REVIEWS IN
AND GEO(

Volume 82 2017

Non-Traditional Stab(
Isotopes

ume of
gures

EDITORS

Fang-Zhen Teng
University of Washington, USA

James Watkins
University of Oregon, USA

Nicolas Dauphas
The University of Chicago, USA

Front-cover: False-color image of a zoned olivine phenocryst (forsterite content) from the Kilauea Iki lava lake, Hawaii. The black marks are spots where the Fe isotopic composition of the olivine was measured by LA-MC-ICPMS and SIMS (Sio et al. 2013, Geochimica et Cosmochimica Acta 123, 302–321). In situ stable isotopic analyses of zoned minerals allow one to tell apart zoning produced by diffusion from zoning produced by growth from an evolving medium.

Series Editor: ***Ian Swainson***

MINERALOGICAL SOCIETY OF AMERICA
GEOCHEMICAL SOCIETY

Non-Traditional Stable Isotopes

82 *Reviews in Mineralogy and Geochemistry* **82**

TABLE OF CONTENTS

1 Non-Traditional Stable Isotopes: Retrospective and Prospective

Fang-Zhen Teng, Nicolas Dauphas, James M. Watkins

2 Equilibrium Fractionation of Non-traditional Isotopes: a Molecular Modeling Perspective

Marc Blanchard, Etienne Balan, Edwin A. Schauble

3 Equilibrium Fractionation of Non-Traditional Stable Isotopes: an Experimental Perspective

Anat Shahar, Stephen M. Elardo, Catherine A. Macris

4 Kinetic Fractionation of Non-Traditional Stable Isotopes by Diffusion and Crystal Growth Reactions

James M. Watkins, Donald J. DePaolo, E. Bruce Watson

5 In Situ Analysis of Non-Traditional Isotopes by SIMS and LA–MC–ICP–MS: Key Aspects and the Example of Mg Isotopes in Olivines and Silicate Glasses

Marc Chaussidon, Zhengbin Deng, Johan Villeneuve, Julien Moureau Bruce Watson, Frank Richter, Frédéric Moynier

6 Lithium Isotope Geochemistry

Sarah Penniston-Dorland, Xiao-Ming Liu, Roberta L. Rudnick

7 Magnesium Isotope Geochemistry

Fang-Zhen Teng

8 Silicon Isotope Geochemistry

Franck Poitrasson

9 Chlorine Isotope Geochemistry

Jaime D. Barnes, Zachary D. Sharp

10 Chromium Isotope Geochemistry

Liping Qin, Xiangli Wang

11 Iron Isotope Systematics

Nicolas Dauphas, Seth G. John, Olivier Rouxel

12 The Isotope Geochemistry of Ni

Tim Elliott, Robert C. J. Steele

13 The Isotope Geochemistry of Zinc and Copper

Frédéric Moynier, Derek Vance, Toshiyuki Fujii, Paul Savage

14 Germanium Isotope Geochemistry

Olivier J. Rouxel, Béatrice Luais

15 Selenium Isotopes as a Biogeochemical Proxy in Deep Time

Eva E. Stüeken

16 *Good Golly, Why Moly?* The Stable Isotope Geochemistry of Molybdenum

Brian Kendall, Tais W. Dahl, Ariel D. Anbar

17 Recent Developments in Mercury Stable Isotope Analysis

Joel D. Blum, Marcus W. Johnson

18 Investigation and Application of Thallium Isotope Fractionation

Sune G. Nielsen, Mark Rehkämper, Julie Prytulak

19 Uranium Isotope Fractionation

Morten B. Andersen, Claudine H. Stirling, Stefan Weyer

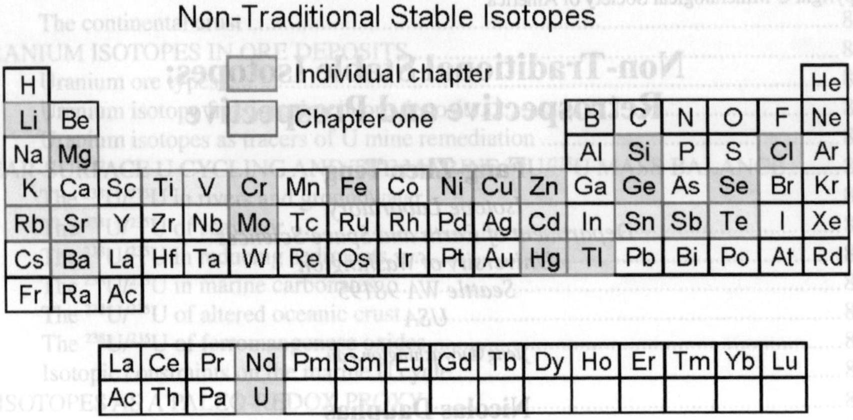

Figure 1. Non-traditional stable isotope systems covered in this volume.

Large variations have been documented in both natural samples and laboratory experiments for non-traditional stable isotopes (Fig. 2). These studies suggest that the following factors control the degree of isotope fractionation in non-traditional stable isotopes during various processes: the relative mass difference between isotopes, change of the oxidation state, biological sensitivity, and volatility. Among these elements, Li displays the largest isotopic variation in terrestrial samples, and since Li is not volatile during geological processes and is not sensitive to redox reactions and biological processes, the large isotope fractionation is controlled mainly by the large relative mass difference (Penniston-Dorland et al. 2017). For many of the other elements, other factors may be equally, if not more, important. For example, Cl exhibits the

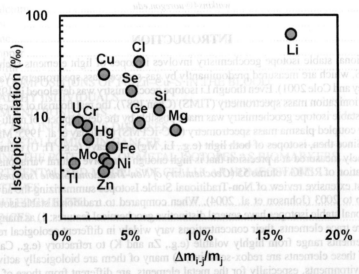

Figure 2. The terrestrial isotopic variation vs. the relative mass difference for non-traditional stable isotopes covered in this volume (Anderson et al. 2017; Barnes and Sharp 2017; Blum and Johnson 2017; Dauphas et al. 2017; Elliott and Steele 2017; Kendall et al. 2017; Moynier et al. 2017; Nielsen et al. 2017; Penniston-Dorland et al. 2017; Poitrasson 2017; Qin and Wang 2017; Rouxel and Luais 2017; Stueken 2017; Teng 2017). The $\Delta m_{i-j} = m_i - m_j$, where i and j represent the two isotopes that are used to report the isotopic variation, with $i > j$.

second largest isotopic variation, which is due to kinetic isotope fractionation during volcanic degassing (Barnes and Sharp 2017). Selenium isotopes also show large isotopic variations, but this reflects redox-controlled Se isotope fractionation (Stueken 2017), while the large Hg isotopic variations are mainly associated with biological processes (Blum and Johnson 2017).

In this chapter, we discuss guidelines and recommendations for reporting non-traditional stable isotopic data and choosing reference materials. We then provide brief introductions to some of the emerging isotope systems that are not covered as individual chapters in this volume. As Ca isotope geochemistry has been recently reviewed in the book series on Advances in Isotope Geochemistry (Gussone et al. 2016), and a similar book is also in preparation for B isotope geochemistry, both Ca and B isotopes are not discussed in this volume. For the basics of stable isotope geochemistry, we recommend prior RIMG volumes (Valley et al. 1986; Valley and Cole 2001; Johnson et al. 2004; MacPherson 2008) and stable isotope geochemistry textbooks (Criss 1999; Faure and Mensing 2005; Sharp 2007; Hoefs 2009).

THE δ NOTATION

The isotopic composition of a sample is commonly reported relative to an international standard as defined by the δ notation. For example, for Mg isotopes:

$$\delta^{26}\mathrm{Mg} = \left\{\frac{(^{26}\mathrm{Mg}/^{24}\mathrm{Mg})_{\mathrm{sample}}}{(^{26}\mathrm{Mg}/^{24}\mathrm{Mg})_{\mathrm{DSM3}}} - 1\right\} \times 1000 \qquad (1)$$

where DSM3 is the international standard for reporting Mg isotopic data (Galy et al. 2003). This definition of δ value has been used extensively for both non-traditional stable isotopes (e.g., this volume, Johnson et al. 2004) and traditional stable isotopes (e.g., Valley et al. 1986).

Recently, a new δ notation without the factor of 1000 has been recommended by the IUPAC for expression of stable isotope ratios (Coplen 2011). For example, the new guideline would suggest a definition of δ for Mg isotopes as:

$$\delta^{26}\mathrm{Mg} = \left\{\frac{(^{26}\mathrm{Mg}/^{24}\mathrm{Mg})_{\mathrm{sample}}}{(^{26}\mathrm{Mg}/^{24}\mathrm{Mg})_{\mathrm{DSM3}}} - 1\right\} \qquad (2)$$

The rationale is that when isotopic data are reported, the ‰ symbol is commonly placed following the value. Then, in the case that $(^{26}\mathrm{Mg}/^{24}\mathrm{Mg})_{\mathrm{sample}}/(^{26}\mathrm{Mg}/^{24}\mathrm{Mg})_{\mathrm{DSM3}} = 1.00025$, the $\delta^{26}\mathrm{Mg}$ should be equal to +0.25 in the traditional δ notation i.e., $\delta^{26}\mathrm{Mg} = +0.25$. If the ‰ symbol is added after the value i.e., $\delta^{26}\mathrm{Mg} = +0.25‰$, then mathematically, this means $\delta^{26}\mathrm{Mg} = +0.25‰ = +0.00025$. In other words, $(^{26}\mathrm{Mg}/^{24}\mathrm{Mg})_{\mathrm{sample}}/(^{26}\mathrm{Mg}/^{24}\mathrm{Mg})_{\mathrm{DSM3}} = 1.00000025$. If the IUPAC δ notation is adopted, then when the $(^{26}\mathrm{Mg}/^{24}\mathrm{Mg})_{\mathrm{sample}}/(^{26}\mathrm{Mg}/^{24}\mathrm{Mg})_{\mathrm{DSM3}} = 1.00025$, the $\delta^{26}\mathrm{Mg} = +0.00025 = +0.25‰$.

While mathematically rigorous, the IUPAC recommendation goes against practices in the field. Retaining the factor of 1000 in the δ notation is, in the view of many in the community, critical to distinguish the δ, ε, and μ notations, which can be used concurrently in a paper.

In the literature on traditional and non-traditional stable isotopes, the δ notation (1) with the factor of 1000 still prevails. For the purpose of being mathematically correct, an alternative to the IUPAC recommendation is to keep the traditional δ notation with the factor of 1000 and omit the ‰ symbol after the δ value, e.g., $\delta^{26}\mathrm{Mg} = +0.25$, which means $(^{26}\mathrm{Mg}/^{24}\mathrm{Mg})_{\mathrm{sample}}/(^{26}\mathrm{Mg}/^{24}\mathrm{Mg})_{\mathrm{DSM3}} = 1.00025$. This expression is also consistent with the usage of the ε notation where the part per 10000 is not added after any ε value.

Another alternative is to adopt the original δ notation as defined by Craig (1957). For Mg isotopes:

$$\delta^{26}\text{Mg}(‰) = \left\{ \left[\frac{(^{26}\text{Mg}/^{24}\text{Mg})_{\text{sample}}}{(^{26}\text{Mg}/^{24}\text{Mg})_{\text{DSM3}}} - 1 \right] \times 1000 \right\} \quad (3)$$

which would read "δ^{26}Mg expressed in ‰ is equal to $[(^{26}\text{Mg}/^{24}\text{Mg})_{\text{sample}}/(^{26}\text{Mg}/^{24}\text{Mg})_{\text{DSM3}} - 1] \times 1000$". Under this definition, it is also mathematically correct to place the ‰ symbol after the value, e.g., δ^{26}Mg = +0.25‰, which means $(^{26}\text{Mg}/^{24}\text{Mg})_{\text{sample}}/(^{26}\text{Mg}/^{24}\text{Mg})_{\text{DSM3}} = 1.00025$.

Although both notations (2) and (3) are mathematically correct, we recommend the notation (3) as it ensures that new publications are consistent with past practices. More important, it is the original definition of δ notation since the beginning of stable isotope geochemistry (Craig 1957). We did not enforce this notation in the present volume but recommend it be used in future publications.

Regardless of which notation is used, it is still correct, as often done by the community, to write sentences like "the ^{26}Mg/^{24}Mg varies 1‰ in water samples from Chicago and Seattle" or "The water sample from Chicago is enriched in heavy Mg isotopes by 1‰ relative to that in Seattle", or "The water sample from Chicago is 1‰ heavier than that from Seattle".

GUIDELINES FOR SELECTING REFERENCE MATERIALS

The field of non-traditional stable isotope geochemistry is confronted with the same issues as traditional stable isotope geochemistry regarding the selection of isotope reference materials (IRMs; Carignan et al. 2004; Vogl and Pritzkow 2010). The recurring problem is that standards used in laboratories either run out or are not readily available. This is the case for reference materials distributed by official organizations as well as in-house standards. For instance, *IRMM-014*, which is used in iron isotope geochemistry and distributed by *the Institute for Reference Materials and Measurements (IRMM),* is no longer available. Similarly, *JMC Lyon Zn* and *DSM3*, which are used in Zn and Mg isotope geochemistry, are not readily available. These issues need to be addressed because systematic errors arise when measurements are carried out against secondary standard solutions. Below, we propose some guidelines for selecting isotopic reference materials used in isotope geochemistry. These are informed by discussions with members of the community as well as our analysis of best practices in traditional and non-traditional stable isotope geochemistry.

Guideline #1: *The isotopic composition of the reference material should be demonstrably homogeneous given current analytical precision and its preparation should be such that future isotopic analyses will be unlikely to reveal isotopic heterogeneities when the precision improves.*

We, as a community, should be forward-thinking in designing preparation protocols that minimize the possibility that reference materials will prove to be heterogeneous as analytical capabilities improve. The purified Mg metal isotopic standard NIST SRM 980, which was made and distributed by *National Institute of Standards and Technology (NIST),* was deemed to be isotopically homogeneous when it was created in 1966 (Catanzaro et al. 1966). However, subsequent higher precision measurements showed that it was isotopically heterogeneous, and it was replaced by a Mg solution made from dissolution of pure Mg metal from Dead Sea Magnesium Ltd, i.e., *DSM3* (Galy et al. 2003). Some steps can be taken to ensure that the maximum level of homogeneity is achieved (i.e., stable solutions, quenched glasses, stable metal sheets or bars are likely to be isotopically homogeneous to a high degree). The processes that can potentially cause heterogeneities in isotopic composition and should be avoided are evaporation, chemical diffusion, the Soret effect, and precipitation/crystallization (e.g., Richter et al. 2009).

Guideline #2: *The reference material should be pure elements or chemical compounds that are either in diluted acids or can be easily dissolved into diluted acids.*

Any unnecessary processing performed in the lab has the potential to induce systematic errors. For example, incomplete digestion and precipitation can induce isotope fractionation, as can chemical purification of analytes in the lab. Although some of these issues can be mitigated using a double-spike approach, it is advantageous to have the reference material in a pure form or a form in which the other elements can be quantitatively removed (e.g., by drying after digestion).

Guideline #3: *The reference material should have an isotopic composition that falls within the range of natural variability, and ideally is representative or similar to a major geological reservoir.*

For non-traditional stable isotope systems, Guideline #2 imposes the reference material be purified by a third party, often at an industrial scale. A benefit is that it ensures that the reference material is available in large quantities and is unlikely to be exhausted. The process of purification can, however, induce significant isotope fractionation, such that the synthetic material can have extreme isotopic composition and may not be representative of any geochemically relevant reservoir. This can lead to all natural isotopic compositions being systematically shifted either to the negative or positive side. Therefore, the reference material should have an isotopic composition that falls within the range of natural variability and ideally is representative of a major geological reservoir for the element investigated. This can be achieved by first surveying aliquots of industrially purified material to find a batch whose isotopic composition approaches that of a geologically relevant reservoir. For example, IRMM-014 coincidentally has an iron isotopic composition that is indistinguishable from chondrites (Craddock and Dauphas 2011).

Guideline #4: *The isotopic composition of the reference material should be characterized at high-precision for all its isotopes to ensure that no measurable anomalies are present that would complicate studies of mass-independent effects and mass-fractionation laws.*

To first order, isotopic variations follow the rules of mass-dependent fractionation, meaning that the variations in isotopic ratios scale as the differences in mass of the isotopes involved. However, it is now possible to discern clear departures from mass-dependent fractionation produced by nuclear field-shift or magnetic effects, and it is also possible to precisely define the laws of mass dependent-fractionation (Dauphas and Schauble 2016). Large-scale purification processes that would fractionate isotopes would impart a certain mass-fractionation law that would complicate comparison between naturally occurring mass fractionation laws. Furthermore, it has been shown previously that some purification processes, notably the Mond-process used in purifying Ni, can create spurious isotopic anomalies (Steele et al. 2011). Characterization of mass-independent effects and mass-fractionation laws is a growing field in geochemistry (Dauphas and Schauble 2016) and it is essential that attention be paid to this issue by documenting whether the material displays spurious isotopic anomalies.

Guideline #5: *The choice of a reference material against which to report isotopic analyses should be a community-led effort and should be consistent.*

Once a standard has reached a certain level of acceptance (e.g., when more than 20 publications report isotopic compositions relative to that standard; if two reference materials have achieved this status, whichever has been more extensively used), that standard should be used in subsequent publications. If it is exhausted, a secondary reference material may be defined and an offset be applied to the isotopic analyses such that the original reference material can still be used to report isotopic compositions. For example, JMC Zn-Lyon is not readily available but future Zn isotopic analyses should still be reported relative to this standard even if new reference materials are used such as IRMM-3702 (Cloquet et al. 2006a; Ponzevera et al. 2006; Moeller et al. 2012) or NIST SRM-683 (Tanimizu et al. 2002; Chen et al. 2016). The secondary reference material used during the measurement should be specified when the δ-notation is defined. For example, one would write: "The Zn isotopic composition is expressed as, $\delta^{66}Zn = [(^{66}Zn/^{64}Zn)_{std} / (^{66}Zn/^{64}Zn)_{JMC\,Lyon} - 1] \times 1000$, where the isotopic analyses were measured relative to IRMM-3702 to which a systematic offset of −0.29 was applied (Moeller et al. 2012) for conversion to the JMC Lyon $\delta^{66}Zn$ scale".

The exception to this guideline is that if the standard is proven to be isotopically heterogeneous as analytical capabilities improve, then a new reference material should be used to report isotopic compositions (see guideline # 1). Ideally, this new standard should have an isotopic composition that is similar to the original reference material used to define the δ scale.

Guideline #6: *The reference materials should be widely available. This implies that lab-defined and owned reference materials should be transferred to organizations that do not have a conflict of interest.*

The role of certification institutes such as IRMM or NIST is to characterize the reference materials that they distribute and ensure that they are available. In many instances, there has been a disconnect between the needs of the geochemical community and what these certification institutes can provide. This is probably due, in part, to the fact that non-traditional stable isotope geochemistry has grown at a rapid pace while preparation of certified materials can take a long time (Vogl and Pritzkow 2010). Part of the certification involves characterization of absolute isotopic abundances using gravimetrically prepared isotope mixtures. In non-traditional stable isotope geochemistry, knowing the absolute ratios is not particularly useful. To cope with the shortage or unavailability of isotopic reference materials, in-house isotopic standards have become the reference materials against which isotopic analyses are reported (DSM3 for Mg, Zn-Lyon for Zn, OL-Ti for Ti). An issue with this practice is the availability of those materials, and conflicts of interest may arise that are detrimental to the advancement of science. Investigator-controlled distribution systems do not work. Moving forward, some organizations/companies could take over that role and distribute (perhaps against a modest fee) the reference materials created by the community. Taking titanium as an example, the bar of pure Ti used to define the OL-Ti standard (Millet and Dauphas 2014; Millet et al. 2016) will be transferred to SARM (Nancy, France), where it will be available to end-users upon request.

Another aspect regarding availability is the preparation of a large enough stock so that the reference material can be used for decades and the material remains stable in time. Aqueous solutions are appealing as they ensure homogeneity (Vogl and Pritzkow 2010) but the concentration is limited by solubility constraints and long-term stability may be an issue. Solids alleviate this issue and are more cost effective for the end user.

Several isotopic systems are, or will be, in crisis if no action is taken to remediate the shortage of isotopic reference materials that can be used by newcomers to the field. There is no committee or working group overseeing the important issues related to standards and we take the opportunity of writing this chapter to propose the establishment of such a working group.

EMERGING ISOTOPE SYSTEMS

The recent advances in instrumentation have made high-precision isotopic analysis possible for almost all elements on the periodic table. Besides those systems reviewed in individual chapters of this volume, there are numerous emerging systems that show great potential as briefly summarized for some of them below.

Stable potassium isotope geochemistry

Table 1. K (atomic number = 19) isotopes and typical natural abundance.

Isotope	Abundance (%)
^{39}K	93.2581
^{40}K	0.0117
^{41}K	6.7302

Isotopic abundance data are from Berglund and Wieser (2011).

Potassium (K) is a volatile, lithophile, incompatible and fluid-mobile element (McDonough and Sun 1995). It is a major cation in both seawater and river water (Pilson 2013), and is well mixed in the ocean because of its long residence time of ~7 Myr (Li 1982). Potassium has three isotopes, ^{39}K, ^{40}K and ^{41}K (Table 1). Among them, ^{40}K is radioactive and exhibits a branched decay scheme to ^{40}Ca and ^{40}Ar, with a half-life of 1.25×10^9 years (Faure and Mensing 2005). The stable ^{39}K and ^{41}K isotopes have > 5% mass difference, which can potentially lead to large K isotope fractionations. Indeed, fractionation of K isotopes during chemical processes has been well known and was documented as early as 1938. Taylor and Urey (1938) found a 10% variation in $^{41}K/^{39}K$ when K was incompletely eluted by an aqueous solution from a zeolite ion exchange column, with ^{39}K preferentially eluted from the exchange medium. This indicates that natural processes such as water-rock interactions could potentially fractionate K isotopes, and generate isotopically distinct reservoirs. Humayun and Clayton (1995a,b) found that both extraterrestrial samples (chondrites, eucrites, SNC meteorites, ureilites, and some lunar highland and mare igneous samples) and terrestrial samples (peridotites, basalts, granites, carbonatites, biotite schists and seawater) have similar K isotopic compositions within ±0.5‰, despite variable levels of volatile element depletion among those bodies. This has been explained by vaporization under a high vapor pressure, as opposed to free evaporation. Indeed, under such conditions, evaporation would take place in an equilibrium rather than a kinetic regime (Richter et al. 2009 and references therein). Chondrules also revealed limited K isotope fractionation, again suggesting that evaporation took place under equilibrium conditions, presumably because chondrule melting and vaporization took place when the density of chondrules was high enough for a high partial pressure of K to build up around them (Alexander et al. 2000; Alexander and Grossman 2005).

The advent of high-resolution mass spectrometry has made it possible to measure K isotopes with higher precision (from ±0.1‰ to < ±0.05‰, 2SE by MC-ICPMS) (Morgan et al. 2014; Li et al. 2016; Wang and Jacobsen 2016a,b) and TIMS (Wielandt and Bizzarro 2011; Naumenko et al. 2013).

Stable K isotopic compositions are reported in the δ notation:

$$\delta^{41}K(‰) = \left\{ \frac{(^{41}K/^{39}K)_{sample}}{(^{41}K/^{39}K)_{std}} - 1 \right\} \times 1000$$

where the standard (std) is either NIST SRM 3141a (Li et al. 2016) or commercial ultrapure potassium nitrate (Wang and Jacobsen 2016a,b). There is a slight difference between these two standards based on the same geostandard and seawater data published from these two groups. Further studies are needed to better quantify the difference and to select one standard for reporting high-precision K isotopic data.

To date, >1.4‰ variation in $^{41}K/^{39}K$ has been reported for terrestrial and extraterrestrial samples (Fig. 3). Morgan et al. (2014) found 0.4‰ K isotopic variation in a diverse range of whole rocks and mineral separates that formed at high temperatures, which is 10 times greater than the analytical uncertainty of < ±0.05‰ (2SE). Wang and Jacobsen (2016a,b) found that

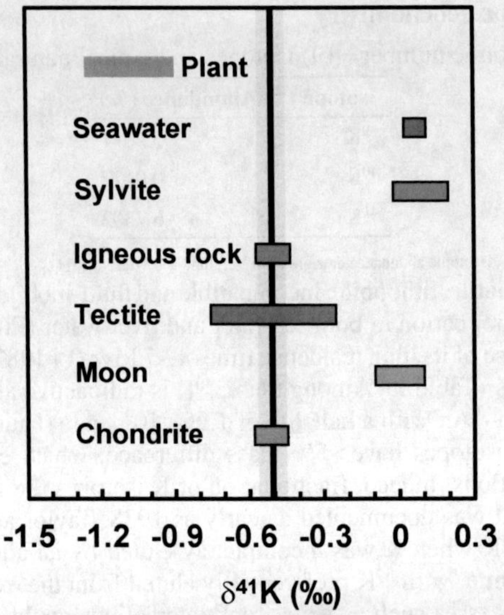

Figure 3. Natural variations in the stable K isotopic composition relative to NIST SRM 3141a. The vertical line and shaded area represent the bulk Earth as represented by igneous rock standards, with $\delta^{41}K = 0.50 \pm 0.04$‰ (2SD). Data from Wang and Jacobsen (2016a,b) are normalized to NIST SRM 3141a by subtracting 0.58. Data are from the literature (Li et al. 2016; Wang and Jacobsen 2016a,b).

seawater samples are ~0.6‰ heavier than terrestrial basalts whereas sylvite samples from two evaporate deposits have heterogeneous K isotopic compositions and overall are also heavier than basalts. Wang and Jacobsen (2016b) reported for the first time that lunar samples are enriched in heavy K isotopes relative to the Earth and chondrites. The isotopically heavy Moon was interpreted as a result of evaporation-driven kinetic K isotope fractionation during the Moon-forming giant impact. Li et al. (2016) documented the largest K isotopic variation in natural samples, with $\delta^{41}K$ (relative to NIST SRM 3141a) ranging from −1.3 in plants to ~0‰ in seawater. Though the underlying processes for large K isotopic variations require further investigation, the large fractionations make K isotope geochemistry a promising avenue for tracing geological and biological processes.

Stable K isotope geochemistry might also play a significant role in refining the ^{40}K–^{40}Ar and ^{40}K–^{40}Ca dating as these methods are based on the assumption that K isotopic composition does not vary in nature to an important degree. As discussed in Naumenk et al. (2013), the $^{40}K/K$ isotopic abundance is the largest contributor to the total K–Ar age uncertainty.

Titanium isotope geochemistry

Table 2. Ti (atomic number = 22) isotopes and typical natural abundance.

Isotope	Abundance (%)
^{46}Ti	8.25
^{47}Ti	7.44
^{48}Ti	73.72
^{49}Ti	5.41
^{50}Ti	5.18

Isotopic abundance data are from Berglund and Wieser (2011).

Titanium (Ti) shares many geochemical similarities with other high field strength elements Hf, Zr, Nb, and Ta. In particular, it has very low solubility in aqueous medium, such that it is largely insusceptible to low temperature aqueous alteration. Titanium is also a highly refractory element that condensed from solar nebula gas early in the condensation sequence (Lodders 2003). For example, refractory inclusions in meteorites (also known as Calcium-Aluminum-rich Inclusions—CAIs) contain significant amounts of Ti. In natural settings, Ti is usually present as Ti^{4+} but significant amounts of Ti^{3+} can be present under low oxygen fugacity conditions, such as those that prevailed during condensation of CAIs.

Titanium has 5 stable isotopes, ^{46}Ti, ^{47}Ti, ^{48}Ti, ^{49}Ti and ^{50}Ti (Table 2). In meteorites and their components, significant departures from mass-dependent fractionation (i.e., isotopic anomalies) have been documented (Dauphas and Schauble 2016). Hibonite inclusions from CM chondrites and CAIs display large enrichments in ^{50}Ti that correlate with anomalies in another neutron-rich isotope, ^{48}Ca (see Fig. 9 and its caption in Dauphas and Schauble 2016 for details). The anomalies measured in hibonite grains can reach ~2500 for ε^{50}Ti (25%). Bulk meteorites also display isotopic anomalies affecting primarily ^{50}Ti but the effects are subtler, ranging from ~−2 to +4 ε-units (Trinquier et al. 2009; Zhang et al. 2012). Zhang et al. (2012) measured the ε^{50}Ti value of lunar rocks and found that they were affected by cosmogenic effects, more specifically neutron capture effects arising from interactions between galactic cosmic rays and lunar rocks. They were able to correct for these effects by using collateral neutron capture effects on Sm and Gd. After correction, they found that lunar and terrestrial rocks have identical ε^{50}Ti values within ±0.04, despite the 6 ε-unit span of the meteorites. This suggests that either most of the Moon came from the protoEarth or the Moon-forming impactor had very similar isotopic composition to the Earth.

The mass-dependent Ti isotopic variations, similar to other non-traditional stable isotopes, are also reported in the traditional δ notation. Nonetheless, no officially certified Ti isotopic standard is available and the measurements published thus far are reported relative to a bar of high-purity metal Ti: OL-Ti (Millet and Dauphas 2014). Millet and Dauphas (2014) showed that different part of the OL-Ti reference material had indistinguishable Ti isotopic compositions. Furthermore, the Ti isotopic compositions of basalts measured by a double spike technique yield a δ^{49}Ti value of +0.004 ± 0.062‰, relative the OL-Ti reference material. Millet et al. (2016) used this technique to measure a more extensive array of mantle-derived magmas and found that they have approximately uniform isotopic compositions (variations can be found but they span less than 0.05‰). Lunar rocks also have Ti isotopic compositions very similar to the Earth. The

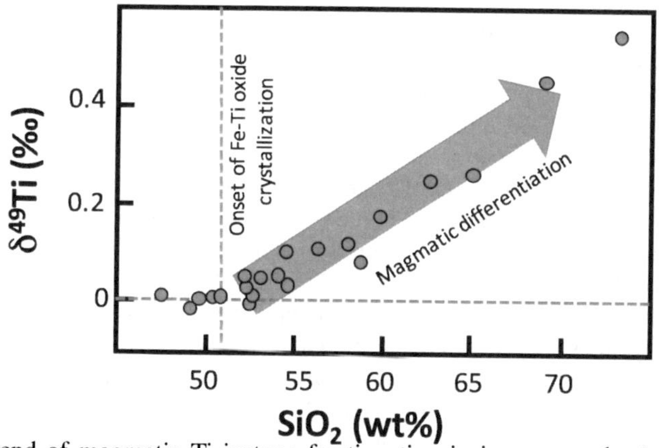

Figure 4. Trend of magmatic Ti isotope fractionation in igneous rocks (modified from Millet et al. 2016).

isotopic composition of OL-Ti is thus representative of an important geochemical reservoir. This reference material fulfills the guidelines highlighted above and we recommend that this standard continue to be used in future studies to facilitate inter-laboratory comparisons.

The available high-precision Ti isotopic data are still limited but suggest a great potential for understanding magmatic differentiation (Miller et al. 2016). Millet et al. (2016) discovered that magmatic differentiation of silicic rocks can fractionate Ti isotopes, producing rocks that have δ^{49}Ti values as high as ~+0.3‰ (Fig. 4). Such enrichment could reflect Ti isotope fractionation during magmatic differentiation, possibly associated with equilibrium fractionation between oxides (e.g., ilmenite) and melt.

Vanadium isotope geochemistry

Table 3. V (atomic number = 23) isotopes and typical natural abundance.

Isotope	Abundance (%)
^{50}V	0.250
^{51}V	99.750

Isotopic abundance data are from Berglund and Wieser (2011).

Vanadium (V) is a biologically active trace element with multiple redox states. The ratio of V to another non-redox element of similar geochemical behavior (e.g., Sc) can provide direct information on the oxygen fugacity during magmatic processes. This approach has been used to examine the redox state of the mantle in the Archean (Canil 1997; Li and Lee 2004) and various other settings (the source of mid-ocean ridge basalts; the sub-arc mantle; Lee et al. 2005). The conclusion is that the redox conditions of melting did not change drastically with time (Canil 1997; Li and Lee 2004). Similarly, there does not appear to be much spatial heterogeneity in the present mantle (Lee et al. 2005). Those results have been questioned recently (Kelley and Cottrell 2009; Aulbach and Stagno 2016) and it is important to develop new proxies of the redox condition of Earth's mantle. Vanadium isotope fractionation could provide clues on that. Indeed, it has been shown that in magmas and minerals relevant to mantle petrology, significant isotope fractionation was present between the two redox states of iron (Dauphas et al. 2014; Roskosz et al. 2015). Similar equilibrium isotope fractionation may be present for V between for example V^4+ and V^{5+}.

Not much work has been done on V isotope geochemistry because it possesses only two stable isotopes, ^{50}V and ^{51}V (Table 3), so a double spike technique cannot be applied and quantitative recovery is required during purification. This element is also not straightforward to purify from other elements. An additional complication arises from the fact that there is a large contrast between the abundances of ^{51}V and ^{50}V (^{51}V/^{50}V = ~400), so that it is difficult to measure its isotopic composition by mass spectrometry. Nielsen et al. (2011) addressed those difficulties and designed a protocol to measure the V isotopic composition of natural rocks. Prytulak et al. (2011) applied this technique to the analysis of igneous geostandards PCC-1, BHVO-2, BCR-2, BIR-1a, GSP-2, and AGV-2 as well as the Allende chondrite and found ~1.2‰ isotopic variation. A later comprehensive study of 64 mafic and ultramafic rocks revealed 1.6‰ V isotopic variation, and yielded a silicate Earth δ^{51}V value of -0.7 ± 0.2‰, where the reference material is a pure V solution distributed by Alfa Aesar (Prytulak et al. 2013). In this δ reference frame, all samples measured so far have markedly negative values. Nielsen et al. (2014) measured the V isotopic composition of meteorites (various chondrites, some HED meteorites, and one martian meteorite). They found that those meteorites have a δ^{51}V value of around -1.7‰, meaning that they are lower than the terrestrial composition by 1‰. The cause for this shift is uncertain but could be related to irradiation from the young Sun, which can produce enrichments in ^{50}V relative to ^{51}V (i.e., it can shift δ^{51}V towards lower values).

More recently, Wu et al. (2015) calculated V isotope fractionation factors using first-principle techniques for V species in solution or adsorbed on goethite surface. They suggested that V isotopes could record past redox conditions in seawater. Wu et al. (2016) reported on a new measurement protocol for V isotopes that yields a precision of better than ±0.1 ‰. Schuth et al. (2016) presented measurements by femtosecond laser ablation MC-ICPMS of V-rich minerals and found a range in $^{51}V/^{50}V$ of ~1.5‰. Overall, V isotopes show a lot of promise as a tracer of metabolic pathways as well as redox conditions in the mantle and possibly in the ocean.

Stable rubidium isotope geochemistry

Table 4. Rb (atomic number = 37) isotopes and typical natural abundance.

Isotope	Abundance (%)
^{85}Rb	72.17
^{87}Rb	27.83

Isotopic abundance data are from Berglund and Wieser (2011).

Rubidium (Rb), similar to K, belongs to the alkali metal group and is a volatile, lithophile, incompatible, and fluid-mobile element. Rb has two isotopes, ^{87}Rb and ^{85}Rb (Table 4). Between them, ^{85}Rb is stable and ^{87}Rb is radioactive and decays to ^{87}Sr, with a half-life of 48.8 billion years (Faure and Mensing 2005). In ^{87}Rb–^{87}Sr dating, the isotopic composition of Rb is often measured to determine the concentration of Rb in the rocks or minerals by isotope dilution but the demand on the precision of those isotopic analyses is not very stringent because other sources of error contribute to the overall error (Waight et al. 2002). To date, high precision Rb isotopic measurements in the literature are still limited. This is presumably due to two difficulties. One is that Rb behaves very similarly to K during most chromatography procedures, so that separating the two can be difficult. The second difficulty is that Rb possesses only two stable isotopes, so that high yield is needed to avoid isotope fractionationation during purification. Achieving both requirements (effective separation from K and high yield) is difficult. Nebel et al. (2005) developed a protocol to purify Rb and analyzed its isotopic composition. The publication does not give details on the degree of separation between Rb and K, other than after processing the K/Rb ratio is < 5. The isotopic analyses are corrected for instrumental mass bias using Zr doping and the precision of the analyses is on the order of ±0.5‰. Nebel et al. (2011) applied this technique to the Rb isotopic analysis of chondrites and found that if any, the isotopic variations were limited to ±1‰ around the terrestrial value. Rb is significantly more volatile than K but improvements in precision are needed to tell whether its isotopic composition was affected by vaporization in the solar system.

Stable strontium isotope geochemistry

Table 5. Sr (atomic number = 38) isotopes and typical natural abundance.

Isotope	Abundance (%)
^{84}Sr	0.56
^{86}Sr	9.86
^{87}Sr	7.00
^{88}Sr	82.58

Isotopic abundance data are from Berglund and Wieser (2011).

Strontium (Sr) is an alkaline earth element, with four isotopes (Table 5). Of them, ^{87}Sr is the decay product of ^{87}Rb, with a half-life of 48.8 billion years (Faure and Mensing 2005). The ^{87}Rb–^{87}Sr decay scheme has been extensively used as a chronometer and fingerprint of mantle source regions (Faure and Powell 1972; Faure and Mensing 2005). Because of the small relative mass difference, the stable Sr isotope ratios are often treated as invariable, with $^{88}Sr/^{86}Sr = 8.375209$. This value has been used to calibrate instrumental fractionations for high-precision Sr isotopic analyses.

With the developments in high precision mass spectrometry using MC-ICPMS and TIMS, significant stable isotope fractionation has been found for Sr in natural samples. The large fractionation has important implications for the measurement and application of radiogenic Sr isotopes, as well as great potential to trace various geological processes.

Stable Sr isotopic compositions are often reported in the δ notation:

$$\delta^{88}Sr(‰) = \left\{ \frac{(^{88}Sr/^{86}Sr)_{sample}}{(^{88}Sr/^{86}Sr)_{std}} - 1 \right\} \times 1000$$

where the standard (std) is the NIST SRM 987.

Sable Sr isotopic compositions have been measured on both TIMS and MC-ICPMS. Double spike methods are usually used for TIMS (DS-TIMS) (Krabbenhoft et al. 2009; Shalev et al. 2013) and some MC-ICPMS (Shalev et al. 2013), though more MC-ICPMS protocols have adopted the standard-sample bracketing method (SSB-MC-ICPMS) (Fietzke and Eisenhauer 2006; Moynier et al. 2010; Charlier et al. 2012; Ma et al. 2013a). In general, the DS-TIMS yield the highest precision and accuracy (±0.02 for $\delta^{88}Sr$, 1SD), followed by DS-MC-ICPMS and SSB-MC-ICPMS. To date, significant stable Sr isotope fractionation has been observed for both high-T and low-T geological processes, with the overall variation in $\delta^{88}Sr$ greater than 1.1‰ in terrestrial materials (Fig. 5).

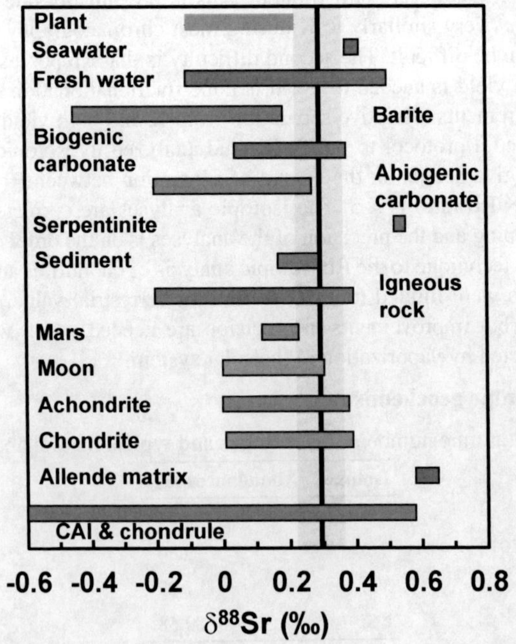

Figure 5. Natural variations in the stable Sr isotopic composition relative to the standard SRM987. The $\delta^{88}Sr$ range of CAI and chondrule extends down to −1.73‰. The vertical line and shaded area represent the mantle, with $\delta^{88}Sr = 0.29 \pm 0.07$‰ (2SD) (Charlier et al. 2012). Data are from literature (Fietzke and Eisenhauer 2006; Halicz et al. 2008; Ohno et al. 2008; Ruggeberg et al. 2008; de Souza et al. 2010; Krabbenhoft et al. 2010; Moynier et al. 2010; Bohm et al. 2012; Charlier et al. 2012; Ma et al. 2013a; Raddatz et al. 2013; Wei et al. 2013; Stevenson et al. 2014, 2016; Vollstaedt et al. 2014; Widanagamage et al. 2014, 2015).

Studies of extraterrestrial and terrestrial materials reveal significant Sr isotope fractionation. Chondrites have heterogeneous Sr isotopic compositions, with δ^{88}Sr ranging from +0.12 to +0.35‰, reflecting the mixing of different proportions of isotopically light CAIs and chondrules with the isotopically heavy matrix materials during nebular processes (Moynier et al. 2010; Charlier et al. 2012). The terrestrial mantle has an average δ^{88}Sr value of +0.29 ± 0.07‰ (2SD), similar to carbonaceous chondrites, angrites, eucrites and martian meteorites (Moynier et al. 2010; Charlier et al. 2012). By contrast, differentiated igneous rocks from the Earth and Moon have light Sr isotopic compositions, likely resulting from the crystallization of isotopically heavy plagioclase during magmatic differentiation (Charlier et al. 2012).

Large stable Sr isotopic variation has also been observed in biogenic and inorganic carbonates (Fietzke and Eisenhauer 2006; Ruggeberg et al. 2008; Krabbenhoft et al. 2010; Bohm et al. 2012; Raddatz et al. 2013; Stevenson et al. 2014; Vollstaedt et al. 2014), barite precipitation (Widanagamage et al. 2014; 2015) and weathered residues (Halicz et al. 2008; de Souza et al. 2010; Krabbenhoft et al. 2010; Wei et al. 2013; Stevenson et al. 2016), indicating the potential of using Sr isotopes as a tracer of continental weathering, marine Sr cycle, and paleoceanographic studies. In general, carbonates tend to have light Sr isotopic compositions relative to seawater, which has a homogenous δ^{88}Sr of 0.387 ± 0.002‰ (2SE mean) (Shalev et al. 2013). The large Sr isotope fractionation correlates with factors such as growth rate and temperature (e.g., Ruggeberg et al. 2008; Bohm et al. 2012; Stevenson et al. 2014). River waters draining carbonates and silicates tend to have distinct Sr isotopic compositions, reflecting the control of source lithologies (isotopically light carbonates vs. heavy silicates) (e.g., Wei et al. 2013). Stable Sr isotopes have also been used in archaeology for paleodietary studies, as there are systematic mass-dependent variations with increasing trophic level (Knudson et al. 2010).

Cadmium isotope geochemistry

Table 6. Cd (atomic number = 48) isotopes and typical natural abundance.

Isotope	Abundance (%)
^{106}Cd	1.25
^{108}Cd	0.89
^{110}Cd	12.49
^{111}Cd	12.80
^{112}Cd	24.13
^{113}Cd	12.22
^{114}Cd	28.73
^{116}Cd	7.49

Isotopic abundance data are from Berglund and Wieser (2011).

Cadmium (Cd) is a highly volatile, chalcophile, moderately incompatible and biologically active trace element with a single oxidation state (Cd2+) in the Earth (McDonough and Sun 1995). Cadmium, together with Hg and Zn, belongs to the group-IIB element, and is toxic and widely distributed in the hydrosphere, biosphere, crust, mantle and extraterrestrial rocks. Cd has eight stable isotopes (Table 6), with a large (>9%) relative mass difference, which could lead to large mass-dependent isotope fractionation.

Stable Cd isotopic compositions are reported in either δ notation:

$$\delta^{114/x}\mathrm{Cd}(‰) = \left\{\frac{(^{114}\mathrm{Cd}/^{x}\mathrm{Cd})_{\mathrm{sample}}}{(^{114}\mathrm{Cd}/^{x}\mathrm{Cd})_{\mathrm{std}}} - 1\right\} \times 1000$$

or ε notation:

$$\varepsilon^{114/x}\text{Cd} = \left\{ \frac{(^{114}\text{Cd}/^{x}\text{Cd})_{\text{sample}}}{(^{114}\text{Cd}/^{x}\text{Cd})_{\text{std}}} - 1 \right\} \times 10000$$

or εCd/amu notation:

$$\varepsilon^{y/x}\text{Cd/amu} = \left\{ \frac{(^{y}\text{Cd}/^{x}\text{Cd})_{\text{sample}}}{(^{y}\text{Cd}/^{x}\text{Cd})_{\text{std}}} - 1 \right\} \times \frac{10000}{(m_y - m_x)}$$

where x = mass 110, 111, 112 or 113, and y = mass 114, 113, 112 or 111. The Cd isotopes are typically measured by MC-ICPMS, and the ratio of ^{114}Cd/^{110}Cd is the one used for reporting purposes. The standard (std), however, is quite different for different labs. Before 2013, different labs used their own in-house Cd isotopic standards (Carignan et al. 2004; Wombacher and Rehkaemper 2004; Cloquet et al. 2005; Schmitt et al. 2009a), which makes it difficult to compare Cd isotopic data reported from different labs, as those in-house standards have very different Cd isotopic compositions. Recently, Abouchami et al. (2013) examined the purity, homogeneity and Cd isotopic composition of NIST SRM 3108 and found that it is pure, homogenous and has a Cd isotopic composition close to the best estimates for the bulk silicate Earth. Thus, NIST SRM 3108 is recommended as the common reference material for reporting high-precision Cd isotopic data for future studies.

To date, >4‰ Cd isotopic variation in ^{114}Cd/^{110}Cd has been reported for terrestrial samples (Ripperger et al. 2007) and >24‰ Cd isotopic variation in extraterrestrial samples (Wombacher et al. 2008). As reviewed in Rehkaemper et al. (2011), these large Cd isotopic variations mainly result from evaporation and condensation during both natural and anthropogenic processes as well as biological processes.

Some ordinary chondrites, type 3 carbonaceous and most enstatite chondrites have highly fractionated Cd isotopic compositions, which reflect kinetic Cd isotope fractionation by volatilization and redistribution of Cd during open system thermal metamorphism on the parental bodies (Wombacher et al. 2003, 2008). The Cd produced during ore refineries has very different isotopic compositions relative to the natural Cd because of the evaporation-driven kinetic isotope fractionation during smelting of sulfide ores, which makes Cd isotopes a great tracer for anthropogenic sources of Cd into soils (Cloquet et al. 2006b; Shiel et al. 2010; Wen et al. 2015) and oceans (Shiel et al. 2012, 2013). Nonetheless, recent studies found significant Cd isotope fractionation during soil weathering, which may compromise the Cd isotopic signatures from pollution sources (Chrastny et al. 2015).

Seawater displays the largest Cd isotopic variation in terrestrial samples because of the large Cd isotope fractionation during biological uptake and utilization of Cd in the seawater column. During Cd uptake into marine carbonates at the ocean surface (Boyle et al. 1976; Boyle 1988), light Cd isotopes are preferentially enriched in carbonates, which leads to a concomitant enrichment of heavy Cd isotopes (> +2‰) in the surface water. The Cd isotopic composition of deep seawater (>1000 m) seems homogenous, with δ^{114}Cd = ~+0.3‰ (relative to NIST SRM 3108; Conway and John 2015, and the references therein). In addition to the vertical heterogeneity, surface seawaters from different oceans also display a large isotopic variation, which reflects both isotope fractionations and mixing among the oceans. Therefore, Cd isotope geochemistry can be used to trace not only global Cd cycle but also changes in marine biological activity in the past (Lacan et al. 2006; Ripperger et al. 2007; Schmitt et al. 2009b; Abouchami et al. 2011; Xue et al. 2013; Abouchami et al. 2014; Conway and John 2015; Georgiev et al. 2015).

Tin isotope geochemistry

Table 7. Sn (atomic number = 50) isotopes and typical natural abundance.

Isotope	Abundance (%)
^{112}Sn	0.97
^{114}Sn	0.66
^{115}Sn	0.34
^{116}Sn	14.54
^{117}Sn	7.68
^{118}Sn	24.22
^{119}Sn	8.59
^{120}Sn	32.58
^{122}Sn	4.63
^{124}Sn	5.79

Isotopic abundance data are from Berglund and Wieser (2011).

Tin (Sn) is a chalcophile and highly volatile trace element with two oxidation states (Sn^{2+} and Sn^{4+}). Natural Sn is mainly associated with sulfide minerals though Sn oxide (cassiterite, SnO_2) is also widespread. Tin has 10 stable isotopes with a large relative mass difference (> 10%) (Table 7). Sn isotope fractionation was considered negligible until 2002 when Clayton et al. (2002) developed a method for high-precision isotopic analysis of Sn based on MC-ICPMS. Since then, tin isotopes have been widely applied in archaeology to trace Sn provenance of artifacts. This is because Sn was a vital commodity in the past and Sn from different ore deposits from different locations has different isotopic compositions (Haustein et al. 2010; Balliana et al. 2013; Yamazaki et al. 2013, 2014; Mason et al. 2016). Nonetheless, the application of Sn isotopes in geochemistry is still limited. A remarkable feature of Sn is that like Fe, it possesses a Mössbauer isotope ^{119}Sn, so it is possible to predict its equilibrium fractionation factor using the synchrotron technique of Nuclear Resonant Inelastic X-ray Scattering (NRIXS) or conventional Mössbauer spectroscopy. Polyakov et al. (2005) used these techniques to predict Sn isotope fractionation for different phases and found a large control of its different oxidation states. Malinovskiy et al. (2009) showed that UV irradiation of methyltin can cause large mass-dependent and mass-independent Sn isotope fractionations. Moynier et al. (2009) observed mass-independent Sn isotope fractionation in chemical exchange reactions by using dicyclohexano-18-crown-6, which they argued are due to nuclear field shift effects.

Antimony isotope geochemistry

Table 8. Sb (atomic number = 51) isotopes and typical natural abundance.

Isotope	Abundance (%)
^{121}Sb	57.21
^{123}Sb	42.79

Isotopic abundance data are from Berglund and Wieser (2011).

Antimony (Sb) is a siderophile, moderately incompatible trace element in the Earth (McDonough and Sun 1995), with multiple oxidation states (dominated by Sb^{3+}, Sb^{5+} and Sb^{3-}). It is toxic, with Sb^{3+} more toxic than Sb^{5+}. Antimony has two stable isotopes (Table 8). At present, the literature on high-precision Sb isotopic data is still limited. Rouxel et al. (2003) characterized the Sb isotopic compositions of seawater, silicate rocks, various environmental samples, deep-sea sediments and hydrothermal sulfides. The continental and oceanic crustal rocks have relatively restricted and lighter Sb isotopic composition than seawater, which has a homogenous Sb isotopic composition. The hydrothermal sulfides display the largest (up to 1.8‰) Sb isotopic variation. The large Sb isotopic variation reflects both heterogeneous Sb sources and isotope fractionation during redox changes in aqueous solutions at low temperatures (Rouxel et al. 2003). Recently, Resongles et al. (2015) have found that two rivers in France, which drained different mining sites, have distinct Sb isotopic compositions, reflecting Sb isotope fractionation during complexed Sb transfer from rocks, mine wastes and sediments to the river water. Overall, Sb isotope systematics may be a useful tool for tracing redox processes, pollution sources and biogeochemical processes in riverine and oceanic systems (Rouxel et al. 2003; Resongles et al. 2015).

Stable tellurium isotope geochemistry

Table 9. Te (atomic number = 52) isotopes and typical natural abundance,

Isotope	Abundance (%)
^{120}Te	0.09
^{122}Te	2.55
^{123}Te	0.89
^{124}Te	4.74
^{125}Te	7.07
^{126}Te	18.84
^{128}Te	31.74
^{130}Te	34.08

Isotopic abundance data are from Berglund and Wieser (2011).

Tellurium (Te) belongs to the same group as S and Se, and is a volatile, chalcophile trace element with multiple oxidation states (Te^{6+}, Te^{4+}, Te^{2+}, and Te^{2-}). Tellurium has eight isotopes with > 8% relative mass difference (Table 9). Among them, ^{126}Te is the decay product of ^{126}Sn, with a half-life of 0.2345 million years (Oberli et al. 1999). In addition, the eight Te isotopes were produced by *r*-, *s*-, and *p*- processes during stellar nucleosynthesis. Hence, Te isotopes have been mainly used in cosmochemistry for searching the short-lived ^{126}Sn in order to use the ^{126}Sn–^{126}Te chronometer to understand early solar system processes, or for searching the nucleosynthetic anomaly in order to understand the solar nebular processes (Fehr et al. 2004, 2005, 2006, 2009; Moynier et al. 2009; Fukami and Yokoyama 2014). To our knowledge, there is only one study that focuses on stable Te isotope geochemistry. Fornadel et al. (2014) developed a method for high-precision (~±0.10‰, 2SD for $^{130}Te/^{125}Te$) analysis of Te isotopes in tellurides and native tellurium from various ore deposits and documented an over 1.6‰ Te isotopic variation. Though the processes responsible for the large fractionation is unknown, Te isotope geochemistry, could become an excellent complement to S and Se isotope geochemistry.

Barium isotope geochemistry

Barium (Ba) is an alkaline earth element with seven stable isotopes (Table 10). Most of the work in Ba stable isotope geochemistry has focused on Ba isotope anomalies in extraterrestrial materials (Eugster et al. 1969; McCulloch and Wasserburg 1978; Hidaka et al. 2003; Savina et al. 2003; Ranen and Jacobsen 2006; Andreasen and Sharma 2007; Carlson et al. 2007; Hidaka and Yoneda 2011). The past several years have seen a rapidly growing literature on Ba isotopes in terrestrial materials.

The Ba isotopic compositions are most often reported using isotopes ^{137}Ba and ^{134}Ba:

$$\delta^{137}\mathrm{Ba}(‰) = \left\{ \frac{(^{137}\mathrm{Ba}/^{134}\mathrm{Ba})_{\mathrm{sample}}}{(^{137}\mathrm{Ba}/^{134}\mathrm{Ba})_{\mathrm{std}}} - 1 \right\} \times 1000$$

where $\delta^{137}Ba$ has been reported relative to one of three different standards in the recent literature: (1) a $Ba(NO_3)_2$ solution from Fluka Aldrich (Von Allmen et al. 2010; Miyasaki et al. 2014; Pretet et al. 2016), (2) a $BaCO_3$ solution (IAEA-CO-9), and (3) a $Ba(NO_3)_2$ solution from the National Institute of Standards and Technology (SRM3104a; Nan et al. 2015, Horner et al. 2015).

Table 10. Ba (atomic number = 56) isotopes and typical natural abundance.

Isotope	Abundance (%)
^{130}Ba	0.106
^{132}Ba	0.101
^{134}Ba	2.417
^{135}Ba	6.592
^{136}Ba	7.854
^{137}Ba	11.232
^{138}Ba	71.698

Isotopic abundance data are from Berglund and Wieser (2011).

The total natural variation in $^{137}Ba/^{134}Ba$ of natural terrestrial materials is ~1.5‰ (Fig. 5). Although this range is quite restricted (owing to the small relative mass difference between Ba isotopes), modern MC-ICPMS methods are able to yield a reproducibility of about ±0.03‰ to ±0.05‰ (Miyazaki et al. 2014; Nan et al. 2015), which is more than an order of magnitude smaller than the natural variations.

The most recent applied studies have focused primarily on the Ba cycle in seawater because the concentration of Ba closely tracks the concentration of dissolved $Si(OH)_4$ and other nutrients (von Allmen et al. 2010; Horner et al. 2015; Pretet et al. 2015; Cao et al. 2016). The thinking behind these studies is that Ba isotopes coupled with Ba concentration measurements may be useful to probe nutrient cycling, biologic productivity, and water mass mixing (Horner et al. 2015; Cao et al. 2016). Ba enters the oceans from rivers with a $\delta^{137}Ba$ of about 0 to +0.3‰ (Fig. 6). From there, precipitation of $BaSO_4$ in organic microenvironments or adsorption of Ba onto organic particulates leads to light isotope enrichment in the solid phase, and consequently, seawater becomes isotopically heavy with respect to Ba (Fig. 6; Horner et al. 2015; Cao et al. 2016). This general process leads to a strong correlation between Ba^{2+} concentrations and $\delta^{137}Ba/^{134}Ba$, and imparts a distinct chemical and isotopic signature on different water masses. Hence, coupled Ba–$\delta^{137}Ba/^{134}Ba$ systematics can be used to trace the mixing of different water masses (Horner et al. 2015; Cao et al. 2016). An open question is whether other factors such as hydrothermal activity, submarine groundwater discharge, or atmospheric inputs, are important within the oceanic Ba cycle or simply have effects that cancel out (Cao et al. 2016).

Like particulate matter, corals are isotopically lighter than seawater (Fig. 6; Pretet et al. 2015). Most of the corals measured so far were analyzed without information regarding the isotopic composition of the host seawater, but the expectation from laboratory cultures is that coral is lighter than seawater by 0.01 to 0.26 ± 0.14‰ (Pretet et al. 2015).

To aid the interpretation of $\delta^{137}Ba/^{134}Ba$ in nature, several studies have investigated isotope fractionation by diffusion (Van Zuilen et al. 2016) and between experimentally grown *inorganic* minerals and their host solution. Von Allmen et al. (2010) grew $BaCO_3$ and $BaSO_4$ from aqueous solutions at 21 and 80 °C. Like many previous experiments using other isotopic systems, they observed that the solid phase was isotopically light relative to the aqueous solution. For $BaCO_3$, the fractionation ranges from 0.1 (fast growth) to 0.3‰ (slow growth). An interesting result for both $BaCO_3$ and $BaSO_4$ is that the fractionations are insensitive to temperature between 21 and 80 °C. Mavromatis et al. (2016) precipitated $BaCO_3$ at 25 °C and documented light isotope enrichment in the solid phase by 0.07 ± 0.04‰, a value that is in

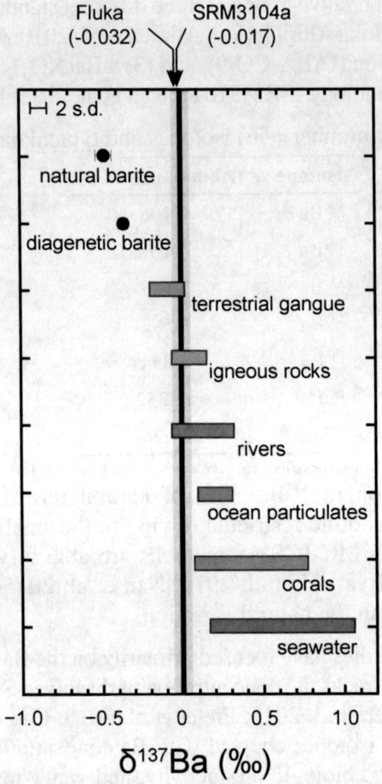

Figure 6. Natural variations in the Ba isotopic composition of terrestrial materials relative to the standard IAEA-CO-9. The vertical line and shaded area represent the standard value with $\delta^{137}Ba/^{134}Ba = 0.00 \pm 0.02$ (2SD). A different choice of standard, shown by the two arrows, would not shift δ values significantly (Miyazaki et al. 2014; Nan et al. 2015). Data are from the literature (Von Allmen et al. 2010; Miyazaki et al. 2014; Horner et al. 2015; Nan et al. 2015; Cao et al. 2016; Pretet et al. 2016).

good agreement with the "fast growth" experiments of Von Allmen et al. (2010). Mavromatis et al. (2016) went a step further and monitored the Ba isotope composition of the aqueous solution over time after chemical (but not isotopic) equilibrium between $BaCO_3$ and fluid was established. After about 6–8 days, the solution had evolved isotopically to a composition that was indistinguishable from the solid phase, indicating a $BaCO_3$–fluid equilibrium fractionation factor of ~0‰. Such a small equilibrium effect is consistent with similarities in the Ba–O bond length of witherite (2.80 Å) compared to aqueous Ba (2.79 Å) (Mavromatis et al. 2016). One other experimental study by Böttcher et al. (2012) showed that $BaMn(CO_3)_2$ grown at 21 ºC yields a similar light isotope enrichment in the solid phase of 0.11 ± 0.06‰.

Stable neodymium isotope geochemistry

Neodymium (Nd) is a rare earth element, with seven isotopes (Table 11). Of them, ^{142}Nd is the decay product of ^{146}Sm, with a half-life of 68 million years (Kinoshita et al. 2012) and ^{143}Nd is the decay product of ^{147}Sm, with a half-life of 106 billion years (Faure and Mensing 2005). The ^{147}Sm–^{143}Nd and ^{146}Sm–^{142}Nd have been extensively used as chronometers and fingerprints of mantle source regions (DePaolo 1988; Boyet and Carlson 2005; Faure and Mensing 2005). Because of the small relative mass difference, the stable Nd isotope ratios are often treated as invariable, with $^{146}Nd/^{144}Nd = 0.7219$. This value has been used to calibrate instrumental fractionations for high-precision Nd isotopic analyses.

Table 11. Nd (atomic number=60) isotopes and typical natural abundance.

Isotope	Abundance (%)
^{142}Nd	27.152
^{143}Nd	12.174
^{144}Nd	23.798
^{145}Nd	8.293
^{146}Nd	17.189
^{148}Nd	5.756
^{150}Nd	5.638

Isotopic abundance data are from Berglund and Wieser (2011).

With the developments in high precision mass spectrometry using MC-ICPMS and TIMS, significant stable isotope fractionation has been found for Nd in natural samples. The large fractionations have important implications for the measurement and application of radiogenic Nd isotopes, as well as great potential to trace various geological processes.

The stable Nd isotopic compositions are often reported in the ε notation:

$$\varepsilon^{x}\text{Nd}(‰) = \left\{ \frac{(^{x}\text{Nd}/^{144}\text{Nd})_{\text{sample}}}{(^{x}\text{Nd}/^{144}\text{Nd})_{\text{std}}} - 1 \right\} \times 10000$$

where the standard (std) is the JNd_{i-1} and x=mass 145, 146, 148 and 150. When x=142 and 143, the ε^{142}Nd and ε^{143}Nd represent the radiogenic Nd isotopic compositions.

The literature on stable Nd isotope geochemistry is still very limited. Nonetheless, the data so far reveal significant Nd isotope fractionation (Fig. 7). Wakaki and Tanaka (2012) reported high-

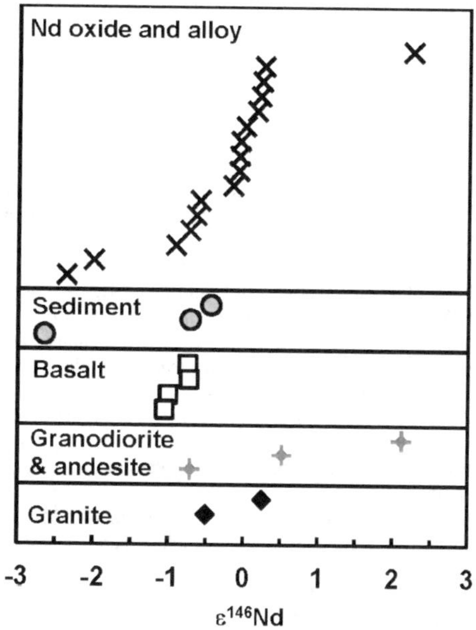

Figure 7. Natural variations in the stable Nd isotopic composition of terrestrial materials relative to the standard JNd_{i-1}. The La Jolla Nd has different δ^{146}Nd value than the JNd_{i-1}. Data are from Wakaki and Tanaka (2012), and Ma et al. (2013b).

precision (±0.3 for ε^{146}Nd, 2SD) stable Nd isotope analysis by using DS-TIMS method and found large mass-dependent Nd isotope fractionation during column chemistry, with ε^{146}Nd varying from +4.60 at the early stage of elution to −9.44 at the late stage of elution. They also found significant Nd isotopic variation in commercial high-purity Nd reagents (ε^{146}Nd = −2.36 to +0.23). In particular, the La Jolla Nd (ε^{146}Nd = −1.97 ± 0.23) has different stable Nd isotopic composition than JNd_{i-1}. Ma et al. (2013b) developed a method of using MC-ICPMS to measure stable Nd isotopes and achieved similar precision (better than ±0.4 for ε^{146}Nd, 2SD) to DS-TIMS. They found a large Nd isotopic variation in 15 rocks and polymetallic nodules (mainly geostandards), with ε^{146}Nd ranging from −2.65 in a stream sediment (JSD-1) to +2.12 in a granodiorite (JG-1a). The in-house Nd standard (Nd-GIG) has the heaviest Nd isotopic composition, with ε^{146}Nd=+2.25. Overall, relative to basalts, river and marine sediments tend to have light Nd isotopic compositions while granites and granodiorites have heavy Nd isotopic compositions.

CONCLUSIONS

Our knowledge of the behavior and utility of non-traditional stable isotopes has expanded greatly since the publication of the RIMG volume 55: *Geochemistry of Non-Traditional Stable Isotopes* in 2004. A testament to the rapid progress in this field is that many of the stable isotope systems that were in their infancy in 2004 have a dedicated chapter in this volume. We anticipate that the isotope systems summarized in this introductory chapter or not even mentioned (e.g., Ga, Br and Zr) will warrant their own dedicated chapter in the not too distant future.

ACKNOWLEDGEMENTS

FZT was supported by NSF (EAR-0838227 and EAR-1340160). ND was supported by NASA (NNX14AK09G, OJ-30381-0036A and NNX15AJ25G) and NSF (EAR144495 and EAR150259). JMW was supported by the NSF (EAR-1249404 and EAR-1050000). FZT thanks Jinlong Ma, Lie-Meng Chen, Yong-Sheng He, Zhuang Ruan, Kwan-Nang Pang, Wang-Ye Li, Yan Hu, Ben-Xun Su, Yang Sun, Shui-Jiong Wang for their helpful comments and discussion.

REFERENCES

Abouchami W, Galer SJG, de Baar HJW, Alderkamp AC, Middag R, Laan P, Feldmann H, Andreae MO (2011) Modulation of the Southern Ocean cadmium isotope signature by ocean circulation and primary productivity. Earth Planet Sci Lett 305:83–9

Abouchami W, Galer SJG, Horner TJ, Rehkamper M, Wombacher F, Xue ZC, Lambelet M, Gault-Ringold M, Stirling CH, Schonbachler M, Shiel AE, Weis D, Holdship PF (2013) A common reference material for cadmium isotope studies–NIST SRM 3108. Geostand Geoanal Res 37:5–17

Abouchami W, Galer SJG, de Baar HJW, Middag R, Vance D, Zhao Y, Klunder M, Mezger K, Feldmann H, Andreae MO (2014) Biogeochemical cycling of cadmium isotopes in the Southern Ocean along the Zero Meridian. Geochim Cosmochim Acta 127:348–367

Alexander C M O'D, Grossman JN, Wang J, Zanda B, Bourot-Denise M, Hewins RH (2000) The lack of potassium-isotopic fractionation in Bishunpur chondrules. Meteorit Planet Sci 35:859–868

Alexander CMDO'D, Grossman JN (2005) Alkali elemental and potassium isotopic compositions of Semarkona chondrules. Meteorit Planet Sci 40:541–556

Andersen MB, Stirling CH, Weyer S (2017) Uranium isotope fractionation. Rev Mineral Geochem 82:799–850

Aulbach S, Stagno V (2016) Evidence for a reducing Archean ambient mantle and its effects on the carbon cycle. Geology 44:751–754

Balliana E, Aramendia M, Resano M, Barante C, Vanhaecke F (2013) Copper and tin isotopic analysis of ancient bronzes for archaeological investigation: development and validation of a suitable analytical methodology. Anal Bioanal Chem 405:2973–2986

Barnes JD, Sharp ZD (2017) Chlorine isotope geochemistry. Rev Mineral Geochem 82:345–377

Berglund M, Wieser ME (2011) Isotopic compositions of the elements 2009 (IUPAC Technical Report). Pure Appl Chem 83:397–410

Blum JD, Johnson MW (2017) Recent developments in mercury stable isotope analysis. Rev Mineral Geochem 83:733–757

Bohm F, Eisenhauer A, Tang JW, Dietzel M, Krabbenhoft A, Kisakurek B, Horn C (2012) Strontium isotope fractionation of planktic foraminifera and inorganic calcite. Geochim Cosmochim Acta 93:300–314

Böttcher ME, Geprägs P, Neubert N, Von Allmen K, Pretet C, Samankassou E, Nägler TF (2012) Barium isotope fractionation during experimental formation of the double carbonate $BaMn[CO_3]_2$ at ambient temperature. Isotopes Environ Health Studies 48:457–46

Boyet M, Carlson RW (2005) ^{142}Nd evidence for early (>4.53 Ga) global differentiation of the silicate Earth. Science 309:576–581

Boyle EA (1988) Cadmium: Chemical tracer of deepwater paleoceanography. Paleoceanography 3:471–489

Boyle EA, Sclater F, Edmond JM (1976) Marine goechemistry of cadmium. Nature 263:42–44

Canil D (1997) Vanadium partitioning and the oxidation state of Archean komatiite magmas. Nature 389:842–845

Cao Z, Siebert C, Hathorne EC, Dai M, Frank M (2016) Constraining the oceanic barium cycle with stable barium isotopes. Earth Planet Sci Lett 434:1–9

Carignan J, Cardinal D, Eisenhauer A, Galy A, Rehkamper M, Wombacher F, Vigier N (2004) A reflection on Mg, Cd, Ca, Li and Si isotopic measurements and related reference materials. Geostand Geoanal Res 28:139–148

Carlson R W, Boyet M, Horan M (2007) Chondrite barium, neodymium, and samarium isotopic heterogeneity and early earth differentiation. Science 316:1175–117

Catanzaro EL, Murphy TJ, Garner EL, Shields WR 1966. Absolute isotopic abundance ratios and atomic weight of magnesium. J Res. Natl. Bur. Stand. 70A:453–458

Chan LH 1987. Lithium isotope analysis by thermal ionization mass- spectrometry of lithium tetraborate. Anal Chem 59:2662–2665

Charlier BLA, Nowell GM, Parkinson IJ, Kelley SP, Pearson DG, Burton KW (2012) High temperature strontium stable isotope behaviour in the early solar system and planetary bodies. Earth Planet Sci Lett 329:31–40

Chen S, Liu Y, Hu J, Zhang Z, Hou Z, Huang F, Yu H (2016) Zinc isotopic compositions of NIST SRM 683 and whole-rock reference materials. Geostand Geoanal Res 40:417–432.

Chrastny V, Cadkova E, Vanek A, Teper L, Cabala J, Komarek M (2015) Cadmium isotope fractionation within the soil profile complicates source identification in relation to Pb-Zn mining and smelting processes. Chem Geol 405:1–9

Clayton R, Andersson P, Gale NH, Gillis C, Whitehouse MJ (2002) Precise determination of the isotopic composition of Sn using MC-ICP-MS J Anal At Spectrom 17:1248–1256

Cloquet C, Rouxel O, Carignan J, Libourel G (2005) Natural cadmium isotopic variations in eight geological reference materials (NIST SRM 2711, BCR 176, GSS-1, GXR-1, GXR-2, GSD-12, Nod-P-1, Nod-A-1) and anthropogenic samples, measured by MC-ICP-MS Geostand Geoanal Res 29:95–106

Cloquet C Carignan J Libourel G (2006a) Isotopic composition of Zn and Pb atmospheric depositions in an urban/ periurban area of northeastern France. Environmental Science and Technology 40:6594–6600

Cloquet C, Carignan J, Libourel G, Sterckeman T, Perdrix E (2006b) Tracing source pollution in soils using cadmium and lead isotopes. Environ Sci Technol 40:2525–2530

Conway TM, John SG (2015) Biogeochemical cycling of cadmium isotopes along a high-resolution section through the North Atlantic Ocean. Geochim Cosmochim Acta 148:269–283

Coplen TB (2011) Guidelines and recommended terms for expression of stable-isotope-ratio and gas-ratio measurement results. Rapid Commun Mass Spectrom 25:2538–2560

Craddock PR, Dauphas N (2011) Iron isotopic compositions of geological reference materials and chondrites. Geostand Geoanal Res 35:101–123.

Craig (1957) Isotopic standards for carbon and oxygen and correction factors for mass-spectrometric analysis of carbon dioxide. Geochim Cosmochim Acta 12:133–149

Criss RE (1999) Principles of Stable Isotope Distribution. Oxford University Press, New York

Dauphas N, Roskosz M, Alp EE, Neuville DR, Hu MY, Sio CK, Tissot FLH, Zhao J, Tissandier L, Médard E, Cordier C (2014) Magma redox and structural controls on iron isotope variations in Earth's mantle and crust. Earth Planet Sci Lett 398:127–140

Dauphas N, Schauble EA (2016) Mass fractionation laws, mass-independent effects, and isotopic anomalies. Ann Rev Earth Planet Sci 44 709–783

Dauphas N, John SG, Rouxel O (2017) Iron isotope systematics. Rev Mineral Geochem 82:415–510

DePaolo DJ (1988) Neodymium Isotope Geochemistry: An Introduction. Springer, Berlin

de Souza GF, Reynolds BC, Kiczka M, Bourdon B (2010) Evidence for mass-dependent isotopic fractionation of strontium in a glaciated granitic watershed. Geochim Cosmochim Acta 74:2596–2614

Elliott T, Steele RCJ (2017) The isotope geochemistry of Ni. Rev Mineral Geochem 82:511–542

Faure G, Powell JL (1972) Strontium Isotope Geology. Springer-Verlag, Berlin

Faure G, Mensing TM (2005) Isotopes: Principles and Applications. John Wiley & Sons, Inc., Hoboken, New Jersey

Fehr MA, Rehkamper M, Halliday AN (2004) Application of MC-ICPMS to the precise determination of tellurium isotope compositions in chondrites, iron meteorites and sulfides. Inter J Mass Spectrom 232:83–94

Fehr MA, Rehkamper M, Halliday AN, Wiechert U, Hattendorf B, Gunther D, Ono S, Eigenbrode JL, Rumble D (2005) Tellurium isotopic composition of the early solar system–A search for effects resulting from stellar nucleosynthesis ^{126}Sn decay, and mass-independent fractionation. Geochim Cosmochim Acta 69:5099–5112

Fehr MA, Rehkamper M, Halliday AN, Schonbachler M, Hattendorf B, Gunther D (2006) Search for nucleosynthetic and radiogenic tellurium isotope anomalies in carbonaceous chondrites. Geochim Cosmochim Acta 70:3436–3448

Fehr MA, Rehkamper M, Halliday AN, Hattendorf B, Gunther D (2009) Tellurium isotope compositions of calcium-aluminum-rich inclusions. Meteorit Planet Sci 44:971–984

Fietzke J, Eisenhauer A (2006) Determination of temperature-dependent stable strontium isotope (^{88}Sr/^{86}Sr) fractionation via bracketing standard MC-ICP-MS Geochem Geophys Geosyst 7

Fornadel AP, Spry PG, Jackson SE, Mathur RD, Chapman JB, Girard I (2014) Methods for the determination of stable Te isotopes of minerals in the system Au–Ag–Te by MC-ICP-MS J Anal At Spectrom 29:623–637

Fukami Y, Yokoyama T (2014) Precise tellurium isotope analysis by negative thermal ionization mass spectrometry (N-TIMS). J Anal At Spectrom 29:520–528

Galy A, Yoffe O, Janney PE, Williams RW, Cloquet C, Alard O, Halicz L, Wadhwa M, Hutcheon ID, Ramon E, Carignan J (2003) Magnesium isotope heterogeneity of the isotopic standard SRM980 and new reference materials for magnesium-isotope-ratio measurements. J Anal At Spectrom 18:1352–1356

Georgiev SV, Horner TJ, Stein HJ, Hannah JL, Bingen B, Rehkaemper M (2015) Cadmium-isotopic evidence for increasing primary productivity during the Late Permian anoxic event. Earth Planet Sci Lett 410:84–96

Gussone N, Schmitt A-D, Heuser A, Wombacher F, Dietzel M, Tipper E, Schiller M (2016) Calcium Stable Isotope Geochemistry. Springer, Berlin

Halicz L, Segal I, Fruchter N, Stein M, Lazar B (2008) Strontium stable isotopes fractionate in the soil environments? Earth Planet Sci Lett 272:406–411

Halliday A, Lee DC, Christensen JN, Walder AJ, Freedman PA, Jones CE, Hall CM, Yi W, Teagle D 1995. Recent developments in inductively coupled plasma magnetic sector multiple collector mass spectrometry. Int J Mass Spectrom Ion Processes 146–147:21–33

Haustein M, Gillis C, Pernicka E (2010) Tin isotopy: A new method for solving old questions. Archaeometry 52:816–832

Hidaka H, Yoneda S (2011) Diverse nucleosynthetic components in barium isotopes of carbonaceous chondrites: Incomplete mixing of *s*- and *r*-process isotopes and extinct ^{135}Cs in the early solar system. Geochim Cosmochim Acta 75:3687–369

Hoefs J (2009) Stable Isotope Geochemistry. Springer-Verlag, Berlin

Horner TJ, Kinsley CW, Nielsen SG (2015) Barium-isotopic fractionation in seawater mediated by barite cycling and oceanic circulation. Earth Planet Sci Lett 430:511–52

Humayun M, Clayton RN (1995a) Potassium isotope cosmochemistry: Genetic implications of volatile element depletion. Geochim Cosmochim Acta 59:2131–2148

Humayun M, Clayton RN (1995b) Precise determination of the isotopic composition of potassium: Application to terrestrial rocks and lunar soils. Geochim Cosmochim Acta 59:2115–2130

Johnson CM, Beard BL, Albarede F (2004) Geochemistry of Non-Traditional Stable Isotopes. Rev Mineral Geochem 55. The Mineralogical Society of America, Washington DC

Kelley KA, Cottrell E (2009) Water and the oxidation state of subduction zone magmas. Science 325:605–607

Kendall B, Dahl TW, Anbar AD (2017) Good golly, why moly? The stable isotope geochemistry of molybdenum. Rev Mineral Geochem 82:683–732

Kinoshita N, Paul M, Kashiv Y, Collon P, Deibel CM, DiGiovine B, Greene JP, Henderson DJ, Jiang CL, Marley ST, Nakaaishi T, Pardo RC, Rehm KE, Robertson D, Scott R, Tang XD, Vondrasek R, Yokoyama A (2012) A shorter ^{146}Sm half-life measured and implications for ^{146}Sm-^{142}Nd chronology in the solar system. Science 335 1614–1617

Knudson KJ, Williams HM, Buikstra JE, Tomczak PD, Gordon GW, Anbar AD (2010) Introducing $\delta^{88/86}$Sr analysis in archaeology: a demonstration of the utility of strontium isotope fractionation in paleodietary studies. J Archaeolog Sci 37:2352–2364

Kondoh A, Oi T, Hosoe M 1996. Fractionation of barium isotopes in cation-exchange chromatography. Separation Science and Technology 31:39–4

Krabbenhoft A, Fietzke J, Eisenhauer A, Liebetrau V, Bohm F, Vollstaedt H (2009) Determination of radiogenic and stable strontium isotope ratios (^{87}Sr/^{86}Sr, $\delta^{88/86}$Sr) by thermal ionization mass spectrometry applying an ^{87}Sr/^{84}Sr double spike. J Anal At Spectrom 24:1267–1271

Krabbenhoft A, Eisenhauer A, Bohm F, Vollstaedt H, Fietzke J, Liebetrau V, Augustin N, Peucker-Ehrenbrink B, Muller MN, Horn C, Hansen BT, Nolte N, Wallmann K (2010) Constraining the marine strontium budget with natural strontium isotope fractionations (^{87}Sr/^{86}Sr*, $\delta^{88/86}$Sr) of carbonates, hydrothermal solutions and river waters. Geochim Cosmochim Acta 74:4097–4109

Lacan F, Francois R, Ji Y, Sherrell RM (2006) Cadmium isotopic composition in the ocean. Geochim Cosmochim Acta 70:5104–5118

Lee CTA, Leeman WP, Canil D, Li ZXA (2005) Similar V/Sc systematics in MORB and arc basalts: implications for the oxygen fugacities of their mantle source regions. J Petrol 46:2313–2336
Li YH (1982) A brief discussion on the mean oceanic residence time of elements. Geochim Cosmochim Acta 46:2671–2675
Li ZXA, Lee CTA (2004) The constancy of upper mantle fO_2 through time inferred from V/Sc ratios in basalts. Earth Planet Sci Lett 228:483–493
Li W, Beard BL, Li S (2016) Precise measurement of stable potassium isotope ratios using a single focusing collision cell multi-collector ICP-MS J Anal At Spectrom 31:1023–1029
Lodders K (2003) Solar system abundances and condensation temperatures of the elements. Astrophys J 591:1220–1247
Ma JL, Wei GJ, Liu Y, Ren ZY, Xu YG, Yang YH (2013a) Precise measurement of stable ($\delta^{88/86}$Sr) and radiogenic (^{87}Sr/^{86}Sr) strontium isotope ratios in geological standard reference materials using MC-ICP-MS Chinese Science Bulletin 58:3111–3118
Ma JL, Wei GJ, Liu Y, Ren ZY, Xu YG, Yang YH (2013b) Precise measurement of stable neodymium isotopes of geological materials by using MC-ICP-MS J Anal At Spectrom 28:1926–1931
MacPherson GJ (2008) Oxygen in the Solar System. Rev Mineral Geochem 68. The Mineralogical Society of America, Washington D
Malinovskiy D, Moens L, Vanhaecke F (2009) Isotopic fractionation of Sn during methylation and demethylation reactions in aqueous solution. Environ Sci Technol 43:4399–4404
Mason AH, Powell WG, Bankoff HA, Mathur R, Bulatovic A, Filipovic V, Ruiz J (2016) Tin isotope characterization of bronze artifacts of the central Balkans. J Archaeolog Sci 69:110–117
Mavromatis V, van Zuilen K, Purgstaller B, Baldermann A, Nägler TF, Dietzel M (2016) Barium isotope fractionation during witherite ($BaCO_3$) dissolution, precipitation and at equilibrium. Geochim Cosmochim Acta 190:72–78
McDonough WF, Sun SS 1995. The composition of the Earth. Chem Geol 120:223–253
Millet MA, Dauphas N (2014) Ultra-precise titanium stable isotope measurements by double-spike high resolution MC-ICP-MS J Anal At Spectrom 29:1444–1458
Millet MA, Dauphas N, Greber ND, Burton KW, Dale CW, Debret B, Macpherson CG, Nowell GM, Williams HM (2016) Titanium stable isotope investigation of magmatic processes on the Earth and Moon. Earth Planet Sci Lett 449 197–205
Miyazaki T, Kimura JI, Chang Q (2014) Analysis of stable isotope ratios of Ba by double-spike standard-sample bracketing using multiple-collector inductively coupled plasma mass spectrometry. J Anal At Spectrom 29:483–49
Moeller K Schoenberg R Pedersen R-B Weiss D Dong S (2012) Calibration of the new certified reference materials ERM-AE633 and ERM-AE647 for copper and IRMM-3702 for zinc isotope amount ratio determinations. Geostand Geoanal Res 36 177–199
Morgan L, Higgins JA, Davidheiser-Kroll B, Lloyd N, Faithfull J, Ellam R (2014) Potassium isotope geochemistry and magmatic processes. Goldschmidt Conference Abstracts
Moynier F, Fujii T, Albarede F (2009) Nuclear field shift effect as a possible cause of Te isotopic anomalies in the early solar system—An alternative explanation of Fehr et al. (2006 and 2009). Meteorit Planet Sci 44:1735–1742
Moynier F, Fujii T, Telouk P (2009) Mass-independent isotopic fractionation of tin in chemical exchange reaction using a crown ether. Analytica Chimica Acta 632:234–239
Moynier F, Agranier A, Hezel DC, Bouvier A (2010) Sr stable isotope composition of Earth, the Moon, Mars, Vesta and meteorites. Earth Planet Sci Lett 300:359–366
Moynier F, Vance D, Fujii T, Savage PS (2017) The isotope geochemistry of zinc and copper. Rev Mineral Geochem 82:543–600
Nan X, Wu F, Zhang Z, Hou Z, Huang F, Yu H (2015) High-precision barium isotope measurements by MC-ICP-MS J Anal At Spectrom. 30:2307–231
Naumenko MO, Mezger K, Naegler TF, Villa IM (2013) High precision determination of the terrestrial ^{40}K abundance. Geochim Cosmochim Acta 122:353–362
Nebel O Mezger K, Scherer EE, Münker C (2005) High precision determinations of ^{87}Rb/^{85}Rb in geologic materials by MC-ICP-MS, Inter J Mass Spectrom 246:10–18
Nebel O, Mezger K, van Westrenen W (2011) Rubidium isotopes in primitive chondrites: constraints on Earth's volatile element depletion and lead isotope evolution. Earth Planet Sci Lett 305:309–316
Nielsen SG, Prytulak J, Halliday AN (2011) Determination of precise and accurate ^{51}V/^{50}V isotope ratios by MC-ICP-MS, Part 1: Chemical separation of vanadium and mass spectrometric protocols. Geostand Geoanal Res 35:293–30
Nielsen SG, Prytulak J, Wood BJ, Halliday AN (2014) Vanadium isotopic difference between the silicate Earth and meteorites. Earth Planet Sci Lett 389:167–175
Nielsen SG, Rehkaemper M, Prytulak J (2017) Investigation and application of thallium isotope fractionation. Rev Mineral Geochem 82:759–798

Oberli F, Gartenmann P, Meier M, Kutschera W, Suter M, Winkler G 1999. The half-life of ^{126}Sn refined by thermal ionization mass spectrometry measurements. Int J Mass Spectrom 184:145–152

Ohno T, Komiya T, Ueno Y, Hirata T, Maruyama S (2008) Determination of ^{88}Sr/^{86}Sr mass-dependent isotopic and radiogenic isotope variation of ^{87}Sr/^{86}Sr, in the Neoproterozoic Doushantuo Formation. Gondwana Res 14:126–133

Penniston-Dorland S, Liu X-M, Rudnick RL (2017) Lithium isotope geochemistry. Rev Mineral Geochem 82:165–217

Pilson MEQ (2013) An Introduction to the Chemistry of the Sea. Cambridge University Press, Cambridge 529 pp

Poitrasson F (2017) Silicon isotope geochemistry. Rev Mineral Geochem 82:289–344

Polyakov VB, Mineev SD, Clayton RN, Hu G, Mineev KS (2005) Determination of tin equilibrium isotope fractionation factors from synchrotron radiation experiments. Geochim Cosmochim Acta 69:5531–5536

Ponzevera E Quétel CR Berglund M Taylor PDP Evans P Loss RD Fortunato G (2006) Mass discrimination during MC-ICP-MS isotopic ratio measurements: Investigation by means of synthetic isotopic mixtures (IRMM-007 series) and application to the calibration of natural-like zinc materials (including IRMM-3702 and IRMM-651). J Am Soc Mass Spectrom 17 1412–1427

Pretet C, Zuilen K, Nägler TF, Reynaud S, Böttcher ME, Samankassou E (2015) Constraints on barium isotope fractionation during aragonite precipitation by corals. Depositional Rec 1:118–12

Prytulak J, Nielsen SG, Halliday AN (2011) Determination of precise and accurate ^{51}V/^{50}V isotope ratios by MC-ICP-MS, Part 2: Isotopic composition of six reference materials plus the Allende chondrite and verification tests. Geostand Geoanal Res 35:307–318

Prytulak J, Nielsen SG, Ionov DA, Halliday AN, Harvey J, Kelley KA, Niu YL, Peate DW, Shimizu K, Sims KWW (2013) The stable vanadium isotope composition of the mantle and mafic magmas. Earth Planet Sci Lett 365:177–189

Qin L, Wang X (2017) Chromium isotope geochemistry. Rev Mineral Geochem 82:379–414

Raddatz J, Liebetrau V, Ruggeberg A, Hathorne E, Krabbenhoft A, Eisenhauer A, Bohm F, Vollstaedt H, Fietzke J, Lopez Correa M, Freiwald A, Dullo WC (2013) Stable Sr-isotope, Sr/Ca, Mg/Ca, Li/Ca and Mg/Li ratios in the scleractinian cold-water coral Lophelia pertusa. Chem Geol 352:143–152

Ranen MC, Jacobsen SB (2006) Barium isotopes in chondritic meteorites: implications for planetary reservoir models. Science 314:809–81

Rehkaemper M, Wombacher F, Horner TJ, Xue Z (2011) Natural and anthropogenic cadmium isotope variations. *In:* Baskaran M (Ed.), Handbook of Environmental Isotope Geochemistry. Springer, Berlin, pp. 124–154

Resongles E, Freydier R, Casiot C, Viers J, Chmeleff J, Elbaz-Poulichet F (2015) Antimony isotopic composition in river waters affected by ancient mining activity. Talanta 144:851–861

Richter FM, Dauphas N, Teng F-Z (2009) Non-traditional fractionation of non-traditional isotopes: Evaporation, chemical diffusion and Soret diffusion. Chem Geol 258 92–103

Ripperger S, Rehkamper M, Porcelli D, Halliday AN (2007) Cadmium isotope fractionation in seawater–A signature of biological activity. Earth Planet Sci Lett 261:670–684

Roskosz M, Sio CK, Dauphas N, Bi W, Tissot FL, Hu MY, Zhao J and Alp EE (2015) Spinel–olivine–pyroxene equilibrium iron isotopic fractionation and applications to natural peridotites. Geochim Cosmochim Acta 169:184–199

Rouxel O, Ludden J, Fouquet Y (2003) Antimony isotope variations in natural systems and implications for their use as geochemical tracers. Chem Geol 200:25–40

Rouxel O, Luais B (2017) Germanium isotope geochemistry. Rev Mineral Geochem 82:601–665

Ruggeberg A, Fietzke J, Liebetrau V, Eisenhauer A, Dullo WC, Freiwald A (2008) Stable strontium isotopes ($\delta^{88/86}$Sr) in cold-water corals–A new proxy for reconstruction of intermediate ocean water temperatures. Earth Planet Sci Lett 269:569–574

Schmitt AD, Galer SJG, Abouchami W (2009a) High-precision cadmium stable isotope measurements by double spike thermal ionisation mass spectrometry. J Anal At Spectrom 24:1079–1088

Schmitt AD, Galer SJG, Abouchami W (2009b). Mass-dependent cadmium isotopic variations in nature with emphasis on the marine environment. Earth Planet Sci Lett 277:262–272

Shalev N, Segal I, Lazar B, Gavrieli I, Fietzke J, Eisenhauer A, Halicz L (2013) Precise determination of $\delta^{88/86}$Sr in natural samples by double-spike MC-ICP-MS and its TIMS verification. J Anal At Spectrom 28:940–944

Sharp Z (2007) Principles of Stable Isotope Geochemistry. Pearson Prentice Hall, Upper Saddle River, N

Shiel AE, Weis D, Orians KJ (2010) Evaluation of zinc, cadmium and lead isotope fractionation during smelting and refining. Sci Total Environ 408:2357–2368

Shiel AE, Weis D, Orians KJ (2012) Tracing cadmium, zinc and lead sources in bivalves from the coasts of western Canada and the USA using isotopes. Geochim Cosmochim Acta 76:175–190

Shiel AE, Weis D, Cossa D, Orians KJ (2013) Determining provenance of marine metal pollution in French bivalves using Cd, Zn and Pb isotopes. Geochim Cosmochim Acta 121:155–167

Steele RC, Elliott T, Coath CD, Regelous M (2011) Confirmation of mass-independent Ni isotopic variability in iron meteorites. Geochim Cosmochim Acta 75 7906–7925

Stevenson EI, Hermoso M, Rickaby REM, Tyler JJ, Minoletti F, Parkinson IJ, Mokadem F, Burton KW (2014) Controls on stable strontium isotope fractionation in coccolithophores with implications for the marine Sr cycle. Geochim Cosmochim Acta 128:225–235

Stevenson EI, Aciego SM, Chutcharavan P, Parkinson IJ, Burton KW, Blakowski MA, Arendt CA (2016) Insights into combined radiogenic and stable strontium isotopes as tracers for weathering processes in subglacial environments. Chem Geol 429:33–43

Stüeken EE (2017) Selenium isotopes as a biogeochemical proxy in deep time. Rev Mineral Geochem 82:657–682

Tanimizu M, Asada Y, Hirata T (2002) Absolute isotopic composition and atomic weight of commercial zinc using inductively coupled plasma mass spectrometry. Anal Chem 74 5814–5819

Taylor TI, Urey HC (1938) Fractionation of the Li and K isotopes by chemical exchange with zeolites. J Chem Phys 6:429–438

Teng F-Z (2017) Magnesium isotope geochemistry. Magnesium isotope geochemistry. Rev Mineral Geochem 82:219–285

Trinquier A, Elliott T, Ulfbeck D, Coath C, Krot AN, Bizzarro M (2009) Origin of nucleosynthetic isotope heteorogeneity in the solar protoplanetary disk. Science 324:374–376

Valley JW, Cole DRE (Eds.) (2001) Stable Isotope Geochemistry. Rev Mineral Geochem 43, Mineralogical Society of America and the Geochemical Society, Washington DC

Valley JW, Taylor HP, O'Neil JR (1986) Stable Isotopes in High-Temperature Geological Processes. Rev Mineral 16. The Mineralogical Society of America, Washington DC

van Zuilen K, Müller T, Nägler TF, Dietzel M, Küsters T (2016) Experimental determination of barium isotope fractionation during diffusion and adsorption processes at low temperatures. Geochim Cosmochim Acta 186:226–24

Vogl J, Pritzkow W (2010) Isotope reference materials for present and future isotope research. J Anal At Spectrom 25:923–932

Vollstaedt H, Eisenhauer A, Wallmann K, Bohm F, Fietzke J, Liebetrau V, Krabbenhoft A, Farkas J, Tomasovych A, Raddatz J, Veizer J (2014) The Phanerozoic $\delta^{88/86}$Sr record of seawater: New constraints on past changes in oceanic carbonate fluxes. Geochim Cosmochim Acta 128:249–265

Von Allmen K, Böttcher ME, Samankassou E, Nägler TF (2010) Barium isotope fractionation in the global barium cycle: First evidence from barium minerals and precipitation experiments. Chem Geol 277:70–7

Waight T, Baker J, Willigers B (2002) Rb isotope dilution analyses by MC-ICPMS using Zr to correct for mass fractionation: towards improved Rb-Sr geochronology? Chem Geol 186:99–116

Wakaki S, Tanaka T (2012) Stable isotope analysis of Nd by double spike thermal ionization mass spectrometry. Inter J Mass Spectrom 323:45–54

Wang K, Jacobsen SB (2016a) An estimate of the Bulk Silicate Earth potassium isotopic composition based on MC-ICPMS measurements of basalts. Geochim Cosmochim Acta 178:223–232

Wang K, Jacobsen SB (2016b) Potassium isotopic evidence for a high-energy giant impact origin of the Moon. Nature 538:487–490

Wei GJ, Ma JL, Liu Y, Xie LH, Lu WJ, Deng WF, Ren ZY, Zeng T, Yang YH (2013) Seasonal changes in the radiogenic and stable strontium isotopic composition of Xijiang River water: Implications for chemical weathering. Chem Geol 343:67–75

Wen HJ, Zhang YX, Cloquet C, Zhu CW, Fan HF, Luo CG (2015) Tracing sources of pollution in soils from the Jinding Pb-Zn mining district in China using cadmium and lead isotopes. Appl Geochem 52:147–154

Widanagamage IH, Schauble EA, Scher HD, Griffith EM (2014) Stable strontium isotope fractionation in synthetic barite. Geochim Cosmochim Acta 147:58–75

Widanagamage IH, Griffith EM, Singer DM, Scher HD, Buckley WP, Senko JM (2015) Controls on stable Sr-isotope fractionation in continental barite. Chem Geol 411:215–227

Wielandt D, Bizzarro M (2011) A TIMS-based method for the high precision measurements of the three-isotope potassium composition of small samples. J Anal At Spectrom 26:366–377

Wombacher F, Rehkaemper M, Mezger K, Munker C (2003) Stable isotope compositions of cadmium in geological materials and meteorites determined by multiple-collector ICPMS Geochim Cosmochim Acta 67:4639–4654

Wombacher F, Rehkaemper M (2004) Problems and suggestions concerning the notation of cadmium stable isotope compositions and the use of reference materials. Geostand Geoanal Res 28:173–178

Wombacher F, Rehkamper M, Mezger K, Bischoff A, Munker C (2008) Cadmium stable isotope cosmochemistry. Geochim Cosmochim Acta 72:646–667

Wu F, Qin T, Li X, Liu Y, Huang JH, Wu Z and Huang F (2015) First-principles investigation of vanadium isotope fractionation in solution and during adsorption. Earth Planet Sci Lett 426:216–224

Wu F, Qi Y, Yu H, Tian S, Hou Z and Huang F (2016) Vanadium isotope measurement by MC-ICP-MS Chem Geol 421:17–25

Xue ZC, Rehkamper M, Horner TJ, Abouchami W, Middag R, van de Flierdt T, de Baar HJW (2013) Cadmium isotope variations in the Southern Ocean. Earth Planet Sci Lett 382:161–172

Yamazaki E, Nakai S, Yokoyama T, Ishihara S, Tang HF (2013) Tin isotope analysis of cassiterites from Southeastern and Eastern Asia. Geochem J 47:21–35

Yamazaki E, Nakai S, Sahoo Y, Yokoyama T, Mifune H, Saito T, Chen J, Takagi N, Hokanishi N, Yasuda A (2014) Feasibility studies of Sn isotope composition for provenancing ancient bronzes. J Archaeolog Sci 52:458–467

Zhang J, Dauphas N, Davis AM, Leya I, Fedkin A (2012) The proto-Earth as a significant source of lunar material. Nat Geosci 5:251–255

Reviews in Mineralogy & Geochemistry
Vol. 82 pp. 27-63, 2017

Equilibrium Fractionation of Non-traditional Isotopes: a Molecular Modeling Perspective

Marc Blanchard, Etienne Balan

Institut de Minéralogie, de Physique des Matériaux, et de Cosmochimie (IMPMC)
Sorbonne Universités
UPMC Univ Paris 06
UMR CNRS 7590
Muséum National d'Histoire Naturelle
UMR IRD 206
4 place Jussieu
F-75005 Paris
France

marc.blanchard@impmc.upmc.fr; etienne.balan@impmc.upmc.fr

Edwin A. Schauble

Department of Earth
Planetary and Space Sciences
University of California, Los Angeles
Los Angeles
California 90095-1567
U.S.A.

schauble@g.ucla.edu

INTRODUCTION

The isotopic compositions of natural materials are determined by their parent reservoirs, on the one hand, and by fractionation mechanisms, on the other hand. Under the right conditions, fractionation represents isotope partitioning at thermodynamic equilibrium. In this case, the isotopic equilibrium constant depends on temperature, and reflects the slight change of free energy between two phases when they contain different isotopes of the same chemical element. The practical foundation of the theory of mass-dependent stable isotope fractionation dates back to the mid-twentieth century, when Bigeleisen and Mayer (1947) and Urey (1947) proposed a formalism that takes advantage of the Teller–Redlich product rule (Redlich 1935) to simplify the estimation of equilibrium isotope fractionations. In this chapter, we first give a brief introduction to this isotope fractionation theory. We see in particular how the various expressions of the fractionation factors are derived from the thermodynamic properties of harmonically vibrating molecules, a surprisingly effective mathematical approximation to real molecular behavior. The central input data of these expressions are vibrational frequencies, but an approximate formula that requires only force constants acting on the element of interest can be applied to many non-traditional isotopic systems, especially at elevated temperatures. This force-constant based approach can be particularly convenient to use in concert with first-principles electronic structure models of vibrating crystal structures and aqueous solutions. Collectively, these expressions allow us to discuss the crystal chemical parameters governing the equilibrium stable isotope fractionation.

1529-6466/17/0082-0002$05.00 http://dx.doi.org/10.2138/rmg.2017.82.2

Since the previous volume of *Reviews in Mineralogy and Geochemistry* dedicated to non-traditional stable isotopes, the number of first-principles molecular modeling studies applied to geosciences in general and to isotopic fractionation in particular, has significantly increased. After a concise introduction to computational methods based on quantum mechanics, we will focus on the modeling of isotopic properties in liquids, which represents a bigger methodological challenge than small molecules in gas phase, or even minerals. Our ability to produce reliable theoretical mineral–solution isotopic fractionation factors is essential for many geosciences problems. The main modeling approaches used in recent studies of fractionation in liquids are molecular cluster models and molecular dynamics with periodic boundary conditions. Their relative advantages and drawbacks will be discussed. So far, the vast majority of theoretical studies applied to isotopic fractionation have been based on the harmonic approximation; in most cases anharmonic effects will be smaller than uncertainties associated with other imperfections in the models (especially in calculated vibrational frequencies), but in some cases (e.g., liquid phases with light elements) it will be important to be able to go beyond the harmonic approximation. More sophisticated methods, such as thermodynamic integration coupled to path integral molecular dynamics, can account for anharmonic effects as well as quantum nuclear effects. We will introduce the basic concepts of this technique and will give some examples of their application.

Among the non-traditional isotopes, the iron isotope system has probably developed the richest and most methodologically varied theoretical literature. This is partly due to the fact that isotope fractionation factors of Mössbauer-active elements (including iron, via ^{57}Fe) can be independently determined using Mössbauer spectroscopy and nuclear resonant inelastic X-ray scattering, which are closely related techniques that probe the vibrational properties of the target element. Expressions used to derive fractionation factors from these spectroscopic techniques are introduced, the accuracy of each method will be discussed, and the results are compared with first-principles calculations.

The discovery of mass-independent isotope fractionation of non-traditional stable isotope systems including Hg, Tl, and U over the past decade has expanded the scope of "stable" isotope geochemistry to include a long-lived radioactive element and almost the whole range of naturally occurring atomic numbers. It has also created a need for theoretical studies of new fractionation mechanisms. Nuclear field shift effects, first proposed to explain laboratory isotope enrichment experiments, including uranium, are now thought to play an important role in driving natural fractionation in uranium and thallium, and a secondary role in mercury isotope geochemistry. Large photochemically induced mass-independent fractionation effects in the mercury isotope system are yet to be explained beyond a qualitative level, and remain an important challenge for isotope fractionation theory. In light isotope systems (particularly ^{16}O–^{17}O–^{18}O) it is now possible to measure variability of mass dependence for different types of fractionation, ranging from equilibrium partitioning to kinetic fractionation. The potential for using variations in mass dependence to identify the types of fractionation affecting non-traditional elements is also a topic of emerging interest for theoretical studies.

THEORETICAL FRAMEWORK

Equilibrium fractionation theory

This section is largely inspired by the articles of Bigeleisen and Mayer (1947), and Ishida (2002).

Let's consider an isotopic exchange reaction between two molecules *A* and *B*, involving a single atomic position:

$$AX' + BX \rightleftarrows AX + BX' \qquad (1)$$

The prime symbol refers to the light isotope of the element *X*. As with any chemical reaction, the equilibrium constant, K_{eq}, can be determined from the free energies of the reactants and products. Isotopic exchange reactions do not, in general, involve significant pressure-volume work because the number of molecules on both sides of the reaction is the same, and because isotope substitution has a negligible effect on the molar volumes of the phases under normal conditions. The above assumption is not true for a complete substitution of hydrogen by deuterium, for instance, or for certain solid–gas equilibria (e.g., Jancso et al. 1993; Horita et al. 2002). Under these general conditions, the standard Gibbs free energy of the exchange reaction can be related to the difference in the Helmholtz free energy of the pure isotopomers (*AX'*, *AX*, *BX'* and *BX*):

$$\Delta F = F(AX) + F(BX') - F(AX') - F(BX)$$

The Helmholtz free energy is related to the molecular partition function, *Q by*:

$$F = -\mathrm{N_a} \mathrm{k} T \ln\left(Q / \mathrm{N_a}\right)$$

where k is Boltzmann's constant, $\mathrm{N_a}$ the Avogadro number, and *T* is the absolute temperature. It is thus possible to express the equilibrium constant of the exchange reaction as:

$$K_{\mathrm{eq}} = \frac{Q(AX) \times Q(BX')}{Q(AX') \times Q(BX)} \tag{2}$$

The molecular partition function is given by the following expression:

$$Q = \sum_n \exp(-E_n / \mathrm{k}T)$$

where the sum spans all the quantum states of the molecule, referred to by their index *n* and their corresponding energies E_n.

A classical partition function, Q_{cl} can be obtained by integration over continuous momenta and position variables that relate to the kinetic energy and potential energy of the molecule, respectively. Its expression is also a function of the symmetry number of the molecule (i.e. the number of equivalent ways to orient a molecule in space); see Equation (5) of Bigeleisen and Mayer (1947). Importantly, atomic masses are only involved in the definition of the kinetic energy term; whereas the configurational integral, obtained from the potential energy of the system, is assumed to be mass-independent. Considering the ratio of the partition functions of the two isotopically substituted molecules, the configurational integrals and the contribution arising for atoms other than the exchanged isotopes cancel out, leading to the following expression:

$$\left(\frac{Q}{Q'}\right)_{\mathrm{cl}} = \frac{s'}{s}\left(\frac{m}{m'}\right)^{3/2}$$

where *s* is the symmetry number and *m*, the atomic mass of isotopes. By inserting this expression into Equation (2), the atomic masses cancel out and one obtains the classical value of the equilibrium constant:

$$(K_{\mathrm{eq}})_{\mathrm{cl}} = \frac{s_{AX'} \; s_{BX}}{s_{AX} \; s_{BX'}} \tag{3}$$

This ratio of symmetry numbers will not lead to an isotopic fractionation as it merely represents the relative probabilities of forming symmetric and antisymmetric molecules. This corresponds to a perfectly random distribution of isotopes; a situation found at $T = \infty$. This demonstrates that isotopic fractionation is a purely quantum effect that cannot be explained by classical statistical mechanics.

The harmonic quantum molecular partition function of a molecule in gas phase, can be written as a product of translational, rotational, vibrational and electronic partition functions:

$$Q_{\mathrm{qm}} = (Q_T)_{\mathrm{qm}} \times (Q_R)_{\mathrm{qm}} \times (Q_V)_{\mathrm{qm}} \times (Q_E)_{\mathrm{qm}}$$

The electronic structure of a molecule is usually assumed to be isotope-independent, so the electronic term is neglected. In some cases, isotope effects will have measurable electronic contributions: isotope mass shift, nuclear field shift effect and nuclear spin effect (e.g. Bigeleisen 1996). Fractionations caused by the nuclear field shift and spin effects will be discussed in a later section. The mass shift arises from the coupling of the motion of the nuclei and the electrons. Mass shift can be important in reaction involving hydrides, i.e. up to a small percentage of H–D fractionation factors (e.g. Kleinman and Wolfsberg 1973), but becomes quickly negligible for heavier atoms since it scales with $\delta M/M^2$. Partition functions for translational and rotational motions are formally quantum mechanical, and sensitive to isotope substitution, but in practice the quanta for both types of motion are so small and closely spaced for most molecules that they do not deviate significantly from their classical equivalents at temperatures relevant to geochemistry (see Schauble 2004, for additional details). This will hold in all cases except hydrogen, where a more sophisticated treatment of rotation may be needed at low temperatures. The classical expressions are not given here since most of their parameters will cancel out in the next step, but they can be found in literature (e.g., Richet et al. 1977; Schauble 2004; Liu et al. 2010). As a result, it is generally the case that only vibrational motions need a quantum-mechanical treatment, and it is vibrational energy that plays the central role in controlling the distribution of isotopes between two phases in thermodynamic equilibrium. The harmonic vibrational partition function is defined by:

$$(Q_V)_{\mathrm{qm}} = \prod_{i=1}^{3N-6} \frac{\exp(-\mathrm{h}\nu_i / 2\mathrm{k}T)}{1-\exp(-\mathrm{h}\nu_i / \mathrm{k}T)}$$

where h is Planck's constant and ν_i is the frequency of the vibrational mode *i*. A molecule with N atoms will have $3N-6$ vibrational degrees of freedom (in addition to 3 rotational and 3 translational degrees of freedom) while a linear molecule will have $3N-5$ vibrational degrees of freedom.

As classical contributions to the partition functions only play a bookkeeping role in the isotopic fractionation, it is useful to define a reduced partition function by ratioing the quantum partition function to its classical counterpart ($Q_{\mathrm{qm}}/Q_{\mathrm{cl}}$). The ratio of reduced partition function, commonly referred to as β, can be written as:

$$\beta_{AX} = \frac{Q_{\mathrm{qm}}(AX)/Q_{\mathrm{cl}}(AX)}{Q_{\mathrm{qm}}(AX')/Q_{\mathrm{cl}}(AX')} \tag{4}$$

The equilibrium isotope fractionation factor of the exchange reaction (1), i.e. $\alpha_{AX\text{-}BX}$, can thus be expressed as a function of the reduced partition function ratios and is related to the equilibrium constant through the following relation (using Eqns. (2) and (3)):

$$\begin{aligned} a_{AX\text{-}BX} &= \frac{\beta_{AX}}{\beta_{BX}} = \frac{Q_{\mathrm{qm}}(AX)\times Q_{\mathrm{qm}}(BX')}{Q_{\mathrm{qm}}(AX')\times Q_{\mathrm{qm}}(BX)} \times \frac{Q_{\mathrm{cl}}(AX')\times Q_{\mathrm{cl}}(BX)}{Q_{\mathrm{cl}}(AX)\times Q_{\mathrm{cl}}(BX')} \\ &= \frac{(K_{\mathrm{eq}})_{\mathrm{qm}}}{(K_{\mathrm{eq}})_{\mathrm{cl}}} \\ &= \frac{s_{\mathrm{AX'}}}{s_{\mathrm{AX}}} \times \frac{s_{\mathrm{BX}}}{s_{\mathrm{BX'}}} \times (K_{\mathrm{eq}})_{\mathrm{qm}} \end{aligned}$$

The above relation shows that, when $(K_{eq})_{qm}$ is equal to $(K_{eq})_{cl}$, there will be no isotopic fractionation (i.e. $\alpha_{AX\text{-}BX} = 1$). This situation occurs, for instance, at very high temperatures when isotopes are randomly distributed. Conversions between fractionation factors and equilibrium constant, written here for the simple isotopic exchange reaction (1), can be more complicated, depending on molecular stoichiometry (Schauble 2004; Liu et al. 2010). By analogy with the isotopic fractionation factor α, we can also see the reduced partition function ratio β_{AX} as the isotopic fractionation factor between the substance *AX* and an ideal atomic gas of *X*. This formulation is a convenient way to tabulate the theoretical fractionations with a simple point of comparison. Fractionations are typically very small, on the order of parts per thousand for non-traditional stable isotopes, so it is common to use the notation 1000 ln α or 1000 ln β expressing the result in permil (‰). The β-factor being the central quantity of theoretical studies, we report below the expressions that apply to situations commonly encountered (i.e. molecules or condensed phases, complete or site by site isotopic substitution).

By inserting the expressions of the quantum and classical partition functions into Equation (4), we obtain the following expression for a molecule in gas phase having only one exchangeable atom:

$$\beta_{AX} = \left(\frac{m'}{m}\right)^{3/2} \times \left(\frac{M}{M'}\right)^{3/2} \times \left(\frac{I_x I_y I_z}{I'_x I'_y I'_z}\right)^{1/2} \times \left(\prod_{i=1}^{3N-6} \frac{\exp(-h\nu_i/2kT)}{1-\exp(-h\nu_i/kT)} \times \frac{1\text{-}\exp(-h\nu'_i/kT)}{\exp(-h\nu'_i/2kT)}\right) \tag{5}$$

where *M* is the mass of molecule *AX*, *M′* is the mass of molecule *AX′*, and I_x, I'_x, etc. are the moments of inertia along each cartesian axis.

Alternatively, for a molecule having *n* exchangeable atoms, the β_{AX} can be determined from the β_i related to a specific atomic site *i*:

$$\beta_{AX} = \frac{1}{n}\sum_{i=1}^{n}\beta_i = \frac{1}{n}\sum_{i=1}^{n}\frac{Q_{qm}(AX'_{n-1}X_i)}{Q_{qm}(AX'_n)} \times \left(\frac{m'}{m}\right)^{3/2} \tag{6}$$

In this "site by site" approach, $Q_{qm}(AX'_{n-1}X_i)$ corresponds to the partition function of the molecule having the atom X′ on the site *i* substituted with X while $Q_{qm}(AX'_n)$ represents the partition function of the molecule with no substituted atoms.

If we further use the Teller–Redlich product rule (e.g., Redlich 1935; Wilson et al. 1955):

$$\left(\frac{I_x I_y I_z}{I'_x I'_y I'_z}\right)^{1/2} \times \left(\frac{M}{M'}\right)^{3/2} \times \left(\frac{m'}{m}\right)^{3N/2} \times \prod_{i=1}^{3N-6}\frac{\nu'_i}{\nu_i} = 1$$

then Equation (5) transforms into a more general expression applicable to any molecule in gas phase undergoing a complete isotopic substitution (i.e. all X′ atoms are substituted with X):

$$\beta_{AX} = \left[\prod_{i=1}^{3N-6}\frac{\nu_i}{\nu'_i} \times \frac{\exp(-h\nu_i/2kT)}{1-\exp(-h\nu_i/kT)} \times \frac{1-\exp(-h\nu'_i/kT)}{\exp(-h\nu'_i/2kT)}\right]^{1/n} \tag{7}$$

This more convenient form involving only the vibrational frequencies before and after full isotope substitution, assumes that: (i) the free energy change associated to the isotopic substitution of an atom *X* does not depend on the isotopic nature of the surrounding *X* atoms, and (ii) the β-factor of each atomic site weakly depends on the site. These assumptions are generally valid except in some specific cases, like for instance, when deuterium is substituted for hydrogen in a water molecule.

Crystalline materials differ from gaseous molecules by their spatial extension involving the presence of long-range interactions. This implies that their vibrational spectra do not exhibit a finite number of vibrational frequencies but rather correspond to a continuum. In crystals, a vibrational mode is defined by a frequency of vibration, the atomic displacement pattern in a given cell and a wave-vector q that describes the phase relation of the atomic displacements in the other cells of the crystal. The wave-vector is defined in the reciprocal space and belongs to the first Brillouin zone. The vibrational frequency thus depends on the wave-vector. It is possible to build dispersion curves by reporting the frequency along specific directions in the reciprocal space, and vibrational density of states by integration over the whole Brillouin zone. A more detailed description of the crystal vibrational properties applied to isotope fractionation can be found in Young et al. (2015). This feature of crystals can be taken into account by modifying the partition function. The energy differences associated with translation motions cancel at equilibrium and the rotational term disappears in crystals. The partition function, in the harmonic approximation, is thus defined by:

$$Q_{\text{qm}} = \left(\prod_{i=1}^{3N} \prod_{\{q\}} \frac{\exp\left(-h\nu_{q,i} / 2kT\right)}{1-\exp\left(-h\nu_{q,i} / kT\right)} \right)^{1/N_q} \tag{8}$$

where $\nu_{q,i}$ is now the frequency of the vibrational mode i, along the wave-vector q. N corresponds to the number of atom in the crystal unit cell. The second product is performed on a uniform grid of N_q q-vectors in the Brillouin zone. In practice, the number of frequencies used is still finite but beyond a sufficiently large number of q-vectors, results are properly converged.

By combining Equations (4) and (8), we obtain the general expression of the reduced partition function ratio (i.e. β-factor) for a crystal undergoing a complete isotopic substitution (i.e. all n atoms X' of the unit-cell are substituted with X):

$$\beta_{AX} = \left(\frac{m'}{m}\right)^{3/2} \times \left[\prod_{i=1}^{3N} \prod_{\{q\}} \frac{\exp\left(-h\nu_{q,i} / 2kT\right)}{1-\exp\left(-h\nu_{q,i} / kT\right)} \times \frac{1-\exp\left(-h\nu'_{q,i} / kT\right)}{\exp\left(-h\nu'_{q,i} / 2kT\right)} \right]^{1/nN_q} \tag{9}$$

Alternatively, Equation (6) for a "site by site" isotopic substitution is still valid for crystals. The rotational and translational terms also disappear from the Teller–Redlich product rule, yielding the high-temperature product rule of Kieffer (1982):

$$\left(\frac{m'}{m}\right)^{3/2} \times \left(\prod_{i=1}^{3N} \prod_{\{q\}} \frac{\nu'_{q,i}}{\nu_{q,i}} \right)^{1/nN_q} = 1$$

This high-temperature product rule imposes the isotope fractionation to be nil at very high-temperatures. If we take advantage of this rule, Equation (9) then becomes:

$$\beta_{AX} = \left[\prod_{i=1}^{3N} \prod_{\{q\}} \frac{\nu_{q,i}}{\nu'_{q,i}} \times \frac{\exp\left(-h\nu_{q,i} / 2kT\right)}{1-\exp\left(-h\nu_{q,i} / kT\right)} \times \frac{1-\exp\left(-h\nu'_{q,i} / kT\right)}{\exp\left(-h\nu'_{q,i} / 2kT\right)} \right]^{1/nN_q} \tag{10}$$

Liquid phases exhibit a higher degree of complexity, in particular because of the absence of long-range translational order and because of their dynamical behavior. More approximations are needed. The isotopic properties of liquid phases can be determined by adopting either the same approach as for crystals (i.e. building a periodic model), or the approach developed for gaseous molecules (i.e. building an isolated molecular cluster of variable size). The latter method can be justified for dissolved molecules that remain more or less intact in solution or for aqueous complexes where intra-complex bonds are probably much stronger than interactions with bulk solvent. The additional complexities that arise in dealing with liquid phases will be discussed in a later section.

Approximate formula based on force constants

The above equations relate conveniently the isotopic fractionation factor to the vibrational frequencies but Bigeleisen and Mayer (1947) also derived a series of approximate formulae that are useful when all vibrational frequencies are not available and also for improving our understanding of the parameters that control equilibrium isotopic fractionation between two phases. Thus, if the frequency shift associated with the isotopic substitution is small and if the reduced energy is small (i.e. $h\nu/kT \lesssim 2$) then Equations (7) or (10) become:

$$\beta_{AX} = 1 + \sum_i \frac{h^2 \Delta\nu_i^2}{24(kT)^2}$$

Treating the vibrations as harmonic, squared vibrational frequencies can be related to the force constants and masses. By doing so, the reduced partition function ratio can be expressed as a function of the force constants acting on the element of interest:

$$\beta_{AX} = 1 + \left(\frac{m - m'}{mm'}\right)\left(\frac{h}{2\pi}\right)^2 \frac{F}{24(kT)^2} \tag{11}$$

F is the sum of force constants in three orthogonal directions opposing displacement of the atom X from its equilibrium position. If atoms X are located in more than one crystallographic site, the force constant for all sites must be averaged. The full derivation of the approximate formula (11) from Equations (7) or (10) can be found for instance in Young et al. (2015). This expression is a valid approximation for Equation (10): (i) at relatively high temperature ($h\nu/kT < 2$ implies, for instance, a temperature higher than 360 K if the vibrations involving atom X in phase A have wavenumbers, $\omega=\nu/c$, smaller than 500 cm^{-1} (1 cm^{-1} is equivalent to 30.0 GHz), or a temperature higher than 720 K if the relevant vibrations extend up to 1000 cm^{-1}), (ii) when the difference in mass between the two isotopes is sufficiently small relative to the average atomic mass (this assumption excludes the very light elements such as hydrogen), (iii) assuming isotope-independent force constants; it is also worth noting that Equation (11) still assumes a harmonic vibrational partition function.

The expression (11) clearly shows that, under the conditions of validity just mentioned, equilibrium isotopic fractionation varies proportionally to the reciprocal of the square of the temperature. When the reduced energy $h\nu/kT$ becomes much higher than 2 (i.e. at low temperature or for high-frequency vibrations), it can be shown that the temperature dependence of fractionation factor weakens and tends to a $1/T$ behavior. This leads to a concave-down curvature of β-factors when plotted against $1/T^2$ and this curvature is more pronounced as one moves away from the conditions of validity of Equation (11). The mass-dependence of the same equation indicates that isotopic fractionations become smaller for heavy elements. Equation (11) also predicts that, at equilibrium, the heavy isotopes of an element will concentrate in the phase where the force constants are the greatest, i.e. in the phase where the element of interest involves the stiffest bonds (e.g., element in higher oxidation state, with lower coordination number). The two first points, i.e. temperature dependence and mass dependence, are well illustrated by the iron and oxygen β-factors of the goethite (Fig. 1). In this iron oxyhydroxide, vibrations involving iron atoms correspond to wavenumbers smaller than 600 cm^{-1}, which dictates that Equation (11) is valid above ~ 400 K. Even at lower temperatures (i.e. in the stability field of goethite), Figure 1 shows that the departure of this approximate formula from Equation (10) is smaller than 0.5‰. Because oxygen is much lighter that iron, the equilibrium isotope fractionation factors are much larger (i.e. oxygen β-factor is ~6 times larger than iron β-factor). In goethite, half oxygen atoms are hydroxylated. The bending and stretching vibrational modes of these OH groups (observed at ~ 800 cm^{-1} and above 3000 cm^{-1}, respectively) contribute

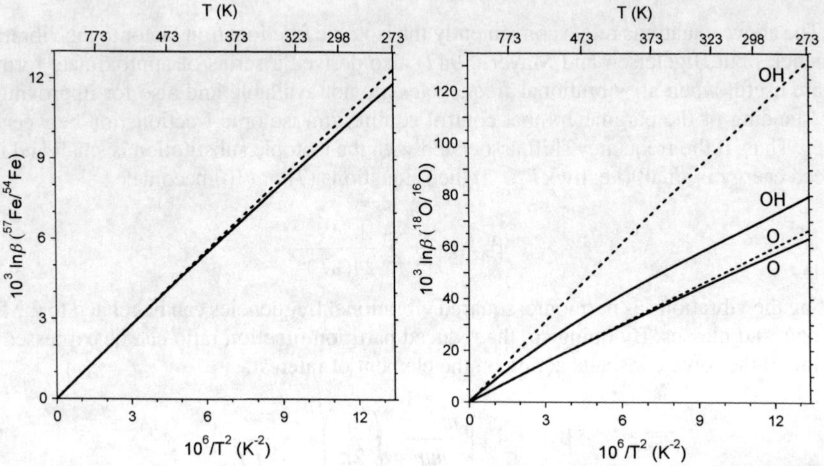

Figure 1. Temperature dependence of the iron (left) and oxygen (right) β-factors of goethite (α-FeOOH). Results obtained using the approximate formula (11) based on force constants (dashed lines, unpublished data) are compared with the results given by Equation (10) using all vibrational frequencies (solid lines, Blanchard et al. 2015).

to the oxygen β-factor. This explains why, in the case of the hydroxylated oxygen atoms, Equation (11) is not a valid approximation and meets the "correct" β-factor curve only at very high temperature whereas the approximation still holds for the other oxygen atoms.

In first-principles calculations based on quantum-mechanics, the determination of β-factors using Equations (7) or (10) requires computing all vibrational frequencies, which is computationally expensive, whereas only electronic energy calculations performed for a limited number of positions of the atom of interest in the vicinity of its equilibrium position are needed to apply the approximate formula (11). When the conditions of validity stated above are met, the use of Equation (11) allows consideration of more phases, and phases with greater structural complexity. This includes liquids and crystal defects, which may require large model systems (i.e. more than a hundred of atoms). This approach has been applied for discussing the Cr isotope fractionation in conditions relevant to the differentiation of the Earth's core (Moynier et al. 2011), the Li isotope fractionation between minerals and aqueous solutions at high pressure and temperature (Kowalski and Jahn 2011), and the S isotope fractionation of sulfate groups incorporated in major calcium carbonates (Balan et al. 2014). Without performing first-principles calculations, a qualitative estimation of the fractionation factors between phases where the element of interest involves contrasted bonding schemes, can be obtained by determining the force constants from an ionic model. In this model based on Pauling's rules, the force constant is assessed mainly from the valences and ionic radii of the central element and its first neighbors. A description and application of the method for non-traditional isotopes is presented in Young et al. (2015). This ionic model derives from earlier studies that demonstrated and used the correlation existing between bond types and oxygen isotope fractionation in silicate minerals (e.g., Taylor and Epstein 1962; Garlick 1966; Schütze 1980; Richter and Hoernes 1988; Smyth and Clayton 1988). Even if this type of ionic model does not show a great accuracy, it highlights the basic crystal chemical parameters that govern the equilibrium stable isotope fractionation, by affecting the stiffness of interatomic bonds.

MODELING APPROACHES

Quantum-mechanical molecular modeling

As shown in the previous section, the determination of the equilibrium isotope fractionation factors can be related to the change of vibrational frequencies associated to the isotopic substitution. These vibrational frequencies can be calculated from empirical force fields built using experimental measurements like structural parameters, elastic properties or known vibrational frequencies. A review of these methods can be found in Schauble (2004). The alternative approach that spread out during the last decade thanks to the advances in processor speed and memory size, consists in using quantum-mechanical molecular modeling. We will give here only a short introduction aiming at helping experimental geochemists to approach these theoretical tools but many general or specialized publications are available elsewhere, like for instance in previous *Reviews in Mineralogy and Geochemistry* volumes (Cygan and Kubicki 2001; Perdew and Ruzsinszky 2010).

The properties of any material can in principle be obtained from the laws of Quantum Mechanics by solving the equations describing the interactions between nuclei and electrons. In practice, a number of approximations are needed to address this complex problem. The first is the Born–Oppenheimer approximation, which considers that the rapid motion of electrons is decoupled from the slower motion of nuclei. Electronic wavefunctions are obtained by solving the Schrödinger equation for a system to which the positions of the nuclei are fixed external parameters. The energy of the system is then a function of the nuclei positions and the nuclei dynamics can be described by considering their motions on a potential energy surface. In comparison, empirical force fields use analytic functions to approximately describe the potential energy surface in terms of interatomic distance, oxidation state, effective ionic charge, and similar parameters that can be fitted by detailed examination of either experimental data or theoretical calculations. This parameterization of the potential energy surface may be assumed to be transferable, meaning that interatomic interactions in a group of related structures can be calculated using parameters fit to data from only one, or a subset of them. The reliability of empirical force fields will then be highly dependent on the quality of data available for parameter fitting, on the correct choice of structural variables to fit, the correct choice of suitable analytic functional forms, and on the consistency of electronic structure in the group of substances to which the force field is applied. Because of this chain of assumptions, and the use of empirical data, it can be difficult to assess the suitability of an empirical force field for calculating isotope fractionation factors. In addition, spectroscopic-quality force fields are not always available for substances of interest, especially for compounds and molecules containing heavy elements, unusual structures, or less common oxidation states. For these reasons, the information obtained using first-principles calculations is often more straightforward to generate, and easier to test against known vibrational and structural properties, than the outputs of analytic potentials. Against this caution, however, it should be noted that typical force-field parameterizations are much more mathematically efficient than electronic structure calculations, making it possible to probe systems with large numbers of atoms and/or dynamical disorder (such as liquids or trace-element substituted crystals) with relatively modest computational effort.

In the quantum mechanical treatment, the Schrödinger equation of a multiple-electron system is most often solved using one of two different schemes. The Hartree–Fock (HF) method (Roothaan 1951) aims at determining the best multi-electronic wavefunction by combining mono-electronic wavefunctions (the so-called orbitals). The multi-electronic wavefunction exactly obeys the Pauli exclusion principle, whereas Coulombic interactions between different electrons are treated in a mean field approximation. It can be shown that the exact system energy is always lower than the Hartree–Fock energy, the difference being often referred to as the correlation energy. Instead of focusing on wavefunctions, density functional theory (DFT)

(Hohenberg and Kohn 1964) is based a theorem requiring that all the ground state properties of a system of electrons moving under the influence of an external potential are uniquely determined by its electron density. Therefore, the ground state energy is a functional of the electronic density. Hohenberg and Kohn (1964) also demonstrated that the ground state energy can be obtained variationally because only the exact ground-state density minimizes this functional. Kohn and Sham (1965) proposed a practical scheme to build this functional by showing that a system of N interacting electrons can be treated as a fictitious system of N electrons that do not interact with each other but operate in an effective external potential taking into account an exchange-correlation term. The corresponding mono-electronic equations, called Kohn–Sham equations, can be solved via an iterative and self-consistent procedure starting from an arbitrary electron density. This procedure should lead to the density that minimizes the energy (i.e. the exact ground state electronic density). Unfortunately, the exact expression of the exchange-correlation potential is unknown and approximate expressions have to be used. Two commonly used approximations are the local density approximation (LDA, based for example on a homogeneous electron gas) and the generalized gradient approximation (GGA) taking partial account of non-homogeneous effects. DFT is popular in part because it provides a description of the electronic ground state of many systems that is more accurate than standard HF methods, at a similar computational cost (for molecules). In this way, DFT makes it possible to efficiently model the static or dynamic properties of relatively complex systems, such as periodic systems containing up to few hundred atoms per unit cell. Unlike HF-based methods, however, there is not (or at least not yet) a well-defined hierarchy of post-GGA theories that can be used to systematically improve the accuracy of DFT models. Partial corrections for some known shortcomings in standard DFT functionals are well established and effective, such as the "DFT+U" technique for improving DFT models of transition-element oxides (Anisimov et al. 1991; Cococcioni and de Gironcoli 2005) or the various methods for including the dispersion interactions into DFT (e.g. Grimme 2011), however.

In practice, electronic wavefunctions are represented using a finite set of fixed functions. These functions can be localized on the atomic positions (as are the atomic orbitals) or can consist of plane waves (which correspond to solutions of the Schrödinger equation for a free particle). Although not a stringent rule, localized basis sets are well suited for isolated molecules or clusters of molecules; whereas plane-waves are more appropriate to treat extended and periodic systems, such as crystalline solids. Localized basis sets make it possible to use hybrid DFT–HF methods such as the Becke three parameter Lee–Yang–Parr (B3LYP) method (Lee et al. 1988; Becke 1993), designed to emphasize the best features of each both theories. Hybrid methods are most commonly used to model the structure and vibrational frequencies of molecules. In order to reduce the computation cost without losing accuracy, it is also possible to restrict the explicit electronic structure calculations to the valence electrons because chemical properties mostly involve changes in the distribution of valence electrons. In this simplified treatment, the potential created by the atomic nucleus and core electrons is replaced by a pseudopotential. Pseudopotentials are most commonly used in conjunction with plane-wave basis sets for elements with $Z > 2$, or in localized basis function calculations involving elements with $Z \geq 20$. Many different types of pseudopotentials have been developed, and high-quality public libraries of basis sets and pseudopotentials for almost all naturally occurring elements are now available online (e.g., GBRV, http://www.physics.rutgers.edu/gbrv/, Garrity et al. 2014; SSSP, http://materialscloud.org/sssp/; EMSL Basis Set Exchange, bse.pnl.gov/bse/portal, Schuchardt et al. 2007). These theoretical methods are implemented in numerous commercial and open-source software packages such as ABINIT (Gonze et al. 2002), CASTEP (Clark et al. 2005), CRYSTAL (Dovesi et al. 2014), GAMESS (Schmidt et al. 1993), Gaussian (Frisch et al. 2009), NWChem (Valiev et al. 2010), Quantum ESPRESSO (Giannozzi et al. 2009), or VASP (Kresse and Furthmüller 1996).

There are typically three steps in first-principles calculations for obtaining the vibrational frequencies needed for the determination of isotope fractionation factors. In the first step, the minimum-energy static structure is determined via geometric relaxation. From an initial guess geometry, often the experimentally determined structure, the forces on each atom and the stress over the cell are calculated, and a refined guess structure is determined. This procedure continues iteratively until the residual forces and stress are sufficiently small. Once the minimum-energy configuration has been calculated, the second step is the determination of force constants for displacements of the atomic nuclei from their equilibrium positions. Finally, vibrational frequencies are determined by a calculation with model force constants and appropriate isotopic masses (Baroni et al. 2001). Isotope substitution is expected to have a negligible effect on electronic structure, so a matrix of force constants for the common isotope in a molecule or a crystal can be recycled to estimate vibrational frequencies of uncommon isotope-substituted species. This means that frequencies corresponding to isotopically substituted species can be calculated very rapidly (i.e., in a few seconds on a personal computer), even for very complex substances, once the force constant matrix has been determined.

In first-principles methods, uncertainty in calculated frequencies is typically the main factor limiting the accuracy of calculated fractionation factors. As mentioned above, isotope effects on vibrational frequencies can be calculated self-consistently, using a single set of force constants for each system. The errors on the vibrational frequencies are expected to be highly systematic and largely cancel when calculating isotope frequency shifts. Méheut et al. (2009) showed that a systematic correction of $n\%$ on the frequencies induces a relative systematic correction on the logarithmic β-factors (ln β) varying between $n\%$ (at low temperatures) to $2n\%$ (at high temperatures). The commonly used generalized gradient approximation is for instance associated with a systematic underestimation of ~5% of the harmonic vibrational frequencies. This would lead to a relative uncertainty of ~0.5‰ on a β-factor of 10‰. Two approaches are sometimes adopted for correcting this systematic frequency error. In some studies, calculated frequencies are re-scaled to experimental ones in order to improve the accuracy of the calculated fractionation factors (e.g., Schauble et al. 2006; Black et al. 2007; Blanchard et al. 2009; Li et al. 2009; Méheut et al. 2009). We must however keep in mind that calculated frequencies are harmonic, as they should be when using equations based on the harmonic approximation (e.g., Eqns. (5) to (10)), while experimental frequencies are influenced by anharmonicity (Liu et al. 2010). In addition, the value of the scaling parameters may be associated with significant uncertainty, depending on the quality and precision of spectroscopic data available for the compound of interest. Some studies that focus on crystals choose to correct the theoretical results by fixing the unit cell parameters to the experimental values and by optimizing only the atomic positions (e.g., Kowalski and Jahn 2011; Blanchard et al. 2015; Pinilla et al. 2015), but this procedure will usually not completely correct systematic errors in the electronic structure method, and it will of course not be applicable in materials where unit cell parameters are not known a priori.

Theoretical studies of non-traditional stable isotope fractionation

A big advantage to quantum-mechanical molecular modeling is the ability to derive a wide range of electronic, structural, energetic, vibrational properties from the same model. These properties can often be directly compared with observations to test the accuracy of the model. First-principles calculations also represent efficient tools to tackle crystal chemical parameters and mechanisms controlling isotopic fractionations. Over the past decade or so DFT studies have been applied to theoretical studies of stable isotope fractionation spanning most of the non-traditional stable isotopes systems represented in this volume. The results of these theoretical studies might best be discussed within the perspective of each system, considering isotopic measurements on natural and synthetic samples as well, and this will be done in the following chapters. Here we present a brief annotated bibliography in order of increasing atomic mass, highlighting some of these works:

- ***Lithium.*** Theoretical studies focused on the equilibrium fractionation of lithium isotopes in aqueous solution (Yamaji et al. 2001) and between aqueous fluids and various Li-bearing minerals such as staurolite, spodumene and mica (Jahn and Wunder 2009; Kowalski and Jahn 2011). Isotopic results were discussed in light of the speciation change of the aqueous lithium at high temperature and pressure.
- ***Boron.*** Most of the first-principles studies investigated the equilibrium distribution of ^{10}B and ^{11}B isotopes between boric acid and borate in aqueous solution at ambient conditions, motivated by the application of boron isotope composition of marine carbonates as paleo-pH proxy (Liu and Tossel 2005; Zeebe 2005; Rustad and Bylaska 2007; Rustad et al. 2010b). Tossel (2006) studied the isotopic fractionation associated with the boric acid adsorption on humic acids, and more recently Kowalski et al. (2013) investigated the B isotope fractionation between minerals, such as tourmaline and micas, and boron aqueous species at high pressure and temperature.
- ***Magnesium.*** Black et al. (2007) studied the equilibrium Mg isotope fractionation in chlorophylls. This and several later studies made efforts to improve methods to determine isotopic fractionation in liquids, with a particular focus on the fractionation between aqueous Mg^{2+} and Mg-bearing carbonate minerals (Rustad et al. 2010a; Schauble 2011; Pinilla et al. 2015, Schott et al. 2016). Mg isotopes in mantle silicates were treated in Schauble (2011), Huang et al. (2013) and Wu et al. (2015b).
- ***Silicon.*** Méheut et al. (2007, 2009, 2014) computed the equilibrium Si isotope fractionation factors in various silicate minerals, including phyllosilicates. Their data analysis enabled to identify the key structural and chemical parameters controlling the isotopic signatures. Huang et al. (2014) and Wu et al. (2015b) applied the DFT method to silicate minerals of the Earth's mantle. Some DFT calculations coupled with isotopic measurements on meteorite and terrestrial samples focused on Si isotope fractionation between metal and silicates, in order to discuss the composition of the Earth's core and the Earth formation (e.g., Georg et al. 2007; Ziegler et al. 2010). The equilibrium fractionation in silicic acid and its potential application as proxies for paleo-pH were investigated in Dupuis et al. (2015) and Fujii et al. (2015). He and Liu (2015), He et al. (2016) complemented equilibrium Si isotope fractionation factors among minerals, organic molecules and the H_4SiO_4 solution. Javoy et al. (2012) determined the Si isotope properties of small gaseous molecules and crystalline compounds in the cosmochemical context of the solar nebula.
- ***Calcium.*** Theoretical Ca isotope fractionation factors between minerals and solution are presented in Rustad et al. (2010a), in Colla et al. (2013), and among pyroxenes in Feng et al. (2014). Griffith et al. (2008) estimated fractionation factors between barite and calcite.
- ***Vanadium.*** Wu et al. (2015a) explored how V isotope fractionation depends on crystal-chemical parameters such as valence, bond length and coordination number. They considered several inorganic V aqueous species and the adsorption of V^{5+} to goethite, by adopting a cluster model with explicit solvation shells.
- ***Chromium.*** Schauble et al. (2004), and Ottonello and Zuccolini (2005) computed the equilibrium Cr isotope fractionation factors of some molecules in the system Cr–H–O–Cl as well as in magnesiochromite (Ottonello et al. 2007). Moynier et al. (2011) extended these theoretical predictions to additional Cr-bearing minerals. These latter data associated with isotopic measurements on a range of meteorites suggest that Cr depletion in the bulk silicate Earth relative to chondrites results from its partitioning into Earth's core.

- ***Iron.*** Several DFT studies focused on isotopic fractionation among Fe species in aqueous solution (Anbar et al. 2005; Domagal-Goldman and Kubicki 2008; Hill and Schauble 2008; Ottonello and Zuccolini 2008, 2009; Hill et al. 2010; Fujii et al. 2014), others looked at iron-bearing minerals such as hematite, goethite, pyrite and siderite (Blanchard et al. 2009, 2010, 2015). Rustad and Yin (2009) investigated the isotopic properties of ferropericlase and ferroperovskite in lower-mantle conditions to discuss the Earth accretion and differentiation. Fe isotope fractionation between mineral and aqueous solution was the object of the studies by Rustad and collaborators (Rustad and Dixon 2009; Rustad et al. 2010a). Moynier et al. (2013) estimated the magnitude of the isotopic fractionation between different Fe species relevant to the transport and storage of Fe in higher plants. In addition to all these first-principles calculations, Mössbauer spectroscopy (e.g., Polyakov 1997; Polyakov and Mineev 2000) and nuclear resonant inelastic X-ray scattering, NRIXS (e.g., Polyakov et al. 2005; Dauphas et al. 2012) represent alternative techniques for obtaining Fe reduced partition function ratios.
- ***Copper.*** Seo et al. (2007) determined the equilibrium isotope fractionation of Cu^{+} complexes relevant of hydrothermal ore-forming fluids. Sherman (2013) modeled Cu-bearing minerals and various aqueous Cu^{+} and Cu^{2+} complexes to predict the equilibrium isotopic fractionation of Cu resulting from oxidation of Cu^{+} to Cu^{2+} and by complexation of dissolved Cu. Additional Cu complexes were considered in Fujii et al. (2013a, 2014).
- ***Zinc.*** Several theoretical works studied the isotope fractionation of Zn between various aqueous zinc complexes including aqueous sulfide, chloride, and carbonate species relevant to hydrothermal conditions (Fujii et al. 2010, 2011, 2014; Black et al. 2011). Other complexes were modeled to discuss the Zn isotope fractionation in roots and leaves of plants (Fujii and Albarède 2012).
- ***Germanium.*** Li et al. (2009) determined equilibrium fractionation factors for a range of Ge-bearing compounds (aqueous species and minerals) simulated using cluster models. Li and Liu (2010) investigated the fractionation associated with Ge adsorption onto Fe(III)-oxyhydroxide surfaces. A cluster model was used to model the adsorption complex. Such adsorption processes occur in many environments, and thus may influence significantly the Ge isotope global budged.
- ***Selenium.*** Equilibrium Se isotope fractionation factors of inorganic and organic Se-bearing species in gaseous, aqueous and condensed phases were computed (Li and Liu 2011).
- ***Strontium.*** A combined theoretical and experimental study focused on the Sr isotope fractionation during inorganic precipitation of barite, where several Sr-bearing minerals and crystalline strontium hydrates were modeled (Widanagamage et al. 2014). This work was preceded by the determination of the isotopic fractionation between SrO_2 and a Sr^{2+} aqueous species (Fujii et al. 2008).
- ***Molybdenum.*** Mo isotope fractionation factors were determined using a cluster approach, for many aqueous species including several forms of molybdic acid and polymolybdate complexes (Tossel 2005; Weeks et al. 2007, 2008; Wasylenki et al. 2008, 2011). These results confronted with experimental data aim at identifying the molecular mechanisms responsible to the Mo isotope fractionation during adsorption to manganese oxyhydroxides, which is a primary control on the global ocean Mo isotope budget.
- ***Cadmium.*** Yang et al. (2015) computed using DFT the equilibrium isotopic fractionation factors for Cd species relevant to hydrothermal fluids.

- ***Rhenium.*** Theoretical Re isotope fractionation has recently been investigated by Miller et al. (2015). They especially assessed the magnitude of nuclear volume fractionation with respect to mass dependent fractionation.
- ***Mercury and Thallium.*** Schauble (2007) performed first-principles calculations on these very heavy elements and could show that isotopic variation in nuclear volume is the dominant cause of equilibrium fractionation, exceeding mass-dependent fractionations. This is supported by two more recent works by Fujii et al. (2013b) and Yang and Liu (2015). Wiederhold et al. (2010) performed additional theoretical calculations, quantifying the relationship between ionic bonding and equilibrium mercury isotope fractionation.
- ***Uranium.*** Abe and his collaborators (Abe et al. 2008a,b, 2010, 2014) investigated the uranium isotope fractionations caused by nuclear volume effects.

Modeling isotopic properties of liquid phases

Many natural processes involve the participation of fluids. Isotopic signatures of minerals are very often related to fluid–rock interactions. Understanding the isotope fractionation processes between minerals and fluids is then of great importance. This understanding will include our ability to produce reliable theoretical mineral–solution isotopic fractionation factors. However calculations of fractionation properties of liquids and solvated elements under thermodynamic equilibrium represent a bigger challenge than for gaseous molecules or minerals. In an aqueous solution, an ion or molecule dissolved in water will interact with water molecules and other dissolved species in a continuously changing arrangement of hydrogen bonds and ion pairs. This disordered and dynamic character complicates significantly the problem. Determining the vibrational frequencies of such systems from first-principles calculations has a computation cost far greater than for minerals. Additional approximations are needed and can include the use of molecular clusters of finite size or the use of relaxed configurations from molecular dynamics simulations.

Most of the theoretical predictions of isotope fractionation in aqueous species are based on the cluster approximation (e.g., Yamaji et al. 2001; Anbar et al. 2005; Black et al. 2007, 2011; Seo et al. 2007; Domagal-Goldman and Kubicki 2008; Hill and Schauble 2008; Ottonello and Zuccolini 2009; Fujii et al. 2010, 2014, 2015; Li and Liu 2010; Rustad et al. 2010a,b; Sherman 2013). In this case, the ion or molecular complex of interest is surrounded by water molecules forming a solvation shell and the whole is sometimes immersed in a continuum approximating the dielectric properties of the solvent. The stable structure of this isolated nanodroplet is obtained at $T = 0\,\mathrm{K}$ by minimizing the forces acting on the atoms and the reduced partition function ratio (β-factor) is computed from the vibrational frequencies obtained in the harmonic approximation. The inclusion of the first solvation shell around the considered species is a first step towards the consideration of the solvation effect, i.e. effect explaining that most gases exhibit measurable isotopic fractionations between the vapor phase and solution. This approach is however hindered by several difficulties, such as the number of water molecules that must be included, the symmetry of the cluster, and the consistency between different aqueous species or between the aqueous species and the mineral. Let's take the example of iron isotopes. First-principles calculations performed on small clusters give equilibrium isotopic fractionation between $Fe(H_2O)_6^{3+}$ and $Fe(H_2O)_6^{2+}$ of 2.5–3‰ at 22 °C for the isotopic ratio $^{56}Fe/^{54}Fe$ (Anbar et al. 2005; Domagal-Goldman and Kubicki 2008; Hill and Schauble 2008). These values are in good agreement with the experimental value of 3.00 ± 0.23‰ (Welch et al. 2003), even if the theoretical value depends on the cluster symmetry chosen. Here the two iron species only differ from the charge and are treated in a consistent way, which allows a cancellation of errors. However when the same theoretical data are combined with mineral β-factors (Polyakov and Mineev 2000; Polyakov et al. 2007; Blanchard et al.

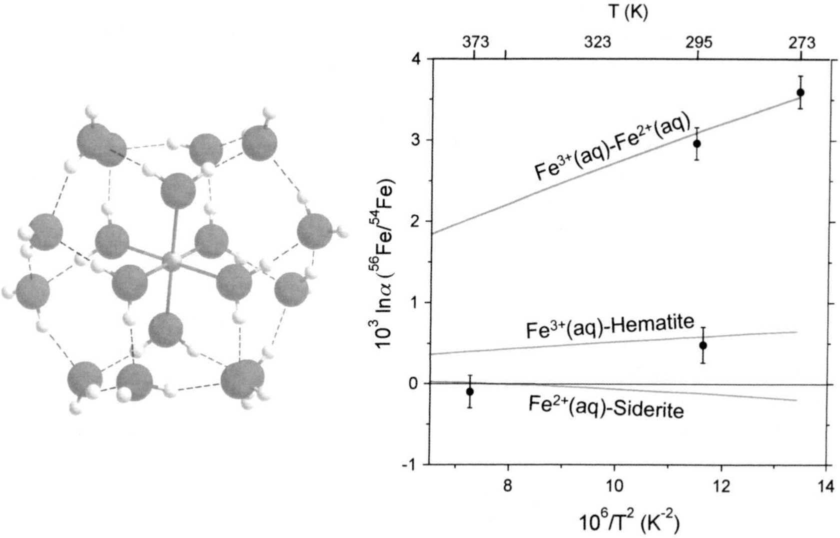

Figure 2. *Left:* Example of molecular cluster used by Rustad et al. (2010a) to model aqueous Fe^{2+} and Fe^{3+}. *Right:* Calculated (curves) and measured (circles) fractionations for the pairs Fe^{3+}(aq)–Fe^{2+}(aq), Fe^{3+}(aq)–hematite and Fe^{2+}(aq)–siderite. Theoretical β-factors are from Rustad et al. (2010a) for aqueous ions, from Polyakov and Mineev (2000) for siderite, and from Polyakov et al. (2007) for hematite. These two latter Mössbauer-derived β-factors are consistent with DFT results (Blanchard et al. 2009). Experimental data are from Skulan et al. (2002), Welch et al. (2003) and Wiesli et al. (2004).

2009), the calculated fractionations for Fe^{3+}-hematite and Fe^{2+}-siderite are in disagreement with experimental data. Preceded by few theoretical works emphasizing the importance of explicitly treating secondary solvation shells (e.g., Schauble et al. 2004; Liu and Tossell 2005), Rustad et al. (2010a) could show that the β-factors can be reliably computed from systems as small as $M(H_2O)_6^{2+}$ but when they are embedded in a set of fixed atoms representing at least the second solvation shell (Fig. 2). Furthermore their results suggest that the aqueous cluster is much more sensitive to improvements in the basis set than the calculations on the mineral systems. By applying these results, Rustad et al. (2010a) obtained more accurate β-factors for aqueous Fe^{2+} and Fe^{3+} and could reconcile theory and experiment for the mineral–solution fractionations (Fig. 2). Obviously, an observed disagreement between theory and experiment may have other reasons, such as kinetic effects during nucleation and crystal growth that could make the equilibrium assumption invalid. For example, if minerals form via oligomers or clusters as an intermediate step between the aqueous species and the mineral, then using a model of the bulk aqueous species will never reproduce the observed fractionation (e.g. Domagal-Goldman et al. 2009). More generally this cluster approach can be justified for dissolved molecules that remain more or less intact in solution (e.g., $[ClO_4]^-$, $B(OH)_3$ and CCl_4) or for aqueous complexes where intra-complex bonds are probably much stronger than interactions with bulk solvent (e.g., $[Cr(H_2O)_6]^{3+}$, $[FeCl_4]^-$, and Mg^{2+} in chlorophyll). On the other hand, this method has the disadvantage of neglecting the constant exchange of particles within the solvation shells and other effects, such as the formation of chemical bonds and structural rearrangements as a function of temperature and pressure considered to be important for the calculation of the isotope fractionation.

To account for these dynamical phenomena, i.e. frequent particle exchange in the hydration shell and structural evolution of the fluid with pressure and temperature like it is for instance the case in Li aqueous solution (Jahn and Wunder 2009) one can go beyond the static calculations on molecular clusters by employing molecular dynamic simulations. In this case, a first-principles molecular dynamics is run at finite temperature where the condensed phase of the fluid is described through periodic boundary conditions. The equilibrated trajectory thus provides a representative distribution of the configurational environments of the species of interest in the fluid. The average fractionation factor is then estimated from the harmonic vibrational frequencies computed on a set of uncorrelated snapshots taken from the molecular dynamic trajectory. Vibrational frequencies can be computed directly from each snapshot without relaxing the atomic positions but this raises a problem because the dynamic structures are often statically unstable, meaning that some calculated frequencies are imaginary numbers. It is not clear how to make reliable thermodynamic calculations of fractionation factors when imaginary frequencies are encountered. A way around this problem is to further process each snapshot structure by allowing atomic positions to relax into the nearest local energy minimum by performing geometry optimization at $T = 0\,K$ (giving the so-called inherent structures, Stillinger and Weber 1983) before computing vibrational frequencies. This approach is more satisfactory for determining the fractionation properties from the equations based on the harmonic approximation, but on the other hand this approach erases some of the desired dynamical sampling, evident for instance in the more homogeneous bond lengths found after snapshot relaxations. An alternative approach was proposed by Kowalski and Jahn (2011) and consists in relaxing only the position of the element of interest before determining the fractionation properties from the high-temperature approximation based on force constants. When this approximation is valid, it reduces significantly the computational cost, which is the major drawback of the molecular dynamics method. In order to keep the calculations as tractable as possible, all parameters must be chosen carefully, including the size of the simulation cell and the snapshot sampling. The simulation cell should be large enough to avoid significant interaction between atoms and their periodic images. The first studies of this kind used simulation cells containing typically 32 or 64 water molecules (Rustad and Bylaska 2007; Kowalski and Jahn 2011; Pinilla et al. 2014, 2015; Dupuis et al. 2015). However Kowalski and Jahn (2011) have shown that for a dissolved Li^+ ion a cell containing only 8 water molecules is enough to get a converged result within the accuracy of the calculations. This highlights the local character of fractionation properties, i.e. isotopic fractionation is mainly controlled by the bonds formed with the first atomic neighbors. The snapshot sampling should be large enough to get a representative distribution of the fluid configurations but small enough to keep the computation time under reasonable limits. Dupuis et al. (2015) tested thoroughly this sampling by considering a random, periodic or selected extraction of the snapshots. Results suggested that the extraction of only 10 snapshots is statistically representative of the whole solution, and that this number can even be decreased by taking advantage of the correlation between the fractionation value and the mean bond length (in cases where such correlation is evidenced).

The first study taking into account dynamical effects on isotope fractionation factors for non-traditional elements is by Rustad and Bylaska (2007). They calculated first the velocities correlation of exchanging isotopes and through its Fourier transform found the vibrational density of states to predict the boron isotope fractionation between $B(OH)_3$ and $B(OH)_4^-$ in aqueous solution. This led to a discrepancy between the calculated fractionation factor and the experimental one (Byrne et al. 2006; Klochko et al. 2006), which was solved after computing the harmonic frequencies of inherent structures taken from the molecular dynamics trajectory. Kowalski and coworkers took advantage of computing partial vibrational properties to investigate the lithium and boron isotope fractionation between aqueous fluids and minerals at high pressure and temperature (Kowalski and Jahn 2011; Kowalski et al. 2013). More recently, Pinilla et al. (2015) studied the equilibrium isotope fractionation between aqueous Mg^{2+} and carbonate minerals, and Dupuis et al. (2015) focused on silicon isotope fractionation in dissolved silicic acid. In conclusion, first-principles molecular dynamics simulations represent an efficient way to take into account the dynamical aspect of the fluid and their compressibility. By employing periodic boundary conditions, this approach also allows to treat minerals and fluids in a consistent manner; a prerequisite for reliable isotope fractionation factors between mineral and solution. All methods mentioned so far are based on the harmonic approximation. For many substances, uncertainties associated with calculated vibrational frequencies are likely to be larger than anharmonic effects. In liquids, anharmonicity effects are expected to have stronger impacts on fractionation properties. Generally anharmonicity will tend to decrease the vibrational frequencies and consequently the reduced partition function ratios (e.g., Richet et al. 1977; Balan et al. 2007; Méheut et al. 2007). To go beyond the harmonic approximation, more sophisticated techniques exist and are presented in the next section.

Beyond harmonic approximation: Path integral molecular dynamics

As already pointed out, isotopic fractionation is a quantum effect. Nuclear quantum effects (e.g., zero-point energy, quantum tunneling) whose relative contribution increases with decreasing temperature, influence significantly the properties of many systems, especially those containing lighter elements. Moreover it is also known that anharmonicity can be substantial especially for light elements and for liquid phases. A method of choice to include quantum nuclear effects without using the harmonic approximation is the method of thermodynamic integration coupled to path integral molecular dynamics (PIMD).

The reduced partition function ratio can be written using the Helmholtz free energy instead of the partition function:

$$\ln\beta_{AX} = \frac{F(AX)-F(AX')}{\mathrm{k}T} + \left(\frac{F(AX)-F(AX')}{\mathrm{k}T}\right)_{cl} \tag{12}$$

where $F(AX)$ and $F(AX')$ are the free energy of a single molecule of the two isotopologues AX and AX', and the subscript *cl* refers as before to quantities calculated using classical mechanics. Unfortunately, the absolute value of the free energy is not a quantity that can be directly obtained for any arbitrary system. Relating the free energy to another physical property, such as the kinetic energy, can circumvent this problem. On this line, it can be shown that the free energy of an isotopic species depends on its kinetic energy and mass (Landau and Lifshitz 1980):

$$\frac{\partial F}{\partial m} = -\frac{\langle K\rangle}{m} \tag{13}$$

where $\langle\rangle$ represents a thermodynamic average in the canonical ensemble (i.e. thermodynamic ensemble *NVT* corresponding to a system in thermal equilibrium: the number of particles, the volume and the temperature of the system are fixed). Inserting Eqn. (13) into Eqn. (12) and taking into account that in the classical limit the kinetic energy of an atom is $\langle K\rangle = 3\mathrm{k}T/2$, the β-factor is then given by:

$$\ln\beta_{AX} = \frac{1}{\mathrm{k}T}\int_{m'}^{m} d\mu \frac{\langle K\langle\mu\rangle\rangle}{\mu} - \frac{3}{2}\ln\left(\frac{m}{m'}\right) \tag{14}$$

where $\langle K(\mu)\rangle$ is the average kinetic energy of the atom X of mass μ in phase AX. In this expression, the β-factor is thus obtained by thermodynamic integration from mass m' to mass m. Here, we stress that the kinetic energy used in the thermodynamic integration is that of the quantum system. It differs from the kinetic energy determined using standard molecular dynamic methods. These latter methods solve the classical equation of atomic motions in a force field, which can be defined either empirically or using *ab initio* electronic structure calculations. In the present case, the determination of the kinetic energy has to take into account the fact that, in a quantum system, the atomic trajectories are not defined. The atoms display some degree of delocalization (i.e. some uncertainty on their position); which is inversely related to their mass. Path integral methods enable the treatment of such effect by replacing the standard classical system by a larger number of replicated classical systems (Fig. 3). The replicated systems interact through harmonic springs connecting a given atom to its counterpart in adjacent replicas. Based on the exact isomorphism between a quantum particle and a classical ring polymer, the quantum thermodynamic averages can be calculated exactly for any force field using path integral methods. PIMD methods are implemented in several codes such as the freely available program CP2K, CPMD (Marx and Hutter 2000), i-PI (Ceriotti et al. 2014) or PINY_MD (Tuckerman et al. 2000). A description of the PIMD methods and their implementations is out of the scope of this chapter but can be found elsewhere (Feynman and Hibbs 1965; Ceperley 1995; Tuckerman 2010). The drawback of PIMD methods is the computational cost that is almost prohibitive for treating at the *ab initio* level most of the systems relevant in geosciences. To address this issue, several studies report new developments that improve the efficiency of the methods concerning isotopic applications (e.g., Ceriotti and Markland 2013; Cheng and Ceriotti 2014; Marsalek et al. 2014).

Regarding the investigation of isotopic effects, many studies focused on small molecules or molecular clusters, including water molecule and ions, hydrated chloride ions, carbon dioxide, organic molecules (e.g., Tachikawa and Shiga 2005; Vanicek and Miller 2007; Suzuki et al. 2008; Mielke and Truhlar 2009; Pérez and von Lilienfeld 2011; Webb et al. 2014). Other studies have modeled condensed phases, like for instance Chialvo and Horita (2009), Ramírez and Herrero (2010), Markland and Berne (2012), Zeidler et al. (2012), and Pinilla et al. (2014) for the water system. Among these studies, Pinilla et al. (2014) determined the H and O isotope equilibrium fractionation between water ice, liquid and vapor, and compared the exact result obtained from PIMD with those of the more common modeling strategies, which involve the use of the harmonic approximation. The same approach was then applied to the aqueous Mg^{2+} (Pinilla et al. 2015). Results show the importance of including configurational disorder for the estimation of isotope fractionation in liquid phases, by using molecular dynamics simulations. In the case of D/H fractionation, neglecting the anharmonic effects leads to an overestimation of the fractionation factor. In other words, the harmonic approximation will overestimate the concentration of heavy isotopes in the aqueous phase. For heavier atoms, like magnesium and to some extent oxygen, methods based on the harmonic approximation give reliable results and in the same time reduce significantly the computational cost.

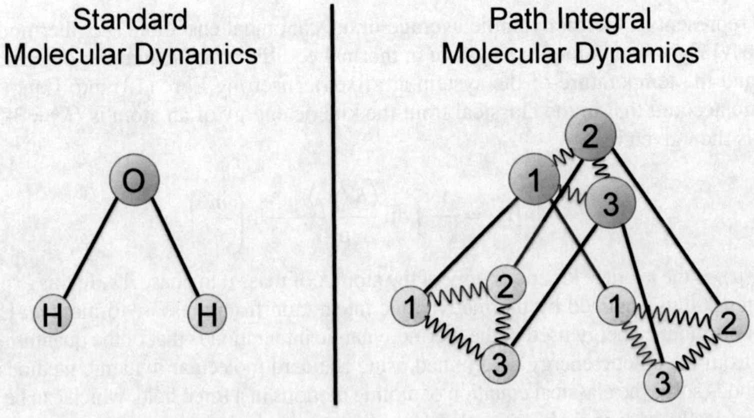

Figure 3. Schematic representations of a water molecule in standard molecular dynamics and path integral molecular dynamics. The straight lines joining the replicas (also called "beads") with the same number represent the interatomic interactions that can be modeled using either empirical or ab initio force fields. Replicas belonging to the same atom interact through harmonic springs. Only three replicas are represented here for clarity reasons but a large number of replicas (several tens) are actually needed to capture the quantum behavior of the system.

MÖSSBAUER AND NRIXS SPECTROSCOPY

In addition to the equilibrium fractionation factors derived experimentally by isotopic composition measurements, we have seen that these equilibrium constants can also be determined theoretically from the computation of the vibrational properties. An additional approach for Mössbauer-active elements (like iron, which is the most commonly studied element) consists in using Mössbauer spectroscopy (e.g., Polyakov 1997, 2000) and nuclear resonant inelastic X-ray scattering, NRIXS (e.g., Polyakov et al. 2005; Dauphas et al. 2012). These two latter techniques probe the vibrational properties of the target element and are thus ideally suited to study complex materials. Let's start again from an expression relating the reduced partition function ratio (β-factor) to the kinetic energy. Using the first-order thermodynamic perturbation theory (Landau and Lifshitz 1980), Equation (14) becomes:

$$\ln\beta_{AX} = \frac{m-m'}{m}\left(\frac{K}{\mathrm{R}T}-\frac{3}{2}\right) \tag{15}$$

In Mössbauer spectroscopy, the kinetic energy K of the active isotope (e.g. ^{57}Fe) is related to the second-order Doppler shift, $S(T)$:

$$S(T) = -\frac{K(T)}{mc}$$

where c is the light velocity. Substituting $S(T)$ for $K(T)$ into Equation (15) leads to:

$$\ln\beta_{AX} = \frac{m-m'}{m}\left(\frac{m'\mathrm{c}S(T)}{\mathrm{R}T}+\frac{3}{2}\right)$$

The second-order Doppler shift $S(T)$ can be determined experimentally from the temperature dependence of the isomer shift because both quantities only differ by a constant value that reflects the fact that the isomer shift is measured relative to a reference spectrum of metallic iron at room temperature. Experimental data are conveniently fitted using a Debye function:

$$S(T) = -\frac{9\mathrm{R}\theta_M}{16mc}\left[1+8\left(\frac{T}{\theta_M}\right)^4\int_0^{\theta_M/T}\frac{x^3}{e^x-1}dx\right]$$

where m is the mass of the resonant isotope (^{57}Fe), and θ_M is a characteristic Mössbauer temperature. However, the SOD shift is not the only factor controlling the temperature shift in the Mössbauer spectra, so model assumptions about the temperature dependence of the Mössbauer isomeric shift are needed. In practice, the prediction of Mössbauer-derived fractionation factors involves an extensive data processing and requires high quality data to achieve a reasonable accuracy. This explains the few cases of conflicting results and revised data reported in the literature (e.g., Polyakov et al. 2007; Rustad et al. 2010a; Blanchard et al. 2012).

In the NRIXS method, kinetic energy is calculated from the measured vibrational density of states of the element of interest using the following expression. The vibrational density of states can also be obtained by *ab initio* calculations.

$$K = \frac{3}{2}\int_0^{e_{\max}} E(e/\mathrm{k}T)g(e)\mathrm{d}e \tag{16}$$

where $g(e)$ is the vibrational density of states of ^{57}Fe, for instance, normalized to unity, and $E(e/\mathrm{k}T)$ is the Einstein function for the vibrational energy of a single harmonic oscillator at frequency $\nu = e/\mathrm{h}$. $e_{\max}$ corresponds to the maximal energy of the vibrational spectrum, and the Einstein function is given by:

$$\frac{E(e/\mathrm{k}T)}{\mathrm{k}T} = \frac{e/\mathrm{k}T}{\exp(e/\mathrm{k}T)-1}+\frac{e}{2\mathrm{k}T} \tag{17}$$

Equations (16) and (17) are valid in the harmonic approximation and Equation (16) takes into account the virial harmonic relation $K = E_{vib}/2$, where E_{vib} is the vibrational energy of the harmonic oscillator. For a given system, the β-factor calculated from the vibrational density of states (Eqns. 15 and 16) is identical to the β-factor calculated directly from the vibrational frequencies (Eqn. 10). When the highest energy of the vibrational density of states is smaller than 2πkT, another expression can be used, based on a Bernoulli expansion of the β-factor (Dauphas et al. 2012):

$$\ln\beta_{AX} \approx \left(\frac{m'}{m}-1\right)\left(\frac{m_2^g}{8\mathrm{k}^2T^2}-\frac{m_4^g}{480\mathrm{k}^4T^4}+\frac{m_6^g}{20160\mathrm{k}^6T^6}\right) \quad (18)$$

where m_i^g is the ith moment of the vibrational density of states $g(e)$, given by:

$$m_i^g = \int_0^{+\infty} g(e)e^i \mathrm{d}e$$

Dauphas et al. (2012) have also shown that the even moments of $g(e)$ can be obtained directly from the moments of the NRIXS spectrum $S(e)$ and Equation (18) can be rewritten as:

$$\ln\beta_{AX} \approx \left(\frac{m'}{m}-1\right)\frac{1}{E_R}\left(\frac{R_3^S}{8\mathrm{k}^2T^2}-\frac{R_5^S-10R_2^SR_3^S}{480\mathrm{k}^4T^4}+\frac{R_7^S+210\left(R_2^S\right)^2R_3^S-35R_3^SR_4^S-21R_2^SR_5^S}{20160\mathrm{k}^6T^6}\right)$$

where E_R is the free recoil energy and R_i^S is the ith moment of $S(e)$ centered on E_R, given by:

$$R_i^S = \int_{-\infty}^{+\infty} S(e)(e-E_R)^i \mathrm{d}e$$

The NRIXS-based method probing directly the vibrational properties of the target element is expected to provide better accuracy for the β-factors than that based on Mössbauer spectroscopy. However NRIXS spectra have only recently been measured specifically for applications to isotope geochemistry (Dauphas et al. 2012). A difficulty that had been unappreciated before, was encountered concerning the baseline at low and high energies (the details of the spectrum at the low- and high-energy ends heavily influence the treatment of experimental data and therefore the value of the β-factor). To address this issue, Dauphas and collaborators have developed a software (SciPhon) that reliably corrects for non-constant baseline (Dauphas et al. 2014; Blanchard et al. 2015; Roskosz et al. 2015).

Figure 4 displays a comparison of the iron β-factors derived from NRIXS, Mössbauer measurements and first-principles calculations (DFT) for various iron-bearing minerals, i.e. sulfides, an oxide and a carbonate. For siderite ($FeCO_3$), DFT results (Blanchard et al. 2009) are in excellent agreement with Mössbauer data (Polyakov et al. 2007). The same kind of agreement is found between the DFT and NRIXS resuls of chalcopyrite ($CuFeS_2$, Polyakov et al. 2013). For hematite (Fe_2O_3) DFT-derived iron β-factors (Blanchard et al. 2009) are very close to NRIXS-derived values (Dauphas et al. 2012) while the results from Polyakov et al. (2007) are slightly above. In the case of pyrite (FeS_2), the apparent discrepancy between DFT and Mössbauer results that was reported in Blanchard et al. (2009) and in Polyakov and Soultanov (2011), could be resolved by using a better constrained temperature dependence of the Mössbauer spectra (Blanchard et al. 2012). The value of the iron β-factor in pyrite was confirmed by NRIXS data (Polyakov et al. 2013) and also appears consistent with experimental measurements of equilibrium isotopic fractionation between pyrite and dissolved Fe^{2+} (Syverson et al. 2013). These comparisons exhibit that DFT, NRIXS and Mössbauer spectroscopy should lead to statistically indistinguishable β-factors, when high-quality measurements are performed followed by a careful data processing. Therefore, the comparison of results from these independent techniques provides reliable isotope fractionation factors. Figure 4 also shows that

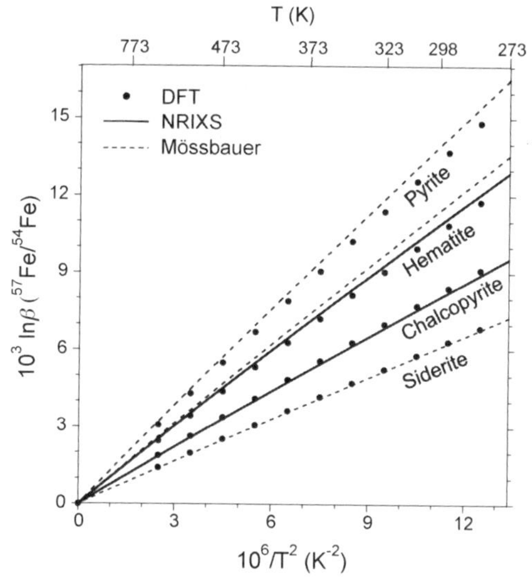

Figure 4. Comparison of iron β-factors derived from first-principles calculations (DFT), NRIXS and Mössbauer measurements for siderite ($FeCO_3$), chalcopyrite ($CuFeS_2$), hematite (Fe_2O_3) and pyrite (FeS_2). Data were taken from Blanchard et al. (2009, 2012), Dauphas et al. (2012), Polyakov and Mineev (2000), Polyakov and Soultanov (2011), Polyakov et al. (2007). In these minerals where iron atoms are always in octahedral coordination, the β-factor is mainly controlled by the iron oxidation state and the degree of covalence of the chemical bonds involved.

iron β-factors of these minerals are noticeably different; pyrite displaying the highest value while siderite has the lowest one. In all minerals except chalcopyrite iron atoms are in octahedral sites, the observed order can be discussed in terms of oxidation state (Fe^{3+} in hematite vs. Fe^{2+} in siderite) and degree of covalence of the interatomic bonds (low-spin, strongly covalent *d*-orbitals in pyrite vs. high-spin, ion-like *d*-orbitals in almost all other minerals).

MASS-INDEPENDENT FRACTIONATION AND VARIATIONS IN MASS LAWS

Typical fractionating processes, including equilibrium isotope partitioning, activation-energy and transport-controlled disequilibrium reactions, and even gravitational and centrifugal isotope separation, almost always impart a characteristic signature in which the magnitude of isotope fractionation scales in close proportion to the difference in isotopic mass (e.g., Hulston and Thode 1965; Matsuhisa et al. 1978; Weston 1999; Young et al. 2002). A typical example is oxygen, where $^{17}O/^{16}O$ fractionation is usually 0.5–0.53 times as large as $^{18}O/^{16}O$ fractionation, very close to the ratio of mass differences (≈0.501). In high temperature igneous and metamorphic rocks the mass-fractionation relationship for oxygen is remarkably consistent, with $\delta^{17}O \approx 0.528 \pm 0.001 \times \delta^{18}O$ (e.g., Rumble et al. 2007). Subtle variations in mass dependence are observed in light stable isotope systems, including oxygen and sulfur (Farquhar et al. 2003; Barkan and Luz 2012; Hofmann et al. 2012), and are of increasing interest as potential tools to unravel the nature of fractionation in the hydrological cycle, in the precipitation of low-temperature minerals from solution, and in biochemical reactions. Young et al. (2002) pointed out that variations in mass-fractionation relationships could also help distinguish equilibrium from kinetic fractionations in non-traditional elements such as magnesium, although such measurements are likely to require very high precision in systems where fractionations of only a few per mil are observed.

Various notations have been developed to describe variations in mass dependence. Here we have followed the basic formulation of Mook (2000), which has been widely adopted. In a stable isotope fractionation involving element X with stable isotopes, ^{1}X, ^{2}X, and ^{3}X that have masses m_1, m_2, and m_3, there are two distinct fractionation factors α:

$$^{3/1}\alpha_{AX/BX} = \frac{\left([^{3}X]/[^{1}X]\right)_{AX}}{\left([^{3}X]/[^{1}X]\right)_{BX}}$$

and

$$^{2/1}\alpha_{AX/BX} = \frac{\left([^{2}X]/[^{1}X]\right)_{AX}}{\left([^{2}X]/[^{1}X]\right)_{BX}} \tag{19}$$

In such a system it is convenient to express the mass dependence of the fractionation as a mass-fractionation exponent, θ, the ratio of the natural logarithms of the fractionation factors:

$$\theta = \frac{\ln\left(^{2/1}\alpha_{AX/BX}\right)}{\ln\left(^{3/1}\alpha_{AX/BX}\right)} \tag{20}$$

Formally, mass-independent fractionation refers to any deviation of an observed fractionation from a reference mass fractionation exponent, typically expressed in delta notation:

$$\Delta'^{2/1} X_{AX/BX}(‰) = 10^3\left(\theta - \theta_{\text{reference}}\right)\ln\left(^{3/1}\alpha_{AX/BX}\right) \tag{21}$$

Note that the prime indicates that logarithmic delta is being used here. Alternative expressions using conventional delta units may also be used, but for a discussion of fractionation factors the logarithmic delta makes the math simpler. The reference exponent might be a theoretical law or an empirically observed trend (an empirical exponent may be indicated by using λ instead of θ). There is generally not a well-accepted consensus in favor of a particular mass law exponent, so it can be tricky to compare Δ' values reported by different labs.

Variability in mass laws for common fractionations

Detailed derivations of mass law exponents for various fundamental fractionation processes have been published elsewhere (e.g., Young et al. 2002; Dauphas and Schauble 2016), and will not be reproduced here. Several of the most significant and/or common mass law exponents are listed in Table 1, below, along with calculated exponents for some traditional and non-traditional stable isotope systems. Among the most important of these is the high-temperature equilibrium mass law exponent, which can be derived from the simplified formula for isotope fractionation in Equation (11), above (Matsuhisa et al. 1978):

$$\theta_{\text{Eq.,Hi-T}} \approx \frac{\frac{1}{m_1} - \frac{1}{m_2}}{\frac{1}{m_1} - \frac{1}{m_3}} \tag{22}$$

As noted above for Equation (11), the constant mass dependence implied by this relationship is a surprisingly good approximation at low temperatures, even for materials with high-frequency vibrations such that $h\nu/kT > 2$ (Matsuhisa et al. 1978). In part, this occurs because the low-temperature equilibrium exponent is *identical* in the limit where the element of interest is much more massive than other atoms in the molecule (Cao and Liu 2011; Dauphas and Schauble 2016). These light-atom molecules tend to have the highest vibrational frequencies

Table 1. Theoretical mass-fractionation exponents.

Type of fractionation	**θ exponent**	^{16,17,18}**O**	^{24,25,26}**Mg**	^{28,29,30}**Si**	^{32,33,34}**S**	^{54,56,57}**Fe**	^{198,200,202}**Hg**
Equilibrium, high-T	$\dfrac{\frac{1}{m_1}-\frac{1}{m_2}}{\frac{1}{m_1}-\frac{1}{m_3}}$	0.5305	0.5210	0.5178	0.5159	0.6780	0.5049
Equilibrium, low-T, light partner	$\dfrac{\frac{1}{m_1}-\frac{1}{m_2}}{\frac{1}{m_1}-\frac{1}{m_3}}$	0.5305	0.5210	0.5178	0.5159	0.6780	0.5049
Equilibrium, low-T, heavy partner†	$\dfrac{\sqrt{\frac{1}{m_1}}-\sqrt{\frac{1}{m_2}}}{\sqrt{\frac{1}{m_1}}-\sqrt{\frac{1}{m_3}}}$	0.5232	0.5160	0.5135	0.5121	0.6750	0.5037
Graham's law (pinhole) effusion, atomic	$\dfrac{\ln\left(\frac{m_1}{m_2}\right)}{\ln\left(\frac{m_1}{m_3}\right)}$	0.5158	0.5110	0.5092	0.5083	0.6720	0.5024
Graham's law effusion, high-mass molecule	$\dfrac{\ln\left(\frac{M_1}{M_2}\right)}{\ln\left(\frac{M_1}{M_3}\right)}$	0.5010	0.5010	0.5006	0.5007	0.6660	0.4999
Kinetic, transition state theory, jump limited	$\dfrac{\ln\left(\frac{\mu_1^*}{\mu_2^*}\right)}{\ln\left(\frac{\mu_1^*}{\mu_3^*}\right)}$	Intermediate between high-T equation and high-mass molecular diffusion					
Gravitational/ centrifugal	$\dfrac{m_2-m_1}{m_3-m_1}$	0.5010	0.5010	0.5006	0.5007	0.6660	0.4999
Calculated variability		*0.0295*	*0.0200*	*0.0172*	*0.0151*	*0.0119*	*0.0050*

Note: m_i are isotopic masses, M_i are masses of isotopically substituted molecules, and μ_i* are reduced masses of a reaction coordinate at the transition state. Based on Matsuhisa et al. (1978), Young et al. (2002), and Dauphas and Schauble (2016).

†This equation only applies to the reduced partition function ratio β_{AX}, and thus to fractionation relative to atomic vapor. Actual β_{AX} exponents for non-traditional elements will rarely, if ever approach this limit because $h\nu/kT$ and/or the masses of bond partners are too small.

and hν/kT, especially when considering non-traditional (typically high atomic mass) isotope systems. This mass law is thus a common and sensible choice as a theory-based reference exponent (e.g., Young et al. 2002).

The mass-dependent relationships described above indicate that there will be a narrow range of variability in mass dependence for typical fractionating processes. For equilibrium fractionations of non-traditional isotopes, it is expected that the variability will be quite small. Indeed, commonly occurring sulfur species ($Z=16$) show very little change in mass dependence at equilibrium at relevant temperatures (Hulston and Thode 1965; Farquhar et al. 2003; Otake et al. 2008). There has not been a focused theoretical effort to quantify the variability in elements heavier than sulfur. However, studies of oxygen and sulfur suggest that even fairly crude electronic structure models (including DFT) can give an accurate picture of mass dependence variations at equilibrium (Cao and Liu 2011), and this seems likely to be an area of future development, as measurement precision continues to improve for many non-traditional elements.

Mass-independent fractionation in light elements (O and S)

In addition to the subtle variations in mass dependence discussed above, there are some natural and laboratory environments that give rise to fractionations that deviate strongly from a proportional relationship with mass differences. These are called mass-independent fractionations, even though they are usually driven, ultimately, by differences in isotopic mass. The best-known examples of mass-independent fractionation are in oxygen and sulfur isotopes, and are thought to be associated with reactions between molecules in the gas phase. Large, approximately 1:1 variation in $^{17}O/^{16}O$ vs. $^{18}O/^{16}O$ is observed in primitive meteoritic oxides and silicates (Clayton et al. 1973). Although the cause of this fractionation is not yet settled, the most common explanation is that it represents a self-shielding effect in carbon monoxide, in which the common isotopologue $^{12}C^{16}O$ is optically thick to incoming light with the right energy to break it apart, while $^{12}C^{17}O$ and $^{12}C^{18}O$ are optically thin, and thus more prone to react in the interior of the solar nebula or a molecular cloud (Clayton 2002; Lyons and Young 2005). In the stratosphere, a large, ~1:1 fractionation of $^{17}O/^{16}O$ and $^{18}O/^{16}O$ is found in ozone, and in gases that exchange oxygen with ozone. It is thought that this fractionation reflects an isotopic effect on the lifetime of excited ozone molecules (Heidenreich and Thiemens 1986; Mauersberger 1987; Gao and Marcus 2001). Mass-independent sulfur isotope fractionation has been found widely in Archean and earliest Proterozoic samples (Farquhar et al. 2000). These samples show a range of $^{33}S/^{32}S$ vs. $^{34}S/^{32}S$ relationships, thought to be caused by photochemical reactions of SO_2 in the early atmosphere, before O_2 became a major constituent of air (Pavlov and Kasting 2002; Lyons 2007).

Mass-independent fractionation in non-traditional elements (Hg, Tl, and U)

For non-traditional stable-isotope systems, at least two different fractionation mechanisms seem to be responsible for mass-independent fractionation effects. Most dramatic are large (> 1‰) mass-independent mercury isotope fractionations that are mainly photochemical (e.g., Bergquist and Blum 2007), and appear to be magnetic isotope effects dependent on the non-zero spins of the odd numbered mercury isotopes ^{199}Hg and ^{201}Hg. Introductory reviews of the magnetic isotope effect have been presented elsewhere (Turro 1983; Buchachenko 1995, 2013). The effect is apparent only in a subset of disequilibrium reactions. As yet, there are not any quantitative theoretical models that can reproduce the observed mass-independent signatures, and this is an area where more work is clearly needed.

Another type of fractionation is observed in the uranium and thallium isotope systems (Stirling et al. 2007; Rehkämper et al. 2002). Variation in isotope abundances in these elements in nature appears to mainly result from an equilibrium mass-independent phenomenon: the nuclear field shift effect. This effect also acts to fractionate mercury isotopes (e.g., Schauble 2007; Estrade et al. 2009; Wiederhold et al. 2010; Ghosh et al. 2013), but the mass-independent signature is much more subtle than the largest MIFs observed in natural samples, e.g, Blum et al. (2014). This effect has been the subject of a number of first-principles theoretical studies.

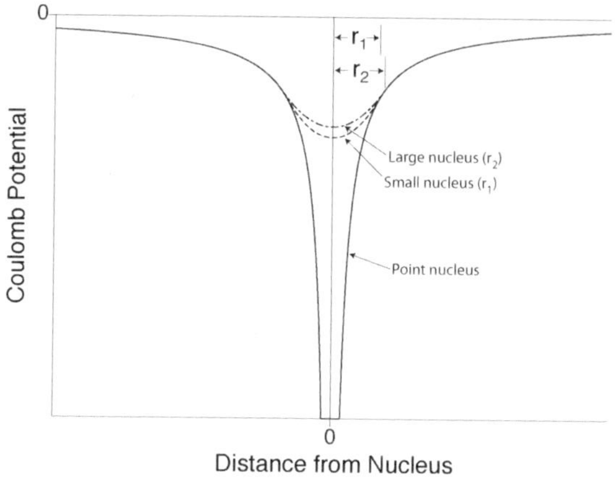

Figure 5. Nuclear field shift fractionations depend on the effect of the size and shape of a nucleus on the binding energy of electrons. The simplified example shown here assumes a single electron attracted to a spherical nucleus with a uniform charge density. The solid line shows the electrostatic potential binding an electron to an infinitesimally small nucleus, which goes to negative infinity as the electron approaches the nuclear center. For finite nuclei, the Coulomb potential does not go to negative infinity, but instead approaches a finite minimum inside the nucleus because the net electrostatic attraction to the shell of nuclear charge farther from the center than the electron is zero. The minimum is higher (the binding potential is weaker) for a large nucleus than for a small nucleus. Here the radius difference is assumed to be 10%, which is much larger than the difference between stable isotopes of most elements. Adapted from Schauble (2007).

Bigeleisen (1996) and Nomura et al. (1996) proposed that equilibrium isotopic fractionation in elements with very high atomic numbers could be driven by differences in the shape and size of nuclei, in addition to differences in mass. Hints of this effect were also described in an earlier experimental study of strontium isotope fractionation (Nishizawa et al. 1995). This is the nuclear field shift effect, and it is caused by overlap of electron density with the spatial volume occupied by the positive charge of a nucleus (Fig. 5). The general effect is to reduce the binding energy of electrons around large nuclei. This nuclear volume effect appears to be the most important component of the field shift, but non-spherical shapes may also be important for field shift effects in some nuclei; this shape-dependent part of the field shift is not as well studied as the volume component (e.g., Knyazev et al. 1999). Bigeleisen, Nomura, and their collaborators used the field shift effect to explain laboratory uranium isotope fractionation experiments in which the oxidized form of uranium, U(VI), had lower $^{238}U/^{235}U$ than coexisting reduced species at equilibrium. Such inverted redox/fractionation relationships are rare. A key observation was that $^{236}U/^{238}U$, $^{238}U/^{234}U$, $^{238}U/^{235}U$, and $^{238}U/^{233}U$ fractionations did not obey a consistent mass dependent relationship, with the magnitude of $^{238}U/^{235}U$ fractionation for instance being very similar to $^{238}U/^{234}U$, despite a 3:4 mass difference ratio (Nomura et al. 1996).

Isotope fractionations caused by the field shift effect can be quantified if field shift energies are known:

$$\ln \alpha_{FS} = \frac{\left[E^0(AX) - E^0(AX^*)\right] - \left[E^0(BX) - E^0(BX^*)\right]}{kT} \tag{23}$$

where α_{FS} is the field shift fractionation factor, and $E^0(AX)$, etc. are the ground state electronic energies of isotopic forms of AX and BX. Approximate expressions for the field shift energies have been derived in the optical spectroscopy literature (e.g., King 1984); these formulations capture the dependence of the field shift effect on the size of the nucleus (i.e., spatial nuclear volume) and on the electron density that overlaps it, e.g.,

$$\ln \alpha_{FS} \approx \frac{2\pi Z e^2}{3kT}\left(\left|\Psi(0)_{AX}\right|^2 - \left|\Psi(0)_{BX}\right|^2\right)\Delta\left\langle r^2\right\rangle \tag{24}$$

where Z is the nuclear charge, e is the charge of an electron, $|\Psi(0)_{AX}|^2$ and $|\Psi(0)_{BX}|^2$ are the electron densities at the center of the nucleus of the atom of interest in substances AX and BX, and $\Delta\langle r^2\rangle$ is the difference in mean-squared nuclear charge radius between the fractionating isotopes. This expression is approximate because it assumes a simplified model of the distribution of charge density in nuclei, that the nuclei are spherical (nuclei with odd numbers of neutrons and/or protons are aspherical) and that the electronic structure does not change as the nuclear size changes. The assumption of sphericity means that only the nuclear volume component of the field shift is considered, and potential shape effects are ignored.

Given these assumptions, however, it is clear that large nuclear field shift fractionations are most likely to occur between substances where the electron densities at the nucleus are very different, and where the difference in nuclear charge radius is large. Because the wavefunctions *d*- and *f*-orbital electrons (and most *p*-orbital electrons) do not overlap significantly with nuclei, variations in *s*-orbital electron population and structure control field shift fractionations. Based on Equation (24), and theoretical studies made so far, it is possible to list some qualitative rules of thumb about the chemical and physical properties that control field shift isotope fractionations:

1. Field shift isotope effects scale with nuclear charge radius, not with isotopic mass. Nuclear charge radii usually (but not always) increase with increasing neutron number, but tend to be smaller for nuclei with odd numbers of neutrons than one would expect from the radii of neighboring even-neutron number nuclei. For this reason, field shift fractionations will often generate a characteristic odd-even fractionation pattern (e.g., Bigeleisen 1996; Nomura et al. 1996). Not all elements show this pattern, however. For platinum, radii are almost perfectly linear for both odd and even numbered nuclei. In contrast, the ^{52}Cr nucleus, with a "magic" number of 28 neutrons, is notably smaller than stable chromium isotopes with fewer or greater numbers of neutrons.
2. Changes in *s*-electron occupation and *s*-orbital shapes control field shift fractionation.
3. Species with more *s*-electrons, and more compact s-orbitals, will tend to attract smaller nuclei. Examples include Hg(0) vs. Hg (II) and Tl(I) vs. Tl(III)—in each case the reduced form has two 6*s* electrons while the oxidized form has none, so Hg(0) and Tl(I) species preferentially incorporate small (neutron poor) isotopes. This pattern is likely for the 0 to +2 oxidation states of most elements in groups 1–12 of the periodic table, for the +1 to +3 oxidation states of group 13 elements, and for the +2 to +4 oxidation states of the group 14 elements. Field shift fractionation and mass-dependent fractionation will often tend to reinforce each other for elements and species in this category, leading to larger observed fractionations. *p*-, *d*-, and *f*-electronic orbitals affect field shifts indirectly; more electrons in these orbitals, and more compact orbital structures, tend to push *s*-electrons away from the nucleus. So species with more *p*-, *d*-, and/or *f*-electrons will tend to attract larger nuclei. U(IV) vs. U(VI) is an example of this behavior, with the two valence 5*f*-electrons in U(IV) species leading to higher $^{238}U/^{235}U$ than in 5*f*-depleted U(VI) species. This pattern is likely to occur in oxidation states higher than +2 for elements in groups 3–11, lanthanides, and actinides, as well as for the –4 to +2 oxidation states of group 14 elements, the –3 to +3 oxidation states of group 15 elements, the –2 to +4 oxidation sates of group 16 elements, and the –1 to +5 oxidation states of group 17 elements. In these systems, field shift effects and mass-dependent fractionation may tend to oppose each other and partially cancel.

4. Valence *s*-orbital electron densities vary much more widely in heavy (high-*Z*) elements than light (low-*Z*) elements. So field shift isotope fractionation effects will be much larger for elements with high atomic numbers (Knyazev and Myasoedov 2001; Schauble 2007). Isotopic variation in nuclear charge radii also varies a lot from one isotope pair to another, but it does not show a strong general trend with atomic number (Knyazev and Myasoedov 2001). It is not yet clear what the minimum atomic number needs to be for significant field shift effects to occur.
5. Field shift fractionation factors scale with T^{-1}, whereas equilibrium mass-dependent fractionations tend to scale as T^{-2} (Eqn. 11). As temperature increases, field shift effects may overwhelm mass-dependent fractionation for some elements.

Mass-independent fractionation signatures in heavy elements, versus light elements

In light elements (e.g., oxygen and sulfur), mass-independent fractionations (despite the terminology) ultimately follow from effects of mass differences on reaction rates and photochemical cross sections. In contrast, mass-independent fractionation effects in elements with high atomic numbers appear to be determined mainly by nuclear properties other than mass, including nuclear spin, volume, and shape. This can lead to some confusion, because mass-independent phenomena in heavy elements may or may not lead to observable departures from proportionality to isotopic mass differences. Illustrative examples can be found in the thallium and mercury isotope systems. Thallium isotope fractionation in nature is likely dominated by the field shift effect (Rehkämper et al. 2002; Schauble 2007; Nielsen et al. 2015), but there are only two stable thallium isotopes, ^{203}Tl and ^{205}Tl, making observation of mass-disproportionate fractionation impossible in natural samples and impractical in laboratory experiments. Among the four common even-numbered isotopes of mercury (^{198}Hg, ^{200}Hg, ^{202}Hg, and ^{204}Hg), isotope fractionations caused by field shift, magnetic, and mass-dependent isotope effects cannot be distinguished solely on the basis of apparent mass-fractionation relationships, because nuclear volume increases by an almost constant increment with each additional pair of neutrons and the magnetic isotope effect is limited to the odd-numbered isotopes ^{199}Hg and ^{201}Hg.

***Ab initio* methods for calculating field shift fractionation factors**

Bigeleisen (1996) and Nomura et al. (1996) identified the nuclear field shift effect in uranium isotope fractionations based on the inverse relationship between $^{238}U/^{235}U$ and oxidation state, and the close correlation between the magnitude of fractionation in other uranium isotopes and nuclear charge radii variations inferred from optical spectra of uranium vapor. Because of the characteristic pattern of isotopic charge radii, which deviates from pattern of mass differences, they were able to draw conclusions without a quantitative, *ab initio* theoretical model of the species present in the experiments. Such charge radius pattern matching has been widely used to search for evidence of field shift isotope fractionations in laboratory experiments and natural samples (Fujii et al. 2006a,b, 2009), and it has been useful for ruling out the field shift effect as the main cause of mass-independent signatures in mercury (e.g., Blum et al. 2014). However, it is not well suited to predict how large field shift effects will be in previously unstudied reactions and isotope systems. The ability to make forward models is important in these situations, and may also be necessary in systems where processes that mimic field shift fractionation patterns might be active, such as during nucleosynthesis. Reasonably accurate compilations of nuclear charge radii for almost all stable and long-lived radioactive nuclei are available in the literature (Nadjakov et al. 1994 and updates; Fricke and Heilig 2004; Angeli and Marinova 2013), so the main goal of theoretical models is to determine electron densities in the species of interest.

Accurate calculations of electron densities bound to high atomic number nuclei must take account of relativity effects. This can be understood by noting that the kinetic energy of a loosely bound *s*-orbital electron, when it is momentarily near the center of a highly charged nucleus, will be of the same order of magnitude as its rest mass energy. Even the *average* kinetic energies of inner-shell *1s*-orbital electrons are ~100 keV (vs. 511 keV rest mass energy) in elements such as Hg, Tl and U. That implies a velocity that is a significant fraction of the speed of light. For this reason, most *ab initio* studies of field shift effects to date have been based on the Dirac equation, a relativistic counterpart to the more familiar Schrödinger equation.

Some early theoretical studies of field shift fractionation are based on atomic and ionic models of electronic structure (Knyazev and Myasoedov 2001; Abe et al. 2008a). These calculations are easily performed on modern personal computers, but they assume a purely ionic bonding environment. More recently, Dirac–Fock and related model chemistries have been used to directly model molecules (e.g., Schauble 2007; Abe et al. 2008b, 2010; Wiederhold et al. 2010; Fujii et al. 2010). Comparisons to atomic spectra and laboratory measurements indicate that such models are usefully accurate (Schauble 2007; Wiederhold et al. 2010). The basic procedure for constructing a relativistic electronic structure model for calculating field shift fractionation is similar to the initial steps in creating a vibrational model for predicting mass-dependent fractionation, which was outlined in the preceding sections: first an initial structure is selected, and an electronic structure calculation is used to estimate the static energy of the structure; this is often followed up by performing a structural optimization. Vibrational frequencies need not be calculated—if mass-dependent fractionation factors are desired it is usually easier to construct a separate model based on non-relativistic theory, such as conventional DFT. The energy associated with isotope substitution can either be determined using an expression like Equation (24), or (even better) directly by manipulating the size of the nucleus in the first-principles model. Much like Hartree–Fock theory, Dirac–Fock theory has been extended to improve model accuracy by considering electron correlation effects and excited electronic states (see Wiederhold et al. 2010 and Nemoto et al. 2015 for comparisons of model results at various levels of theory). These high-accuracy calculations are much more demanding of computation time and memory, however.

Although theoretical studies based on *ab initio* relativistic electronic structure models have shown good agreement with measurements, the calculations are notably more complex, memory-intensive, and slow than typical non-relativistic methods. As a result, only fairly small molecules (no more than ~20 non-hydrogen atoms) can be modeled easily. Solid and liquid phases must be approximated using small clusters, which is likely to increase model uncertainties. It is difficult to confidently formulate a model of substances with long-range bonding interactions, such as metals or strongly solvated aqueous species, within these constraints. Several different ways around this limitation have been proposed. The first takes advantage of strong correlations between field shift effects in mercury compounds and the effective ionic charges of mercury atoms in those structures. This makes it possible to interpolate fractionations involving complex materials, such as liquid mercury, based on effective ionic charges computed with simpler electronic structure models (Wiederhold et al. 2010; Ghosh et al. 2013). This method may be best suited for group 1, 2 and 12 elements, where the field shift effect is dominated by a single valence *s* orbital. The second method, introduced by Nemoto et al. (2015), involves the use of a simplified relativistic modeling approach (the Douglas–Kroll–Hess method) that accurately reproduces variations in electron density near high atomic number nuclei with an order of magnitude less computational effort. This raises the possibility of directly modeling larger, more complex molecules while retaining enough accuracy to be useful. The third method, proposed by Schauble (2013) uses fully relativistic Dirac–Fock models (including some electron correlation effects) of simple molecules to calibrate a corresponding set of DFT models that are built using Projector Augmented Wave (PAW) data sets (Blöchl 1994). PAW is closely related to the pseudopotential methods described earlier in this chapter, and is also typically used in conjunction with plane-wave basis sets and periodic boundary conditions. Roughly the same computational effort is required for PAW methods as for standard pseudopotential-based DFT. But PAW has the advantage that information about the structure of core electronic orbitals is preserved, so that it is possible to calculate the influence of different chemical bonding environments on electron densities at the nucleus with reasonable accuracy (e.g., Zwanziger 2009). Both PAW and pseudopotential basis sets can be constructed using a partial correction for relativistic effects near the nucleus. Like the simplified relativistic approach proposed by Nemoto et al. (2015), the calibrated PAW method can be applied to larger, more complex materials and molecules, and it can even be applied to metals (such as metallic mercury) and semi-conducting materials where long-range bonding interactions are important. However, the PAW method probably loses some accuracy due to its dependence on a limited calibration set of small molecules.

Isomer shifts from Mössbauer spectroscopy

A final method of calculating nuclear field shift fractionations, suggested originally by Knyazev and Myasoedov (2001), uses isomer shifts measured with Mössbauer spectroscopy to determine changes in electron density from one species to another. Isomer shifts and field shifts arise from the same interaction between nuclear charge and electron density, with the main difference being that in Mössbauer spectroscopy the nuclear charge radius changes spontaneously as the Mössbauer nucleus is excited and then decays. In terms of data processing, the isomer shift is distinct from the second-order Doppler shift and NRIXS vibrational spectroscopy—it is typically one of the fundamental parameters measured in a standard Mössbauer experiment (along with quadrupole and magnetic splitting), and does not require measurements at multiple temperatures or a synchotron X-ray source. Like second-order Doppler shift and NRIXS measurements, isomer shifts can be measured in complex materials, selectively and directly probing the Mössbauer isotope's chemical environment. Knyazev and Myasoedov (2001) showed a promising correlation between calculated electron density variations in vapor-phase neptunium ions with varying charge and measured ^{237}Np-isomer shifts in crystals where Np has the same formal oxidation states. Schauble (2013) went a step farther by comparing ^{119}Sn-isomer shifts with electron densities calculated in the same substances using the DFT-PAW approach. The excellent correlation suggests that Mössbauer spectroscopy will be a powerful tool to predict field shift isotope fractionation in elements with Mössbauer-active isotopes.

CONCLUSIONS

While the theory of stable isotope fractionation has been developed in the middle of the twentieth century, the last decade was marked by the growing use of first-principles calculations to apply the theory to non-traditional stable isotopes. The aim of these calculations was first to determine the isotope fractionation factors when the considered phases are in thermodynamic equilibrium in order to identify the factors controlling these equilibrium fractionations. Quantitatively, the theoretical studies mentioned in this chapter but also the others dealing with the traditional stable isotopes show that calculated fractionation factors are reliable enough to be directly compared to experimental values and to values derived from spectroscopic techniques such as Mössbauer and NRIXS. To reach such level of accuracy, high-quality calculations are necessary (the quality of the model can be tested by comparing the calculated structural, electronic and vibrational properties with available experimental data) and most importantly the considered phases must be treated in a consistent manner. In stable isotope geochemistry, first-principles molecular modeling now represents numerical experiments that fully complement laboratory experiments for contributing to the interpretation of isotopic data collected on natural samples.

The advances in computation power enable to model systems of increasing complexity. Among the future directions of research that deserve special efforts, we can cite the investigation of isotopic fractionations associated with complex crystal chemical processes (e.g., solid solutions, chemical impurities, crystal defects, adsorption complexes), and the exploration of the mechanisms producing mass-independent fractionation, not to mention kinetic fractionation.

ACKNOWLEDGEMENTS

This chapter was improved by the thoughtful suggestions from T. Fujii, J. Kubicki and J. Wiederhold.

REFERENCES

Abe M, Suzuki T, Fujii Y, Hada M (2008a) An ab initio study based on a finite nucleus model for isotope fractionation in the U(III)–U(IV) exchange reaction system. J Chem Phys 128:144309

Abe M, Suzuki T, Fujii Y, Hada M, Hirao K (2008b) An ab initio molecular orbital study of the nuclear volume effects in uranium isotope fractionations. J Chem Phys 129:164309

Abe M, Suzuki T, Fujii Y, Hada M, Hirao K (2010) Ligand effect on uranium isotope fractionations caused by nuclear volume effects: An ab initio relativistic molecular orbital study. J Chem Phys 133:044309

Abe M, Hada M, Suzuki T, Fujii Y, Hirao K (2014) Theoretical study of isotope enrichment caused by nuclear volume effect. J Comp Chem, Japan 13:92–104

Anbar AD, Jarzecki AA, Spiro TG (2005) Theoretical investigation of iron isotope fractionation between $Fe(H_2O)_6^{3+}$ and $Fe(H_2O)_6^{2+}$: implications for iron stable isotope geochemistry. Geochim Cosmochim Acta 69:825–837

Angeli I, Marinova KP (2013) Table of experimental nuclear ground state charge radii: An update. Atom Data Nucl Data Tables 99:69–95

Anisimov VI, Zaanen J, Andersen OK (1991) Band theory and Mott insulators: Hubbard *U* instead of Stoner *I*. Phys Rev B 44:943

Balan E, Lazzeri M, Delattre S, Meheut M, Refson K, Winkler B (2007) Anharmonicity of inner-OH stretching modes in hydrous phyllosilicates: Assessment from first-principles frozen-phonon calculations. Phys Chem Minerals 34:621–625

Balan E, Blanchard M, Pinilla C, Lazzeri M (2014) First-principles modeling of sulfate incorporation and $^{34}S/^{32}S$ isotopic fractionation in different calcium carbonates. Chem Geol 374–375:84–91

Barkan E, Luz B (2012) High precision measurements of $^{17}O/^{16}O$ and $^{18}O/^{16}O$ ratios in CO_2. Rapid Commun Mass Spectrom 26:2733–2738

Baroni S, de Gironcoli S, Dal Corso A, Giannozzi P (2001) Phonons and related crystal properties from density-functional perturbation theory. Rev Mod Phys 73:515–561

Becke AD (1993) Density-functional thermochemistry. III. the role of exact exchange. J Chem Phys 98:5648–5652

Bergquist BA, Blum JD (2007) Mass-dependent and -independent fractionation of Hg isotopes by photoreduction in aquatic systems. Science 318:417–420

Bigeleisen J (1996) Nuclear size and shape effects in chemical reactions. Isotope chemistry of the heavy elements. J Am Chem Soc 118:3676–3680

Bigeleisen J, Mayer MG (1947) Calculation of equilibrium constants for isotopic exchange reactions. J Chem Phys 15:261–267

Black JR, Yin Q-Z, Rustad JR, Casey WH (2007) Magnesium isotopic equilibrium in chlorophylls. J Am Chem Soc 129:8690–8691

Black JR, Kavner A, Schauble EA (2011) Calculation of equilibrium stable isotope partition function ratios for aqueous zinc complexes and metallic zinc. Geochim Cosmochim Acta 75:769–783

Blanchard M, Poitrasson F, Méheut M, Lazzeri M, Mauri F, Balan E (2009) Iron isotope fractionation between pyrite (FeS_2), hematite (Fe_2O_3) and siderite ($FeCO_3$): A first-principles density functional theory study. Geochim Cosmochim Acta 73:6565–6578

Blanchard M, Morin G, Lazzeri M, Balan E (2010) First-principles study of the structural and isotopic properties of Al- and OH-bearing hematite. Geochim Cosmochim Acta 74:3948–3962

Blanchard M, Poitrasson F, Méheut M, Lazzeri M, Mauri F, Balan E (2012) Comment on "New data on equilibrium iron isotope fractionation among sulfides: Constraints on mechanisms of sulfide formation in hydrothermal and igneous systems" by VB Polyakov and DM Soultanov. Geochim Cosmochim Acta 86:182–195

Blanchard M, Dauphas N, Hu MY, Roskosz M, Alp EE, Golden DC, Sio CK, Tissot FLH, Zhao J, Gao L, Morris RV, Fornace M, Floris A, Lazzeri M, Balan E (2015) Reduced partition function ratios of iron and oxygen in goethite. Geochim Cosmochim Acta 151:19–33

Blöchl PE (1994) Projector augmented-wave method. Phys Rev B 50:17953–17979

Blum JD, Sherman LS, Johnson MW (2014) Mercury isotopes in earth and environmental sciences. Ann Rev Earth Planet Sci 42:249–269

Buchachenko AL (1995) Magnetic isotope effect. Theor Exp Chem 31:118–126
Buchachenko AL (2013) Mass-independent isotope effects. J Phys Chem B 117:2231–2238
Byrne RH, Yao W, Klochko K, Kaufman AJ, Tossell JA (2006) Experimental evaluation of the isotopic exchange equilibrium $^{10}B(OH)_3 + {}^{11}B(OH)_4^- = {}^{11}B(OH)_3 + {}^{10}B(OH)_4^-$ in aqueous solution. Deep Sea Res 1:684–688
Cao X, Liu Y (2011) Equilibrium mass-dependent fractionation relationships for triple oxygen isotopes. Geochim Cosmochim Acta 75:7435–7445
Ceperley D (1995) Path integrals in the theory of condensed helium. Rev Mod Phys 67:279–355
Ceriotti M, Markland TE (2013) Efficient methods and practical guidelines for simulating isotope effects. J Chem Phys 138:014112
Ceriotti M, More J, Manolopoulos DE (2014) i-PI: A Python interface for ab initio path integral molecular dynamics simulations. Comput Phys Commun 185:1019–1026
Cheng B, Ceriotti M (2014) Direct path integral estimators for isotope fractionation ratios. J Chem Phys 141:244112
Chialvo A, Horita J (2009) Liquid–vapour equilibrium fractionation of water: how well can classical water models predict it? J Chem Phys 130:094509
Clark SJ, Segall MD, Pickard CJ, Hasnip PJ, Probert MJ, Refson K, Payne MC (2005) First principles methods using CASTEP. Zeit Kristallogr 220:567–570
Clayton RN, Grossman L, Mayeda TK (1973) A component of primitive nuclear composition in carbonaceous meteorites. Science 182:485–488
Clayton RN (2002) Solar system: self-shielding in the solar nebula. Nature 415:860–861
Cococcioni M, de Gironcoli S (2005) Linear response approach to the calculation of the effective interaction parameters in the LDA+U method. Phys Rev B 71:035105
Colla CA, Wimpenny J, Yin Q-Z, Rustad JR, Casey WH (2013) Calcium-isotope fractionation between solution and solids with six, seven or eight oxygens bound to Ca(II). Geochim Cosmochim Acta 121:363–373
Cygan RT, Kubicki JD (2001) Molecular modeling theory: Applications in the geosciences. Rev Mineral Geochem 42
Dauphas N, Schauble EA (2016) Mass fractionation laws, mass independent effects, and isotopic anomalies. Ann Rev Earth Planet Sci 44: 709–783, doi:10.1146/annurev-earth-060115-012157
Dauphas N, Roskosz M, Alp EE, Golden DC, Sio CK, Tissot FLH, Hu MY, Zhao J, Gao L, Morris RV (2012) A general moment NRIXS approach to the determination of equilibrium Fe isotopic fractionation factors: Application to goethite and jarosite. Geochim Cosmochim Acta 94:254–275
Dauphas N, Roskosz M, Alp EE, Neuville D, Hu M, Sio CK, Tissot FLH, Zhao J, Tissandier L, Medard E, Cordier C (2014) Magma redox and structural controls on iron isotope variations in Earth's mantle and crust. Earth Planet Sci Lett 398:127–140
Domagal-Goldman SD, Kubicki JD (2008) Density functional theory predictions of equilibrium isotope fractionation of iron due to redox changes and organic complexation. Geochim Cosmochim Acta 72:5201–5216
Domagal-Goldman SD, Paul KW, Sparks DL, Kubicki JD (2009) Quantum chemical study of the Fe(III)-desferrioxamine B siderophore complex—Electronic structure, vibrational frequencies, and equilibrium Fe-isotope fractionation. Geochim Cosmochim Acta 73:1–12
Dovesi R, Orlando R, Erba A, Zicovich-Wilson CM, Civalleri B, Casassa S, Maschio L, Ferrabone M, De La Pierre M, D'Arco P, Noel Y, Causa M, Rerat M, Kirtman B (2014) CRYSTAL14: A Program for the Ab Initio Investigation of Crystalline Solids. Int J Quantum Chem 114:1287–1317
Dupuis R, Benoit M, Nardin E, Méheut M (2015) Fractionation of silicon isotopes in liquids: The importance of configurational disorder. Chem Geol 396:239–254
Estrade N, Carignan J, Sonke JE, Donard OFX (2009) Mercury isotope fractionation during liquid-vapor evaporation experiments. Geochim Cosmochim Acta 73:2693–2711
Farquhar J, Bao H, Thiemens M (2000) Atmospheric influence of Earth's earliest sulfur cycle. Science 289:756–758
Farquhar J, Johnston DT, Wing BA, Habicht KS, Canfield DE, Airieau S, Thiemens MH (2003) Multiple sulphur isotopic interpretations of biosynthetic pathways: implications for biological signatures in the sulphur isotope record. Geobiol 1:27–36
Feng C, Qin T, Huang S, Wu Z, Huang F (2014) First-principles investigations of equilibrium calcium isotope fractionation between clinopyroxene and Ca-doped orthopyroxene. Geochim Cosmochim Acta 143:132–142
Feynman R, Hibbs A (1965) Quantum mechanics and path integrals. McGraw-Hill, New York
Fricke G, Heilig K (2004) Group I: Elementary Particles, Nuclei and Atoms. Nuclear Charge Radii (Landolt-Börnstein: Numerical Data and Functional Relationships in Science and Technology, New Series) (Springer, Berlin), Vol 20
Frisch MJ, Trucks GW, Schlegel HB, Scuseria GE, Robb MA, Cheeseman JR, Scalmani G, Barone V, Mennucci B, Petersson GA, Nakatsuji H, Caricato M, Li X, Hratchian HP, Izmaylov AF, Bloino J, Zheng G, Sonnenberg JL, Hada M, Ehara M, Toyota K, Fukuda R, Hasegawa J, Ishida M, Nakajima T, Honda Y, Kitao O, Nakai H, Vreven T, Montgomery Jr JA, Peralta JE, Ogliaro F, Bearpark M, Heyd JJ, Brothers E, Kudin KN, Staroverov VN, Kobayashi R, Normand J, Raghavachari K, Rendell A, Burant JC, Iyengar SS, Tomasi J, Cossi M, Rega N, Millam NJ, Klene M, Knox JE, Cross JB, Bakken V, Adamo C, Jaramillo J, Gomperts R, Stratmann RE, Yazyev O, Austin AJ, Cammi R, Pomelli C, Ochterski JW, Martin RL, Morokuma K, Zakrzewski VG, Voth GA, Salvador P, Dannenberg JJ, Dapprich S, Daniels AD, Farkas O, Foresman JB, Ortiz JV, Cioslowski J, Fox DJ (2009) Gaussian 09, Revision B.01, Gaussian Inc, Wallingford, CT

Fujii T, Albarède F (2012) Ab initio calculation of the Zn isotope effect in phosphates, citrates, and malates and applications to plants and soil. PLoS ONE 7:e30726

Fujii T, Moynier F, Albarède F (2006a) Nuclear field vs. nucleosynthetic effects as cause of isotopic anomalies in the early Solar System. Earth Planet Sci Lett 247:1–9

Fujii T, Moynier F, Telouk P, Albarède F (2006b) Mass-independent isotope fractionation of molybdenum and ruthenium and the origin of isotopic anomalies in Murchison. Astrophys J 647:1506–1516

Fujii T, Fukutani S, Yamana H (2008) Isotope fractionation of strontium in a precipitation reaction of SrO_2. J Nucl Sci Technol Suppl 6:15–18

Fujii T, Moynier F Albarède F (2009) The nuclear field shift effect in chemical exchange reactions. Chem Geol 267:139–156

Fujii T, Moynier F, Telouk P, Abe M (2010) Experimental and theoretical investigation of isotope fractionation of zinc between aqua, chloro, and macrocyclic complexes. J Phys Chem A 114:2543–2552

Fujii T, Moynier F, Pons M-L, Albarède F (2011) The origin of Zn isotope fractionation in sulfides. Geochim Cosmochim Acta 75:7632–7643

Fujii T, Moynier F, Abe M, Nemoto K, Albarède F (2013a) Copper isotope fractionation between aqueous compounds relevant to low temperature geochemistry and biology. Geochim Cosmochim Acta 110:29–44

Fujii T, Moynier F, Agranier A, Ponzevera E, Abe M, Uehara A, Yamana H (2013b) Nuclear field shift effect in isotope fractionation of thallium. J Radioanal Nucl Chem 296:261–265

Fujii T, Moynier F, Blichert-Toft J, Albarède F (2014) Density functional theory estimation of isotope fractionation of Fe, Ni, Cu, and Zn among species relevant to geochemical and biological environments. Geochim Cosmochim Acta 140:553–576

Fujii T, Pringle EA, Chaussidon M, Moynier F (2015) Isotope fractionation of Si in protonation/deprotonation reaction of silicic acid: A new pH proxy. Geochim Cosmochim Acta 168:193–205

Gao YQ, Marcus RA (2001) Strange and unconventional isotope effects in ozone formation. Science 293:259–263

Garlick GD (1966) Oxygen isotope fractionation in igneous rocks. Earth Planet Sci Lett 1:361–368

Garrity KF, Bennett JW, Rabe KM, Vanderbilt D (2014) Pseudopotentials for high-throughput DFT calculations. Comput Mater Sci 81:446–452

Georg RB, Halliday AN, Schauble EA, Reynolds BC (2007) Silicon in the Earth's core. Nature 447:1102–1106

Ghosh S, Schauble EA, Lacrampe Couloume G, Blum JD, Bergquist BA (2013) Estimation of nuclear volume dependent fractionation of mercury isotopes in equilibrium liquid–vapor evaporation experiments. Chem Geol 336:5–12

Giannozzi P, Baroni S, Bonini N, Calandra M, Car R, Cavazzoni C, Ceresoli D, Chiarotti GL, Cococcioni M, Dabo I, Dal Corso A, de Gironcoli S, Fabris S, Fratesi G, Gebauer R, Gerstmann U, Gougoussis C, Kokalj A, Lazzeri M, Martin-Samos L, Marzari N, Mauri F, Mazzarello R, Paolini S, Pasquarello A, Paulatto L, Sbraccia C, Scandolo S, Sclauzero G, Seitsonen AP, Smogunov A, Umari P, Wentzcovitch RM (2009) Quantum ESPRESSO: a modular and open-source software project for quantum simulations of materials. J Phys Condens Matter 21:395502

Gonze X, Beuken J-M, Caracas R, Detraux F, Fuchs M, Rignanese G-M, Sindic L, Verstraete M, Zerah G, Jollet F, Torrent M, Roy A, Mikami M, Ghosez Ph, Raty J-Y, Allan DC (2002) First-principles computation of material properties : the ABINIT software project. Comput Mat Sci 25:478–492

Griffith EM, Schauble EA, Bullen TD, Paytan A (2008) Characterization of calcium isotopes in natural and synthetic barite. Geochim Cosmochim Acta 72:5641–5658

Grimme S (2011) Density functional theory with London dispersion corrections. WIREs Comput Mol Sci 1:211–228

He H, Liu Y (2015) Silicon isotope fractionation during the precipitation of quartz and the adsorption of H_4SiO_4(aq) on Fe(III)-oxyhydroxide surfaces. Chin J Geochem 34:459–468

He H, Zhang S, Zhu C, Liu Y (2016) Equilibrium and kinetic Si isotope fractionation factors and their implications for Si isotope distributions in the Earth's surface environments. Acta Geochim 35:15–24

Heidenreich JE, Thiemens MH (1986) A non-mass-dependent oxygen isotope effect in the production of ozone from molecular oxygen: the role of molecular symmetry in isotope chemistry. J Chem Phys 84 :2129

Hill PS, Schauble EA (2008) Modeling the effects of bond environment on equilibrium iron isotope fractionation in ferric aquo-chloro complexes. Geochim Cosmochim Acta 72:1939–1958

Hill PS, Schauble EA, Young ED (2010) Effects of changing solution chemistry on Fe^{3+}/Fe^{2+} isotope fractionation in aqueous Fe–Cl solutions. Geochim Cosmochim Acta 74:6669–6689

Hofmann MEG, Horváth B, Pack A (2012) Triple oxygen isotope equilibrium fractionation between carbon dioxide and water. Earth Planet Sci Lett 319–320:159–164

Hohenberg P, Kohn W (1964) Inhomogeneous electron gas. Phys Rev 136:B864–871

Horita J, Cole DR, Polyakov VB, Driesner T (2002) Experimental and theoretical study of pressure effects on hydrogen isotope fractionation in the system brucite-water at elevated temperatures. Geochim Cosmochim Acta 66:3769–3788

Huang F, Chen L, Wu Z, Wang W (2013) First-principles calculations of equilibrium Mg isotope fractionations between garnet, clinopyroxene, orthopyroxene, and olivine: Implications for Mg isotope thermometry. Earth Planet Sci Lett 367:61–70

Huang F, Wu Z, Huang S, Wu F (2014) First-principles calculations of equilibrium silicon isotope fractionation among mantle minerals. Geochim Cosmochim Acta 140:509–520

Hulston JR, Thode HG (1965) Variations in the S^{33}, S^{34}, and S^{36} contents of meteorites and their relation to chemical and nuclear effects. J Geophys Res 70:3475–3484

Ishida T (2002) Isotope effect and isotopic separation: A chemist's view. J Nucl Sci Tech 39:407–412

Jahn S, Wunder B (2009) Lithium speciation in aqueous fluids at high *P* and *T* studied by ab initio molecular dynamics and consequences for Li-isotope fractionation between minerals and fluids. Geochim Cosmochim Acta 73:5428–5434

Jancso G, Rebelo LPN, Van Hook WA (1993) Isotope effects in solution thermodynamics: Excess properties in solutions of isotopomers. Chem Rev 93:2645–2666

Javoy M, Balan E, Méheut M, Blanchard M, Lazzeri M (2012) First-principles investigation of equilibrium isotopic fractionation of O- and Si-isotopes between refractory solids and gases in the solar nebula. Earth Planet Sci Lett 319/320:118–127

Kieffer SW (1982) Thermodynamics and lattice vibrations of minerals: 5. Applications to phase equilibria, isotopic fractionation, and high-pressure thermodynamic properties. Rev Geophys Space Phys 20:827–849

King WH (1984) Isotope shifts in atomic spectra. Plenum Press, New York

Kleinman LI, Wolfsberg M (1973) Corrections to the Born–Oppenheimer approximation and electronic effects on isotopic exchange equilibria. J Chem Phys 59:2043–2053

Klochko K, Kaufman AJ, Yao W, Byrne RH, Tossell JA (2006) Experimental measurement of boron isotope fractionation in seawater. Earth Planet Sci Lett 248:276–285

Knyazev DA, Myasoedov NF (2001) Specific effects of heavy nuclei in chemical equilibrium. Separation Sci Technol 36:1677–1696

Knyazev DA, Semin GK, Bochkarev AV (1999) Nuclear quadrupole contribution to the equilibrium isotope effect. Polyhedron 18:2579–2582

Kohn W, Sham LJ (1965) Self-consistent equations including exchange and correlation effects. Phys Rev 140:A1133–A1138

Kowalski PM, Jahn S (2011) Prediction of equilibrium Li isotope fractionation between minerals and aqueous solutions at high *P* and *T*: An efficient ab initio approach. Geochim Cosmochim Acta 75:6112–6123

Kowalski PM, Wunder B, Jahn S (2013) Ab-initio prediction of equilibrium boron isotope fractionation between minerals and aqueous fluids at high *P* and *T*. Geochim Cosmochim Acta 101:285–301

Kresse G, Furthmüller J (1996) Efficient iterative schemes for ab initio total-energy calculations using a plane wave basis set. Phys Rev B 54:11169–11186

Landau L, Lifshitz E (1980) Course of Theoretical Physics. Statistical Physics. Part 1. Pergamon 5

Lee C, Yang W, Parr RG (1988) Development of the Colle–Salvetti correlation-energy formula into a functional of the electron density. Phys Rev B 37:785–789

Li X, Zhao H, Tang M, Liu Y (2009) Theoretical prediction for several important equilibrium Ge isotope fractionation factors and geological implications. Earth Planet Sci Lett 287:1–11

Li XF, Liu Y (2010) First-principles study of Ge isotope fractionation during adsorption onto Fe(III)-oxyhydroxide surfaces. Chem Geol 278:15–22

Li XF, Liu Y (2011) Equilibrium Se isotope fractionation parameters: A first-principles study. Earth Planet Sci Lett 304:113–120

Liu Y, Tossell JA (2005) Ab initio molecular orbital calculations for boron isotope fractionations on boric acids and borate. Geochim Cosmochim Acta 69:3995–4006

Liu Q, Tossell JA, Liu Y (2010) On the proper use of the Bigeleisen–Mayer equation and corrections to it in the calculation of isotopic fractionation equilibrium constants. Geochim Cosmochim Acta 74:6965–6983

Lyons JR (2007) Mass-independent fractionation of sulfur isotopes by isotope-selective dissociation of SO_2. Geophys Res Lett 34:L22811

Lyons JR, Young ED (2005) CO self-shielding as the origin of oxygen isotope anomalies in the early solar nebula. Nature 435:317–320

Markland TE, Berne B (2012) Unraveling quantum mechanical effects in water using isotopic fractionation. PNAS 109:7988–7991

Marsalek O, Chen P-Y, Dupuis R, Benoit M, Méheut M, Bacic Z, Tuckerman ME (2014) Efficient calculation of free energy differences associated with isotopic substitution using path-integral molecular dynamics. J Chem Theory Comput 10:1440–1453

Marx D, Hutter J (2000) Ab initio molecular dynamics: theory and implementation. *In:* Modern Methods and Algorithms of Quantum Chemistry, vol. 1 (ed. J. Grotendorst). NIC, FZ Jülich, pp. 301–449; CPMD Code: J. Hutter et al. Available from: www.cpmd.org/

Matsuhisa Y, Goldsmith JR, Clayton RN (1978) Mechanisms of hydrothermal crystallization of quartz at 250°C and 15 kbar. Geochim Cosmochim Acta 42:173–182

Mauersberger K (1987) Ozone isotope measurements in the stratosphere. Geophys Res Lett 14:80–83

Méheut M, Schauble EA (2014) Silicon isotope fractionation in silicate minerals: Insights from first-principles models of phyllosilicates, albite and pyrope. Geochim Cosmochim Acta 134:137–154

Mielke SL, Truhlar DG (2009) Improved methods for Feynman path integral calculations of vibrational-rotational free energies and application to isotopic fractionation of hydrated chloride ions. J Phys Chem A 113:4817–4827

Méheut M, Lazzeri M, Balan E, Mauri F (2007) Equilibrium isotopic fractionation between kaolinite, quartz and water: prediction from first-principles density functional theory. Geochim Cosmochim Acta 71:3170–3181

Méheut M, Lazzeri M, Balan E, Mauri F (2009) Structural control over equilibrium silicon and oxygen isotopic fractionation: a first-principles density-functional theory study. Chem Geol 258:28–37

Miller CA, Peucker-Ehrenbrink B, Schauble EA (2015) Theoretical modelling of rhenium isotope fractionation, natural variations across a black shale weathering profile, and potential as a paleoredox proxy. Earth Planet Sci Lett 430:339–348

Mook WG (2000) Environmental Isotopes in the Hydrological Cycle Principles and Applications, V, I: Introduction–Theory, Methods, Review. UNESCO/IAEA, Geneva

Moynier F, Yin Q-Z, Schauble E (2011) Isotopic evidence of Cr partitioning into Earth's core. Science 331:1417–1420

Moynier F, Fujii T, Wang K, Foriel J (2013) Ab initio calculations of the Fe(II) and Fe(III) isotopic effects in citrates, nicotianamine, and phytosiderophore, and new Fe isotopic measurements in higher plants. CR Géosci 245:230–240

Nadjakov EG, Marinova KP, Gangrsky YuP (1994) Systematics of nuclear charge radii. At Data Nucl Data Tables 56:133–157

Nemoto K, Abe M, Seino J, Hada M (2015) An ab initio study of nuclear volume effects for isotope fractionations using two-component relativistic methods. J Comput Chem 36:816–820

Nielsen SG, Klein F, Kading T, Blusztajn J, Wickham K (2015) Thallium as a tracer of fluid–rock interaction in the shallow Mariana forearc. Earth Planet Sci Lett 430:416–426

Nishizawa K, Satoyama T, Miki T, Yamamoto T, Hosoe M (1995) Strontium isotope effect in liquid–liquid extraction of strontium chloride using a crown ether. J Nucl Sci Technol 32:1230–1235

Nomura M, Higuchi N, Fujii Y (1996) Mass dependence of uranium isotope effects in the U(IV)–U(VI) exchange reaction. J Am Chem Soc 118:9127–9130

Otake T, Lasaga AC, Ohmoto H (2008) Ab initio calculations for equilibrium fractionations in multiple sulfur isotope systems. Chem Geol 249:357–376

Ottonello G, Zuccolini MV (2005) Ab-initio structure, energy and stable Cr isotopes equilibrium fractionation of some geochemically relevant H–O–Cr–Cl complexes. Geochim Cosmochim Acta 69:851–874

Ottonello G, Zuccolini MV (2008) The iron-isotope fractionation dictated by the carboxylic functional: An ab-initio investigation. Geochim Cosmochim Acta 72:5920–5934

Ottonello G, Zuccolini MV (2009) Ab-initio structure, energy and stable Fe isotope equilibrium fractionation of some geochemically relevant H–O–Fe complexes. Geochim Cosmochim Acta 73:6447–6469

Ottonello G, Civalleri B, Zuccolini MV, Zicovich-Wilson (2007) Ab-initio thermal physics and Cr-isotopic fractionation of $MgCr_2O_4$. Am Mineral 92:98–108

Pavlov AA, Kasting JF (2002) Mass-independent fractionation of sulfur isotopes in Archean sediments: Strong evidence for an anoxic Archean atmosphere: Astrobiology 2:27–41

Pérez A, von Lilienfeld OA (2011) Path integral computation of quantum free energy differences due to alchemical transformations involving mass and potential. J Chem Theory Comput 7:2358–2369

Perdew JP, Ruzsinszky A (2010) Density functional theory of electronic structure: A short course for mineralogists and geophysicists. Rev Mineral Geochem 71:1–18

Pinilla C, Blanchard M, Balan E, Ferlat G, Vuilleumier R, Mauri F (2014) Equilibrium fractionation of H and O isotopes in water from path integral molecular dynamics. Geochim Cosmochim Acta 135:203–216

Pinilla C, Blanchard M, Balan E, Natarajan SK, Vuilleumier R, Mauri F (2015) Equilibrium magnesium isotope fractionation between aqueous Mg^{2+} and carbonate minerals: insights from path integral molecular dynamics. Geochim Cosmochim Acta 163:126–139

Polyakov VB (1997) Equilibrium fractionation of the iron isotopes: Estimation from Mössbauer spectroscopy data. Geochim Cosmochim Acta 61:4213–4217

Polyakov VB, Mineev SD (2000) The use of Mössbauer spectroscopy in stable isotope geochemistry. Geochim Cosmochim Acta 64:849–865

Polyakov VB, Soultanov DM (2011) New data on equilibrium iron isotope fractionation among sulfides: constraints on mechanisms of sulfide formation in hydrothermal and igneous systems. Geochim Cosmochim Acta 75:1957–1974

Polyakov VB, Mineev SD, Clayton RN, Hu G, Mineev KS (2005) Determination of tin equilibrium isotope fractionation factors from synchrotron radiation experiments. Geochim Cosmochim Acta 69:5531–5536

Polyakov VB, Clayton RN, Horita J, Mineev SD (2007) Equilibrium iron isotope fractionation factors of minerals: reevaluation from the data of nuclear inelastic resonant X-ray scattering and Mössbauer spectroscopy. Geochim Cosmochim Acta 71:3833–3846

Polyakov V, Osadchii E, Chareev D, Chumakov A, Sergeev I (2013) Fe β-factors for sulfides from NRIXS Synchrotron Experiments. Mineral Mag 77:1985

Ramírez R, Herrero CP (2010) Quantum path integral simulation of isotope effects in the melting temperature of ice Ih. J Chem Phys 133:144511

Redlich O (1935) Eine allgemeine Beziehung zwischen den Schwingungsfrequenzen isotoper Molekeln. Z Phys Chem B 28:371–382

Rehkämper M, Frank M, Hein JR, Porcelli D, Halliday A, Ingri J, Liebetrau V (2002) Thallium isotope variations in seawater and hydrogenetic, diagenetic, and hydrothermal ferromanganese deposits. Earth Planet Sci Lett 197:65–81

Richet P, Bottinga Y, Javoy M (1977) A review of hydrogen, carbon, nitrogen, oxygen, sulphur, and chlorine stable isotope fractionation among gaseous molecules. Ann Rev Earth Planet Sci 5:65–110

Richter R, Hoernes S (1988) The application of the increment method in comparison with experimentally derived and calculated O-isotope fractionations. Chem Erde 48:1–18

Roothaan CCJ (1951) New developments in molecular orbital theory. Rev Modern Phys 23:69–89

Roskosz M, Sio CKI, Dauphas N, Bi W, Tissot FHL, Hu MY, Zhao J, Alp EE (2015) Spinel–olivine–pyroxene equilibrium iron isotopic fractionation and applications to natural peridotites. Geochim Cosmochim Acta 169:184–199

Rumble D, Miller MF, Franchi IA, Greenwood RC (2007) Oxygen three-isotope fractionation lines in terrestrial silicate minerals: an inter-laboratory comparison of hydrothermal quartz and eclogitic garnet Geochim Cosmochim Acta 71:3592–3600

Rustad JR, Bylaska EJ (2007) Ab initio calculation of isotopic fractionation in $B(OH)_3(aq)$ and $B(OH)_4^-(aq)$. J Am Chem Soc 129:2222–2223

Rustad JR, Dixon DA (2009) Prediction of iron-isotope fractionation between hematite (α-Fe_2O_3) and ferric and ferrous iron in aqueous solution from density functional theory. J Phys Chem A 113:12249–12255

Rustad JR, Yin Q-Z (2009) Iron isotope fractionation in the Earth's lower mantle. Nature Geoscience 2:514–518

Rustad JR, Casey WH, Yin Q-Z, Bylaska EJ, Felmy AR, Bogatko SA, Jackson VE, Dixon DA (2010a) Isotopic fractionation of Mg^{2+}, Ca^{2+}, and Fe^{2+} with carbonate minerals. Geochim Cosmochim Acta 74:6301–6323

Rustad JR, Bylaska EJ, Jackson VE, Dixon DA (2010b) Calculation of boron-isotope fractionation between $B(OH)_3(aq)$ and $B(OH)_4^-(aq)$. Geochim Cosmochim Acta 74:2843–2850

Schauble EA (2004) Applying stable isotope fractionation theory to new systems. Rev Mineral Geochem 55:65–111

Schauble EA (2007) Role of nuclear volume in driving equilibrium stable isotope fractionation of mercury, thallium, and other very heavy elements. Geochim Cosmochim Acta 71:2170–2189

Schauble EA (2011) First-principles estimates of equilibrium magnesium isotope fractionation in silicate, oxide, carbonate and hexaaquamagnesium(2+) crystals. Geochim Cosmochim Acta 75:844–869

Schauble EA (2013) Modeling nuclear volume isotope effects in crystals. Proc Nat Acad Sci (USA) 110:17714–17719

Schauble EA, Rossman GR, Taylor HP Jr (2004) Theoretical estimates of equilibrium chromium-isotope fractionations, Chem Geol 205:99–114

Schauble EA, Ghosh P, Eiler JM (2006) Preferential formation of ^{13}C–^{18}O bonds in carbonate minerals, estimated using first-principles lattice dynamics. Geochim Cosmochim Acta 70:2510–2529

Schmidt MW, Baldridge KK, Boatz JA, Elbert ST, Gordon MS, Jensen JH, Koseki S, Matsunaga N, Nguyen KA, Su S, Windus TL, Dupuis M, Montgomery JA (1993) General Atomic and Molecular Electronic Structure System. J Comput Chem 14:1347–1363

Schott J, Mavromatis V, Fujii T, Pearce CR, Oelkers EH (2016) The control of magnesium aqueous speciation on Mg isotope composition in carbonate minerals: Theoretical and experimental modeling. Chem Geol doi: 10.1016/j.chemgeo.2016.03.011

Schütze H (1980) Der Isotopenindex—eine Inkrementenmethode zur näherungsweise Berechnung von Isotopenaustauschgleichgewichten zwischen kristallinen Substanzen. Chem Erde 39:321–334

Schuchardt KL, Didier BT, Elsethagen T, Sun L, Gurumoorthi V, Chase J, Li J, Windus TL (2007) Basis Set Exchange: A Community Database for Computational Sciences. J Chem Inf Model 47:1045–1052

Seo JH, Lee SK, Lee I (2007) Quantum chemical calculations of equilibrium copper (I) isotope fractionations in ore-forming fluids. Chem Geol 243:225–237

Sherman DM (2013) Equilibrium isotopic fractionation of copper during oxidation/reduction, aqueous complexation and ore-forming processes: Predictions from hybrid density functional theory. Geochim Cosmochim Acta 118:85–97

Skulan JL, Beard BL, Johnson CM (2002) Kinetic and equilibrium Fe isotope fractionation between aqueous Fe(III) and hematite. Geochim Cosmochim Acta 66:2995–3015

Smyth JR, Clayton RN (1988) Correlation of oxygen isotope fractionation and electrostatic site potentials in silicates. Eos 69:1514 (abstr.)

Stillinger FH, Weber TA (1983) Inherent structure in water. J Phys Chem 87:2833–2840

Stirling CH, Andersen MB, Potter E-K, Halliday AN (2007) Low-temperature isotopic fractionation of uranium. Earth Planet Sci Lett 264:208–225

Suzuki K, Shiga M, Tachikawa M (2008) Temperature and isotope effects on water cluster ions with path integral molecular dynamics based on the forth order Trotter expansion. J Chem Phys 129:144310

Syverson DD, Borrok DM, Seyfried Jr WE (2013) Experimental determination of equilibrium Fe isotopic fractionation between pyrite and dissolved Fe under hydrothermal conditions. Geochim Cosmochim Acta 122:170–183

Tachikawa M, Shiga M (2005) Ab initio path integral simulation study on $^{16}O/^{18}O$ isotope effect in water and hydronium ion. Chem Phys Lett 407:135–138

Taylor HP Jr, Epstein S (1962) Relationships between $^{18}O/^{16}O$ ratios in coexisting minerals of igneous and metamorphic rocks, Part 2. Application to petrologic problems. Geol Soc Am Bull 73:675–694

Tossell JA (2005) Calculating the partitioning of the isotopes of Mo between oxidic and sulfidic species in aqueous solution. Geochim Cosmochim Acta 69:2981–2993

Tossell JA (2006) Boric acid adsorption on humic acids: Ab initio calculation of structures, stabilities, B-11 NMR and B-11, B-10 isotopic fractionations of surface complexes. Geochim Cosmochim Acta 70:5089–5103

Tuckerman M (2010) Statistical mechanics: theory and molecular simulation. Oxford University Press

Tuckerman ME, Yarne DA, Samuelson SO, Hughes AL, Martyna GJ (2000) Exploiting multiple levels of parallelism in molecular dynamics based calculations via modern techniques and software paradigms on distributed memory computers. Comp Phys Commun 128:333–376

Turro NJ (1983) Influence of nuclear spin on chemical reactions: Magnetic isotope and magnetic field effects (a review). Proc Natl Acad Sci USA 80:609–621

Urey HC (1947) The thermodynamic properties of isotopic substances. J Chem Soc (Lond.) 562–581

Valiev M, Bylaska EJ, Govind N, Kowalski K, Straatsma TP, van Dam HJJ, Wang D, Nieplocha J, Apra E, Windus TL, de Jong WA (2010) NWChem: a comprehensive and scalable open-source solution for large scale molecular simulations. Comput Phys Commun 181:1477

Vanicek J, Miller W (2007) Efficient estimators for quantum instanton evaluation of the kinetic isotope effects: application to the intramolecular hydrogen transfer in pentadiene. J Chem Phys 127:114309

Wasylenki LE, Rolfe BA, Weeks CL, Spiro TG, Anbar AD (2008) Experimental investigation of the effects of temperature and ionic strength on Mo isotopic fractionation during adsorption to manganese oxide. Geochim Cosmochim Acta 72:5997–6005

Wasylenki LE, Weeks CL, Bargar JR, Spiro TG, Hein JR, Anbar AD (2011) The molecular mechanism of Mo isotope fractionation during adsorption to birnessite. Geochim Cosmochim Acta 75:5019–5031

Webb MA, Miller TF (2014) Position-specific and clumped stable isotope studies: comparison of the Urey and path-integral approaches for carbon dioxide, nitrous oxide, methane, and propane. J Phys Chem A 118:467–474

Weeks CL, Anbar AD, Wasylenki LE, Spiro TG (2007) Density functional theory analysis of molybdenum isotope fractionation. J Phys Chem A 111:12434–12438

Weeks CL, Anbar AD, Wasylenki LE, Spiro TG (2008) Density functional theory analysis of molybdenum isotope fractionation (correction to v111, 12434, 2007) J Phys Chem A 112:10703

Welch SA, Beard BL, Johnson CM, Braterman PS (2003) Kinetic and equilibrium Fe isotope fractionation between aqueous Fe(II) and Fe(III). Geochim Cosmochim Acta 67:4231–4250

Weston RE Jr. (1999) Anomalous or mass-independent isotope effects. Chem Rev 99:2115–2136

Widanagamage IH, Schauble EA, Scher HD, Griffith EM (2014) Stable strontium isotope fractionation in synthetic barite. Geochim Cosmochim Acta 147:58–75

Wiederhold JG, Cramer CJ, Daniel K, Infante I, Bourdon B, Kretzschmar R (2010) Equilibrium mercury isotope fractionation between dissolved Hg(II) species and thiol-bound Hg. Environ Sci Technol 44:4191–4197

Wiesli RA, Beard BL, Johnson CM (2004) Experimental determination of Fe isotope fractionation between aqueous Fe(II), siderite and "green rust" in abiotic systems. Chem Geol 211:343–362

Wilson EBJ, Decius JC, Cross PC (1955) Molecular Vibrations: The Theory of Infrared and Raman Vibrational Spectra. Dover, New York

Wu F, Qin T, Li X, Liu Y, Huang J-H, Wu Z, Huang F (2015a) First-principles investigation of vanadium isotope fractionation in solution and during adsorption. Earth Planet Sci Lett 426:216–224

Wu ZQ, Huang F, Huang S (2015b) Isotope fractionation induced by phase transformation: First-principles investigation for Mg_2SiO_4. Earth Planet Sci Lett 409:339–347

Yamaji K, Makita Y, Watanabe H, Sonoda A, Kanoh H, Hirotsu T, Ooi K (2001) Theoretical estimation of lithium isotopic reduced partition function ratio for lithium ions in aqueous solution. J Phys Chem A 105:602–613

Yang S, Liu Y (2015) Nuclear volume effects in equilibrium stable isotope fractionations of mercury, thallium and lead. Sci Reports 5:12626

Yang JL, Li YB, Liu SQ, Tian HQ, Chen CY, Liu JM, Shi YL (2015) Theoretical calculations of Cd isotope fractionation in hydrothermal fluids. Chem Geol 391:74–82

Young ED, Galy A, Nagahara H (2002) Kinetic and equilibrium mass-dependent isotope fractionation laws in nature and their geochemical and cosmochemical significance. Geochim Cosmochim Acta 66:1095–1104

Young ED, Manning CE, Schauble EA, Shahar A, Macris CA, Lazar C, Jordan M (2015) High-temperature equilibrium isotope fractionation of non-traditional stable isotopes : Experiments, theory, and applications. Chem Geology 395:176–195

Zeebe RE (2005) Stable boron isotope fractionation between dissolved $B(OH)_3$ and $B(OH)_4^-$. Geochim Cosmochim Acta 69:2753–2766

Zeidler A, Salmon PS, Fischer HE, Neuefeind JC, Simonson JM, Markland TE (2012) Isotope effects in water as investigated by neutron diffraction and path integral molecular dynamics. J Phys: Condens Matter 24:284126
Ziegler K, Young ED, Schauble EA, Wasson JT (2010) Metal–silicate silicon isotope fractionation in enstatite meteorites and constraints on Earth's core formation. Earth Planet Sci Lett 295:487–496
Zwanziger J (2009) Computation of Mössbauer isomer shifts from first principles. J Phys Condens Matter 21:195501

Reviews in Mineralogy & Geochemistry
Vol. 82 pp. 65-83, 2017

Equilibrium Fractionation of Non-traditional Stable Isotopes: an Experimental Perspective

Anat Shahar, Stephen M. Elardo

Geophysical Laboratory
Carnegie Institution for Science
Washington, DC 20015
USA

ashahar@carnegiescience.edu; selardo@carnegiescience.edu

Catherine A. Macris

Indiana University – Purdue University Indianapolis
Indianapolis, IN 46202
USA

camacris@iupui.edu

INTRODUCTION

In 1986, O'Neil wrote a Reviews in Mineralogy chapter on experimental aspects of isotopic fractionation. He noted that in order to fully understand and interpret the natural variations of light stable isotope ratios in nature, it was essential to know the magnitude and temperature dependence of the isotopic fractionation factor amongst minerals and fluids. At that time it was difficult to imagine that this would become true for the heavier, so called non-traditional stable isotopes, as well. Since the advent of the multiple collector inductively coupled plasma-source mass spectrometer (MC–ICP–MS), natural variations of stable isotope ratios have been found for almost any polyisotopic element measured. Although it has been known that as temperature and mass increase, isotope fractionation decreases very quickly, the MC–ICP–MS has revolutionized the ability of a geochemist to measure very small differences in isotope ratios. It was then that the field of experimental non-traditional stable isotope geochemistry was born. As O'Neil (1986) pointed out there are three ways to obtain isotopic fractionation factors: theoretical calculations, measurements of natural samples with well-known formation conditions, and laboratory calibration studies. This chapter is devoted to explaining the techniques involved with laboratory experiments designed to measure equilibrium isotope fractionation factors as well as the best practices that have been learned. Although experimental petrology has been around for a long time and basic experimental methods have been well-refined, there are additional considerations that must be taken into account when the goal is to measure isotopic compositions at the end of the experiment. It has been only about ten years since these initial studies were published, but much has been learned in that time about how best to conduct experiments aimed at determining equilibrium fractionation factors. We will not focus on the scientific results that have been determined by such experiments, as each chapter in this book will focus on a different element of interest. Instead we will provide a *how-to* for those interested in conducting these experiments in the future.

1529-6466/17/0082-0003$05.00 http://dx.doi.org/10.2138/rmg.2017.82.3

FACTORS INFLUENCING EQUILIBRIUM FRACTIONATION FACTORS

Equilibrium isotope fractionation is driven by the effects of atomic mass on bond vibrational energy. The relationships are easily understood using a simple molecule as an example. All molecules have a zero-point vibrational energy (ZPE) = ½ hν, where ν is the vibrational frequency and h is Planck's constant. The vibrational frequency of a particular mode for a molecule can be approximated using Hooke's Law, $\nu = ½\pi\sqrt{(k/\mu)}$, where k is the force constant and μ is the reduced mass of the molecule (e.g. $\mu = m_a m_b / [m_a + m_b]$, where m_a and m_b are the atomic weights of two atoms in a diatomic molecule). When a light isotope in a molecule is substituted by a heavy isotope the potential energy curve does not (to a first approximation) change shape, hence the force constant does not change, but the vibrational frequency does change. When a more massive isotope is substituted, the reduced mass of the molecule increases, which decreases the vibrational frequency and the energy. Therefore, equilibrium stable isotope fractionations are quantum-mechanical effects that depend on the ZPE of the molecule being investigated.

The same principle applies to crystalline materials. To a first approximation, the most important factor that determines the magnitude of isotopic fractionation is differences in bond strength; stiffer bonds concentrate the heavy isotope. Bond strength (stiffness) determines vibrational frequency and vibrational frequency determines internal energy. Stiffer bonds tend to correlate with high oxidation state, covalent bonding, and low coordination number. Therefore, all else being equal, the heavy isotopes of an element will partition preferentially into phases with these characteristics. Furthermore, increasing the pressure experienced by a given phase stiffens the bonds while increasing the temperature weakens the bonds. Therefore, the main variables that influence the equilibrium fractionation factor between two phases are temperature, pressure, oxygen fugacity and composition, as these are the variables that will most notably change the bond strength of the isotope of interest. For a thorough explanation of how temperature, pressure, composition and oxygen fugacity affect isotope fractionation the interested reader can reference Young et al. (2002, 2015) and Schauble (2004).

It has been known since the work of Urey (1947) and Bigeleisen and Mayer (1947) that temperature is crucial for the determination of equilibrium constants. These seminal works included calculations of equilibrium constants for isotopic exchange reactions as a function of temperature. At high temperature, the equilibrium constant becomes proportional to the inverse square of temperature (i.e., isotope fractionation decreases proportional to $1/T^2$). Thus, at the high temperatures involved in many Earth systems isotope fractionation was considered negligible and ignored for elements heavier than the traditional stable isotopes (C, H, S, O, N). Similarly, the effect of pressure on isotope fractionation was also ignored.

Joy and Libby (1960) first calculated the effect of pressure on isotope fractionation. They suggested that oxygen isotope fractionation might be pressure dependent at low temperatures. However, in 1961, Hoering did not observe a pressure effect on oxygen isotope partitioning between water and bicarbonate at 43.5 °C and between 0.1 and 400 MPa. Then in 1975, Clayton et al. found no pressure effect on the calcite-water fractionation at 500 °C, 0.1–2 GPa and at 700 °C, 50–100 MPa. Due to these initial studies, the effect of pressure on isotope fractionation was assumed to be negligible for all elements (except for hydrogen) at all pressures. However, in 1994, a study again predicted that pressure should have an effect on isotopic fractionation (Polyakov and Kharlashina 1994).

There are at least two ways in which pressure can affect isotope fractionation: differential molar volume decrease and force constant stiffening. The first cause is due to a molar volume isotope effect in which the heavy isotopes make slightly shorter bonds and therefore pack more tightly than lighter isotopes, and has been discussed in the literature extensively (e.g., Polyakov 1998). This effect comes mainly from the quantum vibrations of the nuclei inside their surrounding electronic cloud. The second cause is due to an increase of the force constants and, correspondingly, of the vibrational frequencies due to the stiffening of the bonds as the volume decreases due to increasing pressure.

The effect of oxygen fugacity on isotope fractionation was largely untested until the redox-active transition elements and main group elements could be measured with adequate precision. Iron was the first element to be studied in great detail and since 1999 many studies have shown that, in both low and high temperature environments, the oxidation state (Fe^{3+}, Fe^{2+}) exerts strong control on the magnitude of the Fe isotope fractionation (e.g., Williams et al. 2004; Shahar et al. 2008; Dauphas et al. 2014). Likewise, theoretical and experimental (e.g., Hill et al. 2009) results at low temperature have shown that speciation (bond partner and coordination in a solution) also regulates mass fractionation of Fe. In high-temperature solids or melts, the ligand to which Fe is bound also should affect the Fe isotopic fractionation (Schauble 2004). One of the main reasons that experiments are so critical to the study of equilibrium fractionation factors is that each of these variables can be studied individually in an experiment in order to determine which has the largest effect and how the variables influence each other. As mentioned above, Fe isotopes have been studied in the greatest detail experimentally, so many of the examples in this chapter will focus on Fe. However there are many elements that have been, or have the potential to be experimentally studied.

PROOF OF EQUILIBRIUM IN ISOTOPE EXPERIMENTS

The goal of typical stable isotope fractionation experiments is to determine the equilibrium fractionation factor between two phases at a set of temperatures, pressures, oxygen fugacities, and/or compositions of interest. The initial challenge for the experimentalist is to design the simplest, most elegant experiment that will achieve their goal, given the myriad constraints imposed by the laboratory. These issues include (but are not limited to) sample size, time, *P–T* capabilities of the experimental apparatus, availability and usefulness of appropriate starting materials and oxygen buffers, and the ability to separate phases for analysis after an experiment. For example, experimental apparati usually limit sample sizes to milli- or microgram levels, and realistic laboratory time scales impose challenges with systems involving diffusion as the method of isotope exchange. Additionally, and perhaps most importantly, is the ability to prove that the fractionation measured in an experiment is, in fact, the equilibrium fractionation factor between the phases of interest. This point is nontrivial and must be given significant thought when planning a set of experiments. This section will highlight some of the more effective experimental approaches and techniques employed in the past and speak to best practices for achieving a very close approach to equilibrium in stable isotope fractionation experiments.

Time series

The use of a series of experiments conducted over a range of durations to assess the approach to equilibrium is one of the most commonly used methods in experimental petrology. The concept is straightforward: a series of experiments (≥ 3) with the same starting materials is conducted at the same pressure and temperature (usually the lowest of interest) for increasingly longer durations. As the isotopes exchange between phases, the system moves closer to having an equilibrium distribution of isotopes with increasing experimental duration. The equilibrium fractionation between the two phases is estimated by running long enough experiments so that the system appears to have reached a steady state, i.e., there is no further detectable net isotope exchange at progressively longer run times. The run products are then analyzed to ensure that they yield consistent results, which can be used to argue for a close approach to equilibrium (or at minimum a steady state). The time series results then guide future experiments.

A time series should also be used to assess chemical equilibrium independent of isotopic equilibrium. Chemical zoning in experimentally grown phases is always an indicator of disequilibrium and should lead to longer duration experiments whenever practical. An exception to this is dynamic cooling and/or decompression experiments designed to assess kinetic effects at controlled cooling and/or decompression rates. Some chemical zoning is tolerable in more traditional phase equilibrium experiments, as microbeam analyses can be made at phase boundaries where a local equilibrium may exist. However, most isotopic fractionation experiments require the physical separation of phases and analysis of the bulk phase. In this case chemical zoning is less tolerable as it will likely be accompanied by isotopic zoning. For this reason, the durations of phase equilibrium experiments designed to investigate isotopic fractionation may need to be longer than is typical for studies of element partitioning or the like (Shahar et al. 2008). *In situ* isotopic analyses using laser ablation MC–ICP–MS (LA–MC–ICP–MS) or secondary ion mass spectrometry (SIMS) are ways to determine if isotopic zoning has occurred in an experiment, and to obtain spatially oriented measurements of the experimental product. However, extreme care must be taken when treating the resulting data to show the measured values represent a close approach to equilibrium.

Multi-direction approach

The multi-direction approach is also discussed in O'Neil (1986) and has been termed "reverse reactions" in the experimental petrology community (Fig. 1). It is an extension of the time series described above with the added benefit of approaching the equilibrium isotope value from two sides, which results in a more precise determination of the equilibrium value by bracketing. This is done by conducting at least two time series in which the difference between the isotopic compositions of the starting materials lie on either side of the equilibrium isotope fractionation value of the system (line labeled 'Equilibrium' in Fig. 1). Over time, the two sets of starting materials (differing only in isotopic value) evolve towards the equilibrium isotopic state of the system, thereby bracketing the true equilibrium fractionation factor of the two phases. Although this idea seems straightforward there are subtleties associated with non-ideal conditions and solid phases that could complicate the system which are discussed in great detail by Pattison (1994). Although the multi-direction approach is more rigorous than a simple time series, its accuracy in determining the equilibrium fractionation factor is limited by the ability to come close to equilibrium in laboratory time scales. For some systems, this technique may not be practical due to sluggish isotope exchange. In such cases, a different strategy is required to determine equilibrium fractionation factors.

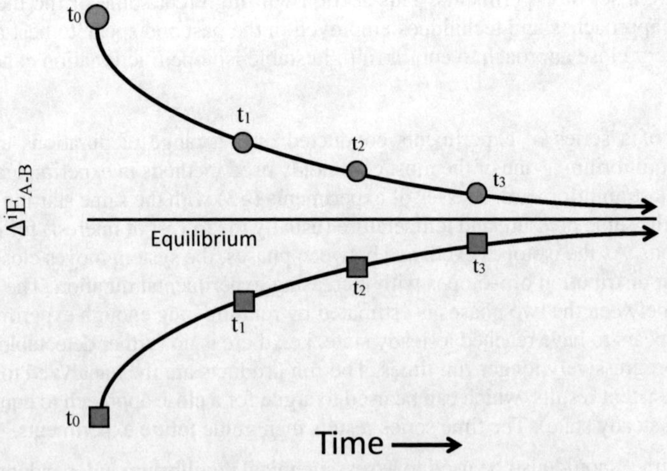

Figure 1. A schematic of a multi-direction approach experiment showing two phases ('A' and 'B') that have starting fractionation values (Δ^iE_{A-B}) lying on opposite sides of the equilibrium value. As the experimental duration progresses the values bracket the true equilibrium value. Here 'i' represents an arbitrary isotope of element 'E', and $\Delta^iE_{A-B} = \delta^iE_A - \delta^iE_B$.

The Northrop–Clayton method is a variation of the multi-direction approach that allows an estimation of the equilibrium fractionation between two phases by extrapolation. This is a strategy that can be employed when attaining equilibrium is not possible on time scales relevant to laboratory experiments. This approach, termed the "partial exchange technique" in O'Neil (1986), was first described by Northrop and Clayton (1966) and later by Deines and Eggler (2009). The partial exchange method involves two phases that each contain at least one site for an element of interest that exchanges isotopes throughout an experiment. A series of experiments are conducted with identical solids and several isotopically variable fluids that bracket the equilibrium isotopic fractionation factor. Just like in the multi-direction approach, the evolution of the fractionation with time will bracket the equilibrium value.

The benefit of the Northrop–Clayton method is that a close approach to equilibrium is not required to estimate the fractionation factor. This is true because this method assumes that rates of isotopic exchange for 'companion' experiments (runs that are the same in every respect except the isotopic composition of starting materials) are identical. This assumption allows one to extrapolate to the equilibrium value of the system based on the 'percentage of isotopic exchange' in two or more companion experiments (O'Neil 1986). For this method, the extent of isotopic exchange is directly proportional to the accuracy of the fractionation factor. In other words, if the isotopes are not largely exchanged, the fractionation might appear to be larger than the true equilibrium value. A combination of this partial exchange technique and the classical time series approach was used successfully by Schuessler et al. (2007) in obtaining the equilibrium Fe isotope fractionation between pyrrhotite and rhyolitic melt.

Three-isotope exchange method

Another method of determining equilibrium fractionation factors is termed the three-isotope exchange method. This method was pioneered at the University of Chicago (Matsuhisa et al. 1978; Matthews et al. 1983) for oxygen isotopes between a mineral and an aqueous phase and later modified by Shahar et al. (2008) to determine direct mineral–mineral fractionation of Fe isotopes. This method requires that the element of interest has at least three stable isotopes in measurable abundances and utilizes the addition of a known amount of isotopic 'spike' (addition of a specific isotope to a phase in excess of its natural abundance) to one of the starting materials containing the element of interest. The choice of which isotope to use as the spike is made obvious by looking at a three-isotope plot (Fig. 2). By adding an excess amount of the isotope present in the denominator of both axes (represented by an 'x' in the axes labels on Fig. 2A, B) to one of the starting materials, the spiked phase will be displaced from the terrestrial fractionation line (TFL) in three-isotope space by a line with a slope of unity.

The principle of the three-isotope exchange method is to replace the TFL, which has a zero intercept on a three-isotope plot, with a secondary fractionation line (SFL) with the same slope as the TFL, but a non-zero intercept determined by the bulk composition of the isotopically spiked system. Figure 2A depicts the typical trend to equilibrium expected in these experiments. As the spiked (Phase 2 Initial) and unspiked (Phase 1 Initial) starting materials exchange isotopes at the pressure and temperature of interest for increasingly long times (t_1, t_2), their isotopic compositions migrate towards the SFL. When isotope exchange is complete and the system reaches isotopic equilibrium, all phases containing the element of interest will lie on the SFL (Phase 1 & 2 Equilibrium), and the equilibrium fractionation factor will be defined by the distance between the final measured values along the x- and y-axes corresponding to isotopes 'j' and 'i', respectively. The three-isotope exchange method can also be used to extrapolate to the equilibrium isotopic value in cases where exchange is slow. In this case it is quite important that the experiment not be 'over' spiked so as to limit the amount of extrapolation that is necessary. A good rule of thumb is to incorporate no more (and sometimes much less) than 1% of the element as a spike.

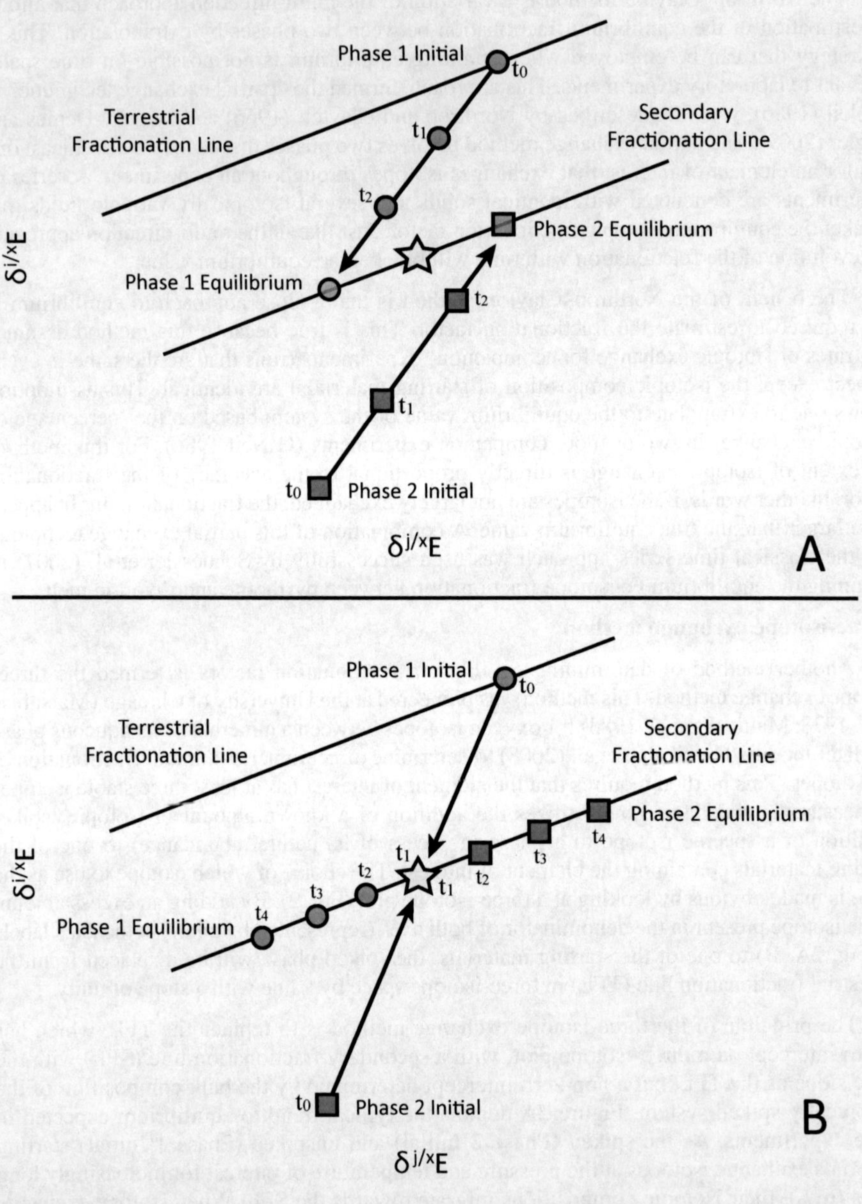

Figure 2. A schematic of the three-isotope exchange method shown for arbitrary isotopes 'i', 'j', and 'x', of element 'E'. Figure 2A depicts the more traditional path during a dissolution and precipitation experiment. Phase 1 (red circles) is unspiked and starts on the terrestrial fractionation line. Over time, as the system evolves towards equilibrium, Phase 1 moves to lighter values, while Phase 2 (blue squares), which is spiked so that is starts off of the terrestrial fractionation line, moves towards heavier values. The yellow star represents the bulk value of the system. Equilibrium is reached when both phases reach the secondary fractionation line, though in some cases the phases may not fully reach the equilibrium values, as depicted by the transparent symbols. In such cases, the equilibrium fractionation is determined by extrapolation. Figure 2B depicts a higher temperature experiment where all the isotopes first mix at the bulk value for the system and then 'un-mix' along the secondary fractionation line. In this scenario a time series is also needed to prove equilibrium.

By way of practical example, we turn to the three Fe isotopes, a system for which the three-isotope method has been executed successfully for mineral–mineral, mineral–fluid, and metal–silicate exchange experiments (e.g., Shahar et al. 2008, 2015; Beard et al. 2010). For Fe, the axes of the three-isotope plot are defined by two isotope ratios with the same denominator: $^{56}Fe/^{54}Fe$ on the *y*-axis and $^{57}Fe/^{54}Fe$ on the *x*-axis (see Fig. 2 of Shahar et al. 2008). These ratios are compared to those of a standard and reported in per mil (‰) using the conventional delta notation:

$$\delta^{i}\text{Fe} = 10^{3}\left(\frac{\left({}^{i}\text{Fe}/{}^{54}\text{Fe}\right)_{\text{Sample}}}{\left({}^{i}\text{Fe}/{}^{54}\text{Fe}\right)_{\text{Standard}}} - 1\right)$$

where $i = 56$ or 57. If the goal is to derive the equilibrium isotope distribution by extrapolation, one can take advantage of the known equilibrium mass fractionation relationship between the two isotopes of interest:

$$\delta^{56}\text{Fe} = (10^{3} + \delta^{56}\text{Fe}_{\text{bulk}})\left(\frac{10^{3} + \delta^{57}\text{Fe}}{10^{3} + \delta^{57}\text{Fe}_{\text{bulk}}}\right)^{\gamma} - 10^{3}$$

This slightly concave fractionation relationship can be approximated by a straight line in three-isotope space with a slope of approximately γ and an intercept defined by the bulk isotopic composition. The intersections of the line defined by the above equation for the SFL and the two lines derived from the trajectories of the isotopically evolving starting materials represent extrapolation to the equilibrium values of the system. The exponent, γ, is 0.67795 (for Fe) at equilibrium based on the equation for equilibrium mass fractionation:

$$\gamma = \frac{\dfrac{1}{m_{56}} - \dfrac{1}{m_{54}}}{\dfrac{1}{m_{57}} - \dfrac{1}{m_{54}}}$$

where m is the mass of subscripted Fe isotope. Although both methods offer extrapolation to equilibrium, an advantage of the three-isotope technique over the Northrop–Clayton method is the ability to know when isotopic exchange is complete and equilibrium has been reached, i.e., when the final isotope values lie on the SFL.

An entirely different path towards equilibrium using the three-exchange isotope method is presented schematically in Figure 2B. Here, the spiked (Phase 2 Initial, t_0) and unspiked (Phase 1 Initial, t_0) starting materials are thoroughly chemically and isotopically mixed before equilibrating because either the sample is completely melted or dissolved during the experiment. When the system homogenizes at the beginning of the experiment (t_1), the isotopes mix and collapse to the bulk isotopic value of the system (yellow star), which by definition lies on the secondary fractionation line (SFL). Following mixing (t_1), there is an 'unmixing' (t_2, t_3) among the isotopes along the secondary fractionation line (SFL) as the system equilibrates isotopically as well as chemically over time. In progressively longer experiments, the isotopic values of the phases move away from each other along the SFL until they reach their equilibrium values (Phase 1 & 2 Equilibrium, t_4). In this case, equilibrium is determined by time series during unmixing; i.e., when the values stop changing (reach a steady state) along the SFL, the system is thought to be in equilibrium and the fractionation factors can be determined. This is critically important and considered a rigorous way to prove equilibrium in these experiments.

The utility of the three-isotope method is in the ability to trace the trajectories of isotopic exchange between phases in three-isotope space as the experiments progress through a time series. Without an isotope tracer (the isotopic spike) it could be difficult to prove that isotopic equilibrium had been reached, know whether the different phases had exchanged isotopes, and be certain that the exchange occurred in a closed system. By spiking one of the phases in the experiments with a known amount of spike, there is better control on mass balance and one can determine if the experiment was truly a closed system. Open system behavior was successfully diagnosed by Lazar et al. (2012) in three-isotope exchange experiments involving Ni alloying with Au capsules.

Proving that the isotopes have come to equilibrium is a crucial part of conducting high temperature experiments. If the isotopes are mixed but have not equilibrated yet then the fractionation would either be underestimated or determined as zero. In silicate–metal experiments, the majority of the Fe in the experiment is in the metallic phase, so any contamination from the metal to the silicate (which cannot be seen by the eye) will cause the apparent fractionation factor to be underestimated or appear to be zero. However, with the three-isotope exchange method this can be easily seen and a duplicate experiment would be performed and more carefully separated to obtain the correct result. Also, it is important to remember that chemical equilibrium is not equivalent to isotopic equilibrium and must be independently assessed. Shahar et al. (2008) found that while chemical equilibrium between fayalite and magnetite occurred within 6 hours, isotopic equilibrium required 48 hours at 800 °C. This is extremely important for experiments where equilibrium is assumed to have occurred based on electron probe measurements showing equilibrium textures and chemical compositions.

The use of a classical time series to assess the approach to equilibrium in isotopic fractionation experiments is valuable even in conjunction with other methods and we highly recommend its use in all experimental studies of isotope fractionation. Furthermore, the three-isotope exchange method may not be available for use in all systems of interest. In such cases, the use of a time series is essential and is highly recommended in conjunction with a multi-directional approach. Frierdich et al. (2014) argued that even the three-isotope technique is not sufficient on its own to guarantee that there are no possible kinetic contributions to the data. The authors argued that adding a multi-direction approach to the three-isotope technique unambiguously demonstrates that equilibrium has been reached during the experiment. Investigators should thoughtfully develop an experimental plan involving some combination of the techniques presented above that will most rigorously yield the equilibrium isotope fractionation factors sought for their system.

Kinetic effects

Kinetic isotope effects are commonly observed in nature and in the lab, making it crucial to be able to distinguish between isotopic equilibrium and disequilibrium. Both states are important to understand and quantify for isotopic systems, but the experimentalist must be able to discern which of the two is represented in their experiments. Kinetic isotope effects are usually associated with fast, unidirectional, or incomplete processes (O'Neil 1986). Some of the pathways that might lead to kinetic effects in isotope exchange experiments are evaporation, diffusion, and rapid crystallization. No matter how carefully an experiment is designed there is always the possibility of a kinetic isotope effect that has not been considered. From quenching effects in a piston cylinder to isotopic zoning within crystals, it is nearly impossible to avoid all kinetic isotope effects. However, this does not mean that the equilibrium value cannot be determined if the kinetic processes are properly identified and understood, and the data are treated accordingly. For example, Shahar et al. (2011) noted that when measuring Si isotopes *in situ* by laser ablation MC–ICP–MS, the isotopic values within each phase showed small deviations thought to be due to a kinetic effect during the quenching of the experiment. In order to determine the true isotopic ratio of the phases, the values were averaged. Had the experiment been measured by solution instead of by laser ablation these local variations would not have been seen and the kinetic behavior would have gone unnoticed.

In some cases kinetic processes are not an impedance to equilibrium, but a means to achieve it. For example, diffusion may be the dominant mechanism by which isotopes exchange between phases in an experiment. In this case, it is important to understand the rates of diffusion corresponding to the isotopes and materials of interest. Diffusion coefficients for many systems have been experimentally determined and can be used in some cases to make predictions as to how long it should take for two phases to chemically and isotopically equilibrate at the pressure and temperature of interest. This information can be used to guide the experimentalist when deciding on experimental durations in a time series.

In an experiment depending on diffusion (or another kinetic process) to exchange isotopes, it is critically important to allow enough time for a steady state to be reached so that a reasonable estimation of equilibrium can be made. However, it is also possible to achieve a spurious result by running an experiment for too long. For example, Lazar et al. (2012) suggested that there is a divergence in the isotope ratios *after* an experiment had reached equilibrium when kinetic effects can start to play a role again, especially when there is diffusive loss to the capsule. Therefore it is important to remember that while a time series is a good way to prove that isotopic equilibrium has been reached, there is a limit to the duration of the experiment that should not be passed. These are just two of many examples of how kinetic effects can be found within experimental charges even when isotopic equilibrium has been reached.

EXPERIMENTAL METHODS

Since the advent of MC–ICP–MS and associated chemical purification methods that allow for the accurate measurement of small isotopic fractionations between co-existing materials, researchers have been exploring pressure, temperature, and composition space in non-traditional stable isotope exchange experiments. Drawing from the well-established experimental methods used successfully by geochemists and petrologists for decades, as well as newer methods developed by mineral physicists and beam line scientists, the community is making great advances in our understanding of how non-traditional stable isotopes fractionate in geo- and biogeochemical processes. In order to reproduce the range of conditions occurring from Earth's surface to its core, researchers are utilizing a variety of experimental apparatuses, including simple flasks and low-temperature ovens, cold-seal hydrothermal vessels, piston–cylinders, multi-anvils, and diamond anvil cells. In the following sections we provide a brief (non-exhaustive) review of these methods, starting with low temperature experiments and their applications, and then moving progressively higher in temperature and pressure.

Low temperature experiments

Early experiments on the non-traditional stable isotopes focused on low temperature Fe isotope fractionation in aqueous fluids (e.g., Johnson et al. 2002; Welch et al. 2003; Hill et al. 2009) and more commonly between an aqueous fluid and a mineral (e.g., Johnson et al. 2002; Skulan et al. 2002). At this time in Fe isotope geochemistry history, it was unclear if abiotic systems could fractionate isotopes as efficiently as biologic systems, so a baseline was needed in order to determine if Fe isotopes could be used as a biosignature. These experiments employed the use of an isotopic spike (excess ^{57}Fe) in combination with classical time series and multi-direction approaches to assess Fe isotope fractionation between phases at low temperatures. Johnson et al. (2002) and Welch et al. (2003) investigated fractionation in aqueous Fe(II) and Fe(III) solutions at room temperature and in an ice bath. These studies have implications for the fractionation of Fe isotopes between aquo and hydroxy complexes in oxygenated natural waters.

If isotope fractionation associated with a redox change exists it would be most noticeable at low temperature and amongst two phases with strikingly different bonding environments, such as a fluid and a solid. As explained in detail in Skulan et al. (2002), low-temperature mineral–fluid exchange experiments rely on a steady state of dissolution and precipitation so that the isotopes can move around within the experiment and tend towards equilibration. This study utilized a range of experimental strategies to ultimately distinguish kinetic effects from equilibrium fractionation between hematite and aqueous Fe(III). One approach they used was simply mineral synthesis. The premise is that a mineral, such as hematite, will be thermodynamically stable at the conditions of the experiment, but will not be present in the starting materials, thus facilitating precipitation. In this case, Skulan et al. (2002) reacted aqueous Fe(III) with dilute HNO_3 in sealed Pyrex flasks at 98 °C to precipitate hematite from solution. The first hematite crystals will not be in isotopic equilibrium, but as the experiment progresses the system will reach a steady state of dissolution and precipitation in which the hematite crystals will then record the equilibrium fractionation between the crystal and the fluid.

The main disadvantage to these experiments is that at low temperature it is much more difficult to attain isotopic equilibrium as the rates of diffusion and nucleation are slow. There are also several kinetic effects that must be sorted out such as isotopic inhomogeneity in the mineral (i.e., zoning during precipitation) or other kinetic effects associated with the rates of dissolution and precipitation. In order to speed up diffusion rates, hydrothermal experiments can also be done using a cold seal technique at slightly elevated temperatures. For example, Li et al. (2015) determined the equilibrium Mg isotope fractionation between dolomite and aqueous Mg in multi-direction mineral synthesis experiments using a ^{25}Mg spike to track isotope exchange. These experiments were conducted at 130–220 °C using Parr bombs and an internal sealed container. By using this technique the experimentalist can avoid possible kinetic effects due to energy barriers that could inhibit crystal growth or dissolution.

Another subject of interest in low temperature experimental work is the isotopic effect of adsorption onto mineral surfaces (e.g., Barling and Anbar 2004; Icopini et al. 2004; Johnson et al. 2004; Beard et al. 2010; Nakada et al. 2013; Wasylenki et al. 2014, 2015). These experiments are aimed at understanding the isotope fractionation between aqueous metal species and sorbed metal complexes. In many of these experiments it has been shown that the isotope fractionation is constant throughout the experiment even as the amount of the element sorbed increases. To prove isotopic equilibrium in these experiments a time series and/or the three-isotope technique has been used; we recommend the use of both whenever possible. Additionally, in these types of experiments, the fractionation of interest in nature might be due to kinetic processes or open-system equilibrium-driven fractionation. Therefore, the approach to treating and interpreting the data may include modeling by constant offset as a function of reactant/product ratio (equilibrium closed-system), or as Rayleigh behavior (equilibrium open-system, or kinetic). For a more thorough explanation of these different treatments, see Johnson et al. (2004).

High temperature, low pressure experiments

The least amount of experimental work to date has focused on experiments at high-temperature and ambient or low pressure (≤ 0.5 GPa). This is due in part to the assumption that pressure was not an important parameter in isotope fractionation, and the more practical concern that it is typically easier to achieve a closed system when conducting high-pressure experiments. However, there have been a few key studies conducted in low-pressure experimental apparatuses in order to study the effect of oxygen fugacity and temperature on isotope fractionation. As far as we are aware, the first experiment on fractionation between a silicate melt and metal was conducted in this fashion in a 1 atm gas-mixing furnace (Roskosz et al. 2006). Figure 3A depicts a cutaway of the interior of such a furnace, exposing a molten sample suspended by a metal wire loop (e.g., Pt, Re) in a sealed alumina or mullite tube, which is flushed with an appropriate gas mixture (e.g., CO–CO_2, CO_2–H_2) to impose the chosen oxygen fugacity.

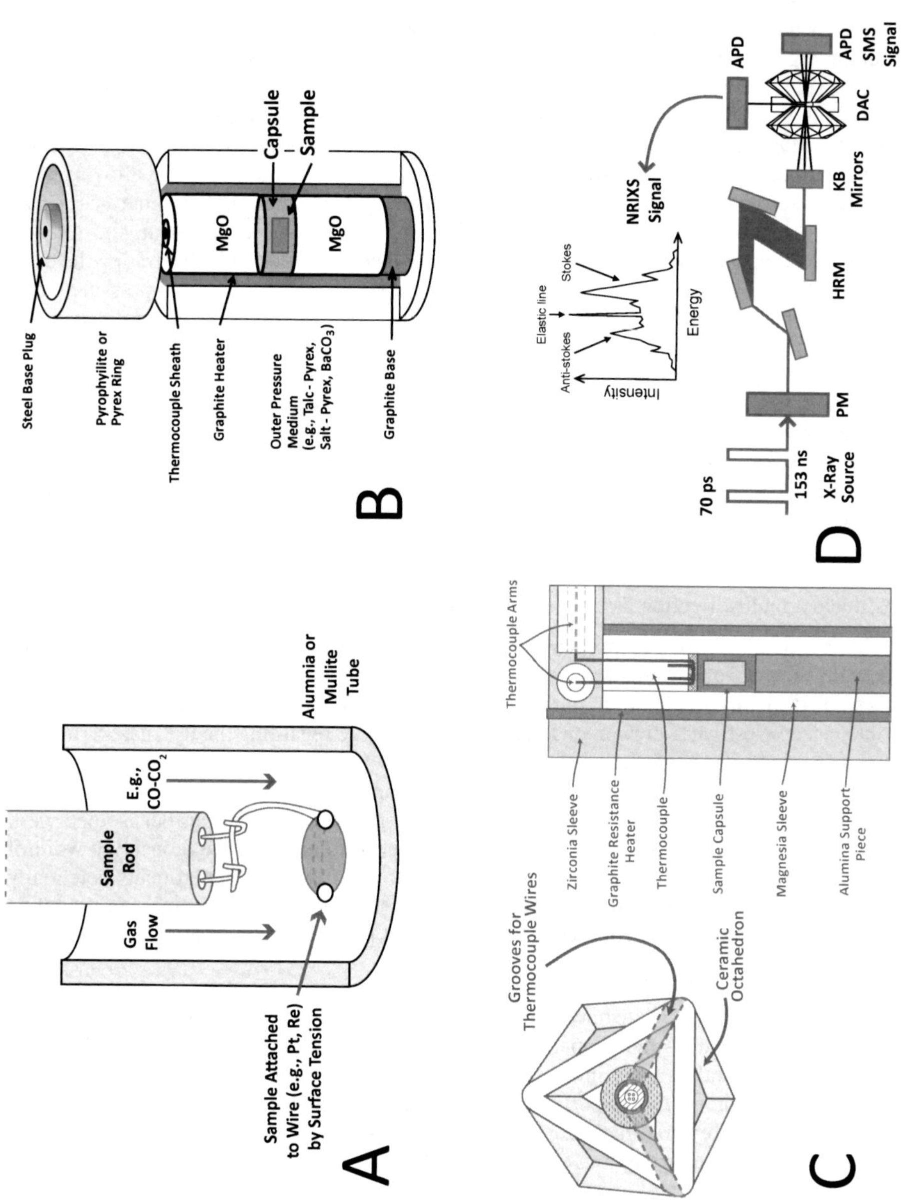

Figure 3. Schematics of A) a controlled-atmosphere metal loop experiment, B) a piston cylinder experiment (after Young et al. 2015), C) a multi-anvil experiment (after Bennett et al. 2015) and D) an NRIXS experiment (after Dauphas et al. 2012). A) A cross-sectional view of the internal arrangement of a typical ambient pressure controlled-atmosphere wire loop experiment. Typical gas mixtures consist of CO–CO_2 and H_2–CO_2, though other mixtures are sometimes used. Wire loops are typically Pt or Re. B) A cross-sectional view of a typical piston cylinder experiment showing where the capsule resides and typical assembly materials. C) An external view of a typical multi-anvil ceramic octahedron and a cross-sectional view of the internal arrangement of the octahedron. D) A typical NRIXS experimental set-up. Incoming synchrotron X-ray pulses (separated by 153 ns) first pass through the diamond pre-monochromator (PM), and then through the Si channel-cut high resolution monochromator (HRM). The beam is focused by small horizontal (20 cm) and vertical (10 cm) Kirkpatrick–Baez (KB) mirrors onto the DACTA sample. The forward synchrotron Mössbauer spectroscopy (SMS) signal is collected through the diamond axis on an avalanche photodiode (APD). Three additional side APDs collect the NRIXS signal through the X-ray transparent gasket.

In the study by Roskosz et al. (2006), it was shown that at 1500 °C there was a resolvable and quite large fractionation of Fe isotopes between silicate melt and metal in a series of Pt loop experiments, although it proved difficult to demonstrate isotopic equilibrium. These experiments used the propensity for Fe to alloy with Pt as a way to segregate the metal from the silicate at a fixed fO_2 and temperature. The authors noted that kinetic fractionation was rampant during the first ~300 minutes of an experiment. Initially, the Fe alloying with the Pt was isotopically light compared to that in the silicate melt and became heavier as the experiment approached equilibrium. In the longest duration experiment, the relative fractionation flipped such that the silicate was lighter than the metal, thus illustrating the need for, and utility of, conducting a time series in such experiments. Another low-pressure study by Schuessler et al. (2007) used an internally heated pressure vessel (IHPV) to investigate Fe isotope fractionation between a rhyolitic melt and pyrrhotite between 840 and 1000 °C at 0.5 GPa. A rigorous assessment of the approach to equilibrium was employed in this study. Three sets of experiments were done to distinguish kinetic effects from equilibrium fractionation for the system. The first was a classical time series in which ^{57}Fe spiked glass was reacted with natural pyrrhotite to determine the time required for a close approach to equilibrium. The second set of experiments employed the Northrop–Clayton method of partial isotope exchange discussed above to extrapolate to the equilibrium value (Northrop and Clayton 1966). In the third set, pyrrhotite was crystallized from the melt instead of being present in the starting materials. This combination of experimental approaches essentially constituted a multi-direction assessment of equilibrium supported by time series.

Other challenges for low-pressure experiments may include (but are not limited to) the effects of slow diffusion creating chemical and isotopic zoning in solid mineral or metal phases and difficulty in obtaining pure phase separates post-experiment. We recommend that the interested investigator draw upon the extensive literature regarding the use of experimental techniques such as controlled-cooling rates (e.g., Donaldson 1976; Corrigan 1982), temperature cycling (e.g., Mills and Glazner 2013), and crystal seeding (e.g., Larsen 2005; Dalou et al. 2012) to develop methodology to overcome these issues. Further isotope work using a controlled-atmosphere furnace or IHPV will need to build on the pioneering work of Roskosz et al. (2006) and Schuessler et al. (2007), as these types of experiments potentially have huge advantages over high pressure experiments, such as long-term temperature stability, large sample volumes, and precise control over oxygen fugacity.

High temperature and pressure experiments

Many experiments to date have been carried out in a piston cylinder apparatus (Fig. 3B) and a few in the multi-anvil apparatus (Fig. 3C). Both of these instruments allow the experimental materials to stay in a sealed container while at the pressure and temperature of interest throughout the experiment. The benefit is that at elevated pressures and temperatures, equilibrium is more easily attained, and with closed capsules, the composition is less likely to change throughout the experiment. Most experiments are carried out using a ½-inch piston–cylinder cell assembly (Fig. 3B), which is typically capable of maintaining pressures up to ~4 GPa. The cell assembly varies by laboratory but a typical assembly, as illustrated by Figure 3B, consists of a graphite tube heater surrounded by a pressure medium, the most common of which are $BaCO_3$, talc or NaCl with an inner Pyrex sleeve, and occasionally CaF_2. A thermocouple is inserted axially into the cell through a small hole in the stainless steel base plug, which is typically surrounded by a thin insulating pyrophyllite or Pyrex sleeve. Capsule material also varies according to the specific needs of the experiment (e.g., graphite, Au, Pt), and capsules are typically centered within the cell. Hotspots are typically large (i.e., on the order of millimeters) in piston–cylinder experiments, so the thermal gradient over a sample is not expected to exceed ~15 °C in most cases. The first experiment to use the piston–cylinder apparatus to directly determine mineral–mineral fractionation was a precipitation–dissolution experiment to measure the Fe isotope fractionation between fayalite and magnetite (Shahar et al. 2008).

Multi-anvil cells have smaller sample sizes but can extend the accessible pressure range to approximately 25 GPa (Fig. 3C). Multi-anvil assemblies typically consist of a ceramic octahedron with a cylindrical sample chamber drilled through the center. A heater, typically made of graphite, Re-foil, or La-chromite surrounds the sample capsule and confining spacers. A thermocouple can be inserted either axially or radially depending on the assembly. Although the multi-anvil apparatus can access a far greater pressure range than a piston–cylinder, there is a significant tradeoff with sample volume. The maximum pressure achievable by a particular multi-anvil cell assembly is inversely proportional to the maximum sample size within that assembly. For an in-depth explanation and visualization of piston–cylinder and multi-anvil experimental setups, the interested reader is referred to Bennett et al. (2015).

One of the most important choices an experimentalist has to make is what type of capsule to use in an experiment. The choice of capsule material is crucial to the success of the experiment, as in order to assess the approach to isotopic equilibrium and ensure that the possibility of kinetic processes is limited, a closed system must be maintained. An open system, with respect to isotopes, will in all likelihood result in non-equilibrium conditions during the experiment. Therefore, a capsule that alloys or reacts with the element of interest should not be used for the experiment (e.g., metal capsules in metal–silicate experiments simulating core formation). This would lead to a loss of the element of interest into the capsule or the creation of a mineral on the capsule wall that has an isotope fractionation of its own (possibly kinetic) and would therefore change the mass balance of the system and the fractionation throughout the experiment.

There is no perfect capsule material in experimental petrology, but acceptable choices can be found. Graphite capsules are the most popular choice. Graphite acts as an inert phase in most systems and is well-suited for many investigations, provided that the condition of carbon saturation can be tolerated. Noble metal and alloy capsules are another popular choice. Platinum, Re, Au, Ag, and Au–Pd are the most commonly used metal capsules; each having its own strengths and drawbacks. The most popular metal capsule, Pt, will cause Fe depletion from silicate melts (Merrill and Wyllie 1973). Although this behavior is well-characterized in the literature and such capsules can be pre-saturated with Fe to minimize this issue, the Pt will provide an additional "sink" for Fe that must come to isotopic equilibrium with the melt in order to avoid kinetic fractionations. Other potential capsule materials include olivine, magnesia, alumina, boron nitride, and silica glass/fused quartz. When the parameters of an experiment require a compromise on capsule material, it is best to test the conditions of interest with more than one capsule material to confirm the results.

Isotope fractionation experiments present an additional challenge over more traditional high *P*–*T* experiments in that two phases of interest typically need to be physically separated and analyzed in solution by MC–ICP–MS. As the achievable pressure range of an experimental apparatus increases, sample sizes necessarily decrease and the ability of a researcher to truly separate the phases, purify the element of interest, and obtain a precise isotope ratio becomes more limited. A potential work around is to use a selective leaching process (Williams et al. 2012) to chemically separate the phases of interest. However, the possibility of isotopic fractionation during the leaching process exists. For this reason though there are not as many published studies of multi-anvil experiments as piston–cylinder experiments. However, it is expected that more researchers will push the current boundaries with this technique, especially if using a large volume multi-anvil assembly.

Selective leaching (also called preferential dissolution) as a strategy to separate phases post-experiment was also used successfully by Macris et al. (2013). This study determined the equilibrium Mg isotope fractionation between spinel and forsterite in piston–cylinder experiments at 1 GPa and 600–800 °C. Two Au capsules were placed side by side in the cell assembly, one containing spinel spiked with excess ^{24}Mg and 'normal' magnesite, and the other containing forsterite spiked with excess ^{24}Mg and 'normal' magnesite. The three-isotope technique was used to determine the spinel–magnesite fractionation and the forsterite-magnesite fractionation, then the spinel–forsterite fractionation was obtained by difference. The magnesite in these experiments served two purposes: it acted as the Mg isotope exchange medium and it allowed for complete separation of phases after the experiment by preferential dissolution. The use of carbonate as an exchange medium has been well known since the pioneering oxygen isotope fractionation experiments of Clayton et al. (1989) and Chiba et al. (1989). It is thought to be superior to water as an isotopic exchange medium and reference material because of potential 'isotope salt effects' that can affect the observed fractionation in an experiment due to mineral dissolution changing the fluid composition (Hu and Clayton 2003).

Experiments in the piston–cylinder have relied on a time series approach or utilized the three-isotope technique to address isotopic equilibrium. Whenever possible both methods should be used to demonstrate a close approach to equilibrium. The Fe isotope system has been the most-well studied system thus far with the piston–cylinder (Poitrasson et al. 2009; Hin et al. 2012; Shahar et al. 2015). However the piston cylinder has proven useful for studies of many other isotope systems such as S (Labidi et al. 2016), Mo (Hin et al. 2013), Cr (Bonnand et al. 2016), Si (Shahar et al. 2009; Kempl et al. 2013; Hin et al. 2014), Mg (Richter et al. 2008; Huang et al. 2009; Macris et al. 2013), Ni (Lazar et al. 2012) and more. Multi-anvil experiments have been conducted mostly for Fe isotopes (Poitrasson et al. 2009; Williams et al. 2012), but also for Si isotopes (Shahar et al. 2011). It is our belief that these experiments will begin to become more commonplace, leading to a more complete database of experimentally determined equilibrium fractionation factors.

NRIXS and diamond anvil cell experiments

One might imagine that if multi anvil experiments have sample sizes that are almost too small to be useful, then diamond anvil cell experiments would be all but impossible. However, it is important to know if pressure truly is a negligible variable within isotope geochemistry (e.g., Clayton et al. 1975). Therefore, a new technique has emerged in order to measure isotope fractionation factors at high pressure. Polyakov and co-workers (Polyakov et al. 2005, 2007; Polyakov 2009) have pioneered the use of nuclear resonant inelastic X-ray scattering (NRIXS) synchrotron data to obtain vibrational properties of minerals from which isotopic fractionation factors can be calculated. The technique is based on the fact that certain isotopes possess a low-lying nuclear excited state that can be populated by X-ray photons of a particular energy and therefore directly probe the vibrational properties of the solid.

As shown in Figure 3D, in nuclear resonant X-ray spectroscopy an in-line high-resolution monochromator is used to narrow down the photon energy to meV resolution and fine-tune the monochromatic X-ray near the exceedingly narrow nuclear resonant (elastic) line. Avalanche photodiodes (APD) are used to collect only the signal from nuclear resonance absorption and reject non-resonantly scattered radiation. The APD directly downstream of the sample collects nuclear resonant forward scattering, or synchrotron Mössbauer spectroscopy (SMS) signal, which can provide a precise determination of the hyperfine interaction parameter and Lamb–Mössbauer factor. The APDs surrounding the sample radially collect the NRIXS signal, which is a result of creation (Stokes) or annihilation (anti-Stokes) of phonons as the incident X-ray beam is scanned over a small range (approximately ±100 meV) around the resonant energy. The phonon density of states (DOS) is extracted from this phonon excitation spectrum associated with the nuclear resonant isotope (Hu et al. 2003). The integration of the whole DOS provides values for key thermodynamic properties. In addition, NRIXS data can also be used to derive reduced partition function ratios (β-factors) from which equilibrium isotopic fractionation factors can be calculated: $\delta_A - \delta_B = 1000 \times (\ln \beta_A - \ln \beta_B)$, where A and B are two different phases of interest. This technique has been widely used by the mineral physics community to investigate seismic velocities and phonon density of states of high-pressure minerals (Sturhahn et al. 1995) but is relatively new to the isotope geochemistry community.

The most commonly used software for processing NRIXS data is the PHOENIX program. The PHOENIX program (Sturhahn 2000) analyzes the data obtained by NRIXS and exports three different force constants. The first is the determination of the force constant using the moments approach in which the normalized excitation probability is computed using the raw data and the experimental resolution function. The outcome is based on the raw data and therefore multiphonon contributions are included in the analysis. The second determination of the force constant is also based on this moments approach; however it is based on data that are first 'refined,' that is, the multiphonon contribution is corrected for and *S(E)* is extrapolated. Dauphas et al. (2012) argued that this moments approach is the best method for determining the force constant as it has smaller uncertainties, is not as dependent on the background, and is not as sensitive to asymmetric scans. The third determination of the force constant is based on the partial density of states that is computed from the raw data. The force constant is then calculated based on the density of states and is therefore only based on the one phonon contribution of the data and does not need to be corrected for the possible multiple phonon contributions in the data. Murphy et al. (2013) suggested that this last force constant calculation is the most robust. The two methods produce slightly different force constants thereby changing the beta factor calculation as well.

Dauphas et al. (2012) also suggested that the high-pressure data used by Polyakov (2009) were inaccurate due to the truncation of the high-energy tail during the data acquisition. These data were originally collected by different groups to derive seismic velocities and were not specifically intended for the novel applications pioneered by Polyakov (2009) regarding β-factors. However, since this type of data has proven effective in calculating equilibrium isotope fractionation factors, its collection and treatment has been improved. For example, Dauphas et al. (2012) further suggested that longer acquisition times are necessary to determine the partial phonon density of states (PDOS) of the phases at high pressure with particular emphasis on the high energy tail. To aid in this, Dauphas et al. (2014) introduced a new program, SciPhon, which determines the background on each side of the elastic peak and subtracts them individually instead of using one background value for the whole spectral range. The broadest spectral energy ranges allow for the most accurate determination of the background correction, apparently reducing overall uncertainty. A comparison of the force constant obtained from PHOENIX and SciPhon indicate that SciPhon uses more accurate background subtraction, gives values closer to theoretical calculations, and is more consistent over differing energy ranges (Shahar et al. 2016). Therefore, SciPhon should be used to calculate force constants when conducting NRIXS experiments. Roskosz et al. (2015) successfully applied SciPhon to NRIXS data to describe the equilibrium Fe isotope fractionation between spinels and silicates, which is in agreement with the direct experimental determination by Shahar et al. (2008).

Once the force constant is calculated, the β-factor determination is straightforward (e.g., Polyakov 1998; Murphy et al. 2013). These calculations are then used to obtain an estimate of the equilibrium isotope fractionation as described above: the β-factors are calculated for each phase that is being probed and then subtracted from one another to get the isotopic fractionation between the two phases. In this way, each phase is analyzed *in situ* and does not need to be separated post-experiment. To date, this technique has been used solely for room temperature experiments, both at atmospheric pressure and high pressure (Dauphas et al. 2012; Shahar et al. 2016). It is expected that high temperature experiments will follow when the temperature gradients in the diamond anvil cell can be better controlled and reduce uncertainties in the analyses.

POST-EXPERIMENT ANALYSIS

The characterization of run products for experiments designed to investigate isotopic fractionations is typically more in-depth than for traditional petrologic experiments, and in most instances requires more sample mass. For low-temperature experiments, both the phases involved and experimental techniques employed are more varied, and are often specific to the system being investigated. Here we will focus on the process for high-temperature experiments by way of example.

Ambient-pressure, piston–cylinder, and multi-anvil experimental run products are all similar in their analytical requirements. Once the samples have been removed from the experimental assembly, the samples are broken or cut in half. For piston–cylinder and especially multi-anvil assemblies, it is best practice to cut the sample capsules axially rather than radially in order to assess any compositional gradients that may be present due to temperature gradients in the cell. The first portion of the run product is mounted and polished for electron microprobe analysis. Geochemical characterization of the run product via electron microbeam analyses is essential and should always be conducted in tandem with isotopic analyses.

The rest of the sample is then processed for isotopic analyses. The most effective method for phase separation will likely vary with the phases of interest. Fortunately, in the case of experiments simulating the process of core–mantle differentiation, Fe metal alloys and quenched silicate melt have extremely different visual properties and are often easily separated by gently crushing the run product, followed by hand picking separates with the aid of a binocular microscope. The metal phase is typically very easy to mechanically separate, whereas hand picking a pure quenched silicate melt phase is more difficult. The portions of sample that appear to be pure are then placed under a strong magnet and screened thoroughly to ensure that the silicate portions are not at all magnetic. Only the silicate portions that make it through that test are dissolved in strong acids for isotopic analyses. After acid-digestion, the element of interest is chemically separated and purified by column chemistry for isotopic analyses, usually by MC–ICP–MS. Chemical purification procedures will vary for different isotopic systems and phases of interest, as will the analytical instrumentation and measurement procedures. Reviewing these procedures is beyond the scope of this chapter; however excellent reviews for isotopic systems of interest are available in the following chapters of this volume.

These post-experiment steps (phase separation and dissolution) are not necessary if the isotopic analyses are conducted *in situ* by LA-MC–ICP–MS. In this case, the second portion of the sample capsule will be mounted in epoxy and polished prior to laser ablation. To our knowledge, no study has yet physically separated quenched silicate melt from a mineral phase of interest to measure the isotopic fractionation. Ensuring a "clean" mineral phase may prove difficult for experimental run products of this kind. The crystallization kinetics during the experiment will undoubtedly prove important in growing crystals that are both free of melt inclusions or embayments and large enough to mechanically separate from the quenched melt. In cases such as this, *in situ* analysis may be the best option to ensure that the measured isotopic values are accurate.

CONCLUSIONS

The field of experimental non-traditional stable isotope geochemistry has expanded rapidly in the past decade and is likely to continue to grow as more and more natural fractionations in these isotopic systems are uncovered. Indeed, discoveries of fractionations in geologic and planetary materials thus far have greatly outpaced experimental efforts to quantify fractionation factors capable of explaining such fractionations. The current state of the non-traditional stable isotope field can be understood by considering the difficulties of using trace element geochemistry in a time when almost no partition coefficients had been determined. Therefore, there is a huge opportunity for creative and careful experimentalists to push this new field forward at a rapid pace. But in order to understand what these experiments can tell us we need to be cautious with interpreting the results. Not all experiments can be compared and often times should not be compared. Changing one variable in the experiment is enough to cause a difference in the equilibrium isotope fractionation and users of the data should be aware of this before extrapolating such data beyond its useful range. Therefore, the most useful path forward is to adopt the approach taken by the experimental trace element partitioning community, wherein the effects of pressure, temperature, solvent and solute composition, oxygen fugacity, and other potential variables are independently evaluated, and parameterized expressions for isotopic fractionation factors are derived. In this way, the community will move toward the goal of understanding the factors that influence equilibrium isotopic fractionations at the relevant conditions in all systems of geologic interest.

ACKNOWLEDGEMENTS

A.S. acknowledges NSF Grants EAR1321858 and EAR1464008. We thank the editors, James Watkins, Nicolas Dauphas and Fang-Zhen Teng, for the opportunity to contribute to this volume. The chapter was greatly improved by thoughtful reviews from Laura Wasylenki and Mathieu Roskosz.

REFERENCES

Barling J, Anbar AD (2004) Molybdenum isotope fractionation during adsorption by manganese oxides. Earth Planet Sci Lett 217:315–29

Beard BL, Handler RM, Scherer MM, Wu LL, Czaja AD, Heimann A, Johnson CM (2010) Iron isotope fractionation between aqueous ferrous iron and goethite. Earth Planet Sci Lett 295:241–250

Bennett NR, Brenan JM, Fei YW (2015) Metal–silicate partitioning at high pressure and temperature: experimental methods and a protocol to suppress highly siderophile element inclusions. J Visualized Exp:e52725

Bigeleisen J, Mayer MG (1947) Calculation of equilibrium constants for isotopic exchange reactions. J Chem Phys 15:261–267

Bonnand P, Williams H, Parkinson I, Wood B, Halliday A (2016) Stable chromium isotopic composition of meteorites and metal–silicate experiments: Implications for fractionation during core formation. Earth Planet Sci Lett 435:14–21

Chiba H, Chacko T, Clayton RN, Goldsmith JR (1989) Oxygen isotope fractionations involving diopside, forsterite, magnetite, and calcite: application to geothermometry. Geochim Cosmochim Acta 53:2985–2995

Clayton RN, Goldsmith JR, Karel KJ, Mayeda TK, Newton RC (1975) Limits on effect of pressure on isotopic fractionation. Geochim Cosmochim Acta 39:1197–1201

Clayton RN, Goldsmith JR, Mayeda TK (1989) Oxygen isotope fractionation in quartz, albite, anorthite and calcite. Geochim Cosmochim Acta 53:725–733

Corrigan GM (1982) Cooling rate studies of rocks from two basic dykes. Mineral Mag 46:387–394

Dalou C, Koga KT, Shimizu N, Boulon J, Devidal JL (2012) Experimental determination of F and Cl partitioning between lherzolite and basaltic melt. Contrib Mineral Petrol 163:591–609

Dauphas N, Roskosz M, Alp E, Golden D, Sio C, Tissot F, Hu M, Zhao J, Gao L, Morris R (2012) A general moment NRIXS approach to the determination of equilibrium Fe isotopic fractionation factors: application to goethite and jarosite. Geochim Cosmochim Acta 94:254–275

Dauphas N, Roskosz M, Alp EE, Neuville DR, Hu MY, Sio CK, Tissot FL, Zhao J, Tissandier L, Médard E, Cordier C (2014) Magma redox and structural controls on iron isotope variations in Earth's mantle and crust. Earth Planet Sci Lett 398:127–140

Deines P, Eggler DH (2009) Experimental determination of carbon isotope fractionation between $CaCO_3$ and graphite. Geochim Cosmochim Acta 73:7256–7274

Donaldson CH (1976) Experimental investigation of olivine morphology. Contrib Mineral Petrol 57:187–213

Frierdich AJ, Beard BL, Scherer MM, Johnson CM (2014) Determination of the Fe $(II)_{aq}$–magnetite equilibrium iron isotope fractionation factor using the three-isotope method and a multi-direction approach to equilibrium. Earth Planet Sci Lett 391:77–86

Hill PS, Schauble EA, Shahar A, Tonui E, Young ED (2009) Experimental studies of equilibrium iron isotope fractionation in ferric aquo-chloro complexes. Geochim Cosmochim Acta 73:2366–2381

Hin RC, Schmidt MW, Bourdon B (2012) Experimental evidence for the absence of iron isotope fractionation between metal and silicate liquids at 1 GPa and 1250–1300 °C and its cosmochemical consequences. Geochim Cosmochim Acta 93:164–181

Hin RC, Burkhardt C, Schmidt MW, Bourdon B, Kleine T (2013) Experimental evidence for Mo isotope fractionation between metal and silicate liquids. Earth Planet Sci Lett 379:38–48,

Hin RC, Fitoussi C, Schmidt MW, Bourdon B (2014) Experimental determination of the Si isotope fractionation factor between liquid metal and liquid silicate. Earth Planet Sci Lett 387:55–66

Hoering TC (1961) The physical chemistry of isotopic substances: The effect of physical changes on isotope fractionation. Carnegie Inst Wash 60:201–204

Hu GX, Clayton RN (2003) Oxygen isotope salt effects at high pressure and high temperature and the calibration of oxygen isotope geothermometers. Geochim Cosmochim Acta 67:3227–3246

Hu MY, Sturhahn W, Toellner TS, Mannheim PD, Brown DE, Zhao JY, Alp EE (2003) Measuring velocity of sound with nuclear resonant inelastic X-ray scattering. Phys Rev B 67:094304

Huang F, Lundstrom C, Glessner J, Ianno A, Boudreau A, Li J, Ferré E, Marshak S, DeFrates J (2009) Chemical and isotopic fractionation of wet andesite in a temperature gradient: experiments and models suggesting a new mechanism of magma differentiation. Geochim Cosmochim Acta 73:729–749

Icopini GA, Anbar AD, Ruebush SS, Tien M, Brantley SL (2004) Iron isotope fractionation during microbial reduction of iron: The importance of adsorption. Geology 32:205–208

Johnson CM, Skulan JL, Beard BL, Sun H, Nealson KH, Braterman PS (2002) Isotopic fractionation between Fe(III) and Fe(II) in aqueous solutions. Earth Planet Sci Lett 195:141–153

Johnson CM, Beard BL, Roden EE, Newman DK, Nealson KH (2004) Isotopic constraints on biogeochemical cycling of Fe. Rev Mineral Geochem 55:359–408

Joy HW, Libby WF (1960) Size effects among isotopic molecules. J Chem Phys 33:1276–1276

Kempl J, Vroon PZ, Zinngrebe E, van Westrenen W (2013) Si isotope fractionation between Si-poor metal and silicate melt at pressure-temperature conditions relevant to metal segregation in small planetary bodies. Earth Planet Sci Lett 368:61–68

Labidi J, Shahar A, Le Losq C, Hillgren V, Mysen B, Farquhar J (2016) Experimentally determined sulfur isotope fractionation between metal and silicate and implications for planetary differentiation. Geochim Cosmochim Acta 175:181–194

Larsen JF (2005) Experimental study of plagioclase rim growth around anorthite seed crystals in rhyodacitic melt. Am Mineral 90:417–427

Lazar C, Young ED, Manning CE (2012) Experimental determination of equilibrium nickel isotope fractionation between metal and silicate from 500 °C to 950 °C. Geochim Cosmochim Acta 86:276–295

Li W, Beard BL, Li C, Xu H, Johnson CM (2015) Experimental calibration of Mg isotope fractionation between dolomite and aqueous solution and its geological implications. Geochim Cosmochim Acta 157:164–181

Macris CA, Young ED, Manning CE (2013) Experimental determination of equilibrium magnesium isotope fractionation between spinel, forsterite, and magnesite from 600 to 800 °C. Geochim Cosmochim Acta 118:18–32

Matsuhisa Y, Goldsmith JR, Clayton RN (1978) Mechanisms of hydrothermal crystallization of quartz at 250 °C and 15 kbar. Geochim Cosmochim Acta 42:173–184

Matthews A, Goldsmith JR, Clayton RN (1983) Oxygen isotope fractionation between zoisite and water. Geochim Cosmochim Acta 47:645–654

Merrill RB, Wyllie PJ (1973) Absorption of iron by platinum capsules in high pressure rock melting experiments. Am Mineral 58:16–20

Mills RD, Glazner AF (2013) Experimental study on the effects of temperature cycling on coarsening of plagioclase and olivine in an alkali basalt. Contrib Mineral Petrol 166:97–111

Murphy CA, Jackson JM, Sturhahn W (2013) Experimental constraints on the thermodynamics and sound velocities of hcp–Fe to core pressures. J Geophys Res: Solid Earth 118:1999–2016

Nakada R, Takahashi Y, Tanimizu M (2013) Isotopic and speciation study on cerium during its solid–water distribution with implication for Ce stable isotope as a paleo-redox proxy. Geochim Cosmochim Acta 103:49–62

Northrop DA, Clayton RN (1966) Oxygen-isotope fractionations in systems containing dolomite. J Geol 74:174-196

O'Neil JR (1986) Theoretical and experimental aspects of isotopic fractionation. Rev Mineral 16:1–40

Pattison DRM (1994) Are reversed Fe-Mg exchange and solid-solution experiments really reversed? Am Mineral 79:938–950

Poitrasson F, Roskosz M, Corgne A (2009) No iron isotope fractionation between molten alloys and silicate melt to 2000 °C and 7.7 GPa: Experimental evidence and implications for planetary differentiation and accretion. Earth Planet Sci Lett 278:376–385

Polyakov VB (1998) On anharmonic and pressure corrections to the equilibrium isotopic constants for minerals. Geochim Cosmochim Acta 62:3077–3085

Polyakov VB (2009) Equilibrium iron isotope fractionation at core–mantle boundary conditions. Science 323:912–914

Polyakov VB, Kharlashina NN (1994) Effect of pressure on equilibrium isotopic fractionation. Geochim Cosmochim Acta 58:4739–4750

Polyakov VB, Mineev SD, Clayton RN, Hu G, Mineev KS (2005) Determination of tin equilibrium isotope fractionation factors from synchrotron radiation experiments. Geochim Cosmochim Acta 69:5531–5536

Polyakov VB, Clayton RN, Horita J, Mineev SD (2007) Equilibrium iron isotope fractionation factors of minerals: Reevaluation from the data of nuclear inelastic resonant X-ray scattering and Mossbauer spectroscopy. Geochim Cosmochim Acta 71:3833–3846

Richter FM, Watson EB, Mendybaev RA, Teng FZ, Janney PE (2008) Magnesium isotope fractionation in silicate melts by chemical and thermal diffusion. Geochim Cosmochim Acta 72:206–220

Roskosz M, Luais B, Watson HC, Toplis MJ, Alexander CMO, Mysen BO (2006) Experimental quantification of the fractionation of Fe isotopes during metal segregation from a silicate melt. Earth Planet Sci Lett 248:851–867

Roskosz M, Sio CKI, Dauphas N, Bi WL, Tissot FLH, Hu MY, Zhao JY, Alp EE (2015) Spinel–olivine–pyroxene equilibrium iron isotopic fractionation and applications to natural peridotites. Geochim Cosmochim Acta 169:184–199

Schauble EA (2004) Applying stable isotope fractionation theory to new systems. Rev Mineral Geochem 55:65–111

Schuessler JA, Schoenberg R, Behrens H, von Blanckenburg F (2007) The experimental calibration of the iron isotope fractionation factor between pyrrhotite and peralkaline rhyolitic melt. Geochim Cosmochim Acta 71:417–433

Shahar A, Young ED, Manning CE (2008) Equilibrium high-temperature Fe isotope fractionation between fayalite and magnetite: An experimental calibration. Earth Planet Sci Lett 268:330–338

Shahar A, Ziegler K, Young ED, Ricolleau A, Schauble EA, Fei YW (2009) Experimentally determined Si isotope fractionation between silicate and Fe metal and implications for Earth's core formation. Earth Planet Sci Lett 288:228–234

Shahar A, Hillgren VJ, Young ED, Fei YW, Macris CA, Deng LW (2011) High-temperature Si isotope fractionation between iron metal and silicate. Geochim Cosmochim Acta 75:7688–7697

Shahar A, Hillgren VJ, Horan MF, Mesa-Garcia J, Kaufman LA, Mock TD (2015) Sulfur-controlled iron isotope fractionation experiments of core formation in planetary bodies. Geochim Cosmochim Acta 150:253–264

Shahar A, Schauble EA, Caracas R, Gleason AE, Reagan MM, Xiao Y, Shu J, Mao W (2016) Pressure-dependent isotopic composition of iron alloys. Science 352:580–582, DOI: 10.1126/science.aad9945

Skulan JL, Beard BL, Johnson CM (2002) Kinetic and equilibrium Fe isotope fractionation between aqueous Fe(III) and hematite. Geochim Cosmochim Acta 66:2995–3015,

Sturhahn W (2000) CONUSS and PHOENIX: Evaluation of nuclear resonant scattering data. Hyperfine Interact 125:149–172

Sturhahn W, Toellner TS, Alp EE, Zhang X, Ando M, Yoda Y, Kikuta S, Seto M, Kimball CW, Dabrowski B (1995) Phonon density-of-states measured by inelastic nuclear resonant scattering. Phys Rev Lett 74:3832–3835

Urey HC (1947) The thermodynamic properties of isotopic substances. J Chem Soc:562–581

Wasylenki LE, Swihart JW, Romaniello SJ (2014) Cadmium isotope fractionation during adsorption to Mn oxyhydroxide at low and high ionic strength. Geochim Cosmochim Acta 140:212–26

Wasylenki LE, Howe HD, Spivak-Birndorf LJ, Bish DL (2015) Ni isotope fractionation during sorption to ferrihydrite: Implications for Ni in banded iron formations. Chem Geol 400:56–64

Welch SA, Beard BL, Johnson CM, Braterman PS (2003) Kinetic and equilibrium Fe isotope fractionation between aqueous Fe(II) and Fe(III). Geochim Cosmochim Acta 67:4231–4250

Williams HM, McCammon CA, Peslier AH, Halliday AN, Teutsch N, Levasseur S, Burg JP (2004) Iron isotope fractionation and the oxygen fugacity of the mantle. Science 304:1656–1659

Williams HM, Wood BJ, Wade J, Frost DJ, Tuff J (2012) Isotopic evidence for internal oxidation of the Earth's mantle during accretion. Earth Planet Sci Lett 321:54–63

Young ED, Galy A, Nagahara H (2002) Kinetic and equilibrium mass-dependent isotope fractionation laws in nature and their geochemical and cosmochemical significance. Geochim Cosmochim Acta 66:1095–1104

Young ED, Manning CE, Schauble EA, Shahar A, Macris CA, Lazar C, Jordan M (2015) High-temperature equilibrium isotope fractionation of non-traditional stable isotopes: Experiments, theory, and applications. Chem Geol 395:176–195

Reviews in Mineralogy & Geochemistry
Vol. 82 pp. 85-125, 2017

Kinetic Fractionation of Non-Traditional Stable Isotopes by Diffusion and Crystal Growth Reactions

James M. Watkins

Department of Geological Sciences
University of Oregon
Eugene, OR
USA

watkins4@uoregon.edu

Donald J. DePaolo

Earth Sciences Division
Lawrence Berkeley National Laboratory
Berkeley, CA
USA
and
Department of Earth and Planetary Science
University of California
Berkeley, CA
USA

DJDepaolo@lbl.gov

E. Bruce Watson

Department of Geology
Rensselaer Polytechnic Institute
Troy, NY
USA

watsoe@rpi.edu

INTRODUCTION

Natural variations in the isotopic composition of some 50 chemical elements are now being used in geochemistry for studying transport processes, estimating temperature, reconstructing ocean chemistry, identifying biological signatures, and classifying planets and meteorites. Within the past decade, there has been growing interest in measuring isotopic variations in a wider variety of elements, and improved techniques make it possible to measure very small effects. Many of the observations have raised questions concerning when and where the attainment of equilibrium is a valid assumption. In situations where the distribution of isotopes within and among phases is not representative of the equilibrium distribution, the isotopic compositions can be used to access information on mechanisms of chemical reactions and rates of geological processes. In a general sense, the fractionation of stable isotopes between any two phases, or between any two compounds within a phase, can be ascribed to some combination of the mass dependence of thermodynamic (equilibrium) partition coefficients, the mass dependence of diffusion coefficients, and the mass dependence of reaction rate constants.

1529-6466/17/0082-0004$10.00 http://dx.doi.org/10.2138/rmg.2017.82.4

Many documentations of kinetic isotope effects (KIEs), and their practical applications, are described in this volume and are therefore not reviewed here. Instead, the focus of this chapter is on the measurement and interpretation of mass dependent diffusivities and reactivities, and how these parameters are implemented in models of crystal growth within a fluid phase. There are, of course, processes aside from crystal growth that give rise to KIEs among non-traditional isotopes, such as evaporation (Young et al. 2002; Knight et al. 2009; Richter et al. 2009a), vapor exsolution (Aubaud et al. 2004), thermal diffusion (Richter et al. 2009a, 2014b; Huang et al. 2010; Dominguez et al. 2011), mineral dissolution (e.g., Brantley et al. 2004; Wall et al. 2011; Pearce et al. 2012; Druhan et al. 2015), and various biological processes (e.g., Zhu et al. 2002; Weiss et al. 2008; Nielsen et al. 2012; Robinson et al. 2014). Coverage of these topics, and how they give rise to KIEs, can be found throughout this volume and in the recent literature.

Organization of the article

In the first part of this review, we provide a compilation of the mass dependence of diffusion coefficients in low-temperature aqueous solutions, high-temperature silicate melts, solid metals and silicate minerals. The reader will appreciate both the complexity of isotope diffusion in condensed media as well as the simplicity of the systematic relationships that have emerged, which allow for general predictions regarding the sign and magnitude of isotope fractionation by diffusion in solids and liquids. The second part of this review covers isotope fractionation during crystal growth. We start with models that involve isotope mass dependent diffusion of impurities to a growing crystal. The impurities could be compatible or incompatible elements, provided that they do not affect the growth rate of the crystal itself. We then discuss kinetic isotope fractionation of the stoichiometric constituents of a mineral (e.g., Ca isotopes in $CaCO_3$) due to diffusion as well as surface reaction controlled kinetics, followed by consideration of isotope fractionation of impurities that affect growth rate itself. The presentation includes discussion of three different types of "surface entrapment models," the underlying mechanisms of mass-dependent reaction rates, and whether isotope fractionation occurs on the aqueous side or the mineral side of the solid–liquid interface.

Throughout the chapter, we rely heavily on trace element and stable isotope data for the mineral calcite because of our own familiarity with this mineral and because it is perhaps the best studied phase in the KIE and crystal growth contexts. We note at the outset that while many of the principles developed herein can be transferred to other minerals with similar (desolvation rate-limited) surface reaction mechanisms or growth pathways, additional work is required to adequately describe KIEs for crystals precipitated via non-classical, particle mediated pathways, as described at the end of this chapter. Along the way, it will be seen that molecular dynamics simulations are playing a key role in drawing connections between nano-scale processes and macro-scale observables related to KIEs.

ISOTOPE FRACTIONATION BY DIFFUSION

The recognition that diffusion is capable of generating measurable (sub-‰) to large (tens of ‰) isotopic fractionations has catalyzed efforts over the past decade towards figuring out how, when and where diffusion is responsible for isotopic variations in nature (e.g., Ellis et al. 2004; Lundstrom et al. 2005; Beck et al. 2006; Roskosz et al. 2006; Teng et al. 2006, 2011; Dauphas 2007; Jeffcoate et al. 2007; Marschall et al. 2007; Parkinson et al. 2007; Rudnick and Ionov 2007; Bourg and Sposito 2008; Gallagher and Elliott 2009; Dauphas et al. 2010; Sio et al. 2013; Müller et al. 2014; Richter et al. 2014a, 2016; Oeser et al. 2015). In this section, we focus on the progress towards a predictive theory for the mass dependence of diffusion coefficients in aqueous solutions, silicate melts, silicate minerals, and metallic alloys at high temperature.

Expressions for diffusive fluxes

Fick's first law states that the flux of a chemical species i is directly proportional to the concentration gradient:

$$J_i = -D_i \nabla C_i, \tag{1}$$

where J_i is the flux (moles m^{-2} s^{-1}), D_i is the diffusion coefficient (m^2 s^{-1}), and C_i is the concentration (moles m^{-3}). Note that concentration can be expressed as $C_i = \rho w_i / M_i$, where ρ is the density of the liquid, w_i is the weight fraction of i, and M_i is the molecular weight of i. If the density of the liquid is constant, then concentration gradients can equivalently be expressed in units of wt%. Equation (1) applies to diffusion of a solute in dilute aqueous solution or diffusion of a trace species in an otherwise homogeneous silicate melt. In concentrated solutions or in cases where the diffusing species are unknown, it is customary to define a basis set of chemical components and to recognize that the flux of a component can be driven by concentration gradients in any of the other components. A more general form of Fick's first law is (Onsager 1945; De Groot and Mazur 1963):

$$J_i = -\sum_{j=1}^{n-1} D_{ij} \nabla C_j, \tag{2}$$

where D_{ij} is a matrix of diffusion coefficients. If diffusion of each component i is independent of all other n components, then D_{ij} is a diagonal matrix and each component obeys Equation (1). Generally, this is not the case and the off-diagonal elements of D_{ij} specify the extent of diffusive coupling between the chosen components.

The diffusion coefficients D_i or D_{ij} are where most of the complication arises in problems involving chemical diffusion. The diffusivity of an element or species depends on the physical properties of the diffusing medium; it may vary spatially in an anisotropic material and it may depend on variables such as temperature, pressure and chemical composition.

Isotopic mass dependence of diffusion in "simple" systems

The kinetic theory of gases gives the diffusivity of molecules in an ideal gas as:

$$D = \frac{1}{3}\lambda \bar{v}, \tag{3}$$

where $\bar{v}$ is the mean molecular velocity of particles and λ is the mean free path between collisions. Inserting expressions for $\bar{v}$ and λ, the tracer diffusion coefficient for a molecular species in a dilute gas is (see Lasaga 1998):

$$D = \frac{RT}{3\sqrt{2}\pi PNd^2}\sqrt{\frac{8RT}{\pi M}}, \tag{4}$$

where T, P, N, d, and M are temperature, pressure, Avogadro's number, the molecular diameter, and the molecular weight of the gas. This expression is the basis for the square-root-of-mass law, which states that the ratio of diffusion coefficients of two gaseous species is proportional to the inverse square root of their mass; i.e.,

$$\frac{D_2}{D_1} = \left(\frac{m_1}{m_2}\right)^{1/2}, \tag{5}$$

which is only valid for systems in which the assumptions of kinetic theory (point masses, low pressure such that collisions are infrequent and intermolecular forces are negligible) are approximately valid.

In condensed systems the effect of mass on diffusion coefficients is considerably more complicated, primarily because the diffusing species have non-negligible potential interactions with their nearest neighbors. Intermolecular potentials are theoretically complex because they depend on the shape and rotation of molecules whose identities are often unknown or are not well defined in systems such as aqueous solutions and silicate melts, as discussed in Watkins et al. (2009, 2011). It has therefore become customary to express the ratio of diffusion coefficients of solute isotopes (D_2/D_1) as an inverse power-law function of the ratio of their masses, m_2 and m_1 (Richter et al. 1999):

$$\frac{D_2}{D_1}=\left(\frac{m_1}{m_2}\right)^{\beta}, \tag{6}$$

where β is a dimensionless empirical parameter. In this review, m will refer to isotopic mass and not the mass of isotopically substituted molecules such as, for example, CO_2 or CH_4. The β factor is a convenient means of reporting the ratio of isotopic diffusion coefficients because it allows for direct comparison between different elements that have different fractional mass differences between the isotopes (e.g., $^7Li/^6Li$, which differ in mass by about 14%, versus $^{29}Si/^{28}Si$, which differ in mass by about 3%).

Isotopic mass dependence of diffusion in aqueous solution

A compilation of β factors for diffusion in aqueous solution is provided in Table 1 and Figure 1. The first takeaway is that isotope fractionation by diffusion in aqueous solution is not nearly as efficient as isotope fractionation by diffusion in a dilute gas, as all but two of the β factors (one measurement for He and one for Ar; see Table 1) are significantly less than the kinetic theory value of 0.5. The second takeaway is that the noble gas elements have larger β factors than the rest of the solutes and their β factors correlate with atomic mass (or radius) such that lighter noble gas elements exhibit greater mass discrimination by diffusion.

The low βs for charged species relative to uncharged species, and the dependence of β on atomic size, can be rationalized by considering solute–solvent interactions and the plausible physical mechanisms of diffusion in aqueous solution. The horizontal axis of Figure 1a is the diffusivity of the species normalized by that of H_2O. Most of the ionic species diffuse more

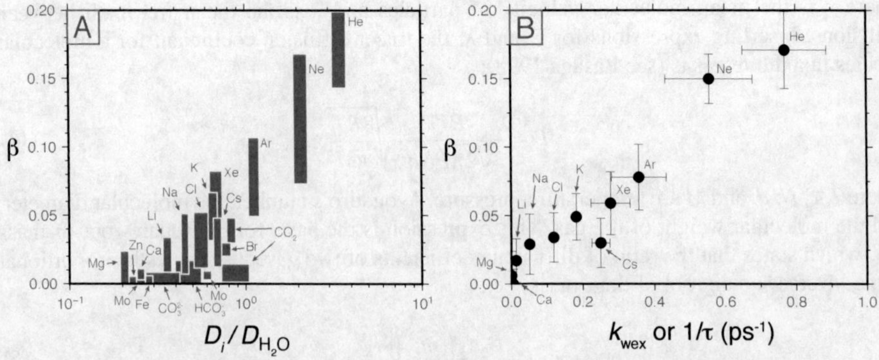

Figure 1. The efficiency of isotope fractionation by diffusion (β) versus metrics for the strength of solute–solvent interactions in aqueous solutions. (a) The β factors correlate with the solvent-normalized diffusivity. Most of the charged species diffuse more slowly than H_2O because they interact strongly with their surrounding H_2O molecules. (b) Molecular dynamics simulation show that the β factors also correlate with the water exchange rate, k_{wex}, which is equivalent to $1/\tau$, where τ is the residence time of water molecules in the first hydration sphere surrounding the cation (Bourg et al. 2010). See Table 1 for references.

Table 1. Isotopic mass dependence of diffusion in aqueous solutions

T	Description	Isotopic system	D_{solute} (10^{-11} m²/s)	$D_{solvent}$ (10^{-11} m²/s)	β	References
				Noble gases		
25 °C	MD simulation	He	785 ± 54	230 ± 10	0.171 ± 0.028	Bourg and Sposito (2008)
25 °C	Experiment	He	722 ± 27	230 ± 10	0.492 ± 0.122	Jahne et al. (1987)
25 °C	MD simulation	Ne	478 ± 37	230 ± 10	0.150 ± 0.018	Bourg and Sposito (2008)
20 °C	Experiment	Ne	–	204 ± 10	0.108 to 0.145	Tempest and Emerson (2013)
Room?	Experiment	Ne	–	–	0.104 ± 0.031	Tyroller et al. (2014)
25 °C	MD simulation	Ar	257 ± 15	230 ± 10	0.078 ± 0.029	Bourg and Sposito (2008)
20 °C	Experiment	Ar	–	–	0.055 to 0.074	Tempest and Emerson (2013)
Room?	Experiment	Ar	N/A	N/A	0.508 ± 0.036	Tyroller et al. (2014)
25 °C	MD simulation	Xe	157 ± 11	230 ± 10	0.059 ± 0.023	Bourg and Sposito (2008)
				Ionic species		
75 °C	Experiment	Ca^{2+}	–	590 ± 10	0.0045 ± 0.0005	Richter et al. (2006)
75 °C	MD simulation	Ca^{2+}	150 ± 3	590 ± 10	0.0000 ± 0.0108	Bourg et al. (2010)
75 °C	Experiment	Mg^{2+}	–	590 ± 10	0.0000 ± 0.0015	Richter et al. (2006)
75 °C	MD simulation	Mg^{2+}	121 ± 6	590 ± 10	0.006 ± 0.018	Bourg and Sposito (2007)
75 °C	Experiment	Li^{+}	–	590 ± 10	0.015 ± 0.002	Richter et al. (2006)
75 °C	MD simulation	Li^{+}	212 ± 8	590 ± 10	0.0171 ± 0.0159	Bourg and Sposito (2007)
25 °C	Experiment	Na^{+}	–	–	0.023 ± 0.023	Pikal (1972); Richter et al. (2006)
75 °C	MD simulation	Na^{+}	269 ± 9	590 ± 10	0.029 ± 0.0.022	Bourg et al. (2010)
75 °C	Experiment	K^{+}	–	590 ± 10	0.042 ± 0.002	Bourg et al. (2010)
75 °C	MD simulation	K^{+}	385 ± 17	590 ± 10	0.049 ± 0.017	Bourg et al. (2010)
75 °C	MD simulation	Cs^{+}	404 ± 20	590 ± 10	0.030 ± 0.018	Bourg et al. (2007)
75 °C	Experiment	Cl^{-}	–	590 ± 10	0.0258 ± 0.0144	Richter et al. (2006)
21 °C	Experiment	Cl-	114 ± 14	209 ± 10	0.0296 ± 0.0027	Eggenkamp and Coleman (2009)
75 °C	MD simulation	Cl^{-}	327 ± 14	590 ± 10	0.034 ± 0.018	Bourg and Sposito (2007)
21 °C	Experiment	Br^{-}	153 ± 17	209 ± 10	0.025 ± 0.005	Eggenkamp and Coleman (2009)
20 °C	Experiment	Fe	57.8 ± 2.3	204 ± 10	0.0025 ± 0.0003	Rodushkin et al. (2004)
20 °C	Experiment	Zn	47.1 ± 0.7	204 ± 10	0.0019 ± 0.0003	Rodushkin et al. (2004)
				Molecular species		
20 °C, pH ~0.5	Diffusion of $Mo_7O_{24}^{6-}$ and $Mo_8O_{26}^{4-}$	Mo	52 ± 10	204 ± 10	0.0000 ± 0.0010	Malinovsky et al. (2007)
20 °C, pH ~7	Diffusion of MoO_4^{2-}	Mo	124 ± 2	204 ± 10	0.0058 ± 0.0019	Malinovsky et al. (2007)
25 °C	Experiment	CO_2 (^{13}C)	191 ± 7	230 ± 10	0.039 ± 0.009	Jähne et al. (1987)
25 °C	MD simulation	CO_2 (^{13}C)	204 ± 35	230 ± 10	0.01 to 0.14	Zeebe (2011)
25 °C	MD simulation	HCO_3^- (^{13}C)	110 ± 18	230 ± 10	−0.04 to 0.17	Zeebe (2011)
25 °C	MD simulation	CO_3^{2-} (^{13}C)	80 ± 18	230 ± 10	−0.04 to 0.13	Zeebe (2011)
20 °C	Gas–water exchange exp.	H_2	N/A	204 ± 10	0.067 to 0.090	Knox et al. (1992)
20 °C	Experiment	CH_4 (^{13}C)	N/A	204 ± 10	0.020 to 0.026	Fuex (1980)
20 °C	Experiment	O_2	N/A	204 ± 10	0.069 to 0.093	Benson and Krause (1980)
20 °C	Experiment	N_2	N/A	204 ± 10	0.050 to 0.066	Benson and Krause (1980)

Notes

1. Water diffusivity comes from Bourg and Sposito (2007) (their Fig. 1)
2. For gas-water exchange experiments, the range in b represents the range of n = 0.5 to 0.67, where n is related to the dynamics of the air-water interface (cf. Tempest and Emerson 2013)

slowly than H_2O ($D_i / D_{H_2O} < 1$) because the polar water molecules interact strongly with ions in solution, forming a hydration sphere. If a number of water molecules are strongly bound to the ion, the hydrodynamic radius becomes larger, and as a result, the diffusivity as well as the fractional mass difference between isotopically substituted hydrated ions is smaller, leading to an attenuated mass discrimination.

The nature of solute-solvent interactions can now be probed in more detail using molecular dynamics simulations. For example, the metal–water exchange frequency (k_{wex}), which is the inverse of the average residence time of water molecules in the first solvation shell, has been shown to correlate with the calculated mass dependence on diffusivity (Fig. 1b). A low k_{wex} implies that water molecules are more strongly affixed to the diffusing cation. The correlations between β and proxies for the strength of solute–solvent interactions are evidence that whenever the hydration shell is massive enough, or alternatively, the lifetime of the water molecules in the hydration shell long enough, diffusion is sluggish and the isotopic effect on diffusion tends toward zero, as seems to be the case for aqueous Mg^{2+}. An important side note is that k_{wex} itself has a mass dependence (Hofmann et al. 2012), which has implications for the origin of kinetic isotope effects due to desolvation at a mineral surface, as discussed in a later section.

Isotopic mass dependence of diffusion in silicate melts

Diffusion data in silicate liquids is usually presented and modeled in terms of simple oxide components even though diffusion does not occur through the motion of long-lived or 'indestructible' molecules. Instead, the evidence from nuclear magnetic resonance (NMR) studies is that diffusion occurs through rapid chemical exchange between the constituent molecular structures (Stebbins 1995). The main structural units in silicate materials are silica and alumina tetrahedra, $(Si,Al)O_4^{4-}$, that are linked together by bridging oxygen atoms to form chains, sheets, and three-dimensional networks. The degree of interlinking, or polymerization, is dependent on a number of factors, including the presence and abundance of other cations (Fe, Mg, Ca, Na, and K), making it difficult to classify the diffusing species. A loose analogy to aqueous solutions can be drawn by considering silicate melts as "concentrated solutions," with the $(Si,Al)O_4^{4-}$ tetrahedra units being the "solvent" molecules. Such an analogy is useful insofar as the mass dependence on diffusivity seems to vary systematically with the strength of cation–aluminosilicate ("solute–solvent") interactions, as seen above for diffusion in dilute aqueous solutions.

Richter et al. (1999) were the first to investigate the mass dependence of diffusion in silicate melts. They measured Ca isotope fractionation by diffusion in molten $CaO–Al_2O_3–SiO_2$ melts and Ge isotope fractionation by diffusion in molten GeO_2. Those experiments, and the ones that have followed, involve juxtaposing two silicate glass cores or powders of different composition but the same (or nearly the same) stable isotope ratio for the element of interest. When heated above the liquidus for a specified duration, diffusion leads to stable isotope fractionation because light isotopes tend to diffuse faster than heavier isotopes. The diffusion couple experiments of Richter et al. (1999) showed that measurable isotopic fractionations can arise even at the high temperatures of molten silicates.

Richter et al. (2003, 2008, 2009b) and Watkins et al. (2009) reported β factors for Ca, Li, Mg, and Fe in diffusion-couple experiments involving natural silicate liquid compositions (e.g., basalt–rhyolite and ugandite–rhyolite). A key observation is that the major elements (Ca, Mg, and Fe) exhibit less mass discrimination ($\beta \approx 0.05 \pm 0.05$) than elements that are present in minor quantities such as Li ($\beta \approx 0.22$). This has been attributed to the cooperative nature of diffusion in dense systems, either due to diffusion of multi-atom complexes or other factors such as mass balance; a diffusive flux of a major component must be accommodated by a concomitant flux of the other components (Watkins et al. 2009). Subsequently, Watkins et al. (2011) studied diffusive isotopic fractionation of Ca and Mg in diffusion-couple experiments involving simplified liquids along the albite–anorthite and albite–diopside join. In these experiments, the cations Mg and Ca were present in relatively minor quantities (<5 wt%) and it was shown that β can be highly variable for a given cation depending on liquid composition. To further understand the relationship between isotope fractionation and diffusive coupling between components, Watkins et al. (2014) investigated isotope fractionation by multicomponent diffusion in molten $CaO–Na_2O–SiO_2$. Their experiments show that diffusive isotopic fractionation can depend on the direction of diffusion in composition space, even for a given bulk composition, and that large diffusive isotope fractionations do not require large concentration gradients in the diffusing element.

Despite the complexity of isotope diffusion in silicate melts, some generalizations can be made from the few experiments that have been carried out thus far. The overall range of measured β factors is 0–0.22, considerably less than the value of 0.5 for dilute gases and comparable to the range observed for diffusion in aqueous solutions. As shown in Figure 2a, the β factors vary systematically with the solvent-normalized diffusivity, here taken to be the diffusivity of the cation normalized by that of Si. Since Si is strongly bound in multi-atom complexes with O (as well as Al and other Si atoms), it diffuses more slowly than other elements, and hence the ratio D_i/D_{Si} is generally greater than 1. This is opposite the case for diffusion in aqueous solutions, where the solvent H_2O molecules interact weakly with one another and hence are more mobile than the solute ions. In cases where the elements are present in major quantities, the ratio D_i/D_{Si} is close to unity. For minor or trace species, faster diffusion implies a decoupling between the diffusing component and the rest of the components in the liquid, and corresponds to greater mass discrimination. The regime where $D_i/D_{Si} \rightarrow 1$ tends to occur in more silica-rich liquids and with decreasing temperature (Dingwell 1990). This is borne out in Figure 2b, which indeed suggests a direct link between the β factors for Ca and the $SiO_2+Al_2O_3$ content of the liquid.

Just as MD simulations have been valuable for probing the mechanisms of mass discrimination by diffusion in aqueous solution, MD simulations involving simplified silicate melt compositions are being used to complement the experimental studies. Goel et al. (2012) used MD simulations to determine β factors for Mg and Si in $MgSiO_3$ and SiO_2 liquids at 4000–4500 K. Their results are in good agreement with experimental data (Fig. 2a), and suggest that large diffusive isotope effects persist at extreme temperatures. The details of their study, and their discussion about the cooperative nature of diffusion in dense liquids, is exemplary for how MD simulations can be used to probe solute–solvent interactions in ways that are not accessible to spectroscopic and experimental diffusion studies.

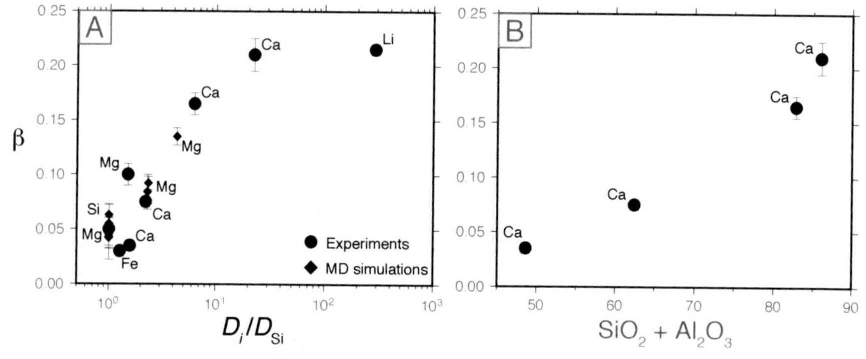

Figure 2. The efficiency of isotope fractionation by diffusion (β) versus metrics for the strength of solute-solvent interactions in silicate melts. (a) The β factors correlate with the solvent-normalized diffusivity. Cations that diffuse much faster than silicon (e.g., lithium) move in a way that is decoupled from the motion of melt matrix and therefore exhibit a larger mass discrimination. Other cations, such as iron, that diffuse only slightly faster than silicon are inferred to be coupled to the motion of larger aluminosilicate complex and therefore exhibit much smaller mass discrimination. (b) Higher values of β (as well as D_i/D_{Si}) tend to occur in higher silica liquids. See Table 2 for references.

Isotopic mass dependence of diffusion in minerals and metals

The classical theory of the isotopic mass dependence of diffusion rates in crystalline solids was developed in the 1950s to 1960s (Vineyard 1957; Schoen 1958; Tharmalingam and Lidiard 1959; Mullen 1961; Le Claire 1966; Rothman and Peterson 1969). The primary goal of these studies was to use β factors alongside crystal lattice models to infer diffusion mechanisms (e.g., interstitial, vacancy, divacancy, collinear interstitialcy, direct exchange, etc.; see Rothman and Peterson 1969) as well as the properties of point defects (Schüle and Scholz 1979).

Richter et al. (2009a) compiled β factors for diffusion of cations in metals and metalloids (their Fig. 8 and grey symbols in Fig. 3). All of the values are less than 0.5, with many between 0.15 and 0.35, demonstrating that mass discrimination is generally larger for diffusion in solids than in aqueous solutions and silicate melts. The scatter of β factors plotted against temperature reflects the sensitivity of mass discrimination by diffusion to the nature of the cation as well as the composition of the solid host. Diffusion in solids is further complicated by diffusion anisotropy, wherein the diffusion mechanism or rate of diffusion by a single mechanism depends on the direction of the diffusive flux in relation to, for example, the crystallographic axes.

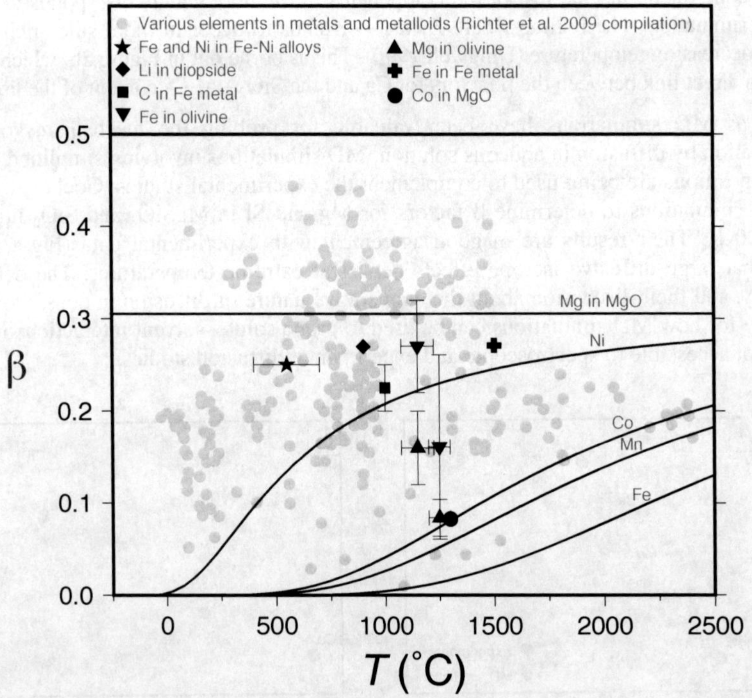

Figure 3. Compilation of β factors in metals and minerals. Gray symbols are data from the materials science literature compiled by Richter et al. (2009a). Isotope fractionation by diffusion is more efficient in ideal gases ($\beta = 0.5$) than in solids ($\beta \sim 0.25 \pm 0.15$) and liquids ($\beta < 0.2$; Figs. 1 and 2). The scatter in the data for solids reflects the sensitivity of β factors to the diffusing cation, the host composition, purity of the host, and temperature. Black symbols are data from the relatively recent geoscience literature (Roskosz et al. 2006; Dauphas 2007; Sio et al. 2013; Müller et al. 2014; Richter et al. 2014a; Oeser et al. 2015; Van Orman and Krawczynski 2015). Curves are calculated βs for different cations diffusing by a vacancy mechanism in periclase (MgO) (Van Orman and Krawczynski 2015).

In the materials science literature, isotopic diffusion coefficients are often discussed in terms of the number and masses of other atoms whose motions are correlated with the movement of the atom of interest (e.g., Vineyard 1957; Schoen 1958; Mullen 1961; Mundy et al. 1966; Rothman and Peterson 1969). Such information is folded into the so-called correlation and coupling coefficients:

$$\beta \approx \frac{\left(\dfrac{D^{\mathrm{L}}}{D^{\mathrm{H}}}-1\right)}{2\left[\left(\dfrac{m^{\mathrm{H}}}{m^{\mathrm{L}}}\right)^{1/2}-1\right]}=\frac{1}{2}f\kappa, \tag{7}$$

where f (<1) is the correlation coefficient, representing the degree to which the diffusion process deviates from a random walk, and the κ is the coupling coefficient, representing the degree to which the motion of an atom during a single jump is coupled to that of other nearby atoms (Van Orman and Krawczynski 2015). The values of f and κ depend on the diffusion mechanism(s) as well as the type of lattice (Mullen 1961; Le Claire 1966). For example, f is close to unity for diffusion by an interstitial mechanism but deviates from unity for diffusion by a vacancy mechanism (e.g., Le Claire 1966). The parameter κ is related to the deformation (dilation/relaxation) of the crystal lattice that accompanies the diffusive jump of an atom (Le Claire 1966; Müller et al. 2014). Historically, β factors have been used to infer either f or κ given independent knowledge of one or the other, but in principle, if f and κ are known independently, they can be used to predict β.

Van Orman and Krawczynski (2015) provide a comprehensive review of the classical theory applied to diffusion in silicate minerals. They point out that κ values range from 0.5 to 1.0 whereas f values range from 0 to 1.0. They show how to calculate f for (1) diffusion of trace elements by a vacancy, interstitial, or interstitialcy mechanism, (2) diffusion of major elements (e.g., Mg in MgO) and (3) diffusion along grain boundaries. In most cases there are no experimental data on the mass dependence of the diffusivity to compare to the theoretical predictions, and most of the theoretical predictions are not purely theoretical, but are derived from experimental diffusion data by applying the classical theory. In some instances, there is enough information to make predictions. The curves in Figure 3 show the expected behavior of Fe, Mn, Co and Ni diffusion by a vacancy mechanism in MgO using diffusion parameters from density functional theory calculations (Crispin et al. 2012; Van Orman and Krawczynski 2015). An encouraging result is that the curve for Co intersects the only experimental data point for Co at 1300 °C. It is noteworthy that the β factors are expected to increase with temperature for these cations because they exchange with vacancies more readily than Mg. The temperature-dependence of β has the opposite sense for trace elements that exchange with vacancies less readily than the solvent Mg atoms. For olivine, the prediction is that $\beta_{\mathrm{Fe}}<\beta_{\mathrm{Ni}}<\beta_{\mathrm{Mg}}<\beta_{\mathrm{Ca}}$, but the experimental data suggest that $\beta_{\mathrm{Fe}}>\beta_{\mathrm{Mg}}$ (Fig. 3; Sio et al. 2013; Oeser et al. 2015). Some possible reasons for this discrepancy are discussed by Van Orman and Krawczynski (2015) and include uncertainty in the binding energy between Fe and vacancies and local ordering in olivine that is not accounted for in the theory. For magnetite, the prediction is that the slowly diffusing cations Ni, Cr, Al and Ti have β factors of about 0.4–0.45 that do not vary with Ti content. Other cations such as Fe, Co, and Mn are predicted to have smaller β factors (~0.2 to 0.3) that are sensitive to Ti content. For rutile, the β factors for Co, Mn, Fe, Ni and Li are expected to exhibit anisotropy because these cations diffuse rapidly in the *c*-direction and slowly in the *a*-direction. In general, faster diffusion in solids implies less mass discrimination, opposite the relationship observed in aqueous solutions and silicate melts. All of these predictions are subject to caveats, as discussed by Van Orman and Krawczynski (2015), but offer much in the way of testable hypotheses moving forward.

The motivation to interpret stable isotope variations in nature has led to renewed efforts to determine β factors in silicate minerals and metals relevant to geochemistry and cosmochemistry (black symbols in Fig. 3). The β factors measured over the past decade have been used in models of Fe and Ni isotope fractionation during growth of kamacite (α-Fe) lamellae in taenite (γ-Fe) (Dauphas 2007), diffusion of Fe into metal from a silicate liquid reservoir (Roskosz et al. 2006), C isotope fractionation during growth of γ-Fe in meteorites (Müller et al. 2014), and Fe, Mg, and Li isotope exchange between olivine or pyroxene and the host rock/melt during subsolidus cooling (Beck et al. 2006; Sio et al. 2013; Richter et al. 2014a; Oeser et al. 2015). The experimental studies are also guiding theoretical developments, leading to a deeper understanding of how β factors are related to the physical mechanisms of diffusion in solids (see, in particular, Müller et al. 2014 and Richter et al. 2014a).

DIFFUSIVE BOUNDARY LAYERS IN THE GROWTH MEDIUM

In this section we describe the effects of competition between mineral growth rate and isotope fractionation by diffusion in the fluid phase, with growth rate being a specified dependent variable. A common assumption, which is adopted here for now, is that the surface of the mineral is in equilibrium with the composition of the fluid at the solid–fluid interface. It is also assumed, for now, that diffusion in the solid is negligible. Hence, any kinetic effects described in this section are due to diffusion in the fluid phase. These effects are placed in a broader context in the next section, where growth rate, diffusion in the growth medium and the kinetics of chemical exchange at the mineral surface are addressed concurrently.

In situations where crystal growth outpaces diffusion in the growth medium, incompatible impurities must accumulate, and compatible impurities must be depleted, in advance of the crystal interface. This "pile up" or "draw down" of impurity concentrations is inevitable when diffusion in the fluid is too slow to maintain uniform concentration (Fig. 4). Drawing upon the work of Tiller et al. (1953) and Smith et al. (1955), Albarède and Bottinga (1972) introduced the phenomenon of diffusive boundary layers (DBLs) to the geoscience community in the context of trace-element uptake in rapidly growing phenocrysts. They invoked the equations of Smith et al. (1955) to show that the uptake of trace elements in phenocrysts could deviate significantly from equilibrium, or more broadly, from control by surface reaction. The equation for the concentration profile in the solid is:

$$\frac{C_s(x_s)}{C_0}=\frac{1}{2}\left\{\begin{aligned}&1+\operatorname{erf}\left(\frac{\sqrt{(v/D)x_s}}{2}\right)\\&+(2K_{eq}-1)\exp\left[-K_{eq}(1-K_{eq})\sqrt{(v/D)x_s}\right]\operatorname{erfc}\left(\frac{(2K_{eq}-1)\sqrt{(v/D)x_s}}{2}\right)\end{aligned}\right\}, \quad (8)$$

where $C_s(x_s)$ is the concentration (mole fraction or weight fraction) in the solid after growth amount x_s ($x_s=0$ when growth begins at $t=0$), C_0 is the initial (uniform) concentration in the growth medium and also at infinite distance from the interface, v is the linear growth rate (cm s^{-1}), D is the diffusivity of the element of interest in the growth medium (cm^2 s^{-1}), and K_{eq} is the equilibrium partition coefficient (no units) between the crystal and the growth medium. Equation (8) describes the response of the impurity concentration in a growing crystal (lower panel of Fig. 4) to the boundary layer that develops in the growth medium ahead of the moving interface. The impurity profile in the boundary layer itself is given by:

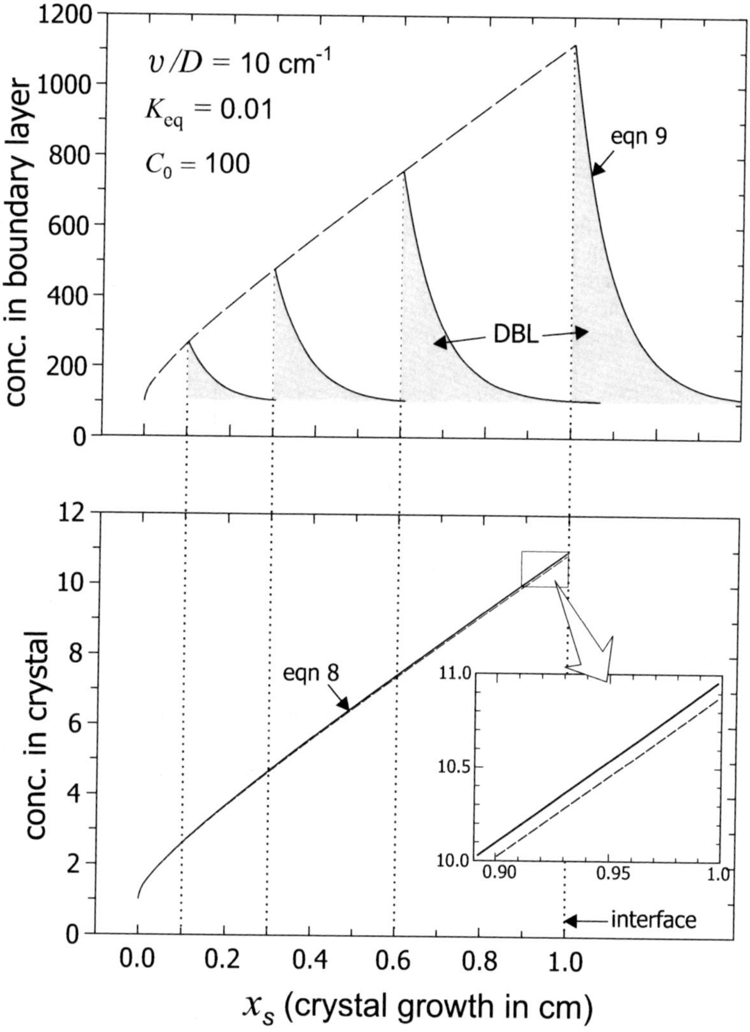

Figure 4. Development of a diffusive boundary layer (DBL) during 1-D crystal growth. *Top*: Four snapshots in time of a DBL (shaded) in the growth medium against a crystal interface advancing from left to right, where the curves above the shaded areas are given by Equation (9). Assuming equilibrium at the immediate interface, the concentration of the element of interest in the crystal follows that in the immediately contacting growth medium ($C_s = K_{eq} \cdot C_l$), where K_{eq} is the equilibrium partition coefficient between the crystal and the growth medium. *Bottom*: The concentration profile in the crystal, which is given by Equation (8). The inset shows the slight difference in concentration between two isotopes whose diffusivities in the growth medium differ by 1%. The partition coefficient is assumed to be identical for the two isotopes (i.e., no solid–liquid equilibrium fractionation), so the observed fractionation is attributable entirely to the difference in diffusivity.

$$\frac{C_l(x_l)}{C_0}=1+\frac{1-K_{eq}}{2K_{eq}}\exp\left(-\frac{v}{D}x_l\right)\mathrm{erfc}\left(\frac{(x_l-vt)}{2\sqrt{Dt}}\right)-\frac{1}{2}\mathrm{erfc}\left(\frac{(x_l+vt)}{2\sqrt{Dt}}\right)$$
$$+\frac{1-K_{eq}}{2}\left(\frac{1}{1-K_{eq}}-\frac{1}{K_{eq}}\right)\exp\left[-(1-K_{eq})\frac{v}{D}(x_l+K_{eq}vt)\right]\times \tag{9}$$
$$\mathrm{erfc}\left\{\frac{1}{2\sqrt{Dt}}\left[x_l+(2K_{eq}-1)vt\right]\right\},$$

where $C_l(x_l)$ is the concentration of the impurity in the growth medium at time t and distance x_l from the interface (upper panel of Fig. 4). Tiller et al. (1953) showed that for sustained crystal growth, the concentration profile given by Equation (9) (i.e., the diffusive boundary layer) eventually reaches a steady state in which growth and diffusion are balanced and the effective bulk partition coefficient becomes unity. At this stage, the steady-state concentration profile in the growth medium is given by

$$\frac{C_l(x_l)}{C_0}=1+\frac{1-K_{eq}}{K_{eq}}\exp\left(-\frac{v}{D}x_l\right). \tag{10}$$

The simplicity of this equation is appealing, but for the range of growth rates and diffusivities pertinent to natural systems, this steady-state condition is never likely to be achieved. It can be seen by inspection of Equations (8) and (9) that the ratio v/D (along with K_{eq}) is key in determining the amplitude of the concentration perturbation in the growth medium and the consequent deviation from the equilibrium in the crystal. Albarède and Bottinga (1972) emphasized the importance of v/D, but they were handicapped in making specific predictions about non-equilibrium behavior of individual trace elements by the lack of data constraining the growth rate v of phenocrysts and the diffusivity D of elements of interest in the molten silicate growth medium. For phenocryst growth in molten basalt, much more is known today than was known in 1972 about appropriate values of v and D, but the usefulness of the equations of Smith et al. (1955) nevertheless remains limited by the fact that they pertain to growth and diffusion in one dimension only. In petrological systems, the length scales of diffusion in fluid growth media are comparable to the crystal sizes under consideration, which means that multidimensional aspects of diffusion are probably important. In addition, the width of the diffusive boundary layer and the magnitude of any non-equilibrium effects resulting from its presence will be influenced by the dynamics of the liquid phase.

In the context of the present review of *isotopic* fractionations during crystal growth, we can use the β factors in Table 2 to calculate non-equilibrium isotopic profiles in crystals and their surrounding host medium. The lower panel of Figure 4 shows the subtle difference in the concentration profiles in a hypothetical crystal of two isotopes of a given element whose abundances in the growth medium are the same but whose diffusivities differ by 1%. The slower-diffusing isotope is enriched in the crystal because its concentration rises to a slightly higher level in the diffusive boundary layer (for an incompatible element). The net result is that the isotope ratio in the crystal is different from that in the bulk growth medium, amounting to a kinetic fractionation. The example in Figure 4 serves as a useful illustration of the potential for rapid crystal growth to fractionate isotopes, but the large magnitude of the isotopic separation (~10‰) is unrealistic for three reasons: (1) nearly a centimeter of sustained rapid growth is required to produce it, (2) the 1% difference in diffusivity is on the high side of the plausible range for natural systems (see below), and (3) the 1-D equation of Smith et al. (1955) was used to generate the diagram (Eqn. 8). Crystal growth in 3-D produces more subdued effects, but these may nevertheless be significant in some instances.

Table 2. Isotopic mass dependence of diffusion in silicate melts

Description	Composition	Element	D_x/D_{Si}	Isotope masses: heavy, light	β	Diffusivity ratio	References
Natural volcanic liquids	basalt–rhyolite	Ca	1.6	44, 40	0.075	0.993	Richter et al. (2003)
	basalt–rhyolite	Ca	2.2	44, 40	0.035	0.997	Watkins et al. (2009)
	basalt–rhyolite	Mg[1]	–	26, 24	0.050	0.996	Richter et al. (2008)
	basalt–rhyolite	Li	290	7, 6	0.215	0.967	Richter et al. (2003)
	basalt–rhyolite	Fe	1.3	56, 54	0.030	0.999	Richter et al. (2009b)
Simple silicate liquids	albite + anorthite	Ca	23	44, 40	0.210	0.980	Watkins et al. (2011)
	albite + diopside	Ca	6.3	44, 40	0.165	0.984	Watkins et al. (2011)
	albite + diopside	Mg	1.5	26, 24	0.100	0.992	Watkins et al. (2011)
	$CaO–Na_2O–SiO_2$	Ca	–	44, 40	0.100	0.991	Watkins et al. (2014)
	$CaO–Na_2O–SiO_2$	Ca	–	44, 40	0.060	0.994	Watkins et al. (2014)
Molecular dynamics simulations	SiO_2	Si	1	30, 28	0.055	0.996	Goel et al. (2012)
	SiO_2	Si	1	30, 28	0.042	0.997	Goel et al. (2012)
	SiO_2	Si	1	30, 28	0.063	0.996	Goel et al. (2012)
	SiO_2	Si	1	30, 28	0.045	0.997	Goel et al. (2012)
	$MgSiO_3$	Si	1	30, 28	0.047	0.997	Goel et al. (2012)
	$MgSiO_3$	Si	1	30, 28	0.043	0.997	Goel et al. (2012)
	$MgSiO_3$	Si	1	30, 28	0.047	0.997	Goel et al. (2012)
	$MgSiO_3$	Mg	4.3	26, 24	0.135	0.989	Goel et al. (2012)
	$MgSiO_3$	Mg	2.3	26, 24	0.092	0.993	Goel et al. (2012)
	$MgSiO_3$	Mg	2.3	26, 24	0.084	0.993	Goel et al. (2012)

Notes:
1. In Figure 2 it is assumed that $D_{Mg}/D_{Si} \approx 1$.

Watson and Müller (2009) extended the approach of Smith et al. (1955) and Albarède and Bottinga (1972) to spherical geometries using finite-difference numerical approaches incorporating a moving boundary. Here, the growth rate v for a spherical crystal is considered to be a linear increase in radius with time, although other growth rate laws are just as worthy of consideration (see Gardner et al. 2012). For the present purposes, we consider three hypothetical pairs of isotopes whose diffusivities differ by 0.1%, 0.5% and 1%. These differences can be related to those between actual isotope pairs using Table 2, which lists diffusivity ratios for relevant isotopes of Li, Mg, Ca, and Fe based on measured or estimated β values. In Figure 5, the isotopic consequences of a diffusive boundary layer in the growth medium are shown as radial isotope profiles in the crystal for both incompatible and compatible elements; i.e., $K_{eq}=0.01$ and 100. These partition coefficients are arbitrary, but the results for all elements with $K_{eq}<0.1$ are very similar, as are those for all elements with $K_{eq}>10$.

Figure 5 shows model isotope profiles in crystals grown from 0.001 to 0.5 cm in radius for v/D ratios varying from 0.2 to 10 cm^{-1}, which is considered a plausible range for phenocryst growth in which diffusive isotope fractionation in a boundary might occur. Values of $v/D<0.2\,cm^{-1}$ produce minimal isotope fractionation, and sustained v/D values $>10\,cm^{-1}$

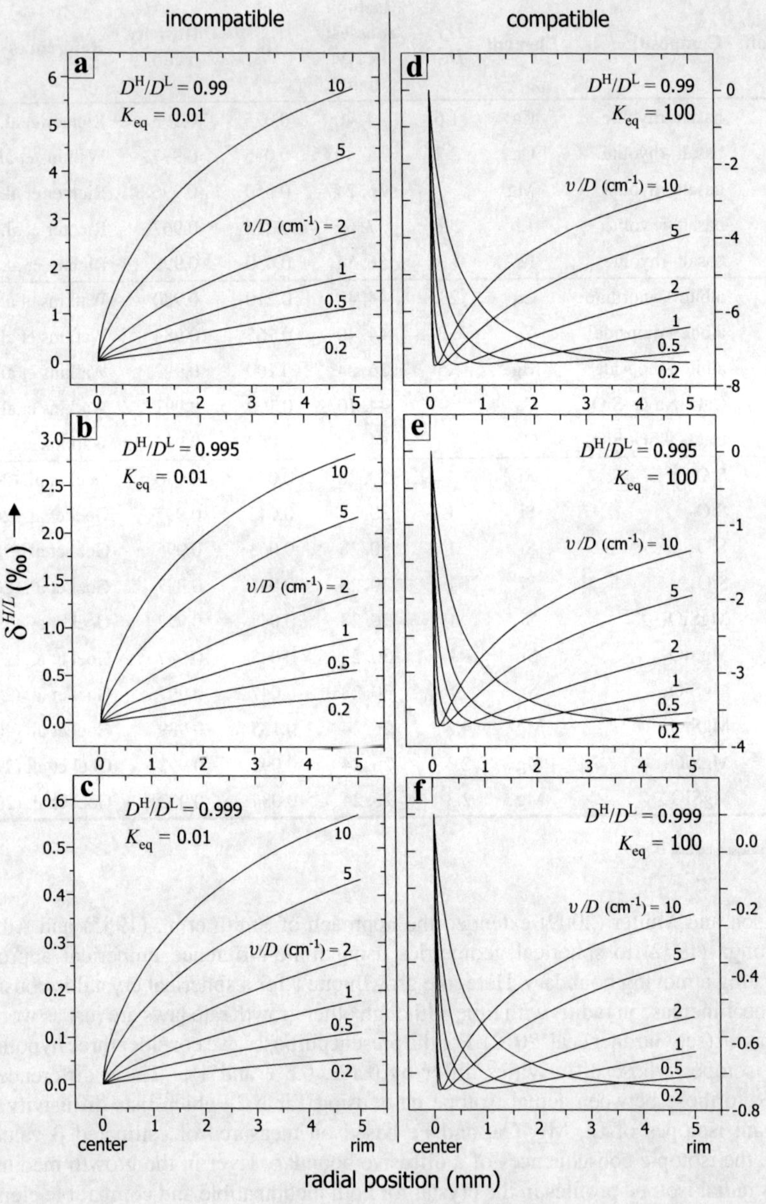

Figure 5. Radial isotope fractionation profiles resulting from development of diffusive boundary layers in the growth medium of a spherical crystal. The curves are outputs of numerical simulations in which the crystal was allowed to grow from 0.01 to 5 mm in radius, with v/D varying from 0.2 to 10 cm^{-1} (see Watson and Müller 2009). Isotope fractionation is expressed in conventional geochemical δ notation, referenced to the isotope ratio in the growth medium (Eqn. 11). Superscripts H and L refer to the heavy and light isotope, respectively; the ratio D^H/D^L is indicated in each panel (compare with specific isotope pairs listed in Table 2). Panels a–c show behavior of the isotopes of an incompatible element ($K_{eq} = 0.01$); panels d–f are for compatible elements ($K_{eq} = 100$). Note that the magnitude of the isotopic fractionation scales linearly with D^H/D^L.

are probably uncommon given typical rates of phenocryst growth (e.g., Maaløe 2011). Isotopic fractionation is portrayed in Figure 5 using geochemical δ notation:

$$\delta^{H/L} = \left[\frac{\left(C^{H}/C^{L}\right)^{int}}{\left(C^{H}/C^{L}\right)^{\infty}} - 1 \right] \times 1000, \tag{11}$$

where $(C^H/C^L)^{int}$ is the concentration ratio of the heavy (H) to light (L) isotope in the melt at the interface with the crystal, and $(C^H/C^L)^{\infty}$ is the "far-field" isotope ratio in the melt. Assuming no equilibrium fractionation of isotopes between crystal and melt (i.e., assuming K_{eq} is not dependent on mass), $(C^H/C^L)^{int}$ is acquired by the crystal as it grows, leading to the radial isotopic profiles shown in Figure 5. An important generalization from this figure is that fractionation of the isotopes of incompatible elements leads to enrichment of the heavy isotope in the crystal, with the isotopic fractionation increasing monotonically as the crystal grows (Fig. 5a–c). For compatible elements, on the other hand, the heavy isotope is depleted in the crystal and the isotopic fractionation is greatest (i.e., $\delta^{H/L}$ reaches a minimum) after a relatively small amount of growth and lessens thereafter (Fig. 5d–f).

The radial isotope profiles in crystals shown in Figure 5 were produced by numerically monitoring the development of diffusive boundary layers around spherical crystals during growth. After 5 mm of growth, the resulting diffusion fields ("haloes") extend from the crystal interface to a distance of ~0.2 cm to ~2 cm into the growth medium for v/D values of 10 and 0.2 cm^{-1}, respectively (see Fig. 5 of Watson and Müller 2009). It is important to ask under what natural circumstances such large diffusion fields could develop without disruption by physical factors. The low viscosity of mafic magmas means that growing crystals are likely to be in motion relative to their growth medium, which would lead to erosion of a developing diffusion field. For this reason, even though the fractionation curves in Figure 5 are quantitatively accurate, they may overestimate isotopic fractionation during phenocryst growth in mafic systems. The purely diffusive model represented by Figure 5 may be more relevant to crystal growth in relatively viscous, silicic melts in which crystal motion relative to the growth medium might be unimportant, and perhaps also to porphyroblast growth in metamorphic systems.

In addition to static environments in which mass transport occurs solely by diffusion, Watson and Müller (2009) also discussed dynamic crystal-growth environments in which the width of the diffusive boundary layer is regulated by fluid dynamics. Despite the seeming greater complexity of dynamic systems, isotope fractionation during crystal growth can be modeled in such systems without resorting to complex numerical simulations. A simple analytical expression can be used, provided the width of the boundary layer in the growth medium can be specified on the basis of fluid dynamical considerations:

$$\delta(‰) = 1000\left(1 - \frac{D^{H}}{D^{L}}\right)\left(\frac{v x_{BL}}{D^{H}}\right)\left(1 - K_{eq}\right). \tag{12}$$

This general equation of Watson and Müller (2009) describes the expected steady-state (maximum) fractionation of two isotopes in a growing crystal in terms of the ratio of their diffusivities (D^H/D^L) in the growth medium and the thickness (x_{BL}) of the physically regulated boundary layer. As in the case of Equation (10), the steady-state described by Equation (12) is established only after a certain amount of growth, which in this case depends critically on the width of the physical boundary layer. If the boundary is thin, as in the case of low viscosity silicate melts (e.g., 10–100 μm; see Kerr 1995), the steady-state is closely approached after a relatively small amount of growth (Fig. 6). However, this figure also shows that for physical boundary layers substantially wider than ~100 μm, the steady state may not be approached for

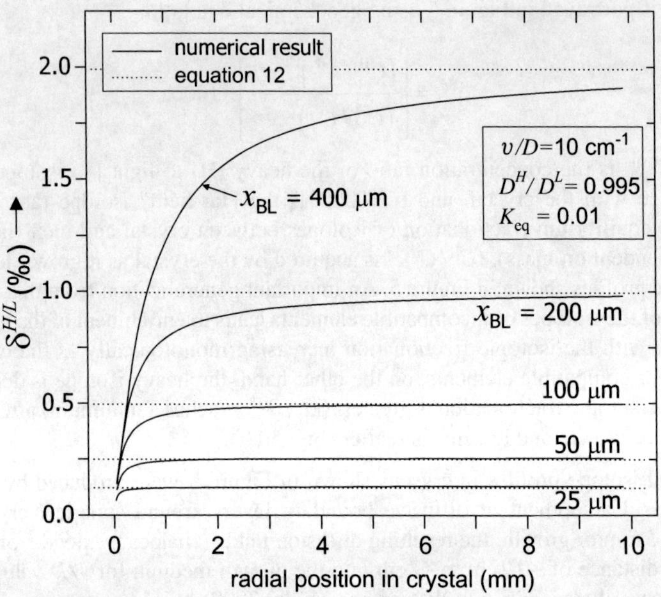

Figure 6. Isotope fractionation as a consequence of crystal growth with $v/D = 10\,\text{cm}^{-1}$ in a dynamic medium where the diffusive boundary layer (DBL) is limited to a fixed width (x_{BL}) by fluid dynamical considerations. The smooth curves represent numerical results for radial crystal growth in which x_{BL} widens by diffusion until the pre-specified maximum is reached and a steady state established. The dotted lines show the fractionation given by Equation (12), which was derived for linear growth with the steady state DBL present from the start (see Watson and Müller 2009). For x_{BL} values that are small relative to the crystal size, Equation (12) provides a good estimate of $\delta^{H/L}$. For phenocryst growth in magmas, x_{BL} is probably <100 mm.

typical crystal sizes of a few millimeters. As noted by Watson and Müller (2009), Equation (12) provides an accurate prediction of the level of steady-state fractionation for highly incompatible elements ($K_{eq} \lesssim 0.1$), but accuracy decreases as the element of interest becomes more compatible. For $K_{eq} \sim 0.4$, for example, the fractionation is underestimated by ~14‰.

A fundamental assumption in the development and use of Figures 5 and 6 and Equation (12) is that the radius of the crystal of interest increases linearly with time (dr/dt is constant). Other possibilities are illustrated in Figure 7; these include an inverse square-root dependence of v upon time and a linear increase in v with time (the specifics are shown in Fig. 7b). In the first of these two models, the initially high growth rate decelerates sharply at the outset and more slowly as growth continues (Fig. 7c), which is the expected pattern for diffusion-controlled crystal growth (e.g., Zhang 2008, p. 277). The linearly accelerating growth rate, on the other hand, might be realistic for interface reaction-controlled growth under conditions of strong undercooling; i.e., with $\Delta G_{crystallization}$ becoming increasingly negative with increasing time. In Figure 7, these two markedly different growth scenarios are compared with the constant-dr/dt model used in generating Figures 5 and 6. The specific growth laws were chosen to produce the same total amount of growth (5 mm) in a given time (6.7×10^6 s) (Fig. 7b–c). The potential of these different growth scenarios to fractionate isotopes is illustrated in Figure 7e using $D^H/D^L = 0.993$, which would apply to diffusion of ^{44}Ca versus ^{40}Ca in basaltic melt, for example. The partition coefficient K_{eq} between the crystal and the growth medium was assumed to be 0.01, but the results would be similar for any moderately to strongly incompatible element. A specific value of D ($10^{-8}\,\text{cm}^2\,\text{s}^{-1}$) was used to produce Figure 7, but because it is v/D that determines outcome, the ratio is plotted in Figures 7c–d. The simple conclusion from

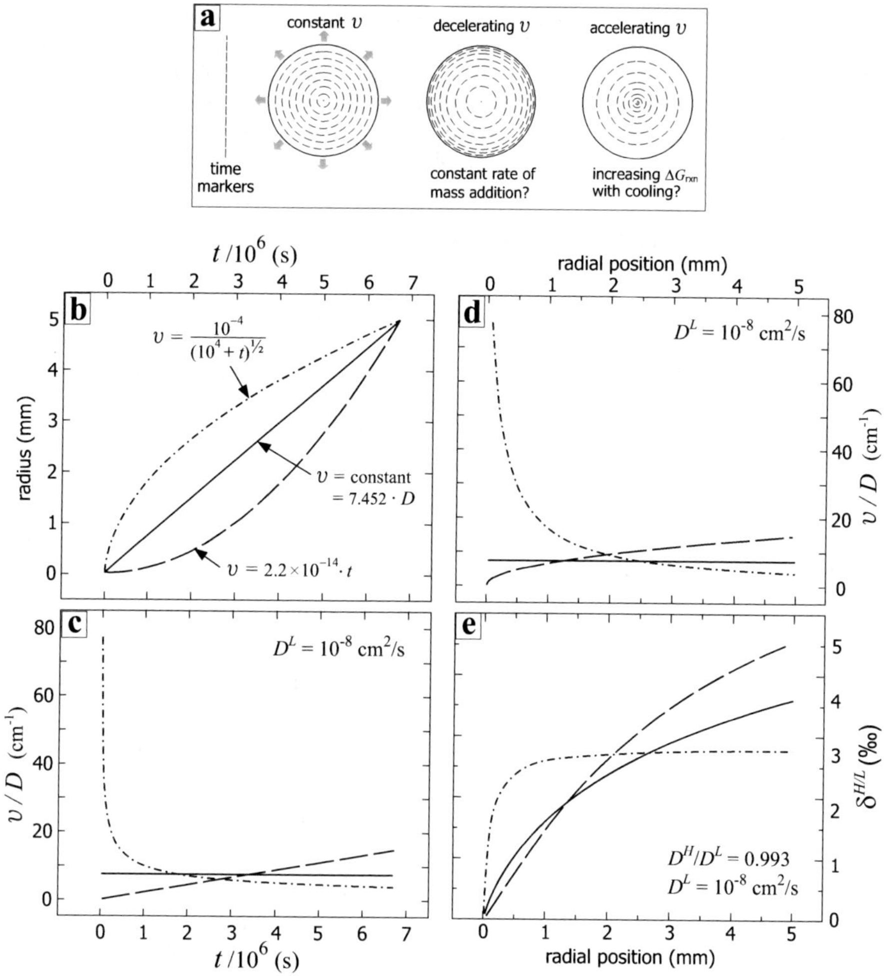

Figure 7. Numerical model output illustrating the effects of various growth laws on isotope fractionation in a growing crystal due to the presence of a diffusive boundary layer (DBL) in the growth medium. Cartoons in (a) show progressive crystal growth with equivalent time intervals indicated by the dashed markers. The reference case is constant v (left), but decelerating or accelerating v is also possible (middle and right; see text). The curves in (b) show three specific growth laws used in numerical models, corresponding to the cartoons in (a). The equations describing v as a function of time are shown on the graph (v in cm/s); these were chosen so that all three growth laws yield the same total crystal growth (5 mm) in the same time (6.7×10^6 s), thus allowing direct comparison. The diffusivity was also assumed to be the same (10^{-8} cm^2 s^{-1}) in all cases. Panel (c) shows v/D as a function of time for the three growth laws; in (d), v/D is shown as a function of radial position. Panel (f) compares the ultimate consequences of the three growth laws for isotopic fractionation in the crystal when $D^H/D^L = 0.993$. Note that the line symbols (solid, dot–dash, and dash) are used consistently in b–e for the three growth laws shown in (b).

Figure 7e is that the constant v and $v \propto t$ models produce broadly similar $\delta^{H/L}$ versus radial position curves. However, the rapid initial growth scenario causes an abrupt deviation from equilibrium (rise in $\delta^{H/L}$) near the center of the crystal that is maintained throughout the 5 mm of growth. The lesson seems to be that an early perturbation from equilibrium caused by fast initial growth ($v/D \sim 80$ cm^{-1}) can be maintained by only modest subsequent v/D (~5 cm^{-1}).

In closing this section on isotope fractionation caused by diffusive boundary layers (DBLs) of trace elements in the growth medium, we note that in general the rate of crystal growth will itself be limited by diffusion of structural constituents. This more general case is treated in the following two sections. Trace elements likely to experience deviations from isotopic equilibrium are those whose diffusivities in the growth medium are similar to or lower than those of the structural constituents of the mineral under consideration.

ISOTOPE FRACTIONATION BY COMBINED REACTION AND DIFFUSION

In the previous section, we considered the special case of isotopic fractionation by diffusion of a trace element where the growth rate is unaffected by the incorporation or presence of the tracer. It was further assumed that the local equilibrium partition coefficient K_{eq} is independent of mass. Here we review a more general model applicable to diffusion and reaction of the stoichiometric constituents of a mineral. The kinetic isotope effects may arise from diffusive transport to the mineral surface as well as the mass dependence on reaction rates at the mineral–fluid interface.

General framework for crystal growth from an infinite solution

The physical framework of the surface reaction kinetic model (SRKM) that we will use for discussion (Fig. 8), is generalized from the discussions in DePaolo (2011) and includes:

- the mineral surface, advancing as a consequence of precipitation at a velocity v,
- a surface liquid layer that represents that region immediately above the mineral surface from which ions are transferred onto the mineral surface and to which ions are transferred from the mineral surface,
- a stagnant liquid layer of thickness h through which dissolved ions must diffuse to the mineral surface from the bulk fluid reservoir, and
- a stirred or well-mixed fluid reservoir that constitutes the ultimate source of ions that precipitate to grow the crystal.

The advancement of the mineral surface as a consequence of precipitation (or growth) can be described in terms of a velocity (v) expressed in units of $\mathrm{m\,s^{-1}}$ or a precipitation flux $R_p = v\rho/M$, in units of $\mathrm{moles\,m^{-2}\,s^{-1}}$. In this expression M is molar mass of the crystal ($\mathrm{kg\,mol^{-1}}$) and ρ is density ($\mathrm{kg\,m^{-3}}$). At the mineral surface there is in general an exchange of ions between the mineral surface and the fluid surface layer. The exchange can be separated into a forward flux (R_f) of ions from the fluid to the mineral surface, and a backward flux (R_b) from the mineral surface to the fluid surface layer. The growth rate is the difference between the forward flux and the backward flux, or:

$$R_p = \frac{v\rho}{M} = R_f - R_b. \tag{13}$$

In situations where diffusion through the stagnant boundary layer limits growth, then R_p must be equal to the flux through the boundary layer, or:

$$R_p^d = \frac{v\rho}{M} = R_f - R_b = D\frac{C^* - C_s}{h}, \tag{14}$$

where C^* is the concentration in the bulk solution, C_s is the concentration in the immediate vicinity of the mineral surface, and h is the diffusive boundary layer thickness. This relationship is sufficient to describe the various limiting cases for crystal growth that are discussed below and in other chapters of this volume.

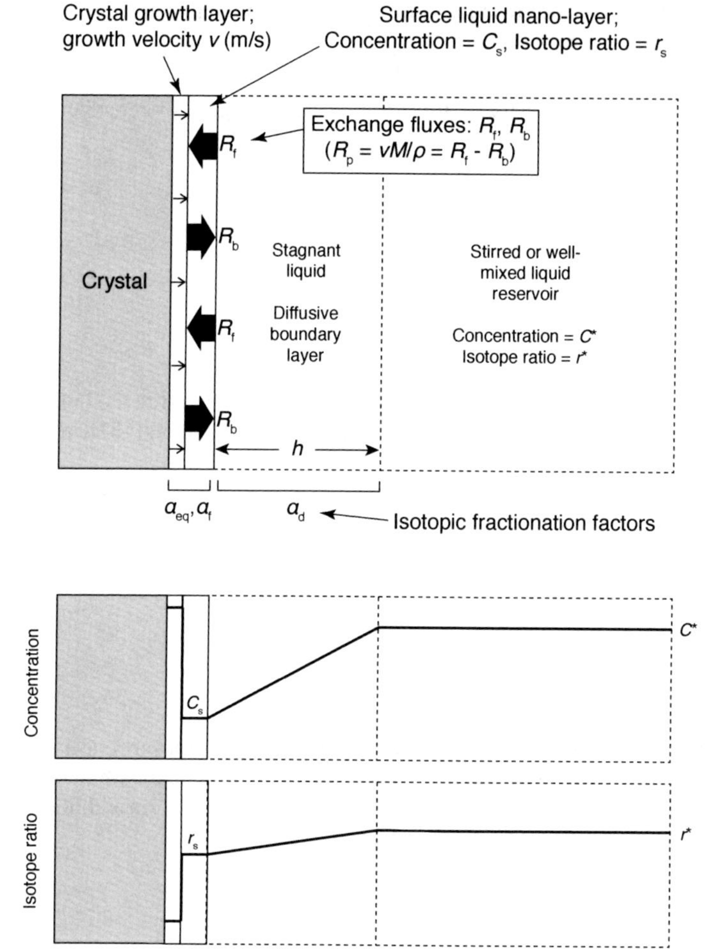

Figure 8. Schematic of the surface reaction kinetic model (SRKM) of DePaolo (2011), where stable isotope fractionation arises during diffusion- and reaction-controlled mineral growth. This model differs from the one presented in Figures 4–7 in that the reaction rate R is mass dependent and the model applies to structural elements of the crystal and not just passive tracers. The α's are fractionation factors associated with diffusion to, and reaction at, the mineral surface. The bottom panels depict steady-state boundary layer concentration and isotope ratio profiles. Note the distinction between the surface concentration and isotopic ratio (C_s and r_s) versus the bulk concentration and isotopic ratio (C^* and r^*).

When diffusion through the boundary layer does not limit growth, then the precipitation rate R_p is determined by the saturation state of the surface layer, which is effectively the same as the saturation state of the bulk solution. This case is referred to as surface reaction-controlled growth. As we will describe further below, the rate of growth is dependent on the extent of oversaturation of the solution in the surface layer, and the behavior for this case can be divided into two limiting regimes. The first corresponds to small degrees of oversaturation, and translates to $R_f - R_b \ll R_b$, or $R_p \ll R_b$. The other regime corresponds to high degrees of oversaturation, and translates to $R_f - R_b \gg R_b$, or $R_p \gg R_b$. The crossover between these two limiting cases occurs at $R_p = R_b$ (DePaolo 2011).

If we represent the surface reaction controlled precipitation rate as

$$R_p^s = k_p\left(C_s - C_{eq}\right), \tag{15}$$

then we can derive an expression for the diffusion-controlled regime by equating Equations (14) and (15):

$$k_p\left(C_s - C_{eq}\right) = \frac{D}{h}\left(C^* - C_s\right). \tag{16}$$

Solving for C_s and substituting into Equation (15) gives:

$$R_p = \frac{k_p D\left(C^* - C_{eq}\right)}{D + hk_p}. \tag{17}$$

This expression gives the growth rate in terms of the concentration in the bulk solution, the equilibrium concentration and the parameters D, h, and k_p. Equation (17) can be rearranged to the form:

$$\frac{1}{R_p} = \frac{1}{k_p\left(C^* - C_{eq}\right)} + \frac{h}{D\left(C^* - C_{eq}\right)}, \tag{18a}$$

which can also be written as

$$\frac{1}{R_p} = \frac{1}{R_p^{s*}} + \frac{1}{R_p^{d*}}. \tag{18b}$$

where R_p^{s*} is the surface-controlled rate assuming that the surface layer concentration is equal to the bulk–fluid concentration C^*, and R_p^{d*} is the diffusion-limited growth rate assuming that the surface layer concentration is the equilibrium concentration, C_{eq}. An additional useful way to write Equation (18b) is in terms of an effective rate constant k_{eff}:

$$R_p = k_{eff}(C^* - C_{eq}) \tag{19}$$

where

$$k_{eff} = \left(\frac{1}{k_p} + \frac{h}{D}\right)^{-1}. \tag{20}$$

Equation (18) provides a description of the transition from surface-reaction-controlled growth to diffusion-limited growth. When $h = 0$, the growth rate is exactly the surface-controlled rate. Diffusion to the mineral surface begins to have an effect when h/D is a significant fraction of $1/k_p$, and dominates when $h/D > 1/k_p$. The crossover in behavior between surface reaction controlled and diffusion-controlled growth is at $D/h = k_p$, or when $D/hk_p = 1$.

The above analysis leads to the framework described by DePaolo (2011) and reproduced in slightly modified form as Figure 9. The conditions that describe the controls on kinetic isotope effects, under both near-equilibrium and far-from-equilibrium conditions, are the same as those that describe surface reaction and diffusion-controlled crystal growth. Crystal growth at isotopic equilibrium can only occur when $R_p \ll R_b$, and in the absence of diffusion control. The extent to which diffusion control affects KIEs is determined by the dimensionless parameter D/hk_p, although for most mineral growth reactions the elemental concentration in solution that controls growth rate may not be the element of interest for isotopic composition. For this reason it may be more useful to rewrite Equation (18a) in the form:

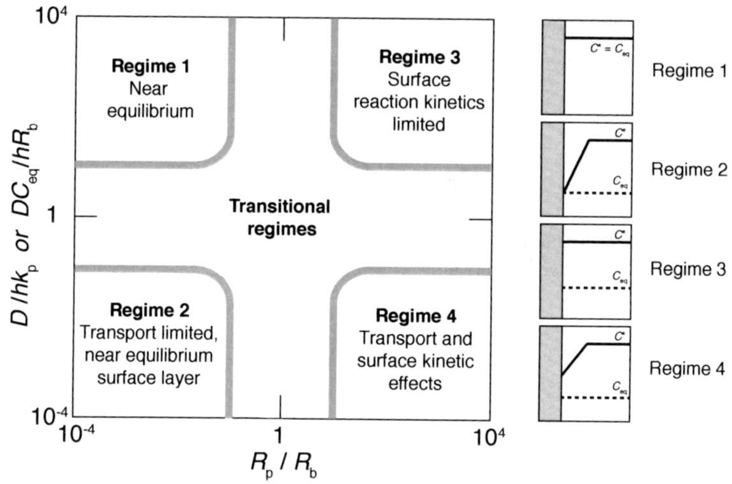

Figure 9. Regime diagram showing when mineral growth rates are controlled by surface reaction kinetics versus diffusion kinetics. The x-axis is the ratio of the net mineral precipitation rate ($R_p = R_f - R_b$) to the rate of ion detachment from the mineral surface (R_b). The y-axis is a dimensionless parameter that describes the relative importance of diffusive transport versus surface-reaction kinetics in determining the overall growth rate. The four panels on the right depict the steady-state concentration profiles for the four regimes. Note that the assumption of local equilibrium at the crystal-liquid interface only applies when $R_p / R_b \ll 1$ (Regimes 1 and 2). Hence the model presented in Figures 4–7 is effectively a "Regime 2" model. Modified from DePaolo (2011).

$$\frac{1}{R_p} = \frac{1}{k_p C_{eq}(S-1)} + \frac{h}{DC_{eq}(S-1)} = \frac{1}{R_b(S-1)} + \frac{h}{DC_{eq}(S-1)}, \tag{21}$$

where S is the saturation of the bulk solution defined as $S = C^*/C_{eq}$. This leads to a form for the dimensionless parameter of $DC_{eq} = h/R_b$, which is similar to the form used in DePaolo (2011). Large D and/or small boundary layer thickness are associated with fast diffusive transport and thus push the system towards a surface reaction controlled regime. A smaller ion exchange rate at the mineral surface (related to R_b) also pushes the system towards a surface reaction controlled regime. Finally, an increase in C_{eq} independent of C^* decreases $C^* - C_{eq}$, which is effectively the definition of moving away from transport control.

Crystal growth and kinetic isotope effects

Although it would be best to have a completely general formulation for KIEs during crystal growth, there are many variations for different cations and anions, and the details are just as important as the generalities. For this reason we start here with a formulation for KIEs for two isotopes of Ca (^{44}Ca and ^{40}Ca) in calcite, which is a fairly simple example. The first step, following DePaolo (2011), is to write equations for the forward (R_f) and backward (R_b) reaction rates for each isotope in the surface reaction controlled regime. It is expected that the aqueous isotopic species $^{44}Ca^{2+}$ and $^{40}Ca^{2+}$ will have slightly different k_f and k_b values and hence different rates of reaction. The simplified equations for the rate of attachment (precipitation) of the two Ca isotopic species as mineral crystal growth proceeds can be written:

$$^{40}R_f = {}^{40}k_f \left[{}^{40}Ca^{2+}\right]_s \left[CO_3^{2-}\right]_s \tag{22a}$$

and

$$^{44}R_f = {}^{44}k_f \left[{}^{44}Ca^{2+}\right]_s \left[CO_3^{2-}\right]_s . \tag{22b}$$

The dissolution rates are:

$$ {}^{40}R_b = {}^{40}k_b\left[{}^{40}\mathrm{CaCO_3}\right] \tag{23a}$$

and

$$ {}^{44}R_b = {}^{44}k_b\left[{}^{44}\mathrm{CaCO_3}\right]. \tag{23b}$$

Implicit in these expressions are the two kinetic isotopic fractionation factors associated with the forward and backward reactions:

$$\alpha_f = \frac{{}^{44}k_f}{{}^{40}k_f} \tag{24a}$$

and

$$\alpha_b = \frac{{}^{44}k_b}{{}^{40}k_b}. \tag{24b}$$

The equilibrium fractionation factor is:

$$\alpha_{eq} = \frac{\alpha_f}{\alpha_b} = \frac{{}^{44}K_{eq}}{{}^{40}K_{eq}} = \left(\frac{r_{solid}}{r_{fs}}\right)_{eq}, \tag{25}$$

where r is shorthand for the isotopic ratio $^{44}Ca/^{40}Ca$, and 'fs' refers to the fluid at the mineral surface. If the above expressions are substituted into the previous equations, two equations can be derived that relate the isotope-specific forward and backward rates:

$$ {}^{44}R_f = \alpha_f r_{fs}\, {}^{40}R_f \tag{26a}$$

and

$$ {}^{44}R_b = \alpha_b r_{solid}\, {}^{40}R_b = \frac{\alpha_f}{\alpha_{eq}} r_{solid}\, {}^{40}R_b. \tag{26b}$$

These equations can be used to derive a general equation for the fractionation attending mineral precipitation under steady state, surface reaction controlled conditions. It should be noted that the isotopic effects are not dependent on the exact form of the kinetic rate expression; they are only dependent on the rates.

The effective isotopic fractionation factor for steady state precipitation can be derived starting first with the rate of change of the isotopic ratio of the solid surface layer:

$$r_{solid} = \frac{N_{^{44}Ca}}{N_{^{40}Ca}}, \tag{27}$$

where N designates the number of atoms (or moles of atoms), and at steady state:

$$\frac{dr_{solid}}{dt} = 0 = \frac{1}{N_{^{40}Ca}}\left(\frac{dN_{^{44}Ca}}{dt} - r_{solid}\frac{dN_{^{40}Ca}}{dt}\right) = \frac{1}{N_{^{40}Ca}}\left({}^{44}R_p - r_{solid}\,{}^{40}R_p\right). \tag{28}$$

Steady state in this case means that the isotopic composition of the surficial layer of the solid is not changing with time as the crystal grows. After substitution of Equations (26a) and (26b) into (28) and some algebraic manipulation, the following expression is obtained for the steady condition:

$$\alpha_p^s = \left(\frac{r_{solid}}{r_{fluid}}\right)_{ss} = \frac{\alpha_f}{1+\frac{R_b}{R_f}\left(\frac{\alpha_f}{\alpha_{eq}}-1\right)} = \frac{\alpha_f}{1+\frac{R_b}{R_p+R_b}\left(\frac{\alpha_f}{\alpha_{eq}}-1\right)}. \quad (29)$$

Note that when the precipitation rate $R_p \to 0$, we have $\alpha_p^s \to \alpha_{eq}$, and hence the equilibrium condition is recovered.

Interpreting the model parameters

The expression in Equation (29) is generally applicable, but requires knowledge of the parameters R_p, R_b, a_{eq} and a_f. DePaolo (2011) argued that R_b can be estimated from the far from equilibrium dissolution rate of calcite, which is dependent on pH and temperature. This proposal fares well when the model is compared to data on a_p versus growth rate (R_p) reported by Tang et al. (2008a) (Fig. 10a). The inflection point of Equation (29) occurs at $R_p = R_b$, which at 25 °C is very close to the dissolution rate of 6×10^{-7} moles m^2 s^{-1} measured by Chou et al. (1989). This correspondence supports the theory that underlies Equation 29, and the data are best fit with Equation (29) using $a_{eq}=0.9995$ and $a_f=0.9984$. However, as also noted by DePaolo (2011), these parameters do not extrapolate to $a_p = 1.0000$ as R_p goes to zero (equilibrium). Based on analysis of precipitation rate data from the literature, DePaolo (2011) hypothesized that R_b is proportional to $R_p^{1/2}$, when $R_p < R_b$. This is the basis for "Model 2" of DePaolo (2011), which is shown for Ca isotope fractionation at 25 °C in Figure 10a. The values of a_{eq} and a_f need to be inferred from comparison of model to data. For "Model 2," the inferred values are $a_{eq}=0.9998$ and $a_f=0.9984$.

A next step in developing the model is to obtain a better representation of R_b and its relationship to growth rate and solution composition. This has been done by Nielsen et al. (2012), who adapted the ion-by-ion crystal growth models of Zhang and Nancollas (1998) to predict both precipitation rate and Ca isotopic fractionation in calcite (Fig. 10a–b). In Figure 10b, the equilibrium and kinetic fractionation factors are assumed to be independent of temperature, but the data could also accommodate a slight temperature dependence to these parameters. The model begins with the recognition that calcite growth proceeds by ion attachment and detachment on kink sites along growth steps on the crystal surface (Fig. 11). The growth rate is determined largely by the rate constants for attachment and detachment of Ca^{2+} and CO_3^{2-}, which are directly related to the forward and backward reaction rates. With additional thermodynamic and geometric parameters, a formulation can be made for the growth rate in terms of solution chemistry, and R_p is found to depend not only on solution saturation state, but also on the ratio of $[Ca^{2+}]$ to $[CO_3^{2-}]$ in solution.

Some of the predictions for R_b from the ion-by-ion growth model are shown in Figure 12. Because there is a dependence on Ca^{2+}/CO_3^{2-} of the solution, the values of R_b are plotted for different values of $[Ca^{2+}]_{aq}$. The DePaolo (2011) Model 2 values for R_b vs. R_p behave in a manner similar to that predicted by the ion-by-ion model, with $R_b = f(R_p^{1/2})$ as is indicated by literature data on calcite precipitation kinetics. The ion-by-ion model predicts that the values of R_b will be dependent on $[Ca^{2+}]_{aq}$, and that the values are a different function of S and R_b at high supersaturations and growth rates. These additional features in the ion-by-ion model make prediction of Ca isotope fractionation somewhat more complex, but also more interesting. The ion-by-ion model predicts that, for the same growth rate, Ca isotope fractionation should be larger when the solution Ca^{2+}/CO_3^{2-} is low and smaller when Ca^{2+}/CO_3^{2-} is high (Fig. 13). This implies that Ca isotope fractionation can be dependent on pH and total dissolved carbon, and this has been verified to some extent with measurements of carbonate formed in an alkaline lake (Nielsen and DePaolo 2013). It is also noteworthy that the dependence of Ca isotope fractionation on solution Ca^{2+}/CO_3^{2-} is one way to differentiate the surface kinetic models for isotope fractionation from those based on the diffusion in the near surface of the mineral discussed in a later section.

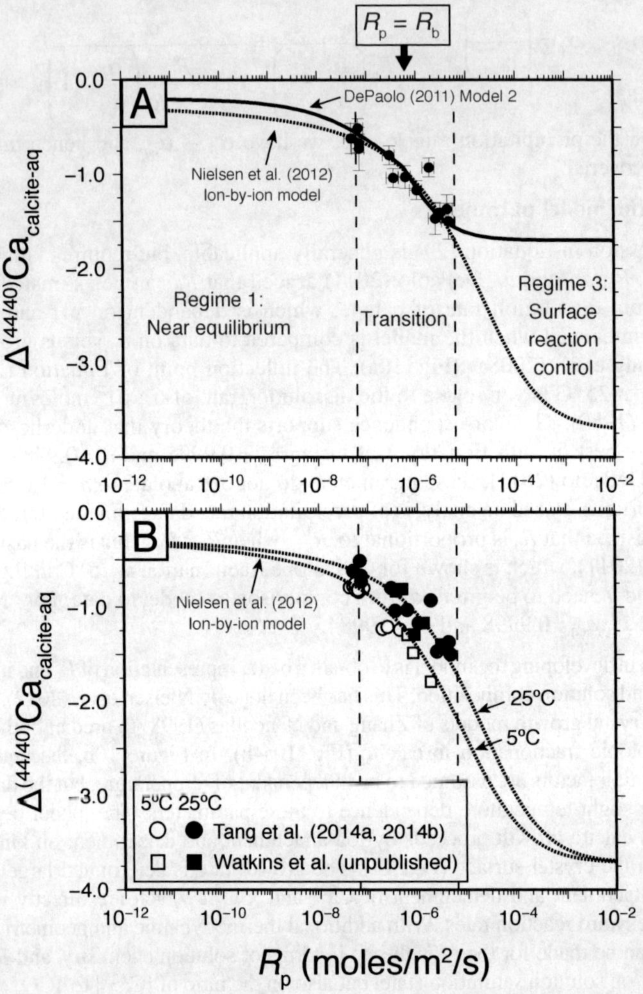

Figure 10. Ca isotope fractionation between calcite and aqueous solution as a function of crystal growth rate. (a) Results from two different modeling approaches. The DePaolo (2011) curve is based on the equations presented herein, and the crossover point from Regime 1 to Regime 3 occurs at $R_p=R_b$. DePaolo (2011) noted that this crossover corresponds to the measured far-from-equilibrium dissolution rate of calcite at 25 °C (Chou et al. 1989), suggesting that R_b can be estimated from dissolution experiments. The Nielsen et al. (2012) curve is based on an ion-by-ion growth model where the rate constants for ion attachment and detachment are mass-dependent, and R_b is a function of solution chemistry. Both models can explain the data but the ion-by-ion model predicts a larger kinetic fractionation factor, a_f, which corresponds to larger fractionations in the fast growth limit. (b) The temperature dependence of kinetic isotope effects for calcium in calcite. In the ion-by-ion model, temperature shifts R_b via the temperature dependence of kink density. According to this model, the equilibrium and kinetic end member fractionation factors, a_{eq} and a_f, are relatively insensitive to temperature.

A further issue with isotope effects due to surface reaction kinetics is to identify the controls on the parameters (k_i and ν_i; Fig. 11) that determine the rates of attachment to and detachment from the crystal surface. These parameters are key to understanding calcite precipitation kinetics, but it is their dependence on isotopic mass that controls the kinetic isotope effect.

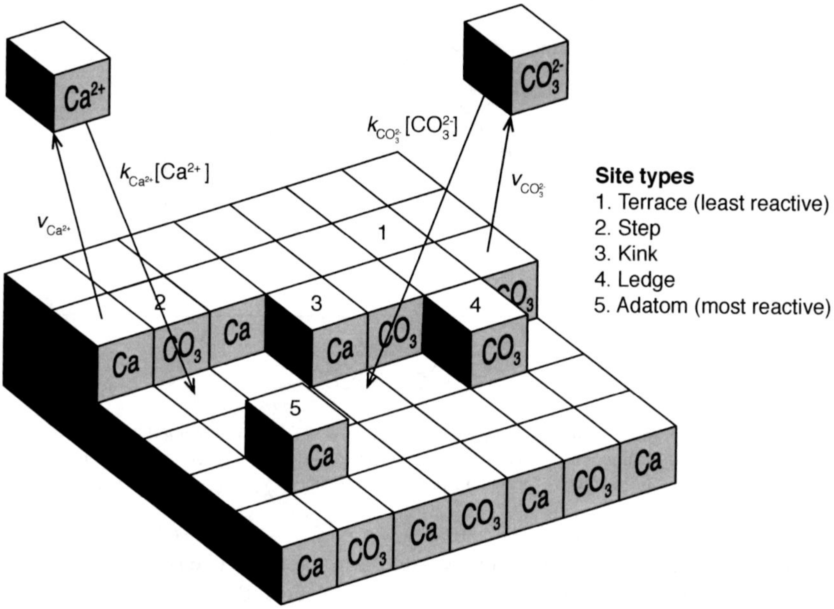

Figure 11. Schematic for the ion-by-ion approach to modeling crystal growth (see Zhang and Nancollas 1998), which was adapted by Nielsen et al. (2012) to explain KIEs during crystal growth. The mass-dependent parameters are the attachment and detachment rate constants, k_i (M^{-1} s^{-1}) and n_i (s^{-1}) to the kink sites and step edges of a cubic crystal. Although kink sites react more slowly than ledge and adatom sites, the kink sites and step edges control the overall growth and dissolution rate owing to their higher concentration.

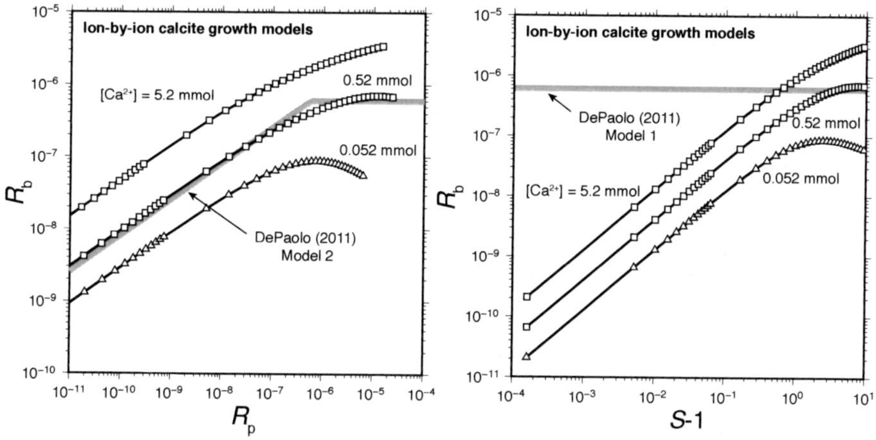

Figure 12. A key parameter in modeling KIEs during crystal growth is the rate of ion detachment, R_b, from the mineral surface relative to the net growth rate (R_p) or degree of supersaturation ($S = [Ca^{2+}][CO_3^{2-}]/K_{sp}$). Based on analysis of precipitation rate data from the literature, DePaolo (2011) hypothesized that R_b is proportional to $R_p^{1/2}$, when $R_p < R_b$. This functional form is in good agreement with the rate dependence of R_b in the ion-by-ion model of Nielsen et al. (2012). Note that R_b is an input of the DePaolo (2011) model and an output of the ion-by-ion model.

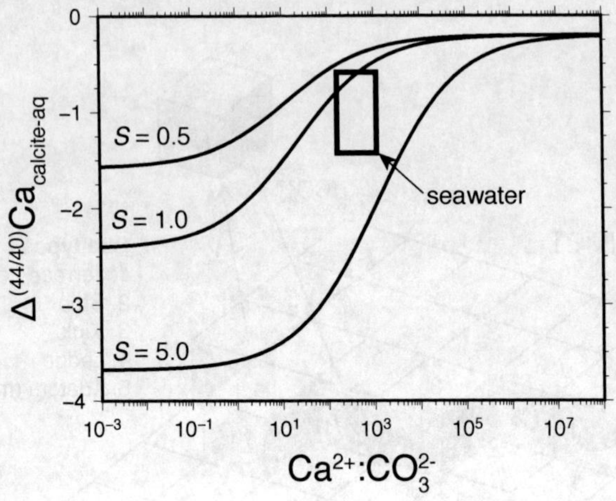

Figure 13. Calculations from the ion-by-ion model of Nielsen et al. (2012) showing the expected calcium isotope fractionation as a function of solution composition for different values of the saturation index. Predictions such as these are attractive targets for future experiments.

One possibility is that the desolvation rate of Ca^{2+} ions in solution is mass dependent. To investigate whether this mass dependence would be in the correct sense and of appropriate magnitude, Hofmann et al. (2012) carried out MD simulations that produce estimates of the residence time of water molecules in the first hydration shell of various cations in a solution of pure water. Their findings show that water molecules have smaller residence times in the solvation shells around lighter isotopes, meaning that lighter isotope cations are more easily desolvated. When converted to estimates of possible kinetic isotopic fractionation factors, the MD results correspond to a_f values of a few permil favoring the light isotope (Fig. 14), close to the values observed for a_f in experiments (Fig. 10). The Hofmann et al. (2012) results suggest that it may be possible to gain a molecular level understanding of, and predictive capability for, the KIEs of major cations through a combination of modeling and experiments. The result also implies that any aspect of solution chemistry that changes the stability of the hydration shells around cations could also change the KIEs of those cations during crystal growth.

Stable isotope fractionation during electroplating

Over the past decade, electrochemistry experiments have been used to investigate kinetic isotope effects for Fe, Zn, and Li during redox reactions (Fig. 15; Kavner et al. 2005, 2008; Black et al. 2009, 2010a,b, 2014). The observations that have been made, and the extremely well-controlled nature of the experiments, present an opportunity to develop and test models for combined reaction and diffusion (i.e., Regimes 3 and 4) in a "simple" system.

In a typical experiment, a rotating disc electrode is immersed in aqueous solution and controls the driving force of the reaction as well as the transport of dissolved ions to the metal surface. The dependence of isotopic fractionation between electro-deposited metal and aqueous solution has been studied as a function of temperature, overpotential, composition of the aqueous solution, composition of the electrode and rotation rate of the electrode. The key findings from these studies were recently summarized by Black et al. (2014) and are repeated here: (1) the electro-deposited metal has thus far been observed to be isotopically lighter than the dissolved species in aqueous solution, (2) the fractionation does not depend on electrode composition, (3) the fractionation varies with solution composition, (4) the fractionation generally increases with increasing temperature, (5) the fractionation decreases

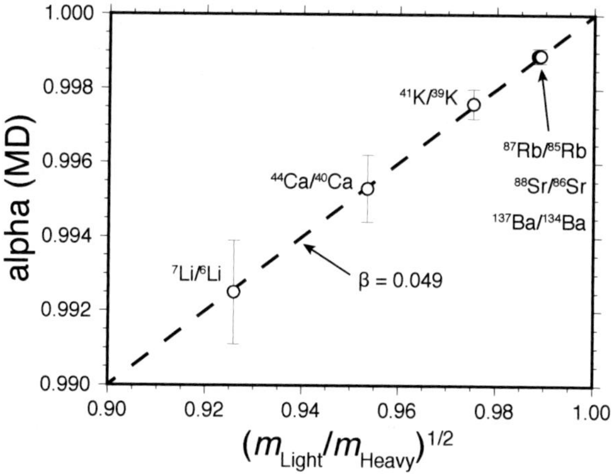

Figure 14. Estimates of possible kinetic isotopic fractionation factors (a_f) based on mass dependent ion desolvation rates (i.e., mass dependent k_{wex}; see Fig. 1). These estimates were made using MD simulations and calculating the residence time of water molecules in the first hydration shell of various cations in a solution of pure water (Hofmann et al. 2012).

with increasing overpotential or deposition rate, (6) the fractionation varies depending on the cation of interest, (7) the fractionation increases with increasing rotation rate of the electrode, (8) the sensitivity to rotation rate increases with overpotential and (9) the KIEs are more scattered when electroplating efficiency is low (e.g., Fe in $FeSO_4$ solutions).

The overall rate of an electrochemical reaction under simultaneous control of mass-transport to the electrode and electrochemical processes at the electrode is given by the Koutecky–Levitch equation (Bard and Faulkner 1980):

$$\frac{1}{i_{\text{total}}} = \frac{1}{i^*_{\text{reaction}}} + \frac{1}{i^*_{\text{transport}}}, \tag{30}$$

where i is the current (Coulombs $m^{-2} s^{-1}$) and the asterisks are a reminder that $C_s = C^*$ for the reaction term and $C_s = 0$ for the transport term when i_{total} is expressed in this form. This expression is exactly analogous to Equation (18b) for the crystal growth problem. For electrochemistry problems, the far-from-equilibrium rate of reaction under non-transport-limited conditions (Regime 3) is given by the far-from-equilibrium limit of the Butler–Volmer equation (also known as the Tafel equation):

$$i^*_{\text{reaction}} = nFC_s k^0 e^{\frac{-\kappa nF\eta}{RT}}, \tag{31}$$

where n is the number of electrons transferred in the reaction, F is Faraday's constant (96485 C mol^{-1}), C_s is the concentration at the electrode surface (moles cm^{-3}), k^0 is a standard rate constant, κ is a transfer coefficient (Bard and Faulkner 1980), η is the electrochemical driving force ($V_{\text{applied}} - V_{\text{equilibrium}}$), R is the gas constant (8.3145 J mol^{-1} K^{-1}) and T is temperature (K).

In the limit where electroplating takes place as fast as ions can be supplied to the surface (Regime 4), the current at the electrode is equal to the rate of mass transport by advection and diffusion as given by the Levitch equation (Bard and Faulkner 1980):

$$i^*_{\text{transport}} = 0.62nFC^* D^{2/3} \nu^{-1/6} \omega^{1/2}, \tag{32}$$

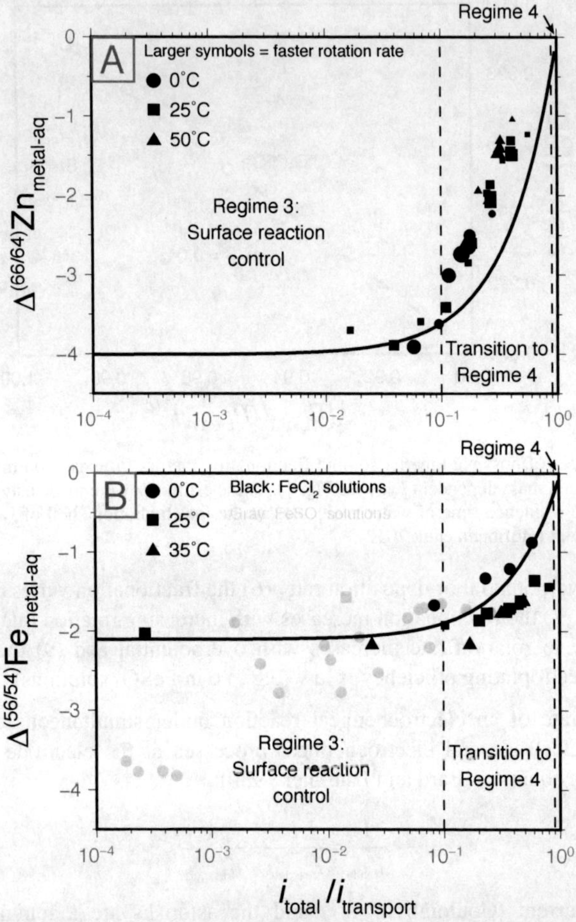

Figure 15. Kinetic isotope effects in electroplating experiments where the temperature, rotation rate of the electrode, and overpotential are controlled (see text references and additional details). The curves are model results based on the Koutecky–Levitch equation (30) with mass dependent diffusivities and reaction rate constants. (a) For Zn, we use ${}^{66}\Delta/{}^{64}\Delta = 0.99993$ ($\beta = 0.0019$) and $a_f = {}^{66}k^0/{}^{64}k^0 = 0.996$. (b) For Fe, we use ${}^{56}\Delta/{}^{54}\Delta = 0.99993$ ($\beta = 0.0025$) and $a_f = {}^{56}k^0/{}^{54}k^0 = 0.998$.

where C^* is the concentration in the bulk solution (moles cm^{-3}), D is the diffusion coefficient (cm^2 s^{-1}), ω is the rotation rate (s^{-1}), and v is the kinematic viscosity (cm s^{-1}). Lower viscosity, faster rotation and faster diffusion all serve to increase the rate of ion transport to the surface of the electrode.

For isotopic fractionation, it is expected the rate constant k^0 and the diffusion coefficient D are mass dependent. The isotopic ratio can be written as (Black et al. 2010b):

$$\frac{{}^{H}i_{total}}{{}^{L}i_{total}} = \frac{1}{\dfrac{{}^{L}i^*_{reaction}}{{}^{H}i^*_{reaction}} + \dfrac{{}^{L}i^*_{reaction}}{{}^{H}i^*_{transport}}} + \frac{1}{\dfrac{{}^{L}i^*_{transport}}{{}^{H}i^*_{transport}} + \dfrac{{}^{L}i^*_{transport}}{{}^{H}i^*_{reaction}}}, \tag{33}$$

where

$$\frac{{}^{\mathrm{L}}i^*_{\mathrm{reaction}}}{{}^{\mathrm{H}}i^*_{\mathrm{reaction}}} = \frac{{}^{\mathrm{L}}k^0}{{}^{\mathrm{H}}k^0}\left(\frac{{}^{\mathrm{L}}C^*}{{}^{\mathrm{H}}C^*}\right), \tag{34}$$

$$\frac{{}^{\mathrm{L}}i^*_{\mathrm{transport}}}{{}^{\mathrm{H}}i^*_{\mathrm{transport}}} = \frac{{}^{\mathrm{L}}D^{2/3}}{{}^{\mathrm{H}}D^{2/3}}\left(\frac{{}^{\mathrm{L}}C^*}{{}^{\mathrm{H}}C^*}\right), \tag{35}$$

$$\frac{{}^{\mathrm{L}}i^*_{\mathrm{reaction}}}{{}^{\mathrm{H}}i^*_{\mathrm{transport}}} = \frac{{}^{\mathrm{L}}k^0}{{}^{\mathrm{H}}D^{2/3}}\left(\frac{{}^{\mathrm{L}}C^*}{{}^{\mathrm{H}}C^*}\right)\frac{e^{\frac{-\kappa nF\eta}{RT}}}{0.62\nu^{-1/6}\omega^{1/2}}, \tag{36}$$

and

$$\frac{{}^{\mathrm{L}}i^*_{\mathrm{transport}}}{{}^{\mathrm{H}}i^*_{\mathrm{reaction}}} = \frac{{}^{\mathrm{L}}D^{2/3}}{{}^{\mathrm{H}}k^0}\left(\frac{{}^{\mathrm{L}}C^*}{{}^{\mathrm{H}}C^*}\right)\frac{0.62\nu^{-1/6}\omega^{1/2}}{e^{\frac{-\kappa nF\eta}{RT}}}. \tag{37}$$

The model curves in Figure 15 were constructed by substituting Equations (34–37) into Equation (33) and plotting the result against $i_{\mathrm{total}}/i_{\mathrm{transport}}$. In the diffusion-controlled limit ($i_{\mathrm{total}}/i_{\mathrm{transport}} \rightarrow 1$), the magnitude of isotopic fractionation is dictated by the β factor, or ratio of isotopic diffusion coefficients. For both Zn^{2+} and Fe^{2+}, the β factor is near zero (Table 1). In the surface reaction-controlled limit, the magnitude of isotopic fractionation is dictated by the ratio of reaction rate constants. Hence, for each curve the Regime 4 limit is constrained by independent experimental data on β factors, but the Regime 3 limit is a fitting parameter that is not yet constrained by independent data.

The fact that the data for Zn in $ZnCl_2$, and also perhaps Fe in $FeCl_2$, collapse to a single trend suggests that β and a_{f} are relatively insensitive to temperature over the temperature range of the experiments. This is significant because the insensitivity of kinetic isotope fractionation factors to temperature might be a common feature, as it is seen in the available data for calcium, carbon, and oxygen isotopes in calcite (e.g., Fig. 10b; Tang et al. 2008a; Dietzel et al. 2009; Watkins et al. 2013, 2014; Baker 2015). The systematic misfit in the data for Zn could be taken as evidence that a_{f} is sensitive to factors other than temperature. For example, a much better fit can be obtained by casting a_{f} as function of electro-deposition rate (results not shown). For Fe in $FeSO_4$ (gray symbols in Fig. 15), the data are more scattered, especially in Regime 3 where surface reaction kinetics dominate. Black et al. (2010a) noted that the electrodeposition efficiency is low in these experiments, meaning that the redox reaction of interest accounts for only a small fraction of the current registered by the electrode (Kavner, pers. comm.). The low efficiency could be taken as an indication that the complex surface chemistry responsible for the deposition of charge (in excess of that produced by the redox reaction) affects a_{f} and β in ways that are not captured by the model.

Stable isotope fractionation of trace elements

Many of the factors that govern kinetic isotope effects of major structural constituents also apply to the partitioning of trace elements and their isotopes. Trace element abundances are widely used to infer the temperatures of mineral formation, but many recent experimental studies have shown that factors beyond temperature affect trace element uptake under non-equilibrium conditions. This leads to correlations between trace element partition coefficients and kinetic isotope effects, as shown in Figure 16 for the mineral calcite.

Different impurities are incorporated into minerals in different ways, presenting an opportunity to probe the micro-scale processes that underlie KIEs from multiple angles. Both Mg^{2+} and Sr^{2+} behave like Ca^{2+} in calcite insofar as they occupy the same structural site and the light isotope of each of these elements is enriched in the solid phase. For Sr^{2+}, the degree

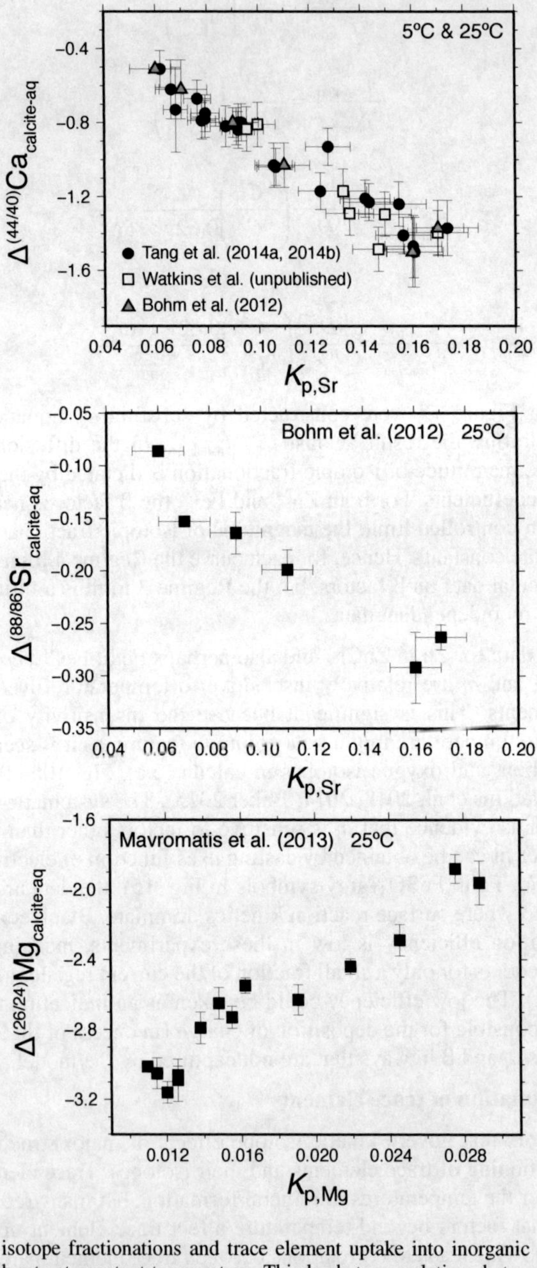

Figure 16. Stable isotope fractionations and trace element uptake into inorganic calcite varies systematically with growth rate at constant temperature. This leads to correlations between (a) calcium isotope fractionations and strontium partitioning, (b) strontium isotope fractionation and strontium partitioning and (c) magnesium isotope fractionation and magnesium partitioning. Strontium and magnesium behave differently in that strontium isotope fractionation increases as more strontium is incorporated into calcite whereas magnesium isotope fractionation decreases as more magnesium is incorporated. This may be related to differences in the desolvation rates of Sr^{2+} (fast) versus Mg^{2+} (slow).

of light isotope enrichment increases with growth rate and strontium uptake (Tang et al. 2008b; Böhm et al. 2012; Fig. 16b). For Mg^{2+}, the degree of light isotope enrichment decreases with growth rate and magnesium uptake, leading to the opposite relationship (Mavromatis et al. 2013; Saenger and Wang 2014; Fig. 16c). An important difference between these two elements is that Sr^{2+} desolvates as readily as Ca^{2+} ($k_{wex} \sim 10^9\ s^{-1}$) whereas Mg^{2+} desolvates ~3 orders of magnitude more slowly (Nielsen 1984). Mavromatis et al. (2013) compared infrared spectra from calcite grown at different rates and found evidence for incomplete desolvation of Mg^{2+} ions at higher growth rates based on the presence of water in the infrared spectra of fast-grown calcite. The conclusion is that with less complete desolvation, the kinetic fractionation factor (a_f) tends toward unity—a finding that supports the hypothesis that ion desolvation is the primary physical mechanism underlying a_f for free aqueous metal cations (Hofmann et al. 2012).

The growing dataset provides exciting opportunities to develop process-based models for trace element uptake and isotope discrimination. For the relatively simple case of strontium uptake into calcite, DePaolo (2011) wrote an expression for trace element partitioning that is analogous to Equation (29), with Sr replacing ^{44}Ca and Ca replacing ^{40}Ca:

$$K_{p,Sr} = \left(\frac{(Sr/Ca)_{solid}}{(Sr/Ca)_{fluid}} \right)_{ss} = \frac{K_f}{1 + \frac{R_b}{R_p + R_b}\left(\frac{K_f}{K_{eq}} - 1 \right)}, \tag{38}$$

where the $K_f = K_{Sr}/K_{Ca}$ is the forward kinetic fractionation factor for Sr/Ca and K_{eq} is the equilibrium Sr/Ca partition coefficient. While this expression can explain the rate-dependence of Sr partitioning into calcite as well as the observed correlation between Sr/Ca and $\Delta^{(44/40)}Ca$ (Fig. 16a), it does not account for the isotopes of trace elements and the feedbacks between impurity uptake and calcite growth kinetics.

Nielsen et al. (2013) used the ion-by-ion model to derive expressions relating strontium and magnesium concentrations in calcite to ion detachment frequencies and overall crystal growth rate. They obtained attachment and detachment rate coefficients from fits to atomic force microscopy (AFM) measurements of step velocity versus impurity concentration in solution. Their Figure 9 shows how the detachment rate coefficient for calcium is directly correlated with the mole fraction of $SrCO_3$ but inversely correlated with the mole fraction of $MgCO_3$. The conclusion is that Sr^{2+} impedes calcite growth by straining the local crystal lattice and increasing calcite solubility (note that Sr^{2+} is larger than the Ca^{2+} ion) whereas Mg^{2+} impedes calcite growth by blocking Ca^{2+} from active growth sites (recall that Mg^{2+} desolvates slowly). The theory has yet to be extended to include the isotopes of strontium and magnesium, in part because data on the growth rate dependence of stable isotope fractionation were not available until recently (Immenhauser et al. 2010; Böhm et al. 2012; Li et al. 2012; Mavromatis et al. 2013). Additional complexity arises when one considers a trace element such as boron, which is delivered to calcite from multiple, isotopically distinct dissolved species ($B(OH)_3$ and $B(OH)_4^-$) that can be incorporated into structural sites as well as non-structural (defect) sites (Gabitov et al. 2014; Mavromatis et al. 2015; Uchikawa et al. 2015). Even for cations such as Ca^{2+} and Mg^{2+}, the formation of ion pairs in solution (e.g., $MgHCO_3^+$ or $MgCO_3^0$) can perturb the coordination sphere and contribute to isotopic fractionation in ways that have not been investigated until recently (Schott et al. 2016).

THE ROLE OF THE NEAR SURFACE OF CRYSTALS

The surface reaction kinetic model (SRKM) of DePaolo (2011) and the ion-by-ion models are, in effect, "surface entrapment models" that involve a competition between growth rate (R_p) and the gross ion detachment rate (R_b) from the surface. Up to this point, the interpretation of kinetic isotope effects (KIEs) has focused on processes operating on the liquid side of the solid–liquid interface; i.e., mass-dependent diffusion in the liquid and mass-dependent solvation–desolvation kinetics. In some circumstances, it may be possible for isotope ratios to be influenced by the properties of, and atomic mobility within, the immediate near-surface of the crystal itself.

High-resolution X-ray reflectivity measurements reveal that the outermost few monolayers of crystals are structurally different from the bulk lattice (Fenter et al. 2000a,b, 2003; Schlegel et al. 2002). Recent MD simulations have confirmed that the outermost layers of crystals are characterized by changes in bond angles relative to the bulk lattice, as well as a general lengthening of bonds and consequent reduction in the binding energies (vacancy formation energies) of cations (Fig. 17; Lanzillo et al. 2014). These characteristics suggest that the equilibrium isotopic composition of the near-surface region could be distinct from that of the bulk lattice and that atomic mobility may be enhanced relative to the "deep" lattice. The equilibrium isotopic composition of, and atomic mobility within, the near-surface is difficult to measure experimentally, but efforts have been made using isotope-exchange experiments involving calcite and CO_2 at 200 °C (Hamza and Broecker 1974).

The growth entrapment model (GEM)

Watson and Liang (1995) used the concept of a chemically anomalous surface region to develop a growth entrapment model (GEM) that differs conceptually from the SRKM in that it deals exclusively with the solid side of the solid–liquid interface. The essence of the growth entrapment model (GEM) is shown schematically in Figure 18a–b, where growth is depicted as overplating of new atomic layers onto the pre-existing surface. This aspect of the model is essentially identical to what is shown in Figure 8a for the SRKM. At the time of deposition, each new layer has the equilibrium composition of the structurally anomalous near-surface region, which is given by the surface concentration factor F (Fig. 18a,c). This part of the model is different from what is shown in Figure 8a, and effectively specifies that, for all cases, $R_b \gg R_p$ (i.e., local equilibrium). If the assumption of local equilibrium in the GEM is relaxed, however, there is more room to reconcile the SRKM and GEM, as discussed further below. In the GEM, the entrapped layer will adjust composition insofar as atomic mobility or "near surface diffusion" allows this to happen. As depicted in Figure 18b, atomic mobility may be only partially effective, so the composition of the resulting crystal (i.e., the concentration plateau) lies between the equilibrium value and the anomalous near-surface concentration.

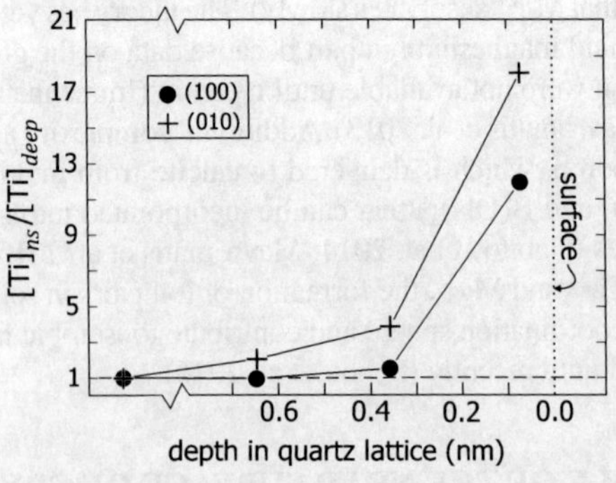

Figure 17. Estimated near-surface (ns) equilibrium Ti concentration relative to the "deep-lattice" value as a function of proximity to the [100] and [010] surfaces of alpha quartz. These estimates were obtained by assuming $[Ti]_{ns}/[Ti]_{deep} \approx \exp(\Delta E_{ex}/RT)$, where ΔE_{ex} is the energy change associated with the exchange reaction $Si_{ns} + Ti_{deep} = Si_{deep} + Ti_{ns}$. The value for ΔE_{ex} was obtained from *ab initio* molecular dynamics simulations (Lanzillo et al. 2014). Note the similarity of this plot with the description of the near-surface anomaly shown in Figure 18c.

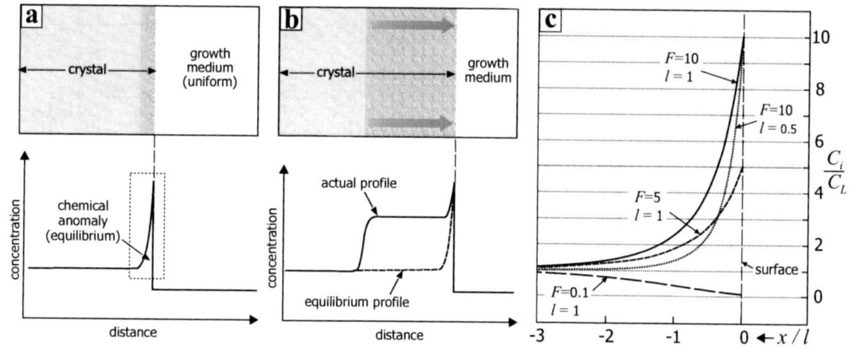

Figure 18. Illustration of some aspects of the growth entrapment model (GEM). The schematic in (a) shows the structurally and chemically anomalous near surface region present at equilibrium in most if not all crystals (see text). During crystal growth, this region is "overplated" with new material and is no longer in equilibrium. If growth is fast, diffusion may not be able to fully restore equilibrium, as suggested in (b). Panel (c) shows mathematical representations of the near-surface equilibrium chemical anomaly as postulated by Watson and Liang (1995), illustrating the surface concentration factor F and the half-width l of the anomalous near-surface region. C_i is the concentration of impurity i; C_L is the equilibrium concentration at depth in the crystal.

In the GEM, trace element uptake depends on the competition between mineral growth rate, expressed in terms of a growth velocity (v) normal to the crystal surface, and the redistribution of the trace constituent back to the mineral surface. This competition is quantified by the dimensionless quantity $\mathrm{Pe}=v\cdot l/D$, where l is the half-width of the anomalous near-surface layer and D is an effective diffusivity meant to describe the sum of all transport processes in the near surface of the mineral. When growth is slow relative to atomic transport ($\mathrm{Pe} \ll 1$), the crystal is able to relax to the true lattice equilibrium composition. When growth is fast ($\mathrm{Pe} \gg 1$), the crystal completely entraps and preserves the composition of the anomalous surface.

In this formulation, a growth rate dependence to trace element uptake should be observed whenever the growth rate spans $\mathrm{Pe}=v\cdot l/D \approx 0.1$ to 10 (Fig. 19a). The growth velocity, v, is determined from laboratory experiments, and the width of the structurally relaxed near-surface region, l, is about 1–2 nm in a variety of minerals (including silicates, carbonates and sulfates) based on the X-ray reflectivity measurements of Fenter and colleagues (Fenter et al. 2000a,b, 2003; Schlegel et al. 2002). This length scale was also returned in the MD simulations of Ti in quartz (Lanzillo et al. 2014) (see Fig. 17). For a typical calcite precipitation experiment, $R_p \approx 10^{-6}$ moles m^{-2} s^{-1}, which translates to a growth velocity of $v=3.7\times10^{-11}$ m s^{-1}. In order for Pe to be on the order of unity, the diffusivity must be on the order of 10^{-19}–10^{-20} m^2 s^{-1}, which is about 16 orders of magnitude greater than the diffusivity of Ca or Sr in the calcite lattice and 10 orders of magnitude less than the diffusivity of Ca or Sr in the aqueous phase. This analysis led DePaolo (2011) to conclude that diffusion within the solid phase was far too slow to allow the competition between solid state diffusion and growth rate to be the process responsible for the growth rate-dependence of trace element partitioning into calcite. The same analysis led Watson and Liang (1995) to hypothesize that the ionic diffusivity was much faster in the nm-scale surface layer of calcite, and in particular, had values that caused the Peclet number to fall in the range of 0.1–10 so that the solid state diffusion could cause a growth rate dependence to the observed Sr/Ca partitioning. To match the data from other trace elements such as Mn/Ca, Co/Ca, and Cd/Ca in calcite, which show a more gradual change in partitioning over four or five orders of magnitude change in Pe, Watson (2004) modified the GEM by expressing D as a function of depth within the crystal:

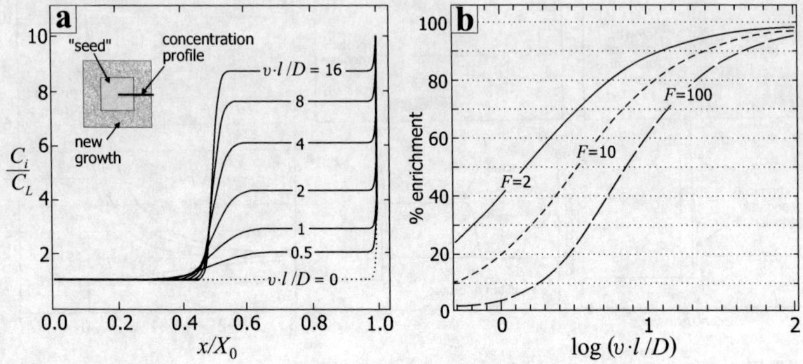

Figure 19. (a) Numerically computed curves illustrating the efficiency of growth entrapment for several values of the dimensionless quantity $v{\cdot}l/D$. As shown in the inset, the concentration profiles are "traverses" across new growth on a pre-existing seed. The surface concentration factor F was assumed to be 10 for these calculations; the "trapped" concentration is that represented by the plateaus identified with specific $v{\cdot}l/D$ values. As in the previous figure, C_i is the concentration of impurity i and C_L is the equilibrium concentration at depth in the crystal. In (b), percent enrichment (i.e., the plateau level relative to the equilibrium concentration) is shown as a function of $v{\cdot}l/D$ for various values of F.

$$D(x) = D_{\text{lattice}}\left(D_{\text{surface}} / D_{\text{lattice}}\right)^{\exp[x/(m \cdot l)]}, \tag{39}$$

where $D(x)$ is the diffusivity at some distance x from the surface, D_{lattice} is the diffusivity in the normal (deep) lattice, and D_{surface} is the diffusivity at the immediate surface. The parameter m is a multiplier relating the width of the diffusively anomalous region to that of the chemically anomalous region. The ad hoc postulate of a spatially variable diffusivity added more adjustable parameters to the GEM, which gives the model a great deal of flexibility, but the model parameters (D_{surface}, m, and l) are too poorly known for the model to be used in a predictive way.

Future molecular dynamics simulations may shed light on F values (Fig. 17), but quantification of near-surface diffusion may not yield so readily to the same approaches. Lanzillo et al. (2014) were able to show that the activation energy for diffusion of Ti in the near-surface of the quartz lattice is strongly depth-dependent in the outermost 2–3 polyhedral layers, but they were not able to place actual values on the diffusivity in this region. The near-surface diffusivity in the GEM may be likened to the grain boundary diffusivity in polycrystals, perhaps limited at the high end by the diffusivity of atoms on the actual surface. In general, the Arrhenius laws for lattice and grain boundary diffusion tend to converge at high temperature (Dohmen and Milke 2010), so D_{lattice} may be the relevant diffusivity in the GEM applied to phenocryst growth in magmas (with $m = 1$). This is the assumption made by Watson et al. (2015) in applying the GEM to phosphorus uptake during growth of olivine phenocrysts.

The surface reaction kinetic model (SRKM), growth entrapment model (GEM), and isotopes

The GEM has also been used to model kinetic *isotope* effects by specifying the equilibrium composition of the bulk lattice (a_{eq}) and the distorted surface relative to the bulk lattice (mass-dependent F). This yields curves that are similar, but not identical, to those produced by the SRKM (Fig. 20), as discussed by DePaolo (2011). In both the SRKM and GEM, there is a slow growth limit and fast growth limit to the trace element concentration and stable isotope composition. The slow growth limit corresponds to the equilibrium composition and is a specified input. The fast growth limit corresponds to the composition of the surface in the absence of a detachment or outward diffusive flux and is also a specified input. These similarities

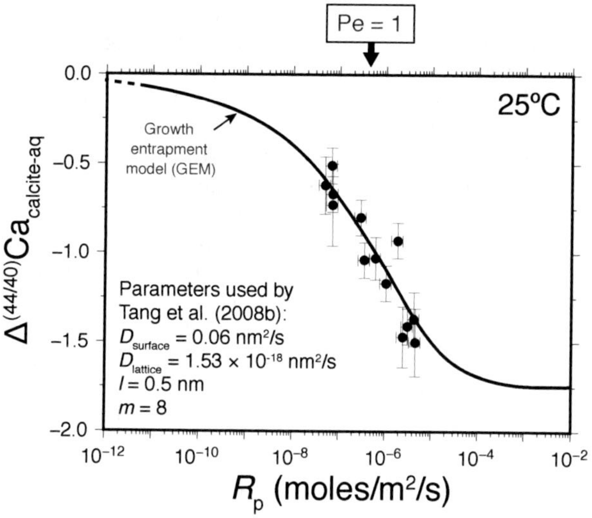

Figure 20. The GEM can match the growth rate dependence of calcium isotope fractionation between calcite and aqueous solution. The crossover between equilibrium and kinetic effects occurs at Pe = 1, which requires a surface diffusivity that is 16 orders of magnitude greater than the lattice diffusivity and 10 orders of magnitude less than the aqueous diffusivity. In the GEM, the surface composition is attributed to local equilibrium and the thermodynamic properties of the distorted near surface of the solid phase. The actual "entrapped" composition is determined by the efficiency of near surface, solid-state diffusion relative to crystal growth rate. This is conceptually different from the SRKM and ion-by-ion models (Fig. 10), where the surface is inferred to be out of equilibrium with the solution (perhaps due to ion desolvation kinetics) and the "entrapped" composition is determined by the efficiency of ion detachment relative to crystal growth rate.

led Thien et al. (2014) to suggest that R_b is similar to $D/(m \cdot l)$ and K_f is similar to F', where F' is the surface/bulk solution partition coefficient. For trace element partitioning, they wrote:

$$K_p = \frac{K_f}{1 + \frac{R_b}{R_p + R_b}\left(\frac{K_f}{K_{eq}} - 1\right)} = \frac{F'}{1 + \frac{\frac{D}{m \cdot l}}{v + \frac{D}{m \cdot l}}(F - 1)}. \tag{40}$$

This expression is mathematically identical to the SRKM but with a detachment flux R_b that varies depending on the values of D, m, and l. Thien et al. (2014) referred to diffusion in the near surface as a "generic homogenization process" that could include dissolution and reprecipitation. Thien et al. (2014) also relaxed the assumption that the surface composition represents an equilibrium surface composition, which allows for solution hydrodynamics and ion desolvation kinetics to be part of the consideration (Hofmann et al. 2012). This helps to reconcile the conceptual differences that set the SRKM and GEM apart, namely (a) what processes set the surface composition (kinetics versus local equilibrium), and (b) what processes modulate that composition through ion detachment from the surface (detachment of adsorbed ions versus solid-state diffusion)?

As discussed throughout this chapter, the composition of the surface depends on many factors, including surface roughness at the atomic scale and the thermodynamic properties of the mineral surface on both sides of the solid–liquid interface. While these properties may be a key factor influencing ion attachment and detachment kinetics, there are situations (Regimes 3 and 4 of Fig. 9) where the assumption of local equilibrium is not tenable. Consider, for example, the desolvation rates of various impurities relative to the desolvation rate of Ca^{2+}, which is likely the rate-limiting step for calcite growth because Ca^{2+} desolvates 1–2 orders of magnitude more slowly than the carbonate anions (Nielsen 1984; Larsen et al. 2010). As summarized by Mavromatis et al. (2013), Sr^{2+} and Ba^{2+} desolvate as readily as Ca^{2+} ($k_{wex} \sim 10^9\ s^{-1}$) whereas Mg^{2+}, Fe^{2+}, Co^{2+} and Ni^{2+} desolvate 3–5 orders of magnitude slower. If calcite growth proceeds as fast as the Ca^{2+} ions can desolvate, it is difficult to imagine how Mg^{2+}, Fe^{2+}, Co^{2+} and Ni^{2+} could establish an equilibrium distribution of desolvated species on the mineral surface, let alone within the outer layers of the solid phase. For rapidly desolvating species like HCO_3^- and CO_3^{2-}, an equilibrium distribution of adsorbed species can possibly be established even when calcite growth is fast. This line of reasoning was used by Watkins et al. (2014) to explain how a_f could plausibly be greater than or less than one for carbon or oxygen isotopes in calcite, depending on the bonding properties on, near, or perhaps within the mineral surface. As noted by Thien et al. (2014), application of the GEM does not require the surface to be in equilibrium with the solution, only that the surface composition be different from the bulk lattice equilibrium value.

PERSPECTIVES

The fact that kinetic isotope effects (KIEs) are sensitive to the pathways of chemical reactions presents both challenges and opportunities for decoding the isotopic composition of minerals to infer conditions of mineral formation. Many of the advances toward understanding KIEs have been made by investigating minerals precipitated from aqueous solution, which presents opportunities to extend the insights gained to problems involving crystal growth from a silicate melt. Further advances in our understanding of kinetic fractionation of non-traditional stable isotopes by diffusion and crystal growth reactions will come from a combination of new experiments designed to isolate specific kinetic processes, continued advances in *in situ* isotopic measurements, and the continued application of computational methods to simulate what happens at phase boundaries. Not only can these tools be used to probe the underlying mechanisms of KIEs, but KIEs can be used as a tool to probe nano-scale processes that are inaccessible to direct observation.

There are many possible diffusion and growth pathways that are not explicitly accounted for in the classical theories of crystal growth covered in this chapter. The classical theories are based on diffusion and attachment of monomers to an isolated crystal, but it is now recognized that biogenic and inorganic crystals can grow by addition of particles that range in size from multi-ion complexes (Schott et al. 2016) to fully formed nanoparticles (Gebauer et al. 2008, 2014; Li et al. 2012; De Yoreo et al. 2015; Gal et al. 2015). These pathways and the concomitant isotope effects may not be adequately described by the surface reaction kinetic model (SRKM), ion-by-ion models, or the growth entrapment model (GEM) described herein. Further complexity arises when one considers biogenic minerals, which are often characterized by unusual, sometimes convoluted, shapes—perhaps reflecting growth by attachment of amorphous or quasi-crystalline nanosphere particles to a pre-existing template or preformed space (Gal et al. 2013, 2015). In such cases, the isotopic composition may be the relict of an initial disordered phase that is no longer present. Bearing these factors in mind, one thing is clear: the real world of crystal growth in nature involves multiple nano-scale phenomena that may discriminate on the basis of mass, and an important challenge moving forward is to better understand how the different nano-scale processes contribute to the kinetic fractionation factors (KFFs) that are deduced from diffusion and crystal growth experiments.

ACKNOWLEDGMENTS

We acknowledge with admiration Frank Richter's leadership of a nearly two-decade effort to characterize diffusive isotope fractionation in silicate melts and aqueous solutions. The authors appreciate the valuable feedback from James Van Orman, Vasileios Mavromatis, Abby Kavner, and an anonymous reviewer. JMW thanks Frederick Ryerson, Frank Richter, Yan Liang, Ian Bourg, Christian Huber, and Laura Lammers (nee Nielsen) for numerous discussions on the themes of this chapter over the past several years. This work was supported by (1) the National Science Foundation through grant nos. EAR-0337481 and EAR-0738843 to EBW and grant no. NSF-EAR-1249404 to JMW, and (2) the U.S. Department of Energy, Office of Science, Office of Basic Energy Sciences, Chemical Sciences, Geosciences, and Biosciences Division, under Award Number DE-AC02-05CH11231.

REFERENCES

Albarède F, Bottinga Y (1972) Kinetic disequilibrium in trace element partitioning between phenocrysts and host lava. Geochim Cosmochim Acta 36:141–156

Aubaud C, Pineau F, Jambon A, Javoy M (2004) Kinetic disequilibrium of C, He, Ar and carbon isotopes during degassing of mid-ocean ridge basalts. Earth Planet Sci Lett 222:391–406

Baker E (2015) Carbon and Oxygen Isotopes in Laboratory-Grown Inorganic Calcite. Master's thesis, University of Oregon

Bard AJ, Faulkner LR (1980) Electrochemical methods: Fundamentals and applications. Vol. 2 Wiley New York

Beck P, Chaussidon M, Barrat JA, Gillet P, Bohn M (2006) Diffusion induced Li isotopic fractionation during the cooling of magmatic rocks: the case of pyroxene phenocrysts from nakhlite meteorites. Geochim Cosmochim Acta 70:4813–4825

Benson BB, Krause D (1980) The concentration and isotopic fractionation of gases dissolved in freshwater in equilibrium with the atmosphere. 1. Oxygen Limnol Oceanogr 25:662–671

Black JR, Umeda G, Dunn B, McDonough WF, Kavner A (2009) Electrochemical isotope effect and lithium isotope separation. J Am Chem Soc 131:9904–9905

Black JR, John S, Young ED, Kavner A (2010a) Effect of temperature and mass transport on transition metal isotope fractionation during electroplating. Geochim Cosmochim Acta 74:5187–5201

Black JR, Young ED, Kavner A (2010b) Electrochemically controlled iron isotope fractionation. Geochim Cosmochim Acta 74:809–817

Black JR, John SG, Kavner A (2014) Coupled effects of temperature and mass transport on the isotope fractionation of zinc during electroplating. Geochim Cosmochim Acta 124:272–282

Böhm F, Eisenhauer A, Tang J, Dietzel M, Krabbenhöft A, Kisakürek B, Horn C (2012) Strontium isotope fractionation of planktic foraminifera and inorganic calcite. Geochim Cosmochim Acta 93:300–314

Bourg IC, Sposito G (2007) Molecular dynamics simulations of kinetic isotope fractionation during the diffusion of ionic species in liquid water. Geochim Cosmochim Acta 71:5583–5589

Bourg IC, Sposito G (2008) Isotopic fractionation of noble gases by diffusion in liquid water: Molecular dynamics simulations and hydrologic applications. Geochim Cosmochim Acta 72:2237–2247

Bourg IC, Richter FM, Christensen, JN, Sposito G (2010) Isotopic mass dependence of metal cation diffusion coefficients in liquid water. Geochim Cosmochim Acta 74:2249–2256

Brantley SL, Liermann LJ, Guynn RL, Anbar A, Icopini GA, Barling J (2004) Fe isotopic fractionation during mineral dissolution with and without bacteria. Geochimica et Cosmochimica Acta 68:3189–3204

Chou L, Garrels RM, Wollast R (1989) Comparative study of the kinetics and mechanisms of dissolution of carbonate minerals. Chem Geol 78:269–282

Crispin KL, Saha S, Morgan D, Van Orman JA (2012) Diffusion of transition metals in periclase by experiment and first-principles, with implications for core-mantle equilibration during metal percolation. Earth Planet Sci Lett 357:42–53

Dauphas N (2007) Diffusion-driven kinetic isotope effect of Fe and Ni during formation of the Widmanstätten pattern. Meteoritics Planet Sci 42:1597–1613

Dauphas N, Teng F-Z, Arndt NT (2010) Magnesium and iron isotopes in 2.7 Ga Alexo komatiites: mantle signatures, no evidence for Soret diffusion, and identification of diffusive transport in zoned olivine. Geochim Cosmochim Acta 74:3274 3291

De Groot SR, Mazur, P (1963) Non-Equilibrium Thermodynamics. Dover Publications

De Yoreo JJ, Gilbert PU, Sommerdijk NA, Penn RL, Whitelam S, Joester D, Zhang H, Rimer JD, Navrotsky A, Banfield JF, Wallace AF, Michel FM, Meldrum FC, Cölfen H, Dove PM (2015) Crystallization by particle attachment in synthetic, biogenic, and geologic environments. Science 349(6247),aaa6760

DePaolo DJ (2011) Surface kinetic model for isotopic and trace element fractionation during precipitation of calcite from aqueous solutions. Geochim Cosmochim Acta 75:1039–1056
Dietzel M, Tang J, Leis A, Köhler SJ (2009) Oxygen isotopic fractionation during inorganic calcite precipitation: effects of temperature, precipitation rate and pH. Chem Geol 268:107–115
Dingwell DB (1990) Effects of structural relaxation on cationic tracer diffusion in silicate melts. Chem Geol 82:209–216
Dohmen R, Milke R (2010) Diffusion in polycrystalline materials: grain boundaries, mathematical models, and experimental data. Rev Mineral Geochem 72:921–970
Dominguez G, Wilkins G, Thiemens MH (2011) The Soret effect and isotopic fractionation in high-temperature silicate melts. Nature 473:70–73
Druhan JL, Brown ST, Huber C (2015) Isotopic gradients across fluid–mineral boundaries. Rev Mineral Geochem 80:355–391
Eggenkamp H, Coleman ML (2009) The effect of aqueous diffusion on the fractionation of chlorine and bromine stable isotopes. Geochim Cosmochim Acta 73:3539–3548
Ellis AS, Johnson TM, Bullen TD (2004) Using chromium stable isotope ratios to quantify Cr(VI) reduction: lack of sorption effects. Environ Sci Technol 38:3604–3607
Fenter P, Geissbühler P, DiMasi E, Srajer G, Sorensen L, Sturchio N (2000a) Surface speciation of calcite observed *in situ* by high-resolution X-ray reflectivity. Geochim Cosmochim Acta 64:1221–1228
Fenter P, Teng H, Geissbühler P, Hanchar J, Nagy K, Sturchio N (2000b) Atomic-scale structure of the orthoclase (001)-water interface measured with high-resolution X-ray reflectivity. Geochim Cosmochim Acta 64:3663–3673
Fenter P, Cheng L, Park C, Zhang Z, Sturchio N (2003) Structure of the orthoclase (001)- and (010)-water interfaces by high-resolution X-ray reflectivity. Geochim Cosmochim Acta 67:4267–4275
Fuex A (1980) Experimental evidence against an appreciable isotopic fractionation of methane during migration. Phys Chem Earth 12:725–732
Gabitov R, Sadekov A, Leinweber A (2014) Crystal growth rate effect on Mg/Ca and Sr/Ca partitioning between calcite and fluid: An *in situ* approach. Chem Geol 367:70–82
Gal A, Habraken W, Gur D, Fratzl P, Weiner S, Addadi L (2013) Calcite crystal growth by a solid-state transformation of stabilized amorphous calcium carbonate nanospheres in a hydrogel. Angew Chem Int Ed 52:4867–4870
Gal A, Weiner S, Addadi L (2015) A perspective on underlying crystal growth mechanisms in biomineralization: solution mediated growth versus nanosphere particle accretion. CrystEngComm 17:2606–2615
Gallagher K, Elliott T (2009) Fractionation of lithium isotopes in magmatic systems as a natural consequence of cooling. Earth Planet Sci Lett 278:286–296
Gardner JE, Befus KS, Watkins JM, Hesse M, Miller N (2012) Compositional gradients surrounding spherulites in obsidian and their relationship to spherulite growth and lava cooling. Bull Volcanol 74:1865–1879
Gebauer D, Völkel A, Cölfen H (2008) Stable prenucleation calcium carbonate clusters. Science 322:1819–1822
Gebauer D, Kellermeier M, Gale JD, Bergström L, Cölfen H (2014) Pre-nucleation clusters as solute precursors in crystallisation. Chem Soc Rev 43:2348–2371
Goel G, Zhang L, Lacks DJ, Van Orman JA (2012) Isotope fractionation by diffusion in silicate melts: Insights from molecular dynamics simulations. Geochim Cosmochim Acta 93:205–213
Hamza M, Broecker W (1974) Surface effect on the isotopic fractionation between CO_2 and some carbonate minerals. Geochim Cosmochim Acta 38:669–681
Hofmann A, Bourg I, DePaolo D (2012) Ion desolvation as a mechanism for kinetic isotope fractionation in aqueous systems. PNAS 109:18689–18694
Huang F, Chakraborty P, Lundstrom C, Holmden C, Glessner J, Kieffer S, Lesher C (2010) Isotope fractionation in silicate melts by thermal diffusion. Nature 464:396–400
Immenhauser A, Buhl D, Richter D, Niedermayr A, Riechelmann D, Dietzel M, Schulte U (2010) Magnesium-isotope fractionation during low-Mg calcite precipitation in a limestone cave - Field study and experiments. Geochim Cosmochim Acta 74:4346–4364
Jähne B, Heinz G, Dietrich W (1987) Measurement of the diffusion coefficients of sparingly soluble gases in water. J Geophys Res: Oceans 92:10767–10776
Jeffcoate A, Elliott T, Kasemann S, Ionov D, Cooper K, Brooker R (2007) Li isotope fractionation in peridotites and mafic melts. Geochim Cosmochim Acta 71:202–218
Kavner A, Bonet F, Shahar A, Simon J, Young E (2005) The isotopic effects of electron transfer: An explanation for Fe isotope fractionation in nature. Geochim Cosmochim Acta 69:2971–2979
Kavner A, John S, Sass S, Boyle E (2008) Redox-driven stable isotope fractionation in transition metals: Application to Zn electroplating. Geochim Cosmochim Acta 72:1731–1741
Kerr RC (1995) Convective crystal dissolution. Contrib Mineral Petrol 121:237–246
Knight, KB, Kita NT, Mendybaev RA, Richter FM, Davis AM, Valley JW (2009) Silicon isotopic fractionation of CAI-like vacuum evaporation residues. Geochim Cosmochim Acta 73:6390–6401

Knox M, Quay P, Wilbur D (1992) Kinetic isotopic fractionation during air-water gas transfer of O_2, N_2, CH_4, and H_2. J Geophys Res: Oceans 97:20335–20343

Lanzillo N, Watson E, Thomas J, Nayak S, Curioni A (2014) Near-surface controls on the composition of growing crystals: Car-Parrinello molecular dynamics (CPMD) simulations of Ti energetics and diffusion in alpha quartz. Geochim Cosmochim Acta 131:33–46

Larsen K, Bechgaard K, Stipp SLS (2010) The effect of the Ca^{2+} to activity ratio on spiral growth at the calcite surface. Geochim Cosmochim Acta 74:2099–2109

Lasaga AC (1998) Kinetic Theory in the Earth Sciences. Princeton University Press

Le Claire A (1966) Some comments on the mass effect in diffusion. Philos Mag 14:1271–1284

Lemarchand D, Wasserburg G, Papanastassiou D (2004) Rate-controlled calcium isotope fractionation in synthetic calcite. Geochim Cosmochim Acta 68:4665–4678

Li W, Chakraborty S, Beard BL, Romanek CS, Johnson CM (2012) Magnesium isotope fractionation during precipitation of inorganic calcite under laboratory conditions. Earth Planet Sci Lett 333:304–316

Lundstrom C, Chaussidon M, Hsui AT, Kelemen P, Zimmerman M (2005) Observations of Li isotopic variations in the Trinity Ophiolite: Evidence for isotopic fractionation by diffusion during mantle melting. Geochim Cosmochim Acta 69:735–751

Maaløe S (2011) Olivine phenocryst growth in Hawaiian tholeiites: Evidence for supercooling. J Petrol 52:1579–1589

Malinovsky D, Baxter DC, Rodushkin I (2007) Ion-specific isotopic fractionation of molybdenum during diffusion in aqueous solutions. Environ Sci Technol 41:1596–1600

Marschall, HR, von Strandmann PAP, Seitz H-M, Elliott T, Niu Y (2007) The lithium isotopic composition of orogenic eclogites and deep subducted slabs. Earth Planet Sci Lett 262:563–580

Mavromatis V, Gautier Q, Bosc O, Schott J (2013) Kinetics of Mg partition and Mg stable isotope fractionation during its incorporation in calcite. Geochim Cosmochim Acta 114:188–203

Mavromatis V, Montouillout V, Noireaux J, Gaillardet J, Schott J (2015) Characterization of boron incorporation and speciation in calcite and aragonite from co-precipitation experiments under controlled pH, temperature and precipitation rate. Geochim Cosmochim Acta 150:299–313

Müller T, Watson EB, Trail D, Wiedenbeck M, Van Orman J, Hauri EH (2014) Diffusive fractionation of carbon isotopes in γ-Fe: Experiment, models and implications for early solar system processes. Geochim Cosmochim Acta 127:57–66

Mullen JG (1961) Isotope effect in intermetallic diffusion. Phys Rev 121:1649

Mundy J, Barr L, Smith F (1966) Sodium self-diffusion and the isotope effect. Philos Mag 14 (130), 785{802

Nielsen AE (1984) Electrolyte crystal growth mechanisms. J Cryst Growth 67:289–310

Nielsen LC, DePaolo DJ (2013) Ca isotope fractionation in a high-alkalinity lake system: Mono Lake, California. Geochim Cosmochim Acta 118:276–294

Nielsen LC, Druhan JL, Yang W, Brown ST, DePaolo DJ (2012) Calcium isotopes as tracers of biogeochemical processes. *In:* Handbook of Environmental Isotope Geochemistry. Springer, p 105–124

Nielsen LC, De Yoreo JJ, DePaolo DJ (2013) General model for calcite growth kinetics in the presence of impurity ions. Geochim Cosmochim Acta 115:100–114

Oeser M, Dohmen R, Horn I, Schuth S, Weyer S (2015) Processes and time scales of magmatic evolution as revealed by Fe–Mg chemical and isotopic zoning in natural olivines. Geochim Cosmochim Acta 154:130–150

Onsager L (1945) Theories and problems of liquid diffusion. Ann NY Acad Sci 46:241–265

Parkinson IJ, Hammond SJ, James RH, Rogers NW (2007) High-temperature lithium isotope fractionation: Insights from lithium isotope diffusion in magmatic systems. Earth Planet Sci Lett 257:609–621

Pearce CR, Saldi GD, Schott J, Oelkers EH (2012) Isotopic fractionation during congruent dissolution, precipitation and at equilibrium: Evidence from Mg isotopes. Geochim Cosmochim Acta 92:170–183

Pikal MJ (1972) Isotope effect in tracer diffusion. Comparison of the diffusion coefficients of $^{24}Na^+$ and $^{22}Na^+$ in aqueous electrolytes. J Phys Chem 76:3038–3040

Richter FM, Liang Y, Davis AM (1999) Isotope fractionation by diffusion in molten oxides. Geochim Cosmochim Acta 63:2853–2861

Richter FM, Davis AM, DePaolo DJ, Watson EB (2003) Isotope fractionation by chemical diffusion between molten basalt and rhyolite. Geochim Cosmochim Acta 67:3905–3923

Richter F, Mendybaev R, Christensen J, Hutcheon I, Williams R, Sturchio N (2006) Kinetic isotope fractionation during diffusion of ionic species in water. Geochim Cosmochim Acta 70:277–289

Richter FM, Watson EB, Mendybaev RA, Teng F-Z, Janney PE (2008) Magnesium isotope fractionation in silicate melts by chemical and thermal diffusion. Geochim Cosmochim Acta 72:206–220

Richter FM, Dauphas N, Teng F-Z (2009a) Non-traditional fractionation of non-traditional isotopes: evaporation, chemical diffusion and Soret diffusion. Chem Geol 258:92–103

Richter FM, Watson EB, Mendybaev R, Dauphas N, Georg B, Watkins J, Valley J (2009b) Isotopic fractionation of the major elements of molten basalt by chemical and thermal diffusion. Geochim Cosmochim Acta 73:4250–4263

Richter FM, Watson EB, Mendybaev R, Dauphas N, Georg B, Watkins J, Valley J (2009c) Isotopic fractionation of the major elements of molten basalt by chemical and thermal diffusion. Geochim Cosmochim Acta 73:4250–4263

Richter F, Watson B, Chaussidon M, Mendybaev R, Ruscitto D (2014a) Lithium isotope fractionation by diffusion in minerals. Part 1: Pyroxenes. Geochim Cosmochim Acta 126:352–370

Richter FM, Watson EB, Chaussidon M, Mendybaev R, Christensen JN, Qiu L (2014b) Isotope fractionation of Li and K in silicate liquids by Soret diffusion. Geochim Cosmochim Acta 138:136–145

Richter F, Chaussidon M, Mendybaev R, Kite E (2016) Reassessing the cooling rate and geologic setting of Martian meteorites MIL 03346 and NWA 817. Geochim Cosmochim Acta 182:1–23

Robinson LF, Adkins JF, Frank N, Gagnon AC, Prouty NG, Roark EB, van de Flierdt T (2014) The geochemistry of deep-sea coral skeletons: A review of vital effects and applications for palaeoceanography. Deep Sea Res Part II 99:184–198

Rodushkin I, Stenberg A, Andrén H, Malinovsky D, Baxter DC (2004) Isotopic fractionation during diffusion of transition metal ions in solution. Anal Chem 76:2148–2151

Roskosz M, Luais B, Watson HC, Toplis MJ, Alexander CM, Mysen BO (2006) Experimental quantification of the fractionation of Fe isotopes during metal segregation from a silicate melt. Earth Planet Sci Lett 248:851–867

Rothman S, Peterson N (1969) Isotope effect and divacancies for self-diffusion in copper. Phys Stat Solidi (b) 35:305–312

Rudnick RL, Ionov DA (2007) Lithium elemental and isotopic disequilibrium in minerals from peridotite xenoliths from far-east Russia: product of recent melt/fluid–rock reaction. Earth Planet Sci Lett 256:278–293

Saenger C, Wang Z (2014) Magnesium isotope fractionation in biogenic and abiogenic carbonates: implications for paleoenvironmental proxies. Quat Sci Rev 90:1–21

Schlegel ML, Nagy KL, Fenter P, Sturchio NC (2002) Structures of quartz (100)- and (101)-water interfaces determined by X-ray reflectivity and atomic force microscopy of natural growth surfaces. Geochim Cosmochim Acta 66:3037–3054

Schoen A (1958) Correlation and the isotope effect for diffusion in crystalline solids. Phys Rev Letters 1:138

Schott J, Mavromatis V, Fujii T, Pearce C, Oelkers E (2016) The control of magnesium aqueous speciation on Mg isotope composition in carbonate minerals: Theoretical and experimental modeling. Chem Geol in press

Schüle W, Scholz R (1979) Properties of vacancies and divacancies in fcc metals. Phys Stat Solidi (B) 93:K119–K123

Sio CKI, Dauphas N, Teng F-Z, Chaussidon M, Helz RT, Roskosz M (2013) Discerning crystal growth from diffusion profiles in zoned olivine by in situ Mg–Fe isotopic analyses. Geochim Cosmochim Acta 123:302–321

Smith VG, Tiller WA, Rutter J (1955) A mathematical analysis of solute redistribution during solidification. Can J Phys 33:723–745

Stebbins JF (1995) Dynamics and structure of silicate and oxide melts; nuclear magnetic resonance studies. Rev Mineral Geochem 32:191–246

Tang J, Dietzel M, Böhm F, Köhler S, Eisenhauer A (2008a) Sr^{2+}/Ca^{2+} and $^{44}Ca/^{40}Ca$ fractionation during inorganic calcite formation: II. Ca isotopes. Geochim Cosmochim Acta 72:3733–3745

Tang J, Köhler S, Dietzel M (2008b) Sr^{2+}/Ca^{2+} and $^{44}Ca/^{40}Ca$ fractionation during inorganic calcite formation: I. Sr incorporation. Geochim Cosmochim Acta 72:3718–3732

Tempest KE, Emerson S (2013) Kinetic isotopic fractionation of argon and neon during air–water gas transfer. Marine Chem 153:39–47

Teng F-Z, McDonough WF, Rudnick RL, Walker RJ (2006) Diffusion-driven extreme lithium isotopic fractionation in country rocks of the Tin Mountain pegmatite. Earth Planet Sci Lett 243:701–710

Teng F-Z, Dauphas N, Helz RT, Gao S, Huang S (2011) Diffusion-driven magnesium and iron isotope fractionation in Hawaiian olivine. Earth Planet Sci Lett 308:317–324

Tharmalingam K, Lidiard A (1959) Isotope effect in vacancy diffusion. Philos Mag 4 :899–906

Thien BM, Kulik DA, Curti E (2014) A unified approach to model uptake kinetics of trace elements in complex aqueous–solid solution systems. Appl Geochem 41:135–150

Tiller W, Jackson K, Rutter J, Chalmers B (1953) The redistribution of solute atoms during the solidification of metals. Acta Metallurgica 1:428–437

Tyroller L, Brennwald MS, Mächler L, Livingstone DM, Kipfer R (2014) Fractionation of Ne and Ar isotopes by molecular diffusion in water. Geochim Cosmochim Acta 136:60–66

Uchikawa J, Penman DE, Zachos JC, Zeebe RE (2015) Experimental evidence for kinetic effects on B/Ca in synthetic calcite: Implications for potential and $B(OH)_3$ incorporation. Geochim Cosmochim Acta 150:171–191

Van Orman JA, Krawczynski MJ (2015) Theoretical constraints on the isotope effect for diffusion in minerals. Geochim Cosmochim Acta 164:365–381

Vineyard GH (1957) Frequency factors and isotope effects in solid state rate processes. J Phys Chem Solids 3:121–127

Wall AJ, Mathur R, Post JE, Heaney PJ (2011) Cu isotope fractionation during bornite dissolution: an in situ X-ray diffraction analysis. Ore Geol Rev 42:62–70

Watkins JM, DePaolo DJ, Huber C, Ryerson FJ (2009) Liquid composition-dependence of calcium isotope fractionation during diffusion in molten silicates. Geochim Cosmochim Acta 73:7341–7359

Watkins JM, DePaolo DJ, Ryerson FJ, Peterson BT (2011) Influence of liquid structure on diffusive isotope separation in molten silicates and aqueous solutions. Geochim Cosmochim Acta 75:3103–3118

Watkins JM, Nielsen LC, Ryerson FJ, DePaolo DJ (2013) The influence of kinetics on the oxygen isotope composition of calcium carbonate. Earth Planet Sci Lett 375:349–360

Watkins JM, Liang Y, Richter F, Ryerson FJ, DePaolo DJ (2014) Diffusion of multi-isotopic chemical species in molten silicates. Geochim Cosmochim Acta 139:313–326 Watson EB (2004) A conceptual model for near-surface kinetic controls on the trace-element and stable isotope composition of abiogenic calcite crystals. Geochim Cosmochim Acta 68:1473–1488

Watson E (2004) A conceptual model for near-surface kinetic controls on the trace-element and stable isotope composition of abiogenic calcite crystals. Geochim Cosmochim Acta 68:1472–1488

Watson EB, Liang Y (1995) A simple model for sector zoning in slowly grown crystals: Implications for growth rate and lattice diffusion, with emphasis on accessory minerals in crustal rocks. Am Mineral 80:1179–1187

Watson EB, Müller T (2009) Non-equilibrium isotopic and elemental fractionation during diffusion-controlled crystal growth under static and dynamic conditions. Chem Geol 267:111–124

Watson EB, Cherniak D, Holycross M (2015) Diffusion of phosphorus in olivine and molten basalt. American Mineralogist 100:2053–2065

Weiss DJ, Rehkdmper M, Schoenberg R, McLaughlin M, Kirby J, Campbell PG, Arnold T, Chapman J, Peel K, Gioia S (2008) Application of nontraditional stable-isotope systems to the study of sources and fate of metals in the environment. Environ Sci Technol 42:655–664

Young E, Galy A, Nagahara H (2002) Kinetic and equilibrium mass-dependent isotope fractionation laws in nature and their geochemical and cosmochemical significance. Geochim Cosmochim Acta 66:1095–1104

Zhang Y (2008) Geochemical Kinetics. Princeton University Press

Zhang J, Nancollas GH (1998) Kink density and rate of step movement during growth and dissolution of an AB crystal in a nonstoichiometric solution. J Colloid Interface Sci 200:131–145

Zhu X, Guo Y, Williams R, Onions R, Matthews A, Belshaw N, Canters G, De Waal E, Weser U, Burgess B, Salvado B (2002) Mass fractionation processes of transition metal isotopes. Earth Planet Sci Lett 200:47–62

Zeebe R (2011) On the molecular diffusion coefficients of dissolved CO_2, HCO_3^- and CO_3^{2-} and their dependence on isotopic mass. Geochim Cosmochim Acta 75:2483–2498

Reviews in Mineralogy & Geochemistry
Vol. 82 pp. 127-163, 2017

In Situ Analysis of Non-Traditional Isotopes by SIMS and LA–MC–ICP–MS: Key Aspects and the Example of Mg Isotopes in Olivines and Silicate Glasses

Marc Chaussidon[1], Zhengbin Deng[1]

[1] Institut de Physique du Globe de Paris
CNRS UMR 7154, Université Sorbonne Paris Cité (USPC)
1 rue Jussieu 75238 Paris Cedex 05
France

Johan Villeneuve[2], Julien Moureau[1]

[2] Centre de Recherches Pétrographiques et Géochimiques
Université de Lorraine, CNRS UMR 7358
15 rue Notre Dame des Pauvres, Vandoeuvre-lès-Nancy, 54501
France

Bruce Watson

Department of Earth and Environmental Sciences,
Rensselaer Polytechnic Institute
Troy, NY 12180
USA

Frank Richter

The University of Chicago
5734 South Ellis Avenue
Chicago, IL, 60637
USA

Frédéric Moynier[1,5]

[5] Institut de Physique du Globe de Paris
Insitut Universitaire de France, Université Paris Diderot
Sorbonne Paris Cité, CNRS UMR 7154
1 rue Jussieu, 75238 Paris Cedex 05
France

INTRODUCTION

Isotopic variation for traditional elements (H, C, N, O and S) has been widely used in the past 40 years in Earth and planetary sciences to study many processes with an emphasis on environments where fluids are present (e.g., Valley and Cole 2011). More recent developments have allowed high-precision measurements of isotope ratios of what has been called non-traditional elements (i.e., Mg, Si, Fe, Zn, Cu, Mo), which are usually less fractionated than traditional elements by at least an order of magnitude (see this volume). These non-traditional stable isotopes can give insights on processes where fluids are not present (e.g., metal–silicate fractionation, e.g., Georg

1529-6466/17/0082-0005$05.00 http://dx.doi.org/10.2138/rmg.2017.82.5

et al. 2007 and review by Poitrasson et al. 2017, this volume), evaporation processes during planetary formation (e.g., Paniello et al. 2012, Wang and Jacobsen 2016, and review by Moynier et al. 2017 this volume), igneous differentiation (e.g., Williams et al. 2009; Sossi et al. 2012; and review by Dauphas et al. 2017, this volume), and on biological processes (e.g., Walczyk and von Blanckenburg 2002, and review by Albarède et al. 2017 this volume).

Among all these non-traditional isotopic systems, Mg isotopes are of major importance because (i) Mg is a major constituent of the silicate portion of planetary bodies, (ii) Mg has more than two isotopes (^{24}Mg, ^{25}Mg and ^{26}Mg) allowing to study processes leading to various types of mass fractionation (Young et al. 2002; Young and Galy 2004; Davis et al. 2015) and (iii) ^{26}Mg excesses produced by the radioactive decay of short-lived ^{26}Al ($T_{1/2}$=0.73 Ma) (Lee et al. 1976) are a key tool for early Solar system chronology (see reviews by Dauphas and Chaussidon 2011; Chaussidon and Liu 2015). Note that in addition, significant Mg isotopic anomalies (^{26}Mg deficits or excess not related to ^{26}Al decay) have been found in hibonite-bearing refractory inclusions from CM chondrites, these inclusions being presumably very early solar nebula condensates (Ireland 1988; Liu et al. 2009, 2012).

In this paper we review recent developments of in-situ techniques towards high-precision isotopic measurement of non-traditional isotopes and detail their limitations. To illustrate these limitations, we report in-situ Mg isotopic data acquired by different techniques on a set of natural and synthetic olivines and basaltic glasses.

Notations used for Mg isotopes

In the following, Mg isotopic compositions are reported using the classical delta notation (δ^{25}Mg and δ^{26}Mg) which gives the variation in per mil of the ^{25}Mg/^{24}Mg or ^{26}Mg/^{24}Mg in a sample *x* relative to that of the DSM3 standard (Young and Galy 2004):

$$\delta^{25,26}\mathrm{Mg}^{x}_{\mathrm{DSM3}} = \left[\frac{\left({}^{25,26}\mathrm{Mg}/{}^{24}\mathrm{Mg}\right)_x}{\left({}^{25,26}\mathrm{Mg}/{}^{24}\mathrm{Mg}\right)_{\mathrm{DSM3}}} - 1\right] \times 1000 \qquad (1)$$

The Mg isotopic composition of DSM3 relative to the SRM 980 standard is $\delta^{25}\mathrm{Mg}^{\mathrm{DSM3}}_{\mathrm{SRM980}} = -1.744‰$ and $\delta^{26}\mathrm{Mg}^{\mathrm{DSM3}}_{\mathrm{SRM980}} = -3.405‰$ (Young and Galy 2004). The absolute Mg isotopic composition of SRM 980 is ^{25}Mg/^{24}Mg = 0.12663 and ^{26}Mg/^{25}Mg = 0.13932 (Catanzaro et al. 1966). These ratios were re-determined recently to be ^{25}Mg/^{24}Mg = 0.126896 ± 0.000025 and ^{26}Mg/^{25}Mg = 0.139652 ± 0.000033 (Bizzarro et al. 2011). Deviation of the Mg isotopic composition from a reference mass fractionation line (see last section) are generally given in ^{26}Mg excesses or deficits in the Δ^{26}Mg notation according to:

$$\Delta^{26}\mathrm{Mg} = \delta^{26}\mathrm{Mg} - \delta^{25}\mathrm{Mg} / 0.521 \qquad (2)$$

When the ^{26}Mg excesses are considered to be of radiogenic origin (due to the decay of short-lived ^{26}Al) they are generally noted as $\delta^{26}\mathrm{Mg}^*$. It is important to stress here that this definition of Δ^{26}Mg (and postulating a value of 0.521) is a first order approximation that cannot be used when high precision (0.01‰ level) is required for radiogenic ^{26}Mg excesses (this is discussed in the last section).

INSTRUMENTATION FOR IN-SITU STABLE ISOTOPE ANALYSIS

Two approaches are dominant in cosmochemistry and geochemistry to get high-precision in situ isotopic analysis: multicollector secondary ion mass spectrometry (MC–SIMS, often referred to as ion microprobe) and laser ablation multiple-collector inductively-coupled-plasma mass-spectrometry (LA–MC–ICP–MS). Other in situ isotopic techniques, such as time-of-flight SIMS (TOF-SIMS) or resonance ionization mass spectrometry (RIMS), exist and are continuously developed, but they are generally used to make isotopic analysis with precision limited in the ‰–% range on very small amount of material, either surface analysis or analysis of sub-micrometer grains. Of specific interest for cosmochemistry are the recent developments made in the University of Chicago on the CHILI RIMS instrument (Stephan et al. 2016).

MC–SIMS analysis

In SIMS analysis a solid sample is sputtered by a beam of primary ions accelerated to high-energy (4–12 keV positive or negative primary ions) and the secondary ions produced in the sample by the sputtering process are accelerated and analyzed in a magnetic sector mass spectrometer. In order to reach high precision for isotopic ratio measurements, precisions close to those obtained by MC–ICP–MS for solutions, the only SIMS instruments that are suitable are those with a transmission optimized to refocus the beams of ions produced by a large (≈20 μm) primary beam. This excludes the NANOSIMS type of instrument that is optimized for isotopic imaging and analysis at the 10nm–1 μm scales. The two types of magnetic SIMS instruments used in Earth and planetary sciences for high-precision isotopic measurements of non-traditional isotopes are thus (i) the CAMECA IMS 1270–1280 series based on the original idea of Castaing and Slodzian (1962) and developed by CAMECA as the ims 3–6f series and later on (in 1996) as the large geometry ims 1270, and (ii) the SHRIMP I-II-RG series developed at the Australian National University (and manufactured by ANUTECH) following the original idea of Prof W. Compston (Clement et al. 1977). Different reviews exist that allow following the development of analytical techniques with SIMS towards the goal of reaching high-precision isotopic analysis in geochemistry and cosmochemistry (e. g. Lovering 1975; Shimizu and Hart 1982; Zinner 1988; Ireland 2004). We take examples in the following of data acquired with the large radius Cameca ims 1270–1280 instruments of the CRPG-CNRS Nancy ion microprobe laboratory, but the general observations reported here are also valid for SHRIMP type instruments (and the effects are theoretically also the same for NANOSIMS type of instruments even if it is not obvious that they could be observed because of the limitation in precision due to the very small volume of sample sputtered).

The unique advantages offered by SIMS for isotopic analysis of non-traditional isotopes are (i) 10–20 μm spatial resolution (up to a few atomic layers in case of analysis by depth profiling, but limited precision), (ii) high sensitivity (variable but very high for some elements) allowing to carry out analysis at high mass resolution thus removing most isobaric interferences, and (iii) low background (ppm range) since surface contamination can be cleaned up by pre-sputtering.

LA–MC–ICP–MS analysis

For laser ablation analysis, an intense photon beam is used to ablate the solid samples. The generated aerosols are transported to mass spectrometry by argon or helium carrier gases. The different laser ablation cells which exist are not specific to a given ICP–MS or MC–ICP–MS. Depending on the excitation sources, the commonly used laser systems for LA-ICP–MS nowadays can be classified into Nd:YAG, Ti:sapphire and Excimer lasers (Günther and Heinrich 1999; Horn and von Blanckenberg 2007; Shaheen et al. 2012). Since 1990s, the development of laser ablation has been facilitated thanks to coupling with ICP–MS to measure major and trace elements (Jackson et al. 1992; Norman et al. 1998; Günther et al. 1999; Liu et al. 2008) or to analyze U–Pb isotopes in zircon or monazite (Hirata and Nesbitt 1995; Poitrasson et al. 2003; Jackson et al. 2004; d'Abzac et al. 2010). This decade, more approaches have been made for coupling laser ablation with MC–ICP–MS to study isotopes of boron (Fietzke et al. 2010), strontium (Christensen et al. 1995; Fietzke et al. 2008), hafnium (Griffin et al. 2000; Woodhead et al. 2004), silicon (Chmeleff et al. 2008; Steinhoefel et al. 2010; Janney et al. 2011; Schuessler and von Blanckenburg 2014), magnesium (Norman et al. 2006; Janney et al. 2011; Xie et al. 2011; Oeser et al. 2015), iron (Horn et al. 2006; Horn and van Blanckenburg 2007;

Nishizawa et al. 2010; Czaja et al. 2013; d'Abzac et al. 2013; Sio et al. 2013; Oeser et al. 2015) and copper (Jackson and Günther 2003), or with both ICP–MS and MC–ICP–MS (so-called "split-stream") to simultaneously measure trace elements, Lu–Hf and U–Pb isotopes (Yuan et al. 2008; Kylander-Clark et al. 2013) or U–Pb and Sm–Nd isotopes (Goudie et al. 2014).

The quality of LA-ICP–MS analysis depends primarily on three factors, i.e., ablation performance, sample transportation and ionization on mass spectrometry (Shaheen et al. 2012). Of them, ablation performance is the most important, because either delivery efficiency or ionization rate is sensitive to the size distribution of the generated aerosols (Guilong and Günther 2002; Kuhn and Günther 2004; Kroslakova and Günther 2006), which, in turn, is dominantly controlled by the wavelength and pulse width of the ablation laser (Günther and Heinrich 1999; Horn 2008; Shaheen et al. 2012). We want to stress that in this paper we take as a an example LA–MC–ICP–MS Mg isotopic data obtained by nanosecond (≤4 ns) ablation laser, a commercial laser system that is commonly and easily accessible in most Earth and planetary science laboratories. As discussed in the following, such a nanosecond laser has some limitations compared to a femtosecond laser.

LIMITATIONS FOR IN-SITU STABLE ISOTOPES ANALYSIS

The reason why high-precision isotopic analysis has always been challenging to reach for in-situ analysis is that there are three major limitations that are inherent to the in-situ techniques:

(i) the small volume sputtered or ablated limits the possible precision from considerations of simple statistics on the total number of ions measured.

(ii) different elements have different ion yields (see below for definition) and these depend on the nature and/or composition of the sample (matrix effects, see below),

(iii) strong instrumental isotopic fractionations exist for all elements and large matrix effects are present, requiring as for ion yields a number of mineral and glass standards to determine and calibrate these effects (no physical model exists that would allow to predict these effects). In addition, because of these large instrumental fractionations, the laws for instrumental mass fractionations must be determined very precisely if one wants to measure non-mass dependent isotopic effects.

These three limitations are defined and discussed in more details in the following. We concentrate on matrix effects for Mg isotopes by MC–SIMS and LA–MC–ICP–MS in a set of matrices (natural and synthetic olivines and silicate glasses) chosen to match the variability in composition observed for the major components of primitive chondritic meteorites. We review the results of the very few studies available and present a set of consistent data that allow investigating the systematic of Mg ion yield and isotopic fractionation during MC–SIMS and LA–MC–ICP–MS. The reason to study the ion yield of Mg is not to develop an in situ technique to measure precisely the Mg contents from the Mg intensity (electron probe is much more appropriate for that) but to gain insights into the processes which control the production of Mg ions and the associated matrix effects, in order to try to get insights into the origin of instrumental isotopic fractionations. In addition, knowing the Mg ion yield of different phases is a prerequisite in order to use the Mg intensity (and that of other major elements) as a monitor of the fraction of the different phases under the ion beam or the laser beam. As will be explained below, a precise correction of matrix effects on instrumental isotopic fractionation during the sputtering of complex samples made of different phases will require an estimate of the fractions in the spot of these different phases.

Limitations due to the small amount of sample analyzed

Bulk and in situ analyses are by nature fundamentally different: much less material is available for an in-situ analysis. In a few extreme situations, the precision for in-situ analysis can be simply limited by the total number of atoms present in the analyzed volume. This was the case, for instance, for the analysis by SIMS of the $^{7}Li/^{6}Li$ isotopic ratio of solar wind implanted in silicates (olivine and pyroxene) from lunar soils (Chaussidon and Robert 1999). Using a primary beam intensity of 7 nA and sputtering a large surface of $250 \times 250\,\mu m$, the sputtering rate of a silicate in these conditions (4.2×10^{-3} nm/sec/nA) makes that 5.5×10^{-12} g of sample are sputtered by second, giving 8.1×10^{-22} mole of Solar wind Li sputtered per second (for a Solar wind Li content of 1 ppb), and thus 15 atoms of ^{6}Li sputtered per second for a Solar $^{7}Li/^{6}Li = 31$ (Chaussidon and Robert 1999). If the total yield of the SIMS is 10% (i.e., 10% of all the atoms sputtered are counted on the detectors), an integration time of 656 s (11 min) would be necessary to count the ≈ 10000 ^{6}Li ions required to get a $\pm 1\%$ statistical error on the $^{7}Li/^{6}Li$ ratio. Thus, for 1 ppb Solar wind Li, the best depth-resolution that can be achieved by SIMS for measuring Li isotopes at $\pm 1\%$ is theoretically of ≈ 3 nm.

However, counting statistic is not the major limiting factor for in situ SIMS analyses of major elements in spots of several µm sizes. Taking the case of Mg, ≈ 5 ng of Mg is present in a typical ion microprobe spot of 20 µm in diameter and 5 µm in depth (requiring ≈ 5–10 min sputtering with a primary beam intensity of ≈ 20 nA) in an olivine with 40 wt% MgO. This corresponds to $\approx 10^{13}$ ^{26}Mg atoms. Because > 1 out of every hundred Mg atoms sputtered from the sample are ionized, transmitted through the mass spectrometer and detected, more than 10^{11} $^{26}Mg^{+}$ ions are measured on the Faraday cup of the MC–SIMS, allowing to get internal errors at the level of 0.02–0.05‰ on the $^{25}Mg/^{24}Mg$ and $^{26}Mg/^{25}Mg$ ratios.

Solution MC–ICP–MS that is generally used to get high-precision Mg isotopic measurements requires purification of magnesium from matrix elements with chromatography before loading samples in the mass spectrometer. However, due to the procedure blank of 2–10 ng, the amount of Mg processed for each analysis by chromatography would typically range from 2 µg to 10 µg (Teng et al. 2010; Bouvier et al. 2013). For San Carlos olivine (50 wt% MgO) as an example, 10 µg Mg will correspond to a 200 µm diameter microdrilling spot with a drill depth of approximately 250 µm, as used in a recent study of Mg isotopic zoning in olivine due to Mg diffusion (Sio et al. 2013). Such a big sampling volume limits the resolution to which possible isotopic heterogeneity within samples can be studied. In addition, sampling by either microdrilling or physical mineral separation may incorporate adjacent minerals. By comparison, laser ablation in a two-volume HELEX sample chamber optimized by Eggins et al. (2003, 2005) can lower analytical blank to picogram levels when using 99.999% helium delivery gas. For San Carlos olivine, only a 50 µm sputtering spot with a depth of 20 µm, i.e., ca. 50 ng Mg, is required to obtain sample/blank $^{24}Mg^{+}$ intensity ratios of 1000–6000 and internal errors at the level of ± 0.1‰ on the $^{25}Mg/^{24}Mg$ and $^{26}Mg/^{25}Mg$ ratios.

Limitations due to matrix effects on ion yield

Definitions. In SIMS analysis, the efficiency of ion generation during sputtering and analysis in the mass spectrometer is quantified by a parameter called the relative ion yield (it is often defined relative to Si that is a major component of silicates) and defined as:

$$\text{Mg ion yield relative to Si} = \frac{Mg^{+}/Si^{+}}{Mg/Si} \qquad (3)$$

with Mg^{+}/Si^{+} the ratio measured for secondary Mg and Si ions and Mg/Si the atomic ratio in the sample. This formalism is used because it has the advantage to eliminate several instrumental parameters that control the generation of secondary ions in the same way for

Mg and Si. The determination of the relative yield does not require measuring the volume of sample sputtered or the sputtering yield (possible variations in sputtering yield between different samples would cancel out when comparing Mg/Si ratios). Mg contents can thus be determined by SIMS for an unknown sample if the Mg relative ion yield has been previously determined from the analyses of standards and if the Si content of the sample is measured independently. However, the use of relative ion yields for SIMS analysis is complicated by the fact that they depend strongly on the energy of secondary ions (Shimizu and Hart 1982; Hinton 1990). Hinton (1990) determined positive ion yields of Mg relative to Si in the NBS 610 glass (assuming a SiO_2 content of 72 wt% and a Mg content of 500 ppm, more recent analyses give 463 ± 11 ppm Mg, Gao et al. 2002) of 1.88 for high-energy ions and 4.99 for low-energy ions.

There is no fundamental reason not to use relative ion yields for LA–MC–ICP–MS (except the fact that MC–ICP–MS analyses are made by sample bracketing, see below) but relative sensitivity factors (RSF) are often preferred for quantification (e.g., Zhang et al. 2015).

RSF just compare the intensity measured on a sample to that in a standard and is for Mg, for instance, defined

$$RSF_{Mg} = \left(Mg^{+}_{sample} / Mg^{+}_{standard}\right) / \left(Mg_{sample} / Mg_{standard}\right) \tag{4}$$

The fact that relative ion yields, or relative sensitivity factors, show significant variations between different chemical elements results in what is often called elemental fractionation: any elemental ratio measured by SIMS or LA-ICP–MS is a priori different from the true value of this ratio in the sample.

In the example of Mg studied in details in the following, ion yields relative to Si are not used, simply because Si ions were not measured together with Mg ions. The analytical approach followed is optimized for the measurement of the Mg isotopic compositions. Scanning the magnetic field from mass 25 (where the three Mg isotopes ^{24}Mg, ^{25}Mg and ^{26}Mg are measured in multi-collection) to mass 28 for Si, carries the risk to introduce instabilities for Mg isotopic measurements both in MC–SIMS and MC–ICP–MS analysis. The efficiency with which Mg atoms present in a given matrix are sputtered away, ionized and analyzed in the mass spectrometer is defined in the following as the Mg ion yield (Storms et al. 1977) with $^{24}Mg^{+}$ the count rate of ^{24}Mg ions and [MgO] the MgO content of the sample:

$$\text{Ion yield}_{Mg} = \frac{^{24}Mg^{+}}{[MgO]} \tag{5}$$

For MC–SIMS the count rate of $^{24}Mg^{+}$ is normalized to the primary beam intensity (in nA) and thus given in counts/sec/nA (Table 3 in electronic annex). For LA–MC–ICP–MS the count rate is in V ($1V = 6.25 \times 10^{7}$ counts/s for a $10^{11}\Omega$ resistor) normalized to 1V of $^{24}Mg^{+}$ on the San Carlos olivine used for bracketing. The MgO contents are in wt% (Table 2 in electronic annex). Dividing the ion yield observed for one element in one sample by that observed in a standard gives the relative sensitivity factor for that element in this sample as defined in Equation (4).

Matrix effect on ion yield. For a given element and a given secondary energy, SIMS ion yields vary with the concentration of the element and the composition of the sample (Deline et al. 1978a,b; Shimizu et al. 1978). This effect is called the matrix effect. Matrix effects are also observed in LA-ICP–MS (Jackson et al. 1992; Norman et al. 2006). Previous studies of elemental ion yields in SIMS and LA-ICP–MS have been mostly motivated by the measurement of concentrations of minor and trace elements since major element concentration can be routinely measured at the micrometer scale by electron beam techniques. Matrix effects exist for major elements but also for trace elements (e.g., for H in silicate glasses, Sobolev and Chaussidon 1996, or for Ti in silicate matrix, Behr et al. 2011). However, because the precision to which trace elements such as REE elements have been routinely measured by ion probe is from ±15% to ±25‰ relative (e.g., Shimizu et al. 1978; Zinner and Crozaz 1986; Fahey et al. 1987; Gurenko and Chaussidon 1995), matrix effects of typically ±10–20% relative have been often ignored.

Principles of calibration. In SIMS, ion yields show variations of several orders of magnitude between different chemical elements depending on various parameters such as the ionization potential of the element, the mineralogical or chemical nature of the matrix, or the energy of the secondary ion produced (e.g., Shimizu et al. 1978; Williams 1985; Reed 1989; Hinton 1990). Despite attempts, there is no simple physical model allowing an accurate prediction of the ion yield of a given element in a wide range of matrices (Shimizu and Hart 1982; Benninghoven 1987). Thus, quantitative analysis of trace elements by SIMS requires the development of standards, either natural or synthetic minerals and glasses, to determine ion yields and detection limits for each element. In LA-ICP–MS, variations in ion yields are also the rule between different elements in a given matrix (e.g., Stix et al. 1995; Gaboardi and Humayun 2009). Several possible sources of chemical fractionation are present in the chain of processes taking place during ablation, transport of the aerosol, its vaporization and the atomization and ionization of the elements in the plasma source (e.g., Guillong and Gunther 2002; d'Abzac et al. 2013; Zhang et al. 2015). These elemental fractionations can be corrected for using matrix matched standards, and most trace, minor and major element concentrations are accurately measured by LA-ICP–MS in different matrices (e.g., Jochum et al. 2005, 2012).

Limitations due to instrumental isotopic fractionation

Definitions. Isotopic analysis by MC–SIMS or MC–ICP–MS is always accompanied by instrumental isotopic fractionation that is visible from the fact that the rough isotopic ratios calculated from the Mg isotopes intensities measured on the Faraday cups (after correction for backgrounds and gains) is always different from the true isotopic ratios. From mass balance consideration, no instrumental isotopic fractionation would exist for a given element if it was ionized and transmitted through the mass spectrometer with 100% efficiency. We consider here only the real isotopic fractionations in the instrument (e.g., during ionization, mass analysis in the spectrometer, ...) but not effects due to isobaric interferences. The lack of significant isobaric interference can be demonstrated for an element with at least three isotopes such as Mg by showing that the measured rough $^{26}Mg/^{24}Mg$ and $^{25}Mg/^{24}Mg$ ratios follow a mass fractionation line. This is the reason why instrumental isotopic fractionation is generally called "Instrumental Mass Fractionation" (IMF) even if the fact that it is a mass dependent process is not demonstrated when measuring two isotopes only. Note that IMF is not obligatory equilibrium fractionation, so that the slope of IMF in the three Mg isotopes diagram (see next section on instrumental mass fractionation laws) can vary.

As any isotopic fractionation (α), IMF is defined for e.g., the $^{26}Mg/^{24}Mg$ ratios by:

$$\alpha_{\text{inst}}^{26/24} = \frac{\left(^{26}\text{Mg}/^{24}\text{Mg}\right)_{\text{measured}}}{\left(^{26}\text{Mg}/^{24}\text{Mg}\right)_{\text{true}}} \tag{6}$$

which can also be approximated to first order by:

$$\Delta_{\text{inst}}^{26/24} = \delta^{26}\text{Mg}_{\text{measured}} - \delta^{26}\text{Mg}_{\text{true}} \tag{7}$$

with

$$\Delta_{\text{inst}}^{26/24} \approx 1000 \times \ln\left(\alpha_{\text{inst}}^{26/24}\right) \tag{8}$$

In order to reach high precision (5–10 ppm level) on the three Mg isotopes measurement, IMF must be determined very precisely for the $^{26}Mg/^{24}Mg$ and $^{25}Mg/^{24}Mg$ ratios and errors should not be introduced by manipulating delta values instead of isotopic ratios (mass fractionation laws must be determined from the $\alpha_{inst}^{25/24}$ and $\alpha^{26/24}$ values and not from the $\Delta_{inst}^{25/24}$ and $\Delta_{inst}^{26/24}$ values). However, in the following we first look for a systematic of IMF in olivines, basaltic and CMAS glasses with chemical composition, and for that, manipulating delta values does not introduce significant errors. Thus, for simplicity we use delta values in the next sections.

During MC–ICP–MS analysis, IMF is generally corrected for by using the "bracketing technique" which consists in analyzing a pure solution of an internal standard of the element to analyze in between each solution of this element purified from the sample. This allows to determine by interpolation the IMF in between two analyses of the standard solution and to use this value of IMF to correct the measurement of the sample. Doing so IMF is "invisible": typically for instance, $\delta^{26}Mg$ values measured for the San Carlos olivine bracketed by itself during the present study have a two sigma standard deviation of ±0.22‰ while IMF varied by ≈ 1.1‰. This is also the classical approach used in gas source mass spectrometry for H, C, N, O, and S isotopes.

For in situ analysis, additional isotopic fractionations exist. In the case of MC–SIMS, isotopic fractionation takes place during sputtering and ionization in the sample and/or between the sample and the immersion lens, and during ion analysis in the mass spectrometer (Slodzian et al. 1980; Shimizu and Hart 1982; McKeegan et al. 1985; Ireland 1995, 2004). IMF generally decreases with atomic mass (Shimizu and Hart 1982), being maximal for D/H ratio (up to several hundred ‰, Deloule et al. 1991) and of only a few‰ for Pb isotopes (Deloule et al. 1986). IMF always results in an enrichment of the extracted ions in the light isotopes ($\alpha_{inst} < 1$) (Zinner 1988) except in the case of Li isotopes for which the extracted ions are enriched in ^{7}Li relative to ^{6}Li (Chaussidon et al. 1997). Much less is known for instrumental isotopic fractionations by LA–MC–ICP–MS in comparison with MC–SIMS, but their existence has been demonstrated, e.g., for Hf isotopes (Thirlwall and Walder 1995), Si (Janney et al. 2011), Mg isotopes (Young et al. 2002; Norman et al. 2006; Janney et al. 2011) or Fe isotopes (Horn et al. 2006; Sio et al. 2013). At variance with MC–SIMS where the isotopic fractionations due to ion sputtering are unavoidable (the goal with MC–SIMS is to control and calibrate them), LA–MC–ICP–MS is still in a phase of rapid development with efforts going on to reduce instrumental fractionations or even to suppress them by improving laser ablation and the homogeneity in size distribution of the aerosols (see Horn and von Blanckenburg 2007, and discussions in next section).

Matrix effect on instrumental isotopic fractionation. Variation of instrumental isotopic fractionation due to change in sample composition is the limiting factor for increasing the precision of in situ isotopic measurements. These effects are generally called matrix effects on IMF. Matrix effects are well known for MC–SIMS and for LA–MC–ICP–MS, but because of the predominance of SIMS as an in situ technique for isotopic measurements in the past 30 years, much more is known in the case of SIMS.

Previous studies of matrix effects during SIMS analysis have shown that IMF can be assessed from empirical calibrations observed with various parameters describing the changes in chemical composition of the matrix. The physical laws behind these empirical calibrations have remained enigmatic though it has been shown for instance that in the case of oxygen isotopic analysis of silicates, IMF was related to the mean atomic mass of the matrix, its sputter rate, its chemical composition and the amount of network-forming cations relative to network-modifying cations (Eiler et al. 1997). These observations led Eiler et al. (1997) to propose that matrix effects for SIMS analysis of O isotopes in silicates were related to (i) variations in the efficiency of transfer of kinetic energy to the secondary ions and to (ii) variations in the fraction of sputtered ions which are ionized. Examples of empirical calibrations are for instance the fact that IMF for D/H analysis in micas and amphiboles has been shown to vary linearly with the mass over charge ratio of octahedral cations (≈ 200‰ change among biotite, muscovite and amphibole showing a large compositional range, Deloule et al. 1991, 1992), that IMF for $^{34}S/^{32}S$ analysis in iron–nickel sulfides depends on the Ni/Fe ratio (≈20‰ range between millerite and pyrhotite, Chaussidon et al. 1987; Deloule et al. 1992), that IMF for $^{11}B/^{10}B$ ratios in tourmaline depends linearly on the mass/charge ratio of octahedral cations (≈10‰ range from elbaite to shorl and dravite, Chaussidon and Albarède 1992), that IMF for $^{18}O/^{16}O$ analysis in garnets can be calibrated in function of their concentration in the three major elements (≈ 6‰ variations for varying Ca–Fe–Mg contents, Vielzeulf et al. 2005), and that IMF for $^{18}O/^{16}O$ analysis in silicates depends primarily on the SiO_2 content (≈8‰ range for wt% SiO_2 from 40 to 100, Chaussidon et al. 2008). Matrix effects for $^{18}O/^{16}O$ analysis in olivine have been shown to be significant only for large variation in their fayalitic content: ≈0.5‰ change for a change of 10% of the fayalite content (Leshin et al. 1997) or more recently with data acquired by MC–SIMS equipped with Faraday cups ≈0.0075‰ per each Fo number (Bindeman et al. 2008; Gurenko et al. 2011).

In the case of LA–MC–ICP–MS with nanosecond laser ablation, non-ideal ablation performance will result in matrix effects due to (i) preferential evaporation of volatile elements during ablation, (ii) particle size drift with increasing ablation depths and (iii) preferential evaporation of lighter isotopes for large particles in the ICPMS (Horn and von Blanckenburg 2007). Recent studies have shown that these matrix effects during LA–MC–ICP–MS (ns laser) are lower than during MC–SIMS with for instance less than 0.8‰ change in IMF for $^{56}Fe/^{54}Fe$ ratios in olivines ranging in compositions from $Fo^{\#}50$ to $Fo^{\#}95$, while it is ≈8% during MC–SIMS analyses (Sio et al. 2013). For both techniques, matrix effects exist and show clear relationships with chemical composition with, in the case of $^{56}Fe/^{54}Fe$ ratios in olivines linear trends relating IMF (for MC–SIMS and for LA–MC–ICP–MS) and $Fo^{\#}$ in the range $Fo^{\#}50$ to $Fo^{\#}95$ (Sio et al. 2013). Similarly, Janney et al. (2011) showed that IMF for LA–MC–ICP–MS isotopic analysis of Mg in CMAS glasses was at the level of a few per mil at maximum and was linearly related to the Al/Mg ratios.

A possible way to reduce matrix effects in LA–MC–ICP–MS is to optimize the laser system to reduce elemental fractionations (Hu et al. 2011; Diwakar et al. 2014), to filter the aerosols to cut down large particles loaded in the ICP–MS (Guilong and Günther 2002; Guillong et al. 2003) or to decrease the average size of the particles by adding a collision cell between the ablation cell and the ICP–MS. By comparison with ns laser, femtosecond laser with ultra short pulse width (ca. 100 fs) produces sputtered particle of smaller sizes and result in less melting of samples (Horn 2008; Shaheen et al. 2012), thus increasing and stabilizing the ionization rate of particles in the ICP. A series of fs-LA–MC–ICP–MS studies showed that ablation with a fs laser could strongly reduce (or perhaps even eliminate) matrix effects commonly observed with ns laser ablation (at least for Mg, Si and Fe isotopes, Horn et al. 2006; Horn and van Blanckenburg 2007; Chmeleff et al. 2008; Nishizawa et al. 2010; Schuessler and von Blanckenburg 2014; Oeser et al. 2015). However, whether all matrix effects are systematically removed using fs laser is not totally clear yet. d'Abzac et al. (2013, 2014) reported that size-dependent inter-particle elemental and Fe isotopic fractionation could take place due to variable condensation of sample vapors produced by fs laser ablation of natural magnetite, siderite, pyrrhotite and pyrite. In addition, resolvable matrix-induced Mg isotopic fractionations can be observed on MC–ICP–MS by doping matrix elements into analyte solutions or aerosols (up to 1‰ in this study see below; Young et al. 2002; An et al. 2014; Teng et al. 2014).

Instrumental mass fractionation laws. Any high-precision measurement of isotopic ratios in a system with more than three isotopes will be confronted with the difficulty of correcting properly for IMF while minimizing errors due to the correction. This comes from the fact that IMF must often be determined for sample compositions that are not exactly matched by the standards. In such a case, the instrumental mass fractionation law has to be precisely known if interpolating the IMF values determined by several standards to unknown samples.

Errors can be minimized if several standards are used to determine the instrumental mass fractionation law to high precision, allowing to reach ppm level precision for the measurement of the deviation of the isotopic compositions from mass fractionation (without introducing additional errors due to the correction of matrix effects on IMF). This is typically what is required for instance for Mg isotopes to determine the excesses of ^{26}Mg (noted $^{26}Mg^{*}$) due to the decay of short-lived ^{26}Al. As explained in details in the next section of this paper, the precise determination of $^{26}Mg^{*}$ requires determination of both instrumental and natural mass fractionation laws (see Fig. 5 in Luu et al. 2013, and discussion therein).

Principles of calibration. Despite the existence of some systematics for matrix effects on ion yields (Eiler et al. 1997), the only possible approach to correct for matrix effect is to determine them by analyzing standard minerals and glasses with chemical compositions similar to that of the analyzed samples. In case of analysis of a mixture of two (or more) phases, an average matrix effect can be calculated if the proportions of the different phases in the analytical spot are determined (from the measurement, together with the isotope ratio of interest, of selected major elements). Standard minerals and glasses can be either natural or experimental. The bulk composition of these standards must be analyzed precisely by "classical techniques", and in situ analyses are also required to test their homogeneity. Standards with fixed isotopic composition can be produced by implanting a molecule such as HD of mass 3 that gives a D/H ratio of 1 (Williams et al. 1983) but such a standard is not suitable for all applications since its isotopic composition is very far from natural ones. Implantation can also be used to develop standards for measuring concentrations (Burnett et al. 2015).

The procedure for a given analytical session must thus be similar for either MC–SIMS or LA–MC–ICP–MS: (i) determination of laws for matrix effects using a set of appropriate standards, (ii) analyses of samples bracketed by one or two standards (the bracketing must be in between each spot for LA–MC–ICP–MS while it can be only several times during a session for SIMS where IMF is much more stable with time than in ICP–MS), and (iii) correction of IMF taking into account matrix effects. In the following sections, as an example to better understand all these effects, we study whether there is a common systematics for matrix effects on IMF for Mg in olivines and silicate glasses and for MC–SIMS and LA–MC–ICP–MS, in light of what we have observed for ion yields. The existence of such a systematics would strengthen the validity of empirical calibrations of IMF and could show how to extend them to compositions for which standards are not easily available.

STANDARDS AND ANALYTICAL APPROACH USED FOR Mg IN THE PRESENT STUDY

Set of standards studied

We have studied by MC–SIMS and laser-ablation MC–ICP–MS a set of 11 basaltic glasses made of 10 international standards (Jochum et al. 2005, 2010) and of one an in-home standard of MORB glass (Chaussidon et al. 1989), 7 experimental glasses of Fe-free CMAS compositions (Richter et al. 2007) and 21 olivines spanning nearly the whole range of composition from fayalite to forsterite. Nine of these olivines are synthetic olivines (see next paragraph for details) and 12 are natural olivines (Sio et al. 2013), including one meteoritic olivine from the Eagle Station pallasite (Luu et al 2014). All these samples can be regarded as standards for Mg because their chemical compositions and Mg isotopic compositions have been determined precisely (see details and references in Table 2 of electronic annex: Richter et al. 2007; Teng et al. 2010; Janney et al. 2011; Oeser et al. 2014; Davis et al. 2015). The Mg isotopic composition of the MgO (ceramic insulator material from Saint-Gobain SA) used to synthetize the olivines was measured by solution MC–ICP–MS at IPGP after digestion by concentrated HNO_3 and dilution in 0.1 N HNO_3: its $\delta^{26}Mg$ is of $-1.19 \pm 0.06‰$ relative to DSM3 international standard. We focused on CMAS compositions, basaltic glasses and olivines because this range of composition is the one required for in situ analysis of Mg isotopic compositions in meteoritic Ca-, Al-rich inclusions, chondrules and matrices.

The synthetic olivines used in this study belong to a series of 11 Li-doped olivines with major-element compositions ranging from ~Fo_{90} to ~Fo_{12} synthesized in the experimental geochemistry laboratory at Rensselaer Polytechnic Institute. These olivines were prepared by solid-state reaction of oxide mixes consisting of finely-ground amorphous SiO_2 (99.8% pure; Aesar lot no. D13Y012), FeO (99.5% pure; Aesar lot no. Y17A024), and ground MgO ceramic insulator material from Saint-Gobain S.A. The oxide mixtures were thoroughly ground in agate under methanol, air-dried at 110°C, and stored in air at room temperature prior to loading into high-purity graphite containers (inside dimensions ~3 mm diameter × ~9 mm length). The containers were then placed in standard piston-cylinder assemblies used in the RPI laboratory (NaCl + borosilicate glass + MgO) and run for 2–3 days at 1.8 GPa and temperatures ranging from 1260° for the Fo_{12} composition to 1400 °C for the Fo_{90} composition. The resulting run products were polycrystalline olivine with typical grain sizes ranging from 100 to 400 µm. The crystals contain occasional small (< 10 µm) inclusions of graphite and (separate) CO_2–N_2–H_2O fluid, both identified by micro-Raman spectroscopy. Occasional Fe-oxide inclusions were also present in two of the more Fe-rich compositions. Because of their dispersed nature and small size, inclusions of all types were avoidable during the SIMS analyses. Reconnaissance elemental analyses were performed at RPI by electron microprobe (Mg, Fe, Al, Ti) and laser-ablation ICP–MS (Li, Al, Ti). The crystals are homogeneous (both intra- and inter-grain) within analytical uncertainty with respect to Li, Mg and Fe; however, Al and Ti show some variation outside the analytical uncertainty.

To try to better assess the instrumental effects specific to LA–MC–ICP–MS, solutions having chemical compositions similar to the compositions of the different standards studied (olivines and basaltic glasses) were analyzed. These solutions were prepared by mixing and diluting 1000 ppm mono-elemental standard solutions (Si, Al, Mg, Ca, Na, Fe). Two sets of solutions, one with fixed Si concentrations (0.750 or 0.250 ppm) and the other set with fixed Mg concentration (0.300 ppm) were analyzed. These concentrations were chosen to mimic the range of ^{24}Mg intensities observed during LA–MC–ICP–MS analyses of olivines and basaltic glasses. Each set of solutions consisted of 11 solutions with "basaltic" composition and 6 with "olivine" composition (Table 5 in electronic annex).

MC–SIMS for Mg isotopic analysis

MC–SIMS Mg isotopic analyses were performed on a multi-collector Cameca ims 1280 instrument at CRPG-CNRS (Nancy, France) using procedures previously described in detail (Villeneuve et al. 2009; Luu et al. 2013). Samples are sputtered with a primary O^- beam accelerated to 13 kV. The secondary positive Mg ions are accelerated at 10 kV and analyzed in multi-collection at a mass resolution $M/\Delta M = 2500$ (with ΔM calculated as the mass difference between the two flanks of the $^{24}Mg^+$ peak at 10% of its height). Potential interference of $^{24}MgH^+$ that is not totally resolved at this mass resolution is kept below a few counts per second if the vacuum is better than 3×10^{-9} Torr and the studied samples are anhydrous phases. Automatic centering of the transfer deflectors, and automatic control of any charge build-up on the insulating sample are implemented in the analysis routine. The three Mg isotopes are measured on three Faraday cups having their yields and backgrounds calibrated at the beginning of each analytical session. One measurement is typically made of 2 min pre-sputtering followed by 25 cycles of 10 s integration time each, separated by 2 s waiting time. This means that each measurement takes around 7 min per spot. The advantage to make 25 cycles of 10 s instead of 1 cycle of 250 s is to be able to follow the stability of the isotope measurement, eventually to reject one anomalous cycle, and also to determine the internal error from the 2σ standard error of the 25 isotopic ratios. Typical spot sizes (Fig. 1) are of 25–40 μm in diameter and a few micrometers in depth, depending on the primary beam intensity (typically 25 nA), which is adjusted to get the best internal errors for samples having variable MgO contents. Repeated analyses of the San Carlos olivine internal standard shows an external reproducibility of ± 0.15‰ (2σ standard deviation, 0.03‰, 2σ standard error for 25 data) for δ^{26}Mg, ±0.08‰ (2σ standard deviation, 0.02‰ 2σ standard error) for δ^{25}Mg and ± 0.036‰ (2σ standard deviation, 0.007‰ 2σ standard error) for Δ^{26}Mg. Errors reported for delta values measured on the present olivines and silicate glasses are 2σ standard deviation of 3–5 different spots.

LA–MC–ICP–MS for Mg isotopic analysis

LA–MC–ICP–MS Mg isotopic analyses were carried out by coupling the 193 nm ultra-short ATLex 300si excimer laser system (≤4 ns pulse width) with Neptune MC–ICP–MS housed at Institut de Physique du Globe de Paris (IPGP) in France. Instrumental fractionation drift was corrected by bracketing each spot in a sample (olivines and glasses) with two spots (between and after) in San Carlos olivine. Each sample had three to five duplicates to monitor the repeatability. After the cease of ablation, intensity of Mg isotopes generally reached background values within 10 s, and the wash time was set to be 60–90 s. Analyses were conducted on 50 μm spots with laser energy output of 3.37–5.06 J/cm^2 and helium gas flow rates of 1.0–1.6 L/min. Each ablation spot was made of 200–400 laser bursts with repetition rates of 3–4 Hz. In order to compare matrix effects among different samples, same laser settings were used for the measurements in each batch (Table 1 in electronic annex). The typical ablation depth for an ablated spot was ca. 20 μm (Fig. 1). On MC–ICP–MS, the analyses were done under medium resolution mode ($M/\Delta M \approx 5000$ with ΔM calculated as the mass difference between 5% and 95% intensity of the ^{24}Mg peak on its low mass side) in order to get rid of molecular interferences such as $^{12}C^{14}N^+$ on $^{26}Mg^+$. Instead of staying on a narrow plateau (ΔM = ca. 0.004 AMU) to escape $^{48}Ca^{2+}$ interferences on $^{24}Mg^+$, peak jump was made for batch 2–4 to monitor intensity on mass 22 ($^{44}Ca^{2+}$). Data reduction was implemented by integrating all the cycles of individual ablation spots, except for the transient cycles at the beginning or the end of ablation. After background subtraction, an instrumental $^{44}Ca^{2+}/^{48}Ca^{2+}$ ratio of 9.6 was used to correct $^{48}Ca^{2+}$ interferences on $^{24}Mg^+$. Repeated analyses of the San Carlos olivine (bracketed against itself) show an external reproducibility of ± 0.22‰ (2SD, $n = 22$) for δ^{26}Mg and ±0.14‰ (2SD) for Δ^{26}Mg. Errors reported for delta values (relative to San Carlos olivine) measured on the present olivines and silicate glasses are 2σ standard errors of 3–5 different spots. To study Mg emissivity by laser ablation, analyses for multiple elements, including Mg, Al, Si, Ca and Fe, were conducted by coupling laser ablation under laser energy output of 2.77–6.00 J/cm^2 with Agilent quadrupole ICP–MS installed at IPGP (Table 7 in electronic annex).

Solution MC–ICP–MS for Mg isotopic analysis

In order to identify matrix effects due to ionization in the torch on MC–ICP–MS, composed solutions that were chemically the same as the glass and olivine standards were analyzed by MC–ICP–MS via Apex inlet system. The analyses were done under medium resolution mode. The sample uptake rate for Apex inlet system was fixed to be 50 μL/min, with which typical ^{24}Mg sensitivity on the IPGP Neptune MC–ICP–MS was c.a. 600 V/ppm. Each analysis of sample or standard contained 20 cycles (8 s per cycle) for data integration (two duplicates for each sample), and the wash time was set to be 3 min (Table 1 in electronic annex). Due to the high sample/blank Mg intensity ratios (ca. 400–2000), background interferences were negligible in our measurements. Mg isotopic compositions of the diluted solutions were measured on MC–ICP–MS by bracketing the mono-elemental Mg solution or the DSM3 solution with matched HNO_3 molarity and Mg concentration (Galy et al. 2003).

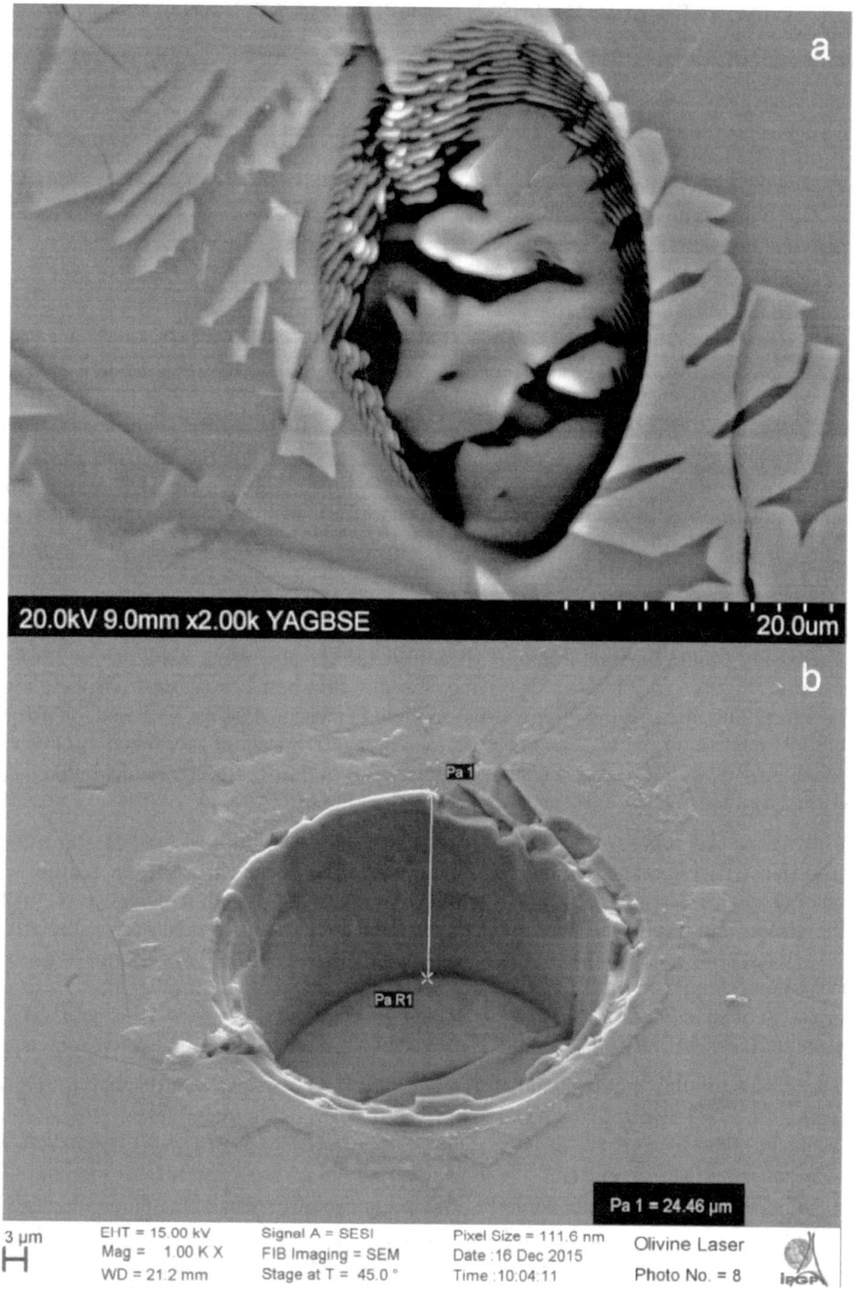

Figure 1. Secondary electron microscope images of typical spots made by SIMS (Fig. 1a) and LA–MC–ICP–MS (Fig. 1b). These two pictures illustrate the fundamental difference in rate of energy deposition in the sample during ion sputtering where preferential sputtering of mixed phases occurs (here preferential sputtering of the glassy matrix in which dendritic pyroxene is present) and laser ablation (here nanosecond laser) where local melting of the sample can occur due to the much higher rate of energy deposition. Note that local melting can be avoided with femtosecond laser ablation. The elongated shape of the crater for SIMS is due to the fact that the primary ion beam, which is circular, arrives on the sample at an angle of 45°.

MAGNESIUM ION EMISSION DURING IN SITU ANALYSIS

Fundamental differences for Mg ion yield between SIMS and laser ablation ICP–MS

Variations observed for Mg ion yields and for Mg ion yields relative to Si in the different silicate matrices studied (olivines, basaltic glasses, CMAS glasses) are shown in Figure 2a,b for MC–SIMS and in Figure 3a,b for LA-Q-ICP–MS (all data are available in the electronic annex). A constant yield would give a linear correlation in these diagrams. Comparison of Figure 2a,b with Figure 3a,b clearly shows that the systematics of production of Mg ions is very different between ion sputtering and laser ablation.

Mg ion yields by ion sputtering show: (i) strong differences between the three different types of standards studied, and (ii) variations for a given type of matrix. At a given MgO content, the $^{24}Mg^{+}$ intensity is the highest in olivine, lower for basaltic glasses and the lowest for CMAS glasses (Fig. 2a). These strong differences in ion yield between different matrices are not surprising for SIMS analysis of major elements (e.g., Shimizu and Hart 1982). For olivines, the increase in $^{24}Mg^{+}$ intensity versus MgO contents is not linear: it levels off with increasing MgO contents until approximately 39–40 wt% MgO (i.e., Fo$^{\#}\approx$75) where it starts to decrease abruptly to stay nearly constant for higher MgO contents. The exact same systematic (at the exception of constant Mg ion yield for very high Mg contents which were not investigated) was observed previously by Steele et al. (1981) who noted that Mg ion yield in olivine was increasing up to Fo$^{\#}\approx$80 and decreasing after. For basaltic glasses the increase of $^{24}Mg^{+}$ intensity versus MgO contents seems linear, but for CMAS glasses it is in fact not linear showing, at the opposite of olivines, a slight but significant increase of ion yield with increasing MgO contents. Relative Mg ion yields are also not constant (Fig. 2b) for a given matrix, this is very obvious for olivine, and different between two matrices at a given MgO content. The fact that the relative Mg ion yield is not constant for olivine demonstrates that the first control on the yield is not variations in sputtering yield as a function of olivine composition. If it were the case, a change by e.g., 20% of the sputtering yield would change the Mg and Si intensities by 20% and would not affect the Mg / Si ratio.

The Mg ion yields observed during laser ablation analysis of olivines and basaltic glasses (Fig. 3a) show (i) systematic variations in function of laser energy, the yield increasing with laser energy, and (ii) a simple common non-linear variation as a function of MgO contents for basaltic glasses and olivines. As obvious in Fig. 3b, a single Mg ion yield of 5.36 (±0.61 for 2 sd, ±0.10 for 2 se) relative to Si exists for all our data whatever the laser energy or the type of matrix or its composition. This is at odds with ion sputtering and implies that the in situ measurement of concentrations for a given element is much more simple by laser ablation than by ion sputtering since only one standard can be used (no matrix effects on relative ion yields).

The reason of this fundamental difference observed between laser ablation and ion sputtering is most likely due to the fact that the energy delivered to the sample is several orders of magnitude higher in the case of laser ablation, as also evidenced by the different types of craters produced on the sample (Fig. 1). Ion sputtering breaks chemical bonds at the nm level in the sample and produces ions while laser ablation produces small fragments (aerosols of ≤0.1–1 μm sizes; Guillong et al. 2003). Assuming that a primary ion beam of 20 nA (accelerated at 13 kV) is used for SIMS analysis with a 40×20 μm ellipse spot, and a laser fluence of 5 J / cm^2 for laser ablation with a pulse width of 3 ns, the power density loaded by laser ablation (1.66 GW / cm^2) will be approximately 4×10^7 times of that delivered by SIMS (41.4 W / cm^2). This difference could explain the distinctive systematics for Mg emissivity between laser ablation and primary ion beam sputtering, although the two processes are physically quite different. In the case of extremely high-energy load on the sample, effects on Mg ion yield of possible variations in structure and composition of the sample (e.g., bond energies, see in next

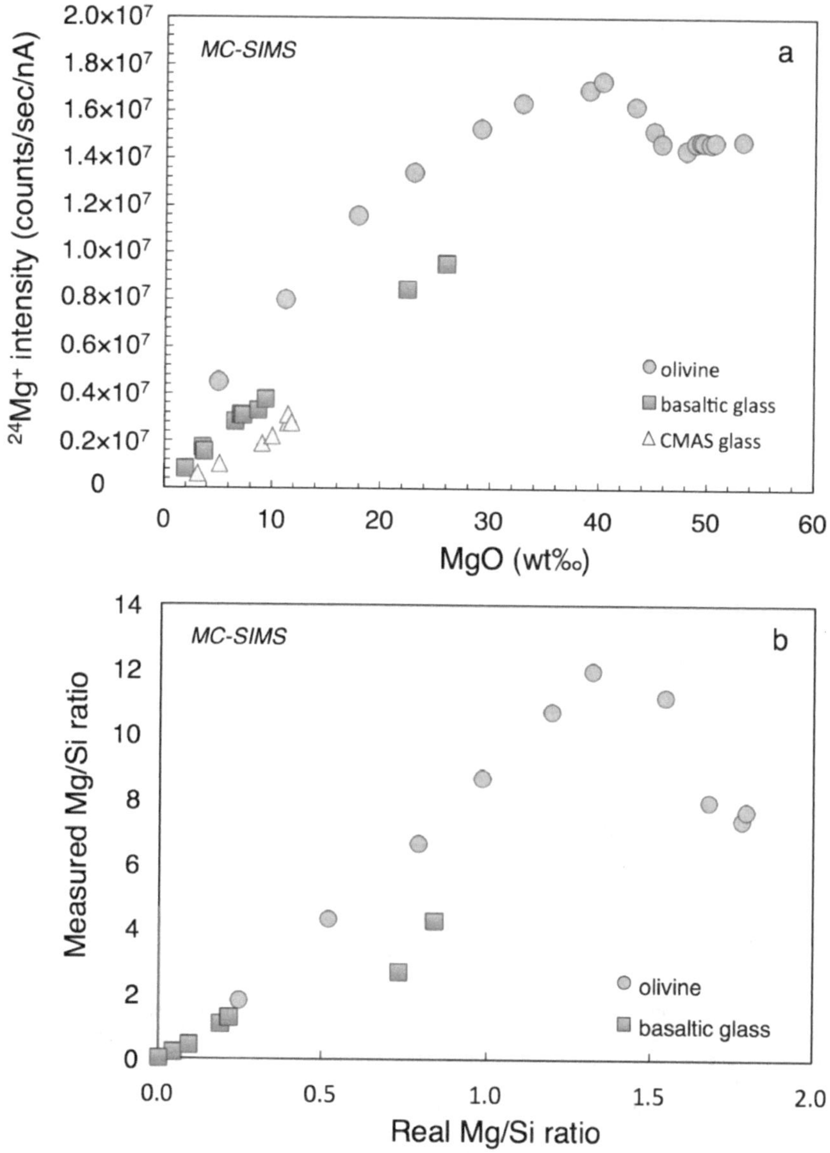

Figure 2. Variations of Mg ion yield ($^{24}Mg^+$ ion intensity versus MgO content in wt%, Fig. 2a) and Mg ion yield relative to Si (observed Mg/Si versus real Mg/Si, Fig. 2b) by MC–SIMS for olivines, basaltic glasses and CMAS glasses. (Data in Fig. 2b were taken in a different analytical session from data in Fig. 2a but their $^{24}Mg^+$ versus MgO content systematics is the same as shown in Fig. 2a). Note that the Mg ion yield is very different at a given MgO content between olivines, basaltic glasses and CMAS glasses and also shows strong non-linear variations among olivines depending on their forsterite content. The Mg ion yield relative to Si shows the same complex variations implying that the variations observed are not due to different sputtering yields between different matrices (if due to differences in sputtering yields these differences would cancel out when looking to the Mg/Si ratios).

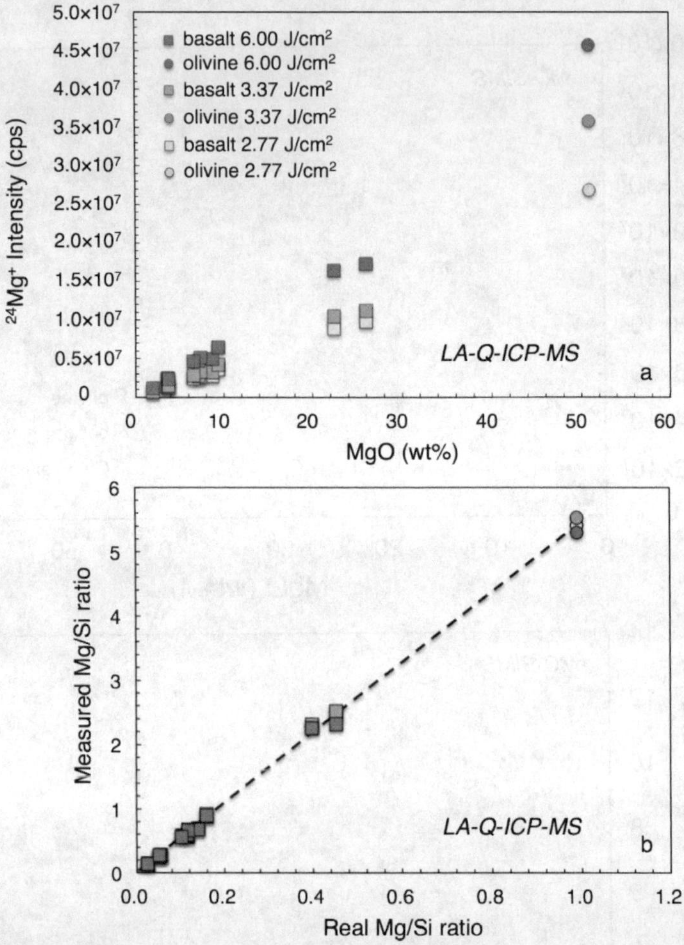

Figure 3. Variations of Mg ion yield ($^{24}Mg^+$ ion intensity versus MgO content in wt%, Fig. 3a) and Mg ion yield relative to Si (observed Mg/Si versus real Mg/Si, Fig. 3b) by LA-Q-ICP–MS for different ablation energies (2.77–6.00 J/cm^2). A quadrupole ICP–MS was used here instead of a magnetic sector ICP–MS to speed up the measurements since the variations observed are due to laser ablation and not to the type of mass spectrometer used after the ICP source. The Mg ion yield shows strong non-linear variations depending on ablation energy but for a given energy all matrices plot on the same curve. Whatever the matrix and the ablation energy, the Mg ion yields relative to Si plot on a single line demonstrating that, at variance with SIMS, all the variations in Mg ion yield are likely due to variations in sputtering yield.

section discussion of matrix effects on IMF) can be expected to be of second order. The small deviation from linearity of Mg ion intensity with MgO content observed for LA-ICP–MS (Fig. 3a) is likely due to the transient melting of the sample during ablation (melting can be significant when using a laser fluence of 10 J/cm^2, Horn and von Blanckenburg 2007), or to incomplete ionization of sputtered particles in the torch of the ICPMS (Guilong and Günther 2002). Analyses of solutions having chemical compositions similar to the compositions of the different standards studied (olivines and basaltic glasses) show a constant Mg ion yield (with a 2σ variability of ≈ 8% relative) for all compositions (Table 5 in electronic annex, Fig. 4). This suggests that, if variations in ionization exist in the torch of the ICP, they are primarily due to variations in the size of the aerosol particles and not to variations in its chemical composition.

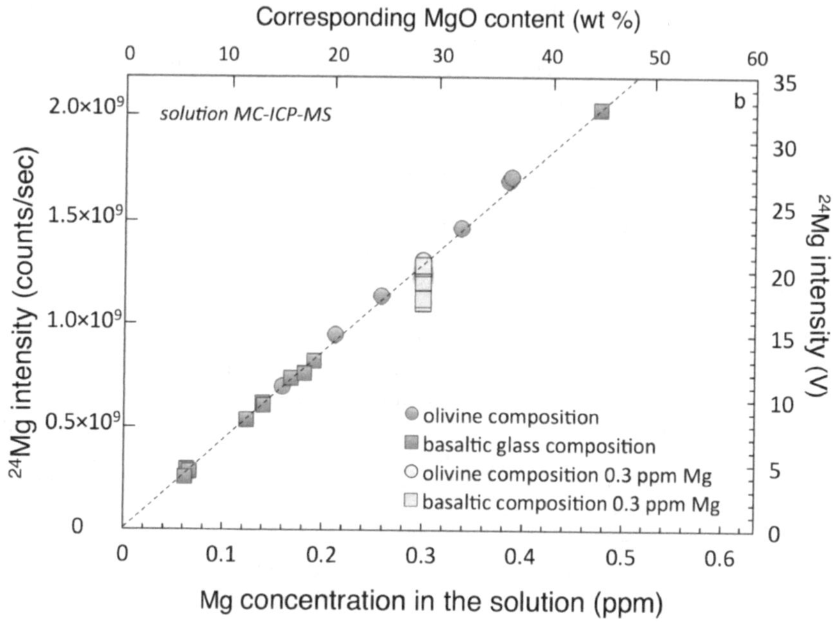

Figure 4. Mg ion yield (shown as $^{24}Mg^+$ intensity versus MgO contents) for solution MC–ICP–MS analysis of solutions having chemical compositions (see Table 5 in electronic annex) similar to that of the aerosols produced by laser ablation of olivines and basaltic glasses. Two sets of data are shown for olivine and basaltic compositions, either with fixed Si concentrations or with a constant Mg concentration (0.3 ppm). $^{24}Mg^+$ intensities are given in cps ($1V = 6.25 \times 10^7$ counts per second for a $10^{11}\Omega$ resistor on the amplifier of the Faraday cup). In order to facilitate the comparison between LA–MC–ICP–MS and solution MC–ICP–MS, the MgO contents (in wt%) corresponding to the Mg content of the solutions are also given. Note that all data imply the existence of a common value of Mg ion yield for all solutions, whatever their composition.

Possible origin of the complex matrix effects on ion yield for SIMS

The complex variations observed for Mg ion yields, and for Mg ion yields relative to Si, as a function of the MgO content of olivines (Fig. 2a, b) are very surprising since there is no chemical or structural property of olivine which shows a singularity at $Fo^{\#} \approx 75$ (Deer et al. 1982). As shown below, this value of $Fo^{\#} \approx 75$ also corresponds to a strong change in the instrumental isotopic fractionation. Thus, it is quite obvious that something fundamental is changing in the production of Mg ions at around $Fo^{\#} \approx 75$. This must be in some way related to the change in Fe content of olivine because it is in fact the only parameter (except the Mg content) that is changing. Several previous studies have found that the presence of Fe in clinopyroxene or in olivine was enhancing the production of Mg, Ca and Ni ions (Shimizu at al 1978; Reed et al. 1979; Steele et al. 1981).

Fe and Mg have structurally the same role in the olivine structure: they are both octahedral cations connecting the $(SiO_4)^{4-}$ tetrahedra by ionic bonds. One process that could be anticipated to take place during sputtering of olivine would be a sort of competition between Mg and Fe for ionization or bond breaking. Shimizu and Hart (1982) proposed that the binary system behavior of Mg and Fe in olivine could be treated like that of Cu–Ni binary alloy for which it was observed that the ratios of secondary intensities of Cu^+ and Ni^+ ions can be related to the Cu/Ni ratio of the alloy and to the ratio of ionization energies of Ni and Cu. The problem is obviously more complicated when the sample is not simply a binary component. Similarly to Shimizu and Hart (1982), Benninghoven et al. (1987) proposed a rule that could describe the variations in the production of ions of A and B from an alloy A–B. They proposed: "in a binary alloy A–B, when A forms a stronger oxide bond than B, the presence of B suppresses the ionization of A, while the presence of A enhances the ionization of B".

In the case of olivine, the difference in bond strengths between Mg and Fe in the olivine structure can be approximated to first order by the difference in enthalpies of atomization between the two elements. In fact, Fe has a much higher enthalpy of atomization than Mg (415 kJ / mole and 146 kJ / mole, respectively), so that breaking Fe bonds is more demanding energetically than breaking Mg bonds. This difference (4.3 eV for a Fe–O bond and 1.5 eV for a Mg–O bond) is small (0.1‰ level) compared to the energy of a primary ion (23 keV here) but during the sputtering process most of the ions emitted from the sample do not result from a direct collision between the primary ion and an atom in the target. The sputtering process is understood as a cascade process during which a primary ion looses all its energy and transfers it to the target via multiple collisions and recoil atoms in turn transfer part of their energy to other atoms in the target, eventually setting into motion out of the target one atom from the surface (Sigmund 1969; Williams 1979). The typical range of energy distribution for emitted secondary ions is of $\approx$ 150 eV, most of the ions being emitted with an energy band pass of $\approx$ 10 eV. It is thus possible that the difference in energy between Mg–O and Fe–O bonds plays a significant role in the yields of Mg and Fe ions from an olivine crystal. One simple idea would then be to consider, by analogy with the rules proposed for binary alloys, a qualitative rule saying that when the Fe content of olivine is high, the presence of Fe enhances the production of Mg ions. Reversely, when the Fe content of olivine is low, the lack of Fe suppresses the production of Mg ions.

In the following we tentatively try to parameterize this rule using enthalpies of atomization of Mg and Fe, to fit the observed variations of Mg ion yield in olivine. It is important to stress that this approach is just an experimental fit to the data, and does not rely on any theoretical consideration of ion production in olivine by sputtering. We define a parameter Ea that describes the relative variations of enthalpy of atomization for Mg and Fe in olivine with varying Fo content

$$\mathrm{Ea} = 1 - \frac{x_{\mathrm{Mg}} \times \mathrm{Ea}_{\mathrm{Mg}} - x_{\mathrm{Fe}} \times \mathrm{Ea}_{\mathrm{Fe}}}{x_{\mathrm{Mg}} \times \mathrm{Ea}_{\mathrm{Mg}} + x_{\mathrm{Fe}} \times \mathrm{Ea}_{\mathrm{Fe}}} \tag{9}$$

with, $\mathrm{Ea}_{\mathrm{Mg}}$ = 146 kJ / mole, $\mathrm{Ea}_{\mathrm{Fe}}$ = 415 kJ / mole, xMg and xFe the molar fractions of Mg and Fe in olivine. For forsterite Ea = 0, Ea = 1 when $x\mathrm{Mg} \times \mathrm{Ea}_{\mathrm{Mg}} = x\mathrm{Fe} \times \mathrm{Ea}_{\mathrm{Fe}}$ and Ea = 2 for fayalite. Strikingly, Ea turns from higher to smaller than 1 exactly at the value of $\mathrm{Fo}^{\#} = 74$ (Fig. 5) for which the ion yield of Mg is showing a singularity (Fig. 2a). Assuming that Mg ionization is function of Ea allows to fit the variation of Mg ion yield observed for $\mathrm{Fo}^{\#} < 74$ but a second term to the fit is required to explain that the Mg ion yield stabilizes for high $\mathrm{Fo}^{\#}$ when Ea is strongly decreasing. We consider simply that for this second term, the ion yield increases as a function of the ratio $(x\mathrm{Mg} \times \mathrm{Ea}_{\mathrm{Mg}}) / (x\mathrm{Fe} \times \mathrm{Ea}_{\mathrm{Fe}})$. In the following, these two components of the ionization are named "enhanced ionization" and "simple ionization". The ion yield of Mg can then be parameterized as the sum of these two processes according to:

$$^{24}\mathrm{Mg}^{+} = \mathrm{K}^{\mathrm{ol}}_{\mathrm{Mg}} \times [\mathrm{MgO}] \times \left(f^{\mathrm{olivine}}_{\mathrm{enhanced}} \times \left(1 - \frac{x_{\mathrm{Mg}} \times \mathrm{Ea}_{\mathrm{Mg}} - x_{\mathrm{Fe}} \times \mathrm{Ea}_{\mathrm{Fe}}}{x_{\mathrm{Mg}} \times \mathrm{Ea}_{\mathrm{Mg}} + x_{\mathrm{Fe}} \times \mathrm{Ea}_{\mathrm{Fe}}} \right) + \left(1 - f^{\mathrm{olivine}}_{\mathrm{enhanced}}\right) \times \frac{x_{\mathrm{Mg}} \times \mathrm{Ea}_{\mathrm{Mg}}}{x_{\mathrm{Fe}} \times \mathrm{Ea}_{\mathrm{Fe}} + k} \right) \tag{10}$$

with $^{24}\mathrm{Mg}^{+}$ the number of ^{24}Mg ions emitted per second and per nA of primary beam, $f^{\mathrm{olivine}}_{\mathrm{enhanced}}$ the relative strength of enhanced ionization and $(1 - f^{\mathrm{olivine}}_{\mathrm{enhanced}})$ the relative strength of simple ionization, [MgO] the MgO content, $\mathrm{K}^{\mathrm{ol}}_{\mathrm{Mg}}$ and k two constants. Fig. 6a is showing the best fit that can be obtained using equation (10) for Mg emissivity in olivine ($\mathrm{K}^{\mathrm{ol}}_{\mathrm{Mg}} = 4.4 \times 10^{5}$, $f^{\mathrm{olivine}}_{\mathrm{enhanced}} = 0.92$, $k = 2$, giving in average $^{24}\mathrm{Mg}^{+}_{\mathrm{modeled}} / {}^{24}\mathrm{Mg}^{+}_{\mathrm{observed}} = 0.99 \pm 0.08$ at 2σ). It appears

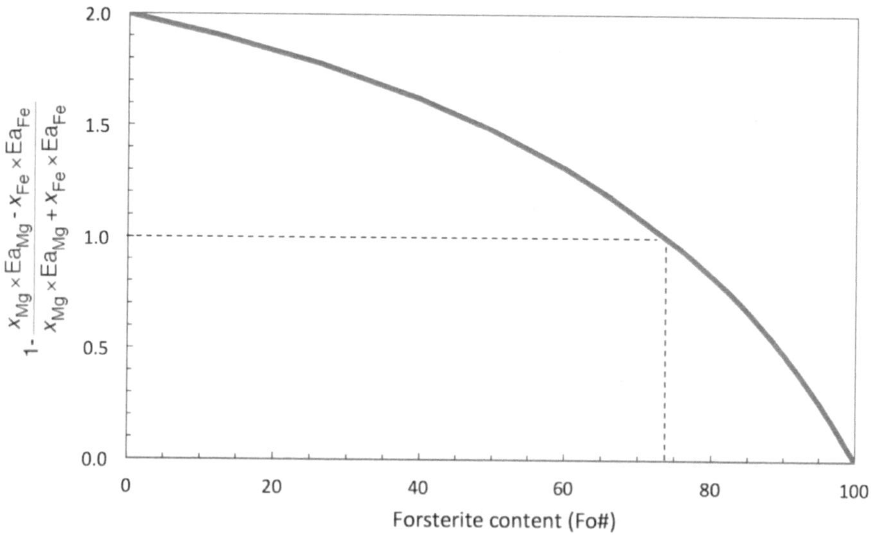

Figure 5. Relative change in enthalpies of atomization (parameter *Ea* along the *Y* axis, see text for definition) for Mg and Fe in olivine as a function of the forsterite content. Note that the enthalpy of atomization due to Fe is dominant over that due to Mg until $Fo^{\#}=74$.

that for glasses the relative strength of enhanced ionization and simple ionization must be different from that in olivine ($f^{glass}_{enhanced}=0.7$ instead of $f_{ol}=0.92$). The two coefficients $K^{basalticglass}_{glass}$ and $K^{CMAS\ glass}_{Mg}$ are also different from K^{ol}_{glass} in order to reproduce the differences in $^{24}Mg^{+}$ intensity between the three phases at the same MgO content ($K^{basalticglass}_{glass}=3.9\times10^{5}$ and $K^{CMAS\ glass}_{Mg}=2.0\times10^{5}$), and k is the same. These values also reproduce the observation that the Mg ionization yield is nearly constant whatever the MgO content in basaltic glasses (Fig. 6b) and that it is slightly increasing with MgO contents (concave shape of the curve in Fig. 6c) in CMAS glasses (for which $x_{Fe}=0$). The linear relationship between $^{24}Mg^{+}$ and [MgO] in basaltic glass is particularly misleading since, in the absence of the evidence from olivines, it would not be obvious to invoke two ionization processes to explain it. For basaltic glasses the model explains perfectly the data (for $K^{basalticglass}_{glass}=3.9\times10^{5}$, $f^{glass}_{enhanced}=0.70$, $k=2$, giving in average $^{24}Mg^{+}_{modeled}/^{24}Mg^{+}_{observed}=1.02\pm0.08$ at 2σ). For CMAS glasses, though the model reproduces the concave shape of the curve, the fit is not as good as for olivines and basaltic glasses (for $K^{CMAS\ glass}_{Mg}=2.0\times10^{5}$, $f^{glass}_{enhanced}=0.70$, $k=2$, giving in average $^{24}Mg^{+}_{modeled}/^{24}Mg^{+}_{observed}=0.94\pm0.29$ at 2σ). As shown below, the combination of these two ionization processes can fit the variations observed for matrix effect on instrumental Mg isotopic fractionation as a function of MgO content.

MAGNESIUM INSTRUMENTAL ISOTOPIC FRACTIONATION

Similarities and differences for Mg instrumental isotopic fractionation between SIMS and laser ablation ICP–MS

A comparison between MC–SIMS and LA–MC–ICP–MS is shown in Figure 7 for instrumental isotopic fractionation of the $^{26}Mg/^{24}Mg$ ratio ($\Delta^{26/24}_{inst}$) in the present set of olivines and basaltic glasses. All values of $\Delta^{26/24}_{inst}$ are normalized to the San Carlos olivine, by sample bracketing during the analyses for LA–MC–ICP–MS, and by external normalization during a

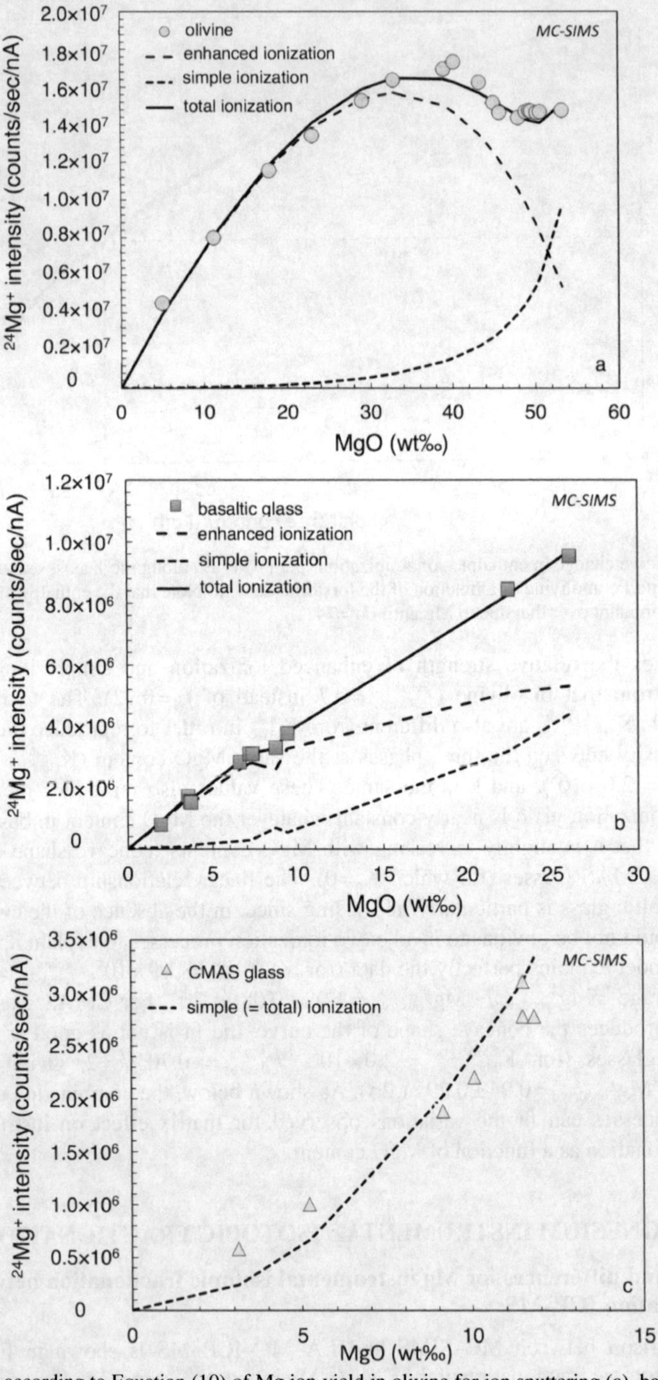

Figure 6. Fit according to Equation (10) of Mg ion yield in olivine for ion sputtering (a), basaltic glasses (b) and CMAS glasses (c). The parameters used in Equation (10) to fit the Mg ion yield of the three types of matrix are given in the text. The solid lines in (a) and (b) are the sum of the two types (enhanced and simple) of ionization. Because CMAS glasses are devoid of Fe, Mg ionization from CMAS glasses is just explained by the simple ionization.

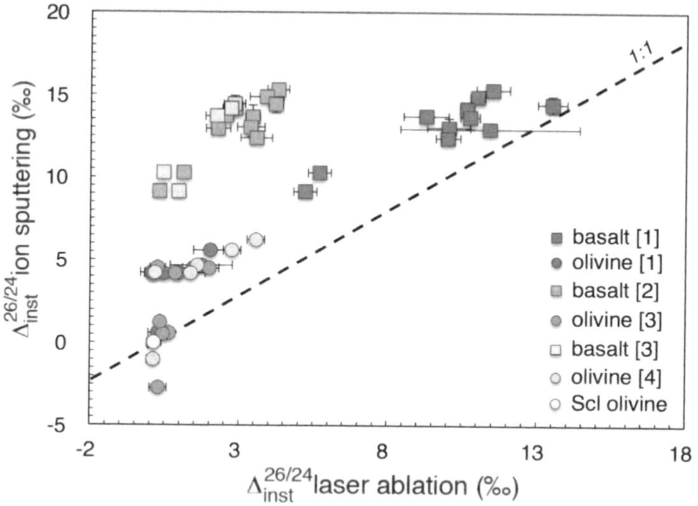

Figure 7. Comparison between the variations of instrumental isotopic fractionation ($\Delta^{26/24}_{inst}$) during ion sputtering (MC–SIMS, *Y*-axis) and during laser ablation (LA–MC–ICP–MS, X axis). $\Delta^{26/24}_{inst}$ is shown in per mil notation relative to San Carlos olivine for the two techniques. Symbols are the same than in Figure 3a. Values of $\Delta^{26/24}_{inst}$ show a similar range for the two techniques but a several ‰ offset, with the exception of the LA–MC–ICP–MS data obtained with the laser energy lower than 5 J/cm^2 and of the MC–SIMS data obtained for Mg-rich olivines.

given analytical session for MC–SIMS (San Carlos olivine is measured several times during a given session but there is no systematic sample bracketing with San Carlos olivine). Most values observed for $\Delta^{26/24}_{inst}$ are similar between MC–SIMS and LA–MC–ICP–MS: they show a similar range of ≈ 13‰ over the compositional range of the present olivines and basaltic glasses and an offset of a few per mil only.

The two exceptions are (i) olivines with $Fo^{\#}>74$ which show larger instrumental fractionation by MC–SIMS while no composition-related effects have been observed for LA–MC–ICP–MS, and (ii) LA–MC–ICP–MS data obtained on basaltic glasses with laser energies lower than 5 J/cm^2. The first of these two differences is not surprising since for olivines with $Fo^{\#}>74$ the suppression of Mg ionization due to the lack of Fe that is observed in MC–SIMS (see Figs. 2a,b and 6a) does not exist in LA–MC–ICP–MS. A decrease of ionization efficiency must increase instrumental isotopic fractionation, simply from mass balance considerations if ionization is taking place with some isotopic fractionation. However the same kind of reasoning does not apply to the lower $\Delta^{26/24}_{inst}$ observed for laser ablation with laser energy lower than 5 J/cm^2 that results in lower ion yields of Mg (see Fig. 3a).

The striking similarity observed for matrix effects on Mg instrumental isotopic fractionation between MC–SIMS and LA–MC–ICP–MS is however surprising since ionization is taking place in different conditions in the two techniques. For SIMS, ionization takes place during electron transfer in the solid target undergoing ion bombardment. For LA-ICP, the solid particles produced by laser ablation are evaporated in the torch of the ICP and the produced gases are ionized through inductive heating under high temperature.

Matrix effects during ionization of solutions in MC–ICP–MS

In order to try to identify matrix effects on Mg isotopic instrumental fractionation in a gas of variable chemical composition we prepared solutions having chemical compositions similar to that of the olivines and basaltic glasses. The solutions were analyzed using the starting solution of pure Mg as the bracketing solution. Thus the δ^{25}Mg and δ^{26}Mg given in the annex (Table 5 in electronic annex) are directly the values of $\Delta^{25/24}_{\text{inst}}$ and $\Delta^{26/24}_{\text{inst}}$ for this type of analyses. Note that because the δ^{25}Mg and δ^{26}Mg values fall on a mass fractionation line (see next section), their variations are not due to the presence of interferences. Thus matrix effects are clear for $\Delta^{26/24}_{\text{inst}}$ (or $\Delta^{25/24}_{\text{inst}}$) with however a small range of variation of up to ≈ 1‰ over the whole compositional range. These matrix effects show a clear systematic with a parameter (Ie) that we defined to describe the relative variations of the ionization energy of Mg due to changes in matrix composition (Fig. 8):

$$\text{Ie} = \frac{x_{\text{Mg}} \times Ie_{\text{Mg}}}{\sum x_i \times Ie_i} \tag{11}$$

with x_i the molar fractions of element i and Ie_i its ionization energy in eV (6 eV for Al, 7.7 eV for Mg, 3.2 eV for Si, 6.1 eV for Ca, 7.9 eV for Fe and 5.1 eV for Na). Note that significant matrix effects were also previously observed (An et al. 2014) for Mg solutions doped with various elements (Na, Al, K, Ca, Rb, Cs) under medium mass-resolution mode and that they show the same systematic with Ie as the present samples (Fig. 8). The regression line calculated from the present data set (Table 5 in electronic annex) is: $\Delta^{26/24}_{\text{inst}} = 0.258\,\text{Ln}(Ie)$. Using this relationship the matrix effect on $\Delta^{26/24}_{\text{inst}}$ can be predicted with an uncertainty of ± 0.12‰ (2σ) for a given chemical composition in the range studied here. In addition, IMF appears to be independent, to first order, of the Mg content of the solution. A physical modeling of this relationship between $\Delta^{26/24}_{\text{inst}}$ and Ie is beyond the scope of this work but two simple hypotheses can be proposed. One possibility is that the ionization of Mg is slightly modified by the presence of other elements that ionize preferentially, but this is a priori not suggested by the constant relative ion yield (at the percent level) observed for Mg in Figure 3b. Another possibility is that the presence of other elements in the ion beam modifies the spatial distribution of ^{24}Mg, ^{25}Mg and ^{26}Mg ions in the plasma (Barling

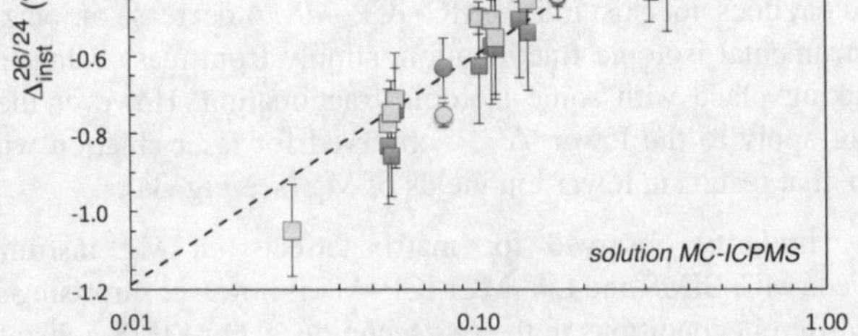

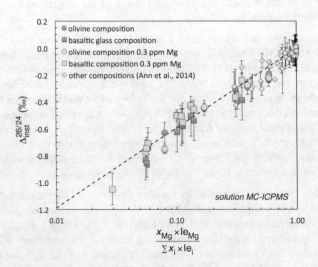

Figure 8. Matrix effects for ^{26}Mg/^{24}Mg ($\Delta^{26/24}_{\text{inst}}$) for solution MC-ICPMS. Data for other solutions with very different compositions (An et al. 2014) are shown as yellow diamonds. Variations of $\Delta^{26/24}_{\text{inst}}$ appear linearly correlated with ln(*Ie*), Ie being a parameter describing the relative variations of the ionization energy of Mg due to changes in matrix composition (see text). Using the regression line (dashed line with a slope of 0.258) allows determination of $\Delta^{26/24}_{\text{inst}}$ at ± 0.12‰ (2σ) for any solution in the compositional range investigated here.

and Weis 2012), thus inducing isotopic fractionations for Mg transmitted in the cones of the MC–ICP–MS. Even if ionization and thus IMF could be slightly different in a wet plasma compared to a dry plasma (Chan et al. 2001), the IMF observed for solutions show that the major component of matrix effects observed for LA–MC–ICP–MS is specific to ablated aerosol particles.

Matrix effects specific to in situ analysis

There is no simple experimental law that could describe perfectly the matrix effects observed on IMF during MC–SIMS and LA–MC–ICP–MS. In both cases $\Delta^{26/24}_{\text{inst}}$ are strongly correlated with the MgO content (Fig. 9a,b) but olivines and basaltic glasses follow different trends. These trends are more linear in the case of LA–MC–ICP–MS and would intersect at a MgO content of ≈50 wt%, a MgO content at which olivine and a basaltic glass would have the same composition, suggesting minor effects of structure on $\Delta^{26/24}_{\text{inst}}$. Because $\Delta^{26/24}_{\text{inst}}$ observed during ionization of solution by ICP depends on the chemistry of the solution, there are also clear correlations between the values of $\Delta^{26/24}_{\text{inst}}$ observed during solution-MC–ICP–MS and LA–MC–ICP–MS (Fig. 10).

Following previous studies of instrumental isotopic fractionation during LA–MC–ICP–MS (see refs in previous section), it can be anticipated that the strong change of $\Delta^{26/24}_{\text{inst}}$ with laser energy most likely results from a change in aerosol mean size: smaller aerosol particles

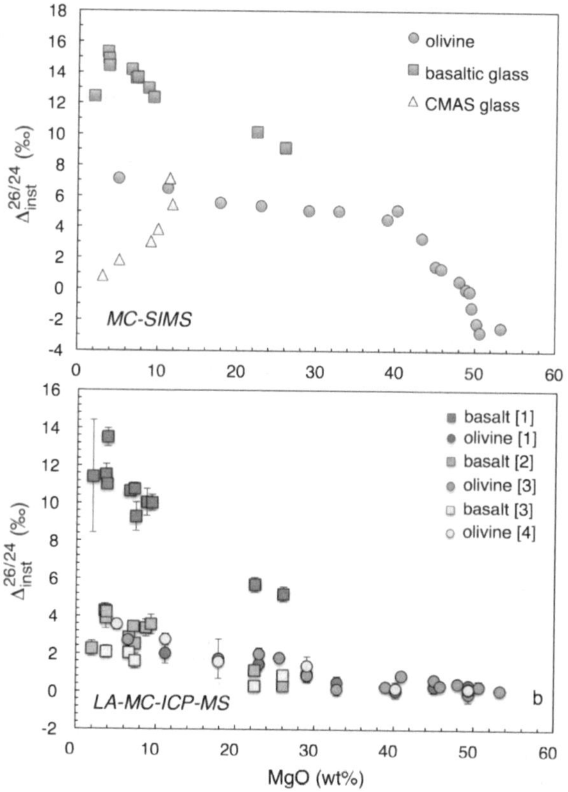

Figure 9. matrix effects observed on IMF during MC–SIMS and LA–MC–ICP–MS. In both cases, $\Delta^{26/24}_{\text{inst}}$ are correlated to the MgO content (a,b) but olivines and basaltic glasses follow different trends. These trends are more linear in the case of LA–MC–ICP–MS.

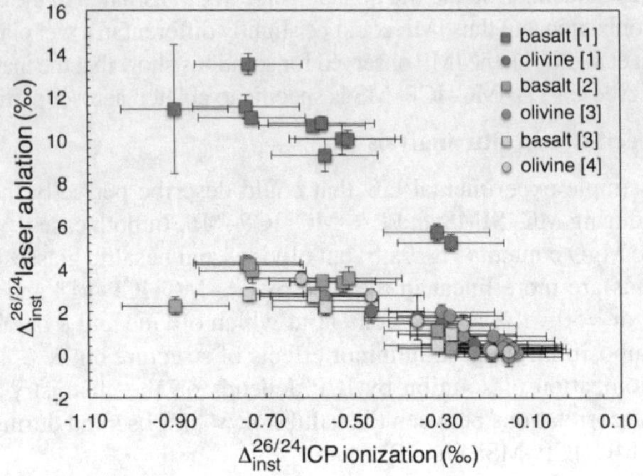

Figure 10. Comparison between the values of $\Delta^{26/24}_{inst}$ observed during solution-MC–ICP–MS and LA–MC–ICP–MS. The two sets of values of $\Delta^{26/24}_{inst}$ are correlated to the chemical composition and thus correlated together. The effects of laser energy on $\Delta^{26/24}_{inst}$ are strong on basaltic glasses (see text)

produced with optimized laser energy can be ionized more efficiently than larger particles ejected by melting at too high laser energy. For a Mg diffusion coefficient D of 10^{-15} m^2/s, a diffusion profile of 50 nm can be formed within ca. 0.025 s $\left(x = 4\sqrt{D \times t}\right)$. This timescale is comparable to the travelling time of ablated particles in plasma sources (ca. 0.02 s for a helium gas flow rate of 1 L/min), implying that, before total evaporation, the particles can undergo a partial diffusion loss of Mg and associated isotopic fractionation. Incomplete evaporation of the large particles fractionated in Mg by diffusion will induce drifts of the measured isotopic ratios. Note that this is not contradictory with the lower Mg ion yield observed at lower energy: this is simply due to higher ablation rate at higher energy (see Fig. 3).

Because the variations in Mg ion yield in olivines by MC–SIMS suggest the existence of two different processes that concur to the production of Mg^+ ions (simple and Fe-enhanced ionization, see Figs. 2, 6), it seems logical to consider the possibility that these two processes have different IMF and different IMF laws. Because CMAS glasses are devoid of Fe, Mg ion yields from CMAS matrix are controlled only by the simple ionization. Values of $\Delta^{26/24}_{inst}$ in CMAS glasses appear to be linearly correlated with optical basicity (Fig. 11). Optical basicity (Duffy 1993) is a parameter that reflects variations of the chemical bonding between Lewis acid-base pairs in, for instance, a silicate framework. It has been shown to describe variations in the chemical composition of a silicate and to be related to various physical properties. In the following, we will use optical basicity (Δ) to describe the variations in the chemical environment of Mg atoms in the different matrix studied. Δ is defined by:

$$\Delta = \frac{\sum_i x_i \times n_i \times \Delta_i}{\sum_i x_i \times n_i} \qquad (12)$$

with x_i the molar fraction of oxide i, n_i the number of oxygen atoms in oxide i and Δ_i the optical basicity of oxide i. We use the following values of optical basicity (Duffy 2004): $\Delta_{SiO_2} = 0.48$, $\Delta_{Al_2O_3} = 0.6$, $\Delta_{MgO} = 0.78$, $\Delta_{CaO} = 1.0$, $\Delta_{Na_2O} = 1.15$, and $\Delta_{FeO} = 1.0$. The linear relationship observed in Fig. 11 between $\Delta^{25/24}_{inst}$ and $\Delta^{26/24}_{inst}$ and optical basicity for CMAS glasses can be tentatively taken to be the law of matrix effects for IMF during the process of simple ionization.

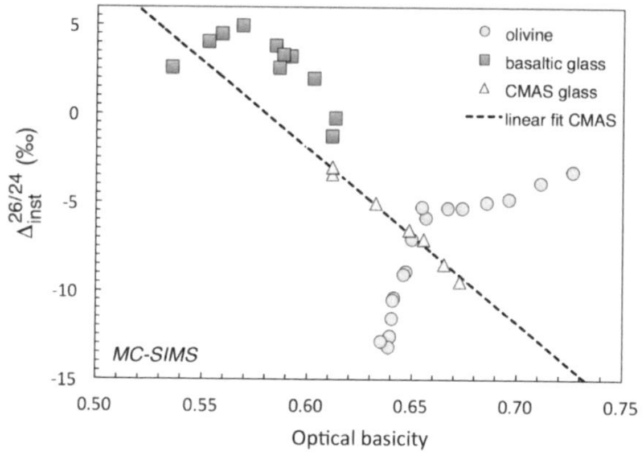

Figure 11. Instrumental isotopic fractionation for $^{26}Mg/^{24}Mg$ ($\Delta^{26/24}_{inst}$) by MC–SIMS for olivine, basaltic glasses and CMAS glasses as a function of the optical basicity. Note that the CMAS glasses which are Fe-free define a linear trend while basaltic glasses and olivines show systematic changes which are not linear. Note the kink in the trend shown by olivine at optical basicity Λ=0.66 which corresponds to $Fo^{\#}=74$.

Note that this linear law intersects the variations of $\Delta^{26/24}_{inst}$ in olivine at a value for optical basicity of ≈0.66, corresponding exactly to $Fo^{\#}=74$. For higher values of optical basicity (i.e., $Fo^{\#}<74$) $\Delta^{26/24}_{inst}$ in olivines show a very different systematics than for lower values of optical basicity ($Fo^{\#}74$) where they go to very negative values. This change corresponds exactly to the change in Mg emissivity in olivines, enhanced ionization becoming dominant for $Fo^{\#}<74$. Similarly to olivines, basaltic glasses dominated by enhanced Mg ionization show $\Delta^{26/24}_{inst}$ above the line defined by the CMAS glasses.

The fractions of Mg ions produced by the enhanced ionization (noted hereafter $f^{Mg}_{\text{enchanced ionization}}$) and the simple ionization ($f^{Mg}_{\text{simple ionization}} = 1 - f^{Mg}_{\text{enchanced ionization}}$) can be calculated from Equation (10). From mass balance considerations the IMF for enhanced ionization ($\Delta^{26/24}_{\text{inst-enhanced ionization}}$) and for simple ionization ($\Delta^{26/24}_{\text{inst-simple ionization}}$)must obey:

$$\Delta^{26/24}_{inst} = f^{Mg}_{\text{enhanced ionization}} \times \Delta^{26/24}_{\text{inst-enhanced ionization}} + \left(1 - f^{Mg}_{\text{enhanced ionization}}\right) \times \Delta^{26/24}_{\text{inst-simple ionization}} \quad (13)$$

Assuming that the values of $\Delta^{26/24}_{\text{inst-simple ionization}}$ for olivine and basaltic glasses are given by the law determined by CMAS glasses, the values of $\Delta^{26/24}_{\text{inst-enhanced ionization}}$ can be calculated for olivines and basaltic glasses from equation (13). At first order the values of $\Delta^{26/24}_{\text{inst-enhanced ionization}}$ appear to vary linearly with the fraction of Mg emitted by the enhanced emission (Fig. 12). It is striking to see that the value of $\Delta^{26/24}_{\text{inst-enhanced ionization}}$ tends towards ≈−36‰ when the fraction of Mg ions produced by enhanced ionization approaches zero. This limit for $\Delta^{26/24}_{\text{inst-enhanced ionization}}$ is closed to the kinetic Mg isotopic fractionation (= −39.3‰ for $(24/26)^{0.5}$; e.g., Davis et al. 1990) which could be expected for Mg ions kinetically fractionated during their extraction from the matrix olivine.

All these considerations are still purely experiment-based and no model can fit all the observations yet, but the clear systematics observed, even though complex, allows a precise experimental calibration using appropriate standards to monitor matrix effects on IMF for in situ Mg isotopic analysis.

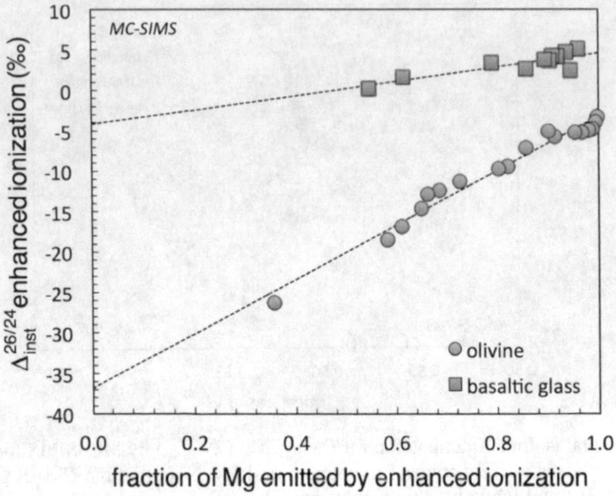

Figure 12. SIMS instrumental isotopic fractionation for $^{26}Mg/^{24}Mg$ ($\Delta^{26/24}_{inst}$) due to the enhanced emission process: versus the fraction of Mg emitted by the enhanced emission. Note that (i) $\Delta^{26/24}_{inst}$ is different between olivine and basaltic glasses, being always more negative for olivine, and that (ii) $\Delta^{26/24}_{inst}$ appears to vary linearly with the fraction of Mg emitted by the enhanced process, both for olivines and basaltic glasses.

MEASUREMENT OF THE THREE MAGNESIUM ISOTOPES

The need for high-precision in situ three Mg isotopes analysis in cosmochemistry

Recent developments towards high-precision and high spatial-resolution Mg isotopic analysis by LA–MC–ICP–MS (Young ct al. 2002), solution MC–ICP–MS (Bizzarro et al. 2011; Jacobsen et al. 2008) and Multiple Collector Secondary Ion Mass Spectrometry (MC–SIMS) (Luu et al. 2013; Villeneuve et al. 2009) have led to significant advances in dating very early solar system events. The analysis of bulk Ca-, Al-rich inclusions (CAIs) allowed to determine an initial Mg and Al isotopic composition ($^{26}Al/^{27}Al_0 = 5.25 \times 10^{-5}$ and $\delta^{26}Mg^*_0$=-0.034 ± 0.032‰) for the Solar system (Jacobsen et al. 2008; Larsen et al. 2011) with a much higher precision than from previous data acquired with mono-collector SIMS (see review by MacPherson et al. 1995). These new data of unprecedented precision made it possible to look for the scale of heterogeneity of Al and Mg isotopes in the early accretion disk (Villeneuve et al. 2009; Larsen et al. 2011; Mishra and Chaussidon 2014) and for the implications on the dynamic of the disk, the origin of ^{26}Al and the significance of ^{26}Al relative ages (Liu and McKeegan 2009; Larsen et al. 2011; MacPherson et al. 2012; Wasserburg et al. 2012; Holst et al. 2013).

The major results obtained for CAIs and chondrules, which are key components of chondritic meteorites, are the following. (i) CAIs formed over a very short time interval of ≈ 10,000–40,000 years (Thrane et al. 2006; Jacobsen et al. 2008; Larsen et al. 2011) from refractory precursors formed at the same time (Larsen et al. 2011) or at maximum 100,000 years before (Mishra and Chaussidon 2014). (ii) Most CAIs experienced a complex history in the accretion disk with episodes of reheating (or remelting) over a period of ≈ 300,000years after their formation, a few of them being reheated much later, up to ≈ 2 My after their formation (MacPherson et al. 2012; Kita et al. 2013; Mishra and Chaussidon 2014). (iii) Most chondrules have ^{26}Al relative formation ages that extend from ≈ 1 to ≈ 3 Myr after CAIs (Villeneuve et al. 2009; Kita et al. 2013) but their precursors formed over an interval from ≈ 0 to ≈ 1.5 Myr after CAIs (Villeneuve et al. 2009; Luu et al. 2015).

All these results open new windows into the understanding of the first few million years of the Solar system but they also open new questions. Evidence exist for a level of ^{26}Al heterogeneity in the disk (Larsen et al. 2011; Makide et al. 2011; Holst et al. 2013) but observations also exist to show that for some objects ^{26}Al chronology is consistent with other radioactive systems like ^{182}Hf–^{182}W (Kruijer et al. 2014), or that heterogeneity is limited to less than 10% relative for ^{26}Al/^{27}Al in most chondrules and CAIs (Villeneuve et al. 2009; Mishra and Chaussidon 2014). Absolute U/Pb ages of single chondrules are apparently older in average than ^{26}Al relative ages (Connelly et al. 2012). There is as yet no simple model of the accretion disk and of the formation of the first planetesimals that would be able to reconcile all these observations. Making progress on these key questions implies being able to get high-precision Mg isotopic compositions at different spatial scales in meteoritic matter, combining information obtained by bulk analysis and in situ analyses, as exemplified in Luu et al. (2015).

The question of potential isobaric interferences

The situation for the resolution of potential isobaric interferences is very different between MC–SIMS and LA–MC–ICP–MS because of the different mass resolutions used in routine mode on the two types of mass spectrometers. The first difficulty comes from the fact that different definitions exist for mass resolution on the two instruments. Mass resolution ($M/\Delta M$) is defined as the difference of mass (ΔM) that can be resolved at a given mass (M).

For MC–SIMS the mass resolution of the instrument ($M/\Delta M_{\text{SIMS}}$) is defined from the mass spectrum as the width (ΔM) of a peak at mass M at 10% of its height. Thus, if the instrument mass resolution is $M/\Delta M_{\text{SIMS}}$, two peaks having a difference of mass of ΔM are resolved. With this definition, the lowest resolution possible on the multicollector ims 1270–1280 series ion probe is $M/\Delta M_{\text{SIMS}} = 2500$. Potential isobaric interferences of $^{48}Ti^{2+}$ ($M/\Delta M = 2167$ relative to $^{24}Mg^{+}$) or of $^{48}Ca^{2+}$ ($M/\Delta M = 2731$ relative to $^{24}Mg^{+}$) are resolved at low mass resolution on MC–SIMS while the interference of $^{24}MgH^{+}$ on $^{25}Mg^{+}$ ($M/\Delta M = 3559$) is not. There is thus no significant isobaric interference for anhydrous samples, and high vacuum (better than 3×10^{-9} Torr) is required when traces of water are present in the sample (Luu et al. 2013).

For MC–ICP–MS instruments, $M/\Delta M_{\text{MC-ICP}}$ is defined as the mass interval between 5% and 95% of maximal height on the shoulder of a peak in the mass spectrum. On Neptune-series MC–ICP–MS, low, medium and high mass resolution modes corresponding to $M/\Delta M_{\text{MC-ICP}}$ of ≈ 3000, ≈ 5000 and ≈ 8000, respectively, are available. The width of the plateau on peak at mass 26 corresponding only to $^{26}Mg^{+}$ions (without $^{12}C^{14}N^{+}$ ions, $M/\Delta M = 1269$ between $^{12}C^{14}N^{+}$ and $^{26}Mg^{+}$) is ≈ 0.015 AMU under medium resolution mode. Under the same resolution mode, the plateau at mass 24 free of $^{48}Ca^{2+}$ interference on $^{24}Mg^{+}$ is only ≈ 0.004 AMU. If switching to high mass resolution ($M/\Delta M_{\text{MC-ICP}} \approx 8000$), the plateau can be ≈ 0.006 AMU, but such a tuning will sacrifice sensitivity on the spectrometer.

A possible way to correct interferences from $^{48}Ca^{2+}$ during LA–MC–ICP–MS is to determine the $^{48}Ca^{2+}$ intensity from monitoring mass 22 ($^{44}Ca^{2+}$) by peak jumping from the masses of the Mg isotopes (Young et al. 2002). Another potential solution is to depress the formation of $^{48}Ca^{2+}$ by introducing solutions together with the ablated aerosols into the plasma source (Young et al. 2002; Janney et al. 2011). Rather than monitoring doubly charged Ca ions directly, Young et al. (2005) used terrestrial samples as calibration standards to correct interferences from $^{48}Ca^{2+}$ (the correction was of 0.04‰ at mass 24 per unit of Ca/Mg under their analytical conditions). In our study, wet plasma can significantly reduce secondary ionization of Ca, but it cannot totally eliminate the formation of Ca^{2+} (e.g., a correction of c.a. 0.3‰ per unit of Ca/Mg on $\Delta^{26}Mg^{*}$ was still necessary). For dry plasma, the Ca^{2+} effect can be stronger (c.a. 0.6‰ correction per unit Ca/Mg on $\Delta^{26}Mg^{*}$). Long-term analyses imply that the secondary ionization rate of Ca really depends on the tuning conditions, and, at least for MC–ICP–MS housed at IPGP, the formation of Ca^{2+} differs among different analytical sessions. This means that it is necessary to monitor Ca^{2+} directly together with Mg isotopes. Improper correction of $^{48}Ca^{2+}$ interferences can lead to significant analytical artifacts on Mg isotope data, especially the ^{26}Mg excesses for silicate glasses (up to 0.63‰ shift on $\Delta^{26}Mg^*$ without correction of Ca^{2+} effect). In addition, because ca. 100–200‰ instrumental isotopic fractionations take place during MC–ICP–MS, the natural Ca isotopic ratios are not appropriate for correction of Ca isobaric interferences, and an instrumentally fractionated Ca isotopic ratio has to be determined. In our study, we determined an instrumental $^{44}Ca^{2+}/^{48}Ca^{2+}$ ratio of 9.6 by adjusting the Mg isotope data of silicate glasses to plot along the same mass fractionation line defined by olivines (Fig. 13). It appears that this $^{44}Ca^{2+}/^{48}Ca^{2+}$ ratio of 9.6 is fractionated relative to the natural $^{44}Ca/^{48}Ca$ ratio by a factor of ≈ 1.7 according to $^{44}Ca^{2+}/^{48}Ca^{2+} = {}^{44}Ca/^{48}Ca \times (44/48)^{1.7}$. This fractionation factor of 1.7 may differ among different laboratories and different analytical conditions.

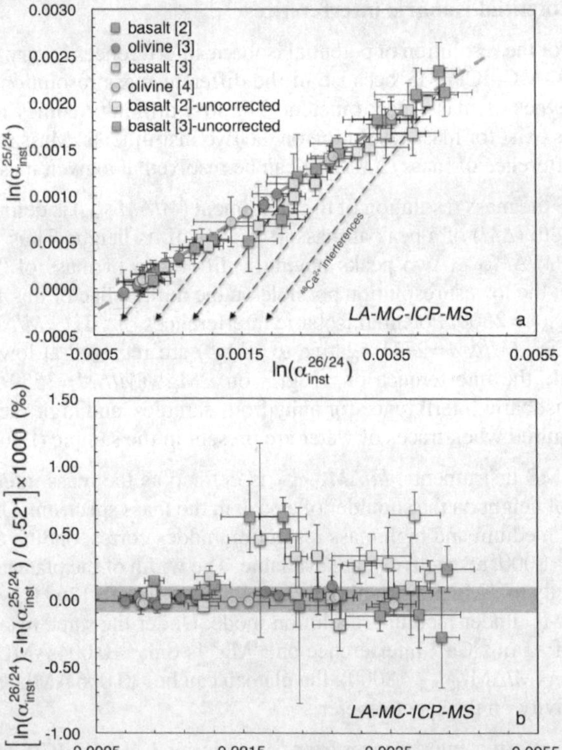

Figure 13. Correction of $^{48}Ca^{2+}$ interference at mass 24: three Mg isotope diagram (a) and ^{26}Mg excesses relative to a mass fractionation law (Eqn. 14) with an exponent $\beta = 0.521$. Symbols for olivine and basaltic glasses are the same as in previous diagrams, the orange and yellow squares corresponding to data without correction of $^{48}Ca^{2+}$ interference. Note that this interference is significant only for basaltic glasses. Because of the magnitude of the correction $\Delta^{26}Mg$ values on Mg-poor basaltic glasses cannot be determined at better than ≈ ±0.1‰.

The question of the mass fractionation law used to correct for instrumental isotopic fractionation

It is perfectly possible de determine ^{26}Mg excesses ($\Delta^{26}Mg$ values) to ppm level precision, without correcting for matrix effect on IMF, and thus without adding the error due to the correction of the matrix effect. This is however not a trivial task because $\Delta^{26}Mg$ values must be calculated relative to the correct mass fractionation law which is responsible of the variations of δ^{25}Mg and of the non-radiogenic variations of δ^{26}Mg (Davis et al. 2005; Villeneuve et al. 2009; Wasserburg et al. 2012; Luu et al. 2013). Several laws have been proposed to describe at best mass dependent Mg isotopic fractionations, initially based on observations made for Ca isotopic fractionation during thermal ionization mass spectrometry (Russel et al. 1978). These different laws (the power law, the Rayleigh law or the exponential law) can be expressed for Mg isotopes according to same formalism (Esat 1984; see also Dauphas and Shauble 2016):

$$\alpha^{25/24} = \left(\alpha^{26/24}\right)^{\beta} \tag{14}$$

with β the exponent defining the mass fractionation law (β=0.501, 0.506, 0.511 for the power law, Rayleigh law and exponential law, respectively). It is considered that β can vary significantly depending on the type of process at the origin of isotopic fractionation, from $\approx$0.511 for kinetic processes to 0.521 for equilibrium processes (Young et al. 2002; Young and Galy 2004). Experimental studies of the vacuum evaporation of CMAS liquids yielded a value of $\beta\approx0.5128$ (Davis et al. 2015) showing that the value of β must be determined and cannot be predicted theoretically to high precision for a given process. This uncertainty on the nature of the processes at the origin of Mg isotopic fractionations in various chondritic components such as CAIs and chondrules, introduces a significant error on the ^{26}Mg excesses (and by consequence the intercept and slopes of the ^{26}Al isochrons) that can be calculated from the ^{25}Mg/^{24}Mg and ^{26}Mg/^{24}Mg ratios (Davis et al. 2005, 2015; Wasserburg et al. 2012, Luu et al. 2013).

A similar difficulty exists for the correction of the ^{25}Mg/^{24}Mg and ^{26}Mg/^{24}Mg ratios for IMF during MC–SIMS and LA–MC–ICP–MS analyses. It is clear from previous sections that in situ Mg isotopic measurements are associated with large instrumental isotopic fractionations ($\alpha_{inst}^{25/24}$ and $\alpha_{inst}^{26/24}$). The lack of isobaric interferences during the measurement of the three Mg isotopes for MC–SIMS is demonstrated by the fact that the rough ^{25}Mg/^{24}Mg and ^{26}Mg/^{24}Mg ratios (before correction for instrumental isotopic fractionation) follow to first order a mass fractionation line (the reason why it is called instrumental mass fractionation or IMF). However, to second order, the instrumental fractionation law is not exactly of the form given in Equation (14): it is a common observation for high-precision MC–SIMS studies to have an instrumental fractionation law of the form:

$$\ln\left(\alpha_{inst}^{25/24}\right) = \beta \times \ln\left(\alpha_{inst}^{26/24}\right) + b \tag{15}$$

with an intercept which can be positive or negative and a slope β which is very often in between 0.511 and 0.521 (Villeneuve et al. 2009; Luu et al. 2013), but can be lower. This instrumental fractionation law must be determined with high precision if precisions at the level of a few ppm are targeted for the measurement of the ^{26}Mg excesses. No explanation has yet been given to the existence of an instrumental fractionation law of the form given in Equation (15). It was tentatively suggested that this could be related to an improper inter-calibration of the gains of the different Faraday cup systems used in the multi-collection. We show below that this linear law between ln $\alpha_{inst}^{25/24}$ and ln $\alpha_{inst}^{26/24}$ is in fact the natural consequence of the existence of two different instrumental mass fractionation processes for MC–SIMS (enhanced and simple ionization as shown above) and for laser ablation MC–ICP–MS (ionization in the torch and possibly evaporation during sputtering or in the torch).

Mg instrumental mass fractionation law for MC–SIMS analyses

We consider in the following that Mg is ionized during sputtering of the sample by two processes (1 and 2), the fraction of ^{24}Mg emitted by process 1 being f_1. We further consider that each process is a mass fractionation process following a law of the form given by Equation (14) with exponent b_1 and b_2 for processes 1 and 2, respectively, and that the true isotopic composition of the sample is $(^{25}Mg/^{24}Mg)_0$ and $(^{26}Mg/^{24}Mg)_0$. This can be schematized as:

$$\left(\frac{^{25}Mg}{^{24}Mg}\right)_0 \xleftrightarrow{\alpha_1^{25/24}} \left(\frac{^{25}Mg}{^{24}Mg}\right)_1 \text{and} \left(\frac{^{25}Mg}{^{24}Mg}\right)_0 \xleftrightarrow{\alpha_2^{25/24}} \left(\frac{^{25}Mg}{^{24}Mg}\right)_2 \quad (16)$$

$\alpha_1^{25/24}$ is the instrumental mass fractionation of the $^{25}Mg/^{24}Mg$ ratio during process 1 with:

$$\alpha_1^{25/24} = \frac{\left(^{25}Mg/^{24}Mg\right)_1}{\left(^{25}Mg/^{24}Mg\right)_0} \quad (17)$$

The total instrumental fractionation for the $^{25}Mg/^{24}Mg$ ratio ($\alpha_1^{25/24}$) is given by:

$$\alpha_{tot}^{25/24} = \frac{\left(^{25}Mg/^{24}Mg\right)_0 \times f_1 \times \alpha_1^{25/24} + \left(^{25}Mg/^{24}Mg\right)_0 \times (1-f_1) \times \alpha_2^{25/24}}{\left(^{25}Mg/^{24}Mg\right)_0} \quad (18)$$

$$= f_1 \times \alpha_1^{25/24} + (1-f_1) \times \alpha_2^{25/24}$$

which can be written:

$$\alpha_{tot}^{25/24} = f_1 \times \left(\alpha_1^{26/24}\right)^{\beta_1} + (1-f_1) \times \left(\alpha_2^{26/24}\right)^{\beta_2} \quad (19)$$

Similarly:

$$\alpha_{tot}^{26/24} = f_1 \times \alpha_1^{26/24} + (1-f_1) \times \alpha_2^{26/24} \quad (20)$$

If f_1, $\alpha_1^{26/24}$ and $\alpha_2^{26/24}$ vary independently from an ion probe spot to another one, there is no simple mathematic law which can describe the relationship between $\alpha_{tot}^{25/24}$ and $\alpha_{tot}^{26/24}$. Rearranging Equations (19) and (20) and considering that to first order

$$\alpha^{25/24} = \left(1 + \frac{\Delta^{26/24}}{1000}\right)^{\beta} \approx 1 + \frac{\beta \times \Delta^{26/24}}{1000} \quad (21)$$

it is obvious that the law relating $\alpha_{tot}^{25/24}$ and $\alpha_{tot}^{26/24}$ cannot be of the form given by Equation (14), even to first order, but should be close to:

$$\alpha_{tot}^{25/24} \approx \left(\alpha_{tot}^{26/24}\right)^{\frac{(\beta_1+\beta_2)}{2}} + \frac{(\beta_1-\beta_2)}{2} \times \left[f_1 \times \left(\alpha_1^{26/24} - 1\right) - (1-f_1) \times \left(\alpha_2^{26/24} - 1\right)\right] \quad (22)$$

In fact, even if $\beta_1 = \beta_2$, the real law is not exactly of the form given by Equation (14), if $\alpha_1^{26/24}$ and $\alpha_2^{26/24}$ are totally independent. Simulating real values of $\alpha_{tot}^{25/24}$ and $\alpha_{tot}^{26/24}$ with randomly varying values of f_1 (from 0 to 1) and of $\alpha_1^{26/24}$ and $\alpha_2^{26/24}$ (from 0.99 to 1.01) shows that $\ln(\alpha_1^{26/24})$ and $\ln(\alpha_2^{26/24})$ are related by a linear law:

$$\ln\left(\alpha_{inst}^{25/24}\right) = \beta \times \ln\left(\alpha_{inst}^{26/24}\right) + b \quad (23)$$

with the slope β ranging from 0.5207 to 0.5211 and the intercept b from -2×10^{-7} to -2.3×10^{-6} when $\beta_1 = \beta_2 = 0.521$, and with the slope β ranging from ≈0.512 to ≈0.519 and the intercept *b* from 6×10^{-6} to -3×10^{-5} when $\beta_1 = 0.511$ and $\beta_2 = 0.521$.

The present MC–SIMS set of data (Table 3 in electronic annex) follows a linear law between $\ln(\alpha_2^{26/24})$ and $\ln(\alpha_1^{26/24})$ with a slope $\beta = 0.5070 \pm 0.0038$ and an intercept $\beta = -8.6(\pm2.4)\times10^{-5}$ (Fig. 14a). Forcing the correlation to a zero intercept (i.e., to be of the form given by Equation (14))

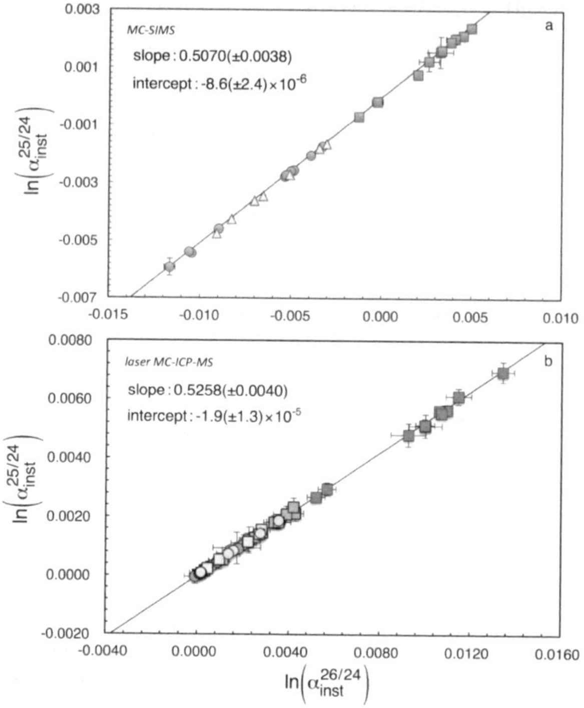

Figure 14. Mg mass fractionation laws observed for MC–SIMS (Fig. 14a) and LA–MC–ICP–MS (Fig. 14b). For MC–SIMS a law for IMF of the form $\ln(\alpha_{inst}^{25/24}) = \beta \times \ln(\alpha_{inst}^{26/24}) + b$ gives a slightly better fit to the data than a law forced to zero (with $\beta = 0.5145$) while for LA–MC–ICP–MS the fit is as good with a law forced to zero (with $\beta = 0.5210$), i.e., of the form $(\alpha_{inst}^{25/24}) = (\alpha_{inst}^{26/24})^{\beta}$.

gives a slope $\beta = 0.5145$. Note that because of the errors associated to the determination of the relative fractions of enhanced ionization and simple ionization (see previous sections), it is not possible to demonstrate with the present data that each process follows the same form of law. The average ^{26}Mg excess calculated for all the standards using the linear law given by Equation (23) is of 0.003 ± 0.037‰ while it is of -0.113 ± 0.047 using the law given by Equation (14). This shows that, effectively, IMF for MC–SIMS follows a law of the type given by Equation (23). It also shows that, in order to reach the best possible precision on ^{26}Mg excess by MC–SIMS, the IMF must be determined with standards having IMF as closest as possible to the samples. Initially high-precision MC–SIMS analysis of ^{26}Mg excesses was developed using a set of four standards (San Carlos olivine, Burma spinel, a glass of pyroxene composition and a MORB glass) and this gave a linear law between $\ln(\alpha_2^{26/24})$ and $\ln(\alpha_1^{26/24})$ with a slope $\beta = 0.5149 \pm 0.0056$ and an intercept $\beta = -6.8(\pm 2.3) \times 10^{-6}$ (Villeneuve et al. 2009). Further studies using this set of standards showed that the highest precision that can be reached on Mg-rich samples like olivines was on the order of ± 0.010‰ (2 sigma standard error for 20 analyses of the standards, Luu et al. 2014). Because this fractionation law will be different depending on the analytical tunings or on the standards used (e.g., different relative fractions of enhanced emissivity or simple emissivity depending on the composition of the sample), its determination is really the limitation to very high-precision for the in situ MC–SIMS analysis on ^{26}Mg excess even on Mg-rich phases (high Mg olivines) where there is no limitation due to counting statistics.

Mg instrumental mass fractionation law for LA–MC–ICP–MS analyses

We consider here that Mg during laser ablation MC-ICPMS is submitted to two successive isotopic fractionations, the first one taking place during evaporation (either of the sample during ablation or of the aerosol in the torch) and the second one during ionization in the torch. This is the simplest assumption that can be made. This can be schematized by:

$$\left(\frac{^{25}Mg}{^{24}Mg}\right)_0 \xleftrightarrow{\alpha_1^{25/24}} \left(\frac{^{25}Mg}{^{24}Mg}\right)_1 \xleftrightarrow{\alpha_2^{25/24}} \left(\frac{^{25}Mg}{^{24}Mg}\right)_2 \tag{24}$$

Assuming as in the previous section that processes 1 and 2 are mass dependent and follow a law as given by Equation (14) with exponent β_1 and β_2 for processes 1 and 2, respectively, the total isotopic fractionation for the $^{25}Mg/^{24}Mg$ ratio is:

$$\alpha_{tot}^{25/24} = \alpha_1^{25/24} \times \alpha_2^{25/24} = \left(\alpha_1^{26/24}\right)^{\beta_1} \times \left(\alpha_2^{26/24}\right)^{\beta_2} = \left(\alpha_{tot}^{26/24}\right)^{\frac{(\beta_1+\beta_2)}{2}} \times \left(\frac{\alpha_1^{26/24}}{\alpha_2^{26/24}}\right)^{\frac{(\beta_1-\beta_2)}{2}} \tag{25}$$

This gives:

$$\ln\left(\alpha_{tot}^{25/24}\right) = \frac{(\beta_1+\beta_2)}{2} \times \ln\left(\alpha_{tot}^{26/24}\right) + \frac{(\beta_1-\beta_2)}{2} \times \ln\left(\frac{\alpha_1^{26/24}}{\alpha_2^{26/24}}\right) \tag{26}$$

Thus, the law relating $\alpha_{tot}^{25/24}$ and $\alpha_{tot}^{26/24}$ is of the simple form given by Equation (14) only when $\beta_1 = \beta_2$, otherwise $\ln(\alpha_1^{26/24})$ and $\ln(\alpha_2^{26/24})$ are related by a linear law with a non-zero intercept.

The present LA–MC–ICP–MS set of data (Table 6 in electronic annex) follows a linear law between $\ln(\alpha_1^{26/24})$ and $\ln(\alpha_2^{26/24})$ with a slope $\beta = 0.5258 \pm 0.0040$ (2σ) and an intercept $\beta = -1.9(\pm 1.3) \times 10^{-5}$ (Fig. 14b). Forcing the correlation to a zero intercept (i.e., to be of the form given by Eqn. (14)) gives a slope $\beta = 0.5210$. The difference between the two regression lines is not significant: the average ^{26}Mg excesses are of -0.01 ± 0.19‰ for the regression with the zero intercept and of -0.02 ± 0.21‰ with the other one. The error on the mass fractionation law observed for LA–MC–ICP–MS is too large to eliminate the possibility of the existence of two IMF processes (following equation (14)), one during evaporation and one during ionization. Note, however, that the data would be consistent with most of the IMF taking place during ionization of the aerosol with a mass fractionation law following Equation (14) and with $\beta = 0.521$.

SUMMARY AND PERSPECTIVES

High-precision in-situ isotopic measurements in silicate samples can be carried out by MC–SIMS and LA–MC–ICP–MS. The limiting steps common to the two techniques are (i) the calibration of matrix effects on IMF and (ii) the determination of the instrumental mass fractionation laws. Matrix effects on IMF can be as strong for both MC–SIMS and LA–MC–ICP–MS, but perspectives exist in the case of LA–MC–ICP–MS to decrease IMF and matrix effects by decreasing and controlling the size of the aerosol particles produced by laser ablation. However it is difficult to suppress totally matrix effects during LA–MC–ICP–MS because all chemical elements present in the sample are introduced in the MC–ICP–MS together with Mg. In the case of MC–SIMS there is no perspective to reduce IMF and matrix effects, but they can be calibrated very precisely because the conditions of sputtering and ionization are fixed for given analytical parameters (primary and secondary high voltage, intensity and type of primary beam, ...). Typical errors, including external reproducibility and corrections for IMF and matrix effects, at the level of ±0.2‰ (2σ deviation), can be reached for the determination of mass-dependent isotopic variations (e.g., $\delta^{26}Mg$) by MC–SIMS and LA–MC–ICP–MS, provided that a full set of standards (having composition as close as possible to the samples) is used.

Instrumental mass fractionation laws are not systematically simple power laws for MC–SIMS contrary to LA–MC–ICP–MS. This is clear in the case of Mg isotopes (as shown above) and might be related to the complex ionization process,. However, very high precisions at the level of ≈5–10 ppm can be reached by MC–SIMS on the determination of non mass-dependent isotopic variations (e.g., radiogenic ^{26}Mg excesses), after calibration of the instrumental mass fractionation law. Interferences on ^{24}Mg are a possible limiting factor for high-precision determination of ^{26}Mg excesses by LA–MC–ICP–MS, especially in the case of samples with moderate Mg content. This implies either to use high mass-resolution, with however reduces sensitivity on the MC–ICP–MS and by consequence requires to ablate large volumes of samples, or to develop a procedure for correction of the interference. The major advantage of MC–SIMS relative to LA–MC–ICP–MS remains that MC–SIMS requires at least one order of magnitude less sample for the same level of precision, being thus much more in situ than LA–MC–ICP–MS. However, LA–MC–ICP–MS might be more appropriate in case of analysis of mixed matrices (different mineral phases) where the simpler laws for ionization and IMF are a potential advantage. It is clear that a number of scientific questions in cosmochemistry and Earth sciences would greatly benefit from the association of the two techniques on the same samples. This conclusion is not only valid for Mg (chosen as an example here) but also for other "non-traditional" isotope systems such as Si and Fe with potentially Ca, Ti and Ni.

ACKNOWLEDGMENTS

We thank Pascale Louvat and Nordine Bouden for heir help. This work was supported by the UnivEarthS Labex program (ANR-10-LABX-0023 and ANR-11-IDEX-0005–02) and by ANR grant ANR grant CRADLE (ANR-15-CE31-0004-01). This is IPGP contribution #3802 and CRPG contribution #2477.

REFERENCES

Albarède et al. (2017, this volume) Emerging applications of non-traditional isotopes in biomedical research. Rev Mineral Geochem 82:xx-xx

An YJ, Wu F, Xiang YX, Nan XY, Yu X, Yang JH, Yu HM, Xie LW, Huang F (2014) High-precision Mg isotope analyses of low-Mg rocks by MC-ICPMS. Chem Geol 390:9–21

Barling J, Weis D (2012) An isotopic perspective on mass bias and matrix effects in multi-collector inductively-coupled-plasma mass spectrometry. J Anal At Spectrom 27:653–662

Behr WM, Thomas JB, Hervig RL (2011) Calibrating Ti concentrations in quartz for SIMS determination using NIST silicate glasses and application to the TitaniQ geothermobarometer. Am Mineral 96:1100–1106

Benninghoven A, Rudenauer FG, Werner HW (1987) Secondary Ion Mass Spectrometry, John Wiley and Sons, New York

Bindeman IN, Gurenko AA, Sigmarsson O, Chaussidon M (2008) Oxygen isotope heterogeneity and disequilibria of olivine phenocrysts in large volume basalts from Iceland: evidence for magmatic digestion and erosion of Pleistocene hyaloclastites. Geochim Cosmochim Acta 72:4397–4420

Bizzarro M, Paton C, Larsen KK, Schiller M, Trinquier A, Ulfbeck D (2011) High-precision Mg-isotope measurements of terrestrial and extraterrestrial material by HR-MC-ICPMS-implications for the relative and absolute Mg-isotope composition of the bulk silicate Earth. J Anal At Spectrom 26:565–577

Burnett DS, Jurewicz AJG, Woolum DS, Wang J, Paque JM, Nittler LR, McKeegan KD, Humayun M, Hervig R, heber VS, Guan Y (2015) In implants as matrix-appropriate calibrators for geochemical ion probe analyses. Geostandard Geoanal Res 39:265–276.

Castaing R, Slodzian G (1962) First attempts at microanalysis by secondary ion emission. Compt Rend 255:1893–1895

Catanzaro EJ, Murphy TJ, Garner EL, Shields WR (1966) Absolute isotopic abundance ratios and atomic weights of magnesium. J Res Nat Bur Stand 70A:453–458

Chakraborty S (2010) Diffusion coefficients in olivine, wadsleyite and ringwoodite. Rev Mineral Geochem 72:603–639

Chan GCY, Chan WT, Mao X, Russo RE (2001) Comparison of matrix effects in inductively coupled plasma using laser ablation and solution nebulization for dry and wet plasma conditions. Spectrochim Acta B 56:1375–1386

Chaussidon M, Albarède F (1992) Secular boron isotope variations in the continental crust: an ion microprobe study. Earth Planet Sci Lett 108:229–241

Chaussidon M, Liu M-C (2015) Timing of nebula processes that shaped the precursors of the terrestrial planets. *In*: The Early Earth: Accretion and Differentiation. James B & Walter MJ (ed) Wiley, New York, p 1–26

Chaussidon M, Albarède F, Sheppard SMF (1987) Sulphur isotope heterogeneity in the mantle from ion microprobe measurements of sulphide inclusions in diamonds. Nature 330:242–244

Chaussidon M, Albarède F, Sheppard SMF (1989) Sulphur isotope variations in the mantle from ion microprobe analyses of micro-sulphide inclusions. Earth Planet Sci Lett 92:144–156

Chaussidon M, Robert F, Mangin D, Hanon P, Rose E (1997) Analytical procedures for the measurement of boron isotope compositions by ion microprobe in meteorites and mantle rocks. Geostand News 21:7–17

Chaussidon M, Libourel G, Krot AN (2008) Oxygen isotopic constraints on the origin of magnesian chondrules and on the gaseous reservoirs in the early solar system. Geochim Cosmochim Acta 72:1924–1938

Chmeleff J, Horn I, Steinhoefel G, von Blanckenburg F (2008) In situ determination of precise stable Si isotope ratios by UV-femtosecond laser ablation high-resolution multi-collector ICP–MS. Chem Geol 249:155–166

Christensen JN, Halliday AN, Lee DC, Hall CM (1995) In situ Sr isotopic analysis by laser ablation. Earth Planet Sci Lett 136:79–85

Clement SWJ, Compston W, Newstead G (1977) Design of a large, high resolution ion microprobe. *In:* Proceedings of the International Secondary Ion Mass Spectroscopy Conference. A Benninghoven (ed.), Munster, Springer Verlag

Connelly JN, Bizzarro M, Krot AN, Nordlund Å, Wielandt D, Ivanova MA (2012) The absolute chronology and thermal processing of solids in the solar protoplanetary disk. Science 338:651–655

Czaja AD, Johnson CM, Beard BL, Roden EE, Li W (2013) Biological Fe oxidation controlled deposition of banded iron formation in the ca. 3770 Ma Isua Supracrustal Belt (West Greenland). Earth Planet Sci Lett 363:192–203

d'Abzac FX, Poitrasson F, Freydier R, Seydoux-Guillaume AM (2010) Near infra red femtosecond laser ablation: the influence of energy and pulse width on the LA-ICP–MS analysis of monazite. J Anal At Spectrom 25:681–689

d'Abzac FX, Beard BL, Czaja AD, Konishi H, Schauer JJ, Johnson CM (2013) Iron isotope composition of particles produced by UV-femtosecond laser ablation of natural oxides, sulfides, and carbonates. Anal Chem 85:11885–11892

d'Abzac FX, Czaja AD, Beard BL, Schauer JL, Johnson CM (2014) Iron distribution in size-resolved aerosols generated by UV-femtosecond laser ablation: Influence of cell geometry and implications for in situ isotopic determination by LA–MC–ICP–MS. Geostand Geoanal Res 38:293–309

Dauphas N, Chaussidon M (2011) A perspective from extinct radionuclides on a young stellar object: the Sun and its accretion disk. Ann Rev Earth Planet Sci 39:351–386

Dauphas N, Schauble EA (2016) Mass fractionation laws, mass-independent effects, and isotopic anomalies. Ann Rev Earth Planet Sci 44:709–783

Dauphas N, John SG, Rouxel O (2017) Iron isotope systematics. Rev Mineral Geochem 82:415–510

Davis AM, Hashimoto A, Clayton RN, Mayeda TK (1990) Isotope mass fractionation during evaporation of Mg_2SiO_4. Nature 347:655–658

Davis AM, Richter FM, Mendybaev RA, Janney PE, Wadhwa M, McKeegan (2005) Isotopic mass fractionation laws and the initial solar system $^{26}Al/^{27}Al$ ratio. 36th Lunar Plan Sci Conf #2334

Davis AM, Richter FM, Mendybaev RA, Janny PE, Wadhwa M, McKeegan KD (2015) Isotopic mass fractionation laws for magnesium and their effects on ^{26}Al–^{26}Mg systematics in solar system materials. Geochim Cosmochim Acta 158:245–261

Deer WA, Howie RA, Zussman J (1982) Orthosilicates. Longman, London and New York

Deline VR, Evans CA, Williams P (1978a) A unified explanation for secondary ion yields. Appl Phys Lett 33:578–580

Deline VR, Katz W, Evans CA, Williams P (1978b) Mechanism of the SIMS matrix effect. Appl Phys Lett 33:832–835

Deloule E, Allegre CJ, Doe B (1986) Lead and sulfur isotope microstratigraphy in galena crystals from Mississippi-valley type deposits. Econ Geol 81:1307–1321

Deloule E, Chaussidon M, Allé P (1992) Instrumental limitations for isotope measurements with a Caméca® ims-3f ion microprobe: Example of H, B, S and Sr. Chem Geol 101:187–192

Deloule E, France-Lanord C, Albarède F (1991) D/H analysis of minerals by ion probe. Geochem Soc Spec Publ 3:53–62

Diwakar PK, Gonzalez JJ, Harilal SS, Russo RE, Hassanein A (2014) Ultrafast laser ablation ICP–MS: Role of spot size, laser fluence, and repetition rate in signal intensity and elemental fractionation. J Anal At Spectrom 29:339–346

Duffy JA (1993) A review of optical basicity and its applications to oxidic systems. Geochim Cosmochim Acta 57:3961–3970

Eggins SM, Grün R, McCulloch MT, Pike AWG, Chappell J, Kinsley L, Mortimer G, Shelley M, Murray-Wallace CV, Spötle C, Taylor L (2005) In situ U-series dating by laser ablation multi-collector ICP–MS: new prospects for Quaternary geochronology. Quat Sci Rev 24:2523–2538

Eiler JM, Graham C, Valley JW (1997) SIMS analysis of oxygen isotopes: matrix effects in complex minerals and glasses. Chem Geol 138:221–244

Esat MT (1984) A 61 cm radius multi-detector mass spectrometer at the Australian National University. Nucl Instr Meth Phys Res B5:545–553.

Fahey AJ, Goswami JN, McKeegan KD, Zinner E (1987) ^{26}Al, ^{244}Pu, ^{50}Ti, REE and trace element abundances in hibonite grains from CM and CV meteorites. Geochim Cosmochim Acta 51:329–350

Fietzke J, Liebetrau V, Günther D, Gürs K, Hametner K, Zumholz K, Hansteen TH (2008) An alternative data acquisition and evaluation strategy for improved isotope ratio precision using LA–MC–ICP–MS applied to stable and radiogenic strontium isotopes in carbonates. J Anal At Spectrom 23:955–961

Gaboardi M, Humayun M (2009) Elemental fractionation during LA-ICP–MS analysis od silicate glasses: implications for matric independent standardization. J Anal At Spectrom 24:1188–1197

Galy A, Yoffe O, Janney PE, Williams RW, Cloquet C, Alard O, Halicz L, Wadhwa M, Hutcheon ID, Ramon E, Carignan J (2003) Magnesium isotope heterogeneity of the isotopic standard SRM980 and new reference materials for magnesium-isotope-ratio measurements. J Anal At Spectrom 18:1352–1356

Georg RB, Halliday AN, Schauble EA, Reynolds BC (2007) Silicon in the Earth's core. Nature 447:1102–1106

Goudie DJ, Fisher CM, Hanchar JM, Crowley JL, Ayers JC (2014) Simultaneously in situ determination of U–Pb and Sm–Nd isotopes in monazite by laser ablation ICP–MS. Geochem Geophys Geosyst 15:2575–2600

Griffin WL, Pearson NJ, Belousova E, Jackson SE, van Achterbergh E, O'Reilly SY, Shee SR (2000) The Hf isotope composition of cratonic mantle LAM-MC–ICP–MS analysis of zircon megacrysts in kimberlites. Geochim Cosmochim Acta 64:133–147

Guillong M, Günther D (2002) Effect of particle size distribution on ICP-induced elemental fractionation in laser ablation-inductively coupled mass-spectrometry. J Anal At Spectrom 17:831–837

Guillong M, Kuhn HR, Günther D (2003) Application of a particle separation device to reduce inductively coupled plasma-enhanced elemental fractionation in laser ablation-inductively coupled plasma-mass spectrometry. Spectrochim Acta B 58:211–220

Günther D, Heinrich C (1999) Comparison of the ablation behavior of 266 nm Nd:YAG and 193 nm ArF excimer lasers for LA-ICP–MS analysis. J Anal At Spectrom 14:1369–1374

Günther D, Jackson SE, Longerich HP (1999) Laser ablation and arc spark solid sample introduction into inductively coupled plasma mass spectrometers. Spectrochim Acta B 54:381–409

Gurenko AA, Chaussidon M (1995) Enriched and depleted primitive melts included in olivine from Icelandic tholeiites: origin by continuous melting of a single mantle column. Geochim Cosmochim Acta 59:2905–2917

Gurenko AA, Bindeman IN, Chaussidon M (2011) Oxygen isotope heterogeneity of the mantle beneath the Canary Islands: insights from olivine phenocrysts. Contrib Mineral Petrol 162:349–363

Hinton RW (1990) Ion microprobe trace-element analysis of silicates: Measurement of multi-element glasses. Chem Geol 83:11–25

Hirata T, Nesbitt RW (1995) U–Pb isotope geochronology of zircon: Evaluation of the laser probe-inductively coupled plasma mass spectrometry technique. Geochim Cosmochim Acta 12:2491–2500

Holst JC, Olsen MB, Paton C, Nagashima K, Schiller M, Wielandt D, Larsen KK, Connelly JN, Jorgensen JK, Krot AN, Nordlund A, Bizzarro M (2013) ^{182}Hf–^{182}W age dating of a ^{26}Al-poor inclusion and implications for the origin of short-lived radioisotopes in the early Solar System. PNAS 110:8819–8823

Horn I (2008) Comparison of femtosecond and nanosecond laser interactions with geologic matrices and their influence on accuracy and precision of LA-ICP–MS data. *In*: Mineralogical Association of Canada Short Course 40, p 53–65

Horn I, von Blanckenburg F (2007) Investigation on elemental and isotopic fractionation during 196 nm femtosecond laser ablation multiple collector inductively coupled plasma mass spectrometry. Spectrochim Acta B 62:410–422

Horn I, von Blanckenburg F, Schoenberg R, Steinhoefel G, Markl G (2006) In situ iron isotope ratio determination using UV-femtosecond laser ablation with application to hydrothermal ore formation processes. Geochimica 70:3677–3688

Hu ZC, Liu YS, Chen L, Zhou L, Li M, Zong KQ, Zhu LY, Gao S (2011) Contrasting matrix induced elemental fractionation in NIST SRM and rock glasses during laser ablation ICP–MS analysis at high spatial resolution. J Anal At Spectrom 26:425–430

Ireland TR (1988) Correlated morphological, chemical, and isotopic characteristics of hibonites from the Murchison carbonaceous chondrite. Geochim Cosmochim Acta 52:2827–2839

Ireland TR (1995) Ion microprobe mass spectrometry: techniques and applications in cosmochemistry, geochemistry, and geochronology. *In*: Advances in Analytical Geochemistry. Vol 2. Rowe MW & Hyman M (ed) JAI Press Limited, Greenwich, p 1–118

Ireland TR (2004) SIMS measurement of stable isotopes. *In*: Handbook of Stable Isotope Analytical Techniques. Vol 1. de Groot PA (ed) Elsevier, Philadelphia, p 652–691
Jackson SE, Günther D (2003) The nature and sources of laser induced isotopic fractionation in laser-multicollector-inductively coupled plasma-mass spectrometry. J Anal At Spectrom 18:205–212
Jackson SE, Longerich HP, Dunning GR, Fryer BJ (1992) The application of laser-ablation microprobe-inductively coupled plasma-mass spectrometry (LAM-ICP–MS) to in situ trace-element determinations in minerals. Can Mineral 30:1049–1064
Jackson EJ, Pearson NJ, Griffin WL, Belousova EA (2004) The application of laser ablation-inductively coupled plasma-mass spectrometry to in situ U–Pb zircon geochronology. Chem Geol 211:47–69
Jacobsen B, Yin Q-Z, Moynier F, Amelin Y, Krot AN, Nagashima K, Hutcheon ID, Palme H (2008) ^{26}Al–^{26}Mg and ^{207}Pb–^{206}Pb systematics of Allende CAIs: canonical solar initial ^{26}Al/^{27}Al ratio reinstated. Earth Planet Sci Lett 272:353–364
Janney PE, Richter FM, Mendybaev RA, Wadhwaa M, Georg RB, Watson BE, Hines RR (2011) Matrix effects in the analysis of Mg and Si isotope ratios in natural and synthetic glasses by laser ablation-multicollector ICPMS: A comparison of single- and double-focusing mass spectrometers. Chem Geol 281:26–40.
Jochum KP, Scholz D, Stoll B, Weis U, Wilson SA, Yang Q, Schwalb A, Börner N, Jacob DE, Andrea MO (2012) Accurate trace element analysis of speleothems and biogenic calcium carbonates by LA-ICP–MS. Chem Geol 318:31–44
Jochum KP, Willbold M, Raczek I, Stoll B, Herwig K (2005) Chemical characterisation of the USGS reference glasses GSA-1G, GSC-1G, GSD-1G, GSE-1G, BCR-2G, BHVO-2G and BIR-1G using EPMA, ID-TIMS, ID-ICP–MS and MA-ICP–MS. Geostand Geoanal Res 29:285–302
Kita NT, Yin QZ, MacPherson GJ, Ushikubo T, Jacobsen B, Nagashima K, Kurahashi E, AKrot AN, Jacobsen SB (2013) ^{26}Al–^{26}Mg isotope systematics of the first solids in the early Solar System. Meteor Planet Sci 48:1383–1400
Kroslakova I, Günther D (2007) Elemental fractionation in laser ablation-inductively coupled plasma-mass spectrometry: evidence for mass load induced matrix effects in the ICP during ablation of a silicate glass. J Anal At Spectrom 22:51–62
Kruijer TS, Kleine T, Fischer-Göge M, Burkhardt C, Wieler R (2014) Nucleosynthetic W isotope anomalies and the Hf-W chronometry of Ca-Al-rich inclusions. Earth Planet Sci Lett 403:317–327
Kuhn HR, Günther D (2004) Laser ablation-ICP–MS: particle size dependent elemental composition studies on filter-collected and online measured aerosols from glass. J Anal At Spectrom 19:1158–1164
Kylander-Clark ARC, Hacker BR, Cottle JM (2013) Laser-ablation split-stream ICP petrochronology. Chem Geol 345:99–112
Larsen KK, Trinquier A, Paton C, Schiller M, Wielandt D, Ivanova MA, Connelly JN, Nordlund O, Krot AN, Bizzarro M (2011) Evidence for magnesium isotope heterogeneity in the solar protoplanetary disk. Astrophys J 735:37–43
Lee T, Papanastassiou DA, Wasserburg GJ (1976) Demonstration of Mg-26 excess in Allende and evidence for Al-26. Geophys Res Lett 3:41–44
Leshin LA, Rubin AE, McKeegan KD (1997) The oxygen isotopic composition of olivine and pyroxene from CI chondrites. Geochim Cosmochim Acta 61:835–845
Liu M-C, Chaussidon M, Göpel C, Lee T (2012) A heterogeneous solar nebula as sampled by CM hibonite grains. Earth Planet Sci Lett 327:75–83
Liu M-C, McKeegan KD (2009) On an irradiation origin for magnesium isotope anomalies in meteoritic hibonite. Astrophys J 697:145–148
Liu M-C, McKeegan KD, Goswami JN, Marhas KK, Sahijpal S, Ireland TR, Davis AM (2009) Isotopic records in CM hibonites: implications for timescales of mixing of isotope reservoirs in the solar nebula. Geochim Cosmochim Acta 73:5051–5079
Liu Y, Hu Z, Gao S, Günther D, Xu J, Gao C, Chen H (2008) In situ analysis of major and trace elements of anhydrous mineral by LA-ICP–MS without applying an internal standard. Chem Geol 257:34–43
Lovering JF (1975) Application of SIMS microanalysis techniques to trace element and isotopic studies in geochemistry and cosmochemistry: Nat Bur Stand Spec Publ 427, 135–178
Luu T-H, Chaussidon M, Mishra RK, Rollion-Bard C, Villeneuve J, Srinivasan G, Birck JL (2013) High precision Mg isotope measurements of meteoritic samples by secondary ion mass spectrometry. J Anal Atom Spec 28:67–76
Luu TH, Chaussidon M, Birck JL (2014) Timing of metal–silicate differentiation in the Eagle Station pallasite parent body. C R Geoscience 346:75–81
Luu T-H, Young ED, Gounelle M, Chaussidon M (2015) A short time interval for condensation of high temperature silicates in the Solar accretion disk. PNAS112:1298–1303
MacPherson GJ, Davis AM, Zinner EK (1995) The distribution of aluminum-26 in the early Solar System—A reappraisal. Meteoritics 30:365–386
MacPherson GJ, Kita NT, Ushikubo T, Bullock ES, Davis AM (2012), Well-resolved variations in the formation ages for Ca-Al-rich inclusions in the early Solar System. Earth Planet Sci Lett 331:43–54

Makide K, Nagashima K, Krot AN, Huss GR, Ciesla FJ, Hellebrand EG, Yang L (2011) Heterogeneous distribution of ^{26}Al at the birth of the Solar System. Astrophys J 733:31–35

McKeegan KD, Walker RM, Zinner E (1985) Ion microprobe isotopic measurements of individual interplanetary dust particles. Geochim Cosmochim Acta 49:1971–1987

Mishra RK, Chaussidon M (2014) Timing and extent of Mg and Al isotopic homogenization in the early inner Solar System. Earth Planet Sci Lett 390:318–326

Moynier F, Vance D, Fujii T, Savage PS (2017) The isotope geochemistry of zinc and copper. Rev Mineral Geochem 82:543–600

Nishizawa M, Yamamoto H, Ueno Y, Tsuruoka S, Shibuya T, Sawaki Y, Yamamoto S, Kon Y, Kitajima K, Komiya T, Maruyama S, Hirata T (2010) Grain-scale iron isotopic distribution of pyrite from Precambrian shallow marine carbonate revealed by a femtosecond laser ablation multicollector ICP–MS technique: Possible proxy for the redox state of ancient seawater. Geochim Cosmochim Acta 74:2760–2778

Norman MD, Griffin WL, Pearson NJ, Garcia MO, O'Reilly SY (1998) Quantitative analysis of trace element abundances in glasses and minerals: a comparison of laser ablation inductively coupled plasma mass spectrometry, solution inductively coupled plasma mass spectrometry, proton microprobe and electron microprobe data. J Anal At Spectrom 13:477–482

Norman MD, McCulloch MT, O'Neill HSC, Yaxley GM (2006) Magnesium isotopic analysis of olivine by laser-ablation multi-collector ICP–MS: composition dependent matrix effects and a composition of the Earth and Moon. J Anal At Spectrom 21:50–54

Oeser M, Dohmen R, Horn I, Schuth S, Weyer S (2015) Processes and time scales of magmatic evolution as revealed by Fe–Mg chemical and isotopic zoning in natural olivines. Geochim Cosmochim Acta 154:130–150

Oeser M, Weyer S, Horn I, Schuth S (2014) High-Precision Fe and Mg isotope ratios of silicate reference glasses determined in situ by femtosecond LA–MC–ICP–MS and by solution nebulisation MC–ICP–MS. Geostand Geoanal Res 38:311–328

Paniello R, Day J, Moynier F (2012) Zn isotopic evidence for the origin of the Moon. Nature. 490:376–379

Poitrasson F (2017) Silicon isotope geochemistry. Rev Mineral Geochem 82:289–344

Poitrasson F, Mao X, Mao SS, Freydier R, Russo RE (2003) Comparison of ultraviolet femtosecond and nanosecond laser ablation inductively coupled plasma mass spectrometry analysis in glass, monazite, and zircon. Anal Chem 75:6184–6190

Reed SJB (1989) Ion microprobe analysis-a review of geological applications. Mineral Mag 53:3–24

Reed SJB, Scott ERD, Long JVP (1979) Ion microprobe analysis of olivine in pallasite meteorites for nickel. Erath Planet Sci Lett 43:5–12

Richter FM, Janney PE, Mendybaev RA, Davis AM, Wadhwa M (2007) Elemental and isotopic fractionation of Type B CAI-like liquids by evaporation. Geochim Cosmochim Acta 71:5544–5564

Russell WA, Papanastassiou DA, Tombrello TA (1978) Ca isotope fractionation on the Earth and other solar system materials. Geochim Cosmochim Acta 42:1075–1090

Schuessler JA, von Blanckenburg F (2014) Testing the limits of micro-scale analyses of Si stable isotopes by femtosecond laser ablation multicollector inductively coupled plasma mass spectrometry with application to rock weathering. Spectrochim Acta B 98:1–18

Shaheen ME, Gagnon JE, Fryer BJ (2012) Femtosecond (fs) lasers coupled with modern ICP–MS instruments provide new and improved potential for in situ elemental and isotopic analyses in the geosciences. Chem Geol 330:260–273

Shimizu N, Hart SR (1982) Applications of the ion microprobe to geochemistry and cosmochemistry. Ann Rev Earth Planet Sci 10:483–526

Shimizu N, Semet MP, Allègre CJ (1978) Geochemical applications of quantitative ion-microprobe analysis. Geochim Cosmochim Acta 42:1321–1334

Sigmund P (1969) Theory of sputtering I. Sputtering yield of amorphous and polycrystalline targets. Phys Rev 184:383–416

Sio CKI, Dauphas N, Teng F-Z, Chaussidon M, Helz RT, Roskosz M (2013) Discerning crystal growth from diffusion profiles in zoned olivine by in situ Mg–Fe isotopic analyses. Geochim Cosmochim Acta 123:302–321

Slodzian G, Lorin JC, Havette A (1980) Isotopic effect on the ionization probabilities in secondary ion emission. J Physique Lett 41:555–558

Sossi P, Foden F, Halverson G (2012) Redox-controlled iron isotope fractionation during magmatic differentiation: an example from the Red Hill intrusion, S. Tasmania. Contrib Mineral Petrol 164:757–772

Steele IM, Hervig RL, Hutcheon ID, Smith JV (1981) Ion microprobe techniques and analyses of olivine and low-Ca pyroxene. Am Mineral 66:526–541

Steinhoefel G, von Blanckenburg F, Horn I, Konhauser KO, Beukes NJ, Gutzmer J (2010) Deciphering formation processes of banded iron formations from the Transvaal and the Hamersley successions by combined Si and Fe isotope analysis using UV femtosecond laser ablation. Geochim Cosmochim Acta 74:2677–2696

Stephan T, Trappitsch R, Davis AM, Pellin MJ, Rost D, Savina MR, Yokochi R, Liu N (in press) CHILI–the Chicago Instrument for Laser Ionization—a new tool for isotope measurements in cosmochemistry. Int J Mass Spectrom

Stix J, Gauthier G, Ludden JN (1995) A critical look at quantitative laser-ablation ICP–MS analysis of natural and synthetic glasses. Can Mineral 33:435–444

Teng F-Z, Li WY, Ke S, Marty B, Dauphas N, Huang S, Wu F-Y, Pourmand A (2010) Magnesium isotopic composition of the Earth and chondrites. Geochim Cosmochim Acta 74:4150–4166

Teng F-Z, Yang W (2014) Comparison of factors affecting the accuracy of high-precision magnesium isotope analysis by multi-collector inductively coupled plasma mass spectrometry. Rapid Commun Mass Spctrom 28:19–24

Thirlwall MF, Walder AJ (1995) In situ hafnium isotope ratio analysis of zircon by inductively coupled plasma multiple collector mass spectrometry. Chem Geol 122:241–247

Thrane KB, Bizzarro M, Baker JA (2006) Extremely brief formation interval for refractory inclusions and uniform distribution of ^{26}Al in the early solar system. Astrophys J 646:159–162

Valley JW, Cole DR (eds) (2011) Stable Isotope Geochemistry. Rev Mineral Geochem Vol 43. Mineralogical Society of America

Vielzeuf D, Champenois M, Valley JW, Brunet F, Devidal JL (2005) SIMS analyses of oxygen isotopes: Matrix effects in Fe–Mg–Ca garnets. Chem Geol 223:208–226.

Villeneuve J, Chaussidon M, Libourel G (2009) Homogeneous distribution of ^{26}Al in the solar system from the Mg isotopic composition of chondrules. Science 325:985–988

Walczyk T, von Blanckenburg F (2002) Natural iron isotope variations in human blood. Science 295:2065–2066

Wang K, Jacobsen S (2016) Potassium isotopic evidence for a high energetic giant impact origin for the Moon. Nature. doi:10.1038/nature19341

Wasserburg GJ, Wimpenny J, Yin QZ (2012) Mg isotopic heterogeneity, Al–Mg isochrons, and canonical ^{26}Al/^{27}Al in the early solar system. Meteorit Planet Sci 47:1980–1997

Williams HM, Nielsen SG, Renac C, Griffin WL, O'Reilly SY, McCammon CA, Pearson N, Viljoen F, Alt JC, Halliday AN (2009) Fractionation of oxygen and iron isotopes by partial melting processes: Implications for the interpretation of stable isotope signatures in mafic rocks. Earth Planet Sci Lett 283:156–166

Williams P (1979) The sputtering process and sputtered ion emission. Surf Sci 90:588–643

Williams P (1985) Limits of quantitative microanalysis using secondary ion mass spectrometry. *In*: Scanning Electron Microscopy. O'Hare AMF (Ed), SEM Inc, Chicago, p 553–561

Williams P, Stika KM, Davies JA, Jackman TE (1983) Quantitative SIMS analysis of hydrogenated amorphous silicon using superimposed deuterium implant standards. Nuclear Instr. Methods B 218:299–302

Woodhead J, Hergt J, Shelley M, Eggins S, Kemp R (2004) Zircon Hf-isotope analysis with an excimer laser, depth profiling, ablation of complex geometries, and concomitant age estimation. Chem Geol 209:121–135

Young ED, Ash RD, Galy A, Belshaw NS (2002) Mg isotope heterogeneity in the Allende meteorite measured by UV laser ablation-MC-ICPMS and comparisons with O isotopes. Geochim Cosmochim Acta 66:683–698

Young ED, Galy A (2004) The isotope geochemistry and cosmochemistry of magnesium. Rev Mineral Geochem 55:197–230

Young ED, Galy A, Nagahara H (2002) Kinetic and equilibrium mass-dependent isotope fractionation laws in nature and their geochemical and cosmochemical significance. Geochim Cosmochim Acta 66:1095–1104

Young ED, Simon JI, Galy A, Russell SS, Tonui E, Lovera O (2005), Supra-canonical ^{26}Al/^{27}Al and the residence time of CAIs in the solar protoplanetary disk. Science 308:223–227

Yuan HL, Gao S, Dai MN, Zong CL, Günther D, Fontaine GH, Liu XM, Diwu CR (2008) Simultaneously determinations of U–Pb age, Hf isotopes and trace element compositions of zircon by excimer laser-ablation quadrupole and multiple-collector ICP–MS. Chem Geol 247:100–118

Zhang SD, He MH, Yin ZB, Zhu EY, Hang W, Huang BL (in press) Elemental fractionation and matrix effects in laser sampling based spectrometry. J Anal At Spectrom

Zinner E (1989) Isotopic measurements with the ion microprobe. *In*: New Frontiers in Stable Isotopic Research: Laser Probes, Ion Probes, and Small-sample Analysis. W. C. III Shanks & R. E. Criss (eds.). U.S. Geol. Surv. Bull. 1890:145–162

Zinner E, Crozaz G (1986) A method for quantitative measurement of rare earth elements in the ion microprobe. Int J Mass Spectrom Ion Phys 69:17–38

Reviews in Mineralogy & Geochemistry
Vol. 82 pp. 165-217, 2017

Lithium Isotope Geochemistry

Sarah Penniston-Dorland[1]

[1]Department of Geology
The University of Maryland
College Park, MD 20742
USA

sarahpd@umd.edu

Xiao-Ming Liu

Department of Geological Sciences
University of North Carolina
Chapel Hill, NC 27599
USA

xiaomliu@email.unc.edu

Roberta L. Rudnick[1, 2]

[2]Now at: Department of Earth Science
University of California at Santa Barbara
Santa Barbara, CA 93106
USA

rudnick@geol.ucsb.edu

INTRODUCTION

The lithium isotope system is increasingly being applied to a variety of Earth science studies, as the burgeoning literature attests; over 180 papers have been published in the last twelve years that report lithium isotope data, including five review papers that cover different aspects of lithium isotope applications (Elliott et al. 2004; Tomascak 2004; Tang et al. 2007b; Burton and Vigier 2011; Schmitt et al. 2012), and a book (Tomascak et al. 2016). The upswing in lithium isotope studies over the past decade reflects analytical advances that have made Li measurements readily obtainable. These include the use of multi-collector inductively coupled plasma mass spectrometry (MC-ICP-MS) for relatively precise solution measurements (Tomascak et al. 1999a) and secondary ion mass spectrometry (SIMS) for high spatial resolution measurements (Chaussidon and Robert 1998; Kasemann et al. 2005; Bell et al. 2009). In addition, lithium isotope studies are motivated by the large variety of problems for which they may provide insight, including crust–mantle recycling, silicate weathering, fluid–rock interaction, as well as geospeedometry.

The great interest in the Li system that spurred the development of these new analytical methods was initiated by the pioneering work of Lui-Heung Chan, who demonstrated not only that Li isotopic fractionation can be very large at or near the Earth's surface (Chan and Edmond 1988), but also that Li isotopes are strongly fractionated during seawater-basalt interaction (Chan et al. 1992). This discovery naturally led to the search for a recycled slab signature in Li isotopes of arc lavas (some of the earlier studies include Moriguti and Nakamura 1998a; Chan et al. 1999, 2002b; Tomascak et al. 2000, 2002; Leeman et al. 2004;

1529-6466/17/0082-0006$10.00 http://dx.doi.org/10.2138/rmg.2017.82.6

Moriguti et al. 2004), as well as more deeply derived intraplate basalts (e.g., Chan and Frey 2003; Kobayashi et al. 2004; Ryan and Kyle 2004). It also spurred the development of Li isotopes as tracers of continental weathering. Thus, Li isotopes have been added to the geochemical toolbox employed to trace the fate of recycled materials in the Earth's mantle, as well as weathering fluxes through time (Figs. 1 and 2).

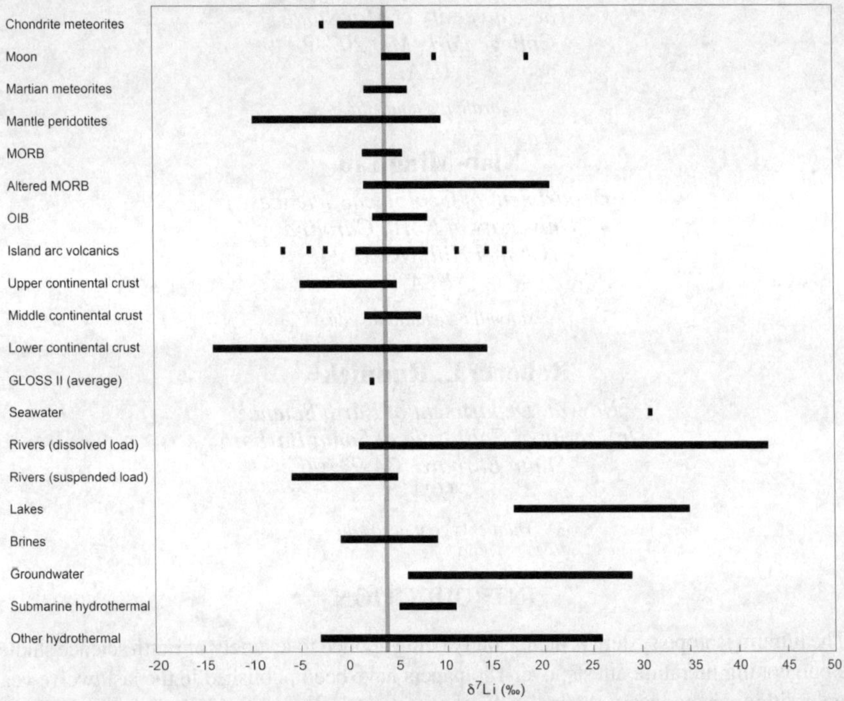

Figure 1. Compilation of whole-rock Li isotopic compositions of various rock types and water reservoirs. Gray vertical line is the composition of bulk silicate Earth. References: *Chondrite:* Seitz et al. (2007); Pogge von Strandmann et al. (2011); *Moon:* Seitz et al. (2006); Magna et al. (2006b); Day et al. (2016); *Mars:* Magna et al. (2006b, 2015); Seitz et al. (2006); Filiberto et al. (2012); *Mantle peridotites:* Brooker et al. (2004); Seitz et al. (2004); Magna et al. (2006b, 2008); Jeffcoate et al. (2007); Ionov and Seitz (2008); Kil (2010); Gao et al. (2011); Pogge von Strandmann et al. (2011); Ackerman et al. (2013); *Mid-ocean ridge basalts (MORB):* Chan et al. (1992); Moriguti and Nakamura (1998a); Elliott et al. (2006); Nishio et al. (2007); Tomascak et al. (2008); *Altered MORB:* Chan et al. (1992, 2002a); Bouman et al. (2004); Brant et al. (2012); *Ocean island basalts (OIB):* Chan and Frey (2003); Kobayashi et al. (2004); Ryan and Kyle (2004); Nishio et al. (2005); Chan et al. (2009); Vlastelic et al. (2009); Magna et al. (2011); Krienitz et al. (2012); Harrison et al. (2015); *Island Arc volcanic rocks:* Moriguti and Nakamura (1998a); Tomascak et al. (2000, 2002); Chan et al. (2002b); Leeman et al. (2004); Moriguti et al. (2004); Magna et al. (2006a); Agostini et al. (2008); Walker et al. (2009); M. Tang et al. (2014); *Upper continental crust:* Teng et al. (2004); Sauzeat et al. (2015); *Lower and middle continental crust:* Teng et al. (2008); Qiu et al. (2011b); *Global Oceanic Subducting Sediment (GLOSS-II):* Plank (2014); *Seawater:* Chan and Edmond (1988); You and Chan (1996); Moriguti and Nakamura (1998b); James and Palmer (2000); Millot et al. (2004); *Rivers:* Huh et al. (1998, 2001); Kisakürek et al. (2005); Pogge von Strandmann et al. (2006, 2010); Vigier et al. (2009); Lemarchand et al. (2010); Dellinger et al. (2015); Liu et al. (2015); Pogge von Strandmann and Henderson (2015); *Suspended river loads:* Dellinger et al. (2014); Liu et al. (2015); *Lakes:* Chan et al. (1988); Tomascak et al. (2003); Witherow et al. (2010); *Brines:* Godfrey et al. (2013); Araoka et al. (2014); *Groundwater:* Hogan and Blum (2003); Negrel et al. (2010); Nishio et al. (2010); Millot et al. (2011); Meredith et al. (2013); Liu et al. (2015); *Submarine hydrothermal:* Chan and Edmond (1988); Chan et al. (1993, 1994); Foustoukos et al. (2004); Mottl et al. (2011); Verney-Carron et al. (2011); *Other hydrothermal:* Millot and Negrel (2007); Millot et al. (2007, 2010a, 2012); Bernal et al. (2014).

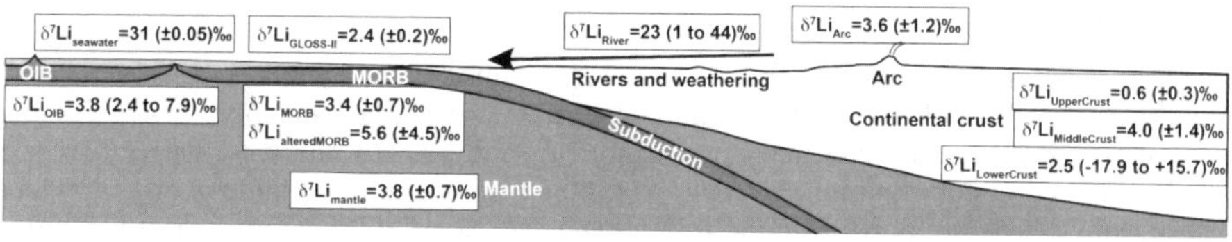

Figure 2. Geochemical reservoirs and cycling of Li isotopes. Values of δ^7Li indicated are average values. Values in parentheses are ±1 standard deviations or else ranges of measured compositions. References: *Mantle:* Lai et al. (2015); *Arc:* Liu and Rudnick (2011); *Upper continental crust:* Sauzeat et al. (2015); *Middle and lower continental crust:* Teng et al. (2008); *MORB (Mid-Ocean Ridge Basalt)*: Tomascak et al. (2008); *altered MORB*: compiled from Chan et al. (1992, 2002b); Bouman et al. (2004); Brant et al. (2012); *OIB (Ocean Island Basalts):* compiled from Nishio et al. (2004); Ryan and Kyle (2004); Vlastelic et al. (2009); Chan et al. (2009); Magna et al. (2011); *GLOSS-II (Global Oceanic Subducting Sediment):* Plank (2014); *Seawater:* Millot et al. (2004).

During the course of writing this chapter, a new book appeared that provides a very readable and comprehensive overview of lithium isotope geochemistry (*Advances in Lithium Isotopes*, by Tomascak, Magna, and Dohmen 2016). Not wishing for Li to be left out of the second *Reviews in Mineralogy and Geochemistry* volume on non-traditional stable isotopes, we continued writing this chapter, which can be viewed as a 'Reader's Digest Condensed' version of the book, albeit from a somewhat different perspective. Anyone interested in Li isotope geochemistry should certainly access the Tomascak et al. (2016) book, which, due to the luxury of space, goes into much greater detail than can be covered in a single chapter and provides very interesting historical background.

LITHIUM SYSTEMATICS

Lithium is the third element on the periodic table—with an atomic number of three and an atomic mass of 6.94. It is an alkali metal, falling in group one on the periodic table, and therefore its common valence state is +1. It has two stable isotopes: ^{7}Li (92.4%) and ^{6}Li (7.6%).

The ratio of Li isotopes is reported using the standard delta notation:

$$\delta^7 Li(‰) = \left[\frac{\left({}^7Li / {}^6Li \right)_{\text{unknown}} - \left({}^7Li / {}^6Li \right)_{\text{standard}}}{\left({}^7Li / {}^6Li \right)_{\text{standard}}} \times 10^3 \right]$$

where the standard material has been L-SVEC, a Li carbonate, NIST standard reference material 8545 (Flesch et al. 1973). This material is no longer produced, so the standard IRMM–016 which has an almost identical isotope ratio (Qi et al. 1997) is used as a replacement standard.

The large relative difference in mass between ^{6}Li and ^{7}Li (~17%) can lead to significant isotopic fractionation. This is particularly evident in low temperature environments where δ^7Li values span a range of over 50‰, but can also be significant in high temperature environments, due to differences in diffusivity between the two isotopes. These properties, together with the fact that Li is highly soluble, monovalent, and is not utilized in biological systems, make Li isotopes potentially useful tracers of continental weathering, fluids, recycled oceanic crust and magmatic cooling histories. However, in order to understand the behavior of Li isotopes, it is essential to understand the site occupancy and behavior of Li in minerals, to establish equilibrium partition coefficients, as well as equilibrium isotope fractionation factors (α), and to understand the disequilibrium behavior of Li.

Li in minerals

Lithium has a similar ionic radius to Mg, and commonly substitutes for Mg in minerals (radii for $Li^+ = 0.76$ Å, $Mg^{2+} = 0.72$ Å, in octahedral coordination, Shannon 1976). For this reason it is partitioned preferentially into many Mg-rich silicate minerals, where Mg is in six-fold, octahedral coordination. Exceptions to this generalization include quartz, in which Li is in either two- or four-fold coordination (Larsen et al. 2004; Sartbaeva et al. 2004), staurolite, in which Li is in four-fold coordination (e.g., Dutrow 1991) and amphibole. Lithium in amphibole is in a strongly distorted octahedral site, in which one of the Li–O bonds has a long bond length, such that the coordination around Li in amphibole can be considered to be five-fold (e.g., Wenger and Armbruster 1991; see also Wunder et al. 2011). The uncommon hydrous silicate mineral nambulite [$(Li,Na)Mn_4Si_5O_{14}OH$] has Li in eight-fold coordination (Narita et al. 1975; Wenger and Armbruster 1991). In aqueous fluids at standard temperature and pressure, Li^+ is coordinated with H_2O dominantly in four-fold hydration shells (Lyubartsev et al. 2001; Wunder et al. 2011). Ab initio molecular dynamics modeling at 1000 K shows that the coordination of Li appears to increase with fluid density and pressure (Jahn and Wunder 2009), ranging from three-fold to five-fold coordination.

Equilibrium fractionation of isotopes between phases depends on the energetics of the bonding environment. It has been demonstrated for many systems that the lighter isotope preferentially fractionates into a site with a higher coordination number due to energetic considerations (see Schauble 2004). However, in the case of solid–fluid interactions, bond valence considerations (Brown 2009) have been shown to explain Li isotopic fractionation (Wunder et al. 2011). Since the coordination of Li in fluids is lower than in most solids, isotopic fractionation of Li in fluid–rock interactions generally results in preferential fractionation of ^7Li into fluids.

Li partitioning

Experimentally derived partition coefficients (K_D) for Li between mineral–silicate melt (Hart and Dunn 1993; Icenhower and London 1995; Latourrette et al. 1995; Blundy et al. 1998; Brenan et al. 1998a; Taura et al. 1998; Schmidt et al. 1999; Evensen and London 2003; McDade et al. 2003; Adam and Green 2006; Aigner-Torres et al. 2007; Ottolini et al. 2009), mineral–aqueous fluid (Brenan et al. 1998b; Caciagli et al. 2011) and mineral–mineral (Coogan et al. 2005; Ottolini et al. 2009; Caciagli et al. 2011; Yakob et al. 2012) are tabulated in Table 1. These demonstrate several interesting features. First, Li is a moderately incompatible element in basaltic systems (most $^{xl-melt}K_D < 0.4$, where xl = crystal) and probably in most igneous systems up to the point at which Li-rich minerals form in highly differentiated magmas (e.g., Marks et al. 2007). Secondly, Li partitions preferentially into aqueous fluids over minerals (most $^{xl-fluid}K_D < 0.3$), consistent with the high solubility of Li deduced from empirical studies. Finally, inter-mineral partitioning of Li between plagioclase (plag) and clinopyroxene (cpx) is strongly temperature-dependent, allowing for the use of Li concentrations in coexisting plagioclase and clinopyroxene to determine cooling histories (Coogan et al. 2005).

High-pressure (4–13 GPa), high-temperature (1100–1400 °C) multi-anvil experiments have shown a pressure-dependent partitioning of Li between garnet and pyroxene in both a simple Ca–Mg–Al–Si system and in rocks with a natural eclogitic composition (Hanrahan et al. 2009a,b). These results suggest that Li partitioning between garnet and pyroxene can be used as a geobarometer for eclogites.

In addition to the experimental investigations, there have been a number of empirical studies of intermineral Li partitioning, particularly for peridotites. There are several difficulties in using real rocks in this endeavor. One is determining whether the minerals are in equilibrium, which can be especially difficult in slowly cooled rocks (see mantle section, below). Another is deducing any temperature-dependent partitioning behavior. If Li partitioning is strongly temperature dependent, then the apparent distribution coefficient determined from real minerals will vary as a function of cooling rate (e.g., Coogan et al. 2005). Nevertheless,

Table 1. Experimentally determined equilibrium partition coefficients for Li from the literature.

	K_D	*P* (GPa)	*T* (°C)	Reference
		Mineral/melt partition		
olivine–basaltic andesite	0.13–0.15	0.0001	1349	Brenan et al. (1998a)
olivine–basalt	0.35	1.0	1320	Brenan et al. (1998a)
olivine–basalt	0.20–0.43	0.0001	1150–1245	Ryan (1989)
olivine–basalt	0.21–0.56	3–14	1600–1900	Taura et al. (1998)
olivine–basalt	0.42–0.44	1	1330	Ottolini et al. (2009)
olivine–basalt	0.29	1.5	1315	McDade et al. (2003)
olivine–ne–basanite	0.29–0.42	1–2	1075–1100	Adam and Green (2006)
cpx–basaltic andesite	0.14–0.20	0.0001	1275	Brenan et al. (1998a)
cpx–basalt	0.59	3	1380	Hart and Dunn (1993)
cpx–basalt	0.27	1.0	1320	Brenan et al. (1998a)
cpx–basalt	0.25	1.5	1255	Blundy et al. (1998)
cpx–basalt	0.11–0.18	0.0001	1150–1245	Ryan (1989)
cpx–basalt	0.23–0.26	1	1330	Ottolini et al. (2009)
cpx–basalt	0.23	1.5	1315	McDade et al. (2003)
cpx–ne basanite	0.14–0.35	1–3.5	1025–1190	Adam and Green (2006)
cpx–leucite lamproite	0.17	1.5	1040–1175	Schmidt et al. (1999)
opx–basalt	0.20	1.0	1320	Brenan et al. (1998a)
opx–basalt	0.17–0.26	1	1330	Ottolini et al. (2009)
opx–basalt	0.17	1.5	1315	McDade et al. (2003)
opx–ne basanite	0.28	2.7	1160	Adam and Green (2006)
plag–basalt	0.22–0.24	0.0001	1150–1245	Ryan (1989)
plag–basalt	0.15	1.5	1255	Blundy et al. (1998)
plag–basalt	0.3–0.7	0.0001	1180–1220	Aigner-Torres et al. (2007)
amph–basaltic andesite	0.13–0.19	1.5	1000	Brenan et al. (1998a)
amph–basanite	0.12	1.5	1092	LaTourrette et al. (1995)
amph–ne basanite	0.09–0.10	1–2	1025–1050	Adam and Green (2006)
phlog–basanite	0.064	1.5	1092	LaTourrette et al. (1995)
phlog–leucite lamproite	0.2	1.5	1040–1175	Schmidt et al. (1999)
phlog–ne basanite	0.19–0.21	2–2.7	1050–1160	Adam and Green (2006)
bio–peralum granite	0.8–1.7*	0.2	650–750	Icenhower and London (1995)
musc–peralum granite	0.8	0.2	650	Icenhower and London (1995)
garnet–ne basanite	0.04	3.5	1190	Adam and Green (2006)
cord–peralum granite	0.12–0.44*	0.2	700–850	Evensen and London (2003)
		Mineral/fluid partition		
cpx	0.16	2.0	900	Brenan et al. (1998b)
[1]cpx	0.07–0.3	1.0	800–1100	Caciagli et al. (2011)
garnet	0.0083	2.0	900	Brenan et al. (1998b)
[2]plagioclase	0.09–0.3	1.0	800–1100	Caciagli et al. (2011)
[3]olivine	0.17–0.57	1.0	800–1100	Caciagli et al. (2011)
chrysotile	0.03–0.06	0.4	400	Wunder et al. (2010)
lizardite, antigorite	0.007–0.015	0.2–4	200–500	Wunder et al. (2010)
		Mineral/mineral partition		
plag–cpx	0.05–0.58*	0.0001	800–1200	Coogan et al. (2005)
ol–opx	2.02	1	1330	Ottolini et al. (2009)
ol–cpx	1.74	1	1330	Ottolini et al. (2009)
ol–cpx	1.2–4	1.0	800–1100	Caciagli et al. (2011)
ol–cpx	2	1.5	700–1100	Yakob et al. (2012)
cpx–opx	1.17	1	1330	Ottolini et al. (2009)

Notes:

amph = amphibole, bio = biotite, cord = cordierite, cpx = clinopyroxene, musc = muscovite, ne = nepheline, ol = olivine, opx = orthopyroxene, plag = plagioclase

*Strongly dependent on temperature

[1] $\ln D_{\text{cpx/fluid}} = (-7.3 \pm 0.5) + (7.0 \pm 0.7) \times 1000/T$

[2] $RT \ln D_{\text{plag/fluid}} = (162{,}000 \pm 26{,}000) - (188{,}000 \pm 28{,}000)X_{an}$

[3] $\ln D_{\text{ol/fluid}} = (-6 \pm 2) + (6 \pm 2) \times 1000/T$

on the basis of apparent equilibrium distribution of Li between minerals in peridotite xenoliths and basalt phenocrysts the following distribution order has been established: olivine > clinopyroxene ≥ orthopyroxene » garnet ~ spinel (Ryan and Langmuir 1987; Eggins et al. 1998; Seitz and Woodland 2000), leading to the following partitioning values: ${}^{ol/cpx}K_D = 1.1$ to 2.0, ${}^{ol/opx}K_D = 1.5$ to 3.5 and ${}^{opx/cpx}K_D = 0.5$ to 1.1. Considering that the equilibration temperatures of most peridotite xenoliths are high (>900–1200 °C in most cases), this is the appropriate temperature range over which these values apply. Seitz and Woodland (2000) concluded that Li partitioning between peridotite minerals is independent of both P and T, a conclusion supported by the recent experimental work of Caciagli et al. (2011) and Yakob et al. (2012).

Marschall et al. (2006) found systematic partitioning of Li between minerals in high-pressure (eclogite and blueschist) metabasaltic rocks from Syros, Greece. The concentration of Li in coexisting minerals follows the order: chlorite > glaucophane > clinopyroxene > paragonite > phengite > talc > tourmaline > Ca-amphibole > garnet > clinozoisite > quartz > titanite > albite > chloritoid > lawsonite.

Equilibrium Isotope Fractionation

Both equilibrium thermodynamics and experimental studies can be used to quantify the sense and temperature dependence of equilibrium isotopic fractionation. For example, the degree of isotopic fractionation, reported as the fractionation factor between two phases, a and b (or the α value, where $\alpha_{a-b} = ({}^7Li/{}^6Li)_a/({}^7Li/{}^6Li)_b$), is close to one at high temperatures and deviates progressively from one with falling temperature (Urey 1947). The nature of the deviation, i.e., whether α is above or below one, depends on the relative partitioning of ${}^7Li$ and ${}^6Li$ between the two phases at equilibrium, which is a function of bond energy. In general, the heavy isotope is preferentially partitioned into the site with the highest bond energy (i.e., strong and short, Schauble 2004). In the case of Li, the bond energy is a function of coordination number: the lower the coordination number, the higher the bond energy. In ferro-magnesian minerals (e.g., olivine, pyroxenes, biotite, chlorite, and cordierite), which form the main hosts for Li in most rocks, Li most commonly substitutes for Mg–Fe in octahedrally coordinated sites (Li and Peacor 1968; Lumpkin and Ribbe 1983; Robert et al. 1983; Brigatti et al. 2003; Bertoldi et al. 2004). By contrast, Li is thought to reside in tetrahedral coordination in aqueous fluids (Yamaji et al. 2001). Thus, ${}^7Li$ preferentially partitions into aqueous fluids relative to mafic minerals, giving waters higher δ^7Li values compared to mafic rocks (e.g., Chan et al. 1992). In a few minerals Li may occur in low coordination number sites as described above for quartz and staurolite. In these cases, the minerals may be isotopically heavier than coexisting water (Teng et al. 2006b; Wunder et al. 2007). Ab initio molecular dynamic modeling successfully predicts mineral-fluid fractionation of Li isotopes within 1‰ of measured experimental values (Kowalski and Jahn 2011). Such modeling of Li speciation in aqueous fluids at $P = 0.3$ to 6 GPa and $T = 1000$ K (Jahn and Wunder 2009) demonstrates an increase in mean coordination number with increasing pressure (and fluid density) from 3.2 up to 5. These results are consistent with the experimental results showing that ${}^7Li$ is preferentially released into fluids from minerals in which Li is octahedrally coordinated, but which show the reverse fractionation for minerals in which Li is tetrahedrally coordinated (e.g., staurolite or quartz). In general however, it appears that fluids are isotopically heavier than coexisting rocks based on field studies (e.g., Chan et al. 1992). This sense of fractionation (heavy water, lighter rock) is also seen in the H (e.g., Chacko et al. 2001) and B isotope systems (e.g., Wunder et al. 2005), but is the opposite of that observed in the O (e.g., Chacko et al. 2001) and Mg isotope systems (Tipper et al. 2006; Teng et al. 2010). Experimentally determined Li isotopic fractionation factors are compiled in Table 2.

High-temperature fractionation. There are a limited, but growing number of experimental determinations of the isotopic fractionation of Li between phases as a function of temperature. To date, there have been no determinations of α in igneous systems, but there have been several studies focused on mineral–fluid partitioning.

Table 2. Experimentally determined mineral(rock)–fluid fractionation factors for Li from the literature.

	α	*P* (GPa)	*T* (°C)	Reference	Notes
quartz	1.010–1.012	0.06–0.1	500	Lynton et al. (2005)	
muscovite	1.019–1.022	0.06–0.1	500	Lynton et al. (2005)	
quartz	1.005	0.05–0.1	400	Lynton et al. (2005)	
muscovite	0.998–0.999	2	300–500	Wunder et al. (2007)	$\Delta^7Li_{Li\text{-}mica\text{-}fluid} = -4.52(1000/T) + 4.74$
staurolite	1.001	3.5	670–880	Wunder et al. (2007)	fractionation not *T* dependent
spodumene	0.996–0.999	2	500–900	Wunder et al. (2006)	$\Delta^7Li_{cpx\text{-}fluid} = -4.61(1000/T) + 2.48$
spodumene	0.997	1,4	500–550	Wunder et al. (2011)	
spodumene	0.998	8	625	Wunder et al. (2011)	
Li amphibole	0.998	2	700	Wunder et al. (2011)	
antigorite, lizardite	0.996–0.999	0.2, 0.4, 4	200–500	Wunder et al. (2010)	α ~ micas $\Delta^7Li_{Li\text{-}mica\text{-}fluid} = -4.52(1000/T) + 4.74$
chrysotile	1.001	0.4	400	Wunder et al. (2010)	
smectite	0.998	0.0001	200–250	Vigier et al. (2008)	α correlates systematically with *T*
smectite	0.99	0.0001	90	Vigier et al. (2008)	
smectite	1	0.0001	22	Pistiner and Henderson (2003)	sorption experiments
gibbsite	0.986	0.0001	22	Pistiner and Henderson (2003)	sorption experiments
illite	0.989	0.1	300	Williams and Hervig (2005)	
calcite	0.9915	0.0001	5–30	Marriott et al. (2004a)	pure water, no *T* dependence of fractionation
calcite	0.997	0.0001	25	Marriott et al. (2004b)	artificial seawater, no dependence on salinity
aragonite	0.989	0.0001	25	Marriott et al. (2004b)	artificial seawater, no dependence on salinity
aragonite	0.9895–0.9923	0.0001	25	Gabitov et al. (2011)	seawater
Na-zeolite	0.978	0.0001	25	Taylor and Urey (1938)	
basalt	0.981	0.0001	25	Millot et al. (2010a)	diluted seawater, $\Delta_{solution\text{-}solid} = 7847/T - 8.093$
basalt	0.993	0.0001	250	Millot et al. (2010a)	diluted seawater, $\Delta_{solution\text{-}solid} = 7847/T - 8.093$
basalt	1	0.0001	22–55	Wimpenny et al. (2010a)	flow–through dissolution experiments

Lynton et al. (2005) reported that both quartz and muscovite take in ^7Li in preference to aqueous fluids (at 500 °C $\alpha_{qtz\text{-}fluid} = 1.010\text{–}1.012$; $\alpha_{musc\text{-}fluid} = 1.019\text{–}1.022$), with muscovite having higher δ^7Li than quartz. Whereas $\alpha_{qtz\text{-}fluid} > 1$ is consistent with the sense of isotopic fractionation predicted on the basis of the relative bond energies reviewed above, the finding that muscovite is heavier than fluid and even heavier than coexisting quartz runs counter to the expectation that octahedrally bound Li in mica should be isotopically lighter than that in coexisting fluids. Another curious finding of their study is that α appears to become closer to one at lower temperature (at 400 °C $\alpha_{qtz\text{-}fluid} = 1.005$, 1.010 to 1.012 at 500 °C), again, the opposite of prediction.

Wunder et al. (2007) also determined isotopic fractionation factors between muscovite and fluid, but in contrast to Lynton et al. (2005), found $\alpha_{musc\text{-}fluid} = 0.998$ to 0.999 at 350–400 °C, 2.0 GPa, consistent with predictions that the fluid should be isotopically heavier than muscovite. Wunder et al. (2007) note the contradictory findings of the two studies and suggest that the results of Lynton et al. (2005) may reflect incomplete equilibration, coupled with kinetic isotopic fractionation, between the starting minerals and fluids. However, arrested diffusive ingress of Li from the fluid into the muscovite should result in low δ^7Li, rather than the high mineral δ^7Li compositions observed (kinetic isotopic fractionation is discussed in detail below). Regardless of the reason for the discrepancies, the muscovite-fluid partitioning results of Lynton et al. (2005) remain enigmatic. Wunder et al. (2007) also investigated isotopic partitioning between staurolite and fluid and found a slight preference for ^{7}Li in staurolite ($\alpha_{staur-fluid} \sim 1.001$ at 670–880 °C, with no obvious temperature dependence), consistent with the tetrahedral coordination of Li in staurolite. In an earlier study Wunder et al. (2006) measured α between spodumene and aqueous fluids as a function of temperature at 2.0 GPa in both alkaline and acidic fluids and confirmed that ^{7}Li preferentially partitions into fluids over pyroxene ($\alpha_{spod-fluid} = 0.996$–$0.999$ at 500–900 °C) and that α correlates strongly with temperature, as predicted. A study by Wunder et al. (2010) determined partitioning and fractionation of Li between serpentine minerals and fluids. This study found that antigorite and lizardite fractionated Li similar to Li-mica (Wunder et al. 2007), however chrysotile fractionated Li in the opposite sense (heavier Li entering the chrysotile), suggesting that chrysotile nanotubes might play a role in hosting ^{7}Li within subduction zones (Wunder et al. 2010).

Using the experimentally determined fractionation factors for fluid–spodumene of Wunder et al. (2006), Marschall et al. (2007) predicted that fractionation of Li isotopes between subducting rock and released aqueous fluid should be limited to < 3‰. This amount of fractionation is significantly lower than that inferred by Zack et al. (2003) for slab dehydration based on the low δ^7Li values observed in some orogenic eclogites from the Alps. Marschall et al. (2007) suggested that the low δ^7Li values reported in Zack et al. (2003) may have been produced by kinetic isotopic fractionation, leading to the conclusion that subducted slabs are unlikely to represent a low δ^7Li component in the mantle, and, in fact, may have *high* δ^7Li values, because of their history of low temperature seawater–rock alteration.

The possible fractionation of Li between liquid and vapor has been addressed in only one study. Liebscher et al. (2007) observed small fractionations (< 0.5‰) between liquid and vapor at low *P* (20–28 MPa) and a temperature of 400 °C, with a very slight preference for ^{7}Li for the liquid relative to the vapor.

The effect of pressure on Li isotopic fractionation between minerals and fluids was investigated by Wunder et al. (2011), who measured Li partitioning between spodumene and fluid at pressures of 1, 4, and 8 GPa at temperatures from 500 to 625 °C. They found that ^{7}Li partitioned preferentially into the fluid ($\Delta^7\text{Li}_{spod-fluid} \approx -3.5‰$; $\alpha_{spod-fluid} = 0.997$) without dependence on pressure at 1 and 4 GPa, but with a slight decrease in fractionation ($\Delta^7\text{Li}_{spod-fluid} \approx -1.9‰$; $\alpha^{spod-fluid} = 0.998$) at 8 GPa. Additional experiments between Li-bearing amphibole and aqueous fluid at 700 °C and 2 GPa resulted in a $\Delta^7\text{Li}_{Li-amph-fluid} = -1.7‰$ ($\alpha_{Li-amph-fluid} = 0.998$).

Low temperature fractionation. There have been significantly more studies of Li isotope fractionation at lower temperatures. While several studies have demonstrated that lithium isotopes are not significantly fractionated during mineral dissolution (e.g., Pistiner and Henderson 2003; Wimpenny et al. 2010a, 2015; Verney-Carron et al. 2011), secondary mineral formation produces large isotopic fractionations, with the secondary minerals preferentially adsorbing or incorporating ^{6}Li, leaving the equilibrium fluid enriched in ^{7}Li (e.g., Taylor and Urey 1938; Chan et al. 1992, 1993, 1994; Huh et al. 1998; Zhang et al. 1998; Pistiner and Henderson 2003; Kisakurek et al. 2004, 2005; Rudnick et al. 2004; Chan and Hein 2007; Vigier et al. 2008; Wimpenny et al. 2010a, 2015). Likewise, experimental studies of carbonates demonstrate significant isotopic fractionation relative to water (Marriott et al. 2004a,b; Rollion-Bard et al. 2009; Gabitov et al. 2011). The results of these isotopic fractionation studies are detailed below.

Vigier et al. (2008) found that the degree of isotopic fractionation in smectite approaches 1 at high temperature, from $\alpha_{smectite-fluid} = 0.990$ at 90 °C to 0.998 at 200–250 °C, which reflects exchange of Li^+ and Mg^{2+} between the mineral's octahedral site and fluid. At lower temperature, from 25 to 90°C, there is little change in α, which they speculate may reflect the poor crystallinity of the clays in their experiments at these low temperatures. Such poorly crystalline materials have a large proportion of edge-octahedra which may fractionate Li differently. They conclude that natural clays, which develop greater crystallinity over time, would likely follow the extrapolation of their $\alpha_{smectite-fluid}$–T trend down to 0.983 at 25°C or 0.981 at 0°C, which is in reasonable agreement with the $\alpha_{clay-seawater}=0.983$ determined from observations on altered MORB by Chan et al. (1992). Sorption experiments performed on smectite at 22°C (Pistiner and Henderson 2003) produced no noticeable fractionation of Li isotopes.

Millot et al. (2010a) similarly observe a decrease in fractionation between basalt and water with increasing temperature in their exchange experiments. In their study, at 25°C, $\Delta^7Li_{solution-solid}=+19.4$‰ and at 250°C, $\Delta^7Li_{solution-solid}=+6.7$‰. This study compared experimental results with measurements of geothermal fluids from several different geothermal reservoirs and found that geothermal fluids were relatively homogeneous at a single site, but there were variations in isotopic composition between sites, which they attributed to different temperatures of fluid–rock interaction and also to different reservoir lithologies.

Isotopic fractionation during the transformation from smectite to illite was investigated by Williams and Hervig (2005). Smectite crystals were loaded with Li- (and B-) rich water into Au capsules in hydrothermal reaction vessels. The materials were K saturated in order to promote illite crystallization. Experimental products were washed, shaken and then exchanged with NH_4Cl to remove adsorbed Li from the interlayer in order to analyze only the isotope ratio of Li in the octahedral site. Reaction products of different sizes had different Li isotopic compositions, but a steady state composition was achieved between 45 and 150 days duration of the experiments. The coarse fraction (>2 mm) was interpreted by Williams and Hervig (2005) to represent new growth (since all starting material was < 0.2 mm). The isotopic composition of the coarse fraction showed an isotopic fractionation of water-clay of 11‰ ($\alpha_{illite-fluid}=0.989$).

Sorption experiments performed on gibbsite at 22°C (Pistiner and Henderson 2003) produced a fluid with δ^7Li that was 8 to 12‰ higher than the source fluid, suggesting a fractionation of $\alpha_{gibbsite-fluid}=0.986$. Wimpenny et al. (2015) suggested that Li isotopic fractionation in gibbsite is complex and that Li uptake may include at least two steps. First, the expansion of the gibbsite structure causes lithium to diffuse into the interlayer and fill octahedral vacancies, resulting in octahedral δ^7Li at least 16‰ lower than in the fluid phase ($\alpha_{gibbsite-fluid}\sim 0.984$). When octahedral vacancies are completely filled, uptake of lithium from solution occurs via relatively weak bonding at the surface and into the interlayer. This uptake of weakly bonded or exchangeable Li occurs with no isotopic preference. Because the δ^7Li of fluids is usually considerably higher than Li in mineral phases, such uptake of Li can reduce the bulk isotopic fractionation in the secondary phase. Thus, isotopic fractionation of Li might occur during diagenesis and the expulsion of water from clay mineral structures, creating shales and mudrocks that are isotopically lighter than the original secondary phases.

Inorganic calcite precipitation experiments (Marriott et al. 2004a) over the temperature range from 5 to 30°C found that calcite δ^7Li is ~8.5‰ lower than the solution from which it was precipitated, resulting in a fractionation factor of $\alpha_{calcite-fluid}=0.9915$. The study found no significant temperature-dependence of isotopic fractionation over this relatively narrow range in temperature. Marriott et al. (2004a) suggested that Li is not readily incorporated into the calcite structure, but rather enters interstitial sites ($D_{(Li/Ca)}=0.0092$ to 0.0030) and that less Li enters calcite with increasing temperature. This study also measured Li concentrations and isotopic compositions of natural coral aragonite. Lower $D_{(Li/Ca)}$ values were measured in the aragonite (0.0022–0.0028) compared to calcite, although a systematic anti-correlation of Li with temperature (as determined by O isotopes in this sample) is consistent with the results from the experimentally precipitated calcite (Marriott et al. 2004a). The lack of temperature-dependent isotopic fractionation in both synthetic and natural carbonates suggests it may directly record δ^7Li of natural seawater, and has the potential to shed light on ancient weathering processes (Marriott et al. 2004a). Further investigation of carbonate–fluid partitioning and fractionation (Marriott et al. 2004b) demonstrated that Li/Ca of calcite precipitated from solution increases as the salinity of the solution increases, however salinity has no significant effect on Li/Ca in aragonite. The authors suggest this may be due to different site occupancies of Li in calcite vs. aragonite and suggest that Li substitutes into the Ca site in aragonite, making Li/Ca of the solution the dominant control on Li/Ca in aragonite. On the other hand, Li is thought to be incorporated into calcite in interstitial sites, in which case the absolute concentration of Li in the solution may be the dominant control on Li/Ca in calcite. Fractionation of Li isotopes relative to the growth solution at 25 °C is observed in both aragonite and calcite, with $\alpha_{\text{calcite–fluid}} = 0.997$ and $\alpha_{\text{aragonite–fluid}} = 0.989$ (Marriott et al. 2004b).

Gabitov et al. (2011) investigated the Li isotopic composition of synthetic aragonite precipitated from seawater at 25 °C at different precipitation rates. Although the goal of the study was to determine the effects of precipitation rate on Li isotopic fractionation, they observed more variability within aragonite from a given experiment than from one experiment to another. The study found $\alpha_{\text{aragonite–fluid}} = 0.9895$ to 0.9923.

No dependence of δ^7Li on temperature, pH, pCO_2, or coral microstructure was found in an experimental study of coral growth by Rollion-Bard et al. (2009). This study found that growth zones within individual corals recorded homogeneous δ^7Li (as measured by SIMS), but they did observe a difference in δ^7Li in deep-sea coral species compared to shallow-water coral species. These results suggest that corals may provide a proxy for paleo-δ^7Li of seawater (Rollion-Bard et al. 2009).

Despite the significant isotopic fractionation observed between aragonite and water in the experiments described above, Misra and Froelich (2009) found no isotopic fractionation between three different species of foraminifera (*O. universa*, *G. menardi*, and *G. triloba*), as well as bulk foraminifera samples, and seawater. They then went on to trace the Li isotopic composition of seawater through the Cenozoic using the data from carefully cleaned foraminifera (Misra and Froelich 2012).

Diffusion and kinetic isotopic fractionation

The second way in which isotopes become fractionated in nature is through kinetic fractionation. In this case, one isotope is transported more rapidly than another in incomplete or uni-directional processes (e.g., evaporation, diffusion along a chemical potential gradient). The difference in diffusion coefficients between two isotopes is characterized by an empirical term β, defined as $\frac{D_{^6\text{Li}}}{D_{^7\text{Li}}} = \left(\frac{m_{^7\text{Li}}}{m_{^6\text{Li}}}\right)^{\beta}$ (Richter et al. 1999). In experiments, kinetic isotopic fractionations of Li so far documented are produced by arrested diffusion, where ^{6}Li diffuses up to 3% faster than ^{7}Li in both silicate melt and water (Richter et al. 2003, 2006) (Fig. 3). Experiments in which diffusion was induced within pyroxenes showed site-dependent mobility of Li, in some cases Li exhibited typical smoothly varying diffusion profiles, attributed to Li occupying fast-diffusing interstitial sites, but in other cases Li concentrations exhibited step-like profiles attributed to partitioning of Li into metal sites with more limited diffusivity (Richter et al. 2014a).

In olivine, Li exhibits complex diffusive behavior suggesting two mechanisms of diffusion operating simultaneously (Dohmen et al. 2010). Time series experiments showed that diffusion of Li within olivine crystals resulted in three regions across the profiles—an outermost region in which (in most cases) the Li concentration exhibits a concave upward shape, a plateau region, and an innermost region in which Li dropped down to the initial Li concentration of the olivine. These profiles are interpreted to be the result of Li partitioning between two different sites in olivine: an octahedral site, and an interstitial site. Diffusion of Li

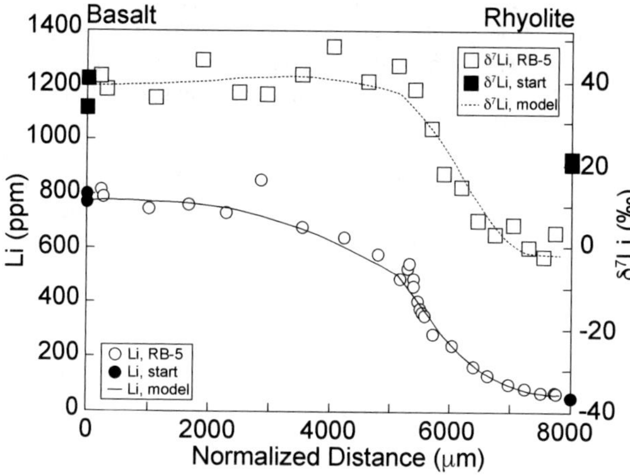

Figure 3: Kinetic isotopic fractionation of Li in basalt–rhyolite experimental couple (after Richter et al. 2003).

in the interstitial site is faster than the diffusion of Li in the octahedral site. Data for peridotites (e.g., Jeffcoate et al. 2007; Parkinson et al. 2007; Rudnick and Ionov 2007) suggest that the slower Li diffusion mechanism dominates in nature, while the experiments suggest that the diffusion rate is, on average, an order of magnitude faster than diffusion of Fe and Mg in olivine (Dohmen et al. 2010). By contrast, correlation of Mg and Li isotopes in peridotite xenoliths is best explained by co-diffusion of Li and Mg, with Mg diffusivity within an order of magnitude of Li diffusivity (Pogge von Strandmann et al. 2011). Concentration profiles through an olivine xenocryst within dioritic magma suggest that Li diffusivity and Fe–Mg interdiffusion are within a factor of three and that diffusivity varies with crystallographic orientation (Qian et al. 2010). Diffusion experiments involving olivine at 1300 °C confirm that there is variation of Li diffusion with olivine orientation, but also found that Li diffusion rates were similar to those of major cations in olivine (Spandler and O'Neill 2010).

Measurements of Li concentrations in stranded diffusion profiles at the edges of garnets from the Makhavinekh Lake Pluton provide estimates for the intracrystalline diffusion of Li in garnet at T=700–900°C and P=0.53 GPa, where $\log_{10}D^*$ ranges from −22.14 to −23.36 m²/s (Cahalan et al. 2014). These diffusivities are slower than those for divalent cations in the same samples, and are comparable to diffusivities of Y and Yb, suggesting that the diffusion of Li in these garnets is controlled by coupled substitution with trivalent cations (Cahalan et al. 2014). These results suggest that garnet may retain Li zoning, making it a useful indicator of Li behavior in crustal rocks.

Elemental redistribution in response to a temperature gradient (Soret 1879) may also cause isotopic fractionation. Bindeman et al. (2013) performed piston cylinder experiments in which they measured the redistribution of O, H, Li, Mg and Fe in wet partially molten andesite in response to a thermal gradient. The experiment was performed at 0.5 GPa for 66 days, with the hot end held at 952 °C and the cold end at 411 °C. There was a Li isotopic fractionation of 18.5‰ from the hot (low δ^7Li) to the cold (high δ^7Li) end. Likewise, Richter et al. (2014b) performed a piston cylinder experiment at 1.4 GPa for 24 hours on basaltic melt in which one end of the capsule was held at 1440°C and the other end at 1320°C. They found ~ 8‰ fractionation in Li isotopes, with high δ^7Li measured at the cold end and low δ^7Li at the hot end of the experiment.

Molecular dynamic simulations of the behavior of Li in water indicate that the self-diffusion coefficients of Li follow an inverse power-law dependence on the ionic mass (Bourg and Sposito 2007; Bourg et al. 2010), where the β value indicates the exponential dependence of the diffusivity on the ionic mass. The results of these simulations are in agreement with β=0.023 (Richter et al. 2006) for Li diffusion in water at 348 K.

Large kinetic isotopic effects have been observed at a variety of scales in natural samples: on the micron scale in minerals from peridotites (Jeffcoate et al. 2007; Rudnick and Ionov 2007; Tang et al. 2007a; Aulbach et al. 2008; Ionov and Seitz 2008; Gao et al. 2011; Pogge von Strandmann et al. 2011; Lai et al. 2015; Xiao et al. 2015) and phenocrysts in lavas (Barrat et al. 2005; Beck et al. 2006; Jeffcoate et al. 2007; Parkinson et al. 2007); at the centimeter scale in subduction-related metamorphic rocks at a blueschist–eclogite contact (Penniston-Dorland et al. 2010) and in an eclogite alteration envelope surrounding a vein in blueschist (John et al. 2012); and at the meter to tens of meters scale in peridotites adjacent to magmatic conduits (Lundstrom et al. 2005) and in contact metamorphic aureoles adjacent to plutons (Teng et al. 2006a; Marks et al. 2007; Liu et al. 2010; Ireland and Penniston-Dorland 2015). The magnitude of the fractionations can be huge (up to 20‰), and the diffusion length scales can be substantial (tens of meters), but the fractionations are ephemeral and will only survive if the system cools relatively quickly.

METHODS

Tomascak (2004) provides a historical perspective on the development of methods for Li isotope analyses, and Tomascak et al. (2016) provide a comprehensive review of all methods that have been employed in Li isotope analyses. Lithium isotopes are analyzed by both whole rock and *in situ* methods. Whole rock analyses require sample digestion followed by separation of Li from the solution matrix, while *in situ* methods (SIMS or laser ablation inductively coupled plasma mass spectrometry (LA-ICP-MS)) generally require use of matrix-matched standard reference materials to act as external calibrants.

Whole rock analyses

Dissolution and column chemistry. Whole rock dissolution is generally accomplished using high purity, concentrated acids (e.g., HNO_3, HCl, HF $\pm$ $HClO_4$) to achieve complete dissolution. High-pressure Teflon bombs may also be employed in order to dissolve refractory accessory phases such as zircon. Ushikubo et al. (2008) reported the surprising result that zircons can contain appreciable Li, up to 100 ppm, though the generally low abundance of zircon in rocks means that failure to dissolve zircon should not adversely influence the Li whole rock result. Perchloric acid ($HClO_4$) can be used to help dissolve fluorides created following HF digestion and is required to achieve complete dissolution of samples containing organic carbon. Flux-dissolution methods cannot be used in Li isotope analyses, as most procedures use Li-compounds such as lithium metaborate and lithium bromide as flux and wetting agents. Indeed, it is advisable to carry out all sample preparation for Li isotopes in a laboratory where such fluxes are not in use, as even a small amount of flux making its way into sample powders can have large effects on the Li isotope results (see the deposited Appendix for an example of this, as well as Košler 2014).

Once the sample is completely dissolved it is passed through cation exchange columns to achieve separation of Li from other constituents. This purification process is arguably the most important step in the analytical procedure, as other ionic constituents can influence the mass spectrometry, particularly the fellow alkali metal Na (e.g., Tomascak et al. 1999a; Nishio and Nakai 2002; Jeffcoate et al. 2004) and ^{7}Li passes more quickly through the exchange resin than ^{6}Li, requiring ~100% yield in order to avoid inducing isotopic fractionation in the eluent (Taylor and Urey 1938; Oi et al. 1997; Moriguti and Nakamura 1998b). Various recipes for Li purification can be found in the literature (e.g., Moriguti and Nakamura 1998a; Jeffcoate et al. 2004); these may involve use of alcohol (either methanol or ethanol, e.g., Moriguti and Nakamura 1998a; Tomascak et al. 1999a) or dilute HCl (James and Palmer 2000b) at some point in the procedure in order to achieve suitable separation from Na. Because Li release from columns may change with the bulk composition of the dissolved sample (James and Palmer 2000b; Chan et al. 2002b), column calibrations need to be performed using a wide variety of possible sample types in order to assure 100% yield. A safe way to assure 100% yield is to analyze the Li concentrations of pre- and post-Li column cuts (e.g., Marks et al. 2007).

Mass spectrometry. With the untimely death of lithium pioneer Dr. Lui-Heung Chan (Louisiana State University) in 2007, and the adaptation of multi-collector inductively coupled plasma mass spectrometry (MC-ICP-MS) methods by the Misasa and Bristol labs, to our knowledge there are no labs actively publishing Li isotope measurements by TIMS. Thus, we review here only ICP-MS protocols. The interested reader is referred to Tomascak et al. (2016) for a review of TIMS methods used in Li isotope analyses.

Lithium isotope analyses by either single or multi-collector ICP-MS employ the standard-sample-standard bracketing method to correct for instrumental mass bias (Tomascak et al. 1999b), where analyses of the unknown are bracketed by analyses of a standard. Final isotopic compositions are calculated as the relative difference between the $^7Li/^6Li$ ratio in the sample and the $^7Li/^6Li$ in a standard (either NIST 8545 (aka L-SVEC), see the earlier section *Lithium Systematics*; or isotopically similar IRMM–016, Qi et al. 1997) rather than absolute $^7Li/^6Li$ ratios.

Long-term reproducibility of Li isotope analysis by this method depends on the type of mass spectrometer (single vs. multi-collector), as well as the ability to achieve adequate purification of Li from other elements, as described above. Single collector ICP-MS (either quadrupole or magnetic sector) can obtain long-term reproducibilites of 1.5 to 2.0‰ (2σ) on solutions containing 1–10 ppb Li (Košler et al. 2001; Carignan et al. 2007; Misra and Froelich 2009). Better long-term reproducibilities are generally obtained by MC-ICP-MS, where 2σ estimates fall over a broad range from ±0.24‰ (Millot et al. 2004, using a Thermo-Finnigan *Neptune*), ±0.3‰ (Jeffcoate et al. 2004, using a Thermo-Finnigan Neptune), ±0.4‰ (Simons et al. 2010, using an VG Elemental Axiom), ±1.0‰ (Rudnick et al. 2004, using a Nu Plasma), to ±1.2‰ (Seitz et al. 2012, using a Thermo-Finnigan Neptune). The reasons for these different levels of long-term reproducibility, often determined on the same instruments of similar generation (cf., Jeffcoate et al. 2004; Seitz et al. 2012, who both used a Neptune), are not apparent, though may relate to the quality of the reagents, and/or the time period over which "long-term" is assessed; short-term reproducibility, for analyses obtained over the course of a project, can be better than long-term reproducibility. For example, Kasemann et al. (2005), using the same MC-ICP-MS method of Jeffcoate et al. (2004) report that "Repeats of full duplicate sample dissolutions [of standard reference materials, SRMs] were usually reproduced within 0.3‰ and always within 0.6‰ (2σ)", implying that the long-term reproducibility of the Bristol group's method may be greater than 0.3‰. Marschall et al. (2007) also found larger within session uncertainty (0.4‰) for the same method, but attributed it to inadvertent use of 'dirty' methanol.

The reproducibility (and accuracy) of MC-ICP-MS methods can also be assessed from data published for SRMs (see supplemental file, as well as GeoREM: http://georem.mpch-mainz.gwdg.de/). The δ^7Li of many well characterized SRMs show only small differences between published values from different laboratories (e.g., BCR–1 δ^7Li values range from +2.0 to +3.0‰, with $2\sigma=0.8$, $n=8$; BHVO–2 δ^7Li values range from +3.6 to +4.9‰, with $2\sigma=0.5$, $n=27$; JA–1 δ^7Li values range from +5.0 to +5.8‰, $2\sigma=0.8$, $n=4$; JR–2 δ^7Li range from +3.8 to +4.0‰, with $2\sigma=0.1$, $n=7$). However, published δ^7Li for other well-characterized SRMs can be much more variable, with δ^7Li varying by more than 2‰. For example, δ^7Li values reported for BHVO–1 range from +4.0 to +5.8‰ ($2\sigma=1.0$, $n=22$), those for AGV–1 range from +4.6 to +6.7‰ ($2\sigma=1.8$, $n=8$), those for AGV–2 range from +5.6 to +8.1‰ ($2\sigma=2.7$, $n=4$), and those for JB–2 range from +3.6 to +6.8‰ ($2\sigma=1.1$, $n=39$).

In situ analyses

To date, *in situ* analyses of Li isotopes have been carried out primarily by SIMS, and there is a significant amount of SIMS data now published for minerals and glasses (for a general review of SIMS methods see Stern 2009). By contrast, Li isotope analyses by LA-ICP-MS have only been presented in four papers (Marks et al. 2008; le Roux 2010; Xu et al. 2013; Martin et al.

2015), and the precision of measurement is generally poorer than for SIMS at comparable spatial resolution (e.g., Marks et al. 2008). Although le Roux (2010) achieved a 2σ precision of 1‰ for glasses having 3–35 ppm Li, this required analyses over large areas (500 μm long line with a spot size of 150 μm). Later studies using spot sizes between 40 and 150 μm obtained 2σ precisions of ~2 to 3‰ on glasses and minerals containing between 3 and 200 ppm Li (Xu et al. 2013; Martin et al. 2015). Thus, both the spatial resolution and precision of the measurement is currently poorer than for SIMS, and for this reason we focus here on SIMS methods.

Lithium isotope analyses by SIMS began in earnest in the late 1990s through the work of Chaussidon and colleagues at CNRS Nancy (Chaussidon and Robert 1998, 1999). As large mass fractionation occurs during the sputtering process, the $^7Li/^6Li$ ratio of the analyte must be corrected. This is usually done based on the mass bias observed in a reference material (e.g., Chaussidon and Robert 1998), though the matrix of the materials may also influence the resulting ratio (e.g., Marks et al. 2008; Bell et al. 2009). Thus, a first-order goal in SIMS analyses is the creation and characterization of homogeneous reference materials of variable bulk compositions (Decitre et al. 2002; Kasemann et al. 2005; Jochum et al. 2006, 2011; Marks et al. 2008; Ushikubo et al. 2008; Bell et al. 2009; Li et al. 2011; Ludwig et al. 2011; Su et al. 2015).

Early analyses of Li isotopes by small radius Cameca ion microprobes (e.g., Cameca 3f or 4f) found no evidence of matrix effects in the analyses of a variety of ferro-magnesian minerals (e.g., biotite, clinopyroxene, olivine, Decitre et al. 2002; Parkinson et al. 2007) or basaltic glasses (Kasemann et al. 2005; Parkinson et al. 2007). However, Kasemann et al. (2005) documented a significant matrix effect in the Si- and Na-rich NIST600 series glass reference materials relative to Fe–Mg-rich minerals. Later, Marks et al. (2008) documented matrix effects for Na-rich pyroxene and amphibole analyzed relative to basaltic glasses, and Bell et al. (2009) documented a matrix effect for olivine as a linear function of Fo content (between Fo_{74-94}) using a Cameca 3f and 6f ion microprobe, highlighting the need to employ matrix-matched standards (or make corrections, e.g., Xiao et al. 2015). Su et al. (2015) later confirmed the matrix effect for olivine using a Cameca 1280 ion microprobe, but found no obvious matrix effect for clinopyroxene or orthopyroxene (though the range of compositions they examined was small).

EXTRATERRESTRIAL LITHIUM RESERVOIRS

Studies of meteorites, lunar, and martian materials provide information about the Li isotopic composition of the solar system, the chondritic reservoir and the composition of the bulk silicate Earth. Additionally, these materials provide evidence for extraterrestrial processes such as galactic cosmic ray bombardment and the production and destruction of Li in stars, as well as interaction with solar wind. In some cases Li isotopes may preserve evidence of processes occurring in the early history of the solar system. A comprehensive review of the cosmology of Li is found in Tomascak et al. (2016).

The interstellar medium and the Sun

Interstellar medium. The interstellar medium contains Li from several sources. One source is pure ^7Li produced by nucleosynthesis during the Big Bang and also by subsequent nucleosynthesis in stars. Added to this is Li produced in spallation reactions. These reactions take place in the solar atmosphere when solar flares interact with ^{16}O and ^{12}C, producing Li with a $^7Li/^6Li$ of 2. Lithium is destroyed in stars by a reaction called astration, which destroys ^6Li more rapidly than ^7Li (Chaussidon and Robert 1999). Because of astration, solar wind is enriched in ^7Li. Measurements of absorption spectra made by training telescopes onto the double star cluster Rho Ophiuchi determined the $^7Li/^6Li$ of the interstellar medium to be 12.5 +2.8/−3.2 (Lemoine et al. 1993). Subsequent spectral measurements confirm this ratio for the composition of most of the gas in the solar neighborhood and suggest that this value has not changed significantly over the last 4.5 billion years (Knauth et al. 2003).

It is hypothesized that ^7Li formed along with other light elements such as H, He, and Be in the first few minutes of the Big Bang. The exact amount is debated because Big Bang Nucleosynthesis predictions do not match observations of cosmic microwave background radiation and observations of absorption spectra from metal-poor halo stars. This so-called "lithium problem" may indicate a new type of particle physics (different from the Standard Model) at work in the early universe (Cyburt et al. 2008).

Solar wind and the Sun. Chaussidon and Robert (1999) studied lunar soil particles to infer the Li isotopic composition of the solar wind. Solar wind particles can become implanted in the outermost 0.03 mm of lunar soil grains. Depth profile measurements of the Li isotopic composition of lunar soil grains showed elevated $^7Li/^6Li$ at grain edges. Using three-component mixing calculations, the compositions of the lunar grains were modeled as a mixture of lunar Li, lunar spallogenic Li, and solar wind Li. Based on these calculations, solar wind $^7Li/^6Li$ was estimated to be 31 ± 4, a value that is consistent with measurements of $^7Li/^6Li > 33$ of the solar photosphere (Ritzenhoff et al. 1997). The solar value of $^7Li/^6Li$ thus deviates significantly from meteoritic $^7Li/^6Li$, and since meteoritic Li isotopic compositions appear to be constant over the life of the solar system, this suggests the solar Li isotopic composition may have varied over time (Tomascak et al. 2016). Analyses of the δ^7Li of different size fractions of lunar soils showed that finer-grained fractions and soil with a longer exposure age have elevated δ^7Li (+8.4 and +8.3‰ respectively), but coarser grained fractions have lower δ^7Li that is similar to that of bulk rock lunar measurements ($\delta^7Li = +5$‰, Magna et al. 2006b). This study confirms the higher δ^7Li of solar wind, which affects the fine-grained soil fractions, but does not affect measurements of coarse-grained soil fractions or measurements of coarse-grained lunar rocks.

Meteorites and their components

Bulk and *in situ* analyses of chondritic and achondritic meteorites have revealed significant heterogeneities in Li isotopic compositions. Measurements of different components of these meteorites, such as Ca–Al-rich inclusions (CAIs), chondrules, and dark inclusions show further that this heterogeneity exists not just between different samples but also within individual samples.

Chondrites. Chondritic δ^7Li is between +3 and +4‰ (averages from Seitz et al. 2007, $\delta^7Li = +3.2 \pm 1.9$‰; Pogge von Strandmann et al. 2011, $\delta^7Li = +3.5 \pm 0.5$‰). Chondritic δ^7Li is likely the product of mixing of Li produced during Big Bang nucleosynthesis with Li produced by spallation reactions during the evolution of the galaxy. Pogge von Strandmann et al. (2011) found lower δ^7Li in enstatite chondrites compared to carbonaceous and ordinary chondrites and suggested this might be due to formation of enstatite chondrites closer to the Sun where there may be greater influence of spallation reactions.

Ion microprobe measurements of δ^7Li in chondrules from the Semarkona meteorite range from −12.7 to +34.8‰ (Chaussidon and Robert 1998) and are interpreted as the product of mixing of different sources. Another study (Seitz et al. 2012) in which ion microprobe measurements of chondrules from carbonaceous (Allende) and five ordinary chondrites (Semarkona, Bishunpur, Saratov, Bjurböle, and Bremervörde) were compared to bulk measurements of the same meteorites also demonstrated intra-sample heterogeneity. *In situ* measurements of chondrules in this study found δ^7Li ranging from −8.5 to +10‰, and the mean δ^7Li of chondrules from a given sample had lower δ^7Li when compared to associated bulk rock measurements. This study inferred the variability in δ^7Li to be the result of heterogeneities on a small scale within the chondrule-forming reservoir (Seitz et al. 2012). Differences between the bulk δ^7Li composition of carbonaceous and ordinary chondrites were interpreted to be a result of mixing different components (i.e., chondrules, CAIs, and matrix) in varying proportions. Heterogeneity of δ^7Li within individual samples was also observed by Pogge von Strandmann et al. (2011) in measurements of the carbonaceous chondrites Allende (CV3), Orgeuil (CI), and Murchison (CM2), and the ordinary chondrite Parnallee (LL3.6).

Lithium–7 is also the product of decay of ^{7}Be, which is a short-lived radionuclide with a half-life of 53 days (Chaussidon et al. 2006a). Correlation of ^{7}Li/^{6}Li with ^{9}Be/^{6}Li in an Allende CAI suggests the former presence of ^{7}Be (Chaussidon et al. 2006a), although this interpretation is debated (see also Desch and Oullette 2006; Chaussidon et al. 2006b; Lundstrom et al. 2006; Leya 2011). The significance of live ^{7}Be is that it would have formed within the early solar system due to irradiation of parts of the nebula and accretion disk by energetic particles from the young Sun (Chaussidon et al. 2006a). A study by Liu et al. (2009) of the Murchison chondritic meteorite found no evidence for live ^{7}Be during its formation.

Phyllosilicates, carbonates, and dark inclusions in chondritic meteorites are interpreted to be the product of aqueous alteration and heating in planetesimals (e.g., Sephton et al. 2004). As described in the section on low temperature fractionation, ^{7}Li partitions preferentially into aqueous fluids during low-temperature fluid-mineral interactions, producing a ^{7}Li-rich fluid. Sephton et al. (2004, 2006) analyzed different constituents of the carbonaceous chondrites Allende and Murchison in order to look for systematic variation in the Li isotopic composition of those parts of meteorites that experienced differing degrees of fluid–rock interaction. Sephton et al. (2004) found an increase in δ^7Li from unaltered chondrules (δ^7Li = –1.9‰) to phyllosilicate-rich matrices (δ^7Li = +5.8 to +6.2‰) to carbonate-rich samples (δ^7Li = +12.6 to +13.0‰) in the Murchison meteorite. Dark inclusions in Allende, however, do not have a resolvably different Li isotopic composition than the whole rock meteorite, suggesting that the history of the Allende dark inclusions is similar to that of the rest of the meteorite (Sephton et al. 2006). Sephton et al. (2013) compared the Li isotopic composition of low-temperature alteration products, such as carbonates and poorly crystalline Fe-(oxy)hydroxides removed from whole rock samples by acetic acid leaching, to the associated unleached whole rock meteorite samples. They found that the alteration products overall had higher δ^7Li relative to the whole rock δ^7Li, and that the type 1 chondrite had a lesser enrichment in ^{7}Li in the alteration products compared to the type 2 chondrites. Sephton et al. (2013) hypothesized that this difference might reflect a higher temperature of alteration for the type 1 chondrite (producing more crystalline phases that do not leach in acetic acid) relative to the temperature of alteration for the type 2 chondrites. Maruyama et al. (2009) found a wide range of Li isotopes in a SIMS study of the Allende meteorite. These *in situ* analyses showed that δ^7Li in chondrules ranged from –32 to +21‰. They found no consistent spatial variation with respect to cores and rims in Li concentrations or isotopic compositions in chondrule olivines. They therefore interpreted this heterogeneity to be the product of heterogeneous chondrule precursor compositions. Variable concentrations of Li in the mesostasis of Allende were interpreted by Maruyama et al. (2009) to be the result of multi-stage fluid-related processes.

Martian Meteorites (SNC: Shergottites, Nahklites and Chassigny). Bulk rock analyses of martian meteorites show a relatively restricted range of Li isotopic compositions. Pristine, high-MgO martian shergottites have an average δ^7Li of +4.2 ± 0.9‰, which is interpreted to represent the Li isotopic composition of bulk silicate Mars (Magna et al. 2015). Most martian meteorites have δ^7Li that falls between +2.1 to +6.2‰ (Magna et al. 2006b, 2015; Seitz et al. 2006; Filiberto et al. 2012). Processes that have been suggested to create the variability of δ^7Li observed in martian meteorites include terrestrial weathering and alteration, crustal contamination on Mars, and assimilation of a Cl-rich, low δ^7Li fluid during cooling. Theoretical calculations (Fairen et al. 2015) suggest that low-temperature weathering on the surface of Mars has the potential to produce variations in δ^7Li and that analysis of sedimentary materials from the surface of Mars could provide information about the nature and extent of weathering, and characterize the martian paleoenvironment, including duration and intensity of pH fluctuations and by association, the density and temporal evolution of the martian atmosphere (Fairen et al. 2015).

Despite the limited range in δ^7Li for martian whole rocks, significant spatial variations in δ^7Li (–17 to +46‰), are found in pyroxenes from martian meteorites measured *in situ* using

SIMS. These variations were initially interpreted as the product of degassing (Beck et al. 2004) during melt crystallization. Evidence for degassing in martian meteorites is significant because it has implications for the water content of primitive martian magmas, which is highly debated. Further *in situ* work on martian meteorite pyroxenes demonstrated that such variations may be better explained by diffusion-induced kinetic fractionation of Li within pyroxene crystals during magmatic cooling (Beck et al. 2006; Udry et al. 2016), although pyroxene crystals from a relatively young, depleted shergottite (QUE 94201) exhibit Li systematics that may indicate degassing (Udry et al. 2016). The measurements of δ^7Li in two pyroxenes from nakhlite meteorites revealed isotopic compositions ranging from −11 to +20‰ with spatial variations suggesting subsolidus diffusion of Li (Beck et al. 2006). Diffusion modeling of these profiles suggested cooling rates of 75 and 50 K/hr (Beck et al. 2006). These cooling rates are one to two orders of magnitude higher than other estimates of cooling rates available in the literature (based on Fe/Mg and Ca exchange between pyroxene and olivine and also on results from experimental petrography), however uncertainties in cooling rates estimated in this fashion are also typically one or two orders of magnitude due to the propagation of uncertainties in measured activation energies (Beck et al. 2006).

Moon

Bulk-rock analyses of lunar meteorites and basalts have revealed a relatively restricted range of δ^7Li, between +3.5 and +6.6‰ (Magna et al. 2006b; Seitz et al. 2006; Day et al. 2016). These values are similar to chondritic and bulk silicate Earth Li isotopic compositions, suggesting a chondritic origin for both Earth and its moon and also a uniform inner solar system Li isotopic reservoir. Exceptions include a ferroan anorthosite that has δ^7Li of +8.9‰ and a KREEP-rich highland breccia with δ^7Li of +18.6‰. SIMS analyses of olivines, pyroxenes, and magmatic inclusions in lunar meteorite NWA 479 (Barrat et al. 2005) exhibit a wide range of δ^7Li (from +2.4 to +15.1‰), which the authors attribute to diffusive fractionation. The cores of the olivines had high δ^7Li, which the authors interpreted to represent the δ^7Li composition of the melt (~+15‰). Magna et al. (2006b) hypothesized that variations in δ^7Li may be due to fractionation during late stages of fractional crystallization of the lunar magma ocean. Day et al. (2016) found that high-Ti lunar basalts had significantly higher δ^7Li (+5.2 ± 1.2‰, 2σ) than low-Ti basalts (+3.8 ± 1.2‰, 2σ). These results suggest that isotopic fractionation of Li occurs after extensive large-scale magma ocean crystallization (Day et al. 2016).

TERRESTRIAL LITHIUM RESERVOIRS

Mantle peridotites

Equilibrium partitioning of Li between peridotitic minerals has been determined in experiments where the different minerals equilibrated with a melt (Brenan et al. 1998a; Blundy and Dalton 2000; Ottolini et al. 2009) or fluid (Caciagli et al. 2011; Yakob et al. 2012) at variable pressure and temperature conditions. These data for olivine–clinopyroxene are summarized in Figure 4, where it can be seen that $D_{Li}^{ol/cpx}$ is ~2. Olivine–orthopyroxene partitioning is similar to olivine–cpx partitioning, with olivine having about twice the Li concentration as orthopyroxene. Equilibrium partitioning inferred from Li concentrations in minerals from 'equilibrated' peridotites and pyroxenites yields similar results to the experimental data (Seitz and Woodland 2000), though the very large range in natural apparent partitioning (small symbols in Fig. 4) make this method more challenging. These partitioning relationships appear to be independent of both temperature and pressure (Seitz and Woodland 2000; Yakob et al. 2012).

In comparison to elemental partitioning, determining isotopic equilibrium fractionation factors for mantle minerals has proved more difficult. Such fractionation factors have yet to be measured experimentally at any temperature, and the only available information comes

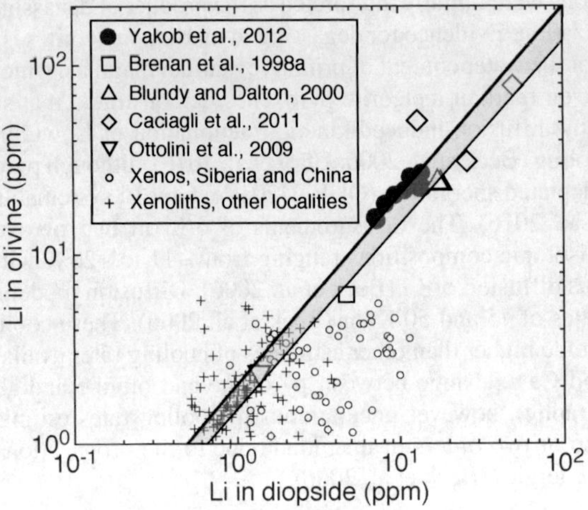

Figure 4. Equilibrium partitioning of Li between clinopyroxene and olivine, after Yakob et al. (2012). Large symbols are experimentally derived partitioning where both olivine and clinopyroxene were in equilibrium with a melt or fluid. Small symbols are data for peridotite or pyroxenite xenoliths. Diagonal shaded field is the range of equilibrium partitioning deduced by Seitz and Woodland (2000) on the basis of Li contents in 'equilibrated' peridotites and pyroxenites. Neither Seitz and Woodland (2000) or Yakob et al. (2012) found evidence for partitioning relationships changing with changing temperature or pressure.

from observations made on natural samples. A subset of xenolithic peridotites studied by Seitz et al. (2004) displaying equilibrium elemental partitioning of Li between minerals show $\delta^7Li_{ol} > \delta^7Li_{opx} > \delta^7Li_{cpx}$, with Δ^7Li_{ol-cpx} (i.e., $\delta^7Li_{ol} - \delta^7Li_{cpx}$) up to ~4‰. Seitz et al. (2004) observed that Δ^7Li_{ol-opx} of these samples correlated negatively with equilibration temperatures calculated from two-pyroxene thermometers, and they hypothesized that the Li isotopic fractionations reflected equilibrium partitioning, with higher δ^7Li in olivine than in clinopyroxene. In contrast to these findings, Jeffcoate et al. (2007) suggested the opposite sense of isotopic fractionation. They hypothesized that, by analogy with the oxygen and magnesium stable isotope systems, olivines should have *lower* δ^7Li than coexisting pyroxenes at equilibrium. The peridotite they measured having the heaviest clinopyroxene, in which $\Delta^7Li_{ol-cpx} = -3.6$‰, was suggested to be closest to equilibrium. These very different estimates of equilibrium isotopic fractionation, coupled with the prevalent isotopic disequilibria observed between and within minerals in many peridotites, described below, highlights the need for experimental determination of isotopic fractionation factors for mantle minerals.

Natural peridotites exhibit a huge range in δ^7Li in both whole rocks and coexisting minerals; δ^7Li varies from −9.7 to +9.6‰ for whole rocks (Figs. 1, 5), whereas δ^7Li varies from −7.9 to +15.1‰ for olivine; −21.4 to +6.8‰ for clinopyroxene, −26.0 to +6.5‰ for orthopyroxene, +2.7 to +4.4‰ for garnet, and −18‰ for phlogopite separates (see Fig. 6 for olivine and pyroxenes). *In situ* data are even more variable, with δ^7Li varying from −42 to +24‰ in olivine, −37 to +15‰ in clinopyroxene, −56 to +14‰ in orthopyroxene and −6 to +17‰ in phlogopite analyzed by SIMS. This great variability in Li isotopes between and within peridotites has been attributed to both kinetic isotopic fractionation due to Li diffusion (Lundstrom et al. 2005; Jeffcoate et al. 2007; Rudnick and Ionov 2007; Tang et al. 2007a, 2011; Aulbach et al. 2008; Ionov and Seitz 2008; Kaliwoda et al. 2008; Magna et al. 2008; Aulbach and Rudnick 2009; Kil 2010; Gao et al. 2011; Pogge von Strandmann et al. 2011; Su et al. 2012, 2016; Lai et al. 2015) and to addition of isotopically distinct Li from metasomatizing melts and fluids (Nishio et al. 2004; YJ Tang et al. 2007a, 2014; Aulbach et

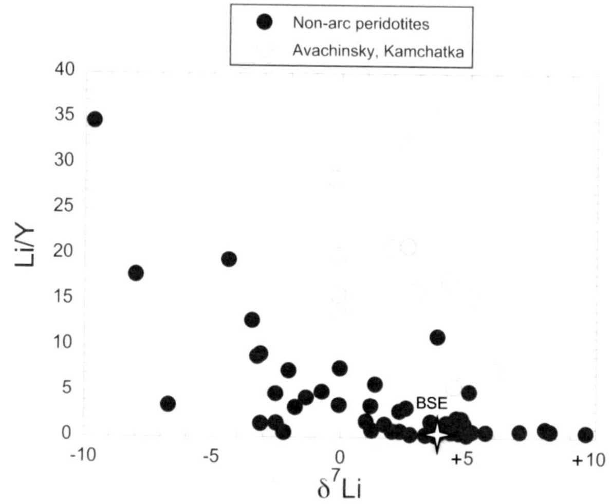

Figure 5. Whole rock δ^7Li vs. Li/Y ratio for mantle peridotites. Avachinsky samples, shown in open symbols (Ionov and Seitz 2008; Pogge von Strandmann et al. 2011) are from an arc setting. Remaining samples (solid symbols) are either xenolithic peridotites carried in alkali basalts erupted in intraplate or rift settings (Magna et al. 2006b, 2008; Jeffcoate et al. 2007; Pogge von Strandmann et al. 2011; Ackerman et al. 2013), or are from the Red Sea rift (Brooker et al. 2004). Arc peridotites are systematically enriched in Li (relative to Y). For the remaining samples there is a general trend of decreasing δ^7Li with increasing Li/Y. Open cross encompasses all estimates for δ^7Li of fertile mantle or bulk silicate earth (BSE) (from Seitz et al. 2004; Magna et al. 2006b; Jeffcoate et al. 2007; Pogge von Strandmann et al. 2011) and Li/Y from McDonough and Sun (1995).

al. 2008; Su et al. 2016). In principle, it may be possible to distinguish between these two very different processes based on the spatial variation in δ^7Li within a mineral or whole rock, as early stages of Li ingress through diffusion generates a characteristic 'dip' in δ^7Li, since ^{6}Li diffuses faster than ^{7}Li (e.g., Richter et al. 2003; Gallagher and Elliott 2009). However, as diffusion progresses, this dip will diminish and may leave a profile with a smooth gradient in δ^7Li, which would be difficult to distinguish from addition of isotopically distinct Li. We review both kinetic fractionation and addition of isotopically distinct Li below.

Kinetic fractionation. The majority of Li data for peridotite minerals exhibits isotopic disequilibrium between coexisting phases. Exceptions are some xenoliths, erupted mainly in pyroclastic units, for which δ^7Li in different phases are within ± 2‰ of one another (Fig. 6, areas within diagonal lines). However, even for these apparently equilibrated samples, absolute δ^7Li varies far beyond what are considered typical mantle values (e.g., δ^7Li varies from ~−10 to +14‰ in 'equilibrated' olivine and orthopyroxene, and over a similar range for olivine and clinopyroxene, Fig. 6). This isotopic variability in whole rocks could also reflect kinetic fractionation, if Li diffused over meters, as documented by Lundstrom et al. (2005) for the mantle, and by many others for the crust (Teng et al. 2006a; Marks et al. 2007; Liu et al. 2010; Ireland and Penniston-Dorland 2015). The general trend of decreasing δ^7Li with increasing Li/Y seen in non-arc peridotites (Fig. 5) is consistent with kinetic fractionation within the whole rocks due to diffusive Li transport over meter scales (e.g., Lundstrom et al. 2005; Tang et al. 2011). The alternative interpretation, that unusually light Li was added to the peridotites, possibly derived from recycled slabs (Nishio et al. 2004; Tang et al. 2012; Ackerman et al. 2013; Su et al. 2016), seems at odds with the fact that Li-enriched arc peridotites do not have unusually low δ^7Li (Fig. 5, Ionov and Seitz 2008; Halama et al. 2009; Pogge von Strandmann et al. 2011).

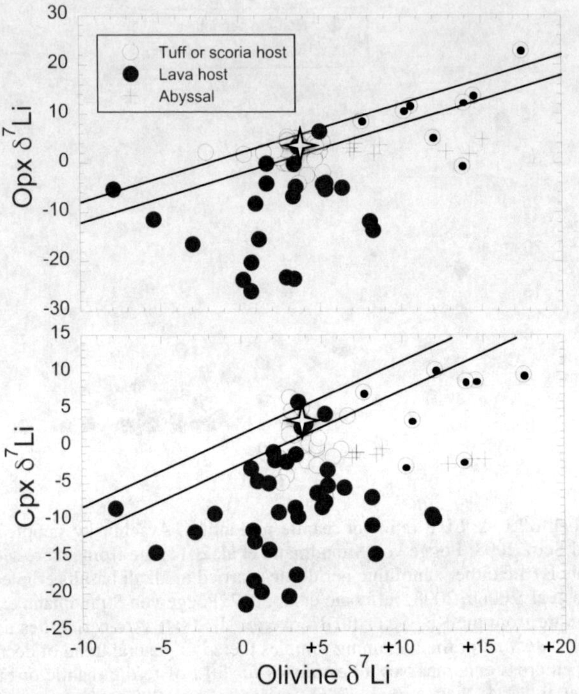

Figure 6. Lithium isotopic compositions of coexisting minerals in peridotite xenoliths carried in tuffs or scoria [open symbols, data from Nishio et al. (2004); Magna et al. (2006b); Jeffcoate et al. (2007); Wagner and Deloule (2007); Aulbach et al. (2008); Ionov and Seitz (2008); Aulbach and Rudnick (2009); Halama et al. (2009)] or lavas [closed symbols, data from Nishio et al. (2004); Seitz et al. (2004); Magna et al. (2006b, 2008); Jeffcoate et al. (2007); Rudnick and Ionov (2007); Tang et al. (2007a, 2011)] as well as data for abyssal peridotites from the Gakkel Ridge (Gao et al. 2011) (crosses). Data determined by ion microprobe are shown with a dot at the center of the symbol (Wagner and Deloule 2007). Most papers that report *in situ* δ^7Li measurements do not provide average values for a given mineral and are not plotted here. White crosses are the δ^7Li estimate for the bulk silicate earth (from Seitz et al. 2004; Magna et al. 2006b; Jeffcoate et al. 2007; Pogge von Strandmann et al. 2011). Diagonal lines mark region where minerals are within ±2‰ of each other, which may represent equilibration between the phases. Most peridotitic minerals exhibit disequilibria, suggesting kinetic isotopic fractionation associated with lithium diffusion and/or recent addition of isotopically distinct lithium from a melt or fluid.

There are two ways to generate isotopic disequilibria between minerals due to diffusion. First, Li could diffuse into the minerals from a grain boundary melt/fluid (Jeffcoate et al. 2007; Rudnick and Ionov 2007; YJ Jang et al. 2007a, 2011). Second, Li may diffuse between phases due to changing partition coefficients during slow cooling (Ionov and Seitz 2008; Kaliwoda et al. 2008; Gao et al. 2011). Lundstrom et al. (2005) were the first to document kinetic isotopic fractionation in nature. They found that δ^7Li changed systematically in clinopyroxenes in harzburgite and lherzolite with distance from discordant dunites in the Josephine ophiolite. The δ^7Li first decreased then increased with increasing distance to the contact, consistent with kinetic fractionation of Li isotopes due to diffusion of Li from the dunite channel into the surrounding peridotite (see section on kinetic fractionation). Rudnick and Ionov (2007) observed up to a 23‰ difference in δ^7Li between coexisting olivine and clinopyroxenes in mantle xenoliths from far east Russia, with the olivines moderately enriched in Li and systematically heavier than normal mantle and the clinopyroxenes strongly enriched in Li and systematically lighter than normal mantle. They attributed these features to fractionation of Li diffusing from a Li-rich grain boundary melt, with faster diffusivity in clinopyroxene than olivine. Jeffcoate et al. (2007)

observed up to a 13‰ difference in δ^7Li between coexisting olivine and clinopyroxene in a peridotite xenolith from San Carlos, Arizona. *In situ* isotopic profiles across minerals from this xenolith revealed large and systematic isotopic fractionations (Fig. 7), which they attributed to kinetic fractionation associated with Li diffusion into the minerals from a grain boundary melt that had an unusually light Li isotopic composition due to diffusion within the melt.

Although inter-mineral isotopic fractionation may be generated by diffusion from a grain boundary, not all features of inter-mineral isotopic disequilibria can be simply explained through kinetic fractionation. For example, even with a two-stage model such as that proposed by Jeffcoate et al. (2007), it is difficult to explain the increase in both Li concentration and δ^7Li towards the rim of the clinopyroxene (Fig. 7), as these two parameters should behave antithetically during diffusion. That is, if Li concentration increases towards the rim, δ^7Li should decrease, and vice versa. Indeed, increases in Li concentration and δ^7Li towards the rims of clinopyroxene have been observed in other xenolithic peridotites where coexisting orthopyroxene and olivine show evidence for kinetic isotopic fractionation (Tang et al. 2007a; Xiao et al. 2015). These authors also explained these observations through a two-stage process of kinetic fractionation followed either by metasomatic overprinting (Tang et al. 2007a) or slow cooling (Xiao et al. 2015).

Some of the most compelling evidence for generation of low δ^7Li by diffusion was presented by Pogge von Strandmann et al. (2011), who observed an excellent correlation between δ^7Li and δ^{24}Mg in whole rock xenolithic peridotites, which they attributed to kinetic fractionation associated with diffusion of both Li and Mg from the host into the xenoliths. This is a compelling argument, as the light isotope in any system will diffuse faster than the heavy isotope. By contrast, equilibrium fractionation between minerals and fluids produces light Li but heavy Mg in the solid, unlike the correlation observed in the xenoliths.

In contrast to diffusion of Li from a Li-enriched grain boundary melt, Ionov and Seitz (2008) suggested that the isotopic disequilibria observed in some xenolithic peridotites could simply reflect Li diffusion driven by changing partitioning behavior during slow cooling of the rock (see also Jeffcoate et al. 2007; Gallagher and Elliott 2009). Specifically, Ionov and Seitz (2008) suggested that Li will diffuse from olivine into clinopyroxene as temperature drops (note that Kaliwoda et al. 2008, suggested the opposite sense of partitioning behavior

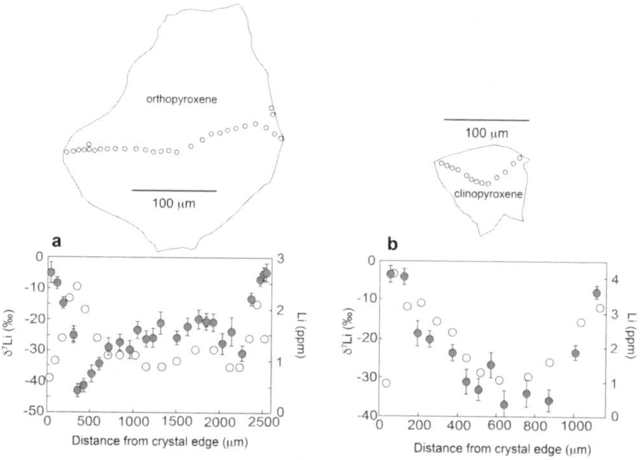

Figure 7. Lithium concentration (open symbols) and δ^7Li (closed symbols) across orthopyroxene (**a**) and clinopyroxene (**b**) in minerals from a xenolithic peridotite from San Carlos, AZ, analyzed by SIMS (modified from Jeffcoate et al. 2007). The symmetrical dip in δ^7Li on either side of the orthopyroxene is accompanied by increase in [Li] and is best explained by diffusion of Li into the orthopyroxene from surroundings. By contrast, the pattern of high δ^7Li and [Li] on towards the rims of the clinopyroxene are difficult to explain through Li diffusion.

as a function of temperature). In support of this hypothesis, Ionov and Seitz (2008) noted that isotopic disequilibrium is mainly developed in xenoliths contained within lavas, which would cool more slowly than xenoliths carried in tuffs or other pyroclastic deposits (e.g, scoria, breccia) that quench upon eruption. The more extensive data now available (Fig. 6) generally support this dichotomy, as most of the peridotites that plot within the equilibrium field (between the diagonal lines in Fig. 6) tend to be from pyroclastic deposits. There are also a number of pyroclastic-hosted peridotites that show disequilibria in Figure 6, and some of these samples show evidence for Li diffusion on the grain scale (e.g., Tang et al. 2012), though one could argue that some pyroclastic units may cool more slowly than others. The changing partition coefficient hypothesis, however, has foundered on recent experimental partitioning data. Both Caciagli et al. (2011) and Yakob et al. (2012) showed that the partitioning of Li between olivine and clinopyroxene does not change significantly as a function of temperature (between 700 and 1,100°C), consistent with the previous observations of Seitz and Woodland (2000) on natural samples. Thus, it seems that if the prevalent isotopic disequilibria observed between mantle minerals is due to diffusion, then it was likely generated from Li diffusion from a Li-rich grain boundary melt or fluid. In this light, it is interesting that Li isotopic disequilibrium has also been observed in abyssal peridotites (Gao et al. 2011) and in massif peridotites (Lai et al. 2015). Although the abyssal peridotite data were ascribed to a cooling effect, the disequilibrium observed in the Horomon Massif was attributed to Li ingress from grain boundary melts that infiltrated shortly before their emplacement. The latter hypothesis is in better keeping with the experimental partitioning data cited above. Finally, it is worth noting that cooling cannot explain whole rock data, described above, that show evidence for kinetic fractionation.

Metasomatic overprinting. An alternative interpretation of the wide variability of δ^7Li commonly observed in peridotites is that it results from recent addition of isotopically distinct Li from a metasomatizing agent (fluid or melt). In most cases, metasomatic addition is thought to have occurred in addition to kinetic fractionation (Tang et al. 2007a, 2012; Aulbach et al. 2008). Given the dramatic changes generated in Li isotopes due to diffusion, which can happen at the scales of millimeters to meters, distinguishing between kinetic fractionation and metasomatic addition of exotic Li can be challenging. In general, though, there have been a few observations that are difficult to ascribe to kinetics and these provide the strongest evidence for addition of exotic Li to peridotites.

Aulbach et al. (2008) observed correlations between olivine composition (forsterite number (Fo), Ni, and Li concentrations) and δ^7Li in Fe-enriched mantle xenoliths from Labait, Tanzania, suggesting addition of isotopically heavy Li associated with Fe enrichment (Fig. 8). As Re enrichment accompanied Fe enrichment, and there was little evidence for ingrowth of ^{187}Os in the samples, they suggested that the enrichment occurred during development of the East African Rift in the Cenozoic, which has been suggested to be fed by a plume with HIMU radiogenic isotopic characteristics and HIMU lavas are associated with heavy Li (see section on OIB). Correlations between δ^7Li and radiogenic isotopes (^{87}Sr/^{86}Sr and ^{143}Nd/^{144}Nd) were cited as further evidence in support of this hypothesis. Similar correlations between FeO, Li and δ^7Li have been observed in peridotites from two other locations: 1) xenolithic peridotites from the Avachinsky Volcano, which were attributed to metasomatism of the subarc mantle by heavy, slab-derived fluids (Pogge von Strandmann et al. 2011), and 2) in olivines from mantle xenoliths from Eastern China, which were also explained by reaction of the peridotites with Li-enriched, isotopically heavy, Fe-rich melts (Xiao et al. 2015).

Additional evidence for metasomatic overprinting by isotopically exotic melts is observed at the mineral scale. As highlighted above, SIMS profiles through peridotitic minerals that show a positive correlation between Li abundance and δ^7Li (e.g., Fig. 7, from Jeffcoate et al. 2007; also Tang et al. 2007a; Xiao et al. 2015) are difficult to explain by kinetic fractionation, which should instead produce a negative correlation. In cases where such positive correlations exist, the addition of Li is accompanied by increased δ^7Li, consistent with addition of heavier Li from a

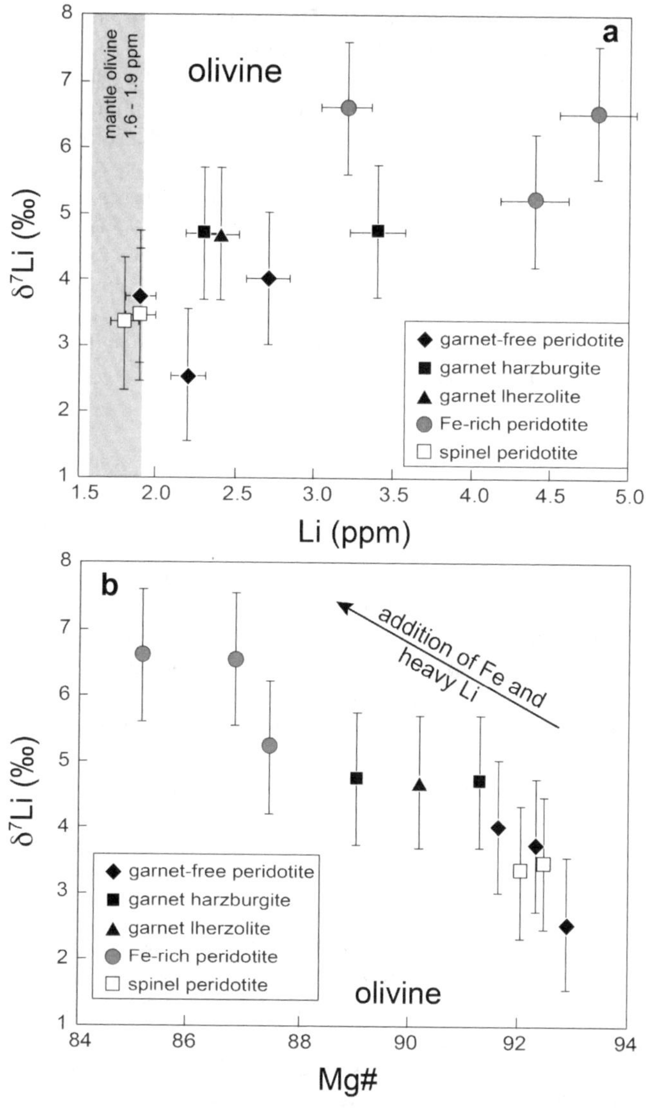

Figure 8. Evidence for addition of isotopically heavy Li in olivines from xenolithic peridotites from Labait, Tanzania (modified from Aulbach et al. 2008). a) Li concentration vs. δ^7Li in olivines; b) olivine Mg# vs. δ^7Li in olivine. The addition of heavy Li was ascribed to overprinting from a lava with HIMU characteristics, likely associated with development of the East African Rift.

grain boundary melt or fluid. Although a number of studies have appealed to isotopically light metasomatic agents, possibly derived from dehydrated slabs (e.g., Nishio et al. 2004; Ackerman et al. 2013; YJ Tang et al. 2014), the signature of metasomatic addition of low δ^7Li, i.e., decreasing δ^7Li with increasing Li concentrations is the same as would be produced by kinetic fractionation and, thus, would be difficult to assign unambiguously to a metasomatic origin. Finally, one might expect to see a correlation between δ^7Li and radiogenic isotopes if metasomatism is an important process affecting δ^7Li in peridotites. Such correlations are rare, but have been observed in whole rock peridotites from Labait, Tanzania (Aulbach et al. 2008). Unfortunately, *in situ* analyses of radiogenic isotopes in mantle minerals are beyond current analytical capabilities.

Average upper mantle from peridotites? Despite the huge isotopic variability seen in peridotites, several studies have sought to define a mantle value through analyses of fertile, equilibrated samples. Seitz et al. (2004) were the first to suggest a bulk silicate Earth (BSE) δ^7Li value of +4‰ based on analyses of minerals from equilibrated, fertile peridotite xenoliths. Jeffcoate et al. (2007) estimated a mantle δ^7Li of +3.5‰ based on reconstructed whole rock compositions for lherzolite xenoliths, whereas Magna et al. (2006b) estimated δ^7Li of ~+3.7‰ based on olivine separates from spinel lherzolites. Likewise, Pogge von Strandmann et al. (2011) found an average δ^7Li value of $+3.5 \pm 0.5$‰ (2σ) for equilibrated, fertile peridotites, and suggested these represent Earth's primitive mantle. These estimates are similar to the prevalent δ^7Li signature of peridotites from the Horoman Massif, which has an average δ^7Li of $+3.8 \pm 1.4$‰ (2σ) for unmetasomatized peridotites (Lai et al. 2015). All of the above studies converge on a bulk silicate Earth δ^7Li of +3.5 to +4.0‰, consistent with the isotopic compositions of mantle-derived basalts (see next section).

Basalts

MORB and altered MORB. Measurements of unaltered mid-ocean ridge basalts (MORB) from multiple ridge segments around the world reveal a relatively restricted range in Li concentration and isotopic composition. The Li concentrations of fresh MORB range from 2.9 to 34 ppm with most samples having concentrations of less than 8 ppm (Ryan and Langmuir 1987; Niu et al. 1999, 2001; Regelous et al. 1999; Danyushevsky et al. 2000; Tomascak et al. 2008), and show negative correlations with MgO and positive correlations with Y, demonstrating Li enrichment during differentiation. The data suggest that Li is a moderately incompatible element with a bulk melt-rock distribution coefficient between 0.25 and 0.35 (Ryan and Langmuir 1987). Measurements of δ^7Li in unaltered MORB range from +1.6 to +5.6‰ with no discernible difference among the different ridges (Chan et al. 1992; Moriguti and Nakamura 1998a; Elliott et al. 2006; Nishio et al. 2007; Tomascak et al. 2008). Some of the variation in isotopic compositions of MORB may be related to varying enrichment in MORB sources. MORB samples with high K_2O/TiO_2 (which is a characteristic of E-MORB) have a slightly higher mean δ^7Li (+4.0‰) compared to N-MORB samples with lower K_2O/TiO_2 ($+3.4 \pm 1.4$‰, Tomascak et al. 2008), though the two populations overlap within uncertainty.

Interactions between MORB and heated seawater at mid-ocean ridges alter the Li concentration and isotopic composition of MORB. There is a dramatic contrast in Li isotopic composition and concentration of MORB compared to seawater. Seawater has a significantly higher δ^7Li of +31.0‰ compared to a δ^7Li of +3.4‰ for MORB, and seawater has a much lower Li concentration of 0.18 ppm compared to ~6 ppm Li in MORB (Ryan and Langmuir 1987; Millot et al. 2004; Tomascak et al. 2008). Hydrothermally altered MORB can be either depleted or enriched in Li, depending on whether leaching or precipitation of secondary phases dominates, respectively.

The earliest studies of hydrothermally altered MORB found that Li is enriched in MORB during alteration (Chan and Edmond 1988; Chan et al. 1992, 2002a). Two of these studies focused on ~5.9 Ma ocean crust sampled in ODP drill holes 896A (Chan et al. 1992) and 504B (Chan et al. 2002a) and documented the formation of alteration products of basalt including sheet silicates such as smectite and celadonite. Lithium is thought to partition preferentially into these minerals relative to water, most likely substituting for Mg. Lithium concentrations in these altered basalts range up to 75 ppm. Smectite separated from rocks in which this type of alteration has been documented (Chan et al. 2002a) contains higher Li concentrations (2–10 ppm) than would be expected from simple exchange with a fluid during a single fluid–rock interaction (expected Li concentration of about 0.4 ppm), suggesting that the exchange of Li between seawater and alteration minerals in basalt is an open-system process (Chan et al. 2002a). These studies also found that δ^7Li increased in MORB due to alteration. The correlation between Li isotopic composition and Li concentration (Fig. 9) suggests incorporation of seawater Li into these new minerals (clays and zeolites) resulting in an increase in δ^7Li.

Various DSDP sites (278, 319A, 332B, 417/418) sample altered MORB ranging in age from 3.5 Ma to 120 Ma. Lithium concentrations and δ^7Li of altered MORB from these cores were generally higher than those of unaltered MORB, similar to the findings of Chan et al. (1992, 2002a) (Fig. 9; Bouman et al. 2004). However, several samples have lower δ^7Li for a given Li concentration than are predicted by this relationship. These samples were from the oldest (120 Ma) basalts, and the deviation in isotopic composition may have resulted from isotopic exchange with fluids derived from overlying sediments (Bouman et al. 2004).

Depletion in Li during seawater–MORB interaction has also been observed in a study of ~1.2 Ma basaltic rocks from Hess Deep (Brant et al. 2012). These samples, obtained by submersible from the seafloor fall at the low Li side of the correlation between Li and Y seen in altered MORB, suggesting Li is depleted by about ~43% due to leaching from plagioclase. The Li concentrations of the Hess Deep samples do not correlate with their δ^7Li values (Fig. 9; Brant et al. 2012). The interpretation of leaching of Li from the Hess Deep rocks hinges on the assumption that Y is immobile. However it has been demonstrated experimentally that Y can be mobile in hydrothermal fluids (Bach and Irber 1998; Bao et al. 2008).

Ocean Island Basalts (OIB) and other intraplate igneous rocks. The Li systematics of intraplate basalts have been determined at a number of localities in order to test the hypothesisthat some OIB derive from a mantle source containing recycled altered oceanic crust. Since low temperature seafloor weathering of oceanic crust imparts an elevated δ^7Li composition relative to asthenospheric mantle, this signal could be preserved in mantle reservoirs that eventually emerge via mantle plumes at hotspot locations. However, analysis of the Li concentrations and isotopic compositions of most OIB are similar to those of unaltered MORB. The only exceptions are some HIMU basalts (high μ, where μ is the $^{238}U/^{204}Pb$ ratio inferred on the basis of Pb isotopes), which can have elevated δ^7Li relative to MORB (Nishio et al. 2004; Chan et

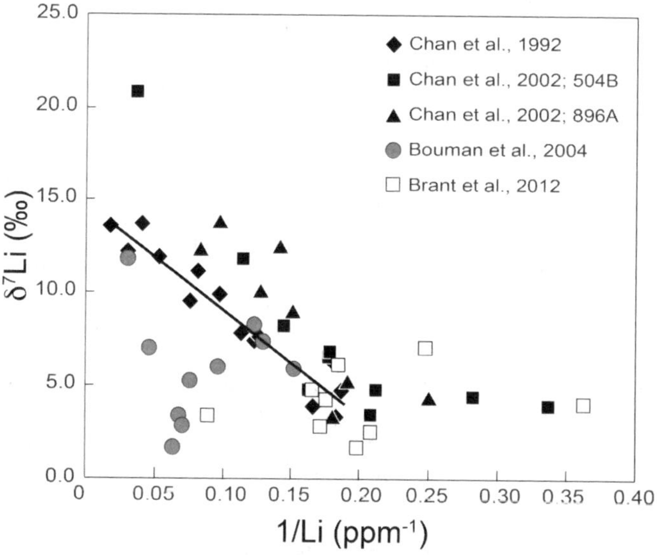

Figure 9. Variation in Li isotopic composition and concentrations in hydrothermally altered MORB (from Chan et al. 1992, 2002; Bouman et al. 2004; Brant et al. 2012).

al. 2009; Vlastelic et al. 2009). There are, however, difficulties in interpreting the HIMU Li data, including changes to whole rock δ^7Li due to post-eruption alteration and weathering (e.g, Chan and Frey 2003; Chan et al. 2009), assimilation of hydrothermally altered oceanic crust upon ascent (e.g., Genske et al. 2014), and kinetic fractionation of Li isotopes within phenocrysts due to diffusion (e.g., Chan et al. 2009).

Most OIB samples are heterogeneous, even for lavas from a single location, and a few OIB δ^7Li values fall outside the range of δ^7Li in MORB (e.g., Chan and Frey 2003; Kobayashi et al. 2004; Ryan and Kyle 2004; Nishio et al. 2005; Chan et al. 2009; Vlastelic et al. 2009; Magna et al. 2011; Krienitz et al. 2012; Harrison et al. 2015). Correlations between δ^7Li and other indicators of subducted oceanic crust such as Pb, Sr, Nd or He isotopic compositions provide the most compelling evidence for mixing of altered oceanic crust in the source region of OIB (e.g., Chan and Frey 2003; Kobayashi et al. 2004; Chan et al. 2009), though the range in δ^7Li is typically within the range observed in MORB and, in some studies, the correlations are not very significant (e.g., Fig. 5 of Krienitz et al. 2012).

Olivines from OIB samples from the Cook–Austral islands and Hawaii have similar δ^7Li values when compared to the associated basaltic whole rocks (Chan and Frey 2003; Jeffcoate et al. 2007; Chan et al. 2009). Clinopyroxenes from OIBs of the Cook–Austral islands, on the other hand, have consistently lower, but variable, δ^7Li values compared to the associated whole rock measurements (Chan et al. 2009). Such isotopically light values have also been observed in clinopyroxene from mantle xenoliths (see previous section), which is likely produced by Li ingress via diffusion (e.g., Jeffcoate et al. 2007; Rudnick and Ionov 2007). Since diffusive processes may alter isotopic compositions at the mineral scale, the possibility exists that such processes also have affected bulk rock isotopic compositions if diffusion occurs during magmatic differentiation (e.g., Halama et al. 2008), although in the case of the Cook–Austral islands, this seems an unlikely scenario (Chan et al. 2009).

Continental intraplate igneous rocks may also provide insight into mantle Li isotopic compositions. Most of these rocks have δ^7Li values that overlap with MORB and OIB compositions (e.g., Marks et al. 2007; Kim et al. 2011). Carbonatites, carbonate-rich, mantle-derived intraplate magmas, mostly have δ^7Li values overlapping with MORB (Halama et al. 2007, 2008; Tian et al. 2015). Some carbonatites from a collisional setting in western Sichuan, SW China display more variability than MORB, with some high δ^7Li values (up to +10.8‰) and some low (down to −4.5‰) which are interpreted to represent variations in the mantle source due to metasomatism by Li-rich fluids rising from a subducting slab prior to collision (Tian et al. 2015). Investigation of carbonatites crystallizing over a wide range of geologic time demonstrated no variability in mantle Li isotopic composition from 2.7 Ga to the present (Halama et al. 2008). In other studies of intraplate igneous rocks, δ^7Li values for samples were interpreted to represent variation in mantle reservoirs (e.g., Gurenko and Schmincke 2002; Hamelin et al. 2009).

The degree to which heterogeneities in Li isotopes can survive at mantle temperatures ($>1000\,^\circ$C) is the subject of debate. Halama et al. (2008) used a simple 1-D diffusion calculation and experimentally determined Li diffusivity in clinopyroxene (Coogan et al. 2005) to suggest that homogenization of Li (and by inference its isotopic composition) occurs within the mantle over relatively short time periods, on the order of millions or tens of millions of years. By contrast, Vlastelic et al. (2009) showed that isotopic heterogeneities can persist over several billions of years after elemental equilibrium is achieved. In these calculations, uncertainties exist in the values used for diffusivity and diffusive fractionation of Li within the mantle (i.e., the β value). Homogenization timescales will also depend on the size and shape of the heterogeneity.

Arc lavas

Island arc lavas were one of the first targets for Li isotope geochemists, as these lavas were expected to record a signature of high δ^7Li slab-derived Li released in fluids derived from altered oceanic basaltic crust. However, most studies have found that arc lavas have δ^7Li that overlap those of MORB (e.g., Chan et al. 2002b; Tomascak et al. 2002; Ryan and Kyle 2004), and the reasons for this are still uncertain, as detailed below. More recent Li isotope studies of arcs have been undertaken to better understand processes of subduction, arc magma genesis and global geochemical cycling.

Relatively uniform δ^7Li values in arc lavas that are similar to MORB compositions have been interpreted as representing mantle compositions seemingly unaffected by subduction or arc magma genesis (e.g., Leeman et al. 2004; Moriguti et al. 2004). This may be because slab-derived Li partitions into and is sequestered in mantle wedge olivine, so that it never makes it to the arc volcanic rocks (Tomascak et al. 2002), and/or diffusive re-equilibration of Li with the mantle (Parkinson et al. 2007; Halama et al. 2009). Alternatively, slab-derived Li, which may be a combination of a high δ^7Li signature from altered basaltic crust and a low δ^7Li signature from subducted terrigenous sediments may have an average isotopic composition that is indistinguishable from the mantle (M Tang et al. 2014). When δ^7Li values in arc lavas deviate from MORB-like values, they have been interpreted as providing evidence for incorporation of isotopically distinct, slab-derived Li in the mantle wedge (e.g., Moriguti and Nakamura 1998a; Chan et al. 2002b; Agostini et al. 2008; Bouvier et al. 2008, 2010; M Tang et al. 2014).

Two studies have observed systematic changes in δ^7Li across arcs. Moriguti and Nakamura (1998a) observed high δ^7Li values (+7.6‰) in the forearc of the Izu arc, which decreased systematically along a traverse across the arc to low values (+1.1‰) in the backarc. They attributed this change to a decrease in heavy, slab-derived Li across the arc. A later study in the northeastern Japan arc found no systematic changes in δ^7Li across the arc, which was attributed to a differing thermal structure of the arc between Izu and northeastern Japan (Moriguti et al. 2004). Magna et al. (2006a) also observed a decrease in δ^7Li values from forearc to backarc in the Cascades of northern California, which they attributed to Rayleigh distillation as Li was progressively released from the dehydrating slab, with ^{7}Li liberated early, on, causing δ^7Li to decrease as the slab continues to dewater.

Variation in the Li isotopic composition in the subducting slab from one arc to the next (e.g., sedimentary rocks, altered oceanic crust) may be responsible for some of the variations in δ^7Li observed in volcanic arcs. M Tang et al. (2014) found that the δ^7Li of lavas from Martinique in the Lesser Antilles island arc are significantly lower than those typical of MORB. They attributed these lower values to incorporation of Li from the subducted terrigenous sediments, which were shed from the highly weathered Guyana Shield and thus have low δ^7Li (see the Section *Continental Weathering Processes).*

Slab processes have also been invoked to explain the variability observed in δ^7Li of arc rocks. A study of age-varying volcanic rocks from Panama showed variations in δ^7Li that were interpreted to represent differences between the slab fluid signature and a slab melting signature (Tomascak et al. 2000). Some studies suggest that variations in arc δ^7Li may be the result of isotopic fractionation that occurred during dehydration reactions occurring in subducting oceanic crust (e.g., Magna et al. 2006a; Agostini et al. 2008), although the degree of this fractionation is debated (e.g, Zack et al. 2003; Marschall et al. 2007). Additionally, the effect of diffusive fractionation between magma and crust upon ascent and emplacement was considered a possible cause for variation in δ^7Li observed in the northern Central American volcanic arc (Walker et al. 2009).

Continental crust

Upper continental crust. Estimates of the Li composition of the upper continental crust have utilized fine-grained sedimentary rocks exposed on the Earth's surface, which provide wide-scale natural averages (e.g., Taylor and McLennan 1985), as well as granites, which are major constituents of the upper continental crust. Teng et al. (2004) investigated the Li composition of shales, loess, granites and upper crustal composites to estimate the average Li composition of upper continental crust. They found that the shales, loess and granites are uniformly lighter than primary mantle-derived igneous rocks (a conclusion supported by a more extensive study of A-type granites by Teng et al. 2009), suggesting loss of ^{7}Li during weathering from the upper crust. The average Li concentration of the upper continental crust was estimated to be 35 ± 11 ppm and the average δ^7Li to be 0 ± 4‰ (2σ). The average δ^7Li of upper continental crust is lower than the average mantle composition (~+3.4‰) and reflects the loss of ^{7}Li due to weathering processes (Teng et al. 2004).

Sauzeat et al. (2015) investigated the Li composition of loess deposits globally to gain a more precise estimate of the Li composition of the upper continental crust, since loess deposits form by mechanical erosion and therefore are more likely to avoid the effects of chemical weathering. They found that desert loess deposits are relatively homogeneous and more suitable than periglacial loess for recording average Li compositions of the upper continental crust. Variations in δ^7Li in glacial loess were found to be a function of mineralogical sorting (correlated with grain size) and thus were inferred to result from varying degrees of weathering. Therefore the average δ^7Li of upper continental crust was inferred, based on correlations between δ^7Li and concentrations of the relatively insoluble REE. The average Li concentration of upper continental crust was estimated to be 30.5 ± 3.6 ppm and the average δ^7Li was estimated to be +0.6 ± 0.6‰ (2σ) (Sauzeat et al. 2015).

Lower and middle continental crust. Studies of materials from the lower continental crust have found highly heterogeneous δ^7Li (Teng et al. 2008; Qiu et al. 2011a). An investigation of granulite-facies xenoliths from Australia and China revealed a large range in δ^7Li (−17.9 to +15.7‰) with inter-mineral variability in some samples suggesting isotopic disequilibrium (Teng et al. 2008). The range of compositions for equilibrated samples (−14 to +14.3‰) is also quite large, however, which was attributed to a number of factors including protolith heterogeneity, diffusion-driven fractionation, and metamorphism. The average lower crustal Li concentration is estimated to be ~8 ppm and the average δ^7Li (using a concentration weighted average) is estimated to be +2.5‰ (Teng et al. 2008). Rocks from high-grade metamorphic terranes in China had lesser variability in δ^7Li (ranging from +1.7 to +7.5‰) with an average δ^7Li of +4.0‰, suggesting that the composition of middle continental crust is possibly less heterogeneous than the lower crust and is similar in composition to the upper mantle (Teng et al. 2008).

A comparison of the Li composition of kinzigites and granulitic stronalites (amphibolite-facies and granulite-facies metapelites, respectively) of the Ivrea Zone show variation in Li concentrations between the lower-grade, biotite-rich kinzigites and the higher-grade, biotite-poor stronalites (Qiu et al. 2011a). These variations reflect the loss of Li during biotite breakdown reactions. The δ^7Li of the two rock types are variable but indistinguishable, suggesting that the variations in δ^7Li reflect protolith heterogeneity. Based on these samples, the average Li concentration of the lower crust is estimated to be ≤ 8 ppm and the average δ^7Li is +1.0‰ (Qiu et al. 2011b).

Correlations between δ^7Li and indicators of assimilation (e.g., Sr, Nd, Pb isotopic compositions) observed in alkalic intraplate volcanic rocks of the Chaîne des Puys allowed Hamelin et al. (2009) to estimate the Li isotopic composition of the lower crustal contaminant for these magmas. The assimilated material had a low δ^7Li ~ −5‰ and a Li concentration of ~40 ppm, consistent with a metapelite (Hamelin et al. 2009).

Seawater

The high solubility of Li means that it is a conservative element in the oceans (i.e., its concentration does not vary from place to place). Its concentration in the modern ocean is relatively low, averaging ~ 0.18 ppm (Riley and Tongudai 1964). The Li isotopic composition of modern seawater is homogeneous and heavy (δ^7Li = +31‰) (Chan and Edmond 1988; You and Chan 1996; Moriguti and Nakamura 1998a; James and Palmer 2000a; Millot et al. 2004) due to its long residence time of ~ 1.5 Ma (Huh et al. 1998) compared to ocean mixing time (~1 kyr). The lithium isotopic composition of seawater is probably maintained at steady state by the balance between inputs from rivers (+23‰, Huh et al. 1998), groundwaters (+15‰, Bagard et al. 2015), and hydrothermal fluids (+7‰, Chan et al. 1993, 1994; Foustoukos et al. 2004) and outputs to clays formed in oceanic basalts (which has a fractionation factor of around 0.98, Chan et al. 1992) and authigenic clay formation in sedimentary rocks.

Rivers

Lithium concentrations in global rivers show great variation (~0.2 ppb to 20 ppb) with a flow-weighted mean Li concentration of ~1.5 ppb (Huh et al. 1998). The global mean riverine δ^7Li at the present day is about +23‰ (Huh et al. 1998), with significant variations for individual rivers, ranging from +1 to +44‰ (Huh et al. 1998, 2001; Kisakurek et al. 2005; Pogge von Strandmann et al. 2006, 2010; Vigier et al. 2009; Lemarchand et al. 2010; Dellinger et al. 2015; Liu et al. 2015; Pogge von Strandmann and Henderson 2015). The suspended loads in rivers have lower Li isotopic compositions compared to those of associated dissolved loads and display much less variation, with δ^7Li mostly ranging from −6 to +5 ‰ (e.g., Huh et al. 2001; Kisakurek et al. 2005; Pogge von Strandmann et al. 2006; Dellinger et al. 2014; Liu et al. 2015). Lithium isotopic studies of rivers are discussed in greater detail in the weathering process section below.

Lakes

The early work of Chan and Edmond (1988) showed that lake water δ^7Li varies from +32 to +34‰ based on studies of Lake Tanganyika, and the Caspian and Dead Seas. The authors speculated that seawater-like δ^7Li values must be related to sedimentary processes in these lakes, which may be a result of incorporation of Li into evaporites or through "reverse weathering" reactions. Later, Tomascak et al. (2003) found lower δ^7Li, ~+20‰ in Mono Lake, California and suggested that sorption onto clays results in the removal of ^{6}Li and enrichment of the water in ^{7}Li. A δ^7Li lower than seawater is common in other Great Basin lakes, which have δ^7Li ranging from +17 to +24‰ (Tomascak et al. 2003). A Li isotope study of lakes in the McMurdo Dry Valleys, Antarctica, showed a large range in Li concentration of 0.11 ppb to 22 ppm and δ^7Li ranging from +17 to +35‰, demonstrating the complexity of lake water sources, including stream water, seawater, deep-seated brine, as well as groundwater (Witherow et al. 2010).

Early work reported relatively high δ^7Li values in brines, up to +47‰, with an average of +35‰ (Bottomley et al. 1999, 2003). However, these values were reported before common usage of external standards to document accuracy of Li isotopic measurement and the seawater values reported by these studies were 2‰ heavier compared to the recognized modern seawater value (δ^7Li = +31‰). More recent studies have reported lower δ^7Li values for brines. For example, Godfrey et al. (2013) show that δ^7Li in brines in the Salar del Hombre Muerto, central Andes, vary from +3‰ to +9‰, with a significant contribution from geothermal springs (δ^7Li = +1‰ to +18‰). In addition, they support the mechanism of surface sorption rather than Li incorporation into octahedral structural sites of clays for Li isotopic fractionation in brines. Araoka et al. (2014) found very high Li concentrations (50–1020 ppm) in playas from Nevada, with relatively small δ^7Li variations of −1 to +8‰.

Like rivers and groundwater, the Li concentrations and δ^7Li in lakes and brines are mainly controlled by the incorporation of Li into secondary minerals, such as clays, resulting in the removal of ^{6}Li from solution and thereby enriching water in ^{7}Li. The Li isotopic composition of lakes and brines are also influenced by fluid–rock interactions at different temperatures (deep springs vs. shallow springs), and mixing from various water reservoirs, such as groundwater, stream water, seawater, and brines.

Groundwater

Compared to rivers, there are more limited studies on Li isotopes in groundwater due to difficulty in sampling them and characterizing the subsurface flow paths. Nonetheless, a few case studies shed light on the factors affecting Li in these waters. The δ^7Li in groundwater generally represents a mixture from different sources, including seawater (Hogan and Blum 2003; Millot et al. 2011), meteoric water (Millot et al. 2011), and deep-seated fluids (which produces δ^7Li as low as −5‰) (Nishio et al. 2010). In addition, anthropogenic contamination can cause a significant increase in δ^7Li (up to more than 1000‰) in groundwater (Negrel et al. 2010). However, mixing from different sources cannot explain all of the variations of δ^7Li observed in uncontaminated groundwater. For example, Meredith et al. (2013) and Negrel et al. (2012) suggested that varying degrees of Li isotopic fractionation occurring during water–rock interaction is the main reason for significant δ^7Li variation in groundwater. Pogge von Strandmann et al. (2014) also suggested that δ^7Li in groundwater is controlled by the extent of isotope fractionation during secondary mineral formation. Liu et al. (2015) observed that all of the shallow groundwaters from a single lithology, the Columbia River Basalts, have lower and more variable δ^7Li (+7 to +21‰) compared to their riverine counterparts, which was also confirmed by Bagard et al. (2015) through a systematic study of river waters and groundwaters in the Ganges-Brahmaputra floodplain. Liu et al. (2015) used a transport–reaction model to quantify the importance of fluid–rock interactions, and showed that δ^7Li in rivers increases relative to groundwater as clays in the suspended load react with the river water during transport. Overall, Li concentrations in groundwater are generally higher than those of rivers and δ^7Li in groundwater is low compared to that of rivers, mostly ranging from +6 to +29‰ (for those without the significant influence of deep/hydrothermal fluids).

Hydrothermal fluids

Submarine hydrothermal vents. Compared to seawater, Li concentrations in hydrothermal fluids, the other important source of water in the oceans besides rivers, vary by two orders of magnitude, from 0.1 to 10 ppm (Chan and Edmond 1988; Chan et al. 1993, 1994; Foustoukos et al. 2004; Mottl et al. 2011). In contrast to Li concentrations, δ^7Li values in subaerial and submarine hydrothermal fluids vary over a small range (mostly +5 to +11‰), and have an average value of +8 ± 4‰ (2σ, $n = 33$) (Verney-Carron et al. 2015).

These high-temperature hydrothermal fluids have higher δ^7Li compared to their source rocks, especially relative to fresh MORB (δ^7Li$_{MORB}$ = +4 ± 2‰ (2σ, $n = 53$), Tomascak et al. 2008). However, the processes responsible for the elevated and rather homogeneous signatures in seafloor hydrothermal fluids are not well known. Parameters that may control the Li concentration and isotopic composition of hydrothermal fluids include the primary minerals they interact with, water/rock ratios, degrees of water–rock interactions, and temperature (Chan et al. 1994; Foustoukos et al. 2004).

Geothermal waters. Lithium concentrations in geothermal waters vary from 0.2–150 ppm (e.g., Millot and Negrel 2007; Millot et al. 2007, 2010a, 2012; Bernal et al. 2014). Compared to the relatively homogeneous Li isotopic compositions in submarine hydrothermal fluids, δ^7Li values in geothermal waters vary greatly. For example, geothermal waters from the Massif Central, France, have Li concentrations that vary from 3 to 150 ppm, and δ^7Li values are between −0.1 and +10.0‰ (Millot and Negrel 2007), suggesting that Li isotopes may track the intensity of water–rock interactions. Millot et al. (2007) reported measurements of additional geothermal waters from France and showed that reservoir rock types also influence the Li isotopic signature. Millot et al. (2010a) later investigated geothermal systems from Guadeloupe, Martinique and found that δ^7Li values vary from +4 to +26‰. Although Li isotopic compositions of the geothermal fluids collected from deep reservoirs were found to be homogeneous for a given site, δ^7Li values from each of these reservoirs were different, likely a result of temperature differences during water–rock interactions. In addition, they suggested that mixing with seawater and Li uptake by secondary minerals may help to explain the large δ^7Li variations (Millot et al. 2010a).

Most recently, Verney-Carron et al. (2015) investigated geothermal fluids from Iceland, where they applied a mass balance model to MOR high-temperature hydrothermal data and suggested that the large range of Li concentrations associated with a small range of δ^7Li values observed in high-temperature hydrothermal fluids can be explained by a combination of high Li leaching rates, low water/rock ratios, and relatively small amounts of secondary phases.

By contrast, geothermal fluids from the Taupo Volcanic Zone (TVZ), New Zealand, display relatively homogeneous and low δ^7Li, ranging from +0.5 to +1.4‰ reported by Millot et al. (2012) with a slightly larger range (−3 to +2‰) reported in a later study of Bernal et al. (2014), where they explored more samples from a wider area within the TVZ. Millot et al. (2012) suggested that temperature might not be the main factor responsible for the Li isotopic composition of the geothermal fluids. Instead, they attributed the observed homogeneous and low δ^7Li values in the TVZ geothermal fluids to leaching of Li from the same source rock. Bernal et al. (2014) confirmed that Li isotopes are greatly influenced by water–rock interactions and that the low δ^7Li in the Taupo geothermal fluids suggests that the fluids have interacted with, and Li may be leached from low δ^7Li source rocks. Similarly, Pogge von Strandmann et al. (2016) investigated geothermal springs at high temperature (200–300 °C) from Lake Myvatn in Iceland, where secondary mineral formation is limited, resulting in little δ^7Li fractionation of Li in the springs compared to that of the bedrock.

IGNEOUS PROCESSES

Differentiation

Mass-dependent isotope fractionation between phases is an inverse function of temperature; the amount of fractionation generally decreases with increasing temperature. Therefore, while partitioning of Li between minerals and melt during fractional crystallization may result in an increase in Li concentration in the melt during differentiation (e.g., Schuessler et al. 2009), fractionation of Li isotopes at magmatic temperatures is expected to be minimal. This concept was tested by Tomascak et al. (1999b), by analyzing samples from the differentiating Kilauea Iki lava lake, which was filled in a single eruption in 1959. Following its formation, the lava lake experienced closed-system fractionation, and its subsequent thermal structure and crystallization sequence are well-characterized. The MgO content of samples ranges from high (~27 wt. %) in olivine-cumulates, to as low as ~2.6% in late-stage silicic veins. The original lava had an average MgO content of 15.5%. The δ^7Li of these whole-rock samples ranged between +3.0 and +4.8‰, i.e., within the uncertainty of measurement of 1.1‰ (2σ) and the most primitive sample was within 0.4‰ of the most evolved sample, indicating that there was no measurable fractionation of Li isotopes during magmatic differentiation at $T > 1050$ °C (Tomascak et al. 1999b).

Later work on lavas documented disequilibrium between phenocrysts and host magmas due to Li diffusion following mineral crystallization (Jeffcoate et al. 2007; Parkinson et al. 2007; Weyer and Seitz 2012). In most cases the phenocrysts have lower δ^7Li compared to coexisting matrix, with differences between phenocrysts and associated matrix ranging up to 30‰. Preferential partitioning of Li into the melt phase during crystallization leads to enrichment of Li in the melt and Li diffusion into the earlier formed phenocrysts, imparting a low δ^7Li to the crystals since ^{6}Li diffuses faster than ^{7}Li (Weyer and Seitz 2012).

Studies of a wide variety of other types of igneous rocks have reached a similar conclusion—that there is little to no isotopic fractionation during magmatic differentiation. Teng et al. (2006b) found no systematic correlation between δ^7Li and SiO_2 content of Harney Peak granitic rocks in the Black Hills, South Dakota. Only at the most extreme extents of differentiation in the nearby pegmatitic rocks of Tin Mountain are variations in δ^7Li associated with elevated Rb concentrations, suggesting that significant isotopic fractionation occurs only at the very lowest-temperature, latest stages of crystallization of granitic bodies (Teng et al. 2006b). Bryant et al. (2004) and Teng et al. (2004, 2009) also found no systematic fractionation of Li isotopic compositions due to differentiation within suites of granitic igneous rocks. A suite of carbonatites from Oldoinyo Lengai (Halama et al. 2007) showed no correlation between δ^7Li and indicators of igneous differentiation (i.e., Mg#). Likewise, a study of the alkaline to peralkaline Ilimaussaq plutonic complex (Marks et al. 2007) found relatively uniform δ^7Li of amphiboles within the inner part of the complex, reflecting a homogeneous melt reservoir. Samples from the margin of the complex show variations in δ^7Li that may have been produced by a combination of isotopic fractionation due to diffusion between the syenite and surrounding country rock and circulation of fluid with high δ^7Li along the contact between these lithologies (Marks et al. 2007).

There are only two studies that have suggested significant Li isotopic fractionation has occurred due to equilibrium partitioning within igneous rocks. Day et al. (2016) found a difference in δ^7Li between high and low Ti lunar basalts and suggested that Li isotopic fractionation may occur during extensive high-temperature fractional crystallization, such as that likely experienced during crystallization of a lunar magma ocean (see section on the Moon). Most recently, Su et al. (2016) attributed differences in δ^7Li between Jurassic leucogranites (δ^7Li = +4 to +9‰) and their entrained garnet–biotite-rich enclaves (δ^7Li = +0.6 to +8‰) to reflect fractionation of Li isotopes at magmatic temperatures. However, their interpretation hinges critically on the assumption that the enclaves represent unmodified residua from granite melting, a link that has traditionally proven difficult to establish (e.g., Yang et al. 2007).

Eruptive processes

The rapid diffusivity of Li lends itself to use as a geospeedometer for short duration processes, and this approach has been applied to igneous rocks in order to understand relatively short-duration cooling events. Rapid diffusivity, along with the dramatic change in partitioning of Li between plagioclase and clinopyroxene, was used by Coogan et al. (2005) to investigate the cooling rates of ocean floor basalts and gabbros, which varied from quenching of the lavas in the uppermost 1 km, to slower cooling rates (2–7 °C/hour) deeper in the sheeted dike complex.

Zoning in Li concentrations and Li isotopic compositions within phenocrysts in igneous rocks provides additional constraints on the duration of relatively rapid cooling events. Ion microprobe measurements of Li concentrations and δ^7Li along traverses within olivine and clinopyroxene phenocrysts in arc volcanic rocks from the Solomon Islands revealed up to 29‰ variation within a single phenocryst (Parkinson et al. 2007). Lithium concentrations also showed zoning, with the greatest effects found in a clinopyroxene in which the rim had ~8× the Li compared to the core. The spatial patterns of δ^7Li within the phenocrysts were systematic and generally produced the "trough" pattern that is predicted by the different diffusivities of ^{6}Li and ^{7}Li. Calibration of the Li isotope profiles with associated Mg-Fe zoning allowed for a comparison of the diffusion rates of Mg and Li. The Li diffusion in olivine is calculated to be 4–8 times slower than Mg diffusion, but it was 20–30 times faster than Mg diffusion in clinopyroxene.

In situ measurements of Li and Li isotopes in phenocrysts in volcanic rocks have also been used to infer the movement of magmatic volatiles and vapor enrichment of magmas. Variable lithium concentrations in amphibole relative to plagioclase phenocrysts in dacites from Mt. St. Helens were attributed to Li partitioning into fluids and its loss from the magma during differentiation (Rowe et al. 2008). Measurements of δ^7Li in plagioclase phenocrysts from Palaea and Nea Kameni, two islands in the center of the flooded Santorini caldera, show that some crystals exhibit extremely low δ^7Li (−24.0 to −30.2‰) in crystal centers along with elevated δ^7Li at rims (+2.8 to +9.0‰) consistent with kinetic fractionation due to degassing and Li loss from the ascending magma during growth of the crystals (Cabato et al. 2013).

METAMORPHIC PROCESSES

Not surprisingly, metamorphic rocks exhibit a wide range of Li concentrations and isotopic compositions, reflecting both variability in protolith compositions and the effects of metamorphic processes. Both lithium elemental and isotopic variations in metamorphic rocks can tell us a lot about metamorphic fluid–rock interactions, including products of dehydration reactions during metamorphism, effects of fluid infiltration, and diffusive fractionation through an intergranular metamorphic fluid. Investigated settings include contact metamorphic aureoles, subduction zone metamorphic complexes, and progressively metamorphosed rocks in regional metamorphic terrains.

Dehydration

One of the earliest studies of the lithium isotopic composition of subduction-related metamorphic rocks (Zack et al. 2003) revealed some extremely low δ^7Li eclogites (as low as −11‰). The interpretation at the time was that metamorphic dehydration reactions generated these very low δ^7Li eclogites, with preferential partitioning of ^{7}Li into the fluid. Additional measurements of metabasaltic rocks from subduction zone metamorphic complexes (e.g., Marschall et al. 2007) unearthed even lower values of δ^7Li in eclogites (as low as −21.9‰). Modeling of partitioning of lithium and lithium isotopic fractionation due to metamorphic dehydration reactions (Marschall et al. 2006, 2007) using experimentally determined fractionation factors (Wunder et al. 2006, 2007), however, demonstrates that dehydration reactions are unlikely to produce such dramatic fractionation (Marschall et al. (2007) calculated a maximum of ~3‰ change). Thus, Marschall et al. (2007) suggested that the low δ^7Li in eclogites is likely produced by Li diffusion during metamorphic fluid–rock interactions. There has been a recent focus on investigating metamorphic rocks to find evidence for diffusive fractionation (see section on diffusion during metamorphism) and for its use as a geospeedometer to constrain the duration of fluid–rock interactions.

Studies of metamorphic rocks in other settings have shown little change in lithium isotopic composition with extent of dehydration (Teng et al. 2007; Qiu et al. 2009, 2011b). These studies investigated the variation in Li concentrations and isotopic compositions as a function of metamorphic grade. Teng et al. (2007) investigated rocks of the Onawa contact aureole and found that, while Li concentrations decreased dramatically with increasing metamorphic grade, there was only limited evidence for variation in δ^7Li. Similarly Qiu et al. (2009) investigated low-grade mudrocks of the British Caledonides and found no evidence for variation in δ^7Li with increasing metamorphic grade. Instead, Qiu et al. (2009) found that variations in Li concentrations and Li isotopic composition correlated with the chemical index of alteration (CIA), reflecting the degree of weathering of the provenance of the original sedimentary rock (Fig. 10). A similar correlation was observed in metamorphosed sedimentary rocks of the Catalina Schist, a subduction zone complex, suggesting there that fluid infiltration was limited among the metasedimentary rocks (Fig. 10; Penniston-Dorland et al. 2012). Finally, Qiu et al. (2011a) found no change in δ^7Li in subgreenschist to greenschist-facies metasedimentary rocks from the accretionary wedge of New Zealand.

Investigation of high-grade metapelitic rocks of the Ivrea-Verbano Zone of NW Italy (Qiu et al. 2011b) revealed a correlation between biotite content and Li concentration, supporting a relationship between Mg and Li in partitioning of Li, and the formation of granulite-facies stronalites by biotite breakdown reactions resulted in significantly lower Li concentrations (8 ± 6 ppm, 2σ) compared to amphibolite facies kinzigites (79 ± 69 ppm, 2σ). No significant isotopic fractionation accompanied the biotite breakdown (average δ^7Li = −1.4 ± 2.0‰, 2σ (amphibolite facies) vs. +0.9 ± 2.9‰, 2σ (granulite facies)). The average Li concentration (≤ 8 ppm) and δ^7Li (+1.0‰) correlates well with previous estimates for lower continental crust (8 ppm, +2.5‰).

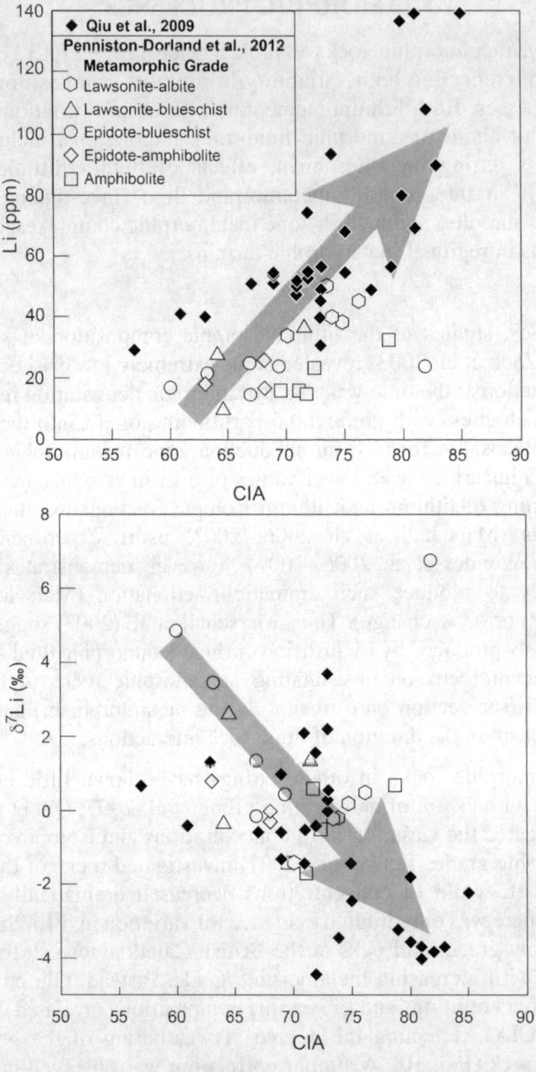

Figure 10. Correlation between **a.** [Li] and CIA and **b.** δ^7Li and CIA in metasedimentary rocks. Data from the British Caledonides (Qiu et al. 2009) and the Catalina Schist (Penniston-Dorland et al. 2012).

Redistribution of Li through fluid infiltration

There is abundant evidence for fluid flow within subduction zone metamorphic rocks, as witnessed by fluid inclusions, variable stable isotopic compositions, and the presence of veins and reaction zones that show evidence for metasomatism (e.g., Sorensen and Barton 1987; Bebout and Barton 1989; Giaramita and Sorensen 1994).

Many if not all subduction zone metamorphic complexes have zones of mixed lithologies referred to as mélange zones. Mélange zones typically contain boulders of a range of sizes (up to hundreds of meters in diameter) and varying lithologies (including metasedimentary, metabasalt, and metaultramafic rocks) that in some cases record different peak metamorphic conditions. These boulders are surrounded by a fine-grained matrix (e.g., shale, serpentinite), which can be a different lithology from the boulders. Such juxtaposition of rocks of different lithologies along with focused fluid transport through the fine-grained mélange matrix provides the opportunity for soluble elements such as lithium to be redistributed among the rocks within subduction zone metamorphic complexes. Comparison of the Li concentration and δ^7Li of subduction-related mélange metabasalts to that of MORB and altered MORB (their presumed protoliths) reveals that metabasalts are more Li-rich and, in some cases, have lower δ^7Li (Penniston-Dorland et al. 2010, 2012; Halama et al. 2011). This phenomenon can be explained by infiltrative transport of Li-rich fluids within the subduction zone mélange. The source of Li-rich fluids in some metamorphic rocks may be Li-rich sedimentary rocks (Marschall et al. 2007; Penniston-Dorland et al. 2010, 2012; Simons et al. 2010), or fluids derived from dehydrating metamorphosing basalt (Halama et al. 2011).

By contrast, a study of the Otago Schist, an accretionary wedge in which there is evidence for abundant fluid infiltration, suggests that infiltration of Li-poor high δ^7Li slab-derived fluids left little to no traces in the overlying metasedimentary units (Qiu et al. 2011a). The Li isotopic composition of the metamorphosed graywackes of the Otago Schist correlates with CIA, reflecting the protolith Li composition, as has been observed for other metasedimentary rocks (Qiu et al. 2011a; Penniston-Dorland et al. 2012).

Diffusion

Kinetic fractionation of Li isotopes has been proposed to explain variations of Li concentrations and isotopic compositions in some metamorphic rocks where fluids or melts of different Li concentration and/or isotopic composition are juxtaposed (e.g., Teng et al. 2006a; Marks et al. 2007; Liu et al. 2010; Penniston-Dorland et al. 2010; John et al. 2012; Ireland and Penniston-Dorland 2015). In many of these natural examples, Li is thought to move through pore spaces within an intergranular fluid, and in some cases, the distances affected can be quite large, up to 30 m (Teng et al. 2006a). Traverses across fluid-related features such as veins (e.g., John et al. 2012), igneous–country rock contacts (e.g., Teng et al. 2006a; Marks et al. 2007; Liu et al. 2010; Ireland and Penniston-Dorland 2015) and rock features that show evidence for infiltration of fluids (e.g., eclogite altered to blueschist, Penniston-Dorland et al. 2010) show variations in Li concentration and δ^7Li consistent with what is predicted by diffusion. The observed isotopic fractionations range up to ~30‰ and the distances over which Li compositions vary range from ~0.5 cm up to tens of meters.

A direct comparison of variations in oxygen and lithium isotopic compositions across a contact between the Bushveld Complex and metasedimentary rocks of the Phepane Dome demonstrates the greater diffusivity of Li relative to O (Ireland and Penniston-Dorland 2015). Along this traverse, O isotopes vary over a distance of < 1 m, while Li isotopes vary over a distance of ~100 m (Fig. 11). The distance over which lithium concentrations and isotopic compositions can be affected by diffusive fractionation depends on the time available for diffusive exchange, and conditions affecting the diffusivity of Li, such as permeability, temperature, and tortuosity. The dependence on time means that Li can be used as a geospeedometer to record the duration of fluid–rock interactions, if the other factors can be estimated.

CONTINENTAL WEATHERING PROCESSES

Previous investigations of the behavior of Li isotopes during weathering mainly focused on two targets: weathered regolith and rivers. These studies found that ^{6}Li partitions preferentially into secondary minerals formed during chemical weathering of silicate rocks (e.g., clays and oxides/hydroxides), and ^{7}Li partitions preferentially into the associated water (Huh et al. 1998, 2001, 2004; Kisakurek et al. 2004, 2005; Rudnick et al. 2004; Pogge von Strandmann et al. 2006, 2010; Vigier et al. 2009; Teng et al. 2010; Liu et al. 2013, 2015). Thus, continental weathering produces low δ^7Li regolith in the upper continental crust, and high δ^7Li waters. This fractionation means that Li isotopes have the potential to be used as tracers of weathering (Liu and Rudnick 2011; Misra and Froelich 2012). Moreover, Li contents in carbonate rocks are very low compared

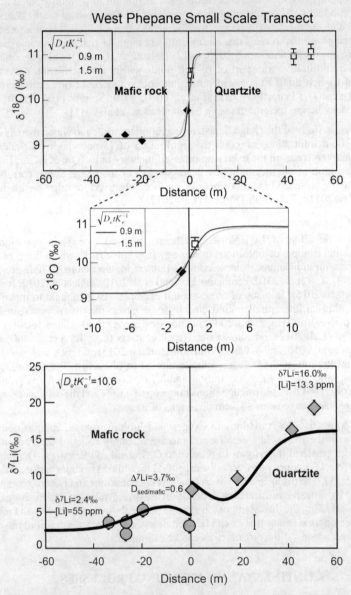

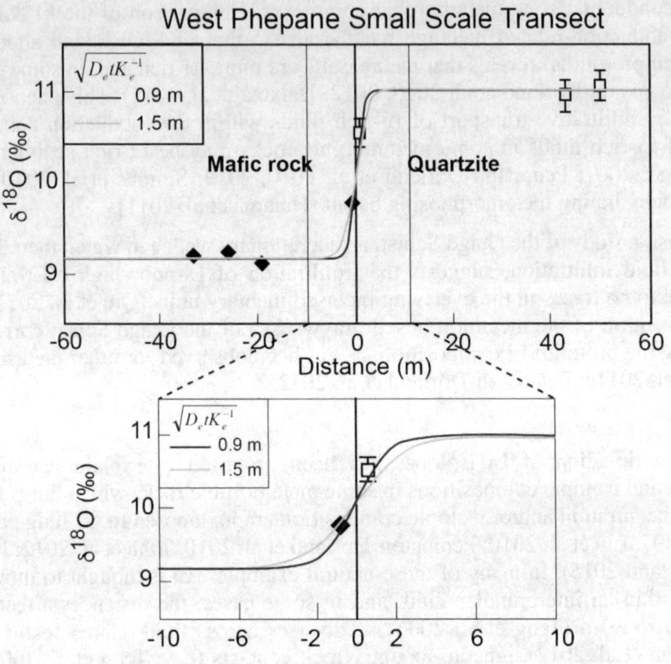

Figure 11. Profiles of δ^7Li and δ^{18}O across the contact between igneous rocks of the Bushveld Complex and adjacent Lakenvalei quartzite (modified from Ireland and Penniston-Dorland 2015). Lines represent results of 1-D quantitative diffusion models. The diffusive distance ($\sqrt{D_e t K_e^{-1}}$) is ten times greater for Li than it is for O. Partitioning and fractionation of Li between minerals in sedimentary quartzite and mafic rock ($D_{\text{sed/mafic}}$ and Δ^7Li) are required to explain difference at contact. Values modeled are the best fits to the data.

to the concentrations in silicate rocks; therefore, Li isotopes are controlled by silicate weathering, and not carbonate weathering (e.g., Kisakurek et al. 2005; Millot et al. 2010b). In addition, Li isotopes are not influenced by plants or primary productivity (Lemarchand et al. 2010; Pogge von Strandmann et al. 2016). All of these features make Li isotopes excellent tracers of silicate weathering with an advantage over other isotopic systems that are also sensitive to carbonate weathering (e.g., Sr isotopes) and biological activity (e.g., Si isotopes).

Weathering profiles

Studies of weathering profiles and soils revealed that the amount of Li isotopic fractionation primarily depends on the type of secondary minerals formed, which may, in turn, reflect the weathering intensity experienced by the rocks (e.g., Pistiner and Henderson 2003; Lemarchand et al. 2010; Pogge von Strandmann et al. 2012; Ryu et al. 2014). However, the lithium isotopic composition of the regolith can also be influenced by the addition of eolian dust (Kisakurek et al. 2004; Liu et al. 2013; Clergue et al. 2015), marine aerosol in coastal regions (Huh et al. 2004; Pogge von Strandmann 2012; Clergue et al. 2015), and even kinetic fractionation, if Li diffuses through water-saturated soil (Teng et al. 2010). All of these factors could complicate our understanding of the behavior of Li during chemical weathering.

Pistiner and Henderson (2003) observed large lithium isotopic fractionation associated with the formation of certain secondary minerals during weathering, such as gibbsite (+14‰), but little Li isotopic fractionation associated with formation of some other secondary minerals, such as smectite. Li isotopic fractionation is also observed in soil profiles developed on Hawaiian basalts, where the influence of high δ^7Li marine aerosols on Li isotopes has been documented (Huh et al. 2004). Similarly, Pogge von Strandmann et al. (2012) examined the Li concentrations and isotopic compositions in soils and pore waters from a soil profile in western Iceland and found that marine aerosol addition can significantly shift the Li isotopic composition of the soil. Analysis of saprolite profiles developed on the Deccan Traps and a meta-diabase further confirmed that secondary mineral formation (e.g., kaolinite, goethite, hematite) plays an important role in determining lithium isotopic distributions during weathering (Kisakurek et al. 2004; Rudnick et al. 2004). A negative correlation between Li concentrations and δ^7Li in the Deccan saprolites was explained by addition of eolian dust to the top of the profile, in addition to Rayleigh fractionation associated with Li removal. Liu et al. (2013) confirmed the significance of eolian addition in a study of bauxite weathering profiles developed on the Columbia River Basalts, where they combined mineralogic evidence with Li and Nd isotopes to demonstrate the addition of dust to the top of the weathering profiles. In addition to the effects of Li isotopic fractionation due to secondary mineral formation, recent work by Clergue et al. (2015) confirmed the significance of Li addition from sea salts and eolian dust in a study based on a highly weathered catchment in Guadeloupe, Lesser Antilles. Finally, Rayleigh fractionation of Li isotopes during leaching could be overprinted by kinetic fractionation associated with diffusion across a paleo-water table, such as that reported in a saprolite profile developed on a diabase dike (Rudnick et al. 2004; Teng et al. 2010).

In summary, studies of weathering profiles have concluded that the most important control on Li isotope fractionation is the formation of secondary minerals. In addition, eolian addition in form of either marine aerosol or dust and kinetic processes may also affect the Li isotope composition of weathering profiles.

Rivers

Riverine studies complement the weathering profile/soil studies in understanding the behavior of Li during chemical weathering (Huh et al. 1998). In a pioneering study, Huh et al. (2001) measured Li isotopes in suspended and dissolved loads from the Orinoco River. They found that Li isotope signatures in the dissolved load reflect chemical weathering and are not simply inherited from rainwater or rocks, and that δ^7Li in the suspended load is homogeneous and lighter compared to that in dissolved loads. In addition, they found that the greatest Li isotopic fractionation between river water and rock occurs in relatively less weathered regions, such as the Andes. A later study of Himalayan rivers confirmed these findings and proposed that Li concentrations in the dissolved load reflect the silicate weathering flux, while δ^7Li is correlated with weathering intensity, as an indication of how much previous weathered material is present in the catchments (Kisakurek et al. 2005).

Iceland has been the target of several Li isotope studies of weathering because of the relative uniformity of the lithology there and because weathering of Icelandic and other basalts contributes disproportionately high weathering fluxes. A study of rivers that drain Icelandic basalts with only minor sediments show that the δ^7Li difference between the dissolved load and suspended load is correlated with weathering intensity, with greater weathering intensity corresponding to larger Li isotopic fractionation between the suspended and dissolved load (Pogge von Strandmann et al. 2006). Later, a systematic study of major rivers in Iceland found a negative correlation between silicate weathering rate and δ^7Li (Vigier et al. 2009), suggesting that greater isotopic fractionation occurs in regions that are more heavily weathered and where erosion is lower. Vigier et al. (2009) suggested that this correlation may be global, but noted the need for investigations in areas with crystalline rocks that are more typical of upper continental crust.

Such an investigation was provided by Millot et al. (2010b), who reported Li isotopic compositions for river waters and suspended sediments from the Mackenzie River Basin, northwest Canada, which has highly variable lithologies and topography. By contrast with the empirical relationship reported by Vigier et al. (2009) for the Icelandic basalts, Millot et al. (2010b) found, like Huh et al. (2001) before them, that the weathering regime is the most important factor in controlling the Li isotopic signature in rivers, rather than weathering fluxes. In incipient weathering regimes, where Li is released from relatively fresh rocks in topographically high or recently glaciated areas, isotopic fractionation is greatest, likely due to partitioning between newly formed oxyhydroxides and water. By contrast, in low-lying, more intensively weathered regimes, Li isotopic fractionation is more subdued as Li partitions between clays and water. They suggested two end member processes that are responsible for Li isotopic fractionation in river waters: formation of secondary minerals, which preferentially take ^{6}Li into their structure, leaving heavy ^{7}Li in river waters (e.g., Zhang et al. 1998; Pistiner and Henderson 2003; Vigier et al. 2008; Wimpenny et al. 2010a); and preferential sorption of ^{6}Li onto Fe and Mn oxyhydroxides, also leading to ^{7}Li-enriched waters, similar to that seen in rivers from Greenland (Wimpenny et al. 2010b). Pogge von Strandmann et al. (2010) compared weathering of basalts from two climatic conditions by studying Li isotopes in rivers from Iceland and the Azores. Their study concluded that neither weathering intensity nor climate has a direct influence on Li isotope behavior; rather the formation of secondary minerals, which is indirectly controlled by climate, seems to play the dominant role.

Yoon (2010) showed through analyses of waters from the Lena River, Russia, that evaporites can be important sources of Li in some rivers (representing ~20% of the dissolved Li). He proposes a 'mineralogy-specific view' of Rayleigh fractionation, where not only is the degree of leaching (associated with weathering intensity) important, but the different fractionation factors between the different kinds of secondary minerals and dissolved loads of river waters influences δ^7Li values in dissolved loads during weathering via Rayleigh fractionation.

Several studies have shown that the δ^7Li composition of rivers is not only dependent upon input water composition, but also on reactions that occur during transport within the rivers between dissolved and suspended loads. Wanner et al. (2014) developed a global reactive transport model that allows reaction between newly formed secondary minerals during flow and showed that the larger the δ^7Li in a specific river, the more water–rock interactions (primary mineral dissolution + secondary mineral precipitation) had occurred. Later, Wang et al. (2015) studied one of the largest river systems, the Changjing River in China, and found that δ^7Li of the dissolved load increased from upper to lower streams, while Li concentration decreased, suggesting that δ^7Li of the dissolved load is a balance between dissolution of primary minerals and formation of and exchange with secondary minerals. In a study of streams and groundwater from single lithology catchments (the Columbia River Basalts), Liu et al. (2015) demonstrated that dissolved riverine δ^7Li is negatively correlated with Li/Na (a proxy for Li abundance), suggesting that δ^7Li vs. Li/Na may be a sensitive indicator of the extent of chemical weathering occurring in streams and groundwater reservoirs. Their reaction transport modeling showed that dissolved δ^7Li is a function of water–rock interaction time. The negative correlations between δ^7Li and Li/Na seen for the streams draining the Columbia River Basalts is observed globally for river waters worldwide that drain only or mainly basalts (Fig. 12), suggesting that these processes occur globally. Finally, Pogge von Strandmann and Henderson (2015) found that δ^7Li in rivers in areas of rapid uplift in New Zealand have low δ^7Li compared to those in areas of slow uplift. They suggested that the higher δ^7Li values reflect a longer sediment transport duration in flood plains.

In large, complex multi-lithology river systems, however, Li isotopes show much more complicated behavior. For instance, Dellinger et al. (2015) suggested that low δ^7Li in rivers can come from both low and high weathering intensities via a study of one of the world's largest river systems—the Amazon River, where the low δ^7Li observed in low intensity weathering catchments is due to primary mineral dissolution, and at high intensity it is due to secondary clay dissolution. Overall, weathering and erosion are dynamic and open system processes, and therefore we need a more fundamental understanding of tectonic, climatic, hydrologic, and mineralogical changes and interrelationships as well as transport and reactive processes before we can fully explain the Li isotopic behavior in continental weathering.

In summary, the main process controlling Li isotopes in rivers is the balance between primary mineral dissolution and secondary mineral formation, which is also referred as the degree of water–rock interaction.

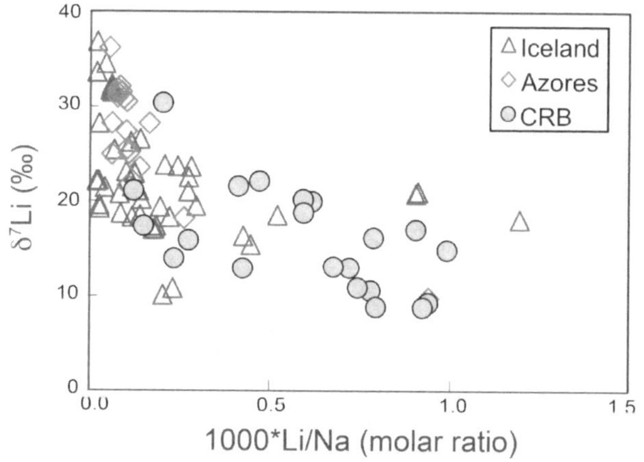

Figure 12. δ^7Li vs. 1000 × Li/Na (molar) for river waters worldwide that drain only or mainly basalts (modified from Liu et al. 2015).

LITHIUM AS A TRACER OF CONTINENTAL WEATHERING THROUGH TIME

We currently have a limited understanding of how the Li isotopic composition in various reservoirs may have changed through Earth's history. Most efforts have focused on the secular evolution of δ^7Li in seawater and its implications for changes in the nature of the continental weathering input to the oceans through time. Hathorne and James (2006) and Misra and Froelich (2012) analyzed Li isotopic compositions in planktonic foraminifera, and demonstrated that the δ^7Li of seawater increased by 9‰ during the Cenozoic. In particular, Misra and Froelich (2012) found that this increase in δ^7Li appeared to mimick, but not exactly parallel, an increase in seawater ^{87}Sr/^{86}Sr compositions over time that was documented decades ago. They attributed this change to an increase in incongruent weathering, resulting in increased clay formation, which could elevate δ^7Li in the riverine input into the ocean that was associated with uplift of the Himalaya around 40 Ma. However, the interpretation of this δ^7Li curve is complicated by the many variables that influence δ^7Li in seawater. The Misra and Froelich (2012) data generated new waves of modeling and observational studies to understand the rise in seawater δ^7Li over the Cenozoic. For example, Bouchez et al. (2013) presented a mass balance box model that includes Li-bearing primary mineral dissolution and precipitation of secondary minerals involved in Li isotope fractionation (but excludes kinetic and thermodynamic properties of mineral phases) to explain the Cenozoic δ^7Li curve. They concluded that the denudation rates change dramatically over the Cenozoic, causing an increase in Li isotope fractionation. Additional mass balance models have been generated by Li and West (2014), who, in agreement with Misra and Froelich (2012) suggested that Li outputs from the ocean were unlikely to account for all δ^7Li changes and therefore substantial changes in seawater δ^7Li over the Cenozoic must be significantly influenced by inputs from riverine Li derived from continental weathering. By contrast, Vigier and Godderis (2015), employing a combined Li and C cycle model, suggested that the δ^7Li increase in seawater through the Cenozoic could be explained by Li storage on the continents in secondary minerals, with a global flux decrease towards the present, while the export of Li to the ocean by weathering increased. In addition to mass balance modeling approaches, Wanner et al. (2014) developed a reactive transport model and the simulation results suggested that the Cenozoic seawater δ^7Li increase was unlikely due to an increase in congruent silicate weathering nor climate change, but rather to an increasing amount of primary silicate mineral dissolution associated with lower weathering intensity, but high riverine suspended loads, ultimately caused by tectonic uplift. This conclusion was supported by a combined observational and modeling study from streams draining the Columbia River Basalts (Liu et al. 2015). Through a study of Li isotopes in rivers draining catchments with variable uplift rates from New Zealand, Pogge von Strandmann and Henderson (2015) suggested that floodplains associated with mountain formation caused clay formation (instead of mountain weathering) and therefore also the Cenozoic seawater δ^7Li increase.

In summary, the observed increase in δ^7Li in seawater through the Cenozoic most likely reflects increasing δ^7Li in rivers and/or increasing riverine fluxes, although it is still very difficult to explain the detailed causes. These increases were probably related to tectonic uplift, which caused increased weathering, while climate may also have played a role in controlling the secular evolution of the Li isotopic composition of seawater.

Several recent studies have attempted to use Li isotopes in carbonate rocks to trace continental weathering during key transitions in Earth history, such as the Mesozoic Ocean Anoxic Events (OAEs). Pogge von Strandmann et al. (2013) measured lithium isotopes in carbonates from the Ocean Anoxic Event 2 (OAE2, ~93.5 Ma) intervals to assess the role of continental silicate weathering. They find the lowest δ^7Li values occurred during OAE2, suggesting that it was likely caused by increased silicate weathering rates and a shift to a transport-limited, intense weathering regime. Lechler et al. (2015) also observed that δ^7Li values decrease to a minimum coincident with the negative carbon-isotope excursion that defines the OAE 1a (~120 Ma). They concluded that weathering of an oceanic Large Igneous Province (increase in silicate weathering) and changes in weathering congruencies might be responsible for the observed low δ^7Li values at the OAE1a. In addition, Ullmann et al. (2013) investigated the Li isotopic composition of a well-preserved Late Jurassic (~155–148 Ma) belemnite from New Zealand and concluded that the average δ^7Li value in Late Jurassic seawater was +27‰, which is slightly lower than its present day value (+31‰).

Several studies have focused on the secular evolution of continental weathering using Li isotopes in proxies other than marine carbonates. For example, Ushikubo et al. (2008) studied the Li isotopic composition of detrital Hadean zircons from the Jack Hills, Western Australia and discovered low δ^7Li in some of these zircons, which they attributed to intense weathering in the Hadean. Although later experimental documentation of rapid Li diffusivity in zircon (Cherniak and Watson 2010; Trail et al. 2016), as well as *in situ* determinations of Li isotopic fractionation within zircons consistent with kinetic fractionation due to diffusion (Gao et al. 2015) has called this interpretation into question, intense weathering early in Earth history is consistent with the results of two later studies. Liu and Rudnick (2011) developed a mass balance model based on Li isotope systematics during continental weathering that suggests that chemical weathering and subsequent subduction of soluble elements have had a major impact on both the mass and the compositional evolution of the continental crust, where they also predicted that the accumulated weathering rates may have been much faster in Earth's history than present-day weathering rates. Li et al. (2016) investigated Li isotope systematics in glacial diamictites as proxies of the upper continental crust and concluded that post-Paleoproterozoic upper continental crust experienced less intense chemical weathering compared to Archean and Paleoproterozoic upper continental crust.

FUTURE DIRECTIONS

Since the publication of the lithium isotope chapter in the original volume on *Non-Traditional Stable Isotopes* in the Reviews in Mineralogy and Geochemistry series (Tomascak 2004), Li isotope geochemistry has emerged as being a particularly powerful tool for tracing continental weathering and for geospeedometry. The following paragraphs outline future promising directions for Li isotope geochemistry in these realms.

Weathering processes and Li fractionation experiments

Although Li isotopic fractionation factors between water and secondary minerals such as gibbsite, kaolinite, and smectite, have previously been determined experimentally (see Table 2 and references therein), there have been no systematic studies with realistic and controlled experimental conditions comparable to Earth's surface conditions. In addition, detailed understanding of how Li is incorporated into these secondary minerals is needed. For example, adsorption onto mineral surfaces, absorption into interlayers, and incorporation into mineral structures have all been proposed to cause Li isotope fractionation during weathering, but how each of them contributes to the isotopic fractionation of Li isotopes is not well understood.

Continental weathering through time

Lithium isotopes are important for tracing continental silicate weathering given the very low Li concentrations in carbonate rocks compared to their silicate counterparts. Recently, many studies have focused on interpretation of the Cenozoic Li isotope seawater curve that used planktonic foraminifera as paleo-ocean proxies in order to understand how continental silicate weathering may have changed over time. However, details of what Li isotopes mean in these marine proxies and how well Li reflects the changes in continental silicate weathering and its association with tectonics and climate are not entirely clear.

For example, we need to understand better how well different marine proxies can record secular variations of the Li isotope composition of seawater. In addition, the interpretation that δ^7Li in Cenozoic seawater reflects changes in continental weathering depends on the assumption that hydrothermal input throughout the Cenozoic did not change, which may not be the case. Therefore, more effort is needed to increase our understanding of the global hydrologic cycle and its impact on Li isotopic fractionation. Additional proxies need to be developed in time periods other than the Cenozoic in order to study the secular evolution of continental weathering and its influence on climate change, relationship to tectonic uplift, and the overall mass and compositional evolution of the continental crust.

Geospeedometry

The rapid diffusivity of Li, and the difference in diffusivities of its two isotopes allow for the use of Li and δ^7Li to enhance our understanding of the duration of relatively short-lived geologic events. Measurements of Li concentrations and isotopic compositions *in situ* in minerals in volcanic rocks have been used to further our understanding of cooling rates in the oceanic crust (e.g., Coogan et al. 2005), the residence time of phenocrysts in magma in arc volcanic rocks (Parkinson et al. 2007), and magma degassing (Cabato et al. 2013). Whole rock measurements of Li and δ^7Li across features indicative of fluid–rock interaction have provided insight into the duration of this interaction, with recent investigations focusing on the duration of pulses of fluid within subduction-related metamorphic rocks (Penniston-Dorland et al. 2010; John et al. 2012). Most recently, Trail et al. (2016) have suggested the use of Li in zircon to document peak metamorphic temperatures and thermal histories of detrital zircons.

In general, a firmer knowledge of the diffusive properties of Li is needed—within minerals, fluids, and melts. Recently it has been shown that Li may have different diffusivities for different diffusive mechanisms in olivine (Dohmen et al. 2010) and in pyroxene (Richter et al. 2014a). More experimental work shedding light on the diffusive behavior of Li in minerals coupled with further investigations of natural rocks will contribute to a better understanding of the meaning of inter-mineral variations in Li concentrations and δ^7Li. Similarly, experimental work shedding light on equilibrium Li isotopic fractionation between minerals and fluids and also between minerals and melts, coupled with experimental investigation of the diffusivity of Li in these media can lead to a better understanding of the duration of short-lived igneous and metamorphic processes.

ACKNOWLEDGEMENTS

We thank Philip Pogge von Strandmann, Paul Tomascak, and Josh Wimpenny for thorough and constructive reviews. We also thank Rick Hervig for providing helpful feedback on portions of the manuscript and Fangzhen Teng for his patience during the writing of this manuscript. We acknowledge the support of the National Science Foundation for our work on Li isotopes, specifically grants EAR−0106719, 0208012, 0609641, 0948549, and 1551388 to RLR and EAR−091110 to SPD.

REFERENCES

Ackerman L, Spacek P, Magna T, Ulrych J, Svojtka M, Hegner E, Balogh K (2013) Alkaline and carbonate-rich melt metasomatism and melting of subcontinental lithospheric mantle: Evidence from mantle xenoliths, NE Bavaria, Bohemian Massif. J Petrol 54:2597–2633

Adam J, Green T (2006) Trace element partitioning between mica- and amphibole-bearing garnet lherzolite and hydrous basanitic melt: 1. Experimental results and the investigation of controls on partitioning behaviour. Contrib Mineral Petrol 152:1–17

Agostini S, Ryan JG, Tonarini S, Innocenti F (2008) Drying and dying of a subducted slab: Coupled Li and B isotope variations in Western Anatolia Cenozoic volcanism. Earth Planet Sci Lett 272:139–147

Aigner-Torres M, Blundy J, Ulmer P, Pettke T (2007) Laser Ablation ICPMS study of trace element partitioning between plagioclase and basaltic melts: an experimental approach. Contrib Mineral Petrol 153:647–667

Araoka D, Kawahata H, Takagi T, Watanabe Y, Nishimura K, Nishio Y (2014) Lithium and strontium isotopic systematics in playas in Nevada, USA: constraints on the origin of lithium. Mineral Deposita 49:371–379

Aulbach S, Rudnick RL (2009) Origins of non-equilibrium lithium isotopic fractionation in xenolithic peridotite minerals: Examples from Tanzania. Chem Geol 258:17–27

Aulbach S, Rudnick RL, McDonough WF (2008) Li–Sr–Nd isotope signatures of the plume and cratonic lithospheric mantle beneath the margin of the rifted Tanzanian craton (Labait). Contrib Mineral Petrol 155:79–92

Bach W, Irber W (1998) Rare earth element mobility in the oceanic lower sheeted dyke complex: evidence from geochemical data and leaching experiments. Chem Geol 151:309–326

Bagard M-L, West AJ, Newman K, Basu AR (2015) Lithium isotope fractionation in the Ganges–Brahmaputra floodplain and implications for groundwater impact on seawater isotopic composition. Earth Planet Sci Lett 432:404–414

Bao S-X, Zhou H-Y, Peng X-T, Ji F-W, Yao H-Q (2008) Geochemistry of REE and yttrium in hydrothermal fluids from the Endeavour segment, Juan de Fuca Ridge. Geochem J 42:359–370

Barrat JA, Chaussidon M, Bohn M, Gillet P, Gopel C, Lesourd M (2005) Lithium behavior during cooling of a dry basalt: An ion-microprobe study of the lunar meteorite Northwest Africa 479 (NWA 479). Geochim Cosmochim Acta 69:5597–5609

Bebout GE, Barton MD (1989) Fluid flow and metasomatism in a subduction zone hydrothermal system: Catalina Schist terrane, California. Geology 17:976–980

Beck P, Barrat JA, Chaussidon M, Gillet P, Bohn M (2004) Li isotopic variations in single pyroxenes from the Northwest Africa 480 shergottite (NWA 480): a record of degassing of Martian magmas? Geochim Cosmochim Acta 68:2925–2933

Beck P, Chaussidon M, Barrat JA, Gillet P, Bohn M (2006) Diffusion induced Li isotopic fractionation during the cooling of magmatic rocks: The case of pyroxene phenocrysts from nakhlite meteorites. Geochim Cosmochim Acta 70:4813–4825

Bell DR, Hervig RL, Buseck PR, Aulbach S (2009) Lithium isotope analysis of olivine by SIMS: Calibration of a matrix effect and application to magmatic phenocrysts. Chem Geol 258:5–16

Bernal NF, Gleeson SA, Dean AS, Liu XM, Hoskin P (2014) The source of halogens in geothermal fluids from the Taupo Volcanic Zone, North Island, New Zealand. Geochim Cosmochim Acta 126:265–283

Bertoldi C, Proyer A, Garbe-Schonberg D, Behrens H, Dachs E (2004) Comprehensive chemical analyses of natural cordierites: implications for exchange mechanisms. Lithos 78:389–409

Bindeman IN, Lundstrom CC, Bopp C, Huang F (2013) Stable isotope fractionation by thermal diffusion through partially molten wet and dry silicate rocks. Earth Planet Sci Lett 365:51–62

Blundy J, Dalton J (2000) Experimental comparison of trace element partitioning between clinopyroxene and melt in carbonate and silicate systems, and implications for mantle metasomatism. Contrib Mineral Petrol 139:356–371

Blundy JD, Robinson JAC, Wood BJ (1998) Heavy REE are compatible in clinopyroxene on the spinel lherzolite solidus. Earth Planet Sci Lett 160:493–504

Bottomley DJ, Katz A, Chan LH, Starinsky A, Douglas M, Clark ID, Raven KG (1999) The origin and evolution of Canadian Shield brines: evaporation or freezing of seawater? New lithium isotope and geochemical evidence from the Slave craton. Chem Geol 155:295–320

Bottomley DJ, Chan LH, Katz A, Starinsky A, Clark ID (2003) Lithium isotope geochemistry and origin of Canadian shield brines. Ground Water 41:847–856

Bouchez J, von Blanckenburg F, Schuessler JA (2013) Modeling novel stable isotope ratios in the weathering zone. Am J Sci 313:267–308

Bouman C, Elliott T, Vroon PZ (2004) Lithium inputs to subduction zones. Chem Geol 212:59–79

Bourg IC, Sposito G (2007) Molecular dynamics simulations of kinetic isotope fractionation during the diffusion of ionic species in liquid water. Geochim Cosmochim Acta 71:5583–5589

Bourg IC, Richter FM, Christensen JN, Sposito G (2010) Isotopic mass dependence of metal cation diffusion coefficients in liquid water. Geochim Cosmochim Acta 74:2249–2256

Bouvier AS, Metrich N, Deloule E (2008) Slab-derived fluids in the magma sources of St. Vincent (Lesser Antilles arc): Volatile and light element imprints. J Petrol 49:1427–1448

Bouvier AS, Metrich N, Deloule E (2010) Light elements, volatiles, and stable isotopes in basaltic melt inclusions from Grenada, Lesser Antilles: Inferences for magma genesis. Geochem Geophys Geosyst 11:Q09004

Brant C, Coogan LA, Gillis KM, Seyfried WE, Pester NJ, Spence J (2012) Lithium and Li-isotopes in young altered upper oceanic crust from the East Pacific Rise. Geochim Cosmochim Acta 96:272–293

Brenan JM, Neroda E, Lundstrom CC, Shaw HF, Ryerson FJ, Phinney DL (1998a) Behaviour of boron, beryllium, and lithium during melting and crystallization: Constraints from mineral–melt partitioning experiments. Geochim Cosmochim Acta 62:2129–2141

Brenan JM, Ryerson FJ, Shaw HF (1998b) The role of aqueous fluids in the slab-to-mantle transfer of boron, beryllium, and lithium during subduction: experiments and models. Geochim Cosmochim Acta 62:3337–3347

Brigatti MF, Kile DE, Poppi L (2003) Crystal structure and chemistry of lithium-bearing trioctahedral micas–3*T*. Euro J Mineral 15:349–355

Brooker RA, James RH, Blundy JD (2004) Trace elements and Li isotope systematics in Zabargad peridotites: evidence of ancient subduction processes in the Red Sea mantle. Chem Geol 212:179–204

Brown ID (2009) Recent developments in the methods and applications of the bond valence model. Chem Rev 109:6858–6919

Bryant CJ, Chappell BW, Bennett VC, McCulloch MT (2004) Lithium isotopic compositions of the New England Batholith: correlations with inferred source rock compositions. Trans R Soc Edinburgh-Earth Sci 95:199–214

Burton KW, Vigier N (2011) Lithium Isotopes as Tracers in Marine and Terrestrial Environments. Handbook of Environmental Isotope Geochemistry, Vols. 1 and 2, 41–59 pp

Cabato J, Altherr R, Ludwig T, Meyer HP (2013) Li, Be, B concentrations and δ^7Li values in plagioclase phenocrysts of dacites from Nea Kameni (Santorini, Greece). Contrib Mineral Petrol 165:1135–1154

Caciagli N, Brenan JM, McDonough WF, Phinney D (2011) Mineral–fluid partitioning of lithium and implications for slab-mantle interaction. Chem Geol 280:384–398

Cahalan RC, Kelly ED, Carlson WD (2014) Rates of Li diffusion in garnet: Coupled transport of Li and Y plus REEs. Am Mineral 99:1676–1682

Carignan J, Vigier N, Millot R (2007) Three secondary reference materials for lithium isotope measurements: Li^7–N, Li^6–N and LiCl–N solutions. Geostandard Geoanal Res 31:7–12

Chacko T, Cole DR, Horita J (2001) Equilibrium oxygen, hydrogen and carbon isotope fractionation factors applicable to geologic systems. Rev Mineral Geochem 43:1–81

Chan LH, Edmond JM (1988) Variation of lithium isotope composition in the marine environment: a preliminary report. Geochim Cosmochim Acta 52:1711–1717

Chan LH, Frey FA (2003) Lithium isotope geochemistry of the Hawaiian plume: Results from the Hawaii Scientific Drilling Project and Koolau volcano. Geochem Geophys Geosyst 4:8707

Chan LH, Hein JR (2007) Lithium contents and isotopic compositions of ferromanganese deposits from the global ocean. Deep-Sea Res Part II 54:1147–1162

Chan LH, Edmond JM, Thompson G, Gillis K (1992) Lithium isotopic composition of submarine basalts: implications for the lithium cycle in the oceans. Earth Planet Sci Lett 108:151–160

Chan LH, Edmond JM, Thompson G (1993) A lithium isotope study of hot springs and metabasalts from Mid-Ocean Ridge Hydrothermal Systems. J Geophys Res-Solid Earth 98(B6):9653–9659

Chan LH, Gieskes JM, You CF, Edmond JM (1994) Lithium isotope geochemistry of sediments and hydrothermal fluids of the Guaymas Basin, Gulf of California. Geochim Cosmochim Acta 58:4443–4454

Chan LH, Leeman WP, You CF (1999) Lithium isotopic composition of Central American Volcanic Arc lavas: implications for modification of subare mantle by slab-derived fluids. Chem Geol 160:255–280

Chan LH, Alt JC, Teagle DAH (2002a) Lithium and lithium isotope profiles through the upper oceanic crust: a study of seawater-basalt exchange at ODP Sites 504B and 896A Earth Planet Sci Lett 201:187–(201)

Chan LH, Leeman WP, You CF (2002b) Lithium isotopic composition of Central American volcanic arc lavas: implications for modification of subarc mantle by slab-derived fluids: correction. Chem Geol 182:293–300

Chan LH, Lassiter JC, Hauri EH, Hart SR, Blusztajn J (2009) Lithium isotope systematics of lavas from the Cook–Austral Islands: Constraints on the origin of HIMU mantle. Earth Planet Sci Lett 277:433–442

Chaussidon M, Robert F (1998) Li-7/Li-6 and B-11/B-10 variations in chondrules from the Semarkona unequilibrated chondrite. Earth Planet Sci Lett 164:577–589

Chaussidon M, Robert F (1999) Lithium nucleosynthesis in the Sun inferred from the solar-wind Li-7/Li-6 ratio. Nature 402:270–273

Chaussidon M, Robert F, McKeegan KD (2006a) Li and B isotopic variations in an Allende CAI: Evidence for the in situ decay of short-lived Be-10 and for the possible presence of the short-lived nuclide Be-7 in the early solar system. Geochim Cosmochim Acta 70:224–245

Chaussidon M, Robert F, McKeegan KD (2006b) Reply to the Comment by Desch and Ouellette on "Li and B isotopic variations in an Allende CAI: Evidence for the in situ decay of short-lived Be-10 and for the possible presence of the short-lived nuclide Be-7 in the early solar system. Geochim Cosmochim Acta 70:5433–5436

Cherniak DJ, Watson EB (2010) Li diffusion in zircon. Contrib Mineral Petrol 160:383–390

Clergue C, Dellinger M, Buss HL, Gaillardet J, Benedetti MF, Dessert C (2015) Influence of atmospheric deposits and secondary minerals on Li isotopes budget in a highly weathered catchment, Guadeloupe (Lesser Antilles). Chem Geol 414:28–41

Coogan LA, Kasemann SA, Chakraborty S (2005) Rates of hydrothermal cooling of new oceanic upper crust derived from lithium-geospeedometry. Earth Planet Sci Lett 240:415–424

Cyburt RH, Fields BD, Olive KA (2008) An update on the big bang nucleosynthesis prediction for Li-7: the problem worsens. J Cosmol Astropart Phys 11:012

Danyushevsky LV, Eggins SM, Falloon TJ, Christie DM (2000) H2O abundance in depleted to moderately enriched mid-ocean ridge magmas; Part I: Incompatible behaviour, implications for mantle storage, and origin of regional variations. J Petrol 41:1329–1364

Day JMD, Qiu L, Ash RD, McDonough WF, Teng F-Z, Rudnick RL, Taylor LA (2016) Evidence for high-temperature frationation of lithium isotopes during differentiation of the Moon. Meteorit Planet Sci 51:1046–1062

Decitre S, Deloule E, Reisberg L, James R, Agrinier P, Mevel C (2002) Behavior of Li and its isotopes during serpentinization of oceanic peridotites. Geochem Geophys Geosyst 3:1–20

Dellinger M, Gaillardet J, Bouchez J, Calmels D, Galy V, Hilton RG, Louvat P, France-Lanord C (2014) Lithium isotopes in large rivers reveal the cannibalistic nature of modern continental weathering and erosion. Earth Planet Sci Lett 401:359–372

Dellinger M, Gaillardet J, Bouchez J, Calmels D, Louvat P, Dosseto A, Gorge C, Alanoca L, Maurice L (2015) Riverine Li isotope fractionation in the Amazon River basin controlled by the weathering regimes. Geochim Cosmochim Acta 164:71–93

Desch SJ, Oullette N (2006) Comment on "Li and Be isotopic variations in an Allende CAI: Evidence for the in situ decay of short-lived ^{10}Be and for the possible presence of the short-lived nuclide ^{7}Be in the early solar system," by M Chaussidon F Robert, and KD McKeegan. Geochim Cosmochim Acta 70:5426–5432

Dohmen R, Kasemann SA, Coogan L, Chakraborty S (2010) Diffusion of Li in olivine. Part I: Experimental observations and a multi species diffusion model. Geochim Cosmochim Acta 74:274–292

Dutrow B (1991) The effects of Al and vacancies on Li substitution in iron staurolite—A synthesis approach. Am Mineral 76:42–48

Eggins SM, Rudnick RL, McDonough WF (1998) The composition of peridotites and their minerals: A laser-ablation ICP-MS study. Earth Planet Sci Lett 154:53–71

Elliott T, Jeffcoate A, Bouman C (2004) The terrestrial Li isotope cycle: light-weight constraints on mantle convection. Earth Planet Sci Lett 220:231–245

Elliott T, Thomas A, Jeffcoate A, Niu YL (2006) Lithium isotope evidence for subduction-enriched mantle in the source of mid-ocean-ridge basalts. Nature 443:565–568

Evensen JM, London D (2003) Experimental partitioning of Be, Cs, and other trace elements between cordierite and felsic melt, and the chemical signature of S-type granite. Contrib Mineral Petrol 144:739–757

Fairen AG, Losa-Adams E, Gil-Lozano C, Gago-Duport L, Uceda ER, Squyres SW, Rodriguez JAP, Davila AF, McKay CP (2015) Tracking the weathering of basalts on Mars using lithium isotope fractionation models. Geochem Geophys Geosyst 16:1172–1(197)

Filiberto J, Chin E, Day JMD, Franchi IA, Greenwood RC, Gross J, Penniston-Dorland SC, Schwenzer SP, Treiman AH (2012) Geochemistry of intermediate olivine–phyric shergottite Northwest Africa 6234, with similarities to basaltic shergottite Northwest Africa 480 and olivine–phyric shergottite Northwest Africa 2990. Meteorit Planet Sci 47:1256–1273

Flesch G, Anderson A, Svec H (1973) A secondary isotopic standard for ^{6}Li/^{7}Li determinations. Int J Mass Spectrom Ion Phys 12:265–272

Foustoukos DI, James RH, Berndt ME, Seyfried WE (2004) Lithium isotopic systematics of hydrothermal vent fluids at the Main Endeavour Field, Northern Juan de Fuca Ridge. Chem Geol 212:17–26

Gabitov RI, Schmitt AK, Rosner M, McKeegan KD, Gaetani GA, Cohen AL, Watson EB, Harrison TM (2011) In situ delta Li-7, Li/Ca, and Mg/Ca analyses of synthetic aragonites. Geochem Geophys Geosyst 12:Q03001

Gallagher K, Elliott T (2009) Fractionation of lithium isotopes in magmatic systems as a natural consequence of cooling. Earth Planet Sci Lett 278:286–296

Gao YJ, Snow JE, Casey JF, Yu JB (2011) Cooling-induced fractionation of mantle Li isotopes from the ultraslow-spreading Gakkel Ridge. Earth Planet Sci Lett 301:231–240

Gao YY, Li XH, Griffin WL, Tang YJ, Pearson NJ, Liu Y, Chu M-F, Li QL, Tang GQ, O'Reilly SY (2015) Extreme lithium isotopic fractionation in three zircon standards (Plešovice, Qinghu and Temora). Sci Rep 5:16878

Genske FS, Turner SP, Beier C, Chu M-F, Tonarini S, Pearson NJ, Haase KM (2014) Lithium and boron isotope systematics in lavas from the Azores islands reveal crustal assimilation. Chem Geol 373:27–36

Giaramita MJ, Sorensen SS (1994) Primary fluids in low-temperature eclogites: evidence from two subduction complexes (Dominican Republic, and California, USA). Contrib Mineral Petrol 117:279–292

Godfrey LV, Chan LH, Alonso RN, Lowenstein TK, McDonough WF, Houston J, Li J, Bobst A, Jordan TE (2013) The role of climate in the accumulation of lithium-rich brine in the Central Andes. Appl Geochem 38:92–102

Gurenko AA, Schmincke HU (2002) Orthopyroxene-bearing tholeiites of the Iblean Plateau (Sicily): constraints on magma origin and evolution from glass inclusions in olivine and orthopyroxene. Chem Geol 183:305–331

Halama R, McDonough WF, Rudnick RL, Keller J, Klaudius J (2007) The Li isotopic composition of Oldoinyo Lengai: Nature of the mantle sources and lack of isotopic fractionation during carbonatite petrogenesis. Earth Planet Sci Lett 254:77–89

Halama R, McDonough WF, Rudnick RL, Bell K (2008) Tracking the lithium isotopic evolution of the mantle using carbonatites. Earth Planet Sci Lett 265:726–742

Halama R, John T, Herms P, Hauff F, Schenk V (2011) A stable (Li, O) and radiogenic (Sr, Nd) isotope perspective on metasomatic processes in a subducting slab. Chem Geol 281:151–166

Halama R, Savov IP, Rudnick RL, McDonough WF (2009) Insights into Li and Li isotope cycling and sub-arc metasomatism from veined mantle xenoliths, Kamchatka. Contrib Mineral Petrol 158:197–222

Hamelin C, Seitz HM, Barrat JA, Dosso L, Maury RC, Chaussidon M (2009) A low δ^7Li lower crustal component: Evidence from an alkalic intraplate volcanic series (Chaine des Puys, French Massif Central). Chem Geol 266:205–217

Hanrahan M, Brey G, Woodland A, Altherr R, Seitz HM (2009a) Towards a Li barometer for bimineralic eclogites: experiments in CMAS. Contrib Mineral Petrol 158:169–183

Hanrahan M, Brey G, Woodland A, Seitz HM, Ludwig T (2009b) Li as a barometer for bimineralic eclogites: Experiments in natural systems. Lithos 112:992–1001

Harrison L, Weis D, Hanano D, Barnes E (2015) Lithium isotopic signature of Hawaiian basalts. *In:* Carey R, Cayol V, Poland M, Weis D (Eds.), Hawaiian Volcanoes: From Source to Surface. Geophysical Monograph. John Wiley & Sons, Inc., American Geophysical Union pp. 79–104

Hart SR, Dunn T (1993) Experimental cpx/melt partitioning of 24 trace elements. Contrib Mineral Petrol 113:1–8

Hathorne EC, James RH (2006) Temporal record of lithium in seawater: A tracer for silicate weathering? Earth Planet Sci Lett 246:393–406

Hogan JF, Blum JD (2003) Boron and lithium isotopes as groundwater tracers: a study at the Fresh Kills Landfill, Staten Island, New York, USA. Appl Geochem 18:615–627

Huh Y, Chan LH, Zhang L, Edmond JM (1998) Lithium and its isotopes in major world rivers: Implications for weathering and the oceanic budget. Geochim Cosmochim Acta 62:2039–2051

Huh Y, Chan LH, Edmond JM (2001) Lithium isotopes as a probe of weathering processes: Orinoco River. Earth Planet Sci Lett 194:189–199

Huh Y, Chan LH, Chadwick OA (2004) Behavior of lithium and its isotopes during weathering of Hawaiian basalt. Geochem Geophys Geosyst 5:Q09002

Icenhower J, London D (1995) An experimental study of element partitioning among biotite, muscovite, and coexisting peraluminous silicic melt at 200MPa (H_2O). Am Mineral 80:1229–1251

Ionov DA, Seitz H-M (2008) Lithium abundances and isotopic compositions in mantle xenoliths from subduction and intra-plate settings: Mantle sources vs. eruption histories. Earth Planet Sci Lett 266:316–331

Ireland RHP, Penniston-Dorland SC (2015) Chemical interactions between a sedimentary diapir and surrounding magma: Evidence from the Phepane Dome and Bushveld Complex, South Africa. Am Mineral 100:1985–2000

Jahn S, Wunder B (2009) Lithium speciation in aqueous fluids at high *P* and *T* studied by ab initio molecular dynamics and consequences for Li-isotope fractionation between minerals and fluids. Geochim Cosmochim Acta 73:5428–5434

James RH, Palmer MR (2000a) The lithium isotope composition of international rock standards. Chem Geol 166:319–326

James RH, Palmer MR (2000b) Marine geochemical cycles of the alkali elements and boron: The role of sediments. Geochim Cosmochim Acta 64:3111–3122

Jeffcoate AB, Elliott T, Thomas A, Bouman C (2004) Precise, small sample size determinations of lithium isotopic compositions of geological reference materials and modern seawater by MC-ICP-MS. Geostandard Geoanal Res 28:161–172

Jeffcoate AB, Elliott T, Kasemann SA, Ionov D, Cooper K, Brooker R (2007) Li isotope fractionation in peridotites and mafic melts. Geochim Cosmochim Acta 71:202–218

Jochum KP, Stoll B, Herwig K, Willbold M, Hofmann AW, Amini M, Aarburg S, Abouchami W, Hellebrand E, Mocek B, Raczek I, Stracke A, Alard O, Bouman C, Becker S, Ducking M, Bratz H, Klemd R, de Bruin D, Canil D, Cornell D, de Hoog CJ, Dalpe C, Danyushevsky L, Eisenhauer A, Gao YJ, Snow JE, Goschopf N, Gunther D, Latkoczy C, Guillong M, Hauri EH, Hofer HE, Lahaye Y, Horz K, Jacob DE, Kassemann SA, Kent AJR, Ludwig T, Zack T, Mason PRD, Meixner A, Rosner M, Misawa KJ, Nash BP, Pfander J, Premo WR, Sun WD, Tiepolo M, Vannucci R, Vennemann T, Wayne D, Woodhead JD (2006) MPI-DING reference glasses for in situ microanalysis: New reference values for element concentrations and isotope ratios. Geochem Geophys Geosyst 7:Q02008

Jochum KP, Wilson SA, Abouchami W, Amini M, Chmeleff J, Eisenhauer A, Hegner E, Iaccheri LM, Kieffer B, Krause J, McDonough WF, Mertz-Kraus R, Raczek I, Rudnick RL, Scholz D, Steinhoefel G, Stoll B, Stracke A, Tonarini S, Weis D, Weis U, Woodhead JD (2011) GSD-1G and MPI-DING reference glasses for in situ and bulk isotopic determination. Geostandard Geoanal Res 35:193–226

John T, Gussone N, Podladchikov YY, Bebout GE, Dohmen R, Halama R, Klemd R, Magna T, Seitz H-M (2012) Volcanic arcs fed by rapid pulsed fluid flow through subducting slabs. Nat Geosci 5:489–492

Kaliwoda M, Ludwig T, Altherr R (2008) A new SIMS study of Li, Be, B and δ^7Li in mantle xenoliths from Harrat Uwayrid (Saudi Arabia). Lithos 106:261–279

Kasemann SA, Jeffcoate AB, Elliott T (2005) Lithium isotope composition of basalt glass reference material. Analytical Chemistry 77:5251–5257

Kil Y (2010) Lithium isotopic disequilibrium of minerals in the spinel lherzolite xenoliths from Boeun, Korea. J Geochem Explor 107:56–62

Kim T, Nakaii S, Gasperini D (2011) Lithium abundance and isotope composition of Logudoro basalts, Sardinia: Origin of light Li signature. Geochem J 45:323–340

Kisakurek B, Widdowson M, James RH (2004) Behaviour of Li isotopes during continental weathering: the Bidar laterite profile, India. Chem Geol 212:27–44

Kisakurek B, James RH, Harris NBW (2005) Li and δ7Li in Himalayan rivers: Proxies for silicate weathering? Earth Planet Sci Lett 237:387–401

Knauth DC, Federman SR, Lambert DL (2003) An ultra high-resolution survey of the interstellar $^{7}Li/^{6}Li$ isotope ratio in the solar neighborhood. Astrophys J 586:268–285

Kobayashi K, Tanaka R, Moriguti T, Shimizu K, Nakamura E (2004) Lithium, boron, and lead isotope systematics of glass inclusions in olivines from Hawaiian lavas: evidence for recycled components in the Hawaiian plume. Chem Geol 212:143–161

Košler JMT (2014) Developments in clean lab practices. *In:*McDonough WF (Ed.) Analytical Methods. Treatise on Geochemistry. Elsevier Ltd., Oxford UK, p. 111–122

Košler J, Kučera M, Sylvester P (2001) Precise measurement of Li isotopes in planktonic foraminiferal tests by quadrupole ICPMS. Chem Geol 181:169–179

Kowalski PM, Jahn S (2011) Prediction of equilibrium Li isotope fractionation between minerals and aqueous solutions at high *P* and *T*: An efficient ab initio approach. Geochim Cosmochim Acta 75:6112–6123

Krienitz MS, Garbe-Schonberg CD, Romer RL, Meixner A, Haase KM, Stroncik NA (2012) Lithium isotope variations in ocean island basalts—implications for the development of mantle heterogeneity. J Petrol 53:2333–2347

Lai YJ, von Strandmann P, Dohmen R, Takazawa E, Elliott T (2015) The influence of melt infiltration on the Li and Mg isotopic composition of the Horoman Peridotite Massif. Geochim Cosmochim Acta 164:318–332

Larsen RB, Henderson I, Ihlen PM, Jacamon F (2004) Distribution and petrogenetic behaviour of trace elements in granitic pegmatite quartz from South Norway. Contrib Mineral Petrol 147:615–628

LaTourrette T, Hervig RL, Holloway JR (1995) Trace element partitioning between amphibole, phlogopite, and basanite melt. Earth Planet Sci Lett 135:13–30

le Roux PJ (2010) Lithium isotope analysis of natural and synthetic glass by laser ablation MC-ICP-MS. J Anal At Spectrom 25:1033–1038

Lechler M, Pogge von Strandmann P, Jenkyns HC, Prossera G, Parente M (2015) Lithium-isotope evidence for enhanced silicate weathering during OAE1a (Early Aptian Selli event). Earth Planet Sci Lett 432:210–222

Leeman WP, Tonarini S, Chan LH, Borg LE (2004) Boron and lithium isotopic variations in a hot subduction zone—the southern Washington Cascades. Chem Geol 212:101–124

Lemarchand E, Chabaux F, Vigier N, Millot R, Pierret MC (2010) Lithium isotope systematics in a forested granitic catchment (Strengbach, Vosges Mountains, France). Geochim Cosmochim Acta 74:4612–4628

Lemoine M, Ferlet R, Vidalmadjar A, Emerich C, Bertin P (1993) Interstellar lithium and the Li^{7}/Li^{6} ratio toward Rho OPH. Astron Astrophys 269:469–476

Leya I (2011) Cosmogenic effects of $^{7}Li/^{6}Li$ $^{10}B/^{11}B$, and $^{182}W/^{184}W$ in CAIs from carbonaceous chondrites. Geochim Cosmochim Acta 75:1507–1518

Li CT, Peacor DR (1968) Crystal structure of $LiAlSi_2O_6$–II (beta spodumene). Z Kristallogr Kristallgeom Kristallphys Kristallchem 126:46–65

Li GJ, West AJ (2014) Evolution of Cenozoic seawater lithium isotopes: Coupling of global denudation regime and shifting seawater sinks. Earth Planet Sci Lett 401:284–293

Li X-H, Li Q-L, Liu Y, Tang G-Q (2011) Further characterization of M257 zircon standard: A working reference for SIMS analysis of Li isotopes. J Anal At Spectrom 26:352–358

Li S, Gaschnig RM, Rudnick RL (2016) Insights into chemical weathering of the upper continental crust from the geochemistry of ancient glacial diamictites. Geochim Cosmochim Acta 176:96–117

Liebscher A, Meixner A, Romer RL, Heinrich W (2007) Experimental calibration of the vapour–liquid phase relations and lithium isotope fractionation in the system H_2O–LiCl at 400 °C/20–28 MPa. Geofluids 7:369–375

Liu XM, Rudnick RL (2011) Constraints on continental crustal mass loss via chemical weathering using lithium and its isotopes. PNAS 108:20873–20880

Liu MC, McKeegan KD, Goswami JN, Marhas KK, Sahijpal S, Ireland TR, Davis AM (2009) Isotopic records in CM hibonites: Implications for timescales of mixing of isotope reservoirs in the solar nebula. Geochim Cosmochim Acta 73:5051–5079

Liu XM, Rudnick RL, Hier-Majumder S, Sirbescu MLC (2010) Processes controlling lithium isotopic distribution in contact aureoles: A case study of the Florence County pegmatites, Wisconsin. Geochem Geophys Geosyst 11:Q08014

Liu XM, Rudnick RL, McDonough WF, Cummings ML (2013) Influence of chemical weathering on the composition of the continental crust: Insights from Li and Nd isotopes in bauxite profiles developed on Columbia River Basalts. Geochim Cosmochim Acta 115:73–91

Liu XM, Wanner C, Rudnick RL, McDonough WF (2015) Processes controlling delta Li–7 in rivers illuminated by study of streams and groundwaters draining basalts. Earth Planet Sci Lett 409:212–224

Ludwig T, Marschall HR, Pogge von Strandmann PAE, Shabaga BM, Fayek M, Hawthorne FC (2011) A secondary ion mass spectrometry (SIMS) re-evaluation of B and Li isotopic compositions of Cu-bearing elbaite from three global localities. Mineral Mag 75:2485–2494

Lumpkin GR, Ribbe PH (1983) Composition, order-disorder and lattice-parameters of olivines—relationships in silicate, germanate, beryllate, phosphate and borate olivines. Am Mineral 68:164–176

Lundstrom CC, Chaussidon M, Hsui AT, Kelemen P, Zimmerman M (2005) Observations of Li isotopic variations in the Trinity Ophiolite: Evidence for isotopic fractionation by diffusion during mantle melting. Geochim Cosmochim Acta 69:735–751

Lundstrom CC, Sutton AL, Chaussidon M, McDonough WF, Ash R (2006) Trace element partitioning between type B melts and melilite and spinel: Implications for trace element distribution during CAI formation. Geochim Cosmochim Acta 70:3421–3435

Lynton SJ, Walker RJ, Candela PA (2005) Lithium isotopes in the system Qz–Ms–fluid: An experimental study. Geochim Cosmochim Acta 69:3337–3347

Lyubartsev A, Laasonen K, Laaksonen A (2001) Hydration of Li^+ ion. An ab initio molecular dynamics simulation. J Chem Phys 114:3120–3126

Magna T, Wiechert U, Grove TL, Halliday AN (2006a) Lithium isotope fractionation in the southern Cascadia subduction zone. Earth Planet Sci Lett 250:428–443

Magna T, Wiechert U, Halliday AN (2006b) New constraints on the lithium isotope compositions of the Moon and terrestrial planets. Earth Planet Sci Lett 243:336–353

Magna T, Ionov DA, Oberli F, Wiechert U (2008) Links between mantle metasomatism and lithium isotopes: Evidence from glass-bearing and cryptically metasomatized xenoliths from Mongolia. Earth Planet Sci Lett 276:214–222

Magna T, Wiechert U, Stuart FM, Halliday AN, Harrison D (2011) Combined Li–He isotopes in Iceland and Jan Mayen basalts and constraints on the nature of the North Atlantic mantle. Geochim Cosmochim Acta 75:922–936

Magna T, Day JMD, Mezger K, Fehr MA, Dohmen R, Aoudjehane HC, Agee CB (2015) Lithium isotope constraints on crust-mantle interactions and surface processes on Mars. Geochim Cosmochim Acta 162:46–65

Marks MAW, Rudnick RL, McCammon C, Vennemann T, Markl G (2007) Arrested kinetic Li isotope fractionation at the margin of the Ilimaussaq complex, South Greenland: Evidence for open-system processes during final cooling of peralkaline igneous rocks. Chem Geol 246:207–230

Marks MAW, Rudnick RL, Ludwig T, Marschall H, Zack T, Halama R, McDonough WF, Rost D, Wenzel T, Vicenzi EP, Savov IP, Altherr R, Markl G (2008) Sodic pyroxene and sodic amphibole as potential reference materials for in situ lithium isotope determinations by SIMS Geostandard Geoanal Res 32:295–310

Marriott CS, Henderson GM, Belshaw NS, Tudhope AW (2004a) Temperature dependence of δ^7Li, $\delta^{44}Ca$ and Li/Ca during growth of calcium carbonate. Earth Planet Sci Lett 222:615–624

Marriott CS, Henderson GM, Crompton R, Staubwasser M, Shaw S (2004b) Effect of mineralogy, salinity, and temperature on Li/Ca and Li isotope composition of calcium carbonate. Chem Geol 212:5–15

Marschall HR, Altherr R, Ludwig T, Kalt A, Gmeling K, Kasztovszky Z (2006) Partitioning and budget of Li, Be and B in high-pressure metamorphic rocks. Geochim Cosmochim Acta 70:4750–4769

Marschall HR, Pogge von Strandmann PAE, Seitz HM, Elliott T, Niu YL (2007) The lithium isotopic composition of orogenic eclogites and deep subducted slabs. Earth Planet Sci Lett 262:563–580

Martin C, Ponzevera E, Harlow G (2015) In situ lithium and boron isotope determinations in mica, pyroxene, and serpentine by LA-MC-ICP-MS. Chem Geol 412:107–116

Maruyama S, Watanabe M, Kunihiro T, Nakamura E (2009) Elemental and isotopic abundances of lithium in chondrule constituents in the Allende meteorite. Geochim Cosmochim Acta 73:778–793

McDade P, Blundy JD, Wood BJ (2003) Trace element partitioning on the Tinaquillo Lherzolite solidus at 1.5 GPa. Phys Earth Planet Inter 139:129–147

McDonough WF, Sun SS (1995) The Composition of the Earth. Chem Geol 120:223–253

Meredith K, Moriguti T, Tomascak P, Hollins S, Nakamura E (2013) The lithium, boron and strontium isotopic systematics of groundwaters from an arid aquifer system: Implications for recharge and weathering processes. Geochim Cosmochim Acta 112:20–31

Millot R, Negrel P (2007) Multi-isotopic tracing (δ^7Li, $\delta^{11}B$, $^{87}Sr/^{86}Sr$) and chemical geothermometry: evidence from hydro-geothermal systems in France. Chem Geol 244:664–678

Millot R, Guerrot C, Vigier N (2004) Accurate and high-precision measurement of lithium isotopes in two reference materials by MC-ICP-MS Geostandard Geoanal Res 28:153–159

Millot R, Negrel P, Petelet-Giraud E (2007) Multi-isotopic (Li, B, Sr, Nd) approach for geothermal reservoir characterization in the Limagne Basin (Massif Central, France). Appl Geochem 22:2307–2325

Millot R, Scaillet B, Sanjuan B (2010a) Lithium isotopes in island arc geothermal systems: Guadeloupe, Martinique (French West Indies) and experimental approach. Geochim Cosmochim Acta 74:1852–(1871)

Millot R, Vigier N, Gaillardet J (2010b) Behaviour of lithium and its isotopes during weathering in the Mackenzie Basin, Canada. Geochim Cosmochim Acta 74:3897–3912

Millot R, Guerrot C, Innocent C, Negrel P, Sanjuan B (2011) Chemical, multi-isotopic (Li–B–Sr–U–H–O) and thermal characterization of Triassic formation waters from the Paris Basin. Chem Geol 283:226–241

Millot R, Hegan A, Negrel P (2012) Geothermal waters from the Taupo Volcanic Zone, New Zealand: Li, B and Sr isotopes characterization. Appl Geochem 27:677–688

Misra S, Froelich PN (2009) Measurement of lithium isotope ratios by quadrupole-ICP-MS: application to seawater and natural carbonates. J Anal At Spectrom 24:1524–1533

Misra S, Froelich PN (2012) Lithium isotope history of Cenozoic seawater: changes in silicate weathering and reverse weathering. Science 335:818–823

Moriguti T, Nakamura E (1998a) Across-arc variation of Li isotopes in lavas and implications for crust/mantle recycling at subduction zones. Earth Planet Sci Lett 163:167–174

Moriguti T, Nakamura E (1998b) High-yield lithium separation and the precise isotopic analysis for natural rock and aqueous samples. Chem Geol 145:91–104

Moriguti T, Shibata T, Nakamura E (2004) Lithium, boron and lead isotope and trace element systematics of Quaternary basaltic volcanic rocks in northeastern Japan: mineralogical controls on slab-derived fluid composition. Chem Geol 212:81–100

Mottl MJ, Seewald JS, Wheat CG, Tivey MK, Michael PJ, Proskurowski G, McCollom TM, Reeves E, Sharkey J, You CF, Chan LH, Pichler T (2011) Chemistry of hot springs along the Eastern Lau Spreading Center. Geochim Cosmochim Acta 75:1013–1038

Narita H, Koto K, Morimoto N, Yoshii M (1975) The crystal structure of nambulite (Li, Na)$Mn_4Si_5O_{14}$(OH). Acta Crystallogr Sect B:31:2422–2426

Negrel P, Millot R, Brenot A, Bertin C (2010) Lithium isotopes as tracers of groundwater circulation in a peat land. Chem Geol 276:119–127

Negrel P, Millot R, Guerrot C, Petelet-Giraud E, Brenot A, Malcuit E (2012) Heterogeneities and interconnections in groundwaters: Coupled B, Li and stable-isotope variations in a large aquifer system (Eocene Sand aquifer, Southwestern France). Chem Geol 296:83–95

Nishio Y, Nakai S (2002) Accurate and precise lithium isotopic determinations of igneous rock samples using multi-collector inductively coupled plasma mass spectrometry. Anal Chim Acta 456:271–281

Nishio Y, Shun'ichi N, Yamamoto J, Sumino H, Matsumoto T, Prikhod'ko VS, Arai S (2004) Lithium isotopic systematics of the mantle-derived ultramafic xenoliths: implications for EM1 origin. Earth Planet Sci Lett 217:245–261

Nishio Y, Nakai S, Kogiso T, Barsczus HG (2005) Lithium, strontium, and neodymium isotopic compositions of oceanic island basalts in the Polynesian region: constraints on a Polynesian HIMU origin. Geochem J 39:91–103

Nishio Y, Nakai Si., Ishii T, Sano Y (2007) Isotope systematics of Li, Sr, Nd, and volatiles in Indian Ocean MORBs of the Rodrigues Triple Junction: Constraints on the origin of the DUPAL anomaly. Geochim Cosmochim Acta 71:745–759

Nishio Y, Okamura K, Tanimizu M, Ishikawa T, Sano Y (2010) Lithium and strontium isotopic systematics of waters around Ontake volcano, Japan: Implications for deep-seated fluids and earthquake swarms. Earth Planet Sci Lett 297:567–576

Niu YL, Collerson KD, Batiza R, Wendt JI, Regelous M (1999) Origin of enriched-type mid-ocean ridge basalt at ridges far from mantle plumes: The East Pacific Rise at 11°20'N. J Geophys Res-Solid Earth 104(B4): 7067–7087

Niu YL, Bideau D, Hekinian R, Batiza R (2001) Mantle compositional control on the extent of mantle melting, crust production, gravity anomaly, ridge morphology, and ridge segmentation: a case study at the Mid-Atlantic Ridge 33–35° N. Earth Planet Sci Lett 186:383–399

Oi T, Odagiri T, Nomura M (1997) Extraction of lithium from GSJ rock reference samples and determination of their lithium isotopic compositions. Anal Chim Acta 340:221–225

Ottolini L, Laporte D, Raffone N, Devidal JL, Le Fevre B (2009) New experimental determination of Li and B partition coefficients during upper mantle partial melting. Contrib Mineral Petrol 157:313–325

Parkinson IJ, Hammond SJ, James RH, Rogers NW (2007) High-temperature lithium isotope fractionation: Insights from lithium isotope diffusion in magmatic systems. Earth Planet Sci Lett 257:609–621

Penniston-Dorland SC, Sorensen SS, Ash RD, Khadke SV (2010) Lithium isotopes as a tracer of fluids in a subduction zone melange: Franciscan Complex, CA. Earth Planet Sci Lett 292:181–190

Penniston-Dorland SC, Bebout GE, Pogge von Strandmann PAE, Elliott T, Sorensen SS (2012) Lithium and its isotopes as tracers of subduction zone fluids and metasomatic processes: Evidence from the Catalina Schist, California, USA. Geochim Cosmochim Acta 77:530–545

Pistiner JS, Henderson GM (2003) Lithium-isotope fractionation during continental weathering processes. Earth Planet Sci Lett 214:327–339

Plank T (2014) The chemical composition of subducting sediments. *In:* Holland, H.D., Turekian, K.K. (eds.) Treatise on Geochemistry, 2nd ed., 4: 607-629. Oxford: Elsevier.

Pogge von Strandmann PAE, Henderson GM (2015) The Li isotope response to mountain uplift. Geology 43:67–70

Pogge von Strandmann PAE, Burton KW, James RH, van Calsteren P, Gislason SR, Mokadem F (2006) Riverine behaviour of uranium and lithium isotopes in an actively glaciated basaltic terrain. Earth Planet Sci Lett 251:134–147

Pogge von Strandmann PAE, Burton KW, James RH, van Calsteren P, Gislason SR (2010) Assessing the role of climate on uranium and lithium isotope behaviour in rivers draining a basaltic terrain. Chem Geol 270:227–239

Pogge von Strandmann PAE, Elliott T, Marschall HR, Coath C, Lai YJ, Jeffcoate AB, Ionov DA (2011) Variations of Li and Mg isotope ratios in bulk chondrites and mantle xenoliths. Geochim Cosmochim Acta 75:5247–5268

Pogge von Strandmann PAE, Opfergelt S, Lai Y-J, Sigfusson B, Gislason SR, Burton KW (2012) Lithium, magnesium and silicon isotope behaviour accompanying weathering in a basaltic soil and pore water profile in Iceland. Earth Planet Sci Lett 339:11–23

Pogge von Strandmann PAE, Jenkyns HC, Woodfine RG (2013) Lithium isotope evidence for enhanced weathering during Oceanic Anoxic Event 2. Nat Geosci 6:668–672

Pogge von Strandmann PAE, Porcelli D, James RH, van Calsteren P, Schaefer B, Cartwright I, Reynolds BC, Burton KW (2014) Chemical weathering processes in the Great Artesian Basin: Evidence from lithium and silicon isotopes. Earth Planet Sci Lett 406:24–36

Pogge von Strandmann PAE, Burton KW, Opfergelt S, Eiríksdóttir ES, Murphy MJ, Einarsson A, Gislason SR (2016) The effect of hydrothermal spring weathering processes and primary productivity on lithium isotopes: Lake Myvatn, Iceland. Chem Geol, doi:10.1016/j.chemgeo.2016.02.026

Qi HP, Taylor PDP, Berglund M, De Bièvre P (1997) Calibrated measurements of the isotopic composition and atomic weight of the natural Li isotopic reference material IRMM–016. Int J Mass Spectrom Ion Processes 171:263–268

Qian Q, O'Neill HSC, Hermann J (2010) Comparative diffusion coefficients of major and trace elements in olivine at ~950 ºC from a xenocryst included in dioritic magma. Geology 38:331–334

Qiu L, Rudnick RL, McDonough WF, Merriman RJ (2009) Li and δ^7Li in mudrocks from the British Caledonides: Metamorphism and source influences. Geochim Cosmochim Acta 73:7325–7340

Qiu L, Rudnick RL, Ague JJ, McDonough WF (2011a) A lithium isotopic study of sub-greenschist to greenschist facies metamorphism in an accretionary prism, New Zealand. Earth Planet Sci Lett 301:213–221

Qiu L, Rudnick RL, McDonough WF, Bea F (2011b) The behavior of lithium in amphibolite- to granulite-facies rocks of the Ivrea-Verbano Zone, NW Italy. Chem Geol 289:76–85

Regelous M, Niu YL, Wendt JI, Batiza R, Greig A, Collerson KD (1999) Variations in the geochemistry of magmatism on the East Pacific Rise at 10° 30' N since 800 ka. Earth Planet Sci Lett 168:45–63

Richter FM, Liang Y, Davis AM (1999) Isotope fractionation by diffusion in molten oxides. Geochim Cosmochim Acta 63:2853–2861

Richter FM, Davis AM, DePaolo DJ, Watson EB (2003) Isotope fractionation by chemical diffusion between molten basalt and rhyolite. Geochim Cosmochim Acta 67:3905–3923

Richter FM, Mendybaev RA, Christensen JN, Hutcheon ID, Williams RW, Sturchio NC, Beloso AD (2006) Kinetic isotopic fractionation during diffusion of ionic species in water. Geochim Cosmochim Acta 70:277–289

Richter F, Watson B, Chaussidon M, Mendybaev R, Ruscitto D (2014a) Lithium isotope fractionation by diffusion in minerals. Part 1: Pyroxenes. Geochim Cosmochim Acta 126:352–370

Richter FM, Watson EB, Chaussidon M, Mendybaev R, Christensen JN, Qiu L (2014b) Isotope fractionation of Li and K in silicate liquids by Soret diffusion. Geochim Cosmochim Acta 138:136–145

Riley JP, Tongudai M (1964) The lithium content of sea water. Deep Sea Res Abstr 11:563–568

Ritzenhoff S, Schroter EH, Schmidt W (1997) The lithium abundance in sunspots. Astron Astrophys 328:695–701

Robert JL, Volfinger M, Barrandon JN, Basutcu M (1983) Lithium in the interlayer space of synthetic trioctahedral micas. Chem Geol 40:337–351

Rollion-Bard C, Vigier N, Meibom A, Blamart D, Reynaud S, Rodolfo-Metalpa R, Martin S, Gattuso JP (2009) Effect of environmental conditions and skeletal ultrastructure on the Li isotopic composition of scleractinian corals. Earth Planet Sci Lett 286:63–70

Rowe MC, Kent AJR, Thornber CR (2008) Using amphibole phenocrysts to track vapor transfer during magma crystallization and transport: An example from Mount St. Helens, Washington. J Volcanol Geotherm Res 178:593–607

Rudnick RL, Ionov DA (2007) Lithium elemental and isotopic disequilibrium in minerals from peridotite xenoliths from far-east Russia: Product of recent melt/fluid–rock reaction. Earth Planet Sci Lett 256:278–293

Rudnick RL, Tomascak PB, Njo HB, Gardner LR (2004) Extreme lithium isotopic fractionation during continental weathering revealed in saprolites from South Carolina. Chem Geol 212:45–57

Ryan JG (1989) The systematics of lithium, beryllium and boron in young volcanic rocks. PhD dissertation, Columbia University, New York, NY.

Ryan JG, Kyle PR (2004) Lithium abundance and lithium isotope variations in mantle sources: insights from intraplate volcanic rocks from Ross Island and Marie Byrd Land (Antarctica) and other oceanic islands. Chem Geol 212:125–142

Ryan JG, Langmuir CH (1987) The systematics of lithium abundances in young volcanic rocks. Geochim Cosmochim Acta 51:1727–1741

Ryu JS, Vigier N, Lee SW, Lee KS, Chadwick OA (2014) Variation of lithium isotope geochemistry during basalt weathering and secondary mineral transformations in Hawaii. Geochim Cosmochim Acta 145:103–115

Sartbaeva A, Wells SA, Redfern SAT (2004) Li^+ ion motion in quartz and beta-eucryptite studied by dielectric spectroscopy and atomistic simulations. J Phys-Condens Matt 16:8173–8189

Sauzeat L, Rudnick RL, Chauvel C, Garcon M, Tang M (2015) New perspectives on the Li isotopic composition of the upper continental crust and its weathering signature. Earth Planet Sci Lett 428:181–192

Schauble EA (2004) Applying stable isotope fractionation theory to new systems. Rev Mineral Geochem 55:65–111

Schmidt K, Bottazzi P, Vannucci R, Mengel K (1999) Trace element partitioning between phlogopite, clinopyroxene and leucite lamproite melt. Earth Planet Sci Lett 168:287–299

Schmitt AD, Vigier N, Lemarchand D, Millot R, Stille P, Chabaux F (2012) Processes controlling the stable isotope compositions of Li, B, Mg and Ca in plants, soils and waters: A review. Comptes Rendus Geoscience 344:704–722
Schuessler JA, Schoenberg R, Srsson O (2009) Iron and lithium isotope systematics of the Hekla volcano, Iceland—Evidence for Fe isotope fractionation during magma differentiation. Chem Geol 258:78–91
Seitz HM, Woodland AB (2000) The distribution of lithium in peridotitic and pyroxenitic mantle lithologies—an indicator of magmatic and metasomatic processes. Chem Geol 166:47–64
Seitz HM, Brey GP, Lahaye Y, Durali S, Weyer S (2004) Lithium isotopic signatures of peridotite xenoliths and isotopic fractionation at high temperature between olivine and pyroxenes. Chem Geol 212:163–177
Seitz HM, Brey GP, Weyer S, Durali S, Ott U, Munker C, Mezger K (2006) Lithium isotope compositions of Martian and lunar reservoirs. Earth Planet Sci Lett 245:6–18
Seitz HM, Brey GP, Zipfel J, Ott U, Weyer S, Durali S, Weinbruch S (2007) Lithium isotope composition of ordinary and carbonaceous chondrites, and differentiated planetary bodies: Bulk solar system and solar reservoirs. Earth Planet Sci Lett 260:582–596
Seitz HM, Zipfel J, Brey GP, Ott U (2012) Lithium isotope compositions of chondrules, CAI and a dark inclusion from Allende and ordinary chondrites. Earth Planet Sci Lett 329:51–59
Sephton MA, James RH, Bland PA (2004) Lithium isotope analyses of inorganic constituents from the Murchison meteorite. Astrophys J 612:588–591
Sephton MA, James RH, Zolensky ME (2006) The origin of dark inclusions in Allende: New evidence from lithium isotopes. Meteorit Planet Sci 41:1039–1043
Sephton MA, James RH, Fehr MA, Bland PA, Gounelle M (2013) Lithium isotopes as indicators of meteorite parent body alteration. Meteorit Planet Sci 48:872–878
Shannon RD (1976) Revised effective ionic-radii and systematic studies of interatomic distances in halides and calcogenides. Acta Crystallogr Sect A 32:751–767
Simons KK, Harlow GE, Brueckner HK, Goldstein SL, Sorensen SS, Hemming NG, Langmuir CH (2010) Lithium isotopes in Guatemalan and Franciscan HP–LT rocks: Insights into the role of sediment-derived fluids during subduction. Geochim Cosmochim Acta 74:3621–3641
Sorensen SS, Barton MD (1987) Metasomatism and partial melting in a subduction complex Catalina Schist, southern California. Geology 15:115–118
Soret C (1879) Sur l'état d'équilibre que prend au point de vue de sa concentration une dissolution saline primitivement homogène dont deux parties sont portées à des températures différentes. Arch Sci Phys. Nat 2:48
Spandler C, O'Neill HSC (2010) Diffusion and partition coefficients of minor and trace elements in San Carlos olivine at 1300 °C with some geochemical implications. Contrib Mineral Petrol 159:791–818
Stern RA (2009) An Introduction to Secondary Ion Mass Spectrometry (SIMS) in Geology. Mineral Assoc Canada Short Course 41:1–18
Su BX, Zhang HF, Deloule E, Sakyi PA, Xiao Y, Tang YJ, Hu Y, Ying JF, Liu PP (2012) Extremely high Li and low δ^7Li signatures in the lithospheric mantle. Chem Geol 292:149–157
Su BX, Gu XY, Deloule E, Zhang HF, Li QL, Li XH, Vigier N, Tang YJ, Tang GQ, Liu Y, Pang KN, Brewer A, Mao Q, Ma YG (2015) Potential orthopyroxene, clinopyroxene and olivine reference materials for in situ lithium isotope determination. Geostandard Geoanal Res 39:357–369
Su B-X, Zhou M-F, Robinson PT (2016) Extremely large fractionation of Li isotopes in a chromitite-bearing mantle sequence. Sci Rep 6:22370
Tang M, Rudnick RL, Chauvel C (2014) Sedimentary input to the source of Lesser Antilles lavas: A Li perspective. Geochim Cosmochim Acta 144:43–58
Tang YJ, Zhang HF, Nakamura E, Moriguti T, Kobayashi K, Ying JF (2007a) Lithium isotopic systematics of peridotite xenoliths from Hannuoba, North China Craton: Implications for melt–rock interaction in the considerably thinned lithospheric mantle. Geochim Cosmochim Acta 71:4327–4341
Tang YJ, Zhang HF, Ying JF (2007b) Review of the lithium isotope system as a geochemical tracer. Int Geol Rev 49:374–388
Tang YJ, Zhang HF, Nakamura E, Ying JF (2011) Multistage melt/fluid–peridotite interactions in the refertilized lithospheric mantle beneath the North China Craton: constraints from the Li–Sr–Nd isotopic disequilibrium between minerals of peridotite xenoliths. Contrib Mineral Petrol 161:845–861
Tang YJ, Zhang HF, Deloule E, Su BX, Ying JF, Xiao Y, Hu Y (2012) Slab-derived lithium isotopic signatures in mantle xenoliths from northeastern North China Craton. Lithos 149:79–90
Tang YJ, Zhang HF, Deloule E, Su BX, Ying JF, Santosh M, Xiao Y (2014) Abnormal lithium isotope composition from the ancient lithospheric mantle beneath the North China Craton. Sci Rep 4:4274
Taura H, Yurimoto H, Kurita K, Sueno S (1998) Pressure dependence on partition coefficients for trace elements between olivine and the coexisting melts. Phys Chem Mineral 25:469–484
Taylor SR, McLennan S (1985) The Continenal Crust: its Composition and Evolution. Blackwell Oxford
Taylor TI, Urey HC (1938) Fractionation of the lithium and potassium isotopes by chemical exchange with zeolites. J Chem Phys 6:429–438

Teng FZ, McDonough WF, Rudnick RL, Dalpe C, Tomascak PB, Chappell BW, Gao S (2004) Lithium isotopic composition and concentration of the upper continental crust. Geochim Cosmochim Acta 68:4167–4178
Teng FZ, McDonough WF, Rudnick RL, Walker RJ (2006a) Diffusion-driven extreme lithium isotopic fractionation in country rocks of the Tin Mountain pegmatite. Earth Planet Sci Lett 243:701–710
Teng FZ, McDonough WF, Rudnick RL, Walker RJ, Sirbescu MLC (2006b) Lithium isotopic systematics of granites and pegmatites from the Black Hills, South Dakota. Am Mineral 91:1488–1498
Teng FZ, McDonough WF, Rudnick RL, Wing BA (2007) Limited lithium isotopic fractionation during progressive metamorphic dehydration in metapelites: A case study from the Onawa contact aureole, Maine. Chem Geol 239:1–12
Teng FZ, Rudnick RL, McDonough WF, Gao S, Tomascak PB, Liu YS (2008) Lithium isotopic composition and concentration of the deep continental crust. Chem Geol 255:47–59
Teng FZ, Rudnick RL, McDonough WF, Wu FY (2009) Lithium isotopic systematics of A-type granites and their mafic enclaves: Further constraints on the Li isotopic composition of the continental crust. Chem Geol 262:370–379
Teng F-Z, Li W-Y, Rudnick RL, Gardner LR (2010) Contrasting lithium and magnesium isotope fractionation during continental weathering. Earth Planet Sci Lett 300:63–71
Tian SH, Hou ZQ, Su AN, Qiu L, Mo XX, Hou KJ, Zhao Y, Hu WJ, Yang ZS (2015) The anomalous lithium isotopic signature of Himalayan collisional zone carbonatites in western Sichuan, SW China: Enriched mantle source and petrogenesis. Geochim Cosmochim Acta 159:42–60
Tipper ET, Galy A, Bickle MJ (2006) Riverine evidence for a fractionated reservoir of Ca and Mg on the continents: Implications for the oceanic Ca cycle. Earth Planet Sci Lett 247:267–279
Tomascak PB (2004) Developments in the understanding and application of lithium isotopes in the earth and planetary sciences. Rev Mineral Geochem 55:153–195
Tomascak PB, Carlson RW, Shirey SB (1999a) Accurate and precise determination of Li isotopic compositions by multi-collector sector ICP-MS Chem Geol 158:145–154
Tomascak PB, Tera F, Helz RT, Walker RJ (1999b) The absence of lithium isotope fractionation during basalt differentiation: New measurements by multicollector sector ICP-MS Geochim Cosmochim Acta 63:907–910
Tomascak PB, Ryan JG, Defant MJ (2000) Lithium isotope evidence for light element decoupling in the Panama subarc mantle. Geology 28:507–510
Tomascak PB, Widom E, Benton LD, Goldstein SL, Ryan JG (2002) The control of lithium budgets in island arcs. Earth Planet Sci Lett 196:227–238
Tomascak PB, Hemming NG, Hemming SR (2003) The lithium isotopic composition of waters of the Mono Basin, California. Geochim Cosmochim Acta 67:601–611
Tomascak PB, Langmuir CH, le Roux PJ, Shirey SB (2008) Lithium isotopes in global mid-ocean ridge basalts. Geochim Cosmochim Acta 72:1626–1637
Tomascak PB, Magna TS, Dohmen R (2016) Advances in Lithium Isotope Geochemistry. Springer International Publishing
Trail D, Cherniak DJ, Watson EB, Harrison TM, Weiss BP, Szumila I (2016) Li zoning in zircon as a potential geospeedometer and peak temperature indicator. Contrib Mineral Petrol 17:1–5
Udry A, McSween HY, Jr., Hervig RL, Taylor LA (2016) Lithium isotopes and light lithophile element abundances in shergottites: Evidence for both magmatic degassing and subsolidus diffusion. Meteorit Planet Sci 51:80–104
Ullmann CV, Campbell HJ, Frei R, Hesselbo SP, Pogge von Strandmann PAE, Korte C (2013) Partial diagenetic overprint of Late Jurassic belemnites from New Zealand: Implications for the preservation potential of delta Li–7 values in calcite fossils. Geochim Cosmochim Acta 120:80–96
Urey HC (1947) The thermodynamic properties of isotopic substances. J Chem Soc 562–581
Ushikubo T, Kita NT, Cavosie AJ, Wilde SA, Rudnick RL, Valley JW (2008) Lithium in Jack Hills zircons: Evidence for extensive weathering of Earth's earliest crust. Earth Planet Sci Lett 272:666–676
Verney-Carron A, Vigier N, Millot R (2011) Experimental determination of the role of diffusion on Li isotope fractionation during basaltic glass weathering. Geochim Cosmochim Acta 75:3452–3468
Verney-Carron A, Vigier N, Millot R, Hardarson BS (2015) Lithium isotopes in hydrothermally altered basalts from Hengill (SW Iceland). Earth Planet Sci Lett 411:62–71
Vigier N, Godderis Y (2015) A new approach for modeling Cenozoic oceanic lithium isotope paleo-variations: the key role of climate. Clim Past 11:635–645
Vigier N, Decarreau A, Millot R, Carignan J, Petit S, France-Lanord C (2008) Quantifying Li isotope fractionation during smectite formation and implications for the Li cycle. Geochim Cosmochim Acta 72:780–792
Vigier N, Gislason SR, Burton KW, Millot R, Mokadem F (2009) The relationship between riverine lithium isotope composition and silicate weathering rates in Iceland. Earth Planet Sci Lett 287:434–441
Vlastelic I, Koga K, Chauvel C, Jacques G, Telouk P (2009) Survival of lithium isotopic heterogeneities in the mantle supported by HIMU-lavas from Rurutu Island, Austral Chain. Earth Planet Sci Lett 286:456–466
Wagner C, Deloule E (2007) Behaviour of Li and its isotopes during metasomatism of French Massif Central lherzolites. Geochim Cosmochim Acta 71:4279–4296

Walker JA, Teipel AP, Ryan JG, Syracuse E (2009) Light elements and Li isotopes across the northern portion of the Central American subduction zone. Geochem Geophys Geosyst 10:Q06S16

Wang QL, Chetelat B, Zhao ZQ, Ding H, Li SL, Wang BL, Li J, Liu XL (2015) Behavior of lithium isotopes in the Changjiang River system: Sources effects and response to weathering and erosion. Geochim Cosmochim Acta 151:117–132

Wanner C, Sonnenthal EL, Liu XM (2014) Seawater δ^7Li: A direct proxy for global CO_2 consumption by continental silicate weathering? Chem Geol 381:154–167

Wenger M, Armbruster T (1991) Crystal chemistry of lithium; oxygen coordination and bonding. Euro J Mineral 3:387–399

Weyer S, Seitz HM (2012) Coupled lithium- and iron isotope fractionation during magmatic differentiation. Chem Geol 294:42–50

Williams LB, Hervig RL (2005) Lithium and boron isotopes in illite–smectite: The importance of crystal size. Geochim Cosmochim Acta 69:5705–5716

Wimpenny J, Gislason SR, James RH, Gannoun A, Pogge Von Strandmann PAE, Burton KW (2010a) The behaviour of Li and Mg isotopes during primary phase dissolution and secondary mineral formation in basalt. Geochim Cosmochim Acta 74:5259–5279

Wimpenny J, James RH, Burton KW, Gannoun A, Mokadem F, Gislason SR (2010b) Glacial effects on weathering processes: New insights from the elemental and lithium isotopic composition of West Greenland rivers. Earth Planet Sci Lett 290:427–437

Wimpenny J, Colla CA, Yu P, Yin QZ, Rustad JR, Casey WH (2015) Lithium isotope fractionation during uptake by gibbsite. Geochim Cosmochim Acta 168:133–150

Witherow RA, Lyons WB, Henderson GM (2010) Lithium isotopic composition of the McMurdo Dry Valleys aquatic systems. Chem Geol 275:139–147

Wunder B, Meixner A, Romer RL, Wirth R, Heinrich W (2005) The geochemical cycle of boron: Constraints from boron isotope partitioning experiments between mica and fluid. Lithos 84:206–216

Wunder B, Meixner A, Romer RL, Heinrich W (2006) Temperature-dependent isotopic fractionation of lithium between clinopyroxene and high-pressure hydrous fluids. Contrib Mineral Petrol 151:112–120

Wunder B, Meixner A, Romer RL, Feenstra A, Schettler G, Heinrich W (2007) Lithium isotope fractionation between Li-bearing staurolite, Li-mica and aqueous fluids: An experimental study. Chem Geol 238:277–290

Wunder B, Deschamps F, Watenphul A, Guillot S, Meixner A, Romer RL, Wirth R (2010) The effect of chrysotile nanotubes on the serpentine–fluid Li-isotopic fractionation. Contrib Mineral Petrol 159:781–790

Wunder B, Meixner A, Romer RL, Jahn S (2011) Li-isotope fractionation between silicates and fluids: Pressure dependence and influence of the bonding environment. Euro J Mineral 23:333–342

Xiao Y, Zhang HF, Deloule E, Su BX, Tang YJ, Sakyi PA, Hu Y, Ying JF (2015) Large lithium isotopic variations in minerals from peridotite xenoliths from the Eastern North China Craton. J Geol 123:79–94

Xu R, Liu Y, Tong X, Hu Z, Zong K, Gao S (2013) In-situ trace elements and Li and Sr isotopes in peridotite xenoliths from Kuandian, North China Craton: Insights into Pacific slab subduction-related mantle modification. Chem Geol 354:107–123

Yakob JL, Feineman MD, Deane JA, Eggler DH, Penniston-Dorland SC (2012) Lithium partitioning between olivine and diopside at upper mantle conditions: An experimental study. Earth Planet Sci Lett 329:11–21

Yamaji K, Makita Y, Watanabe H, Sonoda A, Kanoh H, Hirotsu T, Ooi K (2001) Theoretical estimation of lithium isotopic reduced partition function ratio for lithium ions in aqueous solution. J Phys Chem A 105:602–613

Yang JH, Wu FY, Wilde SA, Xie LW, Yang YH, Liu XM (2007) Tracing magma mixing in granite genesis: In situ U–Pb dating and Hf-isotope analysis of zircons. Contrib Mineral Petrol 153:177–190

Yoon J (2010) Lithium as a silicate weathering proxy: Problems and perspectives. Aquatic Geochemistry 16:189–206

You CF, Chan LH (1996) Precise determination of lithium isotopic composition in low concentration natural samples. Geochim Cosmochim Acta 60:909–915

Zack T, Tomascak PB, Rudnick RL, Dalpe C, McDonough WF (2003) Extremely light Li in orogenic eclogites: The role of isotope fractionation during dehydration in subducted oceanic crust. Earth Planet Sci Lett 208:279–290

Zhang LB, Chan LH, Gieskes JM (1998) Lithium isotope geochemistry of pore waters from Ocean Drilling Program Sites 918 and 919, Irminger Basin. Geochim Cosmochim Acta 62:2437–2450

Reviews in Mineralogy & Geochemistry
Vol. 82 pp. 219-287, 2017

Magnesium Isotope Geochemistry

Fang-Zhen Teng

Isotope Laboratory
Department of Earth and Space Sciences
University of Washington
Seattle, WA 98195
USA

fteng@u.washington.edu

INTRODUCTION

Magnesium (Mg) has an atomic number of 12 and belongs to the alkaline earth element (Group II) of the Periodic Table. The pure Mg is a silvery white metal and has a melting point of 650 °C and boiling point of 1090 °C at 1 standard atmosphere (Lide 1993–1994). The electronic configuration of Mg is [Ne]3s^2, with low ionization energies, which makes Mg ionic in character with a common valance state of 2+ and a typical ionic radius of 0.72 Å (Shannon 1976).

Magnesium is a major element and widely distributed in the silicate Earth, hydrosphere and biosphere (Fig. 1a). It is the fourth most abundant element in the Earth (after O, Fe and Si, MgO = 25.5 wt%) (McDonough and Sun 1995), the fifth most abundant element in the bulk continental crust (MgO = 4.66 wt%) (Rudnick and Gao 2003) and the second most abundant cation in seawater (after Na, Mg = 0.128 wt%) (Pilson 2013). Nonetheless, the mantle has >99.9% of Mg in the Earth because of its high MgO content (37.8 wt%, McDonough and Sun 1995) and mass fraction. The high abundance of Mg in the silicate Earth makes it a major constituent of minerals (e.g., olivine, pyroxene, garnet, amphibole, mica, spinel, carbonate, sulfate, and clay minerals) in igneous, metamorphic and sedimentary rocks.

Magnesium has three stable isotopes, with mass numbers of 24, 25 and 26, and typical abundances of 78.99%, 10.00% and 11.01%, respectively (Berglund and Wieser 2011) (Fig. 1b), and a standard atomic weight of 24.305 (CIAAW 2015). Because of the limitations in the mass spectrometry, many previous Mg isotopic studies have concentrated on either mass independent isotope anomalies to look for the radiogenic ^{26}Mg produced by the decay of short-lived ^{26}Al (Gray and Compston 1974; Lee and Papanastassiou 1974) or large kinetic mass-dependent isotope fractionation during evaporation (Davis et al. 1990; Goswami et al. 1994; Russell et al. 1998; Richter et al. 2002). The recent advent of multi-collector inductively coupled plasma mass spectrometry (MC-ICP-MS) has made it possible to measure Mg isotopes with unprecedented high-precision (Galy et al. 2001). To date, >7‰ isotope fractionation in ^{26}Mg/^{24}Mg has been observed in terrestrial samples, with carbonates at the light end (δ^{26}Mg low to −5.6, Wombacher et al. 2011) and weathered silicates at the heavy end (δ^{26}Mg up to +1.8, Liu et al. 2014b) (Fig. 2). Similar to other stable isotopes, Mg isotopes are fractionated the greatest during low-temperature processes and fractionated less during high-temperature processes.

This review summarizes the recent advance, important application and future direction of Mg isotope geochemistry. It starts with a brief summary on nomenclature and analytical method, followed by the distribution of Mg isotopes in extraterrestrial and terrestrial reservoirs. In the section on the terrestrial reservoirs, the mantle and carbonates will be discussed separately from the other terrestrial reservoirs. Then the magnitude and mechanisms of Mg

1529-6466/17/0082-0007$10.00 http://dx.doi.org/10.2138/rmg.2017.82.7

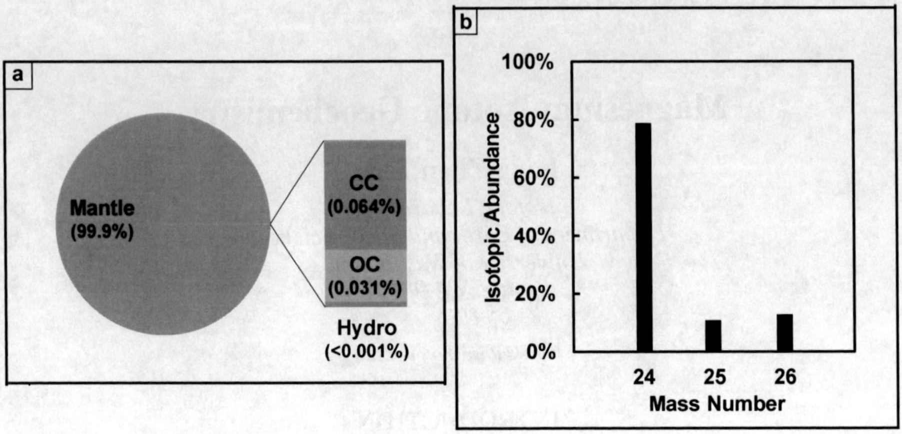

Figure 1. (a) Mass fraction of Mg in the mantle, continental crust (CC), oceanic crust (OC) and hydrosphere (Hydro). (b) Mass number and natural abundance of Mg isotopes. See text for references.

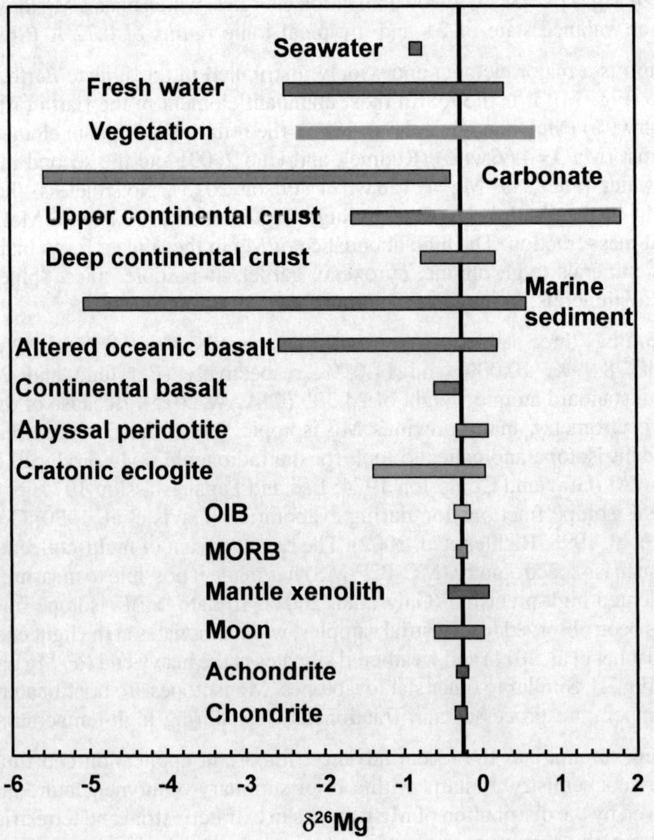

Figure 2. Magnesium isotopic distribution in major extraterrestrial and terrestrial reservoirs. The vertical line represents the average Mg isotopic composition of the mantle, i.e., bulk Earth. See text for details.

isotope fractionation during biological, weathering, magmatic differentiation and metamorphic dehydration processes, as well as equilibrium and kinetic Mg isotope fractionations at high temperatures are summarized. Finally the application and future work that needs to be done to better understand Mg isotope systematics are highlighted.

MAGNESIUM ISOTOPIC ANALYSIS

Magnesium isotope fractionation was clearly documented in 1974 after following the technique developed in Schramm et al. (1970) by using thermal ionization mass spectrometry (TIMS). Meteoritic, lunar and terrestrial feldspars are shown to have identical Mg isotopic composition at ± 1‰ ($^{26}Mg/^{24}Mg$, 2 SD) (Schramm et al. 1970). By contrast, chondrules and Ca–Al rich inclusions (CAIs) in Allende chondrites clearly display large excess in ^{26}Mg (up to 1.3% and 10% respectively), which correlates with $^{27}Al/^{24}Mg$ (Gray and Compston 1974; Lee and Papanastassiou 1974; Lee et al. 1976, 1977). This correlation indicates that ^{26}Mg was produced by the decay of ^{26}Al and provides definitive evidence for the presence of ^{26}Al in the early solar system. Since then, the Al–Mg chronometer has been widely used to study the evolution of the early solar system (MacPherson et al. 1995; Young and Galy 2004; Jacobsen et al. 2008).

At the same time, the large relative mass difference between Mg isotopes also motivated searches for mass-dependent stable Mg isotope fractionation. Daughtry et al. (1962) found large (>4%, $^{24}Mg/^{26}Mg$) mass-dependent Mg isotope fractionation between primary and secondary dolomite during dolomitization. However, this large isotopic variation was not verified by later work that investigated more samples (Shima 1964; Catanzar and Murphy 1966; Catanzar et al. 1966), and likely resulted from the analytical artefact. One of the challenges in Mg isotopic analysis by TIMS is to separate mass-dependent isotope fractionation in nature from that produced in the mass spectrometer since the instrumental fractionation is in many cases larger than natural fractionation. With the development of new instrumentation, especially MC-ICPMS, high-precision measurements of mass-dependent Mg isotope fractionation becomes routine (Galy et al. 2001).

Nomenclature

The early studies of Mg isotope cosmochemistry have used various notations to report Mg isotopic variations in nature (see the review of Young and Galy 2004). As suggested by Young and Galy (2004), the traditional δ notion is used in this study to describe mass-dependent stable Mg isotope fractionation, which is also consistent with other stable isotopic systematics.

Instrumental mass bias is larger and more variable for isotopic ratios measured by MC-ICPMS than for TIMS (Rehkamper et al. 2001). To correct the drift of the instrumental mass bias during Mg isotope analysis by MC-ICPMS, the sample-standard bracketing method is commonly used (Galy et al. 2001). The standards and samples are measured in a sequence of $standard_1$–$sample_1$–$standard_2$–$sample_2$–$standard_3$…. The average value of the two bracketing standards is used to correct the sample analysis for instrumental mass fractionation. The isotopic ratios are then reported in δ-notation:

$$\delta^{X}Mg_{sample_i} = 10^3 \times \left\{ \frac{(^{X}Mg/^{24}Mg)_{sample_i}}{\dfrac{(^{X}Mg/^{24}Mg)_{standard_i} + (^{X}Mg/^{24}Mg)_{standard_{i+1}}}{2}} - 1 \right\} \quad (1)$$

where X refers to mass 25 or 26 and the standard usually refers to DSM3 (Galy et al. 2003). The standard–sample sequences were repeated n times. The reported isotopic composition for a given sample is the average of n repeat analyses. The $\delta^{26}Mg$ values for standards were also computed by considering each $standard_i$ as a sample bracketed by two nearby standards ($standard_{i-1}$ and $standard_{i\ 1}$):

$$\delta^{X}\mathrm{Mg}_{\mathrm{standard}_i} = 10^3 \times \left\{ \frac{({}^{X}\mathrm{Mg}/{}^{24}\mathrm{Mg})_{\mathrm{standard}_i}}{\frac{({}^{X}\mathrm{Mg}/{}^{24}\mathrm{Mg})_{\mathrm{standard}_{i-1}} + ({}^{X}\mathrm{Mg}/{}^{24}\mathrm{Mg})_{\mathrm{standard}_{i+1}}}{2}} - 1 \right\} \quad (2)$$

The dispersion of the δ^{26}Mg values for standards computed by the standard-bracketing method was used to estimate the uncertainty from instrumental instability by assuming that the standard deviation (SD) of the sample measurements equals the standard deviation of the standard measurements. The advantage of using standards rather than samples to quantify the instrumental uncertainty is that the standard deviation of the standard measurements is better known as there are more standard analyses than samples during a batch run. The standard deviation derived from standards, in most cases, is larger than that from samples, and hence is more conservative. Some studies also reported uncertainties by standard error (SE) of the mean ($2SE = 2SD/\sqrt{n}$, where n is the number of replicate analyses) to improve the precision of data. However, caution should be taken for uncertainties on averages of large sample populations as provisional until the accuracy has been tested down to the level of the long-term external precision.

Standard and reference materials

Pure Mg isotopic standard: DSM3 and Cambridge−1. The Mg isotopic standard used extensively before 2003 was Standard Reference Material 980, or SRM 980, which is developed by the National Institute of Standards and Technology (NIST) and distributed by small metal chips. However, Galy et al. (2003) found that SRM 980 has heterogeneous Mg isotopic composition thus is not a suitable reference material; instead, they developed a new liquid Mg isotopic standard that was demonstrated to be homogenous. This new standard solution, which was made from dissolution of 10 g of pure Mg metal from Dead Sea Magnesium Ltd., Israel, in a liter of 0.3 N HNO_3, is called DSM3 (Galy et al. 2003). Aliquots of this concentrated solution have been distributed to different laboratories. Thus the homogeneity of DSM3 is guaranteed. Most of the publications hereafter report mass-dependent Mg isotope data relative to this standard. In addition to DSM3, another mono element nitric solution of Mg called Cambridge −1 was also distributed and measured together with DSM3 in most laboratories (Galy et al. 2003). Cambridge −1 is 2.623 ± 0.030‰ (2 SD) lower than DSM3 (Galy et al. 2003; Teng et al. 2015a).

Mineral and rock reference materials. Cambridge −1 has been analyzed in different laboratories by using different types of instrumental analysis as a way to validate the analytical protocol, and all analyses yielded consistent values (Galy et al. 2003). Nonetheless, this pure Mg standard does not need sample preparation, hence does not test for uncertainties associated with sample dissolution and column chemistry that are required for natural samples. More importantly, sample preparation processes are the key steps to monitor since they have the greatest potential for introducing analytical uncertainties and artefacts. Thus, analysis of well-characterized geostandards that match the chemical composition of samples is needed to ensure data quality and avoid analytical artefacts.

Early studies of some geostandards yield inconsistent results. For example, basalt geostandards BCR−1 and BCR−2 measured from different laboratories yielded significantly different Mg isotopic compositions with δ^{26}Mg value ranges of seven and three times, respectively, of typical reported 2SD uncertainties—although both standards should have similar and homogeneous Mg isotopic composition (Young and Galy 2004; Baker et al. 2005; Bizzarro et al. 2005; Teng et al. 2007b; Wiechert and Halliday 2007; Huang et al. 2009a; Wombacher et al. 2009; Bourdon et al. 2010; Chakrabarti and Jacobsen 2010). Furthermore, Mg isotopic composition of San Carlos olivine was also widely analyzed but the results vary greatly in different laboratories (Pearson et al. 2006; Teng et al. 2007a; Wiechert and Halliday 2007; Handler et al. 2009; Huang et al. 2009a), in particular, analyses of different aliquots of the same powder made by homogenizing 50 g of

San Carlos olivine also yielded different $\delta^{26}Mg$ values, ranging from −0.55 to −0.19 (Young et al. 2009; Chakrabarti and Jacobsen 2010; Liu et al. 2010). These inconsistencies clearly reflect analytical artefacts. A more recent comprehensive study confirms that San Carlos olivine has indeed a homogenous Mg isotopic composition (Hu et al. 2016a).

Recently, a number of international geostandard rock and mineral powders have been widely analyzed by different laboratories and the results demonstrate good agreement (Teng et al. 2015a,b) (Table 1). These standards vary greatly in matrices and include igneous rocks ranging from ultramafic to felsic in composition, metamorphic and sedimentary rocks as well as minerals. This dataset thus lays the foundation for using these geostandards for quality assurance and inter-laboratory comparison of high-precision Mg isotopic data.

Table 1. Recommended magnesium isotopic compositions of geological reference materials and seawater

Standard	Description	δ^{26}Mg	2SD	δ^{25}Mg	2SD	References
		Ultramafic Rocks				
DTS −1	Dunite	−0.30	0.01	−0.13	0.01	Teng et al. (2015a)
		−0.28	0.04	−0.15	0.00	An et al. (2014)
		−0.25	0.08	−0.12	0.02	Wombacher et al. (2009)
		−0.32	0.02	−0.13	0.01	Huang et al. (2009a)
		−0.33	0.14			Huang et al. (2011)
		−1.03	0.28	−0.52	0.11	Chakrabarti and Jacobsen (2010)
		−0.30	**0.01**	**−0.13**	**0.01**	**Recommended value**
DTS −2	Dunite	−0.32	0.06	−0.17	0.04	Teng et al. (2015a)
		−0.23	0.03	−0.12	0.02	Bizzarro et al. (2011)
		−0.32	**0.06**	**−0.17**	**0.04**	**Recommended value**
PCC −1	Peridotite	−0.23	0.06	−0.10	0.01	Teng et al. (2015a)
		−0.22	0.10	−0.13	0.07	Huang et al. (2009a)
		−0.51	0.32	−0.26	0.16	Chakrabarti and Jacobsen (2010)
		−0.24	0.05	−0.12	0.03	An et al. (2014)
		−0.23	**0.06**	**−0.10**	**0.01**	**Recommended value**
JP −1	Peridotite	−0.24	0.01	−0.12	0.01	Pogge von Strandmann et al. (2011)
		−0.23	0.03	−0.12	0.02	Handler et al. (2009)
		−0.24	**0.01**	**−0.12**	**0.01**	**Recommended value**
GBW07101	Ultramafic	−0.28	0.02	−0.12	0.01	Teng et al. (2015b)
		−0.28	**0.02**	**−0.12**	**0.01**	**Recommended value**
GBW07102	Ultramafic	−0.14	0.02	−0.08	0.01	Teng et al. (2015b)
		−0.14	**0.02**	**−0.08**	**0.01**	**Recommended value**

Standard	Description	$\delta^{26}Mg$	2SD	$\delta^{25}Mg$	2SD	References
		Mafic Rocks				
BHVO	Basalt	−0.207	0.07	−0.123	0.033	Teng et al. (2015a)
		−0.207	**0.07**	**−0.123**	**0.033**	**Recommended value**
BHVO −1	Basalt	−0.217	0.067	−0.125	0.026	Teng et al. (2015a)
		−0.12	0.05	−0.06	0.03	Bizzarro et al. (2005)
		−0.09	0.02	−0.04	0.02	Baker et al. (2005)
		−0.3	0.08	−0.14	0.04	Huang et al. (2009a)
		−0.59	0.27	−0.29	0.12	Chakrabarti and Jacobsen (2010)
		−0.16	0.23	−0.08	0.1	Thrane et al. (2006)
		−0.3	0.08			Huang et al. (2011)
		−0.217	**0.067**	**−0.125**	**0.026**	**Recommended value**
BHVO −2	Basalt	−0.24	0.08	−0.12	0.05	Teng et al. (2015a)
		−0.19	0.02	−0.1	0.01	Bizzarro et al. (2011)
		−0.14	0.21	−0.06	0.11	Wiechert and Halliday (2007)
		−0.25	0.11	−0.13	0.08	Pogge von Strandmann et al. (2008a)
		−0.26	0.06	−0.14	0.04	Pogge von Strandmann et al. (2011)
		−0.2	0.08	−0.1	0.03	Bouvier et al. (2013)
		−0.28	0.04	−0.14	0.02	Huang et al. (2015a)
		−0.25	0.01	−0.12	0.01	Huang et al. (2015b)
		−0.24	**0.08**	**−0.12**	**0.05**	**Recommended value**
BR	Basalt	−0.41	0.09	−0.21	0.07	Teng et al. (2015a)
		−0.41	**0.09**	**−0.21**	**0.07**	**Recommended value**
JB −1	Basalt	−0.28	0.10	−0.15	0.04	Teng et al. (2015a)
		−0.28	**0.10**	**−0.15**	**0.04**	**Recommended value**
JB −2	Basalt	−0.22	0.02	−0.11	0.01	Pogge von Strandmann et al. (2011)
		−0.24	0.12	−0.12	0.08	Pogge von Strandmann et al. (2008a)
		−0.18	0.04	−0.09	0.02	Wiechert and Halliday, (2007)
		−0.15	0.13	−0.08	0.07	Bizzarro et al. (2005)
		−0.21	**0.02**	**−0.11**	**0.01**	**Recommended value**

Standard	Description	$\delta^{26}Mg$	2SD	$\delta^{25}Mg$	2SD	References
		−0.34	0.06	−0.18	0.04	Teng et al. (2007b)
		−0.37	0.11	−0.19	0.07	Young et al. (2004)
BCR −1	Basalt	−0.09	0.27	−0.06	0.15	Wiechert and Halliday (2007)
		−0.34	0.12	−0.17	0.08	Huang et al. (2009a)
		−0.19	0.02	−0.10	0.01	An et al. (2014)
		−0.30	0.11			Huang et al. (2011)
		−0.16	0.11	−0.09	0.05	Tipper et al. (2008b)
		−0.26	0.01	−0.13	0.00	Pogge von Strandmann et al. (2011)
		−0.34	0.12			Huang et al. (2011)
		−0.12	0.04	−0.06	0.02	Bourdon et al. (2010)
		−0.14	0.11	−0.07	0.06	Wombacher et al. (2009)
		−0.30	0.08	−0.16	0.09	Teng et al. (2007b)
BCR −2	Basalt	−0.17	0.35	−0.09	0.17	Bizzarro et al. (2005)
		−0.33	0.04	−0.18	0.01	Teng et al. (2007b)
		−0.30	0.11	−0.15	0.07	Huang et al. (2009a)
		−0.22	0.05			Wimpenny et al. (2014a)
		−0.30	0.19	−0.16	0.11	Opfergelt et al. (2012)
		−0.16	0.01	−0.08	0.02	An et al. (2014)
		−0.26	0.08	−0.13	0.05	Lee et al. (2014)
		−0.22	0.06	−0.10	0.02	An et al. (2014)
BIR −1	Basalt	−0.28	0.03	−0.14	0.03	Lee et al. (2014)
		−0.27	0.33	−0.18	0.18	Opfergelt et al. (2012)
		−0.27	**0.03**	**−0.12**	**0.02**	**Recommended value**
DNC −1	Basalt	−0.22	0.05	−0.11	0.03	An et al. (2014)
		−0.22	**0.05**	**−0.11**	**0.03**	**Recommended value**
GBW07105	Basalt	−0.41	0.01	−0.20	0.02	Teng et al. (2015b)
		−0.41	**0.01**	**−0.20**	**0.02**	**Recommended value**
QMC I3	Dolerite	−0.23	0.07	−0.10	0.06	Teng et al. (2015a)
		−0.23	**0.07**	**−0.10**	**0.06**	**Recommended value**
		−0.16	0.08	−0.07	0.10	Teng et al. (2015a)
W −1	Diabase	−0.16	0.10	−0.09	0.06	Wang et al. (2011)
		−0.13	0.13	−0.07	0.06	Huang et al. (2009a)
		−0.16	**0.08**	**−0.07**	**0.10**	**Recommended value**

Standard	Description	$\delta^{26}Mg$	2SD	$\delta^{25}Mg$	2SD	References
W–2	Diabase	–0.16	0.01	–0.08	0.03	Teng et al. (2015a)
		–0.16	0.10	–0.09	0.06	Wang et al. (2011)
		–0.154	0.01	–0.077	0.02	An et al. (2014)
		–0.17	0.08	–0.07	0.05	Lee et al. (2014)
		–0.16	**0.01**	**–0.08**	**0.03**	**Recommended value**
GBW07123	Diabase	–0.19	0.01	–0.09	0.02	Teng et al. (2015b)
		–0.19	**0.01**	**–0.09**	**0.02**	**Recommended value**
GBW07112	Gabbro	–0.18	0.02	–0.07	0.01	Teng et al. (2015b)
		–0.18	**0.02**	**–0.07**	**0.01**	**Recommended value**
		Intermediate Rocks				
DR-N	Diorite	–0.16	0.06	–0.09	0.01	Teng et al. (2015a)
		–0.50	0.04	–0.30	0.04	Brenot et al. (2008)
		–0.52	0.04	–0.25	0.04	Bolou-Bi et al. (2009)
		–0.16	**0.06**	**–0.09**	**0.01**	**Recommended value**
AGV–1	Andesite	–0.12	0.06	–0.06	0.04	Teng et al. (2015a)
		–0.32	0.08	–0.16	0.05	Huang et al. (2009a)
		–0.03	0.10	–0.01	0.01	Wang et al. (2011)
		–0.12	**0.06**	**–0.06**	**0.04**	**Recommended value**
AGV –2	Andesite	–0.15	0.02	–0.07	0.03	Huang et al. (2015a)
		–0.24	0.24	–0.14	0.13	Opfergelt et al. (2012)
		–0.12	0.03	–0.06	0.03	An et al. (2014)
		–0.21	0.09	–0.12	0.07	Lee et al. (2014)
		–0.15	**0.01**	**–0.07**	**0.02**	**Recommended value**
GBW07104	Andesite	–0.66	0.02	–0.35	0.02	Teng et al. (2015b)
		–0.66	**0.02**	**–0.35**	**0.02**	**Recommended value**
JG –1	Granodiorite	–0.31	0.07	–0.16	0.03	Teng et al. (2015a)
		–0.31	**0.07**	**–0.16**	**0.03**	**Recommended value**
GSP –2	Granodiorite	0.042	0.02	0.03	0.011	An et al. (2014)
		0.03	0.09	0	0.06	Huang et al. (2015a)
		–0.02	0.31	–0.05	0.21	Opfergelt et al. (2012)
		0.04	**0.02**	**0.03**	**0.01**	**Recommended value**
GBW07111	Granodiorite	–0.25	0.02	–0.15	0.01	Teng et al. (2015b)
		–0.25	**0.02**	**–0.15**	**0.01**	**Recommended value**

Standard	Description	δ^{26}Mg	2SD	δ^{25}Mg	2SD	References
		Felsic rocks				
G–2	Granite	−0.148	0.071	−0.078	0.057	Teng et al. (2015a)
		−0.22	0.25	−0.07	0.14	Huang et al. (2009a)
		−0.129	0.045	−0.067	0.044	An et al. (2014)
		−0.148	**0.071**	**−0.078**	**0.057**	**Recommended value**
GA	Granite	−0.23	0.065	−0.109	0.052	Teng et al. (2015a)
		−0.75	0.14	−0.36	0.08	Bolou-Bi et al. (2009)
		−0.34	0.15	−0.17	0.11	Huang et al. (2009a)
		−0.165	0.038	−0.084	0.027	An et al. (2014)
		−0.23	**0.065**	**−0.109**	**0.052**	**Recommended value**
GS-N	Granite	−0.23	0.07	−0.12	0.03	Teng et al. (2015a)
		−0.24	0.23	−0.12	0.13	Huang et al. (2009a)
		−0.20	0.06	−0.11	0.04	An et al. (2014)
		−0.23	**0.07**	**−0.12**	**0.03**	**Recommended value**
GSR –1	Granite	−0.23	0.02	−0.12	0.01	An et al. (2014)
		−0.23	**0.02**	**−0.12**	**0.01**	**Recommended value**
GBW07103	Granite	−0.24	0.02	−0.14	0.02	Teng et al. (2015b)
		−0.24	0.02	−0.14	0.02	**Recommended value**
RGM –1	Rhyolite	−0.19	0.03	−0.09	0.02	An et al. (2014)
		−0.19	**0.03**	**−0.09**	**0.02**	**Recommended value**
RGM –2	Rhyolite	−0.18	0.04	−0.09	0.03	An et al. (2014)
		−0.18	**0.04**	**−0.09**	**0.03**	**Recommended value**
GBW07113	Rhyolite	−0.46	0.02	−0.24	0.02	Teng et al. (2015b)
		−0.46	0.02	−0.24	0.02	**Recommended value**
SY –2	Syenite	−0.43	0.10	−0.22	0.04	Teng et al. (2015a)
		−0.43	**0.10**	**−0.22**	**0.04**	**Recommended value**
GBW07109	Syenite	−0.27	0.02	−0.15	0.01	Teng et al. (2015b)
		−0.27	0.02	−0.15	0.01	**Recommended value**
GBW07110	Trachyte	0.08	0.02	0.03	0.01	Teng et al. (2015b)
		0.08	0.02	0.03	0.01	**Recommended value**
		Sedimentary, Metamorphic Rocks and Minerals				
MAG –1	Marine mud	−0.25	0.07	−0.11	0.04	Teng et al. (2015a)
		−0.23	0.09	−0.14	0.04	Wombacher et al. (2009)
		−0.34	0.08	−0.20	0.07	Lee et al. (2014)
		−0.25	**0.07**	**−0.11**	**0.04**	**Recommended value**

Standard	Description	$\delta^{26}Mg$	2SD	$\delta^{25}Mg$	2SD	References
SCo–1	Shale	–0.89	0.08	–0.47	0.05	Teng et al. (2015a)
		–0.91	0.04	–0.48	0.03	Ma et al. (2015)
		–0.81	0.07			Wimpenny et al. (2014)
		–0.89	**0.08**	**–0.47**	**0.05**	**Recommended value**
SGR–1	Shale	–1.00	0.08	–0.51	0.03	Teng et al. (2015a)
		–0.98	0.12	–0.50	0.06	Wombacher et al. (2009)
		–1.00	**0.08**	**–0.51**	**0.03**	**Recommended value**
SDC–1	Mica Schist	–0.11	0.03	–0.06	0.05	Teng et al. (2015a)
		–0.11	**0.03**	**–0.06**	**0.05**	**Recommended value**
UB-N	Serpentine	–0.19	0.04	–0.09	0.02	Teng et al. (2015a)
		–0.12	0.08	–0.06	0.06	Wombacher et al. (2009)
		–0.16	0.09	–0.08	0.07	Wang et al. (2011)
		–0.19	**0.04**	**–0.09**	**0.02**	**Recommended value**
Phlogopite	Phlogopite	–1.371	0.067	–0.719	0.047	Teng et al. (2015a)
		–1.371	**0.067**	**–0.719**	**0.047**	**Recommended value**
JCp–1	Aragonite	–2.01	0.22	–1.05	0.12	Wombacher et al. (2009)
		–1.96	0.07	–1.03	0.02	Hippler et al. (2009)
		1.89	0.10	–0.98	0.06	Yoshimura et al. (2011)
		–2.02	0.11	–1.05	0.06	Planchon et al. (2013)
		–1.96	**0.05**	**–1.03**	**0.02**	**Recommended value**
JCt–1	Aragonite	–3.39	0.09	–1.74	0.04	Yoshimura et al. (2011)
		–2.80	0.11	–1.45	0.08	Planchon et al. (2013)
Jdo –1	Dolomite	–2.36	0.06	–1.25	0.06	Mavromatis et al. (2014)
		–2.37	0.08	–1.25	0.06	Mavromatis et al. (2013)
		–2.38	0.08	–1.23	0.05	Beinlich et al. (2014)
		–2.37	**0.04**	**–1.24**	**0.03**	**Recommended value**
GBW07122	Amphibolite	–0.18	0.02	–0.10	0.02	Teng et al. (2015b)
		–0.18	0.02	–0.10	0.02	**Recommended value**
			Seawater			
Seawater		–0.83	0.09	–0.43	0.06	Teng et al. (2015a)
		–0.83	**0.09**	**–0.43**	**0.06**	**Recommended value**

Note: The recommended values in bold were calculated as a weighted average of independent replicate analysis, taken from either Teng et al. (2015a) or Teng et al. (2015b). In case that those recommended values are not available from Teng et al. (2015a,b), a new weighted average of independent replicate analysis is calculated as the recommended values. When the values by independent replicate analysis yielded inconsistent values, no recommended value is reported.

Seawater. Magnesium is a conservative element in the ocean and has a residence time of ~13 Ma (Li 1982) that is much longer than the ocean mixing time of 1–2 kya (Garrison 2006). Therefore, Mg is homogeneously distributed in the oceans, with a global average content of ~0.13 wt% (Brown et al. 1989) and δ^{26}Mg of -0.83 ± 0.09 (2SD, $n = 90$) (Foster et al. 2010; Ling et al. 2011 and references therein). The matrix elements to Mg ratios in seawater are much more similar to natural samples when compared to the mono-element standard e.g., Cambridge–1. Therefore, analysis of seawater provides a means to test the suitability of the entire chemical and mass spectrometric procedure. This, together with the fact that seawater is readily accessible in large amounts, can be easily processed for isotopic analysis, and has an isotopic composition near the middle of the natural isotopic variation, makes seawater an excellent geostandard for quality control and inter-laboratory accuracy assessment on high-precision Mg isotopic analysis.

Instrumental Analysis

High-precision Mg isotopic ratios have been measured in geological materials by mainly two types of mass spectrometric techniques: *in situ* analysis by laser ablation (LA) MC-ICPMS or secondary ion mass spectrometry (SIMS), and solution-method MC-ICPMS. In-situ analysis by LA-MC-ICPMS or SIMS suffers from greater uncertainties and significant matrix effects but allows for *in situ* determination of Mg isotopes, especially for isotopic zoning in minerals produced by diffusion (Young et al. 2002; Norman et al. 2006b; Pearson et al. 2006; Janney et al. 2011; Xie et al. 2011; Sio et al. 2013; Oeser et al. 2014). By contrast, the solution-method MC-ICPMS techniques require tedious sample dissolution and column chemistry but usually yield at least 3 times better precision than *in situ* analysis.

Galy et al. (2001) carried out the first high-precision analysis of Mg isotopes by using the sample-standard bracketing method on a *Nu Plasma* MC-ICPMS, which since then has rapidly become widespread (Teng et al. 2007b; Wiechert and Halliday 2007; Pogge von Strandmann 2008; Bolou-Bi et al. 2009; Handler et al. 2009; Huang et al. 2009a; Wombacher et al. 2009; Yang et al. 2009; Chakrabarti and Jacobsen 2010; Wang et al. 2011; Choi et al. 2012; Teng and Yang 2014). These techniques have the common feature that the samples are measured relative to bracketing standards in order to correct the instrumental mass bias. Thus this method relies on the following assumptions: (1) the instrumental mass bias changes linearly within a short time period and (2) the sample and standard have the same instrumental mass bias.

As summarized in Teng and Yang (2014), many factors can affect instrumental mass bias and deteriorate the precision and accuracy of Mg isotopic analyses such as the presence of matrix elements, spectral interferences, mismatched Mg concentration between samples and standards and acid molarity, and instrumental settings. In addition, the degree and direction of offsets caused by these factors vary considerably for different instruments, different instrument settings or different laboratories. Overall, wet plasma seems less sensitive to Mg concentration mismatch and matrix effects than dry plasma, hence is recommended for high-precision Mg isotope analysis (Teng and Yang 2014). Furthermore, many of these factors such as matrix effects and spectral interferences can be minimized by separation of Mg from matrix elements through column chemistry, which is commonly used and discussed below.

Sample preparation

Separation of Mg from matrix elements was mainly achieved by cation exchange chromatography with Bio-Rad AG50W-X8 resin in HNO_3 media (Lee and Papanastassiou 1974; Teng et al. 2007b; Huang et al. 2009a; Spivak-Birndorf et al. 2009; Yang et al. 2009), Bio-Rad AG50W-X12 resin in HCl media (Chang et al. 2003; Pogge von Standmann et al. 2008a; Tipper et al. 2008b; Hippler et al. 2009; Immenhauser et al. 2010), Bio-Rad AG50W-X12 resin in HNO_3 media (Black et al. 2006; Pogge von Strandmann 2008; Chakrabarti and Jacobsen 2010; Higgins and Schrag 2010; Wang et al. 2011), Bio-Rad AG MP −50 resin in HCl media, and multiple types of resin or acid (Wiechert and Halliday 2007; Bolou-Bi et al. 2009; Handler et al. 2009; Shen et al. 2009; Wombacher et al. 2009; Young et al. 2009; Schiller et al. 2010; Bizzarro et al. 2011; Pogge von Strandmann et al. 2011; Li et al. 2012). Since the matrix effects vary significantly in different laboratories, the degree to which matrix elements must be eliminated from the Mg cuts after column chemistry differs among different analytical methods, and should be assessed in each laboratory individually (see more details in Teng and Yang 2014).

Magnesium isotopes, similar to other non-traditional stable isotopes, can be significantly fractionated during ion exchange reactions, with heavy Mg isotopes eluted from column first (Chang et al. 2003; Teng et al. 2007b) (Fig. 3). It is, therefore, essential to obtain close to ~100% Mg yield from the column chemistry. Nonetheless, it has been shown that Mg elution curves were not shifted for different matrix compositions, which is different from other isotopes such as Li (Chan et al. 2002) and makes it easier to calibrate the elution curves. Recent studies, however, show that the procedure for cleaning resins can affect the accuracy of Mg isotopic analysis. A cleaning procedure involving diluted HF (e.g., 0.5 N HF) seems to help clean the resin better and eliminate the artefact produced by the resin (Chang et al. 2003; Teng et al. 2015a).

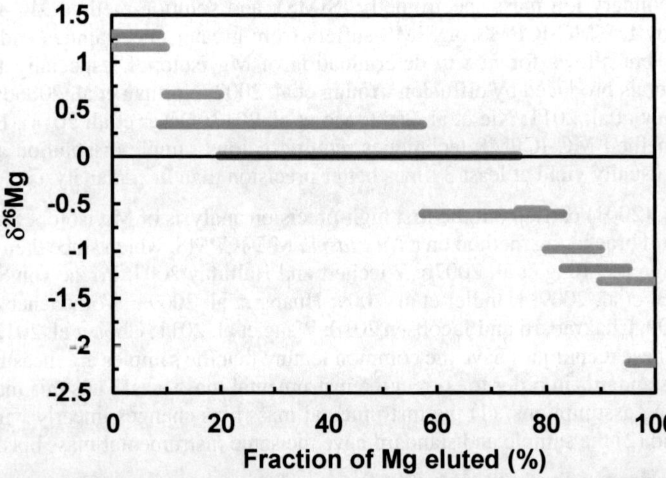

Figure 3. Magnesium isotope fractionation during column chemistry as shown by the deviation of $\delta^{26}Mg$ of each Mg cut of pure Mg solution passed through the column. The original pure Mg solution has $\delta^{26}Mg = 0$ (the horizontal line). Heavy Mg isotopes are always eluted first from the Bio-Rad AG50W-X12 resin in HCl media (green lines, Chang et al. 2003) and Bio-Rad AG50W-X8 resin in HNO_3 media (red lines, Teng et al. 2007b).

MAGNESIUM ISOTOPIC SYSTEMATICS OF EXTRATERRESTRIAL RESERVOIRS

Extraterrestrial materials available for laboratory isotopic analysis consist of meteorites (chondrites, achondrites, stony-iron meteorites and iron meteorites) and samples returned by various missions (e.g., the Apollo missions). Of these samples, chondrites still contain primordial chemical and isotopic signatures in the early solar system because of the lack of high-temperature processes. Therefore, their compositions can provide information on nebular processes (e.g., Scott and Krot 2003). Furthermore, chondrites represent the building blocks of terrestrial planets and provide the best estimate of solar-system abundances for non-volatile elements (McDonough and Sun 1995; Palme and Jones 2003). Therefore, studies of chondrites can place tight constraints on the bulk composition of the Earth. By contrast, the differentiated meteorites (achondrites, stony-iron meteorites, and iron meteorites) are derived from parental bodies having already experienced various degrees of differentiation; hence they can provide information on the differentiation history of the early solar system (e.g., Mittlefehldt 2003).

Finally, lunar meteorites and lunar samples returned by missions vary largely in petrology, mineralogy, and composition, reflecting great differentiation of the Moon. Thus, their compositions can help to constrain the origin and differentiation of the Moon and provide insights into the early evolution of the Earth.

Magnesium isotopic composition of chondrites

Magnesium isotopic compositions of chondrites have been measured from the beginning of the development of MC-ICPMS (Galy et al. 2000; Young and Galy 2004; Baker et al. 2005; Teng et al. 2007b; Wiechert and Halliday 2007; Yang et al. 2009; Young et al. 2009). However, only 11 chondrite samples were analyzed by various laboratories before 2010. The total variation in δ^{26}Mg is >0.5 and the δ^{26}Mg values for even the same meteorite vary >0.5. Some of these large variations could reflect sample heterogeneity since different laboratories likely analyzed a small fragment of the meteorite sample that contains different amounts of constitute chondrules and CAIs, which may have different Mg isotopic compositions (Clayton et al. 1988; Richter et al. 2002). Analytical artefact, however, cannot be ruled out as the same laboratory has reported different values for the same meteorite sample (Young and Galy 2004; Young et al. 2009).

The first comprehensive study of chondrites was carried out by Teng et al. (2010a), which reported Mg isotopic compositions for a total of 38 chondrites, with most of them being falls. These 38 samples come from all three major chondrite classes (carbonaceous, ordinary and enstatite), belong to nine chondrite groups (CI, CM, CO, CV, L, LL, H, EH and EL), vary from 1 to 6 in petrologic grade and cover the whole range of CAI content (from <0.01 vol. % in CI chondrites up to ~13 vol. % in CO chondrites), chemical and isotopic compositions, and oxidation state found in undifferentiated meteorites. Magnesium isotopic compositions of all these chondrites are indistinguishable within uncertainties, with δ^{26}Mg values ranging from −0.35 to −0.23 and an average value of −0.28 ± 0.06 (2SD) (Fig. 4) and δ^{25}Mg values ranging from −0.19 to −0.11 and an average value of −0.15 ± 0.04 (2SD). The δ^{26}Mg and d^{25}Mg values in all major types of chondrites are constant and do not vary with chemical compositions (e.g., Mg/Si and Al/Mg ratios) (Teng et al. 2010a). The homogenous Mg isotopic composition of chondrites observed suggests that Mg isotopes do not fractionate during processes in the solar system parent molecular cloud, in the protosolar nebula or on the parent bodies of chondrites.

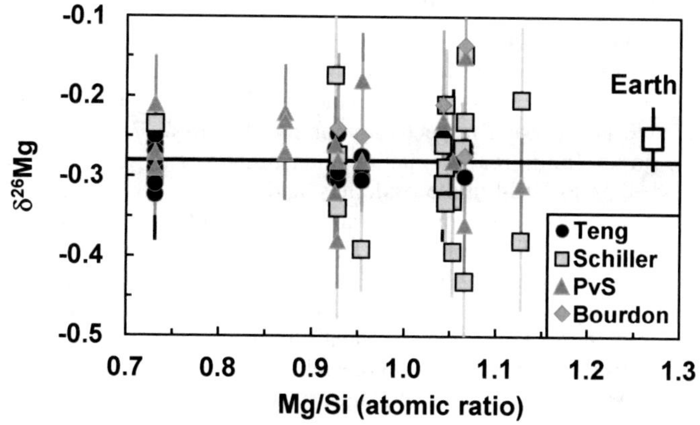

Figure 4. δ^{26}Mg vs. average atomic Mg/Si ratio for chondrites and the Earth. δ^{26}Mg values of chondrites are from Teng et al. (2010a) [Teng], Schiller et al. (2010) [Schiller], Pogge von Strandmann et al. (2011) [PvS] and Bourdon et al. (2010) [Bourdon]. δ^{26}Mg value of the Earth is −0.25 ± 0.04 (2SD) based on peridotite data from Teng et al. (2010a). The Mg/Si ratios of chondrites are from Hutchison (2004), and of the Earth are from McDonough and Sun (1995). The horizontal line and bar represent the average δ^{26}Mg of −0.28 and 2SD of 0.06 for chondrites (Teng et al. 2010a).

Later studies reveal similar results that chondrites have similar Mg isotopic compositions. For example, Schiller et al. (2010) measured Mg isotopes for 18 chondrites including all major classes (carbonaceous, enstatite and ordinary) and 11 chondrite groups (CI, CM, CO, CV, CR, CH, CK, L, LL, H and EH). They found the δ^{26}Mg values vary between −0.43 and −0.15, with an average value of −0.29 ± 0.16 (2SD) (Fig. 4). Correspondingly, δ^{25}Mg values vary from −0.22 to −0.09, with an average value of −0.15 ± 0.08 (2SD). Bourdon et al. (2010) reported δ^{26}Mg for 5 chondrites (CI, CM, CV, LL and H) ranging from −0.27 to −0.14, falling within the range of Teng et al. (2010a) and Schiller et al. (2010). Pogge von Strandmann et al. (2011) reported Mg isotopic data for 17 chondrites including all three classes and 11 groups (CI, CM, CO, CV, CR, CK, L, LL, H, EH and EL). The δ^{26}Mg values range from −0.38 to −0.15, with an average value of −0.27 ± 0.12 (2SD) (Fig. 4). The average values of chondrites in these studies are identical to those of Teng et al. (2010a), although the origins of the slightly large isotopic variation in these chondrites are unknown. It may likely reflect sample heterogeneity since all the three laboratories analyzed different fragments of these meteorite samples.

One study, however, reported different values for both terrestrial and extraterrestrial rocks. Chakrabarti and Jacobsen (2010) analyzed four chondrites (CM, CV, L6 and H3), seven pallasites, three Martian meteorites, and five lunar samples, five terrestrial basalts and four peridotites. They found that all these samples have similar Mg isotopic compositions and hence suggested that the inner solar system has a homogenous Mg isotopic composition, a conclusion consistent with other studies. However, their data are systematically ~0.2‰ lower than the data reported from all other groups. The reason for the systematic offset is still not clear but later measurements of igneous rock standards by this group yielded consistent results with others (Teng et al. 2015b), suggesting that the preveous systematic offset may reflect analytical artefacts.

Magnesium isotopic composition of differentiated meteorites

Early studies of Mg isotopic compositions of achondrites yielded conflicting results. Wiechert and Halliday (2007) reported that Mg isotopic compositions of eucrites and diogenites are similar to those of the Earth and Martian meteorites, but are different from chondrites. By contrast, Chakrabarti and Jacobsen (2010) found achondrites have similar Mg isotopic compositions to the Earth and Moon. Nonetheless, the values for the Earth and chondrites reported in both studies are not consistent with those reported by other groups (Teng et al. 2007b; 2010a; Handler et al. 2009; Yang et al. 2009; Bourdon et al. 2010; Schiller et al. 2010; Pogge von Strandmann et al. 2011).

Sedaghatpour and Teng (2016) carried out a comprehensive Mg isotopic study of 22 well-characterized differentiated meteorites including 7 types of achondrites and pallasites. They found an overall 0.21‰ Mg isotopic variation, with δ^{26}Mg being −0.24 and −0.19 for acapulcoite-lodranite and angrite, respectively, ranging from −0.27 to −0.22 in winonaite-IAB-iron silicate group, from −0.37 to −0.29 in aubrites, −0.27 to −0.16 in HEDs, −0.30 to −0.21 in ureilites, −0.31 to −0.24 in mesosiderites, and −0.30 to −0.24 in pallasites (Fig. 5). Magnesium isotopic compositions of most achondrites are similar to chondrites except angrite and some HEDs, which are slightly heavier. Their slightly higher values likely result from impact-induced evaporation or high abundance of isotopically heavy clinopyroxene. Overall, the average Mg isotopic composition of achondrites (δ^{26}Mg = −0.25 ± 0.08, 2SD, n = 22) is indistinguishable from those of the Earth and chondrites reported by the same laboratory (Teng et al. 2010a), which further supports the homogeneity of Mg isotopes in the solar system and the lack of Mg isotope fractionation during planetary accretion processes and impact events.

Magnesium isotopic composition of the Moon

Early studies on Mg isotopic compositions of the Moon, similar to those of meteorites, were debated. LA-MC-ICPMS analyses revealed homogenous Mg isotopic compositions in lunar mare basalts, glasses and impact melts, similar to the Earth and chondrites at ~±0.2 ‰ amu^{-1} (2SD) (Norman et al. 2006b). By contrast, Wiechert and Halliday (2007), by using solution method MC-ICPMS with a higher precision (~±0.05 ‰ amu^{-1}, 2SD), found that the δ^{26}Mg of the

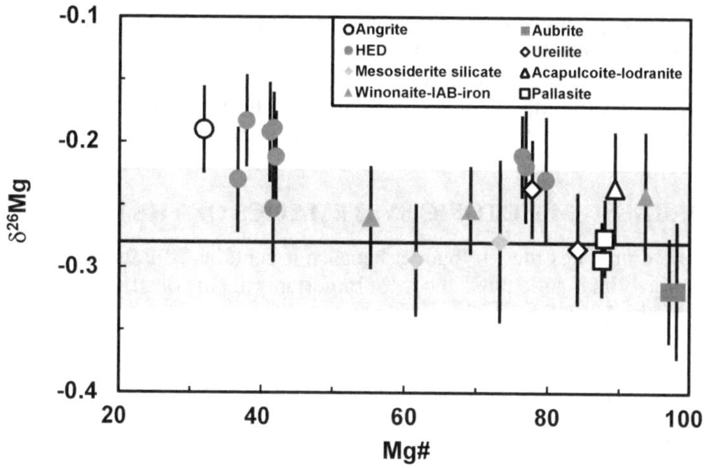

Figure 5. Magnesium isotopic composition of achondrites (Sedaghatpour and Teng 2016). The horizontal line and bar represent the average δ^{26}Mg of –0.28 and 2SD of 0.06 for chondrites (Teng et al. 2010a).

Moon is ~0.17 higher than that of the Earth, and both the Earth and Moon are isotopically heavier than chondrites. By using a similar method, Chakrabarti and Jacobsen (2010) reported that the Moon has an identical Mg isotopic composition to the Earth and chondrites. Nonetheless, the values for most terrestrial and chondrite samples reported in both studies (Wiechert and Halliday 2007; Chakrabarti and Jacobsen 2010) are different from those reported by other groups (Teng et al. 2007b, 2010a; Handler et al. 2009; Yang et al. 2009; Bourdon et al. 2010; Schiller et al. 2010; Pogge von Strandmann et al. 2011), which indicates possible analytical artefacts.

Sedaghatpour et al. (2013) systematically measured Mg isotopes for 47 well-characterized samples from all major types of lunar rocks returned by the Apollo 11, 12, 14, 15, 16 and 17 missions. They found Mg isotopic compositions of lunar samples display a dichotomy between low- and high-Ti samples (Fig. 6). The δ^{26}Mg values of lunar highland rocks, breccias, soil samples and low-Ti basalts vary from –0.41 to –0.14 and are similar to terrestrial basalts and chondrites. By contrast, high-Ti basalts tend to have light Mg isotopic compositions, with δ^{26}Mg ranging from –0.61 to –0.27, which may reflect the source heterogeneity produced during differentiation of the lunar magma ocean (LMO) (Sedaghatpour et al. 2013). Overall,

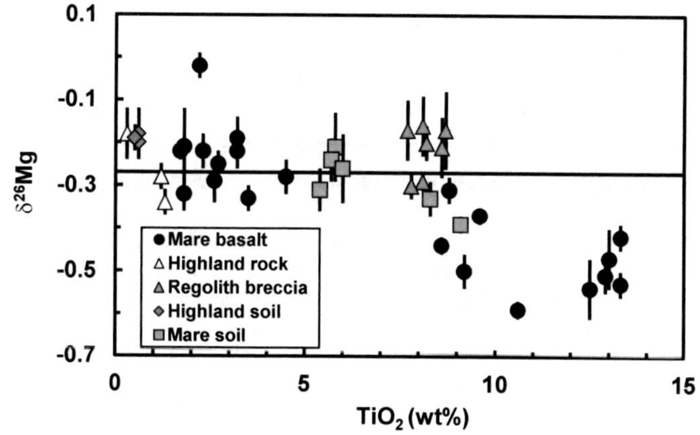

Figure 6. Magnesium isotopic composition of lunar samples (Sedaghatpour et al. 2013). The horizontal line and bar represent the average δ^{26}Mg of –0.28 and 2SD of 0.06 for chondrites (Teng et al. 2010a).

the Mg isotopic composition of the Moon is estimated to be $\delta^{26}Mg = -0.26 \pm 0.16$ (2SD), which is indistinguishable from the Earth, chondrites (Teng et al. 2010a), and achondrites (Sedaghatpour and Teng 2016) reported from the same laboratory. Hence, both the Moon and Earth have chondritic Mg isotopic compositions, reflecting extensive early disk mixing processes and homogeneity of Mg isotopes in the solar system.

MAGNESIUM ISOTOPIC SYSTEMATICS OF THE MANTLE

The mantle dominates the Mg budget because it contains >99.9% of Mg in the Earth (Fig. 1). Therefore, the mantle plays the most important role in constraining the Mg isotopic composition of the bulk Earth. The common samples for studying compositions of the mantle are mantle xenoliths, oceanic basalts, abyssal peridotites, ophiolites, komatiites and continental basalts. Below we review the Mg isotopic systematics in these different mantle and mantle-derived rocks in turn.

Mantle xenoliths

Mantle xenoliths, especially peridotite xenoliths entrained in alkaline lavas, are transported from their mantle source regions, i.e., subcontinental lithospheric mantle, to the surface through volcanic eruption, and have been extensively used to study chemical compositions of the mantle (e.g., McDonough and Sun 1995). Magnesium isotopic measurements have been carried out for a large amount of mantle xenoliths. Although most peridotite xenoliths have a homogenous Mg isotopic composition, some strongly metasomatised xenoliths such as wehrlites and pyroxenites have different isotopic compositions, likely reflecting isotope fractionation during mantle metasomatism.

Peridotite xenoliths. Early attempts to constrain Mg isotopic composition of the mantle were mainly done by studies of peridotite minerals and the results were conflicting. Wiechert and Halliday (2007) carried out the first solution study on Mg isotopic composition of the Earth and found that the Earth, as represented by olivine and pyroxene in a few peridotite xenoliths, has Mg isotopic compositions up to 0.4‰ heavier than chondrites. A later study, based on olivine, pyroxene and spinel from two San Carlos peridotites and two chondrites, also suggested that the Earth has a non-chondritic Mg isotopic composition and is ~0.1‰ heavier than chondrites (Young et al. 2009). By contrast, other studies of olivine and pyroxene in peridotite xenoliths from a range of locations and ages show that mantle minerals cluster around chondrites and the Earth has a chondritic Mg isotopic composition (Handler et al. 2009; Yang et al. 2009). The reason for these conflicting results is still unclear and could reflect mantle metasomatism, which modified Mg isotopic compositions of mantle xenoliths (see details in the next section); nonetheless, analytical artefacts must have played a role. For example, the difference in $\delta^{26}Mg$ of San Carlos olivine is >0.5 based on results from different laboratories (Teng et al. 2007b; Wiechert and Halliday 2007; Handler et al. 2009; Young et al. 2009), and $\delta^{26}Mg$ of a chondrite sample (Orgueil) reported in Young et al. (2009) is >0.3 lower than that reported in Young and Galy (2004), although all laboratories claimed a precision of $\leq \pm 0.1$ ‰ (2SD).

To better characterize the Mg isotopic composition of the mantle i.e., bulk Earth, Teng et al. (2010a) measured Mg isotopes for a globally distributed, geochemically diverse set of 29 well-characterized peridotite xenoliths from Australia, China, France, Tanzania and USA. They found these peridotite xenoliths have identical Mg isotopic compositions, with an average $\delta^{26}Mg = -0.25 \pm 0.04$ (2SD) (Fig. 7), which is used to represent the average Mg isotopic composition of the mantle and bulk Earth. This value is similar to those of olivine and pyroxene separates investigated by Handler et al. (2009) and Yang et al. (2009), and suggests the Earth, Moon, chondrites and achondrites have similar Mg isotopic compositions. Later studies show that most peridotites have similar Mg isotopic compositions whereas some, especially those strongly metasomatised peridotites and pyroxenites do have significantly different values (Bourdon et al. 2010; Huang et al. 2011; Pogge von Strandmann et al. 2011; Xiao et al. 2013; Lai et al. 2015; Hu et al. 2016b; Wang et al. 2016), which will be discussed below.

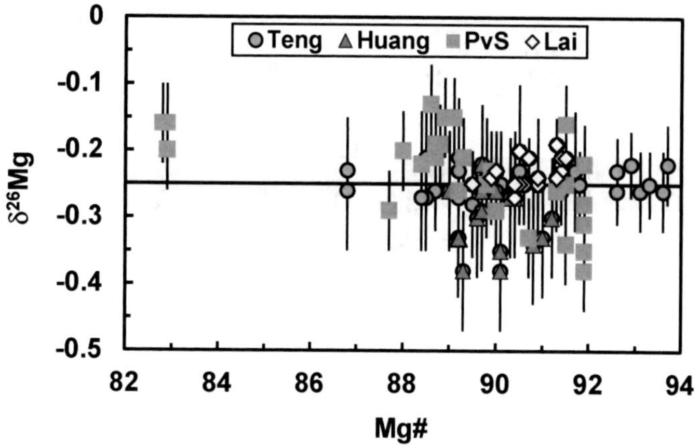

Figure 7. Magnesium isotopic composition of peridotite xenoliths and Massif peridotites. Peridotite xenolith data are from Teng et al. (2010a) [Teng], Huang et al. (2011) [Huang], Pogge von Strandmann et al. (2011) [PvS]; Massif peridotite data are from Lai et al. (2015) [Lai]. The xenolith data from Bourdon et al. (2010); Xiao et al. (2013), Hu et al, (2016b) and Wang et al. (2016) are not plotted because of lack of Mg# data. The horizontal line and bar represent the average δ^{26}Mg of −0.25 and 2SD of 0.04 for the normal mantle, as defined by the data from Teng et al. (2010a).

Metasomatised mantle xenoliths. Studies of moderately metasomatised mantle xenoliths (wehrlites and pyroxenites) have found large Mg isotopic variations at both bulk rock and mineral scales, which has been interpreted to reflect Mg isotope fractionation during mantle metasomatism or during transport of xenoliths in the host magma (Yang et al. 2009; Huang et al. 2011; Liu et al. 2011; Pogge von Strandmann et al. 2011; Xiao et al. 2013; Hu et al. 2016b; Wang et al. 2016).

Wehrlite and pyroxenite xenoliths entrained in alkaline basalts from the North China craton have typical petrographic, geochemical and isotopic features of mantle metasomatism, such as glass veins, presence of metasomatic minerals (phlogopite and apatite), melt and fluid inclusions, and mineral reaction textures, as well as the enrichment in LREE and LILE and positive Pb and Sr anomalies (Xiao et al. 2010, 2013; Hu et al. 2016b). Therefore, they are products of variably high-degree melt-rock interactions and represent the "frozen" metasomatised lithospheric mantle. Both wehrlites and pyroxenites have heterogeneous Mg isotopic compositions relative to the normal mantle (Fig. 8). Furthermore, coexisting minerals (olivine, pyroxene, spinel, and garnet) in these xenoliths display both equilibrium and disequilibrium inter-mineral Mg isotope fractionation (see details in the later section on kinetic Mg isotope fractionation). Finally, phlogopites from clinopyroxenite and lherzolite samples, which are a typical metasomatic mineral, also have heavier Mg isotopic compositions than coexisting olivines and pyroxenes (Liu et al. 2011). These bulk-rock and mineral Mg isotopic signatures cannot be produced by interactions between xenoliths and host basaltic magma during the transport as such interactions should produce kinetic, not equilibrium inter-mineral Mg isotope fractionation. Instead, these isotopically heterogeneous wehrlites and pyroxenites reflect Mg isotopic heterogeneity in their mantle sources. The most likely process responsible for the mantle heterogeneity was metasomatism of subcontinental lithospheric mantle by reactions with melts derived from the subducted slabs. These samples therefore provide the direct evidence for heterogeneous Mg isotopic compositions of the North China craton.

Large Mg isotopic variations were also reported for mantle xenoliths from the SE Siberian craton, from off-craton sites in Mongolia and southern Siberia as well as arc peridotites from Avacha (Fig. 7), with δ^{26}Mg positively correlated with δ^{7}Li for the bulk xenoliths (Pogge von Strandmann et al. 2011). The large Li and Mg co-variations were interpreted as a result of isotope fractionation induced by diffusion between mantle xenoliths and host magma during

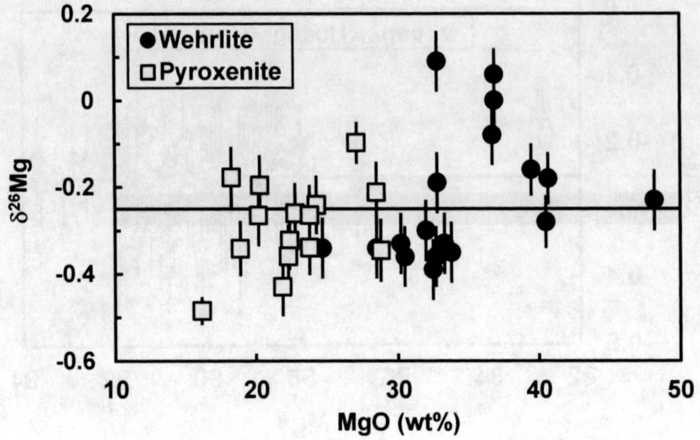

Figure 8. Magnesium isotopic composition of wehrlite and pyroxenite xenoliths (Yang et al. 2009; Pogge von Strandmann et al. 2011; Xiao et al. 2013; Hu et al. 2016b). The horizontal line and bar represent the average δ^{26}Mg of −0.25 and 2SD of 0.04 for the mantle (Teng et al. 2010a).

transport of the xenoliths from the mantle to the surface. The diffusion is mainly triggered by degassing as H is lost from nominally anhydrous minerals during degassing. Consequently, Mg diffused into the xenoliths as a result of charge balance and shifted the xenoliths towards light Mg isotopic composition because light isotopes diffuse faster than heavy ones (Richter et al. 2009a). This kind of kinetic isotope fractionation therefore should not occur in xenoliths hosted in pyroclastic lavas as degassing is limited, which explains why many mantle xenoliths still preserve their mantle Mg isotopic signatures (Pogge von Strandmann et al. 2011).

Cratonic eclogites. Cratonic eclogites are xenoliths entrained by kimberlites and are considered as samples of the subcontinental lithospheric mantle. Although their origins are still highly debated, most studies suggest they are ancient subducted oceanic crust that were stacked beneath old continental crust (Lee et al. 2011, and references therein). Therefore, cratonic eclogites can provide information on compositions of both subcontinental lithospheric mantle and the subducted oceanic crust.

Magnesium isotopic studies of low- and high-MgO cratonic eclogites from the West and South Africa show that they have very heterogeneous Mg isotopic compositions and on average lighter than the mantle; furthermore, both low-MgO and high-MgO eclogites display similar (~1‰) isotopic variations (Wang et al. 2012, 2015c) (Fig. 9). The effects of kimberlite infiltration during transport on the Mg isotopic systematics of cratonic eclogites are considered negligible as the bulk-rock δ^{26}Mg values constructed from garnet and clinopyroxene data are similar to those analyzed on rock powders (Wang et al. 2015c). Therefore, the large Mg isotopic variation reflects source signatures and is consistent with their derivation from the subducted altered oceanic crust. These observations not only further confirm the subducted oceanic crust as the origins of cratonic eclogites, but also indicate that altered oceanic crust after subduction still has very heterogeneous Mg isotopic compositions. This is also consistent with the fact that prograde metamorphism does not shift Mg isotopic compositions of metamorphic rocks (Li et al. 2011a, 2014; Teng et al. 2013; Wang et al. 2014b, 2015a,b), hence the heterogeneous Mg isotopic signature of the subducting oceanic crust (Huang 2013; Hu et al. 2015) can still be preserved after subduction.

Oceanic basalts

Oceanic basalts (e.g., mid-ocean ridge basalts (MORBs) and ocean island basalts (OIBs)) have been extensively used to study chemical and isotopic, especially radiogenic isotopic heterogeneity in the mantle that is produced by crustal recycling (Hofmann 2003). Oceanic basalts are produced by partial melting of the mantle and experience various degrees of differentiation.

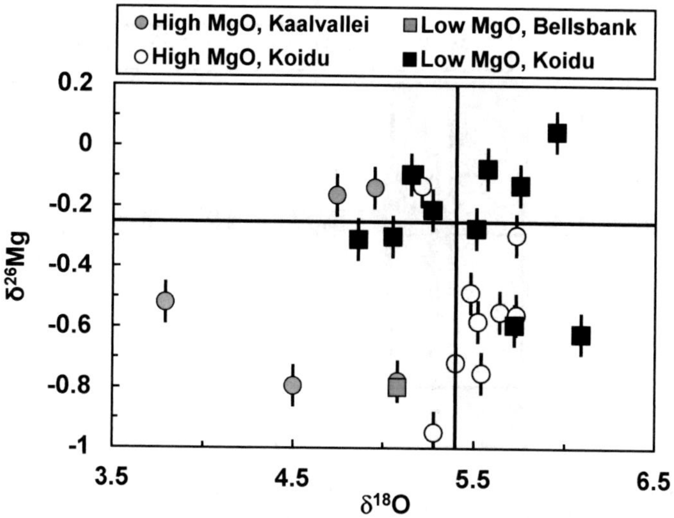

Figure 9. Magnesium and oxygen isotopic compositions of cratonic eclogites from Koidu kimberlite complex, West Africa (Williams et al. 2009; Wang et al. 2012) and Kaalvallei and Bellsbank kimberlite pipes, South Africa (Barth et al. 2001, 2002; Wang et al. 2015c). The lines and bars represent the mean and 2SD of δ^{26}Mg and δ^{18}O values of the normal mantle (Mattey et al. 1994; Teng et al. 2010a).

Therefore, the behavior of stable isotopes during partial melting and magmatic differentiation has to be constrained before basalts can be used to trace their source isotopic compositions. Teng et al. (2007b) carried out a systematic study on the behavior of Mg isotopes during basalt differentiation by investigating a set of well-characterized samples from the Kilauea Iki lava lake, Hawaii. These samples were produced by closed-system crystal-melt fractionation and range from olivine-rich cumulates to highly differentiated basalts with MgO contents ranging from 26.9 to 2.37wt%. They found that the highly differentiated basalts and olivine-rich cumulates have similar Mg isotopic composition to the primitive magma, which indicates a lack of Mg isotope fractionation during crystal–melt fractionation at temperatures of ≥1055 °C. Thus Mg isotopic compositions of oceanic basalts can be used to constrain those of their mantle sources.

Comprehensive studies of a globally distributed, geochemically diverse set of well-characterized MORBs ($n = 47$) and OIBs from Hawaiian and French Polynesian islands ($n = 63$) show that MORBs and OIBs have similar Mg isotopic composition, with δ^{26}Mg ranging from −0.31 to −0.19 (average = −0.25 ± 0.06, 2SD) in MORBs (Fig. 10) and ranging from −0.35 to −0.18 (average = −0.26 ± 0.08, 2SD) in OIBs (Fig. 11). Overall, these oceanic basalts have an average δ^{26}Mg = −0.26 ± 0.07 (2SD), which is similar to global peridotites (Teng et al. 2010a). Later studies of more MORB and OIB samples yield similar results and show that oceanic basalts and peridotite xenoliths have similar Mg isotopic compositions (Bourdon et al. 2010). The similar Mg isotopic compositions among MORBs, OIBs and peridotites further suggest a lack of significant Mg isotope fractionation during partial melting of mantle peridotite and differentiation of basaltic magma. It also suggests that subducted crustal materials cannot produce significant mantle Mg isotopic heterogeneity at global scales as sampled by MORBs and OIBs, although local mantle Mg isotopic heterogeneity has been observed in cratonic eclogites and metasomatised peridotite and pyroxenite xenoliths.

Abyssal peridotites and ophiolites

Abyssal peridotites are partial melting residues of the asthenosphere at oceanic ridges while ophiolites represent an exposed cross section of the oceanic crust and the underlying upper mantle. Thus, they are great proxies for studying compositions of the mantle. Nonetheless, both abyssal peridotites and ophiolites commonly experienced various degrees of post-magmatic

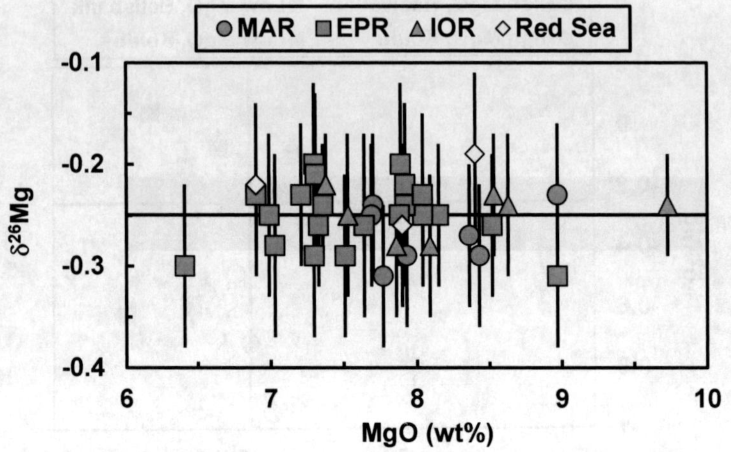

Figure 10. Plot of δ^{26}Mg vs. MgO for MORBs (Teng et al. 2010a). The data from Bourdon et al. (2010) are not plotted because of lack of MgO data. The horizontal line and bar represent the average δ^{26}Mg of −0.25 and 2SD of 0.04 for the normal mantle (Teng et al. 2010a).

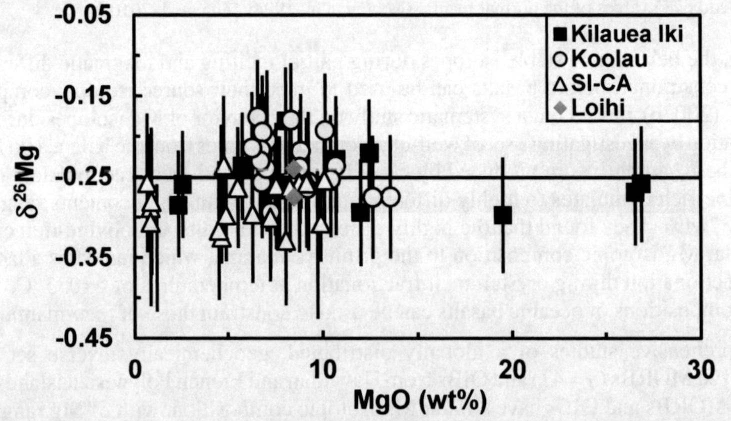

Figure 11. Plot of δ^{26}Mg vs. MgO for OIBs (Teng et al. 2010a). The data from Bourdon et al. (2010) are not plotted because of lack of MgO data. The horizontal line and bar represent the average δ^{26}Mg of −0.25 and 2SD of 0.04 for the normal mantle (Teng et al. 2010a).

hydrothermal alteration and marine weathering (Snow and Dick 1995; Niu 2004; Dilek and Furnes 2011). It is, therefore, important to evaluate the behavior of Mg isotopes during fluid–peridotite interactions before using them to constrain Mg isotopic composition of the mantle.

Magnesium isotopic compositions of abyssal peridotites reported so far are significantly heavier than the mantle because of the Mg isotope fractionation induced by hydrothermal processes during accretion and residence in the deep ocean (Wimpenny et al. 2012; Liu et al. 2016). Wimpenny et al. (2012) reported that abyssal peridotites from the Mid Atlantic Ridge that were altered to various degrees during hydrothermal processes displayed a significant variation in δ^{26}Mg ranging from −0.25 to −0.02, which was interpreted as a result of marine weathering of peridotites. Liu et al. (2016) investigated a suite of 32 abyssal peridotite samples from the Gakkel Ridge and Southwest Indian Ridge (SWIR). Most of these abyssal peridotite samples have been intensively altered, dominated by serpentinization and/or weathering and have $\delta^{26}Mg$ varying from −0.24 to 0.03, with an average value of −0.12 ± 0.13 (2SD, n = 32) (Fig. 12). In particular, $\delta^{26}Mg$ increases with increasing loss on ignition (LOI), suggesting that low-T seafloor weathering and the formation of clays are the main causes for the enrichment of heavy Mg isotopes in the abyssal peridotites (Fig. 12). The significance of these studies is that Mg isotopic compositions of altered peridotites are not directly controlled by Mg isotopic composition of the reacting melt/fluids, but are strongly dependent on the secondary minerals formed. In addition, abyssal peridotites with heavy Mg isotopic signatures can be recycled into the mantle in subduction zones and may thus be the potential sources for arc magmas with heavy Mg isotopes (e.g., Teng et al. 2016).

Magnesium isotopic data for ophiolites are still very limited. Su et al. (2015) carried out the first study on Mg isotopic systematics of ophiolites. They found $\delta^{26}Mg$ values of the ophiolitic peridotite, dunite and gabbro samples from Purang ophiolite, southwestern Tibet, varied from −0.28 to −0.14, with an average $\delta^{26}Mg$ of −0.20 ± 0.10 (2SD, n = 19), a value slightly higher than the mantle value (−0.25 ± 0.04, 2SD). The subtle difference in Mg isotopic composition between ophiolites and the mantle is not clear and might reflect Mg isotope fractionation during hydrothermal alteration.

Continental basalts

Continental basalts are derived from the subcontinental lithosphere and/or asthenosphere, and may have been modified during their ascent by interactions with the continental crust. Thus, chemical and isotopic studies of continental basalts may shed light on the composition and evolution of the subcontinental mantle as well as the crust–mantle interactions (Farmer 2014).

Yang et al. (2012) first investigated Mg isotope systematics of continental basalts from the North China craton and found these basalts display dichotomy Mg isotopic compositions with ages. Yixian lavas, with ages > 120 Ma, have a typical mantle-like Mg isotopic composition but low Ce/Pb, Nb/U ratios and lower-crust-like Sr–Nd–Pb isotopic compositions. By contrast Fuxin and Taihang basalts, formed after 110 Ma, have light Mg isotopic compositions, with $\delta^{26}Mg$ down to −0.60, whereas have mantle-like Ce/Pb, Nb/U ratios and Sr–Nd–Pb isotopic compositions (Fig. 13). These temporal isotopic variations cannot result from crustal contamination as indicated by their mantle-like trace elemental and radiogenic isotopic

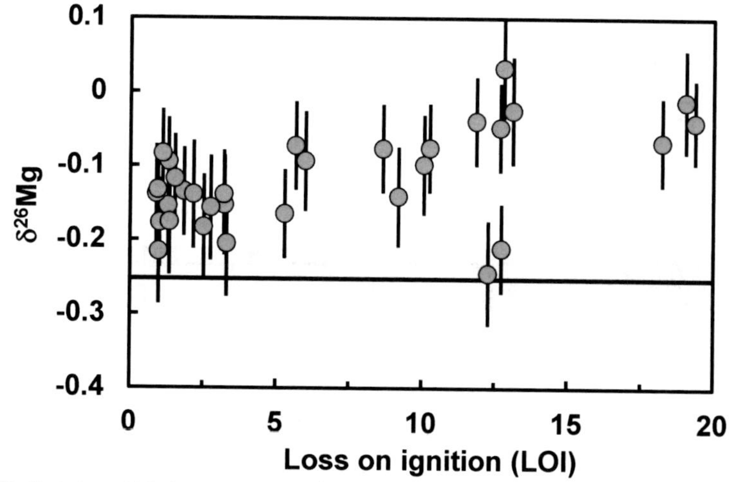

Figure 12. Variation of Mg isotopic composition of abyssal peridotites as a function of loss on ignition (LOI). The horizontal line and bar represent the average $\delta^{26}Mg$ of −0.25 and 2SD of 0.04 for mantle peridotites (Teng et al. 2010a).

compositions (Yang et al. 2012). Instead, they reflect the change in Mg isotopic composition of the mantle sources at ~120 Ma, likely induced by the subduction of the Pacific oceanic crust under the North China craton. The upper mantle beneath the North China craton was metasomatised by melts derived from the subducting slabs (Yang et al. 2012). Melting of this heterogeneous and isotopically light mantle produced the Mg isotopic variations in continental basalts from North China craton. A later study of continental basalts from South China craton also reveals a large and overall light Mg isotopic variation, with δ^{26}Mg negatively correlated with total alkalis, incompatible trace element contents and elemental ratios (Huang et al. 2015a). The isotopically light basalts were interpreted as a result of incongruent partial melting of an isotopically light carbonated mantle, which was also proposed as a result of subduction of the Pacific slabs (Huang et al. 2015a).

Significant Mg isotopic variation was also observed in the HIMU-like continental basalts from Antipodes Volcano in New Zealand (Wang et al. 2016). Magnesium isotopic compositions of these lavas are different from those of spinel-facies peridotite xenoliths that represent the lithospheric mantle of Zealandia, with δ^{26}Mg varying from −0.47 to −0.06 and negatively correlating with the $(Gd/Yb)_N$ ratio (Fig. 14). The difference in Mg isotopic composition between the Zealandia peridotite xenoliths and Antipodes continental basalts suggests that the former cannot be the only source to produce the latter, and another isotopically light end member is required. The negative correlation between δ^{26}Mg and $(Gd/Yb)_N$, together with trace elemental and radiogenic isotopic studies, suggests the most likely source for these

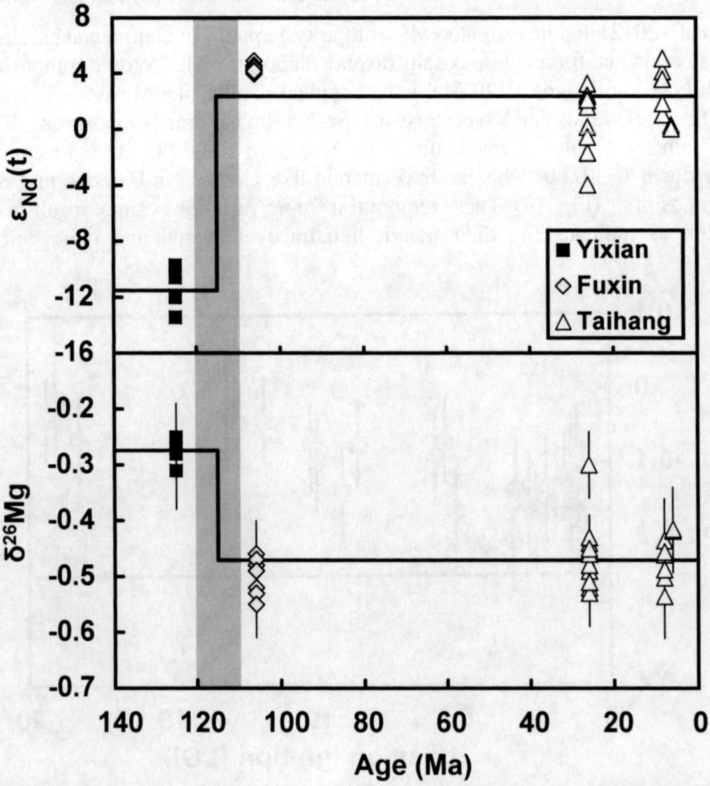

Figure 13. Magnesium and Nd isotopic compositions of continental basalts from North China craton (Yang et al. 2012). The grey vertical bar indicates the time period when the abrupt changes in δ^{26}Mg and ε_{Nd}(t) occurred, which also marked the start of the Pacific slab subduction under the North China craton.

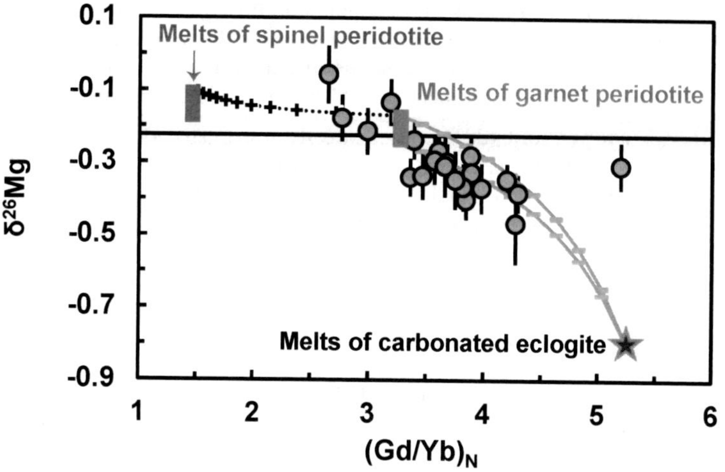

Figure 14. Correlation of $\delta^{26}Mg$ with $(Gd/Yb)_N$ in HIMU-like continental basalts from New Zealand (Wang et al. 2016). The data can be modeled by mixing of melts derived from spinel peridotite, garnet peridotite and carbonated eclogite. The horizontal solid line represents the average $\delta^{26}Mg$ of the Zealandia mantle.

HIMU-like continental basalts is a mixture of melts derived from the peridotites and recycled carbonated eclogites formed by the subduction of oceanic crust (Wang et al. 2016).

MAGNESIUM ISOTOPIC SYSTEMATICS OF THE OCEANIC CRUST, CONTINENTAL CRUST AND HYDROSPHERE

The outermost layers of the Earth (hydrosphere, continental and oceanic crust) contain <0.1% of the total Mg in the bulk Earth (Fig. 1), however, display the largest Mg isotopic variation (Fig. 2). Therefore, they are important for understanding Mg isotopic balance of the bulk Earth and global Mg cycling. Overall, the hydrosphere is isotopically light and the crust is heavy relative to the mantle, reflecting the large Mg isotope fractionation during low-temperature water–rock interactions, during which light isotopes prefer hydrosphere to the crustal residue.

Magnesium isotopic composition of the oceanic crust

The oceanic crust is generally consisted of three layers, beginning at the top: layer 1 = thin layer of pelagic sediment, layer 2 = (altered and fresh) basalts and sheeted dike complex and layer 3 = gabbros (White and Klein 2014). Since fresh oceanic basalts have $\delta^{26}Mg$ of -0.25 ± 0.07 (2SD) (Teng et al. 2010a), this value can represent Mg isotopic compositions of the layer 3 as well as fresh basalts and sheeted dike complex in the layer 2 (see section on Mg isotopic composition of the mantle). The most significant unknowns thus are the Mg isotopic compositions of pelagic sediments and altered basalts. These components are also the most important end members for producing arc magmatism and mantle heterogeneity (Hofmann 2003; Plank 2014; Ryan and Chauvel 2014; White and Klein 2014). Recent studies summarized below suggest that marine sediments and altered oceanic crust (AOC) have very heterogeneous Mg isotopic composition and subduction of the oceanic crust thus can recycle isotopically distinct Mg into the mantle.

Subducting sediments. Marine sediments are the most compositionally heterogeneous component in the subducting oceanic crust. Early studies of Mg isotopes in marine sediments mainly focused on carbonates and carbonate-rich sediments, and found these sediments have very low $\delta^{26}Mg$, down to −5.12 (Higgins and Schrag 2010, 2012; Rose-Koga and Albarede 2010; Blattler et al. 2015). Nonetheless, carbonates only account for approximately 7% of the Global

Subducted Sediments (GLOSS) (Plank and Langmuir 1998). More comprehensive studies on different types of marine sediments, particularly biosiliceous or alumino-silicate marine sediments that make up the bulk of GLOSS, are thus needed to better characterize the sedimentary Mg inputs to subduction zones and understand the behavior of Mg isotopes in marine environments.

Hu et al. (2015) and Teng et al (2016) systematically investigated Mg isotopic compositions of 94 well-characterized marine sediments from 12 drill sites outboard of the world's major subduction zones. The $\delta^{26}Mg$ values in these subducting sediments vary greatly from −3.65 to +0.52 (Fig. 15). The detritus-dominated sediments have $\delta^{26}Mg$ (−0.57 to +0.52) comparable to the weathered crustal materials (e.g., −0.52 to +0.92) (Shen et al. 2009; Li et al. 2010; Liu et al. 2010; Huang et al. 2012, 2013b; Ling et al. 2013; Wimpenny et al. 2014b). By contrast, the calcareous oozes yield low $\delta^{26}Mg$ (−3.65 to −0.32), falling in the range of previous studies of marine carbonates (Higgins and Schrag 2010, 2012; Rose-Koga and Albarede 2010; Blattler et al. 2015). Furthermore, the $\delta^{26}Mg$ is negatively correlated with CaO/TiO_2 in these sediments, which indicates the mineralogical control on Mg isotopic distribution among sediment types. This is consistent with the fact that carbonate-rich sediments are generally enriched in light Mg isotopes and silicate-rich ones are enriched in heavy ones (Fig. 2). Overall, the marine sediments have very heterogeneous Mg isotopic compositions.

Altered oceanic basalts. Seafloor alteration occurs widely and heterogeneously, involves different types of water–rock interactions, and is important for understanding global elemental cycle (Staudigel 2014). Although studies of altered oceanic basalts are still limited, the available data suggest they have very heterogeneous Mg isotopic compositions, comparable to the degree of isotope fractionation observed in chemical weathering of continental basalts (see later sections for details on continental weathering).

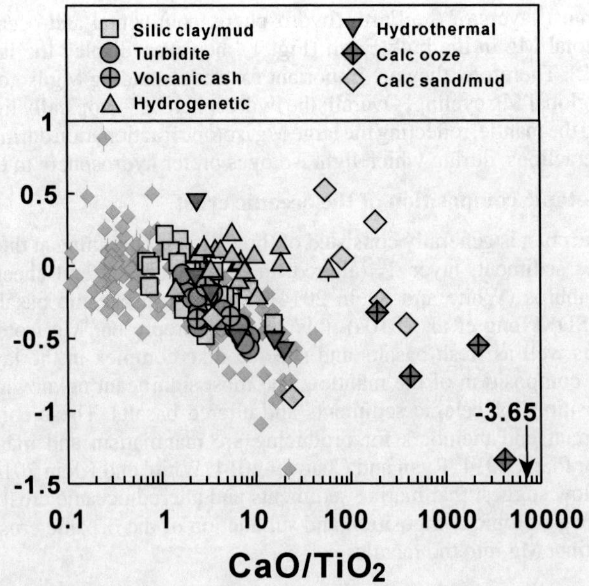

Figure 15. Correlation of $\delta^{26}Mg$ with CaO/TiO_2 for different types of subducting sediments (Hu et al. 2015; Teng et al. 2016). The negative correlation indicates a mineralogical control of Mg isotopic composition of subducting sediments by mixing of different proportions of isotopically light carbonates with isotopically heavy silicates. The shaded diamonds in the background represent clastic sedimentary rocks (see Fig. 18 for more details).

Huang (2013) systematically investigated Mg isotopic compositions of well-characterized altered basalts from the ODP Site 801 outboard of the Mariana trench in the western Pacific. The ODP Hole 801C represents an ideal natural laboratory to evaluate the elemental and isotopic behavior during low-temperature alteration of the oceanic crust, and the subduction input flux in the AOC. This is because altered oceanic basalts recovered from the ODP Site 801 are representatives of materials subducting in the western Pacific, and have been demonstrated to have undergone extensive low-temperature seawater alteration (Plank et al. 2000; Kelley et al. 2003, 2005; Rouxel et al. 2003). Magnesium isotopic compositions of these AOC samples are highly variable, with $\delta^{26}Mg$ ranging from −2.76 to +0.21 (Fig. 16), suggesting large Mg isotope fractionation during seafloor alteration. Both MgO contents and $\delta^{26}Mg$ values of the AOC samples positively correlate with the Fe_2O_3/CaO ratio, indicating that the formation of iron-rich secondary minerals (e.g., saponite) and carbonates accounts for the large Mg isotope fractionation during low-temperature alteration of the oceanic crust.

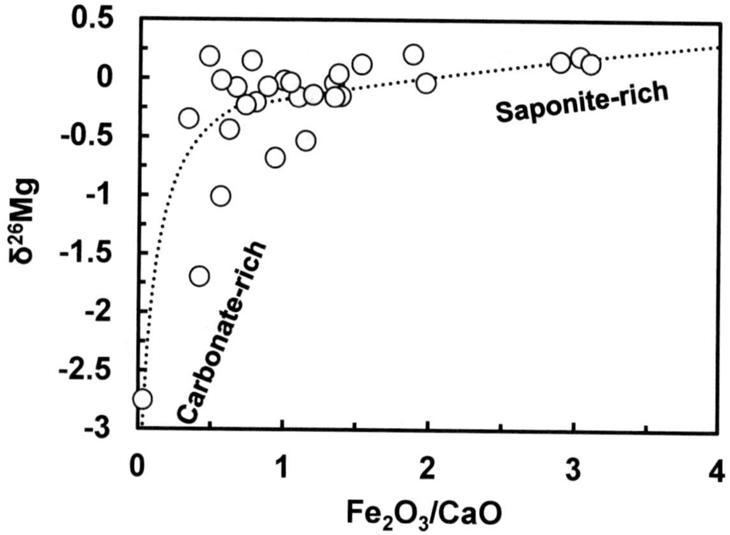

Figure 16. Correlation of $\delta^{26}Mg$ with F_2O_3/CaO for altered oceanic basalts from the ODP site 801 (Huang 2013). The dashed line represents mixing between saponite-rich and carbonate-rich endmembers.

Magnesium isotopic composition of the continental crust

Magnesium is the fifth most abundant element in the bulk continental crust, with an average MgO concentration of 4.66 wt%, and increases with depth from the upper to the lower continental crust (Rudnick and Gao 2003). Magnesium isotopic composition is also highly heterogeneous in the continental crust (Fig. 17).

Upper continental crust. The upper continental crust (between surface and 10–15 km depth) has a very heterogeneous chemical composition, with an average MgO of 2.48 wt% (Rudnick and Gao 2003), which is the lowest among the different reservoirs in the silicate Earth. Because of the compositional heterogeneity, two types of samples are commonly used to estimate the composition of the upper crust: composites that represent weighted averages of the compositions of rocks exposed at the surface and fine-grained clastic sedimentary rocks (e.g., shale) or glacial deposits (e.g., loess) (Rudnick and Gao 2003).

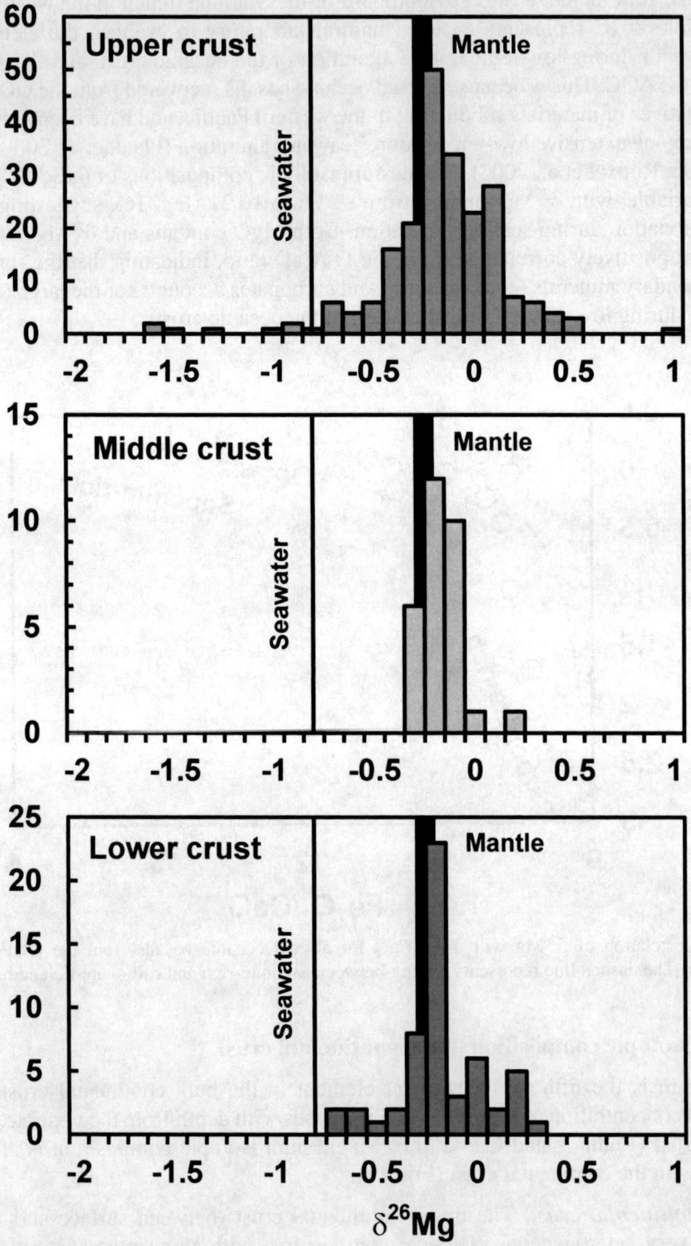

Figure 17. Magnesium isotopic composition of the upper, middle and lower continental crust. The mantle (shaded bar, Teng et al. 2010a) and seawater (vertical solid line, Ling et al. 2011) values are plotted for comparison. Data sources for the upper crust (Shen et al. 2009; Li et al. 2010, 2014; Liu et al. 2010; Huang et al. 2013b; Ling et al. 2013; Wimpenny et al. 2014b; Wang et al. 2015b), middle crust (Yang et al. 2016) and lower crust (Teng et al. 2013; Wang et al. 2015a; Yang et al. 2016).

Granite is a major component of the upper continental crust; hence early attempts to study Mg isotopic composition of the upper crust focused on granites and associated Mg-rich minerals. For example, Shen et al. (2009) found ~0.8‰ Mg isotopic variation in biotites from I-type granitites from southern California. The large Mg isotopic variation is coupled with major elements (MgO and SiO_2) and O, Sr, Pb isotopic variations, which was interpreted as mixing of pre-existing heterogeneous crustal materials in the source of the granitic magma (Shen et al. 2009). In another study, Liu et al. (2010) carried out Mg isotopic analyses for a suite of well-characterized I-type granitoids and constitute hornblende and biotite minerals from the Dabie orogen in central China. Although these granitoids formed through different degrees of partial melting and fractional crystallization with large variations in elemental and mineral compositions, their Mg isotopic compositions are indistinguishable and overall similar to the mantle composition. Furthermore, coexisting hornblende and biotite in these granitoids display limited inter-mineral fractionation (Liu et al. 2010). Therefore, both whole-rock and mineral separate data suggest limited Mg isotope fractionation during I-type granite differentiation; hence Mg isotopic composition of I-type granites can be used to trace source heterogeneity. Nonetheless, later studies of metapelites from Ivrea Zone, NW Italy show that Mg isotopes can be significantly fractionated during the production of S-type granites by partial melting of metapelites in the deep crust when garnet is a stable phase (Wang et al. 2015a and see below for details).

The most comprehensive study of upper crust rocks were from Li et al. (2010), who systematically measured ~100 major types of upper crustal rocks including I (igneous)-, S (sedimentary)-, and A (anorogenic)-type granites, loess and shale samples, as well as upper crustal composites (greywacke, pelite, tillite, granite, granodiorite and diorite). They found that Mg isotopic composition of the upper continental crust is highly heterogeneous, with δ^{26}Mg ranging from −0.52 to +0.92, and overall has a mantle-like weighted average δ^{26}Mg of −0.22 (Li et al. 2010). Later studies of more granite and loess samples found that the upper crust is even more heterogeneous, with δ^{26}Mg down to −1.64 (Telus et al. 2012; Huang et al. 2013b; Ling et al. 2013; Wimpenny et al. 2014b; Wang et al. 2015b; Ke et al. 2016). In detail, A-type granites from northeastern China have heterogeneous Mg isotopic compositions, with δ^{26}Mg ranging from −0.28 to +0.34 (Li et al. 2010), which was interpreted as a result of source heterogeneity produced by incorporation of old crustal components with variable Mg isotopic compositions. By contrast, Mg isotopes display the largest (>2.6‰) variation in sedimentary rocks (loess, shale and mudrock) (Li et al. 2010; Huang et al. 2013b; Wimpenny et al. 2014b; Wang et al. 2015b), which is controlled by mixing of different proportions of isotopically light carbonates with isotopically heavy silicates (Fig. 18). Leaching experiments on loess and mudrock further confirm that carbonates have light whereas silicates have heavy Mg isotopic compositions relative to bulk sediments (Wimpenny et al. 2014a; Wang et al. 2015b). Overall, the large Mg isotopic variation in the upper crust rocks reflects the extreme Mg isotope fractionation during continental weathering.

Deep continental crust. The deep continental crust (below ~10–15 km depth) can be divided into two heterogeneous layers: the middle and lower crust (Rudnick and Fountain 1995; Rudnick and Gao 2003). The middle crust (between 10–15 km and 20–25 km depth) is dominated by amphibolite-facies to lower granulite-facies metamorphic rocks, with an average MgO content of 3.59 wt% (Rudnick and Gao 2003). The lower crust (below 20–25 km depth) is mainly made of granulite-facies rocks and has the highest MgO content (=7.24 wt%) in the continental crust (Rudnick and Gao 2003).

Teng et al. (2013) first investigated Mg isotopic compositions of the lower continental crust by studying granulite xenoliths from Chudleigh and McBride, North Queensland. Chudleigh is a set of exclusively mafic xenoliths whereas McBride varies from felsic to mafic in composition (Rudnick et al. 1986; Rudnick and Taylor 1987). These samples are well-characterized and have average compositions matching the lower continental crust. Their Mg isotopic compositions are very heterogeneous, with δ^{26}Mg ranging from −0.72 to

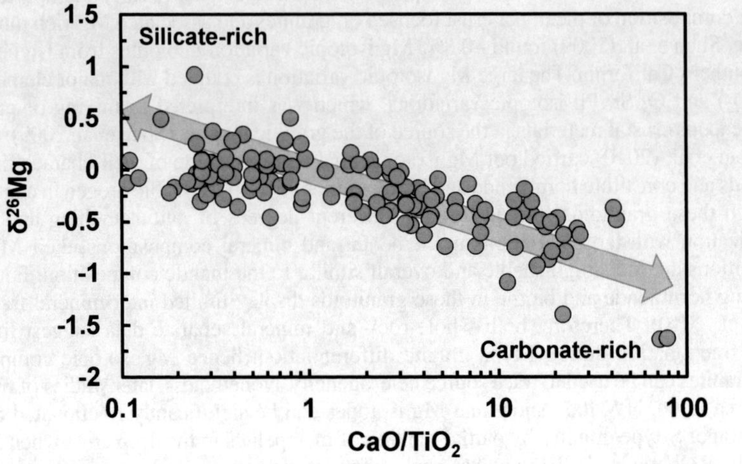

Figure 18. Correlation of δ^{26}Mg with CaO/TiO_2 for clastic sedimentary rocks (Li et al. 2010, 2014; Huang et al. 2013b; Wimpenny et al. 2014b; Wang et al. 2015b). The negative correlation indicates a mineralogical control on Mg isotopic composition of clastic sedimentary rocks by mixing of different proportions of isotopically light carbonates with isotopically heavy silicates.

+0.19 (Fig. 17), and likely reflect their distinct source compositions that involve preexisting, isotopically heterogeneous crustal materials (Teng et al. 2013). In a later study, Wang et al. (2015a) measured Mg isotopes for a suite of amphibolite- to granulite-facies metapelites as well as the biotite and garnet minerals therein from the Ivrea Zone, NW Italy. The mineralogy of metapelites changes from biotite-dominated in the amphibolite-facies to garnet-dominated in the granulite-facies. The δ^{26}Mg of bulk rocks varies from –0.23 to +0.20 and does not correlate with metamorphic grade. By contrast, biotite δ^{26}Mg values vary widely from –0.08 to +1.10, and increase with increasing metamorphic grade. Correspondingly, coexisting garnets become isotopically heavy (–1.22 to +0.10) as metamorphism proceeds, in order to equilibrate with the biotite. These results indicate a nearly closed system for Mg isotopes of metapelites during metamorphism, and the bulk Mg isotopic compositions are therefore reconciled by the shifting garnet and biotite modes accompanied by increasing mineral δ^{26}Mg values. The systematic Mg isotopic variation in the biotite implies a possible Mg isotope fractionation between melts and residues during biotite dehydration melting, which makes Mg isotopes a potential monitor of crustal melting and a tracer of granite petrogenesis. More recently, Yang et al. (2016) systematically studied 30 composite samples from high-grade metamorphic terranes and 18 granulite xenoliths from eastern China to better constrain Mg isotopic compositions of the middle and lower continental crust respectively. The middle continental crust, as represented by the composite samples, has heterogeneous δ^{26}Mg values, varying from –0.40 to +0.12, reflecting heterogeneity of their protoliths. The deep crust, as sampled by granulite xenoliths, is highly heterogeneous, with a large δ^{26}Mg variation from –0.76 to –0.24 (Fig. 17), which reflects both source heterogeneity (by incorporation of chemical weathered upper crustal materials) and deep-crustal fluid metasomatism (Yang et al. 2016).

Overall, the bulk continental crust has an average δ^{26}Mg of –0.24 (Yang et al. 2016), similar to the mantle. This similarity reflects the fact that Mg in the bulk continental crust is mainly controlled by igneous and metamorphic rocks (Wedepohl 1995); because of the limited Mg isotope fractionation during high-temperature silicate magmatic differentiation and metamorphism, igneous and metamorphic rocks still have similar Mg isotopic compositions to oceanic basalts. Therefore, the bulk crust has a mantle-like Mg isotopic composition even though large Mg isotope fractionation occurs during low-temperature water–rock interactions.

Magnesium isotopic composition of the hydrosphere

Magnesium is fluid mobile and its concentration is highly variable in the hydrosphere. It is one of the major elements in fresh waters (Livingstone 1963; Holland 1978) and the second most abundant cation in seawater (after Na) (Pilson 2013). Similar to Mg concentration, Mg isotopic composition of the hydrosphere is highly variable (Fig. 19), and is important for understanding the behavior of Mg isotopes during continental weathering and seafloor alteration as well as the global Mg cycle.

Seawater. Seawater accounts for ~97% of water in the hydrosphere and has the highest Mg concentration (0.13 wt%) in all reservoirs of the hydrosphere (Pilson 2013). Thus, seawater controls the Mg isotopic budget of the hydrosphere. Analyses of seawater samples from various depth and global locations yielded identical Mg isotopic compositions, with $\delta^{26}Mg = -0.83 \pm 0.09$ (2SD), indicating a homogenous Mg isotopic composition of the oceans (Foster et al. 2010; Ling et al. 2011 and references therein).

River water. Rivers on the surface of the continents constitute <1% of water in the hydrosphere, but play a key role in transferring crustal materials from the continents to the oceans and controlling the seawater composition. Magnesium concentration in different rivers varies significantly, ranging from <1 to 50 ppm, with an average concentration of 4 ppm, and mainly depends on the climate and weathered bedrocks in the drainage areas (Livingstone 1963; Holland 1978). Magnesium isotopic compositions of rivers are highly variable, with $\delta^{26}Mg$ ranging from −2.50 to +0.64 (Fig. 19), and reflect climate, bedrock lithology and Mg isotope fractionation during water–rock interactions (de Villiers et al. 2005; Tipper et al. 2006a,b, 2008a, 2012a,b; Brenot et al. 2008; Pogge von Strandmann et al. 2008a,b; Jacobson et al. 2010; Wimpenny et al. 2011; Lee et al. 2014; Mavromatis et al. 2014; Ma et al. 2015; Dessert et al. 2015).

Tipper et al. (2006b) systematically measured Mg isotopes for 45 rivers including 16 of the largest rivers in the world. These rivers are globally, lithologically, tectonically and climatically representative and make up 30% of the global Mg riverine flux to the oceans. They found an overall >2‰ Mg isotopic variation ($\delta^{26}Mg = -2.5$ to -0.3) that mainly results from Mg isotope fractionation during continental weathering. The flux-weighted $\delta^{26}Mg$ of global runoff was estimated to be −1.09 (Tipper et al. 2006b), which is significantly lower than that of seawater. This difference mainly reflects isotope fractionation associated with seawater-ocean floor interactions such as carbonate precipitation, low-temperature seafloor alteration and high-temperature hydrothermal alteration.

Magnesium isotopic compositions of small rivers, different from large rivers, are more affected by bedrock lithology and climate, in addition to Mg isotope fractionation during weathering. Tipper et al. (2006b) found that small rivers draining silicate, dolomite and limestone bedrocks from various locations have different Mg isotopic compositions that are similar to their bedrocks, i.e., small Himalayan rivers draining dolomite have very light Mg isotopic composition, identical to dolomite rocks. Later studies of rivers draining mono or mixed bedrock lithology reveal similar results; rivers draining carbonates usually have light Mg isotopic compositions than those draining silicates, although in many cases the riverine $\delta^{26}Mg$ is different from bedrock $\delta^{26}Mg$, which reflects Mg isotope fractionation during weathering (Tipper et al. 2006a, 2008a; Brenot et al. 2008; Pogge von Strandmann et al. 2008a; Jacobson et al. 2010; Lee et al. 2014; Dessert et al. 2015). Nonetheless, many small rivers that drain mono lithology display large seasonal Mg isotopic variation, which was mainly caused by the annual changes in the input and the type of weathering (Wimpenny et al. 2011; Tipper et al. 2012b; Dessert et al. 2015).

Groundwater. Magnesium isotopic data for groundwater are limited. The available data display a large variation, with $\delta^{26}Mg$ ranging from −1.70 to +0.23 and an average value of

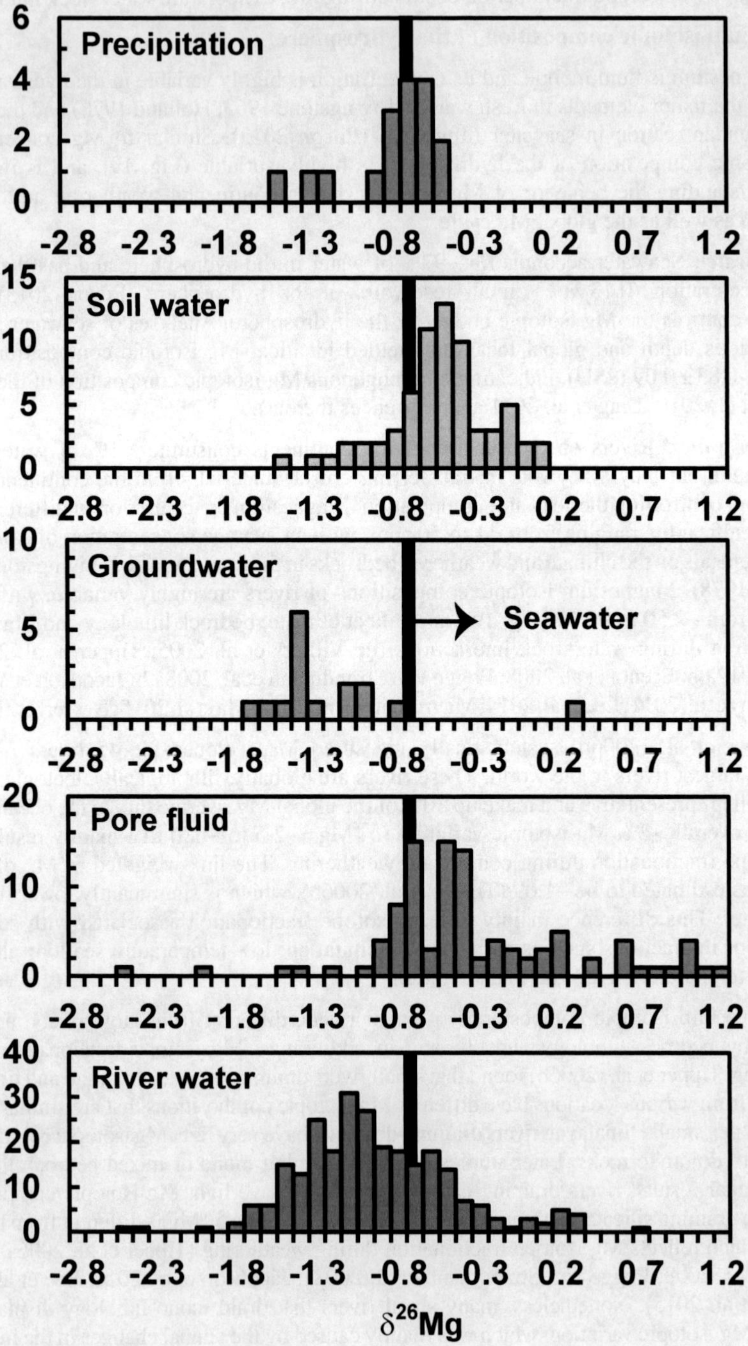

Figure 19. Magnesium isotopic composition of the hydrosphere. The vertical bar represents the average δ^{26}Mg of −0.83 for the seawater. See text for references.

−1.23 ± 0.89 (2SD, $n = 16$) (Fig. 19). The large Mg isotopic variation was mainly caused by source Mg isotopic heterogeneity and isotope fractionation during mineral dissolution and formation of secondary minerals.

The groundwater sample from a basaltic catchment in Iceland has the highest δ^{26}Mg of 0.23, which was interpreted as a result of secondary mineral formation, during which light Mg isotopes were removed, leaving isotopically heavy water (Pogge von Strandmann et al. 2008a). The groundwater samples from carbonate-dominated environment tend to have low δ^{26}Mg ranging from −1.70 to −1.18, which reflects the inputs of light Mg isotopes from carbonates (Galy et al. 2002; Tipper et al. 2006a). Groundwater samples collected from one well but at the different time yielded different δ^{26}Mg of −1.50 and −0.83, respectively, likely reflecting a change in the proportion of Mg inputs between weathering of granite and precipitation of dust (Tipper et al. 2012b). Two groundwater samples collected from a shale catchment have δ^{26}Mg of −1.15 and −1.05 respectively, and are 1.5‰ lower than the bedrock, which likely results from the dissolution of isotopically light ankerite with little contribution from silicate weathering (Ma et al. 2015). Finally, eight groundwater samples collected along a 236 km flow path in Madison Aquifer, South Dakota have δ^{26}Mg ranging from −1.63 to −1.10, which was produced by Mg isotope fractionation during Mg-for-Na ion exchange during dedolomitization (Jacobson et al. 2010).

Soil water (soil pore fluid). Magnesium isotopic compositions of soil water samples vary significantly from −1.50 to +0.06, with an average δ^{26}Mg of −0.62 ± 0.69 (2SD, $n = 53$) (Fig. 19). The large isotopic variation, similar to river water, mainly reflects different Mg sources (rainwater, carbonate, silicate and organic materials) and Mg isotope fractionation during weathering (Immenhauser et al. 2010; Tipper et al. 2010; Pogge von Strandmann et al. 2012; Riechelmann et al. 2012b; Mavromatis et al. 2014; Geske et al. 2015b; Ma et al. 2015).

Tipper et al. (2010) found soil water samples in a smectite soil from Santa Cruz, California, have δ^{26}Mg values that continuously increase from the surface to the base of the weathering profile, with δ^{26}Mg being intermediate between the rain and the smectite. The Mg isotopic variation is broadly consistent with inputs of Mg from both smectite and rainwater. Nonetheless, Mg isotope fractionation between water and soils is still needed in order to completely reconcile Mg isotope data of soil water (Tipper et al. 2010). Pogge von Strandmann et al. (2012) also found significant Mg isotopic variation with depth in soil water in a basaltic soil from Iceland, which falls between basalts and river waters. In addition to source difference, Mg isotopic composition of soil water was also controlled by Mg isotope fractionation during adsorption of isotopically heavy Mg to the soils (Pogge von Strandmann et al. 2012). Similarly, other studies also found large Mg isotopic variations in soil water samples from the Shale Hills catchment, Pennsylvania (Ma et al. 2015) and from the Bunker Cave, Germany (Immenhauser et al. 2010; Riechelmann et al. 2012b), which likely reflects different contribution of Mg from carbonate vs. silicate as well as Mg isotope fractionation during continental weathering.

Seasonal variation in Mg isotopic composition of soil waters was also observed and mainly reflects the change in contribution of Mg from different sources such as carbonate vs. silicate and different type of plants grown under different climatic conditions (Bolou-Bi et al. 2012; Riechelmann et al. 2012b). For example, Bolou-Bi et al. (2012) found δ^{26}Mg of soil varies from −1.44 to +0.06, which does not correlate with soil depth, instead correlates with various seasons. Soil water samples collected in May (high rainfall and start of plant growth season) have low δ^{26}Mg values whereas those collected in September (low rainfall period) have the highest values, which suggests a biological and hydrological control on soil Mg isotopic composition. Riechelmann et al (2012b) also found a cyclical Mg isotopic variation in soil water collected over a period of eight months, which was interpreted as a result of difference in carbonate vs. silicate weathering rates at different seasons.

Deep-sea pore fluid. Magnesium content of pore fluids in marine sediments in general decreases with depth, reflecting low-temperature alteration of oceanic crust, precipitation and dissolution of carbonate and silicate minerals, and changes in seawater composition or pore-fluid advection (Higgins and Schrag 2010). Magnesium isotopic composition of pore fluids also varies significantly, with δ^{26}Mg ranging from −2.56 to +1.13 (Fig. 19), but displays different trends with depth, depending on the type of minerals formed (Higgins and Schrag 2010, 2012; Mavromatis et al. 2014; Geske et al. 2015b). For example, Mg isotopic compositions increase with depth by as much as 2‰ in some ODP sites, because of the removal of light Mg isotopes caused by the precipitation of dolomite. By contrast, Mg isotopic compositions decrease with depth by up to 2‰ in other ODP sites due to the incorporation of isotopically heavy Mg into clay minerals (Higgins and Schrag 2010).

Precipitation (rainwater, snow and ice). Magnesium isotopic data for precipitation samples are still limited and only available for a few sites. The overall variation is, however, large, with δ^{26}Mg ranging from −1.59 to −0.51 and an average value of −0.85 ± 0.58 (2SD, $n = 17$) (Fig. 19).

Rainwater samples from Santa Cruz, CA, have an average δ^{26}Mg of −0.79 ± 0.05 (2SD, $n = 5$), identical to seawater (Tipper et al. 2010); similarly, one glacial ice sample from Iceland also has seawater-like Mg isotopic composition (Pogge von Strandmann et al. 2008a). Both sites are close to the oceans, therefore, Mg isotopic compositions of the precipitation reflect oceanic origins. Rainwater and snow samples from Vosges Mountains, France have slightly heavy Mg isotopic composition, with δ^{26}Mg ranging from −0.75 to −0.51 and an average value of −0.65 ± 0.16 (2SD, $n = 4$), which may result from interactions with isotopically heavy silicate dust as this site is 600 km away from the ocean (Bolou-Bi et al. 2012). Rainwater and snow samples from Rhenish Slate Mountains in NW Germany display a large variation, with δ^{26}Mg ranging from −1.35 to −0.59, reflecting different inputs from dust and aerosols (Riechelmann et al. 2012b). The lowest δ^{26}Mg value for rainwater samples comes from Swiss Alps (−1.59 and −1.29), likely resulting from the Mg input from carbonate dusts (Tipper et al. 2012b).

MAGNESIUM ISOTOPIC SYSTEMATICS OF CARBONATES

Carbonates (biogenic and abiogenic) display the largest Mg isotopic variation among all types of terrestrial rocks, with δ^{26}Mg ranging from −5.57 to −1.04 in calcite-rich carbonates and −3.25 to −0.38 in dolomite-rich ones (Fig. 20). When compared to seawater, carbonates tend to have light Mg isotopic compositions and silicates have heavy ones, which is consistent with the bonding strength, where Mg bonding strength is generally stronger in silicates than water, and is the weakest in carbonates (Schauble 2011). Of the many factors controlling Mg isotope fractionation in carbonates, mineralogy plays the most important role, followed by other factors such as temperature, precipitation rate, kinetic isotope fractionation and vital effects. Below I first summarize Mg isotopic variations in natural abiogenic carbonates, followed by biogenic carbonates, and finally discuss the experimentally calibrated and theoretically calculated fractionation factors between carbonates and fluids at various temperatures.

Abiogenic carbonates

Large (~4.8‰) Mg isotope fractionation has been observed in abiogenic carbonates (Fig. 20). Overall, dolomite-rich samples are enriched in heavy Mg isotopes than calcite-rich ones, with δ^{26}Mg varying from −5.14 to −1.42 in calcite speleothem and limestone, and from −3.25 to −0.38 in dolostone, clearly indicating a strong mineralogical control on Mg isotope fractionation in carbonates.

Calcite speleothem samples. The interest in Mg isotopes in carbonates begins with the potential application of Mg isotopes in speleothems as a new proxy for studies of climate change in continental settings. Overall, calcite speleothems have very light and heterogeneous Mg isotopic composition, with δ^{26}Mg ranging from −4.84 to −3.56 (Fig. 20). This large Mg isotopic variation likely reflects kinetic Mg isotope fractionation during calcite precipitation, the change of Mg isotopic composition of drip water, and to a lesser extent, temperature.

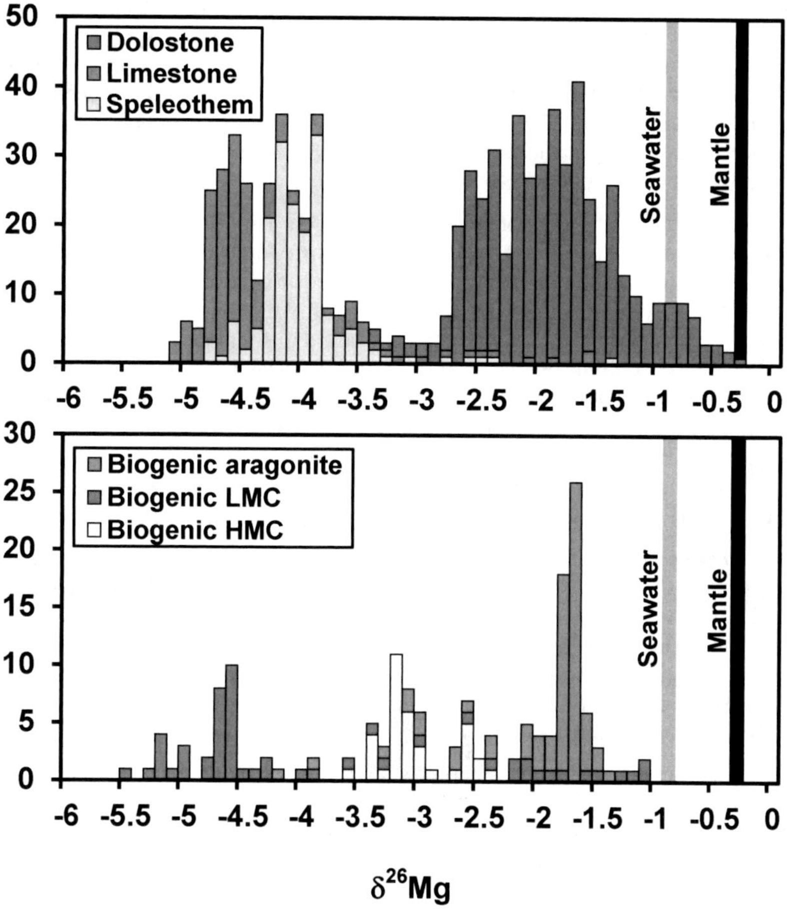

Figure 20. Magnesium isotopic composition of biogenic and abiogenic carbonates. The average δ^{26}Mg values of the seawater and mantle are also plotted for comparison. See text for references.

Galy et al. (2002) first investigated Mg isotopic systematics of calcite speleothems from four caves from French Alps and Israel under various climate conditions. Overall, δ^{26}Mg of these 12 calcite speleothems varies from −4.84 to −4.16, does not vary as a function of temperature, and instead reflects the change of Mg isotopic composition of the local water. Furthermore, these calcites are >2‰ lighter than the drip water, suggesting a large Mg isotope fractionation during the precipitation of calcite speleothems.

Buhl et al. (2007) carried out the first Mg isotopic study for a set of high-resolution time-series speleothem samples from a limestone cave in Morocco. The δ^{26}Mg in these 18 calcite speleothem samples varies from −4.39 to −4.17 and correlates with other proxies such as C, O and Sr isotopic data. The small Mg isotopic variation was not controlled by the change of temperature, instead likely results from the change of Mg isotopic composition of drip water. The composition of the drip water is further controlled by the change in either the carbonate precipitation rate from the meteoric water or the silicate vs. carbonate weathering rates (Buhl et al. 2007). Although this study suggests that Mg isotopic systematics of speleothems may not be a great tracer for temperature change, it may help to trace the rates of carbonate vs. silicate weathering under different climate conditions.

Immenhauser et al. (2010) systematically measured Mg isotopic compositions of soils above the cave, soil water, carbonate host rock, drip water, runoff water, speleothem and cave loam from a monitored cave in Germany. The δ^{26}Mg in these 19 calcite speleothem samples varies from −4.62 to −3.56, which are >2‰ lower than the drip water. As the temperature in the cave is nearly constant at 10.6 °C, the large Mg isotopic variation in these speleothem samples mainly reflects the change in the precipitation rate, with smaller Mg isotope fractionation occurring at faster precipitation rate (Immenhauser et al. 2010).

Riechelmann et al. (2012a) reported the most comprehensive time-series Mg isotopic data for eight speleothems collected in six caves from Germany, Morocco, Peru and Austria on three continents under different climate conditions. They found an overall 3‰ Mg isotopic variation with δ^{26}Mg ranging from −4.39 to −1.42, which reflects different Mg isotopic sources, soil zone biogenic activity, changes in silicate vs. carbonate weathering and the residence time of water in the soil and karst zone. In particular, the isotopically heavy speleothem samples mainly result from the contamination of Mg from detrital materials. This comprehensive study shows that it is still premature to unambiguously use Mg isotopes in speleothems as a high-resolution proxy for continental climate change because of the complicated processes that can affect Mg isotopic systematics of cave carbonates. Nonetheless, Mg isotopes seem very sensitive to changes in soil-zone parameters and may be a powerful tool for tracing soil zone processes.

Limestone and calcite-rich carbonate samples. Galy et al. (2002) first investigated Mg isotopic systematics of limestone samples from the Himalaya and found the δ^{26}Mg varies from −4.47 to −2.68 in five limestone samples, falling within the range of calcite speleothems. Later studies of random limestone samples and standards from various locations reveal similarly large isotopic variations (from −4.38 to −1.69) (Tipper et al. 2006a; Buhl et al. 2007; Brenot et al. 2008; Wombacher et al. 2009; Immenhauser et al. 2010; Azmy et al. 2013; Fantle and Higgins 2014; Kasemann et al. 2014).

Higgins and Schrag (2012) systematically studied Mg isotopic compositions of calcite-rich carbonates from the ODP site 807 and found δ^{26}Mg ranges from −5.12 to −3.92. The Mg isotopic variation can be explained by a secular rise in the seawater Mg and the recrystallization of low-Mg biogenic carbonate to a higher-Mg diagenetic calcite. In a later study, Higgins and Schrag (2015) reported Mg isotopic data for ~100 pelagic calcite-rich carbonates from the ODP sites 1265 and 807, with δ^{26}Mg ranging from −5.14 to −3.89. The large Mg isotopic variation mainly reflects that of foraminiferal carbonates, modified by recrystallization, and to a lesser extent, contributions from silicates, organics, and coccolith calcites.

Dolostone and dolomite-rich carbonate samples. Dolostone samples have been extensively measured for Mg isotopes because of their high Mg contents, which makes it easier to carry out high-precision isotopic analysis, and more importantly, because of their great potentials for tracing global Mg cycle and paleoenvironments of the Earth surface. To date, ~3‰ Mg isotopic variation has been documented in dolostones, with δ^{26}Mg ranging from −3.25 to −0.38 (Fig. 20).

Galy et al. (2002) first reported δ^{26}Mg values of −2.29 to −1.09 for five dolostone samples from the Himalaya, which are significantly heavier than the limestone samples and calcite speleothems, indicating a strong mineralogical control on Mg isotope fractionation in carbonates. Later studies reported Mg isotopic compositions for a few more dolostone samples from various geological settings (Chang et al. 2003; Brenot et al. 2008; Wombacher et al. 2009; Jacobson et al. 2010), all of which fall within the range of Himalaya dolomites (Galy et al. 2002).

The potential applications of Mg isotopes in marine sediments as a tracer of ancient seawater chemistry and geochemical cycle of Mg stimulate extensive analyses of marine dolomites from various drill cores. Higgins and Schrag (2010) first analyzed six marine dolomite samples from the ODP sites 1082 and 1012 and found their $\delta^{26}Mg$ values range from −2.52 to −1.72 and are 2.0 to 2.7 lower than the pore fluids, suggesting large Mg isotope fractionation between dolomite and fluids ($\Delta^{26}Mg_{dolomite-fluid}$ = −2.0 to −2.7) during the dolomite precipitation. Later analyses of seven dolomite-rich marine carbonates from the ODP Site 1196A found slightly lower $\delta^{26}Mg$ values from −2.80 to −2.59 (Fantle and Higgins 2014). Geske et al. (2012) reported the heaviest dolomite so far from short cores in Abu Dhabi, with $\delta^{26}Mg$ ranging from −1.09 to −0.38. Moreover, the apparent fractionation factor between dolomite and pore water ($\Delta^{26}Mg_{dolomite-fluid}$ = −0.7 to 0.1) is also much smaller than that reported by Higgins and Schrag (2010) though the reasons are still unclear. Recently, Blattler et al. (2015) systematically measured Mg and other isotopes for a set of 80 authigenic dolomite samples from a marine drill core recovered from the Miocene-age Monterey Formation of offshore California. The $\delta^{26}Mg$ values display a large range from −2.86 to −0.52 and vary with stratigraphic depth and Ca isotopes, reflecting the change in Mg isotopic composition of pore fluids during early diagenetic processes in addition to the >2‰ Mg isotope fractionation between dolomite and pore fluids during dolomite formation (Blattler et al. 2015).

Besides dolomite samples from drill cores, ancient dolostone samples of various ages and from various geological settings have also been extensively measured for Mg isotopes in order to better constrain the origins of dolomites, solve the 'dolomite problem' and reconstruct paleoenvironments of the Earth surface. To date, 2.8‰ Mg isotopic variation has been reported for these ancient dolostone samples. Overall, the Mg isotopic variation does not correlate with ages or dolostone types, mainly reflecting multiple processes such as changes in Mg isotopic composition of fluids, dolomite–fluid fractionation factors and diagenesis (Pokrovsky et al. 2011; Geske et al. 2012, 2015a,b; Azmy et al. 2013; Lavoie et al. 2014; Liu et al. 2014a; Wang et al. 2014a; Huang et al. 2015b; Husson et al. 2015).

Geske et al. (2012) measured Mg isotopic compositions of 38 dolomites that were affected by diagenesis and low-grade metamorphism (T = 100 to >350 °C). The $\delta^{26}Mg$ varies significantly from −2.27 to −1.41 but the average $\delta^{26}Mg$ does not vary as a function of temperature though the data are less scattered at elevated temperatures. This study thus suggests diagenesis and metamorphism do not significantly modify Mg isotopic composition of dolomites; hence dolomites can still preserve their pristine $\delta^{26}Mg$ values, at least for the rock-buffered system. Nonetheless, later studies show that diagenesis plays a complicated role in affecting Mg isotopic compositions of dolomites, depending on the local environments and water/rock ratios (e.g., Azmy et al. 2013; Fantle and Higgins 2014; Geske et al. 2015b).

Pokrovsky et al. (2011) studied 32 late Precambrian dolostones from various parts of the Siberian Platform and found an overall ~2‰ Mg isotopic variation, with $\delta^{26}Mg$ fluctuating between −2.43 and −0.47 from ~1500 Ma to 530 Ma. Four dolostone samples display a positive correlation between Mg and C isotopes, which likely reflects isotope fractionations during diagenesis. By contrast, the other 28 dolostone samples display a strong negative correlation between Mg and C isotopes, which cannot be caused by diagenesis. Instead, the large Mg isotopic variation in these Precambrian dolostones was interpreted as a result of changes in Mg isotopic composition of the oceans by assuming the Mg isotope fractionation during dolostone precipitation does not change significantly during the sample period. Furthermore, as the seawater Mg isotopic composition is mainly controlled by the riverine input, the large change in dolomite Mg isotopic composition thus suggests a change of riverine Mg isotopic input, which is further controlled by weathering different proportions of carbonate vs. silicate rocks under different climates and sea levels (Pokrovsky et al. 2011). Similarly, Kasemann et al. (2014) interpreted the large Mg and Ca isotopic variations in carbonates from the Neoproterozoic Otavi Group in Namibia as a result of changes in the continental weathering flux, from a mixed carbonate and silicate character to a silicate-dominated character.

Liu et al. (2014a) reported Mg isotopic data for Neoproterozoic cap dolostone samples from Nuccaleena Formation in South Australia and Ol Formation in Mongolia. They applied a 15 step-leaching procedure to minimize the potential contamination and identify the least-altered Mg isotopic compositions of dolostone samples. The δ^{26}Mg of the least-altered dolostone samples starts with a low value of −2.22 at the bottom, rises to −1.79 in the middle and decreases to −1.97 at the top of the Nuccaleena Formation. By contrast, the dolostone samples from the Ol Formation have similar Mg isotopic composition through the whole profile, with δ^{26}Mg ranging from −1.84 to −1.64. The overall Mg isotopic variation in these cap dolostone is interpreted as a result of isotopically different Mg sources: glacial meltwater plume vs. saline ocean residue.

Geske et al. (2015a) comprehensively investigated Mg isotopic compositions of four types of dolomites (marine evaporative dolomite, marine altered/mixing zone dolomite, Lacustrine/palustrine dolomite and hydrothermal dolomite) from various geological settings and ages. The δ^{26}Mg varies largely from −2.49 to −0.45, with one hydrothermal dolomite having the highest value and one marine evaporative dolomite the lowest value. Nonetheless, Mg isotopic compositions of these four groups overlap each other and do not vary as either ages or precipitation environments. The overall large Mg isotopic variations in these dolomites reflect various processes such as different diagenetic settings, alteration and secular variations of source Mg etc. Similarly, Huang et al. (2015b) measured Mg isotopic compositions for Mesoproterozoic dolostones from the Wumishan Formation in North China and also found they have similar Mg isotopic compositions (δ^{26}Mg = −1.72 to −1.35) to the other types of dolomites formed at different ages. Lavoie et al. (2014) reported Mg isotopic data for hydrothermal saddle dolomite samples in the lower Paleozoic of Canada and found large variation in δ^{26}Mg (−3.25 to −0.78), which likely reflects the change in the isotopic composition of the diagenetic fluids.

Biogenic carbonates

Magnesium isotopic variation in marine biogenic carbonates may be used as a proxy for studying paleoceanographic changes. Therefore, significant efforts have been made to investigate Mg isotopic systematics of different types of biogenic carbonates from different environments (Chang et al. 2003, 2004; Pogge von Strandmann 2008; Hippler et al. 2009; Ra et al. 2010a; 2010b; Muller et al. 2011; Wombacher et al. 2011; Yoshimura et al. 2011; Planchon et al. 2013; Pogge von Strandmann et al. 2014; Rollion-Bard et al. 2016). Saenger and Wang (2014) comprehensively reviewed Mg isotopic variations in biogenic carbonates and their potential implication for paleoenvironmental proxies. Hence only a brief summary is provided here.

To date, an overall > 4.5‰ Mg isotopic variation has been observed in biogenic carbonates (Fig. 20), with δ^{26}Mg ranging from −5.57 in one low Mg calcite (LMC) planktonic foraminifera to −1.04 in one LMC coccolith ooze (Wombacher et al. 2011). In detail, biogenic LMC tends to have lighter Mg isotopic composition than biogenic high Mg calcite (HMC) whereas biogenic aragonite has on average the heaviest Mg isotopic composition (Fig. 20). When compared to seawater (δ^{26}Mg = −0.83, Ling et al. 2011), marine biogenic carbonates are significantly ($\Delta^{26}Mg_{\text{biogenic carbonate–seawater}}$ = −4.74 to −0.21) lighter, suggesting large Mg isotope fractionation during the precipitation of biogenic carbonates. Particularly, Mg isotope fractionations between HMC and seawater ($\Delta^{26}Mg_{\text{HMC–seawater}}$) appears most likely to be equilibrium as $\Delta^{26}Mg_{\text{HMC–seawater}}$ values for some species are temperature-dependent and fall on the fractionation lines calibrated by laboratory precipitation experiments. By contrast, Mg isotope fractionation involving LMC shows the largest variation ($\Delta^{26}Mg_{\text{LMC-seawater}}$), with the majority of $\Delta^{26}Mg_{\text{LMC-seawater}}$ values displaying little or no temperature dependence, which indicates the isotope fractionation is most likely kinetic and is related to the biomineralization strategy of different species. Finally, when compared to biogenic calcite, Mg isotope fractionation in biogenic aragonite is typically smaller. Although the direction of the isotope fractionation is consistent with experimental studies, the magnitude of the Mg isotope fractionation in biogenic aragonite in many cases is much larger, suggesting large kinetic isotope fractionation. Overall, the large Mg isotope fractionations in these different types of biogenic carbonates likely result from multiple factors (see review of Saenger and Wang 2014): 1) Mineralogy: Aragonite is expected to be enriched in heavy Mg isotopes relative to calcite when all else being equal; 2) Temperature: larger isotope fractionation occurs at lower growth temperature; 3) Precipitation rate: Faster precipitation rates usually favor smaller isotope fractionations; 4) Aqueous Mg concentration and pH: Higher Mg concentration and pH usually favor temperature-dependent Mg isotope fractionation; and 5) biological vital effects. These different processes can affect Mg isotope fractionation differently in different species, and therefore, compromise the applications of Mg isotope geochemistry of biogenic carbonates as a paleoenvironmental proxy.

Carbonate precipitation experiments and theoretical calculations

Studies of natural carbonate samples have found large Mg isotope fractionation between speleothem and drip water ($\Delta^{26}Mg_{calcite-fluid}$) with speleothem 1.6 to 3.2‰ lighter than the dripping water, between dolomite and pore fluids ($\Delta^{26}Mg_{dolomite-fluid}$) with dolomite 2 to 2.7‰ lighter than the pore fluids, and between different types of biogenic carbonates and seawater ($\Delta^{26}Mg_{biogenic\ carbonate-seawater}$) with biogenic carbonates up to 4.7‰ lighter than seawater (see sections above). These large variations may reflect either equilibrium or kinetic Mg isotope fractionation. Experimentally calibrated and theoretically calculated Mg isotope fractionation factors between carbonate and fluid are therefore needed and will lay the foundation for interpreting Mg isotopic variations in natural carbonates.

Precipitation of carbonates under controlled laboratory conditions indeed found large Mg isotope fractionation between calcite/aragonite/magnesite/dolomite and solution, with the direction of fractionation consistent with studies of natural systems i.e., carbonates are lighter than the solution. However, the degrees and mechanisms responsible for the large Mg isotope fractionation are still controversial. For example, it is still unclear whether temperature and precipitation rate affect equilibrium Mg isotope fractionation between calcite and solution. Immenhauser et al. (2010) and Mavromatis et al. (2013) observed a correlation between Mg isotope fractionation and calcite precipitation rate, with smaller Mg isotope fractionation ($\Delta^{26}Mg_{calcite-fluid}$) occurring at faster precipitation rates. By contrast, Li et al. (2012) found that the isotope fractionation between calcite and solution ($\Delta^{26}Mg_{calcite-fluid}$) does not correlate with the precipitation rate. Instead, the $\Delta^{26}Mg_{calcite-fluid}$ is temperature-dependent and changes from −2.7 at 4 °C to −2.2 at 45 °C. Nonetheless, in a similar calcite precipitation experiment, Saulnier et al. (2012) didn't find measurable change in $\Delta^{26}Mg_{calcite-fluid}$ when temperature changes from 25.5 to 16.2 °C or pH changes from 7.41 to 8.51. Their estimated $\Delta^{26}Mg_{calcite-fluid}$ equals −2.09 ± 0.23 at 25.5 °C, which is slightly lower than that obtained by Li et al. (2012). Overall, these $\Delta^{26}Mg_{calcite-fluid}$ values are close to those equilibrium fractionation factors derived from slow precipitated speleothems and cave waters (Galy et al. 2002; Immenhauser et al. 2010).

Experimental studies of the precipitation of aragonite, magnesite, and dolomite, though limited, also reveal large Mg isotope fractionation. Wang et al. (2013) found temperature-dependent Mg isotope fractionation during precipitation of aragonite, with $\Delta^{26}Mg_{aragonite-fluid}$ varying from −1.1 at 25 °C to −0.8 at 55 °C, which is smaller than the fractionation during calcite precipitation. Pearce et al. (2012) found large Mg isotope fractionation between magnesite and fluid during congruent dissolution, precipitation and at equilibrium. The equilibrium fractionation between magnesite and fluid ($\Delta^{26}Mg_{magnesite-fluid}$) changes from −1.2 at 150 °C to −0.88 at 200 °C. Li et al. (2015) for the first time calibrated Mg isotope fractionation between dolomite and aqueous solution through hydrothermal experiments at 130, 160 and 220 °C. Near-complete isotope exchange was assured by using Sr isotopes and ^{25}Mg tracers. The fractionation factor ($\Delta^{26}Mg_{dolomite-fluid}$) decreases with increasing temperature from −0.93, −0.84 to −0.65 at 130, 160 and 220 °C, respectively. The results clearly show large Mg isotope fractionation between dolomite and aqueous solution, with heavy Mg isotopes preferring fluids to dolomites.

Different from the experimental studies summarized above, Mavromatis et al. (2012) investigated Mg isotope fractionation during hydrous Mg carbonate precipitation with and without cyanobacteria. Light Mg isotopes are always preferentially incorporated into the solid phase in all experiments, with $\Delta^{26}Mg_{solid-liquid}$ ranging from −1.55 to −1.17. More importantly, the presence of cyanobacteria in the precipitation experiments does not affect the $\Delta^{26}Mg_{solid-liquid}$ values, suggesting that *Gloeocapsa* sp. Cyanobacterium does not cause additional Mg isotope fractionation during carbonate precipitation.

Though significant progress has been made (Rustad et al. 2010; Schauble 2011), the theoretically calculated fractionation factors between carbonate and fluid are consistent with some experiments but not all. Furthermore, the fractionation factors calculated by using different methods yield different values, sometimes even different directions of fractionation. The inconsistency between laboratory studies and theoretical calculations may reflect either equilibrium was not reached during laboratory experiments or over simplification of the complicated Mg bonding environments in carbonates and fluids during theoretical calculations. Nonetheless, the following enrichment of ^{26}Mg in various carbonate species can be concluded based on natural samples, experimental calibration and theoretical calculation: aragonite > dolomite > magnesite > calcite (Wang et al. 2013).

BEHAVIOR OF MAGNESIUM ISOTOPES DURING MAJOR GEOLOGICAL PROCESSES

The large Mg isotopic variation in major reservoirs summarized above indicates significant Mg isotope fractionation during various geological processes. This section will briefly summarize the magnitude and mechanisms of Mg isotope fractionation during biological processes, chemical weathering, magmatic differentiation and metamorphic dehydration. The following section will focus on high-temperature inter-mineral equilibrium fractionation and diffusion-driven kinetic fractionation in various geological processes. Overall, knowledge on these processes helps to understand Mg isotopic variation in major reservoirs and eventually lays the foundation for applying Mg isotopes to trace both high- and low-temperature geological processes.

Behavior of Mg isotopes during biological processes

Magnesium is an essential element in the biosphere and Mg isotopes can be largely fractionated in marine biogenic carbonates (see the section above), chlorophylls and during plant growth. Understanding the behavior of Mg isotopes during biological processes thus is the prerequisite for using Mg isotopes as a potential tracer of biological processes and is also important for constraining Mg isotopic systematics during continental weathering.

Magnesium plays an important role in photosynthesis as it is the center of all chlorophylls. Studies of chlorophylls from both laboratory experiments and natural samples have found large Mg isotope fractionation. Chlorophylls have lighter Mg isotopic composition than the growth media for experimentally grown cyanobacteria chlorophyll, coccolithophores and wheat (Black et al. 2006, 2008; Ra et al. 2010b). By contrast, chlorophylls have heavier Mg isotopic composition than the leaves of English ivy from which they were extracted (Black et al. 2007). Furthermore, chlorophyll of marine red algae is also enriched in heavy Mg isotopes relative to the growth media i.e., seawater (Ra and Kitagawa 2007). Therefore, Mg isotope fractionation in chlorophyll depends on both plant species and environmental conditions.

Laboratory studies found large Mg isotope fractionation during plant growth. Black et al. (2008) reported that plants were enriched in heavy Mg isotopes than the nutrient solutions. Within the plant, seeds and exudates are enriched in heavy Mg isotopes relative to leaves, shoots, and roots, suggesting Mg isotope fractionation during uptake and transport within the plants. Bolou-Bi et al. (2010) investigated the behavior of Mg isotopes during growth of rye grass and clover on two substrates (solution and phlogopite). They found both plants prefer heavy Mg isotopes to the substrates during growth although the magnitude of the isotope fractionation between plant and source depends on the plant species and nutrient substrate. Within one plant, roots are enriched in heavy Mg isotopes than leaves, similar to Black et al. (2008), which suggests that different biological processes such as organic molecule formation, Mg transport, and internal cycles affect leaves and roots differently and lead to the internal Mg isotope fractionation (Bolou-Bi et al. 2010).

Studies of natural samples confirmed the above laboratory studies and show that grass and spruce are indeed enriched in heavy Mg isotopes compared to their source Mg in soils, and shoots are enriched in light isotopes than roots within the grass (Tipper et al. 2010, 2012b; Bolou-Bi et al. 2012; Opfergelt et al. 2014). Within the spruce, needles always have lighter Mg isotopic composition than the bole wood at the same height and both of them are isotopically lighter than the roots (Fig. 21), further confirming significant Mg isotope fractionation during Mg transport within plants and trees (Bolou-Bi et al. 2012).

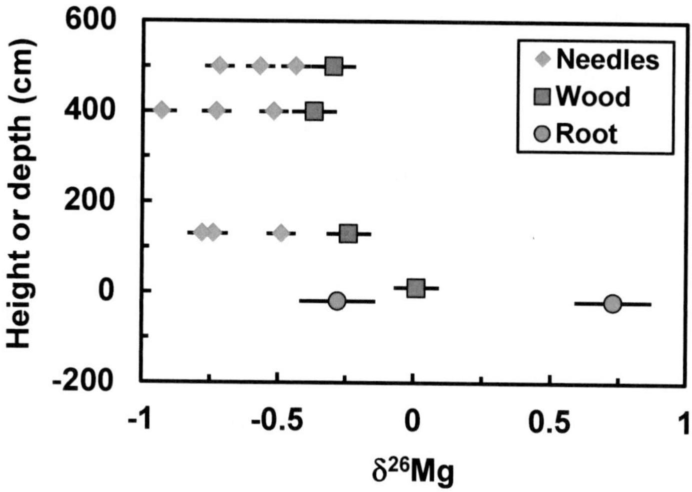

Figure 21. Magnesium isotope fractionation within a Norway spruce (Bolou-Bi et al. 2012).

Behavior of Mg isotopes during continental weathering

Continental weathering links the outer parts of the Earth, i.e., atmosphere, hydrosphere, and continents, and is one of the most important processes that produce the largest stable isotope fractionation. Studies on mechanism and magnitude of Mg isotope fractionation during continental weathering can provide constraints on global Mg cycling and lay the foundation for using Mg isotopes to trace crust–mantle interactions. To date, the behavior of Mg isotope fractionation during continental weathering has been extensively investigated, through studies of river waters, weathering profiles and laboratory experiments. These studies found large Mg isotope fractionation during different weathering processes such as mineral dissolution, formation of secondary mineral, adsorption and desorption, and biological activities. The direction and magnitude of Mg isotope fractionation vary and largely depend on the exact weathering process.

River systems. Studies of Mg isotopic compositions of river systems (e.g., river water, suspended load, soil, bedrock, and vegetation) can help to constrain the direction and magnitude of Mg isotope fractionation during continental weathering. The general observations that river waters, suspended load, soil, and bedrocks have different Mg isotopic compositions (Fig. 22) indicate significant Mg isotope fractionation during continental weathering.

Tipper et al. (2006b) studied 45 rivers including 16 of the largest rivers in the world and found that the first-order control on δ^{26}Mg for small rivers is the bedrock lithology whereas that for large rivers is isotope fractionation during weathering. Even for small rivers, isotope fractionation has to play an important role. This is because small rivers draining silicate rocks have lighter Mg isotopic composition whereas those draining carbonate rocks have heavier Mg isotopic composition than corresponding bedrocks, clearly indicating Mg isotope fractionation during continental weathering.

Studies of small rivers draining largely monolithological bedrocks further confirm large Mg isotope fractionation during weathering and also reveal detailed mechanisms of Mg isotope fractionation. River water is usually lighter whereas the weathered residue (silicate soil) is heavier than the bedrocks, regardless of basaltic or granitic, indicating that light Mg isotopes were released into water during weathering, leaving an isotopically heavy clay-rich residue (Tipper et al. 2006a, 2012b; Brenot et al. 2008; Lee et al. 2014; Ma et al. 2015). By contrast, river waters usually have similar Mg isotopic composition to bedrock dolomites and are, however, isotopically heavier than the limestone bedrock (Tipper et al. 2006b, 2008a; Lee et al. 2014), suggesting a preference of light Mg isotopes in limestone to water during weathering of limestone and limited Mg isotope fractionation during weathering of dolomite.

Pogge von Strandmann et al. (2008a) found that river waters and suspended loads had δ^{26}Mg values both higher and lower than the basaltic bedrocks. Glacier-fed and some direct runoff river waters tend to have heavy Mg isotopic composition, which suggests that the secondary minerals prefer light Mg isotopes. This interpretation is supported by the isotopically light suspended loads in these rivers. By contrast, the other direct runoff river waters tend to

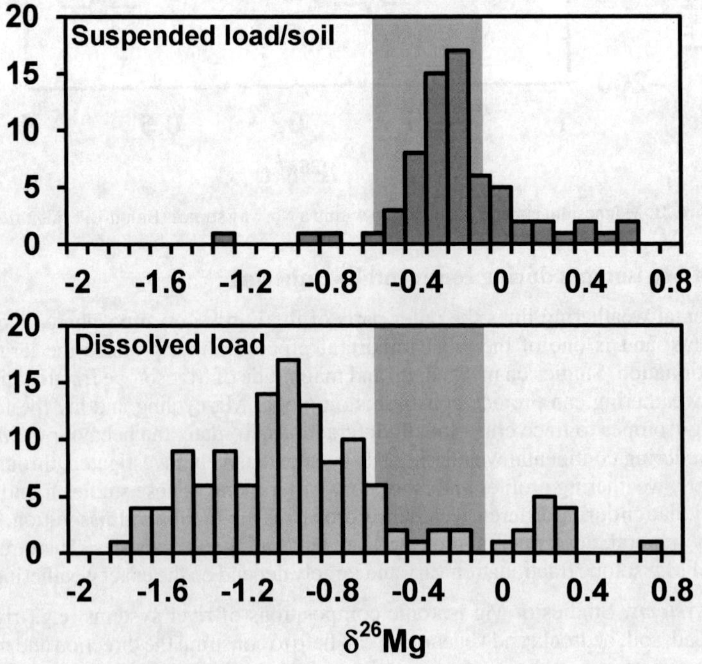

Figure 22. Magnesium isotope fractionation during continental weathering as revealed by studies of dissolved load, suspended load and soil in river systems (Tipper et al. 2006a, 2008a; Brenot et al. 2008; Pogge von Strandmann et al. 2008a,b, 2012; Wimpenny et al. 2011; Opfergelt et al. 2012). The vertical bars represent bedrocks.

have lighter Mg isotopic composition than bedrocks, consistent with the preference of heavy Mg isotopes in the secondary minerals, which is also supported by the isotopically heavy suspended load in these rivers. This study thus highlights the important role of secondary mineral formation on Mg isotope fractionation during continental weathering. Similarly, Tipper et al. (2012a) also found river waters draining mixed carbonate and silicate bedrocks display a positive correlation between Li and Mg isotopic compositions. Such a positive correlation cannot be produced by mixing between carbonate and silicate rocks; instead it likely reflects clay-controlled Mg and Li isotope fractionations.

Weathering profiles: Saprolite. Different from river systems that integrate over the whole catchment area, weathering profiles developed on different types of rocks under different climates can be used to study how chemical weathering affects mineral dissolution, secondary phase crystallization, elemental transport and isotope fractionation as a function of protolith mineralogy, fluid property, temperature, humidity and other chemical and physical conditions (e.g., Goldich 1938). Thus, studies of weathering profiles can provide complementary information to studies of river systems.

Teng et al. (2010b) measured Mg isotopic composition of a weathering saprolite profile developed on a diabase dike from South Carolina and found large Mg isotope fractionation during weathering. The δ^{26}Mg values increase as the weathering progresses from −0.22 in the unweathered diabase in the base to +0.65 in the most weathered saprolite at the top of the profile. The isotopic variation is coupled with Mg concentration, clay mineral proportion and density of the saprolite (Fig. 23). These observations are consistent with the fact that Mg is fluid mobile and is released into water during continental weathering when Mg-rich minerals break down and secondary minerals form. During this process, light Mg isotopes were released to the hydrosphere and heavy Mg isotopes were enriched in the weathered products.

Studies of extremely weathered basalt profiles further reveal large Mg isotope fractionation associated with different types of secondary minerals. Huang et al. (2012) reported large Mg isotope fractionation in a set of clay-rich saprolites developed on the Neogene tholeiitic basalt from Hainan Island in southern China. These saprolites are strongly (93–99%) depleted in Mg and have highly variable but heavier δ^{26}Mg (−0.49 to +0.40) than the unweathered basalts (−0.36) (Fig. 24). Both Mg concentration and δ^{26}Mg of saprolites increase upwards in the lower part of the profile, but decrease towards the surface in the upper part. The contrasting behavior of Mg isotopes between the upper and lower parts was produced by Mg isotope fractionation through adsorption and desorption processes.

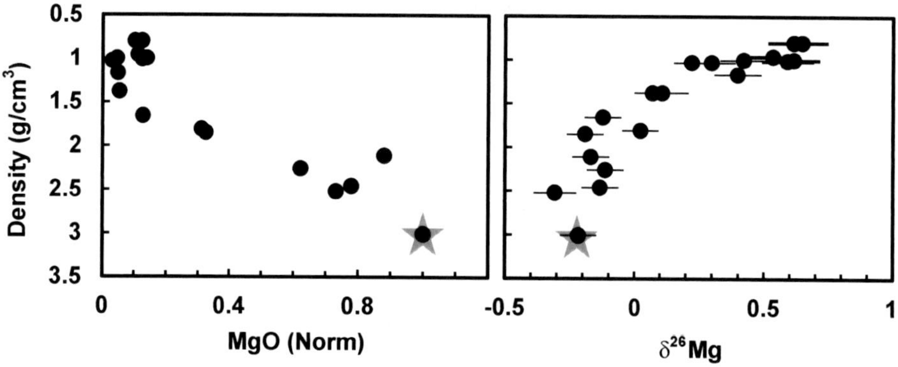

Figure 23. Magnesium isotope fractionation during continental weathering of diabase from South Carolina (Teng et al. 2010b). The Mg concentration was normalized to titanium and density of sample is used as an index of degree of weathering. The star represents the unweathered diabase.

Adsorption preferentially uptakes heavy Mg isotopes onto kaolin minerals in saprolites in the lower part whereas desorption of Mg through cation exchange of Mg with the relatively lower hydration energy cations in the upper profile releases isotopically light Mg to the hydrosphere. This study thus highlights that adsorption and desorption of Mg on kaolin minerals play an important role in behavior of Mg isotopes during extreme weathering.

Another study of ~10 m deep drill cores through bauxites developed on Columbia River Basalts found the heaviest isotopic composition ever reported for Mg (Liu et al. 2014b). These bauxite samples have near complete (>99%) loss of Mg and have variable but heavy Mg isotopic composition (δ^{26}Mg = −0.1 to +1.8) relative to the unweathered basalts (δ^{26}Mg = −0.23). The most highly weathered bauxites at the tops of the profiles have the lowest δ^{26}Mg values, reflecting the addition of isotopically light eolian dust as indicated by their quartz content. Excluding these samples, δ^{26}Mg in the bauxites displays a positive correlation with gibbsite abundance (Fig. 24), suggesting that gibbsite preferentially adsorbs heavy Mg isotopes in the bauxites (Liu et al. 2014b).

Weathering profiles: Soil. Studies of top soil profiles (< 2 m deep) developed on igneous (andesite and basalt) and sedimentary (shale) rocks also reveal large Mg isotope fractionation, associated with sorption (adsorption and desorption) of Mg onto secondary minerals and biological activities (Bolou-Bi et al. 2012; Opfergelt et al. 2012, 2014; Pogge von Strandmann et al. 2012; Ma et al. 2015).

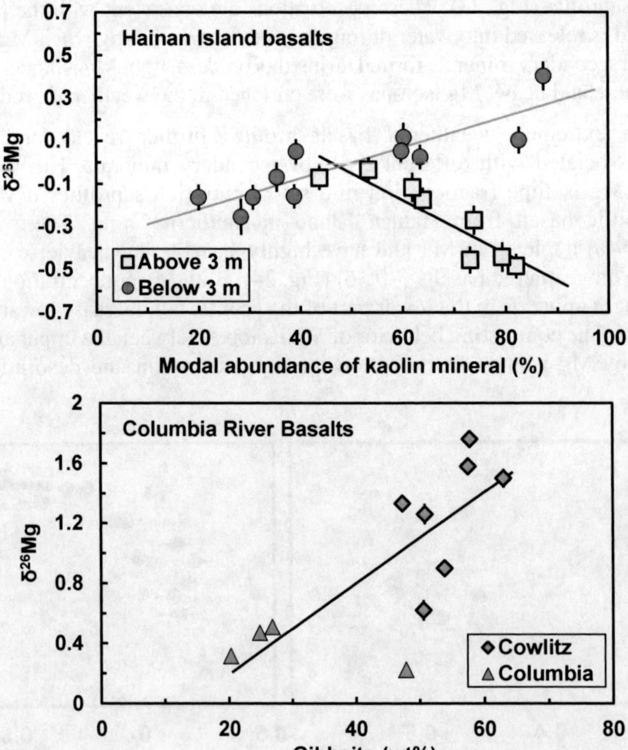

Figure 24. Magnesium isotope fractionation during continental weathering as revealed by studies of extremely weathered basalts developed on Hainan Island basalts (Huang et al. 2012) and Columbia River Basalts (Liu et al. 2014b). The correlations between δ^{26}Mg and mineral abundances reflect mineralogical-control Mg isotope fractionation during adsorption and desorption of Mg on various secondary minerals.

Opfergelt et al. (2012) measured bulk soils and clay fractions in andesitic topsoil sequences from Guadeloupe and found that clays had heavier Mg isotopic composition than the parent andesite, consistent with previous studies of river systems. More importantly, Mg isotopic composition of the bulk soil is negatively correlated with the amount of exchangeable Mg (Fig. 25), which suggests that Mg adsorption on the soil exchange complex prefers light Mg isotopes to heavy ones (Opfergelt et al. 2012). Thus, Mg isotopic compositions of the clays are likely controlled by a mixture of isotopically heavy structural Mg and light exchangeable Mg. Later studies of soil system developed on Icelandic basalts show that clay fractions, exchangeable Mg and soil solutions are all isotopically lighter than the unweathered basalt, which reflects isotope fractionation associated with adsorption and desorption of heavy Mg from the soil profile, producing isotopically light clay fractions (Opfergelt et al. 2014). By contrast, Pogge von Strandmann et al. (2012) observed a negative correlation between Mg isotopic composition of soil pore water and the amount of exchangeable Mg in a basaltic soil and pore water profile developed on Icelandic basalts (Fig. 25). This correlation is interpreted as a result of preference of isotopically heavy Mg on the soil exchange complexes, which leaves pore waters isotopically light. Regardless of the direction of the isotope fractionation, all these studies suggest that adsorption and desorption of Mg onto/from secondary minerals can significantly fractionate Mg isotopes therefore play an important role in controlling Mg isotopic systematics of the soil system.

Ma et al. (2015) recently showed that chemical weathering of shales can also significantly fractionate Mg isotopes. They found stream and soil pore waters are ~0.5‰ to 1‰ lighter than the shale bedrock in the Shale Hills catchment in central Pennsylvania, consistent with previous observations that lighter Mg isotopes are preferentially released to water during silicate weathering. By contrast, Mg isotopic compositions of the soil samples display a slight but systematic decreasing trend with increasing weathering duration, and

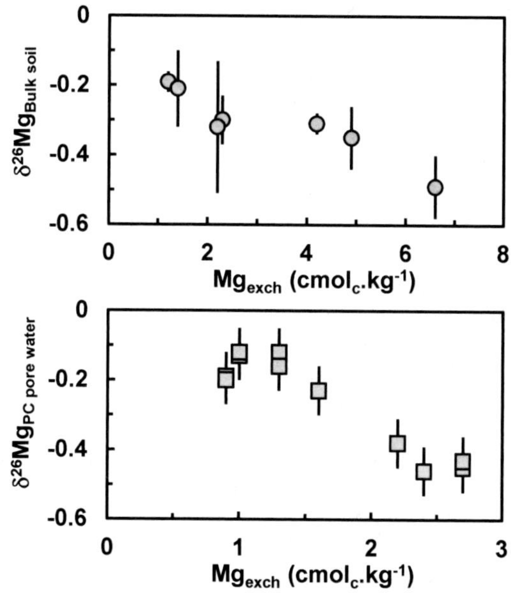

Figure 25. Magnesium isotope fractionation during continental weathering as revealed by studies of soil profiles. The top panel displays a negative correlation between Mg isotopic composition of bulk soils and the amount of exchangeable Mg in bulk soils developed on andesites from Guadeloupe (Opfergelt et al. 2012). The bottom panel shows a negative correlation between Mg isotopic composition of pore waters and the amount of exchangeable Mg in the soils developed on basalts from Iceland (Pogge von Strandmann et al. 2012).

are either similar or up to ~0.2‰ lighter than the bedrock. This isotopically light residue likely reflects the adsorption of light Mg onto secondary clay minerals i.e., vermiculite. The heavy Mg isotopic reservoir, which is complementary to the isotopically light pore water and stream water, is believed to be lost as suspended loads in the stream, which is supported by the isotopically heavy stream sediments accumulated on the valley floor. This study thus provides the first field evidence that changes in clay mineralogy lead to accumulation of light Mg isotopes in residual bulk soils, and that transport of isotopically distinct fine particles from clay-rich systems could be a new and important mechanism to modify Mg isotope compositions of residues produced during silicate weathering (Ma et al. 2015).

Experimental studies. Behaviors of Mg isotope fractionation during continental weathering have also been investigated in laboratory studies through well-controlled experiments such as mineral dissolution, clay mineral formation, and precipitation of carbonate and sulfate minerals. Most of these experiments found large Mg isotope fractionation between fluid and solid and the direction of fractionation is consistent with the observation on natural samples. In this section, only experiments associated with silicates and sulfates are discussed. Those related with biogenic and inorganic carbonates are covered in the section on carbonates.

Wimpenny et al. (2010) carried out a series of experiments to characterize the behavior of Mg isotopes during primary phase dissolution and secondary mineral formation during basalt weathering. Magnesium isotopic compositions of the solutions are always lighter than the primary basalt glass or olivine (Fig. 26), indicating the preferential loss of light Mg isotopes into solutions during primary phase dissolution. By contrast, Mg isotopic composition of the solutions becomes heavy during the formation of secondary mineral chrysotile, suggesting the secondary phase selectively incorporates light Mg isotopes from the solution. This observation is opposite to previous studies of weathered silicate residues and suggests the direction of Mg isotope fractionation during the formation of secondary minerals may be mineral-specific (Wimpenny et al. 2010).

The recent comprehensive study of natural and synthesized clay minerals confirms that secondary minerals preferentially incorporate isotopically heavy Mg into their structure, but also show that the exchangeable Mg adsorbed to the interlayer and surface sites is isotopically light (Fig. 26) (Wimpenny et al. 2014a). Therefore, Mg isotopic composition of bulk clays depends on the proportion of the isotopically heavy structural Mg and light exchangeable Mg, and can be lighter than the unweathered rocks if the exchangeable Mg significantly dominates the Mg budget of the bulk clays. This is consistent with the observations of natural samples (Huang et al. 2012; Opfergelt et al. 2012, 2014; Pogge von Strandmann et al. 2012; Ma et al. 2015)

Similar to mineral-fluid reactions, laboratory studies of rock weathering also reveal a large Mg isotopic variation during granite weathering. Ryu et al. (2011) found that Mg isotopic compositions of the solutions display ~1.4‰ variation during the whole experiment. At the same time, granite minerals also display large Mg isotopic variation with chlorite ~1.5‰ heavier than coexisting biotite and hornblende. Therefore, the Mg isotopic variation of the solutions likely reflects the release of isotopically different Mg from different minerals at different dissolution stages, hence are controlled by the mineral Mg isotopic composition, not by isotope fractionation (Ryu et al. 2011). The reason for the different behavior of Mg isotopes during dissolution (e.g., olivine/basaltic glass dissolution with isotope fractionation vs. granite mineral dissolution without isotope fractionation) is not clear and more studies on dissolution of granitic minerals are needed.

Large Mg isotope fractionation was also reported between solid (epsomite) and aqueous phases of $MgSO_4$ in recrystallization experiments (Li et al. 2011b). The equilibrium fractionation factor $\Delta^{26}Mg_{epsomite-solution}$ (= $\delta^{26}Mg_{epsomite} - \delta^{26}Mg_{solution}$) is 0.6 and does not vary when temperature changes from 7 20 to 40 ºC . Thus, Mg isotopic composition of sulfate can reflect that of source solutions and could be used to constrain evaporative history.

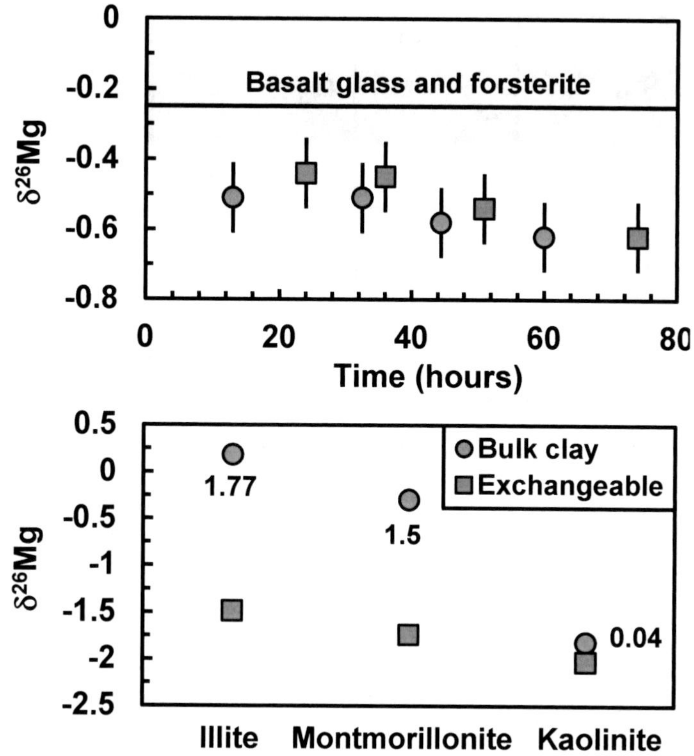

Figure 26. Magnesium isotope frac tionation during weathering as revealed by experimental studies. The top panel shows Mg isotopic composition of experimental fluids during dissolution of basalt glass (circle) and forsterite (square) (Wimpenny et al. 2010). The bottom panel displays Mg isotopic compositions of bulk clays and their exchangeable Mg (Wimpenny et al. 2014a). The Mg concentrations (wt%) of bulk clays are also labelled.

Behavior of Mg isotopes during magmatic differentiation

Magnesium is compatible during mantle melting and magmatic differentiation as it is a major element in rock-forming minerals such as olivine, pyroxene, hornblende, and biotite. Therefore, MgO content in magma generally decreases with magmatic differentiation. Magnesium isotopic compositions, however, do not change significantly in whole rocks during partial melting of the mantle and differentiation of basaltic magma (Teng et al. 2007b, 2010a) (Fig. 11) or granitic magma (Li et al. 2010; Liu et al. 2010; Ke et al. 2016) (Fig. 27). The absence of Mg isotope fractionation during silicate magmatic differentiation thus suggests that basalts and granites can be used to study isotopic variations in their source region. However, as discussed in the later section on high-temperature isotope fractionation, spinel and garnet in mantle peridotites, eclogites, and deep crust granulites have different Mg isotopic composition than coexisting olivine, pyroxene and biotite (Li et al. 2011a; Liu et al. 2011; Wang et al. 2012, 2014a,b, 2015a,c, 2016; Xiao et al. 2013; Hu et al. 2016b). Partial melting or magmatic differentiation involving spinel and garnet could potentially fractionate Mg isotopes and produce melts with different isotopic composition from the source rocks (e.g., Wang et al. 2015a, 2016).

In contrast to silicate magmatism, large Mg isotope fractionation during carbonatite magmatism has been recently reported. Li et al. (2016b) investigated natrocarbonatites and

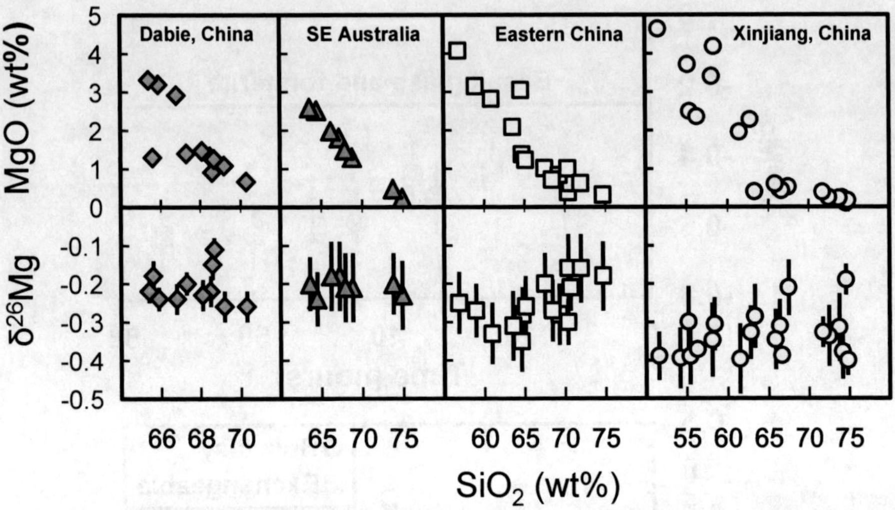

Figure 27. Lack of magnesium isotope fractionation during the differentiation of granitic magma as shown by the lack of correlation between δ^{26}Mg and SiO_2 (Li et al. 2010; Liu et al. 2010; Ke et al. 2016).

peralkaline silicate rocks from Oldoinyo Lengai, Tanzania and found that the silicate rocks not related to silicate–carbonatite liquid immiscibility have δ^{26}Mg (from −0.25 to −0.10) clustered around the mantle value. By contrast, samples derived from the silicate melts that were presumably generated by liquid immiscibility, have heavier Mg isotopic composition (δ^{26}Mg of −0.06 to +0.09) (Fig. 28). Such a difference suggests Mg isotope fractionation during liquid immiscibility and implies, based on mass-balance calculations, that the original carbonatite melts at Lengai were isotopically light. The variable and positive δ^{26}Mg values of natrocarbonatites (from +0.13 to +0.37) hence require a change of their Mg isotopic compositions subsequent to liquid immiscibility. The negative correlations between δ^{26}Mg values and contents of alkali and alkaline earth metals of natrocarbonatites suggest Mg isotope fractionation during fractional crystallization of carbonatite melts, with heavy Mg isotopes enriched in the residual melts relative to fractionated carbonate minerals. Collectively, significant Mg isotope fractionation may occur during both silicate–carbonatite liquid immiscibility and fractional crystallization of carbonatite melts, making Mg isotopes a potentially useful tracer of processes relevant to carbonatite petrogenesis (Li et al. 2016a).

Behaviors of Mg isotopes during metamorphic dehydration

Metamorphic dehydration widely occurs in the continental crust and subduction zones, and can significantly fractionate fluid-mobile elements and their isotopes (Elliott 2003). Magnesium is fluid mobile and can be significantly fractionated during fluid–rock interactions. Studies of Mg isotope fractionation during metamorphism thus can shed light on Mg isotopic composition and evolution of the crust, the behavior of Mg isotopes during subduction processes and global Mg cycle.

Magnesium isotope fractionation during metamorphic dehydration of silicate rocks is limited. Li et al. (2011a) found that orogenic eclogites have similar Mg isotopic compositions to their potential protoliths–basalts/gabbros, and concluded that limited Mg isotope fractionation occurred during eclogite-facies metamorphism of mafic rocks. Similarly, Teng et al. (2013) show that mafic granulites have retained similar Mg isotopic compositions to their inferred basaltic protoliths, suggesting Mg isotopes do not fractionate during granulite-facies metamorphism of mafic rocks. Nonetheless, both studies focused on high-grade metamorphic rocks and their

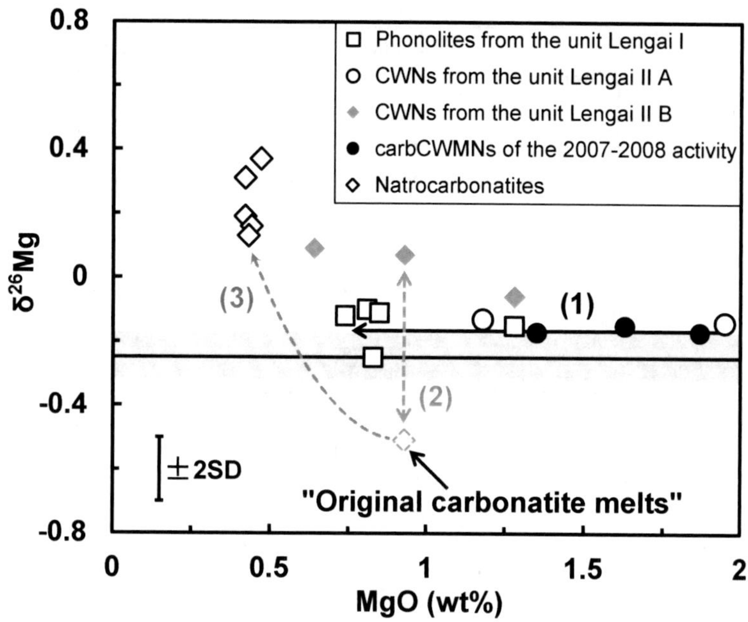

Figure 28. Magnesium isotope fractionation during carbonatite magmatism at Oldoinyo Lengai (Li et al. 2016a). The three main magmatic evolution processes are: (1) silicate magma differentiation, (2) silicate-carbonatite liquid immiscibility, and (3) carbonatite magma differentiation.

protoliths were not directly measured. To directly assess the behavior of Mg isotopes during prograde metamorphism, Li et al. (2014) analyzed a set of well-characterized metapelites from the Onawa contact aureole and found Mg isotopic compositions of metapelites do not change even though significant amount of fluids was lost during prograde metamorphism (Fig. 29). A similar conclusion is also reached for metamorphic dehydration of subduction zone rocks. Wang et al. (2014b) systematically measured Mg isotopes for a set of genetically related prograde metamorphosed meta-basaltic rocks from the Dabie orogen, with metamorphic grade ranging from greenschist, amphibolite to eclogite. All these samples have similar Mg isotopic composition, indicating limited Mg isotope fractionation during prograde metamorphism of mafic rocks (Fig. 29). Overall, these studies suggest that metamorphic dehydration does not produce measurable Mg isotopic changes and metamorphic rocks still preserve Mg isotopic compositions of their protoliths. The lack of isotope fractionation during prograde metamorphism mainly results from the conservative behavior of Mg as Mg is always hosted in major rock-forming minerals during metamorphic reactions and little Mg was lost into fluids.

Behaviors of Mg isotopes during prograde metamorphism of carbonate-silicate mixtures, however, are more complicated (Shen et al. 2013; Wang et al. 2014a). Shen et al. (2013) measured Mg isotopic compositions of endoskarns formed by contact metamorphism between a granodioritic pluton and dolomitic wallrocks. The outer pyroxenite zone of the endoskarn is significantly (>0.8‰) lighter than the inner plagioclase-quartz dominated zone (δ^{26}Mg = −0.82), both of which are lighter than typical mantle rocks. The large spatial Mg isotopic variation most likely reflects mixing of Mg between Mg-rich, isotopically light dolomitic wallrock and Mg-poor, isotopically heavy granodiorite magma (Shen et al. 2013). Thus, carbonate–silicate interactions during contact metamorphism can significantly lower Mg isotopic compositions of silicate rocks. Wang et al. (2014a) investigated Mg isotopes

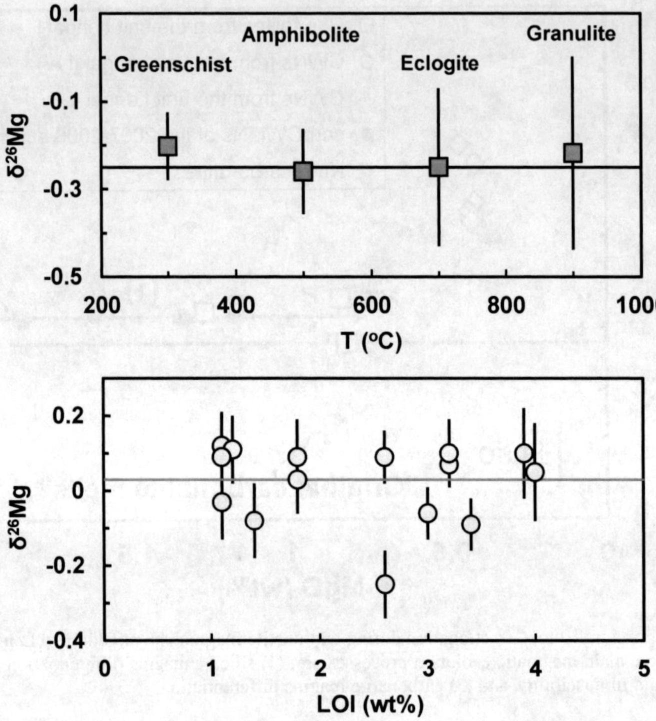

Figure 29. Limited magnesium isotope fractionation during metamorphic dehydration. The top panel reports the average and 2SD of Mg isotopic compositions of regionally metamorphosed mafic rocks ranging from greenschist-facies to granulite-facies (Teng et al. 2013; Wang et al. 2014b, 2015b; Yang et al. 2016). The horizontal line represents the typical Mg isotopic composition of mafic protoliths e.g., oceanic basalts. The bottom panel displays Mg isotopic compositions of metapelites from the Onawa contact aureole (Li et al. 2014). The horizontal line represents the average Mg isotopic composition of the protolith pelites.

in ultrahigh pressure metamorphic marbles and enclosed carbonated eclogites from Dabie orogen, China. Different from normal eclogites that are enclosed in gneiss and generally have mantle-like δ^{26}Mg, carbonated eclogites have anomalously light Mg isotopic composition, with δ^{26}Mg down to −1.93. The marbles also display elevated δ^{26}Mg values when compared to their protoliths (Fig. 30). These unusual Mg isotopic signatures reflect large, differential Mg isotopic exchanges between marbles and the enclosed carbonated eclogites during prograde metamorphism, which is also supported by experimental studies of Mg isotope fractionation between silicates and carbonates (Macris et al. 2013). One important implication of this study is that silicates can partially inherit the light Mg isotopes released from ambient carbonates during subduction and form an isotopically light silicate Mg endmember in the mantle.

HIGH-TEMPERATURE MAGNESIUM ISOTOPE FRACTIONATION

The degrees of high-temperature Mg isotope fractionation were highly debated in early studies. Pearson et al. (2006) analyzed Mg isotopic compositions of olivines in mantle-derived peridotite xenoliths and megacrysts by using laser-ablation (LA) MC-ICPMS and found >4‰ Mg isotopic variation in olivines. These large variations were interpreted as results of diffusion-driven kinetic Mg isotope fractionation produced during mantle metasomatism. By

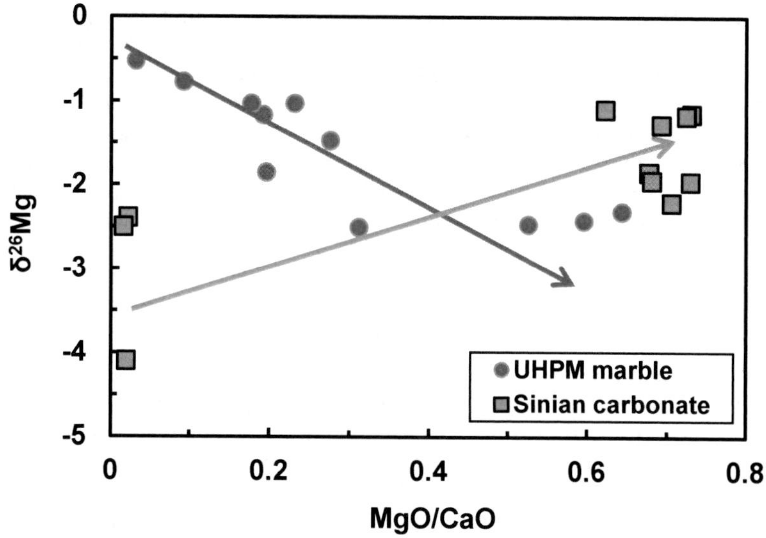

Figure 30. Plot of $\delta^{26}Mg$ vs. MgO/CaO. The UHPM marbles display an opposite trend to their protoliths (Sinian carbonates), suggesting significant Mg isotope exchange between carbonates and silicates during prograde metamorphism (Wang et al. 2014a). Calcite-rich marbles are modified more significantly from their protolith $\delta^{26}Mg$ than dolomite-rich ones during the carbonate-silicate interaction.

contrast, another study, using the same technique, i.e., LA-MC-ICPMS, found that olivines in oceanic basalts and peridotite xenoliths with a range of ages and tectonic settings have similar Mg isotopic compositions (Norman et al. 2006b). The conflicting results from these two studies mainly result from the analytical artefacts associated with matrix effects in LA-MC-ICPMS. This is because matrix effects can easily produce > 1‰ mass bias (Norman et al. 2006a; Janney et al. 2011; Xie et al. 2011; Sio et al. 2013), especially if chemical compositions between samples and standards are significantly different. Thus, matching sample and standard compositions is required for high-precision and high-accuracy Mg isotopic analysis by LA-MC-ICPMS. Handler et al. (2009) analyzed olivine samples from peridotite xenoliths compositionally similar to those measured by Pearson et al. (2006) by solution method MC-ICPMS with higher precision and accuracy. They found all olivine samples have the same Mg isotopic composition, further confirming that the large variation observed in Pearson et al. (2006) was likely produced by matrix effects in LA-MC-ICPMS. Nonetheless, as disused below, later studies do find large Mg isotopic variations at high temperatures though the mechanism and direction of fractionations are different from those proposed in Pearson et al. (2006).

High-temperature equilibrium inter-mineral Mg isotope fractionation

Magnesium is a major element in the silicate Earth and is a major cation in all major mafic rock-forming minerals: olivine (Ol), orthopyroxene (Opx), clinopyroxene (Cpx), hornblende (Hbl) and biotite (Bt). Measurements of coexisting olivine, Opx and Cpx in mantle xenoliths (Handler et al. 2009; Yang et al. 2009; Huang et al. 2011; Liu et al. 2011; Pogge von Strandmann et al. 2011; Xiao et al. 2013; Lai et al. 2015; Hu et al. 2016b; Wang et al. 2016) (Fig. 31), and coexisting hornblende and biotite in granitoids (Liu et al. 2011) suggest limited (<0.2‰) equilibrium inter-mineral fractionations. As these five minerals contain most Mg in the Earth and control Mg budget from ultramafic to felsic rocks, the limited inter-mineral fractionations among these minerals indicate small Mg isotope fractionation during magmatic differentiation, be it basaltic or granitic, as shown in previous studies (Teng et al. 2007b, 2010a; Liu et al. 2010). The limited inter-mineral fractionations reflect the similar coordination of Mg in these minerals. Equilibrium isotope fractionation is qualitatively controlled by the coordination number of Mg in minerals, with heavy isotopes preferring minerals with high-bonding energy sites (i.e., low coordination sites) (Urey 1947). The coordination number of Mg in Ol, Opx, Cpx, Hbl and Bt is the same of 6. Therefore, limited equilibrium inter-mineral fractionation among these minerals is expected. However, the coordination of Mg is not 6 in two important rock-forming minerals as it is 4 in spinel (Spl) and 8 in garnet (Grt) (Deer et al. 1992). Thus spinel is expected to be isotopically heavier whereas garnet is expected to be isotopically lighter than coexisting Ol, Opx, Cpx, Hbl or Bt. The following sequence of ^{26}Mg enrichment can be expected: Spl $\gg$ Bt > Hbl $\approx$ Cpx > Opx > Ol $\gg$ Grt. This qualitative estimate agrees well with the quantitative calculation and experimental calibration (Schauble 2011; Huang et al. 2013a; Macris et al. 2013). Measurements of these minerals from igneous and metamorphic rocks indeed found large equilibrium inter-mineral fractionation involving spinel and garnet.

Studies of mantle xenoliths found large equilibrium Mg isotope fractionation between spinel and olivine (Fig. 32). Young et al. (2009) first reported up to 0.8‰ Mg isotope fractionation between coexisting spinel and olivine in two peridotite xenoliths from San Carlos and interpreted the large isotope fractionation as a result of equilibrium isotope partitioning. Nonetheless, these two peridotite samples formed at similar temperatures thus it is difficult to check whether or not the fractionation is temperature-dependent, which is a key feature of equilibrium isotope partitioning. Liu et al. (2011) systematically measured coexisting Ol, Opx, Cpx, phlogopite (Phl), and Spl in 13 mantle xenoliths in the North China craton that formed over a wide (>300 °C) temperature range. They found small detectable (up to 0.2‰) equilibrium Mg isotope fractionations between coexisting Cpx, Phl and Ol, and unmeasurable fractionation between coexisting Opx and Ol in mantle xenoliths. In particular, they found large temperature-dependent spinel-olivine Mg isotope fractionation with $\Delta^{26}Mg_{Spl-Ol} = (\delta^{26}Mg_{Spl} - \delta^{26}Mg_{Ol})$ decreasing from +0.55 to +0. 25 when the equilibrium temperature increasing from 807 to 1150 °C (Fig. 32). This temperature-dependent correlation, together with the lack of intra-mineral isotope fractionation and compositionally dependent inter-mineral fractionation, suggests an equilibrium isotope fractionation. Nonetheless, the degree of Spl–Ol fractionation is lower than the expected from theoretical calculation (Schauble 2011), which was interpreted as a result of compositional effects on the Spl–Ol equilibrium fractionation (Liu et al. 2011).

Studies of eclogites, granulites and mantle xenoliths show that garnets are always isotopically lighter than coexisting olivine, biotite and pyroxene, reflecting the control of coordination difference between garnet and other coexisting minerals (Li et al. 2011a; Wang et al. 2012, 2014a,b, 2015a,c; Pogge von Strandmann et al. 2015; Hu et al. 2016b). Li et al. (2011a) for the first time reported Mg isotopic data for 10 whole rocks and 13 mineral separates from a set of orogenic eclogites from Bixiling in the Dabie orogen and found omphacite (a kind of Na-rich high-pressure clinopyroxene) is >1‰ heavier in δ^{26}Mg than coexisting garnet. The large inter-mineral Mg isotope fractionation $\Delta^{26}Mg_{omphacite-garnet}$ ($= \delta^{26}Mg_{omphacite} - \delta^{26}Mg_{garnet} = 1.14 \pm 0.04$), together with homogeneous mineral chemistry and equilibrium oxygen isotopic partitioning between omphacite and garnet, suggests an equilibrium Mg isotope fractionation. Later studies of coexisting garnet and pyroxene in cratonic eclogites (Wang et al. 2012, 2015c), orogenic eclogites (Wang et al. 2014a,b; Li et al. 2016b) and mantle pyroxenites (Hu et al. 2016b) as well as of coexisting garnet and biotite in lower crust granulites (Wang et al. 2015a) also found large inter-mineral Mg isotope fractionation. Most of the inter-mineral fractionation reflects equilibrium partitioning as evidenced by the lack of elemental diffusion profiles between coexisting minerals, equilibrium O isotope fractionation, temperature-dependent fractionation, and more or less falls within the equilibrium fractionation lines defined by theoretical calculations (Huang et al. 2013a) (Fig. 33). Nonetheless, some of the inter-mineral fractionations involving garnets vary significantly within a narrow range of temperatures, clearly indicating disequilibrium fractionation, which will be discussed below.

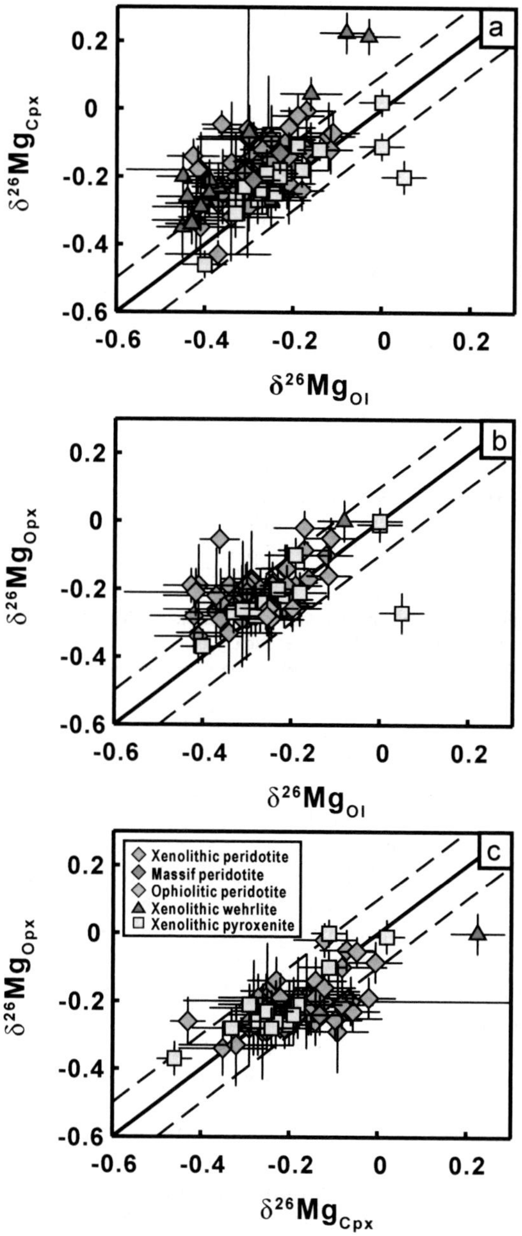

Figure 31. Limited Mg isotope fractionation among Ol, Cpx and Opx in mantle xenoliths, massif peridotites and ophiolitic peridotites (Handler et al. 2009; Yang et al. 2009; Huang et al. 2011; Liu et al. 2011; Pogge von Strandmann et al. 2011; Xiao et al. 2013, 2016; Lai et al. 2015; Hu et al. 2016b; Wang et al. 2006). (a) Ol–Cpx fractionation; (b) Ol–Opx fractionation; (c) Cpx–Opx fractionation.

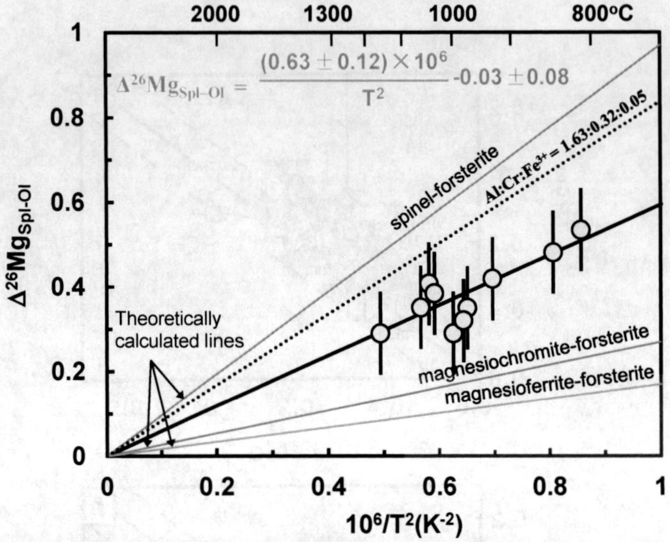

Figure 32. Equilibrium Mg isotope fractionation between spinel and olivine as calibrated by natural samples (Liu et al. 2011) and theoretical calculation (Schauble 2011). The natural samples define a temperature-dependent inter-mineral fractionation line with $10^3 \ln a_{Spl-Ol} \approx \Delta^{26}Mg_{Spl-Ol} = 0.63\,(\pm 0.12) \times 10^6/T^2 - 0.03(\pm 0.08)$ (Liu et al. 2011). The spinel composition plays an important role in controlling the equilibrium fractionation factors as shown by the different lines (Schauble 2011). The dotted line represents the expected $\Delta^{26}Mg_{Spl-Ol}$ values calculated based on the relative proportions of three different spinel species. The small difference between natural samples and the theoretically calculated values reflects an additional effect of Mg coordination numbers in spinel on its Mg isotopic composition (Liu et al. 2011).

Diffusion-driven kinetic Mg isotope fractionation

Recent experimental studies have found large kinetic isotope fractionation associated with diffusion of Mg along chemical and temperature gradients (Richter et al. 2008; Huang et al. 2009b, 2010; Watkins et al. 2009, 2011; Chopra et al. 2012). Diffusion occurs when there is a potential difference in a system. The chemical diffusivity (D) between isotopes is generally inversely correlated to their mass (m) by the following equation: $D_2/D_1 = (m_1/m_2\beta)$, $\beta = 0.5$ for ideal gas and <0.5 for liquid and solid (Richter et al. 2009a). Therefore, lighter isotopes always diffuse faster than heavy ones in all media and the difference in diffusivity between isotopes can cause large kinetic fractionation. For example, gas molecules have the same kinetic energy at the same temperature: $KE_{ave} = ½\ m_i v_i^2$. The velocities (v) of two molecules can be written as $v_2/v_1 = (m_1/m_2)^{0.5}$, e.g., $^{16}O_2$ diffuses $\sqrt{36/32} = 1.06$ times faster than $^{18}O_2$. Richter et al. (2008) first showed that Mg isotopes can be fractionated (up to 7‰ in $^{26}Mg/^{24}Mg$) during diffusion of Mg from a basaltic melt to a rhyolitic melt at a constant temperature with $\beta = 0.05$ (Fig. 34). In addition, they also showed that isotopes can be largely fractionated (up to 8‰ in $^{26}Mg/^{24}Mg$) along a temperature gradient, with lighter isotopes diffusing towards the hot ends, in a way similar to elemental fractionation along a temperature gradient (Soret diffusion) (Richter et al. 2008) (Fig. 35). More experimental studies were conducted later to explore the exact mechanisms driving the large isotope fractionation during chemical and Soret diffusion (Huang et al. 2009b, 2010; Watkins et al. 2009, 2011; Chopra et al. 2012) though the results are still highly debated.

To date, no isotope fractionation caused by Soret diffusion has been found in natural samples. This is partially because of the extreme fast diffusion of heat, which makes it difficult for the kinetic isotope fractionation to preserve for a long time. Nonetheless, Lundstrom (2009)

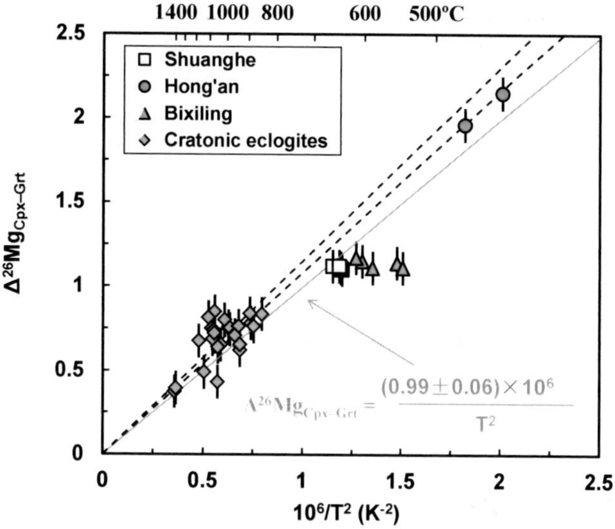

Figure 33. Equilibrium Mg isotope fractionation between garnet and clinopyroxene as calibrated by natural samples (Li et al. 2016b) and theoretical calculations (Huang et al. 2013a). The dashed lines represent the theoretical calculations at 1.5 and 3 GPa respectively (Huang et al. 2013a). The natural samples define a temperature-dependent inter-mineral fractionation line with $10^3 \ln \alpha_{Cpx-Grt} \approx \Delta^{26}Mg_{Cpx-Grt} = (0.99 \pm 0.06) \times 10^6 / T^2$ (Li et al. 2016b).

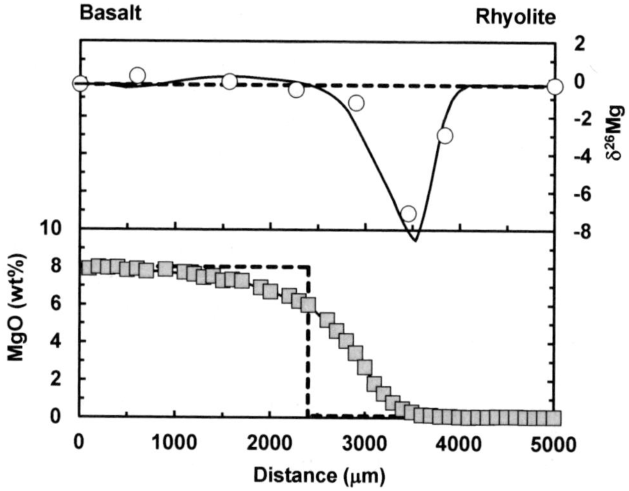

Figure 34. Magnesium isotope fractionation during chemical diffusion of Mg from molten basalt to molten rhyolite (Richter et al. 2008). The dashed lines show the initial MgO concentration and Mg isotopic composition of the basalt and rhyolite used in the diffusion couple. The solid curves are the calculated diffusion profiles (See Richter et al. 2008 for details).

proposed a hypothesis to explain the origin of convergent margin granitoids and the continental crust by thermal migration. In this model, the granitoids are formed by a top-down thermal migration zone refining process in the thick volcanic pile. Continuous injection of magma at the base of the pile provides heat to keep the near-steady-state top-down temperature gradient

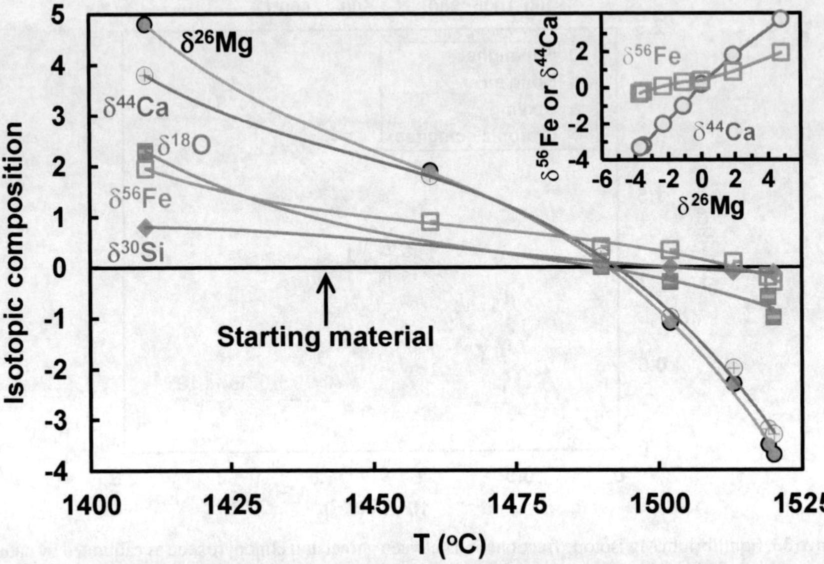

Figure 35. Magnesium isotope fractionation during Soret diffusion. Light isotopes always prefer hot ends to the cold ends, which results in a positive correlation between isotopic compositions of different elements (the inset). Data are from Richter et al. (2008, 2009b).

in the volcanic pile. Felsic materials i.e., granitoids migrate to the top, cold part whereas mafic materials i.e., hornblende gabbros, form at the bottom, hot part. The thermal migration thus led to the magmatic differentiation. As heavy isotopes of one element always diffuse towards the cold ends, granitoids are expected to be enriched in heavy isotopes of non-traditional stable isotopes whereas the hornblende gabbros are enriched in light ones. Studies of Mg isotopes do show some granitoids have heavier Mg isotopic compositions (Li et al. 2010) but were interpreted as a result of incorporation of old crustal materials in their sources. More thorough studies of Mg, in a combination of other isotopes, are needed to test this hypothesis.

Different from Soret diffusion, chemical diffusion-driven Mg isotope fractionation has been recently found in natural samples at both mineral and outcrop scales (Teng et al. 2011; Chopra et al. 2012; Sio et al. 2013; Oeser et al. 2015; Pogge von Strandmann et al. 2015). The corresponding chemical and isotopic variations have been used to infer the thermal history of the rocks and the nature of mineral reactions.

Diffusion-driven Mg isotope fractionation at the mineral scale. Coupled Mg and Fe isotopic studies have revealed a classic example of chemical diffusion-driven isotope fractionation through studies of zoned olivines (Teng et al. 2011; Sio et al. 2013). Olivine zoning is very common in igneous rocks and can be produced by two mechanisms: crystal growth and diffusion. Mineral zoning caused by crystal growth reflects changing chemical composition during magmatic differentiation. Because equilibrium isotope fractionation during magmatic differentiation of basalt is limited (Teng et al. 2007b, 2010a), olivine grown at different times should have similar Mg isotopic composition but different MgO concentration. In this case, elemental zoning is not associated with isotopic zoning. By contrast, mineral zoning produced by diffusion results from inter-diffusion exchange of Mg and Fe, which should produce large Mg and Fe isotope fractionation (Richter et al. 2008, 2009b). In addition, since lighter isotopes of elements always diffuse faster than the heavy ones, a negative correlation between δ^{26}Mg and δ^{56}Fe is expected because the lighter Fe isotopes will diffuse in and lighter Mg isotopes will diffuse out of the olivine during Mg–Fe inter-diffusion. Therefore, Mg–Fe inter-diffusion is expected to produce coupled elemental and isotopic zonings in olivines. Though in theory, Soret diffusion could also occur and potentially produce large Mg and Fe isotope fractionation, Mg and Fe isotopes should be fractionated in the same direction as the lighter isotopes of an element always prefer hot ends, thus a positive correlation between δ^{26}Mg and δ^{56}Fe is expected for Soret diffusion-driven isotope fractionations (Fig. 35).

Teng et al. (2008, 2011) found large (up to 0.45 for δ^{26}Mg and 1.6 for δ^{56}Fe) Mg and Fe isotopic variations in olivine fragments from the Kilauea Iki lava lake, with δ^{26}Mg lineally and negatively correlated with δ^{56}Fe. Furthermore, Mg and Fe isotopic variations are coupled with Fo contents of olivine fragments. These observations, together with the decrease of forsterite (Fo) contents in olivine during cooling of the lava lake (Helz 1987), suggest the large Mg and Fe isotope fractionation was caused by diffusive exchange of Mg and Fe between olivines and melts during cooling. In-situ analyses of Mg and Fe isotopes in the same types of zoned olivines from the Kilauea Iki lava lake by both LA-MC-ICPMS and SIMS reveal large, negatively coupled Mg and Fe elemental and isotopic variations from the rim to the core and directly confirm that the fractionations of Mg and Fe isotopes were caused by chemical diffusion (Sio et al. 2013). The elemental zoning was used to model the cooling rates of the Kilauea Iki lava lake and the results agree well with the documented cooling timescale, further confirming that the large elemental and isotopic zoning was produced by Mg–Fe inter-diffusion (Teng et al. 2011; Sio et al. 2013). Therefore, coupled Mg and Fe isotopic studies of zoned minerals can be used to unambiguously tell the role of Mg–Fe inter-diffusion vs. crystal growth as both can create zoned crystals but only diffusion-controlled zoning is accompanied with large and negatively correlated Mg and Fe isotopic compositions (Fig. 36).

The large kinetic Mg and Fe isotope fractionation is also documented in olivine xeno- and phenocrysts from intra-plate volcanic regions (Oeser et al. 2015). Oeser et al. (2015) applied *in situ* Mg and Fe isotopic analyses by using femtosecond laser ablation (fs-LA) MC-ICPMS on zoned olivine samples from Massif Central (France), Vogelsberg (Germany) and Tenerife (Canary Islands). They found large (up to 0.7 for δ^{26}Mg and 1.7 for δ^{56}Fe) Mg and Fe isotope fractionations, coupled with Mg and Fe elemental zoning. Both chemical and isotopic zonings can be modeled by inter-diffusion of Mg and Fe between olivine and melts. The large elemental zoning in olivines is then used to model the cooling rates of the host magma.

Diffusion-driven inter-mineral disequilibrium Mg isotope fractionation. Large disequilibrium inter-mineral Mg isotope fractionation induced by diffusion was also observed in mantle rocks, reflecting kinetic processes during mantle metasomatism or subsolidus Mg–Fe exchange between coexisting minerals (Fig. 37) (Xiao et al. 2013, 2016; Hu et al. 2016b).

Xiao et al. (2013) systematically investigated Mg isotope fractionation between Ol, Opx, Cpx, and Spl in a suit of variably metasomatised lherzolites, Cpx-rich lherzolites and wehrlites from the North China craton. Although most of the coexisting minerals display equilibrium inter-mineral isotope fractionations in these peridotites, Cpx and Opx are clearly out of equilibrium with coexisting olivine in most Cpx-rich lherzolites ($\Delta^{26}Mg_{Opx\text{-}Ol}$ = +0.16 ~+0.32; $\Delta^{26}Mg_{Cpx\text{-}Ol}$ = +0.04 ~+0.34), and Spl is clearly out of equilibrium with coexisting olivine in one Cpx-rich lherzolite and two wehrlites (Xiao et al. 2013). The large disequilibrium inter-mineral isotope fractionations in Cpx-rich lherzolites and wehrlites likely reflect kinetic isotope fractionation during melt-rock interactions. This also agrees with the geochemical and petrographic studies showing various degrees of interactions between lherzolites and melts, and incomplete mineral reactions between Spl, Cpx and Opx (Xiao et al. 2013). In addition, most strongly metasomatised wehrlites, with equilibrium inter-mineral isotope fractionation between Cpx and Ol, have lower δ^{26}Mg (−0.39 ~−0.30) than the normal mantle. The existence of both equilibrium and disequilibrium inter-mineral isotope fractionations also reflects different degrees of mantle metasomatism. The isotopically light wehrlites with equilibrium

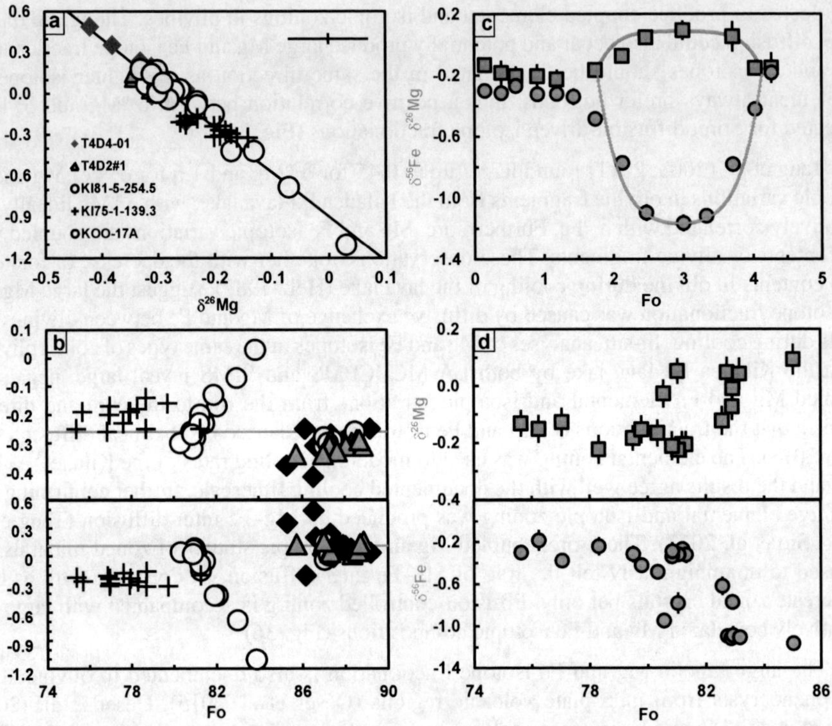

Figure 36. Magnesium and iron isotope fractionation in zoned Hawaiian olivines. Panels a) and b) display the coupled Mg and Fe isotope fractionations as well as Fo content in olivine fragments from Hawaiian lavas. The data in panel a) define a linear relationship with a slope of -3.3 ± 0.3 (Teng et al. 2011). Panels c) and d) show the coupled Mg and Fe isotope fractionations as well as Fo content measured by microdrilling a zoned olivine profile (KI81 −5−254.5, same as the one reported in panels a and b) (Sio et al. 2013).

inter-mineral isotope fractionation reflect the complete melt-peridotite reactions whereas those with disequilibrium fractionation result from incomplete melt-peridotite reactions. Therefore, the large disequilibrium inter-mineral fractionation indicates that the mantle metasomatism occurred not long before the entrainment of the xenoliths in the host magma.

Studies of strongly metasomatised garnet pyroxenite xenoliths from the North China craton further revealed large disequilibrium Mg isotope fractionation between garnet and coexisting olivine and pyroxene (Hu et al. 2016b). Although garnets, as expected from equilibrium isotope fractionation, are systematically lighter than coexisting olivine and pyroxene, the magnitudes of fractionation between garnet and olivine/pyroxene are reduced and not correlated with equilibrium temperature, suggesting disequilibrium Mg isotopic partitioning. Especially, δ^{26}Mg values of olivine and pyroxene are relatively constant and around the normal mantle value whereas those of coexisting garnets vary greatly. Therefore, the reduced degrees of inter-mineral isotope fractionation between garnet and olivine/pyroxene are mainly caused by the variably higher-than-equilibrium δ^{26}Mg in garnets. The most likely process for the disequilibrium Mg isotopic variation in garnets is the rapid and incomplete metasomatic reaction, during which garnets were formed at the expense of isotopically heavier co-existing minerals, particularly spinels. This interpretation is supported by the petrographic observations of incomplete replacement of spinel by garnet and the highly variable grain sizes of garnet. As spinel has much higher δ^{26}Mg than coexisting garnet, pyroxene, and olivine, garnets formed

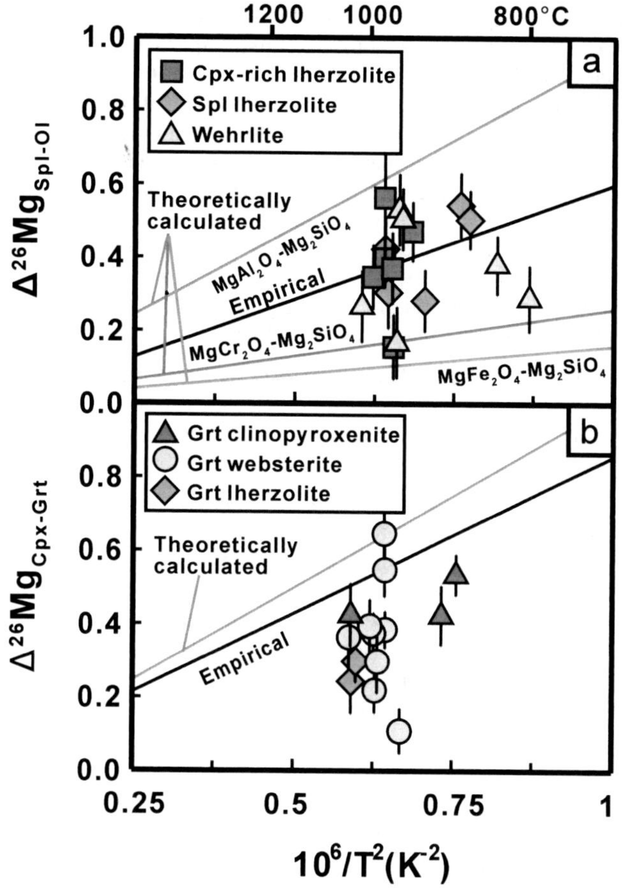

Figure 37. a) Disequilibrium inter-mineral Mg isotope fractionations between spinel and olivine in Cpx-rich lherzolites and wehrlites (Xiao et al. 2013). The black line represents equilibrium fractionation line calibrated by natural samples (Liu et al. 2011). The three colored lines represent theoretically calculated equilibrium fractionation lines between forsterite (Mg_2SiO_4) and three endmembers of the spinel group minerals, including spinel ($MgAl_2O_4$), magnesiochromite ($MgCr_2O_4$) and magnesioferrite ($MgFe_2O_4$) (Schauble 2011). (b) Disequilibrium inter-mineral Mg isotope fractionation between clinopyroxene and garnet in garnet-bearing lherzolites and pyroxenites (Hu et al. 2016b). The lower line represents equilibrium fractionation line calibrated by natural samples (Li et al. 2016b). The upper line represents theoretically calculated equilibrium fractionation line between diopside ($CaMgSi_2O_6$) and pyrope ($Mg_3Al_2Si_3O_{12}$) (Huang et al. 2013a).

by the replacement of spinels could partially inherit their heavy Mg isotopic compositions, resulting in the disequilibrium between garnet and coexisting olivine and pyroxene. On the other hand, olivine and pyroxene have much higher modal abundance and Mg content than garnet, therefore, the kinetic effects associated with garnet formation are much less evident in these two phases. Collectively, these observations suggest that mantle metasomatism plays an important role in producing the inter-mineral Mg isotopic disequilibrium and local Mg isotopic variation in the lithospheric mantle (Hu et al. 2016b).

Studies of chromite, olivine and orthopyroxene in harzburgites and chromitites from the mantle section of the Luobusa ophiolite, Southern Tibet, also reveal large disequilibrium Mg and Fe isotope fractionations (Xiao et al. 2016). In contrast to olivine and orthopyroxene in the harzburgites, which have typical mantle Mg and Fe isotopic compositions, chromite and olivine in the chromitites display large Mg and Fe isotopic variations, with δ^{26}Mg ranging from −0.41 to 0.14 in chromite and −0.29 to −0.03 in olivine, and δ^{56}Fe ranging from −0.25 to 0.04 in chromite and −0.15 to 0.22 in olivine. The large Mg and Fe isotopic variations between chromite and olivine reflect disequilibrium inter-mineral fractionations because they fall off the equilibrium fractionation lines defined by theoretical studies and analyses of natural samples (Xiao et al. 2016). The most likely process for the disequilibrium fractionations is Mg–Fe exchange between chromite and olivine at subsolidus temperature. This process fractionates Mg and Fe isotopes the largest in disseminated chromitites where chromite has lower abundance than olivine and the smallest in the massive chromitites where chromite dominates. Therefore, the subsolidus Mg–Fe exchange between chromite and olivine plays an important role in controlling chemical and isotopic compositions of chromite and olivine in chromitites from the Luobusa ophiolite (Xiao et al. 2016).

Diffusion-driven Mg isotope fractionation at large scales. Diffusion-driven kinetic isotope fractionation not only occurs at mineral (e.g., mm) scale but also was reported recently at hand specimen (e.g., cm) and outcrop (e.g., m) scales, suggesting that chemical diffusion can transport elements, e.g., Mg at large scales (Chopra et al. 2012; Pogge von Strandmann et al. 2015).

Chopra et al. (2012) report ~3‰ Mg isotope fractionation along two 1 cm traverses from felsic to mafic rocks at Vinal Cove in the Vinalhaven Intrusive Complex, Maine. The Mg concentrations and isotopic variations along both profiles can be well modeled by chemical diffusion by using parameters derived from laboratory diffusion couples. The diffusion profiles were then used to calculate the cooling rate of the sample, which is ~ 1.5 ºC per day. This fast cooling rate indicates that chemical diffusion occurred when the system was still in a molten state. Therefore, the kinetic Mg isotope fractionation provides important constraints on the transport mechanisms and time scale in the system where a different type of melts interacted.

Pogge von Strandmann et al. (2015) found ~1‰ Mg isotope fractionation along a 1-m profile across an exhumed contact between subducted crustal rocks and serpentinite in the Syros mélange zone, Greece. The large isotopic variation along the contact was interpreted as a result of kinetic isotope fractionation driven by diffusion of Mg through fluid-infiltrated grain boundary. This study thus suggests that Mg can diffuse to a large scale (e.g., meters) under the presence of grain boundary fluids, in a way similar to the grain boundary diffusion of Li (Teng et al. 2006). Magnesium isotopes may thus help to trace the fluid–rock interactions during metamorphism.

APPLICATIONS AND FUTURE DIRECTIONS

The research summarized here demonstrates significant and systematic Mg isotope fractionation in geological and biological materials at both low and high temperatures. The distribution of Mg isotopes in major reservoirs and the mechanism of Mg isotope fractionation during major geological processes are defined well enough to enable potential applications in several areas. For example, the large equilibrium inter-mineral Mg isotope fractionation associated with garnet and spinel in igneous and metamorphic rocks could be developed as a high-precision geothermometry. The large kinetic isotope fractionation induced by diffusion can be used to tell the nature of mineral zoning, i.e., crystal growth or diffusion, which lays the foundation for using mineral zoning as a geospeedometry. The large Mg isotope fractionation during low-temperature water–rock interactions can be used to constrain continental weathering. Furthermore, because of the limited Mg isotope fractionation during silicate magmatic differentiation and metamorphic dehydration, the large Mg isotopic variation produced during low-temperature processes (especially in carbonates) can survive during subduction and form a distinct mantle end member, and is later sampled in mantle/mantle-derived rocks. Thus Mg isotopes can be used as a tracer of mantle metasomatism, subducted carbonates and origins of cratonic eclogites and continental basalts. The large Mg isotope fractionation in carbonates also has the potential to provide additional constraints on the evolution of seawater chemistry, paleoclimate environments, the origin of dolostone and global Mg cycles. Finally, biological processes clearly fractionate Mg isotopes; hence Mg isotopes may be used to trace biological processes in the geologic record.

Nonetheless, Mg isotope geochemistry is still in its childhood and there are many areas should benefit from future research. Below is a list of the most significant outstanding problems in the field of Mg isotope geochemistry:

1. Origins of mantle Mg isotopic heterogeneity still need further research. In particular, if the large Mg isotopic variation in continental basalts was caused by partial melting of peridotites that were metasomatised by subducted carbonates, then carbonated peridotites, $\delta^{26}Mg$ of which are still rare, should have heterogeneous and light Mg isotopic compositions.

2. Some end-members of OIBs, such as HIMU and EM II, which are known to incorporate altered oceanic crust and subducted sediments in their sources, may have different Mg isotopic composition than MORBs and other OIBs. Their $\delta^{26}Mg$ values will depend on the nature of subducted sediments with silicate-rich ones isotopically heavier and carbonate-rich ones lighter than MORBs.

3. Some island arc lavas should have different Mg isotopic compositions than MORBs depending on their origins. For example, adakites are expected to have high $\delta^{26}Mg$ values if the residual garnet plays a significant role in controlling Mg budget during non-modal partial melting because garnet has lighter Mg isotopic composition than melts and pyroxene. Nonetheless, most island arc lavas should have similar Mg isotopic composition to MORBs unless they incorporate a large amount of Mg from the subducted slab such as those recently studied from the Lesser Antilles Arc (Teng et al. 2016).

4. The Mg isotopic composition of high-Ti continental basalts should be characterized and compared to high-Ti lunar basalts. The results may help better understand the magmatic differentiation in both terrestrial and lunar basalts

5. The average Mg isotopic composition of the continental crust should change secularly as a result of changes in both crustal composition and intensity of continental weathering. Nonetheless, samples that can represent the average Mg isotopic composition of the crust need to be characterized first before this hypothesis can be tested.

6. Although preliminary studies show that seafloor alteration of oceanic basalts leads to large Mg isotope fractionation, the exact direction and magnitude of Mg isotope fractionation are still not well understood and need further studies.

7. Analysis of Mg isotopic composition of well-characterized marine carbonates may help to define the chemical evolution of the oceans and indirectly constrain the river contributions from continental weathering, which is ultimately linked to climate change. Nonetheless, weathering and diagenesis effects on Mg isotope geochemistry need to be quantified first.

8. Biogenic Mg isotope fractionation is well documented but the detailed mechanism and magnitude of the isotope fractionation during different biologic systems need to be investigated.

9. To use equilibrium Mg isotope fractionation as a geothermometry, studies of inter-mineral equilibrium Mg isotope fractionation factors in crust and mantle rocks through laboratory experiments, theoretical calculation and analysis of well-characterized natural samples are needed, especially for systems involving spinel and garnet.

10. To apply kinetic Mg isotope fractionation to constrain magmatic and metamorphic processes, diffusivity of Mg and diffusive fractionation factor of Mg isotopes in major Mg-rich rock-forming minerals, calibrated by laboratory experiments, theoretical calculation and analysis of natural samples are needed, especially for minerals that tend to form and preserve chemical zoning such as olivine and garnet.

11. Methods for high-precision *in situ* Mg isotopic analysis, by SIMS or LA-MC-ICPMS need to be set up to rapidly check isotopic zoning in chemically zoned minerals, though micro mill can partially fulfill this purpose for large zoned minerals, before using zoned minerals to understand their crystallization and cooling histories.

12. Though routine, high precision Mg isotopic measurements are now possible by a number of techniques at a large number of laboratories, the analytical artefacts derived from sample preparation and instrumental analysis could significantly compromise the accuracy of the data. Thus assurance of the accuracy for high-precision data by analysis of well-characterized standards or laboratory doping experiments is required.

ACKNOWLEDGEMENTS

Many thanks to Yan Hu, Shui-Jiong Wang, Zhe Yang, Casey Saenger, Wang-Ye Li, Kang-Jun Huang, Shan Ke, Sheng-Ao Liu, Yan Xiao, Xiao-Ming Liu, and Wei Yang for sharing figures and tables, Wang-Ye Li, Shui-Jiong Wang, Yan Hu, Yan Xiao, Ben-Xun Su, Yongsheng He, Lie-Meng Chen, Fata Sedaghatpour, Kang-Jun Huang, and Hui Huang for comments on an early version of the manuscript, Fred Moynier, Philip Pogge von Strandmann and an anonymous reviewer for constructive reviews, and Jim Watkins for handling. This study was supported by National Science Foundation (EAR –0838227 and EAR –1340160).

REFERENCES

An Y-J, Wu F, Xiang Y, Nan X-Y, Y X, Yang J-H, Yu H-M, Xie L-W and Huang F (2014) High-precision Mg isotope analyses of low-Mg rocks by MC-ICP-MS Chem Geol 390:9–21

Azmy K, Lavoie D, Wang Z, Brand U, Al-Aasm I, Jackson S, Girard I (2013) Magnesium-isotope and REE compositions of Lower Ordovician carbonates from eastern Laurentia: Implications for the origin of dolomites and limestones. Chem Geol 356:64–75

Baker JA, Bizzarro M, Wittig N, Connelly J, Haack H (2005) Early planetesimal melting from an age of 4.5662 Gyr for differentiated meteorites. Nature 436:1127–1131

Barth MG, Rudnick RL, Horn I, McDonough WF, Spicuzza MJ, Valley JW, Haggerty SE (2001) Geochemistry of xenolithic eclogites from West Africa, Part I: a link between low MgO eclogites and Archean crust formation. Geochim Cosmochim Acta 65:1499–1527

Barth MG, Rudnick RL, Horn I, McDonough WF, Spicuzza MJ, Valley JW, Haggerty SE (2002) Geochemistry of xenolithic eclogites from West Africa, Part II: origins of the high MgO eclogites. Geochim Cosmochim Acta 66:4325–4346

Beinlich A, Mavromatis V, Austrheim H, Oelkers EH (2014) Inter-mineral Mg isotope fractionation during hydrothermal ultramafic rock alteration—Implications for the global Mg-cycle. Earth Planet Sci Lett 392:166–176

Berglund M, Wieser ME (2011) Isotopic compositions of the elements 2009 (IUPAC Technical Report). Pure Appl Chem 83:397–410

Bizzarro M, Baker JA, Haack H, Lundgaard KL (2005) Rapid timescales for accretion and melting of differentiated planetesimals inferred from ^{26}Al–^{26}Mg chronometry. Astrophys J 632:L41–L44

Bizzarro M, Paton C, Larsen K, Schiller M, Trinquier A, Ulfbeck D (2011) High-precision Mg-isotope measurments of terrestrial and extraterrestrial material by HR-MC-ICPMS-implications for the relative and absolute Mg-isotope composition of the bulk silicate Earth. J Anal At Spectrom 26:565–577

Black JR, Yin QZ, Casey WH (2006) An experimental study of magnesium-isotope fractionation in chlorophyll-a photosynthesis. Geochim Cosmochim Acta 70:4072–4079

Black JR, Yin QZ, Rustad JR, Casey WH (2007) Magnesium isotopic equilibrium in chlorophylls. J Am Chem Soc 129:8690–8691

Black JR, Epstein E, Rains WD, Yin Q-Z, Casey WH (2008) Magnesium-isotope fractionation during plant growth. Environ Sci Technol 42:7831–7836

Blattler CL, Miller NR, Higgins JA (2015) Mg and Ca isotope signatures of authigenic dolomite in siliceous deep-sea sediments. Earth Planet Sci Lett 419:32–42

Bolou-Bi EB, Vigier N, Brenot A, Poszwa A (2009) Magnesium isotope compositions of natural reference materials. Geostand Geoanal Res 33:95–109

Bolou-Bi EB, Poszwa A, Leyval C, Vigier N (2010) Experimental determination of magnesium isotope fractionation during higher plant growth. Geochim Cosmochim Acta 74:2523–2537

Bolou-Bi EB, Vigier N, Poszwa A, Boudot J-P, Dambrine E (2012) Effects of biogeochemical processes on magnesium isotope variations in a forested catchment in the Vosges Mountains (France). Geochim Cosmochim Acta 87:341–355

Bourdon B, Tipper ET, Fitoussi C, Stracke A (2010) Chondritic Mg isotope composition of the Earth. Geochim Cosmochim Acta 74:5069–5083

Bouvier A, Wadhwa M, Simon SB, Grossman L (2013) Magnesium isotopic fractionation in chondrules from the Murchison and Murray CM2 carbonaceous chondrites. Meteorit Planet Sci 48:339–353

Brenot A, Cloquet C, Vigier N, Carignan J, France-Lanord C (2008) Magnesium isotope systematics of the lithologically varied Moselle river basin, France. Geochim Cosmochim Acta 72:5070–5089

Brown J, Colling A, Park D, Phillips J, Rothery D, Wright J (1989) Seawater: its composition, properties and behaviour (First edition). Pergamon, Oxford

Buhl D, Immenhauser A, Smeulders G, Kabiri L, Richter DK (2007) Time series δ^{26}Mg analysis in speleothem calcite: Kinetic versus equilibrium fractionation, comparison with other proxies and implications for palaeoclimate research. Chem Geol 244:715–729

Catanzar EJ, Murphy TJ (1966) Magnesium isotopic ratios in natural samples. J Geophys Res 71:1271–1274

Catanzar EJ, Murphy TJ, Garner EL, Shields WR (1966) Absolute isotopic abundance ratios and atomic weight of magnesium. J Res Nat Bureau Stand Sect A 70:453–458

Chakrabarti R, Jacobsen SB (2010) The isotopic composition of magnesium in the Solar System. Earth Planet Sci Lett 293: 349–358

Chan LH, Leeman WP, You CF (2002) Lithium isotopic composition of Central American volcanic arc lavas: implications for modification of subarc mantle by slab- derived fluids: correction. Chem Geol 182:293–300

Chang VT-C, Makishima A, Belshaw NS, O'Nions RK (2003) Purification of Mg from low-Mg biogenic carbonates for isotope ratio determination using multiple collector ICP-MS. J Anal At Spectrom 18:296–301

Chang VT-C, Williams RJP, Makishima A, Belshawl NS, O'Nions K (2004) Mg and Ca isotope fractionation during $CaCO_3$ biomineralisation. Biochem Biophys Res Commun 323:79–85

Choi MS, Ryu J-S, Lee S-W, Shin HS, Lee K-S (2012) A revisited method for Mg purification and isotope analysis using cool-plasma MC-ICP-MS J Anal At Spectrom 27:1955–1959

Chopra R, Richter FM, Watson EB, Scullard CR (2012) Magnesium isotope fractionation by chemical diffusion in natural settings and in laboratory analogues. Geochim Cosmochim Acta 88:1–18

CIAAW (2015) Atomic Weights of the Elements 2015, http://ciaaw.org/atomic-weights.htm

Clayton RN, Hinton RW, Davis AM (1988) Isotopic variations in the rock-forming elements in meteorites. Philos Trans R Soc London, Ser A 325:483–501

Daughtry AC, Perry D, Williams M (1962) Magnesium isotopic distribution in dolomite. Geochim CosmochimActa 26:857–866

Davis AM, Hashimoto A, Clayton RN, Mayeda TK (1990) Isotope mass fractionation during evaporation of Mg_2SiO_4. Nature 347:655–658

de Villiers S, Dickson JAD, Ellam RM (2005) The composition of the continental river weathering flux deduced from seawater Mg isotopes. Chem Geol 216:133–142

Deer WA, Howie RA, Zussman J (1992) An Introduction to the Rock-Forming Minerals. Pearson Prentice Hall, London 695 pp

Dessert C, Lajeunesse E, Lloret E, Clergue C, Crispi O, Gorge C, Quidelleur X (2015) Controls on chemical weathering on a mountainous volcanic tropical island: Guadeloupe (French West Indies). Geochim Cosmochim Acta 171:216–237

Dilek Y, Furnes H (2011) Ophiolite genesis and global tectonics: Geochemical and tectonic fingerprinting of ancient oceanic lithosphere. GSA Bulletin 123:387–411

Elliott E (2003) Tracers of the slab. *In:*Eiler J (Ed.) Inside the Subduction Factory. Geophysical Monograph Series. American Geophysical Union, Washington, DC, pp. 23–45

Fantle MS, Higgins J (2014) The effects of diagenesis and dolomitization on Ca and Mg isotopes in marine platform carbonates: Implications for the geochemical cycles of Ca and Mg. Geochim Cosmochim Acta 142:458–481

Farmer GL (2014) Continental basaltic rocks. *In:*Rudnick RL (Ed.), The Crust. Treatise on Geochemistry. Elsevier, Oxford, pp. 75–110

Foster GL, Pogge von Strandmann PAE, Rae JWB (2010) Boron and magnesium isotopic composition of seawater. Geochem Geophys Geosyst 11

Galy A, Young ED, Ash RD, O'Nions RK (2000) The formation of chondrules at high gas pressures in the solar nebula. Science 290:1751–1753

Galy A, Belshaw NS, Halicz L, O'Nions RK (2001) High-precision measurement of magnesium isotopes by multiple-collector inductively coupled plasma mass spectrometry. Inter J Mass Spectrom 208:89–98

Galy A, Bar-Matthews M, Halicz L, O'Nions RK (2002) Mg isotopic composition of carbonate: insight from speleothem formation. Earth Planet Sci Lett 201:105–115

Galy A, Yoffe O, Janney PE, Williams RW, Cloquet C, Alard O, Halicz L, Wadhwa M, Hutcheon ID, Ramon E, Carignan J (2003) Magnesium isotope heterogeneity of the isotopic standard SRM980 and new reference materials for magnesium-isotope-ratio measurements. J Anal At Spectrom 18:1352–1356

Garrison T (2006) Oceanography: an Invitation to Marine Science (6th edition). Brook and Cole Publishers 608 pp

Geske A, Zorlu J, Richter DK, Buhl D, Niedermayr A, Immenhauser A (2012) Impact of diagenesis and low grade metamorphosis on isotope (δ^{26}Mg, δ^{13}C, δ^{18}O and ^{87}Sr/^{86}Sr) and elemental (Ca, Mg, Mn, Fe and Sr) signatures of Triassic sabkha dolomites. Chem Geol 332:45–64

Geske A, Goldstein RH, Mavromatis V, Richter DK, Buhl D, Kluge T, John CM, Immenhauser A (2015a) The magnesium isotope (δ^{26}Mg) signature of dolomites. Geochim Cosmochim Acta 149:131–151

Geske A, Lokier S, Dietzel M, Richter DK, Buhl D, Immenhauser A (2015b) Magnesium isotope composition of sabkha porewater and related (Sub-) recent stoichiometric dolomites, Abu Dhabi (UAE). Chem Geol 393–394:112–124

Goldich SS (1938) A study in rock weathering. J Geol 46:17–58

Goswami JN, Srinivasan G, Ulyanov AA (1994) Ion microprobe studies of Efremovka CAIs: Magnesium isotope composition. Geochim Cosmochim Acta 58:431–447

Gray CM, Compston W (1974) Excess ^{26}Mg in the Allende meteorite. Nature 251:495–497

Handler MR, Baker JA, Schiller M, Bennett VC, Yaxley GM (2009) Magnesium stable isotope composition of Earth's upper mantle. Earth Planet Sci Lett 282:306–313

Helz RT (1987) Diverse olivine types in lava of the 1959 eruption of Kilauea volcano and their bearing on eruption dynamics. USGS Prof. Paper 1350:691–722

Higgins JA, Schrag DP (2010) Constraining magnesium cycling in marine sediments using magnesium isotopes. Geochim Cosmochim Acta 74:5039–5053

Higgins JA, Schrag DP (2012) Records of Neogene seawater chemistry and diagenesis in deep-sea carbonate sediments and pore fluids. Earth Planet Sci Lett 357–358:386–396

Higgins JA, Schrag DP (2015) The Mg isotopic composition of Cenozoic seawater—evidence for a link between Mg-clays, seawater Mg/Ca, and climate. Earth Planet Sci Lett 416:73–81

Hippler D, Buhl D, Witbaard R, Richter DK, Immenhauser A (2009) Towards a better understanding of magnesium-isotope ratios from marine skeletal carbonates. Geochim Cosmochim Acta 73:6134–6146

Hofmann AW (2003) Sampling mantle heterogeneity through oceanic basalts: Isotopes and trace elements. *In:*Carlson RW (Ed.), The Mantle and Core. Treatise on Geochemistry. Elsevier-Pergamon, Oxford, pp. 61–101

Holland HD (1978) The Chemistry of the Atmosphere and Oceans. John Wiley & Sons

Hu Y, Teng F-Z, Plank T, Huang K-J (2015) Magnesium isotopic composition of the subducting marine sediments. American Geophysical Union Fall Meeting, San Francisco

Hu Y, Harrinton M D, Sun Y, Yang Z, Konter J, Teng F-Z (2016a) Magnesium isotopic homogeneity of San Carlos olivine: a potential standard for Mg isotope analysis by multi-collector inductively coupled plasma mass spectrometry. Rapid Commun Mass Spectrom 30 2123–2132

Hu Y, Teng F-Z, Zhang H-F, Xiao Y, Su B-X M (2016b) Metasomatism-induced mantle magnesium isotopic heterogeneity: Evidence from pyroxenites. Geochim Cosmochim Acta 185:88–111

Huang K-J (2013) The behavior of magnesium isotopes during low-temperature water–rock interactions. PhD dissertation. China University of Geosciences, Wuhan, China.

Huang F, Glessner JJ, Ianno A, Lundstrom CC, Zhang Z (2009a) Magnesium isotopic composition of igneous rock standards measured by MC-ICP-MS Chem Geol 268:15–23

Huang F, Lundstrom CC, Glessner J, Ianno A, Boudreau A, Li J, Ferre EC, Marshak S, DeFrates J (2009b) Chemical and isotopic fractionation of wet andesite in a temperature gradient: Experiments and models suggesting a new mechanism of magma differentiation. Geochim Cosmochim Acta 73:729–749

Huang F, Chakraborty P, Lundstrom CC, Holmden C, Glessner JJG, Kieffer SW, Lesher CE (2010) Isotope fractionation in silicate melts by thermal diffusion. Nature 464:396–400

Huang F, Zhang Z, Lundstrom CC, Zhi X (2011) Iron and magnesium isotopic compositions of peridotite xenoliths from Eastern China. Geochim Cosmochim Acta 75:3318–3334

Huang K-J, Teng F-Z, Wei G-J, Ma J-L, Bao Z-Y (2012) Adsorption- and desorption-controlled magnesium isotope fractionation during extreme weathering of basalt in Hainan Island, China. Earth Planet Sci Lett 359–360:73–83

Huang F, Chen L, Wu Z, Wang W (2013a) First-principles calculations of equilibrium Mg isotope fractionations between garnet, clinopyroxene, orthopyroxene, and olivine: Implications for Mg isotope thermometry. Earth Planet Sci Lett 367:61–70

Huang K-J, Teng F-Z, Elsenouy A, Li W-Y, Bao Z-Y (2013b) Magnesium isotopic variations in loess: Origins and implications. Earth Planet Sci Lett 374:60–70

Huang J, Li S-G, Xiao Y, Ke S, Li W-Y, Tian Y (2015a) Origin of low δ^{26}Mg cenozoic basalts from South China Block and their geodynamic implications. Geochim Cosmochim Acta 164:298–317

Huang K-J, Shen B, Lang X-G, Tang W-B, Peng Y, Ke S, Kaufman AJ, Ma H-R, Li F-B (2015b) Magnesium isotopic compositions of the Mesoproterozoic dolostones: Implications for Mg isotopic systematics of marine carbonates. Geochim Cosmochim Acta 164:333–351

Husson JM, Higgins JA, Maloof AC, Schoene B (2015) Ca and Mg isotope constraints on the origin of Earth's deepest δ^{13}C excursion. Geochim Cosmochim Acta 160:243–266

Hutchison R (2004) Meteorites: a petrologic, chemical and isotopic synthesis. Cambridge Planetary Science 2. Cambridge University Press, Cambridge 506 pp

Immenhauser A, Buhl D, Richter D, Niedermayr A, Riechelmann D, Dietzel M, Schulte U (2010) Magnesium-isotope fractionation during low-Mg calcite precipitation in a limestone cave—Field study and experiments. Geochim Cosmochim Acta 74:4346–4364

Jacobsen B, Yin QZ, Moynier F, Amelin Y, Krot AN, Nagashima K, Hutcheon ID, Palme H (2008) ^{26}Al–^{26}Mg and ^{207}Pb–^{206}Pb systematics of Allende CAIs: Canonical solar initial ^{26}Al/^{27}Al ratio reinstated. Earth Planet Sci Lett 272:353–364

Jacobson AD, Zhang Z, Lundstrom C, Huang F (2010) Behavior of Mg isotopes during dedolomitization in the Madison Aquifer, South Dakota. Earth Planet Sci Lett 297: 446–452

Janney PE, Richter FM, Mendybaev RA, Wadhwa M, Georg RB, Watson EB, Hines RR (2011) Matrix effects in the analysis of Mg and Si isotope ratios in natural and synthetic glasses by laser ablation-multicollector ICPMS: A comparison of single- and double-focusing mass spectrometers. Chem Geol 281:26–40

Kasemann SA, Pogge von Strandmann PAE, Prave AR, Fallick AE, Elliott T, Hoffmann K-H (2014) Continental weathering following a Cryogenian glaciation: Evidence from calcium and magnesium isotopes. Earth Planet Sci Lett 396:66–77

Ke S, Teng F-Z, Li S, Gao T, Liu S-A, He Y, Mo XX (2016) Mg, Sr and O isotope geochemistry of syenites from northwest Xinjiang, China: Tracing carbonate recycling during Tethyan oceanic subduction. Chem Geol 437:109–119

Kelley KA, Plank T, Ludden J, Staudigel H (2003) Composition of altered oceanic crust at ODP Sites 801 and 1149. Geochem Geophys Geosyst 4, doi:10.1029/2002GC000435

Kelley KA, Plank T, Farr L, Ludden J, Staudigel H (2005) Subduction cycling of U, Th, and Pb. Earth Planet Sci Lett 234:369–383

Lai Y-J, Pogge von Strandmann PAE, Dohmen R, Takazawa E, Elliott E (2015) The influence of melt infiltration on the Li and Mg isotopic composition of the Horoman Peridotite Massif. Geochim Cosmochim Acta 164:318–332

Lavoie D, Jackson S, Girard I (2014) Magnesium isotopes in high-temperature saddle dolomite cements in the lower Paleozoic of Canada. Sediment Geol 305:58–68

Lee T, Papanastassiou DA (1974) Mg isotopic anomalies in the Allende meteorite and correlation with O and Sr effects. Geophys Res Lett 1:225–228

Lee C-TA, Luffi P, Chin EJ (2011) Building and destroying continental mantle. Ann Rev Earth Planet Sci 39:59–90

Lee S-W, Ryu J-S, Lee K-S (2014) Magnesium isotope geochemistry in the Han River, South Korea. Chem Geol 364:9–19

Lee T, Papanastassiou DA, Wasserburg GJ (1976) Demonstration of ^{26}Mg excess in Allende and evidence for ^{26}Al. Geophys Res Lett 3:109–112

Lee T, Papanastassiou DA, Wasserburg GJ (1977) Aluminum–26 in early solar system—fossil or fuel. Astrophys J 211: L107–L110

Li YH (1982) A brief discussion on the mean oceanic residence time of elements. Geochim Cosmochim Acta 46:2671–2675

Li W-Y, Teng F-Z, Ke S, Rudnick RL, Gao S, Wu F-Y, Chappell BW (2010) Heterogeneous magnesium isotopic composition of the upper continental crust. Geochim Cosmochim Acta 74:6867–6884

Li W-Y, Teng F-Z, Xiao Y, Huang J (2011a) High-temperature inter-mineral magnesium isotope fractionation in eclogite from the Dabie orogen, China. Earth Planet Sci Lett 304: 224–230

Li W, Beard BL, Johnson CM (2011b) Exchange and fractionation of Mg isotopes between epsomite and saturated $MgSO_4$ solution. Geochim Cosmochim Acta 75:1814–1828

Li W, Chakraborty S, Beard BL, Romanek CS, Johnson CM (2012) Magnesium isotope fractionation during precipitation of inorganic calcite under laboratory conditions. Earth Planet Sci Lett 333:304–316

Li W-Y, Teng F-Z, Wing BA, Xiao Y (2014) Limited magnesium isotope fractionation during metamorphic dehydration in metapelites from the Onawa contact aureole, Maine. Geochemistry, Geophysics, Geosysts 15:408–415

Li W, Beard BL, Li C, Xu H, Johnson CM (2015) Experimental calibration of Mg isotope fractionation between dolomite and aqueous solution and its geological implications. Geochim Cosmochim Acta 157:164–181

Li W-Y, Teng F-Z, Halama R, Keller J, Klaudius J (2016a) Magnesium isotope fractionation during carbonatite magmatism at Oldoinyo Lengai, Tanzania. Earth Planet Sci Lett 444:26–33

Li W-Y, Teng F-Z, Xiao Y, Gu H-O, Zha X-P, Huang J (2016b) Empirical calibration of the clinopyroxene-garnet magnesium isotope geothermometer and implications. Contrib Mineral Petrol: 171:1–14

Lide DR (1993–1994) CRC Handbook of Chemistry and Physics 74th Edition. CRC Press, Boca Raton

Ling M-X, Sedaghatpour F, Teng F-Z, Hays PD, Strauss J, Sun W (2011) Homogenous magnesium isotopic composition of seawater: an excellent geostandard for Mg isotope analysis. Rapid Commun Mass Spectrom 25:2828–2836

Ling M-X, Li Y, Ding X, Teng F-Z, Yang X-Y, Fan W-M, Xu Y-G, Sun W (2013) Destruction of the North China craton induced by ridge subudctions. J Geol 121:197–213

Liu S-A, Teng F-Z, He Y, Ke S, Li S (2010) Investigation of magnesium isotope fractionation during granite differentiation: Implication for Mg isotopic composition of the continental crust. Earth Planet Sci Lett 297: 646–654

Liu S-A, Teng F-Z, Yang W, Wu FY (2011) High-temperature inter-mineral magnesium isotope fractionation in mantle xenoliths from the North China craton. Earth Planet Sci Lett 308:131–140

Liu C, Wang Z, Raub TD, Macdonald FA, Evans DAD (2014a) Neoproterozoic cap-dolostone deposition in stratified glacial meltwater plume. Earth Planet Sci Lett 404:22–32

Liu X-M, Teng F-Z, Rudnick RL, McDonough WF, Cummings M (2014b) Massive magnesium depletion and isotopic fractionation in weathered basalts. Geochim Cosmochim Acta 135:336–349

Liu P-P, Teng F-Z, Zhou M-F, Dick HJB, Chung S-L (2016) Magnesium isotopic composition of abyssal peridotite. Goldschmidt Conference Abstracts, Yokohama

Lundstrom C (2009) Hypothesis for the origin of convergent margin granitoids and Earth's continental crust by thermal migration zone refining. Geochim Cosmochim Acta 73:5709–5729

Livingstone DA (1963) Chemical composition of rivers and lakes. USGS Professional Paper

Ma L, Teng F-Z, Jin L, Ke S, Yang W, Gu H-O, Brantley SL (2015) Magnesium isotope fractionation during shale weathering in the Shale Hills Critical Zone Observatory: Accumulation of light Mg isotopes in soils by clay mineral transformation. Chem Geol 397:37–50

MacPherson GJ, Davis AM, Zinner EK (1995) The distribution of aluminum–26 in the early Solar System—a reappraisal. Meteoritics 30:365–386

Macris CA, Young ED, Manning CE (2013) Experimental determination of equilibrium magnesium isotope fractionation between spinel, forsterite, and magnesite from 600 to 800 °C Geochim Cosmochim Acta 118:18–32

Mattey D, Lowry D, Macpherson C (1994) Oxygen isotope composition of mantle peridotite. Earth Planet Sci Lett 128:231–241

Mavromatis V, Pearce CR, Shirokova LS, Bundeleva IA, Pokrovsky OS, Benezeth P, Oelkers EH (2012) Magnesium isotope fractionation during hydrous magnesium carbonate precipitation with and without cyanobacteria. Geochim Cosmochim Acta 76:161–174

Mavromatis V, Gautier Q, Bosc O, Schott J (2013) Kinetics of Mg partition and Mg stable isotope fractionation during its incorporation in calcite. Geochim Cosmochim Acta 114:188-(203)

Mavromatis V, Meister P, Oelkers EH (2014) Using stable Mg isotopes to distinguish dolomite formation mechanisms: A case study from the Peru Margin. Chem Geol 385:84–91

McDonough WF, Sun SS (1995) The composition of the Earth. Chem Geol 120(3–4): 223–253

Mittlefehldt DW (2003) Achondrites. *In:*Davis AM (Ed.), Meteorites, Comets, and Planets. Elsevier-Pergamon, Oxford, pp. 291–324

Muller MN, Kisakuerek B, Buhl D, Gutperlet R, Kolevica A, Riebesell U, Stoll H, Eisenhauer A (2011) Response of the coccolithophores *Emiliania huxleyi* and *Coccolithus braarudii* to changing seawater Mg^{2+} and Ca^{2+} concentrations: Mg/Ca, Sr/Ca ratios and $\delta^{44/40}$Ca, $\delta^{26/24}$Mg of coccolith calcite. Geochim Cosmochim Acta 75:2088–2102

Niu Y (2004) Bulk-rock major and trace element compositions of abyssal peridotites: Implications for mantle melting, melt extraction and post-melting processes beneath Mid-Ocean Ridges. J Petrol 45:2423–2458

Norman MD, McCulloch MT, O'Neill HS, Yaxley GM (2006a) Magnesium isotopic analysis of olivine by laser-ablation multi-collector ICP-MS: composition dependent matrix effects and a comparison of the Earth and Moon. J Anal At Spectrom 21:50–54

Norman MD, Yaxley GM, Bennett VC, Brandon AD (2006b) Magnesium isotopic composition of olivine from the Earth, Mars, Moon, and pallasite parent body. Geophys Res Lett 33: L15202, doi:10.1029/2006GL026446

Oeser M, Weyer S, Horn I, Schuth S (2014) High-precision Fe and Mg isotope ratios of silicate reference glasses determined in situ by femtosecond LA-MC-ICP-MS and by solution nebulisation MC-ICP-MS Geostand Geoanal Res 38:311–328

Oeser M, Dohmen R, Horn I, Schuth S, Weyer S (2015) Processes and time scales of magmatic evolution as revealed by Fe–Mg chemical and isotopic zoning in natural olivines. Geochim Cosmochim Acta 154:130–150

Opfergelt S, Georg RB, Delvaux B, Cabidoche YM, Burton KW, Halliday AN (2012) Mechanisms of magnesium isotope fractionation in volcanic soil weathering sequences, Guadeloupe. Earth Planet Sci Lett 341:176–185

Opfergelt S, Burton KW, Georg RB, West AJ, Guicharnaud RA, Sigfusson B, Siebert C, Gislason SR, Halliday AN (2014) Magnesium retention on the soil exchange complex controlling Mg isotope variations in soils, soil solutions and vegetation in volcanic soils, Iceland. Geochim Cosmochim Acta 125:110–130

Palme H, Jones A (2003) Solar system abundances of the elements. *In:*Davis AM (Ed.), Meteorites, Comets, and Planets. Elsevier-Pergamon, Oxford, pp. 41–61

Pearce CR, Saldi GD, Schott J, Oelkers EH (2012) Isotopic fractionation during congruent dissolution, precipitation and at equilibrium: Evidence from Mg isotopes. Geochim Cosmochim Acta 92:170–183

Pearson NJ, Griffin WL, Alard O, O'Reilly SY (2006) The isotopic composition of magnesium in mantle olivine: Records of depletion and metasomatism. Chem Geol 226:115–133

Pilson MEQ (2013) An Introduction to the Chemistry of the Sea. Cambridge University Press, Cambridge 529 pp

Planchon F, Poulain C, Langlet D, Paulet Y-M, Andre L (2013) Mg-isotopic fractionation in the manila clam (*Ruditapes philippinarum*): New insights into Mg incorporation pathway and calcification process of bivalves. Geochim Cosmochim Acta 121:374–397

Plank T (2014) The chemical composition of subducting sediments. *In:*Rudnick RL (Ed.) The Crust. Treatise on Geochemistry. Elsevier, Oxford, pp. 607–629

Plank T, Ludden JN, Escutia C, Party SS (2000) Leg 185 summary; inputs to the Izu–Mariana subduction system. Ocean Drilling Program Proceedings, Initial Reports, Leg. 185:1–63

Pogge von Strandmann PAE (2008) Precise magnesium isotope measurements in core top planktic and benthic foraminifera. Geochem. Geophys. Geosyst. 9: Q12015, doi:10.1029/2008GC002(209)

Pogge von Strandmann PAE, Burton KW, James RH, Van Calstern P, Gislason SR, Sigmarsson O (2008a) The influence of weathering processes on riverine magnesium isotopes in a basaltic terrain. Earth Planet Sci Lett 276:187–197

Pogge von Strandmann PAE, James RH, Van Calstern P, Gislason SR, Burton KW (2008b) Lithium, magnesium and uranium isotope behaviour in the estuarine environment of basaltic islands. Earth Planet Sci Lett 274:462–471

Pogge von Strandmann PAE, Elliott E, Marschall HR, Coath C, Lai Y-J, Jeffcoate AB, Ionov DA (2011) Variations of Li and Mg isotope ratios in bulk chondrites and mantle xenoliths. Geochim Cosmochim Acta 75:5247–5268

Pogge von Strandmann PAE, Opfergelt S, Lai Y-J, Sigfusson B, Gislason SR, Burton KW (2012) Lithium, magnesium and silicon isotope behaviour accompanying weathering in a basaltic soil and pore water profile in Iceland. Earth Planet Sci Lett 339:11–23

Pogge von Strandmann PAE, Forshaw J, Schmidt DN (2014) Modern and Cenozoic records of seawater magnesium from foraminiferal Mg isotopes. Biogeosciences 11:5155–5168

Pogge von Strandmann PAE, Dohmen R, Marschall HR, Schumacher JC, Elliott E (2015) Extreme magnesium isotope fractionation at outcrop scale records the mechanism and rate at which reaction fronts advance. J Petrol 56:33–58

Pokrovsky BG, Mavromatis V, Pokrovsky OS (2011) Co-variation of Mg and C isotopes in late Precambrian carbonates of the Siberian Platform: A new tool for tracing the change in weathering regime? Chem Geol 290:67–74

Ra K, Kitagawa H (2007) Magnesium isotope analysis of different chlorophyll forms in marine phytoplankton using multi-collector ICP-MS J Anal At Spectrom 22:817–821

Ra K, Kitagawa H, Shiraiwa Y (2010a) Mg isotopes and Mg/Ca values of coccoliths from cultured specimens of the species *Emiliania huxleyi* and *Gephyrocapsa oceanica*. Mar Micropaleontol 77:119–124

Ra K, Kitagawa H, Shiraiwa Y (2010b) Mg isotopes in chlorophyll-a and coccoliths of cultured coccolithophores (*Emiliania huxleyi*) by MC-ICP-MS. Mar Chem 122:130–137

Rehkamper M, Schonbachler M, Stirling CH (2001) Multiple collector ICP-MS: Introduction to instrumentation, measurement techniques and analytical capabilities. Geostand Geoanal Res 25:23–40

Richter FM, Davis AM, Ebel DS, Hashimoto A (2002) Elemental and isotopic fractionation of Type B calcium-, aluminum-rich inclusions: Experiments, theoretical considerations, and constraints on their thermal evolution. Geochim Cosmochim Acta 66:521–540

Richter FM, Watson EB, Mendybaev RA, Teng F-Z, Janney PE (2008) Magnesium isotope fractionation in silicate melts by chemical and thermal diffusion. Geochim Cosmochim Acta 72:206–220

Richter FM, Dauphas N, Teng F-Z (2009a) Non-traditional fractionation of non-traditional isotopes: Evaporation, chemical diffusion and Soret diffusion. Chem Geol 258:92–103

Richter FM, Watson EB, Mendybaev RA, Dauphas N, Georg RB, Watkins J, Valley JW (2009b) Isotopic fractionation of the major elements of molten basalt by chemical and thermal diffusion. Geochim Cosmochim Acta 73:4250–4263

Riechelmann S, Buhl D, Schroder-Ritzrau A, Riechelmann DFC, Richter DK, Vonhof HB, Wassenburg JA, Geske A, Spotl C, Immenhauser A (2012a) The magnesium isotope record of cave carbonate archives. Clim Past 8:1849–1867

Riechelmann S, Buhl D, Schroeder-Ritzrau A, Spoetl C, Riechelmann DFC, Richter DK, Kluge T, Marx T, Immenhauser A (2012b) Hydrogeochemistry and fractionation pathways of Mg isotopes in a continental weathering system: Lessons from field experiments. Chem Geol 300:109–122

Rollion-Bard C, Saulnier S, Vigier N, Schumacher A, Chaussidon M, Lecuyer C (2016) Variability in magnesium, carbon and oxygen isotope compositions of brachiopod shells: Implications for paleoceanographic studies. Chem Geol 423:49–60

Rose-Koga EF, Albarede F (2010) A data brief on magnesium isotope compositions of marine calcareous sediments and ferromanganese nodules. Geochem Geophys Geosyst 11:Q03006, doi:10.1029/2009GC002899

Rouxel O, Dobbek N, Ludden J, Fouquet Y (2003) Iron isotope fractionation during oceanic crust alteration. Chem Geol 202:155–182

Rudnick RL, McDonough WF, McCulloch MT, Taylor SR (1986) Lower crustal xenoliths from Queensland, Australia—Evidence for deep crustal assimilation and fractionation of continental basalts. Geochim Cosmochim Acta 50:1099–1115

Rudnick RL, Taylor SR (1987) The composition and petrogenesis of the lower crust: A xenolith study. J Geophys Res-Solid Earth and Planets 92(B13):13981–14005

Rudnick RL, Fountain DM (1995) Nature and composition of the continental-crust—a lower crustal perspective. Rev Geophys 33:267–309

Rudnick RL, Gao S (2003) Composition of the continental crust. *In:*Rudnick RL (Ed.) The Crust. Treatise on Geochemistry. Elsevier-Pergamon, Oxford, pp. 1–64

Russell SS, Huss GR, Fahey AJ, Greenwood RC, Hutchison R, Wasserburg GJ (1998) An isotopic and petrologic study of calcium–aluminum-rich inclusions from CO3 meteorites. Geochim Cosmochim Acta 62:689–714

Rustad JR, Casey WH, Yin Q-Z, Bylaska EJ, Felmy AR, Bogatko SA, Jackson VE, Dixon DA (2010) Isotopic fractionation of $Mg^{2+}{}_{(aq)}$, $Ca^{2+}{}_{(aq)}$, and $Fe^{2+}{}_{(aq)}$ with carbonate minerals. Geochim Cosmochim Acta 74:6301–6323

Ryan JG, Chauvel C (2014) The subduction-zone filter and the impact of recycled materials on the evolution of the mantle. *In:*Carlson RW (Ed.) The Mantle. Treatise on Geochemistry. Elsevier, Oxford, pp. 1727–1741

Ryu J-S, Jacobson AD, Holmden C, Lundstrom C, Zhang Z (2011) The major ion, $\delta^{44/40}$Ca, $\delta^{44/42}$Ca, and $\delta^{26/24}$Mg geochemistry of granite weathering at pH = 1 and T = 25 °C: power-law processes and the relative reactivity of minerals. Geochim Cosmochim Acta 75:6004–6026

Saenger C, Wang Z (2014) Magnesium isotope fractionation in biogenic and abiogenic carbonates: implications for paleoenvironmental proxies. Quat Sci Rev 90:1–21

Saulnier S, Rollion-Bard C, Vigier N, Chaussidon M (2012) Mg isotope fractionation during calcite precipitation: An experimental study. Geochim Cosmochim Acta 91:75–91

Schauble EA (2011) First-principles estimates of equilibrium magnesium isotope fractionation in silicate, oxide, carbonate and hexaaquamagnesium^{2+} crystals. Geochim Cosmochim Acta 75:844–869

Schiller M, Handler MR, Baker JA (2010) High-precision Mg isotopic systematics of bulk chondrites. Earth Planet Sci Lett 297:165–173

Schramm DN, Tera F, Wasserburg GJ (1970) The isotopic abundance of 26Mg and limits on 26Al in the early solar system. Earth Planet Sci Lett 10:44–59

Scott ERD, Krot AN (2003) Chondrites and their components. *In:*Davis AM (Ed.), Meteorites, Comets and Planets. Treatise on Geochemistry. Elsevier-Pergamon, Oxford, pp. 143-(200)

Sedaghatpour F, Teng F-Z, Liu Y, Sears DW, Taylor LA (2013) Magnesium isotopic composition of the Moon. Geochim Cosmochim Acta 120:1–16

Sedaghatpour F, Teng F-Z (2016) Magnesium isotopic composition of achondrites. Geochim Cosmochim Acta 174:167–179

Shannon RD (1976) Revised effective ionic radii and systematic studies of interatomic distances in halides and chalcogenides. Acta Crystallogr A32:751–767

Shima M (1964) The isotopic composition of magnesium in terrestrial samples. Bull Chem Soc Jpn 37:284–285

Shen B, Jacobsen B, Lee CTA, Yin QZ, Morton DM (2009) The Mg isotopic systematics of granitoids in continental arcs and implications for the role of chemical weathering in crust formation. PNAS 106: 20652-(20657)

Shen B, Wimpenny J, Lee C-TA, Tollstrup D, Yin Q-Z (2013) Magnesium isotope systematics of endoskarns: Implications for wallrock reaction in magma chambers. Chem Geol 356:209–214

Sio CK, Dauphas N, Teng F-Z, Chaussidon M, Helz RT, Roskosz M (2013) Discerning crystal growth from diffusion profiles in zoned olivine by in situ Mg–Fe isotopic analyses. Geochim Cosmochim Acta 123:302–321

Snow JE, Dick HJB (1995) Pervasive magnesium loss by marine weathering of peridotite. Geochim Cosmochim Acta 59:4219–4235

Spivak-Birndorf L, Wadhwa M, Janney PE (2009) ^{26}Al–^{26}Mg systematics in D'Orbigny and Sahara 99555 angrites: Implications for high-resolution chronology. Geochim Cosmochim Acta 73:5202–5211

Staudigel H (2014) Chemical fluxes from hydrothermal alteration of the oceanic crust. *In:*Rudnick RL (Ed.), The Crust. Treatise on Geochemistry. Elsevier, Oxford, pp. 583–606

Su B-X, Teng F-Z, Hu Y, Shi R-D, Zhou M-F, Zhu B, Liu F, Gong X-H, Huang Q-S, Xiao Y, Chen C, He Y-S (2015) Iron and magnesium isotope fractionation in oceanic lithosphere and sub-arc mantle: Perspectives from ophiolites. Earth Planet Sci Lett 430:523–532

Telus M, Dauphas N, Moynier F, Tissot FL, Teng F-Z, Nabelek PI, Craddock PR, Groat LA (2012) Iron, zinc and magnesium isotopic fractionation during continental crust differentiation: The tale from migmatites, granites and pegmatites. Geochim Cosmochim Acta 97:247–265

Teng F-Z, Yang W (2014) Comparison of factors affecting the accuracy of high-precision magnesium isotope analysis by multi-collector inductively coupled plasma mass spectrometry. Rapid Commun Mass Spectrom 28:19–24

Teng F-Z, McDonough WF, Rudnick RL, Walker RJ (2006) Diffusion-driven extreme lithium isotopic fractionation in country rocks of the Tin Mountain pegmatite. Earth Planet Sci Lett 243:701–710

Teng F-Z, McDonough WF, Rudnick RL, Wing BA (2007a) Limited lithium isotopic fractionation during progressive metamorphic dehydration in metapelites: A case study from the Onawa contact aureole, Maine. Chem Geol 239:1–12

Teng F-Z, Wadhwa M, Helz RT (2007b) Investigation of magnesium isotope fractionation during basalt differentiation: Implications for a chondritic composition of the terrestrial mantle. Earth Planet Sci Lett 261:84–92

Teng F-Z, Dauphas N, Helz RT (2008) Iron isotope fractionation during magmatic differentiation in Kilauea Iki lava lake. Science 320:1620–1622

Teng F-Z, Li W-Y, Ke S, Marty B, Dauphas N, Huang S, Wu F-Y, Pourmand A (2010a) Magnesium isotopic composition of the Earth and chondrites. Geochim Cosmochim Acta 74:4150–4166

Teng F-Z, Li W-Y, Rudnick RL, Gardner LR (2010b) Contrasting behavior of lithium and magnesium isotope fractionation during continental weathering. Earth Planet Sci Lett 300:63–71

Teng F-Z, Dauphas N, Helz RT, Gao S, Huang S (2011) Diffusion-driven magnesium and iron isotope fractionation in Hawaiian olivine. Earth Planet Sci Lett 308:317–324

Teng F-Z, Yang W, Rudnick RL, Hu Y (2013) Heterogeneous magnesium isotopic compostion of the lower crust: A xenolith perspective. Geochem. Geophys. Geosyst. 14:3844–3856, doi: 10.1002/ggge.(20238)

Teng F-Z, Li W-Y, Ke S, Yang W, Liu S-A, Sedaghatpour F, Wang S-J, Huang K-J, Hu Y, Ling M-X, Xiao Y, Liu X-M, Li X-W, Gu H-O, Sio CK, Wallace DA, Su B-X, Zhao L, Chamberlin J, Harrington M, Brewer A (2015a) Magnesium isotopic compositions of international geological reference materials. Geostand Geoanal Res 39:329–339

Teng F-Z, Yin Q-Z, Ullmann CV, Chakrabarti R, Pogge von Strandmann PAE, Yang W, Li W-Y, Ke S, Sedaghatpour F, Wimpenny J, Meixner A, Romer RL, Wiechert U, Jacobsen SB (2015b) Interlaboratory comparison of magnesium isotopic composition of 12 felsic to ultramafic igneous rock standards analyzed by MC-ICPMS Geochem Geophys Geosyst 16:3197–3209, doi:1002/2015GC005939

Teng F-Z, Hu Y Chauvel, C (2016) Magnesium isotope geochemistry in arc volcanism. PNAS 113:7082–7087, doi:10.1073/pnas.1518456113

Thrane K, Bizzarro M, Baker, JA (2006) Extremely brief formation interval for refractory inclusions and uniform distribution of ^{26}Al in the early solar system. Astrophys J 646:L159–L162

Tipper ET, Galy A, Bickle MJ (2006a) Riverine evidence for a fractionated reservoir of Ca and Mg on the continents: Implications for the oceanic Ca cycle. Earth Planet Sci Lett 247:267–279

Tipper ET, Galy A, Gaillardet J, Bickle MJ, Elderfield H, Carder EA (2006b) The magnesium isotope budget of the modern ocean: Constraints from riverine magnesium isotope ratios. Earth Planet Sci Lett 250:241–253

Tipper ET, Galy A, Bickle MJ (2008a) Calcium and magnesium isotope systematics in rivers draining the Himalaya–Tibetan-Plateau region: Lithological or fractionation control? Geochim Cosmochim Acta 72:1057–1075

Tipper ET, Louvat P, Capmas F, Galy A, Gaillardet J (2008b) Accuracy of stable Mg and Ca isotope data obtained by MC-ICP-MSusing the standard addition method. Chem Geol 257:65–75

Tipper ET, Gaillardet J, Louvat P, Capmas F, White AF (2010) Mg isotope constraints on soil pore-fluid chemistry: Evidence from Santa Cruz, California. Geochim Cosmochim Acta 74:3883–3896

Tipper ET, Calmels D, Gaillardet J, Louvat P, Capmas F, Dubacq B (2012a) Positive correlation between Li and Mg isotope ratios in the river waters of the Mackenzie Basin challenges the interpretation of apparent isotopic fractionation during weathering. Earth Planet Sci Lett 333:35–45

Tipper ET, Lemarchand E, Hindshaw RS, Reynolds BC, Bourdon B (2012b) Seasonal sensitivity of weathering processes: Hints from magnesium isotopes in a glacial stream. Chem Geol 312:80–92

Urey HC (1947) The thermodynamic properties of isotopic substances. J Chem Soc (London):562–581

Wang G, Lin Y, Liang X, Liu Y, Xie L, Yang Y, Tu X (2011) Separation of magnesium from meteorites and terrestrial silicate rocks for high-precision isotopic analysis using multiple collector-inductively coupled plasma-mass spectrometry. J Anal At Spectrom 26:1878–1886

Wang S-J, Teng F-Z, Williams HM, Li S (2012) Magnesium isotopic variations in cratonic eclogites: Origins and implications. Earth Planet Sci Lett 359–360:219–226

Wang S-J, Teng F-Z, Li S-G (2014a) Tracing carbonate-silicate interaction during subduction using magnesium and oxygen isotopes. Nat Commun 5:5328 doi: 10.1038/ncomms6328

Wang S-J, Teng F-Z, Li S-G, Hong J-A (2014b) Magnesium isotope systematics of mafic rocks during continental subduction. Geochim Cosmochim Acta 143:34–48

Wang S-J, Teng F-Z, Bea F (2015a) Magnesium isotopic systematics of metapelite in the deep crust and implications for granite petrogenesis. Geochem Perspect Lett 1:75–83

Wang S-J, Teng F-Z, Rudnick RL, Li S-G (2015b) The behavior of magnesium isotopes in low-grade metamorphosed mudrocks. Geochim Cosmochim Acta 165:435–448

Wang S-J, Teng F-Z, Rudnick RL, Li S-G (2015c) Magnesium isotope evidence for a recycled origin of cratonic eclogites. Geology 43:1071–1074

Wang S-J, Teng F-Z, Scott J (2016) Tracing the origin of continental HIMU-like intraplate volcanism using magnesium isotope systematics. Geochim Cosmochim Acta 185:78–87

Wang Z, Hu P, Gaetani G, Liu C, Saenger C, Cohen A, Hart S (2013) Experimental calibration of Mg isotope fractionation between aragonite and seawater. Geochim Cosmochim Acta 102:113–123

Watkins JM, DePaolo DJ, Huber C, Ryerson FJ (2009) Liquid composition-dependence of calcium isotope fractionation during diffusion in molten silicates. Geochim Cosmochim Acta 73:7341–7359

Watkins J, DePaolo DJ, Ryerson FJ, Peterson B (2011) Influence of liquid structure on diffusive isotope separation in molten silicates and aqueous solutions. Geochim Cosmochim Acta 75:3103–3118

Wedepohl KH (1995) The composition of the continental crust. Geochim Cosmochim Acta 59:1217–1232

White WM, Klein EM (2014) Composition of the oceanic crust. *In:*Rudnick RL (Ed.), The Crust. Treatise on Geochemistry. Elsevier, Oxford, pp. 457–496

Wiechert U, Halliday AN (2007) Non-chondritic magnesium and the origins of the inner terrestrial planets. Earth Planet Sci Lett 256:360–371

Williams HM, Nielsen SG, Renac C, Griffin WL, O'Reilly SY, McCammon CA, Pearson N, Viljoen F, Alt JC, Halliday AN (2009) Fractionation of oxygen and iron isotopes by partial melting processes: implications for the interpretation of stable isotope signatures in mafic rocks. Earth Planet Sci Lett 283:156–166

Wimpenny J, Gislason SR, James RH, Gannoun A, Pogge Von Strandmann PAE, Burton KW (2010) The behaviour of Li and Mg isotopes during primary phase dissolution and secondary mineral formation in basalt. Geochim Cosmochim Acta 74:5259–5279

Wimpenny J, Burton KW, James RH, Gannoun A, Mokadem F, Gislason SR (2011) The behaviour of magnesium and its isotopes during glacial weathering in an ancient shield terrain in West Greenland. Earth Planet Sci Lett 304: 260–269

Wimpenny J, Harvey J, Yin Q (2012) The effects of serpentinization on Mg isotopes in Mid-Atlantic ridge peridotite. American Geophysical Union Fall Meeting, San Francisco

Wimpenny J, Colla CA, Yin Q-Z, Rustad JR, Casey WH (2014a) Investigating the behaviour of Mg isotopes during the formation of clay minerals. Geochim Cosmochim Acta 128:178-194

Wimpenny J, Yin Q-Z, Tollstrup D, Xie L-W, Sun J (2014b) Using Mg isotope ratios to trace Cenozoic weathering changes: A case study from the Chinese Loess Plateau. Chem Geol 376:31–43

Wombacher F, Eisenhauer A, Heuser A, Weyer S (2009) Separation of Mg, Ca and Fe from geological reference materials for stable isotope ratio analyses by MC-ICP-MS and double-spike TIMS J Anal At Spectrom 24:627–636

Wombacher F, Eisenhauer A, Bohn M, Gussone N, Regenberg M, Dullo W-C, Ruggeberg A (2011) Magnesium stable isotope fractionation in marine biogenic calcite and aragonite. Geochim Cosmochim Acta 75:5797–5818

Xiao Y, Zhang H-F, Fan W-M, Ying JF, Zhang J, Zhao X-M, Su B-X (2010) Evolution of lithospheric mantle beneath the Tan-Lu fault zone, eastern North China Craton: Evidence from petrology and geochemistry of peridotite xenoliths. Lithos 117:229–246

Xiao Y, Teng F-Z, Zhang H-F, Yang W (2013) Large magnesium isotope fractionation in peridotite xenolithis from eastern North China craton: Product of melt–rock interaction. Geochim Cosmochim Acta 115:241–261

Xiao Y, Teng F-Z, Su B-X, Hu Y, Zhou M-F, Zhu B, Shi R-D, Huang Q-S, Gong X-H, He Y-S (2016) Iron and magnesium isotopic constraints on the origin of chemical heterogeneity in podiform chromitite from the Luobusa ophiolite, southern Tibet. Geochem Geophys Geosyst 17:940–953, doi:10.1002/2015GC006223

Xie L-W, Yin Q-Z, Yang J-H, Wu F-Y, Yang Y-H (2011) High precision analysis of Mg isotopic composition in olivine by laser ablation MC-ICP-MS J Anal At Spectrom 26:1773–1780

Yang W, Teng F-Z, Zhang H-F (2009) Chondritic magnesium isotopic composition of the terrestrial mantle: A case study of peridotite xenoliths from the North China craton. Earth Planet Sci Lett 288: 475–482

Yang W, Teng F-Z, Zhang H-F, Li S (2012) Magnesium isotopic systematics of continental basalts from the North China craton: Implications for tracing subducted carbonate in the mantle. Chem Geol 328:185–194

Yang W, Teng F-Z, Li W-Y, Liu S-A, Ke S, Liu Y-S, Zhang H-F, Gao S (2016) Magnesium isotopic composition of the deep continental crust. Am Mineral 101:243–252

Yoshimura T, Tanimizu M, Inoue M, Suzuki A, Iwasaki N, Kawahata H (2011) Mg isotope fractionation in biogenic carbonates of deep-sea coral, benthic foraminifera, and hermatypic coral. Anal Bioanal Chem 401:2755–2769

Young ED, Ash RD, Galy A, Belshaw NS (2002) Mg isotope heterogeneity in the Allende meteorite measured by UV laser ablation-MC-ICPMS and comparisons with O isotopes. Geochim Cosmochim Acta 66:683–698

Young ED, Galy A (2004) The isotope geochemistry and cosmochemistry of magnesium. Rev Mineral Geochem 55:197–230

Young ED, Tonui E, Manning CE, Schauble EA, Macris C (2009) Spinel–olivine magnesium isotope thermometry in the mantle and implications for the Mg isotopic composition of Earth. Earth Planet Sci Lett 288:524–533

Reviews in Mineralogy & Geochemistry
Vol. 82 pp. 289-344, 2017

Silicon Isotope Geochemistry

Franck Poitrasson

Laboratoire Géosciences Environnement Toulouse
CNRS UMR 5563 – UPS – IRD
14–16, Avenue Edouard Belin
31400 Toulouse
France

Franck.Poitrasson@get.obs-mip.fr

In contrast to many other stable isotopes of the elements discussed in this book, those of silicon are not strictly speaking "Non-Traditional Stable Isotopes" because they have been studied for more than 60 years. After the pioneering works of Reynolds and Verhoogen (1953) and Allenby (1954), a steady increase in silicon isotope studies of geological materials has led to a substantial corpus of data. These data were compiled by Ding et al. (1996) alongside new measurements that, collectively, included over a thousand samples of rocks, minerals, waters and biological materials. Most of these data were produced using the well established method of gas source mass spectrometry after sample decomposition and silicon purification via fluorination techniques.

As for many non-traditional stable isotopes, silicon isotope research has flourished with the advent of second generation of multicollector plasma source mass spectrometers (MC–ICP–MS). These instruments eliminated the requirement of hazardous gaseous fluorine sample preparation methods while permitting improved analytical precision in both wet plasma (De La Rocha 2002) and in dry plasma (Cardinal et al. 2003). Subsequent analytical developments involving high mass resolution MC–ICP–MS combined with improved silicon purification methods (Georg et al. 2006) made this analytical technique more robust and precise enough to study even the subtle silicon isotope variations produced during high temperature geological processes (Savage et al. 2014).

ELEMENT PROPERTIES

Silicon is the fourteenth element of the Periodic Table. Its atomic mass was precisely determined to be 28.08553 ± 0.00039 in atomic mass units (a.m.u.) on a pure silicon reference material (NIST SRM–990, Barnes et al. 1975). This 95% confidence limit error includes the overall natural isotopic variation range for $^{30}Si/^{28}Si$ known by the time, estimated to be about 5‰ from the analysis of biological, meteoritic and terrestrial materials (Tilles 1961). As detailed below, the current database suggests that the actual natural range of mass-dependent isotopic variations on Earth is rather more than twice this value.

Silicon has three naturally occurring stable isotope with the following mean abundances (Debievre and Taylor 1993):

^{28}Si: 92.23%

^{29}Si: 4.67%

^{30}Si: 3.10%.

1529-6466/17/0082-0008$05.00 http://dx.doi.org/10.2138/rmg.2017.82.8

Again, the exact values of these relative proportions will depend on the isotopic abundances of the material under study. As a result, the uncertainties of these values were estimated by De Bièvre and Taylor (1993) to be of one unit on the last digit. Silicon has 19 reported radioactive isotopes, of which only ^{31}Si and ^{32}Si have half-lives of 2.62 h or longer (Holden 2007). The latter, with an estimated half-life of ~140 years (Fifield and Morgenstern 2009), is produced by cosmic ray induced spallation of atmospheric argon. Like other cosmogenic nuclides, it may be used to date sediments, ice and groundwater. It has also been proposed to trace seawater masses (Geyh and Schleicher 1990).

Silicon mainly occurs in the tetravalent oxidation state in nature as it generally combines with oxygen to form silicate minerals or amorphous forms of silica in rocks. In aqueous solutions, it occurs as orthosilicic acid and its dissociated species. Under low oxygen fugacities in interplanetary space, it may occur as the divalent gaseous species SiO and/or SiS. Lastly, it is found alloyed in metallic inclusions in meteorites and iron meteorites (Ringwood 1961; Keil 1968; Pack et al. 2011) and this Si^0 form presumably occurs in the metallic core of terrestrial planets.

NOMENCLATURE, REFERENCE MATERIALS AND ANALYTICAL TECHNIQUES

Because stable isotope mass spectrometric methods allow measuring much more precisely the difference of an isotopic ratio of a sample relative to another rather than their absolute abundance, "stable isotope compositions" of an element are commonly reported using the delta notation. For silicon isotopes, the delta value that gives the deviation of, for example, the $^{30}Si/^{28}Si$ of a sample relative to that of a standard may be expressed as:

$$\delta^{30}Si(‰) = \left(\frac{^{30}Si/^{28}Si_{sample}}{^{30}Si/^{28}Si_{standard}} - 1 \right) \times 1000 \qquad (1)$$

The same definition can be applied when ^{29}Si is reported instead of ^{30}Si. These delta notations involving silicon isotopes may be manipulated using the expressions common to the stable isotopic literature that may be found elsewhere in this book. A particular mention should be made to the computation of isotopic differences between two substances A and B. It is simply a subtraction between two delta values to make:

$$\delta^{30}Si_A - \delta^{30}Si_B = \Delta^{30}Si_{A\text{-}B} \qquad (2)$$

Because the isotopic fractionation factor α between substances A and B defined as:

$$\alpha_{A\text{-}B} = \frac{\left(^{30}Si/^{28}Si\right)_A}{\left(^{30}Si/^{28}Si\right)_B} \qquad (3)$$

is close to unity in most situations, the following approximations can be made:

$$\Delta^{30}Si_{A\text{-}B} \approx 1000 \cdot (\alpha_{A\text{-}B} - 1) \approx 1000 \cdot \ln \alpha_{A\text{-}B} \qquad (4)$$

Because this is an approximation, some authors in the silicon isotope literature prefer to use the epsilon notation:

$$\varepsilon = 1000 \cdot (\alpha_{A\text{-}B} - 1) \qquad (5)$$

However, this does not bring anything useful except confusion with the radiogenic εNd isotopic notation since the big delta notation approximation is sufficiently accurate given the small silicon isotopic variations in nature. It may also be confused with the epsilon notation used to define mass-independent isotopic fractionation. This epsilon notation expressed in Equation (5) should therefore rather be avoided (Criss 1999).

Various silicon isotope reference materials have been produced through time (Coplen et al. 2002). Among these, the two reference materials most often used as a standard against which silicon isotopic data are reported are the Caltech Rose Quartz and the National Bureau of Standards (NBS) 28 Sand Quartz. The latter, subsequently renamed National Institute of Standard and Technology Reference Material (NIST RM) 8546, has been almost universally adopted since Molini-Velsko et al. (1986). It is therefore the 0‰ reference point for $\delta^{30}Si$ and $\delta^{29}Si$, by definition. To compare delta values among studies using different standards, it is essential to know the value of one standard relative to another. Fortunately for silicon isotopes, recent precise measurements suggest that the Caltech Rose Quartz and the NBS–28 Sand Quartz yield $\delta^{30}Si$ indistinguishable within ±0.1‰ (2 standard deviation, Georg et al. 2007a; Chmeleff et al. 2008). This indicates that early silicon isotope studies (e.g., Epstein and Taylor Jr 1971; Douthitt 1982) provided $\delta^{30}Si$ directly comparable to other data reported in the literature since.

It is also essential for inter-laboratory comparisons and to assess the quality of analytical methods to have access to other reference materials that may be used as secondary standards and check the accuracy of the $\delta^{30}Si$ and $\delta^{29}Si$ values produced. For silicate rocks, the database is growing (e.g., Abraham et al. 2008; Zambardi and Poitrasson 2011) and some of the USGS reference rock samples, like the Hawaiian basalt (BHVO–1 and 2) have now been particularly well characterized by many laboratories independently (Savage et al. 2014). For those working on samples unrelated to igneous rocks, an interesting laboratory intercalibration exercise has been conducted on three silicon elemental samples, including a biologically-derived diatomite (Reynolds et al. 2007). Unfortunately, the commercially available silicon isotopic reference material from the European Institute for Reference Materials and Measurements (IRMM) 18 appeared contaminated or heterogeneous among batches, which was subsequently confirmed (Chmeleff et al. 2008). However, the homogeneous "Big batch" and "Diatomite" samples should still be available until at least 2025 according to Mark Brzezinski, from the University of California at Santa Barbara, who may provide these samples upon request (personal communication).

There are four main analytical methods reported in the literature for silicon isotopic determination that are still used today. They are, by order of appearance in the literature, Gas source mass spectrometry, Secondary Ion Mass Spectrometry (SIMS), MultiCollector – Inductively Coupled Plasma – Mass Spectrometry (MC–ICP–MS) and Laser Ablation MC–ICP–MS.

Gas source mass spectrometry was the first method to measure silicon isotope compositions precisely enough to be of interest in geochemistry (Reynolds and Verhoogen 1953; Allenby 1954). This method involves sample decomposition using HF followed by several chemical conversion steps, varying among the authors, to produce SiF_4 that can be introduced in the gas source mass spectrometer for isotopic measurements. Since sample decomposition and subsequent purification frequently generates isotopic fractionation, it is essential to find a method leading to a silicon purification yield as close as possible to 100%. Subsequent improvements came with the lunar sample return Apollo program with reported $\delta^{30}Si$ values that were duplicated or triplicated and quoted to fall within an "average deviation" of ±0.11‰ or better using sample fluorination with the F_2 gas (Epstein and Taylor Jr 1970). However, a more comprehensive study using a similar approach leads to the conclusion that the long term reproducibility assessed on full duplicates of both the chemical purification and mass spectrometry of the Caltech Rose Quartz is rather of ±0.3‰ for $\delta^{30}Si$ at the "two sigma level" (Douthitt 1982). Alternative sample fluorination methods using BrF_5 led to a comparable long term reproducibility of ±0.4‰ at the "two sigma level", presumably two standard deviations (Molini-Velsko et al. 1986). Recent automated methods using modern gas source mass spectrometry combined with sample decomposition by HF under vacuum, as done in early studies, yielded excellent reproducibility of ±0.15‰, 2 standard deviations, on $\delta^{30}Si$ (Brzezinski et al. 2006). However, this was demonstrated on "Big batch" and "Diatomite" standard samples that have a relatively simple matrix, essentially consisting of silica, compared to common silicate rocks with more chemical components.

Secondary Ion Mass Spectrometry (SIMS) has been applied since the mid–1980s to perform *in situ* silicon isotope measurements of presolar grains found in certain type of primitive chondrites (e.g., Zinner 1997). The analysis of these silicon carbide (SiC) grains of the micron to submicron size resolved very large, percent-level mass-independent variation in silicon isotopic compositions produced in stars during nucleosynthesis. The main requirement of the ion probe used (CAMECA IMS3f) for this application is its spatial resolution rather than analytical precision. The advent of larger, multicollector SIMS like the CAMECA IMS1270 that have high mass resolution capabilities significantly reduced analytical uncertainties. With sample bombardment by a primary Cs^+ ion beam of diameter 10 to 20 μm on the sample surface, subsequent energy filtering and mass separation from the interferences (like $^{29}SiH^-$) of the secondary $^{28}Si^-$ and $^{30}Si^-$ beams allowed reproducibilities of ~±0.7‰ (2 standard deviation) to be achieved on $\delta^{30}Si$ analyses of silcretes (Basile-Doelsch et al. 2005). Recent progress using the latest generation of SIMS instruments (CAMECA IMS1280) yielded a long term reproducibility of ±0.3‰ on $\delta^{30}Si$ on quartz and Precambrian cherts (Heck et al. 2011). This is comparable to the analytical performances achieved by gas source mass spectrometry after silicon fluorination, but at a spatial resolution of 10 μm on the sample surface.

The use of MultiCollector – Inductively Coupled Plasma – Mass Spectrometry (MC–ICP–MS) for silicon isotope measurements started a little more than a decade ago (De La Rocha 2002). However, this analytical methodology has clearly boosted the field of silicon isotope geochemistry, which justifies their inclusion in a "Non-Traditional Stable Isotopes" volume. What is striking when making a review of the analytical literature on the topic is how innovative the community has been, given the variety of different methodologies used to decompose the samples, purify silicon and measure its isotopic composition by MC–ICP–MS. The first steps performed on relatively pure silica samples ("Diatomite", "Big Batch" and marine sponge spicules) yielded promising $\delta^{29}Si$ results (De La Rocha 2002). Improvements were obtained in doping silicon solutions aspirated to the MC–ICP–MS with magnesium to better correct for the mass bias (Cardinal et al. 2003) instead of only using the sample-standard bracketing approach previously used. Cardinal et al. (2003) also used a desolvator to work in the dry plasma mode to improve uncertainties with the instrument used. Assuming the instrumental mass bias follows an exponential law (Russell et al. 1978), the relationship between the true (t) and measured (m) silicon isotopic ratio of, for example $^{30}Si/^{28}Si$, may be written as:

$$\left(\frac{^{30}\mathrm{Si}}{^{28}\mathrm{Si}}\right)_t = \left(\frac{^{30}\mathrm{Si}}{^{28}\mathrm{Si}}\right)_m \cdot \left(\frac{M_{^{30}\mathrm{Si}}}{M_{^{28}\mathrm{Si}}}\right)^{f_{\mathrm{Si}}} \tag{6}$$

where, e.g., $M_{^{30}Si}$ is the mass of isotope ^{30}Si and f_{Si} the mass bias fractionation factor for Si. A similar expression may be written for the $^{25}Mg/^{24}Mg$ isotopic ratio and rearranged as:

$$f_{\mathrm{Mg}} = \frac{\ln\dfrac{\left(\dfrac{^{25}\mathrm{Mg}}{^{24}\mathrm{Mg}}\right)_t}{\left(\dfrac{^{25}\mathrm{Mg}}{^{24}\mathrm{Mg}}\right)_m}}{\ln\left(\dfrac{M_{^{25}\mathrm{Mg}}}{M_{^{24}\mathrm{Mg}}}\right)} \tag{7}$$

Assuming that the mass bias is the same for silicon and magnesium, i.e., $f_{Mg} = f_{Si}$, then the measured silicon isotope ratio may be corrected from the measurement of the isotopic composition of the magnesium added to the purified silicon solution using Equations (6) and (7). This assumption is not strictly valid, however, and for iron isotope determinations using mass bias correction with nickel, it was found useful to use the daily regression method devised by Maréchal et al. (1999) which does not require such an assumption. When combined with the sample standard bracketing method, it was possible to obtain precise and accurate values within the reproducibility achieved (Poitrasson and Freydier 2005). For silicon isotope measurements though, this assumption of $f_{Mg} = f_{Si}$ does not induce a bias outside uncertainties and it was therefore found that the use of Equations (6) and (7) combined with sample standard bracketing is sufficient as long as the same magnesium solution is used to dope both samples and NBS–28 standards (Cardinal et al. 2003; Zambardi and Poitrasson 2011). Note that the isotopic composition of the magnesium dopant does not need to be precisely known and it can be approximated to that found in any table of naturally occurring isotopes (e.g., Platzner 1997).

This approach resulted into a reproducibility of ±0.08‰, 2 standard deviations, on $\delta^{29}Si$ (Cardinal et al. 2003). This was half the value obtained by gas source mass spectrometry after laser-assisted fluorination of the same samples (±0.18‰, 2SD). But overall, these two different analytical techniques proved to be accurate within those uncertainties on biogenic silica samples "Big Batch" and "Diatomite". These early MC–ICP–MS analytical works, however, were plagued by the low mass resolution capability of the instrument used that prevented the measurement of ^{30}Si due to an interference with $^{14}N^{16}O$. This issue was solved with the use of MC–ICP–MS instruments having high mass resolution capabilities (Georg et al. 2006; van den Boorn et al. 2006) to produce both $\delta^{29}Si$ and $\delta^{30}Si$ accurate values, like gas source mass spectrometry. Outside the need to measure ^{28}Si, ^{29}Si and ^{30}Si isotopes to study mass independent fractionation in extra-terrestrial materials, three isotope plots involving $^{30}Si/^{28}Si$ and $^{29}Si/^{28}Si$ ratios are essential to assess the quality of the mass spectrometric measurements and to evaluate the remaining interferences, at least on ^{29}Si and ^{30}Si. Sample decomposition and silicon purification in more complex matrixes like silicate igneous rocks through wet chemistry is not trivial either, however, because the standard methodology employing HF decomposition may lead to silicon loss via gaseous SiF_4 if not performed in strictly closed vessel or with sufficiently dilute HF. Further, the presence of HF in the nebulization solution relative to a simple HCl matrix was found by Georg et al. (2006) to degrade the instrument sensitivity by 30–40% and the silicon isotope measurement precision by a factor of two. For this reason, these authors developed a NaOH fusion method in Ag crucibles followed by cation exchange chromatography with a BioRad DOWEX AG50-X12 cationic resin that has become the standard method adopted by the silicon isotope community measuring by MC–ICP–MS since. They were some remaining accuracy issues with this approach, though (Georg et al. 2007a). Although the exact reasons are unclear (see review in Savage et al. 2014), it was found that the combination of high mass resolution MC–ICP–MS, the rock sample fusion and silicon purification procedure of Georg et al. (2006), along with the mass bias correction using Mg-doping (Cardinal et al. 2003) proved to yield accurate and reproducible $\delta^{30}Si$ data to better than ±0.08‰ (2SD, Zambardi and Poitrasson 2011). The magnesium doping approach, although not beneficial to all types of instruments (see, e.g., De La Rocha 2002), appears to strengthen the robustness of the method relative to the simple sample-standard bracketing approach. This is because it provides a further correction for remaining mass bias variations produced by drifts due to imperfectly purified samples or if, for example, air conditioning of the room in which the instrument is located is uneven. For river water samples, Georg et al. (2006) found that they can be directly processed through column chemistry after a simple acidification, although Hughes et al. (2011a) recommended a prior UV irradiation step to decompose the organic matter not separated by the column chemistry of Georg et al. (2006) to minimize matrix effects during MC–ICP–MS measurements. Marine water samples are also difficult to handle because of their low silicon and high NaCl contents.

For these, preconcentration schemes are required, such as $Mg(OH)_2$ co-precipitation, but this should be carefully done because these may lead to silicon isotope fractionation if the silicon recovery from seawater is not quantitative (Reynolds et al. 2006; de Souza et al. 2012b). For plants, an initial calcination to 450 °C in an oven (e.g., Delvigne et al. 2009) before fusion also allows minimizing remaining organic molecules in the silicon fraction and therefore limits deleterious matrix effects during MC–ICP–MS isotopic analysis.

The latest analytical development is the coupling of high mass resolution MC–ICP–MS instruments with laser ablation systems for *in situ* silicon isotope analyses in solids. Relative to the pre-existing ion microprobes, LA–MC–ICP–MS systems offer in principle mass spectra with less interferences given the more destructive, more energetic nature of the laser ablation system, and a lower sensitivity to matrix effects. The first attempt proved very satisfactory (Shahar and Young 2007), with an estimated precision of ±0.4‰ (2SD) for both δ^{29}Si and δ^{30}Si, although measurement accuracy was not easy to assess due to limited available solid reference material that could be analyzed with the methodology developed. The use of a 193 nm nanosecond Excimer laser was indeed a limitation since the NBS–28 Quartz Sand reference material could not be analyzed due to poor laser-matter coupling (Shahar and Young 2007). This issue was rapidly overcome by the use of a femtosecond (fs) laser instead that can ablate quartz (Chmeleff et al. 2008). Furthermore, it has been found that LA–MC–ICP–MS is much less sensitive to matrix matching for calibration than SIMS (Janney et al. 2011). This is especially true if a femtosecond laser is used instead of a nanosecond (ns) laser. This is clearly a big advantage given the difficulty to find natural solid samples, like (a)biogenic silica or minerals, with chemical and isotopic compositions homogeneous enough to be used to calibrate *in situ* techniques. Using fs LA–MC–ICP–MS analysis, Chmeleff et al. (2008) were able to demonstrate a long term reproducibility of ±0.15‰ and ±0.24‰ (2SD) on δ^{29}Si and δ^{30}Si, respectively, over 6 months. This is only three times the uncertainties currently achieved by MC–ICP–MS after fusion and silicon purification by chromatography (Zambardi and Poitrasson 2011; Savage et al. 2014). This also compares well with uncertainties achieved with the latest generation of SIMS (IMS1280, Heck et al. 2011), although the latter work was conducted at a much higher spatial resolution (spot sizes of 10 μm instead of a square of 150 μm side length), which was certainly less favorable in terms of counting statistics. Both *in situ* approaches based on high mass resolution instruments remain used by a few laboratories around the world, however, because they require highly skilled operators and expensive instruments. We may nevertheless expect that fs LA–MC–ICP–MS will become more widespread as femtosecond laser ablation systems are becoming more routine.

ELEMENTAL AND ISOTOPIC ABUNDANCES IN MAJOR RESERVOIRS

Extraterrestrial reservoirs

Meteorites and components. The most exotic silicon isotope compositions measured so far have been found in the so-called "presolar grains" that are the residues remaining after the acid dissolution of primitive carbonaceous chondrites, like the CM2 Murchison or Murray. After diamonds, silicon carbide (SiC) grains are the most abundant minerals found in these acid decomposition residues, yet they represent only about 0.001% of the initial meteorite mass (Zinner 1998). These SiC grains are also big enough (from 0.3 to 20 μm; Zinner, 1998) to permit SIMS analyses that revealed huge silicon isotopic fractionation not found anywhere else in nature (Zinner et al. 1987). The range is close to 2000‰ in δ^{30}Si (Fig. 1) and SiC grains also yield large mass-independent fractionation trends easily observable in δ^{29}Si versus δ^{30}Si plots (Fig. 1). Besides silicon carbide, such mass-independent silicon isotopic variations are also found in the less abundant primitive meteorite leach residues consisting of silicon nitride (Si_3N_4) and

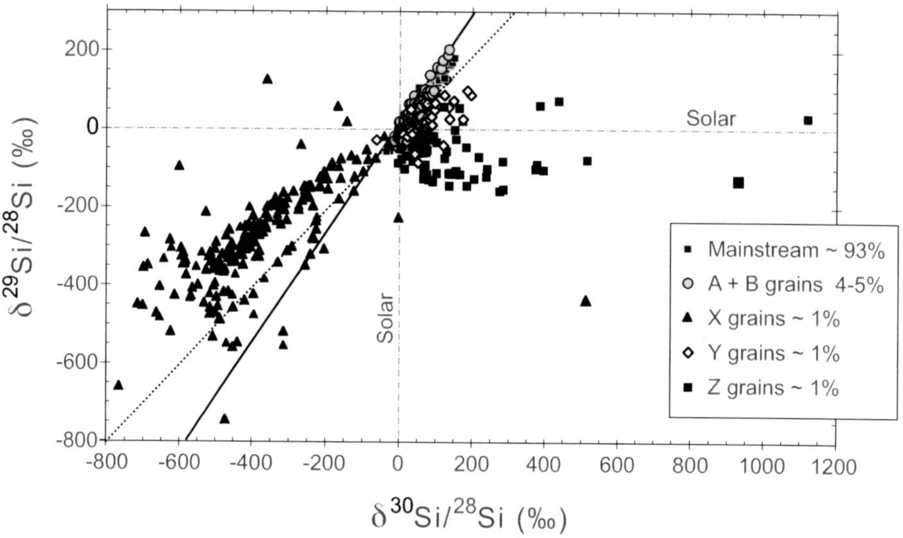

Figure 1. Silicon isotope composition of presolar SiC grains. The different types of grains correspond to different nucleosynthetic production processes and their abundance in primitive meteorites is given in % (Inset). Note the strong deviation from the mass-dependent isotopic fractionation line (dashed line with a slope of 1) of the rare X- and Z-type SiC grains. After Zinner et al. (2006).

graphite (collectively also called "refractory residues"). Lastly, they also occur in a few refractory Calcium–Aluminium-rich Inclusions (CAI) known for their mass independent isotopic variations in many elements resulting from a combination of Fractionation and Unknown Nuclear effects (FUN) found in the CV2 Allende chondrite (Clayton et al. 1978, 1984). No mass-independent Si isotope compositions have been detected so far in other CAIs (Shahar and Young 2007), bulk meteorites (Pringle et al. 2013), or in any other natural materials to date.

Mass independent isotopic variations found in meteorite inclusions are interpreted since the 1970's as presolar grains that record nucleosynthesis in stars that were present before the sun (see reviews in Zinner 1998; Birck 2004). This opened a new branch of astrophysics allowing for direct constraints on the thermonuclear fusion processes occurring at the end of the life of stars larger than the Sun. Silicon carbide grains have been classified into various types (Fig. 1) according to their different isotopic compositions of silicon but also of C, N and Ca that reflect different thermonuclear processes occurring in large stars forming silicon (Zinner et al. 2006). This happens from the oxygen-burning shell of stars having at least eight solar masses that should reach minimal temperatures of 1.5 billons K for the silicon production to occur (Seeds 2001). While early generations of supernovae produced mainly the most abundant ^{28}Si isotope, later generations of stars produced increasing amounts of the more neutron-rich ^{29}Si and ^{30}Si so that throughout the galactic evolution, $^{29}Si/^{28}Si$ and $^{30}Si/^{28}Si$ ratios continuously increased (Fig. 2). Deviation from the mass-dependent line in the three isotopes plot result from heterogeneities introduced depending on the exact reaction involved into the different silicon isotope production processes. For instance, neutron capture in Asymptotic Giant Branch (AGB) stars are expected to particularly enrich silicon into its most neutron-rich stable isotope ^{30}Si (Fig. 2; Zinner et al. 2006).

Silicon concentrations vary a lot among meteorites. They go from a few hundred parts per billions (ppb) in iron meteorites to 25 wt% Si in Howardites–Eucrites–Diogenites (HED) meteorites, which corresponds to 53.5 wt% SiO_2. In the following, silicon concentrations will be reported as the SiO_2 oxide component. Iron meteorites have long been difficult to analyze

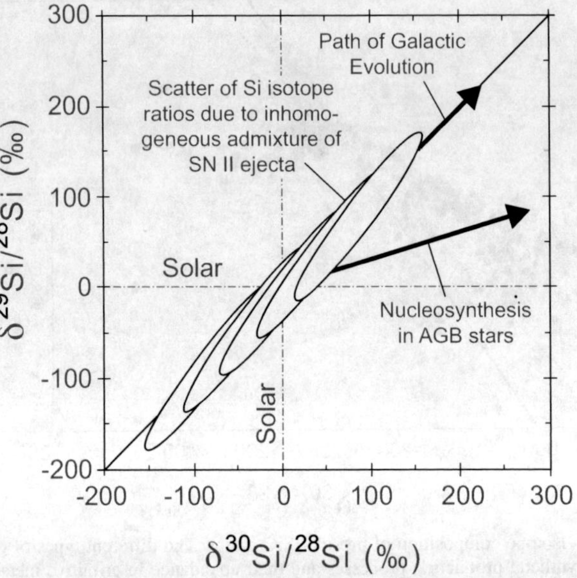

Figure 2. Nucleosynthetic model proposed to explain the range of silicon isotope compositions measured in presolar SiC grains. SN II: Supernovae of Type-II. AGB: Asymptotic Giant Branch stars. The solar isotopic composition is reported for reference. See text for explanations. After Zinner et al. (2006).

accurately for their silicon concentrations. Wai and Wasson (1969) reported values below their detection limits (< 25 part per million (ppm) Si) for all but two iron meteorites, Tucson and Horse Creek. These low silicon concentrations were confirmed by SIMS analyses (Pack et al. 2011) which yielded values ranging from 0.10 to 0.46 ppm for a set of 15 magmatic and non-magmatic iron meteorites. Among meteorite groups consisting of silicates, carbonaceous chondrites display the lowest concentrations, from 22.5 to 34.0 wt% SiO_2 (Jarosewich 1990; Lodders and Fegley Jr 1998; Hutchison 2004). Ordinary and enstatite chondrites have higher SiO_2 contents, going from 35.5 to 40.4 wt%, with increasing values going from H (High total Fe contents) to L (low total Fe contents) and LL (Low metallic Fe relative to total Fe and low total Fe contents) for ordinary chondrites and from EH (High Fe contents) to EL (Low Fe contents) for enstatite chondrites. This essentially reflects the decreasing amount of iron metal in meteorite modes. Concerning less common achondrites for which silicon isotopes have been already analyzed, ureilites, aubrites and angrites show SiO_2 ranges of 36.8–45.2 wt%, 56.5–59.5 wt% and 33.4–43.7 wt%, respectively (Jarosewich 1990; Keil 2010; Armytage et al. 2011; Keil 2012). Lastly, the most abundant achondrites, HEDs, display SiO_2 values going from 46.5 to 53.5% (Jarosewich 1990; Hutchison 2004; Armytage et al. 2011) with diogenites showing the highest values.

In contrast to presolar grains uncovered in primitive chondrites (Fig. 1), no mass-independent silicon isotopic variation has been found so far at the bulk meteorite scale (Pringle et al. 2013). And, the mass-dependent isotopic fractionation range found so far is three orders of magnitude smaller, on the order of ~0.65‰, in δ^{30}Si (Fig. 3). This is commensurate with the range uncovered from the first systematic survey conducted on meteorites using gas source mass spectrometry (Molini-Velsko et al. 1986). However, and in contrast to the conclusions reached in this early study, current techniques now allow for distinction of different meteorite groups according to their δ^{30}Si values (Fig. 3). This was not obvious from the beginning since early MC–ICP–MS studies showed inconsistent results among different laboratories.

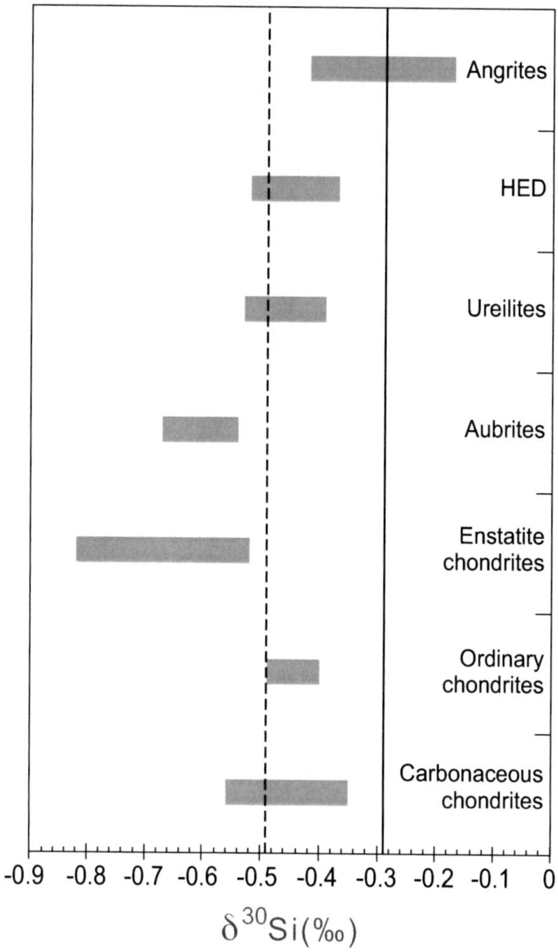

Figure 3. Range of silicon isotope compositions against the NBS-28 reference standard of various types of bulk meteorite samples reported in the literature. Data for *carbonaceous chondrites* are from Fitoussi et al. (2009), Pringle et al. (2013), Savage and Moynier (2013) and Zambardi et al. (2013); Data for *ordinary chondrites* are from Fitoussi et al. (2009), Armytage et al. (2011), Pringle et al. (2013, 2014), Savage and Moynier (2013), Zambardi et al. (2013) and Dauphas et al. (2015); Data for *enstatite chondrites* are from Fitoussi et al. (2009), Armytage et al. (2011), Fitoussi and Bourdon (2012), Savage and Moynier (2013) and Zambardi et al. (2013); Data for *aubrites* are from Savage and Moynier (2013); Data for *ureilites* are from Armytage et al. (2011); Data for *Howardites, Eucrites and Diogenites (HED)* are from Fitoussi et al. (2009), Armytage et al. (2011), Pringle et al. (2013, 2014) and Zambardi et al. (2013); Data for *angrites* are from Pringle et al. (2014) and Dauphas et al. (2015). The bulk silicate Earth (-0.29 ± 0.07‰ δ^{30}Si, continuous line) and bulk silicate Mars (-0.49 ± 0.03‰ δ^{30}Si, dashed line) estimates are shown for comparison. See text for details.

This was particularly evident for carbonaceous chondrites like Orgueil and Murchison for which discrepancies of up to ~0.3‰ δ^{30}Si were noticed among different laboratories (see reviews in Zambardi et al. 2013; Savage et al. 2014). While isotopic studies of extra-terrestrial matter are sometimes plagued by the limited amount of material available (small powder aliquots made out of a few tens to hundred of mg of sample may not be representative of coarse grained samples), this is not the case for Murchison and Orgueil because they are fine-grained (see, e.g., Norton 2002). This probably reveals analytical problems possibly linked with the extra-terrestrial organic matter that may sometime lead to purification problems and/or enhanced MC–ICP–MS mass biases. Nevertheless, using the data compilations of Zambardi et al. (2013) and Savage et al. (2014), along with the latest results reported in the literature, it is now possible to find clear δ^{30}Si differences among the isotopic ranges of different meteorite groups (Fig. 3). While carbonaceous and ordinary chondrites are isotopically lighter than the bulk silicate Earth, enstatite chondrites, and particularly those of EH types are particularly light (Savage and Moynier 2013). In showing that the silicate fraction of these enstatite chondrites have δ^{30}Si undistinguishable from other chondrite values, these authors concluded that these light silicon isotope composition come from the metal inclusions that are particularly abundant in EH chondrites. This is supported by the very light δ^{30}Si measured in iron metal (~6.5‰ in δ^{30}Si) inclusions of aubrites (Ziegler et al. 2010), which are achondrites thought to be resulting from asteroidal processing of enstatite chondrites. This would suggest enstatite chondrite formation under very reducing conditions (with $fO_2 \sim IW - 8$), thereby allowing percent levels of isotopically light silicon partitioning into iron metal (Ziegler et al. 2010; Javoy et al. 2012).

Aside aubrites discussed above, other achondrites analyzed so far for their silicon isotope composition (Ureilites and HED) yield δ^{30}Si undistinguishable from those of carbonaceous and ordinary chondrites (Fig. 3). This suggests a homogeneous source in terms of silicon isotope composition in the protoplanetary accretion disk and/or minimal δ^{30}Si fractionation during asteroidal formation that led to the ureilite and HED parent bodies, assuming ordinary and/or carbonaceous chondrites can tell us something on their formation. Recent work has shown that angrites are the only achondrite meteorites found so far that have a silicon isotope composition that can be as heavy as the bulk silicate Earth (Pringle et al. 2014; Dauphas et al. 2015). This angrite heavy stable isotope composition was previously noticed also for Fe isotopes (Wang et al. 2012). Such heavy δ^{30}Si angrite values was interpreted by Pringle et al. (2014) as resulting from an impact related-light Si isotope loss in space, as was previously proposed to explain the heavy Fe and Zn isotope composition of the Earth–Moon system (Poitrasson et al. 2004; Paniello et al. 2012). However, Dauphas et al. (2015) modeled planetary impact vaporization and concluded that this process is unlikely to leave a measurable isotopic imprint for bodies smaller than 0.2 Earth mass, like the putative angrite parent body. Rather, they explained the heavy angrite δ^{30}Si in terms of isotopic fractionation occurring during high temperature gas condensation in the protoplanetary solar nebula. Whatever the exact interpretation, this discussion clearly shows the important information silicon isotope signatures of meteorites can provide on planet formation and differentiation processes, as further detailed below.

Moon. In contrast to all other extra-terrestrial bodies, geochemical research on lunar rocks has benefitted greatly from the manned Apollo missions that returned over 340 kg of rocks for precise analyses. The study of these rocks has revealed a relatively limited range of SiO_2 contents compared to rocks from other planets like Mars and the Earth (see below), varying from about 36 wt% in some high-Ti basalts to ~52 wt% in highland norites and potassium, Rare Earth Elements and phosphorous-rich (KREEP) basalts (Meyer 2008). The values measured for an individual sample may vary by several percent among authors, however, as a result of the coarse-grained and heterogeneous nature of some lunar rocks returned by the Apollo missions. This makes it sometimes difficult to yield representative SiO_2 values for those rocks. High-Ti basalts are rich in titanium and iron oxides and relatively poor in silicate minerals, which explains their low silica content. However, other Mare basalts (low-Ti Basalts, Al-rich basalts) have SiO_2 contents comparable to terrestrial basalts (Meyer 2008). Highland rocks that make up 83% of the lunar surface (Papike et al. 1998) consist of anorthosites, KREEP basalts and the Mg-plutonic suite. They yield SiO_2 going from ~43 to 52 wt%. Representative analyses of the main petrologic type found on the Moon are given in Lodders and Fegley (1998), whereas a recent compilation of bulk lunar composition estimates is given by Elardo et al. (2011), with SiO_2 values ranging from 44.3 to 46.1 wt%.

Despite early claims (Yeh and Epstein 1978), Molini-Velsko et al. (1986) found the $\delta^{30}Si$ of lunar samples not being significantly different from meteorites. However, the growing database of precise and accurate lunar silicon isotopic data has revealed that, in fact, the Moon is isotopically heavier than most meteorites, with the exception of angrites. In fact, current accurate bulk Moon $\delta^{30}Si$ estimate yield values ranging from -0.29 ± 0.05‰ (Armytage et al. 2012; Fitoussi and Bourdon 2012) to -0.27 ± 0.04‰ (Zambardi et al. 2013), which is indistinguishable from the current BSE estimate.

While the silicon isotope homogeneity of lunar low- and high-Ti Mare basalts is now well established with the current database, a recent study has uncovered a positive relationship between $\delta^{30}Si$ and the tectosilicate content of the lunar rocks (Fig. 4). Such a relationship also occurs among terrestrial igneous rocks, but it is much more scattered, as will be discussed below. This $\delta^{30}Si$ was interpreted by Poitrasson and Zambardi (2015) as resulting from the increasing degree of melt polymerization on the basis of the relationship between silicon isotope composition and the Non Bridging Oxygen/Tetrahedral cation (NBO/T) index defined by Mysen et al. (1985).

Mars. Most of what we know on the Martian rock chemistry comes from the Chassigny–Nahkla–Shergotty (SNC) meteorite group, thought to come from Mars, and from landers and rovers that studied the surface of the Red Planet (See review in McSween et al. 2009). The latest results from the CHEMical CAMera, based on Laser Induced Breakdown Spectroscopy (LIPS) onboard the Curiosity Rover currently studying the Gale Crater on Mars has considerably expanded the range in SiO_2 content of Martian rocks uncovered so far, with estimated values above 60 wt% (Sautter et al. 2015). This leads Martian crustal composition going up to the trachyte domain in a Harker diagram (Fig. 5). The SNC meteorite SiO_2 range is much narrower, however. Besides the low 37.4 wt% in Chassigny, a dunite, shergottites and nakhlites display SiO_2 values ranging from 42.4 to 52.8 wt% (Lodders 1998), which confine these rocks into the basaltic field of a Harker diagram (Fig. 5).

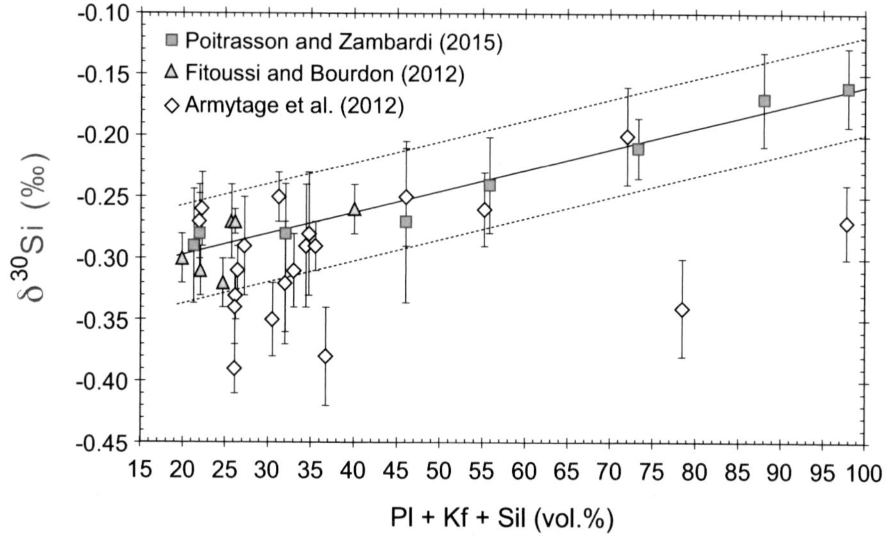

Figure 4. Silicon isotope compositions relative to the NBS-28 reference of bulk lunar samples plotted against their felsic mineral modal composition. The solid line represents the linear fit of the Poitrasson and Zambardi (2015) data and the dashed lines its 2SE envelope. Note that some samples deviate markedly from the trend, possibility because of too small aliquot size that make them unrepresentative of such coarse-grained rocks. Plagioclase (Pl), alkali-feldspar (Kf) and silica (Sil) modal abundances have been compiled by Meyer (1998). After Poitrasson and Zambardi (2015).

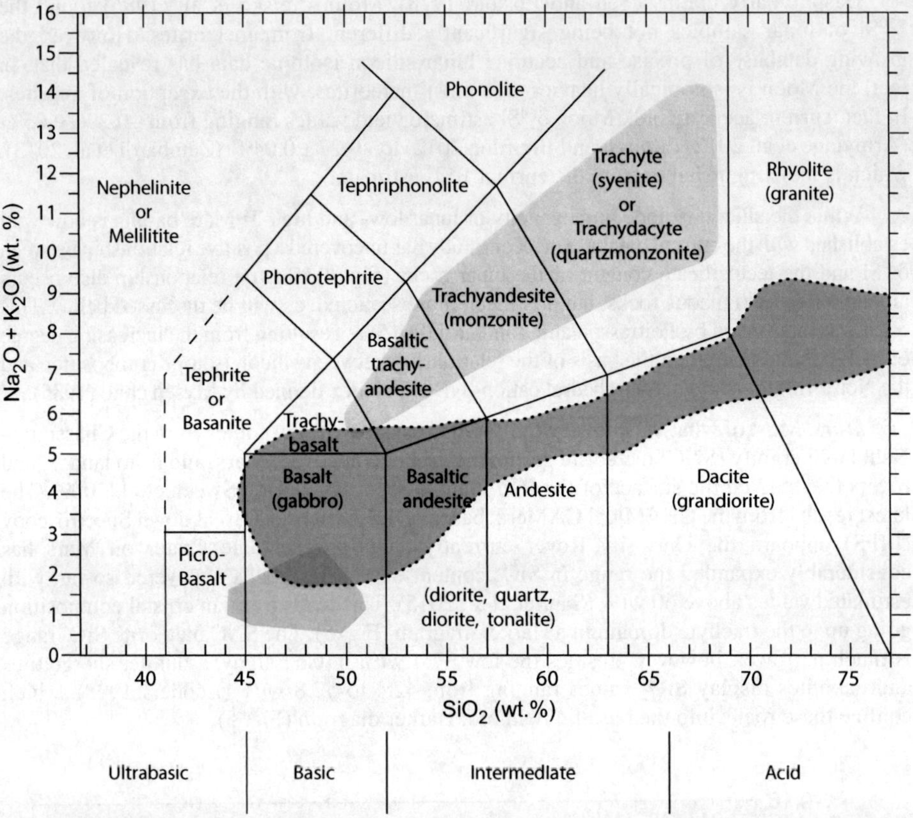

Figure 5. Harker diagram showing the range of silicate rock type defined from their SiO_2 content versus their Na_2O and K_2O contents. The pink field surrounded by a dashed line encloses over 50% of more than 41 000 bulk terrestrial rock analyses compiled in Best (2003). The blue field, occurring mostly in the basalt domain, includes most Martian SNC meteorites while the green field, expanding from trachy-basalts to trachytes corresponds to the silica-rich composition of some Martian outcropping rocks found by the latest NASA Curiosity rover using Light Induced Breakdown Spectroscopy (LIBS), according to Sautter et al. (2015).

However, precise and accurate silicon isotope determination can only be done in the laboratory yet, so we have to rely on data obtained on SNC meteorites, bearing in mind that new findings will certainly arise once more felsic Martian meteorites are uncovered, or when samples of the Martian surface will become available for precise laboratory work from return missions. The database for precise and accurate silicon isotope data is scarce since they essentially come from two studies showing an overall range going from −0.56 to −0.33‰ in $\delta^{30}Si$ (Armytage et al. 2011; Zambardi et al. 2013). Martian igneous rocks, as seen from SNC meteorites, therefore have a silicon isotope composition indistinguishable from those of ordinary and carbonaceous chondrites, ureilites and HED meteorites, but they are lighter than the BSE estimate (Fig. 3). Zambardi et al. (2013) found a hint towards heavier $\delta^{30}Si$ for the most differentiated shergottite Los Angeles, following an igneous differentiation trend uncovered for the Earth. However, this remains limited and overall, these authors propose a Martian mean $\delta^{30}Si = -0.49 \pm 0.03‰$ for Mars on 9 samples, indistinguishable from the mean value previously estimated by Armytage et al. (2011) on 6 other, mostly different SNC meteorites.

Terrestrial reservoirs

Silicon is, after oxygen and iron, the third most abundant element in the Earth (17.1±0.2wt% Si; Allègre et al. 2001). Outside the metallic core for which we are left with inferences and hypotheses discussed below, silicon has a simple chemistry in rocks and waters. In silicate minerals, which make 75% of the Earth's crust (Ding et al. 1996), this element is bonded to four oxygen atoms, making the silica tetrahedron that is the building block of silicate minerals. This translates into silicate mineral classes going from nesosilicates consisting of SiO_4 tetrahedra linked by divalent atoms in six-fold coordination (Fig. 6), like olivines, to tectosilicates like the feldspars, in which tetrahedra are linked to another by shared oxygens in all directions of space (Fig. 6). Between these two extremes are the sorosilicates (Fig. 6), consisting of tetrahedral pairs linked by one oxygen (e.g., epidote), cyclosilicates (Fig. 6) consisting of rings of 3, 4 or 6 SiO_4 tetrahedra (e.g., tourmaline), inosilicates (Fig. 6) forming single (e.g., pyroxenes) or double (e.g., amphiboles) chains of tetrahetra. There are also

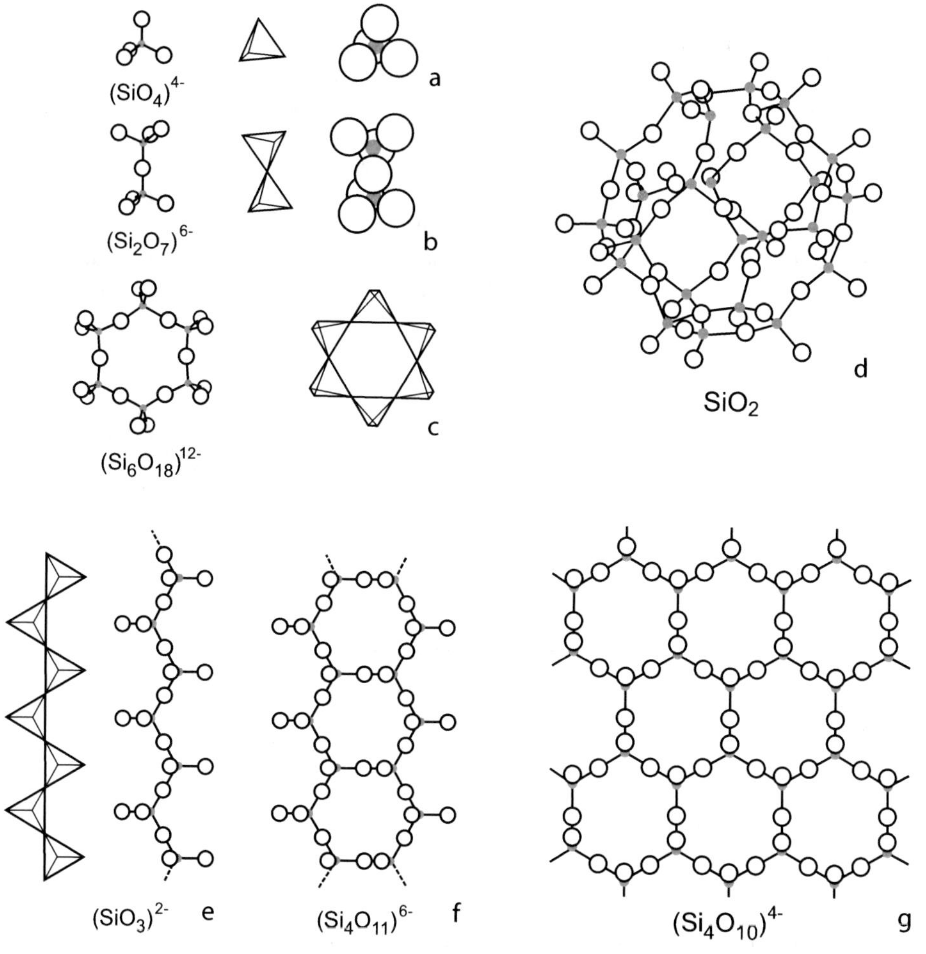

Figure 6. Structural arrangement of oxygen (large spheres) and silicon (smaller red dots) atoms in nesosilicate (a), sorosilicate (b), cyclosilicate (c), tectosilicate (d), single chain (e) and double chain (f) inosilicate and phyllosilicate (g) minerals. After Lameyre (1986).

phyllosilicates (e.g., micas) in which tetrahedra are arranged as sheets that sandwich layers of octahedrally coordinated cations (Fig. 6). A more comprehensive description of the properties of silicate minerals is beyond the scope of this chapter and they may be found in mineralogy textbooks (e.g., Deer et al. 1992). It is nevertheless important to bear in mind this progressive polymerization of SiO_4 tetrahedra from nesosilicates to tectosilicates as this is considered by some authors a key parameter to explain silicon stable isotope compositions, starting from the early theoretical work of Grant (1954). Analyses of mineral separates from igneous rocks has since broadly confirmed that increasing polymerization favors heavier silicon stable isotope compositions (see reviews and new data in Douthitt 1982; Ding et al. 1996; Savage et al. 2014), although more recent theoretical work has shown that the nature of the cations filling the silicate structures also plays a role (Méheut et al. 2009; Méheut and Schauble 2014).

It is also important to consider the various forms of amorphous silica (SiO_2) that is estimated to represent up to 12 wt% of the Earth's crust (Ding et al. 1996). This is a component found in ancient sedimentary rocks as cherts. It also makes up biogenic opaline silica produced by diatoms or plants as phytoliths (e.g., Leng et al. 2009). Like in solids, aqueous silica does not change its redox state, which is normally an important cause for stable isotope fractionation (e.g., Valley and Cole 2001; Johnson et al. 2004). Its concentration in natural waters should in principle be controlled by the speciation of orthosilicic acid (H_4SiO_4) and its dissociated species, depending on the pH (Fig. 7) and other aqueous properties (e.g., Stumm and Morgan 1996). In the following however, we will see that kinetic processes and life play an important role in the aqueous silicon cycle in low temperature terrestrial environments like oceans, river waters and soils. Under higher temperature hydrothermal environments, the prograde solubility of silicate minerals will increase silica concentrations in fluids while temperature rises.

Our knowledge of the silicon isotope composition of the bulk silicate Earth has considerably improved over the past decade since the early estimates of Douthitt (1982)

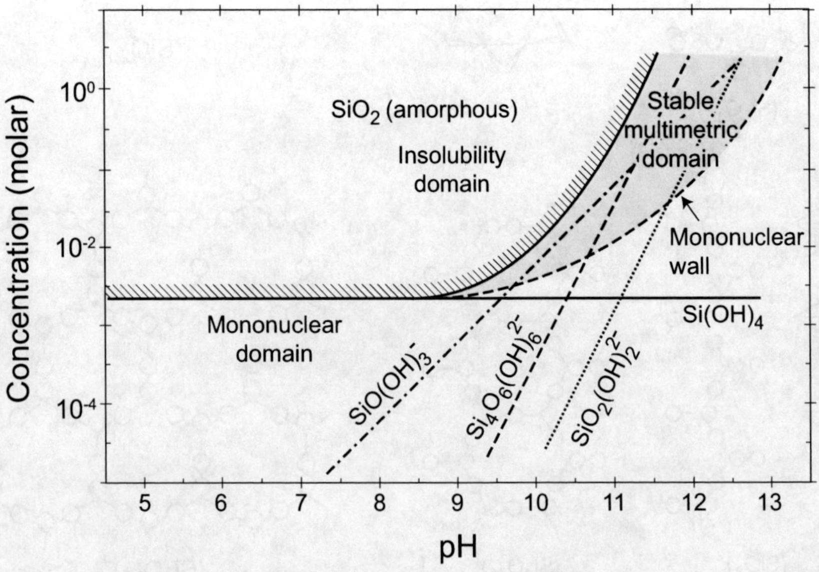

Figure 7. Distribution of aqueous silicon species and solubility of amorphous silica as a function of solution pH. Values are computed from equilibrium constants at 25°C and ionic strength I = 0.05M. The multimeric domain corresponds to the condition where aqueous species like $Si_4O_6(OH)_6^{-2}$ occur. Under the mononuclear wall, only species with one silicon atom are stable. Note that in this graph, aqueous silica ionic species are written in the hydroxyl form, e.g., H_4SiO_4 is expressed as $Si(OH)_4$. After Stumm and Morgan (1996).

and Ding et al. (1996). The first MC–ICP–MS-based estimates were sometimes inaccurate due to analytical issues and/or based on a too-limited number of samples to produce a reliable planetary mean value (See reviews in Zambardi et al. 2013; Savage et al. 2014). A first precise and accurate estimate produced by Fitoussi et al. (2009) of -0.29 ± 0.06‰ (2SD) in δ^{30}Si and -0.15 ± 0.04‰ in δ^{29}Si was subsequently confirmed (Savage et al. 2010; Armytage et al. 2011; Zambardi et al. 2013). Taking these results together, Savage et al. (2014) computed -0.29 ± 0.07‰ in δ^{30}Si (Fig. 8) and -0.15 ± 0.05‰ in δ^{29}Si for the BSE. The isotopic composition of the surficial terrestrial reservoirs (seawater and river waters, soils, plants, cherts, etc.), although representing only a very small fraction of the terrestrial silicon compared to that of igneous rocks, are more highly variable, with a δ^{30}Si ranging by more than 11‰, in a mass-dependent fashion. These will be discussed in the following sections.

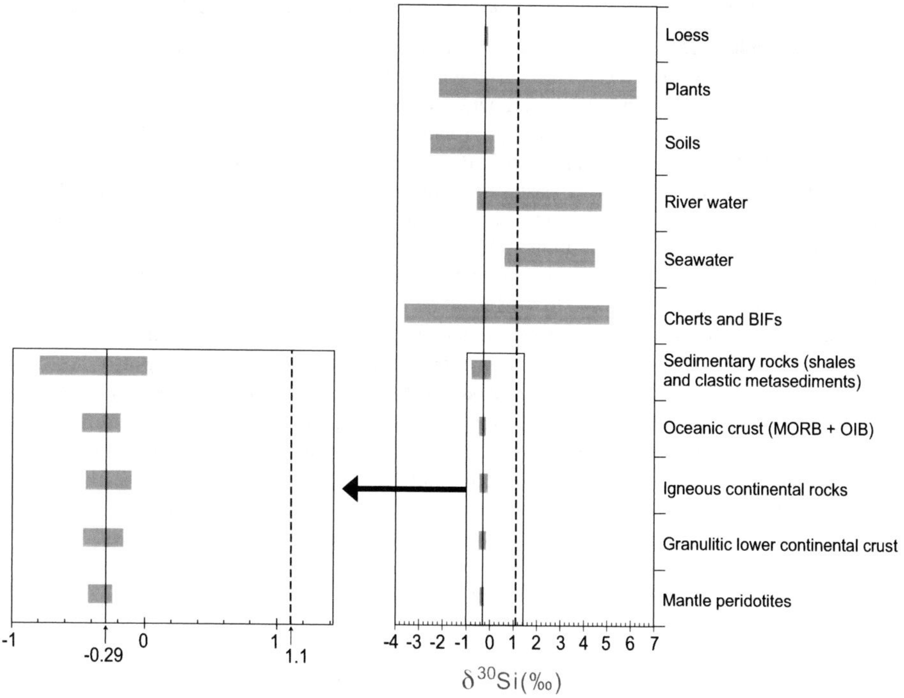

Figure 8. Range of silicon isotope compositions against the NBS-28 reference standard of various types of terrestrial samples reported in the literature. Data are from Fitoussi et al. (2009), Savage et al. (2010) and Zambardi et al. (2013) for *mantle peridotites*; Savage et al. (2013b) for *granulitc lower continental crust*; Savage et al. (2011, 2012); Zambardi and Poitrasson (2011), Zambardi et al. (2014), Poitrasson and Zambardi (2015) for *igneous continental rocks*; Abraham et al. (2008), Fitoussi et al. (2009), Savage et al. (2010, 2011), Zambardi and Poitrasson (2011), Armytage et al. (2012) and Pringle et al. (2016) for *oceanic crust (MORB and OIB)*; André et al. (2006) and Savage et al. (2013a) for *sedimentary rocks (shales and clastic metasediments)*; André et al. (2006), Robert and Chaussidon (2006), van den Boorn et al. (2007, 2010), Steinhoefel et al. (2009, 2010), Heck et al. (2011) and Marin-Carbonne et al. (2012, 2014) for *cherts and BIFs*; De La Rocha et al. (2000, 2011), Varela et al. (2004), Cardinal et al. (2005), Reynolds et al. (2006), Beucher et al. (2008), de Souza et al. (2012a,b) and Grasse et al. (2013) for *seawater*; De La Rocha et al. (2000), Ding et al. (2004, 2011), Georg et al. (2006, 2009), Cardinal et al. (2010), Cockerton et al. (2013), Hughes et al. (2013), Pokrovski et al. (2013) and Frings et al. (2015) for *river water*; Ziegler et al. (2005b), Bern et al. (2010), Steinhoefel et al. (2011), Opfergelt et al. (2012) and Pogge von Strandmann et al. (2012) for *soils*; Ding et al. (2005, 2008a) and Delvigne et al. (2009) for *plants*; Savage et al. (2013a) for *loess*. The bulk silicate Earth (-0.29 ± 0.07‰ δ^{30}Si, continuous line) and global ocean 1.1 ± 0.3‰ δ^{30}Si, dashed line) estimates are shown for comparison. The inset is a blow up for the rocks showing only small variations. See text for details.

Continental crust. In a compilation of over 41,000 igneous rocks of all ages from around the world, R.W. Le Maitre (reported in Best 2003) obtained a SiO_2 range going from 30 to 80 wt%. More than 50% of these data reported in a Harker diagram (Fig. 5) scatter around a path following the silica-saturated igneous trend of magmatic series going from basalts (from ~45 wt% SiO_2) to granites (to ~75 wt%). Taking into account the granulitic lower continental crust, mafic and felsic intrusives and sedimentary rocks of the upper continental crust, Wedepohl (1995) provided chemical composition estimates of whole continental crust, which included silicon concentrations. He obtained 61.5 wt% SiO_2 for the whole continental crust with 64.9 wt% and 58.1 wt% for the upper and lower continental crust, respectively, bearing in mind that the upper continental crust represents ~53% of the continental crust thickness on average according to Wedepohl (1995). Previous composition estimates compiled by Lodders and Fegley (1998) yielded consistent values within ±1 wt%, with exception of two estimates by Taylor and McLennan (1985) that are lower by ~2 wt%.

A considerable effort has been made since 2010 by the Oxford group to update our knowledge on rocky terrestrial reservoirs using modern analytical techniques, with emphasis on high temperature silicon isotope systematics (Savage et al. 2010, 2011, 2012, 2013a,b; Armytage et al. 2012). These results have been reviewed and integrated with other recent results in Savage et al. (2014) and the reader is referred to this publication for more detail on the topic. The five-fold improvement in $\delta^{30}Si$ measurement uncertainties clearly opened the field of high temperature silicon isotope geochemistry that was barely accessible using classical gas source mass spectrometry. For instance, Douthitt (1982) reported a silicon isotope composition range of igneous rock that was four times larger than that uncovered recently by modern methods (with $\delta^{30}Si$ actually going only from −0.40 to −0.10‰, see Fig. 8). Yet his results could barely hint at a difference in $\delta^{30}Si$ means of rocks below and above 55 wt% SiO_2. Similarly, Ding et al. (1996) reported new analyses using the same gas source mass spectrometry method. They also showed a larger scatter than recent MC–ICP–MS values but this time, they could more clearly conclude that granites were isotopically heavier than mafic rocks. However, they could not find $\delta^{30}Si$ differences among various granite types. In contrast, recent MC–ICP–MS studies clearly showed a positive correlation between $\delta^{30}Si$ and SiO_2 among terrestrial igneous rocks (Savage et al. 2011, 2012; Zambardi et al. 2014; Poitrasson and Zambardi 2015). This led Savage et al. (2011, 2014) to define for rocks representing an equilibrium melt assemblage an igneous array of equation:

$$\delta^{30}Si\ (‰) = 0.0056 \times SiO_2\,(wt\%) - 0.567 \tag{8}$$

As will be discussed later, this trend is difficult to attribute solely to simple fractional crystallization. These recent works also revealed that the trend was scattered for some granitoids (S-type, or peraluminous leucogranites) and andesites (Fig. 9), possibly as a result of a source effect. A further study of granulitic xenoliths representative of the lower crust revealed a narrow range of $\delta^{30}Si$, from −0.43 to −0.15‰ (Savage et al. 2013b), comparable to that found for igneous rocks from the upper continental crust (Fig. 8). This led the authors to conclude that the process generating the $\delta^{30}Si$ ranges observed in the lower and upper continental crust were comparable and that the lower crust did not have a distinct isotopic signature. In contrast, silicic sedimentary rocks, as represented by shales and clastic metasediments, exhibit a much larger $\delta^{30}Si$ range compared to that uncovered for igneous rocks, from −0.82‰ to 0.01‰ (Fig. 8). This was interpreted by Savage et al. (2013a) as reflecting, for the light values, the incorporation of clay rich-material, that are known to develop light $\delta^{30}Si$ values when produced through continental rock weathering (Ziegler et al. 2005b; Opfergelt et al. 2012). At the other end of the spectrum, Savage et al. (2013a) suggested that the heavy $\delta^{30}Si$ of some shales, with $\delta^{30}Si$ close to 0‰, could reveal seawater-derived silica that has $\delta^{30}Si$ close to 1‰ at a global scale (De La Rocha et al. 2000). However, these silicic sediments represent only a minor portion of the continental crust silicon budget (Wedepohl 1995), and this reservoir

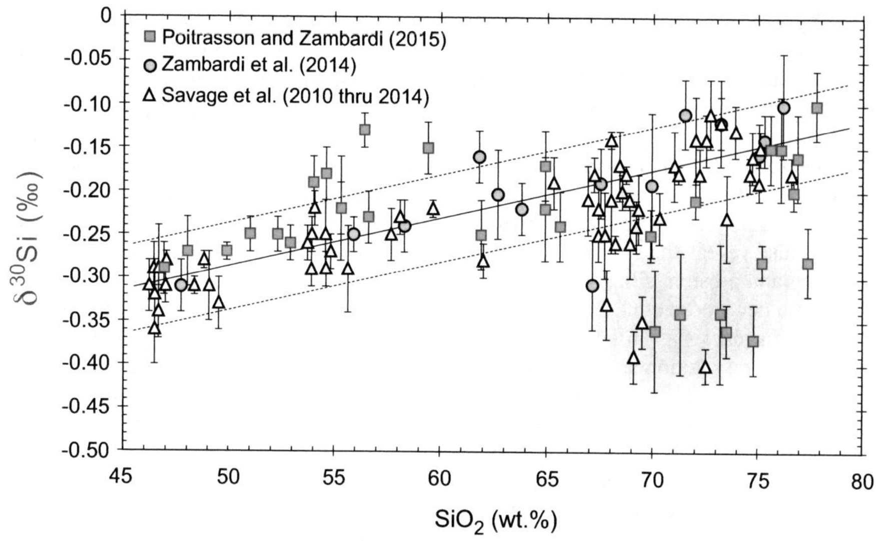

Figure 9. Silicon isotope compositions relative to the NBS-28 reference of bulk terrestrial igneous samples plotted against their SiO_2 concentrations. The solid line represents the linear "igneous array" defined by Savage et al. (2014), of Equation (8) from the text, and the dashed lines its 2SE envelope, corresponding to ±0.05‰ on $\delta^{30}Si$ values. Note that anatectic peraluminous granitoids deviate markedly towards low $\delta^{30}Si$ values at SiO_2 contents of 67 wt% and above (shaded area). Reversely, some andesites yield $\delta^{30}Si$ significantly above the igneous array at SiO_2 contents between 55 to 57 wt%. Both deviations are interpreted in terms different sources having contrasted silicon isotope signatures. Data include those from Savage et al. (2010, 2011, 2012 and 2014). After Poitrasson and Zambardi (2015).

therefore does not affect significantly the upper continental crust mean value computed to be −0.25 ± 0.16‰ (2SD) in $\delta^{30}Si$ (Savage et al. 2013a).

Oceanic crust The igneous portion of the oceanic crust consists essentially of Mid Ocean Ridge Basalts (MORBs), their underlying gabbros and minor ultramafic cumulates. These rocks typically display SiO_2 contents ranging from 47 to 51 wt% (Wilson 1989), with an estimated mean of ~50 wt% (Lodders and Fegley Jr 1998). Of more minor volumetric importance are the Oceanic Island Basalts (OIBs) that show a comparable range of SiO_2 content. This oceanic crust is overlain by detritic and chemical sediments that display a large range of silicon concentrations. They go from ~90% silica in biogenic diatomites and chemically precipitated cherts to very little silicon in carbonate or iron-oxide-rich deposits. The latter may be found in shallow water carbonated platforms or in the iron-rich layers of Precambrian Banded Iron Formations (BIFs).

The precise $\delta^{30}Si$ values reported so far for MORBs and OIBs (Fig. 8) do not differ markedly from continental mafic rocks (e.g., Savage et al. 2010). Except for some OIB from the Northern Rift Zone of Iceland or characterized by an elevated time-integrated $^{238}U/^{204}Pb$ ratio and having $\delta^{30}Si$ going down to −0.46‰ (Pringle et al. 2016), most of these rocks show a narrow range of −0.36 to −0.22‰. They therefore yield a mean value undistinguishable from the BSE. In contrast, cherts and BIFs are among the rocks showing the largest mass-dependent silicon isotope fractionation on Earth (Fig. 8) since they display a $\delta^{30}Si$ range going from −3.7 to 5‰ (e.g., Andre et al. 2006; Robert and Chaussidon 2006; van den Boorn et al. 2010; Heck et al. 2011), excluding the lightest $\delta^{30}Si$ compositions affected by diagenesis (Stefurak et al. 2015). The largest variations are found in studies using *in situ* techniques (SIMS), although reconstructed "bulk-layer" data nevertheless show large ranges of several per mil (e.g., Heck et al. 2011). A recent review on the topic may be found in Chakrabarti (2015).

Interestingly, Marin-Carbonne et al. (2012) showed data suggesting that Proterozoïc cherts appear to have a heavier and narrower range of δ^{30}Si compositions compared to Archean chert silicon isotopic signatures. The latter remain pristine and were not affected by metamorphic and metasomatic resetting according to André et al. (2006). It has been proposed that water temperature of ancient oceans could be estimated based on the δ^{30}Si values of cherts (Robert and Chaussidon 2006). However, the data were scattered and the Archean ocean temperatures so computed (70 °C) seemed high. Subsequent work concluded that the δ^{30}Si signatures of ancient cherts may rather reflect alternative processes. Besides their formation temperature, they may also reveal the isotopic signature of the source that may come from continental weathering and erosion, silicification of clastic sediments or from ocean floor hydrothermal activity (van den Boorn et al. 2007). Furthermore, δ^{30}Si compositions in cherts may not be as resistant to diagenesis as initially thought (Marin-Carbonne et al. 2012; Stefurak et al. 2015), since repeated dissolution and reprecipitation steps of quartz during diagenesis may produce very negative δ^{30}Si values (Basile-Doelsch et al. 2005; Stefurak et al. 2015). This has led to the definition of preservation criteria, in part based on the study of recent (Eocene) cherts drilled on the ocean floor (Marin-Carbonne et al. 2014). Recently, Chakrabarti et al. (2012) highlighted that cherts associated with BIFs tend to yield lighter δ^{30}Si (Fig. 10). Based on experimental work showing that adsorption of silica onto iron oxides tend to favor isotopically light silicon (Delstanche et al. 2009), they concluded that the chert precipitation mechanisms will also have an imprint on δ^{30}Si values. Combined with open-system, Rayleigh-type silicon isotopic fractionation from silica-saturated pore-fluids that leads to up to 3‰ δ^{30}Si variations among peritidal Proterozoic cherts within a single basin (Chakrabarti et al. 2012), this makes paleo-oceanographic reconstructions even more complicated than initially thought. *In situ* studies of the chert layers of Archean BIFs may help to disentangle the various processes at play and yield a better estimate of the silicon isotopic composition of the ocean water of that time, which remains another big challenge (Heck et al. 2011).

Upper mantle. The compositional range observed among the main rocks of the upper mantle, spinel and garnet peridotites, goes from ~42 to 48 wt% SiO_2 (Maaloe and Aoki 1977), although the range is larger when rarer dunites and pyroxenites are taken into account. Bulk present mantle estimates yield ~45 wt% (Lodders and Fegley Jr 1998), that is 5 wt% less than the oceanic crust mean SiO_2 content.

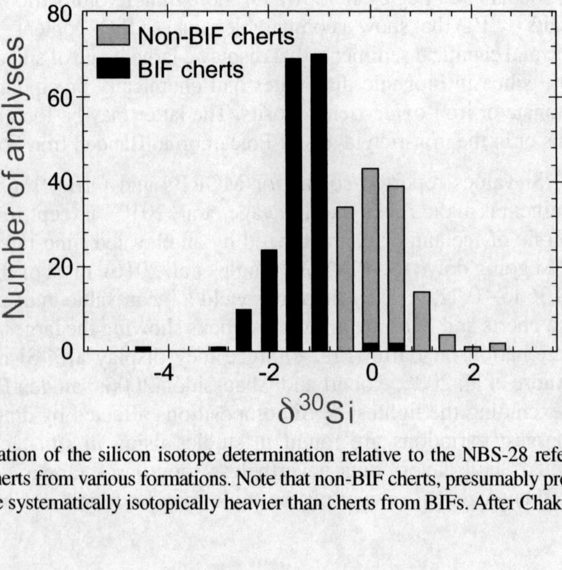

Figure 10. Compilation of the silicon isotope determination relative to the NBS-28 reference measured by MC–ICP–MS on cherts from various formations. Note that non-BIF cherts, presumably precipitated in peritidal environments, are systematically isotopically heavier than cherts from BIFs. After Chakrabarti et al. (2012).

As far as silicon isotope compositions are concerned, three different research groups returned a consistent narrow range of –0.39 to –0.23‰ in δ^{30}Si (Fig. 8) for bulk mantle peridotites of different petrological characteristics and from various locations around the world (Fitoussi et al. 2009; Savage et al. 2010; Zambardi et al. 2013). Collectively, these data yield a mean δ^{30}Si that is indistinguishable from that of the BSE, as well as from basalts. This led Savage et al. (2014) to conclude that mantle partial melting generates only restricted silicon isotopic fractionation, if any, as is also the case for iron isotopes (Poitrasson et al. 2013). However, Huang et al. (2014) raised the question as to whether it is possible to estimate a correct BSE δ^{30}Si value on the basis of upper mantle peridotites and mantle-derived basalts only. They did *ab initio* calculations of the silicon isotope partition functions for various mantle phases and found that if silicon goes from a four-fold coordination, like for olivine in the upper mantle, to a six-fold coordination, as in Mg-perovskite in the lower mantle, this may lead the deep Earth mantle to favor isotopically light silicon. Assuming a primordial magma ocean deep enough to allow equilibration with high pressure phases, and a subsequent preservation of the mantle isotopic zoning throughout the Earth history, Huang et al. (2014) could explain why the Earth has an isotopically heavy δ^{30}Si compared to most other planets and chondrites, as previously proposed for δ^{57}Fe by Polyakov (2009). However, the preservation of a primordial mantle isotopic zoning throughout the Earth's history is questionable and the effect of the change of the silicon coordination chemistry in olivine on δ^{30}Si values has yet to be verified experimentally.

Core. Like the lower mantle, the Earth's core remains mysterious in many respects. While seismology describes it as consisting of a solid inner core and a liquid outer core made of a Fe–Ni alloy, with approximately 10% of lighter elements (Birch 1952), the exact composition remains unknown. Hirose et al. (2013) reviewed the studies addressing this problem and among the three most frequently proposed light element candidates to be present in the Earth's core (O, Si and S), the most recent works suggest that silicon may occur at concentrations ranging from 2.8 to 12.5 wt%. Aside from a few ungrouped iron meteorites (Wai and Wasson 1969), most iron meteorites, taken to be representative of asteroidal cores, contain typically sub-ppm silicon concentrations (Pack et al. 2011). However, the metals of enstatite meteorites and of aubrites, their achondritic equivalent, that both formed under very reducing conditions (with $fO_2 \sim IW-8$), contain percent levels of silicon (Ziegler et al. 2010). This may give clues on the possible conditions required to incorporate silicon in the Earth's core during its formation, although more refined accretion models should also be taken into account (Corgne et al. 2008). Specifically, the progressively increasing temperature and pressure occurring during the terrestrial core segregation may have compensated for likely higher oxygen fugacity conditions relative to the aubrite parent body formation to incorporate silicon in the Earth's metallic core. Georg et al. (2007a) first determined that the Earth and the Moon have heavy δ^{30}Si relative to chondrites Aided by *ab initio* silicon isotope fractionation calculation estimates between the silicate and metallic portions of the Earth, they proposed that this was an indication that silicon entered the core in quantities close to the maximum permissible by geophysical constraints. If correct, this finding would be one of the most spectacular recent applications of silicon stable isotope geochemistry in planetary sciences. Subsequent studies showed that, while the Georg et al. (2007a) isotopic data were fraught with an accuracy issue, the relative planetary differences were correct and they supported the proposed interpretation (Fitoussi et al. 2009; Armytage et al. 2011). However, the light δ^{30}Si composition of enstatite chondrites make these unlikely to be viable building blocks of the Earth as this would yield unrealistically high amount of Si in the core (Fitoussi and Bourdon 2012; Zambardi et al. 2013). Using a more realistic, protracted model for core formation with evolving *P–T–X* conditions and/or the consideration of recently measured angrite heavy δ^{30}Si composition, recent studies (Zambardi et al. 2013; Dauphas et al. 2015; Sossi et al. 2016) have somewhat changed the picture. They concluded that the heavy silicon isotope composition of the Earth and Moon relative to chondrites and most other solar system rock bodies analyzed so far require an additional or alternative mechanisms relative to the unique silicon incorporation in the Earth's core. This applies even when the recent smaller metal–silicate Si isotopic fractionation factor, relative to previous studies, produced experimentally by Hin et al. (2014) is used. Only Savage et al. (2014) seem to find a protracted terrestrial core formation model still compatible with the unique explanation of silicon incorporation in the Earth's metallic core. However, the details given on the modeling are insufficient to compare with other recent studies of this kind and understand what is the key parameter that may explain those differing conclusions. In contrast, other studies require, to quantitatively account for these isotopic measurements, the additional effect of vaporization from a Moon-forming giant impact (Zambardi et al. 2013; Pringle et al. 2014). Alternatively, planetary δ^{30}Si compositions may also in part reflect nebular condensation processes occurring in the proto-planetary accretion disk (Dauphas et al. 2015). If correct, this new idea may lead us to considerably revise our way of interpreting mass-dependent stable isotope composition variations of silicon and iron among planets (Sossi et al. 2016) that were so far interpreted as resulting from either a Moon-forming giant impact, core–mantle differentiation and/or crust extraction. And, as mentioned in the previous section, a heavy upper silicate mantle isotope composition might also be explained by lower mantle minerals favoring isotopically light silicon (Huang et al. 2014). Hence, all these recent works would lead towards the lower-end of geophysical silicon concentration estimates (e.g., 3.6 +6.0/−3.6 wt% Si, according to Dauphas et al, 2015). But overall, they show that the question of the concentration (and therefore corresponding isotopic composition) of silicon in the Earth's metallic core remains largely open.

Seawater. Dissolved silicon concentrations in the oceans range from less than 0.1 to up 180 μM Si (See review and references in de Souza et al. 2014), corresponding to about 0.003 to 5 ppm (or mg/l) of Si. Given the ocean's surface pH~8.2, dissolved silica occurs essentially in the form of orthosilicic acid (H_4SiO_4; see Fig. 7). This concentration varies with the seasons in the ocean's surface sunlit (euphotic) waters because it is controlled by diatom activity that requires this dissolved silicon to make their frustules, i.e., opaline cell walls. As a result, dissolved silica concentrations are particularly low in tropical euphotic waters during planktonic blooms. Dissolved silica profiles show systematically increasing concentrations with depth, although the exact shape of the concentration profiles may vary with locations in the world oceans (e.g., De La Rocha et al. 2000; Cardinal et al. 2005; Reynolds et al. 2006; de Souza et al. 2012a,b; Fripiat et al. 2012; Grasse et al. 2013). According to Tréguer and De La Rocha (2013), a large fraction of the silica input to the ocean is brought by the dissolved load from rivers (6.2 ± 1.8 Tmol/year of Si). Other more minor dissolved silicon inputs come from the seafloor weathering, hydrothermal fluids, aerosols and groundwaters (Fig. 11). Conversely, an even larger output flux is estimated through the burial of diatoms and sponges living on the continental shelves (Treguer and De La Rocha 2013). The balance is compensated by the dissolved silica output from groundwaters to oceans at continental margins (Fig. 11). Its precise quantitative estimate has yet to be made, however (Georg et al. 2009). According to Treguer and De La Rocha (2013), the overall silicon residence time in the oceans, corresponding to the total dissolved Si divided by the net input (or output), is ~10 000 years.

Despite the considerable interest in using stable silicon isotope compositions to infer diatom productivity in order to better constrain the carbon cycle in current and past oceans (see e.g., Holzer and Brzezinski 2015), it has been only 15 years since the first of such silicon isotopic measurements of seawater was published (De La Rocha et al. 2000). The silicon isotope measurement database has since steadily increased (e.g., Varela et al. 2004; Cardinal et al. 2005; Reynolds et al. 2006; Beucher et al. 2008; De La Rocha et al. 2011; de Souza et al. 2012a,b; Grasse et al. 2013). We now have a reasonably good idea of the δ^{30}Si systematics in the world oceans, with dissolved Si having a large δ^{30}Si range relative to igneous rocks, going from 0.5 to 4.4‰ (Fig. 8). De La Rocha et al. (1997) demonstrated by laboratory cultures that diatoms preferentially take up light silicon, thereby leaving residual silicon in water

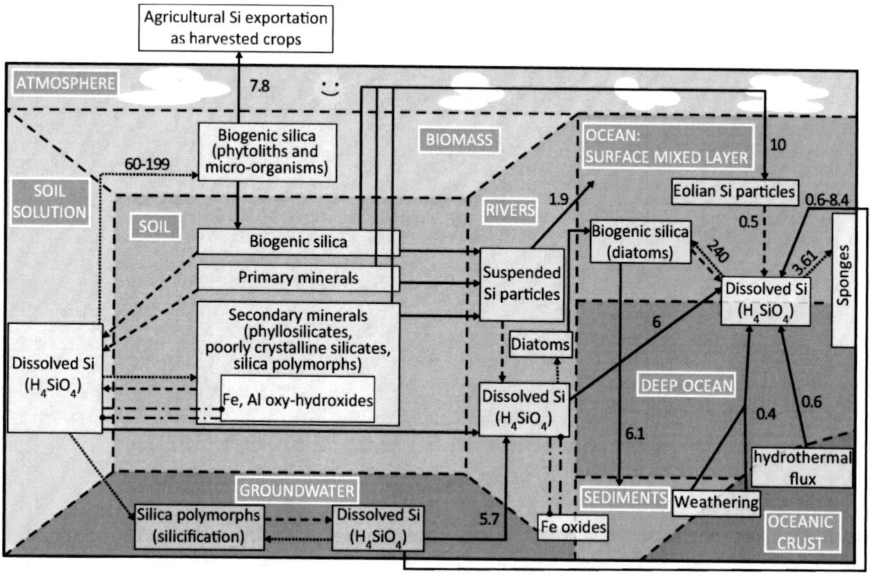

Figure 11. Biogeochemical cycling of silicon after Cornelis et al. (2011), with some Si flux (thick numbers recalculated in teramoles of Si per year and updated from the more recent figures of Tréguer and De La Rocha, 2013), along with more details for the oceanic compartments, when needed. Numbers from Georg et al. (2009) and Frings et al. (2016) were also taken into account. Solid line arrows mean transport; Dashed line: dissolution; Dotted line: precipitation; Point-dotted line: adsorption/desorption. Note the large uncertainty remaining on the dissolved silica uptake from soils solutions by plants and micro-organisms or from the dissolved silicon released to the oceans by groundwaters.

enriched in heavy isotopes. Subsequent open ocean fertilization experiments confirmed this with a similar apparent silicon isotope fractionation factor between diatoms and seawater of $\Delta^{30}Si = 1.36 \pm 0.11‰$ (Cavagna et al. 2011). These experimental results may be used to explain the negative relationship between $\delta^{30}Si$ and silica concentration in shallow seawater (De La Rocha et al. 2000; Reynolds et al. 2006; de Souza et al. 2012a; Grasse et al. 2013). Shallow euphotic zones depleted in dissolved silicon display particularly elevated $\delta^{30}Si$ values. This shows the major role of phytoplankton in the silicon oceanic cycle that can be seen even when silicon concentration and isotopic measurements are made along profiles going from the surface to the ocean floor (Fig. 12). Current models of the silicon cycle integrating measurement data show maps with superficial waters with particularly heavy $\delta^{30}Si$ in tropical oceans as a result of diatom blooms that consume dissolved silicon (Fig. 13). In contrast, deep waters show more constant $\delta^{30}Si$ signatures that are rather controlled by inputs coming from, e.g., the dissolution of the opaline frustules when they sink after the death of diatoms, rivers, and oceanic currents. Hence, whereas silicon is a major nutrient near the ocean surface, silicon isotope compositions are considered as a conservative tracer in the deep ocean (de Souza et al. 2014). This made it possible to propose a current average $\delta^{30}Si$ value of ~1.2‰ for the dissolved silicon in the deep Pacific and Atlantic Oceans in the southern hemisphere, which are less influenced by shallower waters as in the northern hemisphere (de Souza et al. 2012a,b). This therefore supports the initial global ocean estimate of $\delta^{30}Si = 1.1 \pm 0.3‰$ by De La Rocha et al. (2000).

River and lake waters. Rivers typically have lower pH than the oceans, from ~8 down to ~4, and their dissolved silicon is thus also in the form of orthosilicic acid. However, rather than thermodynamic equilibrium, silicon concentrations are more controlled by evaporation rates, weathering mechanisms and rates, along with biological activity (i.e., diatoms, sponges

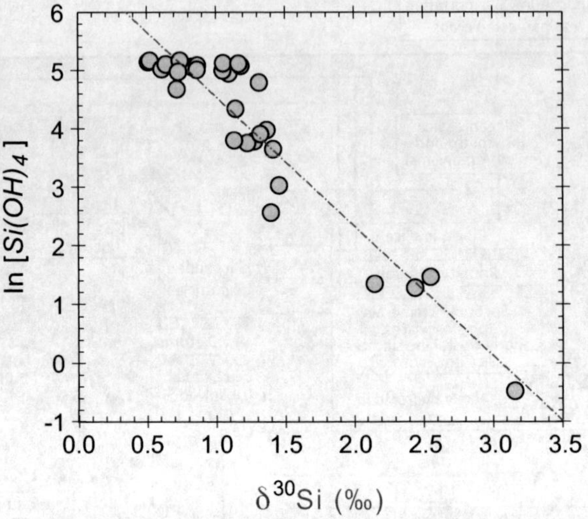

Figure 12. Linear relationship observed in seawater between dissolved silicon isotope compositions expressed relative to the NBS-28 standard against the logarithm of orthosilicic acid concentration, in micromoles, in the Northern Pacific Ocean. This trend is interpreted as reflecting to a first order diatoms activity in surface waters, although it is also impacted by silicon-rich Pacific deep waters because the water samples encompass the whole water column. After Reynolds et al. (2006).

and the riparian vegetation that produces and releases silicon in the form of phytoliths). As a result, dissolved silicon concentration ranges in river waters may be large but in most cases below equilibrium values (of ~120 ppm; Fig. 7). For instance, they have been found to vary from 7 to 673 μM Si in the Nile River and its tributaries (Cockerton et al. 2013), corresponding to 0.2 to 18.9 ppm (or mg/l) of Si. However, the Nile River is the longest in the world and it probably encounters more variable hydroclimatic conditions, weathering contexts and different biological activity as it flows towards the Mediterranean Sea than any other river, which may explain this large range. Accordingly, Ding et al. (2011) showed in a review that rivers usually display average dissolved silicon concentrations ranging from about 60 to 200 μM. These studies were in part motivated by the need to know the silicon output from rivers to the oceans and Georg et al. (2009) pointed out that the groundwater system associated with the rivers at the surface should not be neglected in this context. In the Bay of Bengal, groundwaters showed dissolved silicon concentrations higher than the river waters, even in the dry season (with values going as high as 790 μM), and they accounted for 2/3 of the silicon flux released by the Ganges-Brahmaputra Rivers themselves at the surface to the Indian Ocean (Georg et al. 2009).

Lake waters have pHs that are within the riverine range, although they may occasionally be higher, by up to ~9 or more (Opfergelt et al. 2011) which may imply a higher solubility and more complex silicon aqueous speciation in water (Fig. 7). However, here again, silicon concentrations may be dictated by kinetic and biological processes since dissolved silicon concentration in lake waters are in the range of those found in rivers, and therefore lower than equilibrium values.

Aside from Precambrian silicon chemical deposits found as cherts and banded iron formations and plants, river waters show the largest variations in silicon isotopic composition on Earth: they range from −0.70 to 4.66‰ in δ^{30}Si (Fig. 8). Dissolved silicon generally has positive δ^{30}Si values, with again the heaviest silicon isotopic compositions measured in the Nile River and its tributaries (Cockerton et al. 2013), and they also have been found to exhibit seasonal variations. It has been proposed that this probably results from the combined effect of variable residence time of the waters in their watershed, which enhances weathering processes

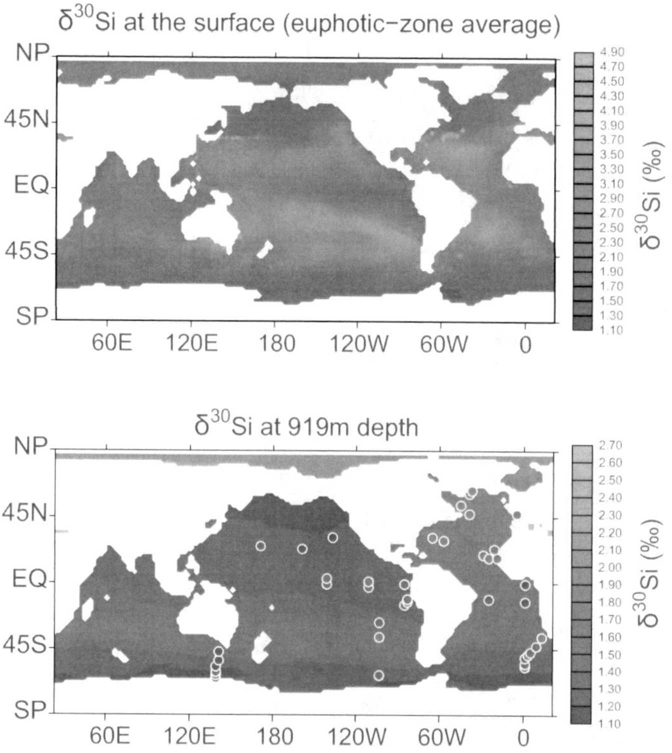

Figure 13a. Silicon isotope maps of the world oceans, expressed relative to the NBS-28 standard. The heaviest $\delta^{30}Si$ are found in the tropical zones whereas the lighter silicon isotope signatures occur at high latitudes and near some equatorial zones of the surface waters. At 919 m deep however, the silicon isotope composition of the oceans is much less variable, and mostly varies between 1.1 to 1.5‰. The dots on the lower pannel represent measurements, with a median uncerainty of ±0.14‰. NP: North Pole; EQ: Equator; SP: South Pole. After Holzer and Brzezinski (2015).

in low water level periods (and thereby increases dissolved silicon $\delta^{30}Si$), and/or changes of biological activity with the seasons (e.g., Engstrom et al. 2010; Ding et al. 2011; Hughes et al. 2011b; Cockerton et al. 2013; Pokrovsky et al. 2013; Frings et al. 2015). These dissolved $\delta^{30}Si$ values have also been shown to increase downstream as a result of lower water flow, and therefore increased water residence time in the river bed, which again increases weathering (e.g., Cockerton et al. 2013; Frings et al. 2015). While congruent mineral dissolution will not significantly fractionate silicon isotopes relative to the continental crust $\delta^{30}Si$ value, weathering accompanied with clay mineral formation will lead to isotopically heavy river waters as a result of light isotope incorporation in clay minerals (Ziegler et al. 2005b; Opfergelt et al. 2012; Savage et al. 2013a). Any subsequent redissolution of these clays-rich soils, as observed in intertropical acidic black waters of the Congo and the Amazon river basins, will lead to isotopically lighter dissolved silicon, with $\delta^{30}Si$ approaching 0‰ (Cardinal et al. 2010; Hughes et al. 2013). Hence, different weathering processes based on the different tropical water chemistries will produce contrasting trends in a dissolved silicon content versus $\delta^{30}Si$ diagram (Fig. 14). Opfergelt and Delmelle (2012) and Frings et al. (2016) also surmised that there should be a relationship between the relative importance of mechanical erosion and chemical weathering on the one hand, and the dissolved silicon concentration and isotopic composition of the river waters on the other hand. In the extreme cases of mechanical erosion (as in mountains, and/or cold streams, like in Iceland, Georg et al. 2007b) or under hot and wet chemical weathering (as

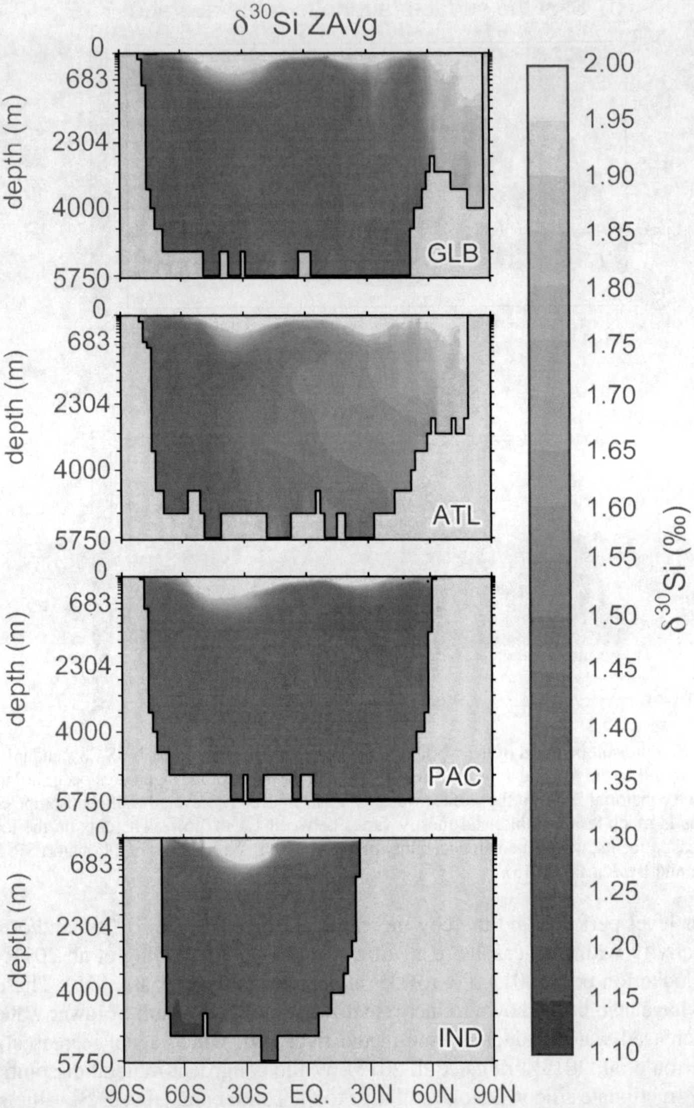

Figure 13b. Silicon isotope cross sections of the world oceans, expressed relative to the NBS-28 standard. The cross sections also show a general tendency of decreasing $\delta^{30}Si$ with increasing depth from 2 to 1.1‰ for the Indian (IND) and Pacific (PAC) Oceans, with an added gradient of decreasing $\delta^{30}Si$ from 2 to 1.1‰ going from the Northern to the Southern latitudes of the Atlantic (ATL) and therefore global (GLB) Ocean. After Holzer and Brzezinski (2015).

under equatorial evergreen forests of Congo or Brazil, Cardinal et al. 2010; Hughes et al. 2013), minimal isotopic fractionation should occur between waters and minerals. In between these two extreme cases, the opposite should happen and a significant silicon isotopic fractionation is expected between surface waters and minerals once the water–rock interaction has occurred. This relationship has yet to be properly demonstrated in nature, however, and it should only be possible to see it clearly in the rare cases where the biological imprint is minimal.

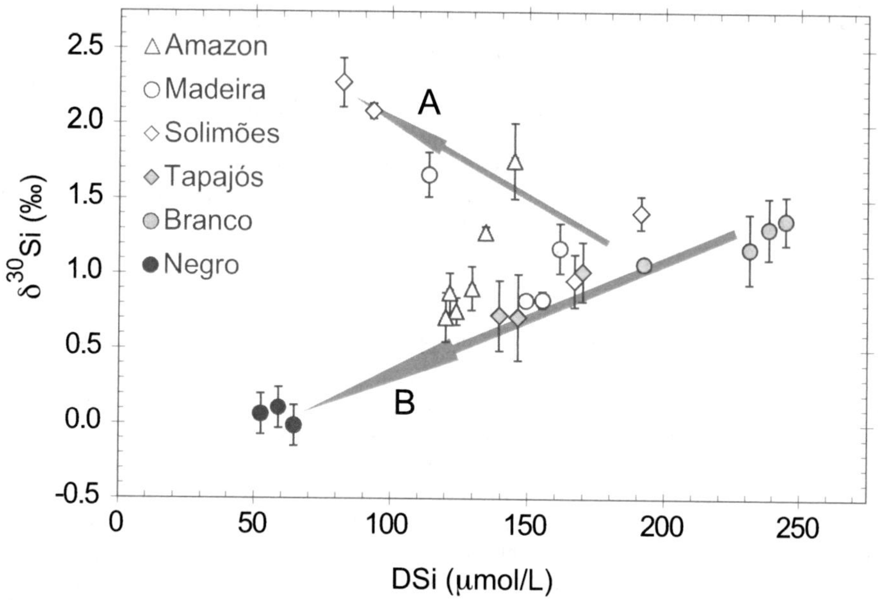

Figure 14. Silicon isotope composition of dissolved silicon (DSi) from rivers of the Amazon Basin. Note the two different trends towards lower Si concentration possibly denoting two different weathering styles: with clay formation, which leads to heavy dissolved silicon in the Solimoes and Madeira River Basins (trend A), or with clay dissolution, which leads towards light dissolved silicon isotope compositions in the Negro River Basin (trend B). After Hughes et al. (2013).

Diatom growth is the biological activity having the strongest imprint on the silicon cycle in rivers. It will also lead to isotopically heavier dissolved riverine silicon as a result of the light silicon isotope uptake by diatoms to make their opaline frustule (e.g., Hughes et al. 2011b, 2013). Hence, whether of weathering or biological origin, the particulate silicon should be isotopically lighter than the dissolved silicon, which is what Ding et al. (2004, 2011) found in the Yangtze and the Yellow rivers from China. They measured a riverine suspended matter δ^{30}Si range going from −0.7 to 0.3‰, thereby encompassing the BSE δ^{30}Si value (Fig. 8) and the upper continental crust estimate of −0.25 ± 0.16‰ (Savage et al. 2013a). However, quantum mechanical calculations show that equilibrium silicon isotope fractionation should lead to aqueous H_4SiO_4 isotopically lighter than minerals like quartz and kaolinite (Dupuis et al. 2015). This therefore supports the idea that the silicon isotope systematics in rivers is not controlled by equilibrium processes occurring between water and minerals, but rather by kinetic reactions involved during weathering and biological processes (e.g., Frings et al. 2015). This view is also supported by silicon adsorption experiments on the solid aluminum hydroxide gibbsite showing that faster reaction will favor preferential adsorption of lighter isotopes (Oelze et al. 2014). As a result of these large dissolved δ^{30}Si variations among major rivers depending on the weathering regime and biological activity, Ding et al. (2011) concluded that the riverine δ^{30}Si output to the oceans is more variable than initially surmised (De La Rocha et al. 2000) as it likely varies among regions, seasons and probably also through geological times. As mentioned above, groundwaters associated with riverine systems may represent a significant proportion of the dissolved silicon released from continental runoff to the ocean (Georg et al. 2009). Although the available data remain scarce, deeper groundwater may be lighter than associated surface riverine waters (e.g., δ^{30}Si down to −0.15‰ for the groundwater of the Ganges–Brahmaputra River System). And this has to be taken into account in estimates of the dissolved silicon isotopic composition delivered to the oceans from the riverine systems (Fig. 11) since these groundwaters may represent an additional 40% of dissolved Si flux in Ganges–Brahmaputra River case.

Silicon isotope studies of lake waters uncovered δ^{30}Si values of dissolved silicon within the range found in river waters (Alleman et al. 2005; Opfergelt et al. 2011; Panizzo et al. 2016). Like in oceans, lake waters show a dissolved silicon concentration decrease towards the top of the water column linked to increasingly heavier silicon isotope composition as a result of light silicon uptake by diatoms (Alleman et al. 2005). Biological activity also imprints a seasonal change on dissolved Si concentration and δ^{30}Si, which will therefore affect the values continental waters deliver to the oceans throughout the year (Opfergelt et al. 2011). In a recent work, Panizzo et al. (2016) confirmed the key role of diatoms on the dissolved silicon concentration and isotopic composition of Lake Baikal. They also found through diatom sampling throughout the water column using a sediment trap, and on lake surface sediments, the lack of silicon isotope fractionation during diatom dissolution. This therefore validates the use of diatom δ^{30}Si from freshwater sediment core for palaeoenvironmental reconstructions.

Soils. Bulk soil silicon concentrations will depend on the parent rock chemistry from which these soils originated through weathering. This will also depend on the weathering degree along with the local climatic conditions. As a result, bulk soil SiO_2 contents have been reported to show a large range, varying between ~2 to 77 wt% (e.g., Ziegler et al. 2005a; Opfergelt et al. 2010, 2012; Steinhoefel et al. 2011). However, soils are complex systems and these and other studies have shown that they may consist of remains of the parent rock minerals such as quartz, newly formed crystalline phases like various clay minerals, amorphous silica that frequently derive from plant opal products found as phytoliths in soils after plant organic compounds have decayed. Furthermore, a full account of the silicon systematics in soils should also consider interstitial soil solutions that carry dissolved silica between soil, plants, groundwater and rivers (Fig. 11), and the quantitative silicon fluxes among these compartments remain poorly known. There have been a number of reviews of the silicon cycling through the plant–soil–river and ocean continuum published, to which the reader is referred for more details (Conley 2002; Sauer et al. 2006; Leng et al. 2009; Cornelis et al. 2011; Opfergelt and Delmelle 2012).

Bulk soils have a δ^{30}Si ranging from −2.7 to 0.1‰ so far (Fig. 8). Like river waters, they encompass the upper continental crust estimate (−0.25±0.16‰; Savage et al. 2013b), but they appear to be somewhat mirroring the riverine δ^{30}Si relative to the continental crust on the light silicon isotopic side (Fig. 8). This likely results from the kinetic silicon isotopic fractionation between water and clay mineral discussed in the previous section that leads to a residual soil containing clay minerals enriched in light δ^{30}Si relative to the starting rock (Ziegler et al. 2005b; Steinhoefel et al. 2011; Opfergelt et al. 2012). Interestingly, and in contrast with silicon concentrations, those bulk soil δ^{30}Si compositions do not seem to be much affected by the nature of the parent rock, or by the weathering style accompanying different climatic conditions (Opfergelt and Delmelle 2012). In contrast, there seems to be a negative relationship between silicon isotope composition and the weathering intensity, i.e., as weathering increases, the soil δ^{30}Si becomes lighter (Ziegler et al. 2005a; Opfergelt and Delmelle 2012; Opfergelt et al. 2012). This trend has been shown using various indicators of weathering intensity, including the Chemical Index of Alteration (CIA; Fig. 15) of Nesbitt and Young (1982), defined by:

$$\text{CIA} = \left(\left(\text{Al}_2\text{O}_3\right)/\left(\text{Al}_2\text{O}_3 + \text{CaO} + \text{Na}_2\text{O} + \text{K}_2\text{O}\right)\right) \times 100 \qquad (9)$$

where oxides are reported in moles. Additional factors also affect soil δ^{30}Si values. Besides clay formation, the adsorption of dissolved silicon onto iron oxyhydroxides (Delstanche et al. 2009; Opfergelt et al. 2009) or aluminium hydroxides (Oelze et al. 2014) will also retain light δ^{30}Si in the soils. Interestingly, silcretes, that are the products of repeated silicon dissolution and reprecipitation in pedogenetic contexts and beyond in the Earth's crust display the lightest δ^{30}Si found so far (−5.7‰; Basile-Doelsch et al. 2005). This illustrates the capability of repeated kinetic dissolution-reprecipitation processes to produce extremely light silicon isotope compositions.

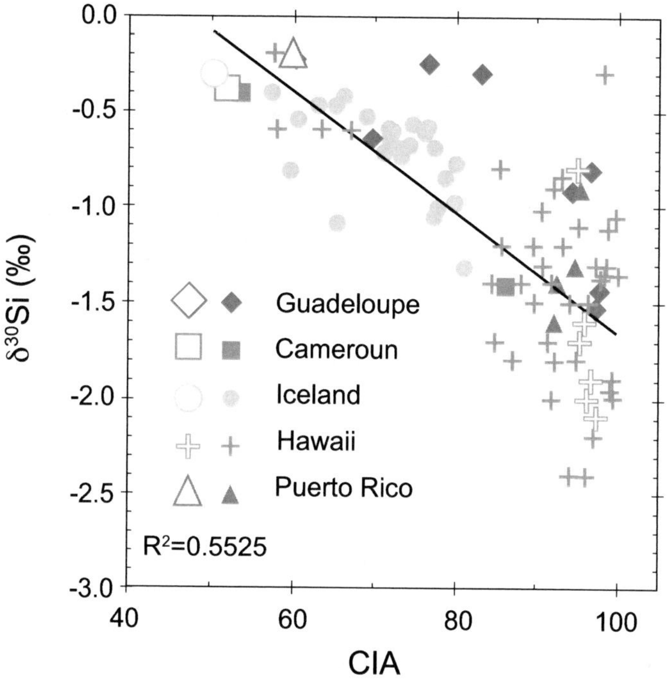

Figure 15. Relationship between the silicon isotope composition of bulk samples relative to the NBS-28 standard as a function of the Chemical Index of Alteration (CIA, see Equation (9) in the text) for various soil profiles (filled symbols) and their parent rocks (open symbols). Note the tendency towards lighter δ^{30}Si as weathering increases, which is interpreted to reflect clay formation and the release of isotopically heavy silicon in surface waters, corresponding to the trend A of Figure 14. After Opfergelt and Delmelle (2012).

The limit of the trend linking δ^{30}Si and weathering index (Fig. 15) occurs when the isotopically light clay minerals themselves dissolve (Cornelis et al. 2010). This is what happens in tropical podzols drained by equatorial acidic black rivers that, as a result, display isotopically light dissolved silicon relative to most other river waters, as discussed above (Fig. 14). In some soils, airborne particulate minerals, such as quartz and mica, deposited on the soil upper horizons should also be taken into account for a correct, well mass-balanced interpretation of the silicon isotope variations in soil profiles. Their input was computed to bring a component with a δ^{30}Si of ~–0.5 to 0.1‰ (Ziegler et al. 2005b, after Ding et al. 1996; Bern et al. 2010).

Plants. A complete understanding of soil formation and evolution cannot be made without the consideration of the biomass, and notably the impact of plants on the shallower soil horizons. Plant activity, characterized by the littering of biogenic silica at the soil surface and dissolved silicon uptake in deeper horizons, will also affect the δ^{30}Si of bulk soils along depth profiles (Ziegler et al. 2005b; Bern et al. 2010). Plants preferentially take up light silicon from interstitial soil solutions to produce opal in their structures during their growth (e.g., Ding et al. 2005; Sun et al. 2008; Delvigne et al. 2009). Plant species vary in the amount of silicon they accumulate and therefore in their phytolith production. For instance, gramineae, like rice, are notable silicon accumulators (Fig. 16). Besides strengthening their structure, plants need to accumulate this element because it enhances their resistance to diseases and pest. It also improve their light-interception ability and minimize transpiration losses and therefore plant resistance to drought (Ma et al. 2006). Subsequent silicon transport towards higher plant organs will translate into phytoliths production via silica polymerization and then silica-gel formation.

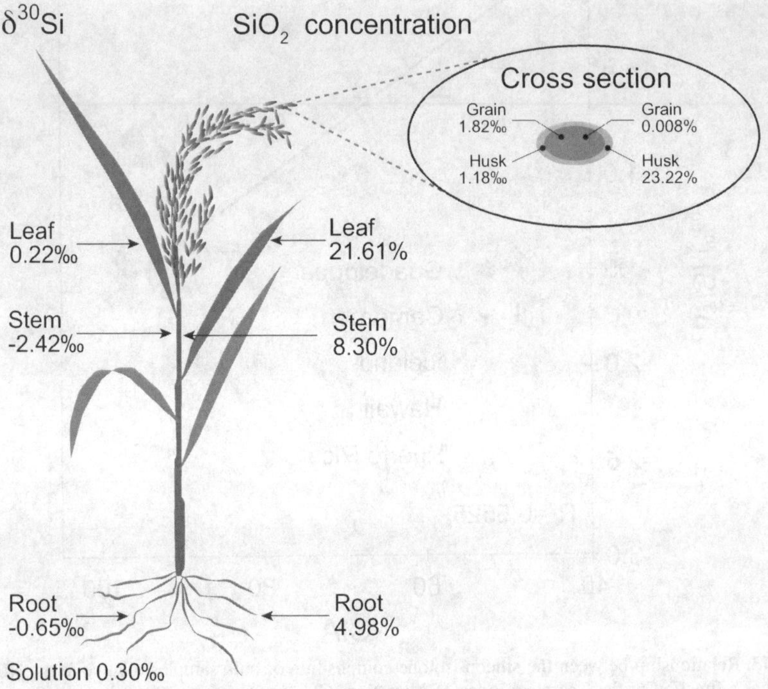

Figure 16. Silicon isotope composition relative to the NBS-28 standard (left) and SiO_2 concentration in wt% (right) of different parts of hydroponic cultivated rice plants. Each value represents the mean of four analyses from four different plants. See text for discussion. After Sun et al. (2008).

Phytoliths can be extremely heavy isotopically in strong silicon accumulators (up to ~6‰ $\delta^{30}Si$ measured in rice grains; Ding et al. 2005). As a result, plant littering may lead into either light (Ziegler et al. 2005b) or heavy silicon isotope additions through their phytoliths to the upper soil horizons (Bern et al. 2010). Whilst a comprehensive review of the silicon isotope systematics in plants is probably beyond the scope of this section (but see Leng et al. 2009, for a more comprehensive account), Ding et al. (2005, 2008a,b) and Opfergelt et al. (2008) have shown that after the preferential uptake of light silicon from soil solutions, subsequent silicon transport will translate into heavier $\delta^{30}Si$ as silicon diffuses through the plant towards the upper organs, possibly with evapotranspiration as the main driving factor (Ding et al. 2008b). There might be exceptions, however, since Delvigne et al. (2009) showed an opposite $\delta^{30}Si$ trend for the most part of laboratory cultivated banana plants. Overall, these authors concluded on the basis of both natural case studies and experimentally grown plants on hydroponia that this process may be modeled through a Rayleigh isotopic fractionation model for at least rice and bamboo (Ding et al. 2005; Ding et al. 2008b; Ding et al. 2008a). The resulting range of $\delta^{30}Si$ measured in plant organs sampled in the field is, together with cherts and BIFs, among the largest found on terrestrial materials, and goes from −2.3 to 6.1‰ (Fig. 8).

Atmosphere. Our current understanding of the silicon cycling in the atmosphere is limited to dust as sampled in loess deposits (Savage et al. 2013a), or computed estimates of $\delta^{30}Si$ of dust inputs in soil profiles discussed in the previous section (Ziegler et al. 2005a; Bern et al. 2010). Ding et al. (2011) also considered that the suspended particulate matter of the

Yellow River draining the Loess plateau essentially sampled this deposit. Silicon concentration of representative loess samples studied by Savage et al. (2013a) range from 52 to 81%, a range somewhat larger than that reported in an earlier global study (Taylor et al. 1983). They consist mainly of quartz, feldspars, sheet silicates and sometimes carbonates, depending on the geology of the eroded continental rocks that lead to the dust which subsequently formed these loess deposits (Taylor et al. 1983). Although the loess were sampled from various continents with different geological histories, the δ^{30}Si values uncovered were extremely narrow, ranging from −0.28 to −0.15‰ (Fig. 8) and thus scattering about the continental crust. These silicon isotope results confirm that loess deposits can provide a sampling of a wide continental area by virtue of the dust transfers and mixing by winds, which is why loess deposits are interesting for studies aiming at global continental estimates (e.g., Taylor et al. 1983). Nevertheless this δ^{30}Si range might expand somewhat as the database will become more comprehensive. It is consistent though with the less precise estimates previously made by Ziegler et al. (2005a) and Bern et al. (2010) from mass balance calculation in soils or with the Yellow River suspended matter analysis from Ding et al. (2011).

ELEMENTAL AND ISOTOPIC BEHAVIORS DURING MAJOR GEOLOGICAL PROCESSES

The aim of the remaining of this chapter is to review the isotopic fractionation mechanisms of silicon isotopes beyond processes common to stable isotopes of every element, such as their dependency on temperature, that may be found elsewhere in the literature (e.g., Criss 1999; Valley and Cole 2001; Johnson et al. 2004; Hoefs 2015) or in previous chapters of this book. We will therefore focus on isotopic effects linked to the specific silicon chemical properties, mineral hosts and/or aqueous speciation summarized at the start of this chapter. These are those specific fractionation mechanisms related to, for example, the general occurrence of silicon as SiO_4 tetrahedra in silicate minerals, its particular aqueous speciation or specific biological utilization that will make the study and subsequent use of silicon isotope fractionation systematics of interest in Earth and environmental sciences. Most of these processes were already touched upon in the previous paragraphs reviewing the silicon stable isotope composition variations in various terrestrial and extra-terrestrial reservoirs.

As reminded by Ding et al. (1996), the relative mass difference between ^{28}Si and ^{30}Si is of similar magnitude (1/14) to those of nitrogen or sulfur isotopes and therefore, one would expect at first sight similar ranges of stable isotope fractionation on the Earth. However, while δ^{15}N and δ^{34}S show overall ranges in excess of 100‰ (Hoefs 2015), δ^{30}Si values barely reach a tenth of this range (Fig. 8). This is because silicon does not share the rich redox chemistries of nitrogen and sulfur, nor does it commonly occur in various physical and chemical forms. These are powerful factors for stable isotope fractionation (e.g., Criss 1999; Valley and Cole 2001; Johnson et al. 2004; Hoefs 2015). Silicon is also less present than nitrogen and sulfur in biological processes, which tend to leave a kinetic imprint on isotopic signatures typically leading to larger variations than equilibrium processes. The common silicon occurrence as SiO_4 tetrahedra in silicate rocks along with the main speciation of this element as H_4SiO_4 in aqueous solution from acidic to pH ~9 allow for only limited stable isotopic fractionation on Earth.

Diffusion, condensation and evaporation

There is a growing body of literature on theoretical studies of stable isotope fractionation of silicon and other elements aimed at investigating the physical processes behind diffusion in silicate melts at elevated temperatures. Two main diffusion processes may generate kinetic isotope effects in igneous rocks: chemically driven and thermally driven diffusion. The latter is generated by a thermal gradient, as opposed to a chemical gradient, and has attracted much attention in the recent theoretical literature (Dominguez et al. 2011; Lacks et al. 2012; Li and Liu 2015). Whereas most authors generally agree that the cold end of a volume of magma will be enriched in heavier isotopes and the hot end will be isotopically lighter, there is a debate on the simplifications taken to make atomistic models of the diffusion process. According to Li and Liu (2015), important parameters have erroneously been ignored in previous modeling efforts. These include vibrational and rotational energies at molecular excited states or intermolecular interaction and inelastic collisions, thereby leading to unrealistically zero point energy differences, for example. Whereas theoretical studies flourish, experimental studies against which they can be tested remain scarce. Nevertheless, the modeling efforts lead to a reasonable reproduction of the only existing experimental data on silicon isotope kinetic fractionation generated by diffusion in a silicic melt under a thermal gradient (Richter et al. 2009). In particular, they were able to reproduce the lack of a compositional effect, along with the slower fractionation of silicon isotopes upon diffusion relative to metals (Li and Liu 2015). This effect is attributed to the nature of the diffusing species, presumably consisting of silicon oxides like SiO_3, SiO_4 or Si_2O_5 instead of free species or monoxides for metals. Li and Liu (2015) proposed the following equation describing the thermal isotopic fractionation of O and Si between the starting (T_0) and final temperature (T) gradient of a silicate melt, for example for silicon:

$$\Delta^{30}\mathrm{Si}_{T-T_0} = -\frac{3}{2}\ln\left(\frac{M^*}{M}\right)\ln\frac{T}{T_0} \tag{10}$$

where M and M^* are the molecular weight of the different diffusing species, * denoting the heavy isotopologue. This equation could, in principle, shed light on the nature of the diffusing molecule and further experimental studies will be needed to assess this. The next question is whether thermally-driven diffusion is geologically significant beyond the few centimeter scale, as advocated by some (Lundstrom 2009; Zambardi et al. 2014). One difficulty is to maintain over a sufficient amount of time (probably of a million years time-scale, Lundstrom 2009) a thermal gradient in a large magma body to allow the process to operate, thus potentially restricting this effect to only slow cooling igneous plutons (Savage et al. 2014). In real geological cases, Zambardi et al. (2014) have proposed that this feature might be enhanced at the km-, pluton-scale through repeated top-down sill injection of a magma providing new inputs of silicic melts to maintain a large-scale thermodiffusion process explaining the zoning observed for silicon and other stable isotopes. However, Zambardi et al. (2014) themselves noted that the silicon isotopic variations in igneous bodies do not unequivocally argue for thermodiffusion as the main responsible factor. This process may be blurred by the occurrence of crystallizing minerals in the melt and the observed isotopic trend may well be a mix of several factors, including the effect of mineral fractional crystallization previously proposed (Savage et al. 2011). Clearly, besides experimental work, there is a strong need for more detailed studies of igneous bodies involving silicon isotope determinations at the bulk-rock and mineral scale in conjunction with structural, petrological and mineralogical data. On the basis of limited silicon concentration gradients found in natural systems, Richter et al. (1999) hypothesized that chemically driven diffusion is unlikely to significantly fractionate silicon isotopes. This view was subsequently backed by more recent diffusion experiments in chemically simplified silicic melts confirming the slower diffusivity of silicon relative to other cations and the putatively much more reduced resulting effect on silicon isotopes, if any, relative to those of Ca, Mg, Fe or Li (Watkins et al. 2011). In meteorites, where larger silicon concentration gradients may occur between silicate and metal, Ziegler et al. (2010) computed that the silicon diffusivities are too slow to allow for a significant isotopic effects at temperatures of 1200 K or lower.

Silicon evaporation from a melt in vacuum and/or condensation are other processes through which notable mass-dependent silicon isotopic fractionation can be generated. A typical example for such processes are early refractory materials found as inclusions in meteorites, or even presolar grains encountered in primitive meteorites. Besides the change in physical state, silicon may also become more reduced due to the very low partial pressure of oxygen in space and form SiO and SiS gases or SiC in presolar grains. These silicon redox changes, not found in the silicate Earth, are the reasons why thousands of permil δ^{30}Si ranges have been uncovered in presolar grains (Fig. 1; Zinner et al. 2006), besides changes in physical state. Furthermore, the nucleosynthetic effects described above, which generate mass-independent isotopic fractionation, may also add an extra component of mass-dependent silicon isotope variations. To explain the large mass-dependent silicon isotope variation encountered in non-FUN CAI inclusion that gave a range in excess of 20‰ in δ^{30}Si, Clayton et al. (1978) proposed a Rayleigh-type evaporation/condensation model to account for the observed isotopic data. This interpretation proved successful according to subsequent vaporization experiments (e.g., Davis et al. 1990; Knight et al. 2009) as long as a liquid melt is being evaporated. Solid sublimation shows no isotopic fractionation effect because silicon diffusion in the solid is much slower than the vaporization process. This prevents isotopic reequilibration of the solid between its margins and its inner portions, in contrast to a liquid that does it at a faster pace (Davis et al. 1990). According to this Rayleigh model, the evolution of the ^{30}Si/^{28}Si ratio of a silicic melt undergoing evaporation may be described by:

$$\frac{\left(\frac{^{30}\mathrm{Si}}{^{28}\mathrm{Si}}\right)}{\left(\frac{^{30}\mathrm{Si}}{^{28}\mathrm{Si}}\right)_0} = f_{^{28}\mathrm{Si}}^{\,\alpha_{\mathrm{Si}}-1} \tag{11}$$

where $\left(\frac{^{30}\mathrm{Si}}{^{28}\mathrm{Si}}\right)_0$ is the isotopic composition of the starting melt, $f_{^{28}\mathrm{Si}}$ is the ^{28}Si fraction remaining in the evaporating melt and α_{Si} is the kinetic isotopic fractionation factor for silicon isotopes. This equation describes a process by which vaporization corresponds to a specific isotopic fractionation between melt and vapor, but the vapor is subsequently isolated from the solid and no longer equilibrates with it. Taking the natural logarithm of both sides of Equation (11) leads to the following convenient expression:

$$\ln\left(\frac{\left(\frac{^{30}\mathrm{Si}}{^{28}\mathrm{Si}}\right)}{\left(\frac{^{30}\mathrm{Si}}{^{28}\mathrm{Si}}\right)_0}\right) = -(1-\alpha_{Si})\cdot \ln f_{^{28}\mathrm{Si}} \tag{12}$$

Using this expression, Knight et al. (2009) compared their silicic melt evaporation residues aiming at reproducing vaporization processes occurring in space during CAI inclusions formation and evolution. They also included previous experimental results from Davis et al. (1990) and Wang et al. (2001). The results showed several interesting properties of the evaporation process on silicon isotopes (Fig. 17). First, and unlike Mg isotopes (Richter et al. 2007), silicon isotope kinetic fractionation upon evaporation does not seem to be affected by temperature, at least within the range (1600–2050 °C) investigated in the three studies. Second, there is a clear melt composition dependence of the kinetic fractionation factors, as illustrated by the three sets of experiments using forsterite, chondrite-like or CAI-like starting compositions (Fig. 17). Lastly, this type of experiment can give an idea of the vaporizing

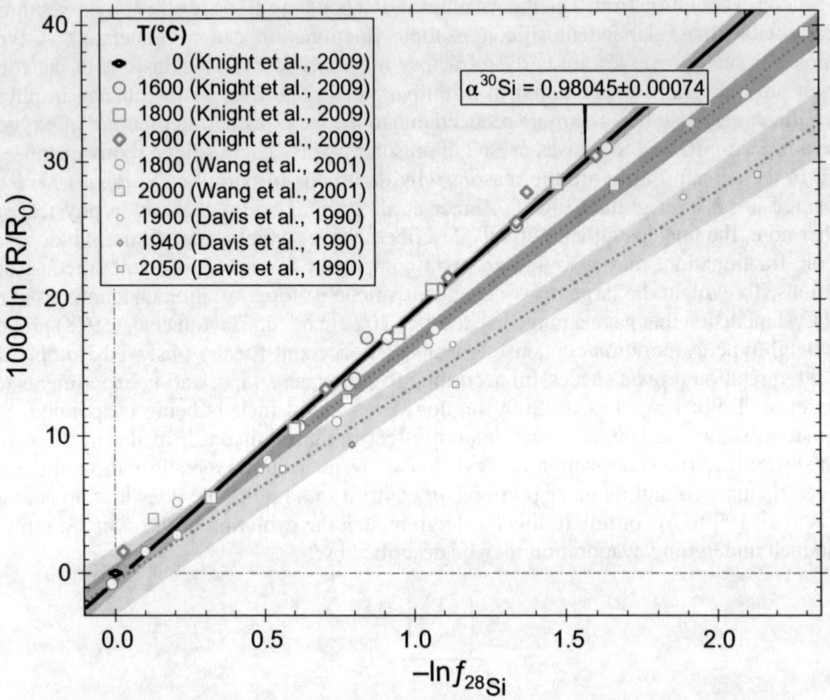

Figure 17. Rayleigh diagram showing the relationship between the $^{30}Si/^{28}Si$ ratio (R) of melt residues as a function of the remaining silicon during evaporation experiments. The corresponding Rayleigh expression is Equation (12) in the main text. This graph shows that whereas different melt compositions (CaO–MgO–Al_2O_3–SiO_2: CMAS, solid line and its 2 sigma envelope (Knight et al. 2009); CI chondrite-like (Wang et al. 2001), middle dotted line and its 2 sigma envelope; Mg_2SiO_4 (Davis et al. 1990), lower dotted line and its 2 sigma envelope) lead to different isotopic fractionation factors, the evaporation temperature has no effect on this parameter within the investigated range of 1600–2050 °C. After Knight et al. (2009).

silicon species since the kinetic isotopic fractionation factor may be expressed as:

$$\alpha_{Si} = \sqrt{M / M^*} \tag{13}$$

where M^* is the isotopologue with ^{30}Si, e.g., in the case of Equations (11) and (12), and M is the light isotopologue with ^{28}Si. Using this expression, Knight et al. (2009) concluded that the silicon-bearing vaporizing species from their experiments was silicon monoxide rather than elemental silicon. It should also be mentioned here that Knight et al. (2009) measured ^{29}Si as well and verified that these processes were mass-dependent. Hence, with this theoretical and experimental background, it is clear that silicon isotopes measured in refractory inclusion of meteorites can give insight into the conditions that occurred in the earliest times of the solar protoplanetary accretion disk. Such experimental results were also useful to assess more quantitatively the hypothesis by which silicon vaporization upon an interplanetary Moon-forming impact could account for the observed $\delta^{30}Si$ values of the Earth–Moon system. Using the kinetic isotopic fractionation factor for silicon of Knight et al. (2009), Zambardi et al. (2013) computed that only ~1% of the bulk Earth's Si loss through volatilization can account for the observed $\delta^{30}Si$ shift between the Earth–Moon system and chondrites. This resolved an issue since the simple incorporation of silicon into the terrestrial metallic core cannot account for the observed interplanetary $\delta^{30}Si$ differences (Zambardi et al. 2013). It should be noted at this point that such a Moon-forming interplanetary impact should have been followed by a thorough homogenization process of the ejecta before re-accretion to form the Earth and Moon to explain their identical $\delta^{30}Si$ compositions (e.g., Pahlevan and Stevenson 2007).

Condensation may also be modeled through the same Rayleigh equation (e.g., Clayton et al. 1978) to reproduce the silicon isotopic fractionation occurring of silicates from a gas. In this case, the $\left(\frac{^{30}Si}{^{28}Si}\right)_0$ ratio of Equation (11) becomes the isotopic composition of the starting gas and $f_{^{28}Si}$ is the ^{28}Si fraction of gas condensed. In this modeling, Clayton et al. (1978) define the kinetic isotopic fractionation factor as:

$$\alpha_{Si} = \frac{\left(\frac{^{30}Si}{^{28}Si}\right)_{solid}}{\left(\frac{^{30}Si}{^{28}Si}\right)_{gas}} \tag{14}$$

It is typically above 1, e.g., 1.002 for a condensing solid heavier by 2‰ in $\delta^{30}Si$ relative to the gas. Again, the solid, once condensed, no longer equilibrated isotopically with the gas. This Rayleigh approach is used in many other contexts, such as upon biotic or abiotic opal precipitation from aqueous silica or during silica uptake by plants from soil solutions, as will be discussed below. The case for equilibrium gaseous condensation has been computed by Javoy et al. (2012) using isotopic partition functions obtained via *ab initio* techniques between some relevant silicon species deemed to be important in the protoplanetary accretion disk, such as silicon monoxide. They found that at 1000 K, a silicate like enstatite will be heavier by ~4‰ in $\delta^{30}Si$ relative to the condensing SiO gas whereas iron metal containing silicon ($Fe_{15}Si$) will be

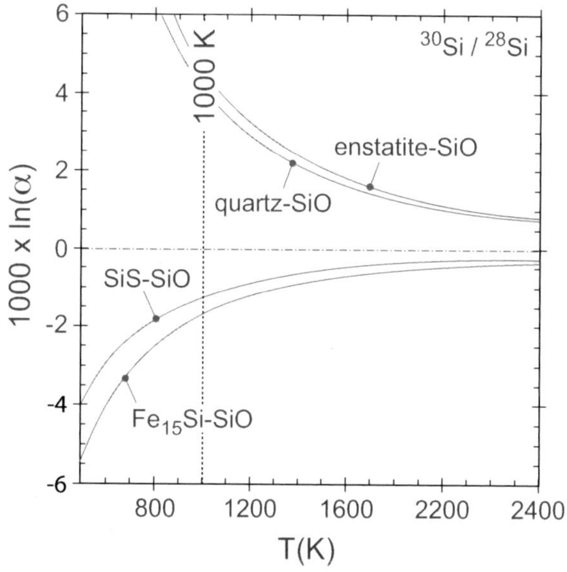

Figure 18. First-principles computed silicon isotopic fractionation factors of various species relative to gaseous SiO, as a function of temperature in Kelvins. Note that whereas silicate partition functions become progressively enriched in heavy silicon isotopes relative to SiO when temperature decreases, the opposite trend is observed for silicon alloyed to metallic iron, or occurring as SiS gas relative to the SiO gas. After Javoy et al. (2012).

lighter by ~−2‰ at the same temperature (Fig. 18), which is consistent with enstatite chondrites and aubrites δ^{30}Si analyses and experimental calibration of Ziegler et al. (2010) and Shahar et al. (2009, 2011). These and other *ab initio* calculations (Méheut et al. 2009) were subsequently used by Dauphas et al. (2015) to explain the δ^{30}Si of angrites found as heavy as those of the Earth. The model involved an equilibrium condensation model of fayalite from SiO from the nebular gas leading to correlated δ^{30}Si and Mg/Si ratio among planetary bodies. As mentioned above, this view may lead us to considerably revise our way of interpreting mass-dependent stable composition variation of silicon and iron among planets (Sossi et al. 2016).

Igneous processes

While bulk rock δ^{30}Si differences between mafic and felsic igneous rocks were hard to see with the analytical uncertainties available more than 40 years ago, it was recognized since the 1970's that minerals from lunar and terrestrial rocks showed systematic differences (see compilation in Douthitt 1982). This was subsequently confirmed by more precise data obtained using modern MC–ICP–MS methods, although the database remains slim and, like for the bulk-rock case discussed above, the range measured among igneous minerals is narrower than initially reported (see compilation in Savage et al. 2014). The general tendency is that mafic minerals like olivine and pyroxenes are lighter than feldspars; in granitoids, micas are lighter than feldspars, themselves lighter than quartz. This therefore seems to support the early view of Grant (1954) who suggested on the basis of theoretical computations that δ^{30}Si should correlate positively with the degree of polymerization of SiO_4 tetrahedra. This means that in igneous rocks, nesosilicates would tend to be lighter than inosilicates, themselves lighter than phylosilicates and the tectosilicates being the heaviest. Subsequent theoretical efforts using quantum mechanical approaches subsequently confirmed this in broad terms, although the key parameter is the mean Si–O interatomic distance of the crystalline structures, not strictly the degree of polymerization of SiO_4 tetrahedra (Fig. 19). These authors also noted that other parameters also play a role, such as the occurrence of certain metals like Al and Mg in the structure (Méheut et al. 2009; Méheut and Schauble 2014). Some minerals, like forsterite, are still problematic as they do not follow the general trend computed between silicon isotope fractionation factors and the average Si–O bond length of silicate minerals (Fig. 19). However, Qin et al. (2016) recently showed that considering the mean volume of the SiO_4 tetrahedron rather than mean Si–O bond length leads to a better fit of forsterite with other silicate minerals in such a graph. Another interesting property revealed by *ab initio* calculations is the sensitivity of silicon fractionation factors for some minerals, like forsterite, and its high pressure polymorph, ringwoodite, to the very high pressures occurring in the deep mantle, beyond the phase change effect (Huang et al. 2014). However, all these computational efforts are facing the very restricted mineral silicon isotope database currently available in the literature, which therefore limits their validation and allows their use in broad terms only.

Georg et al. (2007a) made the initial suggestion that the heavy Si isotope composition of the Earth–Moon system relative to other planets and asteroids could be explained by the incorporation of silicon in the Earth metallic core. Therefore, equilibrium partitioning between an iron alloy and silicate melt under the early Earth conditions should result in the heavy silicon going into the silicate fraction. It was subsequently concluded that additional processes/explanation are required to fully account for the interplanetary δ^{30}Si differences (Zambardi et al. 2013; Dauphas et al. 2015). Nevertheless, such an equilibrium silicon isotope fractionation between metallic core and silicate mantle and its sense has been confirmed by experimental studies (Shahar et al. 2009, 2011; Hin et al. 2014). This effect has now been observed between the metal and silicate fractions of meteorites forming under particularly reducing environments (fO_2=IW−4; Berthet et al. 2009, such as enstatite chondrites (Savage and Moynier 2013) and their achondritic equivalents, aubrites (Ziegler et al. 2010). Such very reducing environment favors silicon incorporation in the metallic phase. These naturally based, experimental and computational calibrations of the metal–silicate δ^{30}Si

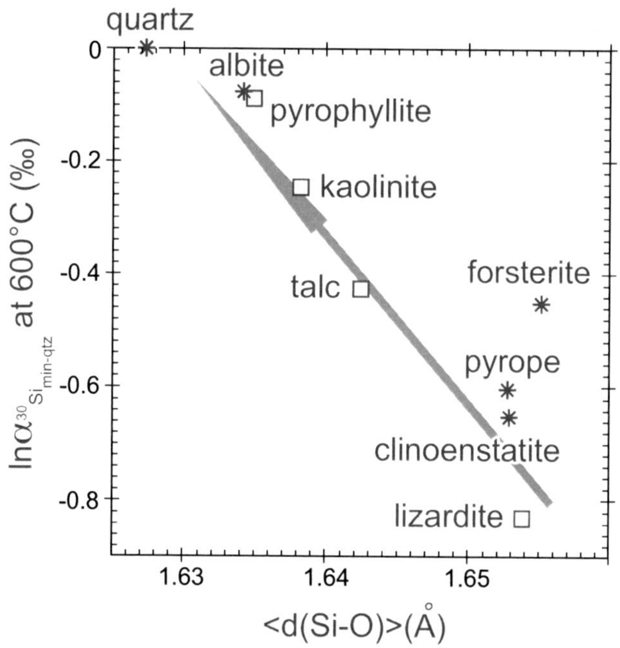

Figure 19. First-principles computed silicon isotopic fractionation factors of various mineral phases relative to quartz (a tectosilicate), as a function of the mean Si–O interatomic distances of the crystalline structures. Note that forsterite, a nesosilicate, deviate off the trend, whereas phyllosilicates (squares) with different structural arrangements and compositions can have vastly different silicon isotopic fractionation factors. All this suggests that the mean Si–O cannot fully account for the relative differences among mineral isotopic fractionation factors. Other computed phases (stars) are albite (tectosilicate), clinoenstatite (inosilicate) and pyrope (nesosilicate). See text for details. After Méheut and Schauble (2014).

fractionation factor (in ‰) lead to the equation:

$$\Delta^{30}Si_{\text{metal-silicate}} = -\frac{B \cdot 10^6}{T^2} \tag{15}$$

where the temperature, T is in Kelvins and the coefficient B varies among the studies from 4.42 ± 0.05 (Hin et al. 2014) to 7.64 ± 0.47 (Ziegler et al. 2010). This is one of the largest (and so far best characterized) equilibrium silicon isotope fractionations occurring at high temperature (Fig. 20) because it is one of the rare cases where the silicon redox changes, from 4^+ in the silicate phase to 0 in the metallic phase. Hence, and even at terrestrial mantle–core differentiation conditions (i.e., from ~1700 to 3300 °C; Wade and Wood, 2005), metal–silicate silicon isotope fractionation effects remain within current analytical reach. In contrast, the chondritic silicon isotope composition of Mars likely result from a smaller planet size that resulted into a too low core formation pressure and temperature given the more oxidizing conditions relative to the Earth to incorporate significant amounts of silicon in the Martian metallic core (Armytage et al. 2011; Zambardi et al. 2013).

The lack of silicon redox changes during mantle melting and subsequent magma differentiation make the detection of $\delta^{30}Si$ changes on peridotites and basalts more difficult, even if occurring at a much lower temperature compared to the conditions of the Earth's core formation. Hence, no silicon isotope fractionation has been detected so far upon mantle melting (Savage et al. 2010, 2014). On the other hand, Savage et al. (2011) could identify a

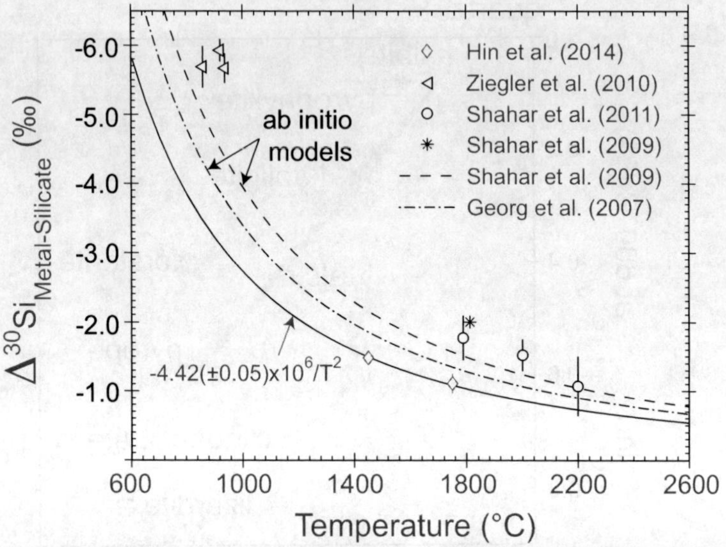

Figure 20. Compilation of the experimental and theoretical silicon isotope fractionation factors between metallic alloys of iron and silicate. This is one of the rare case in nature where silicon changes its redox state, hence the large fractionation factors despite the elevated temperatures. See text for details. After Hin et al. (2014).

general trend of increasing δ^{30}Si with increasing magma differentiation leading to a linear trend termed the "igneous array" (see Equation (8) above). This confirmed earlier hints of Douthitt (1982) and Ding et al. (1996) based on data limited by the analytical uncertainties of that time. Savage et al. (2011) interpreted this positive relationship between δ^{30}Si and SiO_2 of bulk rocks in terms of fractional crystallization of mafic phases like olivine and pyroxene from the evolving melt. This view was subsequently supported by mineral partition function calculations based on first principles that led Méheut and Schauble (2014) to suggest that "light Si isotopes follow electropositive cations", i.e., crystallizing minerals containing elements like Mg and Ca should progressively deplete the silicate melt in light silicon isotopes through fractional crystallization. However, Méheut and Schauble (2014) noted that this view should be considered as "preliminary" because important cations like Fe were not considered in the modeling and because of the anomalous behavior of forsterite in their model (see Fig. 19). Savage et al. (2011, 2014) themselves remained cautious in interpreting their silicon isotope measurements of rocks from the Hekla volcano suite studied in detail. This is because, in contrast to iron isotopes (Schuessler et al. 2009), silicon isotopes could not resolve the contrasting trends of fractional crystallization and magma mixing uncovered earlier using trace elements, oxygen stable isotopes, and strontium radiogenic isotopes (Sigmarsson et al. 1992). They thus concluded that the continuous positive trend between δ^{30}Si and SiO_2 of bulk rocks should rather reveal a "fundamental property of the system". They assigned it to the relative silicon polymerization degree of the phases fractionating versus the melt, i.e., a feldspar crystallizing from a basaltic melt will have a different fractionation factor compared to a feldspar crystallizing from a rhyolite melt, following the initial view of Grant (1954). A further complication arises from the finding that δ^{30}Si of peraluminous granites show a source effect linked to an earlier weathering episode of the protolith, with clay production, leading to lighter isotopic composition (Savage et al. 2012; Poitrasson and Zambardi 2015). There is also a hint that this source effect also occurs for andesites, but this time leading to a heavy silicon composition relative to the "igneous array" (Fig. 9). It was interpreted as a result of isotopically heavy marine silica deposition in the oceanic crust that subsequently affected the δ^{30}Si composition of the andesite source via a subduction zone (Poitrasson and Zambardi 2015). More recently, Pringle et al. (2016) found a slight deviation of δ^{30}Si towards low values for some OIB characterized by their high time-integrated $^{238}U/^{204}Pb$ or from the Iceland Northern Rift Zone. This silicon isotopic shift of opposite direction compared to some andesites was interpreted in terms of altered oceanic crust recycling in the mantle source of these specific basalts coming from Iceland, the Canary Islands and Cape Verde. But overall, a support to the polymerization view is given by the positive relationship between δ^{30}Si and an indicator of the degree melt polymerization, the ratio of non-bridging oxygen over tetrahedral cations (NBO/T; Mysen et al. 1985) in lunar rocks (Fig. 4). Given the lack of notable amounts of water involved in the lunar igneous history, this relationship was not blurred since δ^{30}Si were obviously not affected by a protolith bearing the isotopic signature of water–rock interaction, such as for anatectic peraluminous granitoids or andesites (Poitrasson and Zambardi 2015). Lastly, the effect of thermodiffusion has been proposed to explain the increase of δ^{30}Si with indicators of igneous differentiation (Lundstrom 2009; Zambardi et al. 2014). However, this interpretation is based on experimental observations performed at the cm-scale. Extrapolating it at the intrusion-, km-scale, remains a challenge, as discussed in more detail above. Hence, our exact understanding of the mechanism explaining the positive relationship of δ^{30}Si with igneous differentiation (Fig. 4 and 12) remains elusive. It will require in situ isotopic work by fs LA–MC–ICP–MS to understand the fractionation laws at the mineral-scale on both natural samples and experimental charges to be able to propose a viable mechanism for igneous suites.

Metamorphic processes

There are few studies of the effect of metamorphism on silicon isotope compositions. Ding et al. (1996) investigated various metamorphic rocks. They found no notable silicon isotope imprint of metamorphism within their δ^{30}Si analytical uncertainties for various schists and gneisses with metamorphic grades going up to from the amphibolitic and to the granulitic grade. This conclusion was subsequently confirmed by studies using more precise MC–ICP–MS techniques (Andre et al. 2006; Savage et al. 2013b). In a study investigating a suite of lower crust xenoliths from Australia, Savage et al. (2013b) found no evidence for a granulitic metamorphic imprint on the δ^{30}Si values of these rocks. Their silicon isotope compositions were therefore interpreted to record the isotopic signature of their protolith. André et al. (2006) studied major rock-types from the ~3.8 Ga Isua greenstone belt and their Eoarchean country rocks. They concluded that neither the granutilic metamorphism nor the hydrothermal overprint affected the rock δ^{30}Si values, even at the sub-millimetric scale. This allowed these authors to study the supergene and/or igneous history of the rocks investigated through the metamorphic events they subsequently underwent.

Low temperature processes

As for high temperature geological processes, theoretical calculation of silicon isotope partitioning at equilibrium may provide important benchmarks against which experimental and naturally based studies may be compared. For instance, Fuji et al. (2015) calculated the δ^{30}Si fractionation expected from the orthosilicic acid deprotonation and found that a variation in 0.1 pH unit should translate into a ~0.2‰ change in δ^{30}Si for seawater, which is clearly within the reach of current analytical techniques. They thus proposed that the silicon isotope composition of, for example, ancient seawater records such as cherts could be used as a new paleo-pH proxy. However, given the pKa of the H_4SiO_4 deprotonation reaction (9.46; Fig. 7), the application of this proxy will be limited to the rare natural environments where extremely alkaline pHs prevail, like near some remains of submarine hydrothermal sources, ancient serpentine mud volcanoes or alkaline lakes and streams. Of wider use are the calculation of the equilibrium silicon isotopic fractionation factor between H_4SiO_4 and quartz and between H_4SiO_4 and kaolinite (Dupuis et al. 2015) given their abundance in nature. Such a comparison between minerals and an aqueous species is not trivial, however, since it involves solids and liquids that have to be treated differently in *ab initio* calculations given their contrasted molecular configurations. While the isotopic partition functions of solids are commonly calculated from the density functional theory (DFT), liquids are treated through molecular dynamic simulations. To ensure that the partition functions obtained could be compared whether silicon was an aqueous species or included in a crystalline solid, Dupuis et al. (2015) had to verify that the spectroscopic properties computed matched well experimental values of the same substances. An important outcome of this work is that the computed silicon isotope fractionation between H_4SiO_4 and quartz or between H_4SiO_4 and kaolinite are in opposite sense to those measured in nature or experiments (Fig. 21). In other words, these computations show that many isotopic exchanges observed between dissolved silicon in natural waters and quartz or kaolinite are kinetic in nature and do not reach equilibrium. This is an important step for our understanding of the silicon isotope systematics and exchanges on continental surfaces and in the oceans. Therefore, these theoretical computations (Dupuis et al. 2015) lead to the conclusion that the heavy $\delta^{30}Si$ of dissolved silicon in rivers and the light $\delta^{30}Si$ of soils (Fig. 8) essentially result from the kinetic nature of continental weathering by which igneous rocks are transformed into soils, and which notably involved the alteration of igneous silicate phases into clay minerals.

There is also a growing number of experimental studies substantiating the strong effect of the kinetic nature of many low temperature processes on $\delta^{30}Si$ values. This includes amorphous silica precipitation from aqueous solutions, dissolved silicon adsorption on oxides or the dissolution of biogenic opal. In the case of silica kinetic precipitation, the general principle is that

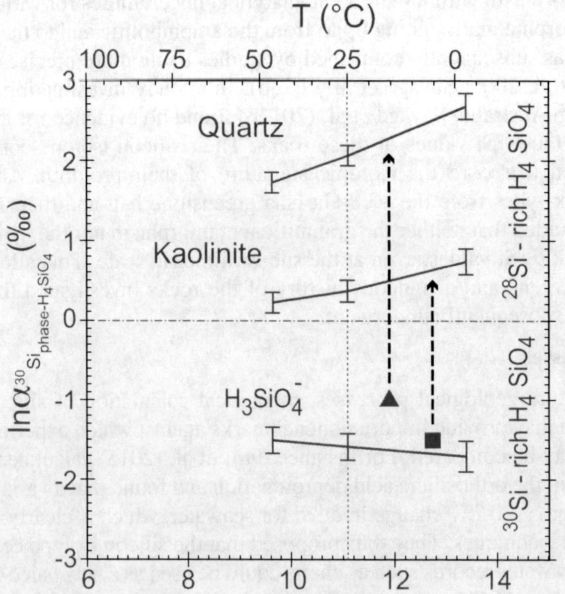

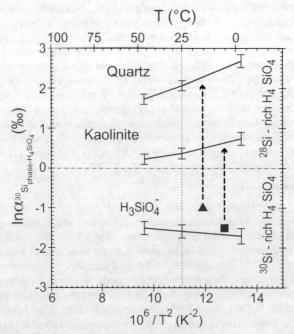

Figure 21. First-principles computed silicon isotopic fractionation factors of quartz, kaolinite and aqueous $H_3SiO_4^-$ against orthosilic acid H_4SiO_4. Note that the sense of isotopic fractionation so calculated at equilibrium (thick lines with error bars) is opposite to what is observed in nature (triangle and square; the thick dashed arrows relate the natural observations to the theoretical fractionation curve). This illustrates the kinetic nature of natural low temperature silicon isotope fractionation between silicate minerals and dissolved silicon in waters. After Dupuis et al. (2015).

the lighter molecular weight of the silica species bearing the lighter ^{28}Si isotope will move and become involved in chemical reactions at a faster rate. Ding et al. (1996) conducted experimental amorphous silica precipitation to illustrate this effect. They obtained a precipitate enriched in light silicon, leaving behind a solution with a higher $\delta^{30}Si$ than the starting solution. Since the precipitate no longer reacts with aqueous silica once formed, this process can be interpreted using the Rayleigh model discussed above. Introducing the delta notation in Equation (11) using:

$$\frac{\left(\frac{^{30}Si}{^{28}Si}\right)}{\left(\frac{^{30}Si}{^{28}Si}\right)_0} = \frac{1000 + \delta^{30}Si}{1000 + \delta^{30}Si_0} \tag{16}$$

leads to:

$$\delta^{30}Si = 1000 \cdot \left(f_{^{28}Si}^{\alpha_{Si}-1} \cdot \left(1 + \frac{\delta^{30}Si_0}{1000}\right) - 1 \right) \tag{17}$$

where the isotopic composition of the silicon remaining in solution ($\delta^{30}Si$ in Eqn. 17) may be computed as a function of the remaining fraction $f_{^{28}Si}$. The isotopic composition of the precipitating silicon may be computed using the kinetic isotopic fractionation factor α_{Si} defined in Equation (14), replacing the silicon isotope composition of the gas by that of the dissolved silicon in solution. These expressions lead to the common graphical representation (Fig. 22) in which it can be clearly seen the progressive $\delta^{30}Si$ increase of the silica remaining in

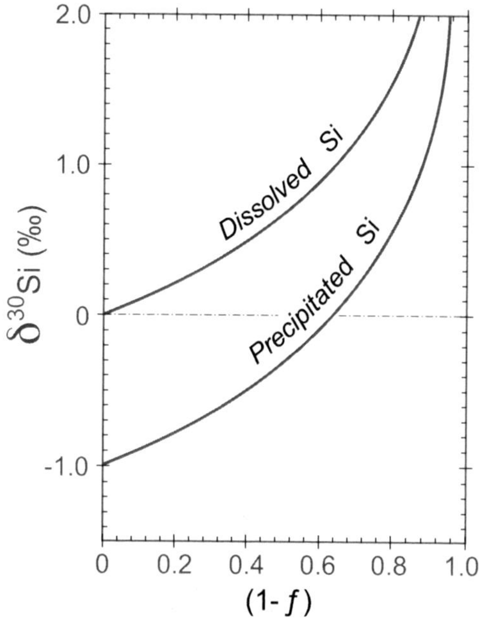

Figure 22. Silicon isotope composition of aqueous and precipitated silicon through a Rayleigh process as a function of the amount of precipitated silica (1–*f*), with *f* being the fraction of silica remaining in solution. This modeling was performed with Equation (17) and an isotopic fractionation factor α_{Si} = 0.999. After Ding et al. (1996).

solution. The precipitating amorphous silica will be initially lighter than the starting solution, but its $\delta^{30}Si$ will progressively increase as a result of the increasing $\delta^{30}Si$ values of the silica remaining in solution and available for further precipitation. More recent experimental investigations in closed reactors have shown that things are further complicated by the fact that the fractionation factor α_{Si} decreases with progressive precipitation and reduced reaction rate (Roerdink et al. 2015). This aspect was subsequently tackled by the same group using flow-through reactors (Geilert et al. 2014). They showed that the apparent fractionation factor was larger at the start of the reaction and it progressively decreased before a steady state was reached, corresponding to conditions approaching chemical equilibrium between amorphous silica and aqueous silicon. Both studies found a direct relationship between silicon isotope fractionation factors and temperature, with fractionation factors decreasing with increasing temperature and nearing 0 when $T \geq 50\,°C$ (Fig. 23). These studies clearly illustrated the sensitivity of abiotic amorphous silica $\delta^{30}Si$ on the precipitation rate and therefore the reaction affinity. It further puts in question earlier paleoenvironmental temperature reconstructions

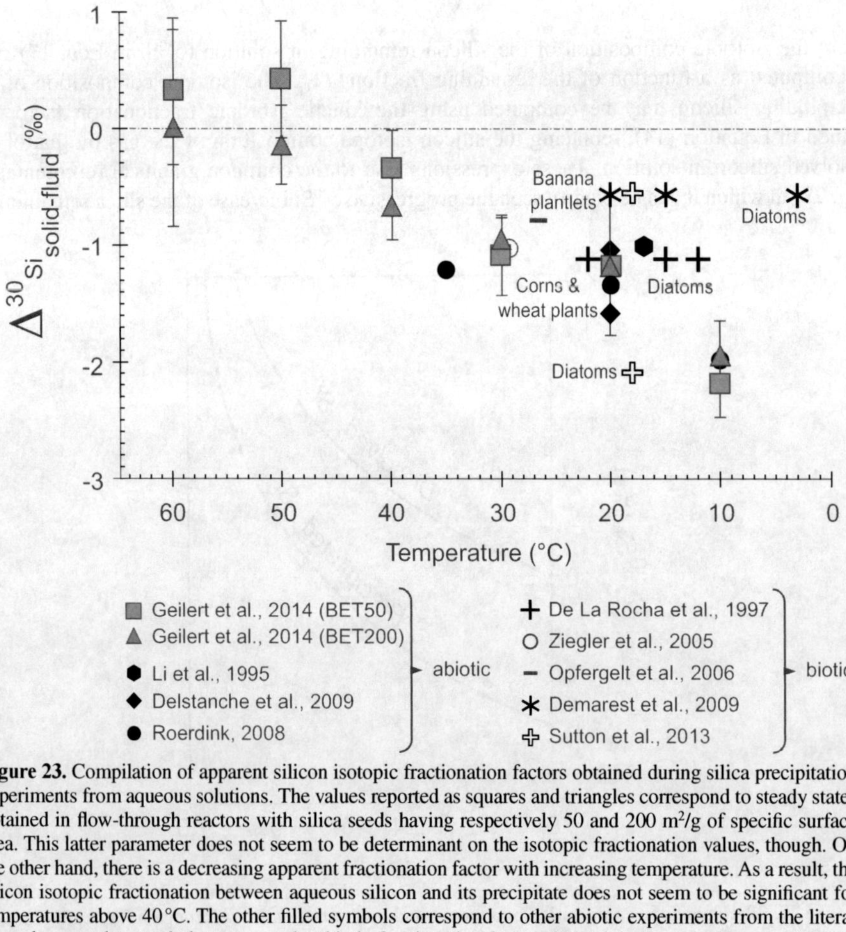

Figure 23. Compilation of apparent silicon isotopic fractionation factors obtained during silica precipitation experiments from aqueous solutions. The values reported as squares and triangles correspond to steady states obtained in flow-through reactors with silica seeds having respectively 50 and 200 m^2/g of specific surface area. This latter parameter does not seem to be determinant on the isotopic fractionation values, though. On the other hand, there is a decreasing apparent fractionation factor with increasing temperature. As a result, the silicon isotopic fractionation between aqueous silicon and its precipitate does not seem to be significant for temperatures above 40 °C. The other filled symbols correspond to other abiotic experiments from the literature whereas other symbols correspond to biotic fractionation factors derived during experiments or natural case studies. Note that the abiotic experiments reproduce well the fractionation factors observed with plants, but diatoms values are more scattered. After a compilation and original results from Geilert et al. (2014).

of past oceans yielding values above 50 °C (Robert and Chaussidon 2006) since under these conditions, aqueous silica precipitation should not lead to an isotopic fractionation and therefore, the silica precipitate $\delta^{30}Si$ value should essentially reflect that of the dissolved silicon. These experiments thus support the view that $\delta^{30}Si$ from ancient cherts yield an indication of the source of the dissolved silicon rather than their precipitation conditions (van den Boorn et al. 2007). Another complication arises from the finding that the silicon isotope fractionation factor between aqueous solution and amorphous silica also depends on the chemical composition of the substrate on which silica precipitate. While confirming the general tendency of having light silicon isotopes precipitating preferentially, Delstanche et al. (2009) have shown that silica precipitation onto a ferrihydrite substrate leads to stronger silica adsorption than goethite for the same experimental time and produces a stronger fractionation factor for ferrihydrite relative to goethite. With the aim to reproduce aqueous silica adsorption onto clay mineral precursors during weathering, Oelze et al. (2014) conducted experiments to study the kinetics of silica precipitation onto gibbsite. As observed by Geilert et al. (2014) and Roerdink et al. (2015) for simple amorphous silica precipitation experiments, they found that the silica precipitation rate increased with increasing silica concentration (Fig. 24) and this effect also went with increasing kinetic silicon isotope fractionation during precipitation (Fig. 25). Oelze et al. (2014) interpreted this as the direct effect of aqueous silica adsorption kinetics onto gibbsite surfaces. Collectively, these experiments also confirm the essentially kinetic nature of silicon isotope fractionation during weathering processes leading to light clay minerals and a heavy aqueous silicon in surface waters (Fig. 8). This conclusion is also in agreement with equilibrium isotope fractionation factors computed through atomistic methods showing an opposite sense of isotopic fractionation between quartz, kaolinite and aqueous silica species (Dupuis et al. 2015). In a recent study of cyclic freezing-thawing experiments aiming at reproducing silica precipitation in low temperature environments, Oelze et al. (2015) confirmed that silicon isotope fractionation factor during precipitation is greater if the reaction

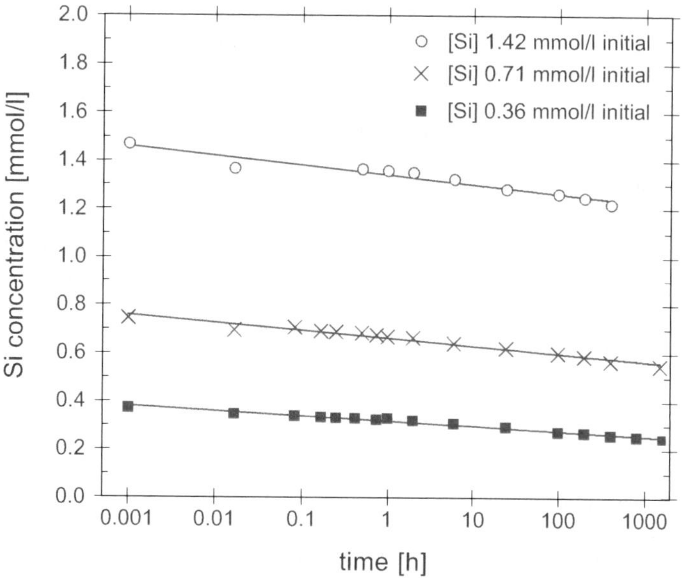

Figure 24. Dissolved silicon concentration through time during adsorption experiments on Al-hydroxides. Note that the higher starting aqueous silicon concentration leads to a steeper slope and therefore higher adsorption rate. After Oelze et al. (2014).

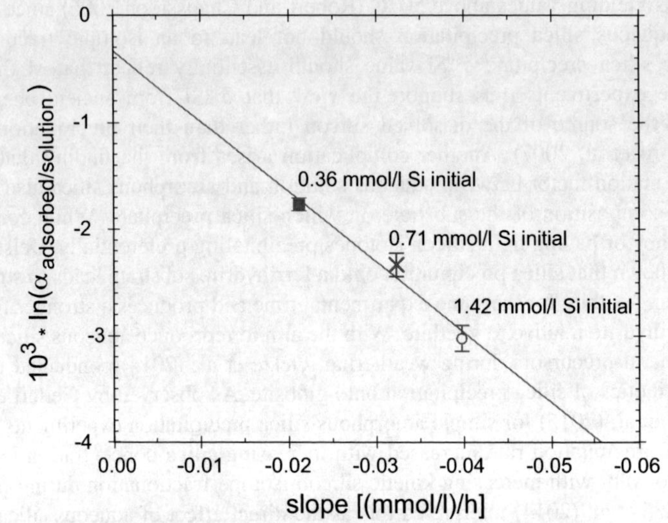

Figure 25. Silicon isotopic fractionation factors derived from the three Si adsorption experiments on gibbsite depicted on Figure 24, as a function of the Si adsorption rate. The apparent kinetic isotopic fractionation factors decrease with decreasing adsorption rate and therefore decreasing starting aqueous silicon concentration. After Oelze et al. (2014).

occurs faster. They also showed that this silicon isotopic fractionation factor will be greater if there is aluminum in the aqueous solution since the coprecipitation with aluminum will enhance the silica precipitation rate. At slow precipitation rate (i.e., over more than 4 months) or without aluminum as a carrier phase, no δ^{30}Si fractionation was observed (Oelze et al. 2015).

Another low temperature, abiotic process generating mass-dependent silicon isotope fractionation involves the dissolution of silica in water. Demarest et al. (2009) performed batch dissolution experiments of opaline frustules from diatoms in seawater at temperatures ranging from 3 to 20 °C and found a systematic fractionation factor of Δ^{30}Si = –0.55‰, with lighter isotopes going preferentially into solution. This value remained the same for diatom frustules of different provenance and even in the case of repeated dissolution experiments of the same frustule lot after removing several times the interacting aqueous solution. The isotopic trend could be either interpreted in terms of a Rayleigh model or as an open, equilibrium fractionation process, and unraveling the potential mechanism at play proved challenging, however (Demarest et al. 2009). Wetzel et al. (2014) conducted further diatom opal dissolution and they could not reproduce this fractionation factor, however. They nevertheless used a completely different dissolution medium (5 mM NaOH instead of silicon-stripped natural seawater) and different diatom opaline material. While Demarest et al. (2009) performed their experiments on freshly harvested plankton using sediment traps from the Southern Oceans, Wetzel et al. (2014) used diatom frustules collected from an Ocean Drilling Program sedimentary core. The distinct chemical and physical properties of diatom opal in terms of higher aluminium content, lower specific surface area and lower reactivity of opal from sediments is the likely explanation for the different conclusions reached in the two studies, according to Wetzel et al. (2014). Further dissolution experiments will be required, accompanied by high precision SIMS surface isotopic characterization of the dissolving solids, to fully understand the mechanisms at play and apply them to interpret observations from natural case studies.

Biological processes

Although present in trace amounts in many organisms (Douthitt 1982), silicon is an essential structural component for plants in the form of cell wall framework and phytoliths.

Phytoplankton (e.g., diatoms), zooplankton (e.g., radiolarian) or siliceous sponges use it to produce opal forming their frustules, tests or spicules. Beyond the interest in understanding the biological silicon cycling in plants and other organisms, the consideration of the biological processes affecting silicon concentrations and isotopic compositions is important in Earth and environmental sciences given that the phytolith pool represent Si fluxes (60–200 Tmoles/year; Conley, 2002) potentially approaching that of siliceous organisms (diatoms and sponges) in the oceans (240 Tmoles/year; Treguer and De La Rocha, 2013 and see Fig. 11). This silicon cycling in surface waters also has an important role in regulating atmospheric CO_2 through carbon burial into the deep ocean upon the sinking of dead diatoms towards the ocean floor since they also carry the carbon-rich organic matter parts of the organisms. Biogenic silica is amorphous and depending on the organism, it may show a variable structure (see review in Leng et al. 2009). This may be expressed through different O–Si–O bond angles and variable Si–O bond length that may be probed using Fourier Transform Infrared Spectroscopy (FTIR). Such changes on silicon and oxygen bonds in silica will obviously impact silicon isotope compositions. Biogenic silica may also show different levels of hydration through the replacement of Si–O by a silanol group (Si–OH), which can be probed by Nuclear Magnetic Resonance (NMR). Those studies revealed that diatom frustules show a higher level of hydration relative to sponge spicules or higher plant phytoliths (Leng et al. 2009). Silica polymerization is mediated in organisms by biomolecules acting either as catalysts and/or aggregation promoting agents. Of related interest for silicon isotope signatures is that those biochemical reactions occur strongly out of equilibrium, as silicic acid within diatoms has been found to be supersaturated by 30–40 times relative to natural concentrations measured in lakes and oceanic waters (See Leng et al. 2009). In plants, the silica biochemistry is less known. Whereas some transporters involved in silicon transfer in roots or for silicon unloading from xylem into shoots have been identified (Ma et al. 2006; Yamaji et al. 2008) or the possible role of hemicellulose in binding silicon in cell walls pointed out (He et al. 2015), the complete biological cycle is far from being clearly understood. Notably, the molecules involved into the silicon transport to form phytoliths remain unknown (Leng et al. 2009).

Only after each biochemical reaction involving silicon metabolism in organisms, from absorption, transport steps and silica polymerization to form opal, is well characterized, will it become possible to predict accurately the silicon isotope fractionation involved during its processing by plants and plankton under various conditions. In the meantime, we are left with the analysis of natural samples and of the products of experimental cultures to provide fractionation trends at the scale of organisms or of their components for plants. In the marine environment, De La Rocha et al. (1997) first demonstrated experimentally the preference of diatoms for light silicon isotopes. Although the exact mechanism for this preference could not be precisely established, they found that the data obtained could be interpreted in terms of a Rayleigh process by which light silicon isotopes are taken up by the organisms without subsequent exchange with the surrounding culture solution. In a more recent study, the same group made further experiments and found that the silicon isotope fractionation by diatoms is species-dependent, with isotope fractionation factors ranging from ~−0.5 to −2‰ in $\delta^{30}Si$ (Sutton et al. 2013). A mechanistic explanation for these species differences was not provided either, however. In a previous experimental work, Milligan et al. (2004) concluded on the basis of mass balance considerations, that the main $\delta^{30}Si$ fractionating step is the diatom H_4SiO_4 uptake from the surrounding water, rather than the subsequent silicon polymerization inside the cell and/or eventual efflux of silicon in excess.

Like diatoms, plants preferentially take up light silicon from interstitial soil solutions to produce opal in their structures during their growth (e.g., Ding et al. 2005; Sun et al. 2008; Delvigne et al. 2009). This feature has been observed in rice, bamboo, banana, corn and wheat (see review in Opfergelt and Delmelle 2012). However, subsequent silicon translocation will translate into heavier $\delta^{30}Si$ as silicon is transported through the plant by mass flow following the transpiration streams towards the upper organs (Fig. 16), possibly with evapotranspiration as the main driving factor (Ding et al. 2005, 2008a,b; Opfergelt et al. 2008; Sun et al. 2008). The argument in favor of such a passive silicon uptake and transport in plants lies with the observation that the higher silicon concentration and heavier isotopic composition are observed near the terminal parts of transpiration streams. It also goes with the limited silicon-bearing organic compounds in higher plants identified so far (Leng et al. 2009). The positive relationship between $\delta^{30}Si$ of bulk plants with that of soluble silicon in soil led Ding et al. (2008b) to favor the uptake of aqueous H_4SiO_4 rather than particulate silicon as previously proposed by some. There might be exceptions to the general trend of increasing $\delta^{30}Si$ in going towards upper plant parts, however, since Delvigne et al. (2009) showed an opposite $\delta^{30}Si$ trend for the most part of laboratory cultivated banana plants. Overall, these authors concluded on the basis of both natural case studies and experimentally grown plants in hydroponia that this process may again be modeled through a Rayleigh isotopic fractionation for at least rice and bamboo (Ding et al. 2005, 2008a,b).

Other biological systems studied for their silicon isotope compositions are marine siliceous sponges that represent a significant silicon reservoir in the ocean (Fig. 11). They show large ranges of $\delta^{30}Si$ compositions in their spicules (Douthitt 1982; De La Rocha 2003), with isotopic fractionation factors going up to $\Delta^{30}Si = -6‰$ relative to seawater (Wille et al. 2010). An interesting feature with sponge is that whereas the silicon isotope fractionation factor is constant during Si uptake from seawater, $\delta^{30}Si$ decreases with increasing seawater dissolved silicon concentration during spicule formation (Fig. 26). This possibly corresponds to an increasing kinetic effect during opaline precipitation in sponge spicules, as illustrated in abiotic experiments discussed above (Geilert et al. 2014; Oelze et al. 2014; Roerdink et al. 2015) and

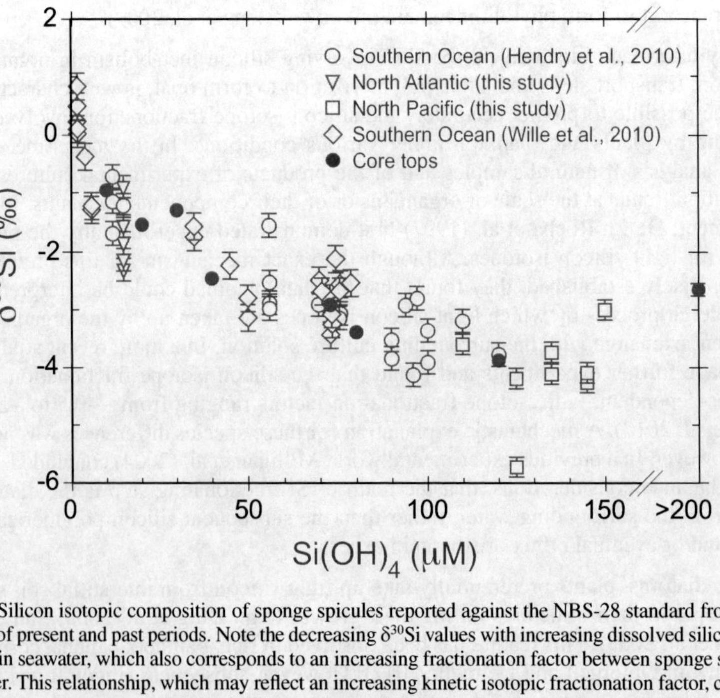

Figure 26. Silicon isotopic composition of sponge spicules reported against the NBS-28 standard from various oceans of present and past periods. Note the decreasing $\delta^{30}Si$ values with increasing dissolved silicon concentrations in seawater, which also corresponds to an increasing fractionation factor between sponge spicules and seawater. This relationship, which may reflect an increasing kinetic isotopic fractionation factor, could be used to reconstruct past oceans dissolved silicon concentrations. After Hendry and Robinson (2012).

illustrated in Figure 25. An interesting property of this relationship observed in sponge spicules is that it seems independent of the oceanic basin, seawater temperature and salinity, thereby opening the way for paleoreconstructions of silicon concentrations in ancient oceans using $\delta^{30}Si$ determination of fossil sponge spicules (Hendry and Robinson 2012). In contrast to diatoms, Hendry and Robinson (2012) interpret their results in terms of the combination of silicon isotope fractionation occurring during silicon uptake, polymerization and subsequent efflux on the basis of our current knowledge of silicon cycling in sponges based on experimental cultures. However, and like for other living organisms bearing significant amounts of silicon, our knowledge of the biochemical reactions involving this element remains too limited to provide a mechanistic interpretation of silicon isotope fractionation variations beyond empirical mass balance and Rayleigh models based on the study of natural and experimentally produced specimens.

IMPORTANT IMPLICATIONS AND FUTURE DIRECTIONS

Although silicon isotope geochemistry has been studied for more than fifty years, significant advances have been made in the past fifteen years into our knowledge of the isotopic ranges and governing factors influencing silicon isotope variations in nature. Much recent progress has been permitted by the advent of plasma source mass spectrometry and the latest generation of ion microprobe that led to notable improvements in terms of measurement accuracy and reproducibility. With an overall mass-dependent $\delta^{30}Si$ variation range exceeding 11‰ found so far on Earth, MC–ICP–MS now allows long-term reproducibility, after sample purification, of better than ±0.08‰, 2SD, and *in situ* techniques yield long-term reproducibility of only a factor of three higher.

At first sight, silicon did not seem such a good candidate for stable isotope geochemistry studies for both low and high temperature. In either case, silicon retains the same redox chemistry (4+) and physical state (solid, or as a dissolved species in water). In outcropping igneous rocks, silicon essentially occurs in silicate tetrahedra while in surface waters, its speciation remains essentially as H_4SiO_4 at $pH \leq 9$, which is the case for the vast majority of rivers, lakes and oceans.

Despite these apparent limitations, silicon isotope geochemistry has been an increasingly active field of research over the past ten years. Much attention has been focused on one of the rare silicon redox change potentially happening at the silicate mantle and metallic core interface of terrestrial planets (Georg et al. 2007a; Shahar et al. 2009, 2011; Ziegler et al. 2010; Hin et al. 2014), assuming that the temperature and oxygen fugacity conditions allow significant amounts of silicon partitioning into metallic cores. At low temperature, silicon isotope geochemistry seems largely driven by kinetic processes according to the latest theoretical and experimental studies (Geilert et al. 2014; Oelze et al. 2014, 2015; Dupuis et al. 2015; Roerdink et al. 2015), whether of biotic or abiotic origin. This yields the large isotopic range uncovered so far at the surface of the Earth (Fig. 8).

While the increase of $\delta^{30}Si$ with indicators of magma differentiation such as bulk rock SiO_2 concentrations or tectosilicate abundances is now well established (Savage et al. 2011, 2014; Zambardi et al. 2014; Poitrasson and Zambardi 2015), the exact mechanisms at play remain obscure. As reviewed above, several hypotheses have been proposed and include fractional crystallization, thermodiffusion or an effect of the melt structure, which may include the degree of silicon polymerization. To investigate this, theoretical efforts aimed at determining the equilibrium silicon isotope fractionation factor among rock-forming minerals (Huang et al. 2014; Méheut and Schauble 2014) should be pursued. This should be accompanied by *in situ* analyses of natural igneous bodies to measure the fractionation factors among rock forming minerals and glasses and detect any zoning due to thermodiffusion, assuming this plays a significant role in this context. Experimental works on silicate melts should also be conducted. They should combine silicon isotope measurements with spectroscopic analyses to determine the atomic environment of silica tetrahedra, depending on the composition of the silicate melts.

High precision measurements found analytically significant variations that seem to trace the protolith of igneous rocks, like some granites and andesites (Savage et al. 2012; Poitrasson and Zambardi 2015). A better understanding of the interaction between seawater and the oceanic crust on the silicon cycle, or during subsequent subduction and interaction between the dehydrating slab and the mantle wedge may help to understand the origin of the slightly heavy δ^{30}Si of some andesites (Poitrasson and Zambardi 2015) relative to the "igneous array" trend previously defined by Savage et al. (2011, 2014).

Beyond the knowledge of the δ^{30}Si ranges in nature and the understanding of the silicon isotope fractionation mechanisms that are topics of geochemical research in themselves, several applications of silicon isotopes in Earth and environmental sciences have already emerged. While still under discussion, the comparison of planetary δ^{30}Si estimates proved to yield insights into the chemistry of inaccessible reservoirs of the Earth (core and/or deep mantle). This led to the proposal that one of the unknown light elements in the Earth's core, besides its main Fe and Ni constituents, could be silicon (Georg et al. 2007a; Fitoussi et al. 2009; Armytage et al. 2011). An alternative interpretation of interplanetary δ^{30}Si difference was sparked by the discovery of the heavy silicon isotope composition of angrites relative to terrestrial planets, chondrite and achondrite parent bodies (Pringle et al. 2014; Dauphas et al. 2015). This led Dauphas et al. (2015) to provide, through a nebular condensation model, more precise estimates than before of Mg/Si ratio of terrestrial planets and asteroids, and therefore of the mineralogical and geophysical properties of planetary interiors. Alternative interpretations of these planetary isotopic differences were used by others to refine our knowledge of the nature of the Earth building blocks (Fitoussi and Bourdon 2012), or provide a more precise protracted Earth's core formation model and infer that traces of the Moon-forming giant impact were apparent with silicon isotopic signatures, as for iron (Zambardi et al. 2013). It is expected that further progresses will be possible through a more detailed modeling of planetary accretion processes and of mantle–core differentiation. These models should incorporate as many observables as possible, such as physical, petrological, mineralogical and geochemical data. While still under investigation, this example illustrates that the study of silicon isotopes from planetary materials goes well beyond the simple determination of the silicon isotope fractionation systematics. Rather, and as other non-traditional stable isotopes like those of iron (Poitrasson 2009), they potentially make up new tracers of inaccessible deep planetary reservoirs and/or on very ancient events confined to the beginning of the solar system history.

On the low temperature side, experimental and theoretical studies of silicon isotopes have clearly illustrated the kinetic nature of the abiotic weathering processes in which silicon is mobilized (Geilert et al. 2014; Oelze et al. 2014, 2015; Dupuis et al. 2015; Roerdink et al. 2015). In doing so, these studies provided further insights into the chemical mechanisms of igneous mineral weathering and replacement by secondary phases. This is particularly significant given the abundance of silicon in silicate rocks that make up the larger mass fraction of the continental crust. This abundance and the systematic comparisons of soils from different parts of the world led Opfergelt and Delmelle (2012) to use δ^{30}Si signatures of soils (Fig. 15) as a new index to assess the degree of continental surface weathering intensity. This silicon isotope evolution, largely driven towards low δ^{30}Si values by kinetically-driven and incongruent silicate mineral alteration leading to clay formation, seems independent of the nature of the parent rock, age of the soil and climatic context. In rivers, silicon isotope compositions give further insights into the river water residence time and resulting riverbed weathering processes (Ding et al. 2011; Hughes et al. 2011b; Cockerton et al. 2013; Pokrovsky et al. 2013; Frings et al. 2015). Seasonal diatom activity may also be inferred through their effect on silicon isotope signatures (Hughes et al. 2011b). Silicon isotopes could also be used to assess the role of plant activity that is still poorly known quantitatively

(Fig. 11), through their phytoliths, or the silicon cycling at the interface between soils and surface waters (Ding et al. 2011; Pokrovsky et al. 2013). Lastly, these silicon isotope studies of rivers and their associated groundwaters may provide further constraints on the silicon delivery to the oceans through time (Georg et al. 2009). This is needed because the amount of silicon delivered by these groundwaters to the ocean remains poorly known (Fig. 11).

However, many unanswered questions remain in low temperature silicon isotope geochemistry as well. For instance, a good knowledge of equilibrium fractionation factors of minerals and aqueous silicon is an essential reference against which measurements performed on natural samples may be compared. Yet, the silicon isotope mineral database is far from being complete (Méheut and Schauble 2014), and additional partition functions of silicate minerals have yet to be computed. A particularly challenging task is the production of isotopic partition functions of mineral phases and aqueous species computed by first principles that can be compared against each other since the calculation methods are different (Dupuis et al. 2015). We are just at the beginning of this approach and it can be expected that more progresses will be conducted in this direction in the future.

The quantitative relative importance on silicon isotope composition of rivers of (1) the interaction kinetics between water and the riverbed, (2) exchanges with the biosphere through plant phytoliths and (3) with diatoms activity or (4) the exchanges with groundwaters through the seasons (e.g., Ding et al. 2011; Hughes et al. 2011b; Cockerton et al. 2013; Pokrovsky et al. 2013; Frings et al. 2015) has yet to be fully evaluated. This can be investigated through further case studies to estimate their relative impact on different river types to completely understand the message riverine δ^{30}Si values can deliver. It is also important for the general issue of continent to ocean chemical transfers and its evolution through environmental change. The finding of a relationship between δ^{30}Si of bulk soils and indexes of weathering intensities (Opfergelt and Delmelle 2012) holds much promise towards a better understanding of continental weathering processes. However, the exact mechanisms at play remain poorly known, and it is expected that further detailed case studies such as the early work of Ziegler et al. (2005a), along with experimental studies, like those of Geilert et al. (2014) or Oelze et al. (2014) will help in this endeavor. A pressing question remains the relative role of biotic and abiotic processes in this context (Opfergelt and Delmelle 2012), and particularly the role of plant phytoliths on the silicon cycling in soils and their quantitative impact on the silicon transfer from soils to rivers. A better understanding of these processes will certainly help in assessing the transformations at play during land use change through time (e.g., from forest to farmland). The work of Georg et al. (2009) revealed the potentially important role of groundwaters associated with major riverine systems in the silicon export from continent to oceans. However, this silicon flux has yet to be properly quantified at the global scale, along with that of other elements, such as iron, also brought to the oceans by these groundwaters (e.g., Rouxel et al. 2008).

In oceans, silicon isotope studies have emphasized the key role of diatoms on the oceanic silicon cycle (De La Rocha et al. 2000; Reynolds et al. 2006; Holzer and Brzezinski 2015). Beyond silicon, this provides a better grasp on the carbon cycle through its drawdown in the deep ocean upon diatom death since sinking organisms towards the ocean floor carry silicon from the diatoms frustules along with carbon from the diatom components consisting of organic matter. However, Demarest et al. (2009) have provided very interesting experimental data showing that diatom dissolution generates a constant silicon isotope fractionation regardless of temperature, nature of the diatom frustule or even after repeated dissolution steps of the same lot. These counterintuitive results remain unexplained and call for further experimental investigation, combined with *in situ* isotope measurements and microscopic imaging techniques to unravel the mechanisms at play. This potentially has important implications for our understanding of the oceanic and freshwater silicon cycling. Other unanswered questions in oceanography that will require further work concern a better quantification of the silicon burial flux in sediments or the role of sponge and diatoms on the silicon oceanic budget (Treguer and De La Rocha 2013).

Such an improved knowledge of the silicon cycling on continents and its impact on the isotopic composition of silicon delivered to the oceans will also help improving paleoproductivity reconstructions. Frings et al. (2016) recently reviewed records of ancient oceanic $\delta^{30}Si$ variations and they concluded that these could not uniquely be interpreted in terms of simple paleoproductivity records as classically done. They notably suggested that glaciation periods may potentially lower by ~1‰ the $\delta^{30}Si$ signature entering the oceans through change in the style and rate of continental silicate weathering. According to these authors, this would be due to evolving temperature and vegetation on continents, change in the seasonal river flux variations and increased dust and groundwater fluxes.

The research on the present day silicon surface cycling is also essential for studies aimed at deciphering past terrestrial conditions on the continents and in the oceans, as potentially recorded in ancient cherts. These display among the largest $\delta^{30}Si$ variations found on Earth (Fig. 8) and they seem to be not so much affected by metamorphic events (Andre et al. 2006). Palaeoenvironmental reconstructions using silicon isotope measurements in cherts have been a topic of much research in the past decade. As reviewed above however, the meaning of the measured $\delta^{30}Si$ remains obscure. Does it represent the past seawater temperature, or trace the silicon origin, or does it rather reflect different seawater adsorption mechanisms and subsequent opal formation depending on the nature of the substrate? In this domain, *in situ* analyses of cherts combined with other isotopic tracers, like oxygen isotopes, along with detailed petrological studies, will certainly lead to progresses, as exemplified by recent studies (Heck et al. 2011; Marin-Carbonne et al. 2012; Stefurak et al. 2015). In fossilized sponge spicules, these may reflect the past seawater silicon concentrations, which is also of use for palaeoenvironmental reconstructions (Hendry and Robinson 2012). In particular, these sponge spicule $\delta^{30}Si$ values may reflect paleoproductivity of biogenic opal in surface ocean waters that are directly linked with the global carbon cycle and global climate. For these applications and others, however, further research is required.

Another important frontier in silicon isotope research concerns the development of a mechanistic understanding of silicon cycling in plants. At present, only a few transporters involved in silicon root uptake (Ma et al. 2006; Yamaji et al. 2008) or hemicellulose silicon ligands in cell wall formation (He et al. 2015) have been identified and most biochemical reactions involving silicon remain unknown (Leng et al. 2009). It is therefore necessary to conduct silicon isotopic research in connection with biomolecular chemistry to go beyond the empirical mass balance and Rayleigh models approach currently used to study plants with silicon isotopes in nature or through culture experiments.

Hence, several fundamental processes underlying stable silicon isotope fractionation in nature still remain to be understood, notably in the context of silicate magma evolution and fluid-mineral interactions, both at high and low temperature. Besides, the field of a proper mechanistic understanding of the biochemical reactions influencing silicon isotope signatures is virtually unexplored. These are only a few among the possible future directions that could be taken. Hence, silicon isotope research is a topic full of promises that will certainly remain a lively geochemical domain in the future.

ACKNOWLEDGEMENTS

This review writing and previous basic research activities on related topics have been made possible to the author thanks to 20 years of CNRS support. Ben Reynolds is thanked for his advice while the author and PhD candidate Thomas Zambardi were setting up the silicon isotope analytical methodology in Toulouse. The author's conceptions on silicon isotope variations in nature benefited from many discussions on the topic with Thomas Zambardi and Merlin Méheut. The volume editors are thanked for the invitation to write this review. It benefited from critical readings and edits by Chuck Douthitt, two anonymous referees and volume editor Jim Watkins. Christiane Cavare-Hester and Anne-Marie Cousin are acknowledged for redrafting all figures.

REFERENCES

Abraham K, Opfergelt S, Fripiat F, Cavagna AJ, de Jong JTM, Foley SF, Andre L, Cardinal D (2008) δ^{30}Si and δ^{29}Si determinations on USGS BHVO–1 and BHVO–2 reference materials with a new configuration on a nu plasma multi-collector ICP-MS. Geostand Geoanal Res 32:193–202, doi:10.1111/j.1751–908X.2008.00879.x

Allègre CJ, Manhès G, Lewin E (2001) Chemical composition of the Earth and the volatility control on planetary genetics. Earth Planet Sci Lett 185:49–69

Alleman LY, Cardinal D, Cocquyt C, Plisnier PD, Descy JP, Kimirei I, Sinyinza D, Andre L (2005) Silicon isotopic fractionation in Lake Tanganyika and its main tributaries. J Great Lakes Res 31:509–519

Allenby RJ (1954) Determination of the isotopic ratios of silicon in rocks. Geochim Cosmochim Acta 5:40–48, doi:10.1016/0016–7037(54)90060–5

Andre L, Cardinal D, Alleman LY, Moorbath S (2006) Silicon isotopes in similar to 3.8 Ga West Greenland rocks as clues to the Eoarchaean supracrustal Si cycle. Earth Planet Sci Lett 245:162–173

Armytage RMG, Georg RB, Williams HM, Halliday AN (2012) Silicon isotopes in lunar rocks: Implications for the Moon's formation and the early history of the Earth. Geochim Cosmochim Acta 77:504–514, doi:10.1016/j.gca.2011.10.032

Armytage RMG, Georg RB, Savage PS, Williams HM, Halliday AN (2011) Silicon isotopes in meteorites and planetary core formation. Geochim Cosmochim Acta 75:3662–3676, doi:10.1016/j.gca.2011.03.044

Barnes IL, Moore LJ, Machlan LA, Murphy TJ, Shields WR (1975) Absolute isotopic abundance ratios and atomic weight of a reference sample of silicon. J Res Nat Bureau Stand Section a-Physics and Chemistry 79:727–735, doi:10.6028/jres.079A.029

Basile-Doelsch I, Meunier JD, Parron C (2005) Another continental pool in the terrestrial silicon cycle. Nature 433:399–402, doi:10.1038/nature03217

Bern CR, Brzezinski MA, Beucher C, Ziegler K, Chadwick OA (2010) Weathering, dust, and biocycling effects on soil silicon isotope ratios. Geochim Cosmochim Acta 74:876–889, doi:10.1016/j.gca.2009.10.046

Berthet S, Malavergne V, Righter K (2009) Melting of the Indarch meteorite (EH4 chondrite) at 1 GPa and variable oxygen fugacity: Implications for early planetary differentiation processes. Geochim Cosmochim Acta 73:6402–6420, doi:10.1016/j.gca.2009.07.030

Best MG (2003) Igneous and Metamorphic Petrology. Blackwell, Malden

Beucher CP, Brzezinski MA, Jones JL (2008) Sources and biological fractionation of Silicon isotopes in the Eastern Equatorial Pacific. Geochim Cosmochim Acta 72:3063–3073, doi:10.1016/j.gca.2008.04.021

Birch F (1952) Elasticity and constitution of the Earth's interior. J Geophys Res 57:227–286

Birck JL (2004) An overview of isotopic anomalies in extraterrestrial materials and their nucleosynthetic heritage. Rev Mineral Geochem 55:25–64

Brzezinski MA, Jones JL, Beucher CP, Demarest MS, Berg HL (2006) Automated determination of silicon isotope natural abundance by the acid decomposition of cesium hexafluosilicate. Anal Chem 78:6109–6114, doi:10.1021/ac0606406

Cardinal D, Alleman LY, de Jeong J, Ziegler K, André L (2003) Isotopic composition of silicon measured by multicollector plasma source mass spectrometry in dry plasma mode. J Anal At Spectrom 18:213–218

Cardinal D, Alleman LY, Dehairs F, Savoye N, Trull TW, Andre L (2005) Relevance of silicon isotopes to Si-nutrient utilization and Si-source assessment in Antarctic. Global Biogeochem Cycles 19:13, doi:10.1029/2004gb002364

Cardinal D, Gaillardet J, Hughes HJ, Opfergelt S, Andre L (2010) Contrasting silicon isotope signatures in rivers from the Congo Basin and the specific behaviour of organic-rich waters. Geophys Res Lett 37, doi:L1240310.1029/2010gl043413

Cavagna AJ, Fripiat F, Dehairs F, Wolf-Gladrow D, Cisewski B, Savoye N, Andre L, Cardinal D (2011) Silicon uptake and supply during a Southern Ocean iron fertilization experiment (EIFEX) tracked by Si isotopes. Limnol Oceanogr 56:147–160, doi:10.4319/lo.2011.56.1.0147

Chakrabarti R (2015) Silicon isotopes: from cosmos to benthos. Curr Sci 108:246–254

Chakrabarti R, Knoll AH, Jacobsen SB, Fischer WW (2012) Si isotope variability in Proterozoic cherts. Geochim Cosmochim Acta 91:187–201, doi:10.1016/j.gca.2012.05.025

Chmeleff J, Horn I, Steinhoefel G, von Blanckenburg F (2008) In situ determination of precise stable Si isotope ratios by UV-femtosecond laser ablation high-resolution multi-collector ICP-MS. Chem Geol 249:155–166

Clayton RN, Mayeda TK, Epstein S (1978) Isotopic fractionation of silicon in Allende inclusions. Proc Ninth Lunar Planet Sci Conf 9:1267–1278

Clayton RN, Macpherson GJ, Hutcheon ID, Davis AM, Grossman L, Mayeda TK, Molinivelsko C, Allen JM (1984) Two forsterite-bearing FUN inclusions in the Allende meteorite. Geochim Cosmochim Acta 48:535–548, doi:10.1016/0016–7037(84)90282–5

Cockerton HE, Street-Perrott FA, Leng MJ, Barker PA, Horstwood MSA, Pashley V (2013) Stable-isotope (H, O, and Si) evidence for seasonal variations in hydrology and Si cycling from modern waters in the Nile Basin: implications for interpreting the Quaternary record. Quat Sci Rev 66:4–21, doi:10.1016/j.quascirev.2012.12.005

Conley DJ (2002) Terrestrial ecosystems and the global biogeochemical silica cycle. Global Biogeochem Cycles 16:8, doi:10.1029/2002gb001894

Coplen TB, Hopple JA, Boehike JK, Peiser HS, Rieder SE (2002) Compilation of minimum and maximum isotope ratios of selected elements in naturally occurring terrestrial materials and reagents. Water Resources Investigations Report. United States Geological Survey

Corgne A, Keshav S, Wood BJ, McDonough WF, Fei YW (2008) Metal–silicate partitioning and constraints on core composition and oxygen fugacity during Earth accretion. Geochim Cosmochim Acta 72:574–589

Cornelis JT, Delvaux B, Cardinal D, Andre L, Ranger J, Opfergelt S (2010) Tracing mechanisms controlling the release of dissolved silicon in forest soil solutions using Si isotopes and Ge/Si ratios. Geochim Cosmochim Acta 74:3913–3924, doi:10.1016/j.gca.2010.04.056

Cornelis JT, Delvaux B, Georg RB, Lucas Y, Ranger J, Opfergelt S (2011) Tracing the origin of dissolved silicon transferred from various soil-plant systems towards rivers: a review. Biogeosciences 8:89–112, doi:10.5194/bg-8-89-2011

Criss RE (1999) Principles of Stable Isotope Distribution. Oxford University Press, Oxford

Dauphas N, Poitrasson F, Burkhardt C, Kobayashi H, Kurosawa K (2015) Planetary and meteoritic Mg/Si and delta Si-30 variations inherited from solar nebula chemistry. Earth Planet Sci Lett 427:236–248, doi:10.1016/j.epsl.2015.07.008

Davis AM, Hashimoto A, Clayton RN, Mayeda TK (1990) Isotope mass fractionation during evaporation of Mg_2SiO_4. Nature 347:655–658, doi:10.1038/347655a0

De La Rocha CL (2002) Measurement of silicon stable isotope natural abundances via multicollector inductively coupled plasma mass spectrometry (MC–ICP–MS). Geochem Geophys Geosyst 3:8, doi:10.1029/2002gc000310

De La Rocha CL, Brzezinski MA, DeNiro MJ (1997) Fractionation of silicon isotopes by marine diatoms during biogenic silica formation. Geochim Cosmochim Acta 61:5051–5056, doi:10.1016/s0016-7037(97)00300-1

De La Rocha CL, Brzezinski MA, DeNiro MJ (2000) A first look at the distribution of the stable isotopes of silicon in natural waters. Geochim Cosmochim Acta 64:2467–2477

De La Rocha CL (2003) Silicon isotope fractionation by marine sponges and the reconstruction of the silicon isotope composition of ancient deep water. Geology 31:423–426, doi:10.1130/0091-7613(2003)031<0423:sifbms>2.0.co;2

De La Rocha CL, Bescont P, Croguennoc A, Ponzevera E (2011) The silicon isotopic composition of surface waters in the Atlantic and Indian sectors of the Southern Ocean. Geochim Cosmochim Acta 75:5283–5295, doi:10.1016/j.gca.2011.06.028

de Souza GF, Reynolds BC, Johnson GC, Bullister JL, Bourdon B (2012a) Silicon stable isotope distribution traces Southern Ocean export of Si to the eastern South Pacific thermocline. Biogeosciences 9:4199–4213, doi:10.5194/bg-9-4199-2012

de Souza GF, Reynolds BC, Rickli J, Frank M, Saito MA, Gerringa LJA, Bourdon B (2012b) Southern Ocean control of silicon stable isotope distribution in the deep Atlantic Ocean. Global Biogeochem Cycles 26:13, doi:10.1029/2011gb004141

de Souza GF, Slater RD, Dunne JP, Sarmiento JL (2014) Deconvolving the controls on the deep ocean's silicon stable isotope distribution. Earth Planet Sci Lett 398:66–76, doi:10.1016/j.epsl.2014.04.040

Debievre P, Taylor PDP (1993) Table of the isotopic compositions of the elements. Int J Mass Spectrom Ion Processes 123:149–166, doi:10.1016/0168-1176(93)87009-h

Deer WA, Howie RA, Zussman J (1992) An introduction to the rock-forming minerals. Longman Scientific and Technical, Harlow

Delstanche S, Opfergelt S, Cardinal D, Elsass F, Andre L, Delvaux B (2009) Silicon isotopic fractionation during adsorption of aqueous monosilicic acid onto iron oxide. Geochim Cosmochim Acta 73:923–934, doi:10.1016/j.gca.2008.11.014

Delvigne C, Opfergelt S, Cardinal D, Delvaux B, Andre L (2009) Distinct silicon and germanium pathways in the soil-plant system: Evidence from banana and horsetail. J Geophys Res-Biogeosci 114:11, doi:10.1029/2008jg000899

Demarest MS, Brzezinski MA, Beucher CP (2009) Fractionation of silicon isotopes during biogenic silica dissolution. Geochim Cosmochim Acta 73:5572–5583, doi:10.1016/j.gca.2009.06.019

Ding T, Jiang S, Wan D, Li Y, Li J, Song H, Liu Z, Yao X (1996) Silicon Isotope Geochemistry. Geological Publishing House, Beijing

Ding T, Wan D, Wang C, Zhang F (2004) Silicon isotope compositions of dissolved silicon and suspended matter in the Yangtze River, China. Geochim Cosmochim Acta 68:205–216

Ding TP, Ma GR, Shui MX, Wan DF, Li RH (2005) Silicon isotope study on rice plants from the Zhejiang province, China. Chem Geol 218:41–50

Ding TP, Zhou JX, Wan DF, Chen ZY, Wang CY, Zhang F (2008a) Silicon isotope fractionation in bamboo and its significance to the biogeochemical cycle of silicon. Geochim Cosmochim Acta 72:1381–1395, doi:10.1016/j.gca.2008.01.008

Ding TP, Tian SH, Sun L, Wu LH, Zhou JX, Chen ZY (2008b) Silicon isotope fractionation between rice plants and nutrient solution and its significance to the study of the silicon cycle. Geochim Cosmochim Acta 72:5600–5615, doi:10.1016/j.gca.2008.09.006

Ding TP, Gao JF, Tian SH, Wang HB, Li M (2011) Silicon isotopic composition of dissolved silicon and suspended particulate matter in the Yellow River, China, with implications for the global silicon cycle. Geochim Cosmochim Acta 75:6672–6689, doi:10.1016/j.gca.2011.07.040

Dominguez G, Wilkins G, Thiemens MH (2011) The Soret effect and isotopic fractionation in high-temperature silicate melts. Nature 473:70–73, doi:10.1038/nature09911

Douthitt CB (1982) The geochemistry of the stable isotopes of silicon. Geochim Cosmochim Acta 46:1449–1458

Dupuis R, Benoit M, Nardin E, Meheut M (2015) Fractionation of silicon isotopes in liquids: The importance of configurational disorder. Chem Geol 396:239–254, doi:10.1016/j.chemgeo.2014.12.027

Elardo SM, Draper DS, Shearer CK (2011) Lunar Magma Ocean crystallization revisited: Bulk composition, early cumulate mineralogy, and the source regions of the highlands Mg-suite. Geochim Cosmochim Acta 75:3024–3045, doi:10.1016/j.gca.2011.02.033

Engstrom E, Rodushkin I, Ingri J, Baxter DC, Ecke F, Osterlund H, Ohlander B (2010) Temporal isotopic variations of dissolved silicon in a pristine boreal river. Chem Geol 271:142–152, doi:10.1016/j.chemgeo.2010.01.005

Epstein S, Taylor Jr HP (1970) The concentration and isotopic composition of hydrogen, carbon, and silicon in Apollo 11 lunar rocks and minerals. Proc Apollo 11 Lunar Sci Conf 2:1085–1096

Epstein S, Taylor Jr HP (1971) $^{18}O/^{16}O$, $^{30}Si/^{28}Si$, D/H and $^{13}C/^{12}C$ ratios in lunar samples. Proc Second Lunar Planet Sci Conf:1421–1441

Fifield LK, Morgenstern U (2009) Silicon-32 as a tool for dating the recent past. Quat Geochron 4:400–405, doi:10.1016/j.quageo.2008.12.006

Fitoussi C, Bourdon B (2012) Silicon isotope evidence against an enstatite chondrite earth. Science 335:1477–1480, doi:10.1126/science.1219509

Fitoussi C, Bourdon B, Kleine T, Oberli F, Reynolds BC (2009) Si isotope systematics of meteorites and terrestrial peridotites: implications for Mg/Si fractionation in the solar nebula and for Si in the Earth's core. Earth Planet Sci Lett 287:77–85

Frings PJ, Clymans W, Fontorbe G, De La Rocha CL, Conley DJ (2016) The continental Si cycle and its impact on the ocean Si isotope budget. Chem Geol 425:12–36, doi:10.1016/j.chemgeo.2016.01.020

Frings PJ, Clymans W, Fontorbe G, Gray W, Chakrapani GJ, Conley DJ, De La Rocha C (2015) Silicate weathering in the Ganges alluvial plain. Earth Planet Sci Lett 427:136–148, doi:10.1016/j.epsl.2015.06.049

Fripiat F, Cavagna AJ, Dehairs F, de Brauwere A, Andre L, Cardinal D (2012) Processes controlling the Si-isotopic composition in the Southern Ocean and application for paleoceanography. Biogeosciences 9:2443–2457, doi:10.5194/bg–9–2443–2012

Fujii T, Pringle EA, Chaussidon M, Moynier F (2015) Isotope fractionation of Si in protonation/deprotonation reaction of silicic acid: A new pH proxy. Geochim Cosmochim Acta 168:193–205, doi:10.1016/j.gca.2015.07.003

Geilert S, Vroon PZ, Roerdink DL, Van Cappellen P, van Bergen MJ (2014) Silicon isotope fractionation during abiotic silica precipitation at low temperatures: Inferences from flow-through experiments. Geochim Cosmochim Acta 142:95–114, doi:10.1016/j.gca.2014.07.003

Georg RB, Reynolds BC, Frank M, Halliday AN (2006) New sample preparation techniques for the determination of Si isotopic compositions using MC-ICPMS. Chem Geol 235:95–104

Georg RB, Halliday AN, Schauble EA, Reynolds BC (2007a) Silicon in the Earth's core. Nature 447:1102–1106

Georg RB, West AJ, Basu AR, Halliday AN (2009) Silicon fluxes and isotope composition of direct groundwater discharge into the Bay of Bengal and the effect on the global ocean silicon isotope budget. Earth Planet Sci Lett 283:67–74, doi:10.1016/j.epsl.2009.03.041

Georg RB, Reynolds BC, West AJ, Burton KW, Halliday AN (2007b) Silicon isotope variations accompanying basalt weathering in Iceland. Earth Planet Sci Lett 261:476–490, doi:10.1016/j.epsl.2007.07.004

Geyh MA, Schleicher H (1990) Absolute age determinations: physical and chemical dating methods and their application. Springer-Verlag, Berlin

Grant FS (1954) The geological significance of variations in the abundances of the isotopes of silicon in rocks. Geochim Cosmochim Acta 5:225–242

Grasse P, Ehlert C, Frank M (2013) The influence of water mass mixing on the dissolved Si isotope composition in the Eastern Equatorial Pacific. Earth Planet Sci Lett 380:60–71, doi:10.1016/j.epsl.2013.07.033

He CW, Ma J, Wang LJ (2015) A hemicellulose-bound form of silicon with potential to improve the mechanical properties and regeneration of the cell wall of rice. New Phytol 206:1051–1062, doi:10.1111/nph.13282

Heck PR, Huberty JM, Kita NT, Ushikubo T, Kozdon R, Valley JW (2011) SIMS analyses of silicon and oxygen isotope ratios for quartz from Archean and Paleoproterozoic banded iron formations. Geochim Cosmochim Acta 75:5879–5891, doi:10.1016/j.gca.2011.07.023

Hendry KR, Robinson LF (2012) The relationship between silicon isotope fractionation in sponges and silicic acid concentration: Modern and core-top studies of biogenic opal. Geochim Cosmochim Acta 81:1–12, doi:10.1016/j.gca.2011.12.010

Hin RC, Fitoussi C, Schmidt MW, Bourdon B (2014) Experimental determination of the Si isotope fractionation factor between liquid metal and liquid silicate. Earth Planet Sci Lett 387:55–66, doi:10.1016/j.epsl.2013.11.016

Hirose K, Labrosse S, Hernlund J (2013) Composition and State of the Core. Ann Rev Earth Planet Sci 41:657–691, doi:10.1146/annurev-earth–050212–124007

Hoefs J (2015) Stable Isotope Geochemistry. Springer-Verlag, Berlin

Holden NE (2007) Table of the isotopes. *In*: CRC Handbook of Chemistry and Physics, Internet version 2007 (87th edition). Lide DR, (ed) Taylor and Francis, Boca Raton, FL, p 2388

Holzer M, Brzezinski MA (2015) Controls on the silicon isotope distribution in the ocean: New diagnostics from a data-constrained model. Global Biogeochem Cycles 29:267–287, doi:10.1002/2014gb004967

Huang F, Wu ZQ, Huang SC, Wu F (2014) First-principles calculations of equilibrium silicon isotope fractionation among mantle minerals. Geochim Cosmochim Acta 140:509–520, doi:10.1016/j.gca.2014.05.035

Hughes HJ, Delvigne C, Korntheuer M, de Jong J, Andre L, Cardinal D (2011a) Controlling the mass bias introduced by anionic and organic matrices in silicon isotopic measurements by MC–ICP–MS. J Anal At Spectrom 26:1892–1896, doi:10.1039/c1ja10110b

Hughes HJ, Sondag F, Cocquyt C, Laraque A, Pandi A, Andre L, Cardinal D (2011b) Effect of seasonal biogenic silica variations on dissolved silicon fluxes and isotopic signatures in the Congo River. Limnology and Oceanography 56:551–561, doi:10.4319/lo.2011.56.2.0551

Hughes HJ, Sondag F, Santos RV, Andre L, Cardinal D (2013) The riverine silicon isotope composition of the Amazon Basin. Geochim Cosmochim Acta 121:637–651, doi:10.1016/j.gca.2013.07.040

Hutchison R (2004) Meteorites: A Petrologic, Chemical and Isotopic Synthesis. Cambridge University Press, Cambridge

Janney PE, Richter FM, Mendybaev RA, Wadhwa M, Georg RB, Watson EB, Hines RR (2011) Matrix effects in the analysis of Mg and Si isotope ratios in natural and synthetic glasses by laser ablation-multicollector ICPMS: A comparison of single- and double-focusing mass spectrometers. Chem Geol 281:26–40, doi:10.1016/j.chemgeo.2010.11.026

Jarosewich E (1990) Chemical analyses of meteorites—A compilation of stony and iron meteorite analyses. Meteoritics 25:323–337

Javoy M, Balan E, Meheut M, Blanchard M, Lazzeri M (2012) First-principles investigation of equilibrium isotopic fractionation of O- and Si-isotopes between refractory solids and gases in the solar nebula. Earth Planet Sci Lett 319:118–127, doi:10.1016/j.epsl.2011.12.029

Johnson CM, Beard BL, Albarède F (eds) (2004) Geochemistry of Non-traditional Stable Isotopes. Rev Mineral Geochem Vol. 55. The Mineralogical Society of America; The Geochemical Society, Washington

Keil K (1968) Mineralogical and chemical relationships among enstatite chondrites. J Geophys Res 73:6945–6976, doi:10.1029/JB073i022p06945

Keil K (2010) Enstatite achondrite meteorites (aubrites) and the histories of their asteroidal parent bodies. Chem Erde-Geochem 70:295–317, doi:10.1016/j.chemer.2010.02.002

Keil K (2012) Angrites, a small but diverse suite of ancient, silica-undersaturated volcanic-plutonic mafic meteorites, and the history of their Parent asteroid. Chem Erde-Geochem 72:191–218, doi:10.1016/j.chemer.2012.06.002

Knight KB, Kita NT, Mendybaev RA, Richter FM, Davis AM, Valley JW (2009) Silicon isotopic fractionation of CAI-like vacuum evaporation residues. Geochim Cosmochim Acta 73:6390–6401, doi:10.1016/j.gca.2009.07.008

Lacks DJ, Goel G, Bopp CJ, Van Orman JA, Lesher CE, Lundstrom CC (2012) Isotope Fractionation by Thermal Diffusion in Silicate Melts. Phys Rev Lett 108:5, doi:10.1103/PhysRevLett.108.065901

Lameyre J (1986) Roches et Minéraux. Doin Editeurs, Paris

Leng MJ, Swann GEA, Hodson MJ, Tyler JJ, Patwardhan SV, Sloane HJ (2009) The potential use of silicon isotope composition of biogenic silica as a proxy for environmental change. Silicon 1:65–77, doi:10.1007/s12633–009–9014–2

Li XF, Liu Y (2015) A theoretical model of isotopic fractionation by thermal diffusion and its implementation on silicate melts. Geochim Cosmochim Acta 154:18–27, doi:10.1016/j.gca.2015.01.019

Lodders K (1998) A survey of shergottite, nakhlite and chassigny meteorites whole-rock compositions. Meteorit Planet Sci 33:183–190

Lodders K, Fegley Jr B (1998) The Planetary Scientist's Companion. Oxford University Press, New York

Lundstrom C (2009) Hypothesis for the origin of convergent margin granitoids and Earth's continental crust by thermal migration zone refining. Geochim Cosmochim Acta 73:5709–5729, doi:10.1016/j.gca.2009.06.020

Ma JF, Tamai K, Yamaji N, Mitani N, Konishi S, Katsuhara M, Ishiguro M, Murata Y, Yano M (2006) A silicon transporter in rice. Nature 440:688–691, doi:10.1038/nature04590

Maaloe S, Aoki KI (1977) Major element composition of upper mantle estimated from composition of lherzolites. Contrib Mineral Petrol 63:161–173, doi:10.1007/bf00398777

Maréchal CN, Télouk P, Albarède F (1999) Precise analysis of copper and zinc isotopic composition by plasma source mass spectrometry. Chem Geol 156:251–273

Marin-Carbonne J, Chaussidon M, Robert F (2012) Micrometer-scale chemical and isotopic criteria (O and Si) on the origin and history of Precambrian cherts: Implications for paleo-temperature reconstructions. Geochim Cosmochim Acta 92:129–147, doi:10.1016/j.gca.2012.05.040

Marin-Carbonne J, Robert F, Chaussidon M (2014) The silicon and oxygen isotope compositions of Precambrian cherts: A record of oceanic paleo-temperatures? Precambrian Res 247:223–234, doi:10.1016/j.precamres.2014.03.016

McSween HY, Taylor GJ, Wyatt MB (2009) Elemental Composition of the Martian Crust. Science 324:736–739, doi:10.1126/science.1165871

Méheut M, Schauble EA (2014) Silicon isotope fractionation in silicate minerals: Insights from first-principles models of phyllosilicates, albite and pyrope. Geochim Cosmochim Acta 134:137–154, doi:10.1016/j.gca.2014.02.014

Méheut M, Lazzeri M, Balan E, Mauri F (2009) Structural control over equilibrium silicon and oxygen isotopic fractionation: A first-principles density-functional theory study. Chem Geol 258:28–37, doi:10.1016/j.chemgeo.2008.06.051

Meyer C (2008) Lunar sample compendium. NASA, Houston, http://curator.jsc.nasa.gov/lunar/compendium.cfm

Milligan AJ, Varela DE, Brzezinski MA, Morel F (2004) Dynamics of silicon metabolism and silicon isotopic discrimination in a marine diatom as a function of pCO_2. Limnology and Oceanography 49:322–329

Molini-Velsko C, Mayeda TK, Clayton RN (1986) isotopic composition of silicon in meteorites. Geochim Cosmochim Acta 50:2719–2726

Mysen BO, Virgo D, Seifert FA (1985) Relationships between properties and structure of aluminosilicate melts. Am Mineral 70:88–105

Nesbitt HW, Young GM (1982) Early Proterozoic climates and plate motions inferred from major element chemistry of lutites. Nature 299:715–717, doi:10.1038/299715a0

Norton OR (2002) The Cambridge encyclopedia of meteorites. Cambridge University Press, Cambridge

Oelze M, von Blanckenburg F, Hoellen D, Dietzel M, Bouchez J (2014) Si stable isotope fractionation during adsorption and the competition between kinetic and equilibrium isotope fractionation: Implications for weathering systems. Chem Geol 380:161–171, doi:10.1016/j.chemgeo.2014.04.027

Oelze M, von Blanckenburg F, Bouchez J, Hoellen D, Dietzel M (2015) The effect of Al on Si isotope fractionation investigated by silica precipitation experiments. Chem Geol 397:94–105, doi:10.1016/j.chemgeo.2015.01.002

Opfergelt S, Delmelle P (2012) Silicon isotopes and continental weathering processes: Assessing controls on Si transfer to the ocean. C R Geosci 344:723–738, doi:10.1016/j.crte.2012.09.006

Opfergelt S, Delvaux B, Andre L, Cardinal D (2008) Plant silicon isotopic signature might reflect soil weathering degree. Biogeochemistry 91:163–175, doi:10.1007/s10533–008–9278–4

Opfergelt S, de Bournonville G, Cardinal D, Andre L, Delstanche S, Delvaux B (2009) Impact of soil weathering degree on silicon isotopic fractionation during adsorption onto iron oxides in basaltic ash soils, Cameroon. Geochim Cosmochim Acta 73:7226–7240, doi:10.1016/j.gca.2009.09.003

Opfergelt S, Cardinal D, Andre L, Delvigne C, Bremond L, Delvaux B (2010) Variations of delta Si−30 and Ge/Si with weathering and biogenic input in tropical basaltic ash soils under monoculture. Geochim Cosmochim Acta 74:225–240, doi:10.1016/j.gca.2009.09.025

Opfergelt S, Eiriksdottir ES, Burton KW, Einarsson A, Siebert C, Gislason SR, Halliday AN (2011) Quantifying the impact of freshwater diatom productivity on silicon isotopes and silicon fluxes: Lake Myvatn, Iceland. Earth Planet Sci Lett 305:73–82, doi:10.1016/j.epsl.2011.02.043

Opfergelt S, Georg RB, Delvaux B, Cabidoche YM, Burton KW, Halliday AN (2012) Silicon isotopes and the tracing of desilication in volcanic soil weathering sequences, Guadeloupe. Chem Geol 326:113–122, doi:10.1016/j.chemgeo.2012.07.032

Pack A, Vogel I, Rollion-Bard C, Luais B, Palme H (2011) Silicon in iron meteorite metal. Meteorit Planet Sci 46:1470–1483, doi:10.1111/j.1945–5100.2011.01239.x

Pahlevan K, Stevenson DJ (2007) Equilibration in the aftermath of the lunar-forming giant impact. Earth Planet Sci Lett 262:438–449

Paniello RC, Day JMD, Moynier F (2012) Zinc isotopic evidence for the origin of the Moon. Nature 490:376–380, doi:10.1038/nature11507

Panizzo VN, Swann GEA, Mackay AW, Vologina E, Sturm M, Pashley V, Horstwood MSA (2016) Insights into the transfer of silicon isotopes into the sediment record. Biogeosciences 13:147–157, doi:10.5194/bg−13−147−2016

Papike JJ, Ryder G, Shearer CK (1998) Lunar samples. Rev Mineral 36:5-1–5-234

Platzner IT (1997) Modern isotope ratio mass spectrometry. John Wiley & sons, Chichester

Pogge von Strandmann PAE, Opfergelt S, Lai YJ, Sigfusson B, Gislason SR, Burton KW (2012) Lithium, magnesium and silicon isotope behaviour accompanying weathering in a basaltic soil and pore water profile in Iceland. Earth Planet Sci Lett 339:11–23, doi:10.1016/j.epsl.2012.05.035

Poitrasson F (2009) Probes of the ancient and the inaccessible. Science 323:882–883

Poitrasson F, Freydier R (2005) Heavy iron isotope composition of granites determined by high resolution MC–ICP–MS. Chem Geol 222:132–147
Poitrasson F, Zambardi T (2015) An Earth–Moon silicon isotope model to track silicic magma origins. Geochim Cosmochim Acta 167:301–312
Poitrasson F, Halliday AN, Lee DC, Levasseur S, Teutsch N (2004) Iron isotope differences between Earth, Moon, Mars and Vesta as possible records of contrasted accretion mechanisms. Earth Planet Sci Lett 223:253–266
Poitrasson F, Delpech G, Grégoire M (2013) On the iron isotope heterogeneity of lithospheric mantle xenoliths: implications for mantle metasomatism, the origin of basalts and the iron isotope composition of the Earth. Contrib Mineral Petrol 165:1243–1258, doi:DOI 10.1007/s00410–013–0856–7
Pokrovsky OS, Reynolds BC, Prokushkin AS, Schott J, Viers J (2013) Silicon isotope variations in Central Siberian rivers during basalt weathering in permafrost-dominated larch forests. Chem Geol 355:103–116, doi:10.1016/j.chemgeo.2013.07.016
Polyakov VB (2009) Equilibrium iron isotope fractionation at core–mantle boundary conditions. Science 323:912–914
Pringle EA, Savage PS, Jackson MG, Barrat JA, Moynier F (2013) Si isotope homogeneity of the solar nebula. Astrophys J 779:123–127, doi:12310.1088/0004–637x/779/2/123
Pringle EA, Moynier F, Savage PS, Badro J, Barrat JA (2014) Silicon isotopes in angrites and volatile loss in planetesimals. PNAS 111:17029–17032, doi:10.1073/pnas.1418889111
Pringle EA, Moynier F, Savage PS, Jackson MG, Moreira M, Day JMD (2016) Silicon isotopes reveal recycled oceanic crust in the mantle sources of Ocean Island Basalts. Geochim Cosmochim Acta 189:282–295
Qin T, Wu F, Wu ZQ, Huang F (2016) First-principles calculations of equilibrium fractionation of O- and Si-isotopes in quartz, albite, anorthite, and zircon. Contrib Mineral Petrol in press
Reynolds JH, Verhoogen J (1953) Natural variations in the isotopic constitution of silicon. Geochim Cosmochim Acta 3:224–234, doi:10.1016/0016–7037(53)90041–6
Reynolds BC, Frank M, Halliday AN (2006) Silicon isotope fractionation during nutrient utilization in the North Pacific. Earth Planet Sci Lett 244:431–443, doi:10.1016/j.epsl.2006.02.002
Reynolds BC, Aggarwal J, Andre L, *et al.* (2007) An inter-laboratory comparison of Si isotope reference materials. J Anal At Spectrom 22:561–568, doi:10.1039/b616755a
Richter FM, Liang Y, Davis AM (1999) Isotope fractionation by diffusion in molten oxides. Geochim Cosmochim Acta 63:2853–2861, doi:10.1016/s0016–7037(99)00164–7
Richter FM, Janney PE, Mendybaev RA, Davis AM, Wadhwa M (2007) Elemental and isotopic fractionation of Type BCAI-like liquids by evaporation. Geochim Cosmochim Acta 71:5544–5564, doi:10.1016/j gca.2007.09.005
Richter FM, Watson EB, Mendybaev R, Dauphas N, Georg B, Watkins J, Valley J (2009) Isotopic fractionation of the major elements of molten basalt by chemical and thermal diffusion. Geochim Cosmochim Acta 73:4250–4263, doi:10.1016/j.gca.2009.04.011
Ringwood AE (1961) Silicon in the metal phase of enstatite chondrites and some geochemical implications. Geochim Cosmochim Acta 25:1–13, doi:10.1016/0016–7037(61)90056–4
Robert F, Chaussidon M (2006) A palaeotemperature curve for the Precambrian oceans based on silicon isotopes in cherts. Nature 443:969–972, doi:10.1038/nature05239
Roerdink DL, van den Boorn S, Geilert S, Vroon PZ, van Bergen MJ (2015) Experimental constraints on kinetic and equilibrium silicon isotope fractionation during the formation of non-biogenic chert deposits. Chem Geol 402:40–51, doi:10.1016/j.chemgeo.2015.02.038
Rouxel O, Sholkovitz E, Charette M, Edwards KJ (2008) Iron isotope fractionation in subterranean estuaries. Geochim Cosmochim Acta 72:3413–3430, doi:10.1016/j.gca.2008.05.001
Russell WA, Papanastassiou DA, Tombrello TA (1978) Ca isotope fractionation on the Earth and other solar system materials. Geochim Cosmochim Acta 42:1075–1090
Sauer D, Saccone L, Conley DJ, Herrmann L, Sommer M (2006) Review of methodologies for extracting plant-available and amorphous Si from soils and aquatic sediments. Biogeochemistry 80:89–108, doi:10.1007/s10533–005–5879–3
Sautter V, Toplis MJ, Wiens RC, Cousin A, Fabre C, Gasnault O, Maurice S, Forni O, Lasue J, Ollila A, Bridges JC (2015) In situ evidence for continental crust on early Mars. Nat Geosci 8:605-+, doi:10.1038/ngeo2474
Savage PS, Moynier F (2013) Silicon isotopic variation in enstatite meteorites: Clues to their origin and Earth-forming material. Earth Planet Sci Lett 361:487–496, doi:10.1016/j.epsl.2012.11.016
Savage PS, Georg RB, Armytage RMG, Williams HM, Halliday AN (2010) Silicon isotope homogeneity in the mantle. Earth Planet Sci Lett 295:139–146
Savage PS, Georg RB, Williams HM, Burton KW, Halliday AN (2011) Silicon isotope fractionation during magmatic differentiation. Geochim Cosmochim Acta 75:6124–6139, doi:10.1016/j.gca.2011.07.043
Savage PS, Georg RB, Williams HM, Turner S, Halliday AN, Chappell BW (2012) The silicon isotope composition of granites. Geochim Cosmochim Acta 92:184–202, doi:10.1016/j.gca.2012.06.017

Savage PS, Georg RB, Williams HM, Halliday AN (2013a) The silicon isotope composition of the upper continental crust. Geochim Cosmochim Acta 109:384–399, doi:10.1016/j.gca.2013.02.004

Savage PS, Georg RB, Williams HM, Halliday AN (2013b) Silicon isotopes in granulite xenoliths: Insights into isotopic fractionation during igneous processes and the composition of the deep continental crust. Earth Planet Sci Lett 365:221–231, doi:10.1016/j.epsl.2013.01.019

Savage PS, Armytage RMG, Georg RB, Halliday AN (2014) High temperature silicon isotope geochemistry. Lithos 190–191:500–519

Schuessler J, Schoenberg R, Sigmarsson O (2009) Iron and lithium isotope systematics of the Hekla volcano, Iceland – Evidence for Fe isotope fractionation during magma differentiation. Chem Geol 258:78–91

Seeds MA (2001) Foundations of astronomy. Brooks/Cole, Pacific Grove, CA, USA

Shahar A, Young ED (2007) Astrophysics of CAI formation as revealed by silicon isotope LA–MC-ICPMS of an igneous CAI. Earth Planet Sci Lett 257:497–510

Shahar A, Ziegler K, Young ED, Ricolleau A, Schauble EA, Fei YW (2009) Experimentally determined Si isotope fractionation between silicate and Fe metal and implications for Earth's core formation. Earth Planet Sci Lett 288:228–234

Shahar A, Hillgren VJ, Young ED, Fei YW, Macris CA, Deng LW (2011) High-temperature Si isotope fractionation between iron metal and silicate. Geochim Cosmochim Acta 75:7688–7697, doi:10.1016/j.gca.2011.09.038

Sigmarsson O, Condomines M, Fourcade S (1992) A detailed Th, Sr And O isotope study of Hekla—Differentiation processes in an Icelandic volcano. Contrib Mineral Petrol 112:20–34, doi:10.1007/bf00310953

Sossi PA, Nebel O, Anand M, Poitrasson F (2016) On the iron isotope composition of Mars and volatile depletion in the terrestrial planets. Earth Planet Sci Lett 449:360–371

Stefurak EJT, Fischer WW, Lowe DR (2015) Texture-specific Si isotope variations in Barberton Greenstone Belt cherts record low temperature fractionations in early Archean seawater. Geochim Cosmochim Acta 150:26–52, doi:10.1016/j.gca.2014.11.014

Steinhoefel G, von Blanckenburg F, Horn I, Konhauser KO, Beukes NJ, Gutzmer J (2010) Deciphering formation processes of banded iron formations from the Transvaal and the Hamersley successions by combined Si and Fe isotope analysis using UV femtosecond laser ablation. Geochim Cosmochim Acta 74:2677–2696, doi:10.1016/j.gca.2010.01.028

Steinhoefel G, Breuer J, von Blanckenburg F, Horn I, Kaczorek D, Sommer M (2011) Micrometer silicon isotope diagnostics of soils by UV femtosecond laser ablation. Chem Geol 286:280–289, doi:10.1016/j.chemgeo.2011.05.013

Stumm W, Morgan JJ (1996) Aquatic Chemistry. Chemical Equilibria and Rates in Natural Waters. Wiley-Interscience, New York

Sun L, Wu LH, Ding TP, Tian SH (2008) Silicon isotope fractionation in rice plants, an experimental study on rice growth under hydroponic conditions. Plant Soil 304:291–300, doi:10.1007/s11104–008–9552–1

Sutton JN, Varela DE, Brzezinski MA, Beucher CP (2013) Species-dependent silicon isotope fractionation by marine diatoms. Geochim Cosmochim Acta 104:300–309, doi:10.1016/j.gca.2012.10.057

Taylor SR, McLennan SM (1985) The Continental Crust: Its Origin and Evolution. Blackwell Science Publishers, Oxford

Taylor SR, McLennan SM, McCulloch MT (1983) Geochemistry of loess, continental crustal composition and crustal model ages. Geochim Cosmochim Acta 47:1897–1905, doi:10.1016/0016–7037(83)90206–5

Tilles D (1961) Natural variations in isotopic abundances of silicon. J Geophys Res 66:3003-&, doi:10.1029/JZ066i009p03003

Treguer PJ, De La Rocha CL (2013) The world ocean silica cycle. *In*: Annual Review of Marine Science, Vol 5. Carlson CA, Giovannoni SJ (eds). Annual Reviews, Palo Alto, p 477–501

Valley JW, Cole DR (eds) (2001) Stable Isotope Geochemistry. Rev Mineral Geochem Vol. 43. The Mineralogical Society of America; The Geochemical Society.

van den Boorn S, Vroon PZ, van Belle CC, van der Wagt B, Schwieters J, van Bergen MJ (2006) Determination of silicon isotope ratios in silicate materials by high-resolution MC–ICP–MS using a sodium hydroxide sample digestion method. J Anal At Spectrom 21:734–742

van den Boorn S, van Bergen MJ, Nijman W, Vroon PZ (2007) Dual role of seawater and hydrothermal fluids in Early Archean chert formation: Evidence from silicon isotopes. Geology 35:939–942

van den Boorn S, van Bergen MJ, Vroon PZ, de Vries ST, Nijman W (2010) Silicon isotope and trace element constraints on the origin of similar to 3.5 Ga cherts: Implications for Early Archaean marine environments. Geochim Cosmochim Acta 74:1077–1103, doi:10.1016/j.gca.2009.09.009

Varela DE, Pride CJ, Brzezinski MA (2004) Biological fractionation of silicon isotopes in Southern Ocean surface waters. Global Biogeochem Cycles 18:8, doi:10.1029/2003gb002140

Wade J, Wood BJ (2005) Core formation and the oxidation state of the Earth. Earth Planet Sci Lett 236:78–95

Wai CM, Wasson JT (1969) Silicon concentrations in metal of iron meteorites. Geochim Cosmochim Acta 33:1465–1468, doi:10.1016/0016–7037(69)90150–1

Wang J, Davis AM, Clayton RN, Mayeda TK, Hashimoto A (2001) Chemical and isotopic fractionation during the evaporation of the FeO–MgO–SiO_2–CaO–Al_2O_3–TiO_2 rare earth element melt system. Geochim Cosmochim Acta 65:479–494

Wang K, Moynier F, Dauphas N, Barrat JA, Craddock P, Sio CK (2012) Iron isotope fractionation in planetary crusts. Geochim Cosmochim Acta 89:31–45, doi:10.1016/j.gca.2012.04.050

Watkins JM, DePaolo DJ, Ryerson FJ, Peterson BT (2011) Influence of liquid structure on diffusive isotope separation in molten silicates and aqueous solutions. Geochim Cosmochim Acta 75:3103–3118, doi:10.1016/j.gca.2011.03.002

Wedepohl KH (1995) The composition of the continental crust. Geochim Cosmochim Acta 59:1217–1232

Wetzel F, de Souza GF, Reynolds BC (2014) What controls silicon isotope fractionation during dissolution of diatom opal? Geochim Cosmochim Acta 131:128–137, doi:10.1016/j.gca.2014.01.028

Wille M, Sutton J, Ellwood MJ, Sambridge M, Maher W, Eggins S, Kelly M (2010) Silicon isotopic fractionation in marine sponges: A new model for understanding silicon isotopic variations in sponges. Earth Planet Sci Lett 292:281–289, doi:10.1016/j.epsl.2010.01.036

Wilson M (1989) Igneous Petrogenesis. A Global Tectonic Approach. Unwin Hyman, London

Yamaji N, Mitatni N, Ma JF (2008) A transporter regulating silicon distribution in rice shoots. Plant Cell 20:1381–1389, doi:10.1105/tpc.108.059311

Yeh HW, Epstein S (1978) $^{29}Si/^{28}Si$ and $^{30}Si/^{28}Si$ of meteorites and Allende inclusions. Lunar Planet Sci 9:1289–1291

Zambardi T, Poitrasson F (2011) Precise silicon isotopes determination in silicate rock reference materials by MC–ICP–MS. Geostand Geoanal Res 35:89–99

Zambardi T, Poitrasson F, Corgne A, Meheut M, Quitte G, Anand M (2013) Silicon isotope variations in the inner solar system: Implications for planetary formation, differentiation and composition. Geochim Cosmochim Acta 121:67–83, doi:10.1016/j.gca.2013.06.040

Zambardi T, Lundstrom CC, Li XX, McCurry M (2014) Fe and Si isotope variations at Cedar Butte volcano; Insight into magmatic differentiation. Earth Planet Sci Lett 405:169–179

Ziegler K, Chadwick OA, Brzezinski MA, Kelly EF (2005a) Natural variations of $\delta^{30}Si$ ratios during progressive basalt weathering, Hawaiian Islands. Geochim Cosmochim Acta 69:4597–4610

Ziegler K, Chadwick OA, White AF, Brzezinski MA (2005b) $\delta^{30}Si$ systematics in a granitic saprolite, Puerto Rico. Geology 33:817–820

Ziegler K, Young ED, Schauble EA, Wasson JT (2010) Metal–silicate silicon isotope fractionation in enstatite meteorites and constraints on Earth's core formation. Earth Planet Sci Lett 295:487–496

Zinner E (1997) Presolar material in meteorites: An overview. *In*: Astrophysical Implications of the Laboratory Study of Presolar Materials. Vol 402. Bernatowicz TJ, Zinner E, (eds). American Institute of Physics, New York, p 3–26

Zinner E (1998) Stellar nucleosynthesis and the isotopic composition of presolar grains from primitive meteorites. Ann Rev Earth Planet Sci 26:147–188, doi:10.1146/annurev.earth.26.1.147

Zinner E, Tang M, Anders E (1987) Large isotopic anomalies of Si, C, N and noble-gases in interstellar silicon carbide from the Murray meteorite. Nature 330:730–732, doi:10.1038/330730a0

Zinner E, Nittler LR, Gallino R, Karakas AI, Lugaro M, Straniero O, Lattanzio JC (2006) Silicon and carbon isotopic ratios in AGB stars: SiC grain data, models, and the galactic evolution of the Si isotopes. Astrophys J 650:350–373, doi:10.1086/506957

Reviews in Mineralogy & Geochemistry
Vol. 82 pp. 345-378, 2017

9

Chlorine Isotope Geochemistry

Jaime D. Barnes

Department of Geological Sciences
University of Texas
Austin, Texas 78712
USA

jdbarnes@jsg.utexas.edu

Zachary D. Sharp

Department of Earth and Planetary Sciences
University of New Mexico
Albuquerque, New Mexico 87131–0001
USA
and
Center for Stable Isotopes
University of New Mexico
Albuquerque, New Mexico, 87122
USA

zsharp@unm.edu

INTRODUCTION

Chlorine played a prominent role in the discovery of isotopes. The famous Cavendish Laboratory scientists were fascinated with the atomic mass of Cl. Most elements have a mass that is a close approximation of the multiple of hydrogen (e.g., Aston 1927). By 1920, it was recognized that the atomic weight of Cl was ~35.5, which appeared to violate Francis Aston's whole number rule. Sir Joseph J. Thomson started the famous "Discussion on Isotopes" (Thomson et al. 1921) with the following: "I will plunge at once into the most dramatic case of the isotopes—the case of chlorine". The discussion that followed between three Nobel Prize winners pitted Thomson against Aston and Frederick Soddy, the latter two in defense of multiple isotopes of a single element. And so the game began.

Aston (1919, 1920) argued that the mass spectra of Cl-bearing compounds (e.g., HCl, COCl) supported the existence of at least two isotopes of Cl, ^{35}Cl and ^{37}Cl. However, Thomson contended that the spectra may be the result of different compounds of Cl and not necessarily different isotopes of Cl (Thomson et al. 1921). Ultimately, Aston was proven correct (e.g., Harkins and Hayes 1921; Harkins and Liggett 1923) and is now credited with the discovery of the two stable isotopes of Cl, which is notable for the unusually large abundance of its "rare" isotope. The relative abundances of ^{35}Cl and ^{37}Cl are currently accepted to be 75.76% and 24.24%, respectively (Berglund and Wieser 2011).

It was not until ~75 years after the discovery of the stable isotopes of Cl that they become more "routinely" analyzed and the chlorine isotope compositions of various chlorine reservoirs were beginning to be determined. Here we summarize the current state of chlorine isotope standards, analytical methods, and fractionation, as well as the isotopic composition of different reservoirs and examples of how Cl isotopes have been used as a tracer of Cl flux. This contribution is meant as an overview; for more details on Cl isotopes, one is referred to the comprehensive book by Eggenkamp (2014).

1529-6466/14/0082-0009$05.00 http://dx.doi.org/10.2138/rmg.2017.82.9

CHLORINE ISOTOPE NOMENCLATURE AND STANDARDS

Chlorine isotope data are reported in the standard per mil notation (‰):

$$\delta^{37}Cl = \frac{\left({}^{37}Cl/{}^{35}Cl\right)_{sample} - \left({}^{37}Cl/{}^{35}Cl\right)_{standard}}{\left({}^{37}Cl/{}^{35}Cl\right)_{standard}} \times 1000 \quad (1)$$

in which the standard is seawater (Standard Mean Ocean Chloride (SMOC)). Kaufmann et al. (1984) proposed seawater to be the international standard due to its isotopic homogeneity (based on seawater from the Gulf of Mexico and Pacific Ocean down to depths of 600 mbsl) and defined its isotopic composition to be 0‰. Subsequent work has confirmed the isotopic homogeneity of seawater (based on seawater from the Atlantic, Pacific, Indian, and Antarctic Oceans, Mediterranean and Red Sea, and depths down to 4560 mbsl) is less than analytical error (±0.1‰) and validated its use as a standard (Godon et al. 2004b). To date, SMOC remains the standard of choice. There are no internationally recognized secondary chlorine isotope standards and inter-laboratory calibrations remain limited.

CHLORINE ISOTOPE ANALYTICAL METHODS

The most common method for Cl isotope analysis is measurement of methyl chloride via isotope ratio mass spectrometry (IRMS). Cl isotope ratios have been measured using thermal ionization mass spectrometry (TIMS) with cesium chloride as the ionizing species. Analytical techniques for secondary ion mass spectrometry (SIMS) have been developed and are being used, particularly for meteoric studies of Cl-rich apatites. Laser ablation inductively coupled plasma mass spectrometry (LA–ICP–MS) methods have also been developed by several groups, but have yet to receive widespread application.

Isotope ratio mass spectrometry (IRMS)

The earliest comprehensive attempt to measure the Cl isotope composition of natural materials ranging from aqueous fluids to silicate rocks was by Hoering and Parker (1961). Their measurements were made with HCl as an analyte using dual inlet isotope ratio mass spectrometry (IRMS) (Hoering and Parker 1961). HCl was used because it is easy to prepare quantitatively; however, it is highly reactive with the vacuum walls if any water is present, resulting in strong memory effects. Hoering and Parker (1961) measured 81 natural samples ranging from seawater, oil brines, igneous rocks to meteorites and noted no variation in the Cl isotope composition among them. This limited variability was largely due to the poor analytical precision (±1.0‰).

Methyl chloride (chloromethane) is now used exclusively as the analyte for Cl isotope measurements in gas source mass spectrometers. The methodology for methyl chloride formation from silver chloride was first described by Langvad (1954) and further improved upon by other studies (Hill and Fry 1962; Taylor and Grimsrud 1969; Kaufmann et al. 1984; Long et al. 1993; Eggenkamp 1994). In brief, an aqueous chloride solution is reacted with silver nitrate to produce silver chloride, which is then reacted with excess methyl iodide to produce methyl chloride (Kaufmann et al. 1984; Long et al. 1993; Eggenkamp 1994; for extensive method details see Eggenkamp 2014). Extraction of chloride from silicate rocks into an aqueous form has been made using a number of different procedures, including sodium hydroxide fusion (Eggenkamp 1994), hydrogen fluoride dissolution (Musashi et al. 1998), and pyrohydrolysis (Magenheim et al. 1994; Schnetger and Muramatsu 1996). Pyrohydrolysis, in which silicate rock powder is melted in a water vapor stream, is the most common extraction method. When a rock is melted in a water vapor stream, Cl^- is released into the vapor and then condensed as aqueous chloride. A number of variations in the technique exist, including heating methods and fusion temperature, the use of a flux (V_2O_5) and the use of a sodium hydroxide solution to capture all aqueous Cl^- (e.g., Bonifacie et al. 2008a; Sharp et al. 2013c).

The next step is to quantitatively convert the Cl^- in solution to solid AgCl. Our own procedure first involves removing dissolved sulfur by adding nitric acid to the solution and allowing sulfur to degas over a 24 hour period. Silver chloride is precipitated by adding $AgNO_3$ to the solution, following standard procedures which are nicely outlined by Eggenkamp (1994). The silver chloride must be protected from light in order to prevent its decomposition and thus loss of Cl. Silver chloride is then reacted with excess methyl iodide by heating to 80 °C for two days in order to cause the following exchange reaction to occur:

$$AgCl + CH_3I \rightleftharpoons AgI + CH_3Cl \tag{2}$$

Excess CH_3I must be added to the solution to drive the reaction strongly to the right (e.g., quantitatively convert AgCl to CH_3Cl).

Once the chloride is converted to methyl chloride, the methyl chloride must be purified of excess methyl iodide prior to introduction into the mass spectrometer. This can be achieved by either cryogenic separation (removal of CH_3I by liquid-solid slush of *n*-pentane at −130 °C) or by gas chromatography (GC) (e.g., Long et al. 1993; Holt et al. 1997; Barnes and Sharp 2006; Eggenkamp 2014). Error on analyses of methyl chloride in dual inlet-IRMS have been reduced to ~±0.1‰ using these methods, and thus useful for detecting small variations in the chlorine isotope composition.

The dual inlet method of IRMS is hampered by the large sample size required for an analysis, as well as a long analysis time. The development of continuous flow IRMS methods has significantly reduced sample size with only a modest reduction in precision (Wassenaar and Koehler 2004; Shouakar-Stash et al. 2005; Sharp et al. 2007). The sample preparation procedure is similar to that described above. The main difference is that the CH_3Cl–CH_3I gas mixture is passed through a GC column to purify the CH_3Cl and then transferred directly into the mass spectrometer where it is measured as a single integrated peak. Sample sizes are in the microgram to 10s of microgram Cl-equivalent range with uncertainties on the order of ±0.2‰.

Thermal ionization mass spectrometry (TIMS)

Prior to the development of continuous flow mass spectrometry for the analysis of chlorine isotope ratios, TIMS methods were developed in order to analyze samples with low chlorine concentrations. Although attempts have been made to measure $\delta^{37}Cl$ values via nTIMS (negative thermal ionization mass spectrometry), the method has been hampered by large errors (≥ ±0.9‰) (Vengosh et al. 1989; e.g., Fujitani et al. 2010). In the 1990s, the interest in Cl isotopic analyses of rocks was reinvigorated (e.g., Magenheim et al. 1994). Analyses were made using pTIMS. CsCl (1 to 6 μg of Cl) is deposited on a Ta filament and measured as Cs_2Cl^+ with typical internal precisions of ~±0.25‰ (e.g., Xiao and Zhang 1992; Magenheim et al. 1994). In recent years, many of the $\delta^{37}Cl$ values of solid material determined via the pTIMS methods could not be reproduced by IRMS methods prompting several researchers to suggest that analytical artifacts from the pTIMS method gave erroneous results (Bonifacie et al. 2007b; Sharp et al. 2007; Sharp and Barnes 2008). Several papers which employ the pTIMS methods state that impurities, particularly CsF, will result in irreproducible fractionation during ionization resulting in artificially high $\delta^{37}Cl$ values (Magenheim et al. 1994; Stewart 2000; Willmore et al. 2002). Stewart (2000) specifically writes "Despite the consistent results of the seawater standard measurements, the determinations of $^{37}Cl/^{35}Cl$ in [solid] samples were not satisfactorily reproducible. We suspect the cation exchange procedure was the cause of the difficulty in obtaining consistent results." To date, there is limited work on cross calibration between laboratories and methods, particularly on solid samples, making evaluation of different techniques difficult.

Secondary ion mass spectrometry (SIMS)

Chlorine isotope ratios with somewhat lower precision can be made using SIMS if Cl concentrations are sufficiently high (several 100 ppm). Layne et al. (2004) outlined the method using a Cs^+ primary beam on a large-radius Cameca IMS 1270. Mass resolution has to be sufficiently high to eliminate potential isobaric interferences, such as $^{34}SH^-$ at mass on $^{35}Cl^-$. Initial SIMS reproducibility for $\delta^{37}Cl$ values was ±1.5‰ (Layne et al. 2004). Subsequent large-radius SIMS work has been made using multicollection mode to simultaneously collect $^{35}Cl^-$ and $^{37}Cl^-$ signals, greatly improving the error to ~±0.4‰ (for samples with ≥400 ppm Cl; errors are larger for lower Cl concentrations) (e.g., John et al. 2010; Sharp et al. 2010b; Kusebauch et al. 2015). Recent Cl isotope analyses have been made on lunar apatites using nanoSIMS and small-radius SIMS to determine $\delta^{37}Cl$ values, with correspondingly larger errors (≳2‰, 2σ) (Tartèse et al. 2014; Treiman et al. 2014; Boyce et al. 2015).

Laser ablation inductively coupled plasma mass spectrometry (LA–ICP–MS)

There have been only a few methods papers describing laser ablation inductively coupled plasma mass spectrometry (LA-ICP-MS) for the measurement of $\delta^{37}Cl$ values (Van Acker et al. 2006; Fietzke et al. 2008; Toyama et al. 2015). Van Acker et al. (2006) measured the Cl isotope composition of chlorinated solvents. Fietzke et al. (2008) determined the $\delta^{37}Cl$ values of silicate rocks. Chloride was extracted from rocks via pyrohydrolysis and precipitated as AgCl. The AgCl was ablated with the laser and introduced into the plasma stream (Fietzke et al. 2008). In order to determine a $^{37}Cl/^{35}Cl$ ratio, $^{37}Cl^+$ must be corrected from interference of $^{36}ArH^+$ from the plasma (Van Acker et al. 2006; Fietzke et al. 2008). For accurate results, Fietzke et al. (2008) report precipitating ≥25 µg of Cl of which about 1 µg of Cl is consumed per spot with an internal error of ±0.06‰. Van Acker et al. (2006) report a minimum of 25 µg of Cl is needed with a larger error than more traditional IRMS methods. To our knowledge, no geological applications of the ICP–MS technique have yet been published.

ISOTOPIC FRACTIONATION

Equilibrium Cl isotope fractionation—theoretical constraints

Chlorine was one of the first stable isotope systems to be investigated theoretically and for which equilibrium fractionation factors were determined (Urey and Greiff 1935; Urey 1947). Urey's 1947 publication lays out the fractionation between Cl in different oxidation states (ClO_4^-, ClO_3^-, ClO_2, Cl_2, and Cl^-). These results were later expanded upon to include more geologically and environmentally relevant species (Schauble et al. 2003). In general, the oxidation state of Cl is the predominant control on isotope fractionation. The higher the oxidation state, the stronger the incorporation of the heavy isotope, ^{37}Cl. The theoretical fractionation between perchlorate–chloride, for example, is on the order of 90‰ at 0 °C (Urey, 1947; Schauble et al., 2003). Chloride bonded to +2 cations (e.g., $FeCl_2$, $MnCl_2$) will be preferentially enriched in ^{37}Cl relative to substances in which Cl^- is bonded to +1 cations (e.g., NaCl) by ~2 to 3‰ at 25 °C (Fig. 1). Cl-bearing organic molecules with Cl-C bonds (+1 oxidation state) should be enriched in ^{37}Cl by ~5 to 9‰ at 25 °C relative to Cl^- (Schauble et al. 2003). These results are supported by more recent theoretical work (Czarnacki and Halas 2012).

Interestingly, both organic and inorganic halogenation undergo significant kinetic isotope effects in which ^{35}Cl is preferentially incorporated into organic matter relative to coexisting aqueous chloride, with $\Delta^{37}Cl_{org\text{-}inorg}$ as large as −12 ‰ (Reddy et al. 2002; Aeppli et al. 2013). This is opposite from the expected equilibrium fractionation.

Equilibrium Cl isotope fractionation—experimental constraints

There are very few experimental data for Cl isotope fractionation. The data that do exist are in good agreement with theoretical calculations of Schauble et al. (2003). The earliest experimental fractionation experiments were made by Howald (1960) and Hoering and Parker (1961). Howald (1960) measured fractionations for different metal chloride solutions. Hoering and Parker (1961) measured the fractionation for four Cl exchange reactions: Cl_2HCl, NH_4Cl–HCl, $NaCl_{(s)}$–$Cl^-_{(aq)}$ and a hexachloroplatinate–chloride exchange. The fractionation between NaCl and a saturated NaCl solution yielded $\alpha = 1.0002 \pm 0.0003$ at 23 °C, where α is the isotopic fractionation factor between two substances, suggesting that NaCl was enriched in ^{37}Cl relative to the aqueous solution. Subsequent work between NaCl and a saturated solution confirms ^{37}Cl enrichment in the alkali salt with a calculated $\alpha = 1.00026 \pm 0.00007$ at 22 ± 2 °C (Eggenkamp et al. 1995), consistent with observations from natural samples (e.g., Eastoe et al. 1999; Eastoe and Peryt 1999). In contrast, K and Mg salts are depleted in ^{37}Cl relative to the saturated solution by -0.09 ± 0.09 and -0.06 ± 0.10‰ at 22 ± 2 °C (Eggenkamp et al. 1995).

The fractionation between $HCl_{(g)}$ and dissolved Cl^- was determined over a range of temperatures by extracting the HCl gas in the headspace of a hydrochloric acid solution (Sharp et al. 2010a). The $1000\ln\alpha$ value of 1.4 to 1.8 (between 50 and 80 °C) is in good agreement with a value of ~1.65 at 50 °C determined from the combined theoretical β factors (reduced partition function ratios) for HCl and NaCl (Schauble et al. 2003) and the measured fractionation between NaCl and aqueous chloride (Eggenkamp et al. 1995) (Fig. 1).

Higher temperature fractionation experiments are lacking, yet are greatly needed. Liebscher et al. (2006) measured the Cl isotope fractionation between vapor and liquid in the NaCl-H_2O system at 400 °C and 450 °C (23 to 28 MPa). Fractionations are small and show an interesting "oscillating" variation with pressure (Fig. 2). These experiments are consistent with empirical data of mid-ocean ridge vent fluids which indicate little to no vapor–liquid Cl isotope fractionation (Bonifacie et al. 2005). However, other studies have proposed large Cl isotope fractionation (several per mil) due in phase separation in sphalerite-hosted fluid inclusions

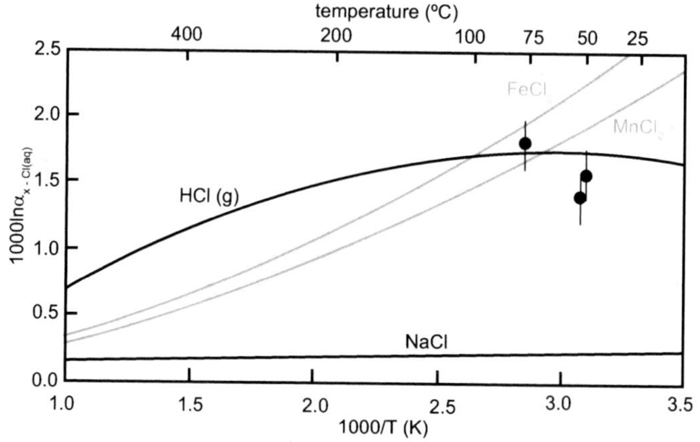

Figure 1. Equilibrium isotope fractionation between HCl, $FeCl_2$, $MnCl_2$, or NaCl and aqueous Cl^- as a function of temperature. Theoretical curves for HCl, $FeCl_2$, $MnCl_2$, and NaCl are from Schauble et al. (2003). $FeCl_2$ and $MnCl_2$ (gray lines) are analogs for Cl-bearing silicate minerals. The equilibrium isotopic fractionation between KCl or RbCl and aqueous Cl^- is negative and therefore not shown. The isotopic composition of aqueous Cl^- is determined from the experimental data between NaCl and Cl^- (aq) from Eggenkamp et al. (1995). Black dots are experimental data for equilibrium fractionation between HCl and Cl^- (aq). Figure reprinted with modifications from *Geochimica Cosmochimica Acta* Vol. 74 Sharp ZD, Barnes JD, Fischer TP, Halick M. A laboratory determination of chlorine isotope fractionation in acid systems and applications to volcanic fumaroles,p 264–273 (2010) with permission from Elsevier.

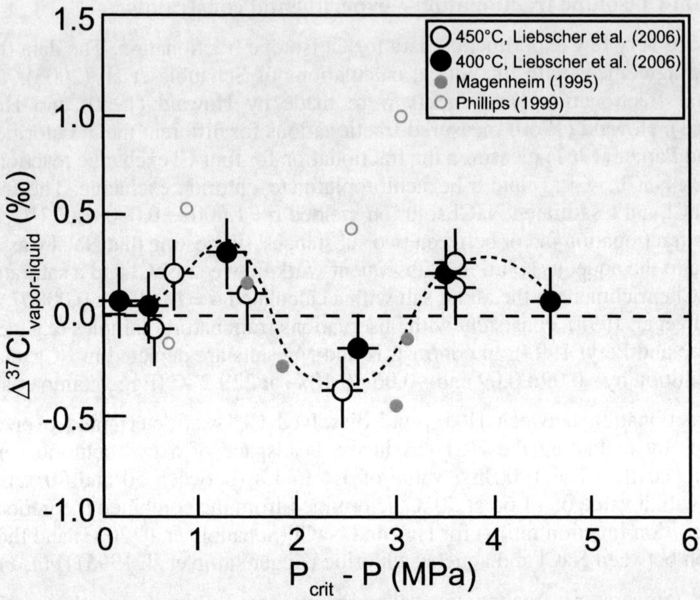

Figure 2. Chlorine isotope fractionation between coexisting vapor and liquid ($\Delta^{37}Cl_{vapor\text{–}liquid} = \delta^{37}Cl_{vapor} - \delta^{37}Cl_{liquid}$) at 450 °C (open black circles) and 400 °C (black solid circles) as a function of the pressure difference to the critical curve at these temperatures (Liebscher et al. 2006). The dashed line highlights zero Cl isotope fractionation. Unpublished thesis data from Magenheim (1995) (solid gray circles) and Phillips (1999) (open gray circles) are shown for comparison. Figure reprinted with minor alterations from *Chemical Geology* Vol. 234 Liebscher A, Barnes JD, and Sharp ZD. Chlorine isotope vapor–liquid fractionation during experimental fluid–phase separation at 400 °C/23 MPa to 450 °C/42 MPa, p. 340–345 (2006) with permission from Elsevier.

from the JADE field and the North Fiji basin and in quartz-hosted fluid inclusions from the Iberian Pyrite Belt (Lüders et al. 2002; Germann et al. 2003). These data indicate preferential incorporation of ^{37}Cl into the brine, counter to theoretical and experimental estimates (Schauble et al. 2003). The only other published high temperature experimental result that we are aware of is the fractionation between sodalite ($Na_8Al_6Si_6O_{24}Cl_2$ and $NaCl_{(l)}$ at 825 °C, with a measured $1000\ln\alpha_{sodalite\text{–}NaCl_{(l)}}$ of –0.3‰ (Sharp et al. 2007). Preliminary experimental work documents a slight positive fractionation ($10^3\ln\alpha_{amphibole\text{–}fluid} = 0.1‰ \pm 0.2‰$) between synthetic amphibole (hastingsite) and a NaCl-bearing solution at 700 °C and 0.2 GPa (Cisneros 2013). Empirical results from seafloor serpentinites and altered oceanic crust (see below) also suggest a small positive fractionation between serpentine and Cl^- and amphibole and Cl^-.

Kinetic Cl isotope fractionation–diffusion and ion filtration

Chlorine species separate from one another by diffusion due to the higher translational velocity of light isotopologues. Cl isotope fractionation due to diffusion along a concentration gradient has been demonstrated both experimentally and empirically by fitting measured data to a diffusion model. Laboratory experiments of aqueous diffusion of Cl through polyacrylamide gel show $\Delta^{35}Cl/\Delta^{37}Cl$ (the ratio of the diffusion coefficients of ^{35}Cl and ^{37}Cl) to be between 1.00128 at 2 °C and 1.00192 at 80 °C (Eggenkamp and Coleman 2009). These values are in agreement with earlier experimental data, ranging from 1.0009 to 1.0021 and averaging 1.0015 (Madorsky and Straus 1947; Konstantinov and Bakulin 1965; Richter et al. 2006). All empirical studies have been made in systems in which a saline fluid diffuses into a more dilute fluid through a sedimentary column (Desaulniers et al. 1986; Eggenkamp et al. 1994; Groen et al. 2000). Empirically determined $\Delta^{35}Cl/\Delta^{37}Cl$ values range from ~1.0010 to ~1.0030 with most data between 1.0015 and 1.0020 (Desaulniers et al. 1986; Eggenkamp et al. 1994; Groen et al. 2000; Beekman et al. 2011).

Ion filtration is another mechanism that fractionates Cl isotopes. A negatively charged membrane (e.g., a clay layer) repels dissolved ions (Cl^-) in an advective flow regime. Because the $^{35}Cl^-$ ion has higher ionic mobility than $^{37}Cl^-$, it tends to become enriched behind the membrane relative to the direction of flow (Phillips and Bentley 1987). Ion filtration is thought to play a role in producing isotopically negative sedimentary pore fluids (Godon et al. 2004a; Bonifacie et al. 2007a) (see *Seawater and seawater-derived chloride* section below).

Kinetic Cl isotope fractionation—Cl loss

Kinetic Cl isotope fractionation can also occur due to Cl loss during volcanic degassing. Unusually high $\delta^{37}Cl$ values (up to +12‰) have been observed in volcanic gases from boiling fumaroles (>100 °C) associated with volcanic lakes and/or hydrothermal systems (Eggenkamp 1994; Barnes et al. 2009a; Sharp et al. 2010a). Theoretical calculations show that $1000\ln\alpha_{HCl\text{-}Cl(aq)}$ is less than 2‰ at all temperatures (Schauble et al. 2003), failing to explain these high values. Fractionation between $HCl_{(g)}$ and $Cl^-_{(aq)}$ has been determined experimentally by measuring the isotopic composition of HCl vapor equilibrated with 1 M HCl solution and confirm these theoretical calculations (Fig. 1, Sharp et al. 2010a). However, this experiment evidently does not mimic the dynamic system of a volcanic fumarole. Flow-through experiments, in which air is bubbled through hydrochloric acid and the removed HCl vapor is allowed to condense along the length of a cool tube, produced fractionations as large as +10‰. These high values are due to kinetic exchange between the solvated Cl^- ion in the aqueous solution condensing along the walls of the tube and the flowing ^{37}Cl-enriched HCl vapor (Sharp et al. 2010a). In contrast to preferential loss of ^{37}Cl into HCl, ^{35}Cl has been shown to be lost to the vapor phase during HCl evaporation experiments due to its higher translation velocity (greater escaping tendency) and vapor pressure compared to ^{37}Cl (Sharp et al. 2010a). Chlorine isotopes can be used as an indicator of volcanic processes such as degassing, but care must be taken to recognize the possibility of Cl loss due to HCl degassing when using Cl isotopes as a tracer of source (Rizzo et al. 2013; Barnes et al. 2014b; Fischer et al. 2015).

The largest kinetic isotope effects measured to date are seen in lunar samples. $\delta^{37}Cl$ values as high as 34‰ have been measured in lunar basalts (Sharp et al. 2010b; Tartèse et al. 2014; Treiman et al. 2014), and are thought to be a result of rapid diffusion of Cl-species to the vapor phase during degassing (see section on the *Moon* for details).

CHLORINE ISOTOPIC COMPOSITION OF VARIOUS GEOLOGIC RESERVOIRS

In general, the chlorine isotope variation in nature is relatively small, ranging from ~ −2 to +2‰, due to small equilibrium fractionation factors. However, large variations are observed, e.g., in extraterrestrial materials and volcanic gases, due to kinetic fractionation (Fig. 3). Below we summarize the variations in $\delta^{37}Cl$ values observed in natural materials. The isotopic composition of many of these Cl reservoirs has only been recently determined and, in some cases, is not universally agreed upon. All $\delta^{37}Cl$ values discussed below were determined by IRMS, unless indicated as pTIMS or SIMS measurements.

Mantle/OIB/mantle derived material

The first systematic Cl isotope analyses of Mid-Ocean Ridge Basalts (MORB) averaged -0.2 ± 0.5‰ ($n = 12$) and showed no variation with Cl content (Sharp et al. 2007). Analysis of sub-continental mantle samples were indistinguishable (-0.03 ± 0.25‰). These results supported previous analyses of mantle-derived carbonatites with $\delta^{37}Cl$ values averaging -0.2 ± 0.4‰ (Eggenkamp and Koster van Groos 1997) and are similar to seawater and evaporites. The following year, Bonifacie et al. (2008a) analyzed MORB samples and found a positive correlation between the $\delta^{37}Cl$ value and Cl content. They concluded that the mantle has a $\delta^{37}Cl$ value

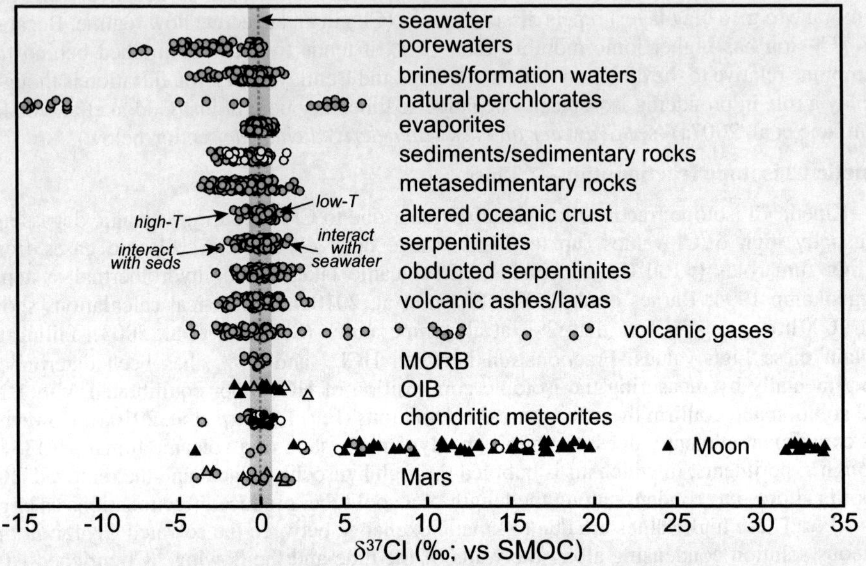

Figure 3. Chlorine isotope variability in different terrestrial and extraterrestrial reservoirs. Dashed line is seawater (SMOC). Wider gray band delineates −0.5 to +0.5‰. Almost all evaporite and MORB values fall within this gray band illustrating that the largest Cl terrestrial reservoirs are near 0‰. Triangles are SIMS analyses; circles are IRMS analyses. Each symbol represents one sample, with the exception of some SIMS analyses which are individual analyses of different mineral grains within the same sample. For multiple analyses of the same sample, only the average is shown. All IRMS data are bulk δ^{37}Cl values, unless indicated otherwise. Sediments/sedimentary rocks: gray circles are marine; open circles are non-marine. Almost all marine sediments are negative; non-marine are negative and positive. OIB: gray triangles = EM1; open triangles = EM2; black triangles = HIMU. Chondritic meteorites: gray circles = ordinary chondrites; open circles = carbonaceous chondrites; black circles = enstatite chondrites. Moon = gray circles = water-soluble chloride; open circles = structurally bound chloride. Data sources: porewaters: Ransom et al. (1995), Hesse et al. (2000), Godon et al. (2004), Bonifacie et al. (2007); brines/formation waters: Kaufmann et al. (1987, 1988, 1993), Eggenkamp (1994), Eastoe et al. (1999, 2001), Zeigler et al. (2001), Shouakar-Stash et al. (2007), Zhang et al. (2007), Stotler et al. (2010); natural perchlorates: Böhlke et al. (2005); Sturchio et al. (2006); Jackson et al. (2010); evaporites: Eastoe et al. (1999, 2001, 2007), Eastoe and Peryt (1999), Eggenkamp et al. (1995), Arcuri and Brimhall (2003); sediments/sedimentary rocks: Acuri and Brimhall (2003), Barnes et al. (2008; 2009), Selverstone and Sharp (2015); metasedimentary rocks: John et al. (2010), Selverstone and Sharp (2013, 2015), Barnes et al. (2014); altered oceanic crust: serpentinites: Barnes and Sharp (2006), Bonifacie et al. (2007, 2008), Barnes et al. (2008, 2009), Barnes and Cisneros (2012), Boschi et al. (2013); obducted serpentinites: Barnes et al. (2006, 2013, 2014), Bonifacie et al. (2008), John et al. (2011), Selverstone and Sharp (2013); volcanic ashes/lavas: Barnes et al. (2008, 2009), Barnes and Straub (2010), Chiaradia et al. (2014), Cullen et al., (2015), Rizzo et al. (2013); volcanic gases: Barnes et al. (2008, 2009), Rizzo et al. (2013), MORB: Sharp et al. (2007); OIB: John et al. (2010); chondritic meteorites: Sharp et al. (2007, 2012); Moon: Sharp et al. (2010), Tartese et al. (2014), Treiman et al. (2014), Boyce et al. (2015); Mars: Sharp et al. (2016).

of ≤ −1.6‰, arguing that the higher δ^{37}Cl values represent seawater contamination. This trend was not observed in the earlier study. Analyses of additional mantle/mantle-derived material (e.g., alkali inclusions in fibrous diamonds) as well as an extensive suite of chondrites were all close to 0‰; the mantle-derived samples were −0.2 ± 0.3‰, analytically indistinguishable from the chondrite value of −0.3 ± 0.3‰ (Sharp et al. 2013c). The authors suggested that the low δ^{37}Cl values analyzed previously may not reflect uncontaminated samples, but rather were an analytical artifact associated with analysis of small samples in dual inlet mode.

Mantle Cl isotope heterogeneities are not in doubt. The mantle likely preserves some heterogeneities due to the subduction of crustal material. Using SIMS analyses of Ocean Island Basalt (OIB) glasses, John et al. (2010) showed that $\delta^{37}Cl$ values range from −1.6 to +1.1‰ for HIMU-type and −0.4 to +2.9‰ for EM-type lavas, and positively correlate with $^{87}Sr/^{86}Sr$ ratios. Sharp et al. (2007) report values for two MORB glasses from a similar location in the East Pacific Rise that are distinctly lower (−1‰) than the near zero values of other MORB samples. They may also represent a contaminated mantle origin.

Seawater and seawater-derived chloride

The oceans and associated ocean-derived Cl in the form of evaporates, sedimentary pore fluids, and marine aerosols are the second largest Cl reservoir on Earth (e.g., Schilling et al. 1978; Jarrard 2003; Sharp and Barnes 2004). As discussed in the *Chlorine Isotope Nomenclature and Standards* section, the Cl isotope composition of seawater is homogeneous (Godon et al. 2004b) due to its extraordinarily long residence time in the ocean, with a defined $\delta^{37}Cl$ value of 0‰ (Kaufmann et al. 1984). The chlorine isotope composition of the ocean has remained relatively constant through time based on the isotopic composition of Phanerozoic halite and Precambrian cherts, barite, and dolomite (Eastoe et al. 2007; Sharp et al. 2007; Sharp et al. 2013c) (Fig. 4).

First-formed halite in equilibrium with seawater at 22±2 °C will have a $\delta^{37}Cl$ value of +0.26±0.07‰ (Eggenkamp et al. 1995). If seawater is isolated and allowed to completely evaporate, the $\delta^{37}Cl$ value of the residual salt will have a bulk $\delta^{37}Cl$ value identical to seawater. As salt is precipitated from the fluid, the fluid itself becomes progressively enriched in ^{35}Cl. If this fluid is then physically removed from the salt, any redeposition of salt from the ^{35}Cl-enriched fluid will also produce a ^{35}Cl-enriched salt. Naturally occurring halite formed from seawater via equilibrium fractionation is reported to have $\delta^{37}Cl$ values ranging from −0.6 to +0.4‰. This range of values is due to multiple precipitation and dissolution events of the evaporite minerals. $\delta^{37}Cl$ values of halite outside of this range require an alternate explanation, typically attributed to the influx of a non-zero ‰ fluid (Eastoe et al. 1999; Eastoe and Peryt 1999).

Sedimentary pore fluids have some of the lowest reported $\delta^{37}Cl$ values, ranging from −7.8 to +0.3‰ with the vast majority of the samples being negative (Ransom et al. 1995; Hesse et al. 2000; Spivack et al. 2002; Godon et al. 2004a; Bonifacie et al. 2007a). Multiple explanations have been proposed for the predominantly negative values, including mineral-fluid interaction, ion filtration, diffusion, and melting of gas hydrates (Ransom et al. 1995; Hesse et al. 2000; Spivack et al. 2002; Godon et al. 2004a; Bonifacie et al. 2007a). Although these processes may, in part, contribute to the low $\delta^{37}Cl$ values of pore waters, none of these mechanisms should lower the $\delta^{37}Cl$ values of seawater-derived fluids to such a large extent. Authigenic clay minerals do not generally have positive $\delta^{37}Cl$ values (Barnes et al. 2008; Barnes et al. 2009a; Selverstone and Sharp 2015), so their formation could not produce a negative residual fluid. No diffusion studies have documented Cl-isotope changes in natural systems greater than ~3‰. Ion filtration has been invoked to explain values of natural samples as low as ~−5‰ (Ziegler et al. 2001; Godon et al. 2004a). However, in most cases, the negative extreme remains unexplained.

An alternative explanation for the low $\delta^{37}Cl$ values may be related to kinetic fractionation associated with the formation of organochlorine compounds. Up to 20% of marine 'natural products' are organohalogens (Gribble 2010). Organohalide microbial communities may play an important role in the carbon and chlorine cycle in marine sediments and have been shown to be extensively distributed in sediments of the Nankai Trough subduction zone (Futagami et al. 2013). High organochlorine concentrations have also been found in sediment traps from the Arabian Penninsula (Leri et al. 2015). Organic matter incorporates Cl in the +1 oxidation state, as opposed to dissolved Cl^- in the −1 oxidation state. As such, it should strongly incorporate ^{37}Cl relative to ^{35}Cl. However, in natural systems kinetic fractionation often results in an enrichment of ^{35}Cl in organic matter in excess of 10‰ (Reddy et al. 2002; Aeppli et al. 2013). Marine organic-rich sediments may therefore have strong negative $\delta^{37}Cl$ values. As they

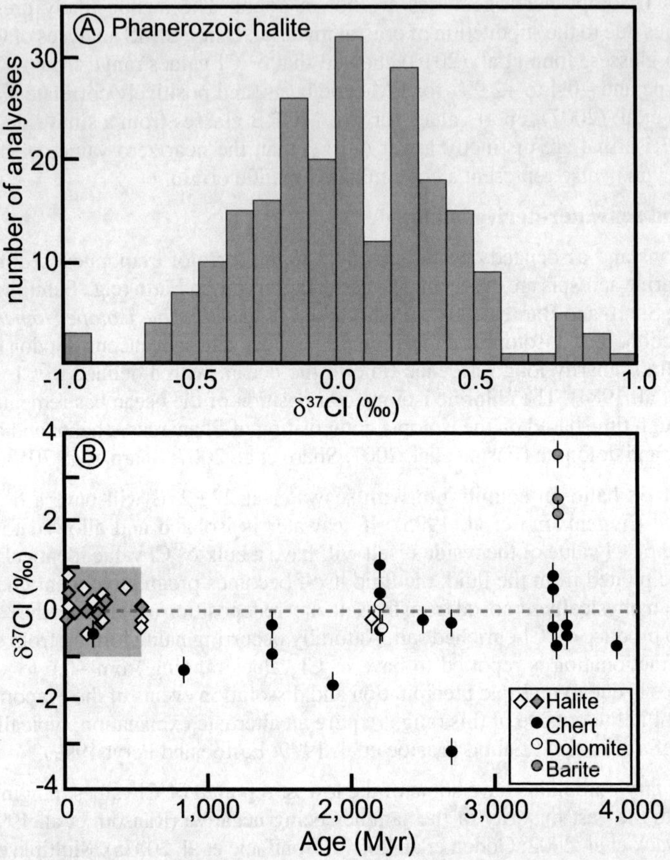

Figure 4. A. Histogram showing the $\delta^{37}Cl$ values of Phanerozoic halite. Data from Eastoe et al. (1999, 2001, 2007), Arcuri and Brimhall (2003), Eastoe and Peryt (1999), and Eggenkamp et al. (1995). B. $\delta^{37}Cl$ values of Phanerozoic halite (open diamond data from Sharp et al. (2007) and gray diamond data are average values from halite facies (Eastoe et al. 2007)) and older halite and sediments (fluid inclusions in chert, barite, and dolomite) (Sharp et al. 2007, 2013b) as a function of age. Gray box delineates $\delta^{37}Cl$ values of Phanerozoic halite shown in A.

are degraded (microbially or through inorganic oxidation), they could release Cl into the porewaters with a very low $\delta^{37}Cl$ value. Organic materials may be responsible for the low $\delta^{37}Cl$ values of marine sedimentary pore fluids and may be the prime driver in the creation of low $\delta^{37}Cl$ values on Earth. The role of organic material is mostly unexplored and may be critical to Cl fractionation in sedimentary systems.

Marine aerosols form from the interaction of sea-spray NaCl with atmospheric sulfuric and nitric acids. Cl^- in the aerosols are replaced by sulfate or nitrate ions, releasing HCl in the process. Marine aerosols have $\delta^{37}Cl$ values, determined by pTIMS, ranging from −0.85 to +2.53‰ (Volpe and Spivack 1994; Volpe et al. 1998). In general, $\delta^{37}Cl$ values of the aerosols increase as chlorine loss increases (Volpe and Spivack 1994). These values are interpreted to be due to fractionation during volatilization of HCl (Volpe and Spivack 1994); however, the fractionation due to HCl exchange has been shown to have the opposite sign (Schauble et al. 2003; Sharp et al. 2010a). Subsequent works show that smaller particles have positive values and larger particles have negative values, suggesting kinetic effects likely play a role in release of Cl from marine aerosols (Volpe et al. 1998).

Sediments

Our knowledge of the Cl isotope composition of sediments is limited to only a handful of bulk analyses. Marine sediments from the Pacific Ocean and sampled by Ocean Drilling Program have δ^{37}Cl values ranging from −2.5 to +0.7‰ with the majority of the samples being negative (average = −0.8 ± 0.8‰; n = 24) (Barnes et al. 2008; Barnes et al. 2009a). Jurassic sedimentary rocks deposited in a shallow marine basin from Chile have also have primarily negative δ^{37}Cl values (−2.6 to +0.5‰; average = −0.9 ± 0.9‰; n = 17) (Arcuri and Brimhall 2003). Marine shales from the Swiss Alps are isotopically negative with bulk δ^{37}Cl values of −3.0 to −0.7‰; whereas, interlayered playa-facies sedimentary rocks range from 0 to +1.8‰ and fluvial/deltaic facies rocks have δ^{37}Cl values between −2 and −1‰ (Selverstone and Sharp 2015). Intra-bed samples are generally isotopically homogeneous, however, great isotopic variably occurs between sedimentary beds. Isotopic differences are interpreted to be due to variations in protolith composition. Protolith variability may reflect biological processes, evaporation of source waters, and input of an external Cl source (e.g., aerosols, formation waters) (Selverstone and Sharp 2015). In sum, there is no unambiguous chlorine isotopic "signature" for sedimentary material; however, all organic-rich marine sediments and their lithified equivalents that have been measured to date have negative δ^{37}Cl values. Non-marine sedimentary material can have either positive or negative δ^{37}Cl values.

In order to further investigate isotopic variability and possible fractionation during subduction, a few studies have analyzed metasedimentary rocks. High-pressure (HP) marine metasedimentary rocks from Ecuador which have δ^{37}Cl values ranging from −2.2 to +2.2‰ (average = 0.0 ± 1.7‰; n = 8), leading the authors to speculate that the high δ^{37}Cl values may be due to preferential loss of ^{35}Cl during prograde metamorphism (John et al. 2010). However, no such high values are reported in a suite of HP and ultra-high-pressure (UHP) metasedimentary rocks of marine origin from the Western Alps—δ^{37}Cl values range from −3.6 to +0.0‰ (average = −2.3 ± 0.9‰; n = 25), overlapping with modern marine sediments and extending to lower values (Selverstone and Sharp 2013). These low values are interpreted to have been derived from interaction with isotopically negative sedimentary pore fluids expelled from the accretionary wedge during plate bending and subduction, rather than isotopic fractionation during subduction (Selverstone and Sharp 2013). In order to address Cl isotope fractionation over a range of metamorphic temperatures, Selverstone and Sharp (2015) determined the Cl isotope composition of unmetamorphosed marine and non-marine sedimentary rocks (see above) and their metamorphosed equivalents through increasing metamorphic grade. The metamorphosed equivalents overlap with the respective protoliths supporting minimal modification of the Cl isotope composition during Alpine metamorphism, despite significant Cl and H_2O loss (Selverstone and Sharp 2015). HP and UHP metamorphosed sedimentary rocks retain their original isotopic composition despite devolatilization unless infiltrated by an externally derived fluid. Future work, particularly on the role of organochlorides, is necessary to fully understand the isotopic variability of sedimentary materials.

Altered Oceanic Crust (AOC)

Hydrothermal alteration of oceanic crust results in the formation of secondary hydrous minerals (e.g., clays, chlorite, amphiboles), many of which can host high concentrations of Cl (amphibole is the dominant Cl host containing up to 4 wt. % Cl) (e.g., Ito et al. 1983; Vanko 1986; Philippot et al. 1998). A detailed study by Barnes and Cisneros (2012) report δ^{37}Cl values and concentrations for 50 altered basalt (extrusive lavas and sheeted dikes) and gabbro samples from seven DSDP/ODP/IODP sites. δ^{37}Cl values range from −1.4 to +1.8‰. In general, chlorine concentrations and δ^{37}Cl values increase with increasing temperature of alteration and amphibole content. Positive δ^{37}Cl values in amphibole-rich samples are consistent with theoretical calculations and preliminary experimental data that indicate amphibole should be slightly enriched in ^{37}Cl compared to co-existing fluid (Schauble et al. 2003; Cisneros 2013). Nevertheless, the very high δ^{37}Cl values cannot be explained solely by equilibrium fractionation at high temperatures. The samples with negative δ^{37}Cl values are dominated by clay minerals (Fig. 5). The correlation between chlorine isotope composition/chlorine concentration and modal mineralogy suggests that chlorine chemistry is a rough indicator of metamorphic grade in altered oceanic crust (Barnes and Cisneros, 2012). Bonifacie et al. (2007b) also analyzed AOC samples from ODP Hole 504B which had δ^{37}Cl values ranging from −1.6 to −0.9‰ (n=3). These values are all lower than the δ^{37}Cl values from Hole 504B (n=9) reported by Barnes and Cisneros (2012), although none of the analyses are on the exact same sample. Until inter-laboratory calibrations are more universal, it is difficult to discuss analytical differences between labs.

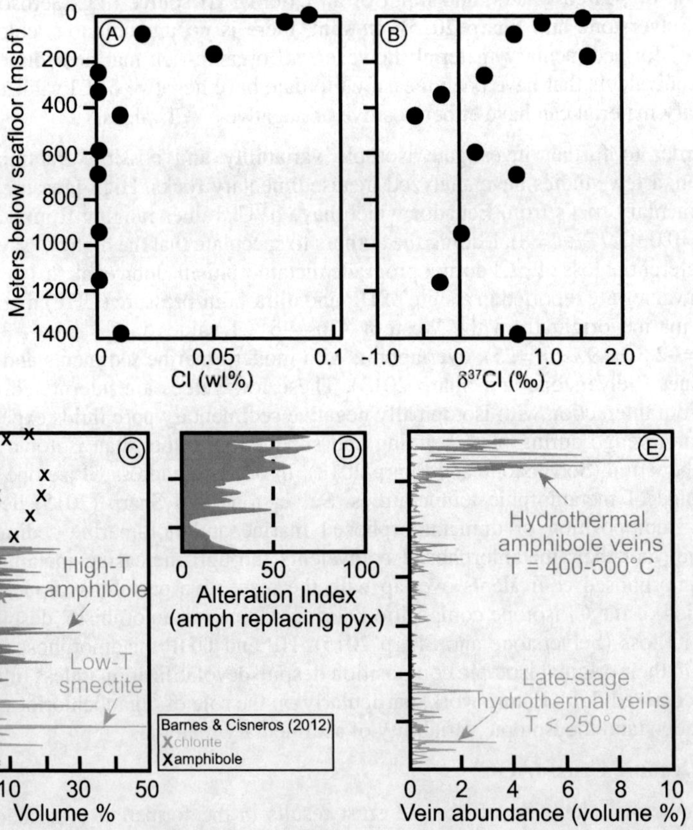

Figure 5. Variation in chlorine geochemistry and mineralogy in Ocean Drilling Program Hole 735B. A) Variation of Cl concentration (wt.%) and B) Cl isotope composition with depth. C) Variation in the modal abundance of amphibole and smectite as bulk volume percent with depth (Dick et al. 2000) and volume percentage of amphibole and chlorite on bulk sample XRD powders (Barnes and Cisneros 2012). D) Alteration index determined by the percentage of amphibole replacing pyroxene (Stakes et al. 1991). E) Volume percent abundance of higher temperature amphibole veins and lower temperature veins (clays, carbonates, zeolites, prehnite) (Bach et al. 2001). Figure reprinted with minor alterations from *Chemical Geology* Vol. 326–327 Barnes JD and Cisneros M. Mineralogical control on the chlorine isotope composition of altered oceanic crust, p. 51–60 (2012) with permission from Elsevier.

Serpentinites

The chlorine isotope composition of serpentinites ($n > 200$) has been better characterized than most of the above discussed reservoirs. The chlorine isotope composition of seafloor serpentinites (sampled by the DSDP/ODP/IODP), including serpentinite clasts extruded at serpentine seamounts, typically have $\delta^{37}Cl$ values ranging from ~ −1.6 to +0.5‰ (Barnes and Sharp 2006; Barnes et al. 2008, 2009b; Bonifacie et al. 2008b; Boschi et al. 2013). Cl is hosted as both a water-soluble phase (thought to be halite precipitated along grain boundaries) and structurally bound within the serpentine structure. Structurally bound Cl is ~ +0.2‰ higher than water-soluble Cl (Sharp and Barnes 2004; Barnes and Sharp 2006), consistent with predicted ^{37}Cl enrichment in silicate hydrous phases relative to salts (Eggenkamp et al. 1995; Schauble et al. 2003). Seafloor serpentinites can be divided into two populations based on tectonic setting and fluid source. The vast majority of analyzed seafloor serpentinities are isotopically positive due to hydration via seawater along faults and fissures, in which the serpentine preferentially incorporates ^{37}Cl into its structure (Barnes and Sharp 2006). Isotopically negative serpentinites are interpreted to have formed via hydration from isotopically negative pore fluids. These serpentinites are typically found juxtaposed to sediments via thrust faults. The isotopically negative pore fluids hydrate the peridotite resulting in an isotopically negative serpentinite (Barnes and Sharp 2006).

Obducted serpentinites have $\delta^{37}Cl$ values overlapping seafloor serpentinite values and some extending to higher values (−1.5 to +2.4‰) (Barnes et al. 2006; Barnes et al. 2013, 2014a; Bonifacie et al. 2008b; John et al. 2011; Selverstone and Sharp 2011, 2013). Despite significant volatile loss, original seafloor serpentinite $\delta^{37}Cl$ values are preserved throughout prograde metamorphism (Barnes et al. 2006; Bonifacie et al. 2008b; John et al. 2011; Selverstone and Sharp 2013), similar to observations in metasedimentary rock sequences (Selverstone and Sharp 2015). However, Cl isotope compositions can be modified from original values due to infiltration of an externally derived fluid, as exemplified by the high $\delta^{37}Cl$ values of some obducted serpentinites. The sources responsible for these high $\delta^{37}Cl$ values are not clear, but in some cases are speculated to derive from nearby sediments (Selverstone and Sharp 2011, 2013; Barnes et al. 2014a).

Perchlorates

Perchlorates are an oxidized form of Cl that form by photochemical reactions in the stratosphere (Bao and Gu 2004). The equilibrium fractionation between chloride (Cl^-) and perchlorate (ClO_4^-, oxidation state of +7) is enormous. Harold Urey calculated the ClO_4^-–Cl^- fractionation to be 80‰ at 23 °C (Urey 1947). Schauble et al. (2003) estimated the fractionation to be 70‰ at 25 °C. Hoering and Parker (1961) first measured the $\delta^{37}Cl$ values of perchlorates and found them to be indistinguishable from the normal range of chlorides. They concluded that perchlorates and chlorides from the same nitrate deposits from Tarpaca, Chile were not in isotopic equilibrium. The formation mechanism of perchlorates was assumed to be one of oxidation of Cl^- to ClO_4^- without isotopic fractionation.

Preservation of natural perchlorates on Earth is rare. Because they are soluble in water, they are found primarily in arid environments (Jackson et al. 2015b). Large deposits are known from several notable locations around the world, including the Antarctic dry valleys, deserts in the American southwest (Death Valley, Amargosa Desert), the Atacama Desert (Chile), and China. The isotopic composition of the Atacama perchlorates range from −15 to −9‰ (Sturchio et al. 2006; Böhlke et al. 2009), whereas perchlorates from the Southwest United States have $\delta^{37}Cl$ values of −3 to +6‰, similar to their presumed chloride source (Sturchio et al. 2012a). Perchlorates from groundwater samples from the Southern High Plains of Texas and New Mexico that are thought to be of natural origin have $\delta^{37}Cl$ values of 3.1 to 5.1‰ (Jackson et al. 2010).

Laboratory experiments involving bacterial reduction of perchlorate by *Azospira suillum* indicates a large kinetic fractionation (Coleman et al. 2003; Sturchio et al. 2003). As is seen in the majority of bacterially mediated reactions, the light isotope is preferentially reduced to Cl^-, resulting in increasing $\delta^{37}Cl$ values of the remaining perchlorate. The $1000 \ln \varepsilon$ value (perchlorate-chloride) value (where ε represents a kinetic fractionation factor given by $\varepsilon = R_{perchlorate}/R_{chloride}$, and $R = {}^{37}Cl/{}^{35}Cl$) determined from these studies span a range of −15.8 to −14.8‰ at 37 °C (Coleman et al. 2003) and −16.6 to −12.7‰ at 22 °C (Sturchio et al. 2003). In a more recent study in which both Cl and O isotope ratios were monitored, a $1000 \ln \varepsilon$ value of −13.2‰ was obtained, independent of temperature or bacterial strain (Sturchio et al. 2007). The high $\delta^{37}Cl$ values measured in the Southern High Plains groundwaters are interpreted as the result of bacterial reduction (Jackson et al. 2010). The low $\delta^{37}Cl$ values of the Atacama samples remains unexplained.

Synthetic perchlorates have a narrow range of $\delta^{37}Cl$ values of −3.1 to +2.3‰ (Ader et al. 2001; Bao and Gu 2004; Böhlke et al. 2005). The similarity to natural salt deposits suggests that the formation of perchlorate by electrolysis of NaCl solutions has a < 1‰ isotopic fractionation from the chloride precursor (Sturchio et al. 2012a).

Natural perchlorates, like nitrates, show large $\Delta^{17}O$ anomalies. Bao and Gu (2004) first identified this anomaly in Atacama perchlorates with $\Delta^{17}O$ values ranging from 4.2 to 9.6‰, where $\Delta^{17}O$ is defined in their publication as $\Delta^{17}O = \delta^{17}O - \delta^{18}O \times 0.52$. The extreme deviation from the Terrestrial Fractionation Line (e.g., Thiemens et al. 2012) was attributed it to photochemical oxidation of volatile chlorine by O_3 in the stratosphere. Böhlke et al. (2005) extended the Atacama range to 10.5‰. Groundwaters from Death Valley have $\Delta^{17}O$ values as high as 18.4‰ (Sturchio et al. 2012a). In contrast to the natural perchlorates, all synthetic perchlorates have $\Delta^{17}O$ values of near 0‰ (Bao and Gu 2004; Böhlke et al. 2005; Sturchio et al. 2012a). The combined $\delta^{37}Cl$–$\Delta^{17}O$ values of perchlorates can be used to easily identify synthetic *vs* natural perchlorates in groundwaters and estimate the likely source of perchlorate to groundwaters (Fig. 6).

Perchlorates are also found on Mars (e.g., Hecht et al. 2009) and in chondritic meteorites (Jackson et al. 2015a). To our knowledge, no Cl isotope analyses have been made of extraterrestrial materials due to sample size limitations.

Extraterrestrial Materials

Hoering and Parker (1961), in their classic study of the Cl isotope compositions of natural materials, found that the total isotopic variation of terrestrial samples was surprisingly small. They attributed the narrow range to an absence of high fundamental vibrational frequencies in naturally occurring chlorine compounds that undergo equilibrium exchange. Perchlorates, $HCl_{(g)}$ and $Cl_{2(g)}$ should fractionate relative to Cl^-, but oxidation to perchlorate in the stratosphere appears not to be an equilibrium process and in any event, HCl and Cl_2 gases have too low an abundance to cause significant fractionation. Indeed, the materials that have $\delta^{37}Cl$ values significantly different from that of seawater are rare. Extreme $\delta^{37}Cl$ values have been measured in Atacama perchlorates (Sturchio et al. 2006) and in $HCl_{(g)}$–bearing volcanic fumaroles (Barnes et al. 2009a; Sharp et al. 2010a), and in some porewaters from marine sediments (Ransom et al. 1995; Godon et al. 2004a; Bonifacie et al. 2007a). Perchlorate formation, HCl degassing and perhaps organic matter production are the major processes for significant chlorine isotope fractionation on Earth. Analysis of extraterrestrial materials could illuminate the importance of these, and other unidentified processes on other bodies. Perchlorates have been found on Mars (Hecht et al. 2009) and in meteorites (Jackson et al. 2015a), which could lead to moderate fractionations. Organic matter is found in some meteorites, and could also result in some isotope fractionation. Isotope fractionation associated with HCl degassing would indicate water presence on the parent body.

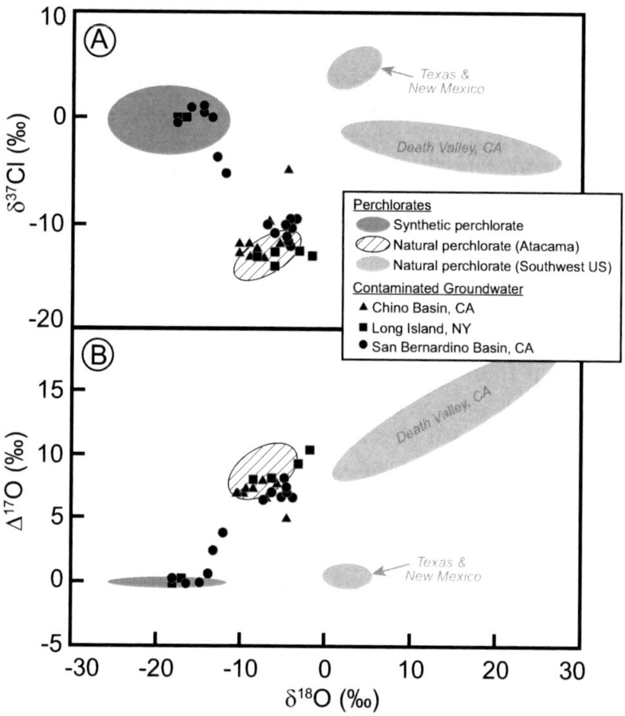

Figure 6. A. $\delta^{18}O$ vs $\delta^{37}Cl$ and B. $\delta^{18}O$ vs $\Delta^{17}O$ of various synthetic and natural perchlorates, as well as, groundwater contaminated by perchlorates. In conjunction, the three isotope systems can be used to uniquely define the source of the perchlorate. Modified from Sturchio et al. (2012a) with data from Böhlke et al. (2005, 2009), Jackson et al. (2010), and Sturchio et al. (2006, 2008, 2012b).

The first ever Cl isotope analyses were made on meteorites for the purposes of defining the atomic weight of Cl (Harkins and Stone 1926). Precision was very low and no differences from terrestrial materials were observed. Hoering and Parker (1961) analyzed four 'stone meteorites' and also found terrestrial-like values. The first high precision analyses of meteorites using gas source mass spectrometry and CH_3Cl as the analyte had $\delta^{37}Cl$ values that were slightly negative, ranging from near 0‰ to slightly less than −1‰ (Sharp et al. 2007; Bonifacie et al. 2008a). The enormous range of $\delta^{37}Cl$ values in extraterrestrial materials was only recognized when samples from the Moon, Mars and selected chondrites were measured.

The Moon. The first analyses of Apollo samples were made on bulk material using gas source mass spectrometry and *in situ* on apatite grains using the ion microprobe (Sharp et al. 2010b). The $\delta^{37}Cl$ values ranged from −1‰ up to a remarkable 24‰ (Fig. 3). Lunar samples with $\delta^{37}Cl$ values up to 34‰ have since been found (Tartèse et al. 2014; Treiman et al. 2014). The authors considered a number of potential processes to explain the high $\delta^{37}Cl$ values. It had long been recognized that solar wind appears to increase the isotopic ratios of sulfur, silicon and even oxygen (Epstein and Taylor 1971; Thode and Rees 1976). Interaction of solar wind with lunar chloride could produce volatile HCl which would then result in an isotopic fractionation. No correlation with surface exposure was observed, however, and a laboratory-based neutron bombardment experiment designed to test the importance of solar wind interaction did not cause any measurable Cl isotope fractionation (Sharp et al. 2007). Instead, the authors suggested that the high $\delta^{37}Cl$ values were due to degassing from an anhydrous melt. On Earth, Cl degasses as $HCl_{(g)}$ in nearly all volcanic systems. In general, vaporization should lead to a preferential loss of the light isotope to the vapor phase (Graham's law). However, degassed terrestrial basalts do not have elevated $\delta^{37}Cl$ values relative to their undegassed equivalents (Sharp et al. 2013c). The proposed explanation for the *lack* of fractionation in terrestrial basalts is that there is a positive Cl isotope fractionation between HCl gas and Cl dissolved in basalt melt, which offsets the negative fractionation associated with vaporization. In other words, HCl incorporates the heavy isotope of Cl (Schauble et al. 2003), which lowers the $\delta^{37}Cl$ value of the degassing basalt. This effect is offset by the preferential fractionation of the light isotope into the vapor phase. Overall, the two processes appear to cancel each other, so that fractionation during degassing of hydrous basalts is negligible.

Degassing of lunar basalts occurred under very different conditions than on Earth. Due to the general anhydrous character of lunar basalts, Cl would have degassed as volatile metal chlorides, such as NaCl, $FeCl_2$, $ZnCl_2$. Such salts have been found as coatings on Apollo samples (McKay et al. 1972; Meyer et al. 1975). The fractionation between a metal chloride and Cl dissolved in a basalt melt is expected to be very small. Preferential incorporation of the light isotope in the vapor phase would occur, as it does in degassing of terrestrial basalts, but the 'opposing' fractionation that occurs on Earth—namely the enrichment of ^{37}Cl in HCl gas—would not occur during metal chloride formation. Depending on the effective fractionation associated with metal chloride degassing, a $\delta^{37}Cl$ value of 24‰ is obtained for a Cl loss of 70 to 99%.

The idea that the Moon is anhydrous is at odds with a number of studies that have documented elevated levels of H in melt inclusions and apatite grains in lunar samples (Saal et al. 2008; McCubbin et al. 2010, 2011; Hauri et al. 2011). This apparent discrepancy is explained when considering the relative diffusion rates of H and Cl in basalt melts. The hydrogen diffusion rate in a silicate melt is orders of magnitude faster than Cl (Sharp et al. 2013a). As a result, it is reasonable that an H-bearing melt could lose all of its H before Cl degassing is complete. Once the melt has lost its H and becomes essentially anhydrous, Cl loss would occur by metal chloride volatilization, and large fractionations would result. In other words, hydrous melts and anhydrous degassing are not incompatible. The major difference between the Earth and Moon is that the latter had low initial volatile concentrations and a low oxygen fugacity that stabilizes H_2 gas over H_2O, allowing for near-complete H_2 loss prior to significant Cl loss.

Boyce et al. (2015) compared the $\delta^{37}Cl$ and δD values of different types of lunar basalts and found a correlation between rock type and $\delta^{37}Cl$ value. The samples rich in the KREEP component (rich in potassium (K), rare earth elements (REE) and phosphorus (P)) have higher $\delta^{37}Cl$ values than KREEP-poor samples. This result is consistent with earlier work (Sharp et al. 2010b). They further noted that samples with high Cl contents tend to have the highest $\delta^{37}Cl$ values. On the basis of this correlation, they suggested that degassing was an early event associated with the lunar magma ocean. They proposed that HCl loss early in Moon's history may explain the enrichment.

There are several problems with this conclusion. First, the argument that the Cl concentrations should correlate with the $\delta^{37}Cl$ values is not valid when samples with different initial Cl concentrations are considered. The increase in $\delta^{37}Cl$ value by degassing of a ^{35}Cl-enriched vapor is a function only to the fraction lost (following Rayleigh fractionation), not the final concentration. A sample with high initial Cl could lose a greater fraction of Cl than a second sample with low initial concentrations and still end up with a higher final Cl content. In fact, it is likely that a sample with high initial Cl concentration will lose a greater proportion of Cl than a low concentration sample simply because it is more likely to be oversaturated during decompression.

A more problematic consideration for the early magma ocean degassing model is the extraordinary variability seen in individual samples. Boyce et al. (2015) suggest that the range of $\delta^{37}Cl$ values is a result of two lunar mantle reservoirs, one with an Earth-like value and the second with a $\delta^{37}Cl$ value in excess of 30‰ (the KREEP reservoir). However, individual samples have *in situ* spot analyses with $\delta^{37}Cl$ values spanning a range of over 10‰ (Tartèse et al. 2014; Boyce et al. 2015). Such heterogeneities on the thin section scale are not compatible with mixing between two mantle reservoirs prior to emplacement. While there could certainly be some fractionation that occurred during the early magma ocean, the heterogeneity of individual apatite grains in a single thin section is best explained by heterogeneous degassing at the local scale.

Why the Moon shows such large variations while the Earth does not is still open to debate. Large variations in δ^{37}Cl values are not seen in terrestrial basalts (Selverstone and Sharp 2013; Sharp et al. 2013c). Anhydrous degassing of metal chlorides remains an attractive explanation for the range in δ^{37}Cl values. To test this, Sharp (unpublished) melted a mixture of basalt and NaCl in a graphite crucible allowing for partial chloride degassing to occur. Following quenching, the glass had a range of δ^{37}Cl values from 0‰ to over 8‰, consistent with ^{37}Cl-enrichment during metal chloride degassing.

Tartèse et al. (2014) noted a positive correlation between δ^{37}Cl and δD values in the matrix of sample NWA 4472 and argued for local HCl degassing. They postulated that Cl isotope fractionation may be significantly different in vacuum *vs.* atmospheric conditions. Simple degassing experiments of NaCl in vacuum and at ambient pressures do not show any enhanced degassing in vacuum (Sharp et al. 2013b), although more work is warranted. Parent body size could also be implicated, as samples from the Martian surface also have elevated (albeit muted relative to the Moon) δ^{37}Cl values. Earth is likely too large to have lost significant HCl to space, whereas the lower gravity of Mars and the Moon would allow for some to extensive Cl loss to occur. Impact induced metasomatism may also be a factor for some samples. The very high δ^{37}Cl values of 32‰ in apatite grains of lunar granulite 79215 are thought to be due to local degassing related to impact (Treiman et al. 2014). While the definitive causes for the high δ^{37}Cl values are debated, it appears clear that the enrichments are related to some combination of low escape velocity, anhydrous degassing and/or lack of an atmosphere.

Meteorites. The δ^{37}Cl values of primitive meteorites (chondrites) range from −4 to +1‰ (Fig. 7). It was originally noted that the chondrites with the lowest degree of aqueous alteration or metamorphism (petrologic type 3) have an average value of −0.3 ± 0.5‰, indistinguishable from bulk Earth (Sharp et al. 2013c). The only exception was Parnallee, an LL 3.6 ordinary chondrite with a very low δ^{37}Cl value (−3.61‰). No correlation was found between the δ^{37}Cl value and Cl concentration or the degree of aqueous alteration or metamorphism. The conclusions of Sharp et al. (2013c) were that chondrites and Earth both record an unmodified, presumably nebular Cl isotope composition, whereas the majority of lunar samples have high

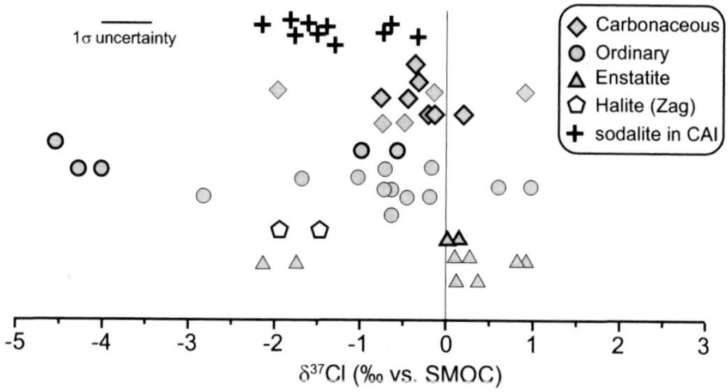

Figure 7. Cl isotope composition of chondritic meteorites. The δ^{37}Cl values of the least processed samples (type 3, shown with a heavy outline) are similar to the bulk Earth with the exception of Parnallee and NWA 8276 (both less than −4‰). Sodalite inclusions in Allende are shown as black crosses. Data from the following sources: Bridges et al. (2004); Bonifacie et al. (2007b); Sharp et al. (2007, 2013c); Williams et al. (2016).

δ^{37}Cl values due to preferential volatilization of ^{35}Cl. The scatter towards low values was difficult to explain. The only identified process that has large enough fractionation to cause a 4 ‰ shift is the formation of $HCl \cdot 3H_2O$ ice at low temperatures (Zolotov and Mironenko 2007; Schauble and Sharp 2011). Partial condensation of isotopically heavy $HCl{\cdot}3H_2O$ would lower the δ^{37}Cl value of the remaining $HCl_{(g)}$. The authors did not see any reasonable mechanism by which the light HCl (*g*) would be preferentially incorporated into the Parnallee parent body, and the low δ^{37}Cl value was left unexplained. A second type 3 chondrite, NWA 8276, has since been analyzed with similarly low δ^{37}Cl values (Williams et al. 2016). Isotopic analyses of Martian meteorites provided an alternative explanation for both the low δ^{37}Cl value of this ordinary chondrite and the low Cl contents of ordinary chondrites in general.

Mars. Mars is an unusual planet. It formed at the outer periphery of the inner protoplanetary disk, incorporating primitive planetesimals scattered beyond the edge of the truncated disk (Hansen 2009; Walsh et al. 2011; Izidoro et al. 2014). It is only 1/10 the mass of Earth and is also thought to be significantly older then the Earth and Moon, with a calculated core formation age of only several million years after collapse of the solar nebula (Dauphas and Pourmand 2011; Tang and Dauphas 2014). Mars has escaped much or all of the extreme energetic collisions experienced by the other Terrestrial planets, such as the Giant Impact thought to be responsible for the Earth–Moon system. Martian meteorites may therefore preserve Cl isotope evidence of the early solar system, in contrast to samples from the Earth and Moon, which may have been modified by later energetic processing.

The δ^{37}Cl values of bulk Martian meteorites range from −3.9 to +1.8‰, with one extreme analysis of a single apatite grain at +8.6‰ (Sharp et al. 2016). The large range is explained in terms of two component mixing. The most primitive olivine phyric shergottites have the lowest values between −4 and −2‰. Samples with petrologic or geochemical evidence for crustal assimilation have positive δ^{37}Cl values whereas basaltic shergottites have intermediate δ^{37}Cl values of −2 to 0‰ (Fig. 8).

The elevated δ^{37}Cl values of the Martian crust are consistent with other isotopic systems and are easily explained in terms of volatile loss to space. Interaction with solar wind (Chassefière and Leblanc 2004), impact erosion, and/or hydrodynamic escape (Hunten 1993) are all thought to have contributed to volatile loss over time. Most atmophilic elements on Mars have extreme heavy isotope enrichment due to preferential loss of the light isotope to space (e.g., Bogard et al. 2001). The Martian atmosphere has an elevated ^{38}Ar/^{36}Ar ratio

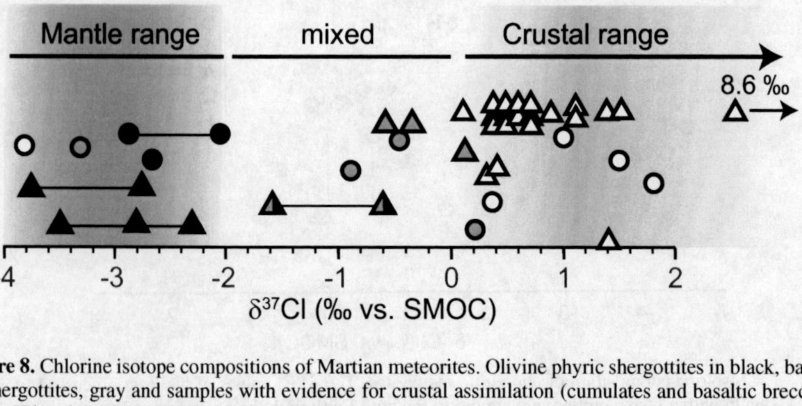

Figure 8. Chlorine isotope compositions of Martian meteorites. Olivine phyric shergottites in black, basaltic shergottites, gray and samples with evidence for crustal assimilation (cumulates and basaltic breccia), white. Triangles are ion probe analyses, circles are structurally bound bulk Cl isotope analyses. The data are explained as a two-component mixing: a mantle value of ~−4‰ and a crustal value with positive δ^{37}Cl values. The basaltic shergottites have a small crustal component (mixed). See Sharp et al. (2016) for details.

relative to chondrites and Earth (Swindle et al. 1986; Bogard 1997). The average $^{15}N/^{14}N$ ratio of Mars is almost double the Earth value (Owen et al. 1977), and CO_2 and H_2O are similarly enriched in the heavy isotopes, presumably due to volatile loss (Webster et al. 2013). Early hydrodynamic escape during intense extreme ultraviolet radiation (EUV) followed by prolonged atmospheric 'erosion' explains the heavy isotope enrichments (Bogard 1997). Cl loss, presumably as HCl, would raise the $\delta^{37}Cl$ value of residual material.

Perchlorates may partly contribute to the high $\delta^{37}Cl$ values of the crustal component. The Martian surface has high perchlorate concentrations (Hecht et al. 2009) and perchlorate has been identified in Martian meteorites (Kounaves et al. 2014). Thermodynamically, perchlorate should concentrate ^{37}Cl relative to chloride, but natural perchlorates with both positive and negative $\delta^{37}Cl$ values have been found on Earth (Coleman et al. 2003; Böhlke et al. 2005; Sturchio et al. 2006; Jackson et al. 2010), suggesting non-equilibrium during oxidation. The $\delta^{37}Cl$ value of Martian perchlorate has not yet been measured.

The low isotope ratios for the mantle component are more difficult to explain. The $\delta^{37}Cl$ values of the olivine–phyric shergottites, Shergotty (a basaltic shergottite) and NWA 2737 (a chassignite) are all lower than any mantle-derived sample from the Earth or Moon. All fractionation processes that could occur in mantle-derived rocks should raise, rather than lower, their $\delta^{37}Cl$ value. Degassing causes a preferential loss of the light isotope and should raise the $\delta^{37}Cl$ value of the residue. Assimilation of the high $\delta^{37}Cl$ crust would obviously raise the $\delta^{37}Cl$ value of a mantle-derived sample. Contamination from meteoritic material should drive the $\delta^{37}Cl$ value towards 0‰, if the measured chondritic data (Sharp et al. 2013c) are representative of meteoric samples. Fractional crystallization should have virtually no effect, as the Cl isotope fractionation between melts and crystals at high temperatures are thought to be negligible. As an alternative to parent-body fractionation, Sharp et al. (2016) suggested that the low $\delta^{37}Cl$ value of the Martian mantle is inherited from the solar nebula and that other bodies, including most chondrites, Earth and Moon have had their Cl isotope ratios modified to higher values.

Mars has avoided the protracted growth experienced by Earth. Mars evolution was arrested very early on during the stage of planetary formation (Brasser 2013). There is no evidence for plate tectonics on Mars, so that the $\delta^{37}Cl$ values of Martian mantle-derived rocks should have been shielded from later events that could have delivered isotopically fractionated Cl to the planet. In the early hot solar nebula, Cl is stable as HCl, and then (partially) incorporated into sodalite at ~1000 K (Fegley and Lewis 1980). There should be little isotopic fractionation during this process (Sharp et al. 2007). Any remaining HCl gas would have persisted metastably below the stability temperature of sodalite. HCl has been observed in galactic star forming regions (Peng et al. 2010), but at a much lower abundance that expected on the basis of constraints from our solar system. HCl is below detection in Oort-family comets, requiring a depletion with respect to solar abundances of a factor of 6 (Bockelée-Morvan et al. 2014). Presumably HCl was not the main Cl reservoir in the nebular regions where these objects formed (Codella et al. 2012; Bockelée-Morvan et al. 2014).

Zolotov and Mironenko (2007) proposed that nebular $HCl_{(g)}$ reacted to form an HCl hydrate ($3H_2O{\cdot}HCl$) during cooling at a temperature of ~150–160 K. This is essentially equivalent to the 'snow line' temperature for H_2O ice formation, thought to be located in the outer region of the asteroid belt. The fractionation between $3H_2O{\cdot}HCl$ and $HCl_{(g)}$ is very large, calculated to be between 3 and 6‰ (Schauble and Sharp 2011). Therefore, partial condensation of HCl will lead to an HCl hydrate component with a $\delta^{37}Cl$ value that is significantly higher than that of the unprocessed solar nebula. Zolotov and Mironenko (2007) further suggested that this acidic ice was incorporated into chondrite parent bodies, a process which would raise their $\delta^{37}Cl$ values. This mechanism could explain the near-zero values of most, but not all, chondrites. The Moon and Earth may have incorporated this processed material in addition to losing isotopically light Cl during volatilization to space. In this scenario, it is only the

Martian interior and the least processed chondrites (e.g., L3.00 chondrite NWA 8276 with a $\delta^{37}Cl$ value of −4.5 ‰; Williams et al. 2016 and Parnallee an LL 3.6 ordinary chondrite) that preserve the initial nebular Cl isotope value. Additional work on the least processed chondrites is necessary to see if the idea of a light solar nebula, with values less than −4 ‰, is a plausible explanation for both the Martian and C3 ordinary chondrite data.

CHLORINE ISOTOPES AS A TRACER

Tracer through subduction zones

A number of studies have used Cl stable isotopes to trace volatile sources in subduction zones by determining the isotopic composition of volcanic outputs (ashes, tephras, lavas, thermals springs) (Barnes et al. 2008, 2009a; Barnes and Straub 2010; Rizzo et al. 2013; Bernal et al. 2014; Chiaradia et al. 2014; Cullen et al. 2015; Li et al. 2015). Due to complications from kinetic fractionation in volcanic gases at the near-surface (see section on *Kinetic isotope fractionation* above), volcanic gases are not used as a tracer of source, but rather to infer near-surface plumbing processes (e.g., Sharp et al. 2010a). The first study using Cl isotopes to trace volatile source was in the Izu–Bonin–Mariana (IBM) subduction zone (Barnes et al. 2008). The IBM is noted for a series of different outputs across the arc which have the potential to trace fluids sourced from different depths within the subduction zone (Fig. 9). Serpentinite clasts from fore-arc seamount tap fluids from depths of ~30 km depth, whereas, volcanic front ash and back-arc lavas are located ~115—130 km and ~200 km above the subducting slab, respectively. $\delta^{37}Cl$ values of ashes along the length of the arc front are negative suggesting fluid contribution from subducted marine sediments; whereas, $\delta^{37}Cl$ values of outputs from both forearc seamounts and back-arc cross-chains are zero to slightly positive suggesting a serpentinite-derived fluid source. Fluids are released at shallow depths from the transition between serpentine polymorphs (chrysotile/lizardite to antigorite) and fluids released behind the back-arc cross-chains are from the breakdown of antigorite (Barnes et al. 2008). Subsequent work on Izu–Bonin arc tephras show no correlation between $\delta^{37}Cl$ values and either ^{207}Pb or Sr isotopes (tracers of fluids derived

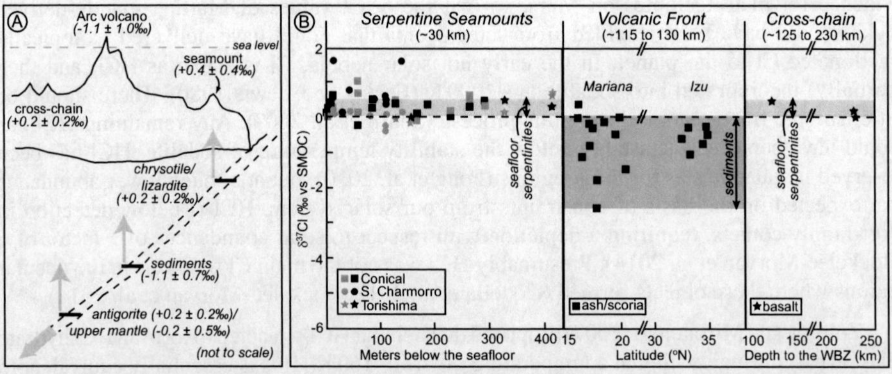

Figure 9. A. Schematic diagram of the Izu–Bonin–Mariana subduction zone showing volatile sources in the subducting slab, outputs across the arc, and their average $\delta^{37}Cl$ values (modified from Barnes et al. 2008). Average bulk $\delta^{37}Cl$ values of seawater-hydrated seafloor serpentinites from Barnes and Sharp (2006); upper mantle mantles from Sharp et al. (2007); and western Pacific sediment $\delta^{37}Cl$ values from Barnes et al. (2008). Gray arrows represent assumed near vertical rise of fluids. B. $\delta^{37}Cl$ values of outputs across Izu–Bonin–Mariana system (modified from Barnes et al., 2008). $\delta^{37}Cl$ values of serpentine mud and serpentinite clasts (gray = water-soluble chloride; black = structurally bound chloride) from the forearc seamounts (Conical, S. Charmorro, Torishima) versus meters below seafloor. $\delta^{37}Cl$ values of ash and/or scoria samples along the length of arc. $\delta^{37}Cl$ values of back-arc cross-chain basalt samples versus depth to subducting slab. Light and dark gray bands show $\delta^{37}Cl$ values for seafloor serpentinites hydrated by seawater (Barnes and Sharp 2006) and western Pacific sediment $\delta^{37}Cl$ values (Barnes et al. 2008). WBZ—Wadati–Benioff Zone.

from subducting sediment and igneous crust, respectively). This lack of correlation was used as an additional argument for Cl sourced from serpentinites (Barnes and Straub 2010). A similar study used Cl isotopes to trace the sources of slab-derived fluids across the Ecuadorian volcanic arc. δ^{37}Cl values of lavas correlate well with other signatures of a slab-derived fluid (e.g., Ba/La, Pb/Ce) and support a decreasing amount of fluid released by the subducted slab away from the trench (Chiaradia et al. 2014). In addition to across arc studies, an along arc study of ashes from the Central America volcanic front record large Cl isotope variations along the length of the arc (−2.6 to +3.0‰) (Barnes et al. 2009a). Nicaraguan ashes from the central arc record contributions of sediment and/or serpentinite-derived fluids, whereas ashes from the northern and southern ends of the arc reflect a more mantle-like signature, in agreement with other geochemical tracers (e.g., δ^{15}N, δ^{18}O, and Ba/La) (e.g., Patino et al. 2000; Fischer et al. 2002; Eiler et al. 2005).

The Cl isotope composition of thermal springs associated with arc volcanism is also a potential tracer of subduction inputs; however, distinguishing the magmatic Cl contribution to spring waters versus Cl enrichment from other processes, such as near surface fluid-rock interaction, can be difficult. Hydrothermal waters from the Taupo Volcanic Zone in New Zealand have δ^{37}Cl values ranging from −0.8‰ to +0.7‰ (Bernal et al. 2014). The authors divide the thermal fluids into two groups: 1) fluids associated with rhyolitic magmas are isotopically positive and have high Cl/Br molar ratios, and 2) fluids associated with andesitic magmas are isotopically negative and have low Cl/Br molar ratios. The authors hypothesize that isotopic variations between the magma sources results from Cl fractionation during magmatic differentiation in the crust as a function of magma water contents (Bernal et al. 2014). However, isotopic work on rhyolitic glasses from Mono Craters, California show that some rhyolitic magmas have distinctly negative δ^{37}Cl values (−1.9 to 0.0‰) (Barnes et al. 2014b). Additional work on isotopic variability of different magma compositions is necessary before assumptions can be made about magma composition based on Cl isotope values. In another recent study, the δ^{37}Cl values of spring waters from the Lesser Antilles were found to vary between −0.65‰ and +0.12‰ (Li et al. 2015). These values are interpreted to reflect contribution of slab-derived Cl to the springs, without any isotopic modification due to limited HCl degassing (Li et al. 2015). Both of these studies discount any contribution from fluid–rock interaction. Cullen et al. (2015) measured the Cl isotope composition of thermal and mineral springs in the Cascade arc and associated volcanic rocks. They conclude that water-rock interaction can explain most, but not all, the isotopic variability of the waters (+0.2 to +1.9‰). Incorporation of some magmatic Cl from degassing of HCl is needed to account for the high δ^{37}Cl values (>+1.0‰). Future work is needed to better constrain Cl isotope behavior during fluid-rock interaction and volcanic degassing in order to assess volatile sources to arc magmatism.

Not all studies of Cl in subduction zones have focused on volcanic and associated spring outputs. A study of metasomatized, suprasubduction-zone mantle (Finero peridotite, Ivrea Zone, Italy) used Cl isotopes to distinguish two slab-derived fluid infiltration events into the overlying mantle wedge: one, with a δ^{37}Cl value ≤ −2‰ and the other with a δ^{37}Cl value ≥ +2‰ (Selverstone and Sharp 2011). A subsequent study on exhumed subduction zone rocks of the Zermatt–Saas ophiolite, Italy, documents infiltration of pore fluids during the initial stages of subduction (Selverstone and Sharp 2013).

Crustal fluids

Chlorine isotopes have been used to trace sources and hence subsurface flow and transport mechanisms of groundwater, primarily basinal brines and formation waters associated with oil fields. These waters have a wide range of δ^{37}Cl values from −4.25 to +1.54‰ (Fig. 10). In many cases these δ^{37}Cl values reflect mixing of multiple fluid sources and evaporation/halite dissolution. As discussed in the *Equilibrium Cl isotope fractionation* section, halite precipitation from seawater produces halite with a δ^{37}Cl value of +0.26‰ (Eggenkamp et al. 1995), lowering the δ^{37}Cl value of the residual fluid. If evaporation proceeds until the formation of K–Mg chlorides (f=0.3), which

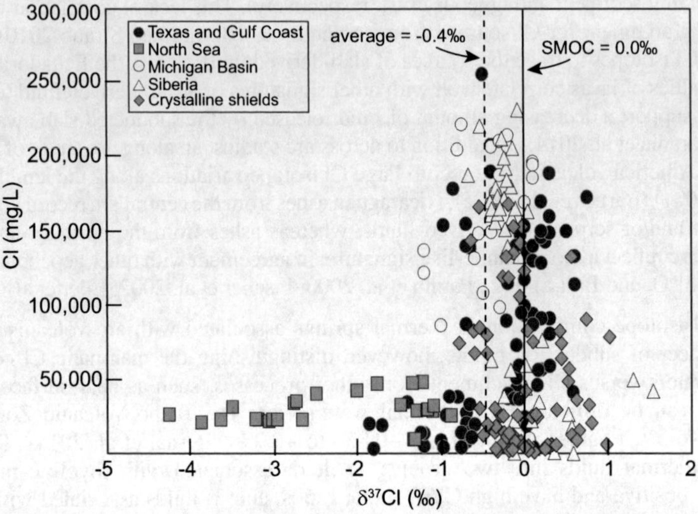

Figure 10. Compilation of basinal brines and formation waters from the Texas and Gulf Coast, United States (Kaufmann et al. 1988; Eastoe et al. 1999; Eastoe et al. 2001), North Sea oil fields (Eggenkamp 1994; Ziegler et al. 2001), Michigan Basin (Kaufmann et al. 1993), Siberia (Shouakar-Stash et al. 2007), and crystalline basement of Canadian and Fennoscandian shields (Kaufmann et al. 1987; Stotler et al. 2010).

are depleted in ^{37}Cl relative to the saturated solution, the δ^{37}Cl values of the solution can then begin to increase (Fig. 11A) (Eggenkamp et al. 1995; Eastoe et al. 1999). Multiple evaporative cycles (e.g., dissolution due to influx of water followed by evaporation) can result in δ^{37}Cl values >+0.3‰, although little of the original halite remains. Removal of about half of the original halite results in a shift of ~0.15‰ (Fig. 11B) (Eastoe and Peryt 1999; Eastoe et al. 2007).

Mixing and dissolution–precipitation processes do not explain the full range of crustal fluid values, which require additional fractionation processes such as diffusion and ion filtration. As discussed in the *Kinetic Cl isotope fractionation* section, diffusion due to large concentration gradients can result in a shift in several per mil due to the greater translational velocity of light isotopologues. The minimum δ^{37}Cl values strongly depends on the initial Cl concentration of the "fresher" fluid (Fig. 11C) (Eggenkamp 1994). Shifts greater than ~3‰ in natural samples cannot be explained solely in terms of diffusion, in part because concentration gradients are typically too small. Ion filtration has been invoked to explain larger shifts in δ^{37}Cl values, up to ~−5‰ (Ziegler et al. 2001; Godon et al. 2004a), in natural samples. An isotopic shift of this magnitude requires a ~20 times increase in Cl concentration behind the membrane (Phillips and Bentley 1987) (Fig. 11D). While not considered in any crustal fluid study, organohalides may partly explain the very negative δ^{37}Cl values measured. As discussed in the *Isotopic Fractionation* section, kinetic effects result in very negative δ^{37}Cl values of organic matter (down to −12‰). Degradation of this material would result in fluids with isotopically negative values. Additional work evaluating the importance of organic matter as a means of fractionating Cl isotopes is warranted. Figure 12 summaries the approximate maximum magnitude of change in crustal fluid δ^{37}Cl values in response to the above discussed mechanisms.

Basinal brines and formation waters from Texas and the Gulf Coast of the United States have δ^{37}Cl values of −1.9 to +0.7‰ (Kaufmann et al. 1988; Eastoe et al. 1999, 2001). Three component mixing between 1) seawater; 2) an isotopically positive brine due to halite dissolution; and 3) an isotopically negative brine (due to halite precipitation and/or diffusion) explains the full range of δ^{37}Cl values (Kaufmann et al. 1988; Eastoe et al. 1999, 2001).

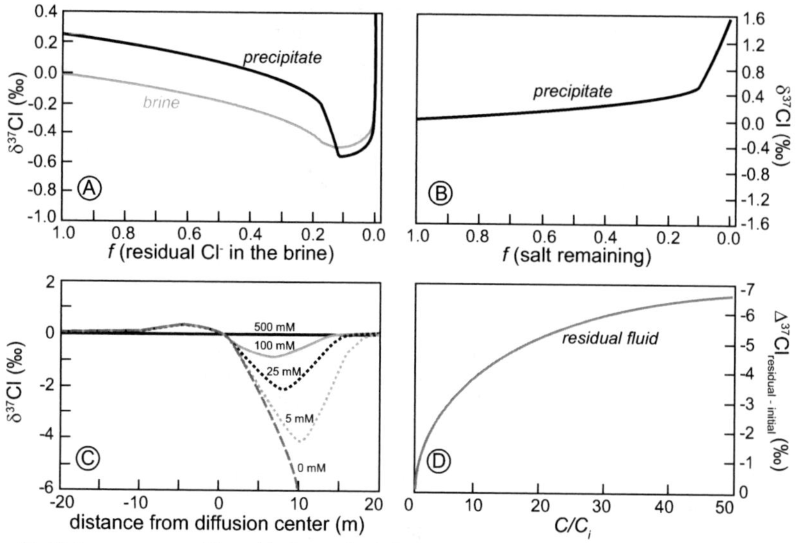

Figure 11. Cl isotope composition of halite and/or brine as a function of various fractionation mechanisms. A. Calculated δ^{37}Cl values of salt precipitate (black line) and residual brine (gray line) as a function of *f*, the residual fraction of chloride in the brine. δ^{37}Cl value of the precipitate is dictated by $\alpha_{\text{halite–Cl}^-} = 1.00026$ (Eggenkamp et al. 1995), until the formation of K–Mg chlorides by which the δ^{37}Cl value of the residual brine can increase. Modified from Eggenkamp et al. (1995). B. δ^{37}Cl value of halite (using $\alpha_{\text{halite–Cl}^-} = 1.00026$) due to multiple evaporative cycles. *f* = residual fraction of salt remaining. Modified from Eastoe et al. (2007). C. The δ^{37}Cl value of a fluid due to diffusion from a constant chloride source. Sediment saturated with seawater (540 mM) at negative distance from the contact with sediment saturated with water of different chloride concentrations (0, 5, 25, 100, and 500 mM). Δ^{35}Cl/Δ^{37}Cl = 1.00245. (Lower modelled Δ^{35}Cl/Δ^{37}Cl ratios would decrease the magnitude of the change in the δ^{37}Cl value.) Modified from Eggenkamp (1994). D. Calculated change in δ^{37}Cl value of the residual fluid ($\Delta^{37}\text{Cl} = \delta^{37}\text{Cl}_{\text{residual}} - \delta^{37}\text{Cl}_{\text{initial}}$) as a function of the change in chloride concentration (C/C_{initial}) due to ion filtration. Modified from Phillips and Bentley (1987).

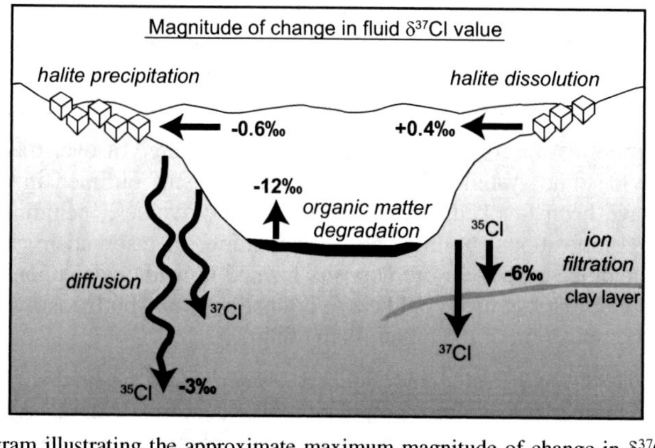

Figure 12. Diagram illustrating the approximate maximum magnitude of change in δ^{37}Cl values of the fluid in response to: 1) halite precipitation, 2) halite dissolution, 3) organic matter degradation, 4) diffusion, and 5) ion filtration.

Formation waters from the Siberian Platform have overlapping δ^{37}Cl values, but extend to more positive values (−0.67 to +1.54‰) (Shouakar-Stash et al. 2007). As with the Gulf Coast waters, negative values are thought to result from evaporation of seawater, with the resulting fluids having negative values due to the precipitation of isotopically positive halite.

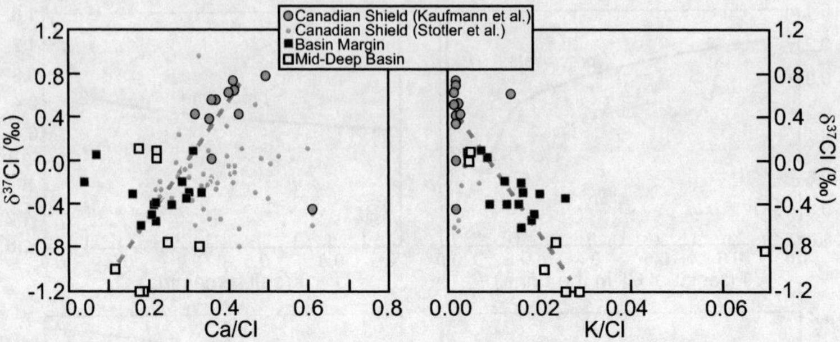

Figure 13. Ca/Cl, K/Cl, and δ^{37}Cl value of Michigan Basin formation waters (open and solid squares) (Kaufmann et al. 1993). The formation waters show possible mixing trends (dashed gray line) between the deeper basin waters and Canadian Shield waters (gray circles) (Kaufmann et al. 1987); however additional, and more isotopically negative shield waters, make the mixing argument less robust (small gray circles) (Stotler et al. 2010).

Some positive values suggest dissolution of halite; however, the authors point out that the high values are above any evaporite samples and instead suggest mixing with some additional heavy reservoir. The source of the very positive values is uncertain and may involve fluid-rock interaction or permafrost freezing effects (Shouakar-Stash et al. 2007). Formation waters from the Michigan Basin have δ^{37}Cl values ranging from −1.2 to +0.1‰ and show nice correlations with Ca/Cl and K/Cl, suggesting mixing between isotopically negative deep basinal brines and isotopically positive Canadian Shield waters (Fig. 13) (Kaufmann et al. 1987, 1993). Diffusional processes have been used to explain isotopically negative (as low as −2.5‰) groundwaters from the Great Artesian Basin in Australia (Zhang et al. 2007); whereas, extreme negative formation waters from the North Sea (as low as −4.25‰; Eggenkamp, 1994) necessitate ion filtration effects (Ziegler et al. 2001) or organochloride contribution.

Tracer in ore deposits

Hydrothermal ore fluids may be a mixture of magmatic, metamorphic, meteoric, and groundwater derived fluids. These hybridized fluids transport metal ions responsible for exploitable deposits; therefore, determining fluid source and percent contribution of each source is of economic importance. Early work recognized the potential for Cl isotopes to be used as a tracer of fluid source in ore deposits, yet its application remains relatively under-utilized.

The δ^{37}Cl variations observed in ore deposits cover a range of over 6‰, with very low values of <−5‰ in some systems. Many of the mechanisms outlined in the earlier parts of this chapter have been invoked to explain the range of values, including inheritance of formation water compositions, boiling, phase separation, metasomatism, and contribution from anomalous mantle sources. In ore deposits formed in acid environments, boiling could lead to the degassing of HCl, which could cause dramatic shifts—both positive and negative—in the Cl isotope composition of the mineralizing fluid.

Eastoe et al. (1989) and Eastoe and Guilbert (1992) report the first δ^{37}Cl values of fluid inclusions trapped in quartz veins from porphyry copper deposits (PCD; Bingham, Silver Bell, Panguna) and in hydrothermal minerals from Mississippi Valley-type deposits (carbonate-hosted lead and zinc deposits). Overall, δ^{37}Cl values range from −1.1 to +0.8‰. In addition, PCD biotite (Cl extracted by leaching in nitric acid) has δ^{37}Cl values of +0.3 to +1.7‰ (Eastoe and Guilbert 1992). Data from hydrothermal minerals fall into different populations depending on mineral host, suggesting an alternating fluid source from different formation waters (Eastoe and Guilbert 1992). Most hydrothermal fluids trapped in fluid inclusions from PCD have slightly positive values and biotites are distinctly positive. The positive δ^{37}Cl values were interpreted as a reflection of the bulk magmatic component or preferential loss of chlorine with a negative

$\delta^{37}Cl$ values to the aqueous vapor phase (Eastoe et al. 1989; Eastoe and Guilbert 1992). However, these interpretations are not consistent with an isotopically negative value of the upper mantle, unless they reflect a subducted component (see the chlorine isotopic composition of the mantle discussion above), and minimal fractionation due to phase separation in the H_2O–NaCl system (Bonifacie et al. 2005; Liebscher et al. 2006). The positive biotite data contrast with those from the Stillwater Complex, Montana, U.S.A., which have slightly negative to near zero pTIMS $\delta^{37}Cl$ values (−0.93 to +0.27‰; average = −0.18 ± 0.46‰; $n = 5$) (Boudreau et al. 1997). These values were interpreted to reflect metasomatism via a crustal fluid (Boudreau et al. 1997) (note, at the time of this contribution the mantle was believed to be isotopically positive. Reinterpretation based on modern data is consistent with a magmatic fluid source).

In the past ~10 years, several other studies have expanded on this early fluid inclusion work to trace metal sources; however, the number of studies is still surprisingly sparse, whereas the range of $\delta^{37}Cl$ values is surprisingly large. Nahnybida et al. (2009) analyzed fluid inclusions from the Butte and Bingham PCD ($\delta^{37}Cl$ values = −2.3 to −0.8‰ and −4.1 to −0.9‰, respectively). The fluid inclusions from this study are from the deepest part of the deposits and are thought to represent supercritical fluid trapped at depth before unmixing into brine and vapor. The values for Bingham are much lower than those previously reported (Eastoe et al. 1989; Eastoe and Guilbert 1992), possibly due to mixing of multiple generations of fluid inclusions in the previous studies. The low $\delta^{37}Cl$ values are interpreted as contribution from slab-derived fluids (Nahnybida et al., 2009). In addition to PCD, $\delta^{37}Cl$ values of fluid inclusions from iron oxide–copper–gold (IOCG) deposits in Sweden and the South American Andes have been analyzed (Chiaradia et al. 2006; Gleeson and Smith 2009). Deposits in Sweden have $\delta^{37}Cl$ values ranging from −5.6 to −1.3‰ and near zero (0 ± 0.5‰ with one outlier). These values are interpreted to reflect input from mantle derived Cl, with very low values due to extreme Rayleigh fractionation. In this process, the incorporation of heavy Cl into newly forming minerals would make the aqueous Cl^- ever lighter. Depending upon the assumed α value (fractionation between mineral and fluid), a $\delta^{37}Cl$ value as low as −10‰ could occur with ~90‰ incorporation in minerals (Gleeson and Smith 2009). This idea is certainly possible, but no experimental or empirical data support a large fractionation between minerals and fluid. Additional work is clearly warranted in this field.

Data from the above studies contrast with mostly isotopically positive pTIMS data ($\delta^{37}Cl$ values = −0.58 to +2.10‰; average = +0.77 ± 0.66‰) from IOCG deposits in the Andes, which were interpreted as a mixture between mantle-derived Cl (at the time thought to be isotopically positive) and basinal brines (Chiaradia et al. 2006). Additional fluid inclusions from iron-oxide-apatite deposits from Sweden have $\delta^{37}Cl$ values of −3.1 to −1.0‰ interpreted to reflect a mantle component as well as modification via fractionation (Gleeson and Smith 2009). Negative $\delta^{37}Cl$ values (−0.6 to 0.0‰) from quartz and carbonate-hosted fluid inclusions are reported from unconformity-related uranium deposits from the Athabasca Basin, Canada suggesting predominantly evaporated seawater as the Cl-source for ore-forming brines (Richard et al. 2011). Fluid inclusions from vein mineralization in the Iberian Pyrite Belt have wide range of pTIMS $\delta^{37}Cl$ values (−1.8 to +3.2‰; $n = 5$). These values correlate with Cl/Br ratios suggesting extreme isotopic fractionation due to phase separation (Germann et al. 2003). Unfortunately, most of the pTIMS data tend to give positive $\delta^{37}Cl$ values which likely reflect analytical problems, and must be treated with caution.

In a survey paper of $\delta^{37}Cl$ values of various minerals, Eggenkamp and Schuiling (1995) show that in contrast to evaporite minerals and most magmatic minerals, metal chlorides and fumarolic minerals (i.e., from late-stage volcanic environments) can have a wide range of $\delta^{37}Cl$ values from −4.88 (sal ammoniac, NH_4Cl) to +5.96‰ (atacamite, $Cu_2(OH)_3Cl$). In an attempt to use Cl isotopes to trace metal sources, Arcuri and Brimhall (2003) determined the Cl isotope composition of atacamite from the Radomiro Tomic PCD in Chile, as well as, associated sediments. $\delta^{37}Cl$ values of the atacamite range from −3.1 to +0.6‰, consistent with derivation of chloride leached from marine sediments with some contribution from halite in the isotopically positive samples. Although the isotopic variation of magmatic minerals is not typically large, Hanley et al. (2011) measured the of $\delta^{37}Cl$ values of magmatic biotite and scapolite associated with Cu–Ni–PGE (platinum group element) sulfide deposits in the Sudbury Igneous Complex to determine the contribution of magmatic fluid to the ore-forming system. Biotite in the Archean country rocks have $\delta^{37}Cl$ values (−0.88 to −0.53‰); whereas, biotite and scapolite in the footwall and main mass of the Sudbury Complex and have positive $\delta^{37}Cl$ values (biotite = +0.98 to +1.61‰; scapolite = +0.16 to +1.34‰) (Hanley et al. 2011). These positive values are interpreted to reflect a magmatic component from crustal melting and the negative values from interaction with local groundwater (see section above on *crustal fluids*) (Hanley et al. 2011).

Environmental

Chlorine stable isotopes, in conjunction with nitrogen, hydrogen, and carbon isotopes, have been used to trace environmental contaminants, such as agrochemicals (herbicides, pesticides, fungicides) (Annable et al. 2007) and chlorinated solvents (e.g., van Warmerdam et al. 1995; Beneteau et al. 1999; Jendrzejewski et al. 2001; Shouakar-Stash et al. 2003), in addition to perchlorates (discussed above). Here we briefly summarize some examples of environmental tracer work.

Chlorinated solvents (e.g., trichloroethylene (TCE), perchloethylene (PCE), dichloromethane (DCM), 1,1,1-trichloroethane (TCA), and chloroform (CFM)) are commonly used by industry as degreasers (e.g., Jendrzejewski et al. 2001; Shouakar-Stash et al. 2003). Because chlorinated solvents are highly soluble in water, they have the potential to contaminant surface and ground water (e.g., Sturchio et al. 1998; Beneteau et al. 1999). The range of $\delta^{37}Cl$ and $\delta^{13}C$ values of chlorinated solvents is large (~−7‰ to ~+4‰ vs SMOC; ~−58 to −23‰ vs PDB). Although different studies report slightly different ranges, it is not clear whether the differences are due to differing analytical methods or variabilities between batches/manufacturers. In general, individual chemicals can be "fingerprinted" by coupled $\delta^{37}Cl$ and $\delta^{13}C$ analyses, showing great promise in tracing source of solvent contamination of water supplies (e.g., van Warmerdam et al. 1995; Beneteau et al. 1999; Jendrzejewski et al. 2001; Shouakar-Stash et al. 2003). Future work regarding possible fractionation during degradation will help to accurately fingerprint polluted water samples (e.g., Jendrzejewski et al. 2001; Shouakar-Stash et al. 2003).

Annable et al. (2007) published the first study to determine the Cl isotope composition of herbicides. $\delta^{37}Cl$ values range from −4.6‰ to +3.4‰. When Cl isotope data are combined with N and C isotope data, individual herbicides are statistically different and identifiable, suggesting a new geochemical tracer of herbicides in the environment.

SUMMARY

The existence of Cl isotopes was first demonstrated by Aston nearly 100 years ago (Aston 1919, 1920); however, almost 40 years would pass until the first systematic study was conducted to determine Cl isotope composition of a range of geological materials. Hoering and Parker (1961) were hampered by poor analytical precision and therefore their data showed no significant isotopic variability among Earth's reservoirs. It was not until 25 years later, with improved methods, that Kaufmann et al. (1984) showed measurable isotopic variability. Starting in the mid-1990s the field grew with further perfection of techniques and improved precision (e.g., Long et al. 1993; Eggenkamp 1994). There have been significant developments in the last ten years: Cl isotope ratios are being routinely analyzed, the isotopic composition of different Cl reservoirs is being continually better defined, and SIMS techniques have been developed for in situ analyses. Despite our better understanding of natural Cl variability, there are still many outstanding questions. For example, what is controlling the isotopic variability observed in sediments and metasediments and what role do organochlorides play in this variability? How variable is the isotopic composition of the mantle and do heterogeneities reflect a deeper subducted component? What are the mechanisms controlling the dramatic isotopic variability seen in extraterrestrial materials and was there an initial isotopically negative solar nebula? There is also a great need for a community effort to define secondary standards and inter-laboratory calibrations to facilitate comparison of data among different laboratories and methods. And finally, studies to determine the magnitude of fractionation between minerals and fluids as a function of temperature are required in order to better explain the wide range of Cl isotope variations seen in natural materials.

ACKNOWLEDGEMENTS

The authors thank H. Eggenkamp and S. Gleeson for helpful and constructive reviews and J. Watkins for editorial handling.

REFERENCES

Ader M, Coleman ML, Doyle SP, Stroud M, Wakelin D (2001) Methods for the stable isotopic analysis of chlorine in chlorate and perchlorate compounds. Anal Chem 73:4946–4950, doi:10.1021/ac010489u

Aeppli C, Bastviken D, Andersson P, Gustafsson O (2013) Chlorine isotope effects and composition of naturally produced organochlorines from chloroperoxidases, flavin-dependent halogenases, and in forest soil. Environ Sci Technol, 47:790–797

Annable WK, Frape SK, Shouakar-Stash O, Shanoff T, Drimmic RJ, Harvey FE (2007) ^{37}Cl, ^{15}N, ^{13}C isotopic analysis of common agro-chemicals for identifying non-point source agricultural contaminants. Appl Geochem 22:1530–1536

Arcuri T, Brimhall G (2003) The chloride source for atacamite mineralization at the Radomiro Tomic porphyry copper deposit, northern Chile. Econ Geol 98:1667–1681

Aston FW (1919) The constitution of the elements. Nature 104:393

Aston FW (1920) Isotopes and atomic weights. Nature 105:617–619

Aston FW (1927) Bakerian lecture. A new mass-spectrograph and the whole number rule. Proc R Soc 115:487–514

Bach W, Alt JC, Niu Y-L, Humphris SE, Erzinger J, Dick HJB (2001) The geochemical consequences of late-stage low-grade alteration of lower ocean crust at the SW India Ridge: results from ODP Hole 735B (Leg 176). Geochim Cosmochim Acta 65:3267–3287

Bao H, Gu B (2004) Natural perchlorate has a unique oxygen isotope signature. Environ Sci Technol 38:5073–5077, doi:10.1021/es049516z

Barnes JD, Cisneros M (2012) Mineralogical control on the chlorine isotope composition of altered oceanic crust. Chem Geol 326–327:51–60

Barnes JD, Sharp ZD (2006) A chlorine isotope study of DSDP/ODP serpentinized ultramafic rocks: insights into the serpentinization process. Chem Geol 228:246–265

Barnes JD, Straub SM (2010) Chlorine stable isotope variations in Izu Bonin tephra: Implications for serpentinite subduction. Chem Geol 272:62–74

Barnes JD, Selverstone J, Sharp ZD (2006) Chlorine chemistry of serpentinites from Elba, Italy, as an indicator of fluid source and subsequent tectonic history. Geochem Geophys Geosyst7:Q08015, doi:08010.01029/02006GC001296

Barnes JD, Sharp ZD, Fischer TP (2008) Chlorine isotope variations across the Izu-Bonin-Mariana arc. Geol 36:883–886

Barnes JD, Sharp ZD, Fischer TP, Hilton DR, Carr MJ (2009a) Chlorine isotope variations along the Central American volcanic front and back arc. Geochem Geophys Geosyst 10:Q11S17, doi:10.1029/2009GC002587

Barnes JD, Paulick H, Sharp ZD, Bach W, Beaudoin G (2009b) Stable isotope (δ^{18}O, δD, δ^{37}Cl) evidence for multiple fluid histories in mid-Atlantic abyssal peridotites (ODP Leg 209). Lithos 110:83–94

Barnes JD, Eldam R, Lee C-TA, Errico JC, Loewy SL, Cisneros M (2013) Petrogenesis of serpentinites from the Franciscan Complex, western California, USA. Lithos 178:143–157

Barnes JD, Beltrando M, Lee C-TA, Cisneros M, Loewy SL, Chin E (2014a) Geochemistry of Alpine serpentinites from rifting to subduction: A view across paleogeographic domains and metamorphic grade. Chem Geol 389:29–47

Barnes JD, Prather T, Cisneros M, Befus K, Gardner JE, Larson TE (2014b) Stable chlorine isotope behavior during volcanic degassing of H_2O and CO_2 at Mono Craters, CA. Bulletin of Volcanology 76:DOI 10.1007/s00445-00014-00805-y

Beekman HE, Eggenkamp HGM, Appelo CAJ (2011) An integrated modelling approach to reconstruct complex solute transport mechanisms–Cl and δ^{37}Cl in pore water of sediments from a former brackish lagoon in The Netherlands. Appl Geochem 26:257–268

Beneteau KM, Aravena R, Frape SK (1999) Isotopic characterization of chlorinated solvents—laboratory and field results. Org Geochem 30:739–753

Berglund M, Wieser ME (2011) Isotopic compositions of the elements 2009 (IUPAC technical report). Pure Appl Chem 83:397–410

Bernal NF, Gleeson SA, Dean AS, Liu XM, Hoskin P (2014) The source of halogens in geothermal fluids from the Taupo Volcanic Zone, North Island, New Zealand. Geochim Cosmochim Acta 126:265–283

Bockelée-Morvan D, Biver N, Crovisier J, Lis DC, Hartogh P, Moreno R, de Val-Borro M, Blake GA, Szutowicz S, Boissier J, Cernicharo J (2014) Searches for HCl and HF in comets 103P/Hartley 2 and C/2009 P1 (Garradd) with the Herschel space observatory. Astron Astrophys 562, doi:10.1051/0004–6361/201322939

Bogard DD (1997) A reappraisal of the Martian ^{36}Ar/^{38}Ar ratio. J Geophys Res: Planets 102:1653–1661, doi:10.1029/96JE02796

Bogard DD, Clayton RN, Marti K, Owen T, Turner G (2001) Martian volatiles: Isotopic composition, origin, and evolution. Space Sci Rev 96:425–458

Böhlke JK, Sturchio NC, Gu B, Horita J, Brown GM, Jackson WA, Batista J, Hatzinger PB (2005) Perchlorate isotope forensics. Anal Chem 77:7838–7842, doi:10.1021/ac051360d

Böhlke JK, Hatzinger PB, Sturchio NC, Gu B, Abbene I, Mroczkowski SJ (2009) Atacama perchlorate as an agricultural contaminant in groundwater: Isotopic and chronologic evidence from Long Island, New York. Environ Sci Technol 43:5619–5625, doi:10.1021/es9006433

Bonifacie M, Charlou J-L, Jendrzejewski N, Agrinier P, Donval JP (2005) Chlorine isotopic compositions of high temperature hydrothermal vent fluids over ridge axes. Chem Geol 221:279–288

Bonifacie M, Monnin C, Jendrzejewski N, Agrinier P, Javoy M (2007a) Chlorine stable isotopic composition of basement fluids of the eastern flank of the Juan de Fuca Ridge (ODP Leg 168). Earth Planet Sci Lett 260:10–22

Bonifacie M, Jendrzejewski N, Agrinier P, Coleman M, Pineau F, Javoy M (2007b) Pyrohydrolysis-IRMS determination of silicate chlorine stable isotope compositions. Application to oceanic crust and meteorite samples. Chem Geol 242:187–201

Bonifacie M, Jendrzejewski N, Agrinier P, Humler E, Coleman M, Javoy M (2008a) The chlorine isotope composition of the Earth's mantle. Science 319:1518–1520

Bonifacie M, Busigny V, Mével C, Philippot P, Agrinier P, Jendrzejewski N, Scambelluri M, Javoy M (2008b) Chlorine isotopic composition in seafloor serpentinites and high-pressure metaperidotites. Insights into oceanic serpentinization and subduction processes. Geochim Cosmochim Acta 72:126–139

Boschi C, Bonatti E, Ligi M, Brunelli D, Dallai L, D'Orazio M, Früh-Green G, Tonarini S, Barnes JD, Bedini R (2013) A 10Ma year old window into deep hydrothermal circulation at the Vema Fracture Zone (11ºN, MAR). Lithos 178:3–23

Boudreau AE, Stewart MA, Spivack AJ (1997) Stable Cl isotopes and origin of high-Cl magmas of the Stillwater Complex, Montana. Geol 25:791–794

Boyce JW, Treiman AH, Guan Y, Ma C, Eiler JM, Gross J, Greenwood JP, Stolper EM (2015) The chlorine isotope fingerprint of the lunar magma ocean. Sci Adv 1, doi:10.1126/sciadv.1500380

Brasser R (2013) The formation of Mars: Building blocks and accretion time scale. Space Sci Rev 174:11–25, doi:10.1007/s11214-012-9904–2

Bridges JC, Banks DA, Smith M, Grady MM (2004) Halite and stable chlorine isotopes in the Zag H3–6 breccia. Meteoritics & Planetary Science 39:657–666, doi:10.1111/j.1945–5100.2004.tb00109.x

Chassefière E, Leblanc F (2004) Mars atmospheric escape and evolution; interaction with the solar wind. Planet Space Sci 52:1039–1058, doi:10.1016/j.pss.2004.07.002

Chiaradia M, Banks D, Cliff R, Marschik R, De Haller A (2006) Origin of fluids in iron oxide–copper–gold deposits: constraints from δ^{37}Cl, $^{87}Sr/^{86}Sr_i$ and Cl/Br. Mineralium Deposita 41:565–573

Chiaradia M, Barnes JD, Cadet-Voisin S (2014) Chlorine isotope variations across the Quaternary volcanic arc of Ecuador. Earth Planet Sci Lett 396:22–33

Cisneros M (2013) An experimental calibration of chlorine isotope fractionation between amphibole and fluid at 700 °C and 0.2 GPa. M.S. Thesis, University of Texas at Austin

Codella C, Ceccarelli C, Bottinelli S, Salez M, Viti S, Lefloch B, Cabrit S, Caux E, Faure A, Vasta M, Wiesenfeld (2012) First detection of hydrogen chloride toward protostellar shocks. Astrophys J 744, doi:10.1088/0004-637X/744/2/164

Coleman ML, Ader M, Chaudhuri S, Coates JD (2003) Microbial isotopic fractionation of perchlorate chlorine. Appl Environ Microbiol 69:4997–5000

Cullen J, Barnes JD, Hurwitz S, Leeman W (2015) Halogen and chlorine isotope composition of thermal springs along and across the Cascadia arc. Earth Planet Sci Lett 426:225–234

Czarnacki M, Halas S (2012) Isotope fractionation in aqua-gas systems: Cl_2–HCl–Cl^-, Br_2–HBr–Br- and H_2S–S_2. Isot Environ Health Stud 45:55–64

Dauphas N, Pourmand A (2011) Hf–W–Th evidence for rapid growth of Mars and its status as a planetary embryo. Nature 473:489–492

Desaulniers DE, Kaufmann RS, Cherry JA, Bentley HW (1986) ^{37}Cl-^{35}Cl variations in a diffusion-controlled groundwater system. Geochim Cosmochim Acta 50:1757–1764

Dick HJ, Natland JH, Alt JC, Bach W, Bideau D, Gee JS, Haggas S, Hertogen JG, Hirth G, Holm PM, Ildefonse B (2000) A long in situ section of the lower ocean crust: results of ODP Leg 176 drilling at the Southwest Indian Ridge. Earth Planet Sci Lett 179:31–51

Eastoe CJ, Guilbert JM (1992) Stable chlorine isotopes in hydrothermal processes. Geochim Cosmochim Acta 56:4247–4255

Eastoe CJ, Peryt T (1999) Stable chlorine isotope evidence for non-marine chloride in Badenian evaporites, Carpathian mountain region. Terra Nova 11:118–123

Eastoe CJ, Guilbert JM, Kaufmann RS (1989) Preliminary evidence for fractionation of stable chlorine isotopes in ore-forming hydrothermal systems. Geol 17:285–288

Eastoe CJ, Long A, Knauth LP (1999) Stable chlorine isotopes in the Palo Duro Basin, Texas: evidence for preservation of Permian evaporite brines. Geochim Cosmochim Acta 63:1375–1382

Eastoe CJ, Long A, Land LS, Kyle JR (2001) Stable chlorine isotopes in halite and brine from the Gulf Coast Basin: brine genesis and evolution. Chem Geol 176:343–360

Eastoe CJ, Peryt TM, Petrychenko OY, Geisler-Cussey D (2007) Stable chlorine isotopes of Phanerozoic evaporites. Appl Geochem 22:575–588

Eggenkamp HGM (1994) The geochemistry of chlorine isotopes. PhD Dissertation, Universiteit Utrecht

Eggenkamp H (2014) The geochemistry of stable chlorine and bromine isotopes. Springer, Berlin

Eggenkamp HGM, Schuiling RD (1995) δ^{37}Cl variations in selected minerals: a possible tool for exploration. J Geochem Explor 55:249–255

Eggenkamp HGM, Middelburg JJ, Kreulen R (1994) Preferential diffusion of ^{35}Cl relative to ^{37}Cl in sediments of Kau bat, Halmahera, Indonesia. Chem Geol 116:317–325

Eggenkamp HGM, Kreulen R, Koster van Groos AF (1995) Chloride stable isotope fractionation in evaporites. Geochim Cosmochim Acta 59:5169–5175

Eggenkamp HGM, Koster van Groos AF (1997) Chlorine stable isotopes in carbonatites: evidence for isotopic heterogeneity in the mantle. Chem Geol 140:137–143

Eggenkamp HGM, Coleman ML (2009) The effect of aqueous diffusion on the fractionation of chlorine and bromine stable isotopes. Geochim Cosmochim Acta 73:3539–3548

Epstein S, Taylor HP, Jr. (1971) O^{18}/O^{16}, Si^{30}/Si^{28}, D/H and C^{13}/C^{12} ratios in lunar samples. Lunar Sci Conf 2:1421–1441

Fegley BJ, Lewis JS (1980) Volatile element chemistry in the solar nebula: Na, K, F, CI, Br, and P. Icarus 41:439–455

Fietzke J, Frische M, Hansteen TH, Eisenhauer A (2008) A simplified procedure for the determination of stable chlorine isotope ratios (δ^{37}Cl) using LA-MC-ICP-MS. J Anal Atom Spectr 23:769–772

Fischer TP, Ramírez C, Mora-Amador RA, Hilton DR, Barnes JD, Sharp ZD, Le Brun M, de Moor JM, Barry PH, Füri E, Shaw AM (2015) Temporal variations in fumarole gas chemistry at Poás volcano, Costa Rica. J Volcanol Geotherm Res 294:56–70

Fujitani T, Yamashita K, Numata M, Kanazawa N, Nakamura N (2010) Measurement of chlorine stable isotopic composition by negative thermal ionization mass spectrometry using total evaporation technique. Geochem J 44:241–246

Futagami T, Morono Y, Terada T, Kaksonen AH, Inagaki F (2013) Distribution of dehalogenation activity in subseafloor sediments of the Nankai Trough subduction zone. Philos Trans R Soc Lond B Biol Sci 368:doi: 10.1098/rstb.2012.0249

Germann K, Lüders V, Banks DA, Simon K, Hoefs J (2003) Late Hercynian polymetallic vein-type base-metal mineralization in the Iberian Pyrite Belt: fluid-inclusion and stable-isotope geochemistry (S–O–H–Cl). Mineral Depos 38:953–967

Gleeson SA, Smith MP (2009) The sources and evolution of mineralising fluids in iron oxide–copper–gold systems, Norrbotten, Sweden: Constraints from Br/Cl ratios and stable Cl isotopes of fluid inclusion leachates. Geochim Cosmochim Acta 73:5658–5672

Godon A, Jendrzejewski N, Castrec-Rouelle M, Dia A, Pineau F, Boulègue J, Javoy M (2004a) Origin and evolution of fluids from mud volcanoes in the Barbados accretionary complex. Geochim Cosmochim Acta 68:2153–2165

Godon A, Jendrzejewski N, Eggenkamp HGM, Banks DA, Ader M, Coleman ML, Pineau F (2004b) A cross-calibration of chlorine isotopic measurements and suitability of seawater as the international reference material. Chem Geol 207:1–12

Gribble GW (2010) Naturally Occurring Organohalogen Compounds—A Comprehensive Update. Springer-Verlag, Vienna

Groen J, Velstra J, Meesters AGCA (2000) Salinization processes in paleowaters in coastal sediments of Suriname: evidence from δ^{37}Cl analysis and diffusion modelling. J Hydrol 234:1–20

Hanley J, Ames D, Barnes J, Sharp Z, Guillong M (2011) Interaction of magmatic fluids and silicate melt residues with saline groundwater in the footwall of the Sudbury Igneous Complex, Ontario, Canada: New evidence from bulk rock geochemistry, fluid inclusions and stable isotopes. Chem Geol 281:1–25

Hansen BMS (2009) Formation of the terrestrial planets from a narrow annulus. Astrophys J 703:1131–1140

Harkins WD, Hayes A (1921) The separation of the element chlorine into isotopes (isotopic elements): The heavy fraction from the diffusion. J Am Chem Soc 43:1803–1825

Harkins WD, Liggett TH (1923) The discovery and separation of the isotopes of chlorine, and the whole number rule. J Phys Chem 28:74–82

Harkins WD, Stone SB (1926) The isotopic composition and atomic weight of chlorine from meteorites and from minerals of non-marine origin: (Papers on atomic stability). J Am Chem Soc 48:938–949

Hauri EH, Weinreich T, Saal AE, Rutherford MJ, Van Orman JA (2011) High pre-eruptive water contents preserved in lunar melt inclusions. Science 333:213–215

Hecht MH, Kounaves SP, Quinn RC, West SJ, Young SM, Ming DW, Catling DC, Clark BC, Boynton WV, Hoffman J, DeFlores LP (2009) Detection of Perchlorate and the Soluble Chemistry of Martian Soil at the Phoenix Lander Site. Science 325:64–67, doi:10.1126/science.1172466

Hesse R, Frape SK, Egeberg PK, Matsumoto R (2000) Stable isotope studies (Cl, O, and H) of interstitial waters from Site 997, Blake Ridge has hydrate field, West Atlantic. *In:* Proceedings of the Ocean Drilling Program, Scientific Results. Vol. 164. Paull CK, Matsumoto R, Wallace PJ, Dillon WP (eds) College Station, TX, p 129–137

Hill JW, Fry A (1962) Chlorine isotope effects in the reactions of benzyl and substituted benzyl chlorides with various nucleophiles. J Amer Chem Soc 84:2763–2769

Hoering TC, Parker PL (1961) The geochemistry of the stable isotopes of chlorine. Geochim Cosmochim Acta 23:186–199

Holt BD, Sturchio NC, T.A. A, Heraty LJ (1997) Conversion of Chlorinated Volatile Organic Compounds to Carbon Dioxide and Methyl Chloride for Isotopic Analysis of Carbon and Chlorine. Anal Chem 69:2727–2733

Howald RA (1960) Ion Pairs. I. Isotope Effects Shown by Chloride Solutions in Glacial Acetic Acid. J Am Chem Soc 82:20–24

Hunten D (1993) Atmospheric evolution of the terrestrial planets. Science 259:915–920

Ito E, Harris DM, Anderson AT, Jr. (1983) Alteration of oceanic crust and geologic recycling of chlorine and water. Geochim Cosmochim Acta 47:1613–1624

Izidoro A, Haghighipour N, Winter OC, Tsuchida M (2014) Terrestrial planet formation in a protoplanetary disk with a local mass depletion: A successful scenario for the formation of Mars. Astrophys J 782, doi:10.1088/0004-637X/782/1/31

Jackson WA, Böhlke JK, Gu B, Hatzinger PB, Sturchio NC (2010) Isotopic composition and origin of indigenous natural perchlorate and co-occuring nitrate in the Southwestern United States. Environ Sci Technol 44:4869–4876, doi:10.1021/es903802j

Jackson WA, Davila AF, Sears DWG, Coates JD, McKay CP, Brundrett M, Estrada N, Böhlke JK (2015a) Widespread occurrence of (per)chlorate in the Solar System. Earth Planet Sci Lett 430:470–476

Jackson WA, Böhlke JK, Andraski BJ, Fahlquist L, Bexfield L, Eckardt FD, Gates JB, Davila AF, McKay CP, Rao B, Sevanthi R (2015b) Global patterns and environmental controls of perchlorate and nitrate co-occurrence in arid and semi-arid environments. Geochim Cosmochim Acta 164:502–522, doi:http://dx.doi.org/10.1016/j.gca.2015.05.016

Jarrard RD (2003) Subduction fluxes of water, carbon dioxide, chlorine, and potassium. Geochem Geophys Geosyst4:8905, doi:8910.1029/2002GC000392

Jendrzejewski N, Eggenkamp HGM, Coleman ML (2001) Characterisation of chlorinated hydrocarbons from chlorine and carbon isotopic compositions: scope of application to environmental problems. Appl Geochem 16:1021–1031

John T, Layne GD, Haase KM, Barnes JD (2010) Chlorine isotope evidence for crustal recycling into the Earth's mantle. Earth Planet Sci Lett 298:175–182

John T, Scambelluri M, Frische M, Barnes JD, Bach W (2011) Dehydration of subducting serpentinite: implications for halogen mobility in subduction zones and the deep halogen cycle. Earth Planet Sci Lett 308:65–76

Kaufmann R, Long A, Bentley H, Davis S (1984) Natural chlorine isotope variations. Nature 309:338–340

Kaufmann R, Frape S, Fritz P, Bentley H (1987) Chlorine stable isotope composition of Canadian Shield brines. *In*: Saline Water and Gases in Crystalline Rocks. Vol 33. Fritz P, Frape SK (eds). Geol Assoc Can Spec Pap p 89–93

Kaufmann RS, Long A, Campbell DJ (1988) Chlorine Isotope Distribution in Formation Waters, Texas and Louisiana: Geologic Note. AAPG Bulletin 72:839–844

Kaufmann RS, Frape SK, McNutt R, Eastoe C (1993) Chlorine stable isotope distribution of Michigan Basin formation waters. Appl Geochem 8:403–407

Konstantinov BP, Bakulin EA (1965) Separation of chloride isotopes in aqueous solutions of lithium chloride, sodium chloride and hydrochloric acid. Russ J Phys Chem 39:315–318

Kounaves SP, Carrier BL, O'Neil GD, Stroble ST, Claire MW (2014) Evidence of martian perchlorate, chlorate, and nitrate in Mars meteorite EETA79001: Implications for oxidants and organics. Icarus 229:206–213

Kusebauch C, John T, Whitehouse MJ, Engvik AK (2015) Apatite as probe for the halogen composition of metamorphic fluids (Bamble Sector, SE Norway). Contrib Mineral Petrol 170:1–20

Langvad T (1954) Separation of chlorine isotopes by ion-exchange chromatography. Acta Chem Scand 8:526–527

Layne GD, Godon A, Webster JD, Bach W (2004) Secondary ion mass spectrometry for the determination of δ37Cl: Part I. Ion microprobe analysis of glasses and fluids. Chem Geol 207:277–289

Leri AC, Mayer LM, Thornton KR, Northrup PA, Dunigan MR, Ness KJ, Gellis AB (2015) A marine sink for chlorine in natural organic matter. Nat Geosci 8:620–624

Li L, Bonifacie M, Aubaud C, Crispi O, Dessert C, Agrinier P (2015) Chlorine isotopes of thermal springs in arc volcanoes for tracing shallow magmatic activity. Earth Planet Sci Lett 413:101–110

Liebscher A, Barnes JD, Sharp ZD (2006) Chlorine isotope vapor-liquid fractionation during experimental fluid-phase separation at 400 °C/23 MPa to 450 °C/42 MPa. Chem Geol 234:340–345

Long A, Kaufmann RS, Martin JG, Wirt L, Finley JB (1993) High-precision measurement of chlorine stable isotope ratios. Geochim Cosmochim Acta 57:2907–2912

Lüders V, Banks DA, Halbach P (2002) Extreme Cl/Br and δ37Cl isotope fractionation in fluids of modern submarine hydrothermal systems. Mineral Depos 37:765–771

Madorsky SL, Straus S (1947) Concentration of isotopes of chlorine by the countercurrent electromigration method. J Res Nat Bur Stand 38:185–189

Magenheim AJ (1995) Oceanic borehole fluid chemistry and analysis of chlorine stable isotopes in silicate rocks. PhD Dissertation, University of California, San Diego

Magenheim AJ, Spivack AJ, Volpe C, Ransom B (1994) Precise determination of stable chlorine isotopic ratios in low-concentration natural samples. Geochim Cosmochim Acta 58:3117–3121

McCubbin FM, Steele A, Hauri EH, Nekvasil H, Yamashita S, Hemley RJ (2010) Nominally hydrous magmatism on the Moon. PNAS 107:11223–11228

McCubbin FM, Jolliff BL, Nekvasil H, Carpenter PK, Zeigler RA, Steele A, Elardo SM, Lindsley DH (2011) Fluorine and chlorine abundances in lunar apatite: Implications for heterogeneous distributions of magmatic volatiles in the lunar interior. Geochim Cosmochim Acta 75:5073–5093, doi:10.1016/j.gca.2011.06.017

McKay DS, Clanton US, Morrison DA, Ladle GH (1972) Vapor phase crystallization in Apoll 14 breccia. Proc Third Lunar Sci Conf 1:739–752

Meyer C, Jr., McKay DS, Anderson DH, Butler P, Jr. (1975) The source of sublimates on the Apollo 15 green and Apollo 17 orange glass samples. Proc Lunar Sci Conf 6:1673–1699

Musashi M, Markl G, Kreulen R (1998) Stable chlorine isotope analysis of rock samples: new aspects of chlorine extraction. Anal Chem Acta 362:261–269

Nahnybida T, Gleeson SA, Rusk BG, Wassenaar LI (2009) Cl/Br ratios and stable chlorine isotope analysis of magmatic–hydrothermal fluid inclusions from Butte, Montana and Bingham Canyon, Utah. Mineral Depos 44:837–848

Owen T, Biemann K, Rushneck DR, Biller JE, Howarth DW, Lafleur AL (1977) The composition of the atmosphere at the surface of Mars. J Geophys Res 82:4635–4639

Peng R, Yoshida H, Chamberlin RA, Phillips TG, Lis DC, Gerin M (2010) A comprehensive survey of hydrogen chloride in the Galaxy. Astrophys J 723, doi:10.1088/0004-637X/723/1/218

Philippot P, Agrinier P, Scambelluri M (1998) Chlorine cycling during subduction of altered oceanic crust. Earth Planet Sci Lett 161:33–44

Phillips FM, Bentley HW (1987) Isotopic fractionation during ion filtration: I. Theory. Geochim Cosmochim Acta 51:683–695

Phillips J (1999) Chlorine isotopic composition of hydrothermal vent fluids from the Juan de Fuca Ridge. Master Thesis, University of North Carolina

Ransom B, Spivack AJ, Kastner M (1995) Stable Cl isotopes in subduction-zone pore waters: Implications for fluid–rock reactions and the cycling of chlorine. Geol 23:715–718

Reddy CM, Xu L, Drenzek NJ, Sturchio NC, Heraty LJ, Kimblin C, Butler A (2002) A chlorine isotope effect for enzyme-catalyzed chlorination. J Am Chem Soc 49:14526–14527

Richard A, Banks DA, Mercadier J, Boiron MC, Cuney M, Cathelineau M (2011) An evaporated seawater origin for the ore-forming brines in unconformity-related uranium deposits (Athabasca Basin, Canada): Cl/Br and δ^{37}Cl analysis of fluid inclusions. Geochim Cosmochim Acta 75:2792–2810

Richter FM, Mendybaev RA, Christensen JN, Hutcheon ID, Williams RW, Sturchio NC, Beloso AD (2006) Kinetic isotopic fractionation during diffusion of ionic species in water. Geochim Cosmochim Acta 70:277–289

Rizzo AL, Caracausi A, Liotta M, Paonita A, Barnes JD, Corsaro RA, Martelli M (2013) Chlorine isotopic composition of volcanic gases and rocks at Mount Etna (Italy) and inferences on the local mantle source. Earth Planet Sci Lett 371:134–142

Saal AE, Hauri EH, Cascio ML, Van Orman JA, Rutherford MC, Cooper RF (2008) Volatile content of lunar volcanic glasses and the presence of water in the Moon's interior. Nature 454:192–195

Schauble EA, Sharp ZD (2011) Modeling isotopic signatures of nebular chlorine condensation. Goldschmidt Conf Abstr 21:1810

Schauble EA, Rossman GR, Taylor HPJ (2003) Theoretical estimates of equilibrium chlorine-isotope fractionations. Geochim Cosmochim Acta 67:3267–3281

Schilling J-G, Unni CK, Bender ML (1978) Origin of chlorine and bromine in the oceans. Nature 273:631–636

Schnetger B, Muramatsu Y (1996) Determination of halogens, with special reference to iodine, in geological and biological samples using pyrohydrolysis for preparation and inductively coupled plasma mass spectrometry and ion chromatography for measurement. Analyst 121:1627–1631

Selverstone J, Sharp ZD (2011) Chlorine isotope evidence for multicomponent mantle metasomatism in the Ivrea Zone. Earth Planet Sci Lett 310:429–440

Selverstone J, Sharp ZD (2013) Chlorine isotope constraints on fluid-rock interactions during subduction and exhumation of the Zermatt-Saas ophiolite. Geochem Geophys Geosyst14:doi:10.1002/ggge.20269

Selverstone J, Sharp ZD (2015) Chlorine isotope behavior during prograde metamorphism of sedimentary rocks. Earth Planet Sci Lett 417:120–131

Sharp ZD, Barnes JD (2004) Water soluble chlorides in massive seafloor serpentinites: a source of chloride in subduction zones. Earth Planet Sci Lett 226:243–254

Sharp ZD, Barnes JD (2008) Comment to "Chlorine stable isotopes and halogen concentrations in convergent margins with implications for the Cl isotopes cycle in the ocean" by Wei et al. A review of the Cl isotope composition of serpentinites and the global chlorine cycle. Earth Planet Sci Lett 274:531–534

Sharp ZD, Barnes JD, Fischer TP, Halick M (2010a) A laboratory determination of chlorine isotope fractionation in acid systems and applications to volcanic fumaroles. Geochim Cosmochim Acta 74:264–273

Sharp ZD, Shearer CKJ, McKeegan KD, Barnes JD, Wang YQ (2010b) The chlorine isotope composition of the moon and implications for an anhydrous mantle. Science 329:1050–1053

Sharp ZD, McCubbin M, Shearer CK (2013a) A hydrogen-based oxidation mechanism relevant to planetary formation. Earth Planet Sci Lett 380:88–97, doi:http://dx.doi.org/10.1016/j.epsl.2013.08.015

Sharp ZD, Shearer CK, McCubbin FM, Agee CB, McKeegan KD (2013b) The effect of vapor pressure on Cl isotope fractionation: Application to δ^{37}Cl value(s) of Mars. Lunar Planet Sci Conf 44:2611

Sharp ZD, Williams J, Shearer CK, Jr., Agee CB, McKeegan KD (2016) The chlorine isotope composition of Martian meteorites 2. Implications for the early solar system and the formation of Mars. Meteor Planet Sci, doi: 10.1111/maps.12591

Sharp ZD, Barnes JD, Brearley AJ, Chaussidon M, Fischer TP, Kamenetsky VS (2007) Chlorine isotope homogeneity of the mantle, crust and carbonaceous chondrites. Nature 446:1062–1065

Sharp ZD, Mercer JA, Jones RH, Brearley AJ, Selverstone J, Bekker A, Stachel T (2013c) The chlorine isotope composition of chondrites and Earth. Geochim Cosmochim Acta 107:189–204

Shouakar-Stash O, Frape SK, Drimmie RJ (2003) Stable hydrogen, carbon and chlorine isotope measurements of selected chlorinated organic solvents. J Contam Hydrol 60:211–228

Shouakar-Stash O, Drimmie RJ, Frape SK (2005) Determination of inorganic chlorine stable isotopes by continuous flow isotope ratio mass spectrometry. Rapid Commun Mass Spectr 19:121–127

Shouakar-Stash O, Alexeev SV, Frape SK, Alexeeva LP, Drimmie RJ (2007) Geochemistry and stable isotopic signatures, including chlorine and bromine isotopes, of the deep groundwaters of the Siberian Platform, Russia. Appl Geochem 22:589–605

Spivack AJ, Kastner M, Ransom B (2002) Elemental and isotopic chloride geochemistry and fluid flow in the Nankai Trough. Geophys Res Lett 29, doi:doi:10.1029/2001GL014122

Stakes DS, Meyer PS, Cannat M, Chaput T (1991) Metamorphic stratigraphy of Hole 735B. *In*: Proc ODP, Sci Results. Vol 118. Von Herzen RP, Robinson PT, et al. (eds). Ocean Drilling Program, College Station, TX, p 153–180

Stewart MA (2000) Geochemistry of dikes and lavas fron Hess Deep: implications for crustal construction processes beneath Mid-Ocean Ridges and the stable-chlorine isotope geochemistry of Mid-Ocean Ridge Basalt glasses. PhD Dissertation, Duke University

Stotler RL, Frape SK, Shouakar-Stash O (2010) An isotopic survey of δ^{81}Br and δ^{37}Cl of dissolved halides in the Canadian and Fennoscandian Shields. Chem Geol 274:38–55

Sturchio NC, Clausen JL, Heraty LJ, Huang L, Holt BD, Abrajano TA (1998) Chlorine isotope investigation of natural attenuation of trichloroethene in an aerobic aquifer. Environ Sci Technol 32:3037–3042

Sturchio NC, Hatzinger PB, Arkins MD, Suh C, Heraty LJ (2003) Chlorine isotope fractionation during microbial reduction of perchlorate. Environ Sci Technol 37:3859–3863, doi:10.1021/es034066g

Sturchio N, Böhlke JK, Gu B, Horita J, Brown G, Beloso A, Jr., Patterson L, Hatzinger P, Jackson WA, Batista J (2006) Stable isotopic composition of chlorine and oxygen in synthetic and natural perchlorate. *In*: Perchlorate. Gu B, Coates J, (eds). Springer US, p 93–109

Sturchio NC, Böhlke JK, Beloso AD, Streger SH, Heraty LJ, Hatzinger PB (2007) Oxygen and chlorine isotopic fractionation during perchlorate biodegradation: Laboratory results and implications for forensics and natural attenuation studies. Environ Sci Technol 41:2796–2802, doi:10.1021/es0621849

Sturchio NC, Beloso Jr AD, Heraty LJ, LeClaire J, Rolfe T, Manning KR (2008) Isotopic evidence for agricultural perchlorate in groundwater of the western Chino Basin, California. *In:* Sixth International Conference on Remediation of Chlorinated and Recalcitrant Compounds. Monterey, CA, p 18–22

Sturchio NC, Böhlke JK, Gu B, Hatzinger PB, Jackson WA (2012a) Isotopic tracing of perchlorate in the environment. *In*: Handbook of Environmental Isotope Geochemistry. Baskaran M (ed) Springer Berlin Heidelberg, p 437–452

Sturchio NC, Hoaglund JR, Marroquin RJ, Beloso AD, Heraty LJ, Bortz SE, Patterson TL (2012b) Isotopic mapping of groundwater perchlorate plumes. Ground Water 50:94–102

Swindle TD, Caffee MW, Hohenberg CM (1986) Xenon and other noble gases in shergottites. Geochim Cosmochim Acta 50:1001–1015, doi:http://dx.doi.org/10.1016/0016–7037(86)90381–9

Tang H, Dauphas N (2014) ^{60}Fe–^{60}Ni chronology of core formation in Mars. Earth Planet Sci Lett 390:264–274, doi:http://dx.doi.org/10.1016/j.epsl.2014.01.005

Tartèse R, Anand M, Joy KH, Franchi IA (2014) H and Cl isotope systematics of apatite in brecciated lunar meteorites Northwest Africa 4472, Northwest Africa 773, Sayh al Uhaymir 169, and Kalahari 009. Meteor Planet Sci 49:2266–2289, doi:10.1111/maps.12398

Taylor JW, Grimsrud EP (1969) Chlorine isotopic ratios by negative ion mass spectrometry. Anal Chem 41:802–810

Thiemens MH, Chakraborty S, Dominguez G (2012) The physical chemistry of mass-independent isotope effects and their observation in nature. Ann Rev Phys Chem 63:155–177, doi:10.1146/annurev-physchem-032511–143657

Thode HG, Rees CE (1976) Sulphur isotopes in grain size fractions of lunar soils. Proc Lunar Sci Conf 7:459–468

Thomson JJ, Aston FW, Soddy F, Merton TR, Lindemann FA (1921) Discussion on isotopes. Proc R Soc 99:87–104

Toyama C, Kimura J-I, Chang Q, Vaglarov BS, Kuroda J (2015) A new high-precision method for determining stable chlorine isotopes in halite and igneous rock samples using UV-femtosecond laser ablation multiple Faraday collector inductively coupled plasma mass spectrometry. J Anal Atom Spectr 30:2194–2207

Treiman AH, Boyce JW, Gross J, Guan Y, Eiler JM, Stolper EM (2014) Phosphate-halogen metasomatism of lunar granulite 79215: Impact-induced fractionation of volatiles and incompatible elements. Am Mineral 99:1860–1870, doi:10.2138/am-2014–4822

Urey HC (1947) The thermodynamic properties of isotopic substances. J Chem Soc:562–581

Urey HC, Greiff LJ (1935) Isotopic exchange equilibria. J Am Chem Soc 57:321–327

Van Acker MR, Shahar A, Young ED, Coleman ML (2006) GC/multiple collector—ICPMS method for chlorine stable isotope analysis of chlorinated aliphatic hydrocarbons. Anal Chem 78:4663–4667

van Warmerdam EM, Frape SK, Aravena R, Drimmie RJ, Flatt H, Cherry JA (1995) Stable chlorine and carbon isotope measurements of selected chlorinated organic solvents. Appl Geochem 10:547–552

Vanko DA (1986) High-chlorine amphiboles from oceanic rocks: product of highly-saline hydrothermal fluids? Am Mineral 71:51–59

Vengosh A, Chivas AR, McCulloch MT (1989) Direct determination of boron and chlorine isotopic compositions in geological materials by negative thermal-ionization mass spectrometry. Chem Geol 79:333–343

Volpe C, Spivack AJ (1994) Stable chlorine isotopic composition of marine aerosol particles in the western Atlantic Ocean. Geophys Res Lett 21:1161–1164

Volpe C, Wahlen M, Pszenny AA, Spivack AJ (1998) Chlorine isotopic composition of marine aerosols: Implications for the release of reactive chlorine and HCl cycling rates. Geophys Res Lett 25:3831–3834

Walsh KJ, Morbidelli A, Raymond SN, O'Brien DP, Mandell.AM (2011) A low mass for Mars from Jupiter's early gas-driven migration. Nature 475:206–209

Wassenaar LI, Koehler G (2004) On-line technique for the determination of the δ37Cl of inorganic and total organic Cl in environmental samples. Anal Chem 76:6384–6388

Webster CR, Mahaffy PR, Flesch GJ, Niles PB, Jones JH, Leshin LA, Atreya SK, Stern JC, Christensen LE, Owen T, Franz H (2013) Isotope Ratios of H, C, and O in CO_2 and H_2O of the Martian Atmosphere. Science 341:260–263, doi:10.1126/science.1237961

Williams JT, Shearer CK, Sharp ZD, Burger PV, McCubbin FM, Santos AR, Agee C, McKeegan KD (2016) The chlorine isotope composition of Martian meteorites 1: Chlorine isotope compostion of the martain mantle, crustal, and atmospheric reservoirs and their interactions. Meteor Planet Sci 51:2092–2110

Willmore CC, Boudreau AE, Spivack A, Kruger FJ (2002) Halogens of the Bushveld Complex, South Africa: δ^{37}Cl and Cl/F evidence for hydration melting of the source region in a back-arc setting. Chem Geol 182:503–511

Xiao YK, Zhang CG (1992) High precision isotopic measurement of chlorine by thermal ionization mass spectrometry of the Cs_2Cl^+ ion. Int J Mass Spectr Ion Processes 116:183–192

Zhang M, Frape SK, Love AJ, Herczeg AL, Lehmann BE, Beyerle U, Purtschert R (2007) Chlorine stable isotope studies of old groundwater, southwestern Great Artesian Basin, Australia. Appl Geochem 22:557–574

Ziegler K, Coleman ML, Howarth RJ (2001) Palaeohydrodynamics of fluids in the Brent Group (Oseberg Field, Norwegian North Sea) from chemical and isotopic compositions of formation waters. Appl Geochem 16:609–632

Zolotov MY, Mironenko MV (2007) Hydrogen chloride as a source of acid fluids in parent bodies of chondrites. Lunar Planet Sci Conf 38:2340

Reviews in Mineralogy & Geochemistry
Vol. 82 pp. 379-414, 2017

Chromium Isotope Geochemistry

Liping Qin

CAS Key Laboratory of Crust–Mantle Materials and Environment
University of Science and Technology of China
Hefei
Anhui, 230026
China

lpqin@ustc.edu.cn

Xiangli Wang

Yale University
Department of Geology and Geophysics
New Haven, CT 06511
USA

xiangli.wang@yale.edu

INTRODUCTION

Chromium consists of four stable isotopes (^{50}Cr, ^{52}Cr, ^{53}Cr and ^{54}Cr) with natural abundances of 4.35%, 83.79%, 9.50% and 2.36%, respectively (Rossman and Taylor 1998). Among these four isotopes, ^{50}Cr, ^{52}Cr and ^{54}Cr are non-radiogenic, whereas ^{53}Cr is a radiogenic product of the extinct nuclide ^{53}Mn, which has a half-life of 3.7 Myr (Honda and Imamura 1971). Chromium isotope systems have a wide range of applications in geochemistry and cosmochemistry. They have been used to study early solar system processes (e.g., Rotaru et al. 1992); the oxidation/reduction (redox) potential of underground systems, which governs the transport and fate of many contaminants (e.g., Ellis et al. 2002); and more recently, the redox evolution of Earth's early ocean-atmosphere system, which is intimately linked to the evolution of life (Frei et al. 2009; Crowe et al. 2013; Planavsky et al. 2014; Cole et al. 2016).

Chemical properties of Cr

Chromium is redox-sensitive. In Earth's near-surface environments, Cr has two main valence states, +3 and +6, which are expressed as Cr(III) and Cr(VI), respectively. The valence state of Cr is controlled by the prevailing redox potential (Eh) and pH conditions (Fig. 1). Cr(VI) is always bound with O^{2-} to form the oxyanion species CrO_4^{2-} (chromate), $HCrO_4^-$ (bichromate), and $Cr_2O_7^{2-}$(dichromate), all of which are water-soluble. In contrast, Cr^{3+} usually forms oxyhydroxides or oxides, which are insoluble and immobile in the natural pH range. During oxidative weathering, Cr(III) in minerals can be oxidized by O_2 to Cr(VI), a process that is catalyzed by manganese oxides (Fendorf and Zasoski 1992; Economou-Eliopoulos et al. 2014). The Cr(VI) migrates to rivers and eventually to the ocean. In the modern ocean, Cr occurs as both Cr(VI) and Cr(III), with the former as the dominant species (Van der Weijden and Reith 1982; Frei et al. 2014). Chromium has a poorly constrained ocean residence time of 9000–40000 years (Van der Weijden and Reith 1982; Campbell and Yeats 1984; Reinhard et al. 2013a) and mean concentrations in the range of 0.05–1 ppb (e.g., Jeandel and Minster 1987; Bonnand et al. 2013; Pereira et al. 2015; Scheiderich et al. 2015).

1529-6466/17/0082-0010$05.00 http://dx.doi.org/10.2138/rmg.2017.82.10

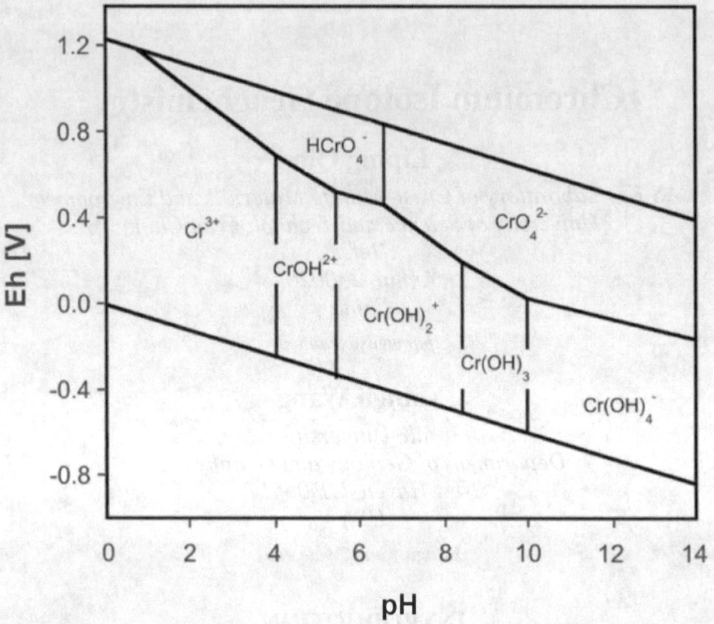

Figure 1. Eh–pH diagram of Cr species.

Chromium is a compatible element during magmatic activity and thus resides mostly in ultra-mafic and mafic rocks (Faure 1991). Compared with its state in Earth's near-surface environment, Cr mainly exists as Cr(III) in igneous rocks. Although Cr(II) has been observed at noticeable abundances in some meteorites (Berry and O'Neill 2004; Berry et al. 2006; Eeckhout et al. 2007; Bell et al. 2014), it is unstable in the oxic near-surface environment. Deep in the mantle where the oxygen fugacities are lower, a small fraction of Cr(II) is possible [in olivine (Burns 1975)], but the dominant Cr species is Cr(III). It is also expected that there is a considerable amount of Cr in the Earth's core present as metallic Cr (Moynier et al. 2011).

Because Cr is widely used in industrial activities, it is also an anthropogenic contaminant in surface and groundwater (e.g., (Rai et al. 1989)). Cr(VI) is highly soluble and is carcinogenic, whereas Cr(III) is sparingly soluble and has limited toxicity because it lacks the oxidizing power that causes damage to cells, unlike Cr(VI). Because of the contrasting geochemical behaviors between Cr(III) and Cr(VI), Cr(VI) contamination can be remediated through reduction induced by either microbial or abiotic mechanisms (Fendorf and Li 1996; Patterson et al. 1997; Pettine et al. 1998; Kim et al. 2001; Wielinga et al. 2001; Sikora et al. 2008; Graham and Bouwer 2009; Han et al. 2012; Kitchen et al. 2012; Xu et al. 2015).

Research History of the Cr isotopic System

Early interest in Cr isotopes has been largely advanced by the application of ^{53}Mn–^{53}Cr short-lived radiochronometers to dating early solar system events; e.g., Birck and Allègre (1985). Although ^{53}Mn became extinct shortly after the formation of the solar system, the variations in the abundance of its daughter product, ^{53}Cr, measured in meteorites reflect "alive" ^{53}Mn in the early solar system, which can be used to date the early solar system history. In addition, anomalous ^{54}Cr abundance was discovered first in refractory inclusions of carbonaceous chondrites and more recently in various types of meteoritic materials; e.g., Birck and Allègre (1984); Trinquier et al. (2007); Qin et al. (2011b). Chromium-54 anomalies are thought to result from nucleosynthetic heterogeneities (incomplete mixing of nuclides from different nucleosynthetic sources) in the solar nebula; e.g., Qin et al. (2011b). Nucleosynthetic anomalies have become powerful tools for tracing the astrophysical environments of solar system formation, the mixing processes of nuclides from different sources, and the generic relationships between different meteorite groups. Both the aforementioned ^{53}Cr and ^{54}Cr anomalies are mass-independent because the anomalies lie outside of the mass-dependent isotopic fractionation line (Young et al. 2002). However, all terrestrial rocks measured so far do not show such mass-independent Cr isotopic effects.

Whereas meteorites show isotopic variations related to ^{53}Mn decay and nucleosynthetic processes, stable Cr isotopic compositions of earth materials vary in response to mass-dependent fractionation. In this process, kinetic and/or thermodynamic differences between Cr isotopes lead to variations in the abundances of heavier versus lighter isotopes (e.g., Schauble 2004), which are quantified by measuring the $^{53}Cr/^{52}Cr$ ratio. Mass-dependent fractionation of Cr isotopes was first used to monitor reduction of Cr(VI) contamination. During reductive immobilization of toxic Cr(VI) in groundwater systems (Ellis et al. 2002, 2004; Johnson and Bullen 2004), CrO_4^{2-} molecules with ^{52}Cr break more easily, and thus react faster, than those with ^{53}Cr. Therefore, as the reduction of Cr(VI) to Cr(III) proceeds, the $^{53}Cr/^{52}Cr$ ratio in the remaining Cr(VI) reservoir progressively increases (Ellis et al. 2002). Hence, isotopically heavy Cr(VI) measured in groundwater samples can be used to indicate, and potentially to quantify, the extent of reduction. To quantify the extent of reduction, the fractionation factor during reduction is needed. In the past decade, numerous experiments (reviewed later in this chapter) have determined isotopic fractionation factors during both biotic and abiotic reduction of Cr(VI) to Cr(III). These experimental results provide the basis for using Cr isotopes to trace Cr biogeochemical cycles in Earth surface processes.

A relatively new research direction is the use of Cr isotopic composition as a paleo-redox proxy. The interest in Earth's redox evolution stems from its intimate relationship with the evolution of life, including oxygenic photosynthesizers and animals together with their subsequent proliferations and extinctions (e.g., Bekker et al. 2004; Lau et al. 2016). Briefly, Cr redox proxy works because under an O_2-rich atmosphere, the Cr(III)-rich minerals contained in crustal rocks can be oxidized and mobilized, leading to isotopically heavy mobile Cr(VI) (e.g., Frei et al. 2009). In contrast, it was assumed that, under an atmosphere devoid of O_2, Cr is transported as solid detrital minerals or as dissolved Cr(III), leaving no opportunities for Cr isotopic fractionation (e.g., Konhauser et al. 2011). Therefore, Cr isotopic compositions in ancient sediments may reveal the oxidation state of the contemporaneous atmosphere (Frei et al. 2009; Crowe et al. 2013; Planavsky et al. 2014). Once the atmosphere is fully oxygenated, the Cr isotopic composition in seawater is mainly controlled by sink fluxes (reviewed later in this chapter), which are largely dependent on the extent of spatial expansion of reducing conditions; for example, an expansion in reducing conditions tends to push seawater to lower Cr inventory and higher $^{53}Cr/^{52}Cr$ ratios (e.g., Reinhard et al. 2013a, 2014; Wang et al. 2016a). Therefore, Cr isotopes may also be used to reflect the expansion of reducing conditions in a largely oxic Phanerozoic ocean (Wang et al. 2016a), although hydrothermal influence must also be considered (Holmden et al. 2016).

In this review, we will focus on the research advances that have been made in the ten years since the last review by Johnson and Bullen (2004). Much of the basis of Cr isotope geochemistry has been discussed in detail in Johnson and Bullen (2004), and the reader is referred to this detailed review for this information.

ANALYTICAL METHODS AND NOTATION

Analytical methods

High-precision measurements of Cr isotope ratios can be obtained on both multi-collector thermal ionization mass spectrometry (TIMS) (Ellis et al. 2002; Trinquier et al. 2008a; Qin et al. 2010a) and multi-collector inductively coupled plasma source mass spectrometry (MC–ICP–MS) (Schoenberg et al. 2008; Schiller et al. 2014).

During ion transmission within a mass spectrometer, relatively heavy isotopes are biased over light isotopes owing to a spatial-charge effect (Maréchal 1999). This instrumental mass bias was corrected in many MC–ICP–MS studies of various elements by performing bracketing standard runs. This method requires that standards and samples be analyzed under identical conditions (same matrix and same concentration); therefore, this approach cannot be used with TIMS because filaments vary. To ensure the accuracy of the method, a high sample recovery rate is essential during sample purification (discussed later in this section) because incomplete sample recovery can potentially lead to isotopic fractionation. Another method of ensuring accuracy involves adding tracers containing two enriched isotopes of the target element (usually called the "double spike" method; in the case of Cr, ^{50}Cr and ^{54}Cr are commonly added) with well-calibrated isotopic compositions before sample preparation. Any artificial isotopic fractionation during sample preparation and instrument analysis can be corrected based on the calibrated and measured ^{50}Cr/^{54}Cr ratio (Dodson 1970).

The double-spike method has several advantages over the standard–sample bracketing method, provided that the spike and the samples are fully equilibrated: (1) Complete recovery during Cr purification is not required because the double spike is added before sample preparation; (2) differences in mass fractionation in the mass spectrometer or sample processing due to different matrices are equally well corrected for by the double spike; (3) short-term drift in mass fractionation in the mass spectrometer can be monitored and corrected for during every integration. The double spike method is essential when performing TIMS measurements because of large and variable isotopic fractionation during ionization.

Sample preparation. Because Cr is a trace element and Cr isotopic variations (both mass-independent and mass-dependent) in both nature and in laboratory conditions are of limited extent, sample purification is required before performing mass spectrometry to achieve the necessary precision to resolve Cr isotopic variability. Two prevailing sample preparation schemes have been used to separate Cr from matrix elements: anion and cation exchange methods. Anion exchange takes advantage of Cr(VI) anion being adsorbed on an anion exchange resin while metal cations pass through. Cation exchange is based on the fact that Cr(III) cation has a unique affinity for cation exchange resin in certain acids.

Anion exchange methods are suitable both for water samples (Ball and Bassett 2000) and for rock sample digests containing Cr(III), after oxidation of Cr(III) to Cr(VI) (Schoenberg et al. 2008). Samples containing Cr(VI) are first passed through columns filled with the anion exchange resin AG 1-X8 (100–200 mesh). Dilute HCl (0.2 N) is used to elute most cations, and Cr(VI) in the form of CrO_4^{2-} anion is retained on the column. Cr(VI) is then converted to Cr(III) and released from the resin by 2 N HNO_3 doped with trace H_2O_2 (Schoenberg et al. 2008; Frei et al. 2009). The eluted solution containing Cr is next passed through an AG 1-X8 (100–200 mesh) microcolumn to remove the interference element Fe and through an AGW 50-X8 (200–400 mesh) microcolumn to remove trace interference elements V and Ti (Yamakawa et al. 2009).

Chromium can also be separated from a sample in the Cr(III) form via a two-step cation exchange method (Trinquier et al. 2008a; Yamakawa et al. 2009). Samples dissolved in 1 N HCl are passed through columns filled with AG 50W-X8 cation exchange resin (200–400 mesh); Cr is immediately collected, and residual Cr is eluted with additional 1 N HCl. This step ensures the separation of Cr from the majority of matrix cations and isobaric interferences. The samples are further purified in a cleanup cation exchange column filled with AG 50W-X8 cation exchange resin (200–400 mesh). Samples are loaded on the columns in 0.5 N HNO_3. Ti and V interference elements are eluted with 0.5 N HF and 1 N HCl. Finally, Cr is released from the resin by 2 N HCl. This cation exchange method can also be used to separate Cr(III) from Cr(VI) from some natural water samples, and it allows Cr(III) and Cr(VI) concentration and isotopic compositions to be measured separately (Ball and McClesky 2003). For Fe-rich samples, Fe columns can be used to remove Fe before the two-step cation exchange method (Qin et al. 2010a; Schoenberg et al. 2016).

The aforementioned anion exchange method typically has higher yields than the cation exchange method (Schoenberg et al. 2008; Trinquier et al. 2008a; Bonnand et al. 2011; Larsen et al. 2016). Low yields offered by cationic exchange methods can be problematic if the double spike technique is not used because it has been shown that the separation step fractionates Cr isotopes if the recovery is incomplete (Larsen et al. 2016). However, Larsen et al. (2016) recently improved the cation exchange method to achieve >90% yield.

Mass spectrometry. For TIMS, the measurements are relatively straightforward. Sub-microgram to a few microgram quantities of Cr are loaded onto Re filaments with silica gel and saturated boric acid, with or without Al. Chromium is steadily ionized at ~1200–1400 °C. Within this temperature range, the ionization efficiency of Fe is very low. Although Fe signals can be detected if there is any Fe left over from the sample preparation procedure, they can usually be corrected by monitoring the ^{56}Fe signal. A study showed that interferences from Fe are correctable even when the ^{56}Fe/^{52}Cr signal ratio is as high as 10^{-4} (Qin et al. 2010a). The other major isobaric interference, Ti, is not ionized in the temperature range where Cr is ionized, generating no signal above background on a Faraday cup, which has a detection limit of 0.02 mV with a 4.2-second integration time. Given the extremely low natural abundance of ^{50}V relative to ^{51}V and removal of the majority of V during column procedures, V is usually not problematic for either TIMS or ICP–MS measurements. Regardless, any potential isobaric interferences from Fe, Ti, and V can be monitored and corrected by measuring ^{56}Fe, ^{51}V, and ^{49}Ti signals.

Because selective ionization is not possible, isobaric interferences from Fe and Ti can be much more severe for MC–ICP–MS than for TIMS. Fe, Ti, and V need to be eliminated by chemistry as much as possible before performing mass spectrometry. In addition to direct isobaric interferences from Fe, Ti, and V, molecular (polyatomic) interferences associated with Ar (the typical primary gas in the plasma source) such as ^{40}Ar^{12}C on ^{52}Cr, ^{40}Ar^{14}N on ^{54}Cr, and ^{40}Ar^{16}O on ^{56}Fe, are also unavoidable. The measurements are usually conducted using desolvating systems to minimize the formation of these molecular interferences. The measurements must be conducted using a high-mass-resolution instrument (resolving power greater than ~6,000) to distinguish the peaks of molecular interferences from those of Cr and Fe (Weyer and Schwieters 2003). MC–ICP–MS instruments built before 2000 could not offer this resolution (see Weyer and Schwieters 2003).

At the present time, new-generation TIMS (Triton and Triton plus) have provided the most precise (Table 1) and efficient mass-independent measurements of Cr isotopes (Trinquier et al. 2008a; Qin et al. 2010a). To achieve a comparable level of precision, a several-fold increase in Cr quantity is required for MC–ICP–MS measurements (Schiller et al. 2014). For mass-dependent measurements, assuming a Cr double spike is used, choosing between TIMS and MC–ICP–MS depends on several factors: (1) TIMS can achieve higher precision using less Cr than MC–ICP–MS; therefore, for samples (such as seawater) where only a small amount (nanogram quantities)

Table 1. Comparison of precision of Cr isotopic measurements from different laboratories.

Reference	Instrument	Quantity of Cr used	Reported precision	Double spike
Ball and Bassett (2000)	VG–336 and MAT–261 (TIMS)	665 ng	—	No
Ellis et al. (2002)	VG–354 and MAT–261 (TIMS)	200 ng	±0.2‰	Yes
Schoenberg et al. (2008)	Neptune(MC–ICP–MS)	2 μg	±0.048‰	Yes
Halicz et al. (2008)	Neptune(MC–ICP–MS)	10 μg	±0.06‰	No
Frei et al. (2009)	IsotopX/GV IsoProbe(TIMS)	2–5 μg	±0.08‰	Yes
Bonnand et al. (2011)	Neptune(MC–ICP–MS)	250 ng	±0.059‰	Yes
Han et al. (2012)	Isoprobe(MC–ICP–MS)	1 μg	±0. 15‰	Yes
Farkaš et al. (2013)	Neptune(MC–ICP–MS)	2 μg	±0.066‰	Yes
Schiller et al. (2014)	Neptune(MC–ICP–MS)	30–60 μg	±0.011‰	No
Shen et al. (2015)	Neptune plus(MC–ICP–MS)	1 μg	±0.04‰	Yes
Bonnand et al. (2016)	Neptune(MC–ICP–MS)	2 μg	±0.022‰	Yes
Zhang et al. (2016) (submitted)	Neptune plus(MC–ICP–MS)	1 μg	±0.05‰	Yes
Wang et al. (2016) (in revision)	Neptune plus (MC–ICP–MS)	10 ng	±0.26‰	Yes
Scheiderich et al. (2015)	Thermo Triton TIMS	20–600 ng	0.026‰	Yes

of Cr can be extracted, TIMS is a better choice. (2) TIMS instruments are less complex than MC–ICP–MS instruments because unlike the latter, they do not have a plasma source, which needs additional devices to power and cool the plasma, which in turn makes MC–ICP–MS more error-prone. (3) High resolution is not needed for TIMS because argon-related molecular interferences are irrelevant, making instrument tuning much easier. (4) The mass bias is more predictable and stable in MC–ICP–MS than in TIMS. However, mass bias can be corrected for each integration on both types of instruments by using the double spike technique, which typically lasts a few seconds. (5) MC–ICP–MS has greater sample throughput because it takes less time to measure one sample on the instrument (~15 minutes for MC–ICP–MS vs. 1–2 hours for TIMS). This advantage, however, may not be as significant as one might think for the following reasons: First, the isotopic results obtained by TIMS are relatively stable (no repeat is need), but repeated analyses are usually needed for MC–ICP–MS, as are more standard measurements. Second, instrument tuning time is much longer for MC–ICP–MS. Third, sample preparation can be a limiting factor. That said, with the development of automatic sample preparation systems (e.g., the prepFAST by Elemental Scientific) and optimization of sample purification procedures (e.g., Larsen et al. 2016), the high throughput offered by MC–ICP–MS could be highly advantageous for samples that do not require very high precision.

The two inter-laboratory Cr isotope standards that have been used in most studies (both mass-independent and mass-dependent effects) are both from the National Institute of Standards and Technology (NIST): Standard Reference Material (SRM) 979, which has certified isotopic composition ($^{50}Cr/^{52}Cr = 0.05186 \pm 0.00010$, $^{53}Cr/^{52}Cr = 0.11339 \pm 0.00015$,

$^{54}Cr/^{52}Cr = 0.02822 \pm 0.00006$), and SRM 3112a, which does not have certified isotopic composition; however, the latter has been measured to have slightly lighter Cr isotopic composition than SRM 979 (Schoenberg et al. 2008; Shen et al. 2015). Both standards are supplied in Cr(III) form in nitric solution.

Notation

Studies of mass-dependent and mass-independent Cr isotopic fractionation use different notation. "Mass-dependent" means that $^{53}C/^{52}Cr$ (*y*-axis) and $^{54}C/^{52}Cr$ (*x*-axis) are plotted on a straight line with a slope of ~0.5 (see Young et al. 2002 for details), whereas "mass-independent" means that $^{53}C/^{52}Cr$ and $^{54}C/^{52}Cr$ deviate from this line.

To express mass-dependent Cr isotopic fractionation relative to a laboratory standard, δ notation is used:

$$\delta^{53}\mathrm{Cr} = \left[\frac{\left(\dfrac{^{53}\mathrm{Cr}}{^{52}\mathrm{Cr}} \right)_{\mathrm{smp}}}{\left(\dfrac{^{53}\mathrm{Cr}}{^{52}\mathrm{Cr}} \right)_{\mathrm{std}}} - 1 \right] 10^3 \qquad (1)$$

where the subscripts smp and std denote the Cr isotopic ratio of the sample and the standard, respectively. In some cases, it is appropriate to express the magnitude of isotopic fractionation between two phases a and b using a fractionation factor α:

$$\alpha = \frac{R_\mathrm{b}}{R_\mathrm{a}} \qquad (2)$$

In the case of isotopic equilibrium, the extent of the equilibrium fractionation between a and b can be conveniently expressed as follows:

$$\Delta_{\mathrm{b-a}} = \delta^{53}\mathrm{Cr_b} - \delta^{53}\mathrm{Cr_a} \approx 10^3 \ln \alpha \qquad (3)$$

In the case of Cr reduction where isotope exchange between Cr(III) and Cr(VI) is limited because of the low solubility of Cr(III), R_b and R_a can be substituted by R_{prod} and R_{reac}, respectively, which are the $^{53}Cr/^{52}Cr$ ratios of the product Cr(III) at an instant in time and of the reactant Cr(VI) pool, respectively. The magnitude of kinetic fractionation can also be more conveniently expressed using ε notation:

$$\varepsilon = (1 - \alpha)10^3 \qquad (4)$$

For mass-independent isotopic measurements, the Cr isotopic composition is typically expressed in ε notation:

$$\varepsilon^{x}\mathrm{Cr} = \left[\frac{\left(\dfrac{^{x}\mathrm{Cr}}{^{52}\mathrm{Cr}} \right)_{\mathrm{smp}}}{\left(\dfrac{^{x}\mathrm{Cr}}{^{52}\mathrm{Cr}} \right)_{\mathrm{std}}} - 1 \right] 10^4 \qquad (5)$$

where $x = 53$, 54, and the subscripts smp and std indicate the Cr isotopic ratio in the sample and the standard, respectively. The isotopic ratios are corrected for the mass-dependent fractionation that occurs during measurements. To make this correction, the ratio $^{50}Cr/^{52}Cr$

is assumed to be natural. The difference between the measured value of $^{50}Cr/^{52}Cr$ and the natural ratio of the two isotopes is then taken as a measure of the instrumental mass fractionation, which is then used to correct $^{53}Cr/^{52}Cr$ and $^{54}Cr/^{52}Cr$ assuming the instrumental fractionation is mass-dependent and obeys exponential fractionation law (Maréchal 1999). We note that the notation ε has been used in both mass-dependent and mass-independent Cr isotopic studies but has completely different meanings in the two contexts.

CHROMIUM ISOTOPE COSMOCHEMISTRY

Chromium isotope cosmochemistry studies have mostly focused on the mass-independent effects of Cr isotopes, namely radiogenic effects on ^{53}Cr (from the decay of short-lived ^{53}Mn) and nucleosynthetic effects on ^{54}Cr (also known as isotope anomalies). These two subjects have recently been reviewed, and readers are referred to recent reviews on short-lived chronometers by McKeegan and Davis (2014) and on nucleosynthetic isotope anomalies by Qin and Carlson (2016) for details. Extensive discussions on mass-independent effects are beyond the scope of this paper, but some of the important aspects are briefly reviewed here.

^{53}Mn–^{53}Cr short-lived chronometer

Short-lived radionuclide chronometers. Short-lived nuclides were produced in supernovae before the formation of the solar system. They were accreted into the solar nebula and then the solar system; thus they were present at the very early stages of the Solar System. Briefly, for a short time (a few to tens of million years) in the early history of the solar system, ^{53}Mn was still "alive". During this time, differentiation of the solar materials led to varying Mn/Cr ratios, which ultimately led to different daughter ^{53}Cr ingrowth. Excesses or deficits (deviation from the mass-dependent fractionation line) of ^{53}Cr observed today provide evidence that differentiation events occurred prior to ^{53}Mn becoming extinct. Various short-lived radionuclides are very useful tools for dating early solar system events that occurred within the first few million years to tens of millions of years of the formation of the solar system. A "fossil" isochron can be constructed for objects that were formed at the same time, and the slope corresponds to the initial parent isotope abundance at the time when the event occurred. By comparing this value with the initial abundance at the beginning of the solar system, the time elapsed after the formation of the solar system can be calculated. A "fossil" isochron can be constructed in a way similar to a traditional isochron except that on the x-axis, the parent isotope abundance is substituted with that of a stable isotope of the parent element because the parent isotope itself is extinct at the present day. It is important to note that unlike long-lived radioactive chronometers, short-lived radio chronometers can only date relative ages. To know absolute ages, one needs to have an age anchor for which both the absolute age (determined by long-lived radiochronometers) and the abundance of the short-lived nuclide at this age are known. Common age anchors are CAIs (calcium–aluminum-rich inclusions in chondrites, widely regarded as the oldest solids in the solar system) and angrites (differentiated achondrites, products of fast cooling; thus, the difference in closure time between various radiogenic chronometries is minimal).

^{53}Mn–^{53}Cr is one of the few short-lived chronometers that have achieved prominence (Birck and Allègre 1985, 1988; Rotaru et al. 1992; Lugmair and Shukolyukov 1998). Other examples are ^{26}Al–^{26}Mg, ^{146}Sm–^{142}Nd, and ^{182}Hf–^{182}W. The prominence of ^{53}Mn–^{53}Cr is partially due to the relatively high abundance of both Mn and Cr in most solar system objects. This system is very useful for dating processes that are related to volatility because of the different condensation temperatures of Mn and Cr [the 50% condensation temperatures for Cr and Mn are 1296 and 1158 K, respectively (Lodders 2003)]. The relatively long half-life of ^{53}Mn of 3.7 Myr makes this chronometer suitable for dating events from nebular processes to the formation and differentiation of early planetesimals.

Evidence of extant ^{53}Mn has been reported for various solar system objects including CAIs, chondrules, and bulk and components of chondrites and achondrites (Birck and Allègre 1985, 1988; Lugmair and Shukolyukov 1998; Nyquist et al. 2001; Trinquier et al. 2008b).

One of the complications of using this chronometer in early days, however, was that there were debates about the homogeneity in the initial distribution of ^{53}Mn in the solar system (e.g. Lugmair and Shukolyukov 1998; Nyquist et al. 2001). However a recent study showed that ^{53}Mn was initially homogeneously distributed in the inner Solar System (Trinquier et al. 2008b). CAIs are the obvious target to obtain initial ^{53}Mn, but they contain very small amounts of Mn and Cr because both are moderately and differentially volatile (Lodders 2003). The most appropriate age anchors for Mn–Cr system are angrites. The initial ^{53}Mn/^{55}Mn ratio was determined for angrite D'Orbigny ($(3.24 \pm 0.04) \times 10^{-6}$; Glavin et al. 2004) and LEW 86010 ($(1.25 \pm 0.07) \times 10^{-6}$; Lugmair and Shukolyukov 1998). Combining the absolute Pb–Pb ages of these two angrites (4563.37 ± 0.025 Ma for D'Orbigny and for 4558.62 ± 0.15 Ma for LEW 86010; Brennecka and Wadhwa 2012), the Mn–Cr and U–Pb ages calculated for various solar system objects are mostly concordant.

^{54}Cr anomalies

Heavy elements (>He) in the solar system were synthesized before the solar system formed. Different nucleosynthetic sources produce elements with distinct isotopic signatures. Isotope anomalies are mass-independent isotopic variations that can be explained by different relative contributions of the nucleosynthetic processes that made all the heavy elements in the solar system. Isotope anomalies recorded in meteorites are very useful for tracing the sources and distributions of nucleosynthetic products in the early solar nebula and for understanding the stellar processes that made the elements. Recently, the discovery of planetary-scale isotope anomalies has provided a viable means for tracing the astrophysical environments of solar system formation, the mixing processes of nuclides from different sources, and the generic relationship between different meteorite groups.

^{54}Cr is a neutron-rich isotope, along with ^{48}Ca, ^{50}Ti, ^{62}Ni, and ^{64}Ni from other iron peak elements, which are thought to be produced in similar astrophysical environments, but the site and method of their production are still somewhat uncertain (Meyer et al. 1996; Woosley and Heger 2007; Wasserburg et al. 2015). Anomalies in ^{54}Cr (i.e. non-zero ε^{54}Cr) were first found in Allende CAIs, with calcium-aluminum-rich inclusions with fractionation and unknown nuclear effects (FUN CAIs) exhibiting diverse ε^{54}Cr values from −151 to 48 parts per ten thousand, compared with ~7 for normal CAIs (Birck and Allègre 1984; Papanastassiou 1986). ^{54}Cr anomalies were also revealed during stepwise acid digestion of chondrites: The fractions dissolved by early, weak acid leaches are characterized by negative ^{54}Cr anomalies. Later, stronger acid leachates and the acid residues have positive anomalies (Rotaru et al. 1992; Podosek et al. 1997; Trinquier et al. 2008b; Qin et al. 2011a); this suggests heterogeneous mixing of products from different nucleosynthetic sources within carbonaceous chondrites.

At the whole-meteorite scale, different groups of meteorites show a range in ^{54}Cr/^{52}Cr of approximately 2 parts per ten thousand. The variability is very systematic and therefore can be used for distinguishing meteorite classes (Table 2).

For instance, carbonaceous chondrites display ε^{54}Cr values ranging from 0.4 to 1.6, compared with enstatite chondrites, which have similar ε^{54}Cr values (0) to those of terrestrial rocks; ordinary chondrites exhibit a constant deficit of ~0.4; achondrites, iron meteorites, and Martian meteorites show deficits of ~0.2 to 0.9 in ε^{54}Cr; lunar rocks have the same ε^{54}Cr values as terrestrial rocks (Shukolyukov and Lugmair 2006; Trinquier et al. 2007; Qin et al. 2010a,b; Yamakawa et al. 2010; Warren 2011). The carrier phases of ^{54}Cr were identified as sub-micron oxide grains from a Type II supernova, and the distribution of ^{54}Cr anomalies among different meteorite groups is thought to result from temporally or spatially heterogeneous distribution of these carrier phases (Dauphas et al. 2010; Qin et al. 2011b).

Table 2. ^{54}Cr Isotopic anomalies in bulk meteorites.

Material type	ε^{54}Cr (parts per ten thousand)
Terrestrial	0
Carbonaceous chondrites	0.4–1.6
Enstatite chondrites	0
Ordinary chondrites	−0.4
Achondrites	
Iron meteorites	~ −0.2 to −0.9
Martian meteorites	
Lunar rocks	0

Data sources: Shukolyukov and Lugmair (2006); Trinquier et al. (2007); Qin et al. (2010a,b); Yamakawa et al. (2010); Warren (2011)

Anomalies of other neutron-rich isotopes have also been found, and these anomalies are more or less correlated with each other at the bulk-meteorite scale (Regelous et al. 2008; Trinquier et al. 2009; Warren 2011; Dauphas et al. 2014), suggesting similar causes for these isotope anomalies.

It is important to note that most terrestrial rocks do not exhibit variations in ε^{53}Cr and ε^{54}Cr. However, if they do, it is taken as evidence of meteoritic impact (Trinquier et al. 2006).

CHROMIUM ISOTOPIC FRACTIONATION IN HIGH-TEMPERATURE SETTINGS

Bulk silicate earth and meteorites

Compared with low-temperature processes (discussed below), the stable Cr isotopic fractionation induced by high-temperature processes is poorly constrained. Schoenberg et al. (2008) first studied the Cr isotopic compositions of a set of mantle-derived rocks from different sources, including mantle xenoliths, ultramafic rocks and cumulates, and oceanic and continental basalts. The investigated samples gave a limited δ^{53}Cr range of −0.211‰ to −0.017‰, with an average of −0.124 ± 0.101‰, implying a relatively homogeneous mantle source in regard to Cr isotopic compositions and limited Cr isotopic fractionation during partial melting and magma differentiation. Farkaš et al. (2013) observed that igneous chromites have slightly heavier Cr isotopic compositions (with an average δ^{53}Cr value of −0.079 ± 0.129‰) than the average value of the bulk silicate earth (BSE) reported by Schoenberg et al. (2008) and then inferred that chromite-bearing mantle might be isotopically heavier than mantle without chromite. This finding was confirmed by Shen et al. (2015), who directly compared the chromite-bearing mantle and chromite-free mantle. Lunar mare basalts show Cr isotopic variation correlated with MgO, likely due to crystallization of spinel (Bonnand et al. 2016).

Moynier et al. (2011) found that chondrites from all subgroups have systematically lower δ^{53}Cr, ranging from −0.4 to −0.2‰ (Moynier et al. 2011), than the silicate earth (Schoenberg et al. 2008). This was interpreted by Moynier et al. (2011) as a result of incorporation of light Cr into the core. This finding was based on first-principle calculation results, which showed that potential Cr-bearing phases (metal or sulfide in the core) have light Cr isotopic composition compared with co-existing Cr-bearing minerals in the silicate, with Cr in the former dominated by Cr^0 and Cr^{2+} and in the latter dominated by Cr^{3+} (Moynier et al. 2011), given that the condition of the proto-Earth is sufficiently oxidizing to contain a large portion of Cr^{3+}. However, the light Cr isotopic data for chondrites have not been reproduced by several recent studies, which showed instead that chondrites have similar δ^{53}Cr values to those of Earth rocks (Qin et al. 2015; Bonnand et al. 2016; Schoenberg et al. 2016).

We note that the first-principle calculation results for Cr isotopes assume that Cr is a major element, likely because of limitations on the computing power of modern computers (Moynier et al. 2011). However, Cr is a trace element in most major mantle minerals. Further theoretical calculations and measurements on major minerals from the mantle will provide important insights into Cr isotopic behavior during partial melting and fractional crystallization. Shen et al. (in preparation) have performed both ionic modeling and measurements on the Cr isotopic composition of major mantle minerals. Both results indicate that the δ^{53}Cr value decreases slightly in the order of spinel, chromite > clinopyroxene, orthopyroxene > olivine.

Serpentinization and metamorphism

Recent experiments and natural observations revealed the high mobility of Cr(III) in Cl-rich fluids (Ottaway et al. 1994; Klein-BenDavid et al. 2009, 2011; Spandler et al. 2011; Watenphul et al. 2014), which are common during serpentinization and oceanic crust dehydrations. High Cr mobility allows the opportunity for Cr isotopic fractionations, which in turn permits the study of subduction and crust–mantle recycling. Recently, several studies have been focused on Cr isotopic fractionations during serpentinization and metamorphic dehydration. Farkaš et al. (2013) first reported extremely high δ^{53}Cr in serpentinites (up to ~+1.22‰ δ^{53}Cr), and positive correlations between δ^{53}Cr values and various alteration indexes. Thus, they proposed that serpentinization could shift altered peridotites to isotopically high δ^{53}Cr, which was interpreted as a result of reduction of isotopically heavier Cr(VI) in serpentinizing fluids. Subsequently, Wang et al. (2016c) presented detailed Cr isotopic investigations of a series of serpentinites and found that the altered peridotites with lower Cr contents had higher δ^{53}Cr values. Two possibilities were proposed: (i) Light Cr isotopes were lost to fluids during serpentinization as a result of kinetic fractionation, leaving isotopically heavy Cr behind; or (ii) Cr was initially lost to fluids without isotopic fractionation, then isotopically heavy Cr was accumulated during later-stage sulfate reduction.

Although the specific Cr isotopic fractionation mechanism remains controversial, the consistent observations of isotopically heavy serpentinites imply potentially significant Cr isotopic fractionation during dehydration accompanying subduction. To balance the isotopically heavy serpentinites, isotopically light-Cr phases are yet to be identified.

Some isotopically heavy metamorphic minerals have been reported by Farkaš et al. (2013), but no systematic Cr isotopic fractionation was observed in subduction-related metamorphosed mafic rocks from the Dabie-Sulu orogen (Shen et al. 2015; Wang et al. 2016c). These metamorphic rocks were thought to have been subducted to ultra-high P–T conditions and then exhumed to low P–T conditions (Zheng 2008; Wang et al. 2013). They span greenschists, amphibolites, and eclogites, yet their δ^{53}Cr values are indistinguishable from the BSE range reported by Schoenberg et al. (2008). The contrasting δ^{53}Cr between serpentinized peridotites and metamorphosed mafic rocks is enigmatic but interesting because it motivates the continued search for the missing isotopically light counterparts in the subduction system. One explanation for the absence of Cr isotopic variation in some metamorphic processes could be a lack of fluid, which hinders Cr mobility and thus limits Cr isotopic fractionation (Shen et al. 2015). Geochemical indices for fluid activities can be utilized to test this hypothesis.

Finally, the Cr isotopic compositions of different terrestrial reservoirs are summarized in Figure 2. The reader will notice that weathered rocks are the only geological reservoir possessing negative δ^{53}Cr. Therefore, our current terrestrial Cr isotope inventory may be skewed to the positive side, and additional future measurements are needed for a more complete picture regarding Cr isotope inventory on Earth.

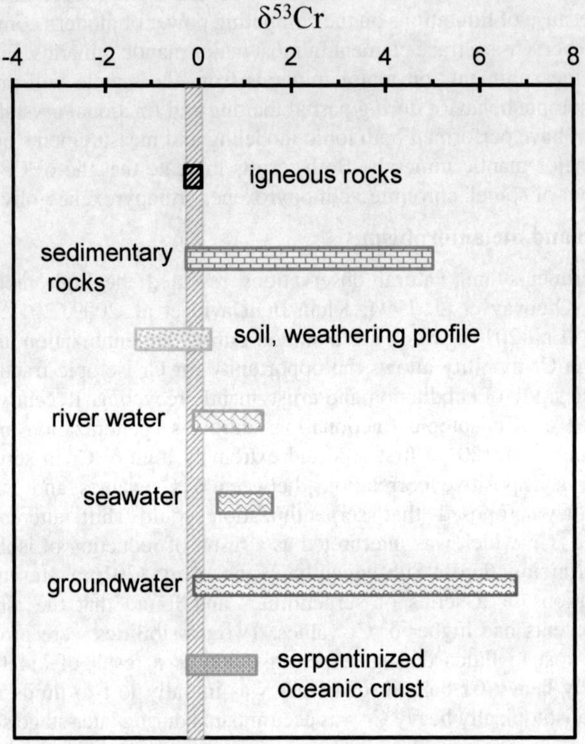

Figure 2. Cr isotopic composition of different terrestrial reservoirs.

MECHANISMS OF Cr ISOTOPIC FRACTIONATION IN LOW-TEMPERATURE SETTINGS

Owing to the contrasting geochemical behavior between insoluble Cr(III) and soluble Cr(VI), reduction of Cr(VI) provides a viable means to remediate Cr(VI) contamination of water sources (Blowes et al. 2000; Crane and Scott 2012). Because substantial isotopic fractionation occurs during reduction, the Cr isotopic composition of groundwater samples is very useful for monitoring and potentially quantifying the extent of reduction. Furthermore, the reduction-associated isotopic fractionation in ancient sediments has been used to track the redox evolution of Earth's surface environment over geological timescales (e.g., Frei et al. 2009; Crowe et al. 2013, Planavsky et al. 2014). Given the great interest in using Cr isotopes in environmental and paleo-redox studies, various reduction mechanisms (described below) have been subjected to extensive studies because they can potentially generate distinctive Cr isotopic fractionations. The results of these experiments are summarized in Table 3; these results can be used for fingerprinting these processes.

Reduction

Abiotic reduction. Sulfides, ferrous iron, organic matter and other organic reductants in sediments and in surface and groundwater can serve as natural Cr(VI) reductants (Fendorf and Li 1996; Patterson et al. 1997; Pettine et al. 1998; Kim et al. 2001; Wielinga et al. 2001; Graham and Bouwer 2009; Kitchen et al. 2012). Hexavalent Cr can also be reduced by granular zero-valent iron (ZVI) particles during remediation (Alowitz and Scherer 2002; Lee et al. 2003). The reduction by ZVI often takes place in permeable reactive barriers, where ZVI is used to reduce

Table 3. Chromium isotopic fractionation ($\delta_{product} - \delta_{reactant}$) induced by different mechanisms determined by laboratory experiments. [continued on next page]

Reagent	Isotopic fractionation	References
	Reduction	
magnetite, sediment slurries	−3.5 ± 0.1‰	Ellis et al. (2002)
Sediment slurry	−2.4 ± 0.1‰ to −3.1 ± 0.1‰	Berna et al. (2010)
Fe(II)-doped goethite	−3.91 ± 0.16‰	Basu and Johnson (2012)
FeS	−2.11 ± 0.04‰	
Green rust	−2.65 ± 0.11‰	
Siderite	−2.76 ± 0.25‰	
Hanford sediments	−3.27 ± 0.12‰	
Geobacter sulfurreducens	−3.03 ± 0.12‰	Basu et al. (2014)
Shewanella sp. (NR)	−2.17 ± 0.22‰	
Pseudomonas stutzeri DCP-Ps1	−3.14 ± 0.13‰	
Desulfovibrio vulgaris	−3.14 ± 0.11‰	
Desulfovibrio vulgaris (aerated)	−2.78 ± 0.27‰	
Pseudomonas stutzeri RCH2 (aerated)	~−2‰	Han et al. (2012)
Pseudomonas stutzeri RCH2 (denitrifying)	~−0.4‰	
Bacillus sp. QH−1 with glucose	−2.00 ± 0.21‰	Xu et al. (2015)
Bacillus sp. QH−1 without glucose	−3.74 ± 0.16‰	
Bacillus sp. QH−1, T = 4 °C	−7.62 ± 0.36‰	
Bacillus sp. QH−1 glucose, T = 15 °C	−4.59 ± 0.28‰	
Bacillus sp. QH−1 glucose, T = 25 °C	−3.09 ± 0.16‰	
Bacillus sp. QH−1 glucose, T = 37 °C	−1.99 ± 0.23‰	
Shewanella oneidensis strain MR−1 low e^- donor	−4.21 ± 0.38‰	Sikora et al. (2008)
Shewanella oneidensis strain MR−1 high e^- donor	−1.8 ± 0.2‰	
Dissolved Fe(II) in batch mode	−3.6‰	Døssing et al. (2011)
Dissolved Fe(II) in constant addition mode with green rust being formed	−1.5‰	
Dissolved Fe(II), pH 4 to 5.3	−4.20 ± 0.11‰	Kitchen et al. (2012)
Organic acids (Elliot fulvic acid, Waskish humic acid, Mandelic acid	−3.11 ± 0.11‰	

Notes:
The "-"and "+" signs of fractionation denote that the product is isotopically lighter and heavier than the reactant, respectively.

Cr(VI) (Blowes et al. 2000; Crane and Scott 2012). ZVI particles can reduce Cr(VI) directly, or they can react with H^+ to form ferrous iron, which then reduces Cr(VI). The reduction of Cr(VI) by organic matter also requires the incorporation of H^+ (Jamieson-Hanes et al. 2012b).

Table 3 (cont'd). Chromium isotopic fractionation ($\delta_{product} - \delta_{reactant}$) induced by different mechanisms determined by laboratory experiments.

Reagent	Isotopic fractionation	References
	Adsorption	
γ-Al_2O_3, goethite	< −0.04‰*	Ellis et al. (2004)
	Oxidation	
Pyrolusite (β-MnO_2)	≤ +1‰	Ellis et al. (2008)
Birnessite (δ-MnO_2)	−2.5‰ to +0.7‰	Bain and Bullen (2005)
Birnessite (δ-MnO_2)	−0.5‰ to 0.0‰	Wang et al. (2010)
H_2O_2	~ +0.2‰ to 0.6‰	Zink et al. (2010)
	Aqueous Cr(III)–Cr(VI) isotope exchange	
No reductant/oxidant	5.8 ± 0.5‰, Cr(VI) isotopically heavier than Cr(III)**	Wang et al. (2015)
	Cr(VI) uptake by abiotic calcite	
Fast precipitation	+0.06‰ to +0.18‰	Rodler et al. (2015)
Slow precipitation	+0.29 ± 0.08‰	
	Cr uptake by coral aragonite	
	~ −1.4‰ to −0.6‰	Pereira et al. (2015)

Notes:

The "-"and "+" signs of fractionation denote that the product is isotopically lighter and heavier than the reactant, respectively.

*This is true only at adsorption–desorption equilibrium. Under disequilibrium conditions, larger but transient isotopic fractionation could occur.

**Experimentally determined equilibrium fractionation is in line with the theoretical estimate (Schauble et al. 2004).

Biotic reduction. Biotic Cr(VI) reduction is one of the most important processes in the Cr biogeochemical cycle. Hexavalent Cr can be naturally reduced by microbes in aquifers. Because Cr(VI) is a toxic species and often does not directly participate in microbial metabolism, it is typically reduced through co-metabolism; that is to say, bacteria do not use Cr for energy, and Cr is reduced by a variety of reducing compounds found in living cells. However, in reducing environments such as some organic-rich sediment packages within aquifers where oxygen is depleted, bacteria such as *Shewanella oneidensis* can use Cr(VI) instead of oxygen for respiration (Middleton et al. 2003). Specifically, *S. oneidensis* oxidizes organic matter to extract energy, a process that produces electrons. The bacteria then transfer the electrons to extracellular Cr(VI) using c-type cytochromes (Myers and Myers 2000). In other scenarios, CrO_4^{2-} could mistakenly enter cells because of the structural similarity between CrO_4^{2-} and SO_4^{2-}. Once within the cell membrane, Cr(VI) is reduced via the sulfate reduction mechanism (Viti et al. 2009). Chromium can also be reduced by chromate reductases, which use NADH or NADPH as cofactors under aerobic conditions (Park et al. 2000).

Because of the Cr(VI)-reducing capabilities of microbes, bio-stimulation—injection of electron donors to stimulate microbial growth—is considered an efficient way to enhance Cr(VI) immobilization. Because many microbes are native to the environment, bio-remediation has proven to be efficient and inexpensive, and it usually causes no secondary pollution to the environment (e.g., Palmer and Wittbrodt 1991; Zayed and Terry 2003). In addition to anaerobic microbes in aquifer systems, aerobic microbes have also been utilized to immobilize Cr(VI) in industrial wastewater (Ohtake et al. 1990; Ackerley et al. 2004; Xu et al. 2015).

Reduction-induced Cr isotopic fractionation. Because Cr(VI) usually exists in the form of CrO_4^{2-} or $HCrO_4^{-}$, the reduction of Cr(VI) involves the breaking of a Cr–O bond. Because light Cr isotopes have higher vibrational frequencies, their bonds with O are easier to break than those of heavy isotopes, resulting in greater reaction rates for lighter Cr isotopes and thus in their enrichment in the reduction product Cr(III) (Schauble 2004). The solid Cr(III) is then physically separated from soluble Cr(VI) to prevent backward reaction. Therefore, during reduction, the Cr(III) produced at each moment in time is constantly offset from the isotopic composition of the remaining Cr(VI) pool by a kinetic isotope fractionation factor ε (a negative value), which varies widely depending on the reaction mechanisms and environmental conditions (Table 3).

In a closed aqueous system, isotope fractionation during reduction of Cr(VI) follows a Rayleigh fractionation model

$$\delta^{53}\mathrm{Cr} = \left[\left(\delta^{53}\mathrm{Cr}_{\mathrm{ini}} + 10^{3}\right) f^{(a-1)}\right] - 10^{3} \qquad (6)$$

where δ^{53}Cr and $\delta^{53}Cr_{ini}$ refer to the unreacted Cr(VI) pool at the time of sampling and at the start of the reaction, respectively; *f* refers to the fraction of Cr(VI) remaining in the solution. To use the Rayleigh fractionation model, three conditions must be met: (i) the system is well mixed and closed; (ii) all of the reduction product Cr(III) is sufficiently removed from the solution, and/or no further isotopic exchange between Cr(VI) and Cr(III) occurs; and (iii) the kinetic isotopic fractionation factor α does not change with time. Although Cr(III) is usually insoluble in the natural pH range, it can be solubilized by complexation with organic matter. Fortunately, as will be shown later, the exchange timescale between Cr(VI) and Cr(III) is relatively long compared with most experimental as well as remediation timescales. Therefore, under controlled laboratory conditions, these three conditions are usually met, and the fractionation factor can be derived from the correlation diagram between Cr isotopic composition and the concentration of Cr(VI) remaining in the solution at different time points (Fig. 3).

Batch and column (flow-through) Cr reduction experiments performed under a wide range of conditions with various abiotic reductants including magnetite, FeS, dissolved Fe(II), and natural sediments obtained ε values ranging from 0.2 to 5, with most values falling within the range of 2‰ to 4‰ (Ellis et al. 2002; Berna et al. 2010; Zink et al. 2010; Døssing et al. 2011;

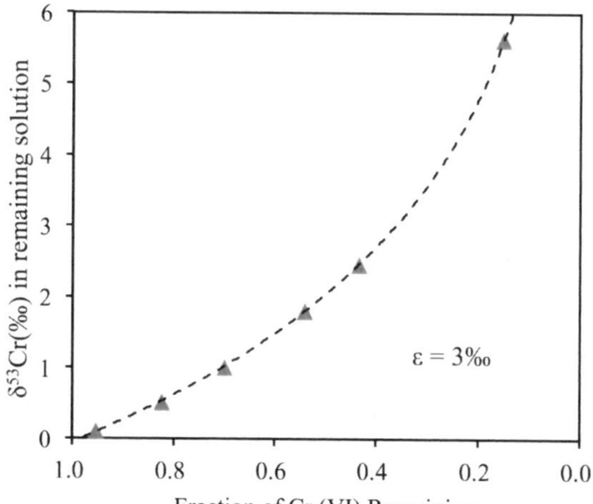

Figure 3. δ^{53}Cr in the remaining Cr(VI) pool vs. the fractionation of Cr(VI) remaining in a Rayleigh fractionation model during Cr(VI) reduction (0‰ at start) with an ε value of 3. The ε value is obtained by fitting the Rayleigh equation (Eqn. 7) to the data. Conversely, if ε is known, one can estimate the extent of reduction (i.e., the fraction of Cr(VI) remaining) by measuring the δ^{53}Cr of Cr(VI) in solution.

Basu and Johnson 2012; Jamieson-Hanes et al. 2012b, 2014; Kitchen et al. 2012) (Fig. 4). Small ε values are usually associated with fast Cr(VI) reduction. The smallest ε values were observed in flow-through experiments with ZVI as the reductant (ε values ranged from 0.2 to 1.5; Jamieson-Hanes et al. 2012b).

Similarly, kinetic isotopic fractionation induced by biotic Cr(VI) reduction varies widely (Table 3). For instance, under anaerobic conditions, higher donor concentrations were linked to faster Cr reduction accompanied by lower ε values, compared with that at low donor concentrations (Sikora et al. 2008), but donor concentration does not seem to have an effect on isotopic fractionation under aerobic conditions (Xu et al. 2015). In addition, ambient redox conditions also seem to have an effect. For instance, Han et al. (2012) found that under denitrifying conditions, there is much less fractionation by enzymatic reduction of Cr(VI) than under aerobic conditions, even though the reduction rates are similar. Furthermore, Xu et al. (2015) also found an inverse relationship between temperature and ε during microbial Cr(VI) reduction. Reduction of Cr under comparable experimental conditions by different types of microbes did not generate a large effect on the isotopic fractionation factor, except in one species (Basu et al. 2014). Chromium isotopic fractionation during plant uptake was investigated by Ren et al. (2015); the authors used an enriched tracer and showed that light Cr isotopes were enriched in the leaves, whereas the roots were enriched in heavy isotopes.

As for the field studies, the ε values derived from field studies are smaller than those obtained in laboratory experiments with the same reductant by a factor of at least 2 (e.g. Berna et al. 2010).

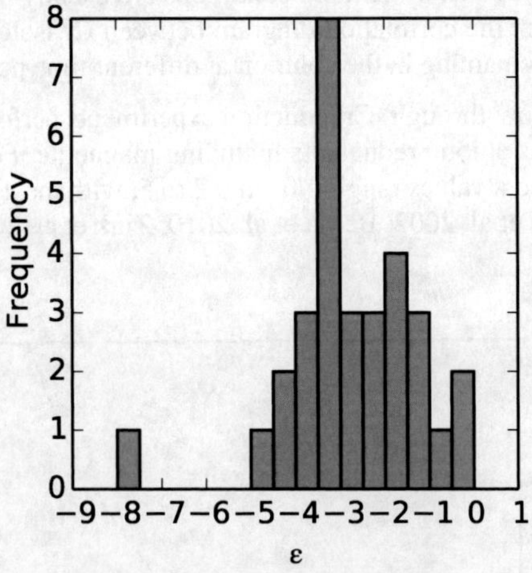

Figure 4. A histogram showing experimentally determined isotopic fractionation during reduction of Cr(VI) to Cr(III) by a range of reductants (Table 3). All reduction mechanisms led to isotopically light Cr(III) and heavy Cr(VI); thus, $\varepsilon < 0$..

Mixed reduction pathways. From the discussions above and the results shown in Table 3, there is a large variation in the Cr isotopic fractionation factor obtained in various laboratory conditions and in field studies. Many authors seek to explain some of the variation in the Cr isotopic fractionation factors either by mixed reduction mechanisms (Døssing et al. 2011; Jamieson-Hanes et al. 2012a,b) or by "reservoir effects" (Berna et al. 2010).

In some cases, it is difficult to use a single isotopic fractionation factor to describe the entire reaction, either because the reaction involves multiple mechanisms or because the reaction rate changes over time. For example, Døssing et al. (2011) explained the small Cr isotopic fractionation factor during the reduction of Cr(VI) by Fe(II) with a combination of two reduction mechanisms: direct reduction by Fe(II), and sorption followed by reduction by green rust; the former induced an isotopic fractionation, but the latter did not. Jamieson-Hanes et al. (2012a,b) observed that the Cr isotopic fractionation factor during Cr(VI) reduction by organic carbon in flow-through experiments was much smaller than those obtained during batch experiments. A transport-reactive model suggests that the batch experimental data can be explained by a dominant Cr(VI) removal reaction generating a large Cr isotopic fractionation, but the explanation of flow-through experimental data requires a second rapid Cr(VI) removal that produces little isotopic fractionation (Jamieson-Hanes et al. 2012a). The specific mechanisms were not identified. In addition, in a series of replicate flow-through experiments with ZVI as the Cr(VI) reductant, Jamieson-Hanes et al. (2014) documented changing ε values during reduction. In the early stage of the experiments when the reducing power of ZVI was high, and therefore the reduction rate was fast, the ε value was −0.2‰, compared with −1.5‰ in the later stage, when reduction became slow likely because of the depletion of reducing power.

Reservoir effects. The "reservoir effect" concept has been invoked to explain the reduced effective/apparent isotopic fractionation factor in systems where isotopic shifts are measured in ocean water in response to O_2 reduction in underlying sediments (Bender 1990; Brandes and Devol 1997). In this model (Fig. 5), below the sediment–water interface, the sediment pile can be divided into an upper, non-reactive diffusion zone and a lower diffusion–reaction zone where O_2 is consumed by respiration. The effective fractionation factor ε_{eff} for O isotopes is inversely correlated with $L\times\sqrt{\frac{R}{D}}$, where L is the thickness of the upper diffusional zone, and R and D are the reaction rate and the diffusion rate, respectively. This "reservoir effect" was later invoked by Clark and Johnson (2008) and Berna et al. (2010) to explain the apparently diminished Se and Cr isotopic fractionation in diffusion-limited systems. Similar effects were modeled by a transport-reactive model (Wanner and Sonnenthal 2013) to explain the diminished ε_{eff} values observed in flow-through experiments with Cr(VI) reduction by Fe(II) (Døssing et al. 2011). The modeling results showed that the apparent ε values decreased with increasing reduction rates and decreasing diffusion rates of Cr(VI).

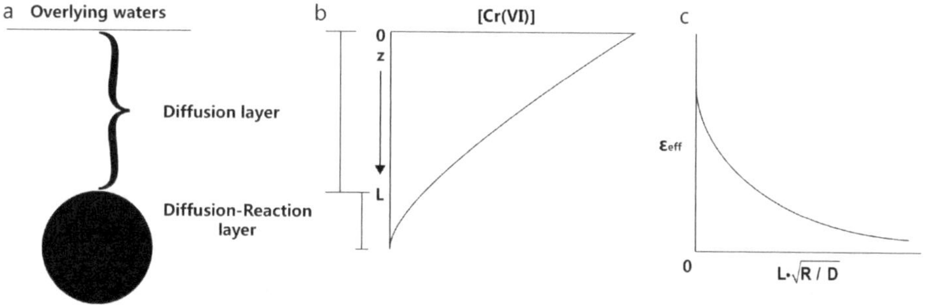

Figure 5. Schematic illustration of the reservoir effect of Cr isotopes in an idealized microzone-sediment oxygen profile. The figure is modified from (Brandes and Devol 1997).

Isotopic fractionation during multi-step reactions: insights from biotic reduction of SO_4^-. As mentioned above, the actual reduction of Cr(VI) likely occur in multiple steps, and the reaction kinetics of different steps will likely affect the observed Cr isotopic fractionation factor. To help explain how different steps in multi-step reactions contribute to the overall Cr isotopic fractionation, insights can be gained from the extensive literature on sulfate reduction (e.g. Farquhar et al. 2003). Insights derived from biotic sulfate reactions also apply to isotopic fractionations in abiotic multi-step reactions. Because the simultaneous transfer of multiple electrons is highly unlikely in nature, the reduction of SO_4^{2-} to S^{2-} and that of CrO_4^{2-} to Cr^{3+} involves multi-step electron transfers. Therefore, the reduction of SO_4^{2-} or CrO_4^{2-} usually proceeds in several steps and with several intermediate species. A conceptual model was proposed by Rees (1973) to explain the S isotopic variations during dissimilatory microbial sulfate reduction. A revised version of this model was put forward later (Brunner and Bernasconi 2005). In a multi-step reaction, each step includes forward (f_i) and backward (b_i) fluxes with a ratio of $X_i = b_i/f_i$; the kinetic isotopic fractionation factors for the f_i and f_b are Δ_{f_i} and Δ_{b_i}, respectively. In general, at a constant reaction rate, the overall fractionation factor can be calculated:

$$\Delta_{\text{total}} = \Delta_{f_1} + \sum_{u=1...(z-2)} \left(\prod_{v=1...u} X_v\right).\Delta_{f_{u+1}} - \sum_{u=1...(z-1)} \left(\prod_{v=1...u} X_v\right).\Delta_{b_u} \quad (7)$$

where $z-1$ is the number of reaction steps (Brunner and Bernasconi 2005). If we assume that all backward reactions induce no isotopic fractionation, the equation can be simplified as follows (a 4-step reaction for example):

$$\Delta_{\text{total}} = \Delta_{f_1} + X_1.\Delta_{f_2} + X_1X_2.\Delta_{f_3} + X_1X_2X_3.\Delta_{f_4} \quad (8)$$

In this example, the theoretical maximum fractionation is

$$\Delta_{\text{total}} = \Delta_{f_1} + \Delta_{f_2} + \Delta_{f_3} + \Delta_{f_4} \quad (9)$$

If thermodynamic equilibrium is reached in a reaction chain, the backward fluxes and forward fluxes become equal, and the total isotopic effect is the equilibrium isotopic effect, which is the sum of all kinetic isotopic effects. Equation (7) can be written as (a 4-step reaction for example):

$$\Delta_{\text{equ}} = \Delta f_1 + \Delta f_2 + \Delta f_3 + \Delta f_4 - \Delta_{b_1} - \Delta_{b_2} - \Delta_{b_3} - \Delta_{b_4} \quad (10)$$

These equations not only can explain the variable S isotopic fractionation during reduction but also can be used to estimate the S isotopic fractionation in a single specific reaction step, if all other steps can be constrained.

Although biotic reduction of Cr(VI) very likely does not consist of the exact same steps, several important implications can be drawn from the equations above to understand the enzyme-meditated Cr isotopic fractionations in a multiple-step reaction chain in general. These implications can be extended to abiotic reduction as well. The overall isotopic effects of multi-step reactions depend on the rate-limiting step, as summarized by Johnson and Bullen et al. (2004), and are re-emphasized here: (1) If a single step of the reaction is the rate-limiting step, the isotopic effect of the total reaction is equal to the kinetic isotopic effect occurring at the rate-limiting step plus the sum of all equilibrium isotopic effects in the preceding reaction steps; (2) kinetic isotopic effects occurring after the rate-limiting step do not contribute to the overall isotopic fractionation of the reaction. These statements can be illustrated as follows: Assuming a reaction chain containing $z-1$ steps among which step w is the rate-limiting step, then X_w is approaching 0, and X_i $[i=1...(w-1)]$ is approaching 1. Thus, Eqn. (7) can be written as follows:

$$\Delta_{\text{total}} = \Delta_{f_1} + \sum_{u=1...(w-1)} \Delta_{f_{u+1}} - \sum_{u=1...(w-1)} \Delta_{b_u} \quad (11)$$

Then, we have:

$$\Delta_{\text{total}} = \sum_{u=1...(w-1)} (\Delta_{f_u} - \Delta_{b_u}) + \Delta_{f_w} \quad (12)$$

Equilibrium Cr isotopic fractionation and Cr(III)–Cr(VI) isotope exchange in aqueous systems

During Cr(VI) reduction, forward reaction is much faster than backward reaction, therefore, isotopic evolution of the remaining Cr(VI) typically conforms to a Rayleigh-type kinetic fractionation model in laboratory experiments lasting hours to days (e.g. Ellis et al. 2002; Basu and Johnson 2012). However, at natural timescales (years to thousands of years), Cr(VI) and Cr(III) can exchange isotopes and reach isotopic equilibrium (Altman and King 1961), overprinting original kinetic signature. Therefore, understanding the rates (isotope exchange kinetics) and maximum impact (equilibrium Cr isotope fractionation) of isotope exchange between Cr(III) and Cr(VI) over hundreds of years or even longer is important for interpreting field Cr isotope data.

Equilibrium Cr isotopic fractionation. Theoretical calculation predicts that the equilibrium isotopic fractionation between two species of the same element depends on the difference between their bonding environments; at isotopic equilibrium, heavier isotopes (e.g. ^{53}Cr) are enriched in species (e.g. CrO_4^{2-}) with stronger bonds, whereas lighter isotopes (e.g. ^{52}Cr) are enriched in species (e.g. $Cr(H_2O)_6^{3+}$) with weaker bonds (Bigeleisen and Mayer 1947; Urey 1947; Schauble et al. 2004). Laboratory experiments confirmed that the equilibrium Cr fractionation factors between coexisting Cr(III) and Cr(VI) species can be as high as 6‰ at room temperature (Wang et al. 2015). Isotopic fractionation could also occur among Cr species that have the same valence states but are complexed with different ligands (Schauble et al. 2004; Larsen et al. 2016). For instance, theoretical calculations by Schauble et al. (2004) predicted that at isotopic equilibrium, Cl-bounded Cr(III) is ~4‰ lower in $\delta^{53}Cr$ than H_2O-bound Cr(III) at room temperature. This finding was also confirmed by later experiments performed on ion exchange resin (Larsen et al. 2016).

Cr(III)-Cr(VI) isotope exchange in aqueous systems. Isotope exchange between Cr(III) and Cr(VI) involves three electron transfers from Cr(III) to Cr(VI), but with no net oxidation or reduction (Altman and King 1961). For example, three electrons can be transferred in three steps (Wang et al. 2015):

$$\text{Step 1: } {}^{53}\mathrm{Cr(III)} + {}^{52}\mathrm{Cr(VI)} = {}^{53}\mathrm{Cr(IV)} + {}^{52}\mathrm{Cr(V)}$$

$$\text{Step 2: } {}^{53}\mathrm{Cr(IV)} + {}^{52}\mathrm{Cr(V)} = {}^{53}\mathrm{Cr(V)} + {}^{52}\mathrm{Cr(IV)}$$

$$\text{Step 3: } {}^{53}\mathrm{Cr(V)} + {}^{52}\mathrm{Cr(IV)} = {}^{53}\mathrm{Cr(VI)} + {}^{52}\mathrm{Cr(III)}$$

Transfer of three electrons at once is considered extremely rare. Combined with the change in bonding environments [e.g., Cr(VI) as CrO_4^{2-} vs. Cr(III) as $Cr(OH)_3$ or $Cr(H_2O)_6^{3+}$], isotope exchange between Cr(III) and Cr(VI) is expected to be slow (Schauble 2004). The timescales for isotopic equilibrium range from months to thousands of years (Wang et al. 2015) under circumneutral pH conditions, depending on the Cr concentrations. However, at circumneutral pH, the impact of isotope exchange on dissolved Cr(VI) is limited because only very limited amounts of Cr(III) on particle surfaces are available for exchange. For comparison, isotopic equilibrium between Fe(II) and Fe(III) is obtained in a few minutes, where only one electron transfer is involved and no bonding environment change is needed (Johnson et al. 2002).

In summary, potential isotope exchange effects should be taken into consideration when interpreting field isotopic data in low-temperature studies, especially when contact time is long, surface area is high, or Cr(III) and/or Cr(VI) concentrations are high (Wang et al. 2015).

Oxidation

Chromium isotopic fractionation during oxidation of Cr(III) to Cr(VI) remains poorly understood, with experimental determinations ranging from approximately −2.5‰ to approximately + 1‰ (Table 3). The size and direction of the fractionation is variable even with the same oxidant [e.g., birnessite, (Bain and Bullen 2005; Wang et al. 2010; Zink et al. 2010)]. One explanation for variable isotopic fractionation during oxidation is the hypothesis of multi-step fractionation as described above. During oxidation, Cr(III) is thought to be oxidized first to Cr(V), followed by disproportionation of Cr(V) to Cr(III) and Cr(VI) (Ardon and Plane 1959; Knoblowitz and Morrow 1976; Banerjee and Nesbitt 1999; Zink et al. 2010). Oxidation of Cr(III) to Cr(V) has been recognized as the rate-limiting step in this case (Knoblowitz and Morrow 1976; Impert et al. 2008). Alternatively, another parallel pathway, Cr(III)–Cr(IV)–Cr(V)–Cr(VI), is also possible (Silvester et al. 1995; Zhao et al. 1995). In addition, the relative importance of kinetic vs. equilibrium fractionation may vary between reaction steps, making it difficult to describe the oxidation-induced fractionation by Rayleigh fractionation alone (Zink et al. 2010). Furthermore, in many natural systems, Cr(III) is present in solid form, which further complicates the situation because fractionation is limited as a solid is dissolved; only the surface of the solid is accessible, and the surface can become quite fractionated. More experimental work is needed to better understand the mechanism of isotopic fractionation during oxidation of Cr(III).

Adsorption

Adsorption of Cr(VI) onto mineral surfaces induces negligible fractionation under equilibrium conditions (Ellis et al. 2004). This is expected because the bonding environment does not differ between dissolved Cr(VI) and adsorbed Cr(VI), both of which are bonded to four oxygen atoms in a tetrahedral geometry (CrO_4^{2-} or $HCrO^{4-}$). However, significant kinetic isotopic fractionation could arise from adsorption under disequilibrium conditions. For example, Ellis et al. (2004) reported up to 3.5‰ difference in δ^{53}Cr between Cr(VI) in solution and Cr(VI) adsorbed to goethite at pH 4, but the isotopic fractionation decreased to zero after 24 hours, when adsorption–desorption reached equilibrium. The experiments conducted by Ellis et al. (2004) were under acidic conditions (pH 4 and 6) and included only two adsorbents: goethite and γ-Al_2O_3. Different adsorbents (e.g., manganese oxides and clays) under neutral pH conditions should be investigated in future experiments.

Coprecipitation

There has been substantial interest in the paleo-redox community in the use of marine carbonate to track seawater δ^{53}Cr. Rodler et al. (2015) investigated Cr isotopic fractionation during calcite precipitation from Cr(VI)-containing solutions. With fast calcite precipitation rates, calcite-Cr(VI) was isotopically heavier than the solution-Cr(VI) by 0.06‰ to 0.18‰, with higher values associated with higher starting Cr(VI) concentrations. With slow calcite precipitation rates, calcite-Cr(VI) was isotopically heavier than the solution-Cr(VI) by 0.29 ± 0.08‰. These results imply that in the marine environment, where Cr(VI) concentration is low and calcite precipitation is fast, inorganic calcite should be able to track seawater δ^{53}Cr. However, most naturally occurring marine carbonate minerals (e.g., calcite, aragonite) are linked with biological processes. Carbonate producers can regulate the chemistry (pH) of the seawater immediately surrounding their cell structures to create favorable calcite/aragonite precipitation conditions (Elderfield et al. 1996; Erez 2003). Therefore, the δ^{53}Cr of biologically produced carbonate minerals may be decoupled from ambient seawater values. Indeed, the δ^{53}Cr values of corals and foraminifera have been shown to be offset from seawater values (Pereira et al. 2015; Wang et al. 2016b). These initial measurements appear to suggest that marine carbonates are currently not straightforward archives for seawater δ^{53}Cr and that further calibration is needed.

CHROMIUM ISOTOPIC VARIATIONS IN SURFACIAL ENVIRONMENTS

Groundwater

Chromium isotopes in groundwater can be used to identify the sources and transport of Cr contamination, to estimate the degree of Cr(VI) reduction, and to distinguish between Cr(VI) mixing and reduction processes (Ellis et al. 2002; Izbicki et al. 2008, 2012; Berna et al. 2010; Raddatz et al. 2010; Wanner et al. 2012, 2013; Basu et al. 2014; Economou-Eliopoulos et al. 2014; Heikoop et al. 2014; Novak et al. 2014). However, in many cases, owing to the complexity of Cr-related reactions as described above, the use of Cr isotopes to identify reduction has limitations.

Experimentally determined Cr isotopic fractionation factors can be applied to groundwater and surface water systems to understand Cr(VI) reduction. Ellis et al. (2002) first reported elevated δ^{53}Cr values (relative to that of NIST SRM 979) for eight samples from California groundwater systems in which Cr(VI) reduction was expected to occur. A later study, by Izbicki et al. (2008), presented a more extensive data set for a Mojave Desert groundwater system containing Cr(VI). Elevated δ^{53}Cr values ranging from 0.7 to 5.2 ‰ were found and were attributed to anaerobic Cr(VI) reduction. However, because of mixing between natural and anthropogenic Cr, the δ^{53}Cr was not correlated with Cr concentrations. Berna et al. (2010) provided an extensive data set from a point-source groundwater plume in Berkeley, CA, collected over five years and at 14 locations that covered the entire plume. For the first time, they showed straightforward relationships between δ^{53}Cr and position, time, and Cr(VI) concentration. A general trend of increasing δ^{53}Cr with decreasing Cr concentration was observed. The data reasonably fit a Rayleigh model, using an effective fractionation factor determined from batch experiments with incubations of site sediments, along with reservoir effects in the restricted subsurface zone. Using the estimated effective fractionation factor, the authors semi-quantitatively estimated the extent of reduction at various locations on the plume. Taking into consideration the uncertainties of the effective fractionation factor, they concluded that Cr(VI) reduction nearly reached completion at the distal end of the plume. Chromium isotopic data were compared with Rayleigh models of Cr reduction and two end-member mixing models to understand the reduction of Cr contamination in a groundwater plume at Idaho National Laboratory (Raddatz et al. 2010). Depth profiles of Cr isotopes in the groundwater further indicated that Cr reduction proceeded more slowly with increasing depth.

A challenge of using Cr isotopes to trace the transport and fate of Cr(VI) in groundwater systems is determining the δ^{53}Cr value of contaminant sources. For anthropogenic sources, the δ^{53}Cr values seem to be very close to that of Cr ore materials (Ellis et al. 2002), which have a silicate-earth-like δ^{53}Cr value of -0.12 ± 0.10‰ (Ellis et al. 2002; Schoenberg et al. 2008). However, geogenic Cr(VI) contamination can also occur near Cr-rich rocks. Oxidative weathering mobilizes Cr(III) from the Cr-rich materials to generate mobile Cr(VI), which is then distributed into aquatic systems. For example, Farkaš et al. (2013) reported $\leq$3.96‰ δ^{53}Cr and $\leq$23 ppm Cr in stream waters draining serpentinized ultramafic rocks characterized by ~1‰ δ^{53}Cr and ~0.4 wt.% Cr_2O_3. Similar ultramafic-sourced, isotopically heavy Cr(VI) contaminant sources have also been reported by Novak et al. (2014) and by Economou-Eliopoulos et al. (2014). Therefore, when interpreting field groundwater δ^{53}Cr data, source rock lithology and δ^{53}Cr must be considered.

Weathering systems

At modern atmospheric oxygen levels, Cr(III) in continental rocks can be oxidized to Cr(VI) by O_2, catalyzed by MnO_2 (Fendorf and Zasoski 1992; Economou-Eliopoulos et al. 2014). Several ancient soil profiles (i.e., paleosols) and modern laterites show negative δ^{53}Cr because of Cr loss (Crowe et al. 2013; Frei and Polat 2013; Berger and Frei 2014; Frei et al. 2014, 2016). However, positive δ^{53}Cr values coupled with Cr enrichment have been found in several weathering profiles; this phenomenon can be explained by re-deposition of mobilized isotopically heavy Cr (Crowe et al. 2013; Frei and Polat 2013; Berger and Frei 2014; Frei et al. 2014; Wang et al. 2016c). In the soil profile, Cr loss and Cr enrichment occur at distinct horizons, which is likely caused by the layered nature of the parent rocks and by a fluctuating water table (e.g., Paulukat et al. 2015). However, in similar soil profiles developed on similar parent rocks in the same region, distinct distribution patterns of δ^{53}Cr can still occur (e.g., Paulukat et al. 2015).

Rivers and seawater

Speciation analyses have revealed that Cr(VI) dominates in modern river waters, which reflects oxidative weathering of continental rocks (Cranston and Murray 1980; Comber and Gardner 2003). Limited measurements of river water have yielded δ^{53}Cr ranging from ~−0.17‰ to 4‰ (Fig. 6A) (Farkaš et al. 2013; Frei et al. 2014; Novak et al. 2014; Paulukat et al. 2015; D'Arcy et al. 2016). The mostly positively fractionated δ^{53}Cr values in river water are consistent with preferential leaching of isotopically heavy Cr(VI) during oxidative weathering.

There is substantial regional variability in Cr isotopes entering the ocean from different drainage systems (Farkaš et al. 2013; Frei et al. 2014; Paulukat et al. 2015). Stream waters draining serpentinized ultramafic rocks show higher δ^{53}Cr values (2.5‰ to 4‰) and Cr concentrations (up to 450 μM, or approximately 22 ppm) than regular river waters (<2‰ and nM) draining typical continental rocks, partly because serpentinites have positively fractionated δ^{53}Cr to begin with (Farkaš et al. 2013; Wang et al. 2016c). However, rivers draining non-serpentinized ultramafic rocks have lower δ^{53}Cr values (up to 1.33 ± 0.04) despite high Cr concentrations (~24 μM, or approximately 1 ppm; Paulukat et al. 2015). The Brahmani River in India (Paulukat et al. 2015) yielded δ^{53}Cr values ranging from BSE-like values to ~1.33‰. Similar values have also been observed at its estuary. However, the δ^{53}Cr value dropped abruptly (0.55 ± 0.08) in the coastal surface water at the Bay of Bengal, whereas the Cr concentration was similar to that of the estuary sample. Because no Cr isotopic data were obtained in the upper stream of the Brahmani River, transportation effects could not be assessed. Similarly, Frei et al. (2014) found that water samples from the Paraná River, its tributaries, and its estuary in Argentina possessed a tight average δ^{53}Cr of 0.35 ± 0.14‰, even though the Cr concentration in the estuary dropped by a factor of 2. This finding suggests minimal Cr isotopic variations during long-distance transport. D'Arcy et al. (2016) showed that both the concentration and the isotopic composition of Cr in weathering fluxes may be affected by the redox conditions of the catchment.

Reported Cr residence times in seawater range from 9000 to 40000 years, compared with the ~1000-year ocean mixing time (Broecker and Peng 1982). This evidence might lead one to predict that seawater δ^{53}Cr is homogeneous. However, seawater samples from global oceans exhibit a large variation in δ^{53}Cr, ranging from ~0.4‰ to 1.6‰ (Fig. 6B) (Bonnand et al. 2013; Pereira et al. 2015; Scheiderich et al. 2015). In the Arctic region, surface seawater has higher δ^{53}Cr values but lower Cr concentrations than deep waters (Scheiderich et al. 2015). The δ^{53}Cr values of Arctic samples decrease sharply with depth in water < 500 m but remain almost unchanged at depths > 500 m (Fig. 6D), and the Cr concentration exhibits the opposite trend (Fig. 6C), providing evidence for Cr reduction in the surface water. However, in the lower-latitude Argentine Basin, the surface versus deep contrast is not apparent (Fig. 6C–D) (Bonnand et al. 2013).

An overall negative correlation between δ^{53}Cr and ln[Cr] was observed for global seawater samples including deep water samples (Rayleigh model) (Fig. 6B). The slope of the correlation line gives an overall ε value of -0.8 ± 0.03‰. This finding indicates that Cr reduction in global oceans occurs faster than mixing of water masses (i.e., water masses are partially closed systems over the timescale of reduction). As shown previously, there is great regional variability in Cr isotopes entering the oceans, likely caused by the nature of the river drainage systems (Scheiderich et al. 2015). However, this variability is likely largely removed once rivers mix with seawater. For example, the estuary sample from the Brahmani River yielded a value of 1.02 ± 0.11‰, but the δ^{53}Cr value dropped significantly (0.55 ± 0.08‰)

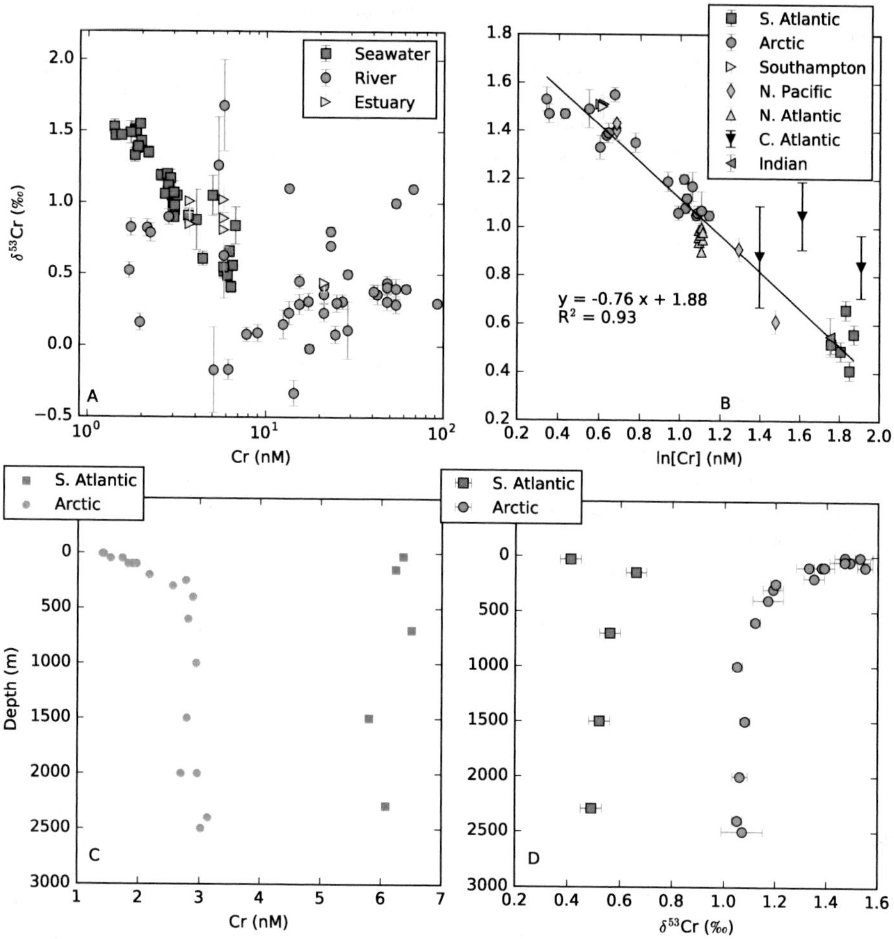

Figure 6. Comparison between seawater and river water δ^{53}Cr (A), detailed seawater δ^{53}Cr (B), depth profile of Cr concentration (C), and δ^{53}Cr (D). Some river or stream water samples have extremely high (> 1 ppm) Cr concentrations because they are from catchments of Cr-rich ultramafic bedrock (Farkaš et al. 2013, Paulukat et al. 2015). These samples are not plotted in panel A. The black line in panel B is the least square linear fit to all data in B except those from the Central Atlantic. Data sources: Bonnand et al. 2013; Farkas et al. 2013; Frei et al. 2014; Scheiderich et al. 2015; Pereira et al. 2015; Paulukat et al. 2015; D'Arcy et al. (2016).

in the surface water at the Bay of Bengal, while the Cr concentration remains unchanged (Paulukat et al. 2015). In contrast to surface seawater, currently limited deep seawater data seem to suggest a relatively homogeneous δ^{53}Cr value of 0.5–0.6‰ (Bonnand et al. 2013; Scheiderich et al. 2015). More deep seawater measurements will tell.

Cr isotope mass balance

Modern global Cr isotope mass balance has not been constrained previously. Reinhard et al. (2013a) presented an oceanic Cr mass balance for the first time, but Cr isotope variation was not incorporated. Reinhard et al. (2013a) reported that Cr in the ocean is supplied mainly by rivers, with an annual flux of 5.8×10^8 mol/yr, and is removed to oxic, anoxic, and reducing sediments with fluxes of 3.7×10^7 mol/yr, 5.8×10^7 mol/yr, and 4.8×10^8 mol/yr, respectively. Chromium sources and sinks are discussed below.

At the present time, rivers are thought to be the main sources of Cr to the ocean. McClain and Maher (2016) recently provided a new estimate of river Cr flux of 1.7×10^9 mol/year, which is approximately three times the estimate by Reinhard et al. (2013a). Hydrothermal input was estimated to be small based on heat flux and chemical anomalies in seafloor hydrothermal systems (Reinhard et al. 2013a). However, hydrothermal Cr input may be important in some local environments (Sander and Koschisky 2000). Because aeolian dust can only supply insoluble Cr to the ocean, it can be ignored. However, reflux of Cr from sediments to overlying seawater may be significant in some locations (Jeandel and Minster 1987).

The groundwater Cr flux to the ocean has not yet been constrained; below, we provide a first-order estimate. Groundwater Cr flux to the ocean can be estimated if the Cr concentration and the groundwater discharging flux are known. McClain and Maher (2016) estimated the annual river water flux to the ocean to be 3.74×10^4 km^3/yr. Moore (1996) estimated that groundwater discharge to coastal sea in the South Africa Bight region is approximately 40% river water discharge. If we extend Moore's result to global ocean and use a global river discharge of 3.73×10^4 km^3/yr (McClain and Maher 2016), the annual groundwater discharge to global ocean is ~1.49×10^{16} L/yr. The Cr concentration in groundwater discharging into the ocean has not been constrained, but we can make some guesses using published groundwater Cr concentrations in uncontaminated areas. For example, Izbicki et al. (2012) reported background groundwater total Cr concentrations of 0.3 ± 0.25 (1sd) nM in Mojave Desert, California, USA. If we use this value as a first-order estimate of the average total dissolved Cr concentration in global groundwater discharge, the groundwater Cr flux to the ocean is ~4.47×10^6 mol/yr, which is ~0.3% of the annual river input (McClain and Maher 2016).

The terms "oxic," "anoxic," and "reducing" are defined as follows: Oxic sediments are deposited in pelagic settings where biological productivity is low in the surface ocean, and the seafloor is therefore well oxygenated. Anoxic sediments are deposited in isolated basins (the Black Sea, the Cariaco Basin, and some fjords) or in continental shelf settings with strong upwelling, where intense oxygen minimum zones (OMZ) impinge on the seafloor (e.g., eastern South Pacific). Reducing sediments are deposited in areas where the biological productivity is relatively high but the OMZs do not touch the seafloor, so that the bottom water stays oxic, but the sediments below become anoxic (e.g., eastern Sub-Tropical North Pacific). Carbonate and Fe–Mn sediments were not considered owing to their low sedimentation rate and low Cr concentration.

With the new input flux estimate and the ocean Cr inventory of 5.48×10^{12} mol (Reinhard et al. 2013a), the residence time of Cr in the ocean is then ~3000 yr (seawater Cr inventory divided by riverine flux). This residence time is significantly lower than previous estimates of 9000–40000 yr (Van der Weijden and Reith 1982; Campbell and Yeats 1984; Reinhard et al. 2013a). This new estimate of ~3000 yr is close to the ocean mixing time of 1000 yr and thus may partly explain the observed heterogeneous Cr concentrations and δ^{53}Cr values (Bonnand et al. 2013; Pereira et al. 2015; Scheiderich et al. 2015).

Using the new global river flux of 1.7×10^9 mol/yr given by McClain and Maher (2016), and assuming the proportion of oxic-anoxic-reducing removal flux remain the same as those used by Reinhard et al. (2013), the new sink fluxes become 1.1×10^8 mol/yr, 1.7×10^8 mol/yr, and 1.42×10^9 mol/yr for oxic, anoxic, and reducing sediments, respectively. The seawater Cr mass balance is provided in Fig. 7.

Even though some seawater and marine sediments have been analyzed for their δ^{53}Cr in the past few years, it is not yet feasible to construct a global Cr isotope mass balance. However, recent developments will be reviewed here. The δ^{53}Cr values of oxic pelagic sediments have been characterized as -0.05 ± 0.10‰ (Gueguen et al. 2016), which is indistinguishable from the BSE value of -0.12 ± 0.10‰. The authigenic δ^{53}Cr values in modern anoxic sediments (0.38 ± 0.10‰) (Reinhard et al. 2014; Gueguen et al. 2016) appear to be consistent with the deep

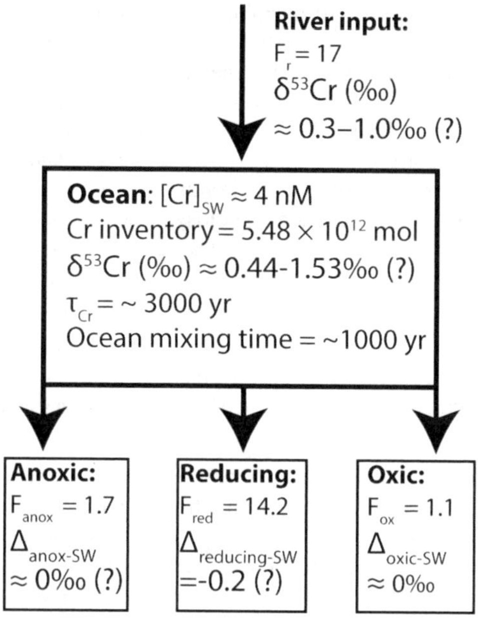

Figure 7. Mass balance of Cr in the modern ocean. Unit of flux (F) is 10^8 mole/yr. "SW" stands for seawater. The $\delta^{53}Cr$ values of river water, seawater, and reducing sediments are marked with question marks and need future research. See the text for definitions of anoxic (anox), reducing (red), and oxic (ox) sediments.

South Atlantic seawater value of ~0.5‰ (Bonnand et al. 2013), although more measurements are needed to test this hypothesis. In addition, authigenic Cr in sediments from the Peru Margin, where the OMZ impinges on the seafloor, yielded a $\delta^{53}Cr$ value of 0.61 ± 0.06‰ (Gueguen et al. 2016), which seems to be consistent with the single published Pacific deep water value (Scheiderich et al. 2015). In reducing settings where Cr(VI) needs to diffuse into the sediments to be reduced, the removal process is likely to fractionate Cr isotopes with a fractionation factor that depends on the diffusional length and reaction kinetics. However, there have been no Cr isotopic data published for these types of sediments, except for the few suboxic sediments reported by Schoenberg et al. (2008), which yielded $\delta^{53}Cr$ values indistinguishable from those of oxic sediments. In summary, more measurements on seawater and reducing sediments are needed to better constrain the currently sketchy global Cr isotope mass balance.

The Cr isotope system as a paleo-redox proxy

How it works. Because of its redox sensitivity, the Cr isotopic system has been used to track the redox evolution of Earth's ocean–atmosphere. The most important application of the Cr isotopic system in this field is tracking the onset of oxygenic photosynthesis, one of the most important biological innovations to have occurred on Earth.

Using $\delta^{53}Cr$ to track Earth's initial oxygenation is relatively relies on the following assumptions: variation of $\delta^{53}Cr$ in sedimentary rocks requires a mobile Cr(VI) pool. Generation of Cr(VI) requires the formation of manganese oxides, the formation of which requires free O_2 (Fendorf and Zasoski 1992; Frei et al. 2009; Crowe et al. 2013; Planavsky et al. 2014; Reinhard et al. 2014). Furthermore, for Cr(VI) to be delivered to the ocean, the surface environment needs to be Fe(II)-free (Crowe et al. 2013). This condition requires atmospheric oxygen levels to be ~3×10^{-4} higher than the present atmospheric level, or PAL (Crowe et al.

2013). This level of O_2 is 7–8 orders of magnitude higher than that obtainable on a prebiotic planet (Liang et al. 2006; Ettwig et al. 2010; Haqq-Misra et al. 2011; Mahaffy et al. 2013). Once the isotopically heavy soluble Cr(VI) is generated and delivered to the ocean, it becomes possible for ocean sedimentary δ^{53}Cr values (which record that of the seawater) to vary outside of the bulk silicate earth range (BSE = −0.12 ± 0.10‰, Schoenberg et al. 2008). In other words, the point at which sedimentary δ^{53}Cr values started to vary outside of the BSE range can be taken as the onset of oxidized surface environments, which serves as a minimum time estimate for the first emergence of oxygenic photosynthesis. This approach is not adversely affected by the heterogeneity of seawater δ^{53}Cr because it is the variation of δ^{53}Cr values, not the absolute values, that is important. However, the heterogeneity issue may complicate the endeavor of tracking the absolute values of seawater δ^{53}Cr (e.g., during the Phanerozoic).

Geological archives. Successful reconstruction of the variation of the δ^{53}Cr in past seawater relies on robust geological archives. Potential archives include iron formations, organic-rich shales, and carbonates. Each archive has its advantages and disadvantages, which are discussed below.

Iron formations are chemical sediments that directly precipitate from seawater and thus can potentially directly track seawater δ^{53}Cr with little terrigenous contamination. Banded iron formations (BIFs) form when soluble Fe(II) is converted to insoluble Fe(III), during which soluble Cr(VI) is reduced to insoluble Cr(III), which then coprecipitates with the formed Fe(III) oxides. The kinetics of the Cr(VI)–Fe(II) reaction are fast (e.g., Buerge and Hug 1997), such that Cr(VI) is quantitatively sequestered, leading to no Cr isotopic fractionation. Based on this framework, Frei et al. (2009) proposed that BIFs can faithfully record ambient seawater δ^{53}Cr.

However, there are two types of iron formations based on their depositional settings: Algoma-type and Superior-type. Algoma-type iron formations are associated with submarine volcanism, whereas Superior-type iron formations are deposited in passive-margin environments and thus are typically not directly linked to submarine hydrothermal systems (Bekker et al. 2010). Therefore, Superior-type iron formations are more likely to record open-ocean δ^{53}Cr, whereas Algoma-type iron formations may only record local hydrothermal signals.

Black shales are another promising archive for reconstructing seawater δ^{53}Cr. During deposition of black shales, the sediment pore water and possibly the overlying water column as well are anoxic and contain reductants such as hydrogen sulfide or ferrous iron. If the water column is anoxic, the reduction is nearly quantitative because of fast reduction kinetics (Buerge and Hug 1997; Kim et al. 2001). Therefore, authigenic Cr in black shales deposited below anoxic water columns can potentially record seawater δ^{53}Cr values (Reinhard et al. 2014). Wang et al. (2016a) recently observed a negative δ^{53}Cr excursion during Cretaceous Ocean Anoxic Event II at Demerara Rise Site 1258. The negative excursion is coincident with a positive δ^{13}C excursion and Cr–Mo depletion. These findings suggest that the Cr isotopic system is a promising redox proxy that can be used to study not only Archean oxygenation but also Phanerozoic marine redox perturbations.

If the bottom water is oxic but the pore water is anoxic, the dissolved Cr(VI) in seawater needs to diffuse into the reducing zones to be reduced. This diffusion–reduction process induces Cr isotopic fractionation, leading to authigenic Cr that is isotopically offset from seawater δ^{53}Cr. The size of the fractionation, however, is smaller than the intrinsic fractionation because of the aforementioned reservoir effects. The size of the effective fractionation depends on the diffusion length, which is typically related to the oxygen penetration depth (Bender 1990; Clark and Johnson 2008; Reinhard et al. 2014). Therefore, when interpreting δ^{53}Cr from black shales, it is necessary to consider the bottom water redox state based on independent proxies (e.g., biomarkers, iron speciation, and multi-trace element approaches (Grice et al. 2005; Zhang et al. 2016).

Detrital contamination. Because of the high concentration of detrital particles in black shales, the measured $\delta^{53}Cr$ is a mixture of detrital and authigenic components. However, it is the authigenic $\delta^{53}Cr$ that has paleo-redox significance. To determine the authigenic $\delta^{53}Cr$, the detrital contribution needs to be corrected for. There are generally two ways to make this correction: Detrital correction and selective leaching, as discussed below.

Measured bulk $\delta^{53}Cr$ values can be corrected for detrital components using the equation

$$\delta^{53}Cr_{auth} = \left(\delta^{53}Cr_{bulk} - \delta^{53}Cr_{det}\left(1 - f_{auth}\right)\right) / f_{auth} \tag{16}$$

where $\delta^{53}Cr_{auth}$, $\delta^{53}Cr_{bulk}$, and $\delta^{53}Cr_{det}$ refer to the Cr isotopic compositions for the authigenic, bulk, and detrital components, respectively. $\delta^{53}Cr_{det}$ can be assumed to be the BSE value of $-0.05 \pm 0.1‰$, based on measurement of pelagic sediments deposited in oxic settings (Schoenberg et al. 2008; Gueguen et al. 2016). f_{auth} refers to the fraction of authigenic Cr and can be calculated according to

$$f_{auth} = \left([Cr]_{bulk} - [Ti]_{bulk}(Cr/Ti)_{det}\right) / [Cr]_{bulk} \tag{17}$$

where $[Cr]_{bulk}$, $[Ti]_{bulk}$, and $(Cr/Ti)_{det}$ correspond to the authigenic Cr concentration, the bulk Cr concentration, the bulk Ti concentration, and the Cr/Ti ratio of the detrital sediments. Ti is typically considered to be an immobile element during weathering; Al can serve the same purpose. Therefore, normalizing Cr to Ti or Al avoids potential complications caused by varying sedimentation rates and dilutions by biogenic components (see Tribovillard et al. 2006). $(Cr/Ti)_{det}$ can be taken as the upper continental crust value: 0.016–0.024 (Condie 1993; McLennan 2001; Rudnick and Gao 2003). Note that $(Cr/Ti)_{det}$ may vary slightly with geography because of different sediment provenances. For this reason, the $(Cr/Ti)_{det}$ value used for detrital correction should ideally be derived from sediments deposited under oxic conditions at the same location.

The second method that can be used to minimize detrital contamination is to selectively leach out the authigenic Cr while leaving the detrital Cr intact. Reinhard et al. (2014) showed that HNO_3 (0.6–6 N) can effectively dissolve the authigenic Cr selectively without detectable isotopic fractionation, and the leaching method yielded the same authigenic $\delta^{53}Cr$ as Cr/Ti-based detrital correction.

Although iron formations and black shales are two promising archives for seawater $\delta^{53}Cr$, they are very rare in the geological history. To understand seawater redox evolution on more continuous geological timescales, more geologically common archives are needed, and marine carbonates are good candidates. However, Cr(VI) uptake mechanisms and Cr isotopic fractionation during carbonate precipitation have been explored only recently (Tang et al. 2007; Bonnand et al. 2011; Frei et al. 2011; Rodler et al. 2015). During inorganic carbonate precipitation, Cr(VI) is incorporated into the crystal lattice (Tang et al. 2007) with up to ~0.3‰ positive isotopic fractionation (Rodler et al. 2015). If these experimental results can be applied to the natural world, inorganic carbonate seems to be a promising archive for seawater $\delta^{53}Cr$.

However, most marine carbonate precipitation involves biological activities, which can induce complicated Cr isotopic fractionation. For instance, coral aragonite has been found to fractionate Cr isotopes, ranging from −0.5 to +0.3‰, depending on the species. The large negative fractionation that occurs during biological carbonate precipitation is likely linked to redox transformations during Cr uptake from seawater (Pereira et al. 2015). In addition, foraminiferal calcification also appears to induce species-dependent Cr isotopic fractionation (Wang et al. 2016b). In summary, until better understanding of Cr isotopic fractionation during biological carbonate precipitation is achieved, marine carbonates cannot be considered to be as a straightforward archive for seawater $\delta^{53}Cr$. One future direction is to explore microbially influenced carbonate as a potential archive for Precambrian seawater $\delta^{53}Cr$.

A sedimentary* δ^{53}*Cr record. The available δ^{53}Cr data from iron formations, shales, carbonates, and paleosols are compiled in Figure 8. The record can be interpreted in terms of the redox evolution of Earth's surface environment. Extensive evidence has shown ≤4.9‰ δ^{53}Cr variation in sedimentary rocks deposited back to the Neoproterozoic (Frei et al. 2009; Planavsky et al. 2014; Reinhard et al. 2014), suggesting highly oxic surface environments during this time. The amount of fractionation is comparable to that observed in modern seawater and river water samples and soil profiles (Bonnand et al. 2013; Farkaš et al. 2013; Berger and Frei 2014; Frei et al. 2014; Novak et al. 2014; Paulukat et al. 2015; Pereira et al. 2015; Scheiderich et al. 2015).

Gilleaudeau et al. (2016) recently reported some fractionated δ^{53}Cr in ~1.1 Ga carbonate rocks. If these data are primary signals, they indicate that the atmospheric oxygen level was higher than 10^{-4}–10^{-3} PAL and was close to the estimated minimum O_2 requirement (10^{-3} PAL) for multi-cellular animals (Sperling et al. 2013; Mills et al. 2014) approximately 300 million years before the first estimated onset of sponges (Erwin et al. 2011). However, because carbonate rocks are prone to digenetic alteration and Cr levels are generally very low (<1 ppm), their Cr isotopic composition should be accepted with caution.

Moderate δ^{53}Cr variation can also be observed, at least sporadically if not continuously, in sedimentary rocks deposited from the early Paleoproterozoic to the Eoarchean (~1.8 Ga to ~3.8 Ga), suggesting mildly oxic surface environments during this time period. δ^{53}Cr fractionation was observed not only in sedimentary rocks but also in paleosols that were in direct contact with the atmosphere, interpreted as relatively high atmospheric oxygen levels (10^{-4}–10^{-3} PAL; Crowe et al. 2013) in the atmosphere as early as 3.0 Ga.

Black shales and iron-rich rocks deposited before 3.0 Ga and between 1.8 Ga and 1.1 Ga are generally unfractionated from the BSE range, indicating the absence, or very limited presence,

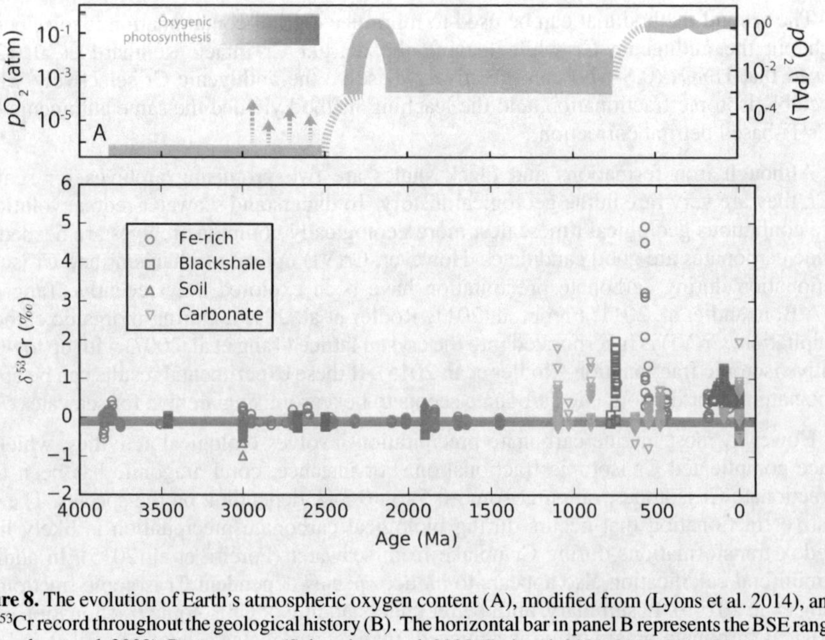

Figure 8. The evolution of Earth's atmospheric oxygen content (A), modified from (Lyons et al. 2014), and the δ^{53}Cr record throughout the geological history (B). The horizontal bar in panel B represents the BSE range (Schoenberg et al. 2008). Data sources: (Schoenberg et al. 2008; Frei et al. 2009, 2011, 2014, 2016; Bonnand et al. 2011, 2013; Crowe et al. 2013; Farkaš et al. 2013; Frei and Polat 2013; Wille et al. 2013; Berger and Frei 2014; Planavsky et al. 2014; Reinhard et al. 2014; Pereira et al. 2015; Rodler et al. 2015; Scheiderich et al. 2015; Cole et al. 2016; Gueguen et al. 2016; Gilleaudeau et al. 2016; Holmden et al. 2016; Wang et al. 2016c).

of a mobile Cr(VI) pool delivered to the ocean, and by inference very low atmospheric O_2 during these periods. The exception is some Fe-rich rocks from the highly metamorphosed Isua Greenstone Belt (Frei et al. 2016), which showed small δ^{53}Cr variations (up to +0.4‰). However, whether these values are primary signals or metamorphic overprint requires further investigation.

The increasing variation of sedimentary δ^{53}Cr from the Archean to the present is generally consistent with records of redox-sensitive detrital grains (e.g., pyrite and uraninite) and red beds (Grandstaff 1980; Holland 1984; Rasmussen and Buick 1999) and mass-independent sulfur isotopes (Farquhar et al. 2000; Pavlov and Kasting 2002; Reinhard et al. 2013b). However, the Cr isotope record provides at least two new insights: (1) Oxygenic photosynthesis likely evolved hundreds of millions of years before the great oxidation event (GOE; ~2.4 Ga), which has profound implications for future phylogenetic studies of the evolution of life on the early Earth, and (2) the unfractionated δ^{53}Cr between 1.8 Ga and 0.8–1.1 Ga indicates that approximately 700 million years after the GOE, atmospheric O_2 seemed to have returned to such low levels that the delivery of mobile Cr(VI) to the ocean was shut off. The low O_2 levels during this period may have contributed to the delayed rise of animal life (Planavsky et al. 2014). Importantly we note that the current interpretation for the Cr isotopic data in sedimentary rocks is based on the assumption of oxidative immobilization of Cr triggered by the rise of AOL. The possibility of Cr (III) immobilization by organic acids generated by bacteria has not been fully explored (Konhauser et al. 2011).

CONCLUDING REMARKS AND OUTLOOK

The chromium isotope system is a very useful redox proxy that can be used to monitor the attenuation of Cr(VI) contamination in groundwater systems, to reconstruct the redox evolution of Earth's surface environment, and to investigate planetary formation, magmatic differentiation and crust–mantle cycling. This chapter has reviewed theoretical and experimental work that provided Cr isotopic fractionation factors in various biological and geochemical processes; these factors are needed for interpreting natural sample data. This review also evaluated various geological archives in terms of their fidelity in recording seawater δ^{53}Cr, and it discussed the existing sedimentary δ^{53}Cr record in terms of Earth's oxygen evolution.

Although an increasing number of studies in the past decade or so have shown that the Cr isotopic system could be a very useful geochemical tracer, many problems remain. First, better understanding of Cr isotopic fractionation in multi-step microbial Cr reduction pathways is needed to better estimate the effective isotopic fractionation factor in the field, and thus allow better use of Rayleigh fractionation models to estimate the extent of Cr(VI) reduction. Second, we need to better understand the heterogeneity of seawater δ^{53}Cr in modern seawater and the factors that control such heterogeneity. This information is important because it helps in gauging the spatial significance of paleocean redox reconstruction. Third, carbonate sedimentary archives could potentially play a significant role in applying the Cr isotopic paleo-redox proxy to broad geological timescales. However, studies to date suggest that foraminiferal calcite and coral aragonite precipitation induce large (likely negative) and species-dependent isotopic fractionation. Therefore, inter-species calibration is needed to better understand the fidelity of carbonate archives in recording seawater δ^{53}Cr. In addition, microbially mediated marine carbonate precipitation (e.g., ooids) could be explored as potential seawater δ^{53}Cr archives. Finally, redox transformation has been assumed to be the major contributor to large Cr isotopic variation, but this assumption is based on theoretical considerations and on limited Cr(VI) adsorption experiments. Potential isotopic fractionations between same-valence species have received insufficient attention. Processes such as ligand solubilization of Cr(III) oxyhydroxides, isotope exchange between same-valence species complexed with different ligands, and Cr(III) precipitation with carbonate and iron oxides are all possible mechanisms to investigate. Although the magnitude of isotopic fractionation during non-redox processes is expected to be limited, natural Cr isotopic variations are also small in some systems such as Archean sediments and magmatic fluids. In these systems, non-redox-related Cr isotopic fractionation could be relevant.

Furthermore, Cr isotopic behavior during magmatic processes needs further characterization. For example, small Cr isotopic variations have been suggested to occur during magmatic differentiation. However, the governing process has not been unambiguously identified. In addition, large Cr isotopic variation has been observed in serpentinites, but the specific mechanisms responsible for such variation continue to be debated. The sharp contrast in the Cr isotopic variability between serpentinites and deep-subduction-related mafic rocks also warrants further investigation because it could shed light on the dehydration processes that occur during subduction and during material recycling between crust and mantle.

ACKNOWLEDGEMENT

This work was supported by the Strategic Priority Research Program (B) of Chinese Academy of Sciences (Grant No. XDB18010203), the National Nature Science Foundation of China (41273076, 41473066, 41571130052), the"111" Project, and Fundamental Research Funds for the Central Universities. XLW would like to thank the Agouron Institute for financial support.

REFERENCES

Ackerley D, Gonzalez C, Park C, Blake R, Keyhan M, Matin A (2004) Chromate-reducing properties of soluble flavoproteins from Pseudomonas putida and Escherichia coli. Appl Environ Microbiol 70:873–882

Alowitz MJ, Scherer MM (2002) Kinetics of nitrate, nitrite, and Cr(VI) reduction by iron metal. Environ Sci Technol 36:299–306

Altman C, King EL (1961) The mechanism of the exchange of chromium(III) and chromium(VI) in acidic solutions. J Am Chem Soc 83:2825–2830

Ardon M, Plane RA (1959) The formation of a dinuclear Cr(III) species by oxidation of chromous solutions 1. J Am Chem Soc 81:3197–3200

Bain DJ, Bullen TD (2005) Chromium isotope fractionation during oxidation of Cr(III) by manganese oxides. Geochim Cosmochim Acta 69:S212

Ball J, Bassett R (2000) Ion exchange separation of chromium from natural water matrix for stable isotope mass spectrometric analysis. Chem Geol 168:123–134

Ball JW, Mccluskey RB (2003) A new cation-exchange method for accurate field speciation of hexavalent chromium. Talanta 61:305–313

Banerjee D, Nesbitt H (1999) Oxidation of aqueous Cr(III) at birnessite surfaces: Constraints on reaction mechanism. Geochim Cosmochim Acta 63:1671–1687

Basu A, Johnson TM (2012) Determination of hexavalent chromium reduction using Cr stable isotopes: Isotopic fractionation factors for permeable reactive barrier materials. Environ Sci Technol 46:5353–5360

Basu A, Johnson TM, Sanford RA (2014) Cr isotope fractionation factors for Cr(VI) reduction by a metabolically diverse group of bacteria. Geochim Cosmochim Acta 142:349–361

Bekker A, Holland H, Wang P-L, Rumble D, Stein H, Hannah J, Coetzee L, Beukes N (2004) Dating the rise of atmospheric oxygen. Nature 427:117–120

Bekker A, Slack JF, Planavsky N, Krapež B, Hofmann A, Konhauser KO, Rouxel OJ (2010) Iron formation: The sedimentary product of a complex interplay among mantle, tectonic, oceanic, and biospheric processes. Econ Geol 105:467–508

Bell A, Burger P, Le L, Shearer C, Papike J, Sutton S, Newville M, Jones J (2014) XANES measurements of cr valence in olivine and their applications to planetary basalts. Am Mineral 99:1404–1412

Bender ML (1990) The $\delta^{18}O$ of dissolved O_2 in seawater: A unique tracer of circulation and respiration in the deep sea. J Geophys Res Oceans 95:22243–22252

Berger A, Frei R (2014) The fate of chromium during tropical weathering: A laterite profile from Central Madagascar. Geoderma 213:521–532

Berna EC, Johnson TM, Makdisi RS, Basu A (2010) Cr stable isotopes as indicators of Cr(VI) reduction in groundwater: a detailed time-series study of a point-source plume. Environ Sci Technol 44:1043–1048

Berry AJ, O'Neill HSC (2004) A XANES determination of the oxidation state of chromium in silicate glasses. Am Mineral 89:790–798

Berry AJ, O'Neill HSC, Scott DR, Foran GJ, Shelley J (2006) The effect of composition on Cr^{2+}/Cr^{3+} in silicate melts. Am Mineral 91:1901–1908

Bigeleisen J, Mayer MG (1947) Calculation of equilibrium constants for isotopic exchange reactions. J Chem Phys 15:261–267

Birck J-L, Allègre CJ (1984) Chromium isotopic anomalies in Allende refractory inclusions. Geophys Res Lett 11:943–946

Birck J-L, Allègre CJ (1985) Evidence for the presence of ^{53}Mn in the early solar system. Geophys Res Lett 12

Birck J-L, Allègre CJ (1988) Manganese chromium isotope systematics and development of the early solar system. Nature 331:579–584

Blowes DW, Ptacek CJ, Benner SG, McRae CW, Bennett TA, Puls RW (2000) Treatment of inorganic contaminants using permeable reactive barriers. J Contam Hydrol 45:123–137

Bonnand P, Parkinson IJ, James RH, Karjalainen A-M, Fehr MA (2011) Accurate and precise determination of stable Cr isotope compositions in carbonates by double spike MC–ICP–MS. J Anal At Spectrom 26:528–535

Bonnand P, James R, Parkinson I, Connelly D, Fairchild I (2013) The chromium isotopic composition of seawater and marine carbonates. Earth Planet Sci Lett 382:10–20

Bonnand P, Williams HM, Parkinson IJ, Wood BJ, Halliday AN (2016) Stable chromium isotopic composition of meteorites and metal–silicate experiments: Implications for fractionation during core formation. Earth Planet Sci Lett 435:14–21

Brandes JA, Devol AH (1997) Isotopic fractionation of oxygen and nitrogen in coastal marine sediments. Geochim Cosmochim Acta 61:1793–1801

Brennecka GA, Wadhwa M (2012) Uranium isotope compositions of the basaltic angrite meteorites and the chronological implications for the early Solar System. PNAS 109:9299–9303

Broecker W, Peng T (1982) Tracers in the Sea. Lamont-Doherty Geol. Obs, Palisades, NY, USA

Brunner B, Bernasconi SM (2005) A revised isotope fractionation model for dissimilatory sulfate reduction in sulfate reducing bacteria. Geochim Cosmochim Acta 69:4759–4771

Buerge IJ, Hug SJ (1997) Kinetics and pH dependence of chromium (VI) reduction by iron (II). Environ Sci Technol 31:1426–1432

Burns RG (1975) On the occurrence and stability of divalent chromium in olivines included in diamonds. Contrib Mineral Petrol 51:213–221

Campbell JA, Yeats PA (1984) Dissolved chromium in the St. Lawrence estuary. Estuar Coast Shelf S 19:513–522

Clark SK, Johnson TM (2008) Effective isotopic fractionation factors for solute removal by reactive sediments: A laboratory microcosm and slurry study. Environ Sci Technol 42:7850–7855

Cole DB, Reinhard CT, Wang X, Gueguen B, Halverson GP, Gibson T, Hodgskiss MS, McKenzie NR, Lyons TW, Planavsky NJ (2016) A shale-hosted Cr isotope record of low atmospheric oxygen during the Proterozoic. Geology 44:555–558

Comber S, Gardner M (2003) Chromium redox speciation in natural waters. J Environ Monit 5:410–413

Condie KC (1993) Chemical composition and evolution of the upper continental crust: contrasting results from surface samples and shales. Chem Geol 104:1–37

Crane R, Scott T (2012) Nanoscale zero-valent iron: future prospects for an emerging water treatment technology. J HAZARD MATER J Hazard Mater 211:112–125

Cranston R, Murray J (1980) Chromium species in the Columbia River and estuary. Limnol Oceanogr 25:1104–1112

Crowe SA, Døssing LN, Beukes NJ, Bau M, Kruger SJ, Frei R, Canfield DE (2013) Atmospheric oxygenation three billion years ago. Nature 501:535–538

D'Arcy J, Babechuk MG, Døssing LN, Gaucher C, Frei R (2016) Processes controlling the chromium isotopic composition of river water: Constrains from basaltic river catchments. Geochim Cosmochim Acta 186:296–315

Dauphas N, Remusat L, Chen JH, Roskosz M, Papanastassiou DA, Stodolna J, Guan Y, Ma C, Eiler JM (2010) Neutron-rich chromium isotope anomalies in supernova nanoparticles. Astrophys J 720:1577–1591

Dauphas N, Chen JH, Zhang J, Papanastassiou DA, Davis AM, Travaglio C (2014) Calcium–48 isotopic anomalies in bulk chondrites and achondrites: Evidence for a uniform isotopic reservoir in the inner protoplanetary disk. Earth Planet Sci Lett 407:96–108

Dodson MH (1970) Simplified equations for double-spiked isotonic analyses: Geochim Cosmochim Acta 34:1241–1244

Døssing LN, Dideriksen K, Stipp SLS, Frei R (2011) Reduction of hexavalent chromium by ferrous iron: A process of chromium isotope fractionation and its relevance to natural environments. Chem Geol 285:157–166

Economou-Eliopoulos M, Frei R, Atsarou C (2014) Application of chromium stable isotopes to the evaluation of Cr(VI) contamination in groundwater and rock leachates from central Euboea and the Assopos basin (Greece). Catena 122:216–228

Eeckhout SG, Bolfan-Casanova N, McCammon C, Klemme S, Amiguet E (2007) XANES study of the oxidation state of Cr in lower mantle phases: Periclase and magnesium silicate perovskite. Am Mineral 92:966–972

Elderfield H, Bertram C, Erez J (1996) A biomineralization model for the incorporation of trace elements into foraminiferal calcium carbonate. Earth Planet Sci Lett 142:409–423

Ellis AS, Johnson TM, Bullen TD (2002) Chromium isotopes and the fate of hexavalent chromium in the environment. Science 295:2060

Ellis AS, Johnson TM, Bullen TD (2004) Using chromium stable isotope ratios to quantify Cr(VI) reduction: lack of sorption effects. Environ Sci Technol 38:3604–3607

Ellis AS, Johnson TM, Villalobos-Aragon A, Bullen T (2008) Environmental cycling of Cr using stable isotopes: kinetic and equilibrium effects. 2008 Fall Meeting, AGU, San Francisco, Calif 15–19 Dec, abstract #H53F–08

Erez J (2003) The source of ions for biomineralization in foraminifera and their implications for paleoceanographic proxies. Rev Mineral Geochem 54:115–149

Erwin DH, Laflamme M, Tweedt SM, Sperling EA, Pisani D, Peterson KJ (2011) The Cambrian conundrum: early divergence and later ecological success in the early history of animals. Science 334:1091–1097

Ettwig KF, Butler MK, Le Paslier D, Pelletier E, Mangenot S, Kuypers MM, Schreiber F, Dutilh BE, Zedelius J, De Beer D (2010) Nitrite-driven anaerobic methane oxidation by oxygenic bacteria. Nature 464:543–548

Farkaš J, Chrastny V, Novak M, Cadkova E, Pasava J, Chakrabarti R, Jacobsen SB, Ackerman L, Bullen TD (2013) Chromium isotope variations ($\delta^{53/52}$Cr) in mantle-derived sources and their weathering products: Implications for environmental studies and the evolution of $\delta^{53/52}$Cr in the Earth's mantle over geologic time. Geochim Cosmochim Acta 123:74–92

Farquhar J, Bao H, Thiemens MH (2000) Atmospheric influence of Earth's earliest sulfur cycle. Science 289:756–758

Farquhar J, Johnston DT, Wing BA, Habicht KS, Canfield DE, Airieau S, Thiemens MH (2003) Multiple sulphur isotopic interpretations of biosynthetic pathways: implications for biological signatures in the sulphur isotope record. Geobiology 1: 27–36

Faure G (1991) Principles and Applications of Geochemistry. Wiley, New York

Fendorf SE, Li G (1996) Kinetics of chromate reduction by ferrous iron. Environ Sci Technol 30:1614–1617

Fendorf SE, Zasoski RJ (1992) Chromium (III) Oxidation by δ-MnO_2. Environ Sci Technol 26:79–85

Frei R, Gaucher C, Poulton SW, Canfield DE (2009) Fluctuations in Precambrian atmospheric oxygenation recorded by chromium isotopes. Nature 461:250–253

Frei R, Gaucher C, Døssing LN, Sial AN (2011) Chromium isotopes in carbonates—a tracer for climate change and for reconstructing the redox state of ancient seawater. Earth Planet Sci Lett 312:114–125

Frei R, Polat A (2013) Chromium isotope fractionation during oxidative weathering-implications from the study of a Paleoproterozoic (ca. 1.9 Ga) paleosol, Schreiber Beach, Ontario, Canada. Precambrian Res 224:434–453

Frei R, Poiré D, Frei KM (2014) Weathering on land and transport of chromium to the ocean in a subtropical region (Misiones, NW Argentina): A chromium stable isotope perspective. Chem Geol 381:110–124

Frei R, Crowe SA, Bau M, Polat A, Fowle DA, Døssing LN (2016) Oxidative elemental cycling under the low O_2 Eoarchean atmosphere. Sci Rep 6:21058

Gilleaudeau G, Frei R, Kaufman A, Kah L, Azmy K, Bartley J, Chernyavskiy P, Knoll A (2016) Oxygenation of the mid-Proterozoic atmosphere: clues from chromium isotopes in carbonates. Geochem Persp Lett 2:178–187

Glavin DP, Kubny A, Jagoutz E, Lugmair GW (2004) Mn–Cr isotope systematics of the D'Orbigny angrite. Meteor Planet Sci 39:693–700

Graham AM, Bouwer EJ (2009) Rates of hexavalent chromium reduction in anoxic estuarine sediments: pH effects and the role of acid volatile sulfides. Environ Sci Technol 44:136–142

Grandstaff DE (1980) Origin of uraniferous conglomerates at Elliot Lake, Canada and Witwatersrand, South Africa: implications for oxygen in the Precambrian atmosphere. Precambrian Res 13:1–26

Grice K, Cao C, Love GD, Böttcher ME, Twitchett RJ, Grosjean E, Summons RE, Turgeon SC, Dunning W, Jin Y (2005) Photic zone euxinia during the Permian-Triassic superanoxic event. Science 307:706–709

Gueguen B, Reinhard CT, Algeo TJ, Peterson LC, Nielsen SG, Wang XL, Rowe H, Planavsky NJ (2016) The chromium isotope composition of reducing and oxic marine sediments. Geochim Cosmochim Acta 184:1–19

Halicz L, Yang L, Teplyakov N, Burg A, Sturgeon R, Kolodny Y (2008) High precision determination of chromium isotope ratios in geological samples by MC-ICP-MS. J Anal Atomic Spec 23:1622–1627

Han R, Qin L, Brown ST, Christensen JN, Beller HR (2012) Differential isotopic fractionation during Cr(VI) reduction by an aquifer-derived bacterium under aerobic versus denitrifying conditions. Appl Environ Microbiol 78:2462–2464

Haqq-Misra J, Kasting JF, Lee S (2011) Availability of O_2 and H_2O_2 on pre-photosynthetic Earth. Astrobiology 11:293–302

Heikoop JM, Johnson TM, Birdsell KH, Longmire P, Hickmott DD, Jacobs EP, Broxton DE, Katzman D, Vesselinov VV, Ding M, Vaniman DT (2014) Isotopic evidence for reduction of anthropogenic hexavalent chromium in Los Alamos National Laboratory groundwater. Chem Geol 373:1–9

Holland HD (1984) The chemical evolution of the atmosphere and oceans. Princeton University Press

Holmden C, Jacobson A, Sageman B, Hurtgen M (2016) Response of the Cr isotope proxy to Cretaceous Ocean Anoxic Event 2 in a pelagic carbonate succession from the Western Interior Seaway. Geochim Cosmochim Acta 186:277–295

Honda M, Imamura M (1971) Half-life of ^{53}Mn. Phys Rev C4:1182–1188

Impert O, Katafias A, Kita P, Woroniecka M (2008) Kinetics of chromate (III) oxidation by hydrogen peroxide in alkaline solutions-revisited. Pol J Chem 82:1121–1126

Izbicki JA, Ball JW, Bullen TD, Sutley SJ (2008) Chromium, chromium isotopes and selected trace elements, western Mojave Desert, USA. Appl Geochem 23:1325–1352

Izbicki JA, Bullen TD, Martin P, Schroth B (2012) Delta chromium–53/52 isotopic composition of native and contaminated groundwater, Mojave Desert, USA. Appl Geochem 27:841–853

Jamieson-Hanes JH, Amos RT, Blowes DW (2012a) Reactive transport modeling of chromium isotope fractionation during Cr(VI) reduction. Environ Sci Technol 46:13311–13316

Jamieson-Hanes JH, Gibson BD, Lindsay MB, Kim Y, Ptacek CJ, Blowes DW (2012b) Chromium isotope fractionation during reduction of Cr(VI) under saturated flow conditions. Environ Sci Technol 46:6783–6789

Jamieson-Hanes JH, Lentz AM, Amos RT, Ptacek CJ, Blowes DW (2014) Examination of Cr(VI) treatment by zero-valent iron using in situ, real-time X-ray absorption spectroscopy and Cr isotope measurements. Geochim Cosmochim Acta 142:299–313

Jeandel C, Minster J (1987) Chromium behavior in the ocean: Global versus regional processes. Global Biogeochem Cycles 1:131–154

Johnson TM, Bullen TD (2004) Mass-dependent fractionation of selenium and chromium isotopes in low-temperature environments. Rev Mineral Geochem 55:289–317

Johnson CM, Skulan JL, Beard BL, Sun H, Nealson KH, Braterman PS (2002) Isotopic fractionation between Fe(III) and Fe(II) in aqueous solutions. Earth Planet Sci Lett 195:141–153

Kim C, Zhou Q, Deng B, Thornton EC, Xu H (2001) Chromium(VI) reduction by hydrogen sulfide in aqueous media: stoichiometry and kinetics. Environ Sci Technol 35:2219–2225

Kitchen JW, Johnson TM, Bullen TD, Zhu J, Raddatz A (2012) Chromium isotope fractionation factors for reduction of Cr(VI) by aqueous Fe(II) and organic molecules. Geochim Cosmochim Acta 89:190–201

Klein-BenDavid O, Logvinova AM, Schrauder M, Spetius ZV, Weiss Y, Hauri EH, Kaminsky FV, Sobolev NV, Navon O (2009) High-Mg carbonatitic microinclusions in some Yakutian diamonds—a new type of diamond-forming fluid. Lithos 112, Supplement 2:648–659

Klein-BenDavid O, Pettke T, Kessel R (2011) Chromium mobility in hydrous fluids at upper mantle conditions. Lithos 125:122–130

Knoblowitz M, Morrow JI (1976) Kinetic study of an intermediate present in the hydrogen peroxide oxidation of chromium (III) to chromium (VI). Inorg Chem 15:1674–1677

Konhauser KO, Lalonde SV, Planavsky NJ, Pecoits E, Lyons TW, Mojzsis SJ, Rouxel OJ, Barley ME, Rosìere C, Fralick PW, Kump LR (2011) Aerobic bacterial pyrite oxidation and acid rock drainage during the Great Oxidation Event. Nature 478:369–373

Larsen K, Wielandt D, Schiller M, Bizzarro M (2016) Chromatographic speciation of Cr (III)-species, inter-species equilibrium isotope fractionation and improved chemical purification strategies for high-precision isotope analysis. J Chromatogr A 1443:162–174

Lau KV, Maher K, Altiner D, Kelley BM, Kump LR, Lehrmann DJ, Silva-Tamayo JC, Weaver KL, Yu M, Payne JL (2016) Marine anoxia and delayed Earth system recovery after the end-Permian extinction. PNAS 113:2360–2365

Lee T, Lim H, Lee Y, Park J-W (2003) Use of waste iron metal for removal of Cr(VI) from water. Chemosphere 53:479–485

Liang M-C, Hartman H, Kopp RE, Kirschvink JL, Yung YL (2006) Production of hydrogen peroxide in the atmosphere of a Snowball Earth and the origin of oxygenic photosynthesis. PNAS 103:18896–18899

Lodders K (2003) Solar System abundances and condensation temperatures of the elements. Astrophys J 591:1220–1247

Lugmair GW, Shukolyukov A (1998) Early solar system timescales according to ^{53}Mn–^{53}Cr systematics. Geochim Cosmochim Acta 62:2863–2886

Lyons TW, Reinhard CT, Planavsky NJ (2014) The rise of oxygen in Earth's early ocean and atmosphere. Nature 506:307–315

Mahaffy PR, Webster CR, Atreya SK, Franz H, Wong M, Conrad PG, Harpold D, Jones JJ, Leshin LA, Manning H (2013) Abundance and isotopic composition of gases in the martian atmosphere from the Curiosity rover. Science 341:263–266

Maréchal (1999) Precise analysis of copper and zinc isotopic compositions by plasma-source mass spectrometry. Chem Geol 156:251–273

McClain C, Maher K (2016) Chromium fluxes and speciation in ultramafic catchments and global rivers. Chem Geol 426:135–157

McKeegan KD, Davis AM (2014) Early Solar System Chronology. *In*: Treatise on Geochemistry. Vol 1. Turekian KK, Holland HD, (eds). Elsevier, Oxford and San Diego, p 431–460

McLennan SM (2001) Relationships between the trace element composition of sedimentary rocks and upper continental crust. Geochem Geophys Geosyst 2:2000GC000109

Meyer BS, Krishnan TD, Clayton DD (1996) ^{48}Ca production in matter expanding from high temperature and density. Astrophys J 462:825–838

Middleton SS, Bencheikh-Latmani R, Mackey MR, Ellisman MH, Tebo BM, Criddle CS (2003) Cometabolism of Cr(VI) by Shewanella oneidensis MR–1 produces cell-associated reduced Cr and inhibits growth. Biotechnol Bioeng 83:627–637

Mills DB, Ward LM, Jones C, Sweeten B, Forth M, Treusch AH, Canfield DE (2014) Oxygen requirements of the earliest animals. PNAS 111:4168–4172

Moore WS (1996) Large groundwater inputs to coastal waters revealed by ^{226}Ra enrichments. Nature 380:612–614

Moynier F, Yin QZ, Schauble E (2011) Isotopic evidence of Cr partitioning into Earth's core. Science 331:1417–1420

Myers JM, Myers CR (2000) Role of the tetraheme cytochrome CymA in anaerobic electron transport in cells of *Shewanella putrefaciens* MR–1 with normal levels of menaquinone. J Bacteriol 182:67–75

Novak M, Chrastny V, Čadková E, Farkas J, Bullen T, Tylcer J, Szurmanova Z, Cron M, Prechova E, Curik J (2014) Common occurrence of a positive δ^{53}Cr shift in Central European waters contaminated by geogenic/industrial chromium relative to source values. Environ Sci Technol 48:6089–6096

Nyquist L, Lindstrom D, Mittlefehldt D, Shih C-Y, Wiesmann H, Wentworth S, Martinez R (2001) Manganese-chromium formation intervals for chondrules from the Bishunpur and Chainpur meteorites. Meteor Planet Sci 36:911–938

Ohtake H, Fujii E, Toda K (1990) Reduction of toxic chromate in an industrial effluent by use of a chromate-reducing strain of *Enterobacter cloacae*. Environ Technol 11:663–668

Ottaway T, Wicks F, Bryndzia L, Kyser T, Spooner E (1994) Formation of the Muzo hydrothermal emerald deposit in Colombia. Nature 369:552–554

Palmer CD, Wittbrodt PR (1991) Processes affecting the remediation of chromium-contaminated sites. Environ Health Perspect 92:25–40

Papanastassiou (1986) Chromium isotopic anomalies in the Allende meteorite. Astrophys J 308:L27–L30

Park C, Keyhan M, Wielinga B, Fendorf S, Matin A (2000) Purification to homogeneity and characterization of a novel *Pseudomonas putida* chromate reductase. Appl Environ Microbiol 66:1788–1795

Patterson RR, Fendorf S, Fendorf M (1997) Reduction of hexavalent chromium by amorphous iron sulfide. Environ Sci Technol 31:2039–2044

Paulukat C, Døssing LN, Mondal SK, Voegelin AR, Frei R (2015) Oxidative release of chromium from Archean ultramafic rocks, its transport and environmental impact—a Cr isotope perspective on the Sukinda Valley ore district (Orissa, India). Appl Geochem 59:125–138

Pavlov AA, Kasting JF (2002) Mass-independent fractionation of sulfur isotopes in Archean sediments: strong evidence for an anoxic Archean atmosphere. Astrobiology 2:27–41

Pereira NS, Vögelin AR, Paulukat C, Sial AN, Ferreira VP, Frei R (2015) Chromium-isotope signatures in scleractinian corals from the Rocas Atoll, Tropical South Atlantic. Geobiology 4:1–13

Pettine M, D'Ottone L, Campanella L, Millero FJ, Passino R (1998) The reduction of chromium (VI) by iron (II) in aqueous solutions. Geochim Cosmochim Acta 62:1509–1519

Planavsky NJ, Reinhard CT, Wang X, Thomson D, McGoldrick P, Rainbird RH, Johnson T, Fischer WW, Lyons TW (2014) Low Mid-Proterozoic atmospheric oxygen levels and the delayed rise of animals. Science 346:635–638

Podosek FA, Ott U, Brannon JC, Neal CR, Bernatowicz TJ (1997) Thoroughly anomalous chromium in Orgueil. Meteor Planet Sci 32:617–627

Qin L, Carlson RW (2016) Nucleosynthetic isotope anomalies and their cosmochemical significance. Geochem J 50:43–65

Qin L, Alexander CMD, Carlson RW, Horan MF, Yokoyama T (2010a) Contributors to chromium isotope variation of meteorites. Geochim Cosmochim Acta 74:1122–1145

Qin L, Rumble D, Alexander CMOD, Carlson RW, Jenniskens P, Shaddad MH (2010b) The chromium isotopic composition of Almahata Sitta. Meteorit Planet Sci 45:1771–1777

Qin L, Carlson RW, Alexander CMOD (2011a) Correlated nucleosynthetic isotopic variability in Cr, Sr, Ba, Sm, Nd and Hf in Murchison and QUE 97008. Geochim Cosmochim Acta 75:7806–7828

Qin L, Nittler LR, Alexander CMOD, Wang J, Stadermann FJ, Carlson RW (2011b) Extreme 54Cr-rich nano-oxides in the CI chondrite Orgueil – Implication for a late supernova injection into the solar system. Geochim Cosmochim Acta 75:629–644

Qin L, Xia J, Carlson RW, Zhang Q (2015) Chromium stable isotope composition of meteorites. 46th Lunar Planet Sci Conf:#2015

Raddatz AL, Johnson TM, McLing TL (2010) Cr stable isotopes in Snake River Plain aquifer groundwater: evidence for natural reduction of dissolved Cr(VI). Environ Sci Technol 45:502–507

Rai D, Eary L, Zachara J (1989) Environmental chemistry of chromium. Sci Total Environ 86:15–23

Rasmussen B, Buick R (1999) Redox state of the Archean atmosphere: evidence from detrital heavy minerals in ca. 3250–2750 Ma sandstones from the Pilbara Craton, Australia. Geology 27:115–118

Patterson RR, Fendorf S, Fendorf M (1997) Reduction of hexavalent chromium by amorphous iron sulfide. Environ Sci Tech 31:2039–2044

Rees C (1973) A steady-state model for sulphur isotope fractionation in bacterial reduction processes. Geochim Cosmochim Acta 37:1141–1162
Regelous M, Elliott T, Coath CD (2008) Nickel isotope heterogeneity in the early Solar System. Earth Planet Sci Lett 272:330–338
Reinhard CT, Planavsky NJ, Robbins LJ, Partin CA, Gill BC, Lalonde SV, Bekker A, Konhauser KO, Lyons TW (2013a) Proterozoic ocean redox and biogeochemical stasis. PNAS 110:5357–5362
Reinhard CT, Planavsky NJ, Lyons TW (2013b) Long-term sedimentary recycling of rare sulphur isotope anomalies. Nature 497:100–103
Reinhard CT, Planavsky NJ, Wang X, Fischer WW, Johnson TM, Lyons TW (2014) The isotopic composition of authigenic chromium in anoxic marine sediments: A case study from the Cariaco Basin. Earth Planet Sci Lett 407:9–18
Ren B-H, Wu S-L, Chen B-D, Wu Z-X, Zhang X (2015) Cr stable isotope fractionation in arbuscular mycorrhizal dandelion and Cr uptake by extraradical mycelium. Pedosphere 25:186–191
Rodler A, Sánchez-Pastor N, Fernández-Díaz L, Frei R (2015) Fractionation behavior of chromium isotopes during coprecipitation with calcium carbonate: implications for their use as paleoclimatic proxy. Geochim Cosmochim Acta 164:221–235
Rossman KJR, Taylor PDP (1998) Isotopic composition of the elements. Pure Appl Geophys 70:217–236
Rotaru M, Birck J-L, Allègre CJ (1992) Clues to early solar system history from chromium isotopes in carbonaceous chondrites. Nature 358:465–470
Rudnick RL, Gao S (2003) Composition of the continental crust. *In*: Treatise on Geochemistry. Vol 3. Turekian K, Holland HD, (eds). Pergamon, Oxford, p 1–64
Sander S, Koschinsky A (2000) Onboard-ship redox speciation of chromium in diffuse hydrothermal fluids from the North Fiji Basin. Mar Chem 71:83–102
Schauble E (2004) Applying stable isotope fractionation theory to new systems. Rev Mineral Geochem 55:65–111
Schauble E, Rossman GR, Taylor HP, Jr (2004) Theoretical estimates of equilibrium chromium-isotope fractionations. Chem Geol 205:99–114
Scheiderich K, Amini M, Holmden C, Francois R (2015) Global variability of chromium isotopes in seawater demonstrated by Pacific, Atlantic, and Arctic Ocean samples. Earth Planet Sci Lett 423:87–97
Schiller M, Van Kooten E, Holst JC, Olsen MB, Bizzarro M (2014) Precise measurement of chromium isotopes by MC–ICPMS. J Anal At Spectrom 29:1406–1416
Schoenberg R, Zink S, Staubwasser M, von Blanckenburg F (2008) The stable Cr isotope inventory of solid Earth reservoirs determined by double spike MC–ICP–MS. Chem Geol 249:294–306
Schoenberg R, Merdian A, Holmden C, Kleinhanns IC, Haßler K, Wille M, Reitter E (2016) The stable Cr isotopic compositions of chondrites and silicate planetary reservoirs. Geochim Cosmochim Acta 183:14–30
Shen J, Liu J, Qin L, Wang SJ, Li S, Xia J, Ke S, Yang J (2015) Chromium isotope signature during continental crust subduction recorded in metamorphic rocks. Geochem Geophys Geosystem 16:3840–3854
Shukolyukov A, Lugmair G (2006) Manganese–chromium isotope systematics of carbonaceous chondrites. Earth Planet Sci Lett 250:200–213
Sikora ER, Johnson TM, Bullen TD (2008) Microbial mass-dependent fractionation of chromium isotopes. Geochim Cosmochim Acta 72:3631–3641
Silvester E, Charlet L, Manceau A (1995) Mechanism of chromium (III) oxidation by Na-buserite. J Phys Chem 99:16662–16669
Spandler C, Pettke T, Rubatto D (2011) Internal and external fluid sources for eclogite-facies veins in the Monviso meta-ophiolite, Western Alps: Implications for fluid flow in subduction zones. J Petrol 52:1207–1236
Sperling EA, Frieder CA, Raman AV, Girguis PR, Levin LA, Knoll AH (2013) Oxygen, ecology, and the Cambrian radiation of animals. PNAS 110:13446–13451
Tang Y, Elzinga EJ, Lee YJ, Reeder RJ (2007) Coprecipitation of chromate with calcite: batch experiments and X-ray absorption spectroscopy. Geochim Cosmochim Acta 71:1480–1493
Tribovillard N, Algeo TJ, Lyons T, Riboulleau A (2006) Trace metals as paleoredox and paleoproductivity proxies: an update. Chem Geol 232:12–32
Trinquier A, Birck J-L, Allègre CJ (2006) The nature of the KT impactor. A ^{54}Cr reappraisal. Earth Planet Sci Lett 241:780–788
Trinquier A, Birck J-L, Allègre CJ (2007) Widespread ^{54}Cr heterogeneity in the inner solar system. Astrophys J 655:1179–1185
Trinquier A, Birck J-L, Allègre CJ (2008a) High-precision analysis of chromium isotopes in terrestrial and meteorite samples by thermal ionization mass spectrometry. J Anal Atom Spectrom 23:1565–1574
Trinquier A, Birck J-L, Allègre CJ, Göpel C, Ulfbeck D (2008b) ^{53}Mn–^{53}Cr systematics of the early Solar System revisited. Geochim Cosmochim Acta 72:5146–5163
Trinquier A, Elliott T, Ulfbeck D, Coath C, Krot AN, Bizzarro M (2009) Origin of nucleosynthetic isotope heterogeneity in the solar protoplanetary disk. Science 324:374–376
Urey HC (1947) The thermodynamic properties of isotopic substances. J Chem Soc (Resumed):562–581

Van der Weijden C, Reith M (1982) Chromium (III)—chromium (VI) interconversions in seawater. Mar Chem 11:565–572

Viti C, Decorosi F, Mini A, Tatti E, Giovannetti L (2009) Involvement of the oscA gene in the sulphur starvation response and in Cr(VI) resistance in Pseudomonas corrugata 28. Microbiology 155:95–105

Wang DT, Fregoso DC, Ellis AS, Johnson TM, Bullen TD (2010) Stable isotope fractionation during chromium(III) oxidation by δ-MnO_2. Abstract H53F-1109 presented at 2010 Fall Meeting, AGU, San Francisco, Calif 13–14 Dec

Wang H, Wu Y-B, Gao S, Liu X-C, Liu Q, Qin Z-W, Xie S-W, Zhou L, Yang S-H (2013) Continental origin of eclogites in the North Qinling terrane and its tectonic implications. Precambrian Res 230:13–30

Wang XL, Johnson TM, Ellis AS (2015) Equilibrium isotopic fractionation and isotopic exchange kinetics between Cr(III) and Cr(VI). Geochim Cosmochim Acta 153:72–90

Wang XL, Reinhard C, Planavsky N, Owens JD, Lyons T, Johnson CM (2016a) Sedimentary chromium isotopic compositions across the Cretaceous OAE2 at Demerara Rise Site 1258. Chem Geol 429:85–92

Wang XL, Planavsky N, Hull P, Tripati A, Reinhard C, Zou H, Elder L, Henehan M (2016b) Chromium isotopic composition of core-top planktonic foraminifera. Geobiology doi: 10.1111/gbi.12198

Wang XL, Planavsky NJ, Reinhard CT, Zou HJ, Ague JJ, Wu YB, Gill BC, Schwarzenbach EM, Peucker-Ehrenbrink B (2016c) Chromium isotope fractionation during subduction-related metamorphism, black shale weathering, and hydrothermal alteration. Chem Geol 423:19–33

Wanner C, Sonnenthal EL (2013) Assessing the control on the effective kinetic Cr isotope fractionation factor: A reactive transport modeling approach. Chem Geol 337–338:88–98

Wanner C, Eggenberger U, Kurz D, Zink S, Mäder U (2012) A chromate-contaminated site in southern Switzerland—Part 1: Site characterization and the use of Cr isotopes to delineate fate and transport. Appl Geochem 27:644–654

Wanner C, Zink S, Eggenberger U, Mäder U (2013) Unraveling the partial failure of a permeable reactive barrier using a multi-tracer experiment and Cr isotope measurements. Appl Geochem 37:125–133

Warren PH (2011) Stable-isotopic anomalies and the accretionary assemblage of the Earth and Mars: A subordinate role for carbonaceous chondrites. Earth Planet Sci Lett 311:93–100

Wasserburg GJ, Trippella O, Busso M (2015) Isotope anomalies in the Fe-Group elements in meteorites and connections to nucleosynthesis in AGB Stars. Astrophys J 805: 7

Watenphul A, Schmidt C, Jahn S (2014) Cr(III) solubility in aqueous fluids at high pressures and temperatures. Geochim Cosmochim Acta 126:212–227

Weyer S, Schwieters J (2003) High precision Fe isotope measurements with high mass resolution MC–ICP–MS. Int J Mass spectrom 226:355–368

Wielinga B, Mizuba MM, Hansel CM, Fendorf S (2001) Iron promoted reduction of chromate by dissimilatory iron-reducing bacteria. Environ Sci Technol 35:522–527

Wille M, Nebel O, Van Kranendonk MJ, Schoenberg R, Kleinhanns IC, Ellwood MJ (2013) Mo–Cr isotope evidence for a reducing Archean atmosphere in 3.46–2.76 Ga black shales from the Pilbara, Western Australia. Chem Geol 340:68–76

Woosley SE, Heger A (2007) Nucleosynthesis and remnants in massive stars of solar metallicity. Phys Rep 442: 269–283

Xu F, Ma T, Zhou L, Hu Z, Shi L (2015) Chromium isotopic fractionation during Cr(VI) reduction by *Bacillus sp.* under aerobic conditions. Chemosphere 130:46–51

Yamakawa A, Yamashita K, Makishima A, Nakamura E (2009) Chemical separation and mass spectrometry of Cr, Fe, Ni, Zn, and Cu in terrestrial and extraterrestrial materials using thermal ionization mass spectrometry. Anal Chem 81:9787–9794

Yamakawa A, Yamashita K, Makashima A, Nakamura E (2010) Chromium isotope systematics of achondrites: chronology and isotopic heterogeneity of the inner solar system. Astrophys J 720:150–154

Young ED, Galy A, Nagahara H (2002) Kinetic and equilibrium mass-dependent isotope fractionation laws in nature and their geochemical and cosmochemical significance. Geochim Cosmochim Acta 66:1095–1104

Zayed AM, Terry N (2003) Chromium in the environment: factors affecting biological remediation. Plant Soil 249:139–156

Zhang S, Wang X, Wang H, Bjerrum CJ, Hammarlund EU, Costa MM, Connelly JN, Zhang B, Su J, Canfield DE (2016) Sufficient oxygen for animal respiration 1,400 million years ago. PNAS:201523449

Zhao Z, Rush JD, Holcman J, Bielski BH (1995) The oxidation of chromium (III) by hydroxyl radical in alkaline solution. A stopped-flow and pre-mix pulse radiolysis study. Radiat Phys Chem 45:257–263

Zheng Y (2008) A perspective view on ultrahigh-pressure metamorphism and continental collision in the Dabie-Sulu orogenic belt. Chin Sci Bull 53:3081–3104

Zink S, Schoenberg R, Staubwasser M (2010) Isotopic fractionation and reaction kinetics between Cr(III) and Cr(VI) in aqueous media. Geochim Cosmochim Acta 74:5729–5745

Reviews in Mineralogy & Geochemistry
Vol. 82 pp. 415-510, 2017

Iron Isotope Systematics

Nicolas Dauphas

Origins Lab
Department of the Geophysical Sciences and Enrico Fermi Institute
The University of Chicago
5734 South Ellis Avenue
Chicago IL 60637
USA

dauphas@uchicago.edu

Seth G. John

University of Southern California
Department of Earth Science
Marine Trace Element Laboratory
3651 Trousdale Pkwy
Los Angeles, CA 90089
USA

sethjohn@usc.edu

Olivier Rouxel

IFREMER
Department of Physical Resources and Deep-Sea Ecosystems
Plouzané
29280
France
University of Hawaii
Department of Oceanography
1000 Pope Road
Honolulu, HI 96822
USA

orouxel@hawaii.edu

INTRODUCTION

Iron is a ubiquitous element with a rich (i.e., complex) chemical behavior. It possesses three oxidation states, metallic iron (Fe^0), ferrous iron (Fe^{2+}) and ferric iron (Fe^{3+}). The distribution of these oxidation states is markedly stratified in the Earth.

- Metallic iron is primarily present in the core, where it is alloyed with Ni, Co, and light elements such as S, Si or O. Some metallic iron may be present at depth in the mantle because Fe^{2+} in bridgmanite can disproportionate into Fe^{3+} and Fe^0 (Frost et al. 2004). Natural metallic iron also exists at the surface of the Earth in rare occurrences in the form of meteorite falls, metallic iron produced by reduction of lavas through interaction with coal sediments as in Disko Island (Greenland), and Josephinite (awaruite) produced by serpentinization reaction in peridotites.

1529-6466/17/0082-0011$10.00 http://dx.doi.org/10.2138/rmg.2017.82.11

- The main repository of ferrous iron is the mantle, where it is present in two spin states (the manner in which electrons fill the orbitals). At low pressure, mantle minerals contain iron in a high spin electronic state. Under the high-pressure conditions of the lower mantle, iron transitions into a low-spin electronic state (Badro et al. 2003; Lin et al. 2013). This spin transition influences the physical, chemical, and rheological properties of minerals.
- The lower mantle presumably contains significant Fe^{3+} produced by Fe^{2+} disproportionation (Frost et al. 2004). At the pressures relevant to the upper mantle, iron is not disproportionated, yet Fe^{3+} represents ~3% of total iron (Canil et al. 1994). Ferric iron is much more common in surface oxygenated environments and in crustal rocks due to the presence of an oxygen-rich atmosphere. Iron is an essential micronutrient and the low solubility of Fe^{3+} in seawater has a significant influence on biomass productivity in the modern oceans (Falkowski 1997; Mills et al. 2004).

Planetary objects other than Earth show similar stratification, including Mars, which has a metallic iron-rich core, a ferrous iron-rich mantle, and a ferric iron-rich surface, giving Mars its red color through the presence of nano-crystalline hematite. Many questions pertaining to the establishment and implications of such stratification remain unanswered, such as what were the P–T–fO_2 conditions that prevailed during core formation, how did the terrestrial mantle get oxidized, what was the timing of Earth's surface oxygenation, how did Earth's atmosphere become oxic, when did iron-based anoxygenic photosynthesis and dissimilatory iron reduction (a form of respiration) begin, and how did these inventions influence the global geochemical cycle of iron? Progress towards answering these fundamental questions and other related ones has been slow for lack of proxies to unravel the riddles of iron's complex cosmochemical, geochemical and biochemical behaviors.

The situation changed drastically over the past 15 years with the bloom of iron isotope geochemistry. Iron possesses four stable isotopes, ^{54}Fe, ^{56}Fe, ^{57}Fe, and ^{58}Fe, which represent 5.845, 91.754, 2.1191, and 0.2919 atom% respectively of the total (Berglund and Wieser 2011). The isotopic composition of iron is usually reported as $\delta^{56}Fe$, which is the deviation in part per mil of the $^{56}Fe/^{54}Fe$ ratio relative to the IRMM–014 reference standard. The $\delta^{58}Fe$ value is almost never reported because ^{58}Fe is a rare isotope and measurements have shown that it was related to $\delta^{56}Fe$ by mass-dependent fractionation (Dauphas et al. 2008; Tang and Dauphas 2012), except in magnetite produced by some magnetotactic bacteria (Amor et al. 2016). The $\delta^{57}Fe$ value is often reported because it allows one to ensure that there are no unresolved analytical artifacts. As with $\delta^{58}Fe$, it is related to $\delta^{56}Fe$ through mass-dependent fractionation. In the present chapter, we will thus focus on $\delta^{56}Fe$, from which $\delta^{57}Fe$ and $\delta^{58}Fe$ can be derived ($\delta^{57}Fe \approx 1.5 \times \delta^{56}Fe$ and $\delta^{58}Fe \approx 2 \times \delta^{56}Fe$).

Before the advent of MC-ICPMS (Multi Collector Inductively Coupled Plasma Mass Spectrometry) (Maréchal et al. 1999), the method of choice for measuring the isotopic composition of iron was TIMS (Völkening and Papanastassiou 1989; Beard and Johnson 1999; Johnson and Beard 1999), which stands for Thermal Ionization Mass Spectrometry. TIMS instruments have a relatively small bias but that bias is highly unstable and cannot be easily corrected for. Better precision and reproducibility was achieved by MC-ICPMS, which are affected by a large instrumental mass bias but this bias is more stable than in TIMS (Belshaw et al. 2000; Dauphas et al. 2009a; Millet et al. 2012). The relative ease with which the isotopic composition of iron can now be measured by MC-ICPMS has contributed to its widespread use to trace geochemical and biochemical processes involving iron.

Several reviews have been published over the past 15 years covering all aspects of iron isotope geochemistry (Beard and Johnson 2004a; Johnson et al. 2004, 2008a; Dauphas and Rouxel 2006; Anbar and Rouxel 2007). Dauphas and Rouxel (2006) reviewed the literature available in the field up to that date and made an effort to cite every single paper published on iron isotope geochemistry before 2006. In the present review, we will therefore put more emphasis on developments in iron isotope geochemistry that took place over the past decade, while still highlighting the important discoveries made before that time. The most important developments in the past 10 years include the recognition that igneous rocks and minerals can display iron isotopic variations, a better understanding of the ancient iron marine cycle, and the first extensive use of iron isotope measurements in modern seawater to better understand the modern marine iron cycle. The past decade has also seen a large increase in the number of laboratory experiments aimed at determining equilibrium and kinetic fractionation factors needed to interpret iron isotope variations in natural samples.

METHODOLOGY

The range of iron isotopic variations in natural samples is a few permil (from ~−4 to +2‰; Dauphas and Rouxel 2006). The method of TIMS, which gives a precision of approximately ±0.2‰ on δ^{56}Fe (see Eqn. 1 for a definition of this notation), opened the field of iron isotope systematics to investigation (Beard and Johnson 1999; Beard et al. 1999; Fantle and DePaolo 2004). Subsequent work by MC-ICPMS, which can reach a precision of ±0.03‰ (Dauphas et al. 2009a; Millet et al. 2012), proved that there were significant iron isotope variations hidden in the ±0.2 ‰ uncertainty of early TIMS measurements. Another advantage of MC-ICPMS relative to TIMS is the high sample throughput, as it is possible to measure up to a few tens of samples in a day, depending on the precision needed. MC-ICPMS has thus been established as the method of choice to measure the isotopic composition of iron.

Iron isotope variations are usually defined using the δ^{56}Fe notation as,

$$\delta^{56}\mathrm{Fe} = \left[\frac{\left({}^{56}\mathrm{Fe} / {}^{54}\mathrm{Fe} \right)_{\mathrm{sample}}}{\left({}^{56}\mathrm{Fe} / {}^{54}\mathrm{Fe} \right)_{\mathrm{standard}}} - 1 \right] \times 10^3 \qquad (1)$$

where the standard is usually IRMM–014; a metallic iron standard distributed by the Institute for Reference Materials and Methods. Although this is a synthetic standard, Craddock and Dauphas (2011a) showed that it has an isotopic composition indistinguishable from chondrites, all of which have a relatively uniform iron isotopic composition regardless of their group or petrologic type, defining a mean value of -0.005 ± 0.006‰ relative to IRMM–014. Early on, the isotopic composition of iron was defined relative to the average of terrestrial igneous rocks (Beard and Johnson 1999) but it was subsequently shown that those rocks displayed significant iron isotopic variations and are not necessarily representative of the composition of the terrestrial mantle (Williams et al. 2004a,b; Schoenberg and von Blanckenburg 2006; Weyer and Ionov 2007; Teng et al. 2008; Dauphas et al. 2009b; Teng et al. 2013), which may have a chondritic composition. Most iron isotope data published in the literature now adopt IRMM–014 as reference standard. An issue is that the stock of IRMM–014 has been exhausted and the Institute for Reference Materials and Measurements no longer sells this standard. Craddock and Dauphas (2011a) showed that IRMM-524a, a reference material for neutron dosimetry, has the same Fe isotopic composition as IRMM–014, which is understandable because IRMM–014 was prepared from IRMM-524a. Quoting the IRMM–014 certificate, "IRMM–014 was made up from the neutron dosimetry reference material EC-NRM 524. The cubes were prepared by melting pieces of foil, rolling into a plate and cutting with a diamond wheel. The wires were taken as such from the EC-NRM 524 stock." Until the shortage of IRMM–014 is addressed, we recommend that IRMM-524a be used for normalization in the lab but that δ^{56}Fe values still be expressed relative to IRMM–014.

Rocks and solid samples

Iron is a major rock-forming element whose chemical separation from rock matrices is relatively straightforward (Dauphas et al. 2004a, 2009a). Because iron is so abundant, blanks are usually not an issue provided that clean fluoropolymer vessels and distilled reagents are used. We describe below the measurement protocol as is used at the University of Chicago (Dauphas et al. 2009a) and elsewhere (e.g., WHOI, Ifremer) to measure rock samples. The procedure usually starts by powdering samples to make sure that the material that is analyzed is representative of the bulk rock. A powder aliquot is then sampled (typically up to a few tens of mg) and transferred in a Teflon beaker to be digested by acids, whose composition depends on the nature of the sample studied. The samples are usually cycled through several evaporations to dryness and acid dissolutions to ensure that insoluble fluorides are eliminated and that iron is present in its 3+oxidation state. The second aspect is usually achieved by using oxidizing reagents in the digestion, such as nitric acid, perchloric acid, hydrogen peroxide, or a combination thereof. The sample is then taken up in acid for passage on the chromatographic column.

All separation chemistries involve anion exchange resins. The nature of the resin can vary, with AG1-X8 200–400 mesh being the most commonly used (Dauphas et al. 2009a) and AG-MP1 coming second when other transition metals such as Cu and Zn have to be recovered for independent isotopic analysis (Liu et al. 2014). The latter resin is used most often with aqueous fluids (Borrok et al. 2007; Conway et al. 2013). The most straightforward chemistry is that based on AG1-X8, which is a rapid and cost-saving stick–non-stick chemical procedure. Its main drawback is that Cu is not separated from Fe, although separation can be achieved by running more matrix-eluting acid through the column (Tang and Dauphas 2012; Sossi et al. 2015). The sample is usually loaded onto the column in concentrated HCl (e.g., 6 M HCl). In that acid, Fe^{3+} is strongly bound to the resin while most of the matrix, except Cu and Mo, is eluted (Strelow 1980). Sometimes H_2O_2 is added to HCl to ensure that iron is not reduced by reaction with the resin, although this may be unnecessary as protocols not involving H_2O_2 still achieve excellent yields and give $\delta^{56}Fe$ values similar to those measured by protocols involving H_2O_2, provided that iron is all oxidized into Fe^{3+} before passage on the column. Iron is then recovered by running dilute HCl (e.g., 0.4 M HCl). The chromatography protocol can be repeated once or more to ensure that a sufficient purity level is achieved. Measurements by MC-ICPMS impose tight constraints on the level of purity required. If the measurements are done by standard-sample bracketing, then two passages through column chemistry may be needed to achieve the highest precision and accuracy (Dauphas et al. 2009a). Double-spike measurements are more forgiving for the presence of matrix elements that can influence instrumental mass bias, as long as there are no direct isobaric interferences on iron isotopes (Millet et al. 2012)

One difficulty intrinsic to iron isotopic analyses by MC-ICPMS is the presence of oxygen and nitrogen argide interferences on iron isotopes. For example, $^{40}Ar^{14}N^+$ can interfere with $^{54}Fe^+$, $^{40}Ar^{16}O^+$ can interfere with $^{56}Fe^+$, and $^{40}Ar^{16}O^1H^+$ can interfere with $^{57}Fe^+$. Several instrumental strategies have been used to deal with these important isobaric interferences: brute force measurements with high iron concentrations (Belshaw et al. 2000), collision cell technology (Beard et al. 2003a; Rouxel et al. 2003; Dauphas et al. 2004a), cold plasma (Kehm et al. 2003), and high-resolution (Weyer and Schwieters 2003; Dauphas et al. 2009a). The second one was implemented on the Micromass Isoprobe and involved filling a hexapole collision cell on the ion beam path with Ar and H_2. That collision cell plays two roles. One is to thermalize the incoming ions, so that their energy dispersion is reduced; the hexapole collision cell plays the role of an energy focusing ion optics component. The second role is to break down and stop molecular ions, so that argide interferences can be reduced to a level that is acceptable for iron isotopic analysis. The Isoprobe and its collision cell have largely been phased out and replaced by high-resolution instruments that partly separate argide interferences from the iron isotope peaks. Those instruments are the Nu Plasma 1700 (a.k.a. Big Nu), Nu Plasma II, and Neptune Plus. The

Nu Plasma 1700 is a large geometry instrument equipped with a large magnet that can achieve high-resolution while maintaining high transmission (Williams et al. 2005). This is seldom used in iron isotope geochemistry; the Nu Plasma II and Neptune Plus are the workhorses of iron stable isotope labs around the world. The discussion hereafter therefore focuses on the operation of those two instruments. The masses of ArN, ArO, and ArOH are sufficiently different from those of Fe isotopes that they can be separated using pseudo medium and high mass resolutions (Weyer and Schwieters 2003; Dauphas et al. 2009a). The word pseudo is used because the interfering isobars are not completely resolved but instead show up as flat top peak shoulders (Fig. 1). A significant difficulty with these measurements is that the width of the flat top peak shoulder is relatively narrow, so high stability is needed for the magnetic field and acceleration voltage. Those are highly sensitive to the temperature of the room, which must be maintained constant. The difficulty is exacerbated when the resolution slit deteriorates, so that the mass resolution decreases and the width of the flat top peak shoulder decreases.

The samples dissolved in dilute nitric acid (e.g., 0.3 or 0.45 M) are introduced into the mass spectrometer using either a desolvating nebulizer like the Apex (ESI) or Aridus II (Cetac), or a standard spray chamber most often made of quartz. Desolvating nebulizers have two advantages; they can significantly increase the overall transmission efficiency by increasing the fraction of atoms in solution that make it into the torch, and they also drastically reduce ArN, ArO, and ArOH peaks by drying down the aerosols and removing the solvent that carries N, O, and H (in the forms of H_2O and HNO_3). The standard quartz spray chamber is less efficient (some aerosols are lost by collision with the walls of the spray chamber) and leads to much higher argide peaks but instrumental mass bias is more stable in a wet plasma than when a desolvating nebulizer is used and wash-out time between samples is also smaller. For standard-sample-bracketing measurements of samples that are relatively rich in iron, the quartz spray chamber gives iron isotopic results that are more precise than when a desolvating nebulizer is used (Dauphas et al. 2009a). The desolvating nebulizer may be better suited for double-spike measurements (Millet et al. 2012) or measurements of low iron concentration samples.

The most commonly used technique for iron isotopic analyses is known as sample-standard-bracketing (SSB). It relies on the fact that while instrumental mass bias (i.e., departure between measured and true ratios) in MC-ICPMS is large, it is relatively stable, so that bracketing sample measurements by standard measurements of known compositions

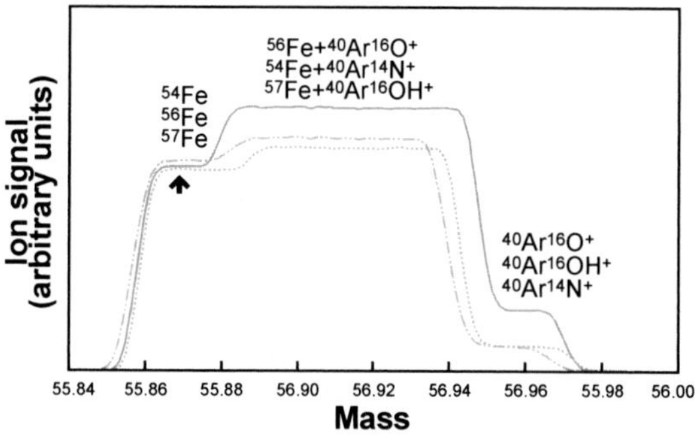

Figure 1. Peak scans of 54, 56, and 57 masses. The measurements were made on a Thermo Scientific Neptune at the University of Chicago. Iron was introduced into the mass spectrometer using the SIS stable introduction system (a dual cyclonic-Scott type spray chamber) and the measurements were made in medium resolution ($m=\Delta m \approx 4000$), where Δm is taken on the peak side between 5 and 95% peak height). The iron concentration was adjusted so that both Fe^+ ions and argide interferences are visible. Iron isotope measurements are performed on the left flat-topped peak shoulders (indicated with an arrow). Iron–58, which is seldom used or reported is not shown.

can correct the measurements for drift in instrumental mass bias. The simplest SSB scheme and the one that seems to give the most reliable results consists in bracketing the sample measurement by two standard measurements in a sequence STD1, SMP_1, STD_2, SMP_2, STD_3, SMP_3…SMP_n, STD_{n+1}. The ith measured δ^{56}Fe value is simply given by,

$$\delta^{56}\mathrm{Fe}_i = \left[\frac{\left({}^{56}\mathrm{Fe}/{}^{54}\mathrm{Fe}\right)_{\mathrm{SMP},i}}{0.5\left({}^{56}\mathrm{Fe}/{}^{54}\mathrm{Fe}\right)_{\mathrm{STD},i}+0.5\left({}^{56}\mathrm{Fe}/{}^{54}\mathrm{Fe}\right)_{\mathrm{STD},i+1}}-1\right]\times 10^3 \quad (2)$$

One could interpolate over more standards using Lagrangian interpolation. However, our experience is that this leads to more noise and simple SSB seems to be the bracketing scheme that gives the best results. The SSB technique has no safeguard to mitigate the influence of matrix elements remaining in solution or other artifacts arising from differences in the composition of the sample and bracketing standards. To achieve the most precise and accurate results, it is important to ensure that (i) iron is well purified, which can involve two passages on ion chromatography columns, (ii) the sample and standard concentrations are well matched, ideally within 5%, and (iii) the acid molarities of the sample and standard solutions are the same, which is most easily achieved by using the same solution to dissolve the standard and purified samples (Dauphas et al. 2009a).

Another approach derived from the SSB technique is the Cu or Ni doping technique (Poitrasson and Freydier 2005; Schoenberg and von Blanckenburg 2005). Copper is more difficult to separate from Fe than Ni, so we will focus below on Ni doping. In this approach, Ni is added to both the sample and standards and the ratio of two Ni isotopes is measured (Poitrasson and Freydier 2005; Rouxel et al. 2005). Iron−58 cannot be analyzed because it has a direct interference from ^{58}Ni. The collector arrays usually allow one to analyze the Ni isotopic ratio ^{62}Ni/^{60}Ni (or ^{61}Ni/^{60}Ni), which we note R_{Ni}. One assumes that the mass fractionation follows the exponential law, $r_{2/1} = R_{2/1}\left(m_2/m_1\right)^{\beta}$, where $r_{2/1}$ and $R_{2/1}$ are the measured and true ratios of isotopes 2 and 1. One can assume a fixed value for the isotopic ratio of the doping Ni solution and calculate the β-exponent of the exponential mass fractionation law given above,

$$\beta_{\mathrm{Ni}} = \ln\left[\left({}^{62}\mathrm{Ni}/{}^{60}\mathrm{Ni}\right)_{\mathrm{measured}} / \left({}^{62}\mathrm{Ni}/{}^{60}\mathrm{Ni}\right)_{\mathrm{reference}}\right] / \ln\left(M_{{}^{62}\mathrm{Ni}} / M_{{}^{60}\mathrm{Ni}}\right) \quad (3)$$

The iron isotopic composition corrected for this instrumental mass fractionation is denoted with a * superscript and calculated as follows,

$$\left({}^{56}\mathrm{Fe}/{}^{54}\mathrm{Fe}\right)^{*}_{\mathrm{measured}} = \left(\frac{{}^{56}\mathrm{Fe}}{{}^{54}\mathrm{Fe}}\right)_{\mathrm{measured}} \left(M_{{}^{56}\mathrm{Fe}} / M_{{}^{54}\mathrm{Fe}}\right)^{\frac{\ln\left[\left({}^{62}\mathrm{Ni}/{}^{60}\mathrm{Ni}\right)_{\mathrm{reference}} / \left({}^{62}\mathrm{Ni}/{}^{60}\mathrm{Ni}\right)_{\mathrm{measured}}\right]}{\ln\left(M_{62_{\mathrm{Ni}}}/M_{60_{\mathrm{Ni}}}\right)}} \quad (4)$$

The corrected iron isotopic ratio is then used in the regular SSB equation given above. The advantage of Ni doping is that it can partially mitigate the influence of matrix elements remaining in the solution that can change the instrumental mass bias. It can also potentially better correct for instrumental drift. Its drawbacks are that ^{58}Fe cannot be measured and the method is sensitive to isobaric interferences on Ni isotopes in addition to Fe. In practice, simple SSB and SSB+Ni-doping can achieve accurate measurements with precisions that are comparable.

A virtue of the SSB bracketing technique is that instrumental stability (and error bars) can be ascribed on the basis of the reproducibility of the standards that are used to bracket the sample measurements. One can, of course, use the standard deviation of the sample measurements to quantify error bars. The number of sample solution measurements is usually limited (typically up to ~10 for high precision measurements but most often just a few), so a large Student *t*-factor has to be applied to calculate the 95% confidence interval and the errors thus calculated are not very robust. In contrast, there are typically tens of standard measurements during a session, which provide a better estimate of the instrument stability. From the n standard measurements in a session, one can calculate $n-2$ δ-values of the standard bracketed by itself, $(1,2,3)\ldots(i-1,i,i+1)$, $(n-2,n-1,n)$, and use the standard deviation of those $n-2$ values σ_{STD} as a measure of error. If a sample solution is measured k times, one can take the mean of those k measurements and ascribe an error of $\pm 2\sigma_{STD}/\sqrt{k}$. In practice, other sources of errors than mass spectrometry limit error bars to ±0.03‰, regardless of the number of replicate analyses (Dauphas et al. 2009a)

The double-spike (DS) technique is the technique of choice in TIMS. It has gained a renewed attention in MC-ICPMS, in particular for elements that are present at trace levels, are difficult to fully purify, and have low chemical yields. A difficulty with the DS technique is that it is a bit more complicated to implement, as a double-spike solution needs to be calibrated and mixed with the sample in the right proportions. Rudge et al. (2009) recently presented a comprehensive study of the DS method, evaluating mixing proportions to minimize error magnification under the assumption that the total quantity of sample+spike must remain constant, so that increasing the quantity of spike used necessitates using less sample and vice versa. Under this assumption, the smallest error propagation is achieved for a 48% ^{57}Fe + 52% ^{58}Fe DS mixture mixed with the sample in a proportion 45% DS + 55% sample. However, the assumption that the total quantity of sample+spike cannot vary does not represent realistic conditions and the values calculated by the double-spike toolbox are not preferable. In practice, the quantities of sample and spike can be varied independently, subject only to the quantity of sample available and the maximum signal which can be measured on the detectors. Using a Monte Carlo method, John (2012) found that the DS compositions suggested by Rudge et al. (2009) were a good starting point, but that sample spike mixtures should generally be chosen to maximize the total quantity of sample+spike analyzed. For δ^{56}Fe, when sample quantity is not limited, a mixture of ~33% sample and ~67% spike is preferred because this produces a mixture with similar quantities of ^{56}Fe, ^{57}Fe, and ^{58}Fe which allows for each of these isotopes to be collected near the maximum voltage of the detector (assuming that each detector has the same range). When sample quantity is limited there may be a decrease in the theoretical error by further increasing the proportion of spike used. The overall error in δ^{56}Fe is, however, dominated by the small amount of ^{54}Fe, so there is little additional gain from increasing the proportion of spike above ~67%. In practice, there is no mathematically optimum DS composition and sample-spike mixture which is best for all conditions but a mixture of sample and spike in a 1:2 ratio with a DS composed of equal proportions of ^{57}Fe and ^{58}Fe is a good practical choice because it yields close to the minimum error under a very wide range of analytical conditions (Fig. 2) (John 2012).

Double-spike is always preferred to triple-spike (Millet and Dauphas 2014) because when adding another isotope, the mixture always resembles more the natural composition and the error propagation factor increases (in the limit of a spike with isotopic ratios similar to the sample composition, the double or triple spike acts as an isotope dilutant and provides no constraint on the isotopic composition of the sample; only on its concentration). Dideriksen et al. (2006) first applied the DS to MC-ICPMS measurements of iron. Millet et al. (2012) showed that double-spike measurements could provide precisions on par with the best SSB measurements, often in fewer replicate analyses. In double-spike analysis, a synthetic mixture of isotopes is mixed with the sample and the isotopic composition is measured (Dodson 1963; Albarède and Beard 2004; Rudge et al. 2009). To very high precision, all iron isotope variations in solar system materials are related to the terrestrial (IRMM–014) composition by mass-dependent fractionation (assumed to follow the exponential law below; but the exact form of the law has little influence) (Dauphas et al. 2004a, 2008; Tang and Dauphas 2012), so we can write that the ratio $R_{2/1}^{SMP} = i_2 / i_1$ in the sample is related to the ratio in the IRMM–014 standard through,

$$R_{2/1}^{SMP} = R_{2/1}^{STD} \left(m_2 / m_1 \right)^{\alpha} \tag{5}$$

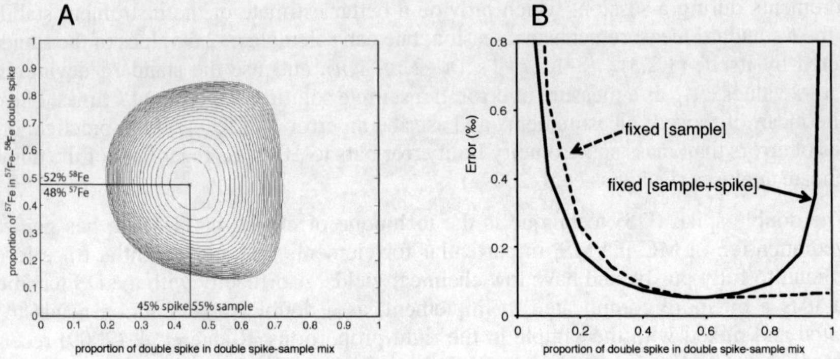

Figure 2. The effect of double-spike composition and mixing proportions on analytical δ^{56}Fe errors. The double-spike toolbox of Rudge et al. (2009) has been used to calculate the optima based on an assumption of a total sample+spike ion beam of 10 V on 10^{11} Ω detectors and isotopically pure spikes (A). The contours represent lines of equal errors which are evenly spaced corresponding to intervals of 1% increments relative to the optimal error (cross). The Monte Carlo method of John (2012) was used to calculate the optimum mixing proportions using a similar DS composition, but different assumptions about sample and spike quantities (B). When the total quantity of sample+spike is fixed, the results are similar to those of Rudge et al. (solid black line). With only the concentrations of sample fixed (for example when only a small amount of sample is available) error is minimized with a ~1:2 mixture of sample and spike, and further increases in spike quantity have very little effect on error. When sample quantity is not limited, a 1:2 mixture of sample and spike is also beneficial because ^{54}Fe, ^{56}Fe, and ^{57}Fe are present in similar quantities so that high concentrations of sample can be run without overloading detectors. When sample+spike voltages approach the detector limitations, increasing the proportion of spike higher than ~1:2 is expected to increase error because it requires using less sample (not shown).

where α quantifies the extent to which the sample is fractionated relative to the standard, which is the unknown that must be estimated in the DS data reduction procedure. The sample is mixed with a spike of non-natural composition. We note f_1^{SP} the fraction of isotope i_1 that comes from the spike ($1-f_1^{SP}$ is the fraction that comes from the sample). We have for the mixture,

$$R_{2/1}^{\text{MIX}} = f_1^{\text{SP}} R_{2/1}^{\text{SP}} + \left(1-f_1^{\text{SP}}\right) R_{2/1}^{\text{SMP}} = f_1^{\text{SP}} R_{2/1}^{\text{SP}} + \left(1-f_1^{\text{SP}}\right) R_{2/1}^{\text{STD}} \left(m_2 / m_1\right)^{\alpha} \quad (6)$$

The mixture is fractionated during chemical processing and isotopic analysis. We note β the exponent of the exponential law corresponding to this fractionation. We have for the measured ratio $r_{2/1}^{\text{MIX}}$,

$$r_{2/1}^{\text{MIX}} = \left[f_1^{\text{SP}} R_{2/1}^{\text{SP}} + \left(1-f_1^{\text{SP}}\right) R_{2/1}^{\text{STD}} \left(m_2 / m_1\right)^{\alpha} \right] \left(m_2 / m_1\right)^{\beta} \quad (7)$$

There are three unknowns in this equation, α, β, and f_1^{SP}. If one assumes that all mass-fractionations can be related to the terrestrial standard composition by a mass-dependent law (here exponential), then one can write the same equation for 3 isotope pairs (e.g., ^{56}Fe/^{54}Fe, ^{57}Fe/^{54}Fe, and ^{58}Fe/^{54}Fe) and solve the system as α, β, and f_1^{SP} are the same for all these ratios. Instrumental mass fractionation in MC-ICPMS is large and the law describing it is not always precisely known, although it is well approximated by the exponential law (Maréchal et al. 1999). The absolute isotopic ratios after double-spike data reduction are variable from session-to-session and can show drifts within a session. In MC-ICPMS, the double-spike approach is therefore often used in tandem with the SSB technique, meaning that samples and standards are spiked at the same level and the δ-values are calculated from the isotopic ratios of the sample and bracketing standards after DS data reduction.

To properly use the double-spike technique, one should dope the sample with the spike as early as possible in the procedure (i.e., before digestion), to ensure that iron in the sample and spike are completely equilibrated, and more importantly to account for any lab-induced isotopic fractionation. Such fractionations could arise during dissolution if some analyte is lost as precipitate and more importantly during chromatographic separation if the yield is not 100%. Iron is a major rock-forming element and even when digesting a few mg of samples, the amount of spike needed to spike the sample at that stage would be prohibitively expensive. Therefore, the double-spike technique as it has been applied in iron isotope geochemistry (Dideriksen et al. 2006; Millet et al. 2012) departs from the golden standard DS approach, as iron is first purified from the sample by chromatography and the DS is then added just before mass spectrometry to correct for instrumental mass fractionation. Iron DS measurements as they have been applied so far thus suffer from some of the same shortcomings as the SSB method, meaning that a high yield is needed to ensure that the measurements are accurate. The DS technique can also suffer from the fact that more isotopes are involved in the reduction and high washout times are needed as the isotopic ratios can vary significantly from sample to sample. Finally, use of the double-spike technique does not allow one to investigate easily mass fractionation laws and potential isotopic anomalies.

To summarize, various strategies exist for measuring the isotopic composition in rocks at high precision that all have advantages and shortcomings. A very reassuring fact is that geostandards are routinely measured in all laboratories practicing iron isotope geochemistry and the results obtained by the various methods agree, even down to precisions of ±0.03‰ on δ^{56}Fe (Fig. 3). This, together with extensive testing performed in laboratories practicing iron isotope geochemistry, gives confidence that iron isotope measurements are accurate at those levels and minute iron isotope variations can be discussed with confidence. New practitioners of iron isotope geochemistry should run well-documented geostandards with a range of δ^{56}Fe values (e.g., BCR–2, BHVO–1, BHVO–2, BIR–1, AC-E, AGV–2, IF-G; Craddock and Dauphas 2011) to ensure that their results are accurate.

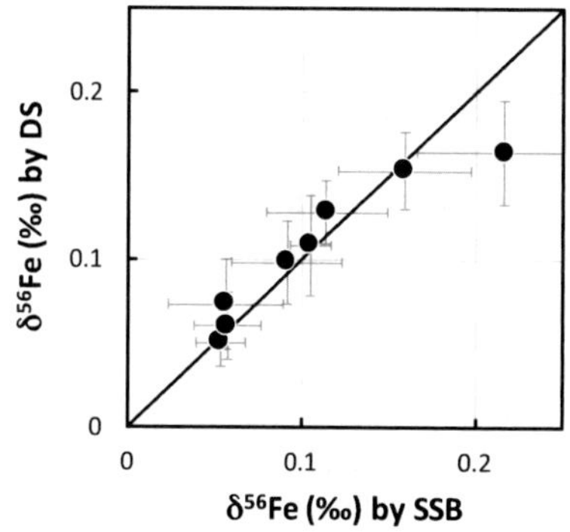

Figure 3. Comparison between Fe isotopic analyses of geostandards (BHVO–2, BCR–2, JB2, BIR–1, AGV–2, JA1, GSP–1, RGM–1) by Sample Standard Bracketing –SSB (Weyer et al. 2005; Craddock and Dauphas 2011a; Liu et al. 2014) and double-spike –DS (Millet et al. 2012). The excellent agreement between SSB and DS, which are sensitive to different analytical pitfalls, supports the accuracy of these measurements. Accurate Fe isotopic analyses can be routinely performed with precisions of ~±0.03‰ on δ^{56}Fe.

Water samples

The analysis of Fe isotopes in water samples presents unique challenges. In liquids where Fe concentrations are high (~μM), such as hydrothermal fluids, sediment porewaters, and river waters, it is generally possible to dry down the liquid and redissolve the residue in concentrated HCl for anion exchange purification using the same procedures as utilized for solid and rock samples. For liquid samples where Fe concentrations are much lower, such as seawater, additional methods are used to concentrate iron and remove it from high concentrations of matrix elements before proceeding with anion exchange purification and analysis.

Seawater $\delta^{56}Fe$ analyses present a great challenge because of the very low concentrations of Fe in seawater and the high concentration of dissolved salts which must be removed from the sample before analysis. Away from particular Fe sources such as hydrothermal vents or reducing sediments, the concentrations of Fe in seawater typically range from as low as 0.02 nM in the surface ocean to around 1 nM in the deep ocean, corresponding to roughly 1–50 ng of Fe per L of seawater. Analytical developments have made it possible to accurately measure $\delta^{56}Fe$ on a few ng of Fe, but it is still necessary to extract iron from liters of seawater to make a measurement. Simply drying down this seawater is not an option because this is impractical, and large amounts of HCl and resin would be needed for redissolution and iron purification, at which point the blank would overwhelm the quantity of Fe present in the sample.

Analyses of $\delta^{56}Fe$ in seawater and other dilute liquid matrices therefore require a preconcentration step before anion exchange purification. Early efforts concentrated Fe from seawater by coprecipitating Fe with $Mg(OH)_2$ after increasing the pH by ammonia addition (de Jong et al. 2007), though this method still results in relatively high salt concentrations after preconcentration which seem to interfere with subsequent analyses. More recently, seawater Fe has been pre-concentrated from seawater onto resins with organic chelating moieties that have a very high affinity for Fe and can bind Fe even at pH as low as 2. A resin with NTA (nitroloacetic acid) functional groups has been used for extracting Fe from seawater in both batch and column processes (John and Adkins 2010; Lacan et al. 2010; Rouxel and Auro 2010). A similar resin with ethylenediaminetriacetic acid (EDTriA) functional groups has been used in batch form and has been found to have slightly lower blank, and has the additional benefit of being able to simultaneously preconcentrate Fe, Zn, and Cd from the same seawater samples (Conway et al. 2013). Further purification and analysis of samples is achieved using methods similar to those for other geological samples, with modifications to increase analytical sensitivity and minimize contamination. For example, anion exchange chromatography is typically performed on smaller columns in order to reduce blank, with some methods using as little as 35 μL of resin (Conway et al. 2013). A typical MC-ICPMS setup includes several changes to increase sensitivity such as the use of an Apex desolvating inlet system, the use of larger hole-diameter cones (e.g., Jet sampler cones and X skimmer cones for the Neptune), and the use of higher impedance resistors (John and Adkins 2010; John 2012; Lacan et al. 2010; Rouxel and Auro 2010; Conway et al. 2013).

In situ analyses

The focus of the first *in situ* (i.e., spatially resolved) stable iron isotope analyzes, by either secondary ion mass spectrometry (SIMS) or MC-ICPMS, was on samples that displayed large fractionations such as banded-iron formations (BIFs). The $\delta^{56}Fe$ value of those samples range between approximately −1.5 and +2‰, so that the precisions of *in situ* techniques (±0.1 to ±0.4‰) were sufficient to detect those variations (Horn et al. 2006; Whitehouse and Fedo 2007). The scope of *in situ* analyses of iron isotopes has significantly expanded over the past 10 years, as the analytical capabilities and methodological approaches have improved.

One important difficulty of *in situ* analyses is the evaluation of accuracy. Indeed, the matrix standards used for correction of instrumental artifacts are ideal model compositions that fail to capture the chemical and structural complexity of a natural system. Micromilling can help bridge the gap between bulk and *in situ* analyses. Sampling by micromilling is done by depositing a drop of water on the sample and drilling the sample surface through that water droplet (Charlier et al. 2006; Sio et al. 2013). The sample slurry thus produced can be retrieved using a fine pipette. The powder retrieved can then be processed through chemistry like regular bulk samples. The hole size made by the tungsten carbide milling tip can be ~300 μm in diameter and ~300 μm depth.

Iron isotopic analyses by SIMS have targeted several matrices, most notably sulfides, magnetite, and olivine. The instrument of choice for *in situ* SIMS iron isotopic analysis has been the Cameca–1270 or 1280. Whitehouse and Fedo (2007) measured the iron isotopic compositions of magnetite and seconday pyrite in >3.7 Ga old BIFs from Isua (southwestern Greenland). The quoted precision of those measurements is ±0.4‰ and the authors reported variations of up to 2‰ in δ^{56}Fe for magnetite grains a few millimeters apart. These BIFs are highly metamorphosed (to amphibolite facies) and no such variation was found in equivalent samples from Isua and Nuvvuaggituq when measured by micromilling, wet chemistry and MC-ICPMS (Dauphas et al. 2007a,b). Kita et al. (2011) reported SIMS analyses of iron isotopes in magnetite for which they achieved a precision of ±0.2‰ on δ^{56}Fe. They found some grain-to-grain variability of 0.6‰ but reported a correlation between $^{56}Fe^+$ ion yield and δ^{56}Fe values. They suggested that it could be a crystallographic orientation effect, meaning that the instrumental mass fractionation depends on the crystallographic orientation of the magnetite grain relative to the incident beam, which could explain the discrepancy between previous SIMS (Whitehouse and Fedo 2007) and micromilling measurements (Dauphas et al. 2007a; 2007b). Marin-Carbonne et al. (2011) presented the most extensive technical study of *in situ* iron isotopic analyses by SIMS. In particular, they explained in detail how an unresolvable ^{54}Cr interference on ^{54}Fe could be corrected for by monitoring the intensities of ^{52}Cr and ^{53}Cr. They reported precisions of ±0.3‰. More emphasis was put on magnetite analyses but instrumental mass fractionation was also reported for metallic iron, siderite, hematite, and pyrite. The technique was applied in a subsequent paper to the analysis of pyrite in 2.7 Gyr old shales (Marin-Carbonne et al. 2014). Through combined δ^{56}Fe and Δ^{33}S analyses, the authors showed that the pyrite nodules could not have formed solely by dissimilatory iron reduction and sulfate reduction (two respiration modes that use oxidized forms of iron and sulfur in place of O_2 as electron acceptors) but must have involved a complex diagenetic history and source mixing.

SIMS measurements were also used to address questions relevant to high-temperature geochemistry and cosmochemistry. In these fields, the temperatures involved are high enough that no equilibrium isotopic fractionation is expected to be measurable by SIMS. Kinetic effects can still impart large fractionation at magmatic temperatures (Richter et al. 2009a). One such process is evaporation/condensation. *In situ* measurements by SIMS of Fe–Ni zoned metal grains from CBb chondrites have revealed large iron isotopic fractionation spanning ~10‰ on δ^{56}Fe that is correlated with Ni isotopic fractionation (Alexander and Hewins 2004; Richter et al. 2014a). Those large effects were also found by laser-ablation MC-ICPMS (Zipfel and Weyer 2007). They are interpreted to reflect kinetic isotopic fractionation associated with partial condensation of metallic Fe and Ni in the aftermath of a vapor-forming impact in the early solar system. Another kinetic process that can impart significant isotopic fractionation at high temperature is diffusion. Sio et al. (2013) measured a zoned olivine xenocryst from Kilauea Iki lava lake by SIMS (Fig. 4). They found large iron isotopic fractionation that corresponds to what is expected for diffusion in olivine, meaning opposite to that measured for Mg isotopes and corresponding to a light iron isotope enrichment in the core of the xenocryst. SIMS instrumental isotopic fractionation for iron in olivine is large and highly sensitive to the forsterite content (spanning almost 10‰ in δ^{56}Fe between Fo_0 and Fo_{100}). Nevertheless, this can be well corrected for, yielding accurate iron isotopic analyses with precisions of ~0.3‰.

Laser-ablation (LA) MC-ICPMS has been used quite extensively in iron isotope systematics. The natural samples that were targeted for isotopic analysis by this method are the same as the ones that were measured by SIMS, namely samples formed in low-*T* aqueous environments and high-*T* samples where transport processes (diffusion, evaporation, condensation) fractionated iron isotopes. The equipment used to carry out those measurements is diverse, as various lasers can be used with various ablation cells to feed different types of MC-ICPMS. The 193 nm ArF Excimer nanosecond lasers are efficient at ablating a large variety of matrices but their main

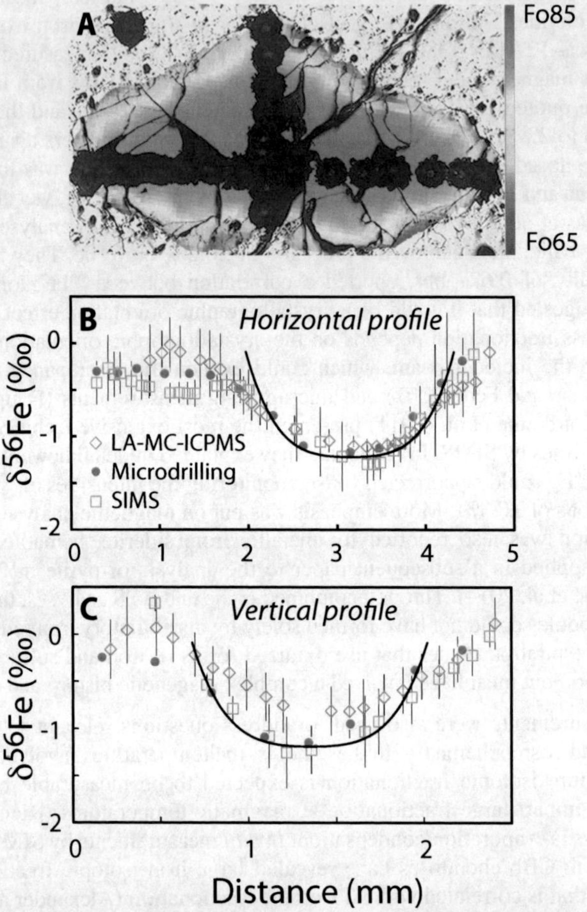

Figure 4. Fe isotopic compositions across a zoned olivine phenocryst from Kilauea Iki lava lake (Sio et al. 2013). The top image (A) is a false color image of the olivine (the colors [online] indicate the Fo content). The middle (B) and bottom (C) panels show comparisons of the measurements made by LA-MC-ICPMS, microdrilling, and SIMS. The continuous curve is a model calculation for isotopic fractionation driven by diffusion. Mg is more compatible than Fe in olivine. During magmatic evolution, Mg will diffuse out while Fe will diffuse in because the melt is becoming progressively more Fe-rich and Mg-poor. The light Fe isotopes diffuse faster than the heavy ones, which explains why the olivine core has negative δ^{56}Fe. Isotopic profiles can help tease apart zoning due to crystal growth from zoning due to diffusion.

drawback is that significant energy is converted into heat, which can effectively vaporize and melt the target sample, inducing significant iron isotopic fractionation that is matrix-dependent. Femtosecond lasers deposit energy over a very brief pulse duration, so the sample ablation is associated with less collateral heating and attendant melting and vaporization. The degree of isotopic fractionation imparted by the laser ablation in femtosecond measurements is consequently smaller than with Excimer lasers. The aerosols produced by laser ablation are carried into the mass spectrometer by a flow of Ar+He. Helium yields the best transport efficiency but it has to be mixed with argon for plasma stability (Günther and Heinrich 1999). The efficiency with which the ablated aerosols are carried to the torch also depends on the ablation cell construction and geometry. The two-volume cell technology (e.g., Laurin Technic) is particularly well suited for MC-ICPMS (Müller et al. 2009). It indeed allows measuring samples exposed over large surface areas while keeping the ablation volume to a minimum, thus promoting fast washout between samples and efficient aerosol transport. The mass-spectrometer plays a lesser role than the nature of the laser, ablation cell, and ablation parameters used. Data acquisition in LA-MC-ICPMS is a bit different than using more standard sample introduction systems because the signal can be recorded constantly, so the beginning and end of a measurement are defined in post-processing by examining (either manually or through an automated algorithm) signal increase and decrease associated with sample introduction and washout.

Horn et al. (2006), and Horn and von Blanckenburg (2007) pioneered the use of LA-MC-ICPMS in iron isotope analyses. They used a 196 nm UV-femtosecond laser with a pulse width of ~100–200 femtosecond on a variety of mineral matrices: iron metal, sulfides, hematite, siderite, goethite, and magnetite. The IRMM–014 metal reference material was used as bracketing standard. They showed that with this setup, matrix matching was not critical and that precisions of better than ~0.1‰ could be achieved for an ablation hole size of 35 μm. Comparison of LA-MC-ICPMS results with values measured by solution MC-ICPMS after chemical purification validated the method. As with SIMS, Cr and Ni isobaric interferences cannot be resolved in mass from iron peaks, so these interferences are corrected for by monitoring masses ^{52}Cr and ^{60}Ni. Steinhoefel et al. (2009a) expanded this work to the analysis of silicates, which can contain significant amounts of Cr and Ni, so that the correction for isobaric interferences can become unreliable and $\delta^{56}Fe$ is calculated as $2 \times \delta^{57/56}Fe$, meaning that ^{54}Fe is not involved in this calculation. The precision achievable for silicates is similar to that reported for metals, oxides, hydroxides, sulfides and carbonates (i.e., ±0.1‰ on $\delta^{56}Fe$) and the measurements appear to be accurate even when a mismatched standard matrix is used for bracketing the samples. Steinhoefel et al. (2009b, 2010) measured the Fe isotopic compositions of minerals in BIFs of low metamorphic grades and found significant Fe isotopic fractionation between layers. They also found that $\delta^{56}Fe$ values of bulk BIF layers correlate coarsely with the Si isotopic composition of those layers, the explanation of which is still unclear. Dziony et al. (2014) applied the same technique to the analysis of ilmenite and magnetite in basalt and gabbros, and found significant fractionation between those two minerals, inconsistent with high temperature equilibrium. Those fractionations must reflect kinetic isotope effects associated with interactions between magmas and surrounding hydrothermal fluids. Oeser et al. (2014) reported *in situ* measurements of reference glasses BIR-1G, BCR-2G, BHVO-2G, KL2-G, ML3B-G, GOR128-G, and GOR132-G, by femtosecond laser mass spectrometry. They showed that the glasses have uniform Fe isotopic compositions, so they can be used as secondary bracketing standards in *in situ* analyses for analytical techniques that require matrix matched standards. Building on ideas of Poitrasson and Freydier (2005) and O'Connor et al. (2006), Oeser et al. (2014) also showed that using Ni as an external mass bias monitor (which is introduced as a Ni standard solution into the plasma along with the ablation aerosol) can significantly improve the accuracy and reproducibility of *in situ* Fe isotope analyses by LA-MC-ICPMS.

Horn et al. (2006) used a UV femtosecond laser but equally good results can be obtained by near-infrared (NIR) femtosecond laser (Nishizawa et al. 2010). Specifically, Nishizawa et al. (2010) used a NIR 780 nm femtosecond (227 fs pulse length 20 Hz repetition rate) to measure the Fe isotopic composition of pyrite grains. The measurements were done by rastering a 40×40 μm surface with a 10 μm laser beam, and excavating sample material over a depth of ~10 μm. The sample measurements were bracketed by standard pyrite measurements and the authors were able to achieve a precision of ±0.3‰ on $\delta^{56}Fe$. They applied this technique to the analysis of pyrite grains associated with shallow marine carbonates.

d'Abzac et al. (2013) found significant isotopic fractionation between different aerosol size fractions produced by femtosecond laser. This fractionation depends on the mineral analyzed and is close to zero for pyrite but can reach 2‰ for magnetite. This probably results from kinetic isotopic fractionation during condensation. To minimize mass fractionation in femtosecond laser ablation, it is thus important to ensure that no size sorting can occur during transport to the plasma torch or that all aerosol sizes are efficiently ionized. Different ablation cells (e.g., single vs. two-volume) and laser wavelengths can influence the extent of this fractionation. A two volume cell seems to yield the most stable and best isotopic results (d'Abzac et al. 2014).

Recently, nanosecond lasers have taken a back seat relative to femtosecond lasers. The reason is that nanosecond lasers impart more variable isotopic fractionation that is matrix dependent (Graham et al. 2004; Košler et al. 2005; Sio et al. 2013; Toner et al. 2016). By exercising some care in matching the matrix well and assessing positional effects, it is nevertheless possible to generate high precision measurements. Sio et al. (2013) measured the isotopic composition of Fe in olivine using a 193 nm Excimer laser ablation system with a relatively large single cell ablation setup. Between Fo_{55} and Fo_{95}, the degree of instrumental mass fractionation changed by ~0.8‰. By bracketing the measurements with standards matched in Fo content, that bias was corrected for. Another bias was present that depended on the position of the laser spot in the ablation cell. To correct for this positional effect, Sio et al. (2013) measured the same profile after rotating the sample. Assuming that the bias introduced by the ablation varies linearly with position in the cell, then doing the same measurement at zero angle and rotating the sample by 180° around the center of the mount should eliminate this effect. Sio et al. (2013) measured the same sample by laser ablation ICPMS, SIMS and microdrilling and the isotopic profile generated by these three independent techniques are in excellent agreement, lending confidence to each of these techniques (Fig. 4).

Isotopic anomalies and mass-fractionation laws

The methodologies outlined above focus on the mass-dependent component of the isotopic variations. By mass-dependent, we mean that the isotopic variations scale, within the precision of the measurements, with the difference in mass of the isotopes involved. For example, $\delta^{56}\text{Fe} \simeq \delta^{57}\text{Fe} \times (56-54)/(57-54)$. Isotopic variations can depart from a reference mass fractionation law (e.g., the exponential law) for two reasons: nucleosynthetic anomalies could be present, or the mass-fractionation law controlling the isotopic variations may differ from the reference law that is used (Dauphas and Schauble 2016). Isotopic anomalies or departures from a reference mass fractionation law are usually calculated by internal normalization. This consists of fixing an isotopic ratio to a constant value and correcting mass fractionation accordingly using the reference law. The ^{57}Fe/^{54}Fe ratio and exponential law are most often used for internal normalization (Dauphas et al. 2004a; 2008; Tang and Dauphas 2012). The exponent parameter of the exponential law is given by (Eqn. 3; $M_{^i\text{Fe}}$ is the mass of isotope ^{i}Fe),

$$\beta^* = \frac{\ln\left[\left({}^{57}\text{Fe}/{}^{54}\text{Fe}\right)_{\text{measured}} / \left({}^{57}\text{Fe}/{}^{54}\text{Fe}\right)_{\text{reference}}\right]}{\ln\left(M_{^{57}\text{Fe}}/M_{^{54}\text{Fe}}\right)} \tag{8}$$

The ^{56}Fe/^{54}Fe and ^{58}Fe/^{54}Fe ratios corrected by internal normalization (noted R^*) are given by,

$$\left(\frac{^{56}\text{Fe}}{^{54}\text{Fe}}\right)^* = \left(\frac{^{56}\text{Fe}}{^{54}\text{Fe}}\right)_{\text{measured}} \left(M_{^{56}\text{Fe}}/M_{^{54}\text{Fe}}\right)^{\frac{\ln\left[\left({}^{57}\text{Fe}/{}^{54}\text{Fe}\right)_{\text{reference}} / \left({}^{57}\text{Fe}/{}^{54}\text{Fe}\right)_{\text{measured}}\right]}{\ln\left(M_{^{57}\text{Fe}}/M_{^{54}\text{Fe}}\right)}} \tag{9}$$

$$\left(\frac{^{58}\text{Fe}}{^{54}\text{Fe}}\right)^* = \left(\frac{^{59}\text{Fe}}{^{54}\text{Fe}}\right)_{\text{measured}} \left(M_{^{58}\text{Fe}}/M_{^{54}\text{Fe}}\right)^{\frac{\ln\left[\left({}^{57}\text{Fe}/{}^{54}\text{Fe}\right)_{\text{reference}} / \left({}^{57}\text{Fe}/{}^{54}\text{Fe}\right)_{\text{measured}}\right]}{\ln\left(M_{^{57}\text{Fe}}/M_{^{54}\text{Fe}}\right)}} \tag{10}$$

The notations used to discuss isotopic anomalies and mass fractionation laws are,

$$\delta' = 1000\ln\left(R / R_{STD}\right) \simeq 1000\left(R / R_{STD} - 1\right)$$

$$\varepsilon' = 10{,}000\ln\left(R^* / R^*_{STD}\right) \simeq 10{,}000\left(R^* / R^*_{STD} - 1\right)$$

ε' is the departure in parts per ten thousand of the internally normalized ratio of a sample relative to the internally normalized ratio of a standard. Usually, to define the internally normalized ratio of the standard, the same bracketing approach is used as when measuring mass-dependent fractionation. A standard (IRMM–014, IRMM-524a) is measured between samples, and the composition of the standard goes through the same internal normalization procedure as the sample. If all samples were derived from the same terrestrial Fe isotopic composition by the exponential law, then their ε' value would be zero. Measurements of mass-independent effects are demanding as the expected effects are small and inaccuracy may arise from isobaric interferences.

Isotopic anomalies have been documented in presolar grains (Marhas et al. 2008) as well as some early solar system condensates known as FUN refractory inclusions (FUN stands for Fractionated and Unknown Nuclear effects). In FUN inclusions, anomalies on ^{58}Fe reach ~+300‰ (Völkening and Papanastassiou 1989). The other isotopes do not display marked departures from the terrestrial mass fractionation line. Some effort has also been expanded towards searching for the presence of nucleosynthetic iron isotopic anomalies at a bulk scale but no such anomaly has been detected so far (Dauphas et al. 2004, 2008; Tang and Dauphas 2012).

Departures from mass-dependent fractionation can also arise because the fractionation is either not mass-dependent or it is mass-dependent but follows a different law than exponential (or the reference law adopted) (Matsuhisa et al. 1978; Luz et al. 1999; Young et al. 2002; Dauphas and Schauble 2016). For example, Steele et al. (2011) found small anomalies in Ni purified by the Mond process (an industrial process that involves evaporation of Ni as a nickel tetracarbonyl). No such effect has been documented thus far for iron. Recently, Nie and Dauphas (2015) and Nie et al. (2016) studied iron isotope mass fractionation in products of partial UV-photo-oxidation. The measurements were made at high precision, allowing identification of the underlying mass fractionation law, which was found to be equilibrium. Banded iron formation must have formed by partial iron oxidation in the water column but the exact mechanism for BIF formation is uncertain and three scenarios are usually considered. The first scenario involves partial oxidation by O_2 in oxygen oases. The second scenario is oxidation by anoxygenic photosynthesis, whereby Fe^{2+} is used as an electron acceptor in place of water in photosynthesis. The third scenario is photo-oxidation, whereby the energetic UV photons from the Sun could have induced photo-oxidation of iron. Evidence for and against these various scenarios is circumstantial. Nie and Dauphas (2015) and Nie et al. (2016) measured the extent of iron isotopic fractionation imparted by photo-oxidation and found that it was similar to other oxidation processes, so photo-oxidation remains a viable explanation for iron oxidation in the Archean oceans.

KINETIC AND EQUILIBRIUM FRACTIONATION FACTORS

Over the past several years, some progress has been achieved on the front of mass spectrometry capabilities. These include (1) development of new plasma interfaces that use different cone geometries (Jet cones) and higher capacity vacuum pumps, which increases instrument sensitivity, and (2) development of femtosecond laser capabilities and two volume cells for isotopic analysis by LA-MC-ICPMS. However, those developments have not been complete game changers for iron isotope geochemistry, where sensitivity is not an issue and most geological questions can be addressed by macro-scale measurements. Where significant progress was made is in documenting iron isotope variations in nature, as well as developing the framework for interpreting iron isotope variations in rock and water samples.

By framework, we mean establishing the kinetic and equilibrium fractionation factors that control iron isotopic variations in natural systems. This was achieved through a combination of ab initio calculations, experimental studies, synchrotron measurements of iron bond strengths, and studies of well-constrained natural case studies.

Kinetic processes

Diffusion. Diffusive processes are very efficient at fractionating isotopes. The light isotopes diffuse faster than the heavier ones, such that in diffusive processes, the source reservoir gets enriched in the heavy isotopes while the sink reservoir gets enriched in the light isotopes. The manner in which diffusion-driven isotopic fractionation is empirically parameterized in isotope geochemistry is,

$$\frac{D_2}{D_1}=\left(\frac{m_1}{m_2}\right)^{\beta} \tag{11}$$

where D_1 and D_2 are the diffusivities of isotopes 1 and 2, and β is a parameter that can vary between 0 (no isotopic fractionation) and 0.5 (equipartition of energy in a monoatomic gas). Note that the β notation here is not the same as the one used to describe the exponential mass fractionation law (e.g., Eqn. 3) or the β-factor (reduced partition function ratio) used in stable isotope geochemistry to calculate equilibrium isotopic fractionation between phases.

Some of the earliest iron isotope diffusion experiments were carried out on metals to learn about the diffusion mechanism (LeClaire and Lidiard 1956; McCracken and Love 1960; Pell 1960; Mullen 1961). Dauphas (2007) and Richter et al. (2009a) provide extensive compilations of β exponents for diffusion in metals. Those early experiments (e.g., Mullen 1961) used radioactive iron isotopes ^{55}Fe and ^{59}Fe as sources, let them diffuse in metal, and measured the $^{55}Fe/^{59}Fe$ ratio in metal after diffusion. All the experiments give β exponents of around 0.25 (Dauphas 2007), with little influence of the crystalline structure (α, γ or δ) or the nature of the metal (Fe in Cu, Ag, V, FeCo). More recently, Roskosz et al. (2006) also measured the β exponent of Fe in Pt by mass spectrometry and reported a value of ~0.27; close to the values estimated in the 60s and 70s in other metals. Such a large β exponent means that diffusion in metal is very efficient at fractionating iron isotopes, a feature that has important consequences for interpreting iron isotopic variations in iron meteorites (Dauphas 2007).

Iron diffusion in silicate melts of relevance to geological processes was studied by mass spectrometry. In that case, one has to specify what redox state is concerned, as Fe^{2+} and Fe^{3+} can be present in any proportions in silicate melts synthesized in the laboratory. Richter et al. (2009b) studied diffusion of iron in silicate by juxtaposing basaltic and rhyolitic melts. Iron in those experiments was most likely present as both Fe^{2+} and Fe^{3+} as the oxygen fugacity was probably between WM and FMQ (Richter, personal communication). Richter et al. (2009b) found a β-exponent of 0.030; much lower than in metal. A controlling factor on the degree of isotopic fractionation imparted by diffusion in silicate melts seems to be the contrast in diffusivity between the solute and the solvent, i.e., D_{Fe}/D_{Si}. Elements with lower solute/solvent contrast tend to display lower isotopic separation by diffusion (they are more strongly bound to the network) (Watkins et al. 2011).

Diffusive fractionation of Mg and Fe isotopes in olivine has been discussed in the context of the establishment of forsterite zoning in that mineral (Fig. 4). No clear laboratory experiment has been performed so far to establish this value but studies of natural olivines provide useful bounds on this fractionation. Dauphas et al. (2010) and Teng et al. (2011) showed that the inverse correlation between Mg and Fe isotopic compositions required that β_{Fe}/β_{Mg} be ~2. Subsequent work by Sio et al. (2013) and Oeser et al. (2015) constrained the β exponent (Eqn. 11) between 0.16 and 0.25; i.e., much higher than the value inferred for silicate melt. Further work is clearly needed to ascertain this value but it is clear that the β_{Mg} and β_{Fe} values for diffusion in olivine are large.

Diffusive fractionation of iron isotopes was also documented for both Fe^{2+} and Fe^{3+} in aqueous medium. Rodushkin et al. (2004) let Fe diffuse in a solution of 0.33 mol/L HNO_3 (pH 0.5). The authors seem to infer that Fe is present as Fe^{2+} but given the relatively oxidizing conditions, it is likely that some Fe would be oxidized into Fe^{3+}. Regardless of this complication, Rodushkin et al. (2004) finds β exponents of around 0.0015 and 0.0025 for Zn and Fe, respectively. Note that Zn is present as 2+ and has similar diffusive properties as Fe^{2+}. The β exponent for diffusion of iron in aqueous solutions is thus very low and it is likely that diffusion in natural aqueous systems played a negligible role in governing iron isotopic fractionation (Dauphas and Rouxel 2006). The reason why the β exponent is so small probably stems from the fact that Fe or Zn do not diffuse as free ions but are instead surrounded by a large solvation shell (e.g., hexaaqua ions; Fe^{2+} or Fe^{3+} surrounded by a first hydration shell composed of 6 water molecules with the oxygen atoms pointing towards iron), so that the difference in mass of the effective diffusing molecule must be small when iron isotopes are substituted. As with silicates, the contrast $D_{\text{solute}}/D_{\text{solvent}}$ ($D_{\text{Fe}}/D_{\text{H}_2\text{O}}$ with $D_{\text{H}_2\text{O}}$ the self-diffusion coefficient of H_2O) seems to correlate with the degree of isotopic fractionation imparted by diffusion (Watkins et al. 2011).

Soret Diffusion. Soret or thermal diffusion is the process by which elements when placed in a thermal gradient can concentrate at the hot or cold ends. Bowen (1928) recognized almost 100 years ago that the timescale for heat transport is always much faster than the timescale for chemical diffusion, such that any thermal gradient would be erased before atoms can move significantly. This means that a thermal gradient must be actively sustained for Soret diffusion to occur in a natural setting. Lundstrom (2009) and Zambardi et al. (2014) have argued that Soret diffusion (or a more sophisticated version of this process that they named thermal migration zone refining) could be responsible for the differentiation of some granitoids. The past 10 years has seen some renewed interest in the Soret process, in large part due to the experimental work that showed that silicate melts placed in a thermal gradient develop chemical zoning, as is expected for Soret effect, but more surprisingly that this is accompanied by large isotopic zoning (Kyser et al. 1998; Richter et al. 2008; Huang et al. 2009). Iron isotopes are indeed fractionated by this process (Huang et al. 2009; Richter et al. 2009b). In Soret diffusion in silicate melts, the cold end gets enriched in the heavy isotopes of iron while the hot end gets enriched in the light isotopes of iron. The same is true for all isotope systems investigated, regardless of whether the elements themselves tend to concentrate at the cold or hot ends. Richter et al. (2009b) found that for each 100 °C contrast between the cold and hot ends, the $\delta^{56}Fe$ value will be fractionated by 2.2‰ (corresponding to 2 amu difference between ^{56}Fe and ^{54}Fe). Huang et al. (2009) reported a somewhat similar value of ~4‰/100 °C. Several attempts have been made to describe those fractionations using parameterized equations or simple models (Huang et al. 2010; Dominguez et al. 2011; Lacks et al. 2012; Li and Liu 2015), some of which have been subsequently questioned (Richter et al. 2014b). Even for gases, describing the Soret effect is quite involved and it depends on details of the interaction potential. Isotopic ratios are a very powerful approach by which one can test whether Soret diffusion is present in natural samples such as komatiites (Dauphas et al. 2010) or granitoids (Telus et al. 2012).

Evaporation/condensation. Evaporation and condensation played important roles in establishing the chemical and isotopic compositions of planets, meteorites, and their constituents. Condensation is difficult to investigate experimentally, so more attention has been paid to iron isotopic fractionation associated with evaporation. During either vaporization or condensation, the degree of isotopic fractionation is influenced by the degree of over- or under-saturation. Applying the Hertz-Knudsen equation to the evaporation and condensation processes, one can show that the isotopic ratio i/j will be fractionated according to (Richter et al. 2002, 2007; Richter 2004; Dauphas et al. 2015),

$$\Delta_{\text{Evaporation}} = \Delta_{\text{Equilibrium}} - \left(1 - \frac{P}{P_{\text{sat}}}\right)\Delta_{\text{Kinetic}} \tag{12}$$

$$\Delta_{\text{Condensation}} = \frac{P_{\text{sat}}}{P}\Delta_{\text{Equilibrium}} + \left(1 - \frac{P_{sat}}{P}\right)\Delta_{\text{Kinetic}} \tag{13}$$

where $\Delta_{\text{Evaporation}}$ is the isotopic fractionation between the condensed phase and the evaporating gas that is leaving the surface, $\Delta_{\text{Condensation}}$ is the fractionation between the solid/liquid that is condensing and the overlying vapor, P is the vapor partial pressure of the element of interest, P_{sat} is the saturation vapor pressure for that element, $\Delta_{\text{Equilibrium}}$ is the equilibrium isotopic fractionation between the condensed phase and the gas, and $\Delta_{\text{Kinetic}} = 1000\left(\frac{\gamma_i}{\gamma_j}\sqrt{\frac{m_j}{m_i}} - 1\right)$ is the kinetic fractionation factor, that depends on the evaporation coefficients γ and masses m of the isotopes involved. The experiments done so far for iron are free evaporation experiments, meaning that $\frac{P}{P_{\text{sat}}} = 0$ (Wang et al. 1994; Dauphas et al. 2004a). At the high temperatures relevant to iron evaporation in the nebula, the equilibrium fractionation factor is negligible and the fractionation, if any, is almost entirely kinetic. If the evaporation coefficients γ of the different isotopes are identical, then the fractionation is given by the inverse square root of the isotopes involved. This assumption is often made when investigating new isotopic systems but experiments have shown that this is not necessarily the case. The square root law for $^{56}Fe/^{54}Fe$ gives a kinetic isotope fractionation factor of 1.835‰ between condensed phase and vapor. During vacuum evaporation experiments, Dauphas et al. (2004a) and Wang et al. (1994) measured a fractionation of 1.877‰ for FeO evaporation, while Dauphas et al. (2004a) reported a value of 1.322‰ for a solar-like mixture of oxides. The evaporation coefficients of ^{54}Fe and ^{56}Fe seem to be approximately identical for FeO but they differ for the solar oxide mixture. Wiesli et al. (2007) also investigated equilibrium iron isotopic fractionation of iron pentacarbonyl between liquid and vapor and found very limited equilibrium iron isotopic fractionation of ~0.05‰ on $\delta^{56}Fe$.

Reaction kinetics. Of the isotope fractionation processes discussed here, chemical reaction kinetics is perhaps the most poorly understood. Firstly, chemical kinetic isotope effects lack a well-developed theoretical framework of the sort which is available for equilibrium processes and other kinetic processes such as diffusion. Secondly, reaction kinetic isotope effects are experimentally challenging to measure because it is difficult to design experiments where reaction occurs in a single step. Nonetheless, fractionation of Fe isotopes driven by reaction kinetics has been studied in a number of different ways.

So far, electron transfer is the only reaction kinetic isotope effect that has been studied theoretically. Using experimental results from the 2-electron reduction of Fe(II) to metallic Fe, Kavner et al. (2005) developed a theoretical framework for isotope fractionation during electroplating based on Marcus's theory for the kinetics of electron transfer (Marcus 1965). While this theory provides a valuable starting point for understanding isotope reaction kinetics, it lags behind theoretical studies of equilibrium isotope effects in two key respects. First, this theory has only been developed for redox kinetics, specifically for electroplating. Second, this theory has not yet been developed to make quantitative predictions of isotope fractionation. Subsequent work on Fe electrochemistry revealed that the observed $\delta^{56}Fe$ fractionations were not based on reaction kinetics alone. Instead, the observed isotope fractionation was the result of an interplay between the isotope effect of the reduction reaction at the electrode, the isotope effect of diffusion towards the electrode, and closed-system isotope effects due to Fe depletion within the boundary layer next to the electrode (Black et al. 2010). While these experiments highlighted the complex way in which various kinetic isotope effects may interact, the simplicity of electroplating means that these processes are relatively well constrained compared to many other experimental conditions. It is likely that most experiments which seek to measure 'pure' reaction kinetic isotope effects are similarly confounded by a mixture of processes such as diffusion limitation, multiple reaction steps, and closed-system isotope fractionation of the reactants.

Still, with careful experimentation it may be possible to measure reaction kinetic isotope effects. Matthews et al. (2001) found large Fe isotope fractions between Fe(III) chloride complexes and Fe(II) bipyridine complexes, which they ascribe to a ~6.6‰ kinetic isotope effect during degradation of the $[Fe(II)(bipy)_3]^{2+}$. Skulan et al. (2002) measured the isotopic offset between aqueous Fe(III) and hematite under both equilibrium and kinetically dominated conditions, finding that while the equilibrium fractionation was close to 0, the kinetic isotope effect was −1.3‰. In addition to these experiments which have sought to explicitly measure reaction kinetic isotope effects, the result of reaction kinetic isotope effects are observed indirectly in many experimental systems such as during biological Fe oxidation (Croal et al. 2004; Balci et al. 2006) and during biological and abiotic mineral dissolution (Brantley et al. 2004; Wiederhold et al. 2006, 2007a; Chapman et al. 2009; Kiczka et al. 2010; Revels et al. 2015), and in natural systems such as the precipitation of Fe from seawater (John et al. 2012a) and the biological uptake of Fe from seawater (Ellwood et al. 2015). In these more complex systems, however, the isotope fractionation associated with a single-step chemical reaction cannot be deconvoluted from the overall observed isotope fractionation.

Equilibrium processes

When two or more phases are juxtaposed, iron isotopes can exchange until the system has reached thermodynamic equilibrium. Under those conditions, isotopes are not distributed uniformly among those phases. The manner in which isotopes are partitioned between coexisting phases is related to the free energy of the isotope exchange reaction, which depends on the strength of the iron bonds (Bigeleisen and Mayer 1947; Urey 1947). Phases that form stronger bonds with iron will be enriched in the heavy isotopes of iron at equilibrium. Because higher valence state and lower coordination tend to be associated with stronger, stiffer bonds, Fe^{3+} in low coordination usually has heavier Fe isotopic composition than Fe^{2+} or Fe^{0} in high coordination. Equilibrium fractionation between phases is usually expressed using the α notation,

$$\alpha_{B-A} = R_B / R_A \tag{14}$$

or in Δ notation,

$$\Delta_{B-A} = \left(\delta_B - \delta_A\right)_{eq} \simeq 1000\left(\alpha_{B-A} - 1\right) \simeq 1000 \ln \alpha_{B-A} \tag{15}$$

The fractionation factors reflect differences in bond strength and can be related to the reduced partition function ratio β through,

$$\Delta_{B-A} = 1000\left(\ln \beta_B - \ln \beta_A\right) \tag{16}$$

β is the equilibrium fractionation factor between a given phase and the reference state of monoatomic vapor Fe. Knowing β-factors is advantageous relative to α values because equilibrium fractionation factors between any coexisting phases can be calculated from β-factors while α is only concerned with the fractionation between two specific phases. Note that the β-factor discussed here has nothing to do with the β exponent of the exponential mass fractionation law (Eqn. 3) or the β exponent of the diffusive law (Eqn. 11). The β-factors cannot directly be measured experimentally (Fe is only gaseous at high temperature), so one relies on other methods such as ab initio calculation or, for iron, the synchrotron method of nuclear resonant inelastic X-ray scattering (NRIXS). Series expansions of the kinetic energy (Polyakov and Mineev 2000) or reduced partition function ratio (Dauphas et al. 2012) give a polynomial expression for the β-factors (and α values),

$$1000\ln\beta = \frac{A_1}{T^2} + \frac{A_2}{T^4} + \frac{A_3}{T^6} \tag{17}$$

where A_1, A_2, and A_3 are constants that do not depend on T but vary between phases or chemical species. This expression will be used hereafter to summarize experimental and theoretical results. At high temperature (above ~400 °C), the higher order terms disappear and the formula is well approximated by the first term $1000\ln\beta = A_1 / T^2$, or written as a function of the mean force constant of iron bonds, F (Herzfeld and Teller 1938; Bigeleisen and Mayer 1947),

$$1000\ln\beta = 1000\left(\frac{1}{m_{54}} - \frac{1}{m_{56}}\right)\frac{\hbar^2}{8k^2T^2}F = 2904\frac{F}{T^2} \tag{18}$$

where m_{54} and m_{56} are the masses of isotopes ^{54}Fe and ^{56}Fe, $\hbar$ is the reduced Planck constant, k is the Boltzmann constant, and T is the temperature. Below, we review those different approaches and try to build a database of β-factors by cross-calibrating the different techniques.

Mineral–mineral and mineral–melt fractionation. To experimentally determine equilibrium fractionation factors involving solids, the main difficulty is that one must ensure that equilibrium is reached or one must correct the measurements for incomplete equilibration. For solids, equilibration most often involves volume diffusion, which can be particularly slow even at high temperature. To determine equilibrium fractionation factors between mineral–mineral or mineral–melt pairs, three strategies can be used.

The first one consists of running time series experiments. Two phases are juxtaposed and the temperature is usually increased to speed up diffusion kinetics. One runs several such experiments, retrieving the run products at different times. As time goes, the two phases approach equilibrium, at which point the isotopic fractionation between the coexisting phases should not change. The equilibrium fractionation factor is defined as the fractionation when longer experiment duration does not induce any change in the isotopic compositions of the coexisting phases.

The second approach consists of using an isotopic label to trace the equilibration process. If one starts with two phases on the terrestrial mass fractionation line in a three-isotope diagram δ^{57}Fe vs. δ^{56}Fe, the run products stay on this fractionation line as the reaction proceeds. However, if one dopes one of the reactants with a specific isotope, the two reactants will plot off the mass fraction line. As the reaction proceeds, they will move towards a mass-dependent isotopic fractionation relationship with each other, so one can tell that equilibrium has been achieved. This is known as the three-isotope technique.

The third approach consists of using natural samples for which the context (closure temperature) is well known and which have all characteristics (chemical, textural, isotopic) consistent with equilibrium. If these conditions are met, one can measure the isotopic compositions of coexisting phases. The virtue of this approach is that the timescales involved in natural systems are much longer than what can be achieved in the laboratory, but the temperature conditions are only indirectly known, so sample selection is critical.

Schuessler et al. (2007a) measured equilibrium iron isotopic fractionation between pyrrhotite (FeS) and peralkaline rhyolitic melt containing 62% Fe^{3+} and 38% Fe^{2+}. The authors investigated the exchange kinetics using an isotope doping experiment and ran a time series, allowing them to conclude that the fractionations at 840 and 1000 °C were both ~+0.35‰.

As discussed above, equilibrium iron isotopic fractionation is expected to scale as $1/T^2$, so that if the fractionation at 840 °C was +0.35‰, then the fractionation at 1000 °C should have been $0.35\times(840+273)^2/(1000+273)^2=0.27$‰. This is at the limit of detection and as a first approximation, the experimental results can be accounted for by a melt-pyrrhotite fractionation of $50\times10^4/T^2$ for the $^{56}Fe/^{54}Fe$ ratio. How much of this fractionation is due to different β-factors between Fe^{2+} in the rhyolite melt and pyrrhotite vs. the presence of large amount of Fe^{3+} in rhyolite melt is unknown.

Metal–silicate equilibrium isotopic fractionation has also been the subject of much work. Poitrasson et al. (2009) studied Fe isotopic fractionation between silicate and metal. By running a time series experiment, the authors were able to estimate the fractionation between ultramafic silicate liquid and molten Fe–Ni alloy at or near equilibrium to be $+0.03\pm0.04$‰ at 2000 °C. Hin et al. (2012) also studied experimentally iron isotopic fractionation between liquid metal and silicate. They used a centrifuge piston cylinder to separate completely metal from silicate, so that their Fe isotopic compositions could be subsequently analyzed by MC-ICPMS. They ran a time series experiment and measured a metal–silicate fractionation factor of $+0.01\pm0.04$ at 1300 °C. Shahar et al. (2015) reported measurements of liquid metal–silicate equilibrium isotopic fractionation in experiments that used the 3 isotope technique and found a fractionation of $+0.08\pm0.03$‰ for $\delta^{56}Fe$ at 1650 °C. They also found that dissolving S in metal had an important effect of iron equilibrium isotopic fractionation, reaching ~+0.3‰ for 25% atomic S. To summarize, 3 studies have investigated experimentally the equilibrium fractionation factor between metal and silicate. Although the measurements were performed at different temperatures, some of these studies seem to disagree. Shahar et al. (2015) give a fractionation that is higher than that given by Hin et al. (2012) at a similar temperature. Further experiments will be needed to tell which value is correct.

Shahar et al. (2008) studied equilibrium fractionation between fayalite and magnetite using the 3-isotope equilibration technique at temperatures between 600 and 800 °C. They found significant equilibrium fractionation that can be described as $\delta^{56}Fe_{magnetite}-\delta^{56}Fe_{fayalite}=(0.20\pm0.016)\times10^6/T^2$. Equilibrium isotopic fractionation was also studied in natural samples for the mineral pair magnetite–pyroxene. Dauphas et al. (2004b, 2007b) studied quartz–pyroxene banded rocks from the island of Akilia (SW Greeland). These rocks are thought to be metamorphosed chemical sediments of BIF affinity. They were metamorphosed to a peak temperature of 750 °C (granulite facies conditions). The magnetite/pyroxene fractionation documented in these rocks is $+0.25\pm0.08$‰ and was interpreted to reflect equilibrium fractionation between these two minerals (Dauphas et al. 2004b, 2007b).

Sossi and O'Neill (2016) used an elegant method to measure equilibrium fractionation between minerals. They equilibrated in piston cylinders at 1073 K and 1 GPa the fluid phase $FeCl_2.4H_2O$ with several minerals: almandine (Fe-bearing garnet), ilmenite, fayalite (Fe-bearing olivine), chromite, hercynite (Fe-bearing spinel) and magnetite. The fluid promotes the formation of large crystals and speeds up mineral equilibration. By taking the differences between pairs of mineral–fluid equilibration experiments, Sossi and O'Neill (2016) were able to estimate mineral–mineral equilibrium fractionation factors, which are difficult to determine otherwise.

Fluid–mineral and fluid–fluid fractionations. Establishing equilibrium fractionation factors at high temperature is fraught with difficulties, as equilibration is difficult to achieve and kinetic isotope effects can also influence the net fractionation. Measuring those values at low temperature in aqueous systems is even more challenging as diffusion is very slow and the exchange process is probably governed by dissolution–precipitation reactions, which may be associated with kinetic isotope effects. Nevertheless, significant work has been done, especially with regard to iron oxides and Fe^{2+} and Fe^{3+} dissolved in water.

Welch et al. (2003) tackled an important and difficult experimental problem, which is to determine the equilibrium fractionation factor between aqueous Fe(II) and Fe(III). The difficulty stems from the fact that those two species are mixed in solution and that to measure how iron isotopes are fractionated between them, one has to separate one from the other, which Welch et al. (2003) achieved by precipitating Fe(III). The difficulties associated with such precipitation is to ensure that no kinetic isotope effects are present during precipitation and that Fe(II) does not exchange isotopically with $Fe(III)_{aq}$ as the latter is precipitating, or one must account and correct for these effects. Two measurements were performed at 0 and 22 °C that yielded an equilibrium fractionation factor for the $^{56}Fe/^{54}Fe$ ratio between aqueous Fe(III) and Fe(II) that can be expressed as $0.334\times10^6/T^2-0.88$. The authors calculated the speciation of the Fe in solution and found that, for the compositions considered, they would be hexaaqua-coordinated, i.e., $[FeII(H_2O)_6]^{2+}$, $[FeIII(H_2O)_6]^{3+}$ and $[FeIII(H_2O)_5(OH)]^{2+}$

Much work has also been done on equilibrium isotopic fractionation factors between aqueous species and minerals. Skulan et al. (2002) found that the isotopic fractionation between $[FeIII(H_2O)_6]^{3+}$ and hematite depended on the rate of precipitation of hematite and by extrapolating the experimental results to a precipitation rate of zero, they inferred that the equilibrium fractionation between aqueous Fe(III) and hematite was -0.1 ± 0.2‰ for the $^{56}Fe/^{54}Fe$ ratio at 98 °C. Saunier et al. (2011) measured equilibrium fractionation between hematite and ferrous/ferric chloride iron complexes under hydrothermal conditions. When the fluid is dominated by Fe(III) chloride complexes, the isotopic fractionation between fluid and hematite is small: $+0.01\pm0.05$‰ for the $^{56}Fe/^{54}Fe$ ratio at 300 °C. When Fe(II) chloride complexes ($FeCl_2$ and $FeCl^+$) are dominant in the fluid, the fractionations are larger, reaching -0.36 ± 0.10‰ at 300 °C and $+0.10\pm0.12$‰ at 450 °C. The experiments of Skulan et al. (2002) and Saunier et al. (2011) were performed at different temperatures and correspond to different fluid speciations, but Saunier et al. (2011) pointed out that they may be difficult to reconcile with each other based on ab initio calculations that can be used to account for differences in the experimental setups. Wu et al. (2010) evaluated the influences of pH, presence of dissolved Si, and $Fe(II)_{aq}$/hematite ratio on $Fe(II)_{aq}$-hematite surface iron isotopic fractionation.

Wu et al. (2011) investigated the equilibrium fractionation between Fe(III)-bearing ferrihydrite (hydrous ferric oxide) and aqueous $Fe(II)_{aq}$. They obtained an equilibrium fractionation factor of $+3.2\pm0.1$‰ for the $^{56}Fe/^{54}Fe$ ratio at 25 °C. Wu et al. (2011, 2012a) found that this fractionation was also influenced by the presence of silica. Beard et al. (2010), Friedrich et al. (2014a), and Reddy et al. (2015) reported a much smaller equilibrium fractionation factor of $+1.05\pm0.08$‰ for the $^{56}Fe/^{54}Fe$ ratio at 22 °C between goethite and Fe(II). The latter value is puzzling because it is not clear what in the crystal chemistry of goethite would impart and equilibrium fractionation factor of ~+2‰ between hematite or ferrihydrite on the one hand, and goethite on the other hand.

Friedrich et al. (2014b) used the three isotope technique to estimate the extent of equilibrium fractionation between $Fe(II)_{aq}$ and magnetite and obtained a value of -1.56 ± 0.20‰ at 22 °C for the $^{56}Fe/^{54}Fe$ ratio.

Wiesli et al. (2004) measured the fractionation between Fe(II) and siderite. Siderite precipitation induced iron isotopic fractionation that followed a Rayleigh distillation model, from which they could estimate an equilibrium fractionation factor between aqueous Fe(II) and siderite of $+0.48\pm0.22$‰ at 20 °C. The extent to which these experiments could have been affected by kinetic effects is unclear.

Dideriksen et al. (2008) measured the equilibrium fractionation factor between aqueous Fe(III) and the siderophore complex Fe(IIII) desferrioxamine B. It is an important fractionation for the modern marine iron cycle as much of iron dissolved in seawater and accessible as a nutrient is in the form of siderophore complexes. They measured what they inferred to be an equilibrium fractionation factor of +0.60±0.15‰ between the siderophore complex and inorganic Fe(III) species Fe^{3+}_{aq} and $Fe(OH)^{2+}$ (those two species are not expected to be significantly fractionated relative to each other).

Guilbaud et al. (2011a) estimated the equilibrium fractionation factor between $Fe(II)_{aq}$ and mackinawite (FeS) at 25 and 2 °C. They obtained fractionations of −0.52±0.16 and −0.33±0.12‰ at 2 and 25 °C, respectively. They used the three isotope method to correct for incomplete fluid–mineral equilibration. Wu et al. (2012b) studied the same system and obtained a similar fractionation of −0.32±0.29‰ for the $^{56}Fe/^{54}Fe$ ratio between $Fe(II)_{aq}$ and mackinawite at 20 °C, indicating that the results are reproducible.

Syverson et al. (2014) investigated iron isotopic fractionation between vapor, halite, and liquid for application to seafloor hydrothermal vents, where phase separation can take place. The experiments were conducted at relatively high temperature (~420 to 470 °C) and the measured fractionations were small, not exceeding 0.1‰ for the most part. Hill and Schauble (2008) measured and predicted theoretically equilibrium iron isotopic fractionation in ferric aquo-chloro complexes. They equilibrated iron dissolved in aqueous solution of various chlorinities with $FeCl_4^-$ dissolved in immiscible diethyl ether. The immiscible diethyl ether played the role of a spectator phase that allowed them to investigate how iron coordination with chlorine affects its equilibrium isotope fractionation factor. At low chlorinity, the dominant species in the aqueous solution is $FeCl^{2+}$. At high chlorinity, the two dominant phases are $FeCl^{2+}$ and $FeCl_3$. Hill and Schauble (2008) found that the fractionation for $\delta^{56}Fe$ between $FeCl^{2+}$ in aqueous medium and $FeCl_4^-$ in ether is ~0.8‰ while the values goes down to near zero at high chlorinity, when the dominant species in the aqueous medium is $FeCl_3$.

Beta factors from NRIXS and ab initio approaches. The experimental methods outlined above can only provide relative fractionation factors, meaning differences in iron isotopic compositions of coexisting phases at equilibrium. Two approaches give access to absolute β-factors (or reduced partition function ratios).

The first approach is computational and it involves calculations grounded in physical chemistry of the vibration modes of molecules or minerals. Discussing the details of those approaches and their uncertainties is beyond the scope of the present review, so we will focus on discussing the modeling results. The reader is referred to the review paper of Schauble (2004), which provides a comprehensive introduction to this topic. A wide variety of phases have been calculated with this approach. To give a more palatable overview of the work done, we have summarized the results of those calculations in Table 1. Sometimes, the β-factor is only given at one or a few temperature values. In those cases, we used a similar approach to Dauphas et al. (2012) to calculate the coefficients of the polynomial that gives the temperature dependence of the β-factor. The β-factor can be written as a function of the force constant of iron bonds as,

$$1000\ln\beta = B_1\frac{F}{T^2} - B_2\frac{F^2}{T^4} + B_3\frac{F^3}{T^6} \tag{19}$$

where $B_1 = 2904\,\mathrm{m.N^{-1}.K^2}$ is a true constant and the two other coefficients B_2 and B_3 depend on the shape of the phonon density of states of the iron sublattice (PDOS; energy distribution of lattice vibrations; through statistical mechanics, this distribution relates to the thermodynamics of the mineral). For a Debye PDOS, $B_2 = 37538\,\mathrm{m^2.N^{-2}.K^4}$. In practice, the coefficients vary little from phase to phase or their variations have little influence on the calculated β-factors because these are higher order terms. To express the temperature dependence of β given a few T values, we can write,

Table 1. Iron β-factors ($^{56}Fe/^{54}Fe$) from ab-initio and NRIXS studies (high-P phases are not included).

			$1000 \times \ln\beta$ ($^{56}Fe/^{54}Fe$) $= A_1/T^2 + A_2/T^4 + A_3/T^6$			$10^3 \ln\beta$				
Phase	**Note**	**Method**	A_1 (×10⁶)	A_2 (×10⁹)	A_3 (×10¹²)	**22 °C**	**100 °C**	**500 °C**	**1200 °C**	**References**
Fe(gas)			0	0	0	0.0	0.0	0.00	0.00	
Magnetite		NRIXS	0.64943	−3.2094	2.76200	7.0	4.5	1.08	0.30	Polyakov et al. (2007)
FeO (wüstite)		NRIXS	0.45569	−1.5357	9.89950	5.0	3.2	0.76	0.21	Polyakov et al. (2007)
Troilite		NRIXS	0.29735	−0.6693	0.30274	3.3	2.1	0.50	0.14	Polyakov et al. (2007)
Fe_3S		NRIXS	0.32109	−0.5830	0.18086	3.6	2.3	0.54	0.15	Polyakov et al. (2007)
Chalcopyrite		NRIXS	0.48888	−1.7631	0.95290	5.4	3.4	0.81	0.22	Polyakov and Soultanov (2011)
Magnetite		NRIXS	0.65756	−3.0366		7.1	4.6	1.09	0.30	Dauphas et al. (2012)
Troilite		NRIXS	0.29101	−0.6353		3.3	2.1	0.49	0.13	Dauphas et al. (2012)
Fe_3S		NRIXS	0.34150	−0.7702		3.8	2.4	0.57	0.16	Dauphas et al. (2012)
Chalcopyrite		NRIXS	0.41654	−1.2068		4.6	2.9	0.69	0.19	Dauphas et al. (2012)
Orthoenstatite		NRIXS	0.50403	−2.0501		5.5	3.5	0.84	0.23	Dauphas et al. (2012)
Hematite		NRIXS	0.69649	−3.2728		7.6	4.8	1.16	0.32	Dauphas et al. (2012)
$(Mg_{0.75}Fe_{0.25})O$		NRIXS	0.51354	−1.7475		5.7	3.6	0.85	0.24	Dauphas et al. (2012)
α−Fe (bcc)		NRIXS	0.49970	−1.1424		5.6	3.5	0.83	0.23	Dauphas et al. (2012)
α−Fe (bcc)		NRIXS	0.52781	−1.2747		5.9	3.7	0.88	0.24	Dauphas et al. (2012)
α−$Fe_{0.525}Cr_{0.475}$ (bcc)		NRIXS	0.44507	−0.8841		5.0	3.2	0.74	0.20	Dauphas et al. (2012)
σ−$Fe_{0.525}Cr_{0.475}$		NRIXS	0.44792	−0.9808		5.0	3.2	0.75	0.21	Dauphas et al. (2012)
γ−Fe (fcc) 80nm particles		NRIXS	0.39657	−0.6982		4.5	2.8	0.66	0.18	Dauphas et al. (2012)

			$1000 \times \ln\beta\ (^{56}Fe/^{54}Fe) = A_1/T^2 + A_2/T^4 + A_3/T^6$			$10^3 \ln\beta$				
Phase	Note	Method	A_1 (×10⁶)	A_2 (×10⁹)	A_3 (×10¹²)	22 °C	100 °C	500 °C	1200 °C	References
Goethite		NRIXS	0.76590	−5.3010	93.18000	8.2	5.3	1.27	0.35	Blanchard et al. (2015)
H–Jarosite		NRIXS	0.79790	−5.4750	72.57000	8.5	5.5	1.32	0.37	Blanchard et al. (2015)
K–Jarosite		NRIXS	0.82870	−6.2890	117.30000	8.9	5.7	1.37	0.38	Blanchard et al. (2015)
Olivine		NRIXS	0.56100	−2.5900	30.40000	6.1	3.9	0.93	0.26	Dauphas et al. (2014)
Fe^{2+} in basalt, andesite, dacite		NRIXS	0.56990	−3.7024	66.68200	6.2	3.9	0.94	0.26	Dauphas et al. (2014)
Fe^{3+} in basalt, andesite, dacite		NRIXS	1.00159	−8.9240	172.49790	10.6	6.8	1.65	0.46	Dauphas et al. (2014)
Fe^{2+} in rhyolite		NRIXS	0.68297	−5.3311	82.01470	7.3	4.7	1.13	0.31	Dauphas et al. (2014)
Fe^{3+} in rhyolite		NRIXS	1.10381	−9.8285	173.90380	11.6	7.5	1.82	0.51	Dauphas et al. (2014)
Fe^{2+} in MgFeAl spinel		NRIXS	0.54341	−3.7432	75.89850	5.9	3.7	0.90	0.25	Roskosz et al. (2015)
Fe^{3+} in MgFeAl spinel		NRIXS	0.86294	−6.3164	102.77160	9.2	5.9	1.43	0.40	Roskosz et al. (2015)
Troilite	*	NRIXS	0.27674	−0.5821	3.47588	3.1	2.0	0.46	0.13	Krawczynski et al. (2014)
γ–Fe–Ni (fcc)	*	NRIXS	0.41939	−1.3368	12.09765	4.7	2.9	0.70	0.19	Krawczynski et al. (2014)
Ilmenite	*	NRIXS	0.44507	−1.5054	14.45849	4.9	3.1	0.74	0.20	Williams et al. (2016)
Pyrite		ab initio	0.82913	−3.2161	17.88867	9.1	5.8	1.38	0.38	Blanchard et al. (2009)
Hematite		ab initio	0.66267	−3.1255	24.75800	7.2	4.6	1.10	0.30	Blanchard et al. (2009)
Siderite		ab initio	0.38180	−1.5765	24.56000	4.2	2.7	0.63	0.18	Blanchard et al. (2009)
Goethite		ab initio	0.68320	−3.9026	45.41467	7.4	4.7	1.13	0.31	Blanchard et al. (2015)
Hematite	*	ab initio	0.67790	−3.4926	51.09130	7.4	4.7	1.12	0.31	Rustad and Dixon (2009)

			$1000 \times \ln\beta$ ($^{56}Fe/^{54}Fe$) $= A_1/T^2 + A_2/T^4 + A_3/T^6$			$10^3 \ln\beta$				
Phase	**Note**	**Method**	A_1 (×10⁶)	A_2 (×10⁹)	A_3 (×10¹²)	**22 °C**	**100 °C**	**500 °C**	**1200 °C**	**References**
Fe^{3+}_{aq}	*	ab initio	0.76983	−4.5128	75.25849	8.4	5.3	1.28	0.35	Rustad and Dixon (2009)
Fe^{2+}_{aq}	*	ab initio	0.46976	−1.6877	17.32170	5.2	3.3	0.78	0.22	Rustad and Dixon (2009)
$[FeII(CN)_6]^{4-}$	*	ab initio	1.72962	−22.7506	850.21972	18.1	11.6	2.83	0.79	Schauble et al. (2004)
$[FeIII(CN)_6]^{3-}$	*	ab initio	1.36547	−14.1740	417.86430	14.4	9.2	2.25	0.63	Schauble et al. (2004)
$[FeIIIBr_4]^-$	*	ab initio	0.59811	−2.7188	35.08997	6.6	4.2	0.99	0.28	Schauble et al. (2004)
α-Fe (bcc)	*	ab initio	0.49377	−1.8531	19.74768	5.5	3.5	0.82	0.23	Schauble et al. (2004)
Fe^{3+}_{aq}	*	ab initio	1.10968	−9.3586	224.09873	11.8	7.6	1.83	0.51	Schauble et al. (2004)
Fe^{2+}_{aq}	*	ab initio	0.57054	−2.4741	30.46468	6.3	4.0	0.95	0.26	Schauble et al. (2004)
$[FeIII(H_2O)_4Cl_2]^+$	*	ab initio	0.83044	−5.2414	93.93647	9.0	5.7	1.38	0.38	Schauble et al. (2004)
$[FeIII(H_2O)_3Cl_3]^0$	*	ab initio	0.72793	−4.0275	63.27410	7.9	5.0	1.21	0.33	Schauble et al. (2004)
$[FeIIICl_4]^-$	*	ab initio	0.66964	−3.4080	49.24720	7.3	4.7	1.11	0.31	Schauble et al. (2004)
$[FeIICl_4]^{2-}$	*	ab initio	0.36518	−1.0135	7.98695	4.1	2.6	0.61	0.17	Schauble et al. (2004)
$[FeIIICl_6]^{3-}$	*	ab initio	0.35579	−0.9626	7.39929	4.0	2.5	0.59	0.16	Schauble et al. (2004)
Fe^{3+}_{aq}	*	ab initio	0.88735	−5.9850	114.63554	9.6	6.1	1.47	0.41	Anbar et al. (2005)
Fe^{2+}_{aq}	*	ab initio	0.62685	−2.9874	40.43608	6.9	4.4	1.04	0.29	Anbar et al. (2005)
$[FeII(H_2O)_6]^{2+}$		ab initio	0.49264	−2.6344	23.86742	5.3	3.4	0.82	0.23	Ottonello and Zuccolini (2009)
$[FeII(H_2O)_{18}]^{2+}$		ab initio	0.52897	−3.9712	60.78421	5.6	3.6	0.87	0.24	Ottonello and Zuccolini (2009)

		$1000 \times \ln\beta$ ($^{56}Fe/^{54}Fe$) = $A_1/T^2 + A_2/T^4 + A_3/T^6$				$10^3 \ln\beta$				
Phase	**Note**	**Method**	**A_1 (×10^6)**	**A_2 (×10^9)**	**A_3 (×10^{12})**	**22 °C**	**100 °C**	**500 °C**	**1200 °C**	**References**
$[FeII(H_2O)_5(OH)]^+$		ab initio	0.56994	−4.8733	65.98762	6.0	3.9	0.94	0.26	Ottonello and Zuccolini (2009)
$[FeII(H_2O)_4(OH)]^+$		ab initio	0.55521	−4.8825	61.46983	5.8	3.8	0.92	0.25	Ottonello and Zuccolini (2009)
$[FeII(H_2O)_4(OH)_2]^0$		ab initio	0.64249	−6.5575	96.42366	6.7	4.3	1.06	0.29	Ottonello and Zuccolini (2009)
$[FeII(H_2O)_2(OH)_2]^0$		ab initio	0.65228	−7.4150	105.04000	6.7	4.3	1.07	0.30	Ottonello and Zuccolini (2009)
$[FeII(H_2O)_3(OH)_3]^-$		ab initio	0.71362	−7.4312	96.50000	7.4	4.8	1.17	0.33	Ottonello and Zuccolini (2009)
$[FeII(OH)_3]^-$		ab initio	0.70670	−7.4079	88.35000	7.3	4.7	1.16	0.32	Ottonello and Zuccolini (2009)
$[FeIII(H_2O)_6]^{3+}$		ab initio	0.84274	−6.8500	84.66667	8.9	5.7	1.39	0.39	Ottonello and Zuccolini (2009)
$[FeIII(H_2O)_5(OH)]^{2+}$		ab initio	0.93316	−11.6355	180.31175	9.5	6.2	1.53	0.43	Ottonello and Zuccolini (2009)
$[FeIII(H_2O)_4(OH)]^{2+}$		ab initio	0.96969	−12.4540	196.00000	9.8	6.4	1.59	0.44	Ottonello and Zuccolini (2009)
$[FeIII(H_2O)_4(OH)_2]^+$		ab initio	0.96685	−12.8868	202.29194	9.7	6.4	1.58	0.44	Ottonello and Zuccolini (2009)
$[FeIII(H_2O)_3(OH)_2]^+$		ab initio	1.00359	−13.2628	202.20000	10.1	6.6	1.64	0.46	Ottonello and Zuccolini (2009)
$[FeIII(H_2O)_3(OH)_3]^0$		ab initio	0.96638	−11.5614	167.14000	9.8	6.4	1.59	0.44	Ottonello and Zuccolini (2009)
$[FeIII(OH)_3]^0$		ab initio	1.01056	−15.3823	237.31089	9.9	6.6	1.65	0.46	Ottonello and Zuccolini (2009)
$[FeIII(H_2O)_2(OH)_4]^-$		ab initio	1.00662	−11.0661	145.07086	10.3	6.7	1.65	0.46	Ottonello and Zuccolini (2009)
$[FeIII(OH)_4]^-$		ab initio	1.06690	−11.8236	147.36411	10.9	7.1	1.75	0.49	Ottonello and Zuccolini (2009)
$[FeIII(H_2O)_6]^{3+}$	*	ab initio	0.87065	−5.7610	108.23681	9.4	6.0	1.44	0.40	Hill and Schauble (2008)
$[FeIIICl(H_2O)_5]^{2+}$	*	ab initio	0.78359	−4.6664	78.90463	8.5	5.4	1.30	0.36	Hill and Schauble (2008)
$FeIIICl(H_2O)_4]^+$	*	ab initio	0.73282	−4.0814	64.54233	8.0	5.1	1.21	0.34	Hill and Schauble (2008)
$FeCl_3(H_2O)_3]^0$ octahedral fac	*	ab initio	0.64815	−3.1928	44.65530	7.1	4.5	1.08	0.30	Hill and Schauble (2008)
$FeCl_3(H_2O)_3]^0$ octahedral mer	*	ab initio	0.66164	−3.3270	47.50126	7.2	4.6	1.10	0.30	Hill and Schauble (2008)

		$1000 \times \ln\beta$ ($^{56}Fe/^{54}Fe$) $= A_1/T^2 + A_2/T^4 + A_3/T^6$				$10^3 \ln\beta$				
Phase	Note	Method	A_1 (×10^6)	A_2 (×10^9)	A_3 (×10^{12})	22 °C	100 °C	500 °C	1200 °C	References
$FeCl_3(H_2O)_3]^0$ trigonal bipyramidal	*	ab initio	0.70194	−3.7447	56.72098	7.7	4.9	1.16	0.32	Hill and Schauble (2008)
$[FeCl_4]^-$	*	ab initio	0.67188	−3.4308	49.74153	7.3	4.7	1.11	0.31	Hill and Schauble (2008)
$[FeCl_5(H_2O)]^{2-}$	*	ab initio	0.45918	−1.6025	15.87837	5.1	3.2	0.76	0.21	Hill and Schauble (2008)
$[FeCl_5]^{2-}$	*	ab initio	0.47729	−1.7313	17.83183	5.3	3.3	0.79	0.22	Hill and Schauble (2008)
$[FeCl_6]^{3-}$	*	ab initio	0.32724	−0.8139	5.74701	3.7	2.3	0.55	0.15	Hill and Schauble (2008)
$[Fe(H_2O)_6]^{2+}$	*	ab initio	0.53168	−2.1484	24.64819	5.9	3.7	0.88	0.24	Hill and Schauble (2008)
$[Fe(H_2O)_6]^{3+}\cdot 12(H_2O)$	*	ab initio	0.94089	−6.7281	136.60294	10.1	6.5	1.56	0.43	Hill and Schauble (2008)
$[FeCl_4]^-.12(H_2O)$	*	ab initio	0.68395	−3.5552	52.47019	7.5	4.7	1.13	0.31	Hill and Schauble (2008)
$[FeIII(H_2O)_6]^{3+}$	*	ab initio	0.80447	−4.9193	85.42416	8.7	5.6	1.33	0.37	Domagal–Goldman and Kubicki (2008)
$[FeII(H_2O)_6]^{2+}$	*	ab initio	0.53143	−2.1466	24.61976	5.9	3.7	0.88	0.24	Domagal–Goldman and Kubicki (2008)
$[FeIII(O_x)_3]^{3-}$	*	ab initio	0.73720	−4.1304	65.71121	8.0	5.1	1.22	0.34	Domagal–Goldman and Kubicki (2008)
$[FeII(O_x)_3]^{4-}$	*	ab initio	0.45825	−1.5959	15.78170	5.1	3.2	0.76	0.21	Domagal–Goldman and Kubicki (2008)
$[FeIII(Cat)_3]^{3-}$	*	ab initio	0.67939	−3.5088	51.46783	7.4	4.7	1.13	0.31	Domagal–Goldman and Kubicki (2008)
$[FeII(Cat)_3]^{4-}$	*	ab initio	0.41094	−1.2835	11.38358	4.6	2.9	0.68	0.19	Domagal–Goldman and Kubicki (2008)
$FeII(H_2O)_6{}^{2+}$	*	ab initio	0.47192	−1.6927	17.23970	5.2	3.3	0.78		Fujii et al. (2014)
$FeIICl(H_2O)_5{}^+$	*	ab initio	0.45038	−1.5417	14.98583	5.0	3.2	0.75		Fujii et al. (2014)
$FeIICl_2(H_2O)_4$	*	ab initio	0.42797	−1.3921	12.85772	4.7	3.0	0.71		Fujii et al. (2014)
$FeIISO_4(H_2O)_5$	*	ab initio	0.50398	−1.9310	21.01328	5.6	3.5	0.84		Fujii et al. (2014)

Phase	Note	Method	A_1 (×10^6)	A_2 (×10^9)	A_3 (×10^{12})	22 °C	100 °C	500 °C	1200 °C	References
		$1000 \times \ln\beta$ ($^{56}Fe/^{54}Fe$) = $A_1/T^2 + A_2/T^4 + A_3/T^6$				$10^3 \ln\beta$				
$FeIIOH(H_2O)_5^+$	*	ab initio	0.51383	−2.0075	22.27952	5.7	3.6	0.85		Fujii et al. (2014)
$FeII(OH)_2(H_2O)_4$	*	ab initio	0.50772	−1.9608	21.52108	5.6	3.6	0.84		Fujii et al. (2014)
$FeII_2(OH)_6^{2-}$	*	ab initio	0.50268	−1.9206	20.83898	5.5	3.5	0.84		Fujii et al. (2014)
$FeIIHCO_3(H_2O)_4^+$	*	ab initio	0.48936	−1.8202	19.22527	5.4	3.4	0.81		Fujii et al. (2014)
$FeIICO_3(H_2O)_4$	*	ab initio	0.54274	−2.2392	26.23650	6.0	3.8	0.90		Fujii et al. (2014)
$FeIIHS(H_2O)_5^+$	*	ab initio	0.40735	−1.2612	11.08847	4.5	2.9	0.68		Fujii et al. (2014)
$FeII(HS)_2(H_2O)_4$	*	ab initio	0.36475	−1.0112	7.96105	4.1	2.6	0.61		Fujii et al. (2014)
$FeII_2S_2(H_2O)_4$	*	ab initio	0.44661	−1.5160	14.61071	4.9	3.1	0.74		Fujii et al. (2014)
$FeIIH_2PO_4(H_2O)_5^+$	*	ab initio	0.49486	−1.8613	19.88031	5.5	3.5	0.82		Fujii et al. (2014)
$FeIIHPO_4(H_2O)_5$	*	ab initio	0.54965	−2.2965	27.24806	6.0	3.8	0.91		Fujii et al. (2014)
$FeIIH_4(PO_4)_2(H_2O)_4$	*	ab initio	0.49161	−1.8369	19.48944	5.4	3.4	0.82		Fujii et al. (2014)
$FeIIH_3(PO_4)_2(H_2O)_4^-$	*	ab initio	0.48697	−1.8027	18.95237	5.4	3.4	0.81		Fujii et al. (2014)
$FeIIH(cit)(H_2O)_3$	*	ab initio	0.48664	−1.8000	18.90732	5.4	3.4	0.81		Fujii et al. (2014)
$FeII(cit)(H_2O)^{3-}$	*	ab initio	0.48896	−1.8174	19.18295	5.4	3.4	0.81		Fujii et al. (2014)
$FeIIH(cit)_2^{3-}$	*	ab initio	0.42467	−1.3707	12.56175	4.7	3.0	0.71		Fujii et al. (2014)
$FeII(cit)_2^{4-}$	*	ab initio	0.35424	−0.9537	7.29000	4.0	2.5	0.59		Fujii et al. (2014)
$FeII(cit)_2OH^{5-}$	*	ab initio	0.62871	−3.0047	40.77936	6.9	4.4	1.04		Fujii et al. (2014)
$FeIII(H_2O)_6^{3+}$	*	ab initio	0.76263	−4.4209	72.77484	8.3	5.3	1.26		Fujii et al. (2014)
$FeIIICl(H_2O)_5^{2+}$	*	ab initio	0.68879	−3.6060	53.60719	7.5	4.8	1.14		Fujii et al. (2014)
$FeIIICl_2(H_2O)_4^+$	*	ab initio	0.64367	−3.1490	43.74452	7.0	4.5	1.07		Fujii et al. (2014)

			$1000\times\ln\beta$ ($^{56}Fe/^{54}Fe$) $= A_1/T^2 + A_2/T^4 + A_3/T^6$			$10^3\ln\beta$				
Phase	**Note**	**Method**	A_1 (×10⁶)	A_2 (×10⁹)	A_3 (×10¹²)	**22 °C**	**100 °C**	**500 °C**	**1200 °C**	**References**
$FeIIISO_4(H_2O)_5^+$	*	ab initio	0.80033	−4.8696	84.15302	8.7	5.5	1.33		Fujii et al. (2014)
$FeIIIOH(H_2O)_5^{2+}$	*	ab initio	0.86462	−5.6852	106.21067	9.3	6.0	1.43		Fujii et al. (2014)
$FeIII(OH)_2(H_2O)_4^+$	*	ab initio	0.89166	−6.0474	116.55276	9.6	6.1	1.48		Fujii et al. (2014)
$FeIII(OH)_3(H_2O)_3$	*	ab initio	0.89990	−6.1582	119.72347	9.7	6.2	1.49		Fujii et al. (2014)
$FeIIIHCO_3(H_2O)_4^{2+}$	*	ab initio	0.73461	−4.1018	65.03534	8.0	5.1	1.22		Fujii et al. (2014)
$FeIIICO_3(H_2O)_4^+$	*	ab initio	0.74190	−4.1840	67.01201	8.1	5.1	1.23		Fujii et al. (2014)
$FeIIIH_3PO_4(H_2O)_5^{3+}$	*	ab initio	0.79978	−4.8619	83.92847	8.7	5.5	1.32		Fujii et al. (2014)
$FeIIIH_2PO_4(H_2O)_5^{2+}$	*	ab initio	0.81651	−5.0680	89.33434	8.8	5.6	1.35		Fujii et al. (2014)
$FeIIIHPO_4(H_2O)_5^+$	*	ab initio	0.86912	−5.7426	107.76645	9.4	6.0	1.44		Fujii et al. (2014)
$FeIIIH_5(PO_4)_2(H_2O)_4^{2+}$	*	ab initio	0.81892	−5.0977	90.11474	8.9	5.7	1.36		Fujii et al. (2014)
$FeIIIH_4(PO_4)_2(H_2O)_4^+$	*	ab initio	0.82966	−5.2326	93.72130	9.0	5.7	1.37		Fujii et al. (2014)
$FeIIIH_3(PO_4)_2(H_2O)_4$	*	ab initio	0.84821	−5.4693	100.15822	9.2	5.8	1.40		Fujii et al. (2014)
$FeIIIH_7(PO_4)_3(H_2O)_3^+$	*	ab initio	0.82181	−5.1337	91.06875	8.9	5.7	1.36		Fujii et al. (2014)
$FeIIIH_6(PO_4)_3(H_2O)_3$	*	ab initio	0.85591	−5.5683	102.86933	9.2	5.9	1.42		Fujii et al. (2014)
$FeIII(cit)(H_2O)_3$	*	ab initio	0.84780	−5.4635	99.98582	9.2	5.8	1.40		Fujii et al. (2014)
$FeIII(cit)OH(H_2O)_2^-$	*	ab initio	0.81552	−5.0560	89.02412	8.8	5.6	1.35		Fujii et al. (2014)
$FeIIIH(cit)_2^{2-}$	*	ab initio	0.80178	−4.8859	84.54333	8.7	5.5	1.33		Fujii et al. (2014)
$FeIII(cit)_2^{3-}$	*	ab initio	0.68018	−3.5161	51.61026	7.4	4.7	1.13		Fujii et al. (2014)
$FeIII(cit)_2OH_4^-$	*	ab initio	0.65511	−3.2623	46.13236	7.2	4.6	1.09		Fujii et al. (2014)

*For studies that report equilibrium fractionation at just a few temperatures or just give a force constant, the coefficients were calculated by writing (with $x = A_1/T^2$): $1000\times\ln\beta\,(^{56}Fe/^{54}Fe) = x - 7.6\times10^{-3}x^2 + 164\times10^{-6}x^3$ where the coefficients were determined by regressing A_2 vs A_1^2 and A^3 vs A_1^3 for the data that have all three coefficients reported.

$$1000\ln\beta = \frac{A_1}{T^2} + C\frac{A_1^{\,2}}{T^4} + D\frac{A_1^{\,3}}{T^6} \quad (20)$$

The coefficients C and D were obtained by regressing A_2 vs. $A_1^{\,2}$ and A_3 vs. $A_1^{\,3}$ from Equation (17), for the phases for which the full polynomial expansion was provided in either NRIXS or ab initio studies. The one-parameter formula to calculate the temperature dependence of $1000\ln\beta$ given as little as one data point is (Fig. 5 ; note that a similar approach can be used for other isotope systems),

$$1000\ln\beta = \frac{A_1}{T^2} - 7.6\times10^{-3}\frac{A_1^{\,2}}{T^4} + 164\times10^{-6}\frac{A_1^{\,3}}{T^6} \quad (21)$$

Much of the early work to calculate β-factors focused on aqueous species. With modern softwares, calculating β-factors is relatively straightforward and a significant fraction of the work involves validating the calculations by comparing the results with existing spectroscopic data and equilibration experiments. Not all studies do that and the reader should exercise caution in using indiscriminately calculated β-factors from the litterature. Schauble et al. (2001) calculated the β-factors of $[FeII(CN)_6]^{4-}$, $[FeIII(CN)_6]^{3-}$, $[FeIII(H_2O)_6]^{3+}$,

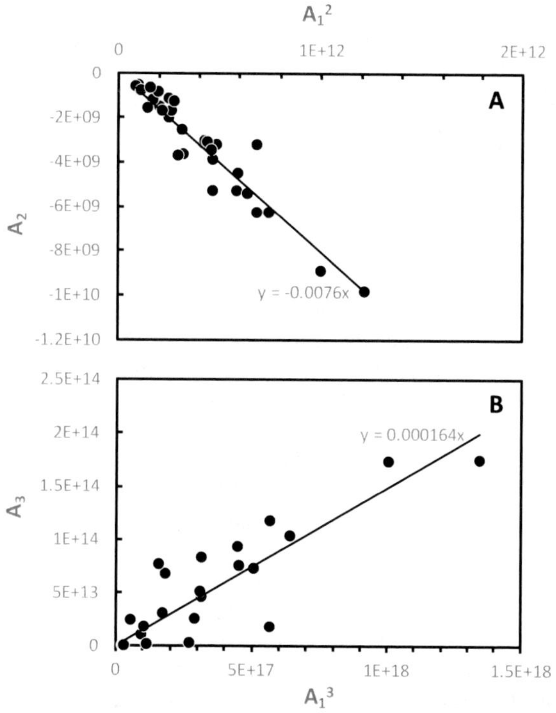

Figure 5. Correlations between the coefficients of the polynomial expansion of β equilibrium fractionation factors. The values of the coefficients of the temperature expansion to the 3rd order are available for some phases from NRIXS and ab initio studies; 1000 $\ln\beta = A_1/T^2 + A_2/T^4 + A_3/T^6$. To a good approximation (see Dauphas et al. 2012 for a theoretical justification), this can be written as 1000 $\ln\beta = A_1/T^2 + CA_1^{\,2}/T^4 + D\,A_1^{\,3}/T^6$ (Eqn. 20), with C and D constant. The values of C and D are calculated by regressing A_2 vs. $A_1^{\,2}$ and A_2 vs. $A_1^{\,3}$, respectively.

$[FeIII(H_2O)_4Cl_2]^+$, $[FeIII(H_2O)_3Cl_3]^0$, $[FeIIICl_4]^-$, $[FeIIIBr_4]^-$, $[FeII(H_2O)_6]^{2+}$, α-Fe metal, $[FeIICl_4]^{2-}$, $[FeIIICl_6]^{3-}$. Anbar et al. (2005) and Jarzecki et al. (2004) calculated the β-factors of $[FeIII(H_2O)_6]^{3+}$ and $[FeII(H_2O)_6]^{2+}$, paying attention to the treatment of the solvation shell. Ottonello and Zuccolini (2009) also studied those two important species as well as many other hydrous complexes, $[Fe(II)(H_2O)_{18}]^{2+}$, $[Fe(II)(H_2O)_5(OH)]^+$, $[Fe(II)(H_2O)_4(OH)]^+$, $[Fe(II)(H_2O)_4(OH)_2]^0$, $[Fe(II)(H_2O)_2(OH)_2]^0$, $[Fe(II)(H_2O)_3(OH)_3]^-$, $[Fe(II)(OH)_3]^-$, $[Fe(III)(H_2O)_6]^{3+}$, $[Fe(III)(H_2O)_5(OH)]^{2+}$, $[Fe(III)(H_2O)(OH)]^{2+}$, $[Fe(III)(H_2O)_4(OH)_2]^+$, $[Fe(III)(H_2O)_3(OH)_2]^+$, $[Fe(III)(H_2O)_3(OH)_3]^0$, $[Fe(III)(OH)_3]^0$, $[Fe(III)(H_2O)_2(OH)_4]^-$, $[Fe(III)(OH)_4]^-$. Hill and Schauble (2008) calculated the β-factors of ferric aquo-chloro compounds $[FeIII(H_2O)_6]^{3+}$, $[FeIIICl(H_2O)_5]^{2+}$, $[FeIII(H_2O)_4Cl_2]^+$, $[FeIII(H_2O)_3Cl_3]^0$, $[FeIII(H_2O)_2Cl_3]^0$, $[FeIIICl_4]^-$, $[FeIIICl_5]^{2-}$, $[FeIIICl_6]^{3-}$ to evaluate the influence of the coordination environment of iron. Hill et al. (2009) compared those results with laboratory experiments that they performed. Domagal-Goldman and Kubicki (2008) studied iron equilibrium fractionation factors involving redox changes and organic complexation. They specifically estimated the β-factors of $[FeIII(H_2O)_6]^{3+}$, $[FeII(H_2O)_6]^{2+}$, $[FeIII(Ox)_3]^{3-}$, $[FeII(Ox)_3]^{4-}$, $[FeIII(Cat)_3]^{3-}$, $[FeII(Cat)_3]^{4-}$, where those last 4 molecules refer to iron trisoxalate and triscatecholate. Moynier et al. (2013) also studied the β-factors of iron bound to organic molecules, namely Fe(II)-citrate, Fe(III)-citrate, Fe(II)-nicotianamine, and Fe(III)-phytosiderophore. Fujii et al. (2006) studied the β-factors of $[Fe(III)Cl_4]$- and $[Fe(III)Cl_2(H_2O)_4]^+$. Rustad and Dixon (2009) calculated the β-factors of $Fe(II)_{aq}$, $Fe(IIII)_{aq}$. Fujii et al. (2014) calculated the β-factors of various Fe(II) iron complexes relevant to geochemical and biological systems, namely $Fe(H_2O)_6{}^{2+}$, $FeCl(H_2O)_5{}^+$, $FeCl_2(H_2O)_4$, $FeSO_4(H_2O)_5$, $FeOH(H_2O)_5{}^+$, $Fe(OH)_2(H_2O)_4$, $Fe_2(OH)_6{}^{2-}$, $FeHCO_3(H_2O)_4{}^+$, $FeHCO_3(H_2O)_4$, $FeHS(H_2O)_5{}^+$, $Fe(HS)_2(H_2O)_4$, $Fe_2S_2(H_2O)_4$, $FeH_2PO_4(H_2O)_5{}^+$, $FeHPO_4(H_2O)_5$, $FeH_4(PO_4)_2(H_2O)_4$, $FeH_3(PO_4)_2(H_2O)_4{}^-$, $FeH(cit)(H_2O)_3$, $Fe(cit)(H_2O)_3{}^-$, $FeH(cit)_2{}^{3-}$, $Fe(cit)_2{}^{4-}$, and $Fe(cit)_2OH^{5-}$. They also studied Fe(III) species, namely $Fe(H_2O)_6{}^{3+}$, $FeCl(H_2O)_5{}^{2+}$, $FeCl_2(H_2O)_4{}^+$, $FeSO_4(H_2O)_5{}^+$, $FeOH(H_2O)_5{}^{2+}$, $Fe(OH)_2(H_2O)_4{}^+$, $Fe(OH)_3(H_2O)_3$, $FeHCO_3(H_2O)_4{}^{2+}$, $FeCO_3(H_2O)_4{}^+$, $FeH_3PO_4(H_2O)_5{}^{3+}$, $FeH_2PO_4(H_2O)_5{}^{2+}$, $FeHPO_4(H_2O)_5{}^+$, $FeH_5(PO_4)_2(H_2O)_4{}^{2+}$, $FeH_4(PO_4)_3(H_2O)_4{}^+$, $FeH_3(PO_4)_2(H_2O)_4$, $FeH_7(PO_4)_3(H_2O)_3{}^+$, $FeH_6(PO_4)_3(H_2O)_3$, $Fe(cit)(H_2O)_3$, $Fe(cit)OH(H_2O)_2{}^-$, $FeH(cit)_2{}^{2-}$, $Fe(cit)_2{}^{3-}$, $Fe(cit)_2OH^{4-}$.

Calculating β-factors of minerals is more involved than for aqueous species, and as a result, the database is much less extensive and the number of independent studies is also much lower. Rustad and Dixon (2009) calculated the β-factors of $Fe(II)_{aq}$, $Fe(III)_{aq}$, and hematite. Blanchard et al. (2009) calculated the β-factors of pyrite, hematite, and siderite. Blanchard et al. (2010) also evaluated the effect of Al_2O_3 substitution in hematite on the β-factor of iron. Blanchard et al. (2015) calculated the β-factor of goethite and compared the predictions with estimates from the NRIXS method. Rustad and Yin (2009) calculated β-factors of ferropericlase and ferroperovskite (bridgmanite) in *P–T* conditions relevant to Earth's lower mantle.

The β-factors can also be measured experimentally by using methods that rely on the fact that iron possesses a Mössbauer isotope. Polyakov and Mineev (2000) used the second order Doppler shift measured in conventional Mössbauer spectroscopy to derive the kinetic energy of the iron atoms, from which they could calculate iron fractionation factors for a variety of minerals based on literature data: aegirine, akakenite, ankerite, bernalite, celadonite, chloritoid, diopside, enstatite, ferrite, ferrochromite, ferrocyanide, goethite, grandidierite, hedenbergite, hematite, ilmenite, lepidocrocite, magnetite, metallic iron, magnesium oxide, nickel sulfide, nitroprusside, olivine, pyrite, and siderite. This study revealed some of the fundamental crystal chemical controls on iron stable isotope variations. Polyakov et al. (2007) subsequently reported β-factors for iron metal, hematite, pyrite, and marcasite. Polyakov and Soultanov (2011) calculated β-factors for mackinawite using conventional Mössbauer data. While sound theoretically, determining the second order Doppler shift is fraught with difficulties and some of the β-factors reported in that publication were later shown to be erroneous when better quality Mössbauer data were acquired (Polyakov et al. 2007). Another, more robust approach to this problem is to use a synchrotron technique known as Nuclear Resonant Inelastic X-ray Scattering (NRIXS) (Polyakov et al. 2007; Dauphas et al. 2012). The Mössbauer isotope ^{57}Fe has a low lying nuclear excited state that can be reached by X-ray photons of 14.4125 keV, the nominal resonance energy. In NRIXS, the energy of the incoming beam is changed in small steps within an energy range as large as −200 to +200 meV around the nominal resonance energy. The energy resolution of the incoming beam can be reduced to 1.3 meV full width at half maximum, which requires the use of sophisticated X-ray monochromators. At each energy, the scattered X-rays induced by the nuclear transition are analyzed. The prompt X-rays scattered by electrons are discarded by imposing some time discrimination, as electronic scattering is almost instantaneous, while nuclear scattering is delayed due to the finite lifetime (141 ns) of the excited ^{57}Fe nucleus.

When the energy of the incoming X-rays is lower than the nominal resonance energy, some nuclear transitions can still occur because lattice vibrations can fill the gap by providing energy in the form of phonons (the particle-like equivalent of interatomic vibrations). Conversely, when the energy is higher than the nominal resonance energy, some of that extra energy can be absorbed by the solid lattice and the transition can still take place. The first process is known as phonon annihilation while the second process is known as phonon creation. By measuring the flux of the scattered X-rays for various values of the incident X-ray energy, one can thus probe the vibrational properties of the target material, which govern equilibrium fractionation factors (Bigeleisen and Mayer 1947; Urey 1947; Kieffer 1982; Dauphas et al. 2012). Two approaches exist to retrieve β-factors from NRIXS spectra. Polyakov et al. (2007) used the kinetic energy and first order perturbation theory to calculate the β-factor from the PDOS of the iron sublattice $g(E)$. The PDOS is calculated using a Log Fourier transform of the raw signal $S(E)$. Dauphas et al. (2012) used a different approach based on the determination of the force constant of iron bonds and other higher order moments of the raw spectrum $S(E)$. The two approaches are mathematically equivalent (Dauphas et al. 2012). A virtue of the second approach is that the error bars are easy to calculate and one knows how data processing affects the results.

Polyakov et al. (2007) originally used published PDOS (sometimes digitized) to calculate β-factors. However, the β-factor calculation in NRIXS depends on high moments of $S(E)$: the 3rd moment for the force constant and higher order moments for the other coefficients of the polynomial expansion (Dauphas et al. 2012; Hu et al. 2013). Most work done on materials of geological relevance was originally done to determine the Debye velocity, from which compression and shear velocities can be derived (Hu et al. 2003; Sturhahn and Jackson 2007). The Debye velocity is determined by the low energy part of the spectrum, where the signal is high. Not much attention was paid to the high energy part of the spectrum, which influences the β-factor dramatically. Published spectra are often very noisy at high energies and sometimes the measurements are truncated before the signal reaches zero. One should therefore exercise caution when using NRIXS data acquired with other objectives than iron isotope geochemistry. Using previously published data, Polyakov et al. (2007) reported β-factors of magnetite, FeO (wüstite), troilite, and Fe_3S. Polyakov (2009) calculated the β-factors at high pressure-temperature for metallic iron, ferropericlase, perovskite and post-perovskite compositions. Those data were acquired in diamond anvil cells and are quite uncertain, with NRIXS spectra that are sometimes truncated. Polyakov and Soultanov (2011), calculated the β-factors of chalcopyrite, troilite, and Fe_3S. Dauphas et al. (2012), also using previously published data calculated the β-factors of myoglobin, cytochrome f, orthoenstatite, hematite, magnetite, troilite, FeS (MbP-type), FeS (monoclinic), chalcopyrite, Fe_3S, $(Mg_{0.75}Fe_{0.25})O$, α-Fe, γ-Fe, ε-Fe, and σ-Fe.

A difficulty with NRIXS data that was not fully appreciated until 2014 is that the signal does not seem to always reach a background value (Dauphas et al. 2014). Sometimes, the low or high energy ends are above background and are different from each other. This has led to the erroneous determination of the β-factors of goethite and jarosite (Dauphas et al. 2012), that were later corrected based on a new data reduction procedure that yielded more reproducible results from session to session (Blanchard et al. 2015). The new data reduction uses a new software, SciPhon, which has a routine to adequately subtract a non-constant baseline. The database of high quality NRIXS data is rapidly expanding and there are now good estimates of iron β-factors for goethite and jarosite (Blanchard et al. 2015), olivine, Fe^{2+} and Fe^{3+} in glasses of basaltic, andesitic, dacitic, and rhyolitic compositions (Dauphas et al. 2014), and spinels (Roskosz et al. 2015). Krawczynski et al. (2014a) reported force constant measurements for troilite and fcc Fe–Ni metal alloy. Williams et al. (2016) measured the force constant of iron in ilmenite and glasses with compositions relevant to lunar petrology.

Shahar et al. (2016) recently reported NRIXS measurements of high-pressure alloys FeO, FeH_x, and Fe_3C. Liu et al. (2016) also measured the mean force constants of basaltic glass, metallic iron, and iron-rich alloys of Fe–Ni–Si, Fe–Si, and Fe–S up to 206 GPa. Those studies were done specifically for the purpose of deriving β-factors and used extended energy ranges and the SciPhon software to do the data reduction. They thus do not suffer from some of the shortcomings of early NRIXS studies whose aims were to derive seismic velocities.

Comparisons of ab initio, NRIXS, and experimental studies. To compare the different methods, in particular to compare α-values determined experimentally with absolute β-factors determined by ab initio calculations or NRIXS, the easiest method is probably to convert all α-values to β-factors (Eqns. 15, 16) using a mineral that has been extensively studied. Polyakov et al. (2007) proposed to use hematite as there is reasonably good agreement between the β-factors estimated by conventional Mössbauer (Polyakov et al. 2007), NRIXS (Polyakov et al. 2007; Dauphas et al. 2012), and ab initio calculations (Blanchard et al. 2009). We concur with this assessment.

In Figure 6, we compare the experimentally determined fractionation factors between $Fe(III)_{aq}$–$Fe(II)_{aq}$ (Welch et al. 2003) and $Fe(III)_{aq}$–hematite (Skulan et al. 2002) with the predictions from ab initio and NRIXS studies (Table 1). The ab initio studies (Anbar et al. 2005; Domagal-Goldman and Kubicki 2008; Hill and Schauble 2008; Ottonello and Zuccolini 2009; Rustad and Dixon 2009; Fujii et al. 2014) can reproduce the equilibrium fractionation factor for $Fe(III)_{aq}$-$Fe(II)_{aq}$ (Welch et al. 2003). However, ab initio (Anbar et al. 2005; Domagal-Goldman and Kubicki 2008; Hill and Schauble 2008; Ottonello and Zuccolini 2009b; Rustad and Dixon 2009b; Fujii et al. 2014; Blanchard et al. 2009) and NRIXS studies (Dauphas et al. 2012) fail to reproduce the measured fractionation between $Fe(III)_{aq}$–hematite (Skulan et al. 2002). The β-factor of hematite agrees well between different techniques (Polyakov et al. 2007; Rustad and Dixon 2009; Blanchard et al. 2009; Dauphas et al. 2012) while the β-factors of $Fe(III)_{aq}$ can be quite variable from study to study, giving values at 22 °C of +8.4‰ (Rustad and Dixon 2009), +11.8‰ (Schauble et al. 2001), +9.6‰ (Anbar et al. 2005), +8.9‰ (Ottonello and Zuccolini 2009), +9.4‰, +10.1‰ (Hill and Schauble 2008), and +8.3‰ (Fujii et al. 2014). Regardless of this complication, all calculations predict a $Fe(III)_{aq}$-hematite isotopic fractionation that is larger than what is measured. This problem was recognized by Rustad and Dixon (2009) and more recent estimates have not solved the discrepancy. It is currently unknown whether the issue is with the calculation or if kinetic effects were present in the experiments.

To try to go around the issue that comparing β-factors for aqueous species and minerals may introduce biases, we have used the fluid–mineral equilibration experiments to recalculate equilibrium fractionation factors between mineral pairs at low temperature. For the reasons outlined above, we use hematite as the reference mineral against which other minerals are compared. The results of NRIXS/ab initio studies and experimentally determined mineral-mineral fractionation factors at 22 °C are compared in Figure 7. The measured ferrihydrite–hematite (Skulan et al. 2002; Welch et al. 2003; Wu et al. 2011) and

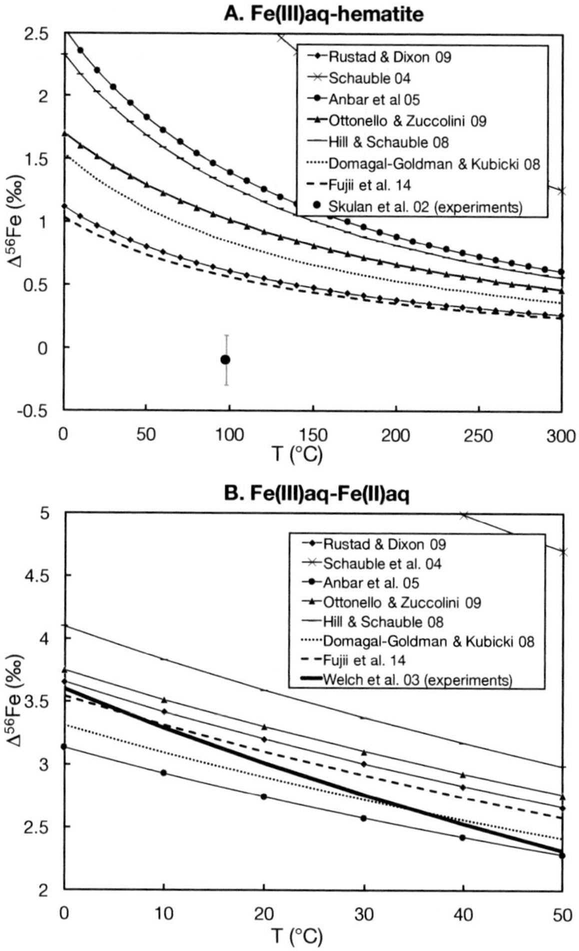

Figure 6. Comparison between experimentally measured and predicted (based on NRIXS data and/or ab initio calculations) equilibrium fractionation factors between $Fe(III)_{aq}$–hematite (A) and $Fe(III)_{aq}$–$Fe(II)_{aq}$ (B). The most recent ab initio calculations can reproduce well the equilibrium fractionation factor between $Fe(III)_{aq}$ and $Fe(II)_{aq}$. However, the calculations predict a larger $Fe(III)_{aq}$–hematite fractionation than is measured. The data sources are indicated on the figure and the β-factors used for the calculation are compiled in Table 1. For hematite in the top panel, the average value of several NRIXS and ab initio studies was used (Rustad and Dixon 2009; Blanchard et al. 2009; Dauphas et al. 2012).

siderite–hematite (Skulan et al. 2002; Welch et al. 2003; Wiesli et al. 2004) fractionation factors agree well with the predictions for goethite–hematite and siderite–hematite (Blanchard et al. 2009, 2014; Rustad and Dixon 2009; Dauphas et al. 2012). Figure 7 also shows a comparison between measured mackinawite–$Fe(II)_{aq}$ fractionation (Guilbaud et al. 2011; Wu et al. 2012b) converted to mackinawite–hematite using the relevant fractionation factors, and predicted troilite–hematite fractionation (Polyakov et al. 2007; Rustad and Dixon 2009; Dauphas et al. 2012; Blanchard et al. 2015). There is a ~1‰ disagreement at 22 °C but it is unsure to what extent troilite can be taken as a proxy for mackinawite. For

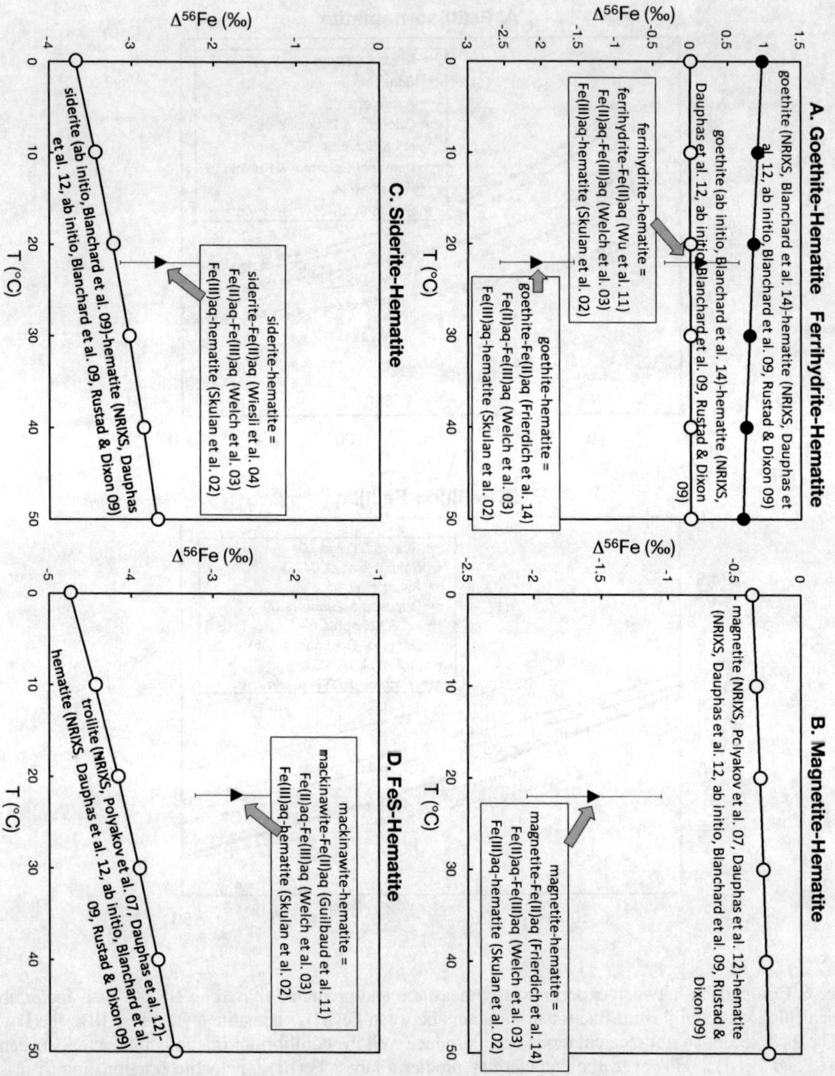

Figure 7. Equilibrium fractionation between mineral pairs based on fluid–mineral equilibration experiments (recalculated to a common temperature of 22 °C) and comparison with predictions from NRIXS and ab initio studies. Hematite was taken as a reference because the equilibrium fractionation factor between $Fe(III)_{aq}$ and hematite was measured experimentally (Skulan et al. 2002) and the β-factor of hematite estimated by NRIXS (Dauphas et al. 2012) and ab initio calculations (Blanchard et al. 2009a; Rustad and Dixon 2009) agree well. As an example of how the mineral–mineral fractionation factors were calculated, goethite–hematite fractionation was calculated as follows: The experimentally determined fractionation for $Fe(III)_{aq}$–hematite at 98 °C is -0.1 ± 0.2‰ (Skulan et al. 2002). At 22 °C, this corresponds to a fractionation of $-0.1\times(273+98)^2/(273+22)^2=-0.16\pm0.32$‰. At 22 °C, laboratory experiments give a $Fe(III)_{aq}$–$Fe(II)_{aq}$ equilibrium fractionation of $+2.95\pm0.38$‰ (Welch et al. 2003). Experiments also give a goethite–$Fe(II)_{aq}$ fractionation of $+1.05\pm0.08$‰ (Frierdich et al. 2014a). The net goethite–hematite fractionation is thus $1.05-2.95-0.15=-2.05\pm0.50$‰. Note that the FeS–hematite panel compares the measured fractionation between mackinawite and hematite against predictions for troilite and hematite. The data sources are indicated on the figure and the β-factors used for the calculation are compiled in Table 1.

the pair magnetite–hematite, the experimentally determined (Skulan et al. 2002; Welch et al. 2003; Frierdich et al. 2014b) and NRIXS predicted (Polyakov et al. 2007; Dauphas et al. 2012) fractionation factors are also off by ~1‰. The β-factor of magnetite by NRIXS is uncertain and Roskosz et al. (2015) suggested that it may need to be revised upwards, which would exacerbate the discrepancy. The most striking discrepancy is for the pair goethite-hematite, where experiments (Skulan et al. 2002; Welch et al. 2003; Beard et al. 2010; Frierdich et al. 2014a) are off from NRIXS and ab initio predictions (Blanchard et al. 2009, 2015; Rustad and Dixon 2009; Dauphas et al. 2012) by ~2‰. Experiments also indicate that there would be a 2‰ equilibrium fractionation between ferrihydrite and goethite (Beard et al. 2010; Wu et al. 2011; Frierdich et al. 2014a), which is difficult to explain from a crystal chemical point of view.

In Fig. 8, we compare measured high-temperature equilibrium fractionation factors with predictions from NRIXS measurements for 3 systems. The fractionation measured by Schuessler et al. (2007) for the system FeS–rhyolitic melt is relatively well reproduced by NRIXS predictions (Polyakov et al. 2007; Dauphas et al. 2012, 2014) (Fig. 8a). The fractionations measured by Hin et al. (2012) and Poitrasson et al. (2009) for silicate–metal are marginally consistent with the fractionation predicted by NRIXS for basaltic glass–bcc iron (Dauphas et al. 2012, 2014) but are lower than the predicted fractionation for basaltic glass–fcc iron (Dauphas et al. 2012, 2014; Krawczynski et al. 2014a) (Fig. 8b). The results of Shahar et al. (2015) are, however, much lower than the NRIXS predictions for both basaltic glass–bcc and basaltic glass–fcc iron (Fig. 8b). The experiments were done on molten systems, and it is unknown to what extent some of these discrepancies are due to the fact that the solid (glass or crystal) systems measured by NRIXS are imperfect proxies for melt β-factors. The measured fractionation between magnetite and olivine (Shahar et al. 2008; Sossi and O'Neill 2016) is significantly higher than the prediction from NRIXS measurements for magnetite-olivine (Polyakov et al. 2007; Dauphas et al. 2012, 2014) (Fig. 8c). This could be due to the fact that the β-factor of magnetite as measured by NRIXS is not accurate (it was not measured for the purpose of estimating β-factors), as a more recent study of the spinel solid solution predicted an equilibrium fractionation factor at a Fe^{3+}/Fe_{tot} relevant to magnetite more in line with what is measured (Roskosz et al. 2015).

IRON ISOTOPES IN COSMOCHEMISTRY

Nucleosynthetic anomalies and iron-60

Chondrites are undifferentiated meteorites whose compositions are thought to represent the composition of the solar system at one point in time and space. During their formation, they incorporated dust formed in the outflows of stars that lived and died before the solar system was formed. The composition of those presolar grains is representative of the star from which they formed. They are thus invaluable tools to study stellar nucleosynthesis and the chemical evolution of the Galaxy. Mahras et al. (2008) measured the Fe isotopic compositions of presolar silicon carbide (SiC) grains of mostly AGB-star (Asymptotic Giant Branch) and type II supernova origins (those origins are ascribed on the basis of isotopic analyses of C, N and Si). Most supernova grains show large excesses in ^{57}Fe while AGB grains have more subdued anomalies. The measurements were done with a NanoSIMS, which does not allow one to resolve the peaks of the low abundance isotope ^{58}Fe and the high abundance ^{58}Ni, so ^{58}Fe was not reported. The compositions of the supernova grains could be reproduced by arbitrarily mixing material from the He/N and He/C zones of supernova simulations. Outstanding questions at the present time are: how was such fine-scale mixing achieved and why did SiC grains preferentially sample those regions? Progress may be achieved by measuring the Fe and Ni isotopic compositions

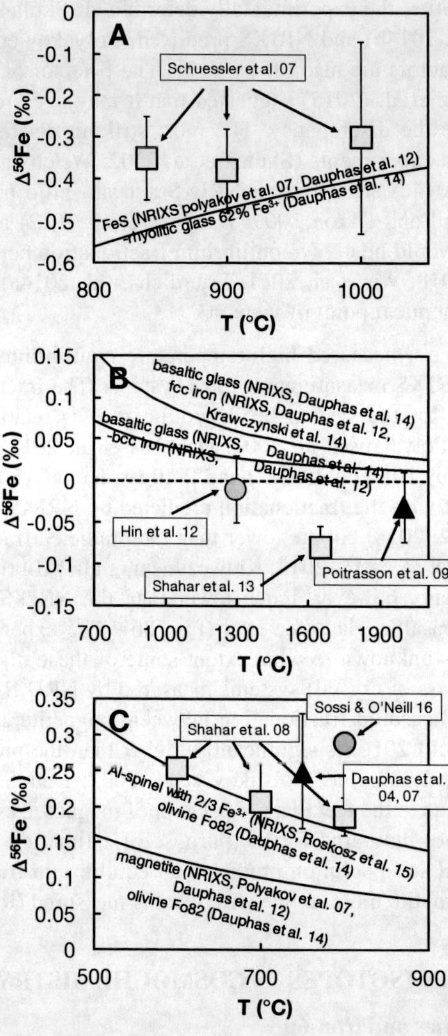

Figure 8. Comparisons between experimentally determined and calculated (from NRIXS measurements) equilibrium fractionation factors between high temperature phases (pyrrhotite–peralkaline rhyolitic melt with 2/3 Fe^{3+} in panel A; silicate–metal in panel B; magnetite–olivine in panel C). The data sources are indicated on the figure and the β-factors used for the calculation are compiled in Table 1.

of X-grains (SiC presolar grains that are thought to come from supernovae) using resonant ionization mass spectrometry, which allows one to measure the isotopic composition of iron, including ^{58}Fe, even in the presence of large quantities of Ni (Trappitsch et al. 2016).

FUN (fractionated and unknown nuclear effects) refractory inclusions are other significant carriers of nucleosynthetic anomalies in meteorites (Dauphas and Schauble 2016, and references therein). Völkening and Papanastassiou (1989) discovered large excesses in the neutron-rich isotope of iron (^{58}Fe) in a FUN inclusion named EK-1-4-1 from the Allende CV chondrite. This nucleosynthetic excess in ^{58}Fe of ~+30‰ is consistent with enrichments in other neutron-rich isotopes such as ^{48}Ca and ^{50}Ti found in the same refractory inclusion. For some elements like Ti, isotopic anomalies are pervasive in all kinds of refractory inclusions, while only one such inclusion has revealed such a large ^{58}Fe excess. One reason for that may be that iron is a relatively volatile element that is not expected to have condensed quantitatively, so most iron found in refractory inclusions is probably of secondary origin, having been introduced by aqueous alteration

The abundance of the short-lived radionuclide ^{60}Fe ($t_{1/2}$ = 2.62 Myr) has been the subject of much discussion over the past couple of years. Measurements of ^{60}Ni, the decay product of ^{60}Fe, in bulk meteorites and mineral separates by MC-ICPMS gives a ^{60}Fe/^{56}Fe ratio of ~1×10^{-8} (Tang and Dauphas 2012, 2014, 2015) (Fig. 9). In situ measurements by SIMS give a much higher initial ^{60}Fe/^{56}Fe ratio of ~7×10^{-7} (Mishra and Goswami 2014; Mishra and Chaussidon 2014; Mishra and Marhas 2016). It was argued that this discrepancy could be due to heterogeneous distribution of ^{60}Fe in solar system materials (Quitté et al. 2010). One way to test that idea is to measure the ^{58}Fe isotopic composition, because ^{58}Fe and ^{60}Fe are produced in the same stellar environments by the same neutron-capture reactions, so those two nuclides should be very well coupled. Therefore, if ^{60}Fe was heterogeneously distributed in meteorites, one would expect to detect collateral isotope effects on ^{58}Fe (Dauphas et al. 2008). Such collateral effects were not detected, which led Tang and Dauphas (2012, 2015) to conclude that the initial abundance of ^{60}Fe was low and uniform (^{60}Fe/^{56}Fe ~1×10^{-8}). This ratio is low compared to the prediction for supernova injection if the short-lived radionuclide ^{26}Al ($t_{1/2}$ = 0.7 Myr) shared such a supernova origin. The reason for this low abundance is debated but one possibility is that ^{26}Al was ejected in the interstellar medium in the form of stellar winds, which contained little ^{60}Fe as this nuclide

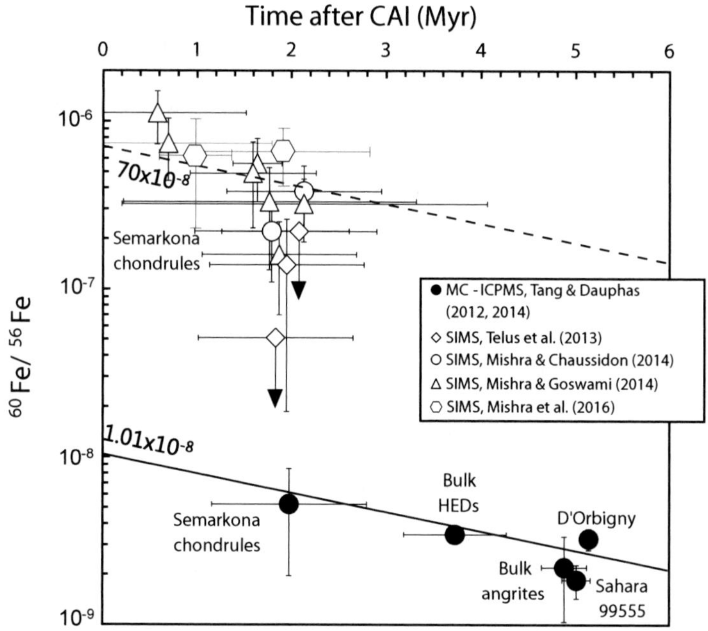

Figure 9. Inferred initial ^{60}Fe/^{56}Fe ratio (^{60}Fe decays into ^{60}Ni with a half-life of 2.6 Myr) of various meteoritic materials (from Tang and Dauphas 2015; updated with the data of Mishra and Marhas 2016). MC-ICPMS measurements detected ^{60}Fe in various materials (Semarkona chondrules, bulk HED meteorites, bulk angrites, mineral separates in D'Orbigny and Sahara 99555 angrites) corresponding to an intial ^{60}Fe/^{56}Fe ratio of ~10^{-8} (Tang and Dauphas 2012a, 2015). In situ measurements by SIMS gave variable initial ^{60}Fe/^{56}Fe ratios that reach 10^{-6} (Mishra and Goswami 2014; Mishra and Chaussidon 2014; Mishra and Marhas 2016). Downward pointing vertical arrows correspond to upper-limits. The weight of evidence supports a low ^{60}Fe/^{56}Fe ratio at solar system formation (see text for details).

was produced in more internal regions of the star. Iron–60 was ejected in the interstellar medium at a later time by the supernova explosion that ended the life of the massive star (Tang and Dauphas 2012). Further work will be needed to assess the likelihood of this scenario.

Overview of iron isotopic compositions in extraterrestrial material

Chondrites and their components. A characteristic feature of most chondrites is the presence of chondrules, which are quenched beads of silicate whose origin is still very much debated. During the heat event that melted the chondrules (shockwaves, planetary collisions, and lightning have been proposed), some of the iron, which is relatively volatile, could have volatilized. An important question in this respect is whether iron-poor type I chondrules could have formed from iron-rich type II chondrule material by iron evaporation. Alexander and Wang (2001) measured the Fe isotopic compositions of chondrules from Chainpur, a LL3.4 chondrite (the 3.4 grading means that the meteorite was not much heated and the Fe isotopic signature was not disturbed by parent-body metamorphism). They found no detectable variations in the isotopic composition of iron with a precision of ~2‰ (the measurements were acquired by ion probe). Mullane et al. (2005), Needham et al. (2009), and Hezel et al. (2010) also measured the Fe isotopic compositions of chondrules from various meteorite groups (CV, H, L, and LL) by MC-ICPMS. The precision of those measurements was significantly better than SIMS and isotopic variations were resolved but the variations were limited. The δ^{56}Fe values of most chondrules fall within ±0.2‰ of the bulk chondrite composition and the average chondrule δ^{56}Fe value is indistinguishable from the bulk chondrite value. If significant Fe was evaporated under vacuum conditions, then large Fe isotopic fractionation would be expected (Wang et al. 1994; Dauphas et al. 2004a). The lack of detectable Fe isotopic fractionation can be explained if iron was not lost by evaporation, or significant iron vapor pressure built up around the chondrules, which suppressed the kinetic fractionation (as P/P_{sat} approaches 1, the system approaches equilibrium between gas and condensed phase; Eqn. 12). Evaporation in a closed system is supported by the fact that highly volatile potassium shows little isotopic fractionation in chondrules (Alexander et al. 2000; Alexander and Grossman 2005). It is also supported by the fact that another volatile element, sodium, is present in the cores of zoned olivines, meaning that even at the highest temperatures reached during chondrule heating, significant Na was still present in molten chondrules, which is only explainable if the chondrule density was sufficiently high for vapor to build up and partially suppress evaporation (Alexander et al. 2008).

The Urey–Craig diagram (Urey and Craig 1953) groups chondrites according to their total iron content and the fraction of iron as metal vs. silicate or sulfide. Despite these variations in the Fe/Si ratio and redox state of iron, chondrites have very constant Fe isotopic compositions, at least given present analytical precision (Fig. 10). Craddock and Dauphas (2011a) measured the Fe isotopic compositions of 10 carbonaceous chondrites, 15 ordinary chondrites, and 16 enstatite chondrites. Except for Kelly (a LL4) and high metamorphic grade EL6 chondrites, all samples have the same Fe isotopic composition, defining an average δ^{56}Fe value of -0.005 ± 0.006‰. This composition is indistinguishable from that of the reference material IRMM–014/524a (δ^{56}Fe$=0$ by definition). This is purely coincidental but is another argument for the general adoption of this reference material as standard in iron isotope geochemistry, as it is representative of an important geochemical reservoir (the chondritic composition).

Mars, Vesta, and the Angrite parent-body. SNC (Shergottite-Nakhlite-Chassignite) and HED (Howardite, Eucrite, Diogenite) achondrite meteorites have Fe isotopic compositions that are almost indistinguishable from the chondritic composition (Poitrasson et al. 2004; Weyer et al. 2005; Anand et al. 2006; Schoenberg and von Blanckenburg 2006; Wang et al. 2012; Sossi et al. 2016) (Fig. 11). SNC and HED meteorites are magmatic rocks that, as several lines of evidence indicate, come from Mars and Vesta, respectively. Among HEDs, there seems to be some isotopic variations, in particular with Stannern group eucrites that are enriched in

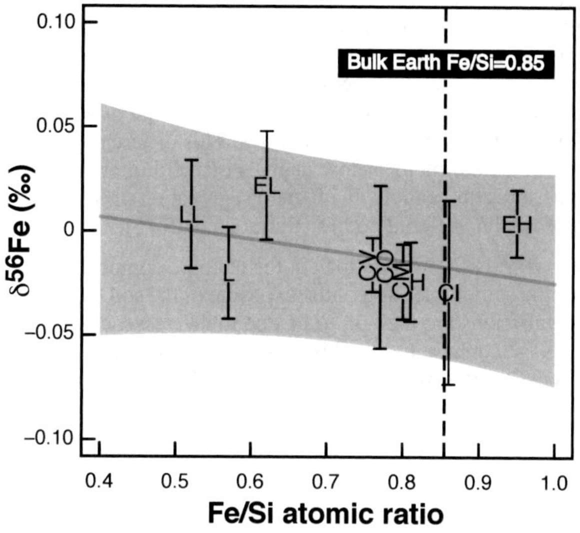

Figure 10. Iron isotopic compositions of chondrite groups as a function of Fe/Si ratios (Dauphas et al. 2009b). Even if chondrites have variable iron contents (Fe/Si ratios) and redox states, the δ^{56}Fe values are relatively constant. The grey band is the 95% confidence for the regression.

heavy Fe isotopes by ~+0.03‰ in δ^{56}Fe (Wang et al. 2012). Stannern eucrites are enriched in incompatible elements relative to main group eucrites, which was explained by smaller degree of partial melting or, more likely, by assimilation of differentiated crustal material (Barrat et al. 2007). The enrichment in the heavy isotopes of iron can be understood in the context of the crust assimilation model if the bulk of Vesta crust has heavy Fe isotopic composition or if assimilation took place in a non-modal manner, so that isotopically heavy minerals such as ilmenite were preferentially incorporated in the parental magma to Stannern group eucrites.

Martian samples also have on average chondritic Fe isotopic composition with a hint for heavier Fe isotopic compositions in Nakhlites but the total range in δ^{56}Fe values is small with most samples within −0.03 and +0.03‰ (Poitrasson et al. 2004b; Weyer et al. 2005a; Anand et al. 2006; Wang et al. 2012). The slightly heavy Fe isotopic composition of nakhlites and evolved shergottites can be explained by magmatic differentiation (Sossi et al. 2016).

Angrites, which are a group of basaltic meteorites that come from a body that was relatively oxidized (~IW+1), have heavy δ^{56}Fe values similar to terrestrial basalts, i.e., δ^{56}Fe values of ~+0.1‰ (Wang et al. 2012) (Fig. 11). On Earth, it was suggested that the presence of Fe^{3+} in significant quantities could affect the Fe isotopic composition of igneous rocks. While angrites are more oxidized than other achondrites, they are significantly more reduced than terrestrial basalts (~IW+1 vs. ~IW+2.5) and Fe^{3+} is not expected to directly influence the Fe isotopic composition of the angrite parent body. Impact volatilization (Poitrasson et al. 2004; Wang et al. 2012) is also unlikely because for bodies of the size of the angrite parent-body, the energy available from collisional accretion is thought to be small compared to the latent heat of vaporization (Dauphas et al. 2015). It is presently unknown why angrites have heavy Fe isotopic composition, one possibility being that this is a feature inherited from nebular processes, as was suggested for Si (Dauphas et al. 2015). Overall, studies of achondrites show that magmatic processes such as magmatic differentiation and assimilation on planetary bodies can fractionate iron isotopes. Nevertheless, those fractionations are subtle and the mantles of Mars and Vesta have Fe isotopic compositions that are indistinguishable from chondrites to very high precision (Fig. 11).

Barrat et al. (2015) analyzed the Fe isotopic compositions of 30 samples from the Ureilite Parent Body (UPB) including 29 unbrecciated ureilites and one ureilitic trachyandesite (ALM-A). Ureilites are ultramafic achondrites, which are thought to be mantle restites formed after extraction of magmas and S-rich metallic melts. The δ^{56}Fe of the whole rocks fall within a restricted range, from +0.01 to +0.11‰, with an average of +0.056 ± 0.008‰, which is significantly higher than that of chondrites. This difference has been ascribed to the segregation of S-rich metallic melts at low degrees of melting at a temperature close to the Fe–FeS eutectic. These results point to an efficient segregation of S-rich metallic melts during the differentiation of small terrestrial bodies (Barrat et al. 2015).

The Moon. Lunar soils show variations in the isotopic composition of iron that correlate with the maturity of the soils, a measure of the exposure of the soil to space weathering (Wiesli et al. 2003). Those variations are thought to be due to the presence of nanophase Fe metal in the most mature soils. Nanophase Fe metal is produced by vapor deposition by micrometeorite impacts and solar-wind sputtering. Thus, to discuss the Fe isotopic composition of the Moon or other airless bodies, one should stay clear of soils.

Of all the lithologies present at the surface of the Moon, the ones that provide the most insights into the composition of the Moon are mare basalts that fill the maria. These formed relatively late (radiometric dating gives ages of ~3 to 4 Ga) and are thought to sample a deep-seated source in the lunar mantle (150–600 km; Lee et al. 2009). A difficulty with estimating the composition of the Moon is that the lack of plate tectonics like on Earth has allowed the survival of significant heterogeneities in the lunar mantle, presumably inherited from lunar magma ocean differentiation. High- and low-Ti mare basalts have distinct Fe isotopic compositions, with the δ^{56}Fe of high-Ti basalts averaging +0.191 ± 0.020‰ compared to +0.073 ± 0.018‰ for low-Ti basalts (Wiesli et al. 2003; Poitrasson et al. 2004; Weyer et al. 2005; Liu et al. 2010) (Fig. 11). A difference between low- and high-Ti basalts was also found for the isotopic compositions of O (Liu et al. 2010) and Mg (Sedaghatpour et al. 2013). For Fe, this contrast is unlikely to reflect shallow magma differentiation processes and reflects instead differences in the source region composition or the mode of melting (Liu et al. 2010). The δ^{26}Mg and δ^{56}Fe values of lunar samples are negatively correlated; the low-Ti basalts have δ^{26}Mg indistinguishable from the terrestrial composition of −0.25 ± 0.10‰ whereas high-Ti basalts have more negative δ^{26}Mg values averaging −0.49 ± 0.14‰ (Sedaghatpour et al. 2013). Low-Ti basalt have δ^{49}Ti isotopic composition similar to terrestrial basalts while high-Ti basalts have slightly elevated δ^{49}Ti values, athough this is at the limit of detection (Millet et al. 2016). Magnesium, iron, and titanium isotopes thus suggest that low-Ti mare basalts are more representative of the bulk silicate Moon. At the present time, the Fe isotopic composition of the bulk Moon is very uncertain but its best estimate is close to the average Fe isotopic composition of low-Ti mare basalts; δ^{56}Fe ≈ +0.084‰ (Liu et al. 2010; Fig. 11)

HIGH-TEMPERATURE GEOCHEMISTRY

Low temperature aqueous processes can impart large iron isotopic fractionation. Because equilibrium fractionation factors decrease as $\sim 1/T^2$ (with T in K), it was initially thought that all igneous rocks should have identical Fe isotopic composition, providing a useful baseline to reference iron isotope variations documented in sediments (Beard and Johnson 1999). At 1200 °C the equilibrium fractionation factor is expected to decrease by a factor $(1473/293)^2 = 25$ compared to that at 20 °C, meaning that variations at the few tenths of permil at most are expected at high temperature in natural systems (at least when equilibrium processes are involved). Once analytical capabilities reached sufficient precision, iron isotope variations in mineral separates and bulk igneous rocks were readily detected (Zhu et al. 2002;

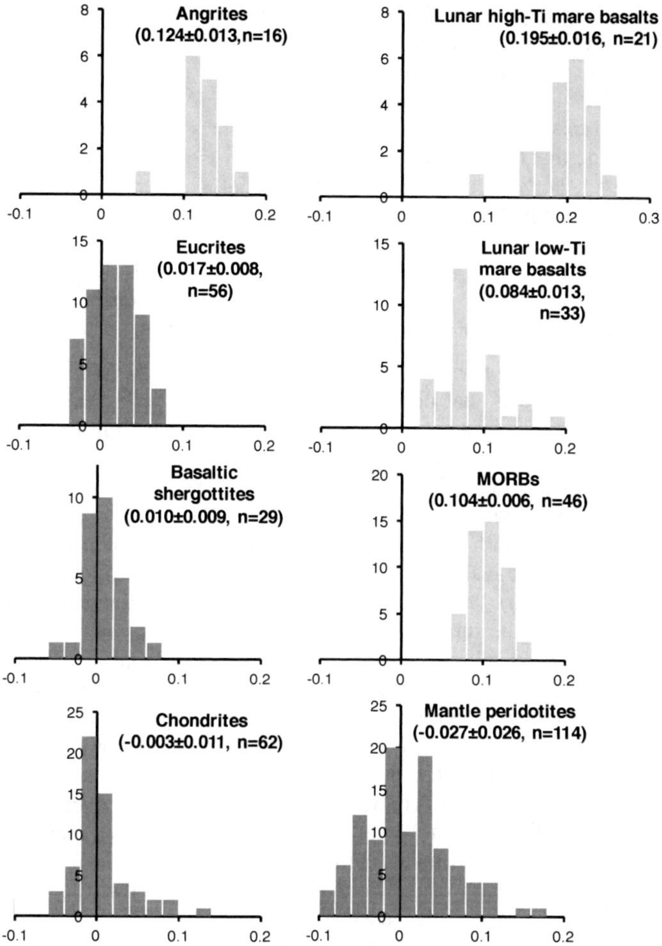

Figure 11. Histograms of δ^{56}Fe values in planetary reservoirs (compilation from the authors; see text for details and references). Eucrites, SNCs (martian meteorites) and mantle peridotites have near chondritic Fe isotopic compositions. MORBs, lunar basalts, and angrites have heavy δ^{56}Fe values relative to chondrites and mantle peridotites. The origins of those enrichments are not well understood. The mean, 95% confidence interval of the mean, and number of samples analyzed are given in parenthesis.

Beard and Johnson 2004b; Poitrasson et al. 2004; Williams et al. 2004a,b; Poitrasson and Freydier 2005; Weyer et al. 2005; Schoenberg and von Blanckenburg 2006; Teng et al. 2011). This is now a very active field of research in iron isotope geochemistry.

Iron isotope variations in mafic and ultramafic terrestrial rocks and the composition of the silicate Earth

Several sample types can be used to estimate the Fe isotopic composition of Earth's mantle. The first samples that were used for that purpose were basalts (Beard and Johnson 1999). One good reason to use those is that they sample large volumes of the Earth. Taking ocean ridges as an example, the yearly production of lavas is ~21 km^3 (Crisp 1984). Given that they form by ~10% partial melting, this means that globally and every year, 210 km^3 of Earth's mantle is sampled by MORBs. Actually, much of the early measurements were done on igneous rocks from oceanic island and continental settings, the reason being that geostandards are readily available for these settings (e.g., BHVO basalt from Hawaii, BCR flood basalt from Columbia river). On the basis of these measurements, it was concluded that the silicate Earth had heavy Fe isotopic composition relative to chondrites (Poitrasson et al. 2004). Teng et al. (2013) subsequently published an extensive survey of the Fe isotopic composition of mid-ocean ridge basalts and showed that MORBs have constant δ^{56}Fe values (MSWD~1) (Fig. 11). It is thus tempting to conclude that the silicate Earth as a whole is isotopically fractionated relative to chondrites. However, Fe^{2+} has an effective partition coefficient of ~1 during partial melting (the FeO concentrations in MORBs and fertile peridotites are similar at ~8 wt%), meaning that when a mantle reservoir melts by 10% partial melting, 90% of iron is left behind in the mantle. While it is true that equilibrium fractionation decreases rapidly with increasing temperature, the fact that 90% of iron is left behind in the mantle provides considerable leverage to fractionate the isotopic composition of iron in the melt.

The samples that should provide the best estimate of the Fe isotopic composition of Earth's mantle are mantle peridotites that directly sample the shallowest portion of Earth's mantle (Fig. 11). Several peridotite types have been studied, including xenoliths from arcs, continental settings, ultramafic massifs, and abyssal peridotites. Zhu et al. (2002), Beard and Johnson (2004b), and Williams et al. (2004, 2005) were among the first to measure significant differences in δ^{56}Fe values between coexisting mantle minerals. Inter-mineral fractionations in peridotites do not vary systematically with the inferred closure temperatures, indicating that they are in disequilibrium, consistent with alteration by metasomatism (Beard and Johnson 2004b; Zhao et al. 2010; Roskosz et al. 2015). Williams et al. (2004, 2005) studied both bulk rocks and mineral separates of samples from the sub-continental margin mantle lithosphere, the sub-continental mantle lithosphere, and the sub-arc mantle. They found large iron isotopic variations, with high fO_2 sub-arc samples carrying lower δ^{56}Fe values than other peridotites. Spinels also show a strong correlation between the recorded oxygen fugacity and Fe isotopic composition. Quantitative modeling of those results is difficult but the pattern of isotopic variations in those rocks is qualitatively consistent with melt depletion under oxidizing conditions in arc samples. In an effort to better understand what controls Fe isotopic variations in mantle rocks and assess the composition of the silicate Earth, Weyer et al. (2005) and Weyer and Ionov (2007) studied mantle peridotites from diverse tectonic settings. Weyer and Ionov (2007) found that most of those samples defined a trend between δ^{56}Fe and Mg/(Mg+Fe), an indicator of melt extraction, such that the samples that have experienced the most extensive melt extraction were isotopically the lightest. The authors could explain the trend that they measured by a fractionation between melt and residue of +0.1 to +0.3‰. Extrapolating the iron isotopic data to a fertile mantle Mg# of 0.894 yielded a fertile mantle δ^{56}Fe value of $+0.02\pm0.03$‰. They also identified several metasomatic trends in metasomatized peridotites that shifted δ^{56}Fe values towards heavy or extremely light values. The heavy Fe isotopic signatures in the metasomatized peridotites can be explained by the fact that the metasomatizing melt had heavy isotopic composition inherited from partial melting and silicate melt–solid interactions. The very low δ^{56}Fe values may reflect kinetic isotope fractionation during diffusion of Fe (the light isotopes tend to diffuse faster; Richter et al. 2009a). Schoenberg and von Blanckenburg (2006), Huang et al. (2011), Zhao et al. (2010, 2012), and Poitrasson et al. (2013) also measured the Fe isotopic compositions of mantle xenoliths and found that on average, those samples have an Fe isotopic composition indistinguishable from chondrites but display significant variations, presumably associated with melt extraction and metasomatic processes. All studies of mantle peridotites have found δ^{56}Fe values that were on average lower than the mean MORB δ^{56}Fe value of ~+0.1‰, suggesting that the Earth may be chondritic. Poitrasson et al. (2013) argued, however, that lithospheric mantle xenoliths are not representative of the Fe isotopic composition of the asthenosphere because they have been affected by complex melt extraction–metasomatic events, so that the silicate Earth may have a δ^{56}Fe value of +0.1‰, as is measured in MORBs.

Volumetrically, MORBs are the most important magmas at the surface of the Earth and they constitute much of Earth's oceanic crust (Crisp 1984). To directly address the question of whether iron isotopes can be fractionated during partial melting, Craddock et al. (2013) measured the Fe isotopic compositions of abyssal peridotites, which represent the mantle residues left behind from MORB generation. The difficulty with these samples is that ultramafic rocks in contact with seawater are extensively serpentinized and weathered. The MgO/SiO_2 ratio is a good proxy for the extent of marine weathering as Mg is more easily mobilized than Si during this process (Snow and Dick 1995). Craddock et al. (2013) found that abyssal peridotites have on average the same Fe isotopic composition as chondrites. The most weathered samples showed more scatter than the least weathered but focusing on the least weathered samples, Craddock et al. (2013) obtained an average $\delta^{56}Fe$ value of $+0.010 \pm 0.007$‰. For 10% partial melting, the Fe isotopic composition is not shifted much but accounting for this effect, Craddock et al. (2013) estimated the composition of the mantle source of MORBs to be $+0.025 \pm 0.025$‰. For comparison, MORBs have an average $\delta^{56}Fe$ value of $+0.105 \pm 0.006$‰. These results therefore suggest that partial melting can fractionate iron isotopes and that the silicate Earth has chondritic Fe isotopic composition.

An important question in modern igneous petrology is the mineralogical nature of Earth's mantle and the extent to which enriched (pyroxenitic) and depleted (peridotitic) lithologies coexist in the mantle and control the chemical composition of magmas sampled at Earth's surface, such as their SiO_2, Ni contents, and Fe/Mn ratios (Humayun et al. 2004; Sobolev et al. 2007). Williams and Bizimis (2014) investigated whether such sources could be traced using iron isotopes. For that purpose, they analyzed the Fe isotopic compositions of mineral separates from peridotite and pyroxenite xenoliths from Hawaii. The bulk rock compositions were recalculated using the known mineral $\delta^{56}Fe$ values, Fe concentrations, and modal mineralogy. Other geochemical proxies indicate that the peridotites have experienced variable extents of melt extraction and refertilization (addition of a melt component). The pyroxenites (garnet clinopyroxenite) are high pressure cumulates. Similarly to what Weyer and Ionov (2007) found for peridotites from various settings, Williams and Bizimis (2014) found that the $\delta^{56}Fe$ values of bulk peridotites are lower in samples that have experienced the greatest extent of melt extraction. The range of variations measured in peridotites (~0.3‰) is impossible to explain by a simple partial melting model because the most depleted peridotites have only experienced ~10% melt extraction, which would be insufficient to impart large isotopic fractionation to the residue (the iron melt/solid partition coefficient is ~1). A complex history of melt extraction and refertilization is therefore suggested by these data. The pyroxenites have heavy Fe isotopic compositions reaching +0.18‰. The pyroxenite samples with the highest $\delta^{56}Fe$ values also have the highest TiO_2 content, presumably because they formed as cumulates from the most evolved magmas and mafic magma differentiation can drive the melt towards heavier Fe isotopic composition through combinations of equilibrium and kinetic processes. This study showed that the mantle may be heterogeneous in its Fe isotopic composition, mirroring the distribution of pyroxenitic vs. peridotitic sources. Such heterogeneity may be visible in lavas in the form of heavy $\delta^{56}Fe$ values in hotspot basalts from the Society, Cook-Austral, and Samoan islands (Teng et al. 2013; Konter et al. 2016).

The weight of evidence from Fe isotope measurements of peridotite supports the view that the Fe isotopic composition of the accessible Earth is close to chondritic (Fig. 11). Unfortunately, the iron isotope record in those samples is complicated by the overprinting of melt extraction and metasomatic events, so the community is still debating to what extent those samples are representative of the whole mantle. As discussed below, other samples have been used in order to attempt to constrain its composition.

Komatiites are products of high degrees of partial melting (up to 50%) in the Archean, when the mantle potential temperatures were significantly higher than at present (~300 °C higher than MORBs for the mantle source of Alexo komatiites). High-degree of partial melting and high temperature should concur to minimize the extent of equilibrium iron isotopic fractionation between the magma and its source, so komatiites may be used as another proxy isotopic composition for the silicate Earth. One difficulty working with komatiites is that they have been serpentinized, which seems to have affected the isotopic composition of Fe (Dauphas et al. 2010). Superimposed on that is the fact that magmatic fractionation of olivine has also induced Fe isotopic fractionation at a bulk rock scale (Dauphas et al. 2010). Using geochemical proxies to avoid these effects, Dauphas et al. (2010) estimated the Fe isotopic composition of the initial lava flow at Alexo to be $\delta^{56}Fe = +0.044 \pm 0.030$‰ (and its $\delta^{25}Mg = -0.138 \pm 0.021$‰; indistinguishable from chondrites). Hibbert et al. (2012) studied the Fe isotopic compositions of komatiites from Belingwe (2.7 Ga), Vetreny (2.4 Ga), and more recent Gorgona (89 Ma). They found correlations with proxies of magmatic differentiation. Correcting for these effects is difficult but Hibbert et al. (2012) proposed that the mantle source may have a lower $\delta^{56}Fe$ value than the value proposed by Dauphas et al. (2010); possibly as low as -0.13 ± 0.05‰. Nebel et al. (2014) measured the Fe isotopic compositions of 3.5 Ga Coonterunah Subgroup and 3.16 Ga Regal Formation komatiites. After correction for crystal fractionation and accumulation, they obtain primary magma $\delta^{56}Fe$ values of −0.06 and ~0‰ for these two komatiite occurrences, consistent with other studies. To summarize, inferring the source Fe isotopic composition of komatiites is fraught with difficulties but available studies hint at a $\delta^{56}Fe$ value lower than MORBs and intraplate basalts. The extent to which those low values reflect prior episodes of melt extraction is uncertain. In addition, several studies point out the importance of immiscible sulfide melt segregation and potentially hydrothermal processes in controlling Fe isotope signatures of komatiites and associated Ni-rich mineralization (Bekker et al. 2009; Hiebert et al. 2013; Hofmann et al. 2014).

Arc basalts and boninites also show rich and complex Fe isotopic variations that have some bearing on Fe isotopic fractionation during partial melting and the composition of the mantle. Boninites are products of high degree flux melting of depleted mantle sources associated with subduction. Dauphas et al. (2009b) measured the Fe isotopic composition of many boninites and found that they have very uniform Fe isotopic composition, clustering around the Fe isotopic composition of chondrites. Those boninites were generated from a highly depleted mantle source that was fluxed with fluids. Hibbert et al. (2012) suggested that the low $\delta^{56}Fe$ value measured in boninites may reflect the fact that their mantle source has low $\delta^{56}Fe$ value inherited from previous partial melting events. This is unlikely to be the case because it would take extraordinary circumstances for melt depletion to shift the Fe isotopic composition of the residue by just the right amount to be offset during a subsequent episode of partial melting to produce lavas that have chondrite-like Fe isotopic compositions. Furthermore, some isotopic scatter would be expected while very little is found. This led Dauphas et al. (2009b) to suggest that Fe isotopic composition of boninites may be representative of the mantle, indicating that it is chondritic. Island arc basalts also show interesting systematics in their iron isotope variations. The samples from the New Britain island arc have Fe isotopic compositions that vary systematically with the distance from the trench and the degree of partial melting (Dauphas et al. 2009b) (Fig. 12). The ones formed by the largest degrees of partial melting with the largest amount of water have Fe isotopic compositions that are almost indistinguishable from the chondritic composition. On the other side of the array, the samples with the lowest degree of partial melting and lowest amount of water fluxing have Fe isotopic compositions similar to MORBs. This is another piece of evidence to support the view that partial melting can fractionate Fe isotopes and that the mantle has chondritic Fe isotopic composition. Nebel et al. (2013) found low $\delta^{56}Fe$ values in the Central Lau Spreading Center located in the Lay back-arc basin similar to signatures found in arcs that they interpreted to reflect the dragging of the proximal arc-front mantle to more distal regions where it can resurface in back arcs. Nebel et al. (2015) presented a study of island arc lavas along the Banda arc, Indonesia. After correction for fractional crystallization and possible crustal contamination, they found that the

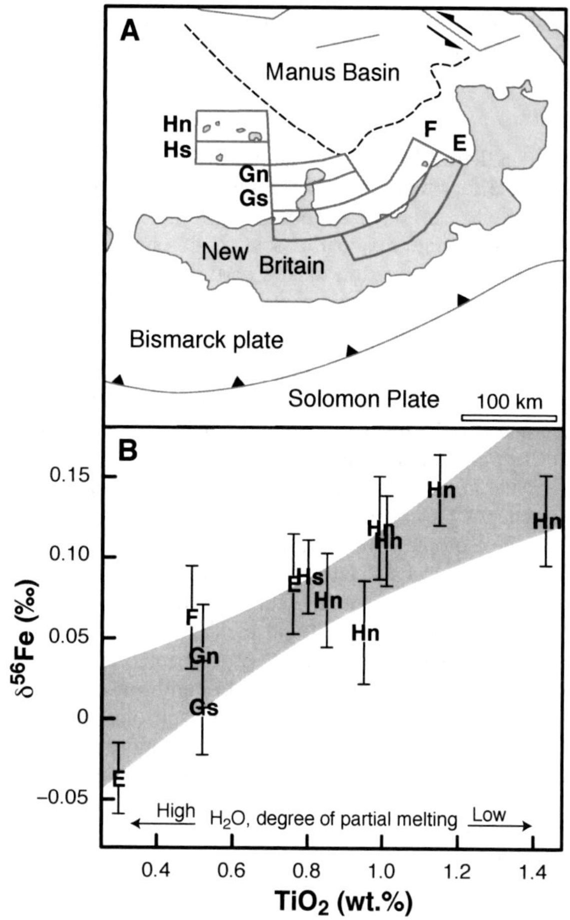

Figure 12. δ^{56}Fe variations in island arc basalts from the New Britain island arc (Dauphas et al. 2009b). The samples farther away from the trench that were fluxed by the smallest amounts of water and correspond to the lowest degrees of partial melting have the heaviest (MORB-like) δ^{56}Fe values. The samples closest to the trench that formed by the highest degrees of partial melting have δ^{56}Fe values similar to chondrites and peridotites.

pristine lavas have distinctly lower δ^{56}Fe values than MORBs, in agreement with the results obtained by Dauphas et al. (2009b). To explain the low δ^{56}Fe values of arc lavas, Dauphas et al. (2009b) suggested that the Fe isotopic fractionation in the source of island arc basalts may be partially suppressed. In this scenario, the starting δ^{56}Fe value of the mantle would be the same in MORBs and IABs but the extent of the fractionation would be reduced in the samples that have experienced the most extensive degree of partial melting and fluxing by subduction fluids. Nebel et al. (2015) instead postulate a different scenario, whereby the Fe isotopic composition would have started the same as mantle source of MORBs (~0‰) but an episode of melt extraction left behind a low δ^{56}Fe mantle, which was subsequently fluxed by oxidizing fluids that generated a second episode of melting. In this scenario, the low δ^{56}Fe value of arc lavas would be inherited from the first melt extraction event. The varied interpretations illustrate the difficulty of inferring the Fe isotopic composition of the mantle from lavas but they also show the strong potential of iron isotopes for tracing petrogenetic processes.

Debret et al. (2016) studied the Fe isotopic composition of serpentinites from Western Alps metaophiolites and found evidence for iron mobility in fluids during subduction, whereby isotopically light iron is lost as Fe(II)–SO_x or Fe(II)–Cl_2 species that could induce oxidation of the mantle wedge.

Partial mantle melting

As discussed above, there still is significant uncertainty on the Fe isotopic composition of the silicate Earth. The fact that all peridotites from various tectonic settings define an average δ^{56}Fe value close to chondritic supports the view that the mantle has a lower δ^{56}Fe value than MORBs and intraplate magmatic rocks. If this is correct, this begs the question of what causes Fe isotopic fractionation during partial mantle melting. A difficulty at the present time is that equilibrium fractionation factors between minerals relevant to Earth's mantle and silicate magmas are not fully established.

Williams et al. (2005) first proposed a quantitative model of iron isotopic fractionation during partial melting. As pointed out by Craddock et al. (2013), some of the calculations violate mass-balance because the residual mantle is calculated to be isotopically fractionated even when the degree of partial melting is ~0, which is impossible (see Fig. 3A of Williams et al. 2005). With this caveat in mind, we can examine some of the parameters and assumptions made by Williams et al. (2005) in calculating melt–solid residue fractionation factors. Williams et al. (2005) used the Fe isotopic compositions of mineral pairs in peridotites to calculate equilibrium fractionation factors between coexisting minerals. They thus used fractionation factors of –0.15‰ between olivine and melt, –0.08‰ between opx and melt, –0.06‰ between cpx and melt, and –0.03‰ between spinel and melt. As discussed by Beard and Johnson (2004b), Zhao et al. (2010), and Roskosz et al. (2015), mineral pairs are often in isotopic disequilibrium in peridotites and it is difficult to infer equilibrium fractionation factors from such samples. Secondly, these fractionations can reflect subsolidus re-equilibration and need to be corrected to mantle solidus temperatures, introducing more uncertainties. Thirdly, there is significant scatter when one plots the δ^{56}Fe values of one mineral against another (olivine vs. pyroxene), so that the inter-mineral fractionation factors are plagued by large uncertainties. With no other data available, Williams et al. (2005) had no other option than to rely on those fractionation factors. They pointed out that the partitioning of more incompatible Fe^{3+} into the melt could in itself impart a heavy Fe isotopic composition to the melt. Williams and Bizimis (2014) built on that early effort. They again used inter-mineral fractionation factors from peridotites as inputs (not corrected for temperature) and left the melt-clinopyroxene as a free adjustable parameter. The melt-residue fractionation factor can be written as,

$$\alpha_{\text{residue}}^{\text{melt}} = \frac{R_{\text{melt}}}{R_{\text{residue}}} = \frac{R_{\text{melt}}}{\sum_i [FeO]_i n_i R_i / \sum_i [FeO]_i n_i} = \frac{\sum_i [FeO]_i n_i R_{\text{melt}} / R_{\text{cpx}}}{\sum_i [FeO]_i n_i R_i / R_{\text{cpx}}} = \frac{\sum_i [FeO]_i n_i \alpha_{\text{cpx}}^{\text{melt}}}{\sum_i [FeO]_i n_i \alpha_{\text{cpx}}^{i}} \tag{22}$$

where R is the ^{56}Fe/^{54}Fe ratio, n is the modal abundance of each mineral, i denotes a particular mineral in the residue (this is Eqn. 3 of Williams and Bizimis 2014). They used a realistic melting model whereby the modal abundance of the various minerals changes as melting progresses, with minerals like clinopyroxene being consumed more rapidly (they did not explicitly track Fe^{3+} but given that Fe^{3+} is concentrated in cpx and this mineral in preferentially consumed during melting, their model indirectly takes into account the more incompatible and heavy Fe isotopic composition of Fe^{3+}). Even when they set the fractionation between melt and cpx to 0‰, they were able to reproduce the fractionation of ~+0.1‰ for 10% partial melting measured in MORBs. However, because most iron is in olivine, the net fractionation between the melt and the residue in Williams and Bizimis (2014) is effectively given by the sum $\Delta^{melt}_{cpx} + \Delta^{cpx}_{ol}$, meaning that their conclusion rests on the assumption that the fractionation Δ^{cpx}_{ol} is ~+0.1‰ when $\Delta^{melt}_{cpx} = 0$‰. Clearly, more work is needed to test those assumptions. Because the partition coefficient of Fe is around unity, when the system has reached 10% partial melting, only ~10% of the initial iron has been removed from the system and little change is expected in the Fe isotopic composition of the melt or the residue over such a melting interval (Dauphas et al. 2009b; Williams and Bizimis 2014).

Weyer and Ionov (2007) also investigated partial melting models and used a very simple formulation to describe the isotopic fractionation. They fitted the measured Fe isotopic composition of the peridotites (mantle residues) with Rayleigh distillation models and concluded that the melt–solid fractionation factor had to be between ~+0.1 and +0.3‰ to explain the variations.

Dauphas et al. (2009b) developed a model of isotopic fractionation that is entirely driven by the redox state of iron, meaning that Fe^{2+} in minerals and melt have the same isotopic compositions but that Fe^{3+} is systematically heavier in both minerals and melts relative to Fe^{2+}. They estimated the Fe^{3+}–Fe^{2+} equilibrium fractionation as a function of temperature using NRIXS and Mössbauer data as well as experimental results on the fractionation between peralkaline rhyolitic melt containing ~2/3 Fe^{3+} and pyrrhotite. An important driver of the melt–solid fractionation in this model is the mild incompatibility of Fe^{3+} relative to Fe^{2+}. While Fe^{2+} has a melt/solid equilibrium partition coefficient of ~1, the partition coefficient of Fe^{3+} is ~5. During melting, the melt has a high Fe^{3+}/Fe^{2+} ratio compared to the residual mantle, and because Fe^{3+} tends to have heavy Fe isotopic composition relative to Fe^{2+}, the melt could also have heavy Fe isotopic composition. Dauphas et al. (2009b) could only explain the heavy Fe isotopic composition of MORBs for an Fe^{3+}/Fe^{2+} ratio in the melt that is ~0.24 at 10% partial melting, whereas the measured value is around 0.12–0.16.

Dauphas et al. (2014) generalized this model by allowing for the presence of equilibrium fractionation between Fe^{3+}–Fe^{2+} in minerals, Fe^{3+}–Fe^{2+} in melt, and Fe^{2+}-melt–Fe^{2+}-minerals. The only simplification remaining in this model is that Fe^{2+} has the same Fe isotopic composition in all minerals (e.g., olivine, pyroxene, spinel). Dauphas et al. (2014) and Roskosz et al. (2015) also measured the force constant of iron bonds using the NRIXS synchrotron techniques (see Dauphas et al. 2012 for details) in olivine as well as basaltic glasses and spinels synthesized under various oxygen fugacities (Fig. 13). The glasses were taken as proxies for melts because the NRIXS technique only works with solids. The authors found that the force constants of Fe^{2+} in olivine (197 ± 10 N/m), Al-bearing spinel (207 ± 14 N/m), and basalt (199 ± 15 N/m) are very similar, so that little equilibrium fractionation is predicted for Fe^{2+} in minerals and melts at magmatic temperatures. In contrast, Fe^{3+} in basaltic glasses (351 ± 29 N/m) and spinel (300 ± 18 N/m) displays much higher force constants. The equilibrium fractionation factors at high temperature are directly proportional to differences in force constants between two phases A and B: $\Delta^{B}_{A} = 2{,}904\left(F_B - F_A\right)/T^2$. At a temperature relevant to MORB generation (~1300 °C), the force constants translate into a fractionation of ~+0.2‰ between melt Fe^{3+} and Fe^{2+} and 0.00 ± 0.02‰ between Fe^{2+} in melt and Fe^{2+} in olivine. Using those fractionation factors, Dauphas et al. (2014) could only explain approximately half of the difference in $\delta^{56}Fe$ values of MORBs and the inferred mantle value. Several explanations can be considered as to why the model does not predict the full extent of the fractionation that is measured: (1) glasses are imperfect proxies for melts and the fractionation factor is actually higher than that measured by NRIXS, (2) the bonds show some anharmonicity so that extrapolation of room temperature measurements to magmatic temperature suffers from inaccuracies, (3) the bonds stiffen with pressure and the melt is more sensitive to this stiffening, or (4) non-equilibrium (kinetic) processes are at play. Further work is needed to document equilibrium fractionation between magmas and minerals relevant to mantle melting before we can fully understand how iron isotopes are fractionated during partial mantle melting.

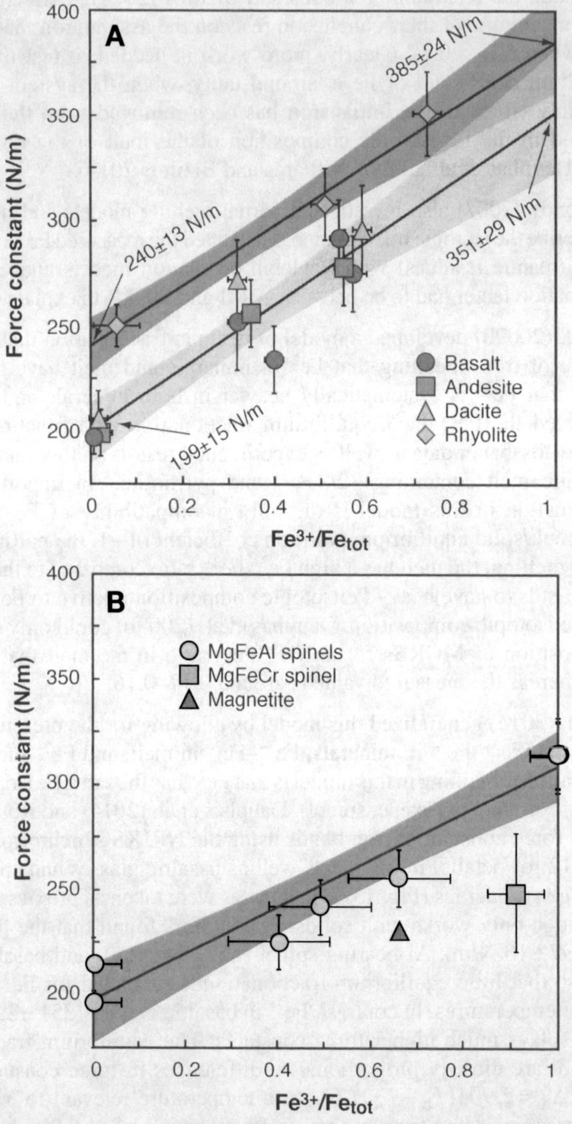

Figure 13. Force constant measurements of basalt, andesite, dacite, rhyolite glasses (A) and spinels (B) as a function of the redox state of iron (Dauphas et al. 2014; Roskosz et al. 2015). At high temperature equilibrium iron isotopic fractionation is directly proportional to the force constant; $1000\ln\beta = 2904\langle F\rangle/T^2$.

Impact evaporation and core formation

One motivation to investigate how iron isotopes are fractionated during partial melting is the observation that unlike MORBS, peridotites (and by inference Earth's mantle) have near chondritic Fe isotopic compositions (Fig. 11). Some, however, argue that peridotites are not necessarily representative of the silicate Earth composition and they posit instead that Earth's mantle as a whole has a heavy Fe isotopic composition. Poitrasson et al. (2004) suggested that impact-driven evaporation could be responsible for such heavy isotope enrichment. The ramifications of this idea have not been tested thoroughly yet but one should note that highly volatile and depleted potassium (Humayun and Clayton 1995a,b) or zinc (Luck et al. 2005; Othman et al. 2006) isotopes are not very fractionated in the Earth relative to chondrites.

Most of iron in the Earth is present in its core (~93%; the rest is in the mantle and crust). If there was any isotopic fractionation between metal and silicate, this offers significant leverage to fractionate iron isotopes in Earth's mantle. High temperature equilibration experiments between metal and silicate (Poitrasson et al. 2009; Hin et al. 2012; Shahar et al. 2015) yield contradictory results (Fig. 8). Shahar et al. (2015) report the largest fractionation beween silicate and metal; −0.08 ± 0.03‰ at 1650 °C. There is some uncertainty on the conditions that prevailed during core formation but Ni and Co impose relatively high pressures of around 40–60 GPa, and a corresponding liquidus temperature of ~3000 K (Li and Agee 1996; Rubie et al. 2011; Siebert et al. 2013). At such elevated temperature, the predicted shift in the Fe isotopic composition of the mantle is divided by a factor of $(3000/1923)^2$, meaning that the fractionation would be at most −0.03‰ between the mantle and core. The results of Hin et al. (2012) suggest that it could be even less. Predicted equilibrium fractionation factors between basaltic glass (Dauphas et al. 2012, 2014) and either fcc iron (Dauphas et al. 2012; Krawczynski et al. 2014b) or bcc iron (Dauphas et al. 2012) based on NRIXS measurements also give different values than the experimental result of Shahar et al. (2015). The system basaltic glass–fcc Fe gives the largest predicted fractionation by NRIXS but even then, at 3000 K, the fractionation between silicate and metal would only be +0.015‰. Starting with a chondritic δ^{56}Fe value of ~0‰, the δ^{56}Fe value of the mantle would be predicted to be between approximately −0.03 and +0.014‰; i.e., much lower than the value of ~+0.1‰ measured in MORBs and intraplate magmatic rocks. Those estimates correspond to relatively low-P conditions and rest on the assumption that the bonds are harmonic, meaning that the fractionation factors can be extrapolated to high-T using a $1/T^2$ relationship, which still needs to be tested. As a side note, basalts from Mars and Vesta have Fe isotopic compositions similar to chondrites, suggesting that even when the core forms at lower temperature than the Earth, metal–silicate fractionation remains small (Fig. 11). This also puts constraints on the amount of sulfur that can be present in the martian core (Shahar et al. 2015).

Polyakov (2009) used NRIXS data on high pressure phases to calculate equilibrium fractionation factors at high pressure-high temperature between metal and high pressure mantle phases ferropericlase and postperovskite. He concluded that at core-mantle boundary conditions, the fractionation between ferropericlase and metal is small (+0.0006 ± 0.0030‰ at 4000 K) but that the fractionation for the system postperovskite and metal could be significant (+0.04 ± 0.01‰ at 4000 K) and possibly sufficient to explain the heavy Fe isotopic composition of the silicate Earth. While intriguing, this conclusion poses two problems. The first one is that postperovskite is irrelevant to core formation conditions (this phase would be unstable; the system most relevant is molten silicate–metal). Furthermore, the NRIXS data on postperovskite used in the calculation are of insufficient quality to reliably derive an equilibrium fractionation factor.

Shahar et al. (2016) and Liu et al. (2016) recently reported NRIXS measurements and ab initio calculations of iron isotopic fractionation between bridgmanite, basaltic glass, pure iron, and Fe-rich alloys of H, C, O Si, S, and Ni. The force constant measurements at high pressure involve the use of diamond anvil cells and are particularly challenging. The conclusion that is emerging from those studies is that the δ^{56}Fe isotopic shift in the mantle induced by core formation is small (<0.03‰ and most likely <0.02‰) because the temperatures involved (~3000–4000 K) were very high, which limited the extent of equilibrium iron isotopic fractionation

Williams et al. (2012) proposed that disproportionation of Fe^{2+} at high pressure in the lower mantle to Fe^0 and Fe^{3+} in bridgmanite could lead to an enrichment in the heavy Fe isotopic composition of the silicate Earth if metal thus formed was removed. However,

Craddock et al. (2013) showed that this does not work from a mass-balance point of view because the mantle only contains ~3% Fe^{3+} (97% Fe^{2+}) and the shift in $\delta^{56}Fe$ value from iron disproportionation would be negligible.

To summarize, some experiments and NRIXS data suggest that the fractionation between silicate and metal at conditions relevant to core formation should be small but more work needs to be done at high pressure on molten compositions to further assess whether this conclusion is correct.

Fractional crystallization, fluid exsolution, immiscibility, and thermal (Soret) diffusion

MORBs and intraplate basalts have a relatively constant $\delta^{56}Fe$ value of ~+0.1‰ (Fig. 11). Poitrasson and Freydier (2005) first showed that granites have heavier Fe isotopic composition than MORB and subsequent work showed that such heavy Fe isotope enrichments are not limited to granite and include rhyolites as well as other rock types such as pegmatites (Heimann et al. 2008; Telus et al. 2012). The $\delta^{56}Fe$ values of silicic rocks correlate broadly with the SiO_2 contents (Fig. 14). Below ~70 wt% SiO_2, igneous rocks have $\delta^{56}Fe$ values that are more or less constant ($\delta^{56}Fe$ between ~+0.08 and +0.14‰) and $\delta^{56}Fe$ then rapidly increases above 70 wt% SiO_2 to reach values as high as +0.4‰. Such large Fe isotopic variations are more commonly encountered in low-*T* environments, begging

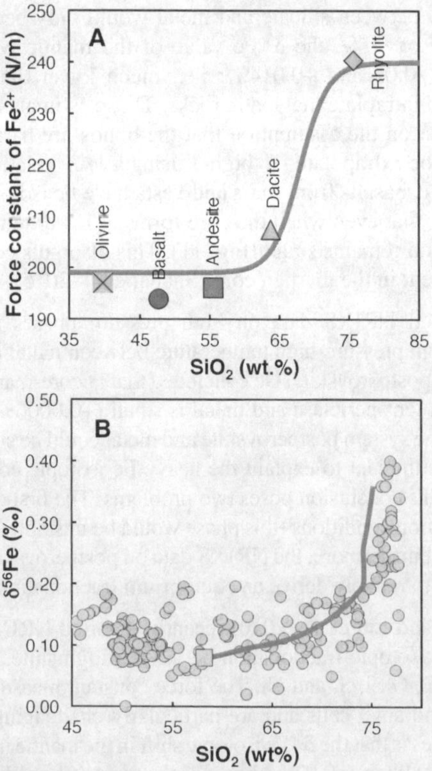

Figure 14. Controls on iron isotopic fractionation in silicic rocks (Dauphas et al. 2014). The force constant of Fe^{2+} shows an abrupt change between 65 and 75 wt% (panel A). This change can partially explain the rapid increase in $\delta^{56}Fe$ values of silicic rocks above 70 wt% SiO_2 (panel B). The circles are data points compiled from the literature and the curve is a fractional crystallization calculation for a starting andesitic melt at fixed fO_2 = FMQ using the Rhyolite–MELTS software and available measured force constant values of glasses, oxides, and silicates.

the question of what causes such large fractionation at magmatic temperatures. Below, we review the scenarios that have been put forward and evaluate their strengths and weaknesses.

Poitrasson and Freydier (2005) favored a scenario of aqueous fluid exsolution. In the later stages of granitic magma body evolution, aqueous fluids rich in chlorine can be exsolved from the magma and mobilize elements such as Zn, Cu, Mo, Au, or Fe that can form porphyry-style mineralizations of economic relevance. For this scenario to work, the fluids that are removed need to have low $\delta^{56}Fe$ values so as to drive the Fe isotopic composition of the residual magma towards heavier values. Heimann et al. (2008) extended the work of Poitrasson and Freydier (2005) and showed that both granitic and volcanic rocks have heavy Fe isotopic compositions. They also interpreted these results in terms of fluid exsolution on the basis that the samples with high $\delta^{56}Fe$ values have sub-chondritic Zr/Hf ratios. They assert that those low ratios are tracers of fluid exsolution. The model that they propose involves exchange between isotopically heavy magnetite and isotopically light $FeCl_2^0$. There are two arguments against the fluid exsolution model. The first one is that some of the granites that show some of the highest $\delta^{56}Fe$ values are of the A-type (anorogenic), which are among the most anhydrous granites encountered and are thought to have formed by partial melting of a dry source. One would not expect those A-type granites to have experienced the most extensive aqueous fluid exsolution (Sossi et al. 2012; Telus et al. 2012). A second argument against the fluid exsolution hypothesis is the study of Zn isotopic compositions and Zn/Fe ratios. Zinc is very mobile in chlorinated fluids and the most silicic magmas show significant Zn/Fe variations that cannot be easily explained by magmatic differentiation, suggesting that indeed Zn was mobilized in those rocks (Telus et al. 2012). If Fe had been mobilized in a similar manner to Zn, one would expect the samples with the highest $\delta^{56}Fe$ values to also exhibit fractionated Zn isotopic compositions. Telus et al. (2012) measured Zn isotopes in granites and did not find any significant correlation with iron (i.e., most samples with fractionated $\delta^{56}Fe$ values have non-fractionated Zn isotopic composition). Schuessler et al. (2009) measured the Li isotopic composition of samples from the Hekla volcano, Iceland, where Fe isotopic fractionation associated with magmatic differentiation was found. The authors did not detect any variation in the isotopic composition of the fluid-mobile element Li. Those studies suggest that aqueous fluid exsolution plays a minor role in fractionating the isotopic composition of iron in silicic rocks. Some pegmatites, however, show fractionated Zn and Fe isotopic compositions for which fluid exsolution likely played a role (Telus et al. 2012).

Another scenario that has been proposed to explain Fe isotopic variations in some silicic rocks is thermal diffusion. The general idea behind this model is that sills are injected by underplating under a thick volcanic pile that is a barrier to the flow. The magma in the sill is cooled from the top and partially differentiates by thermal migration. It was shown relatively recently that thermal gradients could drive Fe isotopic fractionation in magmas (Huang et al. 2009; Richter et al. 2009b), which led Lundstrom (2009) and Zambardi et al. (2014) to suggest that thermal diffusion could be responsible for some of the isotopic fractionations measured for Fe and other elements in silicic rocks. Zambardi et al. (2014) tested one aspect of the thermal migration model, namely that there should be a depth dependence of the isotopic composition (note that magmatic differentiation would also predict such a relationship). A second test of this idea follows from the observation that in thermal diffusion experiments, all elements can be fractionated isotopically, so that isotopic correlations are expected between elements. Telus et al. (2012) measured the stable isotopic compositions of Fe, Zn, Mg and U in the same samples but failed to detect the correlations expected for thermal diffusion, leading them to conclude that thermal/Soret diffusion was not the main driver of Fe isotopic fractionation in silicic rocks. Zambardi et al. (2014) countered that whether or not fractionation would be expressed depends on the leverage given by the melt/solid partition coefficient. In igneous systems, zinc is geochemically similar to iron (somewhat similar partition behaviors in many minerals), yet no clear isotopic correlation was found between Fe and Zn in the rocks studied by Telus et al. (2012).

Zhu et al. (2015) proposed a scenario to explain the heavy Fe isotopic composition of A-type granites that involves diffusion in a silica-rich immiscible melt. As the silica-rich melt grows at the expense of a Fe-rich magma, iron which is preferentially partitioned in the Fe-rich melt accumulates at the interface and diffuses away from the immiscible silicic melt. One would therefore predict that the immiscible melt and the interface should be enriched in the heavy isotopes of iron (during diffusion, source reservoirs are enriched in the heavy isotopes) while the bulk of the Fe-rich melt should be enriched in the light isotopes of iron (sink reservoirs are enriched in the light isotopes during diffusion because those diffuse faster than heavier ones) (Watson and Müller 2009). This is an intriguing possibility but a significant unknown is the extent to which advective transport of iron from the immiscible liquid interface to the far field can play a role to obliterate diffusion in such a system. Carbonatites, which may have formed by liquid immiscibility, define a large span in δ^{56}Fe values (almost 2‰) whose origin is still uncertain and may involve multiple processes (Johnson et al. 2010).

A third scenario, and in our view the most likely one that applies to the majority of magmatic rocks, is fractional crystallization. Obviously, we cannot exclude that thermal migration or fluid exsolution play some roles in some settings, as was suggested for pegmatites. Fractional crystallization is usually thought to be one of the main drivers behind the chemical differentiation of magmatic rocks. Teng et al. (2008) presented the first evidence that fractional crystallization can fractionate Fe and Mg isotopes. They measured samples from the Kilauea Iki lava lake, which is a crater pit that was filled with lava in 1959 and slowly cooled afterwards. The lava lake was drilled at regular intervals by the USGS. The rocks span a significant range in MgO content, from MgO-rich cumulate rocks to MgO-poor segregation veins. The variations in δ^{56}Fe values span 0.2‰, with heavy isotope enrichments in the more differentiated (more MgO-poor) compositions. Teng et al. (2008) could explain their data with a melt–solid fractionation coefficient of +0.1‰. This fractionation factor does not reflect equilibrium as it was subsequently shown by Teng et al. (2011) and Sio et al. (2013) that the fractionation is driven by kinetic fractionation accompanying Mg–Fe diffusion in olivine (this will be discussed more extensively in the next section). Other localities than Kilauea Iki lava lake have shown somewhat similar trends of isotopic fractionation associated with magmatic differentiation, including the Red Hill intrusion (Tasmania; Sossi et al. 2012), the Cedar Butte volcano (USA; Zambardi et al. 2014), the Hekla volcano (Iceland; Schuessler et al. 2009) and the Bergell intrusion (Switzerland, Schoenberg and von Blanckenburg 2006). Teng et al. (2011) showed that the fractionations that they measured in Kilauea Iki lava lake were due to Fe-Mg inter-diffusion in olivine but this interpretation is obviously not applicable to the overall pattern of δ^{56}Fe variations in silicic rocks and other interpretations/models must be sought. Dauphas et al. (2014) and Foden et al. (2015) used the Rhyolite-MELTS software to model the Fe isotopic composition of silicic rocks. Several important factors were identified that control the evolution of δ^{56}Fe values during fractionational crystallization.

Foden et al. (2015) examined the influence on iron isotopic fractionation of the initial redox state of the magma and whether or not the system is open (oxygen fugacity buffered, equilibrium crystallization) or closed (oxygen fugacity non-buffered, fractional crystallization). To model this process, the authors assumed a pyroxene–melt fractionation factor equal to $-0.17\times10^6/T^2$, corresponding to fractionations of −0.07 and −0.11‰ between pyroxene/ilmenite and melt at the relevant magmatic temperatures (Sossi et al. 2012). The assumed fractionation between magnetite and melt is $+0.13\times10^6/T^2$, corresponding to fractionations of +0.07 to +0.09‰ between magnetite and melt at the relevant magmatic temperatures of 900–1000 °C (Sossi et al. 2012). Those fractionation factors are derived semi-empirically to reproduce some of the iron isotope variations that are observed. In a closed system (non-buffered), Fe^{3+} being more incompatible early on than Fe^{2+}, the magma gets progressively more oxidized as crystallization proceeds. Magnetite crystallizes early, which mitigates the incompatible behavior of Fe^{3+} in silicate minerals. Because the amount of Fe^{3+} in the closed system scenario is finite, by the time the system has reached 65–70% SiO_2, the Fe^{3+}/Fe_{tot} ratio of the melt is low and the late crystallizing phases are dominated by Fe^{2+}-bearing silicates, so the melt–mineral fractionation factor is positive. There is also very little Fe left in the melt at that point, so mineral crystallization gives a lot of leverage to enrich the residual magma in the heavy isotopes of iron. While this model can reproduce the heavy Fe isotopic composition of some A-type granites that are characterized by large enrichments in the heavy iron isotopes, the mineral-melt fractionation factors are somewhat arbitrary and do not evolve as the Fe^{3+}/Fe^{2+} ratio in the melt evolves (or only indirectly, through the modal abundances of the crystallizing phases, but not because the β-factor of Fe in the melt changes). The onset of the heavy isotope enrichment also does not match very well observations in natural samples, which see a turning point at ~68 wt%, while the model predictions of Foden et al. (2015) predict variations in $\delta^{56}Fe$ values in rocks with lower SiO_2 content.

Dauphas et al. (2014) modeled fractional crystallization of magmas to explain the trends towards heavy iron isotope values of silicic magmas based on the β-factors that they measured for Fe^{2+} and Fe^{3+} in minerals and glasses (taken as proxies for silicate melts) (Fig. 13). Measurements of olivine, pyroxene, spinels, magnetite and other minerals suggest that the β-factors for Fe^{2+} and Fe^{3+} in silicate minerals and oxides are approximately constant and equal to $0.57\times10^6/T^2$ and $0.73\times10^6/T^2$. Measurements of iron force constants in basalt, andesite, dacite, and rhyolite also show that the β-factor of Fe^{3+} in magmas is approximately constant and equal to $1.06\times10^6/T^2$. In a plot of force constant vs. Fe^{3+}/Fe^{2+} ratio, basalt, andesite and dacite define the same trend, corresponding to a β-factor of $0.57\times10^6/T^2$ for Fe^{2+}. The trend defined by rhyolite glasses is, however, shifted towards higher force constant for Fe^{2+}, corresponding to a β-factor of $0.68\times10^6/T^2$. This difference in the behavior of Fe^{2+} in rhyolitic glass is also seen in XANES (X-ray Absorption Near Edge Structure) spectroscopy, which probes the coordination environment of iron (Dauphas et al. 2014). Both NRIXS and XANES thus suggest that somewhere between dacite and rhyolite, the coordination environment of iron changes, so that ferrous iron forms stronger bonds with surrounding atoms. Dauphas et al. (2014) used those fractionation factors in a model of fractional crystallization of an andesitic starting composition using the Rhyolite-MELTS software, assuming buffered oxygen fugacity. They were able to reproduce the trend of $\delta^{56}Fe$ values vs. SiO_2 content, including the inflexion point at ~70wt% (Fig. 14).

The studies of Dauphas et al. (2014) and Foden et al. (2015) showed that fractional crystallization could explain the isotopic variations in silicic igneous rocks, with no need to invoke more exotic phenomena such as magma immiscibility, aqueous fluid exsolution, or thermal diffusion.

A new tool to improve on geospeedometry reconstructions in igneous petrology

Identifying diffusive processes in igneous petrology is important as it is often the only manner by which one can derive timescales and durations of magmatic processes. One such tool is Mg–Fe interdiffusion in olivine (Costa et al. 2008). It is a simple binary diffusion problem with well-established diffusion coefficients (Dohmen and Chakraborty 2007). Furthermore, the exchange coefficient of Mg and Fe between olivine and melt varies little depending on the melt composition (Toplis 2005), so that it is straightforward to relate trends of magmatic differentiation to crystal boundary conditions. A difficulty with Mg–Fe zoning in olivine (and other chemical profiles in other minerals) is that such zoning cannot always unambiguously be ascribed to diffusion. Let us consider two end-member scenarios to illustrate this point. Zoning in olivine is established in response to magmatic differentiation. Because olivine incorporates preferentially Mg relative to Fe, the magma evolves towards higher Fe/Mg ratio as crystallization proceeds. If the rate of diffusion in olivine is much slower than the rate of magmatic differentiation or cooling, then at each time step, the olivine that grows is in equilibrium with a magma that becomes more Fe rich. The growing olivine will thus develop normal zoning with a composition that becomes more fayalitic from core to rim. In a second end-member scenario, let us consider an early formed olivine that is put into contact with a more evolved magma. Iron will diffuse in the olivine while Mg will diffuse out, creating a zoning pattern with a composition that is more fayalitic from core to rim. In the second scenario, zoning is entirely attributable to diffusion and one can calculate magmatic timescales from it. In the first scenario, zoning is due to crystal growth with no diffusion, so any timescale that one derives from it is meaningless. It is most often impossible to distinguish the two zoning patterns. Trace elements have been used for that purpose (Costa et al. 2008) but they suffer from the fact that the partitioning behavior of those trace elements are not always well constrained, the magma body evolution is uncertain leading to uncertainties in the boundary condition, and the diffusion coefficients are also uncertain. Magnesium and iron isotopes are fractionated during diffusive processes, which led Dauphas et al. (2010) and Teng et al. (2011) to suggest that the two modes of zoning could be distinguished (or their contributions teased apart when both processes are at play). Those authors showed that the Fe isotopic composition of olivine in Alexo komatiite and Kilauea Iki lava lake were affected by diffusion. Sio et al. (2013) and Oeser et al. (2015) built on these early studies and reported *in situ* Fe isotope measurements of zoned olivines (Fig. 4). An important uncertainty remains regarding the β-value of Mg and Fe, with measurements of natural samples suggesting that it should be around 0.084 to 0.16 for Mg and 0.16 to 0.27 for Fe (the ratio βFe/βMg is better constrained to be 1.8 ± 0.3) (Sio et al. 2013; Oeser et al. 2015). As the β-values are refined and *in situ* measurement techniques are improved, it will be possible to inverse Fe isotope data in olivine (or other minerals) to reconstruct the thermal and crystal growth history of a magma (Sio and Dauphas 2016). The study of zoned minerals evolved from characterizing the petrography of the samples by optical microscopy to *in situ* analyses of major, minor and trace elements by microprobe. The next step in studies of zoned minerals is *in situ* stable isotopic analyses, which provide a unique tool to tease apart diffusion from crystal growth.

IRON BIOGEOCHEMISTRY

Microbial cycling of Fe isotopes

Dissimilatory iron reduction (DIR). The importance of Fe(III) oxides as electron acceptors for anaerobic respiration in Fe-rich modern sediments is widely recognized (Lovley et al. 1987; Roden 2004) and may have been associated with some of the earliest forms of metabolism on Earth (Vargas et al. 1998). Mineralogical products of dissimilatory Fe(III) reduction that may be preserved in the rock record include magnetite, Fe carbonates, and sulfides. In pioneering studies, Bullen and McMahon (1998) and Beard et al. (1999b) reported that the δ^{56}Fe value of dissolved Fe(II) produced by dissimilatory Fe-reducing bacteria was fractionated by ~−1.3‰ relative to the ferrihydrite substrate. Subsequent results have been extended to different DIR bacteria, growth conditions, and substrates (Beard et al. 2003a; Icopini et al. 2004; Crosby et al. 2005, 2007; Johnson et al. 2005; Wu et al. 2009; Tangalos et al. 2010; Percak-Dennett et al. 2011). In particular, at high reduction rates, Fe(II) produced by DIR has δ^{56}Fe values that are up to 2.6‰ lower than ferric substrate, possibly reflecting the effect of adsorption of heavy Fe(II) on hydrous ferric oxide. However, Crosby et al. (2005, 2007) have shown that the low δ^{56}Fe values for aqueous Fe(II) produced by DIR reflect isotopic exchange among three Fe inventories: aqueous Fe(II) [$Fe(II)_{aq}$], sorbed Fe(II) [$Fe(II)_{sorb}$], and a reactive Fe(III) component on the ferric oxide surface [$Fe(III)_{reac}$]. The fractionation in $^{56}Fe/^{54}Fe$ ratios between $Fe(II)_{aq}$ and $Fe(III)_{reac}$ was −2.95‰, and independent of the ferric Fe substrate (hematite or goethite) and bacterial species, indicating a common mechanism for Fe isotope fractionation during DIR. Moreover, the $Fe(II)_{aq}$–$Fe(III)_{reac}$ fractionation in $^{56}Fe/^{54}Fe$ ratios during DIR is identical within error of the equilibrium $Fe(II)_{aq}$–ferric oxide fractionation in abiological systems at room temperatures (Johnson et al. 2002; Welch et al. 2003; Wu et al. 2011). This suggests that the role of bacteria in producing Fe isotope fractionations during DIR lies in catalyzing coupled atom and electron exchange between $Fe(II)_{aq}$ and $Fe(III)_{reac}$ so that equilibrium Fe isotope partitioning occurs (Crosby et al. 2007). Other parameters such as the removal or local accumulation of $Fe(II)_{aq}$, presence of dissolved Si, and pH may also affect the iron isotopic record of DIR in sediments (Wu et al. 2010).

Bacterial Fe oxidation. Although chemical oxidation of ferrous iron is thermodynamically favored during the interaction of reduced fluids with oxygenated waters, bacterial Fe(II) oxidation may prevail in acidic, microaerobic or anoxic environments. Microorganisms that oxidize Fe(II) to generate energy for growth include those that couple Fe(II)-oxidation to the reduction of nitrate at neutral pH (Benz et al. 1998), or to the reduction of oxygen at either low (Edwards et al. 2000), or neutral pH (Emerson and Moyer 1997), and the anaerobic Fe(II)-oxidizing phototrophs (Widdel et al. 1993).

Croal et al. (2004) investigated Fe isotope fractionation produced by Fe(II)-oxidizing phototrophs under anaerobic conditions. Among key results, the ferrihydrite precipitate has a $\delta^{56}Fe$ value that is ~+1.5‰ higher than the aqueous Fe(II) source. Since the degree of isotopic fractionation is not correlated with the rate of oxidation (controlled by changing the light intensity), it has been suggested that kinetic isotope effects were not of great importance in controlling the fractionation factor. The fractionation factor, however, is higher than for abiotic Fe(II) oxidation experiments (about 1‰, Bullen et al. 2001) and lower than for equilibrium fractionation between aqueous Fe(II) and Fe(III) of 3‰ at room temperature (Welch et al. 2003).

In fact, it has been difficult to determine Fe isotope fractionation between aqueous Fe(II) and poorly crystalline ferric hydrous oxides (HFO, or ferrihydrite) due to the rapid transformation of the latter to more stable minerals. Wu et al. (2011) experimentally determined the equilibrium Fe(II)–HFO fractionation factor using a three-isotope method. Iron isotope exchange between Fe(II) and HFO was rapid and near complete in the presence of dissolved silica. Equilibrium Fe(II)–HFO $^{56}Fe/^{54}Fe$ fractionation factors of −3.17‰ were obtained for HFO plus silica. In contrast, when coprecipitates of Si–HFO form during the experiment, a smaller fractionation factor of −2.6‰ was obtained, possibly reflecting blockage of oxide surface sites by sorbed silica leading to incomplete isotope exchange.

Magnetotactic bacteria. Magnetotactic bacteria (MB) are prokaryotes participating in the chemical transformation of Fe and S species via both redox and mineral precipitation processes. MB precipitates intracellular single domain ferromagnetic iron oxide (magnetite) or iron sulfide (greigite) minerals, causing them to respond to geomagnetic fields. These bacteria are globally distributed in suboxic to anoxic freshwater (Frankel et al. 1979; Spring et al. 1993) and marine sediments, soils, and stratified marine water columns (Bazylinski et al. 2000; Simmons et al. 2004).

Initial Fe isotope studies of magnetite produced by MB (Mandernack et al. 1999) have shown no detectable fractionation when either an Fe(II) or Fe(III) source was used in the growth media. These results contrast strongly with recent experimental work of Fe isotope fractionation during magnetite formation coupled to dissimilatory hydrous ferric oxide reduction, which shows large isotopic fractionation between Fe in magnetite and Fe in the fluid (Johnson et al. 2005). However, this does not preclude that there is an isotopic effect produced by MB because only two strains were investigated over a restricted range of laboratory conditions. In particular, Fe isotope fractionation might be dependent on the kinetics of Fe uptake by MB, which may vary with Fe concentration, Fe redox state, and the presence of Fe chelators (Schuler and Baeuerlein 1996). In a recent study, Amor (2015) investigated the chemical and isotopic properties of magnetite produced by *Magnetospirillum magneticum* AMB−1 model magnetotactic bacterium. Results suggest that AMB−1 bacteria preferentially incorporate heavy iron isotopes within the cell. Magnetite is then produced from partial reduction of iron accumulated within the cell. This led to magnetite crystal mineralizations that were enriched in light isotopes and displayed δ^{56}Fe values from −1 to −1.5‰ lower than those of the growth medium. Magnetite biomineralizations may therefore have the potential to produce magnetite enriched in light iron isotopes relative to the precipitation solution. Amor et al. (2016) detected deviation from mass-dependent fractionation in ^{57}Fe in growth media and magnetite produced by magnetotactic bacterium *M. magneticum* strain AMB−1. This is the first documented iron isotopic anomaly induced by natural processes. The anomaly was only present when the growth medium was Fe(III)-quinate but was absent from Fe(II)-ascorbate experiments.

Fe isotopes in plants, animals, and humans

Iron is essential for all living organisms as it is used to maintain cellular homeostasis and plays a vital role in oxygen and carbon dioxide shuttling as well as enzymatic reactions required for DNA and hormone synthesis. In plants, Fe is required for iron-sulfur proteins and as a catalyst in enzyme-mediated redox reactions (e.g., Briat and Lobreaux 1997). Although the use of Fe isotopes as robust microbial biosignatures still remains controversial, a rapidly growing number of studies reported that higher organisms, including plants, animals and humans produce in fact the largest isotope fractionations.

In an initial study, Guelke et al. (2007) proposed that Fe isotope fractionation patterns in plants are related to two different strategies that plants have developed to incorporate Fe from the soil:

- strategy I: incorporation and potentially reduction of Fe(III) in soils resulting in the uptake of isotopically light Fe by up to 1.6‰ relative to available Fe in soils;
- strategy II: complexation with siderophores resulting in the uptake of iron that is 0.2‰ heavier than that in soils.

It remains however unclear whether redox-related plant metabolism could be the main cause of isotopic variation in the biogeochemical cycling of Fe. Guelke-Stelling and von Blanckenburg (2012) investigated strategy I and II plants grown in nutrient solutions and proposed a non-reductive translocation process in strategy I plants. Other studies determined the Fe isotopic composition of different plant parts, including the complete root systems, seeds, leaves and stems in order to distinguish between uptake and in-plant fractionation processes (von Blanckenburg et al. 2009; Kiczka et al. 2010; Moynier et al. 2013; Akerman et al. 2014; Arnold et al. 2015). The overall range of fractionation among the different plant tissues and organ systems was up to 4.5‰ (Kiczka et al. 2010), and may result from at least 4 fractionation steps: (1) before active plant uptake, probably during mineral dissolution; (2) during selective uptake of Fe at the plasma membrane; (3) during translocation processes and storage in plants; (4) during remobilization and transfer from old to new plant tissue, further changing the isotopic composition over the season. Ab initio calculations (Moynier et al. 2013) also provided a mechanistic explanation to the enrichment in heavy Fe isotopes in roots of strategy-II plants (by about 1‰ for the δ^{56}Fe value) relative to the upper parts of the plants.

A small number of research groups worldwide are currently using Fe isotopes for biomedical research, with the aim to develop new methods for medical diagnosis on the basis of Fe isotopic analysis of biofluids, either for diseases or metabolism studies (Albarede 2015; Vanhaecke and Costas-Rodriguez 2015; Larner 2016). Following the pioneering study of Walczyk and von Blanckenburg (2002), subsequent studies further improved our understanding of Fe isotope fractionation in the human body (Krayenbuehl et al. 2005; Walczyk and von Blanckenburg 2005; Albarede et al. 2011; Hotz et al. 2011, 2012; Van Heghe et al. 2012, 2013, 2014; Hotz and Walczyk 2013; Jaouen et al. 2013a; von Blanckenburg et al. 2013, 2014). It is now well-established that human blood and muscle tissues are enriched in light Fe isotopes by 1 to 2‰ when compared to the diet intake. It was also found that blood yields the lightest Fe isotopic composition, while the liver is less enriched in light isotopes. The same picture was also observed in other mammals such as mice and sheeps (Balter et al. 2013). Although the exact mechanisms of Fe isotope fractionation during intestinal uptake, binding to the protein transferrin and transport within the blood plasma to various organs and tissues remain under investigation, it appears that whole blood Fe isotopic composition is an indicator of the efficiency of dietary Fe absorption (Hotz and Walczyk 2013). The isotopic composition of whole blood is also related to serum ferritin level, typically seen as a measure for the amount of Fe stored in the liver (Van Heghe et al. 2013). As a consequence, hereditary hemochromatosis, a disease characterized by excessive Fe uptake, is also reflected in the Fe isotopic composition of blood (Krayenbuehl et al. 2005). Finally, recent studies agree that the gender-based difference in whole blood Fe isotopic composition results from the response of the organism to the Fe loss accompanying menstruation rather than differential intestinal absorption between female and male (Van Heghe et al. 2013, 2014; Jaouen and Balter 2014). Considering a simple isotopic mass balance, it is possible that the isotopically heavier Fe in women's blood relative to that of men could be due to higher hepatic mobilization of Fe by women as a result of menstrual Fe loss, which represents up to 40% of the monthly dietary uptake (Harvey et al. 2005). With the assumption that Fe isotopic ratios in bones reflect the patterns observed in blood (Jaouen et al. 2012), Fe isotopes might be used in paleoanthropology as sex indicators for past populations or as proxies for the age at menopause in ancient populations.

Finally, considering the range of Fe isotopic compositions in plants and animals, it is likely that Fe isotopes may also provide important information on trophic levels and food chains, in particular for marine animals. Poigner et al. (2015) measured the Fe isotopic composition in bivalve hemolymph, which represents the product of cutaneous (gills) and intestinal (digestive tract) assimilation, while Emmanuel et al. (2014) measured the Fe isotopic composition of chiton teeth capped with magnetite (chitons are marine molluscs). Overall, δ^{56}Fe ranged from near 0 to −1.9‰, which results from either (i) physiologically controlled processes that lead to species-dependent fractionation; (ii) diet-controlled variability due to different Fe isotope fractionation in the food sources; and (iii) environmentally controlled fractionation that causes variation in the isotopic signatures of bioavailable Fe in the different regions. Jaouen et al. (2013b) found Fe isotopic fractionation in mammal trophic chains between plants, herbivores, and carnivores. Integrated studies, combining Fe isotopic composition of the different environmental compartments (seawater, sediments, porefluid, phytoplankton, marine particles) and the different tissues or organs of the organism are required to further understand what controls iron isotope variations in plants and animals.

FLUID–ROCK INTERACTIONS

In both terrestrial and marine environments Fe cycles between the solid and dissolved phases, and such phase transfers are associated with iron isotope fractionation. In this section we discuss what is known about the fractionation of Fe isotopes during fluid–mineral interactions in hydrothermal systems, soils, and rivers, as well as the application of δ^{56}Fe as a tracer for ore formation processes in the field of economic geology.

High- and low-temperature alteration processes at the seafloor

Past studies have demonstrated the complexity and diversity of seafloor hydrothermal systems and have highlighted the importance of subsurface environments where a variety of chemical reactions between seawater, rocks and hydrothermal deposits is taking place over a wide range of temperatures (German and VonDamm 2003; Hannington et al. 1995; Humphris et al. 1995). Alteration of oceanic crust by seawater is one of the most important processes controlling the global fluxes of many elements (Staudigel and Hart 1983; Wheat and Mottl 2004) and the composition of the aging oceanic crust (Alt 1995).

Leaching of Fe from basalts, either at high- or low-temperature, typically results in the preferential release of the lighter Fe isotopes (Rouxel et al. 2003). High-temperature (>300 °C) vent-fluids yield a range of Fe isotopic compositions that are systematically shifted toward light δ^{56}Fe values compared to igneous and mantle rocks (Sharma et al. 2001; Beard et al. 2003b; Severmann et al. 2004; Rouxel et al. 2008a, 2016; Bennett et al. 2009). Values as low as −0.67‰ and as high as −0.09‰ have been measured in hydrothermal vent fluids along the Mid-Atlantic Ridge and East Pacific Rise (EPR). The heaviest values have been reported for high-temperature hydrothermal fluids from ultramafic-hosted systems (e.g., Rainbow field; Severmann et al. 2004) while lighter values were reported for Fe-depleted vents from basaltic-hosted vent sites (e.g., Bio-vent, Rouxel et al. 2008a). In general, high-temperature hydrothermal fluids from basaltic-hosted fields have a restricted range from −0.3 down to −0.5‰ (Rouxel et al. 2008a, 2016; Bennett et al. 2009).

Potential processes controlling the variability of Fe isotopes in hydrothermal fluids include phase separation, high-temperature basalt alteration, and subsurface processes leading to Fe precitation or remobilization below seafloor. Although phase separation is one of the fundamental processes controlling mid-ocean ridge vent fluid chemistry (Von Damm 1988; Von Damm et al. 1995), several lines of evidence suggest only limited Fe-isotope fractionation during this process. First, Beard et al. (2003b) measured δ^{56}Fe values of both the vapor and brine phases from the Brandon Vent at EPR 21.5°S and found less than 0.15‰ difference between these two fluids. Secondly, although not spatially related, Fe isotopic compositions of the high salinity fluid at K-vent at EPR 9°30'N (i.e., Na above seawater) does not differ significantly from lower salinity, vapor-rich fluids at Tica vent at EPR 9°50'N (Rouxel et al. 2008a). Thirdly, a recent experimental study of Fe isotope fractionation during phase separation in the NaCl–H_2O system (Syverson et al. 2014) yielded a maximum Fe isotope fractionation between the vapor and liquid of 0.15 ± 0.05‰ with, in most cases, variations of δ^{56}Fe values indistinguishable within analytical uncertainties.

Hence, the general enrichment in light Fe isotopes in vent fluids relative to volcanic rocks should be explained by two alternative mechanisms (Rouxel et al. 2003, 2004, 2008a):

(1) high-temperature alteration of basalt and the formation of isotopically heavy secondary minerals (e.g., Mg–Fe amphibole) in the high-temperature reaction zone. This mechanism has been already observed during low-temperature alteration of basalts at the seafloor (Rouxel et al. 2003). In particular, highly altered basalts that are depleted in Fe by up to 80% from their original Fe concentration displayed an increase in δ^{56}Fe values relative to fresh values (up to 1.3‰), which suggests preferential leaching of light Fe isotopes (between −0.5 and −1.3‰) during alteration.

(2) precipitation of isotopically heavy pyrite in subsurface environments or in the reaction zone. Using first-principle methods based on density-functional theory (DFT) and Mössbauer spectroscopy methods, previous theoretical studies have demonstrated that pyrite should be enriched in heavy Fe isotopes under equilibrium conditions (Blanchard et al. 2009; Polyakov and Soultanov 2011; Blanchard et al. 2012). Using the reduced isotopic partition function ratios of FeS_2 and Fe(II)-aquo-chloro complexes, the isotope fractionation between FeS_2 and $Fe(II)_{aq}$ is estimated to be +1.0 to +1.5 at 350 °C. The experimentally determined equilibrium pyrite–hydrothermal fluid Fe isotopic fractionation also agrees with theoretical and spectrally based predictions (Syverson et al. 2013). Under hydrothermal conditions (300–350 °C, 500 bars) in NaCl- and sulfur-bearing aqueous fluids, the fractionation between FeS_2 and $Fe(II)_{aq}$ was determined to be +0.99 ± 0.29‰ (Syverson et al. 2013). Hence, in the case of pyrite precipitation in subsurface environments due to conductive cooling of the fluids, near equilibrium Fe-isotope fractionation is expected which should result in the preferential partitioning of isotopically light Fe in the hydrothermal fluids. This suggests that pyrite acts as important mineral buffer in the composition of high-temperature hydrothermal fluids.

Rivers and soils

In soils and river systems, isotopic fractionations are generally small (often less than ~0.5‰), and isotopically light Fe tends to be preferentially leached from source rocks into the dissolved phase. An early observation found that rivers with a high suspended load were similar to continental δ^{56}Fe, while rivers with a higher proportion of Fe in the dissolved phase typically had light δ^{56}Fe signatures (Fantle and DePaolo 2004). In the Amazon River, dissolved and suspended loads have similar δ^{56}Fe values to continental crust, while in the organic-rich "black" Negro River, there is a preferential concentration of lighter isotopes in the dissolved phase and heavier isotopes in the particulate phase (Bergquist and Boyle 2006). A more extensive subsequent study of δ^{56}Fe values in the Amazon River and its tributaries found similar results: δ^{56}Fe was isotopically light compared to continental material in organic rich rivers with a low suspended load, and similar to continental sources in rivers with more suspended material (Poitrasson et al. 2014). Direct measurements of the suspended material in the Amazon river system demonstrated that it also had an Fe isotope signature similar to the continental source (dos Santos Pinheiro et al. 2013).

While it is more common to find that rivers are isotopically light compared to continental material, many exceptions have been found. The North River in Massachusetts, for example, had dissolved δ^{56}Fe up to +0.3‰ (Escoube et al. 2009). Ingri et al. (2006) focused on the colloidal size fraction within the dissolved phase and found that δ^{56}Fe was dependent on the type of colloids present, with isotopically lighter organic colloids as low as −0.13‰ and heavier oxyhydroxide colloids up to +0.3‰. Escoube et al. (2015) and Ilina et al. (2013) found that larger Arctic rivers had δ^{56}Fe similar to continental material, while smaller rivers could vary dramatically from −1.7 to +1.6‰, which they attributed to active redox cycling and colloid formation. Conversely, a glacial outflow river on Svalbard had nearly continental δ^{56}Fe even over large changes in dissolved Fe concentrations, pointing to the lack of active redox cycling in this environment (Zhang et al. 2015).

Anthropogenic contamination may also influence riverine δ^{56}Fe values. River and lake samples from the South China Karst region show suspended material with δ^{56}Fe values ranging from −2.0 to +0.4‰, with the most negative values being attributed to biological activity and contaminated coal drainages (Song et al. 2011). Chen et al. (2014a) found that anthropogenic Fe could be traced in the Seine river because of its isotopically light δ^{56}Fe signature.

Much of the water in rivers, and nearly all of the Fe, originates in the surrounding soils when meteoric water dissolves iron-containing soil minerals. Studies of redoximorphic soils showed a depletion of light isotopes during anoxic soil dissolution, and a subsequent precipitation of that isotopically light Fe under oxic conditions (Wiederhold et al. 2007b). In contrast, soils from a rainforest in Cameroon did not have variable δ^{56}Fe (Poitrasson et al. 2008). Guelke et al. (2010) found that the most mobile phases in soil were isotopically light, while the residual silicate fraction was isotopically heavy, suggesting the preferential release of lighter Fe isotopes during silicate weathering. Similarly, Kiczka et al. (2010) found an enrichment in lighter Fe isotopes in Fe oxyhydroxides, which they attributed to the preferential leaching of light Fe from silicates in an alpine glacier and subsequent precipitation of this Fe as oxyhydroxides. Recently, field observations of the preferential release of lighter Fe isotopes during weathering has been reproduced under experimental conditions where the redox conditions of a soil were artificially manipulated (Schuth et al. 2015).

Mineral deposits

One motivation to study the behavior of Fe isotopes in the natural world is to better understand how economically important iron ore deposits are formed. Many studies have therefore characterized the distribution of δ^{56}Fe within various different rocks and minerals collected in active mining regions in order to better understand the processes by which these deposits were formed. Recent examples include the Xishimen (Chen et al. 2014b), Han-Xing (Zhu et al. 2016) and Gaosong (Cheng et al. 2015) deposits in China, the Grangesberg Mining District in Sweden (Weis et al. 2013), nickel deposits in Zimbabwe (Hofmann et al. 2014), the Schwarzwald region in Germany (Horn et al. 2006; Markl et al. 2006), Sn–W deposits in Tasmania (Wawryk and Foden 2015), and iron oxide–apatite ore deposits in Chile (Bilenker et al. 2016). The study of Fe isotope systematics in modern systems such as hydrothermal vents and marine and terrestrial sediments can also inform our understanding of how economically important deposits (e.g., volcanogenic massive sulfide) form.

Rouxel et al. (2004, 2008b) and Toner et al. (2016) investigated coupled Fe- and S-isotope systematics of sulfide deposits from the East Pacific Rise at 9–10°N and Lucky Strike vent fields to better constrain processes affecting Fe-isotope fractionation during the formation and aging of sulfide deposits at the seafloor. The results showed systematically lower δ^{56}Fe and δ^{34}S values in marcasite/pyrite relative to chalcopyrite and hydrothermal fluids within a single chimney and suggest isotope disequilibrium in both Fe- and S-isotopes. The concomitant Fe and S-isotope fractionations during pyrite/marcasite precipitation are explained by (1) isotopic S-exchange between fluid H_2S and SO_4^{2-} during precipitation of pyrite from FeS precursors by reaction with thiosulfate and (2) rapid formation of pyrite from FeS, thus preserving negative Fe-isotope fractionation factors during FeS precipitation. In contrast, δ^{56}Fe and δ^{34}S values of pyrite precipitated in massive sulfides, either in the subsurface during conductive cooling of the fluid (i.e., slow rate of precipitation) or during multiple stages of remineralization, are expected to be similar to the δ^{56}Fe and δ^{34}S values of the hydrothermal fluid. This hypothesis is consistent with the limited range of δ^{56}Fe values between high-temperature, Fe-rich black smokers and lower temperature, Fe-poor vents suggesting minimal Fe-isotope fractionation during subsurface sulfide precipitation. It is also consistent with previous work showing opposite Fe-isotope fractionation factors during kinetic Fe-sulfide (mackinawite) precipitation (Butler et al. 2005) and equilibrium pyrite precipitation (Polyakov et al. 2007).

Although still not widely applied, Fe isotopes may provide useful application to distinguish hydrothermal vs. magmatic formation pathways, and address potential reservoir effects due to sulfide precipitation during subsurface cooling of the hydrothermal fluid. Furthermore, the effects of equilibrium or kinetic precipitation of sulfide pairs (pyrite–chalcopyrite) and the temperature of precipitation may be traced using Fe isotopes. Both reservoir effects and partial Fe equilibrium may result in contrasted Fe isotopic compositions in co-existing chalcopyrite and pyrite from the Grasberg Cu–Au porphyry and its associated skarn deposits (Irian Jaya, West New Guinea) (Graham et al. 2004).

Combined Δ^{33}S and δ^{56}Fe analyses have been also applied to determine the source(s) of sulfur in Archean komatiite-hosted Fe–Ni sulfide deposits. Samples were collected from Ni-rich sulfide deposits from (1) the ~2.7-Ga Hart komatiite, Abitibi greenstone belt, Ontario, Canada (Hiebert et al. 2016); (2) the ~2.71-Ga Agnew–Wiluna and Norseman–Wiluna greenstone belts of Western Australia and the time-equivalent Abitibi greenstone belt (Bekker et al. 2009); (3) the ~2.7-Ga volcano–sedimentary sequences of the Zimbabwe craton (Trojan and Shangani mines) (Hofmann et al. 2014); (4) the ~2.6-Ga Tati greenstone belt and the Phikwe Complex of Eastern Botswana (Fiorentini et al. 2012); (5) the ~1.3-Ga Voisey's Bay deposit, Labrador, Canada (Hiebert et al. 2013). While Δ^{33}S values suggest that sulfur in Archean komatiite-hosted Fe–Ni sulfide deposits comes from mixing of hydrothermally remobilized magmatic and sedimentary sulfur, Fe isotopes of sulfides from these deposits show a relatively small range of negative δ^{56}Fe values, consistent with high-temperature fractionations in magmatic systems at high silicate magma / sulfide melt ratios.

Iron isotopes have also been used to decipher the origins of detrital pyrite in Archean sedimentary rocks. Rounded grains of pyrite are a common component of conglomerate hosted gold and uranium deposits of the Mesoarchaean Witwatersrand basin of South Africa. Different sources for detrital pyrite have been discussed, including sedimentary, igneous, and various hydrothermal origins (Barton and Hallbauer 1996; England et al. 2002), while the source of gold remains poorly constrained in this model (Robb and Meyer 1990). The placer model (deposit of pyrite-bearing sand or gravel in the bed of a river or lake) has been challenged by several workers who ascribe these grains to post-depositional pyritisation of non-sulphidic (e.g., Fe-oxide) detrital grains during hydrothermal alteration (e.g., Barnicoat et al. 1997; Phillips and Law 2000). Multiple S ($\delta^{34}S$ and $\delta^{33}S$) and Fe ($\delta^{56}Fe$) isotope analyses of rounded pyrite grains from 3.1 to 2.6 Ga conglomerates of southern Africa (Hoffmann et al. 2009) confirm their detrital origin, which supports anoxic surface conditions in the Archaean.

IRON BIOGEOCHEMICAL CYCLING IN THE MODERN OCEAN

The biogeochemical cycling of Fe in the modern ocean is of particular interest because Fe limits the growth of phytoplankton in so much of the surface ocean, meaning that Fe has a large impact on marine biology and carbon cycling. The sources of Fe to the ocean are not well constrained, but Fe isotope 'fingerprinting' of Fe from various sources may provide a new tool to trace Fe as it mixes into the global ocean. Additionally, Fe stable isotopes may be used in order to better understand the internal cycling of Fe by biological and chemical processes in the marine realm.

The importance of iron in the global ocean

In high-nutrient low-chlorophyll (HNLC) regions of the world ocean, Fe is the element primarily responsible for limiting the growth of phytoplankton. Before the importance of Fe was fully understood, a wide variety of hypotheses had been considered about why there was an abundance of major nutrients (N, P, and Si) in HNLC regions such as the Southern Ocean, subarctic North Pacific, and the Equatorial Pacific. Prior hypotheses included light limitation and lack of grazing pressure (e.g., Landry et al. 1997). With the advent of trace-metal clean sampling techniques, however, it became clear that Fe concentrations in HNLC regions were very low (<1 nM) and the possibility of Fe limitation was considered. Work by John Martin and others at Monterey Bay Aquarium Research Institute was crucial to understanding the importance of Fe, including early experiments showing that cultures of phytoplankton from HNLC regions responded to Fe additions (Martin and Fitzwater 1988), and advancing the so-called 'iron hypothesis' that changes in dust Fe flux to the oceans was responsible for glacial–interglacial changes in atmospheric carbon dioxide (Martin 1990). The work of Martin and colleagues culminated with one of the most dramatic experiments in the history of oceanography, the addition of Fe to a large patch of the HNLC equatorial Pacific which resulted in a massive Fe-fertilized phytoplankton bloom (Martin et al. 1994).

Iron limitation plays a crucial role in marine biogeochemistry in both the modern and the past ocean. Modeling studies suggest that primary productivity in about a quarter of the global ocean is limited by Fe (Moore et al. 2002).There is also strong evidence of a correlation between dust flux to the Southern Ocean and glacial-interglacial climate cycles (Martinez-Garcia et al. 2011; Murray et al. 2012), although the full 'iron hypothesis' that Fe is the primary cause of glacial-interglacial cycles has not been supported by paleoceanographic research.

While there is no doubt about the importance of Fe to marine biogeochemical cycles, important questions remain. Iron stable isotopes may help to address two key questions about the marine biogeochemical cycling of Fe: What are the sources of Fe to the oceans? What are the chemical and biological processes that cycle Fe within the oceans?

Sources and sinks for Fe in the ocean

One of the most promising applications of Fe isotopes in the marine realm is to trace various sources of Fe to the ocean (Fig. 15). Often the measurement of Fe concentrations alone is not sufficient to determine what was the original source of Fe into the ocean, but because different sources often have a unique δ^{56}Fe signature, analysis of seawater δ^{56}Fe may be used to constrain these sources. Four different sources of Fe have been proposed as contributing significantly to global marine productivity; 1) atmospheric dust, 2) hydrothermal vents, 3) reducing sediments along continental margins, and 4) oxic seafloor sediments. Below we discuss the evidence that these sources contribute significant amounts of Fe to the ocean, and evidence regarding the isotope 'fingerprint' of each source.

Atmospheric dust. The importance of dust as a source of Fe to the ocean has been recognized since Fe concentrations were first accurately measured in seawater. The importance of dust can be easily appreciated from the global distribution of iron concentrations. Relatively high Fe concentrations in the North Atlantic coincide with a significant input of atmospheric dust from the Saharan Desert, while iron concentrations are lower in regions with less dust input such as the HNLC Southern Ocean and Equatorial Pacific (Fung et al. 2000; Jickells et al. 2005; Mahowald et al. 2005; Moore et al. 2002). Similarly, Fe concentrations in the surface ocean vary seasonally in response to seasonal changes in dust deposition, and even on shorter timescales in response to individual dust deposition events (Boyle et al. 2005; Fitzsimmons et al. 2015; Sedwick et al. 2005). Dust was considered to be such a dominant source of iron to the marine realm that early models of the global iron cycle considered it as the only source of Fe to the global ocean (Moore et al. 2002; Parekh et al. 2004).

While the importance of dust as a source of bioavailable Fe to the surface ocean is unquestionable, there are great uncertainties in the size of the dust-Fe flux to the ocean. Estimates of the total amount of Fe deposited in the surface ocean vary by nearly an order of magnitude from 1×10^{11} to 6×10^{11} mol.y^{-1} (Fung et al. 2000). This uncertainty is compounded by the even larger uncertainties about how much of that dust Fe dissolves in seawater. Typical estimates for the solubility of mineral dust range from 1% to 10%, and solubility estimates for Fe from less common sources such as biomass burning range up to 100% (Fung et al. 2000; Jickells and Spokes 2001; Luo et al. 2008; Mahowald et al. 2005). With the uncertainties in total dust

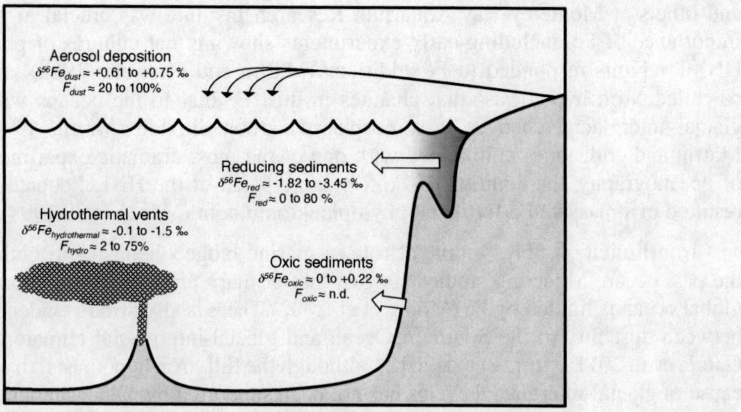

Figure 15. A pictoral representation of the major fluxes of Fe to the oceans including aerosol deposition to the surface ocean, hydrothermal vent input, and sources from both reducing and oxic sediments. The commonly observed range in δ^{56}Fe for each of these sources is taken from Conway and John (2014). Approximate ranges in the fraction of global surface-ocean Fe attributable to each of these sources is based on modeling studies, highlighting the large uncertainties in the relative magnitude of these various fluxes.

fluxes compounded by uncertainties in dust solubilities, there is very little constraint on the total amount of dissolved iron in the oceans that is delivered from dust. Of course in some cases it is clear dust is the source of dissolved Fe based on its distribution, for example when surface Fe maxima are observed away from other obvious Fe sources. But to better understand how much of the Fe is from dust in more ambiguous regions such as the global deep ocean, it may be useful to trace the isotopic signature of dust-derived Fe.

While the total δ^{56}Fe of most natural aerosols is similar to that of continental material, this may be different from the δ^{56}Fe of the Fe released from dust into seawater. The total δ^{56}Fe of most natural aerosols and loess samples falls within a small range near +0.1‰ (Beard et al. 2003; Majestic et al. 2009; Waeles et al. 2007). There is evidence, however, that industrial or anthropogenic aerosols may have a lighter δ^{56}Fe signature (Majestic et al. 2009). While the global importance of anthropogenic aerosols to the ocean is not well known, there is evidence that they can be transported far into the ocean. Fine size fraction aerosols from Bermuda have anomalously lower δ^{56}Fe values, generally between 0 and −0.5‰, during seasons when the aerosols are less influenced by Saharan dust and more influenced by North American sources, leading the authors to suggest that biomass burning may contribute isotopically light aerosol δ^{56}Fe (Mead et al. 2013). In addition to small variability in total dust δ^{56}Fe, there may be much larger fractionations between total dust δ^{56}Fe and soluble δ^{56}Fe. Leaching of Arizona Test Dust with an oxalate-EDTA solution meant to imitate the action of dissolved organic ligands in seawater resulted in the preferential release of slightly isotopically lighter Fe (roughly 0 to −0.5‰) (Revels et al. 2015).

However, neither the observed variability in aerosol δ^{56}Fe nor the observed fractionation of Fe isotopes during leaching with organic ligands can explain the observed δ^{56}Fe value of dust-influenced surface seawater in the North Atlantic. It was observed that surface ocean seawater dissolved δ^{56}Fe in the regions which showed the most obvious signs of dust Fe input (surface waters with high concentrations of Fe and low concentrations of other nutrients) were about +0.7‰, or roughly 0.6‰ heavier than expected total dust δ^{56}Fe (Conway and John 2014). While this observation appears robust at several locations across the North Atlantic, the reasons for this effect are unknown. Perhaps the dissolved seawater ligands are siderophores which bind iron with a very high affinity constant (Butler 2005; Reid et al. 1993), consistent with both theoretical and experimental evidence that stronger ligands preferentially bind heavier isotopes (Dideriksen et al. 2008).

While the evidence for a positive isotope effect during dust dissolution in the North Atlantic is strong, it is not yet known whether this is a universal 'fingerprint' which can be used to trace dust Fe throughout the global ocean. Future studies assessing the impact of dust deposition on seawater δ^{56}Fe in other locations, and a more complete understanding of the mechanisms that fractionate Fe during dust dissolution will help to answer these questions.

Hydrothermal input and plume dispersal. As with dust, there are large uncertainties about the amount of Fe delivered to the global ocean by hydrothermal vents. As hydrothermal fluids pass through the seafloor, they leach Fe from the surrounding rocks, yielding up to millimolar concentrations of Fe, many orders of magnitude higher than deep ocean seawater. If all of this Fe were to remain dissolved in seawater it would have an extraordinarily large impact on global ocean Fe concentrations. However, when hot and reducing hydrothermal fluids mix with the surrounding colder and more oxygenated seawater, nearly all of the hydrothermal iron is precipitated close to the hydrothermal vents as iron oxyhydroxide minerals (e.g., German and Von Damm 2003). While early studies suggested that iron was nearly quantitatively precipitated near hydrothermal vents, more recent work has suggested that a small portion of hydrothermal Fe may be stabilized within hydrothermal plumes through binding with dissolved and particular organic ligands (Bennett et al. 2008; Toner et al. 2009), and that this hydrothermal Fe may be transported great distances in the ocean. Ultimately, hydrothermal Fe is hypothesized to contribute significantly to global deep ocean Fe and surface ocean productivity.

The first evidence that hydrothermal vents might be a globally important source of Fe was based on the positive correlation between Fe concentrations and concentrations of excess ^3He in the North and South Pacific oceans (Boyle et al. 2005; Fitzsimmons et al. 2014), where excess ^3He is a tracer for deep sea hydrothermalism. These correlations were combined with global estimates of hydrothermalism to suggest that hydrothermal Fe contributes significantly to Fe concentrations and surface ocean productivity, particularly in the Southern Ocean which is far from other sources of Fe and where phytoplankton are known to be Fe-limited (Tagliabue et al. 2014, 2010). Modeling efforts are complemented by sampling efforts such as the GEOTRACES program, which have led to basin-scale sections of dissolved Fe concentrations across the world oceans. Elevated Fe concentrations in hydrothermal plumes have been found to extend hundreds, or thousands (in the case of the East Pacific rise) of kilometers from hydrothermal vents (Saito et al. 2013; Conway and John 2014; Resing et al. 2015). While such efforts have led to dramatic figures demonstrating the qualitative impact of hydrothermal vents on Fe concentrations, uncertainty remains about how quantitatively important vents are to the global ocean Fe inventory. Estimates of the hydrothermal Fe flux range from a few percent of the global dust flux up to 65%, with significant uncertainties derived from the roughly 2-orders of magnitude variability in Fe/^3He ratios measured at different vent sites (Tagliabue et al. 2010, 2014; Carazzo et al. 2013; Fitzsimmons et al. 2014; Saito et al. 2013).

While the $\delta^{56}Fe$ signature of most primary hydrothermal fluids falls within a narrow range, the $\delta^{56}Fe$ value of the Fe that eventually becomes stabilized within the hydrothermal plume is less well known. Primary hydrothermal fluid $\delta^{56}Fe$ is typically lighter than that of igneous rocks by about −0.2 to −0.3‰ (Beard et al. 2003; Rouxel et al. 2004, 2008b, 2016; Severmann et al. 2004; Sharma et al. 2001). Hydrothermal plume particulates enriched in Fe-oxyhydroxide have been found to be heavier than the original hydrothermal fluids (Severmann et al. 2004) while (buoyant) plume particulates dominated by Fe-sulfide minerals were isotopically lighter (Bennett et al. 2009; Rouxel et al. 2016). Those results led the authors to conclude that hydrothermal vents might be a source of either isotopically lighter or heavier Fe to the oceans depending on the end-member hydrothermal fluid composition (i.e., Fe/H_2S ratios) and geological setting.

In the North Atlantic ocean near the TAG (Trans-Atlantic Geotraverse) hydrothermal vent site, $\delta^{56}Fe$ has been measured in the hydrothermal plume but not in the immediate vicinity of hydrothermal venting (Conway and John 2014). These samples from within the neutrally buoyant hydrothermal plume provide the first look at how hydrothermal $\delta^{56}Fe$ might be expressed in the global ocean. Here, $\delta^{56}Fe$ value of dissolved iron within the plume ranges from −0.1 to −1.35‰. Because the −0.1‰ value represents mixing of hydrothermal Fe with ambient seawater, the data suggest that, at least for this particular TAG site, hydrothermal vents are a source of isotopically light Fe to the oceans.

Reducing sediments. The low solubility of Fe(III) in seawater is a major factor in defining the low Fe concentrations observed throughout the oxygenated oceans. Under certain conditions, however, Fe(III) may be reduced to Fe(II) which is very soluble in seawater. In the ocean, such reducing conditions typically occur at continental margins. Here, physical circulation brings nutrients into the upper ocean leading to high biological productivity in the surface, and a high flux of reduced organic carbon to the sediments. Within these sediments, particulate Fe(III) can be directly reduced by organisms which use it as an electron acceptor for metabolism of organic carbon, or it can be inorganically reduced by other reduced species in sediment porewater. Either way, the end result is sedimentary porewaters which contain extraordinarily high (μM to mM) concentrations of dissolved Fe(II). These porewaters may eventually diffuse into the overlying water column where Fe(II) may be reoxidized and precipitate back to the sediments, or where some portion of the Fe(II) may be stabilized and contribute to global ocean dissolved Fe.

As with other fluxes of Fe to the ocean, estimates vary considerably about the importance of reducing sediments to the global Fe pool. While early models of the global Fe cycle neglected sedimentary Fe inputs entirely, more recent models suggest that sediments may in fact be a dominant source of Fe to the oceans. A version of the BEC (Biogeochemical Elemental Cycling) model suggests that sediments contribute roughly half of the global ocean Fe, with the greatest impacts on productivity in Arctic regions (Moore and Braucher 2008). A study with the NEMO-PISCES (NEMO=Nucleus for European Modelling of the Ocean) model suggests an even greater impact of sediments, supplying roughly 80% of global Fe (Tagliabue et al. 2014). Scaling up observations from benthic landers to the global ocean would suggest that 98% of the Fe flux into the oceans is supported by reducing continental margin sediments (Elrod et al. 2004).

The δ^{56}Fe signature of reducing sediments is typically very isotopically light compared to other sources. The reason for this unique δ^{56}Fe signature is understood to be the isotopic equilibration of dissolved Fe(II), and Fe(III) either dissolved or attached to sediment particles. Both theoretical calculations and experimental observations suggest that Fe(II) is about −3‰ lighter than Fe(III) when the two species are at chemical equilibrium (Anbar et al. 2005; Johnson et al. 2002; Welch et al. 2003). Consequently, many studies of porewaters in reducing environments show porewater δ^{56}Fe values which are several permil lighter than bulk sediments (Bergquist and Boyle 2006; Severmann et al. 2006, 2010). This characteristically light signature has also been observed just above the sediment-water interface in benthic landers (Severmann et al. 2010), and in the water column tens to hundreds of meters above the bottom in the Santa Barbara and San Pedro basins and in the Peru upwelling region (John et al. 2012a; Chever et al. 2015). While Fe from reducing sediments is commonly isotopically light, there is significant variability in the exact δ^{56}Fe, with values almost always between roughly −3‰, which is the maximum fractionation expected between Fe(II) and Fe(III), and roughly 0‰ corresponding to a continental δ^{56}Fe value. Values of δ^{56}Fe higher than −3‰ could be due to closed-system isotope fractionation, where the first Fe(II) produced is roughly −3‰ lighter than the bulk particle δ^{56}Fe but continued dissolution of those particles drives the porewater δ^{56}Fe closer to the continental value (Chever et al. 2015). Alternatively, within sulfidic sediments the precipitation of isotopically light FeS has been shown to increase porewater δ^{56}Fe (Roy et al. 2012; Severmann et al. 2006; Sivan et al. 2011).

Non-reductive sediments. Oxic sediments have not traditionally been considered an important source of Fe to the oceans, though recent work has revived interest in this subject (Jeandel et al. 2011) and suggested that it may be a globally important source. The isotopic signature of Fe released from oxic sediments appears to be similar to that of continental material. In the first study which used seawater δ^{56}Fe as evidence for input of Fe from oxic sediments, it was suggested that the δ^{56}Fe signature of non-reductive sedimentary dissolution was about +0.3‰ (Radic et al. 2011). Subsequent studies have suggested that oxic sediments may have a δ^{56}Fe value close to that of continental material (Homoky et al. 2013).

Other sources. Most other sources of Fe to the oceans are not thought to contribute significantly to the global ocean Fe pool. However, point sources of Fe to the oceans may have a large impact on Fe concentrations and δ^{56}Fe locally. Such potentially important sources of Fe to the oceans include glaciers (Bhatia et al. 2013; Hawkings et al. 2014; Raiswell et al. 2006; Zhang et al. 2015), rivers (Escoube et al. 2009, 2015), and submarine groundwater discharge (Windom et al. 2006; Rouxel et al. 2008a; Roy et al. 2012).

The potential importance of coastal groundwater discharge in producing both highly negative and positive δ^{56}Fe values in local seawater has been hypothesized by Roy et al. (2012) and Rouxel et al. (2008a). The δ^{56}Fe value of groundwater discharge ultimately depends on mixing processes within subterranean estuaries. In Waquoit Bay (Massachussetts, USA), the groundwater source has high Fe(II) concentrations and δ^{56}Fe values between 0.3 and −1.3‰ (Rouxel et al. 2008a). Within the coastal sandy sediments, pore waters from the mixing zone of the subterranean estuary have even lower δ^{56}Fe values down to −5‰ as a result of partial Fe(II) oxidation and precipitation of isotopically heavy Fe-oxhydroxides. Hence, the input of groundwater Fe-sources at Waquoit Bay is the most likely explanation for both negative δ^{56}Fe values and high Fe-concentrations observed in surface seawater (Rouxel and Auro 2010).

In contrast to Waquoit Bay, the Indian River Lagoon subterranean estuary (Florida, USA) is characterized by organic-rich sediments leading to SO_4^{2-} reduction and S_2-rich porewaters (Roy et al. 2010). In this system, Fe-oxide reduction produces dissolved Fe(II) below the zone of Fe-sulfide precipitation and this dissolved Fe(II) flows upward. The total range of Fe isotope fractionations, of about 2.5‰, results from in situ diagenetic reactions in the subterranean estuary. The near-surface Fe-sulfide precipitation ultimately delivers dissolved Fe with slightly positive δ^{56}Fe values, averaging about +0.24‰, via submarine groundwater discharge (SGD). Consequently, we suggest that Fe-sulfide precipitation in subterranean estuaries could be a previously unidentified source of isotopically heavy Fe to the coastal waters

Using Fe isotopes to trace sources of Fe in the oceans

A key motivation for efforts to analyze the δ^{56}Fe value of various marine Fe sources is to constrain the relative contributions of various different sources of Fe to the oceans. With orders of magnitude uncertainty in the flux of Fe from various sources to the ocean based on other methods, δ^{56}Fe provides a valuable tool for tracing these sources as they mix into the global ocean. Given certain assumptions, the δ^{56}Fe value of seawater can be used to quantify the amount of Fe coming from various sources. The general isotope mass balance equation:

$$\delta^{56}\mathrm{Fe}_{\mathrm{total}} = \delta^{56}\mathrm{Fe}_{\mathrm{A}}\Delta f_{\mathrm{A}} + \delta^{56}\mathrm{Fe}_{\mathrm{B}}\Delta f_{\mathrm{B}}$$

can be applied in the ocean to calculate the relative fraction of Fe from different sources where $\delta^{56}\mathrm{Fe}_{\mathrm{total}}$ is the measured δ^{56}Fe of seawater, and $\delta^{56}\mathrm{Fe}_{\mathrm{A/B}}$ and $f_{A/B}$ are the end-member isotopic composition and fraction of Fe in the seawater coming from source A or B. This technique has been widely applied to quantify the amount of Fe coming from different sources in many locations in the global ocean (Conway and John 2014; John and Adkins 2012; Lacan et al. 2008; Radic et al. 2011; Revels et al. 2014).

The weaknesses of this approach should be carefully considered. The calculated fraction of Fe coming from different sources depends on the choice of end-member δ^{56}Fe values, and studies have shown significant variability in the source δ^{56}Fe value from different locations. Also, such equations are only valid when assuming that only two sources of Fe are mixing at any individual location. Incorporating Fe stable isotopes into global models of the Fe cycle, combined with new datasets on the global distribution of δ^{56}Fe from GEOTRACES and other sampling efforts, presents an opportunity to combine the strengths of δ^{56}Fe as a tracer for Fe sources with the more complex formulations of mixing and circulation possible in global 3D models. Finally, the application of δ^{56}Fe to trace various Fe sources in the ocean relies on the assumption that δ^{56}Fe is not modified by chemical reaction as it travels away from the source.

Internal cycling of Fe isotopes within the ocean

While the assumption that δ^{56}Fe can be used as a passive tracer for mixing of Fe from different sources in the deep ocean, away from regions of biological productivity and strong gradients in chemistry, there are other locations where seawater δ^{56}Fe has been shown to be modified by chemical reaction.

Biological uptake of Fe appears to preferentially remove the heavier isotopes from seawater. In the North Atlantic, this is inferred from the fact that a minimum in δ^{56}Fe is observed in the upper-ocean, coincident with the Fe concentration minimum and the chlorophyll maximum (Conway and John 2014). In the Southern Ocean, the biological uptake of heavier Fe isotopes is observed as a decrease in δ^{56}Fe over the course of the spring phytoplankton bloom (Ellwood et al. 2015). So-called 'inverse' isotope effects, where heavier isotopes react more quickly or are preferentially assimilated during biological processes, are very rare. Most known reactions of C, N, and O isotopes result in the preferential biological assimilation of lighter isotopes, as does the assimilation of Zn and Cd by phytoplankton (John and Conway 2013; John et al. 2007; Lacan et al. 2006). In contrast to these other nutrients, culture data on Fe isotope fractionation by phytoplankton are scarce (John et al. 2012b), limiting our ability to form a mechanistic explanation for the isotope fractionations observed in nature.

The fractionation of Fe isotopes has also been observed during Fe precipitation, though the magnitude and sign of this isotope effect changes under different conditions. Hydrothermal particles have been found to be heavier, lighter, and similar to the dissolved phase in various locations (Bennett et al. 2009; Conway and John 2014; Revels et al. 2014; Severmann et al. 2004; Rouxel et al. 2016). Altogether, these studies suggest that both the initial Fe isotope composition of the high-temperature vent fluids and its initial Fe/H_2S ratio (i.e., isotopically light Fe sulfide precipitation versus isotopically heavy Fe-oxyhydroxide precipitation) should impose characteristic Fe isotope "fingerprints" for hydrothermally derived Fe in the deep ocean.

Particles precipitating near reducing continental margins have been found to be similar to or heavier than the dissolved phase (Chever et al. 2015), while the weakly bound 'ligand leachable' phase within particles is generally isotopically lighter than seawater (Revels et al. 2014).

THE GEOLOGICAL RECORD AND PALEOCEANOGRAPHIC APPLICATIONS

The ferromanganese crust record

Since Fe is actively involved in key biogeochemical processes and has a variety of potential sources to the oceans (e.g., atmospheric deposition, sediment input, rivers, hydrothermal vents) (Hutchins et al. 1999; Johnson et al. 1999; Chase et al. 2005), the record of Fe isotopes in Fe–Mn deposits has attracted significant interest (Zhu et al. 2000; Beard et al. 2003b; Levasseur et al. 2004; Chu et al. 2006; Horner et al. 2015; Marcus et al. 2015). Ferromanganese (Fe–Mn) crusts and nodules have long been characterized for their elevated concentrations of transition metals (e.g., Ni Co, and Cu) and their very slow growth rates (1–6 mm/Myr). Long-lived radiogenic isotope (e.g., Os, Pb, Nd, Hf, Be) and more recently non-traditional stable isotope (e.g., Tl, Mo, Cd, Ni) compositions of hydrogenous Fe–Mn deposits have been used to reconstruct the connection between plate tectonics, climate change, weathering processes and (1) metal sources, (2) carbon and metal cycles, and (3) mixing of water masses in the ocean through the Cenozoic (Frank 2002; Siebert et al. 2003; Horner et al. 2010; Nielsen et al. 2011; Gall et al. 2013)

Modern marine sediments, such as deep-sea clays, terrigenous sediments, turbidite clays, and volcanoclastites, have a restricted range of δ^{56}Fe values clustered around average crust (Beard et al. 2003b; Rouxel et al. 2003; Homoky et al. 2013), suggesting minor Fe isotope fractionation during continental weathering and particle transport in seawater (Radic et al. 2011). In contrast, marine sediments in which the Fe budget is controlled by authigenic or diagenetic precipitates are characterized by a range of δ^{56}Fe values generally enriched in light Fe isotopes (Severmann et al. 2006; Homoky et al. 2013). This is particularly true for hydrogenous (i.e., seawater-derived) precipitates, including Fe–Mn crusts and nodules whose δ^{56}Fe values vary widely from −0.05 to −1.13‰, averaging -0.41 ± 0.49 (2SD, $n = 41$) in modern oceans (Fig. 16). Beard et al. (2003b) initially suggested that δ^{56}Fe variations in Fe–Mn deposits are controlled by the relative flux of Fe from aerosols (with δ^{56}Fe ~ 0‰) and Fe from mid-oceanic ridge hydrothermal fluids (with δ^{56}Fe ranging from −0.5 to −1‰). This model assumes no Fe isotope fractionation during crust growth nor modification of the primary Fe isotope signature, which is unlikely considering the variety of Fe isotope fractionation processes in hydrothermal plumes (Severmann et al. 2004; Bennett et al. 2009). Levasseur et al. (2004) reported the global variations of δ^{56}Fe of the surface scrapings of hydrogenetic Fe–Mn crusts (i.e., near modern values), and found no significant basin to basin trends or relationships with expected hydrothermal contributions (Fig. 16). Hence, although hydrothermal Fe may contribute significantly to some Fe–Mn crusts in the west Pacific (Chu et al. 2006), other marine sources characterized by light δ^{56}Fe values, such as dissolved Fe derived from shelf sediments (Severmann et al. 2006; Homoky et al. 2009; Conway and John 2014; Chever et al. 2015) may also contribute to these paleoceanographic records.

More recently, Horner et al. (2015) reported a range of δ^{56}Fe values from −1.12‰ to +1.54‰ along a 76 Ma-old Fe–Mn crust from the central Pacific (crust CD29–2). This range encompasses the range of δ^{56}Fe values measured for dissolved and particulate Fe from open seawater and oxygen minium zones—OMZs (Radic et al. 2011; John et al. 2012; Conway and John 2014; Chever et al. 2015). By considering a fractionation factor during crust uptake of Δ^{56}Fe–FeMn–SW = −0.77 ± 0.06‰, Horner et al. (2015) proposed that heavy δ^{56}Fe values of seawater (up to 2.2‰) may result from the modification of hydrothermally sourced Fe by precipitation of isotopically light Fe sulfides. Marcus et al. (2015) also investigated the Fe isotopic compositions over 1–3 mm increments across a nodule from the South Pacific Gyre. The δ^{56}Fe values showed limited range from −0.16 to −0.07‰ (Fig. 16), suggesting constant Fe isotope values over a period of 4 Ma despite the diversity of Fe mineral phases identified in the nodule layers (e.g., feroxyhite, goethite, lepidocrocite, and poorly ordered ferrihydrite-like phases). Hence, the results indicate that mineral alteration (i.e., recrystallization) did not affect the primary Fe isotopic composition of the nodule.

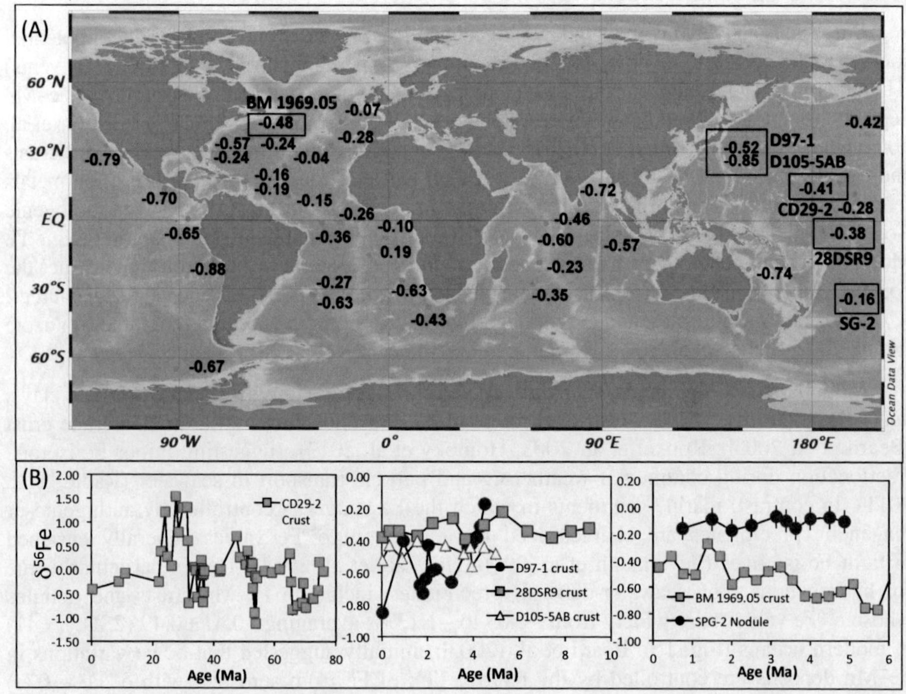

Figure 16. (A) Map of ferromanganese (FeMn) crust sample locations and δ^{56}Fe values of surface scrapings from each crust and (B) temporal δ^{56}Fe record from selected Fe–Mn crusts and nodules (location highlighted as framed number in panel A. Data from Zhu et al. (2000), Levasseur et al. (2004), Chu et al. (2006), Marcus et al. (2015).

Oceanic Anoxic Events

Extensive work in the Black Sea, the world's largest modern euxinic basin, reveals that additional reactive Fe, derived from benthic fluxes out of oxic-suboxic sediments in the shallow margin, is deposited as pyrite in the deep basin sediments (Raiswell and Canfield 1998; Wijsman et al. 2001; Anderson and Raiswell 2004; Lyons and Severmann 2006). Hence, sulfidic sediments often show pronounced enrichments in reactive Fe relative to oxic shelf sediments, leading to an increase in Fe/Al ratios and significant Fe isotope variability. Sediments from the oxic shelf of the Black Sea have bulk $\delta^{56}Fe$ values averaging +0.16 ± 0.02‰ that are slightly elevated relative to values for the average crust, whereas sulfidic sediments from the deep basin have lower values (mean $\delta^{56}Fe = -0.13 \pm 0.04$‰) (Severmann et al. 2008). The source of isotopically light Fe in bulk sediment is likely related to a shelf to basin Fe shuttle, whereby reactive Fe is sequestered nearly quantitatively during Fe sulfide precipitation in the euxinic water column (Severmann et al. 2008). This model is well supported by the fact that diagenetic fluids in anoxic and suboxic marine sediments have isotopically light Fe(II) (down to −2‰) (Severmann et al. 2006; Homoky et al. 2009). In contrast, Fe isotope data for a sediment core transect across the Peru upwelling area, which hosts one of the ocean's most pronounced OMZs (Scholz et al. 2014), show that the heaviest $\delta^{56}Fe$ values of the surface sediments coincide with the greatest Fe enrichment. The observed trend is the opposite of expected results (i.e., transfer of isotopically light Fe to the sediments below the OMZ) but could be explained by partial Fe(II) oxidation in the water column and precipitation of isotopically heavy Fe-oxyhydroxides (Chever et al. 2015).

Iron isotopic compositions of Phanerozoic organic-rich sediments studied so far display a range of $\delta^{56}Fe$ values consistent with the Black Sea model (Fig. 17). Jenkyns et al. (2007) examined the Fe isotopic compositions of shales formed during the Cenomanian–Turonian oceanic anoxic event (OAE). Black shales deposited before the onset of the OAE had light $\delta^{56}Fe$ values (−1.08 ± 0.28‰), unlike those found in the black shale deposited during the oceanic anoxic

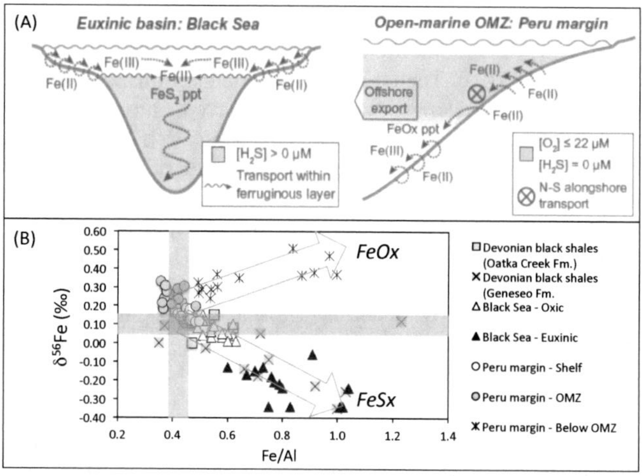

Figure 17. Cross plot of Fe/Al (in g/g) vs. $\delta^{56}Fe$ (B) of Devonian black shales (data from Duan et al. 2010) and sediment cores from the Black Sea (data from Severmann et al. 2008); the Peru margin (data from Scholz et al. 2014). The gray shaded areas correspond to Fe/Al and $\delta^{56}Fe$ of the lithogenic components while the two arrows illustrates the addition of authigenic Fe as Fe-sulfides and Fe-oxyhyroxide respectively showing the Fe isotope fingerprint for the Fe shuttle. The sketches at the top (A) are from Scholz et al. (2014) and illustrate mechanistic differences between Fe shuttles in euxinic basins and open-marine OMZs (FeS_2 ppt = pyrite precipitation; FeO_x ppt = Fe (oxyhydr)oxide precipitation). The gray shaded areas (see legend) refer to the redox state of the water column and do not provide information about the redox state of the sediments.

event itself (–0.07 ± 0.27‰). By comparison, Upper Jurassic pyrite- and siderite-bearing organic carbon-rich shales of the Kimmeridge Clay Formation (UK) yielded $\delta^{56}Fe = 0.08 \pm 0.13$‰ (Matthews et al. 2004). Duan et al. (2010) measured Fe isotopic compositions of bulk samples and chemically extracted pyrite in two black shale units deposited during the mid-to-late Devonian. Samples yielded light and variable bulk $\delta^{56}Fe$ values ranging from –0.53 to –0.06‰ and inversely correlated with Fe/Al. From these studies, it appears that the enrichment of isotopically light Fe in marine sediments can be used as diagnostic of enrichment mechanism (e.g., shelf to basin shuttle) in the geological record at times of widespread oxygen deficiency in the ocean.

The Precambrian record

A host of available geochemical and geological indicators demonstrate that the transition from an atmosphere with almost no oxygen to a fully oxygenated one—called the Great Oxidation Event (GOE)—occurred between 2.45 and 2.22 Ga (Holland 1984; see Lyons et al. 2014 for a recent review). It is now recognized that the rise of O_2 by ~2.3 Ga has been directly linked to an increase in the ocean's sulfate content and variable concentrations of redox-sensitive elements due to the combined effect of oxidative continental weathering and efficient trapping due to the development of local sulfidic environments (Canfield 1998; Scott et al. 2008; Poulton et al. 2010). Despite the recent progress, our understanding of ancient Fe biogeochemical cycling is still limited—if not biased—by the available rock archive and our ability to establish robust paleooceanographic proxies from rock compositions. The Banded Iron Formation (BIF) rock record, spanning every continent and encompassing sediments from as young as 0.55 Ga to as far back as Earth's earliest known marine deposit (~3.8 Ga BIF in Greenland) has been extensively used in the recent years. An alternative approach, which provides better temporal resolution than the BIF record, has also involved black shales. Considering that Fe sources (e.g., continental vs. hydrothermal) as well as Fe redox cycling mechanisms (biotic vs. abiotic) in Precambrian oceans remain controversial, the iron isotope geochemistry of Precambrian sedimentary rocks remains an active field of study, as presented in the following sections.

The archive of iron formations

The key to understanding the origin of Iron Formations (IFs) is to identify the mechanism that could have caused the oxidation of ferrous iron in an ocean that was globally anoxic (see the review by Bekker et al. 2010). There are three mechanisms that are generally considered: (i) oxygenic photosynthesis, releasing O_2 which could have then oxidized Fe(II) into Fe(III); (ii) anoxygenic photosynthesis, involving organisms using Fe(II) as an electron donor; (iii) UV photo-oxidation, promoted by an atmosphere devoid of ozone. The first two processes require the involvement of life, while photo-oxidation is a purely abiotic process. Recently, experiments have been performed to characterize the isotopic fractionation of Fe during photo-oxidation (Nie and Dauphas 2015; Nie et al. 2016). Those studies showed that Fe(II) photo-oxidation follows a Rayleigh fractionation model, with Fe isotope fractionation between Fe(III) precipitates and aqueous Fe(II) of about +1.2‰ at 45 °C. This fractionation is similar to that of anoxygenic photosynthetic oxidation and of O_2-mediated oxidation (Bullen et al. 2001; Welch et al. 2003; Croal et al. 2004; Balci et al. 2006; Beard et al. 2010a; Swanner et al. 2015), which therefore cannot be used to rule out possible pathways to BIF formation.

There have been a number of Fe isotope studies of IF, including Fe-oxide and carbonate facies IF, with the end goal of tracking the biogeochemical cycling of Fe on the early Earth (Beard et al. 1999; Johnson et al. 2003, 2008a,b; Dauphas et al. 2004b; Planavsky et al. 2009, 2012; Steinhoefel et al. 2009b; Heimann et al. 2010; Craddock and Dauphas 2011b; Li et al. 2015). Bulk samples of IF show a large range of $\delta^{56}Fe$ values (Fig, 18) which provide insights into Fe enrichment mechanisms and Fe sources. The two most commonly proposed iron sources, hydrothermal and benthic/diagenetic, have both negative (sub-crustal) $\delta^{56}Fe$ values (Rouxel et al. 2008b; Severmann et al. 2008; Homoky et al. 2013; Chever et al. 2015). Hence, the presence of heavy δ^{56}Fe values in oxide-facies IF must reflect the fractionation during partial Fe(II) oxidation and Fe(III) mineral precipitation. This enrichment in heavy Fe isotopes contrasts with the isotope fractionations associated with Fe(II)-bearing siderite, ankerite, and green rust precipitation, which are depleted in the heavy Fe isotopes relative to the ambient $Fe(II)_{aq}$ pool (Wiesli et al. 2004). Although Fe silicates have been recently proposed to form the primary sediments (i.e., precursor) of Banded Iron Formations (BIFs), before being silicified upon deposition and diagenesis (Rasmussen et al. 2013, 2015), the exact fractionation factor during precipitation of Fe silicates in anoxic seawater is currently not known. Therefore, positive Fe isotope values in IF indicate that Fe(III) delivery was the main process driving the deposition of IF. Additionally, the expression of the Fe isotope fractionation implies partial Fe(II) oxidation, pointing towards oxidation at low Eh conditions (Dauphas et al. 2004b; Planavsky et al. 2009, 2012). If oxidation took place during mixing of anoxic Fe-rich and fully oxic marine waters, as was commonly envisaged in the past (Cloud 1973), oxidation would have been essentially quantitative given the rapid oxidation kinetics of iron at neutral to alkaline pH. This rapid and quantitative oxidation would have prevented any significant expression of iron isotope fractionations as is the case with modern hydrothermal plume fall-outs (Severmann et al. 2004).

A compilation of bulk-rock and mineral-specific δ^{56}Fe values for Archean and Paleo-proterozoic IFs (Fig. 18) reveals heavier δ^{56}Fe values in Archean and early Paleo-proterozoic IF, in contrast to later Proterozoic and Phanerozoic Fe oxide-rich rocks. Notably, the lowest δ^{56}Fe values are typical for the ca. 2.22 Ga Hotazel Formation of South Africa and especially for manganese-rich samples (Planavsky et al. 2012; Tsikos et al. 2010). This unique feature may reflect deposition of iron and manganese from hydrothermal fluids depleted in heavy Fe isotopes by progressive Fe oxidation and precipitation in the deeper part of the redox-stratified basin that was at a redox state intermediate between that required for iron and manganese oxidation (Tsikos et al. 2010). In general, BIF-hosted siderite has negative δ^{56}Fe values, which is expected given the isotopic fractionation during siderite precipitation and expected Fe isotope values for seawater. Therefore, the rare case of siderite with positive δ^{56}Fe values must have been derived from reductive dissolution of iron oxides rather than having precipitated directly from seawater. In most cases, microbial iron reduction can be assumed to be driving the reductive oxide dissolution. This also implies that the IF carbonates do not reflect seawater compositions, but instead record extensive diagenetic Fe cycling in the soft sediment prior to lithification (Heimann et al. 2010; Craddock and Dauphas 2011b; Johnson et al. 2013). A dichotomy exists among BIF associated carbonates (i.e., within nominal BIF units or stratigraphically close to BIF units) between those that are iron-rich (siderite, ankerite), have low δ^{13}C values, and high δ^{56}Fe values, and those that are iron-poor (calcite, dolomite), have normal δ^{13}C values, and low δ^{56}Fe values (Heimann et al. 2010; Craddock and Dauphas 2011b) (Fig. 19). The isotopic and chemical signatures of the second group of carbonates are consistent with precipitation from seawater. The first group of carbonates, however, is best explained by cycling of organic carbon and hydrous ferric iron oxide during diagenesis, possibly mediated by biological activity (DIR). Iron isotopes may point towards microbial iron reduction in the rock record, possibly dating back to the earliest sedimentary rocks at ca. 3.8 Ga (Craddock and Dauphas 2011b). An alternative is that the iron-rich carbonates formed entirely abiotically by reaction between ferric iron precipitate and organic carbon during metamorphism (Perry et al. 1973; Köhler et al. 2013; Halama et al. 2016).

Li et al. (2015) reported an extensive dataset of high-precision *in situ* Fe isotope variations in BIF from the Dales Gorge member (core DDH-47A). Fe isotope analysis performed along and across a magnetite microband showed limited variability at the grain scale, with δ^{56}Fe clustering at -0.49 ± 0.22‰ (2SD, $n = 136$). The homogeneity of Fe isotopic composition at centimeter scale argues either for constancy in supply and iron isotopic fractionation

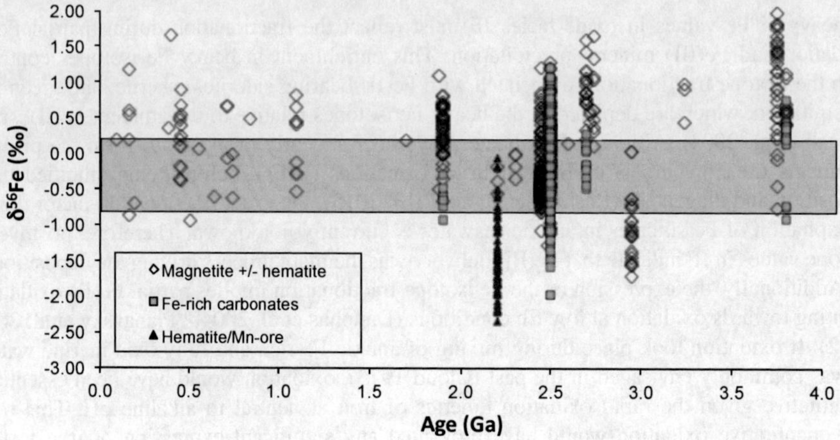

Figure 18. Secular variations in δ^{56}Fe values for Archean and Paleoproterozoic iron formations and younger sediments for comparison. Data include mineral separate analyses (magnetite, hematite, Fe-rich carbonates: ankerite and siderite, pyrite, and metamorphic Fe-silicates) and *in situ* analysis (magnetite) using laser-ablation MC-ICPMS and ion-microprobe. The gray area is bounded by the average δ^{56}Fe values for igneous rocks and δ^{56}Fe values for hydrothermal sources (Beard et al. 2003b; Rouxel et al. 2008a). A total of 663 datapoints is shown including: (a) 83 datapoints for the ca. 1.88 Ga GIF from the Animikie basin, North America (Frost et al. 2007; Hyslop et al. 2008; Planavsky et al. 2009); (b) 138 datapoints for the ca. 2.5 Ga Brockman Iron Formation, Western Australia (Johnson et al. 2008b); (c) 154 datapoints for the 2.5 Ga Kuruman IF and Gomahaan Fm., Transvaal supergroup, South Africa (Johnson et al. 2003; Heimann et al. 2010); (d) 27 datapoints for the Shurugwi and Belingwe greenstone belt IF, Zimbabwe (Rouxel et al. 2005; Steinhoefel et al. 2009a); (e) 137 datapoints for the Eoarchean Isua, Akilia, and Innersuartuut IF and metamorphic rocks (Dauphas et al. 2004a, 2007a,b; Whitehouse and Fedo 2007b); (f) 12 datapoints for the Neoproterozoic Rapitan IF (Halverson et al. 2011); (g) 52 datapoints for IF and Mn formation from the Paleoproterozoic Hotazel Formation, Transvaal Supergroup, South Africa (Tsikos et al. 2010); (h) 45 datapoints for other Paleoproterozoic and Archean IF (Planavsky et al. 2012); (i) 263 datapoints for the Dales Gorge member of the ca. 2.5 Ga Brockman IF (Hamersley Basin, Western Australia) (Craddock and Dauphas 2011; Li et al. 2015) (j) 42 datapoints for the Joffre Member of the Brockman Iron Formation (Hamersley Basin, Western Australia) (Haugaard et al. 2016); (k) 20 datapoints for 2.95 Ga Sinqeni Formation, Pongola (Planavsky et al. 2014); (l) 8 datapoints for the Ordovician Jasper beds from the Lokken ophiolite complex (Norway, Moeller et al. 2014) and additional data from Phanerozoic jasper (Planavsky et al. 2012).

for ~1000 yr or significant homogenization during early BIF diagenesis, consistent with the lines of evidence showing that magnetite and quartz are not primary minerals (Bekker et al. 2012; Rasmussen et al. 2015). Using coupled Fe and Nd isotope systems, Li et al. (2015) also evaluated potential variations in Fe sources during BIF deposition. The results suggested mixing of seawater masses with distinct Fe- and Nd-isotopic compositions: one is continentally sourced and has near-zero to negative δ^{56}Fe values, whereas the other is mantle/hydrothermally sourced and has slightly to strongly positive δ^{56}Fe values. The origin of the isotopically light Fe source in Archean seawater, however, remains a matter of controversy. On the one hand, negative δ^{56}Fe values in BIFs have been explained by the progressive oxidation of hydrothermal fluids following a Rayleigh fractionation-type model (Rouxel et al. 2005; von Blanckenburg et al. 2008; Steinhoefel et al. 2009b; Tsikos et al. 2010; Planavsky et al. 2012; Busigny et al. 2014). In this scenario, anoxygenic phototrophic oxidation could have established significant water column Fe concentration gradients—and therefore Fe isotope gradients—through ferric Fe removal during upwelling. On the other hand, the source of light Fe in IF could result from the near-quantitative oxidation of Fe(II) produced through dissimilatory Fe reduction (DIR) (Severmann et al. 2008; Heimann et al. 2010; Li et al. 2015). In fact, both models may be reconciled considering an Fe-rich deep ocean globally affected by strong hydrothermal input, and shallow water masses dominated by continentally derived Nd and isotopically light

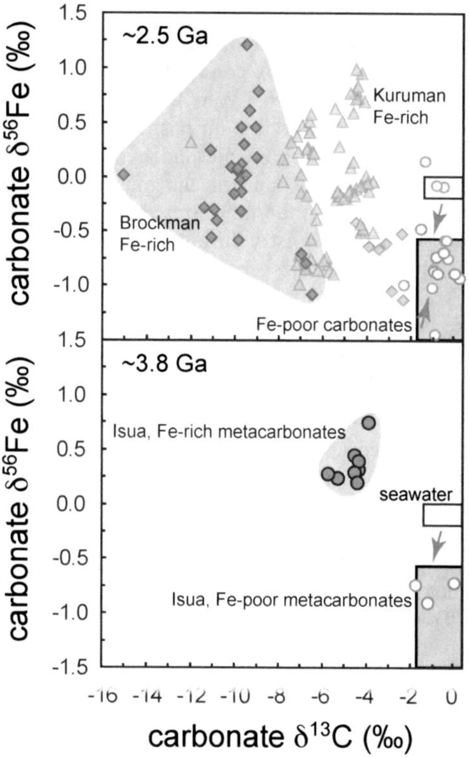

Figure 19. Iron and carbon isotopic compositions of carbonates associated with iron formations (Craddock and Dauphas 2011b and references therein). Iron-poor carbonates have low δ^{56}Fe values and high δ^{13}C values consistent with precipitation from seawater. Iron-rich carbonates have high δ^{56}Fe values and low δ^{13}C values consistent with cycling during diagenesis between low-δ^{13}C organic matter and high-δ^{56}Fe ferric iron oxide precipitate. Such cycling may have been mediated by bacterial dissimilatory iron reduction.

Fe (and Fe-depleted) due to efficient oxidation of hydrothermally affected Fe in the photic zone. This model is consistent with the generally accepted genetic models for Archean and Paleoproterozoic IF (Bau and Dulski 1996; Sumner 1997; Bekker et al. 2012).

The average Fe isotopic composition of different types of IFs remains a major unresolved question. In some cases IF have been estimated to have near crustal average δ^{56}Fe values (Johnson et al. 2008b) or slightly heavier (Planavsky et al. 2012). Recently, Haugaard et al. (2016) conducted detailed petrologic and geochemical analyses of a core section drilled through the entire 355 m of stratigraphic depth of the ca. 2.45 Ga Joffre Member from the Brockman Iron Formation. As a whole, δ^{56}Fe ranges from −0.74 to +1.21‰ with no obvious relationships with either lithology or stratigraphic levels. Although the average δ^{56}Fe value of +0.11‰ for the entire section is close to crustal values defined at +0.09‰, this value may still be affected by sampling bias considering the large variability and still limited sample size (i.e., 42 samples for about 350 m of stratigraphic unit). Furthermore, the Joffre Member was likely deposited in a restricted basin and Algoma-type iron formations are more likely to have positive δ^{56}Fe values (Planavsky et al. 2012). More work is needed to better understand the Fe isotope mass-balance of BIFs and in particular to test whether deposition of isotopically heavy IFs in the deeper parts of the basins could create pools of isotopically light dissolved iron that were buried in shallow-water environments in the Archean.

Black Shales and Sedimentary Pyrite Archives

Trace-metal concentrations of laminated, organic-rich shale facies have long been used to draw inferences concerning paleoredox conditions as well as metal inventory in ancient oceans (Algeo et al. 2004; Brumsack 2006; Lyons and Severmann 2006; Lehmann et al. 2007; Scott et al. 2008) with potential constraints on past atmospheric oxygenation, weathering intensity, marine productivity, global volcanic and hydrothermal events and ocean redox structure. Past studies of non-mass dependent and mass dependent sulfur isotope records of sedimentary pyrite have placed important constraints on the biogeochemical cycle of sulfur and the evolution of ocean chemistry during the rise in atmospheric oxygen (Cameron 1982; Farquhar et al. 2000, 2007; Mojzsis et al. 2003; Ono et al. 2003; Bekker et al. 2004; Johnston et al. 2006). Rouxel et al. (2005) applied a similar time-record approach to explore potential changes in Fe isotope values of sedimentary pyrite in black shale. They identified a direct link between the rise in atmospheric oxygen and changes in the Fe ocean cycle that provides new insights into past ocean redox states.

The general pattern of this record divides Earth's history into three stages (Fig. 20) which are strikingly similar to the stages defined by the $\delta^{34}S$ and $\Delta^{33}S$ records as well as other indicators of the redox state of the atmosphere and ocean (Holland 1984; Bau and Moller 1993; Karhu and Holland 1996; Farquhar et al. 2000; Bekker et al. 2004, 2005). The first stage in >2.3 Ga black shales corresponds to highly variable (from approximately −3.5 to +0.5‰) and overall negative $\delta^{56}Fe$ values in diagenetic pyrite. The second stage between 2.3 and 1.8 Ga shows more subdued variations in $\delta^{56}Fe$ values that extend from ~−0.5 to +1‰. The last stage after 1.8 Ga shows limited variations in $\delta^{56}Fe$ values and an average that is near zero or slightly negative. Although several interpretations of this Fe isotope record were proposed in subsequent studies (Archer and Vance 2006; Rouxel et al. 2006; Severmann et al. 2008; Guilbaud et al. 2011b, 2012), there is a general consensus that the shift from high $\delta^{56}Fe$ variability in >2.3 Ga black shales to little variability <1.8 Ga reflects redox-related changes in the global oceanic Fe cycle. The variable and light $\delta^{56}Fe$ values in pyrites older than about 2.3 Ga suggest that an iron-rich global ocean was strongly affected by the deposition of Fe oxides in a globally anoxic and Fe-rich ocean. The Fe oxide pool provides a means to generate isotopically light dissolved Fe through (1) reservoir effects during partial Fe(II) oxidation, and (2) dissimilatory Fe reduction. Between 2.3 and 1.8 Ga, positive Fe isotope values of pyrite up to +1‰ likely reflect an increase in the precipitation of

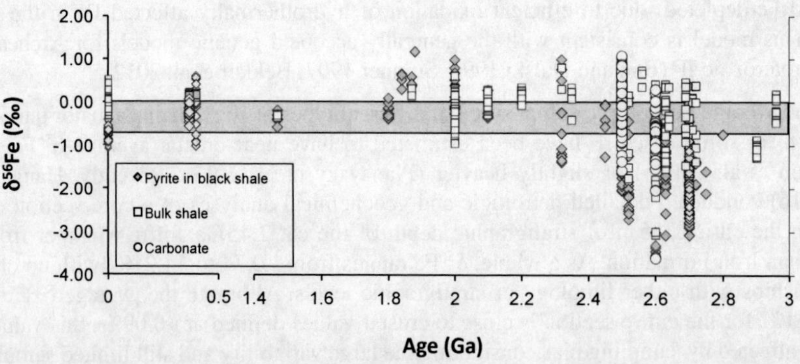

Figure 20. Secular trend of $\delta^{56}Fe$ values of pyrite from black shales (data from Rouxel et al. 2005; Duan et al. 2010 and Rouxel, unpublished), and bulk samples of organic-rich shales (data from Rouxel et al. 2005; Yamaguchi et al. 2005; Archer and Vance 2006; Czaja et al. 2010; Duan et al. 2010) and carbonate (data from Czaja et al. 2010; Heimann et al. 2010). Organic matter-rich sediments from the Black Sea and Peru Margin (Severmann et al. 2008; Scholz et al. 2014) are shown for comparison. *In situ* $\delta^{56}Fe$ values of pyrite, ranging from ca. −4 to +4‰ (Yoshiya et al. 2012; Tahata et al. 2015) are not displayed. The gray area is bounded by the average $\delta^{56}Fe$ values for igneous rocks and $\delta^{56}Fe$ values for hydrothermal sources (Beard et al. 2003b; Rouxel et al. 2008a).

isotopically light Fe sulfides relative to Fe oxides in a redox stratified ocean (Rouxel et al. 2005). Regardless of the exact process leading to Fe isotope variations in sedimentary pyrite, it apprears that Fe isotopes are particularly sensitive to the concentration of dissolved Fe(II) and H_2S (i.e., the size of seawater Fe and sulfate reservoir) and can be used to place important constraints on the sources to and sinks from this Fe(II) reservoir in past oceans.

In order to better assess intra-grain δ^{56}Fe variability, several studies applied *in situ* techniques such as laser ablation coupled to MC-ICPMS (Nishizawa et al. 2010b; Yoshiya et al. 2012; Tahata et al. 2015) or ion microprobe (Marin-Carbonne et al. 2014). Nishizawa et al. (2010) performed *in situ* Fe isotope analyses of pyrite within 2.7 Ga shallow marine carbonates from the Fortescue Group, and showed that the δ^{56}Fe values range from −3.0 to +2.2‰ with a bimodal distribution pattern. Pyrites from the upper Mingah Member of the Fortescue Group showed the widest variation of δ^{56}Fe values from −4.18 to +2.10‰ (Yoshiya et al. 2012). The extremely light δ^{56}Fe values are best explained by the combination of Fe isotope fractionation during precipitation of Fe-sulfide precursors (Guilbaud et al. 2011b) and a globally isotopically light Fe(II) pool in seawater (Rouxel et al. 2005). The pyrite grains with positive δ^{56}Fe values may originate from Fe(III)-bearing minerals like Fe oxide or ferrihydrite (Nishizawa et al. 2010; Yoshiya et al. 2012), or equilibrium pyrite precipitation/recrystallization (Polyakov and Soultanov 2011). The high δ^{56}Fe variability observed at the grain scale or between pyrite grains also suggests strong diagenetic signals. Marin-Carbonne et al. (2014) further adressed the importance of diagenetic overprinting by investigating coupled Fe and S isotope signatures of two pyrite nodules in a ca. 2.7 Ga black shale from the Bubi greenstone belt (Zimbabwe). Spatially resolved analysis across the nodules shows a large range of variation at micrometer-scale for both Fe and S isotopic compositions that cannot be explained by combined DIR and Bacterial Sulfate Reduction (BSR) as previously proposed (Archer and Vance 2006), but rather by the contributions of different Fe and S sources during a complex diagenetic history.

Because of the complication in interpreting the Fe isotope record in pyrite, the indentification of Fe isotope biosignatures in Late Archean black shales remains a matter of debate (Rouxel et al. 2005; Yamaguchi et al. 2005; Archer and Vance 2006; Rouxel et al. 2006; Severmann et al. 2006). Since dissimilatory Fe(III) reduction has been suggested to be important on the early Earth (Vargas et al. 1998) and is known to produce significant Fe isotope fractionation that may be preserved in organic-rich sediments (Johnson et al. 2004a) and carbonates associated with IFs (Heimann et al. 2010; Craddock and Dauphas 2011), it has been hypothesized that the extreme Fe isotope fractionations in the Archean were produced by this metabolic activity (Archer and Vance 2006; Yamaguchi et al. 2007; Johnson et al. 2008a; Nishizawa et al. 2010; Yoshiya et al. 2012). In contrast, Guilbaud et al. (2011a) suggested that the strongly negative δ^{56}Fe values in > 2.4 Ga black shales could be produced solely by pyrite precipitation. However, there are numerous examples of negative δ^{56}Fe values in bulk shales (i.e., where Fe budget is not controlled by pyrite) correlating with δ^{56}Fe values in pyrite (Rouxel et al. 2006; Czaja et al. 2010) suggesting that pyrite precipitation is not the sole mechanism generating isotopically light values in black shales. Additionally, most of the low-δ^{56}Fe shales of Neoarchean and Paleoproterozoic age are enriched in Fe (i.e., Fe/Al greater than 0.5, the value of average Archean shale), and there is no correlation between Fe enrichment, δ^{56}Fe value, and the proportion of pyrite in the Fe inventories of these samples (Czaja et al. 2010; Asael et al. 2013).

To summarize, Fe isotope studies in Archean environments confirmed the existence of an isotopically negative, anoxic, Fe-rich seawater pool. This negative pool was fueled by dissimilatory Fe reduction (similar to the Fe-shuttle found in modern redox stratified basins) and/or by hydrothermal input with partial oxidation in the photic zone (process likely dominant in anoxic Archean oceans). Large variations of δ^{56}Fe in sedimentary pyrite may be diagnostic of ferruginous conditions, as recently proposed by Tahata et al. (2015). Fe isotopes may therefore prove important proxies for paleoceanographic reconstructions when used in conjunction with Fe specitation studies (Poulton et al. 2004).

CONCLUSION

Iron isotope geochemistry is among non-traditional stable isotope systems one of the most rapidly growing fields. This is justified by the ubiquity, and rich and complex chemistry of iron in natural systems. Measuring iron's chemical bonding, abundance, ratio to geochemically similar elements Zn and Mn, redox state, and electronic spin is sometimes insufficient to unravel the riddles of iron's complex geochemical and biochemical behaviors. Iron isotopes provide a unique tool to lift these uncertainties and have already found important applications in several fields:

(1) Iron isotope measurements in ancient rocks and mineral separates (oxides, carbonates, and sulfides) provide a record of the geobiochemical cycling of iron in the ocean and sediments. In particular, processes that involved oxidation or reduction of iron seem to have left an imprint in iron isotopic compositions. Iron isotopes have thus been used to trace partial oxidation of iron in the dominantly anoxic Archean ocean and dissimilatory iron reduction (a form of respiration) in banded-iron formations.

(2) Iron availability is an important factor limiting biological productivity and carbon fixation in the ocean. One of the most promising applications of Fe isotopes in the marine realm is to trace various sources of Fe to the ocean (atmospheric dust, hydrothermal vents, reducing sediments along continental margins, and oxic seafloor sediments). For example, iron isotope measurements of seawater samples from the North Atlantic Ocean show that Saharan dust aerosol is the dominant source of dissolved iron.

(3) Differences in iron metabolism between individual and sexes are manifested in the iron isotopic composition of blood. For example, blood samples from individuals affected by hereditary hemochromatosis (excessive intestinal absorption of dietary iron associated with a number of pathologies) tend to have heavy iron isotopic compositions. Iron isotopes are also fractionated along trophic chains and might be used as tracers of paleodiets.

(4) Iron isotopes have found many applications in igneous petrology. They are sensitive to partial melting, metasomatism, and magmatic differentiation. They can also help disentangle mineral zoning that arises from diffusion vs. crystal growth. Iron isotopes may even provide some insights into the conditions of core formation and the redox evolution of Earth's mantle. Extraterrestrial bodies (the Moon, Mars, meteorites) have also been studied and those measurements provide context for understanding what controls iron isotopic fractionation during magmatic processes.

(5) Studies of the iron isotopic compositions of presolar grains provide clues on the conditions of nucleosynthesis in stars. Detection of ^{60}Ni isotope variations in meteorites from the decay of ^{60}Fe ($t_{1/2}$ = 2.62 Myr) constrains the abundance of the short-lived nuclide ^{60}Fe at the formation of the solar system. The low abundance of ^{60}Fe suggests that the solar system was contaminated by the outflows of one or several massive stars.

New applications are emerging, and the iron age of stable isotope geochemistry is just starting.

ACKNOWLEDGMENTS

The extensive reviews of Adrianna Heimann, Noah Planavsky, Martin Oeser, and Fang Huang greatly improved the manuscript. Nicole Nie is thanked for discussions and assistance in preparing figures 1 and 2. This work was supported by grants from NASA (Laboratory Analysis of Returned Samples, NNX14AK09G; Cosmochemistry, OJ-30381-0036A and NNX15AJ25G) and NSF (Petrology and Geochemistry, EAR144495; Cooperative Studies of the Earth's Deep Interior, EAR150259) to N.D. OR acknowledges the support from the National Science Foundation, Europole Mer, LabexMer (ANR-10-LABX-19–01)

REFERENCES

Akerman A, Poitrasson F, Oliva P, Audry S, Prunier J, Braun JJ (2014) The isotopic fingerprint of Fe cycling in an equatorial soil-plant-water system: The Nsimi watershed, South Cameroon. Chem Geol 385:104–116

Albarède F (2015) Metal stable isotopes in the human body: a tribute of geochemistry to medicine. Elements 11:265–269

Albarède F, Beard B (2004) Analytical methods for non-traditional isotopes. Rev Mineral Geochem 55:113–152

Albarède F, Telouk P, Lamboux A, Jaouen K, Balter V (2011) Isotopic evidence of unaccounted for Fe and Cu erythropoietic pathways. Metallomics 3:926–933

Alexander C, Wang J (2001) Iron isotopes in chondrules: Implications for the role of evaporation during chondrule formation. Meteorit Planet Sci 36:419–428

Alexander COD, Grossman J, Wang J, Zanda B, Bourot-Denise M, Hewins R (2000) The lack of potassium isotopic fractionation in Bishunpur chondrules. Meteorit Planet Sci 35:859–868

Alexander CMOD, Hewins R (2004) Mass fractionation of Fe and Ni isotopes in metal in Hammadah Al Hamrah 237. Meteorit Planet Sci 39:5080

Alexander COD, Grossman JN (2005) Alkali elemental and potassium isotopic compositions of Semarkona chondrules. Meteorit Planet Sci 40:541–556

Alexander COD, Grossman J, Ebel D, Ciesla F (2008) The formation conditions of chondrules and chondrites. Science 320:1617–1619

Algeo TJ, Schwark L, Hower JC (2004) High-resolution geochemistry and sequence stratigraphy of the Hushpuckney Shale (Swope Formation, eastern Kansas): implications for climato-environmental dynamics of the Late Pennsylvanian Midcontinent Seaway. Chem Geol 206:259–288

Alt JC (1995) Subseafloor processes in mid-ocean ridge hydrothermal systems. *In:* Seafloor Hydrothermal Systems: Physical, Chemical, Biological, Geological Interactions. Geophysical Monograph 91. Humphris SE, Zierenberg RA, Mullineaux LS, Thomson RE (eds). American Geophysical Union, p 85–114

Amor M (2015) Chemical and isotopic signatures of the magnetite from magnetotactic bacteria. PhD Thesis Institut de Physique du Globe de Paris, Paris

Amor M, Busigny V, Louvat P, Gelabert A, Cartigny P, Durand-Dubief M, Ona-Nguema G, E Alphandery, I Chebbi, F Guyot. Mass-dependent and-independent signature of Fe isotopes in magnetotactic bacteria Science 352:705–708

Anand M, Russell SS, Blackhurst RL, Grady MM (2006) Searching for signatures of life on Mars: an Fe-isotope perspective. Phil Trans R Soc B: Biol Sci 361:1715–1720

Anbar AD, Rouxel O (2007) Metal stable isotopes in paleoceanography. Ann Rev Earth Planet Sci 35:717–746

Anbar AD, Jarzecki AA, Spiro TG (2005) Theoretical investigation of iron isotope fractionation between $Fe(H_2O)_6^{3+}$ and $Fe(H_2O)_6^{2+}$: Implications for iron stable isotope geochemistry. Geochim Cosmochim Acta 69:825–837

Anderson TF, Raiswell R (2004) Sources and mechanisms for the enrichment of highly reactive iron in euxinic black sea sediments. Am J Sci 304:203–233

Archer C, Vance D (2006) Coupled Fe and S isotope evidence for Archean microbial Fe(III) and sulfate reduction. Geology 34:153–156

Arnold T, Markovic T, Kirk GJD, Schonbachler M, Rehkamper M, Zhao FJ, Weiss DJ (2015) Iron and zinc isotope fractionation during uptake and translocation in rice (*Oryza sativa*) grown in oxic and anoxic soils. C R Geosci 347:397–404

Asael D, Tissot FLH, Reinhard CT, Rouxel O, Dauphas N, Lyons TW, Ponzevera E, Liorzou C, Cheron S (2013) Coupled molybdenum, iron and uranium stable isotopes as oceanic paleoredox proxies during the Paleoproterozoic Shunga Event. Chem Geol 362:193–210

Badro J, Fiquet G, Guyot F, Rueff J-P, Struzhkin VV, Vankó G, Monaco G (2003) Iron partitioning in Earth's mantle: toward a deep lower mantle discontinuity. Science 300:789–791

Balci N, Bullen TD, Witte-Lien K, Shanks WC, Motelica M, Mandernack KW (2006) Iron isotope fractionation during microbially stimulated Fe(II) oxidation and Fe(III) precipitation. Geochim Cosmochim Acta 70:622–639

Balter V, Lamboux A, Zazzo A, Telouk P, Leverrier Y, Marvel J, Moloney AP, Monahan FJ, Schmidt O, Albarede F (2013) Contrasting Cu Fe, Zn isotopic patterns in organs and body fluids of mice and sheep, with emphasis on cellular fractionation. Metallomics 5:1470–1482

Barnicoat AC, Henderson IHC, Knipe RJ, Yardley BWD, Napler RW, Fox NPC, Lawrence SR (1997) Hydrothermal gold mineralization in the Witwatersrand basin. Nature 386:820–824

Barrat J, Yamaguchi A, Greenwood R, Bohn M, Cotten J, Benoit M, Franchi I (2007) The Stannern trend eucrites: Contamination of main group eucritic magmas by crustal partial melts. Geochim Cosmochim Acta 71:4108–4124

Barrat JA, Rouxel O, Wang K, Moynier F, Yamaguchi A, Bischoff A, Langlade J (2015) Early stages of core segregation recorded by Fe isotopes in an asteroidal mantle. Earth Planet Sci Lett 419:93–100

Barton EA, Hallbauer DK (1996) Trace-element and U–Pb isotopic compositions of pyrite types in the Proterozoic Black Reef, Transvaal Sequence, South Africa: Implications on genesis and age. Chem Geol 133:173–199

Bau M, Dulski P (1996) Distribution of yttrium and rare-earth elements in the Penge and Kuruman iron-formations, Transvaal Supergroup, South Africa. Precambrian Res 79:37–55

Bau M, Moller P (1993) Rare earth element systematics of the chemically precipitated component in Early Precambrian iron formations and the evolution of the terrestrial atmopshere-hydrosphere-lithosphere system. Geochim Cosmochim Acta 57:2239–2249

Bazylinski DA, Schlezinger DR, Howes BH, Frankel RB, Epstein SS (2000) Occurence and distribution of diverse populations of magnetic protists in a chemically stratified coastal salt pond. Chem Geol 169:319–328

Beard BL, Johnson CM (1999) High precision iron isotope measurements of terrestrial and lunar materials. Geochim Cosmochim Acta 63:1653–1660

Beard BL, Johnson CM (2004a) Fe isotope variations in the modern and ancient earth and other planetary bodies. Rev Mineral Geochem 55:319–357

Beard BL, Johnson CM (2004b) Inter-mineral Fe isotope variations in mantle-derived rocks and implications for the Fe geochemical cycle. Geochim Cosmochim Acta 68:4727–4743

Beard BL, Johnson CM, Cox L, Sun H, Nealson KH, Aguilar C (1999) Iron isotope biosignatures. Science 285:1889–1892

Beard BL, Johnson CM, Skulan JL, Nealson KH, Cox L, Sun H (2003a) Application of Fe isotopes to tracing the geochemical and biological cycling of Fe. Chem Geol 195:87–117

Beard BL, Johnson CM, Von Damm KL, Poulson RL (2003b) Iron isotope constraints on Fe cycling and mass balance in oxygenated Earth oceans. Geology 31:629–632

Beard BL, Handler RM, Scherer MM, Wu L, Czaja AD, Heimann A, Johnson CM (2010) Iron isotope fractionation between aqueous ferrous iron and goethite. Earth Planet Sci Lett 295:241–250

Bekker A, Holland HD, Wang PL, RumbleIII D, Stein HJ, Hannah JL, Coetzee LL, Beukes NJ (2004) Dating the rise of atmospheric oxygen. Nature 427:117–120

Bekker A, Kaufman AJ, Karhu JA, Eriksson KA (2005) Evidence for Paleoproterozoic cap carbonates in North America. Precambrian Res 137:167–206

Bekker A, Barley ME, Fiorentini ML, Rouxel OJ, Rumble D, Beresford SW (2009) Atmospheric sulfur in Archean komatiite-hosted nickel deposits. Science 326:1086–1089

Bekker A, Slack JF, Planavsky N, Krapez B, Hofmann A, Konhauser KO, Rouxel OJ (2010) Iron formation: the sedimentary product of a complex interplay among mantle, tectonic, oceanic, biospheric processes. Econ Geol 105:467–508

Bekker A, Krapez B, Slack JF, Planavsky N, Hofmann A, Konhauser KO, Rouxel OJ (2012) Iron formation: the sedimentary product of a complex interplay among mantle, tectonic, oceanic, biospheric processes—a Reply. Econ Geol 107:379–380

Belshaw N, Zhu X, Guo Y, O'Nions R (2000) High precision measurement of iron isotopes by plasma source mass spectrometry. Inter J Mass Spectrom 197:191–195

Bennett SA, Achterberg EP, Connelly DP, Statham PJ, Fones GR, German CR (2008) The distribution and stabilisation of dissolved Fe in deep-sea hydrothermal plumes. Earth Planet Sci Lett 270:157–167

Bennett SA, Rouxel O, Schmidt K, Garbe-Schönberg D, Statham PJ, German CR (2009) Iron isotope fractionation in a buoyant hydrothermal plume, 5°S Mid-Atlantic Ridge. Geochim Cosmochim Acta 73:5619–5634

Benz M, Brune A, Schink B (1998) Anaerobic and aerobic oxidation of ferrous iron at neutral pH by chemoheterotrophic nitrate-reducing bacteria. Arch Microbiol 169:159–165

Berglund M, Wieser ME (2011) Isotopic Compositions of the Elements 2009 (IUPAC Technical Report). Pure Appl Chem 83:397–410

Bergquist BA, Boyle EA (2006) Iron isotopes in the Amazon River system: Weathering and transport signatures. Earth Planet Sci Lett 248:54–68

Bhatia MP, Kujawinski EB, Das SB, Breier CF, Henderson PB, Charette MA (2013) Greenland meltwater as a significant and potentially bioavailable source of iron to the ocean. Nat Geosci 6:274–278

Bigeleisen J, Mayer MG (1947) Calculation of equilibrium constants for isotopic exchange reactions. J Chem Phys 15:261–267

Bilenker LD, Simon AC, Reich M, Lundstrom CC, Gajos N, Bindeman I, Barra F, Munizaga R (2016) Fe–O stable isotope pairs elucidate a high-temperature origin of Chilean iron oxide-apatite deposits. Geochim Cosmochim Acta 177:94–104

Black JR, John SG, Young ED, Kavner, A (2010) Effect of temperature and mass transport on transition metal isotope fractionation during electroplating. Geochim Cosmochim Acta 74:5187–5201

Blanchard M, Poitrasson F, Meheut M, Lazzeri M, Mauri F, Balan E (2009) Iron isotope fractionation between pyrite (FeS_2), hematite (Fe_2O_3) and siderite ($FeCO_3$): A first-principles density functional theory study. Geochim Cosmochim Acta 73:6565–6578

Blanchard M, Morin G, Lazzeri M, Balan E (2010) First-principles study of the structural and isotopic properties of Al-and OH-bearing hematite. Geochim Cosmochim Acta 74:3948–3962

Blanchard M, Poitrasson F, Meheut M, Lazzeri M, Mauri F, Balan E (2012) Comment on "New data on equilibrium iron isotope fractionation among sulfides: Constraints on mechanisms of sulfide formation in hydrothermal and igneous systems" by VB Polyakov and DM Soultanov. Geochim Cosmochim Acta 87:356–359

Blanchard M, Dauphas N, Hu M, Roskosz M, Alp E, Golden D, Sio C, Tissot F, Zhao J, Gao L (2015) Reduced partition function ratios of iron and oxygen in goethite. Geochim Cosmochim Acta 151:19–33

Borrok D, Wanty R, Ridley W, Wolf R, Lamothe P, Adams M (2007) Separation of copper, iron, zinc from complex aqueous solutions for isotopic measurement. Chem Geol 242:400–414

Bowen N (1928) The Evolution of the Igneous Rocks Princeton University Press. Princeton, New Jersey

Boyle EA, Bergquist BA, Kayser RA, Mahowald, N (2005) Iron, manganese, lead at Hawaii Ocean Time-series station ALOHA: Temporal variability and an intermediate water hydrothermal plume. Geochim Cosmochim Acta 69:933–952

Brantley SL, Liermann LJ, Guynn RL, Anbar A, Icopini GA, Barling, J (2004) Fe isotopic fractionation during mineral dissolution with and without bacteria. Geochim Cosmochim Acta 68:3189–3204

Briat JF, Lobreaux S (1997) Iron transport and storage in plants. Trends Plant Sci 2:187–193

Brumsack H-J (2006) The trace metal content of recent organic carbon-rich sediments: Implications for Cretaceous black shale formation. Palaeogeogr Palaeoclimatol Palaeoecol 232:344– 361

Bullen TD, McMahon PM (1998) Using stable Fe isotopes to assess microbially mediated Fe^{3+} reduction in a jet-fuel contaminated aquifer. Mineral Mag 62A:255–256

Bullen TD, White AF, Childs CW, Vivit DV, Schulz MS (2001) Demonstration of significant abiotic iron isotope fractionation in nature. Geology 29:699–702

Busigny V, Planavsky NJ, Jezequel D, Crowe S, Louvat P, Moureau J, Viollier E, Lyons TW (2014) Iron isotopes in an Archean ocean analogue. Geochim Cosmochim Acta 133:443–462

Butler A (2005) Marine siderophores and microbial iron mobilization. Biometals 18:369–374

Cameron EM (1982) Sulphate and sulphate reduction in early Precambrian oceans. Nature 296:45–148

Canfield DE (1998) A new model for Proterozoic ocean chemistry. Nature 396:450–453

Canil D, O'Neill HSC, Pearson D, Rudnick R, McDonough W, Carswell D (1994) Ferric iron in peridotites and mantle oxidation states. Earth Planet Sci Lett 123:205–220

Carazzo G, Jellinek AM, Turchyn AV (2013) The remarkable longevity of submarine plumes: Implications for the hydrothermal input of iron to the deep-ocean. Earth Planet Sci Lett 382:66–76

Chapman JB, Weiss DJ, Shan Y, Lemburger, M (2009) Iron isotope fractionation during leaching of granite and basalt by hydrochloric and oxalic acids. Geochim Cosmochim Acta 73:1312–1324

Charlier B, Ginibre C, Morgan D, Nowell G, Pearson D, Davidson J, Ottley C (2006) Methods for the microsampling and high-precision analysis of strontium and rubidium isotopes at single crystal scale for petrological and geochronological applications. Chem Geol 232:114–133

Chase Z, Johnson KS, Elrod VA, Plant JN, Fitzwater SE, Pickell L, Sakamoto CM (2005) Manganese and iron distributions off central California influenced by upwelling and shelf width. Mar Chem 95:235–254

Chen J-B, Busigny V, Gaillardet J, Louvat P, Wang Y-N (2014a) Iron isotopes in the Seine River (France): Natural versus anthropogenic sources. Geochim Cosmochim Acta 128:128–143

Chen Y Su, S He, Y Li S, Hou J, Feng S, Gao K (2014b) Fe isotope compositions and implications on mineralization of Xishimen iron deposit in Wuan, Hebei. Acta Petrolog Sinica 30:3443–3454

Cheng Y, Mao J, Zhu X, Wang, Y (2015) Iron isotope fractionation during supergene weathering process and its application to constrain ore genesis in Gaosong deposit, Gejiu district, SW China. Gondwana Res 27:1283–1291

Chever F, Rouxel OJ, Croot PL, Ponzevera E, Wuttig K, Auro, M (2015) Total dissolvable and dissolved iron isotopes in the water column of the Peru upwelling regime. Geochim Cosmochim Acta 162:66–82

Chu NC, Johnson CM, Beard BL, German CR, Nesbitt RW, Frank M, Bohn M, Kubik PW, Usui A, Graham I (2006) Evidence for hydrothermal venting in Fe isotope compositions of the deep Pacific Ocean through time. Earth Planet Sci Lett 245:202–217

Cloud PE (1973) Paleoecological significance of banded iron-formation. Economic Geology 68:1135–1143

Conway TM, John SG (2014) Quantification of dissolved iron sources to the North Atlantic Ocean. Nature 511:212–215

Conway TM, Rosenberg AD, Adkins JF, John SG (2013) A new method for precise determination of iron, zinc and cadmium stable isotope ratios in seawater by double-spike mass spectrometry. Analytica Chimica Acta 793:44–52

Costa F, Dohmen R, Chakraborty S (2008) Time scales of magmatic processes from modeling the zoning patterns of crystals. Rev Mineral Geochem 69:545–594

Craddock PR, Dauphas N (2011a) Iron isotopic compositions of geological reference materials and chondrites. Geostand Geoanal Res 35:101–123

Craddock PR, Dauphas N (2011b) Iron and carbon isotope evidence for microbial iron respiration throughout the Archean. Earth Planet Sci Lett 303:121–132

Craddock PR, Warren JM, Dauphas N (2013) Abyssal peridotites reveal the near-chondritic Fe isotopic composition of the Earth. Earth Planet Sci Lett 365:63–76

Crisp JA (1984) Rates of magma emplacement and volcanic output. J Volcanol Geotherm Res 20:177–211

Croal LR, Johnson CM, Beard BL, Newman DK (2004) Iron isotope fractionation by Fe(II)-oxidizing photoautotrophic bacteria. Geochim Cosmochim Acta 68:1227–1242

Crosby HA, Johnson CM, Roden EE, Beard BL (2005) Coupled Fe(II)–Fe(III) electron and atom exchange as a mechanism for Fe isotope fractionation during dissimilatory iron oxide reduction. Environ Sci Technol 39:6698–6704, doi:10.1021/es0505346

Crosby HA, Roden EE, Johnson CM, Beard BL (2007) The mechanisms of iron isotope fractionation produced during dissimilatory Fe(III) reduction by *Shewanella putrefaciens* and *Geobacter sulfurreducens*. Geobiology 5:169–189

Czaja AD, Johnson CM, Beard BL, Eigenbrode JL, Freeman KH, Yamaguchi KE (2010) Iron and carbon isotope evidence for ecosystem and environmental diversity in the similar to 2.7 to 2.5 Ga Hamersley Province, Western Australia. Earth Planet Sci Lett 292:170–180

d'Abzac F-X, Beard BL, Czaja AD, Konishi H, Schauer JJ, Johnson CM (2013) Iron isotope composition of particles produced by UV-femtosecond laser ablation of natural oxides, sulfides, carbonates. Anal Chem 85:11885–11892

d'Abzac FX, Czaja AD, Beard BL, Schauer JJ, Johnson CM (2014) Iron distribution in size-resolved aerosols generated by UV-femtosecond laser ablation: influence of cell geometry and implications for in situ isotopic determination by LA-MC-ICP-MS Geostand Geoanal Res 38:293–309

Dauphas N (2007) Diffusion-driven kinetic isotope effect of Fe and Ni during formation of the Widmanstatten pattern. Meteorit Planet Sci 42:1597–1613

Dauphas N, Rouxel O (2006) Mass spectrometry and natural variations of iron isotopes. Mass Spectrometry Reviews 25:515–550

Dauphas N, Schauble E (2016) Mass fractionation laws, mass-independent effects, isotopic anomalies. Ann Rev Earth Planet Sci 44:709–783

Dauphas N, Janney PE, Mendybaev RA, Wadhwa M, Richter FM, Davis AM, van Zuilen M, Hines R, Foley CN (2004a) Chromatographic separation and multicollection-ICPMS analysis of iron. Investigating mass-dependent and-independent isotope effects. Anal Chem 76:5855–5863

Dauphas N, van Zuilen M, Wadhwa M, Davis AM, Marty B, Janney PE (2004b) Clues from Fe isotope variations on the origin of early Archean BIFs from Greenland. Science 306:2077–2080

Dauphas N, Cates NL, Mojzsis SJ, Busigny V (2007a) Identification of chemical sedimentary protoliths using iron isotopes in the >3750 Ma Nuvvuagittuq supracrustal belt, Canada. Earth Planet Sci Lett 254:358–376

Dauphas N, Van Zuilen M, Busigny V, Lepland A, Wadhwa M, Janney PE (2007b) Iron isotope, major and trace element characterization of early Archean supracrustal rocks from SW Greenland: Protolith identification and metamorphic overprint. Geochim Cosmochim Acta 71:4745–4770

Dauphas N, Cook DL, Sacarabany A, Frohlich C, Davis AM, Wadhwa M, Pourmand A, Rauscher T, Gallino R (2008) Iron-60 evidence for early injection and efficient mixing of stellar debris in the protosolar nebula. Astrophys J 686:560–569

Dauphas N, Pourmand A, Teng F-Z (2009a) Routine isotopic analysis of iron by HR-MC-ICPMS: How precise and how accurate? Chem Geol 267:175–184

Dauphas N, Craddock PR, Asimow PD, Bennett VC, Nutman AP, Ohnenstetter D (2009b) Iron isotopes may reveal the redox conditions of mantle melting from Archean to Present. Earth Planet Sci Lett 288:255–267

Dauphas N, Teng F-Z, Arndt NT (2010) Magnesium and iron isotopes in 2.7 Ga Alexo komatiites: Mantle signatures, no evidence for Soret diffusion, identification of diffusive transport in zoned olivine. Geochim Cosmochim Acta 74:3274–3291

Dauphas N, Roskosz M, Alp E, Golden D, Sio C, Tissot F, Hu M, Zhao J, Gao L, Morris R (2012) A general moment NRIXS approach to the determination of equilibrium Fe isotopic fractionation factors: application to goethite and jarosite. Geochim Cosmochim Acta 94:254–275

Dauphas N, Roskosz M, Alp E, Neuville D, Hu M, Sio C, Tissot F, Zhao J, Tissandier L, Médard E (2014) Magma redox and structural controls on iron isotope variations in Earth's mantle and crust. Earth Planet Sci Lett 398:127–140

Dauphas N, Poitrasson F, Burkhardt C, Kobayashi H, Kurosawa K (2015) Planetary and meteoritic Mg/Si and δ^{30}Si variations inherited from solar nebula chemistry. Earth Planet Sci Lett 427:236–248

Debret B, Millet M-A, Pons M-L, Bouihol P, Inglis E, Williams H (2016) Isotopic evidence for iron mobility during subduction. Geology 44:215–218

de Jong J, Schoemann V, Tison JL, Becquevort S, Masson F, Lannuzel D, Petit J, Chou L, Weis D, Mattielli, N (2007) Precise measurement of Fe isotopes in marine samples by multi-collector inductively coupled plasma mass spectrometry (MC-ICP-MS). Anal Chim Acta 589:105–119

Dideriksen K, Baker J, Stipp S (2006) Iron isotopes in natural carbonate minerals determined by MC-ICP-MS with a ^{58}Fe–^{54}Fe double-spike. Geochim Cosmochim Acta 70:118–132

Dideriksen K, Baker JA, Stipp SLS (2008) Equilibrium Fe isotope fractionation between inorganic aqueous Fe (III) and the siderophore complex Fe (III)-desferrioxamine B. Earth Planet Sci Lett 269:280–290

Dodson M (1963) A theoretical study of the use of internal standards for precise isotopic analysis by the surface ionization technique: Part I—General first-order algebraic solutions. J Sci Instrum 40:289

Dohmen R, Chakraborty S (2007) Fe–Mg diffusion in olivine II: point defect chemistry, change of diffusion mechanisms and a model for calculation of diffusion coefficients in natural olivine. Phys Chem Mineral 34:409–430

Domagal-Goldman SD, Kubicki JD (2008) Density functional theory predictions of equilibrium isotope fractionation of iron due to redox changes and organic complexation. Geochim Cosmochim Acta 72:5201–5216

Dominguez G, Wilkins G, Thiemens MH (2011) The Soret effect and isotopic fractionation in high-temperature silicate melts. Nature 473:70–73

dos Santos Pinheiro GM, Poitrasson F, Sondag F, Vieira LC, Pimentel MM (2013) Iron isotope composition of the suspended matter along depth and lateral profiles in the Amazon River and its tributaries. J South Am Earth Sci 44:35–44

Duan Y, Severmann S, Anbar AD, Lyons TW, Gordon GW, Sageman BB (2010) Isotopic evidence for Fe cycling and repartitioning in ancient oxygen-deficient settings: Examples from black shales of the mid-to-late Devonian Appalachian basin. Earth Planet Sci Lett 290:244–253

Dziony W, Horn I, Lattard D, Koepke J, Steinhoefel G, Schuessler JA, Holtz F (2014) In-situ Fe isotope ratio determination in Fe–Ti oxides and sulfides from drilled gabbros and basalt from the IODP Hole 1256D in the eastern equatorial Pacific. Chem Geol 363:101–113

Edwards KJ, Bond PL, Gihring TM, Banfield JF (2000) An archaeal iron-oxidizing extreme acidophile important in acid mine drainage. Science 287:1796–1799

Ellwood MJ, Hutchins DA, Lohan MC, Milne A, Nasemann P, Nodder SD, Sander SG, Strzepek R, Wilhelm SW, Boyd PW (2015) Iron stable isotopes track pelagic iron cycling during a subtropical phytoplankton bloom. PNAS 112:E15–E20

Elrod VA, Berelson WM, Coale KH, Johnson KS (2004) The flux of iron from continental shelf sediments: A missing source for global budgets. Geophys Res Lett 31:10.1029/2004GL020216

Emerson D, Moyer C (1997) Isolation and characterization of novel iron-oxidizing bacteria that grow at circumneutral pH Appl Environ Microbiol 61:2681–2687

Emmanuel S, Schuessler JA, Vinther J, Matthews A, von Blanckenburg F (2014) A preliminary study of iron isotope fractionation in marine invertebrates (chitons, Mollusca) in near-shore environments. Biogeosciences 11:5493–5502

England GL, Rasmussen B, Krapez B, Groves DI (2002) Palaeoenvironmental significance of rounded pyrite in siliciclastic sequences of the Late Archaean Witwatersrand. Sedimentology 49:1133–1156

Escoube R, Rouxel OJ, Sholkovitz E, Donard OF (2009) Iron isotope systematics in estuaries: The case of North River, Massachusetts (USA). Geochim Cosmochim Acta 73:4045–4059

Escoube R, Rouxel OJ, Pokrovsky OS, Schroth A, Holmes RM, Donard, OFX (2015) Iron isotope systematics in Arctic rivers. C R Geosci 347:377–385

Falkowski PG (1997) Evolution of the nitrogen cycle and its influence on the biological sequestration of CO_2 in the ocean. Nature 387:272–275

Fantle MS, DePaolo DJ (2004) Iron isotopic fractionation during continental weathering. Earth Planet Sci Lett 228:547–562

Farquhar J, Bao H, Thiemens M (2000) Atmospheric influence of Earth's earliest sulfur cycle. Science 289:756–758

Farquhar J, Peters M, Johnston DT, Strauss H, Masterson A, Wiechert U, Kaufman AJ (2007) Isotopic evidence for *Mesoarchaean anoxia* and changing atmospheric sulphur chemistry. Nature 449:706–709

Fiorentini ML, Bekker A, Rouxel O, Wing BA, Maier W, Rumble D (2012) Multiple sulfur and iron isotope composition of magmatic Ni–Cu–(PGE) sulfide mineralization from Eastern Botswana. Econ Geol 107:105–116

Fitzsimmons JN, Boyle EA, Jenkins WJ (2014) Distal transport of dissolved hydrothermal iron in the deep South Pacific Ocean. PNAS 111:16654–16661

Fitzsimmons JN, Carrasco GG Wu J Roshan S Hatta M Measures CI, Conway TM, John SG, Boyle EA (2015) Partitioning of dissolved iron and iron isotopes into soluble and colloidal phases along the GA03 GEOTRACES North Atlantic Transect. Deep Sea Research Part II 116:130–151

Foden J, Sossi PA, Wawryk CM (2015) Fe isotopes and the contrasting petrogenesis of A-, I-and S-type granite. Lithos 212:32–44

Frank M (2002) Radiogenic isotopes: Tracers of past ocean circulation and erosional input. Rev Geophys 40, doi:100110.1029/2000rg000094

Frankel RB, Blakemore RP, Wolfe RS (1979) Magnetite in freshwater magnetotactic bacteria. Science 203:1355–1356

Frierdich AJ, Beard BL, Reddy TR, Scherer MM, Johnson CM (2014a) Iron isotope fractionation between aqueous Fe (II) and goethite revisited: New insights based on a multi-direction approach to equilibrium and isotopic exchange rate modification. Geochim Cosmochim Acta 139:383–398

Frierdich AJ, Beard BL, Scherer MM, Johnson CM (2014b) Determination of the Fe $(II)_{aq}$–magnetite equilibrium iron isotope fractionation factor using the three-isotope method and a multi-direction approach to equilibrium. Earth Planet Sci Lett 391:77–86

Frost DJ, Liebske C, Langenhorst F, McCammon CA, Trønnes RG, Rubie DC (2004) Experimental evidence for the existence of iron-rich metal in the Earth's lower mantle. Nature 428:409–412

Frost CD, von Blanckenburg F, Schoenberg R, Frost BR, Swapp SM (2007) Preservation of Fe isotope heterogeneities during diagenesis and metamorphism of banded iron formation. Contrib Mineral Petrol 153:211–235

Fujii T, Moynier Fdr, Telouk P, Albarède F (2006) Isotope fractionation of iron(III) in chemical exchange reactions using solvent extraction with crown ether. J Phys Chem A 110:11108–11112

Fujii T, Moynier F, Blichert-Toft J, Albarède F (2014) Density functional theory estimation of isotope fractionation of Fe Ni Cu, Zn among species relevant to geochemical and biological environments. Geochim Cosmochim Acta 140:553–576

Fung IY, Meyn SK, Tegen I Doney SC, John JG, Bishop, JKB (2000) Iron supply and demand in the upper ocean. Global Biogeochem Cycles 14:281–295

Gall L, Williams HM, Siebert C, Halliday AN, Herrington RJ, Hein JR (2013) Nickel isotopic compositions of ferromanganese crusts and the constancy of deep ocean inputs and continental weathering effects over the Cenozoic. Earth Planet Sci Lett 375:148–155

German CR, Von Damm KL (2003) Hydrothermal processes. *In:* The Oceans and Marine Geochemistry. Treatise on Geochemistry Volume 6, pp.181–222

Graham S, Pearson N, Jackson S, Griffin W, O'Reilly S (2004) Tracing Cu and Fe from source to porphyry: in situ determination of Cu and Fe isotope ratios in sulfides from the Grasberg Cu–Au deposit. Chem Geol 207:147–169

Guelke M, von Blanckenburg F (2007) Fractionation of stable iron isotopes in higher plants. Environ Sci Technol 41:1896–1901

Guelke M, von Blanckenburg F, Schoenberg R, Staubwasser M, Stuetzel, H (2010) Determining the stable Fe isotope signature of plant-available iron in soils. Chem Geol 277:269–280

Guelke-Stelling M, von Blanckenburg F (2012) Fe isotope fractionation caused by translocation of iron during growth of bean and oat as models of strategy I and II plants. Plant Soil 352:217–231

Guilbaud R, Butler IB, Ellam RM, Rickard D, Oldroyd A (2011a) Experimental determination of the equilibrium Fe^{2+} isotope fractionation between and FeSm (mackinawite) at 25 and 2 °C Geochim Cosmochim Acta 75:2721–2734

Guilbaud R, Butler IB, Ellam RM (2011b) Abiotic pyrite formation produces a large fe isotope fractionation. Science 332:1548–1551

Guilbaud R, Butler IB, Ellam RM (2012) Response to Comment on "Abiotic pyrite formation produces a large Fe isotope fractionation". Science 335:538

Günther D, Heinrich CA (1999) Enhanced sensitivity in laser ablation-ICP mass spectrometry using helium-argon mixtures as aerosol carrier. J Anal At Spectrom 14:1363–1368

Halama M, Swanner ED, Konhauser KO, Kappler A (2016) Evaluation of siderite and magnetite formation in BIFs by pressure-temperature experiments of Fe(III) minerals and microbial biomass. Earth Planet Sci Lett 450:243–253

Halverson GP, Poitrasson F, Hoffman PF, Nedelec A, Montel JM, Kirby J (2011) Fe isotope and trace element geochemistry of the Neoproterozoic syn-glacial Rapitan iron formation. Earth Planet Sci Lett 309:100–112

Hannington MD, Jonasson IR, Herzig PM, Petersen S (1995) Physical and chemical processes of seafloor mineralization at mid-ocean ridges. *In:* Seafloor Hydrothermal Systems: Physical, Chemical, Biological, Geological Interactions, SE Humphris, RA Zierenberg, LS Mullineaux, RE Thomson (eds), American Geophysical Union, Washington, D C. doi:10.1029/GM091p0115

Harvey LJ, Armah CN, Dainty JR, Foxall RJ, Lewis DJ, Langford NJ, Fairweather-Tait SJ (2005) Impact of menstrual blood loss and diet on iron deficiency among women in the UK Br J Nutr 94:557–564

Haugaard R, Pecoits E, Lalonde SV, Rouxel OJ, Konhauser KO (2016) The Joffre Banded Iron Formation, Hamersley Group, Western Australia: Assessing the palaeoenvironment through detailed petrology and chemostratigraphy. Precambrian Res 273:12–37

Hawkings JR, Wadham JL, Tranter M, Raiswell R, Benning LG, Statham PJ, Tedstone A, Nienow P, Lee K, Telling J (2014) Ice sheets as a significant source of highly reactive nanoparticulate iron to the oceans. Nat Commun 5,:doi: 10.1038/ncomms4929

Heimann A, Beard BL, Johnson CM (2008) The role of volatile exsolution and sub-solidus fluid/rock interactions in producing high Fe–56/Fe–54 ratios in siliceous igneous rocks. Geochim Cosmochim Acta 72:4379–4396

Heimann A, Johnson CM, Beard BL, Valley JW, Roden EE, Spicuzza MJ, Beukes NJ (2010) Fe C, O isotope compositions of banded iron formation carbonates demonstrate a major role for dissimilatory iron reduction in similar to 2.5 Ga marine environments. Earth Planet Sci Lett 294:8–18

Herzfeld KF, Teller E (1938) The vapor pressure of isotopes. Phys Rev 54:912

Hezel DC, Needham AW, Armytage R, Georg B, Abel RL, Kurahashi E, Coles BJ, Rehkämper M, Russell SS (2010) A nebula setting as the origin for bulk chondrule Fe isotope variations in CV chondrites. Earth Planet Sci Lett 296:423–433

Hibbert K, Williams H, Kerr AC, Puchtel I (2012) Iron isotopes in ancient and modern komatiites: evidence in support of an oxidised mantle from Archean to present. Earth Planet Sci Lett 321:198–207
Hiebert RS, Bekker A, Wing BA, Rouxel OJ (2013) The role of paragneiss assimilation in the origin of the voisey's bay Ni–Cu sulfide deposit, labrador: multiple S and Fe isotope evidence. Econ Geol 108:1459–1469
Hiebert RS, Bekker A, Houlé MG, Wing BA, Rouxel OJ (2016) Tracing sources of crustal contamination using multiple S and Fe isotopes in the Hart komatiite-associated Ni–Cu–PGE sulfide deposit, Abitibi greenstone belt, Ontario, Canada. Miner Deposita 51:919–935
Hill PS, Schauble EA (2008) Modeling the effects of bond environment on equilibrium iron isotope fractionation in ferric aquo-chloro complexes. Geochim Cosmochim Acta 72:1939–1958
Hill PS, Schauble EA, Shahar A, Tonui E, Young ED (2009) Experimental studies of equilibrium iron isotope fractionation in ferric aquo-chloro complexes. Geochim Cosmochim Acta 73:2366–2381
Hin RC, Schmidt MW, Bourdon B (2012) Experimental evidence for the absence of iron isotope fractionation between metal and silicate liquids at 1 GPa and 1250–1300 °C and its cosmochemical consequences. Geochim Cosmochim Acta 93:164–181
Hofmann A, Bekker A, Rouxel O, Rumble D, Master S (2009) Multiple sulphur and iron isotope composition of detrital pyrite in Archaean sedimentary rocks: a new tool for provenace analysis. Earth Planet Sci Lett 286:436–445
Hofmann A, Bekker A, Dirks P, Gueguen B, Rumble D, Rouxel OJ (2014) Comparing orthomagmatic and hydrothermal mineralization models for komatiite-hosted nickel deposits in Zimbabwe using multiple-sulfur, iron, nickel isotope data. Mineralium Deposita 49:75–100
Holland HD (1984) The Chemical Evolution of the Atmosphere and Oceans. Princeton Univ. Press., New York
Homoky WB, Severmann S, Mills RA, Statham PJ, Fones GR (2009) Pore-fluid Fe isotopes reflect the extent of benthic Fe redox recycling: Evidence from continental shelf and deep-sea sediments. Geology 37:751–754
Homoky WB, John SG, Conway TM, Mills RA (2013) Distinct iron isotopic signatures and supply from marine sediment dissolution. Nat Commun 4, doi: 10.1038/ncomms3143
Horn I, von Blanckenburg F, Schoenberg R, Steinhoefel G, Markl G (2006) In situ iron isotope ratio determination using UV-femtosecond laser ablation with application to hydrothermal ore formation processes. Geochim Cosmochim Acta 70:3677–3688
Horn I, von Blanckenburg F (2007) Investigation on elemental and isotopic fractionation during 196 nm femtosecond laser ablation multiple collector inductively coupled plasma mass spectrometry. Spectrochim Acta Part B: At Spectr 62:410–422
Horner TJ, Schönbächler M, Rehkämper M, Nielsen SG, Williams H, Halliday AN, Xue Z, Hein JR (2010) Ferromanganese crusts as archives of deep water Cd isotope compositions. G-cubed: doi:10.1029/2009GC002987
Horner TJ, Williams HM, Hein JR, Saito MA, Burton KW, Halliday AN, Nielsen SG (2015) Persistence of deeply sourced iron in the Pacific Ocean. PNAS 112:1292–1297
Hotz K, Walczyk T (2013) Natural iron isotopic composition of blood is an indicator of dietary iron absorption efficiency in humans. J Biol Inorg Chem 18:1–7
Hotz K, Augsburger H, Walczyk T (2011) Isotopic signatures of iron in body tissues as a potential biomarker for iron metabolism. J Anal At Spectrom 26:1347–1353
Hotz K, Krayenbuehl PA, Walczyk T (2012) Mobilization of storage iron is reflected in the iron isotopic composition of blood in humans. J Biol Inorg Chem 17:301–309
Hu MY, Sturhahn W, Toellner TS, Mannheim PD, Brown DE, Zhao J, Alp EE (2003) Measuring velocity of sound with nuclear resonant inelastic X-ray scattering. Phys Rev B 67:094304
Hu MY, Toellner TS, Dauphas N, Alp EE, Zhao J (2013) Moments in nuclear resonant inelastic x-ray scattering and their applications. Phys Rev B 87:064301
Huang F, Lundstrom C, Glessner J, Ianno A, Boudreau A, Li J, Ferré E, Marshak S, DeFrates J (2009) Chemical and isotopic fractionation of wet andesite in a temperature gradient: experiments and models suggesting a new mechanism of magma differentiation. Geochim Cosmochim Acta 73:729–749
Huang F, Chakraborty P, Lundstrom C, Holmden C, Glessner J, Kieffer S, Lesher C (2010) Isotope fractionation in silicate melts by thermal diffusion. Nature 464:396–400
Huang F, Zhang Z, Lundstrom CC, Zhi X (2011) Iron and magnesium isotopic compositions of peridotite xenoliths from Eastern China. Geochim Cosmochim Acta 75:3318–3334
Humayun M, Clayton RN (1995a) Precise determination of the isotopic composition of potassium: Application to terrestrial rocks and lunar soils. Geochim Cosmochim Acta 59:2115–2130
Humayun M, Clayton RN (1995b) Potassium isotope cosmochemistry: Genetic implications of volatile element depletion. Geochim Cosmochim Acta 59:2131–2148
Humayun M, Qin L, Norman MD (2004) Geochemical evidence for excess iron in the mantle beneath Hawaii. Science 306:91–94

Humphris SE, Herzig PM, Miller DJ, Alt JC, Becker K, Brown D, Brugmann G, Chiba H, Fouquet Y, Gemmell JB, Guerin G, Hannington MD, HolmNG, Honnorez JJ, Iturrino GJ, Knott R, Ludwig R, Nakamura K, Petersen S, Reysenbach AL, Rona PA, Smith S, Sturz AA, Tivey MK, Zhao X (1995) The internal structure of an active sea-floor massive sulfide deposit. Nature 377:713–716

Hutchins DA, Witter AE, Butler A, Luther III GW (1999) Competition among marine phytoplankton for different chelated iron species. Nature 400:858–861

Hyslop EV, Valley JW, Johnson CM, Beard BL (2008) The effects of metamorphism on O and Fe isotope compositions in the Biwabik Iron Formation, Northern Minnesota. Contrib Mineral Petrol 155:313–328

Icopini GA, Anbar AD, Ruebush SS, Tien M, Brantley SL (2004) Iron isotope fractionation during microbial reduction of iron: The importance of adsorption. Geology 32:205–208

Ilina SM, Poitrasson F, Lapitskiy SA, Alekhin YV, Viers J, Pokrovsky OS, (2013) Extreme iron isotope fractionation between colloids and particles of boreal and temperate organic-rich waters. Geochim Cosmochim Acta 101:96–111

Ingri J, Malinovsky D, Rodushkin I, Baxter DC, Widerlund A, Andersson P, Gustafsson O, Forsling W, Ohlander, B (2006) Iron isotope fractionation in river colloidal matter. Earth Planet Sci Lett 245:792–798

Jaouen K, Balter V (2014) Menopause effect on blood Fe and Cu isotope compositions. Am J Phys Anthropol 153:280–285

Jaouen K, Balter V, Herrscher E, Lamboux A, Telouk P, Albarede F (2012) Fe and Cu stable isotopes in archeological human bones and their relationship to sex. Am J Phys Anthropol 148:334–340

Jaouen K, Gibert M, Lamboux A, Telouk P, Fourel F, Albareede F, Alekseev AN, Crubeezy E, Balter V (2013a) Is aging recorded in blood Cu and Zn isotope compositions? Metallomics 5:1016–1024

Jaouen K, Pons M-L, Balter V (2013b) Iron, copper and zinc isotopic fractionation up mammal trophic chains. Earth Planet Sci Lett 374:164–172

Jarzecki A, Anbar A, Spiro T (2004) DFT analysis of $Fe(H_2O)_6^{3+}$ and $Fe(H_2O)_6^{2+}$ structure and vibrations; implications for isotope fractionation. J Phys Chem A 108:2726–2732

Jeandel C, Peucker-Ehrenbrink B, Jones M, Pearce C, Oelkers EH, Godderis Y, Lacan F, Aumont O, Arsouze, T (2011) Ocean margins: The missing term in oceanic element budgets? Eos 92:217

Jenkyns HC, Matthews A, Tsikos H, Erel Y (2007) Nitrate reduction, sulfate reduction, sedimentary iron isotope evolution during the Cenomanian–Turonian oceanic anoxic event. Paleoceanography:PA3208,doi:3210.1029/2006PA001355

Jickells TD, Spokes LJ (2001) Atmospheric iron inputs to the oceans. *In:* DR Turner, KA Hunter (Eds.) The Biogeochemistry of Iron in Seawater pp. 85–118. John Wiley, Hoboken, NJ

Jickells TD, An ZS, Andersen KK, Baker AR, Bergametti G, Brooks N, Cao JJ, Boyd PW, Duce RA, Hunter KA, Kawahata H (2005) Global iron connections between desert dust, ocean biogeochemistry, climate. Science 308:67–71

John SG (2012) Optimizing sample and spike concentrations for isotopic analysis by double-spike ICPMS J Anal At Spectrom 27:2123

John SG, Adkins JF (2010) Analysis of dissolved iron isotopes in seawater. Mar Chem 119:65–76

John SG, Adkins JF (2012) The vertical distribution of iron stable isotopes in the North Atlantic near Bermuda. Global Biogeochem Cycles 26, doi: 10.1029/2011GB004043

John SG, Conway TM (2013) A role for scavenging in the marine biogeochemical cycling of zinc and zinc isotopes. Earth Planet Sci Lett 394:159–167

John SG, Geis RW, Saito MA, Boyle EA (2007) Zinc isotope fractionation during high-affinity and low-affinity zinc transport by the marine diatom *Thalassiosira oceanica*. Limnol Oceanogr 52:2710–2714

John SG, Mendez J, Moffett J, Adkins JF (2012a) The flux of iron and iron isotopes from San Pedro Basin sediments. Geochim Cosmochim Acta 93:14–29

John SG, King A, Hutchins D, Adkins JF Fu F, Wasson A, Hodierne, C (2012b) Biological, chemical, electrochemical, photochemical fractionation of Fe isotopes. AGU Fall Meeting Abstracts

Johnson CM, Beard BL (1999) Correction of instrumentally produced mass fractionation during isotopic analysis of Fe by thermal ionization mass spectrometry. Inter J Mass Spectrom 193:87–99

Johnson KS, Chavez FP, Friederich GE (1999) Continental-shelf sediment as a primary source of iron for coastal phytoplankton. Nature 398:697–700

Johnson CM, Skulan JL, Beard BL, Sun H, Nealson KH, Braterman PS (2002) Isotopic fractionation between Fe(III) and Fe(II) in aqueous solutions. Earth Planet Sci Lett 195:141–153

Johnson CM, Beard BL, Beukes NJ, Klein C, O'Leary JM (2003) Ancient geochemical cycling in the Earth as inferred from Fe isotope studies of banded iron formations from the Transvaal Craton. Contrib Mineral Petrol 144:523–547

Johnson CM, Beard BL, Roden EE, Newman DK, Nealson KH (2004) Isotopic constraints on biogeochemical cycling of Fe. Rev Mineral Geochem 55:359–408

Johnson CM, Roden EE, Welch SA, Beard BL (2005) Experimental constraints on Fe isotope fractionation during magnetite and Fe carbonate formation coupled to dissimilatory hydrous ferric oxide reduction. Geochim Cosmochim Acta 69:963–993

Johnston DT, Poulton SW, Fralick PW, Wing BA, Canfield DE, Farquhar J (2006) Evolution of the oceanic sulfur cycle at the end of the Paleoproterozoic. Geochim Cosmochim Acta 70:5723–5739

Johnson CM, Beard BL, Roden EE (2008a) The iron isotope fingerprints of redox and biogeochemical cycling in the modern and ancient Earth. Ann Rev Earth Planet Sci 36:457–493

Johnson CM, Beard BL, Klein C, Beukes NJ, Roden EE (2008b) Iron isotopes constrain biologic and abiologic processes in banded iron formation genesis. Geochim Cosmochim Acta 72:151–169

Johnson CM, Bell K, Beard BL, Shultis AI (2010) Iron isotope compositions of carbonatites record melt generation, crystallization, late-stage volatile-transport processes. Mineral Petrol 98:91–110

Johnson CM, Ludois JM, Beard BL, Beukes NJ, Heimann A (2013) Iron formation carbonates: Paleoceanographic proxy or recorder of microbial diagenesis? Geology 41:1147–1150

Karhu JA, Holland HD (1996) Carbon isotopes and the rise of atmospheric oxygen. Geology 24:867–870

Kavner A, Bonet F, Shahar A, Simon J, Young, E (2005) The isotopic effects of electron transfer: An explanation for Fe isotope fractionation in nature. Geochim Cosmochim Acta 69:2971–2979

Kehm K, Hauri E, Alexander COD, Carlson R (2003) High precision iron isotope measurements of meteoritic material by cold plasma ICP-MS Geochim Cosmochim Acta 67:2879–2891

Kiczka M, Wiederhold JG, Frommer J, Kraemer SM, Bourdon B, Kretzschmar, R (2010) Iron isotope fractionation during proton- and ligand-promoted dissolution of primary phyllosilicates. Geochim Cosmochim Acta 74:3112–3128

Kieffer SW (1982) Thermodynamics and lattice vibrations of minerals: 5. Applications to phase equilibria, isotopic fractionation, high-pressure thermodynamic properties. Rev Geophys 20:827–849

Kita N, Huberty J, Kozdon R, Beard B, Valley J (2011) High-precision SIMS oxygen, sulfur and iron stable isotope analyses of geological materials: accuracy, surface topography and crystal orientation. Surf Interfac Anal 43:427–431

Košler J, Pedersen RB, Kruber C, Sylvester PJ (2005) Analysis of Fe isotopes in sulfides and iron meteorites by laser ablation high-mass resolution multi-collector ICP mass spectrometry. J Anal At Spectrom 20:192–199

Krawczynski MJ, Van Orman JA, Dauphas N, Alp EE, Hu M (2014) Iron isotope fractionation between metal and troilite: a new cooling speedometer for iron meteorites. Lunar Planet Sci Conf 45:2755

Krayenbuehl PA, Walczyk T, Schoenberg R, von Blanckenburg F, Schulthess G (2005) Hereditary hemochromatosis is reflected in the iron isotope composition of blood. Blood 105:3812–3816, doi:10.1182/blood-2004-07–2807

Kyser T, Lesher C, Walker D (1998) The effects of liquid immiscibility and thermal diffusion on oxygen isotopes in silicate liquids. Contrib Mineral Petrol 133:373–381

Lacks DJ, Goel G, Bopp IV CJ, Van Orman JA, Lesher CE, Lundstrom CC (2012) Isotope fractionation by thermal diffusion in silicate melts. Phys Rev Lett 108:065901

Lacan F, Francois R, Ji YC, Sherrell RM (2006) Cadmium isotopic composition in the ocean. Geochim Cosmochim Acta 70:5104–5118

Lacan F, Radic A, Jeandel C, Poitrasson F, Sarthou G, Pradoux C, Freydier, R (2008) Measurement of the isotopic composition of dissolved iron in the open ocean. Geophys Res Lett 35, doi: 10.1029/2008GL035841

Lacan F, Radic A, Labatut M, Jeandel C, Poitrasson F, Sarthou G, Pradoux C, Chmeleff J, Freydier, R (2010) High-precision determination of the isotopic composition of dissolved iron in iron depleted seawater by double-spike multicollector-ICPMS Anal Chem 55:7103–7111

Landry MR, Barber RT, Bidigare RR, Chai F, Coale KH, Dam HG, Lewis MR, Lindley ST, McCarthy JJ, Roman MR, Stoecker DK, Verity PG, White JR (1997) Iron and grazing constraints on primary production in the central equatorial Pacific: An EqPac synthesis. Limnol Oceanogr 42:405–418

Larner F (2016) Can we use high precision metal isotope analysis to improve our understanding of cancer? Anal Bioanal Chem 408:345–349

LeClaire A, Lidiard A (1956) LIII Correlation effects in diffusion in crystals. Phil Mag 1:518–527

Lee C-TA, Luffi P, Plank T, Dalton H, Leeman WP (2009) Constraints on the depths and temperatures of basaltic magma generation on Earth and other terrestrial planets using new thermobarometers for mafic magmas. Earth Planet Sci Lett 279:20–33

Lehmann B, Nägler TF, Holland HD, Wille M, Mao J, Pan J, Ma D, Dulski P (2007) Highly metalliferous carbonaceous shale and Early Cambrian seawater. Geology 35:403–406

Levasseur S, Frank M, Hein JR, Halliday AN (2004) The global variation in the iron isotope composition of marine hydrogenetic ferromanganese deposits: implications for seawater chemistry? Earth Planet Sci Lett 224:91–105

Li J, Agee CB (1996) Geochemistry of mantle–core differentiation at high pressure. Nature 381:686–689

Li X, Liu Y (2015) A theoretical model of isotopic fractionation by thermal diffusion and its implementation on silicate melts. Geochim Cosmochim Acta 154:18–27

Li WQ, Beard BL, Johnson CM (2015) Biologically recycled continental iron is a major component in banded iron formations. PNAS 112:8193–8198

Lin JF, Speziale S, Mao Z, Marquardt H (2013) Effects of the electronic spin transitions of iron in lower mantle minerals: implications for deep mantle geophysics and geochemistry. Rev Geophys 51:244–275

Liu Y, Spicuzza MJ, Craddock PR, Day J, Valley JW, Dauphas N, Taylor LA (2010) Oxygen and iron isotope constraints on near-surface fractionation effects and the composition of lunar mare basalt source regions. Geochim Cosmochim Acta 74:6249–6262

Liu S-A, Li D, Li S, Teng F-Z, Ke S, He Y, Lu Y (2014) High-precision copper and iron isotope analysis of igneous rock standards by MC-ICP-MS J Anal At Spectrom 29:122–133

Liu J, Dauphas N, Roskosz M, Hu MY, Yang H, Bi W, Zhao J, Alp EE, Hu JY, Lin J-F (2016) Iron isotopic fractionation between silicate mantle and metallic core at high pressure. Nat Commun: in press

Lovley DR, Stolz JF, Nord GL, Phillips EJP (1987) Anaerobic production of magnetite by a dissimilatory iron-reducing microorganism. Nature 330:252–254

Luck J-M, Othman DB, Albarède F (2005) Zn and Cu isotopic variations in chondrites and iron meteorites: early solar nebula reservoirs and parent-body processes. Geochim Cosmochim Acta 69:5351–5363

Lundstrom C (2009) Hypothesis for the origin of convergent margin granitoids and Earth's continental crust by thermal migration zone refining. Geochim Cosmochim Acta 73:5709–5729

Luo C, Mahowald N, Bond T, Chuang PY, Artaxo P, Siefert R, Chen Y, Schauer J (2008) Combustion iron distribution and deposition. Global Biogeochem Cycles, 22

Luz B, Barkan E, Bender ML, Thiemens MH, Boering KA (1999) Triple-isotope composition of atmospheric oxygen as a tracer of biosphere productivity. Nature 400:547–550

Lyons TW, Severmann S (2006) A critical look at iron paleoredox proxies: New insights from modern euxinic marine basins. Geochim Cosmochim Acta 70:5698–5722

Lyons TW, Reinhard CT, Planavsky NJ (2014) The rise of oxygen in Earth's early ocean and atmosphere. Nature 506:307–315

Mahowald NM, Baker AR, Bergametti G, Brooks N, Duce RA, Jickells TD, Kubilay N, Prospero JM, Tegen, I (2005) Atmospheric global dust cycle and iron inputs to the ocean. Global Biogeochem Cycles 19

Mandernack KW, Bazylinski DA, Shanks WC, Bullen TD (1999) Oxygen and iron isotope studies of magnetite produced by magnetotactic bacteria. Science 285:1892–1896

Majestic BJ, Anbar AD, Herckes, P (2009) Stable Isotopes as a Tool to Apportion Atmospheric Iron. Environ Sci Technol 43:4327–4333

Marcus RA (1965) On the theory of electron-transfer reactions. VI Unified treatment for homogeneous and electrode reactions. J Chem Phys:43:679–701

Marcus MA, Edwards KJ, Gueguen B, Fakra SC, Horn G, Jelinski NA, Rouxel O, Sorensen J, Toner BM (2015) Iron mineral structure, reactivity, isotopic composition in a South Pacific Gyre ferromanganese nodule over 4 Ma. Geochim Cosmochim Acta 171:61–79

Maréchal CN, Télouk P, Albarède F (1999) Precise analysis of copper and zinc isotopic compositions by plasma-source mass spectrometry. Chem Geol 156:251–273

Marhas KK, Amari S, Gyngard F, Zinner E, Gallino R (2008) Iron and nickel isotopic ratios in presolar SiC grains. Astrophys J 689:622

Marin-Carbonne J, Rollion-Bard C, Luais B (2011) In-situ measurements of iron isotopes by SIMS: MC-ICP-MS intercalibration and application to a magnetite crystal from the Gunflint chert. Chem Geol 285:50–61

Marin-Carbonne J, Rollion-Bard C, Bekker A, Rouxel O, Agangi A, Cavalazzi B, Wohlgemuth-Ueberwasser CC, Hofmann A, McKeegan KD (2014) Coupled Fe and S isotope variations in pyrite nodules from Archean shale. Earth Planet Sci Lett 392:67–79

Markl G, von Blanckenburg F, Wagner T (2006) Iron isotope fractionation during hydrothermal ore deposition and alteration. Geochim Cosmochim Acta 70:3011–3030

Martin JH (1990) Glacial-interglacial CO_2 change: The iron hypothesis. Paleoceanography 5:1–13

Martin JH, Fitzwater SE (1988) Iron-deficiency limits phytoplankton growth in the Northeast Pacific Subarctic. Nature 331:341–343

Martin JH, Coale KH, Johnson KS, Fitzwater SE, Gordon RM, Tanner SJ, Hunter CN, Elrod VA, Nowicki JL, Coley TL, others (1994) Testing the iron hypothesis in ecosystems of the equatorial Pacific Ocean. Nature 371:123–129

Martinez-Garcia A, Rosell-Mele A, Jaccard SL, Geibert W, Sigman DM, Haug GH (2011) Southern Ocean dust—climate coupling over the past four million years. Nature 476:312–315

Matsuhisa Y, Goldsmith JR, Clayton RN (1978) Mechanisms of hydrothermal crystallization of quartz at 250 °C and 15 kbar. Geochim Cosmochim Acta 42:173–182

Matthews A, Zhu XK, O'Nions, K (2001) Kinetic iron stable isotope fractionation between iron (-II) and (-III) complexes in solution. Earth Planet Sci Lett 192:81–92

Matthews A, Morgans-Bell HS, Emmanuel S, Jenkyns HC, Erel Y, Halicz L (2004) Controls on iron-isotope fractionation in organic-rich sediments (Kimmeridge Clay, Upper Jurassic, southern England). Geochim Cosmochim Acta 68:3107–3123

McCracken G, Love H (1960) Diffusion of lithium through tungsten. Phys Rev Lett 5:201

Mead C, Herckes P, Majestic BJ, Anbar AD (2013) Source apportionment of aerosol iron in the marine environment using iron isotope analysis. Geophys Res Lett 40:5722–5727

Millet M-A, Dauphas N (2014) Ultra-precise titanium stable isotope measurements by double-spike high resolution MC-ICP-MS J Anal At Spectrom 29:1444–1458

Millet M-A, Baker JA, Payne CE (2012) Ultra-precise stable Fe isotope measurements by high resolution multiple-collector inductively coupled plasma mass spectrometry with a ^{57}Fe–^{58}Fe double-spike. Chem Geol 304:18–25

Millet M-A, Dauphas N, Greber ND, Burton KW, Dale CW, Debret B, Pacpherson CG, Nowell GM, Williams HM (2016) Titanium stable isotope investigation of magmatic processes on the Earth and Moon. Earth Planet Sci Lett 449:197–205

Mills MM, Ridame C, Davey M, La Roche J, Geider RJ (2004) Iron and phosphorus co-limit nitrogen fixation in the eastern tropical North Atlantic. Nature 429:292–294

Mishra RK, Chaussidon M (2014) Fossil records of high level of ^{60}Fe in chondrules from unequilibrated chondrites. Earth Planet Sci Lett 398:90–100

Mishra R, Goswami J (2014) Fe–Ni and Al–Mg isotope records in UOC chondrules: Plausible stellar source of ^{60}Fe and other short-lived nuclides in the early Solar System. Geochim Cosmochim Acta 132:440–457

Mishra RK, Marhas KK (2016) Abundance of ^{60}Fe inferred from nanoSIMS study of QUE 97008 (L3. 05) chondrules. Earth Planet Sci Lett 436:71–81

Moeller K, Schoenberg R, Grenne T, Thorseth IH, Drost K, Pedersen RB (2014) Comparison of iron isotope variations in modern and Ordovician siliceous Fe oxyhydroxide deposits. Geochim Cosmochim Acta 126:422–440

Mojzsis SJ, Coath CD, Greenwood JP, McKeegan KD, Harrison TM (2003) Mass-independent isotope effects in Archean (2.5 to 3.8 Ga) sedimentary sulfides determined by ion microprobe analysis. Geochim Cosmochim Acta 67:1635–1658

Moore JK, Braucher, O (2008) Sedimentary and mineral dust sources of dissolved iron to the world ocean. Biogeosciences 5:631–656

Moore JK, Doney SC, Glover DM, Fung IY (2002) Iron cycling and nutrient-limitation patterns in surface waters of the World Ocean. Deep Sea Res Part II 49:463–507

Moynier F, Fujii T, Wang K, Foriel J (2013) Ab initio calculations of the Fe (II) and Fe (III) isotopic effects in citrates, nicotianamine, phytosiderophore, new Fe isotopic measurements in higher plants. C R Geosci 345:230–240

Mullane E, Russell S, Gounelle M (2005) Nebular and asteroidal modification of the iron isotope composition of chondritic components. Earth Planet Sci Lett 239:203–218

Mullen JG (1961) Isotope effect in intermetallic diffusion. Phys Rev 121:1649

Müller W, Shelley M, Miller P, Broude S (2009) Initial performance metrics of a new custom-designed ArF excimer LA-ICPMS system coupled to a two-volume laser-ablation cell. J Anal At Spectrom 24:209–214

Murray RW, Leinen M, Knowlton CW (2012) Links between iron input and opal deposition in the Pleistocene equatorial Pacific Ocean. Nature Geosci 5:270–274

Nebel O, Arculus R, Sossi P, Jenner F, Whan T (2013) Iron isotopic evidence for convective resurfacing of recycled arc-front mantle beneath back-arc basins. Geophys Res Lett 40:5849–5853

Nebel O, Campbell IH, Sossi PA, Van Kranendonk MJ (2014) Hafnium and iron isotopes in early Archean komatiites record a plume-driven convection cycle in the Hadean Earth. Earth Planet Sci Lett 397:111–120

Nebel O, Sossi P, Bénard A, Wille M, Vroon P, Arculus R (2015) Redox-variability and controls in subduction zones from an iron-isotope perspective. Earth Planet Sci Lett 432:142–151

Needham A, Porcelli D, Russell S (2009) An Fe isotope study of ordinary chondrites. Geochim Cosmochim Acta 73:7399–7413

Nie N, Dauphas N (2015) Iron isotope constraints on the photo-oxidation pathway to BIF formation. Lunar Planet Sci Conf 46: #2635

Nie N, Dauphas N, Greenwood R (2016) Iron and oxygen isotope fractionation during photo-oxidation. Oxygen and iron isotope fractionation during iron UV photo-oxidation: Implications for early Earth and Mars. Earth Planet Sci Lett, in press

Nielsen SG, Gannoun A, Marnham C, Burton KW, Halliday AN, Hein JR (2011) New age for ferromanganese crust 109D-C and implications for isotopic records of lead, neodymium, hafnium, thallium in the Pliocene Indian Ocean. Paleoceanography 26:PA2213

Nishizawa M, Yamamoto H, Ueno Y, Tsuruoka S, Shibuya T, Sawaki Y, Yamamoto S, Kon Y, Kitajima K, Komiya T (2010) Grain-scale iron isotopic distribution of pyrite from Precambrian shallow marine carbonate revealed by a femtosecond laser ablation multicollector ICP-MS technique: possible proxy for the redox state of ancient seawater. Geochim Cosmochim Acta 74:2760–2778

O'Connor C, Sharp BL, Evans P (2006) On-line additions of aqueous standards for calibration of laser ablation inductively coupled plasma mass spectrometry: theory and comparison of wet and dry plasma conditions. J Anal At Spectrom 21:556–565

Oeser M, Weyer S, Horn I, Schuth S (2014) High-precision Fe and Mg Isotope ratios of silicate reference glasses determined in situ by femtosecond LA-MC-ICP-MS and by solution nebulisation MC-ICP-MS Geostand Geoanal Res 38:311–328

Oeser M, Dohmen R, Horn I, Schuth S, Weyer S (2015) Processes and time scales of magmatic evolution as revealed by Fe–Mg chemical and isotopic zoning in natural olivines. Geochim Cosmochim Acta 154:130–150

Ono S, Eigenbrode JL, Pavlov AA, Kharecha P, Rumble D, Kasting JF, Freeman KH (2003) New insights into Archean sulfur cycle from mass-independent sulfur isotope records from the Hamersley Basin, Australia. Earth Planet Sci Lett 213:15–30

Othman DB, Luck J, Bodinier J, Arndt N, Albarède F (2006) Cu–Zn isotopic variations in the Earth's mantle. Geochim Cosmochim Acta 70:A46

Ottonello G, Zuccolini MV (2009) Ab-initio structure, energy and stable Fe isotope equilibrium fractionation of some geochemically relevant H–O–Fe complexes. Geochim Cosmochim Acta 73:6447–6469

Parekh P, Follows MJ, Boyle, E (2004) Modeling the global ocean iron cycle. Global Biogeochem Cycles 18, doi: 10.1029/2003GB002061

Pell E (1960) Diffusion of Li in Si at high *T* and the isotope effect. Phys Rev 119:1014

Percak-Dennett EM, Beard BL, Xu H, Konishi H, Johnson CM, Roden EE (2011) Iron isotope fractionation during microbial dissimilatory iron oxide reduction in simulated Archaean seawater. Geobiology 9:205–220

Perry EC, Tan FC, Morey GB (1973) Geology and stable isotope geochemistry of the Biwabik iron formation, Northern Minnesota. Economic Geology 68:1110–1125

Phillips GN, Law JDM (2000) Witwatersrand gold fields: geology, genesis, exploration. SEG Rev. 13:439–500

Planavsky N, Rouxel O, Bekker A, Shapiro R, Fralick P, Knudsen A (2009) Iron-oxidizing microbial ecosystems thrived in late Paleoproterozoic redox-stratified oceans. Earth Planet Sci Lett 286:230–242

Planavsky N, Rouxel OJ, Bekker A, Hofmann A, Little CTS, Lyons TW (2012) Iron isotope composition of some Archean and Proterozoic iron formations. Geochim Cosmochim Acta 80:158–169

Planavsky NJ, Asael D, Hofman A, Reinhard CT, Lalonde SV, Knudsen A, Wang X, Ossa FO, Pecoits E, Smith AJ, Beukes N (2014) Evidence for oxygenic photosynthesis half a billiion years before the Great Oxidation Event. Nat Geosci 7:283–286

Poigner H, Wilhelms-Dick D, Abele D, Staubwasser M, Henkel S (2015) Iron assimilation by the clam *Laternula elliptica*: Do stable isotopes (δ^{56}Fe) help to decipher the sources? Chemosphere 134:294–300

Poitrasson F, Freydier R (2005) Heavy iron isotope composition of granites determined by high resolution MC-ICP-MS Chem Geol 222:132–147

Poitrasson F, Halliday AN, Lee D-C, Levasseur S, Teutsch N (2004) Iron isotope differences between Earth, Moon, Mars and Vesta as possible records of contrasted accretion mechanisms. Earth Planet Sci Lett 223:253–266

Poitrasson F, Viers J, Martin F, Braun, J-J (2008) Limited iron isotope variations in recent lateritic soils from Nsimi, Cameroon: Implications for the global Fe geochemical cycle. Chem Geol 253:54–63

Poitrasson F, Roskosz M, Corgne A (2009) No iron isotope fractionation between molten alloys and silicate melt to 2000 degrees C and 7.7 GPa: Experimental evidence and implications for planetary differentiation and accretion. Earth Planet Sci Lett 278:376–385

Poitrasson F, Delpech G, Grégoire M (2013) On the iron isotope heterogeneity of lithospheric mantle xenoliths: implications for mantle metasomatism, the origin of basalts and the iron isotope composition of the Earth. Contrib Mineral Petrol 165:1243–1258

Poitrasson F, Vieira LC, Seyler P, dos Santos Pinheiro GM, Mulholland DS, Bonnet, M-P, Martinez, J-M, Lima BA, Boaventura GR, Chmeleff J, others (2014) Iron isotope composition of the bulk waters and sediments from the Amazon River Basin. Chem Geol 377:1–11

Polyakov VB (2009) Equilibrium iron isotope fractionation at core-mantle boundary conditions. Science 323:912–914

Polyakov VB, Mineev SD (2000) The use of Mössbauer spectroscopy in stable isotope geochemistry. Geochim Cosmochim Acta 64:849–865

Polyakov VB, Soultanov DM (2011) New data on equilibrium iron isotope fractionation among sulfides: Constraints on mechanisms of sulfide formation in hydrothermal and igneous systems. Geochim Cosmochim Acta 75:1957–1974

Polyakov V, Clayton R, Horita J, Mineev S (2007) Equilibrium iron isotope fractionation factors of minerals: reevaluation from the data of nuclear inelastic resonant X-ray scattering and Mössbauer spectroscopy. Geochim Cosmochim Acta 71:3833–3846

Poulton SW, Fralick PW, Canfield DE (2004) The transition to a sulfidic ocean ~1.84 billion years ago. Nature 431:173–177

Poulton SW, Fralick PW, Canfield DE (2010) Spatial variability in oceanic redox structure 1.8 billion years ago. Nat Geosci 3:486–490

Quitté G, Markowski A, Latkoczy C, Gabriel A, Pack A (2010) Iron–60 Heterogeneity and Incomplete Isotope Mixing in the Early Solar System. Astrophys J 720:1215–1224

Radic A, Lacan F, Murray JW (2011) Iron isotopes in the seawater of the equatorial Pacific Ocean: New constraints for the oceanic iron cycle. Earth Planet Sci Lett 306:1–10

Raiswell R, Canfield DE (1998) Sources of iron for pyrite formation in marine sediments. Am J Sci 298:219–245

Raiswell R, Tranter M, Benning LG, Siegert M, De'ath R, Huybrechts P, Payne T (2006) Contributions from glacially derived sediment to the global iron (oxyhydr)oxide cycle: Implications for iron delivery to the oceans. Geochim Cosmochim Acta 70:2765–2780

Rasmussen B, Meier DB, Krapez B, Muhling JR (2013) Iron silicate microgranules as precursor sediments to 2.5-billion-year-old banded iron formations. Geology 41:435–438

Rasmussen B, Krapez B, Muhling JR, Suvorova A (2015) Precipitation of iron silicate nanoparticles in early Precambrian oceans marks Earth's first iron age. Geology 43:303–306

Reddy TR, Frierdich AJ, Beard BL, Johnson CM (2015) The effect of pH on stable iron isotope exchange and fractionation between aqueous Fe(II) and goethite. Chem Geol 397:118–127

Reid RT, Livet DH, Faulkner DJ, Butler, A (1993) A siderophore from a marine bacterium with an exceptional ferric ion affinity constant. Nature 366:455–458

Resing JA, Sedwick PN, German CR, Jenkins WJ, Moffett JW, Sohst BM, Tagliabue, A (2015) Basin-scale transport of hydrothermal dissolved metals across the South Pacific Ocean. Nature 523:200–203

Revels BN, Ohnemus DC, Lam PJ, Conway TM, John SG (2014) The isotope signature and distribution of particulate iron in the north Atlantic ocean. Deep Sea Research Part II: 116:321-31

Revels BN, Zhang R, Adkins JF, John SG (2015) Fractionation of iron isotopes during leaching of natural particles by acidic and circumneutral leaches and development of an optimal leach for marine particulate iron isotopes. Geochim Cosmochim Acta 166:92–104

Richter FM (2004) Timescales determining the degree of kinetic isotope fractionation by evaporation and condensation. Geochim Cosmochim Acta 68:4971–4992

Richter FM, Davis AM, Ebel DS, Hashimoto A (2002) Elemental and isotopic fractionation of Type B calcium-, aluminum-rich inclusions: experiments, theoretical considerations, constraints on their thermal evolution. Geochim Cosmochim Acta 66:521–540

Richter FM, Janney PE, Mendybaev RA, Davis AM, Wadhwa M (2007) Elemental and isotopic fractionation of Type B CAI-like liquids by evaporation. Geochim Cosmochim Acta 71:5544–5564

Richter FM, Watson EB, Mendybaev RA, Teng F-Z, Janney PE (2008) Magnesium isotope fractionation in silicate melts by chemical and thermal diffusion. Geochim Cosmochim Acta 72:206–220

Richter FM, Dauphas N, Teng F-Z (2009a) Non-traditional fractionation of non-traditional isotopes: evaporation, chemical diffusion and Soret diffusion. Chem Geol 258:92–103

Richter FM, Watson EB, Mendybaev R, Dauphas N, Georg B, Watkins J, Valley J (2009b) Isotopic fractionation of the major elements of molten basalt by chemical and thermal diffusion. Geochim Cosmochim Acta 73:4250–4263

Richter FM, Huss GR, Mendybaev RA (2014a) Iron and nickel isotopic fractionation across metal grains from three CBb meteorites. Lunar Planet Sci Conf 45:1346

Richter FM, Watson EB, Chaussidon M, Mendybaev R, Christensen JN, Qiu L (2014b) Isotope fractionation of Li and K in silicate liquids by Soret diffusion. Geochim Cosmochim Acta 138:136–145

Robb LJ, Meyer FM (1990). The nature of the Witwatersrand hinterland; conjectures on the source area problem. Econ Geol 85:511–536

Roden E (2004) Analysis of long-term bacterial vs. chemical Fe(III) oxide reduction kinetics. Geochim Cosmochim Acta 68:3205–3216

Rodushkin I, Stenberg A, Andrén H, Malinovsky D, Baxter DC (2004) Isotopic fractionation during diffusion of transition metal ions in solution. Anal CHem 76:2148–2151

Roskosz M, Luais B, Watson HC, Toplis MJ, Alexander CMD, Mysen BO (2006) Experimental quantification of the fractionation of Fe isotopes during metal segregation from a silicate melt. Earth Planet Sci Lett 248:851–867

Roskosz M, Sio CK, Dauphas N, Bi W, Tissot FL, Hu MY, Zhao J, Alp EE (2015) Spinel–olivine–pyroxene equilibrium iron isotopic fractionation and applications to natural peridotites. Geochim Cosmochim Acta 169:184–199

Rouxel OJ, Auro M (2010) Iron isotope variations in coastal seawater determined by multicollector ICP-MS Geostand Geoanal Res 34:135–144

Rouxel O, Dobbek N, Ludden J, Fouquet Y (2003) Iron isotope fractionation during oceanic crust alteration. Chem Geol 202:155–182

Rouxel O, Fouquet Y, Ludden JN (2004) Subsurface processes at the Lucky Strike hydrothermal field, Mid-Atlantic Ridge: Evidence from sulfur, selenium, iron isotopes. Geochim Cosmochim Acta 68:2295–2311

Rouxel OJ, Bekker A, Edwards KJ (2005) Iron isotope constraints on the Archean and Paleoproterozoic ocean redox state. Science 307:1088–1091

Rouxel OJ, Bekker A, Edward KJ (2006) Response to comment on "Iron isotope constraints on the archean and paleoproterozoic ocean redox state". Science 311, doi:Doi 10.1126/Science.1118420

Rouxel O, Sholkovitz E, Charette M, Edwards KJ (2008a) Iron isotope fractionation in subterranean estuaries. Geochim Cosmochim Acta 72:3413–3430

Rouxel O, Shanks WC, Bach W, Edwards KJ (2008b) Integrated Fe- and S-isotope study of seafloor hydrothermal vents at East Pacific Rise 9–10°N Chem Geol 252:214–227

Rouxel O, Toner B, Manganini S, German C (2016) Geochemistry and iron isotope systematics of hydrothermal plume fall-out at EPR9°50'N Chem. Geol. 441:212–234

Roy M, Martin JB, Cherrier J, Cable JE, Smith CG (2010) Influence of sea level rise on iron diagenesis in an east Florida subterranean estuary. Geochim Cosmochim Acta 74:5560–5573

Roy M, Rouxel O, Martin JB, Cable JE (2012) Iron isotope fractionation in a sulfide-bearing subterranean estuary and its potential influence on oceanic Fe isotope flux. Chem Geol, 300–301:133–142

Rubie DC, Frost DJ, Mann U, Asahara Y, Nimmo F, Tsuno K, Kegler P, Holzheid A, Palme H (2011) Heterogeneous accretion, composition and core–mantle differentiation of the Earth. Earth Planet Sci Lett 301:31–42

Rudge JF, Reynolds BC, Bourdon B (2009) The double-spike toolbox. Chem Geol 265:420–431

Rustad JR, Dixon DA (2009) Prediction of iron-isotope fractionation between hematite (α–Fe_2O_3) and ferric and ferrous iron in aqueous solution from density functional theory. J Phys Chem A 113:12249–12255

Rustad JR, Yin Q-Z (2009) Iron isotope fractionation in the Earth's lower mantle. Nat Geosci 2:514–518

Saito, MA, Noble AE, Tagliabue A, Goepfert TJ, Lamborg CH, Jenkins WJ (2013) Slow-spreading submarine ridges in the South Atlantic as a significant oceanic iron source. Nat Geosci 6:775–779

Saunier G, Pokrovski GS, Poitrasson F (2011) First experimental determination of iron isotope fractionation between hematite and aqueous solution at hydrothermal conditions. Geochim Cosmochim Acta 75:6629–6654

Schauble EA (2004) Applying stable isotope fractionation theory to new systems. Rev Mineral Geochem 55:65–111

Schauble E, Rossman G, Taylor H (2001) Theoretical estimates of equilibrium Fe-isotope fractionations from vibrational spectroscopy. Geochim Cosmochim Acta 65:2487–2497

Schoenberg R, von Blanckenburg F (2005) An assessment of the accuracy of stable Fe isotope ratio measurements on samples with organic and inorganic matrices by high-resolution multicollector ICP-MS Inter J Mass Spectrom 242:257–272

Schoenberg R, von Blanckenburg F (2006) Modes of planetary-scale Fe isotope fractionation. Earth Planet Sci Lett 252:342–359

Scholz F, Severmann S, McManus J, Hensen C (2014) Beyond the Black Sea paradigm: The sedimentary fingerprint of an open-marine iron shuttle. Geochim Cosmochim Acta 127:368–380

Schuessler JA, Schoenberg R, Behrens H, von Blanckenburg F (2007) The experimental calibration of the iron isotope fractionation factor between pyrrhotite and peralkaline rhyolitic melt. Geochim Cosmochim Acta 71:417–433

Schuessler JA, Schoenberg R, Sigmarsson O (2009) Iron and lithium isotope systematics of the Hekla volcano, Iceland—evidence for Fe isotope fractionation during magma differentiation. Chem Geol 258:78–91

Schuler D, Baeuerlein E (1996) Iron-limited growth and kinetics of iron uptake in Magnetospirilum gryphiswaldense. Arch Microbiol 166:301–307

Schuth S, Hurrass J, Muenker C, Mansfeldt, T (2015) Redox-dependent fractionation of iron isotopes in suspensions of a groundwater-influenced soil. Chem Geol 392:74–86

Scott C, Lyons TW, Bekker A, Shen Y, Poulton SW, Chu X, Anbar AD (2008) Tracing the stepwise oxygenation of the Proterozoic ocean. Nature 452:456–459

Sedaghatpour F, Teng F-Z, Liu Y, Sears DW, Taylor LA (2013) Magnesium isotopic composition of the Moon. Geochim Cosmochim Acta 120:1–16

Sedwick PN, Church TM, Bowie AR, Marsay CM, Ussher SJ, Achilles KM, Lethaby PJ, Johnson RJ, Sarin MM, McGillicuddy DJ (2005) Iron in the Sargasso Sea (Bermuda Atlantic Time-series Study region) during summer: Eolian imprint, spatiotemporal variability, ecological implications. Global Biogeochem Cycles 19, doi: 10.1029/2004GB002445

Severmann S, Johnson CM, Beard BL, German CR, Edmonds HN, Chiba H, Green, DRH (2004) The effect of plume processes on the Fe isotope composition of hydrothermally derived Fe in the deep ocean as inferred from the Rainbow vent site, Mid-Atlantic Ridge, 36°14' N Earth Planet Sci Lett 225:63–76

Severmann S, Johnson CM, Beard BL, McManus, J (2006) The effect of early diagenesis on the Fe isotope compositions of porewaters and authigenic minerals in continental margin sediments. Geochim Cosmochim Acta 70:2006–2022

Severmann S, Lyons TW, Anbar A, McManus J, Gordon G (2008) Modern iron isotope perspective on the benthic iron shuttle and the redox evolution of ancient oceans. Geology 36:487–490

Severmann S, McManus J, Berelson WM, Hammond DE (2010) The continental shelf benthic iron flux and its isotope composition. Geochim Cosmochim Acta 74:3984–4004

Shahar A, Young ED, Manning CE (2008) Equilibrium high-temperature Fe isotope fractionation between fayalite and magnetite: An experimental calibration. Earth Planet Sci Lett 268:330–338

Shahar A, Hillgren V, Horan M, Mesa-Garcia J, Kaufman L, Mock T (2015) Sulfur-controlled iron isotope fractionation experiments of core formation in planetary bodies. Geochim Cosmochim Acta 150:253–264

Shahar A, Schauble EA, Caracas R, Gleason AE, Reagan MM, Xiao Y, Shu J, Mao W (2016) Pressure-dependent isotopic composition of iron alloys. Science 352:580–582

Sharma M, Polizzotto M, Anbar AD (2001) Iron isotopes in hot springs along the Juan de Fuca Ridge. Earth Planet Sci Lett 194:39–51

Siebert C, Nagler TF, von Blanckenburg F, Kramers JD (2003) Molybdenum isotope records as a potential new proxy for paleoceanography. Earth Planet Sci Lett 211:159–171

Siebert J, Badro J, Antonangeli D, Ryerson FJ (2013) Terrestrial accretion under oxidizing conditions. Science 339:1194–1197

Simmons SL, Sievert SM, Frankel RB, Bazylinski DA, Edwards KJ (2004) Spatiotemporal distribution of marine magnetotactic bacteria in a seasonally stratified coastal salt pond. Appl Environ Microbiol 70:6230–6239

Sio CKI, Dauphas N (2016) Thermal histories of magmatic bodies by Monte Carlo inversion of Mg—Fe isotopic profiles in olivine. Geology, in press.

Sio CKI, Dauphas N, Teng F-Z, Chaussidon M, Helz RT, Roskosz M (2013) Discerning crystal growth from diffusion profiles in zoned olivine by in situ Mg–Fe isotopic analyses. Geochim Cosmochim Acta 123:302–321

Sivan O, Adler M, Pearson A, Gelman F, Bar-Or I, John SG, Eckert, W (2011) Geochemical evidence for iron-mediated anaerobic oxidation of methane. Limnol Oceanogr 56:1536–1544

Skulan JL, Beard BL, Johnson CM (2002) Kinetic and equilibrium Fe isotope fractionation between aqueous Fe (III) and hematite. Geochim Cosmochim Acta 66:2995–3015

Snow JE, Dick HJB (1995) Pervasive magnesium loss by marine weathering of peridotite. Geochim Cosmochim Acta 59:4219–4235

Sobolev AV, Hofmann AW, Kuzmin DV, Yaxley GM, Arndt NT, Chung S-L, Danyushevsky LV, Elliott T, Frey FA, Garcia MO (2007) The amount of recycled crust in sources of mantle-derived melts. Science 316:412–417

Song L, Liu C, Wang Z, Zhu X, Teng Y, Wang J, Tang, S Li J, LIANG, L (2011) Iron isotope compositions of natural river and lake samples in the karst area, Guizhou Province, Southwest China. Acta Geologica Sinica - English Edition 85:712–722

Sossi PA, Foden JD, Halverson GP (2012) Redox-controlled iron isotope fractionation during magmatic differentiation: an example from the Red Hill intrusion, S Tasmania. Contrib Mineral Petrol 164:757–772

Sossi PA, Halverson GP, Nebel O, Eggins SM (2015) Combined separation of Cu, Fe and Zn from rock matrices and improved analytical protocols for stable isotope determination. Geostand Geoanal Res 39:129–149

Sossi PA, O'Neill HSC (2016) The effect of bonding environment on iron isotope fractionation between minerals at high temperature. Geochim Cosmochim Acta, in press

Sossi PA, Nebel O, Anand M, Poitrasson F (2016) On the iron isotope composition of Mars and volatile depletion in the terrestrial planets. Earth Planet Sci Lett 449:360–371

Spring S, Amann R, Ludwig W, Schleifer K, Petersen N (1993) Dominating role of an unusual magnetotactic bacteria. System Appl Microbiol 15:116–122

Staudigel H, Hart SR (1983) Alteration of basaltic glasses: Mechanisms and significance for the oceanic crust-seawater budget. Geochim Cosmochim Acta 47:337–350

Steele RC, Elliott T, Coath CD, Regelous M (2011) Confirmation of mass-independent Ni isotopic variability in iron meteorites. Geochim Cosmochim Acta 75:7906–7925

Steinhoefel G, Horn I, von Blanckenburg F (2009a) Matrix-independent Fe isotope ratio determination in silicates using UV femtosecond laser ablation. Chem Geol 268:67–73

Steinhoefel G, Horn I, von Blanckenburg F (2009b) Micro-scale tracing of Fe and Si isotope signatures in banded iron formation using femtosecond laser ablation. Geochim Cosmochim Acta 73:5343–5360

Steinhoefel G, von Blanckenburg F, Horn I, Konhauser KO, Beukes NJ, Gutzmer J (2010) Deciphering formation processes of banded iron formations from the Transvaal and the Hamersley successions by combined Si and Fe isotope analysis using UV femtosecond laser ablation. Geochim Cosmochim Acta 74:2677–2696

Strelow F (1980) Improved separation of iron from copper and other elements by anion-exchange chromatography on a 4% cross-linked resin with high concentrations of hydrochloric acid. Talanta 27:727–732

Sturhahn W, Jackson JM (2007) Geophysical applications of nuclear resonant spectroscopy. Geol Soc Am Spec Pap 421:157–174

Sumner DY (1997) Carbonate precipitation and oxygen stratification in Late Archean seawater as deduced from facies and stratigraphy of the Gamohaan and Frisco formations, Transvaal Supergroup, South Africa. Am J Sci 297:455–487

Swanner ED, Wu WF, Schoenberg R, Byrne J, Michel FM, Pan YX, Kappler A (2015) Fractionation of Fe isotopes during Fe(II) oxidation by a marine photoferrotroph is controlled by the formation of organic Fe-complexes and colloidal Fe fractions. Geochim Cosmochim Acta 165:44–61

Syverson DD, Borrok DM, Seyfried WE (2013) Experimental determination of equilibrium Fe isotopic fractionation between pyrite and dissolved Fe under hydrothermal conditions. Geochim Cosmochim Acta 122:170–183

Syverson DD, Pester NJ, Craddock PR, Seyfried WE (2014) Fe isotope fractionation during phase separation in the NaCl–H_2O system: An experimental study with implications for seafloor hydrothermal vents. Earth Planet Sci Lett 406:223–232

Tagliabue A, Bopp L, Dutay JC, Bowie AR, Chever F, Jean-Baptiste P, Bucciarelli E, Lannuzel D, Remenyi T, Sarthou G, Aumont O (2010) Hydrothermal contribution to the oceanic dissolved iron inventory. Nat Geosci 3:252–256

Tagliabue A, Aumont O, Bopp, L (2014) The impact of different external sources of iron on the global carbon cycle. Geophys Res Lett 41:920–926

Tahata M, Sawaki Y, Yoshiya K, Nishizawa M, Komiya T, Hirata T, Yoshida N, Maruyama S, Windley BF (2015) The marine environments encompassing the Neoproterozoic glaciations: Evidence from C, Sr and Fe isotope ratios in the Hecla Hoek Supergroup in Svalbard. Precambrian Res 263:19–42

Tang H, Dauphas N (2012) Abundance, distribution, origin of ^{60}Fe in the solar protoplanetary disk. Earth Planet Sci Lett 359:248–263

Tang H, Dauphas N (2014) ^{60}Fe–^{60}Ni chronology of core formation on Mars. Earth Planet Sci Lett 390:264–274

Tang H, Dauphas N (2015) Low ^{60}Fe abundance in Semarkona and Sahara 99555. Astrophys J 802:22

Tangalos GE, Beard BL, Johnson CM, Alpers CN, Shelobolina ES, Xu H, Konishi H, Roden EE (2010) Microbial production of isotopically light iron(II) in a modern chemically precipitated sediment and implications for isotopic variations in ancient rocks. Geobiology 8:197–208

Telus M, Dauphas N, Moynier F, Tissot FL, Teng F-Z, Nabelek PI, Craddock PR, Groat LA (2012) Iron, zinc, magnesium and uranium isotopic fractionation during continental crust differentiation: The tale from migmatites, granitoids, pegmatites. Geochim Cosmochim Acta 97:247–265

Teng FZ, Dauphas N, Helz RT (2008) Iron isotope fractionation during magmatic differentiation in Kilauea Iki Lava Lake. Science 320:1620–1622

Teng F-Z, Dauphas N, Helz RT, Gao S, Huang S (2011) Diffusion-driven magnesium and iron isotope fractionation in Hawaiian olivine. Earth Planet Sci Lett 308:317–324

Teng F-Z, Dauphas N, Huang S, Marty B (2013) Iron isotopic systematics of oceanic basalts. Geochim Cosmochim Acta 107:12–26

Toner BM, Fakra SC, Manganini SJ, Santelli CM, Marcus MA, Moffett J, Rouxel O, German CR, Edwards KJ (2009) Preservation of iron(II) by carbon-rich matrices in a hydrothermal plume. Nat Geosci 2:197–201

Toner BM, Rouxel OJ, Santelli CM, Bach W, Edwards KJ (2016) Iron transformation pathways and redox micro-environments in seafloor sulfide-mineral deposits: spatially resolved Fe XAS and $\delta^{57/54}$Fe observations. Front Microbiol 7:648

Toplis M (2005) The thermodynamics of iron and magnesium partitioning between olivine and liquid: criteria for assessing and predicting equilibrium in natural and experimental systems. Contrib Mineral Petrol 149:22–39

Trappitsch R, Stephan T, Davis AM, Pellin MJ, Rost D, Savina MR, Kelly CH, Dauphas N (2016) Simultansous analysis of iron and nickel isotopes in presolar SiC grains with CHILI Lunar Planet Sci Conf 47:#3025

Tsikos H, Matthews A, Erel Y, Moore JM (2010) Iron isotopes constrain biogeochemical redox cycling of iron and manganese in a Palaeoproterozoic stratified basin. Earth Planet Sci Lett 298:125–134

Urey HC (1947) The thermodynamic properties of isotopic substances. J Chem Soc (Resumed):562–581

Urey HC, Craig H (1953) The composition of the stone meteorites and the origin of the meteorites. Geochim Cosmochim Acta 4:36–82

Van Heghe L, Engstrom E, Rodushkin I, Cloquet C, Vanhaecke F (2012) Isotopic analysis of the metabolically relevant transition metals Cu, Fe and Zn in human blood from vegetarians and omnivores using multi-collector ICP-mass spectrometry. J Anal At Spectrom 27:1327–1334

Van Heghe L, Delanghe J, Van Vlierberghe H, Vanhaecke F (2013) The relationship between the iron isotopic composition of human whole blood and iron status parameters. Metallomics 5:1503–1509

Van Heghe L, Deltombe O, Delanghe J, Depypere H, Vanhaecke F (2014) The influence of menstrual blood loss and age on the isotopic composition of Cu, Fe and Zn in human whole blood. J Anal At Spectrom 29:478–482

Vanhaecke F, Costas-Rodriguez M (2015) What's up doc?—High-precision isotopic analysis of essential metals in biofluids for medical diagnosis. Spectroscopy Europe 27:11–14

Vargas M, Kashefi K, BluntHarris EL, Lovley DR (1998) Microbiological evidence for Fe(III) reduction on early Earth. Nature 395:65–67

Völkening J, Papanastassiou D (1989) Iron isotope anomalies. Astrophys J 347:L43-L46

von Blanckenburg F, Marnberti M, Schoenberg R, Kamber BS, Webb GE (2008) The iron isotope composition of microbial carbonate. Chem Geol 249:113–128

von Blanckenburg F, von Wiren N, Guelke M, Weiss DJ, Bullen TD (2009) Fractionation of metal stable isotopes by higher plants. Elements 5:375–380

von Blanckenburg F, Noordmann J, Guelke-Stelling M (2013) The iron stable isotope fingerprint of the human diet. J Ag Food Chem 61:11893–11899

von Blanckenburg F, Oelze M, Schmid DG, van Zuilen K, Gschwind HP, Slade AJ, Stitah S, Kaufmann D, Swart P (2014) An iron stable isotope comparison between human erythrocytes and plasma. Metallomics 6:2052–2061

Von Damm KL (1988) Systematics of and postulated controls on submarine hydrothermal solution chemistry. J Geophys Res-Solid Earth and Planets 93:4551–4561

Von Damm KL, Oosting SE, Kozlowski R, Buttermore LG, Colodner DC, Edmonds HN, Edmond JM, Grebmeier JM (1995) Evolution of East Pacific Rise hydrothermal vent fluids following a volcanic eruption. Nature 375:47–50

Waeles M, Baker AR, Jickells T, Hoogewerff, J (2007) Global dust teleconnections: aerosol iron solubility and stable isotope composition. Environ Chem 4:233–237

Walczyk T, vonBlanckenburg F (2002) Natural iron isotope variations in human blood. Science 295:2065–2066

Walczyk T, von Blanckenburg F (2005) Deciphering the iron isotope message of the human body. Inter J Mass Spectrom 242:117–134

Wang J, Davis A, Clayton R, Mayeda T (1994) Kinetic isotopic fractionation during the evaporation of the iron oxide from liquid state. Lunar Planet Sci Conf 25:1459

Wang K, Moynier F, Dauphas N, Barrat J-A, Craddock P, Sio CK (2012) Iron isotope fractionation in planetary crusts. Geochim Cosmochim Acta 89:31–45

Watkins JM, DePaolo DJ, Ryerson FJ, Peterson BT (2011) Influence of liquid structure on diffusive isotope separation in molten silicates and aqueous solutions. Geochim Cosmochim Acta 75:3103–3118

Watson EB, Müller T (2009) Non-equilibrium isotopic and elemental fractionation during diffusion-controlled crystal growth under static and dynamic conditions. Chem Geol 267:111–124

Wawryk CM, Foden JD (2015) Fe-isotope fractionation in magmatic-hydrothermal mineral deposits: a case study from the Renison Sn–W deposit, Tasmania. Geochim Cosmochim Acta 150:285–298

Weis F, Troll VR, Jonsson E, Hogdahl K, Barker A, Harris C, Millet M-A, Nilsson KP (2013) Iron and oxygen isotope systematics of apatite-iron oxide ores in central Sweden. *In:* E Jonsson (Ed.) Mineral Deposit Research for a High-Tech World pp. 1675–1678

Welch SA, Beard BL, Johnson CM, Braterman PS (2003) Kinetic and equilibrium Fe isotope fractionation between aqueous Fe(II) and Fe(III). Geochim Cosmochim Acta 67:4231–4250

Weyer S, Ionov DA (2007) Partial melting and melt percolation in the mantle: The message from Fe isotopes. Earth Planet Sci Lett 259:119–133

Weyer S, Schwieters J (2003) High precision Fe isotope measurements with high mass resolution MC-ICPMS Inter J Mass Spectrom 226:355–368

Weyer S, Anbar AD, Brey GP, Münker C, Mezger K, Woodland AB (2005) Iron isotope fractionation during planetary differentiation. Earth Planet Sci Lett 240:251–264

Wheat CG, Mottl MJ (2004) Geochemical fluxes through mid-ocean ridge flanks. *In:* Hydrogeology of the Oceanic Lithosphere. Davis EE, Elderfield H, (eds). Cambridge University Press, Cambridge

Whitehouse MJ, Fedo CM (2007) Microscale heterogeneity of Fe isotopes in >3.71 Ga banded iron formation from the Isua Greenstone Belt, southwest Greenland. Geology 35:719–722

Widdel F, Schnell S, Heising S, Ehrenreich A, Assmus B, Schink B (1993) Ferrous iron oxidation by anoxygenic phototrophic bacteria. Nature 362:834–836, doi:10.1038/362834a0

Wiederhold JG, Kraemer SM, Teutsch N, Borer PM, Halliday A, Kretzschmar, R (2006) Iron isotope fractionation during proton-promoted, ligand-controlled, reductive dissolution of goethite. Environ Sci Technol 40:3787–3793

Wiederhold JG, Teutsch N, Kraemer SM, Halliday A, Kretzschmar, R (2007a) Iron isotope fractionation in oxic soils by mineral weathering and podzolization. Geochim Cosmochim Acta 71:5821–5833

Wiederhold JG, Teutsch N, Kraemer SM, Halliday A, Kretzschmar, R (2007b) Iron isotope fractionation during pedogenesis in redoximorphic soils. Soil Sci Soc Am J 71:1840–1850

Wiesli RA, Beard BL, Taylor LA, Johnson CM (2003) Space weathering processes on airless bodies: Fe isotope fractionation in the lunar regolith. Earth Planet Sci Lett 216:457–465

Wiesli RA, Beard BL, Johnson CM (2004) Experimental determination of Fe isotope fractionation between aqueous Fe (II), siderite and "green rust" in abiotic systems. Chem Geol 211:343–362

Wiesli RA, Beard BL, Braterman PS, Johnson CM, Saha SK, Sinha MP (2007) Iron isotope fractionation between liquid and vapor phases of iron pentacarbonyl. Talanta 71:90–96

Wijsman JWM, Middelburg JJ, Herman PMJ, Bottcher ME, Heip CHR (2001) Sulfur and iron speciation in surface sediments along the northwestern margin of the Black Sea. Mar Chem 74:261–278

Williams HM, Bizimis M (2014) Iron isotope tracing of mantle heterogeneity within the source regions of oceanic basalts. Earth Planet Sci Lett 404:396–407

Williams HM, McCammon CA, Peslier AH, Halliday AN, Teutsch N, Levasseur S, Burg J-P (2004) Iron isotope fractionation and the oxygen fugacity of the mantle. Science 304:1656–1659

Williams HM, Peslier AH, McCammon C, Halliday AN, Levasseur S, Teutsch N, Burg JP (2005) Systematic iron isotope variations in mantle rocks and minerals: The effects of partial melting and oxygen fugacity. Earth Planet Sci Lett 235:435–452

Williams HM, Wood BJ, Wade J, Frost DJ, Tuff J (2012) Isotopic evidence for internal oxidation of the Earth's mantle during accretion. Earth Planet Sci Lett 321:54–63

Williams KB, Krawczynski MJ, Nie NX, Dauphas N, Couvy H, Hu MY, Alp EE (2016) The role of differentiation processes in mare basalt iron isotope signatures. Lunar Planet Sci Conf 47:2779

Windom HL, Moore WS, Niencheski, LFH, Jahrike RA (2006) Submarine groundwater discharge: A large, previously unrecognized source of dissolved iron to the South Atlantic Ocean. Mar Chem 102:252–266

Wu LL, Beard BL, Roden EE, Johnson CM (2009) Influence of pH and dissolved Si on Fe isotope fractionation during dissimilatory microbial reduction of hematite. Geochim Cosmochim Acta 73:5584–5599

Wu L, Beard BL, Roden EE, Kennedy CB, Johnson CM (2010) Stable Fe isotope fractionations produced by aqueous Fe(II)–hematite surface interactions. Geochim Cosmochim Acta 74:4249–4265

Wu L, Beard BL, Roden EE, Johnson CM (2011) Stable iron isotope fractionation between aqueous Fe (II) and hydrous ferric oxide. Environ Sci Technol 45:1847–1852

Wu L, Percak-Dennett EM, Beard BL, Roden EE, Johnson CM (2012a) Stable iron isotope fractionation between aqueous Fe (II) and model Archean ocean Fe–Si coprecipitates and implications for iron isotope variations in the ancient rock record. Geochim Cosmochim Acta 84:14–28

Wu L, Druschel G, Findlay A, Beard BL, Johnson CM (2012b) Experimental determination of iron isotope fractionations among FeS_{aq}–Mackinawite at low temperatures: Implications for the rock record. Geochim Cosmochim Acta 89:46–61

Yamaguchi KE, Johnson CM, Beard BL, Ohmoto H (2005) Biogeochemical cycling of iron in the Archean-Paleoproterozoic Earth: Constraints from iron isotope variations in sedimentary rocks from the Kaapvaal and Pilbara Cratons. Chem Geol 218:135–169

Yamaguchi KE, Johnson CM, Beard BL, Beukes NJ, Gutzmer J, Ohmoto H (2007). Isotopic evidence for iron mobilization during Paleoproterozoic lateritization of the Hekpoort paleosol profile from Gaborone, Botswana. Earth Planet Sci Lett 256:577–587

Yoshiya K, Nishizawa M, Sawaki Y, Ueno Y, Komiya T, Yamada K, Yoshida N, Hirata T, Wada H, Maruyama S (2012) In situ iron isotope analyses of pyrite and organic carbon isotope ratios in the Fortescue Group: Metabolic variations of a Late Archean ecosystem. Precambrian Res 212:169–193

Young ED, Galy A, Nagahara H (2002) Kinetic and equilibrium mass-dependent isotope fractionation laws in nature and their geochemical and cosmochemical significance. Geochim Cosmochim Acta 66:1095–1104

Zambardi T, Lundstrom CC, Li X, McCurry M (2014) Fe and Si isotope variations at Cedar Butte volcano; insight into magmatic differentiation. Earth Planet Sci Lett 405:169–179

Zhang R, John SG, Zhang J, Ren, J Wu Y, Zhu Z, Liu S, Zhu X, Marsay CM, Wenger, F (2015) Transport and reaction of iron and iron stable isotopes in glacial meltwaters on Svalbard near Kongsfjorden: From rivers to estuary to ocean. Earth Planet Sci Lett 424:201–211

Zhao X, Zhang H, Zhu X, Tang S, Tang Y (2010) Iron isotope variations in spinel peridotite xenoliths from North China Craton: implications for mantle metasomatism. Contrib Mineral Petrol 160:1–14

Zhao X, Zhang H, Zhu X, Tang S, Yan B (2012) Iron isotope evidence for multistage melt–peridotite interactions in the lithospheric mantle of eastern China. Chem Geol 292:127–139

Zhu XK, O'Nions RK, Guo Y, Reynolds BC (2000) Secular variation of iron isotopes in North Atlantic Deep Water. Science 287:2000–200

Zhu D, Bao H, Liu Y (2015) Non-traditional stable isotope behaviors in immiscible silica-melts in a mafic magma chamber. Sci Rep 5:17561

Zhu B, Zhang, H-F, Zhao, X-M He, Y-S (2016) Iron isotope fractionation during skarn-type alteration: Implications for metal source in the Han-Xing iron skarn deposit. Ore Geol Rev 74:139–150

Reviews in Mineralogy & Geochemistry
Vol. 82 pp. 511-542, 2017

12

The Isotope Geochemistry of Ni

Tim Elliott

Bristol Isotope Group
School of Earth Sciences
University of Bristol
Bristol BS8 1RJ
UK

tim.elliott@bristol.ac.uk

Robert C. J. Steele

Institute for Geochemistry and Petrology
ETH Zürich
Zürich 8092
Switzerland

r.steele@uclmail.net

INTRODUCTION

Nickel is an iron-peak element with 5 stable isotopes (see Table 1) which is both cosmochemically abundant and rich in the information carried in its isotopic signature. Significantly, ^{60}Ni is the radiogenic daughter of ^{60}Fe, a short-lived nuclide ($t_{1/2} = 2.62$ Ma; Rugel et al. 2009) of a major element. ^{60}Fe has the potential to be both an important heat source and chronometer in the early solar system. ^{60}Ni abundances serve to document the prior importance ^{60}Fe and this is a topic of on-going debate (see *Extinct ^{60}Fe and radiogenic ^{60}Ni*). The four other stable Ni nuclides span a sizeable relative mass range of ~10%, including the notably neutron-rich nuclide ^{64}Ni. The relative abundances of these isotopes vary with diverse stellar formation environments and provide a valuable record of the nucleosynthetic heritage of Ni in the solar system (see *Nucleosynthetic Ni isotopic variations*). Ni occurs widely as both elemental and divalent cationic species, substituting for Fe and Mg in common silicate structures and forming Fe/Ni metal alloys. The Ni isotope chemistry of all the major planetary reservoirs and fractionations between them can thus be characterized (see *Mass-Dependent Ni isotopic Variability*). Ni is also a bio-essential element and its fractionation during low-temperature biogeochemical cycling is a topic that has attracted recent attention (see *Mass-Dependent Ni isotopic Variability*).

Notation

Much of the work into Ni has been cosmochemical, focussing on the nucleosynthetic origins of different meteoritic components. Such studies have primarily investigated mass-independent isotopic variations, both radiogenic and non-radiogenic, which require choosing a reference isotope pair for normalization. Throughout this work we use ^{58}Ni–^{61}Ni as the normalizing pair, in keeping with current practice in the field. An alternative ^{58}Ni–^{62}Ni normalization scheme has previously been used for bulk analyses (Shimamura and Lugmair 1983; Shukolyukov and Lugmair 1993a,b; Cook et al. 2006, 2008; Quitté et al. 2006, 2011; Chen et al. 2009) and one early study used ^{58}Ni–^{60}Ni (Morand and Allègre 1983). Although the large isotopic variability accessible by *in situ* analyses often makes external normalization a viable option for mass-independent measurements by secondary ionization mass-spectrometry (SIMS), some have employed internal

1529-6466/17/0082-0012$05.00 http://dx.doi.org/10.2138/rmg.2017.82.12

Table 1. Ni isotopic abundances (Gramlich et al. 1989), nuclide masses (Wang et al. 2012) and atomic weight (Wieser et al. 2013). Ni isotopic abundances also expressed as atomic ratios.

	^{58}Ni	^{60}Ni	^{61}Ni	^{62}Ni	^{64}Ni
Atomic Fraction	0.68076886	0.26223146	0.01139894	0.03634528	0.00925546
2 SE	5.92×10^{-5}	5.14×10^{-5}	4.33×10^{-5}	1.14×10^{-5}	5.99×10^{-5}
Nuclide Mass (amu)	57.93534241	59.93078589	60.93105557	61.92834537	63.92796682
IUPAC Atomic weight (amu)			58.6934 ± 0.0004		
		^{60}Ni/^{58}Ni	^{61}Ni/^{58}Ni	^{62}Ni/^{58}Ni	^{64}Ni/^{58}Ni
Atomic Ratio		0.385198965	0.016744215	0.053388576	0.013595598
2 SE		8.27×10^{-5}	6.52×10^{-6}	1.74×10^{-5}	8.88×10^{-6}

normalization in determinations of ^{60}Ni/^{61}Ni (Tachibana and Huss 2003; Tachibana et al. 2006; Mishra and Chaussidon 2014; Mishra et al. 2016). Given unresolvable Fe and Zn interferences on masses 58 and 64, this requires normalizing to ^{62}Ni/^{61}Ni. In this review, all data have been renormalized to ^{58}Ni–^{61}Ni, where possible. Some studies have only reported normalized data and so such conversion cannot be made. Fortunately, the subtle differences resulting from different normalizations do not affect the inferences being made in these cases and we simply indicate the normalization scheme used. To be clear about these potentially important details, we use a notation proposed by Steele et al. (2011), which includes this information. For example:

$$\varepsilon^{60/58}\mathrm{Ni}_{58/61} = \left({}^{60}\mathrm{Ni}/{}^{58}\mathrm{Ni}^{\mathrm{sample}}_{\mathrm{norm}\ 58/61} / {}^{60}\mathrm{Ni}/{}^{58}\mathrm{Ni}^{\mathrm{standard}}_{\mathrm{norm}\ 58/61} - 1 \right) \times 10000 \quad (1)$$

or the parts per ten thousand variation of ^{60}Ni/^{58}Ni (internally normalized to a reference ^{58}Ni/^{61}Ni) relative to a standard measured in the same way. The established isotopic standard for Ni is the National Institute of Standards and Technology Standard Reference Material (NIST SRM) 986, Gramlich et al. (1989). Reference Ni isotope ratios for this standard are reported in Table 1. This NIST SRM has been widely used, providing a valuable common datum in all but the earliest work. If it is necessary to clarify which reference standard has been used, the notation above can be augmented, e.g. $\varepsilon^{60}\mathrm{Ni}_{58/61}$ (NIST 986). For elements such as Ni, however, where the same standard is conventionally used, we feel this additional information can be omitted without too much confusion, provided it is imparted elsewhere (as we do here). We use the epsilon notation solely for mass-independent isotopic data (internally normalized). This approach is typical although not universal and the presence of the subscript in our notation (Eqn. 1) makes the use of internal normalization evident.

We report mass-dependent variations in the delta notation:

$$\delta^{60}\mathrm{Ni} = \left({}^{60}\mathrm{Ni}/{}^{58}\mathrm{Ni}^{\mathrm{sample}} / {}^{60}\mathrm{Ni}/{}^{58}\mathrm{Ni}^{\mathrm{standard}} - 1 \right) \times 1000 \quad (2)$$

As for the mass-independent work, NIST SRM 986 is extensively used as the Ni isotope reference standard in all studies other than Moynier et al. (2007), and is implicit in Equation (2). In Equation (2), we follow another proposal made in Steele et al. (2011) to report the isotope ratio used, i.e. $\delta^{60/58}$Ni instead of δ^{60}Ni. This removes any ambiguity over which nuclide is used as the denominator. For an element such as Ni, with more than two stable isotopes, such qualification is valuable. We suggest this systematic notion could be useful more generally.

Moynier et al. (2007) reported their mass-dependent Ni isotope data as $\overline{\delta}$Ni, an error weighted, average fractionation per unit mass difference, using the three measured ratios (^{60}Ni/^{58}Ni)/2, (^{61}Ni/^{58}Ni)/3 and (^{62}Ni/^{58}Ni)/4. This is an interesting idea (see also Albalat et al. 2012), which reduces the error for the sample-standard bracketing technique by using all measured data. The alternative method for determining mass-dependent isotopic fractionation is by double-spiking (see review by Rudge et al. 2009). Double-spiking requires measurements of four isotopes, thus yielding three independent isotope ratio determinations. Given a troublesome Zn interference on mass 64, all double-spiked studies to date have used a ^{61}Ni–^{62}Ni double spike and employed ^{58}Ni, ^{60}Ni, ^{61}Ni and ^{62}Ni in the data reduction. With ostensible similarity to the sample-standard bracketing approach of Moynier et al. (2007), the combined measurements of these four isotopes yield a single value of natural isotopic fractionation, which is normally converted into a more tangible delta value for a specific but arbitrary isotope ratio (e.g. $\delta^{60/58}$Ni). The key difference in double-spiking is that the additional isotope ratios are used to constrain explicitly instrumental mass-bias. Namely this procedure improves accuracy, whereas in sample-standard bracketing instrumental mass bias is assumed to be identical for sample and standard and the additional isotope measurements are used to improve precision.

Both methods described above use measurements of ^{60}Ni for determining mass-dependent Ni isotope variability. It is germane to consider whether or not it makes good sense to use a radiogenic isotope for such a purpose. For terrestrial samples, there should be no variability in the relative abundance of radiogenic ^{60}Ni, given likely terrestrial isotopic homogenization after parental ^{60}Fe became extinct. For extra-terrestrial samples, this is potentially a consideration, but bulk variations in $\varepsilon^{60}\mathrm{Ni}_{58/61}$ are typically small (~0.1), dominantly nucleosynthetic rather than radiogenic and associated with mass-independent variability of other isotopes (see the section *Nucleosynthetic Ni isotopic variations*). For the most accurate mass-dependent measurements, a second mass-independent isotopic determination is therefore required (e.g. Steele et al. 2012), but such mass-independent variability does not have a significant impact on more typical mass-dependent determinations at the delta unit level (see the section *Magmatic Systems*).

In the section *Extinct ^{60}Fe and radiogenic ^{60}Ni* we address the presence of ^{60}Fe in the early solar system from ^{60}Ni measurements of meteoritic samples. Such determinations yield initial ^{60}Fe/^{56}Fe for the objects analyzed, denoted ^{60}Fe/^{56}Fe°. Given samples that yield precise values of ^{60}Fe/^{56}Fe° may have different ages, it is useful to calculate ^{60}Fe/^{56}Fe° at a common reference time, typically the start of the solar system as marked by calcium aluminium rich inclusion formation. Such a solar system initial value is abbreviated to $^{60}\mathrm{Fe}/^{56}\mathrm{Fe°}_{\mathrm{SSI}}$. Throughout this review, the uncertainties quoted for various average measurements are two standard errors, unless otherwise stated.

NUCLEOSYNTHETIC Ni ISOTOPIC VARIATIONS

Nickel is significant element in stellar nucleosynthesis. Nickel-62 has the highest binding energy per nucleon of any nuclide; no nuclear reaction involving heavier nuclides can produce more energy than it consumes. It is often, incorrectly, said that ^{56}Fe has the highest binding energy per nucleon, likely due to the anomalously high abundance of ^{56}Fe. In fact, ^{56}Fe is dominantly produced in stars as the decay product ^{56}Ni, which is the result of the last energetically favorable reaction during Si burning.

It is thought the Ni isotopes are dominantly produced during nuclear statistical equilibrium (NSE or the *e*-process) in supernovae (Burbidge et al. 1957). There are two main astrophysical environments in which the majority of Ni is thought to be produced, these are the type Ia (SN Ia) and type II (SN II) supernovae. SN Ia are thought to be the violent explosions of carbon–oxygen white dwarves which accrete material from a binary host to reach the Chandrasekhar limit (<1.39 $M_\odot$ non-rotating). SN Ia are highly neutron enriched environments and have been hypothesized to be the source of some important neutron-rich nuclides, including ^{48}Ca, ^{60}Fe, ^{62}Ni and ^{64}Ni. Due to the size of the progenitor (<1.39 $M_\odot$) the stars are old, ~1 Ga, as they have burned their fuel slowly. This means they out-live the stellar nurseries (lifetime ~10 Ma) in which they formed. Therefore, they would make an unlikely, and so very interesting, source for isotope anomalies in the Solar System. SN II are the terminal explosions of much larger stars (>12 $M_\odot$) which consequently have much shorter main sequence lifetimes. Their shorter lifetimes make them much more likely to return material to star forming regions, meaning they are a more probable source for nucleosynthetic anomalies in the Solar System.

Trying to identify nucleosynthetic contributions from specific stellar events, such as those described above, within the average composition of the solar system is very difficult. However, the presence of 'isotopic anomalies' within meteoritic material provides opportunities to try to fingerprint individual sources. Isotopic anomalies are identified by non-zero mass-independent isotopic compositions relative to a 'normal' isotopic composition. The latter is arbitrary, but frequently the anthropocentric datum of Earth is used, or more specifically a particular terrestrial reference material, NIST SRM 986 in the case of Ni. Although mass independent variations can be generated by *in situ* nuclear reactions and certain physico-chemical processes, in cosmochemistry they commonly indicate a blend of nucleosynthetic products different to the Earth. Such heterogeneities reveal either imperfect mixing of contrasting stellar inputs to the solar system (e.g. Wasserburg et al 1977), or else the unmixing of components in a generally well mixed, but locally heterogeneous nebula (e.g. Trinquier et al. 2009). These observations are of great interest to understanding the processes occurring in the early solar nebula and the stellar contributors to our solar system.

Calcium aluminium rich inclusions (CAIs), found in some primitive meteorites, document isotopic anomalies in the elements hosted in the refractory minerals from which they are formed. This was initially evident in the mass-independent oxygen isotopic compositions (Clayton et al. 1973) of the abundant CAIs in the meteorite Allende, although this signature is now largely attributed to gas-phase processes in the nebula (e.g. Yurimoto and Kuramoto 2004). However, isotopic anomalies were also discovered for a range of refractory metals (e.g. McCulloch and Wasserburg 1978a,b) which are still believed to document, at the macroscopic scale (>1 mm), nucleosynthetic mixtures that contrast with the bulk solar system. Following reports of mass independent variations of other iron-peak nuclides (e.g. Lee et al. 1978; Heydegger et al. 1979), the first Ni isotopic analyses of Allende CAIs (Morand and Allègre 1983; Shimamura and Lugmair 1983) failed to resolve signatures that differed from terrestrial values, in all but a single, highly anomalous 'FUN' inclusion (Shimamura and Lugmair 1983). Subsequent improvements in precision allowed Birck and Lugmair (1988) to resolve excesses of ~1 $\varepsilon^{62}Ni_{61/58}$ and ~3 $\varepsilon^{64}Ni_{61/58}$ within Allende CAIs (Fig. 1), which they noted was in keeping with a neutron-rich, equilibrium process nucleosynthesis. These findings were pleasingly compatible with anomalies in neutron-rich isotopes of Ca (Jungck et al. 1984) and Ti (Heydegger et al. 1979; Niederer et al. 1980; Niemeyer and Lugmair 1981) from previous studies and excesses of ^{54}Cr in their own work (Birck and Lugmair 1988). This landmark contribution identified the key mass-independent variations in Ni isotopes that would subsequently become apparent in bulk meteorite analyses.

The CAIs analysed by Birck and Lugmair (1988) also displayed ^{60}Ni enrichments, $\varepsilon^{60}Ni_{61/58} \sim 1$ (Fig. 1a), potentially related to the decay of ^{60}Fe co-produced with the neutron-rich Ni isotopes (see *Extinct ^{60}Fe and radiogenic ^{60}Ni*). If these $\varepsilon^{60}Ni_{61/58}$ values are taken solely as the consequence of *in situ* ^{60}Fe decay, they imply initial $^{60}Fe/^{56}Fe \sim 1 \times 10^{-6}$, but Birck and Lugmair (1988) cautioned against such an inference, given associated nucleosynthetic variations of comparable magnitude. Further analyses of CAIs by Quitté et al. (2007) similarly showed positive $\varepsilon^{60}Ni_{61/58}$ and $\varepsilon^{62}Ni_{61/58}$ (see Fig. 1a); the method used in this study suffered from too large ^{64}Zn interferences to make precise $\varepsilon^{64}Ni_{61/58}$ measurements. The authors attributed their observations to synthesis of ^{60}Fe and ^{62}Ni (and ^{96}Zr) in a neutron burst event, followed by the decay of ^{60}Fe. Birck and Lugmair (1988) had argued against this style of model, given the absence of predicted, associated ^{46}Ca anomalies in CAIs. However, Quitté et al. (2007) tentatively inferred $^{60}Fe/^{56}Fe^{\circ}_{SSI} > 1 \times 10^{-6}$ from their nucleosynthetic model, a two point internal CAI isochron and the interpretation of $\varepsilon^{60}Ni_{61/58}$ excesses in two CAIs without $\varepsilon^{62}Ni_{61/58}$ anomalies.

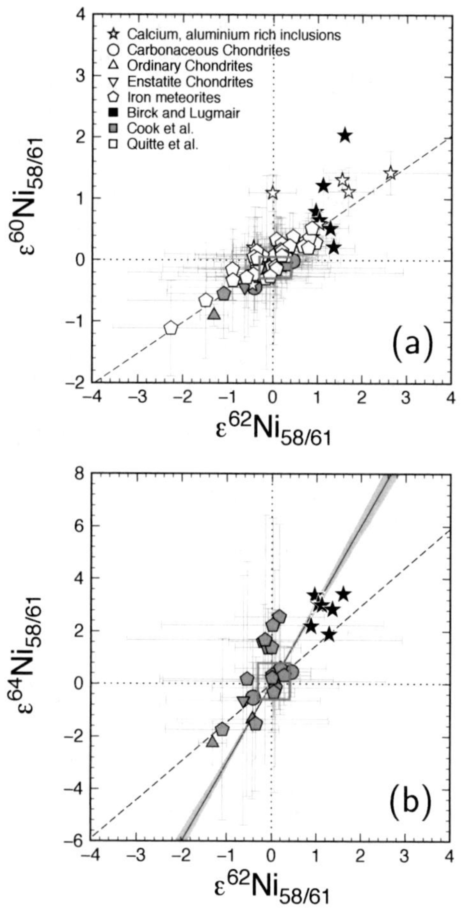

Figure 1. Mass independent nickel isotope data for meteorites and CAIs from earlier studies, (a) $\varepsilon^{60}Ni_{58/61}$ vs. $\varepsilon^{62}Ni_{58/61}$ and (b) $\varepsilon^{64}Ni_{58/61}$ vs. $\varepsilon^{62}Ni_{58/61}$. Data from Birck and Lugmair 1988; Cook et al. 2006, 2008; Quitté et al. 2006, 2007. The dashed lines indicate the vector of change in composition (from the origin) as a result of error or interference (moving to the lower left) in ^{61}Ni. Some scatter appears to be along this trajectory. For ready comparison with the higher precision data, the blue boxes indicate the dimensions of parameter space shown in Figure 4. The black line with grey band shows an extrapolation of a least squares York regression and the 2σ uncertainty error envelope (York 1969; Mahon 1996) through the bulk meteorite and peridotite data of Steele et al. (2012), see Figure 4. This trend is shown in Figure 1b for reference and notably passes through the CAI data.

Better precision was required to investigate mass-independent Ni isotopic variability between bulk meteorite samples. This came with improvements in mass-spectrometry, including multi-collection systems and their coupling with plasma sources (MC–ICPMS). The latter allows intense beams to be runs with relative ease, permitting counting statistical limitations to be overcome given sufficient sample availability. Nonetheless, the technique does require careful monitoring of a wide range of potential sample and plasma related interferences that may be significant at high precision (see common interferences listed in Quitté and Oberlei 2006 and Steele et al. 2011). Some of the earlier MC–ICPMS studies focussed on iron meteorites or the metallic phases of iron-bearing chondrites (Cook et al. 2006; Quitté et al. 2006; Moynier et al.

2007; Dauphas et al. 2008). This approach usefully exploits the natural concentration of Ni in a form readily purified to provide sufficient material for high precision analysis. In comparison, work on silicate samples requires more involved separation of Ni from a wider range of elements, typically accomplished using the highly Ni-specific complexing agent dimethylglyoxime either in solvent extraction (e.g. Morand and Allègre 1983; Shimamura and Lugmair 1983), mobile phase (e.g. Wahlgreen et al. 1970; Victor 1986; Steele et al. 2011; Gall et al. 2012, Chernonozhkin et al. 2015) or stationary phase of ion-chromatography (e.g. Quitté and Oberli 2006; Cameron et al. 2009). Alternatively, cation chromatography using a mixed HCl–acetone eluent (Strelow et al. 1971) has also been successfully used (e.g. Tang and Dauphas 2012).

Initial MC–ICPMS studies on meteoritic metal phases (Cook et al. 2006; Quitté et al. 2006; Moynier et al. 2007; Dauphas et al. 2008) dominantly argued against Ni isotope anomalies in bulk meteorites (Figs. 1–3), as did Chen et al. (2009) for their multi-collector thermal ionization mass-spectrometry (MC–TIMS) analyses (Fig. 3). As exceptions to these overall observations of bulk Ni isotope homogeneity, Quitté et al. (2006) reported correlated, negative values of $\varepsilon^{60}Ni_{61/58}$ and $\varepsilon^{62}Ni_{61/58}$ in many of the sulfide inclusions they analyzed from iron meteorites (Fig. 2a). Subsequently, Cook et al. (2008) reported more modest anomalies in troilites from iron meteorites (Fig. 2). Both studies argued for the preservation of a pre-solar component in these sulfide inclusions, although how this occurred mechanistically was problematic. The MC–TIMS work of Chen et al. (2009) provided a different measurement perspective. This study argued against resolvable differences in $\varepsilon^{60}Ni_{61/58}$ and $\varepsilon^{62}Ni_{61/58}$ in either bulk or sulfide samples, at a level of ±0.2ε and ±0.5ε respectively (Fig. 2a). Since then, no one has further pleaded for the case of anomalous sulfides and the original analyses seem likely to have been measurement artefacts. However, there has been on-going debate about the presence of Ni isotopic anomalies in bulk meteorites.

In striking contrast to the bulk meteorite analyses described above, Bizzarro et al. (2007) reported a dataset with near constant negative $\varepsilon^{60}Ni_{61/58}$ (and $\varepsilon^{62}Ni_{61/58}$) in differentiated meteorites but $\varepsilon^{60}Ni_{61/58}$ ~0 and positive $\varepsilon^{62}Ni_{61/58}$ in chondrites. These data were used to invoke a late super-nova injection of ^{60}Fe into the solar system. Subsequent studies were unable to reproduce these results (Dauphas et al. 2008; Regelous et al. 2008; Chen et al. 2009; Steele et al. 2011; Tang and Dauphas 2012, 2014) and noted that the systematics of the Bizzarro et al. (2007) dataset were consistent with an interference on ^{61}Ni. In reporting the results of further analyses, Bizzarro et al. (2010) commented that their new data were inconsistent with Bizzarro et al. (2007) but agreed with the observations of Regelous et al. (2008). The data from Bizzarro et al. (2007) will thus not be further considered.

Regelous et al. (2008) presented bulk analyses of $\varepsilon^{60}Ni_{61/58}$ and $\varepsilon^{62}Ni_{61/58}$ on a suite of chondrites and iron meteorites with precisions of around ± 0.02ε and ± 0.04ε respectively. By making higher precision measurements, in part by pooling multiple repeats of the same sample and by examining a wider range of meteorites than earlier studies, Regelous et al. (2008) were able to resolve differences in bulk meteorite compositions (Fig. 3a). They illustrated that variablity between different chondrite groups is largely echoed by that in iron meteorites (Fig. 3a). Notably the IVB irons have Ni isotopic compositions similar to carbonaceous chondrites (positive $\varepsilon^{62}Ni_{61/58}$), whilst the other magmatic irons resemble ordinary chondrites (with negative $\varepsilon^{62}Ni_{61/58}$). As for a number of other isotopic systems, enstatite chondrites were largely within error of terrestrial values.

These observations were further refined at higher precision (Fig. 4a) and with the inclusion of $\varepsilon^{64}Ni_{61/58}$ data (Fig. 4b) by Steele et al. (2011, 2012) and Tang and Dauphas (2012, 2014). These data revealed a continuous, well defined array in $\varepsilon^{62}Ni_{61/58}$ vs $\varepsilon^{64}Ni_{61/58}$ from ordinary chondrites and most magmatic irons, through terrestrial values in EH chondrites to carbonaceous chondrites and IVB irons (Fig. 4b). This ordering of meteorite groups is the same as observed in the mass-independent isotopic compositions of other first row, transition elements (Trinquier

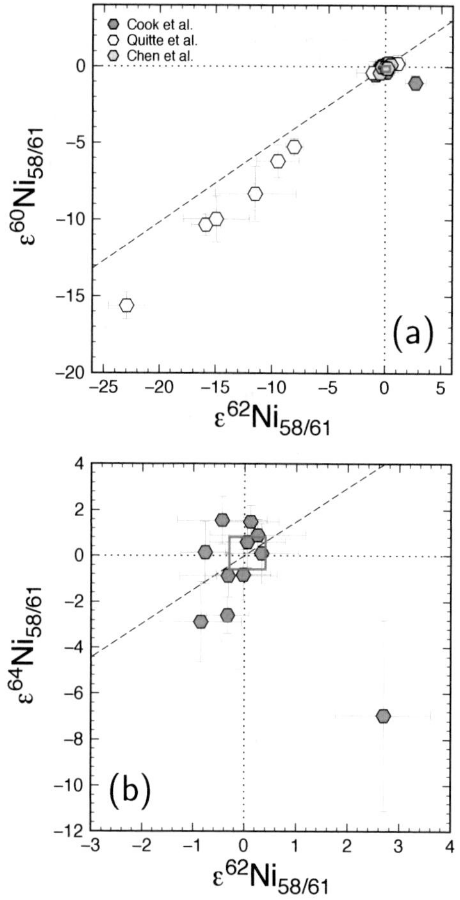

Figure 2. Mass independent nickel isotope data of sulfide inclusions from iron meteorites (Quitté et al. 2006; Cook et al. 2008; Chen et al. 2009), (a) $\varepsilon^{60}Ni_{58/61}$ vs. $\varepsilon^{62}Ni_{58/61}$ and (b) $\varepsilon^{64}Ni_{58/61}$ vs. $\varepsilon^{62}Ni_{58/61}$. Again the dashed lines indicate trajectories caused by error or interference (moving to the lower left) on ^{61}Ni. It appears that ^{61}Ni error may be a dominant (but not sole) cause of variability in the highly anomalous sulfides. The blue boxes indicate the dimensions of parameter space shown in Figure 4.

et al. 2007, 2009). Although the total isotopic variability in Ni is smaller than for Ti or Cr, its notable strength it that both chondrites and iron meteorites can be analyzed to high precision, allowing genetic associations to be made from bulk compositions of iron meteorites rather than from occasional oxygen-bearing inclusions they contain (e.g. Clayton et al. 1983). Strikingly, the bulk Ni isotope array in Figure 4b points towards the CAI values presented by Birck and Lugmair (2008), see Figure 1b. For the same arguments as made by Trinquier et al. (2009), however, the bulk meteorite array is not formed by simple mixing between a single bulk composition and CAI, which would created a strongly curved array (see Fig. 5). Carbonaceous chondrites are enriched, relative to ordinary chondrites, in the same isotopic component that is manifest more strongly in CAIs. Yet, it is not addition of CAIs themselves that causes the trend in Figure 4b but presumably variable abundances of specific pre-solar grains in chondrite matrices (see Dauphas et al. 2010 and Qin et al. 2011 for the Cr isotope case).

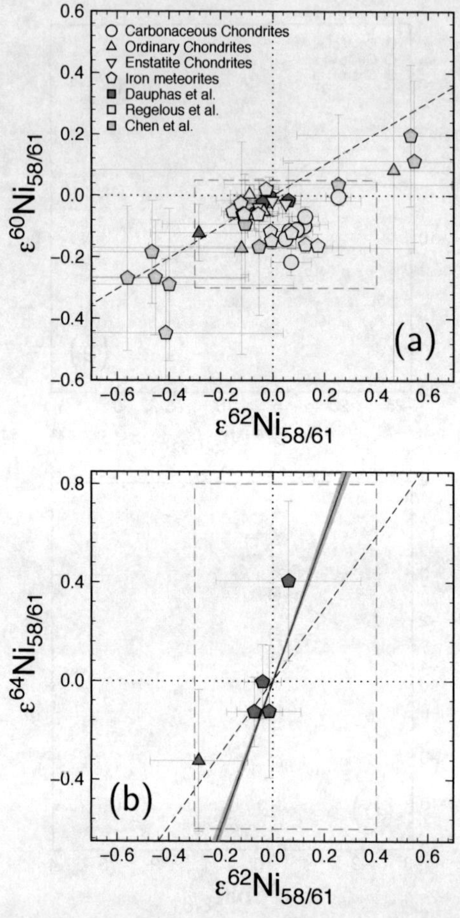

Figure 3. Mass independent nickel isotope data a) $\varepsilon^{60}Ni_{58/61}$ vs. $\varepsilon^{62}Ni_{58/61}$ and b) $\varepsilon^{64}Ni_{58/61}$ vs. $\varepsilon^{62}Ni_{58/61}$ of bulk meteorite analyses from 'second generation' studies (Dauphas et al. 2008; Regelous et al. 2008; Chen et al. 2009). The samples of Regelous et al. (2008) show small but resolved anomalies in both chondritic and iron meteorites. These analyses are consistent with the less precise data of Dauphas et al. (2008), who conversely argued against bulk Ni isotopic variability (as did Chen et al. 2009). Dashed lines represent ^{61}Ni error vectors and solid line with grey band the best fit array of Steele et al. (2012), as discussed in caption to Figure 1. Again (dashed) blue boxes indicate the dimensions of parameter space shown in Figure 4.

The well-defined array in non-radiogenic isotopes (Fig. 4b) provides key constraints on the nucleosynthetic origins of this important nebular component. At face value it represents coupled enrichments in the neutron rich isotopes of ^{62}Ni and ^{64}Ni. However, the 3:1 slope of the array (Fig. 4b) can also be reproduced by variable meteoritic values of $^{58}Ni/^{61}Ni$, the normalizing isotope ratio. Indeed, from high-precision mass-dependent isotopic measurements (see the section *Mass-Dependent Ni isotopic Variability*), Steele et al. (2012) showed that variability in ^{58}Ni best explains all observations. This implies the source of this anomalous Ni is from the Si–S zone of a SNII. This contrasts with material from the O–Ne zone required to account for a similarly constrained component in the Ti isotopic system. Steele et al. (2012) suggested ways in which the different, contributing zones for these different elements might be reconciled by grains from different zones being sorted (homogenized or unmixed) by solar system processes. However, these issues remain unresolved.

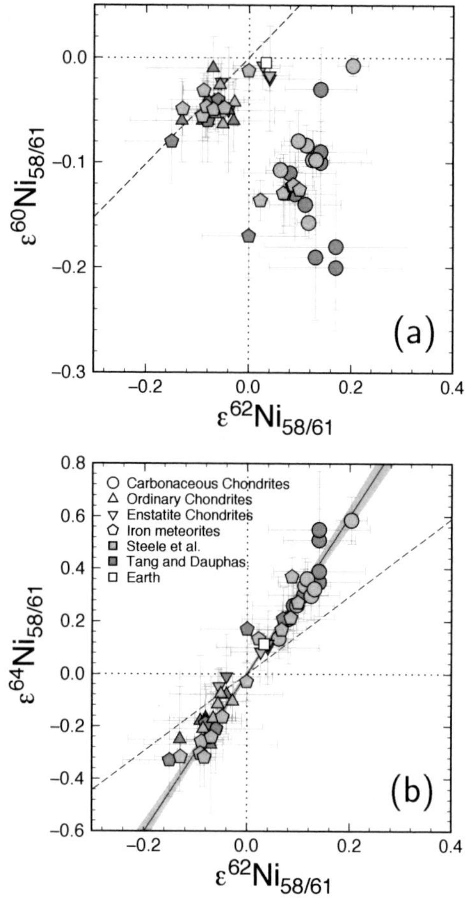

Figure 4. Mass independent Ni isotope measurements a) $\varepsilon^{60}Ni_{58/61}$ vs. $\varepsilon^{62}Ni_{58/61}$ and b) $\varepsilon^{64}Ni_{58/61}$ vs. $\varepsilon^{62}Ni_{58/61}$ on bulk meteorites from the highest precision studies (Steele et al. 2011, 2012; Tang and Dauphas 2012, 2014). Data from these studies show a consistent set of anomalies between carbonaceous, ordinary and enstatite chondrites and different groups of iron meteorites. There is a strong positive correlation between $\varepsilon^{62}Ni_{58/61}$ and $\varepsilon^{64}Ni_{58/61}$, first identified by Steele et al. (2012) and confirmed by Tang and Daphaus (2012). This is array quite distinct from the dashed ^{61}Ni error line. A least squares York regression and the 2σ uncertainty error envelope (York 1969; Mahon 1996) through the bulk meteorite and peridotite data of Steele et al. (2012) is shown with black line and grey band and yields a slope of 3.003 ± 0.116. Due to the bias in least squares regressions when using averages of repeat sample measurements and individual uncertainties, Steele et al. (2012) used each individual measurement of each sample and the homoscedastic uncertainty (see Appendix A.2 of Steele et al. 2012 for a discussion). Since individual measurement of meteorite samples are not available from Tang and Dauphas (2012, 2014) these data have not been included in the regression, but there is clearly excellent agreement between all three datasets.

The correlation of bulk analyses of $\varepsilon^{60}Ni_{61/58}$ with the other isotope ratios ($\varepsilon^{62}Ni_{61/58}$ or $\varepsilon^{64}Ni_{61/58}$) is much less systematic (Regelous et al. 2008; Steele et al. 2012; Tang and Dauphas 2014), see Figure 4a. In general, carbonaceous chondrites have lower $\varepsilon^{60}Ni_{61/58}$ than ordinary chondrites and enstatite chondrites, but CI chondrites are a notable exception. It is tempting to attribute these different relationships to the radiogenic nature of $\varepsilon^{60}Ni_{61/58}$, but given the abundance of ^{60}Fe inferred from bulk meteorite studies (see the section *Extinct ^{60}Fe and radiogenic ^{60}Ni* below) this seems unlikely. Instead, this presumably reflects part of a more complex nucleosythetic

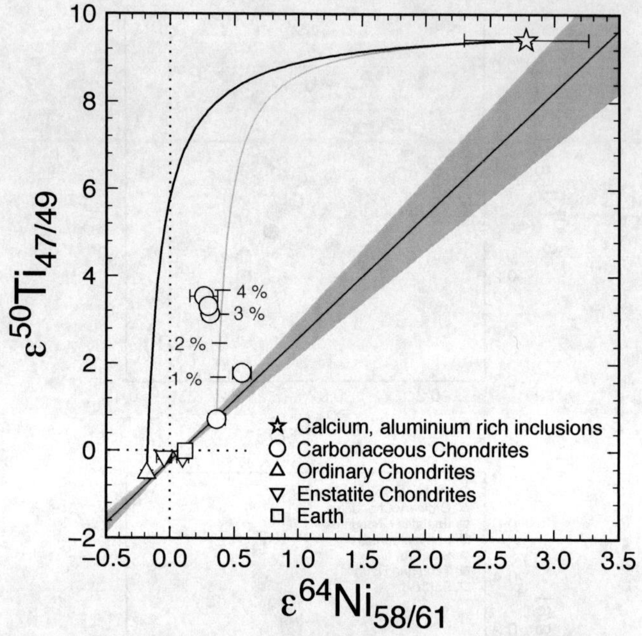

Figure 5. Plot of $\varepsilon^{64}Ni_{58/61}$ vs $\varepsilon^{50}Ti_{47/49}$ for bulk meteorites (Ti isotope data from Trinquier et al. (2009), Ni data from Birck and Lugmair (1988) and Steele et al. (2012)). A linear array is defined by analyses of ordinary, enstatite and CI chondrites and this extends to intersect with CAI compositions (see caption to Fig. 4). However, CAI-rich meteorite groups (CV CO) plot above the array, likely reflecting analyses of unrepresentative sub-samples of these meteorites with an excess of CAI ~ 3–4% by mass (illustrated with a mixing curve from a CI composition), see Hezel et al. (2008). The high Ti/Ni of CAIs means mixing trajectories with bulk chondrite compositions are highly convex up. Thus simple mixing between CAIs and ordinary chondrites (black curve) cannot explain the straight bulk meteorite array.

signature, with several components, potentially one which incorporates high fossil ^{60}Fe, involved in generating the Ni isotopic compositions of ordinary and carbonaceous chondrites.

The carriers of the exotic Ni isotopic components that shape variable bulk meteorite compositions remain to be identified. The highly anomalous isotopic compositions found in separated SiC grains from the CM2 meteorite, Murchison (Marhas et al. 2008) do not readily account for the mass-independent Ni isotope variations seen in bulk samples. These ion-probe analyses show $^{61}Ni/^{58}Ni$ and $^{62}Ni/^{58}Ni$ ratios > 1000ε higher than terrestrial values (for these extreme ratios there is no internal normalization) in X-grains believed to be derived from SNII sources. These Ni isotopic signatures suggest derivation from outer He/N and He/C zones of an SNII event, different to those inferred by Steele et al. (2012) to be necessary to account for bulk isotopic variability. As with other analyses of pre-solar SiC, these measurements bear striking testimony to the diversity of stellar sources that contribute to the bulk composition of the solar system, but do not identify a carrier for the signature that causes variability between different bulk, meteoritic objects (Steele et al. 2012).

EXTINCT ^{60}Fe AND RADIOGENIC ^{60}Ni

As alluded to in the preceding section, much of the initial interest in Ni isotope cosmochemistry was focussed on trying to identify the presence of live ^{60}Fe in the early solar system. A salient property of ^{60}Fe is that it is only made during stellar nucleosynthesis and cannot be produced by particle irradiation in our own solar system, unlike ^{26}Al. Thus an initial solar system value of ^{60}Fe/^{56}Fe (^{60}Fe/^{56}Fe°$_{SSI}$) significantly higher than the calculated steady state background of the interstellar medium (ISM), recently estimated at ~3×10^{-7} by Tang and Dauphas (2012), would provide strong evidence for injection of short-lived nuclides by a proximal stellar explosion. Such an explanation, linked to a trigger for solar system formation, has long been invoked to account for the elevated initial solar system ^{26}Al/^{27}Al (e.g. Cameron and Turan 1977), but alternative means of generating ^{26}Al (e.g. Clayton and Jin 1995) have made a more definitive test using ^{60}Fe highly desirable (see Wasserburg et al. 1998). As explored below, values of ^{60}Fe/^{56}Fe°$_{SSI}$ determined from Ni isotopic measurements of meteorites remain contentious although our own perspective is that values above those of the ISM are questionable.

In a pioneering contribution, Shukolyukov and Lugmair (1993a) reported evidence for live ^{60}Fe from analyses of ^{60}Ni/^{58}Ni on samples of the eucrite Chervony Kut. Eucrites represent a good target for detecting *in situ* decay of ^{60}Fe, given their extremely high Fe/Ni (>10000). Such high Fe/Ni ratios are a consequence of core formation at low pressures (i.e. on a small planetary body) and subsequent magmatic fractionation. Given the formation and differentiation of the eucrite parent body within the first few million years of solar system history (e.g. Papanastassiou and Wasserburg 1969) the resultant Fe/Ni were likely to generate resolvable ^{60}Ni anomalies given significant initial ^{60}Fe. Different components of Chervony Kut showed elevated ε^{60}Ni$_{62/58}$ (up to 50) and three bulk samples with a range of Fe/Ni defined a straight line with a slope that implied ^{60}Fe/^{56}Fe° ~ 4×10^{-9} (Fig. 6a). The authors used an age of 10 ± 2 Ma post CAI, from the similarity of the meteorites' ^{87}Sr/^{86}Sr to dated angrites (Lugmair and Galer 1992), to infer ^{60}Fe/^{56}Fe°$_{SSI}$ ~ 2×10^{-6}. Since then, determinations of the timing of differentiation of the eucrite parent body give older ages, ~3Ma post CAI (Lugmair and Shukolyukov 1998; Bizzarro et al. 2005; Trinquier et al. 2008; Schiller et al. 2011) and the accepted half-life of ^{60}Fe has increased from 1.49 Ma (Kutschera et al. 1984) to 2.62 Ma (Rugel et al. 2009). As a result of these changes a more contemporary interpretation of these results would yield ^{60}Fe/^{56}Fe°$_{SSI}$ ~ 2×10^{-8}.

Shukolyukov and Lugmair (1993a) originally noted that their value of ^{60}Fe/^{56}Fe°$_{SSI}$ was in keeping with ε^{60}Ni$_{61/58}$ measurements made on CAIs by Birck and Lugmair (1988), assuming that such ^{60}Ni excesses were radiogenic. From the discussion of nucleosynthetic variability in the section *Nucleosynthetic Ni isotopic variations*, it should be clear that this assumption is by no means valid, nor is the value thus derived consistent with the revised ^{60}Fe/^{56}Fe°$_{SSI}$. The magnitude of possibly radiogenic ε^{60}Ni$_{61/58}$ in CAIs is no bigger than ε^{62}Ni$_{61/58}$, which must be nucleosynthetic (Fig. 1a). Although Kruijer et al. (2014) demonstrated that it is possible to disentangle nucleosynthetic from radiogenic contributions to ^{182}W/^{184}W in CAIs in order to define ^{182}Hf/^{180}Hf°$_{SSI}$, neither the abundance of initial parent nor parent–daughter fractionation in CAIs are sufficiently large to make this approach viable for determining ^{60}Fe/^{56}Fe°$_{SSI}$. Indeed, most CAIs have sub-chondritic ^{56}Fe/^{58}Ni (e.g. Quitté et al. 2007), as a result of the slightly more refractory cosmochemical behavior of Ni relative to Fe. Thus at best, minor ε^{60}Ni$_{61/58}$ deficits in bulk CAI would be expected if ^{60}Fe/^{56}Fe°$_{SSI}$ were large enough for the bulk nebula to evolve to more radiogenic ε^{60}Ni$_{61/58}$, as is the case for the ^{53}Mn–^{53}Cr system (Birck and Allègre 1985). Moreover, internal CAI isochrons offer little scope for constraining ^{60}Fe/^{56}Fe°$_{SSI}$ given both modest Fe/Ni fractionation between common phases in CAIs and the frequent disturbance of Fe/Ni ratios in CAIs specifically (Quitté et al. 2007) and chondritic meteorites in general (e.g. Telus et al. 2016). In contrast to the ^{26}Al–^{26}Mg system, but similar to the ^{53}Mn–^{53}Cr pair, bulk CAI measurements therefore offer scant opportunity to determine the initial parent abundance of the ^{60}Fe–^{60}Ni system. Instead, basaltic achondrites, with their highly fractionated Fe–Ni, have been the focus of much additional work to constrain ^{60}Fe/^{56}Fe°$_{SSI}$.

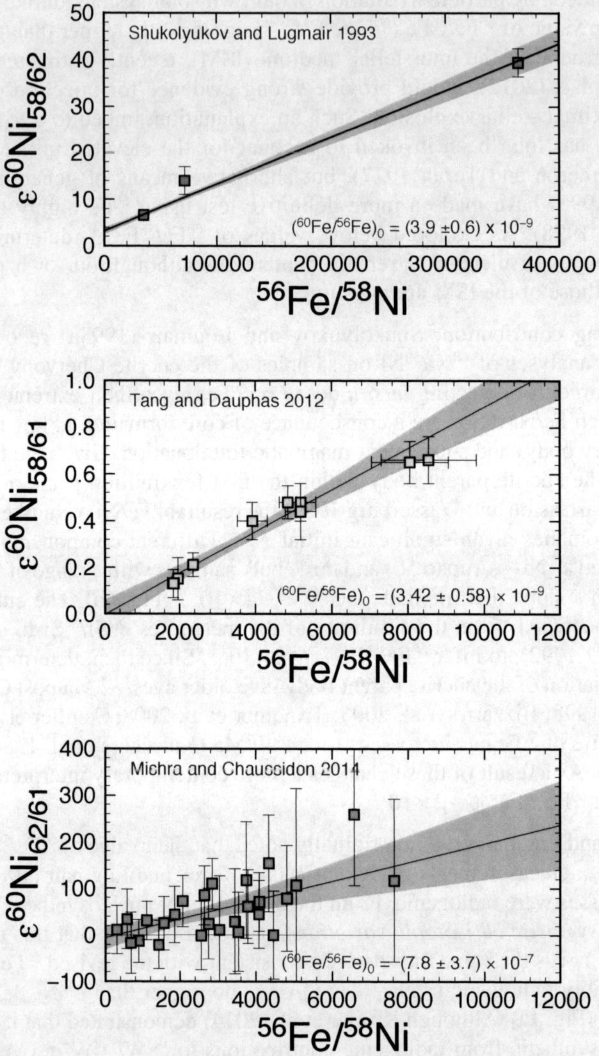

Figure 6. Representative isochron diagrams for the ^{60}Fe–^{60}Ni system measured by different analytical techniques. (a) TIMS isochron of different bulk samples of the eucrite Chervony Kut (Shukolyukov and Lugmair 1993a). (b) MC–ICPMS isochron of mineral separates from the quenched angrite D'Orbigny (Tang and Dauphas 2012). (c) SIMS isochron of a single Efremovka chondrule (Ch 1), Mishra and Chaussidon (2014). Also shown are recalculated least squares linear York regressions (York 1968, Mahon 1995) for each dataset, yielding ^{60}Fe/^{56}Fe°. Note the highly contrasting values obtained for ^{60}Fe/^{56}Fe° by bulk analyses in (a) and (b) versus *in situ* analyses (c) of objects of comparable age.

In a follow-on study, Shukolyukov and Lugmair (1993b) analyzed the eucrite Juvinas and noted an order of magnitude lower ^{60}Fe/^{56}Fe° (~4×10^{-10}), which the authors attributed to ~4 Ma evolution of the eucrite parent body mantle source between generation of the basaltic melts represented by Chervony Kut and Juvinas. Using the newer ^{60}Fe half-life this period would be ~9 Ma, a value which resonates with the age difference of ~11 Ma inferred from Hf–W analyses between a group of eucrites including Juvinas and an older group of eucrites (Touboul et al. 2015); sadly Chervony Kut itself was not analyzed for its Hf–W systematics. As with Chervony Kut (Shukolyukov and Lugmair 1993a), separated mineral phases from Juvinas did not show meaningful isochronous relations and Shukolyukov and Lugmair (1993b) argue that the mobility/diffusivity of Ni made it susceptible to resetting on this shorter length scale, especially given the complex history of eucrites, with well documented thermal metamorphism (e.g. Takeda and Graham 1991).

Subsequent studies of eucrites have been made with increasingly precise analyses using MC–ICPMS, but in essence show comparable features. Quitté et al. (2011) reported data for Bouvante and further analyses of Juvinas. Whole rock sub-samples of Bouvante define two arrays and if these are taken to be isochrons they imply $^{60}Fe/^{56}Fe^{\circ} = 5 \times 10^{-9}$ (but with an implausibly negative intercept of $\varepsilon^{60}Ni_{62/58} = -24 \pm 3$) and $^{60}Fe/^{56}Fe^{\circ} = 5 \times 10^{-10}$ (with a reasonable, near zero $\varepsilon^{60}Ni_{62/58}$ intercept). Instead, the authors argue that the data represent two mixing lines between clasts of different compositions. Whilst the second array gives a similar value of $^{60}Fe/^{56}Fe^{\circ}$ to Juvinas, as obtained by Shukolyukov and Lugmair (1993b), additional measurements of Juvinas by Quitté et al. (2011) are inconsistent with the array of Shukolyukov and Lugmair (1993b). Quitté et al. (2011) obtain a higher $^{60}Fe/^{56}Fe^{\circ}$ (2×10^{-9}), using their two unwashed bulk samples and two unwashed, lower Fe/Ni samples from the Shukolyukov and Lugmair (1993b) array (ignoring the washed samples which they argue may have suffered Fe–Ni fractionation as a result of this preparation). As with the work of Shukolyukov and Lugmair (1993b), the Juvinas mineral analyses of Quitté et al. (2011) are scattered and presumably perturbed by secondary processes. There is clear difficulty in distinguishing primary from secondary signatures in these brecciated, thermally metamorphosed meteorites, but none of the eucrite arrays define $^{60}Fe/^{56}Fe^{\circ}$ greater than 5×10^{-9}.

Tang and Dauphas (2012) comprehensively reassessed this issue using a collection of bulk eucrites and diogenites. This dataset gave a large range in Fe/Ni, which coupled with their high precision measurements (typical $\varepsilon^{60}Ni_{61/58}$ better than ± 0.3), provided a more definitive $^{60}Fe/^{56}Fe^{\circ} = (3.5 \pm 0.3) \times 10^{-9}$ for the eucrite parent body. This result is notably in keeping with the higher values obtained from 'internal isochrons' on sub-samples of meteorites discussed above (e.g. Fig. 6a). Using an age of 2.4 ± 1.1 Ma post CAI for silicate differentiation of 4 Vesta (Trinquier et al. 2008; Connelly et al. 2012), which presumably sets the variable Fe/Ni seen in the eucrites, these data yield a $^{60}Fe/^{56}Fe^{\circ}_{SSI}$ $(6.6 \pm 2.5) \times 10^{-9}$. This work of Tang and Dauphas (2012) also yields a bound of 4 ± 2 Ma on the age of core formation on 4 Vesta, from a two-stage evolution model of its mantle. Namely, the $\varepsilon^{60}Ni_{61/58}$ of the mantle, determined by the intercept of the eucrite–diogenite array with an estimated bulk mantle $^{56}Fe/^{58}Ni \sim 2700$ informs on the time since core formation. Tang and Dauphas (2014) subsequently used a similar approach to constrain the timing of core formation and hence growth of Mars. They argued that the planet reached 44% of its size no earlier than 1.2 Ma post CAI or otherwise the $\varepsilon^{60}Ni_{61/58}$ of the SNC meteorites they measured would be more radiogenic.

Quenched angrites provide a more petrologically robust sample for determining $^{60}Fe/^{56}Fe^{\circ}$, even if their Fe/Ni are not quite as extreme as the eucrites (cf. Figs. 6a,b). Moreover, the well-defined ages for these samples determined using extant isotope chronometry (Amelin 2008a,b; Connelly et al. 2008, Brennecka and Wadhwa 2012), potentially provide more accurate decay correction in calculating $^{60}Fe/^{56}Fe^{\circ}_{SSI}$. Three independent studies (Quitté et al. 2010; Spivak-Birndorf et al. 2011; Tang and Dauphas 2012) obtained consistent values for internal isochrons of d'Orbigny (e.g. Fig. 6b) which give a weighted average $^{60}Fe/^{56}Fe^{\circ} = (3.3 \pm 0.5) \times 10^{-9}$. Two internal isochrons from a second quenched angrite, Sahara 99555, are also in mutual agreement but yield a lower weighted mean $^{60}Fe/^{56}Fe^{\circ} = (1.9 \pm 0.4) \times 10^{-9}$ (Quitté et al. 2010; Tang and Dauphas 2015). Tang and Dauphas (2015) convincingly argue this difference relative to d'Orbigny likely reflects terrestrial weathering experienced by Sahara 99555.

In all, these angrite $^{60}Fe/^{56}Fe°$ are in keeping with those of eucrites, suggesting a similar initial ^{60}Fe and timing of planetary differentiation on these two bodies. So a reassuringly consistent value has emerged from these various TIMS/MC–ICPMS studies which span several decades of work. The data from d'Orbigny provides the best constrained value and using an age of 3.9 Ma post CAI (Amelin 2008a; Brennecka and Wadhwa 2012; Connelly et al. 2012) we calculate $^{60}Fe/^{56}Fe°_{SSI} = (9.8 \pm 4.5) \times 10^{-9}$. We note this value is lower than the equivalent cited by Tang and Dauphas (2012) as a consequence of our using the Pb–Pb ages rather than Mn–Cr chronometry.

Further bulk analyses of a range of meteoritic materials are supportive of the low $^{60}Fe/^{56}Fe°_{SSI}$ determined from angrite and eucrite analyses, albeit from less well constrained scenarios. Shukolyukov and Lugmair (1993b) and Quitté et al. (2010) reported no systematic differences in $\varepsilon^{60}Ni_{62/58}$ for various bulk samples (ureilites) and separated phases (e.g. troilite) with high but variable Fe/Ni ratios. Moynier et al. (2011) placed a maximum upper bound on $^{60}Fe/^{56}Fe°_{SSI}$ of 3×10^{-9} from the absence of ^{60}Ni isotope anomalies in measurements of troilite from the iron meteorite Muonionalusta. These troilites have Pb–Pb model ages as old as the quenched angrites (Blichert-Toft et al. 2010), which coupled with their high Fe/Ni (up to 1500) should result in radiogenic $\varepsilon^{60}Ni_{61/58}$ given sufficiently high $^{60}Fe/^{56}Fe°$. Although appealing targets for analysis, re-equilibration of the troilites with the surrounding Ni-rich metal during parent body during cooling would tend to erase any $\varepsilon^{60}Ni_{61/58}$ anomalies. The authors briefly argue against such an interpretation on the basis of the preservation of ancient Pb–Pb ages, but the potential for diffusional exchange of Ni with the Ni-rich host metal (see Chernonozhkin et al. 2016) seems much greater than for Pb. Whilst the conclusions of Moynier et al. (2011) are thus compatible with other studies, whether or not the measurements represent an independent constraint on $^{60}Fe/^{56}Fe°_{SSI}$ remains open to debate.

Analysis of chondrules from the CB_a meteorite Gujba and ungrouped 3.05 ordinary chondrite NWA 5717 by Tang and Dauphas (2012) form near horizontal arrays that yield $^{60}Fe/^{56}Fe°$ from 1 to 3×10^{-9}. However, elemental mapping by Telus et al. (2016) showed that chondrules in all chondrites they studied had experienced some open system behavior of Fe and Ni. Only the most pristine, LL3.0 meteorite, Semarkona, retained undisturbed chondrules, about ~40% of those studied. Prompted by these findings, Tang and Dauphas (2015) made measurements of single chondrules from Semarkona, to yield a valuable but still relatively poorly defined $^{60}Fe/^{56}Fe° = (5 \pm 3) \times 10^{-9}$. The lack of significant differences between $^{60}Fe/^{56}Fe°$ for these meteorites of different metamorphic grade suggests that they are not unduly compromised by this open system behavior. Given an average chondrule age of 2 Ma post CAI (see recent compilation of data in Budde et al. 2016), the $^{60}Fe/^{56}Fe°_{SSI} \sim 9 \times 10^{-9}$ derived from these individual chondrule measurements is notably compatible with the studies from achondrite meteorites. In a grand compilation of various determinations of $^{60}Fe/^{56}Fe°$ from bulk measurements, Tang and Dauphas (2015) derived a weighted average $^{60}Fe/^{56}Fe°_{SSI} = (1.0 \pm 0.3) \times 10^{-8}$.

In contrast to the work described above on Ni separated from bulk samples, much higher $^{60}Fe/^{56}Fe°$ have been inferred from *in situ* work by SIMS. Initially, the absence of detectable differences in the $^{60}Ni/^{61}Ni$ of olivines from type II chondrules from Semarkona, relative to terrestrial olivines, was used to place an upper limit of 3.4×10^{-7} on their $^{60}Fe/^{56}Fe°$ (Kita et al. 2000). However, later studies documented correlated $\varepsilon^{60}Ni$ and Fe/Ni in matrix sulfides and oxides from primitive ordinary chondrites suggesting $^{60}Fe/^{56}Fe°$ from 1×10^{-7} to 1×10^{-6} (Tachibana and Huss 2003; Mostefaoui et al. 2004, 2005; Guan et al. 2007). As the time of formation of these phases is uncertain, the significance of these data arrays for inferring $^{60}Fe/^{56}Fe°_{SSI}$ was open to question. In an elegant study, Tachibana et al. (2006) subsequently analysed different phases from the chondrules of Semarkona. Although the ranges in correlated Fe/Ni and $\varepsilon^{60}Ni_{62/61}$ were lower than in the sulfide work, interpretation of the arrays as constraining $^{60}Fe/^{56}Fe°_{SSI}$ (5–10) × 10^{-7} seemed less equivocal. Similar results were reported by Mishra et al. (2010) for single chondrule analyses from a wider range of unequilibrated ordinary chondrites. Yet, all such analyses are controlled by the errors on the very small ^{61}Ni and ^{62}Ni beams used for determining $\varepsilon^{60}Ni_{62/61}$ (the larger ^{58}Ni beam cannot be used as it is interfered by ^{58}Fe). These measurements thus critically require accurate background determination and interference free spectra. Moreover, the use of the minor Ni isotope in the denominator of such low intensity measurements can lead to a statistical bias in calculated ratios (Ogliore et al. 2011, see also Coath et al. 2013). This artefact resulted in spurious correlations between Fe/Ni and $\varepsilon^{60}Ni_{62/61}$ in all earlier work (Telus et al. 2012). Yet subsequent work has continued to report high inferred $^{60}Fe/^{56}Fe°$ (see Fig. 6c) from *in situ* analyses of chondrules in studies for which such statistical bias is argued to be insignificant (Mishra and Chaussidon 2014; Mishra and Goswami 2014; Mishra et al. 2016).

Hence, inferred $^{60}Fe/^{56}Fe°$ from TIMS/MC–ICPMS studies and SIMS analyses of individual chondrules are markedly different. This contrast in conclusions from bulk and *in situ* approaches also extends to other MC–ICPMS work. Regelous et al. (2008) and Steele et al. (2012) obtained the loose constraint that the $^{60}Fe/^{56}Fe°$ of carbonaceous chondrites was $< 1 \times 10^{-7}$, given their indistinguishable compositions relative to IVB iron meteorites. Tang and Dauphas (2012) noted that the constant $^{58}Fe/^{54}Fe$ in all their analyses was incompatible with nucleosynthetic models that could account for $^{60}Fe/^{56}Fe°_{SSI} \sim 1 \times 10^{-6}$. Explicit comparison can be made between SIMS and MC–ICPMS data for individual chondrule analyses from Semarkona, reported in the studies of Mishra and Chaussidon (2014) and Tang and Dauphas (2015) respectively. The former inferred $^{60}Fe/^{56}Fe° = (3 \pm 2) \times 10^{-7}$ whereas the latter, as discussed above, argued for $(1 \pm 3) \times 10^{-9}$. Although a detailed comparison requires knowledge of individual chondrule ages (e.g. Mishra et al. 2010) which can be variable (see Connelly et al. 2012), this cannot account for the two orders of magnitude difference in the results of the two studies. It is also worth noting that a SIMS study of basaltic achondrites reported $^{60}Fe/^{56}Fe° = (6 \pm 9) \times 10^{-9}$ for quenched angrites (Sugiura et al. 2006), in good agreement with the MC–ICPMS studies (Quitté et al. 2010; Spivak-Birndorf 2011; Tang and Dauphas 2012, 2015). Thus the divergent results between SIMS and MC–ICPMS seem restricted to analyses of chondrules.

The difference in inferred $^{60}Fe/^{56}Fe°_{SSI}$ between the bulk and *in situ* studies continues to be debated, see Mishra et al. (2016) and Tang and Dauphas (2015). Although we cannot offer an unbiased opinion, we argue strongly for the validity of the interpretations based on MC–ICPMS and TIMS analyses. Not only does this work show consistency in values obtained on a range of materials, from chondrules to bulk achondrites, but the approach removes sample matrix before analysis. SIMS analyses of high Fe/Ni samples typically collect only ~10000 counts of the minor Ni isotopes and apply a background correction determined from a single point on the mass spectrum. Perhaps most critically, the *in situ* approach has not documented accurate ^{60}Ni measurement for materials with the highest Fe/Ni, that define the isochrons (see Fig. 6c).

In this light, we note there may be a possible interference problem for SIMS analyses from the presence of $^{59}CoH^+$. Even for high resolution SIMS measurements, $^{59}CoH^+$ overlaps considerably with $^{60}Ni^+$ (Mishra and Chaussidon 2014) and depending on measurement mass, it may not be resolved at all. The high Fe/Ni portions of chondrules are the result of fractional removal of olivine and/or low-Ca pyroxene and troilite from the cooling melt droplet. Ni is considerably more compatible than Fe in these fractionating phases and so the residual melt (and phases that subsequently crystallize from it) acquire high Fe/Ni. Ni is also more compatible than Co under the same conditions and so as crystallization proceeds, Co/Ni will also increase. Thus high Fe/Ni portions of chondrules inevitably have high Co/Ni. We have quantified this process using the partition coefficients from experiments designed to mimic crystallization of chondritic liquids at low pressure, which provide simultaneously determined Fe, Ni and Co data (Gaetani and Grove 1997). In detail, the evolution of Fe/Ni and Co/Ni depend on the amount of fractionating sulfide. This parameter changes the amount of crystallization required to reach a given Fe/Ni and the magnitude of Co/Ni fractionation. We examine two scenarios using relatively low (0.2%) and high (1%) amounts of sulfide for Semarkona Type II chondrules, taken from the work of Jones (1990). In the latter case, we calculate that an extreme $^{56}Fe/^{58}Ni \sim 30000$ such as reported in SIMS studies (see Fig. 6c) will be associated with $^{59}Co/^{60}Ni \sim 100$, relative to starting chondrule values of 23 and 0.07 respectively. These are model values for a low-Ca pyroxene grown from the residual melt, as this was the main target in the study of Mishra and Chaussidon (2014). Assuming that the fraction of hydride production determined for iron (Mishra and Chaussidon 2014) can be used for cobalt, $^{59}Co/^{60}Ni$ ratios can be used to calculate the magnitude of the hydride interference. This is expressed in terms of apparent ^{60}Ni anomaly, $\varepsilon^{60}Ni_{62/61}$, using the two endmember crystallization scenarios described above (Fig. 7). The model results are compared to the magnitude of observed anomalies and this sensitivty test suggests that $^{59}CoH^+$ could play a substantial role in creating these signatures. This explanation of the divergence of bulk and *in situ* measurements seems promising; it can be tested by measuring Co intensities during SIMS analysis and refining input parameters for the calculations above.

In summary we infer $^{60}Fe/^{56}Fe°_{SSI} \sim 1 \times 10^{-8}$, although this is not a consensus view. This conclusion implies that the solar system did not form with $^{60}Fe/^{56}Fe$ discernibly different from ISM. Thus if the elevated ^{26}Al of the solar system is to be explained by recent injection of stellar material, this must come from a source which does not also generate significant ^{60}Fe. Tang and Dauphas (2012) suggest that winds from the outer portions of a giant, Wolf-Rayet star could constitute an appropriate explanation. These results argue against a supernova trigger of solar system collapse (see Wasserburg et al. 1998).

MASS-DEPENDENT Ni ISOTOPIC VARIABILITY

Magmatic systems

Mass-dependent Ni isotopic variations in magmatic systems have received relatively little interest. This may be due to the lack of redox variability of Ni in common silicate phases on Earth, which has been a prime driver for exploration of several other isotopic systems, based on the notion that redox related changes in bonding environment will lead to large mass-dependent fractionations. However, the notable change in Ni oxidation state from 2^+ to 0 and partitioning during core formation provides a significant point of interest.

More prosaically, but in common with many other isotopic systems recently investigated, commercially available metals can show marked mass-dependent isotopic variability. For example, Tanimizu and Hirata (2006) reported a range of 0.6‰ in the $\delta^{60/58}Ni$ of a selection of high-purity commercial Ni reagents. The authors noted the potential of isotopic fractionation during purification of Ni by the Mond process, which involves separation of Ni in the vapor phase as the volatile $Ni(CO)_4$ species. However, two Ni ores analyzed by Tanimizu and Hirata (2006) showed an even greater variability ($\delta^{60/58}Ni = 0.5$ to -0.4), spanning the compositions of the laboratory Ni reagents and making it hard to distinguish the influences of anthropogenic from natural fractionation. An important consequence of the fractionated values of $\delta^{60/58}Ni$ seen in purified Ni is that NBS SRM 986 is sadly not representative of bulk terrestrial reservoirs, either in mass-dependent (Cameron et al. 2009; Steele et al. 2011; Gall et al. 2012; Gueguen et al. 2013) or mass-independent Ni isotopic compositions (Steele et al. 2011), see Table 2. The latter likely stems from imperfect mass-bias correction using a single exponential form for combined instrumental, natural and industrial fractionations. Separation of Ni as a carbonyl species in the Mond process but measurement as an elemental ion in mass spectrometry offers a tangible mechanism to account for this feature (Steele et al. 2011). Although cosmetically unappealing, the slight deviation of the Ni reference standard from bulk terrestrial values does not diminish its value as a common datum.

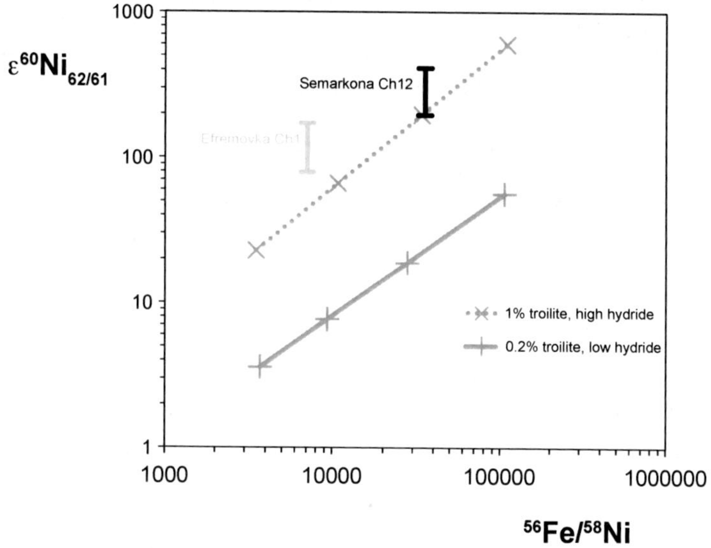

Figure 7. An illustration of the possible influence of $^{59}CoH^+$ formation on the $^{60}Ni/^{61}Ni$ isotope ratio (expressed as $\varepsilon^{60}Ni_{62/61}$, parts per ten thousand anomaly) for model scenarios of SIMS chondrule analysis. The composition of a low-Ca pyroxene crystallized from mesostatis after a given amount of prior olivine crystallization is used as a representation of the samples analyzed by Mishra and Chaussidon (2014). The fractional interference on ^{60}Ni is calculated from the model $^{59}Co/^{60}Ni$, assuming $^{59}CoH^+$ production is equal to $^{56}FeH^+$ production. Values for the latter are estimated from mass scans reported in the supplements of Mishra and Chaussidon (2014) in which $^{56}FeH^+/^{57}Fe^+$ varies from $(1–3)\times10^{-3}$, corresponding to $^{56}FeH^+/^{56}Fe^+=(2–7)\times10^{-5}$. To show a full range of possible outcomes, the higher production ratio is used with the high sulfide crystallization scenario (see below) which generates the highest Co/Ni and the lower value with the low sulfide crystallization scenario. The evolution of Co/Ni and Fe/Ni with fractional (Rayleigh) crystallization of a chondrule is calculated using representative partition coefficients from Gaetani and Grove (1995), $D(Ni)_{olivine–melt}=5.9$, $D(Co)_{olivine–melt}=1.96$, $D(Fe)_{olivine–melt}=0.916$ with two endmember sets of sulfide coefficients $D(Ni)_{sulfide–melt}=4400$, $D(Co)_{sulfide–melt}=164$, $D(Fe)_{sulfide–melt}=16.9$ and $D(Ni)_{sulfide–melt}=510$, $D(Co)_{sulfide–melt}=17.7$, $D(Fe)_{sulfide–melt}=1.53$. We assume the chondrule starts with chondritic elemental abundances [Ni]=10500 ppm, [Co]=500 ppm, [Fe]= 181000 ppm and crystallizes only olivine and sulfide, in keeping with the petrology of the chondrules analyzed. Since sulfide has a strong influence on partitioning, we show cases with masses of (modally) crystallizing sulfide at the low (0.2%) and high (1%) ends of the ranges of troilite abundances in similar chondrules reported by Jones (1990). We use the lower partition coefficients with low modal sulfide and the higher partition coefficients with the higher modal sulfide, again to show a full range of possible outcomes. Tick marks represent 0.12, 0.14, 0.16 and 0.18 solid fractionation for the high sulfide case and 0.65, 0.7, 0.75 and 0.8 for the low sulfide case. The composition of the low-Ca pyroxene to crystallize from our modelled mesostatis is calculated using partition coefficients of $D(Ni)_{px–melt}=2.75$, $D(Co)_{px–melt}=1.31$ $D(Fe)_{px–melt}=0.77$ from the parameterizations of Beattie et al. (1991). To compare observations with our model data we show the error envelopes of two points at the high Fe/Ni end of isochrons from two chondrule analyzed by SIMS (Mishra and Chaussidon 2014). One of the ^{60}Ni anomalies (Ch12), and much of the other (Ch1, see Fig. 6c), can be explained by $^{59}CoH^+$ according to these calculations, although more work is required to constrain better all input values.

Moynier et al. (2007) and Cook et al. (2007) reported Ni isotropic analyses for a wide range of iron meteorites and metal phases in chondrites. Planetary cores dominate the Ni budgets of planets and planetesimals, so iron meteorites should provide a representative mass-dependent Ni isotopic composition of their parent body. Likewise a metal phase will control the Ni bulk composition of host chondrites. These sample-standard bracketing data of Moynier et al. (2007) and Cook et al. (2007) are replotted as $\delta^{60/58}Ni$ in Fig. 8a and vary from –0.1–0.9‰ with an average ~0.3‰ (cited 2SD external reproducibility of $\delta^{60/58}Ni$ for both studies ~±0.15‰). We note that Moynier et al. (2007) reference their data to an Aesar Ni solution, not NIST SRM 986 and that Cook et al (2007) report an Aesar Ni solution with $\delta^{60/58}$Ni 0.2 lower than NIST SRM 986. Until the Aesar solution of Moynier et al (2007) is calibrated against NIST SRM 987, the relationship of the associated meteorite data to other studies is unclear, even though there appears good overlap with the data of Cook et al. (2007) without further correction. Cameron et al. (2009) presented $\delta^{60/58}$Ni for a representative selection of bulk chondrite and iron meteorites by double spiking, which showed less scatter than the Moynier et al. (2007) and Cook et al. (2007) data but with a similar average of 0.27 ± 0.06‰. A set of iron meteorites measured by sample-standard bracketing (Chernonozhkin et al. 2016) and two higher precision double-spiked measurements of chondrites Steele et al. (2012) fall within this range. The iron meteorite and chondrite data summarized in Table 2 are combined to give a planetary reference of $\delta^{60/58}$Ni = 0.25 ± 0.02. Whilst there may be systematic differences in $\delta^{60/58}$Ni between different meteorite types (Moynier et al. 2007), given the limited number of samples currently available and their small variability relative to precision, we have done no more than to compile this average, which implicitly assumes a uniform bulk Ni mass-dependent isotopic composition for different planetary objects in the solar system.

To date, most literature $\delta^{60/58}$Ni analyses of peridotitic material or mantle derived melts, appropriate for estimating the composition of the silicate Earth, are reference materials. In their assessment of a baseline value for silicates, Cameron et al. (2009) found that samples from peridotite to granite, in addition to a range of continental sediments varied little, with $\delta^{60/58}$Ni = 0.15 ± 0.24. Combining two higher precision measurements of fresh peridotites with the data of Cameron et al. (2009), Steele et al. (2011) calculated a weighted average of 0.18 ± 0.04‰ for the silicate Earth. Additional analyses of mafic and ultra-mafic samples by Gall et al. (2012), Gueguen et al. (2013) and Chernonozhkin et al. (2015) support such isotopically lighter values of the terrestrial mantle relative to the meteorite reference (Fig. 8a,b). Gueguen et al. (2013) compiled analyses of silicate samples to calculate $\delta^{60/58}$Ni = 0.05 ± 0.05 for bulk silicate Earth. The latter includes measurements of deep sea clays, as representative of the continental crust. Given the likely pelagic component of some Ni in these sediments and possible associated isotopic fractionations (see below), coupled with the trivial contribution of crustal Ni to bulk silicate Earth, the inclusion of these sediments is unwarranted but does not unduly bias the estimate. Revising this value to include only samples of the mantle and mantle derived melts, a better defined $\delta^{60/58}$Ni = 0.11 ± 0.01 is obtained for the bulk silicate Earth (Table 2).

These data thus imply that the bulk silicate Earth is slightly isotopically light relative to a bulk planetary reference, by ~0.15‰. Most plausibly this is the result of core formation, as Ni is a siderophile element and strongly partitioned into the metal phase. The sense of isotopic fractionation between the silicate Earth and chondrites is consistent with the isotope partitioning experiments of Lazar et al. (2012), who determined a fractionation factor between metal and talc, $\Delta^{62/58}\text{Ni}_{\text{metal–silicate}} = (0.25 \pm 0.02) \times 10^6/T^2$. Using this expression to calculate isotopic fractionation between metal and silicate during core formation predicts a rather more modest than observed difference in $\delta^{60/58}$Ni between bulk and silicate Earth of ~0.02‰ for a single core–mantle equilibrium at 2500 K. Yet, as the authors note, larger net fractionations are possible for a Rayleigh process with a well mixed mantle magma ocean. Clearly, there is also scope for measuring isotopic fractionation under conditions that more closely approximate core formation, although these are difficult experiments. Empirical constraints on high temperature fractionation factors could potentially be gleaned from the analyses of some metal and silicate separates from mesosiderites and pallasites presented by Chernonozhkin et al. (2016). However, the sense of fractionation between metal and silicate in these two different meteorite types is mutually inconsistent and an important role for kinetic (diffusive) fractionation during cooling seems likely (Chernonozhkin et al. 2016, see also Cook et al. 2007).

Table 2. Averages of mass-dependent and mass-independent Ni isotope compositions for calcium aluminium rich meteorite inclusions, meteorite groups and various terrestrial reservoirs. Given the several generations of mass independent data with widely varying precisions (Figs. 1–4) we have only used those studies with sufficient precision to resolve different meteorite groups and we use weighted means. For the mass-dependent data, the variability in precision is less significant and so we do not use weighting in the calculation of means and standard errors. For some reservoirs, in which samples clearly do not represent a homogenous population, we report standard deviations rather than standard errors and these uncertainties are indicated with *. We do not include the Moynier et al. (2007) data, as unlike all other data these are not normalized to NIST SRM 986 and their relation to the other data is uncertain. All uncertainties (2SD and 2SE) have been adjusted by multiplying by the expression $1+\frac{1}{4(n-1)}$ which compensates for the bias in standard deviation of small sample sizes (Gurland and Tripathi 1971). Data sources indicated as follows, where capitals indicate studies dominantly used as sources of mass independent data and lower case for mass dependent measurements: [A] Birck and Lugmair (1988) [b] Cameron et al. (2009) [c] Cameron and Vance (2014) [d] Chernonozhkin et al. (2015) [e] Chernonozhkin et al. (2016) [f] Cook et al. (2007) [G] Dauphas et al. (2008) [h] Estrade et al. (2015) [i] Gall et al. (2013) [j] Gueguen et al. (2013) [k] Porter et al. (2014) [L] Quitté et al. (2007) [m] Ratié et al. (2015) [n] Ratié et al. (2016) [O] Regelous et al. (2008) [P] Steele et al. (2011) [Q] Steele et al. (2012) [R] Tang and Dauphas (2012) [S] Tang and Dauphas (2014) [T] Tang and Dauphas (2015).

Sample	Type	Reference	*n*	$\varepsilon^{60}Ni_{58/61}$	2SE	*n*	$\varepsilon^{62}Ni_{58/61}$	2SE	*n*	$\varepsilon^{64}Ni_{58/61}$	2SE	*n*	$\delta^{60/58}Ni$	2SE
CAI	—	[A,L]	15	0.515	0.069	15	0.599	0.135	7	2.779	0.503	—	—	—
							Chondrites							
CI	CC	[b,O,Q,S]	3	−0.008	0.010	3	0.208	0.027	2	0.558	0.090	3	0.18	0.03
CV	CC	[b,O,Q,R,S]	7	−0.109	0.008	7	0.102	0.014	5	0.296	0.032	1	0.30	0.05
CM	CC	[b,O,Q,R,S]	7	−0.098	0.013	7	0.114	0.023	5	0.308	0.059	1	0.21	0.03
CO	CC	[b,O,Q]	2	−0.074	0.027	2	0.103	0.031	1	0.262	0.107	1	0.31	0.07
CR	CC	[f,O,Q]	2	−0.163	0.018	2	0.113	0.026	1	0.361	0.065	1	0.30	0.06
CB	CC	[f,R]	3	−0.190	0.039	3	0.153	0.081	—	—	—	1	0.41	0.07
EH	EC	[b,f,O,Q,R]	5	−0.016	0.011	5	0.035	0.028	3	0.100	0.052	3	0.18	0.01
EL	EC	[O,Q,R]	3	−0.031	0.019	3	−0.053	0.027	2	−0.044	0.074	—	—	—
H	OC	[e,f,Q,S]	8	−0.052	0.006	8	−0.055	0.012	8	−0.164	0.030	3	0.25	0.06
HL	OC	[O,Q]	2	−0.052	0.021	2	−0.068	0.038	1	−0.253	0.048	—	—	—
L	OC	[e,Q]	2	−0.033	0.019	2	−0.043	0.023	2	−0.108	0.080	1	0.24	0.01
LL	OC	[b,e,f,G,O,Q,R]	6	−0.047	0.012	6	−0.083	0.010	3	−0.204	0.056	4	0.19	0.07

Table 2 (cont'd). Data sources indicated as follows, where capitals indicate studies dominantly used as sources of mass independent data and lower case for mass dependent measurements: [A] Birck and Lugmair (1988) [b] Cameron et al. (2009) [c] Cameron and Vance (2014) [d] Chernonozhkin et al. (2015) [e] Chernonozhkin et al. (2016) [f] Cook et al. (2007) [G] Dauphas et al. (2008) [h] Estrade et al. (2015) [i] Gall et al. (2013) [j] Gueguen et al. (2013) [k] Porter et al. (2014) [L] Quitté et al. (2007) [m] Ratié et al. (2015) [n] Ratié et al. (2016) [O] Regelous et al. (2008) [P] Steele et al. (2011) [Q] Steele et al. (2012) [R] Tang and Dauphas (2012) [S] Tang and Dauphas (2014) [T] Tang and Dauphas (2015).

Sample	Type	Reference	*n*	$\varepsilon^{60}Ni_{58/61}$	2SE	*n*	$\varepsilon^{62}Ni_{58/61}$	2SE	*n*	$\varepsilon^{64}Ni_{58/61}$	2SE	*n*	$\delta^{60/58}Ni$	2SE
							Achondrites							
IC	Iron	[O,P]	4	–0.021	0.013	4	–0.053	0.021	2	–0.144	0.073	—	—	—
IIAB	Iron	[b,e,f,G,O,P,R]	5	–0.029	0.013	5	–0.096	0.028	4	–0.248	0.036	4	0.29	0.08
IIIAB	Iron	[b,f,G,O,P,R]	7	–0.053	0.013	7	–0.093	0.021	5	–0.276	0.040	18	0.22	0.06
IVA	Iron	[b,f,j,O,P,R]	5	–0.057	0.010	5	–0.053	0.017	3	–0.191	0.049	4	0.29	0.06
IVB	Iron	[b,f,O,P,R]	11	–0.128	0.004	11	0.086	0.005	7	0.222	0.027	5	0.26	0.05
IIC	Iron	[e]	—	—	—	—	—	—	—	—	—	1	0.23	0.02
IIE	Iron	[e]	—	—	—	—	—	—	—	—	—	1	0.24	0.02
Angrite	Silicate	[R]	—	—	—	18	0.032	0.035	—	—	—	—	—	—
Aubrite	Silicate	[R]	—	—	—	1	0.050	0.190	—	—	—	—	—	—
HED	Silicate	[R]	—	—	—	9	0.065	0.060	—	—	—	—	—	—
Ureilite	Silicate	[R]	—	—	—	2	–0.070	0.170	—	—	—	—	—	—
SNC	Mars	[S]	5	–0.010	0.023	5	0.036	0.048	—	—	—	—	—	—
							Terrestrial Reservoirs							
Bulk Silicate Earth	Earth	[b,d,j,O,Q,R,S,T]	16	–0.005	0.006	16	0.032	0.008	3	0.114	0.023	15	0.15	0.03
Seawater	Earth	[c]	—	—	—	—	—	—	—	—	—	30	1.44	0.03
River Water	Earth	[c]	—	—	—	—	—	—	—	—	—	16	0.84	0.63*
Organic-Rich Shales	Earth	[k]	—	—	—	—	—	—	—	—	—	18	0.92	1.22*
Oceanic Fe–Mn crusts	Earth	[i]	—	—	—	—	—	—	—	—	—	25	1.58	0.89*
Soils	Earth	[h,m,n]	—	—	—	—	—	—	—	—	—	17	–0.07	0.32*

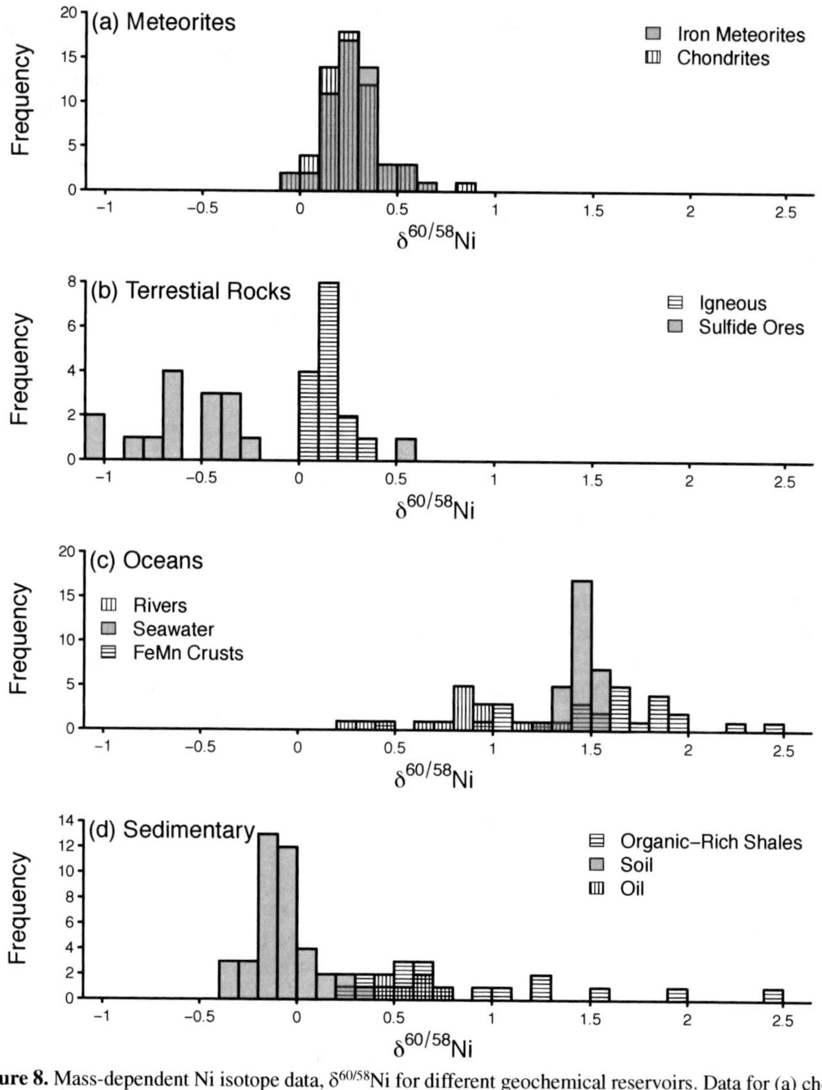

Figure 8. Mass-dependent Ni isotope data, $\delta^{60/58}$Ni for different geochemical reservoirs. Data for (a) chondritic and iron meteorites (Cook et al. 2007; Moynier et al. 2007; Cameron et al. 2009; Steele et al. 2011, 2012; Gueguen et al. 2013; Chernonozhkin et al. 2016), (b) terrestrial igneous rocks and Ni ores, formed by sulfide segregation in igneous systems (Tanimizu and Hirata 2006; Cameron et al. 2009; Steele et al. 2011, 2012; Gueguen et al. 2013; Hofmann et al. 2014; Chernonozhkin et al. 2015) (c) the modern ocean and its major input and output fluxes (Gall et al. 2013; Cameron and Vance 2014) and (d) biosphere processed materials (Porter et al. 2014; Estrade et al. 2015; Ratié et al. 2015, 2016: Ventura et al. 2015).

In contrast to the relative constancy of $\delta^{60/58}$Ni in many silicate samples noted above, Steele et al. (2011), Gueguen et al. (2013) and Hofmann et al. (2014) reported a wide range of isotopically light values ($\delta^{60/58}$Ni = −0.3 to −1) in Ni-bearing sulfides from komatiitic magmas (see Fig. 8b). Such ores are believed to be formed by sulfide saturation following crustal assimilation by these ultra-mafic magmas (Huppert et al. 1984). The Ni in the sulfides is dominantly from the mantle derived melts and so the data imply a significant Ni isotopic fractionation between sulfide and silicate magma, although kinetic effects may also play a role. These low $\delta^{60/58}$Ni echo the value of −0.34‰ measured previously for a millerite ore from the Thompson Ni Belt (Manitoba, Canada) by Tanimizu and Hirata (2006). Despite experiencing some secondary reworking, these Ni deposits were similarly formed by the interaction of ultra-mafic melts with crustal sulfides (e.g. Bleeker and Macek 1996) and so this process generally seems to result in low $\delta^{60/58}$Ni. On the other hand, an isotopically heavy pentlandite from Sudbury ($\delta^{60/58}$Ni ~ 0.5) was reported by Tanimizu and Hirata (2006). These sulfides were formed from a meteorite impact induced melt sheet, with the Ni largely sourced from the molten continental crust (Walker et al. 1991). This different source of Ni for the Sudbury ore relative to the komatiite hosted samples is a plausible explanation of their contrasting $\delta^{60/58}$Ni. Another style of Ni ore deposit, produced from lateritic weathering of ultramafic lithologies in Brazil, exhibits less anomalous compositions, $\delta^{60/58}$Ni = 0.02–0.2 (Ratié et al. 2016), implying little net fractionation during their formation.

Weathering and the hydrological cycle

As discussed above, there is currently only a single experiment to underpin understanding of the controls on Ni isotopic fractionation at magmatic temperatures (Lazar et al. 2012). In contrast, isotopic fractionations between Ni species relevant for the hydrological cycle have received more extensive, theoretical investigation (Fujii et al. 2011, 2014). Despite the absence of redox changes at the Earth's surface, differences in the bonding environment between common, aqueous Ni species result in significant isotopic fractionations (Fig. 9). These first principles calculations for a homogenous aqueous system provide a useful guide to potential magnitudes of Ni isotopic fractionation but do not currently provide direct information on the effects of sorption of Ni onto Fe/Mn oxide surfaces, for example, which likely play an important role in the surface cycle of Ni (e.g. Peacock and Sherman 2007). However, experimental work by Wasylenki et al. (2015) documents that sorption of Ni onto the ferrihydride favours the lighter isotopes, leaving an isotopically heavier residual Ni in solution (Fig. 9).

The processes of weathering, transport to and removal from the ocean can result in major isotopic fractionations, such that the isotopic composition of seawater is a notable end-member for a number of elements (e.g. Li, B, Mg, Mo). In a comprehensive study of seawater, the first to define its Ni isotopic composition, Cameron and Vance (2014) reported $\delta^{60/58}$Ni = 1.44 ± 0.15 which is near constant across all ocean basins and at various depths. Seawater is thus markedly isotopically heavy relative to typical silicate rocks (Fig. 8b,c). Cameron and Vance (2014) further documented that the dissolved Ni loads of rivers are dominantly isotopically heavy relative to unaltered silicates (Fig. 8b,c), showing that weathering processes result in significant Ni isotopic fractionation. Similarly, the $\delta^{60/58}$Ni ~0.5 of organic-rich marine shales (Fig. 8d) is argued to reflect the end product of the weathering process (Porter et al. 2014), as is also likely the case for the more modestly elevated $\delta^{60/58}$Ni of coal and banded ironstone formation (BIF) standards (Gueguen et al. 2013). As a corollary, several studies have shown that soils (Fig. 8d), the residues of weathering, are commensurately isotopically light (Gall et al. 2013; Estrade et al. 2015; Ratié et al. 2015).

A major sink of Ni in the oceans is its incorporation into Fe-Mn crusts (e.g. Peacock and Sherman 2007). A detailed study of these materials by Gall et al. (2013) showed $\delta^{60/58}$Ni that ranged from 0.9–2.5‰, with modern samples yielding an average of 1.6 ± 0.8‰ (Fig. 8c). This work found little systematic variation in the $\delta^{60/58}$Ni of samples between ocean basins or indeed over an ~80 Ma record in the Pacific. In large part, it appears that the $\delta^{60/58}$Ni of Fe–Mn crusts reflects the isotopically heavy composition of seawater and so this sink does not represent a major source of isotopic fractionation as is the case in the Mo isotope cycle (e.g. Barling et al. 2001). Earlier in geological history, sorption of Ni to iron oxide phases, preserved in BIF, may have been a significant sink for oceanic Ni. The near constant

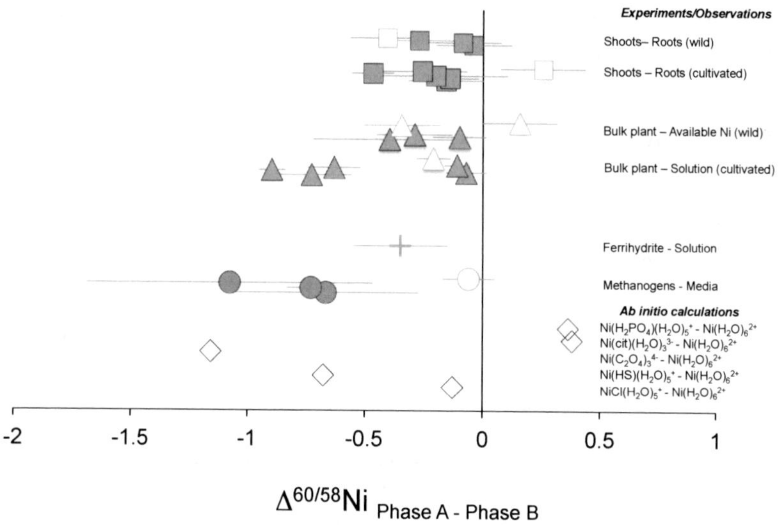

Figure 9. Mass dependent Ni isotope fractionations (in parts per thousand difference in $^{60}Ni/^{58}Ni$) between co-existing phases determined by theory and observation. *Ab initio* calculations of fractionations between different aqueous Ni species (triangles) from Fujii et al. (2011, 2014), cultured methanogens (filled circles) and non-methanogenic Archea (open circle) from Cameron et al. (2009), laboratory sorption onto ferrihydrite (cross) from Wasylenki et al. (2015), cultivated and naturally grown planets from Deng et al. (2014) and Estrade et al. (2015) respectively, with fractionation between bulk plants and available-Ni shown as triangles and shoot–root fractionations as squares (filled symbols indicate Ni hyperaccumulators and open symbols nonaccumulators).

$\Delta^{60/58}Ni_{dissolved-sorbed} = 0.35 \pm 0.10$ determined for Ni sorption onto ferrohydrite over a range of conditions (Wasylenki et al. 2015), however, is in the wrong direction to account for the isotopic difference between seawater and most modern Fe–Mn crusts (Fig. 8c) by such a sorption process. On the other hand, the sense of fractionation could help account for the isotopically heavy composition of Ni in rivers as the product of weathering.

Although the first order observations on the hydrological cycling of Ni isotopes have thus recently been outlined, the isotopic balance of Ni is still not obviously complete (Cameron and Vance 2014) and this further echoes a long identified elemental imbalance between oceanic inputs and outputs for Ni (e.g. Krishmaswami 1976). Whilst rivers are isotopically heavy inputs to the oceans, the oceans are ~0.6‰ heavier than the mean riverine input. There always remains an uncertainty over the representativeness of the river sample, but the magnitude of the Ni isotopic mismatch is troublesome. Since Fe–Mn crusts are isotopically heavier, on average, they do not provide a ready explanation to the problem. A tempting scenario would be to invoke the role of kinetic isotopic fractionation of Ni resulting from its essential role in several biological pathways in the marine environment (see below). Yet the lack of difference in $\delta^{60/58}Ni$ between surface and deep water limits the magnitude of this light isotope sink (Cameron and Vance 2014). Potential solutions to resolve this imbalance are discussed by Cameron and Vance (2014) who tentatively propose the release of Ni carried on Fe–Mn coatings from the suspended load during estuarine processes. The authors argue against the role of serpentinization, given an absence of vertical $\delta^{60/58}Ni$ variation in the water column. Yet, a single, oceanic serpentine sample reported by Gueguen et al. (2013) is notably isotopically light ($\delta^{60/58}Ni = -0.13$), at high [Ni] ~ 1500 ppm and so tantalizingly hints at a plausible reservoir to complement the isotopically heavy oceans.

Most recently, however, measurements of isotopically heavy Ni ($\delta^{60/58}$Ni ~ 2) in euxinic deep waters of the Black Sea (Vance et al. 2016) appear to be the complement to partial removal to the sediment pile of a sorped, Ni sulfide species, which Fujii et al. (2011) predicted to be isotopically light (Fig. 9). If such a process occurred more generally in the sediment interface on upwelling margins of the Pacific, which become euxinic a few centimetres below the surface, this might provide an answer to oceanic Ni isotopic balance (Vance et al. 2016).

Biological systems

As with some other first row transition elements, Ni plays an important role in several biogeochemical cycles. It is an essential trace element for key enzymes in methane production and decomposition of urea. Kinetic fractionation in metabolic pathways will favour the least strongly bound, lightest isotopes and given this basic notion it might be anticipated that Ni incorporated into biological systems would have low $\delta^{60/58}$Ni. On the other hand, the first principles calculations (Fujii et al. 2011, 2014) show substantial equilibrium fractionation of Ni isotopes between aqueous species complexed with different naturally occurring organic ligands (Fig. 9). Fractionation factors at 298 K, calculated between the hexa-aqua ion, the dominant Ni species over a range of conditions in natural waters and complexes with oxalate, phosphate and citrate ions are shown in Figure 9. These examples are indicative of the range of fractionations reported by Fujii and co-workers. Moreover they serve as model representations of possible bonding environments for Ni in soils and plants (Fujii et al. 2011, 2014). If biological pathways efficiently discriminate between such species, macroscopic Ni isotopic variability can result. Yet, some caution needs to be exercised in using simple equilibria to predict the effects of a complex sequence of biochemical reactions. For example, the inorganic oxidation of iron results in a 2.75‰ ^{56}Fe/^{54}Fe isotopic fractionation at room temperature (Johnson et al. 2002), whereas if this processes is biologically mediated through iron oxidizing bacteria, a much smaller 1.5‰ fractionation is observed (Croal et al. 2004).

Given the critical role of Ni in biological methanogenesis, it might be anticipated that Ni isotopic fractionation would be associated with species that had evolved to incorporate environmental Ni to these ends. In a reduced environment, such as the early Earth, methanogens likely represent a major component of the surface Ni cycle and have the potential to impart an isotopic record of their presence. In the first study to look at the effects of biological processes on Ni isotope compositions, Cameron et al. (2009) examined Ni isotopic fractionations during laboratory growth of 3 species of methanogens (2 mesophiles, *Methanosarcina barkeri* and *Methanosarcina acetivorans* and one hyperthermophile, *Methanococcus jannaschii*) and one non-methanogen, a heterotrophic Archea (*P. calidifontis*) as reference. Cameron et al. (2009) showed that Ni incorporated into all three methanogens was isotopically light compared to the media in which they grew, $\Delta^{60/58}\mathrm{Ni}_{\mathrm{cells-media}} \sim -1$ but $\Delta^{60/58}\mathrm{Ni}_{\mathrm{cells-media}} \sim 0$ for the heterotrophic archea, *P. Calidifontis* (Fig. 9).

These substantial fractionations show promise as a means for identifying the active role of methanogens in the rock record of the early Earth. As noted earlier, there is a near constant and smaller fractionation when Ni is sorbed from aqueous solution onto ferrihydrite (Wasylenki et al. 2015) and further experiments suggest that such Ni remains immobile during subsequent diagnesis (Robbins et al. 2015). Hence the isotopic composition of Ni in Banded Ironstone Formations (BIF), originally sorbed from Archean seawater is anticipated to be isotopically heavy for a system in which methanogens dominated the marine Ni cycle. Analyses of BIF of different ages therefore should provide a possible means to investigate the proposed demise of the methanogens prior to the rise in atmospheric oxygen ~2.4 Ga (Konhauser et al. 2009). Yet other fractionations could possibly produce a similar signature (Fig. 9). As discussed above, the partial removal of an isotopically light Ni sulfide species has been invoked to account for modern ocean Ni mass balance. This process can impart a similar sense and magnitude of Ni isotopic fractionation to residual ocean water as Ni sequestration by methanogens (Fig. 9). Such *partial* removal of sulfide species, however, seems unlikely to have occurred under the fully reduced oceanic conditions of the Archean and should have become more important at the onset of oxidation- the converse to that expected from the activity of methanogens.

Ni is an essential micro-nutrient in higher plants (e.g. Brown et al. 1987) given its functional presence in the enzyme urease (Dixon et al. 1975). However, Ni abundances as low as 90 ng/g can be sufficient to maintain this function (Dixon et al. 1975) and so plants do not represent a large Ni reservoir, with Ni concentrations typically less than 10 µg/g. Exceptions are the so-called Ni hyperaccumulators, that have evolved to be tolerant of Ni rich soils that develop on ultramafic lithologies. Many plants deal with the toxicity of metals by sequestering them in their roots, but hyperaccumulators transport large quantities of metals to their leaves. Such Ni hyperaccumulators make interesting subjects for Ni isotopic investigation, as high Ni concentrations make them amenable to analysis and their specialization to cope with high Ni concentrations might lead to characteristic isotopic fractionations.

Laboratory experiments (Deng et al. 2014) and a field study (Estrade et al. 2015) reported isotopic analyses for several Ni hyperaccumulators, together with a nonaccumulator species for comparison. In all but one case, the plants have net isotopically light compositions relative to dissolved Ni in growth media or inferred for soil water ($\Delta^{60/58}Ni_{bulk\ plant-solution} = -0.1$ to -0.9), see Figure 9. In the case of the field study, fractionation factors were derived relative to the bio-available Ni obtained by extraction from the soil with diethylenetriaminepentaacetic acid (Estrade et al. 2015) and this process adds some uncertainty to the values derived. However, the isotopically heavy composition of this extractable Ni would be be consistent with predictions if $Ni(C_2O_4)_3^{2-}$ is used as a proxy for Ni bound to humic acid, as suggested by Fujii et al. (2011), and removed from the bio-available pool. The largest Ni isotopic fractionations are observed in some laboratory experiments with Ni-hyperaccumulators but in general there is not a clear distinction between hyperaccumulators and nonaccumulators (Fig. 9). The overall light isotopic composition of the plants is likely the result of kinetic fractionation during Ni up-take by 'low affinity' transport systems across the cell membranes of plant roots. The dominant role of such ion channels, as opposed to 'high affinity' carried-mediated transport, has previously been invoked by Weiss et al. (2005) to account for the light bulk Zn isotopic compositions of plants they studied.

Isotopically lighter Zn in shoots relative to roots was first reported by Weiss et al. (2005) and similar features are observed in the Ni isotopic studies of plants (Deng et al. 2014; Estrade et al. 2015). In a comparison of shoot-root isotopic differences of Zn and Ni in the same species, Deng et al. (2014) noted a more modest fractionations for Ni. They attributed this to higher Ni mobility in plants relative to Zn. It should be re-emphasized that most of the species analyzed for their Ni isotopic compositions are hyperaccumulators, in which Ni is anomalously enriched in shoots relative to roots. The two nonaccumulators studied so far show contrasting behavior, with both positive, *Thlapsi arvensa* (Deng et al. 2014) and negative, *Euphorbia spinosa* (Estrade et al. 2015) values of $\Delta^{60/58}Ni_{shoots-roots}$, see Figure 9. Some variability in $\Delta^{60/58}$Ni between roots, stems and leaves may be related to the growth stage of the plants (Estrade et al. 2015). A full explanation of the diverse results requires more detailed work on the transport of Ni in plants, which is currently poorly understood. However, these data do provide a means to assess the generality of a hypothesis proposed to explain plant isotopic variability in the Zn system. Fujii and Albarède (2012) invoked isotopic fractionations between Zn phosphate species concentrated in the roots and citrates and malates in the leaves. In the case of Ni, Fujii et al. (2014) calculated that no discernible fractionations are predicted between Ni phosphate and citrate species (Fig. 9) and so this mechanism cannot explain the root-shoot variability in $\delta^{60/58}$Ni.

Not only is Ni incorporated by living plants, but also during diagenesis of plant matter Ni frequently becomes bound by porphyrins, derived from chlorophyll, that originally complexed Mg. Thus, as an important trace element in oils, the abundance of Ni has, especially in relation to V, long been used as a tracer of source and evolution of oil reservoirs (e.g. Hodgson 1954). The additional constraint of the Ni isotopic composition of crude oils is an intriguing new means to understand oil formation. A preliminary study by Ventura et al. (2015) report $\delta^{60/58}$Ni (together with V and Mo isotopic compositions) for a global selection of crude oils. These show a limited range of $\delta^{60/58}$Ni from 0.4 to 0.7, which overlaps the composition of organic-rich shales (Fig. 8c).

OUTLOOK

There are exciting prospects for Ni isotopic measurements across the fields outlined above.

Although the most anomalous macroscopic objects in the solar system, CAIs have not been systematically studied for their Ni isotopic compositions and rather few analyses currently exist. A recent study has reported extreme $\varepsilon^{64}Ni_{61/58}$ (up to 55) in two Allende CAIs (Chen and Papanastassiou 2014). Further investigation of Ni isotopic compositions of CAIs and their relation to anomalies of other neutron rich nuclides would be valuable (e.g. recent discussion by Chen et al. 2015). More generally, finding the carrier that is responsible for bulk meteoritic Ni isotopic variation is a key quest.

It is clearly critical to resolve definitively the value of $^{60}Fe/^{56}Fe°_{SSI}$, given its implications for the seeding of the solar system with exotic isotopic signatures. We strongly favour low values of $^{60}Fe/^{56}Fe°_{SSI}$, consistent with background ISM, but a community consensus is required. We suggest that exploring the potential role of interfering $^{59}CoH^+$ as highlighted in the section *Extinct ^{60}Fe and radiogenic ^{60}Ni* would be a useful step forward.

The measurable abundances of Ni in both silicate and iron meteorites allows valuable comparison of different meteoritic objects. In contrast to the other magmatic irons studied so far, the IVB iron meteorites have affinities with carbonaceous rather than ordinary chondrites. It would be useful to make a more comprehensive investigation of these genetic relationship in other, so far unanalyzed meteorite types.

As with Mg (Teng et al. 2007), for which Ni typically substitutes, there appears to be little mass-dependent Ni isotopic fractionation during silicate partial melting and differentiation. This observation would benefit from more detailed study and unpublished work by Gall (2011) suggests that there might be some systematic silicate melt-solid fractionations at magmatic temperatures. Publication of these and additional empirical data in conjunction with experimental determinations of fractionation factors would be helpful. In the absence of major fractionations during silicate differentiation, the small, but apparently significant fractionation of Ni isotopes during core formation becomes of interest. To substantiate this observation a more robust estimate of bulk silicate Earth is required. The current value relies heavily on a few reference standards. Additional observations of $\delta^{60/58}$Ni on the silicate mantles of smaller planetary bodies, where Ni isotopic fractionation might be larger (due to greater Ni partitioning into the core at low pressure), should prove informative. Interpretation of such results would greatly benefit from further experimental determinations of silicate–metal fractionation factors.

Better understanding the seemingly significant fractionation of Ni isotopes during weathering processes and a clear resolution of the apparent isotope imbalance in the ocean are obvious targets for future research. Comparisons with the better studied, non-redox sensitive, divalent cations, Mg and Ca, would be welcome. The contrast in oceanic sinks for these different elements (Fe-Mn crusts, versus seafloor alteration, versus carbonate production) makes differences in their behavior through time of considerable interest.

As with other novel, isotopic systems, their nascent application to biological systems carries unknown potential. The large Ni isotopic fractionation of methanogens is a specifically appealing prospect, but whether or not such Ni isotopic fractionations remains locked in the rock record and sufficiently distinct over background variability, including abiotic reactions in the weathering environment, remains to be seen.

ACKNOWLEDGEMENTS

In the writing of this review TE was supported by the ERC (AdG 321209 ISONEB) and RCJS by ETH. This funding is gratefully acknowledged. The authors wish to thank the Editor for his patience and two anonymous reviewers for numerous useful suggestions, kindly pointing to some egregiously missed references and encouraging us to expand the final section to speculate in a field in which we have limited expertise. A timely, final sanity check by Derek Vance and access to his unpublished manuscript was appreciated. The authors would like to doff their caps to the many who have helped in the development of their ideas about the Ni system, not least Chris Coath, Marcel Regelous, Hart Chen, Matthias Willbold, Derek Vance, Vyll Cameron and Corey Archer.

REFERENCES

Albalat E, Telouk P, Albarède F (2012) Er and Yb isotope fractionation in planetary materials. Earth Planet Sci Lett 355–356:39–50

Amelin Y (2008a) U–Pb ages of angrites. Geochim Cosmochim Acta 72:221–232

Amelin Y (2008b) The U–Pb systematics of angrite Sahara 99555. Geochim Cosmochim Acta 72:4874–4885

Barling J, Arnold GL, Anbar AD (2001) Natural mass-dependent variations in the isotopic composition of molybdenum. Earth Planet Sci Lett 193:447–457

Beattie P, Ford C, Russell D (1991) Partition coefficients for olivine–melt and orthopyroxene–melt systems. Contrib Mineral Petrol 109:212–224

Birck J-L, Allègre CJ (1985) Evidence for the presence of ^{53}Mn in the early solar system. Geophys Res Lett 12:745–748

Birck J-L, Lugmair GW (1988) Nickel and chromium isotopes in Allende inclusions. Earth Planet Sci Lett 90:131–143

Bizzarro M, Baker JA, Haack H, Lundgaard KL (2005) Rapid timescales for accretion and melting of differentiated planetesimals inferred from ^{26}Al–^{26}Mg chronometry. Astrophys J 632:L41–L44

Bizzarro M, Ulfbeck D, Trinquier A, Thrane K, Connelly JN, Meyer BS (2007) Evidence for a late supernova injection of Fe-60 into the protoplanetary disk. Science 316:1178–1181

Bizzarro M, Ulfbeck D, Boyd JA, Haack H (2010) Nickel isotope anomalies in iron meteorites. Meteorit Planet Sci 45:A15

Blichert-Toft J, Moynier F, Lee C-TA, Telouk P, Albarède F (2010) The early formation of the IVA iron meteorite parent body. Earth Planet Sci Lett 296:469–480

Bleeker W, Macek JJ (1996) Evolution of the Thompson Nickel Belt, Manitoba: setting of Ni–Cu deposits in the western part of the circum Superior boundary zone. Fieldtrip Guidebook Geological Association of Canada Annual Meeting, Winnipeg, Manitoba

Brennecka GM, Wadhwa A (2012) Uranium isotope compositions of the basaltic angrite meteorites and the chronological implications for the early Solar System. PNAS 109 A:9299–9303

Brown PH, Welch RM, Cary EE (1987) Nickel: a micronutrient essential for higher plants. Plant Physiol 85:801–803

Budde G, Kleine T, Kruijer TS, Burkhardt C, Metzler K (2016) Tungsten isotopic constraints on the age and origin of chondrules. PNAS 113:2886–2891

Burbidge EM, Burbidge GR, Fowler WA, Hoyle F (1957) Synthesis of the elements in stars. Rev Mod Phys 29:547–650

Cameron AGW, Truran JW (1977) Supernova trigger for formation of Solar-System. Icarus 30: 447–461

Cameron V, Vance D (2014) Heavy nickel isotope compositions in rivers and the oceans. Geochim Cosmochim Acta 128:195–211

Cameron V, Vance D, Archer C, House CH (2009) A biomarker based on the stable isotopes of nickel. PNAS 106:10944–10948

Chen JH, Papanastassiou DA (2014) Endemic ^{64}Ni effects in Allende Ca–Al-rich inclusions. Lunar Planet Sci 45:2327

Chen JH, Papanastassiou DA, Wasserburg GJ (2009) A search for nickel isotopic anomalies in iron meteorites and chondrites. Geochim Cosmochim Acta 73:1461–1471

Chen H-W, Lee T, Lee D-C, Chen L-C (2015) Correlation of ^{48}Ca, ^{50}Ti and ^{138}La heterogeneity in the Allende refractory inclusions. Astrophys Lett J 806:L21

Chernonozhkin SM, Goderis S, Lobo L, Claeys P, Vanhaecke F (2015) Development of an isolation procedure and MC-ICP-MS measurement protocol for the study of stable isotope ratio variations of nickel. Anal J Atom Spectrom 30:1518–1530

Chernonozhkin SM, Goderis S, Costas-Rodríguez M, Claeys P, Vanhaecke F (2016) Effect of parent body evolution on equilibrium and kinetic isotope fractionation: a combined Ni and Fe isotope study of iron and stony-iron meteorites. Geochim Cosmochim Acta 186:168–188

Clayton DD, Jin L (1995) A new interpretation of ^{26}Al in meteoritic inclusions. Astrophys J 451:L87–L91

Clayton RN, Grossman L, Mayeda TK (1973) A component of primitive nuclear composition in carbonaceous meteorites. Science 182:485–488

Clayton RN, Mayeda TK, Olsen EJ, Prinz M (1983) Oxygen isotope relationships in iron meteorites. Earth Planet Sci Lett 65:229–232

Coath CD, Steele RCJ, Lunnon WF (2013) Statistical bias in isotope ratios. Anal J Atom Spectrom 28:52–58

Connelly JN, Bizzarro M, Thrane K, Baker JA (2008) The Pb–Pb age of angrite SAH99555 revisited. Geochim Cosmochim Acta 72:4813–4824

Connelly JN, Bizzarro M, Krot AN, Nordlund A, Wielandt D, Ivanova MA (2012) The absolute chronology and thermal processing of solids in the protoplanetary disk. Science 338:651–655

Cook DL, Wadhwa M, Janney PE, Dauphas N, Clayton RN, Davis AM (2006) High precision measurements of non-mass-dependent effects in nickel isotopes in meteoritic metal via multicollector ICPMS. Anal Chem 78:8477–8484

Cook DL, Clayton RN, Wadhwa M, Janney PE, Davis AM (2008) Nickel isotopic anomalies in troilite from iron meteorites. Geophys Res Lett 35:L01203

Croal LR, Johnson CM, Beard BL, Newman DK (2004) Iron isotope fractionation by Fe(II)-oxidizing photoautotrophic bacteria. Geochim Cosmochim Acta 68:1227–1242

Dauphas N, Cook DL, Sacarabany A, Frohlich C, Davis AM, Wadhwa M, Pourmand A, Rauscher T, Gallino R (2008) Iron 60 evidence for early injection and efficient mixing of stellar debris in the protosolar nebula. Astrophys J 686:560–569

Dauphas N, Remusat L, Chen JH, Roskosz M, Papanastassiou DA, Stodolna J, Guan Y, Ma C, Eiler JM (2010) Neutron-rich chromium isotope anomalies in supernova nanoparticles. Astrophys J 720:1577–1591

Deng T-H-B, Cloquet C, Tang Y-T, Sterckeman T, Echevarria G, Estrade N, Morel J-L, Qiu R-L (2014) Nickel and zinc isotope fractionation in hyperaccumulating and nonaccumulating plants. Environ Sci Technol 48:11926–11933

Dixon NE, Gazzola C, Blakeley RL, Zerner B (1975) Jack bean urease (EC 3.5.1.5). Metalloenzyme. Simple biological role for nickel. Am J Chem Soc 97:4131–4133

Estrade N, Cloquet C, Echevarria G, Sterckeman T, Deng T, Tang Y, Morel J-L (2015) Weathering and vegetation controls on nickel isotope fractionation in surface ultramafic environments (Albania). Earth Planet Sci Lett 423:24–35

Fujii T, Albarède F (2012) An initio calcuation of the Zn isotope effect in phosphates, citrates and malates and applications to plants and soils. PLoS One 7:e30726

Fujii T, Moynier F, Dauphas N, Abe M (2011) Theoretical and experimental investigation of nickel isotopic fractionation in species relevant to modern and ancient oceans. Geochim Cosmochim Acta 75:469–482

Fujii T, Moynier F, Blichert J-Toft, Albarède F (2014) Density functional theory estimation of isotope fractionation of Fe, Ni Cu and Zn among species relevant to geochemical and biological environments. Geochim Cosmochim Acta 140:553–576

Gaetani GA, Grove TL (1997) Partitioning of moderately siderophile elements among olivine silicate melt and sulfide melt: constraints on core formation in the Earth and Mars. Geochim Cosmochim Acta 61:1829–1846

Gall L (2011) Development and application of nickel stable isotopes as a new geochemical tracer. PhD, University of Oxford

Gall L, Williams H, Siebert C, Halliday AN (2012) Determination of mass-dependent variations in nickel isotope compositions using double spiking and MC–ICPMS Anal J At Spectrom 27:137–145

Gall L, Williams HM, Siebert C, Halliday AN, Herrington RJ, Hein JR (2013) Nickel isotopic compositions of ferromanganese crusts and the constancy of deep ocean inputs and continental weathering effects over the Cenozoic. Earth Planet Sci Lett 375:148–155

Gramlich JW, Machlan LA, Barnes IL, Paulsen PJ (1989) Absolute isotopic abundance ratios and atomic weight of a reference sample of nickel. Res J NIST 94:347–356

Guan Y, Huss GR, Leshin LA (2007) ^{60}Fe–^{60}Ni and ^{53}Mn–^{53}Cr isotopic systems in sulfides from unequilibrated enstatite chondrites. Geochim Cosmochim Acta 71:4082–4091

Gueguen B, Rouxel O, Ponzevra E, Bekker A, Fouquet Y (2013) Nickel isotope variations in terrestrial silicate rocks and geological reference materials measured by MC-ICP-MS Geostand Geoanal Res 37:297–317

Gurland J, Tripathi RC (1971) A simple approximation for unbiased estimation of the standard deviation. Am Stat 25:30–32

Heydegger HR, Foster JJ, Compston W (1979) Evidence of a new isotopic anomaly from titanium isotopic ratios in meteoritic materials. Nature 278:704–707

Hezel DC, Russell SS, Ross AJ, Kearlsey AT (2008) Modal abundances of CAIs: implications for bulk chondrite element abundances and fractionations. Meteorit Planet Sci 43:1879–1894

Hofmann A, Bekker A, Dirks, Gueguen B, Rumble D, Rouxel OJ (2014) Comparing orthomagmatic and hydrothermal mineralization models for komatiite-hosted nickel deposits in Zimbabwe using multiple-sulfur iron and nickel isotope data. Miner Deposita 49:75–100

Hodgson GW (1954) Vanadium, nickel and porphyrins in crude oils of Western Canada. Am Assoc Petrogeol Bull 41:2413–2426

Huppert HE, Sparks RSJ, Turner JS, Arndt NT (1984) Emplacement and cooling of komatiite lavas. Nature 309:19–22

Johnson CM, Skulan JL, Beard BL, Sun H, Nealson KH, Braterman PS (2002) Isotopic fractionation between Fe(III) and Fe(II) in aqueous solutions. Earth Planet Sci Lett 195:141–153

Jones RH (1990) Petrology and mineralogy of Type II, FeO-rich chondrules in Semarkona (LL3.0): origin by closed-system fractional crystallization with evidence for supercooling. Geochim Cosmochim Acta 54:1785–1802

Jungck MHA, Shimamura T, Lugmair GW (1984) Ca isotope variations in Allende. Geochim Cosmochim Acta 48:2651–2658

Kita NT, Nagahara H, Togashi S, Morshita Y (2000) A short duration of chondrule formation in the solar nebula: evidence from ^{26}Al in Semarkona ferromagnesian chondrules. Geochim Cosmochim Acta 64:3913–3922

Konhauser KO, Pecoits E, Lalonde SV, Papineau D, Nisbet EG, Barley ME, Arndt NT, Zahnle K, Kamber BS (2009) Oceanic nickel depletion and a methanogen famine before the Great Oxidation Event. Nature 458:750–753

Krishnaswami S (1976) Authigenic transition elements in Pacific pelagic clays. Geochim Cosmochim Acta 40:425–434

Kruijer TS, Kleine T, Fischer-Gödde M, Burkhardt C, Wieler R (2014) Nucleosynthetic W isotope anomalies and the Hf–W chronometry of Ca-Al-rich inclusions. Earth Planet Sci Lett 403:317–327

Kutschera W, Billquist PJ, Frekers D, Henning W, Jensen KJ, Xiueng M, Pardo R, Paul M, Rehm KE, Smither RK, Yntema JL (1984) Half-life of ^{60}Fe. Nucl Instrum Meth B5:430–435

Lazar C, Young ED, Manning CE (2012) Experimental determination of equilibrium nickel isotope fractionation between metal and silicate from 500 °C to 950 °C. Geochim Cosmochim Acta 86:276–295

Lee T, Papanastassiou DA, Wasserburg GJ (1978) Calcium isotopic anomalies in Allende meteorite. Astrophys J 220:L21-L25

Luck J-M, Othman DB, Barrat JA, Albarède F (2003) Coupled ^{63}Cu and ^{16}O excesses in chondrites. Geochim Cosmochim Acta 67:143–151

Lugmair GW, Galer SJG (1992) Age and isotopic relationships among the angrites Lewis Cliff-86010 and Angra dos Reis. Geochim Cosmochim Acta 56:1673–1694

Lugmair GW, Shukolyukov A (1998) Early solar system timescales according to ^{53}Mn–^{53}Cr systematics. Geochim Cosmochim Acta 62:2863–2886

Mahon K (1996) The new "York" regression: application of an improved statistical method to Geochemistry. Int Geol Rev 38:293–303

Marhas K, Amari S, Gyngard F, Zinner E, Gallino R (2008) Iron and nickel isotopic ratios in presolar SiC grains. Astrophys J 689:622–645

McCulloch MT, Wasserburg GJ (1978a) Barium and neodymium isotopic anomalies in the Allende meteorite. Astrophys J 220:L15–L19

McCulloch MT, Wasserburg GJ (1978b) More anomalies from the Allende meteorite: Samarium. Geophys Res Lett 5:599–602

Mishra RK, Chaussidon M (2014) Fossil records of high level of ^{60}Fe in chondrules from unequilibrated chondrites. Earth Planet Sci Lett 398:90–100

Mishra RK, Goswami JN (2014) Fe–Ni and Al–Mg isotope records in UOC chondrules: plausible stellar source of ^{60}Fe and other short-lived nuclides in the early solar system. Geochim Cosmochim Acta 132:440–457

Mishra RK, Goswami JN, Tachibana S, Huss GR, Rudraswami NG (2010) ^{60}Fe and ^{26}Al in chondrules from unequilibrated chondrites: implications for early solar system processes. Astrophys J 714:L217-L221

Mishra RK, Marhas KK, Sameer (2016) Abundance of ^{60}Fe inferred from nanoSIMS study of QUE 97008 (L3.05) chondrules. Earth Planet Sci Lett 436:71–81

Morand P, Allègre CJ (1983) Nickel isotopic studies in meteorites. Earth Planet Sci Lett 63:167–176

Mostefaoui S, Lugmair GW, Hoppe P, El Gorsey A (2004) Evidence for live ^{60}Fe in meteorites. New Astron Rev 48:155–159

Mostefaoui S, Lugmair GW, Hoppe P (2005) ^{60}Fe: a heat source for planetary differentiation from a nearby supernova explosion. Astrophys J 625:271–277

Moynier F, Blichert J-Toft:Telouk, Luck J-M, Albarède F (2007) Comparative stable isotope geochemistry of Ni, Cu, Zn and Fe in chondrites and iron meteorites. Geochim Cosm Chim Acta 71:4365–4379

Moynier F, Blichert-Toft J, Wang K, Herzog GF, Albarède F (2011) The elusive ^{60}Fe in the solar system. Astrophys J 741:71

Niederer FR, Papanastassiou DA, Wasserburg GJ (1980) Endemic isotopic anomalies in titanium. Astrophys J 240:L73–L77

Niemeyer S, Lugmair GW (1981) Ubiquitous isotopic anomalies in Ti from normal Allende inclusions. Earth Planet Sci Lett 53:211–225

Ogliore RC, Huss GR, Nagashima K (2011) Ratio estimation in SIMS analysis. Nucl Instrum Meth Phys Res B 269:1910–1918

Papanastassiou DA, Wasserburg GJ (1969) Initial strontium isotopic abundances and the resolution of small time differences in the formation of planetary objects. Earth Planet Sci Lett 5:361–376

Peacock C, Sherman DM (2007) Sorption of Ni by birnessite: equilibrium controls on Ni in seawater. Chem Geol 238:94–106

Porter SJ, Selby D, Cameron V (2014) Characterizing the nickel isotopic composition of organic-rich marine sediments. Chem Geol 387:12–21

Qin L, Nittler LR, Alexander CO, Wang J, Staldermann FJ, Carlson RW (2011) Extreme ^{54}Cr-rich nano-oxides in the CI chondrite Orgueil—implication for a late supernova injection into the solar system. Geochim Cosmochim Acta 75:629–644

Quitté G, Oberlei F (2006) Quantitative extraction and high precision isotope measurements of nickel by MC–ICPMS. Anal J At Spectrom 21:1249–1255

Quitté G, Meier M, Latkoczy C, Halliday AN, Günther D (2006) Nickel isotopes in iron meteorites-nucleosynthetic anomalies in sulfides with no effects in metals and no trace of ^{60}Fe. Earth Planet Sci Lett 242:16–25

Quitté G, Halliday AN, Meyer BS, Markowski A, Latkoczy C, Günther D (2007) Correlated iron 60, nickel 62 and zirconium 96 in refractory inclusions and the origin of the solar system. Astrophys J 655:678–684

Quitté G, Markowski A, Latkoczy C, Gabriel A, Pack A (2010) Iron-60 heterogeneity and incomplete isotope mixing in the early solar system. Astrophys J 720:1215–1224

Quitté G, Latkoczy C, Schönbächler M, Halliday AN, Günther D (2011) ^{60}Fe–^{60}Ni systematics in the eucrite parent body: a case study of Bouvante and Juvinas. Geochim Cosmochim Acta 75:7698–7706

Ratié G, Jouvin D, Garnier J, Rouxel O, Miska S, Guimarães E, Cruz Viera L, Sivry Y, Zelano I, Montarges -Pelletier E, Thil F, Quantin C (2015) Nickel isotope fractionation during tropical weathering of ultramafic rocks. Chem Geol 402:68–76

Ratié G, Quantin C, Jouvin D, Calmels D, Ettler V, Sivry Y, Cruz Viera L, Ponzevera E, Garnier J (2016) Nickel isotope fractionation during laterite Ni ore smelting and refining: implications for tracing the sources of Ni in smelter-affected soils. Appl Geochem 64:136–145

Regelous M, Elliott T, Coath CD (2008) Nickel isotope heterogeneity in the early Solar System. Earth Planet Sci Lett 272:330–338

Robbins LJ, Swanner ED, Lalonde SV, Eickhoff M, Parnich ML, Reinhard CT, Peacock CL, Kappler A, Konhauser KO (2015) Limited Zn and Ni mobility during simulated iron formation diagenesis. Chem Geol 402:30–39

Rudge JF, Reynolds BC, Bourdon B (2009) The double spike toolbox. Chem Geol 265:420–431

Rugel G, Faestermann T, Knie K, Korschinek G, Poutivtsev M, Schumann D, Kivel N, Günther Leopold I, Weinreich R, Wohlmuther M (2009) New measurement of the ^{60}Fe half-life. Phys Rev Lett 103:072502

Schiller M, Baker J, Creech J, Paton C, Millet M-A, Irving AJ, Bizzarro M (2011) Rapid timescales for magma ocean crystallization on the Howardite–Eucrite–Diogenite parent body. Astrophys J 740:L22

Shimamura T, Lugmair GW (1983) Ni isotopic compositions in Allende and other meteorites. Earth Planet Sci Lett 63:177–188

Shukolyukov A, Lugmair GW (1993a) Live iron-60 in the early solar-system. Science 259:1138–1142

Shukolyukov A, Lugmair GW (1993b) ^{60}Fe in eucrites. Earth Planet Sci Lett 119:159–166

Sugiura N, Miyazaki A, Yin Q-Z (2006) Heterogeneous distribution of ^{60}Fe in the early solar nebula: achondrite evidence. Earth Planets Space 58:1079–1086

Spivack-Birndorf LJ, Wadhwa M, Janney PE (2011) ^{60}Fe–^{60}Ni chronology of the d'Orbigny angrite: implications for the initial solar system abundance of ^{60}Fe. Lunar Planet Sci 42:2281

Steele RCJ, Elliott T, Coath CD, Regelous M (2011) Confirmation of mass-independent Ni isotopic variability in iron meteorites. Geochim Cosmochim Acta 75:7906–7925

Steele RCJ, Coath CD, Regelous M, Russell S, Elliott T (2012) Neutron-poor nickel isotope anomalies in meteorites. Astrophys J 756:59

Strelow FWE, Victor AH, van Zyl CR, Eloff C (1971) Distribution coefficients and cation exchange behavior of elements in hydrochloric acid-acetone. Anal Chem 43:870–876

Tachibana S, Huss GR (2003) The initial abundance of ^{60}Fe in the solar system. Astrophys J 588:L41–L44

Tachibana S, Huss GR, Kita NT, Shimoda G, Morishita Y (2006) ^{60}Fe in chondrites: debris from a nearby supernova in the early solar system? Astrophys J 639:L87–L90

Takeda H, Graham AL (1991) Degree of equilibration of eucritic pyroxenes and thermal metamorphism of the earliest planetary crust. Meteoritics 26:129–134

Tang H, Dauphas N (2012) Abundance distribution and origin of ^{60}Fe in the solar protoplanetary disk. Earth Planet Sci Lett 359–360:248–263

Tang H, Dauphas N (2014) ^{60}Fe–^{60}Ni chronology of core formation in Mars. Earth Planet Sci Lett 390:264–274

Tang H, Dauphas N (2015) Low ^{60}Fe abundance in Semarkona and Sahara 99555. Astrophys J 802:22

Tanimizu M, Hirata T (2006) Determination of natural isotopic variation in nickel using inductively coupled plasma mass spectrometry. Anal J At Spectrom 21:1423–1426

Telus M, Huss GR, Ogliore RC, Nagashima K, Tachibana S (2012) Recalculation of data for short-lived radionuclide systems using less-biased ratio estimation. Meteorit Planet Sci 47:2013–2030

Telus M, Huss GR, Ogliore RC, Nagashima K, Howard DL, Newville MG, Tomkins AG (2016) Mobility of iron and nickel at low temperature: implications for the ^{60}Fe–^{60}Ni systematics of chondrules from unequilibrated ordinary chondrites. Geochim Cosmochim Acta 178:87–105

Teng F-Z, Wadhwa M, Helz RT (2007) Investigation of magnesium isotope fractionation during basalt differentiation: implications for a chondritic composition of the terrestrial mantle. Earth Planet Sci Lett 261:84–92

Touboul M, Sprung P, Aciego SM, Bourdon B, Kleine T (2015) Hf–W chronology of the eucrite parent body. Geochim Cosmochim Acta 156:106–121

Trinquier A, Birck J-L, Allègre CJ (2007) Widespread ^{54}Cr heterogeneity in the inner solar system. Astrophys J 655:1179–1185

Trinquier A, Birck J-L, Allègre CJ, Göpel C, Ulfbeck D (2008) ^{53}Mn–^{53}Cr systematics of the early Solar System revisited. Geochim Cosmochim Acta 72:5146–5163

Trinquier A, Elliott T, Ulfbeck D, Coath C, Krot AN, Bizzarro M (2009) Origin of nucleosynthetic isotope heterogeneity in the solar protoplanetary disk. Science 324:374–376

Vance D, Little SH, Archer C, Cameron V, Andersen MB, Rijkenberg MJ, Lyons TW (2016) The oceanic budgets of nickel and zinc isotopes: the importance of sulfidic environments as illustrated by the Black Sea. Phil Trans R Soc A 374:20150294

Ventura GT, Gall L, Siebert C, Prytulak J, Szatmari P, Hürlimann M, Halliday AN (2015) The stable isotope composition of vanadium, nickel and molybdenum in crude oils. Appl Geochem 59:104–117

Victor AH (1986) Separation of nickel from other elements by cation-exhange chromatography in dimethylglyoxime/ hydrochloric acid/acetone media. Anal Chim Acta 183:155–161

Wahlgreen M, Orlandini KA, Korkisch J (1970) Specific cation-exchange separation of nickel. Anal Chim Acta 52:551–553

Walker RJ, Morgan JW, Naldrett AJ, Li C, Fassett JD (1991) Re–Os isotope systematics of Ni–Cu sulfide ores, Sudbury Igneous Complex (1991) Ontario: evidence for a major crustal component. Earth Planet Sci Lett 105:416–429

Wang W, Audi G, Wapstra AH, Kondev FG, MacCormick M, Xu XS, Pfeiffer B (2012) The AME2012 atomic mass evaluation (II). Chinese Phys C36:1603–2014

Wasserburg GJ, Lee T, Papanastassiou DA (1977) Correlated O and Mg isotopic anomalies in Allende inclusions: II Magnesium. Geophys Res Lett 4:299–302

Wasserburg GJ, Gallino R, Busso M (1998) A test of the supernova trigger hypothesis with Fe-60 and Al-26. Astrophys J 500:L189–L193

Wasylenki LE, Howe HD, Spivak-Birndorf L, Bish DL (2015) Ni isotope fractionation during sorption to ferrihydrite: implications for Ni in banded iron formations. Chem Geol 400:56–64

Weiss DJ, Mason TFD, Zhao FJ, Kirk JDG, Coles BJ, Horstwood MSA (2005) Isotopic discrimination of zinc in higher plants. New Phytol 165:703–710

Wieser ME, Holden NE, Coplen TB, Böhlke JK, Berglund M, Brand WA, de Bièvre P, Gröning M, Loss RD, Meija J, Hirata T, Prohaska T, Schoenberg R, O'Connor G, Walczyk T, Yoneda S, Zhu X-K (2013) Atomic Weights of the Elements 2011 (IUPAC Technical Report). Pure Appl Chem 85:1047–1078

Yurimoto H, Kuramoto K (2004) Molecular cloud origin for the oxygen isotope heterogeneity in the solar system. Science 305:1763–1766

York D (1969) Least squares fitting of a straight line with correlated errors. Earth Planet Sci Lett 5:320–324

Reviews in Mineralogy & Geochemistry
Vol. 82 pp. 543-600, 2017

The Isotope Geochemistry of Zinc and Copper

Frédéric Moynier[1,2]

[1]Institut de Physique du Globe de Paris
Université Paris Diderot
Université Sorbonne Paris Cité
CNRS UMR 7154, 1 rue Jussieu,
75238 Paris Cedex 05
France
[2]Institut Universitaire de France
Paris
France

moynier@ipgp.fr

Derek Vance

ETH Zürich
Institute for Geochemistry and Petrology
Department of Earth Sciences
Clausiusstrasse 25
8092 Zürich
Switzerland

derek.vance@erdw.ethz.ch

Toshiyuki Fujii

Division of Sustainable Energy and Environmental Engineering
Graduate School of Engineering
Osaka University
2-1 Yamadaoka Suita
Osaka 565-0871
Japan

fujii@see.eng.osaka-u.ac.jp

Paul Savage

Department of Earth and Environmental Sciences
University of St Andrews
Irvine Building
Fife KY16 9AL
United Kingdom

pss3@st-andrews.ac.uk

INTRODUCTION

Copper, a native metal found in ores, is the principal metal in bronze and brass. It is a reddish metal with a density of $8920\,kg\,m^{-3}$. All of copper's compounds tend to be brightly colored: for example, copper in hemocyanin imparts a blue color to blood of mollusks and crustaceans. Copper has three oxidation states, with electronic configurations

1529-6466/17/0082-0013$10.00 http://dx.doi.org/10.2138/rmg.2017.82.13

of Cu^0([Ar]3d^{10}4s^1), Cu^+([Ar]3d^{10}), and Cu^{2+}([Ar]3d^9). Cu^0 does not react with aqueous hydrochloric or sulfuric acids, but is soluble in concentrated nitric acid due to its lesser tendency to be oxidized. Cu(I) exists as the colorless cuprous ion, Cu^+. Cu(II) is found as the sky-blue cupric ion, Cu^{2+}. The Cu^+ ion is unstable, and tends to disproportionate to Cu^0 and Cu^{2+}. Nevertheless, Cu(I) forms compounds such as Cu_2O. Cu(I) bonds more readily to carbon than Cu(II), hence Cu(I) has an extensive chemistry with organic compounds.

In aqueous solutions, Cu^{2+} ion occurs as an aquacomplex. There is no clearly predominant structure among the four-, five-, and six-fold coordinated Cu(II) species (Chaboy et al. 2006). Hydrated Cu(II) ion has been represented as the hexaaqua complex $Cu(H_2O)_6{}^{2+}$, which shows the Jahn–Teller distortion effect (Sherman 2001; Bersuker 2006), whereby the two Cu–O distances of the vertical axial bond (Cu–O_{ax}) are longer than four Cu–O distances in the equatorial plane (Cu–O_{eq}). The Jahn–Teller effect lowers the symmetry of $Cu(H_2O)_6{}^{2+}$ from octahedral T_h to D_{2h}. The sixfold coordination of hydrated Cu(II) species is questioned by a finding of fivefold coordination (Pasquarello et al. 2001; Chaboy et al. 2006; Little et al. 2014b; Sherman et al. 2015). The bond distance related to $Cu(H_2O)_6{}^{2+}$ is considered to reflect a rapid switch between the square pyramid and trigonal bipyramid configurations (Pasquarello et al. 2001; de Almeida et al. 2009). The fivefold coordination is supported by computational (Amira et al. 2005) and spectroscopic (Benfatto et al. 2002) studies.

In aqueous media at elevated temperatures, Cu(I) is thermodynamically more stable than Cu(II). The structures of Cu(I) species are thought to be due to the splitting of degenerate 4$p_{x,y,z}$ orbitals by a ligand field (Kau et al. 1987). Cu(I) complexes possess simple linear structures (Fulton et al. 2004) due to 4p_z and 4$p_{x,y}$ orbitals. The splitting of 4$p_{x,y}$ orbital and/or the formation of degenerate 4$p_{y,z}$ orbitals give the Cu(I) species threefold coordination structures (T-shaped or trigonal planar coordination). For the fourfold tetrahedral coordination (T_d) structure, the $p_{x,y,z}$ orbitals may be close to degenerate.

Zinc is an element of Group 2B, the last column of the *d* block. Zinc is not a transition metal by definition because it has a *d* subshell that is fully occupied. Zinc has two oxidation states, with electronic configurations of Zn^0([Ar]3d^{10}4s^2) and Zn^{2+}([Ar]3d^{10}), where Zn(II) has 3d^{10} with two electrons per orbital. Zinc is sometimes included with the transition metals because its properties are more similar to these than to the post-transition metals, whose properties are determined by partially filled *p* subshells. Fresh zinc has a shiny metallic luster, but it tarnishes easily. It is hard and brittle, becomes malleable with increasing temperature, and melts at 419.53 °C. Metallic zinc is easily oxidized and hence it is used as a reducing agent. Reduction of acids like HCl to H_2 by Zn^0 is well known.

In compounds or complex ions, Zn is present only as Zn(II). Hydrated Zn^{2+} is generally thought to be present as the octahedral $Zn(H_2O)_6{}^{2+}$, this being the most stable structure (Mhin et al. 1992). Besides the marked preference for sixfold coordination, Zn(II) can easily be fourfold or fivefold coordinated. The coordination number is attributable to a balance between bonding energies and repulsions among the ligands.

Zinc and Cu are both moderately volatile elements, with 50% condensation temperatures (T_c) of 726 K and 1037 K, respectively (Lodders 2003). It was long thought that Zn behaved as a lithophile element during planetary (and especially, Earth's) differentiation, hence there is negligible Zn in Earth's core (e.g., McDonough 2003). This assumption was used to place broad bounds on the amount of S (which has a similar T_c to Zn) in Earth's core (around 1.7wt% Dreibus and Palme 1996). However, more recent work indicates that Zn behaves as a moderately siderophile element, with potentially ~30% of terrestrial Zn stored in Earth's core (Siebert et al. 2011), significantly affecting the conclusions of Dreibus and Palme (1996). Zinc is the most abundant lithophile element with a T_c < 750 K, 100 times more abundant than the second-most abundant (Br, $T_c = 546\,K$). Its high abundance relative to other moderately volatile elements (due to the relatively high binding energies per nucleon of its isotopes) makes Zn a good tracer of volatility in rocks and a major application of its isotopes has been related to understanding volatility processes.

Copper is a siderophile and highly chalcophile element (Siebert et al. 2011), with ~2/3 of the terrestrial Cu thought to be stored in Earth's core (Palme and O'Neill 2003). Copper is also moderately volatile, but is the most refractory of the chalcophile elements, meaning that Cu may be a good tracer of the role of sulfides during differentiation and igneous processes.

Zn is comprised of five natural stable isotopes, ^{64}Zn (49.2%), ^{66}Zn (27.8%), ^{67}Zn (4.0%), ^{68}Zn (18.4%) and ^{70}Zn (0.6%) and Cu of two stable isotopes, ^{63}Cu (69.2%), and ^{65}Cu (30.8%) (Shields et al. 1964). Due to their relatively high first ionization potentials (9.4 eV for Zn and 7.7 eV for Cu), the measurement of Zn and Cu isotope ratios by Thermal-Ionization Mass-Spectrometry (TIMS) is very difficult. This explains the very limited amount of Zn and Cu isotopic data produced before the advent of Multiple-Collector Inductively Coupled-Plasma Mass-Spectrometry (MC–ICP–MS). In addition, since Cu has only two stable isotopes it is not possible to use a double spike technique to correct for instrumental bias on TIMS. Since the first commercialized MC–ICP–MS in the late 90s and the first 'high precision' Zn and Cu isotope ratio measurements (Maréchal et al. 1999), more than 500 papers have been published (source: ISI Web of Science) on various geochemical topics associated with Zn and Cu isotopes (e.g., oceanography, cosmochemistry, environmental sciences, medical sciences). With the exception of medical sciences, for which there is a dedicated chapter in this volume, here we review these varied applications and discuss the potential of these isotope systems for future studies.

METHODS

Measurement of Zn and Cu isotope ratio was originally made using TIMS (Shields et al. 1964; Shields et al. 1965; Rosman 1972). As for any element with only two isotopes, it was not possible to properly assess the instrumental isotopic fractionation for Cu and the analytical uncertainty was therefore poor (no better than 2‰/amu; Shields et al. 1964, 1965). With five stable isotopes, for Zn it is possible to correct for instrumental bias and TIMS was originally used with double spike methods to measure Zn isotopic compositions. The earliest measurements, on the older generation of TIMS were associated with analytical precisions of around 1‰/amu (Rosman 1972; Loss et al. 1990), but modern generation TIMS can reach precisions of 0.1–0.2‰/amu (Ghidan and Loss 2011).

The vast majority of recent Cu and Zn isotopic data have been acquired by MC–ICP–MS, either by standard-sample bracketing (e.g., Maréchal et al. 1999; Mason et al. 2004a,b; Weiss et al. 2005; Bermin et al. 2006; Viers et al. 2007; Balistrieri et al. 2008; Peel et al. 2008; Vance et al. 2008; Savage et al. 2015a,b; Sossi et al. 2015) or by the double spike method (e.g., Bermin et al. 2006; Arnold et al. 2010b; Conway and John 2015) for Zn. The pioneering work of Maréchal et al. (1999) showed that instrumental mass bias could be corrected by a combination of elemental doping (Cu for Zn, and Zn for Cu) and standard bracketing, so that it was possible to obtain isotope ratios of both Cu and Zn with precisions better than 0.1‰/amu on the VG Elemental Plasma 54 MC–ICP–MS. Subsequent studies by Zhu et al. (2000, 2002) and Archer and Vance (2002, 2004) have further tested this approach and together with Maréchal et al. (1999) provided the ground work for modern Zn and Cu isotopic studies. An alternative method using Ni doping (instead of Zn) for Cu isotope analyses has also been used (Larner et al. 2011).

More recently, double spike Zn isotopic measurements by MC–ICP–MS have also been employed, providing consistent results with those obtained by standard bracketing techniques. An advantage of the double spike technique is that it provides high precision absolute elemental abundances together with the isotope ratios. The fact that the double spike approach also accounts for mass discrimination during chemical separation means that it has been a key methodology for the analysis of Zn in difficult matrixes such as seawater (e.g., Bermin et al. 2006; Arnold et al. 2010b; Zhao et al. 2014; Conway and John 2015; Vance et al. 2016b). Using a similar approach, the absolute abundance of Zn isotopes were determined by analyzing synthetic isotope mixtures (Tanimizu et al. 2002; Ponzevera et al. 2006).

The precision of Zn and Cu isotopic measurements depends on the quality of the chemical extraction (purity, low blank compared to the amount of Zn and Cu present in the samples, high/quantitative yields) and on the correction of the instrumental bias. The high purity of the final Zn fraction is needed to remove both isobaric interferences and non-isobaric interference that are the cause of so-called 'matrix effects' (see Chaussidon et al. 2017, this volume). As Zn and Cu isotopes can be fractionated during ion-exchange chromatography (Maréchal and Albarède 2002) the chemical procedure requires quantitative yields, unless a double-spike is added pre-column chemistry.

The chemical purification of Cu and Zn is generally made by ion-exchange chromatography in 6–10 N HCl medium on either macro-porous resin such as AG-MP1 or on regular bead resin such as AG1-X8 (e.g., Maréchal et al. 1999; Archer and Vance 2004; Borrok et al. 2007; Conway and John 2015; Sossi et al. 2015). In order to obtain a very pure elution of Cu, many workers (e.g., Savage et al. 2015b; Vance et al. 2016a) repeat the whole procedure. For Zn purification, an alternative method takes advantage of the strong complexation of Zn with bromide, which allows for the use of more dilute acids (HBr/HNO_3 media) on micro-columns (0.1 μl) of anion-exchange resin (AG1-X8; Luck et al. 2005; Moynier et al. 2006; Moynier and Le Borgne 2015).

When analyzing Zn by MC–ICP–MS, the potential nickel interference on mass 64 is normally monitored and corrected for by analyzing the intensity of the ^{62}Ni beam. Typically, ^{70}Zn is not measured (or at least, not reported) due to the low abundance of this isotope, and the potential for overwhelming interference from ^{70}Ge. In most instances, it is not necessary to measure ^{70}Zn (even when using the double-spike method), as terrestrial isotope variations are all mass-dependent. However, the introduction of higher resistance amplifiers attached to Faraday detectors should allow the more accurate measurement of ^{70}Zn in, for example, studies involving mass-independent Zn isotope variations. However, so far the results have been inconclusive (Moynier et al. 2009a; Savage et al. 2014). For Cu, neither masses 63 and 65 have direct elemental interferences, although there is evidence that the formation of $^{23}Na^{40}Ar^{+}$ and $^{25}Mg^{40}Ar^{+}$ in the plasma can create anomalous isotope ratios (Archer and Vance 2004; Larner et al. 2011; Savage et al. 2015b), so that careful monitoring to ensure complete removal of both Na and Mg from each sample aliquot is necessary to ensure accurate data.

The correction of instrumental mass bias by elemental doping (Cu for Zn and Zn for Cu, or Ni for Cu) has been extensively discussed in Maréchal et al. (1999) and further by Larner et al. (2011). The principle is that the instrumental bias can be expressed with an exponential law, for example for the ^{66}Zn/^{64}Zn and ^{65}Cu/^{63}Cu ratios:

$$\left(\frac{^{66}\mathrm{Zn}}{^{64}\mathrm{Zn}}\right)_{\mathrm{Measured}} = \left(\frac{^{66}\mathrm{Zn}}{^{64}\mathrm{Zn}}\right)_{\mathrm{True}} \times \left(\frac{M_{66}}{M_{64}}\right)^{f_{\mathrm{Zn}}} \tag{1}$$

$$\left(\frac{^{65}\mathrm{Cu}}{^{63}\mathrm{Cu}}\right)_{\mathrm{Measured}} = \left(\frac{^{65}\mathrm{Cu}}{^{63}\mathrm{Cu}}\right)_{\mathrm{True}} \times \left(\frac{M_{65}}{M_{63}}\right)^{f_{\mathrm{Cu}}} \tag{2}$$

where M_{63}, M_{64}, M_{65}, M_{66} are the atomic masses of ^{64}Zn, ^{65}Cu, ^{66}Zn, ^{68}Zn, respectively. f_{Zn} and f_{Cu} are mass-independent fractionation factors that depend on the element. Taking the example of Zn measurements, the elemental doping method consists of adding an identical Cu elemental standard to all aliquots to be analyzed, which can then be used to determine the f_{Zn}. Because the ionization behavior of Cu and Zn is not the same, f_{Zn} cannot be assumed to be equal to f_{Cu}, and thus the relation between f_{Cu} and f_{Zn} is estimated by taking the Napierian logarithm of Equations (1) and (2) and ratioing the two equations:

$$\frac{\ln\left(\frac{^{66}\text{Zn}}{^{64}\text{Zn}}\right)_{\text{Measured}} - \ln\left(\frac{^{66}\text{Zn}}{^{64}\text{Zn}}\right)_{\text{True}}}{\ln\left(\frac{^{65}\text{Cu}}{^{63}\text{Cu}}\right)_{\text{Measured}} - \ln\left(\frac{^{65}\text{Cu}}{^{63}\text{Cu}}\right)_{\text{True}}} = \frac{f_{\text{Zn}}}{f_{\text{Cu}}}\frac{\ln\left(\frac{M_{66}}{M_{64}}\right)}{\ln\left(\frac{M_{65}}{M_{63}}\right)} \tag{3}$$

By plotting $\ln\left(\frac{^{66}\text{Zn}}{^{64}\text{Zn}}\right)_{\text{Measured}}$ vs. $\ln\left(\frac{^{65}\text{Cu}}{^{63}\text{Cu}}\right)_{\text{Measured}}$ for the standard solution data generated during a session of analyses, the $f_{\text{Zn}}/f_{\text{Cu}}$ ratio can be estimated from the slope of this diagram (see for example Fig. 9 in Maréchal et al. 1999). The calculated f_{Zn} can then be used to calculate $\left(\frac{^{66}\text{Zn}}{^{64}\text{Zn}}\right)_{\text{true}}$ (and vice-versa for Cu isotopic measurements), provided that there is enough drift in the mass bias during an analytical session. This correction is coupled with a standard bracketing method that consists of measuring a standard before and after each sample, whereby the same correction is applied to both the standard and sample ratios. Once all isotope ratios are corrected for mass discrimination, the data are usually reported using the delta notation:

$$\delta^{65}\text{Cu} = \left[\frac{\left(^{65}\text{Cu}/^{63}\text{Cu}\right)_{\text{sample}}}{\left(^{65}\text{Cu}/^{63}\text{Cu}\right)_{\text{NIST-SRM-976}}} - 1\right] \times 1000 \tag{4}$$

$$\delta^{x}\text{Zn} = \left[\frac{\left(^{x}\text{Zn}/^{64}\text{Zn}\right)_{\text{sample}}}{\left(^{x}\text{Zn}/^{64}\text{Zn}\right)_{\text{JMC-Lyon}}} - 1\right] \times 1000 \tag{5}$$

with $x = 66$, 67, 68 or 70.

The double spike approach to mass discrimination was first described for Zn by Bermin et al. (2006), and involves the addition of a mixture of a tracer solution of known exotic isotopic composition to each sample (e.g., Dodson 1963; Rudge et al. 2009; John 2012). Equations that relate three measured and mass-bias corrected isotope ratios, for ^{66}Zn/^{64}Zn, ^{67}Zn/^{64}Zn and ^{68}Zn/^{64}Zn in terms of mixing, the exponential mass discrimination law, and natural mass-dependent fractionation relative to a standard, are solved to obtain the isotopic composition of the sample. The quality of data obtained with any double spike depends on the spike isotopes used and the sample/spike ratio in the mixture created, which control the magnification of analytical uncertainties propagated through the double spike algebra; hence, optimal spike compositions and abundances need to be deduced, although there is a range for both over which good isotopic data are obtainable. Other considerations include the potential for isobaric interference. Thus, although the optimal Zn spike is a mixture of ^{66}Zn and ^{70}Zn, practical applications have used a ^{64}Zn–^{67}Zn spike (e.g., Bermin et al. 2006; Conway et al. 2013). Bermin et al. (2006) showed that such a spike yields precise and accurate sample isotopic compositions over about a factor 20 range in sample/spike ratios in the mixture. Over a 2 year period, on a Neptune at ETH Zürich, the δ^{66}Zn of IRMM-3702 standard gave +0.300 ± 0.058‰ relative to JMC Lyon (2 SD, $n = 163$). For comparison, standard-sample bracketing standard precisions are of a similar magnitude (i.e., long term 2 SD quoted by Chen et al. 2013, is δ^{66}Zn ± 0.04‰) but this is on samples with relatively high Zn concentrations (~100 ppm) and relies heavily on very stable instrument running conditions.

All of the Zn isotopic variations measured in terrestrial samples that have been analyzed to date follow a mass-dependent law, i.e., $\delta^{70}Zn/3 \approx \delta^{68}Zn/2 \approx \delta^{67}Zn/1.5 \approx \delta^{66}Zn$. By contrast, extra-terrestrial mass-independent isotopic effects on ^{66}Zn of over 1500 ppm have been observed in refractory inclusions (Loss and Lugmair 1990; Völkening and Papanastassiou 1990) and, more recently, these isotope anomalies have been discovered in bulk primitive meteorites, albeit of a much smaller magnitude (20–70 ppm Savage et al. 2014). The survival of these anomalies is perhaps surprising, given the volatile behavior of Zn during solar system condensation. These will be discussed further below.

Because it only has two isotopes, mass independence in Cu isotope variations measured in terrestrial samples cannot be discerned, although there is no reason to assume that such variations occur in this realm. For extra-terrestrial samples, Luck et al. (2003) showed that the Cu isotope variations correlate with $\Delta^{17}O$ anomalies in bulk primitive meteorites, which suggests that the variations measured between solar system materials may not be completely generated by 'mass-dependent' fractionation processes.

A number of different reference standards have been used for Cu and Zn isotopic measurements; NIST SRM 976 for Cu isotopes and the JMC 3-0749C (usually called JMC-Lyon) are the two standards used in the original work of Maréchal et al. (1999). Though neither standard is still commercially available they are still the most commonly cited as references. Following Maréchal et al. (1999) other standards have been developed and used in a routine manner in different laboratories. Data for those that have been measured by a number of different laboratories are summarized in Table 1. Other standards relevant to more specific areas of research have been detailed in Cloquet et al. (2008). We suggest, for the sake of consistency, that future data should always be normalized with respect to NIST SRM 976 for Cu and JMC Lyon for Zn. New reference standards for both Cu and Zn will need to be developed soon, and for Zn a round-robin analysis programme of one such new standard is under way (C. Archer, ETH Zürich, pers. comm.). When these new reference standards come on line, we recommend maintaining the isotopic compositions of the original references at zero, with a reference value for new standards set relative to that, as recently proposed for Mo isotopes (Nägler et al. 2014), so that old data can be directly compared with new.

ZINC AND COPPER ISOTOPE FRACTIONATION FACTORS FROM *AB INITIO* METHODS

A considerable amount of progress has been made in calculating isotope fractionation factors between free metal ions and inorganic complexes in aqueous solution, laying the basis for an understanding of surface terrestrial fluids. Almost all the Cu and Zn in the oceans, in rivers and in soils is organically complexed or sorbed to the surfaces of oxyhydroxides and clays (see section on low temperature processes later in the chapter), rather than being found as a free metal ion. There is therefore an urgent need to build on the existing theoretical work to extend the calculations to species and processes that represent those that are most important at the surface of the Earth.

The equilibrium constant of an isotopic exchange reaction can be theoretically obtained as the Reduced Partition Function Ratio (RPFR or β) of isotopologues (e.g., Schauble 2004). Here we will summarize the isotopic enrichment factors that have been calculated for aqueous solutions and molecules relevant to Zn and Cu in biogeochemistry. We also provide new results for certain molecules that were missing from the published studies (see Tables 2, 3, 4) using the method described in Fujii et al. (2014).

The isotope enrichment factor is evaluated from the reduced partition function ratio $(s/s')f$ (Bigeleisen and Mayer 1947), also denoted β, such that,

$$\ln \frac{s}{s'} f = \sum \left[\ln \beta(u_i') - \ln \beta(u_i) \right] \tag{6}$$

where

$$\ln \beta(u_i) = -\ln u_i + \frac{u_i}{2} + \ln\left(1 - e^{-u_i}\right) \tag{7}$$

and

$$u_i = \frac{\mathrm{h}\nu_i}{\mathrm{k}T} \tag{8}$$

In the latter expression, ν stands for vibrational frequency, *s* for the symmetry number of the considered compound, h for Plank's constant, k for the Boltzmann constant, and *T* for the absolute temperature. The subscript *i* denotes the *i*th normal mode of molecular vibration, and primed variables refer to the light isotopologue. The isotope enrichment factor due to molecular vibrations can be evaluated from the frequencies ν_i summed over all normal modes.

The $\ln\beta$ values of Cu(I) and Cu(II) species (Seo et al. 2007; Fujii et al. 2013, 2014; Sherman 2013; Telouk et al. 2015) are shown in Tables 2 and 3. The inorganic aqueous Cu species are represented in Figure 1. At low pH, positive δ^{65}Cu is found in copper sulfates and carbonates, relative to other inorganic species like hydrated Cu^{2+} and chlorides. At pH ~6, ^{65}Cu is enriched in $CuSO_4$ and $CuHCO_3^+$, while ^{63}Cu is enriched in the other inorganic species like Cu^{2+} and $CuCl^+$ (Fig. 1). With increasing pH, $Cu(OH)_2$ and $CuCO_3$ become the prevalent species (Zirino and Yamamoto 1972). At high pH, δ^{65}Cu is positive in Cu hydroxides and negative in carbonates. At a typical pH of seawater (8.22; Macleod et al. 1994), isotope fractionation among inorganic species favors ^{63}Cu in $CuCO_3$ and ^{65}Cu in $Cu(OH)_2$. However, Cu in soil solutions, rivers and seawater is overwhelmingly complexed to organics (McBride 1981; Coale and Bruland 1988; Moffett and Brand 1996; Shank et al. 2004; Grybos et al. 2007; Vance et al. 2008; Ryan et al. 2014), and the above calculations for inorganic species are relevant only to the tiny inorganic pool of Cu. Little et al. (2014b) and Sherman et al. (2015) show how heavy Cu in simple organic complexes controls seawater Cu isotopes. Note that the interpretation strongly depends on the speciation diagram applied (Fujii et al. 2013). Powell et al. (2007) uses small hydrolysis constants, which depress the role of hydroxides in Cu isotope fractionation; isotope fractionation may therefore not be seen in Cu(II) hydroxide, but Cu in $CuCO_3$ should nevertheless remain isotopically light with respect to the remaining inorganic pool of Cu in seawater, and much lighter than the dominant organically complexed pool.

Since sulfide-bearing euxinic seawater systems are reducing, isotope fractionation of Cu caused by the co-presence of Cu(I) becomes important (Fujii et al. 2013). The $\ln\beta$ value of sulfides is 1–2‰ lower than Cu(II) carbonates, hydroxides, and hydrated Cu^{2+}. This suggests that the dominant organically complexed pool, as well as minor Cu^{2+}, Cu(II) chlorides, carbonates, and hydroxides will all be isotopically heavier than sulfides. The speciation of Cu(I) under hydrothermal conditions (Mountain and Seward 1999) indicates that the prevailing species are CuCl, $CuCl_2^-$, CuHS, and $Cu(HS)_2^-$. Increasing complexation of Cu(I) chlorides and sulfides results in decreasing $\ln\beta$. The $\ln\beta$ values of Cu(II) chlorides and sulfides at 573 K are 0.2–0.5‰ higher than those of corresponding Cu(I) species (Fujii et al. 2013, 2014). Under hydrothermal conditions, the δ^{65}Cu value of Cu(I) may be 0.2–0.5‰ lower than that of Cu(II) with a ±0.1‰ range of variation among Cu(I) species.

Table 1 : $\delta^{66}Zn_{JMC\,Lyon}$ and $\delta^{65}Cu_{SRM\,976}$ of commonly used geological standards and isotopically certified materials.

	$\delta^{66}Zn_{JMC\,Lyon}$	2SE	Ref	$\delta^{65}Cu_{SRM\,976}$	2SE	Ref
IRMM-3702	0.38	0.10	1			
IRMM-3702	0.25	0.09	1			
IRMM-3702	0.30	0.02	2			
IRMM-3702	0.29	0.05	3			
IRMM-3702						
IRMM-3702 recommended value	**0.30**	**0.01**				
ERM-AE633				-0.01	0.05	3
ERM-AE647				0.21	0.05	3
BHVO-2	0.33	0.04	4	0.10	0.08	5
	0.21	0.09	6	0.10	0.04	3
	0.27	0.06	2	0.10	0.07	4
	0.31	0.03	15	0.15	0.05	9
	0.29	0.09	16	0.13	0.03	8
BHVO-2 recommended value	**0.28**	**0.04**		**0.12**	**0.02**	
BCR-1/2	0.20	0.04		0.19	0.07	5
	0.29	0.12	10	0.14	0.05	3
	0.32	0.13	1	0.07	0.08	11
	0.23	0.08	12	0.19	0.08	12
	0.25	0.01	2	0.22	0.04	8
	0.20	0.09	11	0.21	0.04	8
	0.26	0.05	13			
	0.26	0.09	17			
BCR-1/2 recommended value	**0.25**	**0.03**		**0.17**	**0.05**	
BIR-1	0.31	0.04	4	0.00	0.03	2
	0.26	0.09	7	0.08	0.07	3
	0.20	0.04	2	-0.02	0.10	14
BIR-1 recommended value	**0.26**	**0.06**		**0.02**	**0.06**	
AGV1/2	0.32	0.04	4	-0.01	0.03	5
	0.25	0.09	6	-0.01	0.09	5
	0.29	0.03	15	0.01	0.11	5
	0.28	0.05	15	0.11	0.04	3
				0.1	0.11	3
				0.05	0.04	8
AGV1/2 recommended value	**0.29**	**0.03**		**0.04**	**0.04**	
G2	0.34	0.04	15			
	0.30	0.09	18			
	0.32	0.09	17			
G2 recommended value	**0.32**	**0.02**				

Ref: 1 = Cloquet et al. (2006); 2 = Sossi et al. (2014); 3 = Moeller et al. (2012); 4 = Chen et al. (2013); 5 = Savage et al. (2015); 6 = Moynier et al. (2010); 7 = Herzog et al. (2009); 8 = Liu et al. (2015); 9 = Liu et al. (2014); 10 = Chapman et al. (2006); 11 = Archer and Vance (2004); 12 = Bigalke et al. (2010); 13 = Viers et al. (2007); 14 = Li et al. (2009); 15 = S. Chen et al. (2016); 16 = Telus et al. (2012); 17 = Paniello et al. (2012a); 18 = Paniello et al. (2012b).

Table 2. Logarithm of the reduced partition function, ln β, for the pair ^{65}Cu–^{63}Cu. Cu(II) species. Method/Basis set used: B3LYP/TZP for Sherman (2013) and B3LYP/6-311+G(d,p) for Fujii et al. (2013, 2014).

	Species	*N*	Temperature (K)							Ref[b]
			273	298	310	323	373	473	573	
Solid	CuO (Tenorite)	–	6.63	5.62	–	4.81	3.65	2.29	1.57	1
Aquo-ion	$Cu(H_2O)_5^{2+}$	5	5.355	4.546	–	3.905	2.968	1.876	1.290	2
			5.36	4.55	–	3.91	2.97	1.88	1.29	1
	$Cu(H_2O)_6^{2+}$	6	5.053	4.288	–	3.682	2.798	1.767	1.215	2
Chloride	$CuCl(H_2O)_4^+$	5	4.906	4.161	–	3.572	2.712	1.711	1.176	2
	$CuCl(H_2O)_5^+$	6	4.67	3.96	–	3.40	2.58	1.63	1.12	1
	$CuCl_2(H_2O)_3$	5	4.709	3.988	–	3.420	2.592	1.633	1.120	2
	$CuCl_2(H_2O)_4$	6	4.397	3.724	–	3.193	2.421	1.525	1.046	2
	$CuCl_3H_2O^-$	4	3.530	2.985	–	2.556	1.933	1.214	0.832	2
Hydroxide	$CuOH(H_2O)_4^+$	5	5.307	4.517	–	3.889	2.967	1.883	1.298	2
			5.30	4.52	–	3.89	2.97	1.89	1.30	1
	$Cu(OH)_2(H_2O)_3$	5	5.814	4.966	–	4.288	3.286	2.098	1.451	2
Carbonate	$CuCO_3(H_2O)_2$	4	5.091	4.323	–	3.715	2.825	1.787	1.230	2
	$Cu(CO_3)_2^{2-}$	4	6.176	5.239	–	4.498	3.416	2.158	1.483	2
			6.38	5.41	–	4.65	3.53	2.23	1.53	1
	$CuHCO_3(OH)_2^-$	4	5.951	5.075	–	4.376	3.346	2.130	1.471	2
Sulfate	$CuSO_4(H_2O)_4$	5	6.041	5.144	–	4.430	3.381	2.148	1.481	2
Sulfide	$CuHS(H_2O)_4^+$	5	4.002	3.386	–	2.900	2.194	1.377	0.942	2
	$Cu(HS)_2(H_2O)_3$	5	3.855	3.264	–	2.797	2.119	1.333	0.914	2
Phosphate	$CuH_2PO_4(H_2O)_4^+$	5	5.515	4.684	–	4.026	3.063	1.939	1.334	3
	$CuH_4(PO_4)_2(H_2O)_3$	5	5.553	4.714	–	4.050	3.079	1.947	1.339	3
	$CuH_3(PO_4)_2(H_2O)_3^-$	5	5.290	4.492	–	3.861	2.937	1.860	1.280	3
	$CuH_2(PO_4)_2(H_2O)_2^{2-}$	4	6.360	5.403	–	4.645	3.535	2.238	1.540	3
Citrate	$CuH_2(cit)(H_2O)_2^+$	5	5.286	4.486	–	3.852	2.927	1.850	1.272	2
	$CuH(cit)(H_2O)_2$	5	5.622	4.772	–	4.099	3.117	1.972	1.357	2
	$Cu(cit)(H_2O)_2^-$	5	6.092	5.177	–	4.451	3.389	2.147	1.479	2
	$Cu(cit)_2^{4-}$	4	4.998	4.231	–	3.626	2.748	1.730	1.188	2
Oxalate	$CuC_2O_4(H_2O)_2$	4	6.236	5.302	–	4.561	3.474	2.202	1.516	2
Ascorbate	$CuH(L\text{-ascorbate})(H_2O)_4^+$	5	3.924	3.324	–	2.850	2.161	1.362	0.935	2
	$CuH(D\text{-ascorbate})(H_2O)_4^+$	5	3.989	3.380	–	2.899	2.199	1.386	0.951	2
Malonate	$Cu(H_2C_3O_4)_2(H_2O)_2^{2-}$	6	7.00	5.94	–	5.10	3.88	2.45	1.68	1
Amino acid	$Cu(Glu)(H_2O)_3^{2+}$	5	5.230	4.436	4.117	3.808	2.891	-	-	3

Table 2 (cont'd).

	Species	*N*	Temperature (K) 273	298	310	323	373	473	573	Ref[b]
complex	$Cu(Thr)(H_2O)_4^{2+}$	5	5.220	4.429	4.110	3.803	2.889	–	–	3
	$Cu(His)(H_2O)_3^{2+}$	5	5.274	4.470	4.148	3.836	2.911	–	–	3
	$Cu(His)(H_2O)_4^{2+}$	5	5.299	4.492	4.168	3.855	2.926	–	–	3
	$Cu(Cys)(H_2O)_4^{2+}$	5	3.981	3.369	3.124	2.888	2.187	–	–	3
	$Cu(Met)(H_2O)_4^{2+}$	5	4.632	3.932	3.650	3.378	2.568	–	–	3
	$Cu(GS)H^0$	4	4.945	4.194	3.892	3.600	2.734	–	–	3
Lactate	$Cu(L\text{-}lact)(H_2O)_3^+$	5	5.530	4.695	4.359	4.034	3.068	–	–	3
	$Cu(L\text{-}lact)_2$	4	7.110	6.045	5.616	5.199	3.961	–	–	4
	Cu(L-lact)(D-lact)[a]	4	7.125	6.057	5.627	5.210	3.969	–	–	4

N = Coordination Number
[a] Reproduced from Telouk et al. (2015).
[b] 1 = Sherman (2013); 2 = Fujii et al. (2013), 3 = Fujii et al. (2014), 4 = This study.

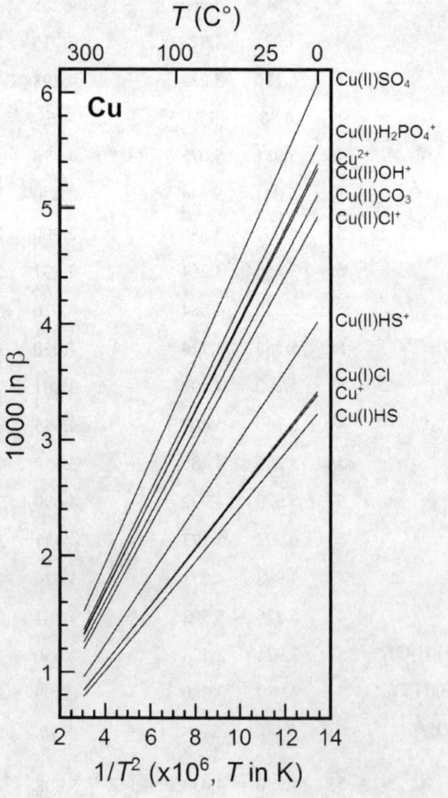

Figure 1. Temperature dependence of ln β. The ln β values of hydrated Cu^{2+} and Cu(II) chlorides, sulfides, phosphates, carbonates and sulfates (see Table 1) are shown as linear functions of T^2.

Table 3. Logarithm of the reduced partition function, ln β, for the pair $^{65}Cu^{63}Cu$. Cu(I) species. Method/Basis set used: B3LYP/TZP for Sherman (2013) and B3LYP/6-311+G(d,p) for Fujii et al. (2013, 2014), Seo et al. (2007) and the present study.

	Species	*N*	Temperature (K) 273	298	310	323	373	473	573	Ref[b]
Solid	Cu_2O (Cuprite)	–	3.99	3.40	–	2.92	2.23	1.41	0.97	1
	$CuFeS_2$ (Chalcopyrite)	–	1.80	1.51	–	1.29	0.97	0.61	0.41	1
Aquo-ion	$Cu(H_2O)_2^+$	2	3.368	2.867	–	2.468	1.882	1.193	0.822	2
Chloride	$CuCl(H_2O)$	2	3.401	2.887	–	2.480	1.885	1.191	0.818	3
			3.40	2.89	–	2.48	1.89	1.19	0.82	4
	$CuCl_2^-$	2	2.775	2.350	–	2.014	1.526	0.960	0.659	3
			2.71	2.29	–	1.97	1.49	0.94	0.64	4
			2.87	2.42	–	2.08	1.57	0.99	0.68	1
	$CuCl_3^{2-}$	3	1.012	0.851	–	0.725	0.545	0.339	0.231	3
			1.02	0.85	–	0.73	0.55	0.34	0.23	4
			1.41	1.19	–	1.02	0.76	0.48	0.33	1
Sulfide	$CuHS(H_2O)$	2	3.208	2.722	–	2.337	1.775	1.121	0.770	3
			2.89	2.45	–	2.10	1.59	1.00	0.69	1
	$Cu(HS)_2^-$	2	2.940	2.489	–	2.133	1.616	1.017	0.697	3
			2.90	2.46	–	2.11	1.60	1.00	0.69	4
			2.69	2.28	–	1.95	1.48	0.93	0.64	1
	$Cu_2S(HS)_2^{2-}$	2	2.648	2.239	–	1.917	1.450	0.911	0.624	3
Lactate	Cu(L-lact) [a]	2	2.195	1.859	1.725	1.595	1.209	–	–	5
	Cu(D-lact) [b]	2	2.202	1.866	1.731	1.600	1.214	–	–	5

N = Coordination Number
[a] Reproduced from Telouk et al. (2015).
[b] 1 = Sherman (2013); 2 = Fujii et al. (2013); 3 = Fujii et al. (2014); 4 = Seo et al. (2007); 5 = This study.

As an application to plant uptake, $\delta^{65}Cu$ for Cu phosphates, citrates, hydroxides, and hydrated Cu^{2+} ions was estimated as a function of pH (Fujii et al. 2014). At neutral pH, the major Cu(II) species are phosphates and citrates, and a range of ~0.5‰ can be expected for $\delta^{65}Cu$. This range overlaps with observations on higher plants (Weinstein et al. 2011; Jouvin et al. 2012). A reduction of Cu^{2+} to Cu^+ by a reductase within roots has also been reported by Jouvin et al. (2012). Since the range of ln β values for Cu(I) species is ~2‰ smaller than those of Cu(II) species at 298 K, a fractionation of −0.84 to −0.11‰ between roots and nutrient solutions (Jouvin et al. 2012) may be expected.

A variety of metabolic processes may induce Cu isotope fractionation. A positive $\delta^{65}Cu$ of 1.5‰ was found in both sheep kidney (Balter and Zazzo 2011) and mouse kidney (Albarède et al. 2011), which may be interpreted in terms of isotope exchange reactions among Cu(I) and Cu(II) species. Oxalic acid is a ubiquitous toxic organic acid in bodily fluids. High oxalate contents in urine and plasma may be correlated with kidney damage. Ascorbate is efficiently converted to oxalate when the coexisting copper concentration is high (Hayakawa et al. 1973). The $\delta^{65}Cu$ of the Cu species relative to the bulk solution as a function of Eh and extent of

Table 4. Logarithm of the reduced partition function, ln β, for the pair ^{66}Zn–^{64}Zn. Zn(II) species. Method/Basis set used: B3LYP/6-311+G(d,p) for Fujii et al. (2013, 2014), Fujii and Albarede (2013), Moynier et al. (2013a), BP86/SVP for Singha Deb et al. (2014), and B3LYP/aug-cc-pVDZ for Black et al. (2012).

	Species	***N***	Temperature (K)							**Ref[d]**
			273	298	310	323	373	473	573	
Aquo-ion	$Zn(H_2O)_4^{2+}$	4	4.539	3.853	3.577	3.310	2.516	–	–	1
			–	5.0	–	–	–	–	–	2
	$Zn(H_2O)_6^{2+}$	6	3.854	3.263	–	2.797	2.119	1.334	0.915	1
			3.61	3.05	–	2.61	1.98	1.25	0.85	3
			–	3.9	–	–	–	–	–	2
	$Zn(H_2O)_{18}^{2+}$	6	–	3.576	–	–	–	–	1.004	4
			3.67	3.11	–	2.66	2.02	1.27	0.87	3
			–	4.3	–	–	–	–	–	2
Chloride	$ZnCl(H_2O)_5^+$	6	3.702	3.136	–	2.689	2.039	1.285	0.882	1
	$ZnCl_2(H_2O)_4$	6	3.486	2.950	–	2.528	1.915	1.205	0.826	1
	$ZnCl_3(H_2O)^-$	4	3.490	2.952	–	2.528	1.913	1.202	0.824	1
	$ZnCl_4^{2-}$	4	2.722	2.293	–	1.957	1.474	0.921	0.629	1
			2.77	2.33	–	1.99	1.50	0.94	0.64	3
Hydroxide	$Zn(OH)_2(H_2O)_4$	6	4.185	3.567	–	3.075	2.350	1.495	1.032	1
Carbonate	$ZnHCO_3(H_2O)_3^+$	5	4.573	3.877	–	3.326	2.525	1.593	1.095	1
	$ZnHCO_3(H_2O)_4^+$ [a]	5	4.579	3.885	–	3.335	2.534	1.602	1.102	5
	$ZnHCO_3(H_2O)_5^+$ [a]	6	4.109	3.482	–	2.988	2.267	1.431	0.983	5
	$ZnCO_3(H_2O)_3$	5	4.940	4.199	–	3.612	2.752	1.745	1.202	1
	$ZnCO_3(H_2O)_4$ [a]	5	4.789	4.076	–	3.509	2.677	1.700	1.172	1
	$ZnCO_3(H_2O)_5$ [a]	6	4.356	3.704	–	3.187	2.429	1.541	1.062	5
Sulfate	$ZnSO_4(H_2O)_6$	5	4.31	3.65	–	3.13	2.38	1.50	1.03	3
	$ZnSO_4(H_2O)_5$	6	4.154	3.527	–	3.031	2.306	1.460	1.006	5
Sulfide	$Zn(HS)_2(H_2O)_4$	6	3.207[b]	2.717	–	2.330[b]	1.766[b]	1.113[b]	0.764	4
	$Zn(HS)_3(H_2O)_2^-$	5	3.580[b]	3.028	–	2.593[b]	1.962[b]	1.233[b]	0.845	4
	$Zn(HS)_4^{2-}$	4	2.598[b]	2.190	–	1.871[b]	1.411[b]	0.883[b]	0.604	4
	$ZnS(HS)H_2O^-$	5	3.112[b]	2.628	–	2.247[b]	1.697[b]	1.064[b]	0.728	4
Phosphate	$ZnH_2PO_4(H_2O)_5^+$	6	4.092	3.468	–	2.975	2.257	1.424	0.978	6
	$ZnH_4(PO_4)_2(H_2O)_4$	6	4.047	3.428	–	2.940	2.229	1.405	0.965	6
	$ZnH_3(PO_4)_2(H_2O)_4^-$	6	5.027	4.268	–	3.667	2.789	1.764	1.214	6
	$ZnHPO_4(H_2O)_5$	6	4.188	3.559	–	3.060	2.330	1.476	1.017	6
	$Zn_2H_2(PO_4)_2(H_2O)_4$	6	5.156	4.380	–	3.765	2.865	1.814	1.249	6
Citrate	$ZnH(cit)(H_2O)_4$	6	4.033	3.419	–	2.934	2.227	1.406	0.967	6
	$Zn(cit)(H_2O)_3^-$	6	4.154	3.523	–	3.024	2.297	1.452	0.999	6
			4.39	3.72	–	3.20	2.43	1.54	1.06	3
	$Zn(cit)_2^{4-}$	6	2.889	2.437	–	2.083	1.572	0.986	0.675	6
	$Zn_2H_{-2}(cit)_2(H_2O)_4^{4-}$	4	5.330	4.523	–	3.884	2.953	1.867	1.284	6

Table 4 (Cont'd).

	Species	N	Temperature (K) 273	298	310	323	373	473	573	Ref
Malate	$ZnH_2(mal)(H_2O)_4^{2+}$	6	3.842	3.250	–	2.784	2.107	1.325	0.909	6
	$ZnH(mal)(H_2O)_4^{+}$	6	3.984	3.376	–	2.896	2.197	1.386	0.952	6
	$Zn(mal)(H_2O)_4$	6	4.103	3.479	–	2.987	2.268	1.433	0.986	6
	$Zn(mal)_2(H_2O)_2^{2-}$	6	3.274	2.771	–	2.376	1.801	1.135	0.780	6
Oxalate	$ZnC_2O_4(H_2O)_2$	4	5.500	4.678	–	4.025	3.068	1.946	1.341	5
	$Zn(C_2O_4)_2^{2-}$	4	5.215	4.421	–	3.794	2.880	1.818	1.250	5
Amino acid complex	$Zn(Glu\text{-}H_{-1})^{+\ c}$	2	1.923	1.633	1.517	1.404	1.070	–	–	7
	$Zn(Glu)(H_2O)_2^{2+}$	4	4.473	3.796	3.524	3.260	2.478	–	–	1
	$Zn(Glu)(H_2O)_4^{2+}$	6	3.888	3.292	3.053	2.822	2.139	–	–	1
	$Zn(Thr)(H_2O)_3^{2+}$	4	4.774	4.056	3.767	3.487	2.654	–	–	1
	$Zn(Thr)(H_2O)_5^{2+}$	6	3.916	3.315	3.075	2.842	2.154	–	–	1
	$Zn(His\text{-}H_{-1})^{+\ c}$	2	4.381	3.728	3.465	3.210	2.448	–	–	7
	$Zn(His)^{2+\ c}$	2	4.223	3.591	3.336	3.090	2.355	–	–	7
	$Zn(His)(H_2O)_2^{2+}$	4	4.670	3.959	3.673	3.397	2.578	–	–	1
	$Zn(His)(H_2O)_4^{2+}$	6	3.541	2.996	2.777	2.566	1.943	–	–	1
	$Zn(His)(H_2O)_3^{2+}$	4	4.635	3.930	3.647	3.373	2.561	–	–	1
	$Zn(His)(H_2O)_5^{2+}$	6	3.724	3.150	2.921	2.699	2.043	–	–	1
	$Zn(Cys\text{-}H_{-1})^{+\ c}$	1	1.417	1.196	1.108	1.023	0.771	–	–	7
	$Zn(Cys)^{2+\ c}$	1	1.545	1.307	1.211	1.119	0.847	–	–	7
	$Zn(Cys)(H_2O)_3^{2+}$	4	3.912	3.313	3.072	2.840	2.152	–	–	1
	$Zn(Cys)(H_2O)_5^{2+}$	6	3.196	2.702	2.504	2.313	1.750	–	–	1
	$Zn(Met)(H_2O)_3^{2+}$	4	4.397	3.733	3.466	3.207	2.438	–	–	1
	$Zn(Met)(H_2O)_5^{2+}$	6	3.478	2.947	2.734	2.528	1.918	–	–	1
	$Zn(GS)^{-}$	4	4.311	3.655	3.392	3.137	2.381	–	–	1

N = Coordination Number

[a] HCO_3^- and CO_3^{2-} were treated as monovalent ligands.

[b] Reproduced from Fujii et al. (2011).

[c] Hydration water molecules were not arranged (anhydrous).

[d] 1 = Fujii et al. (2014); 2 = Singha Deb et al. (2014); 3 = Black et al. (2011); 4 = Fujii et al. (2011); 5 = This Study; 6 = Fujii and Albarède (2012); 7 = Moynier et al. (2013a).

oxalate formation has been estimated (Fujii et al. 2013); $\delta^{65}Cu$ of Cu ascorbate varies from −1.0 to +0.5‰ when Eh increases from −1 V to +1 V, but its mole fraction remains very small, while the heavy isotope is enriched (+0.6 to +2.5‰) in the Cu oxalate relative to total Cu. It is expected that degradation of ascorbate and excretion of oxalate should leave isotopically heavy Cu in the kidney. With respect to food, which has a $\delta^{65}Cu$ value of about 0‰, if even trace amounts of oxalate form it should leave behind copper with a $\delta^{65}Cu$ of ~1.4‰ ($\delta^{65}Cu$ at 0‰ extent of oxalate formation). This value is very close to the $\delta^{65}Cu$ (1.5‰) found in sheep (Balter and Zazzo 2011) and mice (Albarède et al. 2011) kidneys.

Variations in Cu isotopes among Cu^{2+}-amino acid complexes have been estimated (Fujii et al. 2014). The ln β of Cu^{2+} complexes with O and N-donor amino acids is ~1‰ (at the body temperature of 310 K typical for mammals) higher than those with S-donor amino acids. In a same donor amino acid complex, $\delta^{65}Cu$ of ~1‰ may be created via Cu^{2+}/Cu^{+} redox processes in biological activity. This latter study also theoretically estimated the β of Cu lactates. The extent of ^{65}Cu preference over ^{63}Cu in Cu lactates with respect to Cu bound to cysteine is more than 1‰. From a study on the $^{65}Cu/^{63}Cu$ ratios in the serums of cancer patients, a $\delta^{65}Cu$ alarm threshold was found to be at −0.35‰. The decrease of $\delta^{65}Cu$ in the serum of cancer patients is assigned to the extensive oxidative chelation of copper by cytosolic lactate (Telouk et al. 2015).

The ln β values of Zn(II) species (Fujii et al. 2010, 2011, 2014; Black et al. 2011; Fujii and Albarède 2012) are shown in Table 4 and some species are represented in the Figure 2. It is known that Zn sulfate and carbonate create larger ln β. The fivefold and sixfold coordination of Zn in carbonate complexes, for which carbonates are treated as monovalent and divalent ligands, results in large ln β (Fujii et al. 2011, 2014). At circumneutral pH the dominant inorganic species of free Zn^{2+} shows small isotope fractionation relative to the total inorganic pool. Zinc sulfate is enriched in ^{66}Zn, whereas Zn chlorides are enriched in ^{64}Zn with a $\Delta^{66}Zn$ ~0.5‰ being expected between Zn sulfate and chloride. With increasing pH, $Zn(OH)_2$ and $ZnCO_3$ become the dominant species. Small amounts of free Zn^{2+} and $ZnCl^{+}$ still exist at pH = 8.2. In seawater, a fractionation $\Delta^{66}Zn$ of ~1‰ is expected between Zn carbonate and chloride. Zinc hydroxides and sulfates do

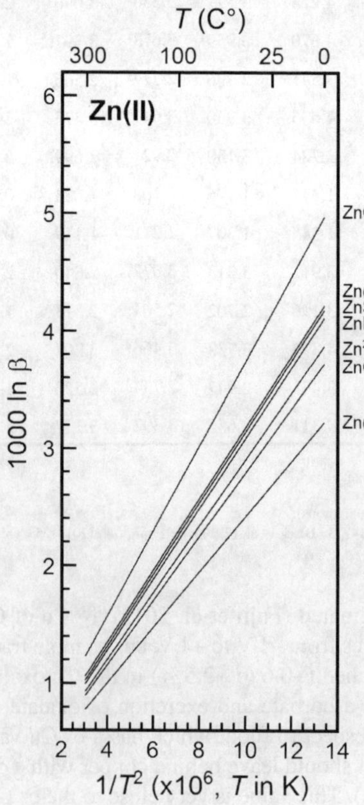

Figure 2. Temperature dependence of ln β. The ln β values of hydrated Zn^{2+} and Zn(II) sulfides, phosphates, carbonates and sulfates (see Table 1) are shown as linear functions of T^2.

not play an important role for Zn isotope fractionation for pH≥8.2. It is again noted, however, that the free and inorganically complexed pool of both Cu and Zn in soil solutions, rivers and seawater is very minor (in seawater, on the of order 2%; McBride 1981; Coale and Bruland 1988; Moffett and Brand 1996; Shank et al. 2004; Grybos et al. 2007; Vance et al. 2008; Ryan et al. 2014) and most is organically complexed. As such, the above discussion is only relevant to the minor free metal and inorganically complexed pool.

The role of sulfides is central to a broad range of geological scenarios. The status of sulfur in ancient oceans in particular is still an outstanding issue (Canfield 1998). Hydrothermal vent solutions discharging either at mid-ocean ridges (Edmond et al. 1979) or along subduction zones (Mott et al. 2004) comprise additional environments dominated by sulfides. Fujii et al. (2011) evaluated the isotope fractionation among the different Zn sulfide species present in geological fluids between 298 and 573 K (Fig. 2). At the high P_{CO_2} conditions of hydrothermal solutions, Zn precipitated as sulfides is isotopically nearly unfractionated with respect to a low-pH parent fluid. In contrast, negative δ^{66}Zn, down to at least –0.6‰, can be expected in sulfides precipitated from solutions with pH>9. Zinc isotopes in sulfides and rocks therefore represent a potential indicator of mid to high pH in ancient hydrothermal fluids (Pons et al. 2011).

Citric acid also plays an important role in the transport of trace metals in the soil-plant system. Citrate is released from the roots of vascular plants and acts as a biological chelating agent for the uptake of metals from soil. Isotope fractionation induced by higher plants has been found for Zn (Weiss et al. 2005; Moynier et al. 2009b). In a pioneering study of isotope fractionation of Zn in the soil-plant system, Weiss et al. (2005) found that Zn was isotopically lighter in the shoots relative to the roots, with a δ^{66}Zn difference of –0.13 to –0.26‰. The origin of this isotope fractionation may be isotopic exchange between Zn(II) phosphates in roots and citrates (or malates) in plants (Fujii and Albarède 2012).

The $\ln\beta$ values for optimized structures of Zn^{2+}-amino acid complexes have been calculated (Fujii et al. 2014). Heavy isotopes tend to bind to O-donor ligands, whereas light isotopes are positively fractionated by S-donor ligands (Balter et al. 2013; Moynier et al. 2013a). This is clearly seen in complexes with identical coordination number (four and six). Isotope fractionation correlated with N-donor ligands may be intermediate between O-donor and S-donor systems or even stronger than with O-donor ligands. Besides the donor type, coordination number is important, implying that four-fold complexation gives larger $\ln\beta$ values relative to complexes with six-fold coordination. The $\ln\beta$ of $Zn(His)^{2+}$ complexes is 0.2 to 0.6‰ larger than that of $Zn(Cys)^{2+}$. This matches the observation that organs rich in proteins with histidine residues show larger δ^{66}Zn than organs in which proteins rich in cysteine residues dominate (Moynier et al. 2013a).

ZINC AND COPPER IN EXTRA-TERRESTRIAL SAMPLES AND IGNEOUS ROCKS

Chondritic reference frame and meteorites. Luck et al. (2003, 2005) were the first to measure Cu and Zn isotopes in a selection of carbonaceous and ordinary chondrites by MC–ICP–MS. The carbonaceous chondrite data show resolvable isotopic variation for both Zn and Cu isotopes between the different groups (e.g., CI, CV, CO, CM…) with the latter system showing the largest variability ($0.16<\delta^{66}\text{Zn}<0.52$ and $-1.44<\delta^{65}\text{Cu}<-0.09$, Fig. 3 and 4). For both systems, each carbonaceous group has a distinct isotope composition (Figs. 3 and 5). There is also a broad positive co-variation between the Cu and Zn isotope compositions of the carbonaceous chondrites, with the CI chondrites defining the heaviest compositions in both systems. The most robust average composition for the CI chondrites has been obtained from the average composition of 6 large chips of the Orgueil meteorite as well as samples from Ivuna and Alais and is $\delta^{65}\text{Cu}=0.05\pm0.16$‰ and $\delta^{66}\text{Zn}=0.46\pm0.08$‰ (Barrat et al. 2012).

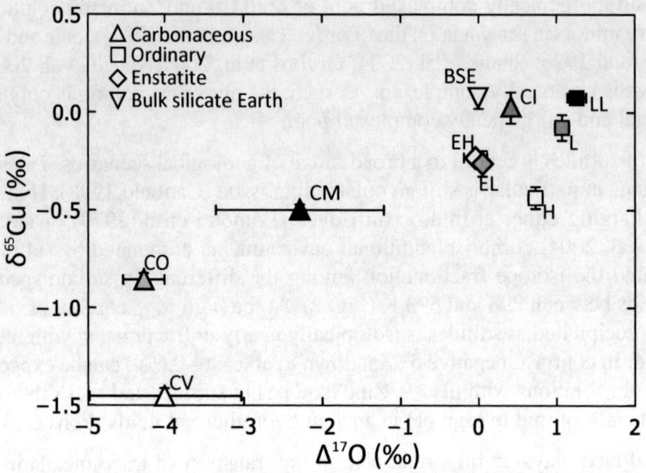

Figure 3. δ^{65}Cu vs. Δ^{17}O for the bulk silicate Earth and the different major groups of chondrites. Meteorite group averages are calculated using data from Luck et al. (2003), Barrat et al. (2012) and Savage et al. (2015b). The estimate of the Bulk Silicate Earth is from Savage et al. (2015b). All error bars are 2 SD of the mean. The trend implies the presence of at least two, and potentially three, distinct Cu isotope reservoirs which then mixed to create the distinct chondritic bodies, as a result of nebula processing.

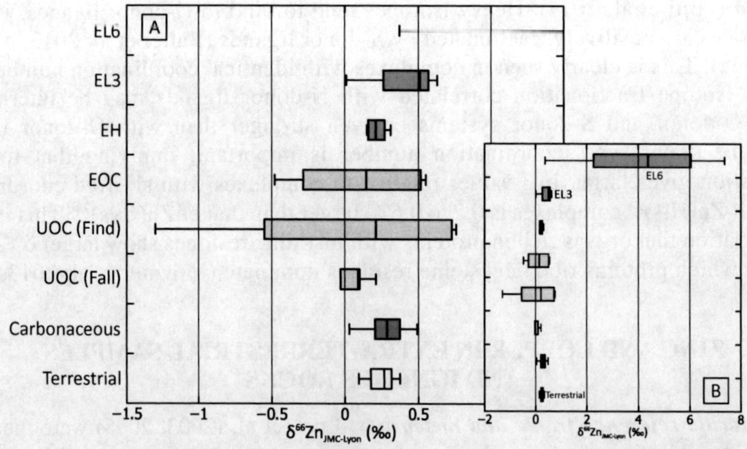

Figure 4. A) Box-and-whisker plot of δ^{66}Zn of the different chondrites groups (UOC=Un-equilibrated ordinary chondrites); whiskers denote the maximum and minimum value of the data, box denotes the first and third quartiles and the band denotes the median. Data are taken from Luck et al. (2005), Moynier et al. (2011) and Barrat et al. (2012). Enstatite chondrites, carbonaceous chondrites, and unequilibrated ordinary chondrites have Zn isotopic composition close to the current estimates of BSE. B) Identical plot as 4A but with increased *x*-axis range. The EL enstatite chondrites of high thermal metamorphic grades (EL6), which are depleted in volatile elements compared to low thermal metamorphic grades (EL3), are highly enriched in the heavier isotopes. This suggests that the origin of the volatile element depletion between EL3 and EL6 chondrites is due to volatilization during the thermal metamorphism.

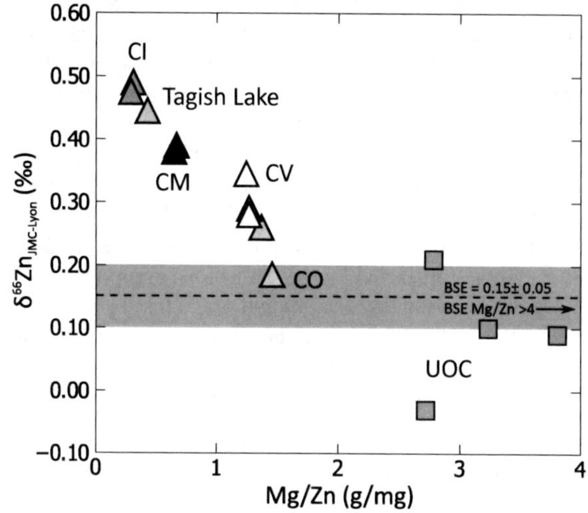

Figure 5. $\delta^{66}Zn$ vs. Mg/Zn ratio for different chondrites groups. The negative correlation between $\delta^{66}Zn$ and Mg/Zn (refractory element/moderately volatile element) suggests that the origin of the Zn depletion in carbonaceous chondrites is not due to evaporation and is of nebular origin. Data from Luck et al. (2005) and Barrat et al. (2012)

For the ordinary chondrites, Zn isotopes define the larger range ($-1.30<\delta^{66}Zn<0.76$, Fig. 4), compared to Cu ($-0.51<\delta^{65}Cu<0.10$, Fig. 3), although the difference between groups (H, L, LL) is much clearer in the Cu system, whereas the variations in Zn isotopes in ordinary chondrites seem to be controlled by degree of parent body metamorphism, as well as subsequent secondary alteration on the Earth's surface (i.e., 'falls' are much less variable than 'finds').

Luck et al. (2003) showed that the $\delta^{65}Cu$ composition of carbonaceous and ordinary chondrites varies systematically with their mass-independent $\Delta^{17}O$ value ($\delta^{18}O-0.52\times\delta^{17}O$, Fig. 3) and $^{58}Ni/^{65}Cu$ ratio, although the two groups fall on distinct trends. They interpreted this phenomenon as revealing the presence of at least two, and potentially three, distinct Cu isotope reservoirs in the early solar system, which subsequently mixed via nebular processing to create the distinct chondritic bodies. They further suggested, based on the relationship with $^{58}Ni/^{65}Cu$ ratio, that the range of Cu isotope compositions was potentially defined early in solar system history by the heterogeneous distribution of a phase enriched in the short-lived radionuclide ^{63}Ni, which decays to ^{63}Cu.

Luck et al. (2005) discovered a negative correlation between $\delta^{66}Zn$ and Mg/Zn in carbonaceous chondrites and un-equilibrated ordinary chondrite falls (Fig. 5). This was taken as robust evidence against evaporation as the origin of the variability of Zn abundance between chondrites groups (with the exception of EL6, see later), and rather suggested that the variation in the volatile element content of chondrite parent bodies was fixed by nebular processes. This argument was later developed by Albarède (2009) to suggest that the volatile element abundance in chondrites was inherited from nebular conditions during accretion, and that the Earth must have accreted "dry" and acquired its volatile elements via later impact events.

Enstatite chondrites are the most reduced type of chondrites and the only group that shares similar isotopic anomaly patterns with Earth for most elements (e.g., O, Cr, Ni, Ti; Moynier and Fegley 2015). For Cu isotopes, the high (EH) and low (EL) iron groups have identical average compositions ($\delta^{65}Cu\approx-0.25$‰; Savage et al. 2015b), falling in the center of the chondritic range, although the more volatile depleted EL define a larger range (Fig. 3). Enstatite chondrites of types EH and EL3 (low thermal metamorphic grade) have a similar Zn isotopic composition (0.15‰$<\delta^{66}$Zn$<$0.31‰, Moynier et al. 2011) to that of carbonaceous chondrites, unequilibrated ordinary chondrites and current estimates of BSE (Fig. 4A). On the other hand, those EL chondrites which experienced strong thermal metamorphic alteration (EL6) are highly enriched in the heavier isotopes (δ^{66}Zn up to 7.35‰, Fig. 4B) and are highly depleted in Zn and other moderately volatile elements. The enrichment in the heavier isotopes of Zn is evidence that the origin of the volatile element depletion between EL3 and EL6 chondrites was due to volatilization during thermal metamorphism (Moynier et al. 2011). Why such large enrichments in heavy Zn are not reflected by the Cu isotope composition of EL6 chondrites is puzzling; however, it should be noted that the amount of Cu loss between EH and EL6 is much less significant than the amount of Zn loss; also there are no EL3 Cu isotopic measurements—it could be that EL3 chondrites have a lighter Cu isotope composition than EH, and that the similarity between EH and EL6 is merely coincidence. This question remains unanswered.

Ureilites are ultramafic achondrites which are widely considered as mantle restites. Ureilites with different shock degrees and volatile element abundances show a positive correlation between δ^{66}Zn (with values up to 1‰) and 1/Zn, whereby samples with the lowest Zn content have the heaviest Zn isotopic compositions (Moynier et al. 2010b). This was taken as evidence that, as with EL6 meteorites and terrestrial tektites (see later), the variations in the abundance of Zn(Isotopes) between ureilite samples is controlled by evaporation processes. In addition, the more depleted samples also exhibited a higher shock state, suggesting an impact may have been responsible for the heating event.

The HED (howardites, eucrite, diogenites) meteorites, presumably derived from the asteroid 4-Vesta, have highly variable δ^{66}Zn (−2 to +1.7‰, Paniello et al. 2012b). On the other hand, unbrecciated eucrites (1.6$<\delta^{66}$Zn$<$6.22‰, $n=4$) and diogenites (0.94‰$<\delta^{66}$Zn$<$1.6‰, n=3) are all isotopically heavy and are all more depleted in Zn (and other moderately volatile elements) than brecciated HED suggesting that some volatile loss by evaporation occurred during the formation of the Vestan crust.

Iron meteorites are mostly composed of Fe and Ni and are enriched in siderophile elements compared to chondrites. The so-called magmatic groups (or fractionally crystallized iron groups) are thought to represent the cores of disrupted asteroids while the silicate-bearing groups have a more complex history, and may have formed as pools of impact-produced melt near the base of a regolith on a chondritic parent body (Wasson and Wang 1986). Luck et al. (2005) found that the IIIAB magmatic iron meteorites show very limited isotopic variations of both Cu and Zn, while silicate-bearing iron meteorites from group IA and IIICD are enriched in the heavier isotopes of Zn by up to 3.7‰ (although their Cu isotopes are mostly unfractionated). Bridgestock et al. (2014) have expanded the set of Zn isotopic data in the silicate bearing IA irons and in the IIAB and IIIAB groups with high precision Zn concentration determined by isotope dilution. They found that, in general, all iron meteorites are isotopically heavy in Zn relative to terrestrial/carbonaceous chondrites, and that the δ^{66}Zn is negatively correlated with 1/Zn for each individual group. They also showed that chromites are Zn-rich and isotopically light (δ^{66}Zn ~ 0) and proposed that the correlation observed between δ^{66}Zn and 1/Zn correspond to the segregation of chromite from metal. Chen et al. (2013a) and Bishop et al. (2012) measured the Cu and Zn isotope compositions of a large set of irons, including Zn-poor iron meteorite groups such as IVA and IVB, and did not find any particular enrichments in the heavier isotopes relative to other iron meteorite groups, suggesting that the low volatile element contents recorded in these meteorites is not related to evaporation during the parent body history. Bishop et al. (2012) found that, as observed in chondrites by Luck et al. (2003), δ^{65}Cu correlates with Δ^{17}O for silicate bearing iron meteorites. Such a correlation between mass-dependent and non-mass dependent isotopic fractionation must reflect mixing between at least two solar nebula components (see above);

hence, the variations in the Cu isotopic composition of the silicate-bearing iron meteorites originates from nebular processes rather than from planetary differentiation effects. Bishop et al. (2012) further proposed that the same may be true for magmatic iron meteorites (which do not contain silicates and so for which it is not possible to determine the $\Delta^{17}O$) and that Cu isotopes could be used to determine genetic connections between meteorite groups. Chen et al. (2016) measured the Cu isotope composition of the IVB (magmatic) iron meteorites and found a very large range of variation ($-5.84‰ < \delta^{65}Cu < -0.24‰$). The IVB irons are the most volatile depleted iron meteorites, with Cu concentration depletions several orders of magnitude larger than other iron meteorites groups and Ni/Cu ratios of $\sim 10^5$. Chen et al. (2016) show that the Cu isotope variations are controlled by neutron capture due to galactic cosmic ray irradiation by a reaction on ^{62}Ni to form ^{63}Ni, which decays to ^{63}Cu ($t_{1/2} \approx 100$ yrs).

Williams and Archer (2011) reported the Cu isotopic composition of phase separates (metal, troilite and silicate) from a variety of iron meteorites and coupled these with Fe isotope composition of the same phases. They found a large range of Cu isotope variations among metals and troilites (~ 10‰ variations) and also within the calculated metal-troilite fractionation factor (≤ ~5‰ variations) suggesting a kinetic control on the isotopic fractionation between the different phases. However, the most equilibrated samples display the smallest metal/troilite fractionation factor of ~0.5‰, with the metal phase being enriched in the heavier Cu isotope.

Mass-independent Zn anomalies in extra-terrestrial material. In terms of nucleosynthesis, Zn is classed as an iron peak (IP) element, along with Ca, Ti, Cr, Ni and of course, Fe. These elements have the highest nuclear binding energies per nucleon, and the 'iron-(abundance)-peak' is defined by the heaviest nuclides for which nuclear fusion becomes is energetically unfavorable during element synthesis (^{56}Ni, which decays to ^{56}Fe). As such, IP elements are only formed in the cores of massive stars or by explosive nucleosynthesis, where formation of some of the nuclides is dominated by nuclear statistical (quasi)equilibrium during explosive nucleosynthesis (NSE/QSE; see Wallerstein et al. 1997, for a review). The measurement of so-called isotope anomalies (identification of isotope reservoirs that do not fall on the terrestrial fractionation line) of the iron peak elements in extra-terrestrial materials have afforded many important insights into the stellar sources of material into the solar system, early solar system processes, and the building blocks of the terrestrial planets (e.g., Birck 2004; Moynier and Fegley 2015).

Of the IP element isotope systems, Zn isotope anomalies are, arguably, the least well constrained with (at the time of writing), less than 10 studies available in the literature. The initial work (using TIMS, Loss and Lugmair 1990; Völkening and Papanastassiou 1990) focused on the analysis of refractory inclusions from primitive meteorites (those phases thought to be the first to condense from a cooling nebula gas). Previously measured Ca, Ni and Cr anomalies in these materials (see Birck 2004 and refs therein) were modelled by Hartmann et al. (1985) in terms of nuclear statistical equilibrium, which also predicted relatively large ^{66}Zn excesses. Although Loss and Lugmair (1990) and Volkening and Pappanastassiou (1990) measured resolvable ^{66}Zn anomalies, these were much smaller than those predicted by the models, and this was explained as a result of the higher volatility of Zn compared to the other IP elements; i.e., by the time Zn began to condense, mixing in the solar nebula had diluted most anomalous Zn. Nevertheless, the presence of anomalous Zn in refractory inclusions is extremely puzzling, because Zn should not condense at all during their formation, and so all Zn in these inclusions must have been introduced by secondary processes. This implies that Zn isotope anomalies survived the hot initial stage of the solar system, potentially as a distinct sulfide phase, but this is a long standing and poorly understood issue (e.g., Chou et al. 1976).

TIMS measurements could not resolve any Zn anomalies at the bulk meteorite scale, and this was apparently confirmed by the first MC–ICP–MS measurements, which showed that bulk chondrites plotted within error of the terrestrial mass fractionation line in $\delta^{66}Zn$ vs. $\delta^{68}Zn$ space (Luck et al. 2005)—seemingly confirming that Zn condensed too late for resolvable anomalies to be detected at this scale. The requirement of such large ^{66}Zn excesses to accompany other neutron rich IP anomalies, particularly ^{48}Ca, was also relaxed with the advancement of nucleosynthesis models into so-called quasi-equilibrium (e.g., Meyer et al. 1998). Nevertheless, the first paper to specifically investigate Zn isotope anomalies on the bulk meteorite scale was Moynier et al. (2009a), which utilized MC–ICP–MS. The advantages of MC–ICP–MS over TIMS, regarding the detection of Zn anomalies, are the much better precision attainable, the ability to accurately measure ^{67}Zn, which always suffered from an unidentified interference on TIMS and, finally, the ability to switch individual amplifier resistances to increase the dynamic range of the instrument. Moynier et al. (2009a) used a smaller resistance amplifier on the ^{64}Zn detector to allow for higher concentration samples to be analyzed, with the specific aim of investigating potential ^{70}Zn heterogeneity in bulk solar system materials—important as this can constrain the distribution of ^{60}Fe (a short-lived radionuclide) in the solar nebula. At the precisions attained in their study (±100 ppm), no resolvable ^{66}Zn, ^{67}Zn or ^{70}Zn anomaly patterns were measured (when normalized to ^{68}Zn/^{64}Zn), which indicated relatively homogeneous distribution of Zn isotopes, and also ^{60}Fe; however, there were some hints in their anomaly patterns that, with further improvements in precision, anomalies in ^{66}Zn or ^{67}Zn may be present and measurable.

With techniques modified from Moynier et al. (2009a) and with analytical precision at the ±10 ppm level, Savage et al. (2014) showed for the first time that carbonaceous chondrites do have resolvable ^{66}Zn and ^{68}Zn excesses (when normalized to ^{67}Zn/^{64}Zn), and, also that enstatite and potentially ordinary chondrites have smaller ^{66}Zn deficits; this is consistent with the sense of ^{48}Ca, ^{50}Ti, ^{54}Cr and ^{62}Ni anomalies measured in the same samples (Trinquier et al. 2009; Steele et al. 2012; Dauphas et al. 2014; Moynier and Fegley 2015; Schiller et al. 2015). The complementary excesses and deficits exhibited by the carbonaceous and enstatite/ordinary chondrites, the more volatile nature of Zn, and the correlations between Zn and other iron-peak anomalies adds further credence to the 'unmixing' hypothesis of solar nebula evolution, where specific phases were remobilized via thermal processing in a previous well-mixed nebula cloud (e.g., Trinquier et al. 2009; Schiller et al. 2015). Sequential leaching experiments show that the Zn anomalies are not limited to one phase, although this is most likely due to post-formation remobilization of Zn. One important insight from this dataset is that Earth is not similar to enstatite chondrites, in terms of Zn isotope budget. Now that Zn isotope anomalies have been discovered in bulk primitive meteorites, there is potential to discover such anomalies in other solar system materials, and the new insights from this system could be hugely important for our understanding of our solar system.

Bulk Silicate Earth composition. The estimation of the isotopic composition of the bulk silicate Earth is not trivial since both Zn and Cu are fractionated during partial melting of the mantle, hence during differentiation processes their isotopes may be fractionated also. In addition, as both Zn and Cu are trace elements, metasomatism or low-temperature alteration could overwhelm any primitive signal that a rock once held. Therefore, in order to estimate the Cu and Zn isotopic compositions of the bulk silicate Earth (BSE) it is necessary to 1) choose pristine samples; 2) constrain the extent to which igneous differentiation processes fractionate the isotopes, and; 3) analyze as wide a variety of mantle derived samples as possible.

The first modern Cu and Zn isotope estimates of BSE were based on the average composition of MORB samples taken from three ocean basins: $\delta^{65}Cu = 0‰$ and $\delta^{66}Zn = 0.25‰$ (Ben Othman et al. 2006). The reason that no precision is given is that these data are given in a conference abstract; nevertheless, for the following decade, these estimates were the accepted values (the abstract was never written up and no further systematic studies were performed).

It is only recently that further investigations into this area have been made. Since Cu is highly incompatible and strongly chalcophile, its behavior during mantle melting is controlled by the fusion of sulfides (Lee et al. 2012); for partial melting degrees <25%, residual sulfides may be retained in the source and could potentially create isotopic fractionation. In order to test the possible effect of partial melting on the isotopic composition of Cu, Savage et al. (2015b) measured komatiites (ultra-mafic lavas formed by >25% partial melting; $\delta^{65}Cu = 0.06 \pm 0.06$‰, 2 SD, $n = 14$) and compared these to fertile orogenic lherzolites (mantle samples that have undergone the least melt depletion: $\delta^{65}Cu = 0.07 \pm 0.09$‰, 2 SD, $n = 16$: Ben Othman et al. 2006; Ikehata and Hirata 2013), as well as a representative selection of both mid-ocean ridge and ocean island basalts. All groups have identical Cu isotope compositions (Fig. 6) which suggests that during mantle melting there is limited Cu isotope fractionation expressed in the melt. Savage et al. (2015a) therefore used all this data to propose a $\delta^{65}Cu = 0.07 \pm 0.10$‰ (2 SD) for the BSE. Furthermore, these authors measured the Cu isotope compositions of two magmatic differentiation sequences, from Kilauea Iki, Hawaii, and Hekla, Iceland, systems. Both suites define a large range of SiO_2 and MgO contents and evolved from a cogenetic source with limited contaminations by crustal materials. The samples from Kilauea Iki showed no variation away from BSE with increasing degree of differentiation, as would be expected given the lack of sulfide fractionation in this system. In comparison, the samples from Hekla show more variability, which seems to be related to the removal of sulfides in the magma chamber. However, these variations are limited (range of compositions from Hekla -0.08‰ $< \delta^{65}Cu < 0.20$‰) and, crucially, the basalts from Hekla are identical to BSE. This indicates that significant igneous differentiation generates only limited Cu isotope fractionation, further confirmed by the similarity of I-type granite ($\delta^{65}Cu = 0.03 \pm 0.15$‰, 2 SD: Li et al. 2009) to BSE.

Liu et al. (2015) reached similar conclusions to Savage et al. (2015b) by comparing a large set of both unmetasomatized cratonic and orogenic peridotites with MORB, and OIB and proposed a $\delta^{65}Cu = 0.06 \pm 0.20$‰ (2 SD) for the BSE. Their data for metasomatized peridotites were much more variable, demonstrating the susceptibility of Cu-depleted rocks to secondary isotope fractionation; this was also seen in large negative Cu isotope excursions in Kilbourne Hole peridotites which correlate with LREE enrichment, and with large positive komatiite Cu isotope enrichments which only occur in those samples whose Cu contents do not plot on olivine control lines (Savage et al. 2015a). This suggests that Cu isotopes could be further utilized as a tracer of recycled materials in the mantle and, for instance, island arc material; indeed, Liu et al. (2015) provide a large set of arc basalt data whose range is much larger than that defined by both MORB and OIB (Fig. 6).

Compared to the Cu isotope system, fewer studies have attempted to address the behavior of Zn isotopes during igneous processes. Because Fe^{2+} and Zn^{2+} have similar ionic radii, the Fe/Zn ratio is not fractionated during partial melting of peridotites and therefore primitive basalts preserve the Fe/Zn ratio of the parent magma (Le Roux et al. 2011). Chen et al. (2013a) evaluated the extent to which Zn isotopes are fractionated during igneous processes by the same set of samples as Savage et al. (2015a) for Cu isotopes, those of Kilauea Iki, USA, and Hekla Volcano, Iceland. Both sets of samples show ~0.1 per mille isotopic variation but only the $\delta^{66}Zn$ of the Kilauea Iki samples vary systematically, correlating with the degree of differentiation (MgO contents) with the most evolved samples enriched in the heavier isotopes (see Fig. 7). These isotopic variations are interpreted as the result of crystallization of isotopically light olivines, and Fe-Ti-oxides at the very end of the differentiation sequence (Chen et al. 2013b). Chen et al. (2013b) combined the data from mafic rocks from Herzog et al. (2009), and their own data to determine the $\delta^{66}Zn$ of the BSE to be 0.28 ± 0.08‰. More recently, Sossi et al. (pers. comm.) have shown that ultramafic rocks comprising unmetasomatized peridotites from the Balmuccia massif and komatiites with ages varying from 3.5 to 2.7 Ga are all isotopically lighter than basalts and complement the $\delta^{66}Zn$ vs. MgO trend defined by Chen et al. (2013b). Sossi et al. (pers. comm.) used the average of these ultramafic samples to determine the most up to date $\delta^{66}Zn$ composition of the BSE to be 0.15 ± 0.05‰. Telus et al. (2012) showed that most granites are not isotopically fractionated in Zn with regards to the

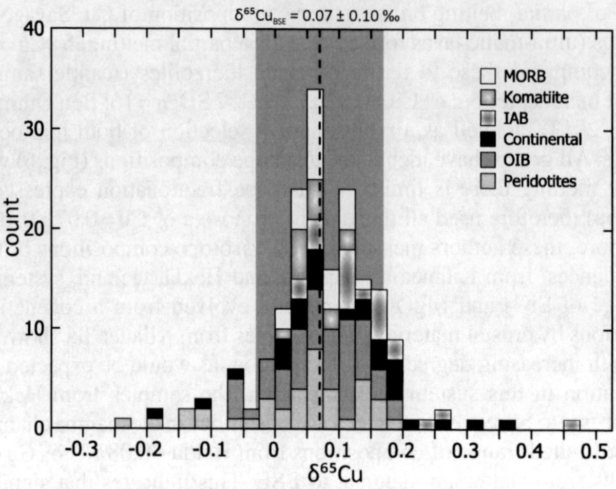

Figure 6. Histogram of the δ^{65}Cu of the various ultramafic and mafic samples analyzed to date (data from Ben Othman et al. 2006; Ikehata and Hirata 2013; Liu et al. 2015; Savage et al. 2015b). The grey box represents the estimate of the BSE composition from Savage et al. (2015b). Komatiites , fertile orogenic lherzolites as well as a representative selection of both mid-ocean ridge and ocean island basalts have identical Cu isotope compositions suggesting that mantle melting produces a limited Cu isotope fractionation.

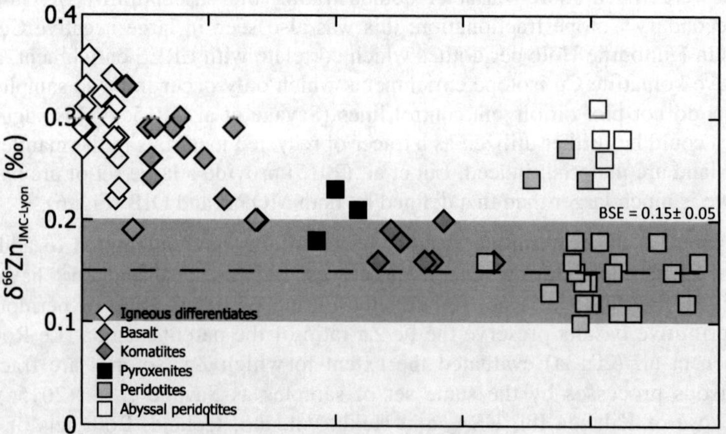

Figure 7. δ^{66}Zn and Zn concentration versus degree of differentiation as represented by MgO content for basalts and their differentiates, komatiites and ultramafic samples (data from Herzog et al. 2011, Chen et al. 2013b and Sossi et al. pers. comm.). The most evolved samples are enriched in the heavier isotopes. This trend is interpreted as the result of crystallization of isotopically light olivines or Ti-oxides (Chen et al. 2013b) and led Sossi et al. (pers. comm.) to propose a BSE δ^{66}Zn composition of 0.15 ± 0.05‰. N.B. the abyssal peridotites underwent metasomatism, most likely affecting their Zn isotope composition.

BSE value, but pegmatites and some granites are isotopically heavy (up to ~0.9‰). Telus et al. (2012) interpreted these heavy isotopic compositions in terms of fluid exsolution and suggest that Zn isotopes can be used to trace fluid exsolution in rocks.

Isotopic fractionation during core formation. Given that both Zn and Cu can partition in measurable quantities into the metal phase during metal–silicate equilibration, there is the potential that both isotope systems could reveal insights into the physiochemical conditions of Earth's differentiation into a planet with a metallic core and silicate mantle. However, for both Zn and Cu, their isotope partitioning behavior is only now being investigated via experimental studies—but there is huge potential for further work.

In the case of Zn, Bridgestock et al. (2014) showed that there appears to be no isotope effect associated with metal silicate equilibration. This indicates that, even if there is measurable Zn in Earth's core, partitioning into this reservoir would not leave its isotopic fingerprints on Earth's mantle. This is consistent with the similarity of the BSE composition with those of carbonaceous, unequilibrated ordinary and EH chondrites (see above).

The case of Cu is more complicated. A series of metal–silicate experiments performed by Savage et al. (2015b) indicated that the heavier Cu isotope prefers to enter the metal phase. This would suggest (given that 2/3 of Earth's Cu is in the core) that bulk Earth is isotopically heavier than the value estimated for BSE (see above). The problem is that there are no primitive meteorites thus far measured which match this isotopically heavy composition (in fact, the Cu isotope composition of BSE is still heavier than most chondrites; Fig. 3). Assuming that the bulk Earth is chondritic, i.e., the bulk Earth has a lighter Cu isotope composition than its mantle, this could imply a number of possibilities:

Cu is also a moderately volatile element, depleted in Earth's mantle relative to chondrites. It could be that impact-driven volatilization created this depletion, which would preferentially lead to the loss of light Cu, leaving a heavy residue. However, as Zn and to a lesser extent Cu isotopes in chondrites have shown, Earth's volatile depletion was more likely caused by nebula processes (see above).

Earth's mantle Cu isotope composition is not in equilibrium with its core: it could be that late addition of Cu to the Earth, during the final stages (i.e., the giant impact) or post-core formation (i.e., the late veneer) may have set the BSE composition. In the case of the late veneer, there is not enough mass delivered by this process to account for all the Cu in the mantle today. In the case of the giant impactor, this is dependent on the composition of the impactor—even if it was CI-like (e.g., Schonbachler et al. 2010), the disruption created by this event would still have led to phase equilibration in the following magma ocean.

Earth contains a hidden, isotopically light Cu reservoir. Savage et al. (2015b) performed sulfide-silicate Cu isotope fractionation experiments, and found that, in this instance, the sulfide phase preferentially takes the lighter isotope of Cu; hence, rather than 2/3 of Cu being held in the core, this reservoir may be split into an isotopically heavy metal and isotopically light sulfide (relative to BSE). One possible sulfide reservoir is that of a 'Hadean Matte' (O'Neill 1991; Lee et al. 2007), an Fe–O–S phase that remains as the final liquid after the crystallization of a magma ocean. Modelling this in terms of [Cu] and δ^{65}Cu suggests that, if such a reservoir formed and eventually was admixed into the core (it would sink through the mantle due to its higher density), it could add up to ~0.8wt.% S to the core (Savage et al. 2015b).

Although there are caveats associated with each of the above models, Cu isotopes could be a powerful tool in tracing the fate of sulfides in various igneous and planetary processes, but the framework to understand this fractionation is still required, and more experimental work is required.

Isotopic fractionation by evaporation on Earth. While the Zn and Cu isotopic database for terrestrial igneous rocks is very limited, tektites are extremely fractionated (Moynier et al. 2009c). Tektites are terrestrial natural glasses produced during a hypervelocity impact of an extraterrestrial projectile onto the Earth's surface, and are extremely depleted in volatile elements, e.g., they are among the driest terrestrial samples (<0.02% of water). Moynier et al. (2009c) found that tektites are extremely enriched in the heavier isotopes of Zn, up to 2.5‰ and attributed this enrichment to kinetic isotopic fractionation during evaporation. Copper can be even more fractionated than Zn, with $\delta^{65}Cu$ up to 12.5‰ found in some European tektite samples (Rodovka et al. pers. comm.). The difference of behavior between Cu and Zn has been explained by isotopic fractionation in a diffusion-limited regime, where the magnitude of the isotopic fractionation is regulated by the competition between the evaporative flux and the diffusive flux at the diffusion boundary layer (Moynier et al. 2010a). Copper diffuses much faster than Zn (due to the difference in ionic charge in silicates of Zn^{2+} vs. Cu^{+}), hence the larger isotopic fractionation in Cu than in Zn in tektites is due to the significant difference in their respective chemical diffusivity.

The Moon. The isotopically heavy lunar regolith (2.2‰ < $\delta^{66}Zn$ < 6.4‰ and 2.6‰ < $\delta^{65}Cu$ < 4.5‰) reflects billions of years of evaporation due to solar wind sputtering and micrometeorite impact gardening (Moynier et al. 2006; Herzog et al. 2009). On the other hand, the low-Ti ($\delta^{66}Zn = 1.31 \pm 0.13$‰) and high-Ti basalts ($\delta^{66}Zn = 1.39 \pm 0.39$) have a more limited isotopic variations and are systematically ~1‰ heavier than the BSE for Zn (Paniello et al. 2012a; Day and Moynier 2014; Kato et al. 2015). The only data available for Cu isotopes ($\delta^{65}Cu = 0.5 \pm 0.1$‰, Herzog et al. 2009) suggests that there is a similar enrichment in the heavy Cu isotope in lunar(?) rocks, but this will need further investigation in the future. Since the isotopically heaviest carbonaceous chondrite group (CI) has a $\delta^{66}Zn = 0.46$‰ and $\delta^{65}Cu = 0.05$‰ (Luck et al. 2003, 2005; Barrat et al. 2012), mixing with chondrites does not explain the Zn or Cu isotopic composition of the lunar basalts. In addition, lunar plutonic rocks (alkali and magnesian suite samples) are isotopically heavier than the mare basalts ($\delta^{66}Zn$ up to 6.27‰) suggesting that the volatile loss could have occurred in two stages: during the proto-lunar disk stage, where a fraction of lunar volatiles accreted onto Earth, and from degassing of a differentiating lunar magma ocean, implying the possibility of isolated, volatile-rich regions in the Moon's interior (Kato et al. 2015).

ZINC AND COPPER IN LOW TEMPERATURE GEOCHEMISTRY

Since the pioneering work of Maréchal et al. (1999, 2000), and Francis Albarède's chapter on Cu and Zn isotopes in the first RiMG volume on non-traditional stable isotopes (Albarède 2004), a considerable amount of effort has gone into understanding and applying isotope variations of these two elements in samples from Earth's surface. As with the development of any relatively new isotope system, documentation of stable isotope variations in nature has been coupled with experimental and theoretical studies aimed at characterizing isotopic fractionations associated with key surface Earth processes. Copper–Zn stable isotope geochemistry of the surface Earth environment has been part of reviews by Cloquet et al. (2008) and Wiederhold (2015). An emerging new interest lies in the application of Zn and Cu isotopes to the study of biological pathways and changes in metabolism associated with diseases. These applications are here treated in Albarède et al. (2017, this volume).

The data obtained to date for important surface Earth reservoirs are summarized in Figure 8. One of the first order features of Cu–Zn isotope geochemistry that this compilation confirms is the contrast between the relative homogeneity of samples whose isotopic characteristics are determined by high temperature processes, versus the variability in materials formed and equilibrated at low temperatures. Thus, igneous rocks (excluding ultramafic rocks) show a very tight distribution, with $\delta^{66}Zn = 0.31 \pm 0.12$‰ ($n = 77$, 1SD) and $\delta^{65}Cu = 0.08 \pm 0.17$‰ ($n = 287$) which overlaps with the BSE estimate presented earlier in this chapter (note that these averages were taken using all igneous rock data available in the literature, without screening for the possibility of secondary alteration/metasomatism—hence they are slightly different, and have poorer precisions, than those defined in the section above). Another first-order feature from the data in Figure 8 is that sediments that have undergone physical, but minimal chemical, processing through the fluid envelopes of the surface Earth (i.e., clastic sediments from rivers, lakes, oceans, as well as aerosols/dust), have average Cu and Zn isotope signatures that are identical

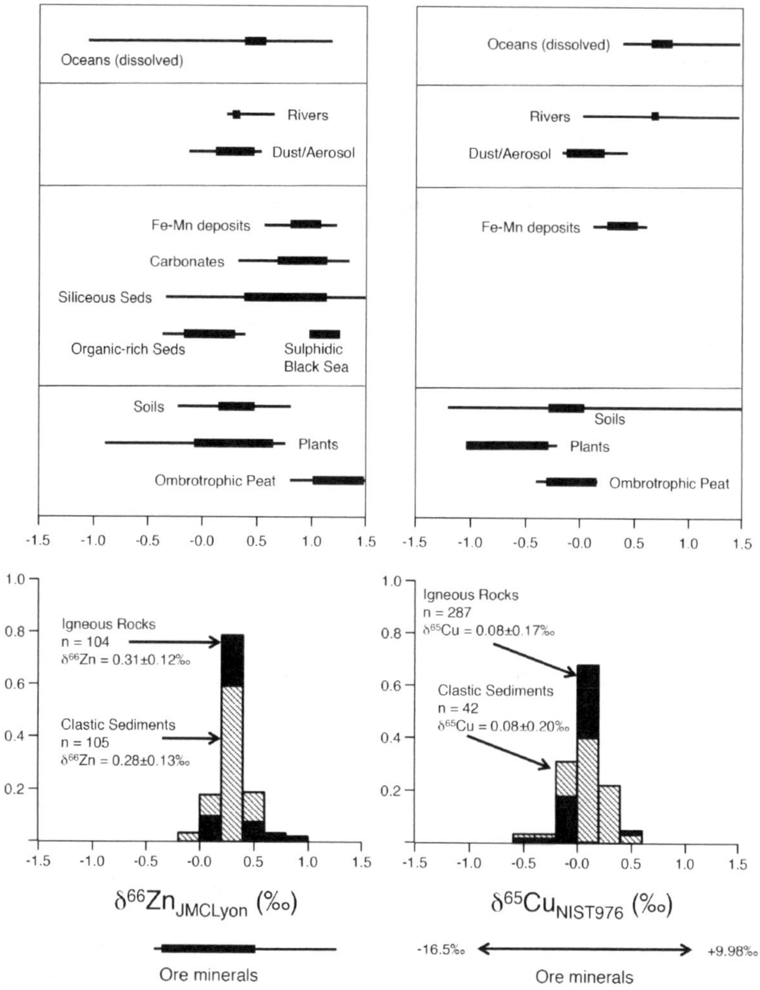

Figure 8. Summary of currently available data for "natural" surface Earth materials. Materials significantly impacted by human activity represent a distinct topic and are treated separately later in this chapter. The range of isotopic compositions found in ore minerals is shown by the black bars at the bottom. Note that Cu isotopes are extremely heterogeneous in Cu-bearing ore minerals, extending well beyond the limits of other Earth surface samples, from δ^{65}Cu −16.5 to +9.98. In the top two panels the thinner lines show the total range of values measured in each type of sample. For the oceanic dissolved pool, the thicker lines show the average and 1 SD for all analyses in the homogeneous deep ocean (beneath 600–800 m). For the river data the square shows the discharge- and [Cu]- or [Zn]-weighted average for the dissolved flux to the oceans as calculated from the large rivers measured to date. For all the other sample types the thicker line shows the average of all the data ±1SD. In the bottom two panels the solid black histograms show the range of variability in all igneous rocks, as relative frequencies, including basalts, andesites, dacites, rhyolites, granites, granodiorites, komatiites and peridotites. Clastic sediments (diagonal pattern), including atmospheric aerosols, show close overlap with the igneous samples. In contrast, surface Earth samples whose genesis involves the partitioning of Cu and Zn between solid and aqueous phases (top two panels) exhibit substantial variability. Note the agreement between the values for igneous rocks and clastic sediments with those for the Bulk Silicate Earth in earlier diagrams.

Figure 8 (cont'd) Data from: *Seawater:* Bermin et al. (2006); Vance et al. (2008); Boyle et al. (2012); Zhao et al. (2014); Conway and John (2014, 2015); Thomson and Ellwood (2014); Takano et al. (2014). *Rivers:* Vance et al. (2008); Ilina et al. (2013); Little et al. (2014a); and including two Zn data for the relatively unpolluted Seine system headwaters from Chen et al. (2008). *Dust/aerosol*: Marechal et al. (2000); Li et al. (2009); Bigalke et al. (2010a); Dong et al. (2013); Little et al. (2014a); and including data for rain (wet deposition?) in Takano et al. (2014). *Fe–Mn deposits:* Marechal et al. (2000); Little et al. (2014a). *Carbonates:* Pichat et al. (2003). *Siliceous sediments:* Andersen et al. (2011); Hendry and Andersen (2013). *Organic-rich sediments:* Little et al. (2016) *Deep*; *sulfidic Black Sea* (dissolved phase):Vance et al. (2016b). *Soils and plants:* Viers et al. (2007); Bigalke et al. (2010a, 2011); Mathur et al. (2012); Liu et al. (2014); Vance et al. (2016a). *Ombrotrophic peat:* Weiss et al. (2007; pre-Anthropocene analyses only). *Igneous rocks:* Marechal et al. (2000); Archer and Vance (2004); Chapman et al. (2006); Cloquet et al. (2006); Bentahila et al. (2008); Toutain et al. (2008); Sonke et al. (2008); Li et al. (2009); Herzog et al. (2008); Chen et al. (2009, 2013); Bigalke et al. (2010); Moynier et al. (2010a;b); Weinstein et al. (2011); Moeller et al. (2012); Telus et al. (2012); Liu et al. (2015); Savage et al. (2015; including data tabulated here from Ben Othman et al. 2006 and Ikehata and Hirata; 2012). *Clastic sediments* (including dust/aersols in refs above): Marechal et al. (1999, 2000); Asael et al. (2007); Bentahila et al. (2008); Sonke et al. (2008; pre-Anthropocene analyses only); Chen et al. (2009; only relatively unpolluted river sediments from the Seine system; with Zn enrichment factors < 2); Bigalke et al. (2010a); Mathur et al. (2012); Gagnevin et al. (2012); Vance et al. (2016a). *Ore minerals*: Marechal et al. (1999); Larson et al. (2003); Mason et al. (2005); Wilkinson et al. (2005) Mathur et al. (2005, 2009, 2010); Markl et al. (2006); Asael et al. (2007); Sonke et al. (2008); Gagnevin et al. (2012).

to high temperature igneous rocks, and are not much more variable. In contrast, environmental samples that *have* seen such biogeochemical processing record a roughly 2.5‰ range in Cu and Zn isotopes, or about 30–40 times the analytical precision. Ore minerals that formed at high temperatures exhibit a range of roughly 1‰ that is also more or less centred on the peak in igneous rocks (Fig. 8, black bars at bottom). In contrast, the isotope compositions of minerals from the supergene environment, containing Cu that has undergone (possibly multiple) oxidation and reduction cycles, exhibit huge (~20‰) variability (e.g., Maréchal et al. 1999; Larson et al. 2003; Mason et al. 2005; Mathur et al. 2005, 2009, 2010; Markl et al. 2006; Asael et al. 2007).

In the following sections we first review the experimental constraints on the size and sign of isotope fractionations of Cu and Zn associated with key surface, low temperature, processes. Secondly, we discuss the origin of, and geochemical constraints available from, Cu and Zn isotope variability in "natural" samples—i.e., those not significantly impacted by human activities. This large subject is separated into sub-sections on (1) the weathering–soil–plant system, and (2) the oceans, their inputs, outputs and internal cycling. In this section we also briefly outline the very few studies that have sought to apply Cu–Zn isotopes in Earth history, and outline the prospects for the future of such a pursuit. Isotopic variations, and their expression in environmental samples, superimposed by human activity on this natural biogeochemical cycling, represent a somewhat distinct topic, and are treated in a third section. For convenience, and since ore minerals often represent the starting material from which pollution of the Anthropocene environment derives, the very large variability seen in ore minerals is also dealt with in this third section.

Experimental constraints on fractionation mechanisms

A survey of the literature reveals four general isotope fractionation mechanisms causing the roughly 2.5‰ variation in both Cu and Zn isotopes in environmental samples, as well as the much greater degree of variability in supergene Cu ore minerals:

1. Equilibrium isotope distributions between Cu in different oxidations states;
2. Equilibrium isotope distribution between dissolved aqueous species;
3. Equilibrium and kinetic effects caused by interactions between solids—abiotic as well as living cells—and aqueous solutions (sorption, precipitation);
4. Kinetic and equilibrium effects related to uptake into the cells of living microbes and higher plants.

The sign and magnitude of these fractionations are summarized in Figure 9 and are discussed in turn below. Though the separate processes bulleted above makes discussion more convenient, distinction between these fractionation mechanisms is not often sharp. To some extent this lack of clarity represents some confusion in the literature. Thus, for example, all metals sorb to the external surfaces of microbial cells, and to some extent the roots of higher plants. This is a somewhat distinct phenomenon from uptake into the cells for metabolic use in enzymes and proteins. When metal uptake is studied in microbiological or hydroponic plant growth experiments, the vast majority of the metals in solution in the media are bound to an added organic complexant, such as EDTA. This leaves only a small pool of free metal ion, which is often regarded as the pool that is available for uptake. The isotopic composition of the metal taken up can be lighter than the bulk experiment, for example where uptake is transport (diffusion)-limited (e.g., John et al. 2007b), or it can be heavier, for example for some plant species that actively bind external metals using phytosiderophores (e.g., Arnold et al. 2010a).

On the other hand, in experiments where organic complexants have not been added, the free metal ion pool is often many orders of magnitude more concentrated. In this case the metals sorb to external surfaces. Though sometimes described as "uptake" in the literature, there is almost certainly no metabolic function of the metals in this case. Such sorption often involves binding to deprotonating functional groups such as carboxyls and amines, so that the fractionation factors measured in these experiments are more relevant to categories 2 and 3 above than 4. On the other hand, there is sometimes genuine uncertainty over whether a pool of metals associated with a cellular experimental product is intra-cellular or sorbed on external surfaces. Some researchers have been able to distinguish between these two mechanisms using experiments with living (active metabolic uptake) versus dead (passive adsorption) cells, or by removing the extra-cellular pool using a desorptive wash prior to analysis (e.g., John et al. 2007b; Navarette et al. 2011). Finally, Cu uptake into microbial cells has been interpreted as involving reduction of external Cu(II) to internal Cu(I)—i.e., a component of 1 above (e.g., Zhu et al. 2002; Navarette et al. 2011; Jouvin et al. 2012; Ryan et al. 2013).

Changes in oxidation state. Zinc does not undergo changes in oxidation state at Earth surface conditions. Thus, although a substantial isotope fractionation has been characterized in an electroplating experiment involving reduction of aqueous Zn(II) to Zn metal (Kavner et al. 2008), this is unlikely to be relevant to natural systems. For copper, on the other hand, the transition between Cu(I) and Cu(II) happens at redox conditions relevant to the Earth's surface, and Cu occurs in both reduced and oxidized forms in Earth materials. Further, it is clear that the redox transition involves large isotope fractionations. This fractionation was first characterized by Zhu et al. (2002) in experiments that found $\Delta^{65}Cu_{Cu(II)\text{-}Cu(I)}$ – 4‰ for the reduction of aqueous Cu(II) to a Cu(I) iodide precipitate at 20 °C (here and throughout $\Delta^{65}Cu_{x\text{-}y} = \delta^{65}Cu_{(\text{phase } x)} - \delta^{65}Cu_{(\text{phase } y)}$). Ehrlich et al. (2004) followed this up with experiments involving the precipitation of Cu(I)S (covellite) from an aqueous Cu(II) solution and found $\Delta^{65}Cu_{Cu(II)aq\text{–}Cu(I)S} = 3.06 \pm 0.14‰$ at 20 °C. Furthermore, and importantly, they contrasted this large fractionation with the small one ($\Delta^{65}Cu_{Cu(II)aq\text{–}Cu(II)(OH)_2} = 0.27 \pm 0.02‰$) for Cu(II) hydroxide precipitation from a $Cu(II)_{aq}$ solution. This finding, as well as that by Maréchal and Sheppard (2002) of small (0.2–0.4‰) isotopic differences between Cu(II) in solution versus malachite, strongly suggests that it is the change in oxidation state, and not the phase change, that causes the large isotopic shift seen in these and other redox experiments. These results were further confirmed by Mathur et al. (2005), who found that Cu(I) in chalcocite (Cu_2S) and chalcopyrite ($CuFeS_2$) was 1.3 and 2.74‰ lighter than aqueous Cu(II) in abiotic batch oxidative leach experiments. In analogous experiments inoculated with *Thiobacillus ferrooxidans* the heavy oxidized Cu was located in amorphous Cu–Fe oxide minerals surrounding bacterial cells.

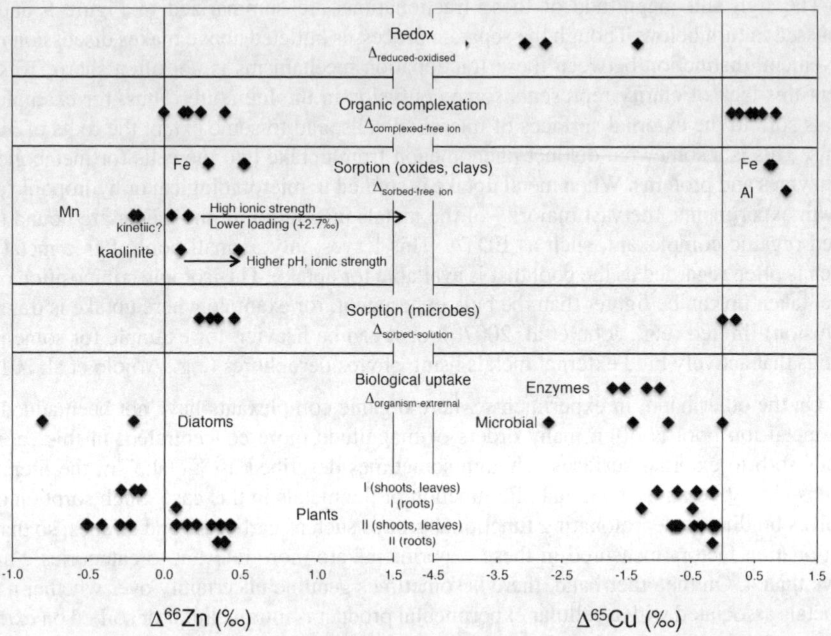

Figure 9. Summary of experimental constraints on the isotopic fractionation of Cu and Zn during important Earth surface processes. Data from: *Redox* (Cu only): Zhu et al. (2002); Ehrlich et al. (2004); Mathur et al. (2005). *Organic complexation*: Ban et al. (2002); Jouvin et al. (2009); Bigalke et al. (2010b); Ryan et al. (2014). *Sorption to oxide; clay and microbial surfaces*: Pokrovsky et al. (2005, 2008); Gélabert et al. (2006); Ballistrieri et al. (2008); Juillot et al. (2008); Navarette et al. (2011); Kafantaris and Borrok (2014); Bryan et al. (2014); Coutaud et al. (2014); Guinoiseau et al. (2016). Biological uptake: Zhu et al. (2002); Weiss et al. (2005); John et al. (2007); Pokrovsky et al. (2008); Arnold et al. (2009); Moynier et al. (2009); Weinstein et al. (2011); Caldelas et al. (2011); Navarette et al. (2011); Jouvin et al. (2012); Tang et al. (2012); Ryan et al. (2013); Conway and John (2014).

Zhu et al. (2002) found that Cu incorporated into proteins expressed in bacteria and yeast cells was 1.0–2.1‰ lighter than in the parent solutions and media, and used these findings to suggest that the biogenic uptake of light Cu also involved reduction. Likewise, Navarette et al. (2011) interpret Cu isotope variations in media from which live bacterial cells remove Cu as due to cellular uptake (as opposed to sorption, which induces a different fractionation in their experiments, as observed in dead cells). This is associated with preferential uptake of the light isotope—by up to 4.4‰—and was also interpreted to involve reduction of Cu(II). As noted in this study, if this reduction occurs within the cell, the changes seen in the media require that there is also efflux of heavy Cu from the cells, allowing equilibration of the two Cu pools. As discussed later in this section, uptake of isotopically light Cu by bacteria and higher plants probably also involves reduction by a reductase protein (e.g., Navarette et al. 2011; Weinstein et al. 2011; Jouvin et al. 2012; Ryan et al. 2013).

Organic complexation. Copper forms very strong inner sphere complexes (conditional stability constants up to 10^{25}) with functional groups in organic matter (McBride 1981; Grybos et al. 2007; Ryan et al. 2014). Virtually all Cu in the operationally defined dissolved phase (that fraction passing through a 0.45 μm filter) of rivers and the oceans is bound in these organic complexes (e.g., McBride 1981; Coale and Bruland 1988; Moffett and Brand 1996; Shank et al. 2004; Grybos et al. 2007; Vance et al. 2008; Ryan et al. 2014), such that inorganically complexed and free Cu^{2+} ion concentrations are 2–5 orders of magnitude lower than total dissolved Cu. Similarly, up to 98% of the "dissolved" Zn in many natural waters is also complexed to organic ligands, though with stability constants that are of order 10^9–10^{11} (e.g., Wells et al. 1998; Bruland 1999; John et al. 2007b). Grybos et al. (2007) suggest that two important processes compete to control transition element behavior in soils: binding to organic complexes, both in condensed organic matter and in an aqueous phase, versus sorption to the surfaces of secondary minerals such as clays and Fe–Mn oxyhydroxides. The last decade of research on Cu–Zn isotopes has revealed that this competition is almost certainly key for isotope distributions, not only in soils but also between the dissolved and particulate phases in rivers and the oceans (e.g., Vance et al. 2008; Bigalke et al. 2010a,b, 2011; Little et al. 2014b; Vance et al. 2016a). In quantifying the isotopic fractionations between an (often) aqueous organically complexed pool and the sorbed pool, the universal approach has thus far been to measure the isotopic separation between each of these and a dissolved free metal ion (Cu^{2+}, Zn^{2+}) pool. Each of these, therefore, are here dealt with separately.

Ban et al. (2002) were the first to quantify the isotopic impact of this important process, finding $\Delta^{66}Zn_{EDTAZn-Zn^{2+}} \sim 0.2‰$ (here and throughout $\Delta^{66}Zn_{x-y} = \delta^{66}Zn_{(phase\ x)} - \delta^{66}Zn_{(phase\ y)}$). Jouvin et al. (2009) used Donnan membranes to separate free Zn from that complexed to humic acid. These authors found no fractionation at pH≤5.4, but $\Delta^{66}Zn_{Humic-Zn^{2+}} = +0.24 \pm 0.06‰$ at pH 6.1–7.2. The variable fractionation as a function of pH was interpreted in terms of the partitioning of the bound Zn between high affinity (HA, bound to phenols) and low affinity (LA, bound to carboxylate groups) sites, and the fact that, at equilibrium, isotopically heavy Zn is partitioned into the strongly bound species. Zinc is increasingly bound to the HA sites at higher pH (50:50 LA and HA at pH around 6.1–6.2). Based on calculated mass balance between the species, Jouvin et al. (2009) proposed a fractionation factor, $\alpha_{HAS-Zn^{2+}}$, of 1.0004. Bigalke et al. (2010a) performed the same experiment for Cu at pH 2–7. In the case of Cu there is no apparent isotopic difference between LA and HA sites, and only 11–35% of the bound Cu is in HA sites. For both, $\Delta^{65}Cu_{Humic-Cu^{2+}} = +0.26 \pm 0.11‰$. More recently, Ryan et al. (2014) measured the Cu isotope fractionation between free Cu and a range of soluble organic ligands. They see a "strong" positive correlation between the isotopic fractionation and the value of the stability constant for each complex. Thus for natural riverine fulvic acid ($\log K = 8$) $\Delta^{65}Cu_{complex-free} = +0.14 \pm 0.11‰$, whereas for desferrioxamine B (DFOB $\log K = 24.7$) $\Delta^{65}Cu_{complex-free} = +0.84 \pm 0.30‰$.

Sorption to abiotic substrates. As with the fractionations associated with organic complexation outlined above, isotopic effects associated with sorption have also been measured experimentally relative to a dissolved free metal ion pool. Pokrovsky et al. (2005) measured Zn isotope fractionations upon sorption from simple aqueous solutions of low ionic strength (0.01M), where Zn is speciated as a hexaquocomplex, and where the sorption equilibrium can often be envisaged as (omitting the solvating waters in the aqueous species):

$$>MeOH^0 + Zn^{2+} \leftrightarrow MeO\text{–}Zn^+ + H^+$$

They found a small preference for the light isotopes of Zn (by about ~0.18–0.23‰) on goethite and birnessite (δ-MnO_2) surfaces, and for the heavy isotopes (by about ~0.11–0.14‰) on pyrolusite (β-MnO_2) and aluminium oxides. Zinc sorbed to hematite exhibited the largest isotopic separation from dissolved free Zn, with $\Delta^{66}Zn_{sorbed-free} = +0.61‰$ when sorption starts at pH 5.5, but decreasing to zero at higher pH as sorption increases. Not all the results of this pioneering study have been reproduced. Indeed in all subsequent studies, sorbed Zn has been found to be universally heavy relative to the aqueous free metal ion pool. Bryan et al. (2015) discuss possible reasons for these discrepancies, including the possibility of kinetic effects in short duration experiments. Moreover, theoretical considerations (e.g., Schauble 2004) suggest that the lower coordination of the metal sorbed on these surfaces (e.g., Peacock and Sherman 2004; Balistrieri et al. 2008; Juillot et al. 2008) should prefer the heavy isotope.

Balistrieri et al. (2008) find heavy Cu and Zn sorbed onto ferrihydrite, in a study incorporating both natural data from streams draining a metal sulfide deposit and experimental results. The experiments were also done with low ionic strength solutions (0.008 M). The sorption experiments only lasted 2–3 hours, but in this case the sorption of heavy isotopes as well as a close-to-linear relationship between the fraction of dissolved metal and isotopic composition rule out a kinetic effect. Again, however, only the aqueous phase was measured so that the mass balance was not confirmed. This study also mixed water draining a mine (acidic and metal-rich) with water from the river (uncontaminated and alkaline) it drains into. Overall, aqueous Cu and Zn concentrations decrease as pH increases and as they are sorbed onto ferrihydrite, and the aqueous phase becomes light. The $\Delta_{\text{sorbed-solution}}$ values are +0.73 ± 0.08‰ for Cu and +0.52 ± 0.04‰ for Zn. Despite potential variation in aqueous speciation with pH, these sorption experiments are well modeled by a single process, which they suggest to be a change in coordination and bond length—octahedral coordination in solution with Me–O bond distances of 2.0–2.4 Å, versus tetrahedral-coordination on ferrihydrite and other Fe oxides and bond lengths of 1.8–2 Å. Pokrovsky et al. (2008) corroborate the finding of sorption of heavy Cu, finding $\Delta^{65}Cu_{\text{sorbed-solution}} = 0.8 \pm 0.2$‰ for goethite and δ of 1.0 ± 0.2‰ for gibbsite.

Juillot et al. (2008) confirmed this result for Zn sorption to ferrihydrite ($\Delta^{66}Zn_{\text{sorbed-solution}}$ = +0.53‰). These authors also obtained $\Delta^{66}Zn_{\text{sorbed-solution}}$ for goethite = +0.29‰. In these experiments ionic strength was kept at 0.1 M using KNO_3 and some solids were measured to close the mass balance. Moreover, time-dependent experiments found light Zn taken up onto ferrihydrite in the first hour, before the observations settled down to a constant heavy value at about 18 hours, possibly explaining the light value found in the Pokrovsky et al. (2005) study. These authors also interpret their results in terms of changes in coordination and bond-length. The smaller fractionation for goethite is ascribed to the fact that bond-lengths are shorter on the surface relative to solution, even though Zn is octahedrally coordinated on goethite, versus tetrahedral coordination on ferrihydrite.

Recently, Bryan et al. (2015) conducted a much more extensive study of Zn isotopic fractionation during sorption to poorly crystalline Mn oxyhydroxide, which dominates the sorption of many metals, including Zn, in the marine environment (e.g., Koschinsky and Hein 2003; Wasylenki et al. 2011; Little et al. 2014b). Isotopic fractionations were monitored as a function of equilibration time, ionic strength of the solution, speciation of inorganic zinc in the aqueous phase, and degree of loading of the Mn oxide surface. The Zn isotopic composition of both solid and dissolved phase were measured, allowing an assessment of overall experimental mass balance as well as of the relative importance of kinetic versus equilibrium fractionation. For low ionic strengths there is a small kinetic effect ($\Delta^{66}Zn_{\text{sorbed-dissolved}} \sim -0.2$‰) for experimental durations up to 48 hours, but for equilibration times greater than 100 hours fractionations are within uncertainty of zero ($\Delta^{66}Zn_{\text{sorbed-dissolved}} = +0.05 \pm 0.08$‰). For high ionic strength solutions heavy isotopes are always preferentially adsorbed, but there is a strong dependence on surface loading, with $\Delta^{66}Zn_{\text{sorbed-solution}} = +2.74$‰ for low surface loadings (8%), reducing to +0.16‰ for high. The authors interpret this variation in terms of a change in coordination as surface loading increases, from tetrahedral for Mn oxide with Zn/Mn = 0.008 to octahedral at Zn/Mn = 0.128 (Manceau et al. 2002). The difference in behavior at different ionic strengths is partially attributed to the fact that in the low ionic strength experiments the surface loadings were also high (Zn/Mn ~ 0.2). There may also be an effect of speciation. The authors suggest that it is free Zn^{2+} that is sorbed. The isotopic composition of free Zn is predicted to change as ionic strength and the proportions of inorganic carbonate and chloride complexes of Zn, with different equilibrium fractionations relative to free Zn (Fujii et al. 2010, 2014; Black et al. 2011), change. For Zn sorption to kaolinite (Guinoiseau et al. 2016), qualitatively similar variation in $\Delta_{\text{sorbed-free aqueous ion}}$ (Fig. 9) has been interpreted as a shift from outer sphere complexation of Zn in basal exchange sites at low pH, when edge sites are protonated, to inner sphere complexation on edge sites, and larger fractionations, at higher pH and ionic strength.

Sorption/binding to biological surfaces. Both elements under consideration here are essential nutrients for plants and animals, but they are toxic at high concentrations in both the terrestrial and marine realm (e.g., Anderson and Morel 1978; Flemming and Trevors 1989; Marschner 1995; Moffett and Brand 1996; Sold and Behra 2000; Peers and Price 2006; Broadley et al. 2007; Yruela 2009; Sinoir et al. 2012; Bruland et al. 2014). Thus, there are important interactions with the cells of living matter that induce significant isotope fractionations. As noted earlier a careful distinction must be made, one that is not always made in the literature (though see John et al. 2007b; Navarette et al. 2011), between metals that are sorbed or bound to the surfaces of microbial cells and plant roots, and those taken up for metabolic utilization. In terms of basic chemical and isotopic mechanisms, the former process is more akin to the binding to the functional groups of both organic matter and inorganic surfaces discussed in previous sections, and usually favours the heavy isotopes. The latter may favour either the light isotope, if governed by a transport-limited, kinetic, process (e.g., John et al. 2007b) or if it involves reduction as may be the case for Cu (e.g., Zhu et al. 2002; Navarette et al. 2011), or the heavy, if it occurs through active uptake by phytosiderophores (e.g., Arnold et al. 2010a).

Gélabert et al. (2004, 2006) conducted experiments that characterized the nature of interactions between marine and freshwater diatoms and aqueous Zn, including isotope fractionation. The aqueous phase (the medium) in these experiments had inorganic Zn concentrations of 0.3–20 μM. With no organic complexant stabilizing Zn in solution, the process studied is sorption, not "uptake". Zinc sorption was strongly controlled by organic layers covering the silica frustule, specifically by carboxylate and silanol groups, with the amount of Zn sorbed to an organic-free silica skeleton being factor five less than cells with organic surface layers. They find $\Delta^{66}Zn_{diatomcell-medium} = +0.1‰$ to $+0.5‰$ in the presence or absence of organic layers. Coutaud et al. (2014) conducted experiments that characterized fractionation upon uptake and release by and from a "phototrophic biofilm" (an aggregate of micro-organisms embedded in an exopolysaccharide matrix) and see adsorption of heavy isotopes to a much greater degree than this—by up to $1.2 \pm 0.4‰$ relative to solution. Some of these fractionations for sorption to cells are very similar to those measured for those outlined earlier for complexation of Zn to organic functional groups, which may be the dominant binding process (Gélabert et al. 2004, 2006). They are also often similar to those found for externally bound Zn in culturing experiments by John et al. (2007b) that were primarily targeted at documenting fractionation upon uptake into the cells themselves.

Three additional studies have been aimed at quantifying and understanding the sorption of Cu and Zn onto cells, but differ in their interpretation of the exact driver of the fractionations observed, specifically whether it was sorption or biological uptake. Pokrovsky et al. (2008), in experiments with Cu sorption onto abiotic metal surfaces and onto bacterial and diatom cells at low ionic strength (0.01–0.1 M), see virtually no fractionation upon sorption to bacterial cells at circumneutral pH (5.1–6.1, $\Delta^{65}Cu$ = mostly $0 \pm 0.3‰$). They see light Cu sorbed onto the cells of soil bacteria at pH 1.8–3.3 (by up to 1.8‰). The rationale given for the sorption of light Cu in this case relates to an outer-sphere monodentate complex likely to form between Cu and phosphoryl groups—with apparently longer bond distances—on bacterial surfaces at low pH.

Navarette et al. (2011), on the other hand, contrast two sets of Cu uptake experiments with live versus dead cells of *E. coli* and *B. subtilis*. When the cells are alive the solution gets much heavier as it loses Cu to the cells, with $\Delta^{65}Cu_{cells-solution}$ as low as −2.6‰ and −4‰ at different pH values. On the other hand, when cells were dead, the solution was lighter, and only by about 0.4‰. As with the small fractionations of Zn in the Gélabert et al. (2004, 2006) experiments, it is likely that the latter process is analogous to the complexation of Cu to organic functional groups outlined earlier (e.g., Bigalke et al. 2010c; Ryan et al. 2014). The uptake of light Cu by live cells, on the other hand, is interpreted in terms of active intra-cellular complexation. Navarette et al. (2011) confirm this finding in experiments where the aqueous Cu is stabilized by organic complexants so that it is not sorbed. They again document large separation factors, with $\Delta^{65}Cu_{cells-solution} = -1.2$ to -4.4‰, depending on species and the nature of bacterial consortia used in each experiment. The authors suggest that the light uptake may be due either to a kinetic fractionation—irreversible incorporation—or to an equilibrium reduction to Cu(I) within the cell. If the latter is important there has to be communication with the outside of the cell to allow the efflux of the oxidized Cu back to the solution. The theme of a paper by Kafantaris and Borrok (2014) is similar, in this case applied to Zn, in that their objective was to try to understand the relative importance of surface complexation versus intracellular incorporation. For experiments with high Zn/bacterial cells ratios, Zn sorption varies with increasing pH in a very similar way to abiotic experiments, presumably due to increased deprotonation of cell surface organic functional groups and consequent binding of Zn. Zinc isotopic data are best fitted by an equilibrium model with a separation factor $\Delta^{66}Zn_{cells-solution}$ of +0.46‰. In contrast, this study found heavy Zn in solutions at low Zn/bacterial cells ratios, with $\Delta^{66}Zn_{cells-solution} = -2.5$‰. This is interpreted in terms of the complexation of Zn in the dissolved phase by organic exudates, generating two pools of Zn, a complexed (heavy) and a free (light) pool, with the light free Zn pool sorbing onto cell surfaces. This would, however, require an isotope separation factor between organically complexed and free Zn of 2–3‰, an order of magnitude greater than that found in experiments to date (Ban et al. 2002; Jouvin et al. 2009). On the other hand, these authors also used an electrolyte wash to remove extra-cellular Zn in an attempt to quantify intra-cellular inventories and isotopic composition, and also found Zn isotopes in cells to be slightly, to very, heavy relative to the aqueous phase. It should be noted, however, that these experiments were conducted at Zn concentrations 3–4 orders of magnitude greater than found in nature, perhaps at levels where Zn is toxic. Moreover, precipitates containing high levels of Zn on cell surfaces, probably not removed by their wash, is likely at these high concentrations.

Metabolic uptake by algae and higher plants. Primary production by photosynthesis on Earth is roughly equally split between higher plants on land and algae in the oceans (Field et al. 1998). As noted earlier, Cu and Zn are both essential micronutrients for photosynthesizing organisms and are indeed required for enzymes and proteins in all organisms, but are also both toxic to plants and algae at very high concentrations. There have been a relatively small number of studies characterizing fractionation of Cu and Zn isotopes during uptake by plants—as opposed to absorption or binding to external surfaces as discussed in the previous section.

John et al. (2007b) report culturing experiments with the diatom *Thalassiosira oceanica* across a range of free Zn ion concentrations, controlled in their media by the addition of complexant EDTA, representative of coastal and open ocean waters. In order to document fractionation during uptake, such culturing studies must remove externally adsorbed Zn by washing prior to analysis, and John et al. (2007b) find that the externally sorbed Zn isolated in this way has an isotopic composition that is 0.1–0.5‰ heavier than the medium, consistent with other studies where binding of Zn to diatom external surfaces has been specifically targeted (Gelabert et al. 2006). In contrast, Zn in washed cells (targeting the internalized cellular Zn pool) is isotopically lighter than the medium. John et al. (2007b) document a range in fractionations, from $\Delta^{66}Zn_{diatom-medium} = -0.2$‰ at low medium free Zn concentrations to -0.8‰ at high, with a step-like transition at free Zn concentrations that are in the range for natural seawater, at around 10^{-10} M. The authors ascribe these two different fractionations to two different Zn uptake systems—high and low affinity. These two systems are well-documented in previous culturing studies (e.g., Sunda and Huntsman 1992), with the high affinity pathway up-regulated when available Zn is low but saturated at high seawater/medium Zn concentrations. The low affinity pathway likely involves diffusive transport across the cell membrane, thus favouring the light isotope. John and Conway (2014) document the same magnitude of fractionation upon uptake into a different kind of phytoplankton—the marine flagellate chlorophyte *Dunaliella tertiolecta*. The $\Delta^{66}Zn_{cells-medium} = -0.76 \pm 0.02‰$ obtained is the same as for the low affinity uptake system in the diatom experiments, consistent with the high free Zn concentrations in their medium.

There have been more studies focusing on fractionations of Cu and Zn isotopes upon uptake into higher plants. Taken as a whole, these studies have documented a number of important features of plant uptake systems for Cu and Zn: (1) all plants have a bulk Cu isotope composition that is lighter than the external pool (cf. bacterial uptake of Cu in Navarette et al. (2011), discussed earlier), leading to the suggestion that Cu reduction upon uptake is an important process; (2) bulk plant Zn isotopic compositions are both lighter and heavier than the external bioavailable pool, perhaps depending on whether free Zn or a complex is taken up; (3) all plants preferentially transfer the lighter isotopes of Zn upwards into stems and leaves, whereas early studies document preferential upward translocation of both light and heavy Cu isotopes.

Before discussing the details of experimental isotopic studies for Cu and Zn in higher plants, it is useful to briefly set the context in terms of plant uptake systems and the constraints that have come from the better studied Fe isotopic system (see Russel et al. 2003 for useful summaries; Jouvin et al. 2012). For Fe, two fundamentally different uptake strategies lead to different isotope fractionations (e.g., Guelke and Von Blanckenburg 2007). Iron acquisition by Strategy I (non-graminaceous) plants involves the uptake of a free metal ion and requires a reduction step that favours uptake of isotopically light Fe. Analogously, uptake of free Cu via transporters such as COPT1 has been described (e.g., Sancenon et al. 2004; Jouvin et al. 2012), and would also require reduction of soil Cu(II) to Cu(I) by a reductase enzyme. For Zn there is no oxidation state change involved. As with the diatom studies outlined above, Strategy I uptake of Zn may involve both a low and high affinity uptake system (see Jouvin et al. 2012). Low affinity uptake, active at external bioavailable Zn concentrations in excess of perhaps 10^{-7}M (e.g., Wang et al. 2009; Jouvin et al. 2012), involves diffusive transport via ion channels and electrogenic pumps, favouring the light isotope. High affinity Strategy I uptake involves zinc–iron–permease (ZIP) proteins that bind free Zn from the external pool at the cell membrane, probably favouring the heavy isotope, and facilitate its uptake and transmembrane transport. In contrast, Strategy II uptake (graminaceous plants), under metal-deficient conditions, can actively complex soil Fe (III) to a phytosiderophore derived from their root—with no reduction step and a small positive isotope fractionation for Fe (e.g., Guelke and Von Blanckenburg 2007; Moynier et al. 2013b). Uptake of Zn and Cu in "phytosiderophores", organic complexes that will favour the heavy isotope as discussed in earlier, have been discussed in the isotopic literature as outlined below.

The pioneering study of Weiss et al. (2005) showed that the roots of tomato, rice and lettuce were all slightly enriched in the heavy isotopes of Zn relative to the bulk nutrient solution in which they were grown—by 0.1–0.2‰. In contrast, the shoots, housing 75–85% of the Zn inventory of the plant, were isotopically light relative to the same nutrient solution—by 0.25–0.5‰, so that the bulk plants contain light Zn relative to the external pool. The simplest explanation of this observation requires the uptake of isotopically light Zn, through ion channels or electrogenic pumps, coupled to the preferential upward transfer of even lighter Zn, leaving the residual root pool heavy. Fujii and Albarède (2012) re-interpreted these observations using *ab initio* calculations and suggested that the fractionation is controlled by the difference in Zn speciation between the root system (isotopically heavy Zn-phosphates) and the upper parts, rich in isotopically light citrates and malates. It is noteworthy that the predicted isotopic signal of Strategy I behavior is found in a Strategy II plant like rice. In a follow up study, however, Arnold et al. (2010a) demonstrate that rice grown in soil rather than hydroponically is isotopically *heavy* relative to the soil, particularly under Zn deficiency. They attribute this to uptake of Zn bound to a "Zn-phytosiderophore". On the other hand, Tang et al. (2012) also observe Zn in plants that is up to ~0.6‰ heavier than in soils, but reject the phytosiderophore hypothesis because the species concerned do not release them. Instead, they favour uptake of heavy isotopes by ZIPs. The upwards transfer of light Zn has been confirmed by later studies (e.g., Moynier et al. 2009b; Caldelas et al. 2011; Jouvin et al. 2012; Tang et al. 2012).

This earlier work on Zn uptake by plants has been followed up by a series of more targeted studies aimed at more detailed investigation of the mechanisms by which plants take up Zn and its isotopes, especially with regard to speciation and including Zn uptake by zinc hyperaccumulators from contaminated soils (Aucour et al. 2011, 2015; Houben et al. 2014; Couder et al. 2015).

Weinstein et al. (2011) first measured the isotopes of Cu in plants, documenting light isotopes in every part of Strategy II plants—by 0.3–0.8‰—relative to the soils in which they were grown. They also document significant transfer of light Cu upwards from the roots, or from the initial stock of Cu in lentils grown from seed without further Cu addition. In all cases, the topmost and youngest leaves contain the lightest Cu. These findings were confirmed by Jouvin et al. (2012). Though the latter study found a difference between Strategy I ($\Delta^{65}Cu_{plant\text{–}nutrient\ solution}$ = −0.84 to −0.47‰) and Strategy II ($\Delta^{65}Cu_{plant\text{–}nutrient\ solution}$ = −0.48 to −0.11‰), all of them took up the light isotope, suggesting that reduction of Cu(II) is an important factor in the uptake of Cu by all plants whether the Cu is complexed or not. Like the previous two studies, Ryan et al. (2013) observe much lighter Cu in plants than the soils in which they were grown, and a very clear difference between Strategy I ($\Delta^{65}Cu_{whole\ plant\text{–}nutrient\ solution}$ = −1.02 ± 0.37‰) and Strategy II ($\Delta^{65}Cu_{whole\ plant\text{–}nutrient\ solution}$ = −0.15 ± 0.11‰) plants. However, in contrast to the previous two studies, their Strategy II plants have a fairly constant isotopic composition in different parts of the plant while for Strategy I the heavier isotope preferentially moves upwards (shoots 0.87–1.35‰ heavier, leaves 0.53–0.98‰ heavier). These authors rationalize their observations in terms of the upward transfer of Cu in organic complexes like nicotinamine, which would indeed preferentially transport the heavy isotope (Ryan et al. 2014), if translocation upwards was not close to quantitative.

Cu–Zn isotopes in the weathering–soil–plant system

Soils represent the interface between the solid Earth and its fluid envelope, the place where chemical weathering of primary minerals and precipitation of secondary minerals begin, the substrate for plant growth, and the locus for the initial partitioning of elements between solid material and the aqueous phase that drains into groundwater, rivers and, eventually, the oceans. In addition, transfer of chemical elements between the atmosphere and soils occurs through the ablation, transport and deposition of dust. As such, soils are sites of complex processes that involve Cu and Zn transfer and isotopic fractionation via all of the mechanisms detailed in the previous section. For ease of discussion here we separate these processes as follows: (1) isotopic effects associated with leaching and dissolution of primary minerals; (2) the partitioning of Cu and Zn and their isotopes between a dissolved pool, often complexed to soluble organics, and a pool sorbed to secondary minerals; (3) overprinting of weathering processes via the addition of atmospheric aerosol to soils; (4) uptake into plants and associated isotope fractionations in the upper organic-rich levels of soils. Anthropogenic addition of Cu and Zn to soils is a fifth important process but is dealt with in the section on the Anthropocene later in this chapter.

Weathering release of Cu and Zn from primary minerals. To our knowledge there are only two studies that have characterized Cu–Zn isotope compositions upon release from primary minerals to an experimental leachate designed to simulate the weathering process. Fernandez and Borrok (2009) measured isotopic compositions of fluids released during oxidative leaching experiments on rocks containing sulfides (pyrite, chalcopyrite, galena, sphalerite). Copper released is 2‰ heavier than the starting rocks at pH 2 and 5. Zinc released is both heavier and lighter than the primary sulfide, depending on the precise rock / mineral being leached, but only by order 0.2‰. For Cu the release of heavy isotopes is almost certainly related to an oxidation state change, from Cu(I) in the sulfides to Cu(II) in the leachates. Weiss et al. (2014) conducted experiments on leaching of biotite granite using 0.5 M HCl and oxalic acid. Zinc mobilized into the aqueous phase in the first hour was as light as −1.2‰ relative to starting material, with 30–40% of the initial rock Zn pool released. The Zn in solution then moved back towards the initial rock, but never got beyond 0.1–0.3‰ lighter after 168 hours (with 45–75% of the original starting Zn mobilized). The early, very negative, fractionations are interpreted as being kinetic.

Cu–Zn isotopes of soils and the impact of sorption and aqueous complexation. Though experimental studies are a useful template for the interpretation of field data, real weathering of rocks in soils is more complex for two main reasons. Firstly, Cu and Zn are not necessarily located in sulfide minerals such as in the Fernandez and Borrok (2009) experimental leaching study. Where this *is* the case the results obtained from field studies are consistent with the experiments. Thus Mathur et al. (2012) studied Cu isotopes in soils and soil waters developed on black shales in Pennsylvania USA, where a very large proportion of the Cu is located in pyrite. Loss or gain of an element of interest (i) during the soil development process is often expressed in terms of a tau (τ) value (Chadwick et al. 1990), which normalizes the concentration (C) of the element in a particular soil horizon (h) to both that in the parent material (p) and to the concentration of an immobile element (j, often Nb, Zr or Ti):

$$\tau_{i/j} = \left[\frac{(C_i / C_j)_h}{(C_i / C_j)_p} - 1 \right]$$

Tau values greater than zero denote addition of the element of interest relative to the immobile element, and values less than zero loss. Unfortunately, not all studies of Cu and Zn isotopes in soils report tau values (Viers et al. 2007; Mathur et al. 2012; Liu et al. 2014; Vance et al. 2016a), but they are essential for identifying net loss or gain of an element given changes in mass that occur during soil development. Tau values for the Mathur et al. (2012) soils are about −0.5, implying loss of 50% of the original Cu in the rock, while δ^{65}Cu is about 0.5–1‰ lighter than the original Cu. In contrast, soil pore waters are all enriched in the heavy isotope, by 0.7–1.7‰. The authors attribute these findings to the preferential mobilization of heavy isotopes due to the oxidative leaching of pyrite, consistent with an abundance of experimental data, including those of Fernandez and Borrok (2009).

Secondly soil solutions, and the interactions between this aqueous phase and the residual solids in the soil, are more complex than those in the experimental studies described above. Mathur et al. (2012) discuss other possible interpretations of their Cu isotopic data, such as organic complexation in solution with a preference for the heavy isotope, but dismiss their relevance to that particular setting given the dominance of pyrite as a reservoir for Cu in the parent rock. However, in many soil settings it is the equilibrium partitioning of both Cu and Zn and their isotopes between dissolved organic complexes in an aqueous phase versus sorption to residual secondary minerals in the soil that appears to dominate trace metal distributions (e.g., Grybos et al. 2007), and the isotopic patterns seen for both Cu–Zn (Bigalke et al. 2010b, 2011; Vance et al. 2016a) and other metals (e.g., Wiederhold et al. 2007).

No soil has yet been studied where conditions are reducing enough for the large isotope fractionations between Cu(I) and Cu(II) to be relevant. However, environmentally relevant redox conditions *do* control the availability of Fe–Mn oxyhydroxides phases as a substrate for sorption. Figure 10a,b documents the isotopic impact of this control in soils distributed across three sites on the island of Maui, Hawaii, that have seen different annual rainfall amounts and in which there is a transition from well-drained conditions that retain Fe oxides to waterlogged conditions that do not (Vance et al. 2016a). If Fe oxides are retained in the soils, depletion of Cu and Zn is accompanied by preferential loss of the heavy isotopes of Cu and slight preferential loss of the heavy isotopes of Zn. When Fe oxides disappear the remaining Cu and Zn is almost completely stripped from the soil and residual isotopic compositions move towards heavy values. Patterns consistent with those seen in Hawaii are also observed by

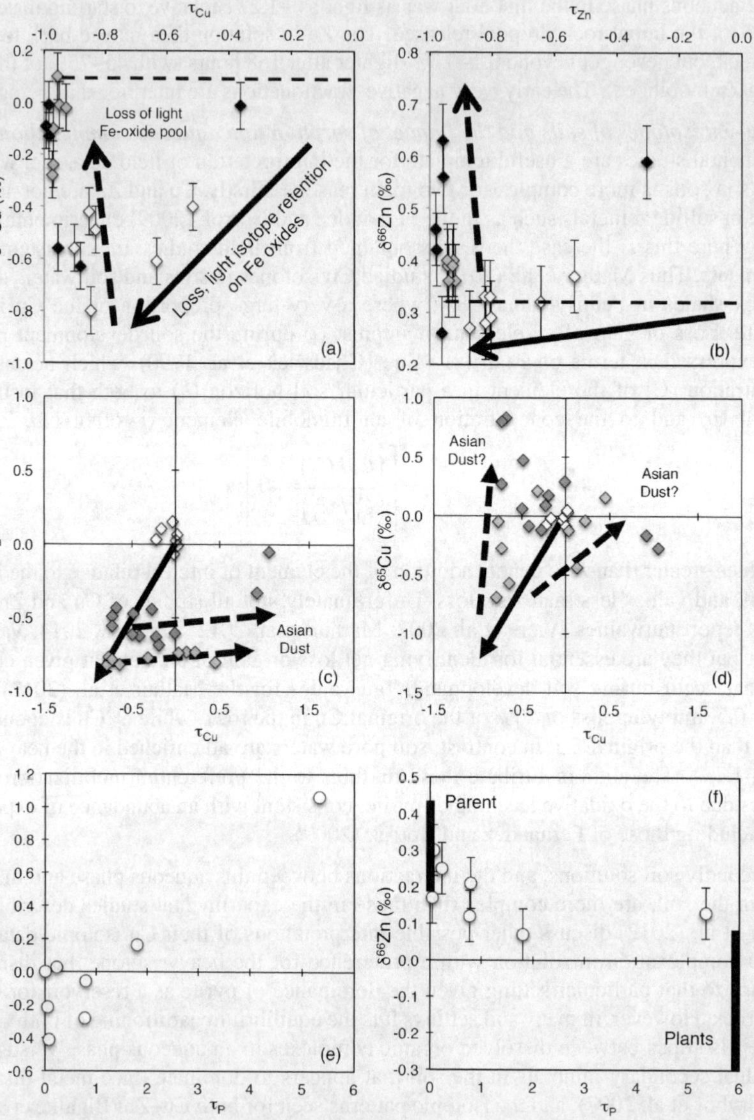

Figure 10. Cu–Zn isotope and tau data from relatively unpolluted soils in Hawaii (basaltic) and Scotland (granitic) to illustrate important soil processes as discussed in the text (from Vance et al. 2016a). *Panels a,b:* data from a sequence of soils, all 400 kyr in age but having seen different annual rainfall, on the island of Maui (Hawaiian Islands). Grey shading represents mean annual precipitation (MAP) from 2500 mm (white), through 3350 mm (grey) to 5050 mm (black). Horizontal dashed lines show the isotopic composition of the parent basalt. At the 2500 mm site Fe is retained in the soil as Fe oxyhydroxides. Cu is depleted with preferential loss of the heavy isotope to aqueous organic complexes and retention of a sorbed isotopically light pool (solid arrow). At higher rainfall these oxyhydroxides are lost by reduction, leading to loss of almost all this residual light Cu and the retention of a very small pool of Cu that is close to the parent material in isotopic composition (dashed arrow). *Panel b* shows the Zn data, which has a similar pattern though the move towards light isotopic compositions at the low rainfall site is barely analytically resolvable and the move back to heavy isotopic compositions overshoots to values 0.3‰ heavier than the original rock. *Panels c,d:* data for soils on the island of Hawaii, all having seen 2500 mm MAP with shading showing [cont'd on next page]

Figure 10 (cont'd), showing different aged soils from 0.3 kyr (white), through 20 kyr (grey), to 150 kyr (black). The solid arrow shows the trajectory for retention of residual light isotopes during chemical weathering in well-drained conditions as in a,b. The dashed arrows in c show the trajectories that would be followed given addition of Asian dust to move the soils away from this trend. In d the dashed arrows are illustrative only because their exact slope depends on how much Cu and Zn the soil has lost when dust is added. *Panels e,f:* tau data for Zn and phosphate in the upper organic-matter-rich horizons of granitic Scottish soils, as well as Zn isotopic data for parent material (bar at left), plants (bar at right) and soils (open circles), to illustrate correlated behaviour between Zn and a major plant nutrient.

Bigalke et al. (2010c, 2011). The most likely interpretation is that heavy Cu is mobilized into aqueous organic complexes, while the oxides in the soil preferentially sorb the light isotopes. As noted earlier, both soluble organic complexes and sorption show a preference for heavy isotopes relative to free Cu ion in aqueous solution, so that this interpretation requires that the preference of the organic complexes for heavy Cu is greater than that of sorption. This in turn, would require the dissolved complexes to bind Cu at least as strongly as the strongest organic ligands in the experiments of Ryan et al. (2014), but there is evidence from the partitioning of Cu isotopes between the aqueous and particulate phases in rivers (Vance et al. 2008, discussed below) that this is indeed the case. The data for Zn in Figure 10a,b document a much more subtle isotopic effect in free-draining soils, mirroring more subtle differences between the dissolved and particulate load of rivers (e.g., Chen et al. 2008, 2009; Little et al. 2014a, also discussion below), and consistent with the fact that the isotopic effects of aqueous complexation versus sorption to mineral surfaces may cancel each other out. Figure 10a,b also show such an effect when Cu and Zn are almost completely stripped away in water-logged conditions. As the Fe-oxyhydroxides are removed, the isotopic composition tends towards heavier values, perhaps reflecting the retention of a very small residual pool on aluminium hydroxides or on condensed organic matter or the addition of dust (see below).

Addition of atmospheric aerosol to soils. There have been few studies of the impact of atmospheric aerosol deposition on soils, and all but one of these concern anthropogenic aerosol deposition (discussed in a later section). Deposition of natural aerosol dust from the atmosphere has the potential to confound and overprint weathering signals. Here we illustrate the impact of such a process again using data from the relatively unpolluted Hawaiian Islands from Vance et al. (2016a). The geochemical impact of the deposition of Asian desert-derived dust on Hawaiian soils is well-documented for many elements (e.g., Kurtz et al. 2001; Vance et al. 2016a and references therein). Its impact on Cu and Zn isotopes is illustrated in Figure 10c,d. Tau data for very young (300 years) soils from the island of Hawaii further help to define the weathering depletion trend illustrated in Figure 10a,b (thick solid black arrow). However, soils with ages in the range 20–150 kyr show deviations from this tau pattern that fall on arrays that are consistent with the addition of Asian dust (thick dashed arrows). The potential impact of this on Cu isotopes is illustrated by the dashed arrows in Figure 10d, but the precise trajectory on this plot depends on the relative concentrations of Cu in the dust versus those in the soil when the dust was added. In general, it might be expected that dust addition would tend to buffer soil Cu and Zn isotopes back to about 0‰ and +0.3‰ respectively, the average Cu and Zn isotope compositions in natural atmospheric aerosol (cf. Fig. 8). However, Weiss et al. (2007) document a heavier Zn isotopic composition for background (un-contaminated) dust deposition in Finland, $\delta^{66}Zn = +0.9‰$, while Dong et al. (2013) found variations of up to 0.5‰ in $\delta^{65}Cu$ among the different size fractions of Asian dust, with some samples of the >63 µm fraction giving isotopic values = 0.4–0.5‰. The fact that natural atmospheric aerosol may be isotopically heavier than the lithogenic values for Cu and Zn isotopes may indicate either a significant contribution from a non-lithogenic source or isotope fractionation during atmospheric processing.

The impact of plants on soil Cu and Zn isotopes. The surface organic-rich layers of soils are often enriched in the light isotopes of both Cu and Zn (e.g., Weiss et al. 2007; Bigalke et al. 2010a, 2011; Liu et al. 2014; Vance et al. 2016a). Though an interpretation in terms of addition from the atmosphere has been discussed (e.g., Bigalke et al. 2010b, 2011), another likely process relates to the concentration of the light isotopes in these surface layers by plant growth and decay. Viers et al. (2007) was the first study to highlight the potential importance of plant cycling for Zn, while Bigalke et al. (2010b, 2011) conclude that light Cu in the upper organic layers of soils is likely attributable to decaying plant material. Likewise, Liu et al. (2014) point to light Cu and high TOC in the upper layers of soils from Hainan, China, as evidence for plant activity. However, in these particular soils Cu is uniformly depleted in the upper relative to deeper soil horizons, whereas the other soils where isotopically light Cu and Zn in the upper organic horizons has been attributed to plants are definitely (Vance et al. 2016a) or probably (Bigalke et al. 2010b, 2011) enriched relative to those underneath. Thus the Liu et al. (2014) data may be more consistent with the loss of heavy Cu by mobilization in aqueous organic complexes, as for the Hawaiian soils discussed above. Schulz et al. (2010) observed the effect of "biolifting" on the distribution and isotopic composition of Fe in soils from Santa Cruz, California. Biolifting is the process by which plant roots and symbiotic fungi (mycorrhizae) transport an element from deep in the regolith to the shallow soil. Vance et al. (2016a) observed increasing τ_{Cu} and τ_{Zn} coupled to increasing τ_P with soil age in the uppermost horizons of Scottish soils, but decreases at depth, suggesting movement of Cu and Zn upwards with increasing soil development (eg. Fig. 10e). As with the experimental studies of plants discussed earlier, these authors document significantly lighter Zn in plant material than in soils, suggesting that biolifting and fractionation by vegetation can also explain some aspects of soil δ^{66}Zn and δ^{65}Cu (e.g., Fig. 10f) for the surface layers of these soils. In contrast, Viers et al. (2015) find little variation in soil Zn isotope compositions related to plant activity in Siberian permafrost soils, which they attribute to the homogenizing impact of seasonal freezing front migration. Plants developed on these latter soils exhibit Zn isotope compositions both lighter and heavier than the bulk soil, possibly due to climate-driven changes in speciation of the plant-available pool.

Summary. Figure 11 presents Cu–Zn isotopes in soils in the form of integrated tau values and isotopic compositions for whole soil profiles (where tau data are also available: Viers et al. 2007; Mathur et al. 2012; Liu et al. 2014; Vance et al. 2016a), in order to make a summary assessment of the degree of loss and isotopic fractionation that occur in this setting. Such an assessment is important for the significance of weathering and other pedogenic process in global biogeochemical budgets, and sets the scene for the discussion of one of the main inputs to the oceans, rivers, in the next section. It is already relatively clear from this still small dataset that soils lose heavy Cu during the weathering process, whether it is because of oxidation of sulfides (e.g., Mathur et al. 2012) or through retention of light Cu isotopes on residual Fe–Mn oxides coupled to the mobilization of heavy Cu in aqueous organic complexes (Bigalke et al. 2010b, 2011; Vance et al. 2016a). In contrast, the isotopic impact of chemical weathering on Zn is much more subdued, with the majority of soils retaining very slightly heavy Zn. It should be noted that the real impact of weathering removal on its own would be more pronounced than these data suggest, given that nearly all these soils will have seen the addition of some dust, buffering the isotopic composition closer to the parent rock than would otherwise be the case. The main point of presenting the summary in Figure 11 is that it predicts that the complementary aqueous reservoir to the residual solids in soils, the dissolved phase of rivers, should be significantly heavier than the average continental crust for Cu, and not very different from the latter for Zn. It will be seen in the next section that this prediction is borne out for estimates of the Cu and Zn isotope composition of the dissolved riverine flux to the oceans obtained from measurements of the dissolved pool in large and small, relatively unpolluted, rivers (Vance et al. 2008; Little et al. 2014a).

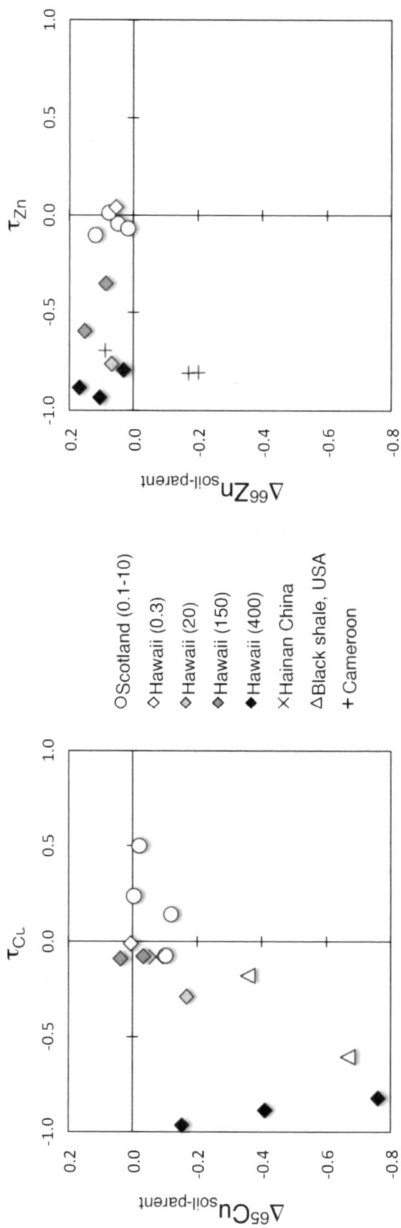

Figure 11. Summary data for soils studied in Viers et al. (2007), Mathur et al. (2012), Liu et al. (2014) and Vance et al. (2016a), in the form of taus and isotopic compositions integrated over the entire soil profile studied, and in order to assess the overall impact of depletion by chemical weathering on Cu and Zn isotopes. The numbers in brackets in the key, and the intensity of shading, indicate the timescale over which soils have developed, where known. All basaltic soils are shown as diamonds, granitic as circles, and soils developed on black shales as triangles. Preferential loss of the heavy Cu isotope during chemical weathering is clearly significant. The Zn data are plotted on the same scale to illustrate the subtlety of isotope fractionation during weathering by comparison with Cu.

The oceans: inputs, outputs and internal cycling of Cu and Zn isotopes

The oceanic dissolved pool and authigenic metals extracted from it to be delivered to sediments represent the ultimate fate of Cu and Zn mobilized on the continents via weathering and erosion, discussed in the previous section. Measurements of the Cu and Zn isotope composition of the dissolved pool of the oceans is extremely challenging due to the low abundances of both metals in seawater (concentrations of order 10^{-10}–10^{-9}M). This presents difficulties related to clean sampling and analysis, in addition to the problem of obtaining large ion currents for the precise measurement of isotope ratios. The challenge is to achieve efficient, low blank, chemical extraction and purification of Cu and Zn from large volumes (of order 0.1–10 L) of seawater, containing up to 8 orders of magnitude more interfering ions such as Mg and Na. The availability of the double spike approach for Zn eliminates concerns over isotope fractionation artefacts during the chemical extraction process, but this approach is not available for Cu. The isotope geochemistry of seawater started with the pioneering work of Bermin et al. (2006) and Vance et al. (2008), but has gained momentum recently and is likely to grow in importance over the next decade for two principal reasons. The first is the inception of the GEOTRACE programme (http://www.geotraces.org), an international collaboration involving many chemical oceanographers worldwide that is now providing large, cleanly collected, seawater samples for a huge body of work aimed at reaching a quantitative understanding of trace elements and their isotopes in seawater. The second is the development of a key new methodology, using Nobias chelate PA-1 resin (e.g., Conway and John 2014, 2015; Takano et al. 2014; Vance et al. (2016b). This new approach is capable of producing a very clean transition metal fraction from seawater, that can then be taken on to the usual anion column for the purification of separate Cu and Zn (as well as Fe, Cd, Mo) fractions.

The data currently available from these endeavours is summarized in Figure 12 and 13. Two principal scientific themes have emerged both from this early work on the dissolved pool of the oceans themselves, as well as from Cu–Zn isotopic characterization of the inputs and outputs: (1) the overall mass balance of Cu and Zn cycling through the oceans as a whole; (2) the cycling of Cu and Zn within the oceans, by biological uptake and regeneration, and through interaction with the surfaces of both biological and abiotic particulates, often termed "scavenging". We discuss each of these in turn below. The work done so far on Cu and Zn isotopes in rivers, atmospheric aerosols, hydrothermal systems, as well as the chemical sediments that represent the outputs from the dissolved pool, are all tied up with the first of these topics and are discussed as part of it.

The overall oceanic budget of Cu and Zn in the oceans. The dissolved pool of the oceans is conventionally regarded as being in steady state with regard to inputs and outputs. Though there are both isotopic and elemental records and models for long residence time elements such as Sr and Mg that suggest the contrary (e.g., Vance et al. 2009; Coggon et al. 2010; Pogge von Strandmann et al. 2014), long-term records of Cu and Zn isotopes in the oceans (Little et al. 2014a) demonstrate a temporal constancy that makes this a useful starting point here.

Vance et al. (2008) and Little et al. (2014a) have characterized the isotopic composition of the dissolved pool of rivers for Cu and Zn isotopes, including relatively unpolluted large and small catchments such as the Amazon and the Kalix (Arctic Circle, Sweden). A key finding for Cu is that the dissolved pool of rivers is isotopically heavier than the continental crust as sampled in high temperature igneous rocks and clastic sediments (Fig. 8), with a discharge- and [Cu]-weighted average δ^{65}Cu of about +0.7‰. This result is common to a number of transition metals, including Mo (Archer and Vance 2008) and Ni (Cameron and Vance 2014), which are also all characterized by weak positive relationships between isotope composition and reciprocal metal concentration. In the case of Cu, at least one small river carries a particulate load with δ^{65}Cu = −0.4 to −0.6‰, and the two pools balance to suggest an estimated total load that is about the same as the rocks being weathered. Vance et al. (2008) attributed this difference between the dissolved and particulate load to a roughly 1.2 ± 0.4‰ equilibrium isotopic fractionation between heavy Cu in dissolved aqueous organic complexes and light Cu in particulate material. Though this suggestion is qualitatively consistent with the fact that the small number of soil systems so far analyzed seem to lose heavy Cu (Fig. 11), and with the experimental finding that organic complexes preferentially sequester the heavy isotope of Cu (Fig. 10: Bigalke et al. 2010a; Ryan et al. 2014), the fractionation seen between the dissolved and particulate phases in rivers is much larger. Little et al. (2014a) found that Zn isotopes in a subset of the same large and small rivers is less variable, and that the discharge- and [Zn]-weighted riverine flux to the oceans, at about +0.33‰, is very close to the continental crust (Fig. 8). This finding is also completely consistent with the very subtle isotopic variations seen in soils (Fig. 11), and with the fact that there is likely to be minimal isotopic difference between dissolved aqueous organic complexes of Zn and that sorbed to surfaces (Fig. 9 and references in the caption).

Other work on rivers has concentrated on rather small catchments, often with a focus on modification of riverine processes due to human activities such as smelting and agriculture (see Anthropocene section). However Ilina et al. (2013), in a study of pristine rivers in subarctic watersheds (NW Russia), also found heavy Cu in the dissolved load (δ^{65}Cu = $+0.46 \pm 0.05$‰), and used ultrafiltration to demonstrate that this isotopic composition characterizes the

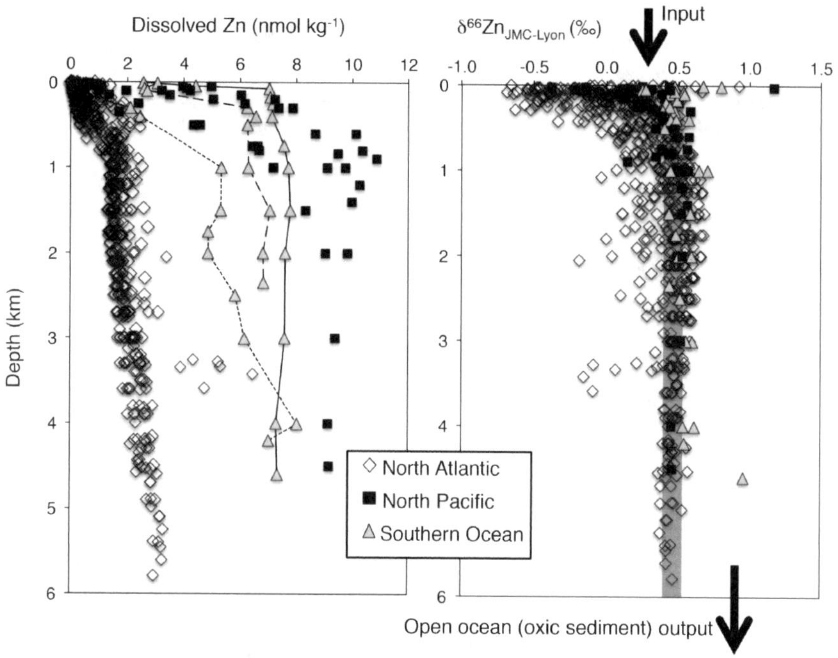

Figure 12. All the Zn isotopic data currently available for the dissolved pool in the oceans (right) with Zn concentrations measured in the same samples as the isotopes (left), plotted versus depth. The middle of the grey bar on the isotope plot marks the average δ^{66}Zn for the deep ocean (beneath 800 m) while its width shows a typical analytical uncertainty (±0.06‰). This deep ocean inventory is generally very homogeneous and has an average δ^{66}Zn ~ +0.47‰. In the deep ocean, the data depart from this ratio locally, such as near hydrothermal vent systems at 3-4 km in the Atlantic (Conway and John 2014). In the surface ocean isotope compositions are also very close to this deep ocean average in the Southern Ocean (Zhao et al. 2014), but the upper ocean in the North Atlantic and North Pacific depart significantly from it (Boyle et al. 2012; Conway and John 2014, 2015; Zhao and Vance, unpublished data). Relative to this average deep ocean value, the estimates input is slightly isotopically light, at about +0.33‰ (arrow at top: Little et al. 2014a), while the dominant outputs in the oxic open ocean are much heavier (arrow at bottom: +0.90‰ in Fe–Mn oxides, carbonates and siliceous sediments; Little et al. 2014a).

riverine load down to <1 kDa, even though 40–60% of the Cu in the rivers they studied is colloidal. Szynkiewicz and Borrok (2015) document a much wider range of Zn isotope compositions (δ^{66}Zn = −0.57 to +0.41‰) than in the global survey of Little et al. (2014a) in streams of the Rio Grande catchment (USA), which they attribute to preferential removal of the light isotope from the dissolved load by adsorption onto particulates. Though part of a study of a river estuary that is at least partially anthropogenically disturbed, particulate and dissolved Cu isotopes in the Gironde estuary show $\Delta^{65}Cu_{particulates-dissolved}$ around +0.4‰ for Cu (Petit et al. 2013). Finally, Chen et al. (2008, 2009), though also in a study primarily aimed at using Zn isotopes to study pollution sources in the River Seine, France, observe rather subtle isotopic differences between the particulate and dissolved load.

Little et al. (2014a) in an assessment of the overall oceanic mass balance of Cu and Zn isotopes, present and summarize further data on the size and isotopic composition of the likely inputs. This paper suggests that the dissolved riverine load is dominant for both, but uncertainties remain. It is conventionally assumed that the metal load of hydrothermal fluids is precipitated and scavenged very close to mid-ocean ridges, and Little et al. (2014a) conclude that the flux that gets past this trap is likely to be very small indeed. However, the recent finding that substantial amounts of iron are transported 1000s of km from hydrothermal systems across the deep Pacific

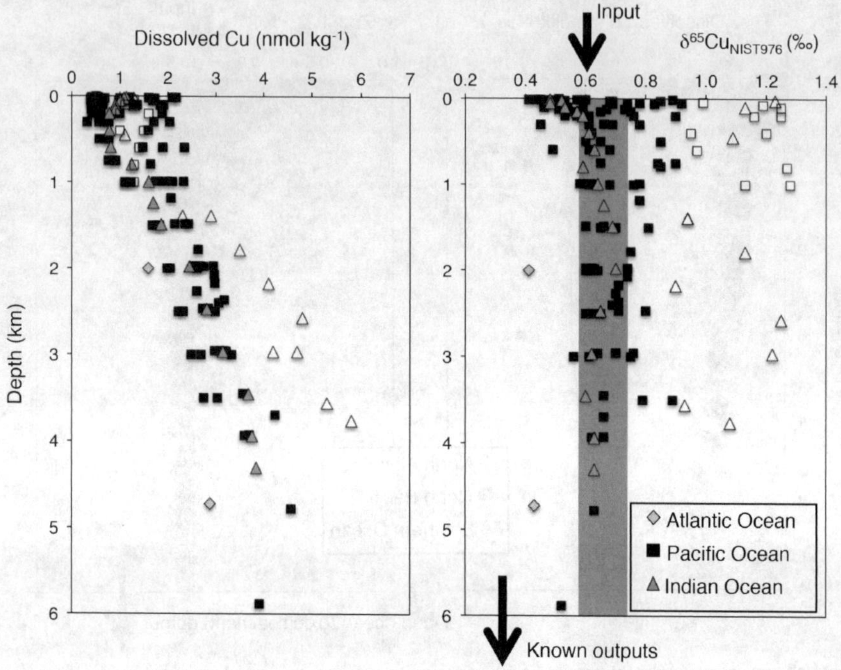

Figure 13. All the Cu isotopic data currently available for the dissolved pool in the oceans (right) with Cu concentrations measured in the same samples as the isotopes (left), plotted versus depth. Open symbols on this plot are for data in the pioneering work of Bermin et al. (2006) and Vance et al. (2008). These data were obtained pre-GEOTRACES, on samples that had been stored for times on the order of 10 years. Though acidified, they stand out as having significantly heavier isotopic compositions than more recent work on new, cleanly-collected, GEOTRACES samples, including new (as yet unpublished) data from the same group, and in Takano et al. (2014) and Thompson and Ellwood (2014). It seems likely that these early measurements are compromised in some way by the long storage. The middle of the grey bar on the isotope plot marks the average δ^{65}Cu for the deep ocean (beneath 800 m, and excluding these older data plotted as open symbols) while its width shows a typical analytical uncertainty (±0.08‰). As with Zn, this deep ocean inventory is generally very homogeneous and has an average δ^{65}Cu ~ +0.66‰. The estimated input is close to this deep ocean average, at about +0.63‰ (arrow at top: Vance et al. 2008; Little et al. 2014a). The only output yet characterized for Cu, though it is probably the most important, is scavenging to Fe–Mn oxide particulates and transfer to sediment (arrow at bottom, +0.31‰: Little et al. 2014a), is significantly lighter than both the input flux and the deep ocean average.

(Resing et al. 2015) may prompt a re-assessment of that conclusion. We know very little about the isotopic composition of end-member hydrothermal fluids, though the [Zn]-weighted δ^{66}Zn in the only study so far (John et al. 2008) is very close to basalts, the continental crust and rivers, at around +0.25‰. Similarly the [Zn]-weighted average δ^{66}Zn of thermal springs and fumaroles from one Caribbean volcano is +0.34‰ (Chen et al. 2014b). Dust, transported through the atmosphere from the continents, was estimated to represent only about 10% of the total input for both Cu and Zn in Little et al. (2014a), whereas Takano et al. (2014) estimate this source to be slightly larger than rivers for Cu. Much of this discussion depends on two inter-related uncertainties. Firstly, the Takano et al. (2014) dust estimate is derived from the Cu concentration of rainwater over Japan, taken to characterize the wet deposition flux of dust Cu to the Earth's surface. However, this rain is likely to be anthropogenically contaminated. Though the residence time of Cu in the oceans is not well-constrained it is certainly of order 10^3–10^4 years, so that it seems unlikely that 100–200 years of this flux is relevant to the whole ocean budget at all. A second uncertainty relates to the size of the dust flux to the oceans itself, and in particular solubility of Cu in that dust. Little et al. (2014a) use a solubility of 27%, likely relevant for mineral dust (Desbouefs et al. 2005), but anthropogenic aerosol may contain a more soluble pool of metals, possibly explaining the high Cu concentrations in Japanese rain.

These uncertainties are important, because the available data for sources and sinks point to missing budget terms (Little et al. 2014a). The isotopic composition of the input to the oceanic dissolved pool for Zn is fairly well-constrained despite large uncertainties on the amount of *total* Zn, simply because the isotopic composition of average rivers, dust and hydrothermal systems are all around +0.25 to +0.33 (Fig. 8). This is significantly lighter than the deep ocean dissolved pool and requires at least one sink from the oceans that is isotopically light. For Cu, the input shown on Figure 13 is that from Little et al. (2014a). If dust is much more important than suggested in that paper, and more like 60% of the total input as in Takano et al. (2014), and with an average δ^{65}Cu in dust of +0.04‰ (Fig. 8) then the total input would be substantially lighter than the dissolved pool of the oceans, so that at least one isotopically *heavy* sink is required.

The current level of knowledge on the isotopic composition of these sinks is summarized in Figure 8 and Figures 12 and 13. For Zn, all the sinks in the open oxic ocean are heavy. One that is likely to be quantitatively important, and the one that we know most about (Little et al. 2014a,b; Bryan et al. 2015), is that which occurs via scavenging of Zn to particulate Fe–Mn oxides and delivery with them to sediment. This sink, as recorded in Fe–Mn crusts, has an isotope composition = +0.94 ± 0.14‰ (Marechal et al. 2000; Little et al. 2014a, see also metalliferous sediments in Dekov et al. 2010), about 0.5‰ heavier than the deep oceans. This is qualitatively consistent with (1) the finding of Little et al. (2014b) that Zn and Cu are both very clearly associated with Mn oxide in these samples and (2) that of Bryan et al. (2015) that Zn sorbed to Mn oxide is heavier than a dissolved pool. However, Zn / Mn ratios in crusts suggest low surface loading, for which Bryan et al. (2015) document an isotopic fractionation upon sorption from a high ionic strength solution in which Zn is inorganically speciated of as great as +2.7‰. The solution to this quantitative discrepancy put forward by Little et al. (2014b) and Sherman et al. (2015)—see schematic in Figure 14—is that, in the dissolved pool, the free metal Zn ion that is likely to be sorbed is actually much lighter than the total dissolved pool because of the fact that most of the oceanic dissolved pool of Zn is organically complexed, which would be heavier than the free ion (Fig. 10; Jouvin et al. 2009).

Organic complexation is likely to be even more important, quantitatively and isotopically, for Cu. Fe–Mn crusts are isotopically lighter than the oceanic dissolved pool by about 0.2‰ (Fig. 13 Little et al. 2014a). No experiments have yet characterized isotope fractionation of Cu upon sorption to Mn oxide surfaces, but experiments for other oxides have universally documented sorption of the heavy isotopes from an aqueous phase containing free Cu(II) (Fig. 10; Balistrieri et al. 2008; Pokrovsky et al. 2008; Navarette et al. 2011). As pointed out by Little et al. (2014b) and Sherman et al. (2015), sorption of the heavy isotope would be consistent with the change in coordination state of Cu (e.g., Schauble 2004) from V in solution to dominantly III–IV on birnessite (δ-MnO_2). But this is inconsistent in sign, never mind magnitude, with the observation of light Cu in natural Fe–Mn crusts. However, Cu again, in all aqueous solutions at the surface of the Earth is ubiquitously complexed to organics (e.g., Coale and Bruland 1988), so that this conundrum probably has a solution similar to the Zn problem. In other words, the free ion that is sorbed is likely to be lighter than the total, as also shown schematically in Figure 14.

Returning to the whole ocean mass balance, at the level of knowledge discussed above there is clearly a substantial budgetary problem for Zn isotopes if the oceans are in steady state. The open ocean outputs are heavier than the dissolved pool while the inputs are light, implying that the oceans should be moving to lighter isotopic compositions through time, which is not seen in records (e.g., Little et al. 2014a). The solution to this problem is likely to lie in a sink for Zn into organic-rich sediments. Very recently, Little et al. (2016) have shown that sediments deposited beneath upwelling continental margins, rich in organic carbon due to high photic zone

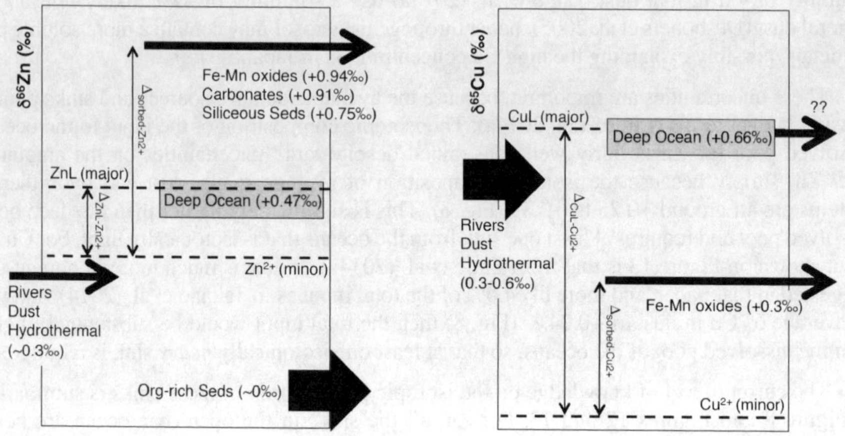

Figure 14. Schematic summary of our current understanding of processes relevant to the overall marine budgets of Cu and Zn isotopes. The inputs are shown as arrows on the left, with the Cu input shown as thick because of the uncertainty over the size of the dust input (Little et al. 2014a; Takano et al. 2014). Within the oceans this input is split into two pools (schematic isotopic compositions shown as horizontal dashed lines): a dominant ligand-bound pool (ZnL or CuL) and a minor free metal ion pool (Zn^{2+} or Cu^{2+}). In both cases the ligand-bound pool is shown as heavy relative to the free metal ion (Jouvin et al. 2009; Ryan et al. 2014). If the oceans are in steady state the isotopic compositions of the outputs (arrows on right) must balance the input. For Zn the outputs to oxic sediments have $\delta^{66}Zn \sim +0.9‰$ (Little et al. 2014a), consistent with a positive $\Delta_{sorbed\text{-}Zn^{2+}}$ (Bryan et al. 2015). This is balanced by a light output to organic-rich sediments (Little et al. 2016) whose isotopic composition is probably controlled by partial sequestration of organic-associated Zn to sulfide in pore waters (Vance et al. 2016b). For Cu the only characterized output is via sorption to Fe–Mn particulates, with $\delta^{65}Cu \sim +0.3‰$ (Little et al. 2014a), shown as being consistent with a positive $\Delta_{sorbed\text{-}Cu^{2+}}$ (e.g., Balistrieri et al. 2008, Pokrovsky et al. 2008). This must be balanced by an output that is heavy relative to the input (question marks on right), possibly quantitative removal of seawater Cu in euxinic settings as speculated by Thompson and Ellwood (2014).

productivity, carry substantial authigenic Zn enrichments and that this Zn is isotopically light. Light Zn isotopes in organic-rich sediment could either be delivered there via uptake of light Zn into phytoplankton in the photic zone, sinking and preservation of the organic matter (e.g., John et al. 2007b; Peel et al. 2009). An alternative is suggested by the data of Vance et al. (pers. comm.) for the sulfidic Black Sea. In this setting Zn behaves very like Mo: sulfidization leads to removal of Zn as particle reactive sulfide species that are isotopically light—by 0.6–0.7‰ (consistent with ab initio calculations: Fujii et al. 2011, see earlier in this chapter)—pushing the residual dissolved pool very heavy (Fig. 8). Pore waters within organic-rich sediments also become sulfidic just beneath the sediment-water interface, due to reduction of pore water sulfate when respiration depletes oxygen and other oxidants (e.g., Froelich et al. 1979). In the Black Sea this process is near-quantitative so that authigenic Zn in sediments is the same as the open ocean dissolved pool, but a non-quantitative version of the same process within the sediments of upwelling continental margins could close the oceanic Zn isotope budget.

As noted earlier, what we currently know about the Cu isotope budget also requires an as yet unidentified sink, in this case isotopically heavy. But we know too little about the oceanic Cu isotope budget to say much more at this stage.

Cycling of Cu and Zn isotopes within the oceans. The discussion in the previous section emphasized the homogeneity of Cu and Zn isotope compositions in the deep ocean (Figs. 12, 13), but there is clearly a great deal of variability *within* the oceanic dissolved pool, in the top 1000 m and particularly for Zn, where the roughly 2‰ range is greater than any other Earth reservoir (Fig. 8). One of the most impressive features of ocean chemistry is the massive drawdown of Zn in the surface ocean, with concentrations there that are sometimes as low as 0.03 nM, more than 2 orders of magnitude lower than the highest concentrations in the deep ocean (Fig. 12 and references therein). For Cu, the surface-deep contrast is smaller, at a maximum of about factor 10 (Fig. 13 and references therein). Thus it is no surprise, if this drawdown is associated with even a small isotope fractionation, that there is substantial variability in the small residual dissolved pool of the upper ocean. The reasons for isotopic variability have focused on two explanations of both the surface ocean drawdown and the isotopic variation: (1) a kinetic fractionation leading to preferential uptake of the light isotope into phytoplankton cells in the photic zone; (2) reversible and non-reversible scavenging of Cu and Zn in both the surface and deep ocean.

For Cu, scavenging is the process that is often regarded as dominant. Though Cu is an important component of enzymes and proteins in phytoplankton (e.g., Peers and Price 2006), it is not limiting to phytoplankton growth and at high concentrations it is toxic (e.g., Moffett and Brand 1996). Indeed, the organic ligands that bind Cu in the dissolved phase of the oceans (and at least to some degree elsewhere on Earth) are probably exuded by phytoplankton to reduce the concentrations of free Cu^{2+} ion beneath the threshold for toxicity at around 10^{-12} M—some 3 orders of magnitude beneath *total* oceanic Cu concentrations (e.g., Moffett and Brand 1996). This dichotomy between the requirement for small amounts of Cu by phytoplankton, coupled to toxicity at high concentrations, has led to Cu being dubbed the "Goldilocks element" of ocean biogeochemistry (e.g., Thomson and Ellwood 2014). The approximately linear increase in Cu concentrations with depth (Fig. 13) is reminiscent of that attributed to reversible scavenging for some other oceanic metals (e.g., Bacon and Anderson 1982), and Little et al. (2013) demonstrated that depth profiles of Cu concentrations are very well modeled by such a process. In support of this, Takano et al. (2014) interpret a good positive relationship between Cu isotope composition and oxygen availability in the deep ocean in terms of preferential scavenging of the light isotope to Fe–Mn oxides, a suggestion that is consistent with what we know of the isotope composition of Cu in Fe–Mn crusts (Little et al. 2014a). Thompson and Ellwood (2014) concur with Vance et al. (2008) and Little et al. (2014a,b) in highlighting the importance of organic ligands in this process: without them, if Cu existed as a free metal ion in seawater, it seems almost certain that sorption would remove the heavy isotope (Balistrieri et al. 2008; Pokrovsky et al. 2008; Little et al. 2014b; Sherman et al. 2015).

Little et al. (2013) showed that, in contrast to Cu, Zn concentrations are not at all well modeled by a reversible scavenging process and conclude that biogeochemical cycling, uptake into phytoplankton at the surface and regeneration by respiration at depth, must be the key process. If this is the case, and if Zn uptake into phytoplankton is associated with a kinetic isotope fractionation that favours uptake of the light isotope (Fig. 10; John et al. 2007b; John and Conway 2014), then the expectation is that drawdown of Zn in the surface ocean should be associated with residual heavy isotopic compositions in the dissolved pool of the upper ocean. An initial examination of Figure 12 seems to imply, if anything, the opposite. John and Conway (2014) suggest that scavenging must play at least some role. These authors conducted a cell degradation experiment in an attempt to simulate regeneration of organic material by respiration, and found that the Zn released was $0.27 \pm 0.11‰$ lighter than the cell inventory. They suggest that this is due to the preferential re-adsorption of the heavy isotopes onto residual organic particles, implying a $\Delta^{66}Zn_{adsorbed-dissolved} = +0.58‰$. However, these experiments contained very high aqueous Zn concentrations and no ligand to complex it, very different from the real ocean where a large portion of the Zn in the dissolved pool is complexed to organic ligands.

We suggest that the main control on oceanic Zn distributions remains to be unequivocally established but that biological uptake and regeneration, with whatever isotope fractionation it is associated with, will turn out to be the key process. In making this suggestion, which will require further work to substantiate, we suggest that the following observations from the data we currently have are going to be key. First, the main oceanic region in which Zn is taken up into cells and exported to the deep is the Southern Ocean. Diatoms dominate the ecology in this region and their cells contain an order of magnitude more Zn than average oceanic phytoplankton (Twining and Baines 2013). Given that diatoms dominate the export of carbon to the deep ocean (e.g., Armbrust 2009), they must completely dominate the export of Zn. Second, the Zn isotopic data for the Southern Ocean (Fig. 12; Zhao et al. 2014) show no isotopic shift in the surface across a 2 order of magnitude drop in Zn concentration away from the locus of upwelling and as diatoms take it up, suggesting the massive uptake of Zn by diatoms in this region causes no isotope fractionation. Third, when the depth profiles in Figure 12 are examined in more detail it is clear that the surface-most point is isotopically heaviest, and the isotopically light Zn that is clearly seen in this figure occupies depths beneath the surface, at 50–200 m (e.g., Zhao et al. 2014). All of these observations are most consistent with very shallow upper ocean recycling of Zn by non-diatom phytoplankton and associated with small isotope fractionations, superimposed on a quantitatively much more important deep export that is controlled by diatoms in the Southern Ocean but that imparts no isotopic variability.

Applications to Earth history. There have been a small number of applications of Zn isotopes in the study of the oceans through Earth history (Pichat et al. 2003; Kunzmann et al. 2013; Pons et al. 2013). Though these have been hampered by a limited understanding of the modern cycle, there is now great potential for such applications given that this understanding has now reached quite an advanced stage through the datasets described in this review. A bridge between attempts to understand the modern cycle and the recent and deep past is provided by the study of the systematics of Zn isotopes in the biological components of core-top sediments, as in Andersen et al. (2011) and Hendry and Andersen (2013). Two studies have sought to harness Zn isotopes in marine carbonate as a monitor of photic zone primary productivity on both kyr timescales in the Quaternary, targeted at understanding upwelling supply of Zn to the photic zone as controlled by climate (Pichat et al. 2003), and on the much longer timescales represented by the recovery from Snowball Earth glaciations, focused on tracking the recovery of the biosphere through the hypothesized burial of isotopically light photosynthesized Zn. Pons et al. (2013), by contrast, link secular changes in Zn isotopes in Banded Iron Formations (BIFS) to relationships between the geochemical cycle of phosphate and isotopic fractionation of Zn isotopes, a phenomenon for which, in all the studies of the modern cycle summarized earlier, evidence has yet to emerge. The burial of light Zn with organic carbon (Little et al. 2016) and the likely control of this burial by sequestration of the light isotope into sulfide in pore waters suggested by the Black Sea data of Vance et al. (pers. comm.) together suggest that the future of successful applications of paleo-Zn isotopes probably lies in the investigation of the links between the biosphere and redox in the ancient oceans. Similarly, a very recent study of Cu isotopes in black shales from across the Great Oxidation Event at 2.7–2.1 Ga (Chi Fru et al. 2016) presents a viable interpretation in terms of removal of the light isotope of Cu into Fe formations before 2.2 Ga and the lack of this removal thereafter. It is not clear, however, whether this interpretation is unique, given what we still have to learn about the controls on other outputs of Cu from the modern ocean, such as to organic-rich sediments and in the sulfidic environments that likely dominated the oceans after the demise of BIFS. Finally, Pons et al. (2011) found that Zn isotopes in 3.8 Ga serpentinites from Isua are depleted in heavy isotopes compared to the BSE ($\delta^{66}Zn$ down to −0.5‰), while serpentinites from modern ophiolites and mid-ocean ridges are isotopically similar to the BSE, at around +0.3‰. Theoretical calculations (Fujii et al. 2011) show that the incorporation of isotopically light Zn in serpentinites requires that the serpentinization reactions occurred at high-pH, with a fluid rich in carbonate at medium temperature (100–300 °C). In addition, Pons et al. (2011) point out that these are the conditions that are found in modern mud volcano environments such as the Mariana's forearc, where the serpentinites are also isotopically light ($\delta^{66}Zn$ down to −0.3‰). Pons et al. (2011) further suggest that Zn isotopes could be used as a pH proxy for ancient hydrothermal fluids.

Cu and Zn isotopes in the Anthropocene

A significant effort has gone into the identification and quantification of human disturbance of the natural Earth surface cycles of Cu and Zn, including via their stable isotopes (see recent review in Fekiacova et al. 2015). Approximately 25% of the Zn and 30% of the Cu released annually to the atmosphere derives directly from human activities such as agriculture, manufacturing and waste management (Rauch and Pacyna 2009). Absolute anthropogenic emissions of Zn are close to double those of Cu (Rauch and Pacyna 2009) and most of the isotopic work on environmental tracing of these emissions has focused on Zn. Virtually all the estimated Cu emitted to the atmosphere annually is thought to derive from non-ferrous metal production (70%) and fossil fuel combustion. For Zn about 70% is also from non-ferrous metal production, 16% from fossil fuel combustion, with 4–5% each via steel and cement production and waste disposal. In terms of riverine transport, Chen et al. (2014a) estimate that excess anthropogenic Cu and Zn in the Seine at Paris represent 15–20% of the total at low water, with little excess over continental crustal concentrations at high water stage.

Zn and Cu isotopes in ore bodies and industrial products. An important starting point is the variability in, and processes responsible for, Cu and Zn isotopes in the mineral ores processed during the smelting that is responsible for about 70% of the emissions to the atmosphere. Though we briefly outline this topic here, it should be emphasized that industrial products and emissions—in the end and at least of Zn—do not reflect the massive variability seen in mineral ores (Fig. 8), implying homogenization of these isotope compositions during processing.

Larson et al. (2003) first documented the large Cu isotope variations in ore minerals in the weathering/supergene environment ($\delta^{65}Cu = -3$ to +2.5‰), while also noting that variability in primary high-temperature minerals was more subdued. These twin results have been confirmed by subsequent studies (Graham et al. 2004; Mason et al. 2005; Markl et al. 2006; Asael et al. 2007, 2009; Mathur et al. 2009, 2010): primary ore minerals predominantly in the 0 ± 0.5‰ range, while the range of Cu isotope compositions for the supergene environment has broadened to −16.5 to +10‰. As first noted by Larson et al. (2003) and confirmed subsequently, redox-induced fractionations almost certainly control the huge variability seen in the supergene environment, where oxidative leaching of high-temperature reduced Cu minerals in the vadose zone is followed by precipitation of reduced minerals beneath the water table (Larson et al. 2003; Ehrlich et al. 2004; Mathur et al. 2005). As noted by Sherman (2013), however, the variability in the natural minerals is much greater than the fractionation factors measured in experiments, so that multiple cycles of oxidation and reduction, or Rayleigh fractionation or open system behavior, or all three, must be in operation. Though redox processes are certainly dominant, Markl et al. (2006) also suggest there may be a component to fractionation controlled by phase changes between fluid and solid mineral. Asael et al. (2009) further suggest that the nature of the ligand-bonding in solution is also important, confirmed by Sherman et al. (2013) and Fujii et al. (2013) using *ab-initio* calculations to show that aqueous Cu complexes have a $\delta^{65}Cu$ range of 1.3‰ (see earlier in this chapter).

There have been fewer studies of Zn isotope variation in ore minerals and, without redox chemistry, the variability is more subdued. Mason et al. (2005) document a range in $\delta^{66}Zn$ of 0.63‰ in a volcanic-hosted massive sulfide ore deposit, attributed to a Zn isotopic difference between co-existing sphalerite and chalcopyrite as well as Rayleigh fractionation during precipitation from a hydrothermal fluid. Wilkinson et al. (2005) and Gagnevin et al. (2012) also explain variability in Zn isotopic data for sphalerite from the Irish ore fields (−0.17 to +1.33‰) as due to a kinetic fractionation and Rayleigh fractionation during progressive precipitation from fluids percolating up from the basement.

To our knowledge, the isotope composition of the final industrial products of these metals has only been measured for Zn. The main result of this study (John et al. 2007a) was that the variability seen in a small number of industrial products, which are mostly −0.4 to +0.2‰, was much less than in the raw ore minerals. In addition, the average δ^{66}Zn of the products measured is identical—at +0.19 ± 11‰ (2SD, n = 14)—to the average for ore minerals in the Wilkinson et al. (2005) and Mason et al. (2005) studies—at +0.15‰. Sivry et al. (2008) note that the extraction yields for Zn processing are higher than 95%, so that a similarity between the final product and average initial ore is perhaps not surprising.

Dispersal of anthropogenic Cu and Zn via the atmosphere. A number of studies have sought to use Zn isotopes (but only one to use Cu: Thapalia et al. 2010) to trace emissions of Zn to the atmosphere in the urban environment (e.g., waste combustors) and in the vicinity of mining and smelting activities through the analysis of lichens and anthropogenic aerosols (Cloquet et al. 2006; Dolgopolova et al. 2006; Gioia et al. 2008; Mattielli et al. 2009), ombrotrophic peat cores (Weiss et al. 2007), and in soils or sediments from rivers and lakes (Sivry et al. 2008; Sonke et al. 2008; Juillot et al. 2011; Aebischer et al. 2015). Though source signatures are not always easily partitioned into natural and anthropogenic (e.g., Cloquet et al. 2006), and though post-depositional processes can induce substantial isotope variability that obscures initial source signatures (e.g., Weiss et al. 2007; Juillot et al. 2011, though cf. Sonke et al. 2008), a systematic feature *has* often emerged from studies of smelting activities (see recent review in Yin et al. 2016). Generally, the slag residues, or tailings, are typically enriched in the heavy isotopes of Zn, while the fine dust and aerosol emitted from smelter chimney stacks tends to be isotopically light (Dolgopolova et al. 2006; Sivry et al. 2008; Sonke et al. 2008; Juillot et al. 2011). In contrast, Ochoa Gonzalez and Weiss (2015) find that heavy isotopes are emitted to the atmosphere during coal combustion while the light isotopes are retained in bottom ashes, an observation that may allow the fingerprinting of sources of Zn pollution from different activities.

Mattielli et al. (2009) provide the most extensive discussion of the causes and consequences of isotopic fractionation during the processing of Zn ores. These authors document a change in the size and Zn isotopic composition of aerosols away from the main chimney, with δ^{66}Zn = +0.01 to +0.19‰ at ≤1 km, and −0.52 to −0.02‰ at 2–5 km. They attribute the light Zn in aerosols to the high temperatures in the smelter (up to 1300 K), leading to fractionation during volatilization/condensation (see above section on Isotopic fractionation by evaporation on Earth). Their main chimney dust samples have the lightest δ^{66}Zn, at −0.67 ± 0.10‰. In contrast Sivry et al. (2008) document Zn in tailings at a smelter with δ^{66}Zn = +0.18 to +1.49‰, a signal also seen in polluted downstream sediments (+0.83 to +1.38‰). Similarly, Juillot et al. (2011) measure δ^{66}Zn = +0.81 ± 0.20‰ for slags at a French smelter and see a shift towards this isotope signature in heavily contaminated top soils within 500 m of the smelter. Consistent with this, Thapalia et al. (2010) document a step towards lighter Zn isotopes in lake sediment ~ 100 km from a smelter after it became active. This is perhaps the only study of a Cu isotope archive of anthropogenic activity, showing a subtle shift to heavier Cu isotopes (δ^{65}Cu from +0.77 ± 0.05 to +0.94 ± 0.10‰) due to human activity.

Processes affecting Cu and Zn isotopes in polluted rivers. A small number of studies have been undertaken of Cu and Zn isotopes in two classes of polluted rivers: (1) rivers that are clearly affected by mining, including acid mine drainage and (2) rivers in areas that have been intensely developed for agriculture and industrial activities more generally. Borrok et al. (2008) studied dissolved Cu and Zn and their isotopes in small streams located in 6 historical mining areas in the US and Europe. It should be noted that the Cu and Zn concentrations in these mining-impacted streams, not unexpectedly, are up to 4 orders of magnitude greater than those in the relatively unpolluted rivers studied in Vance et al. (2008) and Little et al. (2014a). The δ^{66}Zn data covers the range +0.02 to +0.46‰ and exhibits a diel cycle that may be related to uptake by microorganisms. The range of δ^{65}Cu is −0.7 to +1.4‰, similar to that in non-mining-impacted rivers. Kimball et al. (2009) document Cu in acid mine drainage that has $\delta^{65}Cu$ about 1.5‰ heavier than the primary minerals, attributing this fractionation to oxidation of reduced Cu(I) in the minerals. More recently, Wanty et al. (2013, 2015) have documented Zn isotope fractionations associated with biomediated precipitation of hydrozincite in streams draining mining areas.

Chen et al. (2008, 2009) present an array of data for the dissolved and particulate phase of the River Seine, including time-series data in Paris as well as contaminated waters draining roofs and waste water treatment plants. Dissolved Zn concentrations in the Seine increase continuously downstream from 1 to 74 nM, while $\delta^{66}Zn$ decreases from a high in the headwaters of Aube tributary (0.58‰) to +0.09‰ at the estuary. The decreasing $\delta^{66}Zn$ values as anthropogenic Zn is added downstream, particularly in Paris, is consistent with generally light Zn isotope compositions in roadway and roof runoff, and plant-treated and waste water (−0.06 ± 0.05‰). Fertilizers showed heavier Zn isotope values at +0.19 to +0.42‰, leading these authors to suggest that fertilizer Zn is strongly retained in soil. Data for suspended particulate matter presented a similar picture: $\delta^{66}Zn$ decreases from +0.3 to 0.08‰ downstream, associated with a 4-fold increase in concentration. Time-series samples in Paris show an inverse relationship between enrichment factor and isotopes. Chen et al. (2009) interpret the results as due to more or less conservative mixing of silicate and anthropogenic particles, ruling out sorption as an important process generating isotopic variability.

ACKNOWLEDGMENTS

Acknowledgments: FM acknowledges funding from the European Research Council under the H2020 framework program/ERC grant agreement #637503 (Pristine), as well as the financial support of the UnivEarthS Labex program at Sorbonne Paris Cité (ANR-10-LABX-0023 and ANR-11-IDEX-0005-02), and the ANR through a chaire d'excellence Sorbonne Paris Cité. DV is funded by the Swiss National Science Foundation (projects 200021-143262, 200021-153087) and ETH Zürich. We would like to thank Francis Albarède, Morten Andersen, Corey Archer. James Day, Susan Little, Alan Matthews, Emily Pringle, Paolo Sossi and Damien Guinoiseau for discussions. Three anonymous reviewers and editor Fang-Zhen Teng greatly improved the text.

REFERENCES

Aebischer, S, Cloquet, C, Carignan, J, Maurice, C, Pienitz, R (2015) Disruption of the geochemical metal cycle during mining: multiple isotope studies of lake sediments from Schefferville, subarctic Québec. Chem Geol 412:167–178

Albarède F (2004) The stable isotope geochemistry of copper and zinc. Rev Mineral Geochem 55:409–427

Albarède F (2009) Volatile accretion history of the terrestrial planets and dynamic implications. Nature 461:1227–1233, doi:10.1038/Nature08477

Albarède F, Balter V, Jaouen K, Lamboux A (2011) Applications of the stable isotopes of metals to physiology. The 38th Meeting of the Federation of Analytical Chemistry and Spectroscopy Societies (FACSS), Reno (abstract)

Albarède F, Télouk P, Balter V (2017) Medical applications of isotope metallomics. Rev Mineral Geochem 82:851–885

Amira S, Spångberg D, Hermansson K (2005) Distorted five-fold coordination of Cu^{2+}(aq) from a Car–Parrinello molecular dynamics simulation. Phys Chem Chem Phys 7:2874–2880

Anderson D, Morel F (1978) Copper sensitivity of *Gonyaulax tamarensis*. Limnol Oceanogr 23:283–295

Andersen MB, Vance D, Archer C, Anderson RF, Ellwood MJ, Allen CS (2011) The Zn abundance and isotopic composition of diatom frustules, a proxy for Zn availability in ocean surface seawater. Earth Planet Sci Lett 301:137–145, doi:10.1016/j.epsl.2010.10.032

Archer C, Vance D (2002) Mass discrimination correction in multiple collector plasma source mass-spectrometry: an example using Cu and Zn isotopes. J Anal At Spectrom, 64:356–365

Archer C, Vance D (2004) Mass discrimination correction in multiple-collector plasma source mass spectrometry: an example using Cu and Zn isotopes. J Anal At Spectrom 19:656–665, doi:Doi 10.1039/B315853e

Archer C, Vance D (2008) The isotopic signature of the global riverine molybdenum flux and anoxia in the ancient oceans. Nat Geosci 1:597–600

Armbrust EA (2009) The life of diatoms in the world's oceans. Nature 459:185–192

Arnold T, Kirk GJD, Wissuwa M, Frei M, Zhao FJ, Mason TFD, Weiss DJ (2010a) Evidence for the mechanisms of zinc uptake by rice using isotope fractionation. Plant Cell Environ 33:370–381, doi:Doi 10.1111/J.1365-3040.2009.02085.X

Arnold T, Schonbachler M, Rehkamper M, Dong SF, Zhao FJ, Kirk GJD, Coles BJ, Weiss DJ (2010b) Measurement of zinc stable isotope ratios in biogeochemical matrices by double-spike MC–ICP–MS and determination of the isotope ratio pool available for plants from soil. Anal Bio Chem 398:3115–3125, doi:10.1007/S00216-010-4231-5

Asael D, Matthews A, Bar-Matthews M, Halicz L (2007) Copper isotope fractionation in sedimentary copper mineralization (Timna Valley, israel). Chem Geol 243:238–254

Asael D, Matthews A, Oszczepalski S, Bar-Matthews M, Halicz L (2009) Fluid speciation controls of low temperature copper isotope fractionation applied to the Kupferschiefer and Timna ore deposits. Chem Geol 262:147–158

Aucour, AM, Pichat, S, Macnair, MR, Oger, P (2011) Fractionation of stable isotope zinc isotopes in the zinc hyperaccumulator *Arabidopsis halleri* and nonaccumulator *Arabidopsis petrasea*. Environ Sci Technol 45:9212–9217

Aucour, AM, Bedell, J-P, Queyron, M, Magnin, V, Testemale, D, Sarret, G (2015) Dynamics of Zn in an urban wetland soil-plant system: coupling isotopic and EXAFS approaches. Geochim Cosmochim Acta 160:66–69

Bacon MP, Anderson RF (1982) Distribution of thorium isotopes between dissolved and particulate forms in the deep sea. J Geophys Res 87:2045–2056

Balistrieri LS, Borrok DM, Wanty RB, Ridley WI (2008) Fractionation of Cu and Zn isotopes during adsorption onto amorphous Fe(III) oxyhydroxide: Experimental mixing of acid rock drainage and ambient river water. Geochim Cosmochim Acta 72:311–328, doi:DOI 10.1016/j.gca.2007.11.013

Balter V, Zazzo A (2011) An animal model (sheep) for Fe, Cu, and Zn isotopes cycling in the body. Mineral Mag 75:476

Balter V, Lamboux A, Zazzo A, Telouk P, Leverrier Y, Marvel J, Moloney AP, Monahan FJ, Schmidt O, Albarède F (2013) Contrasting Cu, Fe, and Zn isotopic patterns in organs and body fluids of mice and sheep, with emphasis on cellular fractionation. Metallomics 5:1470–1482, doi:10.1039/c3mt00151b

Ban Y, Aida M, Nomura M, Fujii Y (2002) Zinc isotope separation by ligand-exchange chromatography using cation exchange resin. J Ion Exch 14:46–52

Barrat JA, Zanda B, Moynier F, Bollinger C, Liorzou C, Bayon G (2012) Geochemistry of CI chondrites: Major and trace elements, and Cu and Zn Isotopes O. Geochim Cosmochim Acta 83:79–92

Ben Othman D, Luck JM, Bodinier JL, Arndt NT, Albarède F (2006) Cu–Zn isotopic variations in the Earth's mantle. Geochim Cosmochim Acta 70:46

Benfatto M, D'Angelo P, Della Longa S, Pavel NV (2002) Evidence of distorted fivefold coordination of the Cu^{2+} aqua ion from an x-ray-absorption spectroscopy quantitative analysis. Phys Rev B 65:174205

Bentahila Y, Ben Othman D, Luck J-M (2008) Strontium, lead and zinc isotopes in marine cores as tracers of sedimentary provenance: a case study around Taiwan orogeny. Chem Geol 62–82

Bermin J, Vance D, Archer C, Statham PJ (2006) The determination of the isotopic composition of Cu and Zn in seawater. Chem Geol 226:280–297, doi:10.1016/j.chemgeo.2005.09.025

Bersuker IB (2006) The Jahn–Teller effect;. Cambridge Univ. Press, New York

Bigalke M, Weyer S, Wilcke W (2010a) Copper Isotope Fractionation during complexation with insolubilized humic acid. Environ Sci Technol 44:5496–5502, doi:10.1021/Es1017653

Bigalke M, Weyer S, Wilcke W (2010b) Stable copper isotopes: A novel tool to trace copper behavior in hydromorphic soils. Soil Sci Soc Am J 74:60–73, doi:10.2136/sssaj2008.0377

Bigalke M, Weyer S, Kobza J, Wilcke W (2010c) Stable Cu and Zn isotope ratios as tracers of sources and transport of Cu and Zn in contaminated soil. Geochim Cosmochim Acta 74:6801–6813, doi:10.1016/j.gca.2010.08.044

Bigalke M, Weyer S, Wilcke W (2011) Stable Cu isotope fractionation in soils during oxic weathering and podzolization. Geochim Cosmochim Acta 75:3119–3134

Bigeleisen J, Mayer MG (1947) Calculation of equilibrium constants for isotopic exchange reactions. J Chem Phys 15:261–267

Birck JL (2004) An overview of isotopic anomalies in extraterrestrial materials and their nucleosynthetic heritage. Rev Mineral Geochem 55:26–63

Bishop MC, Moynier F, Weinstein C, Fraboulet JG, Wang K, Foriel J (2012) The Cu isotopic composition of iron meteorites. Meteoritics Planet Sci 47:268–276, doi:10.1111/J.1945-5100.2011.01326.X

Black J, Kavner A, Schauble E (2011) Calculation of equilibrium stable isotope partition function ratios for aqueous zinc complexes and metallic zinc. Geochim Cosmochim Acta 75:769–783

Borrok DM, Wanty RB, Ridley WI, Wolf R, Lamothe PJ, Adams M (2007) Separation of copper, iron, and zinc from complex aqueous solutions for isotopic measurement. Chem Geol 242:400–414, doi:10.1016/j.chemgeo.2007.04.004

Borrok DM, Nimick DA, Wanty RB, Ridley WI (2008) Isotopic variations of dissolved copper and zinc in stream waters affected by historical mining. Geochim Cosmochim Acta 72:329–344

Bridgestock LJ, Williams H, Rehkämper M, Larner F, Giscard MD, Hammond S, Coles B, Andreasen R, Wood BJ, Theis KJ, Smith CL (2014) Unlocking the zinc isotope systematics of iron meteorites. Earth Planet Sci Lett 400:153–164

Broadley MR, White PJ, Hammond JP, Zelko I, Lux A (2007) Zinc in plants. New Phytologist 173:677–702

Bruland KW (1999) Complexation of zinc by natural organic ligands in the central North Pacific. Limnol Oceanogr 34:269–285

Bruland KW, Middag R, Lohan MC (2014) Controls of trace metals in seawater. Treatise Geochem 8:19–51

Bryan AL, Dong S, Wilkes EB, Wasylenki LE (2015) Zinc isotope fractionation during adsorption onto Mn oxyhydroxide at low and high ionic strength. Geochim Cosmochim Acta 157:182–197

Caldelas C, Dong SF, Araus JL, Weiss DJ (2011) Zinc isotopic fractionation in Phragmites australis in response to toxic levels of zinc. J Exp Bot 62:2169–2178, doi:10.1093/Jxb/Erq414

Cameron V, Vance D (2014) Heavy nickel isotopic composition of rivers and the oceans. Geochim Cosmochim Acta 129:195–211

Canfield DE (1998) A new model for Proterozoic ocean chemistry. Nature 396:450–453

Chaboy J, Muñoz-Páez A, Merkling PJ, Sánchez Marcos E (2006) The hydration of Cu^{2+}: Can the Jahn–Teller effect be detected in liquid solution? J Chem Phys 124:064509

Chadwick OA, Brimhall GH, Hendricks DM (1990) From a black box to a gray box—a mass balance interpretation of pedogenesis. Geomorphology 3:369–390

Chaussidon M, Deng Z, Villeneuve J, Moureau J, Watson B, Richter F, Moynier F (2017) In situ analysis of non-traditional isotopes by SIMS and LA–MC–ICP–MS: key aspects and the example of mg isotopes in olivines and silicate glasses. Rev Mineral Geochem 82:127–164

Chen H, Nguyem BM, Moynier F (2013a) Zinc isotopic composition of iron meteorites: Absence of isotopic anomalies and origin of the volatile element depletion. Meteorit Planet Sci 48:2441–2450

Chen H, Savage P, Teng FZ, Helz RT, Moynier F (2013b) No zinc isotope fractionation during magmatic differentiation and the isotopic composition of the bulk Earth. Earth Planet Sci Lett 369:34–42

Chen H, Moynier F, Humayun M, Bishop MC, Williams J (2016) Cosmogenic effects on Cu isotopes in IVB iron meteorites. Geochim Cosmochim Acta 182:145–154

Chen J, Gaillardet J, Louvat P (2008) Zinc isotopes in the Seine River waters, France: a probe of anthropogenic contamination. Environ Sci Technol 43:6494–6501

Chen J, Gaillardet J, Louvat P, Huon S (2009) Zn isotopes in the suspended load of the Seine River, France: isotopic variations and source determination. Geochim Cosmochim Acta 73:4060–4076

Chen J, Gaillardet J, Bouchez J, Louvat P, Wang Y-N (2014a) Anthropophile elements in river sediments: overview from the Seine River, France. Geochem Geophys Geosyst 15:4526–2546

Chen J, Gaillardet J, Dessert C, Villemant B, Louvat P, Crispi O, Birck J-L, Wang Y-N (2014b) Zn isotope compositions of the thermal spring waters of La Soufrière volcano, Guadeloupe Island. Geochim Cosmochim Acta 127:67–82

Chen S, Liu Y, Hu J, Zhang Z, Hou Z, Huang F, Yu H (2016) Zinc isotopic composition of NIST683 and whole rock reference materials. Geostand Geoanal Res. doi: 10.1111/j.1751-908X.2015.00377.x

Chou C-L, Baedecker PA, Wasson JT (1976) Allende inclusions: volatile-element distribution and evidence for incomplete volatilization of presolar solids. Geochim Cosmochim Acta 40:85–94

Chi Fru, E, Rodriguez, NP, Partin, CA, Lalonde, SV, Andersson, P, Weiss, DJ, El Albani, A, Rodushkin, I, Konhauser, KO (2016) Cu isotopes in marine black shales record the Great Oxidation Event. PNAS 113:4941–4946, doi: 10.1073/pnas.1523544113

Cloquet C, Carignan J, Libourel G (2006) Isotopic composition of Zn and Pb atmospheric depositions in an urban/periurban area of northeastern France. Environ Sci Technol 40:6594–6600

Cloquet C, Carignan J, Lehmann MF, Vanhaecke F (2008) Variation in the isotopic composition of zinc in the natural environment and the use of zinc isotopes in biogeosciences: a review. Anal Bio Chem 390:451–463

Coale KH, Bruland KW (1988) Copper complexation in the northeast Pacific. Limnol Oceanogr 33:1084–1101

Coggon RM, Teagle DAH, Smith-Duque CE, Alt JC, Cooper MJ (2010) Reconstructing past seawater Mg/Ca and Sr/Ca from mid-ocean ridge flank calcium carbonate veins. Science 327:1114–1117

Conway TM, Rosenberg AD, Adkins JF, John SG (2013) A new method for precise determination of iron, zinc and cadmium stable isotope ratios in seawater by double-spike mass spectrometry. Anal Chim Acta 793:44–52

Conway TM, John SG (2014) The biogeochemical cycling of zinc and zinc isotopes in the North Atlantic Ocean. Glob Biogeochem Cycles 28:1111–1128

Conway TM, John SG (2015) The cycling of iron, zinc and cadmium in the North East Pacific Ocean—Insights from stable isotopes. Geochim Cosmochim Acta 164:262–283, doi:10.1016/j.gca.2015.05.023

Couder, E, Mattielli, N, Drouet, T, Smolders, E, Delvaux, B, Iserentant, A, Meeus, C, Maerschalk, C, Opfergelt, S, Houben, D (2015) Transpiration flow controls Zn transport in *Brassica napus* and *Lolium multiflorum* under toxic levels as evidenced from isotopic fractionation. Comptes Rendus Geoscience 347:386–396

Coutaud A, Mehuet M, Viers J, Rols J-L, Pokrovsky OS (2014) Zn isotope fractionation during interaction with phototrophic biofilm. Chem Geol 390:46–60

Dauphas N, Chen JH, Zhang J, Papanastassiou DA, Davis AM, Travaglio C (2014) Calcium-48 isotopic anomalies in bulk chondrites and achondrites: Evidence for a uniform isotopic reservoir in the inner protoplanetary disk. Earth Planet Sci Lett 407:96–108, doi:10.1016/j.eps1.2014.09.015

Day JM, Moynier F (2014) Evaporative fractionation of volatile stable isotopes and their bearing on the origin of the Moon. Philos Trans R Soc London, Ser A 372:20130259, doi:10.1098/rsta.2013.0259 de Almeida KJ, Murugan NA, Rinkevicius Z, Hugosson HW, Vahtras O, Ågrena H, Cesar A (2009) Phys Chem Chem Phys 11:508–519

Dekov, VM, Cuadros, J, Kamenov, GD, Weiss, D, Arnold, T, Basak, C, Rochette, P (2010) Metalliferous sediments from the H.M.S. Challenger voyage (1872–1876). Geochim Cosmochim Acta 74:5019–5038

Desbouefs KV, Sofikitis A, Losn R, Colin JL, Ausset P (2005) Dissolution and solubility of trace metals from natural and anthropogenic aerosol particulate matter. Chemosphere 58:195–203

Dodson MH (1963) A theoretical study of the use of internal standards for precise isotopic analysis by the surface ionization technique: Part I. General first-order algebraic solutions. J Sci Instrum 40:289–295

Dolgopolova A, Weiss DJ, Seltmann R, Kober B, Mason TFD, Coles BJ, Stanley C (2006) Use of isotope ratios to assess sources of Pb and Zn dispersed in the environment during mining and ore processing within the Orlovka–Spokoinoe mining site (Russia). Appl Geochem 21:563–579, doi:10.1016/j.apgeochem.2005.12.014

Dong S, Weiss DJ, Strekopytov S, Kreissig K, Sun Y, Baker AR, Formenti P (2015) Stable isotope ratio measurements of Cu and Zn in mineral dust (bulk and size fractions) from the Taklimakan Desert and the Sahel and in aerosols from the eastern tropical North Atlantic Ocean. Talanta 114:103–109

Dreibus G, Palme H (1996) Cosmochemical constraints on the sulfur content in the Earth's core. Geochim Cosmochim Acta 60:1125–1130

Edmond JM, Measures C, Mangum B, Grant B, Sclater FR, Collier R, Hudson A, Gordon LI, Corliss JB (1979) On the formation of metal-rich deposits at ridge crests. Earth Planet Sci Lett 46:19–30

Ehrlich S, Butler I, Halicz L, Rickard D, Oldroyd A, Matthews A (2004) Experimental study of copper isotope fractionation between aqueous Cu(II) and covellite, CuS. Chem Geol 209:259–269

Fekiakova S, Cornu S, Pichat S (2015) Tracing contamination sources in soils with Cu and Zn isotope ratios. Sci Tot Environ 517:96–105

Fernandez A, Borrok DM (2009) Fractionation of Cu, Fe and Zn isotopes during weathering of sulfide-rich rocks. Chem Geol 264:1–12

Field CB, Behrenfeld MJ, Randerson JT, Falkowski P (1998) Primary production of the biosphere: integrating terrestrial and oceanic components. Science 281:237–240

Flemming CA, Trevors JT (1989) Copper toxicity and chemistry in the environment—a review. Air Soil Pollut 44:143–158

Froelich PN, Klinkahmmer GP, Bender ML, Luedtke NA, Heath GR, Cullen D, Dauphin P, Hammond D, Hartman B, Maynard V (1979) Early oxidation of organic matter in pelagic sediments of the eastern equatorial Atlantic: suboxic diagenesis. Geochim Cosmochim Acta 43:1075–1090

Fujii T, Albarède F (2012) Ab initio calculation of the Zn isotope effect in phosphates, citrates, and malates and applications to plants and soil. Plos One 7, doi:10.1371/journal.pone.0030726

Fujii T, Moynier F, Telouk P, Abe M (2010) Experimental and theoretical investigation of isotope fractionation of zinc between aqua, chloro, and macrocyclic complexes. J Phys Chem A 114:2543–2552, doi:10.1021/Jp908642f

Fujii T, Moynier F, Pons ML, Albarède F (2011) The origin of Zn isotope fractionation in sulfides. Geochim Cosmochim Acta 75:7632–7643, doi:10.1016/J.Gca.2011.09.036

Fujii T, Moynier F, Abe M, Nemoto K, Albarède F (2013) Copper isotope fractionation between aqueous compounds relevant to low temperature geochemistry and biology. Geochim Cosmochim Acta 110:29–44, doi:10.1016/j.gca.2013.02.007

Fujii T, Moynier F, Blichert-Toft J, Albarède F (2014) Density functional theory estimation of isotope fractionation of Fe, Ni, Cu, and Zn among species relevant to geochemical and biological environments. Geochim Cosmochim Acta 140:553–576, doi:10.1016/j.gca.2014.05.051

Fulton EA, Parslow JS, Smith ADM, Johnson CR (2004) Biogeochemical marine ecosystem models II: the effect of physiological detail on model performance. Ecol Model 173:371–406, doi:10.1016/J.Ecolmodel.2003.09.024

Gagnevin D, Boyce AJ, Barrie CD, Menuge JF, Blakemann RJ (2012) Zn, Fe and S isotope fractionation in a large hydrothermal system. Geochim Cosmochim Acta 88:183–198

Gélabert A, Pokrovsky OS, Schott J, Boudou A, Fertet-Mazel A, Mielczarski J, Mielczarski E, Mesmer-Dudons N, Spalla O (2004) Study of diatoms/aqueous solution interface. I. Acid-base equilibria and spectroscopic observation of freshwater and marines species. Geochim Cosmochim Acta 68:4039–4058

Gélabert A, Pokrovsky OS, Viers J, Schott J, Boudou A, Feurtet-Mazel A (2006) Interaction between zinc and freshwater and marine diatom species: Surface complexation and Zn isotope fractionation. Geochim Cosmochim Acta 70:839–857, doi:10.1016/j.gca.2005.10.026

Gioia, S, Weiss, D, Coles, B, Arnold, T, Babinski, M (2008) Accurate and precise zinc isotope ratio measurements in urban aerosols. Anal Chem 80:9776–9780

Ghidan OY, Loss RD (2011) Isotope fractionation and concentration measurements of Zn in meteorites determined by the double spike, IDMS-TIMS techniques. Meteoritics Planet Sci 46:830–842, doi:0.1111/J.1945–5100.2011.01196.X

Graham S, Pearson N, Jackson S, Griffin W, O'Reilly SY (2004) Tracing Cu and Fe from source to porphyry: in situ determination of Cu and Fe isotope ratios in sulfides from the Grasberg Cu-Au deposit. Chem Geol 207:147–169

Grybos M, Davranche M, Gruau G, Petitjean P (2007) Is trace metal release in wetland soils controlled by organic mobility or Fe-oxyhydroxides reduction? J Colloid Interface Sci 324:490–501

Guelke M, Von Blanckenburg F (2007) Fractionation of stable iron isotopes in higher plants. Environ Sci Technol 41:1896–1901, doi:10.1021/Es062288j

Hartmann D, Woosley SE, El Eid MF (1985) Nucleosynthesis in neutron-rich supernova ejecta. Astrophys J 297:837–845

Guinoiseau, D, Gélabert, A, Moureau J, Louvat, P, Benedetti, MF (2016) Zn isotope fractionation during sorption onto kaolinite. Environ Sci Technol 50:1844–1852

Hayakawa K, Minami S, Nakamura S (1973) Kinetics of the oxidation of ascorbic acid by the copper(II) ion in an acetate buffer solution. Bull Chem Soc Jpn 46:2788

Hendry KR, Andersen MB (2013) The zinc isotopic composition of siliceous marine sponges: investigating nature's sediment traps. Chem Geol:33–41

Herwartz D, Pack A, Friedrichs B, Bischoff A (2004) Identification of the giant impactor Theia in lunar rocks. Science 344:1146–1150

Herzog GF, Moynier F, Albarède F, Berezhnoy AA (2009) Isotopic and elemental abundances of copper and zinc in lunar samples, Zagami, Pele's hairs, and a terrestrial basalt. Geochim Cosmochim Acta 73:5884–5904, doi:10.1016/j.gca.2009.05.067

Houben D, Sonnet P, Tricot G, Matielli N, Couder E, Opfergelt S (2014) Impact of root-induced mobilization of zinc on stable Zn isotope variation in the soil-plant system. Environ Sci Technol 48:7866–73

Ikehata K, Hirata T (2013) Evaluation of UV–fs–LA–MC–ICP–MS for precise in situ copper isotopic microanalysis of cubanite. Anal Sci 29:1213–1217

Ilina SM, Viers J, Lapitsky SA, Mialle S, Mavromatis V, Chmeleff J, Brunet P, Alekhin YV, Isnard H, Pokrovsky OS (2013) Stable (Cu, Mg) and radiogenic (Sr, Nd) isotope fractionation in colloids of boreal organic-rich waters. Chem Geol 434:63–75

John SG, Park JG, Zhan ZT, Boyle EA (2007a) The isotopic composition of some common forms of anthropogenic zinc. Chem Geol 245:61–69

John SG, Geis R, Saito M, Boyle EA (2007b) Zn isotope fractionation during high-affinity zinc transport by the marine diatom *Thalassiosira oceanica*. Limnol Oceanogr 52:2710–2714

John SG, Rouxel OJ, Craddock PR, Engwall AM, Boyle EA (2008) Zinc stable isotopes in seafloor hydrothermal vent fluids and chimneys. Earth Planet Sci Lett 269:17–28

John SG, Conway TM (2014) A role for scavenging in the marine biogeochemical cycling of zinc and zinc isotopes. Earth Planet Sci Lett 394:159–167

Jouvin D, Louvat P, Juillot F, Marechal CN, Benedetti MF (2009) Zinc isotopic fractionation: Why organic matters. Environ Sci Technol 43:5747–5754, doi:10.1021/es803012e

Jouvin D, Weiss D, Mason T, Bravin M, Louvat P, Zhao F, Ferec F, Hinsinger P, Benedetti M (2012) Stable isotopes of Cu and Zn in higher plants: Evidence for Cu reduction at the root surface and two conceptual models for isotopic fractionation processes. Environ Sci Technol 46:2652–2660

Juillot F, Maréchal C, Ponthieu M, Cacaly S, Morin G, Benedetti M, Hazemann JL, Proux O, Guyot F (2008) Zn isotopic fractionation caused by sorption on goethite and 2-Lines ferrihydrite. Geochim Cosmochim Acta 72:4886–4900

Juillot F, Maréchal C, Morin G, Jouvin D, Cacaly S, Telouk P, Benedetti MF, Ildefonse P, Sutton S, Guyot F, Brown GE (2011) Contrasting isotopic signatures between anthropogenic and geogenic Zn and evidence for post-depositional fractionation processes in smelter-impacted soils from Northern France. Geochim Cosmochim Acta 75:2295–2308, doi:10.1016/J.Gca.2011.02.004

Kafantaris F-C, Borrok DM (2014) Zinc isotope fractionation during surface adsorption and intracellular incorporation by bacteria. Chem Geol 366:42–51

Kato C, Moynier F, Valdes MC, Dhaliwal JK, Day JM (2015) Extensive volatile loss during formation and differentiation of the Moon. Nat Commun 6:7617, doi:10.1038/ncomms8617

Kau L-S, Spira-Solomon DJ, Penner-Hahn JE, Hodgson KO, Solomon EI (1987) X-ray absorption edge retermination of the oxidation state and coordination number of copper: Application to the type 3 site in Rhus vernicifera Laccase and its reaction with oxygen. J Am Chem Soc 109:6433–6442

Kavner A, John SG, Sass S, Boyle EA (2008) Redox-driven stable isotope fractionation in transition metals: application to Zn electroplating. Geochim Cosmochim Acta 72:1731–1741

Kimball BE, Mathur R, Dohnalkova AC, Wall AJ, Runkel RL, Brantley SL (2009) Copper isotope fractionation in acid mine drainage. Geochim Cosmochim Acta 73:1247–1263

Koschinsky A, Hein JR (2003) Acquisition of elements from seawater by ferromanganese crusts: Solid phase association and seawater speciation. Mar Geol 198:331–351

Kunzmann N, Halverson GP, Sossi PA, Raub TD, Payne JL, Kirby J (2013) Zn isotope evidence for immediate resumption of primary productivity after snowball Earth. Geology 41:27–30

Kurtz AC, Derry LA, Chadwick OA (2001) Accretion of Asian dust to Hawaiian soils: Isotopic, elemental, and mineral mass balances. Geochim Cosmochim Acta 65:1971–1983

Larner F, Rehkaemper M, Coles B, Kreissig K, Weiss D, Sampson B, Unsworth C, Strekopytov S (2011) A new separation procedure for Cu prior to stable isotope analysis by MC–ICP–MS. J Anal At Spectrom 26:1627–1632

Larson PB, Maher K, Ramos FC, Chang Z, Gaspar M, Meinert LD (2003) Copper isotope ratios in magmatic and hydrothermal ore forming environments. Chem Geol 201:337–350

Le Roux V, Lee CT, Turner SJ (2010) Zn/Fe systematics in mafic and ultramafic systems: Implications for detecting major element heterogeneities in the Earth's mantle. Geochim Cosmochim Acta 74:2779–2796
Lee CTA, Yin Q-Z, Lenardic A, Agranier A, O'Neill C, Thiagarajan N (2007) Trace-element composition of Fe-rich residual liquids formed by fractional crystallization: implications for the Hadean magma ocean. Geochim Cosmochim Acta 71:3601–3615
Lee CTA, Luffi P, Chin EJ, Bouchet R, Dasgupta R, Morton DM, Le Roux V, Yin Q-Z, Jin D (2012) Copper systematics in arc magmas and implications for crust–mantle differentiation. Science 336:64–68, doi:10.1126/science.1217313
Li W, Jackson S, Pearson NJ, Alard O, Chappell BW (2009) The Cu isotopic signature of granites from the Lachlan Fold Belt, SE Australia. Chem Geol 258:38–49
Little SH, Vance D, Siddall M, Gasson E (2013) A modelling assessment of the role of reversible scavenging in controlling oceanic dissolved Cu and Zn distributions. Glob Biogeochem Cycles 27:780–791
Little SH, Vance D, Walker-Brown C, Landing WM (2014a) The oceanic mass balance of copper and zinc isotopes, investigated by analysis of their inputs, and outputs to ferromanganese oxide sediments. Geochim Cosmochim Acta 125:673–693, doi:10.1016/j.gca.2013.07.046
Little SH, Sherman DM, Vance D, Hein JR (2014b) Molecular controls on Cu and Zn isotopic fractionation in Fe–Mn crusts. Earth Planet Sci Lett 396:213–222
Little SH, Vance D, McManus J, Severmann S (2016) Critical role of continental margin sediments in the oceanic mass balance of Zn and Zn isotopes. Geology 44:207–210
Liu S-A, Teng F-Z, Li S, Wei G-J, Ma J-L, Li D (2014) Copper and iron isotope fractionation during weathering and pedogenesis: insights from saprolite profiles. Geochim Cosmochim Acta 146:59–75
Liu S-A, Huang J, Liu J, Woerner G, Yang W, Tang Y-J, Chen Y, Tang L, Zheng J, Li S (2015) Copper isotopic composition of the silicate Earth. Earth Planet Sci Lett 427:95–103, doi:10.1016/j.epsl.2015.06.061
Lodders K (2003) Solar System abundances and condensation temperatures of the elements. Astrophys J 591:1220–1247
Loss RD, Lugmair GW (1990) Zinc isotope anomalies in Allende meteorite inclusions. Astroph J 360:L59-L62
Loss RD, Rosman KJR, Delaeter JR (1990) The isotopic composition of zinc, palladium, silver, cadmium, tin, and tellurium in acid-etched residues of the Allende meteorite. Geochim Cosmochim Acta 54:3525–3536
Luck JM, Othman DB, Barrat JA, Albarède F (2003) Coupled ^{63}Cu and ^{16}O excesses in chondrites. Geochim Cosmochim Acta 67:143–151
Luck JM, Othman DB, Albarède F (2005) Zn and Cu isotopic variations in chondrites and iron meteorites: Early solar nebula reservoirs and parent-body processes. Geochim Cosmochim Acta 69:5351–5363
Macleod G, Mcneown C, Hall AJ, Russel MJ (1994) Hydrothermal and oceanic pH conditions of possible relevance to the origin of life. Biosphere 24:19–41
Manceau A, Lanson B, Drits VA (2002) Structure of heavy metal sorbed birnessite. Part III: results from powder and polarized extended X-ray absorption fine structure spectroscopy. Geochim Cosmochim Acta 66:2639–2663
Maréchal C, Albarède F (2002) Ion-exchange fractionation of copper and zinc isotopes. Geochim Cosmochim Acta 66:1499–1509
Maréchal CN, Sheppard SMF (2002) Isotopic fractionation of Cu and Zn between chloride and nitrate solutions and malachite or smithsonite at 30° and 50 °C (abstr.). Geochim Cosmochim Acta 66:A84
Maréchal C, Télouk P, Albarède F (1999) Precise analysis of copper and zinc isotopic compositions by plasma-source mass spectrometry. Chem Geol 156:251-273
Maréchal CN, Douchet C, Nicolas E, Albarède F (2000) The abundance of zinc isotopes as a marine biogeochemical tracer. Geochem, Geophys, Geosyst 1:1999GC-000029
Markl G, Lahaye Y, Schwinn G (2006) Copper isotopes as monitors of redox processes in hydrothermal mineralization. Geochim Cosmochim Acta 70:4215–4228
Marschner H (1995) Mineral Nutrition of Higher Plants. 2nd edition. Academic Press, London
Mason TFD, Weiss DJ, Horstwood M, Parrish RR, Russell SS, Mullane E, Coles BJ (2004a) High-precision Cu and Zn isotope analysis by plasma source mass spectrometry. Part 1. Spectral interferences and their correction. J Anal At Spectrom 19:209–217
Mason TFD, Weiss DJ, Horstwood M, Parrish RR, Russell SS, Mullane E, Coles BJ (2004b) High-precision Cu and Zn isotope analysis by plasma source mass spectrometry. Part 2. Correcting for mass discrimination effects. J Anal At Spectrom 19:218–226
Mason TFD, Weiss DJ, Chapman JB, Wilkinson JJ, Tessalina SG, Spiro B, Horstwood MSA, Spratt J, Coles BJ (2005) Zn and Cu isotopic variability in the Alexandrinka volcanic-hosted massive sulphide (VHMS) ore deposit, Urals, Russia. Chem Geol 221:170–187
Mathur R, Ruiz J, Titley S, Liermann L, Buss H, Brantley S (2005) Cu isotopic fractionation in the supergene environment with and without bacteria. Geochim Cosmochim Acta 69:5233-5246
Mathur R, Titley S, Barra F, Brantley S, Wilson M, Phillips A, Munizaga F, Maksaev V, Vervoort J, Hart G (2009) Exploration potential of Cu isotope fractionation in porphyry copper deposits. J Geochem Explor 102:1–6
Mathur R, Dendas M, Titley S, Phillips A (2010) Patterns in the copper isotope composition of minerals in porphyry copper deposits in southwestern United States. Econ Geol 105:1457–1467

Mathur R, Jin L, Prush V, Paul J, Ebersole C, Fornadel A, Williams JZ, Brantley S (2012) Cu isotopes and concentrations during weathering of black shale of the Marcellus Formation, Huntingdon County, Pennsylvania (USA). Chem Geol 304:175–184

Mattielli N, Petit JC, Deboudt K, Flament P, Perdrix E, Taillez A, Rimetz-Planchon J, Weis D (2009) Zn isotope study of atmospheric emissions and dry depositions within a 5 km radius of a Pb–Zn refinery. Atmos Environ 43:1265–1272

McBride MB (1981) Forms and distribution of copper in solid and solution phases of soil. *In*: Copper in Soils and Plants. Loneragan J, Robson AD, Graham RD, (eds) Academic Press, NY, p 25–45

McDonough WF (2003) Compositional model for the Earth's core. *In*: Treatise on Geochemistry. Vol 2. Holland HD, Turekian KK (eds) p 547–568

Meyer BS, Krishan TD, Clayton DD (1998) Theory of quasi-equilibrium nucleosynthesis and applications to matter expanding from high temperature and density. Astrophys J 498:808–830

Mhin BJ, Lee S, Cho SJ, Lee K, Kim KS (1992) $Zn(H_2O)_6^{2+}$ is very stable among aqua-Zn(II) ions. Chem Phys Lett 197:77–80

Moffett JW, Brand LE (1996) The production of strong, extracellular Cu chelators by marine cyanobacteria in response to Cu stress. Limnol Oceanogr 41:288–293

Mott lMJ, Wheat CG, Fryer P, Gharib J, Martin JB (2004) Chemistry of springs across the Mariana forearc shows progressive devolatilization of the subducting plate. Geochim Cosmochim Acta 68:4915–4933

Mountain BW, Seward TM (1999) The hydrosulphide/sulphide complexes of copper(I): experimental determination of stoichiometry and stability at 22c and reassessment of high temperature data. Geochim Cosmochim Acta 63:11–29

Moynier F, Le Borgne M (2015) High precision zinc isotopic measurements applied to mouse organs. J Visualized Exp:JoVE:e52479-e52479, doi:10.3791/52479

Moynier F, Fegley B (2015) The Earth's building blocks. *In*: The Early Earth: Accretion and Differentiation. Vol 212. Badro J, Walter MJ, (eds). Wiley, New York, p 27–48

Moynier F, Albarède F, Herzog G (2006) Isotopic composition of zinc, copper, and iron in lunar samples. Geochim Cosmochim Acta 70:6103–6117

Moynier F, Dauphas N, Podosek FA (2009a) A Search for ^{70}Zn Anomalies in Meteorites. Astrophys J 700:L92-L95

Moynier F, Pichat S, Pons ML, Fike D, Balter V, Albarède F (2009b) Isotopic fractionation and transport mechanisms of Zn in plants. Chem Geol 267:125–130, doi:10.1016/j.chemgeo.2008.09.017

Moynier F, Beck P, Jourdan F, Yin Q-Z, Reimold U, Koeberl C (2009c) Isotopic fractionation of zinc in tektites. Earth Planet Sci Lett 277:482–489

Moynier F, Koeberl C, Beck P, Jourdan F, Telouk P (2010a) Isotopic fractionation of Cu in tektites. Geochim Cosmochim Acta 74:799–807, doi:10.1016/j.gca.2009.10.012

Moynier F, Beck P, Yin Q, Ferroir T, Barrat JA, Paniello RC, Telouk P, Gillet P (2010b) Volatilization induced by impacts recorded in Zn isotope composition of ureilites. Chem Geol 276:374–379

Moynier F, Paniello RC, Gounelle M, Albarède F, Beck P, Podosek F, Zanda B (2011) Nature of volatile depletion and genetic relationships in enstatite chondrites and aubrites inferred from Zn isotopes. Geochim Cosmochim Acta 75:297–307, doi:10.1016/J.Gca.2010.09.022

Moynier F, Fujii T, Shaw A, Le Borgne M (2013a) Heterogeneous distribution of natural zinc isotopes in mice. Metallomics 5:693–699

Moynier F, Fujii T, Wang K, Foriel J (2013b) Ab initio calculations of the Fe(II) and Fe(III) isotopic effects in citrates, nicotianamine, and phytosiderophore, and new Fe isotopic measurements in higher plants. C R Geosci 345:230–240, doi:10.1016/j.crte.2013.05.003

Nägler TF, Anbar AD, Archer C, Goldberg G, Gordon GW, Greber ND, Siebert C, Sohrin Y, Vance, D (2014) Proposal for an international molybdenum isotope measurement standard and data representation. Geostand Geoanal Res 38, 149–151

Navarette JU, Borrok DM, Viveros M, Ellzey JT (2011) Copper isotope fractionation during surface adsorption and intracellular incorporation by bacteria. Geochim Cosmochim Acta 75:784–799

Ochoa Gonzalez, R, Weiss, D (2015) Zinc isotope variability in three coal-fired power plants: a predictive model for determining isotopic fractionation during combustion. Environ Sci Technol 49:12560–12567

O'Neill H (1991) The origin of the Moon and the early history of the Earth—A chemical model. Part 2: The Earth. Geochim Cosmochim Acta 55:1159–1172

Pacyna JM, Pacyna EG (2001) An assessment of global and regional emissions of trace metals to the atmosphere from anthropogenic sources worldwide. Environ Rev 9:269–288

Palme H, O'Neill H (2003) Cosmochemical estimates of mantle composition. *In*: Treatise on Geochemistry. Vol 2. Holland HD, Turekian KK (eds). Elsevier, Amsterdam, p 1–38

Paniello RC, Day JM, Moynier F (2012a) Zinc isotopic evidence for the origin of the Moon. Nature 490:376–379, doi:10.1038/naturc11507

Paniello RC, Moynier F, Beck P, Barrat JA, Podosek FA, Pichat S (2012b) Zinc isotopes in HEDs: Clues to the formation of 4-Vesta, and the unique composition of Pecora Escarpment 82502. Geochim Cosmochim Acta 86:76–87, doi:10.1016/J.Gca.2012.01.045

Pasquarello A, Petri I, Salmon PS, Parisel O, Car R, Toth E, Powell DH, Fischer HE, Helm L, Merbach AE (2001) First solvation shell of the Cu(II) aqua ion: evidence for fivefold coordination. Science 291:856–859

Peacock CL, Sherman DM (2004) Copper (II) sorption onto goethite, hematite and lepidocrocite: a surface complexation model based on ab initio molecular geometries and EXAFS spectroscopy. Geochim Cosmochim Acta 68:2623–2637

Peel K, Weiss D, Chapman J, Arnold T, Coles B (2008) A simple combined sample-standard bracketing and inter-element correction procedure for accurate mass bias correction and precise Zn and Cu isotope ratio measurements. J Anal At Spectrom 23:103–110, doi:10.1039/b710977f

Peel K, Weiss D, Sigg L (2009) Zinc isotope composition of settling particles as a proxy for biogeochemical processes in lakes: insights from the eutrophic Lake Greifen, Switzerland. Limnol Oceanogr Methods 54:1699–1708

Peers G, Price NM (2006) Copper-containing plastocyanin used for electron transport by an oceanic diatom. Nature 441:341–344

Petit JC, Schäfer J, Coynel A, Blanc G, Deycard VN, Derriennic H, Lanceleur L, Dutruch L, Cossy C, Mattielli N (2013) Anthropogenic sources and biogeochemical reactivity of particulate and dissolved Cu isotopes in the turbidity gradient of the Garonne River (France). Chem Geol 359:125–135

Pichat S, Douchet C, Albarède F (2003) Zinc isotope variations in deep-sea carbonates from the eastern equatorial Pacific over the last 175 ka. Earth Planet Sci Lett 210:167–178

Pogge von Strandmann PAE, Forshaw J, Schmidt DN (2014) Modern and Cenozoic records of seawater magnesium from foraminiferal Mg isotopes. Biogeosciences 11:5155–5168

Pokrovsky OS, Viers J, Freydier R (2005) Zinc stable isotope fractionation during adsorption on oxides and hydroxides. J Colloid Interface Sci 291:192–200

Pokrovsky OS, Viers J, Emnova EE, Kompantseva EI, Freydier R (2008) Copper isotope fractionation during its interaction with soil and aquatic microorganisms and metal oxy(hyd)oxides: Possible structural control. Geochim Cosmochim Acta 72:1742–1757

Pons ML, Quitte G, Fujii T, Rosing MT, Reynard B, Moynier F, Douchet C, Albarède F (2011) Early Archean serpentine mud volcanoes at Isua, Greenland, as a niche for early life. PNAS 108:17639–17643, doi:10.1073/Pnas.1108061108

Pons M-L, Fujii T, Rosing M, Quitté G, Télouk P, Albarède F (2013) A Zn isotope perspective on the rise of continents. Geobiol 11:201–214

Ponzevera E, Quetel CR, Berglund M, Taylor PDP, Evans P, Loss RD, Fortunato G (2006) Mass discrimination during MC–ICP–MS isotopic ratio measurements: Investigation by means of synthetic isotopic mixtures (IRMM-007 series) and application to the calibration of natural-like zinc materials (including IRMM-3702 and IRMM-651). J Am Soc Mass Spectr 17:1412–1427, doi:10.1016/j.jasms.2006.06.001

Powell KJ, Brown PL, Byrne RH, Gajda T, Hefter G, Sjoberg S, Wanner H (2007) Chemical speciation of environmentally significant metals with inorganic ligands—Part 2: The Cu^{2+}–OH^-, Cl^-, CO_3^{2-}, SO_4^{2-}, and PO_4^{3-} systems. Pure Appl Chem 79:895–950

Rauch, JN, Pacyna, JM (2009) Earth's global Ag, Al, Cr, Cu, Fe, Ni, Pb, and Zn cycles. Glob Biogeochem Cycles 23: GB2001, 10.1029/2008GB003376

Resing JA, Sedwick PN, German CR, Jenkins WJ, Moffett JW, Sohst BM, Tagliabue A (2015) Basin scale transport of hydrothermal dissolved metals across the South Pacific Ocean. Nature 523:200–203

Rosman KJR (1972) A survey of the isotopic and elemental abundances of zinc. Geochim Cosmochim Acta 36:801–819

Rudge JF, Reynolds BC, Bourdon B (2009) The double spike toolbox. Chem Geol 265:420–431

Russel SS, Zhu X, Guo Y, Belshaw N, Gounelle M, Mullane E (2003) Copper isotope systematics in CR, CH-like, and CB meteorites: a preliminary study (abstr.). Meteorit Planet Sci 38

Ryan BM, Kirby JK, Degryse F, Harris H, McLaughlin MJ, Scheiderich K (2013) Copper speciation and isotopic fractionation in plants: uptake and translocation mechanisms. New Phytol 199:367–378

Ryan BM, Kirby JK, Degryse F, Scheiderich K, McLaughlin MJ (2014) Copper isotope fractionation during equilibration with natural and synthetic ligands. Environ Sci Tech 48:862–866

Sancenon V, Puig S, Mateu-Andrés I, Dorcey E, Thiele D, Penarrubia L (2004) The *Arabidopsis* copper transporter COPT1 functions in root elongation and pollen development. J Biol Chem 269:15348–15355

Savage P, Boyet M, Moynier F (2014) Zinc isotope anomalies in bulk chondrites. 77th Meteoritical Soc Meet, Casablanca, Morocco, p 5246

Savage P, Moynier F, Harvey J, Burton K (2015a) The behavior of copper isotopes during igneous processes. AGU conference, San Francisco

Savage P, Moynier F, Chen H, Shofner G, Siebert J, Badro J, Puchtel IS (2015b) Copper isotope evidence for large-scale sulphide fractionation during Earth's differentiation. Geochem Persp Lett 1:53–64

Schauble EA (2004) Applying stable isotope fractionation theory to new systems. Rev Mineral Geochem 55:65–111

Schiller M, Paton C, Bizzarro M (2015) Evidence for nucleosynthetic enrichment of the protosolar molecular cloud core by multiple supernova events. Geochim Cosmochim Acta 149:88–102, doi:10.1016/j.gca.2014.11.005

Schonbachler M, Carlson RW, Horan MF, Mock TD, Hauri EH (2010) Heterogeneous accretion and the moderately volatile element budget of Earth. Science 328:884–887, doi:10.1126/science.1186239

Schulz MS, Bullen TD, White AF, Fitzpatrick JF (2010) Evidence of iron isotope fractionation due to biologic lifting in a soil chronosequence. Geochim Cosmochim Acta 74:A927

Seo JH, Lee SK, Lee I (2007) Quantum chemical calculations of equilibrium copper(I) isotope fractionations in ore-forming fluids. Chem Geol 243:225–237
Shank CG, Ckrabal SA, Whitehead RF, G. BA, Keiber RJ (2004) River discharge of strong Cu-binding ligands to South Atlantic Bight Waters. Mar Chem 88:41–51
Sherman DM (2001) Quantum chemistry and classical simulations of metal complexes in aqueous solutions. Rev Mineral Geochem 42:273–317
Sherman DM (2013) Equilibrium isotopic fractionation of copper during oxidation/reduction, aqueous complexation and ore-forming processes: predictions from hybrid density functional theory. Geochim Cosmochim Acta 118:85–97
Sherman DM, Little SH, Vance D (2015) Reply to comment on "Molecular controls on Cu and Zn isotopic fractionation in Fe–Mn crusts". Earth Planet Sci Lett 411:313–315
Shields WR, Murphy TJ, Garner EL (1964) Absolute isotopic abundance ratio and the atomic weight of a reference sample of copper. J Res NBS 68A:589–592
Shields WR, Goldich SS, Garner EL, Murphy TJ (1965) Natural variations in the abundance ratio and the atomic weight of copper. J Geophys Res:479–491
Shiller AM, Boyle EA (1985) Dissolved zinc in rivers. Nature 317:49–52
Siebert J, Corgne A, Ryerson FJ (2011) Systematics of metal–silicate partitioning for many siderophile elements applied to Earth's core formation. Geochim Cosmochim Acta 75:1451–1489
Singha Deb AK, Ali SkM, Shenoy KT, Ghosh SK (2014) Nano cavity induced isotope separation of zinc: density functional theoretical modeling. J Chem Eng Data, 59:2472–2484
Sinoir M, Butler ECV, Bowie AR, Mongin M, Nesterenko PN, Hassler CS (2012) Zinc marine biogeochemistry in seawater: a review. Mar Freshwater Res 63:644–657
Sivry Y, Riotte J, Sonke JE, Audry S, Schäfer J, Vuers J, Blanc G, Freydier R, Dupré B (2008) Zn isotopes as tracers of anthropogenic pollution from Zn ore-smelters. The Riou Mort-Lot River system. Chem Geol 255:295–304
Sold D, Behra R (2000) Long-term effects of copper on the structure of freshwater periphyton communities and their tolerance to copper, zinc, nickel and silver. Aquatic Toxicol 47:181–189
Sonke JE, Sivry Y, Viers J, Fréydier R, Dejonghe L, André L, Aggarwal JK, Fontan F, Dupré B (2008) Historical variations in the isotopic composition of atmospheric zinc deposition from a zinc smelter. Chem Geol 252:145–157
Sossi PA, Halverson GP, Nebel O, Eggins SM (2015) Combined separation of Cu, Fe and Zn from rock matrices and improved analytical protocols for stable isotope determination. Geostand Geoanaly Res 39:129–149, doi:10.1111/j.1751-908X.2014.00298.x
Steele RCJ, Coath CD, Regelous M, Russell S, Elliott T (2012) Neutron-poor nickel isotope anomalies in meteorites. Astrophys J 758(59) doi:10.1088/0004-637X/758/1/59
Sunda WG, Huntsman SA (1992) Feedback interactions between zinc and phytoplankton in seawater. Limnol Oceanogr Methods 37:25–40
Szynkiewicz A, Borrok DB (2016) Isotope variations of dissolved Zn in the Rio Grande watershed, USA: The role of adsorption on Zn isotope composition. Earth Planet Sci Lett 433:293–302
Takano S, Tanimizu M, Hirata T, Sohrin Y (2014) Isotopic constraints on biogeochemical cycling of copper in the ocean. Nat Commun 5,doi: 10.1038/ncomms6663
Tang Y-T, Cloquet C, Sterckman T, Echevarria G, Carignan J, Qiu R-L, Morel J-L (2012) Fractionation of stable zinc isotopes in the field-grown zinc hyperaccumulator *Noccaea caerulescens* and the zinc-tolerant plant *Silene vulgaris*. Environ Sci Technol 46:9972–9979
Tanimizu M, Asada Y, Hirata T (2002) Absolute isotopic composition and atomic weight of commercial zinc using inductively coupled plasma mass spectrometry. Anal Chem 74:5814–5819
Telouk P, Puisieux A, Fujii T, Balter V, Bondanese VP, Morel A-P, G. C, Lamboux A, Albarède F (2015) Copper isotope effect in serum of cancer patients. A pilot study. Metallomics 7:299–308
Telus M, Dauphas N, Moynier F, Tissot F, Teng FZ, Nebelek PI, Craddock PR, Groat LR (2012) Iron, zinc, magnesium, and uranium isotopic fractionation during continental crust differentiation : The tale from migmatites, granitoids and pegmatites. Geochim Cosmochim Acta 97:247–265
Thapalia A, Borrok DM, van Metre PC, Musgrove M, Landa ER (2010) Zn and Cu isotopes as tracers of anthropogenic contamination in a sediment core from an urban lake. Environ Sci Technol 44:1544–1550
Thomson CM, Ellwood MJ (2014) Dissolved copper isotope geochemistry in the Tasman Sea, SW Pacific Ocean. Mar Chem 165:1–9
Trinquier A, Elliott T, Ulfbeck D, Coath C, Krot AN, Bizzarro M (2009) Origin of nucleosynthetic isotope heterogeneity in the solar protoplanetary disk. Science 324:374–376
Twining BS, Baines SB (2013) The trace metal composition of marine phytoplankton. Ann Rev Earth Planet Sci 5:191–215
Vance D, Archer C, Bermin J, Perkins J, Statham PJ, Lohan MC, Ellwood MJ, Mills RA (2008) The copper isotope geochemistry of rivers and the oceans. Earth Planet Sci Lett 274:204–213, doi:10.1016/j.epsl.2008.07.026
Vance D, Teagle DAH, Foster GL (2009) Variable Quaternary chemical weathering fluxes and imbalances in marine geochemical budgets. Nature 458:493–496

Vance D, Matthews A, Keech A, Archer C, Hudson G, Pett-Ridge J, Chadwick OA (2016a) The behaviour of Cu and Zn isotopes during soil development: controls on the dissolved load of rivers. Chem Geol 445:36–53, http://dx.doi.org/10.1016/j.chemgeo.2016.06.002
Vance D, Little SH, Archer C, Cameron V, Andersen M, Rijkenberg MJA, Lyons TW(2016b) The oceanic budgets of nickel and zinc isotopes: the importance of sulphidic environments as illustrated by the Black Sea. Phil Trans R Soc A 374, DOI: 10.1098/rsta.2015.0294
Viers J, Oliva P, Nonelle A, Gelabert A, Sonke J, Freydier R, Gainville R, Dupre B (2007) Evidence of Zn isotopic frationation in a soil-plant system of a pristine tropical watershed (Nsimi, Cameroon). Chem Geol 239:124–137
Viers, J, Prokushkin, AS, Pokrovsky, OS, Kirdyanov, AV, Zouiten, C, Chmeleff, J, Meheut, M, Chabaux, F, Oliva, P, Dupré, B (2015) Zn isotope fractionation in a pristine larch forest on permafrost-dominated soils in Central Siberia. Geochem Trans 16:10.1186/s12932-015-0018-0
Völkening J, Papanastassiou DA (1990) Zinc isotope anomalies. Astrophys J 358:L29-L32
Wallerstein G, Iben Jr I, Parker P, Boesgaard AM, Hale GM, Champagne AE, Barnes CA, Kaeppeler F, Smith VV, Hoffman RD, Timmes FX, Sneden C, Boyd RN, Meyer BS, Lambert DL (1997) Synthesis of the elements in stars: forty years of progress. Rev Mod Phys 69:995–1084
Wang P, Zhou DM, Luo XS, Li LZ (2009) Effects of Zn-complexes on zinc uptake by wheat (*Triticum aestivum*) roots: a comprehensive consideration of physical, chemical and biological processes on biouptake. Plant Soil 316:177–192
Wanty, RB, Ballistrieri, LS, Wesner, JS, Walsters, DM, Schmidt, TS, Podda, F, De Giudici, G, Stricker, CA, Kraus, J, Lattanzi, P, Wolf, RE, Cidu, R (2015) Isotopic insights into biological regulation of zinc in contaminated systems. Proc Earth Planet Sci 13:60–63
Wanty, RB, Podda, F, De Giudici, G, Cidu, R, Lattanzi, P (2013) Zinc isotope and transition-element dynamics accompanying hydrozincite biomineralization in the Rio Naracauli, Sardinia, Italy. Chem Geol 337–338:1–10
Wasson JT, Wang JM (1986) A non magmatic origin of group-IIE iron-meteorites. Geochim Cosmochim Acta 50:725–732
Wasylenki LE, Weeks CL, Bargar JR, Spiro TG, Hein JR, Anbar AD (2011) The molecular mechanism of Mo isotope fractionation during adsorption to birnessite. Geochim Cosmochim Acta 75:5019–5031
Weinstein C, Moynier F, Wang K, Paniello R, Foriel J, Catalano J, Foriel J (2011) Cu isotopic fractionation in plants. Chem Geol 286:266–271
Weiss DJ, Mason TFD, Zhao FJ, Kirk GJD, Coles BJ, Horstwood MSA (2005) Isotopic discrimination of zinc in higher plants. New Phytologist 165:703–710
Weiss DJ, Rausch N, Mason TFD, Coles BJ, Wilkinson JJ, Ukonmaanaho L, Arnold T, Nieminen TM (2007) Atmospheric deposition and isotope biogeochemistry of zinc in ombrotrophic peat. Geochim Cosmochim Acta 71:3498–3517
Weiss DJ, Boye K, Caldelas C, Fendorf S (2014) Zinc isotope fractionation during early dissolution of biotite granite. Soil Sci Amer 78:171–189
Wells ML, Kozelka PB, Bruland KW (1998) The complexation of "dissolved" Cu, Zn, Cd and Pb by soluble and colloidal organic matter in Narragansett Bay, RI. Mar Chem 62:203–217
Wiederhold JG, Teutsch N, Kraemer SM, Halliday AN, Kretzschmar R (2007) Iron isotope fractionation in oxic soils by mineral weathering and podzolization. Geochim Cosmochim Acta 71:5821–5833, doi:10.1016/J.Gca.2007.07.023
Wiederhold J (2015) Metal stable isotope signatures as tracers in environmental geochemistry. Environ Sci Technol 49:2606–2624
Wilkinson JJ, Weiss DJ, Mason TFD, Coles BJ (2005) Zinc isotope variation in hydrothermal systems: preliminary evidence from the Irish Midlands ore field. Econ Geol 110:583–590
Williams HM, Archer C (2011) Copper stable isotopes as tracers of metal-sulphide segregation and fractional crystallisation processes on iron meteorite parent bodies. Geochim Cosmochim Acta 75:3166–3178
Yin, N-H, Sivry, Y, Benedetti, MF, Lens, PNL, van Hullebusch, ED (2016) Application of Zn isotopes in environmental impact assessment of Zn-Pb metallurgical industries: a mini review. App Geochem 64:128–135
Yruela I (2009) Copper in plants: acquisition, transport and interactions. Funct Plant Biol 26:409–430
Zhao Y, Vance D, Abouchami W, de Baar HJW (2014) Biogeochemical cycling of zinc and its isotopes in the Southern Ocean. Geochim Cosmochim Acta 125:653–672, doi:10.1016/j.gca.2013.07.045
Zhu XK, O'Nions RK, Guo Y, Belshaw NS, Rickard D (2000) Determination of Cu-isotope variation by plasma souce mass spectrometry: implications for use as geochemical tracers. Chem Geol 163:139–149
Zhu XK, Guo Y, Williams RJ, O'Nions RK, Matthews A, Belshaw NS, Canters GW, De Waal EC, Weser U, Burgess BK, Salvato B (2002) Mass fractionation processes of transition metal isotopes. Earth Planet Sci Lett 200:47–62
Zirino A, Yamamoto S (1972) A pH-dependent model for the chemical speciation of copper zinc cadmium, and lead in seawater. Limnol Oceanogr 17:661–671

Reviews in Mineralogy & Geochemistry
Vol. 82 pp. 601-656, 2017

Germanium Isotope Geochemistry

Olivier J. Rouxel

Department of Physical Resources and Deep-Sea Ecosystems
IFREMER, Centre de Brest
Plouzané, 29280
France

Department of Oceanography
University of Hawaii at Mānoa
Honolulu, HI 96822
USA

orouxel@hawaii.edu

Béatrice Luais

Centre de Recherches Pétrographiques et Géochimiques—CRPG
CNRS-UMR 7358, Université de Lorraine
Vandoeuvre-lès-Nancy, 54501
France

luais@crpg.cnrs-nancy.fr

INTRODUCTION

Germanium (Ge) is a trace element in the Earth's crust and natural waters, averaging about 1.6 ppm in rocks and minerals (El Wardani 1957; Bernstein 1985) and 75 picomol/L in seawater (Froelich and Andreae 1981). The naturally occurring oxidation states of Ge are +2 and +4, with the +4 state forming the principal common and stable compounds. Germanium has outer electronic structure $3d^{10}\,4s^2\,4p^2$ and mainly occurs in the quadrivalent state, although in some minerals it is octahedrally coordinated. Germanium is chemically similar to silicon (Si), both belonging to the IVA group in the periodic table, with Ge immediately above Si. Germanium is classified as a semimetal, whereas Si is a nonmetal element. Because of nearly identical ionic radii and electron configurations for Ge and Si, the crustal geochemistry of Ge is dominated by a tendency to replace Si in the lattice sites of minerals (Goldschmidt 1958; De Argollo and Schilling 1978b). These two elements exist in seawater as similar hydroxyacids, i.e., $Ge(OH)_4$ and $Si(OH)_4$ (Pokrovski and Schott 1998a) and the concentration profile of Ge is similar to that of Si (Froelich and Andreae 1981), thus making Ge/Si ratio an interesting tracer for biogenic silica cycling in the ocean. Although Ge and Si are geochemically similar, their behavior is different enough so that decoupling of Ge and Si can occur. Germanium commonly occurs in 4-fold (tetrahedral) coordination but in contrast to Si, Ge has a stronger tendency for the 6-fold coordination. Unlike Si, Ge also forms methylated compounds, and high concentrations of monomethyl- and dimethyl-germanium have been detected in ocean waters, accounting for > 70% of the total Ge (Lewis et al. 1985).

Germanium is a particularly interesting element for geochemists since it exhibits siderophile, lithophile, chalcophile and organophile behaviors in different geologic environments (Bernstein 1985). The siderophile behavior of Ge is well reflected by the fact that it can achieve concentrations of up to 2000 ppm in iron meteorites (Wasson 1966; Wai and Wasson 1979) whereas its lithophile

1529-6466/17/0082-0014$10.00 http://dx.doi.org/10.2138/rmg.2017.82.14

behavior is indicated by the strong geochemical coupling between Si and Ge during partial melting and fractional crystallization (De Argollo and Schilling 1978b). The chalcophile and organophile behaviors of Ge are also well demonstrated by the strong enrichment of Ge in some sphalerite-rich sulfide and coal deposits (Bernstein 1985; Holl et al. 2007; Belissont et al. 2014; Frenzel et al. 2014). In particular, Ge has one of the highest affinities for organic matter of all the elements commonly associated with coal and lignite (Valkovic 1983).

Germanium has five naturally occurring isotopes ^{70}Ge, ^{72}Ge, ^{73}Ge, ^{74}Ge, and ^{76}Ge with relative abundance of 21.2, 27.7, 7.7, 35.9 and 7.5% respectively (Green et al. 1986; Rosman and Taylor 1998; Chang et al. 1999). The possibility of significant Ge isotope fractionation in chemical reactions between various Ge compounds was recognized long ago by Brown and Krouse (1964). Despite the interest, there are less than a dozen of published studies reporting natural variations of Ge isotopes thus far. Early investigations of Ge isotopic compositions using thermal ionization mass spectrometry (TIMS) were limited to an uncertainty of several per mil (Shima 1963; Green et al. 1986) and no significant variations in Ge isotope composition were found. It was only with the advent of multiple collector magnetic sector inductively coupled plasma mass spectrometry (MC-ICP-MS) (Halliday et al. 1995) that progress was made for high precision measurement of Ge isotopic compositions. In a reconnaissance study, Hirata (1997), Xue et al. (1997) and Luais et al. (2000) analyzed the Ge isotopic composition of meteoritic materials and identified the first direct evidence for mass-dependent fractionation of Ge isotopes. Although further analytical developments were made to evaluate inherent instrumental artefacts in the analysis of Ge isotopes (Galy et al. 2003) and their use in cosmochemistry (Luais 2007), the natural variations of Ge isotopes on Earth remained unknown, mainly due to the lack of suitable analytical techniques to analyze silicate matrices and sub-microgram quantities of Ge. Subsequently, new techniques were developed to achieve precise determination of Ge isotopes in geological and aqueous matrices, including silicates and geothermal fluids (Rouxel et al. 2006; Siebert et al. 2006; Escoube et al. 2012; Luais 2012; Meng et al. 2015).

Using these new analytical capabilities, studies have reported Ge isotope variations in low-temperature Earth surface environments (Rouxel et al. 2006; Siebert et al. 2006; Escoube et al. 2015), coal (Qi et al. 2011) and ore deposits (Escoube et al. 2012, 2015; Belissont et al. 2014; Meng et al. 2015), as well as crustal rocks and meteorites (Rouxel et al. 2006; Luais 2007, 2012; Escoube et al. 2012). These studies, together with experimental (Pokrovsky et al. 2014) and theoretical (Li et al. 2009) determination of Ge isotope fractionation factors demonstrate the great potential of Ge isotopes to trace both high- and low-temperature geochemical processes. Although potential mechanisms for fractionation are still poorly kown, Ge isotopes appear particularly sensitive to complexing process with organic matter or sulfide compounds, biological uptake and equilibrium/kinetic isotope effects between fluid and minerals. This chapter summarizes what has been learned thus far in both high- and low-temperature geochemistry of Ge. It also presents some geochemical background on the application of Ge/Si ratios as geochemical tracers, which provide key insights into the isotopic systematics of Ge.

METHODS

Early methods to measure Ge isotope ratios

Early studies using thermal ionisation mass spectrometry (TIMS) were limited by the high ionization potential of Ge (7.899 eV) and a relatively poor precision of several per mil (at one standard deviation, 1 s.d.) precluding the resolution of natural variations of Ge isotope abundances (Reynolds 1953; Shima 1963). Later, Green et al. (1986) improved the TIMS technique using a double spike approach and improved the precision down to 0.5 per mil per mass unit. An ion-enhancing gel loading technique was also developed to enable microgram quantities of Ge to be analyzed. Within the limits of experimental error, no variations in Ge isotopic abundances were found in any of the reagents or minerals analyzed. The analytical capability of TIMS was not further improved, even in very recent studies (Gautier et al. 2012). In rare cases, gas source mass spectrometry (Graham et al. 1951; Kipphardt et al. 1999) and secondary ion mass spectrometry (Nishimura et al. 1988; Onishi et al. 2006) were also employed. Graham et al. (1951) reported the abundances of the Ge isotopes in germanium tetrachloride and tetrafluoride prepared from mineral samples of different geological and geographical origins. Natural variations of several per mil were tentatively found for the $^{76}Ge/^{70}Ge$ ratio. Nearly half a century later, Kipphardt et al. (1999) also used gas isotope mass spectrometry after Ge fluorination and determined the isotopic abundance of Ge in several terrestrial samples. Although no natural variations were reported, this study refined the atomic weight of Ge. Secondary ion mass spectrometry was also used in several studies (Nishimura et al. 1988; Onishi et al. 2006) but not widely applied.

Inductively Coupled Plasma Mass Spectrometry (ICPMS) has further improved the measurement of Ge isotopes over TIMS due to the increased ionisation efficiency of the plasma source. Xue et al. (1997) used Quadrupole ICPMS and found evidences for large variations of Ge isotope ratios during the evaporative loss from oxide rims of Canyon Diabolo spheroids. This technique involved the addition of Ga as internal standard to correct for drifts in instrumental mass bias. However the uncertainty was still too high to resolve sub per mil isotope variations. Multi Collector (MC) ICPMS was subsequently employed by Hirata (1997) and Luais et al. (2000). Instrumental mass bias was monitored using Ga-external correction technique while isobaric interferences were mitigated by purifying Ge from the sample matrix before isotope analysis. Variations in Ge isotopic ratios of three chemical reagents and germanite samples were found to be about 0.1–0.3‰ for $^{70}Ge/^{73}Ge$, whereas isotopic variations for meteoritic samples were about 4‰. An important result of this study was the demonstration that $^{70}Ge/^{73}Ge$ vs. $^{72}Ge/^{73}Ge$ ratios of meteorites and terrestrial samples have mass-dependent relationships (Luais et al. 2000; Luais 2003). Revisiting the measurement of Ge isotopes by MC-ICPMS, Galy et al. (2003) noted that the $^{74}Ge/^{70}Ge$ normalized to $^{72}Ge/^{70}Ge$ of Hirata (1997) was higher by 2‰ than the $^{74}Ge/^{70}Ge$ of Green et al. (1986) (Table 1). The isotope ratio repeatability of a mono-elemental Ge solution was better than 0.06‰ per mass unit at 95% confidence, confirming the superiority of MC-ICPMS over previously used mass spectrometry techniques. Galy et al. (2003) also demonstrated that the addition of Na and K to a standard solution of Ge having a known isotopic composition induces up to 4.4‰ decrease of $^{74}Ge/^{70}Ge$ ratio. This chemical bias is a result of a mass-dependent process and prevents the measurement of natural samples without chemical separation.

State of the art analytical methods

Although significant differences in analytical techniques exist between different research groups (Rouxel et al. 2006; Siebert et al. 2006; Luais 2007), state-of-the-art measurements of Ge isotopes generally involve: (1) chemical purification of Ge through anion-exchange and/or cation-exchange chromatographic columns; (2) hydride generation (HG) and gaseous Ge hydride introduction into the plasma torch; (3) correction of instrumental mass fractionation using either standard-sample bracketing or double spike approaches. The conjunction of HG technique with MC-ICPMS has been previously applied to other stable isotope systems such as Se and Sb isotopes (Rouxel et al. 2002, 2003a) and allowed high-precision isotope ratios for 10- to 100-fold decrease in analyte quantities. Luais et al. (2000) and Luais (2003) also provided high precision measurements of Ge isotopes in iron meteorites (2s.d. less than 0.25‰) using conventional sample introduction system and Ga correction. Rouxel et al. (2006) reported the isotopic results in the same manner as proposed by Galy et al. (2003) and Luais (2003) following the so called "standard-sample bracketing technique" which involves the measurement of the Ge standard solution, before and after each sample. This approach enabled the measurement of some of the first terrestrial Ge isotope ratios, using sample sizes as small as 15 ng and with a

Table 1. Compilation of the natural abundances of Ge isotopes determined by different mass spectrometry techniques. The Ge isotope composition of double spike mixtures used is also shown for comparison

Ge material	$^{74}Ge/^{70}Ge$	1 s.d.	$^{73}Ge/^{70}Ge$	1 s.d.	$^{72}Ge/^{70}Ge$	1 s.d.	Method	Instrument/Lab	Ref.
DS (Ge70:Ge73)	0.07614	0.00005	0.60707	0.00004	0.05626	0.00004	CSC, Ga	Neptune/IFREMER	(1)
NIST3120a	1.76094	0.00003	0.37335	0.00001	1.32901	0.00002	CSC, Ga	Neptune/IFREMER	(1)
NIST3120a	1.76137	0.00014	0.37447	0.00003	1.32934	0.00010	HG, Ga	Neptune/WHOI	(2)
NIST3120a	1.76426	0.00139	0.37484	0.00028	1.33068	0.00096	HG, Cu	Neptune/WHOI	(2)
NIST3120a	1.76738	0.00509	0.37494	0.00065	1.33145	0.00121	CSC, Ga	Isoprobe/CRPG	(2)
NIST3120a	1.76460	0.00021	0.37426	0.00004	1.33073	0.00009	CSC, Ga	Nu Plasma/LCABIE	(2)
NIST3120a	1.76223	0.00057	0.37411	0.00014	1.32979	0.00014	HG, Ga	Nu Plasma/LCABIE	(2)
Aldrich #01704KZ	1.76367	0.00283	0.37370	0.00151	1.33106	0.00070	CSC, Ga	Isoprobe /CRPG	(2)
In house	1.77000	0.00014	0.37520	0.00015	1.33300	0.00070	Ga	Nu Plasma/OSU	(3)
JMC #301230Sn	1.76641	0.00223	0.37499	0.01331	1.33199	0.00092	CSC, Ga	Isoprobe /CRPG	(2)
JMC #301230Sn	1.74337	–	0.37189	–	1.32360	–	Ga	VG–P54	(4)
Ge metal (Aldrich)	1.77937	–	0.37807	–	1.33712	–	CSC, Ga	Neptune	(5)
In house	1.69449	0.00598	0.36383	0.00090	1.30315	0.00276		TIMS	(6)
In house	1.74228	0.00370	0.37107	0.00067	1.31927	0.00491		TIMS- VG-354	(7)
In house	1.73038	–	0.37189	–	1.31833	–		TIMS	(8)
In house	1.78101	–	0.37821	–	1.33699	–		gas MS	(9)
In house	1.71516	–	0.36800	–	1.30938	–		SIMS	(10)

References: (1) Escoube et al. (2012); (2) this study; (3) Siebert et al. (2006); (4) Hirata (1997); (5) Yang and Meija (2010) (6) Green et al. (1986); (7) Chang et al. (1999); (8) Shima (1963); (9) Reynolds et al. (1953); (10) Nishimura et al. (1988)

Notes: SC: cyclonic spray chamber; HG: hydride generation; s.d.: standard deviation; DS (Ge70:Ge73): Ge double spike

precision of less than 0.2‰ on $^{74}Ge/^{70}Ge$ (Rouxel et al. 2006). This represented an important advance as, for the first time, it was possible to measure Ge isotope composition of a range of terrestrial samples, including igneous and sedimentary rocks as well as some marine clays and sponges. Rouxel et al. (2006) also provided a first estimate of the bulk silicate Earth Ge isotope ratio and seawater. In the same year, Siebert et al. (2006) used a double-spike isotope dilution MC-ICPMS technique for the determination of Ge isotope fractionation. Using this technique, they determined Ge isotope compositions of geothermal spring fluids and basalts. Although the double spike approach allowed correcting for instrumental mass bias and isotope fractionation during sample preparation, the external standard reproducibility on $^{74}Ge/^{72}Ge$ ratio was limited to 0.4‰ (2s.d.), corresponding to 4 times the precision reported by Rouxel et al. (2006). Further analytical developments of Ge isotopic measurements were also carried out by Luais (2007, 2012) for the measurement of iron meteorites, sulfides, and silicate rocks. Other groups adapted these available methods for the measurements of sulfide samples (Meng et al. 2015). Escoube et al. (2012) proposed an improved technique by using a double-spike approach, together with HG-MC-ICPMS and chromatographic separation. This study reported an interlaboratory comparison of Ge isotope composition of selected georeference materials and standard solutions. This allowed the definition of a new reference material NIST SRM 3120a to report natural Ge isotope variations. Finally, the most recent progress involved the development of preconcentration methods to measure Ge isotopes in seawater (Baronas et al. 2014; Guillermic et al. 2016).

Sample dissolution issues

Since Ge isotope measurements in rocks and minerals generally require a dissolution step to isolate Ge from the elements constituting the matrix, specific methods must be employed. In the case of sulfides (e.g., sphalerite, pyrite, chalcopyrite), Fe-oxyhydroxides and iron meteorites, the use of concentrated HNO_3 was sufficient to achieve a complete dissolution and quantitative recovery of Ge in solution. After sample digestion, the solutions were evaporated in open Teflon beaker at temperatures ranging from 120 °C (Meng et al. 2015), down to 80 °C (Escoube et al. 2015) or 60 °C (Luais 2007). In all cases, the use of HCl or $HClO_4$ has to be avoided as tests showed that Ge is lost at 85% with $HClO_4$ and at 100% with HCl even at medium temperature 80 °C (Luais 2012) due to the formation of volatile $GeCl_4$ species (Kaya and Volkan 2011). Dissolution of silicate rocks for Ge isotopic measurements is also complicated by the potential volatile behavior of Ge in the presence of hydrofluoric acid (HF) (Chapman et al. 1949), which is commonly used to digest silicate matrices. Previous studies have employed several strategies to address this issue. The method used by Rouxel et al. (2006) involved the dissolution of siliceous rocks in concentrated HF without the critical evaporation step. In short, the sample in powder form was first reacted with concentrated HNO_3 and taken to dryness on hot plate. Total digestion was then achieved using concentrated HF and the solution, along with precipitates, was diluted with water to obtain a solution of ~1 M HF. Insoluble fluorides, containing mostly Ca, Mg and Al and various trace elements (Yokoyama et al. 1999) were separated from the solution by centrifugation. The method used by Siebert et al. (2006) followed standard rock digestions techniques (i.e., mixture of concentrated HF and HNO_3) digestion with the modification of having added double spike to the sample before digestion and evaporation. Luais (2012) used the approach of Ishikawa and Nakamura (1990) for the purification of volatile boron, that consists of dissolving the sample powder in concentrated HF + HNO_3 on hot plate at 60 °C, until obtaining a solution free of any visible particles. The solution was then centrifuged, and the supernatant containing Ge was transferred to another Teflon beaker. Complete Ge extraction from the residual fluorides was achieved through three HF-leaching/centrifugation steps, and then evaporation to dryness at temperatures between 60 and 65 °C of all supernatant fractions. Luais (2012) evaluated in more detail the potential loss of Ge during HF digestion. It was found that Ge is quantitatively recovered after evaporation in HNO_3–HF acid mixture, suggesting that no Ge is lost as GeF_4 species at temperatures between 60 and 65 °C. This confirmed that higher temperatures (i.e., up to 300 °C) are needed to evaporate GeF_4 (Kwasnik 1963). Finally, considering the organophile behavior of Ge, Ge isotope analysis of organic-rich rocks, such as coal and black shales, should also involve the complete breakdown of organic matter. To address this issue, Qi et al. (2011) compared two techniques, one involving repeated evaporation of lignite samples with concentrated HNO_3, and the other involving sample powder ashing at 600 °C for 24 h in a muffle furnace. Both methods yielded similar Ge concentrations and isotope compositions, suggesting that Ge does not form volatile species upon ashing.

Chemical purification of samples

Before undertaking isotopic analysis by MC-ICPMS, it is customary to separate the analyte from (i) isobaric elements that can potentially interfere with the analysis and (ii) matrix elements that can affect the mass bias on the mass spectrometer and can form complex compounds that can also interfere with the element of interest. To achieve this goal, ion exchange chromatography provides the most versatile and convenient technique. Luais (2000, 2003, 2007) adapted the method of Xue et al. (1997) and used cation exchange resin for the separation of Ge from metallic and sulfide matrices in diluted HNO_3 medium. Sample solution in 0.5 mL of 0.5 M HNO_3 was loaded onto 2 mL of AG50W-X8 cationic resin. The extremely low partition coefficient for Ge (occurring as oxyanion) with 0.5 M HNO_3 allows the elution of Ge whereas all the matrix elements (occurring as cations) remain absorbed on the resin. Rouxel et al. (2006) reported a comprehensive chromatography method that is applicable to a range of geological samples and is summarized below:

(1) Samples are dissolved as described above such that a solution of Ge in a matrix of 1 M HF is obtained. The ion exchange resin AG1-X8 (Biorad, Hercules, Ca, USA) is loaded into polypropylene columns (1.8 mL wet volume) and conditioned with repeated elution with 1 M HF, 1.4 M HNO_3 and mMQ water. The sample in 1 M HF is then loaded on the column which allows the binding of Ge to the resin in the form of $(GeF_6)^{2-}$.

(2) Ge is eluted with diluted HNO_3 (between 1.4 M and 0.28 M) and the solution is dried down on hot plate and redissolved in 0.28 M HNO_3.

(3) Although Ge may be further purified through cation-exchange resin as described in Rouxel et al. (2006) and Luais (2000, 2003, 2007, 2012), it is possible to directly analyze the solution after step (2) only if the measurements by MC-ICPMS employ the hydride generation technique. However, because As may be partially eluted together with Ge and could form volatile hydrides, care should be taken to monitor potential isobaric interference from $^{75}AsH^+$ on $^{76}Ge^+$ when analyzed by MC-ICPMS.

Guillermic et al. (2016) recently reported a preconcentration method for the analysis of Ge isotope compositions of inorganic Ge in seawater. Germanium was co-precipitated with iron hydroxide with a yield better than ~70%. Germanium isotopic measurements were performed using a double-spike approach and a hydride generation system coupled to a MC-ICPMS modified from previous methods used for rock analysis (Rouxel et al. 2006; Escoube et al. 2012). An anion exchange resin was used to further purify Ge from Fe and remove potential matrix elements interfering with Ge hydride generation. Variations in $^{74}Ge/^{70}Ge$ ratios were obtained with an external reproducibility better than 0.2‰ (2s.d.). Analytical accuracy was demonstrated using a standard addition method using internal seawater standard. Overall, the analytical method requires minimum Ge amounts of about 2.6 ng, which is sufficient to measure the isotopic composition of inorganic Ge in surface seawater.

Hydride generation (HG) MC-ICPMS

The generation of volatile metalloid hydride has long been the most suitable technique for on-line separation and speciation of ng to pg amounts of Ge, As, Se, Sb, and Sn (Dedina and Tsalev 1995). This method involves the reduction of the element in solution to volatile hydride species using strong reducing agent, such as $NaBH_4$, generating H_2 upon mixing with acidified sample solution. The separation of the evolved gas and remaining solution is performed using a Hydride Generation (HG) system (Fig. 1). The continuous flow HG system has been successfully applied for the high-precision analysis of Ge, Se and Sb isotopes (Rouxel et al. 2002, 2003a, 2006; Layton-Matthews et al. 2006; Zhu et al. 2008).

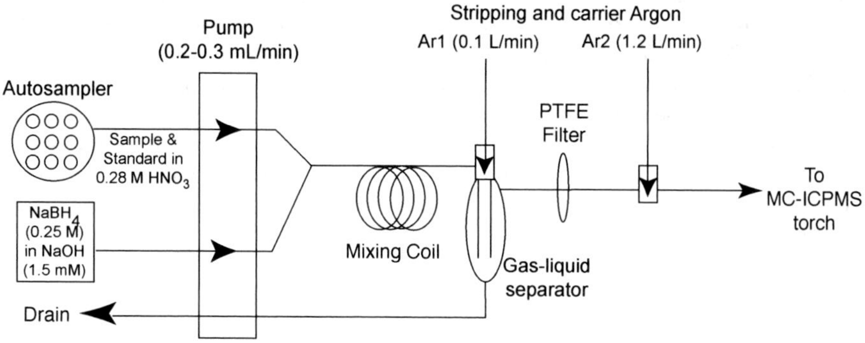

Figure 1. Schematic diagram of the continuous-flow hydride generator system used in Rouxel et al. (2006) and subsequent studies (Qi et al. 2011; Escoube et al. 2012, 2015).

Instrumental mass bias is generally corrected using either the "standard-sample bracketing" or the double-spike method. Important advantages of the use of HG-MC-ICPMS are: (1) higher sensitivity, lowering the total amount of element required for one analysis down to 10 ng or less; (2) further separation of the analyte from its matrix, removing potential isobaric interferences (e.g., Zn). As already applied for Hg isotope analysis (Epov et al. 2008), it is possible to further increase detection limits through the preconcentration of Ge hydrides using either gas chromatography or cold trapping, as done in the past for low level determination of Ge species in seawater (Andreae and Froelich 1981; Hambrick et al. 1984). This new development would also permit the possibility for compound-specific Ge isotope analysis with important prospects for the study of natural environments.

Interference issues

Germanium isotope measurements by MC-ICPMS suffer from molecular interferences, such as $^{35}Cl^{35}Cl$ on ^{70}Ge; $^{40}Ar^{16}O_2$ and $^{36}Ar^{36}Ar$ on ^{72}Ge, $^{58}Ni^{16}O$ and $^{38}Ar^{36}Ar$ on ^{74}Ge and $^{38}Ar^{38}Ar$ and $^{36}Ar^{40}Ar$ on ^{76}Ge, in addition to isobaric interferences of ^{70}Zn on ^{70}Ge. The chemical purification step and hydride generation technique may remove some of these interferences, but it is clear that Ge isotopes overlap with a range of Ar-based interferences. However, it appears that all Ge isotopes, except ^{76}Ge, can be measured without significant correction for interferences (Galy et al. 2003; Rouxel et al. 2006; Luais 2007). Hexapole-collision cell techniques together with soft extraction mode (Isoprobe instruments) have been also successful in eliminating argide interferences, using appropriate H_2 and Ar fluxes in the collision cell (Luais 2007, 2012). Recently, analytical developments at medium mass resolution ($M/\Delta M = 6000$, Neptune*Plus*) consist of Ge isotope measurements on the left part of the Ge spectrum free of any interferences (El Korh et al. 2015, 2016). Since Zn is a ubiquitous source of contamination in laboratory materials, corrections for ^{70}Zn on ^{70}Ge are always recommended. Luais (2007, 2012) showed that a mixture of 25 ppb of Zn + 1 ppm Ge in solution increases the $\delta^{74/70}Ge$ ratios by at least 1‰.

Notation

Germanium isotope composition can be reported using several possible notations, such as:

$$\delta^{x/y}\mathrm{Ge}_{\mathrm{STD}}(‰) = \left(\frac{\left({}^{x}\mathrm{Ge}/{}^{y}\mathrm{Ge}\right)_{\mathrm{sample}}}{\left({}^{x}\mathrm{Ge}/{}^{y}\mathrm{Ge}\right)_{\mathrm{STD}}} - 1 \right) \times 1000 \quad (1)$$

where $x = 74$, 73 or 72; $y = 70$ or 72 and STD corresponds to the normalization to Ge-Standard. Currently, there is no consensus in the way to report Ge-isotope ratios. Siebert et al. (2006) reported $\delta^{74/72}$Ge when using a double spike with ^{73}Ge and ^{70}Ge to correct for mass bias. Galy et al. (2003) reported δGe per mass unit while Rouxel et al. (2006), Escoube et al. (2012) and Luais (2007) reported $\delta^{74/70}$Ge (together with $\delta^{73/70}$Ge and $\delta^{72/70}$Ge). Due to larger abundances and minor isobaric interferences, ^{74}Ge/^{70}Ge ratio seems to be optimum for reporting Ge-isotopes. In order to improve interlaboratory comparisons, Ge isotope ratios should be reported as $\delta^{74/70}$Ge values relative to SRM3120a which is a Ge concentration standard produced in large amount by NIST and readily available.

Analytical precision

In the following, the analytical precision of the Ge isotope measurement is defined as the 2σ (s.d.) of *n* replicate measurements of the same standard solution prepared repeatably and measured over several analytical sessions. This is distinct from the 'internal precision' which is a standard error generated by the instrument for a single isotope ratio measurement. The 'external precision' is determined as the standard deviation (2 s.d.) of *n* replicate measurements of the same sample (e.g., georeference material) processed through the entire chemical procedure. In general, the 'analytical' precision is dependent on the instrumental mass bias correction scheme. Instrumental mass bias may be corrected using several techniques:

(1) The sample-standard bracketing (referred as SSB) technique involves the measurement of a Ge standard solution, before and after each unknown sample (Galy et al. 2003; Rouxel et al. 2006). In general, individual Ge isotope analysis corresponds to the mean of up to three replicate measurements of individual bracketed samples. Standard and sample solutions should be also analyzed within 10% of the same concentration and carefully rinsed to avoid cross contamination. For routine analysis, Rouxel et al. (2006) used Ge concentrations ranging from 10 to 50 ppb, and obtained an overall precision of about 0.14‰, 0.13‰ and 0.09‰ for $\delta^{74/70}$Ge, $\delta^{73/70}$Ge and $\delta^{72/70}$Ge, respectively (84 duplicates). The quantity of Ge used per analysis was between 15 and 150 ng. Compared to a previous investigation using the desolvation system (Galy et al. 2003), the HG technique requires a tenth of the Ge with a similar long-term reproducibility.

(2) The external normalization method involves using Ga isotopes (or Cu isotopes) (Hirata 1997; Galy et al. 2003; Siebert et al. 2006; Luais 2007) to monitor instrumental mass bias. The choice of Ga is based on several criteria, namely that mass 69 and 71 of Ga are within Ge masses (mass 70 to mass 76), the absence of any interferences on Ga masses, and the availability of an international isotopic Ga standard, the NBS SRM 994 Ga reference international isotopic standard having ^{69}Ga/^{71}Ga = 1.50676 (Machlan et al. 1986). The long-term reproducibility on the first generation Nu Plasma instruments (Galy et al. 2003) has been evaluated to be better than 0.06‰/amu (2 s.d.) for 1 µg Ge (i.e., equivalent to 0.24‰ for $\delta^{74/70}$Ge). A similar level of reproducibility, but for only 100 ppb of Ge, has been obtained when using the new generation of MC-ICPMS (Hexapole collision cell Isoprobe: Luais 2007, 2012; and high resolution Neptune Plus: El Korh et al. 2015, 2016). Luais (2012) evaluated several mass bias correction schemes on the reproducibility of Ge isotope measurements: the SSB method as described above; the external correction method using Ga and assuming identical instrumental mass bias isotopic fractionation factors for Ga and Ge (defined as *f*Ga and *f*Ge respectively); and the regression method implying $f\text{Ga} \neq f\text{Ge}$ but constant *f*Ga/*f*Ge, which is similar to the method previously used for Cu and Zn isotope measurements by Maréchal et al. (1999). The later method implies external data reduction to evaluate the accuracy of the measurements that are quantified on the basis of excellent regression lines ($r^2 > 0.99$) and similar slopes in the Ln(^{71}Ga/^{69}Ga)$_{measured}$ vs. Ln(^{x}Ge/^{70}Ge)$_{measured}$ diagram. These three methods have been applied to the JMC and Aldrich Ge reference standards and on

several iron meteorite samples, and they were found to be in excellent agreement (Luais 2012) confirming the robustness of Ga mass fractionation correction for Ge isotopic measurements.

(3) The double spike (referred as DS) correction involves determining simultaneously the instrumental mass bias and natural fractionation factors. The double spike method can be visualized as a three-dimensional diagram where axes represent the three measured isotopic ratios. Line and plane intercepts, defined by isotopic compositions of the double spike, standard solution and unknown sample measured by MC-ICPMS, are used to determine fractionation factors between the measured and corrected isotopic ratios (Siebert et al. 2001; Albarede and Beard 2004). Previous studies have used a double spike prepared from ^{73}Ge and ^{70}Ge spikes and mixed in equal proportions with natural Ge (Siebert et al. 2006; Escoube et al. 2012) (Table 1). The analytical precision reported by Escoube et al. (2012) for $\delta^{74/70}$Ge was about 0.15‰ (2s.d.) over a spike/natural ratio between 0.8 and 3.5 and for 10 ng sample size.

Ge isotope standards and reference materials

Germanium standards used in previous studies include NIST3120a (Lot #000411, 1000 μg/g, Escoube et al. 2012); Spex (Lot #11–160GE, Escoube et al. 2012); Aldrich (Lot #01704 KZ, Luais et al. 2000; Luais 2007), JMC (Johnson Mattey, Karlsruhe, Lot # 301230Sn, Luais et al. 2007) and Aristar (same split used in Rouxel et al. 2006, but incorrectly reported as Aldrich standard solution). The average compositions of Ge isotope standards for different instrumental set-up and research groups are reported in Table 2. 'Aristar' and 'Spex' solutions showed an enrichment in the light isotope in approximately the same proportion ($\delta^{74/70}$Ge = −0.64 ± 0.09‰ and −0.71 ± 0.10‰ respectively). The 'JMC' standard also presented light Ge isotope values, with a $\delta^{74/70}$Ge value of −0.32 ± 0.05‰. The 'Aldrich' standard showed the lightest $\delta^{74/70}$Ge values at −2.01 ± 0.11‰. Several sample introduction systems were tested, using either cyclonic spray chamber (CSC) and hydride generation (HG) for different instruments (Neptune, Nu plasma or Isoprobe), and no systematic differences were found (Table 2). It is also important to note that standard-sample bracketing, Ga normalization and double-spike corrections yielded similar results, both in terms of accuracy and precision.

Several georeference materials were also analyzed in previous studies allowing interlaboratory comparisons (Rouxel et al. 2006; Qi et al. 2011; Siebert et al. 2011a; Escoube et al. 2012; Luais 2012). Those include the following standards: AN–G (anorthosite, Fiskenaesset, Western Greenland), BCR–1 (basalt, Columbia River group USA), BE–N (basalt, Essey-la-Côte, Nancy, France), BHVO–2 (hawaiian basalt), BIR–1 (icelandic basalt), CLB–1 (Lower Bakerstown coal), DNC–1 (Braggtown dolerite, North Carolina, USA), DTS–1 (dunite, Hamilton, Washington), G–2 (granite, Rhode Island USA), GH (granite, Hoggar, Algeria), GLO (glauconite, Normandy France), IF–G (iron formation, West Greenland), Nod-P1 (manganese nodule, Deep Pacific Ocean), PCC–1 (peridotite, a partially serpentinized harzburgite, California USA), SDO–1 (Devonian black shale, Ohio Shale near Morehead, Kentucky), and UB–N (serpentine, Col des Bagenelles, Vosges, France). A compilation of published and unpublished values for these georeference materials is reported in Table 3.

The average $\delta^{74/70}$Ge value of basaltic rocks (BHVO–1&2; BIR–1; BCR–1&2, BE–N) is clustered at 0.56 ± 0.08‰ (2s.d.). Granitic rocks displayed significant heterogeneity with $\delta^{74/70}$Ge ranging from 0.4 to 0.8‰ and ultramafic rocks, although less concentrated in Ge, presented similar composition as granitic and basaltic rocks. We observe that NIST3120a is the standard whose isotope composition is the closest to natural composition compared to other standard solutions used previously. Iron formation IF–G and marine sediment GL–O yielded heavier $\delta^{74/70}$Ge values at 1.03‰ (± 0.1, 2s.d.) and 2.44‰ (± 0.14, 2s.d.) respectively.

Table 2. Average compositions of Ge isotope standards used by different research groups normalized against NIST 3120a isotopic standard.

Lab	**Method**		**nb#**	**$\delta^{74/70}$Ge (‰) NIST3120a**	**2sd**	**$\delta^{73/70}$Ge (‰) NIST3120a**	**2sd**	**$\delta^{72/70}$Ge (‰) NIST3120a**	**2sd**	**$\delta^{74/72}$Ge (‰) NIST3120a**	**2sd**
"ARISTAR" standard				−0.64	0.18	−0.54	0.18	−0.38	0.26	−0.28	0.03
#1	HG	SSB	11	−0.58	0.12	nd		−0.30	0.09	−0.28	0.04
#1	HG	Ga	11	−0.57	0.08	nd		−0.30	0.05	−0.28	0.07
#2	CSC	SSB	4	−0.59	0.02	−0.47	0.04	−0.33	0.01	−0.26	0.01
#2	CSC	DS	4	−0.61	0.11	−0.46	0.08	−0.31	0.06	−0.30	0.06
#3	CSC	Ga	6	−0.76	0.28	−0.61	0.20	−0.41	0.17	nd	
#3	CSC	SSB	6	−0.75	0.28	−0.63	0.20	−0.63	0.17	nd	
"JMC" standard				−0.32	0.10	−0.23	0.12	−0.16	0.07	−0.16	0.05
#1	HG	SSB	4	−0.33	0.33	nd		−0.15	0.27	−0.18	0.06
#1	HG	Ga	4	−0.31	0.29	nd		−0.14	0.25	−0.17	0.06
#2	CSC	SSB	4	−0.23	0.02	−0.14	0.04	−0.11	0.03	−0.12	0.03
#2	CSC	DS	4	−0.32	0.07	−0.24	0.05	−0.16	0.04	−0.16	0.04
#3	CSC	Ga	8	−0.37	0.04	−0.25	0.02	−0.19	0.07	nd	
#3	CSC	SSB	8	−0.37	0.16	−0.28	0.14	−0.20	0.06	nd	
"SPEX" standard				−0.71	0.21	−0.56	0.15	−0.37	0.16	−0.31	0.08
#1	HG	SSB	1	−0.59	nd	nd		−0.28	nd	−0.31	nd
#1	HG	Ga	1	−0.60	nd	nd		−0.28	nd	−0.31	nd
#1	HG	SSB	5	−0.84	0.16	nd		−0.48	0.10	−0.36	0.08
#1	HG	Ga	5	−0.81	0.11	nd		−0.46	0.09	−0.35	0.04
#2	CSC	SSB	3	−0.61	0.04	−0.51	0.09	−0.33	0.03	−0.28	0.05
#2	CSC	DS	4	−0.63	0.13	−0.48	0.10	−0.32	0.06	−0.31	0.06
#3	CSC	Ga	10	−0.81	0.19	−0.62	0.24	−0.41	0.11	nd	
#3	CSC	SSB	10	−0.79	0.18	−0.62	0.16	−0.41	0.12	nd	
#4	CSC	SSB	14	−0.64	0.42	nd		nd	0.00	−0.23	0.26
#4	CSC	Ga	14	−0.79	0.18	nd		nd	0.00	−0.31	0.04
"ALDRICH" standard				−2.01	0.23	−1.54	0.17	−1.03	0.12	−0.97	0.15
#2	CSC	SSB	4	−1.88	0.03	−1.46	0.07	−0.97	0.02	−0.92	0.03
#2	CSC	DS	4	−2.14	0.06	−1.61	0.05	−1.08	0.03	−1.05	0.03
#2	HG	DS	6	−1.90	0.10	−1.44	0.08	−0.96	0.05	−0.94	0.05
#3	CSC	Ga	84	−2.08	0.26	−1.61	0.20	−1.08	0.16	nd	
#3	CSC	SSB	84	−2.05	0.22	−1.60	0.20	−1.06	0.14	nd	

Notes:

Lab 1: WHOI; Lab 2: IFREMER; Lab 3: CRPG; Lab 4: LCBIE. "nd", not determined.

SSB: sample-sample bracketing; DS: double spike correction; Ga: external normalisation to Ga. CSC: cyclonic spray chamber; HG: hydride generation.

Underlined numbers are grand average

Table 3. Compilation of published and unpublished values for georeference materials.

SiO_2 (wt%)	Ref	Ge (ppm)	Ref	$\delta^{74/70}Ge$ (‰) NIST3120a	2sd	Ref	SiO_2 (wt%)	Ref	Ge (ppm)	Ref	$\delta^{74/70}Ge$ NIST3120a	2sd	Ref
BHVO-1 (Hawaiian basalt)							*DTS-1 (dunite)*						
49.9	a	1.64	a	0.55	0.15	g	40.41	a	0.88	a	0.76	0.13	k
		1.55	c						0.84	c	0.65	0.28	g
		1.58	i	0.37	0.10	i			0.83	i	0.50	0.06	i
BHVO-2 (Hawaiian basalt)									0.75	k			
49.9	e	1.52	k	0.55	0.13	k	*PCC-1 (peridotite)*						
		1.63	l	0.54	0.13	k	41.71	a	0.94	a	0.69	0.13	k
		1.61	l	0.52	0.13	k			0.82	k	0.64	0.13	k
		1.64	l	0.43	0.19	h					0.59	0.30	h
				0.53	0.15	h					0.74	0.13	g
				0.55	0.43	l			0.81	i	0.46	0.20	i
				0.64	0.13	l	*UB-N (serpentine)*						
				0.51	0.20	l	39.4	a	0.93	i	0.61	0.16	i
BIR-1 (Icelandic basalt)									0.93	l	0.69	0.24	l
47.96	e	1.50	a	0.74	0.13	k	*AN-G (anorthosite)*						
		1.45	b	0.60	0.13	k	46.3	a	0.80	a	0.67	0.13	k
		1.53	b	0.56	0.30	h			0.93	c	0.66	0.13	g
		1.49	c	0.64	0.33	g			0.84	k			
		1.52	d	0.57	0.39	g	*GL-O (glauconite)*						
		1.52	i	0.57	0.04	i	50.9	a	4.50	a	2.49	0.13	k
		1.40	k						4.02	k	2.34	0.21	h
BCR-1 (Columbia River basalt)											2.51	0.55	g
54.11	e	1.50	a	0.65	0.13	k					2.43	0.60	g
		1.42	b	0.54	0.19	h					2.42	0.26	g
		1.45	c	0.47	0.39	h	*IF-G (iron formation)*						
		1.36	k	0.54	0.22	g	41.2	a	24.00	a	1.11	0.13	k
		1.45	i	0.57	0.12	i			22.00	f	1.03	0.13	k
BE-N (Vosges basalt)											1.01	0.25	j
38.2	a	1.16	i	0.55	0.14	i					1.00	0.27	j
G-2 (granite)									21.80	i	1.01	0.34	i
69.1	a	1.02	b	0.39	0.34	g			23.06	k	1.00	0.15	l
		0.94	b						24.49	l	0.94	0.14	l
		0.92	k				*SDO-1 (Devonian shale)*						
GH (granite)							49.28	a	1.56	l	0.85	0.18	l
75.8	e	2.18	c	0.74	0.13	k			1.64		0.91	0.09	l
		1.63	k	0.55	0.28	g	*CLB-1 (coal)*						
		1.84	i	0.63	0.06	i			n.d.		1.22	0.16	m
DNC-1 (granite)									n.d.		1.44	0.18	m
47.15	e	1.30	a	0.56	0.13	k			n.d.		1.24	0.16	m
47.04	a	1.26	b	0.76	0.13	k			n.d.		1.17	0.20	m
		1.28	k	0.74	0.31	g	*Nod-P1 (Pacific Mn nodule)*						
				0.61	0.42	g	13.9	a	0.54	n	-0.08	0.18	l

References: a: Govindaraju (1994); b: Mortlock and Froelich (1987, 1996); c: Halicz (1990); d: Kurtz et al. (2002); e: USGS; f: Frei and Polat (2007); g: Rouxel et al. (2006); h: Escoube et al. (2012) (double-spike, WHOI); i: Luais (2012); j: Rouxel, unpublished; k: Escoube et al. (2012) (double-spike, Ifremer); l: Lalonde and Rouxel, unpublished; m: Qi et al. (2011); n: Axelsson et al. (2002).

THEORETICAL CONSIDERATIONS AND EXPERIMENTAL CALIBRATIONS

Equilibrium fractionation factors

Li et al. (2009, 2010) provided a thorough theoretical quantification of Ge isotopic equilibrium fractionation between aqueous fluids, mineral phases, and organic compounds. The theoretical quantification of Ge isotopic fractionation is based on the Bigeleisen-Mayer equation that considers isotope exchange reaction between two components A and B:

$$A + B^* \Leftrightarrow A^* + B \quad (2)$$

The isotopic fractionation factor α is defined as a function of the isotope equilibrium constant K_{eq} of the exchange reaction (2), with $\alpha = K^{1/n}$, n equal to the number of each exchanged atom. Li et al. (2009) used $n = 1$ for the calculation of α values for different Ge species. Because K_{eq} is a function of inverse temperature, the isotopic fractionation factor α depends on the temperature of equilibrium between the two phases. The isotopic fractionation factor α is calculated using the reduced partition function ratio (RPFR) or β-factor for each molecule, mainly taking into account the differences in vibrational energies (ΔF_{motion}) between molecules, such as:

$$\Delta G^{\circ}_{reaction} \approx \Delta F_{motion} = -\mathrm{R}T\ln(K_{eq}) \quad (3)$$

$$K_{eq} = \exp\left(-\frac{F_{motion}}{\mathrm{R}T}\right)\frac{RPFR(A)}{RPFR(B)} \quad (4)$$

RPFR values are calculated using *ab initio* quantum chemistry method that allows computing harmonic vibrational frequencies.

Silicate–fluid equilibrium reaction. Li et al. (2009) determined RPFR values for silicate minerals (quartz, albite, K-felsdpar, olivine) and aqueous Ge species. They found a negative correlation between RPFR and Ge–O bond lengths. This enrichment of heavier isotopes in components with shorter Ge–O lengths is in line with the principles of isotopic fractionation given by Schauble (2004). Using Equation (4), $K \approx \alpha$, and the relation $1000 \ln \alpha = \Delta_{A-B} = A \times 10^6/T^2 + B$, the isotopic fractionation Δ_{A-B} between silicate and aqueous species are calculated. Two important conclusions can be drawn from these calculations: (i) The isotopic fractionation between silicate minerals and aqueous fluids indicates that a fluid in equilibrium with olivine should be isotopically heavier than the solid, while fluids in equilibrium with Na-feldspar (albite), K-feldspar, quartz should be isotopically lighter than the solids. Such results may explain the elevated δ^{74}Ge of serpentine compared to unaltered ultramafic rocks and basalts (Luais 2012). (ii) The isotopic fractionation between crust-forming minerals (e.g., quartz–feldspar) and aqueous fluids leads to light isotopic signatures in the fluids. This would predict that the isotopic composition of rivers could be light, although recent studies suggest heavier Ge isotope composition in rivers compared to the crust (Baronas et al. 2014).

Sulfide–fluid equilibrium reaction. Li et al. (2009) gave preliminary Ge isotope fractionation factors $\Delta^{74/70}Ge_{fluid-sulfide}$ at 25 °C for Ge-bearing sulfides having sphalerite-like structures. The calculated Ge isotopic fractionation factors ($\Delta^{74/70}Ge_{fluid-sulfide}$ at 25 °C) of +12.2 to +11.5‰ for Ge(II) and +11.4‰ for Ge(IV) clusters give an important estimation of the direction and the magnitude of Ge isotopic fractionation, with low $\delta^{74/70}$Ge in sulfide in equilibrium with heavy isotope-enriched fluids. No quantification of A and B parameters are given at this stage of the study that could be useful for modelling sulfide isotopic composition from high-T hydrothermal fluids. A first approximation for T determination can however be possible by extracting the α and A parameter values at 25 °C, considering a negligible value for B.

Adsorption processes. Li and Liu (2010) calculated Ge isotopic fractionation factors during adsorption on Fe-oxides and Fe-hydroxides. The calculation was based on cluster models of two main surface structure complexes of$>Fe_2O_nGe(OH)_{4-n}$ and$>Fe_2O_nGeO(OH)^-_{3-n}$, corresponding to adsorption of either $Ge(OH)_4$ or $GeO(OH)_3^-$ species under acidic or basic conditions, respectively. Each surface complex presents specific geometries in term of Ge–O, Fe–Fe, Ge–Fe bond lengths, with Ge remaining in tetrahedral coordination. It is assumed that Ge occurs in tetrahedral coordination and bonded to two corners of adjacent octahedral Fe, consistent with the EXAFS (Extended X-ray Absorption Fine Structure) spectroscopy measurements of Pokrovsky et al. (2006). The equilibrium Ge isotopic fractionation in term of calculated RPFR$\approx\alpha$ values gave 1.02214 and 1.02149 for $Ge(OH)_{4(aq)}$ and $GeO(OH)_3{}^-{}_{(aq)}$ respectively, which translates into a solid-solution fractionation factor $\Delta^{74/70}$Ge of -1.7‰ and -1.6‰ for $>Fe_2O_2Ge(OH)_2(2C)_{(aq)}$–$Ge(OH)_{4(aq)}$ and $>Fe_2O_2GeOOH^-(2C)_{(aq)}$–$GeO(OH)_3{}^-{}_{(aq)}$, respectively. Hence, adsorption onto Fe-oxide and Fe-hydroxide surfaces preferentially favors the light Ge isotopes, yielding a fluid with heavier isotope composition. The *A*, *B* parameters in the equation $1000\ln\alpha = \Delta_{A\text{-}B} = A \times 10^6/T^2 + B$ provide quantitative tools for modelling the isotopic fractionation in terms of temperature of equilibration between Ge in solution and adsorbed Ge.

Quantification of Ge isotopic fractionation during adsorption processes has been further confirmed by the adsorption and co-precipitation experiments of Pokrovsky et al. (2014). These authors show that $Ge(OH)_{4(aq)}$ adsorption on goethite surface forms $>(FeO\text{–}Ge(OH)_3{}^\circ{}_{(aq)}$ or $>(FeO\text{–}Ge(OH)_2^-)$ complexes in acid or alkaline solutions, respectively, and induces a solid-solution isotopic fractionation factor $\Delta^{74/70}$Ge of -1.7 ± 0.1‰. The light Ge isotopic compositions on the surface of Fe-oxy(hydr)oxides is due to changes in the atomic environment of Ge during adsorption-co-precipitation process that increases the distortion and disorder of the complex formed with longer Ge–O bonds, and consequently decreases their stability. In addition, Ge coprecipitation with Fe-oxy(hydr)oxides implies Fe^{2+} oxidation or Fe^{3+} hydrolysis. The Ge/Fe ratio of the precipitated solid will then have a major role on the Fe-oxy(hydr)oxides crystalline structure, and lattice parameters. The X-ray spectroscopy results of Pokrovsky et al. (2006) showed that Ge–Fe complexes with $(Ge/Fe)_{solid}<0.1$ is composed of both tetrahedrally coordinated Ge, and octahedrally coordinated Ge that occurs in substitution with octahedral Fe (Pokrovsky et al. 2006). The configuration of GeO_6 has longer Ge–O bond lengths and is weaker than for GeO_4, thus explaining why $\Delta^{74/70}Ge_{solution\text{-}solid}$ values of 2.0 ± 0.4‰ are higher than for adsorption processes.

Kinetic processes

Isotopic kinetic processes relate to the unidirectional movement of isotopes of a given element from one phase into another one. Kinetic processes describe either *diffusion* between two components of contrasted elemental end isotopic compositions, or *evaporation–condensation* during high- or low-temperature processes. Kinetic isotope fractionation is essentially mass-dependant. According to the kinetic theory of gases, all molecules at the same temperature have the same energy KE, which depends on their mass and velocities: KE = ½ mv^2. Consequently for two isotopes 1 and 2 of masses m_1 and m_2 with similar kinetic energy,

$$KE = \tfrac{1}{2}\, m_1 v_1 = \tfrac{1}{2}\, m_2 v_2 \tag{5}$$

$$\Rightarrow \frac{v_1}{v_2} = \sqrt{\left(\frac{m_1}{m_2}\right)} \tag{6}$$

If $m_2 > m_1$, then $v_2 < v_1$ suggesting that light isotopes will have higher velocities than heavy isotopes. As a result, during diffusion, light isotopes will migrate faster toward the low concentration region than heavy isotopes. Similarly, during evaporation processes, light isotopes will be lost preferentially to the vapor phase, leaving a residue that is enriched in heavy isotopes.

Diffusion of Ge in silicate melts

Jambon (1980) established the law for diffusion between two isotopes of mass m_1 and m_2 and diffusion coefficient D_1 and D_2, of a given element, and Richter et al. (1999) defined the relative diffusivities between isotopes as:

$$D_{rel} = (D_1/D_2) = (m_2/m_1)^{\beta} \quad (7)$$

The isotopic fractionation depends on the relative mass difference between two isotopes, and of β factor, that is determined from experiments. As β values >0, light isotopes will have higher diffusion coefficient (*D*) than heavy isotopes, and will migrate faster than heavy isotopes. Experimentally determined β values are in general lower than ½. Low value for β also reflects the factor that the diffusing molecules are likely to have greater masses than unbonded isotopes (Davis and Richter 2007). Richter et al. (1999) has shown that isotopic diffusion is not only mass-dependent, but also relies on concentration and isotopic contrasts between the two components. They experimentally studied the isotopic fractionation of Ge by diffusion between synthetic GeO_2 melts of contrasting composition, as an analog for Si diffusion in silicate melts. Germanium isotopic interdiffusion was reported on synthetic GeO_2 melts artificially enriched in specific Ge isotopes. The experimental design was to create two end-members with similar $^{76}Ge/^{70}Ge$ but very different $^{74}Ge/^{70}Ge$ allowing an evaluation of the mass differences and isotope concentration contrast on diffusion. The theoretical diffusion profiles indicated an isotopic mobility as inferred from a marked difference between $^{74}Ge/^{70}Ge$ and $^{70}Ge/^{76}Ge$ profiles (Fig. 2). A subtle isotopic fractionation in $^{76}Ge/^{70}Ge$ ratio along the diffusion profile would indicate faster diffusion of ^{70}Ge compared to ^{76}Ge (Fig. 2). β factors between 0.5 and 0.1 were considered. Laboratory experiments were performed on these GeO_2 melts for 3 h at 1400 °C and 0.5 GPa. The $^{74}Ge/^{70}Ge$ diffusion profiles measured by ion microprobe techniques are best fit with β values <0.025, indicating much lower isotopic fractionation for diffusion than prediction based on the simple ratios of isotopic masses.

Diffusion of Ge in metal. Several studies reported the diffusion of Ge in metal or Ge–Si alloys with important application to metallurgy and semi-conductor technologies (McVay and DuCharme 1974; Pike et al. 1974; Vogel et al. 1983). They concluded that Ge diffusion occurs through a lattice vacancy mechanism. In general, isotope diffusion in metal is higher than between silicate melts, as it occurs at lower temperature. Richter et al. (2009) determined β values ranging from 0 to 0.4, regardless of the element, temperature and diffusion type (self-diffusion, or between metal phases). Germanium isotope fractionation for self-diffusion in Ge metal (Campbell 1975), and Ge diffusion in Cu metal (Hehenkamp et al. 1979) were determined using radioactive experiments. Quantification of β values from these experiments indicate significantly higher β values in metal than in silicate melts:

$$\beta_{Ge}\text{-silicate} < 0.025 \ll \beta_{Ge}\text{-Ge} = 0.127 \text{ (self-diffusion)} < \beta_{Ge}\text{-Cu} = 0.215 \text{ (Ge diffusion in Cu)}$$

Diffusion can be an important mechanism constraining the elemental distribution of Ge between the taenite and kamacite phases of iron meteorites. The isotopic effect between these two phases, for example in taenite–kamacite forming Widsmanstätten textures remains unknown. At the bulk scale, kamacite only bearing Fe-meteorites (Ni% <6%) or taenite–kamacite bearing Fe-meteorites (Ni% >6%) have similar Ge isotope composition (Luais 2007, 2012), indicating that diffusion does not contribute to the whole isotopic signature of iron meteorites.

Diffusion and evaporation processes between metal and silicate. Germanium is a moderately siderophile and thus has a large metal–silicate partition coefficient (D_{Ge}) between 10 and 10^4. It is strongly dependant on thermodynamic parameters (fO_2, *P*, *T*). The effect of oxygen fugacity is dominant, with variations in D_{Ge} metal–silicate increasing as a factor 30 with reducing conditions of 3 logfO_2 units (Schmitt et al. 1989). Germanium is volatile at high

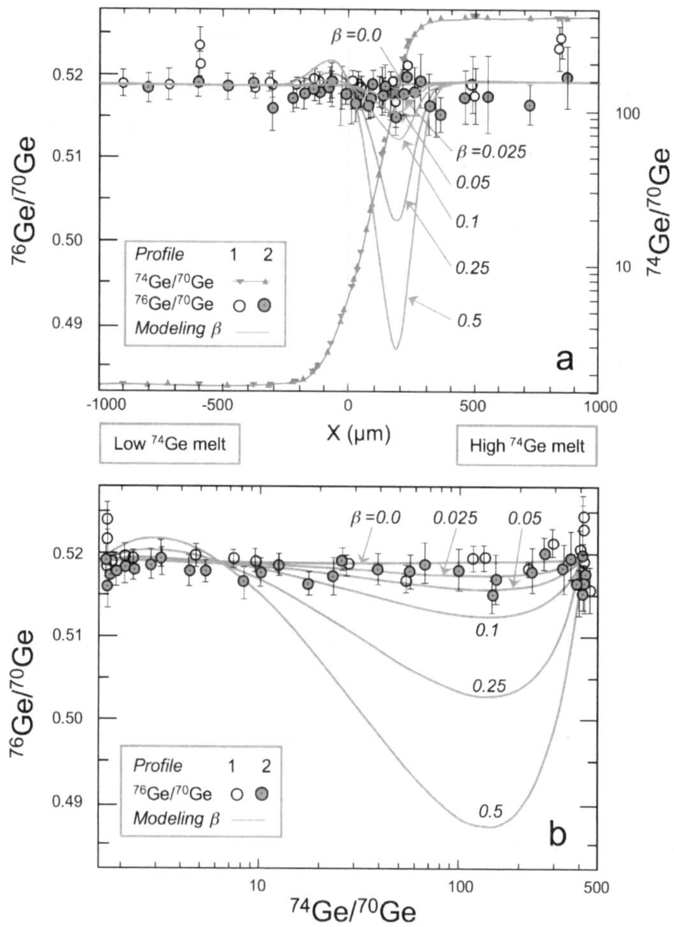

Figure 2. Experimental data of inter-diffusion isotopic fractionation of Ge between two GeO_2 synthetic silicate melts of contrasted initial $^{74}Ge/^{70}Ge$ ratios (low ^{74}Ge melt with $^{74}Ge/^{70}Ge = 1.6969 \pm 0.0013$, high ^{74}Ge melt with $^{74}Ge/^{70}Ge = 408.1 \pm 3.0$), but similar $^{76}Ge/^{70}Ge$ ratios of 0.5193 ± 0.0005 and 0.5188 ± 0.0009, respectively (from Richter et al. 1999). Data are given for two different parallel profiles (profiles 1 and 2) along the length of the experimental glass cylinder. Curves correspond to model results of Ge isotopic ratios for different values of β from 0 to 0.5 for comparison (Richter et al. 1999). (A) Ion probe diffusion profiles are given for $^{74}Ge/^{70}Ge$ and $^{76}Ge/^{70}Ge$ as a function of distance from the interface $\times$ μm = 0. (B) $^{74}Ge/^{70}Ge$ versus $^{76}Ge/^{70}Ge$ variations. Larger Ge isotopic fractionation are observed for $^{74}Ge/^{70}Ge$ than for $^{76}Ge/^{70}Ge$ ratios, highlighting the effect of composition contrast on the degree of diffusive isotopic fractionation between two phases.

temperature, with a condensation temperature T_{cond} of 883 °C at 10^{-4} bar (Lodders et al. 2009), corresponding to the temperature at which 50% of the element is in the condensed phase. Both diffusion and evaporation can induce elemental and isotopic fractionation during high temperature processes. The quantification of Ge isotopic fractionation during partitioning and evaporation are fundamental for understanding the thermodynamic conditions (fO_2, P, T) of core formation and accretion processes in the early history of the Earth and planets.

Laboratory experiments of Ge isotopic fractionation between metal and silicate phases have been undertaken at 1355 °C under various fO_2 (−4 to +2.5 log units above IW buffer) to simulate the Ge transfer from the oxidized silicate phase to a metal, and at different times (up to 60h) to evaluate the effect of chemical and isotopic re-equilibration between phases (Luais et al. 2007). The initial experimental set-up consisted of a pure Ni-capsule and a synthetic silicate phase of eutectic CMAS composition, doped with Ge Aldrich standard solution. Preliminary results showed a negative exponential fit of Ge content with Ge isotopic composition for both fO_2-controlled and time series experiments that highlight two concomitant processes (Luais et al. 2007) :

(i) very reducing conditions (IW–4) or short time duration both result in high Ge content and small enrichment in light isotopes (–0.3‰) in the metal phase, relative to the initial composition. This direction of isotope fractionation is in line with theoretical fractionation for diffusion, with preferential movement of light isotopes from the enriched Ge pole (silicate) toward the depleted pole (pure Ni capsule).

(ii) the progressive increase in fO_2 (i.e., mildly reducing to oxidizing conditions) or longer time experiments produced strong depletion in Ge in the metal phase, associated with an exponential increase in $\delta^{74/70}$Ge values, that can reach +7‰ in the most oxidized conditions. These heavy isotope compositions are indicative of Ge loss due to evaporation.

Future experiments should be devoted to deconvoluting these two processes and to provide parameters for quantification of Ge isotopic fractionation in natural samples.

Evaporation processes in environmental and industrial technologies. The organophile behavior of Ge leads to significant enrichments in coal and lignite. Considering the melting point temperature of metal Ge of 937 °C, coal combustion may represent an important source of Ge in the environment (Froelich and Lesley 2003). In order to investigate possible Ge isotope fractionation during coal combustion, Qi et al. (2011) analyzed soot and cinder samples (i.e., solid waste of coal combustion) as well as various ashes of Ge-rich lignite prepared after ashing at 600 °C for 24 h in a muffle furnace (Fig. 3). Although ashing experiments showed no Ge loss or Ge isotopic fractionation between coal and ash for combustion, industrial coal combustion at $T > 1400$ °C seems to fractionate Ge isotopes, with Ge isotopic compositions of soot being distinctly lighter (up to 2.25‰) than those of cinder (solid residue). The large Ge isotopic fractionation between residue and volatile components is best explained by evaporation during combustion, making Ge isotopes a powerful tool in detecting environmental pollution produced by mining industries.

HIGH-TEMPERATURE GEOCHEMISTRY

Fundamentals of Ge high-temperature geochemistry

Being siderophile, chalcophile and lithophile, Ge incorporates a wide range of crystallographic structures, including Fe-rich metal, sufides, and silicates. Germanium may occur in three oxidation states: Ge^0, Ge^{2+} and Ge^{4+}. Germanium is also a volatile element of particular importance in cosmochemistry. Hence, a significant number of studies have identified the main physical, crystallographic and chemical characteristics of Ge in various compounds, which serve as important guide to identify Ge isotopic fractionation in high-temperature processes leading to the formation of the Earth and planets.

Germanium in metal compounds. Germanium is strongly enriched in metal phase under highly reducing conditions. Only two natural origins for the metal phase are known:

(i) extraterrestrial origin, in particular in iron meteorites that are analogues for cores of planetesimals and planets. Germanium contents can vary from 20 ppb up to 2000 ppm (Wasson 1974).

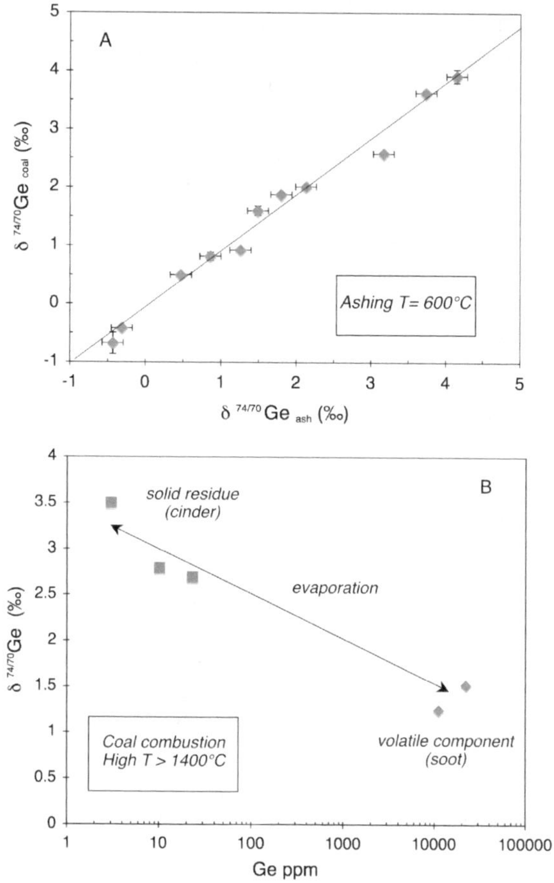

Figure 3. Experimental determination of Ge isotopic fractionation in coal and lignite during ashing and combustion processes (Qi et al. 2010). (A) at low temperature of $T = 600\,°C$: $\delta^{74/70}Ge$ in ash and coal plot on ≈ 1.1 line (slope of 0.963, $R^2 = 0.983$), indicating no isotopic fractionation; (B) at high temperature $T > 1400\,°C$: lighter $\delta^{74/70}Ge$ in volatile component (soot) than in the residue (cinder).

(ii) terrestrial native Fe–Ni metal corresponding to an unusual phase occurring as inclusions in basaltic intrusions. Only three occurrences have been reported worldwide: Disko Island (Greenland) that is the more extensively studied (Ulff-Møllller 1985; Klöck et al. 1986), Siberian trap intrusions (Ryabov and Lapkovsky 2010), and Bühl basalt (Germany) (Medenbach and ElGoresy 1982). Disko Island native iron phases can contain up to 300 ppm of Ge (Klöck et al. 1986). The crystallization of native iron implies strongly reduced conditions ($fO_2 <$ (FMQ) $- 6$) that can be caused by magmatic assimilation of carbonaceous sediments or coal (Iacono-Marziano et al. 2012).

The Fe–Ni phases of extraterrestrial origin are Ni-rich taenite and Ni-poor kamacite (Ni < 6%), the latter can contain high Ge content. Cavell et al. (2004) have studied the incorporation of Ge and Ga in Fe–Ni kamacite phase of Canyon Diablo iron meteorite, using extended X-ray absorption fine structure spectroscopy (EXAFS), and simulate the spectra in a "super lattice" model. The EXAFS patterns describe the kamacite structure as bcc lattice structure (body center cubic) that contains a central atom of Ge, Ga, or Ni surrounded by

Fe atoms. Germanium (and Ga, or Ni) are incorporated in the structure through substitution for Fe atoms. The central atom (e.g., Ge) is surrounded by eight nearest Fe atoms in the first shell. The XAFS quantifies the distances from the central atom to the first shell, for example Ge-Fe length of 2.494 Å (Table 4). Knowledge of bond length parameters is essential for the theoretical prediction of isotopic fractionation between phases.

The taenite and kamacite phases of Fe-meteorites crystallize from a Fe–Ni metallic liquid. Experimental studies show that the incorporation of Ge in metallic crystal structure, which is quantified by the partition coefficient D between coexisting solid metal and liquid metal, is strongly controlled by concentrations of non-metallic elements, such as S, P, C in the metallic liquid (Jones and Walker 1991; Jana and Walker 1997; Liu and Fleet 2001; Chabot et al. 2003; Chabot and Jones 2003; Corrigan et al. 2009). Hence, increasing S, P, C contents in the metallic liquid metal leads to Ge being more compatible, e.g., D(Ge) solid metal/liquid metal increases strongly by two orders of magnitude for an increase in S up to 30%, and to a lesser extent by a factor of 2 for P, and C. These results constitute some basis for the interpretation of elemental trends involving Ge in iron meteorites.

Germanium in silicates. Germanium and silicon have similar crystallographic characteristics, including similar ionic (R_{Ge} = 0.39 Å, R_{Si} = 0.26 Å), and covalent radius (R_{Ge} = 1.20 ± 0.04 Å, R_{Si} = 1.11 ± 0.02 Å). Together with their similar oxidation state (Ge^{4+}, Si^{4+}), this explains the incorporation of Ge in the tetrahedral site of silicate structures by isomorphic substitution for Si. Lattice parameters of Ge–O bonds have been determined by Li et al. (2009) in cluster models for a large range of silicate structures ranging from tectosilicate (quartz, K-feldspar) to nesosilicate (olivine) (Table 4). The Ge–O bond length increases slightly from quartz (1.753 Å), Na- and K-felspars, to olivine having distinctly longer Ge–O bond length (1.79 Å). Variations in Ge–O bond length are negatively correlated with theoretical Ge isotopic fractionation, meaning that distinct Ge isotopic signature can be expected between felsic and mafic natural samples.

Table 4. Lattice parameters for Ge in metal, silicate, oxides and sulfides.

	Oxidation state	Ge–Fe (Å)	Ge–O (Å)	Ge–S (Å)	References
Fe–Ni metal	0	2.494			Cavell et al. (2004)
			Silicates		
olivine	4^+		1.790		Li et al. (2009)
K-felspar	4^+		1.758		Li et al. (2009)
Albite	4^+		1.760		Li et al. (2009)
Quartz	4^+		1.753		Li et al. (2009)
			Oxides		
Brunogierite $(Fe^{2+})_2Ge^{4+}O_4$	4^+		1.771		Cempirek and Groat (2013)
			Sulfides		
$[Ge(II)S_4Zn_2]^{2-}$	2^+			2.257	Li et al. (2009)
$[Ge(II)S_4Zn_4]^2$	2^+			2.267	Li et al. (2009)
$[Ge(IV)S^4Zn_2]^0$	4^+			2.275	Li et al. (2009)
Sphalerite	4^+				Cook et al. (2015) Belissont et al. (2016)

Wickman (1943) initially suggested that Ge⇔Si substitution would govern the Ge content in silicates, with lower Ge contents in the highest polymerized structures such as tectosilicate (quartz, felspar, $[nSiO_{2n}]$) < phyllosilicate (mica, $[Si_4O_{10}(OH)_2]^{6-}$) < inosilicate (pyroxene, $[SiO_3]_2$, amphibole $[Si_4O_{11}(OH)]^{7-}$) and higher Ge contents in less polymerized structures such as nesosilicates (olivine, $[SiO_4]^{4-}$, zircon, garnet). This would imply that ultramafic rocks with high olivine fractions should be enriched in Ge compared to crustal rocks (granite). This does not agree with the fairly homogeneous distribution of Ge in terrestrial silicate rocks (El Korh et al. 2016). The K_d (Ge) for mineral–liquid partition coefficient in high-temperature silicates (Table 5), also indicates, that diopside pyroxene has higher Ge contents than olivine (Malvin and Drake 1987). Thermodynamic parameters (T, P), in addition to Ge–Si substitution, likely control Ge partitioning into silicates (Table 5). The temperature effect seems to be negligible on K_d olivine/liquid (Capobianco and Watson 1982), but possibly significant on K_d cpx/liquid (Hill et al. 2000; Adam and Green 2006). The effect of pressure has been experimentally investigated by Malvin and Drake (1987) who reported lower Ge concentrations in olivine at higher pressure of 2 GPa. Davis et al. (2013) also determined lower D_{Ge} Ol at 3 GPa, which they attributed to a "garnet effect", the only silicate phase in which Ge is compatible (D_{Ge} = 1.51). The K_d (Ge) for mineral–liquid pairs ranged from 0.4 to 0.9 (Table 5, and references therein), Bulk D(Ge) for partial melting was close to 1 (0.8–1.04), for different diopside/olivine ratios of peridotitic source (Malvin and Drake 1987), and a near-solidus of bulk D(Ge) of 0.66 ± 0.01 was calculated for garnet peridotite at 3 GPa (Davis et al. 2013). These experimental results indicate that Ge is moderately incompatible during fractional crystallization process, and partial melting of peridotitic sources.

The latter findings have some important consequences in tracing source composition or source mineralogy. For example, no significant variations in Ge content was observed in volcanic suites from Hawaii, yielding molar Ge/Si = 2.6×10^{-6} (De Argollo and Schilling 1978b). Likewise, variations in Ge/Si ratios in mantle and crustal rocks could trace mantle source compositions from different geodynamic settings (De Argollo and Schilling 1978a,b). New developments of Ge concentration measurements in ocean floor basaltic glass using isotopic dilution (Makishima and Nakamura 2009) and *in situ* laser ICP-MS (Avarelo and McDonough 2010; Jenner and O'Neill 2012; Davis and Humayun 2013) have considerably extended the data set, with higher precision measurements (6%, and 3%, respectively). Avarelo and McDonough (2010) found distinctly higher Ga/Ge ratios up to 20 in tholeiitic basalt from Hawaii compared to MORB (Ga/Ge mean ≈ 10), which would reflect mantle source mineralogy (e.g., garnet vs. spinel).

Germanium in sulfides. Germanium can be strongly and variably enriched in sulfide minerals (Bernstein 1985; Holl et al. 2007; Rosenberg 2009). Besides the rare occurences of sulfides that incorporate Ge at percent level (e.g., germanite, renierite, and briartite), Ge occurs at trace levels in Zn and Fe–Cu sulfides. Germanium is incorporated in the lattice structure through a substitution mechanism with isovalent substitution Ge^{2+} ⇔ Zn^{2+} (Cook et al. 2009) or more complex substitutions involving monovalent elements (Cu^+, Ag^+) and trivalent (Ga^{3+}) with a vacancy (Johan 1988). Recently, on the basis of strong binary elemental correlations, Belissont et al. (2014) proposed coupled substitutions involving Zn and Ag as $3Zn^{2+} \Leftrightarrow Ge^{4+} + 2Ag^+$.

Li et al. (2009) used three sulfide clusters models $[Ge(II)S_4Zn_2]^{2-}$, $[Ge(II)S_4Zn_4]^{2+}$, and $[Ge(IV)S_4Zn_2]^0$ of sphalerite-like lattices that differ in their coordination and oxidation states (Ge^{2+} and Ge^{4+}). These two oxidation states are thought to be present in sulfide minerals. Calculated Ge–S bond lengths differ slightly from 2.257–2.267 Å in Ge(II) structure to 2.275 Å in Ge(IV) structure (Table 4). Cempirek and Groat (2013) have unambiguously identified the occurrence of Ge^{4+} in brunogierite, a spinel-type oxide, with a Ge–O bond length of 1.771 Å, which is shorter than Ge–S bonds.

Table 5. Experimental determination of Ge mineral–liquid partition coefficients in silicate minerals as a function of temperature and pressure.

Mineral	*D*(Ge) mineral/ liquid	Temperature, Pressure	References	*D*(Ge)/*D*(Si)
Anorthite	0.5	1300 °C	Malvin and Drake (1987)	0.58±0.05
Olivine	0.65	1300 °C	Malvin and Drake (1987)	0.73±0.02
	0.68±0.06	1300-1450 °C, 1 atm	Capobianco and Watson (1982)	
	0.54±0.04	20 kbar		
	0.43±0.01	1450-1650 °C, 3 GPa	Davis et al. (2013)	
	0.55	1100 °C 2 GPa	Adam and Green (2006)	
	0.95±0.04	1075 °C 1 GPa	Adam and Green (2006)	
Clinopyroxene		Variable *P*, *T*		
	1.1–1.52	1075–1170 °C 1–3 GPa	Adam and Green (2006)	
Diopside	1.4	1300 °C	Malvin and Drake (1987)	1.4±0.02
	0.87±0.03	1450–1650 °C, 3 GPa	Davis et al. (2013)	
Orthopyroxene	0.87±0.02 0.75±0.28	1450–1650 °C 3 GPa	Davis et al. (2013)	
	1.36±0.07	1160 °C 2.7 GPa	Adam and Green (2006)	
Amphibole	1.61±0.06	1050 °C 2 GPa	Adam and Green (2006)	
	1.109–0.03	1025 °C 1 GPa	Adam and Green (2006)	
Mica	0.51–0.54± 0.02–0.05	1160–1100 °C 2.7–2.5 GPa	Adam and Green (2006)	
	1.08±0.05	1025 °C 1 GPa	Adam and Green (2006)	
Spinel	0.11	1300 °C	Malvin and Drake (1987)	21±12
	0.40±0.04	1450–1650 °C, 3 GPa	Davis et al. (2013)	
Garnet	1.51±0.03	1450–1650 °C, 3 GPa	Davis et al. (2013)	

Note:
D(Ge) mineral-–liquid values for fractionnal crystallisation (Malvin and Drake, 1987; Capobianco and Watson, 1982; Adam and Green 2006) and partial melting (Davis et al. 2013).
D(Ge)/*D*(Si) : exchange coefficient for Ge and Si.

State-of-art investigations of the Ge oxidation state in sulfides, using synchrotron µ-X-ray absorption near edge spectroscopy (µ-XANES), have examined model compounds and samples from Tres Marias Zn deposit, Mexico (Cook et al. 2015) and St-Salvy Zn–Pb deposit, France (Belissont et al. 2016). Sphalerite from Tres Marias was strongly zoned, with (Ge–Fe)-poor and (Ge–Fe)-rich zones, having Ge contents of 252 and 1071 ppm, respectively

(Cook et al. 2015). Sphalerite from St-Salvy displayed complex zoning: (1) rhythmic bands with alternation of Ge-poor (Ge = 140 ppm), Fe-rich dark bands, and Ge-rich (Ge = 270 ppm), Fe-poor light bands, and (2) Ge-rich sector zones with the highest Ge contents (1100 ppm) (Belissont et al. 2014). Whatever the zones (Fig. 4), these two studies established that only Ge^{4+} was present in sulfide (Cook et al. 2015; Belissont et al. 2016).

Cosmochemistry of Ge isotopes

Over the last decade, several studies have emphasized the importance of Ge and its isotopes as novel tracers in planetary sciences. Since the pioneering measurements of elemental Ge in metal–silicate experiments by INAA techniques in the late 60's (Wai et al. 1968), the application of Ge geochemistry has been left aside in favor of other geochemical tracers (e.g., highly siderophile elements). It is only in the recent years that a renewed interest for Ge as an elemental tool for understanding core formation of planetary bodies has been put forward, mainly in experimental cosmochemistry (Righter et al. 2011; Siebert et al. 2011b). Recently, Ge isotopes have added new constraints on metal–silicate segregation, with the identification of distinct Ge isotopic signatures in iron meteorites as analogues of planetesimal cores and silicate reservoirs (Luais 2007, 2012).

Germanium isotope signatures of iron meteorites. Germanium, with Ni and Ir, is a key element for the classification of iron meteorites. Using the correlations between Ge and Ir with Ni and Au, Wasson et al. (1967, 2002) have identified 13 distinct groups, and more than 57 ungrouped meteorites, the minimum of individual members required for defining a group being five. The last identified group was group IIG with six members having the lowest Ni content of 4.2–4.6% (Wasson and Choe 2009). Modelling of trace element variations within the individual groups using solid metal–liquid metal partition coefficients defined two main classes: (1) magmatic irons representing the cooling of metallic liquid core, (2) non-magmatic irons related to the mixing between impact melts formed at the sub-surface of undifferentiated bodies (Choi et al. 1995). Germanium–Ni correlations indicated small within-group variations for magmatic irons, resulting from D_{Ge} solid metal–liquid metal = 0.64, but large variations of 5 orders of magnitude between groups (Ge = 0.01–650 ppm). Wasson and collaborators suggested that each meteorite group represents the core of a distinct parent body. Since metal–silicate partitioning of Ge strongly depends on redox conditions, and variations in Ge/Ni (and Ga/Ni) in iron meteorites may result from loss of Ge through volatility (Davis 2006), Ge isotopes can add strong constraints on thermodynamic conditions during parent body differentiation, and/or formation processes of iron meteorites. Luais (2007) analyzed the Ge isotopic composition of iron meteorites from different groups and classes (Fig. 5) and showed that Ge isotopes bring new constraints on the classification of iron meteorites, with a decoupling between Ge content and its isotopes. Results showed constant $\delta^{74/70}Ge_{NIST3120a}$ of $+1.41 \pm 0.22‰$ for magmatic irons, and $\delta^{74/70}Ge$ ranging from -0.63 to $+1.04‰$ for non-magmatic irons.

Magmatic irons. The similarity in Ge isotopes of magmatic irons implies that they do not fractionate during fractional crystallization of metallic liquid, which is characterized by a large decrease in Ir contents from 70 ppm in IIA irons to less than 0.01 ppm in IIB irons. The lack of correlation between Ge isotopic composition and Ni content (and Ge/Ni), which is a proxy for redox parameter in metallic phase, would indicate similar fO_2 conditions during metal crystallization. However, the specific occurrence of oxygen-bearing phases such as phosphates in IIIAB irons (Olsen et al. 1999), indicated some differences in redox conditions, the IIAB group being more reduced than the IIIAB ones. The constant Ge isotopic composition in magmatic irons, independent of Ge content and redox conditions in the parent body, are best explained by a near-complete partitioning of siderophile Ge into the metal phase that drives Ge isotopic composition close to the initial composition of the parent bodies (Luais 2007). In light of experimental studies (Luais et al. 2007), this implies strong reducing conditions that avoid loss of Ge during iron meteorite formation. Small fO_2 variations would not induce

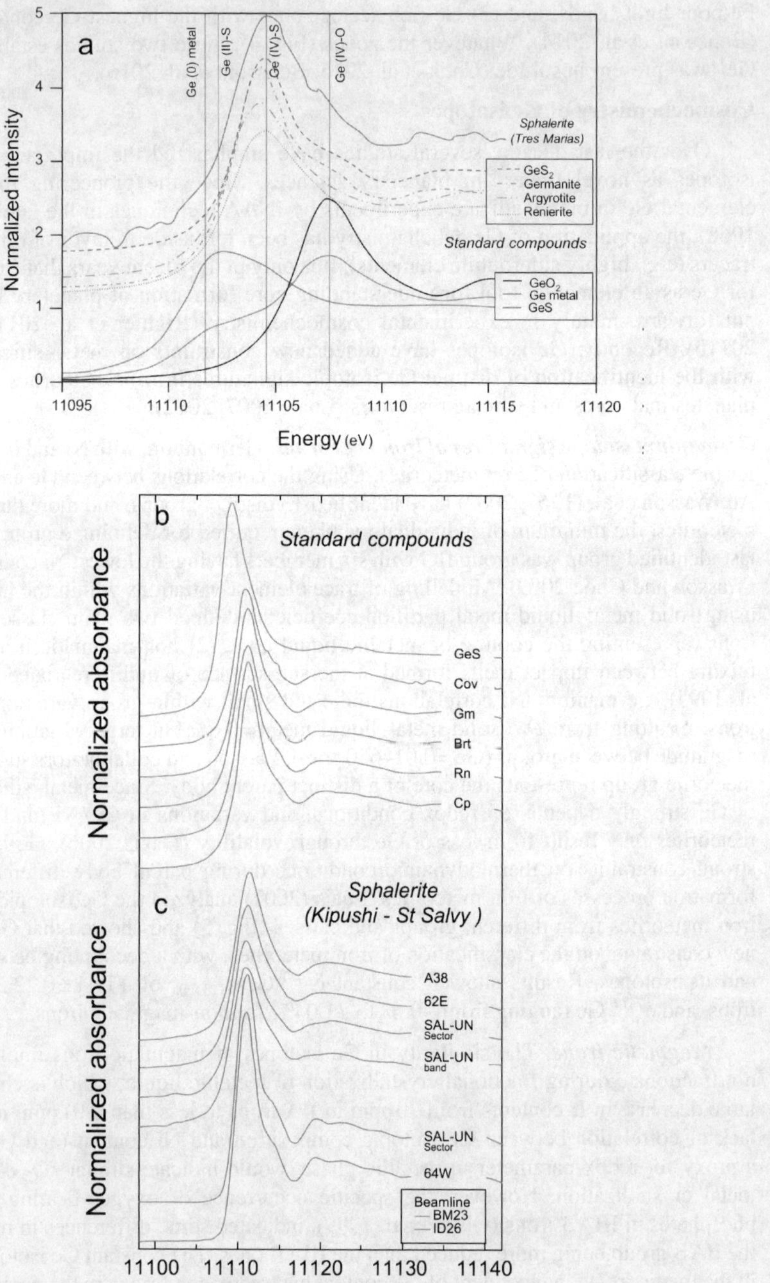

Figure 4. Synchrotron Ge K-edge µ-XANES spectra at the Ge K-edge (a) for standards compounds and Tres Marias sphalerite (Mexico, Cook et al. 2015); (b)(c) from Belissont et al. (2016) : (b) for Ge^{2+} (GeS) compounds, and Ge^{4+} reference materials of renierite (Rn), gernanite (Gm), briartite (Brt) chalcopyrite (Cp), covellite (cv), and (c) St-Salvy–France (64W, 62E, SAL-UN) and Kipushi–D.R. Congo (A38) samples.

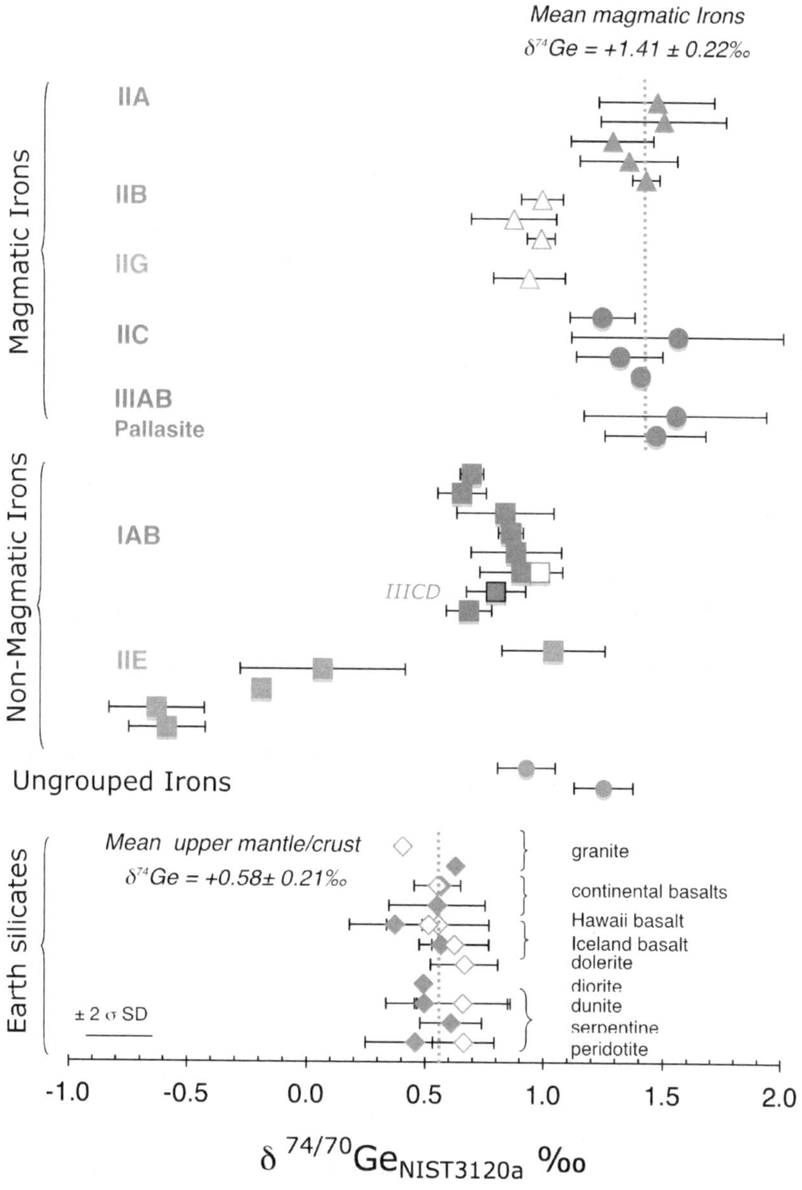

Figure 5. Ge isotopic compositions of magmatic, non-magmatic and ungrouped iron meteorites, pallasite, and Earth silicate samples, relative to NIST3120a Ge standard. Irons and silicate earth (full symbols) data are from Luais (2007, 2012) and Luais et al. (2014); data for Toluca IAB iron and silicates (empty symbols) are from Escoube et al. (2012).

a detectable change in Ge isotopes. The constant Ge isotope compositions of magmatic iron meteorites implied some homogeneity in Ge isotope signatures among parent bodies. The lack of any correlation between $\delta^{74/70}$Ge values with variations in Ge/Ni ratios, the latter quantifying nebular condensation processes at a given oxidation and fractionation state (Kelly and Larimer 1977) led Luais (2007) to propose that the homogeneous $\delta^{74/70}$Ge signature was established early in planet history, through nebular condensation and large-scale isotopic homogenization.

The IIB and IIG magmatic irons have slightly lower Ge isotopic composition than the average for magmatic irons (Fig. 5). Their similar $\delta^{74/70}$Ge values of 0.97±0.16‰ for IIB irons and 0.93±0.02‰ for IIG (Luais 2007, 2012, Luais et al. 2014) confirm their affiliation. Their low Ir contents (Ir IIB=1 to 0.01 ppm, Ir IIG=0.2 to 0.02 ppm), indicate that they represent differentiated metallic liquid. They are also characterised by large inclusions of troilite (FeS) and schreibersite $(FeNi)_3P$, that are thought to be formed by late-stage exsolution of P and S during fractional crystallisation. Measured Ge contents were very low in schreibersite (Ge=0.7–1.8 ppm), and undetectable in troilite (Luais et al. 2014). This was explained by Ge being strongly compatible in metal solid, with $D(\mathrm{Ge})_{\text{solid metal/liquid metal}}$ increasing with increasing S and P in the metallic liquid during the late stage of fractional crystallisation. Thus far, Ge isotopic compositions have been measured in both metal and schreibersite, but not in troilite. $\delta^{74/70}$Ge values in the schreibersite were strongly negative with respect to the metal, giving Δ^{74}Ge metal-schreibersite=+3.5 to+4.2‰ in Santa Luzia (IIB) and Tombigbee River (IIG) (Luais et al. 2014). These results highlight that Ge isotopic fractionation between FeNi liquid and FeNi+S(P) liquid will have to be considered in a global model of early Fe–FeS (and P) metal segregation, preceding metal–silicate differentiation of planetesimals (Kruijer et al. 2014).

Non-magmatic irons. Impacts processes are involved in the formation of non-magmatic iron meteorites (Wasson et al. 1980; Choi et al. 1995). These IAB and IIE groups are characterized by the occurrence of silicate inclusions, which are absent in magmatic irons. Based on petrological, mineralogical and geochemical investigations, the commonly accepted hypothesis of formation is impact processes on chondritic or on already differentiated parent bodies that result in complex processes of parent-body break-up, hit-and-run collision and major silicate–metal mixing (Benedix et al. 2000; Takeda et al. 2000; McDermott et al. 2016). Luais (2007) showed that the Ge isotopic composition of these non-magmatic irons were distinctly lower than those of magmatic irons, and highly variable (Fig. 5).

The IIE irons have highly variable $\delta^{74/70}\mathrm{Ge}_{\mathrm{NIST3120a}}$ ranging from −0.63 to +1.04‰ (Fig. 6) that are inversely correlated with their Ge contents. This trend can be modelled by evaporation processes, leading to Ge loss that preferentially removes light isotopes of Ge, leaving a residue enriched in heavy isotopes. However, loss of light Ge isotopes from IIE irons is smaller than predicted from pure Rayleigh evaporation model ($\beta=0.5$), emphazing that the produced vapor is not simply lost, but continues to interact with particles. The smaller depletion in light isotopes than predicted by kinetic processes, commonly referred as suppressed isotope fractionation, would then reflect gas-residue interaction and back reaction (Cohen et al. 2004). Detailed investigations indicated that $\delta^{74/70}$Ge variations in IIE irons were not random, but were age-correlated: samples with lower $\delta^{74/70}$Ge values, the so called "Miles group", were older than the "Watson group" with higher $\delta^{74/70}$Ge, (Bogard et al. 2000; Snyder et al. 2001). Therefore, increasing $\delta^{74/70}$Ge values from the oldest to the youngest IIE samples possibly reflects progressive and repeated impact events on parent body (Luais 2007). Recently, McDermott et al. (2016) calculated the closure temperature from two-pyroxene geothermometry in silicate inclusions from IIE sample to determine the cooling rate. They found a correlation between the degree of differentiation of silicate inclusions and cooling rates, the less differentiated (chondritic) inclusions in the Watson sample having higher closure temperatures (1120 °C) and faster cooling rates than the more differentiated inclusions in the Miles group with lower closure temperatures (≈1020 °C) and lower cooling rates. It would

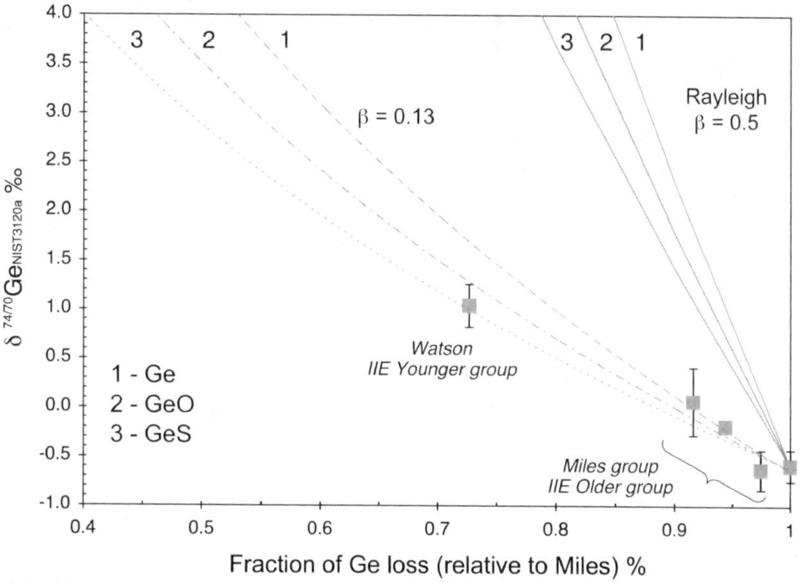

Figure 6. Modelling evaporation processes for Ge isotopic compositions of IIE irons ($\delta^{74/70}Ge_{NIST3120a}$) using Raleigh type fractionation (modified from Luais 2007, 2012). The data can be modelled using $\beta < 0.5$ for the three germanium model compounds: Ge metal, GeO and GeS species. A smaller isotope mass fractionation is shown for heavier molecular species (see sections *Theoretical considerations* and *Diffusion of Ge* for discussion).

be important to decipher if the cooling rate of silicate inclusions can be related to impact events. If so, this would be in line with Ge isotopic data, the high $\delta^{74/70}$Ge–low Ge Watson sample having with faster cooling rates than high $\delta^{74/70}$Ge-high Ge Miles group with slower cooling. Kinetics of isotopic fractionation would then demonstrate that fast cooling rate would produce more extensive evaporation and loss of light Ge isotopes, because of less vapor–residue interactions, thus generating a high $\delta^{74/70}$Ge in the metallic melt.

By contrast, the IAB–IIICD irons have constant $\delta^{74/70}Ge_{NIST3120a}$ of $+0.79 \pm 0.16$‰ (Fig. 5), confirming that former IAB and IIICD groups can be merged into a single group named IAB (Wasson and Kallemeyn 2002). Their constant $\delta^{74/70}$Ge values would indicate they originate from a single melt pod, generated during a single large impact heating event, in full agreement with trace element interpretations of Wasson and Kalleymen (2002).

Germanium isotopes as tracers of metal–silicate segregation. Rouxel et al. (2006), Luais (2012) and Escoube et al. (2012) analyzed Ge isotopes compositions of terrestrial silicates representing oceanic crust, continental crust, and mantle. A detailed discussion of these data is given in the following section. The mean $\delta^{74/70}$Ge composition for Earth silicate reservoir ranged from +0.53 to +0.64‰ (Rouxel et al. 2006; Luais 2012; Escoube et al. 2012, see next section), indicating a light Ge isotopic composition with respect to core reservoir that is reasonably assumed to be that of magmatic Fe meteorites with $\delta^{74/70}$Ge=+1.41‰. The isotopic fractionation factor of $\Delta^{74/70}Ge_{Fe\text{-}met\text{–}Earth\ silicate} \approx 0.85$‰ for the Earth system can be compared with theoretical prediction of equilibrium isotopic fractionation based on crystallographic configuration of Ge in metal and silicate, which depends on Ge oxidation state (Ge^0 in metal, Ge^{4+} in silicate), coordination (octahedral Ge in metal, and tetrahedral Ge in silicate), and length of Ge–Fe (metal) and Ge–O (silicate) bonds (Table 4). Considering that heavy isotopes should prefer shorter (i.e., stronger) bond length, higher oxidation state, and low coordination number, yielding $\Delta^{74}Ge_{Fe\text{-}met\text{–}Earth\ silicate} < 0$, the measured Ge isotopic compositions of iron meteorites and silicate Earth is in contradiction with theoretical considerations. Such a discrepancy raises the question of metal–silicate equilibrium at the time of core formation, although it is known that modern Earth mantle is not in chemical and thermodynamic equilibrium with the Earth's core (e.g., Stevenson 1981; Jones and Drake 1986). While it is conceivable that the Fe-meteorites represent a suitable good proxy for the composition to the Earth's core, the use of accessible silicate Earth samples is more questionable. The influence of pressure on Ge crystallographic configuration in silicate, as well as the volatile behavior of Ge at high temperature must be taken into account in modelling metal–silicate isotopic fractionation of Ge. However, it must be emphasized that the large difference in Ge isotopic composition between Fe-meteorites and Earth mantle makes Ge an interesting tracer to explore the complex processes of accretion and core formation. An important goal for future studies would be to determine where Ge sits in the lower mantle and how it is coordinated. Germanium compatibility and coordination in such phases may influence the Ge isotopic mass balance in Earth mantle.

GERMANIUM ISOTOPE SYSTEMATICS IN IGNEOUS, MANTLE-DERIVED ROCKS, AND METAMORPHIC ROCKS

The Ge isotopic composition of the Earth silicate reservoirs

Germanium isotopic compositions of Earth silicate samples determined by Rouxel et al. (2006), Luais (2012) and Escoube et al. (2012) are detailed in Figure 7. Samples are representative of mantle (peridotite, dunite) and altered lithologies (serpentine), basaltic crust (MORB, OIB : Hawaii, Iceland), continental basalts and continental crust (granite). Independent determinations of average composition for the Earth silicate reservoirs of $\delta^{74/70}$Ge = +0.64 ± 0.16‰ (recalculated to NIST3120a, Rouxel et al. 2006), $\delta^{74/70}$Ge = +0.53 ± 0.16‰ (Luais 2012), $\delta^{74/70}$Ge = +0.56 ± 0.18‰ (Escoube et al. 2012) agreed within analytical errors, and the whole average composition of the bulk silicate Earth (BSE) converges toward $\delta^{74/70}$Ge = +0.58 ± 0.21‰ (2 s.d.). Overall, variations in the Ge isotopic composition of the silicate Earth reservoirs were small.

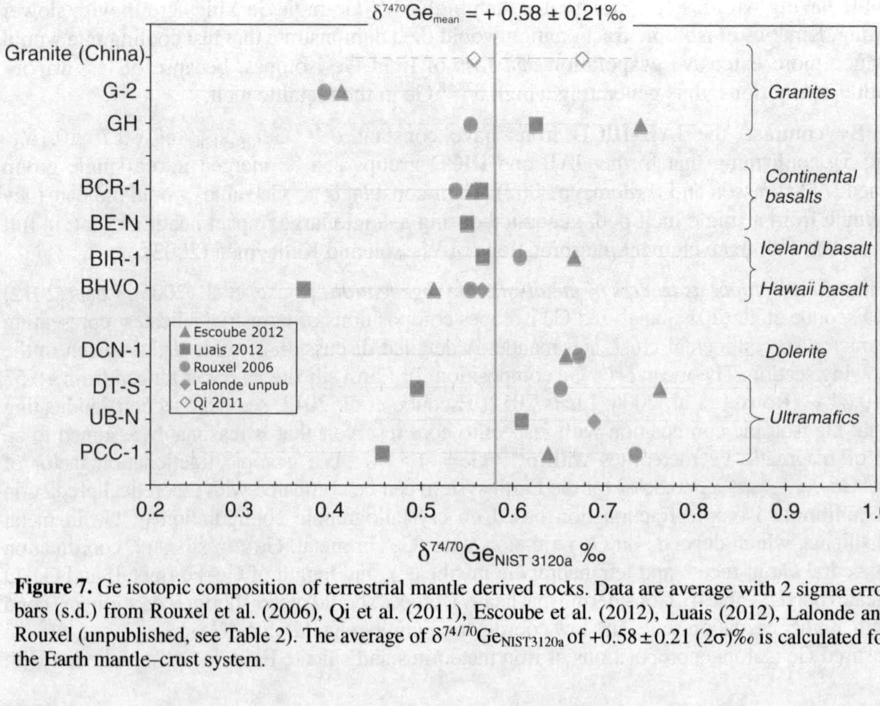

Figure 7. Ge isotopic composition of terrestrial mantle derived rocks. Data are average with 2 sigma error bars (s.d.) from Rouxel et al. (2006), Qi et al. (2011), Escoube et al. (2012), Luais (2012), Lalonde and Rouxel (unpublished, see Table 2). The average of $\delta^{74/70}Ge_{NIST3120a}$ of +0.58 ± 0.21 (2σ)‰ is calculated for the Earth mantle–crust system.

Using this sample set, ultramafic rocks were characterized by Ge/Si ratios of around 1.8×10^{-6} μmol/mol. By comparison, the average Ge/Si ratios of basalts were higher, at around 2.5×10^{-6} μmol/mol which is in good agreement with previous estimates of the oceanic crust based on a much larger sample set (De Argollo and Schilling 1978b; Bernstein 1985). The average Ge/Si ratio of continental crust samples (around 1.8×10^{-6}) was significantly higher than previous estimates (Mortlock and Froelich 1987) which can be explained by the limited number of considered samples and potential lithological heterogeneities. The lack of relationship between the Ge/Si and $\delta^{74/70}$Ge suggests that the estimate of bulk silicate Earth $\delta^{74/70}$Ge value given above was not significantly biased by sampling.

There was no correlation of $\delta^{74/70}$Ge with differentiation index such as SiO_2. Although Luais (2012) identifed some slight distinction in $\delta^{74/70}$Ge between petrologic types, our more recent compilation suggests that $\delta^{74/70}$Ge values of mantle ultramafics, basalts, and granites mainly overlap (Fig. 7). Based on Fe isotope systematics, a distinction between fertile upper mantle and MORB-OIB have been put forwards (Williams et al. 2005; Weyer and Ionov 2007) and interpreted on the basis that Fe^{III} is more incompatible than Fe^{II} during partial melting, resulting in heavier Fe isotopic composition in the melt (Dauphas et al. 2009, 2014). With only Ge (IV) occurring in silicates and silicate melt, a similar valence-dependent process should be excluded, even if the data suggest a progressive increase in $\delta^{74/70}$Ge from mantle ultramafics, to mantle-derived lithologies and igneous rocks through partial melting.

A corrrelation between $\delta^{74/70}$Ge and NBO/T index, where NBO is the number of nonbridging oxygen atoms, was identified (Fig. 8), suggesting a link between Ge isotope fractionation and the degree of polymerization of the silicate (Luais 2012). Basic rocks (tholeiite) have higher NBO/T values (mean 1.8) than felsic rocks (rhyolite) with NBO/T close to 0 (Mysen 2004), integrating the NBO/T values for minerals (NBO/T=4 for olivine, NBO/T=0 for quartz, Mysen 2004). The $\delta^{74/70}$Ge values were found to be lower in rocks having a higher percentage of olivine (NBO/T=4) such as ultramafics, and higher in rocks having higher percentage of tectosilicates (quartz, felsdpar, NBO/T=0) as in granite (Fig. 8). In addition, Ge–O bond lengths are longer in olivine (1.82Å) than in quartz (1.73–1.78Å) (Table 4) (Li et al. 2009). The direction of isotopic fractionation is in line with the theoretical predictions, in that heavy isotopes will be favored with stronger, shorter bonds, then ordering $\delta^{74/70}$Ge quartz>$\delta^{74/70}$Ge olivine. The variation in $\delta^{74/70}$Ge observed for mantle-derived and crustal samples, with $\delta^{74/70}Ge_{granite}>\delta^{74/70}Ge_{basalt}>\delta^{74/70}Ge_{ultramafics}$ is consistent with equilibrium fractionation between mineral and silicate melt. This observation is remarkably similar to Si isotope systematics showing that continental crust is slightly heavy relative to the mantle (Savage et al. 2014). When considering the whole set of data for $\delta^{74/70}$Ge (Fig. 7), this correlation is less clear. One reason is that the range of $\delta^{74/70}$Ge values is small and at the limit of reproducibility (≈0.2‰).

Serpentinite exhibited significantly higher $\delta^{74/70}$Ge value of +0.61‰ than fresh peridotite and dunite with $\delta^{74/70}$Ge values of 0.46 and 0.5‰, respectively. Serpentinite generally formed through hydrothermal interaction with ultrabasics, at high temperature (e.g., $T\approx260\,°C$, Deschamps et al. 2012). Heavier $\delta^{74/70}$Ge value could result from water-rock interactions, with the addition of Ge from hydrothermal fluids with higher $\delta^{74/70}$Ge values, as recently measured in seafloor hydrothermal systems (+1.55±0.3‰; Escoube et al. 2015).

The Hawaiian basalt (BHVO) plots under the $\delta^{74/70}$Ge–NBO/T regression line (Fig. 8). This sample from Kilauea volcano belongs to the "Kea Trend" that is characterised by low $^{87}Sr/^{86}Sr$ and $^{208}Pb/^{206}Pb$ and high $^{143}Nd/^{144}Nd$ ratio typical of the "PREMA" component. According to Weiss et al. (2011), the "Kea trend" could originate from the lower mantle. If it ascended without interaction, it may be sourced from the dense material that is recognized at the core–mantle boundary from geophysical investigations. Alternatively, the presence of garnet in the mantle source with high K_d (Ge) (Table 1), as well as the pressure effect on the equilibrium Ge isotopic fractionation will need to be investigated. If such, it would represent lower mantle Ge isotopic signature. Additional high precision Ge isotopic data are required to confirm this value.

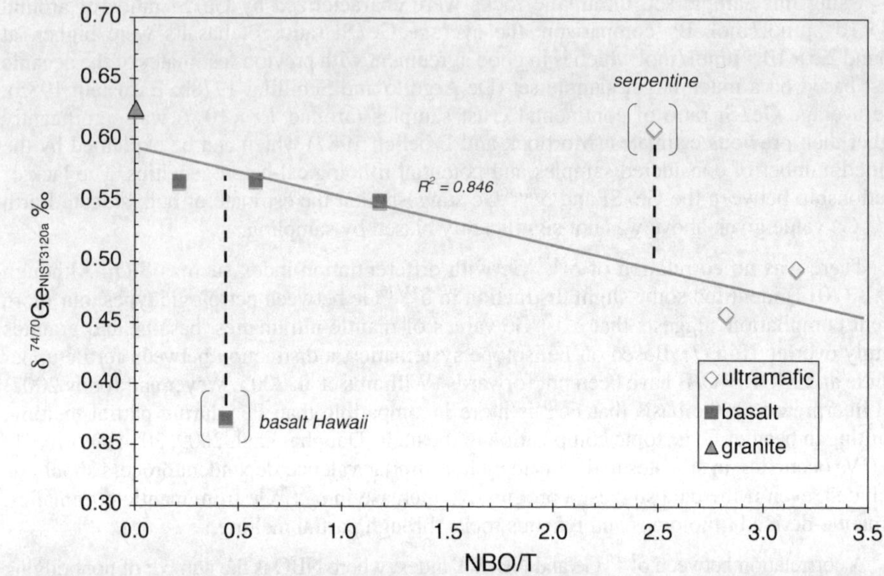

Figure 8. Ge isotopic composition of mantle-derived rocks in function of NBO/T parameter (from Luais 2012). NBO/T quantifies the number of oxygen atoms that are not shared by two tetrahedrons (NBO: nonbridging oxygen), relative to the proportion of tetrahedrally coordinated cations. The serpentine and Hawaiian basalt samples (in brackets) are excluded from the regression line.

Germanium recycling into the mantle: an attempt to evaluate mantle homogeneity

Germanium isotopic data for mantle-derived samples have shown small detectable variations that question the extent of Ge homogeneity in the Earth's mantle. Subduction settings are important zones of intensive fluid–rock interaction driven by metamorphic reactions and dehydration of the subducted slab, then contributing to significant mass transfer into the mantle. Since Ge can be transported with aqueous fluids (Pokrovski and Schott 1998a), it is possible that Ge transport from the subducted slab to the mantle wedge can be associated with significant isotopic fractionation. To address this issue, El Korh et al. (2015, 2016) measured the Ge isotopic composition of metamorphic rocks in a series of HT–LT rocks at peak *P–T* conditions ($P = 1.7$–2.3 GPa and $T = 500$–$600\,°C$) from the Ile de Groix (France). The studied metabasites correspond to former MOR-type basalts that underwent pre-HP hydrothermal alteration, prograde metamorphism in blueschist and eclogite facies, and then retrograde metamorphism in greenschist facies at lower T ($<450\,°C$). El Korh et al. (2015, 2016) reported no Ge isotopic variation in blueschists and eclogites, with $\delta^{74/70}$Ge values (+0.42 to +0.65‰) within the range of tholeiitic basalts (+0.55–0.57‰; Luais 2012). Germanium being mainly hosted in garnet (Ge = 3.3–8.2 ppm) and epidote (2.4–12 ppm) (El Korh et al. 2015), the lack of Ge isotopic fractionation indicated that Ge remains incorporated in different generations of garnet and epidote, that were stable during prograde metamorphism. By contrast, significant Ge isotopic variations were seen in retrograde greenschist samples that were free of garnet, with $\delta^{74/70}$Ge values up to +0.98‰. Hence, Ge isotopic fractionation was likely related to garnet breakdown and fluid interaction during retrogression at lower temperatures, that released Ge in solution. In addition, El Korh et al. (2016) demonstrated that the $\delta^{74/70}$Ge vs. (Fe^{2+}/Fe_{tot}) correlation, was indicative of intensive fluid–rock interaction under reducing conditions. This first study on Ge isotopic variations related to metamorphic processes emphasized that Ge isotopes can fractionate during the breakdown of Ge-bearing minerals, leading to the release of Ge in a fluid phase under reducing conditions. Slab dehydration with prograde metamorphism in the stability field of garnet did not fractionate Ge isotopes, meaning that subduction of oceanic crust would not produce significant mantle heterogeneities.

Ore deposits

Understanding the formation of ore deposits, from crustal or mantle source to deposits, requires deciphering the distinct processes of element dissolution into the fluid phase, fluid transport and mineral precipitation under specific thermodynamics conditions. Whereas Ge is generally present at trace levels in ore deposits (Holl et al. 2007), its concentrations can vary from few ppm to thousands of ppm, with ZnS (sphalerite) being one of the main Ge-bearing mineral. These high concentrations contrast with those found in silicates (1.5–1.6 ppm in the crust and 1.1 ppm in the mantle, e.g., Berstein 1985). The recent analytical development of *in situ* measurements of Ge concentration and associated trace elements by laser ICPMS techniques has opened new fields of research, allowing detection of chemical variations at the mineral scale. It has been shown that optical zonation in sphalerite crystal are related to strong chemical zonings, with variation in Ge contents from hundred to thousands of ppm in single mineral (Belissont et al. 2014). These insights into fine-scale sphalerite structure has led to understanding the incorporation mechanism of Ge in sphalerite through coupled substitutions with other metal trace elements (Ga, In, Cd, Ag, Cu, Fe) (Johan 1988; Yee et al. 2007; Cook et al. 2009), with Cu providing the charge balance for the entire set of substitutions (Belissont et al. 2014). In turn, assessing fine-scale chemical variations provides important constraints, together with thorough statistical tools such as PCA (Belissont et al. 2014; Frenzel et al. 2016) for defining the thermodynamic conditions of sphalerite precipitation and sources.

Li et al. (2009) established the preliminary parameters of theoretical Ge isotope fractionation of sphalerite-like sulfides and calculated a fractionation factor $\Delta^{74/70}Ge_{sph-aqueous}$ at 25 °C of 12.2 to 11.5‰, indicating a very light isotopic composition in sphalerite compared to the fluid in equilibrium. Field data confirmed that sulfides are generally characterized by light Ge isotopic composition, with variable $\delta^{74/70}$Ge ranging from −5 to +2‰ (Fig. 9). Several studies

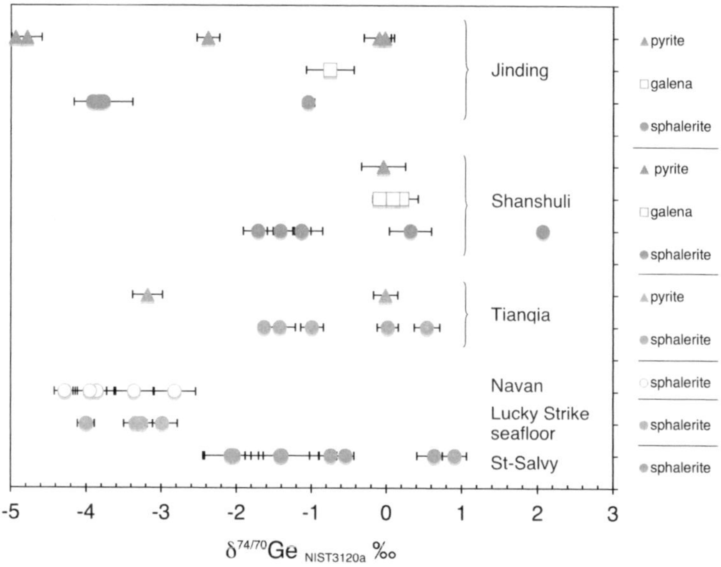

Figure 9. Ge isotopic compositions of sulfides. Data from St-Salvy (Luais, 2007, 2012; Belissont et al. 2014), Navan Irish district and Lucky Strike seafloor hydrothermal system (Escoube et al. 2012), China (Tianqiao, Shanshulin and Jinding sites, Meng et al. 2015).

reported $\delta^{74/70}$Ge values in sphalerites from continental and marine deposits, such as St-Salvy vein-type deposits (France), Irish-type Zn–Pb Navan Deposits (Ireland), Lucky Strike seafloor hydrothermal vent, and in various sulfides (pyrite, sphalerite and galena) from several Zn–Pb deposits in Southwest China (Luais 2007, 2012; Escoube et al. 2012; Belissont et al. 2014; Meng et al. 2015). Overall, there were no systematic differences in $\delta^{74/70}$Ge between sulfide minerals, indicating that local source, or temperature-dependant mineral–mineral equilibrium control the Ge isotopic fractionation. Each case study must be discussed separately.

Sulfides from Zn–Pb deposits in Southwest China (Meng et al. 2015) did not show any correlation between $\delta^{74/70}$Ge and Ge contents as a whole, or within each individual mineral phases. Meng et al. (2015) proposed that Rayleigh-type isotopic fractionation may account for the heavy isotopic composition of Ge in the late galena. However this temporal evolution cannot be confirmed when taking into account $\delta^{74/70}$Ge variations within each sulfide species, because of the large overlap of their $\delta^{74/70}$Ge values. Sphalerites from St-Salvy, Irish and Lucky Strike deposits, when reported together, displayed a well-defined positive trend between Ge contents and $\delta^{74/70}$Ge values, with the Irish and Lucky Strike samples having the lowest values (Fig. 10). Belissont et al. (2014) demonstrated that this trend cannot be modelled by Rayleigh-type fractionation during the cooling of the mineralizing fluid; regardless of the exact α values ($\alpha_{sph-fluid}=0.999–0.995$) and the hydrothermal fluid compositions ($\delta^{74/70}Ge=+1$ to $+2.6$, in the range of data in Escoube et al. 2015), since the different modeling trends all resulted in a negative correlation between Ge concentrations and $\delta^{74/70}$Ge (Fig. 11).

Redox state and temperature are two parameters that can control Ge isotopic fractionation. Li et al. (2009) proposed that redox effects on the theoretical Ge isotopic fractionation in sulfides are small, by calculating a slightly lower $\Delta^{74/70}Ge_{GeIV\ sphalerite-aqueous}$ of 11.4‰ than $\Delta^{74/70}Ge_{GeII\ sphalerite-aqueous}$ of 12.2 and 11.5‰ (25 °C). Belissont et al. (2016) investigated the Ge oxidation state in sphalerites using *K*-edge *μ*XANES synchrotron

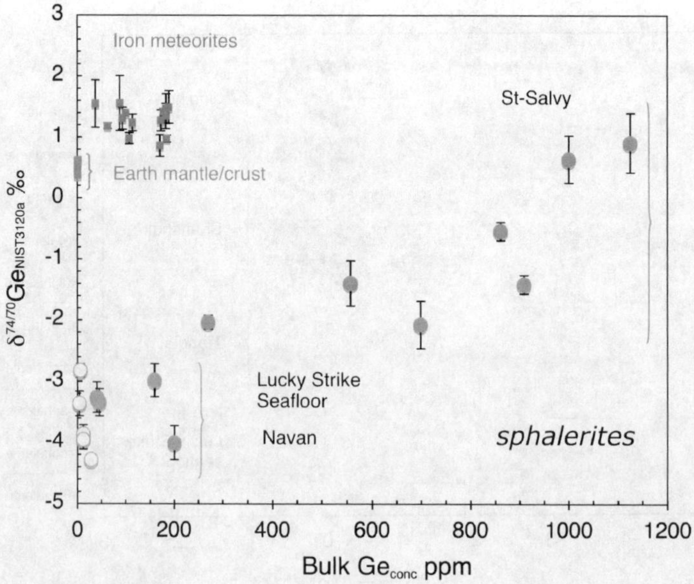

Figure 10. Ge isotopic composition versus Ge content in sphalerites from St-Salvy (Luais 2007, 2012; Belissont et al. 2014) and Navan Irish and Lucky Strike deposits (Escoube et al. 2012). Iron meteorite data from Luais (2007, 2012), and Earth mantle and crust data from Rouxel (2006), Escoube et al. (2012) and Luais (2012).

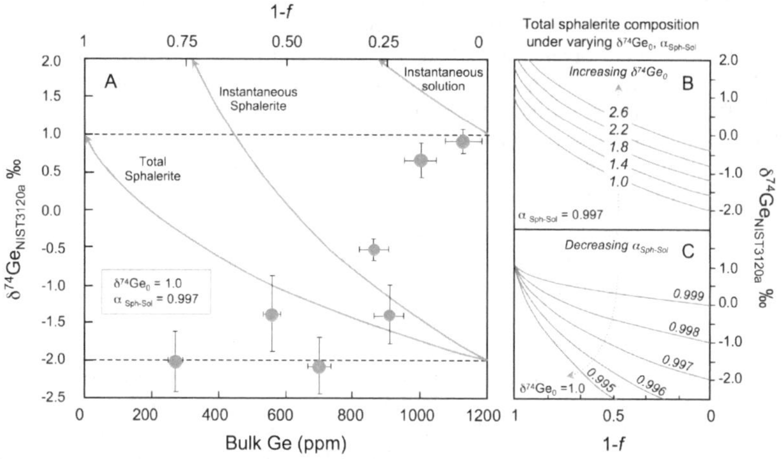

Figure 11. Modelling Ge isotopic fractionation in sphalerite from St-Salvy using Rayleigh laws for kinetic fractionation during sphalerite precipitation, (Belissont et al. 2014), as a function of the fraction f of residual Ge in solution. (A) For given values of $\delta^{74/70}Ge_{fluid}=1$ and $\alpha_{sphalerite–fluid}$ isotopic fractionation factor=0.997; Ge isotopic fractionation curves are given for total sphalerite (i.e. accumulating in the system), instantaneous sphalerite (i.e. produced at in an infinitely short time), and remaining solution, (B) total sphalerite composition for a range of $\delta^{74/70}Ge_{fluid}$ values, (C) total sphalerite for different $\alpha_{sphalerite–fluid}$ values. Modeling curves result in negative $\delta^{74/70}Ge$ versus Ge and (1–f) correlations, and do not fit the St-Salvy sphalerite data (filled circles).

techniques and demonstrated that sphalerites from St-Salvy have incorporated only Ge^{4+}, meaning that redox effect cannot constrain the $\delta^{74/70}Ge$–Ge trends. The authors concluded that Ge isotopic variations were temperature-controlled. A $\Delta^{140\,°C}_{sphalerite\text{-}fluid} - \Delta^{80\,°C}_{sphalerite\text{-}fluid}$ giving $\delta^{74/70}Ge^{140\,°C}_{sphalerite} - \delta^{74/70}Ge^{80\,°C}_{sphalerite}$ of 2.98‰ can be reproduced for a temperature range of 70–150 °C, slightly larger than the T=80–140 °C estimates of fluid inclusions in quartz belonging to the same paragenesis. This means that fluid temperature can partly explain $\delta^{74/70}Ge$ variations in St-Salvy sphalerites.

LOW-TEMPERATURE GEOCHEMISTRY

Fundamentals of Ge low-temperature geochemistry

Speciation in aqueous systems. In natural environments, Ge may occur as both inorganic and organometallic forms. Germanate in aqueous solution includes $Ge(OH)_4$, $GeO(OH)_3^-$, $GeO_2(OH)_2^{2-}$ and H_2GeO_3 species (Ingri 1963), with germanic acid $Ge(OH)_4$ being the predominant inorganic form in natural waters. Organometallic Ge species exist in nature as methyl-Ge (MMGe) and dimethyl-Ge (DMGe) forms (Lewis et al. 1985, 1989), representing more than 70% of the total Ge in the ocean. Although Hambrick et al. (1984) and Hirner et al. (1998) reported the presence of trimethyl-Ge (TMGe) in rainwater and geothermal waters, the occurrence of TMGe in rivers and seawater has not been demonstrated. In contrast to the inorganic form of Ge showing nutrient-type behavior in seawater, much like that of silicic acid, organometallic Ge species are rather conservative throughout the water column, averaging 345±22 pM and 100±5 pM formMGe and DMGe respectively. The sources of methyl-Ge species are still poorly known, but likely involve microbial biomethylation (Lewis et al. 1985).

Unlike Si, Ge forms stable six fold coordination (octahedral) complexes with various organic ligands (Pokrovski and Schott 1998b). This property is particularly important for natural waters with high concentrations of humic substances and organics, as it may explain the strong decoupling of Ge vs. Si during continental weathering and run-off. The ability of Ge to form complexes with different organic ligands (e.g., humic and fulvic acids) is thus an important feature of Ge geochemistry in aqueous systems (Pokrovski and Schott 1998b; Pokrovski et al. 2000), possibly accounting for more than 95% of the total dissolved Ge at $pH \geq 6$–7 in natural river waters. Germanium forms complexes of chelate type with the following functional groups: (a) carboxylic in acid solutions, (b) di-phenolic hydroxyls in neutral and basic solutions, and (c) alcoholic hydroxyls in very basic solutions (Rosenberg 2009).

According to Pokrovski (1998a), Ge does not form complexes with important inorganic ligands like Cl^-, HCO_3^-, H_2S, Na^+ in natural environments, including hydrothermal systems. In hydrothermal systems of near-neutral pH and in the temperature range of 20–350 °C, Ge is mainly transported as the neutral hydroxide $Ge(OH)_4$. According to available experimental data (Ciavatta et al. 1990), Ge-fluoride complexes can occur only in acid and fluor-rich solutions that are rarely found in hydrothermal environments. Although the enrichment of Ge in hydrothermal sulfide deposits may be taken as an indication that S-bearing Ge species may also form in some hydrothermal fluids, there is no evidence for the natural occurrence of thiogermanate complex $[GeS_4]^{4-}$. Finally, the redox potential of most hydrothermal and surficial environments seems too high to cause significant formation of divalent Ge^{2+} species.

Role in biological systems. Germanium has been characterized by Babula et al. (2008) as a non-essential element for human, flora and fauna life. However, Ge can be toxic to plants at high levels (Halperin et al. 1995) and has recognized physiological functions at trace levels (Cakmak et al. 1995) that differ from those of Si (Epstein 1994). Opal phytoliths from higher plants (e.g., graminoids) generally display lower Ge/Si, which underlines a systematic discrimination against Ge during plant uptake (Derry et al. 2005; Blecker et al. 2007) as a likely defense against its toxicity (Puerner et al. 1990). In contrast, significant accumulation of Ge in roots has been observed in both more primitive and modern plants (e.g., horsetails and banana plants) (Delvigne et al. 2009). There is no evidence of any essential function in humans, and no Ge deficiency syndromes have been documented. Although its accute toxicity is low (Tan et al. 2015), there has been several reports of severe human cases linked to prolonged intake of Ge products, leading to renal failure and even death (Tao and Bolger 1997). In both human cases and animal studies, toxicity was associated with ingested Ge products such as GeO_2 dioxide and organic Ge compounds (Schauss 1991). Although it is clear that Ge products present a potential, although poorly investigated, human health hazard, there also have been reports of therapeutic applications (Goodman 1988).

Behavior during weathering processes. Chemical weathering, especially of silicate minerals, has important roles in soil development, nutrient availability in terrestrial and marine ecosystems, buffering of acid rain, and long-term atmospheric CO_2 regulation (Walker et al. 1981; Berner et al. 1983; Gaillardet et al. 1999; Berner 2003). Germanium and Si exhibit substantial geochemical differences during weathering and Ge/Si ratios in stream waters appear to show systematic behavior related to weathering processes. Germanium is preferentially incorporated into weathering products such as clays in soils, leading to a lowering of river water Ge/Si compared to bed-rock (Mortlock and Froelich 1987; Froelich et al. 1992; Kurtz et al. 2002; Anders et al. 2003). Germanium/silicon (Ge/Si) ratios in unpolluted streams vary from ~0.3–1.2 μmol/mol and are almost always lower than Ge/Si of the bedrock they drain (Mortlock and Froelich 1987). The riverine dissolved Ge/Si ratio may exhibit seasonal variations (Mortlock and Froelich 1987) and appears to correlate positively with 1/Si (Murnane and Stallard 1990; Froelich et al. 1992). This suggests that Ge/Si ratios in steam could be approximated by mixing of Si from two sources (Fig. 12), one derived from weathering of primary minerals and the other derived from weathering of secondary clays. The underlying assumption is that incongruent weathering is characterized by a clay/bedrock distribution coefficient > 1 (Murnane and Stallard 1990) resulting in a high Ge/Si component retained by soils and a low Ge/Si component released

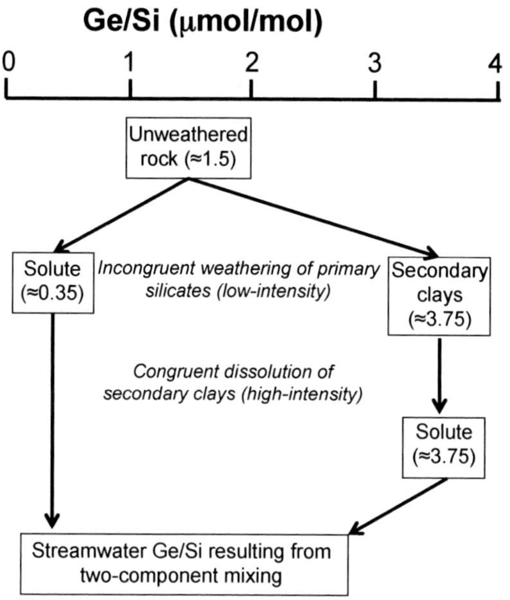

Figure 12. Schematic model of Ge/Si fractionation during weathering (Froelich et al. 1992; Kurtz et al. 2002). In this model, Ge is partitioned into secondary aluminosilicates (clays), resulting in streamwaters with low Ge/Si. More intense weathering releases the Si (and) Ge otherwise stored in clays, contributing higher Ge/Si ratios to streamwater. The Ge/Si ratio of streamwater reflects the balance of these two processes.

to stream waters. According to this scheme, the Si source with high Ge/Si is derived from dissolution of the Ge-enriched secondary minerals under more intense chemical weathering. The Ge/Si ratio in rivers is therefore controlled by the mixing relationships of these two sources and varies as a function of weathering intensity (Kurtz et al. 2002).

Hydrothermal and geothermal systems. Seafloor hydrothermal vent fluids and continental geothermal fluids are typically enriched in Ge relative to Si compared to the source rocks (Arnorsson 1984; Criaud and Fouillac 1986; Mortlock et al. 1993; Evans and Derry 2002; Siebert et al. 2006; Escoube et al. 2015). Oceanic high- and low-temperature systems exhibit Ge/Si ratios that range from 0.9–45 μmol/mol (Mortlock et al. 1993; Escoube et al. 2015), and Ge concentrations which are up to 1000 times seawater concentrations. Continental geothermal systems exhibit an even larger range from ~2–1100 μmol/mol (Arnorsson 1984; Criaud and Fouillac 1986; Evans and Derry 2002; Siebert et al. 2006). The overall enrichment of Ge in even low-temperature hydrothermal fluids contrasts remarkably with the depletion of Ge in rivers (i.e., relative to Si) and may result from the preferential partitioning of Ge vs. Si in hydrothermal fluid, considering the contrasting thermodynamic properties of aqueous $Ge(OH)_4$ and $Si(OH)_4$ in equilibrium with Ge-bearing silicates (Arnorsson 1984; Pokrovski and Schott 1998a). Thermodynamic data showed that Ge/Si ratio in hydrothermal fluid in equilibrium with quartz decreases with increasing temperature from 160 μmol/mol at 50 °C, 43 μmol/mol at 250 °C to about 30 μmol/mol at 400 °C (Pokrovski and Schott 1998a; Evans and Derry 2002). In contrast, Ge/Si ratio in hydrothermal fluid in equilibrium with wollastonite was considered to increase with temperature (Pokrovski and Schott 1998a). The high-temperature results of the equilibrium model with quartz were globally consistent with data from Iceland wells, which have average Ge/Si = 42 μmol/mol at 235 °C (Arnorsson 1984). Based on field data, Evans and Derry (2002) also proposed a model of progressive Si loss via precipitation of Ge-poor quartz (i.e., Rayleigh distillation) in order to explain the extreme increase of the

Ge/Si ratio in cooled hydrothermal fluids. This model, requiring a high level of Si loss along a reaction path in order to produce observed Ge/Si ratios, is however not directly applicable to seafloor hydrothermal systems as discussed in Escoube et al. (2015). Other processes may also influence Ge/Si ratio in hydrothermal fluids including the precipitation or dissolution of sulfides and non-equilibrium behavior.

Germanium isotope systematics in low-temperature marine environments

An estimate of the oceanic crust composition. A crude estimate of the Ge isotopic composition of the Bulk Silicate Earth (BSE) has been estimated at 0.58±0.21‰ (see section *The Ge isotopic composition of the Earth silicate reservoirs*) through the analysis of various mantle-derived rocks such as tholeiitic glasses from mid-ocean ridges, continental and volcanic islands basalts, peridotite and granite (Rouxel et al. 2006; Escoube et al. 2012; Luais 2012). Although slight differences in $\delta^{74/70}$Ge values are possibly observed between ultramafic, basaltic and felsic rocks (Figs. 7–8), the homogeneity of their $\delta^{74/70}$Ge values compared to sedimentary rocks and biogenic materials is remarkable (Fig. 13).

The Ge isotope compositions of deep-sea clays have been reported by Rouxel et al. (2006) using drilled core samples from ODP Site 1149 and 801 in the Western Pacific. The $\delta^{74/70}$Ge values of deep-sea clays (Fig. 13), and to a lesser extent Ge/Si ratios, were homogeneous. The average $\delta^{74/70}$Ge value of deep-sea clays ($\delta^{74/70}$Ge=0.69‰ ± 0.17‰, 2s.d., n=10) overlaps with the bulk crust value ($\delta^{74/70}$Ge=0.58±0.10‰). Because deep-sea sediments are essentially composed of terrigenous materials with only minor biogenic (e.g., siliceous organisms) and authigenic components (e.g., Mn and Fe oxyhydroxides), it can be suggested that, as a whole, secondary clay formations do not significantly fractionate Ge isotopes. As these lithologies also comprise the major part of the deep-sea sediments (Plank and Langmuir 1998), these results suggested that the average $\delta^{74/70}$Ge value for the entire sedimentary section of the oceanic crust is similar to or slightly higher than the igneous value. Mesozoic deep-sea cherts selected by Rouxel et al. (2006) had $\delta^{74/70}$Ge values ranging from +0.04 to +1.31‰ (Fig.13), and had therefore Ge isotope values both lighter and heavier than crustal value, but systematically lighter than seawater value estimated at +3.1‰ (see discussion about seawater value below). Germanium/silicon (Ge/Si) ratios of deep-sea cherts varied between 0.34 to 1.1 μmol/mol, on average slightly higher than found in other Mesozoic cherts, with Ge/Si ratios ranging between 0.19×10^{-6} to 0.72×10^{-6}

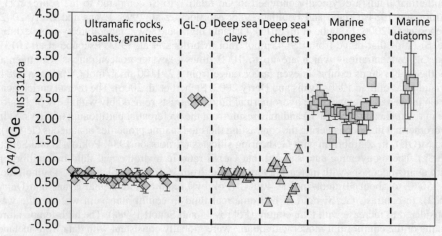

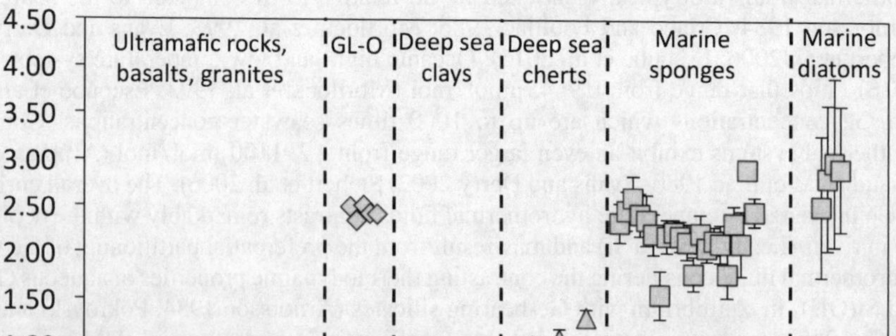

Figure 13. Germanium isotope and Ge/Si ratios of various mantle-derived rocks and deep-sea clays defining a preliminary bulk crustal value (solid horizontal line). $\delta^{74/70}$Ge and Ge/Si values of glauconite (GL-O) other marine sediments and seawater are also shown for comparison. Data from Rouxel et al. (2006), Guillermic et al. (2016) and Rouxel and De La Rocha (unpublished) reported in Table 6.

(Kolodny and Halicz 1988). The calculated amount of clay minerals relative to opal in each sample was generally below 10% and the overall lack of relationship between $\delta^{74/70}$Ge and Ge/Si (or 1/Ge) values for deep-sea cherts rules out simple binary mixing between biogenic and lithogenic pools for the explanation of the observed 1.3‰ range. Instead, Rouxel et al. (2006) proposed that Ge isotope fractionation occurs during opal diagenesis and lithification processes leading to chert formation. One thermal chert was also measured and yielded the highest δ^{74}Ge value of 1.3‰. As this sample was silicified under abnormally high temperature conditions at the proximity of volcanic flows (Rouxel et al. 2003b), $\delta^{74/70}$Ge values might be influenced by hydrothermal fluid input at the sediment-basement interface, which is consistent with recent study of Escoube et al. (2015) and Siebert et al. (2011) showing enrichment of heavy Ge isotopes in low temperature hydrothermal fluids.

One sample of glauconite was analyzed by Rouxel et al. (2006) and showed a $\delta^{74/70}$Ge value at about 2.5‰, close to seawater value (Fig. 13). Glauconite is an Fe-rich authigenic clay mineral, which forms in shallow marine environments at the sediment–water interface within a reducing microenvironment. A part of Ge incorporated in glauconite may be derived from detrital materials or Ge-rich primary mineral (such as biotite). However, the heavy δ^{74}Ge value argues against the predominance of a terrigenous source of Ge, and suggests rather that Ge also derives from seawater with heavy $\delta^{74/70}$Ge value.

Despite their interest for paleoceanography (Frank 2002), no published data are yet available on Ge isotope composition of authigenic or hydrogeneous Mn deposits such as ferromanganese (FeMn) crusts and nodules. Considering their very low growth rate (mm/Ma), FeMn crusts (and nodules) may provide a unique window into Ge biogeochemical cycles throughout the last 10 to 60 Ma of Earth's history. A preliminary analysis of the Ge isotope composition of Nod-P1, a composite Mn nodule from the deep Pacific Ocean with Ge content of 0.54 ppm (Govindaraju 1994) yielded $\delta^{74/70}\text{Ge} = -0.08 \pm 0.09$‰ (S. Lalonde pers. comm.). This value is much lower than the bulk crust and seawater values estimated at about +3.1‰ (Escoube et al. 2012), suggesting that Ge removal in hydrogenous Mn deposits favors the enrichment in the light Ge isotopes. This assumption is further supported by recent experimental and field studies showing the fractionation of Ge isotopes during Ge sorption onto Fe oxyhydroxides (Pokrovsky et al. 2014; Escoube et al. 2015). Hence, unlocking the history of Ge oceanic cycle through would require a good understanding of the mechanisms of Ge isotope fractionation during precipitation on Fe–Mn deposits.

Germanium isotope composition of seawater. Germanium has been long been considered as a "geochemical twin" of silicon. Dissolved inorganic Ge in seawater ranges from less than 2 to about 200 pM and displays a correlation with dissolved silica that is remarkably consistent across the world's oceans, underlying its potential as a complementary tracer for marine silicon cycling (Froelich and Andreae 1981; Froelich et al. 1985a,b; Ellwood and Maher 2003). This coupling is demonstrated by the close correlation of Ge and Si in all oceanic basins, with a Ge/Si of 0.7 µmol/mol. The Ge/Si ratio is also constant through most of the water column. However, there is a positive and significant Ge intercept for the Ge-to-Si relationship. The intercept value varies between 1.7 and 3.6 pM depending on the analytical system used for Ge detection. The positive Ge intercept is thought to result from Ge/Si fractionation during Si uptake (Murnane and Stallard 1988; Froelich et al. 1989; Ellwood and Maher 2003). Likewise, a fractionation factor (K_D) of 0.36 was obtained at low Si concentrations (<6 µM), assuming a Rayleigh distillation-like process (Ellwood and Maher 2003). Profiles for Ge/Si versus depth revealed also a subsurface maximum in the Ge/Si data suggesting either that Ge is being recycled faster than Si from phytoplankton, or that Ge/Si is fractionated at low Ge and Si concentration during uptake (Sutton et al. 2010). Such Ge/Si fractionation during Si and Ge uptake and/or regeneration is the most likely explanation for the positive Ge intercept seen for the global Ge versus Si relationship.

Like Si, natural variation of Ge isotopes may provide additional insights into marine Si biogeochemical cycling and the Ge marine cycle. However, due to the low natural abundance of inorganic Ge in seawater (below 200 pM), the Ge isotope composition of seawater remains largely unconstrained. It was only until recently that new methods were developed for the measurements of Ge isotopes in seawater (Baronas et al. 2014; Guillermic et al. 2016). Guillermic et al. (2016) reported Ge isotope compositions of inorganic Ge across three depth profiles from the Southern Ocean and also from the deep Atlantic and Pacific Ocean. $\delta^{74/70}$Ge values along a water column profiles from the Southern Ocean are isotopically heavier in surface waters compared to deep waters, with a global range of $\delta^{74/70}$Ge from 2.4 to 3.7‰ (relative to NIST3120a). The results may suggest that diatoms, in particular during Ge uptake in soft tissues (Mantoura 2006), take up lighter Ge isotopes leading to surface seawater that is isotopically heavier. However, according to Figure 14, the value for surface seawater ($\delta^{74/70}$Ge up to 3.07 ± 0.09‰) is not isotopically heavier (within error) than deep seawater which has a global $\delta^{74/70}$Ge = 3.14 ± 0.38‰ (2s.d., n = 27). An important result of this preliminary study is that no differences could be identified between the North Atlantic, North Pacific or Southern Ocean.

Unlike Si, a substantial amount of Ge is associated with organic species (Lewis et al. 1985, 1989). The two dominant organic Ge species in the oceans, monomethyl and dimethyl germanium, have concentrations of ~350 and 100 pM, respectively. Both exhibit conservative distributions and appear to be highly stable in the marine environment. Because of their long residence times (>1 million years) these species should not exert a strong influence on the budget for inorganic Ge, which has a residence time of ~10 kyr. The Ge isotope compositions of such species are still unknown.

Germanium isotope composition of biogenic opal. In a reconaissance study, Rouxel et al. (2006) reported the Ge isotope composition of opal sponge spicules obtained from live sponges growing on the seafloor (Table 6, Fig. 13). Two deep-sea specimens (NE Pacific) and

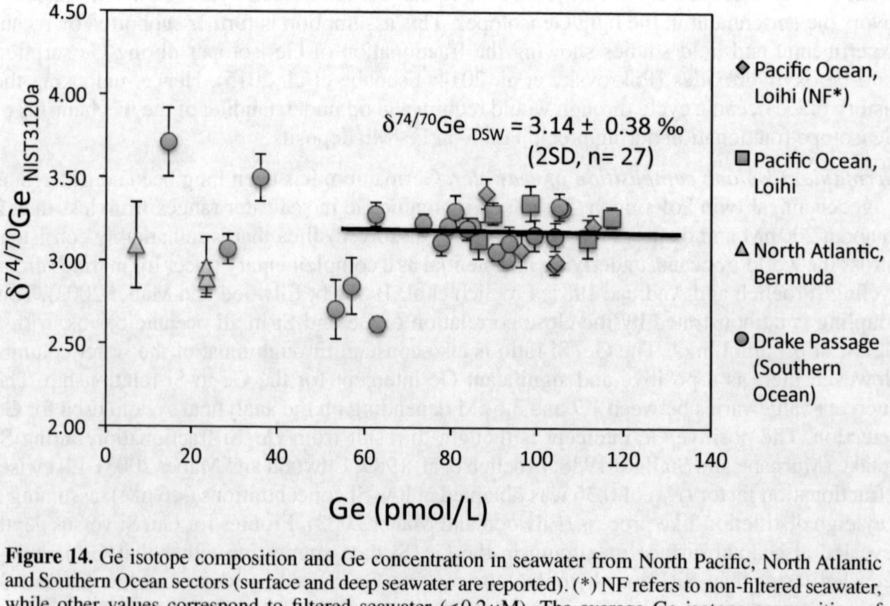

Figure 14. Ge isotope composition and Ge concentration in seawater from North Pacific, North Atlantic and Southern Ocean sectors (surface and deep seawater are reported). (*) NF refers to non-filtered seawater, while other values correspond to filtered seawater (<0.2 μM). The average Ge isotope composition of deep seawater ($\delta^{74/70}Ge_{DSW}$) is also reported, and is calculated using sample recovered below 300 m depth, corresponding typically to samples with Ge concentrations >60 pM for the Pacific and Southern Ocean sectors. Data are from Guillermic et al. (2016).

one coastal marine sponge (California) were analyzed and yielded similar Ge concentrations (between 0.18 and 0.26 ppm) but variable $\delta^{74/70}$Ge values (between 1.6‰ and 2.6‰). These results clearly demonstrate that $\delta^{74/70}$Ge values of modern marine sponges are enriched in heavy isotopes by more than ~1.0‰ relative to the bulk crust. More recently, Guillermic et al. (2016) reported the Ge isotope composition of sponge spicules that were previously examined for Si isotopes (Hendry et al. 2010; Hendry and Robinson 2012). Samples were collected from the Southern Ocean along a transect across the Drake Passage and Scotia Sea to investigate the fractionation of Ge isotopes in sponges as a function of ambient [$Si(OH)_4$]. Although no clear relationships could be derived between $\delta^{74/70}$Ge and oceanographic parameters (e.g., nutrient, water depth, temperature) or sponge species, the results confirmed earlier investigations suggesting heavy Ge isotope composition in biogenic opal. Sponges from the Southern Ocean showed an average $\delta^{74/70}$Ge value of 2.21±0.54‰ (2s.d., n=16) (Table 6) which is about 1.0±0.1‰ lighter than coeval seawater deep water at the same location (Guilllermic et al. 2016). This suggests that sponges fractionate Ge isotopes during biomineralization, and discriminate against heavy isotopes as already observed for Si isotopes (De La Rocha 2003; Hendry and Robinson 2012). Sponges are considered to have a low affinity for silicic acid (Reincke and Barthel 1997) and the inefficient silicon uptake mechanism has been suggested to explain their preferential enrichment in the light Si isotope (De La Rocha 2003) and also Ge isotopes. However, further studies are needed to address whether Ge isotopes (and Ge/Si) in sponges could be used as both tracers of $Si(OH)_4$ utilization and Si sources to the open ocean.

In an unpublished PhD Thesis, Mantoura (2006) carried out laboratory experiments to test whether diatoms fractionate Ge isotopes during biomineralization. Three species were grown at a range of [Ge] and Ge/Si, and no fractionation of Ge isotopes was observed between cleaned diatom opal and the initial culture media (Fig. 15). It was concluded that diatom opal may record the $\delta^{74/70}$Ge value of seawater and therefore may be an accurate recorder of temporal changes in the marine Ge cycle. This is in marked contrast to Si isotopes

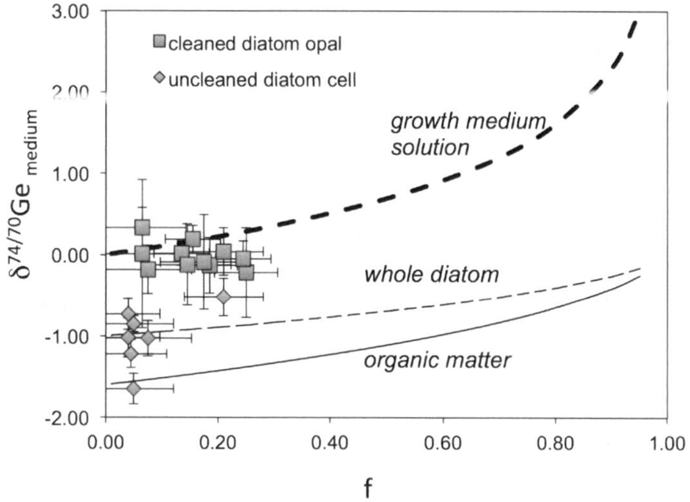

Figure 15. Rayleigh functions are fitted to whole (uncleaned) diatom $\delta^{74/70}$Ge data (diamonds) and cleaned opal $\delta^{74/70}$Ge data (squares). The theoretical $\delta^{74/70}$Ge of the whole diatom cell is represented by the thinner dashed line, which is calculated using α=0.9990. Ge isotope composition of diatom organic matter is determined by mass balance, yielding α=0.9984. Hence, the theoretical accumulating organic matter would have the value of the solid black line. As f (fraction of Ge removed from in solution) increases, the resulting solution would become progressively isotopically heavier (thicker dashed lines corresponding to α=0.9990). Data and isotopic model are from Mantoura (2006).

Table 6. Ge isotope composition and Ge concentration of biogenic opal (sponges and diatoms).

$\delta^{74/70}$Ge (‰) NIST3120a	1sd	Ge (ppm)	Specimen	Ref
North East Pacific (34° 45'N; 123°W); 4000 m depth				
2.33	0.18	0.21	*Hyalonema bianchoratum*	(1)
2.46	0.15	0.21	*Hyalonema bianchoratum*	(1)
2.60	0.12	0.23	*Hyalonema bianchoratum*	(1)
2.42	0.12	0.23	*Hyalonema bianchoratum*	(1)
2.36	0.12	0.27	*Hyalonema bianchoratum*	(1)
2.21	0.12	0.22	*Hyalonema bianchoratum*	(1)
1.56	0.22	0.26	*Bathydorus laevis ssp. spinosus*	(1)
North East Pacific (34°45'N; 123°W); 4000 m depth				
2.19	0.22	0.18	*Axinella mexicana sp.*	(1)
Southern Ocean—Burdwood Bank				
2.24	0.04	0.13	Unidentified Hexactinellid	(2)
2.19	0.05	0.16	Unidentified Demosponge	(2)
2.13	0.04	0.19	*Mycalidae*	(2)
1.79	0.05	0.18	Unidentified Demosponge	(2)
2.22	0.04	0.20	Unidentified Demosponge	(2)
Southern Ocean—Drake Passage				
2.07	0.04	0.15	Unidentified Demosponge	(2)
2.14	0.07	0.15	Unidentified Demosponge	(2)
2.11	0.03	0.20	Unidentified Demosponge	(2)
1.79	0.02	0.25	Unidentified Demosponge	(2)
2.33	0.08	0.18	*Aceolocalyx*	(2)
Southern Ocean—Scotia Sea				
2.11	0.02	0.35	*Acoelocalyx*	(2)
2.91	0.06	0.05	*Acoelocalyx*	(2)
2.45	0.09	0.16	*Acoelocalyx*	(2)
Southern Ocean—Indian Sector				
2.51	0.27	0.76	Holocene diatoms, Core RC11-94	(3)
2.82	0.19	0.78		(3)
2.99	0.05	0.90		(3)
2.92	0.07	0.67		(3)

Reference: (1) Rouxel et al. (2006); (2) Guillermic et al. (2016); (3) Rouxel and De La Rocha, unpublished (pooled cleaned diatoms studied in De La Rocha et al. 1998)

which are significantly fractionated during diatom uptake, allowing Si isotopes in diatom opal to record Si utilisation in the surface ocean (e.g., (De La Rocha et al. 1997, 1998; Sutton et al. 2013). However, there is some indication that fractionation of Ge isotopes may occur in diatom organic matter. Germanium isotopes may be fractionated upon incorporation into diatom organic matter with the preferential enrichment of light Ge isotopes of approximately −1.5‰ (Fig. 15). This suggests that the removal and remineralization of organic matter in the surface ocean may cause some local variation in $\delta^{74/70}$Ge of surface seawater. However, this effect is likely to be diminished due to the rapid remineralization of organic matter, compared to diatom opal, in intermediate waters.

Rouxel and De La Rocha (unpublished data) also measured the $\delta^{74/70}$Ge values for cleaned diatom opal over the penultimate glacial cycle (De La Rocha et al. 1998). The diatom samples were physically and chemically cleaned using a protocol adapted from Shemesh (1988). Since diluted HF was used for opal dissolution, the results may be potentially affected by remaining clay particles. Diatom samples were extracted from Core E50–11 located in the Indian Ocean at different depth below seafloor. Due to limited sample availability, clean diatom samples were pooled to provide sufficient material for Ge isotope analysis. The results in Table 6 and Figure 13 therefore correspond to average values and do not provide a paleoceanographic record. Yet they provide an important record of Holocene seawater, allowing calculating an average $\delta^{74/70}$Ge value of 2.8 ± 0.4‰, confirming that seawater is enriched in heavy Ge isotopes (Baronas et al. 2014; Guillermic et al. 2016).

Ge isotope fractionation during low temperature weathering

In a recent study, Baronas et al. (2014) reported rivers with a wide range of $\delta^{74/70}$Ge values being 1.8–4.8‰ heavier than average crustal values. Based on a relatively limited dataset, river $\delta^{74/70}$Ge values seemed unrelated to Ge/Si or Ge concentrations and bedrock lithology (Baronas et al. 2014). Scavenging or exchange reactions with particulate matter (clays, Fe-oxyhydroxide and organic matter) as well as soil and weathering processes are likely the main contributors to the observed isotopic variability in river water. Pokrovsky et al. (2014) investigated the fractionation of Ge isotopes during Ge adsorption on goethite and its coprecipitation with amorphous Fe oxy(hydr)oxides (see experimental section). Regardless of the pH, surface concentration of adsorbed Ge or exposure time, the solution–solid enrichment factor for adsorption ($\delta^{74/70}Ge_{solution-solid}$) was 1.7 ± 0.1‰. For $(Ge/Fe)_{solid}$ ratio < 0.1, the $\delta^{74/70}$Ge solution–solid increased with the decrease of Ge concentration in the solid phase, with a value as high as 4.4 ± 0.2‰ at $(Ge/Fe)_{solid} < 0.001$, corresponding to the majority of natural settings. These experimental data provide important guides for the cause of heavy Ge isotopes in continental run-off, including river and groundwater systems.

Germanium isotope systematics of hydrothermal waters

In an initial study, Siebert et al. (2006, 2011a) analyzed geothermal spring fluids from the Cascades. These continental high-*T* hydrothermal fluids had a heavy Ge isotope composition relative to crustal value, with a range in $\delta^{74/70}$Ge between 1.1‰ and 1.9‰ (normalized relative to NIST3120a) for Ge concentrations ranging from 20 to 94 nm (Fig. 16). The data were interpreted as reflecting the preferential removal of heavy Ge isotopes out of solution during cooling of the hydrothermal fluid and subsequent precipitation of quartz. However, *ab initio* calculations of Ge isotope fractionation factors reported by Li et al. (2009) suggest that the heavier isotope should be sequestered in the quartz phase, implying that fluids should become lighter as they cool. Hence, alternative mechanisms should be considered, including the sequestration of isotopically light Ge into sulfides as fluids cool or kinetic isotope effects. The existing data sets are insufficient to further speculate about the exact processes driving Ge isotope fractionation in terrestrial hydrothermal fluids. However, these initial observations suggest that continental geothermal springs provide a source of heavy Ge isotopes to rivers and groundwaters and therefore to the oceans.

In a more recent study, Escoube et al. (2015) investigated hydrothermal vents from the Loihi Seamount (Pacific Ocean, 18°54'N, 155°15'W) that were characterized by distinct chemistry with high Fe and Si concentrations and low sulfide concentration (Sedwick et al. 1992; Glazer and Rouxel 2009). Escoube et al. (2015) also investigated Ge isotope signatures

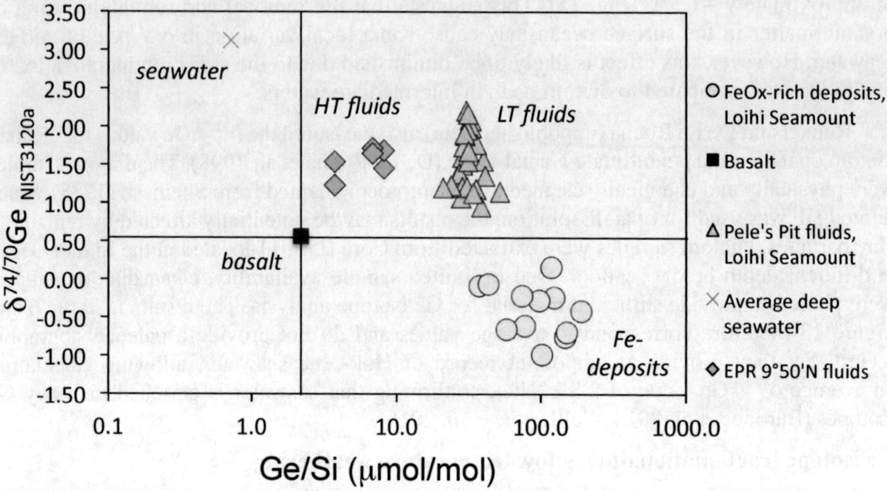

Figure 16. Ge/Si and Ge isotope composition of seafloor hydrothermal fluids from Loihi Seamount (referred as low-temperature (LT) fluids) near Hawaii and East Pacific Rise (EPR) at 9°50'N (referred as high-temperature (HT) fluids). By comparison, Fe oxyhydroxide-rich deposits (FeO_x-rich) at Loihi Seamount are also enriched in Ge (relative to Si and fluid sources) but with significant enrichment in light Ge isotopes. Crustal and seawater values are shown for comparison. Data are from Escoube et al. (2015)

of a well-studied high-temperature hydrothermal system from the East Pacific Rise (EPR) at 9°50'N (e.g., Von Damm 2004). The very young age of the lava flows along the axial summit trough of EPR at 9–10°N offers the opportunity to study Ge isotope systematics during the early stage of high-temperature hydrothermal venting and active formation of hydrothermal sulfide deposits. The high temperature hydrothermal fluids from EPR 9–10°N were enriched in Ge vs. Si compared to source rocks (i.e., basalt with Ge/Si = 2.4 μmol/mol) with Ge/Si ratios between 3.7 and 8.5 μmol.mol^{-1}. Those values were lower than at Loihi Seamount (Ge/Si averaging 29.6 ± 1.6 μmol/mol) but consistent with other high temperature vent fluids from EPR 21°N and Juan de Fuca Ridge (Mortlock et al. 1993) (i.e., Ge/Si between 5 and 15 μmol/mol) (Fig. 16). Loihi vent fluids had $\delta^{74/70}$Ge values ranging from 0.6 to 2.2‰, averaging 1.68 ± 0.70‰ (2s.d.) and were thus systematically heavier than basaltic values at 0.56‰. Those $\delta^{74/70}$Ge values were, however, lighter than seawater with $\delta^{74/70}$Ge about 3‰. High temperature vent fluids from EPR 9–10°N yielded $\delta^{74/70}$Ge values averaging 1.55 ± 0.36‰ (2s.d., $n = 6$) and were similar, albeit slightly lower, than the average $\delta^{74/70}$Ge value of low temperature hydrothermal fluids at Loihi Seamount (Fig. 16). Hence, despite their fundamental differences in temperature and chemical composition, high and low temperature hydrothermal vent fluids showed similar Ge isotope compositions characterized by a global enrichment in heavy Ge isotopes of about 1‰ relative to the basaltic source.

Escoube et al. (2015) further evaluated whether both high Ge/Si ratios and heavy $\delta^{74/70}$Ge values of hydrothermal fluids compared to source rocks may be explained by mineral–fluid partitioning in the reaction zone (i.e., batch fractionation) or upflow zone (i.e., fractional distillation). Because Si concentration in high-temperature hydrothermal fluids is controlled by quartz solubility (Von Damm et al. 1991), higher Ge/Si ratios in hydrothermal fluids venting at the seafloor may be buffered by quartz in the reaction zone (Mortlock et al. 1993; Wheat and McManus 2005). However, the quartz-controlled solubility model cannot fully explain the data as it requires a large fractionation factor $\Delta^{74/70}Ge_{Qz-fluid}$ at about −4‰. Hence both quartz and sulfide precipitation in the reaction zone may control Ge/Si and Ge-isotope signatures in the fluid. Subsurface cooling of the hydrothermal fluid may also affect Ge/Si and $\delta^{74/70}$Ge signatures of the vent fluids if significant Si-rich mineral precipitation occurs along the flow path.

At Loihi Seamount, the distinct fluid chemistry, enriched in Fe and Si and depleted in H_2S, led to the formation of extensive Fe-rich microbial mats and hydrothermal Fe-oxyhydroxide deposits at or below the seafloor (e.g., De Carlo et al. 1983; Karl et al. 1988; Emerson and Moyer 2002; Glazer and Rouxel 2009). The Fe-rich deposits at Loihi Seamount had low $\delta^{74/70}$Ge values from 0.16 down to −0.98‰, which corresponds to a maximum apparent $\Delta^{74/70}Ge_{deposit-fluid}$ of −2.81‰ (Fig. 16). The largest fractionation factor between fluid and Fe-rich deposit was broadly consistent with experimental data for Ge coprecipitation with amorphous Fe-oxyhydroxide up to −4.4‰ (Pokrovsky et al. 2014) (see also the section *Theoretical considerations and experimental calibrations*). Using $\alpha_{sol-liq}$ of about 0.996 as determined by Pokrovsky et al. (2014), Escoube et al. (2015) determined that Ge isotope composition of Fe-oxyhydroxide deposits corresponds to a loss of more than 30% of Ge from the hydrothermal fluid due to precipitation within Fe-rich deposits. This suggests that Fe-rich deposits provide a relatively efficient trapping of reactive elements coming from the diffuse vent fluids. This finding contrasts with high-temperature hydrothermal chimneys that behave as an open system relative to vent fluids; i.e., only a small fraction of elements is trapped within the chimney wall (Tivey 1995).

A preliminary oceanic Ge budget

Rivers supply most of Si to the modern ocean (5.6×10^{12} mol/yr) with minor contributions from aeolian sources, hydrothermal input, and low-temperature basalt weathering, totaling 6.7×10^{12} mol/yr (Treguer et al. 1995; DeMaster 2002; Treguer 2002). The output term for Si is marine biogenic sediments (primarily as diatoms, sponges, and radiolaria) and totals $6.5 - 7.4 \times 10^{12}$ mol/yr (Fig. 17). In contrast to Si, Ge has two primary sources: rivers, which contribute to 3.3×10^6 mol/yr, and hydrothermal inputs, which contribute 4.9×10^6 mol/yr

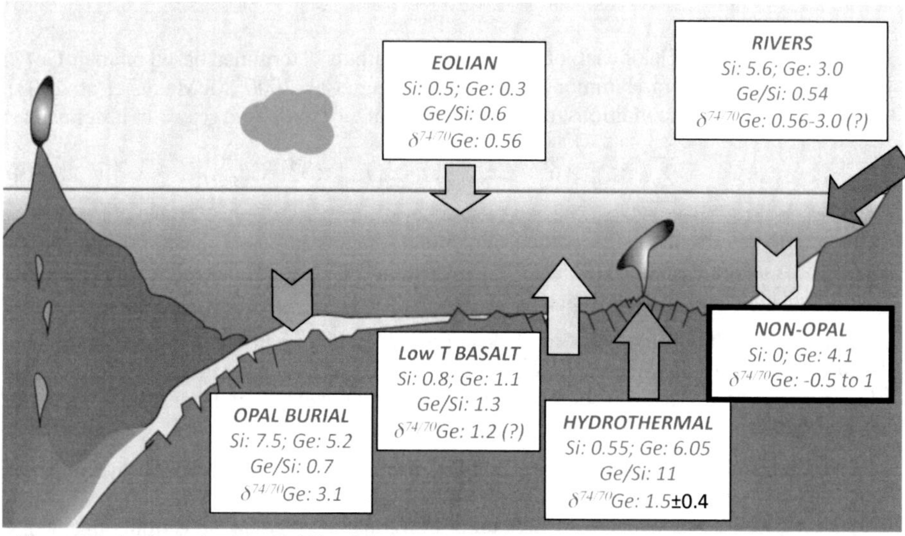

Figure 17. A schematic of the Ge/Si and Ge isotope oceanic budget. The main sources of dissolved Ge into the ocean are wet and dry deposition from the atmosphere, input from rivers, resuspended sediment, and pore water along continental shelves and hydrothermal vents. Silicon fluxes (Si × 10^{12} mol/yr) are after Treguer et al. (1995) while Ge fluxes (Ge × 10^6 mol/yr) are estimated from measured Ge/Si ratios (in μMol/mol) after Mortlock et al. (1993). The hydrothermal flux has been adjusted according to Elderfield and Shultz (1996). Silicon fluxes assume steady state. In order to maintain the current oceanic Ge/Si at 0.72 μMol/mol, 4.1×10^6 mol/yr must be sequestered in a nonopal sink (Froelich and King 1997). Germanium isotope compositions (reported as $\delta^{74/70}$Ge) of Ge sources and sinks are reported in Escoube et al. (2015). Germanium isotope composition of the non-opal sink is determined by mass balance and assuming steady state.

(Mortlock et al. 1993; Elderfield and Schultz 1996; King et al. 2000). Aeolian inputs and low-temperature basalt weathering reactions each produce between 0 to 0.4×10^6 mol/yr. Because the two dominant Si and Ge sources to the ocean carry very different Ge/Si signatures, with $(Ge/Si)_{rivers}$ ~0.4 μMol/mol and $(Ge/Si)_{hydrothermal}$ ~8–14 μmol/mol, it has been proposed that the Ge/Si ratio buried in biogenic siliceous tests on the seafloor reflects the present and past source strength of the river fluxes relative to hydrothermal fluxes (Murnane and Stallard 1988; Froelich et al. 1992; Elderfield and Schultz 1996). However, the use of Ge/Si as a monitor for the relative importance of these two sources through time remains uncertain as the Ge mass balance in the modern ocean is not well understood. Assuming steady state, the contemporary input fluxes from continents and hydrothermal sources require Ge removal with a Ge/Si ratio of 1.6 μMol/mol, which is significantly greater than the observed opal burial ratio of 0.7 μMol/mol (Elderfield and Schultz 1996). Identification and quantification of the so-called "missing Ge sink" has received great interest in the past decade and it is now proposed that Ge may be removed from the ocean in iron-rich reducing sediments of continental margins independently of Si (Hammond et al. 2000; King et al. 2000; McManus et al. 2003). This sink corresponds to about 55±9% of Ge within opal released by dissolution (Hammond et al. 2000). Considering the potential of Ge isotopes to fractionate during marine sediment diagenesis, for example during Ge precipitation with authigenic minerals, Ge isotope systematics should be a powerful tool to further constrain the missing sink through the establishment of a mass balance of Ge isotopes in the oceans.

In order to address these issues, Escoube et al. (2015) established a preliminary mass balance of Ge in seawater, using a one box ocean model based on Ge/Si, $\delta^{74/70}$Ge and the marine Ge source and sink fluxes (Fig. 17). At steady state, the sum of input fluxes and output fluxes of Ge to the ocean are equal:

$$F_{\text{river}}+F_{\text{atmospheric}}+F_{\text{hydrothermal}}+F_{\text{low-}T\text{ basalt}}=F_{\text{opal}}+F_{\text{non-opal}} \tag{8}$$

With, F, the flux of Ge of each of the inputs and outputs determined based on their Ge/Si ratios and Si fluxes (From Hammond et al. 2000; King et al. 2000; McManus et al. 2003). Considering steady-state conditions, the mass balance of Ge isotopes in seawater is defined as:

$$\sum_{\text{Source}}\left(F_{\text{Source}}\times\delta^{74/70}\text{Ge}_{\text{Source}}\right)=\sum_{\text{Sink}}\left(F_{\text{Sink}}\times\delta^{74/70}\text{Ge}_{\text{Sink}}\right) \tag{9}$$

In order to determine the Ge isotope signature of the missing, non-opal Ge sink, Escoube et al. (2015) resolved Equations (8) and (9) using estimated $\delta^{74/70}$Ge and Ge/Si values of other Ge sinks and sources (Fig. 17), such as:

(1) $\delta^{74/70}$Ge value of atmospheric deposition (i.e., aeolian particles) has been estimated to be identical to crustal values, with $\delta^{74/70}$Ge=0.56±0.20‰ as estimated by Escoube et al. (2012). Silicon and Ge fluxes from atmospheric deposition were taken from the estimation of King et al. (2000).

(2) Ridge flanks Ge and Si fluxes, including the sum of warm (40–75 °C) hydrothermal systems and low-temperature seafloor basalt weathering were taken from Wheat and McManus (2005). Because of the large range of Ge/Si ratios in warm hydrothermal fluids from ridge flanks, an uncertainty of 50% should be considered for Ge flux. $\delta^{74/70}$Ge of ridge flank fluxes is also considered similar to the average value obtained at Loihi Seamount ($\delta^{74/70}$Ge=1.2±0.5‰). Despite the relatively large uncertainties, the ridge flank Ge fluxes have a minor influence on global Ge oceanic cycle due to their small contribution representing less than 10% of the total Ge input.

(3) The Si and Ge fluxes from rivers were taken from the estimation of King et al. (2000). $\delta^{74/70}$Ge values of rivers are still poorly known. Germanium/silicon (Ge/Si) in rivers are well

known to be fractionated relative to bulk Earth (with a ratio around 0.58 versus 1.3 for the crust, Mortlock and Froelich 1987). Therefore, the Ge isotope composition of rivers is likely to be fractionated relative to the crust due to Ge sorption on clay or oxide minerals, as previously observed by Kurtz et al. (2002) and Scribner et al. (2006). Because rivers represent 37% of the global input of Ge, it is clear that a better constraint on the riverine flux is important for establishing a reliable isotope mass balance in seawater.

(4) The $\delta^{74/70}$Ge value of the high-temperature (HT) hydrothermal flux is estimated as at 1.55 ± 0.36‰ based on average values obtained from high-temperature vent fluids from the EPR. The Ge/Si of the HT hydrothermal flux is determined as 11 ± 3 μmol/mol based on the compilation of Ge/Si values for high-temperature hydrothermal vents (Mortlock et al. 1993; Escoube et al. 2015). It should also be recognized that Ge (and Si) fluxes from acidic volcanic island arcs, which represent about 10% of the hydrothermal flux (Baker et al. 2008), and back arc spreading centers are poorly known and may contribute differently to Ge fluxes.

(5) The $\delta^{74/70}$Ge values of the opal sink are considered identical to seawater values as demonstrated in previous studies showing a lack of Ge/Si and Ge-isotope fractionation during diatom uptake (Shemesh et al. 1989; Bareille et al. 1998; Mantoura 2006). Mantoura (2006) reported $\delta^{74/70}$Ge values at around 3.3‰ for diatom opal from Holocene sediments while Guillermic et al. (2016) reported Ge isotope composition of deep Pacific and Atlantic waters at 3.14 ± 0.38‰.

(6) The removal of elements in hydrothermal plumes through adsorption onto FeOOH-rich particles is well recognized for seawater oxyanions such as V, As and P (Feely et al. 1990; Feely et al. 1991) but is probably negligible for Ge. In particular, it has been demonstrated that Ge/^{3}He ratios of the dispersing hydrothermal plume remain close to high-temperature hydrothermal vent end-member values, suggesting a near conservative behavior of Ge during seawater-hydrothermal fluid mixing (Mortlock et al. 1993).

The isotopic mass balance is solved using a box model using the parameters defined in Figure 17 and by applying an error propagation scheme (i.e., Monte Carlo simulation). Equations (8–9) are solved simultaneously in order to estimate the $\delta^{74/70}$Ge value of the missing, non-opal Ge sink for the cases of (i) riverine input at crustal values and (ii) riverine input identical to seawater values. In the first case, Escoube et al. (2015) determined the $\delta^{74/70}$Ge value of the non-opal Ge sink to be -0.6 ± 0.4‰, which is about 3.6‰ lighter than seawater. In the second case, Escoube et al. (2015) obtained heavier $\delta^{74/70}$Ge value at 0.9 ± 0.3‰ for the missing Ge sink, which is about 2.1‰ lighter than seawater. These values corresponded well to the expected isotope fractionation produced by Ge adsorption onto Fe-oxyhydroxide. It is however important to note that both the Ge flux estimate and $\delta^{74/70}$Ge value of the non-opal Ge sink were highly sensitive to errors in the estimation of global hydrothermal Ge and Si fluxes. For example, at the lowest estimate of hydrothermal Si flux, the missing (non-opal) Ge sink became insignificant, leading to a very large uncertainty for its Ge isotope signature.

Based on this preliminary mass balance, it can be suggested that the "missing sink" of Ge in the oceans is likely controlled by the adsorption of Ge onto Fe-oxyhydroxide in marine sediments, as already proposed in previous studies (Hammond et al. 2000; King et al. 2000; McManus et al. 2003). It is however important to note that Ge isotope fractionation in authigenic clays remains unknown. Sulfides are also characterized by very light $\delta^{74/70}$Ge values (Escoube et al. 2012; Luais 2012; Meng et al. 2015) which can potentially provide a sink for isotopically light Ge in the ocean. However, the amount of Ge sequestrated in sedimentary sulfides is probably insignificant compared to the opal and authigenic Fe-oxyhydroxide sinks. This assumption is supported by the lack of Ge removal in the sulfidic and anoxic waters of the Baltic Sea (Andreae and Froelich 1984). The importance of Ge removal with organic matter, however, will need to be further assessed considering the large Ge enrichment factors and the range of Ge isotope compositions in organic-rich rocks and compounds (Li et al. 2009; Li and Liu 2010, this study).

The potential for paleoceanography and the rock record

Biogenic Opal. The incorporation of Ge into diatom opal has been previously used to reconstruct changes in the Si water column cycle over time (Shemesh et al. 1988, 1989; Mortlock et al. 1991; Froelich et al. 1992; Bareille et al. 1998). The Ge/Si signature of diatom opal (Ge/Si $_{opal}$) from the past 450,000 years revealed systematic changes that were coherent with glacial–interglacial cycles (Mortlock et al. 1991). Interpretation of the Ge/Si_{opal} record has varied. Initially, it was thought that the decrease in Ge/Si_{opal} from an interglacial value of 0.72 μmol/mol to a glacial value of 0.55 μMol/mol indicated a reduction in Ge uptake relative to Si as a result of lower Si utilisation during glacial times (Mortlock et al. 1991). However, diatom growth experiments have shown that diatoms grown at high Si concentrations (100 μM) did not fractionate Ge/Si significantly leading Froelich et al. (1992) to suggest that glacial ocean Si concentrations were higher than the present day. More recently, the assumption that diatoms do not discriminate Ge versus Si has been challenged by Ellwood and Maher (2003). Because silica is biolimiting in today's ocean, complete consumption of Ge and Si in a Rayleigh distillation surface ocean could camouflage fractionation effects in the opal burial record (Murnane and Stallard 1988). Alternatively, Hammond et al. (2004) demonstrated that the change in Ge/Si for the ocean could be explained by temperature dependent changes in the non-opal sink during glacial times. The fraction of opal dissolution occurring on the seafloor should increase, causing more Ge sequestration in the non-opal sink (Hammond et al. 2004). Clearly, there is a need to understand the processes involved in the uptake and incorporation of Ge into biogenic opal before the Ge/Si_{opal} record can be correctly interpreted.

In an attempt to resolve some of the controversy in the paleorecord of Ge/Si ratios in biogenic opal, Mantoura (2006) investigated Ge isotopes ($\delta^{74/70}$Ge opal pattern) over the penultimate glacial cycle (Fig. 18). The diatom samples were physically and chemically cleaned using a protocol adapted from Shemesh (1988). For opal dissolution, Na_2CO_3 was used instead of diluted HF in order to prevent the dissolution of potentially remaining clay particles. Using samples from the South Atlantic sector of the Southern Ocean (ODP Site 1094), Mantoura (2006) investigated the (Ge/Si) opal and $\delta^{74/70}$Ge opal record and found no difference between average glacial and average interglacial $\delta^{74/70}$Ge opal values. The average $\delta^{74/70}Ge_{opal}$ value over 68 to 178 ka period, was determined at 3.35±0.29‰ (2s.d., $n=29$). $\delta^{74/70}Ge_{opal}$ rises from a glacial value of approximately 3.2‰ to a maximum of 3.6‰ during the termination and then falls again to approximately 3.3‰ later during the interglacial (Fig. 18). Although the changes are barely outside the analytical error, there appears to be a perturbation in $\delta^{74/70}$Ge opal on the termination. Both increased biological productivity and more efficient uptake of upwelled nutrients in high-latitude oceans have been proposed (Martin 1990) as mechanisms responsible for the glacial reduction in atmospheric concentrations of carbon dioxide. Hence, the possibility to combine $\delta^{74/70}Ge_{opal}$, with $\delta^{30}Si_{opal}$ and Ge/Si_{opal} should provide new insights into the variability of Si (and Ge) utilisation of the surface ocean during glacial–interglacial periods.

Iron Formations. As pure chemical sediments largely free of detrital contamination, the Fe- and Si-rich precipitates that form Banded Iron Formations (BIF) are expected to record elemental and isotopic signatures of ancient seawater by sorption and co-precipitation reactions (Bekker et al. 2010). This assumption relies on the predictable nature of metal adsorption reactions occurring at the surface of the authigenic hydrous ferric oxides that would have precipitated from contemporaneous seawater. Although there has been significant work on the sources of silica in BIFs, basic aspects of silica deposition in iron formations and the Precambrian Si cycle remain poorly known (Hamade et al. 2003; Fischer and Knoll 2009; Bekker et al. 2010). Iron formations typically contain 34–56 wt.% SiO_2 (Klein 2005) and, potentially, represent a major sink for dissolved Si in the geological past. Although it is generally accepted that the source of silica in BIF was ambient seawater during most of the Precambrian, when the biological sink for seawater silica was presumably absent (Siever 1992), the source of silica to the oceans of that time period remains uncertain.

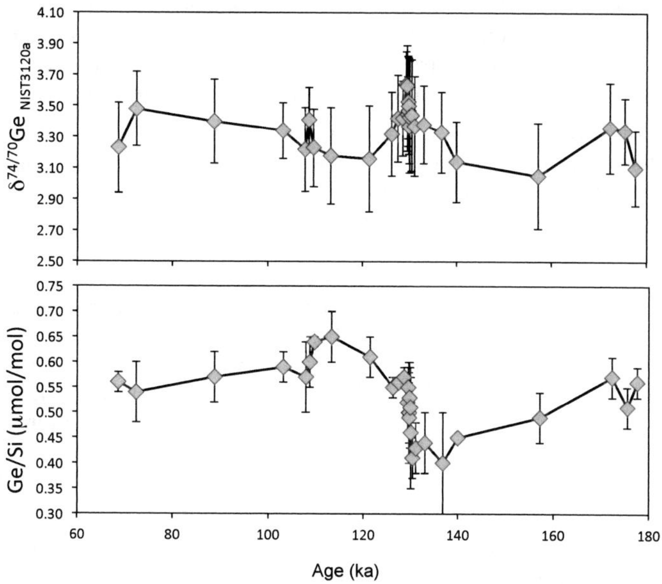

Figure 18. The Ge/Si$_{opal}$ and $\delta^{74/70}Ge_{opal}$ record of samples from site ODP 1094, as reported by Mantoura (2006). Error bars are 95% confidence interval for the data estimated by 2× the standard error of *n* measurements. Age model data are from Mantoura (2006) and established by Schneider-Mor et al. (2005).

In order to better constrain silica sources, a growing number of studies have investigated the Si isotope composition of Precambrian iron formations and cherts (André et al. 2006; Robert and Chaussidon 2006; van den Boorn et al. 2007; Steinhoefel et al. 2009; Marin-Carbonne et al. 2014). Since hydrothermal fluids and rivers have very different Ge/Si ratios (Froelich et al. 1985; Mortlock et al. 1993), Ge/Si systematics in iron formations may provide constraints on the source of Si to the Precambrian oceans. On the basis of co-variation of Ge/Si ratios with silica content in the ca. 2.5 Ga Dales Gorge Member of the Brockman Iron Formation (Hamersley Group, Western Australia), Hamade et al. (2003) proposed a decoupling of Fe and Si sources, with Si being predominantly derived from riverine waters having low Ge/Si ratios due to weathering of continental landmasses. Caution is required, however, due to the strong Ge fractionation relative to Si by sorption onto Fe-oxyhydroxides (Pokrovsky et al. 2006) or quartz precipitation (Evans and Derry 2002). In addition, release of Ge to porewaters may also affect Ge/Si ratios during chert lithification (Rouxel et al. 2006). In either case, Ge/Si ratios in cherts may not reflect seawater composition but instead record multiple, unrelated and geologically protracted processes.

Escoube et al. (2012) and Luais (2012) reported the composition of Ge isotopes in a single BIF sample (IF-G, from Isua, West Greenland) and found $\delta^{74/70}Ge$ values of 1.03 ± 0.09‰, and 1.01 ± 0.13‰, respectively. Germanium was also found enriched relative to crustal values with Ge/Si ratios of 27 μmol/mol. Luais et al. (2011) called attention to the measurements of Ge contents on bulk IF-G sample that exhibit petrographic evidence of fluid-induced metamorphism. They performed in situ LA-ICPMS measurements of Ge concentrations of individual minerals in both quartz and magnetite bands. Germanium/silicon (Ge/Si) ratios of quartz range from 6 to 10 μmol/mol, which is significantly smaller than previously reported data of Ge/Si > 20 μmol/mol on individual bands (Frei and Polat 2007). They found that the occurrence of amphibole with Ge content up to 100 ppm biases the Ge budget and the Ge/Si ratio (and probably the Ge isotopic composition) of individual "bulk" samples. Since Ge adsorbed onto Fe-oxyhydroxide is enriched in light isotopes (Li and Liu 2010; Pokrovsky et al. 2014; Escoube et al. 2015), the heavy $\delta^{74/70}$Ge values for IF-G is best explained by the heavy isotope composition of their Ge sources. Although alternative interpretations are possible, such as metamorphic overprinting, the slightly lighter values for IF-G compared to seawater may result from a higher contribution of hydrothermal Ge to Archaean oceans.

In an unpublished study, Lalonde et al. (2012) reported Ge and Si isotope compositions of Precambrian iron formations. Although the data set is still limited, both $\delta^{74/70}$Ge and $\delta^{30/28}$Si data demonstrated a nearly 3‰ variation over geological time. Archean Iron Formations (2.5–3.2 Ga old) displayed $\delta^{74/70}$Ge values ranging from 0.6‰ to 2.3‰. These Ge isotope data appeared unrelated to Ge/Si and Fe/Si ratios, suggesting that mineralogy or Si sources were not the primary factors controlling Ge isotope composition. Lalonde et al. (2012) also performed a detailed study of $\delta^{74/70}$Ge variations through an entire 355 m core section of the Joffre Member of the Brockman Iron Formation. This ~2.45 billion year old unit is by volume the largest single known banded iron formation (BIF) in the world, providing a window to Paleoproterozoic seawater chemistry (Haugaard et al. 2016). Results showed significant $\delta^{74/70}$Ge variations from 0.97 to 2.27‰, yielding an average of 1.67 ± 0.68‰ (2s.d., $n = 28$). The large variability suggests that local deposition environment and postdepositional effects should be well understood before reconstructing the long-term variations of Ge isotopes across the Precambrian.

Carbonaceous rocks. Trace-metal concentrations of laminated, organic-rich shales have long been used to draw inferences concerning paleoredox conditions as well as metal inventory in ancient oceans (Algeo et al. 2004; Brumsack 2006; Lyons and Severmann 2006; Lehmann et al. 2007) with potential constraints on past atmospheric oxygenation, weathering intensity, marine productivity, global volcanic and hydrothermal events and ocean redox structure. In general, the enrichment of redox-sensitive metals (e.g., Mo, Re, U) in organic-rich sedimentary rocks is highly sensitive to seawater redox conditions, sulfate concentrations, and depositional/post-depositional conditions. As Ge is not redox-sensitive, its behavior in anoxic sedimentary rocks is likely to differ from other elements. However, the utility of Ge enrichment factors in black shales has been the subject of a very limited number of studies. In general, elevated Ge/Si ratios in cherts above ~0.10 µmol/mol are commonly regarded as evidence of a hydrothermal component (Rouxel et al. 2006; Escoube et al. 2015; Slack et al. 2015). In organic-rich black shales, Ge and Si can also fractionate in sulfidic pore fluids during diagenesis (Tribovillard et al. 2011), producing anomalously high Ge/Si ratios unrelated to a hydrothermal signature. Hence, the possiblity of authigenic Ge enrichment in sedimentary rocks deposited under more or less severely reducing conditions open new perspectives for using Ge isotope systematics in paleoceanography.

In contrast to Ge enrichment in black shales, Ge enrichment during coal formation has received significant interest (Holl et al. 2007). Despite the high concentration of Ge in coal, ash and lignified wood (Bernstein 1985), important questions still remain about the origin of Ge in lignite and coal deposits, which is considered to be mainly epigenetic due to migration of Ge-bearing fluids. To adress this issue, Qi et al. (2011) investigated the Ge isotopic composition of Ge-rich lignite samples from China in order to provide important constraints on the mechanisms of Ge enrichment in coal. Result show that Ge-rich lignite samples yield large Ge isotopic fractionation ($\delta^{74/70}$Ge values range from −2.59‰ to 4.72‰). Lignite samples with low Ge concentrations (500 ppm) tend to show heaviest $\delta^{74/70}$Ge values, indicating a preferential enrichment of light Ge isotopes in coal in an open system (i.e., Rayleigh-type fractionation). Considering that Russia and China host the largest Ge-bearing coal deposits, increasing anthropogenic input of Ge in the environments should be considered when investigating natural Ge cycling in surface environments. Hence, Ge isotopes may provide useful tracers of sources of heavy metal pollution caused by high temperature industrial processes (coal combustion and Pb–Zn refining) in the environment.

CONCLUSION

For decades, Ge has been considered as an element of significant interest in Earth Sciences, including cosmochemistry with the characterization of meteorites and Apollo return samples, mantle geochemistry with the study of Iceland and Hawaiian mantle sources, and marine biogeochemistry with important paleoceanographic implications. In this context, Ge isotopes provide new tracers that show great promise to solve fundamental questions related to both high- and low-temperature geochemistry. Although scarce, theoretical and experimental studies have highlighted two distinct modes of enrichment processes leading to significant Ge isotopic fractionations: (i) incorporation in crystal lattice structure of silicate or metal, or (ii) adsorption processes onto Fe-oxides or Fe hydroxides minerals. Being volatile, Ge isotopes are also affected by evaporation/condensation processes. The absolute range of $\delta^{74/70}$Ge values in rocks and minerals is more than 8‰, which is comparable or even larger than most non-traditional stable isotope systems (e.g., Anbar and Rouxel 2007; this volume).

The knowledge of processes that induce high-temperature Ge isotope fractionation is essential for understanding the accretion-collision and differentiation history of planetary bodies. Indeed, the siderophile and volatile behavior of Ge lead to major differences in Ge isotope composition between deep core and mantle-crust reservoirs of planets, unlike other isotopic systems (e.g., Fe, Si). It is expected that further insights into Ge isotopic fractionation mechanisms will provide a key tool for the quantification of metal–silicate segregation in different planetary bodies, when compared to geophysical constraints on core formation of planetary bodies. Initial studies of mantle Ge isotope geochemistry emphasize broad homogeneity in the composition of the bulk silicate Earth. Further investigations at a smaller scale, and in various geodynamic settings will provide constraints into recycling processes into the mantle and crust formation processes. Significant advances have been made in understanding the incorporation of Ge in sulfides and other ore-forming minerals. Germanium isotopes as tracers of kinetic and equilibrium processes of mineral precipitation will undoubtly provide new clues into the mechanisms of formation of large ore deposits.

The promise of low-temperature Ge-isotope geochemistry derives from the close association between Ge and Si biogeochemical cycles. Yet at the same time major Si sources to the ocean (i.e., rivers and hydrothermal sources) have different Ge/Si ratios. Initial work on the Ge-isotope composition of marine environments has revealed striking variations of Ge isotope ratios among Earth surface reservoirs (e.g., seawater, biogenic silica) and sources (e.g., rivers, hydrothermal vents). We hypothesize that coupled Ge/Si, Si isotopes and Ge isotope systematics provide i) a better understanding of the nature of modern marine silica fluxes with the identification of the "missing" Ge sink that prevents closure of the modern marine Ge budget; ii) new approaches to explore the long-term evolution of the marine silica cycle, from ancient silica-rich oceans dominated by inorganic silica cycling to modern oceans effectively Si-stripped by the biological silica pump.

ACKNOWLEDGMENTS

Rouxel's research on Ge isotopes would not have been possible without support from the National Science Foundation, Europole Mer, LabexMer (ANR-10-LABX-19-01), University of Cambridge, Woods Hole Oceanographic Institution and Ifremer. Critical help and inspiration were provided by coworkers, students and colleagues over the years, notably Jotis Baronas, Andrey Bekker, Christina De La Rocha, Olivier Donard, Vesselin Dekov, Katrina Edwards, Harry Elderfield, Raphaelle Escoube, Albert Galy, Maxence Guillermic, Doug Hammond, Stefan Lalonde, Samia Mantoura, Emmanuel Ponzevera, Huawen Qi, Ed Sholkovitz. Béatrice Luais acknowledges fundings from CNRS-INSU research programs

(CESSUR, DYETI-SYSTER, PNP), the French National Research Agency through the national program "Investissements d'avenir" with the reference ANR-10-LABX-21-LABEX RESSOURCES21, the Observatoire Terre Environnement Lorraine (OTELo) program. Germanium isotope research of these last ten years benefited from assistance, interaction, and fruitful discussions with Françoise Ali, Rémi Belissont, Damien Cividini, Etienne Deloule, Afifé El Korh, Alexis Filia, Thibault de Gournay, Aymeric Schumacher, Laurent Tissandier, Mike Toplis, John Wasson. This manuscript benefited from the constructive reviews of Stefan Weyer, Doug Hammond, Nicolas Dauphas and one anonymous reviewer.

REFERENCES

Adam J, Green T (2006) Trace element partitioning between mica-and amphibole-bearing garnet lherzolite and hydrous basanitic melt: 1. Experimental results and the investigation of controls on partitioning behaviour. Contrib Mineral Petrol 152:1–17

Albarede F, Beard B (2004) Analytical methods for non-traditional isotopes. Rev Mineral Geochem 55: 113–152

Algeo TJ, Schwark L, Hower JC (2004) High-resolution geochemistry and sequence stratigraphy of the Hushpuckney Shale (Swope Formation, eastern Kansas): implications for climato-environmental dynamics of the Late Pennsylvanian Midcontinent Seaway. Chem Geol 206:259–288

Anders AM, Sletten RS, Derry LA, Hallet B (2003) Germanium/silicon ratios in the Copper River Basin, Alaska: Weathering and partitioning in periglacial versus glacial environments. J Geophys Res 108(F1):1–9

André L, Cardinal D, Alleman LY, Moorbath S (2006) Silicon isotopes in ~3.8 Ga West Greenland rocks as clues to the Eoarchaean supracrustal Si cycle. Earth Planet Sci Lett 245:162–173

Andreae MO, Froelich PN (1981) Determination of germanium in natural waters by graphite furnace atomic absorption spectrometry with hydride generation. Anal Chem 53:287–291, doi:10.1021/ac00225a037

Andreae MO, Froelich PN (1984) Arsenic, antimony, and germanium biogeochemistry in the Baltic Sea. Tellus Ser B-Chem Phys Meteorol 36:101–117

Arnorsson S (1984) Germanium in Icelandic geothermal systems. Geochim Cosmochim Acta 48:2489–2502

Arevalo R, McDonough WF (2010) Gallium and germanium abundances in MORB and OIB: evidence for pyroxenitic source components. Geochim Cosmochim Acta 74:A32

Babula P, Adam V, Opatrilova R, Zehnalek J, Havel L, Kizek R (2008) Uncommon heavy metals, metalloids and their plant toxicity: a review. Environ Chem Lett 6:189–213, doi:10.1007/s10311-008-0159-9

Baker ET, Embley RW, Walker SL, Resing JA, Lupton JE, Nakamura K, de Ronde CEJ, Massoth GJ (2008) Hydrothermal activity and volcano distribution along the Mariana arc. J Geophys Res-Solid Earth 113, doi:10.1029/2007jb005423

Bareille G, Labracherie M, Mortlock RA, Maier-Reimer E, Froelich PN (1998) A test of (Ge/Si)(opal) as a paleorecorder of (Ge/Si)(seawater). Geology 26:179–182

Baronas JJ, Hammond DE, McManus J, Siebert C, Wheat G (2014) Marine budget for germanium stable isotopes. Ocean Sciences Meeting, Honolulu, 24–28th February 2014:Abstract ID:13978

Bekker A, Slack JF, Planavsky N, Krapez B, Hofmann A, Konhauser KO, Rouxel OJ (2010) Iron formation: the sedimentary product of a complex interplay among mantle, tectonic, oceanic, and biospheric processes. Economic Geology 105:467–508

Belissont R, Boiron MC, Luais B, Cathelineau M (2014) LA-ICP-MS analyses of minor and trace elements and bulk Ge isotopes in zoned Ge-rich sphalerites from the Noailhac–Saint-Salvy deposit (France): Insights into incorporation mechanisms and ore deposition processes. Geochim Cosmochim Acta 126:518–540, doi:10.1016/j.gca.2013.10.052

Belissont R, Muñoz M, Boiron M-C, Luais B, Mathon O (2016) Distribution and oxidation state of Ge, Cu and Fe in sphalerite by μ-XRF and K-edge μ-XANES: insights into Ge incorporation, partitioning and isotopic fractionation. Geochim Cosmochim Acta doi:10.1016/j.gca.2016.01.001

Benedix GK, McCoy TJ, Keil L, Love SG (2000) A petrological study of the IAB iron meteorites: constraints on the formation of the IAB-Winonaite parent body. Meteor Planet Sci 35:1127–1141

Berner RA (2003) The long-term carbon cycle, fossil fuels and atmospheric composition. Nature 426:323–326, doi:10.1038/nature02131

Berner RA, Lasaga AC, Garrels RM (1983) The carbonate–silicate geochemical cycle and its effect on atmospheric carbon dioxide over the past 100 million years. Am J Sci 283:641–683

Bernstein LR (1985) Germanium geochemistry and mineralogy. Geochim Cosmochim Acta 49:2409–2422

Blecker SW, King SL, Derry LA, Chadwick OA, Ippolito JA, Kelly EF (2007) The ratio of germanium to silicon in plant phytoliths: quantification of biological discrimination under controlled experimental conditions. Biogeochemistry 86:189–199, doi:10.1007/s10533-007-9154-7

Bogard DD, Garrison DH, McCoy TJ (2000) Chronology and petrology of silicates from IIE iron meteorites: evidence of a complex parent body evolution. Geochim Cosmochim Acta 64:2133–2154

Brown HM, Krouse HR (1964) Fractionation of germanium isotopes in chemical reactions. Can J Chem 42:1971–1978, doi:10.1139/v64–290

Brumsack H-J (2006) The trace metal content of recent organic carbon-rich sediments: Implications for Cretaceous black shale formation. Palaeogeogr Palaeoclimatol Palaeoecol 232:344– 361

Cakmak I, Kurz H, Marschner H (1995) Short-term effects of boron, germanium and high light intensity on membrane permeability in boron deficient leaves of sunflower. Physiol Plant 95:11–18, doi:10.1034/j.1399–3054.1995.950102.x

Campbell DR (1975) Isotope effect for self-diffusion in Ge. Phys Rev B 12:2318

Capobianco C, Watson E (1982) Olivine/silicate melt partitioning of germanium: an example of a nearly constant partition coefficient. Geochim Cosmochim Acta 46:235–240

Cavell RG, Barne EM, Arboleda PH, Cavell PA, Feng R, Gordon RA, Webb MA (2004) An X-ray and electron microprobe study of Fe, Ni, Ga, and Ge distribution and local structure in a section of the Canyon Diablo iron meteorite. Am Miner 89:519–526

Cempirek J, Groat LA (2013) Note on the formula of brunogeierite and the first bond-valence parameters for Ge^{2+}. J Geosci 58:71–74

Chabot NL, Jones JH (2003) The parameterization of solid metal–liquid metal partitioning of siderophile elements. Meteor Planet Sci 38:1425–1436

Chabot NL, Campbell AJ, Jones JH, Humayun M, Agee CB (2003) An experimental test of Henry's Law in solid metal–liquid metal systems with implications for iron meteorites. Meteor Planet Sci 38:181–196

Chang T-L, Li W-J, Qiao G-S, Qian Q-Y, Chu Z-Y (1999) Absolute isotopic composition and atomic weight of germanium. Int J Mass Spectrom 189:205–211

Chapman FW, Marvin GG, Tyree SY (1949) Volatilization of elements from perchloric and hydrofluoric acid solutions. Anal Chem 21:700–701

Choi B, Ouyang X, Wasson J (1995) Classification and origin of IAB and IIICD iron meteorites. Geochim Cosmochim Acta 59:593–612

Ciavatta L, Iuliano M, Porto R, Vasca E (1990) Fluorogermanate(IV) equilibria in acid-media. Polyhedron 9:1263–1270, doi:10.1016/s0277–5387(00)86762–5

Cohen BA, Hewins RH, Alexander CMOD (2004) The formation of chondrules by open-system melting of nebular condensates. Geochim Cosmochim Acta 68:1661–1675

Cook NJ, Ciobanu CL, Pring A, Skinner W, Shimizu M, Danyushevsky L, Saini-Eidukat B, Melcher F (2009) Trace and minor elements in sphalerite: A LA-ICPMS study. Geochim Cosmochim Acta 73:4761–4791

Cook NJ, Etschmann B, Ciobanu CL, Geraki K, Howard DL, Williams T, Rae N, Pring A, Chen G, Johannessen B, Brugger J (2015) Distribution and substitution mechanism of Ge in a Ge-(Fe)-bearing sphalerite. Minerals 5:117–132, doi:10.3390/min5020117

Corrigan CM, Chabot NL, McCoy TJ, McDonough WF, Watson HC, Saslow SA, Ash RD (2009) The iron, nickel, phosphorus system: effects on the distribution of trace elements during the evolution of iron meteorites. Geochim Cosmochim Acta 73:2674–2691

Criaud A, Fouillac C (1986) Study of CO_2-rich thermomineral waters from the French Massif Central. 2. Behavior of some trace-metals, arsenic, antimony and germanium. Geochim Cosmochim Acta 50:1573–1582, doi:10.1016/0016–7037(86)90120–1

Dauphas N, Craddock PR, Asimow PD, Bennett VC, Nutman AP, Ohnenstetter D (2009) Iron isotopes may reveal the redox conditions of mantle melting from Archean to Present. Earth Planet Sci Lett 288:255–267

Dauphas N, Roskosz M, Alp EE, Neuville DR, Hu MY, Sio CK, Tissot FL, Zhao J, Tissandier L, Médard E, Cordier C (2014) Magma redox and structural controls on iron isotope variations in Earth's mantle and crust. Earth Planet Sci Lett 398:127–140, doi:10.1016/j.epsl.2014.04.033

Davis AM (2006) Volatile evolution and loss. Meteorites and the Early Solar System II. DS Lauretta, HY McSween (eds), p. 295–307

Davis FA, Humayun M, Hirschmann M, Cooper RS (2013) Experimentally determined mineral/melt partitioning of first-row transition elements (FRTE) during partial melting of peridotite at 3 GPa. Geochim Cosmochim Acta 104:232–260

De Argollo RM, Schilling JG (1978a) Ge/Si and Ga/Al variations along the Reykjanes Ridge and Iceland. Nature 276:24–28

De Argollo RM, Schilling JG (1978b) Ge–Si and Ga–Al fractionation in Hawaiian volcanic rocks. Geochim Cosmochim Acta 42:623–630

De Carlo EH, McMurtry GM, Yeh H-W (1983) Geochemistry of hydrothermal deposits from the Loihi submarine volcano, Hawaii. Earth Planet Sci Lett 66:438–449

De La Rocha C (2003) Silicon isotope fractionation by marine sponges and the reconstruction of the silicon isotope composition of ancient deep water. Geology 31:423–426

De La Rocha C, Brzezinski MA, De Niro MJ (1997) Fractionation of silicon isotopes by marine diatoms during biogenic silica formation. Geochim Cosmochim Acta 61:5051–5056

De La Rocha CL, Brzezinski MA, De Niro MJ, Shemesh A (1998) Silicon-isotope composition of diatoms as an indicator of past oceanic change. Nature 295:680–683

Dedina J, Tsalev DL (1995) Hydride Generation Atomic Absorption Spectrometry. Wiley
Delvigne C, Opfergelt S, Cardinal D, Delvaux B, Andre L (2009) Distinct silicon and germanium pathways in the soil–plant system: Evidence from banana and horsetail. J Geophys Res–Biogeosci 114, doi:10.1029/2008jg000899
DeMaster DJ (2002) The accumulation and cycling of biogenic silica in the Southern Ocean: revisiting the marine silica budget. Deep-Sea Res 49:3155–3167
Derry LA, Kurtz AC, Ziegler K, Chadwick OA (2005) Biological control of terrestrial silica cycling and export fluxes to watersheds. Nature 433:728–731, doi:10.1038/nature03299
El Korh A, Luais B, Boiron M-C, Deloule E, Cividini D, Yeguicheyan D (2015) Ge and Ga abundances and Ge isotope ratios in high-pressure metamorphic rocks: a Q-ICPMS and MC-ICPMS study. European Winter Conference February 22–26, Münster (Germany)
El Korh A, Luais B, Boiron M-C, Deloule E (2016) Investigation of Ge and Ga exchange behaviour and Ge isotopic fractionation during subduction zone metamorphism. Chem Geology (in revision)
El Wardani SA (1957) On the geochemistry of germanium. Geochim Cosmochim Acta 13:5–19
Elderfield H, Schultz A (1996) Mid-ocean ridge hydrothermal fluxes and the chemical composition of the ocean. Ann Rev Earth Planet Sci 24:191–224
Ellwood MJ, Maher WA (2003) Germanium cycling in the waters across a frontal zone: the Chatham Rise, New Zealand. Mar Chem 80:145–159
Emerson D, Moyer CL (2002) Neutrophilic Fe-oxidizing bacteria are abundant at the Loihi seamount hydrothermal vents and play a major role in Fe oxide deposition. Appl Environ Microbiol 68:3085–3093
Epov VN, Rodriguez-Gonzalez P, Sonke JE, Tessier E, Amouroux D, Bourgoin LM, Donard OFX (2008) Simultaneous determination of species–specific isotopic composition of Hg by gas chromatography coupled to multicollector ICPMS. Anal Chem 80:3530–3538
Epstein E (1994) The anomaly of silicon in plant biology. PNAS 91:11–17, doi:10.1073/pnas.91.1.11
Escoube R, Rouxel OJ, Luais B, Ponzevera E, Donard OFX (2012) An intercomparison study of the germanium isotope composition of geological reference materials. Geostand Geoanal Res 36:149–159, doi: 10.1111/J.1751–908x.2011.00135.X
Escoube R, Rouxel O, Edwards K, Glazer B, Donard O (2015) Coupled Ge/Si and Ge isotope ratios as geochemical tracers of seafloor hydrothermal systems: case studies at Loihi Seamount and East Pacific Rise 9°50'N. Geochim Cosmochim Acta 167:93–112
Evans MJ, Derry LA (2002) Quartz control of high germanium/silicon ratios in geothermal waters. Geology 30:1019–1022
Feely RA, Massoth GJ, Baker ET, Cowen JP, Lamb MF, Krogslund KA (1990) The effect of hydrothermal processes on midwater phosphorus distributions in the northeast Pacific. Earth Planet Sci Lett 96:305–318, doi:10.1016/0012–821x(90)90009-m
Feely RA, Trefry JH, Massoth GJ, Metz S (1991) A comparison of the scavenging of phosphorus and arsenic from seawater by hydrothermal iron oxyhydroxides in the Atlantic and Pacific Oceans. Deep Sea Res (I Oceanogr Res Pap) 38:617–623, doi:10.1016/0198–0149(91)90001-v
Fischer WW, Knoll AH (2009) An iron shuttle for deepwater silica in Late Archean and early Paleoproterozoic iron formation. Geol Soc Am Bull 121:222–235, doi:10.1130/b26328.1
Frank M (2002) Radiogenic isotopes: Tracers of past ocean circulation and erosional input. Rev Geophys 40:1-1-1-38, doi:1001 10.1029/2000rg000094
Frei R, Polat A (2007) Source heterogeneity for the major components of similar to 3.7 Ga Banded Iron Formations (Isua Greenstone Belt, Western Greenland): Tracing the nature of interacting water masses in BIF formation. Earth Planet Sci Lett 253:266–281, doi:10.1016/j.epsl.2006.10.033
Frenzel M, Ketris MP, Gutzmer J (2014) On the geological availability of germanium. Miner Deposita 49:471–486, doi:10.1007/s00126–013–0506-z
Frenzel M, Hirsch T, Gutzmer J (2016) Gallium, germanium, indium, and other trace and minor elements in sphalerite as a function of deposit type—A meta-analysis. Ore Geol Rev 76:52–78
Froelich PN, Andreae MO (1981) The marine geochemistry of germanium—Ekasilicon. Science 213:205–207
Froelich PN, Lesley P (2003) Tracing germanium contamination from coal-fired power plants down the Chattahoochee–Apalachicola River: implications for the toxic metalloids arsenic and selenium. Proc 2001 Georgia Water Res Conf, University of Georgia, Athens, p. 488–491
Froelich PN, Hambrick GA, Andreae MO, Mortlock RA, Edmond JM (1985a) The geochemistry of inorganic germanium in natural waters. J Geophys Res–Oceans 90:1133–1141, doi:10.1029/JC090iC01p01133
Froelich PN, Hambrick GA, Kaul LW, Byrd JT, Lecointe O (1985b) Geochemical behavior of inorganic germanium in an unperturbed estuary. Geochim Cosmochim Acta 49:519–524
Froelich PN, Mortlock RA, Shemesh A (1989) Inorganic germanium and silica in the Indian Ocean: Biological fractionation during (Ge/Si)OPAL formation. Global Biogeochem Cycles 3:79–88

Froelich PN, Blanc V, Mortlock RA, Chillrud SN, Dunstan W, Udomkit A, Peng TH (1992) River fluxes of dissolved silica to the ocean were higher during glacials: Ge/Si in diatoms, rivers, and oceans. Paleoceanogr 7:739–767, doi:10.1029/92pa02090

Gaillardet J, Dupre B, Louvat P, Allegre CJ (1999) Global silicate weathering and CO_2 consumption rates deduced from the chemistry of large rivers. Chem Geol 159:3–30, doi:10.1016/s0009–2541(99)00031–5

Galy A, Pomiès C, Day JA, Pokrovsky OS, Schott J (2003) High precision measurement of germanium isotope ratio variations by multiple collector–inductively coupled plasma–mass spectrometry. J Anal At Spectrom 18:115–119

Gautier E, Garavaglia R, Lobo A, Fernandez M, Farach H (2012) Isotopic analysis of germanium by thermal ionization mass spectrometry. J Anal At Spectrom 27:881–883, doi:10.1039/c2ja10305b

Glazer BT, Rouxel OJ (2009) Redox speciation and distribution within diverse iron-dominated microbial habitats at Loihi Seamount. Geomicrobiol J 26:606–622, doi: 10.1080/01490450903263392

Goldschmidt VM (1958) Geochemistry. Oxford Univ. Press, London

Goodman S (1988) Therapeutic effects of organic Germanium. Medical Hypotheses 26:207–215, doi:10.1016/0306–9877(88)90101–6

Govindaraju K (1994) compilation of working values and sample description for 383 geostandards. Geostandard Newsl 18:158

Graham RP, MacNamara J, Crocker IH, MacFarlane RB (1951) The isotopic constitution of germanium. Can J Chem 29:89–102

Green MD, Rosman KJR, deLaeter JR (1986) The isotopic composition of germanium in terrestrial samples. Int J Mass Spectrom Ion Processes 68:15–24

Guillermic M, Lalonde SV, Hendry KR, Rouxel O (2016) The isotopic composition of inorganic Germanium in seawater and deep sea sponges. Geochim Cosmochim Acta (in revision)

Halliday AN, Lee DC, Christensen JN, Walder AJ, Freedman PA, Jones CE, Hall CM, Yi W, Teagle D (1995) Recent developments in inductively coupled plasma magnetic sector multiple collector mass spectrometry. Int J Mass Spectrom Ion Processes 146:21–33, doi:10.1016/0168–1176(95)04200–5

Halperin SJ, Barzilay A, Carson M, Roberts C, Lynch J (1995) Germanium accumulation and toxicity in barley. J Plant Nutr 18:1417–1426, doi:10.1080/01904169509364991

Hamade T, Konhauser KO, Raiswell R, Goldsmith S, Morris RC (2003) Using Ge/Si ratios to decouple iron and silica fluxes in Precambrian banded iron formations. Geology 31:35–38

Hambrick GA, Froelich PN, Andreae MO, Lewis BL (1984) Determination of methylgermanium species in natural waters by graphite furnace atomic absorption spectrometry with hydride generation. Anal Chem 56:421–424, doi:10.1021/ac00267a027

Hammond DE, McManus J, Berelson WM, Meredith C, Klinkhammer GP, Coale KH (2000) Diagenetic fractionation of Ge and Si in reducing sediments: The missing Ge sink and a possible mechanism to cause glacial/interglacial variations in oceanic Ge/Si. Geochim Cosmochim Acta 64:2453–2465

Hammond DE, McManus J, Berelson WM (2004) Oceanic Germanium/silicon ratios: Evaluation of the potential overprint of temperature on weathering signals. Paleoceanogr 19:doi:10.1029/2003PA000940

Haugaard R, Pecoits E, Lalonde SV, Rouxel OJ, Konhauser KO (2016) The Joffre banded iron formation, Hamersley Group, Western Australia: Assessing the palaeoenvironment through detailed petrology and chemostratigraphy. Precambrian Res 273:12–37

Hehenkamp T, Lodding A, Odelius H, Schlett V (1979) Isotope effect in the diffusion of the stable germanium isotopes in copper. Acta Metall 27:829–832

Hendry KR, Robinson LF (2012) The relationship between silicon isotope fractionation in sponges and silicic acid concentration: Modern and core-top studies of biogenic opal. Geochim Cosmochim Acta 81:1–12, doi:10.1016/j.gca.2011.12.010

Hendry KR, Georg RB, Rickaby REM, Robinson LF, Halliday AN (2010) Deep ocean nutrients during the Last Glacial Maximum deduced from sponge silicon isotopic compositions. Earth Planet Sci Lett 292:290–300, doi:10.1016/j.epsl.2010.02.005

Hill E, Wood BJ, Blundy JD (2000) The effect of Ca-Tschermaks component on trace element partitioning between clinopyroxene and silicate melt. Lithos 53:203–215

Hirata T (1997) Isotopic variations of germanium in iron and stony iron meteorites. Geochim Cosmochim Acta 61:4439–4448

Hirner AV, Feldmann J, Krupp E, Grumping R, Goguel R, Cullen WR (1998) Metal(loid)organic compounds in geothermal gases and waters. Org Geochem 29:1765–1778, doi:10.1016/s0146–6380(98)00153–3

Holl R, Kling M, Schroll E (2007) Metallogenesis of germanium—A review. Ore Geol Rev 30:145–180, doi:10.1016/j.oregeorev.205.07.034

Iacono-Marziano G, Gaillard F, Scaillet B, Polozov AG, Marecal V, Pirre M, Arndt NT (2012) Extremely reducing conditions reached during basaltic intrusion in organic matter-bearing sediments. Earth Planet Sci Lett 357:319–326

Ingri N (1963) Equilibrium studies of polyanions and polygermanates in NaCl medium. Acta Chem Scand 17:597–616

Ishikawa T, Nakamura E (1990) Suppression of boron volatilization from a hydrofluoric acid solution using a boron–mannitol complex. Anal Chem 62:2612–2616, doi:10.1021/ac00222a017

Jambon A (1980) Isotopic fractionation: A kinetic model for crystals growing from magmatic melts. Geochim Cosmochim Acta 44:1373–1380

Jana D, Walker D (1997) The influence of sulfur on partitioning of siderophile elements. Geochim Cosmochim Acta 61:5255–5277

Jenner FE, O'Neill HSC (2012) Analysis of 60 elements in 616 ocean floor basaltic glasses. Geochem Geophys Geosyst 13

Johan Z (1988) Indium and germanium in the structure of sphalerite : an example of coupled substitution with copper. Mineral Petrol 39:211–229

Jones JH, Drake MJ (1986) Geochemical constraints on core formation in the Earth. Nature 322:221–228, doi:10.1038/322221a0

Jones JH, Walker D (1991) Partitioning of siderophile elements in the Fe–Ni–S system: 1 bar to 80 kbar. Earth Planet Sci Lett 105:127–133

Karl DM, McMurtry GM, Malahoff A, Garcia MO (1988) Loihi Seamount, Hawaii: a mid-plate volcano with a distinctive hydrothermal system. Nature 335:532–535, doi:10.1038/335532a0

Kaya M, Volkan M (2011) Germanium determination by flame atomic absorption spectrometry: An increased vapor pressure-chloride generation system. Talanta 84:122–126, doi:10.1016/j.talanta.2010.12.029

Kelly WR, Larimer JW (1977) Chemical fractionations in meteorites—VIII. Iron meteorites and the cosmochemical history of the metal phase. Geochim Cosmochim Acta 41:93–111

King SL, Froelich PN, Jahnke RA (2000) Early diagenesis of germanium in sediments of the Antartic South Atlantic: In search of the missing Ge sink. Geochim Cosmochim Acta 64:1375–1390

Kipphardt H, Valkiers S, Henriksen F, De Bievre P, Taylor PDP, Tolg G (1999) Measurement of the isotopic composition of germanium using GeF_4 produced by direct fluorination and wet chemical procedures. Int J Mass Spectrom 189:27–37, doi:10.1016/s1387-3806(99)00047-0

Klein C (2005) Some Precambrian banded iron-formations (BIFs) from around the world: their age, geologic setting, mineralogy, metamorphism, geochemistry, and origin. Am Miner 90:1473–1499, doi:10.2138/am.2005.1871

Klöck W, Palme H, Tobschall HJ (1986) Trace elements in natural metallic iron from Disko Island, Greenland. Contrib Mineral Petrol 93:273–282

Kolodny Y, Halicz L (1988) The geochemistry of germanium in deep-sea cherts. Geochim Cosmochim Acta 52:2333–2336

Kruijer TS, Touboul M, Fischer-Gödde M, Bermingham KR, Walker RJ, Kleine T (2014) Protracted core formation and rapid accretion of protoplanets. Science 344.1150–1154

Kurtz AC, Derry LA, Chadwick OA (2002) Germanium–silicon fractionation in the weathering environment. Geochim Cosmochim Acta 66:1525–1537

Kwasnik W (1963) Section 4. Fluorine compounds. *In*: Handbook of Preparative Inorganic Chemistry. Brauer G, (ed) Academic Press Inc., New York, p 150

Lalonde S, Konhauser K, Rouxel O (2012) Germanium and silicon isotopic evolution of seawater inferred from Precambrian Iron Formations. Goldschmidt 2012 Conference, June 24-29, 2012, Montréal, Canada

Layton-Matthews D, Leybourne MI, Peter JM, Scott SD (2006) Determination of selenium isotopic ratios by continuous-hydride generation dynamic-reaction-cell inductively coupled plasma-mass spectrometry. J Anal At Spectrom 21:41–49

Lehmann B, Nägler TF, Holland HD, Wille M, Mao J, Pan J, Ma D, Dulski P (2007) Highly metalliferous carbonaceous shale and Early Cambrian seawater. Geology 35:403–406

Lewis BL, Froelich PN, Andreae MO (1985) Methylgermanium in natural waters. Nature 313:303–305, doi:10.1038/313303a0

Lewis BL, Andreae MO, Froelich PN (1989) Sources and sinks of methylgermanium in natural waters. Mar Chem 27:179–200, doi:10.1016/0304-4203(89)90047-9

Li XF, Liu Y (2010) First-principles study of Ge isotope fractionation during adsorption onto Fe(III)-oxyhydroxides surfaces. Chem Geol:10.1016/j.chemgeo.2010.1005.1008.

Li XF, Zhao H, Tang M, Liu Y (2009) Theoretical prediction for several important equilibrium Ge isotope fractionation factors and geological implications. Earth Planet Sci Lett 287:1–11, doi:10.1016/j.epsl.2009.07.027

Liu M, Fleet ME (2001) Partitioning of siderophile elements (W, Mo, As, Ag, Ge, Ga, and Sn) and Si in the Fe-S system and their fractionation in iron meteorites. Geochim Cosmochim Acta 65:671–682

Lodders K, Palme H, Gail HP (2009) Abundances of the elements in the Solar System. *In*: Solar system. Springer, p 712–770

Luais B (2003) Germanium isotope systematics in meteorites. Meteor Planet Sci Suppl 38:5048

Luais B (2007) Isotopic fractionation of germanium in iron meteorites: Significance for nebular condensation, core formation and impact processes. Earth Planet Sci Lett 262:21–36, doi:10.1016/j.epsl.2007.06.031

Luais B (2012) Germanium chemistry and MC-ICPMS isotopic measurements of Fe–Ni, Zn alloys and silicate matrices: Insights into deep Earth processes. Chem Geol 334:295–311, doi:10.1016/j.chemgeo.2012.10.017
Luais B, Toplis MJ, Roskosz M, Tissandier L (2007) Experimental determination of germanium isotopic fractionation during metal–silicate segregation. Eos Trans AGU 88:Fall Meet. Suppl., Abstract V51E–0833
Luais B, Framboisier X, Carignan J, Ludden JN (2000) Analytical development of Ge isotopic analyses using multi-collection plasma source mass spectrometry: Isoprobe MC-Hex-ICP-MS (Micromass). *In:* Geoanalysis 2000, Symposium B 2000, p 45–46
Luais B, Lach P, Tomassot E, Chaussidon M, Boiron MC (2011) Preliminary high-resolution Ge/Si data in early Archaean BIFs. Mineral Mag 75:1362
Luais B, Ali F, Wasson JT (2014) Low germanium isotopic composition of IIG iron meteorites: relationship with IIAB irons and influence of sulfur and phosphorus. Meteor Planet Sci 49, Supp:A239
Lyons TW, Severmann S (2006) A critical look at iron paleoredox proxies: New insights from modern euxinic marine basins. Geochim Cosmochim Acta 70:5698–5722, doi:10.1016/j.gca.2006.08.021
Machlan LA, Gramlich JW, Powell LJ, Lambert GM (1986) Absolute isotope abundance ratio and atomic weight of a reference sample of gallium. J Res Nat Bur Stand 91:323–331
Makishima A, Nakamura E (2009) Determination of Ge, As, Se and Te in silicate samples using isotope dilution-internal standardisation octopole reaction cell ICP-QMS by normal sample nebulisation. Geostand Geoanal Res 33:369–384
Malvin G, Drake M (1987) Experimental determination of crystal-melt partitioning of Ga and Ge in the system forsterite-anorthite-diopside. Geochim Cosmochim Acta 51:2117–2128
Mantoura S (2006) Development and application of opal based paleoceanographic proxies. Doctor of Philosophy University of Cambridge, Cambridge
Maréchal C, Télouk P, Albarède F (1999) Precise analysis of copper and zinc isotopic compositions by plasma-source mass spectrometry. Chem Geol 156:251–273
Marin-Carbonne J, Robert F, Chaussidon M (2014) The silicon and oxygen isotope compositions of Precambrian cherts: A record of oceanic paleo-temperatures? Precambrian Res 247:223–234, doi:10.1016/j.precamres.2014.03.016
Martin JH (1990) Glacial-interglacial CO_2 change: the iron hypothesis. Paleoceanogr 5:1–13, doi:10.1029/PA005i001p00001
McDermott KH, Greenwood RC, Scott ERD, Franchi IA, Anand M (2016) Oxygen isotope and petrological study of silicate inclusions in IIE iron meteorites and their relationship with H chondrites. Geochim Cosmochim Acta 173:97–113
McManus J, Hammond DE, Cummins K, Klinkhammer GP, Berelson WM (2003) Diagenetic Ge–Si fractionation in continental margin environments: Further evidence for a nonopal sink. Geochim Cosmochim Acta 67:4545–4557
McVay GL, DuCharme AR (1974) Diffusion of Ge in Si-Ge alloys. Phys Rev B 9:627
Medenbach O, ElGoresy A (1982) Ulvöspinel in native iron-bearing assemblages and the origin of these assemblages in basalts from Ovifak, Greenland, and Bühl, Federal Republic of Germany. Contrib Mineral Petrol 80:358–366
Meng YM, Qi HW, Hu RZ (2015) Determination of germanium isotopic compositions of sulfides by hydride generation MC-ICP-MS and its application to the Pb–Zn deposits in SW China. Ore Geol Rev 65:1095–1109, doi:10.1016/j.oregeorev.2014.04.008
Mortlock RA, Froelich PN (1987) Continental weathering of germanium: Ge/Si in the global discharge. Geochim Cosmochim Acta 51:2075–2082
Mortlock RA, Charles CD, Froelich PN, Zibello MA, Saltzman J, Hays JD, Burckle LH (1991) Evidence for lower productivity in the Antartic Ocean during the last glaciation. Nature 351:220–223, doi:10.1038/351220a0
Mortlock RA, Froelich PN, Feely RA, Massoth GJ, Butterfield DA, Lupton JE (1993) Silica and germanium in Pacific Ocean hydrothermal vents and plumes. Earth Planet Sci Lett 119:365–378
Murnane RJ, Stallard RF (1988) Germanium/silicon fractionation during biogenic opal formation. Paleoceanography 3:461–469, doi:10.1029/PA003i004p00461
Murnane RJ, Stallard RF (1990) Germanium and silicon in rivers of the Orinoco drainage basin. Nature 344:749–752, doi:10.1038/344749a0
Mysen BO (2004) Element partitioning between minerals and melt, melt composition, and melt structure. Chem Geol 213:1–16
Nishimura H, Takeshi H, Okano J (1988) Isotopic abundances of germanium determined by secondary ion mass spectrometry. J Mass Spectrom Soc Jpn 36:197–202
Olsen E, Kracher A, Davis AM, Steele IM, Hutcheon ID, Bunch TE (1999) The phosphates of IIIAB iron meteorites. Meteor Planet Sci 34:285–300
Onishi F, Inatomi Y, Tanaka T, Shinozaki N, Watanabe M, Fujimoto A, Itoh K (2006) Time-of-flight secondary mass spectrometry analysis of isotope composition for measurement of self-diffusion coefficient. Jpn J Appl Phys Part 1 45:5274–5276, doi:10.1143/jjap.45.5274

Pike GE, Camp WJ, Seager CH, McVay GL (1974) Percolative aspects of diffusion in binary alloys. Phys Rev B 10:4909

Plank T, Langmuir CH (1998) The chemical composition of subducting sediment and its consequences for the crust and mantle. Chem Geol 145:325–394

Pokrovski GS, Schott J (1998a) Thermodynamic properties of aqueous Ge(IV) hydroxide complexes from 25 to 350 °C: Implications for the behavior of germanium and the Ge/Si ratio in hydrothermal fluids. Geochim Cosmochim Acta 62:1631–1642

Pokrovski GS, Schott J (1998b) Experimental study of the complexation of silicon and germanium with aqueous organic species: Implications for germanium and silicon transport and Ge/Si ratio in natural waters. Geochim Cosmochim Acta 62:3413–3428

Pokrovski GS, Martin F, Hazemann J-L, Schott J (2000) An X-ray absorption fine structure spectroscopy study of germanium-organic ligand complexes in aqueous solution. Chem Geol 163:151–165

Pokrovsky OS, Pokrovski GS, Schott J, Galy A (2006) Experimental study of germanium adsorption on goethite and germanium coprecipitation with iron hydroxide: X-ray absorption fine structure and macroscopic characterization. Geochim Cosmochim Acta 70:3325–3341

Pokrovsky OS, Galy A, Schott J, Pokrovski GS, Mantoura S (2014) Germanium isotope fractionation during Ge adsorption on goethite and its coprecipitation with Fe oxy(hydr)oxides. Geochim Cosmochim Acta 131:138–149, doi:10.1016/j.gca.2014.01.023

Puerner NJ, Siegel SM, Siegel BZ (1990) The experimental phytotoxicology of germanium in relation to silicon. Water Air Soil Pollut 49:187–195, doi:10.1007/bf00279520

Qi HW, Rouxel O, Hu RZ, Bi XW, Wen HJ (2011) Germanium isotopic systematics in Ge-rich coal from the Lincang Ge deposit, Yunnan, Southwestern China. Chem Geol 286:252–265, doi: 10.1016/J.Chemgeo.2011.05.011

Reincke T, Barthel D (1997) Silica uptake kinetics of *Halichondria panicea* in Kiel Bight. Mar Biol 129:591–593

Reynolds JH (1953) The isotopic constitution of silicon, germanium, and hafnium. Phys Rev 90:1047–1049

Richter FM, Liang Y, Davis AM (1999) Isotope fractionation by diffusion in molten oxides. Geochim Cosmochim Acta 63:2853–2861

Richter FM, Watson EB, Mendybaev R, Dauphas N, Georg B, Watkins J, Valley J (2009) Isotopic fractionation of the major elements of molten basalt by chemical and thermal diffusion. Geochim Cosmochim Acta 73:4250–4263, doi:10.1016/j.gca.2009.04.011

Righter K, King C, Danielson L, Pando K, Lee CT (2011) Experimental determination of the metal/silicate partition coefficient of Germanium: Implications for core and mantle differentiation. Earth Planet Sci Lett 304:379–388, doi:10.1016/j.epsl.2011.02.015

Robert F, Chaussidon M (2006) A palaeotemperature curve for the Precambrian oceans based on silicon isotopes in cherts. Nature 443:969–972, doi:10.1038/nature05239

Rosenberg E (2009) Germanium: environmental occurrence, importance and speciation. Rev Environ Sci Biotechnol 8:29–57

Rosman KJR, Taylor PDP (1998) Isotopic compositions of the elements 1997 (Technical Report). Pure Appl Chem 70:217–235

Rouxel O, Ludden J, Carignan J, Marin L, Fouquet Y (2002) Natural variations of Se isotopic composition determined by hydride generation multiple collector inductively coupled plasma mass spectrometry. Geochim Cosmochim Acta 66:3191–3199, doi:Pii S0016–7037(02)00918–3

Rouxel O, Ludden J, Fouquet Y (2003a) Antimony isotope variations in natural systems and implications for their use as geochemical tracers. Chem Geol 200:25–40, doi:10.1016/S0009–2541(03)00121–9

Rouxel O, Dobbek N, Ludden J, Fouquet Y (2003b) Iron isotope fractionation during oceanic crust alteration. Chem Geol 202:155–182, doi:10.1016/J.Chemgeo.2003.08.011

Rouxel O, Galy A, Elderfield H (2006) Germanium isotopic variations in igneous rocks and marine sediments. Geochim Cosmochim Acta 70:3387–3400, doi: 10.1016/J.Gca.2006.04.025

Ryabov VV, Lapkovsky AA (2010) Native iron (-platinum) ores from the Siberian Platform trap intrusions. Aust J Earth Sci 57:707–736

Savage PS, Armytage RMG, Georg RB, Halliday AN (2014) High temperature silicon isotope geochemistry. Lithos 190:500–519, doi:10.1016/j.lithos.2014.01.003

Schauble EA (2004) Applying stable isotope fractionation theory to new systems. Rev Mineral Geochem 55:65–111

Schauss AG (1991) Nephrotoxicity and neurotoxicity in humans from organogermanium compounds and germanium dioxide. Biol Trace Elem Res 29:267–280

Schmitt W, Palme H, Wanke H (1989) Experimental determination of metal/silicate partition coefficients for P, Co, Ni, Cu, Ga, Ge, Mo, and W and some implications for the early evolution of the Earth. Geochim Cosmochim Acta 53:173–185

Scribner AM, Kurtz AC, Chadwick OA (2006) Germanium sequestration by soil: Targeting the roles of secondary clays and Fe-oxyhydroxides. Earth Planet Sci Lett 243:760–770, doi:10.1016/j.epsl.2006.01.051

Sedwick PN, McMurtry GM, Macdougall JD (1992) Chemistry of hydrothermal solutions from Pele's Vents, Loihi Seamount, Hawaï. Geochim Cosmochim Acta 56:3643–3667

Shemesh A, Mortlock RA, Smith RJ, Froelich PN (1988) Determination of Ge/Si in marine siliceous microfossils: separation, cleaning and dissolution of diatoms and radiolaria. Mar Chem 25:305–323
Shemesh A, Mortlock RA, Froelich PN (1989) Late Cenozoic Ge/Si record of marine biogenic opal: implications for variations of riverine fluxes to the ocean. Paleoceanogr 4:221–234, doi:10.1029/PA004i003p00221
Shima M (1963) Isotopic composition of germanium in meteorites. J Geophys Res B: Solid Earth 68:4289–4292
Siebert C, Nagler TF, Kramers JD (2001) Determination of molybdenum isotope fractionation by double-spike multicollector inductively coupled plasma mass spectrometry. Geochem Geophy Geosystem 2:art. no.–2000GC000124
Siebert C, Ross A, McManus J (2006) Germanium isotope measurements of high-temperature geothermal fluids using double-spike hydride generation MC-ICP-MS. Geochim Cosmochim Acta 70:3986–3995
Siebert C, Hammond DE, Ross A, McManus J (2011a) Germanium isotope measurements of high-temperature geothermal fluids using double-spike hydride generation MC-ICP-MS (vol 70, pg 3986, 2006). Geochim Cosmochim Acta 75:6267–6269, doi:10.1016/j.gca.2011.07.032
Siebert J, Corgne A, Ryerson FJ (2011b) Systematics of metal–silicate partitioning for many siderophile elements applied to Earth's core formation. Geochim Cosmochim Acta 75:1451–1489
Siever R (1992) The silica cycle in the Precambrian. Geochim Cosmochim Acta 56:3265–3272, doi:10.1016/0016–7037(92)90303-z
Slack JF, Selby D, Dumoulin JA (2015) Hydrothermal, biogenic, and seawater components in metalliferous black shales of the Brooks Range, Alaska: synsedimentary metal enrichment in a carbonate ramp setting. Economic Geology 110:653–675
Snyder GA, Lee DC, Ruzicka AM, Prinz M, Taylor LA, Halliday AN (2001) Hf–W, Sm–Nd, and Rb–Sr isotopic evidence of late impact fractionation and mixing of silicates on iron meteorite parent bodies. Earth Planet Sci Lett 186:311–324
Steinhoefel G, Horn I, von Blanckenburg F (2009) Micro-scale tracing of Fe and Si isotope signatures in banded iron formation using femtosecond laser ablation. Gechim Cosmochim Acta 73:5343–5360
Stevenson DJ (1981) Models of the Earth's Core. Science 214:611–619, doi:10.1126/science.214.4521.611
Sutton J, Ellwood MJ, Maher WA, Croot PL (2010) Oceanic distribution of inorganic germanium relative to silicon: Germanium discrimination by diatoms. Global Biogeochem Cycles 24, doi:10.1029/2009gb003689
Sutton JN, Varela DE, Brzezinski MA, Beucher CP (2013) Species-dependent silicon isotope fractionation by marine diatoms. Geochim Cosmochim Acta 104:300–309, doi:10.1016/j.gca.2012.10.057
Takeda H, Bogard DD, Mittlefehldt DW, Garrison DH (2000) Mineralogy, petrology, chemistry, and ^{39}Ar–^{40}Ar and exposure ages of the Caddo County IAB iron: evidence for early partial melt segregation of a gabbro area rich in plagioclase–diopside. Geochim Cosmochim Acta 64:1311–1327
Tan CJ, Xiao L, Chen WL, Chen SM (2015) Germanium in ginseng is low and causes no sodium and water retention or renal toxicity in the diuretic-resistant rats. Exp Biol Med 240:1505–1512, doi:10.1177/1535370215571874
Tao SH, Bolger PM (1997) Hazard assessment of germanium supplements. Regul Toxicol Pharmacol 25:211–219, doi:10.1006/rtph.1997.1098
Tivey MK (1995) Modeling chimney growth and associated fluid flow at seafloor hydrothermal vent sites. *In*: Seafloor hydrothermal Systems: Physical, Chemical, Biological, and Geological Interactions. Geophysical Monograph 91. Humphris SE, Zierenberg RA, Mullineaux LS, Thomson RE, (eds). American Geophysical Union, p 158–177
Treguer P (2002) Silica and the cycle of carbon in the ocean. C R Geoscience 334:3–11
Treguer P, Nelson DM, Van Bennekom AJ, Demaster DJ, Leynaert A, Queguiner B (1995) The silica balance in the world ocean: A re-estimate. Science 268:375–379
Tribovillard N, Bout-Roumazeilles V, Riboulleau A, Baudin F, Danelian T, Riquier L (2011) Transfer of germanium to marine sediments: Insights from its accumulation in radiolarites and authigenic capture under reducing conditions. Some examples through geological ages. Chem Geol 282:120–130, doi:10.1016/j.chemgeo.2011.01.015
Ulff-Møllller F (1985) Solidification history of the Kitdlit Lens: immiscible metal and sulphide liquids from a basaltic dyke on Disko, central West Greenland. J Petrol 26:64–91
Valkovic V (1983) Trace Elements in Coal. CRC Press
van den Boorn S, van Bergen MJ, Nijman W, Vroon PZ (2007) Dual role of seawater and hydrothermal fluids in Early Archean chert formation: Evidence from silicon isotopes. Geology 35:939–942, doi:10.1130/g24096a.1
Vogel G, Hettich G, Mehrer H (1983) Self-diffusion in intrinsic germanium and effects of doping on self-diffusion in germanium. J Phys C: Solid Stat Phys 16:6197
Von Damm KL (2004) Evolution of the hydrothermal system at East Pacific Rise 9° 50' N: Geochemical evidence for changes in the Upper Oceanic Crust. *In*: Mid-Ocean Ridges: Hydrothermal Interactions between the Lithosphere and Oceans. Vol 148. German CR, Lin J, Parson LM, (eds). p 285–304
Von Damm KL, Bischoff JL, Rosenbauer RJ (1991) Quartz solubility in hydrothermal seawater: An experimental study and equation describing quartz solubility for up to 0.5 M NaCl solutions. Am J Sci 291:977–1007

Wai CM, Wasson JT (1979) Nebular condensation of Ga, Ge and Sb and the chemical classification of iron-meteorites. Nature 282:790–793, doi:10.1038/282790a0

Wai CM, Wetherill GW, Wasson JT (1968) The distribution of trace quantities of germanium between metal, silicate and sulflde phases. Geochim Cosmochim Acta 32:1269–1278

Walker JCG, Hays PB, Kasting JF (1981) A negative feedback mechanism for the long-term stabilization of Earth's surface temperature. J Geophys Res–Oceans Atmos 86:9776–9782, doi:10.1029/JC086iC10p09776

Wasson JT (1966) Butler, Missouri: An iron meteorite with extremely high germanium content. Science 153:976–978

Wasson JT (1974) Meteorites, Classification and Properties. Springer-Verlag, Berlin Heildelberg New York

Wasson J, Choe W (2009) The IIG iron meteorites: Probable formation in the IIAB core. Geochim Cosmochim Acta 73:4879–4890

Wasson JT, Kallemeyn GW (2002) The IAB iron-meteorite complex: A group, five subgroups, numerous grouplets, closely related, mainly formed by crystal segregation in rapidly cooling melts. Gechimica et Cosmochimica Acta 66:2445–2473

Wasson JT, Kimbeblin J (1967) The chemical classification of iron meteorites–II. Irons and pallasites with germanium concentrations between 8 and 100 ppm. Geochim Cosmochim Acta 31:2065–2093

Wasson JT, Willis J, Wai CM, Kracher A (1980) Origin of iron meteorite groups IAB and IIICD. Z Naturforsch 35a:781–795

Weis D, Garcia MO, Rhodes JM, Jellinek M, Scoates JS (2011) Role of the deep mantle in generating the compositional asymmetry compositional asymmetry of the Hawaiian mantle plume. Nat Geosci 4:831–838, doi:10.1038/NGEO1328

Weyer S, Ionov DA (2007) Partial melting and melt percolation in the mantle: The message from Fe isotopes. Earth Planet Sci Lett 259:119–133

Wheat CG, McManus J (2005) The potential of ridge-flank hydrothermal systems on oceanic germanium and silicon balances. Geochim Cosmochim Acta 69:2021–2029

Wickman FE (1943) Some aspects of the geochemistry of igneous rocks and of differentiation by crystallization. Geol Foeren Stockholm Foerh (GFF) 65:371–396

Williams HM, Peslier AH, McCammon C, Halliday AN, Levasseur S, Teutsch N, Burg J-P (2005) Systematic iron isotope variations in mantle rocks and minerals: The effects of partial melting and oxygen fugacity. Earth Planet Sci Lett 235:435–452

Xue S, Yang Y-L, Hall GS, Herzog GF (1997) Germanium isotopic compositions in Canyon Diablo spheroids. Geochim Cosmochim Acta 61:651–655

Yee D, Grieb T, Mills W, Sedlak M (2007) Synthesis of long-term nickel monitoring in San Francisco Bay. Environ Res 105:20–33, doi:10.1016/j.envres.2007.02.005

Yokoyama T, Makishima A, Nakamura E (1999) Evaluation of the coprecipitation of incompatible trace elements with fluoride during silicate rock dissolution by acid digestion. Chem Geol 157:175–187

Zhu JM, Johnson TM, Clark SK, Zhu XK (2008) High precision measurement of selenium isotopic composition by hydride generation multiple collector inductively coupled plasma mass spectrometry with a Se-74–Se-77 double spike. Chin J Anal Chem 36:1385–1390, doi:10.1016/s1872-2040(08)60075-4

Reviews in Mineralogy & Geochemistry
Vol. 82 pp. 657-682, 2017

Selenium Isotopes as a Biogeochemical Proxy in Deep Time

Eva E. Stüeken

Department of Earth and Space Sciences and Astrobiology Program
University of Washington
Seattle, WA 98195–1310
USA

Department of Earth & Environmental Sciences
University of St. Andrews
St. Andrews, KY16 9AL
UK

evast@uw.edu

INTRODUCTION AND OVERVIEW

Most research on selenium isotopes over the last decade has focused on either one of two avenues in biology and low-temperature geochemistry. Environmental and biological studies of modern systems are primarily concerned with monitoring and controlling the mobility of selenium in terrestrial settings. Selenium is an essential micro-nutrient for many organisms, including humans, but it becomes toxic at high concentrations (Zwolak and Zaporowska 2012). Significant efforts are therefore invested into evaluating the toxicity, mobility and bioavailability of selenium in soils, rivers and agricultural products. Geochemical studies of ancient sedimentary rocks, on the other hand, make use of the redox active nature of selenium to reconstruct atmospheric and marine oxygenation over various spatial and temporal scales (e.g. Rouxel et al. 2004; Mitchell et al. 2012, 2016; Layton-Matthews et al. 2013; Pogge von Strandmann et al. 2015; Stüeken et al. 2015a,b,d). This review will focus on the latter aspect, noting that extensive reviews of environmental problems are provided elsewhere (e.g. Hamilton 2004; Banuelos et al. 2013; Plant et al. 2014)

Selenium (atomic number 34, average mass 78.971 amu, Table 1) is classified as either a metalloid or a non-metal, depending on the allotrope of elemental Se(0) (Fernandez-Marinez and Charlet 2009). It belongs to the chalcophile elements that have a high affinity for sulfur and are enriched in sulfidic ore deposits (Goldschmidt and Strock 1935; Goldschmidt 1937). It is part of the same group as sulfur in the periodic table and shares a number of chemical properties. Like sulfur, selenium has six electrons in its outer shell, two in the 4s subshell and four in the 4*p* subshell, but unlike sulfur, it possess a full 3*d* subshell that lies interior to the outer electrons and provides relatively poor shielding from the nucleus (Greenwood 1984). As a result, the six outer electrons feel a relatively stronger attraction, which leads to a higher energy demand for selenium oxidation compared to sulfur.

Under reducing conditions such as in anoxic sediments, in the Earth's crust or within living cells, both sulfur and selenium are in their −II valence state (Fig. 1) and usually associated with either sulfide minerals or organic compounds. While S(−II) is oxidized to a +VI state at relatively low redox potential, Se(−II) first goes to the thermodynamically stable forms Se(0) and Se(IV) during oxidation; its fully oxidized form (Se(VI)) is only stable at high

1529-6466/14/0082-0015$05.00 http://dx.doi.org/10.2138/rmg.2017.82.15

Table 1. Stable selenium isotopes. Abundances and exact masses are taken from Fernández-Martínez and Charlet (2009), interferences are taken from Stüeken (2013). (a) 82Se decays with a half-life (t1/2) of 1.08·1020 years, which makes it stable for practical purposes. Other unstable isotopes (^{72}Se, $t_{1/2}$ = 8.4 days; ^{75}Se, $t_{1/2}$ = 120 days; ^{79}Se, $t_{1/2}$ = 295,000 years) are omitted. (b) Not included are potential isobaric interferences with doubly-charged heavy metals and metal oxides (cf. Layton-Matthews et al. 2006), for which there is no experimental evidence in MC-ICP-MS, the most common analytical technique (Stüeken et al. 2013).

Isotope	**^{74}Se**	**^{76}Se**	**^{77}Se**	**^{78}Se**	**^{80}Se**	**^{82}Se (a)**
Exact mass [amu]	73.922	75.919	76.92	77.917	79.917	81.917
Natural abundance [%]	0.87	9.36	7.63	23.78	49.61	8.73
Important interferences in MC-ICP-MS (b)	^{74}Ge, ^{36}Ar^{38}Ar	^{76}Ge, ^{36}Ar^{40}Ar, ^{38}Ar^{38}Ar, ^{75}AsH	^{76}SeH, ^{40}Ar^{37}Cl, ^{36}Ar^{40}ArH, ^{38}Ar^{38}ArH	^{77}SeH, ^{38}Ar^{40}Ar	^{40}Ar^{40}Ar, ^{80}Kr, ^{79}BrH	^{82}Kr, ^{81}BrH

Eh, similar to nitrate (N(V) in NO_3^-) (Fig. 1). Elemental Se(0) is a solid phase that is formed by either incomplete reduction or incomplete oxidation and thermodynamically stable over a wide Eh–pH range (Fig. 1). It is found in sediments and soils (Martens and Suarez 1997; Herbel et al. 2002; Kulp and Pratt 2004; Clark and Johnson 2010; Fan et al. 2011), as well as in the colloidal fraction of rivers (Zhang et al. 2004; Doblin et al. 2006). Se(IV) and Se(VI) form oxyanions (selenite, SeO_3^{2-}, hydroselenite or biselenite, $HSeO_3^-$, and selenate, SeO_4^{2-}, respectively) that are highly soluble (Seby et al. 2001). They are the major forms of selenium in the modern deep ocean and in river waters (Conde and Alaejos 1997; Cutter and Cutter 2001). Se(IV) has a relatively higher affinity than Se(VI) for adsorption onto ferromanganese oxides, clay particles and organics, in particular at low pH (e.g. Bar-Yosef and Meek 1987; Balistrieri and Chao 1990; Rovira et al. 2008; Mitchell et al. 2013). Furthermore, trace amounts of both Se(IV) and Se(VI) can be incorporated into carbonate minerals by substitution for CO_3^- (Reeder et al. 1994; Aurelio et al. 2010). Selenium associated with sulfate evaporites appears to be minor (Hagiwara 2000). The major forms of selenium in siliciclastic sediments are organic- and pyrite-bound Se(−II) (Kulp and Pratt 2004; Fan et al. 2011; Schilling et al. 2014b; Stüeken et al. 2015d). Atmospheric selenium gases primarily comprise methylated selenides that are produced by a variety of bacteria, plants, fungi and algae (e.g. Zieve and Peterson 1984; Amouroux et al. 2001; Chasteen and Bentley 2003; Schilling et al. 2011b);

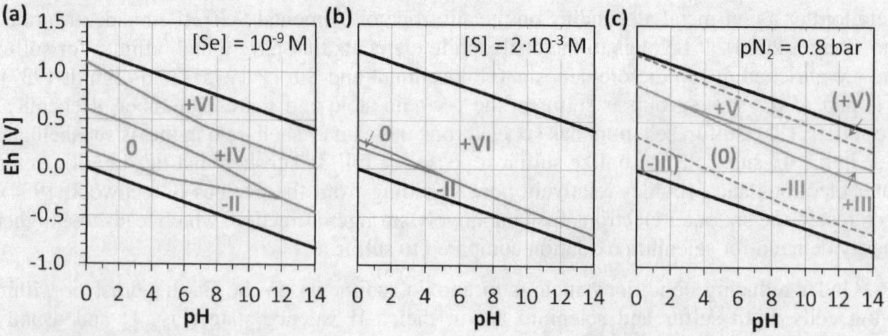

Figure 1. Eh–pH diagrams of (a) selenium, (b) sulfur, and (c) nitrogen. Acid–base transformations are omitted for clarity. Roman numerals indicate oxidation states. Solid black lines mark the boundaries of liquid water stability. For nitrogen, solid lines represent nitrate–nitrite–ammonium equilibria, dashed lines and labels in parentheses represent nitrate–N_2–ammonium equilibria. The figures were constructed with thermodynamic data from Stumm and Morgan (1996).

however, they have a short lifetime of only a few hours because they are rapidly oxidized in the modern oxic atmosphere (Wen and Carignan 2007). Volcanic processes may produce H_2Se gas, but this rapidly oxidizes to Se(0) today (Suzuoki 1965). Similarly, gaseous SeO_2 condenses rapidly below 315 °C and is thus not nearly as volatile as SO_2, the equivalent compound in the sulfur cycle (Wen and Carignan 2007; Floor and Román-Ross 2012). Hence > 80% of volcanic selenium emissions are in particulate form (Mosher and Duce 1987).

Overall, selenium has a complex biogeochemical cycle and undergoes numerous transformations between weathering and burial. Its properties suggest that fluxes, reservoirs and speciation have changed multiple times over the course of Earth's history with the evolution of the atmosphere, oceans and life. Selenium isotopes are a newly emerging proxy for reconstructing the evolution of the global selenium cycle.

NOMENCLATURE, REFERENCE MATERIALS AND ANALYTICAL TECHNIQUES

Some of the first studies of selenium isotopes were conducted by gas-source mass spectrometry where selenium was introduced via fluorination to SeF_6 gas, following similar protocols as for sulfur isotope measurements (Krouse and Thode 1962; Rees and Thode 1966). The prime limitation of this method was the high selenium demand of > 10 mg. Later studies used thermal ionization mass spectrometry (TIMS) which had at least tenfold higher sensitivity (Wachsmann and Heumann 1992; Johnson et al. 1999; Herbel et al. 2000). Traditionally, TIMS operates with cations, but Se^+ production is energetically unfavorable. To circumvent this problem, a negative ion method was developed for TIMS analyses. Nowadays, selenium isotopes are most commonly analyzed by multi-collector inductively-coupled plasma mass spectrometry (MC-ICP-MS) (Rouxel et al. 2002; Elwaer and Hintelmann 2008b; Zhu et al. 2008; Schilling and Wilcke 2011; Mitchell et al. 2012; Stüeken et al. 2013; Pogge von Strandmann et al. 2014). The sample is introduced via cold-vapor hydride-generation (HG), which produces gaseous H_2Se by on-line reduction of aqueous H_2SeO_3 with $NaBH_4$. This process further improves the sensitivity, allowing analyses of as little as 10 ng Se under optimal conditions (Rouxel et al. 2002). MC-ICP-MS allows monitoring all selenium isotopes as well as surrounding masses that may be needed to correct for isobaric interferences.

Interfering elements include residual germanium and arsenic derived from the sample matrix, as well as compounds generated from the argon carrier gas and the hydrochloric acid that contains the dissolved H_2SeO_3 (Table 1). Interferences are most severe for $m/z = 80$, where argon dimers are most abundant. This problem can be ameliorated with a collision cell that reduces dimer production (Rouxel et al. 2002; Layton-Matthews et al. 2006). Nevertheless, results for ^{80}Se are usually not reported. Similarly problematic is ^{74}Se, which has the lowest abundance and is easily masked by traces of ^{74}Ge (Table 1). Measurements of ^{76}Se can be compromised by ^{75}AsH in arsenic-rich samples, requiring accurate correction protocols (e.g. Stüeken et al. 2013). In rare cases, $^{75}AsH_2$ becomes significant and interferes with ^{77}Se (Stüeken et al. 2015a). Data for ^{78}Se and ^{82}Se are comparatively clean.

Two different methods are in use to correct for instrumental mass bias (isotopic fractionation during the transmission of Se ions from the source to the detector) and drift (temporal change of the instrumental mass bias due to slow changes in temperature, vacuum quality, etc.). These methods include double-spiking (e.g. Johnson et al. 1999; Zhu et al. 2008; Schilling and Wilcke 2011; Mitchell et al. 2012; Pogge von Strandmann et al. 2014) and standard-sample bracketing (SSB, e.g. Rouxel et al. 2002; Layton-Matthews et al. 2006; Stüeken et al. 2013). Double-spiking means that at an early stage during sample preparation the sample is spiked with a solution that is artificially enriched in two selenium isotopes. The major advantage of this technique is that isotopic fractionations imparted during sample preparation can be monitored and corrected. SSB, on the other hand, requires close to 100% yields because fractionations during sample preparations cannot be tracked independently. Double-spiking therefore generally leads to higher precision ($2\sigma \approx 0.1$–0.3‰) than SSB ($2\sigma \approx 0.2$–0.4‰) for geological samples. A disadvantage of double spiking is that it requires measurements of at least four isotopes and is thus more prone to isobaric interferences (Table 1). With SSB, accurate isotopic measurements can be made with only the two most interference-free isotopes. Furthermore, if the isotopes enriched in the double spike are affected by any mass independent fractionation, then a second, unspiked analysis is required. Lastly, the quality of interference corrections cannot be monitored with three-isotope diagrams when a double-spike is used. SSB leaves the option of detecting mass-independent fractionation and isobaric interferences, because multiple natural isotope ratios can be monitored. A detailed description of how to implement a selenium double-spike is given by Johnson and Bullen (2004b).

Sample preparation is often done by bulk digestion and column filtration to extract and purify the selenium. Sometimes, the digestion is replaced by a sequential extraction protocol to separate differing selenium phases for isotopic analyses (Clark and Johnson 2010; Schilling et al. 2014b; Stüeken et al. 2015d). Bulk digestion of rock samples usually involves hydrofluoric acid for dissolution of silicates and nitric acids or hydrogen peroxide for oxidation of all selenium to oxyanions. Some studies avoid the use of hydrofluoric acid, assuming that selenium is mobilized quantitatively by oxidation alone (Clark and Johnson 2008; Mitchell et al. 2012). Recalcitrant organics in sedimentary rocks are best oxidized with the addition of perchloric acid (Rouxel et al. 2002; Stüeken et al. 2013). Temperatures need to be kept low to avoid loss of selenium by volatilization, in particular in hydrochloric acid matrices (Rouxel et al. 2002; Johnson 2004; Layton-Matthews et al. 2006); however, in the presence of perchloric acid temperatures up to 150 °C are safe (Stüeken et al. 2013). Insoluble fluoride particles can be removed by centrifugation or filtration (Rouxel et al. 2002; Stüeken et al. 2013). Commercial ion exchange resin are not usually effective for separation of Se from strong acid digests. Instead, researchers use thiolated cotton fibers (TCF) that can be prepared in the laboratory from commercial cotton balls, soaked in a mixture of acetic acid glacial, acetic acid anhydride, mercaptoacetic acid and sulfuric acid (Yu et al. 2002). Some recent studies replaced cotton balls with cellulose powder (Elwaer and Hintelmann 2008a). TCF removes most matrix elements, but it cannot completely remove germanium and arsenic, which cause important isobaric interferences in the mass spectrometer (Table 1, Stüeken et al. 2013). Germanium can be removed effectively by hydride generation (Clark and Johnson 2010) or by treating the sample with aqua regia (Stüeken et al. 2013).

Selenium isotope data are typically normalized to NIST SRM 3149, which is distributed as a solution. Some older reference standards, in particular MERCK, have been calibrated relative to NIST SRM 3149 by Carignan and Wen (2007), where $\delta^{82/78}Se_{NIST3149} \approx \delta^{82/78}Se_{MERCK} + 1.03$‰. Variations in analytical methods dictate different choices of the isotopic ratio that is used to report the data. Some studies report data in terms of $^{80}Se/^{76}Se$ (Johnson et al. 1999; Herbel et al. 2000; Herbel et al. 2002; Clark and Johnson 2010), but most recent work uses either $^{82}Se/^{78}Se$ (Stüeken et al. 2013) or $^{82}Se/^{76}Se$ (Rouxel et al. 2004; Mitchell et al. 2012; Layton-Matthews et al. 2013; Pogge von Strandmann et al. 2014; Schilling et al. 2015). $^{82}Se/^{78}Se$ has the major advantage that isobaric interferences are relatively minor for both isotopes (Stüeken et al. 2013), particularly in sediments with high arsenic concentrations (Stüeken et al. 2015a, unpublished data), where $m/z = 76$ can be compromised. As long as mass-independent fractionation is absent (Stüeken et al. 2015b), $^{82}Se/^{78}Se$ can be converted to $^{82}Se/^{76}Se$ by multiplication with a factor of 1.539 for equilibrium fractionations and 1.519 for kinetic fractionations, assuming atomic rather than molecular masses (Young et al. 2002). The difference between the two factors is less than 0.1‰ for fractionations up to 5‰ and hence within analytical error in most cases. Isotopic data are conventionally presented in delta notation in units of permil (Eqn. 1):

$$\delta^{82/78}\text{Se}\ [\text{‰}] = \left(\frac{\left({}^{82}\text{Se}/{}^{78}\text{Se}\right)_{\text{sample}}}{\left({}^{82}\text{Se}/{}^{78}\text{Se}\right)_{\text{SRM3149}}} - 1 \right) \times 1000 \tag{1}$$

To determine analytical accuracy, recent studies reported measurements of the USGS rock standard SGR-1, which is an Eocene oil shale with 3.51 ± 0.26 ppm selenium (Savard et al. 2009). Results for $\delta^{82/78}$Se range from −0.13‰ to +0.40‰ with a mean between seven studies of +0.14 ± 0.19‰ (1σ), after conversion to the NIST SRM 3149 scale (Rouxel et al. 2002; Layton-Matthews et al. 2006; Schilling et al. 2011a; Mitchell et al. 2012; Pogge von Strandmann et al. 2014; Schilling et al. 2014b; Stüeken et al. 2015b). Other less common reference materials that have been analyzed for selenium isotopes are listed by Rouxel et al. (2002) and Layton-Matthews et al. (2006).

ELEMENTAL AND ISOTOPIC ABUNDANCES IN MAJOR RESERVOIRS

Selenium concentrations have been measured in a variety of natural substrates over the past century (e.g. Goldschmidt and Strock 1935), but accurate isotopic analyses have only been possible since the late 1990s with the establishment of TIMS and associated advances in analytical sensitivity. Our knowledge of isotopic partitioning in natural systems is therefore still severely limited. In compiling information about major geological reservoirs, selenium concentrations and isotopic compositions almost always had to be taken from different sources (Table 2). Selenium concentrations often spread over an order of magnitude within each reservoir and are therefore expressed as geometric means.

Terrestrial and extraterrestrial igneous terrestrial reservoirs

The isotopic compositions in extraterrestrial materials and igneous rocks on Earth have not yet been studied systematically. Iron meteorites (~23 ppm, +0.11 ± 0.34‰, Rouxel et al. 2002) and chondrites (~9.6 ppm, Table 2), the building blocks of the terrestrial planets, are relatively selenium-rich compared to geological reservoirs on Earth (mostly < 1 ppm, Table 2), suggesting that (a) most of Earth's selenium partitioned into the core (Rose-Weston et al. 2009; König et al. 2012), and (b) impacts may have enriched the crust with selenium after core formation (Rose-Weston et al. 2009; Wang and Becker 2013).

Table 2 (overleaf): Concentrations and isotopic compositions of selenium in various reservoirs: blank cells = not determined; *n* = number of samples. (a) concentrations are in ppb [ng/g] unless noted otherwise. (b) isotopic data are in units of permil. (c) average of basalt and diorite. (d) FeMn oxides denotes ferromanganese crusts and nodules from the modern ocean. (e) black smokers reflect high-temperature fluids of marine hydrothermal vents; low-temperature fluids are undetermined. (f) shale includes Precambrian and Phanerzoic samples. Precambrian data may be biased towards more positive $\delta^{82/78}$Se values relative to average crust, as discussed in the text. (g) average selenium concentration of upper crust is calculated following the recipe of Condie (1993) or Wedepohl (1995). *References:* 1. Dreibus et al. (1993); 2. DuFresne (1960); 3. Edgington and Byers (1942); 4. Floor and Roman-Ross (2012); 5. Goldschmidt and Storck (1935); 6. Greenland (1967). 7. Hertogen (1980); 8. Johnson and Bullen (2004a); 9. Koljonen (1973b); 10. Koljonen (1973a); 11. Kulp and Pratt (2004); 12. Lorand et al. (2003); 13. Marin et al. (2001); 14. Mitchell et al. (2012); 15. Schirmer et al. (2014); 16. Shore (2010); 17. Wen et al. (2014); 18. Stüeken et al. (2015b); 19. Stüeken et al. (2015a); 20. Tamari et al. (1990); 21. Tischendorf (1959); 22. Turekian and Wedepohl (1961); 23. Wang et al. (2013); 24. Rouxel et al. (2002); 25. Schilling et al. (2014a); 26. Takematsu et al. (1990); 27. Schilling et al. (2011b); 28. Wedepohl (1995); 29. Conde and Alejos (1997); 30. Plant (2014); 31. Cutter and Cutter (2001); 32. Wen and Carignan (2007); 33. Mosher and Duce (1987); 34. Haygart et al. (1994); 35. von Damm (1990); 36. Condie (1993).

Type	Geom. mean [ppb] (a)	σ [ppb][a]	*n*	Ref. for concentrations	$\delta^{82/78}Se^{(b)}$	$\sigma^{(b)}$	*n*	ref. for isotopes
				Geological Reservoirs				
chondritic meteorites	9,580	+6,548/ −3,889	96	1, 2, 6, 23				
iron meteorites	22,361	+67,264/ −16,782	4	24	0.11	±0.34	4	24
upper mantle	27	+61/−19	140	9, 12, 20, 23	0.15		1	24
gabbro	97	+137/−57	17	9, 13, 20				
basalt	57	+126/−39	47	7, 9, 13, 20	0.36	±0.13	4	24
andesite	7	+23/−6	19	9, 13, 20				
diorite	68	+81/−37	14	9, 13, 20	−0.33		1	24
granodiorite	21	+61/−16	14	9, 13, 20				
granite	18	+46/−13	59	9, 13, 20				
rhyolite	15	+65/−12	14	9, 13, 20				
shale[f]	1,189	+4,765/ −952	646	3, 8, 11, 14, 16, 17, 18, 19, 20, 21	0.09	±0.60	522	14, 16, 17, 18, 19
sandstone	42	+80/−28	16	10, 20, 22				
carbonate	116	+590/−97	23	10, 15, 20, 22	0.14	±0.41	2	24
FeMn oxides[d]	580	+1,176/ −389	71	5, 15, 26	0.32		1	24
Precambrian BIF	79	+77/−39	12	15	< 0?			25
volcanic ash	746	+2,330/ −565	28	4, 9				
gneiss	77	+147/−51	23	10				
serpentine	291	+1,516/ −244	5	10, 13				
amphibolite	217	+163/−93	6	10				
schist	1,007	+7,333/ −885	14	10				
→ upper crust (g)	34 or 59	+137/−41		after 36 or 28	0.01	±0.49		24[c]
				Other Environmental Reservoirs				
seawater		1–2 nM		30, 31	≥ +0.3		0	14, 24
river water		0.1–25.3 nM, avg. 2.17 nM, Amazon 2.66 nM		29, 30	~ 0?		0	18
geothermal water		25–6000 nM, black smokers 73 ± 21 nM[e]		4, 35				
atmospheric particulates		0.0045–1.34 ng/m^3		32, 33, 34	< 0?		0	27
atmospheric gas		0.0003–0.35 ng/m^3		32, 33, 34	< 0?		0	27
marine biomass		1.7–5.6 mmol selenium/mol carbon		14	~ +0.3		1	14

The concentration of selenium in Earth's upper crust has previously been calculated from the concentration of sulfur and an assumed sulfur/selenium ratio of 6000 (Goldschmidt and Strock 1935), yielding a value of 50 ppb (Turekian and Wedepohl 1961; Taylor and McLennan 1995). Wedepohl (1995) used a different approach whereby the concentrations of individual reservoirs were weighted by to their relative mass. According to this scheme, upper crust is composed of 14% sediment (44% shale, 20.9% sandstone and greywacke, 20.3% volcanic ash, 14.6% carbonate), 50% felsic intrusives (50% granite, 40% granodiorite, 10% tonalite), 6% gabbro, and 30% metamorphic rocks (64% gneiss, 15.4% schist, 17.8% amphibolite, 2.6% marble). Using concentration data available at the time, Wedepohl calculated an average concentration of 120 ppb. Adopting the same recipe and combining it with updated estimates of selenium concentrations and their standard deviations in the various reservoirs (Table 2), the new estimate of average upper crust presented here is 59 (+137/−41) ppb. An alternative recipe by Condie (1993), according to which upper Phanerozoic crust is composed of 25% tonalite–trondjamite–granodiorite (TTG), 11% granite, 16% felsic volcanics, 12% basalt, 12% andesite and 24% greywacke, leads to a selenium concentration of 34 (+26/−15) ppb. In this calculation, greywacke was approximated with 50% shale and 50% sandstone, and TTG replaced with granodiorite. The average isotopic composition is less well constrained, because the sedimentary record may be slightly biased (Stüeken et al. 2015b, discussed below) and very few igneous rocks and no high-grade metamorphic rocks have been analyzed to date. Taking the mean of basalts (+0.36 ± 0.13‰) and diorite (−0.33‰) gives a value of +0.01 ± 0.49‰ (Table 2), which may be our current best estimate for bulk crust. This number is consistent with a mass balance of marine sediments presented below.

Reservoirs at the Earth's surface

Selenium is mobilized from the crust by oxidative weathering and transformation into oxyanions. These are bioavailable to organisms and can be subject to re-reduction in the presence of organic or inorganic electron donors (e.g. Stolz et al. 2006; Fernandez-Marinez and Charlet 2009; Schilling et al. 2015; Winkel et al. 2015). The interplay of oxidation and reduction controls the abundance and speciation of selenium in **soils**. To first order, selenium in soil is determined by bedrock composition (e.g. Malisa 2001). Unusually high concentrations (up to 26,000 ppm) and large isotopic fractionations of > 20‰ were reported from a weathering profile through selenium-rich pyritic black shale (Zhu et al. 2014). A slightly smaller isotopic range (up to 7‰) was described from another seleniferous soil with up to 4 ppm selenium (Schilling et al. 2015). On the other hand, soils with concentrations closer to average crust (up to 0.5 ppm) showed a much smaller range of fractionations (±0.25‰, Schilling et al. 2011a). An important factor influencing the isotopic behavior of selenium in soils appears to be the abundance of organic matter and other reductants that re-reduce selenium oxyanions deeper in the soil profile (Zhu et al. 2014; Schilling et al. 2015). This and other processes further affect selenium uptake into plants (Winkel et al. 2015).

Selenium oxyanions that evade re-reduction in soils or plant uptake are ultimately washed into rivers and transported to the ocean. Selenium concentrations in **rivers** are in the nM range (Table 2), but vary widely and are probably affected by anthropogenic activities. Given the scarcity of data from soils, and only a single study of selenium isotopes in river waters (range +0.5‰ to +1‰ for Se(IV), +1.8‰ to +2.5‰ for Se(VI), Clark and Johnson 2010), it is uncertain what the isotopic composition of modern rivers is. The isotopic difference between Se(VI) and Se(IV) found by Clark and Johnson (2010) suggest that at this particular site some Se(VI) is subject to reduction to Se(IV). Hence these data may be affected by local biogeochemical processes. As discussed below, mass balance of marine sediments suggests that the isotopic composition of the global average river flux is close to average continental crust (~0 ± 0.5‰, Table 2).

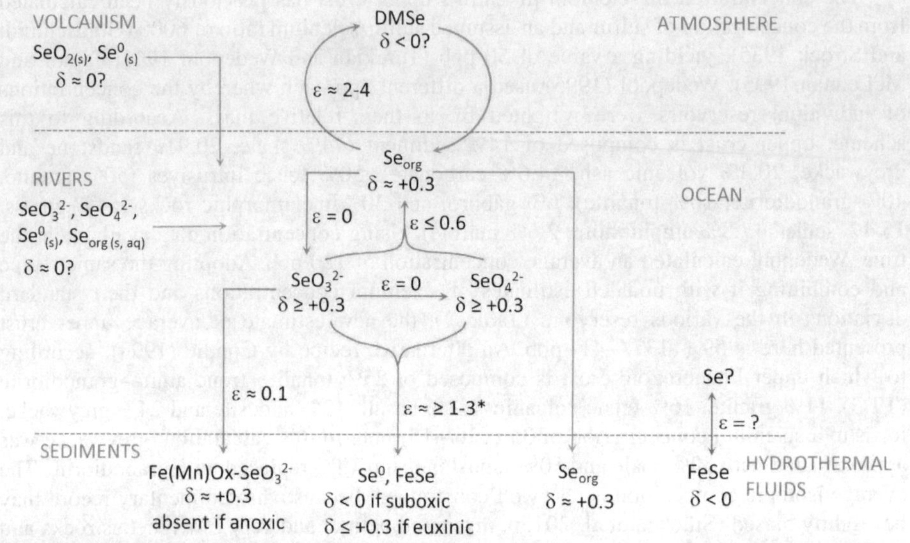

Figure 2. Schematic of the modern marine Se cycle. Red arrows indicate reactions, blue arrows transport. Fractionations and delta values are expressed in terms of $^{82}Se/^{78}Se$ in units of permil. *The fractionation of 1–3‰ for oxyanion reduction to inorganic reduced phases is derived from the mass balance described in the text. FeMn–SeO_3^{2-} denotes Se(IV) adsorption on ferromanganese oxides; FeSe denotes inorganic selenide incorporated in sulfide minerals, in particular pyrite. See text for references.

Selenium concentrations in **seawater** are consistently low (1–2 nM, Table 2). In the modern oxic ocean, Se(IV) and Se(VI) are the most abundant selenium species (Fig. 2). They both show a typical nutrient behavior with depletion in the photic zone and constant concentrations at depth, which reflects assimilation into biomass followed by remineralization of sinking organic matter (Cutter and Cutter 2001). Se(IV), which is produced by oxidation of organics, is thermodynamically unstable in modern oxic seawater, but the oxidation to Se(VI) is kinetically slow (Cutter and Bruland 1984). Therefore, Se(IV) accumulates in the deep ocean. In the photic zone, organic selenide dominates (Cutter and Bruland 1984; Cutter and Cutter 2001). Under anoxic conditions, as in the modern Black Sea, remineralization of organics is limited, such that organic selenide persists throughout the water column, while oxyanions are suppressed (Cutter 1982; 1992). The isotopic composition of dissolved selenium in seawater has so far not been measured directly, because the concentrations are too small. However, data from ferromanganese nodules and marine algae that assimilate selenium with minimal fractionation in the photic zone may provide a lower limit of +0.3‰ (Rouxel et al. 2002; Mitchell et al. 2012). Taking into account potential fractionations during adsorption and assimilation (discussed below), it is probably no higher than +0.9‰ and closer to +0.4‰. Because both Se(VI) and Se(IV) are produced by oxidation of organic matter, which imparts no detectable fractionation (Johnson et al. 1999), their isotopic composition is probably the same today. As noted below, this may not have been the case in the Precambrian when some Se(IV) may have been produced by Se(VI) reduction in the water column.

The nutrient-style behavior is a reflection of the strong accumulation of selenium in **marine biomass**. For illustration, sulfate (~23 mM in seawater, Henderson and Henderson 2009) is about seven orders of magnitude more abundant in the modern ocean than Se(IV) and Se(VI) combined, but only four orders of magnitude more abundant in aquatic microbial biomass (Fagerbakke et al. 1996; Mitchell et al. 2012). Settling of organic matter to the seafloor is therefore a major pathway of selenium export from the ocean into sediments (Fig. 2), especially in anoxic water columns. Furthermore, selenium assimilation can lead to the production of methylated selenide gases that partly escape into the atmosphere.

Dimethyl selenide and dimethyl diselenide are the major **atmospheric selenium gases** (Mosher and Duce 1987; Wen and Carignan 2007), and marine organisms are their main producers, which has led to the idea that this may constitute a transport mechanism of selenium from the ocean to continents (Amouroux et al. 2001). Methylated selenides have a short atmospheric residence time of only a few hours (Wen and Carignan 2007), which probably makes them insignificant on geological timescales. They are rapidly oxidized and converted into particulates, joining the pool of atmospheric particulates that are produced by volcanism and other sources (Mosher and Duce 1987; Wen and Carignan 2007). Mosher and Duce (1987) estimated that nowadays 40% of all selenium emissions into the atmosphere are of anthropogenic origin. The isotopic composition of gaseous selenium has not yet been determined, but methylated selenides are probably isotopically light, given the negative fractionation observed during volatilization experiments (discussed below, Schilling et al. 2011b, 2013).

Apart from settling of organic matter to the seafloor, another important route for selenium into marine sediments is the reduction of oxyanions to solid elemental selenium or inorganic selenide (Fig. 2). The latter can get trapped in sulfide minerals and thus accumulate as a solid phase, together with Se(0) and dead biomass. **Pyritic and organic-rich shales** are thus the most selenium-rich reservoir at the Earth surface (> 1 ppm, Table 2) and the most widely studied substrate for isotopic analyses (Mitchell et al. 2012, 2016; Layton-Matthews et al. 2013; Wen et al. 2014; Pogge von Strandmann et al. 2015; Stüeken et al. 2015a,b,d). The bulk average isotopic composition of all published post-Sturtian (< 700 Myr) marine shale data is −0.14 ± 0.61‰ (1σ, $n = 356$) (Johnson and Bullen 2004a; Shore 2010; Mitchell et al. 2012; Wen et al. 2014; Pogge von Strandmann et al. 2015; Stüeken et al. 2015b,d). For the earlier Precambrian, it is +0.40 ± 0.51‰ (1σ, $n = 247$) (Pogge von Strandmann et al. 2015; Stüeken et al. 2015a,b). As noted below, this difference may be a reflection of changing redox conditions during weathering and transport to the ocean. Sequential extraction studies of a few Phanerozoic shales showed that, in general, most selenium is contained in organic matter, but significant fractions are pyrite-hosted or present as adsorbed Se(IV) or elemental Se(0) (Martens and Suarez 1997; Kulp and Pratt 2004; Clark and Johnson 2010; Fan et al. 2011; Schilling et al. 2014b; Stüeken et al. 2015d). Where isotopic measurements were conducted, the adsorbed Se(IV) fraction often tended to be isotopically lighter than recalcitrant organics by around 1‰, possibly reflecting partial Se(VI) reduction under diagenetic conditions (Clark and Johnson 2010; Schilling et al. 2014b; Stüeken et al. 2015d). Elemental Se(0) was variable (Clark and Johnson 2010), and pyrite-bound selenide was isotopically light (Stüeken et al. 2015d).

Modern **marine ferromanganese oxides** (> 0.5 ppm, Table 2) constitute the second-most selenium-rich reservoir in the ocean after black shales. So far, only one isotopic measurement has been made, producing a value of +0.32‰ (Rouxel et al. 2002). Ferromanganese crusts and nodules are common in deep-sea sediments today. They can also be found as minute particles mixed with pelagic clays. Se(IV) in particular has a high sorption affinity for oxide minerals (e.g. Bar-Yosef and Meek 1987; Balistrieri and Chao 1990; Rovira et al. 2008; Mitchell et al. 2013). This mechanism may thus represent another important exit channel of selenium from the ocean (Fig. 2). Precambrian banded iron formations (BIF) contain comparatively less selenium (< 0.1 ppm, Table 2) and appear to be isotopically light (Schilling et al. 2014a), suggesting a different Se(IV) source (discussed below).

SELENIUM IN BIOLOGY

Organisms can make use of selenium in three different ways: (a) assimilatory reduction, or in short assimilation, which results in incorporation of selenium into proteins, (b) volatilization by conversion of selenium oxyanions into methylated selenides, and (c) dissimilatory reduction for metabolic energy gain or detoxification. Extensive reviews of these processes are provided elsewhere (Birringer et al. 2002; Boeck et al. 2006; Stolz et al. 2006; Gladyshev 2012). The following paragraphs will highlight aspects that are relevant for geobiological studies.

(a) Assimilation. Some but not all organisms have an absolute requirement for selenium as a micro-nutrient (Stadtman 1974). These include vertebrates, protozoa, algae and several groups of prokaryotes (Birringer et al. 2002; Lobanov et al. 2009). Higher plants and fungi do not seem to have any selenium dependence, even though they can contribute to selenium methylation (discussed below). Selenium is used in its reduced form selenide, most commonly as part of the amino acid selenocysteine, sometimes called the 21st amino acid. It is structurally identical to cysteine, with a selenium atom in the place of sulfur, and may have evolved by gradual replacement (Zhang et al. 2006). Other functional selenium compounds include selenomethionine and selenium-bearing nucleosides (Stadtman 1990). Selenocysteine is mostly used in enzymes that catalyze redox reactions. Compared to cysteine, selenocysteine is less susceptible to complete oxidation, which lowers the risk of enzyme deactivation (Ruggles et al. 2012). Important examples of such enzymes include hydrogenases and glutathione peroxidase. Reports of selenocysteine-bearing hydrogenases in methanogens, perhaps the most ancient microbial phylum (Stadtman 1974; Rother et al. 2000), invite speculations about the antiquity and evolutionary trajectory of selenocysteine utilization (Foster 2005). Glutathione peroxidases, on the other hand, which are used by vertebrates to decompose peroxides and relieve oxidative stress (Stadtman 1990; Steinbrenner and Sies 2009), may not have been significant until the 1st or 2nd Proterozoic rise of atmospheric oxygen.

(b) Methylation and volatilization. Methylated selenium compounds primarily include dimethyl selenide and dimethyl diselenide and are produced by a number of bacteria, fungi, algae and plants in both aquatic and terrestrial environments (Chau et al. 1976; Chasteen and Bentley 2003). The evolutionary history of selenium methylation is unclear, but given that eukaryotic organisms are among the main producers of methylated selenides, this process may have gained importance in the late Neoproterozoic when eukaryotes became ecologically dominant (Knoll et al. 2006).

(c) Dissimilatory reduction. A variety of organisms, including Bacteria and Archaea, are capable of reducing selenium oxyanions to Se(0) in exothermic reactions that are sometimes used as sources of metabolic energy (Stolz et al. 2006). Se(VI) and Se(IV) reduction occur at higher Eh than sulfate reduction, i.e. under suboxic conditions. Se(VI) reduction is concurrent with denitrification (Fig 1., cf. Oremland et al. 1990). In fact, the selenate reductase enzyme appears to be phylogenetically related to nitrate reductase (Saltikov and Newman 2003; Watts et al. 2005). Some organisms that do not possess selenate reductase are able to reduce Se(VI) with nitrate reductases (Sabaty et al. 2001). Se(IV) reduction can be carried out by nitrite reductase (DeMoll-Decker and Macy 1993; Basaglia et al. 2007), but may in some cases also have multiple dedicated enzymes (Kessi 2006; Pierru et al. 2006). A specific selenite reductase has, however, not yet been isolated. Some sulfate-reducing bacteria are capable of Se(VI) reduction (Zehr and Oremland 1987), but it is unclear if this reaction is relevant in nature, if Se(VI) is removed at high Eh before sulfate-reducers become environmentally significant. Microbes that reduce selenium oxyanions in dissimilatory reactions usually excrete nanospheres of Se(0) (Stolz et al. 2006). Reduction of Se(0) to Se(−II) has also been described (Herbel et al. 2003), but the enzymatic pathways are still unknown.

ISOTOPIC FRACTIONATION PATHWAYS

Studies of isotopic fractionations have largely focused on low-temperature processes with particular emphasis on redox reactions. As is the case for carbon, sulfur and nitrogen isotopes, kinetic fractionations are likely dominating the selenium biogeochemical cycle and will be the focus of this review. Equilibrium processes can theoretically lead to large fractionations (Li and Liu 2011), but they are probably insignificant under most natural conditions, where various selenium species often coexist in disequilibrium (e.g. Cutter and Bruland 1984; Martens and Suarez 1997; Kulp and Pratt 2004). Adsorption may represent a notable exception (Mitchell et al. 2013).

By far the largest fractionations of up to 23‰ ($\varepsilon \approx \delta^{82/78}Se_{reactant} - \delta^{82/78}Se_{product}$, Table 3) are imparted during abiotic oxyanion reduction to Se(0); biological oxyanion reduction produces slightly smaller fractionations of up to 14‰ in culturing experiments (Krouse and Thode 1962; Rees and Thode 1966; Rashid and Krouse 1985; Johnson et al. 1999; Herbel et al. 2000; Ellis et al. 2003; Johnson and Bullen 2003; Mitchell et al. 2013). As suggested by Johnson and Bullen (2004a), abiotic Se(VI) reduction may be kinetically inhibited in the environment, making biological reduction the major fractionating pathway. Under natural conditions, biological fractionations may be smaller than in laboratory cultures due to differences in microbial physiology and nutrient supply (Ellis et al. 2003; Johnson 2004; Johnson and Bullen 2004a). This is supported by the observation that very few marine shales show fractionations outside the range from −2‰ to +2‰ (Stüeken et al. 2015b).

Table 3. Low-temperature isotopic fractionations. The fractionation between reactant (R) and product (P) of a reaction is defined as $\varepsilon_{R-P} = 1000 \cdot (\alpha_{R-P} - 1)$, where $\alpha_{R-P} = (^{82}Se/^{78}Se)_R/(^{82}Se/^{78}Se)_P$. α_{R-P} approximately equal to $\delta^{82/78}$Se reactant $-\ \delta^{82/78}$Se product; the difference is < 0.1‰ for $\varepsilon_{R-P} \leq 20‰$. Note that most experiments were conducted with much higher selenium concentrations than found in most natural environments, where fractionations may generally be smaller (Ellis et al. 2003). *References*: 1. Johnson et al. (1999); 2. Johnson and Bullen (2003); 3. Johnson and Bullen (2004a); 4. Rees and Thode (1966); 5. Rashid and Krouse (1985); 6. Krouse and Thode (1962); 7. Mitchell et al. (2013); 8. Herbel et al. (2000); 9. Ellis et al. (2003); 10. Herbel et al. (2003); 11. Schilling et al. (2011b); 12. Schilling et al. (2013); 13. Clark and Johnson (2010).

Pathway	$\varepsilon(^{82}Se/^{78}Se)$	References
	Reduction:	
abiotic Se(VI) → Se(IV)	5.6‰ to 11.8‰	1, 2, 4
abiotic Se(IV) → Se(0)	4.6‰ to 11.2‰	4, 5, 6, 7
biotic Se(VI) → Se(IV)	0.2‰ to 5.1‰	8, 9
biotic Se(IV) → Se(0)	1.1‰ to 8.6‰	8, 9, 10
(a-)biotic Se(0) → Se(-II)	<0.5‰	3
	Oxidation:	
(a-)biotic Se(-II) → Se(0)	<0.5‰	1, 3
(a-)biotic Se(0) → Se(IV)	<0.5‰	1, 3
(a-)biotic Se(IV) → Se(VI)	<0.5‰	1, 3
	Adsorption:	
Se(IV) on FeMn-Oxide	<0.1‰	7
Se(VI) on FeMn-Oxide	<0.7‰, average ~0.1‰	7
	Volatilization:	
Se(IV)/(VI) → CH3-Se(-II)	2‰ to 4‰	11, 12
	Assimilation:	
Se(IV)/(VI) → org. Se(-II)	<0.6‰	13

No fractionation beyond analytical uncertainty has been detected during Se(0) reduction to Se(−II) and during oxidation of various selenium phases (both < 0.5‰, Johnson et al. 1999). Adsorption of selenium oxyanions on ferromanganese oxides has a slight preference for the lighter isotopes (range 0.0–0.7‰, average 0.1‰, Mitchell et al. 2013). Higher plants may tend to accumulate isotopically heavy selenium with a fractionation of 1.7‰ to 2.8‰ observed in a recent study (Schilling et al. 2015). Moderate fractionations of up to 2.6‰ were once recorded for oxyanion assimilation into algal biomass (Hagiwara 2000), but subsequent papers raised concerns about the methodology of that study (Johnson 2004; Johnson and Bullen 2004a). More recent work suggests fractionations of <0.6‰ during uptake into aquatic algae (Clark and Johnson 2010). Importantly, this fractionation may not be expressed in the photic zone of the modern ocean where assimilation goes to completion and selenium concentrations are minimal (Johnson 2004). Hence marine phytoplankton may record the composition of seawater (Mitchell et al. 2012). Volatilization of methylated selenium gases is associated with moderate fractionations of 2–4‰ (Schilling et al. 2011b, 2013), but given the short residence time of these gases in the atmosphere, this process is likely insignificant over geological timescales.

Given the range of fractionations quoted above (Table 3), oxyanion reduction can probably be inferred where $\delta^{82/78}$Se values between samples or between selenium sources and sinks differ by more than 1‰. The overall pattern of isotopic behavior in selenium thus resembles that of sulfur, where sulfate reduction imparts by far the largest fractionation. An important difference is the likely absence of Se(0) disproportionation in the selenium cycle. In the case of sulfur, S(0) disproportionation is an exothermic reaction; however, Se(0) is thermodynamically stable (except at very high pH, Fig.1) and would require rather than generate metabolic energy during disproportionation (Johnson 2004). This metabolism may therefore not exist and has not been detected.

Numerous potentially fractionating pathways in the biological and geochemical selenium cycle are yet to be characterized. These include photolysis, volcanic degassing, partial melting, magmatic differentiation, metamorphic reactions, as well as diffusion, condensation and evaporation under natural conditions. Regarding igneous processes, fractionations are likely <1‰, given the small differences between basalts, granite and meteorites (Table 2), and observations of week mass-dependent fractionations at high temperature in other isotopic systems. For the same reason, metamorphic effects on selenium isotopes are expected to be small. Furthermore, selenium concentrations of schists are generally similar to those of shales (Koljonen 1973a, Table 2), suggesting that loss of selenium is minor during metamorphism.

All selenium isotopic fractionations observed to date are mass-dependent within error, even in the Archean when the atmosphere was essentially anoxic and sulfur shows strong mass-independent fractionation (Fig. 3, Farquhar et al. 2000; Stüeken et al. 2015b). Soon after the discovery of mass-independent fractionation in sulfur isotopes during $SO_{2(g)}$ photolysis (Farquhar et al. 2001), researchers at Harvard University hypothesized that selenium may show similar behaviors (A. Bekker, pers. comm). However, isotopic analyses of photolytic reaction products of selenium gases are so far lacking. Photolysis of methylated selenide, the only significant selenium gas in the modern atmosphere (Wen and Carignan 2007), is perhaps the most promising candidate for mass-independent fractionation, given the low volatility of SeO_2 and the discovery of mass-independent fractionation in organo-mercury compounds (Gosh et al. 2008). It is conceivable that mass-independent fractionation in selenium isotopes is preserved in specific environments such as soils and/or only in certain times in Earth's history that have not yet been thoroughly investigated.

Furthermore, it is conceivable that *apparent* mass-independent fractionation could be produced by differing fractionation coefficients in abiotic versus biotic reduction pathways. In particular biological reactions have characteristic intermediates with molecular masses that deviate from atomic masses. Those molecular masses lead to subtle differences in the relative

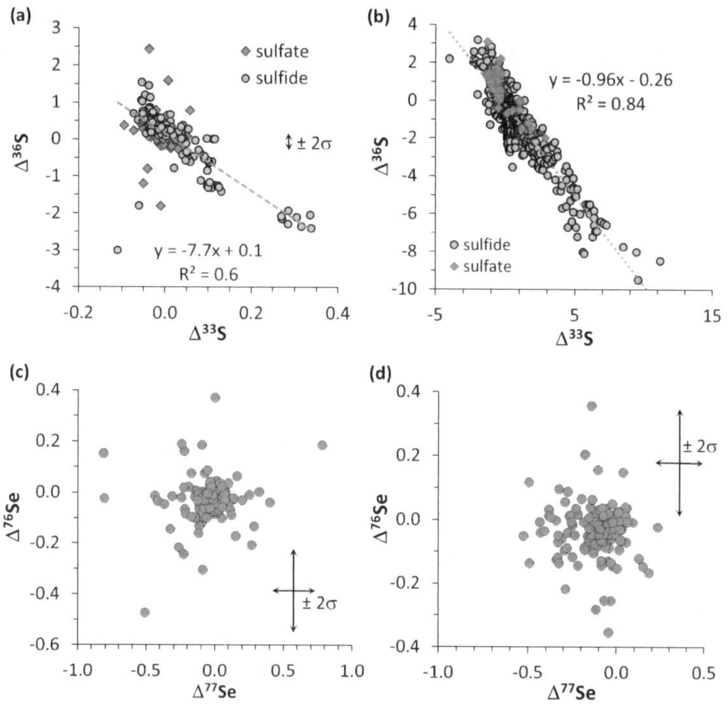

Figure 3. Testing for variations in mass dependence of selenium isotope fractionation. (a) Sulfur isotope data spanning the last 2 billion years, to exclude atmospheric effects. The slope of −7.7 is a biogenic feature (Ono 2008). (b) Sulfur isotope data from the Archean, showing the characteristic atmospheric slope around −1 (Zerkle et al. 2012). Sulfur data are compiled from the literature (Claire et al. 2014). Error bars for $\delta^{33}S$ are 0.02‰. (c) Post-Archean selenium isotope data, $n = 140$ (Stüeken et al. 2015b). (d) Archean selenium isotopic data, $n = 149$ (Stüeken et al. 2015b). Selenium shows no true or apparent mass-independent fractionation beyond analytical uncertainty. Eight data points with As/Se ratios > 200 were omitted due to the possible formation of AsH_2 interferences (Stüeken et al. 2015a).

behavior of heavy and light isotopes. In the case of sulfur, this behavior can be used to detect biological sulfate reduction in the rock record, because it leads to unique correlations between the offsets of the minor sulfur isotopes (^{33}S and ^{36}S) from their expected values (Ono et al. 2006; Johnston 2011). In the sulfur isotope system, first-order fractionations are generally expressed in terms of $^{34}S/^{32}S$ ratios in delta notation ($\delta^{34}S = [(^{34}S/^{32}S)_{sample}/(^{34}S/^{32}S)_{standard} - 1] \times 1000$). With purely atomic masses, i.e. not considering molecular effects and intermediate states of reactions, shifts in $^{33}S/^{32}S$ and $^{36}S/^{32}S$ ratios should theoretically differ from shifts in $^{34}S/^{32}S$ by a factor of 0.515 and 1.89, respectively (Johnston 2011). Deviations from expected values are conventionally quantified as $\Delta^{33}S$ ($= \delta^{33}S - 1000 \cdot [(1 + \delta^{34}S/1000)^{0.515} - 1]$) and $\Delta^{36}S$ ($= \delta^{36}S - 1000 \cdot [(1 + \delta^{34}S/1000)^{1.89} - 1]$). Such deviations occur when the exponents deviate from 0.515 and 1.89, as is the case in biological sulfate reduction due to various intermediate steps with molecular masses that differ from the assumed atomic masses (Johnston et al. 2007). However, this method requires very high analytical precision, which can currently not be attained for selenium isotopes with SSB. Double-spiking, which is more precise, is unsuitable for tracking multiple natural selenium isotopes because of potential problems with isobaric interferences. Furthermore, a theoretical or experimental basis for microbial selenium oxyanion reduction, comparable to equivalent studies on sulfur isotopes (Farquhar et al. 2007; Johnston et al. 2007), is so far lacking. Existing selenium isotope data from sedimentary rocks do not show any relationships, either because the precision is too low or because a significant biological imprint is absent (Fig. 3). Here, deviations from expected values are defined as in Eqns. (2) and (3):

$$\Delta^{82/76}\text{Se} = \delta^{82/76}\text{Se} - 1000\left[\left(1 + \frac{\delta^{82/78}\text{Se}}{1000}\right)^{1.519} - 1\right] \quad (2)$$

$$\Delta^{82/77}\text{Se} = \delta^{82/77}\text{Se} - 1000\left[\left(1 + \frac{\delta^{82/78}\text{Se}}{1000}\right)^{1.258} - 1\right] \quad (3)$$

Whether or not this method can be used to identify microbial selenium reduction in the environment will depend on further methodological improvements as well as a precise calibration through microbial culturing studies. Given the complexity of the selenium cycle, it would provide a valuable additional tool.

GEOBIOLOGICAL APPLICATIONS

Developing a mass balance for the modern ocean

A mass balance of selenium sources and sinks to and from the global ocean has so far not been established. A first-order estimate is attempted here, based on the few existing measurements of selenium isotopes and concentrations and the assumption that the major sinks of selenium are equivalent to those of molybdenum. This assumption is explained below.

The largest selenium flux into the ocean comes from rivers, followed by volcanic eruptions (Table 4, Fig. 4). Their exact isotopic compositions have not yet been determined. The role of hydrothermal circulation is unclear. It may act as a small selenium source, given the small concentrations of selenium in vent fluids (Table 2 and 4). On the other hand, the downdraft of seawater into oceanic crust could be a selenium sink, as evidenced by moderately large isotopic fractionations of several permil in hydrothermal sulfide deposits that likely reflect reduction of selenium oxyanions derived from seawater (Rouxel et al. 2004). Because of this ambiguity, hydrothermal circulation will be omitted from the following mass balance considerations. Not considered as an important sink is oxyanion incorporation into carbonates, because measured concentrations are relatively low and may include traces of pyrite- or organic-bound selenide (Table 2). The same may be true for serpentinite. The major selenium sinks from the ocean (Fig. 4) are therefore probably:

(type i) organic deposition and nearly quantitative oxyanion reduction in restricted euxinic basins. As demonstrated by water-column profiles from the Black Sea (Cutter 1982; 1992) and experimental evidence of Se(IV) reduction by H_2S (Pettine et al. 2012), euxinic environments act as sinks for selenium. Organic matter sinking down from the photic zone is not remineralized efficiently and residual oxyanions are reduced nearly quantitatively.

(type ii) partial oxyanion reduction and minor organic preservation in suboxic regions of the open ocean. Se(VI) reduction occurs in regions of denitrification (Oremland et al. 1990), which is common in upwelling zones and suboxic sediments (Lam and Kuypers 2011; Devol 2015). Furthermore, those environments are conducive to preserving organic matter and hence organic-bound selenium that formed by oxyanion assimilation into biomass.

Table 4. Modern ocean mass balance. Sink fluxes were calculated as fractions of the total input, assuming the same relative proportions as in the molybdenum cycle (Little et al. 2015). Isotopic compositions of sinks were calculated by taking the average of averages of different basins to avoid bias towards larger datasets. References for isotopic data are given in the text. Fluvial and hydrothermal fluxes were calculated by multiplication of average concentrations in mol/liter from Conde and Alaejos (1997, ref. 1) and von Damm (1990, ref. 2), respectively, with the corresponding average water fluxes in liters/year (Emerson and Hedges 2008, ref. 3). Atmospheric input is taken from Mosher & Duce (1987).

Sources [measured]	Average flux	$\delta^{82/78}Se$	Ref. for fluxes
Rivers	$8 \cdot 10^7$ mol/yr	n.d.	1, 3
Volcanoes	$1 \cdot 10^7$ mol/yr	n.d.	4
Hydrothermal vents	$4 \cdot 10^5$ mol/yr	n.d.	2, 3
Total sources	$9 \cdot 10^7$ mol/yr	~ 0?	
Sinks [inferred from steady state assumption and Mo equivalents]:			
FeMnOxide adsorption	$4.3 \cdot 10^7$ mol/yr	+0.32 ± 0.20	48% of total
Restr. euxinic basins	$0.6 \cdot 10^7$ mol/yr	+0.08 ± 0.05	7% of total
Local suboxia	$4.1 \cdot 10^7$ mol/yr	-0.29 ± 0.41	45% of total
Total sinks	$9 \cdot 10^7$ mol/yr	+0.03 ± 0.28	mass balance

(type iii) adsorption of Se(IV) onto ferromanganese oxides. Experiments indicate that Se(IV) has a high affinity for adsorption on various iron and manganese oxide minerals (e.g. Balistrieri and Chao 1990; Rovira et al. 2008), which are common in modern marine sediments. Their moderately high concentrations (Table 2) suggests that they accumulate significant amounts of selenium from seawater.

Examples of the three major sinks are represented in datasets spanning the last 500 kyr of Earth history, when the redox state of the ocean is relatively well known. As discussed by Stüeken et al. (2015b), restricted basins (type i) are captured by data from the Black Sea (Johnson and Bullen 2004a; Mitchell et al. 2012) and the interglacial Cariaco basin (Shore 2010), while locally suboxic or anoxic regions of the global ocean (type ii) are captured by the Arabian Sea (Mitchell et al. 2012), the glacial Cariaco basin (Shore 2010), the mid-Atlantic (Johnson and Bullen 2004a) and the Bermuda Rise (Shore 2010). Partial reduction in these areas probably occurs either in the water column, as in the Arabian upwelling zone, or in sediments under suboxic diagenetic conditions (Stüeken et al. 2015b). The average isotopic composition of the restricted euxinic basins (type i), first averaged by locality then overall, is +0.08 ± 0.05‰, while that of locally suboxic regions in the open ocean (type ii) is −0.29 ± 0.41‰, again first averaged by locality to avoid bias of dataset size (Johnson and Bullen 2004a; Shore 2010; Mitchell et al. 2012). Only one ferromanganese nodule (type iii) has been analyzed for selenium isotopes so far, producing a value of +0.32 ± 0.20‰ (using the analytical uncertainty and conversion from the MERCK to the NIST3149 reference frame) (Rouxel et al. 2002; Carignan and Wen 2007).

The three sinks quoted above are also the most important sinks of molybdenum from the ocean (Anbar 2004). An updated mass balance by Little et al. (2015) suggests that of the total molybdenum output (1.34–$1.9 \cdot 10^8$ mol/yr), 6–8% (mean 7%) are removed in restricted euxinic basins, 13–77% (mean 45%) are removed in anoxic or suboxic parts of continental margins, and 18–77% (mean 48%) are removed in oxic settings, i.e. by adsorption to manganese oxides. These percentages are scaled to add up to 100; the small hydrothermal molybdenum sink (0.04–$0.17 \cdot 10^8$ mol/yr) was omitted. Molybdenum differs from selenium in that it is not as redox active under suboxic conditions. However, like selenium, molybdenum forms oxyanions

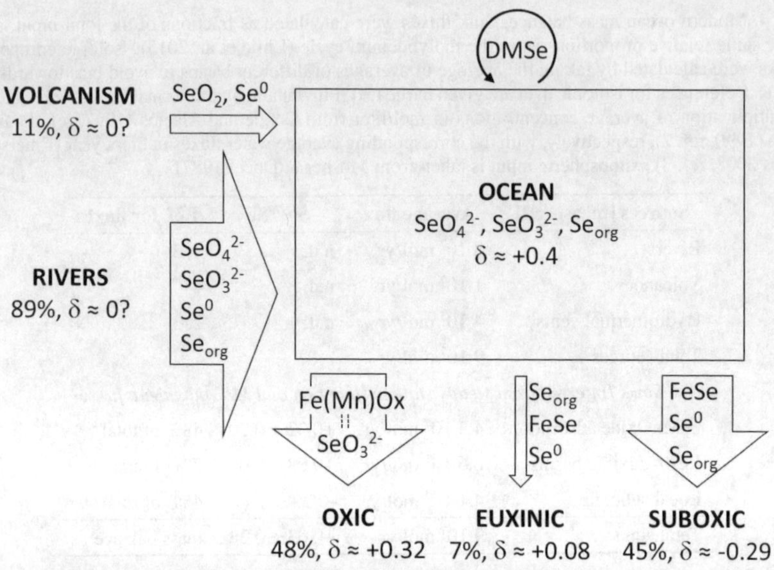

Figure 4. Proposed selenium mass balance for the modern ocean. The schematic was inspired by the molybdenum cycle presented by Anbar (2004). Delta values are expressed in terms of $^{82}Se/^{78}Se$ in units of permil. See text for discussion and references.

with a high affinity for adsorption and it is drawn into sediments by organic matter, especially in the presence of euxinia (Algeo and Lyons 2006). Hence to first order, the proportions of the various sinks may be approximately similar. With this assumption, the average composition of the total output of selenium from the ocean can be calculated as a weighted mean of the values quoted above, which then amounts to +0.03 ± 0.28‰ (Table 4, Fig. 4). This is essentially the same as our current best estimate for bulk average crust (+0.01 ± 0.50‰, Table 2). If this estimate is correct, then at least on the modern Earth, it appears that weathering and transport of selenium to the ocean do, on average, not significantly fractionate selenium isotopes.

The assumed proportions and isotopic compositions of the three main sinks can further be used to derive an approximate average fractionation factor for oxyanion reduction (Fig. 5). If the initial selenium input to the ocean has a composition of 0‰, if modern seawater is close to +0.4‰ (i.e. the composition of manganese nodules with an assumed average adsorption fractionation of ε = 0.1‰, Table 3), and if the suboxic sink removes 13–77% of this dissolved reservoir (i.e. by partial reduction and preservation or organic matter), then one can calculate the minimum isotopic fractionation needed to raise the composition of the dissolved selenium from 0‰ to +0.4‰. Assuming an open-system behavior, the results suggest a minimum fractionation of 3.1‰ to 0.5‰; for a 45% suboxic inorganic sink, it would be 0.9‰ (Fig. 5). The corresponding reduced selenium that is produced during the reduction would fall between −2.7‰ and −0.1‰ (−0.5‰ for a 45% inorganic sink). These are minimum fractionations, because they assume that all of the selenium is inorganic. The data compilation above suggests a composition of −0.3 ± 0.4‰ for the suboxic open marine sink, but those sediments probably contain organic-bound selenide with a composition of perhaps +0.3‰ in addition to the isotopically light Se(0) and Se(-II) in pyrite (Table 2). This organic fraction 'dilutes' the bulk isotopic composition. Hence a fractionation factor equal to or slightly larger than between 0.9‰ and 3‰, corresponding to an inorganic suboxic sink of between 13% and 45% of the total, is perhaps more plausible, because it would generate reduced inorganic selenium with compositions of −2.7‰ to −0.5‰ (or slightly lighter) that could balance the organic selenide. These inferred average fractionation factors fall within the range of values measured in natural microbial consortia (Ellis et al. 2003).

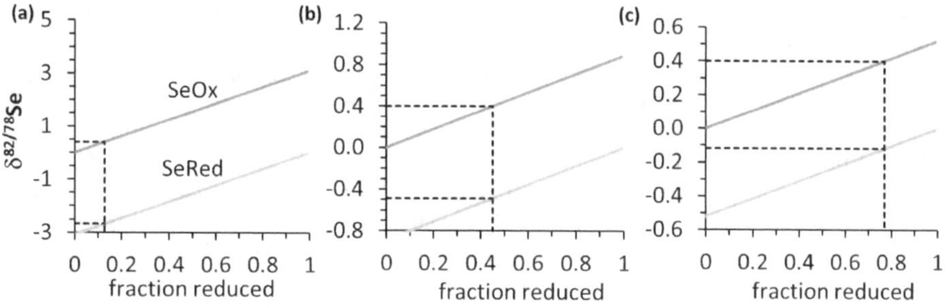

Figure 5. Isotope fractionation model. Shown are the isotopic compositions of reduced inorganic selenium species (SeRed) and residual oxyanions (SeO_x) as a function of the fraction that is subject to reduction. Calculations were done with standard equations for open-system behavior (e.g. Canfield 2001; Stüeken et al. 2015d). (a) ε = 3.09‰, (b) ε = 0.89‰, (c) ε = 0.52‰. Fractionation factors were chosen such that 13% (a), 45% (b) or 77% (c) reduction push residual oxyanions up to +0.4‰. Resulting reduced phases have compositions of −2.68‰ (a), −0.49‰ (b), and −0.12‰ (c). Black dashed lines are for reference.

It is important to note that these calculations are based on a very limited amount of data, but they will hopefully inspire future studies to place more accurate constraints on the importance and internal complexities of the various selenium sources and sinks.

Implications and predictions

A mass balance as presented above (Table 4) allows formulating a number of testable hypotheses. First, it appears that partial oxyanion reduction in suboxic waters and sediments is responsible for raising the isotopic composition of selenium dissolved in seawater compared to its crustal source. This contrasts with the marine molybdenum cycle, where adsorption on ferromanganese oxides imparts the largest fractionation (Anbar 2004). However, it agrees well with observations from other redox-active elements such as sulfur or nitrogen, where the dissolved oxyanions are isotopically heavy due to partial reduction in sediments or oxygen-minimum zones (Canfield 2001; Sigman et al. 2009). If this inference is correct, then the isotopic composition of dissolved selenium should increase with an expansion of suboxic zones, at least up to a threshold where the oxyanion reservoir starts to become depleted.

Another observation that deviates from molybdenum isotope systematics is that sediments from restricted euxinic basins (~ +0.1‰) are apparently not good recorders of the isotopic composition of selenium in seawater (≥ +0.3‰); they appear to be slightly lighter. Even the most sulfidic samples from Mitchell et al. (2012) show this subtle fractionation. This is most likely due to the low concentration (1–2 nM, Table 2) and short seawater residence time (10^4 years) (Henderson and Henderson 2009) of selenium compared to molybdenum (105 nM, 10^5–10^6 years) (Anbar 2004). Hence restricted basins such as the Black Sea not only exchange water with the open ocean, but they are also strongly effected by local selenium sources and redox processes. This implies that euxinic sediments cannot be used as an archive of marine selenium compositions and the extent of oxic or suboxic bottom waters through time, as it is commonly done for molybdenum (e.g. Arnold et al. 2004). However, those sediments will provide a minimum estimate of local seawater values.

The composition of dissolved selenium in modern seawater may be best recorded by ferromanganese oxides, because the fractionation associated with adsorption is small (Table 3). Unfortunately, this proxy fails in the Precambrian, when the ocean was strongly stratified and Se(IV) may have been produced by Se(VI) reduction rather than organic remineralization (Schilling et al. 2014a; Stüeken et al. 2015b).

Despite the possibility of local heterogeneities, marine shales can record useful biogeochemical information, because, as noted above, where fractionations extend over a range larger than about 1‰, they are likely indicative of selenium oxyanion reduction and hence the presence of those ions somewhere in the ocean. This conclusion becomes significant in deep-time studies, because it implies that redox conditions somewhere on the surface of the Earth, either on land or in parts of the ocean or both, were high enough to oxidize selenium to Se(IV) or Se(VI). Given the high Eh of those species (Fig. 1), this reaction requires a strong oxidizer such as nitrate, manganese oxide or molecular oxygen. Evidence of selenium oxyanions can therefore help reconstruct redox changes over Earth's history (e.g. Stüeken et al. 2015a). Furthermore, the absolute value of selenium isotopes in shales can inform about redox conditions in the overlying water column. If $\delta^{82/78}$S values are lighter than seawater, then non-quantitative oxyanion reduction probably occurred in the water column or in sediments at the sampling locality. Conditions were thus probably oxic to suboxic (Stüeken et al. 2015d). If, on the other hand, shale values are generally positive, then partial reduction must have occurred elsewhere, and the residual isotopically heavy oxyanions were drawn down locally at this site, due to either enhanced productivity or strong anoxia or both. In most cases, the composition of local seawater is unknown; however, average crust can be used as a first-order calibration point, because the directions of isotopic fractionations in the selenium cycle make it unlikely that seawater would ever become lighter than the crust.

Lastly, the short marine residence time and possible heterogeneities of isotopic compositions in seawater forbid global extrapolations from single localities. Nevertheless, useful inferences can be made about presence or absence of oxidizing conditions in general (Stüeken et al. 2015a,d).

Selenium isotopes in deep time

A few studies have investigated selenium isotopes in sedimentary rocks spanning the last 3.2 billion years (Rouxel et al. 2004; Mitchell et al. 2012, 2016; Layton-Matthews et al. 2013; Wen et al. 2014; Pogge von Strandmann et al. 2015; Stüeken et al. 2015a,b,d) (Fig. 6). The main findings in chronological order include:

1. A significant increase in selenium abundances from 0.2 (+0.3/−0.1) ppm to 1.4 (+5.6 / −1.1) ppm around 2.75 Gyr (Stüeken et al. 2015b; Mitchell et al. 2016): By analogy to sulfur and molybdenum (e.g. Wille et al. 2007; Kendall et al. 2010; Stüeken et al. 2012), this increase may reflect the onset of mild oxidative weathering on continents, possibly under oxygenic microbial mats (Lalonde and Konhauser 2015). Enhanced volcanism can, however, not be fully ruled out.
2. An isotopic contrast between light non-marine (−0.28 ± 0.67‰) and heavy marine (+0.37 ± 0.27‰) shales in the late Archean Fortescue Group (Stüeken et al. 2015b): Light values down to −1.9‰ in the lacustrine Tumbiana Formation cannot be explained by Se(IV) adsorption or assimilation into biomass; they are most consistent with partial oxyanion reduction in lake sediments or the lake water column. Hence oxyanions were generated in non-marine environments around 2.7 Ga, supporting the interpretation of mild oxidative weathering at this time. Alkaline conditions in volcanic terrains (Stüeken et al. 2015c) could have contributed to oxyanion stability, because they are more soluble and adsorb much less strongly at high pH (Zinabu and Pearce 2003). Partial reduction of selenium oxyanions in these lacustrine and fluvial environments could have raised the isotopic composition of selenium that was transported into the ocean. This would explain the generally positive values found in marine black shales from open marine margins. Hence unlike today, the riverine input to the ocean was likely > 0‰ and the marine mass balance may have shifted towards more positive values. Marine hydrothermal sulfides of similar age are indistinguishable from the marine shales (Stüeken et al. 2015b) and also heavier than Phanerozoic hydrothermal deposits (Rouxel et al. 2004; Layton-Matthews et al. 2013), which is consistent with negligible amounts of dissolved selenium oxyanions in the Archean deep ocean.

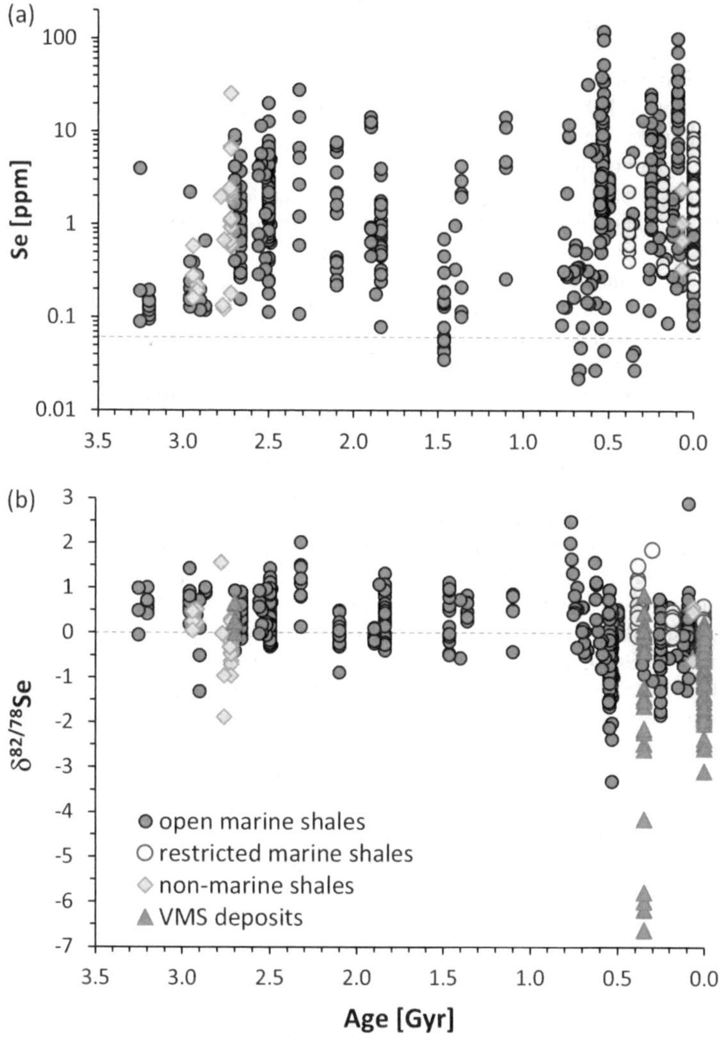

Figure 6. Selenium abundances (a) and isotopes (b) through time. The thin dashed line marks the crustal average. VMS deposits = volcanogenic massive sulfide deposits of hydrothermal marine origin. See text for references.

3. A spike in selenium concentrations and isotopic ratios coincident with the 2.5 Gyr 'whiff' of oxygen (Stüeken et al. 2015a): The selenium spike occurs in marine black shales and coincides with an excursion in molybdenum abundances and isotopes (Anbar et al. 2007), suggesting that both elements were derived from the same source. Multiple lines of evidence point to a pulse of oxidative weathering of the continents and hence an enhanced flux of molybdenum, selenium and other elements to the ocean (Anbar et al. 2007; Reinhard et al. 2009; Kendall et al. 2015). Fractionation of selenium isotopes likely occurred during transport in rivers and estuaries or at the marine chemocline. Hence the complimentary light selenium reservoir that would have balanced the heavy black shales from this outer shelf section may be in shallower facies that were not preserved. It is unlikely to be in the deep ocean, if that was too anoxic to support a selenium oxyanion reservoir.

4. Marine selenium isotopic ratios above crustal average throughout the Proterozoic (Stüeken et al. 2015b; Mitchell et al. 2016): The 'Great Oxidation Event' around 2.4–2.3 Gyr (Lyons et al. 2014) is not obviously reflected in the selenium isotope and abundance record; Proterozoic marine shales are statistically indistinguishable from those of the Neoarchean. As before, partial reduction of selenium oxyanions produced during oxidative weathering may have retained light selenium in soils, rivers and estuaries, such that the average selenium flux into the ocean was isotopically heavy. Perhaps the reason for the lack of response to increasing atmospheric oxygen levels is that a large fraction of the weathered selenium was assimilated by the terrestrial biosphere and transported to the ocean as organic selenides, which did not participate in further redox reactions. However, some negative values from proximal black shales of the Mesoproterozoic Belt Supergroup (Stüeken et al. 2015b), as well as from Paleoproterozoic banded iron formations (Schilling et al. 2014a), indicate that perhaps selenium oxyanions did become more abundant in surface seawater, allowing for partial oxyanion reduction in the ocean. Analyses of non-marine sediments or a greater variety of marine facies may help resolve these questions.

5. A shift towards more negative values in marine shales and hydrothermal deposits of Phanerozoic age (Johnson and Bullen 2004a; Rouxel et al. 2004; Mitchell et al. 2012; Layton-Matthews et al. 2013; Wen et al. 2014; Pogge von Strandmann et al. 2015; Stüeken et al. 2015b): With the oxygenation of the deep ocean in the Neoproterozoic or early Paleozoic (Lyons et al. 2014), selenium oxyanions probably became more stable, and something like a modern selenium cycle (Fig. 2) was established. This included non-quantitative oxyanion reduction in locally suboxic waters and sediments, i.e. in regions that were connected to the oxic selenium reservoir of the global ocean, leading to more negative selenium isotope values in marine shales. In-situ analyses of sedimentary pyrite grains show an increase of pyrite-bound selenium around the Precambrian–Cambrian boundary (Large et al. 2014), which further supports the idea of higher availability of selenium oxyanions for partial reduction to inorganic selenide. Prior to widespread ocean oxygenation, when pyrite-bound selenium levels were lower (Large et al. 2014), most selenium may have been preserved as organic-bound, because remineralization of organic matter was limited in the anoxic Precambrian ocean. Furthermore, a possible 'second rise of oxygen' in the Neoproterozoic atmosphere may have increased the total selenium flux coming from land. Chromium isotopes suggest that the redox state of weathering environments increased around 800 million years ago (Planavsky et al. 2014) to levels that would have been suitable for the production of Se(VI). Oxidative weathering in the earlier Precambrian may have stopped at Se(IV), i.e. the less oxidized form of selenium (Fig. 1). With higher selenium concentrations in seawater from the late Neoproterozoic onwards, hydrothermal oxyanion reduction could have also led to larger net fractionations (Rouxel et al. 2004; Layton-Matthews et al. 2013), which would explain why hydrothermal deposits of Phanerozoic age are much more fractionated than Neoarchean counterparts (Stüeken et al. 2015b) (Fig. 6). Restricted anoxic basins and regions of unusually high productivity, on the other hand, may tend to preserve positive values, presumably due to near quantitative selenium draw-down (Mitchell et al. 2012).

6. Negative excursions in selenium isotopes during the Permian-Triassic extinction and Ocean Anoxic Event 2 (Mitchell et al. 2012; Stüeken et al. 2015b): Net fractionations of selenium isotopes in black shales are usually not as negative as the reduced inorganic phases (Se(0), pyrite-bound Se(–II)) in them, because a large fraction of the selenium is contained in organic matter that does not carry the isotopic signature of dissimilatory oxyanion reduction (Stüeken et al. 2015d). Unusually light selenium isotope ratios down to < –0.5‰ are thus most likely the result of diminished selenium assimilation into biomass relative to the selenium supply. In the case of the Permian–Triassic extinction, selenium assimilation probably decreased as a direct consequence of ecosystem collapse (Stüeken et al. 2015d). As macro-organisms died out, the selenium demand and organic selenium export from the

ocean probably dropped significantly. The 'unused' selenium oxyanions were thus subject to partial reduction in suboxic bottom waters. In the case of Ocean Anoxic Event 2, there is no sign of a productivity collapse, but the selenium supply by volcanism may have been unusually high, far exceeding the selenium demand by living organisms (Mitchell et al. 2012). Hence a relatively greater proportion of Se(IV) and Se(VI) may have been subject to dissimilatory reduction to inorganic reduced phases.

CONCLUSIONS AND FUTURE DIRECTIONS

Multiple studies have shown that selenium isotopes can serve as a useful biogeochemical proxy; however, its interpretation is perhaps more difficult and less unidirectional than for other isotopic systems. This is due to a number of confounding factors, including the complexity of the selenium cycle, ambiguities of local versus global effects, relatively small fractionations in natural systems, and uncertainties about the composition of the crust. The utility of this proxy could be advanced significantly with targeted studies of a greater range of igneous, metamorphic and sedimentary rocks, clarification of the isotopic composition of riverine, estuarine and marine waters, as well as with biological experiments at low selenium concentrations that mimic those in the modern ocean.

Sequential extraction of differing selenium phases from bulk rocks has the potential to reduce several uncertainties and should be pursued and perfected in future work. First, the organic selenium fraction could perhaps serve as a proxy for the composition of selenium dissolved in seawater as phytoplankton does today (Mitchell et al. 2012); second, the separation of organics from inorganic reduced phases such as Se(0) and pyrite-bound Se(−II) allows detecting the full range of fractionations imparted during reduction, thus circumventing the problem of phase mixing (Stüeken et al. 2015d); and third, the relative abundances of organic and inorganic phases can provide a measure of how much selenium was drawn into sediments by dissimilatory reduction versus assimilation versus adsorption. The relative proportions may be another proxy for selenium bioavailability.

Regarding the evolution of the selenium cycle, several topics still need to be addressed, such as the type and magnitude of selenium sources and sinks and their speciation in the Precambrian before the first and second rise of atmospheric oxygen, the bioavailability of selenium to ancient methanogens in the anoxic Archean ocean, the role of banded iron formations and dissolved ferrous iron in the selenium cycle, the importance of the rise of macro-biological productivity at the Precambrian-Cambrian boundary, and the onset of dimethyl selenide production. Studies involving multiple proxies alongside selenium isotopes may help answer some of these and other questions.

ACKNOWLEDGEMENTS

Funding during the compilation of this manuscript was provided by the NASA postdoctoral program. I further than Tom Johnson and an anonymous reviewer for constructive comments.

REFERENCES

Algeo TJ, Lyons TW (2006) Mo–total organic carbon covariation in modern anoxic marine environments: Implications for analysis of paleoredox and paleohydrographic conditions. Paleoceanogr 21:PA1016, doi:10.1029/2004PA001112

Amouroux D, Liss PS, Tessier E, Hamren-Larsson M, Donard OFX (2001) Role of oceans as biogenic sources of selenium. Earth Planet Sci Lett 189:277–283

Anbar AD (2004) Molybdenum stable isotopes: observations, interpretations and directions. Rev Mineral Geochem 55:429–454

Anbar AD, Duan Y, Lyons TW, Arnold GL, Kendall B, Creaser RA, Kaufman AJ, Gordon GW, Scott C, Garvin J, Buick R (2007) A whiff of oxygen before the Great Oxidation Event? Science 317:1903–1906

Arnold GL, Anbar AD, Barling J, Lyons TW (2004) Molybdenum isotope evidence for widespread anoxia in mid-Proterozoic oceans. Science 304:87–90

Aurelio G, Fernandez-Marinez A, Cuello GJ, Roman-Ross G, Alliot I, Charlet L (2010) Structural study of selenium (IV) substitution in calcite. Chem Geol 270:249–256

Balistrieri LS, Chao TT (1990) Adsorption of selenium by amorphous iron oxyhydroxide and manganese dioxide. Geochim Cosmochim Acta 54:739–751

Banuelos GS, Lin Z-Q, Yin X (2013) Selenium in the Environment and Human Health. CRC Press

Bar-Yosef B, Meek D (1987) Selenium sorption by kaolinite and montmorillonite. Soil Sci 144:11–19

Basaglia M, Toffanin A, Baldan E, Bottegal M, Shapleigh JP, Casella S (2007) Selenite-reducing capacity of the copper-containing nitrite reductase of Rhizobium sullae. FEMS Microbiol Lett 269:124–130

Birringer M, Pilawa S, Flohé L (2002) Trends in selenium biochemistry. Natural Prod Rep 19:693–718

Boeck A, Rother M, Leibundgut M, Ban N (2006) Selenium metabolism in prokaryotes. *In*: Selenium - Its molecular biology and role in human health. Hatfield DL, Berry MJ, Gladyshev VN, (eds). Springer, New York

Canfield DE (2001) Biogeochemistry of sulfur isotopes. Rev Mineral Geochem 43:607–636

Carignan J, Wen H (2007) Scaling NIST SRM 3149 for Se isotope analysis and isotopic variations of natural samples. Chem Geol 242:347–350

Chasteen TG, Bentley R (2003) Biomethylation of selenium and tellurium: microorganisms and plants. Chem Rev 103:1–26

Chau YK, Wong PTS, Silverberg BA, Luxon PL, Bengert GA (1976) Methylation of selenium in the aquatic environment. Science 192:1130–1131

Claire MW, Kasting JF, Domagal-Goldman SD, Stüeken EE, Buick R, Meadows VS (2014) Modeling the signature of sulfur mass-independent fractionation produced in the Archean atmosphere. Geochim Cosmochim Acta 141:365–380

Clark SK, Johnson TM (2008) Effective isotopic fractionation factors for solute removal by reactive sediments: a laboratory microcosm and slurry study. Environ Sci Technol 42:7850–7855

Clark SK, Johnson TM (2010) Selenium stable isotope investigation into selenium biogeochemical cycling in a lacustrine environment: Sweitzer Lake, Colorado. J Environ Qual 39:2200–2210

Conde JE, Alaejos MS (1997) Selenium concentrations in natural and environmental waters. Chem Rev 97:1979–2003

Condie KC (1993) Chemical composition and evolution of the upper continental crust: contrasting results from surface samples and shales. Chem Geol 104:1–37

Cutter GA (1982) Selenium in reducing waters. Science 217:829–831

Cutter GA (1992) Kinetic controls on metalloid speciation in seawater. Mar Chem 40:65–80

Cutter GA, Bruland KW (1984) The marine biogeochemistry of selenium: a re-evaluation. Limnol Oceanogr 29:1179–1192

Cutter GA, Cutter LS (2001) Sources and cycling of selenium in the western and equatorial Atlantic Ocean. Deep-Sea Research II 48:2917–2931

DeMoll-Decker H, Macy JM (1993) The periplasmic nitrite reductase of *Thauera selenatis* may catalyze the reduction of selenite to elemental selenium. Arch Microbiol 160:241–247

Devol AH (2015) Denitrification, anammox, and N_2 production in marine sediments. Ann Rev Mar Sci 7:403–423

Doblin MA, Baines SB, Cutter LS, Cutter GA (2006) Sources and biogeochemical cycling of particulate selenium in the San Francisco Bay estuary. Estuarine Coastal Shelf Sci 67:681–694

Dreibus G, Palme H, Spettel B, Waenke H (1993) Sulfur and selenium in chondritic meteorites. Meteoritics 28:343

DuFresne A (1960) Selenium and tellurium in meteorites. Geochim Cosmochim Acta 20:141–148

Edgington G, Byers HG (1942) Geology and biology of North Atlantic Deep-Sea cores between Newfoundland and Ireland - Part 9: Selenium content and chemical analyses. *In*: U.S. Geological Survey Professional Paper 196-F, p 151–155

Ellis AS, Johnson TM, Herbel MJ, Bullen T (2003) Stable isotope fractionation of selenium by natural microbial consortia. Chem Geol 195:119–129

Elwaer N, Hintelmann H (2008a) Selective separation of selenium (IV) by thiol cellulose powder and subsequent selenium isotope ratio determination using multicollector iductively coupled plasma mass spectrometry. J Anal Atom Spectr 23:733–743

Elwaer N, Hintelmann H (2008b) Precise selenium isotope ratios measurement using a multimode sample introduction system (MSIS) coupled with multicollector inductively coupled plasma mass spectrometry (MC-ICP-MS). J Anal Atom Spectr 23:1392–1396

Emerson SR, Hedges JI (2008) Chemical Oceanography and the Marine Carbon Cycle. Cambridge University Press, Cambridge, UK

Fagerbakke KM, Heldal M, Norland S (1996) Content of carbon, nitrogen, oxygen, sulfur and phosphorus in native aquatic and cultured bacteria. Aquatic Microbial Ecology 10:15–27

Fan HF, Wen H, Hu M, Zhao H (2011) Selenium speciation in Lower Cambrian Se-enriched strata in South China and its geological implications. Geochim Cosmochim Acta 75:7725–7740

Farquhar J, Bao H, Thiemens M (2000) Atmospheric influence on Earth's earliest sulfur cycle. Science 289:756–758

Farquhar J, Savarino J, Airieau S, Thiemens MH (2001) Observation of wavelength-sensitive mass-independent sulfur isotope effects during SO_2 photolysis: Implications for the early atmosphere. J Geophys Res 106:32829–32839

Farquhar J, Johnston DT, Wing BA (2007) Implications of conservation of mass effects on mass-dependent isotope fractionations: influence of network structure on sulfur isotope phase space of dissimilatory sulfate reduction. Geochim Cosmochim Acta 71:5862–5875

Fernandez-Marinez A, Charlet L (2009) Selenium environmental cycling and bioavailability: a structural chemist point of view. Rev Environ Sci Biotechnol 8:81–110

Floor GH, Román-Ross G (2012) Selenium in volcanic environments: A review. Appl Geochem 27:517–531

Foster CB (2005) Selenoproteins and the metabolic features of the archaeal ancestor of eukaryotes. Mol Biol Evol 22:383–386

Gladyshev VN (2012) Selenoproteins and selenoproteomes. *In*: Selenium: Its Molecular Biology and Role in Human Health. Hatfield DL, Berry MJ, Gladyshev VN, (eds). Springer Science + Business Media, p 109–123

Goldschmidt VM, Strock LW (1935) Zur Geochemie des Selens II. Nachrichten von der Gesellschaft der Wissenschaften zu Goettingen Fachgruppe IV, Mathematisch–Physikalische Klasse 3:245–252

Goldschmidt VM (1937) The principles of distribution of chemical elements in minerals and rocks. J Chem Soc:655–673, doi:10.1039/JR9370000655

Gosh S, Xu Y, Humayun M, Odom L (2008) Mass-dependent fractionation of mercury isotopes in the environment. Geochem Geophys Geosystems 9:doi:10.1029/2007GC001827

Greenland L (1967) The abundances of selenium, tellurium, silver, palladium, cadmium, and zinc in chondritic meteorites. Geochim Cosmochim Acta 31:849–860

Greenwood NN (1984) Chemistry of the elements. Pergamon Press

Hagiwara Y (2000) Selenium isotope ratios in marine sediments and algae: A reconnaissance study. M.Sc., University of Illinois at Urbana-Champaign, Urbana, IL

Hamilton SJ (2004) Review of selenium toxicity in the aquatic food chain. Sci Total Environ 326:1–31

Haygarth PM, Fowler D, Suerup S, Davison BM, Jones KC (1994) Determination of gaseous and particulate selenium over a rural grassland in the UK. Atmos Environ 28:3655–3663

Henderson P, Henderson GM (2009) The Cambridge Handbook of Earth Science Data. Cambridge University Press, New York

Herbel MJ, Johnson TM, Oremland RS, Bullen T (2000) Fractionation of selenium isotopes during bacterial respiratory reduction of selenium oxyanions. Geochim Cosmochim Acta 64:3701–3709

Herbel MJ, Johnson TM, Tanji KK, Gao S, Bullen TD (2002) Selenium stable isotope ratios in California agricultural drainage water management system. J Environ Qual 31:1146–1156

Herbel MJ, Blum JS, Oremland RS, Borglin SE (2003) Reduction of elemental selenium to selenide: experiments with anoxic sediments and bacteria that respire Se-oxyanions. Geomicrobiol J 20:587–602

Hertogen J, Janssen M-J, Palme H (1980) Trace elements in ocean ridge basalt glasses: implications for fractionations during mantle evolution and petrogenesis. Geochim Cosmochim Acta 44:2125–2143

Johnson TM (2004) A review of mass-dependent fractionation of selenium isotopes and implications for other heavy stable isotopes. Chem Geol 204:201–214

Johnston DT (2011) Multiple sulfur isotopes and the evolution of Earth's surface sulfur cycle. Earth-Sci Rev 106:161–183

Johnson TM, Herbel MJ, Bullen TD, Zawislanski PT (1999) Selenium isotope ratios as indicators of selenium sources and oxyanion reduction. Geochim Cosmochim Acta 63:2775–2783

Johnson TM, Bullen T (2004a) Mass-dependent fractionation of selenium and chromium isotopes in low-temperature environments. Rev Mineral Geochem 55:289–317

Johnson TM, Bullen TD (2004b) Selenium, iron and chromium stable isotope ratio measurements by the double isotope spike TIMS method. *In*: Handbook of Stable Isotope Methods. Vol 1. Elsevier, Amsterdam, Netherlands, p 623–651

Johnson TM, Bullen T (2003) Selenium isotope fractionation during reduction by Fe(II)–Fe(III) hydroxide–sulfate (green rust). Geochim Cosmochim Acta 67:413–419

Johnston DT, Farquhar J, Canfield DE (2007) Sulfur isotope insights into microbial sulfate reduction: when microbes meet models. Geochim Cosmochim Acta 71:3929–3947

Kendall B, Reinhard CT, Lyons TW, Kaufman AJ, Poulton SW, Anbar AD (2010) Pervasive oxygenation along late Archaean ocean margins. Nat Geosci 3:647–652

Kendall B, Creaser RA, Reinhard CT, Lyons TW, Anbar AD (2015) Transient episodes of mild environmental oxygenation and oxidative continental weathering during the late Archean. Sci Adv 1:e1500777, doi: 10.1126/sciadv.1500777

Kessi J (2006) Enzymic systems proposed to be involved in the dissimilatory reduction of selenite in the purple non-sulfur bacteria *Rhodospirillum rubrum* and *Rhodobacter capsulatus*. Microbiol 152:731–743

Knoll AH, Javaux EJ, Hewitt D, Cohen P (2006) Eukaryotic organisms in Proterozoic oceans. Philos Trans R Soc London, Ser B 361:1023–1038

Koljonen T (1973a) Selenium in certain metamorphic rocks. Bull Geol Soc Finland 45:107–117

Koljonen T (1973b) Selenium in certain igneous rocks. Bull Geol Soc Finland 45:9–22

König S, Luguet A, Lorand JP, Wombacher F, Lissner M (2012) Selenium and tellurium systematics of the Earth's mantle from high precision analyses of ultra-depleted orogenic peridotites. Geochim Cosmochim Acta 86:354–366

Krouse HR, Thode HG (1962) Thermodynamic properties and geochemistry of isotopic compounds of selenium. Can J Chem 40:367–375

Kulp TR, Pratt LM (2004) Speciation and weathering of selenium in Upper Cretaceous chalk and shale from South Dakota and Wyoming, USA. Geochim Cosmochim Acta 68:3687–3701

Lalonde SV, Konhauser KO (2015) Benthic perspective on Earth's oldest evidence for oxygenic photosynthesis. PNAS 112:995–1000, doi: 10.1073/pnas.1415718112

Lam P, Kuypers MM (2011) Microbial nitrogen cycling processes in oxygen minimum zones. Ann Rev Mar Sci 3:317–345

Large RR, Halpin JA, Danyushevsky LV, et al. (2014) Trace element content of sedimentary pyrite as a new proxy for deep-time ocean-atmosphere evolution. Earth Planet Sci Lett 389:209–220

Layton-Matthews D, Leybourne MI, Peter JM, Scott SD (2006) Determination of selenium isotopic ratios by continuous-hydride-generation dynamic-reaction-cell inductively coupled plasma-mass spectrometry. J Anal Atom Spectr 21:41–49

Layton-Matthews D, Leybourne MI, Peter JM, Scott SD, Cousens B, Eglington B (2013) Multiple sources of selenium in ancient seafloor hydrothermal systems: compositional and Se, S and Pb isotopic evidence from volcanic-hosted and volcanic-sediment-hosted massive sulfide deposits of the Finlayson Lake District, Yukon, Canada. Geochim Cosmochim Acta 117:313–331

Li X, Liu Y (2011) Equilibrium Se isotope fractionation parameters: A first principle study. Earth Planet Sci Lett 304:113–120

Little SH, Vance D, Lyons TW, McManus J (2015) Controls on trace metal authigenic enrichment in reducing sediments: Insights from modern oxygen-deficient settings. Am J Sci 315:77–119

Lobanov AV, Hatfield DL, Gladyshev VN (2009) Eukaryotic selenoproteins and selenoproteomes. Biochim Biophys Acta (BBA)-General Subjects 1790:1424–1428

Lyons TW, Reinhard CT, Planavsky NJ (2014) The rise of oxygen in Earth's early ocean and atmosphere. Nature 506:307–315

Malisa EP (2001) The behavior of selenium in geological processes. Environ Geochem Health 23:137–158

Martens DA, Suarez DL (1997) Selenium speciation of marine shales, alluvial soils, and evaporation basin soils of California. J Environ Qual 26:424–432

Mitchell K, Mason PRD, Van Cappellen P, Johnson TM, Gill BC, Owens JD, Ingall ED, Reichart G-J, Lyons TW (2012) Selenium as paleo-oceanographic proxy: A first assessment. Geochim Cosmochim Acta 89:302–317

Mitchell K, Couture RM, Johnson TM, Mason PR, Van Cappellen P (2013) Selenium sorption and isotope fractionation: Iron (III) oxides versus iron (II) sulfides. Chem Geol 342:21–28

Mitchell K, Mansoor SZ, Mason PR, Johnson TM, Van Cappellen P (2016) Geological evolution of the marine selenium cycle: Insights from the bulk shale $\delta^{82/76}$Se record and isotope mass balance modeling. Earth Planet Sci Lett 441:178–187

Mosher BW, Duce RA (1987) A global atmospheric selenium budget. J Geophys Res 92:13289–13298

Ono S (2008) Multiple-sulphur isotope biosignatures. Space Sci Rev 135:203–220

Ono S, Wing B, Johnston D, Farquhar J, Rumble D (2006) Mass-dependent fractionation of quadruple stable sulfur isotope system as a new tracer of sulfur biogeochemical cycles. Geochim Cosmochim Acta 70:2238–2252

Oremland RS, Steinberg NA, Maest A, Miller LG, Hollbaugh JT (1990) Measurement of in situ rates of selenate removal by dissimilatory bacterial reduction in sediments. Environ Sci Technol 24:1157–1164

Pettine M, Gennari F, Campanella L, Casentini B, Marani D (2012) The reduction of selenium(IV) by hydrogen sulfide in aqueous solutions. Geochim Cosmochim Acta 83:37–47

Pierru B, Grosse S, Pignol D, Sabaty M (2006) Genetic and biochemical evidence for the involvement of a molybdenum-dependent enzyme in one of the selenite reduction pathways of *Rhodobacter sphaeroides f. sp. denitrificans* IL106. Appl Environ Microbiol 72:3147–3153

Planavsky NJ, Reinhard CT, Wang X, Thomson D, McGoldrick P, Rainbird RH, Johnson T, Fischer WW, Lyons TW (2014) Low Mid-Proterozoic atmospheric oxygen levels and the delayed rise of animals. Science 346:635–638

Plant JA, Bone J, Voulvoulis N, Kinniburgh DG, Smedley PL, Fordyce FM, Klinck B (2014) Arsenic and selenium. Treatise Geochem 11:13–57

Pogge von Strandmann P, Coath CD, Catling DC, Poulton S, Elliott T (2014) Analysis of mass dependent and mass independent selenium isotope variability in black shales. J Anal Atom Spectr 29:1648–1659

Pogge von Strandmann PAE, Stüeken EE, Elliott T, Poulton S, Dehler CM, Canfield DE, Catling DC (2015) Selenium isotope evidence for progressive oxidation of the Neoproterozoic biosphere. Nature Commun 6:doi:10.1038/ncomms10157

Rashid K, Krouse HR (1985) Selenium isotopic fractionation during SeO_3 reduction to Se^0 and H_2Se. Can J Chem 63:3195–3199

Reeder RJ, Lamble GM, Lee JF, Staudt WJ (1994) Mechanism of SeO_4^{2-} substitution in calcite: An XAFS study Geochim Cosmochim Acta 58:5639–5646

Rees CE, Thode HG (1966) Selenium isotope effects in the reduction of sodium selenite and of sodium selenate. Can J Chem 44:419–427

Reinhard CT, Raiswell R, Scott C, Anbar AD, Lyons TW (2009) A late Archean sulfidic sea stimulated by early oxidative weathering of the continents. Science 326:713–716

Rose-Weston L, Brenan JM, Fei Y, Secco RA, Frost DJ (2009) Effect of pressure, temperature, and oxygen fugacity on the metal-silicate partitioning of Te, Se, and S: Implications for earth differentiation. Geochim Cosmochim Acta 73:4598–4615

Rother M, Wilting R, Commans S, Böck A (2000) Identification and characterisation of the selenocysteine-specific translation factor SelB from the archaeon Methanococcus jannaschii. J Molec Biol 299:351–358

Rouxel O, Ludden J, Carignan J, Marin L, Fouquet Y (2002) Natural variations of Se isotopic composition determined by hydride generation multiple collector inductively coupled plasma mass spectrometry. Geochim Cosmochim Acta 66:3191–3199

Rouxel O, Fouquet Y, Ludden JN (2004) Subsurface processes at the Lucky Strike hydrothermal field, Mid-Atlantic Ridge: evidence from sulfur, selenium, and iron isotopes. Geochim Cosmochim Acta 68:2295–2311

Rovira M, Gimenez J, Martinez M, Martinez-Llado X, de Pablo J, Marti V, Duro L (2008) Sorption of selenium (IV) and selenium (V) onto natural iron oxides: Goethite and hematite. J Hazard Mater 150:279–284

Ruggles EL, Snider GW, Hondal RJ (2012) Chemical basis for the use of selenocysteine. *In*: Selenium. Hatfield DL, Berry MJ, Gladyshev VN, (eds). Springer New York, p 73–83

Sabaty M, Avazeri C, Pignol D, Vermeglio A (2001) Characterization of the reduction of selenate and tellurite by nitrate reductases. Applied and Environmental Microbiology 67:5122–5126

Saltikov CW, Newman DK (2003) Genetic identification of a respiratory arsenate reductase. PNAS 100:10983–10988

Savard D, Bedard LP, Barnes S-J (2009) Selenium concentrations in twenty-six geological reference materials: New determinations and proposed values. Geostandards Geoanal Res 33:249–259

Schilling K, Wilcke W (2011) A method to quantitatively trap volatilized organoselenides for stable selenium isotope analysis. J Environ Qual 40:1021–1027

Schilling K, Johnson TM, Wilcke W (2011a) Selenium partitioning and stable isotope ratios in urban topsoil. Soil Sci Soc Am J 75:1354–1364

Schilling K, Johnson TM, Wilcke W (2011b) Isotope fractionation of selenium during fungal biomethylation by Alternaria alternate. Environ Sci Technol 45:2670–2676

Schilling K, Johnson TM, Wilcke W (2013) Isotope fractionation of selenium by biomethylation in microcosm incubations of soil. Chem Geol 352:101–107

Schilling K, Basu A, Johnson TM, Mason PRD, Tsikos H, Mondal SK (2014a) Se isotope signature of Paleoarchean and Paleoproterozoic banded iron formations. Goldschmidt Conference Abstract # 2207

Schilling K, Johnson TM, Mason PRD (2014b) A sequential extraction technique for mass-balanced stable selenium isotope analysis of soil samples. Chem Geol 381:125–130

Schilling K, Johnson TM, Dhillon KS, Mason PR (2015) Fate of Selenium in Soils at a Seleniferous Site Recorded by High Precision Se Isotope Measurements. Environ Sci Technol 49:9690–9698

Schirmer T, Koschinsky A, Bau M (2014) The ratio of tellurium and selenium in geological material as a possible paleo-redox prox. Chem Geol 376:44–51

Seby F, Potin-Gautier M, Giffaut E, Borge G, Donard OFX (2001) A critical review of thermodynamic data for selenium species at 25°C. Chem Geol 171:173–194

Shore AJT (2010) Selenium geochemistry and isotopic composition of sediments from the Cariaco Basin and the Bermuda Rise: a comparison between a restricted basin and the open ocean over the last 500 ka. Ph.D., University of Leicester, Leicester, UK

Sigman DM, Karsh K, Casciotti KL (2009) Ocean process tracers: nitrogen isotopes in the ocean. *In*: Encyclopedia of Ocean Science, 2nd edition. Steele, JH, Thorpe SA, Turekian KK (eds), Elsevier, Amsterdam, Netherlands, p 4138–4152

Stadtman TC (1974) Selenium Biochemistry Proteins containing selenium are essential components of certain bacterial and mammalian enzyme systems. Science 183:915–922

Stadtman TC (1990) Selenium biochemistry. Ann Rev Biochem 59:111–127

Steinbrenner H, Sies H (2009) Protection against reactive oxygen species by selenoproteins. Biochim Biophys Acta (BBA)-General Subjects 1790:1478–1485

Stolz JF, Basu P, Santini JM, Oremland RS (2006) Arsenic and selenium in microbial metabolism. Ann Rev Microbiol 60:107–130

Stüeken EE, Catling DC, Buick R (2012) Contributions to late Archaean sulphur cycling by life on land. Nat Geosci 5:722–725

Stüeken EE, Foriel J, Nelson BK, Buick R, Catling DC (2013) Selenium isotope analysis of organic-rich shales: advances in sample preparation and isobaric interference correction. J Anal Atom Spectr 28:1734–1749

Stüeken EE, Buick R, Anbar AD (2015a) Selenium isotopes support free O_2 in the latest Archean. Geol 43:259–262
Stüeken EE, Buick R, Bekker A, Catling DC, Foriel J, Guy BM, Kah LC, Machel HG, Montanez IP, Poulton SW (2015b) The evolution of the global selenium cycle: Secular trends in Se isotopes and abundances. Geochim Cosmochim Acta 162:109–125
Stüeken EE, Buick R, Schauer AJ (2015c) Nitrogen isotope evidence for alkaline lakes on late Archean continents. Earth Planet Sci Lett 411:1–10
Stüeken EE, Foriel J, Buick R, Schoepfer SD (2015d) Selenium isotope ratios, redox changes and biological productivity across the end-Permian mass extinction Chem Geol 410:28–39
Stumm W, Morgan JJ (1996) Aquatic Chemistry. John Wiley and Sons, Inc
Suzuoki T (1965) A geochemical study of selenium in volcanic exhalation and sulfur deposits. II. On the behavior of selenium and sulfur in volcanic exhalation and sulfur deposits. Bull Chem Soc Jpn 38:1940–1946
Takematsu N, Sato Y, Okabe S, Usui A (1990) Uptake of selenium and other oxyanionic elements in marine ferromanganese concretions of different origins. Mar Chem 31:271–283
Tamari Y, Ogawa H, Fukumoto Y, Tsuji H, Kusaka Y (1990) Selenium content and its oxidation state in igneous rocks, rock-forming minerals, and a reservoir sediment. Bull Chem Soc Jpn 63:2631–2638
Taylor SR, McLennan SM (1995) The geochemical evolution of the continental crust. Rev Geophys 33:241–265
Tischendorf G (1959) Zur Genesis einiger Selenidvorkommen, insbesondere von Tilkerode im Harz. Freiberger Forschungshefte C69:1–168
Turekian KK, Wedepohl KH (1961) Distribution of the elements in some major units of the earth's crust. Geol Soc Am Bull 72:175–192
von Damm KL (1990) Seafloor hydrothermal activity: Black smokers chemistry and chimneys. Ann Rev Earth Planet Sci 18:173–204
Wachsmann M, Heumann KG (1992) Negative thermal ionization mass spectrometry of main group elements Part 2. 6th group: sulfur, selenium and tellurium. Int J Mass Spectr Ion Process 114:209–220
Wang Z, Becker H (2013) Ratios of S, Se and Te in the silicate Earth require a volatile-rich late veneer. Nature 499:328–331
Watts CA, Ridley H, Dridge EJ, Leaver JT, Reilly AJ, Richardson DJ, Butler CS (2005) Microbial reduction of selenate and nitrate: common themes and variations. Biochem Soc Trans 33:173–175
Wedepohl KH (1995) The composition of the continental crust. Geochim Cosmochim Acta 59:1217–1232
Wen H, Carignan J (2007) Reviews on atmospheric selenium: Emissions, speciation and fate. Atmos Environ 41:7151–7165
Wen H, Carignan J, Chu X, Fan H, Cloquet C, Huang J, Zhang Y, Chang H (2014) Selenium isotopes trace anoxic and ferruginous seawater conditions in the Early Cambrian. Chem Geol 390:164–172
Wille M, Kramers JD, Nägler TF, Beukes NJ, Schröder S, Meisel T, Lacassie JP, Voegelin AR (2007) Evidence for a gradual rise of oxygen between 2.6 and 2.5 Ga from Mo isotopes and Re–PGE signatures in shales. Geochim Cosmochim Acta 71:2417–2435
Winkel LH, Vriens B, Jones GD, Schneider LS, Pilon-Smits E, Bañuelos GS (2015) Selenium cycling across soil–plant–atmosphere interfaces: A critical review. Nutrients 7:4199–4239
Young ED, Galy A, Nagahara H (2002) Kinetic and equilibrium mass-dependent isotope fractionation laws in nature and their geochemical and cosmochemical significance. Geochim Cosmochim Acta 66:1095–1104
Yu M, Sun D, Tian W, Wang G, Shen W, Xu N (2002) Systematic studies on adsorption of trace elements Pt, Pd, Au, Se, Te, As, Hg, Sb on thiol cotton fiber. Analytica Chimica Acta 456:147–155
Zehr JP, Oremland RS (1987) Reduction of selenate to selenide by sulfate-respiring bacteria: experiments with cell suspensions and estuarine sediments. Appl Environ Microbiol 53:1365–1369
Zerkle AL, Claire MW, Domagal-Goldman SD, Farquhar J, Poulton SW (2012) A bistable organic-rich atmosphere on the Neoarchaean Earth. Nat Geosci 5:359–363
Zhang Y, Zahir ZA, Frankenberger Jr. WT (2004) Fate of colloidal-particulate elemental selenium in aquatic systems. J Environ Qual 33:559–564
Zhang Y, Romero H, Salinas G, Gladyshev VN (2006) Dynamic evolution of selenocysteine utilization in bacteria: a balance between selenoprotein loss and evolution of selenocysteine from redox active cysteine residues. Genome Biol 7:doi:10.1186/gb-2006-1187-1110-r1194
Zhu J-M, Johnson TM, Clark SK, Zhu X-K (2008) High precision measurement of selenium isotopic composition by hydride generation multiple collector inductively coupled plasma mass spectrometry with a ^{74}Se–^{77}Se double spike. Chin J Anal Chem 36:1385–1390
Zhu JM, Johnson TM, Clark SK, Zhu XK, Wang XL (2014) Selenium redox cycling during weathering of Se-rich shales: A selenium isotope study. Geochim Cosmochim Acta 126:228–249
Zieve R, Peterson PJ (1984) Volatilization of selenium from plants and soils. Sci Total Environ 32:197–202
Zinabu GM, Pearce NJ (2003) Concentrations of heavy metals and related trace elements in some Ethiopian rift-valley lakes and their in-flows. Hydrobiologia 492:171–178
Zwolak I, Zaporowska H (2012) Selenium interactions and toxicity: a review. Cell Biol Toxicol 28:31–46

Reviews in Mineralogy & Geochemistry
Vol. 82 pp. 683-732, 2017

16

Good Golly, Why Moly?

THE STABLE ISOTOPE GEOCHEMISTRY OF MOLYBDENUM

Brian Kendall*

Department of Earth and Environmental Sciences
University of Waterloo
Waterloo, ON
N2L 3G1
Canada

bkendall@uwaterloo.ca

Tais W. Dahl*

Natural History Museum of Denmark
University of Copenhagen
Copenhagen
Denmark

tais.dahl@snm.ku.dk

* Both authors contributed equally to this work.

Ariel D. Anbar

School of Earth and Space Exploration
School of Molecular Sciences
Arizona State University
Tempe, AZ 85287
USA

anbar@asu.edu

INTRODUCTION

"The Answer to the Great Question... Of Life, the Universe and Everything...
Is... Forty-two," said Deep Thought, with infinite majesty and calm...
"I checked it very thoroughly," said the computer, "and that quite definitely is the answer."
— Douglas Adams, *The Hitchhiker's Guide to the Galaxy*

Molybdenum (Mo)—the element with atomic number 42—possesses unique properties that make it the answer to many questions in the geosciences, life sciences, and industry.

In the geosciences, the redox sensitivity of Mo makes it particularly useful for answering questions about environmental redox conditions. In particular, it was first suggested as an ocean paleoredox proxy over 30 years ago (Holland 1984; Emerson and Huested 1991)—an application that finally came to fruition in the late 1990s and 2000s when understanding of Mo geochemical behavior in modern environments improved significantly (e.g., Crusius et al. 1996; Helz et al. 1996, 2011; Morford and Emerson 1999; Erickson and Helz 2000; Barling et al. 2001; Siebert et al. 2003, 2005; Arnold et al. 2004; Vorlicek et al. 2004; Morford et al. 2005; Nägler et al. 2005; Algeo and Lyons 2006; McManus et al. 2006; Poulson et al. 2006; Anbar et al. 2007; Wille et al. 2007; Pearce et al. 2008; Archer and Vance 2008; Neubert et al. 2008; Scott et al. 2008; Gordon et al. 2009; Poulson Brucker et al. 2009).

1529-6466/17/0082-0016$10.00 http://dx.doi.org/10.2138/rmg.2017.82.16

In the life sciences, nature settled on Mo as the answer to the challenge of biological-N_2 fixation at least ~2 billion years ago (Boyd et al. 2011), with the evolution of the Mo-dependent nitrogenase enzyme. Molybdenum is also at the heart of nitrate reductase enzymes, which are essential for assimilatory and dissimilatory nitrate reduction (Glass et al. 2009). Therefore, Mo is central to the nitrogen biogeochemical cycle. This biological role combines with its geochemical behavior in ways that might drive aspects of the coevolution of life and environment (Anbar and Knoll 2002).

Industrially, Mo is variously used as a catalyst, pigment, steel additive, and lubricant. Most of this use is in different types of steel, to improve physical properties like hardness and temperature strength, as well as chemical properties, notably corrosion resistance. Over 230,000 metric tons are used each year, mostly in China (IMOA 2016). Porphyry molybdenum and copper–molybdenum deposits are the most important sources of molybdenite, the ore mineral of Mo.

Isotope geochemists were drawn to Mo because of its biogeochemical importance and economic value, and its seven stable isotopes, all relatively abundant (10–25%) and covering a relatively wide mass range of ~8% (Fig. 1). Beginning in the late 1990s, equipped with new multiple collector inductively coupled plasma mass spectrometers, they began to wonder if Mo isotope compositions varied significantly, and if Mo isotope fractionation could provide new answers to yet more questions.

The subsequent ~15 years of research yielded an emphatic answer of "yes", centered in particular on paleoceanographic applications, but also extending to the solid Earth geosciences and other areas.

This review provides an overview of this maturing isotope system, with an emphasis on paleoredox applications that dominate the literature. It is intended as an update of the reviews written when the Mo isotope system was still emerging (Anbar 2004; Anbar and Rouxel 2007). The first section covers analytical methodology. The next two sections provide the necessary context for Mo isotope studies by reviewing Mo biogeochemistry and Mo isotope fractionation factors. Next, the Mo isotope variations in meteorites and Earth reservoirs are explored, with an emphasis on the large database for marine sediments. In the context of modern observations of the ocean Mo cycle, the use of Mo isotopes as a local and global ocean paleoredox proxy is synthesized in the following section. In the final section, we explore the rapidly growing application of Mo isotopes to ore deposits, oil, and anthropogenic tracing, areas that are expected to see strong growth in the near future.

ANALYTICAL CONSIDERATIONS

Data reporting

Molybdenum stable isotope fractionation is conventionally reported in δ^{98}Mo notation as parts per thousand deviation of the ^{98}Mo/^{95}Mo ratio relative to a universal reference material. Older data were reported relative to in-house reference materials thought to be identical in composition. However, the analytical precision has improved since then and a common reference material is necessary because various in-house reference materials now differ by up to 0.37‰ (Goldberg et al. 2013). The Mo standard solution, NIST–SRM–3134, has been defined as an international reference material, and is assigned a distinct δ^{98}Mo value of 0.25‰ to account for its offset from the most common in-house standards used previously (Nägler et al. 2014). On this scale, the Mo isotope composition of samples can be calculated as follows:

$$\delta^{98}\text{Mo} = \left[\left(\left({}^{98}\text{Mo}/{}^{95}\text{Mo}\right)_{\text{sample}} / \left({}^{98}\text{Mo}/{}^{95}\text{Mo}\right)_{\text{NIST-SRM-3134}}\right) - 1\right] \times 1000 + 0.25\ [‰]$$

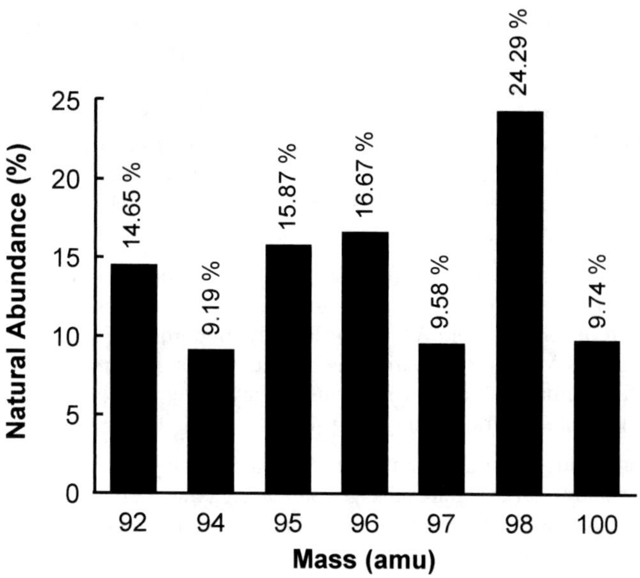

Figure 1. Abundances of the seven stable isotopes of Mo, based on Mayer and Wieser (2014).

If the δ^{98}Mo of the in-house reference material relative to the NIST–SRM–3134 standard is known, then it is possible to re-normalize the Mo isotope composition of a sample from the in-house reference scale to the NIST–SRM–3134 scale. If the isotopic offset between the in-house and NIST–SRM–3134 standards is not known, it is still possible to convert between the two scales by measuring a well-known secondary standard such as seawater (e.g., IAPSO) or the USGS rock reference material SDO–1, which has $\delta^{98}\text{Mo} = 1.05 \pm 0.14‰$ ($2\sigma = 2$ standard deviations) on the NIST–SRM–3134 scale (Goldberg et al. 2013; Nägler et al. 2014).

Hence, the NIST–SRM–3134 scale facilitates the comparison of future work with almost all older data within a reasonable level of precision. On this scale, open ocean water samples have $\delta^{98}\text{Mo} = 2.34 \pm 0.10‰$ irrespective of ocean basin or water depth (Barling et al. 2001; Siebert et al. 2003; Greber et al. 2012; Nakagawa et al. 2012; Goldberg et al. 2013), except in the deep waters of restricted anoxic basins (Nägler et al. 2011; Noordmann et al. 2015) or in highly productive surface ocean waters (Kowalski et al. 2013). This is indistinguishable from the canonical value of 2.3‰ suggested from earlier work. In this review, all values of δ^{98}Mo are reported relative to NIST–SRM–3134 = 0.25‰.

Chemical separation

The Mo isotope composition of molybdenite (MoS_2) can be measured precisely and accurately using mass spectrometry after sample dissolution and dilution because Mo and S are the only major elements in the molybdenite crystal structure (Barling et al. 2001). However, most other natural materials have low Mo abundances (<100 ppm) and much higher concentrations of other elements, and thus require pre-concentration and purification of Mo before the isotope composition can be measured. Doing so minimizes the problem of matrix effects, which arise when the presence of other elements causes the formation of ionic compounds with masses that are similar to those of the Mo isotopes. Such "interferences" on Mo isotope masses can preclude accurate measurement of Mo isotope compositions unless adequately corrected for or minimized.

Removal of Fe and Mn is particularly critical to minimize the formation of argides, which produce polyatomic interferences at masses 94–97. For example, the Fe/Mo ratio in the analyte should be less than 1 to avoid measurable interferences when using multiple collector inductively coupled plasma mass spectrometry (MC–ICP–MS) (Malinovsky et al. 2005). Zirconium has isobaric interferences with Mo on masses 92, 94 and 96, but Mo is efficiently separated from Zr during purification. Both Ru and doubly-charged W interfere on masses 96–100 and 92, respectively, but this has mainly been a concern for synthetic materials and meteorite samples (Burkhardt et al. 2011; Migeon et al. 2015). Other elements including Si may affect the measured isotope ratios, and Si/Mo ratios less than 50 are recommended to avoid such matrix effects (Malinovsky et al. 2005).

For studies exploring Mo isotope variations in meteorites, where nucleosynthetic anomalies may affect the Mo isotope compositions, the measurement of purified Mo without interferences from Zr, Ru, and W is particularly important. Furthermore, comparison of mass-dependent Mo isotope variations in meteorites requires a correction for the nucleosynthetic anomalies found in most meteorite classes except for achondritic, lunar, and Martian meteorites (Burkhardt et al. 2014).

Traditionally, Mo is separated from the matrix elements using ion exchange chromatography. Most schemes deploy both an anion exchange column (e.g., Bio-Rad™ AG1-X8, Dowex™ AG1, Eichrom™ AG1-X8) to separate Mo from Zr and most other matrix elements, and a cation exchange column (e.g., Bio-Rad™ AG50W-X8 or TRU-spec™) to mainly separate Mo from Fe (Anbar et al. 2001; Barling et al. 2001; Siebert et al. 2001; Pietruszka et al. 2006; Migeon et al. 2015). However, purification using a chelating resin (Malinovsky et al. 2005), anion-only resin (Siebert et al. 2001; Pearce et al. 2009; Nagai and Yokoyama 2016), or two distinct cation resins (Archer and Vance 2008; Burkhardt et al. 2011) has also been successfully done.

A key observation is that Mo isotopes are fractionated during elution in anion exchange systems (e.g., Bio-Rad™ AG1-X8, Dowex™ AG1) (Anbar et al. 2001; Siebert et al. 2001). The magnitude of fractionation depends on the column yield, but is large enough (~ 1‰/amu) to completely swamp natural variability (Anbar et al. 2001; Siebert et al. 2001). Therefore, it is necessary to either ensure quantitative yields during purification or to make a correction for isotope fractionation induced by this process. Mixing and equilibrating sample Mo with a double spike of known composition before purification allows for such a correction (discussed further below).

Mass Spectrometry

A fundamental challenge to stable isotope studies (not including mass-independent Mo isotope variations produced by nucleosynthesis; Dauphas et al. 2002a,b, 2004; Yin et al. 2002; Fujii et al. 2006; Burkhardt et al. 2011, 2012) is that mass spectrometry induces mass-dependent isotope fractionation. Therefore, precise determination of the Mo stable isotope composition depends on a precise correction for such fractionation processes.

The magnitude of isotope fractionation differs markedly between MC–ICP–MS and thermal ionization mass spectrometry (TIMS). For MC–ICP–MS, the instrumental mass bias is large (+17‰/amu), but very stable, whereas TIMS produces variable mass bias of smaller magnitude, at −6.4‰/amu and −0.5‰/amu for positively and negatively charged ions, respectively (Wieser et al. 2007; Nagai and Yokoyama 2016). In both cases, the instrumental isotope fractionation exceeds the variability in nature (~ 1‰/amu), and thus a correction for instrumental mass bias is necessary. Wieser et al. (2007) compared the various mass spectrometric techniques and concluded that MC–ICP–MS is the optimal method for accurately measuring the isotope composition of Mo in natural materials.

The earliest Mo isotope measurements were performed using TIMS in positive ion mode (P–TIMS) with a Mo^+ beam, resulting in an analytical precision of 6‰/amu for each Mo isotope ratio, $^{x}Mo/^{100}Mo$ (Murthy 1962, 1963; Wetherill 1964). The large uncertainty was due to the low ionization potential of Mo. Recently, it has been demonstrated that the latest generation TIMS instruments operating in negative ion mode (N–TIMS), measuring MoO_3^-, can yield precisions of <0.01‰/amu for $^{x}Mo/^{100}Mo$ (Nagai and Yokoyama 2016). To achieve highly precise Mo isotope ratios using N–TIMS, it is important to measure and correct for the oxygen isotope composition of the MoO_3^- ions.

Three strategies have been applied to correct for instrumental mass bias during mass spectrometric analysis, including 1) standard-sample bracketing, 2) elemental spiking, and 3) double spiking. All methods are applicable to MC–ICP–MS, whereas double spiking is needed for TIMS analysis.

All three methods are summarized below.

Standard-sample bracketing. The simplest correction for instrumental mass bias is comparison of the sample to a standard run under the same instrumental conditions. Usually, analyses of samples are bracketed by standards to cope with systematic instrumental drift. This correction assumes that instrumental mass bias: a) has a constant drift during analysis, and b) does not vary systematically between samples and standards. In TIMS, instrumental mass bias changes continuously during analysis as a result of isotope enrichment during thermal evaporation and ionization (Murthy 1962, 1963). Therefore, the standard-sample bracketing method is more applicable to MC–ICP–MS, where the instrumental mass bias is not a time-dependent phenomenon (Maréchal et al. 1999). This approach has been successful for some non-traditional isotope systems, including Fe (Beard et al. 2003), and may be suitable for isotopic analysis of molybdenite (Pietruszka et al. 2006). However, an efficient purification protocol is required for a trace metal such as Mo because variation in instrumental mass bias arising from matrix differences between sample and standard solutions cannot be corrected for. If efficient purification cannot be achieved, then other mass bias correction methods must be applied.

Element spike. In MC–ICP–MS, it is possible to dope the purified sample solution with another element immediately before analysis and simultaneously monitor changes in instrumental mass bias and Mo isotope fractionation in the sample. In principle, this correction is applicable without standard-sample bracketing, but typically it is used in combination with bracketing standards doped in an identical fashion as the samples. Some of the first modern observations of Mo isotope fractionation in geological materials employed Zr and Ru element spikes to yield $\delta^{98}Mo$ values with a precision of ~0.3‰ (2σ) (Anbar et al. 2001). Later refinements improved precision to ~0.15‰ (2σ) (e.g., Duan et al. 2010). However, this approach rests on the assumption that the instrumental mass bias of Zr or Ru isotopes varies systematically with the instrumental mass bias of Mo isotopes.

Isotopic double spike. For both MC–ICP–MS and TIMS, a correction for mass-dependent isotope fractionation that occurs during non-quantitative chromatographic purification and mass spectrometric analysis can be made using an isotopic double spike. The spike consists of two Mo isotopes with a known isotopic ratio. The fundamental advantage of this approach is that the spike isotopes follow exactly the same fractionation law as the isotopes of interest. This method can correct for isotope fractionation incurred during both chemical separation and mass spectrometry (Wetherill 1964; Siebert et al. 2001). Therefore, a more pure chemical separation can be prioritized instead of an optimum yield.

Due to its large number of stable isotopes (Fig. 1), Mo is particularly suitable for the double spike method, which thus has become the favored method for correcting isotope fractionation induced in the laboratory (Skierszkan et al. 2015). Several laboratories have calibrated and adopted a ^{97}Mo–^{100}Mo spike to obtain $\delta^{98}Mo$ data on an in-house standard solution that has a long-term external reproducibility of better than ±0.12‰, reaching as low as 0.04‰ (2σ) (Siebert et al. 2001; Goldberg et al. 2013; Willbold et al. 2016). Data from molybdenite samples utilizing TIMS and a ^{94}Mo–^{100}Mo spike with no chemical purification yielded Mo isotope ratios with uncertainties of 0.12‰/amu at the 2σ level (Hannah et al. 2007; Wieser et al. 2007). Recently, Nagai and Yokohama (2016) utilized a ^{92}Mo–^{97}Mo–^{100}Mo triple spike and N–TIMS to determine Mo isotope ratios in a standard solution with a reproducibility of ~0.01‰/amu at the 2σ level (i.e., ~10 ppm on the ^{96}Mo/^{95}Mo ratio).

CHEMICAL AND BIOLOGICAL CONTEXT

Aqueous geochemistry

In the surface environment, interest in Mo has long revolved around its dynamic redox behavior (e.g., Bertine and Turekian 1973; Morford and Emerson 1999). Under oxygenated conditions, Mo is a highly mobile and conservative element that accumulates in seawater to such an extent that it is the most abundant transition metal in the oceans (~10^7 nmol kg^{-1}; Morris 1975; Bruland 1983; Collier 1985). In contrast, in H_2S-bearing waters, Mo is readily removed from solution, leading to pronounced sedimentary enrichments (e.g., Bertine and Turekian 1973; Emerson and Huested 1991; Crusius et al. 1996; Scott and Lyons 2012). This bimodal behavior has made Mo—and its isotopes—particularly powerful for paleoredox investigations.

This bimodality can be understood in terms of chemical speciation. Mo is easily oxidized, so that Mo(VI) species occupy the largest area of Eh–pH phase space, particularly at typical seawater and freshwater conditions (Fig. 2). Mo(VI) readily forms the oxyanion molybdate (MoO_4^{2-}), which coordinates only weakly with other environmentally common inorganic ligands such as Cl^- or OH^-. Thus, the tetrahedrally coordinated oxyanion MoO_4^{2-} is thought to dominate aqueous speciation. However, recent work suggests a significant role for Mo(V) species such as MoO_2^+ (Wang et al. 2011). The potential importance of this species can be seen in Fig. 2, which compares the distribution of Mo species in Eh–pH space (a) with, and (b) without MoO_2^+. This cationic species could be important at pH<8 in dysoxic settings, but the behavior of Mo in oxic surface waters generally fits with the low reactivity of MoO_4^{2-}. Organic complexes also play a role in natural environments, which has been recognized for a long time (Szilagyi 1967; Nissenbaum and Swaine 1976), and remains an active area of investigation (Wichard et al. 2009).

The best analogy for Mo environmental chemistry is S, with MoO_4^{2-} and SO_4^{2-} having similar behaviors and distributions due to similar charges, coordination, and ionic radii as well as element redox behaviors. Not surprisingly, Mo and SO_4^{2-} concentrations are well-correlated in surface water systems (Miller et al. 2011).

Other molybdate species, such as $HMoO_4^-$ and H_2MoO_4 ("molybdic acid"), become quantitatively important only at pH<6 (Fig. 2), but may play a role in Mo adsorption to cationic surfaces. Aqueous polynuclear molybdate species ("polymolybdates") such as $Mo_6O_{19}^{2-}$, $Mo_7O_{24}^{6-}$, or $Mo_8O_{26}^{4-}$ will dominate the solution at pH<8 when Mo concentrations are >1 mM, while MoO_4^{2-} should dominate at all concentrations below 100 μM above a pH of 4 (Baes and Mesmer 1976). While millimolar-level Mo concentrations are rare in the environment, polymolybdates are implicated on some mineral surfaces. Such octahedrally coordinated Mo compounds may play an important role in Mo adsorption to Mn and Fe oxides, reflecting a change in Mo coordination geometry after MoO_4^{2-} has been attracted to protonated oxide mineral surfaces, as discussed further below (Wasylenki et al. 2011).

In sulfidic aqueous solutions, MoO_4^{2-} is progressively transformed into thiomolybdate species ($MoO_{4-x}S_x^{2-}$; Saxena et al. 1968; Diemann and Müller 1973). At $[H_2S]_{aq}$>11 μM, the stable thiomolybdate species is MoS_4^{2-} (Erickson and Helz 2000). This "switchpoint" corresponds to 22–125 μM total sulfide ($\Sigma S^{2-}=H_2S+HS^-+S^{2-}$) at a pH of 7–8, typical of natural sulfidic waters.

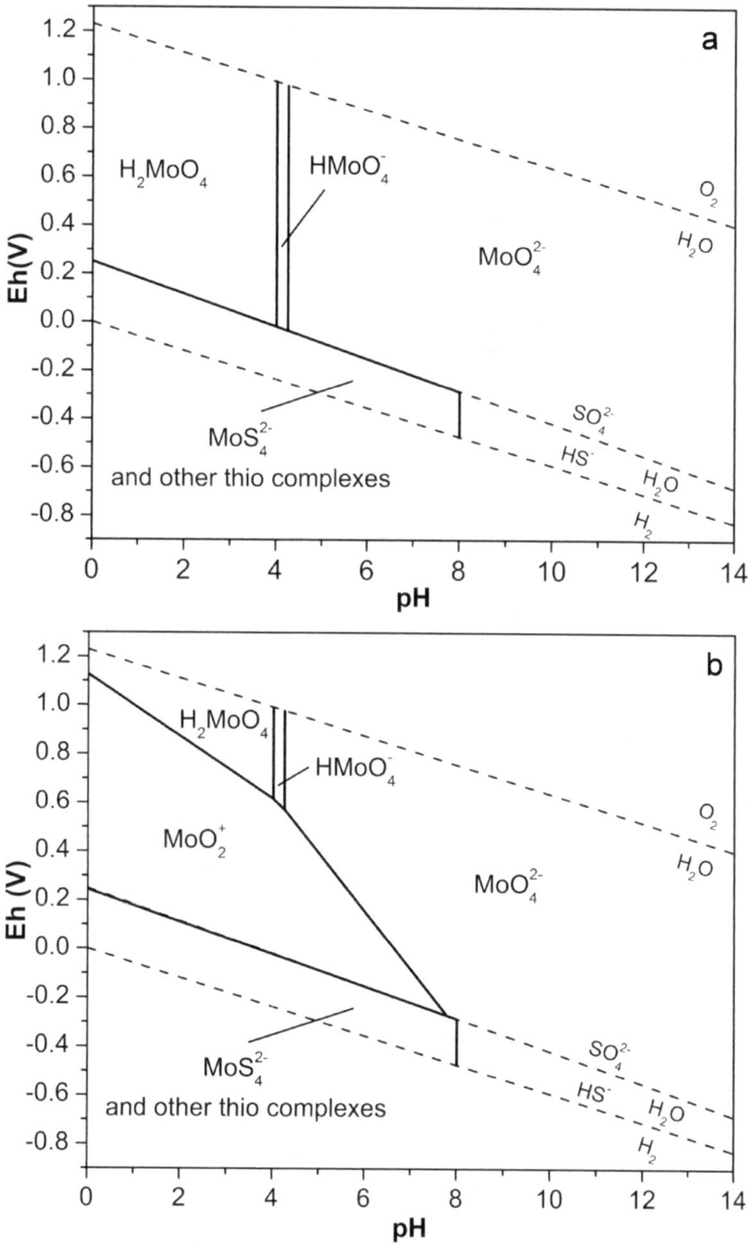

Figure 2. Eh-pH diagram showing dissolved Mo speciation in the system Mo–H_2O–H_2S, assuming that $\Sigma Mo = 10^{-6}$ M and $\Sigma S = 10^{-4}$ M. Molybdate protonation constants from H_2MoO_4 and $HMoO_4^-$ are from Smith and Martell (2004). The Mo speciation below the SO_4^{2-}–H_2S transition is not well known. The boundary between MoO_4^{2-} and MoS_4^{2-} was calculated using equilibrium constants from Erickson and Helz (2000). Other metastable thiomolybdates are not indicated. **a**) Classical diagram that does not include MoO_2^+, modified from Anbar (2004). **b**) Diagram that includes MoO_2^+, recognizing the possible importance of Mo(V) species (Wang et al. 2011).

Polynuclear Mo sulfide species including $Mo_2S_7^{2-}$, $Mo_4S_{15}^{6-}$, and $Mo_4S_{13}^{2-}$, are reported from continuous acidification experiments with molar-level thiomolybdate solutions (Saxena et al. 1968). Ultimately, the hexavalent MoS_3 dominates at pH < 2.4 (Helz et al. 1996). Polynuclear Mo sulfide species have not been observed in sulfidic experiments with 40–350 μM Mo, and are probably irrelevant at the low Mo concentration in sulfidic aqueous environments (< 10 nM) (Vorlicek et al. 2004).

As discussed further below, it is well-documented that Mo is rapidly removed from solution in H_2S-rich waters. Early studies assumed that MoS_2 precipitated via molybdate reduction in natural sulfidic systems (Amrhein et al. 1993):

$$MoS_4^{2-} + 2\,e^- + 2\,HS^- + 6\,H^+ \Leftrightarrow MoS_2 + 4\,H_2O \qquad \Delta G^0 = -314.3\ \text{kJ/mol}$$

However, MoS_2 precipitation is kinetically hindered in most Earth surface environments studied to date (e.g., Helz et al. 1996; Bostick et al. 2003; Chappaz et al. 2008; Dahl et al. 2013a). Instead, the chemistry of thiomolybdate species likely plays a role. In particular, these species are thought to be particle-reactive (Helz et al. 1996) and so may be removed from solution in association with sinking particulates (discussed below in the section *Molybdenum Isotopes in Major Reservoirs*). Yet, there is still a large gap in our understanding of this removal process.

Mo is found as distinct Mo(IV)-sulfide compounds in unknown, submicron, dispersed forms in anoxic muds and organic-rich mudrocks (Helz et al. 1996; Bostick et al. 2003; Dahl et al. 2013a). Hence, post-thiomolybdate reactions involve a Mo reduction step. Zero-valent sulfur present in natural sulfidic environments can reduce thiomolybdate to form highly reactive Mo polysulfide anions (Vorlicek et al. 2004) that, in turn, readily adsorb onto FeS_2, FeS (Bostick et al. 2003; Helz et al. 2004), and clay minerals (i.e., illite and Fe-bearing kaolinite and montmorillonite) (Helz et al. 2004). Scavenging with particulate organic matter is indicated in experiments with sulfate reducing bacteria where Mo precipitation occurs on the periphery of cells (Biswas et al. 2009). This may also explain the general relationship between Mo and organic carbon contents in euxinic sediments (discussed further below).

More recently, it was hypothesized that Mo removal in sulfidic systems is controlled by precipitation of an Fe(II)–Mo(VI) sulfide phase to form nanoscale mineral particles with the chemical formula $Fe_5Mo_3S_{14}$ (Helz et al. 2011). This Mo–Fe–S phase would be consistent with the observed association of Mo with organic matter in sediments, since Mo–Fe-sulfides may be embedded in an organic matrix (Dahl et al. 2013a). The actual removal pathway(s) remain an area for future study.

Biology

Molybdenum is the only second-row transition metal in the periodic table that is required by most living organisms (Hille 2002). Like Fe, Mo is an essential micronutrient required by enzymes catalyzing key reactions in global C, S, and N metabolism (e.g., Mendel and Bittner 2006). This capacity makes Mo an important element in biology despite its scarcity at the Earth's surface (~ 1 ppm), and has presumably led to the evolution of efficient processes for Mo uptake, such as production of siderophore-like binding ligands that target Mo (e.g., Liermann et al. 2005; Bellenger et al. 2008).

The reason for the critical biological role of Mo is probably due to the low reduction potentials of several oxidation states compared with other metals (Fig. 3). The fact that multiple Mo oxidation states can be accessed over a narrow range of voltages makes Mo relatively "redox labile" at low environmental Eh, but it also means that the energy gain from Mo redox transformations is small compared to many other elements. Therefore, unlike Fe and Mn, Mo is not used as a terminal electron acceptor or donor in metabolic pathways.

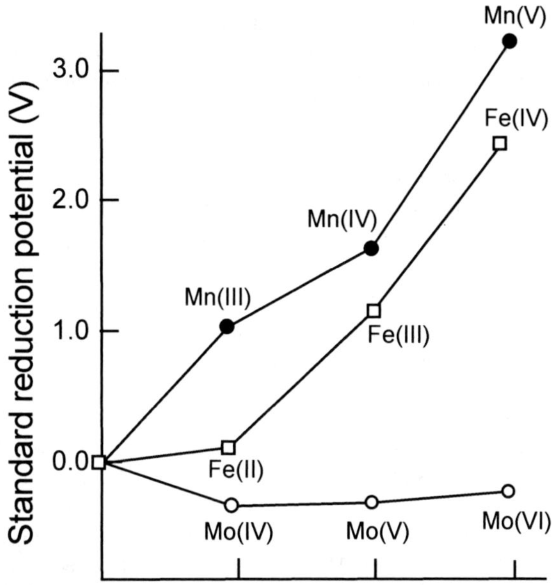

Figure 3. Standard reduction potentials (at pH = 7 relative to the hydrogen electrode) for the different oxidation states of Mo, Fe and Mn. The slope between any two points is equal to the redox potential. In contrast to most metals, Mo has multiple oxidation states that span a small range of potentials. Modified from Frausto da Silva and Williams (2001).

The redox lability makes Mo well-suited as a co-factor in enzymes that catalyze redox reactions. The enzymes that utilize Mo can be grouped into two broad categories: (1) the nitrogenases and (2) the mononuclear Mo enzymes (Stiefel 1997).

Nitrogenase is the enzyme responsible for nitrogen fixation that converts atmospheric N_2 to biologically useful NH_3. Biological nitrogen fixation only occurs in prokaryotes, and is essential for maintaining the nitrogen cycle on Earth. In nitrogenases, Mo sits in a multinuclear Fe–Mo–S cluster known as the FeMo-cofactor, where the six-electron transfer reduction takes place (Rees et al. 2005). Alternative nitrogenases utilizing Fe, W or V in place of Mo do exist, but they are markedly less efficient (Miller and Eady 1988; Eady 1996).

The remaining Mo-containing enzymes include more than 30 distinct enzymes that govern a wide variety of bioessential redox processes of environmental, agronomic, and health relevance. Examples include nitrate reductase, sulfite oxidase, formate dehydrogenase, xanthine oxidase, DMSO reductase, and aldehyde oxidase (Hille 1996; Stiefel 1997). These enzymes are not confined to prokaryotes, but also occur in eukaryotic organisms, including humans. They all contain the Mo cofactor (Moco), which is chemically, biochemically, and genetically distinct from the nitrogenase cofactor (FeMoco). The Moco enzymes all share common structural features with Mo situated at the active center coordinated via S to one or two unusual pterin ligands ("molybdopterin" ligands) and usually one or more oxo groups, depending on the oxidation state of the Mo center. These enzymes carry out two electron transfer (O transfer) reactions (Romao et al. 1997).

Molybdenum deficiency is rare, as are disorders of Mo metabolism, but symptoms may be induced in diets rich in Cu or W, which are Mo antagonists. On the other hand, tetrathiomolybdate has a strong affinity for Cu, and is an active agent for treatment of disorders of copper metabolism (Alvarez et al. 2010).

Molybdenum plays an important role in biology despite its scarcity at the Earth's surface, likely reflecting a combination of the unique chemical character of this element, evolutionary adaptation to higher Mo availability in increasingly more oxygenated oceans, or a legacy of early evolution in Mo-rich environments such as prebiotic chemical evolution in association with sulfide minerals (e.g., Crick and Orgel 1973; Anbar and Knoll 2002).

Molybdenum limitation (<5 nM) in some freshwater lakes can limit rates of nitrogen fixation and nitrate reduction when NH_4^+ is unavailable and biology must rely on N_2 and NO_3^- as sole N sources (Glass et al. 2012). Growth experiments show that N_2 fixation slows down at 1–5 nM Mo in cyanobacteria, presumably due to the expression of high affinity ModABC MoO_4^{2-} uptake systems, which are widely distributed in bacteria and archaea (Zerkle et al. 2006; Glass et al. 2010).

It has been hypothesized that Mo concentrations in Proterozoic oceans were low enough that Mo and N could have co-limited marine primary production (Anbar and Knoll 2002). The Mo concentration in seawater was lower in the Proterozoic, but it is unclear how this influenced marine productivity (Scott et al. 2008; Dahl et al. 2011; Reinhard et al. 2013a). Phylogenetic studies suggest that the Nif proteins necessary for N_2 fixation were not present in the last universal common ancestor (LUCA). Molecular clock estimates suggest a Proterozoic origin, some 2,200–1,500 Myr ago (Raymond et al. 2003; Boyd et al. 2011; David and Alm 2011), although a recent estimate suggests nitrogen fixing cyanobacteria diversified only 850–635 Myr ago (Sánchez-Baracaldo et al. 2014). In contrast, the Moco enzymes are distributed widely amongst extant organisms in the tree of life and could have been present in LUCA (Schoepp-Cothenet et al. 2012). The Mo availability and Mo requirements of early life continue as subjects of scrutiny.

FRACTIONATION FACTORS

Molybdenum isotope fractionation during both abiotic and biotic chemical reactions has been studied in controlled laboratory experiments, in natural systems, and in theoretical *ab initio* calculations. Key conclusions from these studies are reviewed below.

The Mo isotope fractionation observed to date is mass-dependent. Mass-dependent stable isotope fractionation is fundamentally a quantum chemical phenomenon arising from differences in the zero-point energies (ZPEs) between chemical bonds that are identical except for isotopic substitution (Bigeleisen 1947; Urey 1947). The mass dependence of bond strengths leads to differences in reaction rate constants, which give rise to kinetic isotope effects when reactions are unidirectional or incomplete. It also leads to mass dependence of equilibrium constants, so that an isotope offset exists between the reactant and product even for a system that has had infinite time to react (e.g., White 2015).

Adsorption to Mn oxides

The largest Mo isotope fractionation in nature occurs during Mo adsorption onto Mn oxides in oxic seawater. This process has been studied in controlled laboratory experiments, which show that lighter Mo isotopes are preferentially adsorbed onto the mineral surface. Experiments with poorly crystalline potassium birnessite ($\sim K_{0.5}Mn^{3+}Mn^{4+}O_4 \cdot 1.5\,H_2O$) in synthetic seawater yield a fractionation factor $\Delta^{98}Mo_{solution-MnOx} = 2.7 \pm 0.1‰$ at 25 °C (or $\alpha = 1.0027$; $\Delta \sim (\alpha - 1) \times 1000$) (Barling and Anbar 2004; Wasylenki et al. 2008). This finding is in excellent agreement with the isotopic difference between Mo in seawater and natural ferromanganese sediments (Barling et al. 2001; Siebert et al. 2003; Arnold et al. 2004). This fractionation is only weakly dependent on temperature and ionic strength (Wasylenki et al. 2008). It follows the behavior of closed-system equilibrium isotope exchange rather than an open-system with irreversible Rayleigh distillation (Fig. 4), suggesting that the mechanism is a reversible equilibrium isotope effect (Barling and Anbar 2004).

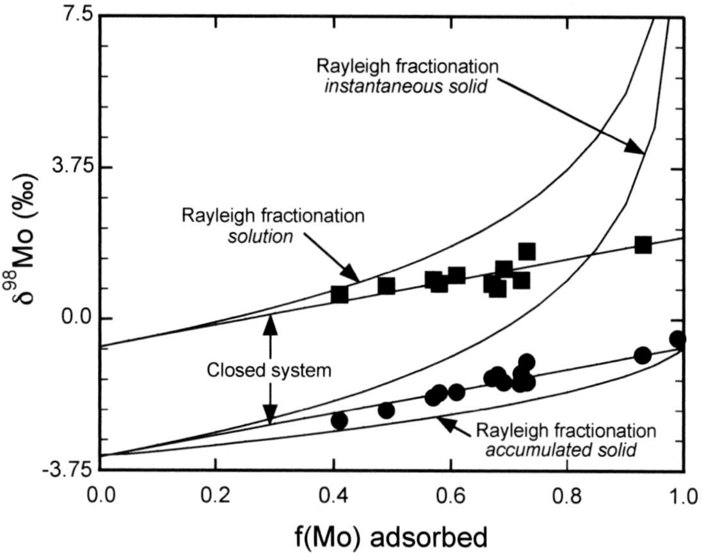

Figure 4. Mo isotope fractionation between Mo-bearing solutions and synthetic Mn oxides (-MnO_2), measured over 2–96 hours at pH = 6.5–8.5. Residual Mo in solution (■) was measured for all experiments. Mo adsorbed to oxide particle surfaces (●) was either measured or inferred from mass balance. Dissolved Mo was systematically heavier than adsorbed Mo with a fractionation factor of 1.0027 ± 0.0008. The data are not consistent with an irreversible Rayleigh-type distillation process, but instead point to closed system equilibrium in which Mo isotopes exchange continuously between Mn oxide surfaces and solution (i.e., a reversible process). Modified from Barling and Anbar (2004).

Ironically, this substantial isotope fractionation appears to be decoupled from the versatile redox chemistry of Mo, and instead results from the change in Mo coordination geometry between MoO_4^{2-} in oxic seawater and Mo adsorbed onto the mineral (Siebert et al. 2003; Wasylenki et al. 2011; Kashiwabara et al. 2011). Whereas MoO_4^{2-} is tetrahedrally coordinated, EXAFS studies reveal that Mo on the mineral surface is present as octahedrally coordinated polymolybdate species (e.g., $Mo_6O_{19}^{2-}$). *Ab initio* calculations show that Mo isotope fractionation between MoO_4^{2-} and polymolybdates in solution produces the observed fractionation factor across a range of temperatures (Wasylenki et al. 2011). Mo may also exist in solution and on surfaces in other octahedrally coordinated compounds, such as $Mo(OH)_6$ and $MoO_3(H_2O)_3$, but these species do not reproduce the observed isotope fractionation (Liu 2008; Oyerinde et al. 2008; Wasylenki et al. 2008).

The predicted concentration of polynuclear Mo species in seawater is $< 10^{-41}$ M, corresponding to < 8000 molecules in the entire ocean. Hence, the mechanism of Mo isotope fractionation on Mn oxide surfaces highlights the unique chemistry possible on mineral surfaces. Most likely, protonated surfaces attract negatively charged MoO_4^{2-} to the mineral surface. Diprotonated molybdic acid at the surface could lead to the formation of polymolybdates (Wasylenki et al. 2011).

Adsorption to Fe oxides and oxyhydroxides

A range of fractionation factors occur during Mo adsorption onto magnetite, ferrihydrite, goethite, and hematite minerals, with lighter Mo isotopes preferentially removed from solution (Goldberg et al. 2009). The isotopic difference between the solid (A) and dissolved (B) phases increases at higher pH, and also varies with mineralogy, increasing in the order magnetite

($\Delta^{98}Mo = 0.83 \pm 0.60‰$) < ferrihydrite ($\Delta^{98}Mo = 1.11 \pm 0.15‰$) < goethite ($\Delta^{98}Mo = 1.40 \pm 0.48‰$) < hematite ($\Delta^{98}Mo = 2.19 \pm 0.54‰$) at 25 °C. The observed isotope behavior is consistent with both adsorption onto the mineral surface and adsorption of different Mo species/structures from solution. For example, both molybdate and an octrahedrally coordinated Mo compound may adsorb onto the mineral with decreasing molybdate affinity for the minerals in the order listed above. The Mo speciation in the Fe-oxyhydroxide minerals has not been directly measured.

Sulfidic species

Molybdate reacts with hydrogen sulfide in anoxic aqueous solutions to form thiomolybdates following the reaction scheme:

$$MoO_4^{2-} \rightarrow MoO_3S^{2-} \rightarrow MoO_2S_2^{2-} \rightarrow MoOS_3^{2-} \rightarrow MoS_4^{2-}$$
$$(\text{mono-}) \rightarrow (\text{di-}) \rightarrow (\text{tri-}) \rightarrow (\text{tetra-}) \rightarrow \text{thiomolybdate}$$

Each step involves a ligand exchange with S donated from H_2S, and O inserted into H_2O. There is a geochemical switchpoint at $[H_2S]_{aq} = 11\ \mu M$, above which Mo exists primarily as tetrathiomolybdate (MoS_4^{2-}) (Erickson and Helz 2000). The intermediate oxythiomolybdates are only minor species in solution. For example, MoS_4^{2-} should account for up to 83% of the total dissolved Mo pool in the deep Black Sea, with $MoOS_3^{2-}$ being the second most abundant species (Nägler et al. 2011). The more S-rich oxythiomolybdate species are considered particle-reactive and so will be removed from solution.

Ab initio calculations indicate that there is a large isotope fractionation associated with each step in this reaction scheme (Tossell 2005). At equilibrium, the isotopic differences calculated for the (MoO_4^{2-}–$MoO_2S_2^{2-}$) pair and the (MoO_4^{2-}–MoS_4^{2-}) pair at 25 °C are −2.4‰ and −5.4‰, respectively (recalculated to $\delta^{98}Mo$; Tossell 2005; Nägler et al. 2011). By interpolation, the four isotope fractionation factors are $\Delta^{98}Mo_{0,1} = \Delta^{98}Mo_{1,2} = 1.20‰$ and $\Delta^{98}Mo_{2,3} = \Delta^{98}Mo_{3,4} = 1.50‰$, where the subscripts (x,y) represent the number of S atoms in the reactant (x) and product (y) species. The magnitude of fractionation is higher in cooler waters, e.g. $\Delta^{98}Mo_{0,1} = \Delta^{98}Mo_{1,2} = 1.40‰$ and $\Delta^{98}Mo_{2,3} = \Delta^{98}Mo_{3,4} = 1.75‰$ in the deep Black Sea (9 °C).

Although the thiomolybdate species have not been measured separately, observations from the Black Sea, Lake Cadagno, and Kyllaren Fjord show that the sulfidic waters are ~0.5‰ heavier than the source waters (Dahl et al. 2010a; Nägler et al. 2011; Noordmann et al. 2015). The muted fractionation relative to that predicted from *ab initio* calculations can be reconciled if multiple oxythiomolybdate species are particle-reactive and scavenged to the sediments (Dahl et al. 2010a; Nägler et al. 2011). Indeed, controlled precipitation experiments with FeS_2 show that both $MoOS_3^{2-}$ and MoS_4^{2-} are particle-reactive (Vorlicek et al. 2004). Although the fractionation factors between consecutive oxythiomolybdate species in solution are large, there is little or no isotope offset expressed between sediments and the Mo source (e.g., seawater) because Mo is quantitatively scavenged from the deep waters in these restricted euxinic basins (Neubert et al. 2008; Dahl et al. 2010a; Nägler et al. 2011; Noordmann et al. 2015).

Biological processes

Molybdenum assimilation in the nitrogen-fixing soil bacterium, *A. Vinelandii*, is associated with the preferential incorporation of lighter Mo isotopes, with a fractionation of $\Delta^{98}Mo = -0.45‰$ (Liermann et al. 2005; Wasylenki et al. 2007). The uptake pathway involves Mo chelation by high-affinity metal-binding ligands, such as the cathecolate "molybdophore" *azotochelin*, where Mo sits in an octahedral coordination geometry (Bellenger et al. 2008). There are several possible fractionating steps, including Mo release from the chelate, conversion to tetrahedrally coordinated MoO_4^{2-}, and uptake in the periplasmic modA transporter protein.

The latter is common among bacteria and archaea. Isotope fractionation could result from: 1) simple kinetic effects associated with irreversible Mo transport; 2) coordination changes during incomplete uptake or release from the chelating ligand and/or the Mo transporter protein; or 3) sorption of Mo onto the cell surface (Liermann et al. 2005; Wasylenki et al. 2007). Molybdenum adsorption onto organic matter of algal origin may cause Mo isotope fractionation with a similar isotope fractionation (–0.3‰) in productive surface waters (Kowalski et al. 2013).

However, Mo isotope fractionation during uptake may not be the only biological story. Studies of the filamentous heterocystous cyanobacterium *Anabaena variabilis* also show isotope fractionation between cells and media (Zerkle et al. 2011). *A. variabilis* is a freshwater species with Mo-dependent enzymes capable of both N_2-fixation and nitrate reduction. Heterocystous cyanobacteria are relatively rare in the modern oceans. However, several lines of evidence point to shared biochemical pathways for Mo uptake and utilization in marine and freshwater cyanobacteria (Zerkle et al. 2011). The isotope fractionation depended on the cell function. During growth on nitrate, *A. variabilis* consistently produced $\Delta^{98}Mo_{cells-media}$ of -0.3 ± 0.1‰. When fixing N_2, *A. variabilis* produced $\Delta^{98}Mo_{cells-media}$ of -0.9 ± 0.2‰ during exponential growth and -0.5 ± 0.1‰ during the stationary phase (very slow metabolic/growth rates). This variability demonstrates that Mo isotope fractionation can be more complex than a simple kinetic effect during Mo uptake because the same uptake system was likely involved in all experiments.

To explain these observations, Zerkle et al. (2011) hypothesized a reaction network model that assumes no isotope fractionation during Mo transport into and out of the cell, and equilibrium isotope fractionation between tetrahedrally bound MoO_4^{2-} in storage proteins and octahedrally bound Mo in the enzymes, applying a fractionation factor $\alpha^{98/95}=0.9982$ derived from *ab initio* calculations. They infer that the isotope fractionation is influenced by the relative proportion of Mo bound to storage proteins vs. Mo bound to enzymes. This model indicates that the largest isotope fractionation was observed during N_2 fixation because at conditions of high Mo demand, less Mo is bound to storage proteins (Zerkle et al. 2011).

High-temperature melt systems

Limited data are available for fractionation factors between mineral-melt pairs and silicate–metal liquid pairs in high-temperature systems. Voegelin et al. (2014) estimated biotite–melt and hornblende–melt fractionation factors at ~700 °C using Mo isotope data from volcanic dacite (representing quenched melt) and single mineral separates. In the two dacite samples they examined, biotite and hornblende had lower δ^{98}Mo than the host rock, with the largest expression of isotope fractionation being 0.4‰ and 0.6‰, respectively, in the sample with the lower abundance of these minerals. Hence, these are minimum fractionation factors for biotite–melt and hornblende–melt pairs, respectively.

Fractionation of Mo isotopes during metal–liquid segregation has also been investigated experimentally at 1400 °C and 1600 °C using a centrifuging piston cylinder, with the goal of exploring the use of Mo isotopes for inferring the temperature of planetary core formation (Hin et al. 2013). These experiments suggest that the fractionation factor between metal and silicate liquids is insensitive to oxygen fugacity at the conditions expected for core formation, as well as silicate melt composition and the C and Sn content of metallic melts. An equilibrium Mo isotope fractionation factor of 0.19 ± 0.03‰ and 0.12 ± 0.02‰ (95% confidence interval), favoring lighter isotopes in the metallic melt, was determined for 1400 °C and 1600 °C, respectively. From these measurements, Hin et al. (2013) inferred the temperature dependence of Δ^{98}Mo to be $\Delta^{98}Mo_{metal-silicate}=-4.70(\pm0.59)\times10^5/T^2$ (2σ). Hence, resolvable Mo isotope fractionation between silicate and metallic liquids is expected to occur up to 2500 °C (>0.06‰).

MOLYBDENUM ISOTOPES IN MAJOR RESERVOIRS

Meteorites

Most iron meteorites and ordinary, enstatite, and carbonaceous chondrites have a narrow range of δ^{98}Mo (average = 0.09 ± 0.02‰; 95% confidence interval, *n* = 12) (Fig. 5; Burkhardt et al. 2014). Higher δ^{98}Mo for some iron meteorites and carbonaceous chondrites may reflect evaporative loss of isotopically light Mo, although isotopic heterogeneity in the region of carbonaceous chondrite formation is also a possibility. Achondrites typically have higher δ^{98}Mo (up to ~ 1.2‰) than chondrites because of the preferential removal of lighter Mo isotopes to metallic liquids during planetary differentiation (Burkhardt et al. 2014), as confirmed by experiments on silicate-metal isotopic partitioning (Hin et al. 2013). The temperature at which silicate and metal phases segregated during planetary differentiation can be estimated using the achondrite δ^{98}Mo and the metal-silicate equilibrium fractionation factor assuming quantitative metal segregation in the core (e.g., 1800 ± 200 °C for the moon). However, some achondrites have δ^{98}Mo that is higher than modeled for planetary core formation. This high δ^{98}Mo may reflect later processes such as high-temperature metamorphism or terrestrial weathering of fallen meteorites on Earth's surface (Burkhardt et al. 2014).

High-precision Mo isotope measurements in meteorites have revealed mass-independent variations in isotope composition arising from nucleosynthetic processes. Heavy elements such as Mo were synthesized in red giant stars (*s*-process) and supernovae (*r*-process and *p*-process) and so bulk meteorites exhibit small, but resolvable, mass-independent nucleosynthetic isotope anomalies in many elements, including Mo, that indicate presolar dust was not isotopically homogenized by high temperatures and mixing during solar system formation (Dauphas et al. 2002a,b, 2004, Yin et al. 2002; Chen et al. 2004; Burkhardt et al. 2011, 2012). With respect to

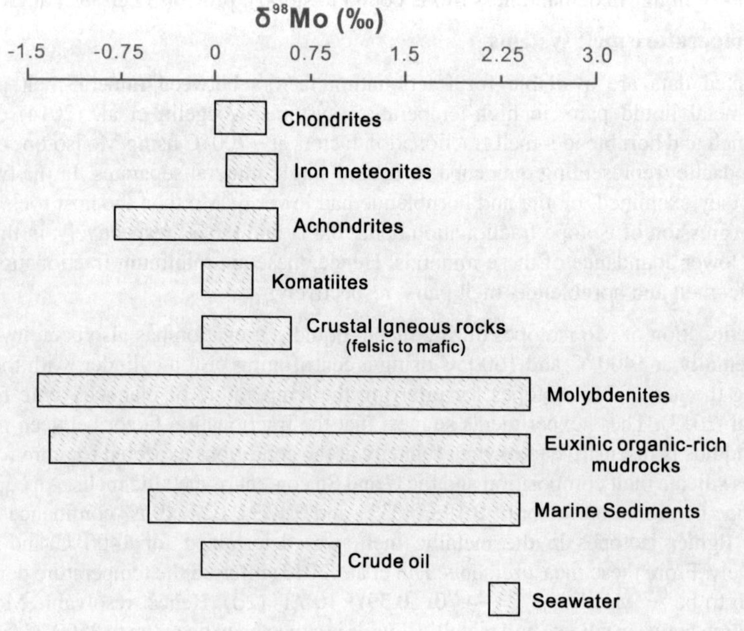

Figure 5. Molybdenum isotope composition of meteorites, the upper mantle (represented by komatiites), various crustal rocks, marine sediments, crude oil, and seawater. See text for sources of data.

tracing isotopic heterogeneity within the early solar system and inferring the source of solar nebula material, the Mo isotope system is a valuable tool because four of the Mo isotopes are produced by only one nucleosynthetic process: ^{92}Mo and ^{94}Mo from the *p*-process; ^{96}Mo from the *s*-process; and ^{100}Mo from the *r*-process (Arlandini et al. 1999).

Early studies demonstrated Mo isotope heterogeneity in solar system materials. Dauphas et al. (2002a,b) reported isotopic evidence from iron meteorites, mesosiderites, pallasites, and chondrites for *s*-process depletion and/or enrichment in *r*- and *p*-process nuclides relative to terrestrial samples. Carbonaceous chondrites were found to have decoupled *p*- and *r*- process anomalies, even though both processes are associated with supernovae, implying that the feeding zone(s) of carbonaceous chondrites contained material from multiple supernova sources that had not been isotopically homogenized (Yin et al. 2002; Chen et al. 2004).

Although one early study did not find nucleosynthetic anomalies in either primitive or differentiated meteorites (Becker and Walker 2003), likely because isotope measurements were being done at the edge of analytical capabilities at the time, recent analyses have confirmed these findings (Burkhardt et al. 2011, 2012). Notable exceptions include angrites, IAB–IIICD irons, and Martian meteorites, which have terrestrial isotopic compositions. Most other bulk meteorites exhibit depletions in Mo produced by the *s*-process. Carbonaceous chondrites such as Murchison seem to have multiple presolar components of variable isotopic composition, including calcium–aluminum-rich inclusions predominantly enriched in *r*-process Mo and SiC grains enriched in *s*-process Mo (Dauphas et al. 2002b; Burkhardt et al. 2011, 2012). By contrast, the Earth is enriched in *s*-process Mo, implying that Earth accreted from material of different isotopic composition compared with the known meteorite classes (Burkhardt et al. 2011).

The Mo isotope anomalies in bulk meteorites for each meteorite class are well-correlated with Ru isotope anomalies as predicted by nucleosynthesis theory, thus confirming that the observed anomalies resulted from variations in *s*-process contributions from low-mass AGB stars (Dauphas et al. 2004; Burkhardt et al. 2011). The magnitude of nucleosynthetic anomalies is generally greater in meteorites that are older and derived from smaller parent bodies, suggesting progressive isotopic homogenization of the solar nebula over time. Because carbonaceous chondrites have even larger nucleosynthetic Mo isotope anomalies than expected given their old age, the material that formed these primitive meteorites may have originated from further out in the solar system (where isotopic homogenization proceeded more slowly at lower temperatures) compared with other meteorites (Burkhardt et al. 2011).

The mantle and crust

The average δ^{98}Mo of the bulk silicate Earth (BSE; crust+mantle; the mantle dominates the mass balance) is estimated to be 0.04 ± 0.12‰ (2σ) using four sets of komatiite samples from widely separated localities (Greber et al. 2015a). Komatiites provide a good estimate of the mantle δ^{98}Mo because the high degree of partial mantle melting necessary to form komatiitic melts results in essentially quantitative melting of sulfide minerals in the mantle source, and thus complete transfer of Mo and its isotope composition from the mantle source to melts. The excellent agreement between the δ^{98}Mo of the BSE and chondritic meteorites indicates that full isotopic equilibrium was attained between the Earth's core and mantle at high temperatures (>2500 °C) during the moon-forming impact (Greber et al. 2015a). At such high temperatures, Mo isotope fractionation between co-existing metal and silicate phases is minimal (Hin et al. 2013).

In contrast to the isotopic homogeneity of most meteoritic and mantle materials, pronounced variability exists in the δ^{98}Mo of the crust. Indeed, the entire range of δ^{98}Mo observed in solid Earth materials is represented by the rocks and minerals of Earth's crust. Significant efforts have thus been devoted to explaining this isotopic variability.

Data from subduction zones reveal that Mo isotope fractionation accompanies crustal formation. In the Mariana island arc, lavas have δ^{98}Mo up to 0.3‰ higher than the average mantle/BSE value, suggesting that continental crust has slightly higher δ^{98}Mo than the mantle (Freymuth et al. 2015; Greber et al. 2015a). The source of the isotopically heavy Mo may be fluids released during dehydration of the subducting slab. In the Aegean continental arc (Kos Island, Greece), fractional crystallization is suggested to have increased the δ^{98}Mo of magmas as they evolved to more silica-rich compositions (Voegelin et al. 2014). The δ^{98}Mo of biotite and hornblende mineral separates suggests minimum melt–crystal fractionation factors of 0.4‰ and 0.6‰, respectively, with lighter isotopes preferentially incorporated into the fractionating crystals. Hence, fractional crystallization may explain the higher δ^{98}Mo of dacites (0.6‰) compared with basalts (0.3‰) at Kos Island. By contrast, negligible Mo isotope fractionation was observed in a suite of basalts to rhyolites in a mid-ocean ridge setting (Hekla volcano, Iceland). At the Icelandic locality, all samples yield an average δ^{98}Mo of 0.10 ± 0.05‰ that is indistinguishable from the mantle (Yang et al. 2015).

These observations indicate that the types of minerals crystallizing from the magma and their associated liquid-crystal fractionation factors exert some control on Mo isotope fractionation during magmatic differentiation. Amphibole and biotite did not crystallize from the largely anhydrous Hekla magmas, thus possibly explaining the lack of Mo isotope fractionation during magmatic differentiation in that mid-ocean ridge setting (Yang et al. 2015). Hence, the tectonic environment (e.g., subduction zone versus mid-ocean ridge) may influence high temperature Mo isotope fractionation via its effect on magmatic chemistry.

Least-altered mid-ocean ridge basalts from near the Mariana arc have δ^{98}Mo similar to the mantle, suggesting that decompression partial melting in the upper mantle is not accompanied by appreciable Mo isotope fractionation (Freymuth et al. 2015). The lack of Mo isotope fractionation in anhydrous systems may thus allow Mo isotopes to serve as a tracer of parent magma composition and possibly depleted versus enriched mantle sources (e.g., from analysis of ocean island basalts; Freymuth et al. 2015; Yang et al. 2015).

Crustal sulfide minerals and organic-rich mudrocks are most likely the major host phases of Mo in Earth's crust and also hold the distinction of having the widest variability in δ^{98}Mo. Significant efforts have been devoted to characterizing the δ^{98}Mo of crustal sulfide minerals, particularly molybdenite, because of their relevance for studies on ore mineralization. Rayleigh distillation, fluid boiling, and redox reactions are thought to be responsible for the wide variation in the δ^{98}Mo of molybdenites (−1.4‰ to +2.5‰; Hannah et al. 2007; Mathur et al. 2010; Greber et al. 2011, 2014; Shafiei et al. 2015; Breillat et al. 2016). Organic-rich mudrocks are characterized by a wide range in δ^{98}Mo (from about −1.3‰ to +2.5‰) that is controlled primarily by local and global ocean redox conditions, as shown by recent papers that compiled Mo isotope data for these rocks (Dahl et al. 2010b; Duan et al. 2010; Wille et al. 2013; Chen et al. 2015; Kendall et al. 2015a; Partin et al. 2015).

The pronounced isotopic variability in crustal rocks makes it difficult to precisely constrain the average δ^{98}Mo of the upper continental crust. Voegelin et al. (2014) calculated an average δ^{98}Mo of ~0.3‰ based on the limited dataset of basalts and granites. A recent compilation of nearly 400 molybdenite samples yielded an average of ~0.3‰, but is associated with a large 2σ (1.04‰) (Breillat et al. 2016). Molybdenites crystallize from hydrothermal fluids that have isotopically heavier Mo than the silica-rich magmas from which they exsolved, and thus the average δ^{98}Mo of molybdenites likely represents a maximum value for the average crust (Greber et al. 2014). Igneous pyrites rather than molybdenites may be the most important Mo reservoir in the crust (Miller et al. 2011), but inadequate data are available to quantify their isotopic composition.

The isotopic distribution of Mo in marine sediments has implications for crustal and mantle cycling of Mo. Deep-ocean pelagic sediments deposited from oxygenated bottom waters are enriched in isotopically light Mo whereas continental margins generally have sediments with isotopically heavier Mo because of reducing conditions in regions of high primary productivity (upwelling) or basin restriction (see the next section on the oceans). Pelagic sediments are preferentially incorporated into subduction zones compared with continental margin sediments, resulting in an upper crust that is isotopically heavier compared with igneous rocks (Neubert et al. 2011; Freymuth et al. 2015).

The isotopically light Mo from subducted pelagic sediments may be returned to Earth's surface via seafloor hydrothermal systems (Neubert et al. 2011) or volcanism (Freymuth et al. 2015). High δ^{98}Mo in Mariana arc lavas may reflect Mo isotope fractionation during dehydration of the subducting slab (Freymuth et al. 2015). If so, this process would cause the subducted slab to have low δ^{98}Mo. Incorporation of subducted oceanic lithosphere into mantle plumes may return this isotopically light Mo to Earth's surface by intraplate volcanism. This hypothesis has yet to be tested rigorously through analysis of ocean island basalts.

The oceans

Global seawater has a uniform δ^{98}Mo of 2.34±0.10‰ (Barling et al. 2001; Siebert et al. 2003; Nakagawa et al. 2012). The uniformity of this value and its magnitude can be understood in terms of the ocean budget of Mo.

Mo is thought to have a comparatively straightforward ocean budget (Fig. 6), entering largely dissolved in river waters and leaving primarily in association with authigenic Fe–Mn oxides and anoxic sediments underlying oxic or anoxic waters, where hydrogen sulfide is present (Crusius et al. 1996; Morford and Emerson 1999; Scott et al. 2008; Scott and Lyons 2012; Reinhard et al. 2013a). The high concentration of Mo in the modern oceans is largely dictated by the high solubility of Mo phases and slow removal rate of MoO_4^{2-} in the presence of dissolved O_2. Essentially, Mo is readily transferred from crust to oceans during oxidative weathering but, because settings in which bottom water $O_2 < 5$ µM represent only ~0.3% of the modern seafloor, Mo is very slowly removed from the oceans.

Quantitatively, the oceanic input is entirely dominated by riverine supply with a small (~5%) contribution from low-temperature hydrothermal systems (Wheat et al. 2002; Miller et al. 2011; Reinhard et al. 2013a). Rivers discharge 3.1×10^{8} mol yr^{-1} to the oceans with an average dissolved concentration of 8.0 nmol kg^{-1} (Miller et al. 2011). Dust and aerosols are negligible fluxes (Morford and Emerson 1999). Anthropogenic Mo contributions may also be low but are not well constrained (Miller et al. 2011). From this, the oceanic residence time for Mo is calculated as ~440 kyr (Miller et al. 2011), which is ~40% lower than previous estimates (Morford and Emerson 1999; Scott et al. 2008). Nevertheless, this is still more than two orders of magnitude higher than the ocean mixing time of ~1.5 kyr (Sarmiento and Gruber 2006). Therefore, the average Mo atom circulates the oceans ~300 times before it comes to rest in sediments. Hence, the oceans are well-mixed with respect to Mo, resulting in a homogeneous elemental and isotopic distribution across almost all ocean basins (Morris 1975; Collier 1985; Nakagawa et al. 2012). The largest variations in the Mo concentration of oxygenated seawater are only ~5% on a salinity-normalized basis (Tuit 2003).

An unusual feature of the Mo isotope system is that seawater represents the isotopically heaviest Mo reservoir on Earth. This observation is readily explained by observations of modern marine sediments (see below), which indicate that any expression of Mo isotope fractionation between seawater and sediments always results in preferential removal of lighter Mo isotopes to sediments, thus driving seawater to higher δ^{98}Mo.

Ocean Inputs. Surface fluids display a linear relationship between Mo and SO_4^{2-} (R^2=0.69), implying that the predominant source of Mo is oxidative weathering of sulfide minerals and that Mo is transported in the form of the hexavalent oxyanion with geochemical behavior similar to that of SO_4^{2-} (Miller et al. 2011).

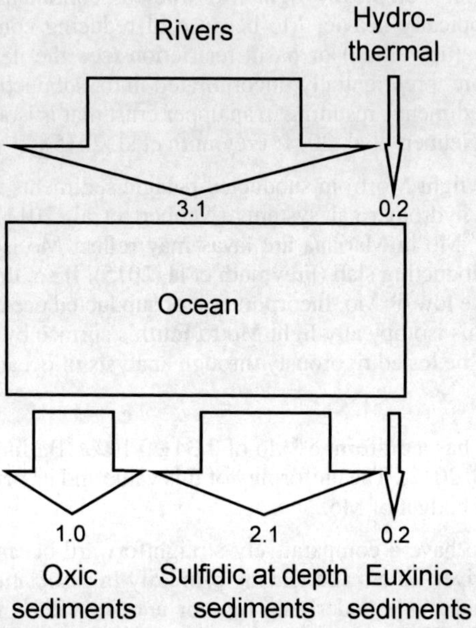

Figure 6. A schematic depiction of the Mo elemental budget in the modern oceans. Rough estimates of the Mo fluxes in 10^8 mol/yr are shown (see text for discussion). Sources of data: rivers: Miller et al. (2011); low-*T* hydrothermal fluids: Wheat et al. (2002); oxic, sulfidic at depth (i.e., dissolved sulfide is confined to sediment pore waters), and euxinic (i.e., dissolved sulfide is in the overlying water column) sediments: Scott et al. (2008) and Reinhard et al. (2013a) (scaled in proportion to the combined river and low-*T* hydrothermal fluid fluxes).

Rivers are characterized by a wide range in δ^{98}Mo values between −0.1‰ and +2.3‰ (Archer and Vance 2008; Pearce et al. 2010a; Neubert et al. 2011; Voegelin et al. 2012; Wang et al. 2015). Archer and Vance (2008) calculated an average riverine δ^{98}Mo of 0.7‰ based on analyses of waters representing ~22% of global riverine discharge. This implies that modern average riverine δ^{98}Mo is higher than the eroding upper continental crust and BSE (Archer and Vance 2008; Neubert et al. 2011).

Multiple mechanisms have been suggested to explain the isotopic fractionation between rivers and the eroding upper crust. During weathering, isotopically light Mo can be adsorbed to residual phases in soils that have experienced net Mo loss relative to the original bedrock (Archer and Vance 2008; Pearce et al. 2010a; Liermann et al. 2011; Siebert et al. 2015; Wang et al. 2015). Organic-rich soils may have a net gain in Mo with higher δ^{98}Mo compared to the original bedrock (Siebert et al. 2015). However, if all Mo in soils is ultimately released to rivers, then long-term Mo isotope fractionation between the eroding upper crust and rivers should not occur (Dahl et al. 2011; Neubert et al. 2011). Adsorption of isotopically light Mo to river particulates is probably of minor importance given that most Mo is dissolved in solution (Archer and Vance 2008; Wang et al. 2015). Desorption of isotopically light Mo from particulates may occur in some estuaries (Pearce et al. 2010a) whereas in others some isotopically light Mo may be retained in estuarine sediments, causing the release of isotopically heavy Mo to the oceans (Rahaman et al. 2014). Catchment lithology may exert significant control on the δ^{98}Mo of individual rivers via incongruent dissolution during weathering of easily oxidized phases like sulfide minerals and organic matter that commonly have higher δ^{98}Mo than crustal silicate minerals (Neubert et al. 2011; Voegelin et al. 2012).

Low-temperature hydrothermal systems provide a subordinate contribution of Mo to the oceans (Wheat et al. 2002; Miller et al. 2011; Reinhard et al. 2013a), but this flux and its isotopic composition are poorly constrained. The lone study for the flank of the Juan de Fuca ridge suggests that Mo is released to the oceans with a δ^{98}Mo of 0.8‰. However, it is not clear whether the isotopic signature truly reflects seawater–basalt reactions or was inherited from Mo diffusion into basaltic rocks from overlying sediments (McManus et al. 2002). High-temperature hydrothermal fluids are not a source of Mo to the oceans (Miller et al. 2011). A terrestrial hydrothermal spring from West Iceland has a δ^{98}Mo of −3.5‰ but the reason for this exceptionally light isotopic signature is not known (Pearce et al. 2010a).

Ocean Outputs. Significant Mo isotope fractionation occurs in the marine environment during removal to sediments (Fig. 7). To first order, the magnitude of Mo isotope fractionation between seawater and sediments correlates with the redox state of the local depositional environment. Well-oxygenated settings are characterized by the largest Mo isotope fractionations, whereas the most reducing conditions (associated with intense water column euxinia in restricted basins) may result in direct capture of seawater δ^{98}Mo by organic-rich sediments. Depositional environments of intermediate redox state have a wide range in δ^{98}Mo. In addition to redox conditions, other factors may affect the δ^{98}Mo of sediments, such as the operation of an Fe–Mn particulate shuttle (Herrmann et al. 2012; Scholz et al. 2013). Careful consideration of local depositional conditions is important for proper application of Mo isotopes in ancient sedimentary rocks as an ocean paleoredox proxy. The three major types of sedimentary sinks, and their isotope systematics, are summarized below.

The euxinic sink. The geochemical behavior of Mo changes sharply in H_2S-bearing systems, so much that it has been likened to a "geochemical switch" (Helz et al. 1996; Erickson and Helz 2000). This change is seen in the concentration depth profiles of these elements in the Black Sea and other restricted sulfidic basins (Fig. 8) (Emerson and Huested 1991; Neubert et al. 2008; Dahl et al. 2010a; Helz et al. 2011; Noordmann et al. 2015). For example, in the Black Sea, oxygenated surface waters give way to deeper anoxic waters at ~100 m, with $[H_2S]_{aq} > 11\ \mu M$ below ~400 m water depth. The total Mo concentration across this redox transition declines from ~40 nmol kg^{-1} at the surface to ~3 nmol kg^{-1} below the chemocline (Emerson and Huested 1991; Nägler et al. 2011).

In euxinic settings, removal of Mo from the water column leads to strong Mo enrichments in the underlying sediments relative to its average crustal abundance of ~1–2 ppm. The magnitude of this enrichment depends on Mo availability in the euxinic water column (Algeo and Lyons 2006). In relatively unrestricted ocean settings, Mo removal to euxinic sediments is readily balanced by Mo recharge to the deep waters, resulting in high Mo enrichments (often > 100 ppm) in sediments (Scott and Lyons 2012). By contrast, euxinic sediments in highly restricted basins with slow rates of deepwater renewal (including the Black Sea), euxinic sediments deposited rapidly (high sedimentation rates), and intermittently euxinic sediments typically have more modest Mo enrichments of ~25–100 ppm (Scott and Lyons 2012).

Particle scavenging in the euxinic water column is widely accepted as an important Mo flux to euxinic sediments. Once the thiomolybdate switch has been achieved, Mo is scavenged by forming bonds with metal-rich particles, organic compounds, and/or iron sulfides. The relative importance of these host phases is not well understood, although pyrite was recently ruled out as a major Mo carrier (Chappaz et al. 2014). Early studies of settling particles caught in sediment traps in the anoxic part of the water column suggested that most Mo removal occurs below the sediment–water interface (Francois 1988; Emerson and Huested 1991; Crusius et al. 1996). However, more recent studies indicate that Mo removal can also occur within euxinic water columns (Dahl et al. 2010a; Helz et al. 2011). The particle affinity of thiomolybdates is also used to explain the general linear relationship between Mo and total organic carbon (TOC) contents in sediments. This may suggest a direct connection between Mo and settling organic particles (e.g., Brumsack and Gieskes 1983; Algeo and Lyons 2006). However, correlation does not mean causation. The Mo–TOC relationships may be indirect, since both organic matter and Mo preferentially accumulate in basins with higher sulfide concentrations (Helz et al. 1996).

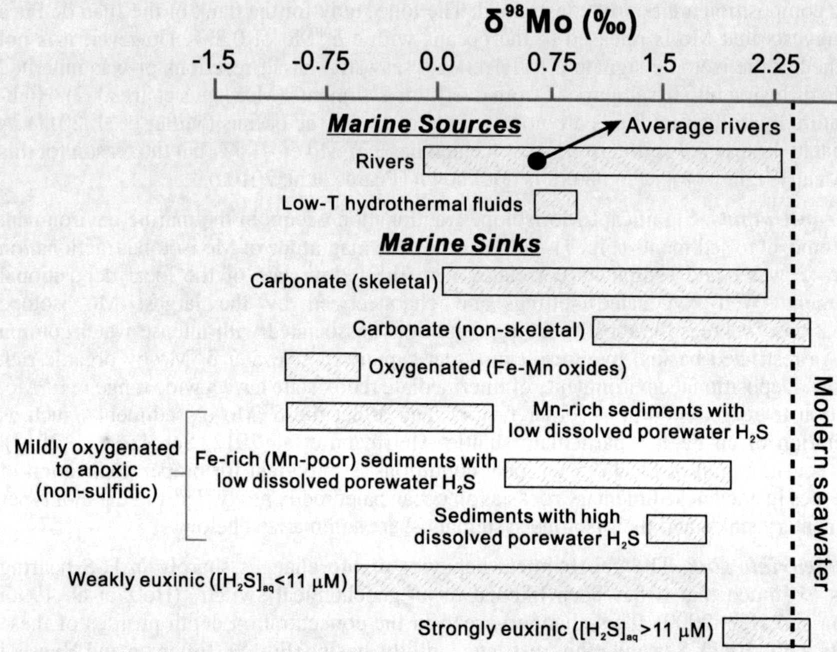

Figure 7. Molybdenum isotope composition of the sources and sinks of Mo in the modern oceans. The Mo isotope system is unusual compared with other isotope systems in that seawater is the isotopically heaviest reservoir, a consequence of the preferential removal of isotopically light Mo to sediments in all redox environments except in some strongly euxinic settings where Mo removal from bottom waters is quantitative. See text for sources of data.

Regardless of the mechanistic details, euxinic sedimentary settings account for removal of ~6–15% of the Mo entering the oceans via rivers each year, despite sulfidic waters only covering ~0.05–0.10% of the seafloor today (Scott et al. 2008; Reinhard et al. 2013a). Paleoredox investigations suggest the euxinic sink was much greater in the past (see section *Application to Ocean Paleoredox*).

Global seawater δ^{98}Mo is recorded by organic-rich sediments in the deep Black Sea and Kyllaren Fjord where bottom waters are strongly euxinic $[H_2S]_{aq} > 11\,\mu M$], MoO_4^{2-} (molybdate) is quantitatively converted to highly reactive $MoOS_3^{2-}$ (trithiomolybdate) and MoS_4^{2-} (tetrathiomolybdate), and Mo is quantitatively removed from sulfidic bottom waters (Erickson and Helz 2000; Barling et al. 2001; Arnold et al. 2004; Vorlicek et al. 2004; Neubert et al. 2008; Noordmann et al. 2015). The long seawater Mo residence time enables the δ^{98}Mo of strongly euxinic sediments in a partially restricted marine basin like the Black Sea to be a proxy for global seawater δ^{98}Mo (Barling et al. 2001; Arnold et al. 2004; Neubert et al. 2008; Noordmann et al. 2015).

Quantitative Mo removal may not be characteristic of all basins with strongly euxinic bottom waters because the rate of Mo removal to sediments depends on other factors such as pH and sulfur speciation as well as $[H_2S]_{aq}$ (Vorlicek et al. 2004; Helz et al. 2011). Non-quantitative removal of dissolved Mo will result in euxinic sediments with a lower δ^{98}Mo than global seawater (and enrichment of overlying euxinic bottom waters in isotopically heavy Mo; Nägler et al. 2011, Noordmann et al. 2015). The Mo isotope fractionation between dissolved MoS_4^{2-} or $MoOS_3^{2-}$ and authigenic solid Mo may be 0.5 ± 0.3‰ (Nägler et al. 2011), which is non-trivial and can lead to an overestimate of the global extent of ocean euxinia if it is incorrectly assumed that ancient euxinic organic-rich mudrocks directly recorded seawater δ^{98}Mo.

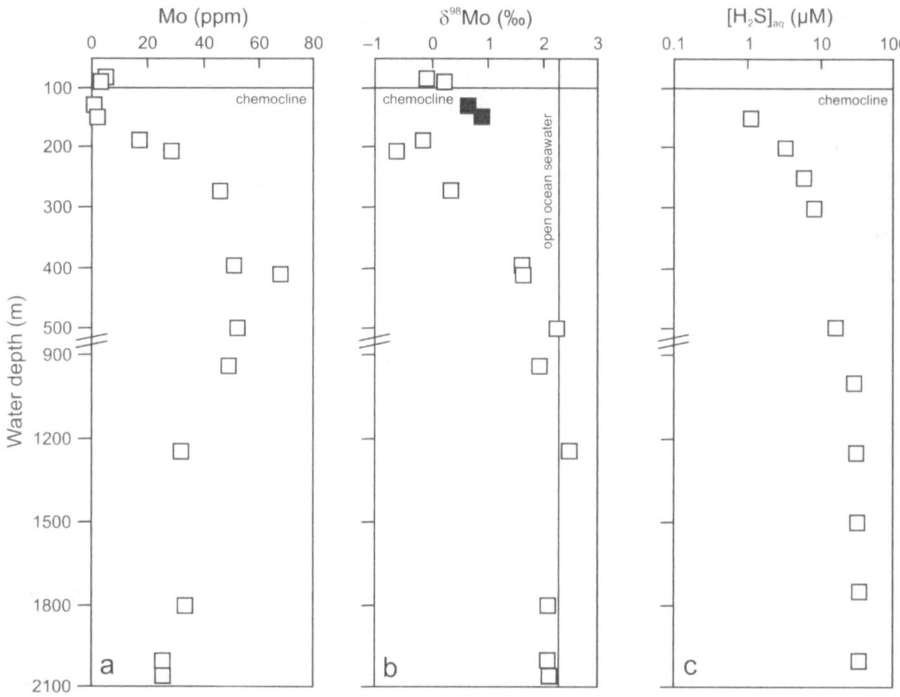

Figure 8. Mo concentration and isotopic composition in sediments at various water depths in the modern Black Sea, illustrating the change in geochemical behavior across the O_2–H_2S chemocline (note the break in scale). **a)** Mo concentrations. **b)** Mo isotope compositions. The two black squares denote samples whose Mo content is significantly influenced by detrital material. **c)** Dissolved hydrogen sulfide concentrations. At $[H_2S]_{aq}$ concentrations greater than 11 μM (below ~400 m water depth), the δ^{98}Mo of the sediments becomes similar to the open ocean seawater δ^{98}Mo. Modified from Neubert et al. (2008).

When bottom waters are intermittently euxinic or contain low $[H_2S]_{aq}$ (< 11 μM), a wide range of δ^{98}Mo (–0.6 to +1.8‰) is observed in the underlying sediments, likely reflecting the slow and incomplete conversion of molybdate to thiomolybdates (Arnold et al. 2004; Nägler et al. 2005; Neubert et al. 2008; Dahl et al. 2010a, Noordmann et al. 2015). Such conditions are characteristic of less restricted continental margin basins (e.g., Baltic Sea and Cariaco Basin) as well as shallower waters proximal to the chemocline along the margins of more restricted basins (e.g., water depths of ~ 100–400 m in the Black Sea). The sediment δ^{98}Mo is not well-correlated with $[H_2S]_{aq}$ at sulfide concentrations below the geochemical switchpoint of Mo. For intermittently euxinic basins, frequent periodic flushing by oxygenated seawater probably has a significant impact on sediment δ^{98}Mo via the formation of Fe–Mn (oxyhydr)oxides and their reductive dissolution in anoxic sediments (Scholz et al. 2013; Noordmann et al. 2015). An Fe–Mn shuttle is likely to be important for efficient transfer of Mo to sediments in less restricted redox-stratified basins and in oxygen minimum zones along upwelling continental margin systems where the redox cline occurs in the water column and deep water renewal times are fast enough to sustain the Fe–Mn shuttle (Algeo and Tribovillard 2009; Scholz et al. 2013).

However, some puzzling observations remain to be explained. For example, weakly euxinic sediments on the shallow Black Sea margin have significantly lighter δ^{98}Mo compared with the weakly euxinic sediments of the deep Cariaco Basin. In modern and ancient environments, distinguishing between the Mo isotope effects of incomplete thiomolybdate formation, the operation of an Fe–Mn shuttle, and periodic ventilation of anoxic basins is not a straightforward task. Careful comparisons with other geochemical redox proxies may narrow the range of possible mechanisms involved (e.g., Herrmann et al. 2012; Azrieli-Tal et al. 2014), but there is still no general approach for this. In such scenarios, the δ^{98}Mo of euxinic sediments is only a minimum estimate for global seawater δ^{98}Mo.

The oxic sink. Surprisingly in view of the stability of MoO_4^{2-} in solution, Mo enrichment to concentrations of 100s – 1000s of ppm, correlated with Mn content, is seen in ferromanganese oxide sediments, especially crusts, nodules, and some oxic pelagic sediments in the abyssal part of the oceans (Cronan and Tooms 1969; Bertine and Turekian 1973; Calvert and Price 1977; Cronan 1980; Calvert and Piper 1984; Shimmield and Price 1986). Such enrichment most likely reflects authigenic accumulation of Mo by adsorption to and/or co-precipitation with Mn oxide phases. This phenomenon is observed in the laboratory (Chan and Riley 1966; Barling and Anbar 2004; Wasylenki et al. 2008, 2011).

This removal process is associated with a large equilibrium isotope fractionation of ~3‰ occurring between Fe–Mn nodules or crusts (–0.7‰) and seawater (2.3‰), in excellent agreement with experimental observations of Mo adsorption to birnessite (Fig. 4; Barling et al. 2001; Siebert et al. 2003; Barling and Anbar 2004; Wasylenki et al. 2008; Poulson Brucker et al. 2009). A similar isotope fractionation was also inferred for hydrothermal Mn crusts (Ryukyu arc; Goto et al. 2015).

Because ferromanganese crusts and nodules accumulate very slowly and the Mo enrichments in widely disseminated pelagic sediments are small (Morford and Emerson 1999), the Mo concentration and isotopic composition of the oceans is much more sensitive to the extent of ocean euxinia than to oxygenated conditions. As Mn oxides are buried into organic-matter containing sediments, they experience reductive dissolution and liberate adsorbed Mo into the pore waters. In the absence of H_2S, Mo will diffuse into the overlying water column and thus the majority of Mo is not permanently buried, particularly in continental margin settings. In this scenario, Mn oxide-rich sediments can be considered failed sinks (e.g., Baja California; Shimmield and Price 1986). Even though deep-sea sediments also leak Mo, these sediments are so widespread that they still constitute an important Mo sink. A range of estimates suggests that some 30–50% of the riverine Mo supply is buried via the Mn oxide pathway in deep-sea sediments (Bertine and Turekian 1973; Morford and Emerson 1999; Scott et al. 2008; Reinhard et al. 2013a). Hence, the oxic sink is disproportionately small compared with the euxinic sink given that >80% and ≤0.1% of the seafloor is covered by well-oxygenated and euxinic waters, respectively (Reinhard et al. 2013a).

The intermediate sink ("sulfidic at depth" —SAD). In the last decade, it has become clear that a substantial portion of Mo removal occurs neither in fully oxic nor in fully euxinic systems. Investigations of Mo in marine sediments and pore waters indicate that Mo is also removed from solution under less intensely reducing conditions (Fig. 9). Authigenic Mo enrichments occur in sediments overlain by waters in which $O_2 < 10\,\mu M$ (Fig. 9c), where both Mn oxides and sulfate are reduced (Emerson and Huested 1991; Crusius et al. 1996; Dean et al. 1999; Zheng et al. 2000; Nameroff et al. 2002). The sedimentary Mo enrichments in these "sulfidic at depth" systems are smaller (typically <25 ppm) than in euxinic settings (Scott and Lyons 2012; Dahl et al. 2013b). Current estimates suggest that ~50–65% of oceanic Mo removal occurs in these environments (Morford and Emerson 1999; McManus et al. 2006; Reinhard et al. 2013a).

In settings with >10 µM of O_2 in the bottom waters, where Mn oxides form in the water column (Shaw et al. 1990), solid-phase Mo enrichment can develop in two redox zones within the sediment (Fig. 9b). First, transient authigenic Mo accumulation occurs at the upper limit of the manganiferous zone, where Mo is released to the pore fluids as Mn oxides undergo reductive dissolution. Secondly, a permanent Mo enrichment is found in the underlying

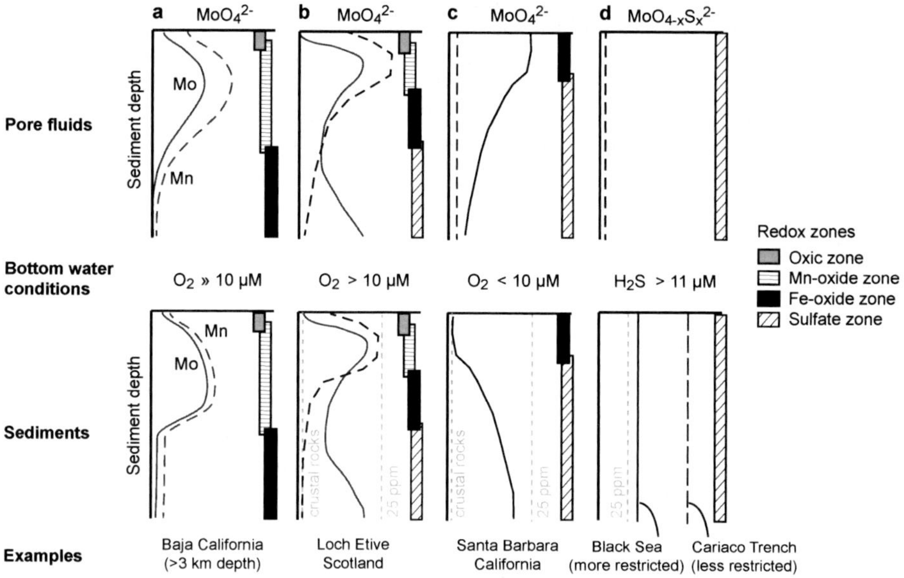

Figure 9. Molybdenum cycling in different redox settings. The relative concentrations of Mo and Mn increase from left to right in each profile and the dissolved Mo species in the bottom waters are shown along the top of each profile. **a)** Non-euxinic sediments with a manganous zone and no sulfidic zone. **b)** Non-euxinic sediments with both manganous and sulfidic zones. **c)** Non-euxinic sediments with a sulfidic zone and no manganous zone. **d)** Euxinic sediments where thiomolybdates are present in bottom waters. Examples of each category are from Baja California (Shimmield and Price 1986), Loch Etive, Scotland (Malcolm 1985), Santa Barbara, California basin (Poulson Brucker et al. 2009), Black Sea (Neubert et al. 2008), and the Cariaco Basin (Dean et al. 1999). The Mo concentration of average crustal rocks is shown. In non-euxinic environments, Mo concentrations in sediments are typically < 25 ppm (the crustal concentration and 25 ppm are shown as grey dashed lines). The heavy dashed line illustrates the higher Mo concentrations in Cariaco Basin euxinic sediments compared with the more restricted Black Sea. Modified from Scott and Lyons (2012).

sulfidic zone, where thiomolybdates can form. This two-fold maximum enrichment is exemplified in the sediments of the fjordic estuary Loch Etive in Western Scotland and in the Gulf of St. Lawrence (Malcolm 1985; Sundby et al. 2004). At many localities in the modern oceans, the Mn-reducing zone is located in the water column and/or the sulfidic capture zone is located at a large enough depth below the sediment-water interface that only small authigenic Mo enrichments (up to ~3 ppm) are expressed in the sediments because most Mo escaped back into the water column (Fig. 9a). This occurs for example in Boston Harbor, USA, Bay of Biscay and Thau lagoon in France, and in the Californian and Mexican border basins (Zheng et al. 2000; Chaillou et al. 2002; Elbaz-Poulichet et al. 2005; Poulson et al. 2006; Siebert et al. 2006; Morford et al. 2007; Poulson Brucker et al. 2009).

The isotopic composition of Mo in anoxic sediments deposited from mildly oxygenated to anoxic (but non-sulfidic) bottom waters depends on a number of factors, including the Fe and Mn content of the (oxyhydr)oxides, the crystallinity of Fe (oxyhydr)oxides, and the amount of dissolved H_2S in sediment pore waters (Poulson Brucker et al. 2009; Goldberg et al. 2009, 2012). Goldberg et al. (2012) identified three groups: 1) Mn-rich sediments with low dissolved porewater H_2S ($\delta^{98}Mo$ = −1.0‰ to +0.4‰); 2) Fe-rich sediments with low dissolved porewater H_2S ($\delta^{98}Mo$ = −0.5‰ to +2.0‰); and 3) sediments with high dissolved porewater H_2S ($\delta^{98}Mo = 1.6 \pm 0.2$‰). The low $\delta^{98}Mo$ of the first group simply reflects the large Mo isotope fractionation between seawater and Mn-rich oxides.

In the second group, the most reactive and poorly crystalline Fe (oxyhydr)oxides (e.g., ferrihydrite) are reduced in the Mn-reducing and upper part of the Fe reduction zones in sediments. The magnitude of isotope fractionation during Mo adsorption to poorly crystalline Fe (oxyhydr)oxides is smaller compared with Mn oxides (Goldberg et al. 2009), resulting in sediments with δ^{98}Mo between 0.5‰ and 2.0‰. By contrast, the lower part of the Fe reduction zone is characterized by sediments with lower δ^{98}Mo between –0.5‰ and + 1.0‰ because of a larger Mo isotope fractionation during Mo adsorption to more crystalline Fe (oxyhydr)oxides such as hematite and goethite (Goldberg et al. 2009).

The third group may be influenced by Mo isotope fractionation during formation of intermediate thiomolybdates, and is represented by open-ocean sediments in continental margin settings where bottom waters are O_2-deficient (< 10 μM) and pronounced microbial H_2S production occurs in sediment pore waters (Poulson et al. 2006; Siebert et al. 2006; Poulson Brucker et al. 2009; Goldberg et al. 2012). This group likely dominates the overall Mo isotope composition of the SAD sink because the higher H_2S concentrations in pore waters promote more efficient removal of Mo to sediments.

Lakes

The Mo isotopic composition of lakes has received less attention compared with marine systems. Molybdenum enrichment processes found in sulfidic marine environments were also recognized in euxinic lake settings (Dahl et al. 2010a; Helz et al. 2011). Smaller Mo enrichments were found in the seasonally dysoxic Castle Lake in California (Glass et al. 2013). Using sediment cores from lakes in Sweden and Russia, Malinovsky et al. (2007) showed that lower δ^{98}Mo in lake sediments is generally associated with deposition from oxygenated bottom waters whereas higher δ^{98}Mo occurs in sediments deposited from anoxic bottom waters. This behavior was also observed in two lakes in eastern Canada (Chappaz et al. 2012). Dahl et al. (2010a) examined in detail the Mo isotope budget of meromictic Lake Cadagno in Switzerland to better understand Mo isotope fractionation in redox-stratified water columns. The oxygenated shallow and sulfidic deep parts of the lake were found to have distinctive δ^{98}Mo (0.8‰ and 1.7‰, respectively) in part because of two different Mo sources to the lake (riverine inputs and groundwater at 0.8‰ and 1.4‰, respectively). The higher δ^{98}Mo of the sulfidic deep waters (1.7‰) compared with the groundwater source (1.4‰) suggests that removal of isotopically light Mo to sediments enriched the sulfidic deep waters in isotopically heavy Mo.

APPLICATION TO OCEAN PALEOREDOX

Observations from modern environments (e.g., Emerson and Huested 1991; Crusius et al. 1996; Helz et al. 1996; Morford and Emerson 1999; Erickson and Helz 2000; Zheng et al. 2000; Morford et al. 2005; Algeo and Lyons 2006; Algeo and Tribovillard 2009; Scott and Lyons 2012; Dahl et al. 2013b) have led to the use of Mo concentrations in sediments as a tracer of local ocean redox conditions and the degree of water mass restriction between a local sedimentary basin and the open ocean during deposition. The Mo concentration of euxinic organic-rich mudrocks (ORMs) deposited in unrestricted or weakly restricted sedimentary basins has been used to obtain a first-order estimate of the global seawater Mo concentration and thus the extent of atmosphere–ocean oxygenation (e.g., Scott et al. 2008; Reinhard et al. 2013a). For similar reasons, it was logical to also explore the use of Mo isotopes in ORMs as an ocean redox proxy (Barling et al. 2001; Siebert et al. 2003; Arnold et al. 2004). This approach has now been extended to chemical sedimentary rocks, notably carbonates, phosphorites, and iron formations (Voegelin et al. 2009; Wen et al. 2011; Baldwin et al. 2013; Planavsky et al. 2014). The discovery that both local and global ocean redox conditions control the δ^{98}Mo of marine sediments has led to ocean paleoredox studies being the most prominent application of the Mo stable isotope system.

Local depositional conditions

Building upon observations of modern environments (described in the section *Molybdenum Isotopes in Major Reservoirs*), the Mo isotope composition of sediments broadly scales with the degree of anoxia in the local depositional environment (Fig. 7). This means that the δ^{98}Mo of ancient ORMs may be used to infer local bottom water redox conditions at different locations in the world if seawater δ^{98}Mo is known. Such an approach is possible for the past ~60 Myr when seawater δ^{98}Mo was generally constant and close to the modern-day value of 2.3‰ as inferred from Pacific and Atlantic Fe–Mn crusts (at a temporal resolution of 1–3 Ma) assuming a constant isotopic offset of ~3‰ between these sedimentary materials and the contemporaneous open ocean (Siebert et al. 2003).

Given that the δ^{98}Mo of ORMs is influenced by both global and local ocean redox conditions, Mo isotopes should not be used alone to infer the redox state of local bottom waters when no constraint on seawater δ^{98}Mo is available. Hence, Mo isotope data for older ORMs can provide insight on local depositional conditions only in combination with independent proxies for local bottom water redox conditions, particularly Mo enrichments, Mo/U and Mo/Re ratios, and sedimentary Fe speciation (Crusius et al. 1996; Morford and Emerson 1999; Morford et al. 2005; Poulton and Canfield 2005, 2011; Tribovillard et al. 2006, 2012; Algeo and Tribovillard 2009; Scott and Lyons 2012).

The usefulness of Mo isotopes as a local redox proxy for Pleistocene-Holocene sediments can be illustrated by recent studies on the Black Sea and eastern Mediterranean Sea. As expected, older oxic-limnic sediments (Unit IIB, III) in the Black Sea record lighter δ^{98}Mo compared with more recent anoxic sediments (Unit I, IIA) (Nägler et al. 2005). Development of strongly euxinic bottom waters in the Bosporus Inlet region around 350–300 B.P. was inferred from an excursion to high δ^{98}Mo (similar to modern seawater) in sediments. Arnold et al. (2012) linked this increase in bottom water sulfide concentrations to shoaling of the chemocline (by more than 65 m) in response to water circulation and temperature changes brought on by the Little Ice Age. The δ^{98}Mo of the overlying sediments declines upsection, reflecting a transition to modern well-oxygenated conditions in the Bosporus Inlet region.

Sapropels from the eastern Mediterranean Sea exhibit more complicated stratigraphic trends in δ^{98}Mo. The youngest organic-rich sapropel (S1) has lighter δ^{98}Mo in its lower part compared with the overlying more oxygenated sediments (Reitz et al. 2007; Azrieli-Tal et al. 2014), a finding that is contrary to modern environments where more oxygenated sediments typically have lower δ^{98}Mo. Reitz et al. (2007) suggested that propagation of an oxidation front into the more reducing sapropel remobilized and transported Mo downwards in the sediment until Mo was co-precipitated with Mn oxides at the oxidation front. In contrast, Azrieli-Tal et al. (2014) used a combination of redox-sensitive metal enrichments and Fe isotope data to show that local bottom waters were euxinic during early sapropel deposition and less reducing during late sapropel deposition, and separated by a transient ventilation event associated with cold climatic conditions at ~8.2 ka. The lightest δ^{98}Mo (< –0.7‰) in the lower sapropel was suggested to reflect weakly euxinic conditions ($[H_2S]_{aq} < 11\ \mu M$) that caused a large Mo isotope fractionation between the sediments and overlying seawater (Azrieli-Tal et al. 2014), similar to that observed in the shallower part of the modern Black Sea (Neubert et al. 2008).

Scheiderich et al. (2010a) also used redox-sensitive metal concentrations and S isotope data from eight Pleistocene Mediterranean sapropels to conclude that euxinic bottom water conditions generally prevailed during sapropel deposition. The range in δ^{98}Mo (0.3–1.8‰) in the sapropels is consistent with deposition from weakly euxinic bottom waters, albeit with a smaller degree of seawater-sediment isotope fractionation compared with lower S1. Hemipelagic sediments beneath the sapropels have high δ^{98}Mo, in some cases exceeding modern seawater δ^{98}Mo, despite trace metal and S isotope evidence for oxygenated bottom water conditions.

These observations suggest that preferential removal of isotopically light Mo to the sapropels enriched pore fluids in isotopically heavy Mo. Downward diffusion of the pore fluids would enable transfer of isotopically heavy Mo to the underlying hemipelagic sediments.

Studies on the Paleocene–Eocene thermal maximum (~ 55.9 Ma) and Eocene Thermal Maximum 2 (~ 54.1 Ma) provide an example of using Mo isotopes and redox-sensitive metal enrichments to reconstruct the development of transient euxinic conditions along ocean margins in response to hyperthermal events (Dickson and Cohen 2012; Dickson et al. 2012). In both cases, the euxinia was fingerprinted by a stratigraphic excursion to higher Mo and Re enrichments and higher δ^{98}Mo in Arctic ocean sediments. The highest δ^{98}Mo (~ 2.0–2.1‰) approaches the modern seawater value, consistent with limited Mo isotope fractionation between seawater and sediments and thus the development of strongly euxinic bottom waters ($[H_2S]_{aq} > 11$ μM). Dickson et al. (2014) further showed that early Eocene anoxic sediments from two continental margin sites in the Tethys Ocean were deposited from non-euxinic or intermittently euxinic bottom waters (based on Fe speciation data) and had a highest δ^{98}Mo that was ~ 0.7‰ lower than the highest δ^{98}Mo observed from the Arctic Ocean euxinic sediments. This 0.7‰ offset is similar to that observed between modern anoxic continental margin sediments and global seawater (Poulson et al. 2006; Poulson Brucker et al. 2009).

The δ^{98}Mo of ORM deposited from euxinic waters (independently verified by trace metal and Fe speciation data) has also been used along with Mo/U ratios to fingerprint the operation of an Fe–Mn particulate shuttle. Specifically, low δ^{98}Mo (< 1‰) and high Mo/U ratios (≥3× the molar Mo/U seawater ratio) in the Late Pennsylvanian Hushpuckney Shale (Midcontinent Sea, USA) and late Ediacaran Doushantuo Formation (South China) raise the possibility that an Fe–Mn particulate shuttle delivered isotopically light Mo to sediments (Herrmann et al. 2012; Kendall et al. 2015a). These examples, along with the Mediterranean sapropels, demonstrate that both weakly euxinic conditions and operation of an Fe–Mn particulate shuttle can compromise the ability of euxinic ORM to record open ocean δ^{98}Mo.

Reconstructing the oceanic Mo isotope mass balance

Global ocean redox conditions can be inferred through mass balance modelling of the oceanic Mo isotope budget. Initial models used a simple isotope mass balance involving two oceanic Mo sinks (oxic and euxinic) (Arnold et al. 2004). Modern studies now typically use more complicated models that take into account both Mo burial fluxes and the isotopic composition of three sinks (oxic, sulfidic at depth, and euxinic; see Section *Molybdenum Isotopes in Major Reservoirs*) as well as the scaling of Mo burial fluxes to the size of the global seawater Mo reservoir (e.g., Dahl et al. 2011; Reinhard et al. 2013a; Chen et al. 2015). Rivers are typically assumed to be the only major source of Mo to the oceans in the Proterozoic and Phanerozoic, as they are today. This is a reasonable assumption for a world with an oxygenated atmosphere (i.e., following the Great Oxidation Event [GOE]) given that subaerial oxidative dissolution of crustal sulfide minerals is efficient even at low O_2 levels (to <0.001% and <0.026–0.046% of present levels in the case of pyrite and molybdenite, respectively; Reinhard et al. 2009, 2013b; Greber et al. 2015b).

From the perspective of the magnitude of Mo isotope fractionation in marine environments, two of the three oceanic Mo sinks are easy to define. The oxic sink (F_{OX}) is typically associated with Mo adsorption onto Mn oxides and Fe–Mn crusts beneath well-oxygenated bottom waters, which is represented by a Mo isotope fractionation factor of ~ 3‰. A euxinic sink (F_{EUX}) has often been used to denote environments where sediments are deposited from highly sulfidic bottom waters ($[H_2S]_{aq} > 11$ μM) and Mo removal from those bottom waters is nearly quantitative, thus enabling preservation of seawater δ^{98}Mo in the sediments. The third sink (F_{SAD}) has traditionally been used to represent all other environments of more intermediate redox character, which range from mildly oxygenated to weakly euxinic bottom waters (e.g., Kendall et al. 2009, 2011; Dahl et al. 2010b, 2011). The magnitude of Mo isotope fractionation in the environments represented by this third sink span the entire range between the oxic and strongly euxinic end-members. An average Mo isotope fractionation of ~0.7‰ is typically chosen to represent this sink because this is the common Mo isotope offset from overlying seawater observed in continental margin sediments where bottom waters are weakly oxygenated, dissolved O_2 penetrates <1 cm below the sediment-water interface, and dissolved sulfide is present in shallow sediment pore fluids (Poulson et al. 2006; Poulson Brucker et al. 2009).

One complication is the weakly euxinic sink (bottom water $[H_2S]_{aq} < 11\ \mu M$), which is characterized by a wide range of Mo isotope fractionations (up to 3‰ in the shallow Black Sea near the chemocline). For mass balance modelling that integrates Mo burial fluxes with the Mo isotope mass balance, it is problematic to assign weakly euxinic settings to the SAD sink because both weakly and strongly euxinic settings have Mo burial fluxes that are significantly higher than in non-euxinic settings (e.g., Scott et al. 2008; Reinhard et al. 2013a). Hence, a compromise is to assign a small Mo isotope fractionation of ~0.5‰ to the euxinic sink such that it represents both strongly and weakly euxinic conditions (e.g., Chen et al. 2015). This Mo isotope fractionation is observed in the deep weakly euxinic Cariaco Basin (Arnold et al. 2004), which may be a good analogue for ancient euxinic environments. In this modelling approach, the assumption that the SAD sink is dominated by the weakly oxygenated settings (Fig. 9c; where Mo isotope fractionation averages 0.7‰; Poulson et al. 2006; Poulson Brucker et al. 2009) is further justified because Mo burial in such settings is more efficient compared with mildly oxygenated settings where dissolved sulfide occurs farther below the sediment-water interface (Fig. 9b). Hence, the average δ^{98}Mo of the oxic, SAD, and euxinic sinks are −0.7‰, 1.6‰, and 1.8‰, respectively, for seawater δ^{98}Mo = 2.3‰.

The oceanic Mo isotope mass balance equation can thus be represented as:

$$\delta_{RIVER} = f_{OX}\delta_{OX} + f_{SAD}\delta_{SAD} + f_{EUX}\delta_{EUX}$$

where f = fraction of each sink flux relative to the total oceanic Mo burial flux ($f_{RIVER} = 1$), $f_{OX} + f_{SAD} + f_{EUX} = 1$, and $\delta = \delta^{98}$Mo. Modern budget estimates (f_{OX} = 30–50%, f_{SAD} = 50–65%, f_{EUX} = 6–15%) yield values for the average riverine input, δ_{RIVER} = 0.5–0.9‰, in good agreement with the observed value of ~0.7‰ (Morford and Emerson 1999; Archer and Vance 2008; Scott et al. 2008, Reinhard et al. 2013a). Each f term in the equation can be linked to that redox setting's average global Mo burial flux, which scales with the size of the global oceanic Mo reservoir. For each F term, this can be expressed as:

$$F = F_0 \times R / R_0$$

where R denotes the size of the global oceanic Mo reservoir, F is the burial flux (g m^{-2} yr^{-1}), and the subscript 0 denotes the modern value. Each f term in the Mo isotope mass balance equation can be replaced by the following expression that relates each sink flux to its areal fraction:

$$f = \left[\left(F_0 \times R / R_0\right) \times \left(A_{TOTAL} \times f_A\right)\right] / F_{RIVER}$$

where f_A = fraction of seafloor represented by the sink and A = total seafloor area covered by the three sinks. In this way, the global seawater δ^{98}Mo can be modelled as a function of the areal extent of each sink (Dahl et al. 2011; Reinhard et al. 2013a; Chen et al. 2015). A limitation of this model is that the average Mo burial flux for each sink is based on observations from continental margin settings (8% of the modern seafloor), where burial fluxes are higher compared with the abyssal seafloor. Hence, the rate at which the global seawater Mo reservoir is drawn down in response to an expansion of ocean anoxia onto the abyssal seafloor will be overestimated (e.g., $f_A > 8\%$). Addressing this weakness would require a more complicated modelling approach that scales burial fluxes from continental margin to abyssal seafloor (cf., Dahl et al. 2011; Reinhard et al. 2013a).

The mass balance model reveals that a combination of high Mo concentrations and high δ^{98}Mo in ancient euxinic ORM is best interpreted as evidence for a large oceanic Mo reservoir and widespread ocean oxygenation (Fig. 10). By contrast, low Mo concentrations and low δ^{98}Mo in euxinic ORM point to a significant extent of ocean euxinia.

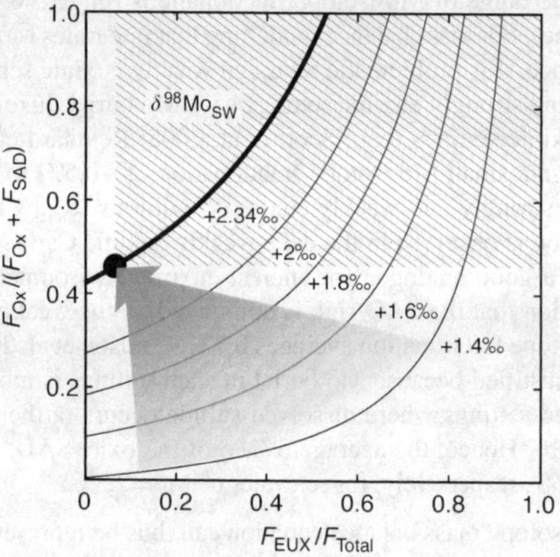

Figure 10. Relationship between the relative sizes of the oxic, sulfidic at depth, and euxinic sinks on the seawater Mo isotope composition, as derived from mass balance modelling. The black dot represents the modern Mo isotope budget. The shaded region encompassed by the arrow represents the overall direction that seawater δ^{98}Mo would take in response to increased deep ocean oxygenation. The hatched area represents mass balance solutions that are unrealistic because it would require that both the oxic and euxinic sinks expand at the expense of the "sulfidic at depth" sink. F = flux; Ox = oxic sink; SAD = sulfidic at depth sink; EUX = euxinic sink. Modified from Chen et al. (2015). Reproduced from Figure 2a of Chen et al. (2015) Nature Comm 6:7142 under a Creative Commons CC-BY license.

Inferring seawater δ^{98}Mo from sedimentary archives

ORMs. Application of Mo isotopes as a global ocean redox proxy depends on knowledge of ancient seawater δ^{98}Mo. The growing database of δ^{98}Mo from modern environments consistently shows that organic-rich sediments deposited from strongly euxinic bottom waters in semi-restricted basins are most likely to directly capture global seawater δ^{98}Mo (Arnold et al. 2004; Neubert et al. 2008; Noordmann et al. 2015). Hence, ORM is the predominant lithology for inferring ancient seawater δ^{98}Mo. Independent indicators are used to establish that ORM were deposited from euxinic bottom waters, especially Mo enrichments and extensive pyritization of biogeochemically highly reactive Fe (Fe-bearing mineral phases that can react with dissolved sulfide in the water column or in sediment pore waters during early diagenesis; e.g., Arnold et al. 2004; Neubert et al. 2008; Gordon et al. 2009; Pearce et al. 2010b; Dahl et al. 2010b; Scott and Lyons 2012). The latter is identified by a combination of high ratios of highly reactive Fe to total Fe (typically >0.38; indicating anoxic bottom waters) and high ratios of pyrite Fe to highly reactive Fe (>0.7; indicating dissolved sulfide in those anoxic waters) (Raiswell and Canfield 1998; Poulton and Raiswell 2002; Poulton and Canfield 2011).

Seawater δ^{98}Mo will be directly captured by ORM if bottom water sulfide concentrations were high enough (i.e., $[H_2S]_{aq} \gg 11\ \mu M$) to enable quantitative conversion of molybdate to highly particle-reactive Mo species, and if Mo removal from bottom waters was quantitative or nearly so. Assessing whether these conditions were met for ancient ORM is not always straightforward because local redox proxies such as Mo concentrations and Fe speciation cannot quantitatively constrain the dissolved sulfide concentration of euxinic bottom waters.

However, careful comparison of elemental and Mo isotope data can provide clues. Positively correlated stratigraphic variations in the δ^{98}Mo and Mo enrichments of euxinic ORM suggest that changes in seawater δ^{98}Mo are being captured because such a correlation is the expected response to changes in the global seawater Mo inventory and ocean redox conditions. By contrast, high Mo enrichments and low δ^{98}Mo (i.e., similar to igneous rocks) in ORM indicates weakly euxinic bottom waters during deposition. High Mo enrichments indicate a sizable oceanic Mo reservoir and thus a significant extent of ocean oxygenation whereas the low δ^{98}Mo portrays a conflicting viewpoint of widespread ocean anoxia. This apparent contradiction can be resolved by invoking a large Mo isotope fractionation between weakly euxinic bottom waters and sediments. A combination of high Mo enrichments and low δ^{98}Mo may also be explained by operation of an Fe–Mn particulate shuttle, particularly if high Mo/U ratios are observed in ORM (Algeo and Tribovillard 2009; Herrmann et al. 2012; Kendall et al. 2015a). In either scenario, another isotope redox proxy that is less sensitive to dissolved sulfide concentrations and the Fe–Mn particulate shuttle is needed to infer the extent of global ocean oxygenation, such as U isotopes (Asael et al. 2013; Kendall et al. 2015a).

Even if bottom waters are strongly euxinic, Mo isotope fractionation between the sediments and seawater will occur if Mo removal from bottom waters is not quantitative. In the deep Black Sea, near-quantitative removal of Mo from bottom waters is indicated by low Mo enrichments compared with TOC contents (average Mo/TOC ratio of 4.5 ppm/wt%) in the euxinic sediments (Algeo and Lyons 2006; Neubert et al. 2008; Scott and Lyons 2012). Higher Mo/TOC ratios in ORM are suggestive of non-quantitative Mo removal, which may be associated with a Mo isotope fractionation of up to ~0.5 ± 0.3‰ between dissolved and authigenic Mo in a strongly euxinic setting (Nägler et al. 2011). Hence, the δ^{98}Mo of euxinic ORM with high Mo/TOC ratios must be regarded as a minimum value for global seawater δ^{98}Mo.

Fe–Mn Crusts. Hydrogenous Fe–Mn crusts have been used to trace the evolution of seawater δ^{98}Mo over the past 60 Myr (Siebert et al. 2003). This approach takes advantage of the constant isotopic offset of ~3‰ that is observed between modern Mn oxides and seawater. The Mo isotope record of hydrogenous Fe–Mn crusts from the Atlantic and Pacific Oceans are homogeneous and similar to modern Mn oxides, suggesting that the global ocean redox conditions during the Cenozoic Era were generally similar to today. However, the poor temporal resolution of Fe–Mn crusts (1–3 Ma) means that short-term variations in global ocean redox conditions will not be well represented. In addition, the possibility of re-equilibration with younger seawater cannot easily be excluded. The use of Fe–Mn crusts to reconstruct seawater δ^{98}Mo is also limited to the recent geological past because subduction of oceanic lithosphere has destroyed the vast majority of this record.

Carbonates. Primary carbonate precipitates and phosphorites may also directly record seawater δ^{98}Mo in some cases (Voegelin et al. 2009; Wen et al. 2011; Romaniello et al. 2016). Molybdenum occurs at sub-crustal abundance in most carbonate rocks ($\ll$ 1 ppm), and carbonates probably constitute a negligible sink for marine Mo. In carbonate rocks, Mo may be bound to detrital silicate minerals, organic matter, sulfide minerals, and carbonate minerals. To avoid detrital material that may have a different Mo isotope composition from authigenic Mo, leaching of carbonate rocks can be done with dilute HCl, which primarily dissolves the carbonate fraction. Otherwise, total digestion techniques can be used and the effect of the detrital component on Mo concentrations and isotopic compositions can be evaluated using immobile elements such as Al or Ti (Voegelin et al. 2009, 2010).

In comparison with fine-grained siliciclastic sediments, little Mo isotope data are available for modern carbonate sediments (Voegelin et al. 2009; Romaniello et al. 2016). However, initial data are encouraging. Most modern skeletal organisms, including bivalves and gastropods, have low and strongly variable Mo contents (0.004–0.120 ppm) and isotope compositions (δ^{98}Mo = 0.07–2.19‰), suggesting a biological Mo isotope fractionation that preferentially incorporates lighter Mo isotopes into shells. Corals, however, display a nearly uniform Mo concentration (0.02–0.03 ppm) and a narrow range of δ^{98}Mo values (2.0–2.2‰) that are slightly lighter than modern seawater. This could mean that MoO_4^{2-} in oxic seawater is principally incorporated directly as an impurity in the crystal lattice, but there are currently no controlled laboratory experiments to confirm this chemical pathway. Nevertheless, corals are a potential archive of δ^{98}Mo in ancient seawater (Voegelin et al. 2009).

Mo isotope fractionation is also observed to be small in some non-skeletal carbonates, including ooids and in bulk carbonate sediments with high sulfide levels in pore waters (Voegelin et al. 2009; Romaniello et al. 2016). Bahamian ooid sands thought to contain pure non-skeletal calcite contain a narrow range of Mo concentrations (0.02–0.04 ppm) and δ^{98}Mo values (2.0–2.2‰) that are only slightly lower than modern seawater (2.3‰). Other ooids contain a detrital Mo component and display lower δ^{98}Mo values (Voegelin et al. 2009). Bulk carbonate sediments from shallow water settings in the Bahamas display low Mo concentrations (< 0.2 ppm) and δ^{98}Mo that is ~ 1‰ lower than seawater when pore water sulfide concentrations are low (i.e., $[H_2S]_{aq} < 20\,\mu M$; Romaniello et al. 2016). By contrast, high Mo concentrations (2–28 ppm) and seawater-like δ^{98}Mo are found in carbonate sediments containing high levels of pore water sulfide ($[H_2S]_{aq} = 20$–$300\,\mu M$). Hence, the ability of non-skeletal carbonates to record coeval seawater δ^{98}Mo may depend on redox conditions in a fashion similar to siliciclastic sediments (Romaniello et al. 2016).

Tracing atmosphere–ocean oxygenation using Mo isotopes

With these caveats in mind, we provide an overview of how the Mo isotope compositions of sedimentary rocks have been used to trace oxygenation of Earth's surface environment. The Mo isotope system is used in two distinct ways depending on atmospheric pO_2 levels. The first is to search for evidence of free O_2 in the Archean environment, with the goal of constraining the onset of oxygenic photosynthesis and the transition from an anoxic to an oxygenated atmosphere. The second is to constrain the global extent of oxygenated seafloor during various intervals in the Proterozoic and Phanerozoic, with the major goals being to infer the magnitude of oceanic anoxic events associated with major Phanerozoic mass extinctions, and to determine when Earth's oceans became predominantly oxygenated.

Part 1: Searching for free O_2 in the Archean surface environment

Molybdenum isotope data from Archean ORMs, carbonates, and iron formations play a prominent role in ongoing efforts to trace the dynamics of initial Earth surface oxygenation leading up to the GOE. In such studies, evidence is sought for Mo isotope fractionation in surface environments (e.g., rivers, oceans), which is manifested in the form of δ^{98}Mo values in sedimentary rocks that are higher or lower than the range observed in crustal igneous rocks. If such δ^{98}Mo values are found, an assessment is made on whether environmental O_2 is likely to explain them. These assessments take into account the range of Mo isotope variations and their correlation with other geochemical redox proxies. Most studies have focused on late Archean sedimentary rocks (2.7–2.5 Ga) deposited in the Hamersley Basin, Western Australia (Duan et al. 2010; Kurzweil et al. 2015a) and the Transvaal Basin and Griqualand West Basin, South Africa (Wille et al. 2007; Voegelin et al. 2010; Czaja et al. 2012; Eroglu et al. 2015) (Fig. 11).

The 2.5 Ga Mt. McRae Shale in drillcore ABDP-9 (Hamersley Basin) has been intensively studied at high stratigraphic resolution using a diverse range of elemental and isotopic (S, Mo, U, N, Se, Os) redox proxies (Anbar et al. 2007; Kaufman et al. 2007; Garvin et al. 2009;

Reinhard et al. 2009; Duan et al. 2010; Kendall et al. 2013, 2015b; Stüeken et al. 2015). In the Mt. McRae Shale, δ^{98}Mo ranges between 0.9‰ and 1.8‰ (Duan et al. 2010). The highest δ^{98}Mo values are found in euxinic ORM (as inferred from sedimentary Fe speciation analyses) characterized by small but distinctive Mo enrichments and isotopic evidence for a dissolved marine Mo reservoir during an episode of mild environmental oxygenation. One explanation for the high δ^{98}Mo values is the removal of isotopically light Mo to oxide minerals, thus leaving behind a dissolved pool of isotopically heavy Mo in seawater that was sequestered into euxinic sediments. Isotopic fractionation during riverine transport and in weakly euxinic settings may also have contributed to the high seawater δ^{98}Mo. Using mass balance calculations, Duan et al. (2010) showed that in a largely anoxic world, the δ^{98}Mo of a small seawater Mo reservoir is susceptible to significant modification by isotope fractionation, thus enabling high seawater δ^{98}Mo to occur without extensive oxygenation.

Building upon these initial efforts, Kurzweil et al. (2015a) measured the δ^{98}Mo of ORMs, carbonates, and iron formations from the underlying 2.6–2.5 Ga stratigraphic units of the Hamersley Group. Although the stratigraphic resolution of this data is low, a general pattern of increasing δ^{98}Mo occurs upsection, peaking in the Mt. McRae Shale. This stratigraphic trend may capture an overall increase of seawater δ^{98}Mo in the Hamersley Basin, but it is also possible there was only a single episode of mild environmental oxygenation during Mt. McRae time (Anbar et al. 2007; Duan et al. 2010; Kendall et al. 2015b). In sedimentary rocks older than the Mt. McRae Shale, δ^{98}Mo typically ranges between 0.5‰ and 1.0‰ and thus is either similar to or only slightly higher than igneous rock compositions, suggesting limited Mo isotope fractionation at low O_2 levels.

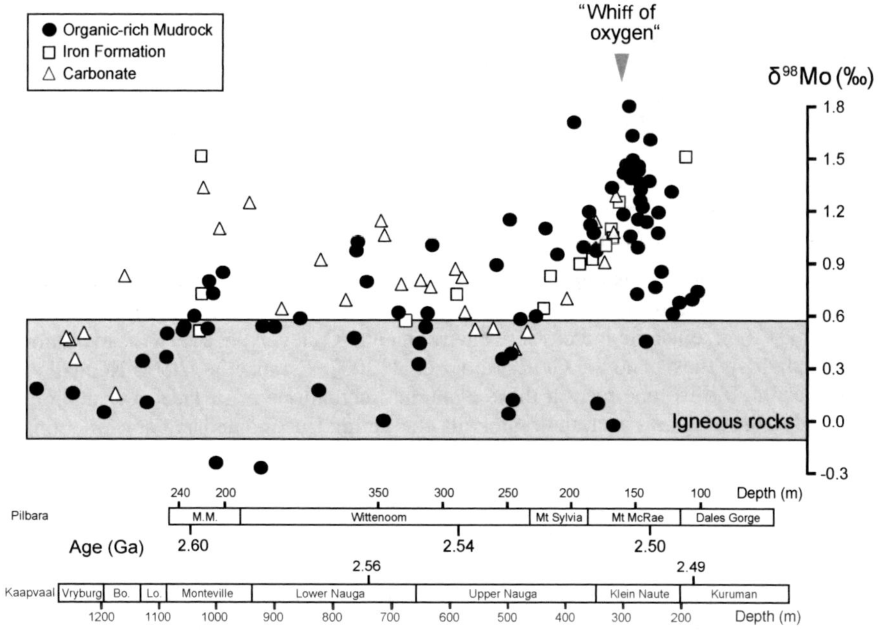

Figure 11. Mo isotope compositions of late Archean sedimentary rocks from the Hamersley Basin (Western Australia), and Griqualand West Basin (South Africa), modified from Kurzweil et al. (2015a). Sedimentary rocks containing higher δ^{98}Mo than the igneous baseline point to fractionation of Mo isotopes in surficial environments, consistent with mild environmental oxygenation. M.M. = Marrra Mamba Formation; Lo = Lokammona Formation; Bo = Boomplaas Formation. Sources of data: Wille et al. (2007), Duan et al. (2010), Voegelin et al. (2010), Kurzweil et al. (2015a).

The Mo isotope data from ca. 2.7–2.5 Ga ORMs and carbonates of the Transvaal Basin and Griqualand West Basin, deposited on the platform and slope of the Campbellrand-Malmani carbonate platform, are also consistent with mild environmental oxygenation but yield a significantly more complex stratigraphic pattern (Wille et al. 2007; Voegelin et al. 2010; Czaja et al. 2012; Eroglu et al. 2015). Appreciable differences were commonly observed between the δ^{98}Mo of ORMs and carbonates in close stratigraphic proximity. These differences may be explained by isotope fractionation associated with non-euxinic bottom water redox conditions during deposition of some ORMs (Voegelin et al. 2010) and by detrital and diagenetic modification of carbonate Mo isotope signatures (Eroglu et al. 2015). Nevertheless, the occurrence of high δ^{98}Mo values (>1.0‰) and the association of negative Fe with positive Mo isotope signatures at multiple stratigraphic levels in the Ghaap Group are consistent with at least episodic environmental oxygenation in the vicinity of the Campbellrand-Malmani carbonate platform (Wille et al. 2007; Voegelin et al. 2010; Czaja et al. 2012; Eroglu et al. 2015). The coupled Fe–Mo isotope data suggest oxidation of Fe^{2+} to Fe^{3+} by photosynthetic O_2, thus producing Fe oxides that adsorbed Mo from seawater (Czaja et al. 2012). Other geochemical data, such as Fe speciation and Re/Mo ratios, from the shallow water and slope sediments are also consistent with the episodic presence of free O_2 in bottom waters (Kendall et al. 2010; Zerkle et al. 2012).

Most older Archean ORMs (3.4–2.7 Ga) have minimal Mo enrichments and δ^{98}Mo values that are similar to or only slightly higher than igneous rocks. The Mo data indicate limited oxidative mobilization of Mo and minimal fractionation of Mo isotopes in the surface environment, and thus low environmental O_2 levels (Siebert et al. 2005; Wille et al. 2007, 2013). High δ^{98}Mo values of up to 1.8‰ were found in ORM at the base of a banded iron formation sequence in the ca. 2.75 Ga Carajás Formation (southern Brazil), but post-depositional potassic metasomatism may have altered the Mo isotope compositions (Cabral et al. 2013).

In contrast to the limited isotopic variation in most pre-2.7 Ga Archean ORMs, a wide range of δ^{98}Mo (spanning ~2.5‰) is observed in the iron formations of the 2.95 Ga Sinqeni Formation (Pongola Supergroup, South Africa) (Planavsky et al. 2014). A positive correlation is observed between Fe/Mn ratios and δ^{98}Mo in these iron formations. This observation suggests that a greater magnitude of Mo isotope fractionation (producing negative δ^{98}Mo) was associated with adsorption of Mo to Mn oxides that formed during local, transient episodes of O_2 production. A similar correlation between Fe/Mn ratios and δ^{98}Mo was also observed for ca. 1.88 Ga iron formations deposited in the Animikie Basin (Lake Superior) after the GOE (Planavsky et al. 2014).

In summary, the Mo isotopic composition of Archean sedimentary rocks, together with other elemental and isotopic redox proxies, are consistent with the emerging notion of "whiffs of O_2" (i.e., episodic increases in environmental O_2 levels) between the evolution of oxygenic photosynthesis and the GOE (Anbar et al. 2007; Lyons et al. 2014; Kendall et al. 2015b). For the Mo isotope record, these dynamic fluctuations in surface oxygenation are manifested in the temporal overlap of intervals containing fractionated and non-fractionated δ^{98}Mo relative to the igneous baseline.

Part 2: Tracing global ocean oxygenation in the post-GOE world

Next, we provide an overview of global ocean redox conditions during the Phanerozoic and Proterozoic Eons from the perspective of the Mo isotope system. The Mo isotope data from each stratigraphic section tells its own story for a specific interval of Earth history. A broader temporal perspective on changes in global ocean redox conditions since the GOE can be obtained from a compilation of Mo isotope data from euxinic ORM (Dahl et al. 2010b; Duan et al. 2010; Wille et al. 2013; Chen et al. 2015; Kendall et al. 2015a; Partin et al. 2015; Fig. 12). The maximum δ^{98}Mo found in ORM for any time interval provides the most conservative estimate of seawater δ^{98}Mo during any particular period of Earth history. Lower δ^{98}Mo values within each interval either indicate that fluctuations in seawater δ^{98}Mo occurred during that interval, or that Mo isotope fractionation occurred locally between seawater and sediments because of weakly euxinic conditions, non-quantitative removal of Mo from bottom waters, or operation of an Fe–Mn particulate shuttle.

Two observations are immediately apparent from the compilation. As expected, the Phanerozoic world overall had higher seawater δ^{98}Mo and thus was more oxygenated compared with the Proterozoic (Fig. 12a), consistent with numerous other types of elemental and isotopic data from sedimentary rocks (e.g., Lyons et al. 2014). Second, the Phanerozoic witnessed oscillations in seawater δ^{98}Mo in response to changes in global ocean redox conditions, including across the Proterozoic–Phanerozoic boundary and in the early Paleozoic (Dahl et al. 2010b; Chen et al. 2015; Kendall et al. 2015a). A moderate positive correlation is observed between the highest δ^{98}Mo and average Mo/TOC ratios of ORM in the compilation (Fig. 12b). Such a correlation is expected because a large seawater Mo inventory, reflected by high Mo/TOC ratios in ORM, should be associated with a more oxygenated ocean floor, resulting in high seawater δ^{98}Mo.

In addition to having low δ^{98}Mo (≤1.4‰), Proterozoic ORM deposited between 2050 and 640 Ma are characterized by Mo/TOC ratios that are intermediate between Archean and Phanerozoic ORM (Arnold et al. 2004; Scott et al. 2008; Kendall et al. 2009, 2011, 2015a; Dahl et al. 2011; Asael et al. 2013; Reinhard et al. 2013a; Partin et al. 2015). Mass balance models suggest that the oceanic Mo reservoir was probably < 20% of today, and that the maximum extent of ocean euxinia was < 1–10% of the seafloor (Dahl et al. 2011; Reinhard et al. 2013a; Chen et al. 2015). These observations are consistent with a redox-stratified ocean structure, specifically oxygenated surface waters, euxinic mid-depth waters along productive ocean margins, and either ferruginous or weakly oxygenated deep waters. The oceanic Mo isotope mass balance model cannot distinguish between weakly oxygenated and ferruginous sinks for Mo (both included in the SAD sink) because the magnitude of Mo isotope fractionation in such settings is similar (Goldberg et al. 2009, 2012; Dahl et al. 2010b; Kendall et al. 2015a).

We emphasize that variations in pre-Ediacaran Proterozoic seawater δ^{98}Mo were likely and that some of the maximum δ^{98}Mo may still only represent minimum values for global seawater. In particular, those intervals with high Mo enrichments (e.g., Velkerri Formation; > 100 ppm Mo; Kendall et al. 2009) likely reflect non-quantitative removal of Mo from bottom waters, suggesting that Mo isotope fractionation was expressed between seawater and sediments. Hence, it is possible that seawater δ^{98}Mo reached higher values at least sporadically between the GOE and Neoproterozoic Oxidation Event. Future work will improve the temporal resolution of the pre-Ediacaran Proterozoic database and better constrain the range of seawater δ^{98}Mo.

The Proterozoic-Phanerozoic transition is currently an interval of intense scrutiny. Excursions to high δ^{98}Mo (≥2‰), similar to modern seawater, are observed in late Ediacaran (Kendall et al. 2015a) and early Cambrian ORM (Wille et al. 2008; Chen et al. 2015; Wen et al. 2015; Cheng et al. 2016), as well as in early Cambrian phosphorite deposits (Wen et al. 2011). Similarly, high δ^{98}Mo is also observed in early Hirnantian ORMs deposited at a time of global cooling and glaciation (Zhou et al. 2012, 2015). However, lower δ^{98}Mo values (<2‰) dominate late Ediacaran and early Phanerozoic (pre-Devonian) ORMs (Lehmann et al. 2007; Wille et al. 2008; Dahl et al. 2010b; Xu et al. 2012; Zhou et al. 2012, 2015; Chen et al. 2015, 2016; Kendall et al. 2015a; Kurzweil et al. 2015b; Wen et al. 2015; Cheng et al. 2016). These low values may reflect a more deoxygenated global ocean state. Alternatively, they can be attributed to Mo isotope fractionation in the local depositional environment because of weakly euxinic or non-euxinic conditions or the operation of an active Fe–Mn particulate shuttle (e.g., Neubert et al. 2008; Gordon et al. 2009; Herrmann et al. 2012). Therefore, it is not clear if the high δ^{98}Mo values represent a permanent transition to a more oxygenated ocean state, episodic oxygenation, or even an episode of expanded Mo burial with large isotope fractionations in reducing settings, specifically the weakly euxinic sink (Wille et al. 2008; Dahl et al. 2010b; Boyle et al. 2014;

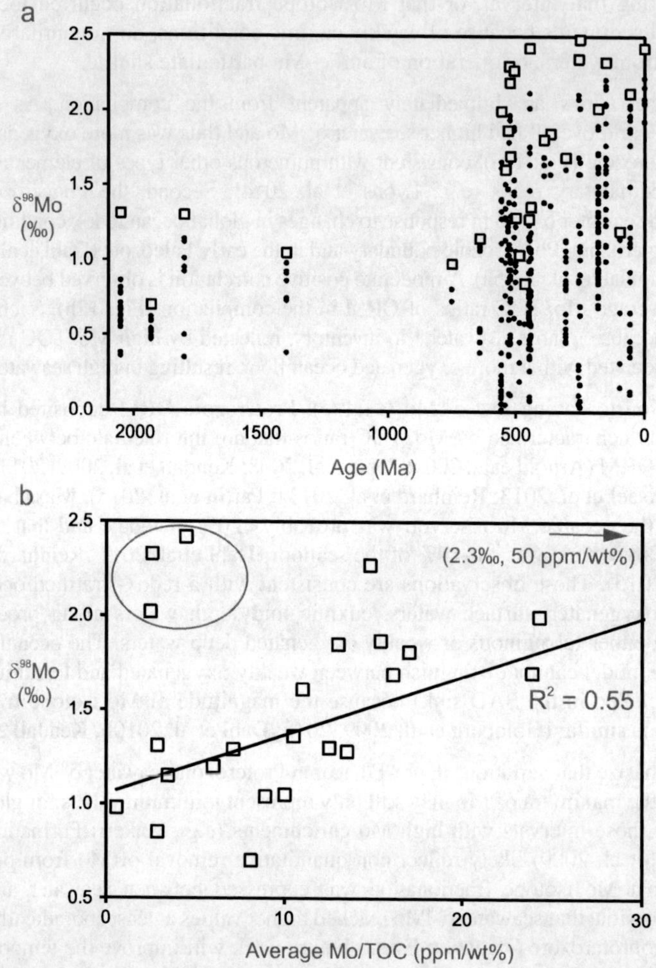

Figure 12. a) Temporal record of Mo isotope compositions in euxinic organic-rich mudrocks. For each time interval, the squares represent the highest δ^{98}Mo, whereas circles represent other data. **b)** Comparison of the highest δ^{98}Mo and associated average Mo/TOC ratios of the time intervals. High δ^{98}Mo and Mo/TOC indicates widespread ocean oxygenation whereas low δ^{98}Mo and Mo/TOC indicates significant ocean anoxia. Exceptions to this trend (upper left circle) are the modern, highly restricted Black Sea and Jurassic oceanic anoxic events (both causing low Mo/TOC). The Jurassic oceanic anoxic events were followed by a return to widespread oxygenation and high seawater δ^{98}Mo. See text for sources of data. Modified from Kendall et al. (2015a).

Chen et al. 2015; Kendall et al. 2015a). Those ORM with high Mo enrichments and high δ^{98}Mo are likely to reflect, at minimum, episodes of widespread oxygenation because such conditions permit both a large oceanic Mo inventory and high seawater δ^{98}Mo.

The Mo isotope composition of ORM has been measured during biotic crises, when expansions of anoxic and sulfidic water masses are thought to have eliminated large portions of the marine fauna. The predicted consequence on seawater δ^{98}Mo during a significant expansion of ocean euxinia is a shift to lower values. Indeed, this behavior is recorded in ORMs both from the Toarcian oceanic anoxic event (~183 Ma, Pearce et al. 2008) and from the Late Cambrian SPICE event (~500 Ma, Gill et al. 2009). However, sediments deposited in basins where the redox conditions of the local bottom waters changed from oxygenated to euxinic can see a positive shift in δ^{98}Mo, due to the smaller fractionation between seawater and sediments expressed in most anoxic environments compared with oxygenated settings (e.g., Zhou et al. 2012; Proemse et al. 2013).

An example of this process is observed in sediments deposited in deep-water slope environments during the Late Permian extinction event at Buchanan Lake in the Sverdrup Basin, Arctic Canada. These sediments show a large positive shift in δ^{98}Mo values from −2.0‰ to 2.2‰, requiring local redox changes. Moreover, the positive δ^{98}Mo trend is associated with a dramatic increase to high Mo enrichments (up to 80 ppm), thus confirming increasingly more reducing conditions in the local basin during peak δ^{98}Mo values (Proemse et al. 2013). A similar scenario was observed in the Shangsi section, Southern China (Zhou et al. 2012). Other parts of the Sverdrup basin remained oxygenated during the mass extinction event, suggesting shallow water anoxia was not a global phenomenon. This observation is consistent with the near-modern seawater δ^{98}Mo values in sediments deposited during the local peak in reducing conditions, which suggest a substantial oxic Mo sink existed at this time (Proemse et al. 2013).

Sediments from the Late Jurassic Kimmeridge Clay Formation (155–148 Ma) show evidence for slightly more widespread euxinia than today (Pearce et al. 2010b), whereas sections from the Cenomanian-Turonian oceanic anoxic event (~94 Ma, OAE2) suggest seawater δ^{98}Mo decreased to ~1‰ at the peak of the event (Westermann et al. 2014; Dickson et al. 2016; Goldberg et al. 2016). Many samples from OAE2 sections have δ^{98}Mo well below the average oceanic input (i.e., < 0.6‰), implying Mo isotope fractionation between seawater and sediments during deposition. This observation illustrates how difficult it is to record seawater δ^{98}Mo through time.

Expansions of anoxic waters during hyperthermal events is observed using local redox proxies at multiple sites during the Paleocene–Eocene thermal maximum and the early Eocene thermal maximum 2 (Dickson and Cohen 2012; Dickson et al. 2012, 2014). The δ^{98}Mo values in these ORMs are persistently high (2.1‰) and close to modern seawater (2.3‰), suggesting that expanded ocean anoxia was limited to the short duration (~100–200 kyr) of the warming events.

In summary, studies of the post-GOE world highlight that the Mo isotope paleoredox proxy can trace variations in the global extent of ocean euxinia, with a greater extent of such conditions suggested by low Mo enrichments and low δ^{98}Mo in ORM deposited from locally euxinic bottom waters. By contrast, high Mo enrichments coupled with high δ^{98}Mo values (i.e., similar to modern seawater) in ORM are a strong indicator of widespread ocean oxygenation. In some cases, it is possible that the δ^{98}Mo of euxinic ORM can be significantly lower than the seawater composition because of deposition from weakly euxinic bottom waters or the operation of an Fe–Mn shuttle in shallower basins where the chemocline is close to the sediment-water interface. In such cases, the Mo data from ORM can provide misleading information. Hence, it is good practice to couple Mo isotope data with other paleoredox proxies to provide the most robust information on global ocean redox conditions.

APPLICATION TO NATURAL RESOURCES

Ore deposits

Application of the Mo isotope system as a process tracer for ore deposits is in its infancy. Initial studies explored the range of Mo isotope compositions for different deposit types, and the relationship between Mo isotope variations, fractionation mechanisms, mineralization processes, and fluid sources for individual deposits.

Predictably, these initial efforts have concentrated on molybdenite (the principal ore mineral of Mo), which is approximately 60% Mo by weight and often dominates the Mo mass balance in mineralizing systems. Rhenium concentrations in molybdenites may range from a few ppm to several weight percent due to the tendency for Re^{4+} to substitute for Mo^{4+}, thus enabling the use of the Re–Os geochronometer to date the timing of molybdenite crystallization and associated mineralization (Stein et al. 2001; Golden et al. 2013). Hence, Mo, S, and Re stable isotope compositions and Re-Os crystallization ages from molybdenites have potential to shed detailed insight on the alteration and mineralization processes responsible for many different types of ore deposit, including porphyry copper(–molybdenum), porphyry molybdenum, lode gold, granite–pegmatite, greisen, skarn, and iron oxide copper–gold deposits (Breillat et al. 2016). A particularly attractive feature of molybdenite is the robustness of this mineral to post-ore events such as granulite facies metamorphism and intense deformation (Stein et al. 2001).

The total range of Mo isotope variation in molybdenites is ~4‰, with isotopic compositions ranging between −1.37‰ and +2.52‰ (Fig. 13; Breillat et al. 2016). The average δ^{98}Mo of molybdenites is 0.29 ± 1.04‰ (2SD). Significant variability in the δ^{98}Mo of molybdenites can occur for specific categories of ore deposits (> 2‰) and even within single deposits (> 1‰), including at the cm-scale (Hannah et al. 2007; Mathur et al. 2010; Greber et al. 2011, 2014; Segato et al. 2015; Shafiei et al. 2015; Breillat et al. 2016). By contrast, minimal Mo isotope variation is observed between the fractions of single coarse grains cut along and across cleavage planes for a number of molybdenites from different porphyry deposits (Segato et al. 2015). No discernible trends are observed for the δ^{98}Mo of molybdenites through time (Hannah et al. 2007; Breillat et al. 2016).

Temperature may exert an influence on the δ^{98}Mo of molybdenites in an ore deposit. For example, molybdenite from porphyry and granite deposits, representing higher temperature crystallization, have lower δ^{98}Mo (average of about 0.1‰ for each type; Shafiei et al. 2015; Breillat et al. 2016). By contrast, higher δ^{98}Mo is observed in molybdenites deposited by lower temperature fluids, such as in greisen and iron oxide copper–gold deposits (average of about 1.25‰ and 1.07‰, respectively; Breillat et al. 2016). However, preliminary studies reveal that Mo isotope fractionation in ore-forming systems is probably also influenced by Rayleigh distillation, fluid boiling, variations in redox conditions, and possibly molybdenite crystal structure (Hannah et al. 2007; Mathur et al. 2010; Greber et al. 2011, 2014; Shafiei et al. 2015). Significant overlap is observed in the δ^{98}Mo of molybdenites from different ore deposit types (Segato et al. 2015; Breillat et al. 2016), indicating that isotopic variations should be interpreted in the context of an individual deposit's geological history rather than the type of ore deposit it represents.

In magmatic-hydrothermal environments, Mo may be transported as a number of different species, such as MoO_3, $MoO_3{\cdot}nH_2O$, MoO_4^{2-}, $HMoO_4^-$, H_2MoO_4, $MoO(OH)Cl_2$, MoO_2Cl_2, K_2MoO_4, $KHMoO_4$, Na_2MoO_4, $NaHMoO_4$, and $NaHMoO_2S_2$ (e.g., Candella and Holland 1984; Cao 1989; Farges et al. 2006; Rempel et al. 2006, 2009; Ulrich and Mavrogenes 2008; Zhang et al. 2012). The dominant species involved and their associated isotope fractionations are poorly understood. Molybdenum may be transported in the vapor state as $MoO_3 \cdot nH_2O$ (Rempel et al. 2006, 2009) and crystallize from the vapor upon reaction with H_2S. If correct, this means that Mo isotope fractionation is possible at high temperatures. For example, Rayleigh distillation associated with molybdenite precipitation along a fracture system would result in different δ^{98}Mo for earlier (proximal) and later (distal) molybdenites (Hannah et al. 2007). The degree of covariation between Mo and S isotope compositions in molybdenites from a single deposit represents one test of this hypothesis because in an ore-forming system with limited Mo and S availability, the isotopic signatures of both elements should be positively correlated if Rayleigh distillation is the main mechanism of isotope fractionation (Hannah et al. 2007). Paired Mo and S isotope analyses have not yet been reported for molybdenite.

Fluid boiling may explain some Mo isotope variations in porphyry systems because of the formation of brine and vapor components with different Mo isotope compositions (Greber et al. 2014; Shafiei et al. 2015). Lighter Mo isotopes may preferentially partition

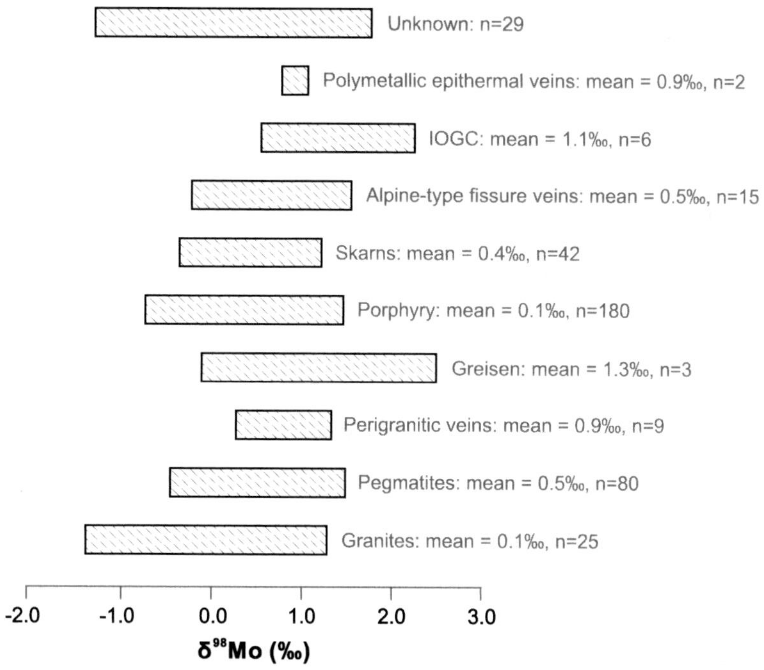

Figure 13. Range and mean of the Mo isotope compositions in molybdenite from different types of ore deposit. The strong overlap in δ^{98}Mo among different ore deposit types indicates that Mo isotopes cannot be used to fingerprint the type of mineralization. IOGC = iron oxide copper–gold deposits. Modified from Breillat et al. (2016).

into the vapor phase whereas heavier Mo isotopes remain in the brine (Shafiei et al. 2015). In the Kerman porphyry copper deposits of Iran, a high-temperature (400–600 °C) brine phase deposited isotopically heavy Mo in the early stages of mineralization, whereas the vapor phase (300–400 °C) crystallized isotopically lighter molybdenite in the hydrothermal fracture system (Fig. 14). Hence, Shafiei et al. (2015) suggested that the δ^{98}Mo of molybdenites in a porphyry system will evolve to lower values over time and with distance from the mineralizing source. The crystal structure of the molybdenite may exert some control on the Mo isotope composition, with heavier Mo isotopes preferentially taken up by the denser 2H polytype compared with the less dense 3R polytype (Shafiei et al. 2015).

Redox reactions and multiple hydrothermal events may also exert a major control on Mo isotope fractionation in ore-forming systems. Molybdenites from Late Paleozoic high temperature (300–600 °C) quartz-molybdenite veins (Aar Massif, Switzerland) have a bimodal distribution in δ^{98}Mo, with peaks at ~0.2‰ and ~1.1‰ (Greber et al. 2011). Single-stage Rayleigh distillation is thus not the main mechanism responsible for Mo isotope fractionation. Isotopic variability in the molybdenites at both small (cm apart) and large (different hand samples) scales suggests Mo isotope fractionation was influenced by redox conditions during precipitation of molybdenite during separate episodes of fluid expulsion from an evolving magma (Greber et al. 2011).

Magmatic evolution and redox reactions may lead to higher δ^{98}Mo of molybdenites in a porphyry system over time (Greber et al. 2014). In the porphyry Questa deposit (New Mexico, U.S.A.), three major fractionation mechanisms were identified by Greber et al. (2014) that

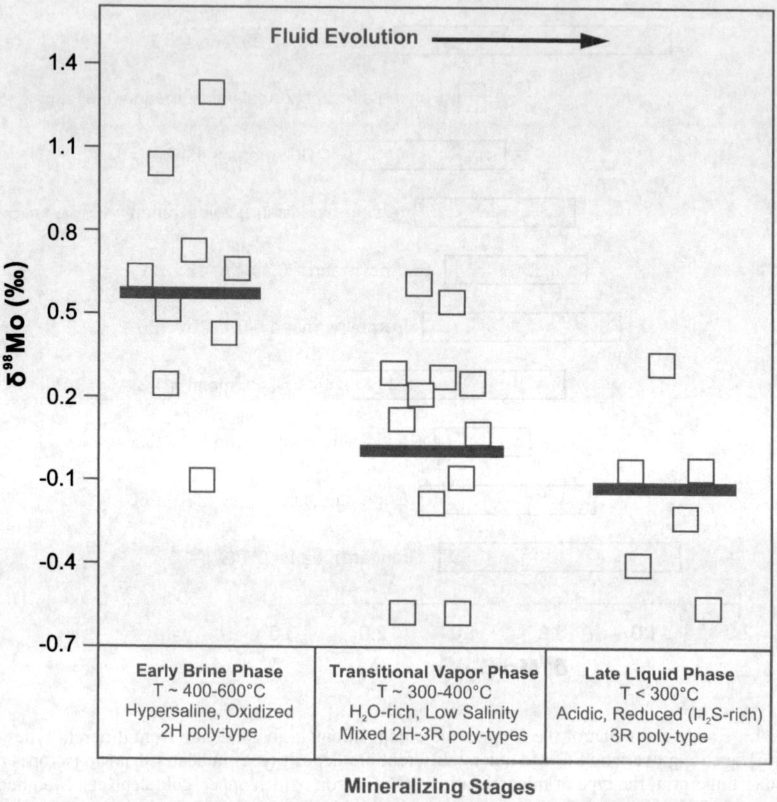

Figure 14. Molybdenum isotope compositions of molybdenite from different stages of mineralization in the Kerman porphyry copper deposits, Iran. The black bars represent the average δ^{98}Mo for each mineralization stage. In this deposit, the molybdenite data suggest an overall evolution of the mineralizing fluid to lower δ^{98}Mo over time. Modified from Shafiei et al. (2015).

operated over a temperature range of ~700 to 350 °C. First, removal of isotopically lighter Mo into minerals during fractional crystallization can enrich the remaining melt in isotopically heavier Mo. Second, fluids exsolved from the magma are preferentially enriched in isotopically heavier Mo isotopes. Third, lighter Mo isotopes are preferentially incorporated into molybdenite during crystallization, causing the remaining fluid to have an isotopically heavier composition. Hence, later-stage molybdenites can have higher δ^{98}Mo than earlier-stage molybdenites. In the Questa deposit, this is reflected by a low δ^{98}Mo for a rhyolite formed after fluid exsolution (~ –0.57‰) and successively higher median δ^{98}Mo for molybdenite in igneous-phase magmatic–hydrothermal breccia (–0.29‰), hydrothermal-phase magmatic–hydrothermal breccia (–0.05‰), and stockwork veins (+0.22‰) (Fig. 15; Greber et al. 2014).

The work of Greber et al. (2014) and Shafiei et al. (2015) on porphyry deposits suggests that the δ^{98}Mo of hydrothermal fluids and molybdenite may evolve to either lower or higher values over time and with distance from the mineralizing source, depending on the relative influence of various processes (fluid boiling, magmatic evolution, fluid exsolution, redox reactions) on the Mo isotope systematics of an ore-forming system. It is also possible that the spatiotemporal variations within a single deposit will be obscured by the interplay of multiple processes operating at different scales, times, and locations within the ore-forming system.

The starting Mo isotope composition of an ore-forming porphyry system can also influence the isotope compositions of molybdenites. Based on the comparison of Nd isotope data from magmatic rocks with Mo isotope data from molybdenites for a number of different deposits, Wang et al. (2016) suggested that porphyry systems with crustal magma sources will precipitate molybdenites with generally higher δ^{98}Mo compared with mantle-derived magmatic systems.

Molybdenum isotope studies point to the importance of redox reactions on the δ^{98}Mo of Mo-bearing mineral phases in low-temperature systems (Ryb et al. 2009; Greber et al. 2011; Song et al. 2011). In a Pliocene low-temperature system (100–160 °C) in Switzerland, molybdate may have been transported by oxidizing surface waters into brecciated rocks (Grimsel breccia) where it was reduced, leading to precipitation of Mo-bearing sulfide phases (Greber et al. 2011; the mineralogy could not be identified by the authors). The larger Mo isotope variation of ~3‰ in the brecciated rocks compared with individual high-temperature systems may reflect a combination of lower temperature crystallization, reduction of MoO_4^{2-} (an uncommon species in high-temperature systems), and multiple stages of re-dissolution and re-precipitation of Mo (Greber et al. 2011). Variable redox conditions and depositional environments (open marine versus restricted) were invoked to explain the range of Mo isotope compositions in the different orebodies of the Dajiangping pyrite deposit in China (Song et al. 2011).

A study of Mo-rich iron oxide veins by Ryb et al. (2009) revealed significant Mo isotopic variation of greater than 4‰ in a low temperature mineralizing system associated with the Dead Sea transform. The isotopic variation likely reflects interaction of dense evaporitic marine brines (δ^{98}Mo ~ 2.3‰) with isotopically lighter igneous and sedimentary rocks, as well as Rayleigh distillation of Mo isotopes along the brine flow path. The latter is suggested to explain Mo isotope compositions in the iron oxide veins that are higher than seawater δ^{98}Mo. This study demonstrates that Mo isotopes have the potential to be used as both a source and process tracer for subsurface fluid migration.

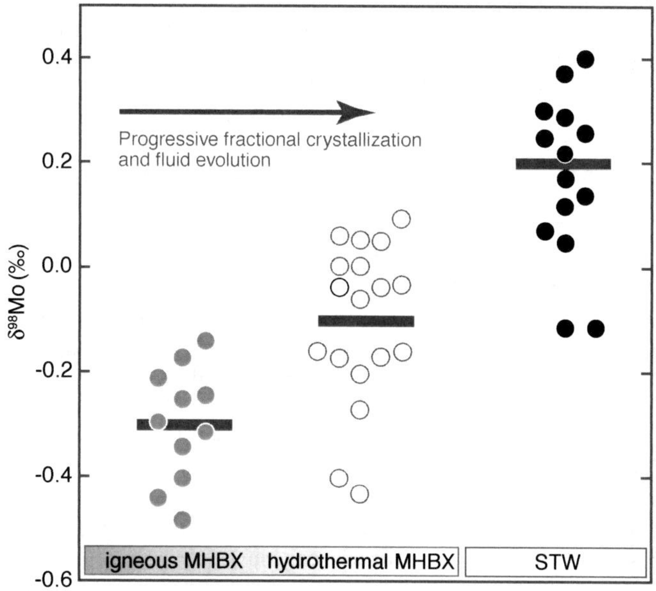

Figure 15. Molybdenum isotope compositions of molybdenite from different stages of mineralization in the Questa porphyry deposit, New Mexico, U.S.A. Black bars represent the median Mo isotope composition of each mineralization stage. This deposit provides an example of possible fluid evolution to higher δ^{98}Mo over time. MHBX = magmatic-hydrothermal breccia; STW = stockwork veins. Modified from Greber et al. (2014).

Petroleum systems

Petroleum metal isotope geochemistry has potential for oil-source rock and oil-oil correlation and tracing petroleum generation and reservoir processes, but has not advanced beyond the exploratory stage. The Mo isotopic analysis of oils is an analytically challenging problem caused by both the highly complex nature of oil matrices as well as the low Mo concentration of oils (typically ppb to low ppm; Ventura et al. 2015). However, as shown by Ventura et al. (2015), it is expected these challenges can be circumvented for the Mo stable isotope system by using the double spike method (to minimize matrix effects) as well as high temperature and pressure microwave digestion of bulk oil samples. Another possible fruitful approach is to develop techniques to isolate the Mo-rich fraction(s) of oils (e.g., similar to asphaltene separation for Re–Os isotope analyses; Selby et al. 2007; Mahdaoui et al. 2013).

Within a single petroleum-producing sedimentary basin, distinctive δ^{98}Mo may be preserved in different petroleum source rocks because of: a) differences in the global seawater δ^{98}Mo associated with variations in global ocean redox conditions; and/or b) differences in the magnitude of Mo isotope fractionation between local seawater and sediments caused by differences in the dissolved O_2 and H_2S concentrations of local bottom waters (Ventura et al. 2015). If the Mo isotope composition of oils is not affected by oil migration or reservoir processes, then it may be possible to infer the relative importance of different source rocks to oil reservoirs by comparing the Mo isotope compositions of oils and source rocks (Archer et al. 2012). This approach would complement traditional methods of oil-source rock correlation using light stable isotopes and biomarkers.

The Mo isotope composition of oil may not be affected by oil maturation, migration, and biodegradation on the scale of a sedimentary basin, thus raising the possibility of using Mo isotopes for oil-source rock correlation (Archer et al. 2012). However, it is not known if source rock Mo isotope compositions are transferred directly to oils. Furthermore, systematic studies are required to assess the impact on oil Mo isotope compositions by other reservoir processes such as thermochemical sulfate reduction, which is known to affect the isotopic composition of other redox-sensitive metals like Re and Os (Lillis and Selby 2013). Ventura et al. (2015) reported a range of ~ 1.1‰ (from –0.1‰ to 1.0‰) for four crude oils from the Campos Basin (Brazil), but did not measure the δ^{98}Mo of the lacustrine source rocks. A total range of ~ 1.5‰ was reported by Archer et al. (2012) for multiple petroleum systems.

Anthropogenic tracing

Application of Mo isotopes as an anthropogenic tracer is confined to a small number of studies. Although anthropogenic Mo is only a small component in most lacustrine and marine settings studied to date (Dahl et al. 2010a; Scheiderich et al. 2010b; Glass et al. 2012), it has been reported from some localities (e.g., Chappaz et al. 2012; Rahaman et al. 2014). Chappaz et al. (2012) used the δ^{98}Mo of sediments to fingerprint the addition of anthropogenic Mo to lakes in eastern Canada from smelting or fossil fuel combustion. In both cases, the anthropogenic source was characterized by a distinct isotope composition of 0.1 ± 0.1‰. Rahaman et al. (2014) calculated that anthropogenic Mo may account for up to 27% of the dissolved Mo load in the Tapi estuary that drains into the Arabian Sea. The δ^{98}Mo of aerosols may also be useful as a tracer of urban anthropogenic emissions (Lane et al. 2013). It is expected that development of Mo isotopes as an anthropogenic tracer will accelerate in the near future.

However, distinguishing isotopically between natural and anthropogenic Mo is not always straightforward because anthropogenic source signatures may be overprinted by natural Mo isotope fractionation in the environment or because of isotopic similarities between the natural and anthropogenic sources of Mo. For example, sediments from the Baltimore Harbor (a site of smelting operations) that are enriched in Mo did not have a different δ^{98}Mo compared with uncontaminated sediments elsewhere in the Chesapeake Bay, in contrast to Os isotope data. Hence, the Mo is either not anthropogenic in origin or the natural and anthropogenic Mo have identical isotopic compositions (Scheiderich et al. 2010b). Neubert et al. (2011) could not find clear evidence for anthropogenic contamination by industry and agriculture in the concentration and isotopic composition of dissolved Mo in small rivers from India, Switzerland, and China.

CONCLUSIONS

The Mo isotope system has matured into a valuable paleoceanographic tracer, as reflected by the large number of studies that seek to characterize local and global ocean redox conditions on the ancient Earth. Although it has long been recognized that rivers are the only major source of Mo to the modern oceans, research efforts over the past decade revealed that there are three major Mo sinks: well-oxygenated settings, sulfidic sediments overlain by weakly oxygenated bottom waters, and euxinic settings characterized by the presence of H_2S in the water column. The δ^{98}Mo of ancient seawater is most commonly inferred from ORM deposited from strongly euxinic bottom waters in semi-restricted marine basins. However, the difficulty of distinguishing between strongly versus weakly euxinic conditions during ORM deposition makes it challenging to confirm that such rocks do indeed record the seawater Mo isotope composition. Chemical sediments including carbonates, phosphorites, and iron formations have recently also been used to infer seawater δ^{98}Mo.

Building from observations of Mo isotope fractionation in modern environments, a wealth of studies have sought to constrain the past extent of global ocean oxygenation from the δ^{98}Mo of Proterozoic and Phanerozoic sedimentary archives, and to look for the Mo isotope expression of initial environmental oxygenation on the Archean Earth. From these studies, it is clear that both local and global conditions affect sedimentary δ^{98}Mo. As a global tracer, the Mo isotope paleoredox proxy is most sensitive to the extent of ocean euxinia, rather than to oxygenated versus anoxic (euxinic and ferruginous) conditions, because the rate of Mo burial into sediments correlates with dissolved H_2S concentrations. Hence, the Mo isotope system should be used in combination with other geochemical proxies to obtain the most reliable information on paleoredox conditions. Refinements in our understanding of the modern oceanic Mo isotope budget, including the significance of biological Mo isotope fractionation and Mo isotope behavior in weakly euxinic settings, will further improve the Mo isotope paleoredox proxy.

New applications to other low-temperature systems (petroleum and anthropogenic tracing) as well as to high-temperature environments (meteorites, magmatic systems, and ore deposits) are rapidly emerging. Many basic questions have yet to be answered. Are Mo isotopes useful for oil-source rock correlation or for tracing oil reservoir processes? Can spatial variations in the Mo isotope composition of molybdenite be used as a vector to mineralization, or for fingerprinting specific processes in ore-forming systems (e.g., fluid boiling, Rayleigh distillation, redox variations, single versus multiple mineralization events)? Will Mo isotopes become a prominent anthropogenic tracer, or does natural fractionation of Mo isotopes limit this application? What more can Mo isotopes tell us about the evolution of magmatic systems, metamorphic prograde-retrograde paths, mantle reservoirs and their relative contributions to volcanism, and early solar system processes?

The Mo isotope system was part of the first wave of non-traditional stable isotope systems to be explored. We fully expect that it will continue to hold a prominent position in studies of low- and high-temperature geochemistry.

ACKNOWLEDGEMENTS

Kendall acknowledges support from an NSERC Discovery Grant RGPIN-435930. Dahl acknowledges a grant from the VILLUM Foundation (VKR023127). Anbar acknowledges support from NSF 1338810. Martin Wille, Ryan Mathur, and two anonymous reviewers provided helpful comments and suggestions. Xinming Chen produced updated Eh–pH diagrams. Alysa Segato is thanked for providing constructive suggestions on Mo isotopes in natural resources. Susan Selkirk and Xinze Lu provided valuable assistance with drafting of the figures.

REFERENCES

Algeo TJ, Lyons TW (2006) Mo–total organic carbon covariation in modern anoxic marine environments: Implications for analysis of paleoredox and paleohydrographic conditions. Paleoceanogr 21:1–23

Algeo TJ, Tribovillard N (2009) Environmental analysis of paleoceanographic systems based on molybdenum–uranium covariation. Chem Geol 268:211–225

Alvarez HM, Xue Y, Robinson CD, Canalizo-Hernandez MA, Marvin RG, Kelly RA, Mondragon A, Penner-Hahn JE, O'Halloran TV (2010) Tetrathiomolybdate inhibits copper trafficking proteins through metal cluster formation. Science 327:331–334

Amrhein C, Mosher PA, Brown AD (1993) The effects of redox on Mo, U, B, V, and As solubility in evaporation pond soils. Soil Science 155:249–255

Anbar AD (2004) Molybdenum stable isotopes: observations, interpretations, directions. Rev Mineral Geochem 55:429–454

Anbar AD, Knoll AH (2002) Proterozoic ocean chemistry and evolution: a bioinorganic bridge? Science 297:1137–1142

Anbar AD, Rouxel O (2007) Metal stable isotopes in paleoceanography. Ann Rev Earth Planet Sci 35:717–746

Anbar AD, Knab KA, Barling J (2001) Precise determination of mass-dependent variations in the isotopic composition of molybdenum using MC-ICPMS. Anal Chem 73:1425–1431

Anbar AD, Duan Y, Lyons TW, Arnold GL, Kendall B, Creaser RA, Kaufman AJ, Gordon GW, Scott C, Garvin J, Buick R (2007) A whiff of oxygen before the Great Oxidation Event? Science 317:1903–1906

Archer C, Vance D (2008) The isotopic signature of the global riverine molybdenum flux and anoxia in the ancient oceans. Nat Geosci 1:597–600

Archer C, Elliott T, van den Boorn S, van Bergen P (2012) Mo and Ni isotope systematics in petroleum fluids across subsurface alteration gradients. Mineral Mag 76:1433

Arlandini C, Käppeler F, Wisshak K, Gallino R, Lugaro M, Busso M, Straniero O (1999) Neutron capture in low-mass asymptotic giant branch stars: cross sections and abundance signatures, Astrophys J 525:886–900

Arnold GL, Anbar AD, Barling J, Lyons TW (2004) Molybdenum isotope evidence for widespread anoxia in Mid-Proterozoic oceans. Science 304:87–90

Arnold GL, Lyons TW, Gordon GW, Anbar AD (2012) Extreme change in sulfide concentrations in the Black Sea during the Little Ice Age reconstructed using molybdenum isotopes. Geology 40:595–598

Asael D, Tissot FLH, Reinhard CT, Rouxel O, Dauphas N, Lyons TW, Ponzevera E, Liorzou C, Chéron S (2013) Coupled molybdenum, iron and uranium stable isotopes as oceanic paleoredox proxies during the Paleoproterozoic Shunga Event. Chem Geol 362:193–210

Azrieli-Tal I, Matthews A, Bar-Matthews M, Almogi-Labin A, Vance D, Archer C, Teutsch N (2014) Evidence from molybdenum and iron isotopes and molybdenum–uranium covariation for sulphidic bottom waters during Eastern Mediterranean sapropel S1 formation. Earth Planet Sci Lett 393:231–242

Baes CF, Mesmer RE (1976) Hydrolysis of Cations. Wiley, New York

Baldwin GJ, Nägler TF, Greber ND, Turner EC, Kamber BS (2013) Mo isotopic composition of the mid-Neoproterozoic ocean: an iron formation perspective. Precambr Res 230:168–178

Barling J, Arnold GL, Anbar AD (2001) Natural mass-dependent variations in the isotopic composition of molybdenum. Earth Planet Sci Lett 193:447–457

Barling J, Anbar AD (2004) Molybdenum isotope fractionation during adsorption by manganese oxides. Earth Planet Sci Lett 217:315–329

Beard BL, Johnson CM, Skulan JL, Nealson KH, Cox L, Sun H (2003) Application of Fe isotopes to tracing the geochemical and biological cycling of Fe. Chem Geol 195:87–117

Becker H, Walker RJ (2003) Efficient mixing of the solar nebula from uniform Mo isotopic composition of meteorites. Nature 425:152–155

Bellenger JP, Wichard T, Kustka AB, Kraepiel AML (2008) Uptake of molybdenum and vanadium by a nitrogen-fixing soil bacterium using siderophores. Nat Geosci 1:243–246

Bertine K, Turekian K (1973) Molybdenum in marine deposits. Geochim Cosmochim Acta 37:1415–1434

Bigeleisen J (1947) Calculation of equilibrium constants for isotopic exchange reactions. J Chem Phys 15:261–267
Biswas KC, Woodards NA, Xu H, Barton LL (2009) Reduction of molybdate by sulfate-reducing bacteria. Biometals 22:131–139
Bostick BC, Fendorf S, Helz GR (2003) Differential adsorption of molybdate and tetrathiomolybdate on pyrite (FeS_2). Environ Sci Technol 37:285–291
Boyd ES, Anbar AD, Miller S, Hamilton TL, Lavin M, Peters JW (2011) A late methanogen origin for molybdenum-dependent nitrogenase. Geobiol 9:221–232
Boyle RA, Dahl TW, Dale AW, Shields-Zhou GA, Zhu M, Brasier MD, Canfield DE, Lenton TM (2014) Stabilization of the coupled oxygen and phosphorus cycles by the evolution of bioturbation. Nature Geosci 7:671–676
Breillat N, Guerrot C, Marcoux E, Négrel P (2016) A new global database of δ^{98}Mo in molybdenites: a literature review and new data. J Geochem Explor 161:1–15
Bruland KW (1983) Trace elements in seawater. *In:* Chemical Oceanography. Riley JP, Chester R (eds) Academic Press, London, p 157–220
Brumsack H, Gieskes J (1983) Interstitial water trace-metal chemistry of laminated sediments from the Gulf of California, Mexico. Mar Chem 14:89–106
Burkhardt C, Kleine T, Oberli F, Pack A, Bourdon A, Wieler R (2011) Molybdenum isotope anomalies in meteorites: constraints on solar nebula evolution and origin of the Earth. Earth Planet Sci Lett 312:390–400
Burkhardt C, Kleine T, Dauphas N, Wieler R (2012) Origin of isotopic heterogeneity in the solar nebula by thermal processing and mixing of nebular dust. Earth Planet Sci Lett 357–358:298–307
Burkhardt C, Hin RC, Kleine T, Bourdon B (2014) Evidence for Mo isotope fractionation in the solar nebula and during planetary differentiation. Earth Planet Sci Lett 391:201–211
Cabral AR, Creaser RA, Nägler T, Lehmann B, Voegelin AR, Belyatsky B, Pašava J, Seabra Gomes Jr AA, Galbiatti H, Böttcher ME, Escher P (2013) Trace-element and multi-isotope geochemistry of Late-Archean black shales in the Carajás iron-ore district, Brazil. Chem Geol 362:91–104
Calvert SE, Piper DZ (1984) Geochemistry of ferromanganese nodules from DOMES site A, Northern Equatorial Pacific: Multiple diagenetic metal sources in the deep sea. Geochim Cosmochim Acta 48:1913–1928
Calvert SE, Price NB (1977) Geochemical variation in ferromanganese nodules and associated sediments from the pacific ocean. Mar Chem 5:43–74
Candela PA, Holland HD (1984) The partitioning of copper and molybdenum between silicate melts and aqueous fluids. Geochim Cosmochim Acta 48:373–380
Cao X (1989) Solubility of molybdenite and the transport of molybdenum in hydrothermal solutions. PhD Dissertation, Iowa State University, Ames, Iowa
Chaillou, G, Anschutz P, Lavaux G, Schafer J, Blanc G (2002) The distribution of Mo, U, and Cd in relation to major redox species in muddy sediments of the Bay of Biscay. Mar Chem 80:41–59
Chan KM, Riley JP (1966) The determination of molybdenum in natural waters, silicates and biological materials. Anal Chim Acta 36:220–229
Chappaz A, Gobeil C, Tessier A (2008) Geochemical and anthropogenic enrichments of Mo in sediments from perennially oxic and seasonally anoxic lakes in Eastern Canada. Geochim Cosmochim Acta 72:170–184
Chappaz A, Lyons TW, Gordon GW, Anbar AD (2012) Isotopic fingerprints of anthropogenic molybdenum in lake sediments. Environ Sci Technol 46:10934–10940
Chappaz A, Lyons TW, Gregory DD, Reinhard CT, Gill BC, Li C, Large RR (2014) Does pyrite act as an important host for molybdenum in modern and ancient euxinic sediments? Geochim Cosmochim Acta 126:112–122
Chen JH, Papanastassiou DA, Wasserburg GJ, Ngo HH (2004) Endemic Mo isotopic anomalies in iron and carbonaceous meteorites. Lunar Planet Sci XXXV:1431
Chen X, Ling H-F, Vance D, Shields-Zhou GA, Zhu M, Poulton SW, Och LM, Jiang S-Y, Li D, Cremonese L, Archer C (2015) Rise to modern levels of ocean oxygenation coincided with the Cambrian radiation of animals. Nat Commun 6:7142
Cheng M, Li C, Zhou L, Algeo TJ, Zhang F, Romaniello S, Jin CS, Lei LD, Feng LJ, Jiang SY (2016) Marine Mo biogeochemistry in the context of dynamically euxinic mid-depth waters: a case study of the lower Cambrian Niutitang shales, South China. Geochim Cosmochim Acta 183:79–93
Collier RW (1985) Molybdenum in the Northeast Pacific-Ocean. Limnol Oceanogr 30:1351–1354
Crick FHC, Orgel LE (1973) Directed panspermia. Icarus 19:341–346
Cronan DS (1980) Underwater Minerals. Academic Press, London
Cronan DS, Tooms JS (1969) The geochemistry of manganese nodules and associated pelagic deposits from the Pacific and Indian Oceans. Deep Sea Res 16:335–359
Crusius J, Calvert S, Pedersen T, Sage D (1996) Rhenium and molybdenum enrichments in sediments as indicators of oxic, suboxic and sulfidic conditions of deposition. Earth Planet Sci Lett 145:65–78
Czaja AD, Johnson CM, Roden EE, Beard BL, Voegelin AR, Nägler TF, Beukes NJ, Wille M (2012) Evidence for free oxygen in the Neoarchean ocean based on coupled iron-molybdenum isotope fractionation. Geochim Cosmochim Acta 86:118–137

Dahl TW, Anbar AD, Gordon GW, Rosing MT, Frei R, Canfield DE (2010a) The behavior of molybdenum and its isotopes across the chemocline and in the sediments of sulfidic Lake Cadagno, Switzerland. Geochim Cosmochim Acta 74:144–163

Dahl TW, Hammarlund EU, Anbar AD, Bond DPG, Gill BC, Gordon GW, Knoll AH, Nielsen AT, Schovsbo NH, Canfield DE (2010b) Devonian rise in atmospheric oxygen correlated to the radiations of terrestrial plants and large predatory fish. PNAS 107:17911–17915

Dahl TW, Canfield DE, Rosing MT, Frei RE, Gordon GW, Knoll AH, Anbar AD (2011) Molybdenum evidence for expansive sulfidic water masses in ~750 Ma oceans. Earth Planet Sci Lett 311:264–274

Dahl TW, Chappaz A, Fitts JP, Lyons TW (2013a) Molybdenum reduction in a sulfidic lake: evidence from X-ray absorption fine-structure spectroscopy and implications for the Mo paleoproxy. Geochim Cosmochim Acta 103:213–231

Dahl TW, Ruhl M, Hammarlund EU, Canfield DE, Rosing MT, Bjerrum CJ (2013b) Tracing euxinia by molybdenum concentrations in sediments using handheld x-ray fluorescence spectroscopy (HH-XRF). Chem Geol 360–361:241–251

Dauphas N, Marty B, Reisberg L (2002a) Molybdenum evidence for inherited planetary scale isotope heterogeneity of the protosolar nebula. Astrophys J 565:640–644

Dauphas N, Marty B, Reisberg L (2002b) Molybdenum nucleosynthetic dichotomy revealed in primitive meteorites. Astrophys J 569:139–142

Dauphas N, Davis AM, Marty B, Reisberg L (2004) The cosmic molybdenum–ruthenium isotope correlation. Earth Planet Sci Lett 226:465–475

David LA, Alm EJ (2011) Rapid evolutionary innovation during an Archaean genetic expansion. Nature 469:93–96

Dean W, Piper D, Peterson L (1999). Molybdenum accumulation in Cariaco basin sediment over the past 24 ky: A record of water-column anoxia and climate. Geology 27:507–510

Dickson AJ, Cohen AS (2012) A molybdenum isotope record of Eocene Thermal Maximum 2: Implications for global ocean redox during the early Eocene. Paleoceanogr 27:PA3230

Dickson AJ, Cohen AS, Coe AL (2012) Seawater oxygenation during the Paleocene–Eocene Thermal Maximum. Geology 40:639–642

Dickson AJ, Cohen AS, Coe AL (2014) Continental margin molybdenum isotope signatures from the early Eocene. Earth Planet Sci Lett 404:389–395

Dickson AJ, Jenkyns HC, Porcelli D, van den Boorn S, Idiz E (2016) Basin-scale controls on the molybdenum-isotope composition of seawater during Oceanic Anoxic Event 2 (Late Cretaceous). Geochim Cosmochim Acta 178:291–306

Diemann E, Müller A (1973) Schefel- und Selenverbindungen von Übergangsmetallen mit d^0-Konfiguration (Thio and Seleno compounds of the transition metals with the d^0 configuration). Coord Chem Rev 10:79–122

Duan Y, Anbar AD, Arnold GL, Lyons TW, Gordon GW, Kendall B (2010) Molybdenum isotope evidence for mild environmental oxygenation before the Great Oxidation Event. Geochim Cosmochim Acta 74:6655–6668

Eady RR (1996) Structure-function relationships of alternative nitrogenases. Chem Rev 96:3013–3030

Elbaz-Poulichet F, Seidel JL, Jézéquel D, Metzger E, Prévot F, Simonucci C, Sarazin G, Viollier E, Etcheber H, Jouanneau J-M, Weber O, Radakovitch O (2005) Sedimentary record of redox-sensitive elements (U, Mn, Mo) in a transitory anoxic basin (the Thau lagoon, France). Mar Chem 95:271–281

Emerson S, Huested SS (1991) Ocean anoxia and the concentrations of molybdenum and vanadium in seawater. Mar Chem 34:177–196

Erickson BE, Helz GR (2000) Molybdenum (VI) speciation in sulfidic waters: stability and lability of thiomolybdates. Geochim Cosmochim Acta 64:1149–1158

Eroglu S, Schoenberg R, Wille M, Beukes N, Taubald H (2015) Geochemical stratigraphy, sedimentology, and Mo isotope systematics of the ca. 2.58–2.50 Ga-old Transvaal Supergroup carbonate platform, South Africa. Precambr Res 266:27–46

Farges F, Siewert R, Ponader CW, Brown Jr GE, Pichavant M, Behrens H (2006) Structural environments around molybdenum in silicate glasses and melts. II. Effect of temperature, pressure, H_2O, halogens, and sulfur. Can Mineral 44:755–773

Freymuth H, Vils F, Willbold M, Taylor RN, Elliott T (2015) Molybdenum mobility and isotopic fractionation during subduction at the Mariana arc. Earth Planet Sci Lett 432:176–186

Francois R (1988) A study on the regulation of the concentrations of some trace metals (Rb, Sr, Zn, Pb, Cu, V, Cr, Ni, Mn and Mo) in Saanich Inlet Sediments, British Columbia, Canada. Mar Geol 83:285–308

Frausto da Silva JJ, Williams RJP (2001) The Biological Chemistry of the Elements: The Inorganic Chemistry of Life. Clarendon Press, Oxford

Fujii T, Moynier F, Telouk P, Albarède F (2006) Mass-independent isotope fractionation of molybdenum and ruthenium and the origin of isotopic anomalies in Murchison. Astrophys J 647:1506

Garvin J, Buick R, Anbar AD, Arnold GL, Kaufman AJ (2009) Isotopic evidence for an aerobic nitrogen cycle in the latest Archean. Science 323:1045–1048

Gill BC, Lyons TW, Dahl T, Saltzman M, Gordon G, Anbar AD (2009) Multiple geochemical proxies reveal a Late Cambrian ocean anoxic event. Geochim Cosmochim Acta 73:A436

Glass JB, Wolfe-Simon F, Anbar AD (2009) Coevolution of metal availability and nitrogen assimilation in cyanobacteria and algae. Geobiol 7:100–123

Glass J, Wolfe-Simon F, Elser J, Anbar A (2010) Molybdenum–nitrogen co-limitation in freshwater and coastal heterocystous cyanobacteria. Limnol Oceanogr 55:667–676

Glass JB, Axler RP, Chandra S, Goldman CR (2012) Molybdenum limitation of microbial nitrogen assimilation in aquatic ecosystems and pure cultures. Front Microbiol 3:1–11

Glass JB, Chappaz A, Eustis B, Heyvaert AC, Waetjen DP, Hartnett HE, Anbar AD (2013) Molybdenum geochemistry in a seasonally dysoxic Mo-limited lacustrine ecosystem. Geochim Cosmochim Acta 114:204–219

Goldberg T, Archer C, Vance D, Poulton SW (2009) Mo isotope fractionation during adsorption to Fe (oxyhydr) oxides. Geochim Cosmochim Acta 73:6502–6516

Goldberg T, Archer C, Vance D, Thamdrup B, McAnena A, Poulton SW (2012) Controls on Mo isotope fractionations in a Mn-rich anoxic marine sediment, Gullmar Fjord, Sweden. Chem Geol 296–297:73–82

Goldberg T, Gordon G, Izon G, Archer C, Pearce CR, McManus J, Anbar AD, Rehkämper M (2013) Resolution of inter-laboratory discrepancies in Mo isotope data: an intercalibration. J Anal At Spectrom 28:724–735

Goldberg T, Poulton SW, Wagner T, Kolonic SF, Rehkämper M (2016) Molybdenum drawdown during Cretaceous Oceanic Anoxic Event 2. Earth Planet Sci Lett 440:81–91

Golden J, McMillan M, Downs RT, Hystad G, Goldstein I, Stein HJ, Zimmerman A, Sverjensky DA, Armstrong JT, Hazen RM (2013) Rhenium variations in molybdenite (MoS_2): evidence for progressive subsurface oxidation. Earth Planet Sci Lett 366:1–5

Gordon GW, Lyons TW, Arnold GL, Roe J, Sageman BB, Anbar AD (2009) When do black shales tell molybdenum isotope tales? Geology 37:535–538

Goto KT, Shimoda G, Anbar AD, Gordon GW, Harigane Y, Senda R, Suzuki K (2015) Molybdenum isotopes in hydrothermal manganese crust from the Ryukyu arc system: Implications for the source of molybdenum. Mar Geol 369:91–99

Greber ND, Hofmann BA, Voegelin AR, Villa IM, Nägler TF (2011) Mo isotope composition in Mo-rich high- and low-T hydrothermal systems from the Swiss Alps. Geochim Cosmochim Acta 75:6600–6609

Greber, ND, Siebert C, Nägler TF, Pettke T (2012) $\delta^{98/95}$Mo values and molybdenum concentration data for NIST SRM 610, 612 and 3134: Towards a common protocol for reporting Mo data. Geostand Geoanal Res 36:291–300

Greber ND, Pettke T, Nägler TF (2014) Magmatic–hydrothermal molybdenum isotope fractionation and its relevance to the igneous crustal signature. Lithos 190–191:104–110

Greber ND, Puchtel IS, Nägler TF, Mezger K (2015a) Komatiites constrain molybdenum isotope composition of the Earth's mantle. Earth Planet Sci Lett 421:129–138

Greber ND, Mäder U, Nägler TF (2015b) Experimental dissolution of molybdenum-sulphides at low oxygen concentrations: a first-order approximation of late Archean atmospheric conditions. Earth Space Sci 2:173–180

Hannah JL, Stein HJ, Wieser ME, de Laeter JR, Varner MD (2007) Molybdenum isotope variations in molybdenite: Vapor transport and Rayleigh fractionation of Mo. Geology 35:703–706

Helz GR, Bura-Nakić E, Mikac N, Ciglenečki I (2011) New model for molybdenum behavior in euxinic waters. Chem Geol 284:323–332

Helz GR, Miller CV, Charnock JM, Mosselmans JFW, Pattrick RAD, Garner CD, Vaughan DJ, (1996) Mechanism of molybdenum removal from the sea and its concentration in black shales: EXAFS evidence. Geochim Cosmochim Acta 60:3631–3642

Helz GR, Vorlicek TP, Kahn MD (2004) Molybdenum Scavenging by Iron Monosulfide. Environ Sci Technol 38:4263–4268

Herrmann AD, Kendall B, Algeo TJ, Gordon GW, Wasylenki LE, Anbar AD (2012) Anomalous molybdenum isotope trends in Upper Pennsylvanian euxinic facies: Significance for use of δ^{98}Mo as a global marine redox proxy. Chem Geol 324–325:87–98

Hille R (1996) The mononuclear molybdenum enzymes. Chem Rev 96:2757–2816

Hille R (2002) Molybdenum and tungsten in biology. Trends Biochem Sci 27:360–367

Hin RC, Burkhardt C, Schmidt MW, Bourdon B, Kleine T (2013) Experimental evidence for Mo isotope fractionation between metal and silicate liquids. Earth Planet Sci Lett 379:38–48

Holland HD (1984) The Chemical Evolution of the Atmosphere and Oceans. Princeton University Press, Princeton

International Molybdenum Association (2016). http://www.imoa.info/

Kashiwabara T, Takahashi Y, Tanimizu M, Usui A (2011) Molecular-scale mechanisms of distribution and isotopic fractionation of molybdenum between seawater and ferromanganese oxides. Geochim Cosmochim Acta 75:5762–5784

Kaufman AJ, Johnston DT, Farquhar J, Masterson AL, Lyons TW, Bates S, Anbar AD, Arnold GL, Garvin J, Buick R (2007) Late Archean biospheric oxygenation and atmospheric evolution. Science 317:1900–1903

Kendall B, Creaser RA, Gordon GW, Anbar AD (2009) Re–Os and Mo isotope systematics of black shales from the Middle Proterozoic Velkerri and Wollogorang Formations, McArthur Basin, northern Australia. Geochim Cosmochim Acta 73:2534–2558

Kendall B, Reinhard CT, Lyons TW, Kaufman AJ, Poulton SW, Anbar AD (2010) Pervasive oxygenation along late Archean ocean margins. Nat Geosci 3:647–652

Kendall B, Gordon GW, Poulton SW, Anbar AD (2011) Molybdenum isotope constraints on the extent of late Paleoproterozoic ocean euxinia. Earth Planet Sci Lett 307:450–460

Kendall B, Brennecka GA, Weyer S, Anbar AD (2013) Uranium isotope fractionation suggests oxidative uranium mobilization at 2.50 Ga. Chem Geol 362:105–114

Kendall B, Komiya T, Lyons TW, Bates SM, Gordon GW, Romaniello SJ, Jiang G, Creaser RA, Xiao S, McFadden K, Sawaki Y, Tahata M, Shu D, Han J, Li Y, Chu X, Anbar AD (2015a) Uranium and molybdenum isotope evidence for an episode of widespread ocean oxygenation during the late Ediacaran Period. Geochim Cosmochim Acta 156:173–193

Kendall B, Creaser RA, Reinhard CT, Lyons TW, Anbar AD (2015b) Transient episodes of mild environmental oxygenation and oxidative continental weathering during the late Archean. Sci Adv 1:e1500777

Kowalski N, Dellwig O, Beck M, Gräwe U, Neubert N, Nägler TF, Badewien TH, Brumsack HJ, van Beusekom JEE, Böttcher ME (2013) Pelagic molybdenum concentration anomalies and the impact of sediment resuspension on the molybdenum budget in two tidal systems of the North Sea. Geochim Cosmochim Acta 119:198–211

Kurzweil F, Wille M, Schoenberg R, Taubald H, Van Kranendonk MJ (2015a) Continuously increasing δ^{98}Mo values in Neoarchean black shales and iron formations from the Hamersley Basin. Geochim Cosmochim Acta 164:523–542

Kurzweil F, Drost K, Pašava J, Wille M, Taubald H, Schoeckle D, Schoenberg R (2015b) Coupled sulfur, iron and molybdenum isotope data from black shales of the Teplá-Barrandian unit argue against deep ocean oxygenation during the Ediacaran. Geochim Cosmochim Acta 171:121–142

Lane S, Proemse BC, Tennant A, Wieser ME (2013) Concentration measurements and isotopic composition of airborne molybdenum collected in an urban environment. Anal Bioanal Chem 405:2957–2963

Lehmann B, Nägler TF, Holland HD, Wille M, Mao J, Pan J, Ma D, Dulski P (2007) Highly metalliferous carbonaceous shale and Early Cambrian seawater. Geology 35:403–406

Liermann LJ, Guynn RL, Anbar A, Brantley SL (2005) Production of a molybdophore during metal-targeted dissolution of silicates by soil bacteria. Chem Geol 220:285–302

Liermann LJ, Mathur R, Wasylenki LE, Nuester J, Anbar AD, Brantley SL (2011) Extent and isotopic composition of Fe and Mo release from two Pennsylvania shales in the presence of organic ligands and bacteria. Chem Geol 281:167–180

Liu Y (2008) Theoretical study on the mechanism of the removal of Mo from seawater in oxic environment. Geochim Cosmochim Acta 72:A564

Lillis PG, Selby D (2013) Evaluation of the rhenium–osmium geochronometer in the Phosphoria petroleum system, Bighorn Basin of Wyoming and Montana, USA. Geochim Cosmochim Acta 118:312–330

Lyons TW, Reinhard CT, Planavsky NJ (2014) The rise of oxygen in Earth's early ocean and atmosphere. Nature 506:307–315

Mahdaoui F, Reisberg L, Michels R, Hauteville Y, Poirier Y, Girard J-P (2013) Effect of the progressive precipitation of petroleum asphaltenes on the Re-Os radioisotope system. Chem Geol 358:90–100

Malcolm S (1985) Early diagenesis of molybdenum in estuarine sediments. Mar Chem 16:213–225

Malinovsky D, Rodushkin I, Baxter DC, Ingri J, Öhlander B (2005) Molybdenum isotope ratio measurements on geological samples by MC-ICPMS. Inter J Mass Spec 245:94–107

Malinovsky D, Hammarlund D, Ilyashuk B, Martinsson O, Gelting J (2007) Variations in the isotopic composition of molybdenum in freshwater lake systems. Chem Geol 236:181–198

Maréchal CN, Telouk P, Albarede F (1999) Precise analysis of copper and zinc isotopic compositions by plasma-source mass spectrometry. Chem Geol 156:251–273

Mathur R, Brantley S, Anbar A, Munizaga F, Maksaev V, Newburry R, Vervoort J, Hart G (2010) Variation of Mo isotopes from molybdenite in high-temperature hydrothermal ore deposits. Min Deposit 45:43–50

Mayer AJ, Wieser ME (2014) The absolute isotopic composition and atomic weight of molybdenum in SRM 3134 using an isotopic double spike. J Anal At Spectrom 29:85–94

McManus J, Nägler TF, Siebert C, Wheat CG, Hammond DE (2002) Oceanic molybdenum isotope fractionation: diagenesis and hydrothermal ridge-flank alteration. Geochem Geophys Geosyst 3:1078, 10.1029/2002GC000356

McManus J, Berelson WM, Severmann S, Poulson RL, Hammond DE, Klinkhammer GP, Holm C (2006) Molybdenum and uranium geochemistry in continental margin sediments: Paleoproxy potential. Geochim Cosmochim Acta 70:4643–4662

Mendel RR, Bittner F (2006). Cell biology of molybdenum. Biochim Biophys Acta 1763:621–635

Migeon V, Bourdon B, Pili E, Fitoussi C (2015) An enhanced method for molybdenum separation and isotopic determination in uranium-rich materials and geological samples. J Anal At Spectrom 30:1988–1996

Miller CA, Peucker-Ehrenbrink B, Walker BD, Marcantonio F (2011) Re-assessing the surface cycling of molybdenum and rhenium. Geochim Cosmochim Acta 75:7146–7179

Miller RW, Eady RR (1988) Molybdenum and vanadium nitrogenases of *Azotobacter chroococcum*. Biochem J 256:429–432

Morford JL, Emerson S (1999) The geochemistry of redox sensitive trace metals in sediments. Geochim Cosmochim Acta 63:1735–1750

Morford JL, Emerson SR, Breckel EJ, Kim SH (2005) Diagenesis of oxyanions (V, U, Re, and Mo) in pore waters and sediments from a continental margin. Geochim Cosmochim Acta 69:5021–5032

Morford JL, Martin WR, Kalnejais LH, François R, Bothner M, Karle I-M (2007) Insights on geochemical cycling of U, Re and Mo from seasonal sampling in Boston Harbor, Massachusetts, USA. Geochim Cosmochim Acta 71:895–917

Morris AW (1975) Dissolved molybdenum and vanadium in the northeast Atlantic Ocean. Deep Sea Res 22:49–54

Murthy VR (1962) Isotopic anomalies of molybdenum in some iron meteorites. J Geophys Res 67:905–907

Murthy VR (1963) Elemental and isotopic abundances of molybdenum in some meteorites. Geochim Cosmochim Acta 27:1171–1178

Nagai Y, Yokoyama T (2016) Molybdenum isotopic analysis by negative thermal ionization mass spectrometry (N–TIMS): effects on oxygen isotopic composition. J Anal At Spectrom 31:948–960

Nägler TF, Siebert C, Lüschen H, Böttcher ME (2005) Sedimentary Mo isotope record across the Holocene fresh—brackish water transition of the Black Sea. Chem Geol 219:283–295

Nägler TF, Neubert N, Böttcher ME, Dellwig O, Schnetger B (2011) Molybdenum isotope fractionation in pelagic euxinia: evidence from the modern Black and Baltic Seas. Chem Geol 289:1–11

Nägler TF, Anbar AD, Archer C, Goldberg T, Gordon GW, Greber ND, Siebert C, Sohrin Y, Vance D (2014) Proposal for an international molybdenum isotope measurement standard and data representation. Geostand Geoanal Res 38:149–151

Nakagawa Y, Takano S, Firdaus ML, Norisuye K, Hirata T, Vance D, Sohrin Y (2012) The molybdenum isotopic composition of the modern ocean. Geochem J 46:131–141

Nameroff TJ, Balistrieri LS, Murray JW (2002). Suboxic trace metal geochemistry in the Eastern Tropical North Pacific. Geochim Cosmochim Acta 66:1139–1158

Neubert N, Nägler TF, Böttcher ME (2008) Sulfidity controls molybdenum isotope fractionation onto euxinic sediments: Evidence from the modern Black Sea. Geology 36:775–778

Neubert N, Heri AR, Voegelin AR, Nägler TF, Schlunegger F, Villa IM (2011) The molybdenum isotopic composition in river water: constraints from small catchments. Earth Planet Sci Lett 304:180–190

Nissenbaum A, Swaine D (1976). Organic matter-metal interactions in recent sediments: the role of humic substances. Geochim Cosmochim Acta 40:809–816

Noordmann J, Weyer S, Montoya-Pino C, Dellwig O, Neubert N, Eckert S, Paetzel M, Böttcher ME (2015) Uranium and molybdenum isotope systematics in modern euxinic basins: case studies from the central Baltic Sea and the Kyllaren fjord (Norway). Chem Geol 396:182–195

Oyerinde OF, Weeks CL, Anbar AD, Spiro TG (2008) Solution structure of molybdic acid from Raman spectroscopy and DFT analysis. Inorg Chim Acta 361:1000–1007

Partin CA, Bekker A, Planavsky NJ, Lyons TW (2015) Euxinic conditions recorded in the ca. 1.93 Ga Bravo Lake Formation, Nunavut (Canada): implications for oceanic redox evolution. Chem Geol 417:148–162

Pearce CR, Cohen AS, Coe AL, Burton KW (2008) Molybdenum isotope evidence for global ocean anoxia coupled with perturbations to the carbon cycle during the Early Jurassic. Geology 36:231–234

Pearce CR, Cohen AS, Parkinson IJ (2009) Quantitative separation of molybdenum and rhenium from geological materials for isotopic determination by MC–ICP–MS. Geostand Geoanal Res 33:219–229

Pearce CR, Burton KW, Pogge von Strandmann PAE, James RH, Gíslason S (2010a) Molybdenum isotope behavior accompanying weathering and riverine transport in a basaltic terrain. Earth Planet Sci Lett 295:104–114

Pearce CR, Coe AL, Cohen AS (2010b) Seawater redox variations during the deposition of the Kimmeridge Clay Formation, United Kingdom (Upper Jurassic): evidence from molybdenum isotopes and trace metal ratios. Paleoceanogr PA4213

Pietruszka A, Walker RJ, Candela PA (2006) Determination of mass-dependent molybdenum isotopic variations by MC–ICP–MS: an evaluation of matrix effects. Chem Geol 225:121–136

Planavsky NJ, Asael D, Hofmann A, Reinhard CT, Lalonde SV, Knudsen A, Wang X, Ossa FO, Pecoits E, Smith AJB, Beukes NJ, Bekker A, Johnson TM, Konhauser KO, Lyons TW, Rouxel OJ (2014) Evidence for oxygenic photosynthesis half a billion years before the Great Oxidation Event. Nat Geosci 7:283–286

Poulson RL, Siebert C, McManus J, Berelson WM (2006) Authigenic molybdenum isotope signatures in marine sediments. Geology 34:617–620

Poulson Brucker RL, McManus J, Severmann S, Berelson WM (2009) Molybdenum behavior during early diagenesis: insights from Mo isotopes. Geochem Geophys Geosyst 10:Q06010

Poulton SW, Canfield DE (2005) Development of a sequential extraction procedure for iron: implications for iron partitioning in continentally derived particulates. Chem Geol 214:209–221
Poulton SW, Canfield DE (2011) Ferruginous conditions: a dominant feature of the ocean through Earth's history. Elements 7:107–112
Poulton SW, Raiswell R (2002) The low-temperature geochemical cycle of iron: from continental fluxes to marine sediment deposition. Am J Sci 302:774–805
Proemse BC, Grasby SE, Wieser ME, Mayer B, Beauchamp B (2013) Molybdenum isotopic evidence for oxic marine conditions during the latest Permian extinction. Geology 41:967–970
Rahaman W, Goswami V, Singh SK, Rai VK (2014) Molybdenum isotopes in two Indian estuaries: Mixing characteristics and input to oceans. Geochim Cosmochim Acta 141:407–422
Raiswell R, Canfield DE (1998) Sources of iron for pyrite formation in marine sediments. Amer J Sci 298:219–245
Raymond J, Siefert JL, Staples CR, Blankenship RE (2003) The natural history of nitrogen fixation. Mol Biol Evol 21:541–554
Rees DC, Akif Tezcan F, Haynes CA, Walton MY, Andrade S, Einsle O, Howard JB (2005) Structural basis of biological nitrogen fixation. Phil Trans Series A Math Phys Eng Sci 363:971–984; discussion 1035–1040
Reinhard CT, Raiswell R, Scott C, Anbar AD, Lyons TW (2009) A late Archean sulfidic sea stimulated by early oxidative weathering of the continents. Science 326:713:716
Reinhard CT, Planavsky NJ, Robbins LJ, Partin CA, Gill BC, Lalonde SV, Bekker A, Konhauser KO, Lyons TW (2013a) Proterozoic ocean redox and biogeochemical stasis. PNAS 110:5357–5362
Reinhard CT, Lalonde SV, Lyons TW (2013b) Oxidative sulfide dissolution on the early Earth. Chem Geol 362:44–55
Reitz A, Wille M, Nägler TF, de Lange GJ (2007) Atypical Mo isotope signatures in eastern Mediterranean sediments. Chem Geol 245:1–8
Rempel K, Migdisov A, Williams-Jones A (2006) The solubility and speciation of molybdenum in water vapour at elevated temperatures and pressures: implications for ore genesis. Geochim Cosmochim Acta 70:687–696
Rempel KU, Williams-Jones AE, Migdisov AA (2009) The partitioning of molybdenum (VI) between aqueous liquid and vapour at temperatures up to 370 °C. Geochim Cosmochim Acta 73:3381–3392
Romaniello SJ, Herrmann AD, Anbar AD (2016) Syndepositional diagenetic control of molybdenum isotope variations in carbonate sediments from the Bahamas. Chem Geol 438:84–90
Romao MJ, Knäblein J, Huber R, Moura JJG (1997) Structure and function of molybdopterin containing enzymes. Progress Biophys Molec Biol 68:121–144
Ryb U, Erel Y, Matthews A, Avni Y, Gordon GW, Anbar AD (2009) Large molybdenum isotope variations trace subsurface fluid migration along the Dead Sea transform. Geology 37:463–466
Sánchez-Baracaldo P, Ridgwell A, Raven JA (2014). A neoproterozoic transition in the marine nitrogen cycle. Current Biol 24:652–657
Sarmiento JL, Gruber N (2006) Ocean Biogeochemical Dynamics. Princeton University Press, New Jersey, USA
Saxena RS, Jain MC, Mittal ML (1968). Electrometric investigations of an acid-thiomolybdate system and the formation of polyanions. Aus J Chem 21:91–96
Schoepp-Cothenet B, van Lis R, Philippot P, Magalon A, Russell MJ, Nitschke W (2012). The ineluctable requirement for the trans-iron elements molybdenum and/or tungsten in the origin of life. Sci Rep 2:263
Scheiderich K, Zerkle AL, Helz GR, Farquhar J, Walker RJ (2010a) Molybdenum isotope, multiple sulfur isotope, and redox-sensitive element behavior in early Pleistocene Mediterranean sapropels. Chem Geol 279:134–144
Scheiderich K, Helz GR, Walker RJ (2010b) Century-long record of Mo isotopic composition in sediments of a seasonally anoxic estuary (Chesapeake Bay). Earth Planet Sci Lett 289:189–197
Scholz F, McManus J, Sommer S (2013) The manganese and iron shuttle in a modern euxinic basin and implications for molybdenum cycling at euxinic ocean margins. Chem Geol 355:56–68
Scott C, Lyons TW (2012) Contrasting molybdenum cycling and isotopic properties in euxinic versus non-euxinic sediments and sedimentary rocks: Refining the paleoproxies. Chem Geol 324–325:19–27
Scott C, Lyons TW, Bekker A, Shen Y, Poulton SW, Chu X, Anbar AD (2008) Tracing the stepwise oxygenation of the Proterozoic ocean. Nature 452:456–459
Segato A, Kendall B, Hanley J (2015) Further insights into Mo isotope variations in molybdenites from different ore deposits. Geol Soc Am Abstr Progr 47:243
Selby D, Creaser RA, Fowler MG (2007) Re–Os elemental and isotopic systematics in crude oils. Geochim Cosmochim Acta 71:378–386
Shafiei B, Shamanian G, Mathur R, Mirnejad H (2015) Mo isotope fractionation during hydrothermal evolution of porphyry systems. Min Deposit 50:281–291
Shaw TJ, Gieskes JM, Jahnke RA (1990) Early diagenesis in differing depositional environments: The response of transition metals in pore water. Geochim Cosmochim Acta 54:1233–1246
Shimmield G, Price N (1986). The behaviour of molybdenum and manganese during early sediment diagenesis offshore Baja California, Mexico. Mar Chem 19:261–280
Siebert C, Nägler TF, Kramers JD (2001) Determination of molybdenum isotope fractionation by double-spike multicollector inductively coupled plasma mass spectrometry. Geochem Geophys Geosyst 2:2000GC000124

Siebert C, Nägler TF, von Blanckenburg F, Kramers JD (2003) Molybdenum isotope records as a potential new proxy for paleoceanography. Earth Planet Sci Lett 211:159–171

Siebert C, Kramers JD, Meisel TH, Morel PH, Nägler TF (2005) PGE, Re–Os, and Mo isotope systematics in Archean and early Proterozoic sedimentary systems as proxies for redox conditions of the early Earth. Geochim Cosmochim Acta 69:1787–1801

Siebert C, McManus J, Bice A, Poulson R, Berelson WM (2006) Molybdenum isotope signatures in continental margin sediments. Earth Planet Sci Lett 241:723–733

Siebert C, Pett-Ridge JC, Opfergelt S, Guicharnaud RA, Halliday AN, Burton KW (2015) Molybdenum isotope fractionation in soils: Influence of redox conditions, organic matter, and atmospheric inputs. Geochim Cosmochim Acta 162:1–24

Skierszkan EK, Amini M, Weis D (2015) A practical guide for the design and implementation of the double-spike technique for precise determination of molybdenum isotope compositions of environmental samples. Anal Bioanal Chem 407:1925–1935

Smith RM, Martell AE (2004) NIST critically selected stability constants of metal complexes database. NIST Standard Reference Database 46, Version 8.0. National Institute of Standards and Technology

Song S, Hu K, Wen H, Zhang Y, Li K, Fan H (2011) Molybdenum isotopic composition as a tracer for low-medium temperature hydrothermal ore-forming systems: A case study on the Dajiangping pyrite deposit, western Guangdong Province, China. Chinese Science Bull 56:2221–2228

Stein HJ, Markey RJ, Morgan JW, Hannah JL, Scherstén A (2001) The remarkable Re–Os chronometer in molybdenite: how and why it works. Terra Nova 13:479–486

Stiefel EI (1997) Chemical keys to molybdenum enzymes. J Chem Soc Dalton Trans 21:3915–3923

Stüeken EE, Buick R, Anbar AD (2015) Selenium isotopes support free O_2 in the latest Archean. Geology 43:259–262

Sundby B, Martinez P, Gobeil C (2004) Comparative geochemistry of cadmium, rhenium, uranium, and molybdenum in continental margin sediments. Geochim Cosmochim Acta 68:2485–2493

Szilagyi M (1967) Sorption of molybdenum by humus preparations. Geochem Internat 4:1165–1167

Tossell JA (2005) Calculating the partitioning of the isotopes of Mo between oxidic and sulfidic species in aqueous solution. Geochim Cosmochim Acta 69:2981–2993

Tribovillard N, Algeo TJ, Lyons T, Riboulleau A (2006) Trace metals as paleoredox and paleoproductivity proxies: an update. Chem Geol 232:12–32

Tribovillard N, Algeo TJ, Baudin F, Riboulleau A (2012) Analysis of marine environmental conditions based on molybdenum–uranium covariation—Applications to Mesozoic paleoceanography. Chem Geol 324–325:46–58

Tuit C (2003) The marine biogeochemistry of molybdenum, Massachussets Institute of Technology and Woods Hole Oceanographic Institution. PhD thesis

Ulrich T, Mavrogenes J (2008) An experimental study of the solubility of molybdenum in H_2O and KCl–H_2O solutions from 500 °C to 800 °C, and 150 to 300 MPa. Geochim Cosmochim Acta 72:2316–2330

Urey HC (1947) The thermodynamic properties of isotopic substances. J Chem Soc 562–581

Ventura GT, Gall L, Siebert C, Prytulak J, Szatmari P, Hürlimann M, Halliday AN (2015) The stable isotope composition of vanadium, nickel, and molybdenum in crude oils. Appl Geochem 59:104–117

Voegelin AR, Nägler TF, Samankassou E, Villa IM (2009) Molybdenum isotopic composition of modern and Carboniferous carbonates. Chem Geol 265:488–498

Voegelin AR, Nägler TF, Beukes NJ, Lacassie JP (2010) Molybdenum isotopes in late Archean carbonate rocks: Implications for early Earth oxygenation. Precambr Res 182:70–82

Voegelin AR, Nägler TF, Pettke T, Neubert N, Steinmann M, Pourret O, Villa IM (2012) The impact of igneous bedrock weathering on the Mo isotopic composition of stream waters: Natural samples and laboratory experiments. Geochim Cosmochim Acta 86:150–165

Voegelin AR, Pettke T, Greber ND, von Niederhäusern B, Nägler TF (2014) Magma differentiation fractionates Mo isotope ratios: evidence from the Kos Plateau Tuff (Aegean Arc). Lithos 190:440–448

Vorlicek TP, Kahn MD, Kasuya Y, Helz GR (2004) Capture of molybdenum in pyrite-forming sediments: role of ligand-induced reduction by polysulfides. Geochim Cosmochim Acta 68:547–556

Wang D, Aller RC, Sañudo-Wilhelmy SA (2011) Redox speciation and early diagenetic behavior of dissolved molybdenum in sulfidic muds. Marine Chem 125:101–107

Wang Y, Zhou L, Gao S, Li JW, Hu ZF, Yang L, Hu ZC (2016) Variation of molybdenum isotopes in molybdenite from porphyry and vein Mo deposits in the Gangdese metallogenic belt, Tibetan plateau and its implications. Min Deposit 51:201–220

Wang Z, Ma J, Li J, Wei G, Chen X, Deng W, Xie L, Lu W, Zou L (2015) Chemical weathering controls on variations in the molybdenum isotopic composition of river water: Evidence from large rivers in China. Chem Geol 410:201–212

Wasylenki LE, Anbar AD, Liermann LJ, Mathur R, Gordon GW, Brantley SL (2007) Isotope fractionation during microbial metal uptake measured by MC–ICP–MS. J Anal At Spectrom 22:905–910

Wasylenki LE, Rolfe BA, Weeks CL, Spiro TG, Anbar AD (2008) Experimental investigation of the effects of temperature and ionic strength on Mo isotope fractionation during adsorption to manganese oxides. Geochim Cosmochim Acta 72:5997–6005

Wasylenki LE, Weeks CL, Bargar JR, Spiro TG, Hein JR, Anbar AD (2011) The molecular mechanism of Mo isotope fractionation during adsorption to birnessite. Geochim Cosmochim Acta 75:5019–5031

Wen H, Carignan J, Zhang Y, Fan H, Cloquet C, Liu S (2011) Molybdenum isotopic records across the Precambrian-Cambrian boundary. Geology 39:775–778

Wen H, Fan H, Zhang Y, Cloquet C, Carignan J (2015) Reconstruction of early Cambrian ocean chemistry from Mo isotopes. Geochim Cosmochim Acta 164:1–16

Westermann S, Vance D, Cameron V, Archer C, Robinson SA (2014) Heterogeneous oxygenation states in the Atlantic and Tethys oceans during Oceanic Anoxic Event 2. Earth Planet Sci Lett 404:178–189

Wetherill GW (1964) Isotopic composition and concentration of molybdenum in iron meteorites. J Geophys Res 69:4403–4408

Wheat CG, Mottl MJ, Rudnicki M (2002) Trace element and REE composition of a low-temperature ridge-flank hydrothermal spring. Geochim Cosmochim Acta 66:3693–3705

White WM (2015) Isotope Geochemistry. Wiley-Blackwell

Wichard T, Mishra B, Myneni SCB, Bellenger J-P, Kraepiel AML (2009). Storage and bioavailability of molybdenum in soils increased by organic matter complexation. Nature Geosci 2:625–629

Wieser ME, De Laeter JR, Varner MD (2007) Isotope fractionation studies of molybdenum. Internat J Mass Spec 265:40–48

Willbold M, Hibbert K, Lai YJ, Freymuth H, Hin RC, Coath C, Vils F, Elliott T (2016) High-precision mass-dependent molybdenum isotope variations in magmatic rocks determined by double-spike MC–ICP–MS. Geostand Geoanal Res, Online Version (Early View)

Wille M, Kramers JD, Nägler TF, Beukes NJ, Schröder S, Meisel TH, Lacassie JP, Voegelin AR (2007) Evidence for a gradual rise of oxygen between 2.6 and 2.5 Ga from Mo isotopes and Re–PGE signatures in shales. Geochim Cosmochim Acta 71:2417–2435

Wille M, Nägler TF, Lehmann B, Schröder S, Kramers JD (2008) Hydrogen sulphide release to surface waters at the Precambrian/Cambrian boundary. Nature 453:767–769

Wille M, Nebel O, Van Kranendonk MJ, Schoenberg R, Kleinhanns IC, Ellwood MJ (2013) Mo–Cr isotope evidence for a reducing Archean atmosphere in 3.46–2.76 Ga black shales from the Pilbara, Western Australia. Chem Geol 340:68–76

Xu L, Lehmann B, Mao J, Nägler TF, Neubert N, Böttcher ME, Escher P (2012) Mo isotope and trace element patterns of Lower Cambrian black shales in South China: Multi proxy constraints on the paleoenvironment. Chem Geol 318–319:45–59

Yang J, Siebert C, Barling J, Savage P, Liang Y-H, Halliday AN (2015) Absence of molybdenum isotope fractionation during magmatic differentiation at Hekla volcano, Iceland. Geochim Cosmochim Acta 162:126–136

Yin Q, Jacobsen SB, Yamashita K (2002) Diverse supernova sources of pre-solar material inferred from molybdenum isotopes in meteorites. Nature 415:881–883

Zerkle AL, Scheiderich K, Maresca JA, Liermann LJ, Brantley SL (2011) Molybdenum isotope fractionation by cyanobacterial assimilation during nitrate utilization and N_2 fixation. Geobiol 9:94–106

Zerkle AL, House CH, Cox RP, Canfield DE (2006). Metal limitation of cyanobacterial N_2 fixation and implications for the Precambrian nitrogen cycle. Geobiol 4:285–297

Zerkle AL, Claire MW, Domagal-Goldman SD, Farquhar J, Poulton SW (2012) A bistable organic-rich atmosphere on the Neoarchaean Earth. Nat Geosci 5:359–363

Zhang L, Audétat A, Dolejš D (2012) Solubility of molybdenite (MoS_2) in aqueous fluids at 600–800 °C, 200 MPa: A synthetic fluid inclusion study. Geochim Cosmochim Acta 77:175–185

Zheng Y, Anderson RF, Geen AV, Kuwabara K. (2000) Authigenic molybdenum formation in marine sediments: A link to pore water sulfide in the Santa Barbara Basin. Geochim Cosmochim Acta 64:4165–4178

Zhou L, Wignall PB, Su J, Feng Q, Xie S, Zhao L, Huang J (2012) U/Mo ratios and $\delta^{98/95}$Mo as local and global redox proxies during mass extinction events. Chem Geol 324–325:99–107

Zhou L, Algeo TJ, Shen J, Hu Z, Gong H, Xie S, Huang J, Gao S (2015) Changes in marine productivity and redox conditions during the Late Ordovician Hirnantian glaciation. Palaeogeogr Palaeoclimatol Palaeoecol 420:223–234

Reviews in Mineralogy & Geochemistry
Vol. 82 pp. 733-757, 2017

17

Recent Developments in Mercury Stable Isotope Analysis

Joel D. Blum and Marcus W. Johnson

Department of Earth and Environmental Sciences
University of Michigan
1100 North University Avenue
Ann Arbor, Michigan 48109
USA

INTRODUCTION

The first *Reviews in Mineralogy* volume on the *Geochemistry of Non-Traditional Stable Isotopes* was compiled before it was appropriate to include a chapter on mercury (Hg) stable isotope geochemistry. At that time there were only a few papers on this new topic (Jackson 2001; Lauretta et al. 2001; Hintelmann and Lu 2003), and there were still some important analytical issues that needed to be resolved. But the field has come a long way in a decade. Now we have a different problem; at our last count there were well over 100 publications utilizing mercury stable isotopes and it is becoming very difficult to synthesize this vast amount of exciting and rapidly developing research. Experimental studies have expanded our knowledge of the mechanisms of mercury isotope fractionation and applications of mercury isotope measurements have touched virtually every area of research in mercury biogeochemistry. There have been a number of previous reviews of the mercury stable isotope literature as it has developed (Ridley and Stetson 2006; Bergquist and Blum 2009; Yin et al. 2010; Blum 2011; Hintelmann 2012; Blum et al. 2014).

It is our view that the field has become too large to comprehensively review the entire literature on mercury stable isotopes. Ten years ago Hg isotope researchers were just beginning to explore the boundaries of natural Hg isotope variation and the mechanisms that cause this variation in the environment. At that time large and relatively easily measured isotope signals were of great interest and mercury isotope researchers were beginning to develop theories to explain mass dependent isotope fractionation (MDF) and mass independent isotope fractionation of the odd mass-numbered isotopes of mercury (odd-MIF). More recently researchers have discovered a wider range of types of isotopic variability (even-MIF), some of which are subtle and require great attention to measurement protocols and quality control measures. Additionally, researchers have continued to refine chemical separation protocols that allow Hg isotope measurements of samples with lower and lower Hg concentrations, which requires more attention to procedural blank effects and matrix interferences.

In writing this chapter our goal was to keep from repeating material presented in previous reviews, in particular our most recent review (Blum et al. 2014). Instead we focus on new observations of mercury isotope variation, new developments in the analysis of mercury isotope ratios, and some of the more recent questions that have arisen from mercury isotope research over the past three to four years. For readers who are new to mercury isotope research we refer you first to Blum et al. (2014), which is a comprehensive compilation of all published mercury isotope data on natural samples as of July 2013. In addition to outlining the boundaries of mercury isotope ratio variation in various types of materials, Blum et al. (2014) provides a summary of experimental studies that help constrain the mechanisms responsible for the observed isotope variation, and also summarizes the basic nomenclature and analytical

1529-6466/17/0082-0017$05.00 http://dx.doi.org/10.2138/rmg.2017.82.17

methods that have been adopted by the mercury isotope community. We have focused our attention in the present chapter on analytical advances that have recently improved research capabilities and on recent studies that are taking the field in new directions.

Element properties

The atomic number of mercury is 80 and the atomic mass based on determination of the isotopic composition of the National Institute of Standards and Technology Standard Reference Material (NIST SRM) 3133 by Blum and Bergquist (2007) is 200.6025 atomic mass units. Mercury has seven stable isotopes ranging from mass ^{196}Hg to mass ^{204}Hg. The longest-lived radioactive isotope is ^{194}Hg, which has a half-life of 520 years. Mercury occurs in three main oxidation states: Hg(0) (elemental mercury), Hg(I) (mercurous mercury), and Hg(II) (mercuric mercury). Hg(II) is the most common oxidation state in natural solid materials and commonly forms sulfides, chlorides, selenides and tellurides.

Hg isotope nomenclature

Although we will not repeat all of the details of mercury isotope nomenclature here, it is appropriate to review some of the basics. Nearly all mercury isotope studies use NIST SRM 3133 as an isotopic reference material. Sample-standard bracketing is used to determine the difference in isotopic compositions between samples and NIST-3133. Hg has seven stable isotopes with the following abundances in NIST-3133 as determined by Blum and Bergquist (2007): ^{196}Hg = 0.16%, ^{198}Hg = 10.04%, ^{199}Hg = 16.94%, ^{200}Hg = 23.14%, ^{201}Hg = 13.17%, ^{202}Hg = 29.73%, ^{204}Hg = 6.83% . The National Research Council of Canada reference material NIMS-1 is a dilution of NIST-SRM-3133 and has been certified for its isotopic composition (Meija et al. 2010).

For most high precision mercury isotope ratio measurements, solutions containing Hg as Hg(II) are reduced online with $SnCl_2$. The resulting Hg(0) is stripped from solution to produce a stream of Hg(0) vapor using a frosted tip phase separator (Lauretta et al. 2001; Blum and Bergquist 2007). A dry thallium (Tl) aerosol is produced from solution using a desolvating nebulizer and this aerosol is mixed with the Hg vapor prior to introduction into a multiple collector inductively coupled plasma mass spectrometer (MC-ICP-MS). It is common practice to use the Tl reference standard NIST-997 (with certified ^{205}Tl/^{203}Tl ratio of 2.38714) to correct the mass bias in the MC-ICP-MS analysis of Hg isotope ratios using an exponential fractionation law. The Hg isotope ratios are reported as delta values in permil (‰) relative to the average ratios measured in the NIST-3133 Hg reference material run before and after each sample following the equation:

$$\delta^{xxx}\mathrm{Hg}(‰) = ([(^{xxx}\mathrm{Hg}/^{198}\mathrm{Hg})_{\mathrm{unknown}} / (^{xxx}\mathrm{Hg}/^{198}\mathrm{Hg})_{\mathrm{NIST\text{-}3133}}] - 1) \times 1000$$

The use of ^{198}Hg as the denominator in δ^{xxx}Hg values was recommended early in the history of Hg isotope studies (Blum and Bergquist 2007) and has been adopted by almost all research groups (see Blum et al. 2014). This follows the convention in stable isotope geochemistry, which is to report isotope fractionation using ratios with the smaller mass in the denominator so that higher delta values always signify a "heavier" average isotopic mass (Johnson et al. 2004; Coplen 2011).

In addition to MDF, almost all studies of Hg isotopes in the environment have observed significant MIF of the odd-mass-number isotopes of Hg and several recent studies suggest that the even-mass-number isotopes of Hg can also undergo MIF in the environment, particularly in samples of Hg derived from the atmosphere. MIF is measured as the difference between a measured δ-value and the δ-value that is predicted based on the measured δ^{202}Hg value and the kinetic MDF law derived from transition state theory. For variations of less than ~10‰, these values are well approximated as (Blum and Bergquist 2007):

$$\Delta^{199}Hg = \delta^{199}Hg - (\delta^{202}Hg \times 0.2520)$$
$$\Delta^{200}Hg = \delta^{200}Hg - (\delta^{202}Hg \times 0.5024)$$
$$\Delta^{201}Hg = \delta^{201}Hg - (\delta^{202}Hg \times 0.7520)$$
$$\Delta^{204}Hg = \delta^{204}Hg - (\delta^{202}Hg \times 1.4930)$$

The odd-mass-number isotopes of Hg undergo the highest degree of MIF due to the magnetic isotope effect, which occurs during aqueous photochemical radical pair reactions. Because only the odd-mass-number isotopes have nuclear magnetic moments and nuclear spin, triplet to singlet and singlet to triplet intersystem crossing is enhanced among these isotopes. This process causes the odd- and even-mass-number isotopes to react at differing rates and often results in high levels of odd-MIF (>1‰). For expression of the magnetic isotope effect, radical pairs need a long enough lifetime for intersystem crossing to compete with dissociation into free radicals. Therefore the magnetic isotope effect is only expressed in aqueous and perhaps surface reactions where radical pairs are in a solvent cage and not in pure gas phase reactions. A small amount of odd-MIF ($<\sim 0.5$‰) can also be produced in equilibrium and kinetic reactions due to the nuclear volume effect, which results from differences in nuclear charge radii of the Hg isotopes that are not linear with mass, and the effect of these variations on bond strengths (Ghosh et al. 2013).

Reference materials

Adoption of a common set of well-characterized and readily available reference materials has aided the rapid development of mercury stable isotope geochemistry. Early in the history of high precision mercury isotope measurement most researchers in this field adopted a common bracketing standard, NIST-SRM-3133, which is defined as having $\delta^{202}Hg = 0.0$‰. Similarly, most labs adopted NIST-SRM-997 as the thallium standard to be used for MC-ICP-MS mass bias correction. For the purposes of quality control a secondary standard solution (UM-Almaden) has been widely distributed and used to determine the precision of Hg isotope ratio analyses and to inter-calibrate isotope ratio measurements between laboratories (Blum and Bergquist 2007). This standard was distributed by the University of Michigan Laboratory between 2007 and 2015 and is now being distributed by NIST. Whereas UM-Almaden is essential for monitoring mass spectrometry measurements, secondary standards with matrices similar to sample unknowns should also be analyzed along with each batch of samples for acceptable quality assurance. This allows comparison of the effects of sample digestion, matrix separation and either dilution or pre-concentration on the accuracy of isotope values and the precision of measurements for each type of sample analyzed.

Research groups have migrated to a common set of secondary standards that are widely distributed by NIST, the National Research Council of Canada (NRC) and the European Commission Institute for Reference Materials and Measurements (IRMM). Although no official values have been certified for the isotopic composition of these secondary standards, reporting of values obtained for each published study allows comparisons of the results of each research laboratory's methods. We have assembled a table with values from the University of Michigan Laboratory of the most commonly used standards and the mean and standard error of isotope values that we have measured (Table 1). Blum and Bergquist (2007) discuss the use of standard error versus standard deviation in the reporting of analytical uncertainties. In order to assure that our analyses are as precise as possible we perform matrix separation for all analyses, we measure the yield of all digestions and matrix separations, and we closely bracket Hg concentrations for all analyses that we report. The importance of matrix separation and its necessity to achieve the highest precision analyses is discussed in detail below.

Analytical advances

As analytical methods have improved, higher precision measurements on smaller samples have become possible. Isotopic differences less than 0.1‰ are now often interpreted as

Table 1. Results of standard reference material analyses performed at University of Michigan

Source	SRM	Material[1]	Analysis Period	n	δ^{202}	2SE	Δ^{199}	2SE	Δ^{200}	2SE	Δ^{201}	2SE	Δ^{204}	2SE	Note
NRC	NIMS-1	Standard solution	Aug 2013–Mar 2014	7	0.01	0.02	0.01	0.02	0.00	0.01	–0.01	0.01	–0.01	0.01	
Inorganic Ventures	IV-MSHG10ppm	Standard solution	Jan–May 2010	16	–0.75	0.02	0.06	0.02	0.01	0.01	0.04	0.01	–0.04	0.03	Lot # 02057
University of Michigan	UM-Almaden	Standard solution	Jan 2010–May 2016	403	–0.57	<0.005	–0.02	<0.005	0.01	<0.005	–0.04	<0.005	–0.01	<0.005	
NRC	DOLT-2	Fish	Jun 2009–Jul 2014	17	–0.51	0.01	0.73	0.01	0.03	0.01	0.60	0.01	–0.03	0.02	
			Apr–Jul 2014	6	–0.49	0.02	0.72	0.01	0.03	0.01	0.60	0.01	–0.05	0.03	Acid Digest (Purged)
NRC	DORM-2	Fish	Jul 2009–Jul 2014	15	0.20	0.02	1.13	0.01	0.05	0.01	0.92	0.01	–0.07	0.01	
			Apr–Jul 2014	9	0.20	0.02	1.14	0.01	0.05	0.01	0.93	0.01	–0.08	0.02	Acid Digest (Purged)
NRC	DORM-3	Fish	Jun 2011–Jan 2016	32	0.48	0.02	1.80	0.01	0.06	0.01	1.48	0.01	–0.09	0.01	
			Mar 2014	1	0.49		1.80		0.05		1.50		–0.05		Acid Digest (Purged)
NRC	DORM-4	Fish	Oct 2014–May 2015	10	0.47	0.03	1.81	0.03	0.07	0.02	1.48	0.02	–0.10	0.02	
IRMM	ERM-CE464	Fish	Oct 2010–Apr 2016	47	0.68	0.01	2.40	0.01	0.08	0.01	1.97	0.01	–0.09	0.01	
			Mar–Apr 2014	3	0.71	0.03	2.40	0.01	0.08	0.01	2.00	0.03	–0.10	0.04	Acid Digest (Purged)
NIST	SRM1947	Fish	Jul 2013–Jan 2015	23	1.19	0.03	5.50	0.01	0.10	0.01	4.30	0.01	–0.14	0.01	
			Mar 2014	5	1.21	0.02	5.47	0.01	0.08	0.02	4.29	0.01	–0.12	0.03	Acid Digest (Purged)
NRC	TORT-2	Lobster	Mar 2011–Mar 2016	89	0.06	0.02	0.75	0.01	0.06	0.01	0.59	0.01	–0.09	0.02	All analyses
			Apr 2011–Nov 2015	36	0.07	0.01	0.75	0.01	0.07	0.01	0.59	0.01	–0.09	0.02	5ppb analyses, only
NRC	TORT-3	Lobster	Dec 2014–May 2015	4	0.07	0.04	0.67	0.05	0.06	0.06	0.55	0.08	–0.08	0.08	UNCG Lot[2]
IRMM	CRM397	Human hair	Jan 2011–Apr 2013	12	0.15	0.01	–0.03	0.02	0.00	0.01	–0.02	0.02	–0.01	0.03	
IRMM	BCR482	Lichen	Aug–Oct 2010	6	–1.67	0.02	–0.67	0.02	0.08	0.01	–0.68	0.01	–0.10	0.01	
NIST	SRM1515	Apple leaves	Oct 2009–Sep 2015	39	–2.61	0.02	0.06	0.02	–0.03	0.01	–0.03	0.01	0.04	0.03	
			Apr–May 2016	10	–2.66	0.04	0.04	0.06	–0.03	0.02	–0.02	0.03	0.05	0.05	UNCG Lot[2]
NIST	SRM1575a	Pine needles	Dec 2012–May 2016	7	–1.32	0.02	–0.37	0.04	0.01	0.02	–0.30	0.03	0.04	0.04	
NRC	MESS-3	Marine sediment	Mar 2011–May 2014	44	–1.88	0.02	0.01	0.01	0.01	0.01	–0.04	0.01	0.01	0.01	
			May 2014–May 2016	14	–2.26	0.05	0.01	0.02	0.01	0.02	–0.04	0.01	0.02	0.04	New Lot (Apr2014)
			Aug 2014–May 2016	7	–1.96	0.07	0.01	0.02	0.01	0.02	–0.03	0.01	–0.04	0.05	UNCG Lot[2]
NIST	SRM1944	Marine sediment	Jan 2010–Mar 2016	62	–0.43	0.01	0.00	0.01	0.01	0.01	–0.02	0.01	–0.01	0.01	
NIST	SRM2711	Contaminated soil	Mar 2014–Oct 2015	11	–0.18	0.05	–0.24	0.01	–0.01	0.01	–0.19	0.02	–0.01	0.03	
NIST	SRM1632c	Coal	Aug 2009–Sep 2010	7	–1.86	0.05	–0.02	0.02	0.01	0.01	–0.04	0.01	–0.03	0.03	

Note 1. Unless otherwise noted, solid SRM material was prepared for isotope analysis by thermal decomposition followed by secondary purge and trap.
Note 2. SRM material supplied by the University of North Carolina, Greensboro.

significant in Hg isotope studies (e.g., Demers et al. 2015; Lepak et al. 2015; Thibodeau et al. 2016). Therefore we think that it is important to review some of the analytical artifacts (or pitfalls) that could potentially compromise the quality of Hg isotope data, and make some recommendations for assuring the highest quality data. We begin by reviewing mass interferences with Hg for MC-ICP-MS analyses and then discuss Hg blanks, sample carry-over and general recommendations for Hg isotope analyses.

Isobaric and molecular interferences

Of the seven stable isotopes of Hg and two stable isotopes of Tl (used for mass bias correction) the only isobaric interferences are from Pt on mass 196 and 198 and from Pb on mass 204. Neither Pt nor Pb readily produce cold vapors and therefore are unlikely to be transported to the MC-ICP-MS as gases. However, aerosol formation and transfer with the Ar carrier gas is possible and should be monitored. The absence of Pt and Pb during analyses can be verified by monitoring the 196/198 ratio for Pt and mass 206 for Pb. ICP-MS cones and guard electrodes are sometimes made from Pt and this material should, of course, be avoided.

Mercury hydride and dihydride can cause interferences on all of the Hg isotopes (except mass 196) and for ^{203}Tl and ^{205}Tl. A recent paper (Georg and Newman 2015) explored the effects of sample matrix composition on Hg hydride and dihydride formation and their effects on Hg isotope ratio measurements. In agreement with the relative stability of the hydrides (Shayesteh et al. 2005), they found HgH_2 to be the dominant hydride. They also found a significant matrix effect whereby volatile hydrocarbon introduction into the plasma enhanced hydride formation as evidenced by an apparent increase in the ^{203}Tl signal with increased Hg concentration. The absence of both HgH_2 and Pb during analyses can be verified by monitoring mass 206 for $^{204}Hg^{1}H_2$.

Oxides and hydroxides of W (WO and WHO) can produce interferences on all of the Hg isotope masses except mass 204. W has a much higher crustal abundance than Hg (~1000-fold) and even small amounts can compromise the quality of Hg concentration data (Guo et al. 2011). In a W isotope study the molecular interference WO was detected at 0.07% of the intensity of the W peak (Shirai and Humayun 2011). W-carbide is a common material used in some types of sample grinders and should be avoided in all sample preparation procedures for Hg isotope analyses. WHO can be monitored by varying the ratio of Tl to Hg and monitoring the 205/203 ratio.

Sample introduction of Hg requires mixture of $SnCl_2$ with solutions containing Hg(II) to reduce Hg(II) to Hg(0) for vapor separation and introduction to the MC-ICP-MS. Sn does not form a vapor under these conditions, but because it is used in very high concentrations and there is the possibility of aerosol formation and transfer with the Ar carrier gas, analysts should monitor Sn-based interferences. Of particular concern are the Sn-bearing molecular interferences $SnCl_2$ (which can interfere with Hg at masses 196 and 198), and SnArAr (which can interfere with Hg at masses 196–200, 202 and 204). $SnCl_2$ can be detected based on the 196/198 ratio and SnArAr can be detected based on mass 197.

As a test of the potential influence of various contaminants on mercury isotope ratios we performed a series of calculations to simulate interferences. We emphasize that we assessed this interference sensitivity as a calculation, not by adding interfering species to actual run solutions. We started with measured peak intensities for 5 ng/g Hg solutions made from the secondary standard UM-Almaden and bracketing standard NIST-SRM-3133. Next, for a particular interference species we limited the maximum contribution for the mass position with the highest relative abundance of the interference (in the mass range from 198 to 204) to a magnitude equal to 10%, 1% and 0.1% of the measured UM-Almaden or NIST-SRM-3133 signal at mass 198, and added it to the appropriate measured signal. Interference contributions for other mass positions with smaller relative abundances of interference were scaled accordingly and added to the measured signals at those mass positions. We repeated this for each interference species. Next we performed an on-peak-zero (OPZ) correction and a mass bias correction based on the Tl peak intensities (allowing the artificially generated interferences at Tl mass positions, if present, to propagate through the mass bias correction and to influence the Hg peak intensities) and calculated small delta and capital delta values for each Hg isotope ratio. This exercise allowed us to A) evaluate which interferences were most significant and B) compare interferences to the small Δ^{200}Hg and Δ^{204}Hg values that we measure in some natural samples to better evaluate whether they could be produced by interferences. Each potential interference (see Table 2 for relative abundances) was evaluated for the following three cases:

1. Interfering species present during measurement of OPZ, bracketing standard and sample solutions.
2. Interfering species present during measurement of bracketing standard and sample solutions, but not during baseline correction measurement.
3. Interfering species present during measurement of sample solution only.

Case 1 accounts for interferences continuously delivered to the MC-ICP-MS, such as: contaminants in the solution delivering Tl to the desolvating nebulizer, the solution delivering $SnCl_2$ to the gas-liquid phase separator, aerosols delivered with the Ar carrier gas, or species entrained from the torch, cone or ion-optic surfaces. This case also includes interferences delivered to the phase separator with the bracketing standard, sample solution and wash solutions, and it assumes equivalent contributions to measurements of the standard and samples. Case 2 limits the sources of contaminants to solutions delivering the bracketing standard and sample to the phase separator. In this case the aerosol stream emanating from the desolvating nebulizer can also contribute interferences if Tl is streamed only during bracketing standard and sample measurement, or if the baseline correction is accomplished by measurement with the ion beams deflected away from the collectors or with the magnetic field set to a half-mass value. Case 3 further limits the source of contaminants to only the sample solution.

For most instances of Case 1, the combination of OPZ correction and standard bracketing completely factors out any effect from the interference, irrespective of the maximum magnitude of the interference (for example, 10% of sample or standard 198 signal), and MDF and MIF results are indistinguishable from those obtained from unmodified source data. Similarly, for most instances of Case 2 assessed at the 10% signal level, standard bracketing is sufficient to eliminate all but very small (0.01 to 0.1‰) interference-related changes to MDF and MIF.

Table 2. The relative abundance of interferences at mercury and thallium isotope mass positions. Bold entries indicate the reference mass for each interference species. Predicted intensities of each interference at other masses are in proportion to the reference mass of 1.00.

Mass	SnArAr	SnClCl	WHO	WO	HgH	HgHH	Pt	Pb
206	0	0	0	0	0.00	0.23	0	0
205	0	0	0.002	0	0.23	0.0004	0	0
204	0.24	0	0.001	0.0019	0	**1.00**	0	**1.00**
203	0	0	0.93	0.0004	**1.00**	0.44	0	0
202	0.19	0	0.002	0.93	0.44	0.78	0	0
201	0	0	**1.00**	0.0013	0.78	0.57	0	0
200	1.34	0	0.47	**1.00**	0.57	0.34	0	0
199	0.35	0	0.86	0.47	0.34	0	0	0
198	**1.00**	**1.00**	0	0.86	0	0.0047	**1.00**	0
197	0	0	0.004	0	0.0047	0	0	0
196	0.61	7.05	0	0.0042	0	0	3.51	0

The exceptions for both Case 1 and Case 2 are the potential interferences present at the Tl mass positions, principally WHO, HgH and HgH_2. These interferences directly affect the mass-bias correction and, at the 10% signal level, have a profound effect on MDF (δ^{202}Hg is changed by more than 1‰), but *not* on MIF. The effect on MDF is much diminished at the 1% level (δ^{202}Hg is changed by less than 0.2‰), and at the 0.1% level the effect on MDF is almost non-existent (δ^{202}Hg is changed by only 0.01 to 0.02‰). For all instances of Case 3 both MDF and MIF are strongly affected by the presence of interfering species, even at the 0.1% interference level, and the MIF patterns reflect the relative magnitude of the interference at each mass position. The remainder of this discussion, therefore, evaluates the fractionation patterns associated with Case 3, where interfering species are present during measurement of the sample solution only.

Because the signal at mass 198 is in the denominator of each isotope ratio, and because δ^{202}Hg is the reference used to define mass-dependency and to determine mass-independent anomalies, interferences that add to the signals at mass 198 and/or 202 can affect the isotopic signature of a sample in very specific ways (see Table 3 for relative fractionation patterns at the 0.1% interference level). Interferences that have no significant abundance at either mass 198 or 202 (WHO, Pb) cause no (or very little) change in δ^{202}Hg. Consequently, apparent positive MIF at one, or more, of masses 199, 200, 201 and 204 is proportional to the magnitude of the interference at each mass position.

Interferences with sub-equal abundances at mass 198 and 202 (WO) cause a significant reduction in δ^{202}Hg because the relative contribution to the signal at 198 is about 3 times larger than the relative contribution to the signal at 202, which causes the 202/198 ratio to be smaller than it should be. Contributions to other masses that are similar to, or relatively smaller than those to masses 198 and 202, result in apparent negative MIF of those masses. Interferences with significant abundance at 198 only ($SnCl_2$, Pt) cause a large reduction in δ^{202}Hg. The ratios of signals at the other Hg masses to the signal at 198 are also smaller than they should be, but because these signals retain their mass dependent relationships with the signal at mass 202, when the ratios are normalized to bracketing standard values δ^{199}Hg has the largest apparent deficit and δ^{204}Hg has an apparent surplus. The consequence is a linear trend on a plot of δ^{xxx}Hg vs. mass position that has a shallower slope than the linear trend defined by predicted mass-dependent values for δ^{xxx}Hg. We did not simulate results for any interference that has a significant abundance at mass 202 only, because we know of no such interference. However, we predict that such an interference would cause δ^{202}Hg to be strongly increased, and apparent MIF at 199, 200, 201 and 204 would be negative. Overall, among the interferences examined, Pt, Pb and the Sn- and W-bearing polyatomic species cause either no change or a reduction in δ^{202}Hg. Only the Hg-hydride interferences cause the apparent δ^{202}Hg to increase.

Table 3. Calculated effects on sample mercury isotope composition attributable to an increase in signal intensity of 0.1% at the reference mass for each interfering species and proportional increases at other masses (see Table 2). Note that these results are based on interference contributions to sample measurements only, and not to measurement of bracketing standard or blank solutions (see text for detailed explanation).

Mass	SnArAr	SnClCl	WHO	WO	HgH	HgHH	HgH+HgHH	Pt	Pb
				‰ Change to Apparent MDF					
δ202	−0.92	−0.98	+0.32	−0.54	+0.45	+0.40	+0.64	−0.98	0
				‰ Contribution to Apparent MIF					
Δ204	+0.72	+0.48	0	0	−0.22	+0.99	+0.89	+0.48	+1.39
Δ201	−0.32	−0.28	+0.70	−0.47	+0.43	+0.19	+0.42	−0.28	0
Δ200	0	−0.51	+0.17	−0.17	+0.14	0	+0.07	−0.51	0
Δ199	−0.59	−0.78	+0.45	−0.48	+0.11	−0.11	0	−0.78	0

Natural samples in which odd-mass MIF is measured always have $\Delta^{199}Hg$ and $\Delta^{201}Hg$ values with the same sign and $\Delta^{199}Hg / \Delta^{201}Hg$ ratios that range from +1 to about +1.7. For most interferences considered here, contributions to odd-mass MIF are consistent with that pattern. Consequently, interference-related changes to the magnitude of measured MIF values and associated subtle changes to $\Delta^{199}Hg / \Delta^{201}Hg$ ratios could be difficult to identify, and could confound the interpretation of results. For example, odd-mass MIF for both $\Delta^{199}Hg$ and $\Delta^{201}Hg$ is reduced by WO, Pt and Sn-bearing interferences, is left unchanged by Pb, and is increased by WHO; but the $\Delta^{199}Hg / \Delta^{201}Hg$ ratios, while variable, are all $\geq +1$. In contrast, the Hg-hydrides, which should be considered in combination, cause a decrease in $\Delta^{199}Hg$ and a much larger increase in $\Delta^{201}Hg$, which causes the $\Delta^{199}Hg / \Delta^{201}Hg$ ratio to be negative, a relationship not observed to date in natural samples.

Studies of even-mass MIF require interpretation of very small $\Delta^{200}Hg$ and $\Delta^{204}Hg$ values and it is particularly important to explore the possible effect of interferences on those values. Natural samples in which even-mass MIF has been measured to date have $\Delta^{200}Hg$ and $\Delta^{204}Hg$ values with opposite sign and $\Delta^{200}Hg / \Delta^{204}Hg$ ratios of about −0.5 (Blum and Johnson 2015). For example, precipitation has $+\Delta^{200}Hg$ and $-\Delta^{204}Hg$, whereas total gaseous mercury (TGM) has $-\Delta^{200}Hg$ and $+\Delta^{204}Hg$. A small subset of interferences ($SnCl_2$ and Pt) investigated here make contributions to even-mass MIF that are similar to the pattern measured for TGM samples, but the $\Delta^{200}Hg / \Delta^{204}Hg$ ratio is about −1 rather than the observed −0.5. Only HgH, which in comparison to HgH_2 is likely significantly less abundant as an interference, can contribute to even-mass MIF in a pattern (with respect to both sign and $\Delta^{200}Hg / \Delta^{204}Hg$ ratio) consistent with that measured for precipitation samples. Other interferences increase $\Delta^{204}Hg$ but leave $\Delta^{200}Hg$ unchanged ($SnAr_2$, HgH_2 and Pb), increase both $\Delta^{200}Hg$ and $\Delta^{204}Hg$ (combined HgH and HgH_2), or leave $\Delta^{204}Hg$ unchanged while $\Delta^{200}Hg$ increases (WHO) or decreases (WO). Therefore, interference contributions to even masses may be more likely to obscure rather than enhance any naturally produced even-MIF signature associated with environmental samples.

We conclude this section with a discussion of aerosol production in frosted tip liquid-gas phase separators. We have monitored the mass range for the most abundant isotopes of Sn while using typical operating conditions for our inlet system and mass-spectrometer for measurement of mercury isotopes. When delivering a 2% solution of $SnCl_2$ in 10% hydrochloric acid combined in equal volume with a 2% hydrochloric acid wash solution we detect a signal of approximately 1 mV for the most abundant isotope of Sn. Two obvious conclusions follow from the magnitude of this measured signal: first, the efficiency of $SnCl_2$ aerosol generation in our liquid-gas phase separator is extremely low, and second, a polyatomic-ion:metal-ion ratio of ~1 is necessary for detection of tin-bearing polyatomic ions at any of the mercury masses. Extrapolation of this observation to other potential interferences in liquid delivered to the phase separator suggests that this mechanism can contribute significant quantities of interfering polyatomic ions only if the ratio of interference-metal:Hg is on the order of 10^7 and production of polyatomic ions incorporating the interference-metal is very efficient. Thus aerosol transfer from frosted tip liquid-gas phase separators to the ICP-MS is not likely to produce significant interferences for trace constituents, but could be significant for reagents used at high concentrations such as Sn.

Reagent blanks and sample carryover

The Hg content of reagents and labware must be carefully monitored throughout all stages of sample preparation for low-level Hg concentration and isotopic analysis. Unlike other trace metals such as Pb, Cd or Cr, Hg exists as a gas in laboratory air and is permeable to most plastics. Reagents and samples stored in plastic may gain or lose Hg from laboratory air. Glass and Teflon are the only universally appropriate materials for storage of reagents and samples used for Hg analyses, and blanks should be verified periodically in reagents. Liquid reagents should be purged with N_2 or Ar gas that has been passed through a gold trap to remove Hg, and solid reagents should be baked at high temperature to minimize Hg blanks.

Hg(II) in solution is notoriously "sticky" (or surface reactive) and thus extensive "washout" and frequent replacement of pump tubing in on-line sample reduction systems is critical. Between periods of data collection for samples and bracketing standards the Hg signal should be monitored until it is reduced to about 0.1% of the analyte signal (<0.1 mv on mass 198 for a 100 mV sample analysis intensity). Additionally, on-peak-zero intensities should be measured for each peak and subtracted from the subsequent analysis to reduce any small interferences from sample carry-over or impurities that are in the Tl aerosol or Ar carrier gas.

Matrix effects and the need to remove matrix materials

In order to minimize potential matrix effects, some of which are discussed above, the University of Michigan Laboratory adopted procedures to purify mercury samples by removing all matrix components prior to mercury isotope measurement by MC-ICP-MS. We found in our earliest studies (Smith et al. 2005, 2008) that simply digesting samples in acid and then diluting them to a constant concentration did not result in the highest quality analyses. We found that matrix components led to higher carry-over of mercury from sample to bracketing standard and also that both the precision and accuracy of the analyses, even in the absence of significant carry-over, were compromised. This is presumably due to very small interferences at the Hg and Tl masses plus the effect of variable matrices on online reduction of Hg(II) by $SnCl_2$. Similarly, we have found that acid digestions of samples with high organic matter content (such as fish muscle tissues, vegetation and coal) can result in incomplete oxidation of organic compounds (even after high pressure and temperature microwave digestion) and lead to compromised analytical performance. This led us to develop the method of thermal decomposition and release of mercury Hg(0) with subsequent trapping as Hg(II) in $KMnO_4$ solution (Biswas et al. 2008). We use a two-stage furnace so that samples can be heated slowly from room temperature to 750 °C (over 4–6 hours) while decomposition products are transferred through a second-stage furnace held at 1000 °C to further decompose organic vapors and aerosols. Although analysis of Hg in these $KMnO_4$ solutions yields less carry-over and higher precision than acid digests, even these results are affected by the presence of incompletely decomposed sample matrix residues in solution with the Hg. In either case (chemical decomposition or thermal decomposition) some of these residues can be filtered from the mercury-bearing solution though a 0.45 μm filter, but that approach will not remove smaller residues. Further improvements are achieved by an additional "purge and trap" cleanup step. To do this we reduce the Hg in these $KMnO_4$ solutions using $SnCl_2$, then purge the Hg(0) with argon or Hg-free air and re-trap the Hg as Hg(II) in another $KMnO_4$ solution. Sample types for which combustion is not appropriate, such as water samples and sediment leachates, are also subjected to purification by purge and trap procedures. Each of the mercury reduction and transfer steps must be done with a high yield so as not to induce mass dependent fractionation of the mercury isotopes in the samples. We routinely obtain yields over 95% and we have found no detectable change in isotopic composition with yields over 85%.

The merits of our sample "cleanup" protocol may be illustrated by comparing the analytical performance of three widely distributed fish muscle tissue standard reference materials used in our laboratory over the past nine years. These SRMs have been prepared for isotopic analysis by the following four methods: 1) acid digestion only, 2) acid digestion followed by purge and trap into a $KMnO_4$ solution, 3) 2-stage furnace combustion into $KMnO_4$, and 4) 2-stage furnace combustion into $KMnO_4$ followed by purge and trap into another $KMnO_4$ solution.

Microwave assisted digests (Method 1) of DORM-2, DOLT-2 and ERM-CE464 were conducted during 2007 and 2008 as part of the QA/QC assessment for research projects in progress at that time (Blum and Bergquist 2007; Senn et al. 2010; Blum 2011). These digest solutions were filtered to 0.45 μm, portions of them were diluted and measured for isotopic composition during that same time period, and then remaining portions were archived in refrigerated storage. Aliquots of one sample each of DORM-2 and DOLT-2 were supplied to another laboratory where a double-spike technique for Hg isotope analysis was being developed, and were measured there (Mead and Johnson 2010). One additional sample of ERM-CE464 was prepared with the same acid digestion technique in 2013, specifically for comparison with the archived preparations of that same material. A small number of these three reference materials were prepared by combustion (Method 3) and were analyzed during 2008 and 2009 as primary sample solutions without additional processing (Gehrke et al. 2011), but the poor quality of the results served as an incentive to adopt secondary trapping as a routine part of the sample preparation protocol. Since 2009, subsequent preparations of DORM-2, DOLT-2 and ERM-CE464 in the University of Michigan Laboratory have utilized combustion followed by secondary purge and trapping (Method 4) as the preferred technique, but only the results from ERM-CE464 and a small subset of DOLT-2 have previously been published (Kwon et al. 2012, 2013, 2014, 2016; Blum et al. 2013; Sherman and Blum 2013; Li et al. 2014). In 2009 and 2014 secondary purge and trapping was applied to portions of archived digestates (Method 2) of DORM-2, DOLT-2 and ERM-CE464 originally prepared in 2007 and 2008 to demonstrate that, in comparison with results from combustion followed by purge and trapping, results from digestion followed by purge and trapping can be of equivalent quality.

The compilation of results (Fig. 1) clearly indicates that preparation techniques employing primary sample solutions for analysis (Methods 1 and 3) yield average values for δ^{202}Hg and Δ^{199}Hg that are lower than those obtained by techniques using secondary purge and trapping (Methods 2 and 4). Moreover, the average precision of results is dramatically better for Methods 2 and 4, improving by at least a factor of 2 in comparison with Methods 1 and 3. Results (both isotopic composition and precision) for samples of DORM-2 and DOLT-2 that were prepared by acid digest and subsequently measured using the double-spike technique are nearly indistinguishable from results obtained using secondary trapping techniques.

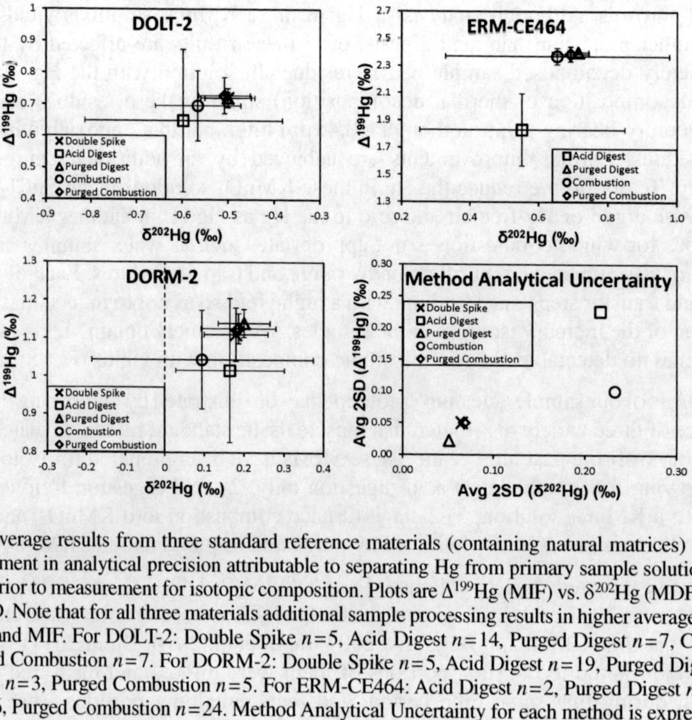

Figure 1. Average results from three standard reference materials (containing natural matrices) illustrating the improvement in analytical precision attributable to separating Hg from primary sample solutions and re-trapping it prior to measurement for isotopic composition. Plots are Δ^{199}Hg (MIF) vs. δ^{202}Hg (MDF) and error bars are 2SD. Note that for all three materials additional sample processing results in higher average values for both MDF and MIF. For DOLT-2: Double Spike $n=5$, Acid Digest $n=14$, Purged Digest $n=7$, Combustion $n=7$, Purged Combustion $n=7$. For DORM-2: Double Spike $n=5$, Acid Digest $n=19$, Purged Digest $n=10$, Combustion $n=3$, Purged Combustion $n=5$. For ERM-CE464: Acid Digest $n=2$, Purged Digest $n=3$, Combustion $n=6$, Purged Combustion $n=24$. Method Analytical Uncertainty for each method is expressed as the average of the 2SD uncertainties determined for the three materials processed by each method.

We suggest the following explanations for why increased sample purity leads to improved measurement precision for Hg stable isotopes by MC-ICP-MS.

1. Residual sample matrix components can contaminate the surfaces of tubing and fittings that carry mercury-bearing solution to the liquid-gas separator and the frosted-post of the phase separator. These contaminants produce active sites that scavenge Hg from both bracketing-standard and sample solutions. Throughout the duration of an analytical session these contaminant sites become more numerous, and once populated with Hg slowly exchange with Hg from subsequent samples or standards, or release Hg into wash solutions. The consequence is two-fold: Hg washout times and practically achievable blank signals increase throughout an analytical session, and Hg delivered to the ICP-MS torch includes a portion that is compositionally a moving average of Hg that was previously delivered. For analysis schemes observing strict standard-sample-standard bracketing the bracketing standard composition dominates that moving average. However, analysis order and the isotopic contrast between samples, or between samples and the bracketing standard, can cause that moving average composition to be variable, which, in turn, contributes to the magnitude of analytical uncertainty associated with either individual analyses or replicate analyses of the same material. In addition, the values for both MDF and MIF can be slightly shifted toward values of the bracketing standard.

2. Primary sample solutions with acid strength above 10 to 20% can cause suppression of analyte signals measured with MC-ICP-MS instruments regardless of sample introduction method (e.g. direct delivery of sample solution to nebulizer-spray chamber-torch, production of analyte aerosol by desolvating nebulizer, or production of analyte by cold-vapor generation). In the case of cold-vapor generation of Hg(0), a significant reduction in signal intensity is likely accompanied by a change in a sample's MDF values, especially if the efficiency of the $SnCl_2$ reduction is impaired. Dilution can minimize or eliminate such effects, but that imposes limits on the abundance of Hg in bulk material prepared by chemical decomposition and the amount of concentrated acid that can be used to extract Hg from the bulk material. For example, our best results are obtained when we can prepare about 30 ng of Hg for measurement. If a sample's bulk material Hg content is 100 ng/g the amount of sample required is about 300 mg and the maximum amount of concentrated acid that can be used is limited to what can be diluted to achieve a run solution of 10% acid strength with enough volume to perform one analysis. Our inlet system is configured to consume about 6 ml per analysis, which limits the amount of concentrated acid to no more than 600 μL. The obvious compromise is to reduce the amount of Hg required for measurement, which can reduce either the amount of sample material needed or the bulk material Hg content required. However, because analytical uncertainty increases nonlinearly as run-solution Hg concentration is reduced (Foucher and Hintelmann 2006; Tsui et al. 2013, 2014; Rua-Ibarz et al. 2015) the practical consequence of that compromise is loss of resolution when evaluating between-sample isotopic contrasts. In addition, variable dilution applied to sample solutions in order to match run-solution concentration between samples, and between samples and bracketing standards, has the potential for introducing analytical artifacts related to measuring solutions of differing acid strengths and matrix, which also contributes to diminished analytical resolution. Although results from standard reference materials can assist in evaluating performance of the sample preparation and analysis processes, many such materials have bulk Hg contents that are much higher (in some cases more than an order of magnitude higher) than the "unknown" environmental samples with which they are analyzed. Such a large difference in

Hg content means that the amount of reference material put into solution is much less than for samples, and the dilution factor used to prepare reference material solutions for measurement is much larger than for samples. Therefore, reference material results are less likely to be impacted by analytical artifacts, and thus can mask the presence of such effects on sample results.

3. Pre-concentration of Hg by purging it from a primary sample solution into a secondary trapping solution maximizes the Hg content in the smallest possible volume, thereby extending the practical working range of the chemical extraction and thermal decomposition preparation methods. For example, where bulk material is limited and the primary sample solution is a strong acid that cannot be diluted sufficiently to ensure an analysis free of artifacts, purging the Hg into a smaller volume of $KMnO_4$ in 10% H_2SO_4 allows a solution matrix of consistent composition to be used for all samples and bracketing standards. Alternatively, when employing thermal decomposition, multiple gram quantities of bulk material containing as little as 10 ng/g can be processed, and, if necessary, Hg from multiple primary sample solutions can be combined into a single secondary solution. The principal limitation in this case is magnification of the process blank, which is dominated by the $KMnO_4$ trapping solution and scales with the number of primary sample solutions.

Recommendations for analyses

The University of Michigan Laboratory has taken an approach to sample preparation for Hg isotope measurement that incorporates matrix separation from samples and is best summarized as "better safe than sorry," and we recommend that other laboratories adopt the same approach. What we mean by this is that measurement protocols should take every feasible precaution to avoid matrix effects, whether they have been demonstrated to be necessary for the analysis that is being done or not. We recognize that many labs have produced reliable Hg isotope data without matrix separation and that other labs have been successful only removing matrices selectively for problematic sample matrices. As an example of why we think our cautious "better safe than sorry" approach is appropriate we review some of our early studies of Hg isotopes, in which we learned "after the fact" that our increased precision yielded interesting scientific information. When we first analyzed rocks and ore deposits for Hg isotope ratios we expected only mass dependent isotope signals, but we also detected small (<0.2‰) mass independent anomalies of $\Delta^{199}Hg$ and $\Delta^{201}Hg$. This led us to perform experiments demonstrating that $\Delta^{199}Hg$ and $\Delta^{201}Hg$ were produced by photochemical reactions and to the recognition of the importance of photochemically reduced mercury in hydrothermal ore deposits, which we would not have otherwise detected (Blum et al. 2014). As another example, our first report of MIF of mercury was based on analyses of fish from New England and Lake Michigan (Bergquist and Blum 2007). We measured $\Delta^{199}Hg$ and $\Delta^{201}Hg$ values of up to 5‰, and it was not apparent that we would have any interest in quantification of very small (<0.1‰) values of $\Delta^{200}Hg$ and $\Delta^{204}Hg$. It wasn't until three years later that we noticed anomalous $\Delta^{200}Hg$ and $\Delta^{204}Hg$ values in precipitation (Gratz et al. 2010; Sherman et al. 2012; Demers et al. 2013; and Donovan et al. 2013). Subsequent re-evaluation of all these results led to the realization that the small, but consistent, $\Delta^{200}Hg$ and $\Delta^{204}Hg$ anomalies are analytically significant and are indicative of atmospherically derived Hg that can be traced into fish and sediment samples (Blum and Johnson 2015; Lepak et al. 2015).

Thus we seek to make the most precise measurements that we can by working to minimize contamination, by making complete matrix separations, and by closely matching concentrations and matrices of samples and standards. We summarize the procedures used in the Michigan Laboratory below. We acknowledge that some excellent laboratories do not follow all of these steps for all samples types and that they obtain reliable data. However, we recommend these steps for the highest precision data possible.

1. ***Sample Preparation:*** Avoid contact of the samples with Hg, W, Pt and Pb during sample collection, storage, preparation and analysis. This includes use of "clean sampling" techniques, acid washing of sample containers and verification of sample collection and storage blank values.
2. ***Mercury Release and Trapping:*** Combust solid samples at 750 °C in a dual stage furnace in an oxygen atmosphere. We trap Hg(0) from combustion in $KMnO_4$, but trapping in a mixture of HNO_3 and HCl can also produce high yields (Sun et al. 2013a). Small volume liquid samples such as urine (Sherman et al. 2013, 2015b) are evaporated to dryness at 60 °C and then combusted. Large volume liquid samples, such as rainwater (Demers et al. 2013) are oxidized with BrCl to release bound Hg(II), reduced with $SnCl_2$ to produce Hg(0), purged with Hg-free air and trapped in $KMnO_4$ solution. Samples that must be digested in solution, such as sequential leaches, are diluted to reduce reagent strength and then processed for further matrix removal by purge and trap methods discussed below.
3. ***Further Matrix Removal:*** All samples including combustions and acid digests of solid samples and liquid samples are reduced with $SnCl_2$ and transferred as Hg(0) to a $KMnO_4$ solution where the Hg is re-oxidized to Hg(II). This step has been considered unnecessary by many laboratories, but we show that it minimizes sample carryover during MC-ICP-MS analyses and results in overall higher precision for Hg isotope ratio measurements, especially for samples rich in organic matter.
4. ***Matrix and Concentration Matching of Samples and Standards:*** Final sample solutions are diluted with neutralized $KMnO_4$ to a constant concentration or concentrations. If there is sufficient Hg for analysis then samples are prepared at 5 ng/g and run multiple times. If there is limited Hg available for analysis samples are prepared at lower concentrations down to 1 ng/g. Bracketing standards and external process standards should also be prepared with the same matrix and the same Hg concentrations as samples. In addition to concentration matching, signal intensities should also be monitored to ensure that there is complete reduction and volatilization of Hg with no suppression of the Hg signal.
5. ***Thallium Mass Bias Correction:*** A dry Tl aerosol is produced by a desolvating nebulizer and added to the Hg cold vapor stream for mass bias correction. The Tl total beam intensity is best maintained at a constant level that is approximately equal to the Hg total beam intensity so that the effect of HgH_2 on ^{203}Tl is minimized. But we have found that over an eight-fold range in Hg/Tl ratio there is no change in the $\delta^{202}Hg$ value or the uncertainty in measurements of the $\delta^{202}Hg$ value for the UM-Almaden standard solution. A clean and matrix-matched Hg sample and standard will result in minimum sample uncertainty due to Hg hydrides. Furthermore, monitoring of mass 206 allows verification that both HgH_2 and Pb are not present at a significant level.

ISOTOPIC ABUNDANCES IN MAJOR RESERVOIRS

The field of Hg isotope geochemistry is quite young and the elemental and isotopic abundances of Hg in major Earth reservoirs have not yet been well documented. Further complicating this goal is the fact that anthropogenic emissions of Hg to the atmosphere are spread globally and deposit into essentially every ecosystem on Earth. Therefore, it is difficult to differentiate the natural isotopic values from the numerous and isotopically variable anthropogenic sources that have overwhelmed most natural reservoirs. Additionally, Hg is a highly redox sensitive element and its isotopic composition is shifted to varying degrees in virtually all dark, biotic, and photochemical reactions that occur in the atmosphere, vegetation, fresh waters, oceans and sediments. In an earlier review (Blum et al. 2014) we compiled isotopic values from all studies that had been published as of July 2013 to show the range of natural and anthropogenic variation and to highlight the processes that were most important in causing different types of Hg isotope fractionation. An updated version of this figure is shown here as Figure 2A. In this chapter we also provide a figure that includes only data for geological materials that are thought to be relatively unaffected by anthropogenic Hg contributions (Fig. 2B). This includes meteorites, crystalline rocks, sedimentary rocks, ore deposits (cinnabar and sphalerite ores), coal deposits (representing ancient vegetation) and marine and freshwater sediments deposited before the year 1800 (see Table 4 for citations to data sources). All atmospheric and aqueous environments and most living organisms on Earth are subjected to some anthropogenic Hg, which is estimated to be three to four times greater in atmospheric abundance than pre-anthropogenic mercury (e.g., Engstrom et al. 2014). Considering only the non-anthropogenically effected geological samples gives a much more limited range in isotopic composition compared to samples affected by anthropogenic Hg (compare Fig. 2B with Fig. 2A).

Carbonaceous chondrites, chondrites and achondrites display a narrow range of Δ^{199}Hg of −0.26 to 0.31 ‰, but a large range in δ^{202}Hg from −7.13 to −0.73 ‰. Crystalline rocks display

Table 4. Sources of data used in Figure 2.

Source	Category				
	Rocks	**Hg Ores and Minerals**	**Coals**	**Sediments**	**Meteorites**
Biswas et al. (2008)			X		
Blum et al. (2014)	X	X			
Cooke et al. (2013)		X		X	
Donovan et al. (2013)				X	
Foucher et al. (2009)		X		X	
Gehrke et al. (2009)				X	
Lefticariu et al. (2011)			X		
Meier et al. (2016)					X
Mil-Homens et al. (2013)				X	
Sherman et al. (2009)		X			
Sherman et al. (2012)			X		
Sial et al. (2016)	X				
Smith et al. (2005)		X			
Smith et al. (2008)	X	X			
Sonke et al. (2010)		X			
Stetson et al. (2009)		X			
Sun et al. (2013b)			X		
Sun et al. (2014)	X		X		
Thibodeau et al. (2016)	X				
Yin et al. (2013)		X			
Yin et al. (2014)			X		
Yin et al. (2016)		X			
Zhang et al. (2014)	X				

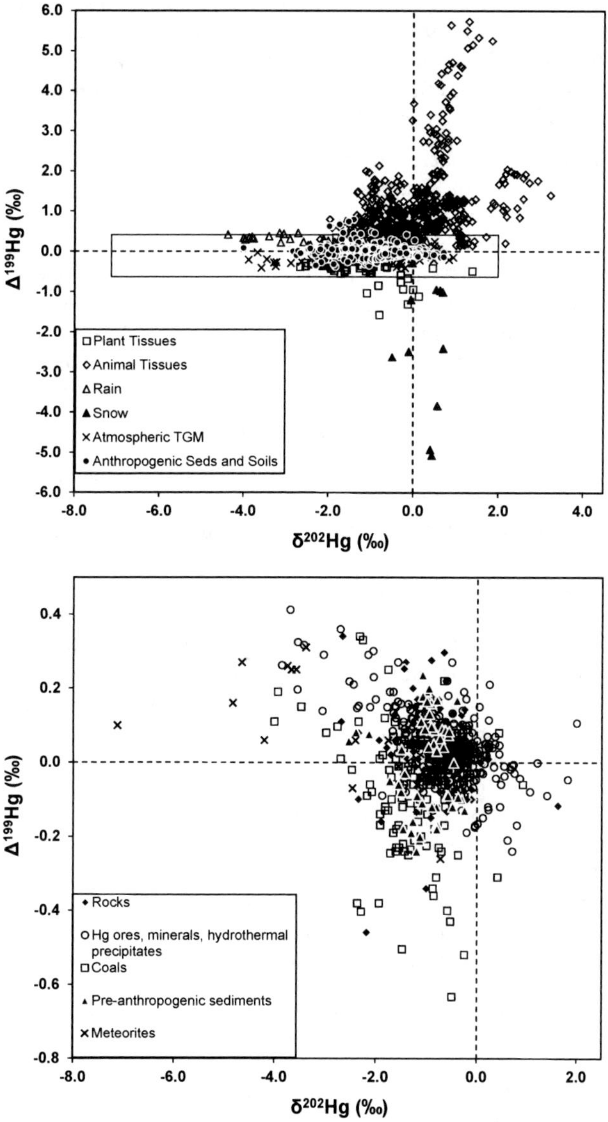

Figure 2. Compilation from published sources of Hg isotopic composition results for atmospheric, biotic, Earth and extraterrestrial materials. Panel A is adapted and updated from Figure 2 in Blum et al. (2014) and illustrates samples that are likely to be influenced by anthropogenic Hg inputs. The rectangle outline delineates the range of isotopic values for Earth and extraterrestrial materials presented in Panel B that are free from contributions of Hg derived from anthropogenic activities (see Table 4 for an updated list of published sources for these results). These plots of $\Delta^{199}Hg$ (MIF) vs. $\delta^{202}Hg$ (MDF) illustrate that for these materials processes that cause mass-dependent fractionation are dominant. Note that the results from pre-anthropogenic sediments (which include materials from both fresh-water and marine settings) are coincident with the dense cluster of points that represents the majority of the results from rocks (both crystalline and sedimentary), Hg ores and minerals, and coals.

$\delta^{202}Hg$ from −1.70 to 1.61 ‰, and $\Delta^{199}Hg$ of −0.13 to 0.13 ‰, sedimentary rocks display $\delta^{202}Hg$ from −2.68 to 0.23 ‰, and $\Delta^{199}Hg$ of −0.46 to 0.34 ‰, and coal deposits display $\delta^{202}Hg$ from −4.00 to 0.91 ‰, and $\Delta^{199}Hg$ of −0.63 to 0.34 ‰. Cinnabar and sphalerite ore deposits display $\delta^{202}Hg$ from −3.85 to 1.99 ‰, and $\Delta^{199}Hg$ of −0.24 to 0.41 ‰. Marine sediments from pre-1800 display $\delta^{202}Hg$ from −2.53 to −0.38 ‰, and $\Delta^{199}Hg$ of −0.09 to 0.17 ‰ and freshwater sediments from pre-1800 (minus a few suspicious outliers that may have post-1800 Hg contributions) display $\delta^{202}Hg$ from −1.72 to −0.25 ‰, and $\Delta^{199}Hg$ of −0.24 to 0.24 ‰.

None of these "non-atmospheric" samples display appreciable even-mass MIF; in marked contrast to modern precipitation samples, which typically have $\Delta^{200}Hg > 0.1$‰ and $\Delta^{204}Hg < -0.2$‰. For the seven categories of "non-atmospheric" materials considered here, average values for $\Delta^{200}Hg$ and $\Delta^{204}Hg$ (Table 5) range from −0.03‰ to 0.03‰, with 1SD uncertainties ≤ 0.05‰ for $\Delta^{200}Hg$ and <0.15‰ for $\Delta^{204}Hg$. However, nearly every category of material has at least one value of $\Delta^{200}Hg$ or $\Delta^{204}Hg$ that is < −0.1‰ or > 0.1‰, and for five of the categories the effect of these potential outlier results on the averages and associated uncertainties is magnified by the very small numbers of samples. Therefore additional high quality Hg isotope measurements of these materials are needed to confirm the absence (or not) of even-mass MIF.

Isotopic behaviors during major geological processes

Mercury isotopes fractionate during virtually every biogeochemical reaction that has been studied (see Blum et al. (2014) and references therein). There are essentially four different types of fractionation behavior that are observed for Hg isotopes. 1) Mass dependent fractionation, which is reported as $\delta^{202}Hg$, and occurs during all biological reactions and abiotic reactions that have been studied. It also accompanies all processes that produce mass independent fractionation. 2) Odd-mass independent fractionation, which is reported as $\Delta^{199}Hg$, with $\Delta^{199}Hg/\Delta^{201}Hg$ = 1.0 to 1.3. This is believed to be caused by the magnetic isotope effect during the kinetic photochemical reduction of Hg(II) and MeHg bonded to organic ligands in aqueous solutions. 3) Odd-mass independent fractionation, which is reported as $\Delta^{199}Hg$, with $\Delta^{199}Hg/\Delta^{201}Hg$ = 1.6. This is believed to be caused by the nuclear volume effect during both equilibrium and kinetic chemical processes including the evaporation of Hg(0) and the dark reduction of Hg(II) by organic matter. 4) Even-mass independent fractionation, which is reported as $\Delta^{200}Hg$ or $\Delta^{204}Hg$, with $\Delta^{200}Hg/\Delta^{204}Hg$ = −0.5 to −0.6. This results from an unknown mechanism in the atmosphere, which appears to be related to photochemical oxidation of Hg(0) exclusively in the upper atmosphere. This mass independent fractionation likely affects both even and odd isotopes, but is termed even-MIF as it is clearly observed on the even isotopes but may be overwhelmed on the odd isotopes by odd-MIF of larger magnitude imparted by other odd-mass independent fractionation mechanisms.

Table 5. Mercury even-MIF in solid-Earth materials.

Material	Δ200					Δ204				
	n	Min	Max	Avg	1SD	*n*	Min	Max	Avg	1SD
Meteorites	16	−0.12	0.07	0.00	0.05	10	−0.4	0.05	−0.06	0.14
Ores and Minerals	329	−0.05	0.13	0.01	0.02	153	−0.07	0.12	0.00	0.02
Crystalline Rocks	24	−0.09	0.07	−0.01	0.04	24	−0.13	0.20	0.02	0.09
Sedimentary Rocks	111	−0.23	0.17	−0.00	0.05	21	−0.13	0.34	0.02	0.11
Coals	168	−0.14	0.19	0.02	0.04	45	−0.13	0.23	−0.01	0.06
Fresh Water Seds, Pre−anthropogenic	28	−0.05	0.08	0.02	0.03	8	−0.06	0.04	0.00	0.03
Marine Seds, Pre−anthropogenic	23	−0.01	0.17	0.03	0.04	6	−0.17	0.06	−0.03	0.08

During geological processes the most important type of fractionation is mass dependent and is due to volatilization and condensation, diffusive transport, and abiotic reduction and oxidation. There are also small odd-MIF effects observed that are most likely caused by deposition of Hg that previously spent time exposed to sunlight and was co-deposited with organic matter into sediments (Gehrke et al. 2009). In some cases this Hg may be recycled from sedimentary materials into crystalline rocks and ore deposits through the geological rock cycle. Laboratory fractionation experiments help us to understand the types of processes that are likely to have affected the Hg isotopic composition of geological materials. Estrade et al. (2009) and Ghosh et al. (2013) demonstrated that the simple volatilization of metallic Hg resulted in odd-MIF due to the nuclear volume effect. Wiederhold et al. (2010) and Jiskra et al. (2012) showed that MDF occurred during sorption of Hg(II) to thiols, goethite and organic ligands, which suggests that Hg dissolved in water and flowing through geological materials could acquire MDF. Smith et al. (2005, 2008), Sherman et al. (2009), and Blum et al. (2014) presented data suggesting that boiling in hydrothermal systems caused MDF and that odd-MIF could be produced when these boiling systems reached the Earth's surface and Hg was reduced in the presence of sunlight. Similarly, sphalerite deposits have been shown to display small amounts of odd-MIF presumably due to recycling of Hg from sedimentary materials into hydrothermal ore deposits (Sonke et al. 2010; Yin et al. 2016). Analyses of crystalline rocks including basalts, schists and serpentinite have been published (Smith et al. 2005, 2008; Blum et al. 2014) and display a range of δ^{202}Hg from −1.70 to 1.61 ‰ but very low Δ^{199}Hg (~ ±0.1 ‰), which suggests minimal mixing of sedimentary Hg into these rocks. Pre-anthropogenic sedimentary rocks have a several permil range in δ^{202}Hg due to a combination of fractionation processes discussed above, and variable Δ^{199}Hg up to ~±0.2 ‰, which most commonly represents contributions of Hg associated with organic matter that was exposed to sunlight in the water column prior to co-deposition with organic matter (Gehrke et al. 2009).

NEW FRONTIERS IN HG ISOTOPE ANALYSIS

Isotopic composition of methyl mercury

Much of the current effort in environmental Hg research is focused on understanding the sources and pathways for formation and biological uptake of methylmercury (MeHg), the most toxic and bio-accumulative form of Hg in the environment. Mercury isotope studies have the potential to address many unanswered questions about MeHg biogeochemistry. For the concentration or isotopic composition of MeHg to be measured directly in a sample it must first be separated from other forms of mercury, and this is usually done by gas chromatography (GC). The earliest attempts to measure the isotopic composition of MeHg (Epov et al. 2008, 2010), while groundbreaking, suffered from two limitations: 1) the amount of MeHg that could be run through standard GC columns was small—and resulted in inadequate signal intensities and 2) MC-ICP-MS measurements were made on transient signals as the Hg derived from MeHg exited the GC, and this compromised the precision of the isotope ratio measurement. Nevertheless, this method has been improved and used successfully to investigate the degradation of MeHg found in marine mammals (Perrot et al. 2016). A recent study addressed the limitations of the above-mentioned method by scaling up the size of the GC separation system and collecting the transient MeHg peak into a $KMnO_4$ trapping solution for off-line analysis (Janssen et al. 2015). This method now allows high precision isotope ratio measurements of MeHg, but is still restricted to sediments with relatively high Hg concentrations. A third approach (Masbou et al. 2013) uses selective extraction techniques followed by back-extraction to isolate MeHg from biological samples and prepare it for isotopic analysis in a mixed HNO_3:HCl solution.

An alternate approach to separating and measuring the isotopic composition of MeHg from samples involves taking advantage of the natural bioaccumulation of MeHg in foodwebs. Several captive feeding experiments have demonstrated that there is no significant trophic fractionation of Hg isotopes in aquatic foodwebs (Kwon et al. 2012, 2013, 2016; Feng et al. 2015). High trophic level fish typically have >95% of their total Hg (THg) as MeHg (%MeHg), and therefore, measurement of the isotopic composition of THg in these organisms yields an isotopic value for MeHg. This approach has been used in many MeHg isotope studies of fish (e.g., Bergquist and Blum 2007; Senn et al. 2010; Blum et al. 2013). It is even more desirable to measure the isotopic composition of both MeHg and inorganic Hg (IHg) in the same foodwebs, which is possible when the foodweb contains organisms with a wide range of %MeHg. In this case the isotopic composition of both MeHg and IHg can be measured across entire foodwebs by analyzing THg in biota from the lowest to the highest trophic levels. Isotopic compositions are then regressed against %MeHg to estimate the Hg isotopic composition of MeHg and IHg by extrapolating isotope values to pure MeHg and IHg endmembers. Four studies to date have been published using this approach; one each in an estuary (Kwon et al. 2014), a lake (Kwon et al. 2015), a relatively uncontaminated river (Tsui et al. 2012) and a river system highly contaminated by Hg use in placer gold mining (Donovan et al. 2016a,b). In each case relationships between %MeHg and Hg isotope values were found to be linear and allowed extrapolation of values for IHg and MeHg, which had different isotopic compositions. It is important to note that the IHg and MeHg in the foodweb and measured by this technique has been bioaccumulated and may differ in Hg isotopic composition from IHg and MeHg in sediment or water from the ambient environment.

Separation of Hg from low concentration waters

An important frontier in Hg isotope studies is the investigation of Hg isotopes in rainfall, surface freshwaters and ocean water. While some very exciting results have been obtained (Gratz et al. 2010; Chen et al. 2012, 2016; Sherman et al. 2015a; Štrok et al. 2015; Yuan et al. 2015) progress has been slow due to the low Hg concentrations in natural waters and the need to separate Hg from large volumes of water with high yield and low blank contribution. The first analyses of Hg isotopes in rainfall were made at the University of Michigan Laboratory by pumping 2 to 4 liters of water into an online $SnCl_2$ mixing cell and then onto a gas-liquid phase separator. Hg(0) was then bubbled into a $KMnO_4$ trap where it was retained in solution as Hg(II). To obtain high yields (>85%) with this method required slow pumping rates and was very time consuming (Gratz et al. 2010). This was followed by the development of two different approaches for separation of Hg from large water samples. The University of Michigan Laboratory optimized batch purge and trapping from 2 L samples into $KMnO_4$ traps (Sherman et al. 2012). The Trent Laboratory (Chen et al. 2010) developed an ion-exchange resin method for separation of Hg from large volumes of water. Studies of MDF and odd-MIF signatures in precipitation have yielded information on the importance of local versus regional sources of Hg in rainfall, atmosphere to surface exchanges of Hg and Hg reduction and oxidation reactions in the atmosphere (Gratz et al. 2010; Chen et al. 2012; Sherman et al. 2012, 2015a; Demers et al. 2013). The recent finding that atmospheric samples often contain even-MIF anomalies in ^{200}Hg and ^{204}Hg (Gratz et al. 2010; Chen et al. 2012; Demers et al. 2013, 2015; Blum and Johnson 2015; Sherman et al. 2015a) and that these anomalies may provide tracers of atmospheric Hg in ecosystems, is further stimulating interest in this area of research.

Separation of Hg from atmospheric gases

Hg is present in the atmosphere predominantly as Hg(0) gas, which can be oxidized to Hg(II) gas and can then dissolve in water droplets or attach to particles. Hg isotopes yield information on the sources of Hg to the atmosphere and on the chemical pathways that lead to deposition from the atmosphere. A small number of studies of Hg isotopes in atmospheric gases have been conducted. Studies by the University of Michigan Laboratory collected Total Gaseous Mercury (TGM) by pumping air through a particulate filter and then through a manifold of eight parallel gold coated bead filters to trap Hg(0) (Gratz et al. 2010; Sherman et al. 2012, 2015a; Demers et al. 2013, 2015). Rolison et al. (2013) used a similar setup but without a pre-filter and collected and analyzed TGM plus particulate Hg together. Collection of TGM has also been accomplished by using a pre-filter followed by either an iodated carbon filter or a chlorinated carbon filter for collection of Hg(0) (Fu et al. 2014, 2016). Hg(0) collected on filters is typically transferred to an oxidizing solution in the laboratory and this solution is analyzed for Hg isotopic composition. Due to the low Hg(0) concentration in the atmosphere and the large amount of Hg needed for high precision isotope analysis, the time duration for sampling has typically been 12 to 24 hours. Recently Demers et al. (2013, 2015) showed that Δ^{200}Hg and Δ^{204}Hg in TGM has values that are complementary to precipitation collected in the same location. This suggests that the photochemical oxidation of Hg(0) in the atmosphere and its removal by precipitation may shift isotopic values of Δ^{200}Hg and Δ^{204}Hg in atmospheric Hg(0) in the opposite directions of that observed in precipitation.

Even-mass MIF

Beginning in the year 2007, MDF and odd-MIF began to be reported in a wide range of environmental and geological samples. In 2010 Gratz et al. (2010) measured a small but consistent even-MIF on mass ^{200}Hg in atmospheric precipitation and TGM, and this anomaly was further confirmed by Chen et al. (2012). More recent papers from the University of Michigan Laboratory have measured Δ^{204}Hg and found that this anomaly is about twice as large and of opposite sign compared to Δ^{200}Hg (Demers et al. 2013, 2015; Donovan et al. 2013; Blum and Johnson 2015; Sherman et al 2015a). The origin of the Hg that displays even-MIF is not well known but several studies shed light on this question. Chen et al. (2012) found a much larger magnitude of Δ^{200}Hg than any other study in winter precipitation and used atmospheric back trajectory analysis and a correlation with temperature to argue that the anomaly was caused by photo-oxidation of Hg(0) in the Arctic tropopause. Mixing of Hg subjected to this process into the troposphere may result in the small Δ^{200}Hg and Δ^{204}Hg anomalies observed in measurements of atmospheric Hg. The observed Δ^{200}Hg and Δ^{204}Hg may also provide a useful tracer of Hg that originated from the atmosphere, has been deposited to the Earth's surface and is now found in sediments and aquatic organisms (Blum and Johnson 2015).

The fractionation mechanism that produces Δ^{200}Hg and Δ^{204}Hg anomalies in the atmosphere is not well understood. Experimental studies of the magnetic isotope effect and nuclear volume fractionation do not produce even-MIF (Bergquist and Blum 2007; Ghosh et al. 2013). However, Mead et al. (2013) observed MIF on all of the Hg isotopes in compact fluorescent light bulbs and pointed out that self-shielding was an isotope fractionation mechanism that might produce even-MIF in the atmosphere. They postulated that UV-light in the stratosphere might be filtered by interaction with ambient Hg(0). Because different isotopes of Hg absorb UV light at slightly different wavelengths, as light passes through the stratosphere wavelengths of light corresponding to the most abundant isotopes of Hg might be attenuated to a greater extent than light corresponding to less abundant isotopes. This intriguing idea has not been tested under conditions relevant to atmospheric processes but remains as one of the most promising explanations of even-MIF in atmospheric Hg.

Gratz et al. (2010) and Chen et al. (2012) pointed out that due to the way Hg isotope ratios are normalized, the Δ^{200}Hg anomaly could be due to fractionation of ^{200}Hg, or alternatively, to the fractionation of one or both of the isotopes ^{198}Hg and ^{202}Hg, which are used to define the baseline for MIF of mass ^{200}Hg. Although some laboratories do not routinely measure the Δ^{204}Hg value due to the limitations of ion beam collector designs, the University of Michigan Laboratory has routinely measured Δ^{204}Hg on all analyses since 2009. Similarly to Δ^{200}Hg, we find anomalous Δ^{204}Hg in all samples for which we expect there to be atmospherically derived Hg. We have also compiled data on Δ^{200}Hg and Δ^{204}Hg for hundreds of samples (Blum et al. 2014) and we have calculated mean values for each data set (ie, a single sample type in a single location). All samples of Hg ores, rocks, marine sediments, soils and coals have both Δ^{200}Hg and Δ^{204}Hg of less than ±0.1‰. In contrast, samples of precipitation, tree lichens, Spanish moss, lake sediments, marine fish and some lake fish have positive values of Δ^{200}Hg and negative values of Δ^{204}Hg. Samples of TGM from the Arctic, Wisconsin, Michigan and Florida plus frost on Artic sea ice have negative values of Δ^{200}Hg and positive values of Δ^{204}Hg. Regression of Δ^{200}Hg against Δ^{204}Hg yields a slope of about −0.5 indicating that Δ^{200}Hg and Δ^{204}Hg anomalies are closely related. The fact that the Δ^{204}Hg values are generally twice the Δ^{200}Hg values means that we can more easily detect these small isotopic signals in natural samples and evaluate whether interferences could cause some of these anomalies. Additionally, the ratio of Δ^{200}Hg to Δ^{204}Hg may provide insight into the mechanism that causes even-MIF.

As discussed earlier, a Δ^{200}Hg anomaly could result from fractionation of any one of the three isotopes ^{198}Hg, ^{200}Hg or ^{202}Hg (Gratz et al. 2010; Chen et al. 2012). Adding the information that Δ^{204}Hg varies with twice the magnitude and the opposite sign as Δ^{200}Hg allows a focusing of the possible reasons for these anomalies. In an earlier section of this chapter we presented calculations of the relative magnitude of Δ^{200}Hg to Δ^{204}Hg anomalies that might be caused by interferences during MC-ICP-MS analyses. These synthetic mass spectra show that the coupled Δ^{200}Hg and Δ^{204}Hg anomalies cannot be caused by any of the likely interferences or combinations of interferences. Interference from HgH can cause coupled Δ^{200}Hg and Δ^{204}Hg anomalies of the correct sign and relative magnitude in precipitation, but cannot produce the values observed in TGM. Precipitation values could match the pattern predicted for interference affected measurement only if HgH is produced in much greater abundance than HgH_2, which is generally not the case (Georg and Newman 2015). The only experiments that we are aware of that have produced coupled Δ^{200}Hg and Δ^{204}Hg anomalies with a slope of about −0.5 are experiments that analyzed Hg(II) embedded in the glass wall of compact fluorescent lights (Mead et al. 2013). This Hg was oxidized under intense UV radiation and although a very different environment than the Earth's upper atmosphere, the similar isotope pattern is suggestive that photochemical oxidation by UV light may be the mechanism creating coupled Δ^{200}Hg and Δ^{204}Hg anomalies in precipitation.

FUTURE DIRECTIONS

The field of mercury stable isotope geochemistry was established only within the past ten years and important new discoveries are being published by an ever-growing group of laboratories and investigators participating in this field of research. There are many exciting future applications of Hg isotopes on the horizon. Although this is an incomplete list, and there are many applications that have not yet been explored, we present our admittedly biased view of some of the most powerful potential applications of Hg stable isotopes.

1. Identification and tracking of isotopically distinct, multiple reservoirs of Hg in sediments and in surface and ground waters. In particular there is a need to understand why only a small portion of Hg in sediments is bioavailable for methylation. There is some promise that multiple Hg reservoirs within individual samples may be identifiable with selective extraction techniques and that these may be traceable through methylation and bioaccumulation.

2. Use of MDF, odd-MIF and even-MIF in precipitation and TGM to identify different sources and reservoirs of Hg in, and deposited from, the atmosphere. This has the potential to aid in monitoring changes in different sources of Hg to the atmosphere through time, and to allow quantification of the relative amounts of Hg(0) versus Hg(II) deposition to ecosystems.

3. Use of odd-MIF and even-MIF measured in samples collected from the marine photic zone to trace the transfer of Hg to the deep ocean and to sediments and to identify

zones of methylation in the ocean. Hg in sediments with even-MIF also provides a tracer of Hg cycling from the ocean surface waters to sediments and finally to the sedimentary rock record.

4. Investigations of the recycling of Hg that was once present at the Earth's surface back into the Earth's mantle through subduction. The odd and even MIF signals imparted on atmospheric Hg by photochemical reactions may provide a unique tracer of crustal Hg recycling.

5. Changes in global volcanism and wildfires over geological timescales. The use of MDF and odd-MIF can differentiate between geologic and biomass inputs of Hg to the global atmosphere and ocean.

The potential applications of Hg isotopes are extremely broad because of the multiple types of isotope fractionation that occur, and their usefulness in fingerprinting important reactions and mass transfers in the Earth system. As a closing thought we point out that while Hg isotopes have mostly been used to better understand the biogeochemistry of the element Hg, they are now finding many uses as tracers of bulk materials— such as air-masses, insects, organic matter and the rock cycle—rather than just as tracers of the Hg within these materials. Future applications of Hg isotopes are limited only by the imaginations of scientists using them, and we are optimistic of the opportunities for Hg isotope studies to advance many areas of the Earth and Environmental Sciences.

ACKNOWLEDGEMENTS

We wish to thank all of the past and current members of the Michigan mercury isotope research team for their hard work and dedication to moving the field of mercury stable isotopes forward. Comments and reviews by BA Bergquist, LS Sherman, MTK Tsui and an anonymous referee were helpful and resulted in an improved manuscript.

REFERENCES

Bergquist BA, Blum JD (2007) Mass-dependent and -independent fractionation of hg isotopes by photoreduction in aquatic systems. Science 318:417–420, doi:10.1126/science.1148050

Bergquist BA, Blum JD (2009) The odds and evens of mercury isotopes: applications of mass-dependent and mass-independent isotope fractionation. Elements 5:353–357, doi:10.2113/gselements.5.6.353

Biswas A, Blum JD, Bergquist BA, Keeler GJ, Xie Z (2008) Natural mercury isotope variation in coal deposits and organic soils. Environ Sci Technol 42:8303–8309, doi:10.1021/es801444b

Blum JD (2011) Applications of stable mercury isotopes to biogeochemistry. *In*: Handbook of Environmental Isotope Geochemistry. Advances in Isotope Geochemistry. Baskaran M (ed.) Springer-Verlag, Berlin, Heidelberg, p 229–245, doi: 10.1007/978-3-642-10637-8_12

Blum JD, Bergquist BA (2007) Reporting the variations in the natural isotopic composition of mercury. Anal Bioanal Chem 388:353–359

Blum JD, Johnson MW (2015) Identification of atmospheric mercury input to ecosystems from precipitation using coupled Δ^{200}Hg and Δ^{204}Hg fractionation, Abstract B13I-02 presented at 2015 Fall Meeting, AGU, San Francisco, Calif., 14–18 Dec

Blum JD, Popp BN, Drazen JC, Anela Choy C, Johnson MW (2013) Methylmercury production below the mixed layer in the North Pacific Ocean. Nat Geosci 6:879–884, doi:10.1038/ngeo1918

Blum JD, Sherman LS, Johnson MW (2014) Mercury isotopes in earth and environmental sciences. Ann Rev Earth Planet Sci 42:249–269, doi:10.1146/annurev-earth-050212-124107

Chen J, Hintelmann H, Dimock B (2010) Chromatographic pre-concentration of Hg from dilute aqueous solutions for isotopic measurement by MC-ICP-MS. J Anal Atom Spectrom 25:1402, doi:10.1039/c0ja00014k

Chen J, Hintelmann H, Feng X, Dimock B (2012) Unusual fractionation of both odd and even mercury isotopes in precipitation from Peterborough, ON, Canada. Geochim Cosmochim Acta 90:33–46, doi:10.1016/j.gca.2012.05.005

Chen J, Hintelmann H, Zheng W, Feng X, Cai H, Wang Z, Yuan S, Wang Z (2016) Isotopic evidence for distinct sources of mercury in lake waters and sediments. Chem Geol 426:33–44, doi:10.1016/j.chemgeo.2016.01.030

Cooke CA, Hintelmann H, Ague JJ, Burger R, Biester H, Sachs JP, Engstrom DR (2013) Use and legacy of mercury in the Andes. Environ Sci Technol 47:4181–4188, doi:10.1021/es3048027

Coplen TB (2011) Guidelines and recommended terms for expression of stable-isotope-ratio and gas-ratio measurement results. Rapid Commun Mass Spectrom 25:2538–2560, doi:10.1002/rcm.5129

Demers JD, Blum JD, Zak DR (2013) Mercury isotopes in a forested ecosystem: Implications for air-surface exchange dynamics and the global mercury cycle. Global Biogeochem Cycles 27:222–238, doi:10.1002/gbc.20021

Demers JD, Sherman LS, Blum JD, Marsik FJ, Dvonch JT (2015) Coupling atmospheric mercury isotope ratios and meteorology to identify sources of mercury impacting a coastal urban-industrial region near Pensacola, Florida, USA. Global Biogeochem Cycles 29:1689–1705, doi:10.1002/2015GB005146

Donovan PM, Blum JD, Yee D, Gehrke GE, Singer MB (2013) An isotopic record of mercury in San Francisco Bay sediment. Chem Geol 349–350:87–98, doi:10.1016/j.chemgeo.2013.04.017

Donovan PM, Blum JD, Singer MB, Marvin-DiPasquale M, Tsui MT (2016a) Methylmercury degradation and exposure pathways in streams and wetlands impacted by historical mining. Sci Total Environ, doi:10.1016/j.scitotenv.2016.04.139

Donovan PM, Blum JD, Singer MB, Marvin-DiPasquale M, Tsui MT (2016b) Isotopic composition of inorganic mercury and methylmercury downstream of a historical gold mining region. Environ Sci Technol 50:1691–1702, doi:10.1021/acs.est.5b04413

Engstrom DR, Fitzgerald WF, Cooke CA, Lamborg CH, Drevnick PE, Swain EB, Balogh SJ, Balcom PH (2014) Atmospheric Hg emissions from preindustrial gold and silver extraction in the Americas: A reevaluation from lake–sediment archives. Environ Sci Technol 48:6533–6543, doi:10.1021/es405558e

Epov VN, Rodriguez-Gonzalez P, Sonke JE, Tessier E, Amouroux D, Bourgoin LM, Donard OFX (2008) Simultaneous determination of species-specific isotopic composition of hg by gas chromatography coupled to multicollector ICPMS. Anal Chem 80:3530–3538, doi:10.1021/ac800384b

Epov VN, Berail S, Jimenez-Moreno M, Perrot V, Pecheyran C, Amouroux D, Donard OFX (2010) Approach to measure isotopic ratios in species using multicollector-ICPMS coupled with chromatography. Anal Chem 82:5652–5662, doi:10.1021/ac100648f

Estrade N, Carignan J, Sonke JE, Donard OFX (2009) Mercury isotope fractionation during liquid–vapor evaporation experiments. Geochim Cosmochim Acta 73:2693–2711, doi:10.1016/j.gca.2009.01.024

Feng C, Pedrero Z, Gentès S, Barre J, Renedo M, Tessier E, Berail S, Maury-Brachet R, Mesmer-Dudons N, Baudrimont M, Legeay A (2015) Specific pathways of dietary methylmercury and inorganic mercury determined by mercury speciation and isotopic composition in zebrafish (*Danio rerio*). Environ Sci Technol 49:12984–12993, doi:10.1021/acs.est.5b03587

Foucher D, Hintelmann H (2006) High-precision measurement of mercury isotope ratios in sediments using cold-vapor generation multi-collector inductively coupled plasma mass spectrometry. Anal Bioanal Chem 384:1470–1478, doi:10.1007/s00216-006-0373-x

Foucher D, Ogrinc, Hintelmann H (2009) Tracing mercury contamination from the Idrija mining region (Slovenia) to the Gulf of Trieste using Hg isotope ratio measurements. Environ Sci Technol 43:33–39, doi:10.1021/es801772b

Fu X, Heimbürger L-E, Sonke JE (2014) Collection of atmospheric gaseous mercury for stable isotope analysis using iodine- and chlorine-impregnated activated carbon traps. J Anal Atom Spectrom 29:841, doi:10.1039/c3ja50356a

Fu X, Marusczak N, Wang X, Gheusi F, Sonke JE (2016) Isotopic composition of gaseous elemental mercury in the free troposphere of the Pic du Midi Observatory, France. Environ Sci Technol 50:5641–5650, doi:10.1021/acs.est.6b00033

Gehrke GE, Blum JD, Meyers PA (2009) The geochemical behavior and isotopic composition of Hg in a mid-Pleistocene western Mediterranean sapropel. Geochim Cosmochim Acta 73:1651–1665, doi:10.1016/j.gca.2008.12.012

Gehrke GE, Blum JD, Slotton DG, Greenfield BK (2011) Mercury isotopes link mercury in San Francisco Bay forage fish to surface sediments. Environ Sci Technol 45:1264–1270, doi:10.1021/es103053y

Georg RB, Newman K (2015) The effect of hydride formation on instrumental mass discrimination in MC-ICP-MS: a case study of mercury (Hg) and thallium (Tl) isotopes. J Anal Atom Spectrom 30:1935–1944, doi:10.1039/c5ja00238a

Ghosh S, Schauble EA, Lacrampe Couloume G, Blum JD, Bergquist BA (2013) Estimation of nuclear volume dependent fractionation of mercury isotopes in equilibrium liquid–vapor evaporation experiments. Chem Geol 336:5–12, doi:10.1016/j.chemgeo.2012.01.008

Gratz LE, Keeler GJ, Blum JD, Sherman LS (2010) Isotopic composition and fractionation of mercury in great lakes precipitation and ambient air. Environ Sci Technol 44:7764–7770, doi:10.1021/es100383w

Guo W, Hu S, Wang X, Zhang J, Jin L, Zhu Z, Zhang H (2011) Application of ion molecule reaction to eliminate WO interference on mercury determination in soil and sediment samples by ICP-MS. J Anal Atom Spectrom 26:1198–1203, doi:10.1039/C1JA00005E

Hintelmann H (2012) Use of stable isotopes in mercury research. *In*: Mercury in the Environment: Pattern and Process. Bank MS (ed.) University of California Press, p 55–71

Hintelmann H, Lu S (2003) High precision isotope ratio measurements of mercury isotopes in cinnabar ores using multi-collector inductively coupled plasma mass spectrometry. Analyst 128:635–639, doi:10.1039/B300451A

Jackson TA (2001) Variations in the isotope composition of mercury in a freshwater sediment sequence and food web. Canadian Journal of Fisheries and Aquatic Sciences 58:185–196, doi:10.1139/f00-186

Janssen SE, Johnson MW, Blum JD, Barkay T, Reinfelder JR (2015) Separation of monomethylmercury from estuarine sediments for mercury isotope analysis. Chem Geol 411:19–25, doi:10.1016/j.chemgeo.2015.06.017

Jiskra M, Wiederhold JG, Bourdon B, Kretzschmar R (2012) Solution speciation controls mercury isotope fractionation of Hg(II) sorption to goethite. Environ Sci Technol 46:6654–6662, doi:10.1021/es3008112

Johnson CM, Beard BL, Albarède F (2004) Overview and general concepts. Rev Mineral Geochem 55:1–24, doi:10.2138/gsrmg.55.1.1

Kwon SY, Blum JD, Carvan MJ, Basu N, Head JA, Madenjian CP, David SR (2012) Absence of fractionation of mercury isotopes during trophic transfer of methylmercury to freshwater fish in captivity. Environ Sci Technol 46:7527–7534, doi:10.1021/es300794q

Kwon SY, Blum JD, Chirby MA, Chesney EJ (2013) Application of mercury isotopes for tracing trophic transfer and internal distribution of mercury in marine fish feeding experiments. Environ Toxicol Chem 32:2322–2330, doi:10.1002/etc.2313

Kwon SY, Blum JD, Chen CY, Meattey DE, Mason RP (2014) Mercury isotope study of sources and exposure pathways of methylmercury in estuarine food webs in the Northeastern U.S. Environ Sci Technol 48:10089–10097, doi:10.1021/es5020554

Kwon SY, Blum JD, Nadelhoffer KJ, Timothy Dvonch J, Tsui MT (2015) Isotopic study of mercury sources and transfer between a freshwater lake and adjacent forest food web. Sci Total Environ 532:220–229, doi:10.1016/j.scitotenv.2015.06.012

Kwon SY, Blum JD, Madigan DJ, Block BA, Popp BN (2016) Quantifying mercury isotope dynamics in captive Pacific bluefin tuna (*Thunnus orientalis*). Elementa 4:15, doi:10.12952/journal.elementa.000088

Lauretta DS, Klaue B, Blum JD, Buseck PR (2001) Mercury abundances and isotopic compositions in the Murchison (CM) and Allende (CV) carbonaceous chondrites. Geochim Cosmochim Acta 65:2807–2818, doi:10.1016/s0016-7037(01)00630–5

Lefticariu L, Blum JD, Gleason JD (2011) Mercury isotopic evidence for multiple mercury sources in coal from the Illinois basin. Environ Sci Technol 45:1724–1729, doi:10.1021/es102875n

Lepak RF, Yin R, Krabbenhoft DP, Ogorek JM, DeWild JF, Holsen TM, Hurley JP (2015) Use of stable isotope signatures to determine mercury sources in the Great Lakes. Environ Sci Technol Lett 2:335–341, doi:10.1021/acs.estlett.5b00277

Li M, Sherman LS, Blum JD, Grandjean P, Mikkelsen B, Weihe P, Sunderland EM, Shine JP (2014) Assessing sources of human methylmercury exposure using stable mercury isotopes. Environ Sci Technol 48:8800–8806, doi:10.1021/es500340r

Masbou J, Point D, Sonke JE (2013) Application of a selective extraction method for methylmercury compound specific stable isotope analysis (MeHg-CSIA) in biological materials. J Anal Atom Spectrom 28:1620, doi:10.1039/c3ja50185j

Mead C, Johnson TM (2010) Hg stable isotope analysis by the double-spike method. Anal Bioanal Chem 397:1529–1538, doi:10.1007/s00216-010-3701-0

Mead C, Lyons JR, Johnson TM, Anbar AD (2013) Unique Hg stable isotope signatures of compact fluorescent lamp-sourced Hg. Environ Sci Technol 47:2542–2547, doi:10.1021/es303940p

Meier MMM, Cloquet C, Marty B (2016) Mercury (Hg) in meteorites: Variations in abundance, thermal release profile, mass-dependent and mass-independent isotopic fractionation. Geochim Cosmochim Acta 182:55–72, doi:http://dx.doi.org/10.1016/j.gca.2016.03.007

Meija J, Yang L, Sturgeon RE, Mester Z (2010) Certification of natural isotopic abundance inorganic mercury reference material NIMS-1 for absolute isotopic composition and atomic weight. J Anal Atom Spectrom 25:384–389, doi:10.1039/B926288A

Mil-Homens M, Blum JD, Canário J, Caetano M, Costa AM, Lebreiro SM, Trancoso M, Richter T, de Stigter H, Johnson M, Branco V (2013) Tracing anthropogenic Hg and Pb input using stable Hg and Pb isotope ratios in sediments of the central Portuguese Margin. Chem Geol 336:62–71, doi:10.1016/j.chemgeo.2012.02.018

Perrot V, Masbou J, Pastukhov MV, Epov VN, Point D, Berail S, Becker PR, Sonke JE, Amouroux D (2016) Natural Hg isotopic composition of different Hg compounds in mammal tissues as a proxy for in vivo breakdown of toxic methylmercury. Metallomics 8:170–178, doi:10.1039/c5mt00286a

Ridley WI, Stetson SJ (2006) A review of isotopic composition as an indicator of the natural and anthropogenic behavior of mercury. Appl Geochem 21:1889–1899, doi:10.1016/j.apgeochem.2006.08.006

Rolison JM, Landing WM, Luke W, Cohen M, Salters VJM (2013) Isotopic composition of species-specific atmospheric Hg in a coastal environment. Chem Geol 336:37–49, doi:10.1016/j.chemgeo.2012.10.007

Rua-Ibarz A, Bolea-Fernandez E, Vanhaecke F (2015) An in-depth evaluation of accuracy and precision in Hg isotopic analysis via pneumatic nebulization and cold vapor generation multi-collector ICP-mass spectrometry. Anal Bioanal Chem 408:417–429, doi:10.1007/s00216-015-9131-2

Senn DB, Chesney EJ, Blum JD, Bank MS, Maage A, Shine JP (2010) Stable Isotope (N, C, Hg) Study of methylmercury sources and trophic transfer in the northern Gulf of Mexico. Environ Sci Technol 44:1630–1637, doi:10.1021/es902361j

Shayesteh A, Yu SS, Bernath PF (2005) Gaseous HgH_2, CdH_2, and ZnH_2. Chem Eur J 11:4709–4712, doi:10.1002/chem.200500332

Sherman LS, Blum JD (2013) Mercury stable isotopes in sediments and largemouth bass from Florida lakes, USA. Sci Total Environ 448:163–175, doi:10.1016/j.scitotenv.2012.09.038

Sherman LS, Blum JD, Nordstrom DK, McCleskey RB, Barkay T, Vetriani C (2009) Mercury isotopic composition of hydrothermal systems in the Yellowstone Plateau volcanic field and Guaymas Basin sea-floor rift. Earth Planet Sci Lett 279:86–96, doi:http://dx.doi.org/10.1016/j.epsl.2008.12.032

Sherman LS, Blum JD, Keeler GJ, Demers JD, Dvonch JT (2012) Investigation of local mercury deposition from a coal-fired power plant using mercury isotopes. Environ Sci Technol 46:382–390, doi:10.1021/es202793c

Sherman LS, Blum JD, Franzblau A, Basu N (2013) New insight into biomarkers of human mercury exposure using naturally occurring mercury stable isotopes. Environ Sci Technol 47:3403–3409, doi:10.1021/es305250z

Sherman LS, Blum JD, Dvonch JT, Gratz LE, Landis MS (2015a) The use of Pb, Sr, and Hg isotopes in Great Lakes precipitation as a tool for pollution source attribution. Sci Total Environ 502:362–374, doi:10.1016/j.scitotenv.2014.09.034

Sherman LS, Blum JD, Basu N, Rajaee M, Evers DC, Buck DG, Petrlik J, DiGangi J (2015b) Assessment of mercury exposure among small-scale gold miners using mercury stable isotopes. Environ Res 137:226–234, doi:10.1016/j.envres.2014.12.021

Shirai N, Humayun M (2011) Mass independent bias in W isotopes in MC-ICP-MS instruments. J Anal Atom Spectrom 26:1414–1420, doi:10.1039/C0JA00206B

Sial AN, Chen J, Lacerda LD, Frei R, Tewari VC, Pandit MK, Gaucher C, Ferreira VP, Cirilli S, Peralta S, Korte C (2016) Mercury enrichment and Hg isotopes in Cretaceous–Paleogene boundary successions: Links to volcanism and palaeoenvironmental impacts. Cretaceous Res 66:60–81, doi:10.1016/j.cretres.2016.05.006

Smith CN, Kesler SE, Klaue B, Blum JD (2005) Mercury isotope fractionation in fossil hydrothermal systems. Geology 33:825–828, doi:10.1130/G21863.1

Smith CN, Kesler SE, Blum JD, Rytuba JJ (2008) Isotope geochemistry of mercury in source rocks, mineral deposits and spring deposits of the California Coast Ranges, USA. Earth Planet Sci Lett 269:399–407, doi:10.1016/j.epsl.2008.02.029

Sonke JE, Schäfer J, Chmeleff J, Audry S, Blanc G, Dupré B (2010) Sedimentary mercury stable isotope records of atmospheric and riverine pollution from two major European heavy metal refineries. Chem Geol 279:90–100, doi:10.1016/j.chemgeo.2010.09.017

Stetson SJ, Gray JE, Wanty RB, Macalady DL (2009) Isotopic variability of mercury in ore, mine-waste calcine, and leachates of mine-waste calcine from areas mined for mercury. Environ Sci Technol 43:7331–7336, doi:10.1021/es9006993

Štrok M, Baya PA, Hintelmann H (2015) The mercury isotope composition of Arctic coastal seawater. CR Geosci 347:368–376, doi:10.1016/j.crte.2015.04.001

Sun R, Enrico M, Heimburger LE, Scott C, Sonke JE (2013a) A double-stage tube furnace—acid-trapping protocol for the pre-concentration of mercury from solid samples for isotopic analysis. Anal Bioanal Chem 405:6771–6781, doi:10.1007/s00216-013-7152-2

Sun R, Heimbürger L-E, Sonke JE, Liu G, Amouroux D, Berail S (2013b) Mercury stable isotope fractionation in six utility boilers of two large coal-fired power plants. Chem Geol 336:103–111, doi:10.1016/j.chemgeo.2012.10.055

Sun R, Sonke JE, Heimburger LE, Belkin HE, Liu G, Shome D, Cukrowska E, Liousse C, Pokrovsky OS, Streets DG (2014) Mercury stable isotope signatures of world coal deposits and historical coal combustion emissions. Environ Sci Technol 48:7660–7668, doi:10.1021/es501208a

Thibodeau AM, Ritterbush K, Yager JA, West AJ, Ibarra Y, Bottjer DJ, Berelson WM, Bergquist BA, Corsetti FA (2016) Mercury anomalies and the timing of biotic recovery following the end-Triassic mass extinction. Nat Commun 7, doi:10.1038/ncomms11147

Tsui MT, Blum JD, Kwon SY, Finlay JC, Balogh SJ, Nollet YH (2012) Sources and transfers of methylmercury in adjacent river and forest food webs. Environ Sci Technol 46:10957–10964, doi:10.1021/es3019836

Tsui MTK, Blum JD, Finlay JC, Balogh SJ, Kwon SY, Nollet YH (2013) Photodegradation of methylmercury in stream ecosystems. Limnol Oceanogr 58:13–22, doi:10.4319/lo.2013.58.1.0013

Tsui MT, Blum JD, Finlay JC, Balogh SJ, Nollet YH, Palen WJ, Power ME (2014) Variation in terrestrial and aquatic sources of methylmercury in stream predators as revealed by stable mercury isotopes. Environ Sci Technol 48:10128–10135, doi:10.1021/es500517s

Wiederhold JG, Cramer CJ, Daniel K, Infante I, Bourdon B, Kretzschmar R (2010) Equilibrium mercury isotope fractionation between dissolved Hg(II) species and thiol-bound Hg. Environ Sci Technol 44:4191–4197, doi:10.1021/es100205t

Yin R, Feng X, Shi W (2010) Application of the stable-isotope system to the study of sources and fate of Hg in the environment: A review. Appl Geochem 25:1467–1477, doi:10.1016/j.apgeochem.2010.07.007

Yin R, Feng X, Chen J (2014) Mercury stable isotopic compositions in coals from major coal producing fields in China and their geochemical and environmental implications. Environ Sci Technol 48:5565–5574, doi:10.1021/es500322n

Yin R, Feng X, Wang J, Li P, Liu J, Zhang Y, Chen J, Zheng L, Hu T (2013) Mercury speciation and mercury isotope fractionation during ore roasting process and their implication to source identification of downstream sediment in the Wanshan mercury mining area, SW China. Chem Geol 336:72–79, doi:10.1016/j.chemgeo.2012.04.030

Yin R, Feng X, Hurley JP, Krabbenhoft DP, Lepak RF, Hu R, Zhang Q, Li Z, Bi X (2016) Mercury isotopes as proxies to identify sources and environmental impacts of mercury in sphalerites. Sci Rep 6:18686, doi:10.1038/srep18686, http://www.nature.com/articles/srep18686#supplementary-information

Yuan S, Zhang Y, Chen J, Kang S, Zhang J, Feng X, Cai H, Wang Z, Wang Z, Huang Q (2015) Large variation of mercury isotope composition during a single precipitation event at Lhasa City, Tibetan Plateau, China. Procedia Earth Planet Sci 13:282–286, doi:10.1016/j.proeps.2015.07.066

Zhang L, Liu Y, Guo L, Yang D, Fang Z, Chen T, Ren H, Yu B (2014) Isotope geochemistry of mercury and its relation to earthquake in the Wenchuan Earthquake Fault Scientific Drilling Project Hole-1 (WFSD-1). Tectonophys 619–620:79–85, doi:http://dx.doi.org/10.1016/j.tecto.2013.08.025

Reviews in Mineralogy & Geochemistry
Vol. 82 pp. 759-798, 2017

Investigation and Application of Thallium Isotope Fractionation

Sune G. Nielsen[1,2]

*[1]NIRVANA laboratories
Woods Hole Oceanographic Institution
Woods Hole, MA
USA*

*[2] Department of Geology and Geophysics
Woods Hole Oceanographic Institution
Woods Hole, MA
USA*

snielsen@whoi.edu

Mark Rehkämper and Julie Prytulak

*Department of Earth Science and Engineering
Imperial College
London
SW7 2AZ
UK*

markrehk@imperial.ac.uk

j.prytulak@imperial.ac.uk

ABSTRACT

This contribution summarizes the current state of understanding and recent advances made in the field of stable thallium (Tl) isotope geochemistry. High precision measurements of Tl isotope compositions were developed in the late 1990s with the advent of multiple collector inductively coupled plasma mass spectrometry (MC-ICPMS) and subsequent studies revealed that Tl, despite the small relative mass difference of the two isotopes, exhibits substantial stable isotope fractionation, especially in the marine environment. The most fractionated reservoirs identified are ferromanganese sediments with $\varepsilon^{205}Tl \approx +15$ and low temperature altered oceanic crust with $\varepsilon^{205}Tl \approx -20$. The total isotopic variability of more than 35 ε^{205}Tl-units hence exceeds the current analytical reproducibility of the measurement technique by more than a factor of 70. This isotopic variation can be explained by invoking a combination of conventional mass dependent equilibrium isotope effects and nuclear field shift isotope fractionation, but the specific mechanisms are still largely unaccounted for.

Thallium isotopes have been applied to investigate paleoceanographic processes in the Cenozoic and there is evidence to suggest that Tl isotopes may be utilized as a monitor of the marine manganese oxide burial flux over million year time scales. In addition, Tl isotopes can be used to calculate the magnitude of hydrothermal fluid circulation through ocean crust. It has also been shown that the subduction of marine ferromanganese sediments can be detected with Tl isotopes in lavas erupted in subduction zone settings as well as in ocean island basalts.

1529-6466/17/0082-0018$05.00 http://dx.doi.org/10.2138/rmg.2017.82.18

Meteorite samples display Tl isotope variations that exceed the terrestrial range with a total variability of about 50 ε^{205}Tl. The large isotopic diversity, however, is generated by both stable Tl isotope fractionations, which reflect the highly volatile and labile cosmochemical nature of the element, and radiogenic decay of extinct ^{205}Pb to ^{205}Tl with a half-life of about 15 Ma. The difficulty of deconvolving these two sources of isotopic variability restricts the utility of both the ^{205}Pb–^{205}Tl chronometer and the Tl stable isotope system to inform on early solar system processes.

INTRODUCTION

The distribution of Tl in natural environments on Earth is controlled in part by its large ionic radius (Tl^+ = 1.50 A), which is akin to the alkali metals potassium (K), rubidium (Rb) and cesium (Cs) (Wedepohl 1974; Shannon 1976; Heinrichs et al. 1980). Thallium's large ionic radius renders it highly incompatible during partial melting and magmatic differentiation, leading to much higher Tl concentrations in the continental crust (Shaw 1952; Wedepohl 1995) compared to the mantle (Fig. 1). However, the electron structure of Tl tends to favor covalent bonding, which makes Tl compatible in some sulfides (Jones et al. 1993; Wood et al. 2008; Nielsen et al. 2011, 2014; Kiseeva and Wood 2013; Genna and Gaboury 2015). In addition to its bonding preferences, Tl has two different valence states: Tl+and Tl^{3+}. Although oxidation of Tl requires a large redox potential (Table 1) the manganese (Mn) oxide birnessite has the ability to adsorb and oxidize Tl at its surface (Bidoglio et al. 1993; Peacock and Moon 2012) and subsequently incorporate Tl^{3+} more firmly in its structure, which leads to very high concentrations of Tl in marine (Fe) Mn oxides (Fig. 1) (Shaw 1952; Hein et al. 2000; Rehkämper et al. 2002; Peacock and Moon 2012; Nielsen et al. 2013). Sorption of Tl onto some clay minerals has also been observed (Matthews and Riley 1970; McGoldrick et al. 1979; Turner et al. 2010) although it is not clear if this behavior is related to the similarity with alkali metals. Thallium's higher particle-reactivity compared to alkali metals results in low concentrations in rivers and the oceans and thus a higher concentration contrast between the ocean and continental crust compared to elements such as Rb, Cs and K (Bruland 1983; Flegal and Patterson 1985; Nielsen et al. 2005b).

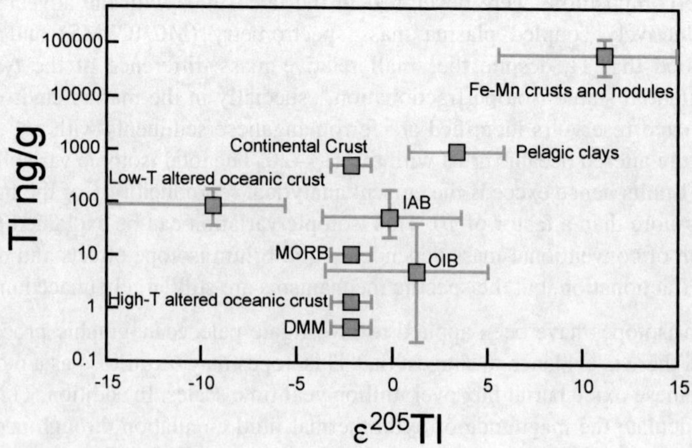

Figure 1. Thallium isotope compositions and concentrations for terrestrial reservoirs relevant to subduction zones and recycled oceanic crust. Note the logarithmic scale for the concentrations. Data sourced from (Rehkämper et al. 2002, 2004; Nielsen et al. 2005a, 2006b,c, 2010, 2014, 2015, 2016; Prytulak et al. 2013).

Thallium has two isotopes with atomic masses 203 and 205 (Table 1) and abundances of ~30% and ~70%, respectively. This equates to a relative mass difference of < 1%. Considering that stable isotope fractionation theory states that the magnitude of isotope fractionation should scale with the relative mass difference of the isotopes (Bigeleisen and Mayer 1947; Urey 1947) one would not expect large stable isotope effects for Tl. Hence, it may be surprising that stable isotope investigations for Tl were commenced in the first place. However, the first attempts to measure Tl isotope ratios were not aimed at terrestrial materials. The main reason behind the first Tl isotope studies was the search for potential radiogenic isotope variations due to decay of the now-extinct radioactive isotope ^{205}Pb to ^{205}Tl with a half-life of 15.1 Ma (Pengra et al. 1978), whereas ^{203}Tl is stable and has no radioactive precursor. A number of studies between 1960 and 1994 failed to register resolvable Tl isotope variation for some selected terrestrial and a large number of extraterrestrial materials (Anders and Stevens 1960; Ostic et al. 1969; Huey and Kohman 1972; Chen and Wasserburg 1987, 1994). These investigations were hampered, however, by the relatively large errors (>2‰) associated with the thermal ionization mass spectrometry (TIMS) measurements used at that time. The breakthrough came in 1999 when the first high precision Tl isotope measurements by MC-ICPMS were published (Rehkämper and Halliday 1999). This technique provided a reduction in the uncertainty of more than an order of magnitude with errors reported at about 0.1–0.2‰ (Rehkämper and Halliday 1999). With the reduced error bars, analysis of various terrestrial samples revealed large Tl isotope variation in excess of 15 times the analytical reproducibility of the original method, which opened up investigations of Tl isotope fractionation on Earth and in meteorites.

In this contribution, we summarize the knowledge that has been accumulated since 1999 on the stable isotope geochemistry of Tl. Three main environments are discussed: 1) extraterrestrial, 2) the solid Earth, and 3) the marine domain. Throughout these sections, current and notable applications of Tl stable isotopes in geochemical research are incorporated. Finally, potential future studies are suggested that are likely to make Tl isotopes a more quantitative tracer in Earth sciences.

Table 1. Physical properties of thallium (Nriagu 1998).

Melting point	**577 K**
Molar weight	204.38 g
Density	11.85 g/cm^3
Valence states	Tl^0, Tl^+, Tl^{3+}
Redox potentials (V)	
$Tl_{(s)} \rightarrow Tl^+ + e^-$	+0.336
$Tl^+ \rightarrow Tl^{3+} + 2e^-$	−1.28
Ionic radii (Å)	
Tl^+	1.50
Tl^{3+}	0.89
Stable isotopes	^{203}Tl, ^{205}Tl

METHODOLOGY

Mass spectrometry

The advent of MC-ICPMS facilitated the development of high-precision Tl isotope ratio measurements. The principal difference to the previous TIMS measurements was the ability to correct for instrumental isotope fractionation that occurs during the measurement (mass bias or mass discrimination). Two-isotope systems are difficult to measure by TIMS because isotope fractionation during volatilization from the filament is both time and mass dependent, which is difficult to correct for. Therefore, precise stable isotope ratios by TIMS are best measured with the use of a double spike (Rudge et al. 2009). However, double spiking can only easily be performed for elements with four or more isotopes (Rudge et al. 2009) and this is the reason why the early Tl isotope studies by TIMS yielded relatively large uncertainties. The great advantage of MC-ICPMS when measuring Tl isotopes (or any two-isotope system) is that, even though the overall magnitude of instrumental mass discrimination is much larger than for TIMS, it can be monitored independently during a measurement and thus corrections can be applied much more precisely than is possible with TIMS. This mass bias correction can be performed because the sample is introduced into the mass spectrometer as a solution (or a desolvated aerosol). Into this solution can be admixed a separate element with a known isotope composition (for Tl this element is Pb) and by assuming that the mass bias incurred for the two elements are proportional, isotope ratios can be determined very accurately and precisely (Rehkämper and Halliday 1999; Nielsen et al. 2004). For Tl isotope measurements this external normalization with Pb is always combined with the more conventional standard-sample bracketing technique that is also very common for stable isotope measurements by MC-ICPMS, which produces the most rigorous measurement stability as both instrumental mass bias and machine drift can be corrected for simultaneously. In practice (as is the case for all other stable isotope systems) isotope compositions are conventionally reported by reference to a standard that is defined as zero. For Tl this standard is the NIST 997 Tl metal, such that:

$$\varepsilon^{205}\text{Tl} = \frac{10^4 \times \left(\dfrac{^{205}\text{Tl}}{^{203}\text{Tl}_{\text{sample}}} - \dfrac{^{205}\text{Tl}}{^{203}\text{Tl}_{\text{NIST 997}}} \right)}{\left(\dfrac{^{205}\text{Tl}}{^{203}\text{Tl}_{\text{NIST 997}}} \right)} \qquad (1)$$

The terminology used for Tl isotope ratios is slightly different from most stable isotope systems, which generally use the δ-notation (variations in parts per 1,000). The reason for this difference is that the Tl isotope system was originally developed as a cosmochemical radiogenic isotope system, which is usually reported using the ε-notation (variations in parts per 10,000). Hence, the original notation was retained in order to facilitate comparisons between cosmochemical and terrestrial data. In addition, the analytical uncertainty on and overall variability of stable Tl isotope ratios make the ε-notation very convenient as most data are thereby shown in whole digits and only a single figure behind the decimal point is needed.

Chemical separation of thallium

Prerequisites to obtaining precise and accurate stable isotope ratios by MC-ICPMS are the complete separation of the element of interest from the sample matrix as well as 100% recovery. This is important because residual sample matrix in the purified sample can result in isotope effects either present as instabilities that lead to large uncertainties or as reproducible isotopic offsets leading to precise but inaccurate data (Pietruszka and Reznik 2008; Poirier and Doucelance 2009; Shiel et al. 2009).

The method for separating Tl from sample matrix when performing isotopic analyses of geologic materials was initially developed by Rehkämper and Halliday (1999). The technique has been modified slightly (Nielsen et al. 2004, 2007; Baker et al. 2009) from the original recipe, but the fundamentals have remained unchanged. Only the elution procedure has been optimized in order to remove matrix elements most efficiently (Fig. 2). All techniques outlined in Figure 2 achieve effective Tl separation from sample matrix. However, methods that use HBr during matrix elution allow for collection of Pb from the same column, although this procedure has a tendency to separate Pb less efficiently from Tl due to the strong partitioning of bromide-complexed Pb onto anion exchange resin. The separation technique relies on the fact that Tl^{3+} produces anionic complexes with the halogens (the technique uses either Cl^- or Br^-) in acidic solutions that partition very strongly to anion exchange resins. Conversely Tl^+ does not form strong anionic complexes and thus does not partition at all to anion exchange resins. Therefore, samples are prepared in oxidizing media by adding small amounts of water saturated in Br_2 to the samples already digested and dissolved in hydrochloric acid. This process ensures that all Tl is in the trivalent state, which will adsorb onto the anion exchange resin prepared in a quartz or teflon column. If the Tl oxidation was only partial during bromine addition this would cause loss of Tl and likely result in Tl isotope fractionation. However, the procedure routinely produces quantitative recovery of sample Tl (Prytulak et al. 2013), which documents that all Tl is oxidized from bromine addition. The sample matrix can then be eluted in various acidic media as long as Br_2 is present. Lastly, Tl is stripped from the resin by elution with a reducing solution that

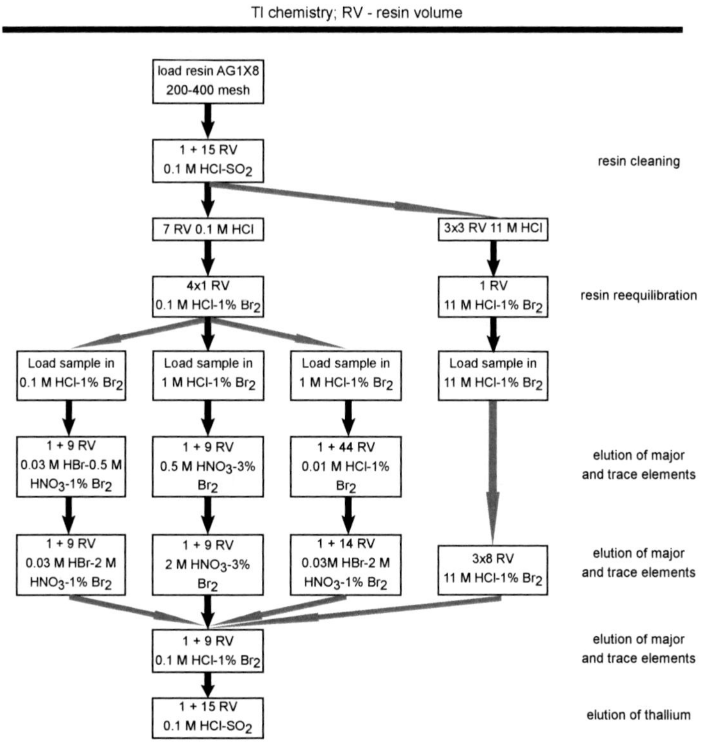

Figure 2. Anion exchange separation procedures for Tl. The recipes can be scaled to any amount of resin (RV – resin volume) depending on sample size, though large samples require a second 100 μl resin column to ensure that Tl is sufficiently pure. Four different elution procedures are outlined, as published by Rehkämper and Halliday (1999), Nielsen et al. (2004) and Baker et al. (2009).

converts Tl to the univalent state. The reducing solution used is 0.1M hydrochloric acid in which 5% by weight of SO_2 gas has been dissolved. As SO_2 is not stable in solution for long periods of time it is important to make this solution fresh before performing the chemical separation of Tl.

Measurement uncertainties and standards

As with most stable isotope measurements by MC-ICPMS, the smallest uncertainties are obtained for pure standard solutions. The most commonly used secondary standard for Tl is a pure 1,000 μg/g standard solution for ICP-MS concentration analyses that was originally purchased from Aldrich. Over more than 10 years this standard has been measured against NIST 997 Tl on seven different mass spectrometers with an average of $\varepsilon^{205}Tl = -0.79 \pm 0.35$ (2sd, $n = 187$). This uncertainty, however, is not necessarily representative of how well samples can be reproduced, mainly because small amounts of sample matrix invariably degrade the measurement precision compared to a pure metal standard even when measuring at the same ion beam intensity. Matrix effects are difficult to quantify and also depend on the sample introduction equipment. However, experiments in which matrix effects were tested by way of doping samples with Tl of a known isotope composition showed no systematic Tl isotope offset due to residual sample matrix (Nielsen et al. 2004). Older studies did find a relationship between sample concentration and measurement uncertainty with the smallest uncertainties obtained for samples with the highest concentrations (Nielsen et al. 2004, 2006a, 2007; Baker et al. 2009), but the most recent studies have reported external reproducibility for real samples and reference materials that are only slightly worse than what can be achieved for the Aldrich standard (Prytulak et al. 2013; Coggon et al. 2014; Kersten et al. 2014; Nielsen et al. 2015) (Table 2).

Table 2. Tl isotope and concentration data for geologic reference materials.

Standard	Description	ε^{205}Tl	*n*	Error[a]	Tl conc (ng/g)	Reference
Nod P1	USGS Ferromanganese nodule	0.5	1	0.5	146000	1
Nod A1[b]	USGS Ferromanganese nodule	10.7	6	0.5	108000	1,2
AGV-2	USGS Andesite	−3.0	8	0.6	269	3,4
BCR-2	USGS Columbia River basalt	−2.5	4	0.4	257	3
BHVO-1	USGS Hawaii Basalt	−3.5	10	0.5	37	3,8
BHVO-2	USGS Hawaii Basalt	−1.8	17	0.3	18	3,4,7
BIR-1	USGS Iceland basalt	1.1	6	1.2	1.3	5
NASS-5	Atlantic surface seawater	−5.0	1	1.0	0.0094	2
Allende	Carbonaceous chondrite	−3.1	8	0.5	55	6
BHVO-2G	USGS Basaltic glass	nd			16	9
BCR-2G	USGS Basaltic glass	nd			234	9
BIR-1G	USGS Basaltic glass	nd			2.5	9

References 1: Rehkämper et al. (2002); 2: Nielsen et al. (2004); 3: Prytulak et al. (2013); 4: Baker et al. (2009); 5: Nielsen et al. (2007); 6: Baker et al. (2010b); 7: Coggon et al. (2014); 8: Nielsen et al. (2015); 9: Nielsen and Lee (2013)

[a] – errors are either 2sd of the population of separate sample splits processed individually ($n \geq 3$) or estimated based on repeat measurements of similar samples ($n = 1$).

[b] – Isotope composition reported for multiple analyses of one large 300 mg aliquot dissolved in 6 M HCl.

nd - not determined

In cases where the amount of sample is limited, it can become an important issue how many ions can be measured for a given amount of Tl. Over the last ten years MC-ICPMS instruments have been developed to achieve increased transmission (i.e. the fraction of the ions introduced into the machine that reach the collector) and the most recent instruments have values of > 1% for Tl and Pb. This level of transmission routinely produces ion beam intensities of ~20nA for a solution containing 1 μg/g Tl and enables Tl isotope analyses on samples as small as 1 ng, without notably compromising counting statistics, and an external precision of better than ±1 ε-unit is achievable (Baker et al. 2009; Nielsen et al. 2004, 2006a, 2007, 2015). Smaller sample sizes down to 200 pg can still be analyzed on regular Faraday collectors, although precision is significantly degraded to ±3 ε-units (Nielsen et al. 2006a, 2007, 2009b).

THALLIUM ISOTOPE VARIATION IN EXTRATERRESTRIAL MATERIALS

With a half-mass condensation temperature (the temperature at which half of the Tl in the solar nebula was condensed) of 532 K (Lodders 2003), Tl is classified as a highly volatile element in cosmochemistry. In addition, Tl has also been termed as highly labile (Lipschutz and Woolum 1988) because it is readily remobilized by processes that are recorded on asteroidal parent bodies and meteroids, including thermal metamorphism and shock heating. Importantly, both properties are conducive to the production of relatively large stable isotope fractionations, and these have been observed for both Tl and other highly volatile elements such as Cd (Wombacher et al. 2003, 2008) and Hg (Lauretta et al. 1999, 2001) in various extraterrestrial materials.

The ^{205}Pb–^{205}Tl decay system

The radiogenic Tl isotope variations recorded in meteorites reflect decay of the short-lived radionuclide ^{205}Pb to ^{205}Tl with a half-life of 15.1 Ma (Pengra et al. 1978). Interest in the ^{205}Pb–^{205}Tl decay system and the initial solar system abundance of ^{205}Pb was responsible for driving the first efforts to precisely determine the Tl isotope compositions of natural materials. In detail, the literature documents six attempts to identify radiogenic Tl isotope variations in iron meteorites and chondrites from 1960 to 1995 (Anders and Stevens 1960; Ostic et al. 1969; Huey and Kohman 1972; Arden and Cressey 1984; Chen and Wasserburg 1987, 1994). Whilst these studies were able to provide an upper limit for the initial solar system ^{205}Pb abundance they were unable to conclusively establish radiogenic Tl isotope variations from the decay of this now-extinct nuclide.

The strong historic interest in the ^{205}Pb–^{205}Tl decay system stems from studies of stellar nucleosynthesis, which indicate that ^{205}Pb is produced primarily or almost exclusively by *s*-process nucleosynthesis (Blake et al. 1973; Yokoi et al. 1985; Wasserburg et al. 1994, 2006), whereby heavier elements are formed by slow neutron capture in the interior of stars. For astrophysicists, precise constraints on the initial solar system abundance of ^{205}Pb, gained from analyses of meteorites, would hence offer unique clues on the site and operation of *s*-process nucleosynthesis and the extent to which this process contributed to the freshly synthesized nucleosynthetic material that was delivered to the nascent solar system. In addition, the ^{205}Pb–^{205}Tl decay system is also of interest as a chronometer of volatile depletion as well as core formation and cooling. The latter applications follow from the observation that Pb/Tl ratios are likely to be fractionated by both processes, as (i) Pb, with a half-mass condensation temperature of 727 K (Lodders 2003), is somewhat less volatile than Tl and (ii) Pb and Tl exhibit different extents of moderately siderophile and chalcophile affinity during metal-sulfide-silicate partitioning (Jones et al. 1993; Wood et al. 2008; Ballhaus et al. 2013).

Despite of the numerous potential applications, only relatively few Tl isotope studies of meteorites have been attempted using the much more precise MC-ICP-MS methods which superseded TIMS measurements following the pioneering study of Rehkämper and Halliday

(1999). These investigations were all motivated by the ^{205}Pb–^{205}Tl chronometer but resolved both radiogenic and stable Tl isotope variations in various stony and iron meteorites. A summary of this work, key results, and cosmochemical implications are presented below.

Chondritic meteorites

Carbonaceous chondrites. Baker et al. (2010b) carried out a comprehensive study of 10 carbonaceous chondrites of groups CI1 (Orgueil), CM2 (incl. Murchison), CR2, CV3 (incl. Allende), and CO3. Two samples were excluded from the evaluation of radiogenic Tl isotope effects as they had fractionated Cd isotope compositions due to thermal processing (Wombacher et al. 2003, 2008), and this may have also affected Tl isotopes, whilst four samples were corrected for terrestrial Pb contributions of between 9 and 77% of total Pb, based on measured Pb isotope data.

Overall, the Tl and Pb concentrations of the meteorites varied by more than a factor of 5, between about 20 ng/g (in CR2 chondrites) to 100 ng/g (in Orgueil CI1) for Tl and 340–2200 ng/g for Pb. Importantly, the results showed a clear co-variation between Tl and Pb, which was used to derive average carbonaceous chondrite and, by inference, solar system Pb/Tl and $^{204}Pb/^{203}Tl$ ratios of 22 ± 2 and 1.43 ± 0.14 respectively (uncertainties are 2sd).

Furthermore, the samples revealed a small but significant correlated variability between Tl isotope compositions, with $\varepsilon^{205}Tl$ values of -4.0 to $+1.2$, and $^{204}Pb/^{203}Tl$ ratios (Fig. 3). A number of observations, in particular unfractionated stable Cd isotope compositions in all the samples, indicate that this correlation is unlikely to be due to stable isotope fractionation from early solar system processes or terrestrial weathering, and is instead most readily explained by in situ decay of ^{205}Pb to ^{205}Tl. Previous ^{53}Mn–^{53}Cr and ^{107}Pd–^{107}Ag studies of bulk carbonaceous chondrites furthermore suggest that the Pb–Tl isochron records volatile fractionation in the solar nebula at close to 4567 Ma (Schönbächler et al. 2008; Shukolyukov and Lugmair 2006). If this interpretation is correct, then the isochron of Figure 3 yields the initial ^{205}Pb abundance and Tl isotope composition of the solar system, with values of $^{205}Pb/^{204}Pb_{SS,0} = (1.0 \pm 0.4) \times 10^{-3}$ and $\varepsilon^{205}Tl_{SS,0} = -7.6 \pm 2.1$, respectively. These results provide clear evidence for the existence of live ^{205}Pb in the early solar system. The inferred $^{205}Pb/^{204}Pb_{SS,0}$ ratio is close to the upper limit of nucleosynthetic production estimates for AGB stars (Wasserburg et al. 2006) and thus in accord with contributions of such stars to the early solar system budget of freshly synthesized nucleosynthetic matter.

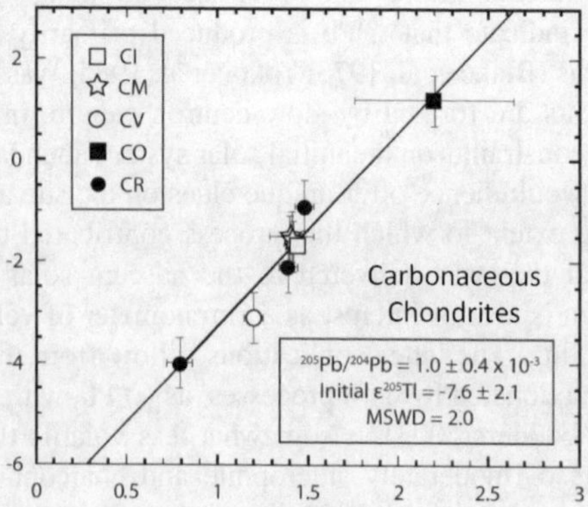

Figure 3. Pb–Tl isochron diagram for carbonaceous chondrites (modified from Baker et al. 2010b). Excluded from the plot and isochron calculation are two meteorites that exhibit stable Cd isotope fractionations as well as a sample of Allende Smithsonian due to contamination with terrestrial Pb (Baker et al. 2010b). All error bars are 2sd.

Enstatite chondrites. Analyses were carried out on a suite of enstatites chondrites, comprised of three less metamorphosed samples from groups EL3, EH4 and three equilibrated meteorites from groups EH5 and EL6 (Palk et al. 2011). In detail, the EL3 and EH4 enstatite chondrites were found to have much higher volatile contents (of about 40 to 100 ng/g Tl) than the intensely metamorphosed samples of groups EH5 and EL6, which had Tl abundances of only about 5 ng/g Tl. Whilst four of the meteorites displayed moderate to extreme Cd isotope fractionations of up to $\varepsilon^{114/110}Cd \approx +70$, only a single sample displayed a clearly fractionated Tl isotope composition with $\varepsilon^{205}Tl \approx +22$. As this sample is known to be highly shocked (with a grade of S5), it is possible that Tl was mobilized and isotopically fractionated during shock metamorphism.

In contrast, all other enstatite chondrites yielded a narrow range of $\varepsilon^{205}Tl$ values, with results of between −2.9 and +0.8. Given the limited isotopic variability, the observation of Tl (and more prevalent Cd) stable isotope fractionation, and the presence of terrestrial Pb contamination (in two meteorites) as revealed by Pb isotopes, a robust Pb–Tl isochron could not be obtained for the enstatite chondrites.

Ordinary chondrites. Only a single Pb–Tl study of ordinary chondrites has been attempted, with results reported for samples of groups L and LL (Andreasen et al. 2009). To minimize the problem of terrestrial Pb contamination, only meteorite falls were selected for analysis. Overall, these meteorites displayed a large spread in $^{204}Pb/^{203}Tl$ ratios from 1.7 to 152, as a result of large variations in Tl concentration (from 0.3 to 19 ng/g), whilst Pb abundances varied by less than a factor of 7. The Tl isotope compositions were also highly variable with $\varepsilon^{205}Tl$ values of between about −20 to +15.

Given the observation of large Cd isotope fractionations in ordinary chondrites (Wombacher et al. 2003, 2008), the Tl isotope variability was assessed to distinguish radiogenic effects from clearly evident mass dependent Tl isotope fractionations, presumably from mobility and redistribution of Tl during thermal metamorphism. Consequently, the equilibrated LL and L chondrites (of type 4–6) were judged to show no evidence of radiogenic ^{205}Tl, indicating that the high Pb/Tl ratios were established by elemental redistribution after ^{205}Pb was extinct. In contrast, the Tl isotope compositions of unequilibrated LL3 chondrites were interpreted to reflect radiogenic ingrowth of ^{205}Tl, with a Pb–Tl age of about 45 Ma after solar system formation.

Iron meteorites

Due to the relatively high volatile element contents, IAB irons are the most straightforward iron meteorites for Tl isotope and Pb–Tl isochron analyses. Such measurements were also conducted for IIAB and IIIAB irons, but are significantly more difficult as much larger samples (generally more than 15 g; Nielsen et al. (2006a)) are required to obtain robust Tl isotope data and because the nearly ubiquitous terrestrial Pb contamination is more problematic due to the low meteoritic Pb contents. No data are currently available for the even more volatile depleted irons of groups IVA and IVB.

Non-magmatic IAB complex iron meteorites. Nielsen et al. (2006a) analyzed seven metal samples and five troilite nodules of the IAB main group irons Toluca and Canyon Diablo. The metal samples had Tl concentrations from about 0.03 ng/g to 16.5 ng/g and variable $^{204}Pb/^{203}Tl$ from about 0.1 to more than 75, equivalent to Pb/Tl ratios of between about 2 and 1200. Notably, the Tl isotope compositions, which ranged between $\varepsilon^{205}Tl$ values of about −3 to +23 were observed to correlate with $^{204}Pb/^{203}Tl$ (Fig. 4). When interpreted as an isochron, this correlation corresponds to initial values of $^{205}Pb/^{204}Pb_0 = (7.3 \pm 0.9) \times 10^{-5}$ and $\varepsilon^{205}Tl_0 = -2.2 \pm 1.5$. Alternative explanations for the correlation, such as mixing of variably mass fractionated meteorite components or terrestrial contamination are difficult to reconcile with the results. Troilite nodules from Toluca and Canyon Diablo contain Tl that is generally significantly less radiogenic than the co-existing metal with isotope compositions that are variable and decoupled from $^{204}Pb/^{203}Tl$ (Fig. 4). These effects were interpreted to result

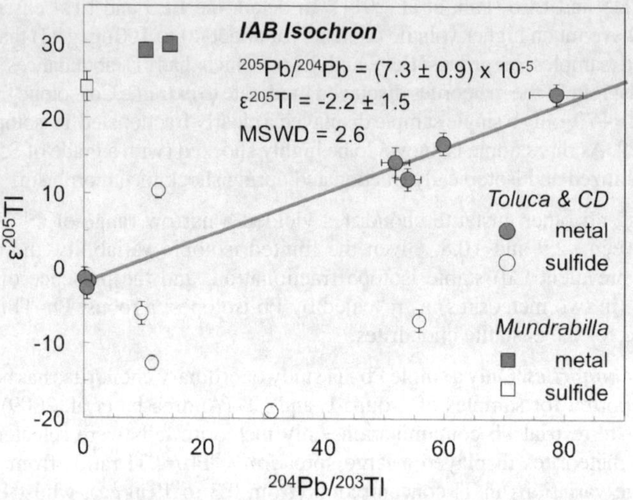

Figure 4. Pb–Tl isochron diagrams for metal samples (filled symbols) and sulfides (open symbols) of IAB complex iron meteorites, based on the results of Nielsen et al. (2006a). The metal samples of Toluca and Canyon Diablo (CD) delineate a well-defined isochron. The isochron slope is slightly revised from Nielsen et al. (2006a), due to improved corrections for terrestrial Pb. The troilite nodules of these meteorites and metal and sulfide from the anomalous IAB iron Mundrabilla do not plot on the IAB main group isochron defined by Toluca and Canyon Diablo, presumably due to stable isotope fractionation of Tl. All error bars are 2sd.

from kinetic stable isotope fractionation during diffusion of Tl between metal and sulfide. Given the relatively low Tl contents of the troilites (with about 1.6 to 27 ng/g Tl) and the low overall abundance of sulfides in IAB irons (≤2% of mass), such processes are unlikely to have significantly affected the Tl isotope compositions of the co-existing metal.

Assuming a solar system initial $^{205}Pb/^{204}Pb_{SS,0} = (1.0 \pm 0.4) \times 10^{-3}$, as derived from carbonaceous chondrites (Fig. 3), the IAB isochron was established 57 +10/−14 Ma after the start of the solar system (Baker et al. 2010b). In comparison, recent Pd–Ag isochron studies of the same meteorite groups indicate that the Pd–Ag fractionation of IAB irons is only about 15–19 Ma younger than carbonaceous chondrites (Carlson and Hauri 2001; Woodland et al. 2005; Schönbächler et al. 2008; Theis et al. 2013). The Pb–Tl and Pd–Ag ages are thus only barely consistent, within the combined uncertainties, hinting at possible discrepancies between the two ages. Most likely, the relatively young Pb–Tl isochron age of the IAB metals reflects late closure of this isotope system, due to a blocking temperature that is significantly lower than the 1100 K blocking temperature of the Pd–Ag chronometer (Sugiura and Hoshino 2003) and possibly similar to that of the K–Ar system at 650 K (Renne 2000; Trieloff et al. 2003). This interpretation is in accord with the highly labile nature of Tl, as opposed to Ag, during thermal processing (Lipschutz and Woolum 1988).

Also not in accord with the IAB isochron are the results obtained for the anomalous IAB complex iron Mundrabilla (Fig. 4; (Nielsen et al. 2006a)). Two metal samples and a sulfide nodule of this meteorite were characterized by ε^{205}Tl values of about +30 and +24, respectively. It is likely that the high ε^{205}Tl found in Mundrabilla is not solely due to rapid radiogenic ingrowth of ^{205}Tl in a high Pb/Tl environment but also reflects, at least in part, significant stable isotope fractionations, most likely associated with volatile loss from or redistribution on the parent body.

Magmatic IIAB and IIIAB iron meteorites. Andreasen et al. (2012) conducted a comprehensive Pb-Tl isochron study of metal samples from six IIAB and six IIIAB iron meteorites. All samples were thoroughly leached prior to dissolution to remove terrestrial Pb contamination. The fraction of primordial Pb was then calculated by assuming that any deviation in the Pb isotope composition of the iron meteorites from primordial Pb is due to residual contributions from terrestrial Pb. The meteorite samples exhibited low and highly variable Tl contents ranging from 0.002 to 0.485 ng/g, with most Tl concentrations at less than 0.02 ng/g. The IIAB and IIIAB iron meteorites are hence, on average, significantly more depleted in Tl than the IAB irons.

The $^{204}Pb/^{203}Tl$ ratios of the meteorites (Fig. 5, Andreasen et al. 2012) varied from 0.05 to 5.8 for the IIAB irons (Pb/Tl ≈ 0.8 to 89) and from 1.6 to 14 for the IIIABs (Pb/Tl ≈ 26 to 215). As such, these magmatic irons have $^{204}Pb/^{203}Tl$ ratios that are noticeably less variable than in IABs (Fig. 4), and with many samples not far removed from the chondritic value of $^{204}Pb/^{203}Tl = 1.43$ (Pb/Tl = 22). The $\varepsilon^{205}Tl$ values of the IIAB and IIIAB irons ranged from −18 to +23 (Fig. 5). The majority of the Tl isotope data, furthermore, display a correlation of $\varepsilon^{205}Tl$ with $^{204}Pb/^{203}Tl$, suggesting that the isotopic variations are primarily governed by decay of ^{205}Pb at variable Pb/Tl ratios.

Further Pb-Tl analyses for metal samples from two IIAB and two IIIAB irons and a IIIAB troilite nodule were reported by Nielsen et al. (2006a). For the IIABs, these data do not support the IIAB errorchron inferred by Andreasen et al. (2012) (Fig. 5). It is possible that this discrepancy arises, at least in part, from the less accurate correction of residual terrestrial Pb contamination that was utilized by Nielsen at al. (2006a) or possibly even minor terrestrial Tl contributions. Alternatively, the non-isochronous behavior of the IIABs analyzed by Nielsen at al. (2006a) may reflect late open system behavior and/or stable isotope fractionation of Tl in some samples. In contrast, the two IIIAB metals and the IIIAB troilite analyzed by Nielsen et al. (2006a) yield results, which are in accord with the IIIAB isochron reported by Andreasen et al. (2012) (Fig. 5).

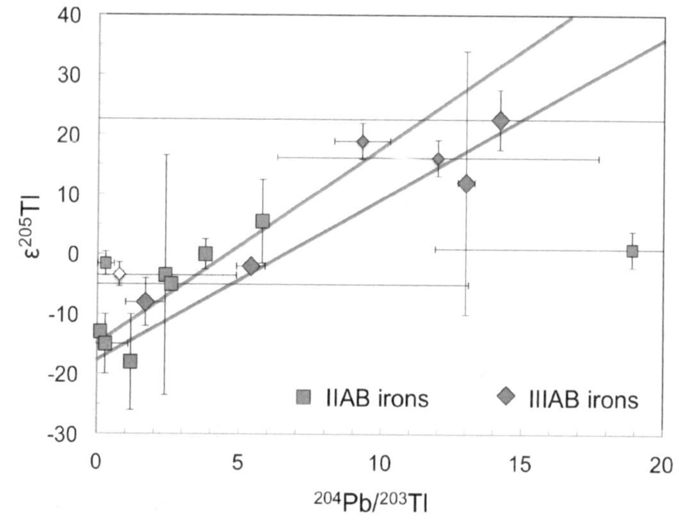

Figure 5. Pb–Tl isochron diagram for metal samples of IIAB (squares) and IIIAB (diamonds) iron meteorites. The large and small symbols, with 2sd error bars, denote the data of Andreasen et al. (2012) and Nielsen et al. (2006a), respectively. The open diamond is for a troilite nodule of Grant IIIAB. Despite of the scatter and some large uncertainties, the results for both groups of magmatic irons display a positive correlation, which is indicative of radiogenic variations in Tl isotope composition from in situ decay of ^{205}Pb. The two errorchron trendlines were calculated from the data of Andreasen et al. (2012) only and correspond to initial $^{205}Pb/^{204}Pb_0$ ratios of $(8 \pm 2) \times 10^{-4}$ and $(8 \pm 4) \times 10^{-4}$ for the IIABs and the IIIABs, respectively.

Assuming that the IIAB data of Andreasen et al. (2012) indeed define a Pb–Tl isochron (Fig. 5), they correspond to an initial $^{205}Pb/^{204}Pb_0$ ratio of $(8.3\pm1.6)\times10^{-4}$. This ^{205}Pb abundance is equivalent to an age of 4+16/−14 Ma, assuming an initial solar system abundance of $^{205}Pb/^{204}Pb_{SS,0}=(1.0\pm0.4)\times10^{-3}$, as defined by carbonaceous chondrites (Fig. 3) (Baker et al. 2010b). A similar but more uncertain initial ^{205}Pb abundance and a less well-constrained age of 4+29/−14 Ma was obtained from the IIIAB correlation (Fig. 5). Both iron meteorite isochrons, however, have y-intercepts with initial $\varepsilon^{205}Tl_0$ values that are more negative (at $\varepsilon^{205}Tl_0=-12\pm1$ for the IIABs) than the initial solar system value of $\varepsilon^{205}Tl_{SS,0}=-7.6\pm2.1$, as defined by the carbonaceous chondrites isochron (Fig. 3). It is conceivable that the observed offset of about 7 ε-units (at the chondritic value of $^{204}Pb/^{203}Tl=1.43$) between the IIAB and the chondrite isochron reflects a nucleosynthetic isotope anomaly or is caused by stable Tl isotope fractionation. However, nucleosynthetic isotope effects of this magnitude are unlikely for Tl, as they are also not observed for other elements of similar atomic mass, such as Hf, Os, and Pt (Yokoyama et al. 2007; Sprung et al. 2010; Walker 2012; Kruijer et al. 2013). Thallium isotope data acquired for metal-silicate and sulfide-silicate partitioning experiments furthermore demonstrate that Tl isotope fractionation during metal or sulfide segregation is either very small or absent (Wood et al. 2008).

As a consequence, Andreasen et al. (2012) suggested an alternative interpretation of the carbonaceous chondrite and magmatic iron meteorites isochrons. By discarding the results of all four carbonaceous chondrites, which featured significant quantities of terrestrial Pb contamination (at about 10 to 80% of total measured Pb), a revised carbonaceous chondrite isochron with a significantly steeper but less well-defined slope was obtained. This revised calculation yields $^{205}Pb/^{204}Pb_{SS,0}=(2\pm1)\times10^{-3}$ and $\varepsilon^{205}Tl_{SS,0}=-13.6$. Using this initial solar system $^{205}Pb_{SS,0}$ abundance, the IIAB and IIIAB isochrons now provide ages of 15+20/−12 Ma and 14+32/−15 Ma, respectively. It is unknown, however, whether these ages date core crystallization or are linked to the breakup of the meteorite parent bodies. Likewise the IAB isochron now gives a younger age of 69+16/−10 Ma after the carbonaceous chondrites, and this may indicate that the Pb-Tl decay system dates the same impact event(s) that are recorded in the ~4.5 Ga Ar–Ar ages of IAB silicate inclusions (Vogel and Renne 2008). Notably, the revised initial solar system ^{205}Pb abundance of $^{205}Pb/^{204}Pb_{SS,0}=(2\pm1)\times10^{-3}$ is somewhat higher than the upper estimate that was obtained in models of *s*-process nucleosynthesis by Wasserburg et al (2006). Whether this discrepancy reflects uncertainties in the analytical data, the stellar models, or in the estimates of recent *s*-process contributions to the solar nebula, remains unclear at present.

Limitations of the ^{205}Pb–^{205}Tl chronometer

In principle, the ^{205}Pb–^{205}Tl decay system offers many promising applications but translation of this potential has been hindered by a number of factors. (i) Due to the low abundances of Tl in many meteorites, large samples of typically more than 1 g but with > 10 g not uncommon, have been used for precise Tl isotope measurements. These requirements make analyses challenging and measurements for highly volatile depleted meteorites (e.g., IVA/B irons and many achondrites) have not been attempted to date. (ii) As a result of the highly labile nature of Tl, isochronous behavior is easily disturbed or fully reset. In addition, remobilization of Tl can be associated with stable isotope fractionation, which hampers unambiguous interpretations. (iii) The pervasive contamination of many meteorites with terrestrial Pb renders the determination of robust primitive $^{204}Pb/^{203}Tl$ ratios (as required for Pb-Tl isochron calculations) and initial $^{205}Pb/^{204}Pb$ values very difficult. Analyses of such contaminated meteorites requires that anthropogenic Pb is removed from samples by thorough cleaning and/or corrections for the effects of contamination must be attempted, based on Pb isotope data. Such corrections, however, can be subject to large uncertainties whilst aggressive cleaning of samples by leaching and/or partial dissolution (as is commonly applied prior to analyses of radiogenic Pb isotope compositions) can lead to partial Tl loss that may be associated with isotope fractionation.

THALLIUM ISOTOPE COMPOSITION OF THE SOLID EARTH

The primitive mantle

The concentration of Tl in the primitive mantle has been estimated as 0.0035 μg/g (McDonough and Sun 1995) and for the depleted mantle as 0.00038 μg/g (Salters and Stracke 2004). The challenge associated with evaluating the Tl concentration and isotope composition of the mantle is largely twofold. First, thallium concentrations are vanishingly low, thus grams of material are required to obtain precise concentration and/or isotope measurements by bulk rock dissolution approaches. Minute inclusions of secondary alteration products such as clays that easily incorporate Tl are much more problematic with very large sample sizes, and any such material can drastically affect the determination of both Tl concentration and isotope composition. Second, even though the required sample sizes are large, there is still worrying potential for 'nugget' effects. For example, Nielsen et al. (2014) have shown by laser-ablation inductively coupled plasma mass spectrometry (LA-ICPMS) that the only phase housing significant Tl in the Lherz peridotite massif, France, are interstitial sulfides, which yielded variable Tl concentrations from 0.023 to 0.430 μg/g. All other investigated mineral phases (olivine, orthopyroxene, clinopyroxene and spinel) had Tl concentrations below detection limit, restricting the Tl concentration in the vast majority of mantle minerals to < 0.001 μg/g (Nielsen et al. 2014). An experimental study by Kiseeva and Wood (2013) further documents Tl partition coefficients between sulfide liquid and a silicate melt with MORB composition of $D_{Tl}^{sulf/sil} = 4.1$ to 18.8, reinforcing the notion that sulfides are the main host for Tl, at least in the upper mantle. To date, only a single measurement of a mantle xenolith has been published (Nielsen et al. 2015). The sample, a harzburgite from the Eifel volcanic field in Germany, revealed an isotope composition of $\varepsilon^{205}Tl = -2.0 \pm 0.8$ and concentration of 1.05 ng/g. The concentration is, thus, slightly higher than what would be expected based on published estimates of Tl in the depleted mantle (Salters and Stracke 2004), but given the potential for nugget effects it is unclear if the concentration is necessarily representative of the upper mantle.

Due to the problems associated with direct measurements, Nielsen et al. (2006b) used five fresh, glassy MORBs from different global ocean basins to estimate the Tl isotope composition of the mantle. The homogeneity of the results ($\varepsilon^{205}Tl_{MORB} = -2 \pm 1$; Fig. 1) strongly indicates that no large-scale geographic Tl isotopic differences exist in the present day upper mantle uncontaminated by crustal components. Given that these lavas are not derived from identical degrees of melting, the restricted range in $\varepsilon^{205}Tl$ strongly suggests that partial melting of the mantle does not fractionate Tl isotopes to an analytically resolvable level. This inference is further supported by the identical Tl isotope composition of these MORBs and the published harzburgite sample (Nielsen et al. 2015) and not unexpected, given the heavy atomic masses of Tl isotopes. However, since stable isotope fractionation has been observed for some lighter elements during mantle melting (e.g., Fe isotopes; Williams et al. 2004; Craddock et al. 2013), further Tl isotope analyses of MORB and peridotites are desirable to validate this assumption.

The continental crust

As discussed, Tl generally follows the alkali metals K, Rb and Cs during melting and fractional crystallization (Heinrichs et al. 1980; Shaw 1952) resulting in much higher Tl concentrations in the continental crust (~0.5 μg/g) versus the primitive mantle (~0.0035 μg/g) (Shaw 1952; Heinrichs et al. 1980; McDonough and Sun 1995; Wedepohl 1995; Rudnick and Gao 2003).

Although the chemical affinity with K, Rb, and Cs suggests lithophile behavior, thallium has been considered both a chalcophile and lithophile element. McGoldrick et al. (1979) suggested that Tl displays chalcophile behavior and follows sulfur in sulfur-saturated magmas. In contrast, a study of sulfur-rich ore deposits (Baker et al. 2010a) found that Tl retained strong lithophile behavior in this setting. Noll et al. (1996) assumed that Tl behaved as a chalcophile element and used chalcophile/lithophile element ratios in subduction-related lavas to determine relative fluid mobility of trace elements. They suggested that Tl has a similar bulk partition coefficient to La. However, Tl showed no significant correlation with 'fluid-mobile' and chalcophile elements such as As, Sb and Pb, again casting doubt on its chalcophile affinity in evolving magmatic systems. A study of lavas from the Mariana arc also suggests lithophile, fluid-immobile behavior, demonstrated by strong co-variation of Tl with La, rather than Ba (Prytulak et al. 2013). Thus the elemental behavior of Tl during magmatic processes and the specific controls on its concentration in the continental crust remain ambiguous. This ambiguity is underlined by the strong affinity Tl has for sulfides formed in aqueous low temperature hydrothermal systems (Xiong 2007), mantle sulfides and early diagenetic pyrite (Nielsen et al. 2011, 2014).

The lack of resolvable isotope differences in global MORB suggests that negligible Tl isotope fractionation occurs during moderate degrees of partial melting (15–20%). Given the evolved bulk composition of the continental crust, it is necessary to consider the impact of further igneous processes on Tl stable isotope fractionation. Though stable isotope fractionation has been documented during igneous processes for relatively heavy elements such as iron (e.g. Williams et al. 2004, 2009), the small relative mass difference between the isotopes of Tl in combination with little or no redox chemistry for Tl in igneous systems, favor negligible Tl isotope fractionation during magmatic processes. Several studies found that the isotope composition of the average upper continental crust, represented by loess, is indistinguishable from MORB, with both exhibiting a value of $\varepsilon^{205}Tl = -2.0 \pm 0.5$ (Nielsen et al. 2005b, 2006b,c, 2007). The uniform isotope composition of the continental crust is further supported by data obtained for an ultrapotassic dike from the Tibetan Plateau ($\varepsilon^{205}Tl = -2.3 \pm 0.5$), which represents a melt originating from the sub-continental lithospheric mantle (Williams et al. 2001). Finally, Nielsen et al., (2016) recently determined $\varepsilon^{205}Tl$ in a suite of co-genetic basalts through andesites from a single volcano on Atka Island in the Aleutian arc. Thallium concentrations display strong positive co-variation with K_2O wt%, as expected for an incompatible element during fractional crystallization. Although the lavas range in $\varepsilon^{205}Tl$ from -2.3 to $+0.7$, the isotope values do not co-vary with Tl concentration and the variability is instead attributed to subduction zone inputs. Hence, on balance, there is little evidence to suggest that melting or fractional crystallization imparts analytically resolvable Tl isotope fractionation and therefore the bulk continental crust can be characterized by $\varepsilon^{205}Tl_{CONTCRUST} = -2 \pm 1$ (Fig. 1).

THALLIUM ISOTOPE COMPOSITION OF SURFACE RESERVOIRS

Volcanic degassing

Due to its low boiling point (Table 1), Tl is significantly enriched in volcanic gasses and particles compared with the geochemically analogous alkali metals (Patterson and Settle 1987; Hinkley et al. 1994; Gauthier and Le Cloarec 1998; Baker et al. 2009). Consequently, volcanic plumes provide a large Tl flux from igneous to surface environments on Earth, with an estimated flux to the oceans of 370 Mg/a (Rehkämper and Nielsen 2004; Baker et al. 2009). The behavior of Tl isotopes during degassing in volcanic systems was investigated by Baker et al. (2009), who identified significant isotope variations. The most likely form of isotope fractionation to occur during degassing is kinetic isotope fractionation, where the light isotope is enriched in the gas phase during evaporation. The extent of kinetic isotope fractionation between two phases is determined by the relative magnitude of the atomic or molecular velocities, where $\alpha = (m_1/m_2)^{\beta}$ and β varies from 0.5 (in the case of a vacuum) down to values approaching 0 (Tsuchiyama et al. 1994). Hence, even though the relative mass difference between the two Tl isotopes is small, the maximum kinetic $\alpha_{liq-vap}$ is 1.0049, equivalent to a fractionation factor of 49 ε-units. An isotopic difference of this magnitude exceeds the stable Tl isotope variability currently known for Earth and should hence be readily detectable. However, contrary to the expectation of a gas phase enriched in light isotopes, Baker et al. (2009) found no systematic enrichment of either ^{203}Tl or ^{205}Tl in volcanic emanations compared with average igneous rocks. Despite large isotopic differences between individual samples, the isotope compositions of 34 samples of gas condensates and particles from six separate volcanoes showed an average of $\varepsilon^{205}Tl = -1.7 \pm 2.0$. This value is indistinguishable from average igneous rocks (Nielsen et al. 2005b, 2006b,c, 2007, 2016; Prytulak et al. 2013) and suggests that degassing as a whole does not significantly alter that Tl isotope composition of degassed lavas. The Tl isotope variation of volcanic emanations was interpreted to reflect the complex evaporation as well as condensation processes that occur in volcanic edifices (Baker et al. 2009), and which eventually produce no net isotope fractionation between Tl in the magma and the Tl transported into the atmosphere and surface environments on Earth.

Weathering and riverine transport of Tl

Stable isotope fractionation during weathering is often monitored via measurements of isotope ratios in the dissolved and particulate phases in rivers. For example, elements like lithium and molybdenum have been shown to display significant isotopic variation between different rivers and rocks of the continental crust (e.g. Huh et al. 1998; Archer and Vance 2008; Pogge von Strandmann et al. 2010). Numerous kinetic and equilibrium processes can potentially affect the stable isotope budgets of rivers, and as such it is difficult to predict isotope fractionation during weathering. In general, Tl is soluble in aqueous solution and should be readily mobilized during weathering. However, the tendency for Tl to partition into potassium rich minerals and manganese oxides (Heinrichs et al. 1980) suggests that the transport of Tl into the ocean may be less efficient. Nielsen et al. (2005b) measured the Tl isotope compositions of dissolved and particulate components for a number of major and minor rivers and found that these generally display values similar to those observed for continental crust. The average value for dissolved riverine Tl is $\varepsilon^{205}Tl = -2.5 \pm 1.0$ (Nielsen et al. 2005b), with particulate matter ($\varepsilon^{205}Tl = -2.0 \pm 0.5$) being indistinguishable from the dissolved phase (Table 3). Although there are clearly resolvable differences in riverine Tl isotope compositions that were inferred to be related to variations in catchment litholigies, the majority of rivers including the worlds largest, the Amazon, display $\varepsilon^{205}Tl$ values similar to average continental crust. These data strongly imply that there is little or no Tl isotope fractionation associated with continental weathering processes.

Table 3. Thallium isotope and concentration data for rivers.

Sample	$\varepsilon^{205}Tl_{diss}$	Tl_{diss} (ng/kg)	$\varepsilon^{205}Tl_{part}$	Tl_{part} (ng/kg)
Amazon	−2.3	16.4		
Danube	−6.7	16.4	−2.9	3.6
Doubs	−5.5	3.36		
Eder	−4.2	1.93		
Kalix	−1.6	1.31	−3.8	0.29
Nahe	−2.5	7.03	−1.2	3.7
Nidda	−2.7	1.67		
Nidder	−2.6	2.85	−1.6	0.26
Nile	0.0	3.13		
Rhine Rueun	−6.4	3.61		
Rhine Laufenbg.	−3.0	4.04		
Rhine Speyer	−2.8	5.35		
Rhine Bingen	−2.9	6.71	−2.1	1.3
Rhone	−2.7	6.54	−2.2	36
Volga	−1.1	1.60		

Uncertainty on Tl isotope measurements is ± 1 $\varepsilon^{205}Tl$-unit
Exact sample locations are given in Nielsen et al. (2005b)

Natural unpolluted Tl abundances in rivers are generally very low and vary between 1 and 10 pg/g (Cheam 2001; Nielsen et al. 2005b). An estimated global average dissolved riverine concentration of 6±4 pg/g (or 30±20 pmol/l) results in a flux to the oceans of 230 Mg/yr. In the study of Nielsen et al. (2005b) three rivers exhibited Tl isotope compositions significantly lighter (at ε^{205}Tl = −6 to −4) than average continental crust (Table 3). The lower ε^{205}Tl values were interpreted to reflect weathering of marine carbonates that are a main constituent of the drainage areas for these rivers. Relatively light Tl isotope compositions are expected for such carbonates based on analyses of modern seawater, which is characterized by ε^{205}Tl = −6.0 ± 0.3 (Rehkämper et al. 2002; Nielsen et al. 2004, 2006c; Owens et al. pers. comm.). However, marine carbonates including carbonate oozes (Rehkämper et al. 2004), corals (Rehkämper, unpublished data) and foraminifera (Nielsen, unpublished data) exhibit very low Tl concentrations. This implies that weathering of marine carbonates will only have a strong impact on the Tl inventories of rivers that predominantly drain such lithologies and will not strongly affect the total global budget of Tl transported by rivers to the oceans.

Anthropogenic mobilization of Tl

Similar to its periodic table neighbors Hg, Cd, and Pb, Tl is also readily volatilized by high-temperature processes, such as combustion. As a consequence, the most important global source of anthropogenic Tl are atmospheric emissions from processes such as pyrite roasting, cement production and coal burning, and associated solid wastes. In addition, Tl release with wastewater from ore processing plants can have a significant impact on terrestrial aquatic system (Nriagu 1998; Peter and Viraraghavan 2005).

Like its elemental neighbors, Tl also has a high acute and chronic toxicity to mammals, including humans. Due to the low concentrations of Tl in most natural and manufactured materials, and its limited use in consumer products or industrial processes, anthropogenic Tl emissions to the environment are fortunately generally low. They only represent a significant health hazard in the vicinity of significant emission sources, particularly when aided by bioaccumulation in the food chain (Nriagu 1998; Peter and Viraraghavan 2005). Such a case of chronic Tl poisoning was reported for the inhabitants of a rural village in the Lanmuchang area of Guizhou Province, China. The village is situated in close proximity to a massive vein of Tl mineralization that has a protracted history of artisanal mining (Xiao et al. 2003, 2004). The health effects seen in the villagers (including muscle and joint pain, hair loss, and disturbance or even loss of vision) were traced to the consumption of local crops (particularly highly Tl-enriched green cabbage), which are grown on soils with elevated Tl concentrations from past mining activities (Xiao et al. 2004, 2007).

The isotopic signature of Pb, which is highly variable due to radiogenic ingrowth from the decay of U and Th isotopes, has long been used to fingerprint anthropogenic Pb emissions to the environment, particularly from the use of leaded gasoline (Settle and Patterson 1982; Rosman et al. 1994; Alleman et al. 1999). Following the advent and more widespread application of MC-ICP-MS, a number of studies demonstrated that similar fingerprinting is also feasible for Hg and Cd emissions, based on analyses of the stable isotope compositions of these elements (Ridley and Stetson 2006; Rehkämper et al. 2012). More recently, an investigation by Kersten et al. (2014) demonstrated, for the first time, that Tl stable isotope compositions can also be used as a tracer of anthropogenic Tl emissions to the environment.

In the study of Kersten et al. (2014), Tl isotope data were used to link the high Tl contents of agricultural soils to past emissions of cement kiln dust (CKD) from a nearby cement plant in Lengerich, northwest Germany. In detail, it was shown that the soils were contaminated by Tl emissions that occurred in the 1970's, when the cement plant utilized pyrite-roasting waste with high Tl contents as a cost-effective, S-rich additive during cement production, a process that involves combustion processes at temperatures of more than 1000 °C. To arrive at this conclusion, contaminated soil samples from three vertical profiles with up to 1 m depth were analyzed for both Tl concentrations and isotope compositions. When viewed in a diagram of $\varepsilon^{205}Tl$ versus inverse Tl concentration (1/[Tl]), the soil data are strongly indicative of a binary mixing relationship (Fig. 6). The mixing endmembers were inferred to be the geogenic background, as defined by isotopically light soils with $\varepsilon^{205}Tl \approx -4$ at depth, and the Tl emissions, represented by Tl-enriched topsoils with a distinctly heavier isotopic signature of $\varepsilon^{205}Tl \approx 0$. This conclusion is further corroborated by Tl isotope compositions of $\varepsilon^{205}Tl \approx \pm 0$ that were obtained for (i) a CKD sample taken at the time of the inferred Tl emissions and (ii) a pyrite, which was sourced from the same Weggen deposit in Germany as the pyrite roasting waste that was added to the cement raw mix prior to combustion in the kiln (Fig. 6).

Additional analyses were carried out by Kersten et al. (2014) for soil and crop samples from the Lanmuchang area (Guizhou Province) in China. These measurements revealed significant isotope fractionation between soils, with a high natural Tl background characterized by $\varepsilon^{205}Tl \approx +0.4$, and locally grown cabbage, which displayed $\varepsilon^{205}Tl$ values of between −2.5 and −5.4. This demonstrates that biological isotope fractionation and subsequent remineralization of Tl from organic material cannot be responsible for the heavier Tl isotope signatures that were found in the Lengerich topsoil. Rather, the high Tl contents and associated heavier isotope signatures seen in the vicinity of the Lengerich cement plant are most reasonably explained by Tl emissions that were released during cement production (Kersten et al. 2014).

The isotope composition of seawater

In the oceans, Tl is a conservative, low-level trace element with an average dissolved concentration of 13 ± 1 pg/g (64 ± 5 pmol/l) (Flegal and Patterson 1985; Schedlbauer and Heumann 2000; Rehkämper and Nielsen 2004; Nielsen et al. 2006c). The seawater average is thereby slightly higher than the dissolved Tl abundances of most rivers (see *Weathering and riverine transport of Tl*). Based on a thorough review of the marine input and output fluxes of Tl, Rehkämper and Nielsen (2004) concluded that the oceans are currently at steady state and that Tl has a residence time of ~21 ka, which is consistent with a number of previous studies (Flegal and Patterson 1985; Flegal et al. 1989). With an inferred marine residence time that

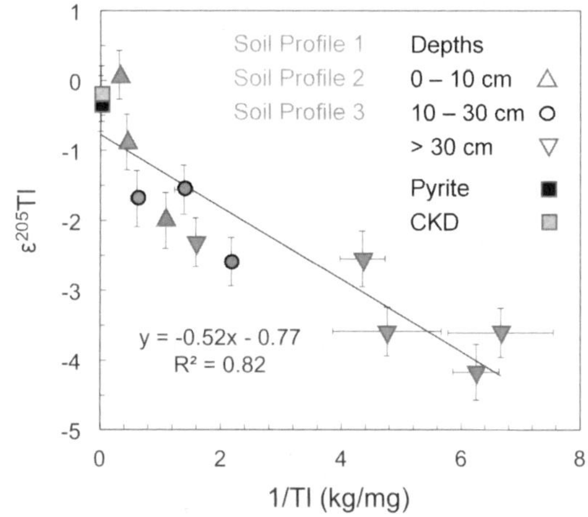

Figure 6. Plot of $\varepsilon^{205}Tl$ vs. 1/total Tl concentration for soil samples from the vicinity of the Lengerich cement plant in Germany (modified from Kersten et al. 2014). Furthermore shown are results for a pyrite from the Meggen deposit (which is also the ultimate origin of the Tl-rich additive that was used in cement production) and the cement kiln dust (CKD). The linear trend and associated correlation coefficient R^2 were calculated from the soil data only. All error bars are 2sd.

is more than an order of magnitude longer than the ocean mixing time and a conservative distribution, Tl should exhibit an invariant isotope composition in seawater. Analyses of Arctic, Atlantic and Pacific seawater confirm this prediction (Rehkämper et al. 2002; Nielsen et al. 2004, 2006c). In particular, a recent data set for about 50 samples covering the GEOTRACES GA10 transect in the Atlantic across 40°S shows no Tl isotope variation, with an overall average of $\varepsilon^{205}Tl = -6.0 \pm 0.3$ (Owens et al. pers. comm.). The invariance of the GA10 data is remarkable because a number of different water masses are present in this transect (Antarctic Bottom Water, North Atlantic Deep Water, Antarctic Intermediate Water) that are sourced from very disparate regions of the ocean. Therefore, these analyses confirm that the open ocean is homogenous with respect to Tl isotopes within current measurement uncertainties.

It may be somewhat surprising that the oceans are significantly enriched in ^{203}Tl compared to the continental crust and the mantle (see the section *Thallium isotope composition of the solid earth*) with an average value of $\varepsilon^{205}Tl = -6.0 \pm 0.3$ (Rehkämper et al. 2002; Nielsen et al. 2004, 2006c; Owens et al. pers. comm.). Hence, it is required either that the marine sources of Tl are isotopically light compared to the continental crust and mantle or that the outputs are fractionated towards heavy isotope compositions relative to seawater. Rehkämper and Nielsen (2004) showed that the most significant marine inputs for Tl are from rivers, high-temperature hydrothermal fluids, mineral aerosols, volcanic emanations and sediment pore water fluxes at continental margins. In contrast, there are only two important marine Tl sinks, namely Tl adsorption by the authigenic phases of pelagic clays and uptake of Tl during low-temperature alteration of oceanic crust. The relative magnitudes and isotope compositions of these fluxes are summarized in Table 4. In the following sections, we outline the main observations that follow from an assessment of these fluxes.

THE MARINE MASS BALANCE OF THALLIUM ISOTOPES

Thallium isotopes in marine input fluxes

As riverine and volcanic input fluxes were already discussed in previous sections, we will here focus on high-temperature hydrothermal fluids, mineral aerosols and sediment pore water fluxes from continental margins.

High temperature hydrothermal fluids. In hydrothermal systems where temperatures exceed ~150 °C it has been shown that Tl behaves much like the alkali metals Rb and K (Metz and Trefry 2000), and is leached from the oceanic crust by circulating fluids. The efficiency of this leaching process is roughly 90%, as shown by Ce/Tl ratios in the sheeted dike complex of ODP Hole 504B that are ~10 times higher than in pristine MORB (Fig. 7). This results in end-member high temperature hydrothermal fluids that exhibit Tl concentrations almost 500 times higher compared to ambient seawater (Metz and Trefry 2000; Nielsen et al. 2006c). We can estimate the flux of Tl into the oceans (M_{Tl}) from high-T hydrothermal fluids by combining the average Tl concentration of MORB with the flux of ocean crust leached by high-T fluids, assuming 90% leaching efficiency:

$$M_{Tl} = F_{oc\ leach} \times [Tl]_{oc} \times f_{Tl\ leach} \qquad (2)$$

Here, $F_{oc\ leach}$ is the annual production rate of ocean crust that is leached by high-T fluids, $[Tl]_{oc}$ is the Tl content of the crust prior to leaching, and $f_{Tl\ leach}$ is the fraction of Tl leached from the rocks during alteration. The annual production rate of ocean crust leached by high-T fluids is comprised of $\sim 1.24 \pm 0.16 \times 10^{16}$ g/a of MORB crust and $\sim 0.76 \pm 0.10 \times 10^{16}$ g/a of cumulate lower ocean crust with Tl concentration estimated at 25% of the fresh MORB value (Mottl 2003; Nielsen et al. 2006c).

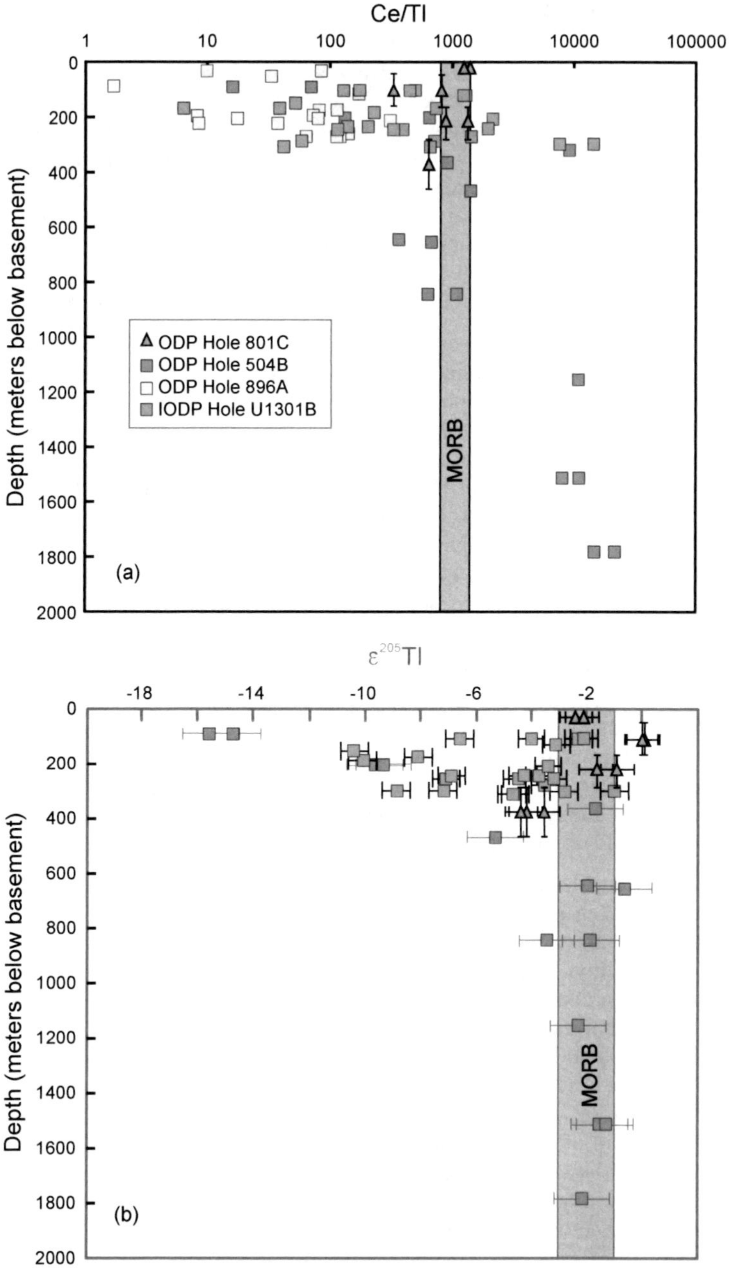

Figure 7. Plots of (a) Ce/Tl and (b) Tl isotopes against depth for different sections of hydrothermally altered oceanic crust. Data compiled from (Teagle et al. 1996; Nielsen et al. 2006c; Prytulak et al. 2013; Coggon et al. 2014). The samples with error bars from ODP 801C are composites prepared from the lithologies encountered in these depth intervals (Kelley et al. 2003).

It was previously thought that Tl partitioned similarly to Cs during mantle melting (Jochum and Verma 1996; Nielsen et al. 2006b), which, based on an assumed Cs/Tl ≈ 6, yielded a Tl content of ~3 ng/g for fresh MORB. However, more recent data has shown that Tl is much more compatible during mantle melting because it partitions into sulfides (Kiseeva and Wood 2013; Nielsen et al. 2014). Although there is a relatively large database for Tl concentrations in fresh MORB (Jenner and O'Neill 2012; Nielsen et al. 2014), the global average Tl concentration for MORB is best determined by combining the constant Ce/Tl = 1110 ± 330 for MORB with > 6% MgO (Nielsen et al. 2014) with the global weighted average Ce concentration for MORB of 14.86 ± 1.26 μg/g (Gale et al. 2013), because Ce data for MORB have a complete global coverage. This combination yields a global average Tl concentration in MORB of 13.4 ± 4.2 ng/g, which is similar to the value obtained by averaging all available Tl data for MORB (12 ± 8ng/g) (Jenner and O'Neill 2012; Nielsen et al. 2014). Inserting the three parameters into equation (2) produces an annual Tl flux into the ocean from high-T hydrothermal fluids of 170 ± 60 Mg/yr.

Due to the relatively high temperatures involved (300–400 °C), isotope fractionation is not expected to be significant. This inference was confirmed by Nielsen et al. (2006c), who determined the Tl isotope composition of hydrothermal fluids from the East Pacific Rise and Juan de Fuca Ridge. All samples had Tl isotope compositions identical to that of average MORB, thus supporting the interpretation that extraction of Tl from the oceanic crust is not associated with isotope fractionation. The chemical and isotopic behavior inferred for Tl from hydrothermal fluids is furthermore in accord with the results of a study of oceanic crust altered by high temperature hydrothermal fluids from ODP Hole 504B. The latter work revealed Tl concentrations for basalts and dikes that were much lower than expected for depleted ocean crust whilst the Tl isotope compositions were identical to average MORB (Nielsen et al. 2006c).

Mineral aerosols. There are no direct investigations of Tl abundances or isotope compositions for mineral aerosols deposited in the oceans. Based on studies of windborne loess sediments deposited on land, Nielsen et al. (2005b) concluded that the average abundance of Tl in dust deposited in the oceans is about 490 ± 130 ng/g. The main uncertainty in determining the Tl flux to the oceans is from estimating the fraction of Tl that is released from the dust into seawater, following deposition and partial dissolution. By comparison with a number of other elements, Rehkämper and Nielsen (2004) concluded that about 5–30% of the Tl transported in aerosol particles would dissolve in seawater, resulting in an annual Tl flux of 10–150 Mg/yr.

It may be reasonable to assume that the bulk Tl isotope composition of the material transported to the ocean is identical to loess, and thereby also the continental crust, with $\varepsilon^{205}Tl = -2$ (Nielsen et al. 2005b). However, it is unknown whether dust dissolution is associated with isotope fractionation. Schauble (2007) has shown that significant equilibrium Tl isotope fractionations will be produced primarily by chemical reactions that involve both valence states, Tl^{+} and Tl^{3+}. Based on the strongly correlated behavior of Tl, Rb, Cs and K in the continental crust, univalent Tl should be dominant in mineral aerosols. Thermodynamic calculations of the valence state of Tl in seawater also predict that all Tl is univalent in this reservoir (Nielsen et al. 2009a). In addition, continental weathering processes, which are ultimately controlled by aqueous dissolution of silicates and are therefore in some ways analogous to the partial dissolution of dust in seawater, induce no detectable Tl isotope fractionations (Nielsen et al. 2005b). Hence, we infer that isotope fractionation is unlikely to be significant during the dissolution of mineral aerosols in seawater.

Benthic fluxes from continental margins. It has long been known that pore waters, which seep into the oceans from reduced continental margin sediments, are rich in Mn (Elderfield 1976; Sawlan and Murray 1983). As Tl has a high affinity to Mn oxides (Koschinsky and Hein 2003) and is more soluble in seawater than Mn, it has been inferred that such pore waters may be an important source of dissolved marine Tl (Rehkämper and Nielsen 2004). However, there are currently no direct data available for sediment pore fluids to constrain either the average Tl concentration or isotope composition of these benthic fluxes. An estimate was therefore derived indirectly, based on (1) Tl/Mn ratios observed in ferromanganese nodules that were known to have precipitated from pore waters (Rehkämper et al. 2002), combined with (2) estimates for benthic Mn fluxes (Sawlan and Murray 1983; Heggie et al. 1987; Johnson et al. 1992). Taken together, these data yield a Tl flux of 5 – 390 Mg/a, and a best estimate of 170 Mg/a (Rehkämper and Nielsen 2004), which constitutes a substantial fraction of the total Tl flux to the oceans (Table 4).

We can use two distinct approaches to estimate the average Tl isotope composition of pore waters. The first utilizes published Tl isotope data for the various components that may supply pore waters. The second relies on the Tl isotope compositions of sediments, which contain a component that was precipitated from pore waters. The application of these approaches is summarized below.

There are three principle components, from which Tl could be mobilized and incorporated into sedimentary pore waters. (1) Labile Tl associated with riverine particles (Nielsen et al. 2005b). (2) Thallium adsorbed onto clay minerals from seawater (Matthews and Riley 1970). (3) Thallium bound to authigenic Mn-oxides that precipitated from seawater as part of pelagic red clays (Rehkämper et al. 2004). All three will be present in continental margin sediments in various proportions depending on sedimentation rate and proximity to estuaries, where high sedimentation rates will tend to dilute the Mn-oxides, as authigenic precipitation should remain fairly constant. Hence, assuming that there is no isotope fractionation associated with the release of Tl into pore waters, the isotope composition of each component provides bounds on the composition of pore waters.

The labile components in riverine particles have been shown to be isotopically similar to the continental crust and river waters and are thus characterized by $\varepsilon^{205}Tl = -2$ (Nielsen et al. 2005b). Matthews and Riley (1970) showed that Tl is readily adsorbed onto some clay minerals (in particular illite) where most Tl is exchanged with K. Because of the required charge balance for adsorption reactions, Tl adsorption onto clays is unlikely to be associated with any Tl reduction/oxidation and should thereby exhibit only minimal or no isotope fractionation (Schauble 2007). Since Tl adsorption by clay minerals will take place primarily within the marine environment, this component should inherit the Tl isotope composition from seawater of $\varepsilon^{205}Tl = -6.0$.

Thallium that is precipitated with Mn oxides onto sedimentary particles has been shown to have a significantly heavier isotope composition than the seawater from which the mineral forms (Rehkämper et al. 2002, 2004). The origin of this isotope fractionation is discussed in *Thallium isotope composition of surface reservoirs*. Theoretically, the pure MnO_2 mineral to which Tl is bound should have approximately the same isotope composition as the surfaces of Fe–Mn crusts, which display $\varepsilon^{205}Tl \approx +13$ (Fig. 8). However, leaching experiments conducted on shelf sediments with HCl and hydroxylamine hydrochloride indicate that the labile Tl on sediment particles features somewhat lower $\varepsilon^{205}Tl$ values of between +3 to +7 (Rehkämper et al. 2004; Nielsen et al. 2005b). These lower values could reflect contributions from components other than Fe–Mn oxides, particularly Tl incorporated into carbonates or adsorbed to clay minerals, with both presumably characterized by negative $\varepsilon^{205}Tl$. Nevertheless, it appears reasonable to infer that the Tl associated with marine authigenic Mn in continental margin sediments exhibits $\varepsilon^{205}Tl \approx +4$ to +10.

In summary, the isotopic data available for these three components indicate that pore waters from continental margin sediments are likely to feature $\varepsilon^{205}Tl$ values of about −4 to +6. This range can be compared with the compositions of two different sedimentary archives that form from pore waters. These are diagenetic ferromanganese nodules (which are inferred to originate primarily from pore waters) and early diagenetic pyrite. Rehkämper et al. (2002) analyzed two Mn nodules from the Baltic Sea and these exhibited $\varepsilon^{205}Tl = -0.2$ and −5.2. Nielsen et al. (2011) measured Tl isotopes in pyrites younger than 10 Ma from continental shelf sediments in the

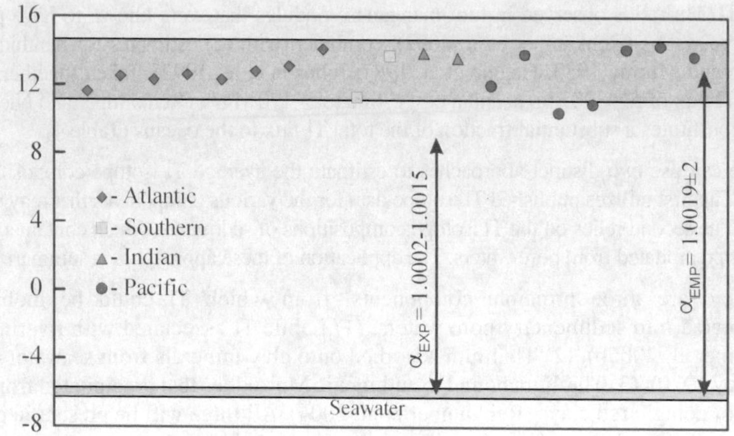

Figure 8. Thallium isotope compositions determined for the growth surfaces of hydrogenetic Fe–Mn crusts and seawater. The Fe–Mn crusts precipitate directly from seawater and hence there is an isotope fractionation of ~19 ε^{205}Tl-units (labeled α_{EMP}) between these two reservoirs. Data from (Rehkämper et al. 2002; Nielsen et al. 2004, 2006c; Owens et al. 2016). Also shown are the isotope fractionation factors (α_{EXP}) determined for experiments in which the manganese oxide birnessite was equilibrated with an aqueous Tl solution (Nielsen et al. 2013).

Northeast Pacific and the Caribbean and found ε^{205}Tl values of between −1 to +2. Importantly, both ferromanganese nodules and pyrites have Tl isotope compositions that are in accord with the range of values inferred for pore waters based on their constituent components. Taken together (and assuming that there is negligible isotope fractionation between pyrite/Fe–Mn nodules and pore waters) these constraints suggests that benthic fluxes from continental shelf sediments are characterized by $\varepsilon^{205}\text{Tl} \sim -3$ to +3, with a best estimate of $\varepsilon^{205}\text{Tl} \approx 0$.

Thallium isotope compositions of marine output fluxes

Thallium adsorption by the authigenic phases of pelagic clays. There is a significant enrichment of Tl in pelagic clays compared with continental shelf sediments (Matthews and Riley 1970; Heinrichs et al. 1980; Rehkämper et al. 2004). This enrichment is caused by the adsorption of Tl onto hydrogenetic Fe–Mn oxy-hydroxides and clay minerals (Matthews and Riley 1970; McGoldrick et al. 1979; Heinrichs et al. 1980). Two independent methods of calculating the Tl flux associated with these authigenic fluxes agree very well and indicate an annual flux of 200–410 Mg/a, with a best estimate of 270 Mg/a (Rehkämper and Nielsen 2004). If the smaller flux of Tl incorporated into pure Fe–Mn deposits of ~40 Mg/a (Rehkämper and Nielsen 2004) is also considered, this yields a total marine authigenic Tl flux of 310 Mg/a (Table 4). As discussed in *Benthic fluxes from continental margins*, the isotope composition of the Tl associated with authigenic phases is best approximated as a mixture of Tl bound to Mn oxides that probably display $\varepsilon^{205}\text{Tl} \approx +13$ and Tl adsorbed onto clay minerals, which likely exhibit the composition of seawater ($\varepsilon^{205}\text{Tl} \approx -6$). Due to the strong enrichment of Mn in pelagic clays, it can be assumed that the majority of authigenic Tl in such sediments originates from Mn oxides and thus may have an isotope composition more akin to Fe–Mn crusts. We can attempt to quantify this effect by performing an isotope mass balance calculation for the modern (core-top) pelagic clays reported in Rehkämper et al. (2004). The bulk isotope composition of these samples is $\varepsilon^{205}\text{Tl} = +3$ to +5. About 50% of the Tl in pelagic clays is thought to be of detrital origin (Rehkämper et al. 2004), which is characterized by $\varepsilon^{205}\text{Tl} = -2$ (Nielsen et al. 2005b). In order to account for the reported bulk isotope compositions, the authigenic component would thus have to display $\varepsilon^{205}\text{Tl} \sim +8$ to +12, and this implies that the authigenic Tl of pelagic clays features an isotope composition that is slightly lighter than pure Fe–Mn deposits. Based on the above considerations we can calculate that ~75–95% of the authigenic Tl in pelagic clays originates from Mn-oxides, whilst the remainder is bound to clay minerals.

Thallium uptake during low temperature hydrothermal alteration. It is well known that alteration minerals produced by interaction between cold (< 100 °C) seawater and MORB are highly enriched in Tl compared to pristine oceanic crust (McGoldrick et al. 1979; Alt 1995; Jochum and Verma 1996), and this leads to a strong Tl enrichment in the upper 500–600 m of the oceanic crust (Fig. 7a). It is unclear, however, which alteration mineral(s) are primarily responsible for accommodating the additional Tl. Palagonitization causes substantial deposition of Tl, which has been attributed to partitioning into smectites and other alkali-rich clay minerals (McGoldrick et al. 1979; Jochum and Verma 1996). In contrast, altered assemblages from IODP Hole U1301B appear to show correlations between bulk Tl and S concentrations, which led to the interpretation that the main carrier phase for Tl during hydrothermal alteration is pyrite (Coggon et al. 2014). Lastly, composites from several depth intervals in Jurassic oceanic crust from ODP Hole 801C showed no clear Tl enrichment and limited isotope fractionation compared with $\varepsilon^{205}Tl_{MORB} = -2 \pm 1$ (Fig. 7). The uncertainties in depositional mechanism, combined with the highly heterogeneous distribution of Tl in low-T altered oceanic crust (Teagle et al. 1996; Nielsen et al. 2006c; Prytulak et al. 2013; Coggon et al. 2014), complicate efforts to estimate the average Tl concentration of low-T altered oceanic crust and determine the annual flux of Tl into this reservoir. Following a conservative approach it was proposed that, on average, the upper 600 m of oceanic crust contains 200 ± 150 ng/g of Tl (Nielsen et al. 2006c), although the number is based mostly on inferred values rather than actual data. The Tl concentration estimate is equivalent to an element flux of 225–1985 Mg/a, with a best estimate of ~1000 Mg/yr. The preferred estimate, however, is similar to the combined flux of all marine Tl inputs (Table 4).

Given the uncertainties, it may thus be more reasonable to apply a mass balance approach (which assumes that marine Tl is at steady state) to estimate the Tl flux into low-T altered ocean crust. With a total input flux of 990 Mg/a (Table 4) and an authigenic output flux of 310 Mg/a, the marine Tl budget can be balanced with a low-T alteration output of 680 Mg/a, which is well within the range of values estimated based on Tl concentration data available for low-T altered ocean crust (Table 4). The reconstructed marine Tl mass balance yields a residence time of ~18.5 ka, which is within error of previous estimates of ~21 ka (Flegal and Patterson 1985; Rehkämper and Nielsen 2004).

The isotope composition of the low-T alteration output is equally difficult to determine. Nielsen et al. (2006c) observed that, in general, the upper oceanic crust is significantly enriched in isotopically light Tl with the shallowest samples displaying the lightest isotope compositions (Fig. 7b). This isotopic signature was interpreted to reflect closed system isotope fractionation during uptake of Tl from seawater, which gradually becomes more depleted in Tl as it penetrated deeper into the oceanic crust. As shown by a Rayleigh fractionation model in which an isotope fractionation factor of $\alpha = 0.9985$ was applied (Nielsen et al. 2006c), the isotope composition of the total Tl deposited during low-T alteration depends critically on the fraction of Tl that is extracted from the seawater before it is re-injected into the oceans as a low-T hydrothermal fluid. Stripping the fluid of all Tl originally present would imply an isotope composition for the low-T alteration flux that is identical to seawater ($\varepsilon^{205}Tl = -6$), whilst lower degrees of depletion result in lighter isotope compositions.

Based on the isotope fractionation observed for rocks from ODP Hole 504B, it was estimated that extraction of ~50% of the original seawater Tl by ocean crust alteration would yield an average isotope composition of $\varepsilon^{205}Tl \sim -18$ for the Tl deposited during low-T alteration. The application of this approach to obtain a global estimate for the average Tl isotope composition of the low-T alteration flux is fraught with many uncertainties, however,

Table 4. The Tl mass balance of the oceans with estimated source and sink fluxes.

	Range of Tl flux estimates (Mg/a)	**Best estimate (Mg/a)**		**ε^{205}Tl**	**Ref.**
		Marine Input Fluxes			
Rivers	76–380	230	23%	−2.5	1, 2
Hydrothermal fluids	110–230	170	17%	−2	3, 9
Subaerial volcanism	42–700	370	37%	−2	6
Mineral aerosols	10–150	50	5%	−2	1, 2
Benthic fluxes from continental margins	5–390	170	17%	0	1, 7
Total Input Flux	465–1850	990	100%	−1.8	
		Marine Output Fluxes			
Pelagic clays	240–450	310	36%	+10	1, 4, 7
Altered ocean crust	225–1985	680	64%	−7.2	1, 3, 7
Total Output Flux	465 –1850	940	100%	−1.8	
	Mass of Tl (Mg)	**Steady–state residence time**			
Global Oceans	1.75 (±0.14) × 10^7*	18,500 a		−6.0	1, 3, 5, 8

References 1: Rehkämper and Nielsen 2004; 2: Nielsen et al. 2005b; 3: Nielsen et al. 2006c; 4: Rehkämper et al. 2004; 5: Rehkämper et al. 2002; 6: Baker et al. 2009; 7: This study; 8: Owens et al. 2016; 9: This work.

* For a global ocean system with 1.348×10^{21} kg, the Tl mass in the oceans is equivalent to an average seawater concentration of 65 ± 5 pmol/kg or 13 ± 1 ng/kg (Rehkämper and Nielsen 2004).

as the database is currently limited to rocks from just three sections of altered ocean crust (504B, 801C, U1301B) and Tl concentration measurements for 3 low-T fluids from the Juan de Fuca ridge (Nielsen et al. 2006c). In principle, average altered upper ocean crust can be characterized by an ε^{205}Tl value of between about −18 to −6. Hence it may again be more appropriate to apply an isotope mass balance approach to obtain a more accurate result. With an isotope composition of $\varepsilon^{205}\text{Tl} \approx -1.8$ for the combined marine Tl input fluxes and an authigenic output flux of $\varepsilon^{205}\text{Tl} \approx +10$, mass balance dictates that the altered basalt flux is characterized by $\varepsilon^{205}\text{Tl} \approx -7.2$ (Table 4). This result is certainly within the range of reasonable values but based on Rayleigh fractionation modeling it requires that more than 95% of seawater Tl is removed by alteration processes, which is not fully supported by the data obtained for ODP Hole 504B (Nielsen et al. 2006c) and IODP Hole U1301B (Coggon et al. 2014). It is unlikely, however, that the results obtained for these two relatively young (< 5Ma) sections of oceanic crust are representative for global average ocean crust alteration processes. On the other hand, the Jurassic oceanic crust at ODP Hole 801C was found to have only limited Tl enrichment. This may reflect the low Tl concentration of ambient seawater at the time, as a consequence of enhanced incorporation of Tl into the abundant euxinic sediments of the Jurassic and Cretaceous, during which most of the alteration at ODP 801C took place (Prytulak et al. 2013).

CAUSES OF THALLIUM ISOTOPE FRACTIONATION

In general, Tl isotope variations on Earth are fairly limited, with only a few environments displaying significant deviations from the MORB and continental crust value of $\varepsilon^{205}\text{Tl} = -2$ (Nielsen et al. 2005b, 2006b,c, 2007). However, the overall magnitude of Tl isotope variations in natural environments on Earth now exceeds 35 ε^{205}Tl-units (Rehkämper et al. 2002, 2004; Nielsen et al. 2006c; Coggon et al. 2014). This variability is substantially larger than what is expected based on classical stable isotope fractionation theory (Bigeleisen and Mayer 1947; Urey 1947) and it is therefore important to understand the fundamental processes responsible for the Tl isotope variability.

There are two principle mechanisms by which most stable isotope fractionations are generated – a kinetic route that is associated with unidirectional processes and an equilibrium pathway, which acts during chemical exchange reactions. Both mechanisms should scale with the relative mass difference between the two isotopes of interest. In principle, kinetic isotope fractionation is capable of generating substantial Tl isotope effects (see *Volcanic degassing*) and there is evidence that such processes are recorded in volcanic fumaroles and some meteorites (Baker et al. 2009, 2010b; Nielsen et al. 2006a). However, the large Tl isotope fractionations observed between seawater and Fe–Mn oxy-hydroxides are more likely to reflect equilibrium isotope effects (Rehkämper et al. 2002; Nielsen et al. 2006c).

The larger-than-expected Tl isotope effects of equilibrium reactions were shown by Schauble (2007) to be partially caused by the so-called nuclear field shift isotope fractionation mechanism (Bigeleisen 1996). In short, the fundamental equilibrium isotope exchange equation of Bigeleisen and Mayer (1947) has five components, of which four were deemed negligible. However, based on unusual isotope fractionation effects observed for uranium (Fujii et al. 1989a,b), it was concluded that the equilibrium term caused by nuclear field shifts may be important in some cases (Bigeleisen 1996). Thus, nuclear field shift isotope fractionation is also an equilibrium isotope fractionation term, with a magnitude that scales broadly with the mass of the isotopes and hence is largest for heavy elements (Knyazev and Myasoedov 2001; Schauble 2007). The calculations that were carried out for Tl isotopes predict that an equilibrium system with aqueous dissolved Tl^+ and Tl^{3+} will feature both regular mass dependent and nuclear field shift isotope effects, with isotope fractionation factors that act in the same direction (Schauble 2007). When combined, these two components can reproduce the approximate magnitude of Tl isotope variation observed on Earth (Schauble 2007). The calculations are particularly relevant for the isotope compositions determined for Fe–Mn crusts and low-T altered basalts as these represent the heaviest and lightest reservoirs, respectively, found to date. The main requirement for substantial equilibrium Tl isotope fractionation to take place, is a chemical exchange reaction that involves two valence states of Tl (Schauble 2007), and these could be Tl^0, Tl^+ or Tl^{3+}. However, the calculations of Schauble (2007) also imply that the largest isotope effects are expected if Tl^{3+} is present.

Based on these theoretical considerations, experimental work was conducted to investigate the mechanism for Tl isotope fractionation during adsorption onto hydrogenetic Fe–Mn crusts (Fig. 8) and other Fe–Mn sediments (Peacock and Moon 2012; Nielsen et al. 2013). The majority of Tl in these deposits is associated with MnO_2 minerals (Koschinsky and Hein 2003; Peacock and Moon 2012) and the isotope fractionation is therefore likely to occur at or in such phases. The EXAFS/XANES spectra for Tl sorbed onto Mn-oxides have shown that the MnO_2 phase hexagonal birnessite has the capacity to oxidize Tl^+ to Tl^{3+}, following Tl sorption as a univalent ion (Bidoglio et al. 1993; Peacock and Moon 2012). This reaction appears to be associated with isotope fractionation whereas sorption of Tl onto other MnO_2 mineral structures (e.g., todorokite) with less oxidation potential, is not associated with significant isotope effects (Nielsen et al. 2013). Several series of experiments revealed that Tl sorbed to birnessite is systematically enriched in ^{205}Tl (Nielsen et al. 2013), which is in agreement with Tl isotope data for Fe–Mn crusts and seawater (Rehkämper et al. 2002; Owens et al. pers. comm.). However, the experimental isotope fractionation factors measured for aqueous univalent Tl versus Tl sorbed to birnessite were more variable and lower than the constant isotopic fractionation observed between Fe–Mn crusts and seawater (Fig. 8).

This discrepancy was interpreted to reflect sorption of Tl to two distinct sorption sites on birnessite: one associated with significant isotope fractionation and one with little or none. If this interpretation is correct, then Tl in natural Fe–Mn crusts either occupies only sorption sites with isotope fractionation or the distribution of Tl between the two types of sites is fairly constant in nature.

In summary, observation (Rehkämper et al. 2002), theory (Schauble 2007) and experiments (Nielsen et al. 2013) provide a relatively consistent picture of the mechanism responsible for Tl isotope fractionation during sorption to Mn oxides. However, it is still unknown what effects, if any, there are on the magnitude of Tl isotope fractionation during sorption to Mn oxides as a function of temperature, pH, ionic strength or other parameters.

The isotope fractionation effects found in low-*T* altered ocean crust are much less well understood. In this environment, Tl is extracted from seawater circulating through the ocean crust with an isotope fractionation factor of about $\alpha = 0.9985$ (Coggon et al. 2014; Nielsen et al. 2006c). It is conceivable that these fractionations reflect kinetic isotope effects, for example as a result of more rapid diffusion of the light isotopes from the hydrothermal fluid to the alteration minerals that concentrate Tl. If an equilibrium reaction is responsible, Tl^{3+} is likely to be involved because large equilibrium isotope effects are not expected for reactions without this species (Schauble 2007). Simple models of Tl speciation in seawater predict that only Tl^{+} is present (Nielsen et al. 2009a) and this implies that the hydrothermal processes, which deposit Tl in the oceanic crust should produce Tl^{3+}. This inference is in accord with the observation that Tl^{3+} has a much lower aqueous solubility than Tl^{+} (Nriagu 1998) and this implies that trivalent Tl should be deposited during hydrothermal alteration. However, these conclusions strongly contradict observations, which demonstrate that low-*T* hydrothermal alteration generally occurs at conditions that are more reducing than those prevalent in the open ocean (Alt et al. 1996). In addition, isotope fractionation calculations indicate that oxidized Tl should be enriched in ^{205}Tl (Schauble 2007), which is at odds with the light Tl isotope signatures observed in altered basalts.

In summary, the large Tl isotope variability observed in the marine environment is most likely produced by a combination of conventional mass dependent and nuclear field shift equilibrium isotope fractionation processes. Contributions from kinetic isotope effects are also possible, but primarily for Tl incorporation during low-*T* ocean crust alteration. In order to reproduce the magnitude of equilibrium isotope fractionation observed for natural samples, reduction-oxidation processes, in which oxidized Tl^{3+} plays a central role, are predicted to be important.

APPLICATIONS OF THALLIUM ISOTOPES

Studies of Tl isotopes in Fe–Mn crusts

Hydrogenetic Fe–Mn crusts grow on hard substrates that experience little or no regular detrital sedimentation, for example on seamounts where ocean currents prevent gravitational settling of particles (Hein et al. 2000). They precipitate directly from the ambient water mass in which they are bathed and feature growth rates of a few mm/Ma (Segl et al. 1984, 1989; Eisenhauer et al. 1992). This implies that samples with a thickness exceeding 10 cm may provide a continuous seawater record for the entire Cenozoic (the last ~65 Ma). Over the last 20 years, extensive investigations of Fe–Mn crusts have been conducted to infer changes in the radiogenic isotope compositions of various elements in deep ocean waters (e.g. Lee et al. 1999; Frank 2002). Depending on the marine residence time for the element investigated, the isotopic variability has mainly been interpreted as reflecting changes in ocean circulation patterns or the marine source and/or sink fluxes (Burton et al. 1997; van de Flierdt et al. 2004).

The globally uniform Tl isotope compositions observed for the surfaces of Fe–Mn crusts (Fig. 8), imply a constant equilibrium isotope fractionation between seawater and Tl incorporated into the Fe–Mn crusts. Time-dependent variations of Tl isotope compositions in Fe–Mn crusts can be interpreted to reflect either changes in the isotope fractionation factor between seawater and Fe–Mn crusts or the Tl isotope composition of seawater. In principle, both interpretations are feasible, but several lines of reasoning currently favor the latter explanation (Rehkämper et al. 2004; Nielsen et al. 2009a).

Two studies have determined Tl isotope depth profiles for several Fe–Mn crusts, and both identified large systematic changes in Tl isotope compositions (Rehkämper et al. 2004; Nielsen et al. 2009a). The first study produced low-resolution depth profiles for a number of samples. The largest Tl isotope variations were observed for the early Cenozoic (Rehkämper et al. 2004), but the low sampling density and uncertainties in the age models of the crusts precluded a precise determination of the timing and duration of the observed changes. It was furthermore argued that the Fe–Mn crusts record variations in the Tl isotope composition of seawater that were caused by changes in the marine input and/or output fluxes of this element (Rehkämper et al. 2004). The second study generated high resolution Tl isotope time series for two Fe–Mn crusts. Improved age models were applied, which resolved a single large shift in Tl isotope composition, which occurred between ~55 and ~45 Ma (Fig. 9; Nielsen et al. 2009a). Based on an improved understanding of the marine input and output fluxes of Tl and their respective isotope compositions, it was proposed that the large shift in the ε^{205}Tl value of seawater reflects a decrease in the amount of authigenic Mn oxides that were deposited with pelagic sediments in the early Eocene (Nielsen et al. 2009a).

It is difficult to assess the underlying mechanism responsible for this global change in Mn oxide precipitation. The strong co-variation of the Tl isotope curve with the sulfur (S) isotope

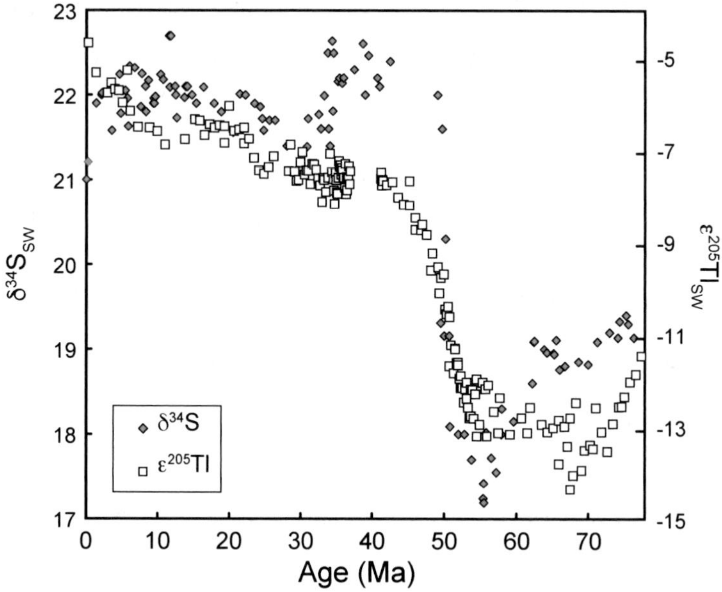

Figure 9. The S and Tl isotope composition of seawater over the last 75Myrs. The Tl isotope curve is based on a Fe–Mn crust from the Pacific Ocean and an inferred constant isotope fractionation factor of $\alpha = 1.0019$. The S isotope data are from Paytan et al. (1998 2004), with ages based on the age model of Kurtz et al. (2003). The chronology of the Tl isotope curve was determined based on Os isotope data (Burton 2006). Figure modified from Nielsen et al. (2009a).

composition of seawater (Fig. 9) may imply, however, that the same mechanism is driving the shift in the isotopic evolution of both stable isotope systems, even though S isotopes are known to be unaffected (at least directly) by changes in Mn oxide precipitation. Baker et al. (2009) proposed that the inferred high Mn oxide precipitation rates for the Paleocene (~65–55 Ma) may be explained by increased deposition of Fe- and Mn-rich volcanic ash particles in the oceans. Such volcanic activity would also supply isotopically light S and this could explain the relatively low $\delta^{34}S$ value for seawater at this time. The changes in the Tl and S isotope compositions of the oceans between ~55 and ~45 Ma (Fig. 9) would then be controlled by diminishing volcanic activity (Wallmann 2001). An alternative model proposes that Mn oxide precipitation is controlled by biological utilization and burial of Mn with organic carbon (Nielsen et al. 2009a). Higher organic carbon burial rates would lead to diminished Mn oxide precipitation rates as less Mn would be available in the water column. Simultaneously, the increased organic carbon burial would result in higher rates of sedimentary pyrite burial, which draws isotopically light S out of seawater (Berner 1984). In summary, the results of initial paleoceanographic studies indicate that it may be possible to utilize Tl isotopes as a proxy for changes in marine Mn sources and/or Mn oxide precipitation rates back in time.

Calculation of hydrothermal fluid fluxes using Tl isotopes in the ocean crust

Hydrothermal fluids are expelled from the seafloor (i) at high temperature on mid ocean ridge axes, as fueled by the magmatic energy from the crystallization and cooling of newly produced ocean crust to ~300–400 °C and (ii) at lower temperatures on the ridge flanks, as the ocean crust cools further over millions of years. These hydrothermal fluxes play pivotal roles in controlling seawater chemistry, but the magnitude of the high temperature water flux at mid-ocean ridge axes remains widely disputed, whilst the volume of low temperature vent fluids expelled at ridge flanks is essentially unconstrained.

As discussed in *High temperature hydrothermal fluids* and *Thallium uptake during low temperature hydrothermal alteration*, Tl displays distinct behavior during high and low temperature hydrothermal alteration of the ocean crust. high-T fluids effectively leach Tl from the cooling rocks whereas low-T fluids deposit Tl into the upper part of the oceanic crust. Following Nielsen et al. (2006c), a mass balance equation can be constructed for the high-T hydrothermal fluid flux (F_{hT}):

$$F_{hT} \times [\mathrm{Tl}]_{hT} = M_{Tl} \tag{3}$$

where $[\mathrm{Tl}]_{hT}$ is the average Tl concentration of the vent fluids and, as defined in Equation (2), M_{Tl} is the annual flux of Tl expelled into the ocean via high-T hydrothermal fluids. When the Tl flux of 170±60 Mg/a determined from Equation (2) is combined with $[\mathrm{Tl}]_{hT}$=6.7±0.7 ng/g (Nielsen et al. 2006c), we obtain a high temperature hydrothermal water flux of $2.5\pm0.9\times10^{13}$ kg/a. This fluid flux corresponds to 50–80% of the heat available at mid-ocean ridge axes from the crystallization and cooling of the freshly formed ocean crust (Mottl 2003). The difference between the available heat at mid-ocean ridge axes and that expelled via hydrothermal fluids requires that some energy at mid ocean ridge axes is lost via conduction and/or through the circulation of intermediate temperature hydrothermal fluids that do not alter the chemical budgets of Tl in the ocean crust (Nielsen et al. 2006c).

For the low-T hydrothermal fluid circulation flux (F_{lT}), the following mass balance equation was shown to apply (Nielsen et al. 2006c):

$$F_{vz} \times \left([\mathrm{Tl}]_{avz} - [\mathrm{Tl}]_{pvz}\right) = F_{lT} \times [\mathrm{Tl}]_{sw} \times f_{upt} \tag{4}$$

where F_{vz} is the mass flux of newly produced ocean crust that is affected by low-T alteration, $[\mathrm{Tl}]_{avz}$, $[\mathrm{Tl}]_{pvz}$, and $[\mathrm{Tl}]_{sw}$ are the Tl concentrations of the altered volcanic zone basalts, their pristine equivalents and seawater, respectively. The fraction of Tl that is removed from seawater by basalt weathering is denoted by f_{upt}. As discussed in *Thallium uptake during low temperature hydrothermal alteration*, $[Tl]_{avz}$ is extremely difficult to assess because the oceanic crust at ODP Holes 504B, 896A and U1301B is younger than 5Ma, which means that hydrothermal alteration is still ongoing at these locations. Analyses of Jurassic oceanic crust at ODP Hole 801C revealed little to no Tl enrichment (Fig. 7), possibly due to much lower Tl concentrations in seawater at the time (Prytulak et al. 2013). A meaningful assessment of Tl accumulation in oceanic crust that experienced complete hydrothermal alteration is hence currently not possible. The Tl enrichment factors observed for rocks from drill holes in young oceanic crust (Fig. 7) are, therefore, likely minimum estimates. The remaining parameters are more easily determined, but the large uncertainty on $[Tl]_{avz}$ generates estimates for the low-T hydrothermal fluid fluxes at ridge flanks that are also highly uncertain at $0.2–5.4 \times 10^{17}$ kg/a (Nielsen et al. 2006c). Using the ridge flank power output of 7.1 TW (Mottl 2003), it was calculated that such fluids have an average temperature anomaly of only about 0.1 to 3.6 °C relative to ambient seawater, which is lower than most flank fluids sampled to date. It is therefore unclear how representative low-T fluids sampled to date are of average low-T ocean crust alteration processes.

In order to improve the utility of Tl mass balance calculations to constrain hydrothermal fluid fluxes it will be essential to obtain more data on altered ocean crust from a number of locations and particularly for older sections of altered oceanic crust. However, Jurassic ocean crust appears to not follow the general pattern of strong Tl enrichment during low-T alteration and this may reflect the particular marine conditions of this era (Prytulak et al. 2013). Analyses of intermediate-age altered oceanic crust (20–100Ma) will provide improved constraints on the behavior of Tl during hydrothermal processes and may thus ultimately yield more reliable Tl-based estimates of global hydrothermal water fluxes.

High temperature terrestrial applications

Mineral Deposits. The accumulation of Tl via hydrothermal activity and the association of Tl and S, suggests obvious potential applications for Tl elemental systematics and isotope signatures in mineral deposit exploration. The significantly elevated concentrations of Tl in many mineral deposits furthermore allows analyses of mineral separates, which are often at the limits of analytical capabilities in barren igneous systems. Such investigations provide useful information on the magnitude and direction of natural mineral fractionation factors. For example, it is of interest to determine if Tl-enriched minerals have a distinctive Tl isotope 'fingerprint' that could dominate bulk rock assays and thus aid exploration activities.

The first economically focused study of thallium and thallium isotopes was undertaken by Baker et al. (2010a), who investigated a porphyry copper deposit hosted in the Collahuasi Formation of northern Chile. These authors examined whole rock andesite, dacite, rhyolite and Cu-porphyry samples. The concentration of thallium was found to vary over an order of magnitude from 0.1 to 3.2 µg/g, correlating with K and Rb, and not Cu, thus indicting that Tl displays lithophile rather than chalcophile behavior in this system (Baker et al. 2010a). The isotope compositions ranged from $\varepsilon^{205}Tl = -5.1$ to $+0.1$, which is not unusual in terms of the range of $\varepsilon^{205}Tl$ seen in barren terrestrial igneous rocks (Fig. 1). The five examined porphyry samples had very restricted $\varepsilon^{205}Tl$ values of -1.9 ± 0.1, identical to $\varepsilon^{205}Tl_{MORB}$. Thus there appears little potential for fingerprinting Cu porphyry rocks with Tl isotopes. Furthermore, there were no clear, systematic relationships between $\varepsilon^{205}Tl$ and major and trace element data or alteration minerals, which also limits the diagnostic potential for Tl isotopes in Cu porphyry systems.

Hettmann et al. (2014) addressed the magnitude of Tl isotope fractionation between different sulfide melts by examining the Pb-As-Tl-Zn deposit at the Lengenbach quarry in Switzerland. The Lengenbach deposit is notable for its abundance of rare Tl-bearing minerals, including hatchite ($AgTlPbAs_2S_5$) with an extreme thallium concentration of up to 24.8 wt%.

The high concentrations of Tl in the ore minerals allowed isotopic analysis of individual mineral phases and thus an evaluation of potential isotope fractionation induced by sulfide melt/mineral partitioning. The overall range of ε^{205}Tl measured in sulfides, sulfosalts and micas spans $\varepsilon^{205}\text{Tl} = -4.1$ to $+1.9$. As for the Collahuasi deposit, the substantial range in Tl isotope compositions over a geographically restricted area was not accompanied by a clear co-variation with other chemical characteristics or mineralogy. Complicated processes including both hydrothermal and sulfide melt inputs and subsequent formation and dissolution of secondary phases are likely responsible for the general lack of coherent behavior. Certainly, the Tl concentrations documented are amongst the highest measured in natural materials, yet the isotopic variability, whilst large, is again within the range of barren igneous rocks.

Given the significant and resolvable variability in ε^{205}Tl for both examined deposits, there is clearly much to be explored in terms of individual mineral controls on isotope signatures and the potential effects of fluid mineral partitioning at different pH, fO_2 and T. The utility of Tl isotopes to fingerprint economically viable deposits, however, remains to be convincingly demonstrated.

The Tl isotope composition of arc lavas. Subduction related magmas are classically considered to be derived from a depleted mantle wedge, chemically flavored by percent-level addition of sediments and/or fluids released from the subducting slab. Due to the large diversity of Tl concentrations and isotope compositions among these three reservoirs (Fig. 1), Tl isotopes appear to be uniquely suited to disentangle these sources and investigate the petrogenesis of subduction-related lavas. Although back arc spreading may cause the mantle wedge to be more depleted than the source of mid-ocean ridge basalts, to a first order, the depleted upper mantle is a reasonable estimate for the Tl content and isotope composition of the mantle wedge. The Tl concentration of the depleted mantle is orders of magnitude lower than possible inputs to the system in the form of sediments and fluids from the altered oceanic crust (Fig. 1). In addition to the large concentration contrast between 'background' mantle and possible inputs, the two most commonly invoked inputs have opposite vectors of isotope fractionation. As discussed in the sections *Thallium adsorption by the authigenic phases of pelagic clays* and *Thallium uptake during low temperature hydrothermal alteration*, pelagic sediments rich in Mn oxides are characterized by positive ε^{205}Tl whilst altered oceanic crust and, by inference, slab fluids derived thereof are expected to yield negative ε^{205}Tl signatures Thus, even minor addition of Mn oxide rich sediments or low-T altered basalts to a mantle wedge can result not only in analytically resolvable Tl isotope signatures, but also identify the type of input a specific lava has experienced. Finally, Tl isotope measurements may be able to contribute to the debate over the petrogenesis of so-called 'adakite' lavas (Kay 1978), whereby one long-standing hypothesis proposes that they are produced by direct melting of the (altered) basaltic portion of the subducting slab.

The prediction is straightforward: lavas with trace element chemistry indicative of fluid contributions are expected to have light Tl isotope signatures, whilst sediment-influenced lavas are expected to be isotopically heavy. However, Tl isotope measurements of lavas from the Mariana subduction zone display very little isotopic variation, with ε^{205}Tl values from -1.8 to -0.4. One lava falls outside this range, with a heavy ε^{205}Tl of $+1.2$, which was suggested to be due to volcanic degassing (Prytulak et al. 2013). Hence, even though volcanic degassing broadly speaking has little impact on the Tl isotope composition of lavas and the continental crust (see section 5.1), individual samples may be affected by this process and must be assessed in studies of subaerial volcanism. The remaining data from the Mariana arc are reasonable, in that the measured inputs to the system, including sediments and altered ocean crust from ODP Hole 801C, did not show the same variability of Tl isotope compositions that was documented in similar lithologies elsewhere. In addition, a subsequent study of Mariana forearc serpentinites revealed that Tl sourced from pelagic sediments was released with fluids from the subducting sediments that caused serpentinization of the forearc peridotites (Nielsen et al. 2015). This result suggests that the Tl isotope signature of subducting sediments may be altered in the forearc before they enter the subarc mantle. This process could be an important mechanism of subduction zone cycling for Tl and other trace metals that concentrate in Mn oxides. In summary, the Tl isotope data for Mariana arc lavas demonstrates a situation where isotopically invariant inputs produce similarly invariant outputs. This is a key finding, as it provides evidence that the subduction process itself does not fractionate Tl isotopes (Prytulak et al. 2013).

The only other subduction zone that has been investigated for Tl isotopes is the Aleutian arc (Nielsen et al. 2016). The Aleutians are notable for along strike variation in subduction parameters, such as slab dip, slab velocity, and variable sediment lithologies dominated by volcaniclastic in the East to pelagic clay in the West. In addition, the Aleutians contain the type locality for so-called adakite lavas, from Adak Island. Nielsen et al. (2016) analyzed both lavas and sedimentary lithologies outboard of the arc and demonstrated that the ε^{205}Tl values of the lavas closely follow the sediment values, whereby the latter change systematically along the arc, with the Central Aleutian sediments and lavas showing enrichments in ^{205}Tl relative to the Eastern Aleutians. This finding further strengthens the conclusion that subduction processes do not fractionate Tl isotopes and which are thus useful for tracing inputs that are isotopically distinct from the background mantle. Nielsen et al. (2016) also determined the Tl isotopic composition of the lava originally used to define adakites (Kay 1978; Defant and Drummond 1990). As expected, the sample has a light ε^{205}Tl of −3.3, in general agreement with the hypothesis that adakites incorporate melts from the altered basaltic ocean crust (Nielsen et al. 2016).

The Tl isotope composition of ocean island basalts (OIB). The recycling of sediments and oceanic crust is a central tenet of mantle geochemistry and has been the subject of intense research over the past 35 years (e.g. Hofmann 2014). Many investigations focus on ocean island basalts, which are argued to be generated at pressures > 2.5 GPa in the mantle, and thus significantly deeper than mid-ocean ridge basalts. Though still debated, there is some consensus that many ocean island basalts are produced by anomalously hot mantle plumes, thus offering a reasonable explanation why their locations are independent of tectonic plate boundaries (Hofmann and White 1982; Hofmann 2014). Traditional approaches to deduce the nature of OIB mantle sources employ a combination of trace element and radiogenic isotope data to distinguish sediment and/or crustal additions. As radiogenic isotopes reflect the time-integrated parent–daughter trace element ratio, there is inescapable ambiguity associated with their employment, with the same isotope signature permissibly generated by markedly different evolution paths.

Stable isotope analyses offer important complementary constraints to measurements that apply radiogenic isotope systems. Some of the earliest applications of stable isotopes to high temperature geochemistry encompass studies of O and Li to trace processes such as ocean crust alteration (Alt et al. 1986; Chan et al. 2002). However, light elements like O and Li are known to record isotope fractionation even during processes that occur at elevated mantle temperatures (Jeffcoate et al. 2007; Marschall et al. 2007; Williams et al. 2009) and their mantle concentrations are relatively high, such that the isotope composition of a mantle source is not readily affected by admixing of sediment and/or ocean crust (Elliott et al. 2004; Thirlwall et al. 2004). However, the concentration contrast and magnitude of isotope fractionation offered by the Tl isotope system is favorable for overcoming these obstacles.

To date, OIB samples from the Azores, Iceland and Hawaii have been investigated (Nielsen et al. 2007; Nielsen et al. 2006b). As it is unclear whether the Azores basalts were affected by post-eruptional alteration (Nielsen et al. 2007), these data will not be discussed in the following. Samples from Hawaii exhibit the most convincing Tl isotope evidence for the presence of sediments in the mantle source (Fig. 10). In detail, about 8 ppm of pure Fe–Mn sediment is sufficient to explain the positive ε^{205}Tl values of up to +4 recorded in these lavas. Whilst it is unlikely that the Tl isotope variation originates from anything else than Fe–Mn sediment, it is uncertain if this component was acquired by the melts during magma ascent via assimilation of modern marine deposits or if it is a feature of the mantle source. The samples with the heaviest Tl isotope compositions are, however,

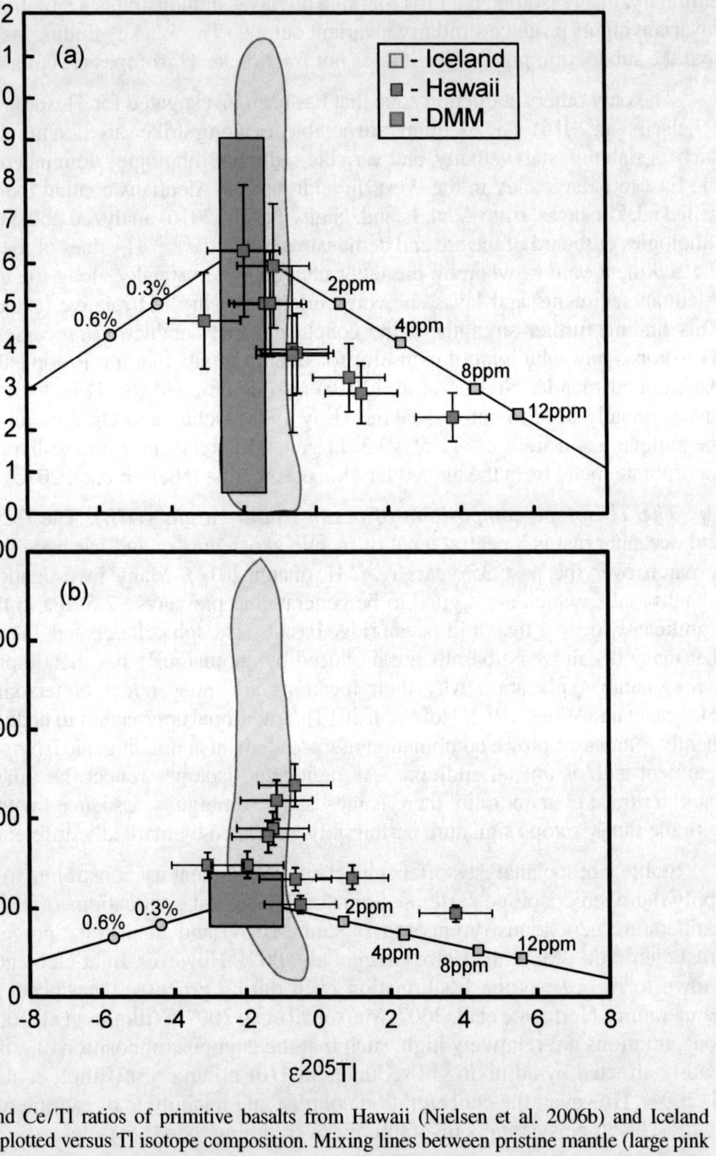

Figure 10. Cs/Tl and Ce/Tl ratios of primitive basalts from Hawaii (Nielsen et al. 2006b) and Iceland (Nielsen et al. 2007) plotted versus Tl isotope composition. Mixing lines between pristine mantle (large pink square), Fe–Mn oxyhydroxides (blue squares) and low-*T* altered MORB (light green circles) are also shown. The mantle is assumed to be characterized by $\varepsilon^{205}Tl = -2$ (Nielsen et al. 2006b) and Ce, Cs and Tl concentrations of 772 ng/g, 4.2 ng/g and 0.7 ng/g, respectively (Salters and Stracke 2004; Nielsen et al. 2014). For the Fe–Mn oxyhydroxides, the Tl concentration and isotope composition are assumed to be 100 µg/g and $\varepsilon^{205}Tl = +10$, akin to values of modern Fe–Mn crusts and nodules (Hein et al. 2000; Rehkämper et al. 2002). The Cs content of Fe–Mn oxyhydroxides is about 500 ng/g (Ben Othmann et al. 1989). Altered MORB is assumed to be characterized by $\varepsilon^{205}Tl = -10$ and Ce, Tl and Cs concentrations of about 13400 ng/g 200 ng/g and 200 ng/g, respectively (Gale et al. 2013; Nielsen et al. 2006c). Error bars denote 2sd uncertainties.

also characterized by the least radiogenic Pb isotope compositions (Nielsen et al. 2006b) which would argue for an old age of the sedimentary component.

The relatively straightforward interpretation of the Tl isotope data for Hawaii could be considered a "smoking gun" for the presence of recycled sediments in the Hawaiian mantle plume and, indeed, agrees with independent constraints from Hf isotopes (Blichert-Toft et al. 1999). However, results obtained for of a suite of lavas from Iceland strongly indicate that there is some way to go before Tl isotopes can be confidently applied as a unique tracer of crustal recycling within the mantle. Seventeen Icelandic samples, including picrites that span all major eruption centers of the island, exhibit an average ε^{205}Tl = −1.6 ± 1.1, completely overlapping with $\varepsilon^{205}Tl_{MORB}$. In contrast to the isotopic homogeneity, Cs/Tl ratios vary from 0.3 to 11. The isotopic invariance is perhaps unsurprising given that the thickness of subducted oceanic lithosphere exceeds 30 km. Thallium isotope anomalies will be situated in only the uppermost ~500 to 1000 m, whilst the remainder of the oceanic crust is expected to be isotopically identical to the ambient mantle. The mantle-like Tl isotope signatures of the Iceland basalts hence do not argue against the presence of recycled ocean crust in the plume source. Of more concern is that the samples display variable Cs/Tl ratios. The large observed Cs/Tl range demonstrates that processes other than the addition of Fe–Mn sediments and low-*T* altered basalts may alter this ratio. It is conceivable that this includes igneous processes, such as partitioning of Tl into sulfides and/or phyllosilicates, Tl mobilization by magmatic fluids (Nielsen et al. 2007) and/or fractionation by accessory phases in the subducted material (e.g., phengite; Prytulak et al. 2013).

Finally, it is prudent to consider three key caveats for the application of Tl isotopes in mantle geochemistry. First, it is likely that the Tl isotope composition of seawater has not remained constant over time (Fig. 9) and the isotope signatures of altered basalts and Fe–Mn sediments are therefore also expected to exhibit temporal variability. Any such variability will obscure the mixing trends that are produced by contamination of the ambient mantle with recycled material. Second, whilst it is unclear when the oceans became sufficiently oxic to support the precipitation of Mn oxides, this probably occurred after ~2.4 Ga (Canfield 1998). Sediments that were recycled more than 2.4 billion years ago are therefore unlikely to be enriched in Tl associated with Mn oxides and are hence probably characterized by ε^{205}Tl ≈ −2. The isotope fractionation mechanism responsible for the highly negative ε^{205}Tl values of modern altered ocean crusts is yet to be determined. Hence, it is unclear whether such basalts are also isotopically fractionated in ancient environments and if past oceanic crust recycling was able to alter the Tl isotope composition of the mantle. Third, it needs to be acknowledged that Tl can also be *too* sensitive as a tracer. Even much less than 1% of subaerial or submarine contamination by secondary clays or precipitation of Mn oxides can have a drastic impact on the Tl concentration and isotope signature of a basalt. Great attention must therefore be paid to sample preparation and selection. For example, the careful leaching experiments for submarine samples conducted by Nielsen et al. (2016) should be considered a minimum requirement to establish confidence in the determined ε^{205}Tl isotope signatures of submarine lavas with low Tl concentrations.

FUTURE DIRECTIONS AND OUTLOOK

High precision measurements of Tl isotope compositions have only been possible for little more than a decade but despite of this, we have already acquired a surprisingly detailed understanding of the diverse isotopic behavior of this element. Significant gaps in knowledge remain, however.

Most meteorite samples display resolvable variations in Tl isotope compositions with an overall isotopic variability of about 50 ε^{205}Tl. In many samples, the isotopic variation appears to be caused by both radiogenic decay of extinct ^{205}Pb and stable isotope fractionations, which reflect the highly volatile and labile nature of Tl. The difficulty of deconvolving these two sources of isotopic variability restricts the utility of both the ^{205}Pb–^{205}Tl chronometer and the Tl stable isotope system to inform on early solar system processes. Nonetheless, further studies of suitable meteorites (including carbonaceous chondrites) are desirable to better constrain the initial solar system abundance of ^{205}Pb, as this is a unique tracer of freshly synthesized *s*-process material that was delivered to the nascent solar system.

For the Earth, it is desirable to further expand the Tl isotope and concentration database for various environments, in order to gain a better understanding of the geochemical distribution and cycling of this element. In addition, there are a few crucial investigations that are needed to advance the utility of Tl isotopes as quantitative tracers of past and present geological processes.

First of all, we must fully understand the mechanisms that govern the two major Tl isotope effects observed on Earth, which produce highly fractionated Tl isotope signatures in marine Mn oxides and low-*T* altered basalts. Experimental studies have shown that Tl oxidation and adsorption to the Mn oxide birnessite is clearly the central process responsible for the heavy isotope ratios recorded in Mn oxide rich marine sediments (Peacock and Moon 2012; Nielsen et al. 2013). However, the magnitude of isotope fractionation and the effect of changes in intensive parameters like T, redox potential and ionic strength are presently unknown. Further studies of both natural and synthetic systems that mimic the conditions of Mn oxide precipitation and low-*T* ocean crust alteration are required to establish the mechanisms controlling Tl isotope fractionation in these two critical reservoirs. This knowledge will not only help expand our appreciation of the physico-chemical processes that cause isotope fractionation in heavy elements, but also enable us to better utilize the Tl isotope system to quantify low-*T* hydrothermal fluid flow (Nielsen et al. 2006c) and help to unravel the causes of the Tl isotope variations observed in the marine environment over time (Baker et al. 2009; Nielsen et al. 2009a).

Furthermore, it will be important to refine the three applications outlined in section 8. For the paleoceanographic studies, this will include a detailed determination of the magnitude and isotope compositions of the most uncertain marine Tl fluxes. These are the fluxes associated with benthic pore waters, adsorption processes on pelagic clays and low-*T* hydrothermal alteration. A complete understanding of the modern marine Tl cycle is a prerequisite for good models of past Tl isotope variations. Another aspect of the oceanic Tl isotope evolution that has yet to be investigated are short-term fluctuations, for example on glacial-interglacial time-scales or ocean anoxic events. Since the marine residence time of Tl is about 18.5 ka (Table 4), it should be feasible to observe perturbations of the Tl cycle that occur on geologically rapid time-scales. To this end, a recent study of Tl isotopes in euxinic sediments has revealed no detectable isotope fractionation between seawater and sediment (Owens et al. pers. comm.). This is an important result because such sediments may now enable reconstruction of marine Tl isotope signatures in deep time.

The use of Tl isotopes as a tool in mantle geochemistry is currently limited by scant data. Thallium isotope and concentration data are presently available only for a few OIB and there is a lack of understanding concerning how Tl is cycled through subduction zones. Recent studies that attempted to address such questions through Tl isotope analyses of cratonic eclogites (Nielsen et al. 2009b) and island arc lavas (Prytulak et al. 2013; Nielsen et al. 2016) were unable to conclusively constrain the behavior of Tl and further investigations of rocks from subduction related environments are thus necessary.

The application of Tl isotopes as a tracer of anthropogenic Tl emissions to the environment has been shown to be viable in an initial study (Kersten et al. 2014) and hence shows promise for future investigations. However, whilst Tl is a highly toxic element, it shows low concentrations in most natural materials and is used to only a limited extent in industrial products and processes. Hence, Tl isotopes are unlikely to develop into a widely used tracer of anthropogenic emission, similar to radiogenic Pb isotope compositions, but will be of utility to trace Tl origin and mobility in particular, localized pollution scenarios.

REFERENCES

Alleman LY, Veron AJ, Church TM, Flegal AR, Hamelin B (1999) Invasion of the abyssal North Atlantic by modern anthropogenic lead. Geophys Res Lett 26:1477–1480

Alt JC (1995) Subseafloor processes in mid-ocean ridge hydrothermal systems. *In:*Humphris SE, Lupton JE, Mullineaux LS, Zierenberg RA (Eds.) Seafloor Hydrothermal Systems, Physical, Chemical, and Biological Interactions. AGU, Washington DC, pp. 85–114

Alt JC, Muehlenbachs K, Honnorez J (1986) An oxygen isotopic profile through the upper kilometer of the oceanic crust, DSDP Hole 504B Earth Planet Sci Lett 80:217–229

Alt JC, Teagle DAH, Bach W, Halliday AN, Erzinger J (1996) Stable and strontium isotopic profiles through hydrothermally altered upper oceanic crust, hole 504B Proc ODP Sci Results 148:57–69

Anders E, Stevens CM (1960) Search for extinct lead 205 in meteorites. J Geophys Res 65:3043–3047

Andreasen R, Schönbächler M, Rehkämper M (2009) The Pb-205-(Tl)-T-205 and Cd isotope systematics of ordinary chondrites. Geochim Cosmochim Acta 73, A43-A43

Andreasen R, Rehkämper M, Benedix GK, Theis KJ, Schönbächler M, Smith CL (2012) Lead-thallium chronology of IIAB and IIIAB iron meteorites and the solar system initial abundance of lead-205, Lunar Planet Sci Conf. Lunar and Planetary Institute, Woodlands, TX, p. Abstract #2902

Archer C, Vance D (2008) The isotopic signature of the global riverine molybdenum flux and anoxia in the ancient oceans. Nature Geoscience 1:597–600

Arden JW, Cressey G (1984) Thallium and Lead in the Allende C3v Carbonaceous Chondrite - a Study of the Matrix Phase. Geochim Cosmochim Acta 48:1899–1912

Baker RGA, Rehkämper M, Hinkley TK, Nielsen SG, Toutain JP (2009) Investigation of thallium fluxes from subaerial volcanism-Implications for the present and past mass balance of thallium in the oceans. Geochim Cosmochim Acta 73:6340–6359

Baker RGA, Rehkämper M, Ihlenfeld C, Oates CJ, Coggon RM (2010a) Thallium isotope variations in an ore-bearing continental igneous setting: Collahuasi Formation, Northern Chile. Geochim Cosmochim Acta 74:4405–4416

Baker RGA, Schönbächler M, Rehkämper M, Williams HM, Halliday AN (2010b) The thallium isotope composition of carbonaceous chondrites - New evidence for live Pb-205 in the early solar system. Earth Planet Sci Lett 291:39–47

Ballhaus C, Laurenz V, Münker C, Fonseca ROC, Albarede F, Rohrbach A, Lagos M, Schmidt MW, Jochum KP, Stoll B, Weis U, Helmy HM (2013) The U/Pb ratio of the Earth's mantle-A signature of late volatile addition. Earth Planet Sci Lett 362:237–245

Ben Othmann D, White WM, Patchett J (1989) The geochemistry of marine sediments, island arc magma genesis, and crust-mantle recycling. Earth Planet Sci Lett 94:1–21

Berner RA (1984) Sedimentary Pyrite Formation - an Update. Geochim Cosmochim Acta 48:605–615

Bidoglio G, Gibson PN, Ogorman M, Roberts KJ (1993) X-Ray-absorption spectroscopy investigation of surface redox transformations of thallium and chromium on colloidal mineral oxides. Geochim Cosmochim Acta 57:2389–2394

Bigeleisen J (1996) Nuclear size and shape effects in chemical reactions. Isotope chemistry of the heavy elements. J Am Chem Soc 118:3676–3680

Bigeleisen J, Mayer MG (1947) Calculation of equilibrium constants for isotopic exchange reactions. J Chem Phys 15:261–267

Blake JB, Lee T, Schramm DN (1973) Chronometer for s-process nucleosynthesis. Natur-Phys Sci 242:98–100

Blichert-Toft I, Frey FA, Albarede F (1999) Hf isotope evidence for pelagic sediments in the source of Hawaiian basalts. Science 285:879–882

Bruland KW (1983) Trace elements in seawater. *In:* Riley JP, Chester R (Eds.) Chemical Oceanography. Academic Press, London, pp. 157–221

Burton KW (2006) Global weathering variations inferred from marine radiogenic isotope records. J Geochem Explor 88:262–265

Burton KW, Ling HF, Onions RK (1997) Closure of the Central American Isthmus and its effect on deep-water formation in the North Atlantic. Nature 386:382–385

Canfield DE (1998) A new model for Proterozoic ocean chemistry. Nature 396:450–453

Carlson RW, Hauri EH (2001) Extending the Pd-107-Ag-107 chronometer to low Pd/Ag meteorites with multicollector plasma-ionization mass spectrometry. Geochim Cosmochim Acta 65:1839–1848

Chan LH, Alt JC, Teagle DAH (2002) Lithium and lithium isotope profiles through the upper oceanic crust: a study of seawater-basalt exchange at ODP Sites 504B and 896A Earth Planet Sci Lett 201:187–201

Cheam V (2001) Thallium contamination of water in Canada. Water Qual. Res J Canada 36:851–877

Chen JH, Wasserburg GJ (1987) A search for evidence of extinct lead 205 in iron meteorites. LPSC XVIII, 165–166

Chen JH, Wasserburg GJ (1994) The abundance of thallium and primordial lead in selected meteorites - the search for ^{205}Pb. LPSC XXV, 245

Coggon RM, Rehkämper M, Atteck C, Teagle DAH, Alt JC, Cooper MJ (2014) Mineralogical and microbial controls on thallium uptake during hydrothermal alteration of the upper ocean crust. Geochim Cosmochim Acta 144:25–42

Craddock PR, Warren JM, Dauphas N (2013) Abyssal peridotites reveal the near-chondritic Fe isotopic composition of the Earth. Earth Planet Sci Lett 365:63–76

Defant MJ, Drummond MS (1990) Derivation of some modern arc magmas by melting of young subducted lithosphere. Nature 347:662–665

Eisenhauer A, Gogen K, Pernicka E, Mangini A (1992) Climatic influences on the growth-rates of Mn crusts during the late Quaternary. Earth Planet Sci Lett 109:25–36

Elderfield H (1976) Manganese fluxes to the oceans. Mar Chem 4:103–132

Elliott T, Jeffcoate A, Bouman C (2004) The terrestrial Li isotope cycle: light-weight constraints on mantle convection. Earth Planet Sci Lett 220:231–245

Flegal AR, Patterson CC (1985) Thallium concentrations in seawater. Mar. Chem 15:327–331

Flegal AR, Sanudo-Wilhelmy S, Fitzwater SE (1989) Particulate thallium fluxes in the northeast Pacific. Mar Chem 28:61–75

Frank M (2002) Radiogenic isotopes: Tracers of past ocean circulation and erosional input. Rev Geophys 40, art. 1001

Fujii Y, Nomura M, Okamoto M, Onitsuka H, Kawakami F, Takeda K (1989a) An anomalous isotope effect of U-235 in U(IV)-U(VI) chemical exchange. Z Naturforsch Sect A 44:395–398

Fujii Y, Nomura M, Onitsuka H, Takeda K (1989b) Anomalous isotope fractionation in uranium enrichment process. J Nucl Sci Technol 26:1061–1064

Gale A, Dalton CA, Langmuir CH, Su YJ, Schilling JG (2013) The mean composition of ocean ridge basalts. Geochem Geophys Geosyst 14:489–518

Gauthier PJ, Le Cloarec MF (1998) Variability of alkali and heavy metal fluxes released by Mt. Etna volcano, Sicily, between 1991 and (1995) J Volcanol Geotherm Res 81:311–326

Genna D, Gaboury D (2015) Deciphering the hydrothermal evolution of a VMS system by LA-ICP-MS using trace elements in pyrite: an example from the Bracemac-McLeodd Deposits, Abitibi, Canada, and implications for exploration. Econ Geol 110 2087–2108

Heggie D, Klinkhammer G, Cullen D (1987) Manganese and copper fluxes from continental margin sediments. Geochim Cosmochim Acta 51:1059–1070

Hein JR, Koschinsky A, Bau M, Manheim FT, Kang J-K, Roberts L (2000) Cobalt-rich ferromanganese crusts in the Pacific. *In:* Cronan DS (Ed.) Handbook of Marine Mineral Deposits. CRC Press, Boca Raton, pp. 239–280

Heinrichs H, Schulz-Dobrick B, Wedepohl KH (1980) Terrestrial geochemistry of Cd, Bi, Tl, Pb, Zn and Rb. Geochim Cosmochim Acta 44:1519–1533

Hettmann K, Kreissig K, Rehkämper M, Wenzel T, Mertz-Kraus R, Markl G (2014) Thallium geochemistry in the metamorphic Lengenbach sulfide deposit, Switzerland: Thallium-isotope fractionation in a sulfide melt. American Mineralogist 99:793–803

Hinkley TK, Lecloarec MF, Lambert G (1994) Fractionation of Families of Major, Minor and Trace-Metals Across the Melt Vapor Interface in Volcanic Exhalations. Geochim Cosmochim Acta 58:3255–3263

Hofmann AW (2014) Sampling Mantle Heterogeneity through Oceanic Basalts: Isotopes and Trace Elements. *In:* Heinrich DHa.KKT (Ed.), Treatise on Geochemistry (Second Edition). Elsevier, Oxford, pp. 67–101

Hofmann AW, White WM (1982) Mantle plumes from ancient oceanic crust. Earth Planet Sci Lett 57:421–436

Huey JM, Kohman TP (1972) Search for extinct natural radioactivity of Pb-205 via thallium-isotope anomalies in chondrites and lunar soil. Earth Planet Sci Lett 16:401–412

Huh Y, Chan LH, Zhang L, Edmond JM (1998) Lithium and its isotopes in major world rivers: Implications for weathering and the oceanic budget. Geochim Cosmochim Acta 62 2039–2051

Jeffcoate AB, Elliott T, Kasemann SA, Ionov D, Cooper K, Brooker R (2007) Li isotope fractionation in peridotites and mafic melts. Geochim Cosmochim Acta 71 202–218

Jenner FE, O'Neill HSC (2012) Analysis of 60 elements in 616 ocean floor basaltic glasses. Geochem Geophys Geosyst 13, article Q02005

Jochum KP, Verma SP (1996) Extreme enrichment of Sb, Tl and other trace elements in altered MORB Chem Geol 130:289–299

Johnson KS, Berelson WM, Coale KH, Coley TL, Elrod VA, Fairey WR, Iams HD, Kilgore TE, Nowicki JL (1992) Manganese flux from continental-margin sediments in a transect through the oxygen minimum. Science 257:1242–1245

Jones JH, Hart SR, Benjamin TM (1993) Experimental partitioning studies near the Fe–FeS eutectic, with an emphasis on elements important to iron meteorite chronologies (Pb, Ag, Pd, and Tl). Geochim Cosmochim Acta 57:453–460

Kay RW (1978) Aleutian magnesian andesites - melts from subducted Pacific ocean crust. J Volcanol Geotherm Res 4:117–132

Kelley KA, Plank T, Ludden J, Staudigel H (2003) Composition of altered oceanic crust at ODP Sites 801 and 1149. Geochem Geophys Geosyst 4, 8910

Kersten M, Xiao TF, Kreissig K, Brett A, Coles BJ, Rehkämper M (2014) Tracing anthropogenic thallium in soil using stable isotope compositions. Environ Sci Technol 48:9030–9036

Kiseeva ES, Wood BJ (2013) A simple model for chalcophile element partitioning between sulphide and silicate liquids with geochemical applications. Earth Planet Sci Lett 383:68–81

Knyazev DA, Myasoedov NF (2001) Specific effects of heavy nuclei in chemical equilibrium. Separ Sci Technol 36:1677–1696

Koschinsky A, Hein JR (2003) Acquisition of elements from seawater by ferromanganese crusts: Solid phase association and seawater speciation. Mar. Geol 198:331–351

Kruijer TS, Fischer-Gödde M, Kleine T, Sprung P, Leya I, Wieler R (2013) Neutron capture on Pt isotopes in iron meteorites and the Hf-W chronology of core formation in planetesimals. Earth Planet Sci Lett 361:162–172

Kurtz AC, Kump LR, Arthur MA, Zachos JC, Paytan A (2003) Early Cenozoic decoupling of the global carbon and sulfur cycles. Paleoceanography 18, 8910

Lauretta DS, Devouard B, Buseck PR (1999) The cosmochemical behavior of mercury. Earth Planet Sci Lett 171:35–47

Lauretta DS, Klaue B, Blum JD, Buseck PR (2001) Mercury abundances and isotopic compositions in the Murchison (CM) and Allende (CV) carbonaceous chondrites. Geochim Cosmochim Acta 65:2807–2818

Lee D-C, Halliday AN, Hein JR, Burton KW, Christensen JN, Günther D (1999) Hafnium isotope stratigraphy of ferromanganese crusts. Science 285:1052–1054

Lipschutz ME, Woolum DS (1988) Highly labile elements. *In:* Kerridge JF, Matthews MS (Eds.), Meteorites and the Early Solar System. Univerisity of Arizona Press, pp. 462–487

Lodders K (2003) Solar system abundances and condensation temperatures of the elements. Astro Phys J 591:1220–1247

Marschall HR, Pogge von Strandmann PAE, Seitz HM, Elliott T, Niu YL (2007) The lithium isotopic composition of orogenic eclogites and deep subducted slabs. Earth Planet Sci Lett 262:563–580

Matthews AD, Riley JP (1970) The occurrence of thallium in sea water and marine sediments. Chem Geol 149:149–152

McDonough WF, Sun S-s (1995) The composition of the Earth. Chem Geol 120:223–253

McGoldrick PJ, Keays RR, Scott BB (1979) Thallium - sensitive indicator of rock–seawater interaction and of sulfur saturation of silicate melts. Geochim Cosmochim Acta 43:1303–1311

Metz S, Trefry JH (2000) Chemical and mineralogical influences on concentrations of trace metals in hydrothermal fluids. Geochim Cosmochim Acta 64:2267–2279

Mottl MJ (2003) Partitioning of energy and mass fluxes between mid-ocean ridge axes and flanks at high and low temperature. *In:* Halbach PE, Tunnicliffe V, Hein JR (Eds.), Energy and Mass Transfer in Marine Hydrothermal Systems. Dahlem University Press, pp. 271–286

Nielsen SG (2010) Potassium and uranium in the upper mantle controlled by Archean oceanic crust recycling. Geology 38:683–686

Nielsen SG, Lee CTA (2013) Determination of thallium in the USGS glass reference materials BIR-1G, BHVO-2G and BCR-2G and application to quantitative Tl concentrations by LA-ICP-MS Geostand Geoanal Res 37:337–343

Nielsen SG, Rehkämper M, Baker J, Halliday AN (2004) The precise and accurate determination of thallium isotope compositions and concentrations for water samples by MC-ICPMS Chem Geol 204:109–124

Nielsen SG, Rehkämper M, Porcelli D, Andersson P, Halliday AN, Swarzenski PW, Latkoczy C, Gunther D (2005a) Thallium isotope composition of the upper continental crust and rivers—An investigation of the continental sources of dissolved marine thallium. Geochim Cosmochim Acta 69 2007–2019

Nielsen SG, Rehkämper M, Porcelli D, Andersson PS, Halliday AN, Swarzenski PW, Latkoczy C, Günther D (2005b) The thallium isotope composition of the upper continental crust and rivers - An investigation of the continental sources of dissolved marine thallium. Geochim Cosmichim Acta 69 2007–2019

Nielsen SG, Rehkämper M, Halliday AN (2006a) Large thallium isotopic variations in iron meteorites and evidence for lead-205 in the early solar system. Geochim Cosmochim Acta 70:2643–2657

Nielsen SG, Rehkämper M, Norman MD, Halliday AN, Harrison D (2006b) Thallium isotopic evidence for ferromanganese sediments in the mantle source of Hawaiian basalts. Nature 439:314–317

Nielsen SG, Rehkämper M, Teagle DAH, Alt JC, Butterfield D, Halliday AN (2006c) Hydrothermal fluid fluxes calculated from the isotopic mass balance of thallium in the ocean crust. Earth Planet Sci Lett 251:120–133

Nielsen SG, Rehkämper M, Brandon AD, Norman MD, Turner S, O'Reilly SY (2007) Thallium isotopes in Iceland and Azores lavas - Implications for the role of altered crust and mantle geochemistry. Earth Planet Sci Lett 264:332–345

Nielsen SG, Mar-Gerrison S, Gannoun A, LaRowe DE, Klemm V, Halliday AN, Burton KW, Hein JR (2009a) Thallium isotope evidence for increased marine organic carbon export in the early Eocene. Earth Planet Sci Lett 278:297–307

Nielsen SG, Williams HM, Griffin WL, O'Reilly SY, Pearson N, Viljoen KS (2009b) Thallium isotopes as a potential tracer for the origin of cratonic eclogites? Geochim Cosmochim Acta 73:7387–7398

Nielsen SG, Goff M, Hesselbo SP, Jenkyns HC, LaRowe DE, Lee CTA (2011) Thallium isotopes in early diagenetic pyrite—A paleoredox proxy? Geochim Cosmochim Acta 75:6690–6704

Nielsen SG, Shimizu N, Lee CTA, Behn M (2014) Chalcophile behavior of thallium during MORB melting and implications for the sulfur content of the mantle. Geochem Geophys Geosyst 15:4905–4919

Nielsen SG, Wasylenki LE, Rehkämper M, Peacock CL, Xue Z, Moon EM (2013) Towards an understanding of thallium isotope fractionation during adsorption to manganese oxides. Geochim Cosmochim Acta 117:252–265

Nielsen SG, Klein F, Kading T, Blusztajn J, Wickham K (2015) Thallium as a Tracer of Fluid-Rock Interaction in the Shallow Mariana Forearc. Earth Planet Sci Lett 430:416–426

Nielsen SG, Yogodzinski GM, Prytulak J, Plank T, Kay SM, Kay RW, Blusztajn J, Owens JD, Auro M, Kading T (2016) Tracking along-arc sediment inputs to the Aleutian arc using thallium isotopes. Geochim Cosmochim Acta 181:217–237

Noll PD, Newsom HE, Leeman WP, Ryan JG (1996) The role of hydrothermal fluids in the production of subduction zone magmas: Evidence from siderophile and chalcophile trace elements and boron. Geochim Cosmochim Acta 60:587–611

Nriagu J (1998) Thallium in the Environment, Advances in Environmental Sciences and Technology. Wiley, New York

Ostic RG, Elbadry HM, Kohman TP (1969) Isotopic composition of meteoritic thallium. Earth Planet Sci Lett 7:72–76

Palk CS, Rehkämper M, Andreasen R, Stunt A (2011) Extreme cadmium and thallium isotope fractionations in enstatite chondrites. Meteorit Planet Sci 46, A183-A183

Patterson CC, Settle DM (1987) Magnitude of lead flux to the atmosphere from volcanos. Geochim Cosmochim Acta 51:675–681

Paytan A, Kastner M, Campbell D, Thiemens MH (1998) Sulfur isotopic composition of Cenozoic seawater sulfate. Science 282:1459–1462

Paytan A, Kastner M, Campbell D, Thiemens MH (2004) Seawater sulfur isotope fluctuations in the cretaceous. Science 304:1663–1665

Peacock CL, Moon EM (2012) Oxidative scavenging of thallium by birnessite: Controls on thallium sorption and stable isotope fractionation in marine ferromanganese precipitates. Geochim Cosmochim Acta 84:297–313

Pengra JG, Genz H, Fink RW (1978) Orbital electron capture ratios in the decay of ^{205}Pb. Nuclear Physics A302, 1–11

Peter ALJ, Viraraghavan T (2005) Thallium: a review of public health and environmental concerns. Environ Int 31:493–501

Pietruszka AJ, Reznik AD (2008) Identification of a matrix effect in the MC-ICP-MS due to sample purification using ion exchange resin: An isotopic case study of molybdenum. Int. J Mass Spectrom. 270:23–30

Pogge von Strandmann PAE, Burton KW, James RH, van Calsteren P, Gislason SR (2010) Assessing the role of climate on uranium and lithium isotope behaviour in rivers draining a basaltic terrain. Chem Geol 270:227–239

Poirier A, Doucelance R (2009) Effective Correction of mass bias for rhenium measurements by MC-ICP-MS Geostand Geoanal Res 33 195–204

Prytulak J, Nielsen SG, Plank T, Barker M, Elliott T (2013) Assessing the utility of thallium and thallium isotopes for tracing subduction zone inputs to the Mariana arc. Chem Geol 345:139–149

Rehkämper M, Nielsen SG (2004) The mass balance of dissolved thallium in the oceans. Mar. Chem 85:125–139

Rehkämper M, Halliday AN (1999) The precise measurement of Tl isotopic compositions by MC- ICPMS: Application to the analysis of geological materials and meteorites. Geochim Cosmochim Acta 63:935–944

Rehkämper M, Frank M, Hein JR, Porcelli D, Halliday A, Ingri J, Liebetrau V (2002) Thallium isotope variations in seawater and hydrogenetic, diagenetic, and hydrothermal ferromanganese deposits. Earth Planet Sci Lett 197:65–81

Rehkämper M, Frank M, Hein JR, Halliday A (2004) Cenozoic marine geochemistry of thallium deduced from isotopic studies of ferromanganese crusts and pelagic sediments. Earth Planet Sci Lett 219:77–91

Rehkämper M, Wombacher F, Horner TJ, Xue Z (2012) Natural and Anthropogenic Cd Isotope Variations. *In:* Baskaran M (Ed.) Handbook of Environmental Isotope Geochemistry: Vol I Springer Berlin Heidelberg, Berlin, Heidelberg, pp. 125–154

Renne PR (2000) Ar-40/Ar-39 age of plagioclase from Acapulco meteorite and the problem of systematic errors in cosmochronology. Earth Planet Sci Lett 175:13–26

Ridley WI, Stetson SJ (2006) A review of isotopic composition as an indicator of the natural and anthropogenic behavior of mercury. Appl GeoChem 21:1889–1899

Rosman KJR, Chisholm W, Boutron CF, Candelone JP, Patterson CC (1994) Anthropogenic lead isotopes in Antarctica. Geophys Res Lett 21:2669–2672

Rudge JF, Reynolds BC, Bourdon B (2009) The double spike toolbox. Chem Geol 265:420–431

Rudnick RL, Gao S (2003) Composition of the continental crust. *In:* Holland HD, Turekian KK (Eds.) Treatise on Geochemistry. Pergamon, Oxford, pp. 1–64

Salters VJM, Stracke A (2004) Composition of the depleted mantle. Geochem Geophys Geosyst 5, Q05004

Sawlan JJ, Murray JW (1983) Trace-metal remobilisation in the interstitial waters of red clay and hemipelagic marine sediments. Earth Planet Sci Lett 64:213–230

Schauble EA (2007) Role of nuclear volume in driving equilibrium stable isotope fractionation of mercury, thallium, and other very heavy elements. Geochim Cosmochim Acta 71:2170–2189

Schedlbauer OF, Heumann KG (2000) Biomethylation of thallium by bacteria and first determination of biogenic dimethylthallium in the ocean. Appl Organometal Chem 14:330–340

Schönbächler M, Carlson RW, Horan MF, Mock TD, Hauri EH (2008) Silver isotope variations in chondrites: Volatile depletion and the initial Pd-107 abundance of the solar system. Geochim Cosmochim Acta 72:5330–5341

Segl M, Mangini A, Beer J, Bonani G, Suter M, Wolfli W (1989) Growth rate variations of manganese nodules and crusts induced by paleoceanographic events. Paleoceanography 4:511–530

Segl M, Mangini A, Bonani G, Hofmann HJ, Nessi M, Suter M, Wolfli W, Friedrich G, Pluger WL, Wiechowski A, Beer J (1984) Be-10-dating of a manganese crust from Central North Pacific and implications for ocean palaeocirculation. Nature 309:540–543

Settle DM, Patterson CC (1982) Magnitudes and sources of precipitation and dry deposition fluxes of industrial and natural leads to the North Pacific at Enewetak. J Geophys Res-Oceans and Atmospheres 87:8857–8869

Shannon RD (1976) Revised effective ionic radii and systematic studies of interatomic distances in halides and chalcogenides. Acta Crystallogr A32, 751–767

Shaw DM (1952) The geochemistry of thallium. Geochim Cosmochim Acta 2:118–154

Shiel AE, Barling J, Orians KJ, Weis D (2009) Matrix effects on the multi-collector inductively coupled plasma mass spectrometric analysis of high-precision cadmium and zinc isotope ratios. Anal Chimica Acta 633:29–37

Shukolyukov A, Lugmair GW (2006) Manganese–chromium isotope systematics of carbonaceous chondrites. Earth Planet Sci Lett 250 200–213

Sprung P, Scherer EE, Upadhyay D, Leya I, Mezger K (2010) Non-nucleosynthetic heterogeneity in non-radiogenic stable Hf isotopes: Implications for early solar system chronology. Earth Planet Sci Lett 295:1–11

Sugiura N, Hoshino H (2003) Mn–Cr chronology of five IIIAB iron meteorites. Meteorit Planet Sci 38:117–143

Teagle DAH, Alt JC, Bach W, Halliday AN, Erzinger J (1996) Alteration of upper ocean crust in a ridge-flank hydrothermal upflow zone: Mineral, chemical, and isotopic constraints from hole 896A Proc ODP Sci Res 148:119–150

Theis KJ, Schönbächler M, Benedix GK, Rehkämper M, Andreasen R, Davies C (2013) Palladium–silver chronology of IAB iron meteorites. Earth Planet Sci Lett 361:402–411

Thirlwall MF, Gee MAM, Taylor RN, Murton BJ (2004) Mantle components in Iceland and adjacent ridges investigated using double-spike Pb isotope ratios. Geochim Cosmochim Acta 68:361–386

Trieloff M, Jessberger EK, Herrwerth I, Hopp J, Fieni C, Ghelis M, Bourot-Denise M, Pellas P (2003) Structure and thermal history of the H-chondrite parent asteroid revealed by thermochronometry. Nature 422:502–506

Tsuchiyama A, Kawamura K, Nakao T, Uyeda C (1994) Isotopic effects on diffusion in MgO melt simulated by the molecular-dynamics (MD) method and implications for isotopic mass fractionation in magmatic systems. Geochim Cosmochim Acta 58:3013–3021

Turner A, Cabon A, Glegg GA, Fisher AS (2010) Sediment–water interactions of thallium under simulated estuarine conditions. Geochim Cosmochim Acta 74:6779–6787

Urey HC (1947) The thermodynamic properties of isotopic substances. J Chem Soc, 562–581

van de Flierdt T, Frank M, Halliday AN, Hein JR, Hattendorf B, Gunther D, Kubik PW (2004) Tracing the history of submarine hydrothermal inputs and the significance of hydrothermal hafnium for the seawater budget—a combined Pb–Hf–Nd isotope approach. Earth Planet Sci Lett 222:259–273

Vogel N, Renne PR (2008) Ar-40-Ar-39 dating of plagioclase grain size separates from silicate inclusions in IAB iron meteorites and implications for the thermochronological evolution of the IAB parent body. Geochim Cosmochim Acta 72:1231–1255

Walker RJ (2012) Evidence for homogeneous distribution of osmium in the protosolar nebula. Earth Planet Sci Lett 351:36–44

Wallmann K (2001) Controls on the Cretaceous and Cenozoic evolution of seawater composition, atmospheric CO_2 and climate. Geochim Cosmochim Acta 65:3005–3025

Wasserburg GJ, Busso M, Gallino R, Raiteri CM (1994) Asymptotic giant branch stars as a source of short-lived radioactive nuclei in the solar nebula. AstroPhys J 424:412–428

Wasserburg GJ, Busso M, Gallino R, Nollett KM (2006) short-lived nuclei in the early solar system: possible AGB sources. Nucl Phys A 777:5–69

Wedepohl KH (1974) Handbook of Geochemistry. Springer

Wedepohl KH (1995) The composition of the continental crust. Geochim Cosmochim Acta 59:1217–1232

Williams H, Turner S, Kelley S, Harris N (2001) Age and composition of dikes in Southern Tibet: New constraints on the timing of east-west extension and its relationship to postcollisional volcanism. Geology 29:339–342

Williams HM, McCammon CA, Peslier AH, Halliday AN, Teutsch N, Levasseur S, Burg JP (2004) Iron isotope fractionation and the oxygen fugacity of the mantle. Science 304:1656–1659

Williams HM, Nielsen SG, Renac C, Griffin WL, O'Reilly SY, McCammon C, Pearson N (2009) Fractionation of oxygen and iron isotopes in the mantle: implications for crustal recycling and the source regions of oceanic basalts. Earth Planet Sci Lett 283:156–166

Wombacher F, Rehkämper M, Mezger K, Münker C (2003) Stable isotope compositions of cadmium in geological materials and meteorites determined by multiple-collector ICPMS Geochim Cosmochim Acta 67:4639–4654

Wombacher F, Rehkämper M, Mezger K, Bischoff A, Münker C (2008) Cadmium stable isotope cosmochemistry. Geochim Cosmochim Acta 72:646–667

Wood BJ, Nielsen SG, Rehkämper M, Halliday AN (2008) The effects of core formation on the Pb- and Tl- isotopic composition of the silicate Earth. Earth Planet Sci Lett 269:325–335

Woodland SJ, Rehkämper M, Halliday A, Lee D-C, Hattendorf B, Günther D (2005) Accurate measurement of silver isotope composition in geological materials including low Pd/Ag meteorites. Geochim Cosmochim Acta 69:2153–2163

Xiao TF, Boyle D, Guha J, Rouleau A, Hong YT, Zheng BS (2003) Groundwater-related thallium transfer processes and their impacts on the ecosystem: southwest Guizhou Province, China. Appl Geochem 18:675–691

Xiao TF, Guha J, Boyle D, Liu CQ, Chen JG (2004) Environmental concerns related to high thallium levels in soils and thallium uptake by plants in southwest Guizhou, China. Sci Total Environ 318:223–244

Xiao TF, Guha J, Liu CQ, Zheng BS, Wilson G, Ning ZP, He LB (2007) Potential health risk in areas of high natural concentrations of thallium and importance of urine screening. Appl Geochem 22:919–929

Xiong YL (2007) Hydrothermal thallium mineralization up to 300 degrees C: A thermodynamic approach. Ore Geol Rev 32:291–313

Yokoi K, Takahashi K, Arnould M (1985) The production and survival of Pb-205 in stars, and the Pb-205- Tl-205 S-process chronometry. Astron AstroPhys 145:339–346

Yokoyama T, Rai VK, Alexander MO, Lewis RS, Carlson RW, Shirey SB, ThiernenS MH, Walker RJ (2007) Osmium isotope evidence for uniform distribution of s- and r-process components in the early solar system. Earth Planet Sci Lett 259:567–580

Reviews in Mineralogy & Geochemistry
Vol. 82 pp. 799-850, 2017

Uranium Isotope Fractionation

Morten B. Andersen[1,2]

[1]ETH-Zürich
Department of Earth Sciences
Institute of Geochemistry and Petrology
Claussiusstrasse 25
8092 Zürich
Switzerland
[2]Cardiff University
School of Earth and Ocean Sciences
Main Place
Cardiff CF10 3AT
United Kingdom

Andersenm1@cardiff.ac.uk

Claudine H. Stirling

University of Otago
Department of Chemistry
PO Box 56
Dunedin 9016
New Zealand

cstirling@chemistry.otago.ac.nz

Stefan Weyer

University of Hannover
Institute of Mineralogy
Calinstrasse 3
D–30167 Hannover
Germany

s.weyer@mineralogie.uni-hannover.de

INTRODUCTION

This review focuses on the rapidly growing field of natural $^{238}U/^{235}U$ variability, largely driven by the technical advances in the measurement of U isotope ratios by mass spectrometry with increasing precision over the last decade. A thorough review on the application of the U-decay series systems within Earth sciences was published in *Reviews in Mineralogy and Geochemistry* (RiMG) volume 52 in 2003, and will not be discussed further within this review. Instead, this article will first focus on the basic chemical properties of U and the evolution of $^{238}U/^{235}U$ measurement techniques, before discussing the latest findings and use of this isotopic system to address questions within geochronology, cosmochemistry and Earth sciences.

Uranium occurrence and properties

Uranium constitutes one of the principal long-lived radioactive elements that was formed over the lifetime of the galaxy, then injected into the solar system and Earth when

1529-6466/17/0082-0019$10.00 http://dx.doi.org/10.2138/rmg.2017.82.19

they formed more than 4.5 billion years ago (Ga; Dicke 1969). The discovery of the three naturally occurring radioactive decay chains of U and Th occurred around the start of the twentieth century (Becquerel 1896). The heat production from U decay, together with the decay of Th and K, provides the major radioactive heat source on Earth (e.g., Jaupart and Mareschal 2010). The ultimate decay of U to stable isotopes of Pb also forms the basis of one of the most important geochronometers for dating the Earth and solar system, namely the U–Pb or Pb–Pb dating systems (e.g., Patterson et al. 1955).

In nature, U commonly occurs in two oxidation states, U^{+4} and U^{+6} (e.g., Langmuir 1978). Intermediate U^{+5} also occurs naturally, but is generally assumed to be unstable through disproportionation and therefore it is short-lived and uncommon in nature (e.g., Grenthe et al. 1992). Chemical species of U^{+4} are generally insoluble and form the dominant U oxidation state in the mantle (e.g., Wood et al. 1999). The estimated abundance of U in the primitive mantle is ~22 ppb with a chondritic estimate of about ~8 ppb (Palme and O'Neill 2003). Because U is refractory and lithophile, differentiation processes on Earth have resulted in the accumulation of U in the continental crust (e.g., Taylor and McLennan 1985). With an average U concentration of ~1.3 ppm, the continental crust contains ~30% of the U on Earth (Rudnick and Gao 2003). Uranium accumulation from magmatic and metamorphic differentiation processes within the continental crust has occurred over the last three billion years (e.g., Cuney 2010).

In the Archean (>2.4 Ga), oxygen-lean atmospheric conditions allowed the accumulation of fluvial-transported detrital U^{+4}, such as in the Witwatersrand Basin placer gold–pyrite–uraninite deposit, formed at ~3 Ga (e.g., Holland 2005). However, under oxidized conditions U occurs as U^{+6} and, in contrast to U^{+4}, is highly soluble and mobile, mainly as $(UO)^{2+}$ carbonate complex forms (Langmuir 1978). The "Great Oxidation Event" (GOE) at ~2.4 Ga, when appreciable amounts of oxygen in the Earth's atmosphere first appeared (Farquhar et al. 2000), therefore resulted in a significant shift in the cycling of U and, as a consequence, no detrital U placer deposits formed thereafter (e.g., Holland 2005). Instead, following the GOE, weathering in an oxygenated surface environment allowed significant hydrological transport of soluble U^{+6} (e.g., Partin et al. 2013a). The majority of this soluble U^{+6} in the hydrological cycle is ultimately transported to the oceans via rivers (e.g., Dunk et al. 2002). In the modern open ocean, U mainly occurs as uranyl carbonate complexes and has a relatively long residence time on the order of a few hundred thousand years (k.y.; e.g., Ku et al. 1977; Henderson and Anderson 2003). The main U removal from the oceans occurs through accumulation in reducing sediments and carbonate deposits (e.g., Dunk et al. 2002; Henderson and Anderson 2003).

Uranium isotopes

Natural U consists of primordial-formed ^{238}U and ^{235}U, which are the two parent isotopes of the ^{238}U- and ^{235}U-series chains. These ultimately decay to stable ^{206}Pb and ^{207}Pb with half-lives of ~4.5 and ~0.7 billion years (b.y.), respectively (see Table 1). A minute fraction of U decays through spontaneous fission, which is negligible compared to the proportion that decays via α-emission, but forms the basis of fission track dating (e.g., Fleischer et al. 1975). A small additional contribution to the abundance of ^{235}U may come from the decay of extinct ^{247}Cm (half-life ~16 million years, m.y., Table 1) in the early solar system. In the ^{238}U-series decay chain, the ^{234}U nuclide (half-life 246 k.y., Table 1) is continuously formed and is the longest-lived daughter of all of the intermediate nuclides of the U-series decay chains. There is a wealth of literature on the ^{234}U–^{238}U system since Chedyntsev et al. (1955) reported the discovery of disequilibrium between these two nuclides from physical α-recoil processes (e.g., see RiMG volume 52). Minute abundances of ^{236}U can also occur naturally ($^{236}U/^{238}U < 10^{-9}$) on Earth from neutron-capture processes within U ores (e.g., Murphy et al. 2015).

Table 1. Curium and uranium isotopes with significantly long half-lives

Isotope	Half-life	Decay-constant (λ)	References
^{247}Cm	15.9 m.y.	4.28×10^{-8}	Tuli (2005)
^{238}U	4.468 b.y.	1.55125×10^{-10}	Jaffey et al. (1971); Villa et al. (2015)
^{236}U	23 m.y.	2.96×10^{-8}	Tuli (2005)
^{235}U	703.7 m.y.	9.85×10^{-10}	Jaffey et al. (1971); Villa et al. (2015)
^{234}U	246 k.y.	2.83×10^{-6}	Cheng et al. (2013); Villa et al. (2015)
^{233}U	159 k.y.	4.35×10^{-6}	Tuli (2005)

The different half-lives of ^{238}U and ^{235}U means that the absolute ^{238}U/^{235}U ratio has changed from an average ratio of ~3 to ~138, over the ~4.56 b.y. timespan of the solar system. In addition, Kuroda (1956) predicted that ^{235}U has the capacity to sustain a natural fissile chain reaction, provided the following three key conditions are met: (i) ^{235}U exceeds ~3% of the total U, (ii) a "critical" U assemblage is present (*"a thickness of a few feet"* Kuroda 1956), and (iii) there is a thermal neutron flux. In 1972 (Bodu et al. 1972; Neuilly et al. 1972) and in subsequent years (IAEA 1975), assemblages of ~2 Ga sandstone-hosted U ores at Oklo and Bangombè in the Republic of Gabon, West Africa, were found to have unusually high ^{238}U/^{235}U values and also revealed a range of ^{235}U fissile elemental products, confirming the existence of fossil natural chain reactors on Earth (e.g., De Laeter et al. 1980; Gauthier-Lafaye et al. 1996). These localities constitute the oldest major redox-controlled sandstone-hosted U ore deposits (e.g., Cuney 2010) suggesting the abovementioned conditions for U accumulation to reach a "critical" stage, may first have been attained on Earth after the GOE at ~2.4 Ga. However, as natural radioactive decay diminished the ^{235}U abundance to <3% of the total U by ~1.7 Ga, the conditions for natural fission chain reactions to occur have not been met during the last billion years of Earth's history (Kuroda 1982). This likely explains the limited occurrence of natural fission chain reactors discovered so far and, apart from these specific cases at Oklo and Bangombè, natural ^{238}U/^{235}U variation in U ores appeared limited and less than 0.1% around the time of discovery (Cowan and Adler 1976). These findings led to the recommendation of a canonical ^{238}U/^{235}U value of 137.88 (Steiger and Jäger 1977) to be used indiscriminately for the entire solar system, including Earth, for the purpose of geochronology (e.g., U–Pb, Pb–Pb, U-series) and this invariant value was used for the next three decades. However, in the last decade, this long-held assumption has been partially invalidated by the discovery of measurable ^{238}U/^{235}U variability in both terrestrial and extraterrestrial materials (Fig. 1).

On Earth, these observed natural ^{238}U/^{235}U variations stem from non-radioactive physico-chemical processes that result in U isotope exchange, but with no net change in the total U abundance, commonly described as "stable isotope fractionation". The term 'stable", however, is problematic for U given its radioactive properties, and, hence, we recommend the term "U isotope fractionation" for this non-radiogenic process. As ^{238}U/^{235}U variations caused by U isotope fractionation are independent of radioactive decay processes, the relative shifts in ^{238}U/^{235}U incurred from U isotope fractionation will be faithfully recorded in the geological record today, irrespective of when they occurred in the past. The U isotope variability caused by these processes are the main focus of this chapter.

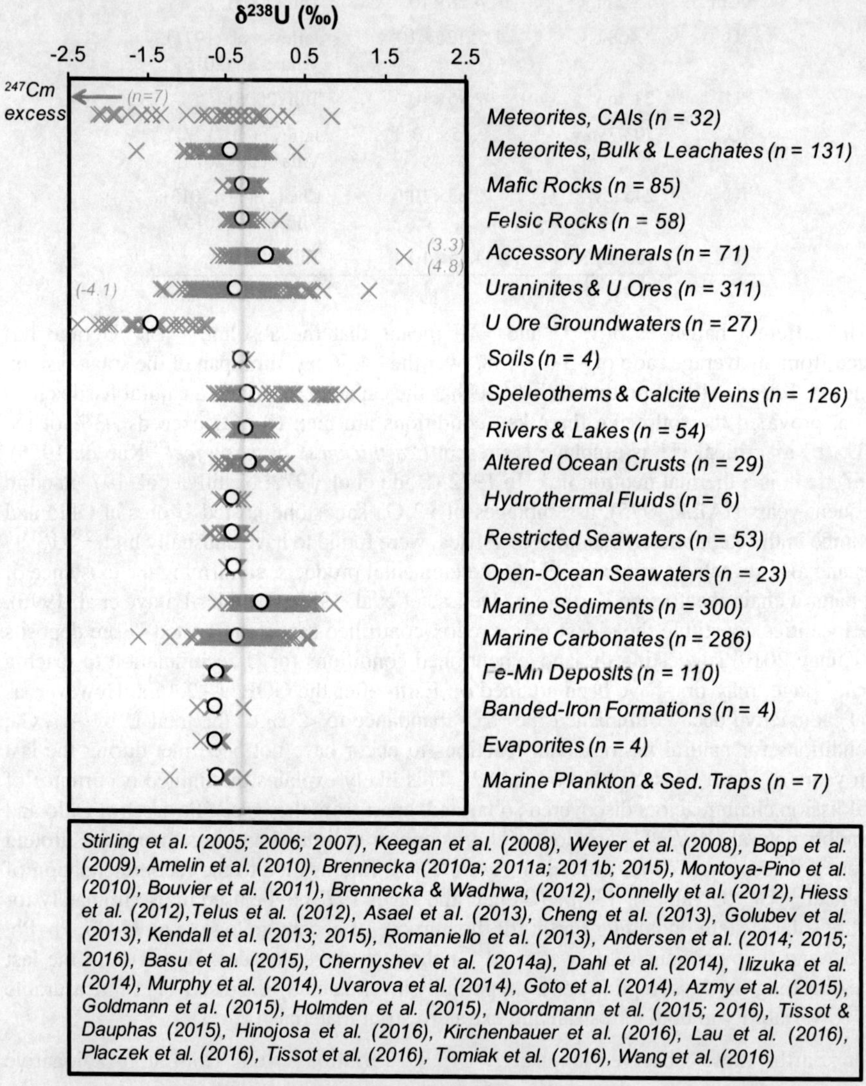

Figure 1. Natural $^{238}U/^{235}U$ compositions (plotted as $\delta^{238}U$ as defined in Eqn. (1)) separated into different reservoirs based on data published from 2005 to May 2016. The 'Bulk Earth' average is marked with a grey vertical line for comparison. Individual data points are shown as grey crosses and the mean values for each of the reservoirs are the plotted as circles. Note that these average values are unweighted and may not accurately represent the best estimated mean of each reservoir. Some of the meteoritic data for calcium-aluminum-rich inclusions (CAIs) may have elevated ^{235}U (low $\delta^{238}U$) from the decay of ^{247}Cm (see Fig. 4). The U ore data from Uvarova et al. (2014) has been renormalized using a CRM-129a–CRM-112a offset of −1.7‰, instead of −0.9‰. Also, U ore groundwaters from IRS mining or test sites (Bopp et al. 2010; Basu et al. 2015; Brown et al. 2016; Placzek et al. 2016; Shiel et al. 2016) are not plotted as these have been effected by manipulation or mining activity.

URANIUM ISOTOPE DETERMINATIONS

Historical overview of $^{238}U/^{235}U$ measurements

A thorough review of measurement techniques and chemical purification methods for U was provided in Goldstein and Stirling (2003), with ^{238}U–^{234}U–^{230}Th–^{232}Th measurements in mind. Here, we focus on analytical developments that have occurred with emphasis on the measurement of $^{238}U/^{235}U$. The first direct mass spectrometric determination of these U isotopes was conducted in the 1930's. Using volatile UF_6, Aston (1931) detected the mass-spectra of the major ^{238}U isotope concluding its abundance to be >97% of total U. Using a spark-source and U electrode technique, Dempster (1935) detected ^{238}U and ^{235}U, concluding that the latter would comprise less than one percent of total U. The first robust estimates of the $^{238}U/^{235}U$ ratio were conducted by Nier (1939) using thermal ionization mass spectrometry (TIMS) with a source of either UCl_4 or UBr_4. Nier (1939) not only estimated the $^{238}U/^{235}U$ of a range of U ores to be 138.9 ± 1.4, but also provided the first direct measurements of the minor ^{234}U isotope. During the following decades, the most precise determinations of $^{238}U/^{235}U$ were performed using gas-source mass spectrometry and UF_6 on U ore samples (e.g., Hamer and Robbins 1960) with precisions down to the ±0.2‰ level (Cowan and Adler 1976) an order of magnitude more precise than those determined previously by thermal ionization mass spectrometry (e.g., Lounsbury 1956). Advances in $^{238}U/^{235}U$ measurement were reported by Chen and Wasserburg (1981b) using the "Lunatic I" TIMS instrument at the California Institute of Technology (Caltech), USA, for applications within cosmochemistry. This method allowed $^{238}U/^{235}U$ measurements to be conducted at the permil level, but using small sample sizes down to sub-ng quantities, orders of magnitude lower than the sample size requirements of gas-source measurements. A major technical advancement in this study was the use of a ^{236}U–^{233}U double spike for reliable internal correction of instrumental mass discrimination of the U isotopes during sample measurement, provided the impurities of natural ^{235}U and ^{238}U in the spike were reliably corrected for (see Chen and Wasserburg 1981b). Further mass spectrometric improvements, including the simultaneous collection capabilities of both multiple-collector thermal ionization (MC-TIMS) and inductively coupled plasma mass spectrometry (MC-ICPMS) allowed the analytical precision of $^{238}U/^{235}U$ measurements, to improve by more than an order of magnitude (e.g., Stirling et al. 2005; Richter et al. 2008).

In particular, the use of ^{236}U–^{233}U double spikes and the high U ionization in the plasma source of MC-ICPMS allowed measurements of $^{238}U/^{235}U$ at the <±0.1‰ level using <100 ng U sample sizes (Stirling et al. 2005, 2006, 2007; Rademacher et al. 2006; Weyer et al. 2008; Bopp et al. 2009; Brennacka et al. 2010a). The prospect for measuring high-precision $^{238}U/^{235}U$ ratios with high accuracy greatly increased with the introduction of the commercially available and gravimetrically calibrated ^{236}U–^{233}U double-spike, IRMM-3636, prepared at the Institute of Reference Materials and Measurements (IRMM), Belgium (Richter et al. 2008, 2010). A range of laboratories now routinely measure $^{238}U/^{235}U$ to the <±0.1‰ level (Fig. 1). Several studies also invoke the simultaneous measurement of the minor ^{234}U isotope, either using Faraday collectors or electron multipliers coupled to ion counting devises (e.g., Hiess et al. 2012; Cheng et al. 2013; Romaniello et al. 2013; Andersen et al. 2014; Tissot and Dauphas 2015). Further measurement improvements in analytical precision down to the <±0.03‰ level for $^{238}U/^{235}U$ have recently been performed, taking advantage of mixed 10^{10} and 10^{11} Ω resistors on Faraday collectors, in place of the standard array of 10^{11} Ω resistors, and MC-ICPMS. This approach allows ^{235}U to be measured at higher intensities, providing better signal to baseline ratios and analytical precision (Andersen et al. 2015). In contrast, for applications focused on samples with very low U concentrations within cosmochemistry, the search for significant (permil to percent level) $^{238}U/^{235}U$ variability has led to the use of 10^{12} Ω resistors (Brennecka et al. 2010b) or electron multipliers coupled to ion counting devises (Snow and Friedrich 2005; Tissot et al. 2016) for lower-precision (permil level), but accurate measurements.

Chemical preparation of U and mass spectrometric corrections

Due to the high mass range of U, molecular interferences, generated in the plasma source of MC-ICPMS instruments, are generally significantly smaller than in the lower mass range (e.g., Wieser and Schwieters 2005). However, these exist, for instance, as $PtAr^+$ and $^{232}ThH^+$, necessitating purification of U from the sample matrix for high precision analysis. Typically, ion exchange resins such as Dowex AG-1x8 have been used for decades to separate U from matrix elements (e.g., Chen and Wasserburg 1981b). Chromatographic resins developed specifically for actinides, such as the TRU (Horwitz et al. 1993) and UTEVA (Horwitz et al. 1992) resins manufactured by Eichrom Technologies (USA) are now widely used for U purification for $^{238}U/^{235}U$ analysis (e.g., Stirling et al. 2005; Weyer et al. 2008; Tissot and Dauphas 2015). Various chromatographic protocols with these resins are generally deemed to provide the necessary purification levels to avoid spectral interferences spanning the U isotope range while analysing in low mass resolution ($\Delta M/M$ ~500) by MC-ICPMS (e.g., Stirling et al. 2005; Weyer et al. 2008; Hiess et al. 2012; Tissot and Dauphas 2015), although some studies have opted to measure in higher resolution modes to avoid potential unidentified polyatomic interferences across the U mass range (Connelly et al. 2012). Both UH^+ and UO^+ may form during mass spectrometric analysis. Using MC-ICPMS, the rate of production of these interferences is highly variable and dependent on operational conditions and the type of sample introduction system. The formation of UO^+ may potentially bias the measured $^{238}U/^{235}U$ ratio as secondary mass instrumental fractionation of UO^+ may not follow the exponential mass fractionation law (Russell et al. 1978) typically used for correcting instrumental mass fractionation. Such an effect has been observed during Nd isotope measurement due to variable generation of NdO^+/Nd^+ ratios (Newman 2012). Also, different mean values have been reported for mass-bias corrected $^{238}U/^{235}U$ ratios of the same U isotope standard, presumably related to the symmetry of the ion beam trajectory into the Faraday cup array (Tissot and Dauphas 2015). Furthermore, tailing of larger U isotopes (e.g., ^{238}U) below minor ion beams (e.g., ^{236}U) may need to be corrected for to obtain high accuracy $^{238}U/^{235}U$ measurements (e.g., Hiess et al. 2012; Cheng et al. 2013). However, the above issues may only be critical for obtaining "absolute" $^{238}U/^{235}U$ ratios, as uncertainties from these effects may be negated if 'unknowns' are compared directly against a standard measured in exactly the same way, using the same spike to sample ratios, to establish "relative" $^{238}U/^{235}U$ variability with respect to the standard.

Anthropogenic U contamination

A concern for measuring natural U isotope variability is potential "anthropogenic" U contamination with isotopically anomalous U. This issue may arise both from laboratory use of commercially available U concentration standards made from depleted uranium and pollution of natural samples from the nuclear industry and weapons. For nuclear fission fuel, ^{235}U is enriched whereas the U residual product, depleted in ^{235}U, may be used in penetrating weapons (e.g., Oliver et al. 2007). The analysis of minute amounts of ^{236}U in the presence of natural U is challenging, but modern accelerator mass spectrometry provides enough sensitivity and background suppression to be able to determine $^{236}U/^{238}U$ atom ratios at the 10^{-12} level (e.g., Steier et al. 2008; Christl et al. 2015). Anthropogenic U is widespread across the globe, as evidenced by ^{236}U above natural backgrounds in some oceanic water masses and this may provide information about the timescales of oceanic mixing and transport processes (e.g., Christl et al. 2012). Such anthropogenic U addition is likely to carry anomalous ^{238}U and ^{235}U as well, although, the very low level of anthropogenic U in seawater does not appreciably effect the estimates for the natural U component in general. However, it may have a more overwhelming effect in more localized areas, such as in sediments near nuclear plants or in soils from war zones and test sites that have been exposed to depleted U (e.g., Oliver et al. 2007; Lloyd et al. 2009). While a spent fuel rod has $^{236}U/^{238}U$ ratios on the order of a few permil (Steier et al. 2008), naturally occurring U ores exhibit ratios below 10^{-9} (e.g., Murphy et al. 2015). Thus, a reliable indicator for anthropogenic U contamination may therefore be the presence of appreciable ^{236}U in natural samples above this threshold.

Table 2. $^{238}U/^{235}U$ measurements of a range of pure U standards

Study	Technique*	CRM-112a** ($^{238}U/^{235}U$)	SRM-950a δ***	IRMM-184 δ***	REIMP–18a δ***
Certified		*137.849 ± 0.079*			
Condon et al. (2010)	T	137.844 ± 0.011	+0.02 ± 0.08	−1.18 ± 0.08	
Richter et al. (2010)	T	137.837 ± 0.023		−1.12 ± 0.16	
Hiess et al. (2012)	T	137.832 ± 0.022			
Hiess et al. (2012)	M	137.829 ± 0.022			
Cheng et al. (2013)	M	137.832 ± 0.015			
Goldmann et al. (2015)	M	137.843 ± 0.008	+0.02 ± 0.01	−1.16 ± 0.04	−0.14 ± 0.01
Tissot and Dauphas (2015)	M	137.842 ± 0.003			
Weyer et al. (2008)	M	"0"	−0.02 ± 0.06		−0.20 ± 0.06
Bopp et al. (2009)	M	"0"			−0.25 ± 0.19
Shiel et al. (2013)	M	"0"			−0.15 ± 0.09
Kendall et al. (2013)	M	"0"	−0.04 ± 0.06		
Chernyshev et al. (2014)	M	"0"		−1.19 ± 0.09	
Noordmann et al. (2016)	M	"0"	+0.04 ± 0.01		
Wang et al. (2016)	M	"0"			−0.17 ± 0.09
Shiel et al. (2016)	M	"0"			−0.14 ± 0.08

* T = TIMS, M = MC-ICPMS

** CRM-112a (or NBL-112a, SRM-960) and CRM-145, the solution form prepared from same material, should be of the same isotopic composition.

***δ numbers are the $^{238}U/^{235}U$ difference of this standard compared to the CRM-112a reported in parts-per-thousand (‰).

$^{238}U/^{235}U$ nomenclature

There is currently no accepted conventional way to report variability between ^{238}U and ^{235}U, and either absolute ratios or nomenclature traditionally used for stable (δ) and radiogenic (ε) isotope systematics have been used (e.g., Cowan and Adler 1976; Chen and Wasserburg 1981b; Stirling et al. 2005; Weyer et al. 2008; Brennecka et al. 2010b; Cheng et al. 2013: Tissot et al. 2016). Absolute ratios are important for geochronology purposes and these are often reported as $^{238}U/^{235}U$ (e.g., Hiess et al. 2012). However, for studies focusing on $^{238}U/^{235}U$ variability in the context of isotope fractionation processes, the direct comparison of measured unknowns to a standard (e.g., δ-notation), eliminates uncertainties arising from the determination of "absolute" ratios, as well as changes in the $^{238}U/^{235}U$ ratio over time through radioactive decay (see *Uranium Isotopes*).

In a similar manner as discussed above for nomenclature, there are no accepted standards to report relative $^{238}U/^{235}U$ ratios. In recent studies, several isotopically pure U standards from the National Institute of Standards and Technology (NIST) and IRMM have been used, including CRM-112a/SRM-960, CRM-129a, CRM-145, SRM-950a, IRMM-184 and REIMP-18a (e.g., Stirling et al. 2007; Weyer et al. 2008; Bopp et al. 2009; Condon et al. 2010; Richter et al. 2010; Brennecka et al. 2011a). Direct estimates of the absolute $^{238}U/^{235}U$ of CRM-112a (Table 2) as well as intercomparison of $^{238}U/^{235}U$ between standards have been reported. Of these, the reported compositions of the CRM-112a, SRM-950a, IRM-184 and REIMP-18a standards compare well between laboratories (Table 2). In contrast, standard CRM-129a has been measured in several studies but variable offsets (0.9 to 1.8‰ lower) compared to the compositions of CRM-112a or SRM-950a (Brennecka et al. 2011a; Kendall et al. 2013; Uvarova et al. 2014; Azmy et al. 2015; Wang et al. 2015b; Lau et al. 2016; Shiel et al. 2016). Different $^{238}U/^{235}U$ mean values (1.5‰ vs. 1.7‰ lower than CRM-112a) for two different splits of this standard were measured by Lau et al. (2016) showing that the available CRM-129a solutions are heterogeneous and, thus, not ideal for use as a primary reference standard.

A large proportion of studies have used the CRM-112a (or CRM-145, which is a solution made of CRM-112 metal) reference standard, which is widely available and also commonly used for $^{238}U/^{234}U$ measurements, reported herein as the activity ratio ($^{234}U/^{238}U$) (e.g., Cheng et al. 2013). We propose that studies should strive to use this standard as a primary normalization standard and, if reporting absolute $^{238}U/^{235}U$ ratios, this standard should be measured and reported to allow for inter-laboratory comparison. In terms of nomenclature, the $\delta^{238}U$ notation (Weyer et al. 2008) has been the most widely used to date. Thus, we suggest reporting $^{238}U/^{235}U$ in δ-notation with respect to the composition of the CRM-112a (or CRM-145) standard as a standardized way to report relative $^{238}U/^{235}U$ variability as follows:

$$\delta^{238}U = \left[\left({}^{238}U/{}^{235}U_{sample}\right) / \left({}^{238}U/{}^{235}U_{CRM\text{-}112a}\right) - 1\right] \times 1000 \quad (1)$$

EXPERIMENTAL EVIDENCE FOR URANIUM ISOTOPE FRACTIONATION PROCESSES

Experimental studies for nuclear ^{235}U enrichment

Following the second world war, attempts to establish equilibrium isotope fractionation factors for U were driven by testing methods for fast redox driven U separation and ^{235}U enrichment for nuclear usage, such as the development of the Japanese Asahi and the French Chemex ion exchange separation processes (e.g., Mioduski 1999 and references therein). These chromatographic uranium isotope separation experiments yielded unexpected findings of both larger U isotope separation factors and a preference for heavier U isotopes in compounds of the lowest oxidation state (e.g., Shimokaua and Kobayashi 1970; Florence et al. 1975; Fukuda et al. 1983). These observations were in complete contrast to the expected magnitude and direction of U isotope shifts from "traditional" mass-dependent isotope fractionation driven by vibrational zero-point energy differences (Bigeleisen and Mayer 1947). Additional work revealed further unanticipated results, most notably, that fractionation of the uneven ^{235}U isotope did not follow the mass-dependent trend predicted by the fractionation of the even ^{234}U and ^{236}U isotopes, with respect to ^{238}U (Fujii et al. 1989a,b; Nomura et al. 1996).

Uranium isotopes and the nuclear field shift

The above findings led Bigeleisen (1996a) to expand the theory of equilibrium isotope fractionation to also include the effects of the nuclear field shift in addition to mass-dependent isotope fractionation based on zero-point vibrational energy differences. The effect of the nuclear field shift on isotope fractionation arises from changes in the nuclear charge density and is particularly important for heavier elements. For U, the nuclear field shift effect involves electron exchange in the s configuration, which are the electrons with significant probability of being near the nucleus, or shielding *f–p* configurations (e.g., King 1984 and references therein). The nuclear field shift effect on isotope fractionation is therefore governed by how variable neutron numbers impact the nuclear binding energy and does not scale evenly with mass for U, but has an "odd-even staggering effect" (King 1984). For species in their electronic ground states (as is usually the case at equilibrium), the isotope fractionation from the nuclear field shift scales with 1/temperature (T) as opposed to $1/T^2$ for mass-dependent isotope fractionation (Bigeleisen 1996a,b). In particular, for the reduction of U^{+6} to U^{+4}, the f electrons are generally exchanged, favouring the heavier U isotopes in the lower oxidation state, which competes against, and often overwhelms, the mass-dependent effect (Bigeleisen 1996a; Fujii et al. 2006; Schauble 2007). This theory could successfully explain the observed U isotope fractionation patterns in previous experimental studies (Bigeleisen 1996a,b; Nomura et al. 1996). Further experimental work (Ismail et al. 2002; Fujii et al. 2006) and *ab initio* modeling (Abe et al. 2008a,b, 2010, 2014; Nemoto et al. 2015) of U, as well as Hg and Tl isotope fractionation behaviour (Schauble 2007), explored the specific combined effects of both mass-dependent and nuclear field shift equilibrium isotope fractionation (Fig. 2). Although the nuclear field shift effect is likely to dominate the U isotope fractionation signature during U reduction, the expected U isotope fractionation for both terms may vary significantly based on specific electron coordination and ligand binding (Abe et al. 2010, 2014). Consequently, it is possible that the observed net U isotope fractionation factors may, at least in principle, become reversed (e.g., Mioduski 1999) or increase over a certain positive temperature range (Ismail et al. 2002).

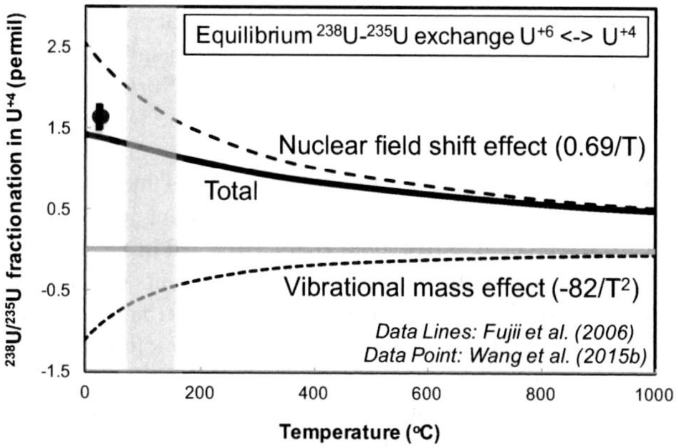

Figure 2. Expected temperature vs. $^{238}U/^{235}U$ fractionation trends based on experimental data of U isotope fractionation at equilibrium during U^{+6}–U^{+4} exchange in a chloride medium (Fujii et al. 2006; Wang et al. 2015b). The observed $^{238}U/^{235}U$ fractionation (Total) is a consequence of both the nuclear field shift and vibrational zero-point energy changes. These fractionate the U isotopes in opposing directions and have different temperature dependence, and the nuclear field shift dominates the total signal. The estimated effect on U isotope fractionation across the range of temperatures shown has been extrapolated from a relatively narrow experimental dataset with a limited temperature range (87–160 °C), which may generate large uncertainties outside of this band (Fujii et al. 2006). These data, however, agree well with experimental data from Wang et al. (2015b) obtained at 25 °C (circle).

Experimental evidence for kinetic and equilibrium $^{238}U/^{235}U$ fractionation

More recent experimental studies have examined U isotope fractionation during various biotic and abiotic processes. These have expanded upon the earlier experimental findings for equilibrium U isotope fractionation driven by the prospect of nuclear usage and moved these towards more natural conditions and applications.

Experimental studies involving no U redox exchange have shown limited U isotope fractionation. One study involving U^{+6} adsorption to birnessite (Mn-oxyhydoxide) from an oxic solution, showed ~0.2‰ lower $^{238}U/^{235}U$ for the adsorbed U, consistent with a change in U-O coordination and the direction of U isotope fractionation expected from mass-dependent isotope fractionation (Brennecka et al. 2011b). Also, at the Rifle Research Field Site, Colorado, USA, groundwaters were measured for $^{238}U/^{235}U$ after stimulated U flow-through in the aquifer and U removal using bi-carbonate induced U^{+6} adsorption (Shiel et al. 2013). Despite significant U removal (>50%), no $^{238}U/^{235}U$ change was observed in the groundwaters, suggesting little net U isotope fractionation during the adsorption process (Shiel et al. 2013). Furthermore, experimental work by Chen et al. (2016) explored potential U isotope fractionation in carbonates precipitated from aqueous solutions. No resolvable U isotope exchange was observed for U incorporation into calcite or aragonite at pH 7.5. However, aragonite precipitation at pH 8.5

results in a small (<0.13‰) but resolvable U isotope fractionation towards heavier values. This latter finding, combined with U-speciation modelling, suggests that small equilibrium U isotope fractionation may occur during U exchange between aqueous pH-dependent U species, with preferential incorporation of these U species into aragonite (Chen et al. 2016).

Experimental studies involving U redox exchange have shown a range of U isotope fractionations. A study by Wang et al. (2015a) found that $^{238}U/^{235}U$ was ~1.1‰ lower in U^{+6} compared to U^{+4} when soluble U^{+6} was oxidized from a U^{+4} solution at low pH (~0.1 M HCl), best described by an equilibrium isotope exchange process. Similarly, Wang et al. (2015b) determined an equilibrium U isotope fractionation factor of ~1.6‰ with higher $^{238}U/^{235}U$ in U^{+4} than U^{+6} during exchange in soluble form (~0.6 M HCl) at ambient temperatures (Fig. 2). A range of kinetically driven experiments using sulfate or metal-reducing bacteria to reduce and remove U from oxidized solutions (Basu et al. 2014; Stirling et al. 2015; Stylo et al. 2015a,b), have shown an enrichment of the heavier U isotope in the reduced phase (Fig. 3). Similarly, at the Rifle Research Field Site, Colorado, USA, groundwaters were measured for $^{238}U/^{235}U$ with U removal stimulated by the activation of metal-reducing bacteria in the aquifer, showing significant $^{238}U/^{235}U$ fractionation consistent with preferential removal of the heavier U isotope (Bopp et al. 2010; Shiel et al. 2016). All of these $^{238}U/^{235}U$ fractionation factors derived for kinetically controlled laboratory experiments are consistently in the ~1‰ range (Basu et al. 2014; Stirling et al. 2015; Stylo et al. 2015a,b), of similar magnitude to those determined experimentally for equilibrium U isotope fractionation (Fig. 2). In contrast, one laboratory study of kinetically driven U reduction using sulfate-reducing bacteria did not follow this trend and instead showed limited enrichment (~0.2‰) of the lighter U isotope in the solid precipitate (Rademarker et al. 2006). It has been suggested that this result is related to an experimental artifact involving adsorption of U^{+6} chloride species onto microbial cells, thereby outcompeting the U reduction process (Basu et al. 2014).

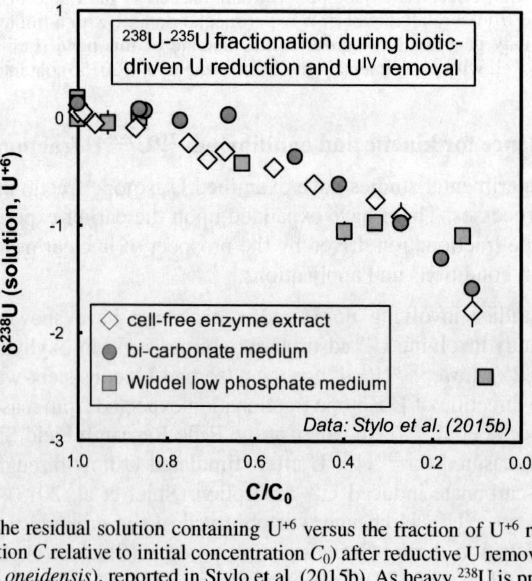

Figure 3. $\delta^{238}U$ for the residual solution containing U^{+6} versus the fraction of U^{+6} remaining in solution (measured concentration C relative to initial concentration C_0) after reductive U removal by metal reducing bacteria (*Shawanella oneidensis*), reported in Stylo et al. (2015b). As heavy ^{238}U is preferentially removed into the solid U^{IV} phase the remaining U^{+6} in the solution is driven isotopically lighter. This pattern can be approximated by a Rayleigh type distillation process and all three experiments here are consistent with a $^{238}U/^{235}U$ fractionation factor of ~0.9‰ for such a process.

Furthermore, experimental studies involving abiotic U reduction via natural reductants (e.g., Fe^{+2} based phases, Zn metal, and aqueous sulfide), have so far shown either limited $^{238}U/^{235}U$ fractionation or lighter U isotope enrichment in the reduced precipitate (Rademacher et al. 2006; Stirling et al. 2007, 2015; Stylo et al. 2015a,b). This highlights that the observed U isotope fractionation is dependent on variable reaction rates and pathways. For instance, laboratory experiments involving the oxidation of solid reduced U (uraninite) only showed limited U isotope fractionation, presumably due to fast reaction rates and limited time for co-existence of the U^{+6} and U^{+4} reservoirs to allow isotopic exchange (Wang et al. 2015a). Also, in an additional series of experiments it was shown that the timescales required for significant U isotope fractionation during U^{+4}–U^{+6} exchange at equilibrium, decreases with increasing U concentration (Wang et al. 2015b). For the biotic experiments discussed above, it has been observed that ^{238}U–^{235}U isotope fractionation occurs during enzymatic U reduction, leading to the tantalizing suggestion that the U isotope systematics may provide a 'fingerprint' of biologically driven U isotope fractionation in nature (Stylo et al. 2015b). It may be the case that enzymatic U reduction is the rate-limiting step in such biological processes and that the environment surrounding the enzyme facilitates the parallel existence of U^{+6} and U^{+4} and isotopic exchange between these two oxidation states (Basu et al. 2014; Stylo 2015b). The parallel existence of U^{+6} and U^{+4} is more unlikely in abiotic macroscopic systems (experimental and natural), as U^{+6} and U^{+4} only coexist in solution at very low pH (Wang et al. 2015a). However, further investigations are necessary to justify that these experimental findings are transferable to natural processes and geological timescales.

$^{238}U/^{235}U$ IN COSMOCHEMISTRY

Our solar system evolved from a giant molecular cloud of dust and gas that collapsed gravitationally more than 4.5 billion years ago. The elemental and isotopic compositions of meteorites, derived from asteroidal-sized objects that formed at the start of the solar system, play a crucial role in constraining its formation and subsequent evolutionary history. Uranium and its short-lived precursor nuclides were formed over the lifetime of the galaxy by rapid neutron capture or '*r*-process' nucleosynthesis, then injected into the protosolar molecular cloud before solar system formation. The relative abundance of uranium's two primordial isotopes, ^{238}U and ^{235}U, recorded by the $^{238}U/^{235}U$ isotopic signature of meteorites, thus provides critical information on the building blocks of the solar system, the exact timing and sequence of events that led to the formation of the Earth and planets from dust-sized objects in the first 100 m.y. of the solar system's history, as well as subsequent planetary differentiation and evolution. The U isotopic composition of meteorites also provides crucial constraints on the processes at work within the highly dynamic environment of the early solar system, and the circumstances leading to the birth of the Sun and planets. Sizeable $^{238}U/^{235}U$ fractionations have been measured in meteorites and their earliest formed mineral aggregate components. In the following sections, this variability and its implications will be discussed in the context of (1) the extinct short-lived nuclide Cm–U *r*-process cosmochronometer, (2) geochemical processes leading to U isotope fractionation in early solar system environments and inherited U isotope anomalies due to heterogeneity in the solar nebula, and (3) the long-lived U–Pb cosmochronometer.

Uranium isotopic anomalies and the search for extant ^{247}Cm

Uranium and the other heavy radioactive elements, collectively referred to as the actinides, can only be produced with a high neutron fluence, as is found in a supernova. Under these astrophysical conditions, the synthesis of the actinides occurs via *r*-process nucleosynthesis. The former existence of many short-lived nuclides (e.g., ^{26}Al, ^{60}Fe, ^{129}I, ^{182}Hf and ^{244}Pu) at the start of the solar system has now been demonstrated (e.g., Wasserburg et al. 1996; Dauphas and Chaussidon 2011). If some of these were produced in a supernova then the first solar system condensates should also have contained short-lived actinides. For example, the short-lived actinide ^{244}Pu, with a half-life of about 80 m.y., appears to have been detected in early solar system objects (Podosek 1970).

A critical short-lived actinide is ^{247}Cm, which has a half-life of 15.6 m.y. (Tuli 2005), considerably shorter than that of ^{244}Pu, and provides an important means of constraining the timing of the last supernova event. The detection of the former existence of ^{247}Cm in meteorites is thus pivotal in determining the presence of supernovae in the pre-solar molecular cloud and the role of a supernova in triggering the birth of our solar system. Because ^{247}Cm is now extinct, this can be achieved by high precision measurements of the abundance of ^{235}U (the daughter product of ^{247}Cm), detectable as small variations in ^{235}U/^{238}U, in early solar system objects. Specifically, evidence for the former existence of "live" ^{247}Cm in the early solar system would be provided if anomalous excesses in its ^{235}U daughter product are observed which correlate with the parent-daughter Cm/U ratio. The magnitude of U isotopic anomalies in meteorites can thus be used to elucidate the timing and character of the last *r*-process nucleo-synthetic event for input into models describing the formation and evolution of the early solar system. These observations can also provide constraints on the timescale of Galactic nucleosynthesis and the *r*-process production curve over the actinide mass range.

This approach relies on three key prerequisites. First, Cm must have been chemically fractionated from U to varying degrees when solids formed in the early nebula environment. Second, knowledge of the magnitude of Cm/U fractionation is required. Third, the magnitude of the inferred Cm/U must be shown to correlate with the extent of U isotope fractionation. The first condition is likely met, as the theoretical calculations of Boynton (1978) indicate that large Cm–U fractionations of up to ~100 are expected in the early solar system as a result of both condensation and vaporization processes due to volatility differences between the actinides, and the differential partitioning of Cm and U between minerals during the formation and thermal processing of meteorites. Ensuring the second condition is met is more problematic; reliable determination of the Cm/U ratio is exceedingly difficult because there are no long-lived isotopes of Cm and a suitable reference nuclide must instead be used as a chemical proxy. The rare earth elements (REEs), and in particular the light REEs (LREEs) Nd, Sm and Pr, are all believed to be good chemical analogues of Cm during nebular fractionation due to their near identical valence states, ionic radii and volatilities (Boynton 1978). Thorium has also been used as a chemical proxy for Cm (Chen and Wasserburg 1981a; Chen and Wasserburg 1981b). The third prerequisite may also be difficult to satisfy when processes not related to ^{247}Cm decay fractionate ^{238}U from ^{235}U, especially if such processes are also capable of generating a positive correlation between ^{235}U/^{238}U and Nd/U, or if the LREEs and/or Th are not reliable chemical proxies of Cm (Stirling et al. 2005, 2006; Tissot et al. 2016), as discussed further in *Uranium isotope fractionation unrelated to ^{247}Cm decay*.

Systematics of the Cm–U r-process chronometer. The systematics of the ^{247}Cm–^{235}U cosmochronometer, described in detail in previous studies (Stirling et al. 2005; Tissot et al. 2016), are based on the decay of ^{247}Cm to stable ^{207}Pb via ^{235}U and the decay of ^{238}U to stable ^{206}Pb, shown in (2) and (3) below. The half-lives ($t_{1/2}$) of ^{247}Cm, ^{238}U and ^{235}U are given in m.y. and their respective decay constants are given in Table 1.

$$^{247}\mathrm{Cm} \xrightarrow{t_{1/2}=16} {}^{235}\mathrm{U} \xrightarrow{t_{1/2}=704} \ldots\ldots \longrightarrow {}^{207}\mathrm{Pb} \quad (2)$$

$$^{238}\mathrm{U} \xrightarrow{t_{1/2}=4468} {}^{234}\mathrm{U} \xrightarrow{t_{1/2}=0.25} \ldots\ldots \longrightarrow {}^{206}\mathrm{Pb} \quad (3)$$

The Cm–U 'isochron' equation is given in Equation (4), assuming that Nd behaves like Cm during nebula processing and adopting ^{144}Nd as the reference element for ^{247}Cm. Similar isotope systematics can be derived using other chemical analogues for ^{247}Cm, such as ^{232}Th.

$$\left(\frac{^{235}\mathrm{U}}{^{238}\mathrm{U}}\right)_P = \left(\frac{^{144}\mathrm{Nd}}{^{238}\mathrm{U}}\right)_P \left(\frac{^{247}\mathrm{Cm}}{^{144}\mathrm{Nd}}\right) e^{-\lambda_{235} t} + \left(\frac{^{235}\mathrm{U}}{^{238}\mathrm{U}}\right)_0 e^{(\lambda_{238}-\lambda_{235})t} \tag{4}$$

In Equation (4), subscript P denotes the present day atomic ratio and subscript 0 represents the initial atomic ratio at the time of closure of the ^{247}Cm–^{235}U system at time $t = 0$. It follows from Equation (4) that a positive correlation between the measured ^{235}U/^{238}U and ^{144}Nd/^{238}U values should be observed if some proportion of the ^{235}U was formed by the decay of live ^{247}Cm in the early solar system. Then, samples of equal age produce a straight line, provided ^{247}Cm/^{144}Nd is constant, as shown schematically in Fig. 4. The slope of the data, given by Equation (5), together with the ^{235}U/^{238}U-axis intercept (Eqn. 6), can be used to calculate the initial abundance of ^{247}Cm at the start of the solar system (i.e. at the time of closure of the ^{247}Cm–^{235}U system when Cm–U fractionation occurred), and in turn, the interval of free decay Δ, representing the time interval between the last actinide-producing *r*-process event and the time the first solid objects formed in the early solar system, provided the present day Nd/U cosmic ratio and the ^{247}Cm/^{238}U production ratio during *r*-process nucleosynthesis is adequately constrained through astrophysical models (e.g., Lingenfelter et al. 2003) and has behaved as a closed system.

$$\mathrm{Slope} = \left(\frac{^{247}\mathrm{Cm}}{^{144}\mathrm{Nd}}\right)_0 e^{-\lambda_{235} t} \tag{5}$$

$$^{235}\mathrm{U}/^{238}\mathrm{U} - \text{axis intercept} = \left(\frac{^{235}\mathrm{U}}{^{238}\mathrm{U}}\right)_0 e^{(\lambda_{238}-\lambda_{235})t} \tag{6}$$

Constraints on the abundance of ^{247}Cm in the early solar system. The potential of ^{247}Cm as a short-lived chronometer was first discussed by Blake and Schramm (1973), who calculated that permil- to percent-level variations in ^{235}U/^{238}U should be present in early solar system condensates. These theoretical calculations immediately stimulated research in the search for uranium isotopic anomalies in meteorite samples using TIMS techniques, although widely discrepant results were obtained. Arden (1977), for example, reported ^{235}U excesses of up to 300‰ in some bulk chondritic meteorites (with respect to the terrestrial value) and ^{235}U-excesses of up to 2400‰ in their acid-etched residues. Furthermore, Tatsumoto and Shimamura (1980) noted ^{235}U-excesses of up to 190‰ and depletions of up to –350‰ in various mineral phases and acid leachates of Allende. These observations, however, were not substantiated by higher precision studies performed by Chen and Wasserburg (1980, 1981a,b) who observed normal ^{235}U/^{238}U at the sub-percent level in all samples, including some of the meteorites considered to have anomalous U isotopic compositions. The implications for Δ from such widely discrepant results are profound. Estimates of the order of 100 m.y. are inferred by assuming no ^{235}U/^{238}U variations, while assuming ^{247}Cm effects at the 2000‰ level gives rise to high initial solar system abundances of ^{247}Cm and estimates for Δ of only a few m.y. The latter scenario offers the possibility of an actinide-producing supernova trigger for the birth of our solar system.

Further exploration of the ^{247}Cm–^{235}U system and resolution of these conflicting data required higher precision measurements than could be attained using the mass spectrometric techniques available at the time of these first measurements. The search for live ^{247}Cm in the early solar system was thus abandoned for more than two decades. The significant advances in instrumentation that have taken place since the late 1990's prompted a renewed search for ^{235}U/^{238}U anomalies in meteorites in the mid-2000's. Uranium isotopic measurements of (1) bulk samples of carbonaceous chondrites, ordinary chondrites and eucrites, for

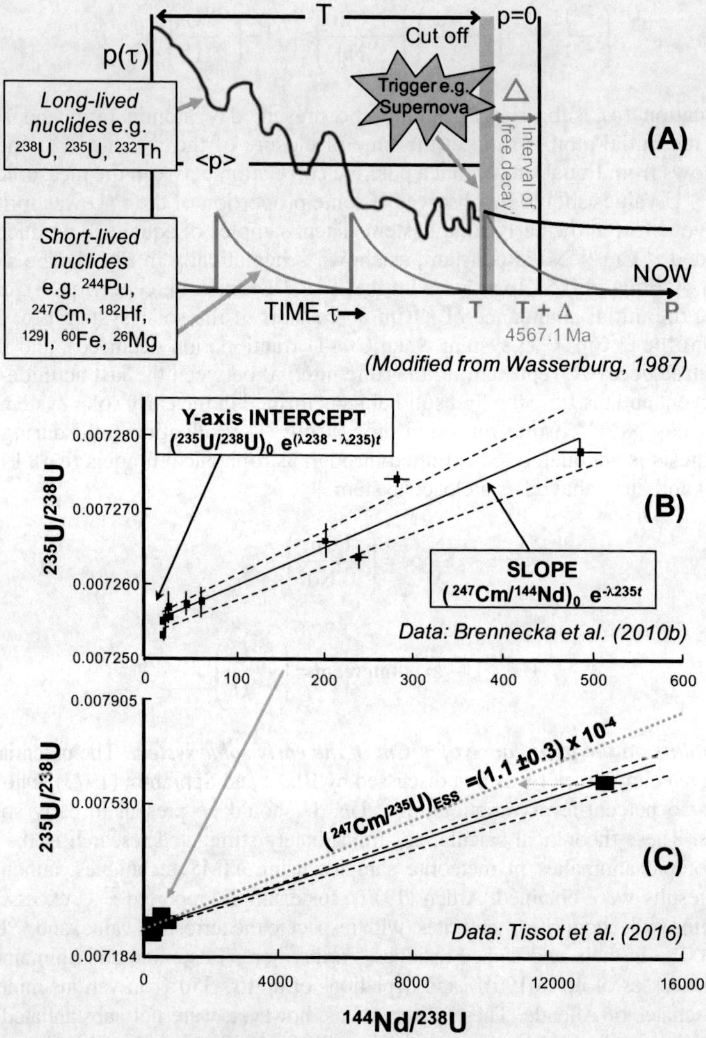

Figure 4. (A) Schematic illustration of the rate of nucleosynthesis p(τ) of some nuclides over galactic history τ; modified from Wasserburg (1987). The abundance of a stable nuclide is the average rate times the timescale. For long-lived nuclei (represented by the black curve), the abundance is decreased over time by decay. For presently extinct, short-lived nuclei (represented by the grey curve), the abundance is given by production near the end of nucleosynthesis and is then decreased by decay when production ceases; P denotes the present day; T represents the time of the last actinide-producing *r*-process event; T + Δ is the time the first solid objects formed in the early solar system; for the ^{247}Cm–^{235}U system, the interval of free decay Δ represents the time interval between the last actinide-producing *r*-process event and the time the first solids formed in the early solar system; the timing of the event which triggered the beginning of the solar system is represented by the grey shaded band, and is shown here to coincide with the timing of the last actinide-forming event. (B) and (C) ^{144}Nd/^{238}U vs. ^{235}U/^{238}U for CAIs extracted from Allende; Brennecka et al. (2010b) and Tissot et al. (2016). A systematic, correlated relationship between Cm/U (Nd/U) and ^{235}U/^{238}U is observed, whereby materials with a high initial Cm/U (Nd/U) contain a higher relative amount of ^{235}U than those with lower initial Cm/U values, consistent with ^{247}Cm decay. The data of Tissot et al. (2016) greatly extends the range of correlated Nd/U and ^{235}U/^{238}U compared to a previous study by Brennecka et al. (2010b). These data give an early solar system ^{247}Cm/^{235}U of 1.1×10^{-4} (grey stippled line), taking into account possible Nd/U fractionation after Solar system formation, see Tissot et al. (2016) for details.

which conflicting results had previously been obtained, as well as (2) acid-etched leachates and mineral assemblages extracted from primitive meteorites showed Cm/U fractionations of up to 75 and no well-resolved excursions in $^{235}U/^{238}U$ away from the terrestrial value at the 0.1–0.2‰ level (Stirling et al. 2005, 2006). These data provided evidence for a solar system initial $^{247}Cm/^{235}U$ of $<1\times10^{-4}$, requiring a long time-scale of 100 to 200 million years between the last actinide producing *r*-process event and the formation of the solar system (Stirling et al. 2005, 2006). These initial studies were followed by further measurements of $^{235}U/^{238}U$ in primitive calcium–aluminum-rich inclusions (CAIs), widely assumed to represent the earliest condensates to have formed in the solar system, extracted from the carbonaceous chondrite Allende (Brennecka et al. 2010b). These data showed larger Cm/U fractionations of up to 500 and U isotopic anomalies of up to 3.5‰ (with the largest ratios occurring in inclusions displaying fractionated 'Group II' REE patterns). Importantly, these observations showed a convincing, well-resolved correlation between $^{235}U/^{238}U$ and the inferred Cm/U parent-daughter ratio (represented by the Nd/U and Th/U), providing strong evidence that ^{247}Cm was "live" in the early solar system. These data constrain the solar system initial $^{247}Cm/^{235}U$ to $1.1–2.4\times10^{-4}$, requiring a time interval of 110 to 140 m.y. between the last actinide producing *r*-process event and the formation of the solar system (Brennecka et al. 2010b). Tissot et al. (2016) extended the database of U isotopic measurements for Allende CAIs, both those displaying Group II and 'normal' REE patterns, and identified extremely large Cm/U fractionations of up to 74,000. In this study, a well-defined positive correlation was also observed between $^{235}U/^{238}U$ and the inferred Cm/U ratio (represented by Nd/U) and ^{235}U excesses of up to 60‰, a factor of twenty larger than those recorded by Brennecka et al. (2010b), were documented. These results constrain the initial solar system $^{247}Cm/^{235}U$ ratio to $\sim1.1\times10^{-4}$ and yield a tightly constrained value for Δ of 87 ± 14 m.y. (Tissot et al. 2016).

All of the above-described estimates for Δ imply that material from the last *r*-process event was injected into the protosolar molecular cloud approximately 100 m.y. before solar system formation, in agreement with independent constraints for Δ from other *r*-process only nuclides, namely extant ^{129}I and ^{244}Pu, of 100 ± 7 m.y. and 158 ± 85 m.y, respectively (Dauphas 2005). Such a long time interval of free decay implies that the supernova event leading to the formation of the actinides and ^{129}I cannot be considered as a viable mechanism for triggering the birth of our solar system.

Uranium isotope fractionation unrelated to ^{247}Cm decay

In addition to the discovery of large U isotope anomalies in some CAIs, resolvable albeit smaller, U isotope variations of up to 2‰ have also been reported in some other meteorites and their mineral components that cannot be explained by ^{247}Cm decay, as no resolvable correlation is observed between $^{238}U/^{235}U$ and Cm/U (or Nd/U) (Amelin et al. 2010; Connelly et al. 2012; Goldmann et al. 2015). Accordingly, several other mechanisms may result in U isotope variations in meteorites that are discussed in turn below. These processes are related to pre-solar system-formed U isotope anomalies, isotope fractionation mechanisms, and terrestrial U contamination. These mechanisms will affect different meteorite classes in varying ways, dependent on the timing of formation (e.g., early formed mineral aggregates, such as CAIs, will record different U isotope variations and fractionating processes than differentiated meteorite classes, such as the angrites and eucrites).

Uranium isotope homogeneity across the proto-planetary disk. Some of the observed U isotope variations in meteorites might be due to heterogeneity in U isotopic composition across the solar nebula. However, the identical average U isotopic compositions of several classes of meteorites, including undifferentiated carbonaceous chondrites and ordinary chondrites as well as differentiated achondrites and terrestrial basalts suggests that the U isotopic composition of the solar system was homogeneous and unfractionated between reservoirs across the proto-planetary disk, at least at the time of parent body formation (Stirling et al. 2005, 2006; Brennecka and Wadhwa 2012; Connelly et al. 2012; Goldmann et al. 2015).

Nucleosynthetic anomalies inherited from pre-solar grains. Some U isotope fractionation in meteorites may potentially reflect inherited nucleosynthetic U isotope anomalies due to heterogeneity in the distribution of pre-solar material within the solar nebula. However, nucleosynthetic heterogeneities of other elements, such as Ti, Cr, Ni, and Mo (e.g., Dauphas et al. 2002; Trinquier et al. 2007; Regelous et al. 2008; Trinquier et al. 2009), have been found more commonly in carbonaceous chondrites, attributed to late-stage addition of materials enriched in pre-solar grains from the outer solar system. The absence of significant U isotope fractionation in carbonaceous chondrites therefore indicates that nucleosynthetic components do not control the U isotopic composition of meteorites (Brennecka et al. 2010b; Goldmann et al. 2015; Tissot et al. 2016).

Uranium isotope fractionation during evaporation and/or condensation. It is possible that U isotope fractionation occurred under kinetic or equilibrium conditions during condensation and evaporation processes within the accreting proto-planetary disk, as a result of the partial transfer of U between the gaseous and solid phases. This could give rise to either heavy or light U isotope compositions in early formed minerals, as well as 'anomalous' chemical signatures, dependent on local physico-chemical conditions. For example, a positive correlation between $^{238}U/^{235}U$ and 1/[U] was observed in some ordinary chondrites, opposite in direction to that expected from ^{247}Cm decay (Goldmann et al. 2015). This might indicate the preferential condensation of early stage U-rich components with light U isotope signatures, followed by the condensation of later-stage U-poor minerals with heavy U isotope compositions from a molecular cloud depleted in U (Goldmann et al. 2015).

Also the highly fractionated $^{238}U/^{235}U$ signatures of some CAIs that have been attributed to ^{247}Cm decay may have been generated instead through kinetic isotope fractionation during condensation processes in the proto-solar disk, resulting in a reservoir with low $^{238}U/^{235}U$. In such a scenario, the correlation between $^{238}U/^{235}U$ and Nd/U observed in Allende CAIs may reflect either progressive condensation and gradual evolution of the reservoir towards lower $^{238}U/^{235}U$ values, or two-component mixing between unfractionated 'average' solar system material and a material with a low content of U (resulting in high Nd/U values) and low $^{238}U/^{235}U$ values (Amelin et al. 2010; Connelly et al. 2012; Goldmann et al. 2015). Theoretical calculations imply that ^{235}U enrichments of up to 6‰ in the solid phase are possible through kinetic isotope effects during partial U condensation (Tissot et al. 2016). It therefore follows that correlated Nd/U and $^{238}U/^{235}U$ variations due to ^{247}Cm decay cannot be easily distinguished from those produced via condensation until Nd/U values are greater than ca. 1000 when ^{235}U enrichments exceed ca. 6‰ (Tissot et al. 2016). To this end, the U isotopic shifts observed in the CAI study of Brennecka et al. (2010b) of up to 3.5‰ fall within the range that can be generated by kinetic isotope effects. However, it is important to bear in mind that the very good agreement between the estimates for Δ derived by Brennecka et al. (2010b) with those of Tissot et al. (2016), despite more limited Nd/U and U isotope variations in the former study, suggests that the U isotopic anomalies of both studies were caused by ^{247}Cm decay. This, in turn, implies that condensation processes in the early solar system may not be a significant driver of U isotope fractionation in the early formed mineral components of meteorites.

Uranium isotope fractionation during planetary differentiation. Uranium isotope fractionation may also occur during the magmatic differentiation of meteorite parent bodies, although ultimately, long-term mantle mixing is likely to homogenize any pre-existing U isotopic variations on a planetary scale. To date, no resolvable differences in the average U isotope composition are observed between a range of meteorite classes spanning the differentiated achondrites (angrites), and undifferentiated chondrites (carbonaceous and ordinary chondrites), and terrestrial materials when their average compositions are compared (Goldmann et al. 2015). Also, these meteorite groups, with the exception of the ordinary chondrites, show almost no resolvable differences in $^{238}U/^{235}U$ between individual bulk samples (Stirling et al. 2005, 2006; Brennecka and Wadhwa 2012; Goldmann et al. 2015). These observations may indicate that large-scale planetary differentiation does not significantly fractionate ^{238}U from ^{235}U, although further studies are required to investigate potential fractionating mechanisms in more detail.

Uranium isotope fractionation during secondary alteration. It is conceivable that U isotope fractionation occurred during later stage secondary alteration processes within the meteorite parent body, such as oxidation-reduction reactions, causing U mobilization during aqueous alteration or thermal metamorphism. In this context, the correlated relationship between $^{238}U/^{235}U$ and 1/[U] as well as Nd/U in some ordinary chondrites may be explained by the preferential uptake of light U isotopes to the U-enriched oxidized species during U oxidation (Goldmann et al. 2015). However, this scenario is considered unlikely, as the undifferentiated carbonaceous chondrites, widely believed to have experienced significant aqueous alteration, display less U isotope variability than the ordinary chondrites. An alternative scenario requiring further investigation invokes thermal metamorphism, rather than aqueous alteration, as a viable mechanism for fractionating ^{238}U from ^{235}U in bulk meteorites.

Terrestrial contamination and U isotope variation. Some of the $^{238}U/^{235}U$ variability in meteorites may have resulted from open-system disturbance of the ^{238}U–^{235}U–^{234}U isotope system during terrestrial weathering. Measurements of ($^{234}U/^{238}U$) disequilibrium may be applied as a tracer of terrestrial disturbance in meteorites over the past ~1.5 m.y., spanning the lifetime of short-lived ^{234}U. For example, ^{234}U excesses of up to 12% in ordinary chondrites relative to radioactive equilibrium have been observed to broadly correlate with U concentration, indicative of oxidative weathering at Earth's surface, which may act to preferentially mobilize ^{235}U as well as ^{234}U with respect to ^{238}U (Andersen et al. 2015). These processes, however, are unlikely to explain the large variations among some ordinary chondrites investigated by Goldmann et al. (2015), as all of these samples were meteorite falls and these samples have therefore only spent a very short time in a terrestrial weathering regime.

Pb–Pb chronometer

Accurate and precise measurement of the U isotopic composition in meteorites is critical for the veracity of the uranium-thorium-lead (U–Th–Pb) chronometers, which have been utilized for decades to provide the most robust 'absolute' constraints on the timing and sequence of events throughout the history of the solar system.

Systematics of the U–Th–Pb chronometer. As discussed earlier, the U–Th–Pb chronometers are based on the ingrowth of stable ^{206}Pb and ^{207}Pb from their respective long-lived parent nuclides, ^{238}U and ^{235}U according to Equations (7) below and (3) above.

$$^{235}\mathrm{U} \xrightarrow{t_{1/2}=704} \ldots\ldots \longrightarrow {}^{207}\mathrm{Pb} \tag{7}$$

Both decay equations can be used as independent chronometers in their own right, provided the Pb isotopic composition and the U/Pb elemental ratio (e.g., $^{238}U/^{206}Pb$ or $^{235}U/^{207}Pb$) can be reliably measured. However, these U–Pb chronometers are not sufficiently precise for resolving early solar system events that are closely separated in time due to the inherent difficulty in reliably measuring the U/Pb elemental ratio. Consequently, the method of choice for high-precision U–Th–Pb dating combines the ^{238}U–^{206}Pb and ^{235}U–^{207}Pb decay series to form a single ^{207}Pb–^{206}Pb chronometer (Eqn. 8), assuming concordancy between the ^{238}U–^{206}Pb and ^{235}U–^{207}Pb systems, and underpinned by the additional requirement that the $^{238}U/^{235}U$ composition of the sample is reliably known. In (Eqn. 8), t is the Pb–Pb age of the sample in years since isotopic closure of the ^{238}U–^{206}Pb and ^{235}U–^{207}Pb systems, and the decay constants of ^{238}U and ^{235}U are given in Table 1.

$$\left(\frac{^{207}Pb^*}{^{206}Pb^*}\right)=\left(\frac{^{235}Ue^{\lambda_{235}t}-1}{^{238}Ue^{\lambda_{238}t}-1}\right) \tag{8}$$

High-precision Pb–Pb dating of meteorites. For several decades $^{207}Pb/^{206}Pb$ (or Pb–Pb) ages were derived making the assumption that the $^{238}U/^{235}U$ ratio was constant and equal to 137.88 across the entire solar system. As discussed in the section *Uranium Isotopes*, this assumption was based on invariant $^{238}U/^{235}U$ compositions in terrestrial materials (Steiger and Jäger 1977) at the ~1‰ limits of analytical precision available at the time of these measurements. In addition, by the mid-1980's, the consensus was that extra-terrestrial meteorites were also characterized by invariant $^{238}U/^{235}U$ compositions at the 10‰-level (Chen and Wasserburg 1980, 1981a). The view of negligible fractionation of $^{238}U/^{235}U$ in high-temperature early solar system environments was further supported by the conventional view that non-radiogenic isotope fractionation was solely a mass-dependent process (Urey 1947) and would be unresolvable in U, even at the 1‰ level, given its heavy mass range.

The assumption of an invariant solar system U isotopic composition equal to 137.88 did not limit the utility of the Pb–Pb chronometer at the time of these early measurements due to the comparatively low levels of analytical precision. However, the advent of modern high-precision Pb–Pb dating in the early 2000's, using state-of-the-art chemical and mass spectrometric methods, offered a significantly improved age resolution of 100,000 years, and the ability to assess models of early solar system formation based on relative age differences of only a few m.y. (Amelin et al. 2006). Therefore, the lack of direct measurement of the U isotope ratio in meteorites, and how this would impact the accuracy of Pb–Pb chronologies, became of increasing concern. Amelin et al. (2002) reported the first high-precision Pb–Pb ages for primitive CAIs from the carbonaceous chondrite Efremovka and chondrules from the carbonaceous chondrite Acfer, adopting the 'canonical' meteoritic $^{238}U/^{235}U$ ratio of 137.88 in all calculations. This unique dataset defined the most precise age of the solar system at 4.5672 ± 0.0006 Ga, and provided an age difference of approximately 2.5 m.y. between the formation of CAIs under high temperature solar nebula conditions as the first solid objects and the flash heating of chondrule formation at lower nebula temperatures. However, despite the significant improvements in Pb–Pb age precision, additional research in this field gave rise to inconsistencies in the chronology of the early solar system, both between different Pb–Pb dating studies and between Pb–Pb ages and those derived using extinct short-lived nuclide chronometers. For example, conventional Pb–Pb ages for CAIs and chondrules extracted from the Allende carbonaceous chondrite, yielded an unresolvable to small age difference of 0.6 ± 1.0 m.y. (Amelin and Krot 2007; Amelin et al. 2010) to 1.8 ± 0.5 m.y. (Connelly et al. 2008). This contrasts with the ca. 2.5 m.y. Pb–Pb age interval reported in Amelin et al. (2002), and with the 2–3 m.y. age interval reported for the ^{26}Al–^{26}Mg and ^{182}Hf–^{182}W extinct short-lived nuclide chronometers (e.g., Kleine et al. 2009; Villeneuve et al. 2009). These, and other, inconsistencies are likely due, at least in part, to variability in the meteoritic $^{238}U/^{235}U$ composition that was not taken into account by the conventional Pb–Pb age calculations.

Towards a high accuracy and high precision Pb–Pb cosmochronometer. The discovery of natural variability in $^{238}U/^{235}U$ in Earth's near surface, low-temperature environments in the mid–2000's (Stirling et al. 2007; Weyer et al. 2008; Bopp et al. 2009) has had a significant impact on the accuracy of Pb–Pb ages. This, together with improved concordance of Pb–Pb ages with those derived using extinct short-lived nuclide chronometers, has led to a revision of the fundamental chronometry of the early solar system, which is ongoing.

An important development occurred with the commercial production of the fully calibrated, SI-traceable, ^{236}U–^{233}U double spike IRMM-3636 (Richter et al. 2008) as discussed in the section *Historical overview of* $^{238}U/^{235}U$ *measurements*. This allowed, for the first time, high accuracy, as well as high precision measurements of $^{238}U/^{235}U$. Subsequent studies of bulk meteorites reinforced previous findings that U isotopic variability is limited to the 0.1–0.2‰ level across all investigated averaged meteorite classes (Stirling et al. 2005; Connelly et al. 2012; Andersen et al. 2015; Brennecka et al. 2015; Goldmann et al. 2015), and the mean of these studies has yielded a value for the average solar system composition of 137.79 ± 0.03. This is significantly lower than the previously adopted value of 137.88 and results in an average Pb–Pb age adjustment of ~0.9 m.y.

Although most meteorite samples display limited $^{238}U/^{235}U$ variability, the discovery of sizeable variations in the $^{238}U/^{235}U$ composition of up to 60‰ in Allende CAIs (Brennecka et al. 2010b; Tissot et al. 2016) has profound implications for the application of Pb–Pb dating in cosmochemistry. For example, the Pb–Pb ages of CAIs, calculated assuming an invariant $^{238}U/^{235}U$ ratio of 137.88 would be overestimated by 5 m.y. if the true ratio is 3.5‰ lower (Stirling et al. 2007; Brennecka et al. 2010b; Tissot et al. 2016). To provide a context, this interval is larger than the apparent age difference between CAIs and chondrules, a crucial time interval for constraining models of solar system formation.

The first study to combine high precision ^{238}U–^{235}U–^{207}Pb–^{206}Pb measurements on the same samples was conducted by Amelin et al. (2010). These authors observed a resolvable difference in the U isotopic composition between Allende CAIs and chondrules, resulting in a significant difference between the Pb–Pb ages of CAIs and chondrules of ca. 2 m.y. compared with the ca. 0.6 m.y. age interval calculated assuming an identical U isotopic composition in all samples (Amelin et al. 2010). Furthermore, this study helps reconcile a major inconsistency between the 'absolute' Pb–Pb chronometer and other 'relative' chronometers based on extinct short-lived nuclides (including the ^{26}Al–^{26}Mg, ^{53}Mn–^{53}Cr and ^{182}Hf–^{182}W systems) regarding the time difference between the formation of the first solid objects (CAIs) and the flash heating associated with chondrule formation, possibly linked to the beginning of planetary accretion. Moreover, in a subsequent combined ^{238}U–^{235}U–^{207}Pb–^{206}Pb study, Connelly et al. (2012) reported revised Pb–Pb ages for a suite of CAIs and chondrules extracted from Allende and Efremovka. Remarkably, all CAI ages were in perfect agreement with each other, and with the CAI Pb–Pb age reported in Amelin et al. (2010), despite highly variable $^{238}U/^{235}U$ in these samples.

Mineral inclusions in undifferentiated chondrites aside, Brennecka and Wadhwa (2012) acquired the first U isotopic measurements for the differentiated angrite meteorites, resulting in revised Pb–Pb ages that are approximately 1 m.y. younger than previously reported ages assuming a constant $^{238}U/^{235}U$ of 137.88. This study included a new age constraint of 4563.37 ± 0.25 Ga for the D'Orbigny angrite, which is widely used as a solar system anchor for mapping the relative 'model' ages of the extinct short-lived nuclide chronometers onto the Pb–Pb absolute timescale.

Ongoing studies continue to demonstrate that precise determination of the $^{238}U/^{235}U$ ratio in meteoritic material is crucial for achieving accurate Pb–Pb ages at the current precision of dating. To this end, a growing number of studies have adopted combined high precision U and Pb isotopic analyses to derive high accuracy Pb–Pb ages and revise the timing and sequence of events in the early solar system (e.g., Iizuka et al. 2014).

URANIUM ISOTOPE SYSTEMATICS IN HIGH TEMPERATURE ENVIRONMENTS ON EARTH

Bulk Earth $^{238}U/^{235}U$

Regarding the U isotope composition of the Earth, extensive investigations of meteorites (see *$^{238}U/^{235}U$ in cosmochemistry*) have revealed two important findings (Connelly et al. 2012; Andersen et al. 2015; Goldmann et al. 2015): 1) despite isotopic heterogeneities among individual meteorites, the solar system seems to be homogeneous at planetary scales with respect to its

U isotope composition, as indicated by the indistinguishable $^{238}U/^{235}U$ of different chondrite groups and achondrites; 2) a solar system $^{238}U/^{235}U$ mean of 137.79±0.03 can be derived from analyses of meteorites which translates to a $\delta^{238}U$ of −0.34‰ relative to the U isotope standard CRM-112a (or CIRM-145). According to these findings it may be assumed that the bulk Earth has a U isotope composition identical to that of the solar system. The U budget of the Earth is almost equally distributed between the mantle and the continental crust (see *Introduction*), assuming that U in the core is negligible (see Wohlers and Wood 2015 for a discussion of this assumption). Thus, if the U isotope composition of these two most important reservoirs is known, the bulk Earth $^{238}U/^{235}U$ may also be estimated from the U isotope compositions of the crust and mantle and their relative U fractions.

The mantle

In a recent study, Andersen et al. (2015) realized that at least the upper mantle is heterogeneous with respect to its U isotope composition. This heterogeneity is reflected in the different U isotope compositions of ocean island basalts (OIB), mid ocean ridge basalts (MORB) and island arc basalts (IAB). As U is dominantly present as U^{+4} in the mantle (e.g., Wood et al. 1999), large U isotope fractionation during partial melting or metasomatic processes is unlikely. Furthermore, U is a highly incompatible element, and as such, it is readily extracted from the mantle source during partial melting. Thus, variable U isotope compositions in melts most likely reflect U isotope variations in the mantle source, as generated by subducted crustal material. Indeed, near-surface processes generate significant U isotope fractionations (see *Near-surface U cycling and the marine* $^{238}U/^{235}U$ *mass balance*), which are largely driven by redox processes and the reduction of U^{+6} to U^{+4}. Accordingly, the variability of $\delta^{238}U$ in marine sinks, including U uptake in altered oceanic crust (AOC), was likely significantly larger during the Phanerozoic, when the oceans were oxidized in contrast to the more reducing Precambrian oceans (Andersen et al. 2015). The comparatively low Th/U values of MORB have been interpreted to reflect a recycled U component from AOC (Elliott et al. 1999) which may have impacted the U isotope composition of the MORB source. A negative correlation between Th/U and $\delta^{238}U$ indicates contamination of the mantle source with a component that is U-rich and has a high $\delta^{238}U$. Indeed, these characteristics are the case for the bulk altered upper oceanic crust (Andersen et al. 2015; Noordmann et al. 2016). Thus subducted oceanic crust may be a likely candidate for the generation of the relatively high $\delta^{238}U$ (−0.26‰) observed for MORB (Fig. 5).

Contrarily, the deeper OIB mantle sources are much older, as indicated by their Pb model ages (Chase 1981; Andersen et al. 2015), and as such, should be significantly less affected by isotopically variable subducted crustal material, as largely reducing oceans in the Archean and early Proterozoic would lead to quantitative U uptake in the subducted sink (Andersen et al. 2015). Indeed, OIB have a fairly constant U isotope composition (independent of Th/U) which is indistinguishable from the U isotope composition of the Solar system (Andersen et al. 2015; Goldmann et al. 2015). We infer from these findings that the U isotope composition of OIB (and intra-continental basalts) is representative of that for the Earth's deeper mantle. Compiling recent data from the literature (n=32: Weyer et al. 2008; Amelin et al. 2010; Brennecka and Whadwa 2012; Cheng et al. 2013; Andersen et al. 2014, 2015; Dahl et al. 2014; Iizuka et al. 2014; Goldmann et al. 2015; Tissot and Dauphas 2015) results in a mean value of $\delta^{238}U$= −0.31‰. The uncertainty associated with this value is only ± 0.01‰, as given by the standard error of the mean at the 95% confidence level of a *Student-t* distribution, referred to herein as "2 SE-mean". However, the uncertainty of the OIB mean value may be better expressed by the level of inter-laboratory agreement, based on analyses of common U isotope and rock standards (e.g., the United States Geological Survey (USGS) reference materials BCR-2, BHVO-2 or seawater) which is typically on the order of ~ ±0.06‰.

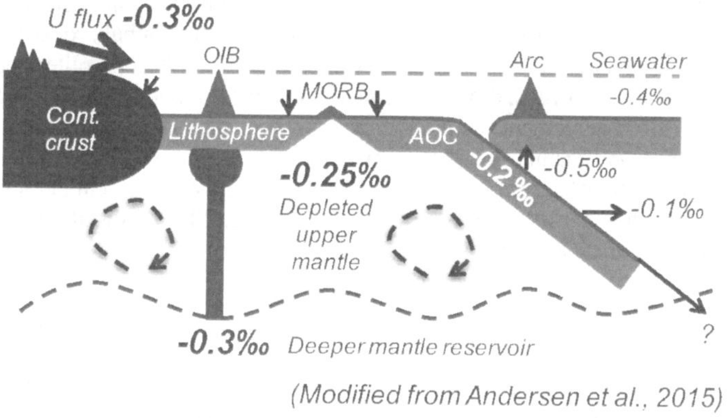

Figure 5. Schematic illustration of the modern cycling of U isotopes ($\delta^{238}U$) in the deep Earth; modified from Andersen et al. (2015). During U uptake in the altered oceanic crust (AOC) $\delta^{238}U$ is fractionated compared to seawater and riverine input. The excess U in the AOC is released into the upper part of the mantle during subduction; an isotopically lighter U fraction is released from the downgoing slab and lost to arc magmatism (e.g., Mariana Arc), while isotopically heavier U is lost from the slab at a deeper level and incorporated into the upper mantle, the source of midocean ridge basalts (MORB). The $\delta^{238}U$ of the deeper mantle source, as recorded in ocean island basalts (OIB), are not effected by this isotopically heavier AOC signature, hypothesized to be due to its older Precambrian origin, when U uptake in AOC was largely isotopically unfractionated (see details in Andersen et al. 2015).

As also revealed by Andersen et al. (2015), IAB from the Mariana Arc have slightly lighter U isotope compositions, on average ($\delta^{238}U = -0.38$‰), than OIB. As arc rocks are the major building blocks of the modern continental crust, this may imply that modern crust is isotopically lighter than the mantle. During subduction, material is lost from the downgoing plate and incorporated into the source of IAB. As shown by Elliott et al. (1997), the Mariana IAB show evidence for two slab-derived components: a melt from the sedimentary section with high Th/U, and a 'fluid' from the mafic oceanic crust with low Th/U. The Mariana IAB with low Th/U also have low $\delta^{238}U$ indicating that the fluid derived from the mafic oceanic crust has a low $\delta^{238}U$ (Andersen et al. 2015). This low $\delta^{238}U$ may originate during dehydration of the slab and, for example, from a leaching process that may have resulted in U isotope fractionation (Stirling et al. 2007; Hiess et al. 2012) or during preferential dehydration of the uppermost AOC that has relatively light U isotope compositions. Independent of the process that generated the light U isotope signatures of basalts from the Mariana Arc, these signatures likely only appeared when the oceans were fully oxidized and became U-rich at approximately 600 Ma (see details in Andersen et al. 2015). More subduction-related basalts should be investigated to verify if low $\delta^{238}U$ are a general feature of such rocks, which may, in turn, impact the U isotope composition of the continental crust.

The continental crust

The U isotope composition of the crust has been estimated by analyses of either granites and other common rocks of the crust (Weyer et al. 2008; Telus et al. 2012; Tissot and Dauphas 2015; Noordmann et al. 2016) or accessory minerals that control the U budget of many crustal rocks (e.g., zircon, monazite and apatite; Hiess et al. 2012). Most of the estimates based on bulk rocks are more representative of the composition of the upper crust, which has been more intensely investigated for its U isotope composition. Tissot and Dauphas (2015) derived an average value for the U isotope composition of the crust of $\delta^{238}U = -0.29$‰ by compiling their own and literature U isotope data of different crustal materials from different reservoirs and applying mass balance constraints. Adding the U isotope data of crustal rocks, reported by Noordmann et al. (2016)

to their compilation, and leaving out any sediments and soils, does not change the average value. There is no significant difference in $\delta^{238}U$ between the different bulk rocks that have been analysed. Specifically, the averages for highly evolved rocks (granites, rhyolites and tonalities), intermediate rocks (granodiorites, diorites, dacites and andesites) and alkali-rich rocks (syenites and trachites) are all in agreement, within the magnitude of their reported uncertainties (Tissot and Dauphas 2015; Noordmann et al. 2016). Likewise, there seems to be no correlation between the U isotope composition and geochemical parameters of crustal differentiation or the age of the crust (Noordmann et al. 2016). Furthermore, Telus et al. (2012) showed, by combining the Fe, Mg and U isotope analyses of granitoids, that thermal Soret diffusion did not dominate the U isotopic variability observed in the investigated crustal rocks.

The estimated $\delta^{238}U$ of the continental crust based on accessory minerals from Hiess et al. (2012) is −0.13‰. This value is based on 44 zircons and excludes the most extreme outlier which is heavier than the average of all other analyzed zircons by ~3‰. This average value agrees with the bulk rock-derived estimate for the continental crust, taking into consideration the spread in the zircon population of Hiess et al. (2012) around the mean (±0.33‰; 2SD). However, considering the 2SD-mean (±0.05‰) instead, the zircon-derived value is readily distinguished from the estimate of the bulk continental crust U isotope composition derived from bulk rock material. Analyses of international reference materials, reported by Hiess et al. (2012), do not indicate that this difference reflects an inter-laboratory bias. More likely, it may be generated by U isotope fractionation among different U-bearing phases or between zircons and a fluid/melt. As recently suggested (Tissot and Dauphas 2015), such processes might be responsible for generating the relatively large range in $\delta^{238}U$ (~0.4‰) observed among differentiated crustal rocks compared with that observed for fresh basalts (~0.2‰). If this is the case, the average $\delta^{238}U$ value of Hiess et al. (2012) should be used for zircon and other accessory minerals, however it provides an unsuitable estimate of the U isotope composition of the bulk continental crust. From the current U isotope database, this hypothesis is difficult to prove, as not enough $\delta^{238}U$ data for other U-bearing accessory mincrals exist.

It may be concluded from the findings above that bulk rocks are likely better suited to providing an estimate for the U isotope composition of the continental crust, although the sample selection thus far is biased towards rocks from the upper crust. Furthermore, sediments and low-temperature ores are not considered here which can have a significant influence on the U isotope composition of individual Earth reservoirs (see *Near-surface U cycling and the marine* $^{238}U/^{235}U$ *mass balance*) but only a small effect on the bulk continental crust U isotope composition. The mean value of $\delta^{238}U = -0.29 \pm 0.03$‰, as compiled in Tissot and Dauphas (2015) and here for bulk continental crust, is indistinguishable from the composition of the deep mantle based on intra-plate basalts ($\delta^{238}U = -0.31 \pm 0.06$‰) and also from that of the average Solar system ($\delta^{238}U = -0.34 \pm 0.07$‰), as derived from meteorites. Accordingly, mass balance prohibits a significantly fractionated U isotope composition of the lower continental crust.

URANIUM ISOTOPES IN ORE DEPOSITS

Geochemical processes, and especially redox reactions, play a significant role in the formation of uranium ore deposits, which occur worldwide in a range of different geologic settings. Uranium ore deposits have been increasingly mined and traded globally as a high-value commodity over the last several decades. In preparation for commercial markets, U is concentrated from the sub-10 wt.% levels found in mined U ore to >65 wt.% through milling processes, resulting in the formation of uranium ore concentrate (UOC), commonly referred to as 'yellowcake'. In recent years, the U isotopic signatures of uranium ore deposits have been shown to offer considerable potential as a tracer of ore body formation processes and remediation efforts in economic geology. Progress to date in this field of research is summarized in the following sections.

Uranium ore types

Uranium ore deposits have conventionally been grouped into two primary categories based on their depositional setting, and specifically the temperature and redox environment in which they formed. Here, we broadly characterise these into two categories of (1) sandstone-type U ore deposits and (2) magmatic-type U ore deposits, described in turn as follows.

Redox-controlled sandstone-type uranium ore deposits. Sandstone-type U deposits are formed at near ambient temperatures, below the water table in permeable sandstone aquifers. In these low-temperature settings, oxidizing groundwaters entering the aquifer from the surface transport aqueous uranyl U^{+6} complexes, leached from the overlying strata, deeper into the aquifer system. When these oxidizing meteoric waters encounter a reductant, such as sulfides, organic matter, or hydrocarbons, the dissolved U^{+6} is reduced, possibly mediated by bacteria, resulting in the precipitation of insoluble U^{+4}-bearing minerals, such as uraninite (UO_2), coffinite ($USiO_4 \cdot nH_2O$) or monomeric U^{+4} species (e.g., Langmuir 1978; Bargar et al. 2013), or the adsorption of relatively insoluble U^{+4} complexes onto the aquifer host lithology or organic materials (Langmuir 1978). These U^{+4}-bearing precipitates therefore accumulate at an oxidation-reduction interface in 'roll-front' or 'tubular sand' ore deposits comprising up to percent levels of uranium oxide concentrate (Dahlkamp 1991). These U^{+4} minerals can undergo subsequent remobilization. For example, continual circulation of groundwater through the aquifer results in the redissolution and reprecipitation of uranium minerals, and hence migration of the oxidation–reduction interface and mineralized 'roll-front' (Hostetler and Garrels 1962). Therefore, the chemical precipitation of U^{+6}- as well as U^{+4}-bearing minerals may occur in this dynamic environment.

Magmatic- and hydrothermal-type uranium ore deposits. Magmatic-type U deposits may form from fractional crystallization, resulting in the accumulation of incompatible U in highly evolved granitic melts until they are sufficiently enriched in U to crystallize U-rich accessory phases, such as uraninite (e.g., Cuney 2010). During the Archean, such magmatic-type uraninites may have survived dissolution during weathering due to low atmospheric oxygen levels (e.g., Holland 2005) and accumulated in fluvial-formed quartz–pebble conglomerate deposits as, for instance, the economically important 3 b.y. old Witwatersrand Supergroup, South Africa (e.g., Cuney 2010).

Hydrothermal-type U deposits involve the precipitation of U-rich minerals under higher temperatures (ca. 200–400 °C degrees) from the circulation of hydrothermal fluids or magmas that are enriched in U due to extensive leaching of source rocks and late-stage fractional crystallization (e.g., Hobday and Galloway 1999; Plant et al. 1999). Although operating under the higher temperatures of igneous and metamorphic processes, the mechanisms of deposition for hydrothermal-type U deposits are similar to those for sandstone-hosted deposits, such that they are also controlled by U^{+6}–U^{+4} reduction. Uranium deposition then occurs when highly soluble, aqueous U^{+6} species are reduced to relatively insoluble U^{+4}-bearing minerals. This category includes some of the richest and largest U deposits worldwide.

Uranium isotope fractionation in ore deposits

To date, most U isotope studies of U ore deposits have focused on gaining an improved understanding of the primary mechanisms responsible for generating fractionation between ^{238}U, ^{235}U, and ^{234}U. Uranium isotope fractionation may occur during any of the above-described stages of mineralization and post-depositional alteration. Chemical milling processes devised to pre-concentrate U and form UOC from mined U ore may also fractionate U isotopes.

In search of natural nuclear fission reactions. In 1976, Cowen and Adler (1976) compiled the available $^{238}U/^{235}U$ observations (e.g., Hamer and Robbins 1960; Smith 1961) for almost ninety U ore deposits from around the world. These authors were motivated by the discovery of several natural 'fossil' fission reactors with very anomalous $^{238}U/^{235}U$ compositions preserved in U ore deposits at the Oklo mine in Gabon (Bosu et al. 1972; Neuilly et al. 1972; IAEA

1975) (see *Uranium Isotopes*). Although the $^{238}U/^{235}U$ ratio was considered to be generally invariant at the time of these measurements, the Oklo discovery raised the possibility that the $^{238}U/^{235}U$ ratio may be slightly variable in U ore environments due to natural nuclear reactions. The compilation of U isotope data revealed a bi-modal distribution and a 0.3‰ difference in composition between high-temperature 'magmatic-type' ores and low-temperature 'sandstone-type' ores. A subsequent compilation of U ores from other sites by Gauthier-Layfaye et al. (1996) also revealed permil-level variations in $^{238}U/^{235}U$. The compiled $^{238}U/^{235}U$ observations, although state-of-the-art at the time, had insufficient precision to adequately resolve whether the apparent difference between the two modal values of $^{238}U/^{235}U$ was caused by natural processes, such as Precambian nuclear fission reactions generating ^{235}U depletion in U ore deposits, or artefacts from the industrial milling process used to chemically pre-concentrate U to form UOC. Later higher precision studies have attempted to 'fingerprint' U ore deposits as natural reactors. In these studies, relatively homogeneous $^{238}U/^{235}U$ values that largely reflect the source rock geology have been obtained, and display a level of variability that only marginally exceeds the ±0.5‰ uncertainties (Richter et al. 1999; Kirchenbaur et al. 2016).

Temperature-dependent U^{+6}–U^{+4} reduction. The advent of high-precision measurements of natural $^{238}U/^{235}U$ using techniques in double spiking coupled to MC-ICPMS in the mid-2000s (Stirling et al. 2005, 2006, 2007; Rademacher et al. 2006; Weyer et al. 2008) prompted a renewed investigation of the U isotopic composition of uranium ore deposits with analytical uncertainties of ±0.05‰. On the basis of such high-precision $^{238}U/^{235}U$ measurements for a suite of the same U ore samples that were previously compiled by Cowen and Adler (1976), Bopp et al. (2009) verified that a resolvable ~1‰ difference in U isotopic composition exists between magmatic-type and sandstone-type U ore deposits, whereby sandstone-type deposits are significantly depleted in ^{235}U and have higher $\delta^{238}U$ values. In a subsequent study, Brennecka et al. (2010a) extended the findings of Bopp et al. (2009) by measuring the U isotopic composition of a suite of UOCs extracted from U ore deposits. These authors observed that the $^{238}U/^{235}U$ compositions of the low temperature sandstone-type UOCs were, on average, 0.4‰ heavier than those of the magmatic-type UOCs. Both studies attributed this offset to U isotope fractionation during U^{+6}–U^{+4} reduction due to volume-dependent nuclear field shift effects (see *Experimental evidence for uranium isotope fractionation processes*) prior to the deposition of U^{+4} in minerals. The difference in U isotopic composition between the low temperature sandstone-hosted ores and the high temperature magmatic-type ores could then be explained by the temperature dependence of the nuclear field shift effect, which predicts significantly larger U isotope fractionation at low temperature than at high temperature. Moreover, both Bopp et al. (2009) and Brennecka et al. (2010a) observed significant 1‰-level variability in the $\delta^{238}U$ signatures of the low temperature sandstone-type U deposits and UOCs, which contrasted with the restricted 0.4‰ variation in the $\delta^{238}U$ values of their high temperature magmatic counterparts. These authors attributed this variability to the non-quantitative removal of U to ore minerals with varying degrees of extraction in the sandstone-hosted deposits, allowing U isotope fractionation to be expressed with varying magnitudes in the U ore deposit record.

A later study by Murphy et al. (2014) investigated U isotopic measurements for cogenetic groundwaters and mineralized sediments formed contemporaneously within the same aquifer system. The U isotope systematics of this system revealed U isotopic shifts of up to 5‰ in the mineralized sediments and 2‰ in the corresponding groundwaters, representing the largest variations observed to date in a single natural redox system (Fig. 6). The $^{238}U/^{235}U$ signatures of the sediments were generally heavier than those of the unreacted groundwaters, recording the preferential removal of ^{238}U from the aqueous phase, and could also be readily explained by nuclear field shift effects during low temperature U^{+6}–U^{+4} reduction. This U isotope system could be described by a Rayleigh fractionation model, yielding an apparent fractionation factor ε of ~0.2‰ (Murphy et al. 2014), in agreement with the range of values obtained from other uranium redox systems.

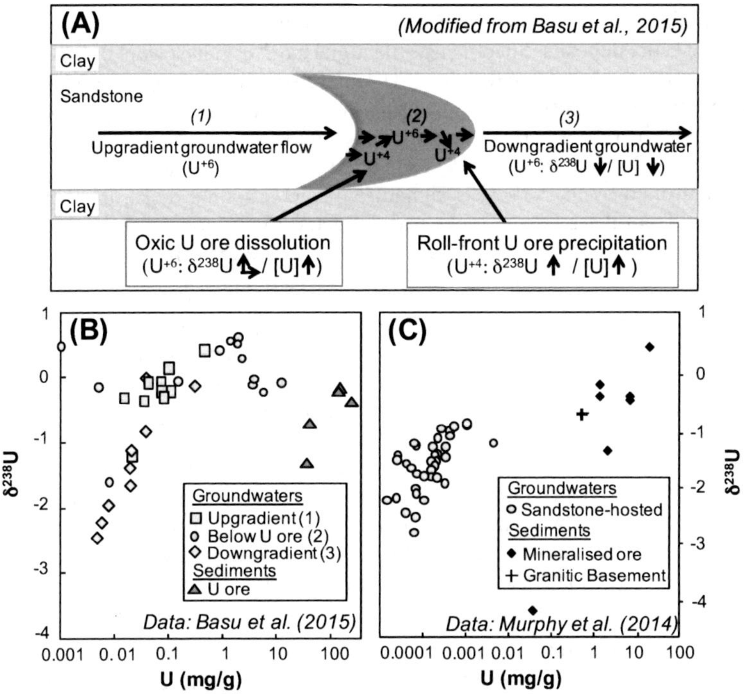

Figure 6. (A) Schematic illustration of $\delta^{238}U$ evolution in a sandstone-hosted roll-front U deposit system, modified from Basu et al. (2015). For a matured roll-front U ore system the evolution can be separated into three-steps: (1) oxidized groundwaters, containing soluble U^{+6}, reaches the reducing redox-barrier of the already developed U ore; (2) the oxidized groundwaters transform the reduced U^{+4} ore into soluble U^{+6}, with little change in $\delta^{238}U$; this U^{+6} moves further into the U ore zone until the redox-potential is lowered enough for U reduction and U^{+4} precipitation with higher $\delta^{238}U$ reoccurs; (3) the downgradient groundwaters show progressively decreasing $\delta^{238}U$ and U concentrations dictated by the degree of U^{+4} precipitation at the roll-front. (B) $\delta^{238}U$ vs. U concentration in groundwaters and U ores from *in situ* recovery (ISR) from the Rosita ISR mining site, USA, where U from roll-front U deposits is mobilized as U^{+6} by oxidative dissolution (Basu et al. 2015) and (C) $\delta^{238}U$ vs. U concentration in groundwaters and mineralized sediments from the Pepegoona high-grade sandstone-hosted U deposit, Australia; from Murphy et al. (2014). Both data-sets (B and C) show trends broadly following those schematically shown in (A) towards lower U^{+6} concentrations and $\delta^{238}U$ values in the groundwaters, consistent with progressive U reduction and removal to concurrent deposition as U-rich minerals. For both (B) and (C) multi-stage U redox cycling processes may cause larger variability in $\delta^{238}U$ and U concentrations, than expected from the simplistic single-stage U roll-front redox cycling process illustrated in (A).

Although U redox processes appear to dominate the U isotope systematics in regions of U ore formation, other potential causes of U isotope fractionation, some of which may depend on local environmental parameters, have been increasingly investigated, and are discussed in turn in the following sections.

Source rock composition, extent of U 'capture' and fluid alteration. Uvarova et al. (2014) investigated the U isotopic compositions of a variety of different U ore deposits and U-bearing phosphates, including those formed in both high temperature and low temperature depositional settings. As observed in previous studies, the U isotopic compositions of the samples investigated by Uvarova et al. (2014) varied widely, spanning a 2‰ range, and could be separated into two distinct groups. However, in contrast to Bopp et al. (2009) and Brennecka et al. (2010a), who observed a clear separation in U isotopic composition between low temperature sandstone-type and high temperature magmatic-type U ores, Uvarova et al.

(2014) found distinctly different $^{238}U/^{235}U$ compositions between high-grade U ore deposits and low-grade U ore deposits. Specifically, the highest $\delta^{238}U$ values were observed in the low-grade ore deposits, furthest from the site of U mineralization. Both low and high temperature deposits were present in both groupings, implying that the temperature of deposition during redox-driven U mineral formation is not the only factor controlling the U isotope signatures of U ore minerals. Uvarova et al. (2014) concluded that the $\delta^{238}U$ values of the U ore deposits were also controlled by the $^{238}U/^{235}U$ signature of the U source rocks, the degree to which U was leached from these source rocks and delivered to ore zone groundwaters, the extent to which U was removed from the U ore zone groundwaters through U reduction (and U adsorption) and its deposition in U-bearing minerals, and later fluid alteration of the U ore.

A suite of U ore samples investigated by Chernyshev et al. (2014a,b) also showed no systematic difference in U isotope composition between low and high temperature U ore deposits, although a greater variability in $\delta^{238}U$ was observed in the low temperature deposits. This reinforces the findings of Uvarova et al. (2014), whereby other factors besides the temperature of deposition control the $\delta^{238}U$ signatures of U ore deposits. Furthermore, Chernyshev et al. (2014a) examined the U isotope systematics within a selection of U ore deposits at the microscale by examining the $\delta^{238}U$ composition of the individual U^{+4}-bearing mineral components uraninite, pitchblende and coffinite. This dataset showed $\delta^{238}U$ variations of up to 0.9‰ between coexisting U^{+4}-bearing minerals, and an apparent shift towards lower values between the earlier- and later-formed growth zones of minerals. The authors attributed this U isotope variability to multiple stages of formation and re-precipitation under differing redox and environmental conditions.

Fluid–mineral interactions. The role of fluid–mineral interactions in controlling U isotope fractionation in U ore deposits has been investigated in several studies by examining the relationship between $\delta^{238}U$ and $(^{234}U/^{238}U)$ in U ore materials. Fluid–mineral interactions include, but are not limited to, leaching from source rocks during chemical weathering prior to U mineral deposition, and post-depositional aqueous alteration of U-bearing minerals. Both of these processes can preferentially remove isotopically light ^{235}U over heavier ^{238}U, and weakly bound ^{234}U (often located in a radiation-damaged site) over lattice-bound ^{238}U from the mineral structure (Stirling et al. 2007; Hiess et al. 2012). Aqueous alteration is believed to be prevalent in sandstone-hosted U ores due to their relatively high porosity that allows for greater re-mobilization of the ^{238}U–^{235}U–^{234}U isotopes. In contrast, magmatic-type U ore deposits are not considered to be susceptible to such open-system behavior due to the low-permeability lithologies of the host rocks.

The suite of U ore deposits and mineralized sediments discussed in Brennecka et al. (2010a), Murphy et al. (2014) and Uvarova et al. (2014) showed no discernible relationship between $\delta^{238}U$ and $(^{234}U/^{238}U)$. Together, these observations indicate that leaching is not the primary control of U isotope fractionation in U ore samples, and that different mechanisms are responsible for the fractionation of the ^{238}U–^{235}U and ^{238}U–^{234}U isotope systems. In contrast, a study of U ore deposits by Golubev et al. (2013) revealed a pronounced correlation between $\delta^{238}U$ and $(^{234}U/^{238}U)$, indicating that a common mechanism controls the preferential leaching of ^{234}U and fractionation of ^{235}U from ^{238}U in the studied samples. These contrasting findings highlight the importance of local environmental controls, in addition to U redox properties, on the U isotopic composition of U ore deposits.

Uranium isotopes as tracers of U mine remediation

Uranium isotopes have shown considerable potential for monitoring the remediation of U ore mines amid serious public health concerns regarding U contamination of waters downstream of mining sites (e.g., Abdelouas 2006). In mining settings, chemically hazardous U can be released to the environment from historic mine tailings or through *in situ* recovery (ISR) mining techniques. Historically, mine tailings containing U-rich materials are often stored in subaerial piles adjacent to the mine and are prone to U^{+4}–U^{+6} oxidation and acid leaching processes that mobilize U from the tailings pile over time (e.g., Abdelouas 2006). More recent ISR techniques recover U from permeable sandstone-hosted U ore deposits by injecting oxidizing solutions into the mine to convert insoluble U^{+4}-minerals to highly mobile U^{+6} (e.g., Mudd 2001; Abzalov 2012), which can continue to be released long after mine operation has ceased. Both processes pose a significant risk to groundwaters downgradient of the mine, which can persist for years to come.

Remediation strategies have been implemented to immobilize U, *in situ*, by injecting amendment reagents directly into the subaerial groundwaters of the mine to stimulate the reduction of mobile U^{+6} to less mobile U^{+4}. Techniques in bioremediation are increasingly used, which take advantage of natural microbial populations within the mine systems to mediate U^{+6}–U^{+4} reduction (e.g., Anderson et al. 2003). Stimulated remediation strategies are not always applied to ISR mines, which commonly rely on a return to the natural reducing conditions of sandstone-hosted U ore zones once U removal through forced oxidation has ceased. Regardless of the remediation strategy, remediated U ore sites may return to more oxic conditions with time, thus increasing the mobility of U^{+6} and its potential for groundwater contamination. Moreover, laboratory-controlled biofilm experiments by Stylo et al. (2015a) demonstrate that noncrystalline U^{+4} species form preferentially over crystalline uraninite during bioremediation amendment procedures. Noncrystalline U^{+4} species are more readily oxidized to U^{+6} and remobilized than uraninite, which raises additional concerns about the stability of the bioremediation of U contaminated sites over time. Therefore, the long-term success of these remediation strategies remains uncertain and continual monitoring of groundwater quality is required to assess their effectiveness.

Several recent studies have investigated the potential of U isotopes to trace the mobility of U in the ore zone groundwater after mining has ceased, taking advantage of the comparatively large and well-characterized isotopic shifts that occur during U^{+6}–U^{+4} reduction. For example, Basu et al. (2015) and Paczek et al. (2016) observed $\delta^{238}U$ variations of more than 3‰ in groundwaters collected from nearby ISR mining sites at Rosita and Kingsville Dome, south Texas, USA (Fig. 6). A similar pattern was seen by Brown et al. (2016) at the ISR mining site of Smith Ranch, Wyoming, USA. At all sites, the respective authors also observed a trend towards lower U^{+6} concentrations and $\delta^{238}U$ values along the hydraulic gradient, consistent with progressive U reduction and removal from the aqueous phase along the direction of groundwater flow (Fig. 6). Estimated $^{238}U/^{235}U$ fractionation factors ε were ~0.48‰ for Rosita (Basu et al. 2015) and ~0.78–1.03‰ for Smith Ranch (Brown et al. 2016). Similar U isotope systematics were observed at the Rifle remediation test site, USA, following stimulated bioremediation through organic carbon amendment (Bopp et al. 2010; Shiel at al. 2016), and at the Pepegoona uranium mineralized system, Australia (Murphy et al. 2014), giving rise to respective ε values of ~0.46–0.85‰ and ~0.20‰ for the progressive U reduction and removal. These variable ε values may potentially be explained by competing processes for U removal, such as concurrent U removal from both adsorption and reduction (Brown et al. 2016; Shiel et al. 2016) or by biotic versus abiotic U reduction (Stylo et al. 2015b). Irrespective of such competing processes, these studies demonstrate the utility of U isotopes as an effective tracer of the zones of active U reduction at remediation sites.

NEAR-SURFACE U CYCLING AND THE MARINE $^{238}U/^{235}U$ MASS BALANCE

Understanding the present isotopic budget of the marine U cycle is important for the use of $\delta^{238}U$ values of sedimentary records to reconstruct environmental changes in the past on a local or even global scale (see *U isotopes as a paleo-redox proxy*). Several studies have been conducted with the aim of quantifying the marine U mass balance in the past (e.g., Barnes and Cochran 1990; Morford and Emerson 1999; Dunk et al. 2002). All of these studies agree that rivers provide 80–100% of the U input to the oceans. A comprehensive compilation of riverine U fluxes is given in Dunk et al. (2002) who concluded that the riverine U input is about 42×10^6 mol/year. The remaining U fluxes to the ocean are very uncertain but comprise input from direct groundwater discharge and a very minor contribution from atmospheric dust. The most important oceanic U sinks are sediments in anoxic basins, sediments at the continental shelves that become anoxic with depth in the sediment pile, biogenic carbonates and hydrothermally altered basalts at mid ocean ridges. Minor sinks represent metalliferous sediments (e.g., ferro-manganese crusts), pelagic clays and opaline silica. A larger amount of U may also be removed at coastal retention zones (Dunk et al. 2002). In the following, we will evaluate the $^{238}U/^{235}U$ ratios of these near-surface U fluxes and potential causes for their variability.

The $^{238}U/^{235}U$ in rivers and groundwater

The U isotope composition of rivers has been investigated by Stirling et al. (2007) and very recently by Tissot and Dauphas (2015), Noordmann et al. (2016), and Andersen et al. (2016). The latter study, provided a compilation of their own data combined with previous literature data, covering 36% of the present U flux to the oceans. They determined a weighted mean riverine $\delta^{238}U$ value, based on the U fluxes of the individual rivers of –0.34‰ (reported without an uncertainty estimate due to the limited knowledge of seasonal variability, see Andersen et al. 2016). However, this average value is strongly influenced by the low $\delta^{238}U$ of the Yangtze river (–0.70‰). Excluding this river from the compilation would lower the mean riverine value to –0.26‰. Nevertheless, in both cases, with or without the Yangtze river, the mean $\delta^{238}U$ is indistinguishable from that of the upper continental crust (–0.29‰, assuming an interlaboratory uncertainty of ±0.06‰; see *Uranium isotope systematics in high temperature environments on Earth*). Despite this conformity between the $\delta^{238}U$ of average crust and average rivers, individual rivers display large variations regarding their $\delta^{238}U$ values (–0.70 to +0.06‰) that exceed the variations observed among granitic rocks. There is no correlation between $\delta^{238}U$ and parameters characterizing the weathering regime, such as the Na/Ca elemental ratio, latitude or the ($^{234}U/^{238}U$) activity (Andersen et al. 2016; Noordmann et al. 2016). Instead, the variable U isotope composition of rivers may largely reflect U isotope variations between different catchments. Although the upper crust has an average granodioritic composition, most of its surface is covered by sediments or meta-sediments, which may have highly variable U isotope compositions. For example, dolomites, evaporites and oxides may have very low $\delta^{238}U$ of <–0.6‰ (Stirling et al. 2007; Weyer et al. 2008; Romaniello et al. 2013; Goto et al. 2014; Tissot and Dauphas 2015; Wang et al. 2016), while black shales and carbonates can be isotopically heavy with $\delta^{238}U \geq 0.0$‰ (e.g., Weyer et al. 2008; Romaniello et al. 2013; Andersen et al. 2014). Notably, the catchment of the Yangtze River, contains significant amounts of isotopically light evaporites and dolomites (Brennecka et al. 2011a) which may explain the very negative $\delta^{238}U$ signature in this river. Contrarily, the catchment of Indus, which has a relatively high $\delta^{238}U$ of –0.18‰ (Noordmann et al. 2016), includes a significant portion of black shales. However, to date, no systematic investigation of the relationship between the U isotope composition of rivers and the respective catchment has been conducted.

At present, no groundwaters have been studied for $\delta^{238}U$ apart from specific settings related to U ore deposits (see *Uranium isotopes in ore deposits*). Although the experimental study of Wang et al. (2015b) indicates that partial oxidation of U^{4+}, a process that may expected to be ubiquitous during weathering, is unlikely to generate significant U isotope fractionation, leaching experiments of Stirling et al. (2007) indicate that weathering may affect the U isotope composition of groundwater. This is also indicated by the highly variable U isotope composition observed for speleothems ($\delta^{238}U$ = –0.70 to +0.44‰) that formed in terrestrial environments by precipitation of carbonates (or sulfates) from sub-surface waters (Stirling et al. 2007). Groundwater is frequently out of secular equilibrium for ($^{234}U/^{238}U$), as α-recoil processes and associated damage to the crystal lattice structure of minerals result in preferential leaching of ^{234}U during weathering. As indicated by a negative correlation between $\delta^{238}U$ and ($^{234}U/^{238}U$) in speleothems, which have precipitated directly from groundwater, weathering may potentially generate ^{238}U–^{235}U fractionation and ($^{234}U/^{238}U$) disequilibrium, as ^{235}U and ^{234}U are preferentially leached from minerals by percolating waters (Stirling et al. 2007). This is supported by the study of Cheng et al. (2013), who also observed preferentially light U isotope signatures in stalagmites ($\delta^{238}U = -0.60$ to -0.30‰). However, the similar average U isotope composition of crust and rivers does not imply large-scale systematic U isotope fractionation during weathering and transport (Andersen et al. 2016; Noordmann et al. 2016). There also seems to be no significant anthropogenic influence on the U isotope composition of rivers, such as from U-rich fertilizers, as no correlation between the U and the P content was observed (Andersen et al. 2016).

The $^{238}U/^{235}U$ of seawater

As U has a long ocean residence time (of ~400,000 years; Ku et al. 1977), exceeding the typical mixing time of ocean water by more than 2 orders of magnitude, the U isotope composition of ocean water can be expected to be invariant. Indeed, several studies investigated the $\delta^{238}U$ of seawater (Stirling et al. 2007; Weyer et al. 2008; Andersen et al. 2014; Tissot and Dauphas 2015) and, with a few exceptions, individual analyses agree within their analytical uncertainties. These data have recently been compiled by Tissot and Dauphas (2015) who suggested a seawater mean value of $\delta^{238}U = -0.39 \pm 0.01$‰. Seawater analyses of more restricted basins, such as the North Sea, the Gulf of Mexico or the Persian Gulf (Tissot and Dauphas 2015), appear to show a larger variability than those of open ocean seawater, however, these do not show any systematic bias towards higher or lower $\delta^{238}U$. The global seawater mean $\delta^{238}U$ is slightly lower than the mean signature of riverine input, indicating that the integrated U sink from the oceans predominantly fractionates U isotopes towards high $\delta^{238}U$. Strongly restricted and stratified basins with an anoxic ($O_2 < 0.2$ mol/L; Berner 1981) deep water column (e.g., the Black Sea, Cariaco Basin, Baltic Sea, Saanich inlet and some fjords), resulting in non-conservative behavior of U, are not considered in the determination of the seawater mean $\delta^{238}U$. In such basins, U isotopes are variable with depth and can be strongly fractionated in the deep water column (Romaniello et al. 2009; Noordmann et al. 2015; Rolison et al. 2015), as a result of coupled U reduction and pore water diffusion.

The $^{238}U/^{235}U$ in reducing sediments. Reducing sediments are the largest U sink in the oceans and can be divided into:

- Sediments of restricted basins with an anoxic or euxinic (anoxic with H_2S) deep water column, referred to herein as anoxic sinks (Fig. 7).
- Sediments that become anoxic or euxinic with depth with an oxic to hypoxic (reduced O_2, e.g., as a result of high primary productivity at continental margins) water column above, referred to herein as hypoxic sinks (Fig. 7).

According to Dunk et al. (2002) these two sedimentary deposits have negative U fluxes of about 11 and 15 Mmol/year, respectively. As U in seawater is present as uranyl $(UO_2)^{2+}$ ions that form stable carbonate complexes, the reaction kinetics for U^{+6}–U^{+4} reduction in the water column are generally slow, even under anoxic conditions (Anderson et al. 1989). Correspondingly, U reduction is assumed to mainly occur in the porewaters within sediment or at the water-sediment interface, likely assisted by microorganisms (Anderson 1987; Anderson et al. 1989; Klinkhammer and Palmer 1991).

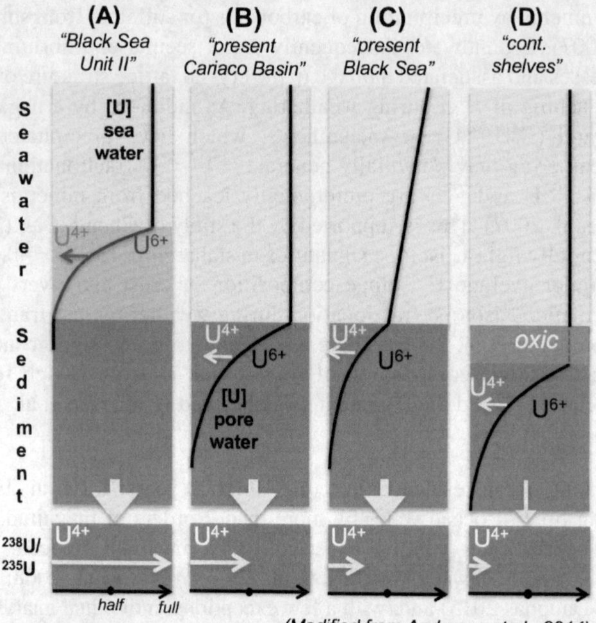

Figure 7. Four schematic scenarios of how different depositional environments may impact $^{238}U/^{235}U$ through reduction of U and authigenic accumulation in sediments; modified from Andersen et al. (2014). The upper panel shows the expected behavior of dissolved U concentrations in the water column and sediments. The downward pointing arrow represents the relative size of expected U accumulation in sediments. The lower panel shows the expected U isotopic shift in the accumulated reduced U^{+4} in the sediment, relative to isotopically unfractionated seawater (white vertical arrow). For comparison, the full length of the lower black vertical arrow indicates the full isotope fractionation expected during an $U^{+6} \rightarrow U^{+4}$ transition. In (A) the U reduction is assumed to occur in the water column (e.g., in the floc layer above the sediments) and not restricted by transport–diffusion, thereby expressing a $^{238}U/^{235}U$ close to the "full" isotope fractionation. In (B), (C) and (D), U reduction occurs within the sediments leading to observed U isotope fractionation that is less than the full shift of $^{238}U/^{235}U$ of the sediments relative to that of seawater; in (B) due to transport–diffusion limitations (Clark and Johnson 2008); in (C) as in (B) but combined with U isotope fractionation in the deep water column of restricted basins; and in (D) as in (B) but with an oxic penetration zone through the sediments. The sediments of the indicated type localities (top) are not expected to completely behave as indicated by the schematic endmembers, but may well be affected by a mixture of different processes.

The $^{238}U/^{235}U$ in sediments from anoxic basins. The U isotope composition of sediments from restricted and permanently anoxic basins, including the Black Sea, Cariaco Basin, Saanich inlet, the Gotland and Landsort deeps of the Baltic Sea, the Kyllaren Fjord (Norway) and four fjords from New Zealand, have been analyzed (Weyer et al. 2008; Montoya-Pino et al. 2010; Andersen et al. 2014; Holmden et al. 2015; Noordmann et al. 2015; Hinojosa et al. 2016). In all of these settings, the sediments typically display $\delta^{238}U$ that are significantly higher than those of seawater and average continental crust (with $\delta^{238}U$ ranging between −0.2 and +0.4‰ in most cases). Likewise, the deep water column of Kyllaren Fjord shows low $\delta^{238}U$ of around −0.7‰ (Noordmann et al. 2015) and similarly low $\delta^{238}U$ have also been reported for the Black Sea (Romaniello et al. 2009; Rolison et al. 2015). These findings are consistent with U isotope fractionation during partial U reduction and accumulation in the sediments, coupled with back-diffusion of isotopically light U into the water column. The direction of U isotope fractionation is also consistent with that observed in many experimental studies (e.g., Nomura et al. 1996; Fujii et al. 2006; Basu et al. 2014; Stirling et al. 2015; Stylo et al. 2015a,b; Wang et al. 2015b) and *ab initio* molecular orbital modeling (e.g., Abe et al. 2008b).

Although anoxic sediments are typically enriched in U, they may also have a substantial contribution of detrital terrigenous U, depending on the authigenic U content (as defined by the ratio of authigenic U flux and the mass accumulation rate) and the detrital U content in the sediment. Even in sediments of restricted basins with highly sulfidic deep water, detrital U may represent up to 50% of the total U budget if sedimentation rates are high (Noordmann et al. 2015). For the determination of the isotopic composition of authigenic U it is thus important to correct for the contribution of detrital U (e.g., Asael et al. 2013; Andersen et al. 2014; Noordmann et al. 2015; Hinojosa et al. 2016). This requires determination of the detrital component of the sediment which may be monitored by the analyses of "non-authigenic" elements, such as Al, Ti, Zr or Th (e.g., Tribovillard et al. 2006). As both the elemental and U isotope composition may be difficult to analyze directly, or are unknown, an average crustal composition (e.g., Rudnick and Gao 2003; Asael et al. 2013; Andersen et al. 2016) is frequently used for the detrital U correction, although detrital components may be variable and may have low $\delta^{238}U$ in some cases (e.g., Saanich Inlet; Holmden et al. 2015). The U isotope composition of the authigenic U may be determined by applying the following equations:

$$[U]_{det} = (U/Al)_{det} \cdot [Al]_{sam} \tag{9}$$

$$[U]_{auth} = [U]_{sam} - [U]_{det} \tag{10}$$

$$\delta^{238}U_{auth} = (\delta^{238}U_{sam} \cdot [U]_{sam} - \delta^{238}U_{det} \cdot [U]_{det}) / [U]_{auth} \tag{11}$$

In these equations, $[U]_{det}$ is the concentration of U that comes from the detritus, $(U/Al)_{det}$ is the U/Al ratio of the detritus, $[Al]_{sam}$ is the measured Al concentration of the sample, $[U]_{auth}$ is the concentration of U that is of authigenic origin, $[U]_{sam}$ is the measured U concentration of the sample, $\delta^{238}U_{auth}$ is the $\delta^{238}U$ of the authigenic U, $\delta^{238}U_{sam}$ is the measured $\delta^{238}U$ of the sample and $\delta^{238}U_{detritus}$ is the $\delta^{238}U$ of the detritus.

The authigenic contribution of U in recent sediment may also be estimated from the deviation of the ($^{234}U/^{238}U$) activity ratio of the sediment sample from that of seawater of 1.147 (Andersen et al. 2010), following the approach of Holmden et al. (2015) and Andersen et al. (2016). Biogenic carbonates may have high U concentrations and thus significantly contribute to the bulk U in sediments. This contribution may be monitored by the analysis of the Ca content of the sediment and a crude first-order correction may be applied assuming an average U concentration for primary carbonate precipitates of ~ 1–1.5 μg/g (Dunk et al. 2002; Romaniello et al. 2013; Andersen et al. 2014).

Assuming partial U reduction occurs at isotopic equilibrium at the sediment–water interface, U removal can be considered a one-directional process from a closed system, because of the sluggish reaction behavior of the U^{+4} reduction products. The resulting U isotope fractionation may then be described by a Rayleigh-type model and approximated by the following equations:

$$\delta_{reactant} = \delta_0 + \varepsilon \ln f \tag{12}$$

$$\delta_{accum.\ product} = \delta_0 + \varepsilon \ln f - (\varepsilon \ln f)/(1-f) \tag{13}$$

In these equations, $\delta_{reactant}$ describes the $\delta^{238}U$ of the U remaining in water column; δ_0 is the original $\delta^{238}U$ of the water column (e.g., seawater); ε is the isotope enrichment factor, derived from the isotope fractionation factor α ($= {}^{238}U/{}^{235}U_{product} / {}^{238}U/{}^{235}U_{reactant}$) using the equation $\varepsilon \equiv (\alpha - 1) \times 1000$ (‰) and which closely approximates $\delta_{reactant} - \delta_{inst.prod}$ (the difference between the $\delta^{238}U$ of the reactant and that of the instantaneous product of U reduction); f is the fraction of U remaining in the reactant, and $\delta_{accum.\ product}$ is the $\delta^{238}U$ of the reduced U that accumulated in the sediment (Fig. 7).

In the case of the present Black Sea, the authigenic U in the deep sediment has a $\delta^{238}U$ of between 0.0‰ and +0.1‰ (Weyer et al. 2008; Montoya-Pino et al. 2010; Andersen et al. 2014), while $\delta^{238}U$ as light as −0.7‰ have been reported for deep waters (Romaniello et al. 2009), significantly lower than the open ocean $\delta^{238}U$ composition (−0.39‰). Thus, the present day difference in the $\delta^{238}U$ signature between the water column and the underlying sediment, given by $\Delta^{238}U_{water-sediment}$, is ~0.7‰, which is similar to estimates for the Saanich inlet (Holmden et al. 2015). Applying these parameters for the Black Sea in the above-described Rayleigh model results in an isotope fractionation factor ε of about 0.6‰. This value is about a factor of 1.3 to 2 lower than that predicted by experimental studies (e.g., Fujii et al. 2006; Basu et al. 2014; Stirling et al. 2015; Stylo et al. 2015a,b; Wang et al. 2015b) and *ab initio* molecular orbital modeling (e.g., Abe et al. 2008b). Very similar U isotope fractionation between the deep water column and authigenic U in the underlying sediment was estimated by Noordmann et al. (2015) for Kyllaren Fjord.

As U reduction is assumed to mainly occur in the sediment, it is likely coupled with U pore water diffusion from the water column into the sediment, which may impact the net isotope effects of the accumulated authigenic U in the sediment (Fig. 7). Andersen et al. (2014) allowed for these effects, applying a diffusion–reaction–transport model based on previous work (Bender 1990; Clark and Johnson 2008). Assuming continuous U reduction from the top to the bottom of a fictive sediment column, the expected net (or effective) isotope fractionation between the average accumulated authigenic U and U in the dissolved phase is expected to be about half of the full isotope fractionation factor for reduction (Fig. 7B). In the case of the Black Sea, this would imply ε ~ 1.2‰ (Andersen et al. 2014) which is within the range given by experimental and *ab initio* studies.

The extent of U isotope fractionation, however, as estimated from the observed $\delta^{238}U$ in the sediment, will ultimately depend on the time-integrated $\delta^{238}U$ of the deep water column during the formation of the sediment (i.e. its diffusive interaction with the overlying water column). If the latter has a low $\delta^{238}U$ that is fractionated relative to open-ocean seawater, as a result of a high U flux into the sediment and low deep water renewal, this will also lower the observed $\delta^{238}U$ in the sediment (Andersen et al. 2014; Noordmann et al. 2015). This is, for example, the case for the present Black Sea. Thus, variable U depletion and $\delta^{238}U$ of the deep water may result in variable observed $\delta^{238}U$ in the sediment, even with the same effective isotope fractionation during U reduction (Fig. 7C). This may potentially explain the relatively low $\delta^{238}U$ of recently deposited sediments from the Cariaco basin (Andersen et al. 2014) which may imply that the overlying anoxic water column displays a stronger time-integrated U depletion than presently observed (~10%; Anderson 1987) and a lower net U isotope fractionation compared to the open-ocean $\delta^{238}U$ during sediment formation.

Some very high $\delta^{238}U$ values of up to ~+0.4‰, as observed in Holocene-aged Unit II sediments of the Black Sea (Weyer et al. 2008; Montoya-Pino et al. 2010), are difficult to explain with variable U isotope compositions of the overlying water column alone. These values more likely indicate variations in the effective isotope fractionation between water and sediment during U reduction. Such variability could be based on variations in the total isotope fractionation factor, which is likely ligand-dependent (Abe et al. 2010) and differences in biotic and abiotic reduction systems (Stylo et al. 2015b). Likewise, variable effective isotope effects could occur depending on whether U reduction takes place in the water column, closer to the sediment–water interface or deeper in the sediment (Fig. 7). If a larger amount of U was reduced in the surface floc which overlies the consolidated sediment (e.g., associated with biomass; Anderson et al. 1989), then the process of U reduction may be less limited by diffusion through the sediment pile, resulting in net U isotope fractionation being closer to the full isotope enrichment factor of ~0.8 to 1.6‰ (e.g., Fujii et al. 2006; Abe et al. 2008b; Basu et al. 2014; Stirling et al. 2015; Stylo et al. 2015a,b; Wang et al. 2015b).

The $^{238}U/^{235}U$ in sediments from hypoxic settings. Hypoxic sediments, derived from the continental margins, are commonly less enriched in U and are thus more affected by detrital components. The U isotope composition of the hypoxic sink has initially been estimated by the analyses of sediments from the Peru continental margin by Weyer et al. (2008) and more recently in studies by Andersen et al. (2014) and (2016) who analyzed sediment samples from the shallower slope of the Black Sea and the continental shelf off Washington State (USA), respectively. These sediments together display a range of $\delta^{238}U$ between −0.42‰ and +0.44‰. However, the Black Sea shelf sediments have been deposited under highly variable redox conditions in the overlying water column due to fluctuations of the chemocline. At present, the water column (and pore waters of the upper few cm of the sediment) is oxic, however it was likely to have been euxinic in the recent past (Lyons et al. 1993; Arnold et al. 2012). These variable conditions may have resulted in U remobilization and associated U isotope fractionation. Likewise, variable $\delta^{238}U$ (−0.39‰ to −0.13‰) have also been observed for sediments of the Gotland and Landsort deeps of the Baltic Sea (Noordmann et al. 2015) which are overlain by sulfidic water at present. However, periodically occurring flushing events have resulted in dramatic changes in the water column redox conditions in the recent past (Huckriede and Meischner 1996).

Thus, the margin sediments off Peru and Washington State may be more suitable than those from the Black Sea shelf for estimating an average authigenic $\delta^{238}U$ for hypoxic sinks. As the overlying water column of these sediments is oxic to suboxic, oxygen is able to penetrate into the uppermost few cm of the sediment and U reduction takes place within a distinct sediment layer, likely mediated by microorganisms (Lovley et al. 1991; McManus et al. 2005; Morford et al. 2009). Under these conditions, coupled pore water diffusion and U reduction results in low effective U isotope fractionation factors (Fig. 7D), as U reduction is largely quantitative (see Clark and Johnson 2008, who quantified these processes in experiments with Se isotopes). Furthermore, variations in the water column redox conditions or bioturbidation (Zheng et al. 2002; McManus et al. 2005) may result in fluctuations of the oxygen penetration zone associated with the remobilization of U, and may affect the net U isotope effect (Andersen et al. 2014; Noordmann et al. 2015). Lastly, a correlation between U concentration and the C flux through the water column in upwelling zones has been highlighted as a potential important source of authigenic U in sediments at continental margins (Zheng et al. 2002; McManus et al. 2005). Such a source of authigenic U would have $\delta^{238}U$ significantly different from U accumulation due to the U reduction process, potentially carrying a low $\delta^{238}U$ signature as observed in plankton traps (Holmden et al. 2015; Hinojosa et al. 2016). Nevertheless, Andersen et al. (2016) estimated a $\delta^{238}U$ of ~+0.1‰ for the seawater-related authigenic endmember of the uppermost 20 cm of the Washington State margin sediments (using a ($^{234}U/^{238}U$)–$\delta^{238}U$ mixing relationship, see *The $^{238}U/^{235}U$ in sediments from anoxic basins*). Combining all available data from the margin sediments of Peru and Washington State and applying a detrital U correction gives an average $\delta^{238}U$ of −0.24‰ for the hypoxic sink, indicating only minor net U isotope fractionation, on average, during the removal of seawater U by marginal sediments.

The $^{238}U/^{235}U$ in marine carbonates

Carbonates may have highly variable U concentrations, depending on their mineralogy and mechanism of carbonate precipitation. Aragonitic carbonates, such as corals and aragonitic slope sediments, commonly have high U concentrations of several µg/g, likely because of their direct incorporation of $UO_2(CO_3)_3^{4-}$ from seawater (Reeder et al. 2000). In contrast, calcitic carbonates commonly have lower U contents as a coordination change is likely necessary to allow U to fit into the calcite structure (Reeder et al. 2000; Kelly et al. 2003, 2006).

Analyses of the $\delta^{238}U$ in recently deposited marine carbonates have been reported by Stirling et al. (2007), Weyer et al. (2008), Andersen et al. (2014) and Tissot and Dauphas

(2015), and in a more detailed study by Romaniello et al. (2013). The overall motivation, in particular for the latter study, was to verify whether carbonates generally record the primary seawater δ^{238}U at the time of their deposition, which would make them an ideal paleo-redox proxy (Brennecka et al. 2011a; Dahl et al. 2014; Lau et al. 2016). Indeed, these studies show that although the investigated primary carbonate precipitates display a large range in [U] ranging between 0.014 and 3.6 µg/g, they have fairly invariant δ^{238}U around the seawater value with a mean of −0.40 ± 0.15‰ (2 SD). These, results are in good agreement with an experimental study (Chen et al. 2016, see *Experimental evidence for kinetic and equilibrium $^{238}U/^{235}U$ fractionation*) showing limited U isotope fractionation during U incorporation into calcium carbonate precipitated from aqueous media. Additionally, fossil carbonates, including those formed several hundred thousand years ago that show clear evidence of alteration with respect to their (^{234}U/^{238}U) compositions, appear to retain 'pristine' δ^{238}U signatures (Stirling et al. 2007; Andersen et al. 2014). Moreover, analyses of ancient carbonates from the late Permian and early Triassic by Brennecka et al. (2011a) and Lau et al. (2016), did not show any evidence for diagenetic alteration and showed either δ^{238}U similar to those of modern primary carbonate or lighter compositions, which were interpreted to reflect global seawater δ^{238}U variations (see *U isotopes as a paleo-redox proxy*).

The above observations of primary δ^{238}U signatures being recorded by carbonate precipitates contrasts with a study of recent carbonates sampled via short (ca. 50 cm long) drill cores from the Bahamas carbonate platform, which show increasing U enrichment with depth (Romaniello et al. 2013). Likewise, these carbonates also display elevated δ^{238}U that ranges between −0.24 and +0.07‰ (with a mean of −0.13‰) and crudely correlates with U concentrations. These observations are similar to those of Andersen et al. (2016) for sediments from the Washington State continental margin and clearly indicate authigenic U enrichment with sediment depth as conditions become more anoxic. Such early diagenetic processes may alter the primary U signal of carbonates. Furthermore, during dolomitization, carbonates may become isotopically light. For example, a δ^{238}U of −0.83‰ was reported for dolomite chimney deposits in Stirling et al. (2007). Thus, the δ^{238}U of biogenic carbonates may be shifted towards both lower and higher values during diagenesis and care must be taken if using carbonates as a recorder for paleo-seawater U isotope compositions. A recent study by Lau et al. (2016) used several indicators for diagenetic alteration, such as oxygen isotopes, Mn/Sr, Mg/Ca, Sr/Ca or TOC to exclude that the observed δ^{238}U were generally affected by diagenetic processes or to eliminate individual samples that may have been affected by post-depositional alteration.

The ^{238}U/^{235}U of altered oceanic crust

Uranium is one of only a few elements that is significantly removed from seawater during hydrothermal alteration processes at midocean ridges (MOR) (e.g., Staudigel et al. 1996) and thus, U is distinctly enriched in alteration products of AOC (e.g., Staudigel et al. 1995; Bach et al. 2003; Kelley et al. 2005). Recent studies have revealed that this process can lead to significant U isotope fractionation (Andersen et al. 2015; Noordmann et al. 2016), as recorded by bulk AOC samples (δ^{238}U = −0.47‰ to +0.27‰), hydrothermal fluids (δ^{238}U = −0.59‰ to −0.28‰) and U-rich calcium carbonate veins within AOC (δ^{238}U = −0.63‰ to +0.11‰).

Andersen et al. (2015) investigated different crustal levels within the upper 500 m of AOC, where most of the U removal during low-temperature hydrothermal cycling is thought to occur, on the basis of δ^{238}U measurements of the Jurassic-aged ODP 801 drill core from the Pacific plate (Plank et al. 2000). Although all investigated samples display significant U-enrichment, on average by a factor of ~5 compared to unaltered MORB, δ^{238}U values show variations that are distinctly depth-dependent. While samples from the uppermost section (0–110 m) display slightly lower δ^{238}U (~ −0.44‰) compared to seawater, samples from two deeper sections (between 110–420 m) show significantly higher δ^{238}U with average values of +0.16‰ and

−0.14‰. These data imply that in the uppermost levels of the oceanic crust, U removal from seawater and incorporation into AOC occurs without a significant redox change (potentially by adsorption), resulting in little U isotope fractionation. Potentially, the conditions in the uppermost oceanic crust are either too oxidizing for U reduction to occur or the high stability of $UO_2(CO_3)_3^{4-}$ complexes prevents U^{+6} from reducing to U^{+4}. The strong shift towards high $\delta^{238}U$ in the lower sections of this core strongly indicates that in the deeper, more reducing oceanic crust, partial reduction of U^{+6} to U^{+4} is the predominant process resulting in U removal from hydrothermal waters and subsequent enrichment in AOC, as no other process seems likely to generate such significant U isotope fractionation.

Noordmann et al. (2016) investigated the U isotope composition of the significantly U-enriched AOC from 3 different drill cores (Pigafetta Basin, Bermuda Rise and Reykjanes Ridge). They observed low $\delta^{238}U$ (−0.46‰) also in deeper crustal levels (~500 m). Potentially, oxygen-rich fluids penetrated into the deeper crust via cracks, or the fluids became isotopically light by previous U reduction. The highest $\delta^{238}U$ (+0.27‰) in this study was observed in a sample from a depth of 823 m that was enriched in U by a factor of ~ 10 compared to unaltered MORB, demonstrating that significant hydrothermal U enrichment can still occur at deeper crustal levels.

From their data, Andersen et al. (2015) determined the U isotope composition of a supercomposite AOC sample, integrated over the full upper 500 m of the ODP 801 AOC, with a $\delta^{238}U = -0.17‰$, that agrees, within uncertainties, with the weighted average for AOC ($\delta^{238}U = -0.25‰$) of Noordmann et al. (2016). This implies that seafloor alteration results in the preferential removal of heavy U isotopes from the oceans. This finding is also consistent with the slightly lower $\delta^{238}U$ of two hydrothermal fluid samples relative to that of seawater (Noordmann et al. 2016). Assuming that the heavy U isotope signatures observed in the AOC were generated by equilibrium U isotope fractionation between U^{+4} and U^{+6}, this requires the coexistence of both U species in solution for a sufficient time to allow for isotopic exchange (Wang et al. 2015b). Potentially, the hydrothermal fluids provide adequate conditions for this to occur at elevated temperature and low pH.

The $^{238}U/^{235}U$ of ferromangenese oxides

Ferromanganese crusts and nodules are moderately enriched in U and provide only a minor sink for U from the oceans (Dunk et al. 2002). Their U isotopic composition was analyzed in several studies (Stirling et al. 2007; Weyer et al. 2008; Goto et al. 2014; Wang et al. 2016) revealing only a limited range of $\delta^{238}U$ (−0.52‰ to −0.71‰) with an average value of −0.62‰. Most analyses have been performed on Fe–Mn crusts from various seamounts (Goto et al. 2014; Wang et al. 2016), however, the limited number of Mn nodules analyzed to date indicates that there is no systematic difference between average $\delta^{238}U$ of Fe–Mn crusts and other metalliferous sediments. These findings imply that fairly constant U isotope fractionation between seawater and Fe–Mn crusts and nodules, represented by $\Delta^{238}U_{\text{seawater–Fe–Mn-oxide}} = -0.23‰$ occurs during the incorporation of U into ferromanganese oxides. An experimental study of Brennecka et al. (2011b) determined similar U isotope fractionation behavior during the adsorption of U onto the Mn oxide birnessite. The latter experimental study attributed the observed U isotope fractionation during adsorption to a subtle difference in U coordination between dissolved and adsorbed U^{+6} species (as determined by EXAFS analyses). Studies by Goto et al. (2014) and Wang et al. (2016) furthermore imply that seawater $\delta^{238}U$ was likely fairly constant throughout the Cenozoic, as indicated by the constant $\delta^{238}U$ of depth profiles through several Fe–Mn crusts. The ($^{234}U/^{238}U$) data for the same Fe–Mn crusts indicate that some post-depositional remobilization of U has affected the uppermost 1–2 mm of the surface layers of the crust (Goto et al. 2014; Wang et al. 2016). This U remobilization was likely the result of diffusional exchange with seawater and may limit the time resolution of the depth profiles (determined with ^{10}Be and Re–Os dating) to about 3 m.y. (Wang et al. 2016). However, as discussed earlier, ^{234}U may have been preferentially mobilized during alteration due to its location in a damaged crystal lattice site, while tightly lattice-bound ^{238}U and ^{235}U may not have been disturbed at all, such that they still record primary ^{238}U/^{235}U signatures.

Isotopic constraints on the marine U cycle

Several recent studies have used the U isotope data of the various oceanic U sources and sinks, combined with previous constraints on the U oceanic mass balance (Goto et al. 2014; Tissot and Dauphas 2015; Andersen et al. 2016; Noordmann et al. 2016; Wang et al. 2016) to: (1) set limits on the mean δ^{238}U of poorly constrained oceanic sinks, (2) assess previous constraints on the relative weight of oceanic U fluxes (based on estimates of the total U fluxes of the diverse range of oceanic sources and sinks and constraints from other parameters, such as (^{234}U/^{238}U)), and (3) assess the likelihood that the present oceanic U budget is at steady state (Broecker 1971; Dunk et al. 2002). Most studies investigating the oceanic U budget, including those based on δ^{238}U variations, assume that the riverine U input to the oceans is by far the dominant source. No other U sources (e.g., dust and groundwater discharge, see Dunk et al. 2002) are isotopically well described and, thus, are not considered further here. The major U sinks display either isotopic fractionation towards heavier U isotope compositions or essentially no U isotope fractionation during U removal. Only the minor U sink of metalliferous sediments is enriched in light U isotopes, although other potential, but currently unconstrained, sinks (notably those in coastal retention zones; Dunk et al. 2002; Holmden et al. 2015; Tissot and Dauphas 2015) may also have low δ^{238}U. These U isotope systematics (Fig. 8) are reflected in the marginally fractionated U isotope composition of modern seawater that is ~0.1‰ lower than the dominant riverine U source.

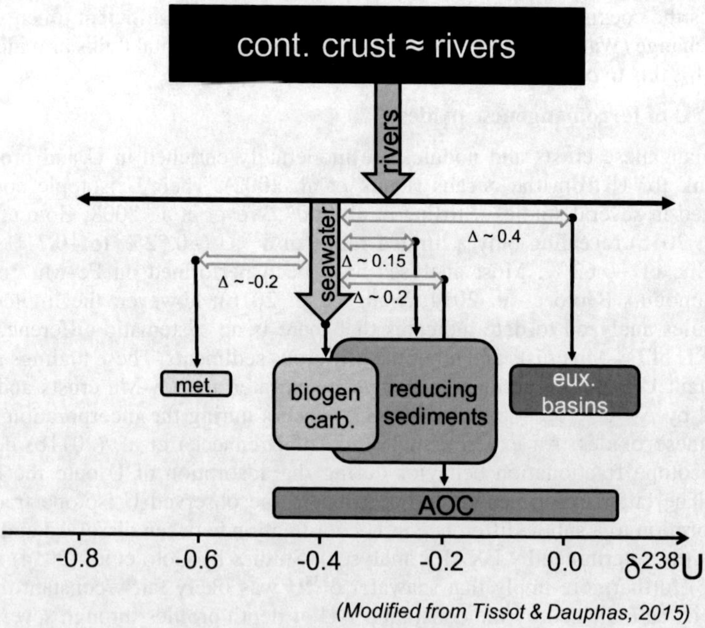

Figure 8. Schematic illustration of the isotopic U mass balance of the modern oceans, similar to Figure 10 of Tissot and Dauphas (2015). The widths of the sink boxes crudely represents their U isotopic range in nature, while the area of the sink boxes crudely represents their relative weights (not scaled). The grey arrows indicate mean U isotope fractionation of the various sinks relative to seawater. The sinks considered are metalliferous sediments (met.), biogenic carbonates (biogen carb.), reducing sediments, anoxic/euxinic sediments from restricted basins (eux. basins) and altered oceanic crust (AOC), taken from Table 3.

The relatively large flux of U into sediments of restricted anoxic basins, where the largest isotope fractionation during U removal from ocean water occurs ($\Delta^{238}U_{seawater-anoxic} \sim 0.4‰$), also has the largest effect on the oceanic U isotope budget. Thus, if the global ocean is currently at steady state regarding its U budget, then there is an upper limit for the size of this sink, even if all other sinks combined do not additionally fractionate U isotopes. However, hypoxic shelf sediments, AOC, and potentially carbonates are also enriched in heavy U isotopes and, thus, their size and $\delta^{238}U$ is also critical to the oceanic U isotope mass balance. We will focus here on the U budget as suggested by Dunk et al. (2002) and combine, for each of the U sinks, the estimated mean $\delta^{238}U$ from above with the respective sizes of the sinks (Table 3). If U in the ocean is at steady state, the following mass balance must be fulfilled:

$$\delta^{238}U_{In} = {}^{238}U_{red} \times f_{red} + {}^{238}U_{anox} \times f_{anox} + {}^{238}U_{carb} \times f_{carb} + {}^{238}U_{hyd} \times f_{hyd} + {}^{238}U_{met} \times f_{met} \quad (14)$$

where f denotes the fraction of the total oceanic U that goes into the individual sinks. Subscripts are "in" for input (i.e. all riverine assumed here), "red" for reducing sediments, excluding those from restricted basins with anoxic/euxinic conditions in the water column above, "anox" for reducing sediments from restricted anoxic/euxinic basins, "carb" for carbonates, "hyd" for products of hydrothermal alteration, and "met" for metalliferous sediments (including Fe–Mn-nodules).

In the U budget of Dunk et al. (2002), the "anoxic" sink includes all sediments where conditions in the water column are anoxic. However, regarding U isotope fractionation, several studies indicate that strong U isotope fractionation only occurs under permanently anoxic conditions, such as in the Black Sea, the Cariaco Basin or some water-flow restricted

Table 3. Modeling of the net $\delta^{238}U$ of all combined U sinks from the oceans

Data used for modeling	**Dunk (2002)**	**Dunk mod***	**Best estimate**	**Lower limit**	**Upper limit**	**References**** ($\delta^{238}U$)**
	U fraction			$\delta^{238}U$		
Input (rivers)	1	1	**−0.29**			1,2,3,4
Sinks						
Anoxic Basins	0.25	0.13	**0**	−0.1	0.1	5,6,7,8
Other reducing	0.33	0.45	**−0.25**	−0.3	−0.2	4,5
Carbonates	0.28	0.28	**−0.4**	−0.4	−0.2	1,2,5,7,9
Hydrothermal AOC	0.12		**−0.2**	−0.4	−0.1	3, 10
Metalliferous (oxic)	0.02		**−0.6**			5,11,12

Modeling results	**Best estimate**		**Lower limit**		**Upper Limit**	
$\delta^{238}U$	Dunk (2002)	Dunk mod	Dunk (2002)	Dunk mod	Dunk (2002)	Dunk mod
	−0.23	**−0.26**				
with alt. anoxic sink			−0.26	−0.27	−0.21	−0.25
with alt. reducing sink			−0.25	−0.28	−0.22	−0.24
with alt. carbonate sink			−0.23	−0.26	−0.18	−0.20
with alt. hydrothermal sink			−0.26	−0.28	−0.26	−0.25

*including anoxic conditions at upwelling zones of continental shelves in "other reducing sinks" (= "suboxic sediments" in Dunk et al. 2002).
** (1) Stirling et al. 2007; (2) Tissot and Dauphas 2015; (3) Noordmann et al. 2016; (4) Andersen et al. 2016; (5) Weyer et al. 2008; (6) Montoya-Pino et al. (2010); (7) Andersen et al. 2014; (8) Noordmann et al. 2015; (9) Romaniello et al. 2013; (10) Andersen et al. 2015; (11) Goto et al. (2014); (12) Wang et al. (2016)

fjords (Weyer et al. 2008; Andersen et al. 2014; Noordmann et al. 2015), while settings with periodic changes from anoxic to oxic/suboxic conditions, such as the oxygen minimum zones of continental slopes, estuaries, or the Baltic Sea (Holmden et al. 2015; Noordmann et al. 2015; Andersen et al. 2016) result in variable, but limited U isotope fractionation. As restricted and permanently anoxic basins account for only half of the anoxic sink as suggested by Dunk et al. (2002), we redefined a modified mass balance, labeled as "Dunk-mod" in Table 3. In our "Dunk-mod" mass balance model, the "reducing" sink, thus, constitutes all reducing settings apart from restricted anoxic basins and represents the "suboxic" sink and half of the "anoxic" sink of Dunk et al. (2002). To explore the effect of the U isotope compositions of each of the individual sinks on the total U isotope budget, we show different model calculations. This includes those based on the "best estimate" $\delta^{238}U$ values of each U sink, as determined above and, furthermore, those based on the lower and upper limits of $\delta^{238}U$ for each U sink that can be accommodated by the uncertainty limits. Regarding the carbonate sink, we assume here that the $\delta^{238}U$ of primary carbonates (~0.4‰; e.g., Romaniello et al. 2013) is the "best estimate". The isotopically heavy U in carbonates at marginal settings is included in the "hypoxic" sink in the mass balance suggested by Dunk et al. (2002) (and in the "other reducing" sink in the "Dunk-mod" mass balance model presented here; see Table 3).

Combining the U fluxes as suggested by Dunk et al. (2002) and our "best estimate" $\delta^{238}U$ for the individual sinks would result in a net $\delta^{238}U$ for all sinks combined that may be slightly higher (−0.23‰) than that of the riverine U input, assuming this reflects the U isotope composition of the continental crust (−0.29‰; Table 3). However, combining the modified U fluxes ("Dunk-mod" as described above, based on Dunk et al. 2002) with the same $\delta^{238}U$ values results in a net $\delta^{238}U$ sink of −0.26‰ that agrees, within uncertainties, with the riverine estimate. Likewise, a slightly lower estimate for the $\delta^{238}U$ of the hypoxic sink ("other reducing" sediments in Table 3) or the hydrothermal sink (AOC) may also result in a net $\delta^{238}U$ that is closer to the riverine estimate. Assuming a modified average carbonate $\delta^{238}U$ of −0.2‰, by including measurements from reducing settings in the compilation (Romaniello et al. 2013), would shift the net $\delta^{238}U$ sink towards values (−0.18‰ to −0.20‰) that are significantly higher than the riverine input. Overall, the estimates for the net oceanic $\delta^{238}U$ sink using variable input data tend to be systematically higher than the $\delta^{238}U$ value of the bulk continental crust, although most of the modeled output values agree with the latter within the magnitude of their uncertainties (Fig. 8, Table 3). This may have the following reasons: (1) the oceanic U budget is not quite at steady state; (2) the riverine $\delta^{238}U$ is slightly higher than that estimated for the bulk continental crust; or (3) the estimate for the $\delta^{238}U$ of the net oceanic sink is too high. This latter possibility could be the case if adsorption processes on oxides or organic material in coastal retention zones (not considered here) result in low $\delta^{238}U$ for this sink and thus lower the $\delta^{238}U$ of the net oceanic sink (Tissot and Dauphas 2015). For example, assuming that costal retention zones have a similarly low $\delta^{238}U$ as metalliferous sediments (~−0.6‰), as suggested by Tissot and Dauphas (2015), then the net oceanic $\delta^{238}U$ lowers by ~0.05‰, resulting in a better agreement with the value for continental crust. Likewise, riverine input might be isotopically heavier than bulk crust if the average catchment of worldwide rivers has a $\delta^{238}U$ that lies closer to that of the major sedimentary U sinks, which are carbonates, black shales and sediments from the continental shelves. Finally, the oceanic budget might be at non-steady state if the modern U isotope budget, as determined and applied here, is not representative of the longer term average relative to the ~400 k.y. U ocean residence time. For instance, glacial-interglacial cycles may influence the relative sizes of the sources and sinks, as has been suggested for other elements with long ocean residence times (e.g., Sr; Vance et al. 2009). This may be particularly relevant for redox-sensitive elements, such as U, as some currently anoxic basins, such as the Cariaco Basin and the Black Sea, had oxic deep waters or were fresh water lakes during the last glacial period (e.g., Haug et al. 1998; Lericolais et al. 2007). Thus, the U isotope budget may be different when averaged over a full glacial cycle compared with the interglacial conditions of the modern ocean, adding uncertainty when using the latter budget for testing steady-state processes over longer timescales.

U ISOTOPES AS A PALEO-REDOX PROXY

According to its redox-sensitive character, the abundance and isotope composition of U recorded in sediments are frequently used as proxies to reconstruct the redox evolution of the oceans and atmosphere (Montoya-Pino et al. 2010; Brennecka et al. 2011b; Asael et al. 2013; Kendall et al. 2013, 2015; Partin et al. 2013a,b; Dahl et al. 2014; Azmy et al. 2015; Satkoski et al. 2015; Lau et al. 2016). The use of U isotopes as a paleo-redox proxy was stimulated by the finding of significant fractionation of U isotopes during redox processes, in particular during U reduction and incorporation into anoxic shales (Weyer et al. 2008) compared to the relatively insignificant fractionation of δ^{238}U observed in other oceanic environments. To date, mainly black shales or carbonates have been used as U isotope archives, assuming that they either directly record the δ^{238}U of ancient seawater, as is likely to be approximately the case for oceanic carbonates (see *The $^{238}U/^{235}U$ in marine carbonates*), or they experienced distinct, well characterized U isotope fractionation during U incorporation, as is inferred for black shales (e.g., Weyer et al. 2008; Montoya-Pino et al. 2010; Andersen et al. 2014). In recent studies, Goto et al. (2014) and Wang et al. (2016) used the δ^{238}U of ferromanganese oxides, which appear to display very constant U isotope fractionation relative to seawater (see *The $^{238}U/^{235}U$ of ferromanganese oxides*), as an archive for reconstructing Cenozoic oceanic paleo-redox conditions.

As U has a long ocean residence time of ca. 400 k.y. (Ku et al. 1977) in the modern oxygenated oceans, oceanic mass balance results in a well-mixed and globally uniform U isotopic signal in seawater. Fluctuations in the relative importance of the individual sinks driven, for example, by the enhanced occurrence of seafloor anoxia, should be recorded in the U isotope composition of the individual sinks, assuming that U isotope fractionation into these sinks remains constant with time, irrespective of changing environmental conditions (Fig. 9). In contrast, in the dominantly anoxic surface Earth of the Archean, favouring particle-reactive U^{+4}, the abundance of dissolved U^{+6} in the ocean was very low and the oceanic residence time of U may have been short relative to the ca. 1,500 year modern ocean mixing time (e.g., Li et al. 2013; Partin et al. 2013a). Thus, Archean δ^{238}U signatures are likely to provide indicators of local or regional, rather than global, redox conditions.

The disappearance of detrital U in terrestrial sediments in the early Proterozoic (at ~2.3 Ga) was one of the striking lines of evidence for a remarkable increase of atmospheric oxygen levels (e.g., Holland 1984). During most of the following Proterozoic era (2.3–0.6 Ga), the oceans were likely stratified, with deep oceans that were still low in oxygen, ferruginous and partially euxinic (Canfield 1998; Anbar and Knoll 2002; Poulton et al. 2010). Under such conditions, the resulting ocean residence time for U may still have been relatively short (in particular between ~2.0 and ~0.6 Ga, see Partin et al. 2013a) and the U isotope signals recorded in the sediments deposited during this era may still provide only limited constraints for the global ocean. However, these signatures may also provide information about changes in atmospheric oxygen levels and the mobilization of U during oxidative weathering. Such conditions may have prevailed until the late Neoproterozoic at 635–550 Ga when atmospheric oxygen may have increased to almost present levels during a "second oxidation event" (e.g., Canfield and Teske 1996). This period is marked by a significant enhancement of redox-sensitive metals in the oceans, expressed, for example, by enhanced variations in S and Cr isotope compositions and increased abundances of Mo and U in the oceanic sediment record (e.g., Canfield and Teske 1996; Scott et al. 2008; Frei et al. 2009; Konhauser et al. 2011; Partin et al. 2013; Planavsky et al. 2014).

After this "second oxidation event", marine U cycling was likely more similar to that of today, with longer oceanic U residence times on the order of several hundred thousand years and U isotopes may be expected to provide constraints on the expansion of oceanic anoxia, for example during 'ocean anoxic events' (e.g., Jenkyns 2010). We thus divide the following discussion of paleo-redox studies that have used U isotopes into two groups, those focused on the Archean and early Proterozoic redox evolution of the ocean–atmosphere system, and studies that investigated redox fluctuations in the oceans of the Ediacaran and Phanerozoic era.

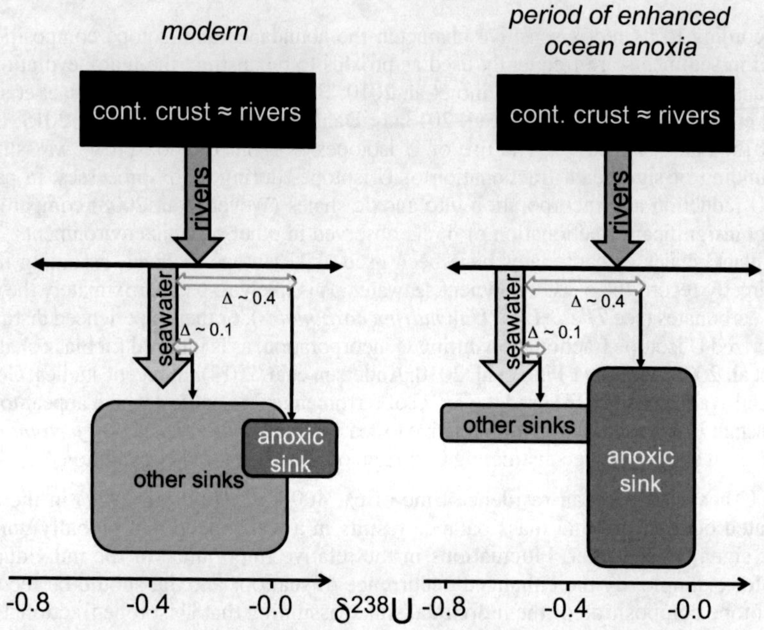

Figure 9. Schematic illustrations of a simplified isotopic U mass balance of the modern oceans (left panel) and an ancient ocean (right panel) during a period of enhanced ocean anoxia (similar to e.g., the illustration in Montoya Pino et al. 2010). The width of the sink boxes crudely represents their U isotopic range in nature. The area of the sink boxes crudely represents their relative weights (not scaled). The grey arrows indicate mean U isotope fractionation of the U sinks relative to seawater, whereas "other sinks" combines all sinks excluding the anoxic sink.

Early Earth redox evolution

During the Archean and Early Proterozoic, U may have been mobilized, at least locally, in sufficiently oxidizing environments, resulting in soluble U in the oceanic water column, and local U enrichment potentially coupled with U isotope fractionation in sediments associated with U reduction. The earliest evidence for such processes were observed in late Archean shales from Mount McRae (Western Australia; Kendall et al. 2013). These shales were deposited at ~2.5 Ga during locally enhanced oxygen levels, as indicated by high accumulation of authigenic Mo and Re (Anbar et al. 2007). They display $\delta^{238}U > -0.1‰$, that is significantly higher than that of present seawater and riverine U input. The correlation of these high $\delta^{238}U$ signatures with high enrichment factors for Mo and Re, indicate that U was mobilized by oxidative weathering and subsequently reduced during deposition in sediments, resulting in U isotope fractionation. As O_2 levels in the atmosphere were likely too low to initiate significant oxidative weathering on land (Pavlov and Kasting 2002), Kendall et al. (2013) concluded that oxidation may have occurred in the uppermost water column in so-called "oxygen oases" along the continental margins. Notably, the high $\delta^{238}U$ values observed in black shales are not correlated with U enrichment (e.g., U/Al or U/Th) in these sediments. However, Kendall et al. (2013) could show that $\delta^{238}U$ is a more sensitive proxy than U/Al for authigenic U enrichment as a result of U reduction in the sediment. Such a scenario would leave a light U isotope composition for residual U remaining in seawater and may explain the very low $\delta^{238}U$ (−0.9 to −0.7‰) observed in some late Archean banded iron formations (BIFs) (Weyer et al. 2008). Assuming that U isotope fractionation during U deposition in BIFs is similar to that observed during U deposition in Fe–Mn oxides (Weyer et al. 2008; Brennecka et al. 2011b; Goto et al. 2014; Wang et al. 2016), this would indicate that Archean seawater had at least locally lower $\delta^{238}U$ than present seawater.

During the Paleoproterozoic and after the 'Great Oxidiation Event' at ca. 2.3 Ga, the atmosphere may have had O_2 levels of up to a few percent of the present atmospheric oxygen level (e.g., Farquhar et al. 2000; Pavlov and Kasting 2002; Bekker et al. 2004), sufficiently high for oxidative weathering. Asael et al. (2013) investigated organic-rich shales and carbonates that were deposited during the Shunga Event at ~ 2.05 Ga. They observed variable and high $\delta^{238}U$ (after correction for an assumed detrital U contribution), with an average value of ~ +0.3‰. As the investigated sediments were likely deposited under euxinic conditions, these authors inferred a seawater $\delta^{238}U$ value of −0.2‰ (assuming $^{238}U/^{235}U$ fractionation of ~0.5‰ during the deposition of the black shales from seawater), which is similar to the $\delta^{238}U$ of present riverine U input to the oceans. Furthermore, $\delta^{98/95}Mo$ (see Kendall et al. 2017, this volume) of these sediments are ~ +0.8‰ (Asael et al. 2013), indicating a low seawater $\delta^{98/95}Mo$ that is similar to the value of modern rivers (Archer and Vance 2008). This is an interesting finding, implying that euxinic sediments, as studied by Asael et al. (2013), were not the dominant sink for U (as they were for Mo), as this should have resulted in seawater with a significantly lower $\delta^{238}U$ than that of rivers and the crust. From this finding, the authors concluded that either carbonates with unfractionated U isotope signatures were the dominant sink for U in the early Proterozoic oceans or light U isotopes were preferentially incorporated into marine iron formations over heavy U isotopes, thereby balancing the $\delta^{238}U$ of the net oceanic sinks.

Neo-Proterozoic and Phanerozoic

After the 'second oxidation event' during the Neoproterozoic (e.g., Canfield and Teske 1996; Scott et al. 2008; Frei et al. 2009; Konhauser et al. 2011; Partin et al. 2013a), the oceans are assumed to be well-oxygenated and the atmosphere had essentially modern O_2 levels (e.g., Berner 2004). Accordingly, U is likely to have displayed generally conservative behavior in the oceans at this time, with a long ocean residence time, similar to that of the modern oceans. Nevertheless, the oceans experienced distinct fluctuations in their redox state during this time-period, resulting in the enhanced occurrence of seafloor anoxia during the Phanerozoic that was, in cases, coupled with extinction events (e.g., Erwin et al. 2002). For redox sensitive trace elements, such as Mo and U, these fluctuations resulted in distinct variations in the relative abundance of the sinks (e.g., Brumsack 1981; Algeo 2004) that may be recorded in the isotope compositions of these elements in sediments (e.g., Arnold et al. 2004; Montoya-Pino et al. 2010). The extent of ocean anoxia may be semi-quantified by a box model, applying a steady state mass balance equation, as shown in Equation (14) above for U (Fig. 9; see Montoya-Pino et al. 2010 and Lau et al. 2016). A similar approach has previously been applied to Mo isotopes (e.g., Arnold et al. 2004; Kendall et al. 2009). This approach assumes (1) that enhanced ocean anoxia simply resulted in a shift in the relative proportions of the individual sinks, such as an enhancement in the magnitude of the anoxic sink at the expense of all other sinks, and (2) that the oceanic U budget was at steady state. For very rapid changes in environmental conditions, resulting in non-steady state conditions, differential equations need to be used (Lau et al. 2016). Other prerequisites for a simple box model approach are that (1) the U isotope composition of the U input to the oceans, solely represented by riverine input at present, has not changed through time, (2) that average U isotope fractionation between seawater and the individual sinks during U removal has not changed with time and 3) that the relative proportions of all other sinks, including suboxic sediments, biogenic carbonates, AOC, and metalliferous sediments, has not significantly changed with time. According to the latter prerequisite, and as we are only interested in variations of the "anoxic" U sink, the mass balance Equation (14) is commonly simplified to the following form:

$$\delta^{238}U_{In} = {}^{238}U_{anox} \times f_{anox} + {}^{238}U_{other} \times f_{other} \qquad (15)$$

In Equation (15), the subscript "anox" represents sediment sinks beneath an anoxic water column and "other" is for all other U sinks combined. The U isotope composition of the "anoxic sink" is either determined directly, by $\delta^{238}U$ analyses of black shales from respective geological periods (Montoya-Pino et al. 2010; Kendall et al. 2015). Alternatively, it is calculated from $\delta^{238}U$ analyses of carbonates (Brennecka et al. 2011a; Dahl et al. 2014; Azmy et al. 2015), assuming that these directly represent the U isotope composition of the ancient seawater, and applying the U isotope fractionation factor between seawater and "anoxic" sediments observed in similar modern settings (e.g., the Black Sea). It can then be assumed that the U isotopic offset between the "anoxic sink" and all other sinks remains constant with time and the fraction of U that goes into sediments of anoxic basins may be calculated by rearranging Equation (15) to the following form:

$$f_{\mathrm{anox}} = \frac{\delta^{238}\mathrm{U}_{\mathrm{In}} - \delta^{238}\mathrm{U}_{\mathrm{other}}}{\delta^{238}\mathrm{U}_{\mathrm{anox}} - \delta^{238}\mathrm{U}_{\mathrm{other}}} \quad (16)$$

According to the necessary assumptions from above, the estimate for f_{anox} will have considerable uncertainties. For example, U isotope fractionation between seawater and the anoxic sink ($\Delta^{238}U_{\text{seawater–anox}}$) may vary significantly, depending on local conditions, such as the extent of basin restriction (see *The* $^{238}U/^{235}U$ *in reducing sediments*). At present, the global average $\Delta^{238}U_{\text{seawater–anox}}$ (if considering only environments with continuously anoxic conditions) is dominated by the U isotope systematics of the Black Sea, which is a highly restricted basin with strong U isotope fractionation in the water column, limiting the net U isotope fractionation between open ocean water and Black Sea sediments (Andersen et al. 2014; Noordmann et al. 2015). If anoxic basins were dominantly more open, this would have resulted in more significant U isotope fractionation between seawater and sediments from such basins (Andersen et al. 2016). Likewise the U isotope fractionation factor between seawater and suboxic sediments, and the relative proportions of the other sinks, may also have changed with time. For example, in times of enhanced seafloor anoxia, hypoxic environments along the continental margins may also have increased in size and extent. Furthermore, hydrothermal activity has also varied significantly with time (Jacobsen and Kaufman 1999). An additional uncertainty arises if the total U flux into euxinic sediments is used to estimate the seafloor area that is covered by euxinic conditions in the water column, as this estimate critically depends on the assumed average U flux (per surface area) in these environments. This, in turn, depends on other environmental factors, such as the degree of restriction of the basin. For example, because of its strong restriction, the Black Sea water column is depleted in U (and other redox-sensitive metals), resulting in a relatively low U flux to the sediments (Anderson et al. 1989). Despite all of these limitations caused by the model assumptions, significant U isotope variations in black shales or carbonates can still provide a solid estimate regarding whether the seafloor area covered by anoxic/euxinic water was significantly enhanced during certain critical geological periods or whether it was similar to that of today. This approach was applied in several case studies, discussed below.

Kendall et al. (2015) investigated black shales from the late Ediacaran period and concluded from their observations of high $\delta^{238}U$ (of $+0.24 \pm 0.16‰$) that at ca. 560–550 Ma, the oceans were governed by similar oxidation conditions as those of today. This finding is consistent with high $\delta^{98/95}Mo$ observed for some of the investigated samples coupled with an extreme increase in Mo concentrations, previously observed by Scott et al. (2008).

Dahl et al. (2014) and Azmy et al. (2015) inferred a significant enhancement of ocean anoxia during the late Cambrian SPICE event (Steptoean Positive Carbon Isotope Excursion at ~500–496 Ma) and at the Cambrian–Ordovician boundary, respectively. These authors draw their conclusions based upon multi-proxy approaches, including the analysis of $\delta^{238}U$, applied to carbonate deposits formed during these periods. Dahl et al. (2014) observed a negative correlation between $\delta^{238}U$ and both carbon and sulfur isotope compositions during the first stage of the SPICE event, indicating that enhanced ocean anoxia was initially associated with increased carbon and pyrite burial. Azmy et al. (2015) observed a crudely positive correlation between $\delta^{238}U$ and carbon isotopes, as well as a negative correlation between $\delta^{238}U$ and nitrogen isotopes, in carbonates from the Cambrian–Ordovician boundary with very low $\delta^{238}U$ in the samples from the early Ordovician. Although local redox variations during the deposition of the carbonates cannot be completely excluded as a cause for the observed U isotope variations, the overall decrease in $\delta^{238}U$ with time is consistent with an enhancement of ocean anoxia as a result of sea level rise.

Brennecka et al. (2011a) and Lau et al. (2016) observed a significant isotope shift from $\delta^{238}U$ values of around −0.4‰ to values of around −0.7‰ in carbonates that were deposited across the end-Permian mass extinction event. This isotopic shift occurred sharply at the extinction horizon and is correlated with a depletion in U concentration in the studied carbonates. As these carbonates did not show any evidence of significant diagenetic alteration, the results have been interpreted to represent a shift in the seawater $\delta^{238}U$ signature, associated with dramatic seawater U drawdown into the underlying sediments as a result of enhanced oceanic anoxia (by a factor of ~ 100; Lau et al. 2016). Lau et al. (2016) additionally analyzed carbonates throughout the Triassic and observed low $\delta^{238}U$ and low U concentrations for a prolonged period of several million years following the end-Permian mass extinction event. They concluded that the oxygen minimum zone (OMZ) was likely expanded to shallow levels, which may have impinged continental shelves, throughout the early Triassic, thus inhibiting the recovery of the marine ecosystem.

Similarly, Montoya Pino et al. (2010) concluded from a general shift in $\delta^{238}U$ towards lower values, observed in black shales deposited across the Cretaceous 'Oceanic Anoxic Event' OAE2, at ~93 Ma, that seafloor anoxia was significantly enhanced during that period. These authors used Equation (16) to estimate that, during OAE2, the U sink into sediments beneath anoxic/euxinic water column conditions was at least a factor of three larger than at present.

OUTLOOK

Earlier efforts in the measurement of $^{238}U/^{235}U$ mainly focused on the search for natural nuclear reactors in terrestrial samples and ^{247}Cm-formed ^{235}-excesses in meteorites. In a review on the state of measured $^{238}U/^{235}U$ on Earth and the "Oklo phenomenon", Kuroda (1982) noted:

> *"Thus the detection of fossil reactors in other uranium ores probably requires the frequent control of the isotopic compositions of uranium with an accuracy better than about 0.1 percent, and it is not certain that such good analyses has been and/or will be conducted in other uranium ore deposits."*

Technical advances in measuring precise and accurate U isotopes over the last decade have indeed allowed these kinds of analyses. Furthermore, the applications of U isotope geochemistry extend well beyond the characterization of U ores and include improved accuracy of the Pb–Pb dating methods, evidence for the presence of now-extinct ^{247}Cm, improved understanding of global U cycling on and within the Earth and a new paleo-redox proxy for application to the surface Earth. It is inevitable that further advances will follow in this rapidly growing research field.

The processes responsible for $^{238}U/^{235}U$ variability that have been reliably identified so far, and discussed in this review, can be categorized into three separate groups:

1. **Cosmogenic nuclear synthetic processes:** *including the relative abundances and distributions of primordial ^{238}U and ^{235}U as well as ^{235}U excesses from ^{247}Cm-decay in the early Solar system*

2. **Terrestrial nuclear synthetic processes:** *including natural nuclear fission chain reactors in the Earth's crust*
3. **Isotope fractionation:** *including both "traditional" mass dependent and volume-dependent nuclear field shift processes in terrestrial and extra-terrestrial environments*

In cosmochemistry, ^{235}U excesses in meteorites have now been identified that appear to be linked to ^{247}Cm decay. However, a detailed understanding of the mechanisms controlling U isotope fractionations that are not related to ^{247}Cm decay, and the extent to which they have been inherited through heterogeneities in the distribution of ^{238}U and ^{235}U, versus U isotope fractionation during physico-chemical processes, such as condensation in the proto-planetary disk, need to be well-characterized in order to fully understand the sequence of events in the early solar system. There is also a pressing need to extend the efforts that have already begun in revising early Solar system chronologies based on the Pb–Pb chronometer that now takes the precise $^{238}U/^{235}U$ of individual samples into account. Such efforts are already showing considerable promise; there is now enhanced coherency between the Pb–Pb chronometer with other independent cosmochronometers based on extinct short-lived nuclides. Additional studies may provide further critical revisions of early solar system chronology and models for its formation.

The observations of the U isotopic composition of U ore deposits offers considerable promise as a powerful tool in economic geology for monitoring U ore deposit type and/or grade and its proximity to mineralization. However, the exact processes controlling U isotope fractionation in U ore deposits remain underconstrained and further efforts are required to explore the relationships between $^{238}U/^{235}U$ and U redox-controlled (biotic and/or abiotic) mineralization processes, temperature of deposition, U source rock composition, and fluid alteration before the full potential of U isotopes as a 'fingerprinting' tool in economic geology can be recognized. A related question for the understanding of U isotope fractionation in U ore deposits is "what is the cause of the very high $^{238}U/^{235}U$ variations observed among some accessory mineral phases, and could such variations be explained by high-temperature U isotope fractionation?"

The exact mechanisms controlling U isotopic shifts in terrestrial environments need to be clearly addressed in order to use ^{238}U–^{235}U isotope systematics to 'fingerprint' specific Earth system processes. Experimental and *ab initio* modeling studies have so far provided invaluable information on the processes driving U isotope fractionation, yet further work is still needed. This includes understanding isotopic mass-balance constraints during different U transfer and reduction pathways, further studies of U isotope fractionation during transfer mechanisms without U reduction (e.g., U–O bond coordination changes and ligand binding) and the effects of kinetic U isotope fractionation. A key question to address is if significant U isotope fractionation in near surface environments is mainly driven by enzymatic U reduction, and, if so, what are the causes of the observed differences, between biotic and abiotic U isotope fractionation? Oxidation-reduction aside, the role of other mechanisms capable of generating isotope fractionation between ^{238}U and ^{235}U also need to be explored in more detail in future studies.

If biotic and abiotic reduction generates different isotopic signatures, are these preserved over geological timescales, therefore providing a tracer of biologically driven U isotope fractionation in the past? An improved understanding of the mechanisms, resulting in U isotope variations on Earth will enable us to more specifically use U isotopes to unravel processes in the geological past. Uranium may provide advantages over other isotope proxies due to its specific chemical and isotopic behavior that differs from other elements. For example, a special feature is that U isotope fractionation in nature is driven by two different mechanisms (mass- and nuclear volume-dependent fractionation) that act in opposing directions. Understanding the contributions of both mechanisms in different environments may help to discriminate between potential reaction pathways or between the existence of isotopic equilibrium or kinetic processes. Such understanding will help explain whether U isotopic differences, as observed in experiments, can be explained by different reduction mechanisms, especially between biotic and abiotic reduction.

In paleo-redox studies, a particularly promising approach is to combine different isotope (and other) proxies, taking advantage of their differing redox response, and allowing the weaknesses of one system to be overcome by combining the particular strengths of all systems together. For example, while other metals are reduced or scavenged (e.g., Mo, Cr, Zn) in the water column, U is mainly reduced in the sediments, likely mediated by microorganisms. Furthermore, U responds isotopically across a range of oxic–anoxic–euxinic environments, whereas other elements may only show a limited isotopic response across some of these environments. These differences in reduction behavior are likely recorded in sediment archives, however correct interpretation of such records relies on more information on the mechanistic controls of metal reduction. With such an improved understanding, the U isotope paleo-redox proxy may become an important piece to complete the puzzle of the evolution of the oceans, atmosphere and life on Earth.

ACKNOWLEDGMENTS

Full reviews from G. Brennecka, X. Wang, F. Tissot and S. Romaniello as well as comments on sections from M. Christl and E. Schauble helped to improve this manuscript.

REFERENCES

Abdelouas A (2006) Uranium mill tailings: Geochemistry, mineralogy, and environmental impact. Elements, 2:355–341

Abe M, Suzuki T, Fujii Y, Hada M (2008a) An ab initio study based on a finite nucleus model for isotope fractionation in the U (III)–U (IV) exchange reaction system. J Chem Phys 128:144309

Abe M, Suzuki T, Fujii Y, Hada M, Hirao K (2008b) An ab initio molecular orbital study of the nuclear volume effects in uranium isotope fractionations. J Chem Phys 129:164309

Abe M, Suzuki T, Fujii Y, Hada M, Hirao K (2010) Ligand effect on uranium isotope fractionations caused by nuclear volume effects: An ab initio relativistic molecular orbital study J Chem Phys 133:044309

Abe M, Hada M, Suzuki T, Fujii Y, Hirao K (2014) Theoretical study of isotope enrichment caused by nuclear volume effect. J Comput Chem Jpn 13:92–104

Abzalov, M.Z (2012) Sandstone-hosted uranium deposits amenable for exploitation by in situ leaching technologies. Trans Inst Min Metall Sect B 121:55–64

Algeo TJ (2004) Can marine anoxic events draw down the trace element inventory of seawater? Geology 32:1057–1060

Amelin Y (2006) The prospcct of high-precision Pb isotopic dating of meteorites. Meteor Planet Sci 41:7–17

Amelin Y, Krot AN (2007) Pb isotopic age of the Allende chondrules. Meteor Planet Sci 42:1321–1335

AmelinY, Krot AN, Hutcheon ID, Ulyanov AA (2002) Lead isotopic ages of chondrules and calcium–aluminum-rich inclusions. Science 297:1678–1683

Amelin Y, Kaltenbach A, Iizuka T, Stirling CH, Ireland TR, Petaev M, Jacobsen SB (2010) U–Pb chronology of the Solar system's oldest solids with variable $^{238}U/^{235}U$. Earth Planet Sci Lett 300:343–350

Anbar AD, Knoll, AH (2002) Proterozoic ocean chemistry and evolution: a bioinorganic bridge? Science 297:1137–1142

Anbar AD, Duan Y, Lyons TW, Arnold GL, Kendall B, Creaser RA, Kaufman AJ, Gordon GW, Scott C, Garvin J, Buick R (2007) A whiff of oxygen before the great oxidation event? Science 317:1903–1906

Anderson RF (1987) Redox Behaviour of Uranium in an Anoxic Marine Basin. Uranium 3:145–164

Anderson RF, Fleischer MQ, LeHurray AP (1989) Concentration, oxidation state, and particulate flux of uranium in the Black Sea. Geochim Cosmochim Acta 53:2215–2224

Anderson RT, Vrionis HA, Ortiz-Bernad I, Resch CT, Long PE, Dayvault R, Karp K, Marutzky S, Metzler DR, Peacock A, White DC (2003) Stimulating the in situ activity of *Geobacter* species to remove uranium from the groundwater of a uranium-contaminated aquifer. Appl Environ Microbiol 69:5884–5891

Andersen, MB, Stirling CH, Zimmermann B, Halliday AN (2010) Precise determination of the open ocean $^{234}U/^{238}U$ composition. Geochem Geophys Geosystems 11:Q12003

Andersen MB, Romaniello S, Vance D, Little SH, Herdman R, Lyons TW (2014) A modern framework for the interpretation of $^{238}U/^{235}U$ in studies of ancient ocean redox. Earth Planet Sci Lett 400:184–194

Andersen MB, Elliott T, Freymuth H, Sims KW, Niu Y, Kelley KA (2015) The terrestrial uranium isotope cycle. Nature 517:356–359

Andersen MB, Vance D, Morford JL, Bura-Nakić E, Breitenbach SF, Och L (2016) Closing in on the marine $^{238}U/^{235}U$ budget. Chem Geol 420:11–22

Archer C, Vance D (2008) The isotopic signature of the global riverine molybdenum flux and anoxia in the ancient oceans. Nat Geosci 1:597–600

Arden JW (1977) Isotopic composition of uranium in chondritic meteorites. Nature 269:788–789

Arnold GL, Anbar AD, Barling J, Lyons TW (2004) Molybdenum isotope evidence for widespread anoxia in Mid-Proterozoic oceans. Science 304:87–90

Arnold GL, Lyons TW, Gordon GW, Anbar, AD (2012) Extreme change in sulfide concentrations in the Black Sea during the Little Ice Age reconstructed using molybdenum isotopes. Geology 40:595–598

Asael D, Tissot FL, Reinhard CT, Rouxel O, Dauphas N, Lyons TW, Ponzevera E, Liorzou C, Chéron S (2013) Coupled molybdenum, iron and uranium stable isotopes as oceanic paleoredox proxies during the Paleoproterozoic Shunga Event. Chem Geol 362:193–210

Aston FW (1931) Constitution of thallium and uranium. Nature 128:725

Azmy K, Kendall B, Brand U, Stouge S, Gordon GW (2015) Redox conditions across the Cambrian–Ordovician boundary: Elemental and isotopic signatures retained in the GSSP carbonates. Palaeogeogr Palaeoclimatol Palaeoecol 440:440–454

Bach, W, Peucker-Ehrenbrink, B, Hart, S.R, Blusztajn, J.S (2003) Geochemistry of hydrothermally altered oceanic crust: DSDP/ODP Hole 504B – Implications for seawater–crust exchange budgets and Sr- and Pb-isotopic evolution of the mantle. Geochem Geophys Geosystems 4:8904

Bargar JR, Williams KH, Campbell KM, Long PE, Stubbs JE, Suvorova EI, Lezama-Pacheco JS, Alessi DS, Stylo M, Webb SM, Davis JA (2013) Uranium redox transition pathways in acetate-amended sediments. PNAS 110:4506–4511

Barnes CE, Cochran JK (1990) Uranium removal in oceanic Sediments and the oceanic-U balance. Earth Planet Sci Lett 97:94–101

Basu A, Brown ST, Christensen JN, DePaolo DJ, Reimus PW, Heikoop JM, Woldegabriel G, Simmons AM, House BM, Hartmann M, Maher K (2015) Isotopic and Geochemical Tracers for U(VI) reduction and U mobility at an in situ recovery U mine. Environ Sci Technol 49:5939–5947

Basu A, Sanford RA, Johnson TM, Lundstrom CC, Löffler, FE (2014) Uranium isotopic fractionation factors during U (VI) reduction by bacterial isolates. Geochim Cosmochim Acta 136:100–113

Becquerel H (1896) Sur les radiations émises par phosphorescence. Comptes Rendus 122:501–503

Bekker A, Holland HD, Wang PL, Rumble DI, Stein HJ, Hannah JL, Coetzee LL, Beukes NJ (2004) Dating the rise of atmospheric oxygen. Nature 427:117–120

Bender ML (1990) The $\delta^{18}O$ of dissolved O_2 in seawater: A unique tracer of circulation and respiration in the deep sea. J Geophys Res Oceans 95(C12): 22243–22252

Berner RA (1981) A new geochemical classification of sedimentary environments. J Sediment Petrol 51:359–365

Berner RA (2004) The Phanerozoic carbon cycle: CO_2 and O_2. Oxford, UK: Oxford University Press

Bigeleisen J, Mayer MG (1947) Calculation of equilibrium constants for isotopic exchange reactions. J Chem Phys 15:261–267

Bigeleisen J (1996a) Nuclear size and shape effects in chemical reactions. Isotope chemistry of heavy elements. J Am Chem Soc 118:3676–3680

Bigeleisen J (1996b) Temperature dependence of the isotope chemistry of the heavy elements. PNAS 93:9393–9396

Blake JB, Schramm DN (1973) ^{247}Cm as a short-lived *r*-process chronometer. Nature, 243:138–140

Bodu R, Bouzigues H, Morin N, Pfiffelmann JP (1972) Sur l'existence d'anomalies isotopiques rencontrées dans l'uranium du Gabon. CR Acad Sci Paris. 275(D):1731.

Bopp CJ, Lundstrom CC, Johnson TM, Glessner JJ (2009) Variations in $^{238}U/^{235}U$ in uranium ore deposits: Isotopic signatures of the U reduction process? Geology 37:611–614

Bopp CJ, Lundstrom CC, Johnson TM, Sanford RA, Long PE, Williams KH (2010) Uranium $^{238}U/^{235}U$ isotope ratios as indicators of reduction: results from an in situ biostimulation experiment at Rifle, Colorado, USA. Environ Sci Technol 44:5927–5933

Boynton WV (1978) Fractionation in the solar nebula, II. Condensation of Th, U Pu, and Cm. Earth Planet Sci Lett 40:63–70

Brennecka GA, Borg LE, Hutcheon ID, Sharp MA, Anbar AD (2010a) Natural variations in uranium isotope ratios of uranium ore concentrates: Understanding the $^{238}U/^{235}U$ fractionation mechanism. Earth Planet Sci Lett 291:228–233

Brennecka GA, Weyer S, Wadhwa M, Janney PE, Zipfel J, Anbar AD (2010b) $^{238}U/^{235}U$ variations in meteorites: extant ^{247}Cm and implications for Pb–Pb dating. Science 327:449–451

Brennecka GA, Herrmann AD, Algeo TJ, Anbar AD (2011a) Rapid expansion of oceanic anoxia immediately before the end-Permian mass extinction. PNAS 108:17631–17634

Brennecka GA, Wasylenki LE, Bargar JR, Weyer S, Anbar AD (2011b) Uranium isotope fractionation during adsorption to Mn-oxyhydroxides. Environ Sci Technol 45:1370–1375

Brennecka GA, Wadhwa M (2012) Uranium isotope compositions of the basaltic angrite meteorites and the chronological implications for the early Solar system. PNAS 109:9299–9303

Brennecka GA, Budde, G, Kleine, T (2015) Uranium isotopic composition and absolute ages of Allende chondrules. Meteor Planet Sci 50:1995–2002

Broecker WS (1971) A kinetic model for the chemical composition of sea water. Quatern Res 1:188–207

Brown ST, Basu A, Christensen JN, Reimus P, Heikoop J, Simmons A, Wolde Gabriel G, Maher K, Weaver K, Clay J, DePaolo DJ (2016) Isotopic evidence for reductive immobilization of uranium across a roll-front mineral deposit. Environ Sci Technol 50:6189–6198

Brumsack HJ (1981) Geochemistry of Cretaceous black shales from the Atlantic Ocean (DSDP Legs 11:14, 36 and 41). Chem Geol 31:1–25

Canfield DE (1998) A new model for Proterozoic ocean chemistry. Nature 396:450–453

Canfield DE, Teske A (1996) Late Proterozoic rise in atmospheric oxygen concentration inferred from phylogenetic and sulphur-isotope studies. Nature 382:127–132

Chase CG (1981) Oceanic island Pb: two-stage histories and mantle evolution. Earth Planet Sci Lett 52:277–284

Chen JH, Wasserburg GJ (1980) A search for isotopic anomalies in uranium. Geophys Res Lett 7:275–278

Chen JH, Wasserburg GJ (1981a) The isotopic composition of uranium and lead in Allende inclusions and meteoritic phosphates. Earth Planet Sci Lett 52:1–15

Chen JH, Wasserburg GJ (1981b) Isotopic determination of uranium in picomole and subpicomole quantities. Anal Chem 53:2060–2067

Chen X, Romaniello SJ, Herrmann AD, Wasylenki LE, Anbar AD (2016) Uranium isotope fractionation during coprecipitation with aragonite and calcite. Geochim Cosmochim Acta 186:189–207

Cheng H, Edwards RL, Shen CC, Polyak VJ, Asmerom Y, Woodhead J, Hellstrom J, Wang Y, Kong X, Spötl C, Wang X (2013) Improvements in ^{230}Th dating, ^{230}Th and ^{234}U half-life values, and U–Th isotopic measurements by multi-collector inductively coupled plasma mass spectrometry. Earth Planet Sci Lett 371–372:82–91

Cherdyntsev VV, Chalov PI, Khitrik ME, Mambetov DM, Khaidarov GZ (1955) Proceedings of the III Session of the Commission on the Determination of the Absolute Ages of Geological Formations, Acad Sci USSR

Chernyshev IV, Golubev VN, Chugaev AV, Baranova AN (2014a) ^{238}U/^{235}U isotope ratio variations in minerals from hydrothermal uranium deposits. Geochem Inter 52:1013–1029

Chernyshev IV, Dubinina EO, Golubev VN (2014b) Fractionation factor of ^{238}U and ^{235}U isotopes in the process of hydrothermal pitchblende formation: A numerical estimate. Geol Ore Deposits 56:315–21

Christl M, Lachner J, Vockenhuber C, Lechtenfeld O, Stimac I, Van der Loeff MR, Synal HA (2012) A depth profile of uranium-236 in the Atlantic Ocean. Geochim Cosmochim Acta 77:98–107

Clark SK, Johnson TM (2008) Effective isotopic fractionation factors for solute removal by reactive sediments: A laboratory microcosm and slurry study. Environ Sci Technol 42:7850–7855

Condon DJ, McLean N, Noble SR, Bowring SA (2010) Isotopic composition ^{238}U/^{235}U of some commonly used uranium reference materials. Geochim Cosmochim Acta 74:7127–7143

Connelly JN, Amelin Y, Krot AN, Bizzarro M (2008) Chronology of the solar system's oldest solids. Astrophys J Lett 675:L121

Connelly JN, Bizzarro M, Krot AN, Nordlund Å, Wielandt D, Ivanova MA (2012) The absolute chronology and thermal processing of solids in the solar protoplanetary disk. Science 338:651–655

Cowan GA, Adler HH (1976) The variability of the natural abundance of ^{235}U. Geochim Cosmochim Acta 40:1487–1490

Cuney M (2010) Evolution of uranium fractionation processes through time: driving the secular variation of uranium deposit types. Econ Geol 105:553–569

Dahl TW, Boyle RA, Canfield DE, Connelly JN, Gill BC, Lenton TM, Bizzarro M (2014) Uranium isotopes distinguish two geochemically distinct stages during the later Cambrian SPICE event. Earth Planet Sci Lett 401:313–326

Dahlkamp FJ (1991) Uranium Ore Deposits. Springer, Berlin, Heidelberg, New York

Dauphas N, Marty B, Reisberg L (2002) Inference on terrestrial genesis from molybdenum isotope systematics. Geophys Res Lett 29:8-1–8-3

Dauphas N (2005) Multiple sources or late injection of short-lived *r*-nuclides in the early solar system? Nucl Phys A 758:757–60

Dauphas N, Chaussidon M (2011) A perspective from extinct radionuclides on a young stellar object: the Sun and its accretion disk. Ann Rev Earth Planet Sci 39:351–386

De Laeter JR, Rosman KJR, Smith CL (1980) The Oklo natural reactor: cumulative fission yields and retentivity of the symmetric mass region fission products. Earth Planet Sci Lett 50:238–246

Dempster A (1935) Isotopic constitution of uranium. Nature 136:180

Dicke RH (1969) The age of the galaxy from the decay of uranium. Astrophys J 155:123

Dunk RM, Mills RA, Jenkins WJ (2002) A reevaluation of the oceanic uranium budget for the Holocene. Chem Geol 190:45–67

Elliott T, Plank T, Zindler A, White W, Bourdon B (1997) Element transport from slab to volcanic front at the Mariana arc. J Geophys Res: Solid Earth (1978–2012) 102(B7):14991–15019

Elliott T, Zindler A, Bourdon B (1999) Exploring the kappa conundrum: the role of recycling in the lead isotope evolution of the mantle. Earth Planet Sci Lett 169:129–145

Erwin DH, Bowring SA, Yugan J (2002) End-Permian mass extinctions: a review. Spec Pap Geol Soc Am: 363–384

Farquhar J, Bao H, Thiemens M (2000) Atmospheric influence of Earth's earliest sulfur cycle. Science 289:756–758

Fleischer RL, Price PB, Walker RM (1975) Nuclear Tracks in Solids: Principles and Applications. University of California Press

Florence TM, Batley GE, Ekstrom A, Fardy JJ, Farrar YJ (1975) Separation of uranium isotopes by uranium (IV)–uranium (VI) chemical exchange. J Inorg Nucl Chem 37:1961–1966

Frei R, Gaucher C, Poulton SW, Canfield, DE (2009) Fluctuations in Precambrian atmospheric oxygenation recorded by chromium isotopes. Nature 461:250–253

Fujii YA, Nomura M, Okamoto M, Onitsuka H, Kawakami F, Takeda K (1989a) An anomalous isotope effect of ^{235}U in U (IV)–U (VI) chemical exchange. Z Naturforsch A 44:395–398

Fujii Y, Nomura M, Onitsuka H, Takeda K (1989b) Anomalous isotope fractionation in uranium enrichment process. J Nucl Sci Technol 26:1061–1064

Fujii Y, Higuchi N, Haruno Y, Nomura M, Suzuki T (2006) Temperature dependence of isotope effects in uranium chemical exchange reactions. J Nucl Sci Technol 43:400–406

Fukuda J, Fujii Y, Okamoto M (1983) A Fundamental study on uranium isotope separation using U (IV)–U (VI) electron exchange reaction. Z Naturforsch A 38:1072–1077

Gauthier-Lafaye F, Holliger P, Blanc PL (1996) Natural fission reactors in the Franceville basin, Gabon: A review of the conditions and results of a "critical event" in a geologic system. Geochim Cosmochim Acta 60:4831–4852

Goldmann A, Brennecka G, Noordmann J, Weyer S, Wadhwa M (2015) The uranium isotopic composition of the Earth and the solar system. Geochim Cosmochim Acta 148:145–158

Goldstein SJ, Stirling CH (2003) Techniques for measuring uranium-series nuclides: 1992–2002. Rev Mineral Geochem 52:23–57

Golubev VN, Chernyshev IV, Chugaev AV, Eremina AV, Baranova AN, Krupskaya VV (2013) U–Pb systems and U isotopic composition of the sandstone-hosted paleovalley Dybryn uranium deposit, Vitim uranium district, Russia. Geol Ore Deposits 55:399–410

Goto KT, Anbar AD, Gordon GW, Romaniello SJ, Shimoda G, Takaya Y, Tokumaru A, Nozaki T, Suzuki K, Machida S, Hanyu T (2014) Uranium isotope systematics of ferromanganese crusts in the Pacific Ocean: Implications for the marine ^{238}U/^{235}U isotope system. Geochim Cosmochim Acta 146:43–58

Grenthe I, Wanner H, Forest I (1992) Chemical Thermodynamics of Uranium. Nuclear Energy Agency

Hamer AN, Robbins EJ (1960) A search for variations in the natural abundance of uranium-235. Geochim Cosmochim Acta 19:143–145

Haug GH, Pedersen TF, Sigman DM, Calvert SE, Nielsen B, Peterson LC (1998) Glacial/interglacial variations in production and nitrogen fixation in the Cariaco Basin during the last 580 kyr. Paleoceanogr 13:427–32

Hiess J, Condon DJ, McLean N, Noble SR (2012) ^{238}U/^{235}U systematics in terrestrial uranium-bearing minerals. Science 335:1610–1614

Hinojosa JL, Stirling CH, Reid MR, Moy CM, Wilsin, GS (2016) Trace metal cycling and ^{238}U/^{235}U in New Zealand's fjords: Implications for reconstructing global paleoredox conditions in organic-rich sediments. Geochim Cosmochim Acta 179:89–109

Henderson GM, Anderson RF (2003) The U-series toolbox for paleoceanography. Rev Mineral Geochem 52:493–531

Hobday DK, Galloway WE (1999) Groundwater processes and sedimentary uranium deposits. Hydrogeol J 7:127–138

Holland HD (1984) The Chemical Evolution of the Atmosphere and Oceans. Princeton, NJ. Princeton University Press

Holland HD (2005) 100th anniversary special paper: sedimentary mineral deposits and the evolution of earth's near-surface environments. Econ Geol 100:1489–1509

Holmden C, Amini M, Francois R (2015) Uranium isotope fractionation in Saanich Inlet: A modern analog study of a paleoredox tracer. Geochim Cosmochim Acta 153:202–215

Horwitz EP, Dietz ML, Chiarizia R, Diamond H, Essling AM, Graczyk D (1992) Separation and preconcentration of uranium from acidic media by extraction chromatography. Anal Chim Acta 266:25–37

Horwitz EP, Chiarizia, R, Dietz ML, Diamond H, Nelson DM (1993) Separation and preconcentration of actinides from acidic media by extraction chromatography. Anal Chim Acta 281:361–372

Hostetler PB, Garrels RM (1962) Transportation and precipitation of uranium and vanadium at low temperatures, with special reference to sandstone-type uranium deposits. Econ Geol 57:137–167

Huckriede H, Meischner D (1996) Origin and environment of manganese-rich sediments within black-shale basins. Geochim Cosmochim Acta 60:1399–1413

IAEA (1975) Proceedings Series: The Oklo Phenomenon. Report, Vienna

Iizuka, T, Amelin, Y, Kaltenbach, A, Koefoed, P, Stirling, CH (2014) U–Pb systematics of the unique achondrite Ibitira: Precise age determination and petrogenetic implications. Geochim Cosmochim Acta 132:259–273

Ismail IM, Nomura M, Fujii Y, Aida M (2002) The effect of temperature on uranium isotope effects studied by cation exchange displacement chromatography. Z Naturforsch A 57:247–254

Jacobsen SB, Kaufman AJ (1999) The Sr, C and O isotopic evolution of Neoproterozoic seawater. Chem Geol 161:37–57

Jaffey A, Flynn K, Glendenin L, Bentley W, Essling A (1971) Precision measurement of half-lives and specific acitivities of ^{235}U and ^{238}U. Phys Rev C4:1889–1906

Jaupart C, Mareschal J-C (2010) Heat Generation and Transport in the Earth. Cambridge University Press

Jenkyns HC (2010) Geochemistry of oceanic anoxic events. Geochem Geophys Geosystems 11: Q03004, DOI: 10.1029/2009GC002788

Keegan E, Richter S, Kelly I, Wong H, Gadd P, Kuehn H, Alonso-Munoz A (2008) The provenance of Australian uranium ore concentrates by elemental and isotopic analysis. Appl Geochem 23:765–77

Kelley KA, Plank T, Farr L, Ludden J, Staudigel H (2005) Subduction cycling of U Th, and Pb. Earth Planet Sci Lett 234:369–383

Kelly SD, Newville MG, Cheng L, Kemner KM, Sutton SR, Fenter P, Sturchio NC, Spötl C (2003) Uranyl incorporation in natural calcite. Environ Sci Technol 37:1284–1287

Kelly SD, Rasbury ET, Chattopadhyay S, Kropf AJ, Kemner KM (2006) Evidence of a stable uranyl site in ancient organic-rich calcite. Environ Sci Technol 40:2262–2268

Kendall B, Brennecka GA, Weyer S, Anbar AD (2013) Uranium isotope fractionation suggests oxidative uranium mobilization at 2.50 Ga. Chem Geol 362:105–114

Kendall B, Creaser RA, Gordon GW, Anbar AD (2009) Re–Os and Mo isotope systematics of black shales from the Middle Proterozoic Velkerri and Wollogorang formations, McArthur Basin, northern Australia. Geochim Cosmochim Acta 73:2534–2558

Kendall B, Komiya T, Lyons TW, Bates SM, Gordon GW, Romaniello SJ, Jiang G, Creaser RA, Xiao S, McFadden K, Sawaki Y (2015) Uranium and molybdenum isotope evidence for an episode of widespread ocean oxygenation during the late Ediacaran Period. Geochim Cosmochim Acta 156:173–193

Kendall B, Dahl TW, Anbar AD (2017) Good golly, why moly? The stable isotope geochemistry of molybdenum. Rev Mineral Geochem 82:683–732

King WH (1984) Isotope Shifts in Atomic Spectra. Springer Science & Business Media

Kirchenbaur M, Maas R, Ehrig K, Kamenetsky VS, Strub E, Ballhaus C, Münker C (2016) Uranium and Sm isotope studies of the supergiant Olympic Dam Cu–Au–U–Ag deposit, South Australia. Geochim Cosmochim Acta 180:15–32

Kleine T, Touboul M, Bourdon B, Nimmo F, Mezger K, Palme H, Jacobsen SB, Yin QZ, Halliday AN (2009) Hf–W chronology of the accretion and early evolution of asteroids and terrestrial planets. Geochim Cosmochim Acta 73:5150–88

Klinkhammer GP Palmer MR (1991) Uranium in the oceans: Where it goes and why. Geochim Cosmochim Acta 55:1799–1806

Konhauser KO, Lalonde SV, Planavsky NJ, Pecoits E, Lyons TW, Mojzsis SJ, Rouxel OJ, Barley ME, Rosìere C, Fralick PW, Kump LR (2011) Aerobic bacterial pyrite oxidation and acid rock drainage during the Great Oxidation Event. Nature 478:369–73

Ku TL, Knauss KG, Mathieu GG (1977) Uranium in open ocean—concentration and isotopic composition. Deep-Sea Res 24:1005–1017

Kuroda PK (1956) On the nuclear physical stability of the uranium minerals. J Chem Phys 25:781–782

Kuroda PK (1982) The Oklo phenomenon. *In:* Origin of the Chemical Elements, Springer-Verlag, Berlin, Heidelberg, New York, pp. 30–55

Langmuir D (1978) Uranium solution–mineral equilibria at low temperatures with applications to sedimentary ore deposits. Geochim Cosmochim Acta 42:547–569

Lau KV, Maher K, Altiner D, Kelley BM, Kump LR, Lehrmann DJ, Silva-Tamayo JC, Weaver KL, Yu M, Payne JL (2016) Marine anoxia and delayed Earth system recovery after the end-Permian extinction. PNAS 113:2360–5

Li W, Czaja AD, Van Kranendonk MJ, Beard BL, Roden EE, Johnson CM (2013) An anoxic, Fe (II)-rich, U-poor ocean 3.46 billion years ago. Geochim Cosmochim Acta 120:65–79

Lingenfelter RE, Higdon JC, Kratz KL, Pfeiffer B (2003) Actinides in the source of cosmic rays and the present interstellar medium. Astrophys J 591:228–237

Lericolais G, Popescu I, Guichard F, Popescu SM, Manolakakis L (2007) Water-level fluctuations in the Black Sea since the last glacial maximum. *In:* The Black Sea Flood Question: Changes in Coastline, Climate, and Human Settlement. Springer Netherlands, p. 437–452

Lloyd NS, Chenery SR, Parrish RR (2009) The distribution of depleted uranium contamination in Colonie NY, USA. Sci Tot Environ 20:408:397–407

Lounsbury M (1956) The natural abundances of the uranium isotopes. Can J Chem 34:259–264

Lovley DR, Phillips EJP, Gorby YA, Landa ER (1991) Microbial reduction of uranium. Nature 350:413–416

Lyons TW, Berner RA, Anderson RF (1993) Evidence for large pre-industrial perturbations of the Black Sea chemocline. Nature 365:538–540

McManus, J, Berelson WM, Klinkhammer GP, Hammond DE, Holm C (2005) Authigenic uranium: Relationship to oxygen penetration depth and organic carbon rain. Geochim Cosmochim Acta 69:95–108
Mioduski, T (1999) Comment to the Bigeleisen's theory of isotope chemistry of the heavy elements. Comments Inorg Chem 21:175–196
Montoya-Pino C, Weyer S, Anbar AD, Pross J, Oschmann W, van de Schootbrugge B, Arz HW (2010) Global enhancement of ocean anoxia during Oceanic Anoxic Event 2: A quantitative approach using U isotopes. Geology 38:315–318
Morford JL, Emerson S (1999) The geochemistry of redox sensitive trace metals in sediments. Geochim Cosmochim Acta 63:1735–1750
Morford JL, Martin WR, François R, Carney CM (2009) A model for uranium, rhenium, and molybdenum diagenesis in marine sediments based on results from coastal locations. Geochim Cosmochim Acta 73:2938–2960
Mudd G (2001) Critical review of acid in situ leach uranium mining: 1. USA and Australia. Environ Geol, 41:390–401
Murphy MJ, Stirling CH, Kaltenbach A, Turner SP, Schaefer BF (2014) Fractionation of $^{238}U/^{235}U$ U by reduction during low temperature uranium mineralisation processes. Earth Planet Sci Lett 388:306–317
Murphy MJ, Froehlich MB, Fifield LK, Turner SP, Schaefer BF (2015) In-situ production of natural ^{236}U in groundwaters and ores in high-grade uranium deposits. Chem Geol 410:213–222
Nemoto K, Abe M, Seino J, Hada M (2015) An ab initio study of nuclear volume effects for isotope fractionations using two-component relativistic methods. J Comput Chem 36:816–20
Neuilly M, Bussac J, Frejacques C, Nief G, Vendryes G, Yvon J (1972) Sur l'existence dans un passe recule dune reaction en chaine naturelle de fissions dans le gisement d'uranium d'Oklo (Gabon). CR Acad Sci Paris 275, Ser D, 1847
Newman K (2012) Effects of the sampling interface in MC-ICP-MS: Relative elemental sensitivities and non-linear mass dependent fractionation of Nd isotopes. J Anal Atom Spectr 27:63–70
Nier AO (1939) The isotopic constitution of uranium and the half-lives of the uranium isotopes. I. Phys Rev 55:150
Nomura M, Higuchi N, Fujii Y (1996) Mass dependence of uranium isotope effects in the U (IV)-U (VI) exchange reaction. J Am Chem Soc 118:9127–9130
Noordmann J, Weyer S, Montoya-Pino C, Dellwig O, Neubert N, Eckert S, Paetzel M, Böttcher ME (2015) Uranium and molybdenum isotope systematics in modern euxinic basins: Case studies from the central Baltic Sea and the Kyllaren fjord (Norway). Chem Geol 396:182–195
Noordmann J, Weyer S, Georg RB, Jöns S, Sharma M (2016) $^{238}U/^{235}U$ isotope ratios of crustal material, rivers and products of hydrothermal alteration: New insights on the oceanic U isotope mass balance. Isotop Environ Health Stud 52:141–163, http://dx.doi.org/10.1080/10256016.(2015)1047449
Oliver IW, Graham MC, MacKenzie AB, Ellam RM, Farmer JG (2007) Assessing depleted uranium (DU) contamination of soil, plants and earthworms at UK weapons testing sites. J Environ Monit 9:740–748
Palme H, O'Neill HS (2003) Cosmochemical estimates of mantle composition. Treatise on Geochemistry 2:1-38
Partin CA, Bekker A, Planavsky NJ, Scott CT, Gill BC, Li C, Podkovyrov V, Maslov A, Konhauser KO, Lalonde SV, Love GD (2013a) Large-scale fluctuations in Precambrian atmospheric and oceanic oxygen levels from the record of U in shales. Earth Planet Sci Lett 369:284–293
Partin CA, Lalonde SV, Planavsky NJ, Bekker A, Rouxel OJ, Lyons TW, Konhauser KO (2013b) Uranium in iron formations and the rise of atmospheric oxygen. Chem Geol 362:82–90
Patterson C, Tilton G, Inghram M (1955) Age of the Earth. Science 121:69–75
Pavlov AA, Kasting JF (2002) Mass-independent fractionation of sulfur isotopes in Archean sediments: strong evidence for an anoxic Archean atmosphere. Astrobiol 2:27–41
Placzek CJ, Heikoop JM, House B, Linhoff BS, Pelizza M (2016) Uranium isotope composition of waters from South Texas uranium ore deposits. Chem Geol 437:44–55
Planavsky NJ, Reinhard CT, Wang X, Thomson D, McGoldrick P, Rainbird RH, Johnson T, Fischer WW, Lyons TW (2014) Low Mid-Proterozoic atmospheric oxygen levels and the delayed rise of animals. Science 346:635–638
Plank T, Ludden JN, Escutia C, Party SS (2000) Leg 185 summary; inputs to the Izu–Mariana subduction system. Ocean Drill Prog Proc, Initial Reports, Leg 185:1–63
Plant JA, Simpson PR, Smith B, Windley BF (1999) Uranium ore deposits; products of the radioactive Earth. Rev Mineral Geochem 38:255–319
Podosek FA (1970) The abundance of ^{244}Pu in the early solar system. Earth Planet Sci Lett 8:183–187
Poulton SW, Fralick PW, Canfield DE (2010) Spatial variability in oceanic redox structure 1.8 billion years ago. Nat Geosci 3:486–490
Rademacher LK, Lundstrom CC, Johnson TM, Sanford RA, Zhao J, Zhang Z (2006) Experimentally determined uranium isotope fractionation during reduction of hexavalent U by bacteria and zero valent iron. Environ Sci Technol 40:6943–6948
Reeder RJ, Nugent M, Lamble GM, Tait CD, Morris DE (2000) Uranyl incorporation into calcite and aragonite: XAFS and luminescence studies. Environ Sci Technol 34:638–644
Regelous M, Elliott T, Coath CD (2008) Nickel isotope heterogeneity in the early solar system. Earth Planet Sci Lett 272:330–338

Richter S, Alonso A, De Bolle W, Wellum R, Taylor PDP (1999) Isotopic "fingerprints" for natural uranium ore samples. Int J Mass Spectr 193:9–14

Richter S, Alonso-Munoz A, Eykens R, Jacobsson U, Kuehn H, Verbruggen A, Aregbe Y, Wellum R, Keegan E (2008) The isotopic composition of natural uranium samples—Measurements using the new $^{233}U/^{236}U$ double spike IRMM-3636. Int J Mass Spectr 269:145–148

Richter S, Eykens R, Kühn H, Aregbe Y, Verbruggen A, Weyer S (2010) New average values for the $^{238}U/^{235}U$ isotope ratios of natural uranium standards. Int J Mass Spectr 295:94–97

Rolison JM, Stirling CH, George E, Middag R, Gault-Ringold M, Rijkenberg MJA, De Baar HJW (2015) Biogeochemical cycling of the uranium, iron and cadmium isotope systems during oceanic anoxia: A case study of the Black Sea. Goldschmidt Abstract

Romaniello SJ, Brennecka GA, Anbar AD, Colman, AS (2009) Natural isotopic fractionation of $^{238}U/^{235}U$ in the water column of the Black Sea. Eos Trans AGU 1:6

Romaniello SJ, Herrmann AD, Anbar AD (2013) Uranium concentrations and $^{238}U/^{235}U$ isotope ratios in modern carbonates from the Bahamas: assessing a novel paleoredox proxy. Chem Geol 362:305–316

Rudnick R, Gao S (2003) Composition of the continental crust. Treatise Geochem 3:1–64

Russell WA, Papanastassiou DA, Tombrello TA (1978) Ca isotope fractionation on the Earth and other solar system materials. Geochim Cosmochim Acta 42:1075–1090

Satkoski AM, Beukes NJ, Li W, Beard BL, Johnson CM (2015) A redox-stratified ocean 3.2 billion years ago. Earth Planet Sci Lett 430:43–53

Schauble EA (2007) Role of nuclear volume in driving equilibrium stable isotope fractionation of mercury, thallium, and other very heavy elements. Geochim Cosmochim Acta 71:2170–2180

Scott C, Lyons TW, Bekker A, Shen YA, Poulton SW, Chu XL, Anbar AD (2008) Tracing the stepwise oxygenation of the Proterozoic ocean. Nature 452:456–459

Shiel AE, Laubach PG, Johnson TM, Lundstrom CC, Long PE, Williams KH (2013) No measurable changes in $^{238}U/^{235}U$ due to desorption–adsorption of U (VI) from groundwater at the Rifle, Colorado, integrated field research challenge site. Environ Sci Technol 47:2535–2541

Shiel AE, Johnson TM, Lundstrom CC, Laubach PG, Long PE, Williams KH (2016). Reactive transport of uranium in a groundwater bioreduction study: Insights from high-temporal resolution $^{238}U/^{235}U$ data. Geochimica et Cosmochimica Acta 187:218-36

Shimokaua J, Kobayashi F (1970) Separation of uranium isotopes by chemical exchange. Isotopenpraxis Environ Health Stud 6:170–176

Smith LA (1961) Variations in the uranium-235 content of fifteen ores. Union Carbide Nuclear Company, Oak Ridge Gaseous Diffusion Plant Rep. No. K-1462, 19 January

Snow JE, Friedrich JM (2005) Multiple ion counting ICPMS double spike method for precise U isotopic analysis at ultra-trace levels. Int J Mass Spectr 242:211–5

Staudigel H, Davies GR, Hart SR, Marchant KM, Smith BM (1995) Large scale isotopic Sr, Nd and O isotopic anatomy of altered oceanic crust: DSDP/ODP sites417/418. Earth Planet Sci Lett 130:169–185

Staudigel H, Plank T, White B, Schmincke H-U (1996) Geochemical fluxes during seafloor alteration of the basaltic upper oceanic crust: DSDP Sites 417 and 418. Geophys Monogr Ser 96:19–38

Steiger RH, Jäger E (1977) Subcommission on geochronology—Convention on use of decay constants in geochronology and cosmochronology. Earth Planet Sci Lett 36:359–362

Stirling CH, Halliday AN, Porcelli, D (2005) In search of live ^{247}Cm in the early solar system. Geochim Cosmochim Acta 69:1059–1071

Stirling CH, Andersen MB, Potter E-K, Halliday AN (2007) Low temperature isotope fractionation of uranium. Earth Planet Sci Lett 264:208–225

Stirling CH, Halliday AN, Potter E-K, Andersen MB, Zanda B (2006) A low initial abundance of ^{247}Cm in the early solar system: Implications for *r*-process nucleo-synthesis. Earth Planet Sci Lett 251:386–97

Stirling CH, Andersen MB, Warthmann R, Halliday AN (2015) Isotope fractionation of ^{238}U and ^{235}U during biologically-mediated uranium reduction. Geochim Cosmochim Acta 163:200–218

Stylo M, Neubert N, Roebbert Y, Weyer S, Bernier-Latmani R (2015a) Mechanism of uranium reduction and immobilization in *Desulfovibrio vulgaris* biofilms. Environ Sci Technol 49:10553–10561

Stylo M, Neubert N, Wang Y, Monga N, Romaniello SJ, Weyer S, Bernier-Latmani R (2015b) Uranium isotopes fingerprint biotic reduction. PNAS 112:5619–5624

Tatsumoto M, Shimamura T (1980) Evidence for live ^{247}Cm in the early solar system. Nature 286:188–192

Taylor SR, McLennan SM (1985) The continental crust: Its composition and evolution. Blackwell Scientific: 1–312

Telus M, Dauphas N, Moynier F, Tissot FL, Teng FZ, Nabelek PI, Craddock PR, Groat LA (2012) Iron, zinc, magnesium and uranium isotopic fractionation during continental crust differentiation: The tale from migmatites, granitoids, and pegmatites. Geochim Cosmochim Acta 97:247–265

Tissot FLH, Dauphas N (2015) Uranium isotopic compositions of the crust and ocean: age corrections, U budget and global extent of modern anoxia. Geochim Cosmochim Acta 167:113–143

Tissot FL, Dauphas N, Grossman L (2016) Origin of uranium isotope variations in early solar nebula condensates. Sci Adv 2:e1501400

Tomiak PJ, Andersen MB, Hendy EJ, Potter EK, Johnson KG, Penkman KE (2016) The role of skeletal micro-architecture in diagenesis and dating of *Acropora palmata*. Geochim Cosmochim Acta 183:153–75

Tribovillard N, Algeo TJ, Lyons T, Riboulleau A (2006) Trace metals as paleoredox and paleoproductivity proxies: an update. Chem Geol 232:12–32

Trinquier A, Birck J-L, Allègre CJ (2007) Widespread ^{54}Cr heterogeneity in the inner solar system. Astrophys J 655:1179

Trinquier A, Elliott T, Ulfbeck D, Coath C, Krot AN, Bizzarro M (2009) Origin of nucleosynthetic isotope heterogeneity in the solar protoplanetary disk. Science 324:374–376

Tuli JK (2005) Nuclear Wallet Cards. Brookhaven National Laboratory, USA

Urey HC (1947) The thermodynamic properties of isotopic substances. J Chem Soc (Resumed) 562–81

Uvarova YA, Kyser TK, Geagea ML, Chipley D (2014) Variations in the uranium isotopic compositions of uranium ores from different types of uranium deposits. Geochim Cosmochim Acta 146:1–17

Villa IM, Bonardi ML, De Bievre P, Holden NE, Renne PR (2016) IUPAC-IUGS status report on the half-lives of ^{238}U, ^{235}U and ^{234}U. Geochim Cosmochim Acta 172:387–392

Vance D, Teagle DAH, Foster GL (2009) Variable Quaternary chemical weathering fluxes and imbalances in marine geochemical budgets. Nature 458:493–496

Villeneuve J, Chaussidon M, Libourel G (2009) Homogeneous distribution of ^{26}Al in the solar system from the Mg isotopic composition of chondrules. Science 325:985–988

Wang X, Johnson TM, Lundstrom CC (2015a) Isotope fractionation during oxidation of tetravalent uranium by dissolved oxygen. Geochim Cosmochim Acta 150:160–170

Wang X, Johnson TM, Lundstrom CC (2015b) Low temperature equilibrium isotope fractionation and isotope exchange kinetics between U(IV) and U(VI). Geochim Cosmochim Acta 158:262–275

Wang X, Planavsky NJ, Reinhard CT, Hein JR, Johnson TM (2016) A Cenozoic seawater redox record derived from ^{238}U/^{235}U in ferromanganese crusts. Am J Sci 1:316:64–83

Wasserburg GJ (1987) Isotopic abundances: inferences on solar system and planetary evolution. Earth Planet Sci Lett 86:129–173

Wasserburg GJ, Busso M, Gallino R (1996) Abundances of actinides and short-lived nonactinides in the interstellar medium: diverse supernova sources for the *r*-processes. Astrophys J Lett 466:L109

Weyer S, Anbar AD, Gerdes A, Gordon GW, Algeo TJ, Boyle EA (2008) Natural fractionation of ^{238}U/^{235}U. Geochim Cosmochim Acta 72:345–359

Wieser ME, Schwieters JB (2005) The development of multiple collector mass spectrometry for isotope ratio measurements. Int J Mass Spectr 242:97–115

Wohlers A, Wood BJ (2015) A Mercury-like component of early Earth yields uranium in the core and high mantle ^{142}Nd. Nature 520:337–340

Wood BJ, Blundy JD, Robinson JAC (1999) The role of clinopyroxene in generating U-series disequilibrium during mantle melting. Geochim Cosmochim Acta 63:1613–1620

Zheng Y, Anderson RF, van Geen A, Fleischer MQ (2002) Preservation of particulate non-lithogenic uranium in marine sediments. Geochim Cosmochim Acta 66:3085-3092

Reviews in Mineralogy & Geochemistry
Vol. 82 pp. 851-885, 2017

Medical Applications of Isotope Metallomics

Francis Albarède[1,2*], Philippe Télouk[1], and Vincent Balter[1]

[1]Ecole Normale Supérieure de Lyon
Université de Lyon 1 and CNRS
69007 Lyon
France
[2] Rice University
Department of Earth Science
Houston TX 77005
USA

[*] *Corresponding author: albarede@ens-lyon.fr*

INTRODUCTION

One may wonder how a paper discussing medical applications of metal isotopes got lost in a review journal dedicated to mineralogy and geochemistry. The justifications are multiple. First, the coming of age of metal isotopic analysis in the mid '90s is largely due to the analytical creativity of the geochemical community and to corporate technical skills allowing the rise of new technologies. Second, many concepts, which can be imbedded in quantitative models testable from their predictions, are common to geochemistry, biochemistry, physiology, and nutrition: a cell, with its organelles, a body with its organ and body fluids, are systems liable to treatments similar to those used to model a lake, the ocean–atmosphere, and the mantle–crust systems. Of course, time scales and length scales differ, the complexity of biology is immense compared to that of the mineral world. Geological systems lack the hallmarks of life, genes and cell signaling. In spite of the overall complexity of the biological systems, pathways, kinetics, and chemical dynamics are better understood than their counterpart in earth sciences. Like in many fields of engineering, comparing the records of inputs and outputs is a powerful tool to identify the internal 'knobs' controlling a given system and learn how to tweak them. Third, although some of the most sophisticated techniques such as *ab initio* calculations of molecular configurations, energetics, and isotopic properties are still limited to molecules with less than a few dozens of atoms, the time is getting closer to when simulations of large molecules will become available for application to 'real' proteins with large molecular weights. The present article reviews some of the basic features of what is now known as *Metallomics* and the preliminary applications of stable isotopes to some medical cases, a discipline for which we suggest the simple term of *Isotope Metallomics*. Stable isotopes are widely used for nutrition studies: enriched stable isotopes are added to the diet of humans or animals to monitor the transit of a particular element (Umpleby and Fielding 2015) or to microbial cultures in ecology (Radajewski et al. 2000). The natural isotope fractionation of ubiquitous elements, such as C, H, O, N, and S, are only rarely used for medical purposes because are usually too unspecific relative to biological processes. In contrast, alkaline earth metals such as Ca and Mg and the transition elements Fe, Cu, and Zn are more promising because of their more specific functional roles in biology and also because their turnover rates in the body are relatively short. Copper plays a major role in oxidizing iron (oxidase) and controlling electron transfer, while hundreds of important enzymes use zinc as a cofactor. Iron is involved in a large number of biological functions and, because of the very large stores contained in red blood cells, muscle, and the

1529-6466/17/0082-0020$05.00 http://dx.doi.org/10.2138/rmg.2017.82.20

liver, its overall turnover time is of several years (Bothwell and Finch 1962; Gropper and Smith 2012). It is an essential component of heme, a cofactor made of large heterocyclic porphyrin rings. Heme is the active component of metalloproteins known as hemoglobin and myoglobin, which are used by the body to shuttle oxygen and carbon dioxide in blood cells and muscle. Since a number of diseases act to disrupt biochemical pathways in which metalloproteins are involved, it is expected that some pathologies may induce a signal on metal isotope compositions observable in easily accessible biological samples, notably the major blood components, serum and red blood cells. The purpose of the present article is to expand the review by Albarède (2015) and Albalat et al. (2016) to some new applications of Ca, Fe, and Cu isotopes to medicine (also see Costas-Rodriguez et al. 2016). It is neither a review of isotope chemistry nor a metallomics compendium. The sections dedicated to isotope chemistry are intended to appeal to biology and medical students, while those with a more biological perspective are directed at geochemists and illustrate some potential applications of stable isotopes to the medical world.

The *isotope effect* (Lindemann 1919; Bigeleisen and Mayer 1947; Urey 1947; Galimov 1985; Criss 1999; Schauble 2004; Wolfsberg et al. 2010) refers to the small energy changes induced by the substitution of one isotope by another. This effect is a straightforward consequence of the Heisenberg uncertainty principle of quantum mechanics, which prescribes that the position and the velocity of an object cannot be both measured exactly at the same time. Even at the minimum of their potential energy, bonds never come to rest and their lowermost energy state is referred to as the zero-point energy. As the kinetic energy of a bond also depends on the mass M of bonding atoms, the zero-point energy and the successive energy levels differ for the isotopes of a same element. This is the origin of mass-dependent stable isotope fractionation in nature. Reporting the relative abundance of one pair of isotopes suffices to describe the strength of isotope fractionation: if an effect is observed with a given intensity between ^{64}Zn and ^{66}Zn (mass difference of two), the effect on ^{64}Zn and ^{68}Zn (mass difference of four) will be twice as strong. By 'intensity' we mean the deviation of isotope abundances relative to a reference value, in general a standard material typically provided by agencies such as the National Institute of Standards and Technology in the USA or the Institute for Reference Materials and Measurements in Europe. The nature and the absolute abundance of isotopes in reference materials are inconsequential to the scale of isotopic variations.

In contrast with organic biomarkers which degrade with time, isotope compositions of metals can be analyzed on biological samples years after the samples have been taken. In some less informed scientific environments, using the term isotope may lead to some confusion, as it is easily perceived as referring to radioactive nuclides, such as ^{14}C, used for labeling products, or ^{99}Tc used to visualize the inside of blood vessels and organs. It may also refer to artificially enriched stable isotopes (spikes) added to the diet of volunteers to monitor nutrition and metabolism. Contrary to C, O, H, and N, metals are not present in the atmosphere. The isotope abundances of metals in the bulk sample therefore are immune to oxidation, as they are unreactive to any chemical or biological reactions taking place in the original container, even if the sample is accidentally heated or transferred to another container. Here, we will review the variations in the abundances of stable isotopes of metals tightly related in cellular and physiological activity and *naturally* present in the body of humans and other organisms. Early metal isotope work on biological samples (Walczyk and von Blanckenburg 2002; Ohno et al. 2004; Krayenbuehl et al. 2005; Ohno et al. 2005; Stenberg et al. 2004, 2005; Albarède et al. 2011; Hotz et al. 2012; Aramendia et al. 2013; Jaouen et al. 2013; Van Heghe et al. 2014; von Blanckenburg et al. 2014; Balter et al. 2015; Costas-Rodríguez et al. 2015a,b; Larner et al. 2015; Télouk et al. 2015) showed promising relationships with age, sex, and pathologies. Although isotopic data on organs will be discussed occasionally, emphasis will be on serum for reason of feasibility: it is a chemically stable liquid medium, more readily available, commonly from bio-banks, than biopsies and resections, even for healthy subjects. Sulfur is not a metal, but is closely related to transition metal biochemistry, and its isotopic abundances in medical materials have not been systematically explored until now (Balter et al. 2015; Albalat et al. 2016). In order to assess the role of sulfur-rich amino acids and proteins, in particular the well-established connection between zinc and sulfur biochemistry through redox control (Maret and Krężel 2007), we will therefore also review these recent observations of sulfur isotope compositions of biological samples (Balter et al. 2015; Albalat et al. 2016).

THE ISOTOPE EFFECT

Isotope fractionation is a general term referring to the variability in the isotopic abundances of a particular element among coexisting species (e.g., sulfide and sulfate for S) or reservoirs (e.g., S in serum and red blood cells) hosting this element. It can be explained in a simple way: (1) vibrational frequencies decrease approximately with $M^{-1/2}$, while bond energy E varies with vibrational frequency ν according to $E = (n + 1/2)\, h\nu$, where h is the Plank constant and n a non-negative integer characterizing the energy 'level'. Favoring heavier isotopes in the lowermost energy levels therefore is a way of reducing the total energy of the system. High temperatures work to randomize the distribution of isotopes across energy levels. At ambient temperatures, however, the total energy is minimized when heavy isotopes concentrate into the 'stiffest' bonds, those with the lowest and therefore most stable energy levels (Bigeleisen and Mayer 1947; Urey 1947; Galimov 1985; Schauble 2004; Kohen and Limbach 2005; Wolfsberg et al. 2010). For a given element, the strength of a particular bond is expected to be higher for the smaller ions with the higher charge and therefore developing the strongest field. It is also higher when the overall binding energy at the site of the metal is shared among fewer partners. The strength of a bond involving metal depends on how easily the metal loses its electrons and how eager the bonding species (O, C, N, S) are to attract the lost electrons. A crude scale of bond stiffness may be guessed from the scale of electronegativity or ionization energy (Fig. 1). Bonds involving high oxidation states (Fe^{3+}, Cu^{2+}) and sites with small coordination numbers therefore prefer heavy over light isotopes. It is worth noting at this stage that isotope variability is a very subtle phenomenon: when differences are noted between 'light' and 'heavy' zinc or copper, shorthand for 'depleted' and 'enriched', respectively, the effects are always within the range of only a few parts in one thousand, which only modern mass spectrometry can resolve.

How do we go from energy and vibrational frequencies to real-life isotope fractionation? It was Bigeleisen and Mayer's great merit in 1947 to demonstrate that the isotope effect at equilibrium (and equilibrium is always a useful reference because it is the state to which systems spontaneous relax) can be predicted by statistical quantum mechanics from the ratio β of the reduced partition function of isotopes of a molecule:

$$\beta = \prod_i \frac{u_i'}{u_i} \frac{e^{u_i/2}}{e^{u_i'/2}} \frac{1-e^{-u_i}}{1-e^{-u_i'}}$$

Here the product extends over all the normal vibrational modes i of the molecule, $u_i = h\nu_i/kT$, and prime/non-prime variables differentiate the two isotopes. A partition function ratio is 'reduced' when configurations equivalent under rotation are counted only once. We will see thereafter that evaluation of these reduced partition functions is at the heart of predicting isotope fractionation factors.

In addition to the effects just described for systems at thermodynamic equilibrium, the smaller activation energy of the lighter isotopes allows them to react faster (Bigeleisen and Wolfsberg 1958; Wolfsberg et al. 2010): the kinetic isotope effect (KIE) has been advocated as a cause of biologically mediated isotope fractionation (Gussone et al. 2003), but it requires either non-steady state conditions (the system grows) or the existence of competing reaction pathways.

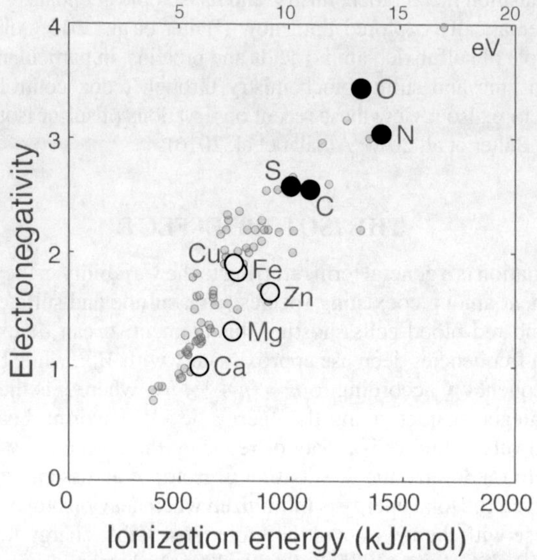

Figure 1. Strong bonds preferentially incorporate heavy isotopes. This is the case of nitrogen, as in the amino acid histidine, and oxygen, as in lactate. In contrast, sulfide bonds, as in amino acid cysteine, tend to preferentially incorporate light isotopes. The scale of bond strength can be crudely estimated using first ionization energy or element electronegativity. The electronegativity difference between bonding elements is a sensitive indicator of isotope fractionation.

Mass spectrometry is the chief method for obtaining precise isotopic abundances.. Except for hydrogen, the isotope effect is usually very small, with variations of isotope abundances rarely exceeding one part per 1000 per unit of mass difference. Measuring such small variations requires a mass spectrometer with high transmission and a magnetic mass filter (sector). Inexpensive quadrupole mass spectrometers do not provide the precision needed for useful applications to natural abundance variations. For decades, only the isotopic abundances of elements that could be introduced into a gas source with electron bombardment, such as H, C, N, O, S, were measured, often as molecular compounds such as CO_2 or SO_2. Mass fractionation in the mass spectrometer itself (instrumental mass bias) would be dealt with by alternating standard material with the unknown samples through calibrated inlet valves (sample-standard bracketing). The actual "true" (absolute) values of isotopic ratios cannot be known with high precision, but this limitation does not really matter: only relative isotopic variations are considered, and they are typically reported on a convenient delta scale, which, for example for ^{65}Cu, is expressed as:

$$\delta^{65}\mathrm{Cu} = \frac{\left(^{65}\mathrm{Cu}/^{63}\mathrm{Cu}\right)_{sample}}{\left(^{65}\mathrm{Cu}/^{63}\mathrm{Cu}\right)_{standard}} - 1$$

It is common practice to place the heavy isotope at the numerator. Most metals are refractory. Gas sources therefore are inefficient for metallic elements, unless volatile components such as SiF_4 are used. Standardization by sample-standard bracketing could not be used on thermal ionization mass spectrometers, which is a method of choice for radiogenic isotope geochemistry. Thermal ionization is also notoriously difficult for metals such as iron. The precision of measured isotope abundances therefore was poor and their subtle variations remained largely unexplored (Shields et al. 1965). Double-spike techniques, in which the abundance dependence of mass fractionation is used, would relieve the constraint for elements with *four* stable isotopes or more (Fe and Zn). This technique is, however, rather time consuming and has found only limited applications (Albarède et al. 2004). In the mid '90s, multiple collector inductively coupled plasma mass spectrometry (MC-ICP-MS) emerged as a game changer for the measurement of metal isotope abundances as the technique, which is based on very efficient ionization and high transmission, combined with sample-standard bracketing, allowed for unprecedented precision (typically 0.01–0.05‰) on metal samples as small as a few tens of nanograms of metal. The metal analyzed represents traces in a matrix of organic material loaded with major elements such as Na, Cl, P, Mg, and Ca. Both isobaric interferences and matrix effects seriously affect the measurement of isotopic abundances when unprocessed samples are run. The MC-ICP-MS technique therefore demands strict purification of the analyzed trace metal with a yield close to 100%. The analytical perspective provided by Costas-Rodríguez et al. (2016) is particularly useful for medical studies.

Why take the trouble to measure metal isotopic abundances, a daunting task, instead of simply relying on the concentrations of the metal in question in various parts of the body? The answer is that changes in metal concentrations in general are not amenable to quantitative predictions, whereas the direction and magnitude of the isotopic effect induced by bonding a metal with a chelate, typically an amino acid such as cysteine or histidine, can be predicted by theoretical methods. In contrast to different elements, which can never truly substitute for one another along all biochemical pathways, the isotopes of a given element behave similarly enough that variations in their relative abundances remain predictable. Decades ago, experimental determination of isotope fractionation of an element between coexisting compounds was the method of choice, but the results are now perceived as much less reliable than those obtained by the so-called *ab initio* or first-principles theories. In addition, the challenge of obtaining results for the very large number of relevant organic compounds is simply daunting. The most commonly used method is the Density Functional Theory or DFT (Parr 1983), a computational quantum mechanical model providing the ground-state orbital geometries and vibrational frequencies of metallic compounds. Each atom is treated as a cloud of electrons orbiting a nucleus. Typically, each calculation is divided into two steps, one step in which atoms are confined in a box and left to drift towards a stable molecular configuration and a subsequent step in which isotopes are substituted to infer the slight thermodynamic changes arising from the substitution. Obtaining results on compounds of biological interest such as the complex environment of Cu in ceruloplasmin, a blood component, and superoxide dismutase 1, an intracellular protein (Fig. 2), is calculation intensive and would require special software and consistent databases. So far, only the bonding of metals with single amino acids ligands has been explored. Even for simple cases, uncertainties on DFT results arise that are due to the difficulty of modeling a quantum mechanical effect known as the exchange and correlation interaction between electrons. Other theoretical approaches calculate the vibrational force constants of the molecular compounds. Methods specific to ^{57}Fe are based on the measure of the kinetic energy of the nucleus of interest by its Mössbauer effect, which is the resonant nuclear fluorescence of γ-rays. The nuclear kinetic energy may be measured by Mössbauer spectroscopy (Polyakov and Mineev 2000; Polyakov and Soultanov 2011), in which the relative velocity between the γ-ray emitter and absorber is determined by the Doppler shift, or by inelastic nuclear resonant X-ray scattering (INRXS) synchrotron experiments, a method giving the phonon density of states (Polyakov et al. 2007; Dauphas et al. 2012).

Large proteins are so far beyond the reach of DFT but efforts to predict fractionation of elements such as Fe, Cu, Zn, Ni, and Ca by smaller molecules have recently been made by a few groups (Seo et al. 2007; Domagal-Goldman and Kubicki 2008; Domagal-Goldman et al. 2009; Black et al. 2011; Fujii and Albarède 2012; Fujii et al. 2013, 2014; Sherman 2013). Isotope fractionation factors for ligand monomers, such as the most common amino acids (histidine, cysteine, methionine), glutathione, and carboxylic acids, such as lactate, oxalate and citrate, have

Table 1. Partition function ratios for $^{66}Zn/^{64}Zn$ and $^{65}Cu/^{63}Cu$ in molecular species relevant to medical studies on a 1000 ln β scale (reduced partition function ratios) ($T = 310\,K$). Isotopic fractionation between two coexisting species 1 and 2 can be computed as $\delta^{65}Cu_2 - \delta^{65}Cu_1 \approx \ln\beta_2 - \ln\beta_1$.

Species	$\ln\beta_{Zn}$	Ref	Species	$\ln\beta_{Cu}$	Ref
$ZnHPO_4(H_2O)_5$	3.309	[1]	Cu(I)L-Lact	1.725	[5]
$ZnH_3(PO_4)_2(H_2O)_4^-$	3.967	[1]	$Cu(I)Cl_2^-$	2.182	[2]
$ZnH_2(PO_4)_2(H_2O)_4$	4.072	[1[	$Cu(I)HS(H_2O)$	2.529	
fourfold			$Cu(I)(H_2O)_2^+$	2.667	[4]
$Zn(Cys)(H_2O)_3^{2+}$	3.072	[2]	$Cu(I)Cl(H_2O)$	2.683	[2]
$Zn(Glu)(H_2O)_2^{2+}$	3.524	[2]			
$Zn(H_2O)_4^{2+}$	3.577	[2]	$Cu(II)H(L\text{-}ascorbate)(H_2O)_4^+$	3.087	[4]
$Zn(His)(H_2O)_3^{2+}$	3.647	[2]	$Cu(II)H(D\text{-}ascorbate)(H_2O)_4^+$	3.139	[4]
$Zn(Met)(H_2O)_3^{2+}$	3.66	[2]	$Cu(II)H_3(PO_4)_2(H_2O)_3^-$	4.176	[2]
$Zn(His)(H_2O)_2^{2+}$	3.673	[2]	$Cu(II)H_2PO_4(H_2O)_4^+$	4.355	[2]
$Zn(Thr)(H_2O)_3^{2+}$	3.767	[2]	$Cu(II)H_4(PO_4)_2(H_2O)_3$	4.382	[2]
			$Cu(II)O_x(H_2O)_2$	4.931	[4]
sixfold			$CuH_2(PO_4)_2(H_2O)_3^{2-}$	5.024	[2]
$Zn(Cys)(H_2O)_5^{2+}$	2.504	[2]			
$Zn(Met)(H_2O)_5^{2+}$	2.734	[2]	$Cu(II)(Cys)(H_2O)_4^{2+}$	3.124	[2]
$Zn(His)(H_2O)_4^{2+}$	2.777	[2]	$Cu(II)(Met)(H_2O)_4^{2+}$	3.650	[2]
$ZnCl(H_2O)_5^{2+}$	2.912	[2]	Cu(II)(GS)H0	3.892	[2]
$ZnSO_4(H_2O)_5^{2+}$	3.279	[2]	$Cu(II)(Thr)(H_2O)_4^{2+}$	4.110	[2]
$Zn(His)(H_2O)_5^{2+}$	2.921	[2]	$Cu(II)(Glu)(H_2O)_3^{2+}$	4.117	[2]
$Zn(H_2O)_6^{2+}$	3.026	[2]	$Cu(II)(His)(H_2O)_3^{2+}$	4.148	[2]
$Zn(Glu)(H_2O)_4^{2+}$	3.053	[2]	$Cu(II)(His)(H_2O)_4^{2+}$	4.168	[2]
$Zn(Thr)(H_2O)_5^{2+}$	3.075	[2]	$Cu(II)(H_2O)_5^{2+}$	4.220	[2]
			$Cu(II)L\text{-}Lact(H_2O)_3^+$	4.359	[2]
anhydrous			$Cu(II)L\text{-}LactH_{-1}(H_2O)_2$	4.969	[5]
$[Zn\text{-}Cys\text{-}H_{-1}]^+$	1.108	[3]	$Cu(II)L\text{-}Lact_2$	5.616	[5]
$[Zn\text{-}Cys]^{2+}$	1.211	[3]	Cu(II)L-Lact D-Lact	5.627	[6]
$[Zn\text{-}Glu\text{-}H_{-1}]^+$	1.517	[3]			
$[Zn\text{-}His]^{2+}$	3.336	[3]			
$[Zn\text{-}His\text{-}H_{-1}]^+$	3.465	[3]			

References: [1] Fujii and Albarède (2012); [2] Fujii et al. (2014); [3] Moynier et al. (2013); [4] Fujii et al. (2013); [5] T. Fujii (pers .comm.); [6] Télouk et al. (2015).

become available for Cu and Zn. As shown in Tables 1 and 2, the data are tabulated as reduced partition functions β (usually as 1000 ln β) and the order and amplitude of isotopic enrichment between two compounds 1 and 2 at equilibrium can be estimated as $\delta^{65}Cu_2 - \delta^{65}Cu_1 \approx \ln\beta_2 - \ln\beta$. For example, the predicted $\delta^{65}Cu$ value in $Cu(II)(His)(H_2O)_4^{2+}$ is $4.168 - 3.124 = 1.044$ ‰ higher than in $Cu(II)(Cys)(H_2O)_4^{2+}$. Table 3 shows some important stability constants for Cu and Zn chelates. Calculation of Fe and Ca fractionation factors are still largely restricted to compounds of environmental interest (Rustad et al. 2010; Fujii et al. 2014).

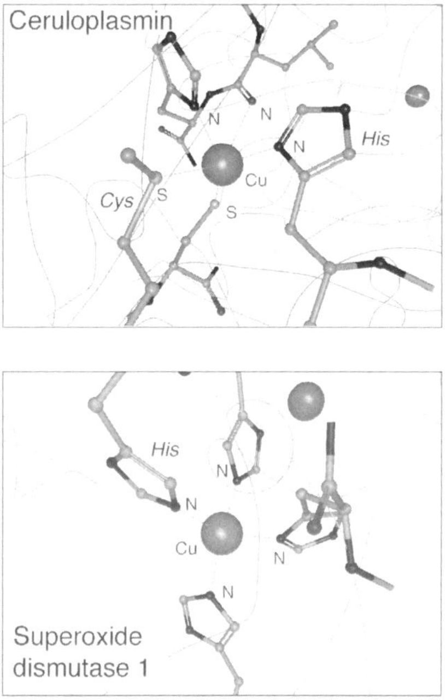

Figure 2. Example of complex coordination encountered in two metalloproteins. The ferroxidase enzyme ceruloplasmin (Cp, PDB ID: 1KCW (Zaitseva et al. 1996)) is the major copper-carrying protein in the blood, and oxidizes seric iron. The large Cu ion in the center is labeled A 1055, while the distant one is labeled A 1049. The intracellular Cu/Zn enzyme superoxide dismutase 1 (SOD1, PDB ID: 1SOS (Parge et al. 1992)) catalyzes the disproportionation of superoxide O_2^- into hydrogen peroxide H_2O_2 or di-oxygen O_2 and therefore protects the cell from free radicals. The large Cu ion in the center is labeled C 154, while the metal sphere in the background is a Zn ion labeled C 155. These two enzymes use the two oxidation states of copper to shuttle electrons. Copper is bound to N from the amino acid histidine (His, hard bond) and S from the amino acid cysteine (Cys, soft bond).

Table 2. Equilibrium $^{34}S/^{32}S$ enrichment in ‰ of different sulfur-bearing inorganic and organic species at 298 K (Albalat et al. 2016). The calculations include the effect of one hydrate shell on sulfate. $^{34}S/^{32}S$ fractionation α between two species may be obtained in ‰ between two coexisting species 1 and 2 and can be computed as $\delta^{34}S_2 - \delta^{34}S_1 \approx \ln\beta_2 - \ln\beta_1$.

HS^-	H_2S	Cysteine	Cystine	Glutathione	Methionine	Taurine	$SO_4^{2-}\,6H_2O$
4.75	11.42	16.11	17.12	15.67	20.21	71.59	73.94

Table 3. Stability constants for the successive chelates of Cu and Zn by relevant carboxylates.

Ion	Species	log β_1	log β_2	log β_3	Ref
Cu^{2+}	pyruvate	2.2	4.9		[1]
	lactate	2.52	3.9	4.28	[2]
	ascorbate	1.57			[1]
Zn^{2+}	pyruvate	1.26	1.98		[1]
	lactate	1.67	2.65	2.94	[2]
	ascorbate	1.0			[1]

References: [1] Smith and Martell (1987); [2] Portanova et al. (2003)

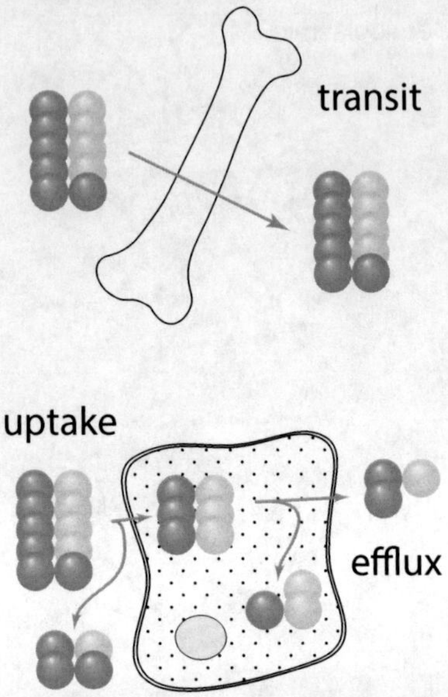

Figure 3. Isotopes of the same element bound to a specific metalloprotein, e.g., ^{63}Cu and ^{65}Cu, are depicted as spheres with two different shades of grey. *Top*: Upon transit at steady-state (no growth), the proportion of each isotope is preserved. The cell or organ is reduced to a single input and a single output. *Bottom*: When the isotopes distribute themselves between two coexisting channels, such as extracellular medium and cytosol, or the cytosol and an organelle, the different pathways allocate different isotope abundances to each channel. Oxidation, biosynthesis, input, storage, and output are expected to result in isotope fractionation.

What comes around goes around: the proportion of isotopes present within a system (cells, organs, body fluids) can only vary if they are imported and exported at different rates (Fig. 3). For a period of time exceeding the mean turnover or residence in the system, input and output must be precisely balanced. When a pathway involves multiple outputs for a single input, the abundances of the different isotopes of a specific element may not be identical in each branch: this is the nature of isotope fractionation,

The best-documented examples so far are the robust trends of Zn and Cu isotope fractionation, two elements for which fractionation by amino acids and other organic ligands have been studied. The following has been observed:

1. Isotope fractionation is less intense for Zn than for Cu
2. Cu(II) compounds are isotopically heavier than Cu(I) compounds
3. Electron donors with a strong electronegativity (N, O) and associated moieties (NH_2, SO_4, PO_4, OH, lactate and pyruvate, two carboxylic acids with a side oxygen or hydroxyl) preferentially bind to heavy isotopes relative to elements with smaller electronegativity, typically S and S-bearing amino acids (cysteine, methionine) (Fig. 1)
4. As demonstrated for zinc by the comparison between four- and sixfold coordination, preference for heavier isotopes decreases with increasing coordination numbers
5. Working out how these results relate to large proteins should attract attention in the future.

AN OVERVIEW OF Ca, Fe, Zn, Cu, AND S BIOCHEMISTRY AND HOMEOSTASIS

We will first take an introductory tour of the biochemistry of these important metals, then summarize a few important facts about sulfur-containing amino acids and proteins. A major trait shared by the ions of all these elements is that their electrostatic field is very strong, which makes their presence as bare cations harmful to many proteins. This is the reason why they are normally bound to metalloproteins. The following subsections are not intended to extensively map the pathways of metal metabolism in the human body, but to highlight some particular reactions, usually associated with transmembrane import, export, and changes in redox states, which are potential mechanisms accounting for isotope fractionation.

Calcium

Calcium is an alkaline earth metal and is always in the divalent state. It has six isotopes at nominal masses 40, 42, 43, 44, 46, and 48. The Ca content of the human body is ~ 1.4 kg and the daily intake recommended for an adult is about 1 g (Gropper and Smith 2012). Of the Ca^{2+} not stored in bones (~1 %), only ~0.1 ‰ resides in the extracellular fluid. The main intracellular Ca^{2+} stores in generic cells are the endoplasmic reticulum and the Golgi apparatus (Brini et al. 2013). The Ca content of the cytosol is particularly low but may be increased through a number of channels. The sodium-calcium exchanger (NCX), an antiporter membrane protein, and the plasma membrane Ca^{2+} ATPase (PMCA) are used to clear Ca^{2+} from the cell. Ninety-nine percent of body calcium (Ca) is in bones in the form of hydroxyapatite. It must be kept in mind that we do not inherit bone from our adolescence and that bones are renewed during our entire life. The feared bone loss with increasing age is therefore a dynamic process. Cartilage collagen is produced by chondrocytes on the outer part the new bone and is gradually replaced by bone. Chondrocytes die out and are replaced by osteoblasts, the bone building cells, which precipitate apatite. Alkaline phosphatase (ALP), an enzyme located in the membrane of osteoblasts liberates phosphate from phosphate esters, such as β-glycerophosphate (Chung et al. 1992), which allows apatite precipitation and mineralization (bone deposition). The bone demolishers, osteoclasts, are responsible for resorption, which returns Ca and phosphate to the blood stream, a process regulated by the parathyroid hormone (PTH). Deficient Ca regulation results in vascular calcification, a rather common disease of the elderly and dialysis patients (Hofmann Bowman and McNally 2012; Evrard et al. 2015). In the blood stream, Ca^{2+} exists in three forms, free ions (50 %), albumin bound (40 %), and protein bound (10 %). Free Ca is one of the most tightly regulated parameters in the body and regulates itself serum concentrations by stimulating bone resorption and Ca reabsorption by the kidney. Cysteine present in albumin, which would favor isotopically light isotopes relative to the free-ion pool, does not seem to bind significantly with Ca^{2+} (Kragh-Hansen and Vorum 1993).

Calcium, which is particularly abundant in the environment, was the first non-conventional element for which the variations of its isotopes were successfully investigated (Skulan et al. 1997). Analysis by TIMS and double-spike is still competing with sample-standard bracketing by MC-ICP-MS. Two main features have emerged: First, Ca becomes isotopically lighter as it moves through the food chain. Second, bone Ca is isotopically light (Skulan and DePaolo 1999; Reynard et al. 2010), which may come as a surprise since, as expected from the electronegativity scale, heavy Ca isotopes should favor PO_4^3 over carboxylate and carbonyl groups. Isotope fractionation inevitably reflects Ca binding with soft ligands since combination of Ca^{2+} with phosphate liberated by phosphate esters must avoid precipitation of the relatively insoluble hydroxyapatite until final delivery of its constituents to the bone.

Iron

The role of iron in human biology is particularly important because Fe(II)-bearing hemoglobin is the prime carrier of oxygen in the blood. More than 99% of Fe in blood (Albarède et al. 2011) and 65% in the body (Gropper and Smith 2012) is accounted for by erythrocytes. An average body of 70 kg contains ~ 3 g of Fe, and the daily requirement is 1–2 mg per day, giving a residence time of Fe in the body of ~5.5 years. Myoglobin is a closely related oxygen-binding protein, which allows the muscle tissue to 'hold its breath' for protracted lengths of time. Other iron stores are present in the liver, the kidney and the spleen, largely as Fe(III) ferritin, a ferrihydrite analog wrapped in a protein shell. A characteristic of iron homeostasis is that excretion is not actively regulated, with body balance being only achieved by the modulation of the intestinal input.

Iron ions exist in two oxidation states, ferrous Fe(II) and ferric Fe(III). Ferric iron binds with many inorganic and organic ligands and ferric hydroxide is highly insoluble. Iron has four major isotopes, 54, 56, 57, and 58 (nominal masses). Iron isotopic abundances are most commonly measured by MC-ICP-MS. The pioneering study of Walczyck and von Blanckenburg (2002) demonstrated the existence of large isotopic variability among organs and body fluids.

Iron homeostasis in mammalian cells has been reviewed by many authors (Fleming and Bacon 2005; Andrews 2008, 2012; Dlouhy and Outten 2013; Winter et al. 2014). After reduction of diet Fe(III) to Fe(II) by the ferric reductase duodenal cytochrome b (Dcytb), the divalent metal transporter 1 (DMT1) allows inorganic iron from the diet across the membrane into intestinal cells. Dietary iron is transported by the transmembrane copper-dependent ferroxidase hephaestin from intestinal enterocytes into the blood stream to facilitate iron loading onto the plasma iron carrier transferrin (Tf). Transferrin (Tf) binds two ferric ions and provides iron for most human cell types (Fig. 4). Intracellular iron is

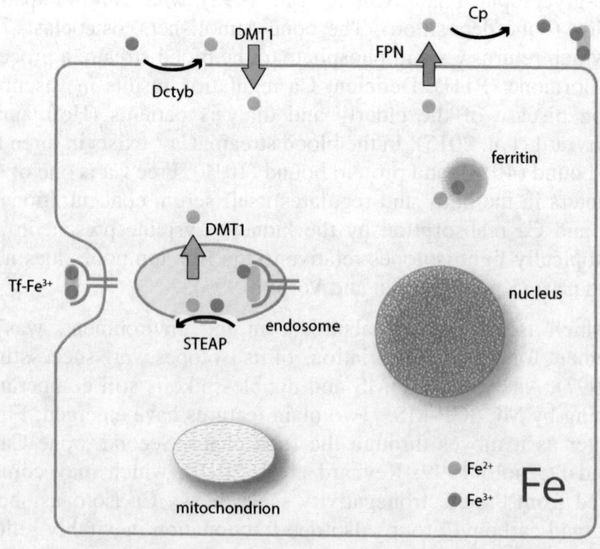

Figure 4. A sketch of iron trafficking in a generalized cell. Abbreviations: transmembrane divalent metal transporter 1 (DMT1, uptake), ferroportin (FPN, efflux), transferrin (Tf, reversible binding). Oxidase and reductase enzymes are shown as curved arrows: duodenal cytochrome B (reductase, dctyb), 6-transmembrane epithelial antigen of prostate (reductase, STEAP), ferritin (FER, oxidase and storage), ceruloplasmin (Cp, oxidase). An endosome is a membrane-bound pocket of extracellular medium created by endocytosis.

either stored in ferritin (see below) or exported by the Fe(II) iron exporter ferroportin (FPN) into the circulation where it is oxidized by ceruloplasmin for binding with transferrin (Tf), which accounts for most of iron transport in the serum. Iron-bound transferrin is imported into the cell with the cell surface Tf receptor 1, and the complex engulfed by the retracting membrane forming an endosome in a process known as endocytosis.

Further reduction of Fe(III) by the STEAP reductase and transport to the cytoplasm via DMT1 makes Fe(II) available for cell needs. Excess Fe(II) is either re-exported back to the blood stream by FPN or stored within the cell after re-oxidation as ferritin (Theil 2011; Linder 2013).

The hypoxia-inducible factor 1 (HIF-1α) links the production of red blood cells (erythropoiesis) with iron homeostasis (Haase 2013). The important observation (Krayenbuehl et al. 2005; Hotz et al. 2012) that blood Fe becomes isotopically heavier after phlebotomy (bloodletting) was interpreted as indicating that Fe is quickly retrieved from ferritin iron stores (liver, kidney) to replace lost iron, which is an important observation for the treatment of anemia. Iron is isotopically heavy in the liver, spleen, and bone marrow, and light in red blood cells (Hotz et al. 2011). A thorough inventory of Fe isotope abundances in human diet has been assembled by von Blanckenburg et al. (2013).

Zinc

Zinc is a transition element with five isotopes, 64, 66, 67, and 68, and the very minor isotope 70. It is always in a divalent state. The Zn content of the human body ranges from 1.5 to 3 g and the daily intake recommended for an adult is about 10 mg (Gropper and Smith 2012) giving a residence (turnover) time in the body of about 225 days. This metal is found in the nucleus and the cytosol of cells in all organs. About 90 % of Zn in blood is accounted for by erythrocytes (Albarède et al. 2011). Excess cytosolic Zn is bound to metallothionein, a short sulfur-rich protein, and then transported to nucleus and organelles for storage. Zinc is a cofactor of carbonic anhydrase, which interconverts carbon dioxide and bicarbonate, and thus regulates the acid-base balance of the cytosol. Zinc is also a cofactor of superoxide dismutase, which controls reactive oxygen species. Zinc further regulates the glutathione metabolism and metallothionein expression (Cruz et al. 2015) and affects signaling pathways and the activity of transcription factors with zinc finger domains.

Zinc homeostasis and its importance in various pathologies have been regularly reviewed (Cousins et al. 2006; Maret and Krężel 2007; Lichten and Cousins 2009; Fukada et al. 2011; Maret 2013; Bonaventura et al. 2015) (Fig. 5). Adjustment of gastrointestinal absorption is the primary control of zinc homeostasis (Lowe et al. 2000). Malnutrition induces cell-mediated immune defects and promotes infections (Golden et al. 1977). Zinc acts on the immune system by potentiating cytokines (Poleganov et al. 2007), a mechanism that may also be controlling chronic inflammation, such as rheumatoid arthritis (Kawashima and Miossec 2004). Zinc in seminal fluid has been suggested as a biomarker of prostate cancer (Costello and Franklin 2008). Albumin is the main transporter of Zn in serum. For a 'generalized' cell, the transmembrane importers consist of 14 'isoforms' of the ZIP family (ZIP1 to ZIP14). Different ZIP transporters are expressed specifically on different cell types (Bonaventura et al. 2015). DMT1 has a lower affinity for Zn (Garrick et al. 2003; Espinoza et al. 2012). No specific chaperone, a dedicated cytosolic transporter, has been identified for the transfer from cytoplasm to organelles, although to some extent metallothionein may be considered one. Zinc efflux from the cell and Zn stockade in organelles is controlled by the ZnT protein family, which consists of 10 isoforms (ZnT1 to ZnT10). The trans-membrane ZnT1 is the only isoform to be ubiquitously expressed on the cell surface, while the expression of other ZnTs depends on the type of cell and organelles where they are localized.

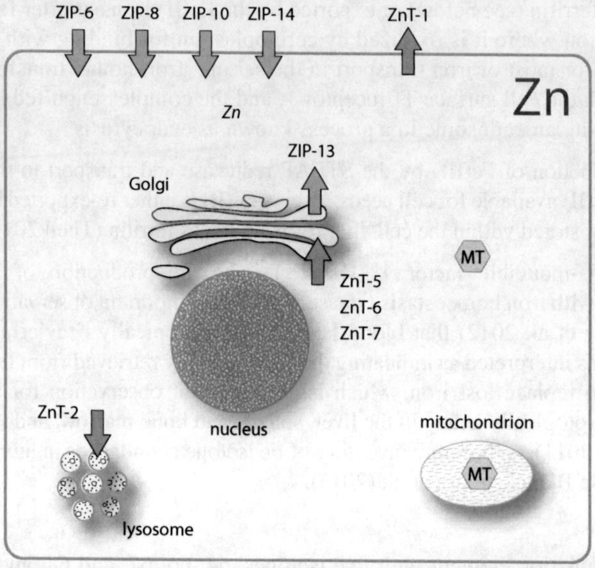

Figure 5. A sketch of zinc trafficking in a generalized cell. Abbreviations: Zn transporter family (ZnTx, uptake), Zn transporter ZIP family (ZIPx, efflux), metallothionein (MT), Cu,Zn-superoxide dismutase 1 (SOD1).

Copper

Copper has two isotopes at nominal masses 63 and 65 and its ions exist in two oxidation states, Cu(I) and Cu(II). The total copper content of the human body ranges from 50 to 150 mg and is found in all tissues and most body fluids and the daily intake recommended for an adult is about 1 mg (Gropper and Smith 2012). The residence (turnover) time in the bulk human body is therefore less than 20 days. About 35 % blood copper is accounted for by erythrocytes (Albarède et al. 2011). Copper is a micronutrient and a catalytic and structural cofactor of many important enzymes involved in tumor development (Linder and Goode 1991; Brewer 2003; Kim et al. 2008; Lutsenko 2010; Vest et al. 2013; Denoyer et al. 2015). Serum ceruloplasmin is a ferroxidase enzyme synthetized in the liver, which allows iron to be transported in the blood as harmless Fe^{3+} hydroxide. It also acts as a modulator of inflammation. A variable fraction of copper is transported by serum albumin. Cytochrome *c* oxidase is a transmembrane protein complex of the mitochondrion associated with the terminal step of electron transport and energy production. Superoxide dismutase 1 (Cu, Zn SOD1) resides mostly in the cytosol. Excess Cu may also be stored in metallothionein.

The dominant Cu importer of cells is hCtr1 (human Cu transporter) (Pope et al. 2011; Kim et al. 2013; Wee et al. 2013), which binds to albumin (Shenberger et al. 2015) and binds both Cu(I) and Cu(II) (Haas et al. 2011) (Fig. 6). Hypoxia-induced DMT1 (divalent metal transporter, notably ferrous iron) has also been invoked in copper transport into mice intestinal cells (Arredondo et al. 2003; Arredondo et al. 2014), but its relevance to other cell types is not established. Depending on the final destination, hCtr1 presents Cu^{+} to chaperones that will deliver it to specific partners: COX17 brings copper to cytochrome c oxidase (CCO) in mitochondria, CCS delivers it to SOD1, while ATOX1 is the chaperone for the copper-transporting ATPases (Cu-ATPases). The latter maintain intracellular Cu(I) levels by regulating its efflux either directly or through the secretory pathway (Lutsenko 2010; Rosenzweig and Argüello 2012). ATP7A and ATP7B differ in their patterns of tissue expression and cellular localization.

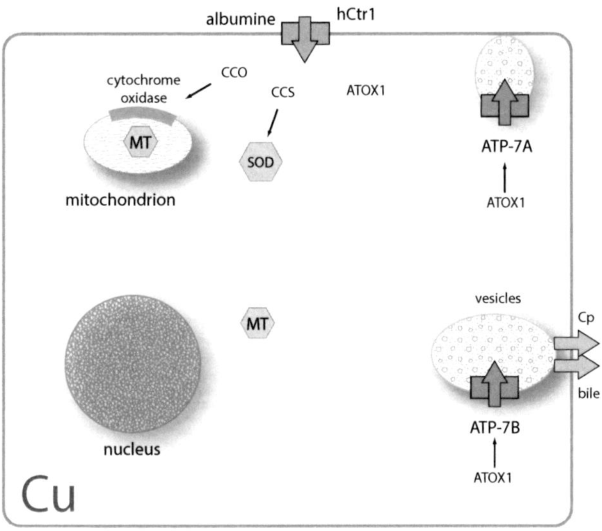

Figure 6. A sketch of copper trafficking in a generalized cell. Abbreviations: human copper transporter (hCtr1, uptake), copper ATPases ATP7A and ATP7B (efflux), cytochrome c oxidase copper chaperone (CCO), copper chaperone for superoxide dismutase (CCS), antioxidant protein 1 chaperone to ATP7A and ATP7B (ATOX1).

Sulfur

The sulfur content of the human body is about 175 g (Gropper and Smith 2012). The daily requirement of 1 g/d suggests a residence time of about 200 days which attests that S is a fairly reactive biological component. Most body sulfur is hosted by two major amino acids, cysteine a thiol ending with an -SH moiety and methionine, an *S*-methyl thioether ending with a -C–S–CH_3 moiety. Methionine is an essential amino acid, which must be obtained from the diet, and is imported by transmembrane importers, notably the Na-independent L-type amino acid transporter 1 (LAT1) (Kanai et al. 1998; Fuchs and Bode 2005; Broer and Palacin 2011). Instead, cysteine can be synthesized from methionine within the cell through the transsulfuration pathway involving methylation by *S*-adenosyl methionine (SAM). Metal binding metallothioneins are rich in cysteine, accounting for up to one third of the amino-acidic sequence (Kojima et al. 1976; Kissling and Kagi 1977).

Sulfur homeostasis is sketchily represented in Figure 7. An essential property of cysteine is the potential of two molecules to bind into cystine by forming a covalent disulfide S–S bridge, which may unlock for metal chelation in a reducing environment, such as the cytosol. Disulfide bridges are very important for the structure and stability of proteins such as serum albumin, the most abundant protein of blood serum and its main sulfur carrier. The properties of the disulfide bridge are at the basis of glutathione's function, a tripeptide essential to the control of cellular redox state by easily switching between its reduced (GSH) and oxidized (GSSG) form. Glutathione is synthetized from cystine imported from the extracellular medium in exchange of glutamate by the x_c^- 'antiporter' (Bridges et al. 2012). Intracellular cysteine is catabolized into either taurine or pyruvate, which is used for energy production, and sulfate (Stipanuk et al. 2006).

Sulfate is associated with membrane proteins known as proteoglycans, such as heparan sulfate, and is also found in heparin, an anticoagulant substance commonly used as an additive to lower the viscosity of blood samples and inhibit blood clotting.

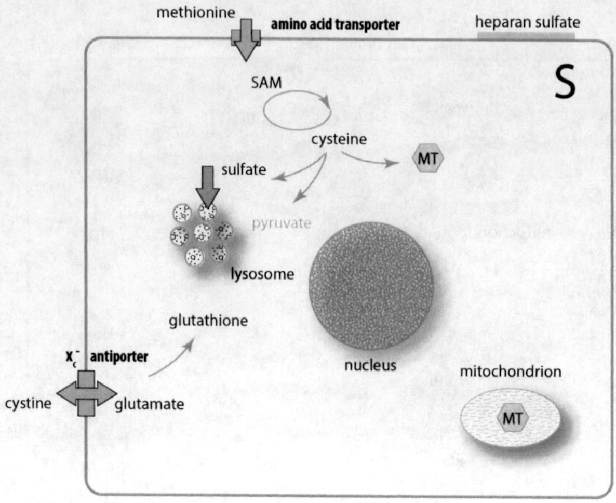

Figure 7. A sketch of sulfur trafficking in a generalized cell. Abbreviations: S-adenosyl methionine (SAM, a methylation agent involved in the intracellular biosynthesis of cysteine from methionin), metallothionein (MT, a cysteine-rich metal store). A lysosome is a membrane-bound cell organelle used to break down unused and toxic molecules and which is eventually eliminated by exocytosis. An antiporter is a two-way transmembrane exchange protein.

ISOTOPE COMPOSITIONS OF Fe–Zn–Cu–S IN THE BLOOD OF HEALTHY INDIVIDUALS

Copper and zinc contents vary in the serum of control individuals in a remarkable way. Figure 8 shows that the Cu content is high and variable among women (Milne and Johnson 1993), whereas Zn tends to be constant. In contrast, male serum tends to have a narrow range of Cu and variable Cu/Zn. The range of overlapping values is, however, relatively large. Although copper is commonly used as a biomarker to assess health status, the Zn/Cu ratio seems to have an even stronger potential (Malavolta et al. 2015; Télouk et al. 2015). Prostate cancer appears to have little effect on serum Zn levels but Cu clearly increases relative to controls. The serum of breast cancer patients seems to plot above the reference line Zn=1200 ppm, which roughly defines the average value of women controls, whereas for colon cancer patients the value plots below this line (higher Cu and/or lower Zn).

How isotope compositions of metals and sulfur vary among the organs and body fluids of a mammal was essentially unknown until a few years ago with the first studies on sheep, mice and minipigs (Balter et al. 2010, 2013; Hotz et al. 2011; Moynier et al. 2013; Tacail et al. 2014). For ethical reasons, access to human material is much more restricted. The first major observation was that, in most cases, the isotope compositions of Cu, Zn, and Fe of each organ falls, for a given species, within a narrow range of values (Fig. 9). In mice, Zn is isotopically heavy in blood and bone and light in liver and brain and independent of genetic background. Copper is specifically light in kidney. This pattern reproduces for sheep except for isotopically light Zn in blood, a feature still awaiting elucidation.

Albarède et al. (2011) conducted a systematic analysis of Zn, Cu, and Fe isotope compositions in human whole blood, serum, and erythrocytes (Table 4). They concluded that, on average, Fe in erythrocytes (red blood cells or RBC) is isotopically light with respect to serum, whereas Zn and Cu are isotopically heavier by ~0.3 and ~0.8‰, respectively. Male-female δ^{66}Zn and δ^{65}Cu differences were less than 0.2% for both serum and RBC. The study found mean values of δ^{66}Zn ~ +0.17‰ and δ^{65}Cu ~ −0.26±0.40‰ for serum

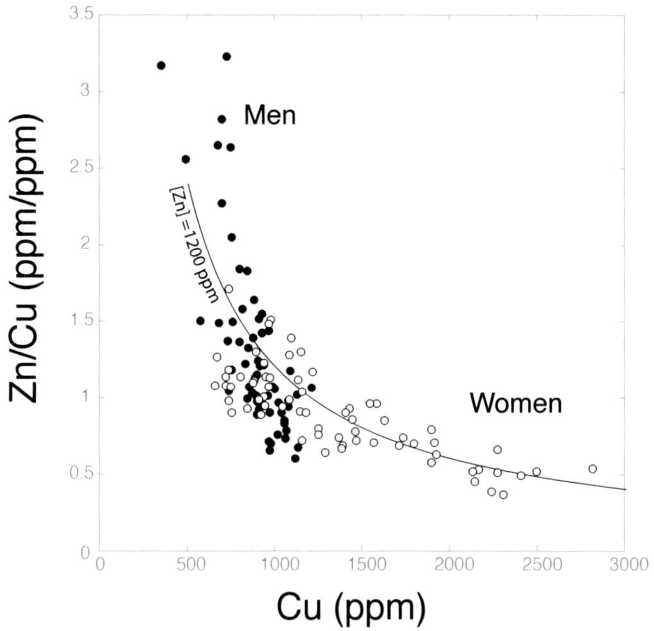

Figure 8. Plot of Zn/Cu vs Cu concentrations in the serum of healthy adults. Data from (Albarède et al. 2011). In this plot, which exacerbates the contrast between men and women, constant Zn concentrations are represented by hyperbolae.

and δ^{66}Zn ~ +0.44±0.26‰ and δ^{65}Cu ~ +0.66‰ for erythrocytes. A similar δ^{65}Cu value 0.29±0.27‰ was obtained by Costas-Rodríguez et al. (2015) on 29 serum samples. The serum-RBC difference is most significant for Cu. δ^{65}Cu is 0.2‰ heavier in men erythrocytes relative to women (Albarède et al. 2011). Total blood values depend on the erythrocyte load (hematocrit) and therefore may vary with sample preparation and should be considered less reliable than serum and RBC values. There is still no consensus on the δ^{56}Fe of human serum, in which iron concentration is very low and hemolysis (rupture of red blood cells) may obscure the results (Albarède et al. 2011; von Blanckenburg et al. 2014).

In a study of *whole-blood* samples on Yakut volunteers aged 18–74, Jaouen et al. (2013) found that the ^{66}Zn/^{64}Zn ratio increases and the ^{65}Cu/^{63}Cu ratio decreases with age. Likewise, the whole blood Fe isotope ratios of postmenopausal women or women on a hormonal anti-conception treatment leading to absence of menstruation are shifting in the direction of those characteristic of the male population (Van Heghe et al. 2013). Van Heghe et al. (2014) observed that ^{65}Cu/^{63}Cu ratios tend to increase in whole blood after menopause but found no age or menstruation effects on Zn isotopes. In contrast, Jaouen and Balter (Jaouen et al. 2013) showed that, while the Fe and Cu isotope compositions of blood of men are steady throughout their lifetime, postmenopausal women exhibit blood δ^{65}Cu values similar to men, and δ^{56}Fe values intermediate between men and menstruating women, an isotopic pattern easily explained by the different residence times of the metals into the body (Fe ~ 5.5 years and Cu ~ 20 days).

Comparison of their results with Albarède et al.'s (2011) study led Jaouen et al. (2013) to emphasize the importance of the ethnic factor. On a fairly small sample set, Van Heghe et al. (2012) observed that δ^{66}Zn in whole blood is about 0.15‰ higher for vegetarian relative to omnivorous volunteers, but the outcome for Cu isotopes was less conclusive. The isotopic

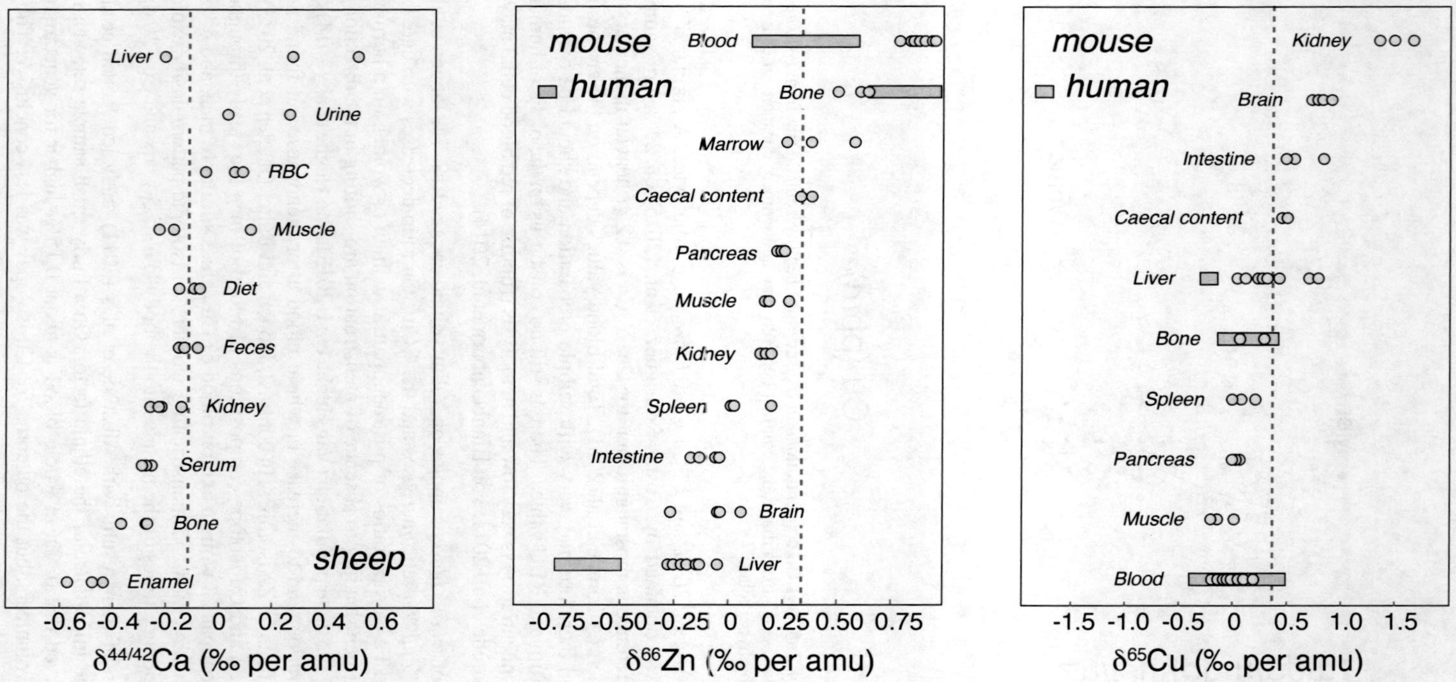

Figure 9. Zinc and copper isotope variability among organs, bones, body fluids, and intestinal content of mice (Balter et al. 2013) (left and center) and for Ca (Tacail et al. 2014) (right). All the data are reported in delta units per mil for Cu and Zn. The shaded bands show the range of variations for humans (Albarède et al. 2011; Jaouen et al. 2013). Typical uncertainties are ±0.05 ‰ (2-sigma error).

Table 4. Average isotope compositions in delta units (‰) and 95% range (2s) for the isotope compositions of Zn and Cu in the serum, erythrocytes, and total blood of 49 blood donors (Albarède 2015). Typical analytical uncertainties are 0.05 ‰. Men–women comparison: p is the probability that the two sets are not identical.

	n	av δ^{66}Zn	2s	av δ^{65}Cu	2s
Serum					
women	28	0.18	0.28	−0.24	0.36
men	21	0.16	0.1	−0.28	0.4
all	49	0.17	0.26	−0.26	0.4
p value (men/women)		0.45		0.3	
Erythrocytes					
women	28	0.46	0.17	0.46	0.47
men	21	0.43	0.45	0.67	0.36
all	49	0.44	0.33	0.56	0.5
p value (men/women)		0.39		0	
Total blood					
women	28	0.41	0.16	0.01	0.16
men	21	0.39	0.41	0.17	0.33
all	49	0.4	0.37	0.09	0.32
p value (men/women)		0.32		0.02	

composition of Zn was shown to be different between food products of plant and of animal origin (Costas-Rodríguez et al. 2014). Clearly, the carrier of assimilated Zn and the effect of diet deserve expanded studies on larger numbers of volunteers and with well-documented blood tests.

The first substantial set of $^{34}S/^{32}S$ values on the blood of healthy individuals were obtained by elemental analysis-isotope ratio mass spectrometry (EA-IRMS; gas source mass spectrometry) by Balter et al. (2015) and an average $\delta^{34}S_{V\text{-}CDT}$ of 5.9±1.5‰ on 11 serum samples and 5.1±1.9‰ on 20 RBC samples were measured. On 25 serum samples of adults, Albalat et al. (2016) obtained a very similar mean value but within a reduced interval 6.0±0.7‰, with the average $\delta^{34}S_{V\text{-}CDT}$ of women being 0.2‰ lower than that of men. On the same samples, both methods agree within one permil. Albalat et al. (2016) show that S in children serum is only slightly heavier but more scattered (6.3±1.0‰) than S in adult serum.

CALCIUM AND BONE LOSS

Calcium isotopic variations in urine and blood of bed rest studies (Heuser and Eisenhauer 2010; Morgan et al. 2012a,b; Channon et al. 2015) were motivated by the observation that astronauts suffer bone loss during space flight (Fig. 10). It was known that Ca in urine increases and becomes isotopically lighter with time, suggesting that this metal is liberated by osteoclasts from the bones into the blood stream (Permyakov and Kretsinger 2010). Increased loss of isotopically lighter Ca may reflect changing input, i.e., bone mineralization,

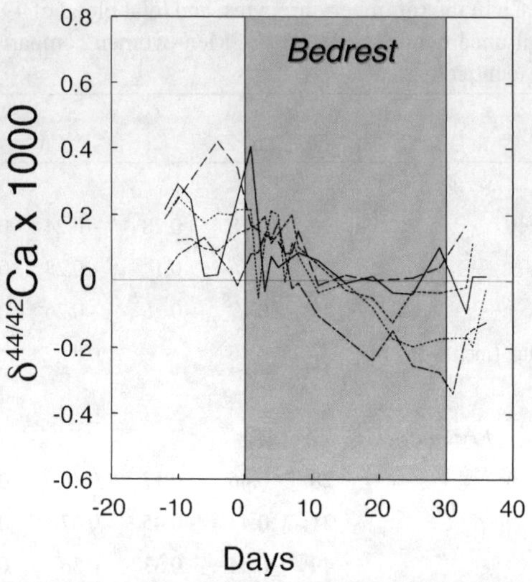

Figure 10. Isotope evolution of Ca in urine of volunteers confined to bed rest for a month. Bone Ca being isotopically light relative to diet, the study by Morgan et al. (2012a) shows that bed rest enhances bone loss.

as a response to changing proportions of free Ca to albumin-bound Ca. It may also signal an increased loss of bone Ca, which is recognized to result from enhanced expression of another calcium-binding protein, sclerostin, during bed rest (Spatz et al. 2012). Calcium isotopes show a strong potential to quantify the Ca fluxes in and out of the bones without having to resort to adding isotope tracers to the diet (Skulan and DePaolo 1999; Heuser and Eisenhauer 2010; Channon et al. 2015). This is of course relevant to the study of Ca in osteoporosis, a condition particularly common with aging women.

GENETIC AND INFECTIOUS DISEASES

Blood of patients with hereditary hemochromatosis contains more of the heavier iron isotopes than blood of healthy individuals (Krayenbuehl et al. 2005). This pathology is characterized by a genetically disrupted function of two critical proteins involved in control of intestinal iron absorption, hepcidin and ferroportin, leading to ineffective control of intestinal iron absorption and progressive iron overload of tissues. Iron in the blood of patients with hemochromatosis is 0.2–0.4 ‰ heavier than in healthy individuals, which suggests that the blood of hemochromatosis patients uses iron stores more heavily than the control cases. An alternative explanation (Van Heghe et al. 2013) is that, due to decreased hepcidin levels, hemochromatosis patients absorb more iron in the intestine.

Wilson's disease is a genetic disorder affecting Cu metabolism, which can lead to severe physiological and neurological symptoms. Aramendia et al. (2013) demonstrated a correlated decrease of both Cu levels and δ^{65}Cu in the serum of Wilson patients. This observation is quite remarkable and calls for an expanded data set.

It is well established that the prion protein (PrP) has an influence on cellular copper metabolism. Buechl et al. (2008) analyzed Zn and Cu isotopes in the brain of wild type and knockout mice and demonstrated that Cu and Zn isotopes in brain tissue are sensitive to prion-related local damage. They concluded that the presence of a form of PrP that does not bind Cu induces the fractionation of Cu isotopes.

ISOTOPE METALLOMICS IN CANCER

We already mentioned that calcium isotopes constitute a biomarker of bone loss. It was therefore suspected that they also have some potential to be affected by multiple myeloma (MM), a cancer of the plasma B-cells (lymphocytes produced in the bone marrow) characterized by bone destruction (Gordon et al. 2014). Radiographs do not demonstrate MM-induced abnormalities until a substantial fraction of the bone has been lost and provide no information about ongoing bone remodeling. Existing MM biomarkers do not provide an estimate of the net bone mineral balance. From the data on a limited number of patients, Gordon et al. (2014) suggested a significant relationship between serum Ca isotope abundances and myeloma activity, likely due to an MM-induced increased level of bone resorption.

Warburg and Krebs (1927) found that serum copper levels increased in various chronic diseases and several types of cancers, resulting in systemic and oncogenic (Ishida et al. 2013) copper accumulation. Anomalously high Cu levels or Cu/Zn ratios were indeed observed in the serum of breast cancer (Gupta et al. 1991; Yücel et al. 1994; Magalova et al. 1996; Piccinini et al. 1996; Koksoy et al. 1997; Wu et al. 2006; Cui et al. 2007) and cervical cancer (Cunzhi et al. 2003) patients. Zowczak et al. (2001) showed significant increase in the mean ceruloplasmin oxidase activity and total Cu concentrations in the serum of 62 patients with breast, lung, gastrointestinal, and cervical cancer relative to a control group. Serum ceruloplasmin was found to be significantly elevated in advanced stages of solid malignant tumors (Senra Varela et al. 1997). A compilation of Cu and Zn concentrations in the serum of cancer patients discussed below confirms that certain types of cancer may affect these parameters (Fig. 11). By themselves, such observations justify that copper and zinc isotopic variability should be investigated in cancer patients. The connection between Cu and Zn through proteins such as superoxide dismutase and metallothionein involved in the control of hypoxia, and therefore of

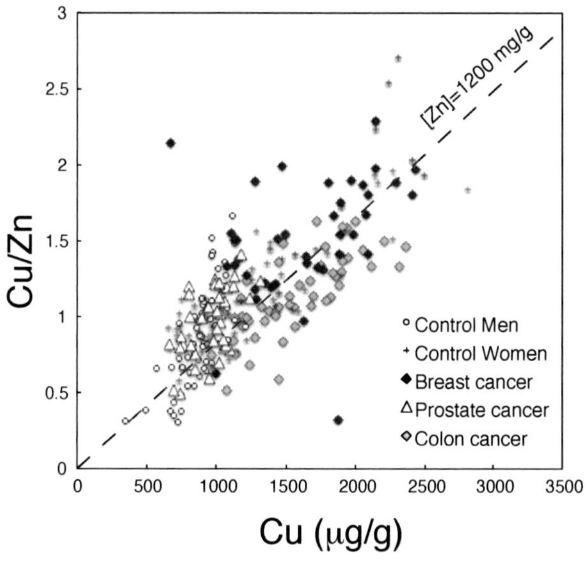

Figure 11. Copper and Cu/Zn in the serum as indicators of cancer status. The control group shows strong correlations, reflecting the tight regulation of Zn concentrations in the body (note that x/y=Zn). The trends for control men and women are different, with men having, on average, less Cu and more Zn than women.. Copper remains stable in prostate cancer patients relative to control men, but increases in colon cancer patients. Zinc in the serum of breast cancer patients is decreased and Cu probably increased relative to healthy subjects.

cancer in particular, gives grounds for interest in the amino acids that control the coordination of both metals, notably histidine, cysteine, and methionine. Hence the importance of exploring the extent of correlated Cu, Zn, and S isotopic variations.

Télouk et al. (2015) measured the $^{65}Cu/^{63}Cu$ ratios in the serum of 20 breast and 8 colorectal cancer patients (Fig. 12). Samples were taken at different times during the treatment, and amount to, respectively, 90 and 49 samples. Phenotypes and molecular biomarker were documented on most of the samples. When compared with the literature data from a control group of 50 healthy blood donors, abundances of Cu isotopes predict mortality in the colorectal cancer group with an error probability $p = 0.018$ (Fig. 13). For the breast cancer patients and the group of control women the probability falls even further to $p = 0.0006$. Most patients considered in this preliminary study and with serum $\delta^{65}Cu$ below the threshold value of −0.35‰ (per mil) did not survive beyond a few months. As a marker, a drop in $\delta^{65}Cu$ precedes molecular biomarkers such as CEA (carcinoembryonic antigen) and CA15.3 (carbohydrate antigen 15.3) by several months (Fig. 14), which is consistent with Cu turnover time in the body. The observed decrease of $\delta^{65}Cu$ in the serum of cancer patients was assigned by Télouk et al. (2015) to the extensive oxidative chelation of copper by cytosolic lactate. The potential of Cu isotope variability as a new diagnostic tool for breast and colorectal cancer seems strong.

Larner et al. (2015) analyzed zinc isotope compositions of various tissues in breast cancer patients. Resections of breast cancers were found to have a significantly lighter Zn isotopic composition than the blood, serum, and healthy breast tissue in both groups. The authors interpret the isotopically light Zn in tumors as attesting to its uptake by metallothionein in breast tissue cells, rather than in Zn-specific proteins. This reveals a possible mechanism of Zn delivery to Zn-sequestering vesicles by metallothionein, and is supported by a similar signature observed in the copper isotope abundances of one breast cancer patient. The number of samples used for this study was rather small and, hence, whether and how cancer may affect the $\delta^{66}Zn$ of serum deserves further attention.

Balter et al. (2015) found that in hepatocellular carcinoma (liver cancer) patients, serum and erythrocyte copper and sulfur are both enriched in light isotopes relative to controls (Fig. 15, bottom). The magnitude of the sulfur isotope effect is similar in red blood cells and serum of hepatocellular carcinoma patients, implying that sulfur fractionation is systemic. In contrast to serum data, the $\delta^{65}Cu$ of tumor resections is notably higher relative to healthy liver tissue (Fig. 15, top). The agreement between sulfur isotope data acquired on the same samples by EA-IRMS and MC-ICP-MS (Albalat et al. (2016) is reasonably good. Balter et al. (2015) concluded that the isotopic shift of either element is not compatible with a dietary origin, but rather reflects the massive reallocation in the body of copper immobilized within cysteine-rich metallothionein. A study by Costas-Rodrıguez et al. (2015b) also found lower $\delta^{65}Cu$ in the serum of patients with end-stage liver disease, with complications such as ascites, encephalopathy, and hepatocellular carcinoma (Fig. 16). These authors pointed out that $\delta^{65}Cu$ was positively correlated with the liver cirrhosis-related parameters, notably aspartate aminotransferase, INR (International Normalized Ratio for prothrombin time), bilirubin, and C-reactive protein, and inversely correlated with albumin and Na. They also found a negative correlation of $\delta^{65}Cu$ with Child-Pugh score based on albumin, bilirubin, and INR and the Mayo Clinic Model for End-stage Liver Disease score (MELD) based on creatinine, bilirubin, and INR.

Albalat et al. (2016) analyzed sulfur isotopes in a large number of pathological samples with emphasis on serum. The serum samples depart from those of healthy volunteers by their much smaller S concentration. This observation echoes the negative correlation between low serum albumin content and mortality (Okuda et al. 1985; Kao et al. 2015). The samples, however, for which the $\delta^{34}S$ departs from the range of healthy individuals are few and correspond to the 'naive' (untreated) patients, in particular those analyzed by Balter et al. (2015). Cancer and

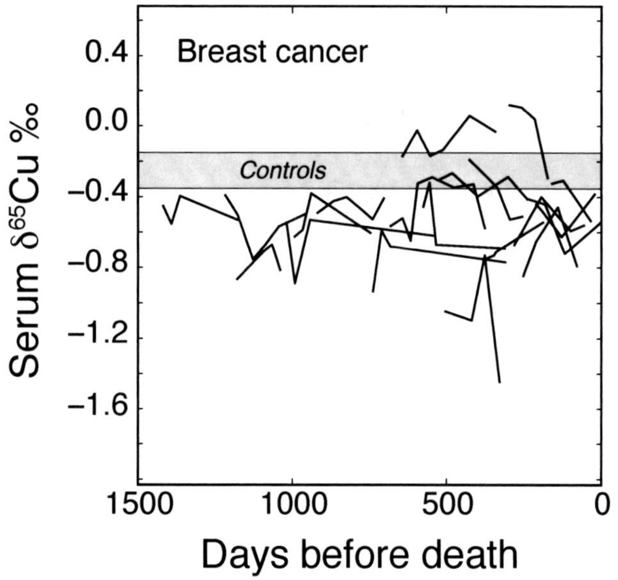

Figure 12. Evolution of serum δ^{65}Cu in 140 samples taken from 20 breast cancer cases until patient death (Télouk et al. 2015). Each line represents a different patient, with patterns used for differentiation purposes. The shaded band (controls) represents the 75 % range of δ^{65}Cu in the serum of healthy donors.

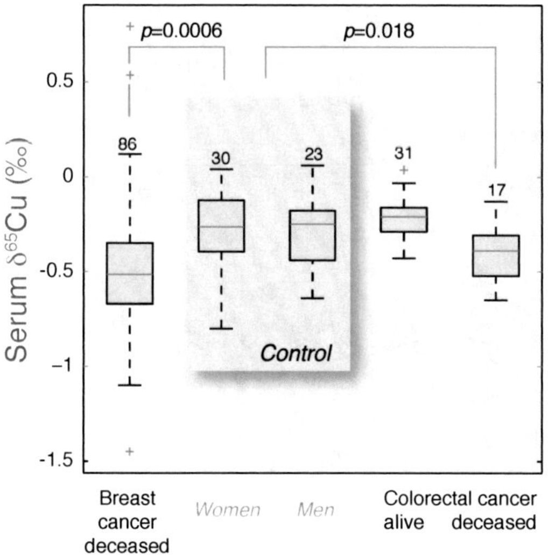

Figure 13. Whisker plots of serum δ^{65}Cu values for healthy men and women compared to breast cancer and colorectal cancer patients (Télouk et al. 2015). Boxes represent the 75 % middle quantiles and the whiskers 95 % quantiles. Parameter p represents the probability that two populations may be identical. Horizontal lines in the box: median; crosses: outliers. Separation between breast cancer patients and healthy women is strong. Separation between breast cancer and colorectal cancer patients and healthy men and women seems to depend on mortality.

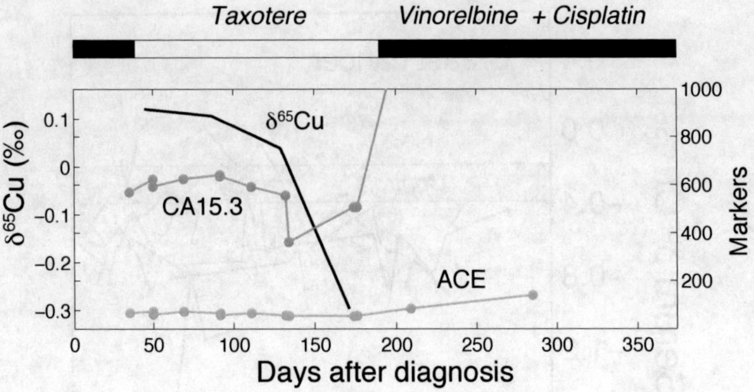

Figure 14. Early alarm by δ^{65}Cu (Télouk et al. 2015). The plot compares the δ^{65}Cu values (left axis, black line) and the molecular biomarkers (right axis): CEA (carcinoembryonic antigen, shade 1) and CA 15.3 (carbohydrate antigens, shade 2). The top bar scale shows the successive therapies received by the patient. The copper isotope signal precedes the other markers by 2–3 months.

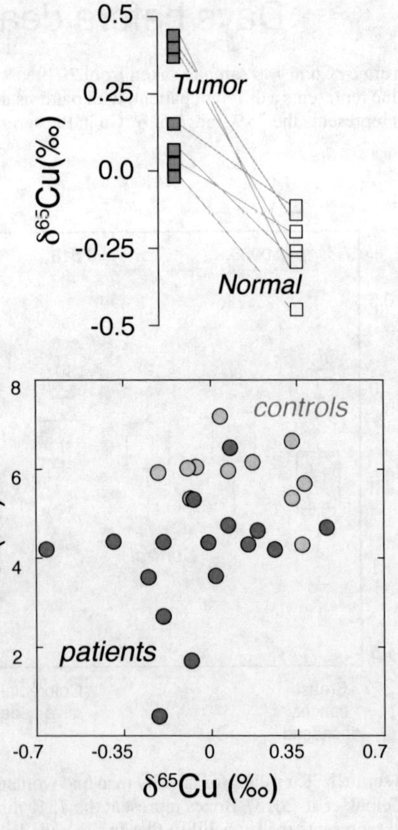

Figure 15. *Bottom*: Isotopically light copper and sulfur in the serum of hepatocellular carcinoma patients relative to controls (Balter et al. 2015). *Top*: Isotopically heavy copper in tumor liver tissue relative to normal tissue. The opposite direction of the changes in Cu isotope abundances in serum and tumor may be explained in different ways (see text).

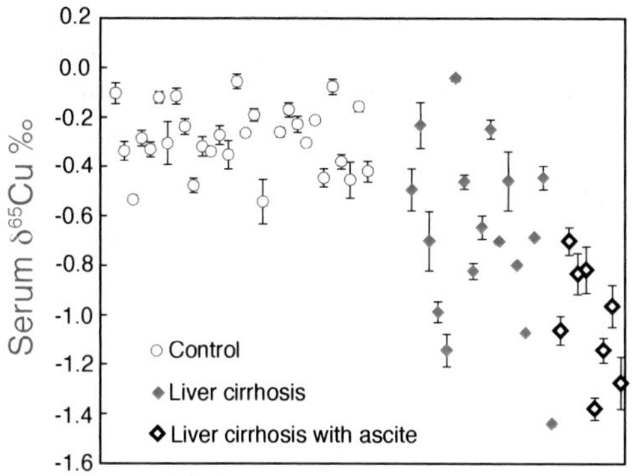

Figure 16. δ^{65}Cu values in the serum of liver cirrhosis patients with and without accumulation of fluid in the peritoneal cavity (ascites) relative to controls (Costas-Rodríguez et al. 2015b). Ascites is often associated with cirrhosis and metastatic cancer.

rheumatoid arthritis conditions increase the scatter of sulfur isotope compositions by up to a factor of two, but with little effect on the mean δ^{34}S values. It has been observed that medication brings δ^{34}S back to normal values but does not change sulfur concentrations in the serum.

Before attempting a biochemical interpretation of isotopic trends in biological samples, let us summarize the observations at hand. Most of the observations so far have been made on serum, on whole blood, occasionally on erythrocytes, and only exceptionally on organ tissues and tumors. Among all the analyzed elements, Cu and Zn isotope compositions seem to show that tumors deviate from healthy tissue (heavy Cu in liver and light Zn in breast neoplastic tissue) (Balter et al. 2015; Larner et al. 2015). In contrast to Cu, which is isotopically lighter in the serum of well over 130 cancer patients relative to a similar number of controls (colon, breast, and liver) (Balter et al. 2015; Télouk et al. 2015), our unpublished Zn isotope serum data show no difference between cancer patients and healthy donors of any age that could be used for medical purposes. (Fig. 17). Likewise, results obtained so far suggest that sulfur isotope compositions in the serum of cancer patients (colon, breast, and liver) could not in general be distinguished from the values in control patients, with the exception of some hepatocellular carcinoma patients (Balter et al. 2015; Albalat et al. 2016), but that the spread of δ^{34}S values is smaller for controls.

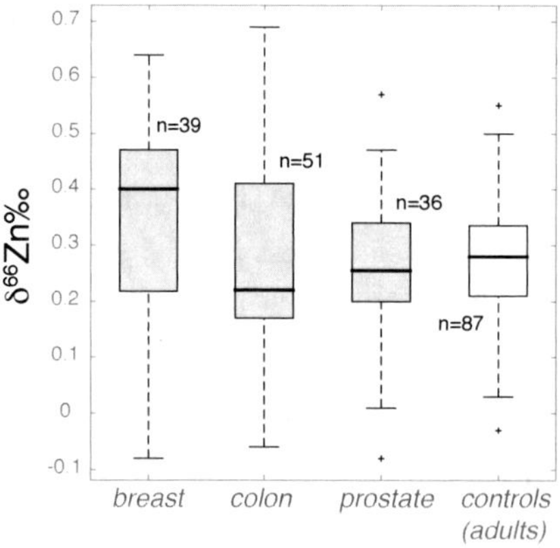

Figure 17. Whisker plot of serum δ^{66}Zn values for healthy men and women compared to breast, colorectal, and prostate cancer patients (unpublished data). The parameter n is the number of samples. Although the database is still rather small, Zn isotopes in serum show little potential as a cancer biomarker.

Both Télouk et al. (2015) and Costas-Rodrıguez et al. (2015b) suggested that low δ^{65}Cu can be used for prognosis in end-stage cancer (liver, colon, breast). Copper isotopes would complement other markers, such as the Child-Pugh score, albumin, or transaminases. The ~6 weeks turnover time (Linder and Goode 1991; Milne 1998) is close enough to the 19 days of albumin (Schaller et al. 2008) that the two parameters may have some biochemical pathways in common, one of them being that albumin is a Cu transporter. Télouk et al. (2015) pointed out that Cu isotopes seem to be reactive over time intervals of weeks to deteriorating health conditions, whereas molecular biomarkers tend to increase, whenever they do, within months.

Among possible reasons for δ^{65}Cu variations, some have been privileged:

1. Télouk et al. (2015) suggested that isotopically heavy Cu chelated by lactate, which cancer cells are known to produce massively, accumulates in the cytoplasm.
2. Balter et al. (2015) proposed that the low δ^{65}Cu value of serum is due to the release of intracellular copper from cysteine clusters in metallothionein.
3. Costas-Rodrıguez et al. (2015b) suggested that low δ^{65}Cu values reveal the hepatocellular and biosynthetic dysfunction of the liver, synergistically with inflammation and water retention.

Isotope abundances add a new 'dimension' to the overall budget of each element in cells and in the organism. None of the Cu isotope studies discussed above have attempted a mass balance evaluation including blood components, healthy tissues, and tumors, simply because the data are not available. Liver accounts for a large fraction of bodily metals such as Cu and Fe. A legitimate concern therefore is that the Cu isotope effects observed in serum and tumors cannot be directly compared until the missing data have been collected.

Three main routes by which Cu interacts with cancer cells are cellular metabolism, angiogenesis, and hypoxia. Copper is a tumor promoter and regulates oxidative phosphorylation in rapidly proliferating cancer cells inside solid tumors (Ishida et al. 2013). In normal cells, glycolysis, the first step of ATP production from glucose, is slow and its end product, pyruvate, is oxidized in mitochondria, where it fuels the much more efficient steps of the citric acid cycle and oxidative phosphorylation. Glucose degradation (glycolysis) is the primary source of ATP, which is achieved by the attachment of inorganic phosphate PO_4 to adenosine diphosphate (ADP). In normal *aerobic glycolysis* ATP is produced by a set of reactions summarized as follows:

$$\text{Glucose} + 2\,\text{ADP} + 2\,\text{NAD}^+ + 2\,\text{PO}_4 \Leftrightarrow 2\,\text{ATP} + 2\,\text{pyruvate}^- + 2\,\text{NADH} + 2\,\text{H}_2\text{O} + 4\,\text{H}^+$$

where nicotinamide adenine dinucleotide (NAD^+) is a ubiquitous electron acceptor. NADH is further re-oxidized at the surface of mitochondrion with consumption of H^+. In cancer cells, in contrast, pyruvate is used as an electron acceptor and aerobic glycolysis is replaced by *anaerobic glycolysis*:

$$\text{Glucose} + 2\,\text{ADP} + 2\,\text{PO}_4 \Leftrightarrow 2\,\text{ATP} + 2\,\text{lactate}^- + 2\,\text{H}_2\text{O} + 2\,\text{H}^+$$

a reaction known as lactic fermentation. Excess protons produced by the latter reaction are pumped out of the cell into the blood stream, which decreases its pH and greatly favors metastasis. The observation that cancer cells show enhanced glycolysis followed by lactate production in the cytosol, even in the presence of O_2, is known as the Warburg effect. Lactate levels are observed to be elevated in the serum of critically ill patients and correlate well with disease severity (Okorie and Dellinger 2011; Kinnaird and Michelakis 2015). Lactate efflux from the cell is regulated by monocarboxylate transporters (MCT) and intracellular and extracellular lactate levels are not simply related (Halestrap and Price 1999; Swietach et al. 2014). Copper is isotopically heavy in both pyruvate and lactate. However, in healthy cells pyruvate is shuttled into mitochondria for further energy processing, whereas lactate is exported from the cell by MCT and is metabolized in the liver. To a large extent, lactate is 'available' in the cell for Cu chelation (Fig. 18), whereas pyruvate is not. This is the substance of Télouk et al.'s (2015) explanation for the accumulation of copper with high δ^{65}Cu in the cell

A major role of copper in cancer is associated with hypoxia, a hallmark of both inflammation and human malignancies. In order to secure delivery of oxygen and nutrients to tumor cells, the growth of cm-sized tumors is accompanied by pervasive neovascularization (Carmeliet and Jain 2000). Several angiogenic factors, notably VEGF, tumor necrosis factor alpha (TNF-α) and interleukin (IL1), are copper activated (Nasulewicz et al. 2004). Under hypoxic conditions, ionic serum copper stabilizes the hypoxia-inducible factor (HIF-1α) (Martin et al. 2005), which is associated with tumor growth, vascularization, and metastasis (Semenza 2007; Wilson and Hay 2011).

Copper transport and uptake are still poorly understood (Lutsenko 2010), as is the mechanism of Cu reduction during uptake. It is likely that albumin is the main serum carrier presenting Cu to the cell and binds both Cu(I) and Cu(II) (Haas et al. 2011; Shenberger et al. 2015). Both Cu^+ and Cu^{2+} are captured by the terminal amino acids of Ctr1, which introduces Cu^+ into the cytosol. Hypoxic stimulation of the HepG2 cells (hepatocellular carcinoma) leads to a down-regulation of albumin (Wenger et al. 1995), supporting a connection between liver, copper, and albumin (Costas-Rodríguez et al. 2015b). Transmembrane Cu uptake is the primary site where isotope fractionation takes place. Fractionation restricted to storage or efflux is unlikely, as it would lead to an open-ended shift in intracellular δ^{65}Cu. Clearly, Cu isotopes may help understand the connections between tumor growth and Cu homeostasis.

ASSESSING THE POTENTIAL OF METAL ISOTOPES AS BIOMARKERS

Will metal isotopes become efficient biomarkers of disease? Efficiency here means reliability, independence relative to pre-existing biomarkers, and a potential for early diagnosis. Much work lies ahead. First, documenting the sensitivity and the specificity of metal isotopes relative to a particular pathology or a family of pathologies is a huge endeavor. Second, promoting Isotope Metallomics as a set of biomarkers will be confronted with the sample throughput capacity of the analytical groups. The medical community expects large datasets to be evaluated before metal isotopic variations can be routinely used as a diagnostic tool or as a therapeutic help. Running thousands of samples, as it is common practice in modern medical cohorts, is labor intensive and costly, and still beyond the analytical reach of existing mass-spectrometry labs.

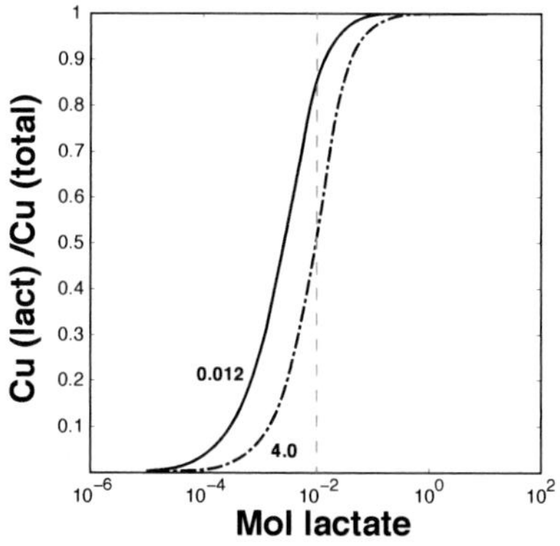

Figure 18. Extent of copper chelation by lactates in the cytosol (Télouk et al. 2015). The numbers on the curves represent the Cu^+/Cu^{2+} ratio for a redox potential of 0.153 V (copper ions) and for a body potential of 0.27 V (Lundblad and Macdonald 2010). The vertical dashed line corresponds to a lactate concentration of 10 mMol typical of tumor cells (Walenta et al. 2000).

Given the expectedly small sample size of the existing isotope abundance datasets, high-profile statistics will have a strong role to play. When facing a decision, geochemists are used to compare two-sigma intervals or, at best, to run t-tests on rather small populations (e.g., patients vs controls). Non-parametric tests, such as the Mann–Whitney and Kruskal–Wallis rank tests, are particularly useful to strengthen a case. When decision comes to human health, the dread of a false positive diagnose adds, however, to that of a failed detection. Receiver Operating Characteristics (ROC) curves (e.g., (Krzanowski and Hand 2009)) are particularly powerful tools of diagnostic medicine. The Y-axis of the ROC curve is the probability of a True Positive decision concluding that a patient with serum δ^{65}Cu ≤ δ^{65}Cu$_c$ died, or that a donor serum δ^{65}Cu exceeded δ^{65}Cu$_c$ for different cutoff values δ^{65}Cu$_c$. The Y-axis is also known as Sensitivity. The X-axis is the probability of a False Positive decision concluding that a patient with serum δ^{65}Cu ≥ δ^{65}Cu$_c$ has died, or that a donor serum δ^{65}Cu is ≤ δ^{65}Cu$_c$. The abscissa is also known as 1– Specificity. A widely used test is the area under the ROC curve (AUC), which varies between 0.5 (pure chance) and 1.0 (fully trustworthy test). The most reliable cutoff value can also be inferred by different techniques (Krzanowski and Hand 2009) from the elevation of the ROC curve above the first diagonal. In the present case of breast cancer diagnostic (Fig. 19), AUC is 0.76, while the optimum cutoff δ^{65}Cu$_c$ value is –0.37‰.

Whether metal isotopes as biomarkers can compete with molecular biomarkers, such as CA 125 for ovarian cancer (Cramer et al. 2011), CA 19–9 for colorectal cancer (Hotakainen et al. 2009), and CA 15–3 for metastatic breast cancer (Fahmueller et al. 2013) seems unlikely in the foreseeable future. Metal isotopes, however, have opened a window on cellular metallomics and therefore show promise of significant return on the biochemical mechanisms associated with each disease.

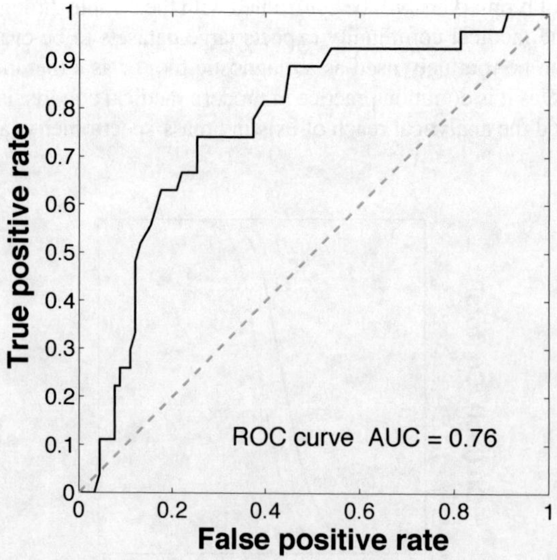

Figure 19. Receiver Operating Characteristics (ROC) curve for δ^{65}Cu in the serum of cancer patients. The ROC curve plots the probability of a True Positive decision vs the probability of False Positive decision. The optimum cutoff value for breast cancer is δ^{65}Cu$_c$ = –0.37‰. The area under the ROC curve (AUC) may vary between 0.5 (pure chance) and 1.0 (fully trustworthy test): the value of 0.76 obtained for the present data set supports the worth of δ^{65}Cu as a diagnostic tool.

COMPARTMENTALIZED MODELS OF CELLULAR HOMEOSTASIS

This field is known as Mathematical Systems Biology (e.g., ref. (Klipp et al. 2013)). Metal concentrations, as all concentrations, are prone to very large uncertainties due in particular to the elusive nature of the amount of material concentrations refer to. Elemental fractionation coefficients between coexisting media, e.g. extracellular medium and membrane, are also known with very large uncertainties. One more time, isotopic ratios help get around these problems. Models of metal homeostasis are useful as they help comparing predictions from fractionation patterns with isotope compositions observed in cells and biological fluids. An element for which modeling is particularly well advanced is zinc (Colvin et al. 2008, 2010), yet the models are still rather complex. For sake of illustration, let us consider a simpler formulation of an arbitrary compartmentalized model that captures anyway the simple features of homeostasis: uptake, efflux, storage, and catabolism. Zinc is absorbed by diffusion by transmembrane proteins from the extra-cellular fluid and stored in metallothionein, which is itself synthesized from cysteine. We leave endocytosis and exocytosis out of the model. The rate of change of the number of moles of ^{64}Zn (n), ^{66}Zn ($\tilde{n}$), cysteine (c), and Zn-loaded methionine (m) within the cytosol changes according to the following set of non-linear differential equations:

$$\frac{dn}{dt} = -k\rho\left(Dn - D^e n^e\right) - \left(Knc^i - K'm\right)$$

$$\frac{d\tilde{n}}{dt} = -\tilde{k}\rho\left(\tilde{D}\tilde{n} - \tilde{D}^e \tilde{n}^e\right) - \left(\tilde{K}\tilde{n}c^i - \tilde{K}'m\right)$$

$$\frac{dc}{dt} = -k_c\rho_c\left(D_c c - D_c^e c^e\right) - i\left(Knc^i - K'm\right) - \lambda_c c$$

$$\frac{dm}{dt} = Knc^i - K'm$$

Variables with superscript e refer to the extracellular medium, tildes (~) to the properties of ^{66}Zn compounds. For simplicity, we assume that $\tilde{n} << n$. The first term on the right-hand side of the first three equations refers to transfer through the membrane in response to a chemical potential gradient (Bonaventura et al. 2016): parameters D are membrane-cytosol and membrane-extracellular medium partition coefficients. We assume that the transmembrane protein density ρ may vary, possibly as a result of inflammation (Bonaventura et al. 2016). For simplicity again, we combined Zn uptake (ZIP) and export (ZnT) in one coefficient. The probability of cysteine catabolism per unit time is denoted l_c. K and K' are the rate constants of bonding and breakdown of Zn-bearing metallothionein. The variable i refers to the cysteine/Zn stoichiometric ratio in metallothionein.

Such a set of first-order non-linear differential equations can be solved by standard numerical packages (e.g., Matlab) or by commercial software (Klipp et al. 2013). A mock-up example is shown in Figure 20 in which the rate K' of metallothionein catabolism increases at time $t = 30$, while the methionine concentration c^e in the extracellular medium decreases. The purpose of this exercise is to show how the isotope composition of a metal in one of the compartments (including the extracellular medium) may respond to changes in biological pathways upon signaling. The actual values of the parameters chosen were not meant to reproduce natural conditions. In this respect, the cell is similar to a chemical reactor, in which the records of input and output variables can be used to document the 'state variables' within the system.

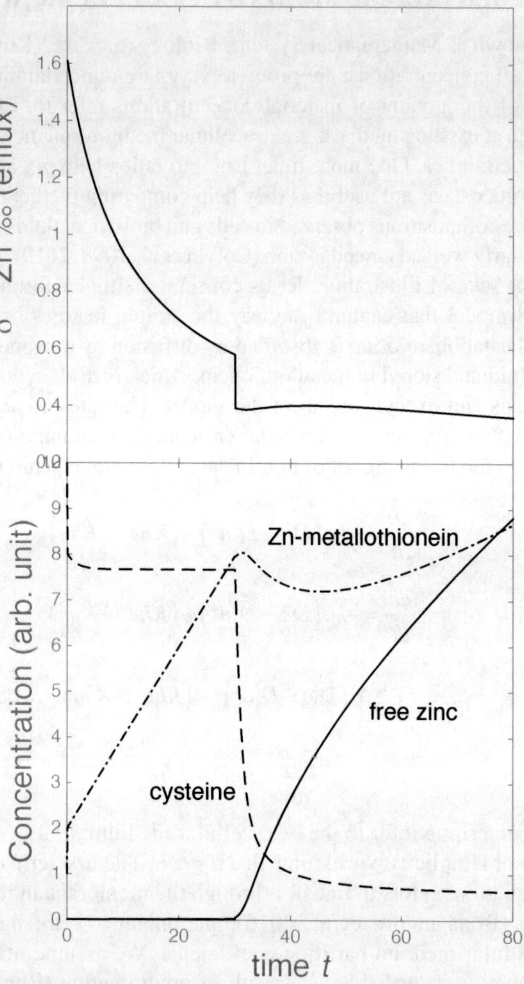

Figure 20. A mockup example of Zn transport through a cell. We assume that Zn is imported and exported through transmembrane proteins. The number of these proteins may be modulated by cell signaling. Transmembrane Zn transfer takes place by diffusion. Excess Zn is stored in Zn-metallothionein, which requires cysteine biosynthesis in the cytosol. Cysteine biosynthesis is dependent on methionine uptake. Cytosolic cysteine is broken down with a first-order rate. All the units and the parameters are arbitrary. It was assumed that at $t = 30$, the rate of metallothionein catabolism increases, while the methionine concentration in the extracellular medium decreases. *Bottom panel:* amounts of cytosolic Zn in each form. *Top panel:* δ^{66}Zn in the efflux.

PERSPECTIVES

So far, Cu has provided the strongest signal associated with a number of diseases, in particular cancer. Zinc, iron, and sulfur have not so far proved to be as informative as copper. The exploratory stage of Cu isotope variations in blood has been very fruitful. Now that this field is becoming mature, descriptive investigations need to be complemented. Data on organs are needed that only animal models can provide. Experiments should be run on cell cultures under hypoxic conditions. Protein expression, notably those controlling metal trafficking, storage, and redox, should be evaluated.

Among the upcoming challenges, several major questions need to be addressed, notably which part of the $\delta^{65}Cu$ signal is due to the cancer itself, and which part is due to other factors, such as age and, even more likely, inflammation. Our preliminary studies of athletes, which shows little or no effect exertion on $\delta^{65}Cu$, and of patients with purely inflammatory diseases, such as rheumatoid arthritis, suggest that if there an effect of inflammation, it only manifests itself on the very long term.

Reduction of copper or ceruloplasmin levels by chelates, without causing clinical copper deficiency, was proposed for therapeutic purposes. Specific copper chelators, such as tetrathiomolybdate, D-penicillamine, and TPEN(Brewer 2001, 2005; Pan et al. 2002, 2003; Lowndes et al. 2008; Fatfat et al. 2014) have been shown to be a potent antiangiogenic and antimetastatic compound possibly through suppression of the NFκB signaling cascade. Recently, Cu-chelation therapy has been proposed as treatment of the broad spectrum of cancers containing the $BRAF^{V600E}$ mutation (Brady et al. 2014). Inhibition of copper Atox1 trafficking has also been investigated (Wang et al. 2015). The isotopic study of copper will certainly add a new dimension to the understanding of chelation pathways and copper mass balance, at the scale of both the cell and the organism, during the treatment.

The prospects of isotope variations of magnesium, which is involved in a large number of biological segments, remain unexplored. Molybdenum plays a role, notably through molybdenum hydroxylase, in a variety of hydroxylation, oxygen atom transfer, and other oxidation–reduction reactions (Hille et al. 1998; Schwarz et al. 2009) and shows promise for future isotopic work.

ACKNOWLEDGEMENTS

We are grateful to Aline Lamboux and Mélanie Simon for collecting the crucial data reported in the original publications. Janne Blichert-Toft is gratefully acknowledged for editing such a long manuscript. The Institut National des Sciences de l'Univers and the Institut des Sciences Biologiques of the Centre National de la Recherche Scientifique, the Ecole Normale Supérieure de Lyon, the Labex Lyon Institute of Origins, the Fondation Mérieux and the Fondation Bullukian, with a particular mention for Jacques Samarut, the late Chantal Rabourdin-Combes, François Juillet and Christian Bréchot, financially supported some of the work reviewed here. Discussions with colleagues, notably Alain Puisieux, Pierre Miossec, and Pierre Hainaut, provided helpful clinical insights. Three quite helpful reviews helped us improve the text and catch some misunderstandings.

REFERENCES

Albalat E, Telouk P, Balter V, Fujii T, Bondanese VP, Plissonnier ML, Vlaeminck-Guillem V, Baccheta J, Thiam N, Miossec P, Zoulim F (2016) Sulfur isotope analysis by MC-ICP-MS and applications to small medical samples. J Anal Atom Spectr 31:1002–11

Albarède F (2015) Metal stable isotopes in the human body: a tribute of geochemistry to medicine. Elements 11:265–269

Albarède F, Télouk P, Blichert-Toft J, Boyet M, Agranier A, Nelson B (2004) Precise and accurate isotopic measurements using multiple-collector ICPMS. Geochim Cosmochim Acta 68:2725–2744

Albarède F, Télouk P, Lamboux A, Jaouen K, Balter V (2011) Isotopic evidence of unaccounted for Fe and Cu erythropoietic pathways. Metallomics 3:926–933

Andrews NC (2008) Forging a field: the golden age of iron biology. Blood 112:219–230, doi:10.1182/blood-2007-12–077388

Andrews NC (2012) Closing the iron gate. N Engl J Med 366:376–377

Aramendia M, Rello L, Resano M, Vanhaecke F (2013) Isotopic analysis of Cu in serum samples for diagnosis of Wilson's disease: a pilot study. J Anal Atom Spectr 28:675–681

Arredondo M, Muñoz P, Mura CV, Núñez MT (2003) DMT1, a physiologically relevant apical Cu^{1+} transporter of intestinal cells. Am J Physiol-Cell Physiol 284:C1525-C1530

Arredondo M, Mendiburo MJ, Flores S, Singleton ST, Garrick MD (2014) Mouse divalent metal transporter 1 is a copper transporter in HEK293 cells. Biometals 27:115–123

Balter V, Zazzo A, Moloney AP, Moynier F, Schmidt O, Monahan FJ, Albarède F (2010) Bodily variability of zinc natural isotope abundances in sheep. Rapid Commun Mass Spectrom 24:605–612, doi:10.1002/rcm.4425

Balter V, Lamboux A, Zazzo A, Télouk P, Leverrier Y, Marvel J, Moloney AP, Monahan FJ, Schmidt O, Albarède F (2013) Contrasting Cu, Fe, and Zn isotopic patterns in organs and body fluids of mice and sheep, with emphasis on cellular fractionation. Metallomics 5:1470–1482, doi:10.1039/c3mt00151b

Balter V, Nogueira da Costa A, Bondanese VP, Jaouen K, Lamboux A, Sangrajrang S, Vincent N, Fourel F, Télouk P, Gigou M, Lécuyer C, Srivatanakul P, Bréchot C, Albarède F, Hainaut (2015) Natural variations of copper and sulfur stable isotopes in blood of hepatocellular carcinoma patients. PNAS 112:982–985, doi:10.1073/pnas.1415151112

Bigeleisen J, Mayer MG (1947) Calculation of equilibrium constants for isotopic exchange reactions. J Chem Phys 15:261–267

Bigeleisen J, Wolfsberg M (1958) Theoretical and experimental aspects of isotope effects in chemical kinetics. Adv Chem Phys 1 1:15–76

Black JR, Kavner A, Schauble EA (2011) Calculation of equilibrium stable isotope partition function ratios for aqueous zinc complexes and metallic zinc. Geochim Cosmochim Acta 75:769–783, doi:10.1016/j.gca.2010.11.019

Bonaventura P, Benedetti G, Albarède F, Miossec P (2015) Zinc and its role in immunity and inflammation. Autoimmun Rev 14:277–285, doi:10.1016/j.autrev.2014.11.008

Bonaventura P, Lamboux A, Albarède F, Miossec P (2016) A feedback loop between inflammation and Zn uptake. PLoS ONE 11:11, doi:10.1371/journal.pone.0147146

Bothwell T, Finch C (1962) Iron Metabolism. 440 pp. Boston: Little, Brown

Brady DC, Crowe MS, Turski ML, Hobbs GA, Yao X, Chaikuad A, Knapp S, Xiao K, Campbell SL, Thiele DJ (2014) Copper is required for oncogenic BRAF signalling and tumorigenesis. Nature 509:492–496

Brewer GJ (2001) Copper control as an antiangiogenic anticancer therapy: Lessons from treating Wilson's disease. Exp Biol Med 226:665–673

Brewer GJ (2003) Copper in medicine. Curr Opin Chem Biol 7:207–212

Brewer GJ (2005) Anticopper therapy against cancer and diseases of inflammation and fibrosis. Drug Discov Today 10:1103–1109

Bridges RJ, Natale NR, Patel SA (2012) System x(c)(-) cystine/glutamate antiporter: an update on molecular pharmacology and roles within the CNS. Br J Pharmacol 165:20–34, doi:10.1111/j.1476 5381.2011.01480.x

Brini M, Calì T, Ottolini D, Carafoli E (2013) Intracellular calcium homeostasis and signaling. *In*: Metallomics and the Cell. Springer, p 119–168

Broer S, Palacin M (2011) The role of amino acid transporters in inherited and acquired diseases. Biochem J 436:193–211

Buechl A, Hawkesworth CJ, Ragnarsdottir KV, Brown DR (2008) Re-partitioning of Cu and Zn isotopes by modified protein expression. Geochem Trans 9, doi:10.1186/1467-4866-9-11

Carmeliet P, Jain RK (2000) Angiogenesis in cancer and other diseases. Nature 407:249–257

Channon MB, Gordon GW, Morgan JLL, Skulan JL, Smith SM, Anbar AD (2015) Using natural, stable calcium isotopes of human blood to detect and monitor changes in bone mineral balance. Bone 77:69–74, doi:http://dx.doi.org/10.1016/j.bone.2015.04.023

Chung CH, Golub EE, Forbes E, Tokuoka T, Shapiro IM (1992) Mechanism of action of beta-glycerophosphate on bone cell mineralization. Calcif Tissue Int 51:305–311, doi:10.1007/bf00334492

Colvin RA, Holmes WR, Fontaine CP, Maret W (2010) Cytosolic zinc buffering and muffling: Their role in intracellular zinc homeostasis. Metallomics 2:306–317, doi:10.1039/b926662c

Colvin RA, Bush AI, Volitakis I, Fontaine CP, Thomas D, Kikuchi K, Holmes WR (2008) Insights into Zn^{2+} homeostasis in neurons from experimental and modeling studies. Am J Physiol-Cell Physiol 294:C726-C742

Costas-Rodríguez M, Van Heghe L, Vanhaecke F (2014) Evidence for a possible dietary effect on the isotopic composition of Zn in blood via isotopic analysis of food products by multi-collector ICP-mass spectrometry. Metallomics 6:139–146

Costas-Rodríguez M, Anoshkina Y, Lauwens S, Van Vlierberghe H, Delanghe J, Vanhaecke F (2015) Isotopic analysis of Cu in blood serum by multi-collector ICP-mass spectrometry: a new approach for the diagnosis and prognosis of liver cirrhosis? Metallomics 7:491–498

Costas-Rodríguez M, Delanghe J, Vanhaecke F (2016) High-precision isotopic analysis of essential mineral elements in biomedicine: natural isotope ratio variations as potential diagnostic and/or prognostic markers. TrAC Trends Anal Chem 76:182–193

Costello L, Franklin R (2008) Prostatic fluid electrolyte composition for the screening of prostate cancer: a potential solution to a major problem. Prostate Cancer Prostatic Dis 12:17–24

Cousins RJ, Liuzzi JP, Lichten LA (2006) Mammalian zinc transport, trafficking, and signals. J Biol Chem 281:24085–24089, doi:10.1074/jbc.R600011200

Cramer DW, Bast RC, Berg CD, Diamandis EP, Godwin AK, Hartge P, Lokshin AE, Lu KH, McIntosh MW, Mor G (2011) Ovarian cancer biomarker performance in prostate, lung, colorectal, and ovarian cancer screening trial specimens. Cancer Prev Res (Phila Pa) 4:365–374

Criss RE (1999) Principles of Stable Isotope Distribution. Oxford Univ. Press, Oxford

Cruz KJC, de Oliveira ARS, Marreiro DdN (2015) Antioxidant role of zinc in diabetes mellitus. World J Diabetes 6:333–337, doi:10.4239/wjd.v6.i2.333

Cui Y, Vogt S, Olson N, Glass AG, Rohan TE (2007) Levels of zinc, selenium, calcium, and iron in benign breast tissue and risk of subsequent breast cancer. Cancer Epidemiol Biomarkers Prev 16:1682–1685

Cunzhi H, Jiexian J, Xianwen Z, Jingang G, Shumin Z, Lili D (2003) Serum and tissue levels of six trace elements and copper/zinc ratio in patients with cervical cancer and uterine myoma. Biol Trace Elem Res 94:113–122

Dauphas N, Roskosz M, Alp EE, Golden DC, Sio CK, Tissot FLH, Hu MY, Zhao J, Gao L, Morris RV (2012) A general moment NRIXS approach to the determination of equilibrium Fe isotopic fractionation factors: Application to goethite and jarosite. Geochim Cosmochim Acta 94:254–275, doi:http://dx.doi.org/10.1016/j.gca.2012.06.013

Denoyer D, Masaldan S, La Fontaine S, Cater MA (2015) Targeting copper in cancer therapy:'Copper That Cancer'. Metallomics 7:1459–76, doi: 10.1039/c5mt00149h

Dlouhy AC, Outten CE (2013) The iron metallome in eukaryotic organisms. *In:* Metallomics and the Cell. Volume 12: Metal Ions in Life Sciences. Bianci L (ed), Springer, p 241–278

Domagal-Goldman SD, Kubicki JD (2008) Density functional theory predictions of equilibrium isotope fractionation of iron due to redox changes and organic complexation. Geochim Cosmochim Acta 72:5201–5216

Domagal-Goldman S, Paul K, Sparks D, Kubicki J (2009) Quantum chemical study of the Fe (III)-desferrioxamine B siderophore complex—electronic structure, vibrational frequencies, and equilibrium Fe-isotope fractionation. Geochim Cosmochim Acta 73:1–12

Espinoza A, Le Blanc S, Olivares M, Pizarro F, Ruz M, Arredondo M (2012) Iron, copper, and zinc transport: inhibition of divalent metal transporter 1 (DMT1) and human copper transporter 1 (hCTR1) by shRNA. Biol Trace Elem Res 146:281–286

Evrard S, Delanaye P, Kamel S, Cristol J-P, Cavalier E (2015) Vascular calcification: from pathophysiology to biomarkers. Clin Chimica Acta 438:401–414

Fahmueller YN, Nagel D, Hoffmann R-T, Tatsch K, Jakobs T, Stieber P, Holdenrieder S (2013) CA 15–3 is a predictive and prognostic biomarker in patients with metastasized breast cancer undergoing Selective Internal Radiation Therapy. Int J Clin Pharmacol Ther 51:63–66

Fatfat M, Merhi RA, Rahal O, Stoyanovsky DA, Zaki A, Haidar H, Kagan VE, Gali-Muhtasib H, Machaca K (2014) Copper chelation selectively kills colon cancer cells through redox cycling and generation of reactive oxygen species. BMC Cancer 14:527

Fleming RE, Bacon BR (2005) Orchestration of iron homeostasis. N Engl J Medicine 352:1741–1744

Fuchs BC, Bode BP (2005) Amino acid transporters ASCT2 and LAT1 in cancer: partners in crime? *In:* Book Amino acid transporters ASCT2 and LAT1 in cancer: partners in crime? Vol 15. Falus, A. (ed). Elsevier, p 254–266

Fujii T, Albarède F (2012) Ab initio calculation of the Zn isotope effect in phosphates, citrates, and malates and applications to plants and soil. PLoS ONE 7, doi:10.1371/journal.pone.0030726

Fujii T, Moynier F, Abe M, Nemoto K, Albarède F (2013) Copper isotope fractionation between aqueous compounds relevant to low temperature geochemistry and biology. Geochim Cosmochim Acta 110:29–44, doi:http://dx.doi.org/10.1016/j.gca.2013.02.007

Fujii T, Moynier F, Blichert-Toft J, Albarède AF (2014) Density functional theory estimation of isotope fractionation of Fe, Ni, Cu, and Zn among species relevant to geochemical and biological environments. Geochim Cosmochim Acta 140:553–576

Fukada T, Yamasaki S, Nishida K, Murakami M, Hirano T (2011) Zinc homeostasis and signaling in health and diseases. J Biol Inorg Chem 16:1123–1134, doi:10.1007/s00775-011-0797–4

Galimov E (1985) The Biological Fractionation of Isotopes. Academic Press, Orlando

Garrick MD, Dolan KG, Horbinski C, Ghio AJ, Higgins D, Porubcin M, Moore EG, Hainsworth LN, Umbreit JN, Conrad ME (2003) DMT1: a mammalian transporter for multiple metals. Biometals 16:41–54

Golden MN, Jackson A, Golden B (1977) Effect of zinc on thymus of recently malnourished children. Lancet 310:1057–1059

Gordon G, Monge J, Channon M, Wu Q, Skulan J, Anbar A, Fonseca R (2014) Predicting multiple myeloma disease activity by analyzing natural calcium isotopic composition. Leukemia 28:2112–2115

Gropper SS, Smith JL (2012) Advanced Nutrition and Human Metabolism. Cengage Learning, Belmont

Gupta SK, Shukla VK, Vaidya MP, Roy SK, Gupta S (1991) Serum trace elements and Cu/Zn ratio in breast cancer patients. J Surg Oncol 46:178–181

Gussone N, Eisenhauer A, Heuser A, Dietzel M, Bock B, Böhm F, Spero HJ, Lea DW, Bijma J, Nägler TF (2003) Model for kinetic effects on calcium isotope fractionation (δ^{44}Ca) in inorganic aragonite and cultured planktonic foraminifera. Geochim Cosmochim Acta 67:1375–1382

Haas KL, Putterman AB, White DR, Thiele DJ, Franz KJ (2011) Model peptides provide new insights into the role of histidine residues as potential ligands in human cellular copper acquisition via Ctr1. J Am Chem Soc 133:4427–4437

Haase VH (2013) Regulation of erythropoiesis by hypoxia-inducible factors. Blood Rev 27:41–53, doi:http://dx.doi.org/10.1016/j.blre.2012.12.003

Halestrap AP, Price NT (1999) The proton-linked monocarboxylate transporter (MCT) family: structure, function and regulation. Biochem J 343:281–299, doi:10.1042/bj3430281

Heuser A, Eisenhauer A (2010) A pilot study on the use of natural calcium isotope ((44)ca/Ca–40) fractionation in urine as a proxy for the human body calcium balance. Bone 46:889–896, doi:10.1016/j.bone.2009.11.037

Hille R, Rétey J, Bartlewski-Hof U, Reichenbecher W, Schink B (1998) Mechanistic aspects of molybdenum-containing enzymes. FEMS Microbiol Rev 22:489–501

Hofmann Bowman MA, McNally EM (2012) Genetic Pathways of Vascular Calcification. Trends Cardiovasc Med 22:93–98, doi:http://dx.doi.org/10.1016/j.tcm.2012.07.002

Hotakainen K, Tanner P, Alfthan H, Haglund C, Stenman U-H (2009) Comparison of three immunoassays for CA 19–9. Clinica Chimica Acta 400:123–127, doi:http://dx.doi.org/10.1016/j.cca.2008.10.033

Hotz K, Augsburger H, Walczyk T (2011) Isotopic signatures of iron in body tissues as a potential biomarker for iron metabolism. J Anal Atom Spectr 26:1347–1353, doi:10.1039/C0JA00195C

Hotz K, Krayenbuehl P-A, Walczyk T (2012) Mobilization of storage iron is reflected in the iron isotopic composition of blood in humans. J Biol Inorg Chem 17:301–309, doi:10.1007/s00775-011-0851-2

Ishida S, Andreux P, Poitry-Yamate C, Auwerx J, Hanahan D (2013) Bioavailable copper modulates oxidative phosphorylation and growth of tumors. PNAS 110:19507–19512

Jaouen K, Gibert M, Lamboux A, Télouk P, Fourel F, Albarède F, Alekseev AN, Crubezy E, Balter V (2013) Is aging recorded in blood Cu and Zn isotope compositions? Metallomics 5:1016–1024, doi:10.1039/c3mt00085k

Kanai Y, Segawa H, Miyamoto K-i, Uchino H, Takeda E, Endou H (1998) Expression cloning and characterization of a transporter for large neutral amino acids activated by the heavy chain of 4F2 antigen (CD98). J Biol Chem 273:23629–23632

Kao W-Y, Chao Y, Chang C-C, Li C-P, Su C-W, Huo T-I, Huang Y-H, Chang Y-J, Lin H-C, Wu J-C (2015) Prognosis of Early-Stage Hepatocellular Carcinoma: The Clinical Implications of Substages of Barcelona Clinic Liver Cancer System Based on a Cohort of 1265 Patients. Medicine (Baltimore) 94:e1929, doi:10.1097/md.0000000000001929

Kawashima M, Miossec P (2004) Decreased response to IL–12 and IL–18 of peripheral blood cells in rheumatoid arthritis. Arthritis Res Ther 6:R39-R45

Kim B-E, Nevitt T, Thiele DJ (2008) Mechanisms for copper acquisition, distribution and regulation. Nat Chem Biol 4:176–185, doi:10.1038/nchembio.72

Kim H, Wu X, Lee J (2013) SLC31 (CTR) family of copper transporters in health and disease. Mol Aspects Med 34:561–570, doi:http://dx.doi.org/10.1016/j.mam.2012.07.011

Kinnaird A, Michelakis E (2015) Metabolic modulation of cancer: a new frontier with great translational potential. J Mol Med 93:127–142, doi:10.1007/s00109-014-1250–2

Kissling MM, Kagi JHR (1977) Primary structure of human hepatic metallothionein. FEBS Letters 82:247–250, doi:http://dx.doi.org/10.1016/0014–5793(77)80594–2

Klipp E, Liebermeister W, Wierling C, Kowald A, Lehrach H, Herwig R (2013) Systems Biology. John Wiley & Sons

Kohen A, Limbach H-H (2005) Isotope Effects in Chemistry and Biology. CRC Press

Kojima Y, Berger C, Vallee BL, Kägi J (1976) Amino-acid sequence of equine renal metallothionein-1B. PNAS 73:3413–3417

Koksoy C, Kavas GO, Akcil E, Kocaturk PA, Kara S, Ozarslan C (1997) Trace elements and superoxide dismutase in benign and malignant breast diseases. Breast Cancer Res Treat 45:1–6, doi:10.1023/a:1005870918388

Kragh-Hansen U, Vorum H (1993) Quantitative-analyses of the interaction between calcium-ions and human serum-albumin. Clinic Chem 39:202–208

Krayenbuehl PA, Walczyk T, Schoenberg R, von Blanckenburg F, Schulthess G (2005) Hereditary hemochromatosis is reflected in the iron isotope composition of blood. Blood 105:3812–3816

Krzanowski WJ, Hand DJ (2009) ROC Curves for Continuous Data. CRC Press

Larner F, Woodley LN, Shousha S, Moyes A, Humphreys-Williams E, Strekopytov S, Halliday AN, Rehkämper M, Coombes RC (2015) Zinc isotopic compositions of breast cancer tissue. Metallomics 7:107–112

Lichten LA, Cousins RJ (2009) Mammalian Zinc Transporters: Nutritional and Physiologic Regulation. Ann Rev Nutr 29:153–176

Lindemann F (1919) XII. Note on the vapour pressure and affinity of isotopes. Philos Mag Ser 6 38:173–181

Linder MC (2013) Mobilization of stored iron in mammals: a review. Nutrients 5:4022–4050

Linder MC, Goode CA (1991) Biochemistry of Copper. Plenum Press

Lowe NM, Woodhouse LR, Matel JS, King JC (2000) Comparison of estimates of zinc absorption in humans by using 4 stable isotopic tracer methods and compartmental analysis. Am J Clin Nutr 71:523–529

Lowndes SA, Adams A, Timms A, Fisher N, Smythe J, Watt SM, Joel S, Donate F, Hayward C, Reich S (2008) Phase I study of copper-binding agent ATN–224 in patients with advanced solid tumors. Clin Cancer Res 14:7526–7534

Lundblad RL, Macdonald F (2010) Handbook of biochemistry and molecular biology. CRC Press

Lutsenko S (2010) Human copper homeostasis: a network of interconnected pathways. Curr Opin Chem Biol 14:211–217, doi:10.1016/j.cbpa.2010.01.003

Magalova T, Bella V, Babinska K, Brtkova A, Kudláčková M, Bederova A (1996) Zinc and copper in breast cancer. *In*: Therapeutic Uses of Trace Elements. Springer, p 373–375

Malavolta M, Piacenza F, Basso A, Giacconi R, Costarelli L, Mocchegiani E (2015) Serum copper to zinc ratio: Relationship with aging and health status. Mech Ageing Dev

Maret W (2013) Zinc and the zinc proteome. *In*: Metallomics and the Cell. Springer, p 479–501

Maret W, Krężel A (2007) Cellular zinc and redox buffering capacity of metallothionein/thionein in health and disease. Mol Med 13:371

Martin F, Linden T, Katschinski DM, Oehme F, Flamme I, Mukhopadhyay CK, Eckhardt K, Tröger J, Barth S, Camenisch G (2005) Copper-dependent activation of hypoxia-inducible factor (HIF)–1: implications for ceruloplasmin regulation. Blood 105:4613–4619

Milne DB (1998) Copper intake and assessment of copper status. Am J Clin Nutr 67:1041S-1045S

Milne DB, Johnson PE (1993) Assessment of copper status - effect of age and gender on reference ranges in healthy-adults. Clin Chem 39:883–887

Morgan JLL, Skulan JL, Gordon GW, Romaniello SJ, Smith SM, Anbar AD (2012a) Rapidly assessing changes in bone mineral balance using natural stable calcium isotopes. PNAS 109:9989–9994, doi:10.1073/pnas.1119587109

Morgan JLL, Skulan JL, Gordon GE, Romaniello SJ, Smith SM, Anbar AD (2012b) Using natural stable calcium isotopes to rapidly assess changes in bone mineral balance using a bed rest model to induce bone loss. FASEB J 26

Moynier F, Fujii T, Shaw AS, Le Borgne M (2013) Heterogeneous distribution of natural zinc isotopes in mice. Metallomics 5:693–699

Nasulewicz A, Mazur A, Opolski A (2004) Role of copper in tumour angiogenesis—clinical implications. J Trace Elem Med Biol 18:1–8

Ohno T, Shinohara A, Kohge I, Chiba M, Hirata T (2004) Isotopic analysis of Fe in human red blood cells by multiple collector-ICP-mass spectrometry. Anal Sci 20:617–621, doi:10.2116/analsci.20.617

Ohno T, Shinohara A, Chiba M, Hirata T (2005) Precise Zn isotopic ratio measurements of human red blood cell and hair samples by multiple collector ICP mass spectrometry. Anal Sci 21:425–428, doi:10.2116/analsci.21.425

Okorie ON, Dellinger P (2011) Lactate: biomarker and potential therapeutic target. Crit Care Clin 27:299–326

Okuda K, Ohtsuki T, Obata H, Tomimatsu M, Okazaki N, Hasegawa H, Nakajima Y, Ohnishi K (1985) Natural history of hepatocellular carcinoma and prognosis in relation to treatment study of 850 patients. Cancer 56:918–928, doi:10.1002/1097–0142(19850815)56:4<918::AID-CNCR2820560437>3.0.CO;2-E

Pan Q, Kleer CG, van Golen KL, Irani J, Bottema KM, Bias C, De Carvalho M, Mesri EA, Robins DM, Dick RD, Brewer GJ (2002) Copper Deficiency Induced by Tetrathiomolybdate Suppresses Tumor Growth and Angiogenesis. Cancer Res 62:4854–4859

Pan Q, Bao LW, Merajver SD (2003) Tetrathiomolybdate Inhibits Angiogenesis and Metastasis Through Suppression of the NFκB Signaling Cascade1 1 NIH grants R01CA77612 (SDM), P30CA46592, and M01-RR00042, Head and Neck SPORE P50CA97248, Susan G. Komen Breast Cancer Foundation, NIH Cancer Biology Postdoctoral Fellowship T32 CA09676 (QP), Department of Defense Breast Cancer Research Program Postdoctoral Fellowship (QP), and Tempting Tables Organization, Muskegon, MI. Mol Cancer Res 1:701–706

Parge HE, Hallewell RA, Tainer JA (1992) Atomic structures of wild-type and thermostable mutant recombinant human Cu, Zn superoxide dismutase. PNAS 89:6109–6113

Parr RG (1983) Density Functional Theory. Ann Rev Phys Chem 34:631–656, doi:doi:10.1146/annurev.pc.34.100183.003215

Permyakov E, Kretsinger RH (2010) Calcium Binding Proteins. Wiley, Hoboken, New Jersey

Piccinini L, Borella P, Bargellini A, Medici CI, Zoboli A (1996) A case-control study on selenium, zinc, and copper in plasma and hair of subjects affected by breast and lung cancer. Biol Trace Elem Res 51:23–30

Poleganov MA, Pfeilschifter J, Mühl H (2007) Expanding extracellular zinc beyond levels reflecting the albumin-bound plasma zinc pool potentiates the capability of IL–1 β, IL–18, and IL–12 to act as IFN-γ-inducing factors on PBMC. J Interferon Cytokine Res 27:997–1002

Polyakov VB, Mineev SD (2000) The use of Mossbauer spectroscopy in stable isotope geochemistry. Geochim Cosmochim Acta 64:849–865, doi:10.1016/s0016–7037(99)00329–4

Polyakov V, Clayton R, Horita J, Mineev S (2007) Equilibrium iron isotope fractionation factors of minerals: reevaluation from the data of nuclear inelastic resonant X-ray scattering and Mössbauer spectroscopy. Geochim Cosmochim Acta 71:3833–3846

Polyakov VB, Soultanov DM (2011) New data on equilibrium iron isotope fractionation among sulfides: Constraints on mechanisms of sulfide formation in hydrothermal and igneous systems. Geochim Cosmochim Acta 75:1957–1974

Pope CR, Flores AG, Kaplan JH, Unger VM (2011) Structure and function of copper uptake transporters. Current topics in membranes 69:97–112

Portanova R, Lajunen LH, Tolazzi M, Piispanen J (2003) Critical evaluation of stability constants for alpha-hydroxycarboxylic acid complexes with protons and metal ions and the accompanying enthalpy changes. Part II. Aliphatic 2-hydroxycarboxylic acids (IUPAC Technical Report). Pure Appl Chem 75:495–540

Radajewski S, Ineson P, Parekh NR, Murrell JC (2000) Stable-isotope probing as a tool in microbial ecology. Nature 403:646–649

Reynard LM, Henderson GM, Hedges REM (2010) Calcium isotope ratios in animal and human bone. Geochim Cosmochim Acta 74:3735–3750, doi:10.1016/j.gca.2010.04.002

Rosenzweig AC, Argüello JM (2012) Toward a molecular understanding of metal transport by P1B-Type ATPases. Curr Top Membr 69:113

Rustad JR, Casey WH, Yin Q-Z, Bylaska EJ, Felmy AR, Bogatko SA, Jackson VE, Dixon DA (2010) Isotopic fractionation of Mg2+(aq), Ca2+(aq), and Fe2+(aq) with carbonate minerals. Geochim Cosmochim Acta 74:6301–6323, doi:http://dx.doi.org/10.1016/j.gca.2010.08.018

Schaller J, Gerber S, Kaempfer U, Lejon S, Trachsel C (2008) Human Blood Plasma Proteins: Structure and Function. John Wiley & Sons

Schauble EA (2004) Applying stable isotope fractionation theory to new systems. Rev Min Geochem 55:65–111

Schwarz G, Mendel RR, Ribbe MW (2009) Molybdenum cofactors, enzymes and pathways. Nature 460:839–847

Semenza GL (2007) Evaluation of HIF–1 inhibitors as anticancer agents. Drug Discov Today 12:853–859

Senra Varela A, Lopez Saez J, Quintela Senra D (1997) Serum ceruloplasmin as a diagnostic marker of cancer. Cancer Lett 121:139–145

Seo JH, Lee SK, Lee I (2007) Quantum chemical calculations of equilibrium copper (I) isotope fractionations in ore-forming fluids. Chem Geol 243:225–237, doi:http://dx.doi.org/10.1016/j.chemgeo.2007.05.025

Shenberger Y, Shimshi A, Ruthstein S (2015) EPR Spectroscopy Shows that the Blood Carrier Protein, Human Serum Albumin, Closely Interacts with the N-Terminal Domain of the Copper Transporter, Ctr1. J Phys Chem B 119:4824–4830, doi:10.1021/acs.jpcb.5b00091

Sherman DM (2013) Equilibrium isotopic fractionation of copper during oxidation/reduction, aqueous complexation and ore-forming processes: Predictions from hybrid density functional theory. Geochim Cosmochim Acta 118:85–97, doi:http://dx.doi.org/10.1016/j.gca.2013.04.030

Shields WR, Goldich SS, Garner EL, Murphy TJ (1965) Natural variations in the abundance ratio and the atomic weight of copper. J Geophys Res:479–491

Skulan J, DePaolo DJ, Owens TL (1997) Biological control of calcium isotopic abundances in the global calcium cycle. Geochim Cosmochim Acta 61:2505–2510, doi:10.1016/s0016–7037(97)00047–1

Skulan J, DePaolo DJ (1999) Calcium isotope fractionation between soft and mineralized tissues as a monitor of calcium use in vertebrates. PNAS 96:13709–13713, doi:10.1073/pnas.96.24.13709

Smith RM, Martell AE (1987) Critical stability constants, enthalpies and entropies for the formation of metal complexes of aminopolycarboxylic acids and carboxylic acids. Sci Total Environ 64:125–147

Spatz J, Fields E, Yu E, Pajevic PD, Bouxsein M, Sibonga J, Zwart S, Smith S (2012) Serum sclerostin increases in healthy adult men during bed rest. J Clin Endocrin Metab 97:E1736-E1740

Stenberg A, Andrén H, Malinovsky D, Engström E, Rodushkin I, Baxter DC (2004) Isotopic variations of Zn in biological materials. Anal Chem 76:3971–3978

Stenberg A, Malinovsky D, Öhlander B, Andrén H, Forsling W, Engström L-M, Wahlin A, Engström E, Rodushkin I, Baxter DC (2005) Measurement of iron and zinc isotopes in human whole blood: preliminary application to the study of HFE genotypes. J Trace Elem Med Biol 19:55–60

Stipanuk MH, Dominy JE, Lee J-I, Coloso RM (2006) Mammalian cysteine metabolism: new insights into regulation of cysteine metabolism. J Nutr 136:1652S-1659S

Swietach P, Vaughan-Jones RD, Harris AL, Hulikova A (2014) The chemistry, physiology and pathology of pH in cancer. Philos Trans R Soc London, Ser B 369:20130099

Tacail T, Albalat E, Télouk P, Balter V (2014) A simplified protocol for measurement of Ca isotopes in biological samples. J Anal Atom Spectr 29:529–535

Télouk P, Puisieux A, Fujii T, Balter V, Bondanese VP, Morel A-P, Clapisson G, Lamboux A, Albarède F (2015) Copper isotope effect in serum of cancer patients. A pilot study. Metallomics 7:299–308, doi:10.1039/C4MT00269E

Theil EC (2011) Ferritin protein nanocages use ion channels, catalytic sites, and nucleation channels to manage iron/oxygen chemistry. Curr Opin Chem Biol 15:304–311

Umpleby M, Fielding BA (2015) Stable Isotopes in Nutrition Research. Nutrition Research Methodologies:250–264

Urey HC (1947) The thermodynamic properties of isotopic substances. J Chem Soc:562–581

Van Heghe L, Engstrom E, Rodushkin I, Cloquet C, Vanhaecke F (2012) Isotopic analysis of the metabolically relevant transition metals Cu, Fe and Zn in human blood from vegetarians and omnivores using multi-collector ICP-mass spectrometry. J Anal Atom Spectr 27:1327–1334, doi:10.1039/c2ja30070b

Van Heghe L, Delanghe J, Van Vlierberghe H, Vanhaecke F (2013) The relationship between the iron isotopic composition of human whole blood and iron status parameters. Metallomics 5:1503–1509

Van Heghe L, Deltombe O, Delanghe J, Depypere H, Vanhaecke F (2014) The influence of menstrual blood loss and age on the isotopic composition of Cu, Fe and Zn in human whole blood. J Anal Atom Spectr 29:478–482, doi:10.1039/C3JA50269D

Vest KE, Hashemi HF, Cobine PA (2013) Chapter 13. The copper metallome in eukaryotic cells. *In:* Metallomics and The Cell. Metal Ions in Life Sciences Vol 12, Bianci, L (ed) Springer, Dordrecht p 1559–0836

von Blanckenburg F, Noordmann J, Guelke-Stelling M (2013) The iron stable isotope fingerprint of the human diet. J Agric Food Chem 61:11893–11899

von Blanckenburg F, Oelze M, Schmid DG, van Zuilen K, Gschwind H-P, Slade AJ, Stitah S, Kaufmann D, Swart P (2014) An iron stable isotope comparison between human erythrocytes and plasma. Metallomics : integrated biometal science 6:2052–2061, doi:10.1039/c4mt00124a

Walczyk T, von Blanckenburg F (2002) Natural iron isotope variations in human blood. Science 295:2065–2066

Walenta S, Wetterling M, Lehrke M, Schwickert G, Sundfør K, Rofstad EK, Mueller-Klieser W (2000) High lactate levels predict likelihood of metastases, tumor recurrence, and restricted patient survival in human cervical cancers. Cancer Res 60:916–921

Wang J, Luo C, Shan C, You Q, Lu J, Elf S, Zhou Y, Wen Y, Vinkenborg JL, Fan J, Kang H. Inhibition of human copper trafficking by a small molecule significantly attenuates cancer cell proliferation. Nat Chem 2015 Nov 9, doi:10.1038/nchem.2381. http://www.nature.com/nchem/journal/vaop/ncurrent/abs/nchem.2381.html - supplementary-information

Warburg O, Krebs H (1927) Über locker gebundenes Kupfer und Eisen im Blutserum. Biochem Z 190:143–149

Wee NK, Weinstein DC, Fraser ST, Assinder SJ (2013) The mammalian copper transporters CTR1 and CTR2 and their roles in development and disease. Int J Biochem Cell Biol 45:960–963

Wenger RH, Rolfs A, Marti HH, Bauer C, Gassmann M (1995) Hypoxia, a novel inducer of acute phase gene expression in a human hepatoma cell line. J Biol Chem 270:27865–27870

Wilson WR, Hay MP (2011) Targeting hypoxia in cancer therapy. Nat Rev Cancer 11:393–410

Winter WE, Bazydlo LA, Harris NS (2014) The molecular biology of human iron metabolism. Lab Medicine 45:92–102

Wolfsberg M, VanHook WA, Paneth P (2010) Isotope Effects in the Chemical, Geological, and Bio sciences. Springer, New York

Wu H-DI, Chou S-Y, Chen D-R, Kuo H-W (2006) Differentiation of serum levels of trace elements in normal and malignant breast patients. Biol Trace Elem Res 113:9–18

Yücel I, Arpaci F, Özet A, Döner B, Karayilanoĝlu T, Sayar A, Berk Ö (1994) Serum copper and zinc levels and copper/zinc ratio in patients with breast cancer. Biol Trace Elem Res 40:31–38

Zaitseva I, Zaitsev V, Card G, Moshkov K, Bax B, Ralph A, Lindley P (1996) The X-ray structure of human serum ceruloplasmin at 3.1 Å: nature of the copper centres. JBIC J Biol Inorg Chemi 1:15–23

Zowczak M, Iskra M, Paszkowski J, Mańczak M, Torliński L, Wysocka E (2001) Oxidase activity of ceruloplasmin and concentrations of copper and zinc in serum of cancer patients. J Trace Elem Med Biol 15:193–196

RiMG Series

HISTORY OF RiMG

Volumes 1–38 were published as "*Reviews in Mineralogy*" (ISSN 0275-0279). Volumes 1-6 originally appared as "*Short Course Notes*" (no ISSN). The name was changed to "*Reviews in Mineralogy & Geochemistry*" (RiMG) (ISSN 1529-6466) starting with Volume 39. Paul Ribbe was sole editor for volumes 1–41. He was joined by Jodi Rosso as series editor for volumes 42–53 in the RiMG series submitted through the Geochemistry Society. With his retirement, Jodi Rosso became sole editor for volumes 54–79 in the RiMG series. With Jodi Rosso's move to Executive Editor of Elements magazine, Ian Swainson became series editor starting with volume 80.

HOW TO PUBLISH IN RiMG

RiMG volumes are based on topics that have been proposed and appoved by the MSA Council or Geochemical Society Board of Directors. If you have an idea for a future RiMG volume, or a short course accompanied by a RiMG volume, you should read the Short Course Guide which describes how to develop and propose a topic for consideration for either case. Proposals should be submitted to the Short Course Committee (http://www.minsocam.org/msa/SC/SCCommittee.html). Contributions to an appoved volume are by invitation only.

A listing of the previous volume numbers and their volume editors is below. Previous volumes can be ordered from https://msa.minsocam.org/orders.html.

Volume 81: *Highly Siderophile and Strongly Chalcophile Elements in High-Temperature Geochemistry and Cosmochemistry*

2015 ISBN 978-0-939950-97-3 J Harvey, JMD Day i-xxiii + 774 pp

Volume 80: *Pore-Scale Geochemical Processes*

2015 ISBN 978-0-939950-96-6 CI Steefel, S Emmanuel, LM Anovitz i-xiv + 491

Volume 79: *Arsenic, Environmental Geochemistry, Mineralogy, and Microbiology*

2014 ISBN 978-0-939950-94-2 RJ Bowell, CN Alpers, HE Jamieson, DK Nordstrom, J Majzlan i-xvi + 635 pp

Volume 78: *Spectroscopic Methods in Mineralogy and Materials Sciences*

2014 ISBN13 978-0-939950-93-5 GS Henderson, DR Neuville, RT Downs i-xviii + 800 pp

Volume 77: *Geochemistry of Geologic CO_2 Sequestration*

2013 ISBN 978-0-939950-92-8 DJ DePaolo, DR Cole, A Navrotsky, IC Bourg i-xiv + 539 pp

Volume 76: *Thermodynamics of Geothermal Fluids*

2013 ISBN 978-0-939950-91-1 A Stefánsson, T Driesner, P Bénézeth i-x + 350 pp

Volume 75: *Carbon in Earth*

2013 ISBN 978-0-939950-90-4 RM Hazen, AP Jones, JA Baross i-xv + 698 pp

Volume 74: *Appied Mineralogy of Cement & Concrete*

2012 ISBN 978-0-939950-88-1 MATM Broekmans, Pöllmann, i-x + 364 pp

Volume 73: *Sulfur in Magmas and Melts: Its Importance for Natural and Technical Processes*

2011 ISBN 978-0-939950-87-4 H Behrens, JD Webster i-xiv + 578 pp

Volume 72: *Diffusion in Minerals and Melts*

2010 ISBN 978-0-939950-86-7 Y Zhang, DJ Cherniak i-xviii + 1036 pp

Volume 71: *Theoretical and Computational Methods in Mineral Physics: Geophysical Appications*

2010 ISBN 978-0-939950-85-0 R Wentzcovitch, L Stixrude i-xviii + 484 pp

Volume 70: *Thermodynamics and Kinetics of Water-Rock Interaction*

2009 ISBN 0-939950-84-7; ISBN13 978-0-939950-84-3 EH Oelkers J Schott i-xvii + 569 pp

Volume 69: *Minerals, Inclusions and Volcanic Processes*

2008 ISBN 0-939950-83-9; ISBN13 978-0-939950-83-6 KD Putirka, FJ Tepley III i-xiv + 674 pages.

Volume 68: *Oxygen In the Solar System*

2008 ISBN 0-939950-80-4; ISBN13 978-0-939950-80-5 GJ MacPherson, DW Mittlefehldt, JH Jones, SB Simon, JJ Papike, S Mackwell i-xx + 598 pages

Volume 67: *Amphiboles: Crystal Chemistry, occurrences, and Health Issues*

2007 ISBN 0-939950-79-0; ISBN13 978-0-939950-79-9 FC Hawthorne, R Oberti, G Della Ventura, A Mottana i-xxv + 545 pages.

Volume 66: *Paleoaltimetry: Geochemical and Thermodynamic Appoaches*

2007 ISBN 0-939950-78-2; ISBN13 978-0-939950-78-2 MJ Kohn i-x + 278 pages

Volume 65: *Fluid-Fluid Interactions*

2007 ISBN 0-939950-77-4; ISBN13 978-0-939950-77-5 A Liebscher CA Heinrich i-xii + 430 pages

Volume 64: *Medical Mineralogy and Geochemistry*

2006 ISBN 0-939950-75-8; ISBN13 978-0939950-75-1 N Sahai, MAA Schoonen i-xi + 332 pp

Volume 63: *Neutron Scattering in Earth Sciences*

2006 ISBN 0-939950-75-8; ISBN13 978-0939950-75-1 HR Wenk i-xx + 471 pp

Volume 62: *Water in Nominally Anhydrous Minerals*

2006 ISBN 0-939950-74-X; ISBN13 978-0-939950-74-4 H Kepper, JR. Smyth i-viii + 478 pp

Volume 61: *Sulfide Mineralogy and Geochemistry*

2006 ISBN 0-939950-73-1; ISBN13 978-0-939950-73-7 DJ Vaughan i-xiii + 714 pp

Volume 60: *New Views of the Moon*

2006 ISBN 0-939950-72-3; ISBN13 978-0-939950-72-0 BL Jolliff, MA Wieczorek, CK Shearer, CR Neal i-xxii + 772 pp

Volume 59: *Molecular Geomicrobiology*

2005 ISBN 0-939950-71-5; ISBN13 978-0-939950-71-3 JF Banfield, J Cervini-Silva, KH Nealson i-xiv + 294 pp

Volume 58: *Low-Temperature Thermochronology: Techniques, Interpretations, and Appications*

2005 ISBN 0-939950-70-7; ISBN13 978-0-939950-70-6 PW Reiners, TA Ehlers i-xxii + 620 pp

Volume 57: *Micro- and Mesoporous Mineral Phases*

2005 ISBN 0-939950-69-3; ISBN13 978-0-939950-69-0 G Ferraris, S Merlino i-xiii + 448 pp

Volume 56: *Epidotes*

2004 ISBN 0-939950-68-5; ISBN13 978-0-939950-68-3 A Liebscher, G Franz i-xviii + 628 pp

Volume 55: *Geochemistry of Non-Traditional Stable Isotopes*

2004 ISBN 0-939950-67-7; ISBN13 978-0-939950-67-6 CM Johnson, BL Beard, F Albarede i-xvi + 454 pp

Volume 54: *Biomineralization*

2003 ISBN 0-939950-66-9; ISBN13 978-0-939950-66-9 PM Dove, JJ De Yoreo, S Weiner i-xiv + 381 pp

Volume 53: *Zircon*

2003 ISBN 0-939950-65-0; ISBN13 978-0-939950-65-2 JM Hanchar, PWO Hoskin i-xviii + 500 pp

Volume 52: *Uranium-Series Geochemistry*

2003 ISBN 0-939950-64-2; ISBN13 978-0-939950-64-5 B Bourdon, GM Henderson, CC Lundstrom, SP Turner, i-xx + 656 pp

Volume 51: *Plastic Deformation of Minerals and Rocks*

2002 ISBN 0-939950-63-4; ISBN13 978-0-939950-63-8 S Karato, H-R Wenk i-xiv + 420 pp

Volume 50: *Beryllium Mineralogy, Petrology, and Geochemistry*

2002 ISBN 0-939950-62-6; ISBN13 978-0-939950-62-1 E Grew i-xii + 691 pp

Volume 49: *Appications of Synchrotron Radiation in Low-Temperature Geochemistry and Environmental Science*

2002 ISBN 0-939950-61-8; ISBN13 978-0-939950-61-4 PA Fenter, ML Rivers, NC Sturchio, SR Sutton, i-xxii + 579 pp

Volume 48: *Phosphates Geochemical, Geobiological, and Materials Importance*

2002 ISBN 0-939950-60-X; ISBN13 978-0-939950-60-7 ML Kohn, J Rakovan, JM Hughes i-xvi + 742 pp

Volume 47: *Nobel Gases in Geochemistry and Cosmochemistry*

2002 ISBN 0-939950-59-6; ISBN13 978-0-939950-59-1 DP Porcelli, CJ Ballentine, R Wieler i-xviii + 844 pp

Volume 46: *Micas: Crystal Chemistry & Metamorphic Petrology*

2002 ISBN 0-939950-58-8; ISBN13 978-0-939950-58-4 A Mottana, FP Sassi, JB Thompson, Jr., S Guggenheim i-xiv + 499 pp

Volume 45:*Naturnal Zeolites: Occurrence, Properties, Appications*

2001 ISBN 0-939950-57-X; ISBN13 978-0-939950-57-7 DL Bish, DW Ming i-xiv + 654 pp

Volume 44: *Nanoparticles and the Environment*

2001 ISBN 0-939950-56-1; ISBN13 978-0-939950-56-0 JF Banfield ,A Navrotsky i-xiv + 349 pp

Volume 43: *Stable Isotope Geochemistry*

2001 ISBN 0-939950-55-3; ISBN13 978-0-939950-55-3 JW Valley, D Cole i-x11 + 531 pp

Volume 42: *Molecular Modeling Theory: Applications in the Geosciences*

2001 ISBN 0-939950-54-5; ISBN13 978-0-939950-54-6 RT Cygan, JB Kubicki i-v +662 pp

Volume 41: *High-Temperature and High-Pressure Crystal Chemistry*

2001 ISBN 0-939950-53-7; ISBN13 978-0-939950-53-9 RM Hazen, RT Downs i-viii + 596 pp

Volume 40: *Sulfate Minerals - Crystallography, Geochemistry, and Environmental Significance*

2000 ISBN 0-939950-52-9; ISBN13 978-0-939950-52-2 CN Alpers, JL Jambor, DK Nordstrom i-xii + 608 pp

Volume 39: *Transformation Processes in Minerals*

2000 ISBN 0-939950-51-0; ISBN13 978-0-939950-51-5 SAT Redfern, MA Carpenter, i-x + 361 pp

Volume 38: *Uranium: Mineralogy, Geochemistry and the Environment*

1999 ISBN 0-939950-50-2; ISBN13 78-0-939950-50-8 PC Burns, R Finch i-xvi + 679 pp

Volume 37: *Ultrahigh-Pressure Mineralogy: Physics and Chemistry of the Earth's Deep Interior*

1998 ISBN 0-939950-48-0; ISBN13 978-0-939950-48-5 — R Hemley — i-xx + 671 pp

Volume 36: *Planetary Materials*

1998 ISBN 0-939950-46-4; ISBN13 978-0-939950-46-1 — JJ Papike — i-xx + 864 pp

Volume 35: *Geomicrobiology: Interaction Between Microbes and Minerals*

1997 ISBN 0-939950-45-6; ISBN13 978-0-939950-45-4 — JF Banfield ,KH Nealson — i-xvi + 448 pp

Volume 34: *Reactive Transport in Porous Media*

1997 ISBN 0-939950-45-6; ISBN13 978-0-939950-45-4 — PC Lichtner, CI Steefel, EH Oelkers — i-xiv + 438 pp

Volume 33: *Boron Mineralogy, Petrology and Geochemistry*

1996 ISBN 0-939950-41-3; ISBN13 978-0-939950-41-6 — LM Anovitz, ES Grew — i-xx + 864 pp

Volume 32: *Structure, Dynamics and Properties of Silicate Melts*

1995 ISBN 0-939950-39-1; ISBN13 978-0-939950-39-3 — JF Stebbins, PF McMillan, DB Dingwell — i-xvi + 616 pp

Volume 31: *Chemical Weathering Rates of Silicate Minerals*

1995 SBN 0-939950-38-3; ISBN13 978-0-939950-38-6 — AF White, SL Brantley — i-xvi + 583 pp

Volume 30: *Volatiles in Magmas*

1994 ISBN 0-939950-36-7; ISBN13 978-0-939950-36-2 — MR Carroll, JR Holloway — i-xviii + 517 pp

Volume 29: *Silica: Physical Behavior, Geochemistry and Materials Appcations*

1994 ISBN 0-939950-35-9; ISBN13 978-0-939950-35-5 — PJ Heaney, CT Prewitt, GV Gibbs — i-xviii + 606 pp

Volume 28: *Health Effects of Mineral Dusts*

1993 ISBN 0-939950-33-2; ISBN13 978-0-939950-33-1 — GD Guthrie, Jr., BT Mossman — i-xvi + 584 pp

Volume 27: *Minerals and Reactions at the Atomic Scale: Transmission Electron Microscopy*

1992 ISBN 0-939950-32-4; ISBN13 978-0-939950-32-4 — PR Buseck — i-xvi + 516 pp

Volume 26: *Contact Metamorphism*

1991 ISBN 0-939950-31-6; ISBN13 978-0-939950-31-7 — DM Kerrick — i-xvi + 672 pp

Volume 25: *Oxide Minerals:Petrologic and Magnetic Significance*

1991 ISBN 0-939950-30-8; ISBN13 978-0-939950-30-0 — DH Lindsley — i-xiv + 509 pp

Volume 24: *Modern Methods of Igneous Petrology: Understanding Magmatic Processes*

1990 ISBN 0-939950-29-4; ISBN13 978-0-939950-29-4 — J Nicholls, JK Russell — i-viii + 314 pp

Volume 23: *Mineral–Water Interface Geochemistry*

1990 ISBN 0-939950-28-6; ISBN13 978-0-939950-28-7 — MF Hochella, Jr., AF White — i-xvi + 603 pp

Volume 22: *The Al_2SiO_5 Polymorphs*

1990 ISBN 0-939950-27-8; ISBN13 978-0-939950-27-0 — DM Kerrick — i-xii + 406 pp

Volume 21: *Geochemistry and Mineralogy of Rare Earth Elements*

1989 ISBN 0-939950-25-1; ISBN13 978-0-939950-25-6 BR Lipin, GA McKay i-x + 348 pp

Volume 20: *Modern Powder Diffraction*

1989 ISBN 0-939950-24-3; ISBN13 978-0-939950-24-9 DL Bish, JE Post i-xii + 369 pp

Volume 19: *Hydrous Phyllosilicates (exclusive of micas)*

1988 ISBN 0-939950-23-5; ISBN13 978-0-939950-23-2 SW Bailey, i-xiii + 725 pp

Volume 18: *Spectroscopic Methods in Mineralogy and Geology*

1988 ISBN 0-939950-22-7; ISBN13 978-0-939950-22-5 FC Hawthorne i-xvi + 512 pp

Volume 17: *Thermodynamic Modeling of Geological Materials: Minerals, Fluids and Melts*

1987 ISBN 0-939950-21-9; ISBN13 978-0-939950-21-8 ISE Carmichael, HP Eugster i-xiv + 499 pp

Volume 16: *Stable Isotopes in High Temperature Geological Processes*

1986 ISBN 0-939950-20-0; ISBN13 978-0-939950-20-1 JW Valley, HP Taylor, Jr., JR O'Neil i-xvi + 570 pp

Volume 15: *Mathematical Crystallography*

1985 ISBN 0-939950-19-7; ISBN13 978-0-939950-19-5 MB Boisen, Jr., GV Gibbs i-xii + 460 pp

Volume 14: *Microscopic to Macroscopic*

1985 ISBN 0-939950-18-9; ISBN13 978-0-939950-18-8 SW Kieffer, A Navrotsky i-x + 428 pp

Volume 13: *Micas*

1984 ISBN 0-939950-17-0; ISBN13 978-0-939950-17-1 SW Bailey i-xii + 584 pp

Volume 12: *Fluid Inclusions*

1984 ISBN 0-939950-16-2; ISBN13 978-0-939950-16-4 Edwin Roedder 1984 i-vi + 646 pp

Volume 11: *Carbonates: Mineralogy and Chemistry*

1983, 1990 ISBN 0-939950-15-4; ISBN13 978-0-939950-15-7 RJ Reeder i-xii + 399 pp

Volume 10: *Characterization of Metamorphism through Mineral Equilibria*

1982 ISBN 0-939950-12-X; ISBN13 978-0-939950-12-6 JM Ferry i-xiv + 397 pp

Volume 9B: *Amphiboles and Other Hydrous Pyriboles—Mineralogy*

1982 ISBN 0-939950-11-1; ISBN13 978-0-939950-11-9 DR Veblen, PH Ribbe i-x + 390 pp

Volume 9A: *Amphiboles: Petrology and Experimental Phase Relations*

1981 ISBN 0-939950-10-3; ISBN13 978-0-939950-10-2 DR Veblen, PH Ribbe i-xii + 372 pp

Volume 8: *Kinetics of Geochemical Processes*

1981 ISBN 0-939950-08-1; ISBN13 978-0-939950-08-9 AC Lasaga, RJ Kirkpatrick i-x + 398 pp

Volume 7: *Pyroxenes*

1980 ISBN 0-939950-07-3; ISBN13 978-0-939950-07-2 CT Prewitt i-x + 525 pp

Volume 6: *Marine Minerals*

1979	ISBN 0-939950-06-5; ISBN13 978-0-939950-06-5	RG Burns	i-x + 380 pp

Volume 5: *Orthosilicates*

1980	ISBN 0-939950-13-8; ISBN13 978-0-939950-13-3	RG Burns	i-xii + 450 pp

Volume 4: *Mineralogy and Geology of Natural Zeolites*

1977	ISBN 0-939950-04-9; ISBN13 978-0-939950-04-1	FA Mumpton	i-xii + 233 pp

Volume 3: *Oxide Minerals*

1976	ISBN 0-939950-03-0; ISBN13 978-0-939950-03-4	D Rumble, III	i-3 + 706 pp

Volume 2: *Feldspar Mineralogy*

1975, 1983	ISBN 0-939950-14-6; ISBN13 978-0-939950-14-0	PH Ribbe	i-vii + 362 pp

Volume 1: *Sufide Mineralogy*

1974	ISBN 0-939950-01-4; ISBN13 978-0-939950-01-0	PH Ribbe	i-v + 301 pp